I0885520

Handbook of Egg Science and Technology

Eggs are one of the most popular foods worldwide due to their great taste and versatility, economical value and high nutritional content. The egg plays an important role in the human diet, both for the nutritional value of its many components (e.g., proteins, vitamins, minerals, choline and specific long chain fatty acids) and its wide range of functional characteristics, including foaming, gelling and emulsifying properties. The egg sector is a vibrant field with many new developments in terms of production, processing and commercialization as well as research. Since the beginning of the 21st century, the global production of eggs has grown by 69.5%, farm production systems have evolved to improve the welfare of laying hens, many shell eggs and egg products have been developed to address the changing demands of consumers and our knowledge of the composition of the egg has been boosted by the latest gene-based technologies. Information on the science and technology of egg and egg processing is essential to governments, academia and industry.

The *Handbook of Egg Science and Technology* aims to be the first book providing a complete source of information about egg science and technology, covering topics such as world egg production, marketing of eggs, chemistry of egg components, functional properties of egg components, egg processing, egg product development, eggshell quality, grading, egg microbiology, egg pasteurization, egg nutrition and bioactive components, egg biotechnology and sustainability of egg production.

Key Features

- This book includes the most current and comprehensive scientific and technical information about egg science and technology
- This book presents an ideal guide for professionals in related food industries, egg business consultants, regulatory agencies and research groups
- This book answers the need for a comprehensive textbook for upper-level undergraduate and graduate courses in food science, animal science and poultry departments

A global panel of experts in the field of egg science was gathered with the aim of providing the most updated information and development on many topics likely to interest readers, ranging from academia and food science students to managers working in the food production and egg processing sectors. This handbook is an excellent resource for the food and poultry industry, R&D sectors, as well as for experts in the field of food and nutrition.

Handbook of Egg Science and Technology

Editor-in-Chief: Yoshinori Mine
Associate Editor: Vincent Guyonnet
Assistant Editors: Hajime Hatta,
Françoise Nau, and Ning Qiu

CRC Press
Taylor & Francis Group
Boca Raton London New York

CRC Press is an imprint of the
Taylor & Francis Group, an **informa** business

First edition published 2024
by CRC Press
6000 Broken Sound Parkway NW, Suite 300, Boca Raton, FL 33487-2742
and by CRC Press

4 Park Square, Milton Park, Abingdon, Oxon, OX14 4RN

CRC Press is an imprint of Taylor & Francis Group, LLC

Library of Congress Cataloging-in-Publication Data

Names: Mine, Yoshinori, editor. | Guyonnet, Vincent, editor. | Hatta, Hajime, 1956- editor. | Nau, Françoise, editor. | Qiu, Ning (Professor of food science), editor.
Title: Handbook of egg science and technology / edited by editor in chief Dr. Yoshinori Mine, co-editor Dr. Vincent Guyonnet, associate editors Dr. Hajime Hatta, Dr. Françoise Nau, Dr. Ning Qiu.
Description: First edition. | Boca Raton : CRC Press, 2023. | Includes bibliographical references.
Identifiers: LCCN 2022060718 (print) | LCCN 2022060719 (ebook) | ISBN 9781032156606 (hardback) | ISBN 9781032184203 (paperback) | ISBN 9781003254430 (ebook)
Subjects: LCSH: Eggs--Handbooks, manuals, etc. | Egg processing--Handbooks, manuals, etc. | Egg trade--Handbooks, manuals, etc. | Egg products industry--Handbooks, manuals, etc.
Classification: LCC SF490 .H36 2023 (print) | LCC SF490 (ebook) | DDC 637/.5--dc23/eng/20230215
LC record available at https://lccn.loc.gov/2022060718
LC ebook record available at https://lccn.loc.gov/2022060719

ISBN: 978-1-032-15660-6 (hbk)
ISBN: 978-1-032-18420-3 (pbk)
ISBN: 978-1-003-25443-0 (ebk)

DOI: 10.1201/9781003254430

Typeset in Times
by KnowledgeWorks Global Ltd.

Contents

Part VIII Egg Nutrition and Human Health

Part IX Egg Biotechnology

Part X Production and Sustainability

Preface

The egg plays an important role in human diets, both for the nutritional value of its many components (e.g., proteins, vitamins, minerals, choline, specific long chain fatty acids) and for its wide range of functional characteristics, including foaming, gelling and emulsifying properties. The egg sector is a vibrant field with many new developments in terms of production, processing and commercialization as well as research. Since the beginning of the 21st century, the global production of eggs has grown by 69.5%, farm production systems have evolved to further improve the welfare of laying hens and many shell egg and egg products have been developed to address the changing demands of consumers. Our knowledge of the composition of the egg has been boosted by the use of the latest gene-based technologies and a wide range of molecules have recently been identified. It is evident that the egg is not only a food rich in important nutrients but also a great source of numerous bioactive molecules, with properties that may lead to a number of commercial applications. The structure and composition of the eggshell may offer great opportunities to better understand the mineralization process and provide some insight into human bone repair mechanisms. Both the yolk and the egg white contain many molecules still to be discovered and their roles still to be identified, offering many avenues for further research and commercial applications. The truth about egg nutrition is now widely recognized by the health community and society at large. With the digestibility of its proteins, their excellent amino acid profile and the high bioavailability of many of its components, the egg is indeed offering great nutrition to all segments of the population: infants and growing children, pregnant and lactating mothers, fitness enthusiasts and senior citizens. Eating eggs regularly may contribute to alleviate many ailments and will certainly contribute to healthier diets and lifestyles. With the lowest impact on the environment of all other terrestrial animal-sourced foods, the egg is also a great sustainable source of nutrients, an important feature highly valued by younger generations of consumers. The sustainability of the egg sector is one of the keys for its continued growth, especially in a number of low-income food deficit countries where the availability of eggs remains scarce. The egg has the potential to be one of the main animal-sourced foods during the remainder of the 21st century and warrants our attention. For this purpose, a global panel of experts in the field of egg science was gathered with the aim to provide the most updated information and development on many topics likely to interest readers ranging from academia and food science students to managers working in the food production and egg processing sectors. The editors wish to acknowledge all of the world-class contributor enthusiasm for this project and thank them for the time they allocated to bring to the readers the most current and pertinent information on a wide range of topics.

Yoshinori Mine and Vincent Guyonnet

About the Editors

Yoshinori Mine, Editor-in-Chief, earned his PhD in biochemistry from Tokyo University of Agriculture and Technology in Japan in 1993. He joined the University of Guelph (Ontario, Canada) in 1996 as a faculty member in the Department of Food Science. Currently, he is a full professor (since 2005). He has also been an associate editor for *Journal of Agricultural and Food Chemistry* (ACS Publications) since 2010 and serves as a member on the editorial advisory board for numerous international journals. He is also actively involved in numerous international food- and health-related symposiums as an invited speaker. His primary research focus is food bioactives and gut health, particularly dietary peptides, amino acids and pro-/prebiotics as well as understanding their molecular mechanism. He has published 5 books, 37 book chapters, 15 review articles and 182 original papers in peer-reviewed international journals.

Vincent Guyonnet, DVM, PhD, Dipl. ACPV, Associate Editor, obtained his veterinary degree from Lyon, France, and his PhD in poultry science from The University of Georgia, USA. During his career, he has worked closely with the poultry sector in various roles and represented the global poultry and egg sectors with intergovernmental organizations (OIE, FAO, World Bank) to address global issues such as animal welfare, climate change, food security and the overall sustainability of animal production. He is regularly invited to teach at universities on topics related to eggs and is currently a lecturer with the World Veterinary Education in Production Animal Health (WVEPAH). He has authored 4 book chapters, 22 refereed articles, presented at 51 international scientific conferences, often as an invited speaker, and writes regular editorials for international poultry magazines.

Françoise Nau, Assistant Editor, obtained her PhD in animal science and technology from ENSA Rennes, France (1991). After three years in the egg product industry, she got an assistant professor position (1993) and since 2003 has a full professor position at the Institut Agro Rennes-Angers, France. Prof. Nau's field of teaching is in food science and technology, including a special focus on ingredient functionalities and egg science. Her research activity is part of the laboratory INRAE-Institut Agro "Dairy and Egg Science and Technology". Her primary area of expertise concerns functionality and fractionation of egg white proteins. Since 2010, she is particularly interested in the understanding of the digestion process in relationship to food processing and food structure and texture, on one hand, and in antimicrobial properties of egg white proteins and peptides on the other.

Hajime Hatta, PhD, Assistant Editor, has been researching eggs for more than 40 years. He worked at a food company for 18 years, then taught food science and processing at Kyoto Women's University for 24 years before retiring in 2022. He is currently a research professor at KWU and the Egg & Chicken Laboratory Director at NBL, Inc. His research interests include basic and applied research on nutrition, cooking and health functions of chicken eggs, especially research on chicken egg antibodies (IgY) and their applications. Several specific-IgY fortified foods and feeds to prevent dental caries and fish diseases were developed for the first time in the world through his IgY project. He was awarded the Technological Research for Industrial Applications of IgY prize from the Japanese Association for Bioscience, Biotechnology, and Biochemistry in 1994.

Ning Qiu, PhD, Assistant Editor, works in research areas including the application of quantitative proteomics to investigate avian egg protein nutrition and functions and modulation of gut microbiota influenced by egg proteins. He obtained his bachelor's in bioengineering from Jiangnan University in 2000 and a PhD in microbiology from Huazhong Agricultural University in 2010. Currently, he is an associate professor at Huazhong Agricultural University. He has a number of honors and scientific publications to his credit including Advisor for Monitoring and Analysis of Egg Processing Industry (Ministry of Agriculture of China), Member of the Youth Committee of the National R&D System for Agricultural Products Processing Technology, External Enterprise Tutor of CP Group, and more.

List of Contributors

Achyut Adhikari
Louisiana State University
Baton Rouge, Louisiana

Tamer A. E. Ahmed
Faculty of Medicine
University of Ottawa
Ottawa, Canada
Faculty of Health Sciences
University of Ottawa
Ottawa, Canada

Dong U. Ahn
Iowa State University
Ames, Iowa

Satoshi Akiniwa
Kewpie Corporation
Tokyo, Japan

Katherine Alvarado-Ramos
Aix-Marseille Université, INRAE, INSERM, C2VN
Marseille, France

Marc Anton
Biopolymères Interactions Assemblages (BIA), INRAE
Nantes, France

Kazuto Arimitsu
Kewpie Corporation
Tokyo, Japan

Xiaoyu Bao
University of Alberta
Edmonton, Canada

Florence Baron
Institut Agro, INRAE, STLO
Rennes, France

Khushwant S. Bhullar
University of Alberta
Edmonton, Canada

Paolo Bonato
Sanovo Technology Italia
Marano Vicentino (VI), Italy

Claire Bordat
Aix-Marseille Université, INRAE, INSERM, C2VN
Marseille, France

Jan Høy Carstens
Sanovo Technology
Odense SØ, Denmark

Giosuè Casagrande
Sanovo Technology Italia
Marano Vicentino (VI), Italy

David Cassin
Groupe Avril
Bruz, France

Yujie Chi
Northeast Agricultural University
Harbin, China

Roberto Colavitti
Sanovo Technology Italia
Marano Vicentino (VI), Italy

Luc Djoussé
Harvard Medical School
Boston, Massachusetts

Xiang Duan
Northwest A&F University
Yangling, China

Pauline Duffuler
University of Alberta
Edmonton, Canada

Xi Feng
San Jose State University
San Jose, California

Joël Gautron
INRAE, Université de Tours, BOA
Nouzilly, France

Fang Geng
Chengdu University
Chengdu, China

Yunsi Guo
Zhejiang Gongshang University
Hangzhou, China

Audrey E. M. Guyonnet
FFI Consulting Ltd.
Brockville, Canada

Vincent Guyonnet
FFI Consulting Ltd.
Brockville, Canada

Nicolas Guyot
INRAE, Université de Tours, BOA
Nouzilly, France

Jae Yong Han
Seoul National University
Seoul, Republic of Korea

Akihiro Handa
Tokyo Denki University
Saitama, Japan

Hajime Hatta
Kyoto Women's University
Kyoto, Japan

Takehiko Hayakawa
DSM Nutritional Products Asia Pacific
Animal Nutrition & Health Division
Tokyo, Japan

Maxwell T. Hincke
Faculty of Medicine
University of Ottawa
Ottawa, Canada

Yasumi Horimoto
University of Guelph
Guelph, Canada

Xi Huang
Huazhong Agricultural University
Wuhan, China

Tasnim F. Imran
Warren Alpert Medical School of Brown University
Providence, Rhode Island

Sophie Jan
Institut Agro, INRAE, STLO
Rennes, France

Mahnoor Khalid
Foundation University Medical College
Islamabad, Pakistan

Yoshimi Kishimoto
Setsunan University
Osaka, Japan

Takashi Kojima
Kewpie Corporation
Tokyo, Japan

Nanase Kubo
Egg & Poultry Research Center, NBL Co., Ltd.
NBL Co., Ltd. could be placed on the same line as Egg & Chicken Research Center
Kyoto, Japan

Garima Kulshreshtha
Faculty of Medicine
University of Ottawa
Ottawa, Canada

Jack N. Losso
Louisiana State University
Baton Rouge, Louisiana

Zihong Ma
Northeast Agricultural University
Harbin City, China

Yaqi Meng
Huazhong Agricultural University
Wuhan, China

Yoshinori Mine
University of Guelph
Guelph, Canada

Tetsuya Miyamoto
Kewpie Corporation
Tokyo, Japan

Takeya Miyashita
Kewpie Corporation
Tokyo, Japan

Françoise Nau
Institut Agro, INRAE, STLO
Rennes, France

Noriaki Nishijima
Kewpie Corporation
Tokyo, Japan

Yoshifumi Oda
Kewpie Corporation
Tokyo, Japan

Jin Se Park
Seoul National University
Seoul, Republic of Korea

Arissara Phosanam
Louisiana State University
Baton Rouge, Louisiana

Ning Qiu
Huazhong Agricultural University
Wuhan, China

Shofiqur Rahman
EW Nutrition Japan K.K.
Gifu, Japan

Hans Chr. D. Rasmussen
Sanovo Technology
Odense SØ, Denmark

Emmanuelle Reboul
Aix-Marseille Université, INRAE, INSERM, C2VN
Marseille, France

Sophie Réhault-Godbert
INRAE, Université de Tours, BOA
Nouzilly, France

Hiroyuki Shidara
Kewpie Corporation
Tokyo, Japan

Yutaka Tahara
Oihama High School
Chiba, Japan

Hidekazu Takeuchi
Kewpie Corporation
Tokyo, Japan

Clarisse Techer
Mixscience
Bruz, France

Manish Thapaliya
Louisiana State University
Baton Rouge, Louisiana

Trevor Toavs
The Warren Alpert Medical School of Brown University
Providence, Rhode Island

Yonggang Tu
Jiangxi Agricultural University
Nanchang, China

Gouta Uematsu
Kewpie Corporation
Tokyo, Japan

Yoshie Ueno
Kyoto Women's University
Kyoto, Japan

Limei Wang
Hubei Provincial Engineering Technology Research Center for Deep Processing of Eggs
Wuhan, China

Zihan Wang
University of Alberta
Edmonton, Canada

Jianping Wu
University of Alberta
Edmonton, Canada

Hui Xue
Hainan University
Haikou, China

Fang Yang
Hubei University of Chinese Medicine
Wuhan, China

Manar Younes
Faculty of Medicine
University of Ottawa
Ottawa, Canada

Luyan Yu
Hubei University of Chinese Medicine
Wuhan, China

Hua Zhang
Jiangxi University of Chinese Medicine
Nanchang, China

Suyun Zhang
Zhejiang Gongshang University
Hangzhou, China

Yue Zhang
Zhejiang Gongshang University
Hangzhou, China

Yan Zhao
Jiangxi Agricultural University
Nanchang, China

Yuting Zheng
Hubei Provincial Engineering Technology Research Center for Deep Processing of Eggs
Wuhan, China

Lingjiao Zhu
Huazhong Agricultural University
Hubei Provincial Engineering Technology Research Center for Deep Processing of Eggs
Wuhan, China

Part I

Introduction

1

Introduction

Yoshinori Mine* **and Vincent Guyonnet**

CONTENT

For centuries, the egg has played a key role in human diets as a great source of nutrients such as proteins, vitamins, minerals, essential fatty acids, and several micronutrients. One of the best ways to demonstrate the great nutritional value of eggs was a study recently conducted with children in a region of Ecuador where stunting is highly prevalent (Iannotti et al., 2017). This randomized controlled study showed that when young infants, 6–9 months of age, were given one egg per day for 6 months, the prevalence of stunting and underweight were decreased significantly, by 47% and 74%, respectively, when compared to infants receiving a standard diet without eggs. However, this appeal of shell eggs to consumers extends well beyond their great nutritional value. The yolk and the egg white, the two main compartments in the egg, provide some valuable functional characteristics such as the foaming and gelling ability, making the egg an ingredient of choice in the preparation of many dishes and meals, both at home and in food processing plants. The egg sector is a vibrant field with many new developments in terms of production, processing, and commercialization, as well as research. Since the beginning of the 21st century, the global production of eggs has grown by 69.5%, farm production systems have evolved to improve the welfare of laying hens, many shell eggs and egg products have been developed to address the changing demands of consumers, and our knowledge of the composition of the egg has been boosted by the use of the latest gene-based technologies. Since the last edition of the Egg science and Technology handbook was published more than 25 years ago (Stadelman & Cotterhill, 1995), it was long overdue to gather the latest information generated by recent studies conducted in various parts of the world. While many egg-related books are available (egg production, egg microbiology, egg nutrition, etc.), most are single-volume books covering either general topics or some very specific topics. Currently, there is no professional comprehensive book covering the complete information about egg science and technology (Board & Fuller, 1993; Yamamoto et al., 1997; Schade et al., 2000; Huopalahti et al., 2007; Mine, 2008 ; Watson, 2010; Nys et al., 2011; Wu, 2019). This book aimed to be the first book providing a complete source of information about egg science and technology, covering topics such as world egg production, marketing of eggs, chemistry of egg components, functional properties of egg components, egg processing, egg product development, shell egg quality, grading, egg microbiology, egg pasteurization, egg nutrition and bioactive, egg biotechnology, and sustainability of egg production. A global panel of experts in the field of Egg Science was gathered with the aim to provide the most updated information and development on many topics likely to interest readers ranging from those in academia and food science students to managers working in food production and egg processing.

The content and format of this textbook were designed to suit the needs of the intended readers, providing numerous tables, figures and graphs to clearly illustrate the key points presented in the text. Following this introduction section along with an overview of the global egg sector (Part I), this handbook of Egg Science and Technology will consists of 42 additional chapters allocated to 9 main themes: PART II: Chemistry and Characterization of Egg Components, PART III: Functional Properties of Egg Components, PART IV: Processing of the Egg and Its Application on the Food System, PART V: Shell Egg Quality, PART VI: Microbiology, Quality, and Safety, PART VII: Bioactives of Egg Components, PART VIII: Egg Nutrition and Human Health, PART IX: Egg Biotechnology, and PART X: Production and Sustainability.

This book will cover comprehensively the most recent advances in fields ranging from egg production to egg science and technology. The breadth of the information provided in this new textbook is so broad that it is anticipated that readers will direct their attention only to specific sections or topics at a time and will return regularly to consult its content. Therefore, a brief overview of the content of each chapter will assist readers in selecting the chapter(s) best addressing their immediate needs for information.

Two chapters serve as an Introduction to this new textbook about Egg Science and Technology (Part I): Chapter 1 highlights the aim and scope of this book and provides a brief introduction for each chapter, providing readers with the means to select the chapters they should read first.

Over the past 20 years, the production of eggs has increased tremendously, and the successful growth

* Corresponding author.

DOI: 10.1201/9781003254430-2

of egg production has been the results of investments by many stakeholders within the egg value chain. Chapter 2 reviews the latest information in terms of egg production trends and figures, the key stakeholders within the egg value chain and the most recent marketing trends for shell eggs and egg products.

The Chemistry and Characterization of Egg Components (Part II) is addressed in a series of five chapters (Chapters 3–7):

Chapter 3 reviews the complex mechanisms involved during the formation of an egg. The folliculogenesis in the ovary, the synthesis of yolk lipoproteins in the liver, the role of the infundibulum in the deposition of the outer layer of yolk membranes, the synthesis of egg white proteins in the magnum, shell membranes in the isthmus, and shell calcification in the shell gland are discussed. This chapter also describes the state of knowledge on the tissues or organs involved in the formation of the egg and on the structure and global composition of the egg constituents.

The unique physicochemical and biological properties of eggshell and associated membrane in terms of composition and mechanical features, along with their abundance, make them promising biomaterials for various biotechnological applications. Chapter 4 aims to summarize the research landscape pertaining to chicken eggshell and membrane constituents over the last three decades, from which much of our knowledge about avian eggshell mineralization, function, and composition are derived.

The egg white contains numerous proteins, with a wide range of physicochemical features and properties. Chapter 5 devotes a large section to these proteins, at least the major ones. This chapter presents the sources of variability in protein composition of the chicken egg white, as well as interspecies variability. Then, the biochemical and structural features of the main proteins are described.

Chapter 6 reviews the composition and the different levels of organizations of the egg yolk fractions and explores the changes induced by external constraints, among which thermo-mechanical and high-pressure treatments. This information allows one to better understand the importance of the different structures of egg yolk for the control of its properties.

The egg displays a perfect balance and diversity in its nutrients, making the egg a desirable dietary source of many micronutrients to help meet daily intake requirements. In Chapter 7, the micronutrients including vitamins, carotenoids, and inorganic minerals, as well as enzymes with biological activity are introduced in detail for the unfertilized chicken eggs.

The Functional Properties of Egg Components (Part III) are reviewed in two chapters (Chapters 8 and 9):

Egg white contributes to multiple functional properties, such as gelling, foaming, binding, and emulsifying, that impart food products with desirable sensory characteristics. Chapter 8 addresses how these functional properties may be improved through proper formulation and processing, and thus improve the overall quality of food products. Elucidating the functional properties of egg components is indeed one of the keys to improving and broadening their application within the food industry.

As natural emulsifiers, egg yolk lipids and proteins both contribute to the egg's functional properties, such as emulsification and gelation. During the processing of egg yolks, their functionalities play important roles in providing unique texture, flavor, color, and mouth feel to food formulations. Both egg yolk plasma and granules have their own functionalities, thus increasing the potential of egg yolk in product development. Given their broad applicability, their processing should be optimized according to the needs of consumers. Chapter 9 discusses research related to the functional properties of egg yolk components to inform their application in the food industry.

A series of five chapters (Chapters 10–14) are included in PART IV: Processing of the Egg and Its Application on the Food System:

The use of egg products is now common in the food service and food industry sector, bringing additional safety and uniformity to the use of eggs. While egg processing is a rather complex process with the presence of egg proteins bringing some challenges not seen with other processed foods, a series of well-defined principles and specifically designed pieces of equipment allow to repeatedly produce quality and safe products. Some key principles and steps involved for the production of liquid egg and egg powder products are reviewed in Chapter 10.

Eggshell makes up about 10% of an egg, and eggshell calcium is the main component of the eggshell, mainly present in the calcite crystalline form of calcium carbonate. Despite being an excellent source of calcium, most of it is discarded or used as a soil conditioner but rarely used as a source of calcium for humans. Chapter 11 discusses in detail the benefits and principles of eggshell calcium intake, including comparisons with other calcium sources and several uses of eggshell calcium other than as a source of calcium.

Eggs may be prepared, processed, and consumed in various forms. Chapter 12 introduces by function several processed egg products mainly consumed in Japan. As an example of the thermal coagulation properties of eggs, the industrial production of boiled eggs and some of its challenges is presented, along with a description of fried eggs and Japanese-style omelets. Next, the foaming properties of eggs and the foods taking advantage of these foaming properties are explained. In order to illustrate the emulsifying properties of shell eggs, the production of mayonnaise is detailed, outlining some of the solutions used to maintain quality during long storage time. The importance of color and the visual aspect of foods, as well their nutritional value concludes this chapter.

The traditional eggs and egg products in Western and Asian countries are introduced in Chapter 13. This chapter describes the history of the expansion of egg consumption in Western Europe, China, and Japan, including the food cultures at different time periods. The traditional egg preservation methods and their uses are presented along with egg dishes typical of each region, including Japan, China, the European Union, and Latin and North America.

The egg product segment is well developed in both the USA and Japan, servicing the food processing industry as well as the retail sector. In fact, Japan is the only country in the world where consumers typically eat more eggs in the form of egg products than shell eggs. Innovative products such as lactic acid fermented egg whites, aged egg yolks, cholesterol-free eggs, and egg white seasonings have been developed to further improved the nutritional value and functionality of shell eggs. Chapter 14 reviews the current market situation for egg products in the USA and Japan, highlighting the different egg products developed in these countries. It also provides some guidance into strategies to consider in the future to develop innovative egg products addressing the new and evolving needs of consumers. Addressing price, taste, convenience, authenticity, and sustainability must be the guiding principles of future product development efforts.

The importance of Shell Egg Quality (Part V) is reviewed in four chapters (Chapters 15–18):

Egg quality is greatly affected by the feed fed to the hens, the husbandry environment, the climate, and the storage conditions and handling of shell eggs after collection on farms. Among them, differences in eggshell aspect and color, along with egg yolk color, are easily noticeable by consumers and will have a great impact on their appeal. Chapter 15 explains in detail the market trends for eggshell and egg yolk color in each country and their impact on egg quality. Nutritional and metabolic considerations involving laying hens are also addressed.

Chapter 16 describes the principles and characteristics of two procedures for measuring the freshness of chicken eggs: one is to examine eggs without breaking them (non-destructive method), and the other is to examine the state of egg white and yolk by breaking them (cracked egg method). In addition, the accuracy of the Haugh unit and Yolk index, well-known freshness indicators, are evaluated using both manual and digital equipment.

Because of nutritional value, eggs are perishable foods prone to spoilage. As eggs lose their freshness, bacteria on the eggshell or certain types of salmonella contaminated inside the egg can multiply and cause food poisoning. The shelf life of eggs is set as the time limit by which eggs are safe and tasty to eat, based on the relationship between the freshness of eggs and the growth of bacteria that contaminate eggs. Therefore, egg shelf-life differs in countries where eggs are eaten raw and eggshells are washed or not. Chapter 17 discusses the shelf life of chicken eggs and the rationale for setting expiration dates for eggs.

As one of the most nutritious and affordable animal-sourced foods, the safety of shell eggs is critical for the continuing growth of the egg sector. Considering that preventing food safety hazards is the most effective and economic way to provide safe foods to consumers, the hygienic production and commercialization of shell eggs is reviewed while considering the different biological, chemical, and physical hazards from the Hazard Analysis Critical Control Point (HACCP) perspective. The regulations in China, the USA, India, the EU27 countries, Japan, Mexico, and South Africa are outlined in detail for similarities and differences in Chapter 18.

PART VI: Microbiology, Quality and Safety is addressed in two chapters (Chapters 19 and 20):

Proteomics, as a large-scale technical strategy to identify and quantify all proteins in a biological sample, has been widely used in the field of egg science in the past two decades. Recent advances in our understanding of the proteome of eggs from different species and the dynamic changes occurring during storage are increasing rapidly, providing new insights for the exploration of egg science. Chapter 19 provides a comprehensive overview of the egg proteome, covering all its components and applications under various conditions, such as marker detection, egg quality evaluation, genetic and biological unknown identification, and embryonic nutritional supplementation.

The protection of eggs from bacterial contamination is ensured by the implementation of appropriate production conditions, egg collection methods, and storage practices. However, egg contamination by *Salmonella* Enteritidis remains a health concern, in particular for un- or under-cooked eggs. This type of contamination is mainly due to the specific ability of this bacterium to sporadically invade the egg through vertical contamination, and to resist the natural antimicrobial defense mechanisms of the egg. Chapter 20 discusses egg and egg product microbiology, underpinned by some antagonist phenomena.

The richness of shell eggs in Bioactives of Egg Components (Part VII) is reviewed in Chapters 21–Chapter 28:

Eggshell membrane (ESM) is made up of a double-layered fibrous meshwork and plays an important role in resistance to microbial invasion, egg protection, and embryonic development. A variety of bioactive components present in the ESM, including proteins (e.g., lysyl oxidases, lysozyme, cysteine-rich eggshell membrane proteins, fibronectin, histones, ovocalyxin-36, and ovotransferrin), bioactive peptides (e.g., avian beta-defensins), glycosaminoglycans (e.g., hyaluronic acid and chondroitin sulfate), and sialic acid are discussed in Chapter 21.

Egg white is regarded as a rich and inexpensive source of bioactive proteins/peptides due to the diverse biological activities that have been revealed. Chapter 22 aims to review the various biological activities of egg white proteins and their derived peptides, including antioxidative, antimicrobial, antiviral, antitumor, immunomodulatory, and angiotensin-converting enzyme-inhibitory (ACEi) activities. The potential applications of egg white proteins and derivative peptides on human health are also discussed.

Egg yolk protein includes lipovitellins, livetins, low-density lipoproteins, and phosvitin. Among these egg yolk proteins, γ-livetin (IgY) and phosvitin are the two most studied, and their functions and potential uses are very well defined. Chapter 23 discusses the possibility of developing industrial-scale sequential separation methods for egg yolk proteins, especially IgY and phosvitin, and the functionality of egg yolk proteins and their potential applications in the nutraceutical, food, and pharmaceutical industries. Future research directions on egg yolk proteins are also discussed. The egg yolk contains lipids, proteins, granules, and several minor components, including carbohydrates, minerals, and vitamins.

Egg white is a good source of water-soluble proteins, most of which have pro-inflammatory properties. The egg yolk is also a good source and carrier of xanthophylls lutein and zeaxanthin that are good for eye health. Chapter 24 first addresses major bioactive compounds in eggs that may contribute to reducing chronic degenerative diseases in clinical nutritional aspects.

The biological activities derived from egg yolk lipids, such as egg yolk oil, unsaturated fatty acids, phospholipids, carotenoids, and fat-soluble vitamins, have anti-inflammatory, anti-hypertension, antioxidant, and antimicrobial biological activities, as well as some beneficial effects on heart health, protecting eyesight, and enhancing memory without side-effects. Additionally, egg yolk lipids regulate cell function, physiological homeostasis, and chronic metabolic disease. The biological functions of egg yolk lipids and other fat-soluble components, as well as their underlying advantages, have been emphasized in Chapter 25.

Lysozyme, a value-added product from hen egg white has the GRAS status and is approved by the FAO/World Health Organization for use as an antimicrobial, antifungal, and antiviral ingredient in minimally processed foods physico-chemical modifications such as hurdle technologies enhancing and expanding the lysozyme antimicrobial activity against Gram-negative bacteria, yeasts, coliforms, and viruses are highlighted in Chapter 26.

Egg yolk constitutes a relevant alternative source of antibodies (IgY). It presents some advantages over mammalian serum immunoglobulins in terms of productivity, animal welfare, and specificity. Chapter 27 aims to review the IgY, its history, avian-non-avian diversifications, physicochemical, immuno-biological characteristics, extraction, purification, basic laboratory investigations, field studies, clinical trials, and extensive applications in the global One Health approach toward addressing in a holistic approach the health of people, animals, and the planet.

The tremendous increase in the discovery of minor proteins in eggs was made possible by applying highly resolutive methods for protein separation and hundreds of minor egg proteins have been regularly identified during the past 20 years. This large number of proteins accounts for a very small percentage of the total amount of protein in the egg and are therefore designated as minor egg proteins. These low abundant proteins are believed to support the harmonious embryonic development of the chick by exhibiting diverse functions including defense, nutrition, and other predicted biological activities, which have been investigated using bioinformatics and experimental investigations (Chapter 28).

The importance of Egg Nutrition and Human Health (Part VIII) is outlined in a series of 10 chapters (Chapters 29–38):

Shell eggs have always played a key role in the human diet. The high digestibility of egg proteins along with their content in indispensable amino acids are just one of the nutritional benefits of shell eggs. With the dietary intake of cholesterol no longer seen as a health risk factor in the general population, shell eggs are now widely recommended by governments and health professionals as an essential component of a healthy diet (Chapter 29).

Although many studies have found no adverse cardiovascular effect of egg consumption, others have reported adverse effects with egg consumption above seven eggs per week. Thus, the association of egg consumption and cardiovascular disease continues to spur some debate. In Chapter 30, the evidence of the effects of egg consumption on cardiovascular disease, cardiovascular mortality, and lipids are reviewed.

In industrialized countries, egg allergy accounts for one of the most prevalent food hypersensitivities, especially in children. The dominant egg allergens are proteins, and are mainly present in the egg white, e.g., ovalbumin, ovomucoid, ovotransferrin and lysozyme. Attempts to produce hypoallergenic egg-containing products through food processing techniques have met some promising results, but the approach is limited due to their potentially undesirable effects on the unique functional and sensory attributes of egg proteins (Chapter 31)

Eggs are a major source of dietary cholesterol; a 50-gram egg contains ~200 mg of cholesterol. The association between eggs and chronic diseases has been discussed mainly in terms of their negative impacts on hypercholesterolemia and cardiovascular disease (CVD), but eggs have recently been recognized as a healthy food with high nutritional value and a variety of functional properties. Chapter 32 discusses whether eggs may offer some protection against chronic diseases such as metabolic syndrome and the related oxidative stress and inflammation.

Today the characterization of intestinal microbiota composition and activity is a hot topic, and more and more studies focus on the effect of diet on gut microbiota. Dietary eggs can modify gut microbiota composition and further affect human health and disease. Chapter 33 summarizes the research investigating the effect of egg components on microbiota that was conducted over the past few decades, aiming to develop a deeper understanding of their potential roles in metabolism and health.

Owing to its broad nutritional profile, low cost, and culinary versatility, chicken egg represents an alternative option as an ingredient used for the production of sport nutritional supplements. The consumption of whole eggs compared to that of egg white has been strongly suggested for a greater stimulation of myofibrillar protein synthesis. But current evidence seems to support that the intake of egg yolk is preferrable to enhance myofibrillar protein synthesis (Chapter 34).

Osteoporosis is characterized by a microarchitectural deterioration of the bone tissue and a low bone mineral density, increasing the risk of fracture. Eggs are an inexpensive, widely available, and highly nutritious food that provide essential proteins and micronutrients (e.g., iron, selenium, vitamin K) for bone health (Chapter 35).

Brain is the control center of the body and the center of the nervous system of human beings. The egg yolk is rich in numerous fatty acids, including phospholipids, choline, polyunsaturated fatty acids (PUFA), and lipophilic vitamins such as carotenoids. Among these fatty acids, phospholipids and PUFAs derived from egg yolks play a key role from infancy to old age in modulating the brain cell membrane structure and its physiological properties. Choline is the essential nutrient for neurodevelopment during the early stage of brain development (Chapter 36).

Carotenoids are vegetal pigments found in significant amounts in the human eye. Eye health is highly dependent on carotenoids status. For instance, Age Related Macular Degeneration (AMD) is one of major causes of blindness in the world, and to date, there is no specific treatment for this pathology. However, carotenoids such as lutein have been shown to play a role in stopping macular degeneration evolution, and their consumption through lifespan contribute to prevent this disease. Egg yolk represents a great source of carotenoids for the human diet. Chapter 37 presents which carotenoids accumulate in eggs, as well as their function and importance for human eye health.

Chapter 38 describes the methods and characteristics of producing high-value-added eggs, which further enhance the nutritional function of chicken eggs through the feed. In particular, the chapter focused on eggs with proven physiological effects on humans, such as eggs fortified with ω3 fatty acids, folic acid, and iodine, and eggs with higher vitamin and mineral content.

The latest development and findings in terms of Egg Biotechnology (Part IX) are presented in Chapters 39–41:

Poultry, especially chickens, are regarded as valuable animals for food resources and are efficient bioreactors for the production of recombinant proteins. Due to their valuable characteristics, genome engineering for industrial purposes has been performed in poultry over the past several decades, and therapeutic proteins generated by transgenic chickens have been commercialized. Due to their distinct physiological characteristics, genome engineering technologies in poultry have been developed independently of those in mammalian species. Chapter 39 introduces the genome engineering methods that have been applied in poultry and describes recent progress in genome editing of poultry, as well as future applications, especially in terms of modulation of egg components.

An avian embryo culture system enables the observation of embryonic development, which is usually enclosed and obscured by an opaque eggshell. Chapter 40 describes in detail the shell-less chick embryo culture method using a transparent film made of gas-permeable polymethylpentene. Therefore, embryonic development processes can be observed from various angles through the entire period.

The chicken egg, as a completely aseptic and self-sufficient biological entity, contains all the components required for the embryonic development. As such, it constitutes not only an excellent model to study the mechanisms of early embryo nutrition and disease origin but can also be used to develop egg-based products with specific applications. Chapter 41 deals with the dynamic molecular processes of the different chicken egg components involved in embryo development.

The final two chapters of this textbook (Chapters 42 and 43) address Part X: Production and Sustainability:

The focus of consumers and various stakeholders on the sustainability of their food choices offers many opportunities to the global egg sector. While having the lowest impact on the planet compared to other livestock productions, the egg sector can adopt various mitigation strategies to further improve its environmental sustainability (Chapter 42).

The access to reliable information is essential for the development of any business and agricultural sector. Technologies to process shell eggs and egg products are also evolving rapidly with the adoption of more automations and innovative technologies. Throughout Chapter 43, a list of valuable websites is provided on a wide range of topics, ranging from national nutritional databases for the nutritional composition of shell eggs to the list of major equipment manufacturers offering equipment and tools for the grading and processing of eggs.

As science evolves rapidly with the use of better, faster, and more accurate technologies, one single individual can no longer keep up with the progress made on all corners of the planet. It is evident that the gathering of the latest information about Egg Science & Technology would not have been possible without the time and dedication of many contributors to this new textbook. The editors wish to acknowledge their enthusiasm for this project and thank them for the time they allocated to bring to the readers the most current and pertinent information on a wide range of topics. The editors wish also to acknowledge the editorial support provided by Drs. F. Nau, H. Hatta, and N. Qiu.

This new textbook on Egg Science and Technology was written with the intention to serve as a reference to a wide range of readers with an association with the egg sector. In addition to Chapter 43, many contributors have been included in the text and in the list of reference of their respective chapters some means for the readers to keep up with future development in these fields. This book will answer the need for a comprehensive textbook for upper-level undergraduate and graduate courses in food science, animal science, nutrition, and poultry departments. It is an ideal guide for professionals in related food industries, egg business consultants, regulatory agencies, and research groups for the food and poultry industry R&D sectors as well as for experts in the field of food and nutrition. This book should therefore be a very valuable resource for many years to come and help various stakeholders ensure that Egg Science continues to thrive and that the egg, an amazing gift of nature, plays an even bigger global role in food security and health.

REFERENCES

Board, R. G., & Fuller, R. (1993). *Microbiology of the avian egg*. Springer.

Huopalahti, R., López-Fandiño, R., Anton, M., & Schade, R. (2007). *Bioactive egg compounds*. Springer.

Iannotti, L. L., Lutter, C. K., Stewart, C. P., Gallegos Riofrío, C. A., Malo, C., Reinhart, G., & Waters, W. F. (2017). Eggs in early complementary feeding and child growth: A randomized controlled trial. *Pediatrics*, *140*(1), e20163459. https://doi.org/10.1542/peds.2016-3459

Mine, Y. (2008). *Egg science and biotechnology*. Wiley-Interscience.

Nys, Y., Bain, M., & Immerseel, F. V. (2011). *Improving the safety and quality of eggs and egg products volume 1: Egg chemistry, production and consumption* (1st ed.). Elsevier.

Schade, R., Behn, I., Erhard, M., Hlinak, A., & Staak, C. (2000). *Chicken egg yolk antibodies, production and application: IgY-technology*. Springer Science & Business Media.

Stadelman, W. J., & Cotterhill, O. J. (1995). *Egg science and technology* (W. J. Stadelman, & O. J. Cotterhill, Eds.; 4th ed.). Food Products Press.

Watson, R. R. (2010). *Eggs and health promotion* (1st ed.). Wiley-Blackwell.

Wu, J. (2019). *Eggs as functional foods and nutraceuticals for human health*. Royal Society of Chemistry.

Yamamoto, T., Juneja, L. R., Hatta, H., & Kim, M. (1997). *Hen eggs: Their basic and applied science*. CRC Press Inc.

2

World Egg Production and Marketing Trends

Vincent Guyonnet

CONTENTS

DOI: 10.1201/9781003254430-3

2.1 Introduction

With the exception of Antarctica, laying hens are present on every continent and provide a great source of nutrition to nearly 8 billion people. Over the years, the production of eggs has increased tremendously, although not at the same pace around the world. With the global population expected to growth to 9.7 billion by 2050, with most of the growth in Asia and Africa, the production of eggs will need to grow faster on these two continents, since the trading of shell eggs is mostly regional in nature and export is limited to countries close to the producing country. Although egg production has grown faster than that of the global population, the promotion of eggs must continue in order to retain consumers but also to gain new ones, especially in countries where the consumption of eggs still remains a rarity. This chapter will review the latest information in terms of egg production figures, the key stakeholders within the egg value chain and the marketing trends — often linked to changes in consumer behaviours — and will conclude by presenting data on egg consumption in various parts of the world, as some significant differences still persist.

2.2 Egg Production

The production of eggs by laying hens has shown some tremendous growth over the past 40 years. During that time, the production of eggs shifted from Europe and North America, towards Asia and Latin America but this growth has been uneven across regions and countries. For the purpose of comparisons, data from the Food and Agriculture Organization of the United Nations (FAO) have been used throughout this section (FAO, 2022a), using the regional classification adopted by the FAO. For instance, the Russian Federation is included in numbers reported for Europe, while egg production in the countries of the Middle East and Turkey are allocated to Asia. North America comprises Canada and the USA while egg production in Mexico, reported by the FAO in Central America, is presented here with the overall results for Latin America.

2.2.1 Global Production

The production of eggs was multiplied by 3 between 1980 and 2020, growing from 26.2 million metric tonnes to 86.7 million metric tonnes (Table 2.1). Since 1980, each decade has registered some impressive growth in egg production: 33.8% (1980–1990), 45.8% (1990–2000), 25.6% (2000–2010) and 35.0% (2010–2020). This growth equated to a compound annual growth rate (CAGR) of +3.03% between 1980 and 2020, below that of chicken meat production (+4.22%) but well above that of pig meat (+1.85%), milk (+1.62%) and cattle meat (+1.00%) productions. In 2020, the global egg production of 86.7 million metric tonnes represented a growth of 2.7% over that of 2019.

While the COVID19 pandemic made animal production more challenging, it is clear that the egg sector was able to overcome many issues to record one of its highest annual growths of the past 10 years with 105 countries registering an increase in production. Unfortunately, 20 countries reported no growth between 2019 and 2020 while 65 countries recorded a decrease in egg production.

2.2.2 Top 10 Egg-Producing Countries (1980–2020)

The changes in the list of the 10 largest egg-producing countries during the past 40 years demonstrate the geographical shift of the global egg production (Table 2.2). Although the USA led the way in 1980, Europe accounted for 49.1% of the global egg production and was well represented in this list with five countries (USSR, West Germany, France, the UK and Spain). China and Japan were the sole representatives from Asia-Oceania, a region accounting for 27.7% of the global output. Brazil and Mexico, ranking 8th and 10th, respectively, represented Latin America, which was responsible for 9.8% of the global production of eggs. By 2000, India and Turkey joined the list of the top 10 egg-producing countries and Asia-Oceania became the dominant region for egg production (57.2% of the global production), represented by four countries (China, Japan, India and Turkey). The Russian Federation, France and Germany remained the sole European countries in this list, while Mexico overtook Brazil as the largest egg-producing country in Latin America. The USA gave up its top position in 1985, when China exceeded US production by 270,000 tonnes, but remained well ahead of Japan. By 2020, Indonesia had joined the list of top 10 egg-producing countries, where it was positioned fourth just behind India and fast approaching the total egg output of the USA. With five representatives in the list, Asia-Oceania accounted for 62.8% of the global output. Along with the Russian Federation, France managed to keep its membership in this illustrious list of egg-producing countries as representative of Europe

TABLE 2.1

Evolution of the Global Egg Production Since 1980

Year	Total Production (Tonnes)	% Growth Over 10 Years	Country with Largest Egg Production	% Global Production
1980	26,204,901		USA	15.7%
1990	35,071,918	33.8%	China	18.1%
2000	51,133,048	45.8%	China	36.3%
2010	64,220,679	25.6%	China	36.6%
2020	86,669,518	35.0%	China	34.4%

Source: From FAO (2022a).

TABLE 2.2

List of the Top 10 Egg-Producing Countries (1980–2000–2020)

1980		2000		2020	
Country	**Production**	**Country**	**Production**	**Country**	**Production**
USA	4,126,100	China	18,547,000	China	29,824,800
USSR	3,795,600	USA	5,017,000	USA	6,607,722
China	2,240,000	Japan	2,535,444	India	6,292,000
Japan	2,001,582	India	2,035,000	Indonesia	5,044,395
W. Germany	1,134,299	Russian Fed.	1,894,600	Brazil	3,260,859
France	853,000	Mexico	1,787,942	Mexico	3,015,959
UK	822,000	Brazil	1,509,464	Japan	2,632,882
Brazil	782,430	France	987,885	Russian Fed.	2,492,190
Spain	678,391	Germany	901,000	Turkey	1,236,754
Mexico	644,427	Turkey	844,287	France	984,638

Source: From FAO (2022a).
Note: Production expressed in metric tonnes.

(12,8% of the global output). While Brazil overtook Mexico by 244,900 tonnes, egg production in Europe remained slightly higher than in Latin America, with 12.8% and 12.0% of the global egg production, respectively.

2.2.3 Regional Evolution (2000–2020)

The regional distribution of egg production, presented in Table 2.3, showed that the production in Latin America (+103.5%), Asia-Oceania (+86.1%) and Africa (+80.9%) has grown between 2000 and 2020 at a faster rate than the global average. Europe recorded the lowest growth rate (17.5%) of all regions, a trend now observed for several decades. In 2020, Asia-Oceania accounted for 62.8% of the global production, followed by Europe (12.8%) slightly ahead of Latin America (12.0%).

2.2.3.1 Asia-Oceania

Consisting of 67 countries and territories, Asia-Oceania became the largest egg-producing region by overtaking Europe in 1988 and has accounted for more than half of the global egg production since 1994. In 2020, the Asia-Oceania region produced 54.44 million tonnes of eggs or 62.8% of the global production. Between 2000 and 2020, this region added 25.19 million tonnes of eggs, accounting for 70.9% of the global growth in egg production and representing a 3.15% CAGR. In 2020, the top 10 egg-producing countries in Asia-Oceania (Table 2.4) represented 90% of the production in the region, as well as 56.5% of the global egg production.

China: For many years, even after becoming the largest egg-producing country in 1985, the production of eggs in China has been dominated by small farms with less than 5,000 hens (Yang, 2011). The first larger commercial farms (>500,000 hens) appeared in the late 1990s-early 2000s and their development has greatly increased over the past 20 years. The contribution of smaller farms (<5,000 hens) decreased from 59.6% to 31% between 2008 and 2014 (Yang et al., 2018). New environmental regulations implemented in 2015 have further contributed to the closure of many small farms and the emergence of large egg producing companies. In 2017, 12 companies had more than one million hens and in 2020, three Chinese companies appeared in the list of global top 25 egg companies, implying a size over 10 million hens (WATT Poultry, 2021). In 2020, the average farm was estimated at 35,000 hens (IEC, 2021).

India: With 6.29 million tonnes produced in 2020, India is fast approaching the level of production in the USA. With

TABLE 2.3

Evolution of the Global Egg Production by Region (2000–2020)

Region	Production 2000	Production 2020	% Growth (2000–2020)	% Global Production 2020
Asia-Oceania	29,244,799	54,438,435	86.1%	62.8%
Europe	9,434,031	11,086,447	17.5%	12.8%
Latin America	5,102,338	10,383,666	103.5%	12.0%
North America	5,389,739	7,212,022	33.8%	8.3%
Africa	1,962,141	3,548,948	80.9%	4.1%

Source: From FAO (2022a).
Note: Production expressed in metric tonnes.

TABLE 2.4
Top 10 Egg-Producing Countries in Asia-Oceania (2020)

Country	Total Production	% Growth Since 2000
China	29,824,800	60.8%
India	6,292,000	209.2%
Indonesia	5,044,395	685.7%
Japan	2,632,882	3.8%
Turkey	1,236,754	46.5%
Pakistan	946,251	175.0%
Malaysia	806,877	106.6%
Iran	755,570	30.3%
Republic of Korea	734,528	53.4%
Thailand	713,242	38.6%

Source: From FAO (2022a).
Note: Production expressed in metric tonnes.

an average egg size slightly lower than that in the USA, the Indian egg sector estimated that they have already overtaken the USA in terms of the number of eggs produced. There are about 10,000 egg farmers in India for an average farm size of slightly less than 30,000 hens, but about 60% of the production is coming from farms with more than 100,000 hens. The production is dominated by a few states of the Indian Union, mostly located in the South and Western parts of the Subcontinent. The Indian government has recently recognized the importance of animal production with the creation of the Ministry for Fisheries, Animal Husbandry and Dairy. More visibility at the highest level of the nation is going to further boost the confidence of local producers to invest in the egg sector (Guyonnet, 2020b).

Indonesia: With a production of 5.04 million tonnes of eggs in 2020, Indonesia entered the list of top 10 egg-producing countries in 2002. Since 2000, egg production has increased by 685.7%, demonstrating the rapid development of the egg sector in Indonesia. However, it should be noted that in 2017, the Central Bureau of Statistics (BPS, 2021) reported an increase of 3.17 million tonnes of eggs over the previous year. This phenomenal growth in a 12-month period was likely the reflection of a revised estimation of the local production, with a greater contribution of small farmers, rather the placement of an additional 175 million hens into production.

Japan: Along with China, Japan (2.64 million tonnes in 2020) is the only Asian country that has been consistently in the list of top 10 egg-producing countries since the first publication of the FAO production data in 1961 (FAO, 2022). Since 2000, the Japanese layer flock has decreased by about 1% to 139 million laying hens and the countryside has lost about 2,500 farms or 53% of its farms. Yet, the Japanese egg sector has managed to continue to deliver growth in production. With about 1% less birds in production, Japan has increased its production by 3.6%, making production efficiencies one of the keys to its success. The consolidation of the sector is also evident when the top 10 companies represent about 40% of the Japanese production, one of the highest concentrations in the world (Guyonnet, 2020d).

Turkey: Entering the list of top 10 egg-producing country in 2003, the production of eggs in Turkey reached 1.24 million tonnes in 2020, an increase of 46.5% since 2000. Unlike most other countries in Asia-Oceania, a large part of this increase in production, especially since 2007, is linked to the growth in export opportunities.

2.2.3.2 Europe

While remaining the second largest region for egg production, European countries had to adapt since 2000 to a number of challenges including a growing number of Avian Influenza outbreaks, new regulations in terms of housing systems (Section 2.2.4) as well as a shift of the production of eggs towards Eastern European countries such as Ukraine and Poland. The production of eggs in Europe was previously reviewed by Magdelaine (2011). In 2020, Europe (39 countries as defined by the FAO classification) produced 11.08 million tonnes of eggs or 12.8% of the global production. Between 2000 and 2020, this region added 1.65 million tonnes of eggs, accounting for only 4.6% of the global growth in egg production and representing a 0.81% CAGR. In 2020, the top 10 egg-producing countries in Europe (Table 2.5) represented 78.7% of the production in the region.

EU27: Made up of 27 countries, the combined production of eggs in the European Union was 6.35 million tonnes in 2020, placing the EU27 just below India in terms of production. Between 2000 and 2020, the production in the EU27 countries increased by 0.31 million tonnes or at a 0.25% CAGR. Most of the growth in production came from Spain and Poland, while countries like Czechia (– 43.2%), Belgium (−40.6%) and Germany (−6.8%) saw some significant reduction in production. In fact, the production of eggs was reduced in 15 countries of the EU27 between 2000 and 2020. In 2020, France remained the only EU27 country as part of the list of the 10 top egg-producing countries (European Union, 2021a).

Russian Federation: With 2.49 million tonnes produced in 2020, the Russian Federation accounted for 22.5% of the European production and 36.2% of the growth in Europe between 2000 and 2020. In 20 years, the Russian Federation

TABLE 2.5
Top 10 Egg-Producing Countries in Europe (2020)

Country	Total Production	% Growth Since 2000
Russian Federation	2,492,190	31.5%
France	984,638	−0.3%
Ukraine	924,325	86.1%
Spain	912,532	38.8%
Germany	839,654	−6.8%
United Kingdom	774,000	36.1%
Italy	692,811	1.0%
Poland	642,354	51.6%
Netherlands	625,000	−6.7%
Romania	260,521	−0.9%

Source: From FAO (2022a).
Note: Production expressed in metric tonnes.

increased its production by 0.598 million tonnes or a 1.38% CAGR. The concentration of the sector has increased over the past 20 years and large farms (>500,000 hens) account for about 80% of the production while a large number of smaller farms and household production still account for the remaining 20%. About 45% of the production in the Russian Federation is concentrated in the central Russia, North-West and Volga regions, shipping eggs across this vast nation to supply major urban centres (Kostenko, 2021). With only 1.2% of the production of shell eggs exported, the increase in production has fuelled a huge rise in consumption (see Section 2.5).

Ukraine: Between 2000 and 2020, the production of eggs increased by 86.1% (3.15% CAGR) to reach 0.924 million tonnes in 2020. Large farms are responsible for 54% of the production while household production accounts for 46%, with little changes since 2015 (Poultry Union of Ukraine, 2020). The production in 2020 represented however a decline of 0.197 million tonnes from its peak production (1.12 million tonnes) in 2013. The growth of egg production in Ukraine has been build on both the increase of local consumption as well as the export of shell eggs (about 14% of the production exported as shell eggs) and egg products.

2.2.3.3 Latin America

In 2020, Latin America was the third largest egg-production region, with 10.38 million tonnes of eggs produced or 12.0% of the global output. Made up of 34 countries located in Central America, South America and the Caribbean, the production of eggs in the region has grown by 103.5% (3.61% CAGR) since 2000, with some significant growth in almost all countries. In 2020, the top 10 egg-producing countries in Latin America (Table 2.6) represented 93.1% of the production in the region.

Brazil: The production of eggs has more than doubled since 2000 to reach 3.36 million tonnes in 2020. With about 125 million hens, the production of eggs is divided into four distinct groups of producers: three main companies with over 10 million hens each, seven producers with 3–5 million hens, many producers with 0.5–1 million hens and many more with less than 0.5 million hens. The production of eggs is located in the South-Eastern part of the country, with slightly more than half of the production located in three states: São Paulo, Minas Gerais and Espiroto Santo. Unlike the broiler meat production, the production of eggs in Brazil is predominantly geared towards local consumption and export accounts for less 0.5% of the local production (ABPA, 2021).

TABLE 2.6
Top 10 Egg-Producing Countries in Latin America (2020)

Country	Total Production	% Growth Since 2000
Brazil	3,260,859	116.0%
Mexico	3,015,959	68.7%
Colombia	982,897	154.4%
Argentina	873,000	167.0%
Peru	497,525	206.6%
Guatemala	274,467	238.8%
Chile	228,044	107.8%
Dominican Republic	199,826	240.4%
Ecuador	168,480	133.5%
Venezuela	165,362	−5.3%

Source: From FAO (2022a).
Note: Production expressed in metric tonnes.

Mexico: With 167 million hens, Mexico is the sixth largest egg production country in the world and host the second largest egg-producing company in the world (Proteina Animal with 36 million hens). Between 2000 and 2020, production has increased by 68.7% to reach 3.01 million tonnes. The production of eggs is highly concentrated in three states: Jalisco (53.3%), Puebla (17.3%) and Somona (4.9%) and the majority of the eggs produced are still sold as unbranded products (CEDRSSA, 2019).

Colombia: Between 2000 and 20202, Colombia has experienced a 154.4% increase in the production of eggs, the third largest growth in Latin America (+0.596 million tonnes), to reach a total production of 0.982 million tonnes of eggs. The strong leadership from the local association of producers (FENAVI) has been one of the main drivers for the significant growth in production, based on the support provided to producers as well as the multi-pronged approach towards the promotion of egg consumption in Colombia (FENAVI, 2022).

Argentina: Ranking fourth in the region in terms of production with 47.5 million laying hens, the production of eggs in Argentina grew by 0.546 million tonnes between 2000 and 2020, or a 5.6% CAGR. With 938 farms and an average of 50,000 hens per farm, the production of eggs in Argentina is principally located in the States of Buenos Aires (40%), Entre Rios (28%) and Córdoba (8%) to service mostly the local consumption of eggs (CAPIA, 2022).

2.2.3.4 North America

USA: Since the first data reported by the FAO in 1961, the USA has always been among the top two largest egg-producing countries, overtaken only by China in 1985 for the number one position. With a production of 6.607 million tonnes in 2020, the US egg sector has grown by 31.7% since 2000, adding 1.59 million tonnes of eggs in spite of various challenges. In addition to a devastating outbreak of Avian Influenza in 2014–2015 (about 50 million hens euthanized), the producers of eggs in the USA have faced since 2008 a tremendous opposition to their housing practices for laying hens (see Section 2.4), leading to the commitment to a supply of eggs only produced in a cage-free environment by 2025. In 2020, about 72.1% of the eggs produced in the USA were consumed as shell eggs while 27.8% of the egg production was used for egg processing and the production of further processed products (UEP, 2022).

Canada: In 2020, the production of eggs reached 0.604 million tonnes of eggs, an increase of 62.3% since 2000. While Canada has only 27.9 million hens (far less than the largest US company, Cal-Maine with 44.3 million hens in 2020), this country has an unusual production system called supply management. Through a restriction on the number of birds placed and a tight control on importation (mainly from the USA), this system offers to producers a guaranteed price for eggs (Guyonnet, 2013).

2.2.3.5 *Africa*

In 2020, the African continent produced 3.548 million tonnes of eggs or about 4.1% of the global production. Made up of 53 countries, the production of eggs in the region has grown by 84.6% (3.11% CAGR) since 2000. While some countries (Mozambique 2827%, Senegal 298%, Togo and Benin 274%) recorded some significant growth, although often from a low base, the production of eggs also declined in nine countries during this 20-year period. In 2020, the top 10 egg-producing countries in Africa (Table 2.7) represented 80.7% of the production in the region. The production of eggs in Africa was reviewed by Tukur (2011), outlining the various challenges faced by the three main egg-producing sectors in African countries: intensive, semi-intensive and scavenging systems. Depending on the countries, the production of eggs by smallholder farmers, also called family poultry, accounts between 20 and 70% of the national production of eggs (Sonaiya, 2007).

Nigeria: With a production of 0.646 million tonnes in 2020, Nigeria ranked as the number one egg-producing country in Africa and the 23rd at the global level. Egg production in Nigeria, a blend of large farms and numerous small farms, has grown by 61.7% since 2000 but is often plagued by egg glut, the chronic problem caused by excess in egg production relative to the demand fluctuating widely by season. This disconnect between supply and demand created in 2020 a drop of 25% in prices and further impacted the economic sustainability of egg production. The production of eggs in Nigeria returned in 2020 to the production level in 2013 while its population increased during the same time by 34.4 million people, implying a lesser contribution of eggs to food security.

South Africa: While the egg sector in South Africa has experienced a number of outbreaks of Avian influenza recently, the production in South Africa has increased by 80.4% since 2000 and reached 0.593 million tonnes in 2020. Supported by the South African Poultry Association established in 1904, one of the oldest poultry associations in the world, the production of eggs is fragmented with 12 major players (representing more than 50% of the national production) and a number of small and medium-sized producers located throughout the country. Given the large number of small producers, about 45% of the total egg production is sold unbranded through informal market channels (DAFF, 2020).

TABLE 2.7

Top 10 Egg-Producing Countries in Africa (2020)

Country	Total Production	% Growth Since 2000
Nigeria	646,667	61.7%
South Africa	593,526	80.4%
Egypt	450,000	154.7%
Morocco	415,000	76.6%
Algeria	308,456	205.4%
Kenya	110,370	81.8%
Tunisia	103,800	26.6%
United Republic of Tanzania	89,967	4.7%
Republic of the Sudan	75,385	New country since 2011
Libya	72,015	20.0%

Source: From FAO (2022a).

Note: Production expressed in metric tonnes.

Egypt: While production in 2020 (0.450 million tonnes) was up by 154.7% since 2000, the production of eggs in Egypt fluctuates often from year to year depending on a number of factors and the production in 2020 was slightly less than in 2012. With a dependence on the importation of a wide range of feed raw materials, Egypt has seen during the 2000–2020 period nine times when the production of eggs was lower than that of a previous year. Egypt is home to the largest egg-producing company on the African continent, the Wadi group, with 10 million hens. As a means to achieve greater food security, the Egyptian government is offering land to promote the production of poultry and eggs (Darwish, 2021).

2.2.4 Housing Systems

Since their introduction in the 1950s, cages have been the dominant housing system for laying hens around the world. Although a few improvements were made to this system over the years, mainly the switch from the A-frame type of cages to the H-frame system with cages directly positioned on top of each other and manure belts located directly under each tier of cages, the biggest driver for changes in housing systems was the promulgation by the European Union Council of the directive 1999/74/EC (European Union, 1999) setting the minimum standards for the protection of laying hens. With a series of intermediate target dates for various requirements, this directive called for the European Member States to prohibit the use of conventional cage systems by January 1, 2012. These new cages, called enriched cages, furnished cages or colony housing, contain perches, a scratching area as well as a dedicated space for nesting, therefore allowing hens to express more natural behaviours. In order to facilitate the transition to these enriched cages, some equipment manufacturers developed systems called pre-enriched or enrich-able, allowing egg producers to convert these cages into fully enriched cages at a later date. Countries in the European Union adjusted at various speeds during this 12-year transition period, with a few countries still housing a large percentage of laying hens in conventional cages in 2011: Belgium (66.8%), Spain (33%), Cyprus (27.4%) and Poland (24%). It was estimated that about 83.6 million hens (23% of the total number of hens in the EU) were still housed in conventional cages at the beginning of 2012 (European Union, 2021b). While many countries simply replaced conventional cages by enriched cages during this transition period, egg producers in some countries saw this transition as an opportunity to invest in alternative housing systems. In Germany, the percentage of laying hens housed in conventional cages decreased from 73.2% in 2005 to 13.2% in 2012, while that of hens in barn systems increased at the same time from 14% to 64%. In the United Kingdom, egg producers opted to replace the cage system (63% in 2005 and 48.6% in 2012) by free-range systems (30% in 2005 and 47.8% in 2012). The directive 1999/74/EC totally changed the housing system landscape in the EU. While 92% of the laying hens were housed in cages in 1996, this percentage was reduced to

TABLE 2.8

Evolution of the Percentage of Hens Housed in Cage System in the Top 6 EU27 Egg-Producing Countries (2004–2020)

Country	2004	2010	2012	2020
France	79%	80%	70%	44%
Spain	99%	96%	94%	78%
Germany	73%	18%	13%	5%
Italy	96%	77%	57%	42%
Poland	>94%	92%	84%	81%
Netherlands	47%	44%	17%	12%

Source: From IEC (2021).

61.13% in 2012. Interestingly, the percentage of hens housed in cages in 2012 was typically higher in Southern European countries than in Northern European countries. By 2020, only 48% of the laying hens kept in EU27 countries were housed in cages, with some significant reduction in the top six egg-producing countries with the exception of Poland (Table 2.8).

More birds in the EU are kept in cage-free systems including barn/aviary system (33.9%), free-range (11.9%) and organic system (6.2%). For each housing system, there is a wide variation from country to country (Table 2.9).

The trend of housing laying hens in cage-free systems will likely continued in the EU27, as seen with the reduction of cage systems between 2018 and 2020 in France (from 61% to 44%) and in Spain (from 83% to 77.6%).

Concurrently, the USA and Canada, to a lesser extend, have experienced a series of challenges to the practice of housing laying hens in cages. In 2008, voters in California approved Proposition 2 by 63%, a measure prohibiting the confinement of laying hens (along with pregnant sows and calves raised for veal) in a manner that does not allow them to turn around freely, lie down, stand up, and fully extend their limbs (Bovay & Sumner, 2019). This initiative in California was quickly followed by a number of initiatives in various US states to ban the use of conventional cages. With the Humane Society of the United States (HSUS) as the driving force behind these initiatives, the United Egg Producers (UEP), the association representing the largest egg farms in the USA, managed in 2011 to secure an agreement with the HSUS related to the phasing out of conventional cages in favour of enriched cages, agreement which was later defeated. A different tactic was then used to pressure large egg buyers (retailers, restaurant chains and other food service companies and food processors) to switch to the sourcing of eggs produced in a cage-free environment. By 2017, more than 229 companies had made this commitment, representing about 72% of the total egg demand in the USA or the equivalent of the total number of eggs produced by about 223 million hens. With a target date of 2025, the US egg sector will need to invest about US$10 billion to switch to cage-free housing systems. In the Spring of 2022, the number of hens housed in cage-free systems had reached 111.1 million, or 33.7% of the total US flock (Agricultural Marketing Service, 2022).

TABLE 2.9

Distribution of Housing Systems Used in the European Union 27 Countries (2020)

Housing System	Combined EU27	Highest % (Country)	Lowest % (Country)
Enriched cages	48%	99.4% (Malta)	0% (Austria/ Luxembourg)
Barn/Aviary	33.9%	76.1% (Sweden)	0.6% (Malta)
Free range	11.9%	43.8% (Eire)	0% (Malta/ Luxembourg)
Organic	6.2%	31% (Denmark)	0% (Belgium/Malta)

Source: From European Union (2021a).

Outside of the European Union and North America, the reduction in the number of hens kept in cages has been also observed in New Zealand (from 86.5% to 56% between 2010 and 2020) and to a lesser extend in Australia (from 68% to 54.5%). Apart from the USA, all top 10 egg-producing countries are still using predominantly conventional cage systems (90% and above) with little changes over the past 10 years and few signs of imminent changes.

In 2020, it was estimated that about 4.8 billion laying hens were kept for the commercial production of eggs. The estimations of the prevalence of the different systems used to house these birds are presented in Table 2.10, showing the global estimates as well as results by region.

Globally, 84.2% of the laying hens are housed in cage systems (mainly conventional cages and only enriched cages in Europe), with the highest % in Latin America (91.5%) and the lowest in Africa (60.7%). The low level of use of cage system in Africa is explained mostly due to the higher cost of cage systems compared to floor systems. For instance, it is estimated that only 40% of the 45 million laying hens in Nigeria are housed in cage systems. The indoor cage-free systems (barn and aviary) housed 12.4% of the global flock of hens, with the highest percentages of use in Africa (36.2%), Europe (20.7%) and North America (18.0%). However, as mentioned previously, the switch to a cage-free supply of eggs in the USA by 2025, will change these percentages over the next few years when the prevalence of indoor cage-free systems should reach 17%. Free range and organic systems are used for about 3.4% of the 4.8 billion hens kept for production, with the highest percentage in Europe (14.6%) and the lowest in Latin America (0.9%) (European Union, 2021a; IEC, 2021; ILH, 2021; FAO, 2022a).

TABLE 2.10

Estimated Distribution of Housing Systems by Regions (2020)

Region	Number of Hens (in millions)	Cage Systems	Barn/Aviary Systems	Free Range/ Organic Systems
Asia-Oceania	3,005	89.2%	9.4%	1.4%
Europe	623.4	64.7%	20.7%	14.6%
Latin America	587.6	91.5%	7.6%	0.9%
North America	352.9	76.9%	18.0%	5.1%
Africa	200.5	60.7%	36.2%	3.1%
World	4,797	84.2%	12.4%	3.4%

Source: From European Union (2021a); IEC (2021); ILH (2021); FAO (2022a).

2.2.5 White versus Brown Eggs

While in some countries there is a choice between eggs from laying hens, quails, ducks, pigeons or even ostriches, consumers in many countries have only really the choice between brown and white chicken eggs when buying eggs at their local grocery store. And not even that if you reside in the United Kingdom (99% brown eggs) or in Mexico (98% white eggs). There is indeed a wide variation in the availability of brown eggs and white eggs between countries, with no real pattern within a region (Figure 2.1).

In Latin America, white eggs are predominant in Mexico (98%), Brazil (79%) and Uruguay (80%) while brown eggs are preferred in Paraguay, Bolivia, Venezuela (100%) and Colombia (96%). In Europe, differences between neighbouring countries are common. White eggs represent 67% of the market in Switzerland while they account for only 8% of the production in Austria. The North American market is dominated by white eggs (Canada, 87% and USA, 93%). Interestingly, the White Leghorn, so popular in North America, originated from Tuscany, a region of Italy where 80% of the consumers purchase brown eggs.

The case of brown eggs versus white eggs is complex as it relates to history, traditions and cultural preferences. The strains of birds first raised commercially in a region have created a consumer habit for shell colour and it is well recognized that consumer habits are extremely hard to change. For instance, the shell colour preference of consumers in Wisconsin, USA, has barely changed since a survey reported by Skala (1965). The colour of the shell has also been used as a way to control the origin of production and prevent illegal imports across borders. For example, in Central America, El Salvador, surrounded by Guatemala and Honduras, produces almost 100% of brown eggs, while the egg production of its neighbours is dominated by white eggs.

With brown eggs, there are also different tones of colour, and consumers in some countries will prefer an intense brown shell colour to a lighter one. The colour of the shell is therefore another criterium evaluated during the genetic selection process of new generations of hens.

There can also be differences in the shell colour preference within a given country. In China, 65% of the market is for brown eggs and 35% for white eggs while the exact opposite is observed with consumers in South Africa. In Argentina and the USA, consumers from specific regions will prefer one colour of shell while the rest of the country will overwhelmingly prefer the other one.

The colour of the shell seems also to create a certain perception in the mind of consumers in terms of healthiness. It is interesting to observe that in countries where consumers predominantly purchase white eggs, they also do view brown eggs more favourably in terms of nutrition. It may be due to the fact that organic and free-range eggs in these countries are brown eggs, for ease of tracking their authenticity and differentiation, and consumers still do believe that these types of eggs are more nutritious (Guyonnet, 2022b).

2.2.6 Egg Products

Although the processing of eggs started in the late 1800s, today the volume of shell eggs transformed still account for less than 10% of the total global egg production. If the egg processing sector initially developed with the production of egg powders as a way to easily ship eggs across oceans, we have seen over the past 40 years many changes in the egg processing sector. From an outlet for eggs in time of oversupply, the processing of eggs through investment and innovation offers now many options to add value to the egg sector. As a proof that egg processing is no longer a sector built on surplus eggs, in several countries, the entire production of farms is solely dedicated to egg processing, where the eggs are automatically conveyed from the layer houses to the processing plant (57% of inline production of processed eggs in the USA in 2020).

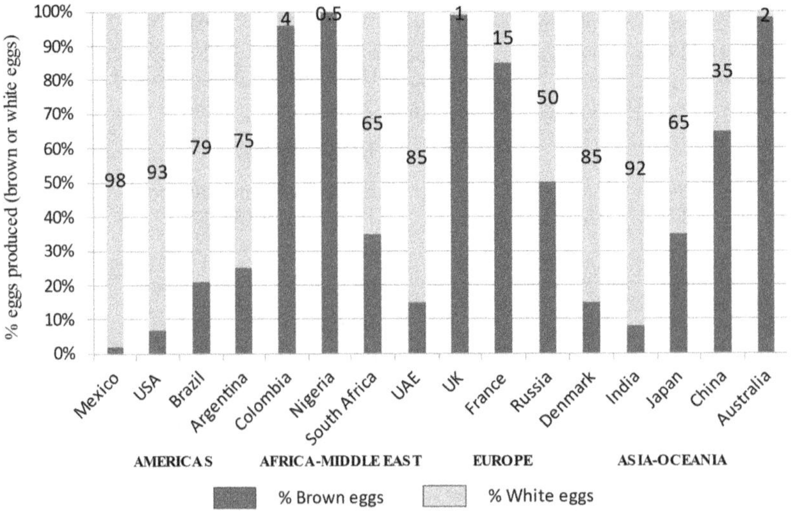

FIGURE 2.1 Percent of brown and white eggs produced in selected countries (2020).

2.2.6.1 *Global Production*

Unlike the production of shell eggs, the FAO does not maintain a global database on the production of egg products (liquid fresh or frozen, powders and further processed products). The best source of information for egg products is provided by the Egg Processors International (EPI), a division of the International Egg Commission, representing egg processing businesses from all over the world. While incomplete from a global standpoint, the data provided by 17 EPI member countries (IEC, 2021) shows that the egg processing sector has been expanding between 2000 and 2020. For instance, the production of liquid eggs in the USA, the largest liquid egg-producing country, increased by about 30% during this 20-year period. Between 2015 and 2019, the processing of liquid eggs increased by 12.6%, with the growth in key producing countries ranging from 5.7% in Japan to 38.9% in China (Table 2.11). However, the growth of the egg processing sector was severely affected by the COVID19 pandemic. Between 2019 and 2020, the global production of liquid eggs declined by about 4.3%, with key markets affected differently (–2.2% in Japan and –23.1% in Canada).

In 2020, the global production of egg products was estimated at 6 million tonnes of egg-equivalent: 3.5 million tonnes for liquid (fresh or frozen) products, 1.5 million tonnes for egg powders and 1 million tonnes for further processed products (e.g., hard-boiled eggs, omelettes). These estimates would indicate that in 2020, about 7% of the global egg production was being processed, slightly decreased during the first year of the COVID19 pandemic. While the percentage of shell eggs processed in Asia and Latin America is relatively low, more than 30% of shell eggs are processed in North America (USA, 33% and Canada, 26%), in the European Union and Japan.

2.2.6.2 *Categories of Egg Products*

For many years, the processing of eggs has grown along with the increasing demand from the food service sector (quick service restaurants) and food manufacturers. Combining convenience with food safety, liquid products (whole egg, egg white, yolk, salted or sugared yolk and any combination of yolk-egg white with additional ingredients like milk or gums), ranging in presentation size from 250 grams to 20,000 kilograms, have been adopted by many and represent today about 60% of the total egg usage by the food service and food manufacturing sectors.

Egg powders (Whole egg, egg white or yolk) present the advantage of a long shelf-life at room temperature and are used frequently by food manufacturers as ingredients in their formulations. Egg powders are also the preferred option in countries lacking a strong refrigerated supply chain.

More recently, egg processing companies have invested in research and development to bring innovative ideas to the market place, targeting both the food service and retail sectors. Hard-boiled eggs appeared as a convenient option for food service companies before reaching the retail shelves in the early 2000s. With hard cooked eggs, the egg processing sector was really able to enter the snack food segment, the rising star of the food sector. Consumers now find in grocery stores a broader range of egg-derived products like quiches, omelette wraps, egg-bite snacks and many other ways to eat eggs as a snack or as a "food on the go". Innovations in the egg processing sector are offering the consumers more opportunities to enjoy eggs at any time of the day (see Chapter 14). Egg products recently launched are targeting the breakfast segment where the consumption of eggs is more prevalent but hampered by the busy lifestyles of consumers. By combining convenience with healthy nutrition, both key buying factors for foods, especially with younger generations of consumers, the egg processing sector will continue to grow and offer new product options to consumers (Guyonnet, 2021).

2.3 Key Players in the Egg Sector

The growth of the egg sector over the years has been achieved with the assistance of key players, each providing some essential inputs into the production and commercialization of eggs. The following section will review different categories of stakeholders and provide a few examples of multinational companies within each category. There are many more companies, often having a national focus, that are supporting the egg sector but could not be presented in this section.

2.3.1 Genetics and Breeder Companies

The outstanding growth of the egg sector over the years was not obtained by simply having more birds producing eggs but truly by obtaining more eggs from each laying hen in

TABLE 2.11

Production of Liquid Eggs in Selected Countries (2015–2020)

Country	2015	2019	% Growth (2015–2019)	2020	% Growth (2019–2020)
USA	1,280,694	1,448,949	13.1%	1,312,154	–9.4%
Japan	469,589	496,300	5.7%	485,380	–2.2%
China	180,000	250,000	38.9%	230,000	–8.0%
Canada	128,907	174,217	35.1%	133,989	–23.1%
Total for 17 IEC member countries	2,496,130	2,811,633	12.6%	2,691,240	–4.3%

Source: From IEC (2021).
Note: Expressed in million tonnes of eggs.

production through an increase in the rate of lay (number of eggs per hen over a given period) and the extension of the egg production cycle (up to 100 weeks of age in a single cycle). Over the years, genetic companies have been able to select birds for optimum performance in terms of egg production, feed conversion, liveability, egg quality and many other traits, working concurrently with male and female lines. The work of genetic companies starts with a few flocks of pure genetic lines used to test various crossing options and obtain great-grand parent and grand-parent flocks. A number of companies, called breeders, manage parent flocks and in some cases also grand-parent flocks, and produced more than 4 billion day-old pullet chicks in 2020. This genetic pyramid makes it possible to obtain close to 100 million commercial laying hens from eight pure line birds. In some countries, companies have specialized in the raising of day-old pullet birds and sell birds at the point of lay (17–19 weeks of age) to egg producers. In other cases, egg-producing companies are fully integrated, from the grand-parent level to the production and commercialization of eggs, and will typically have their own hatcheries and feed mills. The improvements made through genetic selection, combined with a number of other positive factors, were clearly demonstrated by Pelletier et al. (2014) comparing some commercial production parameters in the USA between 1960 and 2010: the number of eggs per hen per year increased from 216 to 276 (+27%) and the feed conversion (kilograms of feed per kilograms of eggs) was reduced from 3.44 to 1.98 (–42%). In no particular order, the main global genetic companies are presented. Hendrix Genetics from the Netherlands, formerly established in 2005, provides strains of laying hens benefiting from over 60 years of experience. In addition to Hisex, a locally developed strain of laying hens, the company has acquired over the years different genetic lines like ISA from France, Shaver from Canada and the Bovans, Dekalb and Babcock lines from the USA, and provides breeds of laying hens benefitting from decades of genetic selection in both Europe and North America. The PHW group from Germany has been involved in the poultry sector since the early 1930s but became involved in the genetic selection of laying hens in the late 1950s through a licensing agreement with H&N Farms from the USA. The company is providing a wide range of breeds through two main entities, Lohmann Breeders GmbH and Hy-Line International. Established in Iowa, USA in 1936, Hy-Line International is considered the oldest modern layer genetic company and was acquired by the PHW group in the late 1970s, bringing together genetic lines addressing the needs of the US and European markets, the two largest centres of egg production at the time. Novogene, part of Groupe Grimaud from France already well known for duck genetics, started its work in layer genetics in the early 2000s with various strains of white-egg and brown-egg-producing hens. In 2022, Novogene, through its acquisition by the PHW group joined H&N, Lohmann Breeders GmbH and Hy-Line International. Although not widely present at the international level, Beijing Yukou Poultry Industry Co., founded in 1975 initially to supply more eggs to Beijing residents, provides various lines of birds to producers in China, the largest egg-producing country in the world, with an estimated 1.4 billion laying hens. These genetic companies offer various lines, meeting the needs of producers in terms of eggshell colour, production performance, ability to cope with a specific environment and body conformation. With the rise of many alternative housing systems, the selection of birds is now focusing also on the ability of laying hens to perform in a cage-free environment, from both a productivity and sociability standpoint.

2.3.2 Housing System Manufacturers

As reviewed in the section presenting the evolution of housing systems over the years, manufacturers have invested a lot of time and resources since the early 2000s to develop new systems meeting the new regulatory requirements or the anticipated expectations of consumers and customers. Since Europe and the USA were initially the main egg-producing markets, a number of global suppliers of housing equipment originate from Germany (Big Dutchman, Hellmann Poultry Equipment and SALMET GmbH & Co KG), Italy (FACCO Poultry equipment and Valli), Spain (Zucami), the Netherlands (Vencomatic Group) or the USA (AGCO, Alaso and Chore-Time). These manufacturers will typically offer a wide range of options for cage and cage-free systems, both for laying hens in production and for the raising of pullet birds. In addition to the housing systems, they provide also systems addressing feeding and drinking, ventilation, lighting, egg collection and manure removal and treatment, all these systems automatically managed and controlled. More recently, housing system manufacturers have integrated a number of sensors and connected Internet-of-Things devices into their offering to facilitate the collection of bird weight and environmental conditions inside the houses and the overall analysis of production data. With the rapid growth of egg production in China, a number of new manufacturers have entered the egg sector such as the Guangzhou Guangxing group, Henan Jinfeng Poultry Equipment Co. and Jiangsu Huali Intelligent Technology Co. Ltd. but at this time, these Chinese manufacturers mostly provide conventional housing systems, the dominant system in China and sell equipment across Asia and Africa.

2.3.3 Feed Ingredients and Feeds

After the birds and the housing systems, feeds and feed ingredients are certainly another key component of the egg value chain. There are a number of trading companies dealing with the sourcing of raw feed ingredients like corn, wheat and soybean. Many international companies also supply vitamins and minerals as well as synthetic amino acids, all ingredients important to provide balanced diets to the hens and therefore optimize egg production and the well-being of birds. While often considered as a commodity, calcium is also an extremely important component in the diet of laying hens, with the provision of the right combination of particle sizes essential for optimum eggshell quality and bird welfare. Feeds are manufactured in feed mills operated by large international corporations, local feed mills as well as feed mills integrated to egg-producing companies. Based on the Global Feed Survey published by Alltech (2021), the production of compounded feeds for the egg production sector in 2020 represented 160.48

million metric tonnes or about 13.6% of the global production of all animal compounded feeds. Between 2016 and 2020, the global tonnage of compounded feeds for layers increased by 17.18 million metric tonnes (+12%), in line with the growth of egg production, confirming that the vast majority of egg producers used compounded feeds.

2.3.4 Animal Health Companies

Along with the provision of balanced diets, the development of effective health programmes has greatly contributed to the growth of the egg sector. The development and adoption of vaccines against the most common diseases affecting laying hens have greatly contributed to the 57% reduction in annual mortality of laying hens between 1960 and 2010 (Pelletier et al., 2014). Vaccines are mainly administered during the pullet stage and even before hatching, using efficient techniques that allow their mass applications. It is common for birds entering production to have received more than 15 different vaccines during their first 15 weeks. In addition to the vaccines, a growing number of animal health companies also supply equipment that will facilitate the administration of vaccines, both at the hatchery and on farms. Antibiotics whether administered in the water, in the feed or through injection, are important to guarantee the health and well-being of birds but should always been used according to local regulation to avoid the presence of residues in eggs and in a prudent manner to avoid the development of drug resistance. Animal health companies supply also vitamins and minerals, often in various presentation forms. There is a large number of companies, both international (e.g., Boehringer Ingelheim, Ceva, DSM, Elanco, Globion India, Hester Biosciences, Phibro Animal Heath, Zoetis), and local, supplying many nutritional and heath products to the egg sector.

2.3.5 Equipment for the Grading of Shell Eggs

The marketing of branded eggs requires the ability to obtain clean eggs, sorted by size and packaged in visually appealing egg cartons. Egg grading equipment are designed to assist the different steps required for the packing of shell eggs. The selection of clean eggs is handled differently around the world (see Chapter 18) since some countries do not allow the washing of eggs (mainly the European Union countries) while others have made the washing of eggs a mandatory requirement (USA, Canada) or an optional step (rest of the world). The inspection of eggs for external and internal defects is performed through candling, the process of shining a bright light through the eggs. The candling of eggs by human operators is now replaced by the use of automated systems to detect micro-cracks at the surface of shell eggs and discard eggs with a range of external and internal defects, allowing to grade eggs at speeds up to 70 eggs per second. Grading machines include also a set of scales, allowing eggs to be sorted by weight, before being automatically placed in the appropriate egg cartons. With the advent of robotics, the entire grading process requires little human intervention, increasing the processing efficiencies (around 250,000 eggs per hour) and reducing the risks of human errors. Globally recognized companies like MOBA (The Netherlands), Nabel (Japan) and SANOVO Technology Group (Denmark) offer a number of options to egg producers and egg graders. With the growth of egg branding in China, a number of Chinese equipment manufacturers have also entered this segment of the egg value chain.

2.3.6 Equipment for Processed Eggs

The processing of shell eggs constitutes one major opportunity for the global egg sector since it is fairly underdeveloped in key egg-producing countries like China, India and Indonesia. The processing of eggs to produce products like egg powders, fresh and frozen liquid eggs and hard cooked eggs require a wide range of pieces of equipment (see Chapter 10), readily available from manufacturers dedicated to the egg sector or manufactured to supply other food processing sectors like the dairy and beverage industries. New type of egg products like omelets and other ready-to-eat or ready-to-drinks products will often require more custom-made types of equipment. Globally recognized companies like Actini (France), MOBA (The Netherlands), Ovobel (Belgium) and SANOVO Technology Group (Denmark) offer various pieces of equipment as well as innovative and customized turnkey solutions.

2.3.7 Packaging Materials

Suppliers of packaging materials played a role in both the egg packing and egg processing sides of the business. The packaging materials for table eggs is rather standard with eggs, ranging in numbers from 4 to 60, packed in egg cartons or trays. Due to their ease of recycling, packaging materials made of paper pulp are rapidly replacing plastic and Styrofoam packaging materials in many countries. Packaging materials for processed products vary from small 250 millilitre containers to large 1,000 kilograms bulk presentations, all designed and manufactured for food contact use. The quality and sterility of the packaging materials used for processed eggs is essential to prevent the risks of contamination of products throughout their shelf-life. Many global suppliers of packaging materials are now based in China.

2.3.8 Chemicals for Cleaning and Sanitation

Suppliers of chemical products play also an important role within the egg value chain. On farms the use of chemical products is an essential component of a good biosecurity programme, applied after a thorough dry-cleaning and removal of all organic materials. For shell eggs, these chemicals may be used during the washing of the eggs (when allowed) or for the cleaning and sanitation of equipment used in egg packing stations. Chemical products in egg processing plants are used in a similar manner to either wash the eggs or clean and sanitize the different pieces of equipment. Companies supplying chemical products will typically assist egg companies to validate their cleaning procedures, both for cleaning-in-place (CIP) and cleaning-out-of-place (COP) situations. A number of chemicals must also be food-grade when in direct or potential contact with products.

2.3.9 National and International Organizations

At the national and international levels, a number of organizations are active within the egg value chain and contribute to the sustainable growth of the egg sector (Table 2.12). These organizations will play various roles: technical support to egg farmers to boost production (P); generic marketing campaigns to increase the consumption of eggs (M); educational programmes for health professionals and consumers to raise the awareness of the value of eggs for human nutrition (N); and educational programmes for field managers and veterinarians to enhance their knowledge and adoption of the latest practices in terms of management, nutrition and health (E).

At the international level, the International Egg Commission (IEC) is the trade organization representing the egg sector since 1964. With members in over 80 countries, the IEC is bringing the voice of egg producers to intergovernmental organizations like the Food and Agriculture Organization of the United Nations (FAO), the World Animal Health Organisation (OIE), the World Bank and the CODEX Alimentarius. The IEC is also an excellent source of statistics about the sector, updated annually on its website. In a number of countries (e.g., Colombia, Mexico, Nigeria, or South Africa), these organizations address the needs of both broiler and egg producers, while in other countries (e.g., Australia, Canada, China, or France), their focus is dedicated to eggs. In the USA, most egg producers belong to the United Egg Producers (UEP) organization while the marketing of eggs is done by the American Egg Board, a branch of the United States Department of Agriculture.

2.4 Marketing Trends

In parallel with the growth in production, the egg sector in many countries was able to grow egg consumption during the past 20 years by adopting new marketing strategies to appeal to younger generations of consumers. The growth in consumption has been achieved both by promoting the value of egg nutrition as well as by creating new market segments and meal opportunities for eggs and egg products.

2.4.1 International Trade

Between 2000 and 2020, the export of shell eggs increased from 0.9 million tonnes to 2.08 million, or an increase of 120.3% (FAO, 2022b). However, the trading of shell eggs represents a very small portion of the global egg production, 1.9% and 2.4% in 2000 and 2020, respectively. At any time, the export of shell eggs has typically represented less than 3% of the global production. During the first year of the COVID19 pandemic (2019–2020), the international trade of eggs experienced a reduction of 8.8%, mainly due to the difficulties encountered by logistic companies at the beginning of the pandemic. In 2020, the main trading zones for shell eggs were within Europe (60.7% of the global trade) and Asia (29.9%). In Europe, the majority of the trade was done between EU27 countries, mainly the Netherlands, Poland, Spain and Germany while the key exporting countries in Asia were Turkey, Uzbekistan, Malaysia and China. While the tonnage of eggs exported has increased during the past 20 years in both Europe and Asia, the trading of eggs within Asia is accelerating but surprisingly, this continent remained a net importer of eggs in 2020 (FAO, 2022b). In a few countries, the percentage of their 2020 national production dedicated to the export market was rather significant: Poland (29.2%), Uzbekistan (28.6%), the Netherlands (24.7%) and Turkey (17.4%), implying a production strategy truly geared towards building an export business. In the future, other countries like Brazil may look at the international trade as an outlet for increased production or local oversupply issues.

2.4.2 Consumer Buying Factors

The attitudes, perceptions and behaviours of consumers towards egg and egg products were previously reviewed (Hansstein, 2011; Guyonnet, 2012; Bertechini, 2017; Rondoni et al., 2020). Price, taste and convenience have typically been the key buying factors for foods and eggs and egg products have always fared rather well for these three key attributes. Whether available as branded products or as a bulk food sold by weight or by unit, eggs represent one of the cheapest sources of quality animal-sourced proteins. At the global level, less than 50% of the eggs are sold as branded products, in nice packaging materials. Since younger generations of consumers do not consider that price is their top buying factor, this will offer an opportunity to switch more consumers from bulk to branded products in large markets like China, Indonesia or China (Guyonnet, 2020c). Taste, as a buying factor, is quite interesting. Eggs are globally recognized for their great taste and for their versatility, that is, their ability to be used for many reasons in many dishes. Yet, the eggs have often a very limited use in our diet. For instance, in the UK, a recent survey showed that 73% of the eggs consumed are either fried, boiled or poached, 3 classic ways to eat eggs for breakfast (Guyonnet, 2012). Convenience is probably best expressed by the rise of hard-boiled eggs ready to be eaten by consumers at any time of the day. Hard-boiled eggs have entered the snack food market and it is a growing segment in many countries (Guyonnet, 2019a).

More recently, with the rise of younger generations of consumers, new buying factors have also emerged, namely food authenticity and food sustainability (Guyonnet, 2020c). Food authenticity relates to the composition of foods, the length of the ingredient list in foods and as well as the healthiness of foods. Since overcoming the myth about the negative health effects of egg consumption due cholesterol (see Chapter 30), eggs are widely recognized as a nutritious and healthy food, produced naturally with no ingredient listed on the packaging. The last key buying factor, sustainability, is extremely important with younger generations of consumers who care more about farming practices, animal welfare, recycling practices and food waste (Guyonnet, 2020c). Depending on the countries, other buying factors such as freshness, remaining shelf-life and food safety considerations are part of the elements taken into account by consumers at the time of purchase (Guyonnet, 2012; Ochs et al., 2018; Rondoni et al., 2020).

TABLE 2.12

Selected International and National Organizations

Country/Region	Name of Organization	Acronym	Main Objectives	Website
International				
	International Egg Commission	IEC	P, N	http://www.internationalegg.com
	Asociación Latino Americana de Avicultural	ALA	P, N	https://www.avicolatina.com/
	Instituto Latin-American del huevo	ILH	N	http://www.ilhala.org/
	World Poultry Science Association	WPSA	E	http://www.wpsa.com/
	World Veterinary Education in Production Animal Health	WVEPAH	E	https://www.wvepah.org
North America				
Canada	Egg Farmers of Canada	EFC	P, M, N	https://www.eggfarmers.ca/
USA	American Egg Board	AEB	M, N	https://www.incredibleegg.org/
	United Egg Producers	UEP	P	https://unitedegg.com/
Latin America				
Argentina	Cámara Argentina de Productores Avícolas	CAPIA	P, M, N	https://www.capia.com.ar/
Brazil	Instituto Ovos Brasil		M, N	https://www.ovosbrasil.com.br/
Chile	Chile Huevos		P, M, N	https://www.chilehuevos.cl/
Colombia	Federación nacional de avicultores de Colombia	FENAVI	P, M, N	https://fenavi.org/
Guatemala	Asociación nacional de avicultores	ANAVI	P	http://www.anaviguatemala.org/
Mexico	Unión nacional de avicultores	UNA	P	https://una.org.mx/
Uruguay	Asociación de Productores Avícolas Sur	APAS	P	http://www.apas.com.uy/
Europe				
France	Comité National pour la Promotion de l'Œuf	CNPO	P, N	https://oeuf-info.fr/
Germany	Zentralverband der Deutschen Geflügelwirtschaft e.V.	ZDG	P	https://zdg-online.de/
Spain	Organización Interprofesional del Huevo y Sus Productos	INPROVO	P, M, N	https://www.inprovo.com/que-es-inprovo/
UK	British Egg Industry Council	BEIC	M, N	https://britisheggindustrycouncil.com/
Africa/Middle East				
Nigeria	Poultry Association of Nigeria	PAN	P	https://panog.org.ng/
Rwanda	Rwanda Poultry Industry Association	RPIA	P, E	https://www.rpiass.rw/
South Africa	South Africa Poultry Association	SAPA	P, N, E	http://www.sapoultry.co.za/
Zambia	Poultry Association of Zambia	PAZ	P	http://www.paoz.org/
Asia/Oceania				
Australia	Australian Eggs		M, N	https://www.australianeggs.org.au/
China	Wonderful Egg Alliance		M	http://www.wonderfulegg.com/
India	National Egg Coordination Committee	NECC	P, N	https://www.e2necc.com/
Japan	Japanese Poultry Association		P, N	https://www.jpa.or.jp/
Korea	Korean Poultry Association		P	https://www.poultry.or.kr/
Malaysia	Federation of livestock farmers associations of Malaysia	FLFAM	P	http://www.flfam.org.my/

Note: P, production; M, marketing; N, nutrition; E, education.

2.4.3 Market Growth–Generic Promotion of Eggs

2.4.3.1 Promoting Health and Nutrition

For many years, the cholesterol content in eggs was considered a risk factor for coronary heart disease (CHD) and a number of countries did recommend a limitation in the daily dietary intake of cholesterol (Griffin, 2011). Through the efforts of national organizations of producers, recruiting leading scientists in the field, new data have clearly demonstrated the flaws of the association of CHD and eggs (Kanter et al., 2012). In 2015, the USDA through the Report of the 2015 Dietary Guidelines Advisory Committee removed the limitation to a 300-mg daily intake of cholesterol since the "available evidence showed no appreciable relationship between consumption of dietary cholesterol and serum cholesterol" (USDA, 2015). While the USA were one of the last countries to remove these restrictions about egg consumption, most national organizations of egg producers had partnered with health professionals and especially dieticians to convey to the medical profession and consumers the health benefits of egg nutrition (Guyonnet & Jing, 2016). With the consequences of the COVID19 pandemic and other crises affecting the price of healthy foods, the promotion of eggs as an affordable source of valuable nutrients is becoming more and more relevant (Walker & Baum, 2022).

2.4.3.2 Promoting Convenience

Convenience has always been one of the key values of shell eggs appreciated by consumers and will continue to be an important characteristic to attract newer generations of consumers (Guyonnet, 2012; Guyonnet, 2020c). Convenience is also a key attribute of hard-boiled eggs and has been a key factor in their growing market share within the booming snack food segment (Guyonnet, 2019a). In addition, recognizing the potential to increase the consumption of eggs during weekdays, companies are now investing in the research and development of new egg products, simple and quick to prepare at home, to accommodate the busy lifestyles of younger generations of consumers (Guyonnet, 2021).

2.4.4 Market Differentiation – Branding of Eggs and Egg Products

2.4.4.1 Regular Eggs

Generic campaigns about eggs and nutrition have been at the inception of the successful growth of egg consumption in many countries. These campaigns are often led by local or national organizations of egg producers and have yield excellent results over the years (Guyonnet, 2018a; Guyonnet, 2020a). For instance, between 2010 and 2020, the consumption of eggs in Brazil by 119 eggs and by 111 eggs in Colombia (IEC, 2021). The significant gains in terms of egg consumption made during the first year of the COVID19 pandemic (globally +5 eggs but + 33 eggs in Colombia, +21 eggs in Brazil and + 20 eggs in Argentina) will need to be sustained through accrued promotion of egg consumption as consumers tend to switch to more expensive animal protein sources during better financial times (Guyonnet, 2022a).

2.4.4.2 Nutritionally Enhanced Eggs

The ease and repeatability of producing nutritionally enhanced eggs have been one of the key factors for the success of eggs in the functional food market segment. Since their first production in the early 1990s, the health benefits of functional eggs are widely promoted, showing some benefits for the entire family: pregnant and lactating mothers, young infants, adults and senior citizens. In many countries, egg companies are promoting functional eggs enriched with polyunsaturated omega-3 fatty acids, selenium, vitamin D, A or E, lutein, iodine or selenium, stressing the additional benefits derived from the consumption of nutritionally enhanced eggs (Guyonnet, 2015).

2.4.4.3 "Welfare" Eggs/Housing Systems

The growing interest of consumers, especially younger generations of consumers, for a more sustainable food supply, including shell eggs, has been demonstrated in many countries (Sánchez-Bravo et al., 2020). This interest of consumers has convinced egg producers to adopt a wide range of production systems claiming better welfare and quality of life for the laying hens. A number of certificating bodies are auditing farms to ensure that egg production meets these different private standards. The analysis of social media communications gives a greater insight into the perceptions of consumers about welfare and in the future should guide the messaging required to attract more consumers to this market segment (Widmar et al., 2020). At this time, it is evident that there remains some confusion in the mind of consumers (Ochs et al., 2018; Sass et al., 2018; Ochs et al., 2019, Widmar et al., 2020) and that there are some wide variations between countries in the willingness of consumers to pay for higher welfare (Lagerkvist & Hess, 2011; Sánchez-Bravo et al., 2021).

2.4.4.4 Other Elements of Differentiation

The COVID19 pandemic has reinforced the desire of consumers to consume local, fresh foods and favoured short distribution channels. Although the egg value chain has always been a typical example of a local, short value chain, it is essential to continue to brand eggs as one of the most local and fresh animal-source proteins, requiring little to no manual handling from the farms to the point of purchase (Guyonnet, 2022a). New egg categories can also appeal to some specific perceptions of consumers. Although hens are omnivorous animals, some people believe that birds should not be fed any animal by-product such as feather meal, bone meal or fish meal. With the growing trend in Western countries of people switching to vegetarian diets, free of any animal protein, some egg companies are now selling "vegetarian eggs", meaning that the eggs are produced by hens fed without the use of any animal by-products including certain vitamins such as vitamin D3 that is often derived from sheep wool. Some claims such as free of hormones or steroids still persist on too many egg cartons, although they are deceptive and contribute to maintain a certain level of confusion in the mind of consumers (Heng et al., 2016).

2.5 Egg Consumption

2.5.1 Global Egg Consumption

The annual consumption of eggs is calculated by adding the number of eggs consumed as shell eggs as well in the form of various egg products, consumed either directly or as an ingredient of processed foods and prepared meals. In 2020, the annual per capita consumption of eggs was estimated at 185 eggs and recorded one of its highest annual growths (+ five eggs) during the first year of the pandemic. Since 2000, in spite of the global population growing by 26.9% and reaching 7.795 billion in 2020, the global sector has been able to make available an additional 46 eggs per person (+29.9%). However, this outstanding growth of the average egg consumption hides some growing differences in egg consumption across the world (Figure 2.2) (FAO, 2022a).

2.5.2 Regional Differences

In 2020, the consumption of eggs in the USA (287 eggs) and Canada (255 eggs) were obtained from shell eggs (73%) and a significant amount of egg products (27%). Overall, the average egg consumption in North America was 284 eggs, the highest average consumption for any region in the world. Between 2010 and 2020, the growth in egg consumption of eggs was higher in Canada (+30.1%) than in the USA (+16.2%), with the consumption of egg products growing faster than that of shell eggs (Table 2.13).

Led by Mexico with an average annual consumption of 380 eggs per person, Latin America recorded in 2020 an average consumption of 265 eggs per person, up by eight eggs during the first year of the COVID19 pandemic. While egg consumption increased only by 4.1% in Mexico between 2010 and 2020, other countries in the region registered increases ranging from 27.6% in Argentina to 90.2% in Brazil, a testimony of the phenomenal growth of egg production in Latin America.

The annual consumption of eggs in Europe, combining shell eggs and egg products, was estimated in 2020 at 230 eggs, ranging from 199 eggs in the UK to 279 eggs in Russia. In most countries, consumption increased moderately between

TABLE 2.13

Egg Consumption in 2010 and 2020 for Selected Countries

Country	2010	2020	Increase	% Increase
North America				
Canada	196	255	59	30.1%
USA	247	287	40	16.2%
Latin America				
Argentina	239	305	66	27.6%
Brazil	132	251	119	90.2%
Colombia	214	325	111	51.9%
Mexico	365	380	15	4.1%
Peru	154	244	90	58.4%
Europe				
France	210	220	10	4.8%
Germany	214	239	25	11.7%
Ireland	165	209	44	26.7%
Italy	210	216	6	2.9%
Russia	260	279	19	7.3%
Spain	214	224	10	4.7%
Switzerland	189	189	0	0%
UK	182	199	17	9.3%
Africa				
Algeria	122	141	19	15.6%
Tunisia	147	157	10	6.8%
South Africa	137	159	22	16.1%
Asia – Oceania				
Australia	198	249	51	25.8%
India	57	78	21	36.8%
Iran	164	226	62	37.8%
Japan	324	340	16	4.9%
Korea	236	281	45	19.1%
New Zealand	230	251	21	9.1%
Turkey	171	194	23	13.5%

Source: From IEC (2021) and FAO (2022a).

Note: Expressed as the total number of eggs consumed per person per year.

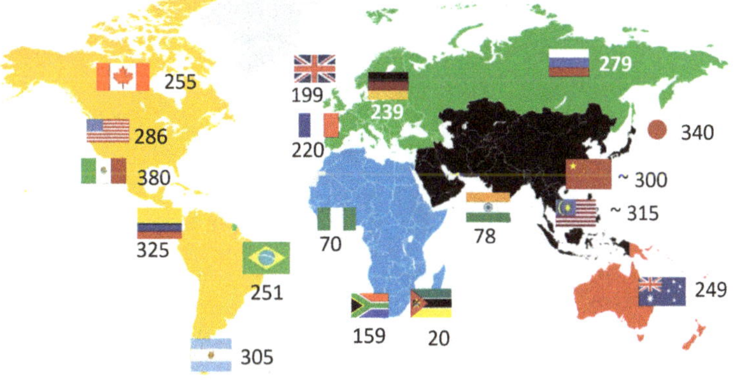

FIGURE 2.2 Egg consumption in selected countries (2020), expressed as the total number of eggs consumed annually per person.

2010 and 2020, with Ireland (26.7%) recording one of the highest growths in the region.

Home to an estimated 4.7 billion people, Asia-Oceania registered in 2020 an average consumption of 194 eggs per person, up by 7 eggs during the first year of the COVID19 pandemic. The consumption of eggs varies widely in the region, from low double digit numbers (South East and Central Asia) to 340 eggs in Japan. Although still low with 78 eggs, India experienced a 36.8% growth in consumption between 2010 and 2020.

While the African continent accounts for 17.2% of the global population, egg production in Africa represents only 4.1% of the global production in 2020, translating into an average annual consumption of 44 eggs. Consumption of eggs is typically well over 140 eggs in countries located in North African countries and in South Africa while eating eggs is often a rare event on the rest of the African continent, with 41 African countries (out of 53 countries) having an average consumption of less than one egg per week.

2.5.3 Factors Limiting Egg Consumption

Limitations of egg consumption are basically linked to two main factors: availability and utilization. In many countries, the egg value chain is well structured, with an efficient and reliable supply of birds, feeds and other inputs, and excellent production results. For them, the main challenge is to attract and retain consumers. Marketing and promotion will play a key role, especially in countries where the rising level of disposable income of consumers will allow them to afford many options in terms of animal-sourced proteins. The growth of egg consumption in Brazil and Colombia is the testimony of the importance of investing in the promotion of egg nutrition. The significant investment in marketing and promotion in the USA and Canada has also yielded some of the highest growth rates among rich nations (Guyonnet, 2019b; Guyonnet, 2020a).

In other countries, mainly in Africa, Central Asia and the Pacific Islands, the production of eggs is constraint by a lack of reliable supply of birds, the fluctuating quality of feeds and high mortality rates due to poor housing systems, lack of strict biosecurity and questionable management practices (Akinola & Essien, 2011; Diarra, 2017). With most of the production managed by smallholder farmers, the economies of scale of intensive production are often absent and the availability and the higher cost of most inputs limit significantly egg production (Guèye, 2002; Tukur, 2011). In addition, smallholders will often not invest sufficient time and funds in promoting egg consumption and rebuking some persisting taboos about eggs (Guyonnet, 2018b).

2.6 Conclusions

The successful growth of egg production over the past 20 years has been the results of investments by many stakeholders within the egg value chain. While individual egg producers have invested to build more production capacity and adopt newer housing systems and management practices, egg companies have also allocated vast resources to develop innovative shell eggs and egg products to give consumers more meal opportunities with eggs. Investment in consumer education and promotion has been the third pillar of the growth of the egg sector. By continuing with this approach, the global egg sector will continue to experience some significant growth during the next 20 years.

REFERENCES

ABPA (2021). Relatório anual 2021. Associação Brasileira de Proteína animal. https://abpa-br.org/wp-content/uploads/2021/04/ABPA_Relatorio_Anual_2021_web.pdf

Agricultural Marketing Service (2022). Monthly USDA cage-free shell egg report. https://www.ams.usda.gov/mnreports/pymcagefree.pdf

Alltech (2021). 2021 Global feed survey. Retrieved December 2, 2021, from https://www.alltech.com/sites/default/files/2021-11/2021_alltech_global_feed_survey-ENG_0.pdf

Akinola, I. A. F., & Essien, A. (2011). Relevance of rural poultry production in developing countries with special reference to Africa. *World's Poultry Science Journal*, *67*(4), 697–705. https://doi.org/10.1017/S0043933911000778

Bertechini, A. G. (2017). Economic and cultural aspects of the table egg as an edible commodity. In P. Y. Hester (Ed.), *Egg innovations and strategies for improvement* (pp. 223–232). Academic Press. http://dx.doi.org/10.1016/B978-0-12-800879-9.00021-4

Bovay, J., & Sumner, D. A. (2019). Animal welfare, ideology, and political labels: Evidence from California's proposition 2 and Massachusetts's question 3. *Journal of Agricultural and Resource Economics*, *44*(2), 246–266. https://10.22004/ag.econ.287970

BPS (2021, December 20). Egg production of layers by province (Tonnes), 2015–2017. https://www.bps.go.id/indicator/24/491/2/produksi-telur-ayam-petelur-menurut-provinsi.html

CAPIA (2022). Avicultura de postura de la Argentina. https://www.capia.com.ar/

CEDRSSA (2019). Reporte – La importancia de la industria avícola en México. http://www.cedrssa.gob.mx/files/b/13/47Industria_Avicola_M%C3%A9xico.pdf

DAFF (2020). A profile of the South African egg market value chain. Department of Agriculture, Forestry & Fisheries. https://www.dalrrd.gov.za/doaDev/sideMenu/Marketing/Annual%20Publications/Egg%20Market%20Value%20Chain%20Profile%202020.pdf

Darwish, N. (2021). Poultry production in Egypt. Poultry middle-East & North Africa yearbook of international suppliers pp. 270–265. http://www.meap.net/news/index.php/en/read-the-magazines/readpoultry

Diarra, S. S. (2017). Poultry industries in the South Pacific region: Issues and future direction. *World's Poultry Science Journal*, *73*(2), 293–300. https://doi.org/10.1017/S0043933916001070

European Union (1999). Council directive 1999/74/EC of 19 July 1999 laying down minimum standards for the protection of laying hens. https://eur-lex.europa.eu/legal-content/EN/TXT/PDF/?uri=CELEX:31999L0074&from=EN

European Union (2021a). Eggs: Dashboard 3 November 2021. https://ec.europa.eu/info/sites/default/files/food-farming-fisheries/farming/documents/eggs-dashboard_en.pdf

European Union (2021b). Laying hens by way of keeping. https://ec.europa.eu/info/food-farming-fisheries/animals-and-animal-products/animal-products/eggs_en

FAO (2022a). FAOSTAT crops and livestock products. https://www.fao.org/faostat/en/#data/QCL

FAO (2022b). FAOSTAT trade – Crops and livestock products. https://www.fao.org/faostat/en/#data/TCL

FENAVI (2022). ¿Qué es Programa Huevo? https://fenavi.org/programa-huevo/que-es-programa-huevo/

Griffin, B. A. (2011). Eggs, dietary cholesterol and disease: Facts and folklore. In Y. Nys, M. Bain, & F. Van Immerseel (Eds.), *Improving the safety and quality of eggs and egg products. volume 2: Egg safety and nutritional quality chemistry* (pp. 237–253). Woodhead Publishing.

Guèye, E. F. (2002). Employment and income generation through family poultry in low-income food-deficit countries. *World's Poultry Science Journal*, *58*(4), 541–557. https://doi.org/10.1079/WPS20020039

Guyonnet, V. (2012). Eggs and egg products: Consumers' attitudes, perceptions and behaviours. World's Poultry Science Congress. August 5–9, http://www.facta.org.br/wpc2012-cd/pdfs/plenary/Vincent_Guyonnet.pdf

Guyonnet, V. (2013). Egg production in Canada. Danish Poultry Congress. Proceedings. https://lf.dk/~/media/lf/for-medlemmer/sektioner-og-udvalg/lf_fjerkrae/kongres-2013/g-vincent-guyonnet-egg-production-in-canada.pdf?la=da

Guyonnet, V. (2015). Chapter 9. Strengths and weaknesses of functional eggs for human health. In R. R. Watson and F. De Meester (Eds.), *Handbook of eggs in human function* (pp. 153–165). Wageningen Academic Publishers. https://doi.org/10.3920/978-90-8686-804-9

Guyonnet, V. (2018a). How egg marketing campaigns increase consumption. *Poultry International*, *57*(6), 10–11. https://www.wattagnet.com/articles/34529-how-egg-marketing-campaigns-increase-consumption

Guyonnet, V. (2018b). How to increase African egg production, consumption. *Poultry International*, *57*(11), 14–15. https://www.wattagnet.com/articles/35352-how-to-increase-african-egg-production-consumption

Guyonnet, V. (2019a). Positioning hard-boiled eggs in the snacking market. *Poultry International*, *58*(6), 30–31. https://www.wattagnet.com/articles/37290-positioning-hard-boiled-eggs-in-the-snacking-market

Guyonnet, V. (2019b). Egg sector must play greater role in consumer education. *Poultry International*, *58*(12), 20–21. https://www.wattagnet.com/articles/38964-egg-sector-must-play-greater-role-in-consumer-education

Guyonnet, V. (2020a). Egg associations key for sector's sustainable growth. *Poultry International*, *59*(1), 30–31. https://www.wattagnet.com/articles/39096-egg-associations-key-for-sectors-sustainable-growth

Guyonnet, V. (2020b). Indian egg production, consumption to keep rising. *Poultry International*, *59*(3), 14–15. https://www.wattagnet.com/egg/egg-processing/article/15529976/india-the-rise-of-a-new-egg-giant-wattagnet

Guyonnet, V. (2020c). How the egg industry can grow its appeal to Gen Z. *Poultry International*, *59*(7), 30–33. https://www.wattagnet.com/articles/40211-how-the-egg-industry-can-grow-its-appeal-to-gen-z

Guyonnet, V. (2020d). Japan consuming more eggs, despite population stagnation. *Poultry International*, *59*(11), 20–21. https://www.wattagnet.com/articles/41113-japan-consuming-more-eggs-despite-population-stagnation

Guyonnet, V. (2021). Value through processing – As simple as boiling an egg! *Poultry International*, *60*(1), 18–19. https://www.wattagnet.com/articles/41640-value-through-processing—as-simple-as-boiling-an-egg

Guyonnet, V. (2022a). The COVID-19 pandemic or a tale of two egg sectors. *Poultry International*, *61*(1), 34–35. https://www.wattagnet.com/articles/43892-the-covid19-pandemic-or-a-tale-of-two-egg-sectors

Guyonnet, V. (2022b). Does egg shell color really matter? *Poultry International*, *61*(3), 22–23. https://www.wattagnet.com/articles/44288-does-eggshell-color-really-matter

Guyonnet, V., & Jing, H. (2016). The eggs and cholesterol: A liability or an opportunity? In R. R. Watson & F. De Meester (Eds.), *Handbook of cholesterol, biology, function and role in health and diseases* (pp. 413–426). Wageningen Academic Publishers. https://doi.org/10.3920/978-90-8686-821-6

Hansstein, F. (2011). Profiling the egg consumer: Attitudes, perceptions and behaviours. In Y. Nys, M. Bain, & F. Van Immerseel (Eds.), *Improving the safety and quality of eggs and egg products. Volume 1: Egg chemistry, production and consumption* (pp. 39–61). Woodhead Publishing.

Heng, Y., Peterson, H. H., & Li, X. (2016). Consumer responses to multiple and superfluous labels in the case of eggs. *Journal of Food Distribution Research*, *47*(2), 62–82. https://doi.org/10.22004/ag.econ.240708

IEC (2021, November 9). International egg commission interactive statistics, Country data. https://marketdata.internationalegg.com/graphs/

ILH (2021, May 14). La industria avícola Latinoamericana u el bienestar animal. Mesa redonda: Producción y comercialización de huevos producidos en sistemas de jaulas en Europa, Estados Unidos u Latinoamérica. Comportamiento del consumidor [Webinar], Instituto Latinoamericano del Huevo. http://www.ilhala.org/presentacioacuten.html

Kanter, M. M., Kris-Etherton, P. M., Fernandez, M. L., Vickers, K. C., & Katz, D..L (2012). Exploring the factors that affect blood cholesterol and heart disease risk: Is dietary cholesterol as bad for you as history leads us to believe? *Advances in Nutrition*, *3*(5), 711–717 https://doi.org/10.3945/an.111.001321

Kostenko, O. (2021). Placement and concentration in agriculture: Latest trends (case study by Russian egg poultry). *E3S Web of Conferences*, *247*, 01022. https://doi.org/10.1051/e3sconf/202124701022

Lagerkvist, C. J., & Hess, S. (2011). A meta-analysis of consumer willingness to pay for farm animal welfare. *European Review of Agricultural Economics*, *38*(2), 55–78. https://doi.org/10.1093/erae/jbq043

Magdelaine, P. (2011). Egg and egg product production and consumption in Europe and the rest of the world. In Y. Nys, M. Bain, & F. Van Immerseel (Eds.), *Improving the safety and quality of eggs and egg products. Volume 1: Egg chemistry, production and consumption* (pp. 3–16). Woodhead Publishing.

Ochs, D., Wolf, C. A., Widmar, N. J. O., & Bir, C. (2018). Consumer perceptions of egg-laying hen housing systems. *Poultry Science*, *97*(10), 3390–3396. https://doi.org/10.3382/ps/pey205

Ochs, D., Wolf, C. A., Widmar, N. J., & Bir, C. (2019). Is there a "Cage-free" lunch in U.S. Egg production? Public views of laying-hen housing attributes. *Journal of Agricultural and Resource Economics*, *44*(2), 345–361.

Pelletier, N., Ibarburu, M., & Xin, H. (2014). Comparison of the environmental footprint of the egg industry in the United States in 1960 and 2010. *Poultry Science*, *93*(2), 241–255. https://doi.org/10.3382/ps.2013-03390

Poultry Union of Ukraine (2020). Current state of the poultry industry in Ukraine. http://www.poultryukraine.com/

Rondoni, A., Asiolia, D., & Millana, E. (2020). Consumer behaviour, perceptions, and preferences towards eggs: A review of the literature and discussion of industry implications. *Trends in Food Science & Technology*, *106*, 391–401. https://doi.org/10.1016/j.tifs.2020.10.038

Sánchez-Bravo, P., Chambers V, E., Noguera-Artiaga, L., López-Liuch, D., Chambers IV, E., Carbonell-Barrachina, A., & Sendra, E. (2020). Consumers' attitude towards the sustainability of different food categories. *Foods*, *9*(11), 1608. https://doi.org/10.3390/foods9111608

Sánchez-Bravo, P., Chambers V, E., Noguera-Artiaga, L., Sendra, E., Chambers IV, E., & Carbonell-Barrachina, A. (2021). Consumer understanding of sustainability concept in agricultural products. *Food Quality and Preference*, *89*, 104136. https://doi.org/10.1016/j.foodqual.2020.104136

Sass, C. A. B., Kuriya, S., da Silva, G. V., Silva, H. L. A., da Cruz, A. G., Esmerino, E. A., & Freitas, M. Q. (2018). Completion task to uncover consumer's perception: A case study using distinct types of hen's eggs. *Poultry Science*, *97*(7), 2591–2599. http://dx.doi.org/10.3382/ps/pey103

Skala, J. H. (1965). Egg preferences of Wisconsin consumers as related to standards for grading. *Poultry Science*, *44*(5), 1185–1192. https://doi.org/10.3382/ps.0441185

Sonaiya, E. B. (2007). Family poultry, food security and the impact of HPAI. *World's Poultry Science Journal*, *63*(1), 132–138. https://doi.org/10.1017/S0043933907001353

Tukur, H. M. (2011). Egg production in Africa. In Y. Nys, M. Bain, & F. Van Immerseel (Eds.), *Improving the safety and quality of eggs and egg products. volume 1: Egg chemistry, production and consumption* (pp. 27–38). Woodhead Publishing.

UEP (2022, February 28). Facts and stats. https://unitedegg.com/facts-stats/

USDA (2015). Scientific report of the 2015 dietary guidelines advisory committee advisory report to the secretary of health and human services and the secretary of agriculture. https://health.gov/sites/default/files/2019-09/Scientific-Report-of-the-2015-Dietary-Guidelines-Advisory-Committee.pdf

Walker, S., & Baum, J. I. (2022). Eggs as an affordable source of nutrients for adults and children living in food-insecure environments. *Nutrition Review*, *80*(2), 178–186. https://doi.org/10.1093/nutrit/nuab019

WATT Poultry (2021). Top world broiler and egg rankings for 2021. Poultry trends 2021. https://www.wattglobalmedia.com/publications/watt-poultry-trends/

Widmar, N., Bir, C., Wolf, C., Lai, J., & Liu, Y. (2020). Eggs: Social and online media-derived perceptions of egg-laying hen housing. *Poultry Science*, *99*(11), 5697–5706. https://doi.org/10.1016/j.psj.2020.07.011

Yang, N. (2011). Social economic aspects of egg production in China. In Y. Nys, M. Bain, & F. Van Immerseel (Eds.), *Improving the safety and quality of eggs and egg products. Volume 1: Egg chemistry, production and consumption* (pp. 17–26). Woodhead Publishing.

Yang, Z., Rose, S. P., Yang, H. M., Pirgozliev, V., & Wang, Z. Y. (2018). Egg production in China. *World's Poultry Science Journal*, *74*(3), 1–10. https://doi.org/10.1017/S0043933918000429

Part II

Chemistry and Characterization of Egg Components

3

Egg Biosynthesis, Structure, and General Composition

Joël Gautron*, Sophie Réhault-Godbert, and Nicolas Guyot

CONTENTS

3.1 Introduction

The egg is an oocyte surrounded by nutritional reserves and a protective envelope (the shell) where the embryo will develop without maternal contact. The hen must therefore anticipate all the needs necessary for the embryo, and deposit in the egg all the resources necessary for its development. These include nutrients, protective systems, regulatory and effectors molecules. To ensure this reproductive function, the constituents of the hen's egg have remarkable properties including a very good digestibility to face the constraints of this closed environment where the elimination of waste is not possible.

The egg also contains biological activities that protect the embryo and/or intervene in the regulation of its development. The egg must remain sterile to allow the development of a healthy embryo. To ensure this protective function, the egg possesses numerous molecules with a wide range of antimicrobial activities. In addition, some egg molecules exhibit complementary biological activities including, among others, anti-hypertensive, anti-cancer, antioxidant, antigenic, cryoprotective, and immunomodulatory activities. The study of the biological activities of egg proteins is therefore of great interest as many new biological activities were recently discovered.

The egg is also characterized by a mineral shell with remarkable mechanical properties to protect the embryo from the external environment. It is a biomaterial manufactured at low temperature and low pressure. When it is intact, the shell is impermeable to all external penetrations with the noticeable exception of gas exchanges that are essential for the embryonic development.

Egg proteins constitute a very high biological value for humans. In particular, egg proteins are of very high nutritional quality, attested by a DIAAS (Digestible Indispensable Amino Acid Score) value higher than 1.00 (Phillips, 2017), because of the perfect nutritional balance of its amino acids for humans and its high digestibility. Therefore, the egg is a low-cost food widely consumed all over the world with more than 1,400 billion units produced each year. It does not suffer from any religious prohibitions and is considered to be the easiest and cheapest animal protein to produce.

The average egg weight is 60 grams. The yolk or vitellus represents 30% of the egg's weight. It is surrounded by an acellular and translucent protein layer, called the vitelline membrane. The yolk, which actually corresponds to an ovulated mature oocyte, includes the germinal disc carrying the female gamete. The white or albumen accounts for 60% of the egg's

* Corresponding author.

DOI: 10.1201/9781003254430-5

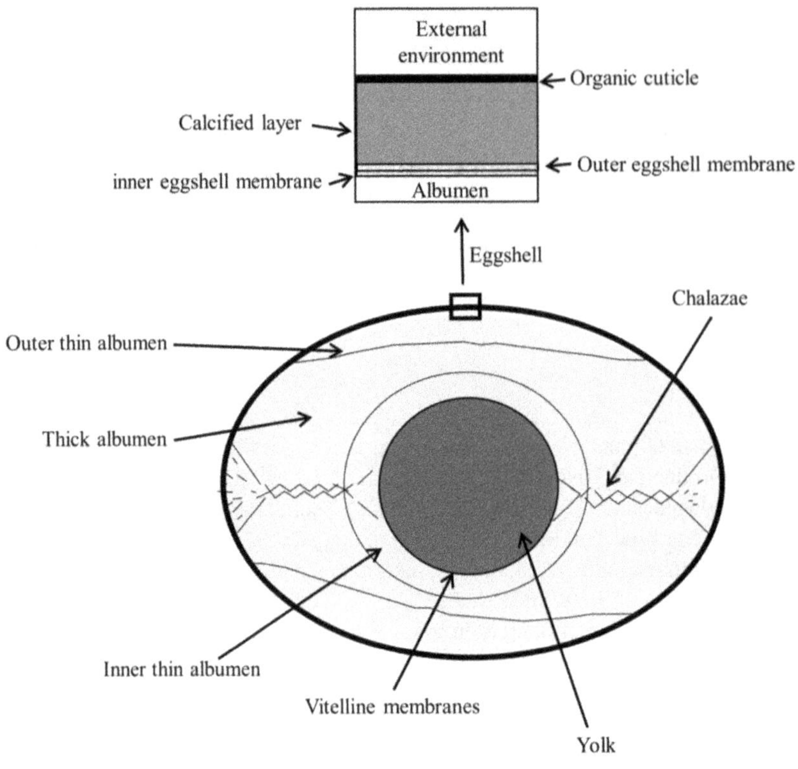

FIGURE 3.1 Schematic representation of the chicken egg. The egg is composed of yolk surrounded by vitelline membranes, the egg white composed of thin and thick textures, and the shell with in the square an enlargement showing the internal and external shell membranes, the calcified layer and the cuticle made of organic matter.

weight. It is not a homogeneous medium since thick and thin whites are distinguished, as well as chalazae, kinds of spiral filaments going from the two extremities of the yolk to the poles of the egg, to maintain the yolk in the center of the egg. The most external layer is the calcified eggshell composed of shell membranes, the calcified part, and the organic cuticle that prevent the contamination of the egg content by microorganisms present in the egg environment (Figure 3.1).

All these characteristics and properties result from physiological mechanisms involving the liver and the reproductive tissues of the hen (ovary and oviduct). In this chapter, we describe the state of knowledge on the tissues or organs involved in the formation of the egg and on the structure and global composition of the egg constituents.

3.2 Development of Reproductive Organs

The organs responsible for egg synthesis of the laying hen encompasses the liver that synthetizes yolk precursors, the ovary that supports steroid synthesis, gametogenesis, folliculogenesis, and the synthesis of the inner layer of the vitelline membrane, and the oviduct that deposits the remaining egg components (outer vitelline membrane, egg white, eggshell membranes, and the eggshell) around the ovulated yolk.

Sexual maturity of domestic pullets (laying of the first egg) occurs around 17 weeks after hatch. During the seven weeks preceding sexual maturity, the weight of the ovary increases from 0.34 to 35 g and that of the oviduct from about 0.12 to 50.5 g (Hrabia et al., 2011). Concomitantly, protein synthesis and lipogenesis are stimulated in the liver to support the incorporation of yolk precursors in the yolky follicles growing in the ovary. The maturation of the female reproductive system depends on orchestrated effects of molecules from the hypothalamus-pituitary-gonads axis and the expression of specific genes, some of them being located on Z and W sex chromosomes (Chue & Smith, 2011), but also microRNA (Cutting et al., 2012). The development of the female reproductive system has been described extensively in many books and reviews (Nys et al., 2011; Réhault-Godbert & Guyot, 2018).

Female layers only have the left ovary and the left oviduct. This particularity results from the asymmetric development of the right and the left oviduct during embryogenesis. It is unique to females in chicken and in most birds, and initiates around day 6.5 of incubation (Cutting et al., 2012). From hatch to 9–12 weeks, the ovary contains only primordial and primary follicles. The presence of white follicles only become visible 3–4 weeks before the first ovulation. It is followed by the development and growth of yolky follicles and the establishment of the pre-ovulatory hierarchy. The hierarchical organization of ovarian yolky follicles results from hormonal stimulation, and allows for the ovulation of the largest follicle into the upper oviduct so that only one yolk is usually enclosed in the laid egg. Follicle growth and ovulation depends on the coordinated effects of hypothalamic, pituitary, and ovarian hormones. The biochemical composition of the follicles was shown to vary significantly during vitellogenesis with, for example, proteins including vitamin-binding proteins and, lipids such as cholesterol, being more abundant in mature yolk follicles compared with pre-hierarchal follicles (small yellow follicles) (Wang et al., 2022).

Similarly to the ovary, the development of the oviduct into a mature reproductive organ depends on the secretion of ovarian steroid hormones. In immature pullets, the oviduct consists of a thin pale, narrow tube attached to the ovary. Its development and cell specialization are concomitant to the development of the ovary between 2 and 3 weeks before the production of the first egg, considering that the development of both organs is controlled by ovarian steroids. The secretory cells responsible of the deposition of egg white, eggshell membranes and calcified layers are tubular shell gland, an epithelium lined by ciliated and non-ciliated cells (Hrabia, 2022). Those cellular layers are forming the mucosa that is folded in the lumen of the organ.

3.3 Egg Formation in the Reproductive Organs

The formation of the egg is characterized by two phases of different duration summarized in Figure 3.2. The process starts with a long phase beginning from the embryonic life with the oogenesis until the development of mature follicles on the ovary. As previously mentioned, this folliculogenesis is supported by the liver, which produces the majority of yolk components, as precursors. Once mature, the yolk, which actually corresponds to a mature oocyte, is ovulated to be captured by the oviduct in the infundibulum. This step is the start of a rapid phase of deposition of the other constituents in the oviduct (Sauveur & De Reviers, 1988; Nys & Guyot, 2011). This oviductal phase lasts approximately 24 h and corresponds to the well-known adage "an egg per day".

3.3.1 Ovary and Folliculogenesis

In total, 12,000 oocytes are formed on the ovary, but only 2,000–3,000 will have the potential to develop into a mature yolk during the whole life of a laying hen. At sexual maturity, a follicular hierarchy is established with a selection of oocytes under the dependence of hormones from the pituitary-hypothalamic axis. A limited number of follicles will grow to give a mature oocyte (yolk). This process is constant and allows the daily ovulation of a yolk, except for the days of pause between clutch series.

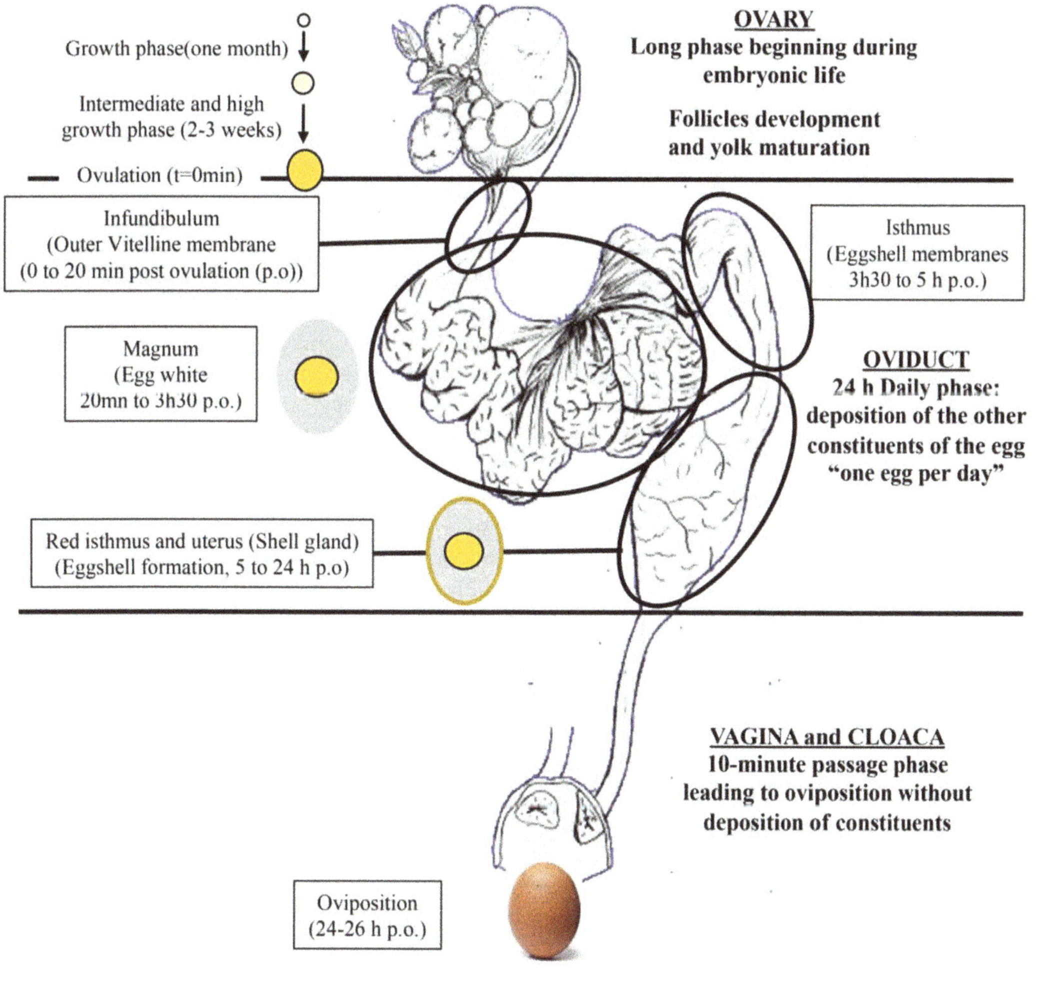

FIGURE 3.2 Schematic drawing representing the spatiotemporal events of egg formation. The drawing includes the long phase on the ovary followed by the daily phase in the oviduct.

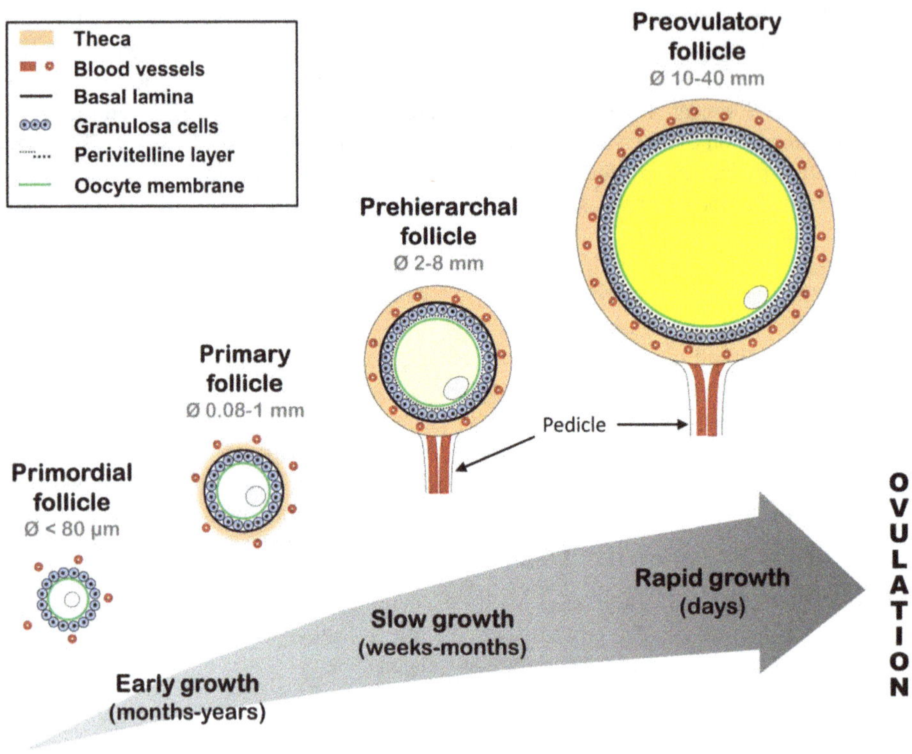

FIGURE 3.3 Follicle growth in the domestic fowl (*Gallus gallus*). (Drawn based on a picture from Johnson and Woods, 2007.)

The adult hen ovary has a "bunch of grapes" appearance due to the numerous follicles present at different phases of growth. The central region of the ovary is composed of a vascularized medulla and a cortex containing the small follicles, which are oocytes covered by the follicular epithelium. The viability and the growth of follicles is controlled by endocrine factors (e.g., gonadotropins) supplied via the bloodstream, paracrine and autocrine factors produced within the ovary, and neurochemical/neurohumoral factors locally expressed by the nervous system (Johnson & Woods, 2007). From the primordial follicle to ovulation, the process of follicle growth comprises three main phases, namely early, slow, and rapid growth phases (Figure 3.3). The early growth of primary follicles is assumed to last months to years, while the slow growth of pre-hierarchal (or pre-vitellogenic) follicles takes weeks to months. During the slow growth phase, the small white follicles (<2 mm) and the large white follicles (2–5 mm) incorporate "white" yolk, and then become small yellow follicles (5–9 mm) concomitantly with the uptake of "yellow" yolk. In comparison with "yellow" yolk, the "white" yolk contains a higher protein to lipid ratio, but only 10–13% solid material (Gilbert, 1979). During the rapid growth of pre-ovulatory follicles, large amounts of "yellow" yolk are transferred into the follicle within six days, and ends 2–3 h before ovulation. Five pre-ovulatory follicles with increasing size (follicular hierarchy) named F1 to F5 (F1 being the largest follicle) are generally present in the ovary. Each pre-ovulatory follicle contains an oocyte filled with vitellus and covered with the inner vitelline membrane that will be expelled into the infundibulum upon ovulation.

In pre-ovulatory follicles, the follicle wall is composed of multiple layers including, from inside to outside: the zona radiata, the inner vitelline membrane, the granulosa cell layer, the basal lamina, and the thecal layers (Rothwell & Solomon, 1977; Perry et al., 1978). The zona radiata corresponds to the interdigitation of processes from granulosa cells and the oolemma (oocyte plasma membrane). The inner vitelline membrane is constituted of protein fibers, and lined by a monolayer of granulosa cells. The onset of formation of the inner vitelline membrane can be observed in oocytes of about 2.5 mm diameter (slow growth phase), while its maximum size is reached in oocytes of 2.5 cm during the rapid growth phase (Bellairs, 1965). The basal lamina (basement membrane) is approximately 1 µm thick and separates the granulosa and the theca layers. Granulosa cells are the main cellular source for progesterone, while theca cells are the cellular site for testosterone and estrogen synthesis (Huang et al., 1979). The thecal layer is anatomically divided into theca interna and theca externa (Perry et al., 1978), each possessing specific functional roles in steroidogenesis. The theca interna and theca externa indeed seem to be involved in the production of testosterone and estradiol, respectively (Porter et al., 1989).

3.3.2 Liver and Maturation of the Yolk

With the exception of immunoglobulins that are produced in the ovary, egg yolk proteins and lipids essentially originate from the liver of laying hens. Indeed, during the 2–3 weeks preceding the onset of egg laying, protein synthesis and lipogenesis are progressively stimulated by 15- to 20-fold in the liver at sexual maturity in hens, to support the inclusion of 0.75 kg proteins and 1.5 kg of triglycerides into the yolk over a year of commercial production (Nys et al., 2011). Once secreted from the liver into the bloodstream, egg yolk precursors are conveyed to the ovarian follicle where they are incorporated in the growing yolky follicle by endocytosis (Perry & Gilbert, 1979).

In the liver, yolk lipids are secreted and transported to the ovary as very low-density lipoproteins (VLDLs). The core of VLDLs consists of triglycerides and cholesteryl esters that are surrounded by a surface layer of phospholipids, cholesterol, and apoproteins (mainly apolipoprotein B and apovitellenin-1). The VLDL concentration in blood plasma ranges from 2 to 5 g/L in immature hens (non-laying hens), whereas it increases up to 20 g/L in mature hens (Noble & Cocchi, 1990). VLDLs are essentially composed of phospholipids (25–29%) and triacylglycerols (65%) (Noble & Cocchi, 1990). This VLDL structure is the circulating form of yolk precursors. It binds to ovarian receptors, undergoes endocytosis, and gives rise to Low Density Lipoproteins (LDLs) and High Density Lipoproteins (HDLs) in the yolk (Nimpf & Schneider, 1998).

The liver of laying hens also expresses many other proteins than VLDLs apolipoprotein B and apovitellenin-1 (Gloux et al., 2019a). These include (i) vitellogenins that are precursors of phosvitins in the yolk, and which are 40-fold overexpressed in the liver of laying hens compared with pullets; (ii) vitamin-binding proteins, such as riboflavin-binding protein, vitamin-D binding proteins, avidin-related protein 6-like; and (iii) proteases that are predicted to participate in proteolytic maturation of yolk protein precursors, such as cathepsin E-A-like or "similar to nothepsin" (Bourin et al., 2012). Interestingly, the liver also overexpresses proteins that are likely to participate in fertilization of the oocyte as components of the vitelline membrane (Zona Pellucida protein 1, ovochymase, Milk fat globule-EGF factor 8 protein) (Lindsay et al., 1992; Stewart et al., 2004; Trigg et al., 2021).

The mature yolk is then captured by the oviduct where the other egg components are deposited. The laying oviduct is characterized by five functionally distinct regions, namely the infundibulum (deposition of the outer layer of the vitelline membrane and site of fertilization), the magnum (deposition of egg white components), the isthmus (formation of eggshell membranes), the uterus (formation of the calcified eggshell and cuticle), and the vagina (passage of the egg before oviposition, without deposition of supplementary constituents) (Figure 3.2).

3.3.3 Role of the Infundibulum

The infundibulum is the most proximal part of the oviduct. It is a 10 cm long thin wall flexible funnel formed by a wide-mouthed (proximal) region and a neck (distal) region. This segment of the oviduct grasps the ovulated oocyte released by the largest follicle and transports it to the next segment (magnum). The infundibulum is involved in the storage of sperm cells (secondary site after the sperm storage glands in the utero-vaginal junction), and in the secretion of the outer layer of the vitelline membrane. It is also the only site where egg fertilization can occur when hens are inseminated. Fertilization occurs before the outer vitelline membrane is deposited. Indeed, sperm cells that have not interacted with the inner vitelline membrane and initiated acrosome reaction are entrapped within the outer layer, thus preventing pathologic polyspermy. The duration of oocyte transit in the infundibulum is approximately 15–30 min.

The infundibular wall comprises outer longitudinal and inner circular layers of muscles, a lamina propria (containing connective tissue and blood vessels), and an epithelium. The luminal surface contains a series of macroscopic ridges forming primary folds that become more prominent (0.5–1 mm high) in the distal region of the infundibulum (neck region), where they are associated with minor folds separated by deep invaginations (crypts) leading to tubular glands (Wyburn et al., 1970). Histologically, the mucosal epithelium exposed to the lumen is composed of non-secretory ciliated cells, and secretory non-ciliated cells (also referred as granular cells). Ciliated cells are more numerous in the proximal part of the infundibulum, while the neck region contains both ciliated and non-ciliated cells alternating at the apex of the minor folds (Wyburn et al., 1970; Chousalkar & Roberts, 2008). In the deeper crypts, however, only non-ciliated cells are observed (Wyburn et al., 1970). Both ciliated and non-ciliated cells possess microvilli at their apical membrane (Chousalkar & Roberts, 2008). Secretory activity is much more obvious in granular and glandular cells than in ciliated cells (Wyburn et al., 1970; Chousalkar & Roberts, 2008).

3.3.4 Magnum and Deposition of the White

Twenty minutes after ovulation, the egg in formation enters the magnum, which is the longest segment of the oviduct. This part of the oviduct continuously synthetizes egg white proteins, and the passage of the developing egg causes their secretion and concomitant deposition. The egg white accumulation lasts between 3 and 3.5 h as the egg evolves in the magnum. The magnum is about 35 cm-long. It is composed of an epithelium consisting of a regular alternation of ciliated and granular cells shown in Figure 3.4. These cells are oriented toward the lumen of the oviduct.

The content of ciliated and non-ciliated cells was shown to be highly glycosylated (Jung et al., 2011; Jones et al., 2021). The nuclei of granular and ciliated cells are essentially found at the apical pole of the cells, and their height increases toward the magnum/isthmus junction ranging from 10 to 90 μm (Wyburn et al., 1970). The granular cells contain granules of low electron density that accumulate at the cell apex, and that are discharged in the lumen when a yolk arrives in the magnum. Ciliated cells are characterized by protuberances resembling microvilli filled with cytoplasm, which project toward the lumen. These cells are supposed to facilitate the movement of the egg and its migration toward the next oviduct segment, namely the isthmus. Beneath the epithelium, there are tubular cells (essentially type A cells) that contain small electron dense components of 0.5–3 μm in diameter. The space between the surface epithelium and the underlying tubular cells hosts fibrous tissue, blood vessels and plasma cells (Figure 3.4). The yolk enclosed in the vitelline membrane spends about 3 h in the magnum section to form the egg white. When the egg leaves the magnum to enter the isthmus, the egg white consists of a thick hydrated gel.

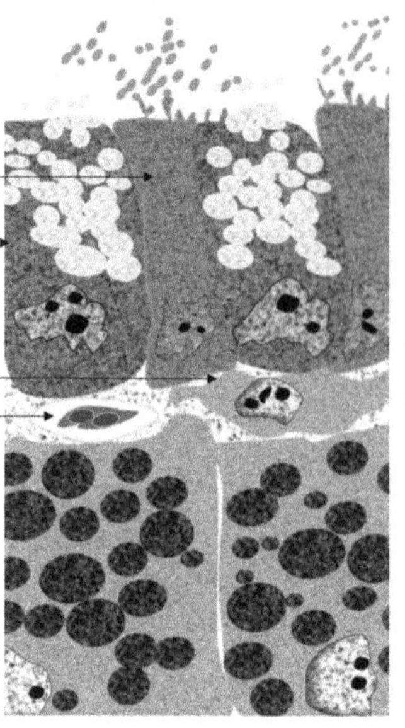

FIGURE 3.4 Schematic representation of the magnum. (Original figure, adapted from micrographs published by Wyburn et al., 1970.)

3.3.5 White Isthmus and Deposition of Shell Membranes

From an anatomical point of view, two regions of different colors can be distinguished in the isthmus, white and red, according to their degree of blood irrigation. As it exits the magnum, the egg penetrates then in the white isthmus where the synthesis of the fibers of the shell membranes takes place. These fibers are strongly intertwined and consist only of proteins. The isthmus is the smallest region of the oviduct (about 8 cm-long). It is separated from the magnum by a thin, transparent, non-glandular band (Hrabia, 2022). This junction between the magnum and the isthmus is involved in the secretion of a thin granular layer, the peri-albumen layer, which is located between the egg white and the shell membranes (Sultana et al., 2003). This layer has the function of covering the irregular surface of the egg white into a smooth layer upon the shell membrane (Mao et al., 2006).

3.3.6 Red Isthmus and Uterus Involvement in Shell Formation

Physiologically, the red region of the isthmus is also called tubular shell gland, and is functionally associated with the uterus (shell gland pouch) to form the shell gland (Solomon, 1991; Nys & Guyot, 2011). The shell gland is the location of two important biological processes, first the hydration of the egg, and second the formation of the shell in the uterine fluid from mineral and organic precursors secreted by the tubular cells and the surface epithelium of the uterus.

When the egg enters the red isthmus (tubular shell gland), organic matter aggregates are deposited randomly on the surface of the shell membranes. These organic clusters constitute the mammillary nuclei, which are the seeding sites of the shell calcification in the uterus.

The uterus (shell gland pouch) is the segment where the egg will stay the longest, about 19 h. The uterus is a sac-like balloon-shaped part of the oviduct, which is about 7 cm-long and 3 cm-wide (Hrabia, 2022). The muscular wall is particularly developed within the uterus compared to the other segments of the oviduct. This particularity is linked to the necessity to maintain in rotation the egg in formation during the calcification of the shell. The lamina propria, which includes the tubular gland cells, is prominent in this organ, and epithelium cells are both ciliated and non-ciliated (Nys & Guyot, 2011; Stapane et al., 2019; Hrabia, 2022). When the egg enters the shell gland, it is weakly hydrated, the shell membranes still have a crumpled appearance, and the egg white is particularly thick. A strong secretion of water then occurs in the uterus to hydrate the egg white in formation and lead the egg in its definitive ovoid shape. The huge hydration of the egg in the early stage in the uterus is called plumping. The water content in the white is 3.5 g of water per g of dry matter at the entry of egg in the uterus. This quantity of water will double during the first 6–7 h of presence in this oviduct segment. This secretion of water associated with the minerals and the organic constituents of the matrix constitutes the uterine fluid, which bathes the egg permanently during the process of biomineralization of the shell. The volume of uterine fluid is particularly important in the first 6 h of mineralization (Etches, 1986). It will remain at about 2 mL for the next 6 h, and will decrease sharply in the last 6 h of stay in the uterus (Nys et al., 1991). Moreover, the rotation of the egg in the uterus will allow the formation of chalazae by twisting of egg white proteins into fibers, as well as the separation of the white in thick and thin by expulsion of water that results from rotation (Nys & Guyot, 2011).

The uterus is also the tissue where the calcification of the eggshell occurs. This formation takes place in the uterine fluid from the mineral and organic precursors secreted by the uterine cells. Mineral secretion is done by tubular cells, while organic matrix proteins can be of tubular or epithelial origin (Nys & Guyot, 2011; Gautron et al., 2021; Nys et al., 2022). The amount of minerals to be supplied is considerable for an animal such as an adult hen weighing about 2 kg. Indeed, the shell weighs 5–6 g, corresponding to an intake of 2–2.5 g per day of calcium and 3–3.5 g of bicarbonates, requiring a considerable metabolic effort and an exacerbated calcium metabolism (Nys & Guyot, 2011; Gautron et al., 2021).

Both components of the shell mineral (Ca^{2+} and CO_3^{2-}) are continuously supplied during eggshell formation via the blood plasma, firstly by trans-epithelial ionic transport through the uterine epithelium, and secondly by vesicular secretion of Amorphous Calcium Carbonate (ACC) mineral particles. A comprehensive and further refined model for calcium and carbonate transport to the mineralization site during eggshell formation was recently proposed (Nys & Le Roy, 2018; Gautron et al., 2021; Nys et al., 2022). Shell bicarbonates can be produced in the blood or in the uterine fluid by hydration of CO_2 by carbonic anhydrases, or actively transferred into uterine cells using the Na^+/HCO_3^- co-transporters and extruded from cells by HCO_3^-/Cl^- exchangers (Brionne et al., 2014; Nys & Le Roy, 2018; Gautron et al., 2021; Nys et al., 2022). The transcellular pathway to secrete calcium and bicarbonate ions into the fluid has been previously described (Jonchere et al., 2012; Brionne et al., 2014; Gautron et al., 2021; Nys et al., 2022). Plasma Ca^{2+} is transferred into uterine cells by active transporters and ion channel exchangers (Nys & Le Roy, 2018; Sah et al., 2018). Intracellular calcium ions are buffered/transferred by calbindin and Ca^{2+} pumps associated with the endoplasmic reticulum. Finally, Ca^{2+}/Na^+ exchangers and the Ca^{2+} pumps are involved in the apical extrusion of calcium into the uterine fluid (Nys & Le Roy, 2018; Sah et al., 2018).

RNA-Seq analysis revealed the expression of several genes involved in the paracellular pathway (Gautron et al., 2021). This paracellular pathway is present in intestine and acts to replenish calcium from dietary sources during eggshell biomineralization, when soluble calcium in the intestinal lumen creates a favorable gradient for passive absorption (Gloux et al., 2019b). The ionic calcium concentration in uterine fluid is six times higher than blood calcium levels (Nys et al., 1991). Consequently, the concentration gradient is not in favor of calcium movement toward the uterine fluid through the paracellular pathway (Nys & Le Roy, 2018). However, it has been suggested that the electrical potential difference could invert this gradient, allowing some paracellular transfer of calcium into the uterine fluid (Bar, 2009). The involvement of this pathway in the movement of calcium is still controversial in uterus. It could participate in maintaining ionic and osmotic homeostasis, notably for sodium and potassium that are exchanged by active transporters to allow the transcellular pathway of calcium, and that have to be removed from the uterine fluid to maintain cell homeostasis.

More recently, a vesicular mechanism was shown to stabilize the transient forms of calcium carbonate necessary for calcite crystal formation (Stapane et al., 2019; Stapane et al., 2020). In avian uterus, vesicles (100 to 500 nm) accumulate at the apical plasma membrane and bud to secrete extracellular vesicles (EVs) (Stapane et al., 2020). The model of vesicular transport includes EDIL3 and MFGE8, which are two major eggshell proteins (Marie et al., 2015). These proteins bind to EVs budding from uterine cells into the uterine fluid, in order to guide vesicular transport of stabilized ACC for delivery to the mineralizing site, and to prevent non-specific precipitation. Annexins are calcium channels proposed to contribute to uptake of calcium for intra-vesicular ACC formation (Stapane et al., 2020). Carbonic anhydrase 4 (CA4), present in the epithelial cells and in EVs, would catalyze the reversible hydration of CO_2 forming HCO_3^-, and might contribute to accumulation of ACC in vesicles.

The second important process in the uterus is the mineralization of the egg, which occurs at different stages in the uterine fluid, where the physicochemical conditions necessary to this calcification are met. The formation of the eggshell results from the development of a mineral structure in a living organism. This phenomenon of mineralization in a biological environment is called biomineralization. Biomineralization is the process by which organisms have the ability to induce and control the formation of a large number of mineralized tissues, such as shell (mollusks, birds, snails), otolith, corals… (Addadi & Weiner, 1992; Marin et al., 2008). Proteins are sequentially secreted and participate in the control of nucleation, polymorphic phase, growth rate, crystal morphology, and thus regulate the formation of biomineralized structures. A number of conditions are necessary for biomineralization:

i. *Existence of a biologically compartmentalized space in which the biomineral will form.* It is in this space that the ionic and organic inputs are controlled to achieve the physicochemical conditions necessary for biomineralization. In the case of the eggshell, it is the uterus that plays this role. It is an extracellular space containing a liquid (uterine fluid) in which all the mineral and organic constituents of the shell are secreted (Nys et al., 1999; Hincke et al., 2012; Gautron et al., 2021).
ii. *An hyper-saturation of the medium.* The constituent ions of the mineral must be in saturating conditions. The egg bathes in the uterine fluid during the whole process of calcification. Calcium carbonate precipitates from the calcium and bicarbonates contained in the uterine fluid. The calcium and carbonate are in such high concentrations that the solubility product of calcite is exceeded by about 100 times, and remains saturated even when calcification stops (Nys et al., 1991).
iii. *Mineral nucleation and growth.* Although essential, hyper-saturation conditions alone are not sufficient to precipitate a salt. Indeed, the initiation of crystallization occurs from particular nucleation sites found in the medium. For the egg, the shell membranes are the support of these sites called mammillary nuclei. The mineralization is initiated from the mammillary nuclei and thus constitute the anchoring sites of the shell. The necessary conditions for the initiation of calcification are gathered on these sites. In a second phase, the growth of the mineral will occur

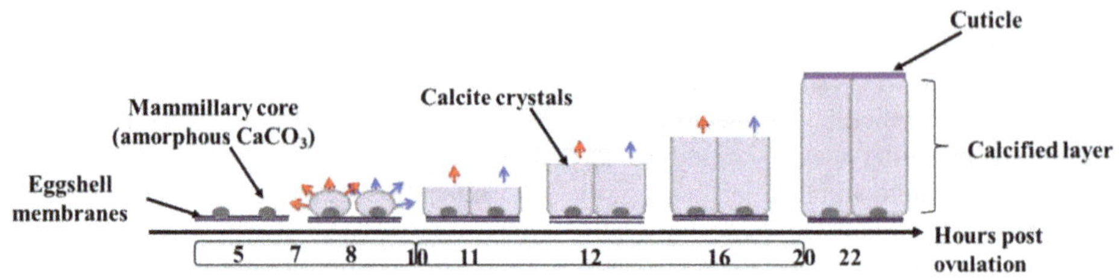

FIGURE 3.5 The 19 h of shell calcification in the uterus to allow the formation of a complete eggshell. Mammillary knobs are deposited 5 h post ovulation (p.o.) simultaneously with a massive deposition of amorphous calcium carbonate that progressively transformed in calcite crystals all oriented in a spherulitic manner (–10 h p.o.). From 10 to 22 h p.o., the fusion of cones allows the formation of the compact calcified layer with a preferred orientation of crystals perpendicular to the surface. Finally, the cuticle is deposited in the last 2 h before the oviposition.

by secondary nucleation on the crystals produced during the previous nucleation. The organic matrix intervenes during this process by inhibiting or promoting the crystal growth according to privileged axes (Nys et al., 1999; Hincke et al., 2012; Gautron et al., 2021).

iv. *The presence of a crystalline structure*. The process of biomineralization results in the deposition of minerals in a given place, which is not random. A biomineral is the result of the development of a controlled and highly ordered crystalline mineral structure. For the same chemical composition, the crystal structure can be different. Calcium carbonate can exist under three different polymorphic types (aragonite, calcite and vaterite). The structure of the biomineral will be different according to the polymorphic type present. The calcium carbonate deposited in the shell is calcite.

The formation of the eggshell is one of the fastest biomineralization phenomenon of the living world. Six grams of shell are formed in about 19 h. The calcification of the eggshell is divided in three phases (Figure 3.5).

The initial phase is initiated 5 h after ovulation in the red isthmus. Organic aggregates (mammillary knobs) are deposited on the surface of the outer shell membrane. Eggshell mineralization begins with the deposition of massive mineral deposits of ACC particles on the mammillary knobs. Mineralization continues within the uterus, in the uterine fluid which contains ionized calcium and bicarbonates in excess of the solubility product of calcite (Nys et al., 1991). ACC progressively dissolves and gives way to calcite crystals (Rodriguez-Navarro et al., 2015). The formation of eggshell calcite crystals through the dissolution of an intermediate metastable amorphous mineral phase maintains a high supersaturation in the uterine fluid that sustains fast and continuous growth of calcite crystals during the linear phase of eggshell biomineralization. Although the uterine fluid is highly supersaturated with respect to all calcium carbonate polymorphs (aragonite, vaterite, calcite and ACC), only calcite is formed from metastable ACC, and ultimately is the sole mineral form observed in the eggshell at laying, which indicates that there is a strict control over eggshell mineralogy (Nys et al., 1991; Rodriguez-Navarro et al., 2015). The growth of the mineral is initially spherulitic. It is inhibited toward the interior of the egg by the shell membranes, and continues toward the exterior to form the cones of the mammillary layer. The mineralization is relatively low during this phase with 0.1 g of shell deposited per hour.

The active growth phase starts 10 h after ovulation, when the cones fuse to form the palisade layer. The mineralization of the shell is fast and linear during this stage (0.33 g/h). This phase is initiated when adjacent mammillary cones become fused, which forms a continuous compact layer (palisade layer) that continues to grow. This phase lasts 12 h, corresponding to deposition of about 4 g of eggshell, which is organized as columnar calcite units with preferred orientation perpendicular to the eggshell surface.

The terminal phase corresponds to the stop of mineralization about 2 h before egg laying. This phase corresponds to the deposition of the outermost layer of the shell, the cuticle, which constitutes a biofilm that covers the shell and obstructs the pore openings to restrict bacterial penetration through the shell.

The uterine wall also secretes organic matrix proteins that represent only 2–2.5% of the calcified part of the shell (Panheleux et al., 2000), corresponding to an amount of 110 to 140 mg of organic matrix for a 5.5 g eggshell. This low quantity does not reflect the great diversity of the constituent proteins of this matrix, which plays an essential role during the calcification of the shell. Indeed, two integrative analyses of shell proteomic data indicate that the organic matrix would be composed of 904 unique proteins when shell membranes are included (Gautron et al., 2019), and 676 if only the mineralized part is considered (Yang et al., 2020). All of these matrix proteins are expressed and secreted by the glandular and epithelial cells of the uterine tissue. They have been mainly identified and characterized by transcriptomic and proteomic approaches reported in numerous review articles (Hincke et al., 2012; Gautron et al., 2019; Gautron et al., 2021; Nys et al., 2022). The complete description of these approaches is available Chapter 4.

The egg is thus completed in one day in the oviduct, and it will then pass quickly in the vagina and the cloaca before being expelled outside. This process is called oviposition. After oviposition, the surface of the egg dries in less than 10 min, and the cuticle then cracks. These small cracks allow gas exchanges between the inside and the outside of the egg, but their size prevents microbial penetration.

3.4 Structure and General Composition of the Egg

The egg is composed of four major components that are, from the outside to the inside, the eggshell and the attached eggshell membranes, the egg white, the vitelline membrane, and the yolk. These components have evolved to supply all nutrients that are necessary to accompany the development of an embryo. Thus, their respective composition is perfectly balanced (Figure 3.6).

The eggshell is mainly composed of minerals, essentially calcium that will support the mineralization of the embryo during incubation. Although the proteins account for only 3–3.5% of the total eggshell, those molecules are essential for eggshell mineralization and ultrastructure, and some of them are supposed to participate in egg defense as antimicrobials (Moreau et al., 2022). The egg white is a source of water, amino acids, vitamin B2 essentially (trace of vitamin K2, and other vitamins B) and some minerals (potassium, sodium, phosphorus, calcium, etc.). Its protein concentration is about 110 g/L. Its viscosity, pH and the diversity of its antimicrobial proteins and peptides also contribute to protect the internal egg components from pathogen contamination. Yolk is the source of lipids, most vitamins, trace elements, and contains minerals including phosphorus, sodium and potassium. It also contains many proteins with binding functions (for vitamins, minerals and trace elements) and immunoglobulin Y. The vitelline membrane is an acellular proteinaceous membrane that separates the egg white from the egg yolk. When the germinal disc is fertilized, the vitelline membrane constitutes a proteinaceous matrix that allows for the development and growth of the yolk sac membrane, an extra-embryonic membrane bearing digestive functions (pre-digestion and assimilation of yolk nutrients).

3.4.1 Structure and Composition of the Yolk

Yolk is a complex medium composed of 55% water, 15% proteins, and 27% lipids (Figure 3.6). It is also a rich source of vitamins (vitamin D, E, B2, B5 etc., except vitamin C), minerals (calcium, phosphorus, potassium and sodium, but also zinc, magnesium and iron) accounting for about 2% ash (Réhault-Godbert et al., 2019). Proteins correspond to 68% low-density lipoproteins (LDLs), 16% high-density lipoproteins (HDLs), 10% livetins and other soluble proteins, and 4% phosvitins. Fifty percent of yolk compounds are recovered in non-soluble protein aggregates termed granules (19–23% of dry matter) while the others are found in clear yellow fluid named plasma, that corresponds to 77–81% of dry matter (Anton & Gandemer, 1997; Anton, 2013). Lipids are part of yolk lipoproteins that are characterized by a core of triglycerides and cholesteryl esters, surrounded by a monolayer of phospholipids and cholesterol in which proteins are embedded (apolipoprotein and apovitellenin). These lipids correspond to saturated fatty acids (8.5 g/100 g of egg yolk), monounsaturated fatty acids (11.9 g/100 g of egg yolk), and polyunsaturated fatty acids (4 g/100 g of egg yolk) (Réhault-Godbert et al., 2019). In addition, egg yolk is a major source of choline (680 mg/100 g of egg yolk). This micronutrient is found as both water-soluble and lipid-soluble forms, and was shown to be critical for brain and eye development (Mun et al., 2019).

The physiological role of egg proteins remains unexplored and is predicted based on their molecular function or similarity

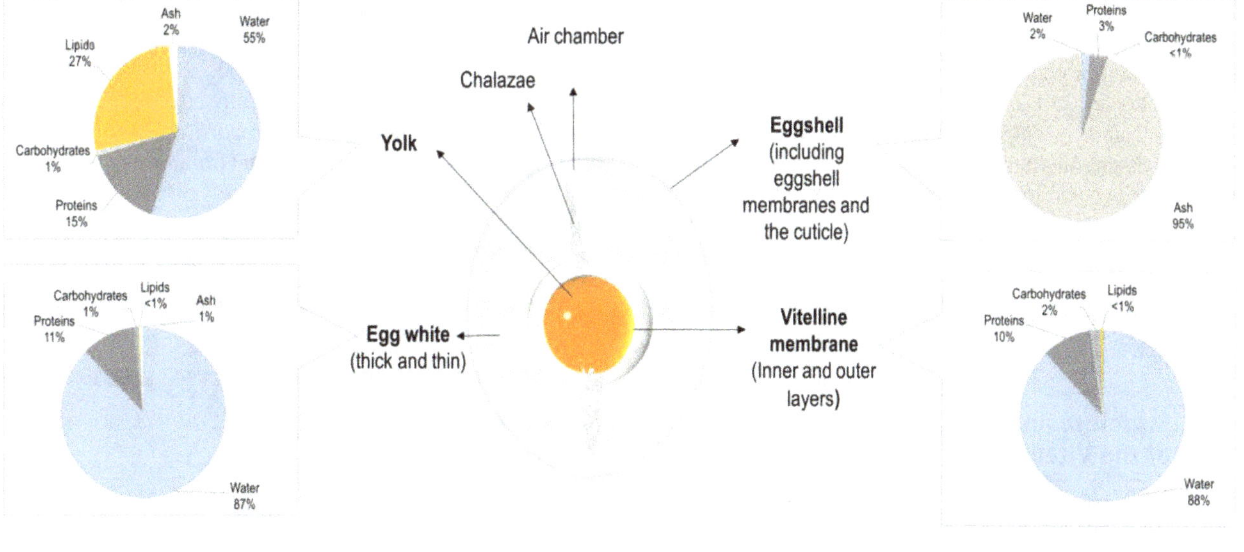

FIGURE 3.6 Overview of the egg's macroscopic structure and composition. The composition of the vitelline membrane was integrated from previous publications (Bellairs et al., 1963; Kido & Doi, 1988), while that of the eggshell (including eggshell membranes) was retrieved from Romanoff's book (Romanoff & Romanof, 1949). The white spot visible on the yolk is the germinal disc corresponding to the oocyte where fertilization occurs. It is to note that the composition shown in this figure for the yolk is likely to include the vitelline membrane (this detail is not mentioned on the CIQUAL website). (Original figure: Egg white and egg yolk general composition was extracted from the CIQUAL website https://ciqual.anses.fr/; French Agency for Food, Environmental and Occupational Health & Safety. ANSES-CIQUAL.)

TABLE 3.1
Egg Yolk Proteins with Highest Abundance (Top-Ten)[a] (Mann & Mann, 2008)

Gene ID	Gene Symbol	Gene Name	Localization (Granule, G; Plasma, P) (Mann & Mann, 2008)	Biological Activity
396197	ALB	Albumin	P>G	Binds water, Ca^{2+}, Na^{+}, K^{+}, fatty acids, hormones, bilirubin, and drugs.
396476	APOV1	Apovitellenin	G>P	Protein component of the very low density lipoproteins (VLDLs) of egg-laying females. Potent lipoprotein lipase inhibitor, preventing the loss of triglycerides from VLDLs on their way from the liver to the growing oocytes.
424533	VTG2	Vitellogenin-2	G>P	Precursor of phosvitin that is believed to be of importance in sequestering calcium, iron and other cations for the developing embryo.
396058	SERPINB14	Ovalbumin	P>G	Non-inhibitory serpin. Storage protein of egg white. Binds calcium.
396535	APOB	Apolipoprotein B	G>P	Heparin binding; lipase binding ; phospholipid binding; cholesterol transfer activity ; low density lipoprotein particle receptor binding.
424547	VTG1	Vitellogenin-1	G>P	Precursor of phosvitin that is believed to be of importance in sequestering calcium, iron and other cations for the developing embryo.
	AVR6	Avidin-related protein 6 like	G>P	Biotin-binding protein.
396241	TF	Ovotransferrin	P>G	Iron-binding protein.
424534	VTG3	Vitellogenin 3	G>P	Precursor of phosvitin that is believed to be of importance in sequestering calcium, iron and other cations for the developing embryo.
396277	TTR	Transthyretin	G>P	Thyroid hormone-binding protein. Probably transports thyroxine from the bloodstream to the brain.

Note: The molecular function was extracted from the Uniprot Protein Database (uniprot.org)
[a] Immunoglobulins Y were also very abundant (but no associated gene ID).

with homologous proteins. Many proteomic approaches on egg yolk have been published (Mann & Mann, 2008; Farinazzo et al., 2009; Geng et al., 2018; Wang et al., 2020). The top-ten most abundant proteins are presented in Table 3.1.

It is noteworthy that five of these major proteins (vitellogenin-1, -2 and -3, apovitellenin, and avidin-related protein 6 like) lack mammalian homologs and are assumed to have specifically evolved to assist egg formation and egg function as an autonomous reproductive chamber (Gloux et al., 2019a). Moreover, the higher abundance of apovitellenin and apolipoprotein B in yolk granules (Table 3.1) is consistent with their known presence as constituents of LDLs (Jolivet et al., 2006). The complete description of the yolk is available Chapter 6.

3.4.2 Structure and Composition of the Vitelline Membrane

The vitelline membrane forms an extracellular protein matrix around the yolk-containing oocyte, thus physically separating it from the egg white. It is approximately 10 µm thick, and consists of two layers, namely the inner and outer vitelline membranes, separated by a thin amorphous "continuous membrane" with 50–100 nm in thickness (Bellairs et al., 1963; Fujii et al., 1972; Li-Chan & Kim, 2008) (Figure 3.7). The vitelline membrane has water content estimated to be about 88% (Bellairs et al., 1963), and a dry weight of about 6 mg per egg (Kido & Doi, 1988; Chung et al., 2010). Outer and inner vitelline membranes have a dry weight of about 3.6 mg and 2.1 mg, respectively (Chung et al., 2010). The dry matter of the vitelline membrane is mainly composed of proteins (81.6–91.5%), but also in smaller proportions of carbohydrates (6.5–16.5%), lipids (0.9–13.3%) and minerals (0.61–0.65%)

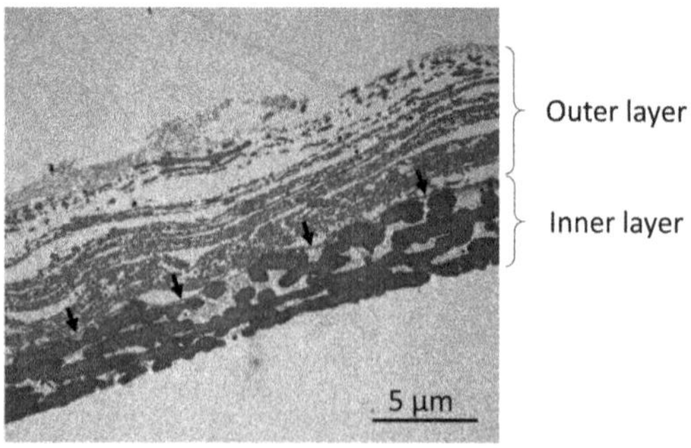

FIGURE 3.7 Chicken egg vitelline membrane in cross-section observed by transmission electron microscopy. The inner and outer layers of the vitelline membrane have different structural organizations and are separated by a very thin "Continuous Membrane" (indicated by black arrows). (© Plateforme IBiSA De Microscopie Electronique, University and CHRU of Tours, France, S. Georgeault.)

TABLE 3.2

Top-Ten Most Abundant Proteins Identified in the Proteome of the Whole Vitelline Membrane (Mann, 2008)

Gene ID	Gene Symbol	Gene Name	Biological Activity
396058	OVAL	Ovalbumin (SERPINB14)	Non-inhibitory serpin. Storage protein of egg white.
396218	LYZ	Lysozyme	Lysozymes have primarily a bacteriolytic function.
396241	TF	Ovotransferrin	Transferrins are iron binding transport proteins which can bind two Fe^{3+} ions in association with the binding of an anion, usually bicarbonate. Ovotransferrin (egg transferrin) has a bacteriostatic function.
420897	OVALY	Ovalbumin-related protein Y (SERPINB14B)	Unknown.
416236	SPINK7	Ovomucoid. serine peptidase inhibitor, Kazal type 7	Serine protease inhibitor. Inhibits trypsin.
418974	VMO1	Vitelline membrane outer layer 1	Exact function unknown. Component of the outer membrane of the vitelline layer of the egg. Seems to be able to synthesize N-acetylchito-oligosaccharides (n = 14–15) from hexasaccharides of N-acetylglucosamine in a manner similar to the transferase activity of lysozyme. Possesses antibacterial activities against *Listeria monocytogenes* and heparin-binding properties (Guyot et al., 2016).
416235	SPINK5	Ovoinhibitor	Serine protease inhibitor involved in antimicrobial egg defense preventing contamination of table eggs (non-fertilized eggs) and protecting the chick embryo (fertilized eggs) (probable). Inhibits trypsin, chymotrypsin, elastase, subtilisin, and a proteinase of fungus *Aspergillus oryzae.*
396260	AVD	Avidin	Biological function unknown. Forms a strong non-covalent specific complex with biotin (one molecule of biotin per subunit of avidin).
420898	OVALX	Ovalbumin-related protein X (SERPINB14C)	Heparin-binding. Defense response to bacteria.
396497	CST3	Cystatin C	This protein binds tightly to and inhibits a variety of thiol proteases including ficin, papain, and cathepsins B, C, H, and L. Although isolated from egg white, it is also present in serum.

Note: The biological functions were extracted from literature and/or the Uniprot Protein Database (uniprot.org)

(Trziszka & Smolińska, 1982; Kido & Doi, 1988; Chung et al., 2010). The inner layer contains about 1.13 times more protein than the outer layer, which in turn contains twice as much carbohydrates compared to the inner layer (Chung et al., 2010). These data are supported by the study from Kido and Doi who showed that the dry matter of the inner vitelline membrane contains 84% proteins and 12.3% sugars, compared to 69% and 24.4% for the outer vitelline membrane (Kido & Doi, 1988). Neutral, amino (hexosamines) and acidic (sialic acid) sugars account for 4.2%, 7.1%, and 1% of the inner layer respectively, while outer layer has 7.7%, 10.5%, and 6.2% of these sugars, respectively.

The first proteomic analysis of the chicken egg vitelline membrane revealed 137 different proteins (Mann, 2008). The protein composition of each layer is different, as evidenced by biochemical analyses and electrophoretic profiles, and suggests that these two layers play different physiological roles. More recently, a study performed on isolated layers obtained after manual separation revealed a total of 412 different proteins (Brégeon et al., 2022). Among them, 239 and 314 proteins were identified in the outer and inner layers, respectively, while 141 proteins were common to both layers. Functional annotation of identified proteins revealed functions associated with fertilization and early development for the inner vitelline membrane, while the outer layer would rather participate in egg defense and embryogenesis (Brégeon et al., 2022). The top-ten most abundant proteins identified by mass spectrometry in the proteomes of the complete vitelline membrane, and isolated outer and inner vitelline membranes, are shown in Tables 3.2, 3.3, and 3.4, respectively.

3.4.2.1 Inner Vitelline Membrane

The inner vitelline membrane is 1–4 μm thick. It consists of an entangled network of fibers (200–600 nm in diameter) (Bellairs et al., 1963) (Figure 3.7). These fibers appear flat in scanning electron microscopy, and join together to form a mesh. Spaces between the fibers ranging from 1.9 to 7.8 μm have been observed by cryo-scanning electron microscopy (Chung et al., 2010).

The inner vitelline membrane actually corresponds to the mammalian Zona Pellucida involved in the oocyte-sperm recognition and fertilization. The early studies identifying inner layer proteins revealed several glycoproteins (GP-I, GP-II, GP-II, GP-IV) (Kido et al., 1975; Back et al., 1982; Kido & Doi, 1988), some of which may correspond to ZP proteins (ZP1, ZP2, ZP3, ZP4, ZPD, ZPAX1 and ZPAX2), identified in later studies (Takeuchi et al., 1999; Bausek et al., 2000; Okumura et al., 2004; Nishio et al., 2014), and especially ZP1, ZP3 et ZPD that have been identified as the most abundant proteins of the inner vitelline membrane (Takeuchi et al., 1999; Okumura et al., 2004).

The interaction between ZP1 and ZP3 proteins, which seems to be the basis of the inner layer matrix, was characterized by biochemical and immuno-histochemical

TABLE 3.3

Top-Ten Most Abundant Proteins Identified in the Proteome of the Isolated Outer Vitelline Membrane (Brégeon et al., 2022)

Gene ID	Gene Symbol	Gene Name	Biological Activity
396218	LYZ	Lysozyme	Lysozymes have primarily a bacteriolytic function.
418974	VMO1	Vitelline membrane outer layer 1	Exact function unknown. Component of the outer membrane of the vitelline layer of the egg. Seems to be able to synthesize N-acetylchito-oligosaccharides (n = 14–15) from hexasaccharides of N-acetylglucosamine in a manner similar to the transferase activity of lysozyme. Possesses antibacterial activities against *Listeria monocytogenes* and heparin-binding properties (Guyot et al., 2016).
414876	AvBD11	Avian beta-defensin 11	Possesses antibacterial, antiparasitic, and antiviral activities and heparin-binding properties (Hervé-Grépinet et al., 2010; Guyot et al., 2016; Guyot et al., 2020).
418125	PTN	Pleiotrophin	Secreted growth factor that mediates its signal through cell-surface proteoglycan and non-proteoglycan receptors (by similarity). Binds cell-surface proteoglycan receptor via their chondroitin sulfate (CS) groups (by similarity). Thereby regulates many processes like cell proliferation, cell survival, cell growth, cell differentiation, and cell migration (by similarity) Possesses antibacterial activities (*Salmonella* Enteritidis, *Listeria monocytogenes*) and heparin-binding properties (Guyot et al., 2016).
101750704	LOC101750704	Beta-microseminoprotein-like (MSMB3 protein)	Possesses antibacterial activities (*Salmonella* Enteritidis, *Listeria monocytogenes*) and heparin-binding properties (Guyot et al., 2016).
419882	OLFML3	Olfactomedin like 3	Secreted scaffold protein that plays an essential role in dorsoventral patterning during early development. Stabilizes axial formation by restricting chordin (CHRD) activity on the dorsal side. Acts by facilitating the association between the tolloid proteases and their substrate chordin (CHRD), leading to enhance chordin (CHRD) degradation (by similarity).
395209	RARRES1	Retinoic acid receptor responder 1 (ovocalyxin-32)	Inhibits bacterial growth (*Bacillus subtilis*) and inhibits bovine carboxypeptidase (Xing et al., 2007).
396058	OVAL	Ovalbumin (SERPINB14)	Non-inhibitory serpin. Storage protein of egg white.
396483	TIMP3	TIMP metallopeptidase inhibitor 3	Complexes with metalloproteinases (such as collagenases) and irreversibly inactivates them by binding to their catalytic zinc cofactor. May form part of a tissue-specific acute response to remodeling stimuli.
420898	OVALX	Ovalbumin-related protein X (SERPINB14C)	Heparin-binding. Defense response to bacteria.

Note: The biological functions were extracted from literature and/or the Uniprot Protein Database (uniprot.org)

approaches (Okumura et al., 2015). These two ZP proteins are linked through their ZP-C domain (C-terminal domain of the ZP module) to form protein fibrils. ZP-C domains are then buried in the core of the fibril, and the N-glycans and N-terminal domains of the ZP proteins are exposed on the surface of the fibril. The fibrils formed by ZP1- ZP3 interaction then assemble into fibers via disulfide bridges between the N-terminal domains of ZP1. Furthermore, ZPD forms homopolymers, and binds to ZP1-ZP3 fibrils to form aggregates (Okumura et al., 2004; Sasanami et al., 2006; Okumura et al., 2007). Thus, ZP1, ZP3 and ZPD are thought to be the major ZP proteins involved in the primary structure of the inner vitelline membrane (Nishio et al., 2018). These proteins are assembled in the ovary although they have different cell sources. Some ZP proteins identified in the inner vitelline membrane are either synthesized locally by granulosa cells (e.g., ZP2, ZP3, ZP4 and ZPD) (Nishio et al., 2014; Brégeon et al., 2022), the oocyte (quail ZP2) (Kinoshita et al., 2010), and possibly theca cells (Nishio et al., 2014; Brégeon et al., 2022). Some other ZP proteins are rather produced by the liver such as ZP1 (Bausek et al., 2000; Brégeon et al., 2022), to be likely secreted into the bloodstream and transported to the growing follicles in the ovary to be assembled.

3.4.2.2 Outer Vitelline Membrane

The outer layer is 3 to 8.5 μm thick and is composed of a varying number of sublayers (Bellairs et al., 1963) (Figure 3.7). Each sublayer consists of a latticework of fine fibrils (as little as 15 nm in diameter) running in straight lines for distances up to 1 μm or more, and forming swellings (300 nm in diameter) at the contact points between fibrils (Bellairs et al., 1963). Polygon-shaped spaces measuring between 1.2 and 7.4 μm in diameter can be observed between the fibrils by cryo-scanning electron microscopy (Chung et al., 2010).

The outer vitelline membrane mainly contains lysozyme, ovomucin, VMO1 (Back et al., 1982), and VMO2/AvBD11 (Kido et al., 1992). Their proportions within this layer remain imprecise and can be very different depending on the study. According to Back and collaborators, it consists of 60% lysozyme, 25% ovomucin, and 15% VMO1 (Back et al., 1982). Kido and Doi showed that it is composed of 43% ovomucin, 32% lysozyme, 20% VMO1, and 5% VMO2 (Kido & Doi, 1988). DeBoeck and Stockx found that lysozyme represents around 37% of the proteins of the outer layer (Deboeck & Stockx, 1986). These differences can be explained, at least in part, by the different techniques used by each of these groups for extraction, identification, and quantification of proteins.

TABLE 3.4

Top-Ten Most Abundant Proteins Identified in the Proteome of the Isolated Inner Vitelline Membrane (Brégeon et al., 2022)

Gene ID	Gene Symbol	Gene Name	Biological Activity
396218	LYZ	Lysozyme	Lysozymes have primarily a bacteriolytic function.
378906	ZP3	Zona Pellucida sperm-binding protein 3	Component of the zona Pellucida, which mediates species-specific sperm binding. Directly binds to sperm. Important for egg fertilization.
395192	SLURP1	Secreted LY6/PLAUR domain containing 1 (Hep21 protein)	Unknown.
395418	ZP1	Zona Pellucida sperm-binding protein 1	Involved in sperm-egg binding (Bausek et al., 2004) and in sperm activation (Okumura et al., 2004).
418974	VMO1	Vitelline membrane outer layer protein 1	Exact function unknown. Component of the outer membrane of the vitelline layer of the egg. Seems to be able to synthesize N-acetylchito-oligosaccharides (n = 14–15) from hexasaccharides of N-acetylglucosamine in a manner similar to the transferase activity of lysozyme. Possesses antibacterial activities against *Listeria monocytogenes* and heparin-binding properties (Guyot et al., 2016).
414876	AvBD11	Avian beta-defensin 11	Possesses antibacterial, antiparasitic and antiviral activities and heparin-binding properties (Hervé-Grépinet et al., 2010; Guyot et al., 2016; Guyot et al., 2020).
396058	OVAL	Ovalbumin	Non-inhibitory serpin. Storage protein of egg white.
418125	PTN	Pleiotrophin	Secreted growth factor that mediates its signal through cell-surface proteoglycan and non-proteoglycan receptors (by similarity). Binds cell-surface proteoglycan receptor via their chondroitin sulfate (CS) groups (by similarity). Thereby regulates many processes like cell proliferation, cell survival, cell growth, cell differentiation and cell migration (by similarity) Possesses antibacterial activities (*Salmonella* Enteritidis, *Listeria monocytogenes*) and heparin-binding properties (Guyot et al., 2016).
101750704	LOC101750704	Beta-microseminoprotein-like (MSMB3 protein)	Possesses antibacterial activities (*Salmonella* Enteritidis, *Listeria monocytogenes*) and heparin-binding properties (Guyot et al., 2016).
404754	UMOD	Uromodulin (Zona Pellucida protein D)	Involved in sperm activation (Okumura et al., 2004).

Note: The biological functions were extracted from literature and/or Uniprot Protein Database (uniprot.org)

Ovomucin, lysozyme, VMO1 and VMO2/AvBD11 are synthesized in the proximal region of the oviduct, not in the liver nor in the ovary. VMO1 is specifically synthesized in the infundibulum-magnum junction (Uyeda et al., 1994; Brégeon et al., 2022), while ovomucin, lysozyme and AvBD11 are mainly produced in this junction and/or in the magnum (Brégeon et al., 2022).

Ovomucin is assumed to form the backbone of the outer layer (Back et al., 1982), and the disulfide bonds of this protein are critical for the structural integrity of the outer layer (Kido & Doi, 1988). The removal of salt-soluble proteins (e.g., lysozyme, VMO1) from the vitelline membrane, following a treatment with 10% NaCl, strongly alters the structure of the outer layer (Back et al., 1982), although it seems to have no significant effect on the membrane resistance (Kido & Doi, 1988). However, the storage of eggs, which is known to weaken the membrane, leads to a loss of VMO1 and VMO2/AvBD11, and to a dimerization of lysozyme in the vitelline membrane (Back, 1984; Schäfer et al., 1998). Lysozyme, VMO1 and VMO2 have high isoelectric points, and are closely associated with ovomucin fibrils, which contain negatively charged sialic acid (Kido & Doi, 1988; Kido et al., 1995; Schäfer et al., 1998). Lysozyme has an important role in the structure of the outer layer through its electrostatic interactions with ovomucin (Deboeck & Stockx, 1986). These interactions are established between the negative charge of the sialic acid of ovomucin and the positive charges of the lysyl ε-amino groups of lysozyme (Kato et al., 1975). Moreover, it has been demonstrated in quail that VMO2 interacts with the ZP1 and ZP3 proteins of the inner layer (Rahman et al., 2009), suggesting that VMO2 would be involved in the adhesion mechanism between the outer and inner layers.

3.4.3 Structure and Composition of the Egg White

The egg white is a gel-like structure mainly composed of water (about 87%) and proteins (Figure 3.6). When the yolk surrounded by the egg white leaves the magnum, the consistence of this latter is very viscous. The completion of its hydration occurs in the uterus where the water content doubles. This process is named the plumping phase (Nys et al., 1999). The egg white structure is not homogenous; it is divided between thick and thin egg whites (Nys et al., 2011). The viscosity of the egg white is assumed to depend on the ovomucin. In freshly laid eggs, the concentration of ovomucin in thick egg white is five to seven times higher compared with thin egg white, depending on the genetic background (Jolivet et al., 2006). The egg white is also characterized by the presence of two spring-like opalescent structures termed chalazae that are anchored to the yolk, thereby stabilizing the egg yolk in the center of the egg (Figures 3.1 and 3.6) (Almquist, 1936). Chalazae contain twisted fibers that result from the continuous rotation of the egg during its formation in the oviduct. They are composed of ovomucin and lysozyme, two proteins that are also abundant proteins of egg white (Baker et al., 1959; Itoh et al., 1987).

TABLE 3.5

Egg White Proteins with Highest Abundance (Top-Ten)

Gene ID	Gene Symbol	Gene Name	Biological Activity
396058	SERPINB14	Ovalbumin (SERPINB14)	Non-inhibitory serpin. Storage protein of egg white.
396218	LYZ	Lysozyme	Lysozymes have primarily a bacteriolytic function.
396241	TF	Ovotransferrin	Transferrins are iron binding transport proteins which can bind two Fe^{3+} ions in association with the binding of an anion, usually bicarbonate. Ovotransferrin (egg transferrin) has a bacteriostatic function.
420897	SERPINB14B	Ovalbumin-related protein Y	Unknown.
416236	SPINK7	Ovomucoid. serine peptidase inhibitor, Kazal type 7	Serine protease inhibitor. Inhibits trypsin.
418974	VMO1	Vitelline membrane outer layer 1	Exact function unknown. Component of the outer membrane of the vitelline layer of the egg. Seems to be able to synthesize N-acetylchito-oligosaccharides (n = 14–15) from hexasaccharides of N-acetylglucosamine in a manner similar to the transferase activity of lysozyme.
416235	SPINK5	Ovoinhibitor	Serine protease inhibitor involved in antimicrobial egg defense preventing contamination of table eggs (non-fertilized eggs) and protecting the chick embryo (fertilized eggs) (probable). Inhibits trypsin, chymotrypsin, elastase, subtilisin and a proteinase of fungus *Aspergillus oryzae*.
396260	AVD	Avidin	The biological function of avidin is unknown. Forms a strong non-covalent specific complex with biotin (one molecule of biotin per subunit of avidin).
420898	OVALX	Ovalbumin-related protein X (SERPINB14C)	Heparin-binding. Defense response to bacteria.
396497	CST3	Cystatin C	This protein binds tightly to and inhibits a variety of thiol proteases, including ficin, papain, and cathepsins B, C, H, and L. Although isolated from egg white, it is also present in serum.

Note: The molecular function was extracted from the Uniprot Protein Database (uniprot.org)

The most abundant egg white protein is ovalbumin that accounts for more than 50% of total egg white proteins (Mann, 2007). Ovalbumin exhibits a three-dimensional structure that is specific to serpin (serine protease inhibitors) proteins, although it lacks inhibitory activities. The physiological function of this protein in egg remains unknown, but ovalbumin is assumed to provide essential amino acids for the chicken embryo growth. Ovalbumin has two paralogs, ovalbumin-related protein Y and X, which are in the top-ten abundant proteins in egg white (Table 3.5) (Silva et al., 2015). Beside ovalbumin, the egg white contains high concentration of lysozyme, an antimicrobial protein (Table 3.5). Remarkably, the egg white is also characterized by the presence of four highly abundant protease inhibitors, ovoinhibitor, cystatin, ovomucoid, and ovostatin (Table 3.5) that prevent the uncontrolled proteolytic degradation of proteins. These proteins are assumed to contribute to its hygienic quality and the high stability of the egg during its storage (Guyot et al., 2017).

Long egg storage is associated with egg white thinning because of the pH increase resulting from the loss of carbon dioxide through eggshell pores. This egg white thinning is characterized by a change in its physiochemical structure that is associated with protein structural modifications and differences in protein-protein electrostatic interactions (Huang et al., 2012; Sheng et al., 2016; Liu et al., 2022;) (see Chapters 5 and 19). These changes have been reported to modify the technological properties of egg for food industries (Jones, 2007; Deleu et al., 2015). The complete description of the egg white is available Chapter 5.

3.4.4 Structure and Composition of the Eggshell

The eggshell is composed of 1.6% water, 95.1% minerals (mainly calcium carbonate in the form of calcite), and 3.3% organic matter. The presence of a perfectly defined and structured eggshell has been highlighted by the work of Von Nathusius in the nineteenth century. This work was taken up and translated by Tyler (1964), and more recently in reviews (Tyler, 1964; Arias et al., 1993; Nys et al., 1999). The eggshell can be divided into five distinct layers (Figure 3.8). From the inner region to the periphery, we distinguish:

i. The shell membranes of which there are two, an inner one of about 20 µm thickness, and an outer one of about 50 µm that is located between the mineralized part of the shell and the inner membrane.
ii. The mammillary layer has a thickness of approximately 70 µm. It corresponds to the most internal part of the calcified layer, and represents less than a third of the thickness of the shell. The mammillary knobs, from which the mineralization is initiated, constitute its base. The mineralization continues toward the outside and gives a structure in the shape of cones or nipples.
iii. The palisade layer represents 2/3 of the shell thickness (about 200 µm). It begins when the growth of the cones of the mammillary layer leads to a fusion of the adjacent cones. The palisade layer is composed of a compact layer of minerals associated with an organic framework, except for the pores, which

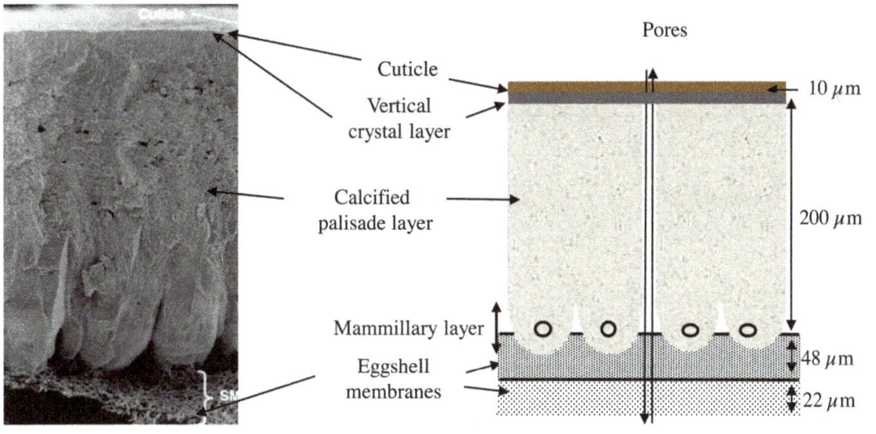

FIGURE 3.8 Scanning electron microscopy photograph and transversal diagram illustrating the highly ordered structure of the eggshell consisting of the shell membranes, the calcified layer, and the organic cuticle; pores cross the shell in a random way. The thickness of the different layers is indicated on the right side of the diagram.

cross the shell from one side to the other to allow the gas exchanges necessary for the development of the embryo. These pores are obtained in the absence of fusions of adjacent cones.

iv. The superficial layer of vertical crystals is deposited on the surface of the palisade layer. It is a thin layer of small calcite crystals, deposited vertically on the surface of the shell under the cuticle.

v. The cuticle is the most external layer of the egg. It is made up of organic matter. It contains most of the pigments responsible for the color of brown eggs. The cuticle is about 10 µm thick. It blocks the pores and thus prevents the penetration of bacteria inside the egg. Gas exchange is made possible by cracks that appear in the cuticle as it dries.

The main function of the shell is to protect the embryo from external aggressions. The shell must resist to the shocks coming from the outside, while having a structure that allows a facilitated rupture from the inside, necessary to the hatching of the embryo. Moreover, its structure must allow the gas exchanges during the development of the embryo. It must also be partially soluble during the embryonic development to provide the calcium needs of the chick. These requirements are fulfilled by the shell, which is a ceramic with an evolution in the organization of the size and orientation of its crystals from the inside to the outside (texture gradient) (Nys et al., 2004). In the outermost zone of the shell, there is a very hard structure made of large crystals. Inter-crystalline organic layers that limit the propagation of intra-crystalline cracks absorb external shocks (Nys et al., 2004). On the other hand, the internal region is composed of spherulitically arranged calcite microcrystals. This organization facilitates the propagation of cracks when they come from the shocks created by the beak of the embryo during its hatching. Moreover, this internal structure facilitates the mobilization of calcium to feed the embryo since the calcite microcrystals are more easily soluble.

The shell is therefore a porous ceramic formed at low temperature and has remarkable mechanical properties (Nys et al., 2004; Hincke et al., 2012; Gautron et al., 2021). Under static pressure, a 0.3 mm thick hen's eggshell resists a pressure of more than 30 N. These properties are the result of a very high degree of control over the size, shape, and orientation of the calcite crystals in the birds' eggshells that are responsible for this unique and particularly strong ultrastructure. This structure is particularly visible when observing cross sections of thin shell layers in polarized light. A nucleation center with a radial structure corresponding to the mammillary nuclei is observed. These nucleation centers are hemispherical, and are made of radially arranged calcite microcrystals. During the growth of the crystals, there is then competition for space. Only crystals oriented perpendicularly to the surface of the shell membranes will be able to grow.

The matrix composition and function of eggshell is widely described in Chapter 4. The ten most abundant proteins in the shell, and their functions as matrix proteins are listed in Table 3.6.

3.4.5 Shared, Common, and Specific Components of the Egg Compartments

The proteomic studies allowed the identification of hundreds of egg proteins, some of them being present in the various egg compartments. One gap in this methodology is the presence of redundancy between proteins, because the proteomic studies used identifiers originated from different databases. Altogether, egg proteomic studies described thousands of different protein identifiers. However, one unique protein often possesses various identifiers. To fix this problem, the entire protein sequences originated from these different identifiers have been loaded and were aligned using a BLAST algorithm to eliminate all redundancies (Gautron et al., 2019).

A total of 1,261 non-redundant protein sequences (gene products) were finally determined using the egg proteomic studies. The resulting distinct proteins were classified in the different compartments of the egg (Figure 3.9). Egg yolk, vitelline membrane, and egg white contain 230, 144, and 273 different proteins, respectively. Thus, the eggshell is the compartment with the lowest amount of proteins, but exhibits the largest variety of proteins, as 904 were reported in this

TABLE 3.6

Eggshell Proteins with Highest Abundance (Top-Ten) (Marie et al., 2015)

Gene ID	Gene Symbol	Gene Name	Biological Activity
396218	LYZ	Lysozyme	Antibacterial egg white proteins able to modify morphology of calcite crystals in eggshell (Hincke et al., 2000).
395256	OC-116	Ovocleidin-116	Protein-core of eggshell dermatan sulfate (Hincke et al., 1999).
100313508	OC-17	Ovocléidin-17	C-type lectin potentially involved in the stabilization of ACC and the formation of crystalline calcite (Hincke et al., 1995; Freeman et al., 2011).
427326	EDIL3	EGF-like repeats and discoidin I-like domains 3	Guidance molecule in extracellular vesicles (Stapane et al., 2019).
396241	OVOT	Ovotransferrin	Antibacterial egg white protein able to modify morphology of calcite crystals in eggshell (Gautron et al., 2001).
396197	ALB	Albumin	Chicken serum albumin.
396058	OVAL	Ovalbumin	Abundant egg white protein known to control the morphology of calcite in the eggshell (Dombre et al., 2017).
396497	CST3	Cystatin C	Protease inhibitor potentially involved in the control of proteins driving mineralization (Marie et al., 2015).
395722	CLU	Clusterin	Chaperone molecule, which could participate to proper folding of the shell matrix (Mann et al., 2003).
396393	LCN8	Lipocalin 8, Extracellular fatty acid-binding protein	Siderocalin-like lipocalin tightly binding a variety of bacterial ferric siderophores, also binds long-chain unsaturated fatty acids such as linoleic acid, oleic acid, arachidonic acid and, with a lower affinity, long chain saturated fatty acids such as stearic acid. May act as an antibacterial factor, through dual ligand specificity, both as a siderophore-sequestrating molecule and a lysophosphatidic acid (LPA) sensor.

Note: The biological functions were extracted from literature or the Uniprot Protein Knowledgebase (uniprot.org)

single compartment. It has to be noted that this integrative analysis was made in 2019, before the publication of the recent proteomic study on the inner and outer vitelline membranes (Brégeon et al., 2022). Consequently, the number of proteins indicated in this compartment is underestimated.

Figure 3.9 illustrates specific, common and shared proteins. The egg proteins only specific to one compartment are representing 82.6% (1,042) of the total number, with 143 proteins specific of the yolk, 41 only found in the vitelline membrane, 149 unique proteins of the egg white, and 709 proteins only observed in the eggshell. Surprisingly, the number of shared proteins between two or more egg compartments is limited to 219 proteins, representing less than 20% of the total number of egg proteins. Only 13 proteins are common to all egg compartments. The other 206 are found within two or three egg compartments.

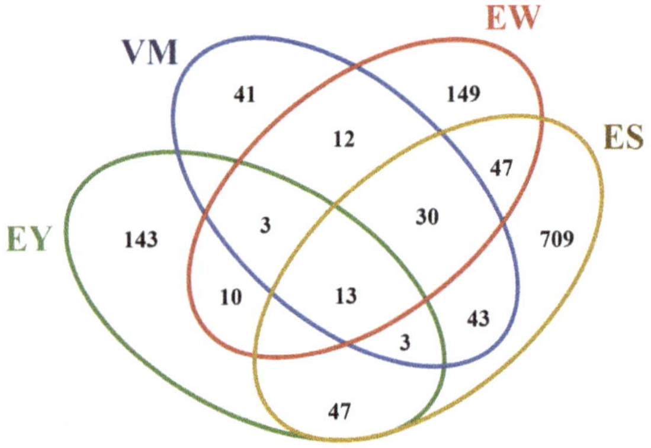

FIGURE 3.9 Venn diagram illustrating specific, common, and shared proteins according to the most recent integrative analysis of egg proteomes (Gautron et al., 2019).

3.5 Conclusion

Eggs have been shaped throughout evolution to support the autonomous development of an embryo outside the mother's body. In this respect, egg nutrients are perfectly balanced and need to be well assimilated by the embryo to avoid the accumulation of toxic metabolites during incubation. Besides their role as a source of nutrients, egg structures also participate in the antibacterial egg defense, either as physical barriers, or through the action of antimicrobial molecules and particular physicochemical conditions (egg white pH, viscosity). Some of their constituents also bear specific biological activities that all assist embryonic development. These characteristics all contribute to the high value of eggs for human nutrition (high nutritional value and high digestibility, especially when cooked/boiled), while being of moderate calorie source. The edible constituents of eggs contain many essential nutrients including macronutrients (proteins, lipids) and micronutrients (vitamins, choline, trace elements), although they lack fibers and vitamin C. Eggs are of low cost and can be stored for several weeks without deleterious degradation, which facilitates egg handling by producers, retailers and consumers. In addition, the diverse technological properties of egg white and egg yolk (emulsifying, foaming, etc., see Chapters 8 and 9) are of great interest for specific culinary preparations.

The biological activity of most egg proteins is unknown, but there is increasing evidence that some of them may have interest as nutraceuticals, considering their antimicrobial, antioxidant, and anti-cancer activities (see Part VII).

Although this aspect has not been addressed in this chapter, it has to be highlighted that there are many factors affecting the intrinsic proportion of egg components of egg, but also its physicochemical characteristics (Réhault-Godbert et al., 2019; Gautron et al., 2022). The first one is genetics. Indeed, laying hens have been selected for decades based on their productivity, the laying persistency and egg quality (eggshell strength and color, egg white quality and yolk percentage), and some major differences in egg quality and overall features can be observed between several genetic backgrounds. There are also some differences in egg composition across avian domestic species (weight, shape, etc.) (Sun et al., 2019). As an example, yolks from duck and goose species contain higher fat, and exhibit a higher percentage of yolk compared to chicken egg (Huang & Lin, 2011). Nutrition of hens (calcium particle size, presence of anti-nutritional factors, carotenoids, etc.) and rearing systems may also modify egg overall physicochemical characteristics (eggshell strength, yolk color, etc.). In addition, the physiological status of hens and more especially, age, stress and health problems can dramatically affect egg quality. Significant changes in egg internal components and quality are also clearly affected once laid, according to storage conditions (high temperature, long storage). Such modifications may also result in changing its technological properties and increase the risk of bacterial contaminations, and thereby consumer's food-borne disease (salmonellosis).

In conclusion, egg is an affordable product of high nutritional value that needs to be properly stored to maintain its nutritional benefits for human health. It is also a very fascinating model of autonomous development that also serves as an animal model to understand the basics for several physiological processes, including embryonic development, digestion, immunity, bone mineralization, etc.

REFERENCES

Addadi, L., & Weiner, S. (1992). Control and design principles in biological mineralization. *Angewandte Chemie International Edition in English*, *31*, 153–169.

Almquist, H. J. (1936). Formation of the chalazae in the Hen's egg. *Poultry Science*, *15*(6), 460–461. https://doi.org/10.3382/ps.0150460

Anton, M. (2013). Egg yolk: Structures, functionalities and processes. *Journal of the Science of Food and Agriculture*, *93*(12), 2871–2880. https://doi.org/10.1002/jsfa.6247

Anton, M., & Gandemer, G. (1997). Composition, solubility and emulsifying properties of granules and plasma of egg yolk. *Journal of Food Science*, *62*(3), 484–487. https://doi.org/10.1111/j.1365-2621.1997.tb04411.x

Arias, J. L., Fink, D. J., Xiao, S. Q., Heuer, A. H., & Caplan, A. I. (1993). Biomineralization and eggshells: Cell-mediated acellular compartments of mineralized extracellular matrix. *International Review of Cytology*, *145*, 217–250.

Back, J. F. (1984). Changes in the proteins of the vitelline membranes of hens eggs during storage. *Biochimica Et Biophysica Acta*, *799*(3), 319–321. https://doi.org/10.1016/0304-4165(84)90277-0

Back, J. F., Bain, J. M., Vadehra, D. V., & Burley, R. W. (1982). Proteins of the outer layer of the vitelline membrane of hen's eggs. *Biochimica Et Biophysica Acta*, *705*(1), 12–19. https://doi.org/10.1016/0167-4838(82)90329-6

Baker, R. C., Hartsell, S. E., & Stadelman, W. J. (1959). Lysozyme studies on chicken egg chalazae *Journal of Food Science*, *24*(5), 529–538. https://doi.org/10.1111/j.1365-2621.1959.tb17304.x

Bar, A. (2009). Calcium transport in strongly calcifying laying birds: Mechanisms and regulation. *Comparative Biochemistry and Physiology Part A: Molecular & Integrative Physiology*, *152*(4), 447–469. https://doi.org/10.1016/j.cbpa.2008.11.020

Bausek, N., Ruckenbauer, H. H., Pfeifer, S., Schneider, W. J., & Wohlrab, F. (2004). Interaction of sperm with purified native chicken ZP1 and ZPC proteins. *Biology of Reproduction*, *71*(2), 684–690. https://doi.org/10.1095/biolreprod.104.028605

Bausek, N., Waclawek, M., Schneider, W. J., & Wohlrab, F. (2000). The major chicken egg envelope protein ZP1 is different from ZPB and is synthesized in the liver. *The Journal of Biological Chemistry*, *275*(37), 28866–28872. https://doi.org/10.1074/jbc.275.37.28866

Bellairs, R. (1965). The relationship between oocyte and follicle in the hen's ovary as shown by electron microscopy. *Journal of Embryology and Experimental Morphology*, *13*, 215–233.

Bellairs, R., Harkness, M., & Harkness, R. D. (1963). The vitelline membrane of the hen's egg: A chemical and electron microscopical study. *Journal of Ultrastructure Research*, *8*(3), 339–359. https://doi.org/10.1016/S0022-5320(63)90012-1

Bourin, M., Gautron, J., Berges, M., Nys, Y., & Réhault-Godbert, S. (2012). Sex- and tissue-specific expression of "similar to nothepsin" and cathepsin D in relation to egg yolk formation in gallus. *Poultry Science*, *91*(9), 2288–2293. https://doi.org/10.3382/ps.2011-01910

Brégeon, M., Tomas, D., Bernay, B., Zatylny-Gaudin, C., Georgeault, S., Labas, V., Réhault-Godbert, S., & Guyot, N. (2022). Multifaceted roles of the egg perivitelline layer in avian reproduction: Functional insights from the proteomes of chicken egg inner and outer sublayers. *Journal of Proteomics*, *258*, 104489. https://doi.org/10.1016/j.jprot.2022.104489

Brionne, A., Nys, Y., Hennequet-Antier, C., & Gautron, J. (2014). Hen uterine gene expression profiling during eggshell formation reveals putative proteins involved in the supply of minerals or in the shell mineralization process. *BMC Genomics*, *15*, 220. https://doi.org/Artn 22010.1186/1471-2164-15-220

Chousalkar, K. K., & Roberts, J. R. (2008). Ultrastructural changes in the oviduct of the laying hen during the laying cycle. *Cell and Tissue Research*, *332*(2), 349–358. https://doi.org/10.1007/s00441-007-0567-3

Chue, J., & Smith, C. A. (2011). Sex determination and sexual differentiation in the avian model. *The FEBS Journal*, *278*(7), 1027–1034. https://doi.org/10.1111/j.1742-4658.2011.08032.x

Chung, W. H., Lai, K. M., & Hsu, K. C. (2010). Comparative study on histological structures of the vitelline membrane of hen and duck egg observed by cryo-scanning electron microscopy. *Journal of Agricultural and Food Chemistry*, *58*(3), 1794–1799. https://doi.org/10.1021/jf903924g

Cutting, A. D., Bannister, S. C., Doran, T. J., Sinclair, A. H., Tizard, M. V. L., & Smith, C. A. (2012). The potential role of microRNAs in regulating gonadal sex differentiation in the chicken embryo. *Chromosome Research*, *20*(1), 201–213. https://doi.org/10.1007/s10577-011-9263-y

Da Silva, Beauclercq, M., Harichaux, S., Labas, G., Guyot, V., Gautron, N., Nys, J., & Rehault-Godbert, Y., S. (2015). The family secrets of avian egg-specific ovalbumin and its related proteins Y and X. *Biology of Reproduction*, *93*(3), 71. https://doi.org/10.1095/biolreprod.115.130856

Deboeck, S., & Stockx, J. (1986). Mode of interaction between lysozyme and teh other proteins of the hens egg vitelline membrane. *International Journal of Biochemistry*, *18*(7), 623–628. https://doi.org/10.1016/0020-711x(86)90292-2

Deleu, L. J., Wilderjans, E., Van Haesendonck, I., Courtin, C. M., Brijs, K., & Delcour, J. A. (2015). Storage induced conversion of ovalbumin into S-ovalbumin in eggs impacts the properties of pound cake and its batter. *Food Hydrocolloids*, *49*, 208–215. https://doi.org/10.1016/j.foodhyd.2015.03.014

Dombre, C., Guyot, N., Moreau, T., Monget, P., Da Silva, M., Gautron, J., & Rehault-Godbert, S. (2017). Egg serpins: The chicken and/or the egg dilemma. *Seminars in Cell & Developmental Biology*, *62*, 120–132. https://doi.org/10.1016/j.semcdb.2016.08.019

Etches, R. J. (1986). *Reproduction in poultry*. CAB International.

Farinazzo, A., Restuccia, U., Bachi, A., Guerrier, L., Fortis, F., Boschetti, E., Fasoli, E., Citterio, A., & Righetti, P. G. (2009). Chicken egg yolk cytoplasmic proteome, mined via combinatorial peptide ligand libraries. *Journal of Chromatography A*, *1216*(8), 1241–1252. https://doi.org/10.1016/j.chroma.2008.11.051

Freeman, C. L., Harding, J. H., Quigley, D., & Rodger, P. M. (2011). Simulations of ovocleidin-17 binding to calcite surfaces and its implications for eggshell formation. *The Journal of Physical Chemistry*, *115*(16), 8175–8183.

Fujii, S., Tamura, T., & Okamoto, T. (1972). Light and scanning electron microscopical studies on the structure of the vitelline membrane of the Hen's egg. *Journal of the Faculty of Fisheries and Animal Husbandry, Hiroshima University*, *11*, 1–13.

Gautron, J., Dombre, C., Nau, F., Feidt, C., & Guillier, L. (2022). Review: Production factors affecting the quality of chicken table eggs and egg products in Europe. *Animal*, *16*, 100425. https://doi.org/10.1016/j.animal.2021.100425

Gautron, J., Guyot, N., Brionne, A., & Réhault-Godbert, S. (2019). CHAPTER 14 bioactive minor egg components. In J. Wu. (Ed.), *Eggs as functional foods and nutraceuticals for human health* (pp. 259–284). The Royal Society of Chemistry. https://doi.org/10.1039/9781788013833-00259

Gautron, J., Hincke, M. T., Panheleux, M., Garcia-Ruiz, J. M., Boldicke, T., & Nys, Y. (2001). Ovotransferrin is a matrix protein of the hen eggshell membranes and basal calcified layer. *Connective Tissue Research*, *42*(4), 255–267. https://doi.org/10.3109/03008200109016840

Gautron, J., Stapane, L., Le Roy, N., Nys, Y., Rodriguez-Navarro, A. B., & Hincke, M. T. (2021). Avian eggshell biomineralization: An update on its structure, mineralogy and protein tool kit. *BMC Molecular and Cell Biology*, *22*(1), 11. https://doi.org/10.1186/s12860-021-00350-0

Geng, F., Xie, Y., Wang, J., Majumder, K., Qiu, N., & Ma, M. (2018). N-glycoproteomic analysis of chicken egg yolk. *Journal of Agriculture and Food Chemistry*, *66*(43), 11510–11516. https://doi.org/10.1021/acs.jafc.8b04492

Gilbert, A. B. (1979). Female genital tract. In A. S. King & J. M. Lelland (Eds.), *Form and function in birds* (pp. 237–360). Academic Press.

Gloux, A., Duclos, M. J., Brionne, A., Bourin, M., Nys, Y., & Réhault-Godbert, S. (2019a). Integrative analysis of transcriptomic data related to the liver of laying hens: From physiological basics to newly identified functions. *BMC Genomics*, *20*(1), 821. https://doi.org/10.1186/s12864-019-6185-0

Gloux, A., Le Roy, N., Brionne, A., Bonin, E., Juanchich, A., Benzoni, G., Piketty, M. L., Prie, D., Nys, Y., Gautron, J., Narcy, A., & Duclos, M. J. (2019b). Candidate genes of the transcellular and paracellular calcium absorption pathways in the small intestine of laying hens. *Poultry Science*, *98*(11), 6005–6018. https://doi.org/10.3382/ps/pez407

Guyot, N., Labas, V., Harichaux, G., Chessé, M., Poirier, J.-C., Nys, Y., & Réhault-Godbert, S. (2016). Proteomic analysis of egg white heparin-binding proteins: Towards the identification of natural antibacterial molecules. *Scientific Reports*, *6*, 27974. https://doi.org/10.1038/srep27974

Guyot, N., Meudal, H., Trapp, S., Iochmann, S., Silvestre, A., Jousset, G., Labas, V., Reverdiau, P., Loth, K., Hervé, V., Aucagne, V., Delmas, A. F., Rehault-Godbert, S., & Landon, C. (2020). Structure, function, and evolution of Gga-AvBD11, the archetype of the structural avian-double-β-defensin family. *Proceedings of the National Academy of Sciences (USA)*, *117*(1), 337–345. https://doi.org/10.1073/pnas.1912941117

Guyot, N., Réhault-Godbert, S., Nys, Y., & Baron, F. (2017). Understanding the natural antibacterial defences of egg white and their regulation. In J. Roberts (Ed.), *Achieving sustainable production of eggs* (Vol. 1, pp. 161–193). Burleigh Dodds Science Publishing.

Hervé-Grépinet, V., Réhault-Godbert, S., Labas, V., Magallon, T., Derache, C., Lavergne, M., Gautron, J., Lalmanach, A.-C., & Nys, Y. (2010). Purification and characterization of avian beta-defensin 11, an antimicrobial peptide of the hen egg. *Antimicrobial Agents and Chemotherapy*, *54*(10), 4401–4409. https://doi.org/10.1128/aac.00204-10

Hincke, M. T., Gautron, J., Panheleux, M., Garcia-Ruiz, J., McKee, M. D., & Nys, Y. (2000). Identification and localization of lysozyme as a component of eggshell membranes and eggshell matrix. *Matrix Biology*, *19*(5), 443–453. https://doi.org/10.1016/S0945-053x(00)00095-0

Hincke, M. T., Gautron, J., Tsang, C. P. W., McKee, M. D., & Nys, Y. (1999). Molecular cloning and ultrastructural localization of the core protein of an eggshell matrix proteoglycan, ovocleidin-116. *Journal of Biological Chemistry*, *274*(46), 32915–32923. https://doi.org/10.1074/jbc.274.46.32915

Hincke, M. T., Nys, Y., Gautron, J., Mann, K., Rodriguez-Navarro, A. B., & McKee, M. D. (2012). The eggshell: Structure, composition and mineralization. *Frontiers in Bioscience*, *17*, 1266–1280. https://www.ncbi.nlm.nih.gov/pubmed/22201802

Hincke, M. T., Tsang, C. P., Courtney, M., Hill, V., & Narbaitz, R. (1995). Purification and immunochemistry of a soluble matrix protein of the chicken eggshell (ovocleidin 17). *Calcified Tissue International*, *56*(6), 578–583.

Hrabia, A. (2022). Chapter 35 – Reproduction in the female. In C. G. Scanes & S. Dridi (Eds.), *Sturkie's avian physiology* (7th ed., pp. 941–986). Academic Press. https://doi.org/10.1016/B978-0-12-819770-7.00002-5

Hrabia, A., Sechman, A., Gertler, A., & Rząsa, J. (2011). Effect of growth hormone on steroid content, proliferation and apoptosis in the chicken ovary during sexual maturation. *Cell and Tissue Research*, *345*(1), 191–202. https://doi.org/10.1007/s00441-011-1187-5

Huang, E. S., Kao, K. J., & Nalbandov, A. V. (1979). Synthesis of sex steroids by cellular components of chicken follicles. *Biology of Reproduction*, *20*(3), 454–461. https://doi.org/10.1095/biolreprod20.3.454

Huang, J. F., & Lin, C. C. (2011). Production, composition and quality of duck eggs. In Y. Nys, M. Bain, & F. Van Immersel (Eds.), *Improving the safety and quality of eggs and egg products* (Vol. 1, pp. 487–508). Woodhead Publishing Limited.

Huang, Q., Qiu, N., Ma, M. H., Jin, Y. G., Yang, H., Geng, F., & Sun, S. H. (2012). Estimation of egg freshness using S-ovalbumin as an indicator. *Poultry Science*, *91*(3), 739–743. https://doi.org/10.3382/ps.2011-01639

Itoh, T., MiyazakiI, J., Sugawara, H., & AdachiI, S. (1987). Studies on the characterization of ovomucin and chalaza of the Hen's egg. *Journal of Food Science*, *52*(6), 1518–1521. https://doi.org/10.1111/j.1365-2621.1987.tb05868.x

Johnson, A. L., & Woods, D. C. (2007). Ovarian dynamics and follicle development. In B. G. M. Jamieson (Ed.), *Reproductive physiology of birds (part A)* (pp. 243–277). Science Publishers.

Jolivet, P., Boulard, C., Beaumal, V., Chardot, T., & Anton, M. (2006). Protein components of low-density lipoproteins purified from hen egg yolk. *Journal of Agricultural and Food Chemistry*, *54*(12), 4424–4429. https://doi.org/10.1021/jf0531398

Jonchere, V., Brionne, A., Gautron, J., & Nys, Y. (2012). Identification of uterine ion transporters for mineralisation precursors of the avian eggshell. *BMC Physiology*, *12*, 10. https://doi.org/10.1186/1472-6793-12-10

Jones, D. R. (2007). Egg functionality and quality during long-term storage. *International Journal of Poultry Science*, *6*, 157–162. https://doi.org/10.3923/ijps.2007.157.162

Jones, C. J. P., Aplin, J. D., Ververs, C., Twink Allen, W. R., Burns, K., & Wilsher, S. (2021). A preliminary study on the glycosylation of the reproductive tract in the ostrich (Struthio camelus camelus). *Reproduction in Domestic Animals*, *56*(1), 130–141. https://doi.org/10.1111/rda.13857

Jung, J. G., Lim, W., Park, T. S., Kim, J. N., Han, B. K., Song, G., & Han, J. Y. (2011). Structural and histological characterization of oviductal magnum and lectin-binding patterns in gallus domesticus. *Reproductive Biology and Endocrinology*, *9*, 62. https://doi.org/10.1186/1477-7827-9-62

Kato, A., Imoto, T., & Yagishita, K. (1975). The binding groups in ovomucin-lysozyme interaction. *Agricultural and Biological Chemistry*, *39*(2), 541–544. https://doi.org/10.1080/00021369.1975.10861614

Kido, S., & Doi, Y. (1988). Separation and properties of the inner and outer layers of the vitelline membrane of Hen's eggs. *Poultry Science*, *67*(3), 476–486. https://doi.org/10.3382/ps.0670476

Kido, S., Doi, Y., Kim, F., Morishita, E., Narita, H., Kanaya, S., Ohkubo, T., Nishikawa, K., Yao, T., & Ooi, T. (1995). Characterization of vitelline membrane outer layer protein-I, VMO-I – Amino acid sequence and structural stability. *Journal of Biochemistry*, *117*(6), 1183–1191. https://doi.org/10.1093/oxfordjournals.jbchem.a124842

Kido, S., Janado, M., & Nunoura, H. (1975). Macromolecular components of the vitelline membrane of hen's egg. I. Membrane structure and its deterioration with age. *The Journal of Biochemistry*, *78*(2), 261–268. https://doi.org/10.1093/oxfordjournals.jbchem.a130903

Kido, S., Morimoto, A., Kim, F., & Doi, Y. (1992). Isolation of a novel protein from the outer layer of the vitelline membrane. *Biochemical Journal*, *286* (Pt 1), 17–22. https://doi.org/10.1042/bj2860017

Kinoshita, M., Rodler, D., Sugiura, K., Matsushima, K., Kansaku, N., Tahara, K., Tsukada, A., Ono, H., Yoshimura, T., Yoshizaki, N., Tanaka, R., Kohsaka, T., & Sasanami, T. (2010). Zona pellucida protein ZP2 is expressed in the oocyte of Japanese quail (coturnix japonica). *Reproduction*, *139*(2), 359–371. https://doi.org/10.1530/rep-09-0222

Li-Chan, E. C. Y., & Kim, H.-O. (2008). Structure and chemical compositions of eggs. In *Egg bioscience and biotechnology* (pp. 1–95). https://doi.org/10.1002/9780470181249.ch1

Lindsay, L. L., Larabell, C. A., & Hedrick, J. L. (1992). Localization of a chymotrypsin-like protease to the perivitelline space of xenopus laevis eggs. *Developmental Biology*, *154*(2), 433–436. https://doi.org/10.1016/0012-1606(92)90081-q

Liu, L., Wang, J., Wang, G., Song, H., Geng, F., Zeng, Q., & Huang, Q. (2022). Quantitative proteomics provides a new perspective on the mechanism of network structure depolymerization during egg white thinning. *Food Chemistry*, *392*, 133320. https://doi.org/10.1016/j.foodchem.2022.133320

Mann, K. (2007). The chicken egg white proteome. *Proteomics*, *7*(19), 3558–3568. https://doi.org/10.1002/pmic.200700397

Mann, K. (2008). Proteomic analysis of the chicken egg vitelline membrane. *Proteomics*, *8*(11), 2322–2332. https://doi.org/10.1002/pmic.200800032

Mann, K., Gautron, J., Nys, Y., McKee, M. D., Bajari, T., Schneider, W. J., & Hincke, M. T. (2003). Disulfide-linked heterodimeric clusterin is a component of the chicken eggshell matrix and egg white. *Matrix Biology*, *22*(5), 397–407.

Mann, K., & Mann, M. (2008). The chicken egg yolk plasma and granule proteomes. *Proteomics*, *8*(1), 178–191. https://doi.org/10.1002/pmic.200700790

Mao, K. M., Sultana, F., Howlider, M. A. R., Iwasawa, A., & Yoshizaki, N. (2006). The magnum-isthmus junction of the fowl oviduct participates in the formation of the avian-type shell membrane. *Zoological Science*, *23*(1), 41–47. https://doi.org/10.2108/zsj.23.41

Marie, P., Labas, V., Brionne, A., Harichaux, G., Hennequet-Antier, C., Rodriguez-Navarro, A. B., Nys, Y., & Gautron, J. (2015). Quantitative proteomics provides new insights into chicken eggshell matrix protein functions during the primary events of mineralisation and the active calcification phase. *Journal of Proteomics*, *126*, 140–154. https://doi.org/10.1016/j.jprot.2015.05.034

Marin, F., Luquet, G., Marie, B., & Medakovic, D. (2008). Molluscan shell proteins: Primary structure, origin, and evolution. *Current Topics in Developmental Biology*, *80*, 209–276.

Moreau, T., Gautron, J., Hincke, M. T., Monget, P., Réhault-Godbert, S., & Guyot, N. (2022). Antimicrobial proteins and peptides in avian eggshell: Structural diversity and potential roles in biomineralization. *Frontiers in*

Immunology, *13*, 946428. https://doi.org/10.3389/fimmu.2022.946428

Mun, J. G., Legette, L. L., Ikonte, C. J., & Mitmesser, S. H. (2019). Choline and DHA in maternal and infant nutrition: Synergistic implications in brain and eye health. *Nutrients*, *11*(5). https://doi.org/10.3390/nu11051125

Nimpf, J., & Schneider, W. J. (1998). The VLDL receptor: An LDL receptor relative with eight ligand binding repeats, LR8. *Atherosclerosis*, *141*(2), 191–202. https://doi.org/10.1016/s0021-9150(98)00172-5

Nishio, S., Kohno, Y., Iwata, Y., Arai, M., Okumura, H., Oshima, K., Nadano, D., & Matsuda, T. (2014). Glycosylated chicken ZP2 accumulates in the egg coat of immature oocytes and remains localized to the germinal disc region of mature eggs1. *Biology of Reproduction*, *91*(5). https://doi.org/10.1095/biolreprod.114.119826

Nishio, S., Okumura, H., & Matsuda, T. (2018). Egg-coat and zona pellucida proteins of chicken as a typical species of aves. *Current Topics in Developmental Biology*, *130*, 307–329. https://doi.org/10.1016/bs.ctdb.2018.02.008

Noble, R. C., & Cocchi, M. (1990). Lipid metabolism and the neonatal chicken. *Progress in Lipid Research*, *29*(2), 107–140. https://doi.org/10.1016/0163-7827(90)90014-c

Nys, Y., Bain, M., & Van Immerseel, F. (2011). *Improving the safety and quality of eggs and egg products. Vol. 1: Egg chemistry, production and consumption* (Vol. 1). Woodhead Publishing. https://www.sciencedirect.com/science/article/pii/B9781845697549500243

Nys, Y., Gautron, J., Garcia-Ruiz, J. M., & Hincke, M. T. (2004). Avian eggshell mineralization: Biochemical and functional characterization of matrix proteins. *Comptes Rendus Palevol*, *3*(6–7), 549–562. https://doi.org/10.1016/j.crpv.2004.08.002

Nys, Y., Gautron, J., Rodriguez-Navarro, A. B., & Hincke, M. (2022). Chapter 32 – Mechanisms and hormonal regulation of shell formation: Supply of ionic and organic precursors, shell mineralization. In C. G. Scanes & S. Dridi (Eds.), *Sturkie's avian physiology* (7th ed., pp. 833–879). Academic Press. https://doi.org/10.1016/B978-0-12-819770-7.00049-9

Nys, Y., & Guyot, N. (2011). Egg formation and chemistry. In Y. Nys, M. Bain, & F. Van Immerseel (Eds.), *Improving the safety and quality of eggs and egg products* (Vol. 1, pp. 83–132). Woodhead Publishing.

Nys, Y., Hincke, M. T., Arias, J. L., Garcia-Ruiz, J. M., & Solomon, S. E. (1999). Avian eggshell mineralization. *Poultry and Avian Biology Reviews*, *10*(3), 143–166.

Nys, Y., & Le Roy, N. (2018). Calcium homeostasis and eggshell biomineralization in female chicken. In D. Feldman (Ed.), *Vitamin D* (4 ed., pp. 361–382). Academic Press.

Nys, Y., Zawadzki, J., Gautron, J., & Mills, A. D. (1991). Whitening of brown-shelled eggs: Mineral composition of uterine fluid and rate of protoporphyrin deposition. *Poultry Science*, *70*(5), 1236–1245. https://doi.org/10.3382/ps.0701236

Okumura, H., Aoki, N., Sato, C., Nadano, D., & Matsuda, T. (2007). Heterocomplex formation and cell-surface accumulation of hen's serum zona pellucida B1 (ZPB1)with ZPC expressed by a mammalian cell line (COS-7): A possible initiating step of egg-envelope matrix construction. *Biology of Reproduction*, *76*(1), 9–18. https://doi.org/10.1095/biolreprod.106.056267

Okumura, H., Kohno, Y., Iwata, Y., Mori, H., Aoki, N., Sato, C., Kitajima, K., Nadano, D., & Matsuda, T. (2004). A newly identified zona pellucida glycoprotein, ZPD, and dimeric ZP1 of chicken egg envelope are involved in sperm activation on sperm-egg interaction. *Biochemical Journal*, *384*(Pt 1), 191–199. https://doi.org/10.1042/bj20040299

Okumura, H., Sato, T., Sakuma, R., Fukushima, H., Matsuda, T., & Ujita, M. (2015). Identification of distinctive interdomain interactions among ZP-N, ZP-C and other domains of zona pellucida glycoproteins underlying association of chicken egg-coat matrix. *FEBS Open Bio*, *5*, 454–465. https://doi.org/10.1016/j.fob.2015.05.005

Panheleux, M., Nys, Y., Williams, J., Gautron, J., Boldicke, T., & Hincke, M. T. (2000). Extraction and quantification by ELISA of eggshell organic matrix proteins (ovocleidin-17, ovalbumin, ovotransferrin) in shell from young and old hens. *Poultry Science*, *79*(4), 580–588. https://doi.org/10.1093/ps/79.4.580

Perry, M. M., & Gilbert, A. B. (1979). Yolk transport in the ovarian follicle of the hen (gallus domesticus): Lipoprotein-like particles at the periphery of the oocyte in the rapid growth phase. *Journal of Cell Science*, *39*, 257–272. https://doi.org/10.1242/jcs.39.1.257

Perry, M. M., Gilbert, A. B., & Evans, A. J. (1978). Electron microscope observations on the ovarian follicle of the domestic fowl during the rapid growth phase. *Journal of Anatomy*, *125*(Pt 3), 481–497.

Phillips, S. M. (2017). Current concepts and unresolved questions in dietary protein requirements and supplements in adults [Review]. *Frontiers in Nutrition*, *4*. https://www.frontiersin.org/articles/10.3389/fnut.2017.00013

Porter, T. E., Hargis, B. M., Silsby, J. L., & Halawani, M. E. (1989). Differential steroid production between theca interna and theca externa cells: A three-cell model for follicular steroidogenesis in avian Species*. *Endocrinology*, *125*(1), 109–116. https://doi.org/10.1210/endo-125-1-109

Rahman, M. A., Moriyama, A., Iwasawa, A., & Yoshizaki, N. (2009). VMO-II mediates the binding of the chalaziferous layer with the vitelline membrane in quail eggs. *The Journal of Poultry Science*, *46*(3), 240–248. https://doi.org/10.2141/jpsa.46.240

Réhault-Godbert, S., & Guyot, N. (2018). Vitellogenesis and yolk proteins, birds. In M. K. Skinner (Ed.), *Encyclopedia of reproduction* (2nd ed., pp. 278–284). Academic Press. https://doi.org/10.1016/B978-0-12-809633-8.20568-2

Réhault-Godbert, S., Guyot, N., & Nys, Y. (2019). The golden egg: Nutritional value, bioactivities, and emerging benefits for human health. *Nutrients*, *11*, 684. https://doi.org/10.3390/nu11030684

Rodriguez-Navarro, A. B., Marie, P., Nys, Y., Hincke, M. T., & Gautron, J. (2015). Amorphous calcium carbonate controls avian eggshell mineralization: A new paradigm for understanding rapid eggshell calcification. *Journal of Structural Biology*, *190*(3), 291–303. https://doi.org/10.1016/j.jsb.2015.04.014

Romanoff, A. L., & Romanof, A. J. (1949). *The avian egg*. John Willey and Sons. https://www.worldcat.org/title/avian-egg/oclc/607757071

Rothwell, B., & Solomon, S. E. (1977). The ultrastructure of the follicle wall of the domestic fowl during the phase of rapid growth. *British Poultry Science*, *18*(5), 605–610. https://doi.org/10.1080/00071667708416409

Sah, N., Kuehu, D. L., Khadka, V. S., Deng, Y., Peplowska, K., Jha, R., & Mishra, B. (2018). RNA sequencing-based analysis of the laying hen uterus revealed the novel genes and biological pathways involved in the eggshell biomineralization. *Scientific Reports*, *8*(1), 16853. https://doi.org/10.1038/s41598-018-35203-y

Sasanami, T., Ohtsuki, M., Ishiguro, T., Matsushima, K., Hiyama, G., Kansaku, N., Doi, Y., & Mori, M. (2006). Zona pellucida domain of ZPB1 controls specific binding of ZPB1 and ZPC in Japanese quail (Coturnix japonica). *Cells Tissues Organs*, *183*(1), 41–52. https://doi.org/10.1159/000094905

Sauveur, B., & De Reviers, M. (1988). *Reproduction des volailles et production d'oeufs*. Institut National de la Recherche Agronomique (INRA).

Schäfer, A., Drewes, W., & Schwägele, F. (1998). Analysis of vitelline membrane proteins of fresh and stored eggs via HPLC. *Zeitschrift für lebensmitteluntersuchung und -forschung A*, *206*(5), 329–332. https://doi.org/10.1007/s002170050267

Sheng, L., Wang, J., Huang, M., Xu, Q., & Ma, M. (2016). The changes of secondary structures and properties of lysozyme along with the egg storage. *International Journal of Biological Macromolecules*, *92*, 600–606. https://doi.org/10.1016/j.ijbiomac.2016.07.068

Solomon, S. E. (1991). *Egg and egg quality*. Wolfe Publishing.

Stapane, L., Le Roy, N., Ezagal, J., Rodriguez-Navarro, A. B., Labas, V., Combes-Soia, L., Hincke, M. T., & Gautron, J. (2020). Avian eggshell formation reveals a new paradigm for vertebrate mineralization via vesicular amorphous calcium carbonate. *Journal of Biological Chemistry*, *22*, 11.

Stapane, L., Le Roy, N., Hincke, M. T., & Gautron, J. (2019). The glycoproteins EDIL3 and MFGE8 regulate vesicle-mediated eggshell calcification in a new model for avian biomineralization. *Journal of Biological Chemistry*, *294*(40), 14526–14545. https://doi.org/10.1074/jbc.RA119.009799

Stewart, S. G., Bausek, N., Wohlrab, F., Schneider, W. J., Janet Horrocks, A., & Wishart, G. J. (2004). Species specificity in avian sperm: Perivitelline interaction. *Comparative Biochemistry and Physiology Part A: Molecular & Integrative Physiology*, *137*(4), 657–663. https://doi.org/10.1016/j.cbpb.2004.01.027

Sultana, F., Yokoe, A., Ito, Y., Ma, K. M., & Yoshizak, N. (2003). The peri-albumen layer: A novel structure in the envelopes of an avian egg. *Journal of Anatomy*, *203*(1), 115–122. https://doi.org/10.1046/j.1469-7580.2003.00197.x

Sun, C., Liu, J., Yang, N., & Xu, G. (2019). Egg quality and egg albumen property of domestic chicken, duck, goose, turkey, quail, and pigeon. *Poultry Science*, *98*(10), 4516–4521. https://doi.org/10.3382/ps/pez259

Takeuchi, Y., Nishimura, K., Aoki, N., Adachi, T., Sato, C., Kitajima, K., & Matsuda, T. (1999). A 42-kDa glycoprotein from chicken egg-envelope, an avian homolog of the ZPC family glycoproteins in mammalian zona pellucida. Its first identification, cDNA cloning and granulosa cell-specific expression. *European Journal of Biochemistry*, *260*(3), 736–742. https://doi.org/10.1046/j.1432-1327.1999.00203.x

Trigg, N. A., Stanger, S. J., Zhou, W., Skerrett-Byrne, D. A., Sipilä, P., Dun, M. D., Eamens, A. L., De Iuliis, G. N., Bromfield, E. G., Roman, S. D., & Nixon, B. (2021). A novel role for milk fat globule-EGF factor 8 protein (MFGE8) in the mediation of mouse sperm-extracellular vesicle interactions. *Proteomics*, *21*(13–14), e2000079. https://doi.org/10.1002/pmic.202000079

Trziszka, T., & Smolińska, T. (1982). Chemical characterisation of the vitelline membrane of hens' eggs. *Food Chemistry*, *8*(1), 61–70. https://doi.org/10.1016/0308-8146(82)90058-9

Tyler, C. (1964). *Wihhelm von Nathusius 1821–1899 on avian eggshells*. University of Reading.

Uyeda, A., Inuzuka, C., Doi, Y., Kido, S., & Kikuchi, M. (1994). Cloning and sequencing of hen magnum cDNAs encoding vitelline membrane outer layer protein I (VMO-I). *Gene*, *144*(2), 311–312. https://doi.org/10.1016/0378-1119(94)90397-2

Wang, H., Qiu, N., Mine, Y., Sun, H., Meng, Y., Bin, L., & Keast, R. (2020). Quantitative comparative integrated proteomic and phosphoproteomic analysis of chicken egg yolk proteins under diverse storage temperatures. *Journal of Agriculture and Food Chemistry*, *68*(4), 1157–1167. https://doi.org/10.1021/acs.jafc.9b06912

Wang, Y., Wang, J., Shi, Y., Ye, H., Luo, W., & Geng, F. (2022). Quantitative proteomic analyses during formation of chicken egg yolk. *Food Chemistry*, *374*, 131828. https://doi.org/10.1016/j.foodchem.2021.131828

Wyburn, G. M., Johnston, H. S., Draper, M. H., & Davidson, M. F. (1970). The fine structure of the infundibulum and magnum of the oviduct of gallus domesticus. *Quarterly Journal of Experimental Physiology and Cognate Medical Sciences*, *55*(3), 213–232. https://doi.org/10.1113/expphysiol.1970.sp002071

Xing, J., Wellman-Labadie, O., Gautron, J., & Hincke, M. T. (2007). Recombinant eggshell ovocalyxin-32: Expression, purification and biological activity of the glutathione s-transferase fusion protein. *Comparative Biochemistry and Physiology Part B: Biochemical and Molecular Biology*, *147*(2), 172–177. https://doi.org/10.1016/j.cbpb.2007.01.015

Yang, R., Geng, F., Huang, X., Qiu, N., Li, S. G., Teng, H., Chen, L., Song, H. B., & Huang, Q. (2020). Integrated proteomic, phosphoproteomic and N-glycoproteomic analyses of chicken eggshell matrix.*Food Chemistry*, *330*, 127167. https://doi.org/10.1016/j.foodchem.2020.127167

4

Eggshell and Membranes

Tamer A. E. Ahmed, Garima Kulshreshtha, Manar Younes, and Maxwell T. Hincke*[*]

CONTENTS

4.1 Introduction

The calcareous egg of birds is a successful reproductive adaptation to the desiccating terrestrial environment, and is often considered as the most advanced amniotic egg in oviparous vertebrates. The eggshell resists external physical and pathogen challenges, and meets the metabolic and nutritional needs of the developing embryo. The eggshell regulates gas and water exchange, serves as a calcium store and provides camouflage and thermal regulation for the egg contents. The shell is a complex bioceramic with properties adapted to the environment of a given species (Hincke et al., 2012). For example, shell permeability depends on the characteristics of its respiratory pores – number/density, caliber, and branching pattern. The shell resists the weight of the brooding hen and provides protection against physical damage, microbial invasion and consummation by small animals. Many physiological, biochemical, nutritional, structural and morphological studies have been conducted on avian eggs and most have utilized the egg of the domestic chicken (*Gallus gallus domesticus*) (Figure 4.1), because of its availability and commercial importance as a nutritious food for human consumption. This chapter will emphasize the constituents of the chicken eggshell, from which much of our knowledge of avian eggshell mineralization, function and composition is derived.

4.2 Structure of Eggshell

In most bird species, the eggshell mass is approximately 10–11% of egg weight; it consists of the calcite crystalline form of calcium carbonate (Nys et al., 2004; Hincke et al., 2012; Gautron et al., 2021). The well-studied chicken eggshell is a thin layer about 0.35–0.4 mm in thickness, containing 95% calcium carbonate and about 3.5% organic matrix when the eggshell membranes are included (Figure 4.2). Minor amounts of Mg and phosphorus are readily detected and are differentially distributed within the shell; however, other elements are

[*] Corresponding author.

DOI: 10.1201/9781003254430-6

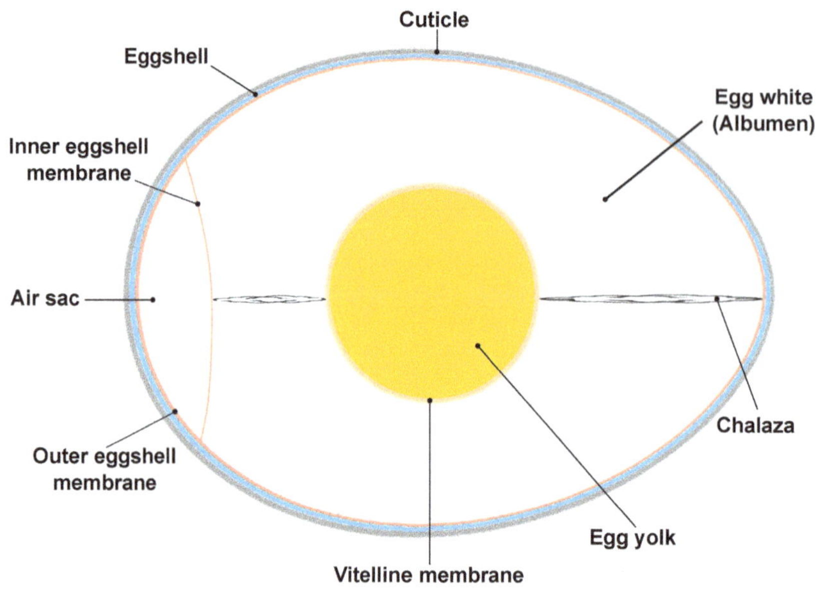

FIGURE 4.1 Longitudinal section to depict the interior contents of a chicken egg.

FIGURE 4.2 Scanning electron micrograph of cross-fractured chicken eggshell showing overall structure and regions of the calcified eggshell. Starting adjacent to the shell membranes (ESM), the eggshell consists of mammillae (ML), palisades (PL), vertical crystal layer (VCL), and cuticle (C). The micrograph also shows the pore plug (PP) and respiratory pore (P).

also present in trace amounts. The eggshell possesses remarkable mechanical strength; the chicken egg is capable of withstanding a static pressure of more than 3 Kg (Gautron et al., 2021). The occluded organic matrix of the eggshell is a complex mixture of proteins, glycoproteins and proteoglycans that regulate mineralization during its formation. Its glycosaminoglycan moieties include hyaluronic acid (48%) and galactosaminoglycans (52%: chondroitin sulfate, dermatan sulfate) (Nakano et al., 2001; Nakano et al., 2002; Nakano et al., 2003).

The avian eggshell is typically subdivided into six layers for descriptive purposes (Cordeiro et al., 2013). The uncalcified inner and outer eggshell membranes are the innermost layers of the ES; each shell membrane layer is composed of a fibrous network of crosslinked proteins. The inner ESM envelops the albumen (Ahmed, Suso et al., 2019), while the outer ESM provides a platform upon which eggshell mineralization is initiated and from which outward calcite crystal growth occurs. The mammillary knob layer is composed of calcite crystals that are distributed pseudo-periodically on the surface of the outer ESM, to which the mineralized shell is anchored by fibers that penetrate the tips of the mammillary knobs. The palisade layer extends outward from the fusion of the bases of the mammillary knobs to form a continuous calcified layer of large columnar units; it corresponds to about two-thirds of the ES mineral thickness. The palisade layer ends with a thin vertical crystal layer, where crystals are aligned perpendicular to the shell surface. The outmost layer of the ES, the cuticle, is a protein-rich, non-calcified film deposited on the surface of the ES mineral; its thickness ranges from 5 to 10 μm in the chicken eggshell (Rose & Hincke, 2009; Hincke et al., 2012). The inner zone of cuticle contains hydroxyapatite crystallites, and the bulk of the superficial eggshell pigments (Dennis et al., 1996; Samiullah & Roberts, 2013; Kulshreshtha et al., 2022a). Numerous respiratory pores penetrate the eggshell; their funnel-shaped outer mouths are blocked by a cuticle plug (Figure 4.2) (Kulshreshtha et al., 2022a).

When thin polished sections of ES are viewed in cross-section under cross-polarized light in optical microscopy, the nucleation centers of the mammillary knobs are observed to be hemispherical in shape and made of microcrystals of calcite with radial arrangement displaying concentric growth banding (Nys et al., 2004; Athanasiadou et al., 2018; Gautron et al., 2021). From the base of these knobs, the dense columnar units called palisades arise, which appear as elongated calcite crystals (70–80 µm wide) and show various degrees of light extinction due to differences in their crystallographic orientation. Under parallel light illumination, considerable amounts of organic matter can be distinguished, which are distributed throughout the palisade region, although the highest concentration is seen in the upper part of the mammillary layer (Gautron et al., 2021).

Analysis of white chicken eggshell microstructure by electron backscatter diffraction (EBSD) allows calcite crystal orientation to be mapped for the eggshell cross-section. Small randomly oriented crystals are visualized at the base from which large columnar calcite crystal units arise. There is an evolution of crystal orientation with eggshell thickness, with an increase in preferential orientation of calcite crystals with their c-axis perpendicular to the outer eggshell surface (Figure 4.3).

The inner and outer eggshell membranes are composed of a fibrous meshwork of highly cross-linked proteins; they are the innermost layer of the ES and lie between the mammillary layer of the ES and the egg white (Cordeiro & Hincke, 2011; Ahmed et al., 2017; Shi et al., 2021). The ESM consists of an outer shell membrane (attached to the ES) and an inner shell membrane, which is lined by the limiting membrane that encloses the egg white. The two ESM layers have different fiber diameters and orientations, with the outermost layer presenting the thickest and the roughest structure. The outer ESM is approximately 50–70 µm thick, which decreases to 15–26 µm in the inner layer (Ahmed et al., 2017; Shi et al., 2021). The fibers of the outer ESM penetrate the tips of the mammillary knobs. (Cordeiro & Hincke, 2016; Shi et al., 2021). The ESM fibers acquire their stability from lysine-derived desmosine and isodesmosine cross-linkages, which are reinforced by multiple disulfide bonds and are responsible for their remarkable insolubility (Ahmed et al., 2017).

FIGURE 4.3 Crystal orientation map obtained by EBSD of a polished eggshell cross-section.

4.3 Assembly of Eggshell and Eggshell Membranes

The hens of highly selected modern strains of chicken can produce 300 eggs per year, whereas wild birds lay eggs seasonally in clutches that are specific in size to each species. Egg formation follows a quasi-circadian rhythm. There is an orderly development of follicles in the hen ovary, and ovulation initiates the release of an ovum (yolk) into the proximal oviduct (Nys et al., 2022). The forming egg progressively acquires its layers as it traverses specialized segments of the oviduct. In the chicken, the vitelline membrane is acquired in the infundibulum in 15 minutes, followed by egg white accumulation in the magnum during about 4 hours (Rose & Hincke, 2009; Hincke et al., 2012; Gautron et al., 2021; Nys et al., 2022). As the yolk and albumen complex pass through the specialized white isthmus segment, the precursors of the eggshell membranes are secreted and assemble during approximately one hour (Rose & Hincke, 2009; Du et al., 2015). The ESM is a meshwork of cross-linked fibers organized into morphologically distinct inner and outer sheets that enclose the albumen (Hincke et al., 2000; Hincke et al., 2012; Cordeiro & Hincke, 2016).

Eggshell calcification is a process which differs from bone or tooth mineralization, in that it occurs in the absence of cell-directed assembly and with precisely synchronized temporo-spatial control of its formation (Nys et al., 2004; Hincke et al., 2012; Stapane et al., 2019, 2020; Gautron et al., 2021). Eggshell calcification is one of the fastest known biomineralization processes in vertebrates, and occurs in approximately 19 hours (Nys et al., 2004; Hincke et al., 2012; Gautron et al., 2021). Once the forming egg enters the distal red isthmus (tubular shell gland), organic aggregated known as the cores of the mammillary knobs are deposited on the surface of the outer ESM at quasi-periodically, but randomly located sites. These subsequently become heterogenous nucleation sites for calcium carbonate polycrystalline aggregation. In the next phase, the egg enters the uterus (shell gland pouch), where it first acquires its final ovoid shape by albumen plumping of the ESM contents (hydration) so that the outer eggshell membranes are pushed toward close contact with the uterine mucosa (Gautron et al., 2021; Nys et al., 2022). Subsequent calcite crystal growth occurs in the acellular uterine fluid, which is supersaturated with ionized calcium and bicarbonate and contains the organic precursors of the shell matrix (Hincke et al., 2000). The concentration of these components is different throughout the sequential process of ES fabrication, i.e., the initiation of mineralization (5–10 hours post-ovulation), active phase with linear deposition (10–22 hours post-ovulation) and finally the arrest of ES calcification (22–24 hours post-ovulation) (Rodríguez-Navarro et al., 2015; Nys et al., 2022). In chickens, ES formation requires deposition of approximately 6 grams of $CaCO_3$ at a sustained rate of about 0.32 g/h during the linear phase. The calcium and carbonate required for ES mineralization are derived from the blood stream via trans-epithelial transport across the uterine gland cells (Rodríguez-Navarro et al., 2015; Nys et al., 2022). Amorphous calcium carbonate (ACC) is an important intermediate in the formation of the highly crystalline eggshell, which is composed of large columnar calcite single crystals. Disk-shaped amorphous calcium carbonate (ACC) particles accumulate at specific sites on the outer eggshell membrane fibers, where nucleation and stabilization of ACC with calcitic short order occurs prior to its transformation to calcite. Rapid calcite crystal growth occurs by addition of ACC to the shell mineralization front (Rodriguez-Navarro et al., 2015). Finally, the cuticle is deposited 1.5 hours before egg expulsion, just after arrest of calcification (Nys et al., 2004; Wellman-Labadie et al., 2010; Gautron et al., 2021; Kushreshtha et al., 2022a).

4.4 Structure – Function Relationships of ES and ESM

The ES is a sophisticated container for the nutrient reserve of the fertilized egg to support embryonic development (Halgrain et al., 2022). It exhibits numerous functions that are vital requirements for the extra-uterine development of the chick embryo within an autonomous chamber (Athanasiadou et al., 2018; Hincke et al., 2019; Stapane et al., 2020). The ES resists the weight of the brooding hen, protects the egg contents against physical aggression and acts as a physical barrier against microbial invasion and predation by small animals (Hincke et al., 2012; Stapane et al., 2019, 2020; Gautron et al., 2021; Halgrain et al., 2022). This protection is accomplished thanks to its unique structure; the outer zone of the ES is made of large crystals where the external impacts are absorbed by thin inter-crystalline organic layers, which hinder intracrystalline crack propagation. During the second half of embryonic development, progressive decalcification of the inner eggshell provides calcium ions for the mineralization of the embryo skeleton (Athanasiadou et al., 2018; Hincke et al., 2019; Halgrain et al., 2022; Ahmed et al., 2022). The solubilized calcium is transferred to the developing embryo via the chorioallantoic membrane (CAM), which is a highly vascularized structure lining the inner ES (Halgrain et al., 2022; Ahmed et al., 2022). The inner region of the eggshell, at the base of the mammillary cones, where microcrystals of calcite are arranged with spherulitic texture, facilitates dissolution of calcium carbonate (calcium reserve body) during embryonic incubation (Dennis et al., 1996). This selective weakening of the eggshell promotes the propagation of cracks during piping, when the embryo breaks out of the eggshell with its beak (Nys et al., 2004). In addition, the ES controls the exchange of water and gases between the external environment and the embryo thanks to its porosity (Nys et al., 2004; Gautron et al., 2021; Halgrain et al., 2022). The cuticle as well as the outer layer of the calcified ES contains pigments, primarily protoporphyrin IX, which modify the appearance of the eggshell surface and serve as in camouflage, temperature control of the egg contents and parental recognition (Rose & Hincke, 2009). The cuticle also has a role in water/gas exchange and prevention of micro-organism entry through water repellent obstruction of the eggshell pores and/or by limiting microbial colonization of the eggshell surface (Rose-Martel et al., 2012; Kulshreshtha et al., 2022a).

The eggshell membranes are a double-layered, mechanically stable, and insoluble cross-linked fibrous meshwork that retains the egg white (Ahmed et al., 2017). The ESM contains the internal egg contents and provides structural support for ES mineralization. The outer and inner ESM layers differ slightly in terms of morphology, chemical composition, and the degree of protection they provide against pathogen invasion (Hincke et al., 2000; Baláž, 2016). Abnormal ESM fibers formation, organization or cross-linking will inhibit normal ES biomineralization, which results in diminished ES quality and strength. The mechanical properties of the eggshell membrane are provided by extensive desmosine, isodesmosine, and disulfide cross-linkages (Kodali et al., 2011). The extensive lysine-mediated crosslinking (i.e., desmosine, isodesmosine) of collagens by lysyl oxidases gives ESM its remarkable insolubility and structural flexibility to enclose the egg white while providing a firm support upon which ES calcification takes place (Ahmed et al., 2017).

4.5 Composition of Eggshell

All avian eggshells are composed mainly of crystalline calcium carbonate in its calcite polymorph. The chicken eggshell (ES) is a well-studied model, from which much information has been derived in the last two decades to elucidate the interaction between the mineral precursors (ACC) and matrix proteins that promote the calcification of this highly resilient biomineral (Gautron, 2019). The chicken eggshell consists of 95% calcium carbonate, 3.3–3.5% organic matrix when the eggshell membranes are included, and 1.6% water (Figure 4.4). ES-specific matrix proteins play major roles that reflect protective functions of ES in avian reproduction including regulation of biomineralization and antimicrobial defense (Rose & Hincke, 2009).

4.5.1 Proteins

Eggshell calcification takes place in the lumen of the uterine segment of the oviduct, within a uterine fluid whose ionic and protein composition changes over three distinct phases (initiation, active calcification, and termination of shell calcification). The eggshell matrix proteins, glycoproteins and proteoglycans that become occluded within the eggshell mineral constitute ~2% of its weight. This organic matrix can be released from eggshell using calcium chelators (ethylenediaminetetraacetic acid [EDTA] or ethylene-bis(oxyethylenenitrilo)tetraacetic acid [EGTA]) or by acid demineralization (acetic acid or HCl), yielding soluble and insoluble constituents for further characterization (Hincke et al., 1992). Moreover, the changing protein

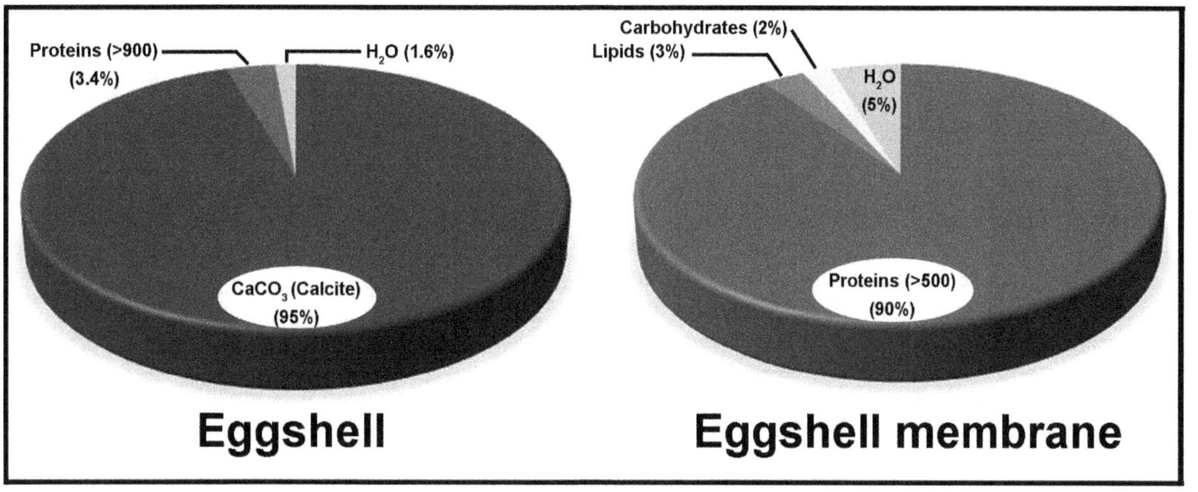

FIGURE 4.4 Chemical composition of eggshell and eggshell membranes.

constituents of the uterine fluid during eggshell formation provide another window into understanding the function of the eggshell organic matrix. Since the uterine fluid constituents are secreted by the epithelial and tubular gland cells of the uterine wall, transcriptomics approaches to characterize differentially expressed genes during eggshell calcification have become a fruitful field of research. Comparative avian genomics, transcriptomics and proteomics have allowed rapid advances in the studies of the organic constituents of avian eggshell and their functional implications (Gautron et al., 2021; Roy et al., 2021).

4.5.1.1 Amino Acid Composition

Bulk measurements of amino acid composition provide insight into the building blocks of the eggshell membranes, ES organic matrix and cuticle. Comprehensive amino acid composition analyses of ES matrix and cuticle are shown in Table 4.1.

TABLE 4.1
Amino Acid Composition (Mol%) of Eggshell Organic Matrix and Cuticle

	Cuticle			ES Matrix			
Amino acid	**Ref. 1**	**Ref. 2**	**Average**	**Ref. 3**	**Ref. 4**	**Ref. 5**	**Average**
Ala	2.9	5.0	4.0 ± 1.4	6.8	6.9	6.9	6.9 ± 0.0
Arg	9.2	8.7	8.9 ± 0.4	5.8	5.7	5.9	5.8 ± 0.1
Asx	14.0	8.4	11.2 ± 4.0	8.3	8.2	8.1	8.2 ± 0.1
Cys/2	2.7	4.3	3.5 ± 1.1	4.8	4.5	-	4.6 ± 0.2
Glx	18.8	10.7	14.8 ± 5.7	11.5	11.4	11.8	11.6 ± 0.2
Gly	9.4	15.3	12.3 ± 4.2	12.8	13.2	13.0	13.0 ± 0.2
His	0.6	1.0	0.8 ± 0.3	3.0	2.9	4.2	3.4 ± 0.6
Hyp	-	-	-	-	-	0.3	0.3
Ile	3.5	4.3	3.9 ± 0.6	2.8	2.8	2.6	2.7 ± 0.1
Leu	6.4	4.0	5.2 ± 1.7	2.6	6.2	6.1	5.0 ± 1.7
Lys	4.6	4.4	4.5 ± 0.1	3.3	3.4	3.6	3.4 ± 0.1
Met	2.8	-	2.8	2.5	2.4	2.0	2.3 ± 0.2
Phe	3.3	2.2	2.7 ± 0.8	2.1	2.3	2.1	2.2 ± 0.1
Pro	3.2	4.5	3.8 ± 1.0	8.3	8.4	8.3	8.3 ± 0.0
Ser	3.3	6.8	5.0 ± 2.5	8.5	8.5	9.7	8.9 ± 0.6
Thr	2.8	6.1	4.4 ± 2.3	6.1	6.1	6.2	6.1 ± 0.0
Tyr	9.1	7.2	8.1 ± 1.3	1.6	1.6	1.8	1.7 ± 0.1
Val	3.5	7.3	5.4 ± 2.7	6.2	6.2	7.3	6.6 ± 0.5

Note: Glx = Glu + Gln; Asx = Asp + Asn; Hyp = hydroyproline. Values are expressed as Mol%. Ref. 1: Wedral et al. (1974), Comp Biochem Physiol B, 47, 631–640; Ref. 2: Baker and Balch (1962), *Biochem J*, 82: 352–361; Ref. 3: Baumgartner et al. (1978), *J. Nutr*, 108, 804–811; Ref. 4: Salevsky and Leach (1980), *Poult Sci*., 59, 438–443; and Ref. 5: Nakano et al. (2003), *Poult Sci*., 82, 510–514.

4.5.1.2 Eggshell Proteomics

Early studies isolated, identified and studied abundant eggshell matrix proteins using classical biochemistry methods: ovocleidin-17 (OC-17), ovalbumin, lysozyme, ovotransferrin (Hincke et al., 1992, 1995; Hincke, 1995; Hincke et al., 2000; Gautron et al., 2001b). Additional advances were made using an expression cloning strategy that identified novel members of the ovocleidin (OC) and ovocalyxin (OCX) families: OC-116, OCX-32, OCX-36 (Hincke et al., 1999; Gautron et al., 2001a; Gautron et al., 2007). More recently, proteomic analyses of the chicken eggshell and its phosphoproteome together identified 528 different proteins as constituents of the soluble eggshell matrix, a forewarning that deciphering their functional role in eggshell function would prove to be a challenging task (Mann et al., 2006; Mann et al., 2007). Major advances from functional genomics-based methods have allowed more than 900 chicken eggshell matrix proteins to be inventoried (Gautron et al., 2021).

Treatment of clean eggshell with acid or EDTA allows removal of the inner eggshell membranes and associated mammillae, and the outer cuticle. Therefore, proteomics to identify the eggshell constituents have generally focused on the soluble proteins obtained after demineralization of the remaining calcified layer (principally the palisade layer) (Mann et al., 2006; Mikšík et al., 2010; Sun et al., 2013). Such studies have identified hundreds of proteins, including those that are relatively eggshell-specific [ovocleidin-17 (OC-17), ovocleidin-116 (OC-116, MEPE), ovocalyxin-32 (OCX32, RARRES1) and ovocalyxin-36 (OCX-36, BPIFB3)], as well as proteins that are most abundant in the egg white [ovalbumin (SERPINB14), ovotransferrin, lysozyme, etc.] and a large number of proteins that are found ubiquitously in many tissues and biological fluids [clusterin (CLU), osteopontin (OPN, SSP1), etc.] (Hincke et al., 2012; Gautron et al., 2021). Specialized approaches have allowed phospho-protein and glycoprotein constituents of the eggshell matrix to be identified. Eggshell matrix phosphopeptides were identified after enrichment by anion-exchange chromatography and reversible binding to titanium oxide. The major phosphorylated eggshell proteins were OPN, OC-17, OC-116, and OCX-32 (Mann et al., 2007). A more recent study identified 676 proteins in the shell matrix after acetic acid-mediated solubilization of ES, including phosphoproteins (69 proteins) and glycoproteins (182 proteins). The most abundant identified phosphoproteins include OC-17, OC-116, OCX-32, and ovalbumin (SERPINB14) (Yang et al., 2020). To identify the insoluble (non-extractable) proteins of ES, EDTA-insoluble proteins were extracted from ES via a gradual decalcification approach that facilitated the identification of 28 proteins including OCX-25, OCX-32, OCX-36, OC-17, OC-116, CLU, lysozyme, and ovotransferrin (Mikšík et al., 2010).

Extraction of ES matrix proteins with acetic acid facilitated the identification of 738 and 600 proteins in pimpled and normal ES, respectively. This study revealed 42 common major proteins in the ES matrix including OC-116, SERPINB14, EGF-like repeats and discoidin domains 3 (EDIL3), and OCX-36 (Liu et al., 2017). In another approach, quantitative proteomics allowed 216 unique matrix proteins to be attributed to four key stages of shell calcification: i. widespread deposition of amorphous calcium carbonate (ACC), ii. ACC transformation into crystalline calcite aggregates, iii. formation of larger calcite crystal units, and iv. development of a columnar structure with preferential calcite crystal orientation. In particular, LYZ, TF, OC-17 and OC-116 were associated with shell calcification, while the calcium binding proteins EDIL3, serum albumin (ALB), milk fat globule-EGF factor 8 protein (MFGE8), and NUCB2 were also identified (Marie et al., 2015a, 2015b).

Proteins associated with the ES mammillary cone layer, hypothesized to regulate the earliest stage of ES calcification, were identified including 49 proteins associated with the ESM fibers, 18 mammillary cone-specific proteins, and 18 proteins enriched in the mammillary cones. The most promising modulators of protein–crystal interactions include ABI family member 3 (NESH) binding protein (ABI3BP), hyaluronan and proteoglycan link protein 3 (HAPLN3), collagen alpha-1(X), collagen alpha-1(II) and fibronectin, in addition to the calcium-binding proteins calumenin, EDIL3, nucleobindin 2 (NUCB2), and secreted protein acidic and cysteine rich (SPARC) (Rose-Martel et al., 2015).

The cuticle is a specialized protein-rich outer zone of the eggshell. The cuticle components are thought to play a major role in microbial resistance, since eggs with incomplete or absent cuticle are more susceptible to bacterial contamination (Kushreshtha et al., 2021). In one study, the extracted insoluble cuticle film was dissected using laser microdissection before proteomics to identify 37 proteins including OC-17, OC-116, lysozyme, SERPINB14, and OCX-32 (Mikšík et al., 2014). In a related study, 47 outer cuticle proteins were identified following solubilization in sodium dodecyl sulfate/dithiothreitol (SDS/DTT). OCX-32 and similar to Kunitz-like protease inhibitor were found to be the most abundant constituents of this cuticle extract, in addition to a number of antimicrobial proteins (lysozyme C, ovotransferrin, cystatin C, ovoinhibitor (SPINK5)) (Rose-Martel et al., 2012). Treatment of the eggshell surface with either EDTA or acetic acid to release the insoluble cuticle identified OCX-32, OCX-36, OC-17 and OC-116 as the most abundant constituents (Bain et al., 2013).

Quantitative proteomics analysis was conducted to determine the candidate proteins and molecular mechanisms underlying variation in chicken ES brownness and identified 147 differentially expressed proteins between dark and light brown eggs. Functional analysis suggested that down-regulation of iron-sulfur cluster assembly protein (Iba57) in light brown eggs would decrease the synthesis of protoporphyrin IX used for heme synthesis, leading to a reduction in the exported protoporphyrin IX available for ES deposition and brownness (Li et al., 2016).

4.5.1.3 Uterine Fluid Proteomics

The uterine fluid displays distinct protein electrophoretic profiles at each phase of shell mineralization (initial, active calcification and termination), suggesting specific roles for different organic constituents during deposition of the mammillary bodies, palisades and the terminal zone (Gautron et al., 1997). The changing protein constituents of the uterine fluid during eggshell formation provide another window into understanding the function of the eggshell organic matrix.

A total of 577 proteins were identified in the uterine fluid from hens laying eggs with strong or weak eggshell. Interestingly, 357 proteins (~80%) of all the proteins identified in uterine fluid associated with strong eggshell (UF-Strong) (447 proteins) were in common with proteins identified in weak eggshell uterine fluid (UF-Weak). Among the 357 common proteins in uterine fluid, 27 proteins in UF-Strong were higher than in UF-Weak including antithrombin-III (SERPINC1), annexin A2 (ANXA2), and syntenin-1 (SDCBP) (Sun et al., 2013).

Proteomic analysis of hen uterine fluid collected at three stages of shell mineralization, Initiation (I), rapid growth (G), and termination (T), allowed the identification of 308 proteins, which could be functionally associated with eggshell formation (Marie et al., 2014; 2015c). A first group of proteins were related to mineralization and mainly present at the onset of calcification (OVOT, OVAL, OC-17, EDIL3, MFGE8). A second group of proteins were mainly secreted during the initiation and the termination phases of shell calcification. OCX-21, protease inhibitors and chaperone proteins predominated in this group. The third group of proteins, including a number with known antimicrobial activity, were present in uterine fluid at all stages of shell calcification (Marie et al., 2014; 2015c).

In a related study, proteomic analysis of extracellular vesicles (EVs) isolated from uterine fluid identified 29 non-redundant proteins proposed to be involved in vesicular transport of amorphous calcium carbonate including carbonic anhydrase-4 (CA4), EDIL3, ezrin (EZR), programmed cell death 6-interacting protein (PDCD6IP), SDCBP, SERPINB14, and LYZ. Also present were several proteins already identified in the ES matrix including ovotransferrin (TF), ovalbumin-related X (SERPINB14C), ovalbumin-related Y (SERPINB14B), ovocalyxin-25 (OCX-25), clusterin (CLU), and ovostatin (OVST) (Stapane et al., 2020).

4.5.1.4 Transcriptomics Analysis

Transcriptome analysis of gene expression by uterine tissue during eggshell calcification has the potential to identify regulatory, signaling and metabolic pathways that control eggshell mineralization, in addition to clarifying the genes encoding the secreted eggshell matrix proteins that become occluded matrix proteins. Moreover, expression of the genes for cell membrane ion transporters, channels, and pumps responsible for movement of Ca and bicarbonate ions into the uterine lumen, as well as those linked to the maintenance of cellular ionic homeostasis during this process, are of great interest (Brionne et al., 2014). The rich datasets available from microarray and RNA-Seq transcriptomic studies provide insight into regulation of eggshell mineralization and may provide new selection tools for improved eggshell quality in commercial flocks. Differences in experimental design for such studies have allowed detailed interpretation of differential gene expression for various comparisons: between oviduct segments, laying hens of different ages, differential sexual maturity of hens, hens laying eggs with strong vs weak shell strength, hens laying normal versus defective eggshell, eggs with good versus poor cuticle, eggs with light versus dark shell color (Jonchère et al., 2010; Liu et al., 2013; Zhang et al., 2015; Guangqi et al., 2017; Sah et al., 2018; Khan et al., 2019; Yin et al., 2019; Zhang et al., 2019; Feng et al., 2020; Poyatos Pertiñez et al., 2020; Yin et al., 2020; Khogali et al., 2021; Chen, Chen et al., 2021; Chen, He et al., 2021).

A transcriptomic study exploring ES proteins involved in the physical defense of the chicken egg identified 605 transcripts that were over-expressed in the uterus compared to the magnum or white isthmus. These transcripts correspond to 469 unique genes that encode 437 different proteins. Gene ontology (GO) enrichment analysis identified terms related to proteins involved in ion transport proteins, calcium binding or biomineralization, proper folding of the ES matrix, and antibacterial properties along with proteases and protease inhibitors (Jonchère et al., 2010). To elucidate the repertoire of genes involved in egg formation, the transcriptomic landscapes of ovary and three oviduct segments including magnum, isthmus and uterus were established. The study identified 547 DEGs in the magnum and 585 in uterus (Yin et al., 2020). Furthermore, in a related transcriptomic study, genes and pathways regulating albumen and ES formation, as well as defense mechanism in oviduct and egg have been investigated 120 (prelaying) and 300 (laying) days old Lohmann layers that allowed the identification of 1,040 and 595 DEGs in the magnum and uterus, respectively. GO enrichment and KEGG pathway analysis showed that gene activities of the magnum and uterus in prelaying chickens were more likely to promote growth and development, and that after egg laying, they increase transmembrane transport and secretory activities (Yin et al., 2019).

A similar transcriptomic study revealed putative proteins involved in the supply of minerals or in the shell mineralization process, and identified a repertoire of 302 over-expressed genes in uterine tissue during shell calcification. A list of 57 proteins involved in the mineralization process was also established and classified into proteins with biomineralization, proteoglycans, molecular chaperone, antimicrobials and proteases/antiproteases functionalities (Brionne et al., 2014). A transcriptomic study to understand genes/proteins and the biological pathways regulating the eggshell formation was conducted and revealed 229 differentially expressed genes (DEGs) that were up-regulated in egg-laying hens compared to non-layers. Of these 229 DEGs, 12 were related to ES mineralization including OC-116 (regulator of mineralization), otopetrin 2 (OTOP2, modulator of cellular calcium influx), calcitonin related polypeptide beta (CALCB, intracellular release of Ca-ions), stanniocalcin 2 (STC2, increases alkaline phosphatase activity), and ATPase secretory pathway Ca^{2+} transporting 2 (ATP2C2, cellular import of Ca-ions) that were significantly higher in the uteri of laying hen at 15–20 hrs post-ovulation (Sah et al., 2018). Similarly, the global gene expression profile of the shell gland of laying hens at two time-points (5 and 15 hrs) of ES formation has been investigated. Out of the 14,334 genes assessed for differential expression in the shell gland tissue, 278 genes were significantly down-regulated, and 413 genes were significantly up-regulated at 15 h relative to the 5 h time-point of eggshell formation. The up-regulated genes were most likely involved in calcium transportation, as well as synthesis and secretory activities of ions and molecules, reflecting the peak stage of eggshell formation (Khan et al., 2019).

Uterine transcriptomic analysis revealed mRNA expression changes associated with eggshell ultrastructural differences. Differentially expressed genes (DEGs; 125 upregulated and 58 downregulated) were identified in uterus of laying hens in the late phase (aged hens) compared to those at peak production (young hens). Genes of several matrix proteins potentially involved in eggshell biomineralization, such as ovalbumin, glypican 3, and versican were differentially expressed between the two hen groups (Feng et al., 2020). Another study performed transcriptome RNA-Seq analysis of the magnum and uterus in 120-d-old (pre-lay) and 300-d-old (egg-laying) hens, revealing that 1,040 genes were differentially expressed in the magnum and 595 genes in the uterus. Gene activities of the magnum and uterus in pre-laying birds were more likely to concentrate on growth and development, whereas, in egg-laying hens, transmembrane transport and secretion activities were enhanced (Yin et al., 2019).

Uterine genes modulating ES strength at the early stage of active calcification were identified by transcriptomics, revealing 1,195 differentially expressed transcripts that corresponded to 941 unique genes in hens laying eggs with strong ES compared to weak ES. These included transcripts encoding ion transporters and secreted extracellular matrix proteins (77 transcripts) and carbohydrate metabolism or post-translation glycosylation modification (26 transcripts), along with 88 signaling transcripts (Liu et al., 2013). In addition, integrating transcriptome and genome re-sequencing data allowed the identification of 14,234 genes expressed in the chicken uterus. A total of 889 DEGs were identified by comparing uterine expression in hens laying eggs with low versus normal ES strength. The DEGs are enriched in calcification-related machinery, including calcium ion transport and calcium signaling pathways (Zhang et al., 2015). Similarly, another transcriptomic study identified 1,777 DEGs in the uterus of hens laying eggs with high breaking strength as compared to eggs with low breaking strength. Among the identified DEGs, 30 DEGs are linked to eggshell calcification including OPN, TF, and SERPINB14 (Zhang et al., 2019). Moreover, a recent transcriptomic analysis of uterine samples collected from 450 days old hens producing normal eggs (NE) and sandpaper-shelled eggs (SE) enabled the identification of 211 DEGs. These DEGs were clustered into 145 GO terms and enriched in 12 KEGG pathways related to organ morphogenesis and development, ion transport, endocrine and cell communication, immune response, and corticotropin-releasing hormones. Corticotropin may be an important factor in SE formation as it alters ion transport (Khogali et al., 2021).

The uterine transcriptome and proteome of laying hens producing eggs with good and poor cuticle (GC and PC, respectively) were compared and identified 53 differentially DEGs (25 up-regulated and 28 down-regulated). Several DEGs, including prostaglandin D2 synthase (PTGDS), phospholipase C gamma 2 (PLCG2), adrenomedullin (ADM), and prolactin receptor (PRLR) might play essential roles in cuticle deposition by regulating the uterine muscular activities and secretion function (Chen, He et al., 2021). Similarly, uterus transcriptomes were compared between hens with naturally good and poor cuticle and, those where the hypothalamo-pituitary-gonadal-oviduct axis was manipulated to produce eggs with or without cuticle. Highly expressed genes were identified encoding ES matrix and cuticle proteins including OC-116, OCX-36, OCX-32, OCX-25 (Poyatos Pertiñez et al., 2020).

Another transcriptomic study of shell gland epithelium was conducted to identify genes responsible for ES brownness. The study revealed the expression of 8461 genes in the shell gland epithelium, of which 34 genes were differentially expressed in hens laying darker brown eggs including TF and heat-shock protein 70 (HSP70) along with the oxidative phosphorylation pathway involved in the synthesis and transport of protoporphyrin IX (Guangqi et al., 2017). RNA-Seq was employed to profile the uterine gland transcriptome (lncRNA and mRNA) of Changshun blue eggshell laying hens, to better understand the genetic mechanisms of deposition of blue eggshell color. Differentially expressed mRNAs, GO terms, and KEGG pathways among pink-eggshell and blue-eggshell hens mainly targeted immune- and transporter-related terms with the SLC family, IgJ, CD family, and MTMR genes. Moreover, lncRNA and mRNA profiles predicted 8 and 22 lncRNA–gene pairs, which were enriched in lipid-related metabolism and predicted functions in color formation by modulating pigment disposition and immune- and lipid-related metabolism (Chen, Chen et al., 2021).

4.5.1.5 Antibody-Based Techniques

The use of antibodies raised to purified ES proteins has facilitated their immunochemical detection and immunolocalization to understand their role in biomineralization and shell function (Rose & Hincke, 2009; Roy et al., 2021). The ES matrix composition of various domesticated bird species (laying and breeder domestic fowl, turkey, Pekin duck, goose, guinea fowl, quail, and pheasant) was verified using Western blotting (WB) to predict the ubiquitous nature of the ES mineralization process across bird species. OC-17 and SERPINB14 were revealed ES from all species (except quail for OC-17). In contrast, LYZ was identified only in hen ES. OVOT showed a positive signal in hens, turkey and quail, while OPN (SPP1) was detected in laying and breeder hens and quail (Panheleux et al., 1999). Likewise, OCX-32, OCX-36, OVOT, LYZ, and SERPINB14 were identified in chicken ES extract using Western blotting (WB) (Wellman-Labadie et al., 2008). In addition, the mineralization-specific and ES matrix proteins annexin 1 (ANXA1), ANXA2, annexin 8 (ANXA8), carbonic anhydrase 4 (CA4), EDIL3, and PDCD6IP have been studied in uterine tissue by immunofluorescence (Stapane et al., 2020). The potential use of OVOT, SERPINB14, and OC-17 proteins as biomarkers of ES quality was assessed by a competitive indirect enzyme-linked immunosorbent assay (ELISA). It has been shown that concentrations of these proteins were high in ES extracts from older hens as compared to young hens. This strategy may be a promising tool for analyzing the origin of ES faults and provide useful information for breeding programs (Panheleux et al., 2000).

Immunocytochemistry at the light and electron microscope levels demonstrated that OCX-32 is limited to the isthmus and uterus. In the ES, OCX-32 is localized to the outer palisade layer, the vertical crystal layer, and the cuticle of the eggshell, in agreement with its demonstration by WB at high levels in the uterine fluid during the termination phase of eggshell formation (Gautron et al., 2001a). Alternatively, OCX-32 which is present at high levels in the uterine fluid, was detected in outermost layer of mineralized shell and cuticle using WB along with antibodies raised to the 32-kDa uterine fluid protein (Hincke et al., 2001). Similarly, OCX-36 protein was identified in the ES extracts and uterine fluid using WB (Gautron et al., 2007). In addition, OCX-36 protein was localized to the calcified ES predominantly in the inner part of the shell using immunochemical techniques (immunofluorescence and colloidal-gold immunocytochemistry) based on antibody raised against sequence 51–65 of the OCX-36 protein. OC-17 protein was localized in ES matrix and shell gland (tubular gland cells) using WB and immunohistochemistry (Hincke et al., 1995). OC-17 was also identified in uterine fluid during the active phases of calcification along with ES extracts using WB. This finding demonstrated the presence of ES matrix precursors in the uterine fluid and supports the hypothesis of their involvement in the process of eggshell mineralization (Gautron et al., 1997). Furthermore, OC-116 was identified in ES extracts and the uterine fluid using WB based on primary antiserum raised to the 116-kDa protein (Hincke et al., 1999).

OPN, OVOT, and SERPINB14 have been shown to play a critical role in the ES calcification process (Zhang et al., 2019). OPN SNPs have been associated with eggshell fracture toughness and unusual patterns of uterine OPN expression may underlie certain defects in eggshell mineralization (Dunn et al., 2008; Arazi et al., 2009). The oviduct expression of OPN is entirely uterine-specific and is temporally associated with eggshell calcification through coupling of uterine wall distension to OPN gene (Sppl) expression (Lavelin et al., 1998). OPN protein was identified in the decalcified palisades using colloidal gold immunolabeling (Hincke et al., 2008; Chien et al., 2009) and ES extracts using WB (Hincke et al., 2008). Moreover, OPN was identified in ES extract using WB and in the decalcified shell samples using immunohistochemistry and immunogold labeling and implicated in the modulation of ES hardness (Athanasiadou et al., 2018). Colloidal-gold immunolabeling allowed the detection of OPN at various organic matrix structures including lamellar sheets, interconnected fine filamentous threads, thin film-like surface coatings of proteins, and among occluded proteins residing preferentially at the internal {104} crystallographic faces of ES calcite (Chien et al., 2008).

Ovotransferrin was identified as an 80 kDa matrix protein in the uterine fluid, uterine tissue extract, and demineralized ES using WB, and its changing concentrations at different stages of ES calcification confirmed by ELISA. OVOT was localized by immunofluorescence in the mammillae (sites of calcite nucleation) and the tubular gland cells of the uterus (Gautron et al., 2001b). Ovalbumin (SERPINB14) was shown by WB of ES extract and immunofluorescence to be an authentic matrix protein, localized to the mammillary bodies, and not distributed throughout the shell matrix (Hincke, 1995). Furthermore, the use of polyclonal antibodies allowed the detection of SERPINB14 (the most abundant egg white protein) protein in ES extracts and in the uterine fluid during the active calcification phases using WB and immunofluorescence, revealing the involvement of this protein in ES mineralization (Hincke, 1995; Gautron et al., 1997). Lysozyme (LYZ) was detected in the ES matrix using WB and immunofluorescence and in the uterine fluid using WB (Hincke et al., 2000).

EDIL3 and MFGE8 were identified in the soluble ES matrix (SEM) using WB, confirming their identification in the uterus and uterine fluid by proteomics (Stapane et al., 2019). In addition, clusterin (CLU, a secretory glycoprotein that acts as an extracellular chaperone) was detected in the ES matrix and uterine fluid using WB. CLU was also detected in the decalcified ES matrix and uterine segment of the oviduct (Mann et al., 2003). Furthermore, follistatin (FST, a member of transforming growth factor-beta (TGF-β) family signaling) was identified in uterus of laying hens at different physiological states using immunohistochemistry (Wasti et al., 2020). In addition, WB with a monoclonal antibody termed Epi2 (which stained approximately 50% of the cells within the uterine pseudostratified columnar epithelium) showed binding affinity to the ES dermatan sulfate proteoglycan (DSPG) which decorates the OC-116 core protein (Hincke et al., 1999; Dennis et al., 2000).

4.5.2 Eggshell Carbohydrates

The ES matrix also contains glycoproteins and proteoglycans, with their associated sugar and glycosaminoglycan (GAG) constituents, respectively. Purified ES GAGs are composed of equimolar amounts of N-acetylglucosamine (GlcNAc) and glucuronic acid. Hyaluronidase-mediated digestion of ES extract revealed the GAGs to be hyaluronic acid (HA) (Heaney & Robinson, 1976). Various strategies have been utilized to comprehensively identify the glycosaminoglycan (GAG) constituents of ES matrix including decalcification, precipitation, proteolysis, and chromatographic techniques (Arias et al., 1992; Carrino et al., 1997; Nakano et al., 2001, 2002, 2003; Ha et al., 2007; Liu et al., 2014; Liu et al., 2016). These strategies allowed the identification of all 4 groups of GAGs including HA, keratan sulfate (KS), chondroitin sulfate (CS), and heparan sulfate (HS) (Liu et al., 2014; Liu et al., 2016). HA was the most abundant GAG in eggshell membranes, while CS was the most enriched in calcified shell (Liu et al., 2014; Liu et al., 2016). Dermatan sulfate (DS) was previously identified in the ES matrix (Arias et al., 1992). In addition, different GAGs in ES matrix were identified, including HA and CS-DS copolymer. Uronic acid, a carbohydrate moiety of GAGs, represented 0.024% of the ES dry weight (Nakano et al., 2001). In comparison to ESM, uronic and sialic acids was five-fold and four-fold higher in the ES organic matrix, respectively (Nakano et al., 2003). In a related study, GAGs were extracted from decalcified ES, and identified as CS-DS copolymers with iduronic acid content between 20% and 59% of the uronic acid (Nakano et al., 2002). Similarly, the uronic acid and sulfated GAGs contents of decalcified ES were measured calorimetrically. Galactose and uronic acid were reported to correlate with eggshell strength (Ha et al., 2007). However, this correlation could not be confirmed in a later study (Liu et al., 2016). Furthermore, acid-mediated demineralization of the palisade matrix allowed the identification of DS and unsulfated chondroitin along with 4-sulfated and 6-sulfated disaccharides (Carrino et al., 1997). Disaccharide compositional analysis revealed that CS in the calcified shell consisted primarily of 6SCS-C, 4SCS-A, and 0SCS-0 disaccharides (Liu et al., 2016). The 4SCS-A content correlated with eggshell strength, and 0SCS-0 correlated with eggshell strength and eggshell thickness.

Although the major ES glycoproteins have been inventoried (OC-116, OC-17, CLU, OVAL, OVOT), and their N-glycosylation sites identified (Yang et al., 2020), much less is known regarding their sugar residues. However, the most abundant ES matrix protein, OC-116, contains two N-glycosylation sites in its sequence (62N-Q-T and 293N-D-S). The glycan structures attached to asparagine (Asn62) were determined by carbohydrate compositional analysis, methylation analysis and MS/MS. This combined strategy allowed the identification of mannose- and fucose-containing glycans with LacdiNAc (GalNAch1–4GlcNAc) as a major motif in the N-glycan structures of OC-116 (Nimtz et al., 2004). Moreover, OC-116 was initially identified as a dermatan sulfate proteoglycan, although the modified residues have not yet been identified (Hincke et al., 1999; Mann et al., 2002). GPC4 and HAPLN3 have also been identified as ES proteoglycan core proteins (Marie et al., 2015a). Finally, different monosaccharides including glucose, galactose, mannose, xylulose, glucose, and sialic acid have been identified in the cuticle (Wedral et al., 1974).

To address the possible role of proteoglycans (PG) and GAGs in mineralization of the ES, immunohistochemistry was carried out using antibodies to various PG and GAG epitopes. The immunostaining approach allowed the identification of DS in ES matrix and the mammillary knobs. Different KS antibodies immunostain distinct regions of the ES suggesting differential distribution of various types of KS (Arias et al., 1992). Different PGs were localized to discrete regions within the ES of various domesticated bird species (laying and breeder domestic fowl, turkey, Pekin duck, goose, guinea fowl, quail, and pheasant) using immunostaining that identified dermatan sulphate (DS) within the matrix of the calcified ES of all species except quail which contained chondroitin-6-sulfate. KS was observed in mammillary bodies of breeder and laying hen, quail, pheasant, and turkey while CS was also present in guinea fowl and duck (Panheleux et al., 1999). In addition, immunostaining of ES facilitated the identification of various GAGs including KS and DS. Multiple different antibodies to KS stained different regions of the ES; one KS epitope is prominent in the calcium reserve assemblies, while DS staining was very intense in the palisade region (Carrino et al., 1996).

4.5.3 Eggshell Minerals

The chicken eggshell is 95% calcium carbonate in the form of calcite (Hincke et al., 2010). The eggshell also contains minor amounts of magnesium and phosphorus, which are differentially distributed, in addition to trace levels of other elements (Table 4.2). Magnesium levels are higher in the inner mammillary layer and at the outer palisade layer, while phosphorus (as inorganic phosphate or associated with phosphoproteins) is mainly found in the outer palisade layer and cuticle (Cusack et al., 2003; Nys et al., 2004; Halgrain et al., 2022). Inorganic material recovered after decalcification of ES with acetic acid allowed the detection of calcium, magnesium, phosphorus (Nakano et al., 2001). The mineral composition of eggshell powder was determined using a variety of methods, detecting Ca, Mg, P, Sr, Zn, Fe, Cu, Cr, F, and Se, while B, V, Pb, Al, Cd, and Hg were below the detection limit (Schaafsma et al., 2000). This observation supports the recommendation that calcium supplements for human consumption derived from eggshell are superior to those derived from geological calcium carbonate (i.e., limestone), due to lower levels of heavy metal contamination. ES powder from another source appear to have a comparable mineral composition except for Mg (Schaafsma et al., 2000). In addition, analysis of ES from fertilized broiler breeder eggs identified B, barium (Ba), Ca, Fe, Mg, P, Sr, and Zn levels that evolved during incubation (Szeleszczuk et al., 2016). Similarily, Ca, Mg, P, sodium (Na), and potassium (K) have been identified in the eggshell of incubated fertilized eggs (Halgrain et al., 2022). The elemental composition of cuticle revealed P, Ca, K, Mg, Na, Zn, manganese (Mn), Fe, Cu, B, and Al as its mineral constituents (Wedral et al., 1974) (Table 4.2).

4.6 Composition of Eggshell Membrane

The ESM surrounds the egg white and provides a structural support for ES mineralization (Cordeiro & Hincke, 2016; Ahmed, Kulshreshtha et al., 2019; Ahmed, Suso et al., 2019; Shi et al., 2021). Defects in ESM fibers formation, organization or cross-linking are reported to hinder normal ES biomineralization, leading to inferior ES quality and strength (Chowdhury, 1990; Chowdhury & Davis, 1995; Du et al., 2015; Gautron et al., 2021). ESM is a highly insoluble structure predominantly composed of proteins (~90%) with low amounts of lipids (~3%) and carbohydrate (~2%) (Hasiak et al., 1970; Vadehra & Rath, 1973, Baker & Balch, 1962; Cooke & Balch, 1970; Chen et al., 2014) (Figure 4.5).

4.6.1 Proteins

4.6.1.1 Amino Acid Composition

Hydroxyproline and hydroxylysine are known posttranslational modifications of collagens, and their presence in the ESM indicates that collagens are major components of ESM (Ahmed, Suso et al., 2019). This is supported by the observation that ESM is digestible by the enzyme collagenase (Nys et al., 2004). A relatively high content of cysteine, approximately 10–12%, is detected in ESM, reflecting the presence of a considerable amount of cysteine-rich eggshell membrane proteins (CREMPs) with multiple disulfide cross-links (Kodali et al., 2011; Ahmed, Kulshreshtha et al., 2019; Ahmed, Suso et al., 2019). Moreover, ESM also contains approximately 10% glycine and the canonical G-X-Y repeats of collagen, correlating with the presence of 25% glycine in avian collagens (Ahmed, Suso et al., 2019). ESM processing by mechanical size reduction techniques and sieving does not affect its amino acid composition; the amino acid profile of Particalized Eggshell Membrane (PEM) in various sizes was similar to that of intact ESM material prior to processing (Kulshreshtha et al., 2020). The amino acid composition

TABLE 4.2

Elemental Compositions of Eggshell Including Cuticle

Reference		Wedral et al. (1974)	Schaafsma et al. (2000)		Nakano et al. (2001)	Szeleszczuk et al. (2016)				Halgrain et al. (2022)	
						Eggshell Powder (Broiler Breeder Eggshell)				Eggshell	
Sample		Cuticle (Dry Ash)	Slovakian & Dutch Eggshell Powder	Japanese Eggshell Powder	Inorganic Material Collected after Decalcification	ED1	ED14	ED18	ED21	ED12	ED16
Element	Ca	20–31**	401 ± 7*	393.0*	24.1†	390 ± 6*	370 ± 12*	345 ± 11*	332 ± 16*	2298 ± 191‡	2172 ± 182‡
	Mg	0.01–0.05**	4.5 ± 0.5*	28.0*	0.23†	5.1 ± 0.3*	4.8 ± 0.2*	4.6 ± 0.1*	4.1 ± 0.2*	20 ± 3.4‡	18 ± 2.9‡
	Na	0.5–0.7**								6.4 ± 0.7‡	6.0 ± 0.7‡
	K	0.02–0.05**								2.7 ± 0.4‡	2.5 ± 0.3‡
	P	10–20**	1 ± 0.2*	0.1*	0.04†	1.1 ± 0.1*	1.2 ± 0.1*	1.5 ± 0.1*	1.7 ± 0.2*	6.8 ± 0.9‡	6.5 ± 0.9‡
	Mn	0.01–0.03**									
	Sr		0.4 ± 0.2*	0.4*		98 ± 2**	106 ± 4**	105 ± 8**	110 ± 5**		
	Zn	0.21–0.50**	5.1 ± 0.5**	41.0**		5.7 ± 0.5**	5.4 ± 0.2**	5.8 ± 0.4**	5.9 ± 0.6**		
	Fe	0.35–5.5**	22.4 ± 1**	27.0**		28.8 ± 1.3**	31.6 ± 2.3**	32.7 ± 3.6**	29.7 ± 5.6**		
	Cu	0.06–1.2**	7.7 ± 1**	9.0**							
	Cr		0.1 ± 0.1**								
	F		3.8 ± 1**								
	B	0.05–1.0**	<0.5**			200 ± 20**	201 ± 13**	197 ± 3**	196 ± 6**		
	V		<0.5**								
	Pb		<0.5**	<0.5**							
	Al	2.0–4.0**	<5.0**	<5.0**							
	Hg		<0.2**	<0.2**							
	Se		23.5 ± 5***								
	Cd		<50.0***	<50.0***							
	Ba					1.4 ± 0.2**	1.7 ± 0.1**	2.2 ± 0.2**	2.5 ± 0.3**		

Note: *: mg/g; **: μg/g; ***: ng/g; †: %; ‡: mg per total dried eggshell weight.

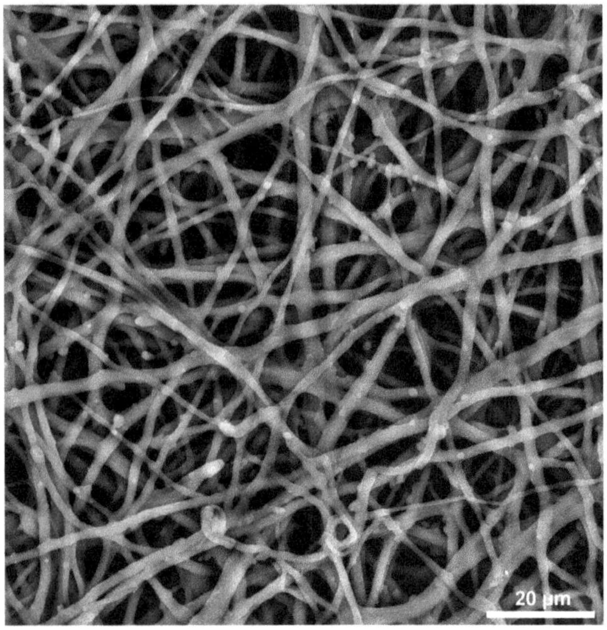

FIGURE 4.5 Scanning electron microscopic view of the inside surface of the inner eggshell membranes.

of eggshell outer and inner membrane as compared to the combined ESM is shown in Table 4.3.

A comparative quantitative determination of the amino acid composition of decalcified ES, ESM, and cuticle is comprehensively presented in (Table 4.4). The contents of most of the amino acids were similar among the decalcified eggshell and ESM (Outer and inner ESM) (Nakano et al., 2003). In contrast, glycine and alanine contents were significantly higher, while proline and hydroxyproline contents were significantly lower, in the decalcified ES as compared to the ESM. The content of leucine was significantly higher in the decalcified eggshell than in the outer ESM (Nakano et al., 2003). Cysteine content was lower in ES matrix as compared to ESM, as previously noted (Abatangelo et al., 1978), reflecting the prominent CREMP content of ESM. Hydroxyproline and hydroxylysine are readily detected in ESM, reflecting its significant collagen content, in contrast to the paucity of collagens in ES and cuticle.

4.6.1.2 *Eggshell Membrane Proteomics/Transcriptomics*

The highly cross-linked fibers of the ESM are mainly constituted by proteins (Figure 4.5) including cysteine-rich eggshell

TABLE 4.3

Amino Acid Composition (Mol%) of ESM (Inner, Outer, and Combined)

Amino Acid	Eggshell Membranes (Combined)								Inner Eggshell Membrane				Outer Eggshell Membrane			
	Ref. 1	Ref. 2	Ref. 3	Ref. 4	Ref. 5	Ref. 6	Ref. 7	Average	Ref. 8	Ref. 5	Ref. 9	Average	Ref. 8	Ref. 5	Ref. 9	Average
Ala	4.0	4.0	3.9	4.1	2.3	4.2	3.5	3.7 ± 0.6	4.6	1.9	4.6	3.7 ± 1.6	4.0	2.8	4.1	3.6 ± 0.7
Arg	5.5	5.1	4.8	5.2	5.9	5.1	5.1	5.2 ± 0.3	5.1	7.7	5.7	6.1 ± 1.3	5.4	4.1	5.8	5.1 ± 0.9
Asx	7.5	8.1	6.1	8.2	11.7	8.2	7.6	8.2 ± 1.6	8.2	9.7	8.4	8.8 ± 0.8	8.7	13.6	8.8	10.4 ± 2.8
Cys/2	10.3	10.0	8.8	10.4	10.9	10.2	12.2	10.4 ± 0.9	5.1	11.3	-	8.2 ± 4.3	5.5	10.6	-	8.1 ± 3.6
Glx	9.7	11.0	10.2	10.5	16.0	10.8	10.1	11.2 ± 2.0	10.0	15.5	11.1	12.2 ± 2.9	11.0	16.6	11.9	13.1 ± 3.0
Gly	9.7	10.3	11.9	10.4	4.0	10.9	9.9	9.6 ± 2.4	11.4	3.7	11.1	8.7 ± 4.4	10.7	4.4	10.6	8.6 ± 3.6
His	3.3	3.1	4.1	2.9	1.3	2.6	3.2	2.9 ± 0.8	2.9	1.8	4.1	2.9 ± 1.2	5.4	0.9	4.3	3.5 ± 2.4
Hyl	0.1	-	3.3	-	-	0.2	0.4	1.0 ± 1.5	0.2	-	-	0.2	0.2	-	-	0.2
Hyp	0.9	-	0.6	-	-	1.3	0.9	0.9 ± 0.3	1.6	-	1.5	1.5 ± 0.0	1.2	-	1.4	1.3 ± 0.1
Ile	3.6	3.2	3.0	3.3	3.8	2.6	3.1	3.2 ± 0.4	2.9	3.7	3.3	3.3 ± 0.4	2.8	4.0	3.4	3.4 ± 0.6
Leu	4.8	4.7	4.8	5.0	6.1	4.4	4.4	4.9 ± 0.5	5.2	6.7	5.6	5.9 ± 0.8	4.4	5.6	4.8	4.9 ± 0.6
Lys	3.4	3.1	3.9	3.0	3.9	2.9	3.1	3.3 ± 0.4	3.3	5.2	3.6	4.0 ± 1.0	3.3	2.6	3.4	3.1 ± 0.4
Met	3.0	3.4	2.0	3.6	4.7	3.6	3.3	3.4 ± 0.7	3.3	4.8	2.3	3.5 ± 1.3	4.0	4.5	2.3	3.6 ± 1.2
Phe	1.5	1.4	1.9	1.7	3.6	0.9	1.7	1.8 ± 0.8	1.6	3.8	1.6	2.3 ± 1.3	1.7	3.4	1.5	2.2 ± 1.1
Pro	11.9	10.5	8.9	9.8	8.4	10.5	9.7	10.0 ± 1.1	10.5	7.5	11.6	9.8 ± 2.1	10.7	9.3	12.0	10.7 ± 1.3
Ser	6.0	6.8	6.4	6.7	1.7	7.4	6.6	5.9 ± 1.8	8.0	2.9	9.2	6.7 ± 3.4	7.4	0.5	9.2	5.7 ± 4.6
Thr	5.7	6.2	6.1	6.2	6.2	6.2	6.0	6.1 ± 0.2	7.0	5.6	6.9	6.5 ± 0.8	6.9	6.7	6.9	6.8 ± 0.1
Tyr	1.4	1.3	1.5	1.5	3.9	1.1	1.3	1.7 ± 0.9	1.9	3.6	2.2	2.6 ± 0.9	1.2	4.1	1.7	2.3 ± 1.5
Val	7.9	7.9	7.8	7.5	5.6	5.6	7.7	7.1 ± 1.0	6.7	4.8	7.2	6.2 ± 1.3	6.9	6.4	7.9	7.1 ± 0.8

Note: Glx = Glu + Gln; Asx = Asp + Asn; Hyp = hydroyproline; Hyl: hydroxylysine. Values are expressed as Mol%. Ref. 1: Leach et al. (1981), *Arch Biochem Biophys*, 207:353–359; Ref. 2: Baumgartner et al. (1978), *J. Nutr*, 108, 804–811; Ref. 3: Blake et al. (1985), Poult Sci, 64, 176–182; Ref. 4: Salevsky and Leach (1980), *Poult Sci*, 59, 438–443; Ref. 5: Wedral et al. (1974), *Comp Biochem Physiol B*, 47, 631–640; Ref. 6: Crombie et al., 1981, *Biochim Biophys Acta*, 640, 365–367; Ref. 7: Kulshreshtha et al. (2020), *Biomater Sci*, 8, 5346–5361; Ref. 8: Candlish & Scougall 1969, *Int J Protein Res*, 1(4):299–302; and Ref. 9: Nakano et al. (2003), *Poult Sci*, 82, 510–514.

TABLE 4.4

A Comparative Quantitative Determination of the Amino Acid Composition of Decalcified ES, ESM, and Cuticle.

	Cuticle	ES Matrix	ESM (Combined)	ESM (Inner)	ESM (Outer)
Amino acid	**Ref. 1 & 2**	**Ref. 3–5**	**Ref. 1 & 3 & 4 & 6–9**	**Ref. 1 & 5 & 10**	**Ref. 1 & 5 & 10**
Ala	4.0 ± 1.4	6.9 ± 0.0	3.7 ± 0.6	3.7 ± 1.6	3.6 ± 0.7
Arg	8.9 ± 0.4	5.8 ± 0.1	5.2 ± 0.3	6.1 ± 1.3	5.1 ± 0.9
Asx	11.2 ± 4.0	8.2 ± 0.1	8.2 ± 1.6	8.8 ± 0.8	10.4 ± 2.8
Cys/2	3.5 ± 1.1	4.6 ± 0.2	10.4 ± 0.9	8.2 ± 4.3	8.1 ± 3.6
Glx	14.8 ± 5.7	11.6 ± 0.2	11.2 ± 2.0	12.2 ± 2.9	13.1 ± 3.0
Gly	12.3 ± 4.2	13.0 ± 0.2	9.6 ± 2.4	8.7 ± 4.4	8.6 ± 3.6
His	0.8 ± 0.3	3.4 ± 0.6	2.9 ± 0.8	2.9 ± 1.2	3.5 ± 2.4
Hyl	-	-	1.0 ± 1.5	0.2	0.2
Hyp	-	0.3	0.9 ± 0.3	1.5 ± 0.0	1.3 ± 0.1
Ile	3.9 ± 0.6	2.7 ± 0.1	3.2 ± 0.4	3.3 ± 0.4	3.4 ± 0.6
Leu	5.2 ± 1.7	5.0 ± 1.7	4.9 ± 0.5	5.9 ± 0.8	4.9 ± 0.6
Lys	4.5 ± 0.1	3.4 ± 0.1	3.3 ± 0.4	4.0 ± 1.0	3.1 ± 0.4
Met	2.8	2.3 ± 0.2	3.4 ± 0.7	3.5 ± 1.3	3.6 ± 1.2
Phe	2.7 ± 0.8	2.2 ± 0.1	1.8 ± 0.8	2.3 ± 1.3	2.2 ± 1.1
Pro	3.8 ± 1.0	8.3 ± 0.0	10.0 ± 1.1	9.8 ± 2.1	10.7 ± 1.3
Ser	5.0 ± 2.5	8.9 ± 0.6	5.9 ± 1.8	6.7 ± 3.4	5.7 ± 4.6
Thr	4.4 ± 2.3	6.1 ± 0.0	6.1 ± 0.2	6.5 ± 0.8	6.8 ± 0.1
Tyr	8.1 ± 1.3	1.7 ± 0.1	1.7 ± 0.9	2.6 ± 0.9	2.3 ± 1.5
Val	5.4 ± 2.7	6.6 ± 0.5	7.1 ± 1.0	6.2 ± 1.3	7.1 ± 0.8

Note: Glx = Glu + Gln; Asx = Asp + Asn; Hyp = hydroyproline; Hyl: hydroxylysine. Values are expressed as Mol% (Mean± Standard deviation). Ref. 1: Wedral et al. (1974), *Comp Biochem Physiol B*, 47, 631–640; Ref. 2: Baker and Balch (1962), *Biochem J*, 82: 352–361; Ref. 3: Baumgartner et al. (1978), *J. Nutr*, 108, 804–811; Ref. 4: Salevsky and Leach (1980), *Poult Sci*, 59, 438–443; Ref. 5: Nakano et al. (2003), *Poult Sci*, 82, 510–514; Ref. 6: Leach et al. (1981), *Arch Biochem Biophys*, 207:353–359; Ref. 7: Blake et al. (1985), *Poult Sci*, 64, 176–182; Ref. 8: Crombie et al. (1981), *Biochim Biophys Acta*, 640, 365–367; Ref. 9: Kulshreshtha et al. (2020), *Biomater Sci*, 8, 5346–5361; and Ref. 10: Candlish & Scougall 1969, *Int J Protein Res*, 1(4): 299–302.

membrane protein (CREMP) with multiple disulfide cross-links; moreover, a variety of lysine-derived desmosine and isodesmosine linkages also stabilize the membranes (Harris et al., 1980; Wong et al., 1984; Arias et al., 1991; Zhao & Chi, 2009; Kodali et al., 2011; Cordeiro & Hincke, 2016). The resulting insolubility of ESM fibers has impeded detailed identification of its protein constituents (Cordeiro et al., 2013; Cordeiro & Hincke, 2016). Nevertheless, constituents of ESM from both fertilized and unfertilized eggs have been identified using combinations of extraction techniques and proteomics analyses. Many proteins identified in the ESM are also present in other egg compartments. For example, egg white proteins such as ovalbumin, ovotransferrin, lysozyme, clusterin, ovoglycoprotein, ovomucoid, and ovoinhibitors are abundant constituents of ESM (Mann, 2007; Boschetti & Righetti, 2008; Rose & Hincke, 2009). Proteomics analysis identified 62 proteins (8 egg white proteins and 53 newly reported) in the ESM of fertilized chicken egg at different times (days 1, 9, 15, and 21) (Kaweewong et al., 2013). This study concluded that some proteins with unique biological functions were present at specific time of embryo development, indicating incubation-dependent evolution of ESM proteins leading to egg hatching (Kaweewong et al., 2013). A comparative analysis of ESM is essential to understand the changes in abundance of proteins during embryogenesis, and to confirm if the proteins are specifically related to embryonic events or, whether the presence of certain proteins is due to thermally induced changes in protein localization during incubation (Cordeiro & Hincke, 2016).

Solubilization of ESM from both fertilized and unfertilized white eggs followed by bioinformatics analysis revealed 228 proteins (Cordeiro & Hincke, 2016). A total of 201 proteins were identified in the ESM from fertilized eggs, while 114 proteins in the ESM proteome from unfertilized eggs and 87 proteins were found to be common to both fertilized and unfertilized models, indicating that many additional proteins (114) are associated with the ESM during embryo development (Cordeiro & Hincke, 2016). Specifically, protease inhibitors were present at all phases of chick development; calcium binding and oxygen transport proteins were present only during the second phase; while extracellular matrix, cell adhesion proteins related to the vascularization of chorioallantoic membrane (CAM), antimicrobial proteins and lipid binding and transport proteins were dominant in the second and third phases of development (Cordeiro & Hincke, 2016). These observations suggest that ESM helps in the development of the chick embryo and provides protection against pathogen invasion from the earliest stages of development until hatching (Cordeiro et al., 2013; Cordeiro & Hincke, 2016). In another study, ESM from unfertilized brown and white eggs were solubilized for analysis by mass spectrometry (Makkar et al., 2015). A total of nine proteins were identified in a methanol extract, including several clusterin peptides and defensins; in particular, two isoforms of gallin were detected at different ratios in brown vs white eggs (Makkar et al., 2015). The analysis suggested the presence of several abundant egg white proteins, as well as other proteins associated with antimicrobial, biomechanical, cytoskeletal organizational, cell signaling, and enzyme activities. Structural proteins such as collagens, agrin and laminin were also identified (Makkar et al., 2015).

Another study solubilized and identified proteins in ESM obtained from hatching eggs (HESM) (Rath et al., 2016). Several previously identified blood (hemoglobin, α-D-globin, annexin, fibronectin) and embryo-associated proteins (vitellogenin, zona pellucida sperm-binding proteins, and vitelline membrane outer layer protein) (Cordeiro et al., 2013; Cordeiro & Hincke, 2016) were identified in the HESM. However, the HESM was differentially populated with several proteins associated with glycolysis, carbohydrate, protein, and nucleic acid metabolisms. Clusterin, a major glycoprotein component, was not detected, and may have been depleted during embryonic development. Similarly, an antimicrobial peptide gallin was absent in the proteomics profile of HESM (Rath et al., 2016). This evolution in the protein profile is likely due to qualitative and quantitative changes occurring during both incubation and passive storage of eggs (Omana et al., 2011; Qiu et al., 2012, 2013; Kaweewong et al., 2013; Liu et al., 2013; Cordeiro & Hincke, 2016). Differences in levels or presence/absence of minor proteins can also be due to issues such as extraction conditions, post-translational modifications, and search parameters.

In order to address the issue of variability arising from ESM solubilization methods, a comprehensive strategic extraction approach identified 472 proteins in the ESM from unfertilized white eggs (Ahmed et al., 2017). Nine different solubilization conditions were progressively utilized, including: water, HCl, 3-mercaptopropionic acid (3-MPA), sodium hydroxide (NaOH) and lithium dodecyl sulfate/dithiothreitol (LDS/DTT), to develop a complete protein inventory. Identified protein superfamilies with distinct functionalities included collagens (III, IV, V, VII, VIII, X, XII and XXII), CREMPs, histones, AvBDs, lysyl oxidase-like 2 (LOXL2), and OCX-36 (Ahmed et al., 2017). The structural proteins identified in this study are likely critical for the mechanical stability and unique physiochemical properties of ESM (Nys et al., 2004; Kodali et al., 2011; Hincke et al., 2012; Kaweewong et al., 2013; Cordeiro & Hincke, 2016). Moreover, antimicrobial proteins such as OCX-36, avian β-defensins and histones provide direct protection against pathogen invasion of the egg interior (Hincke et al., 2000; Sugiarto & Yu, 2004; Zhang & Sunkara, 2014; Yacoub et al., 2015; Zhang et al., 2016). In a follow-up study, proteomics analysis of processed eggshell membrane powder (PEP) obtained from ESM flakes identified 110 proteins including structural proteins which constituted approx. 40% of PEP (Ahmed, Suso et al., 2019). Based on total spectral counts of the unique peptides, ESM mainly contained

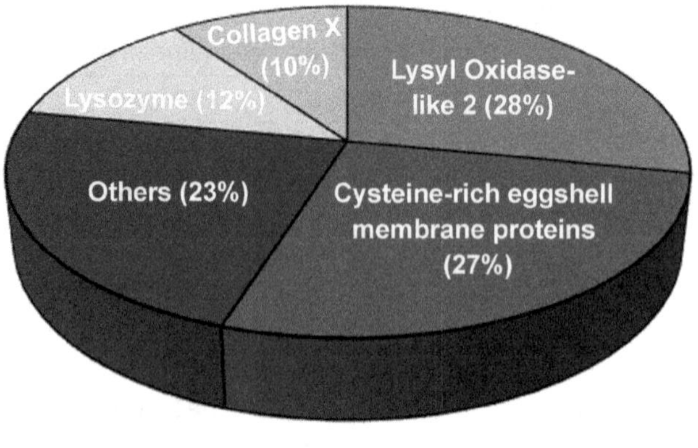

FIGURE 4.6 Pie chart summarizing the protein composition of ESM. Cysteine – rich eggshell membrane proteins (CREMPs) and collagen X constitute about 40% of the ESM proteome.

CREMPs (30%) and collagen X (10%), lysyl oxidase-like protein 2 (LOXL2) (30%) and lysozyme C (LYZ) (10%) (Ahmed, Suso et al., 2019) (Figure 4.6).

Functional annotation clustering predicted wound healing functions of identified proteins mainly associated with response to external stimulus, defense response, inflammatory response, and cell-substrate adhesion (Ahmed, Suso et al., 2019). The wound healing efficacy of PEP was demonstrated *in vivo* using the mouse excisional wound splinting model (Ahmed, Suso et al., 2019).

Accurate gene annotation is essential to support proteomics approaches (Roy et al., 2021). Microarray transcriptomics was used to probe the transcript levels during eggshell membrane fiber synthesis (Du et al., 2015; Yin et al., 2020). Transcriptomic analysis revealed that eggshell membrane synthesis was accompanied by differential gene expression of 107 unique genes in white isthmus (WI) region of the hen oviduct compared to the neighboring magnum and uterine segments (Du et al., 2015). These genes were shown to be associated with processing of collagen X (*COL10A1*), structural proteins including fibrillin-1 (*FBN1*) and cysteine-rich eggshell membrane protein (*CREMP*), and disulfide-mediated cross-linking sulfhydryl oxidase (*QSOX1*), suggesting that coordinated upregulation of genes in the white isthmus is associated with eggshell membrane fiber formation (Du et al., 2015).

4.6.1.3 Antibody-Based Techniques

Immunochemistry including Western blotting and colloidal gold immunocytochemistry has been used to identify shell membrane constituents such as collagen X and lysozyme (Arias et al., 1997; Wang et al., 2002). Immunohistochemistry with specific monoclonal antibodies demonstrated the presence of collagen X in the shell membrane and in the cells of isthmus section of the hen oviduct (Arias et al., 1997; Wang et al., 2002). The ESM functions as a surface, supporting eggshell mineralization that commences at organic matrix nodules on the outer eggshell fibers and giving rise to the mammillary cones (Creger et al., 1976; Stemberger et al., 1977; Hincke et al., 2012). The non-helical domain of type X collagen can function *in vivo* to inhibit mineralization, leading to the hypothesis that its presence in ESM prevents calcite crystals from growing inward toward the albumen (Arias et al., 1997; Wang et al., 2002).

Western blotting and immunofluorescence demonstrated that lysozyme is distinctly distributed within the eggshell, also present in the cuticle, and is particularly abundant in the ESM (Hincke et al., 2000). Colloidal gold immunocytochemistry confirmed its heterogeneous distribution throughout the eggshell membranes and the shell matrix (Hincke et al., 2000). Lysozyme is also an abundant protein constituent of the limiting membrane, which lines the inner ESM and plays an important role in restricting the egg white (Hincke et al., 2000). Lysozyme is a well-known antimicrobial protein with potent activity against both Gram-positive and Gram-negative pathogens (Vadehra et al., 1972; Baker et al., 1983; During et al., 1998). Therefore, its presence in the cuticle and eggshell membranes suggests that it plays an important role in protecting the egg contents against microbial contamination (Hincke et al., 2000).

An eggshell matrix protein, ovotransferrin was identified at high concentration in the uterine fluid at the initial stage of shell mineralization using N-terminal sequencing and western blotting with monoclonal and polyclonal antibodies (Gautron et al., 2001b). Immunofluorescence demonstrated that ovotransferrin is mostly localized in the shell membranes and eggshell mammillae (Gautron et al., 2001b). The presence of ovotransferrin in eggshell and membrane suggests a dual role: 1. during mineralization by influencing nucleation and growth of calcite crystals, 2. as a bacteriostatic filter to inhibit the growth of *Salmonella* in the egg albumen (Gautron et al., 2001b). Another eggshell protein OCX-36 was detected in the ESM by immunofluorescence and is proposed to participate in defense of the developing embryo and in unfertilized table eggs against bacterial pathogens (Gautron et al., 2007; Cordeiro et al., 2013).

4.6.2 Eggshell Membrane Carbohydrates

ESM contains glycosaminoglycans (GAGs) which are linear, highly negative charged polysaccharides containing variable number of repeated disaccharide units (Capila & Linhardt, 2002; Liu et al., 2016). GAGs can be classified into four groups based on the structure of their repeating units: hyaluronan (HA), keratan sulfate (KS), chondroitin sulfate (CS), and heparan sulfate (HS) (Capila & Linhardt, 2002). GAG abundance in ESM has been correlated with shell strength (Ha et al., 2007). A weak correlation between uronic acid content in ESM and eggshell strength has been established. Moreover, galactose content in shell membrane has also been shown to be correlated with eggshell strength, indicating association of GAGs with shell strength (Ha et al., 2007; Liu et al., 2016). In addition, it has been speculated that GAGs or disaccharides perform multiple functions in the eggshell mineralization process including organic matrix organization, mineral deposition, and crystal growth (Liu et al., 2016).

LC-MS analyses demonstrated that HA is most abundant GAG component in the shell membranes followed by CS, KS, and HS (Long et al., 2005; Liu et al., 2016). The biological activity of HA is dependent on its molecular weight (Weigel et al., 1997; Weigel & DeAngelis, 2007). In general, HA is categorized as low molecular weight (LMW-HA) of $0.5–1 \times 10^6$ Da, medium molecular weight (MMW-HA) of $1.2–4.5 \times 10^6$ Da, and high molecular weight (HMW-HA) of $6–7 \times 10^6$ Da (Weigel et al., 1997; Weigel & DeAngelis, 2007). The structure of HA contains sugar monomer units which are connected by beta-1,4 and beta-1,3 alternating glycosidic bonds with hydroxyl and carboxylic groups are sterically linked to the outside and the hydrogen to the inside resulting in a twisting ribbon structure (Necas et al., 2008; Kablik et al., 2009; Stabler Ogawa, 2021). HA plays a crucial role in tissue hydration and various cellular functions including growth, differentiation, and transport (Kulshreshtha et al., 2022b). HA possesses unique properties such as biocompatibility, viscoelasticity, lubricity and immunostimulation, and has diverse health and medical applications for joint injection, ocular surgery, osteoarthritis treatments, plastic surgery, and diverse skin treatments such as major burn healing and anti-aging products (Price et al., 2007; Kulshreshtha et al., 2022b). However, its function in the eggshell/eggshell membranes remains poorly understood.

4.6.3 Eggshell Membrane Minerals

The elemental composition of ESM powder and intact membranes has been identified by energy dispersive X-ray spectroscopy (EDS) analysis. Major elements detected in the ESM were Mg, S, and Ca (Kulshreshtha et al., 2020). The eggshell mineral is strongly adherent to the protein-based eggshell membranes, whose outermost fibers penetrate the tips of the mammillary cones. For various commercial applications, the inner ESM fibers can be removed mechanically, while the outer fibers remain attached to the tips of the mammillary cones and can only be extracted by harsh chemical treatments (Chi et al., 2019; Pasarin & Rovinaru, 2019; Kulshreshtha et al., 2022b).

4.7 Conclusion and Perspectives

This chapter's main aim has been to summarize the research landscape pertaining to chicken eggshell and membrane constituents over the last three decades, from which much of our knowledge of avian eggshell mineralization, function and composition are derived. Many biochemical, nutritional, morphological, physiological and structural studies have been conducted on avian eggs and most have utilized the egg of the domestic chicken because of its availability and commercial importance as a nutritious food for human consumption.

We have summarized various approaches that have been used for the comprehensive identification of eggshell and accompanying membrane constituents, including proteomics and transcriptomics analyses along with antibody-based strategies. In addition, carbohydrate and elemental compositions were determined using chromatographic in-conjunction with spectrometric techniques. The proteomic analysis identified >900 and >500 proteins in the eggshell organic matrix and eggshell membrane, respectively, and proteins and expressed genes have been correlated to a variety of egg and eggshell quality phenotypes.

In addition, insight has been gained into expression of numerous genes coding for cell membrane ion transporters, channels and pumps responsible for movement of Ca and bicarbonate ions into the uterine lumen, which are linked to the maintenance of cellular ionic homeostasis during eggshell mineralization. These rich datasets are expected to provide new selection tools to improve eggshell quality in commercial flocks. Differences in experimental design for such studies have allowed detailed interpretation of differential gene expression according to the experimental design.

All 4 groups of GAGs, including hyaluronic acid, keratan sulfate, chondroitin sulfate, and heparan sulfate, have been identified in eggshell and the eggshell membranes. The elemental composition identified at least 20 elements in the eggshell, while only 4 major elements were identified in the eggshell membrane. Very low levels of heavy metals are detected in ES, and it therefore holds great promise as a nutritional supplement.

Globally, around 100 million tonnes (1.65 trillion table eggs) are produced annually. Chicken egg processing plants produce considerable amounts of ES and ESM waste (~30 Megatonnes), which is a valuable resource as raw materials for various applications and secondary processing (Ahmed, Udenigwe et al., 2021; Ahmed, Wu et al., 2021; Ahmed, Younes et al., 2021). Eggshell organic matrix and its associated membrane contain more than 1,000 proteins with diverse functionalities. The unique physicochemical and biological properties of eggshell and associated membrane in terms of composition and mechanical features, along with their abundance, make them promising biomaterials for various biotechnological applications. Exploration of novel top-down strategies and ES development as a physical platform are new endeavors that are expected to further increase the upscaling of ES waste exploitation. A novel perspective for upcycling ESM waste to commercially viable high-value biomaterials as nutraceutical supplements and cosmetics ingredients has been proposed (Kulshreshtha et al., 2022b). However, innovative processing of ESM into various formats is necessary to diversify its utilization for different applications. Finally, there is a need to prepare micro-or nano-scale formats of both ES and ESM in order to increase the bioavailability of the biologically active components, as well as preserve and enhance their unique structures and bioactivities (Ahmed, Kulshreshtha et al., 2019).

Acknowledgments

This work was supported by funding from the Egg Farmers of Canada (EFC), Livestock Research Innovation Corporation (LRIC) and Canadian Natural Sciences and Engineering Research Council (NSERC, Discovery program RGPIN-2022-04803 and Alliance program ALLRP-556438-2020). The authors would like to thank Jianqun Wang from the Nano Imaging Facility at Carleton University, Ottawa, Canada, for help with energy dispersive X-ray spectroscopy and scanning electron microscopy. Salary support for GK was provided by a MITACS Accelerate Fellowship (Application Ref. IT13653/Funding Request Ref. FR35012).

REFERENCES

Abatangelo, G., Daga-Gordini, D., Castellani, I., & Cortivo, R. (1978). Some observations on the calcium ion binding to the eggshell matrix. *Calcified Tissue Research, 26*(3), 247–252. https://doi.10.1007/BF02013266

Ahmed, T. A. E., Cordeiro, C. M. M., Elebute, O., & Hincke, M. T. (2022). Proteomic analysis of chicken chorioallantoic membrane (CAM) during embryonic development provides functional insight. *BioMed Research International, 2022*, ID, 7813921. https://doi.org/10.1155/2022/7813921

Ahmed, T. A. E., Garima, K., Hincke, M. T., & Hincke (2019). Chapter 19: Value-added uses of eggshell and eggshell membranes. In W. J (Ed.), *Food chemistry, function and analysis no.14; Eggs as functional foods and nutraceuticals for human health* (pp. 359–397). The Royal Society of Chemistry.

Ahmed, T. A. E., Suso, H. P., & Hincke, M. T. (2017). In-depth comparative analysis of the chicken eggshell membrane proteome. *Journal of Proteomics, 155*, 49–62. https://doi.10.1016/j.jprot.2017.01.002

Ahmed, T. A. E., Suso, H.-P., Maqbool, A., & Hincke, M. T. (2019). Processed eggshell membrane powder: Bioinspiration for an innovative wound healing product. *Materials Science and Engineering: C, 95*, 192–203. https://doi.org/10.1016/j.msec.2018.10.054

Ahmed, T. A. E., Udenigwe, C. C., & Gomaa, A. (2021). Editorial: Biotechnology and bioengineering applications for egg-derived biomaterials. *Frontiers Bioengineering and Biotechnology, 9*, 756058. https://doi.10.3389/fbioe.2021.756058

Ahmed, T. A. E., Wu, L., Younes, M., & Hincke, M. T. (2021). Biotechnological applications of eggshell: Recent advances. *Frontiers Bioengineering and Biotechnology, 9*, 675364. https://doi.10.3389/fbioe.2021.675364

Ahmed, T. A. E., Younes, M., Wu, L., & Hincke, M. T. (2021). A survey of recent patents in engineering technology for the screening, separation and processing of eggshell. *Frontiers Bioengineering and Biotechnology, 9*, 677559. https://doi.10.3389/fbioe.2021.677559

Arazi, H., Yoselewitz, I., Malka, Y., Kelner, Y., Genin, O., & Pines, M. (2009) Osteopontin and calbindin gene expression in the eggshell gland as related to eggshell abnormalities, *Poultry Science, 88*, 647–653. https://doi.org/10.3382/ps.2008-00387

Arias, J. L., Carrino, D. A., Fernández, M. S., Rodríguez, J. P., Dennis, J. E., & Caplan, A. I. (1992). Partial biochemical and immunochemical characterization of avian eggshell extracellular matrices. *Archives of Biochemistry and Biophysics, 298*(1), 293–302. https://doi.10.1016/0003-9861(92)90126-h

Arias, J. L., Fernandez, M. S., Dennis, J. E., & Caplan, A. I. (1991). Collagens of the chicken eggshell membranes. *Connective Tissue Research, 26*(1–2), 37–45. https://doi.org/10.3109/03008209109152162

Arias, J. L., Nakamura, O., Fernández, M. S., Wu, J. J., Knigge, P., Eyre, D. R., & Caplan, A. I. (1997). Role of type X collagen on experimental mineralization of eggshell membranes. *Connective Tissue Research, 36*(1), 21–33. https://doi.org/10.3109/03008209709160211

Athanasiadou, D., Jiang, W., Goldbaum, D., Saleem, A., Basu, K., Pacella, M. S., Böhm, C. F., Chromik, R. R., Hincke, M. T., Rodríguez-Navarro, A. B., Vali, H., Wolf, S. E., Gray, J. J.,

Bui, K. H., & McKee, M. D. (2018). Nanostructure, osteopontin, and mechanical properties of calcitic avian eggshell. *Science Advances*, *4*(3), eaar3219. https://doi.10.1126/sciadv.aar3219

Bain, M. M., McDade, K., Burchmore, R., Law, A., Wilson, P. W., Schmutz, M., Preisinger, R., & Dunn, I. C. (2013). Enhancing the egg's natural defence against bacterial penetration by increasing cuticle deposition. *Animal Genetics*, *44*, 661–668. https://doi.org/10.1111/age.12071

Baker, J. R., & Balch, D. A. (1962). A study of the organic material of hen's-egg shell. *The Biochemical Journal*, *82*(2), 352–361. http://doi.10.1042/bj0820352

Baker, R. C., Hogarty, S., Poon, W., & Vadehra, D. V. (1983). Survival of salmonella typhimurium and Staphylococcus aureus in eggs cooked by different methods. *Poultry Science*, *62*(7), 1211–1216. https://doi.org/10.3382/ps.0621211

Baláž, M. (2016). Eggshell membrane biomaterial as a platform for applications in materials science. *Acta Biomaterialia*, *10*(9), 3827–3843. https://doi.org/10.1016/j.actbio.2014.03.020

Baumgartner, S., Brown, D. J., Salevsky, E. Jr, & Leach, R. M. Jr (1978). Copper deficiency in the laying hen. *The Journal of Nutrition*, *108*(5), 804–811. http://doi.10.1093/jn/108.5.804

Blake, J. P., Kling, L. J., & Halteman, W. A. (1985). The relationship of the amino acid composition of a portion of the outer eggshell membrane to eggshell quality. *Poultry Science*, *64*(1), 176–182. https://doi.org/10.3382/ps.0640176

Boschetti, E., & Righetti, P. G. (2008). The ProteoMiner in the proteomic arena: A non-depleting tool for discovering low-abundance species. *Journal of Proteomics*, *71*(3), 255–264. http://doi.10.1016/j.jprot.2008.05.002

Brionne, A., Nys, Y., Hennequet-Antier, C., & Gautron, J. (2014). Hen uterine gene expression profiling during eggshell formation reveals putative proteins involved in the supply of minerals or in the shell mineralization process. *BMC Genomics*, *15*, 220. https://doi.10.1186/1471-2164-15-220

Candlish, J. K., & Scougall, R. K. (1969). L-5-hydroxylysine as a constituent of the shell membranes of the hen's egg. *International Journal of Protein Research*, *1*(4), 299–302. http://doi.10.1111/j.1399-3011.1969.tb01654.x

Capila, I., & Linhardt, R. J. (2002). Heparin-protein interactions. *Angewandte Chemie International Edition*, *41*(3), 390–412. https://doi.org/10.1002/1521-3773(20020201)41:3<390::AID-ANIE390>3.0.CO;2-B

Carrino, D. A., Dennis, J. E., Wu, T. M., Arias, J. L., Fernandez, M. S., Rodriguez, J. P., Fink, D. J., Heuer, A. H., & Caplan, A. I. (1996). The avian eggshell extracellular matrix as a model for biomineralization. *Connective Tissue Research*, *35*(1–4), 325–329. https://doi.10.3109/03008209609029207

Carrino, D. A., Rodriguez, J. P., & Caplan, A. I. (1997). Dermatan sulfate proteoglycans from the mineralized matrix of the avian eggshell. *Connective Tissue Research*, *36*(3), 175–193. http://Doi.10.3109/03008209709160219

Chen, S., Chen, K., Xu, J., Li, F., Ding, J., Ma, Z., Li, G., & Li, H. (2021) Insights into mRNA and long non-coding RNA profiling RNA sequencing in uterus of chickens with pink and blue eggshell colors. *Frontiers in Veterinary Science*, *8*, 736387. https://doi.org/10.3389/fvets.2021.736387

Chen, X., He, Z., Li, X., Song, J., Huang, M., Shi, X., Li, X., Li, J., Xu, G., & Zheng, J. (2021). Cuticle deposition duration in the uterus is correlated with eggshell cuticle quality in white leghorn laying hens. *Scientific Reports*, *11*(1), 22100. https://doi.10.1038/s41598-021-01718-0

Chen, L., Kang, J., & Sukigara, S. (2014). Preparation and characterization of polyurethane/soluble eggshell membrane nanofibers. *Bio-Medical Materials and Engineering*, *24*(6), 1979–1989. https://doi.org/10.3233/BME-141007

Chi, Y., Wang, Y., Li, M. F., Ren, J., & Chi, Y. J. (2019). Numerical simulation and experimental study on eggshell membrane separation device. *International Journal of Agricultural and Biological Engineering*, *12*(2), 173–183. https://doi.org/10.25165/j.ijabe.20191202.3058

Chien, Y. C., Hincke, M. T., & McKee, M. D. (2009). Avian eggshell structure and osteopontin. *Cells Tissues Organs*, *189*(1–4), 38–43. https://doi.10.1159/000151374

Chien, Y. C., Hincke, M. T., Vali, H., & McKee, M. D. (2008). Ultrastructural matrix-mineral relationships in avian eggshell, and effects of osteopontin on calcite growth in vitro. *Journal Structural Biology*, *163*(1), 84–99. https://doi.10.1016/j.jsb.2008.04.008

Chowdhury, S. D. (1990, July). Shell membrane protein system in relation to lathyrogen toxicity and copper deficiency. *World's Poultry Science Journal*, *46*(2), 153–169. https://doi.org/10.1079/WPS19900018

Chowdhury, S. D., & Davis, R. H. (1995) Influence of dietary osteolathyrogens on the ultrastructure of shell and membranes of eggs from laying hens. *British Poultry Science*, *36*, 575–583. https://doi.org/10.1080/00071669508417803

Cooke, A. S., & Balch, D. A. (1970). Studies of membrane, mammillary cores and cuticle of the hen egg shell. *British Poultry Science*, *11*(3), 345–352. http://doi.10.1080/00071667008415825

Cordeiro, C. M. M., Esmaili, H., Ansah, G., & Hincke, M. T. (2013). Ovocalyxin-36 is a pattern recognition protein in chicken eggshell membranes. *PLoS One*, *8*(12), e84112. https://doi.org/10.1371/journal.pone.0084112

Cordeiro, C. M. M., & Hincke, M. T. (2011). Recent patents on eggshell: Shell and membrane applications. *recent patents on food. Nutrition & Agriculture*, *3*(1), 1–8. https://doi.10.2174/2212798411103010001

Cordeiro, C. M. M., & Hincke, M. T. (2016). Quantitative proteomics analysis of eggshell membrane proteins during chick embryonic development. *Journal of Proteomics*, *130*, 11–25. https://doi.org/10.1016/j.jprot.2015.08.014

Creger, C. R., Phillips, H., & Scott, J. T. (1976). Formation of an egg shell. *Poultry Science*, *55*(5), 1717–1723. https://doi.org/10.3382/ps.0551717

Crombie, G., Snider, R., Faris, B., & Franzblau, C. (1981). Lysine-derived cross-links in the egg shell membrane. *Biochimica Et Biophysica Acta*, *640*(1), 365–367. http://doi.10.1016/0005-2736(81)90560-5

Cusack, M., Fraser, A. C., & Stachel, T. (2003) Magnesium and phosphorus distribution in the avian eggshell. *Comparative Biochemistry and Physiology – B Biochemistry and Molecular Biology*, *134*, 63–69. https://doi.org/10.1016/S1096-4959(02)00185-9

Dennis, J. E., Xiao, S.-Q., Agarwal, M., Fink, D. J., Heuer, A. H., & Caplan, A. I. (1996). Microstructure of matrix and mineral components of eggshells from white leghorn chickens *Gallus gallus*. *Journal of Morphology*, *228*, 287–306.

Dennis, J. E., Carrino, D. A., Yamashita, K., & Caplan, A. I. (2000). Monoclonal antibodies to mineralized matrix molecules of the avian eggshell. *Matrix Biology*, *19*(7), 683–692. https://doi.10.1016/s0945-053x(00)00118-9

Du, J., Hincke, M. T., Rose-Martel, M., Hennequet-Antier, C., Brionne, A., Cogburn, L. A., Nys, Y., & Gautron, J. (2015). Identifying specific proteins involved in eggshell membrane formation using gene expression analysis and bioinformatics. *BMC Genomics*, *16*(1), 792. https://doi.org/10.1186/s12864-015-2013-3

Dunn, I. C., Joseph, N. T., Bain, M., Edmond, A., Wilson, P. W., Milona, P., Nys, Y., Gautron, J., Schmutz, M., Preisinger, R., & Waddington, D. (2008) Polymorphisms in eggshell organic matrix genes are associated with eggshell quality measurements in pedigree Rhode Island red hens. *Animal Genetics*, *40*, 110–114. https://doi.org/10.1111/j.1365-2052.2008.01794.x

During, K., Porsch, P., Mahn, A., Brinkmann, O., & Gieffers, W. (1998) The non-enzymatic microbicidal activity of lysozymes. *FEBS Letters*, *449*, 93–100. https://doi.org/10.1016/S0014-5793(99)00405-6

Feng, J., Zhang, H. J., Wu, S. G., Qi, G. H., & Wang, J. (2020) Uterine transcriptome analysis reveals mRNA expression changes associated with the ultrastructure differences of eggshell in young and aged laying hens. *BMC Genomics*, *21*(1), 770. https://doi.10.1186/s12864-020-07177-7

Gautron, J. (2019). Proteomics analysis of avian eggshell matrix proteins: Toward new advances on biomineralization. *Proteomics*, *19*(13), e1900120. https://doi.10.1002/pmic.201900120

Gautron, J., Hincke, M. T., Mann, K., Panheleux, M., Bain, M., McKee, M. D., Solomon, S. E., & Nys, Y. (2001a). Ovocalyxin-32, a novel chicken eggshell matrix protein. Isolation, amino acid sequencing, cloning, and immunocytochemical localization. *Journal of Biological Chemistry*, *276*(42), 39243–39252. https://doi.10.1074/jbc.M104543200

Gautron, J., Hincke, M. T., Panheleux, M., Garcia-Ruiz, J. M., Boldicke, T., & Nys, Y. (2001b). Ovotransferrin is a matrix protein of the hen eggshell membranes and basal calcified layer. *Connective Tissue Research*, *42*(4), 255–267. https://doi.10.3109/03008200109016840

Gautron, J., Hincke, M. T., & Nys, Y. (1997). Precursor matrix proteins in the uterine fluid change with stages of eggshell formation in hens. *Connective Tissue Research*, *36*(3), 195–210. https://doi.10.3109/03008209709160220

Gautron, J., Murayama, E., Vignal, A., Morisson, M., McKee, M. D., Réhault, S., Labas, V., Belghazi, M., Vidal, M. L., Nys, Y., & Hincke, M. T. (2007). Cloning of ovocalyxin-36, a novel chicken eggshell protein related to lipopolysaccharide-binding proteins, bactericidal permeability-increasing proteins, and plunc family proteins. *Journal of Biological Chemistry*, *282*(8), 5273–5286. https://doi.10.1074/jbc.M610 294200

Gautron, J., Stapane, L., Le Roy, N., Nys, Y., Rodriguez-Navarro, A. B., & Hincke, M. T. (2021). Avian eggshell biomineralization: An update on its structure, mineralogy And protein tool kit. *BMC Molecular and Cell Biology*, *22*(1), 11. http://doi.10.1186/s12860-021-00350-0

Guangqi, L., Congjiao, S., Guiqin, W., Fengying, S., Aiqiao, L., Hao, S., & Ning, Y. (2017). Transcriptome sequencing identifies potential regulatory genes involved in chicken eggshell brownness. *Yi Chuan*, *39*(11), 1102–1111. https://doi.10.16288/j.yczz.17-111

Ha, Y. W., Son, M. J., Yun, K. S., & Kim, Y. S. (2007). Relationship between eggshell strength and keratan sulfate of eggshell membranes. *Comparative Biochemistry and Physiology. Part A, Molecular & Integrative Physiology*, *147*(4), 1109–1115. https://doi.10.1016/j.cbpa.2007.03.022

Halgrain, M., Bernardet, N., Crepeau, M., Même, N., Narcy, A., Hincke, M., & Réhault-Godbert, S. (2022) Eggshell decalcification and skeletal mineralization during chicken embryonic development: Defining candidate genes in the chorioallantoic membrane. *Poultry Science*, *101*(2), 101622. http://doi.10.1016/j.psj.2021.101622

Harris, E. D., Blount, J. E., & Leach, R. M. (1980). Localization of lysyl oxidase in hen oviduct: Implications in egg shell membrane formation and composition. *Science*, *208*(4439), 55–56. https://doi.org/10.1126/science.6102412

Hasiak, R. J., Vadehra, D. V., & Baker, D. V. (1970). Lipid composition of the egg exteriors of the chicken (*Gallus gallus*). *Comparative Biochemistry and Physiology*, *37*(3), 429–435. http://doi.10.1016/0010-406x(70)90571-2

Heaney, R. K., & Robinson, D. S. (1976). The isolation and characterisation of hyaluronic acid in egg shell. *Biochimica et Biophysica Acta*, *451*(1), 133–142. https://doi.10.1016/0304-4165(76)90265-8

Hincke, M. T. (1995). Ovalbumin is a component of the chicken eggshell matrix. *Connective Tissue Research*, *31*(3), 227–233. https://doi.10.3109/03008209509010814

Hincke, M. T., Bernard, A. M., Lee, E. R., Tsang, C. P., & Narbaitz, R. (1992) Soluble protein constituents of the domestic fowl's eggshell. *British Poultry Science*, *33(3)*, 505–516. https://doi.org/10.1080/00071669208417489

Hincke, M. T., Chien, Y. C., Gerstenfeld, L. C., & McKee, M. D. (2008). Colloidal-gold immunocytochemical localization of osteopontin in avian eggshell gland and eggshell. *Journal of Histochemistry and Cytochemistry*, *56*(5), 467–476. https://doi.10.1369/jhc.2008.950576

Hincke, M. T., Da Silva, M. T., Guyot, N., Gautron, J., McKee, M. D., Guabiraba-Brito, R., & Réhault-Godbert, S. (2019) Dynamics of structural barriers and innate immune components during incubation of the avian egg: Critical interplay between autonomous embryonic development and maternal anticipation. *Journal of Innate Immunity*, *11(2)*, 111–124. https://doi.org/10.1159/000493719

Hincke, M. T., Gautron, J., Mann, K., Panhéleux, M., McKee, M. D., Bain, M., Solomon, S. E., & Nys, Y. (2001). Purification of ovocalyxin-32, a novel chicken eggshell matrix protein. *Connective Tissue Research*, *44*(Suppl 1), 16–19. https://doi.10.1080/03008200390152025

Hincke, M. T., Gautron, J., Panheleux, M., Garcia-Ruiz, J., McKee, M. D., & Nys, Y. (2000). Identification and localization of lysozyme as a component of eggshell membranes and eggshell matrix. *Matrix Biology*, *19*(5), 443–453. https://doi.10.1016/s0945-053x(00)00095-0

Hincke, M. T., Gautron, J., Tsang, C. P., McKee, M. D., & Nys, Y. (1999). Molecular cloning and ultrastructural localization of the core protein of an eggshell matrix proteoglycan, ovocleidin-116. *Journal of Biological Chemistry*, *274*(46), 32915–32923. https://doi.10.1074/jbc.274.46.32915

Hincke, M. T., Nys, Y., & Gautron, J. (2010). The role of matrix proteins in eggshell formation. *The Journal of Poultry Science*, *47*(3), 208–19.

Hincke, M. T., Nys, Y., Gautron, J., Mann, K., Rodriguez-Navarro, A. B., & McKee, M. D. (2012). The eggshell: Structure, composition and mineralization. *Frontiers Bioscience (Landmark Ed.)*, *17*(4), 1266–1280. https://doi.10.2741/3985

Hincke, M. T., Tsang, C. P., Courtney, M., Hill, V., & Narbaitz, R. (1995). Purification and immunochemistry of a soluble matrix protein of the chicken eggshell (ovocleidin 17). *Calcified Tissue International*, *56*(6), 578–583. https://doi.10.1007/BF00298593

Jonchère, V., Réhault-Godbert, S., Hennequet-Antier, C., Cabau, C., Sibut, V., Cogburn, L. A., Nys, Y., & Gautron, J. (2010). Gene expression profiling to identify eggshell proteins involved in physical defense of the chicken egg. *BMC Genomics*, *11*, 57. https://doi.10.1186/1471-2164-11-57

Kablik, J., Monheit, G. D., Yu, L., Chang, G., & Gershkovich, J. (2009). Comparative physical properties of hyaluronic acid dermal fillers. *Dermatologic Surgery*, *35*(Sup 1), 302–312. https://doi.org/10.1111/j.1524-4725.2008.01046.x

Kaweewong, K., Garnjanagoonchorn, W., Jirapakkul, W., & Roytrakul, S. (2013). Solubilization and identification of hen eggshell membrane proteins during different times of chicken embryo development using the proteomic approach. *The Protein Journal*, *32*(4), 297–308. https://doi.org/10.1007/s10930-013-9487-0

Khan, S., Wu, S. B., & Roberts, J. (2019). RNA-sequencing analysis of shell gland shows differences in gene expression profile at two time-points of eggshell formation in laying chickens. *BMC Genomics*, *20*(1), 89. https://doi.10.1186/s12864-019-5460-4

Khogali, M. K., Wen, K., Jauregui, D., Liu, L., Zhao, M., Gong, D., & Geng, T. (2021). Uterine structure and function contributes to the formation of the sandpaper-shelled eggs in laying hens. *Animal Reproduction Science*, *232*, 106826. https://doi.10.1016/j.anireprosci.2021.106826

Kodali, V. K., Gannon, S. A., Paramasivam, S., Raje, S., Polenova, T., & Thorpe, C. (2011). A novel disulfide-rich protein motif from avian eggshell membranes. *PLoS One*, *6*(3), e18187. https://doi.org/10.1371/journal.pone.0018187

Kulshreshtha, G., Ahmed, T. A. E., Wu, L., Diep, T., & Hincke, M. T. (2020). A novel eco-friendly green approach to produce particalized eggshell membrane (PEM) for skin health applications. *Biomaterials Science*, *8*(19), 5346–5361. https://doi.org/10.1039/d0bm01110j

Kulshreshtha, G., Benavides-Reyes, C., Rodriguez-Navarro, A. B., Diep, T., & Hincke, M. T. (2021). Impact of different layer housing systems on eggshell cuticle quality and salmonella adherence in table eggs. *Foods*, *10*(11), 2559. http://doi.10.3390/foods10112559

Kulshreshtha, G., D'Alba, L., Dunn, I. C., Rehault-Godbert, S., Rodriguez-Navarro, A. B., & Hincke, M. T. (2022a) Properties, genetics and innate immune function of the cuticle in egg-laying species. *Frontiers in Immunology*, *13*, 838525. https://doi.org/10.3389/fimmu.2022.838525

Kulshreshtha, G., Diep, T., Hudson, H.-A., & Hincke, M. T. (2022b). High value applications and current commercial market for eggshell membranes and derived bioactives. *Food Chemistry*, *382*, 132270. https://doi.org/10.1016/j.foodchem.2022.132270

Lavelin, I., Yarden, N., Ben-Bassat, S., Bar, A., & Pines, M. (1998). Regulation of osteopontin gene expression during egg shell formation in the laying hen by mechanical strain. *Matrix Biology*, *17*, 615–623. https://doi.org/10.1016/S0945-053X(98)90112-3

Le Roy, N., Stapane, L., Gautron, J., & Hincke, M. T. (2021). Evolution of the avian eggshell biomineralization protein toolkit – New insights from multi-omics. *Frontiers Genetics*, *12*, 672433. https://doi.10.3389/fgene.2021.672433

Leach, R. M. Jr, Rucker, R. B., & Van Dyke, G. P. (1981). Egg shell membrane protein: A nonelastin desmosine/isodesmosine-containing protein. *Archives of Biochemistry and Biophysics*, *207*(2), 353–359. http://doi.10.1016/0003-9861(81)90042-4

Li, G., Sun, C., Wu, G., Shi, F., Liu, A., & Yang, N. (2016). iTRAQ-based quantitative proteomics identifies potential regulatory proteins involved in chicken eggshell brownness. *PLoS One*, *11*(12), e0168750. https://doi.10.1371/journal.pone.0168750

Liu, Y., Qiu, N., & Ma, M. (2013). Comparative proteomic analysis of hen egg white proteins during early phase of embryonic development by combinatorial peptide ligand library and matrix-assisted laser desorption ionization-time of flight. *Poultry Science*, *92*(7), 1897–1904. https://doi.org/10.3382/ps.2012-02986

Liu, Z., Song, L., Lu, L., Zhang, X., Zhang, F., Wang, K., & Linhardt, R. J. (2017). Comparative proteomics of matrix fractions between pimpled and normal chicken eggshells. *Journal of Proteomics*, *167*, 1–11. https://doi.10.1016/j.jprot.2017.07.015

Liu, Z., Sun, X., Cai, C., He, W., Zhang, F., & Linhardt, R. J. (2016). Characteristics of glycosaminoglycans in chicken eggshells and the influence of disaccharide composition on eggshell properties. *Poultry Science*, *95*(12), 2879–2888. https://doi.10.3382/ps/pew179

Liu, Z., Zhang, F., Li, L., Li, G., He, W., & Linhardt, R. J. (2014). Compositional analysis and structural elucidation of glycosaminoglycans in chicken eggs. *Glycoconjugate Journal*, *31*(8), 593–602. https://doi.org/10.1007/s10719-014-9557-3

Liu, Z., Zheng, Q., Zhang, X., & Lu, L. (2013). Microarray analysis of genes involved with shell strength in layer shell gland at the early stage of active calcification. *Asian-Australasian Journal of Animal Sciences*, *26*(5), 609–624. https://doi.10.5713/ajas.2012.12398

Long, F. D., Adams, R. G., & DeVore, D. P. (2005). Preparation of hyaluronic acid from eggshell membrane: US-6946551-B2 Available at https://patents.google.com/patent/US6946551B2/en (verified 27 June 2022).

Makkar, S., Liyanage, R., Kannan, L., Packialakshmi, B., Lay, J. O., & Rath, N. C. (2015). Chicken egg shell membrane associated proteins and peptides. *Journal of Agricultural and Food Chemistry*, *63*(44), 9888–9898. https://doi.org/10.1021/acs.jafc.5b04266

Mann, K. (2007). The chicken egg white proteome. *Proteomics*, *7*(19), 3558–3568. https://doi.org/10.1002/pmic.200700397

Mann, K., Gautron, J., Nys, Y., McKee, M. D., Bajari, T., Schneider, W. J., & Hincke, M. T. (2003). Disulfide-linked heterodimeric clusterin is a component of the chicken eggshell matrix and egg white. *Matrix Biology*, *22*(5), 397–407. https://doi.10.1016/s0945-053x(03)00072-6

Mann, K., Hincke, M. T., & Nys, Y. (2002). Isolation of ovocleidin-116 from chicken eggshells, correction of its amino acid sequence and identification of disulfide bonds and glycosylated Asn. *Matrix Biology*, *21*, 383–387. https://doi.org/10.1016/S0945-053X(02)00031-8

Mann, K., Macek, B., & Olsen, J. V. (2006). Proteomic analysis of the acid-soluble organic matrix of the chicken calcified eggshell layer. *Proteomics*, *6*(13), 3801–3810. https://doi.10.1002/pmic.200600120

Mann, K., Olsen, J. V., Macek, B., Gnad, F., & Mann, M. (2007). Phosphoproteins of the chicken eggshell calcified layer. *Proteomics*, *7*(1), 106–115. https://doi.10.1002/pmic.200600635

Marie, P., Labas, V., Brionne, A., Harichaux, G., Hennequet-Antier, C., Nys, Y., & Gautron, J. (2014). Data set for the proteomic inventory and quantitative analysis of chicken uterine fluid during eggshell biomineralization. *Data in Brief*, *1*, 65–69. https://doi.10.1016/j.dib.2014.09.006

Marie, P., Labas, V., Brionne, A., Harichaux, G., Hennequet-Antier, C., Rodriguez-Navarro, A. B., Nys, Y., & Gautron, J. (2015a). Quantitative proteomics provides new insights into chicken eggshell matrix protein functions during the primary events of mineralisation and the active calcification phase. *Journal of Proteomics*, *126*, 140–154. https://doi.10.1016/j.jprot.2015.05.034

Marie, P., Labas, V., Brionne, A., Harichaux, G., Hennequet-Antier, C., Rodriguez-Navarro, A. B., Nys, Y., & Gautron, J. (2015b). Data set for the proteomic inventory and quantitative analysis of chicken eggshell matrix proteins during the primary events of eggshell mineralization and the active growth phase of calcification. *Data in Brief*, *4*, 430–436. https://doi.10.1016/j.dib.2015.06.019

Marie, P., Labas, V., Brionne, A., Harichaux, G., Hennequet-Antier, C., Nys, Y., & Gautron, J. (2015c). Quantitative proteomics and bioinformatic analysis provide new insight into protein function during avian eggshell biomineralization. *Journal of Proteomics*, *113*, 178–193. https://doi.10.1016/j.jprot.2014.09.024

Mikšík, I., Ergang, P., Jiri, & Pácha, J. (2014). Proteomic analysis of chicken eggshell cuticle membrane layer. *Analytical and Bioanalytical Chemistry*, *406*(29), 7633–7640. https://doi.10.1007/s00216-014-8213-x

Mikšík, I., Sedláková, P., Lacinová, K., Pataridis, S., & Eckhardt, A. (2010). Determination of insoluble avian eggshell matrix proteins. *Analytical and Bioanalytical Chemistry*, *397*(1), 205–214. https://doi.10.1007/s00216-009-3326-3

Nakano, T., Ikawa, N., & Ozimek, L. (2001). Extraction of glycosaminoglycans from chicken eggshell. *Poultry Science*, *80*(5), 681–684. https://doi.10.1093/ps/80.5.681

Nakano, T., Ikawa, N., & Ozimek, L. (2002). Galactosaminoglycan composition in chicken eggshell. *Poultry Science*, *81*(5), 709–714. https://doi.org/10.1093/ps/80.5.681

Nakano, T., Ikawa, N., & Ozimek, L. (2003). Chemical composition of chicken eggshell and shell membranes. *Poultry Science*, *82*(3), 510–514. https://doi.10.1093/ps/82.3.510

Necas, J., Bartosikova, L., Brauner, P., & Kolar, J. (2008). Hyaluronic acid (hyaluronan): A review. *Veterinarni Medicina*, *53*(8), 397–411.

Nimtz, M., Conradt, H. S., & Mann, K. (2004). LacdiNAc (GalNAcbeta1-4GlcNAc) is a major motif in N-glycan structures of the chicken eggshell protein ovocleidin-116. *Biochimica et Biophysica Acta*, *1675*(1–3), 71–80. https://doi.10.1016/j.bbagen.2004.08.007

Nys, Y., Gautron, J., Garcia-Ruiz, J. M., & Hincke, M. T. (2004). Avian eggshell mineralization: Biochemical and functional characterization of matrix proteins. *Comptes Rendus Palevol*, *3*(6–7), 549–562. https://doi.org/10.1016/j.crpv.2004.08.002

Nys, Y., Gautron, J., Rodriguez-Navarro, A. B., & Hincke, M. T. (2022) Mechanisms and hormonal regulation of shell formation: Supply of ionic and organic precursors, shell mineralization. In C. Scanes & S. Dridi (Eds.), *Sturkie's avian physiology* (7th ed., Chapter 32, pp. 833–879). Elsevier. https://doi.org/10.1016/B978-0-12-819770-7.00049-9

Omana, D. A., Liang, Y., Kav, N. N., & Wu, J. (2011). Proteomic analysis of egg white proteins during storage. *Proteomics*, *11*(1), 144–153. https://doi.org/10.1002/pmic.201000168

Panheleux, M., Bain, M., Fernandez, M. S., Morales, I., Gautron, J., Arias, J. L., Solomon, S. E., Hincke, M., & Nys, Y. (1999). Organic matrix composition and ultrastructure of eggshell: A comparative study. *British Poultry Science*, *40*(2), 240–252. https://doi.10.1080/00071669987665

Panheleux, M., Nys, Y., Williams, J., Gautron, J., Boldicke, T., & Hincke, M. T. (2000). Extraction and quantification by ELISA of eggshell organic matrix proteins (ovocleidin-17, ovalbumin, ovotransferrin) in shell from young and old hens. *Poultry Science*, *79*(4), 580–588. https://doi.10.1093/ps/79.4.580

Pasarin, D., & Rovinaru, C. (2019). Separation methods of the eggshell membranes from eggshell. *Proceedings*, *29*(1), 122. https://doi.org/10.3390/proceedings2019029122

Poyatos Pertiñez, S., Wilson, P. W., Icken, W., Cavero, D., Bain, M. M., Jones, A. C., & Dunn, I. C. (2020). Transcriptome analysis of the uterus of hens laying eggs differing in cuticle deposition.*BMC Genomics*, *21*(1), 516. https://doi.10.1186/s12864-020-06882-7

Price, R. D., Berry, M. G., & Navsaria, H. A. (2007). Hyaluronic acid: The scientific and clinical evidence. *Journal of Plastic, Reconstructive & Aesthetic Surgery*, *60*(10), 1110–1119. https://doi.org/10.1016/j.bjps.2007.03.005

Qiu, N., Liu, W., Ma, M., Zhao, L., & Li, Y. (2013). Differences between fertilized and unfertilized chicken egg white proteins revealed by 2-dimensional gel electrophoresis-based proteomic analysis. *Poultry Science*, *92*(3), 782–786. https://doi.org/10.3382/ps.2012-02565

Qiu, N., Ma, M., Zhao, L., Liu, W., Li, Y., & Mine, Y. (2012). Comparative proteomic analysis of egg white proteins under various storage temperatures. *Journal of Agricultural and Food Chemistry*, *60*(31), 7746–7753. https://doi.org/10.1021/jf302100m

Rath, N. C., Liyanage, R., Makkar, S. K., & Lay, J. O. (2016). Protein profiles of hatchery egg shell membrane. *Proteome Science*, *15*(1), 4. https://doi.org/10.1186/s12953-017-0112-6

Rodríguez-Navarro, A. B., Marie, P., Nys, Y., Hincke, M. T., & Gautron, J. (2015). Amorphous calcium carbonate controls avian eggshell mineralization: A new paradigm for understanding rapid eggshell calcification. *Journal of Structural Biology*, *190*(3), 291–303. http://doi.10.1016/j.jsb.2015.04.014

Rose, M. L. H., & Hincke, M. T. (2009). Protein constituents of the eggshell: Eggshell-specific matrix proteins. *Cellular and Molecular Life Sciences*, *66*(16), 2707–2719. https://doi.org/10.1007/s00018-009-0046-y

Rose-Martel, M., Du, J., & Hincke, M. T. (2012). Proteomic analysis provides new insight into the chicken eggshell cuticle. *Journal of Proteomics*, *75*(9), 2697–2706. https://doi.10.1016/j.jprot.2012.03.019

Rose-Martel, M., Smiley, S., & Hincke, M. T. (2015). Novel identification of matrix proteins involved in calcitic biomineralization. *Journal of Proteomics*, *116*, 81–96. https://doi.10.1016/j.jprot.2015.01.002

Sah, N., Kuehu, D. L., Khadka, V. S., Deng, Y., Peplowska, K., Jha, R., & Mishra, B. (2018). RNA sequencing-based analysis of the laying hen uterus revealed the novel genes and biological pathways involved in the eggshell biomineralization. *Scientific Reports*, *8*(1), 16853. https://doi.10.1038/s41598-018-35203-y

Salevsky, E., & Leach, R. M. (1980). Studies on the organic components of shell gland fluid and the hen's egg shell. *Poultry Science*, *59*(2); 438–443. https://doi.org/10.3382/ps.059043

Samiullah, S., & Roberts, J. R. (2013) The location of protoporphyrin in the eggshell of brown-shelled eggs. *Poultry Science*, *92*, 2783–2788. https://doi.org/10.3382/ps.2013-03051

Schaafsma, A., Pakan, I., Hofstede, G. J., Muskiet, F. A., Van Der Veer, E., & De Vries, P. J. (2000). Mineral, amino acid, and hormonal composition of chicken eggshell powder and the evaluation of its use in human nutrition. *Poultry Science*, *79*(12), 1833–1838. https://doi.10.1093/ps/79.12.1833

Schroeder, K. M., & McRae, S. B. (2018). Variable laying times among king rails (Rallus elegans). *The Wilson Journal of Ornithology*, *130*(4), 908–914. https://doi.org/10.1676/1559-4491.130.4.1036

Shi, Y., Zhou, K., Li, D., Guyonnet, V., Hincke, M. T., & Mine, Y. (2021). Avian eggshell membrane as a novel biomaterial: A review. *Foods*, *10*(9), 2178. https://doi.10.3390/foods10092178

Stabler Ogawa, J. L. (2021). *Extraction and Characterization of Hyaluronic Acid and Collagen from Eggshell Membrane Waste: An Industrial Recycling Process*. Ph.D. Thesis, Trent University, Peterborough, Ontario, Canada. http://digitalcollections.trentu.ca/objects/etd-951

Stapane, L., Le Roy, N., Ezagal, J., Rodriguez-Navarro, A. B., Labas, V., Combes-Soia, L., Hincke, M. T., & Gautron, J. (2020). Avian eggshell formation reveals a new paradigm for vertebrate mineralization via vesicular amorphous calcium carbonate. *Journal of Biological Chemistry*, *295*(47), 15853–15869. https://doi.10.1074/jbc.RA120.014542

Stapane, L., Le Roy, N., Hincke, M. T., & Gautron, J. (2019). The glycoproteins EDIL3 and MFGE8 regulate vesicle-mediated eggshell calcification in a new model for avian biomineralization. *Journal of Biological Chemistry*, *294*(40), 14526–14545. https://doi.10.1074/jbc.RA119.009799

Stemberger, B. H., Mueller, W. J., & Leach, R. M. (1977). Microscopic study of the initial stages of egg shell calcification. *Poultry Science*, *56*(2), 537–543. https://doi.org/10.3382/ps.0560537

Sugiarto, H., & Yu, P.-L. (2004). Avian antimicrobial peptides: The defense role of β-defensins. *Biochemical and Biophysical Research Communications*, *323*(3), 721–727. https://doi.org/10.1016/j.bbrc.2004.08.162

Sun, C., Xu, G., & Yang, N. (2013). Differential label-free quantitative proteomic analysis of avian eggshell matrix and uterine fluid proteins associated with eggshell mechanical property. *Proteomics*, *13*, 3523–3536. https://doi.10.1002/pmic.201300286

Szeleszczuk, L., Kuras, M., Pisklak, D. M., & Wawer, D. M. (2016). Analysis of the changes in elemental composition of the chicken eggshell during the incubation period. *The Journal of Animal & Plant Sciences*, *26*(3), 583–587.

Vadehra, D. V., & Rath, K. R. (1973). Eggs as a source of protein. *CRC Critical Reviews in Food Technology*, *4*, 193–309. https://doi.org/10.1080/10408397309527158

Vadehra, D. V., Baker, R. C., & Naylor, H. B. (1972). Distribution of lysozyme activity in the exteriors of eggs from *Gallus gallus*. *Comparative Biochemistry and Physiology Part B: Comparative Biochemistry*, *43*(3), 503–508. https://doi.org/10.1016/0305-0491(72)90133-2

Wang, X., Ford, B., Praul, C., & Leach, R. (2002). Collagen X expression in oviduct tissue during the different stages of the egg laying cycle. *Poultry Science*, *81*(6), 805–808. https://doi.org/10.1093/ps/81.6.805

Wasti, S., Sah, N., Kuehu, D. L., Kim, Y. S., Jha, R., & Mishra, B. (2020). Expression of follistatin is associated with egg formation in the oviduct of laying hens. *Animal Science Journal*, *91*(1), e13396. http://doi.10.1111/asj.13396

Wedral, E. M., Vadehra, D. V., & Baker, R. C. (1974). Chemical composition of the cuticle, and the inner and outer shell membranes from eggs of *Gallus gallus*. *Comparative Biochemistry and Physiology Part B: Comparative Biochemistry*, *47*(3), 631–640. https://doi.org/10.1016/0305-0491(74)90011-X

Weigel, P. H., & DeAngelis, P. L. (2007). Hyaluronan synthases: A decade-plus of novel glycosyltransferases. *Journal of Biological Chemistry*, *282*(51), 36777–36781. https://doi.org/10.1074/jbc.R700036200

Weigel, P. H., Hascall, V. C., & Tammi, M. (1997). Hyaluronan synthases. *Journal of Biological Chemistry*, *272*(22), 13997–14000. https://doi.org/10.1074/jbc.272.22.13997

Wellman-Labadie, O., Lemaire, S., Mann, K., Picman, J., & Hincke, M. T. (2010). Antimicrobial activity of lipophilic avian eggshell surface extracts. *Journal of Agricultural and Food Chemistry*, *58*(18):10156–10161. http://doi.10.1021/jf101954g

Wellman-Labadie, O., Picman, J., & Hincke, M. T. (2008). Comparative microbial toxicity of cuticle and outer eggshell protein from three species of domestic bird. *British Poultry Science*, *49*(2), 133–143. https://doi.10.1080/00071660802001722

Wong, M., Hendrix, M. J. C., von der Mark, K., Little, C., & Stern, R. (1984). Collagen in the egg shell membranes of the hen. *Developmental Biology*, *104*(1), 28–36. https://doi.org/10.1016/0012-1606(84)90033-2

Yacoub, H. A., Elazzazy, A. M., Abuzinadah, O. A. H., Al-Hejin, A. M., Mahmoud, M. M., & Harakeh, S. M. (2015). Antimicrobial activities of chicken β-defensin (4 and 10) peptides against pathogenic bacteria and fungi. *Frontiers in Cellular and Infection Microbiology*, *5*. https://doi.org/10.3389/fcimb.2015.00036

Yang, R., Geng, F., Huang, X., Qiu, N., Li, S., Teng, H., Chen, L., Song, H., & Huang, Q. (2020). Integrated proteomic, phosphoproteomic and N-glycoproteomic analyses of chicken eggshell matrix. *Food Chemistry*, *330*, 127167. https://doi.10.1016/j.foodchem.2020.127167

Yin, Z., Lian, L., Zhu, F., Zhang, Z. H., Hincke, M., Yang, N., & Hou, Z. C. (2020). The transcriptome landscapes of ovary and three oviduct segments during chicken (*Gallus gallus*) egg formation. *Genomics*, *112*(1), 243–251. https://doi.10.1016/j.ygeno.2019.02.003

Yin, L., Yu, L., Zhang, L., Ran, J., Li, J., Yang, C., Jiang, X., Du, H., Hu, X., & Liu, Y. (2019). Transcriptome analysis reveals differentially expressed genes and pathways for oviduct development and defense in prelaying and laying hens. *American Journal of Reproductive Immunology*, *82*(3), e13159. https://doi.10.1111/aji.13159

Zhang, L., Lu, L., Li, S., Zhang, G., Ouyang, L., Robinson, K., Tang, Y., Zhu, Q., Li, D., Hu, Y., & Liu, Y. (2016). 1,25-dihydroxyvitamin-D3 induces avian β-defensin Gene expression in chickens. *PLoS One*, *11*(5), e0154546. https://doi.org/10.1371/journal.pone.0154546

Zhang, G., & Sunkara, L. (2014). Avian antimicrobial host defense peptides: From biology to therapeutic applications. *Pharmaceuticals*, *7*(3), 220–247. https://doi.org/10.3390/ph7030220

Zhang, J., Wang, Y., Zhang, C., Xiong, M., Rajput, S. A., Liu, Y., & Qi, D. (2019). The differences of gonadal hormones and uterine transcriptome during shell calcification of hens laying hard or weak-shelled eggs. *BMC Genomics*, *20*(1), 707. https://doi.10.1186/s12864-019-6017-2

Zhang, Q., Zhu, F., Liu, L., Zheng, C. W., Wang, de, H., Hou, Z. C., & Ning, Z. H. (2015). Integrating transcriptome and genome re-sequencing data to identify key genes and mutations affecting chicken eggshell qualities.*PLoS One*, *10*(5), e0125890. https://doi.10.1371/journal.pone.0125890

Zhao, Y.-H., & Chi, Y.-J. (2009). Characterization of collagen from eggshell membrane. *Biotechnology (Faisalabad)*, *8*(2), 254–258. https://doi.org/10.3923/biotech.2009.254.258

5

Egg White

Françoise Nau

CONTENTS

DOI: 10.1201/9781003254430-7

5.1 Introduction

Egg white, also called albumen, constitutes the largest part of the chicken egg (~two-thirds in weight). This biological liquid fulfills several roles during the embryo development, the first being a nutritional role, in particular as a source of amino acids, but also minerals and vitamins. The second biological role of egg white is the protection of the embryo against contaminants from the external environment, at least during the first days after oviposition. Hence, in addition to the eggshell, that is a physical barrier and protection, the egg white is usually considered as the chemical protector of the embryo. Indeed, its protective function against microorganisms is related to some physicochemical properties of this biological medium, namely its viscosity and its pH, becoming alkaline within a few days after laying, and to its protein composition, including a lot of antimicrobial proteins, making the egg white not very conducive to the growth of microorganisms (see Chapter 20).

Moreover, the egg white is also an amazing food ingredient, due to its excellent and various functional properties. The egg white continues to be the reference as a foaming agent for home cooking as well as in the food industry, but it is also a very efficient gelling agent, and can be used as an emulsifier as well. The egg white proteins are the main constituents responsible for these properties (see Chapter 8).

Since the proteins play the major roles of the egg white, whether biological or functional, a large portion of this chapter will deal with proteins. Moreover, proteins account for most of the dry matter of egg white. However, some overall information on the composition and structure of the egg white will be first provided, including the different factors that can modulate these features.

5.2 Egg White Composition and Structure

5.2.1 Overall Composition

The egg white is basically an aqueous solution (87–89% water) of proteins (9.7–10.6%). Hence, the dry matter of egg white is made of nearly 90% proteins, emphasizing further the importance of proteins in this biological liquid. Therefore, the second section of this chapter will be exclusively devoted to egg white proteins. Nevertheless, the egg white also contains carbohydrates (0.4–0.9%), of which half is free glucose while the rest are incorporated in glycoproteins. The egg white contains also some minerals (0.5–0.6%) and hydrophilic vitamins (see Chapter 7), while lipids are present only in trace quantities.

The variability in dry matter and protein contents in freshly laid eggs depends mainly on the strain and age of the hens (Powrie & Nakai, 1986), but the rearing system has also an impact, with higher dry matter and protein contents reported in eggs produced in a free-range system (Nistor Cotfas et al., 2015). Nonetheless, the protein-to-dry matter ratio is very close to 90% in all cases, a rather outstanding feature among all animal food products.

5.2.2 Water Content Decrease and Alkalization during Egg Storage

The moisture content of albumen decreases with the length of egg storage, more rapidly in higher temperature environment. During storage at 37°C, the moisture dropped from around 87.3% at the first day after laying to 85.3% after only 12 days (Huang et al., 2022), due to water evaporation through the porous eggshell, and water transfer into the yolk due to osmotic pressure differences between these two egg compartments.

Simultaneously, CO_2 diffusion through the pores of the shell results in a very rapid alkalization of the albumen, from generally accepted pH value of 7.5~8.0 for in freshly laid eggs (Hayakawa et al., 1983). When eggs were stored at 37°C, the pH increased to 8.8 within 24 h after laying, and reached 9.5 after only 2 days. Thereafter, the pH continued to rise, reaching its highest value (around 9.75) after 10 days of storage before decreasing slightly by about 0.1 pH unit within two days, likely as a consequence of protein aggregation and changes in protein surface charge (Huang et al., 2022).

5.2.3 Egg White Is Made of Two Fractions: Thin and Thick Egg White

Beyond the overall composition mentioned above, it is noteworthy that the egg white is actually a heterogeneous medium, made of a thick fraction and a thin one. Both are distributed within four distinct layers that are, from outside to inside of the egg: the outer thin egg white, the thick egg white, the inner thin egg white, and the chalaziferous layer, which includes the chalazae. The percentages for each layer dramatically vary (Table 5.1), depending on many factors described below.

The thick and thin egg white fractions strongly differ in rheological properties. The viscosity of thin egg white is comparable to that of water, with no significant change during egg storage time. However, thick egg white is more viscous, although the accurate measurement of thick egg white viscosity is practically infeasible (Kemps et al., 2010). Indeed, thick egg white is itself a heterogeneous gel-like solution composed of microscopic fibers dispersed in a liquid phase very similar to thin egg white (Romanoff & Romanoff, 1949); these fibers can be separated by using ultracentrifugation (Hiidenhovi et al., 2016). Measurements of the mobility of small ions confirmed that thick albumen may be regarded as a gel network of hydrated proteins held by intermolecular disulfide bridges, surrounded by water pools and channels, whose viscosity is very close to that of the bulk (Krise & Milosavljevic, 2011). This may explain the heterogeneous rheological properties displayed by the thick albumen, its pseudoplastic behavior (apparent viscosity decreases when shear rate increases), and also the hysteresis phenomenon observed (apparent viscosity depends on the previous shearing history of the product) (Kemps et al., 2010).

Because of the difficulty in collecting and analyzing separately the different albumen layers, great variations are reported in the literature for the composition of each albumen layer. Furthermore, hen genetics and age, freshness of the eggs, and analytical strategies may also explain some of the variability observed. However, egg scientists agree that

TABLE 5.1
Percentage, Water Content, and Ovomucin Content of Each Albumen Layer

Albumen Layer		% Albumen		% Moisture	Ovomucin Content	
		Mean Values	Value Range		mg/100 g	% β-ovomucin
Outer thin albumen		23.2	10–60	88.8	125	7
Thick albumen	*Gel-like fraction*	57.3	30–80	87.6	554	29
	Liquid fraction				142	15
Inner thin albumen		16.8	1–40	86.4	186	25
Chalaziferous layer		2.7	-	84.3	857	19

Source: From Powrie and Nakai (1986), and Hiidenhovi et al. (2016).

differences in viscosity between the four albumen layers mainly result from changes in concentration and composition of a specific protein fraction, namely ovomucin (Omana et al., 2010). Actually, ovomucin is an essential component in maintaining the viscoelastic properties of albumen (Wang et al., 2019). A much higher carbohydrate content, especially sialic acid, was measured in ovomucin from the thick albumen than that of the thin egg white (Kato et al., 1970). For their part, Hiidenhovi et al. (2016) measured an ovomucin content about three times higher in the chalaziferous layer (857 mg/100 g of egg layer) than in the thick egg white (276 mg/100g), and about five and seven times higher than in the inner (186 mg/100g) and outer (125 mg/100g) thin egg white (Table 5.1). Moreover, when considering separately the gel-like and liquid fractions of the thick albumen layer, these authors also found a much higher concentration of ovomucin in the gel-like fraction (554 mg/100g) than in the liquid fraction (142 mg/100g). In addition to these huge quantitative differences from one layer to another, qualitative differences were highlighted, namely different distributions of both ovomucin subunits (α-ovomucin or Mucin-5B, and β-ovomucin or Mucin-6) depending on the layers. Hence, Hiidenhovi et al. (2016) confirmed the positive correlation between the percentage of β-ovomucin and high viscosity: the highest percentages of β-ovomucin are actually observed in thick albumen (26%) and in the gel-like fraction of this layer (29%). However, β-ovomucin content was only 19% in the chalaziferous layer, *i.e.* very close to that of the liquid fraction of the thick albumen (15%), underlining the need to consider both the distribution of ovomucin subunits and the ovomucin amount, which is the highest in the chalaziferous layer (see above). Moreover, these differences in protein composition between the albumen layers, and in particular the different ovomucin content, might explain the differences in gel properties and microstructure obtained after heating. Hence, heat-gelation of thick egg white results in a soft and cohesive gel, whereas thin egg white results in a hard and springy, but less cohesive, gel (Liu et al., 2020).

5.2.4 Egg White Thinning during Egg Storage

The most common, albeit controversial indicator to measure the thick-to-thin albumen ratio, and consequently the egg white quality, is expressed in Haugh Unit (HU), which is a logarithmic expression of the height of thick egg white corrected for the egg weight (Silversides & Villeneuve, 1994). The temperature and length of storage of eggs are among the main parameters that influence the height of thick albumen (Silversides & Scott, 2001), and more generally the viscoelastic properties of the egg white (Wang et al., 2019). Thus, Kemps et al. (2010) reported a significant decrease of the mean viscosity (measured at 20°C) of thick albumen during egg storage at 18°C, from around 30 mPa.s for fresh eggs to 18 mPa.s after 24 days. During storage at 0°C, mean values of the HU decreased from around 76 after 15 days to 70 after 7 months. When eggs were stored for an additional week at 18°C, the HU values dropped to 62 (Berardinelli et al., 2008). Sheng et al. (2018) reported a gradual decrease of the HU, from 94 to 60, during a 50-day storage period at 22°C and 65% relative humidity.

The decrease in HU values during egg storage is a well-known phenomenon described as egg white thinning, which reflects the loss of gel structure of thick albumen (Beveridge & Nakai, 1975). It occurs along with the strong pH increase due to CO_2 diffusion, considered, for a long time, as the cause of the strong modifications experienced by ovomucin. In particular, ovomucin depolymerization by alkaline hydrolysis of disulfide bonds and disruption of the ovomucin-lysozyme complex due to the decrease of electrostatic interactions were suggested (Beveridge & Nakai, 1975). The assumption of a noncovalent disaggregation of ovomucin without disulfide cleavage, but with some deglycosylation, was proposed by others (Kato et al., 1981). More recently, the depolymerization hypothesis was confirmed, as the conversion of large ovomucin aggregates into smaller ones was observed (Shan et al., 2020). These authors also demonstrated the progressive decrease in the percentage of β- and α1-subunits of ovomucin in parallel with the HU values decrease, and also the reduction of the molecular weight of each subunit. All these results taken together thus confirmed the gradual degradation of ovomucin. Moreover, infrared spectroscopy analysis strongly suggested some structural changes in the glycan moieties of the protein (Shan et al., 2020), and structural transitions from an ordered to an unordered structure (Huang et al., 2022). This was confirmed by circular dichroism analysis that highlighted an increase in β-turn and random coils, and a decrease in β-sheet and α-helix contents (Shan et al., 2020).

5.2.5 Zootechnical Factors Influencing the Egg White Structure

Some zootechnical parameters also have an impact on the egg white structure. In particular, the percentage of thick egg white decreases as hens get older (Silversides & Scott, 2001; Wan et al., 2018) or suffer from specific diseases (*e.g.* infectious bronchitis) (Roberts, 2004).

It might be also modulated by some nutritional factors, although nutrition has a minor impact (Silversides & Scott, 2001). A decrease of thick albumen height was reported when protein and amino acid content in the hen diet increased, or when soybean was replaced by cottonseed. Conversely, it may increase with increasing lysine, acid ascorbic, or vitamin E concentration in the diet, and with tea polyphenol supplementation (Roberts, 2004; He et al., 2017; Ding et al., 2020).

Genetics is another factor that influences the structure of the egg white. With a heritability estimated to 0.45, the thick-to-thin egg white ratio can be regarded as moderately to highly heritable, suggesting that genetic selection could contribute to increasing this ratio, an important indicator of albumen quality (Wan et al., 2019). Using Rhode Island White strains of laying hens selected for high *vs* low thick-to-thin albumen ratio, and based on RNA sequencing and functional analysis, seven promising candidate genes favoring a high proportion of thick albumen were recently identified (Wan et al., 2017).

5.3 Egg White Proteins

5.3.1 General Points

5.3.1.1 An Increasing Number of Proteins Identified in the Albumen

Egg white proteins, which are the essential part of the albumen dry matter (around 90%), play a major biological role in providing a protective environment and nutritional resources to the chicken embryo (see Chapter 3). They also give the specific and various functional properties making egg white a major food ingredient (see Chapter 8).

For a long time, only six main egg white proteins had been described, namely ovalbumin, ovotransferrin, ovomucoid, lysozyme, ovomucin and ovoinhibitor, along with a few minor proteins. The development of proteomic analysis techniques has allowed over the last few decades to increase significantly the number of proteins identified. From a total of 13 proteins identified in 1991 (Stevens, 1991), 2D-electrophoresis coupled with LC-MS/MS, used for the first time in 2006 for albumen analysis, allowed to raise this number to 16 proteins, including several isoforms (Guérin-Dubiard et al., 2006). In 2011, 158 proteins were isolated, including 79 proteins identified for the first time in egg white (Mann & Mann, 2011). Finally, this number increased to 378 proteins quantified in the most complete proteome established to date (Wang et al., 2019). However, 12 proteins make up more than 97% of the total abundance of all identified egg white proteins (Bílková et al., 2018).

A broad description of the peptides naturally present in freshly laid eggs was reported for the first time by Arena et al. (2020). Nano-LC-ESI-Q-Orbitrap-MS/MS analysis enabled these authors to identify a total of 504 peptides, stemming from 80 proteins. However, just three proteins, namely ovocleidin-116, clusterin and sulfhydryl oxidase 1, were associated with more than 50% of the peptides characterized. Moreover, bioinformatics tools suggest that 359 out of all the identified sequences may be bioactive; antihypertensive, antioxidative/anti-inflammatory, and antimicrobial activities would be the most common ones. These peptides, naturally present in the albumen of unfertilized eggs, likely result from the activity of endoproteases that may migrate into the egg during egg formation.

5.3.1.2 Egg White Proteins Are Mostly Glycosylated and/or Phosphorylated

Many egg white proteins possess post-translational modifications. More specifically, the egg white is rich in glycoproteins that may contribute to the defense of the embryo against microbes, and be responsible for egg white allergies. The four major egg white proteins, ovalbumin, ovotransferrin, ovomucoid, and ovomucin, are all glycoproteins with glycosylation sites identified for a long time (see below), whereas little is known for many minor proteins. Hence, in 2017, 71 N-glycosites were identified in 26 egg white glycoproteins, including many low-abundance proteins that were isolated in egg white for the first time (Geng et al., 2017). Moreover, the extraordinary large degree of glycosylation of some egg white proteins makes their complete characterization very complex. As an example, 15 N-glycosites have been experimentally confirmed in α-ovomucin (Offengenden et al., 2011), and 12 N-glycosites in ovomacroglobulin (Geng et al., 2015). Many minor proteins recently identified in egg white are still to be characterized from this perspective. Furthermore, each glycosylation site may be occupied with different glycan moieties, making the characterization even more complicated. Hence, up to 25 glycoforms have been recently identified for one of the two glycosylation sites of ovotransferrin (N^{492}) (Cavallero et al., 2020).

Another major post-translational modification is phosphorylation, which also involves many egg white proteins. Some of them are registered as phosphoproteins for a long time, but others were identified as such only recently, thanks to proteomics tools combined with original digestion approaches using both trypsin and trypsin/Glu-C enzymes. This led in 2020 to the identification of 189 phosphorylation sites, mainly on serine residues, distributed on 96 phosphoproteins (Sun et al., 2020). Most of the phosphoproteins identified (60) have a single phosphorylation site, but some revealed many sites. For instance, 10 phosphorylation sites were identified in the ovalbumin sequence.

5.3.1.3 A Wide Panel of Biological Functions

Egg white proteins contribute to a wide range of biological functions, such as binding activities, immunity, redox homeostasis, various metabolic activities, chaperone functions… But many of them, including the minor proteins recently identified, have demonstrated or are suggested to have antimicrobial properties (Réhault-Godbert et al., 2011; Guyot et al., 2016). Actually, six out of the 12 most abundant proteins quantified in the egg white exhibit an antimicrobial function. In total, antimicrobial proteins would be more than 50% of the total albumen protein content (Bílková et al., 2018).

Enzymatic proteins are also widely present within the egg white, including in poultry species other than chicken. Hydrolases, primarily glycosylases and peptidases, account for the most important part of these enzymatic proteins (Sun et al., 2017).

5.3.1.4 Sources of Variability in Protein Composition of Chicken Egg White

Differential expression of some egg white proteins has been highlighted between fertilized and unfertilized eggs, with ovotransferrin, cystatin, ovomucoid, lysozyme, TENP and chondrogenesis associated lipocalin (CAL) downregulated in fertilized eggs (Zhang, Chen et al., 2020). Conversely, the concentration of ovalbumin-Y protein was increased in fertilized eggs (Qiu et al., 2013). In addition, egg white proteins logically undergo some highly dynamic and extensive changes during the development of the chick embryo. In particular, from the early stages, high-molecular-weight proteins are hydrolyzed into low-molecular-weight molecules that may play a role in embryo nutrition since more easily absorbed, and immune defense since functionalized with antibacterial activities (Meng et al., 2021), and likely compensating for the gradual decline of antimicrobial proteins observed during embryo growth (Wang et al., 2019). Moreover, a significant increase in abundance was observed during embryo development for some proteins (including ovomucoid, flavoprotein, TENP), whereas a decrease was observed for others (including ovotransferrin, Ex-FABP, serum albumin precursor, hemopexin) (Liu et al., 2015; Meng et al., 2021; Wang et al., 2019).

Variations related to genetics have also been detected. Thus, among the 378 proteins quantified in the egg white by Wang et al. (2019), 102 were differentially regulated in fresh eggs laid by two groups of Rhode Island Red laying hens differing in hatchability (high *vs* low). Similarly, while the qualitative proteome composition of egg white (presence or absence of a given protein) is very similar regardless of the breeds, 15 proteins were detected with significantly different abundances in small and large traditional breeds (Bílková et al., 2018). It is noteworthy that a high number of the proteins differing in abundance are antimicrobial proteins, and were present at higher concentrations in the large breeds included in this study, suggesting that selection led to the acquisition of better egg defense capacities.

Some nutritional factors may also modulate the protein composition of the egg white. For instance, 31 proteins were shown to be differentially accumulated in egg white after supplementation of the ration with tea polyphenol. A total of 19 proteins, including ovalbumin-Y protein and ovoinhibitor were upregulated, while 12 were downregulated (Ding et al., 2020). Similarly, 15 proteins involved in various biological functions (*e.g.* antimicrobial activity, immunity, reproduction), including those present abundantly in the egg white, were downregulated when soybean meal was replaced with cottonseed meal in the diet fed to the laying hens. Conversely, two minor proteins were upregulated (He et al., 2017).

Lastly, environmental conditions may have an impact on egg white protein composition. Thus, the storage of eggs at 37°C for 15 days led to more changes in the concentration of 32 proteins, clustered into six families, in comparison to a storage at 4°C, or even 20°C. More specifically, ovalbumin and ovotransferrin degradation into peptides was accelerated, clusterin almost disappeared, while proteins of the lipocalin family also decreased (Omana et al., 2011; Qiu et al., 2012). Moreover, an environmental stress experienced by hens may also modify the protein composition in egg white, as suggested by the differential abundance measured after dietary corticosterone was used to model stress. This treatment resulted in a decreased concentration for four egg white proteins, including TENP and Ex-FABP, and an increased concentration for two others (ovoinhibitor and OvaX) (Kim & Choi, 2014).

5.3.1.5 Interspecies Variability in Albumen Protein Composition

As expected from evolution over the past million years and domestication for nearly several thousands of years, interspecies variability has been observed with the protein composition of egg white. First, overall protein concentrations vary, between 61% in the ostrich egg white and more than 80% in the quail and duck egg white; moreover, some qualitative and quantitative differences are observed in electrophoresis profiles of egg white depending on the species (Miguel et al., 2005). Similar conclusions regarding the interspecific variability were reported by Hu et al. (2016) using two-dimensional electrophoresis maps of chicken, quail and duck egg whites. However, these authors were able to identify less than 10 proteins since high-abundance proteins were not eliminated before electrophoresis. Thereafter, and using tandem mass tags quantification technology, Sun et al. (2017) completed an exhaustive analysis comparing the egg white proteome of six main domestic poultry species, namely chicken, duck, goose, turkey, pigeon and quail. Among the 138–183 proteins identified (different numbers depending on the species), 52 were common among the six species. However, even the main proteins in chicken egg white, namely ovalbumin, ovotransferrin, ovomucoid, and lysozyme, while being shared with the other five species, displayed some quantitative and qualitative inter-species differences. In particular, the lysozyme content and the ratio between the two lysozyme types (C- and G-types) differed greatly from one species to another. The highest lysozyme abundance was measured in goose egg white, consisting almost exclusively of lysozyme G in this species, while lysozyme C is the main type in all other species excepting quail (G-type to C-type 2:1); C-type is almost exclusive in chicken egg white. Moreover, ovotransferrin was found only in chicken and turkey, while the egg white of the other four species contains a homologous protein named serotransferrin. Furthermore, ovomucin, which is also a major chicken egg white protein, was not detected in the other five domestic species. In summary, the similarity in protein composition and abundance between these species was logically consistent with their birds order affiliation. Hence, strong similarities were reported among chicken, turkey, and quail (Galliformes) and between duck and goose (Anseriformes). However, pigeon (Neoaves) showed a significantly divergent proteome (Sun et al., 2017).

TABLE 5.2
Main Egg White Proteins and Their Physicochemical Features

Protein Family	Protein	Proportion[a] (%)	MW[b] (kDa) Th. / Exp.	pI[c] Th. / Exp.	Specificities
Serpin family	Ovalbumin	54	42.7 / 43.8–46.2	5.2 / 4.4–4.7	Glycosylated and phosphorylated
	OvaY	7	43.6 / 53	5.2 / 5.3–5.5	Glycosylated and phosphorylated
	OvaX	0.3–0.5	45.3 / 53	6.3 / 6.0–6.7	Glycosylated and phosphorylated
Transferrin family	Ovotransferrin	12	75.8 / 75–85	6.7 / 6.2–6.7	Glycosylated, possibly phosphorylated
Kazal-type protease inhibitor family	Ovomucoid	11	20.1 / 37–43	4.8 / 5.0–5.3	Highly glycosylated (25%)
	Ovoinhibitor	1.5	49.4 / 63.6–69.5	6.1 / 6.2–6.4	Glycosylated (10%)
Cystatin family	Cystatin	0.08	13.1 / 17.0	6.5 / 5.6 & 6.5	Two isoforms, one of which phosphorylated
α2-macroglobulin family	Ovomacrogolbulin (ovostatin)	0.6	649 / 780	5.6 / 5.0–6.0	Homo-tetramer; glycosylated (11.5%)
Glycoside hydrolase family 22	Lysozyme	3.4	14.3 / 15 & 17.6	9.3 / nd	A low abundant glycosylated form
Mucin family	Ovomucin α-ovomucin β-ovomucin	3.5	~5,000–8,000 231 / nd 132 / ~370	 5.35 / nd 5.8 / nd	Glycosylated 15% 60%
Folate receptor family	Riboflavin-binding protein (flavoprotein)	0.7	25.3 / 29.2	5.05 / 4.0	Glycosylated (15%) and highly phosphorylated
Avidin / streptavidin family	Avidin	0.05	57.3 / 67.0	9.7 / nd	Homo-tetramer; glycosylated (10%)
Lipocalin family	α1-acid glycoprotein (ovoglycoprotein)	5	20.3 / 37–43	5.0 / 5.0–5.4	Highly glycosylated (30%)
	Ex-FABP (Ch21)	0.1	18.1 / 21	5.4 / 5.7	Putative glycosylation
	CAL-γ	0.1	19 / 22	6.2 / 5.6–6.0	Putative glycosylation
BPI fold-containing family B	TENP (ovoglobulin G2)	0.1–0.5	46.1 / 48.9–50.2	5.7 / 5.9–6.3	Glycosylated
Clusterin family	Clusterin α subunit β subunit	nd	49.4 / 70 24.5 / nd 25.0 / nd	5.5 / 4.5–6.7 7.08 / nd 4.9 / nd	Hetero-dimer; highly glycosylated (30%)
uPAR/CD59/Ly-6/ snake toxin family	HEP21	nd	10 /21	6.7 / 6.3–6.4	Putative glycosylation
	Glutamyl-amino peptidase	nd	nd / 320	nd / 4.2	Dimer; glycosylated
	Methionine aminopeptidase	nd	nd / 180	nd / 4.15–4.40	Glycosylated
	Matrix metalloprotease 2 (MMP-2)	nd	nd/72	nd	-

Note: nd: not determined.
[a] Proportion as % of total egg white proteins.
[b] Molecular weight, either calculated from the polypeptide backbone only (*Th.*), or experimentally determined (*Exp.*).
[c] Isoelectric point, either theoretical, or experimentally determined. For each protein characteristics, references are given in the corresponding section.

In the following sections, only the major chicken egg white proteins will be described, with a specific focus on their main biochemical and structural features, as highlighted in Table 5.2.

5.3.2 Serpin Family Proteins

Serpin family proteins, mostly serine protease inhibitors, share common structural features in their metastable tertiary structure (Whisstock et al., 1998). Six serpins have been identified in egg white, with three of them present in much higher quantity, namely ovalbumin and its related proteins Y and X. The homologous genes of these three proteins (*Serpinb14*, *Serpinb14b*, and *Serpinb14c*, respectively) are located within a 46 kb region on chromosome 2 of *Gallus gallus*. Their expression occurs under estrogen control, but with different hormonal responsiveness resulting in different concentrations in the egg (Da Silva et al., 2015). These three serpins are bird species specific, and they play mainly a nutritional role as a source of amino acids for the embryo (Dombre et al., 2017). Indeed, ovalbumin and ovalbumin-related protein X exhibit no anti-protease activity, while the anti-trypsin like activity of ovalbumin-related protein Y is only putative, based on a predicted reactive site (Benarafa & Remold-O'Donnell, 2005). Ovalbumin and related proteins undergo quantitative and structural changes during incubation of fertilized eggs, likely related to their nutritional function for the embryo (Duan et al., 2017).

5.3.2.1 Ovalbumin

Ovalbumin is the most abundant protein in egg white, representing around 54% of the total proteins (Stevens, 1991). First deduced from the mRNA sequence established by McReynolds et al. (1978), the amino acid sequence of chicken ovalbumin was determined by Nisbet et al. (1981). It consists of a single polypeptide chain of 385 amino acid residues, with a theoretical molecular weight of 42,750 Da and a theoretical pI of 5.2, both predicted using only the polypeptide backbone (UniProtKB-P01012, OVAL_Chick), that is without considering glycosylation and phosphorylation that result in a great number of isoforms (Füssl et al., 2019).

Unlike most secreted proteins, ovalbumin sequence does not include a signal peptide, but only an initiator methionine residue, which is quickly removed after the initiation of protein synthesis. The first amino acid residue of the mature sequence is then an acetylated glycine residue (Palmiter et al., 1978), although a minor proteoform lacking the N-acetylation was observed (Yang et al., 2013). Moreover, the substitution of Glu^{289} to Gln^{289}, and of Asn^{311} to Asp^{311} was reported by Ishihara et al. (1981) and Wiseman et al. (1972), respectively, and the deamidation of Asn^{94} was also established (Füssl et al., 2019).

Ovalbumin is glycosylated on a single amino acid residue (Asn^{292}) located in the loop between a α helix and a β strand (Stein et al., 1991). However, many different glycan structures have been identified on this unique glycosylation site. Since the 1960's, a huge number of studies using various analytical techniques investigated the glycan moiety of ovalbumin, making this protein one of the best characterized on that basis. For instance, by analyzing the intact ovalbumin instead of the glycan moieties previously released from the protein as common done prior to that time, Yang et al. (2013) could detect and identify unambiguously 45 different glycan structures. These glycosylated moieties are composed of hexose (mannose, and likely galactose in very low abundance) and N-acetylglucosamine units, sometimes associated with one or two N-acetylneuraminic acid (sialic acid) units; polylactosamine type structures (repeated Gal(β1-4)GlcNAc units) may also be present. Hence, the glycan structures identified contained between 6 and 19 monosaccharide units, for a mass ranging from 1,074 up to 3,467 Da, providing a mass for ovalbumin ranging from 43,848 up to 46,241 Da. In a more recent study, fucose, a deoxyhexose, was also identified in the glycan moieties linked to ovalbumin (Cavallero et al., 2020).

Ovalbumin also contains two phosphorylation sites, on Ser^{68} and Ser^{344} (Nisbet et al., 1981). Both sites may be phosphorylated or not, thus leading to four isoforms: in addition to the di-phosphorylated (A_1) and non-phosphorylated (A_3) forms, two mono-phosphorylated (A_2) forms have been observed, depending on the phosphate group attaches to Ser^{68} or Ser^{344}. However, in mono-phosphorylated ovalbumin, the phosphorylation on Ser^{68} tends to be more than seven time more frequent than on Ser^{344} (Kinoshita-Kikuta et al., 2012). Nevertheless, the di-phosphorylated species is the major ovalbumin isoform in chicken egg white (77.3%), and the non-phosphorylated species is the less abundant (1.1%), while both mono-phosphorylated species are an intermediate abundance (21.6%) (Yang et al., 2013). The level of phosphorylation of ovalbumin may be upregulated during the embryo development, leading to a potential increase in the ability of the protein to combine with minerals, likely assisting the nutrition of the developing embryo (Sun et al., 2020). Phosphorylation makes the effective isoelectric point of ovalbumin much lower (4.4–4.7) than that predicted on the sole basis of the polypeptide backbone (5.2) (Liu et al., 2012).

The ovalbumin sequence contains four free cysteine residues (Cys^{11}, Cys^{30}, Cys^{367} and Cys^{382}) and a single disulfide bridge (Cys^{73} and Cys^{120}) (Fothergill & Fothergill, 1970). This latter has been proven to contribute directly to the structural stability of ovalbumin against thermal denaturation, and also against proteolysis by pancreatic elastase (Ishimaru et al., 2011). As to the four free thiol groups, increasing temperature results in their reaction with the disulfide Cys^{73}-Cys^{120} to create non-native disulfide bonds, and subsequent gelation above 60°C (Savadkoohi et al., 2016).

In 1991, Stein et al. published the first complete crystal structure of ovalbumin refined at 1.95 Å resolution, which was, at the same time, the first crystal structure of an uncleaved member of the serpin family ever solved (Stein et al., 1991). It revealed a shape approximately ellipsoidal of this globular protein (70 Å x 45 Å x 50 Å), with a major portion of the polypeptide chain involved in secondary structures, namely three β-sheets, nine α-helices, and three short helical segments of three to four residues. Despite common structural features and about 30% sequence identity with several functional inhibitors of the serpin family, ovalbumin lacks any anti-protease activity, likely due to a failure to undergo the appropriate structural change after cleavage within the reactive center loop, characteristic of this protein family (Huntington & Stein, 2001). Using quantitative analysis by Fourier transform infrared spectroscopy (FTIR), Savadkoohi et al. (2016) determined that the structure of native ovalbumin consists of 35% β-sheets, 15% β-turns, 40% α-helices, and 9% random coils. After heating for 20 min at 85°C, at 10% in a water solution and neutral pH, ovalbumin underwent an increase of β-sheets and β-turns (43% and 22%, respectively), and a decrease of α-helices (21%); random coils did not change (10%). However, ovalbumin was shown to be completely denatured when high pressure (600 MPa for 15 min) was applied on diluted solutions (10–20%), whereas increasing amounts of secondary conformation were preserved when the ovalbumin concentration increased above 30%; for instance, 80% of the native structure was maintained after pressurization of 70–80% ovalbumin solutions. Savadkoohi et al. (2016) explained this surprising pressure stability of ovalbumin at intermediate to high concentrations by its high hydrophobicity (around 40% hydrophobic zones along the sequence), resulting in water exclusion from the molecule interior, and therefore preventing its denaturation.

For the first time, Smith and Back identified in 1962 two forms of ovalbumin naturally present in the egg white and different in their heat stability (Smith & Back, 1962). Defined by these authors as the fraction soluble at pH 4.7 after heating at 55°C for 90 min at pH 3.0, the modified ovalbumin was more abundant as egg storage duration and the egg white pH increased. In shell eggs, the modified ovalbumin represented 2% of total ovalbumin after a one-day storage at 20°C, *vs* 86% after three days when the pH reached 9.2. Since the conversion of ovalbumin to the modified form strongly correlates with the duration of egg storage, some researchers proposed to assess the modified ovalbumin content in shell eggs to evaluate egg freshness, using rapid and non-destructive technologies such as hyper-spectral imaging (Fu et al., 2019). Smith and Back (1965) reproduced *in vitro* similar ovalbumin conversion, with an increasing rate when temperature increased, and when pH increased within the 9.0–10.0 range. They named this modified ovalbumin "S-ovalbumin" for its increased stability against heating, exposure to urea and guanidine chloride. The conversion of ovalbumin to S-ovalbumin proceeds through an intermediate species (I-ovalbumin). The same conversion mechanism was observed in fertile eggs (Hatta et al., 2001), with the accumulation of the intermediate species in the embryonic tissues (Shinohara et al., 2005). The thermal denaturation temperature as determined by differential scanning calorimetry (DSC) at pH 8.8 is 84.5°C for ovalbumin, 88.5°C for I-ovalbumin, and 92.5°C for S-ovalbumin, both ovalbumin and S-ovalbumin with the same enthalpy of denaturation (3.6 cal/g, that is 12.55 J/g) (Donovan & Mapes, 1976). While heat-denatured ovalbumin and S-ovalbumin may correctly refold into their respective original non-denatured forms, the conversion of ovalbumin to S-ovalbumin is an irreversible event (Takahashi et al., 2005). Many models have been successively proposed to explain this conversion mechanism, involving either covalent modification or conformational change. In the end, the assumption of a covalent change was discarded. First, the loss of amino acids could not be substantiated since the molecular weight of both forms is the same (Smith & Back, 1965). Secondly, the comparison between ovalbumin and SS-mutated ovalbumin, in which the Cys residues involved in the disulfide bridge were replaced by Ala or Ser residues, enabled to exclude the role of the SS bridge in the conversion of ovalbumin to S-ovalbumin (Ishimaru et al., 2011). However, a conformational modification is supported by several studies, although limited to a small change in secondary structures and no significant change in the overall tertiary structure (Huntington et al., 1995). Hence, α-helix content is slightly lower (3–4% reduction) in S-ovalbumin in comparison with ovalbumin, while antiparallel β-sheet content is increased by an equivalent proportion (Kint & Tomimatsu, 1979). Thanks to the crystal structure of S-ovalbumin solved for the first time in 2003, it was demonstrated that the conversion of ovalbumin into S-ovalbumin resulted from the chemical inversion of the residues Ser^{164}, Ser^{236} and Ser^{320} in D-configuration (Yamasaki et al., 2003), further confirmed using deuterium labeling combined with LC-MS/MS (Miyamoto et al., 2015). This change leads to a decreased solvent accessibility of the hydrophobic core of the protein, thus increasing its structural stability. Unlike Ser^{164} and Ser^{320}, racemized Ser^{236} may not contribute to the increased stability of S-ovalbumin (Takahashi et al., 2010).

5.3.2.2 Ovalbumin-Related Protein Y (OvaY)

Ovalbumin-related protein Y (OvaY), which gene sequence was previously established (Heilig et al., 1982), has been identified in chicken egg white as a new minor protein by Nau et al. (2005). Based on IEF gel image analysis, OvaY concentration in egg white was estimated to 13% of ovalbumin, that is around 7 g/L, consistent with the value deduced from the differential hormonal responsiveness of ovalbumin and OvaY genes, *i.e.* 100:10 (Colbert et al., 1980).

The amino acid sequence of OvaY, determined from the genomic sequence (Heilig et al., 1982), consists of 388 residues including the initiator methionine residue, which is, similarly to ovalbumin, likely cleaved during the secretion of the protein. Likewise, the acetylation of the newly exposed N-terminal residue (Asp^2) was reasonably assumed (Nau et al., 2005).

The polypeptide sequence of OvaY, sharing 57.5% identity with ovalbumin sequence (Da Silva et al., 2015), results in a theoretical molecular weight of 43,641 Da and a theoretical pI of 5.2, both predicted only from the polypeptide backbone (UniProtKB-P01014, OVALY_Chick). This theoretical pI is almost consistent with that estimated from 2D-PAGE analysis, where OvaY appeared as three spots ranging from pI 5.5 to pI 5.3. However, the theoretical molecular weight is much lower than that experimentally estimated, which was around 53 kDa (Nau et al., 2005). This discrepancy was assumed to result from posttranslational modifications, especially glycosylation.

Using monoclonal antibodies, Hirose et al. (2006) confirmed that, among the four putative glycosylation sites predicted on OvaY (Asn^{94}, Asn^{214}, Asn^{292}, Asn^{311}), the first one and at least one of the other three would actually be glycosylated. Moreover, a striking result was that the carbohydrate chains carried by this protein homologous of ovalbumin are actually of ovomucoid type. Later, an omics strategy enabled to establish unambiguously that Asn^{94}, Asn^{214}, and Asn^{311} are glycosylated (Geng et al., 2017).

Phosphorylation of OvaY could not be experimentally established by Nau et al. (2005) with enzymatic and Western blotting technology. However, five phosphorylation sites have been recently discovered by using an innovative approach, which combined two digestion strategies, an enrichment step by metal affinity chromatography, and lastly an identification step by nano-LC-MS/MS (Sun et al., 2020).

Considering the conservation of Cys^{73} and Cys^{120} between ovalbumin and OvaY, it was hypothesized that a disulfide bridge between these residues exists in OvaY, as in ovalbumin (Da Silva et al., 2015). However, to date, there is no data available about the possibility of OvaY conversion to a S-form, similarly to ovalbumin (Dombre et al., 2017).

5.3.2.3 Ovalbumin-Related Protein-X (OvaX)

Ovalbumin-related protein X (OvaX) was detected for the first time in chicken egg white by Guérin-Dubiard et al. (2005). Réhault-Godbert et al. (2013) estimated the concentration of

OvaX in chicken egg white around 0.3 g/L, consistent with the value estimated from the differential transcription rate of ovalbumin and OvaX genes (100:1), *i.e.* 0.5 g/L (Colbert et al., 1980).

Initially, only a 232 amino acid-long fragment of OvaX was known, predicted from the gene sequence (Heilig et al., 1980). However, the whole protein sequence consists of 402 amino acid residues, including the initiating methionine residue. Like ovalbumin and OvaY, OvaX sequence lacks a classical signal peptide, and all three proteins may undergo the same mechanism of secretion, involving the cleavage of the initiator methionine residue. However, unlike the other two homologous proteins, the N-terminal amino acid residue is not predicted to be acetylated. Sequence identity with ovalbumin is 61.9%, and 73.45% with OvaY (Da Silva et al., 2015).

The OvaX sequence corresponds to a theoretical molecular weight of 45,299 Da and a theoretical pI of 6.31, both predicted only from the polypeptide backbone (UniProtKB-R9TNA6, R9TNA6_Chick). OvaX appeared in 2D PAGE as several isoforms of around 53 kDa, ranging from pI 6.0 to 6.7, thus suggesting some posttranslational modifications (Guérin-Dubiard et al., 2006). Because of a longer polypeptide chain, and especially due to five additional residues close to one of the two Cys residues involved in the disulfide bond in ovalbumin (Cys^{73}), it has been assumed that such a disulfide bond could not exist in OvaX (Da Silva et al., 2015).

Unlike ovalbumin, but similarly to OvaY, several glycosylation sites have been predicted in OvaX (Asn^{110}, Asn^{228}, Asn^{306}, Asn^{325}, and Asn^{387}). Among them, Asn^{110} and Asn^{228} were confirmed, although both sites were also identified as non-glycosylated, thus suggesting a certain heterogeneity regarding this glycosylation (Réhault-Godbert et al., 2013). Asn^{110} and Asn^{325} were later confirmed as actually glycosylated sites as well (Geng et al., 2017). Moreover, numerous putative phosphorylation sites exist in OvaX sequence, four of which could be actually experimentally confirmed (Ser^{11}, Ser^{77}, Ser^{358}, and Ser^{368}) (Sun et al., 2020).

Circular dichroism (CD) spectra enabled to conclude that OvaX and ovalbumin exhibit very similar secondary structures, with α-helix contents of 34% for both, and a β-sheet content of 19% and 16%, respectively (Réhault-Godbert et al., 2013). However, OvaX was shown to be more heat-unstable, as indicated by a denaturation temperature of 58.9°C *vs* 71.2°C for ovalbumin. Moreover, unlike ovalbumin and OvaY, OvaX exhibits a specific positively charged domain at the surface of the molecule, which is assumed to be involved in the heparin-binding property of this protein (Réhault-Godbert et al., 2013). It is noticeable that although the heparin-binding property of OvaX slightly decreases during embryogenesis, OvaX incorporated into the embryo tissues retains more or less this activity (Akazawa et al., 2019). This decrease may result from the slight conformational changes of the protein under alkaline conditions, leading to the disorganization of the basic cluster domain involved in the interaction with heparin (Akazawa et al., 2018). Such a decrease in the heparin-binding property may have a physiological role during embryo development.

5.3.3 Ovotransferrin

Ovotransferrin makes up around 12% of the total chicken egg white proteins (Stevens, 1991). The primary sequence of chicken ovotransferrin, translated from the nucleotide sequence (Jeltsch & Chambon, 1982), consists of 686 amino acid residues, for a theoretical molecular weight of 75,828 Da and a theoretical pI of 6.69, both predicted only from the polypeptide backbone (UniProtKB-P02789, TRFE_Chick). The polypeptide sequence is the same as that of chicken serum transferrin, which differs only in the attached carbohydrates (Lee et al., 1980). Sequence variations in chicken ovotransferrin were demonstrated at least in nine positions (Ser/Ala^{33}, Asp/Asn^{292}, Arg/Lys^{438}, Lys/Arg^{512}, Glu/Gln^{519}, Ile/Val^{540}, Lys/Arg^{605}, Gln/Glu^{624}, and Ser/Asn^{667}) (Williams et al., 1982). Inter-species variations were identified as well between chicken and some other avian species, such as turkey (90.7% identity with chicken) and ostrich (Ciuraszkiewicz et al., 2006).

Using a classical glycomic approach, Hwang et al. (2014) identified a single glycosylated residue (Asn^{473}), but 17 different carbohydrate chains, mainly consisting of mannose (42.6%) and N-acetylglucosamine (53.9%), plus some galactose (3.5%), but no sialic acid. The molecular weight of these glycan structures ranged from 891 to 2,272 Da, and 11 out of the 17 carbohydrate chains identified were common with ovomucoid. Thanks to a glycoproteomics approach based on LC-MS/MS analysis, Cavallero et al. (2020) confirmed the glycosylation of Asn^{473}, on which 25 glycoforms were found, ranging from $HexNAc_5Hex_3$ to $HexNAc_7Hex_6$; small amounts of fucose and sialic acid were also detected. Moreover, these authors identified an additional glycosylated site (Asn^{618}), on which four glycoforms were found ($HexNAc_{5-7}Hex_{3-4}$). Both glycosylated sites were previously identified by Geng et al. (2017), in addition to a third one (Asn^{366}). Unlike the amino acid sequence of ovotransferrin, which is highly conserved between bird species, the structure of the carbohydrate moieties is much more species-specific (Jiang et al., 2014). Similarly to the glycoproteomics approach recently developed, LC-MS/MS has also been applied for phosphoproteomics studies, enabling to identify three phosphorylated sites (Ser^{517}, Tyr^{191}, and Tyr^{524}) never previously identified (Sun et al., 2020). Both types of post-translational modifications lead to experimental molecular weight in a range 75–85 kDa, higher than the theoretical one, and pI in a range of 6.2 to 6.7 (Guérin-Dubiard et al., 2006).

Ovotransferrin is part of the transferrin family proteins, all showing high sequence homology. Homology is particularly high between ovotransferrin and human serum transferrin (51%), and human lactoferrin (49%) (Mazurier et al., 1983). Transferrins are all made up of a single polypeptide chain folded in two lobes (N- and C-lobes) connected by a short loop region, with each lobe further divided into two similar-sized sub-domains (N1 and N2 in the N-lobe; C1 and C2 in the C-lobe). Like all transferrins, ovotransferrin is able to bind reversibly two Fe^{3+} ions per molecule, as well as many other metal ions (Garratt et al., 1991; Aramini & Vogel, 1998; Mizutani et al., 2005; Saponja & Vogel, 2011). Ovotransferrin also exhibits a typical secondary structure, consisting of 29% α-helix in its native state (pH 7.0). Its hydrodynamic radius is 5.1 nm (Rabbani et al., 2011).

Crystal structure of the di-ferric form of chicken ovotransferrin has been solved in 1995 (Kurokawa et al., 1995), four years before that of the apo-protein (Kurokawa et al., 1999). More detailed structures of the N-lobe were then proposed

(Mizutani et al., 1999, 2000). The N-lobe includes amino acid residues 1–332, and 343–386 for the C-lobe, both showing very similar structures, resulting from a high sequence identity (37.4%). The folded structure is stabilized by six homologous intra-chain disulfide bridges in each lobe (Cys^{10}-Cys^{45}, Cys^{20}-Cys^{36}, Cys^{115}-Cys^{197}, Cys^{160}-Cys^{174}, Cys^{171}-Cys^{182}, and Cys^{228}-Cys^{242} in the N-lobe, homologous to Cys^{348}-Cys^{380}, Cys^{358}-Cys^{371}, Cys^{454}-Cys^{530}, Cys^{488}-Cys^{502}, Cys^{499}-Cys^{513}, and Cys^{570}-Cys^{584} in the C-lobe), plus three extra bridges in the C-lobe only (Cys^{405}-Cys^{680}, Cys^{421}-Cys^{643}, and Cys^{478}-Cys^{671}) (Williams et al., 1982). Sub-domain N1 includes residues 1–91 and 247–332 *vs* 92–246 for sub-domain N2; sub-domain C1 contains residues 343–429 and 589–686 *vs* 430–588 for sub-domain C2; the loop region between both lobes consists of residues 333–342 (Mizutani et al., 2012). Each lobe contains an iron-binding site, located within the inter-domain cleft between the two sub-domains. These sub-domains are linked by two antiparallel β-strands that allow them to adopt either an open configuration in the apo-form of the protein, or a closed configuration in its holo-form (Kurokawa et al., 1999; Mizutani et al., 2001).

In the holo-protein, iron is hexa-coordinated to four amino acid residues (Asp^{60}, Tyr^{92}, Tyr^{191}, and His^{250} in the N-lobe; Asp^{395}, Tyr^{431}, Tyr^{524}, and His^{592} in the C-lobe) and to a bidentate carbonate anion, which is linked to another residue (Arg^{121} in the N-lobe, Arg^{460} in the C-lobe) (Mizutani et al., 2012). Ovotranferrin tightly binds iron, with the affinity for iron of the C-lobe being greater (K_a ~ 10^{18} M) than that of the N-lobe (K_a ~ 10^{14} M) at neutral pH (Giansanti et al., 2012). With regard to iron release which occurs at acidic pH, it starts with the decarbonation of the holo-protein, after which iron is released first from the N-lobe, and then from the C-lobe. This involves the protonation of the amino acid side-chains engaged in iron coordination, and maybe those involved in inter-domain H-bonds. Iron release is made possible by the conformational changes of the binding lobes, from the closed to the open configuration (Abdallah & Chahine, 1999; Mizutani et al., 2001).

Iron binding increases the heat-stability of ovotransferrin, as shown by the higher denaturation temperatures determined by DSC: 63°C for the apo-form, 68°C and 76°C for the two singly liganded species, and 84°C for the doubly liganded species (holo-form) (Brandts & Lin, 1990). Aluminium binding raises the denaturation temperature of ovotransferrin as well, by around 7°C (Mizutani et al., 2005).

Ovotransferrin plays multiple physiological roles, with a primary one in the avian immune defense (Giansanti et al., 2012). In addition to its antibacterial activities extensively described (Legros et al., 2021), ovotransferrin might also exhibit an antiviral activity, in particular against Marek's disease virus (Giansanti et al., 2002), and an antifungal activity, in particular against *Candida spp* (Valenti et al., 1986). Another physiological function of ovotransferrin, likely related to its iron-binding capacity, is to promote growth during the embryo development; the protein may also participate in the control of inflammation process (Giansanti et al., 2012). All these properties make ovotransferrin a good candidate for many potential applications, for instance as a natural food preservative or as an ingredient in functional foods (Wu & Acero-Lopez, 2012; Wang et al., 2021), especially for iron supplementation in human (Galla et al., 2021).

5.3.4 Antiproteases

Chicken egg white contains four protein inhibitors that are, together, effective against a wide range of proteases. Ovomucoid and ovoinhibitor inhibit serine proteases, cystatin inhibits cysteine proteases, while ovostatin inhibits proteases belonging to the four mechanistic classes, *i.e.* serine proteases, cysteine proteases, acidic proteases, and metalloproteases (Saxena & Tayyab, 1997). The biological role of such an arsenal to overcome protease activity in egg white is not yet known, although several theories have been proposed. In particular, defense against the microorganism growth and control of the tissue differentiation during embryo development are the hypotheses most frequently put forward.

5.3.4.1 Ovomucoid

Ovomucoid represents around 11% of the total chicken egg white proteins (Stevens, 1991). Its primary sequence, determined from the nucleotide sequence, consists of 186 amino acid residues, where tryptophan is totally absent (Catterall et al., 1980). A theoretical molecular weight of 20,098 Da and a theoretical pI of 4.82 are predicted from the polypeptide backbone (UniProtKB-P01005, IOVO_Chick). However, the experimental molecular weight of ovomucoid is much higher (37–43 kDa) than this theoretical value (Guérin-Dubiard et al., 2006), due to its high glycosylation level (20–25%) (Saxena & Tayyab, 1997). Moreover, as with many glycoproteins, ovomucoid exists in many isoforms separated by IEF (Désert et al., 2001) or 2D-PAGE (Guérin-Dubiard et al., 2006). One phosphorylated site (Ser^{55}) was reported for the first time by Sun et al. (2020). Lastly, a certain polymorphism has been observed in chicken ovomucoid, with a predominant form with all the 186 amino acid residues, and a minor form where Val^{134} and Ser^{135} residues are missing (Kato et al., 1987).

Ovomucoid is regarded as the most important allergen in chicken egg white (Urisu et al., 1997). For this reason, highly sensitive and specific methods have been developed for the detection of traces of eggs in foods, some of which by means of ovomucoid determination, whether with the usual ELISA tests, or with more recent electrochemical immunosensors (Benedé et al., 2018).

The amino acid sequence of all avian ovomucoids is organized into three homologous domains. With regard to chicken ovomucoid, and based on the strict amino acid identity, the percentage of homology is 49% between domains I (residues 1–64) and II (residues 65–130), 35% between domains II and III (residues 131–186), and 36% between domains I and III. However, when the homology is calculated based on amino acid groups according to the physicochemical features of their side chains, homology is 71%, 57%, and 70% between domains I and II, II and III, and I and III, respectively (Catterall et al., 1980; Kato et al., 1987). Despite the high homology of primary sequences between domains I and II, their predicted secondary structures are only partially homologous, and domain III is even more different from the two others on that basis (Matsuda et al., 1981). Each domain of ovomucoid is cross-linked by three intra-domain disulfide bridges that stabilize their structure. In chicken ovomucoid, the disulphide bonds are Cys^{5}-Cys^{44}, Cys^{22}-Cys^{41}, and Cys^{30}-Cys^{62} in domain I; Cys^{70}-Cys^{109}, Cys^{87}-Cys^{106}, and Cys^{95}-Cys^{127} in domain II; Cys^{138}-Cys^{168}, Cys^{146}-Cys^{165}, and Cys^{154}-Cys^{186} in domain III. There is no free Cys residue, and no inter-domain disulfide bonds (Kato et al., 1987).

Avian ovomucoids are inhibitors of serine proteases, in which each domain contains one actual or putative reactive site of Kazal-type, which consists of a single peptide bond acting as a substrate. Contact between the inhibitor and the cognate enzyme occurs over short sequences, although fitting perfectly and specifically (Ibrahim & Pattabhi, 2004). As a result, the k_{cat}/K_m value for the hydrolysis of the reactive peptide bond by the cognate enzyme is particularly high for inhibitors of serine proteases, from 10^4 to 10^6 $M^{-1}s^{-1}$ *vs* 10^3 $M^{-1}s^{-1}$ for classical substrates. However, k_{cat} and K_m are very low, leading to a very slow hydrolysis, and a very stable inhibitor-enzyme complex (Laskowski & Kato, 1980). While the amino acid residues that determine the 3D-structure of the protein are highly conserved from one avian species to another, the variability of the residues involved in enzyme-inhibitor contact is very high. Hence, 11 different amino acid residues have been identified at P1 position of the active site in the third domain of ovomucoids from 153 avian species (Apostol et al., 1993). This results in strongly variable inhibitory specificity and affinity between ovomucoid and serine proteases. For instance, chicken ovomucoid strongly inhibits bovine trypsin but not human trypsin, whereas Japanese quail ovomucoid is highly effective toward human trypsin (Asao et al., 1998). However, the strength of interaction with serine proteases is not only determined by the residue at P1 position; the physicochemical features of the residues around the reactive site are also decisive (Kojima et al., 1999).

Chicken ovomucoid, similarly to ostrich and cassoway ovomucoids, contains only one reactive site effective against bovine trypsin in the second domain (Arg^{89}-Ala^{90} peptide bond) (Kato et al., 1987). In golden pheasant, ovomucoid contains only one effective reactive site as well, but effective against chymotrypsin. However, in some other avian species, ovomucoid contains either two (turkey, penguin, guinea) or three (duck, emu) effective reactive sites, making these proteins able to combine simultaneously with two or three different serine proteases, respectively (Saxena & Tayyab, 1997). The anti-trypsin activity of chicken ovomucoid has been suggested as a natural solution to protect bioactive peptides from pancreatic enzymes in some therapeutic strategies (Marchbank et al., 2021).

Among the six potential glycosylation sites Asn-X-Thr/Ser of chicken ovomucoid, one was reported initially as not glycosylated in the third domain (Asn^{158}), similarly with many other equivalent avian ovomucoids. Hence, domain I contains two glycosylation sites (Asn^{10} and Asn^{53}), as well as domain II (Asn^{77} and Asn^{69}), while domain III contains only one (Asn^{175}), which is actually glycosylated or not. It is notable that Asn^{10} and Asn^{77} are at homologous positions, and that the glycosylation site of the third domain is in a highly conserved sequence over avian species (Kato et al., 1987). However, LC-MS/MS analyses enabled recently the detection of N-glysosylation on Asn^{158}, and on Asn^{39}, Asn^{92}, and Asn^{104} as well (Geng et al., 2017). Consequently, there should be nine actual glycosylation sites along the chicken ovomucoid sequence. The carbohydrate chains of chicken ovomucoid are multi-branched and highly heterogeneous structures. The overall composition of the carbohydrate components was first estimated to 55% N-acetylglucosamine, 44% hexose, and 1% sialic acid (Watanabe et al., 1981). Later, Egge et al. (1983) identified 20 different oligosaccharide moieties made up of mannose, N-acetyl-galactosamine, galactose, and N-acetyl-glucosaminol units. More recently, 32 glycan structures have been identified using LC-MS/MS analysis, which in particular, confirmed the minor presence of sialic acid (Cavallero et al., 2020). The glycosylated chains of chicken ovomucoid may protect the protein from proteolysis and heat denaturation. A 2–4°C increase of the melting temperature (T_m) was measured for the glycosylated form (59.0–69.3°C in a pH range from 3.25 to 5.50) of the first domain of chicken ovomucoid, in comparison with its deglycosylated form (54.2–67.4°C) (DeKoster & Robertson, 1997). However, it might not be essential for the biological function of the protein, since deglycosylation does not change its inhibitory activity toward trypsin (Gu et al., 1989). Nonetheless, the consequences of deglycosylation strongly depend on the avian species: it decreases the efficacy of turkey ovomucoid toward trypsin, but it enhances its inhibitory activity toward chymotrypsin (Dabich et al., 1993). Unlike conformational epitopes that play a major role in allergenic properties of ovomucoid, the glycosylated moieties of the protein may have less importance for this property (Matsuda et al., 1982; Cooke & Sampson, 1997).

The chicken ovomucoid sequence was predicted to consist of 33% β–structure, 27% α-helix, 23% β-turn, and 17% random coil (Matsuda et al., 1981), in quite good agreement with CD data (46%, 26%, 10%, and 18%, respectively) (Watanabe et al., 1981). The structure of ovomucoid is particularly thermally stable, since severe heat treatments (90°C for 90 min, and even 100°C for 120 min) do not completely cancel its antigenicity. However, the inhibitory activity of the protein decreases more rapidly than antigenicity: the residual activity is less than 5% after 120 min at 90°C, and completely disappears after 120 min at 100°C (Matsuda et al., 1982). The unusual heat-stability of chicken ovomucoid may be due to the third domain, which is the most stable (Das et al., 1991). The heat-induced structure changes are more significant at alkaline pH (9.5) than at acidic (4.5) or neutral pH (7), and may involve a three-state thermal transition *vs* an all-or-none process in latter cases (Stănciuc et al., 2018). High stability toward heating and pH was reported also for Japanese quail ovomucoid. No loss of inhibitory activity was observed after 24 h at 37°C and pH range from 1 to 12; 70–75% activity was retained after 1 h at 100°C from pH 2 to pH 6, and even 25% after 30 min at 121°C from pH 3 to pH 5 (Takahashi et al., 1994). However, superheated steam, an emerging high temperature technology, resulted in structure changes and antigenicity decrease after heating at 120–200°C for 2–10 min (Wen et al., 2022).

5.3.4.2 Ovoinhibitor

Ovoinhibitor is structurally close to ovomucoid, as indicated by the immunological cross-reactivity between both proteins (Matsuda et al., 1983). However, it is present in chicken egg white at a much lower concentration than that of ovomucoid, around 1.5% of total egg white proteins (Stevens, 1991); different concentrations are reported depending on the avian species (Liu et al., 1971; Saxena & Tayyab, 1997). Chicken ovoinhibitor consists of a single polypeptide chain of 449 amino acid residues, deduced from the nucleotide sequence (Scott et al., 1987). The theoretical pI predicted from this sequence (UniProtKB-P10184, IOV7_Chick) is 6.07, and the theoretical molecular weight is 49,376 Da. In fact, this protein exists in up to nine isoforms of different pI (6.2–6.4) and molecular weights (63.6 – 69.5 kDa), as observed in 2D-PAGE (Guérin-Dubiard et al., 2006). Kim and Choi (2014) reported six isoforms (possibly nine), ranging from 53 to 63 kDa.

Glycosylation (around 10% of the molecule) and the variable sugar composition are responsible for these experimental molecular weights higher than the theoretical one, and for the polymorphism (Saxena & Tayyab, 1997). Chicken ovoinhibitor contains three glycosylation sites on which very similar, but not identical, glycan structures are attached. Glycan composition determination revealed complex type structures made of mannose, N-acetylglucosamine, and minor content of galactose (Yet & Wold, 1990); 2% sialic acid have been also quantified in the carbohydrate moiety of domain I (Galzie et al., 1996).

Like ovomucoids, ovoinhibitors are Kazal-type inhibitors, effective against several serine proteases such as trypsin, chymotrypsin, elastase, subtilisin, alkaline proteases from *Aspergillus oryzae*, *Streptomyces griseus*, and *Aspergillus soyae*, with specificity species-dependent. Hence, chicken, duck, turkey and quail ovoinhibitors exhibit similar inhibiting specificity, unlike ostrich and penguin ovoinhibitors (Liu et al., 1971; Gertler & Ben-Valid, 1980). It is remarkable and quite unique that ovoinhibitor is effective against the esterolytic, proteolytic and elastolytic activities of porcine elastase (Saxena & Tayyab, 1997).

Chicken ovoinhibitor sequence is organized in seven homologous Kazal-type domains, each cross-linked by three intra-domain disulfide bridges (Scott et al., 1987; Yet & Wold, 1990). At least five of the seven domains contain effective reactive sites: two for trypsin, two for chymotrypsin, and one for porcine pancreatic elastase; all five binding sites may be occupied simultaneously (Gertler & Ben-Valid, 1980). The amino acid residues at P1 position are Arg in the first four domains (Arg^{24}, Arg^{89}, Arg^{155}, and Arg^{221}), Phe in the fifth domain (Phe^{286}), and Met in the two last domains (Met^{352} and Met^{411}). This suggests that the cognate enzymes should be trypsin for the first four sites, chymotrypsin for the fifth, and both chymotrypsin and elastase for the two last (Laskowski & Kato, 1980). The specificity of domain I for trypsin was experimentally established later (Galzie et al., 1996). The first six domains of chicken ovoinhibitor are more similar to each other than to the seventh domain, and also more similar to domains I and II of ovomucoid, while the seventh domain is more closely related to the third domain of ovomucoid (Scott et al., 1987; Saxena & Tayyab, 1997).

Like ovomucoid, ovoinhibitor is heat-stable with regard to its inhibitory activity. However, heat-stability dramatically decreases when pH increases. Hence, anti-trypsin and anti-chymotrypsin activities remained after heating at 80°C for 30 min at pH 8.0, but completely disappeared at pH 9.5 (Nakamura & Matsuda, 1984).

5.3.4.3 Cystatin

The cystatin superfamily gathers proteins that are cysteine protease inhibitors, and classified into three structural subfamilies (types 1, 2, 3) on the basis of molecular weight, sequence homology, presence and position of disulfide bridges, and subcellular localization (Kupaj et al., 2020). Cystatin from chicken egg white was the first identified member of this superfamily (Fossum & Whitaker, 1968). It is a type 2 cystatin, *i.e.* a non-glycosylated protein, including four cysteine residues involved into two disulfide bridges (Cys^{71}-Cys^{81}, and Cys^{95}-Cys^{115}) (Grubb et al., 1984).

Cystatin concentration in chicken egg white is low, estimated as 80 μg/mL (Anastasi et al., 1983), *i.e.* around 0.08% of total proteins. Its primary sequence was determined by Schwabe et al. (1984) and Turk et al. (1983), consisting of 116 amino acid residues (UniProtKB-P01038, CYT_Chick), from which the predicted molecular weight is 13,148 Da, and pI is 6.54. However, the protein exists in two forms that differ in pI (6.5 and 5.6 for forms I and II, respectively), but similar in inhibitory activity and in immunological identity (Anastasi et al., 1983). The difference between both forms is due to the phosphorylation of one residue (Ser^{80}) in form 2, which represents 37% of native chicken cystatin (Laber et al., 1989).

The predicted 2D-structure of cystatin from chicken egg white contains 21% α-helix, 42% β-sheet, 24% β-turn, and 12% random coil (Schwabe et al., 1984). The 3D-structure of the protein, determined by X-ray crystallography, consists of a five-stranded antiparallel β-pleated sheet wrapped around a straight five-turn α-helix, with an appending segment of partially α-helical geometry (Bode et al., 1988).

The great majority of cysteine proteases of the papain superfamily are inhibited by cystatin from chicken egg white (Saxena & Tayyab, 1997), which inhibits in particular papain, ficin, and cathepsins B, H and L. Cystatin forms equimolar and tight, but reversible complexes with the cognate enzymes, with K_i values ranging from around 10^{-8} M to 10^{-12} M (Anastasi et al., 1983). The binding site comprises three regions of the inhibitor molecule, namely the segment around Gly^{9}, the Gln^{53}-Gly^{57} sequence, and the loop around Trp^{104} (Bode et al., 1988; Machleidt et al., 1989; Björk & Ylinenjaervi, 1990; Nycander & Björk, 1990).

Cystatin is extremely heat stable, at least at acidic pH, since boiling for 30 min at pH 4 reduces its inhibitory activity by only 10%; at pH 6 and pH 9, the loss of activity reaches 25% and 60%, respectively (Fossum & Whitaker, 1968). Anastasi et al. (1983) reported no loss in activity after storage overnight at pH 11 and 4°C.

Although cystatins are used or are considered for a wide range of medical applications (antimicrobial, antiviral, cancer control…), the use of egg white cystatin would be limited due to its high cost (Kupaj et al., 2020).

5.3.4.4 Ovostatin (Ovomacroglobulin)

Ovostatin, also referred to as ovomacroglobulin, is the last identified protease inhibitor in egg white, at a concentration estimated to be 0.5 mg/mL, *i.e.* around 0.6% of total egg white proteins. This protein is a member of the α_2-macroglobulin family, and exhibits in particular some functional, structural and composition similarities with the α_2-macroglobulin identified in chicken plasma, despite being very different proteins. In particular, there is no immunological cross-reactivity between them (Nagase et al., 1983).

Ovostatin is a high molecular weight (730–780 kDa) tetrameric protein, which consists of four identical subunits of 1437 amino acid residues each (Nielsen et al., 1994). The theoretical molecular weight of each subunit is 162.2 kDa, calculated from the amino acid sequence (UnirProtKB-P20740, OVOS_Chick), while experimental measurements, *i.e.* including the carbohydrate moiety, indicate 195 kDa by sedimentation equilibrium ultracentrifugation (Nagase et al., 1983), and 183.3 kDa using MALDI-TOF MS (Geng et al., 2015). The theoretical pI of the protein is 5.58.

Each subunit is predicted to contain 12 intra-chain disulfide bridges, while subunits are cross-linked in pairs by three inter-chain bridges (Nielsen et al., 1994). The resulting dimers are supposed to be associated by non-covalent interactions (Nagase et al., 1983; Saxena & Tayyab, 1997).

Ovostatin sequence contains 12 effective N-glycosylation sites, only one of which (Asn^{1221}) existing in both glycosylated and non-glycosylated forms. Moreover, Asn^{342} is a putative glycosylation site. The carbohydrate moiety was calculated as 21.1 kDa for each subunit, representing 11.5% of the molecule. However, ovostatin glycosylation is heterogeneous, resulting in many isoforms separated on 2D-PAGE with apparent pI ranging from 5.03 to 6.03 (Geng et al., 2015). N-acetylhexoses, hexoses, fucose, and sialic acid have been identified as components of the carbohydrate chains of ovostatin (Cavallero et al., 2020).

Ovostatin is able to inhibit proteases of all the four mechanistic classes, and in particular trypsin, subtilisin, papain, and thermolysin (Saxena & Tayyab, 1997). However, it exhibits a greater activity toward metalloproteases than proteases of other classes, in a manner such as only the activity of proteases toward high molecular weight substrates is inhibited. The mechanism is initiated by the hydrolysis of a specific peptide bond in ovostatin sequence by the cognate protease, leading to conformational changes of ovostatin, which, in turn, entraps the protease in such a way that the latter can no longer access large substrates. A different mechanism occurs against serine and cysteine proteases (Nagase & Harris, 1983; Saxena & Tayyab, 1997). The association of four subunits of ovostatin is important for its inhibitory activity toward pepsin and rennin (Kato et al., 1991).

Ovostatin is relatively stable at alkaline pH, as suggested by the remaining inhibitory activity (90%) at pH 10.5 after 16 h at 4°C, while it completely disappeared at pH 4.0 after a similar storage. However, ovostatin is heat-labile, since it lost 20% and 95% activity after heating at 55°C for 30 min, and 80°C for 5 min, respectively (Nagase et al., 1983).

5.3.5 Lysozyme

Lysozyme is probably one of the most extensively studied protein worldwide. This protein, discovered in human nasal mucus at the beginning of the 20th century (Fleming, 1922), has later shown to be widely distributed throughout the human body, in many organs and biological fluids (tears, saliva, urine, milk…), and in animal tissues and secretions (Callewaert & Michiels, 2010). Egg white exhibits one of the highest lysozyme concentrations, around 3.4% of total proteins in chicken egg white (Stevens, 1991). This explains why, still today, all the lysozyme commercialized for industrial applications is extracted from chicken egg white (Shahmohammadi, 2018). Lysozymes exist not only in animal organisms, but also in plants, fungi, and bacteriophages. Moreover, animal lysozymes can be further classified in several types: c-type, the most common, including chicken egg white lysozyme; g-type, including goose egg white lysozyme; and i-type, including lysozymes of invertebrates. Homology in primary sequences among these different types is quite low, but strong similarities exist in their 3D-structure (Callewaert & Michiels, 2010).

Initially, the interest in lysozyme was for its lytic activity. Indeed, lysozyme is a hydrolytic enzyme, also called muramidase or N-acetylmuramoyl-hydrolase, able to hydrolyze the β-1,4 glycosidic bond between N-acetylmuramic acid and N-acetylglucosamine in peptidoglycan. The substrate-binding site of chicken egg white lysozyme consists of six sub-sites, able to complex six carbohydrate residues, while Glu^{35} and Asp^{52} play a major role in the enzymatic activity (Kawaguchi et al., 2015). The enzymatic activity toward peptidoglycan, which is the major component of bacterial cell wall, explains, at least in part, the antimicrobial properties of lysozyme. Lysozyme has shown to exhibit other antibacterial mechanisms, making this protein active against a much broader bacteria spectrum than initially described, and an essential component of biological innate immunity (Ragland & Criss, 2017). This explains the high number of lysozyme applications in pharmacology, cosmetics, and food manufacturing (Wu et al., 2019; Khorshidian et al., 2022). Moreover, structure modifications may confer new antimicrobial properties to lysozyme (Ibrahim et al., 2001; Derde et al., 2014, 2017), and native lysozyme exhibits also anti-fungal, antiviral, anti-inflammatory, immunomodulatory, anticancer, and anti-oxidative properties, making this protein already widely used, and still very promising (Jiang et al., 2021).

The primary sequence of chicken egg white lysozyme consists of 129 amino acid residues (Canfield, 1963), resulting in a small molecular weight of 14,313 Da and a predicted pI of 9.32 (UniProtKB-P00698, LYSC_Chick). Natural variants have been reported, at positions 31 (Ala *vs* Val) and 53 (Tyr *vs* Phe or Ser) of the primary sequence (Jung et al., 1980; Downing et al., 2010). It is mainly a non-glycosylated protein, despite a low abundant glycosylated form has been detected (around 0.3% of total lysozyme molecules) (Trudel & Asselin, 1995). Glycosylation sites are Asn^{39} and Asn^{44}, which could not be glycosylated simultaneously. Moreover, it is notable that both sites do not correspond to any known N-glycosylation-specific sequence motif, and that the N-glycan profile is particularly heterogeneous, including at least 26 different carbohydrate chains composed of hexoses and N-acetyl-hexoses (Asperger et al., 2015; Geng et al., 2017).

X-ray crystallography analysis of chicken lysozyme highlights a shape roughly ellipsoidal (Blake et al., 1965). Moreover, four disulfide bonds (Cys^{6}-Cys^{127}, Cys^{30}-Cys^{115}, Cys^{64}-Cys^{80}, Cys^{76}-Cys^{94}) result in a very compact globular protein. These bonds play an essential role in forming and stabilizing the secondary structure of the protein, estimated to be 31–33% α-helix structure, and 11–12% β-structure (White, 1982).

5.3.6 Ovomucin

Ovomucin is a sulfated glycoprotein responsible for the gel-like aspect of fresh egg white (see Section 5.2.3 and Table 5.1). Its concentration is higher in thick albumen and chalazae than in the thin albumen fraction, and the changes it undergoes during egg storage largely explains egg white liquefaction (see Section 5.2.4). Overall, ovomucin represents about 3.5% of total proteins in chicken egg white (Omana et al., 2010).

The analysis and biochemical characterization of ovomucin have long been limited because of its difficult solubilization after precipitation by pH and ionic strength decrease, which is the usual protocol for its extraction (Donovan et al., 1970). Moreover, this protocol results in co-precipitation of other albumen proteins, contaminating the ovomucin fraction. Ovomucin is not yet completely characterized, as it exists as a very high molecular weight and filamentous complex (5,000–8,000 kDa), composed of two types of subunits, namely α-ovomucin and β-ovomucin, which molecular weights are around 220 kDa and 370 kDa, respectively (Hiidenhovi et al., 2016). The structure of the complex is supposed to be similar to that of mammalian mucins, namely highly polymerized macromolecules cross-linked by disulfide bridges (Omana et al., 2010).

The primary sequence of α-ovomucin consists of 2,087 amino acid residues, for a pI predicted to be 5.35, and a theoretical molecular weight of 231 kDa, calculated from the polypeptide backbone (UniProtKB-Q98UI9, MUC5B_Chick). Sixteen intra-chain disulfide bridges are predicted. With regard to β-ovomucin, the primary sequence should consist of 1,185 amino acid residues, corresponding to a predicted pI of 5.78, a theoretical molecular weight of 132 kDa, and eight predicted disulfide bridges (UniProtKB-F1NBL0, MUC6_Chick).

α-ovomucin is a poorly glycosylated subunit (15% carbohydrates), while β-ovomucin is a carbohydrate rich subunit (60%). Carbohydrate chains consist of N-acetylglucosamine, N-acetylgalactosamine, galactose, mannose, sialic acid and sulfate in both types of subunits (Robinson & Monsey, 1971). Besides a high amount of O-linked glycosylation characteristics of mucin family protein (Strecker et al., 1992; Dekker et al., 2002), 18 potential N-glycosylation sites have been determined in the α-subunit, 15 of which were actually glycosylated, and one was either glycosylated or not; only two N-glycosylation sites have been identified in β-ovomucin sequence (Offengenden et al., 2011).

Ovomucin, either as entire protein or peptides obtained by hydrolysis, exhibits many biological functions. For example, anti-haemagglutination, antiviral and anti-tumor activities, as well as a hypocholesterolemic action have been reported (Omana et al., 2010).

5.3.7 Riboflavin-Binding Protein (RfBP, Flavoprotein)

Riboflavin-binding protein (RfBP) also referred to as flavoprotein, represents less than 1% of total proteins in chicken egg white. An average concentration of 0.7 g/L (23.9 μM) has been determined by CD and fluorescence spectroscopy, but with some surprising individual variation (Zandomeneghi & Zandomeneghi, 2009). A similar protein exists in the egg yolk, and in the plasma of laying hens (White & Merrill, 1988). The three proteins are synthesized from the same gene, either in the oviduct for the protein present in the egg white, or in the liver for that present in the egg yolk and plasma. Egg white and yolk RfBP differ in their glycosylation patterns, and in their polypeptide chain; the yolk RfBP lacks the last 11–13 amino acid residues (Hamazume et al., 1984; Norioka et al., 1985). All are globular and monomeric phospho-glycoproteins with the same glycosylation sites (Asn^{36} and Asn^{147}).

The polypeptide chain of the chicken egg white RfBP is composed of 219 amino acid residues, the sequence of which was determined either directly (Hamazume et al., 1984), or from the DNA sequence (Zheng et al., 1988). The molecular weight of RfBP, including carbohydrates and phosphate, is around 29.2 kDa, and its pI around 4.0 (White & Merrill, 1988). Hence, it is the most acidic protein in chicken egg white, making easy its extraction and purification by ionic exchange chromatography (Guérin-Dubiard et al., 2005).

The primary sequence of chicken egg white RfBP contains 18 Cys residues, all of them involved in nine disulfide bridges (Hamazume et al., 1987). Crystal structure indicates a protein consisting of six α-helices, including about 30% of the amino acid residues, and four series of discontinuous areas of β-structure, accounting for less than 15% of the residues (Monaco, 1997). This molecule is organized in two domains. The N-terminal domain, from Glu^{1} to Cys^{169}, contains the riboflavin-binding site, and its 3D-structure is strongly conditioned by a large number of disulfide bridges. The C-domain, extending from Leu^{170} to the C-terminal end, consists of two helices on both sides of a flexible and highly anionic segment, due to eight phosphoserine and ten Glu residues (Monaco, 1997). All the phosphoserine residues are gathered on a polypeptide segment of 11 amino acid residues only (from Ser^{186} to Ser^{197}), thus constituting a cluster (Mega et al., 1986).

Carbohydrate chains, which represent about 15% of the molecular weight of chicken egg white RfBP (Hamazume et al., 1984), consist of fucose, mannose, galactose, N-acetylglucosamine and sialic acid. Both glycosylation positions differ in that the carbohydrates attached to Asn^{36} have a lower degree of branching, while those attached to Asn^{147} exhibit very heterogeneous and complex structures, with triantennary- and tetra-antennary-type structures being the most abundant species. This heterogeneity as well as the presence of sialic acid may explain the existence of many isoforms differing in their pI. Nine isoforms have been identified (Amoresano et al., 1999).

Due to its ability to bind riboflavin (vitamin B2), egg white RfBP is essential for the development and survival of the embryo (White, 1996). The protein has a single high-affinity site that binds riboflavin in a molar ratio 1:1, with a dissociation constant (K_d) of the complex estimated to be 1.3 nM at 25°C and pH 7.0, nearly constant between pH 6 and pH 9 (Becvar & Palmer, 1982). RfBP binds riboflavin at pH values above 4.5, but quickly releases its ligand at pH lower than 4.0, because of structure modifications of the protein (Kumosinski et al., 1982). The interaction between both molecules involves the isoalloxazine ring of riboflavin (Nishina et al., 1978; Kitagawa et al., 1979), which inserts into the binding pocket of the protein molecule between the rings of Tyr^{75} and Trp^{156} (Monaco, 1997). Only one disulfide bridge, maybe involved in maintaining the structure of the binding site but located out of the latter, may be essential for the binding capacity of the protein (Kozik, 1982).

Besides its affinity for riboflavin, RfBP is able to bind copper(II) in a molar ratio 1:1. Thus, RfBP could play a role in the transport and storage of copper in egg (Smith et al., 2006). Another original property of RfBP lies in that it suppresses the sweetness of protein sweeteners. Sweet suppression may stem from the interaction between RfBP and a sweet taste receptor (Maehashi et al., 2006). Similarly, RfBP lowers the bitterness of various substances such as quinine, caffeine, and theobromine, also likely due to the ability of the protein to interact with taste receptors (Maehashi et al., 2008).

CD analysis suggests that holo- and apo-RfBP have very similar secondary structures (Allen et al., 1992). However, even if RfBP is generally heat-stable, it is particularly true with regard to its holo-form. Hence, denaturation temperature was estimated by DSC to be 60.8°C for apo-RfBP *vs* 72.8°C for RfBP-riboflavin complex (Wasylewski, 2000, 2004). Previously, Rhodes et al. (1959) did not observe any loss of binding capacity of the protein after 15 min at 100°C, pH 7.0.

5.3.8 Avidin

Avidin represents only 0.05% of total proteins in chicken egg white, *i.e.* around 50 mg/L (Green, 1975). However, this protein has been extensively studied for a long time, because of its extremely strong binding capacity of biotin (vitamin H or vitamin B8). Very early, dermatosis-type pathologies due to biotine deficiency were reported in rats after the ingestion of raw egg white (György et al., 1941). This biological property may also contribute to the role of avidin as a defense protein in egg (Ahlroth et al., 2000).

Avidin is a tetrameric protein composed of four identical subunits of 128 amino acid residues each, with a carbohydrate chain attached to Asn^{17}, a disulfide bridge between Cys^{4} and Cys^{83}, and a natural variation at position 34 (Ile *vs* Thr) (DeLange & Huang, 1971; Huang & DeLange, 1971). It is synthesized from the only member of the avidin gene family that is expressed at a high level (Ahlroth et al., 2000). The primary sequence of each subunit indicates a theoretical pI of 9.7, making avidin, like lysozyme, a strongly basic protein in egg white. The theoretical molecular weight is 14,343 Da for each subunit, and therefore 57,372 Da for the tetramer, which is lower than the experimental value of 67 kDa (Green, 1975). This discrepancy stems from the carbohydrate moieties that account for roughly 10% of the molecular weight of the protein. Carbohydrate chains, consisting of mannose and N-acetylglucosamine units in a ratio from 1.2 to 2.0, are highly heterogeneous (Bruch & White, 1982). A non-glycosylated form may naturally exist in egg white (Nau et al., 2007a). However, the carbohydrate chains may not be involved in the biotin-binding capacity of the protein (Hiller et al., 1987), and elsewhere do not affect its overall three-dimensional structure (Rosano et al., 1999).

The avidin-biotin interaction is the strongest non-covalent binding between a protein and a ligand known in nature, with a K_d estimated to be around 10^{-15} M (Green, 1963). Each subunit is organized in an eight-stranded antiparallel β-barrel, which defines a binding site for biotin. Therefore, each tetramer is able to bind up to four vitamin molecules. The binding pocket consists of six hydrophobic residues, among which Trp plays a major role, and into which the ureido ring of biotin inserts, before hydrogen bonds reinforce the protein-ligand interaction. Thus, several factors explain this exceptional affinity of avidin for biotin. First, the open structure of the binding pocket in the apo-protein makes easy and fast the access of the vitamin to the binding site. Secondly, the three-dimensional structure of the binding pocket perfectly fits the vitamin shape and volume. Thirdly, once bound, biotin is entrapped by conformational readjustments of the protein (Pugliese et al., 1994; Rosano et al., 1999). The secondary structure of avidin, predominantly consisting of β-sheet with some turns and bends, but no significant content of α-helices, is not significantly modified after biotin binding (Wang & Polavarapu, 2001).

Avidin is a thermostable protein, as evidenced by a transition temperature (T_{tr}) estimated by DSC to be 85°C at pH ranging from 7 to 9, and biotin binding highly increases the protein T_{tr} (up to 132°C) (Donovan & Ross, 1973). Consistently, Fourier-transform infrared spectroscopic studies concluded that the secondary structure of apo-avidin is unchanged up to around 75°C, while thermal stability dramatically increases for the avidin-biotin complex, since no change was detected up to 93°C (Swamy et al., 1996). Treatments as high as 121°C for 25 min were necessary to reduce by 90% the avidin binding activity, while high pressure (700 MPa) combined with heating at 50°C for 10 min resulted in a complete inactivation (Zhu et al., 2018). Moreover, biotin binding protects avidin from proteolysis by proteinase K, due to a loop that is constrained in holo-avidin, while mobile in apo-avidin (Ellison et al., 1995).

The rapid and almost irreversible biotin-binding capacity of avidin is at the foundation of many applications generically called "avidin-biotin technology". This term describes experimental systems in which a target molecule is labeled with a biotinylated binder molecule, which is, in turn, labeled either with an avidin-conjugated probe, or with free avidin before addition of a biotinylated probe. Biochemical applications such as immunoassays, blotting, signal amplification, cytochemistry…, as well as medical applications, such as immune-localization, imaging, drug delivery, and tumor-targeting, have been extensively investigated (Bailly et al., 2017; Jain & Cheng, 2017; Zhang, He et al., 2020). However, avidin creates in some cases the problem of non-specific binding (*i.e.* biotin-independent), in particular due to its glycosylated moieties. Avidin deglycosylation has been proposed to avoid this interference (Bayer et al., 1995). Another way to get around these technical limitations is its substitution by streptavidin, which is a non-glycosylated homologous protein isolated from bacterial culture of *Streptomyces avidinii* (Nau et al., 2007b).

5.3.9 Lipocalin Proteins

The lipocalin family gathers small (160–180 amino acid residues) and extracellular proteins sharing several properties, including the binding of small hydrophobic molecules (lipids, steroid hormones, retinoids…). For this reason, they are commonly regarded as transport proteins, even though they fulfill many other properties and functions. Despite highly diverse primary sequences, lipocalins share conserved short sequence motifs and structural features. In particular, lipocalins are organized in eight-stranded antiparallel β-sheets folded in a highly symmetrical manner and closed on itself, creating a β-barrel which encloses a binding site (Flower et al., 2000). In chicken egg white, three lipocalin proteins have been identified, namely the α1-acid glycoprotein, the extra fatty acid binding protein (Ex-FABP, or Ch21), and the chondrogenesis-associated lipocalin (Cal-γ, or prostaglandin D synthase).

5.3.9.1 α1-Acid Glycoprotein (Ovoglycoprotein)

α1-acid glycoprotein, also referred to as ovoglycoprotein, was identified and isolated from egg white in the early 1960's, and was regarded as a minor protein with a concentration estimated to be 1% of total egg white proteins (Ketterer, 1965). However, this value was likely underestimated because of an incomplete isolation by alcohol precipitation before quantification. Indeed, semi-quantitative immune-detection allowed recently to reassess the average concentration of this protein in chicken egg white around 5 g/L, *i.e.* 5% of total proteins (Julien et al., 2020).

The primary sequence of α1-acid glycoprotein, deduced from cDNA sequence, consists of 183 amino acid residues, and exhibits 31–32% sequence homology with rabbit and human α1-acid glycoproteins (Sadakane et al., 2002). The polypeptide sequence (UniProtKB, Q8JIG5_Chick) corresponds to a theoretical pI of 5.03, and a theoretical molecular weight of 20,317 Da. However, its experimental molecular weight is much higher, around 40 kDa (Guérin-Dubiard et al., 2006), due to a high glycosylation rate.

Carbohydrate moieties of α1-acid glycoprotein account for about 30% of the molecular weight of this protein, consisting of 13.6% hexoses (mannose and glucose at a ratio 2:1), 13.8% glucosamine, and 3% sialic acid (Ketterer, 1965). The N-type carbohydrate chains are attached to five glycosylation sites, namely Asn^{16}, Asn^{70}, Asn^{77}, Asn^{87}, and Asn^{62}, the last one being only partially glycosylated (Matsunaga et al., 2004). Glycosylation may be responsible for the polymorphism of α1-acid glycoprotein, as evidenced by the detection of eight spots on 2D-PAGE gels, different in experimental molecular weight (37–43 kDa) and pI (5.0–5.4) (Guérin-Dubiard et al., 2006). Eleven different carbohydrate chains have been identified recently by LC-MS/MS, distributed onto Asn^{70}, Asn^{77}, and Asn^{62} (Cavallero et al., 2020). In addition to the disulfide bridge experimentally evidenced (Cys^{73}-Cys^{163}), chicken α1-acid glycoprotein may also contain another one, highly probable of occurring (Cys^{6}-Cys^{146}) (Matsunaga et al., 2004). Lastly, a phosphorylation site (Ser^{56}) was recently reported for the first time (Sun et al., 2020).

The biological function of α1-acid glycoprotein in egg is still unknown. However, it may be noted that chicken α1-acid glycoprotein is part of the same subgroup as human α1-acid glycoprotein, which is known to have a variety of immunological functions, and one of the most-significant drug binders in human blood (Flower et al., 2000). As a matter of fact, the ability of chicken α1-acid glycoprotein for the recognition and separation of drug enantiomers has been widely documented, and even used to prepare, for instance, HPLC chiral stationary phases for analysis purposes (Haginaka, 2000). The protein domain of α1-acid glycoprotein, and not the carbohydrate moieties, is responsible for the chiral recognition (Sadakane et al., 2002).

5.3.9.2 Extracellular Fatty Acid-Binding Protein (Ex-FABP, Ch21, Quiescence Specific Protein)

Ex-FABP, also called Ch21 or Quiescence-specific protein, was first detected in chicken egg white by Larsen et al. (1999). It was later confirmed by Désert et al. (2001) and Guérin-Dubiard et al. (2006). Ex-FABP concentration in chicken egg white has been recently estimated by immunoblotting to be around 0.1 g/L, *i.e.* 0.1% of total proteins (Julien et al., 2020).

The primary sequence of Ex-FABP (UniProtKB-P21760, EXFAB_Chick), determined from the cDNA sequence (Giannoni et al., 2004), consists of 158 amino acid residues and corresponds to a theoretical pI of 5.39, and a theoretical molecular weight of 18,065 Da. However, the experimental molecular weight, determined by SDS-PAGE (Larsen et al., 1999) or 2D-PAGE (Guérin-Dubiard et al., 2006) is higher, estimated to be 21 kDa, thus suggesting some degree of glycosylation. Ex-FABP sequence is also predicted to contain only one disulfide bridge (Cys^{60}-Cys^{153}) (Correnti et al., 2011).

Ex-FABP may contribute to antibacterial defense in egg white, due to its ability to sequester the enterobactins produced by bacteria, such as *Escherichia coli* (Garénaux et al., 2013), *Bacillus subtilis* (Correnti et al., 2011), and *Salmonella* Enteritidis (Julien et al., 2020). Ex-FABP is the only one, among the three lipocalins identified in egg white, able to bind this kind of siderophores, required to supply iron to bacteria under iron-limited conditions. However, Ex-FABP was initially identified in chick embryo skeletal tissues, muscle fibers, and blood granulocytes, and is suspected to be a stress protein expressed in tissues under intensive remodeling or during inflammatory phase due to pathological conditions (Descalzi Cancedda et al., 2000, 2001). Injection of Ex-FABP has shown to decrease the damages due to dyschondroplasia in chicken (Jahejo et al., 2022). Ex-FABP is present in serum as well, where it preferentially binds and transports long-chain unsaturated fatty acids (Descalzi Cancedda et al., 1996), which would be essential for the embryo survival (Gentili et al., 2005).

5.3.9.3 Chondrogenesis-Associated Lipocalin γ (CAL-γ)

CAL-γ was identified for the first time in chicken egg white in 2003 (Nau et al., 2003a). Its concentration in chicken egg white was estimated to be around 0.1 g/L, *i.e.* equivalent to that of Ex-FABP (Julien et al., 2020). CAL-γ and Ex-FABP form, with another protein (CAL-β) previously identified and sequenced (Pagano et al., 2002), a genomic cluster coordinately regulated and acting synergistically in bone formation process and inflammatory response (Pagano et al., 2003).

The primary sequence of CAL-γ (UniProtKB, Q8QFM7_Chick), deduced from the cDNA sequence (Pagano et al., 2003), consists of 166 amino acid residues. Its theoretical pI is 6.18, and its theoretical molecular weight is 18,955 Da, *i.e.* slightly lower than that experimentally determined on 2D-PAGE (22 kDa), thus suggesting CAL-γ may be glycosylated. Moreover, CAL-γ appeared into two spots on 2D-PAGE, corresponding to significantly different pI (5.6 and 6.0) (Guérin-Dubiard et al., 2006), suggesting polymorphism the nature of which remains unknown.

5.3.10 Protein TENP

The expression of *tenp*, a gene transiently expressed in neural precursors, was reported in developing brain and retina, and had been sequenced (Yan & Wang, 1998; Whenham et al., 2014), before the related protein TENP was also detected in unfertilized chicken egg white, at an estimated concentration between 0.1 and 0.5% of total proteins (Guérin-Dubiard et al., 2006). It is now established that *tenp* is a typical oviduct and egg specific gene (Whenham et al., 2014). TENP is the same protein than that previously referred to as ovoglobulin G2 (Kinoshita et al., 2016).

The primary sequence (UniProtKB-O42273, TENP_Chick), determined from the cDNA sequence, consists of 427 amino acid residues, considering that the first 13 N-terminal residues are a signal peptide sequence (Kinoshita et al., 2016). The polypeptide chain of TENP corresponds to a theoretical pI of 5.67 and a theoretical molecular weight of 46,115 Da. However, when analyzed by 2D-PAGE, TENP appeared as three spots corresponding to pI 5.9–48.9 kDa, pI 6.1–48.9 kDa, and pI 6.3–50.2 kDa, respectively (Guérin-Dubiard et al., 2006). Some high cDNA sequence polymorphism was reported, with eleven non-synonymous substitutions, but only one of which (Arg^{110}/Gln^{110}) was related to two major polymorphic forms of the protein, separated by native-PAGE. Moreover, a predicted glycosylation site at Asn^{252} position (Kinoshita et al., 2016) was recently experimentally confirmed (Geng et al., 2017).

TENP is a member of the BPI (bactericidal permeability-increasing protein) fold-containing family B (BPIFB), and classified as BPIFB7 (member 7). However, because of its high homology with mammalian BPIFB2, TENP is also referred to as BPIFB2 (Kinoshita et al., 2016). BPI proteins share the property of binding and neutralizing lipopolysaccharides of the outer membrane of Gram-negative bacteria, thus leading to a bacteriostatic, and eventually a bactericidal action (Bingle & Craven, 2004). While the antibacterial activity of chicken TENP still has to be demonstrated experimentally, its protective role in oviduct and in egg white is very likely (Whenham et al., 2014; Guyot et al., 2016). Interestingly, in emu egg white, TENP is a major protein (16% of total proteins), and exhibited activity against Gram-positive bacteria, but not against Gran-negative bacteria (Maehashi et al., 2014). Emu TENP has 62.4% homology with chicken TENP (Kinoshita et al., 2016). Moreover, TENP is involved in oviduct remodeling during the regression and recrudescence phases of molting in laying hens (Lee et al., 2020), and is supposed to protect and promote the development of the nervous system in the embryo (Wang et al., 2019).

5.3.11 Clusterin

Clusterin is a glycoprotein highly conserved across species and widely distributed in biological fluids of mammalian and birds. In particular, clusterin is present in several tissues of the hens, in particular in the oviduct (Mahon et al., 1999), and also in the eggshell matrix and egg white (Mann et al., 2003; Guérin-Dubiard et al., 2006).

Like other clusterins, chicken clusterin is a heterodimer composed of two subunits, derived from a precursor polypeptide by proteolytic cleavage between Arg^{211} and Glu^{212} (Mann et al., 2003), and linked by five disulfide bridges. The primary sequence of the precursor (UniProtKB, Q9YGP0_Chick), deduced from the cDNA sequence (Mahon et al., 1999), consists of 428 amino acid residues, corresponding to a theoretical molecular weight of 49,412 Da. The sequence shows 45 and 95% identity to human and quail clusterins, respectively, including in particular the conservation of all 10 cysteine residues in the three species. The α subunit, the N-terminal domain, corresponds to a theoretical pI and molecular weight of 7.08 and 24,467 Da, respectively, *vs* pI 4.9 and 24,963 Da for the β subunit, the C-terminal domain. However, the experimental molecular weight of clusterin is around 70 kDa in SDS-PAGE, decreased to 35 kDa under reducing conditions (Mann et al., 2003). This larger than predicted molecular weight is consistent with the high glycosylation rate (30%) reported for human clusterin (Wilson & Easterbrook-Smith, 2000). Chicken clusterin sequence contains six potential glycosylation sites, four of which are in common with human clusterin (Mahon et al., 1999). Several spots are revealed on 2D-PAGE gels, within a wide range of pI (4.54–6.67), indicating some heterogeneity in glycosylation (Geng et al., 2017). Glycosylation of Asn^{79}, Asn^{121}, Asn^{255}, Asn^{332}, and Asn^{352} has been experimentally demonstrated (Geng et al., 2017; Cavallero et al., 2020).

Clusterin is involved in numerous biological processes, and in particular, it is a chaperone protein, namely a protein that binds to other proteins thus preventing their aggregation and precipitation (Wilson & Easterbrook-Smith, 2000). For instance, clusterin exerts a powerful chaperone-like activity with lysozyme, inhibiting the low precipitation of lysozyme induced by heat or reduction (Poon et al., 2002). Hence, it may contribute to control the folding and association of proteins serving as the organic matrix of the eggshell, thus regulating mineralization and biomechanical properties of this structure (Mann et al., 2003). Clusterin may have also a role during embryo development by stabilizing several proteins, as suggested by the significantly increasing abundance of this protein during the hatching period (Liu et al., 2015).

5.3.12 HEP21

HEP21 is a small minor egg white protein recently identified, and for which knowledge is still very scarce. The primary sequence (UniProtKB, Q8AV77_Chick), deduced from the cDNA sequence (Nau et al., 2003b), consists of 107 amino acid residues, including a predicted signal peptide of 19 residues. The theoretical molecular weight, *i.e.* 10,001 Da, is much lower than that measured by electrophoresis (21 kDa), likely due to glycosylation, consistently with the predicted glycosylation site at Asn^{20}. Various levels of glycosylation could also explain the polymorphism observed on 2D-PAGE gels, with two spots of an estimated pI of 6.3 and 6.4 (Guérin-Dubiard et al., 2006).

HEP21 is expressed along the length of the oviduct, but at higher intensity in the magnum and white isthmus (Nau et al., 2003b). Its biological role remains unknown, and cannot easily be predicted due to the diverse range of biological functions played by proteins within this family, namely the uPAR/CD59/Ly-6/snake toxin family. HEP21 may be involved in sexual maturity, in ovary and oviduct development, and in egg production and formation (Lim & Song, 2014; Chen et al., 2020).

5.3.13 Proteases and Peptidases

Up until now, only three proteases and peptidases have been identified in the chicken egg white. The first one is a glutamyl aminopeptidase, isolated and characterized in 1990 (Petrović & Vitale, 1990). This enzyme is a dimer; its apparent molecular weight is around 320 kDa, and its pI 4.2. The molecular weight of each subunit, determined by electrophoresis under denaturing conditions, is 180 kDa. This acidic and glycosylated enzyme is strictly substrate-specific; it hydrolyzes peptide bonds of α-glutamyl residue, and to a lesser extent α-aspartyl residue, at the NH_2-end. Proteolysis is optimal at pH 7.6 and 60°C, and is activated by Ca^{2+} and Mg^{2+}. However, it is inhibited by Zn^{2+}, Co^{2+}, Ni^{2+}, and by calcium chelating agents such as EDTA. Glutamyl aminopeptidase remains active after storage for 60 min in pH ranging from 5 to 10, under freezing conditions and up to 45°C (Petrović & Vitale, 1990). This enzyme is also stable during long periods of egg storage (up to four months), characteristic also demonstrated by the second aminopeptidase isolated from chicken egg white, namely methionine aminopeptidase (Tripic et al., 1998).

Methionine aminopeptidase (MAP) is a smaller glycoprotein with an estimated molecular weight around 180 kDa in the native form, and around 120–130 kDa in denaturing conditions. Its pI is 4.15–4.40 (Škrtić & Vitale, 1994). This enzyme preferentially hydrolyzes peptide bonds of methionine at the NH_2-end, and basic or hydrophobic residues at the penultimate position, but also accepts many other amino acid residues. Accordingly, this enzyme has been classified as a methionine-preferring broad specificity aminopeptidase. MAP activity is optimal at pH 7.0–7.5, and at 50°C, and strongly depends on the presence of several metallic ions. MAP remains active after incubation at pH 5–9 for 1 h, but more than 50% irreversible inactivation occurs at acidic pH (<pH 4) and at alkaline pH (>pH 10). No loss of activity was measured after 30 min at 40°C when incubated at pH 8.0, but a 30% decrease was observed at 60°C.

The protease most recently identified in the chicken egg white, and in the yolk as well, is the matrix metalloprotease 2 (MMP-2), or, to be more precise, its 72 kDa precursor, found in the form of a complex with one of its inhibitors, the tissue inhibitor of metallo-proteinases TIMP-2. Indeed, the processing of pro-MMP-2 into the mature form MMP-2 will not occur *in vivo* in fresh eggs. The gelatinolytic activity of pro-MMP-2 in the egg white did not change during egg storage at 4°C and 20°C. However, it drastically and rapidly decreased at 37.8°C, leading to no detectable activity only after four days, likely due to the inhibition or proteolytic degradation by yolk components (Réhault-Godbert et al., 2008). MMP-2 proteins are highly conserved across species, mainly involved in tissue remodeling. Hence, chicken MMP-2 plays essential role during the embryo development (Cai & Brauer, 2002). Although its role in the egg remains unclear, MMP-2 may have a local action in the early embryonic development (Réhault-Godbert et al., 2008).

5.3.14 Other Minor Proteins

In view of the steady increase in the performances of proteomics analysis techniques, many new minor proteins and peptides have been identified in the egg white while it is likely that many still remain to be discovered. For instance, and among the most recent studies, Arena et al. (2020) identified 506 peptides, many associated with various hypothetical bioactivities. Guyot et al. (2016) identified three new potential antibacterial proteins (vitelline membrane outer layer protein 1, β-microseminoprotein-like, and pleiotrophin), while Mann (2007) identified for the first time in the egg white 54 proteins, among which ovosecretoglobin and gallin. Seventy-nine new proteins were similarly identified a few years later (Mann & Mann, 2011). Many minor phosphoproteins, some of them never identified before, have been also reported by Sun et al. (2020), who used an innovative analysis strategy.

5.4 Minerals and Vitamins

The mineral and vitamin content of the egg white highly varies with the authors, thus reflecting the impact of many factors on these components, particularly zootechnical factors such as the nutrition of laying hens. Therefore, mineral and vitamin concentrations may highly differ depending on the eggs analyzed, excluding eventual differences due to the variability in analytical methods. As an illustration of this point, the average, minimal and maximal contents for minerals and vitamins, gathered from the French ANSES-Ciqual homepage, are presented in Tables 5.3 and 5.4, respectively.

The data indicate that chicken egg white is rich in sodium, potassium, phosphorus, magnesium and calcium (Table 5.3), in accordance with data reported by Réhault-Godbert et al. (2019), and available from the USDA National Nutrient Database. The values reported by these authors are actually very close to those presented in Table 5.3. Egg white also contains some trace minerals of interest from a nutrition point of view, such as copper, iron, zinc and selenium.

The egg white, as well as the yolk, totally lacks vitamin C. Lipophilic vitamins are also absent in the egg white since it is a lipid-free medium. However, it contains several hydrophilic vitamins, some of which at high concentrations (Vitamins B2, B3, and B5) (Table 5.4). As for minerals, the data presented here are consistent with values reported by Réhault-Godbert et al. (2019), who in addition reported the presence of 5.7–7.9 μg/100 g of vitamin B8, or biotin in the egg white.

More details on the minerals and vitamins content of the egg white content is presented in Chapter 7.

TABLE 5.3
Mineral Content in Chicken Egg White

	Content in Raw Egg White (mg/100 g)		
Mineral	**Average Value**	**Minimum Value**	**Maximum Value**
Calcium	6	5	9
Copper	0.023	0.019	0.037
Iodine	0.0019	0.0005	0.0052
Iron	0.05	0.02	0.13
Magnesium	11	10	12
Manganese	0.011	-	-
Phosphorus	15	13	17
Potassium	163	148	173
Selenium	0.006	-	-
Sodium	166	159	179
Zinc	0.03	0.01	0.07

Source: Retrieved on May 26, 2022, from the Ciqual homepage: https://ciqual.anses.fr/.

TABLE 5.4
Vitamin Content in Chicken Egg White

	Content in Raw Egg White (µg/100 g)		
Vitamin	**Average Value**	**Minimum Value**	**Maximum Value**
Vitamin B1 or thiamin	4	-	-
Vitamin B2 or riboflavin	440	290	670
Vitamin B3 or niacin	93	70	110
Vitamin B5 or pantothenic acid	200	190	-
Vitamin B6	3.5	2	5
Vitamin B9 or folate	5.5	2.6	11.8
Vitamin B12 or cobalamin	0.09	-	-

Source: Retrieved on May 26, 2022, from the Ciqual homepage: https://ciqual.anses.fr/.

5.5 Conclusion

In this chapter, we highlighted that the apparent simplicity of the egg white (an aqueous solution of proteins) hides actually a considerable variety of proteins with some remarkably rich and diverse physicochemical, structural, and biological features. In particular, the number of proteins with antimicrobial properties, either proven or putative, is noteworthy, and many other properties have also been reported. Hence, beyond its natural protective role for the chick embryo, the egg white should be considered also as a natural source of many bioactive components (see Chapter 22). Moreover, despite the tremendous progress made over recent years to identify and characterize more minor egg white proteins, thanks not least to the development of the "omics" technologies, it is obvious that new proteins will continue to be discovered in the future, some with potential new bioactivities. In that regard, the egg white remains a promising object of studies, whether to improve our understanding of the chick embryogenesis process, or to identify molecules with applications in the medical or biological fields.

REFERENCES

Abdallah, F. B., & Chahine, J.-M. E. H. (1999). Transferrins, the mechanism of iron release by ovotransferrin. *European Journal of Biochemistry, 263*(3), 912–920. https://doi.org/10.1046/j.1432-1327.1999.00596.x

Ahlroth, M. K., Kola, E. H., Kulomaa, M. S., Ewald, D., Masabanda, J., Sazanov, A., & Fries, R. (2000). Characterization and chromosomal localization of the chicken avidin gene family: Characterization of the chicken avidin gene family. *Animal Genetics, 31*(6), 367–375. https://doi.org/10.1046/j.1365-2052.2000.00681.x

Akazawa, T., Ogawa, M., & Hayakawa, S. (2019). Migration of chicken egg-white protein ovalbumin-related protein X and its alteration in heparin-binding affinity during embryogenesis of fertilized egg. *Poultry Science, 98*(10), 5100–5108. https://doi.org/10.3382/ps/pez335

Akazawa, T., Ogawa, M., Hayakawa, S., Hirata, M., & Niwa, T. (2018). Structural change of ovalbumin-related protein X by alkali treatment. *Poultry Science, 97*(5), 1730–1737. https://doi.org/10.3382/ps/pey024

Allen, S., Stevens, L., Duncan, D., Kelly, S. M., & Price, N. C. (1992). Unfolding and refolding of hen egg-white riboflavin binding protein. *International Journal of Biological Macromolecules, 14*(6), 333–337. https://doi.org/10.1016/S0141-8130(05)80074-1

Amoresano, A., Brancaccio, A., Andolfo, A., Perduca, M., Monaco, H. L., & Marino, G. (1999). The carbohydrates of the isoforms of three avian riboflavin-binding protein. *European Journal of Biochemistry, 263*(3), 849–858. https://doi.org/10.1046/j.1432-1327.1999.00570.x

Anastasi, A., Brown, M. A., Kembhavi, A. A., Nicklin, M. J. H., Sayers, C. A., Sunter, D. C., & Barrett, A. J. (1983). Cystatin, a protein inhibitor of cysteine proteinases. Improved purification from egg white, characterization, and detection in chicken serum. *Biochemical Journal, 211*(1), 129–138. https://doi.org/10.1042/bj2110129

Apostol, I., Giletto, A., Komiyama, T., Zhang, W. L., & Laskowski, M. J. (1993). Amino acid sequences of ovomucoid third domains from 27 additional species of birds. *Journal of Protein Chemistry, 12*(4), 419–433. https://doi.org/10.1007/BF01025042

Aramini, J. M., & Vogel, H. J. (1998). Quadrupolar metal ion NMR studies of metalloproteins. *Biochemistry and Cell Biology, 76*, 210–222. https://doi.org/10.1139/o98-037

Arena, S., Renzone, G., & Scaloni, A. (2020). A multi-approach peptidomic analysis of hen egg white reveals novel putative bioactive molecules. *Journal of Proteomics, 215*, 103646. https://doi.org/10.1016/j.jprot.2020.103646

Asao, T., Takahashi, K., & Tashiro, M. (1998). Interaction of second and third domains of Japanese quail ovomucoid with ten mammalian trypsins. *Biochimica et Biophysica Acta (BBA) - Protein Structure and Molecular Enzymology, 1387*(1–2), 415–421. https://doi.org/10.1016/S0167-4838(98)00155-1

Asperger, A., Marx, K., Albers, C., Molin, L., & Pinato, O. (2015). Low abundant *N* -linked glycosylation in hen egg white lysozyme is localized at nonconsensus sites. *Journal of Proteome Research*, *14*(6), 2633–2641. https://doi.org/10.1021/acs.jproteome.5b00175

Bailly, C., Bodet-Milin, C., Rousseau, C., Faivre-Chauvet, A., Kraeber-Bodéré, F., & Barbet, J. (2017). Pretargeting for imaging and therapy in oncological nuclear medicine. *EJNMMI Radiopharmacy and Chemistry*, *2*, 6. https://doi.org/10.1186/s41181-017-0026-8

Bayer, E. A., De Meester, F., Kulik, T., & Wilchek, M. (1995). Preparation of deglycosylated egg white avidin. *Applied Biochemistry and Biotechnology*, *53*(1), 1–9. https://doi.org/10.1007/BF02783477

Becvar, J., & Palmer, G. (1982). The binding of flavin derivatives to the riboflavin-binding protein of egg white. A kinetic and thermodynamic study. *Journal of Biological Chemistry*, *257*(10), 5607–5617. https://doi.org/10.1016/S0021-9258(19)83821-8

Benarafa, C., & Remold-O'Donnell, E. (2005). The ovalbumin serpins revisited: Perspective from the chicken genome of clade B serpin evolution in vertebrates. *Proceedings of the National Academy of Sciences*, *102*(32), 11367–11372. https://doi.org/10.1073/pnas.0502934102

Benedé, S., Ruiz-Valdepeñas Montiel, V., Povedano, E., Villalba, M., Mata, L., Galán-Malo, P., & Pingarrón, J. M. (2018). Fast amperometric immune platform for ovomucoid traces determination in fresh and baked foods. *Sensors and Actuators B*, *265*, 421–428. https://doi.org/10.1016/j.snb.2018.03.075

Berardinelli, A., Ragni, L., Giunchi, A., Gradari, P., & Guarnieri, A. (2008). Physical-mechanical modifications of eggs for food-processing during storage. *Poultry Science*, *87*(10), 2117–2125. https://doi.org/10.3382/ps.2007-00216

Beveridge, T., & Nakai, S. (1975). Effects of sulphydryl blocking on the thinning of egg white. *Journal of Food Science*, *40*(4), 864–868. https://doi.org/10.1111/j.1365-2621.1975.tb00577.x

Bílková, B., Świderská, Z., Zita, L., Laloë, D., Charles, M., Beneš, V., & Vinkler, M. (2018). Domestic fowl breed variation in egg white protein expression: Application of proteomics and transcriptomics. *Journal of Agricultural and Food Chemistry*, *66*(44), 11854–11863. https://doi.org/10.1021/acs.jafc.8b03099

Bingle, C. D., & Craven, C. J. (2004). Meet the relatives: A family of BPI- and LBP-related proteins. *Trends in Immunology*, *25*(2), 53–55. https://doi.org/10.1016/j.it.2003.11.007

Björk, I., & Ylinenjaervi, K. (1990). Interaction between chicken cystatin and the cysteine proteinases actinidin, chymopapain a, and ficin. *Biochemistry*, *29*(7), 1770–1776. https://doi.org/10.1021/bi00459a016

Blake, C. C. F., Koenig, D. F., Mair, G. A., North, A. C. T., Phillips, D. C., & Sarma, V. R. (1965). Structure of hen egg-white lysozyme. a three-dimensional fourier synthesis at 2 a Resolution. *Nature*, *4986*, 757–761. https://doi.org/10.1038/206757a0

Bode, W., Engh, R., Musil, D., Thiele, U., Huber, R., Karshikov, A., & Turk, V. (1988). The 2.0 a X-ray crystal structure of chicken egg white cystatin and its possible mode of interaction with cysteine proteinases. *The EMBO Journal*, *7*(8), 2593–2599. https://doi.org/10.1002/j.1460-2075.1988.tb03109.x

Brandts, J. F., & Lin, L. N. (1990). Study of strong to ultralight protein interactions using differential scanning calorimetry. *Biochemistry*, *29*(29), 6927–6940. https://doi.org/10.1021/bi00481a024

Bruch, R. C., & White, H. B. (1982). Compositional and structural heterogeneity of avidin glycopeptides. *Biochemistry*, *21*(21), 5334–5341. https://doi.org/10.1021/bi00264a033

Cai, D. H., & Brauer, P. R. (2002). Synthetic matrix metalloproteinase inhibitor decreases early cardiac neural crest migration in chicken embryos. *Developmental Dynamics*, *224*(4), 441–449. https://doi.org/10.1002/dvdy.10129

Callewaert, L., & Michiels, C. W. (2010). Lysozymes in the animal kingdom. *Journal of Biosciences*, *35*(1), 127–160. https://www.ias.ac.in/article/fulltext/jbsc/035/01/0127-0160

Canfield, R. E. (1963). The amino acid sequence of egg white lysozyme. *Journal of Biological Chemistry*, *238*(8), 2698–2707. https://doi.org/10.1016/S0021-9258(18)67888-3

Catterall, J. F., Stein, J. P., Kristo, P., Means, A. R., & O'Malley, B. W. (1980). Primary sequence of ovomucoid messenger RNA as determined from cloned complementary DNA. *Journal of Cell Biology*, *87*(2), 480–487. https://doi.org/10.1083/jcb.87.2.480

Cavallero, G. J., Landoni, M., & Couto, A. S. (2020). In depth N-glycoproteomics shows glyco-features of chicken egg white. *Food Bioscience*, *35*, 100590. https://doi.org/10.1016/j.fbio.2020.100590

Chen, B., Liang, G., Zhu, X., Tan, Y., Xu, J., Wu, H., & Liu, S. (2020). Gene Expression profiling in ovaries and association analyses reveal HEP21 as a candidate Gene for sexual maturity in chickens. *Animals*, *10*(2), 181. https://doi.org/10.3390/ani10020181

Ciuraszkiewicz, J., Olczak, M., & Wątorek, W. (2006). Isolation, cloning and sequencing of transferrins from red-eared turtle, African ostrich, and turkey. *Comparative Biochemistry and Physiology Part B: Biochemistry and Molecular Biology*, *144*(3), 301–310. https://doi.org/10.1016/j.cbpb.2006.03.003

Colbert, D. A., Knoll, B. J., Woo, S. L. C., Mace, M. L., Tsai, M.-J., & O'Malley, B. W. (1980). Differential hormonal responsiveness of the ovalbumin gene and its pseudogenes in the chick oviduct. *Biochemistry*, *19*(24), 5586–5592. https://doi.org/10.1021/bi00565a020

Cooke, S. K., & Sampson, H. A. (1997). Allergenic properties of ovomucoid in man. *The Journal of Immunology*, *159*(4), 2026–2032. http://www.jimmunol.org/content/159/4/2026

Correnti, C., Clifton, M. C., Abergel, R. J., Allred, B., Hoette, T. M., Ruiz, M., & Strong, R. K. (2011). Galline ex-FABP is an antibacterial siderocalin and a lysophosphatidic acid sensor functioning through dual ligand specificities. *Structure*, *19*(12), 1796–1806. https://doi.org/10.1016/j.str.2011.09.019

Da Silva, M., Beauclercq, S., Harichaux, G., Labas, V., Guyot, N., Gautron, J., & Rehault-Godbert, S. (2015). The family secrets of avian egg-specific ovalbumin and its related proteins Y and X. *Biology of Reproduction*. https://doi.org/10.1095/biolreprod.115.130856

Dabich, D., Yurewicz, E. C., & Battel, V. A. (1993). Deglycosylation with trifluoromethanesulfonic acid differentially affects inhibitor activities of turkey ovomucoid. *Biochimica et Biophysica Acta (BBA) - Protein Structure and Molecular Enzymology*, *1164*(1), 47–53.

Das, B. K., Agarwal, S. K., & Khan, M. Y. (1991). Unfolding-refolding behaviour of chicken egg white ovomucoid and its correlation with the three domain structure of the protein.

Biochimica et Biophysica Acta (BBA) - Protein Structure and Molecular Enzymology, 1076(3), 343–350.

Dekker, J., Rossen, J. W. A., Büller, H. A., & Einerhand, A. W. C. (2002). The MUC family: An obituary. *Trends in Biochemical Sciences, 27*(3), 126–131. https://doi.org/10.1016/S0968-0004(01)02052-7

DeKoster, G. T., & Robertson, A. D. (1997). Thermodynamics of unfolding for kazal-type serine protease inhibitors: Entropic stabilization of ovomucoid first domain by glycosylation. *Biochemistry, 36*(8), 2323–2331. https://doi.org/10.1021/bi962580b

DeLange, R. J., & Huang, T.-S. (1971). Egg white avidin. *Journal of Biological Chemistry, 246*(3), 698–709. https://doi.org/10.1016/S0021-9258(18)62469-X

Derde, M., Guérin-Dubiard, C., Lechevalier, V., Cochet, M.-F., Jan, S., Baron, F., & Nau, F. (2014). Dry-heating of lysozyme increases its activity against Escherichia coli membranes. *Journal of Agricultural and Food Chemistry, 62*(7), 1692–1700. https://doi.org/10.1021/jf405155p

Derde, M., Vié, V., Walrant, A., Sagan, S., Lechevalier, V., Guérin-Dubiard, C., & Nau, F. (2017). Antimicrobial activity of lysozyme isoforms: Key molecular features. *Biopolymers, 107*(12), e23040. https://doi.org/10.1002/bip.23040

Descalzi Cancedda, F., Dozin, B., Zerega, B., Cermelli, S., & Cancedda, R. (2000). Ex-FABP: A fatty acid binding lipocalin developmentally regulated in chicken endochondral bone formation and myogenesis. *Biochimica et Biophysica Acta (BBA) - Protein Structure and Molecular Enzymology, 1482*(1–2), 127–135. https://doi.org/10.1016/S0167-4838(00)00159-X

Descalzi Cancedda, F., Dozin, B., Zerega, B., Cermelli, S., & Cancedda, R. (2001). Extracellular fatty acid binding protein (ex-FABP) is a stress protein expressed during chondrocyte and myoblast differentiation. *Osteoarthritis Cartilage, 9*(Suppl A), S118–S122.

Descalzi Cancedda, F., Malpeli, M., Gentili, C., Di Marzo, V., Bet, P., Carlevaro, M., & Cancedda, R. (1996). The developmentally regulated avian Ch21 lipocalin is an extracellular fatty acid-binding protein. *Journal of Biological Chemistry, 271*(33), 20163–20169. https://doi.org/10.1074/jbc.271.33.20163

Désert, C., Guérin-Dubiard, C., Nau, F., Jan, G., Val, F., & Mallard, J. (2001). Comparison of different electrophoretic separations of hen egg white proteins. *Journal of Agricultural and Food Chemistry, 49*(10), 4553–4561. https://doi.org/10.1021/jf001423n

Ding, X., Du, J., Zhang, K., Bai, S., Zeng, Q., Peng, H., & Wang, J. (2020). Tandem mass tag-based quantitative proteomics analysis and gelling properties in egg albumen of laying hens feeding tea polyphenols. *Poultry Science, 99*(1), 430–440. https://doi.org/10.3382/ps/pez523

Dombre, C., Guyot, N., Moreau, T., Monget, P., Da Silva, M., Gautron, J., & Réhault-Godbert, S. (2017). Egg serpins: The chicken and/or the egg dilemma. *Seminars in Cell & Developmental Biology, 62*, 120–132. https://doi.org/10.1016/j.semcdb.2016.08.019

Donovan, J. W., Davis, J. G., & White, L. M. (1970). Chemical and physical characterization of ovomucin, a sulfated glycoprotein complex from chicken eggs. *Biochimica et Biophysica Acta (BBA) - Protein Structure, 207*(1), 190–201. https://doi.org/10.1016/0005-2795(70)90151-0

Donovan, J. W., & Mapes, C. J. (1976). A differential scanning calorimetric study of conversion of ovalbumin to S-ovalbumin in eggs. *Journal of the Science of Food and Agriculture, 27*(2), 197–204. https://doi.org/10.1002/jsfa.2740270220

Donovan, J. W., & Ross, K. D. (1973). Increase in the stability of avidin produced by binding of biotin. Differential scanning calorimetric study of denaturation by heat. *Biochemistry, 12*(3), 512–517. https://doi.org/10.1021/bi00727a024

Downing, T., O'Farrelly, C., Bhuiyan, A. K., Silva, P., Naqvi, A. N., Sanfo, R., & Bradley, D. G. (2010). Variation in chicken populations may affect the enzymatic activity of lysozyme. *Animal Genetics, 41*(2), 213–217. https://doi.org/10.1111/j.1365-2052.2009.01974.x

Duan, X., Li, M., Ji, B., Liu, X., & Xu, X. (2017). Effect of fertilization on structural and molecular characteristics of hen egg ovalbumin. *Food Chemistry, 221*, 1340–1345. https://doi.org/10.1016/j.foodchem.2016.11.021

Egge, H., Peter-Katalinić, J., Paz-Parente, J., Strecker, G., Montreuil, J., & Fournet, B. (1983). Carbohydrate structures of hen ovomucoid: A mass spectrometric analysis. *FEBS Letters, 156*(2), 357–362. https://doi.org/10.1016/0014-5793(83)80528-6

Ellison, D., Beynon, R. J., Hinton, J., & Hubbard, S. J. (1995). Limited proteolysis of native proteins: The interaction between avidin and proteinase K. *Protein Science, 4*(7), 1337–1345. https://doi.org/10.1002/pro.5560040709

Fleming, A. (1922). On a remarkable bacteriolytic element found in tissues and secretions. *Royal Society Proceedings B, 93*, 306–317. https://doi.org/10.1098/rspb.1922.0023

Flower, D. R., North, A. C. T., & Sansom, C. E. (2000). The lipocalin protein family: Structural and sequence overview. *Biochimica et Biophysica Acta (BBA) - Protein Structure and Molecular Enzymology, 1482*(1–2), 9–24. https://doi.org/10.1016/S0167-4838(00)00148-5

Fossum, K., & Whitaker, J. R. (1968). Ficin and papain inhibitor from chicken egg white. *Archives of Biochemistry and Biophysics, 125*(1), 367–375. https://doi.org/10.1016/0003-9861(68)90672-3

Fothergill, L. A., & Fothergill, J. E. (1970). Thiol and disulphide contents of hen ovalbumin. C-terminal sequence and location of disulphide bond. *Biochemical Journal, 116*(4), 555–561. https://doi.org/10.1042/bj1160555

Fu, D. D., Wang, Q. H., Ma, M. H., Ma, Y. X., & Vong, C. N. (2019). Prediction and visualisation of S-ovalbumin content in egg whites using hyperspectral images. *International Journal of Food Properties, 22*(1), 1077–1086. https://doi.org/10.1080/10942912.2019.1628775

Füssl, F., Criscuolo, A., Cook, K., Scheffler, K., & Bones, J. (2019). Cracking proteoform complexity of ovalbumin with anion-exchange Chromatography–High-Resolution mass spectrometry under native conditions. *Journal of Proteome Research, 18*(10), 3689–3702. https://doi.org/10.1021/acs.jproteome.9b00375

Galla, R., Grisenti, P., Farghali, M., Saccuman, L., Ferraboschi, P., & Uberti, F. (2021). Ovotransferrin supplementation improves the iron absorption: An in vitro gastro-intestinal model.*Biomedicines, 9*(11), 1543. https://doi.org/10.3390/biomedicines9111543

Galzie, Z., Abul Qasim, M., & Salahuddin, A. (1996). Isolation and characterization of domain I of ovoinhibitor.

Biochimica et Biophysica Acta (BBA) - Protein Structure and Molecular Enzymology, *1293*(1), 113–121. https://doi.org/10.1016/0167-4838(95)00234-0

Garénaux, A., Houle, S., Folch, B., Dallaire, G., Truesdell, M., Lépine, F., & Dozois, C. M. (2013). Avian lipocalin expression in chickens following Escherichia coli infection and inhibition of avian pathogenic Escherichia coli growth by ex-FABP. *Veterinary Immunology and Immunopathology*, *152*(1–2), 156–167. https://doi.org/10.1016/j.vetimm.2012.09.018

Garratt, R. C., Evans, R. W., Hasnain, S. S., Lindley, P. F., & Sarra, R. (1991). X.a.f.s. Studies of chicken dicupric ovotransferrin. *Biochemical Journal*, *280*(1), 151–155. https://doi.org/10.1042/bj2800151

Geng, F., Huang, X., Majumder, K., Zhu, Z., Cai, Z., & Ma, M. (2015). Mass spectrometry and two-dimensional electrophoresis to characterize the glycosylation of hen egg white ovomacroglobulin. *Journal of Agricultural and Food Chemistry*, *63*(37), 8209–8215. https://doi.org/10.1021/acs.jafc.5b02618

Geng, F., Wang, J., Liu, D., Jin, Y., & Ma, M. (2017). Identification of *n* -glycosites in chicken egg white proteins using an omics strategy. *Journal of Agricultural and Food Chemistry*, *65*(26), 5357–5364. https://doi.org/10.1021/acs.jafc.7b01706

Gentili, C., Tutolo, G., Zerega, B., Di Marco, E., Cancedda, R., & Descalzi Cancedda, F. (2005). Acute phase lipocalin ex-FABP is involved in heart development and cell survival. *Journal of Cellular Physiology*, *202*, 683–689. https://doi.org/10.1002/jcp.20165

Gertler, A., & Ben-Valid, I. (1980). Stoichiometry of interaction of chicken ovoinhibitor with pancreatic trypsin, chymotrypsin and elastase I. *European Journal of Biochemistry*, *110*(2), 571–577. https://doi.org/10.1111/j.1432-1033.1980.tb04900.x

Giannoni, P., Zambotti, A., Pagano, A., Cancedda, R., & Dozin, B. (2004). Differentiation-dependent activation of the extracellular fatty acid binding protein (ex-FABP) gene during chondrogenesis. *Journal of Cellular Physiology*, *198*(1), 144–154. https://doi.org/10.1002/jcp.10405

Giansanti, F., Leboffe, L., Pitari, G., Ippoliti, R., & Antonini, G. (2012). Physiological roles of ovotransferrin. *Biochimica et Biophysica Acta (BBA) - General Subjects*, *1820*(3), 218–225. https://doi.org/10.1016/j.bbagen.2011.08.004

Giansanti, F., Rossi, P., Massucci, M. T., Botti, D., Antonini, G., Valenti, P., & Seganti, L. (2002). Antiviral activity of ovotransferrin discloses an evolutionary strategy for the defensive activities of lactoferrin. *Biochemistry and Cell Biology*, *80*(1), 125–130. https://doi.org/10.1139/o01-208

Green, N. M. (1963). Avidin. 3. The nature of the biotin-binding site. *Biochemical Journal*, *89*, 599–609.

Green, N. M. (1975). Avidin. *Advances in Protein Chemistry*, *29*, 85–133.

Grubb, A., Lofberg, H., & Barrett, A. J. (1984). The disulphide bridges of human cystatin c (y-trace) and chicken cystatin. *FEBS Letters*, *170*(2), 370–374.

Gu, J., Matsuda, T., Nakamura, R., Ishiguro, H., Ohkubo, I., Sasaki, M., & Takahashi, N. (1989). Chemical deglycosylation of hen ovomucoid: Protective effect of carbohydrate moiety on tryptic hydrolysis and heat denaturation. *The Journal of Biochemistry*, *106*(1), 66–70. https://doi.org/10.1093/oxfordjournals.jbchem.a122821

Guérin-Dubiard, C., Pasco, M., Hietanen, A., Quiros del Bosque, A., Nau, F., & Croguennec, T. (2005). Hen egg white fractionation by ion-exchange chromatography. *Journal of Chromatography A*, *1090*(1–2), 58–67. https://doi.org/10.1016/j.chroma.2005.06.083

Guérin-Dubiard, C., Pasco, M., Mollé, D., Désert, C., Croguennec, T., & Nau, F. (2006). Proteomic analysis of hen egg white. *Journal of Agricultural and Food Chemistry*, *54*(11), 3901–3910. https://doi.org/10.1021/jf0529969

Guyot, N., Labas, V., Harichaux, G., Chessé, M., Poirier, J.-C., Nys, Y., & Réhault-Godbert, S. (2016). Proteomic analysis of egg white heparin-binding proteins: Towards the identification of natural antibacterial molecules. *Scientific Reports*, *6*(1), 27974. https://doi.org/10.1038/srep27974

György, P., Rose, C. S., Eakin, R. E., Snell, E. E., & Williams, R. J. (1941). Egg white injury as the result of non-absorption or inactivation of biotin. *Science*, *93*, 477–478. https://doi.org/10.1126/science.93.2420.477

Haginaka, J. (2000). Enantiomer separation of drugs by capillary electrophoresis using proteins as chiral selectors. *Journal of Chromatography A*, *875*(1–2), 235–254. https://doi.org/10.1016/S0021-9673(99)01168-1

Hamazume, Y., Mega, T., & Ikenaka, T. (1984). Characterization of hen egg white- and yolk-riboflavin binding proteins and amino acid sequence of egg white-riboflavin binding protein. *The Journal of Biochemistry*, *95*(6), 1633–1644. https://doi.org/10.1093/oxfordjournals.jbchem.a134776

Hamazume, Y., Mega, T., & Ikenaka, T. (1987). Positions of disulfide bonds in riboflavin-binding protein of hen egg white. *Journal of Biochemistry*, *101*(1), 217–223. https://doi.org/10.1093/oxfordjournals.jbchem.a121894

Hatta, H., Nomura, M., Takahashi, N., & Hirose, M. (2001). Thermostabilization of ovalbumin in a developing egg by an alkalinity-regulated, two-step process. *bioscience, Biotechnology, and Biochemistry*, *65*(9), 2021–2027. https://doi.org/10.1271/bbb.65.2021

Hayakawa, S., Kondo, H., Nakamura, R., & Sato, Y. (1983). Effect of β -ovomucin on the solubility of α -ovomucin and further inspection of the structure of ovomucin complex in thick egg white. *Agricultural and Biological Chemistry*, *47*(4), 815–820. https://doi.org/10.1080/00021369.1983.10865728

He, T., Zhang, H., Wang, J., Wu, S., Yue, H., & Qi, G. (2017). Proteomic comparison by iTRAQ combined with mass spectrometry of egg white proteins in laying hens (*Gallus gallus*) fed with soybean meal and cottonseed meal. *PLoS One*, *12*(8), e0182886. https://doi.org/10.1371/journal.pone.0182886

Heilig, R., Mursaskowsky, R., Kloepfer, C., & Mandel, J. L. (1982). The ovalbumin gene family: Complete sequence and structure of the Y gene. *Nucleic Acids Research*, *10*(14), 4363–4382. https://doi.org/10.1093/nar/10.14.4363

Heilig, R., Perrin, F., Gannon, F., Mandel, J. L., & Chambon, P. (1980). The ovalbumin Gene family: Structure of the x Gene and evolution of duplicated split genes. *Cell*, *20*(3), 625–637. https://doi.org/10.1016/0092-8674(80)90309-8

Hiidenhovi, J., Ek-Kommonen, C., Järvenpää, E., Huopalahti, R., & Ryhänen, E.-L. (2016). Characterization of crude ovomucins obtained from various egg white layers. *European Food Research and Technology*, *242*(2), 233–246. https://doi.org/10.1007/s00217-015-2535-6

Hiller, Y., Gershoni, J. M., Bayer, E. A., & Wilchek, M. (1987). Biotin binding to avidin. Oligosaccharide side chain not required for ligand association.*Biochemical Journal*, *248*(1), 167–171. https://doi.org/10.1042/bj2480167

Hirose, J., Doi, Y., Kitabatake, N., & Narita, H. (2006). Ovalbumin-related Gene y protein bears carbohydrate chains of the ovomucoid type. *bioscience, Biotechnology, and Biochemistry*, *70*(1), 144–151. https://doi.org/10.1271/bbb.70.144

Hu, S., Qiu, N., Liu, Y., Zhao, H., Gao, D., Song, R., & Ma, M. (2016). Identification and comparative proteomic study of quail and duck egg white protein using 2-dimensional gel electrophoresis and matrix-assisted laser desorption/ionization time-of-flight tandem mass spectrometry analysis. *Poultry Science*, *95*(5), 1137–1144. https://doi.org/10.3382/ps/pew033

Huang, T.-S., & DeLange, R. J. (1971). Egg white avidin. *Journal of Biological Chemistry*, *246*(3), 686–697. https://doi.org/10.1016/S0021-9258(18)62468-8

Huang, Q., Liu, L., Wu, Y., Huang, X., Wang, G., Song, H., & Luo, P. (2022). Mechanism of differences in characteristics of thick/thin egg whites during storage: Physicochemical, functional and molecular structure characteristics analysis. *Food Chemistry*, *369*, 130828. https://doi.org/10.1016/j.foodchem.2021.130828

Huntington, J. A., Gettins, P. G. W., & Patston, P. A. (1995). S-ovalbumin, an ovalbumin conformer with properties analogous to those of loop-inserted serpins. *Protein Science*, *4*(4), 613–621. https://doi.org/10.1002/pro.5560040403

Huntington, J. A., & Stein, P. E. (2001). Structure and properties of ovalbumin. *Journal of Chromatography B: Biomedical Sciences and Applications*, *756*(1–2), 189–198. https://doi.org/10.1016/S0378-4347(01)00108-6

Hwang, H. S., Kim, B. S., Park, H., Park, H.-Y., Choi, H.-D., & Kim, H. H. (2014). Type and branched pattern of N -glycans and their structural effect on the chicken egg allergen ovotransferrin: A comparison with ovomucoid. *Glycoconjugate Journal*, *31*, 41–50. https://doi.org/10.1007/s10719-013-9498-2

Ibrahim, B. S., & Pattabhi, V. (2004). Crystal structure of trypsin–turkey egg white inhibitor complex.*Biochemical and Biophysical Research Communications*, *313*, 8–16. https://doi.org/10.1016/j.bbrc.2003.11.082

Ibrahim, H. R., Thomas, U., & Pellegrini, A. (2001). A helix-loop-helix peptide at the upper lip of the active site cleft of lysozyme confers potent antimicrobial activity with membrane permeabilization action. *Journal of Biological Chemistry*, *276*(47), 43767–43774. https://doi.org/10.1074/jbc.M106317200

Ishihara, H., Takahashi, N., Ito, J., Takeuchi, E., & Tejima, S. (1981). Either high-mannose-type or hybrid-type oligosaccharide is linked to the same asparagine residue in ovalbumin. *Biochimica et Biophysica Acta (BBA) - Protein Structure*, *669*(2), 216–221. https://doi.org/10.1016/0005-2795(81)90243-9

Ishimaru, T., Ito, K., Tanaka, M., Tanaka, S., & Matsudomi, N. (2011). The role of the disulfide bridge in the stability and structural integrity of ovalbumin evaluated by site-directed mutagenesis. *Bioscience, Biotechnology, and Biochemistry*, *75*(3), 544–549. https://doi.org/10.1271/bbb.100772

Jahejo, A. R., Bukhari, S. A. R., Rajput, N., Kalhoro, N. H., Leghari, I. H., Raza, S. H. A., & Tian, W. (2022). Transcriptome-based biomarker gene screening and evaluation of the extracellular fatty acid-binding protein (ex-FABP) on immune and angiogenesis-related genes in chicken erythrocytes of tibial dyschondroplasia. *BMC Genomics*, *23*(1), 323. https://doi.org/10.1186/s12864-022-08494-9

Jain, A., & Cheng, K. (2017). The principles and applications of avidin-based nanoparticles in drug delivery and diagnosis. *Journal of Controlled Release*, *245*, 27–40. https://doi.org/10.1016/j.jconrel.2016.11.016

Jeltsch, J.-M., & Chambon, P. (1982). The complete nucleotide sequence of the chicken ovotransferrin mRNA. *European Journal of Biochemistry*, *122*(2), 291–295. https://doi.org/10.1111/j.1432-1033.1982.tb05879.x

Jiang, L., Li, Y., Wang, L., Guo, J., Liu, W., Meng, G., & Sun, M. (2021). Recent insights into the prognostic and therapeutic applications of lysozymes. *Frontiers in Pharmacology*, *12*, 767642. https://doi.org/10.3389/fphar.2021.767642

Jiang, K., Wang, C., Sun, Y., Liu, Y., Zhang, Y., Huang, L., & Wang, Z. (2014). Comparison of chicken and pheasant ovotransferrin *n* -glycoforms via electrospray ionization mass spectrometry and liquid chromatography coupled with mass spectrometry. *Journal of Agricultural and Food Chemistry*, *62*(29), 7245–7254. https://doi.org/10.1021/jf501352j

Julien, L. A., Fau, C., Baron, F., Bonnassie, S., Guérin-Dubiard, C., Nau, F., & Andrews, S. C. (2020). The three lipocalins of egg-white: Only ex-FABP inhibits siderophore-dependent iron sequestration by salmonella enteritidis. *Frontiers in Microbiology*, *11*, 913. https://doi.org/10.3389/fmicb.2020.00913

Jung, A., Sippel, A. E., Grez, M., & Schütz, G. (1980). Exons encode functional and structural units of chicken lysozyme. *Proceedings of the National Academy of Sciences*, *77*(10), 5759–5763. https://doi.org/10.1073/pnas.77.10.5759

Kato, A., Kanemitsu, T., & Kobayashi, K. (1991). Inhibitory activity of ovomacroglobulin for pepsin and rennin. *Journal of Agricultural and Food Chemistry*, *39*(1), 41–43. https://doi.org/10.1021/jf00001a007

Kato, A., Nakamura, R., & Sato, Y. (1970). Studies on changes in stored shell eggs: Part V the difference in the chemical and physicochemical properties of ovomucin(B) between the thick and thin white. *Agricultural and Biological Chemistry*, *34*(6), 854–859. https://doi.org/10.1080/00021369.1970.10859697

Kato, A., Ogato, S., Matsudomi, N., & Kobayashi, K. (1981). Comparative study of aggregated and disaggregated ovomucin during egg white thinning. *Journal of Agricultural and Food Chemistry*, *29*(4), 821–823. https://doi.org/10.1021/jf00106a035

Kato, I., Schrode, J., Kohr, W. J., & Laskowski, M. (1987). Chicken ovomucoid: Determination of its amino acid sequence, determination of the trypsin reactive site, and preparation of all three of its domains. *Biochemistry*, *26*(1), 193–201. https://doi.org/10.1021/bi00375a027

Kawaguchi, Y., Yoneda, K., Torikata, T., & Araki, T. (2015). Asp48 function in the hydrogen-bonding network involving Asp52 of hen egg-white lysozyme. *Bioscience, Biotechnology, and Biochemistry*, *79*(2), 196–204. https://doi.org/10.1080/09168451.2014.963502

Kemps, B. J., Bamelis, F. R., Mertens, K., Decuypere, E. M., De Baerdemaeker, J. G., & De Ketelaere, B. (2010). The assessment of viscosity measurements on the albumen of consumption eggs as an indicator for freshness. *Poultry Science*, *89*(12), 2699–2703. https://doi.org/10.3382/ps.2008-00520

Ketterer, B. (1965). Ovoglycoprotein, a protein of hen's-egg white. *Biochemical Journal*, *96*(2), 372–376. https://doi.org/10.1042/bj0960372

Khorshidian, N., Khanniri, E., Koushki, M. R., Sohrabvandi, S., & Yousefi, M. (2022). An overview of antimicrobial activity of lysozyme and its functionality in cheese. *Frontiers in Nutrition*, *9*, 833618. https://doi.org/10.3389/fnut.2022.833618

Kim, J., & Choi, Y.-H. (2014). Differential abundance of egg white proteins in laying hens treated with corticosterone. *Journal of Agricultural and Food Chemistry*, *62*(51), 12346–12359. https://doi.org/10.1021/jf504469t

Kinoshita, K., Shimogiri, T., Ibrahim, H. R., Tsudzuki, M., Maeda, Y., & Matsuda, Y. (2016). Identification of TENP as the Gene encoding chicken egg white ovoglobulin G2 and demonstration of its high genetic variability in chickens. *PLoS One*, *11*(7), e0159571. https://doi.org/10.1371/journal.pone.0159571

Kinoshita-Kikuta, E., Kinoshita, E., & Koike, T. (2012). Separation and identification of four distinct serine-phosphorylation states of ovalbumin by Phos-tag affinity electrophoresis. *Electrophoresis*, *33*, 849–855. https://doi.org/10.1002/elps.201100518

Kint, S., & Tomimatsu, Y. (1979). A Raman difference spectroscopic investigation of ovalbumin and S-ovalbumin. *Biopolymers*, *18*, 1073–1079.

Kitagawa, T., Nishina, Y., Kyogoku, Y., Yamano, T., Ohishi, N., Takai-Suzuki, A., & Yagi, K. (1979). Resonance Raman spectra of carbon-13- and nitrogen-15-labeled riboflavin bound to egg-white flavoprotein. *Biochemistry*, *18*(9), 1804–1808. https://doi.org/10.1021/bi00576a026

Kojima, S., Takagi, N., Minagawa, T., Fushimi, N., & Miura, K. (1999). Effects of amino acid replacements around the reactive site of chicken ovomucoid domain 3 on the inhibitory activity toward chymotrypsin and trypsin. *Protein Engineering, Design and Selection*, *12*(10), 857–862. https://doi.org/10.1093/protein/12.10.857

Kozik, A. (1982). Disulfide bonds in egg-white riboflavin-binding protein - chemical reduction studies. *European Journal of Biochemistry*, *121*, 395–400.

Krise, K. M., & Milosavljevic, B. H. (2011). Mobility of molecules and ions solubilized in protein gels: Diffusion in the thick fraction of hen egg white. *Biomacromolecules*, *12*(6), 2351–2356. https://doi.org/10.1021/bm200417t

Kumosinski, T. F., Pessen, H., & Farrell, H. M. (1982). Structure and mechanism of action of riboflavin-binding protein: Small-angle X-ray scattering, sedimentation, and circular dichroism studies on the holo- and apoproteins. *Archives of Biochemistry and Biophysics*, *214*(2), 714–725. https://doi.org/10.1016/0003-9861(82)90078-9

Kupaj, P., Korus, M., Korzeniowska, M., & Kopec, W. (2020). Egg white cystatin – A review. *Acta Veterinaria*, *70*(4), 401–422. https://doi.org/10.2478/acve-2020-0031

Kurokawa, H., Dewan, J. C., Mikami, B., Sacchettini, J. C., & Hirose, M. (1999). Crystal structure of hen apo-ovotransferrin. *Journal of Biological Chemistry*, *274*(40), 28445–28452. https://doi.org/10.1074/jbc.274.40.28445

Kurokawa, H., Mikami, B., & Hirose, M. (1995). Crystal structure of diferric hen ovotransferrin at 2.4 Å Resolution. *Journal of Molecular Biology*, *254*(2), 196–207. https://doi.org/10.1006/jmbi.1995.0611

Laber, B., Krieglstein, K., Henschen, A., Kos, J., Turk, V., Huber, R., & Bode, W. (1989). The cysteine proteinase inhibitor chicken cystatin is a phosphoprotein. *FEBS Letters*, *248*(1–2), 162–168. https://doi.org/10.1016/0014-5793(89)80453-3

Larsen, L. B., Hammershoj, M., & Rasmussen, J. T. (1999). *Identification of Ch21, a developmentally regulated chicken embryo protein, in egg albumen* [Poster session] (Vol. II, pp. 61–67). In L. G. Cavalchini and D. Baroli (eds.), Proceedings of the VIII European Symposium on the Quality of Eggs and Egg Products (Vol. II), Associazione Italiana di Avicoltura Scientifica, September 19–23, Milano, Italy.

Laskowski, M., & Kato, I. (1980). Protein inhibitors of proteinases. *Annual Review of Biochemistry*, *49*(1), 593–626. https://doi.org/10.1146/annurev.bi.49.070180.003113

Lee, D. C., McKnight, G. S., & Palmiter, R. D. (1980). The chicken transferrin gene. Restriction endonuclease analysis of gene sequences in liver and oviduct DNA. *Journal of Biological Chemistry*, *255*(4), 1442–1450. https://doi.org/10.1016/S0021-9258(19)86050-7

Lee, M., Yang, C., Song, G., & Lim, W. (2020). Alterations in egg white-related genes expression in response to hormonal stimulation. *Reproduction*, *160*(5), 793–801. https://doi.org/10.1530/REP-20-0342

Legros, J., Jan, S., Bonnassie, S., Gautier, M., Croguennec, T., Pézennec, S., & Baron, F. (2021). The role of ovotransferrin in egg-white antimicrobial activity: A review. *Foods*, *10*, 823. https://doi.org/10.3390/foods10040823

Lim, W., & Song, G. (2014). Pivotal roles for hormonally regulated expression of the HEP21 gene in the reproductive tract of chickens for oviduct development and in ovarian carcinogenesis. *Domestic Animal Endocrinology*, *48*, 136–144. https://doi.org/10.1016/j.domaniend.2014.03.003

Liu, J., Liu, J., & Zhang, X. (2012). Production and characterization of ovalbumin and ovotransferrin from chicken egg white. *International Journal of Food Engineering*, *8*(2), 14. https://doi.org/10.1515/1556-3758.2742

Liu, W.-H., Means, G. E., & Feeney, R. E. (1971). The inhibitory properties of avian ovoinhibitors against proteolytic enzymes. *Biochimica et Biophysica Acta (BBA) - Protein Structure*, *229*(1), 176–185. https://doi.org/10.1016/0005-2795(71)90331-X

Liu, Y., Qiu, N., & Ma, M. (2015). Comparative proteomic analysis of egg white proteins during the rapid embryonic growth period by combinatorial peptide ligand libraries. *Poultry Science*, *94*(10), 2495–2505. https://doi.org/10.3382/ps/pev176

Liu, X., Wang, J., Huang, Q., Cheng, L., Gan, R., Liu, L., & Geng, F. (2020). Underlying mechanism for the differences in heat-induced gel properties between thick egg whites and thin egg whites: Gel properties, structure and quantitative proteome analysis. *Food Hydrocolloids*, *106*, 105873. https://doi.org/10.1016/j.foodhyd.2020.105873

Machleidt, W., Thiele, U., Laber, B., Assfalg-Machleidt, I., Esterl, A., Wiegand, G., & Bode, W. (1989). Mechanism of inhibition of papain by chicken egg white cystatin: Inhibition constants of N-terminally truncated forms and cyanogen bromide fragments of the inhibitor. *FEBS Letters*, *243*(2), 234–238. https://doi.org/10.1016/0014-5793(89)80135-8

Maehashi, K., Matano, M., Kondo, A., Yamamoto, Y., & Udaka, S. (2006). Riboflavin-binding protein exhibits selective sweet suppression toward protein sweeteners. *Chemical Senses*, *32*(2), 183–190. https://doi.org/10.1093/chemse/bjl048

Maehashi, K., Matano, M., Nonaka, M., Udaka, S., & Yamamoto, Y. (2008). Riboflavin-binding protein is a novel bitter inhibitor. *Chemical Senses*, *33*(1), 57–63. https://doi.org/10.1093/chemse/bjm062

Maehashi, K., Ueda, M., Matano, M., Takeuchi, J., Uchino, M., Kashiwagi, Y., & Watanabe, T. (2014). Biochemical and functional characterization of transiently expressed in neural precursor (TENP) protein in emu egg white. *Journal of Agricultural and Food Chemistry*, *62*(22), 5156–5162. https://doi.org/10.1021/jf5008117

Mahon, M. G., Lindstedt, K. A., Hermann, M., Nimpf, J., & Schneider, W. J. (1999). Multiple involvement of clusterin in chicken ovarian follicle development. *Journal of Biological Chemistry*, *274*(7), 4036–4044. https://doi.org/10.1074/jbc.274.7.4036

Mann, K. (2007). The chicken egg white proteome. *Proteomics*, *7*(19), 3558–3568. https://doi.org/10.1002/pmic.200700397

Mann, K., Gautron, J., Nys, Y., McKee, M. D., Bajari, T., Schneider, W. J., & Hincke, M. T. (2003). Disulfide-linked heterodimeric clusterin is a component of the chicken eggshell matrix and egg white. *Matrix Biology*, *22*(5), 397–407. https://doi.org/10.1016/S0945-053X(03)00072-6

Mann, K., & Mann, M. (2011). In-depth analysis of the chicken egg white proteome using an LTQ orbitrap velos. *Proteome Sci*, *9*(7), 10–1186.

Marchbank, T., Ten Bruggencate, S. J. M., & Playford, R. J. (2021). Protease inhibitors protect bovine colostrum or chicken egg growth factors from pancreatic enzyme digestion in AGS cells or colitic rats. *The Journal of Nutrition*, *151*(10), 3036–3044. https://doi.org/10.1093/jn/nxab197

Matsuda, T., Watanabe, K., & Nakamura, R. (1982). Immunochemical studies on thermal denaturation of ovomucoid. *Biochimica et Biophysica Acta (BBA) - Protein Structure and Molecular Enzymology*, *707*(1), 121–128. https://doi.org/10.1016/0167-4838(82)90404-6

Matsuda, T., Watanabe, K., & Nakamura, R. (1983). Ovomucoid and ovoinhibitor isolated from chicken egg white are immunologically cross-reactive. *Biochemical and Biophysical Research Communications*, *110*(1), 75–81. https://doi.org/10.1016/0006-291X(83)91262-7

Matsuda, T., Watanabe, K., & Sato, Y. (1981). Secondary structure prediction of chicken egg white ovomucoid. *Agricultural and Biological Chemistry*, *45*(2), 417–423. https://doi.org/10.1080/00021369.1981.10864526

Matsunaga, H., Sadakane, Y., & Haginaka, J. (2004). Identification of disulfide bonds and site-specific glycosylation in chicken α1-acid glycoprotein by matrix-assisted laser desorption ionization time-of-flight mass spectrometry. *Analytical Biochemistry*, *331*(2), 358–363. https://doi.org/10.1016/j.ab.2004.04.041

Mazurier, J., Metz-Boutigue, M.-H., Jollès, J., Spik, G., Montreuil, J., & Jollès, P. (1983). Human lactotransferrin: Molecular, functional and evolutionary comparisons with human serum transferrin and hen ovotransferrin. *Experientia*, *39*(2), 135–141. https://doi.org/10.1007/BF01958861

McReynolds, L., O'Malley, B. W., Nisbet, A. D., Fothergill, J. E., Givol, D., Fields, S., & Brownlee, G. G. (1978). Sequence of chicken ovalbumin mRNA. *Nature*, *273*(5665), 723–728. https://doi.org/10.1038/273723a0

Mega, T., Hamazume, Y., Nong, Y.-M., & Ikenaka, T. (1986). Studies on the methods for the determination of phosphorylation sites in highly phosphorylated peptides or proteins: phosphorylation sites of hen egg white riboflavin binding protein. *The Journal of Biochemistry*, *100*(5), 1109–1116. https://doi.org/10.1093/oxfordjournals.jbchem.a121814

Meng, Y., Qiu, N., Guyonnet, V., & Mine, Y. (2021). Omics as a window to unravel the dynamic changes of egg components during chicken embryonic development. *Journal of Agricultural and Food Chemistry*, *69*(44), 12947–12955. https://doi.org/10.1021/acs.jafc.1c05883

Miguel, M., Manso, M. A., López-Fandiño, R., & Ramos, M. (2005). Comparative study of egg white proteins from different species by chromatographic and electrophoretic methods. *European Food Research and Technology*, *221* (3–4), 542–546. https://doi.org/10.1007/s00217-005-1182-8

Miyamoto, T., Takahashi, N., Sekine, M., Ogawa, T., Hidaka, M., Homma, H., & Masaki, H. (2015). Transition of serine residues to the d-form during the conversion of ovalbumin into heat stable S-ovalbumin. *Journal of Pharmaceutical and Biomedical Analysis*, *116*, 145–149. https://doi.org/10.1016/j.jpba.2015.04.030

Mizutani, K., Aibara, S., & Hirose, M. (2005). Structure of aluminium-bound ovotransferrin at 2.15 Å resolution. *Acta Crystallographica*, *D61*, 1636–1642. https://doi.org/10.1107/S090744490503266X

Mizutani, K., Mikami, B., & Hirose, M. (2001). Domain closure mechanism in transferrins: New viewpoints about the hinge structure and motion as deduced from high Resolution crystal structures of ovotransferrin N-lobe. *Journal of Molecular Biology*, *309*(4), 937–947. https://doi.org/10.1006/jmbi.2001.4719

Mizutani, K., Muralidhara, B. K., Yamashita, H., Tabata, S., Mikami, B., & Hirose, M. (2001). Anion-mediated Fe3+ release mechanism in ovotransferrin C-lobe. *Journal of Biological Chemistry*, *276*(38), 35940–35946. https://doi.org/10.1074/jbc.M102590200

Mizutani, K., Toyoda, M., & Mikami, B. (2012). X-ray structures of transferrins and related proteins. *Biochimica et Biophysica Acta (BBA) - General Subjects*, *1820*(3), 203–211. https://doi.org/10.1016/j.bbagen.2011.08.003

Mizutani, K., Yamashita, H., Kurokawa, H., Mikami, B., & Hirose, M. (1999). Alternative structural state of transferrin. *Journal of Biological Chemistry*, *274*(15), 10190–10194. https://doi.org/10.1074/jbc.274.15.10190

Mizutani, K., Yamashita, H., Mikami, B., & Hirose, M. (2000). Crystal structure at 1.9 Å Resolution of the apoovotransferrin n-lobe bound by sulfate anions: Implications for the domain opening and iron release mechanism. *Biochemistry*, *39*(12), 3258–3265. https://doi.org/10.1021/bi992574q

Monaco, H. L. (1997). Crystal structure of chicken riboflavin-binding protein. *The EMBO Journal*, *16*(7), 1475–1483.

Nagase, H., & Harris, E. D. (1983). Ovostatin: A novel proteinase inhibitor from chicken egg white. II. Mechanism of inhibition studied with collagenase and thermolysin. *Journal of Biological Chemistry*, *258*(12), 7490–7498. https://doi.org/10.1016/S0021-9258(18)32204-X

Nagase, H., Harris, E. D., Woessner, J. F., & Brew, K. (1983). Ovostatin: A novel proteinase inhibitor from chicken egg white. I. Purification, physicochemical properties, and tissue distribution of ovostatin. *Journal of Biological Chemistry*, *258*(12), 7481–7489. https://doi.org/10.1016/S0021-9258(18)32203-8

Nakamura, R., & Matsuda, T. (1984). Heat inactivation of ovoinhibitor in the alkaline pH region. *Journal of Agricultural and Food Chemistry*, *32*(3), 483–486. https://doi.org/10.1021/jf00123a016

Nau, F., Guérin-Dubiard, C., & Croguennec, T. (2007a). Avidin. In R. Huopalahti, R. Lopez-Fandino, M. Anton, & R. Schade (Eds.), *Bioactive egg compounds* (pp. 75–80). Springer. https://doi.org/10.1007/978-3-540-37885-3_11

Nau, F., Guérin-Dubiard, C., & Croguennec, T. (2007b). Avidin-biotin technology. In R. Huopalahti, R. Lopez-Fandino, M. Anton, & R. Schade (Eds.), *Bioactive egg compounds* (pp. 287–292). Springer. https://doi.org/10.1007/978-3540 378839

Nau, F., Pasco, M., Désert, C., Mollé, D., Croguennec, T., & Guérin-Dubiard, C. (2005). Identification and characterization of ovalbumin Gene Y in hen egg white. *Journal of Agricultural and Food Chemistry*, *53*(6), 2158–2163. https://doi.org/10.1021/jf0483691

Nau, F., Guérin-Dubiard, C., Désert, C., Gautron, J., Carré, W., & Lagarrigue, S. (2003a). *Identification of three minor hen egg white proteins* [Conference session]. Xth European Symposium on the Quality of Eggs and Egg Products, Vol. III (pp. 36–42), Ploufragan, France.

Nau, F., Guérin-Dubiard, C., Désert, C., Gautron, J., Bouton, S., Gribonval, J., & Lagarrigue, S. (2003b). Cloning and characterization of HEP21, a new member of the uPAR/Ly6 protein superfamily predominantly expressed in hen egg white. *Poultry Science*, *82*(2), 242–250. https://doi.org/10.1093/ps/82.2.242

Nielsen, K. L., Sottrup-Jensen, L., Nagase, H., & Etzerodt, M. (1994). The primary structure of ovomacroglobulin. *Annals of the New York Academy of Sciences*, *737*(1), 476–479. https://doi.org/10.1111/j.1749-6632.1994.tb44344.x

Nisbet, A. D., Saundry, R. H., Moir, A. J. G., Fothergill, L. A., & Fothergill, J. E. (1981). The complete amino-acid sequence of hen ovalbumin. *European Journal of Biochemistry*, *115*(2), 335–345. https://doi.org/10.1111/j.1432-1033.1981.tb05243.x

Nishina, Y., Kitagawa, T., Shiga, K., Horiike, K., Matsumura, Y., Watari, H., & Yamano, T. (1978). Resonance Raman spectra of riboflavin and its derivatives in the bound state with egg riboflavin binding proteins. *The Journal of Biochemistry*, *84*(4), 925–932. https://doi.org/10.1093/oxfordjournals.jbchem.a132205

Nistor Cotfas, L. I., Albu, A., Nistor, A. C., & Usturoi, M. G. (2015). Aspects of eggs quality provided from free range and conventional systems. *Journal of Microbiology, Biotechnology and Food Sciences*, *05*(02), 186–189. https://doi.org/10.15414/jmbfs.2015.5.2.186-189

Norioka, N., Okada, T., Hamazume, Y., Mega, T., & Ikenaka, T. (1985). Comparison of the amino acid sequences of hen plasma-, yolk-, and white-riboflavin binding proteins. *Journal of Biochemistry*, *97*, 19–28. https://doi.org/10.1093/oxfordjournals.jbchem.a135044

Nycander, M., & Björk, I. (1990). Evidence by chemical modification that tryptophan-104 of the cysteine-proteinase inhibitor chicken cystatin is located in or near the proteinase-binding site. *Biochemical Journal*, *271*(1), 281–284. https://doi.org/10.1042/bj2710281

Offengenden, M., Fentabil, M. A., & Wu, J. (2011). N-glycosylation of ovomucin from hen egg white. *Glycoconjugate Journal*, *28*(3–4), 113–123. https://doi.org/10.1007/s10719-011-9328-3

Omana, D. A., Liang, Y., Kav, N. N. V., & Wu, J. (2011). Proteomic analysis of egg white proteins during storage. *Proteomics*, *11*(1), 144–153. https://doi.org/10.1002/pmic.201000168

Omana, D. A., Wang, J., & Wu, J. (2010). Ovomucin – A glycoprotein with promising potential. *Trends in Food Science & Technology*, *21*(9), 455–463. https://doi.org/10.1016/j.tifs.2010.07.001

Pagano, A., Crooijmans, R., Groenen, M., Randazzo, N., Zerega, B., Cancedda, R., & Dozin, B. (2003). A chondrogenesis-related lipocalin cluster includes a third new gene, CALγ. *Gene*, *305*(2), 185–194. https://doi.org/10.1016/S0378-1119(03)00382-2

Pagano, A., Giannoni, P., Zambotti, A., Randazzo, N., Zerega, B., Cancedda, R., & Dozin, B. (2002). CALß, a novel lipocalin associated with chondrogenesis and inflammation. *European Journal of Cell Biology*, *81*, 264–272. https://doi.org/10.1078/0171-9335-00243

Palmiter, R. D., Gagnon, J., & Walsh, K. A. (1978). Ovalbumin: A secreted protein without a transient hydrophobic leader sequence. *Proceedings of the National Academy of Sciences*, *75*(1), 94–98. https://doi.org/10.1073/pnas.75.1.94

Petrović, S., & Vitale, L. (1990). Purification and properties of glutamyl aminopeptidase from chicken egg-white. *Comparative Biochemistry and Physiology Part B: Comparative Biochemistry*, *95*(3), 589–595. https://doi.org/10.1016/0305-0491(90)90026-P

Poon, S., Treweek, T. M., Wilson, M. R., Easterbrook-Smith, S. B., & Carver, J. A. (2002). Clusterin is an extracellular chaperone that specifically interacts with slowly aggregating proteins on their off-folding pathway. *FEBS Letters*, *513*(2–3), 259–266. https://doi.org/10.1016/S0014-5793(02)02326-8

Powrie, & Nakai, S. (1986). The chemistry of eggs and egg products. In W. J. Stadelman & O. J. Cotterill (Eds.), *Egg science and technology* (3rd ed., pp. 97–139). The Avi Publishing Company, Inc.

Pugliese, L., Malcovati, M., Coda, A., & Bolognesi, M. (1994). Crystal structure of apo-avidin from hen egg-white. *Journal of Molecular Biology*, *235*(1), 42–46. https://doi.org/10.1016/S0022-2836(05)80010-5

Qiu, N., Liu, W., Ma, M., Zhao, L., & Li, Y. (2013). Differences between fertilized and unfertilized chicken egg white proteins revealed by 2-dimensional gel electrophoresis-based proteomic analysis. *Poultry Science*, *92*(3), 782–786. https://doi.org/10.3382/ps.2012-02565

Qiu, N., Ma, M., Zhao, L., Liu, W., Li, Y., & Mine, Y. (2012). Comparative proteomic analysis of egg white proteins under various storage temperatures. *Journal of Agricultural and Food Chemistry*, *60*(31), 7746–7753. https://doi.org/10.1021/jf302100m

Rabbani, G., Ahmad, E., Zaidi, N., & Khan, R. H. (2011). pH-dependent conformational transitions in conalbumin (Ovotransferrin), a metalloproteinase from hen egg white.

Cell Biochemistry and Biophysics, *61*, 551–560. https://doi.org/10.1007/s12013-011-9237-x

Ragland, S. A., & Criss, A. K. (2017). From bacterial killing to immune modulation: Recent insights into the functions of lysozyme. *PLoS Pathogens*, *13*(9), e1006512. https://doi.org/10.1371/journal.ppat.1006512

Réhault-Godbert, S., Gautron, J., Labas, V., Belghazi, M., & Nys, Y. (2008). Identification and characterization of the precursor of chicken matrix metalloprotease 2 (pro-MMP-2) in hen egg. *Journal of Agricultural and Food Chemistry*, *56*(15), 6294–6303. https://doi.org/10.1021/jf8003948

Réhault-Godbert, S., Guyot, N., & Nys, Y. (2019). The golden egg: Nutritional value, bioactivities, and emerging benefits for human health. *Nutrients*, *11*(3), 684. https://doi.org/10.3390/nu11030684

Réhault-Godbert, S., Hervé-Grépinet, V., Gautron, J., Cabau, C., & Nys, Y. (2011). Molecules involved in chemical defence of chicken egg. In Y. Nys, M. Bain, & F. Van Immerseel (Eds.), *Improving the safety and quality of eggs and egg products* (Vol. 1, pp. 183–208). Woodhead Publishing Ltd.

Réhault-Godbert, S., Labas, V., Helloin, E., Hervé-Grépinet, V., Slugocki, C., Berges, M., & Nys, Y. (2013). Ovalbumin-related protein X is a heparin-binding ov-serpin exhibiting antimicrobial activities. *Journal of Biological Chemistry*, *288*(24), 17285–17295. https://doi.org/10.1074/jbc.M113.469759

Rhodes, M. B., Bennett, N., & Feeney, R. E. (1959). The flavoprotein-apoprotein system of egg white. *Journal of Biological Chemistry*, *234*(8), 2054–2060. https://doi.org/10.1016/S0021-9258(18)69866-7

Roberts, J. R. (2004). Factors affecting egg internal quality and egg shell quality in laying hens. *Journal of Poultry Science*, *41*, 161–177. https://doi.org/10.2141/jpsa.41.161

Robinson, D. S., & Monsey, J. B. (1971). Studies on the composition of egg-white ovomucin. *Biochemical Journal*, *121*(3), 537–547. https://doi.org/10.1042/bj1210537

Romanoff, A. L., & Romanoff, A. J. (1949). Structure. In *The avian egg* (pp. 113–173). John Wiley & Sons, Inc.

Rosano, C., Arosio, P., & Bolognesi, M. (1999). The X-ray three-dimensional structure of avidin. *Biomolecular Engineering*, *16*(1–4), 5–12. https://doi.org/10.1016/S1050-3862(99)00047-9

Sadakane, Y., Matsunaga, H., Nakagomi, K., Hatanaka, Y., & Haginaka, J. (2002). Protein domain of chicken alpha(1)-acid glycoprotein is responsible for chiral recognition. *Biochemical and Biophysical Research Communications*, *295*, 587–590. https://doi.org/10.1016/s0006-291x(02)00716-7

Saponja, J. A., & Vogel, H. J. (2011). Quadrupolar central transition (QCT) and ^{13}C NMR competition studies of metal ion binding to ovotransferrin. *Canadian Journal of Chemistry*, *89*(7), 779–788. https://doi.org/10.1139/v11-019

Savadkoohi, S., Bannikova, A., Mantri, N., & Kasapis, S. (2016). Structural properties of condensed ovalbumin systems following application of high pressure. *Food Hydrocolloids*, *53*, 104–114. https://doi.org/10.1016/j.foodhyd.2014.09.021

Saxena, I., & Tayyab, S. (1997). Protein proteinase inhibitors from avian egg whites: *Cellular and Molecular Life Sciences*, *53*(1), 13–23. https://doi.org/10.1007/PL00000575

Schwabe, C., Anastasi, A., Crow, H., McDonald, J. K., & Barrett, A. J. (1984). Cystatin. Amino acid sequence and possible secondary structure. *Biochemical Journal*, *217*(3), 813–817. https://doi.org/10.1042/bj2170813

Scott, M. J., Huckaby, C. S., Kato, I., Kohr, W. J., Laskowski, M., Tsai, M. J., & O'Malley, B. W. (1987). Ovoinhibitor introns specify functional domains as in the related and linked ovomucoid gene. *Journal of Biological Chemistry*, *262*(12), 5899–5907. https://doi.org/10.1016/S0021-9258(18)45659-1

Shahmohammadi, A. (2018). Lysozyme separation from chicken egg white: A review. *European Food Research and Technology*, *244*(4), 577–593. https://doi.org/10.1007/s00217-017-2993-0

Shan, Y., Tang, D., Wang, R., Tu, A., Yi, Y., Wang, X., & Lü, X. (2020). Rheological and structural properties of ovomucin from chicken eggs with different interior quality. *Food Hydrocolloids*, *100*, 105393. https://doi.org/10.1016/j.foodhyd.2019.105393

Sheng, L., Huang, M., Wang, J., Xu, Q., Hammad, H. H. M., & Ma, M. (2018). A study of storage impact on ovalbumin structure of chicken egg. *Journal of Food Engineering*, *219*, 1–7. https://doi.org/10.1016/j.jfoodeng.2017.08.028

Shinohara, H., Iwasaki, T., Miyazaki, Y., Matsuo, K., Aoki, T., Matsumoto, M., & Sugimoto, Y. (2005). Thermostabilized ovalbumin that occurs naturally during development accumulates in embryonic tissues. *Biochimica et Biophysica Acta (BBA) - General Subjects*, *1723*(1–3), 106–113. https://doi.org/10.1016/j.bbagen.2005.02.016

Silversides, F. G., & Scott, T. A. (2001). Effect of storage and layer age on quality of eggs from two lines of hens. *Poultry Science*, *80*(8), 1240–1245. https://doi.org/10.1093/ps/80.8.1240

Silversides, F. G., & Villeneuve, P. (1994). Is the Haugh unit correction for egg weight valid for eggs stored at room temperature? *Poultry Science*, *73*(1), 50–55. https://doi.org/10.3382/ps.0730050

Škrtić, I., & Vitale, L. (1994). Methionine-preferring broad specificity aminopeptidase from chicken egg-white. *Comparative Biochemistry and Physiology Part B: Comparative Biochemistry*, *107*(3), 471–478. https://doi.org/10.1016/0305-0491(94)90213-5

Smith, M. B., & Back, J. F. (1962). Modification of ovalbumin in stored eggs detected by heat denaturation. *Nature*, *4818*, 878–879. https://doi.org/10.1038/193878a0

Smith, M. B., & Back, J. F. (1965). Studies on ovalbumin. II. The formation and properties of S-ovalbumin, a more stable form. *Australian Journal of Biological Sciences*, *18*, 365–377. https://doi.org/10.1071/bi9650365

Smith, S. R., Pala, I., & Benore-Parsons, M. (2006). Riboflavin binding protein contains a type II copper binding site. *Journal of Inorganic Biochemistry*, *100*(11), 1730–1733. https://doi.org/10.1016/j.jinorgbio.2006.06.015

Stănciuc, N., Crețu, A. A., Banu, I., & Aprodu, I. (2018). Advances on the impact of thermal processing on structure and antigenicity of chicken ovomucoid: Effects of heat on chicken ovomucoid. *Journal of the Science of Food and Agriculture*, *98*, 3119–3128. https://doi.org/10.1002/jsfa.8813

Stein, P. E., Leslie, A. G. W., Finch, J. T., & Carrelll, R. W. (1991). Crystal structure of uncleaved ovalbumin at 1-95 a Resolution. *Journal of Molecular Biology*, *221*(3), 941–959. https://doi.org/10.1016/0022-2836(91)80185-W

Stevens, L. (1991). Egg white proteins. *Comparative Biochemistry and Physiology, 100B*(1), 1–9. https://doi.org/10.1016/0305-0491(91)90076-P

Strecker, G., Wieruszeski, J., Cuvillier, O., Michalski, J., & Montreuil, J. (1992). 1H and 13C-NMR assignments for sialylated oligosaccharide-alditols related to mucins. Study of thirteen components from hen ovomucin and swallow nest mucin. *Biochimie, 74*(1), 39–51. https://doi.org/10.1016/0300-9084(92)90182-e

Sun, Y., Jin, H., Sun, H., & Sheng, L. (2020). A comprehensive identification of chicken egg white phosphoproteomics based on a novel digestion approach. *Journal of Agricultural and Food Chemistry, 68*(34), 9213–9222. https://doi.org/10.1021/acs.jafc.0c03174

Sun, C., Liu, J., Li, W., Xu, G., & Yang, N. (2017). Divergent proteome patterns of egg albumen from domestic chicken, duck, goose, Turkey, quail and pigeon. *Proteomics, 17* (17–18), 1700145. https://doi.org/10.1002/pmic.201700145

Swamy, M. J., Heimburg, T., & Marsh, D. (1996). Fourier-transform infrared spectroscopic studies on avidin secondary structure and complexation with biotin and biotin-lipid assemblies. *Biophysical Journal, 71*(2), 840–847. https://doi.org/10.1016/S0006-3495(96)79285-8

Takahashi, K., Kitao, S., Tashiro, M., Asao, T., & Kanamori, M. (1994). Inhibitory specificity against various trypsins and stability of ovomucoid from Japanese quail egg white. *Journal of Nutritional Science and Vitaminology, 40*(6), 593–601. https://doi.org/10.3177/jnsv.40.593

Takahashi, N., Maeda, M., Yamasaki, M., & Mikami, B. (2010). Protein-engineering study of contribution of conceivable d-serine residues to the thermostabilization of ovalbumin under alkaline conditions. *Chemistry & Biodiversity, 7*(6), 1634–1643. https://doi.org/10.1002/cbdv.200900305

Takahashi, N., Onda, M., Hayashi, K., Yamasaki, M., Mita, T., & Hirose, M. (2005). Thermostability of refolded ovalbumin and *S* -ovalbumin. *Bioscience, Biotechnology, and Biochemistry, 69*(5), 922–931. https://doi.org/10.1271/bbb.69.922

Tripic, T., Skrtic, I., & Vitale, L. (1998). Activity of various hydrolytic enzymes in chicken egg white during egg storage. *Food Technology and Biotechnology, 36*(3), 215–218.

Trudel, J., & Asselin, A. (1995). Detection of a glycosylated form of hen egg white lysozyme. *Biochemistry and Cell Biology, 73*(5–6), 307–309. https://doi.org/10.1139/o95-038

Turk, V., Brzin, J., Longer, M., Ritonja, A., Eropkin, M., Borchart, U., & Machleidt, W. (1983). Protein inhibitors of cysteine proteinases. III. Amino-acid sequence of cystatin from chicken egg white. *Hoppe-Seyler´s Zeitschrift Für Physiologische Chemie, 364*(2), 1487–1496. https://doi.org/10.1515/bchm2.1983.364.2.1487

Urisu, A., Ando, H., Morita, Y., Wada, E., Yasaki, T., Yamada, K., & Wakamatsu, T. (1997). Allergenic activity of heated and ovomucoid-depleted egg white. *Journal of Allergy and Clinical Immunology, 100*(2), 171–176. https://doi.org/10.1016/S0091-6749(97)70220-3

Valenti, P., Visca, P., Antonini, G., & Orsi, N. (1986). Interaction between lactoferrin and ovotransferrin and *candida* cells. *FEMS Microbiology Letters, 33*(2–3), 271–275. https://doi.org/10.1111/j.1574-6968.1986.tb01285.x

Wang, F., & Polavarapu, P. L. (2001). Temperature influence on the secondary structure of avidin and avidin–Biotin complex: A vibrational circular dichroism study. *The Journal of Physical Chemistry B, 105*(32), 7857–7864. https://doi.org/10.1021/jp0104177

Wang, Y., Wang, Z., & Shan, Y. (2019). Assessment of the relationship between ovomucin and albumen quality of shell eggs during storage. *Poultry Science, 98*(1), 473–479. https://doi.org/10.3382/ps/pey349

Wang, X., Wei, Z., & Xue, C. (2021). The past and future of ovotransferrin: Physicochemical properties, assembly and applications. *Trends in Food Science & Technology, 116*, 47–62. https://doi.org/10.1016/j.tifs.2021.07.003

Wang, X., Xu, G., Yang, N., Yan, Y., Wu, G., & Sun, C. (2019). Differential proteomic analysis revealed crucial egg white proteins for hatchability of chickens. *Poultry Science, 98*(12), 7076–7089. https://doi.org/10.3382/ps/pez459

Wan, Y., Jin, S., Ma, C., Wang, Z., Fang, Q., & Jiang, R. (2017). RNA-seq reveals seven promising candidate genes affecting the proportion of thick egg albumen in layer-type chickens. *Scientific Reports, 7*(1), 18083. https://doi.org/10.1038/s41598-017-18389-5

Wan, Y., Jin, S., Ma, C., Wang, Z., Fang, Q., & Jiang, R. (2018). Effect of strain and age on the thick-to-thin albumen ratio and egg composition traits in layer hens. *Animal Production Science, 59*(3), 416. https://doi.org/10.1071/AN17231

Wan, Y., Wang, Z., Wang, J., Su, H., Guo, X., Wu, J., & Jiang, R. (2019). Genetic parameters of the thick-to-thin albumen ratio and egg compositional traits in layer-type chickens. *British Poultry Science, 60*(5), 517–521. https://doi.org/10.1080/00071668.2019.1631446

Wasylewski, M. (2000). binding study of riboflavin-binding protein with riboflavin and its analogues by differential scanning calorimetry. *Journal of Protein Chemistry, 19*(6), 523–528. https://doi.org/10.1023/A:1026509717268

Wasylewski, M. (2004). Evaluation of riboflavin binding protein domain interaction using differential scanning calorimetry. *Biochimica Et Biophysica Acta (BBA) - Proteins and Proteomics, 1702*(2), 137–143. https://doi.org/10.1016/j.bbapap.2004.07.004

Watanabe, K., Matsuda, T., & Sato, Y. (1981). The secondary structure of ovomucoid and its domains as studied by circular dichroism. *Biochimica Et Biophysica Acta (BBA) - Protein Structure, 667*(2), 242–250. https://doi.org/10.1016/0005-2795(81)90189-6

Wen, P.-W., Tu, Z.-C., Hu, Y.-M., & Wang, H. (2022). Effects of superheated steam treatment on the allergenicity and structure of chicken egg ovomucoid. *Foods, 11*(2), 238. https://doi.org/10.3390/foods11020238

Whenham, N., Wilson, P. W., Bain, M. M., Stevenson, L., & Dunn, I. C. (2014). Comparative biology and expression of TENP, an egg protein related to the bacterial permeability-increasing family of proteins. *Gene, 538*(1), 99–108. https://doi.org/10.1016/j.gene.2013.12.065

Whisstock, J., Skinner, R., & Lesk, A. M. (1998). An atlas of serpin conformations. *Trends in Biochemical Sciences, 23*(2), 63–67. https://doi.org/10.1016/s0968-0004(97)01172-9

White, F. H. (1982). Studies on the relationship of disulfide bonds to the formation and maintenance of secondary structure in chicken egg white lysozyme. *Biochemistry, 21*(5), 967–977. https://doi.org/10.1021/bi00534a023

White, H. B. (1996). Sudden death of chicken embryos with hereditary riboflavin deficiency. *Journal of Nutrition,*

126(4 Suppl), 1303S–1307S. https://doi.org/10.1093/jn/126.suppl_4.1303s

White, H. B., & Merrill, A. H. (1988). Riboflavin-binding proteins. *Annual Review of Nutrition*, *8*, 279–299. https://doi.org/10.1146/annurev.nu.08.070188.001431

Williams, J., Elleman, T. C., Kingston, I. B., Wilkins, A. G., & Kuhn, K. A. (1982). The primary structure of hen ovotransferrin. *European Journal of Biochemistry*, *122*(2), 297–303. https://doi.org/10.1111/j.1432-1033.1982.tb05880.x

Wilson, M. R., & Easterbrook-Smith, S. B. (2000). Clusterin is a secreted mammalian chaperone. *Trends in Biochemical Sciences*, *25*(3), 95–98. https://doi.org/10.1016/s0968-0004(99)01534-0

Wiseman, R. L., Fothergill, J. E., & Fothergill, L. A. (1972). Replacement of asparagine by aspartic acid in hen ovalbumin and a difference in Immunochemical reactivity. *Biochemical Journal*, *127*(5), 775–780. https://doi.org/10.1042/bj1270775

Wu, J., & Acero-Lopez, A. (2012). Ovotransferrin: Structure, bioactivities, and preparation. *Food Research International*, *46*(2), 480–487. https://doi.org/10.1016/j.foodres.2011.07.012

Wu, T., Jiang, Q., Wu, D., Hu, Y., Chen, S., Ding, T., & Chen, J. (2019). What is new in lysozyme research and its application in food industry? A review. *Food Chemistry*, *274*, 698–709. https://doi.org/10.1016/j.foodchem.2018.09.017

Yamasaki, M., Takahashi, N., & Hirose, M. (2003). Crystal structure of s-ovalbumin as a non-loop-inserted thermostabilized serpin form. *Journal of Biological Chemistry*, *278*(37), 35524–35530. https://doi.org/10.1074/jbc.M305926200

Yan, R.-T., & Wang, S.-Z. (1998). Identification and characterization of tenp, a gene transiently expressed before overt cell differentiation during neurogenesis. *Journal of Neurobiology*, *34*(4), 319–328. https://doi.org/10.1002/(SICI)1097-4695(199803)34:4<319::AID-NEU3>3.0.CO;2-9

Yang, Y., Barendregt, A., Kamerling, J. P., & Heck, A. J. R. (2013). Analyzing protein micro-heterogeneity in chicken ovalbumin by high-Resolution native mass spectrometry exposes qualitatively and semi-quantitatively 59 proteoforms. *Analytical Chemistry*, *85*(24), 12037–12045. https://doi.org/10.1021/ac403057y

Yet, M.-G., & Wold, F. (1990). The distribution of glycan structures in individual n-glycosylation sites in animal and plant glycoproteins. *Archives of Biochemistry and Biophysics*, *278*(2), 356–364. https://doi.org/10.1016/0003-9861(90)90271-Y

Zandomeneghi, G., & Zandomeneghi, M. (2009). Determination of holo- and apo-riboflavin binding protein in avian egg whites through circular dichroism and fluorescence spectroscopy. *Journal of Agricultural and Food Chemistry*, *57*(15), 6510–6517. https://doi.org/10.1021/jf901079n

Zhang, L., Chen, J., Fan, B., Fu, M., Sun, Y., Wang, Y., & Wang, F. (2020). Label-free proteomic analysis reveals the differentiation between unfertilized and fertilized Beijing-You chicken eggs. *International Journal of Biological Macromolecules*, *152*, 1020–1026. https://doi.org/10.1016/j.ijbiomac.2019.10.189

Zhang, C., He, T., Vedadghavami, A., & Bajpayee, A. G. (2020). Avidin-biotin technology to synthesize multi-arm nano-construct for drug delivery. *MethodsX*, *7*, 100882. https://doi.org/10.1016/j.mex.2020.100882

Zheng, D. B., Lim, H. M., Pène, J. J., & White, H. B. (1988). Chicken riboflavin-binding protein. cDNA sequence and homology with milk folate-binding protein. *Journal of Biological Chemistry*, *263*(23), 11126–11129. https://doi.org/10.1016/S0021-9258(18)37931-6

Zhu, Y., Vanga, S. K., Wang, J., & Raghavan, V. (2018). Impact of food processing on the structural and allergenic properties of egg white. *Trends in Food Science & Technology*, *78*, 188–196. https://doi.org/10.1016/j.tifs.2018.06.005

6

Egg Yolk Constituents and Structures

Marc Anton

CONTENTS

6.1 Introduction

At the industrial level, egg yolk is used for its functional properties and comes in different forms: liquid, powdered or frozen. It is incorporated into many food formulations for its exceptional emulsifying and gelling properties and also because it provides food with the desired taste and color. It contains also proteins of high biological value and other nutrients such as vitamins, minerals, essential fatty acids, phospholipids and other lipids. Egg yolk is thus an essential ingredient in the production of cold (mayonnaises, salad dressings) and hot (béarnaise, hollandaise) emulsions. It participates to the formation and stabilization of emulsions by constituting an interfacial film between oil and water. This film makes it possible, on the one hand, to reduce the interfacial tension between oil and water and, on the other hand, to form a barrier protecting the oil droplets from rupture, stabilizing thus the system (Anton, 2013). Despite its intensive use for formulated food products, and this since the invention of mayonnaise three centuries ago, it has not been easy to clarify the role of its major constituents due to its unwell elucidated structures in relation with its properties.

Egg yolk is a mixture of proteins and lipids forming natural assemblies at different size scale levels. These natural supramolecular assemblies contribute to the nano- and macro-structures of the yolk. Thus, the clarification of the emulsifying properties of egg yolk lies in the understanding of these different levels of structure.

Egg yolk can be fractionated into its two main fractions, namely plasma and granules, by a mild centrifugation process without causing any denaturation to the proteins. If plasma is the main contributor to the emulsifying properties of whole

DOI: 10.1201/9781003254430-8

egg yolk, granules bring also interesting emulsifying properties provided that they are in micelle state after disruption due to the addition of salts (Anton, 2013). These two fractions exhibit large differences of structures and compositions, resulting in proper functionalities and sensibilities to thermomechanical treatments. Precisely, as plasma and granules are not yet used separately in the current industrial production of oil/water emulsions, a poor knowledge is available on the effect of processing the fractions on their structure and functionality. These differences observed between plasma and granules indicated that there is a track for a more efficient use of these fractions compared to whole egg yolk.

Consequently, this chapter proposes to make the point and up-date data concerning the composition of yolk and its constituents, the different structures of which it is composed from the macro- to the nano- scale, the different strategies of extraction of its constituents, and finally the effect of some thermomechanical treatments on these structures.

6.2 Egg Yolk Contents and Constituents

Egg yolk, or vitellus, refers to the content of the yolk sac present in the eggs of oviparous animals, especially birds. Unlike mammals, the mother does not feed bird embryos during their development and they have no possibility of eliminating metabolic waste. Thus, the egg parts (yolk and albumen) provide essential nutrients to the embryo, which are extremely well metabolized. Egg is therefore a very interesting source of nutrients for humans. In particular, egg proteins have no limiting amino acid and a DIAAS score (digestible indispensable amino acid score) over 100%, around the same level as whole milk proteins (Phillips, 2017).

The dry matter content of freshly laid chicken egg yolks ranges from 50 to 52% depending on the age of the layer and shelf life, with water transfer taking place from white to yolk during egg storage (Kiosseoglou, 1989; Thapon & Bourgeois, 1994; Li-Chan et al., 1995). The compositions of fresh and dry yolk are presented in Table 6.1: egg yolk has a solid content close to 50% consisting mainly of lipids (about 62.5% of dry matter) and proteins (33%). It also contains a significant amount of minerals including iron and phosphorus. Lipids of egg yolk are found to be fully associated with lipoprotein complexes. These lipids consist of 65% triglycerides, 29% phospholipids and less than 5% cholesterol. Carotenoids represent less than 1% of the yolk lipids, and they are the ones who give it its color. Proteins are divided into free proteins and apoproteins constituting lipoprotein complexes. The interactions between lipids and proteins result in the formation of lipoproteins (low and high density), which represent the main constituents of yolk.

TABLE 6.1
Composition of Hen Egg Yolk

	% of Fresh Yolk	% of Dry Yolk
Water	51.1	-
Lipids	30.6	62.5
Proteins	16.0	33.0
Carbohydrates	0.6	1.2
Minerals	1.7	3.5

Source: Data from Powrie and Nakai (1986).

6.2.1 Plasma and Granules

From a physicochemical point of view, egg yolk is a dispersion of particles (Low-density lipoproteins (LDL) and granules) in equilibrium in an aqueous solution of proteins (Figure 6.1). LDL are assembled structures from 12 nm to 48 nm in diameter whereas granules have the shape of more or less flattened spheres with a diameter of 0.3–2 μm (Anton, 2007a; Laca et al., 2010).

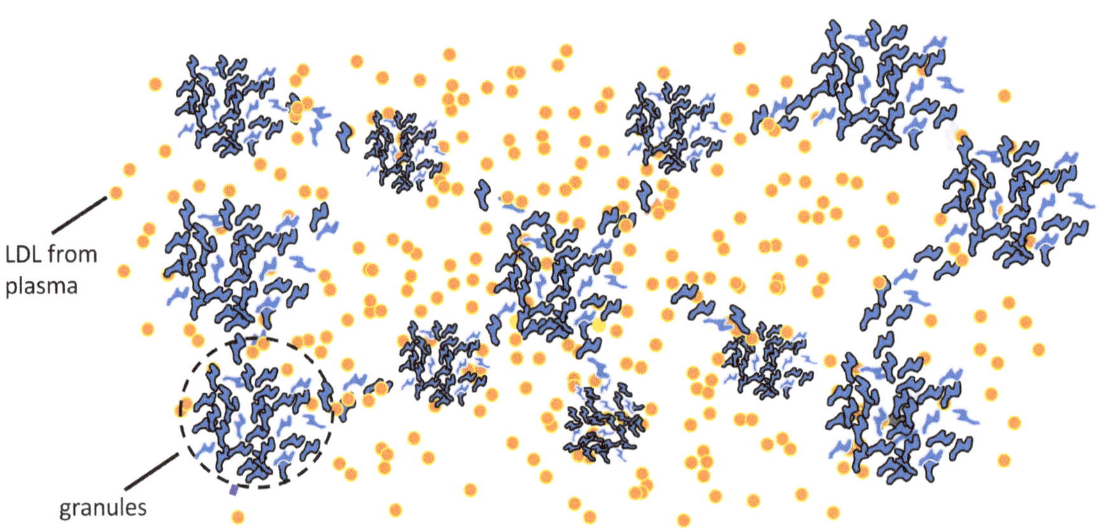

FIGURE 6.1 Macro- and nano-structures of hen egg. (From Anton, 2013.)

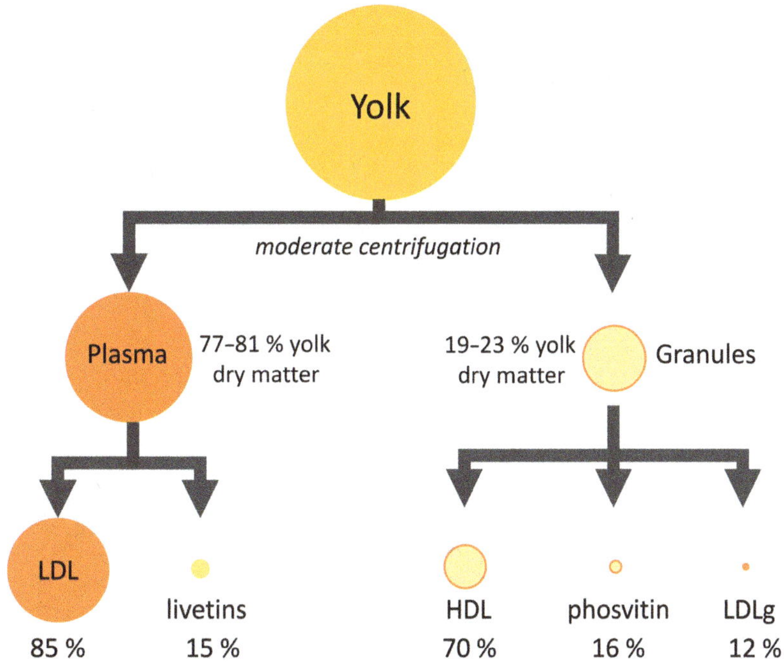

FIGURE 6.2 Constituents of plasma and granules from hen egg yolk (LDL: low-density lipoprotein, HDL: high-density lipoprotein).

By dilution in NaCl (0.17 M) and centrifugation (10,000 g) (McBee & Cotterill, 1979), the egg yolk can be separated into two fractions. Plasma (orange supernatant) accounts for 77–81% of the yolk dry matter and includes LDL (85%) and 15% of soluble proteins named livetins. Granules (pale pellets) constitute 19–23% of the yolk dry matter and contain high-density lipoprotein (HDL) (70%), phosvitin (16%) and residual LDLg (12%) (Figure 6.2).

The plasma contains about 73% of lipids and 25% of proteins, and represents 93% of yolk lipids (out of which 85% of yolk phospholipids) and 53% of yolk proteins. The granules contain 31% of lipids and 64% of proteins, and represent about 47% of yolk proteins and 7% of yolk lipids.

It may be specified that all the lipids are incorporated into LDL and HDL structures. Into granules, HDL and phosvitin contain a high proportion of phosphorylated serine residues, which allow their association *via* divalent calcium ions forming the granular structure. The numerous phosphocalcic bridges make this structure compact, slightly hydrated, poorly accessible to enzymes and protect the proteins against denaturation and thermal gelling (Anton, 2007a; Laca et al., 2014). The repartition of egg yolk constituents into plasma and granules is indicated Table 6.2.

6.2.2 Plasma Constituents

6.2.2.1 LDL Structure

LDL are spherical nanoparticles (17–60 nm diameter with a mean of about 35 nm) with a lipid core of triglycerides and cholesterol esters in a liquid state surrounded by a mono-film of phospholipids and proteins (called apoproteins) (Figure 6.3) (Cook & Martin, 1969; Evans et al., 1973; Anton et al., 2003; Jolivet et al., 2006, 2008). LDL are the major constituent of egg yolk (2/3 of its dry matter) and are composed of 11–17% proteins and 83–89% lipids, out of which latter 74% are neutral lipids and 26% are phospholipids. They are mainly located in plasma but a small portion is included in granules (LDLg).

Egg yolk LDL are derived from lipoproteins synthesized in the liver of the hen, the VLDL (very low density lipoproteins). VLDL are transported by the blood flow to the oocyte, the place where the egg yolk is formed. Specific receptors in the oocyte membrane allow the fixation of the VLDL, which are then secreted in the egg yolk by endocytosis and then modified into LDL. Two types of LDL have been identified: LDL1 (20% of LDL) whose molecular weight is 10×10^3 kDa and LDL2 (80% of LDL) whose molecular weight is 3×10^3 kDa (Cook & Martin, 1969).

TABLE 6.2

Repartition of Hen Egg Yolk Constituents

	% Yolk Dry Matter	% Yolk Lipids	% Yolk Proteins	% of Lipids	% of Proteins
Yolk	100	100	100	64	32
Plasma	78	93	53	73	25
LDL	66	61	22	88	10
Livetins	10	-	30	-	96
Others	2	-	1	-	90
Granules	22	7	47	31	64
HDL	16	6	35	24	75
Phosvitin	4	-	11	-	95
LDLg	2	1	1	88	10

Source: Data from Powrie and Nakai (1986).

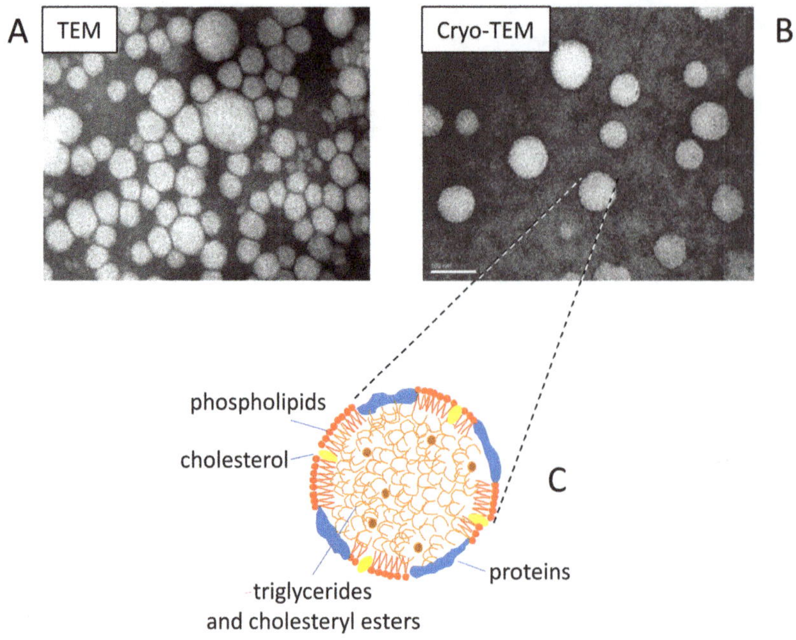

FIGURE 6.3 Transmission electron microscopy (TEM) (A) and cryo electron microscopy (Cryo-TEM) using negative staining (B), and schematic representation of yolk LDL (C). (From Anton et al., 2003.)

LDL are soluble in aqueous solutions (whatever the pH and ionic conditions) due to their low density (0.982). Phospholipids take an essential part in the stability of the LDL structure because association forces are essentially hydrophobic and some cholesterol is included in the phospholipid film, increasing its rigidity (Saari et al., 1964). Electron microscopy and X-ray diffraction strongly support the existence of a mosaic structure at the LDL surface, in which lipids and proteins occupy distinct areas and do not interact tightly together. Electron micrographs (TEM and Cryo-TEM) of LDL negatively stained with 2% sodium phospho-tungstate are shown in Figure 6.3. Particles of LDL appear spherical, with heterogeneous sizes between 20 nm and 60 nm diameter (average 35–40 nm). Contiguous particles have flattened edges; these polygonal shapes could be the result of particle aggregation due to dehydration on grids or to a vacuum in the electron microscope during analysis.

The same sample of LDL was analyzed by photon-correlating spectroscopy (results not shown) and gave an average particle diameter of 35 nm, in agreement with the average value obtained from the electron micrograph. However, we observed some structures of about 200 nm diameter surrounded by aggregated LDL (Anton et al., 2003). It is possible that these structures are VLDL as observed by Martin et al. (1964), or merged LDL as noted for human plasma LDL (Ala-Korpela et al., 1998).

From a techno-functional point of view, it has been largely proven that LDLs are the main contributors to the exceptional emulsifying properties of egg yolk (Anton, 2013, for a review). This capacity is clearly due to the LDL structure, through interactions between amphiphilic apoproteins and phospholipids. This assembly of phospholipids and apoproteins allows transport of these non-soluble amphiphilic molecules through the aqueous phase and until to release them at the oil-water or air-water interface. Using Langmuir film balance (air-water interface), three phase transitions (19, 41 and 54 mN/m) have been detected in compression isotherms. They have been attributed respectively to neutral lipids, proteins and phospholipids by comparison with films of neutral lipids, phospholipids and total lipids extracted from LDL (Anton et al., 2003; Martinet el al., 2003; Dauphas et al., 2006; Anton, 2007b; Dauphas et al., 2007a, 2007b). These different transitions evidenced that LDL actually break down when they come into contact with the interface. Neutral lipids, phospholipids and proteins are then released from the lipoprotein core and they can spread at the interface. It is obvious that LDL serve as vectors of surfactant constituents (proteins and phospholipids) that could not be soluble in water, until the interface. Furthermore, comparing interfacial behavior of LDL with that of synthetic liposomes, constructed by a double phospholipid layer protein-free, it has been revealed that the proteins situated on the LDL surface initiate the LDL disruption mechanism by their initial anchorage. This anchorage provokes an unfolding of the protein leading to the destabilization of the external layer of the LDL. Conversely, the liposomes remained intact at the interface and thus did not release the surfactant molecules for adsorption (Anton et al., 2008). LDL are also effective to deliver hydrophobic bioactive molecules such as curcumin as a nanoscale oral delivery vehicle (Zhou et al., 2016).

6.2.2.2 LDL Apoproteins

LDL proteins (apoproteins) are particularly difficult to study due to their low solubility and their high proportion of hydrophobic amino acids (40%), which places the LDL apoproteins among the most hydrophobic proteins that are listed

(Damodaran & Xu, 1996). They are also glycosylated: 2.45% sugars including 1.3% hexose, 0.67% hexosamine, and 0.38% sialic acid. The pI of these proteins is in a pH range of 6.5–7.3 (Nakamura et al., 1977; Kojima & Nakamura, 1985). Their structures are largely unknown. These would be proteins containing a majority of disordered and helix α domains (Mine, 1997; Anton et al., 2003). Using circular dichroism, Anton et al. (2003) have observed, for the majority of the apoproteins, a double minimum at 208 nm and 222 nm, and a maximum at 191–193 nm, consistent with a considerable proportion of α-helix chains. In the same study, sodium dodecyl sulfate (SDS) was used to solubilize apoproteins, and it is suggested that SDS could lead to an induction of α-helix structures (Montserret et al., 2000). However, in the case of membrane proteins or amphiphilic proteins like LDL apoproteins, SDS is known to mimic the hydrophobic environment existing in biological membranes without changing protein conformation. Apoproteins extracted from LDL with ether-ethanol are very difficult to dissolve in aqueous buffers. A basic pH (pH 12) associated with a low ionic strength helps to attain 30% solubility. For a deeper understanding, an interesting overview compares the different studies around LDL apoproteins, and plasma proteins (Laca et al., 2015).

There are six major LDL apoproteins. The main one accounts for more than 70% of the total and its molecular weight is estimated at 130 kDa. The second apoprotein, with a molecular weight of 15 kDa, represents about 20% of the total. Four other minor apoproteins with molecular weights between 55 and 80 kDa have been identified (Anton et al., 2003). VLDLs of hen blood are the precursors of egg yolk LDL (Evans & Burley, 1987). While VLDLs exist in the blood of immature pullets, their production in hen liver is considerably increased with sexual maturity due to estrogen secretion. VLDLs contain mainly two apoproteins, namely apo-VLDL II and apo-B.

Apo-B is a 500 kDa protein constituted by a single polypeptide chain and highly similar to the human apolipoprotein B-100 precursor. The structural homology of hen apo-B with human apoB-100 and sequence alignment could be an interesting tool for furthering the characterization of hen apo-B. During its transfer into the yolk, hen apo-B is enzymatically cleaved, resulting in the production of apo-B fragments. Hen apo-B is known to be cleaved into several protein fragments by cathepsin D.

Apo-VLDL II is the only apoprotein from blood lipoproteins to be transferred to yolk in large amount without any modification. It is called apovitellenin I in yolk. Apovitellenin I is a small homodimer with disulfide-linked subunits of 9 kDa. Jolivet et al. (2006) have confirmed the existence in egg yolk of apovitellenin I, either as a monomer, or a homodimer of disulfide-linked subunits that is resistant to SDS. The monomer has a molecular weight of 9.331 kDa in spite of a slight heterogeneity, and its amino acid sequence has been confirmed through the analysis of trypsin peptides.

Apart from apovitellenin I, there is a lack of knowledge concerning the exact identification of the other apoproteins of LDL, and especially about the correspondence between LDLs and blood lipoproteins. Furthermore, gaps still exist in the knowledge of the exact maturation mechanism of apoprotein precursors.

6.2.2.3 Lipids

Main components of the yolk (62.5%), namely lipids, are distributed exclusively in lipoproteins (LDL and HDL). They are composed of triglycerides (65%), phospholipids (29%), cholesterol (5%), free fatty acids (<1%), and other lipids including carotenoids (<0.1%).

The fatty acid composition of lipids, based on a standardized hen feeding, is about 30–35% saturated fatty acids (SFA), 40–45% monounsaturated fatty acids (MUFA), and 20–25% polyunsaturated fatty acids (PUFA). The main fatty acids are oleic acid (C18:1, 40–45%), palmitic acid (C16:0, 20–25%), and linoleic acid (C18:2, 15–20%) (Kuksis, 1992). However, this composition is subject to strong variations, in particular according to the nature of fatty acids ingested by the hen (Posati et al., 1975; Anton & Gandemer, 1997; Bruneel et al., 2013; Hammershøj & Johansen, 2016). Dietary fatty acids particularly modify the proportions of PUFA and MUFA, whereas SFA proportions are but slightly affected.

The major triglyceride fatty acids comprise most of the fatty acid total. The glycerol of the triglycerides is mainly esterified by palmitic acid in position 1, by oleic and linoleic acids in position 2, and by oleic, palmitic, and stearic acids in position 3 (Kuksis, 1992).

The phospholipids of the yolk are very rich in phosphatidylcholine (PC): 76% of the phospholipids. PC is a zwitterionic phospholipid on a pH scale from acid to base. Phosphatidylethanolamine (PE) accounts for 22% of the phospholipids in egg yolk. Phospholipids are amphiphilic molecules that contain one hydrophilic head group, which is phosphoric acid + alcohol, amino acid or polyol, and one hydrophobic group, which consists of two fatty acids. Phosphatidylinositol (PI), phosphatidylserine (PS), sphingomyelin (SM), cardiolipins (CL), lysoPC, and lysoPE are present in very low amounts. PUFA represent 30–40% of the fatty acids, whereas SFA account for 45%, and MUFA for 20–25% (Kuksis, 1992). Phosphatidylcholine contains important quantities of ω3 fatty acids as the nonpolar part, choline the polar part. Choline is an important nutrient in brain development, liver function, and cancer prevention (Gutierrez et al., 1997), and is routinely added to commercial infant formulas as an essential nutrient. Consumption of phosphatidylcholine increases plasma and brain choline levels and accelerates neuronal acetylcholine synthesis. It has been demonstrated that consumption of yolk phospholipids tends to alleviate the symptoms of Alzheimer's disease (Juneja, 1997).

Cholesterol is the sterol found in egg yolk. It results partly from the hen feed and partly from synthesis in the liver during the elaboration of lipoproteins. It represents approximately 5% of total lipids in free (85–90%) or in esterified (10–15%) form (Bitman & Wood, 1980). The free cholesterol participates in the structure of LDL. Cholesterol esters are present in the lipid core of LDL and contain 35% oleic acid, 33% palmitic acid, 12% linoleic acid, and 11% stearic acid (Kuksis, 1992).

Carotenoids are the natural pigments of hen egg yolk. They confer it its yellow color, which can go from a very pale yellow to a dark brilliant orange. They are far from plentiful in eggs, but economically important because the color represents a quality criterion. They are mainly carotene and xanthophylls (lutein, cryptoxanthin, and zeaxanthin). The total concentration of lutein and zeaxanthin is ten times greater than that of cryptoxanthin and carotene combined (Shenstone, 1968). Carotene is the main carotenoid found in common corn and alfalfa fed hens. After ingestion, this compound is largely oxidized to form the xanthophylls.

Finally, yolk is a plentiful source of lipids, some of which are particularly suitable for nutrition and health. Particularly, omega 3 fatty acids (ω3), mainly found in phospholipids, are considered an essential nutrient for brain function and visual acuity in humans (Maki et al., 2003). Eicosapentaenoic acid (EPA) and docosahexaenoic acid (DHA) are the main ω3 fatty acids. They are particularly useful for pregnant and nursing mothers for optimizing growth of their infants.

6.2.2.4 Livetins

Livetins are the non-lipid globular proteins that are present in yolk plasma. These proteins represent 10% of yolk dry matter and 30% of its proteins (Kiosseoglou, 1989). These are blood proteins, deposited in the yolk. Their pI is between 4.3 and 5.5 (Ternes, 1989).

They represent a relatively inhomogeneous fraction including α-, β- and γ-livetin in a 2:5:3 ratio. The main component of α-livetin (80 kDa) is an albumin, whereas α-2-glycoprotein is the main component of β-livetin (45 kDa) and IgY (immunoglobulin Y) is the predominant fraction of γ-livetin (150 kDa) (Schade et al., 2005; Schade & Chacana, 2007; Chalamaiah et al., 2017).

We propose a focus on γ-livetin as this protein acts as antibody to perform immunity of embryo into the yolk. Furthermore, potential applications in foods as immunological supplement can be developed for this protein (Yang et al., 2014).

The molecular weight of IgY was found by mass spectrometry to be 167 kDa (that of IgG is about 160 kDa) (Sun et al., 2001). The general structure of the IgY molecule, consists of two light and two heavy chains and is comparable to the mammalian immunoglobulins (Laca et al., 2015). The heavy chain is indicated by the Greek letter Y or v. It has a MW of 65 kDa, and contains one variable domain and four constant domains, in contrast to the three constant domains of the IgG heavy chain. The light chain consists of one variable and one constant domain (18.6 kDa), and the Fab fragment has a MW of 45.3 kDa (Shimizu et al., 1992; Schade & Chacana, 2007). The less developed hinge region of IgY may result in a reduced flexibility of the Fab moiety, which, in turn, may be the reason for some differences between IgY and IgG concerning antigen epitope recognition (Warr et al., 1995). As for IgG, the Fc part of IgY is the site of most biological effector functions due to its capacity to fix the membrane receptors of immune system cells. It contains two carbohydrate side-chains, in contrast to only one in IgG. The isoelectric point of IgY is in the range of 5.7–7.6, whereas that of IgG lies between 6.1 and 8.5 (Davalos-Pantoja et al., 2000; Sun et al., 2001; and Chapter 27 of this book for more details).

6.2.3 Granule Constituents

6.2.3.1 Granules Structure

Granules consist of circular complexes ranging in diameter from 0.3 μm to 2 μm (Chang et al., 1977). At low ionic strength, granules mainly form non-soluble HDL-phosvitin complexes linked by phosphocalcic bridges as HDL and phosvitin contain both a high proportion of phosphoserin residues able to bind divalent calcium (Causeret et al., 1991; Castellani et al., 2004; Guilmineau & Kulozik, 2006). It is by the precipitation of granules that yolk can be easily separated into two fractions after dilution (two times) with 0.3 M NaCl and a centrifugation at 10,000 g (30 min) according to the method of McBee and Cotterill (1979). The numerous phosphocalcic bridges make the granular structure very compact, poorly hydrated, not very accessible to enzymes, and lead to an efficient protection against thermal denaturation and heat gelation. These specificities confer also a strong capacity of iron chelation by phosvitin that could be used for antioxidant purposes (Castellani et al., 2004; Anton et al., 2006).

At an ionic strength over 0.3 M NaCl, the phosphocalcic bridges are disrupted, because monovalent sodium replaces divalent calcium. In such conditions, the solubility of granules reaches 80% because phosvitin is a soluble protein and HDL behave like soluble proteins (Cook & Martin, 1969; Anton & Gandemer, 1997). Complete disruption of granules occurs when ionic strength reaches 1.71 M NaCl.

Acidification or alkalization likewise cause the disruption of granules and the solubilization of these constituents. At acidic pH the increase in number of the positive charges (NH^{3+}) induces electrostatic repulsion and results in disruption of the granules constituents. At basic pH, carboxylate groups are deprotonated leading to the repulsion of negative charges (COO^-), again causing disruption.

Recently a review made the point about the potentialities of granules for food applications, and particularly emulsifying properties. It stressed the fact that properties of granules could be improved by subtilisin treatments to enhance emulsifying properties, and by delipidation to develop film properties or encapsulation applications (Marcet et al., 2022).

A phase diagram drawing the different states of granules as function of pH and ionic strength has been established (Figure 6.4) (Sirvente, 2007; Anton et al., 2008). In their soluble state (high ionic strength or neutral pH), phosvitin and HDL are not in a molecular form but are assembled into soluble micelles of 100–200 nm diameter that can spread and adsorb at oil-water interfaces. The similarity with casein micelles must be pointed out here. On the other hand, in low salt conditions and acidic pH, granules are in their native aggregated state (aggregates of 1–8 μm). Between these two extreme conditions, a large intermediate state includes a continuum between granules aggregates (1–8 μm) and granules micelles (100–200 nm) at different ratio depending on the pH/ionic strength values.

Recently, Li et al. (2022) explored the associated effect of pH and NaCl on granules structure, thus revealing the Janus effect of NaCl on granular structure at varied pH values. This effect could be explained by the counter-ions adsorbed on granules surface. When pH is above 5.0, dissociated granules

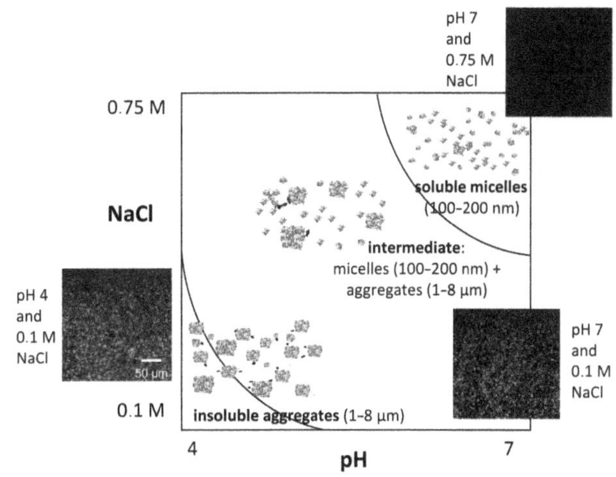

FIGURE 6.4 Physical state of granules as function of pH and ionic strength; and confocal micrographs showing the structure of granules dispersions (30 mg of protein/ml, pH 4, 0.1 M NaCl, pH7 0.1 M NaCl, and pH7 0.75 M NaCl). (From Sirvente et al., 2007; Anton et al. 2008.)

are negatively charged with adsorption of highly hydrated Na^+ and Ca^{2+} as counterions. Due to the strong hydration repulsion of Na^+ and Ca^{2+}, dispersions with good colloidal stability are formed. When pH is below 4.0, Cl^- are adsorbed on granules surface, which fail to provide sufficient hydration repulsion due to their low degree of hydration, leading to phase separation.

6.2.3.2 HDL

HDL are the second group of lipoproteins of yolk and account for about the sixth of its dry matter and 35% of its proteins (Anton, 2007c). They are specifically localized in granules and are designated as lipovitellin. HDL are the result of the proteolytic cleavage of a precursor (vitellogenin), which is synthesized in the liver under the regulation of estrogens. It can be noticed that phosvitin comes from the same precursor.

Vitellogenin interacts with lipids and two calcium ions, and one zinc ion (Montorzi et al., 1995), and is secreted into the blood circulation, from where it reaches the oocyte. During the transfer into oocyte by endocytosis, HDL undergo a proteolytic cleavage generating several polypeptides that preserve its structure (Banaszak et al., 1990). Protein fragments coming from this cleavage form the HDL-phosvitin complexes, basic element of granules.

In native egg yolk, HDL are complexed to phosvitin to form the granular structure through phosphocalcic bridges. HDL are made up of 75–80% proteins and 20–25% lipids. Hence, their density is close to that of proteins (1.120 g/ml). Their lipids are composed of 65% phospholipids, 29% triglycerides, and 5% cholesterol (Cook & Martin, 1969).

HDL have a molecular weight of about 400 kDa and a diameter from 7 nm to 20 nm (Burley & Cook, 1961). Unlike LDL, HDL have no spherical micelle-like structure but rather a pseudo-molecular structure close to that of globular proteins. Consequently, we call them lipoproteins but they are not formally lipoproteins, bearing more of a resemblance to lipid transfer proteins. HDL are present in the form of a dimer of two monomers of about 200 kDa. Each monomer of HDL is a very structured globular protein containing a cavity in the form of a funnel composed essentially of two β-sheets predominantly constituted by hydrophobic amino acids (Anderson et al., 1998). The cavity has a volume of 68 nm^3 large enough to fix about 35 molecules of phospholipids concentrated on a monolayer interacting with the hydrophobic amino acid residues. Triglycerides can then be enclosed in this hydrophobic cavity through new hydrophobic zones created by phospholipids (Banaszak et al., 1990; Timmins et al., 1992). Thus, phospholipids are entrapped and protected by the HDL structure, and phospholipase C can liberate only 6% of the lipid phosphorus of HDL. Four other domains (three β-sheets and one helix) form the rest of the polypeptide. Five disulfide bridges maintain the structure, but the different domains of HDL complex interact through numerous salt and ionic bridges, hydrophobic interactions occurring also (Anderson et al., 1998).

Two sub-groups of HDL are separated by ion exchange chromatography: α- and β-HDL in a ratio of 1:1.5 (Radomski & Cook, 1964). α-HDL contain six times more sialic acid and two times more phosphorus than β-HDL (0.5% *vs* 0.27%) (Kurisaki et al., 1981). Consequently, α-HDL are more acidic than β-HDL while the latter are more water-soluble. Except for these differences, the two types of HDL have a similar chemical composition. About 30% of amino acid residues of both types of HDL are hydrophobic.

At pH values below 7.0, a reversible dimerization of HDL occurs. Dimers of β-HDL are dissociated more rapidly than those of α-HDL. The dissociation reaches 50% at pH 10.5 and 7.8, respectively, for α- and β-HDL. The presence of 1 M NaCl hinders the dissociation in alkaline medium (Bernardi & Cook, 1960), and temperature increases slows dissociation. Sulfhydryl and protein phosphorus groups are not directly involved in the dissociation reactions. All these data indicate that the dimerization proceeds by both electrostatic and hydrophobic interactions. The dissociation process unquestionably alters the organization, hydration, and release of phospholipids in the cavity. HDL precipitate in water and become soluble when ionic strength is higher than 0.3 M NaCl.

Each monomer of HDL is composed of about five main apoproteins, with molecular weights ranging from 35 to 110 kDa (Anton, 2007c). Their respective quantities are 21% for the 110 kDa protein; 16% for the 100 kDa protein; 20% for the 80 kDa protein; 14% for the 50 kDa protein; 28% for the 35 kDa protein. Apoproteins of HDL are glycosylated (0.75% carbohydrate), with carbohydrate chains made up of mannose, galactose, glucosamine, and sialic acid.

6.2.3.3 Phosvitin

Phosvitin is synthesized from a precursor, vitellogenin, from which yolk HDL also comes (Taborsky, 1983). After cleavage of the precursor, the two polypeptides are liberated, but they interact through phosphocalcic bridges to form the granular structure.

Phosvitin from hen egg yolk is a phosphoglycoprotein that represents 4% of yolk dry matter. Of the known proteins, it is the most phosphorylated: almost 50% of its amino acid residues are serine, out of which 96% are phosphorylated (Clark,

1985). The high phosphorus content of phosvitin is responsible for its strong affinity for metal ions, particularly for iron, 95% of yolk iron being bound to phosvitin. Consequently, due to its peculiar physical and chemical characteristics, phosvitin possesses a variety of functional and biological properties (Anton et al., 2007). Particularly, its phospholipid antioxidant activity was demonstrated in both egg yolk emulsion and meat model systems (Jung et al., 2013).

Phosvitin represents about 11% of yolk proteins. It is a mixture of two proteins: α-phosvitin (160 kDa) and β-phosvitin (190 kDa) (Itoh et al., 1983). Both phosvitins are aggregates of many different polypeptides: α-phosvitin contains three or four sub-units of 35–40 kDa, while β-phosvitin contains four or five sub-units of 45 kDa. Amino acid residues of these sub-units consist of about 50% serine. There are very few or no sulfur-containing amino acids. Alpha-phosvitin has about 6% carbohydrate: 2.5% hexose, 1% hexosamine, and 2% sialic acid; whereas β-phosvitin has only 2% carbohydrate (mainly hexose) (Ito & Fujii, 1962). The two phosvitins contain 3% (α) and 10% (β) phosphorus (present as phosphoric acid bound to serine residues), which accounts for about 80% of total protein phosphorus of yolk (Joubert & Cook, 1958). The phosphoserines are arranged in a singular way, forming blocks that can carry up to 15 consecutive residues (Byrne et al., 1984). This confers to the protein a very large and central hydrophilic area surrounded by two small hydrophobic areas at the N- (9 residues) and C- (3 residues) terminal parts. Phosvitin is therefore an elongated molecule of 28 nm × 1.4 nm.

Phosvitin is soluble in pure water and easily precipitated with 0.1 M magnesium sulfate. The isoelectric point of phosvitin is 4.0 (Ternes, 1989), and at neutral pH it is highly charged: -179 mV. At neutral pH, phosvitin predominantly exists in a random coil conformation, whereas a β-type conformation predominates at acidic pH (67% β-sheet) (Taborsky, 1974; Damodaran & Xu, 1996). Owing to its unordered structure, phosvitin is poorly modified by heat treatment: heating at 110°C for 20 minutes does not cause its aggregation or insolubility (Albright et al., 1984).

Phosvitin possesses a very strong metal chelating property due to its poly-anionic character. This unique primary structure makes this protein one of the strongest metal chelating agents (Hegenauer et al., 1979). Whole egg contains about 2 mg of iron per 100 g, and 95% of yolk iron is bound to phosvitin (Albright et al., 1984), which contains about 6 mg of iron per gram. One phosvitin molecule can anchor 113 ions of Mn^{2+} or 120 of Co^{2+}, and the binding is not affected by temperature up to 60°C (Grizzuti & Perlmann, 1973). These authors have also described the interaction with Mg^{2+} and Ca^{2+}: at pH 6.5, phosvitin binds up to 103 and 127 ions/molecule, respectively. However, at pH 4.5, it can only bind 40 and 32 ions of Mg^{2+} and Ca^{2+}, respectively. By equilibrium dialysis, Hegenauer et al. (1979) described a great affinity between ferric ions and phosvitin ($K_f = 10^{18}$ M). Furthermore, this protein can chelate ferric ions better than molecules like citrate or nitrile-triacetate. This capacity of phosvitin to capture easily Ca^{2+} and Fe^{2+} ions confers it good preservative and antioxidant properties (Lesnierowski & Stangierski, 2018).

Phosvitin in egg yolk is not saturated with iron and additional iron can be fixed. At saturation, two moles of organic phosphorus of phosvitin can bind one mole of iron (the maximum Fe/P ratio is then 0.5) (Taborsky, 1974). Iron is bound in ferric form in a tetrahedral coordination structure with the oxygen of the phosphoserine residues.

The best conditions for iron binding capacity of purified phosvitin are almost neutral pH (6.5) and low ionic strength (0.15 M) (Castellani et al., 2004). This is probably because at this pH phosphates are completely ionized, and there are not enough salt ions to prevent the iron binding by phosvitin. The maximum iron binding capacity was found to be 115 mg of iron/mg of phosvitin, which corresponds to one iron per two phosphate groups. In more acidic conditions (pH values lower than 3.5), there is no significant fixation (Castellani et al., 2004). It is likely that precipitation occurs at pH values lower than 2.9, where the secondary structure of phosvitin changes. The weak fixation obtained at acidic pH is not related to an inhibition of iron autoxidation because, at this pH value and for a time of 1 hour, this process is largely allowed. Thus, the presence of partially protonated phosphates could inhibit the formation of an iron/phosvitin complex.

Complexes formed under favorable conditions (pH 6.5 and ionic strength 0.15) are very stable. The decrease of pH to values where the formation of iron/phosvitin complex is normally prevented, does not release the iron bound by phosvitin (Castellani et al., 2004). Some modifications are expected when the complex is formed. Castellani et al. (2004) hypothesized that, in an initial step, the iron ions interact with only one phosphate group, and once their interaction is formed, another phosphate completes the process (tetrahedral stoichiometry), or in the case where the complex has an octahedral stoichiometry, two other molecules (i.e. water molecules) could complete the system. Once this condition is reached, the activation energy required to dissociate the complex could become high. Consequently, once the complex is formed, its dissociation by protons is not observed, even in the presence of a high proton activity and in conditions where the complex formation is prevented. Only a chelating agent (o-phenanthroline or EDTA) with stronger iron binding capacity than phosvitin is able to release iron from the complex (Albright et al., 1984).

6.3 Constituent Extraction

6.3.1 LDL Extraction

Initially, ultracentrifugation has been favored for LDL extraction because the low density of these assemblies (0.98) allowed an efficient separation. However, this type of technique is extremely time consuming (several days of centrifugation) and the quantity of extract is very low (only few grams of LDL), prohibiting a scale-up of the method. A method of extraction of LDL from plasma, transferable at a pilot scale, has been then finalized (Moussa et al., 2002). The principle of separation is the addition of ammonium sulfate (40% saturation) to eliminate livetins from the solution of plasma extracted from egg yolk. After 1 hour under stirring at 4°C, the blend is centrifuged at 10,000 g. The pellet obtained containing γ and β livetins is discarded. The supernatant is then dialyzed against distilled water for 20 hours at 4°C with frequent changes of baths. The dialysis, desalting samples, provokes the aggregation of LDL, which creams. An orange-colored floating dough containing 97% LDL is obtained. The yield of extraction is about 60%. A patent (Anton, Martinet, et al., 2001) protects this process. The purity of 97% obtained can be increased further by the use of gel filtration chromatography. Thirty grams of LDL extract are dispersed in 50 ml Tris-HCl 50 mM pH 7 buffer, and 12 ml of this dispersion are injected into the column of Ultrogel AcA 34 (Pall, East Hills, NY, USA). This additional chromatography step permits the separation of the LDL from contaminating livetins, and the recovery of pure LDL. This last step is efficient for refining LDL purification in the laboratory, but scaling-up is not feasible because of the low quantities treatable by this technique.

6.3.2 Livetin γ Extraction (IgY)

Various IgY extraction methods were reviewed in detail by De Meulenaer and Huyghebaert (2001). Overall, these methods can be divided into three groups: (i) Precipitation methods: mainly by ammonium or sodium sulphate (*e.g.* Akita & Nakai, 1993), and by polyethyleneglycol (PEG)) (Polson et al., 1985); (ii) Chromatographic methods: principally affinity chromatography (*e.g.* Verdoliva et al., 2000); and (iii) Ultrafiltration (Kim & Nakai, 1998). Polson's PEG precipitation method has become the most commonly used and most effective protocol. A simple and frequently used procedure is the water dilution method, described by Akita and Nakai (1992). IgY purification by means of protein A (*Staphylococcus aureus*) or protein G (*Streptococcus* species, strains C and G) affinity columns is not possible, since these proteins have no binding site for IgY.

Among liquid-liquid extraction techniques, the use of thermo-responsive aqueous micellar two-phase systems (AMTPS) appears to be an attractive approach since these comprise large amounts of water, thus being able to maintain the native conformation and biological activities of biomolecules if properly designed, including proteins and antibodies (Vicente et al., 2016). In addition, these systems only require a surfactant, water and temperature to be formed. Upon phase separation, two distinct environments are created with different polarities, namely a hydrophobic surfactant-rich phase and a hydrophilic surfactant-poor phase. The different characteristics of the phases formed are also pointed out as advantageous features, especially when working with real matrices. Some interesting developments were recently carried out by combining AMTPS comprising non-ionic surfactants and surface-active ionic liquids (SAILs) as co-surfactants to improve the extractive performance and selectivity of these processes, features achieved by the tailoring ability of ionic liquids (ILs) (Vicente et al., 2022).

6.3.3 HDL Extraction

Literature reports several methods of extraction based on salt precipitation. Joubert and Cook (1958) suggested that egg yolk be diluted in a solution of $MgSO_4$ 0.4 M in order to disrupt HDL-phosvitin complexes. The solution is then diluted two times, and centrifuged at 10,000 g for 30 min. The pellet (principally phosvitin, because this protein forms aggregates in the presence of low $MgSO_4$ concentrations) is discarded, the supernatant diluted two times with distilled water, and centrifuged again as before. The supernatant is collected, and diluted again with distilled water to attain a final concentration of $MgSO_4$ of 0.05 M. The pellet is dissolved in 50 ml of a 10% NaCl solution and then ultracentrifuged at 105,000 g for 2 hours. The solution is then centrifuged again to obtain a pellet rich in HDL. After this extraction, the HDL extract is purified on a hydroxyapatite column. The separation is based on the affinity of proteins for calcium ions.

6.3.4 Phosvitin Extraction

Depending on the isolation method used, egg yolk phosvitin contains between three and six atoms of iron per molecule (Mecham & Olcott, 1949; Albright et al., 1984). However, the study of iron binding activity of phosvitin requires that the protein at first be free of metal. So Wallace and Morgan (1986) have developed a method of extraction, starting from granules, employing precipitation of non-phosvitin proteins by $(NH_4)_2SO_4$ and recovery of phosvitin from the supernatant by dialyzing against water at pH 1.5. They described the presence of α and β-phosvitin polypeptides and of three additional phosphoproteins with high electrophoretic mobility: phosvettes. The phosvettes were not present in the more-often-used phosvitin preparations. In fact, the protein yield for their isolation method was only 0.96% of egg yolk dry matter, considerably less than the total phosvitin present in egg yolk.

A valuable purification method includes an extraction phase based on the insolubility of Mg^{2+}/phosvitin salt, and further ion-exchange chromatographic fractionation was developed by Castellani et al. (2003). Yolk is fractionated into granules and plasma following the method of McBee and Cotterill (1979). Granules are diluted at 10% (w/v) with NaCl 1.74 M. This provokes the disruption of the phosphocalcic bridges binding phosvitin and HDL, and their replacement with sodium ions. The solution is then dialyzed against distilled water for 24 hour. Calcium ions and excess sodium ions are eliminated by the dialysis. The pH of the solution is adjusted to 7.25 and the solution is centrifuged at 10,000 g for 30 min at 4°C.

The resulting supernatant is then diluted with a solution of magnesium sulfate to obtain a final concentration of magnesium of 0.2 M. This provokes the precipitation of phosvitin after a centrifugation at 10,000 g for 30 min. The pellet enriched in phosvitin contains about 15% impurities (β-livetin, α-HDL and β-HDL). This enriched pellet can be completely purified by the use of anion exchange chromatography through MonoQ resin (Castellani et al., 2003). Furthermore, the phosvitin obtained is free of all metals that are retained by the anionic resin. It is a necessary preliminary to the study of cation interactions with phosvitin.

Jiang et al. (2019) isolated phosvitin from egg yolk granules by dissolving them in $(NH_4)_2SO_4$ solution, heating at 80°C for several and different reaction times and treating the solution with ultrasound at 600 W for 10 min. After dialysis and centrifugation, the phosvitin was isolated in the supernatant. Under the best set of conditions assessed, the purity of the phosvitin was 80% and no significant differences between the properties of the extracted phosvitin and those of the phosvitin standard were detected. Bearing in mind that the different phosvitin extraction procedures may exert different effects on the structure and functionality of the remaining granular proteins, these proteins may show a wide variety of functional properties.

Recently, Giarratano et al. (2020) extracted phosvitin from egg yolk granules by high hydrostatic pressure at 400 MPa for 5 min. After applying such pressure, the supernatant was concentrated by ultrafiltration with a 10 kDa membrane, obtaining a concentrate with a phosvitin content of 26.00 ± 4.12% w/w.

6.4 Effects of Processes on Yolk Structures and Functionalities

Egg yolk multi-scale structures in relation with their functionalities have been detailed above in this chapter. However, these structures are affected by different thermo-mechanical processes used to modify egg yolk (pre-treatments) or to formulate food products (post-treatments). Consequently, it is of primary importance to understand the effect of transformation, formulation and/or stabilization processes on these structures and the functionalities of yolk and its fractions. Different classical and innovative processes directly applied on egg yolk or its fractions, plasma and granules, or on emulsions made with yolk, are reviewed below.

6.4.1 Dynamic High-Pressure Pre-Treatment

Mechanical processing is an effective physical method to modify or standardize the properties of egg yolk constituents (Li et al., 2021, for a review). The development of high-pressure homogenizers allows having technology available for this kind of treatment.

The effect of rotor-stator and high-pressure homogenization (30, 100 and 200 bar at 20°C) on yolk, plasma and granules dispersions at pH 4 or 7 and 0.1 or 0.75 M NaCl has been followed by Sirvente et al. (2007), to understand how the emulsification process impacts their structures. Dispersion characteristics (protein solubility, rheological behavior, aggregation and protein denaturation) and subsequent emulsifying properties (oil/water 60/40 emulsions: droplet size and flocculation) of non-treated and treated dispersions were compared.

Concerning plasma (and obviously LDL that form 85% of plasma dry matter), it appears clearly that the effect of dynamic pressure on LDL structure depends on environmental conditions. All levels of treatment do not affect LDL structure at pH 7 regardless the ionic strength and at pH 4 and 0.1 M NaCl (Figure 6.5). In these conditions, pressure does not modify structure of LDL. Conversely, at pH 4 with 0.75 M NaCl, which are conditions endured during industrial mayonnaises formulation, LDL are disrupted and reorganized in big heaps with a non-spherical shape from 100 bar treatment (Figure 6.5).

Up this threshold reached, the constituents released (neutral lipids and proteins) reorganize and aggregate probably through hydrophobic interactions. These changes in LDL structure and organization are associated with a decrease in LDL solubility,

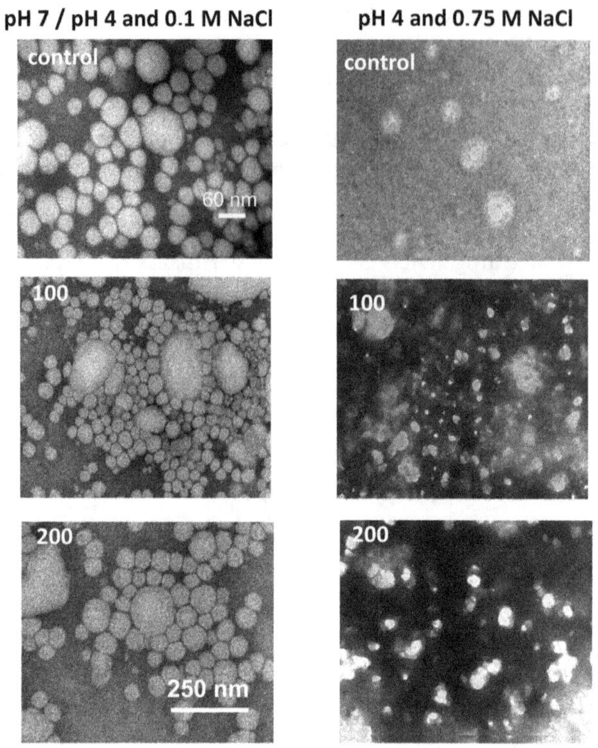

FIGURE 6.5 Transmission electron microscopy of solutions of plasma after treatment at 100 or 200 bars in different conditions. (From Sirvente et al., 2007.)

an increase in LDL apparent viscosity, and a trouble in the suspension of LDL (Figure 6.6).

However, these modifications only slightly affect the emulsifying properties of plasma as the same particle diameter and bridging flocculation were observed for emulsions made with both untreated and treated dispersions. In this experiment, a high viscosity obtained after an intense shear of the emulsion and a subsequent time equilibration is the evidence of droplet flocculation due to depletion (dependent on size and number of droplet aggregates). The stress value corresponding to the drop of viscosity gives an estimation of the strength to apply to break the droplet aggregates. It can only be observed a decrease in the shear stress level to attain a separation of oil droplets for 100 bar-treated yolks, traducing a less important flocculation level when plasma is pre-treated at 100 bar.

Concerning granules, a very limited increase of viscosity and a decrease of solubility and very slight structural changes in the dispersions are visualized by confocal microscopy under dynamic high-pressure. Under experimental conditions used here (pH 4 and 0.75 M NaCl), the constituents of granules (HDL and phosvitins) are in form of micro-particles (1–8 µm) bound together by phosphocalcic bridges between their serine residues to form insoluble complexes. It appears that dynamic high-pressure pre-treatment (up to 200 bar) increased the density of the protein network joining the granules subcomponents without altering more the microstructure. At 100 bar pressure, confocal micrographs (not shown) revealed an open microstructure with large protein clusters and big pores. When

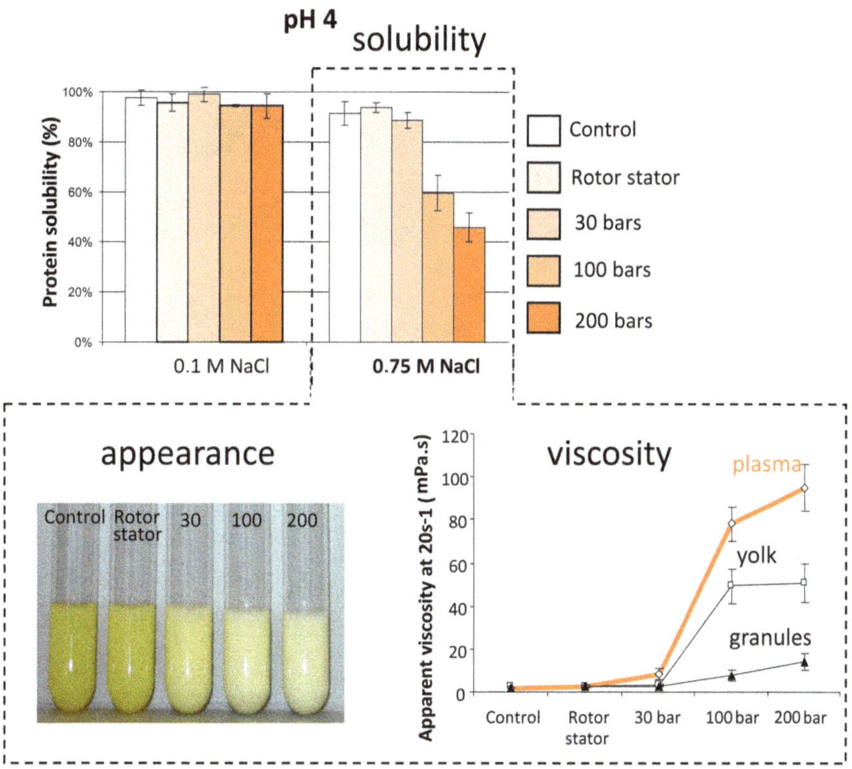

FIGURE 6.6 Changes in solubility, appearance, and viscosity of plasma solutions after treatment with rotor-stator, 30, 100, or 200 bars under different conditions. (From Sirvente et al., 2007.)

granules dispersions were homogenized at 200 bar, the protein network became finer and the microstructure became considerably denser with a decrease in pore size and an increase in the degree of interconnectivity between clusters. The effects of pre-treatments of granules on their emulsifying properties led to a slight decrease in bridging and depletion flocculation. These results suggest that the constituents of pre-treated dispersions can adsorb at the oil-water interface as whole aggregates or as fragments disrupted during the homogenization process.

More recently, Suhag et al. (2021) investigated the effect of micro-fluidization treatment (103, 138, 172, and 207 MPa) on pH, refractive index (RI), color, rheology, particle size, emulsifying properties, and thermal denaturation of egg yolk. Color parameters were significantly affected by micro-fluidization treatment. Rheological measurements showed the conversion of Newtonian nature to the shear-thinning nature of egg yolk upon micro-fluidization treatment. Further, yield stress (τ_o), consistency constant (K), and flow behavior index (n) were affected by the process. Micro-fluidization caused a reduction in particle size of yolk with an increase in pressure up to 172 MPa and particle size increased at 207 MPa. Emulsifying activity index (EAI) increased from 30.49 m^2/g (control) to 32.41 m^2/g (138 MPa) but decreased as micro-fluidization pressure was increased. Emulsifying stability index (ESI) increased at 103 MPa but decreased as micro-fluidization pressure increased further. Thermal analysis revealed an increase in denaturation temperature as micro-fluidization temperature increased.

6.4.2 Static High-Pressure Treatment

6.4.2.1 Pre-Treatment

High-pressure treatments, approved as an effective innovative mean to eradicate microorganisms in foods, is known to have a weaker effect on protein structures than thermal treatments like pasteurization or sterilization. In that sense, it is a promising technology to stabilize food products, and it could represent an alternative to eliminate gently microorganisms. It is of interest to assess the effect of static high-pressure treatments (up to 6000 bar) on the particular structure of LDL, main contributor of egg yolk emulsifying properties (Speroni et al., 2005). At pH 3, high-pressure treatment (between 2000 and 6000 bar) has no significant effect on LDL structure and solubility, whereas aggregation (destructuration-merging-restructuration) happens in the same treatment conditions at pH 8. At this latter pH, a progressive shift of the size of the first population from 30 nm to about 200 nm was observed when the pressure increased up to 6000 bar. It appears that the modifications of LDL observed at pH 8 do not alter the capacity of protein adsorption at the oil-water interfaces: the same percentage of adsorbed proteins and the same size of droplets were found with or without treatment. But the emulsions prepared with these treated samples show a serious decrease of depletion and bridging flocculation, whereas there is no differences of droplet size between all samples (treated or not regardless the pH). In conclusion, a static high-pressure treatment of LDL dispersions combined with an alkaline pH is a very interesting pre-treatment that can lead to the fabrication of more stable oil/water emulsions.

6.4.2.2 *Post-Treatment*

Another way to use static high-pressure treatments is to apply it directly on the emulsions already prepared. This was realized on yolk emulsions at pH 3 and 7 (Anton, Chapleau, et al., 2001). Oil droplet size, interfacial protein concentration and composition are not altered by the high-pressure treatments (2000–5000 bar). The rheological study demonstrated that high-pressure treatment induces a significant increase of viscosity in emulsions prepared at pH 7, whereas no obvious thickening was noticed in emulsions prepared at pH 3. Furthermore, as the protein concentration increases, the effect of high-pressure is more pronounced. Microbial analysis revealed that high-pressure treatment allows an efficient reduction of the microbial charge of emulsions and that this effect is greater at acidic pH than at neutral pH.

In this context, it is interesting to notice that at neutral or quietly basic medium (pH 8) yolk emulsions are flocculated and a pre-treatment by high-pressure performs modifications that decrease their tendency to flocculate oil droplets. In another way, at the same pH, the treatment of flocculated emulsions made with yolk increases more this phenomenon.

6.4.3 Thermal Treatment

6.4.3.1 *Pre-Treatment*

Globally, yolk undergoes a gelation when it is subjected to severe heat treatments or freezing-thawing, and LDL are responsible for this gelation, while the other constituents of yolk do not directly participate (Le Denmat et al., 1999; Anton et al., 2008). It can be specified that we are not talking here about pasteurization of egg yolk, a treatment which, although it can modify the emulsifying performances of the yolk (Sakai & Ikeda, 2022), does not cause drastic modifications in its structures and functional properties.

So, considering drastic heating, LDL solution (4% w/v) starts denaturing at 70°C and form gels at 75°C, more stable than ovalbumin or BSA gels. Unlike ovalbumin and BSA, LDL present a heat-induced gelation in a large range of pH (4–9) with a minimal value around their pI. Between pH 6 and 9, LDL solutions form coagulum gels (opaque) whereas, at extreme pH (4–6 and 8–9), LDL solutions form translucent gels.

Heat-induced gelation of yolk is governed by the unfolding of proteins during heating. Then, the functional groups are exposed and attracted to one another through hydrophobic interactions resulting in a gel. The primary stage of these phenomena (freezing-thawing and heat-induced gelation) is the disruption of the LDL structure.

The critical concentrations (Cg) for heat gelation (90°C/2 min) of dispersions of yolk, plasma and granules were measured in pH and NaCl ranges of, respectively, 3–7 and 0.15–0.55 M (Anton et al., 2000) (Figure 6.7). In the whole domain, Cg range for plasma is 12–28 mg protein/ml, whereas it is 26–120 mg protein/ml for granules. For yolk solution, Cg is comprised between 16 and 39 mg protein/ml.

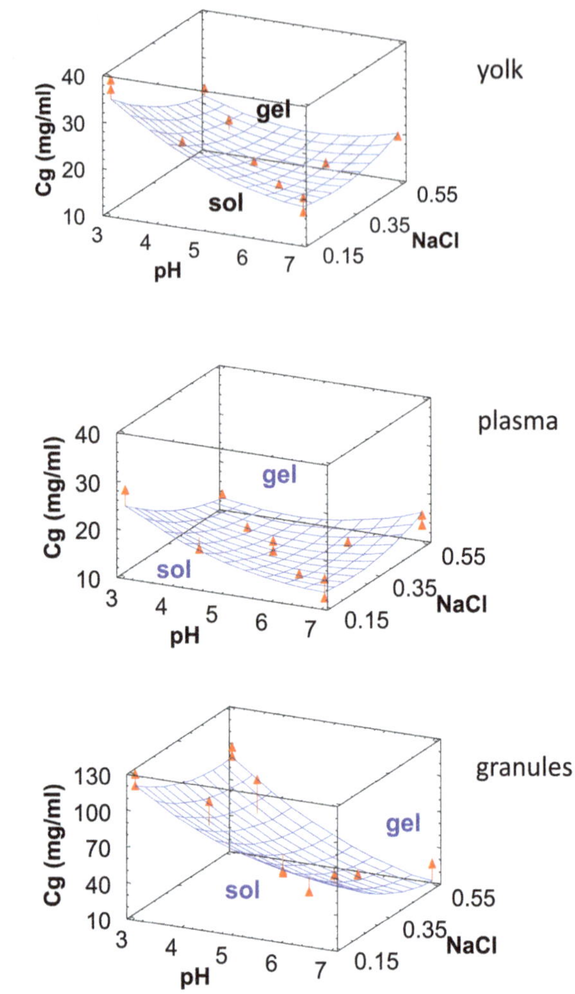

FIGURE 6.7 Critical gelation concentrations as function of pH and ionic strength for yolk, plasma or granules solutions treated by temperature (90°C/2 min). (From Denmat et al., 1999.)

This confirms the preponderant influence of LDL, the major compound of plasma, in the gelation of yolk under heating. This underlines also the excellent capacity of granules to resist to heat treatments, and this could be used for industrial applications. In reality the native structure of granules does not protect proteins against denaturation during heating, but the phosphocalcic bridges, maintaining the structure and resisting to heat treatment, restrict the aggregation of granules constituents (phosvitin and HDL), limiting the increase in apparent viscosity and avoiding gelation of granules solutions and the dramatic decrease in protein solubility. This suggests that granules, in their native form, can be heated to pasteurization temperature without losing their emulsifying properties.

In a same way, it has been observed that pre-heating of egg yolk (74°C for 0–40 min) alters the physico-chemical properties of subsequently formulated emulsions, depending on time of heating and environmental conditions (Guilmineau & Kulozik, 2006).

Furthermore, it has been established that destructuration of LDL by heating resulted in a leakage of lipids conducting to their strong oxidation due to contact of oxygen in water (Ren et al., 2022). Moreover, the free lipids were oxidized and their oxidative products (hexanal, (E)-2-octenal, (E,E)-2,4-decadienal, 1-octen-3-ol, and 2-pentylfuran) were components of off-odor in egg yolk. This stresses the fact that the integrity of native LDL structure is clearly a protection against lipid oxidation as it maintains separated unsaturated fatty acid chains protected inside the LDL lipid core, from external oxygen and catalyzers responsible of oxidation. We can also suppose that an observed oxidation in a LDL solution is the sign of degradation of the integrity of LDL structure.

6.4.3.2 *Post-Treatment*

Heat gelation of oil-in-water emulsions prepared with egg yolk may differ from that of egg yolk solutions because of the influence of oil droplets on network formation. The influence of oil droplets on the gelation of oil-in-water emulsions made with yolk was investigated at 3 pH: 3, 5 and 7 with a constant NaCl concentration: 0.55 M (Anton, Le Denmat, et al., 2001). Emulsions made with yolk and plasma have a similar gelation process with oil droplets acting as inactive fillers. Moreover, acidic conditions provided the fastest gelation of yolk solutions and emulsions. In contrast, oil droplets behave as active filler particles in emulsions prepared with granules, and reinforced the gel strength.

The contribution of oil droplets (or filler particles) to the viscoelasticity of the heat-set emulsion gel depends on the filler particle nature, size and concentration, and on the filler-matrix interactions. Since this study was realized at constant oil volume fraction and similar droplet diameter (d_{32}), only interactions involving the interfacial film of the droplets are expected as a source of variations. The presence of emulsion droplets did not cause severe changes in the gelation process compared to solutions, and observation of TEM micrographs revealed that oil droplets coated with yolk proteins do not act principally as active fillers (Figure 6.8). In conclusion, to the results of rheology and TEM, it can be assumed that heat-set emulsion gels prepared with yolk have a mechanism of gel formation and characteristics of network structure, similar to those of plasma. Consequently, plasma proteins exert control on the formation of the matrix in the aqueous phase, and as well on the interactions between protein-coated droplets and this matrix.

Inversely, with granules, the value of the elastic modulus (G') of the heat-set emulsion gel is much higher than that of the corresponding protein gel (Figure 6.8). It is obvious that the dispersed oil droplets behave as strong active filler particles and hence, dramatically enhance the gel strength. It is conceivable that cross-links between oil droplets in emulsions prepared with granules are due to interactions between granules particles, as proposed these hypothetical schemes (Figure 6.9).

Otherwise, this capacity of granules to connect in all the volume can be used to stabilize air bubbles introduced in granules-coated mayonnaises (Anton et al., 2007). This process can permit to conceive reduced-fat mayonnaises (<30% fat) with the same texture as traditional one.

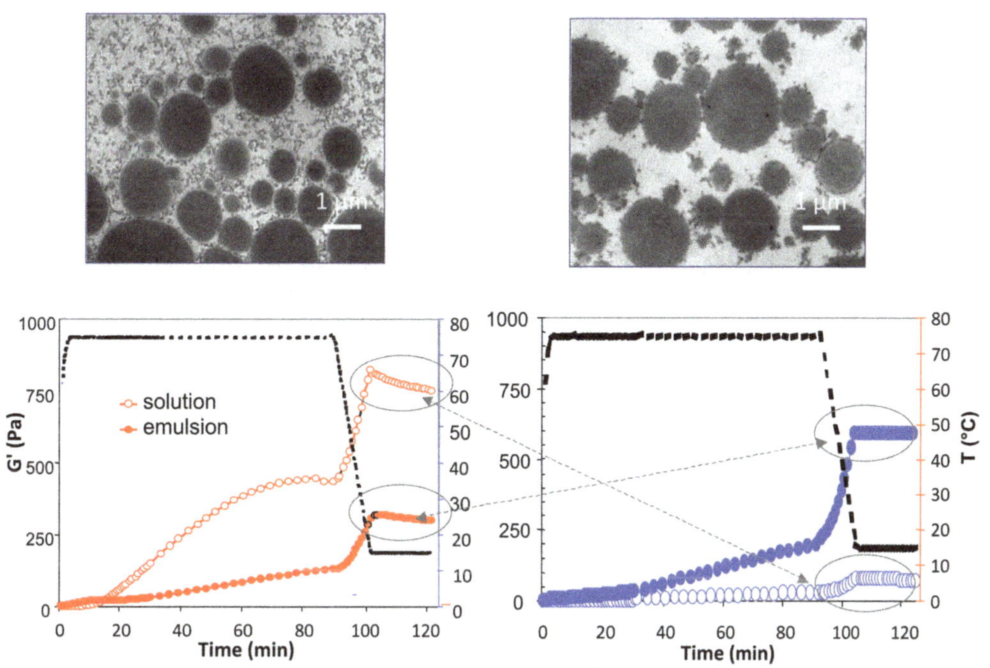

FIGURE 6.8 Development of storage modulus G' of plasma (or yolk) (A) and granules (B) solutions (55 g L^{-1} protein) and emulsions (55 g L^{-1} protein, 30 wt% oil) at 1 Hz during heating/cooling cycle, at pH7, and corresponding electron micrographs (TEM) of heat-set emulsion gels. (From Anton et al., 2001.)

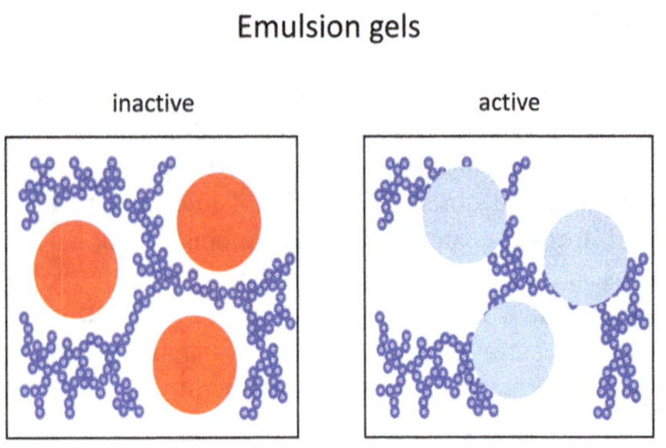

FIGURE 6.9 Scheme explaining the different behavior between inactive and active fillers for emulsion gels.

6.4.4 Freeze-Thawing

In industrial applications, egg yolks may be frozen to facilitate transportation and storage, and extend their shelf life. The benefits of freezing egg yolks also include preventing microbial growth and spoilage, retaining flavor and color, and inhibiting various degradative chemical reactions (Fei et al., 2021). However, when the temperature is less than −6°C, egg yolks lose fluidity, and gelation occurs. The occurrence of this phenomenon increases the egg yolk viscosity, making it difficult to combine with other ingredients. Hence, it limits the egg yolk functionality, including its emulsifying ability and dispersibility, and reduces the utility of frozen-thawed egg yolks. From a recent study of Wang et al. (2021), the gelation mechanism of frozen-thawed egg yolk may include two aspects. First, freezing promotes the formation of small aggregates, and this aggregation contributes to the interaction of egg yolk protein with lipids and water. Consequently, the content of bound water and immobile water increases, and the size of egg yolk particles increases, resulting in a decrease in fluidity of the egg yolk. Second, ice crystal formation during the freezing process mechanically destroys the conformation of egg yolk proteins. It results in the enhancement of hydrogen bonds and the exposure of hydrophobic amino acids, which both increase lipoprotein interactions and promotes aggregation of egg yolk protein and lipid release. This study may provide theoretical support for practical gel strength and a new perspective to control the application of frozen-thawed egg yolks in food processing.

6.5 Conclusion

Hen egg yolk consists of complex multi-scale structures of proteins and lipids with different organization levels (plasma and granules), largely depending on medium conditions and processes applied for extraction, transformation, stabilization and/or formulation of food products. These structures are the key of the remarkable techno-functional properties of egg yolk and we know, for instance, that emulsifying properties of egg yolk are largely due to the peculiar structure of LDL contained in the plasma. For the future, the use of egg yolk, and specially its fractions, for industrial purpose, has then to be conducted by the knowledge of this fourfold relation: structures-environment-processes-functionalities. This is the unique way for taking better advantages of the functionalities of plasma and granules fractions, presently not used at the industrial scale.

REFERENCES

Akita, E. M., & Nakai, S. (1992). Immunoglobulins from egg yolks: Isolation and purification. *Journal of Food Science*, *57*(3), 629–634. https://doi.org/10.1111/j.1365-2621.1992.tb08058.x

Akita, E. M., & Nakai, S. (1993). Comparison of four purification methods for the production immunoglobulins from eggs laid by hens immunized with an enterotoxigenic E. coli strain. *Journal of Immunological Methods*, *160*(2), 207–214. https://doi.org/10.1016/0022-1759(93)90179-b

Ala-Korpela, M., Pentikäinen, M., Korhonen, A., Hevonoja, T., Lounila, J., & Kovanen, P. T. (1998). Detection of low-density lipoprotein particle fusion by proton nuclear magnetic resonance spectroscopy. *Journal of Lipid Research*, *39*(8), 1705–1712. https://doi.org/10.1016/S0022-2275(20)32201-X

Albright, K. J., Gordon, D. T., & Cotterill, O. J. (1984). Release of iron from phosvitin by heat and food additives. *Journal of Food Science*, *49*(1), 78–81. https://doi.org/10.1111/j.1365-2621.1984.tb13674.x

Anderson, T. A., Levitt, D. G., & Banaszak, L. J. (1998). The structural basis of lipid interactions in lipovitellin, a soluble lipoprotein. *Structure*, *6*(7), 895–909. https://doi.org/10.1016/s0969-2126(98)00091-4

Anton, M. (2007a). Composition and structure of hen egg yolk. In R. Huopalathi, R. Lopez-Fandino, M. Anton, & R. Schade (Eds.), *Bioactive egg compounds* (pp. 1–6). Springer. https://doi.org/10.1007/978-3-540-37885-3_1

Anton, M. (2007b). Low-density lipoproteins (LDL) or lipovitellenin fraction. In R. Huopalathi, R. Lopez-Fandino, M. Anton, & R. Schade (Eds.), *Bioactive egg compounds* (pp. 7–12). Springer. https://doi.org/10.1007/978-3-540-37885-3_2

Anton, M. (2007c). High-density lipoproteins (HDL) or lipovitellin fraction. In R. Huopalathi, R. Lopez-Fandino, M. Anton, & R. Schade (Eds.), *Bioactive egg compounds* (pp. 13–16). Springer. https://doi.org/10.1007/978-3-540-37885-3_3

Anton, M. (2013). Egg yolk structures, functionalities and processing: A review. *Journal of the Science of Food and Agriculture*, *93*(12), 2871–2880. http://dx.doi.org/10.1002/jsfa.6247

Anton, M., Castellani, O., & Guérin-Dubiard, C. (2007). Phosvitin. In R. Huopalathi, R. Lopez-Fandino, M. Anton, & R. Schade (Eds.), *Bioactive egg compounds* (pp. 17–24). Springer. https://doi.org/10.1007/978-3-540-37885-3_4

Anton, M., Chapleau, N., Beaumal, V., Delépine, S., & de Lamballerie, M. (2001). Effect of high-pressure treatment on rheology of oil-in-water emulsions prepared with hen egg yolk. *Innovative Food Science and Emerging Technologies*, *2*(1), 9–21. https://doi.org/10.1016/S1466-8564(00)00036-9

Anton, M., & Gandemer, G. (1997). Composition, solubility and emulsifying properties of granules and plasma of hen egg yolk. *Journal of Food Science*, *62*(3), 484–487. https://doi.org/10.1111/j.1365-2621.1997.tb04411.x

Anton, M., Le Denmat, M., Beaumal, V., & Pilet, P. (2001). Filler effects of oil droplets on the rheology of heat-set emulsion gels prepared with egg yolk and egg yolk fractions. *Colloids and Surfaces B: Biointerfaces*, *21*(1–3), 137–147. https://doi.org/10.1016/s0927-7765(01)00167-9

Anton, M., Le Denmat, M., & Gandemer, G. (2000). Thermostability of hen egg yolk granules: Contribution of granular structure. *Journal of Food Science*, *65*(4), 581–584. https://doi.org/10.1111/j.1365-2621.2000.tb16052.x

Anton, M., Martinet, V., Dalgalarrondo, M., Beaumal, V., David-Briand, E., & Rabesona, H. (2003). Chemical and structural characterisation of low-density lipoproteins purified from hen egg yolk. *Food Chemistry*, *83*(2), 175–183. https://doi.org/10.1016/S0308-8146(03)00060-8

Anton, M., Martinet, V., Tainturier, D., & Moussa, M. (2001). Procédé de production d'une fraction purifiée de lipoprotéines de faible densité du jaune d'œuf et milieu de conservation incorporant de telles lipoprotéines. French Patent n° 0100292. EP02290065B1.

Anton, M., Nau, F., & Lechevalier, V. (2008). Egg proteins. In G. O. Phillips & P. A. Williams (Eds.), *Handbook of hydrocolloids* (2nd ed.). Woodhead Publishing Limited.

Anton, M., Nau, F., & Nys, Y. (2006). Bioactive egg components and their potential uses. *World's Poultry Science Journal*, *62*(3), 429–438. https://doi.org/10.1017/S004393390600105X

Banaszak, L. J., Timmins, P., & Poliks, B. J. (1990). The crystal structure of a lipoprotein: Lipovitellin. In *Molecular biology of atherosclerosis: Proceedings of the 20th Steenbock symposium* (pp. 135–140). University of Wisconsin.

Bernardi, G., & Cook, W. H. (1960). Separation and characterization of the two high density lipoproteins of egg yolk, α- and β-lipovitellin. *Biochimica et Biophysica Acta*, *44*(1), 86–96. https://doi.org/10.1016/0006-3002(60)91527-4

Bitman, J., & Wood, D. L. (1980). Cholesterol and cholesteryl esters of eggs from various avian species. *Poultry Science*, *59*(9), 2014–2023. https://doi.org/10.3382/ps.0592014

Bruneel, C., Lemahieu, C., Fraeye, I., Ryckebosch, E., Muylaert, K., Buyse, J., & Foubert, I. (2013). Impact of microalgal feed supplementation on omega-3 fatty acid enrichment of hen eggs. *Journal of Functional Foods*, *5*(2), 897–904. http://dx.doi.org/10.1016/j.jff.2013.01.039

Burley, R. W., & Cook, W. H. (1961). Isolation and composition of avian egg yolk granules and their constituents α- and β-lipovitellins. *Canadian Journal of Biochemical Physiology*, *39*(8), 1295–1307. http://dx.doi.org/10.1139/o61-136

Byrne, B. M., van Het Schip, A. D., van de Klundert, J. A. M., Arnberg, A. C., Gruber, M., & Geert, A. B. (1984). Amino acid sequence of phosvitin derived from the nucleotide sequence of part of the chicken vitellogenin gene. *Biochemistry*, *23*(19), 4275–4279. http://dx.doi.org/10.1021/bi00314a003

Castellani, O., Guérin-Dubiard, C., David-Briand, E., & Anton, M. (2004). Influence of physico-chemical conditions and technological treatment on the iron binding capacity of hen egg phosvitin. *Food Chemistry*, *85*(4), 569–577. http://dx.doi.org/10.1016/j.foodchem.2003.08.002

Castellani, O., Martinet, V., David-Briand, E., Guérin-Dubiard, C., & Anton, M. (2003). Egg yolk phosvitin: Preparation of metal-free purified protein by fast protein liquid chromatography using aqueous solvents. *Journal of Chromatography B: Biomedical Sciences and Applications*, *791*(1–2), 273–284. http://dx.doi.org/10.1016/S1570-0232(03)00230-7

Causeret, D., Matringe, E., & Lorient, D. (1991). Ionic strength and pH effects on composition and microstructure of yolk granules. *Journal of Food Science*, *56*(6), 1532–1536. http://dx.doi.org/10.1111/j.1365-2621.1991.tb08634.x

Chalamaiah, M., Esparza, Y., Temelli, F., & Wu, J. (2017). Physicochemical and functional properties of livetins fraction from hen egg yolk. *Food Bioscience*, *18*, 38–45. https://doi.org/10.1016/j.fbio.2017.04.002

Chang, C. M., Powrie, W. D., & Fennema, O. (1977). Microstructure of egg yolk. *Journal of Food Science*, *42*(5), 1193–1200. http://dx.doi.org/10.1111/j.1365-2621.1977.tb14458.x

Clark, R. C. (1985). The primary structure of avian phosvitins. Contributions through the Edman degradation of methylmercaptovitins prepared from the constituent phosphoproteins. *International Journal of Biochemistry*, *17*(9), 983–988. http://dx.doi.org/10.1016/0020-711X(85)90243-5

Cook, W. H., & Martin, W. G. (1969). Egg lipoproteins. In E. Tria & A. M. Scanu (Eds.), *Structural and functional aspects of lipoproteins in living systems* (pp. 579–615). Academic Press.

Damodaran, S., & Xu, S. (1996). The role of electrostatic forces in anomalous adsorption behavior of phosvitin at the air/water interface. *Journal of Colloid and Interface Science*, *178*(2), 426–435. http://dx.doi.org/10.1006/jcis.1996.0137

Dauphas, S., Beaumal, V., Gunning, P., Mackie, A., Wilde, P., Vie, V., & Anton, M. (2007a). Structure modification in hen egg yolk low density lipoproteins layers between 30 and 45 mN/m observed by AFM. *Colloids and Surfaces B: Biointerfaces*, *54*(2), 241–248. http://dx.doi.org/10.1016/j.colsurfb.2006.10.027

Dauphas, S., Beaumal, V., Gunning, P., Mackie, A., Wilde, P., Vié, V., & Anton, M. (2007b). Structures and rheological properties of hen egg yolk low density lipoproteins layers spread at the air-water interface at pH 3 and 7. *Colloids and Surfaces B: Biointerfaces*, *57*(1), 124–133. http://dx.doi.org/10.1016/j.colsurfb.2007.01.017

Dauphas, S., Beaumal, V., Riaublanc, A., & Anton, M. (2006). Hen egg yolk low density lipoproteins film spreading at the air-water and oil-water interfaces. *Journal of Agricultural and Food Chemistry*, *54*(10), 3733–3737. http://dx.doi.org/10.1021/jf053174e

Davalos-Pantoja, L., Ortega-Vinuesa, J. L., Bastos-Gonzalez, D., & Hidalgo-Alvarez, R. (2000). A comparative study between the adsorption of IgY and IgG on latex particles. *Journal of Biomaterial Science, Polymer Edition*, *11*(6), 657–673. http://dx.doi.org/10.1163/156856200743931

De Meulenaer, B., & Huyghebaert, A. (2001). Isolation and purification of chicken egg yolk immunoglobulins: A review. *Journal of Agricultural and Food Chemistry*, *13*(4), 275–288. http://dx.doi.org/10.1080/09540100120094537

Evans, R. J., Bauer, D. H., Bandemer, S. L., Vaghefi, S. B., & Flegal, C. J. (1973). Structure of egg yolk very low density lipoprotein. Polydispersity of the very low density lipoprotein and the role of lipovitellenin in the structure. *Archives of Biochemistry and Biophysics*, *154*(2), 493–500. http://dx.doi.org/10.1016/0003-9861(73)90001-5

Evans, A. J., & Burley, R. W. (1987). Proteolysis of apoprotein B during the transfer of very low density lipoproteins from hens' blood to egg yolk. *Journal of Biological Chemistry*, *262*(2), 501–504.

Fei, T., Leyva-Gutierrez, F. M. A., Wan, Z., & Wang, T. (2021). Gelation inhibiting additives and freezing impact rheological, thermal, and microstructural properties of yolk. *LWT - Food Science and Technology*, *144*, 111160. https://doi.org/10.1016/j.lwt.2021.111160

Giarratano, M., Duffuler, P., Chamberland, J., Brisson, G., House, J. D., Pouliot, Y., & Doyen, A. (2020). Combination of high hydrostatic pressure and ultrafiltration to generate a new emulsifying ingredient from egg yolk. *Molecules*, *25*(5), 1184. https://doi.org/10.3390/molecules25051184

Grizzuti, K., & Perlmann, G. E. (1973). Binding of magnesium and calcium ions to the phosphoglycoprotein phosvitin. *Biochemistry*, *12*(22), 4399–4403. https://doi.org/10.1021/bi00746a016

Guilmineau, F., & Kulozik, U. (2006). Impact of a thermal treatment on the emulsifying properties of egg yolk. Part. 1: Effect of the heating time. *Food Hydrocolloids*, *20*(8), 1105–1113. http://dx.doi.org/10.1016/j.foodhyd.2005.12.005

Gutierrez, M. A., Takahashi, H., & Juneja, L. R. (1997). Nutritive evaluation of hen eggs. In T. Yamamoto, L. R. Juneja, H. Hatta, & M. Kim (Eds.), *Hen eggs, their basic and applied science* (pp. 25–35). CRC Press.

Hammershøj, M., & Johansen, N. F. (2016). Review: The effect of grass and herbs in organic egg production on egg fatty acid composition, egg yolk colour and sensory properties. *Livestock Science*, *194*, 37–43. https://doi.org/10.1016/j.livsci.2016.11.001

Hegenauer, J., Saltman, P., & Nace, G. (1979). Iron (III)-phosphoprotein chelates: Stoichiometric equilibrium constant for interaction of iron (III) and phosphorylserine residues of phosvitin and casein. *Biochemistry*, *18*(18), 3865–3879. http://dx.doi.org/10.1021/bi00585a006

Ito, Y., & Fujii, T. (1962). Chemical compositions of the egg yolk lipoproteins. *Journal of Biochemistry*, *52*(3), 221–222. http://dx.doi.org/10.1093/oxfordjournals.jbchem.a127600

Itoh, T., Abe, Y., & Adachi, S. (1983). Comparative studies on the α and β-phosvitins from hen's egg yolk. *Journal of Food Science*, *48*(6), 1755–1757. http://dx.doi.org/10.1111/j.1365-2621.1983.tb05077.x

Jiang, B., Wang, L., Wang, X., Wu, S., Li, D., Liu, C., & Feng, Z. (2019). Ultrasonic thermal-assisted extraction of phosvitin from egg yolk and evaluation of its properties. *Polymers*, *11*(8), 1353. https://doi.org/10.3390/polym11081353

Jolivet, P., Boulard, C., Beaumal, V., Chardot, T., & Anton, M. (2006). Protein components of low-density lipoproteins purified from hen egg yolk. *Journal of Agricultural and Food Chemistry*, *54*(12), 4424–4429. http://dx.doi.org/10.1021/jf0531398

Jolivet, P., Boulard, C., Chardot, T., & Anton, M. (2008). New insights into the structure of apolipoprotein B from low-density lipoproteins and identification of a novel YGP-like protein in hen egg yolk. *Journal of Agricultural and Food Chemistry*, *56*(14), 5871–5879. http://dx.doi.org/10.1021/jf800321m

Joubert, F. J., & Cook, W. H. (1958). Preparation and characterization of phosvitin from hen egg yolk. *Canadian Journal of Biochemistry and Physiology*, *36*(4), 399–408. http://dx.doi.org/10.1139/o58-045

Juneja, L. R. (1997). Egg yolk lipids. In T. Yamamoto, L. R. Juneja, H. Hatta, & M. Kim (Eds.), *Hen eggs, their basic and applied science*, (pp. 73–98). CRC Press.

Jung, S., Nam, K. C., Ahn, D. U., Kim, H. J., & Jo, C. (2013). Effect of phosvitin on lipid and protein oxidation in ground beef treated with high hydrostatic pressure. *Meat Science*, *95*(1), 8–13. https://doi.org/10.1016/j.meatsci.2013.04.005

Kim, H., & Nakai, S. (1998). Simple separation of immunoglobulin from egg yolk by ultrafiltration. *Journal of Food Science*, *63*(3), 485–490. http://dx.doi.org/10.1111/j.1365-2621.1998.tb15769.x

Kiosseoglou, V. D. (1989). Egg yolk. In G. Charalambous & G. Doxastakis (Eds.), *Food emulsifiers: Chemistry, technology, functional properties and applications* (pp. 63–85). Elsevier.

Kojima, E., & Nakamura, R. (1985). Heat gelling properties of hen's egg yolk low density lipoprotein (LDL) in the presence of other protein. *Journal of Food Science*, *50*(1), 63–66. http://dx.doi.org/10.1111/j.1365-2621.1985.tb13278.x

Kuksis, A. (1992). Yolk lipids. *Biochimica et Biophysica Acta*, *1124*(3), 205–222. http://dx.doi.org/10.1016/0005-2760(92)90132-F

Kurisaki, J. K., Yamauchi, H., Isshiki, H., & Ogiwara, S. (1981). Differences between α and β-lipovitellin from hen egg yolk. *Agricultural and Biological Chemistry*, *45*(3), 699–704. http://dx.doi.org/10.1080/00021369.1981.10864576

Laca, A., Paredes, B., & Diaz, M. (2010). A method of egg yolk fractionation. Characterization of fractions. *Food Hydrocolloids*, *24*(4), 434–443. http://dx.doi.org/10.1016/j.foodhyd.2009.11.010

Laca, A., Paredes, B., Renduels, B., & Diaz, M. (2014). Egg yolk granules: Separation, characteristics and applications in food industry. *LWT - Food Science and Technology*, *59*(1), 1–5. https://doi.org/10.1016/j.lwt.2014.05.020

Laca, A., Paredes, B., Renduels, B., & Diaz, M. (2015). Egg yolk plasma: Separation, characteristics and future prospects. *LWT - Food Science and Technology*, *62*(1), 7–10. https://doi.org/10.1016/j.lwt.2015.01.048

Le Denmat, M., Anton, M., & Gandemer, G. (1999). Protein denaturation and emulsifying properties of plasma and granules of egg yolk as related to heat treatment. *Journal of Food Science*, *64*(2), 194–197.

Lesnierowski, G., & Stangierski, J. (2018). What's new in chicken egg research and technology for human health promotion? - A review. *Trends in Food Science and Technology*, *71*, 46–51. https://doi.org/10.1016/j.tifs.2017.10.022

Li-Chan, E. C. Y., Powrie, W. D., & Nakai, S. (1995). The chemistry of eggs and egg products. In W. J. Stadelman & O. J. Cotterill (Eds.), *Egg science and technology* (4th ed.). Food Product Press.

Li, T., Su, H., Zhu, J., & Fu, Y. (2022). Janus effects of NaCl on structure of egg yolk granules. *Food Chemistry*, *371*, 131077. https://doi.org/10.1016/j.foodchem.2021.131077

Li, J., Zhai, J., Gu, L., Su, Y., Gong, L., Yang, Y., & Chang, C. (2021). Hen egg yolk in food industry - A review of emerging functional modifications and applications. *Trends in Food Science and Technology*, *115*, 46–51. https://doi.org/10.1016/j.tifs.2021.06.031

Maki, K. C., Van Elswyk, M. E., McCarthy, D., Seeley, M. A., Veith, P. E., Hess, S. P., & Davidson, M. H. (2003). Lipid responses in mildly hypertriglyceridemic men and women to consumption of docosahexaenoic acid-enriched eggs. *International Journal of Vitamins and Nutrition Research*, *73*(5), 357–368. https://doi.org/10.1024/0300-9831.73.5.357

Marcet, I., Saez-Orviz, S., Rendueles, M., & Díaz, M. (2022). Egg yolk granules and phosvitin. Recent advances in food technology and applications. *LWT - Food Science and Technology*, *153*, 112442. https://doi.org/10.1016/j.lwt.2021.112442

Martin, W. G., Augustyniak, J., & Cook, W. H. (1964). Fractionation and characterization of the low-density lipoproteins of hen's egg yolk. *Biochimica et Biophysica Acta*, *84*(6), 714–720. https://doi.org/10.1016/0926-6542(64)90029-0

Martinet, V., Saulnier, P., Beaumal, V., Couthaudon, J., & Anton, L. M. (2003). Surface properties of hen egg yolk low-density lipoproteins spread at the air-water interface. *Colloids and Surfaces B: Biointerfaces*, *31*(1–4), 185–194. https://doi.org/10.1016/S0927-7765(03)00139-5

McBee, L. E., & Cotterill, O. J. (1979). Ion-exchange chromatography and electrophoresis of egg yolk proteins. *Journal of Food Science*, *44*(3), 656–660.

Mecham, D. K., & Olcott, H. S. (1949). Phosvitin, the principal phosphoprotein of egg yolk. *Journal of American Chemical Society*, *71*(11), 3670–3679.

Mine, Y. (1997). Structural and functional changes of hen's egg yolk low-density lipoproteins with phospholipase A(2). *Journal of Agricultural and Food Chemistry*, *45*(12), 4558–4563. https://doi.org/10.1021/jf9704064

Montorzi, M., Falchuk, K. H., & Vallee, B. L. (1995). Vitellogenin and lipovitellin: Zinc proteins of xenopus laevis oocytes. *Biochemistry*, *34*(34), 10851–10858. https://doi.org/10.1021/bi00034a018

Montserret, R., McLeish, M. J., Böckmann, A., Geourjon, C., & Penin, F. (2000). Involvement of electrostatic interactions in the mechanism of peptide folding induced by sodium dodecyl sulfate binding. *Biochemistry*, *39*(29), 8362–8373. https://doi.org/10.1021/bi000208x

Moussa, M., Martinet, V., Trimeche, A., Tainturier, D., & Anton, M. (2002). Low density lipoproteins extracted from hen egg yolk by an easy method: Cryoprotective effect on freeze-thaw bull semen. *Theriogenology*, *57*(6), 1695–1706. https://doi.org/10.1016/S0093-691X(02)00682-9

Nakamura, R., Hayakawa, R., & Sato, Y. (1977). Isolation and fractionation of the protein moiety of egg yolk low density lipoprotein. *Poultry Science*, *56*(4), 1148–1152. https://doi.org/10.3382/ps.0561148

Phillips, S. M. (2017). Current concepts and unresolved questions in dietary protein requirements and supplements in adults. *Frontiers in Nutrition*, *4*(13). https://doi.org/10.3389/fnut.2017.00013

Polson, A., Coetzer, T., Kruger, J., von Maltzahn, E., & van der Merwe, K. J. (1985). Improvements in the isolation of IgY from the eggs laid by immunized hens. *Immunological Investigations*, *14*(4), 323–327. https://doi.org/10.3109/08820138509022667

Posati, L. P., Kinsella, J. E., & Watt, B. K. (1975). Comprehensive evaluation of fatty acids in foods: Eggs and egg products. *Journal of American Diet Association*, *67*(2), 111–115.

Radomski, M. W., & Cook, W. H. (1964). Fractionation and dissociation of the avian lipovitellins and their interaction with phosvitin. *Canadian Journal of Biochemistry*, *42*(3), 395–406. https://doi.org/10.1139/o64-047

Ren, L., Ma, J., Xu, W., Lv, Y., & Tong, K. (2022). Stability of low density lipoprotein particles affect the formation of off-flavor in thermal egg yolk. *Food Research International*, *154*, 111029. https://doi.org/10.1016/j.foodres.2022.111029

Saari, A., Powrie, W. D., & Fennema, O. (1964). Isolation and characterization of low-density lipoproteins in native egg yolk plasma. *Journal of Food Science*, *29*(3), 307–315. https://doi.org/10.1111/j.1365-2621.1964.tb01736.x

Sakai, S., & Ikeda, N. (2022). A numerical analysis to evaluate the emulsifying activity of pasteurized egg yolk. *Food Hydrocolloids*, *123*, 107087. https://doi.org/10.1016/j.foodhyd.2021.107087

Schade, E. G., Calzado, R., Sarmiento, P. A., Chacana, J., Porankiewicz-Asplund, H., & Terzolo, R. (2005). Chicken egg yolk antibodies (IgY-technology): A review of progress in production and use in research and human and veterinary medicine. *Alternative Laboratory Animals*, *33*(2), 129–154. https://doi.org/10.1177/026119290503300208

Schade, R., & Chacana, P. H. (2007). Livetin fraction (IgY). In R. Huopalathi, R. Lopez-Fandino, M. Anton, & R. Schade (Eds.), *Bioactive egg compounds* (pp. 25–32). Springer. https://doi.org/10.1007/978-3-540-37885-3_5

Shenstone, F. S. (1968). Egg quality. In TD. Carter (Ed.), *A study of the avian egg*. Oliver & Boyd.

Shimizu, M., Nagashima, H., Sano, K., Hashimoto, K., Ozeki, M., Tsuda, K., & Hatta, H. (1992). Molecular stability of chicken and rabbit immunoglobulin G. *Bioscience, Biotechnology and Biochemistry*, *56*(2), 270–274. https://doi.org/10.1271/bbb.56.270

Sirvente, H., Beaumal, V., Bialek, L., & Anton, M. (2007). Structuring and functionalization of egg yolk plasma and granules solutions induced by mechanical treatments. *Journal of Agricultural and Food Chemistry*, *55*(23), 9537–9544. https://doi.org/10.1021/jf0719398

Sirvente, H. (2007). *Fonctionnalités des fractions du jaune d'oeuf dans les emulsions alimentaires huile-dans-eau allégées en matière grasse: impact des différents niveaux de structuration*, Ph D thesis, University of Nantes, France.

Speroni, F., Puppo, M. C., Chapleau, N., de Lamballerie, M., Castellani, O., Añón, M. C., & Anton, M. (2005). High pressure induced physicochemical and functional modifications of low-density lipoproteins from hen egg yolk. *Journal of Agricultural and Food Chemistry*, *53*(14), 5719–5725. https://doi.org/10.1021/jf0502808

Suhag, R., Dhiman, A., Thakur, D., Kumar, A., & Upadhyay, A. (2021). Physico-chemical and functional properties of microfluidized egg yolk. *Journal of Food Engineering*, *294*, 110416. https://doi.org/10.1016/j.jfoodeng.2020.110416

Sun, S., Mo, W., Ji, Y., & Liu, S. (2001). Preparation and mass spectrometric study of egg yolk antibody (IgY) against rabies virus. *Rapid Communication on Mass Spectrometry*, *15*(9), 708–712. https://doi.org/10.1002/rcm.271

Taborsky, G. (1974). Phosphoprotein. *Advances in Protein Chemistry*, *28*, 1–210. https://doi.org/10.1016/s0065-3233(08)60230-2

Taborsky, G. (1983). Phosvitin. *Advances on Inorganic Biochemistry*, *5*, 235–279.

Ternes, W. (1989). Characterization of water soluble egg yolk proteins with isoelectric focusing. *Journal of Food Science*, *54*(3), 764–765. https://doi.org/10.1111/j.1365-2621.1989.tb04704.x

Thapon, J.-L., & Bourgeois, C.-M. (1994). *L'œuf et les ovoproduits*. Lavoisier TEC & DOC.

Timmins, P., Poliks, B., & Banaszak, L. (1992). The location of bound lipid in the lipovitellin complex. *Science*, *257*(5070), 652–655. https://doi.org/10.1126/science.1496377

Verdoliva, A., Basile, G., & Fassina, G. (2000). Affinity purification of immunoglobulins from chicken egg yolk using a new synthetic ligand. *Journal of Chromatography B: Biomedical Science Applied*, *749*(2), 233242. https://doi.org/10.1016/S0378-4347(00)00426-6

Vicente, F. A., Castro, L. S., Mondal, D., Coutinho, J. A. P., Tavares, A. P., Ventura, S. P. M, & Freire, M. G. (2022). Purification of immunoglobulin Y from egg yolk using thermoresponsive aqueous micellar two-phase systems comprising ionic liquids. *Separation and Purification Technology*, *288*, 120589. https://doi.org/10.1016/j.seppur.2022.120589

Vicente, F. A., Lario, L. D., Pessoa, A., & Ventura, S. P. M. (2016). Recovery of bromelain from pineapple stem residues using aqueous micellar two-phase systems with ionic liquids as co-surfactants. *Process Biochemistry*, *51*(4), 528–534, https://doi.org/10.1016/j.procbio.2016.01.004

Wallace, P. A., & Morgan, J. P. (1986). Isolation of phosvitin: Retention of small molecular weight species and staining characteristics on electrophoretic gels. *Analytical Biochemistry*, *157*(2), 256–261. https://doi.org/10.1016/0003-2697(86)90623-8

Wang, R., Ma, Y., Zhang, L., Zhang, Z., Chi, Y., & Chi, Y. (2021). Changes in egg yolk gelation behaviour and mechanisms during freezing. *LWT - Food Science and Technology*, *151*, 112223. https://doi.org/10.1016/j.lwt.2021.112223

Warr, G. W., Magor, K. E., & Higgins, D. A. (1995). IgY clues to the origins of modern antibodies. *Immunology Today*, *16*(8), 392–398. https://doi.org/10.1016/0167-5699(95)80008-5

Yang, Y., Wen, J., Zhao, S., Zhang, K., & Zhou, Y. (2014). Prophylaxis and therapy of pandemic H1N1 virus infection using egg yolk antibody. *Journal of Virological Methods*, *206*, 19–26. https://doi.org/10.1016/j.jviromet.2014.05.016

Zhou, M., Wang, T., Hu, Q., & Luo, Y. (2016). Low density lipoprotein/pectin complex nanogels as potential oral delivery vehicles for curcumin. *Food Hydrocolloids*, *57*, 20–29. https://doi.org/10.1016/j.foodhyd.2016.01.010

7

Egg Pigments, Minerals, Enzymes, and Vitamins

Yue Zhang[*]**, Suyun Zhang, and Yunsi Guo**

CONTENTS

7.1 Introduction

Hen egg contains various micronutrients (such as vitamins, minerals) and enzymes, which are stored for the metabolism of chicken in embryonic development. These micronutrients in egg are also believed to be very valuable to supplement the diets of infants and adults. Compared with plant-based resources, these nutrients in eggs are highly bioavailable. However, unlike egg white, the contents of micronutrients including vitamins, minerals and carotenoids are far from plentiful in eggs and greatly rely on the feeding (Réhault-Godbert et al., 2019). The pigments, mainly carotenoids, are responsible for the color in egg yolk, which strongly influences consumer's perception on the quality of eggs. They also have many health benefits, and therefore, egg producers are controlling the carotenoids very carefully so that the fortification of carotenoids in hen's feed is now a very common strategy. The minerals including calcium, phosphorus, zinc, iron, magnesium, selenium and copper are abundant in egg, especially egg yolk. However, egg intake accounts for only 1–3% of the dietary mineral intake of human beings (Huopalahti et al., 2007). The enrichment of minerals in eggs should be carefully regulated to avoid the possible toxicity caused by some elements at high dose level. Egg is also considered as a main resource of vitamins in the human diet. All the vitamins essential for humans could be found in hen egg, mainly in egg yolk, except vitamin C (Sirri & Barroeta, 2007). Hence, egg yolk, as the great carrier of vitamins, is always further enriched with different kinds of vitamins to better meet the daily intake requirement. Numerous enzymes are present in the albumen, yolk, and vitelline member of unfertilized eggs, which are mainly categorized as metabolic enzymes. Many enzymes' activities are strongly regulated by several inorganic minerals. However, unlike other micronutrients, studies on these egg enzymes except lysozyme have been quite limited in the last decades. Thus, it would be of great interests to review all these nutrients in details, to summarize their biological functions, current stage of their enrichments in eggs, and future research opportunities.

7.2 Egg Yolk Color

The color of hen egg yolk can go from a very pale yellow to a dark brilliant orange, conferred by the natural pigments of carotenoids. The contents of carotenoids are very variable in

[*] Corresponding author.

DOI: 10.1201/9781003254430-9

eggs, but are believed to be economically important since the color of egg yolk is perceived as one of the priority criteria of egg quality and market preference.

Carotenoids are a class of naturally occurring fat-soluble pigments synthesized by photosynthetic organisms including plants, algae, and bacteria. In hen eggs, carotenoids mainly exist in yolk. And based on the chemical structures, carotenoids could be divided into carotenes and xanthophylls. Carotene does not contain an oxygen atom in its structure and gives an orange color, whereas xanthophyll contains oxygen as functional group and gives a yellow color.

7.2.1 Carotene

Carotenes are tetraterpenes-related unsaturated hydrocarbon substances with a general formula of $C_{40}H_x$. The main carotenes in hen eggs are α-carotene and β-carotene, while the predominant carotene in fruits and vegetables is lycopene (Nabi et al., 2020). The rich unsaturated C=C structures make carotenes effective quenchers of singlet oxygen with scavenging abilities, and therefore are responsible for the strong activities of carotenes as both antioxidants and prooxidants (Zaheer, 2017). α-Carotene and β-carotene are considered as provitamins A, precursors of vitamin A that can be converted by the body to retinol (with 100% vitamin A biological activity). And β-carotene has higher conversion efficacy than α-carotene and β-cryptoxanthin, one type of main xanthophylls in egg yolk (Tang, 2014).

7.2.2 Xanthophyll

The hen egg yolk contains ten times more xanthophyll molecules than the carotenes. The primary xanthophyll molecules in egg yolk are lutein and zeaxanthin, with a general formula of $C_{40}H_{56}O_2$. Two hydroxyl groups present at the terminal rings of the hydrocarbon chain on the basis of $C_{40}H_{56}$ carotene structure. The two molecules are stereoisomers of each other, with only one difference in the location of a double bond in one ring with the hydroxyl group (Carpentier et al., 2009). Unlike α-carotene and β-carotene, these two main xanthophylls are non-provitamins A, which cannot be converted to retinol. However, the exist of hydroxyl group in these two xanthophylls provides some additional biological functions. For example, these two xanthophylls are promising on prevention of adult macular degeneration (AMD) and age-related cataract formation (Schlatterer & Breithaupt, 2006). In addition, their activities on prevention of cardiovascular disease (CVD), stroke, cancer, and neurodegenerative disorders have also been noticed (Zaheer, 2017). Like carotenes, fortification of hen's feed with commercial xanthophylls could greatly increase their content in eggs. Therefore, eight xanthophylls, including both natural and synthetic xanthophylls, have been used to the feed of poultry or laying hens. For example, besides lutein and zeaxanthin, many other xanthophylls such as β-cryptoxanthin, *cis* isomers of lutein and zeaxanthin, and synthetic xanthophylls (canthaxanthin, β-apo-8′-carotenoic acid ethyl ester and citranaxanthin) have also been found in commercial egg yolk (Schlatterer & Breithaupt, 2006).

7.2.3 Sources and Consumption of Carotenoids

As animals including hens do not have the ability to synthesize carotenoids, all carotenoids exported into the egg during formation come from the hens' diet. In poultry industry, α-carotene, β-carotene, β-cryptoxanthin, lutein, zeaxanthin, and lycopene are the most common dietary carotenoids. The first three are provitamins A while the latter three are non-provitamins. Different carotenoid species have different digestion and absorption abilities. It was suggested that xanthophylls in food matrix are more easily to be released and micellized during digestion than the carotenes while carotenes are more easily to be absorbed by intestinal enterocytes (Yonekura & Nagao, 2007). For hen feeds with corn and alfalfa, carotene is the main carotenoid. It is easily oxidized to xanthophyll after ingestion and accumulated in the body fat or egg yolk lipid. However, in order to produce egg yolk with desirable golden color, the carotenoids should not be used as antioxidants or vitamin precursors. Therefore, carotenes with higher vitamin A activity have lower pigmenting properties, in which case β-carotene with the highest provitamin A activity is not considered as a good choice for coloring egg (Hencken, 1992). In a recent study, corn fortified with zeaxanthin and β-cryptoxanthin showed a deposition efficiency of 25.52 and 26.05%, respectively, which were about three times and five times more efficient than β-cryptoxanthin (8.30%) and β-carotene (5.65%), respectively (Kljak et al., 2021). Thus, xanthophylls are considered to be more important in egg coloration than carotenes.

Besides coloration, the health benefits and nutritional values of carotenoids such as the promoting effect on visual performance and protective impact on AMD and age-related macular degeneration are also important reasons for poultry industry to fortify egg yolk. For example, the lutein content in whole egg is about 0.3–1.0 mg/100g, only provides 10% of the required intake of lutein to achieve the protective effect of lutein against AMD (based on a daily intake of 8 mg). The lutein content in hen eggs could be enriched to 5 to 8 times above regular concentration with a supplement diet of laying hens (Leeson & Caston, 2004). In addition, a study also found that lutein from enriched eggs showed higher bioavailability than lutein from spinach or supplements for adults (Chung et al., 2004).

Many factors could also affect the absorption, conversion of carotenoids in hens and finally their deposition in egg yolk. A higher fat content in diet could result a higher absorption rate of these pigments. Secondary factors such as age, health status that affect digestive health also show a noticeable impact on the absorption of these pigments. Thus, maintaining the healthy intestinal physiological status of hen is very important throughout the laying phase. Recent studies have found that the plant-based carotenoid resources could improve the production performance of poultry birds, and therefore boost the utilization of these natural carotenoids as alternative feed ingredients directly (Nabi et al., 2020). Tomato powder, marigold extract, lucerne meal, and algae are the main natural carotenoid sources (Marounek & Pebriansyah, 2018). However, compared with commercial carotenoid supplementary, the consistency is the main concern for the enriched egg quality based on natural carotenoid extract feeding (Jang et al.,

2013). This requires more research efforts from the academy and poultry industry for the improvement.

7.3 Minerals

Minerals, the essential elements cannot be formed by living lives including human beings, are abundant in eggs. As shown in Table 7.1, egg contains all major minerals such as phosphorus, calcium (Ca), potassium (K), and moderate amounts of sodium (Na), and all essential trace elements including copper (Cu), iron (Fe), magnesium (Mg), manganese (Mn), selenium (Se), and zinc (Zn).

Egg yolk contains 1% minerals, with a very high level of phosphorous. More than 61% phosphorous in egg yolk is contained in phospholipids, with around 80–90% in the form of phosphoprotein, i.e., phosvitin (Samaraweera et al., 2011). Egg yolk also contains most of the Ca and Fe in whole egg. However, their bioavailability is low due to the high binding capacity with phosvitin (Yilmaz & Ağagündüz, 2020). Take Fe as an example, about 95% of Fe in egg yolk exists in the ferric form and forms a complex with the phosphate groups in phosvitin. It is very stable except a stronger chelating agent such as EDTA to be used. The bioavailability of these divalent and trivalent cations could also be improved by fermentation process due to the production of small phosphopeptides instead of the formation of insoluble complexes (Choi et al., 2005).

Egg albumen contains abundant minerals essential for the development of embryo such as chlorine, Mg, K, Na, and sulfur, while Ca and phosphorus are present in trace amounts (Hrabia, 2022). Very little Fe is found in albumen, where it easily forms complex with another protein, ovotransferrin. Unlike Na and K, which contents are very stable in egg white, the Fe content in albumen shows a quite high variability and is affected significantly by dietary (Huyghebaert et al., 2006). Ca and Mg, in the form of divalent cations, vary significantly from the thick egg albumen to the thin egg albumen because of their binding affinity to proteins (Immerseel et al., 2011). These cations play important role in the viscosity of egg white, which is an indicator for consumers to judge the quality of eggs. The level of these trace elements in egg white is critical for the egg quality.

The inorganic minerals account for 94–97% of the weight of whole eggshell, which are mainly made of calcium carbonate crystals. Eggshell powder, containing 385–401 mg/g calcium (300–350 times above the egg yolk content), therefore, is considered as an excellent calcium source for nutrition of human, animals, and plants. Consumption of half an eggshell or 2.7 g eggshell powder could fulfill the daily calcium intake for adults (Milbradt et al., 2015). Many functional foods with eggshell powder incorporated have been developed to improve the calcium intake and therefore reduce the risk of osteoporosis and other diseases (Arnold et al., 2021). The phosphorus is also found in eggshell, which is mainly concentrated in the outer part of eggshell and in the cuticle. Mg and some trace minerals, including Cu, Fe, Se, and Zn, are also present throughout the shell, which contents are much higher than those in egg yolk. It is worth mentioning that eggshell contains a high amount of strontium (Sr), which does not exist in egg yolk or white. This micronutrient is believed to play the key role in the prevention of osteoporosis. Keeping Sr at a certain trace

TABLE 7.1
The Mineral Content in Whole Egg, Egg White, Egg Yolk, and Egg Shell

Nutrients	Whole Egg, Without Shell, Raw mg / 100 g[a]	Egg Yolk, Raw mg / 100 g[a]	Egg White, Raw mg / 100 g[a]	Egg Shell Powder, mg/g[b]
Major minerals				
Calcium	56	129	7	385–401
Magnesium	12	5	11	3.5–4.5
Phosphorus	198	390	15	0.2–1.9
Potassium	138	109	163	
Sodium	142	48	166	
Trace minerals				
Copper	0.072	0.077	0.023	0.007–0.013
Iodine	0.021	0.18	0.002	
Iron	1.75	2.73	0.08	0.022–0.023
Manganese	0.028	0.055	0.011	
Selenium	0.030	0.056	0.020	0.023–0.053
Zinc	1.29	2.30	0.03	0.005
Strontium				0.32–0.88

[a] Retrieved on 06/28/2022 from Department of Agriculture, Agricultural Research Service, USDA FoodData Central (FDC) with the Standard Reference data released in April 2018: https://fdc.nal.usda.gov/fdc-app.html#/food-details/171287/nutrients; https://fdc.nal.usda.gov/fdc-app.html#/food-details/172184/nutrients; https://fdc.nal.usda.gov/fdc-app.html#/food-details/172183/nutrients.

[b] Schaafsma, Pakan, Hofstede, Muskiet, Van Der Veer, and De Vries (2000).

amount and a relatively low content of heavy metals is essential for using eggshell powder as a superior calcium source in food applications (Singh et al., 2021).

Whereas the content of major minerals is quite stable, the trace minerals (i.e., Fe, Se, I) can be significantly influenced by hens' feeding. Therefore, the enrichment of these elements could be achieved through hen's dietary fortification. To ensure the safety, Cu, I and Se could be added at levels between 50 and 100% of the European recommended dietary allowance (EC RDA), while Zn, Fe, Ca, Mg, and phosphorus could be added at 10–40% (Huopalahti et al., 2007). The average hen's egg contains about 5 μg of Se and 4–10 μg of I, which could be enriched to 20–40 μg/egg and up to 26 μg/egg, respectively (Schiavone & Barroeta, 2011). Unlike these two trace elements, the deposition rate of zinc in egg yolk is less dependent on the feeding dose. For example, an average Zn egg content of 70 mg/kg could be achieved by feeding laying hen with additional 1000 mg/kg Zn, which only increased 10% Zn content in egg (Kim & Patterson, 2005). Supplementation with 3000 mg/kg Zn could bring the average Zn egg value to 112 mg/kg but also result reduced body weight, egg production and poor egg quality. Most of the trace element enrichments are achieved by deposition in egg yolk. However, because their content in egg white is relatively lower, a more significant increase of their content in egg white can be achieved. For example, the concentration of Se and Mn could be increased to 428.66% in the albumen and 174.47% in the yolk, upon supplementation (Gravena et al., 2011). To supplement these minerals, additives based on their salts of sulfates, phosphates, and carbonates are used for laying hen feeding. However, currently, the trend in the usage of inorganic materials has switched to organic forms such as organic acid-based chelates, amino acid complexes, and proteinates which show greater mineral bioavailability (Ghasemi et al., 2022).

7.4 Enzymes

About 30 kinds of enzymes have been found in the unfertilized hen egg. They are proteolytic enzymes, phosphatase, glycosidase, etc., which are list in Table 7.2, and discussed as below. Many of them have been purified and characterized, especially with the new advanced technology incorporated into the identification and purification process. However, most of the studies were conducted in last century, no particular progress on these enzymes has been reported recently.

7.4.1 Proteolytic Enzymes

Proteolytic enzymes, also called proteases, peptidases, proteinases, with the main function to hydrolyze peptidyl bonds, are found in all biological kingdoms because of their importance in all physiological processes. However, the identification of proteases in egg was very difficult in the past, due to the relative low concentration and also the presence of protease inhibitors (Huopalahti et al., 2007). There are numerous protease inhibitors contained in the egg, such as ovostatin and ovomucoid, which may inhibit the protease activity and hence hinder the identification of these proteases in egg. Only

TABLE 7.2
The Enzymes Found in Unfertilized Hen's Egg

Catalog	Enzyme	Source
Protease, peptidase		White, yolk
	Glutamyl aminopeptidase	White
	Methionine preferring broad specificity aminopeptidase	White
Phosphatase		
	Acid phosphatase	Vitelline membrane, white, yolk
	Alkaline phosphatase	Vitelline membrane, white, yolk
Glycosidase		White, yolk
	Lysozyme	White, shell
	α-mannosidase	White, yolk
	α-amylase	White, yolk
	β-*N*-acetylglucosaminidase	Vitelline membrane, white
	β-D-glucosidase	Yolk
	β-D-galactosidase	Yolk
	α-D-glucosidase	Yolk
	β-D-glucuronidase	Yolk
Nucleoside triphosphatase		Vitelline membrane, thick white, yolk and shell membrane
	ATPase	Plasma, vitelline membrane
	Rho-family GTPase	Vitelline membrane
Ribonucleic acid-degrading enzymes		
	Ribonuclease	White, yolk
	Deaminase	Yolk
	Nucleosidase	Yolk
	Purine N_1-C_6 hydrolase	Yolk
	5′-nucleotidase	Vitelline membrane, yolk
	Deoxyribonuclease	Yolk
Pyruvate kinase and glycolytic enzymes		
	Pyruvate kinase	Vitelline membrane, yolk
	Hexokinase	Vitelline membrane, yolk
Other enzymes		
	Catalase	White, yolk, shell membrane
	Lysyl oxidase	Shell membrane
	Triacylglycerol lipase	White, yolk
	Lactate dehydrogenase	Vitelline membrane, yolk

Source: Mainly adapted from Yamamoto et al., (1996).

two proteases, namely glutamyl aminopeptidase (302kDa) and methionine preferring broad specificity aminopeptidase (180kDa) with comprehensive characterizations were reported in the 1990s. These two proteases mainly exist in egg white, with their pIs around 4.1–4.4. Later, with more advanced and sensitive techniques developed, numerous proteases have been identified from hen eggs. For example, a chicken matrix metalloprotease, along with its tissue inhibitor of metalloproteinases, has been identified as a complex in egg white and yolk by using zymography and mass spectrometry (Réhault-Godbert et al., 2008). In a recent study, 86 glycoproteins have been identified from egg yolk using a shotgun glycoproteomics strategy, with 22 of them were recognized as proteases and protease inhibitors, confirming that a proteinase/inhibitor regulation system exists in egg yolk (Geng et al., 2018).

7.4.2 Phosphatases

Active acid and alkaline phosphatases, along with ATPase, have been observed in hen eggs. ATPase, the enzyme catalyzing the decomposition of ATP, is also classified into the group of nucleoside triphosphatases. It has been mainly found in the vitelline membrane, which is strongly bound with the membrane function and increases markedly at the time of ovulation (Haaland et al., 1971). Therefore, it plays an important role on the quality of hen's eggshell. The enzyme activity is strongly affected by Mg^{2+} and therefore decreases significantly by adding chelators such as EDTA (Belkameh et al., 2021). It is also responsible for the transportation of Ca. It could be isolated from the unfertilized egg.

Acid and alkaline phosphatases, the enzymes accelerating the hydrolysis of ester linkage of phosphate, have been found in vitelline, yolk, and white. The alkaline phosphatase in egg yolk is identified as a dimeric metalloenzyme with a strong binding ability to Zn^{2+}, followed by Mg^{2+}(Debruyne, 1982). It is used frequently as a major target to understand causative disease mechanism, therefore, has been purified by many conventional methods (Debruyne & Stockx, 1977). The recovery and purification of this enzyme from egg yolk could be further improved by magnetic nanocarrier technology (Farzi-Khajeh et al., 2017).

7.4.3 Glycosidases

Many glycosidases have been found in both egg white and yolk. Since lysozyme, the most well-known glycosidase in egg white, has already been discussed in Chapter 5, and more details of its properties and functions will be introduced in Chapter 26, it is not discussed in this section. There are also other glycosidases such as α-mannosidase and β-*N*-acetylglucosaminidase contained in egg white (Yamamoto et al., 1996). In egg yolk, the presence of α-mannosidase, β-D-glucosidase, β-D-galactosidase, α-D-glucosidase, and β-D-glucuronidase has been reported. These glycosidases are responsible for hydrolytic degradation of complex polysaccharides, glycoproteins and even glycolipids. β-*N*-acetylglucosaminidase, existing in both egg white and vitelline membrane, shows great antibacterial activity, which might be purified and used as a bioactive additive (Ahlborn et al., 2006). It is inactivated by heating at 60°C for 2.5 min, and therefore could be also used as an indicator for heat pasteurization of egg white. α-amylase has also been found in the white and yolk and has been suggested to evaluate the adequacy of pasteurization (Vadehra et al., 1973). Its activity also shows stage dependence after fertilization. But the practicality of α-amylase in egg has not been demonstrated yet.

7.4.4 Nucleoside Triphosphatases

Nucleoside triphosphatases (NTPase), also called nucleoside triphosphate hydrolase, specifically catalyze the cleavage of the β–γ phosphodiester bond in phosphorus-containing anhydrides to yield nucleoside diphosphate and phosphate. NTPases, including the most frequently encountered ATPase, have been found in yolk, thick white, vitelline membrane and shell membrane. In a proteomic analysis study, Na^+/K^+-ATPase, some proteins similar to Rho-family GTPase have been found in the egg vitelline membrane after a salt washing enrichment (Mann, 2008). However, their concentration is believed to be very low.

7.4.5 Ribonucleic Acid-Degrading Enzymes

The ribonuclease (RNase) activity has also been observed in both egg white and yolk. And it was suggested that the RNase in egg yolk would be mainly concentrated in yolk granules (Seko et al., 2018). An adenosine deaminase, with M_w of 14kDa, has been identified in egg yolk, which shows different final structure compared with the adenosine deaminases of chicken liver and blood plasma (De Boeck et al., 1975). The deamination activity on adenosine, deoxyadenosine, guanosine, and cytidine has also been found in egg yolk (Yamamoto et al., 1996). The adenosine deaminase is currently used as a diagnostic marker for T-lymphocyte proliferation, tuberculous pleuritis, and many types of cancers. The egg yolk adenosine deaminase could be purified 4686-fold with a specific activity about 81.4mU/mg (Lopez et al., 1990).

7.4.6 Pyruvate Kinase and Glycolytic Enzymes

Pyruvate kinase is the enzyme playing the most important role in the regulation of glycolysis, which directly accelerates the transfer of phosphate from phosphoenolpyruvate to ADP, resulting the generation of ATP and one pyruvate. Pyruvate kinases type M_2 were purified from vitelline membrane and yolk, with different kinetic properties (Noda et al., 1986). It is believed that the pyruvate kinase from egg were related to pyruvate kinase type M_1 from chicken breast muscle. It is also the substrate of protein kinase C (Siwko & Mochly-Rosen, 2007).

7.4.7 Others

Some other types of enzymes with more or less activities have also been identified in hen eggs. The activities of catalase and cholinesterase were found in egg white and water-soluble fraction of the yolk in 1970s. The catalase activity is also used to reflect the egg quality and the degree of heat pasteurization

(Yamamoto et al., 1996). The activity of lysyl oxidase was found in egg shell membrane, and the H_2O_2 released by this oxidase could be completely decomposed by the catalase coexisted in shell membrane (Akagawa et al., 1999). It believes that this coupling enzyme system is involved in the protection of embryo against H_2O_2.

7.5 Vitamins

Egg is a rich and balanced source of vitamins with almost all the fat-soluble vitamins and most water-soluble vitamins found in egg yolk (Huopalahti et al., 2007). As shown in Table 7.3, the whole egg contains high amount of fat-soluble vitamins (A, D, E, K) and water-soluble vitamin B group, only except for vitamin C (ascorbic acid). Compared to egg yolk, which is the main contributor of vitamins in hen egg, egg white contains low levels of vitamins except for water-soluble niacin(B_3) and riboflavin(B_2) (Huopalahti et al., 2007).

It is believed that eating two medium-size eggs per day covers almost up to 1/3 of the vitamin requirements for humans (Réhault-Godbert et al., 2019). The consumption of one egg daily could greatly increase the daily nutrition intake of human by raising 20–30% of daily intake of vitamins A, E, and B_{12}, 10–20% of folate and <10% of vitamin B_6, respectively (Song & Kerver, 2000).

TABLE 7.3

The Average Amount of Major Vitamins in Whole Egg, Egg White, and Egg Yolk

Nutrients	Whole Egg Without Shell, Raw / 100 g[a]	One Egg Expressed as % of RDI[b]
Vitamin A or retinol	160 μg	15.0
Vitamin D_3 or cholecalciferol	2.46 μg	30.8
Vitamin E or alpha-tocopherol	1.05 mg	8.5
Vitamin K or phylloquinone	0.3 μg	38.5
Vitamin C	0	0
Vitamin B_1 or thiamin	0.04 mg	4.2
Vitamin B_2 or riboflavin	0.457 mg	11.4
Vitamin B_3 or niacin	0.075 mg	11.0
Vitamin B_5 or pantothenic acid	1.53 mg	14.2
Vitamin B_6	170 μg	4.6
Vitamin B_8 or biotin		29.4
Vitamin B_9 or folate	47 μg	8.5
Vitamin B_{12} or colabamin	0.89 μg	35.3

[a] Retrieved on 06/28/2022 from Department of Agriculture, Agricultural Research Service, USDA FoodData Central (FDC) with the Standard Reference data released in April 2018: https://fdc.nal.usda.gov/fdc-app.html#/food-details/171287/nutrients

[b] Seuss-Baum (2007).

7.5.1 Major Vitamins in Egg

Egg, especially egg yolk, is one of the few foods that contain naturally high amounts of vitamin D (D_3). A serving of two eggs can provide more than 60% of human recommended daily intake (RDI). The content of D_3 in egg yolk is lower than fish liver, but higher than in meat (Schmid & Walther, 2013). Although data on the factors affecting vitamin D absorption efficiency are still limited, it is believed that no difference was found on the vitamin D accumulation status from the consumption of egg, cod liver oil, and multivitamin tables. Therefore, egg is considered as a good carrier for vitamin D_3. The content of D_3 in egg yolk could be increased by supplementary feeding, exposing the hens or even the egg yolk to UVB light directly (Barnkob et al., 2020).

Vitamin E (α-tocopherol), is the most abundant fat-soluble vitamin in hen egg. It plays a very fundamental role in the normal functioning of the organism. It is also the main antioxidant system of fresh egg and its consumption could prevent many metabolic disorders. The RDA value for vitamin E is about 15 mg. As a result, production of vitamin E-enriched egg is popular worldwide to better fit the daily requirement. And the deposition of vitamin E in egg yolk could be significantly increased by supplementary of hen's diet.

Egg is also a rich source of vitamin A (retinol). The vitamin A found in one egg could supply around 15% of the vitamin A requirement of an adult. The bioavailability of vitamin A from egg yolk is very high (more than 95%) in humans. Additionally, egg yolk contains many vitamin A precursors such as β-carotene, as discussed above, which could also convert into retinol after ingestion. No significant difference was found between the vitamin A content in free-range and caged hen egg, however, cooking could cause a 17–20% loss of this vitamin due to its thermo-instability of unsaturated structure (Ramalho et al., 2006).

Although the content of vitamin K in egg is the lowest, eating one egg per day could cover 38.5% of the recommended intake requirement for humans. This vitamin has restricted functions, which is believed to be mainly involved in the liver function, bone mineralization, and skin metabolism. Unlike many other fat soluble vitamins, the concentration of vitamin K in modified egg is similar to the regular egg (Leeson & Caston, 2003).

It is remarkable, however, that the content of fat-soluble vitamins in egg yolk, similar as the content of carotenoids, is highly dependent on the hen's diet and living conditions. The vitamins database of egg nutrient penal from different countries can vary a bit due to different culture conditions and standards. Take the vitamin D as an example, it was determined to be 1.7, 2.2, and 2.9 μg/100 g raw egg in France, Finland, and Germany, respectively (Huopalahti et al., 2007). Since vitamin D is the "sunshine vitamin", excluding the dietary supplement, the differences of vitamin D amount could also be a reflection of the degree of intensification of the laying hen industry in these countries.

Egg yolk is considered as a great source for all vitamin B group. Pantothenic acid (B5) is the water-soluble vitamin with the highest concentration in egg yolk, which serves as regulator for several metabolic processes. Biotin, playing a critical role in human function, is also rich in egg yolk. However, eating raw eggs continuously may cause biotin insufficiency because of the binding of biotin with avidin in egg white. This problem is more common in Japan with its traditional raw egg eating culture. From this point, eating cooked egg white with relatively raw egg yolk is recommended not only for microbiological safety but also for better nutrition uptake.

Vitamin B_{12} shows a particularly high value in egg. It plays an important role on brain development and immune system function, and the current recommended dietary allowance (RDA) value for people with age ≥14 is 2.4 μg/d while a recommended adequate intake value of 4.0 μg/d for adults is suggested by European Food Safety Authority (Obeid et al., 2019). The deficiency of B_{12} is common, especially in the elderly and people with low consumption of animal foods. A serving of one egg (~53g) per day could satisfy 35.3% of the daily requirement of B_{12} based on the nutritional facts (Seuss-Baum, 2007). However, the bioavailability of B_{12} from egg seems to be slightly lower compared with fish, beef, and other meat sources (Watanabe, 2007).

7.5.2 Vitamin Enrichment

With the enrichment strategy, much higher content of vitamins could be achieved for table eggs. The enrichment of vitamins in eggs first appeared in Belgium in 1997 starting with the supplementary of vitamin E (Surai & Sparks, 2001). The vitamin enrichment is mainly achieved by hen nutrition regulation, which is commonly used by industry to modify the contents of both water-soluble and fat-soluble vitamins in eggs. For example, hens supplied with extra retinol or vitamin D_3 can produce vitamin A or D_3 rich eggs, with the magnitude of increase up to 10-fold and 15-fold, respectively (Réhault-Godbert et al., 2019). The vitamin E content could be increased 2- to 20-fold to its original value based on the diet content and the nutrient supply. However, a significantly lower vitamin E in egg yolk was found in hens feed with vitamin A and E together (Grobas et al., 2002). For water-soluble vitamins, the magnitude of increase by dietary supply is much lower, for only about 2-fold (Réhault-Godbert et al., 2019), which can be well explained by the fundamental metabolism rules.

However, high dose vitamin as supplementary dietary can be toxic for hens. The sources of vitamin may differ a lot on toxicity. For example, 25-hydroxyvitamin D_3, a synthetic vitamin D_3 is used to be partially or totally substitute vitamin D_3 in hens' feed. However, there are still many studies concerned about the toxicity risk of high dose and long-term feeding of both 25-hydroxyvitamin D_3 and vitamin D_3 (Mattila et al., 2011; Warren et al., 2021). More research efforts should be provided to evaluate the safety and bioavailability of vitamins from eggs with different feeding practices.

7.6 Conclusions

Egg, as a rich resource of all these micronutrients required to support the development of embryo, confers a multitude of health benefits to human beings. Certain micronutrients, including carotenoids, vitamins and inorganic minerals could be significantly enriched through hen's dietary fortification to optimize their health benefits. Their contents could also be affected by various factors such as the housing condition, co-supplementary materials, and the feeding dose of these supplementary nutrients should be carefully determined to ensure hens' health condition and the quality of produced eggs. Unlike other nutrients, many active enzymes identified in unfertilized eggs have yet to be fully exploited, with many interesting subjects awaiting researchers to delve into in greater detail.

REFERENCES

Ahlborn, G., Clare, D., Sheldon, B., & Kelly, R. (2006). Identification of eggshell membrane proteins and purification of ovotransferrin and β-NAGase from hen egg white. *The Protein Journal*, *25*(1), 71–81. https://doi.org/10.1007/s10930-006-0010-8

Akagawa, M., Wako, Y., & Suyama, K. (1999). Lysyl oxidase coupled with catalase in egg shell membrane. *Biochimica et Biophysica Acta (BBA)-Protein Structure and Molecular Enzymology*, *1434*(1), 151–160. https://doi.org/10.1016/S0167-4838(99)00169-7

Arnold, M., Rajagukguk, Y. V., & Gramza-Michałowska, A. (2021). Functional food for elderly high in antioxidant and chicken eggshell calcium to reduce the risk of osteoporosis—a narrative review. *Foods*, *10*(3), 656. https://doi.org/10.3390/foods10030656

Barnkob, L. L., Argyraki, A., & Jakobsen, J. (2020). Naturally enhanced eggs as a source of vitamin D: A review. *Trends in Food Science & Technology*, *102*, 62–70. https://doi.org/10.1016/j.tifs.2020.05.018

Belkameh, M. M., Sedghi, M., & Azarfar, A. (2021). The effect of different levels of dietary magnesium on eggshell quality and laying hen's performance. *Biological Trace Element Research*, *199*(4), 1566–1573. https://doi.org/10.1007/s12011-020-02259-9

Carpentier, S., Knaus, M., & Suh, M. (2009). Associations between lutein, zeaxanthin, and age-related macular degeneration: An overview. *Critical Reviews in Food Science and Nutrition*, *49*(4), 313–326. https://doi.org/10.1080/10408390802066979

Choi, I., Jung, C., Choi, H., Kim, C., & Ha, H. (2005). Effectiveness of phosvitin peptides on enhancing bioavailability of calcium and its accumulation in bones. *Food Chemistry*, *93*(4), 577–583. https://doi.org/10.1016/j.foodchem.2004.10.028

Chung, H.-Y., Rasmussen, H. M., & Johnson, E. J. (2004). Lutein bioavailability is higher from lutein-enriched eggs than from supplements and spinach in men. *The Journal of Nutrition*, *134*(8), 1887–1893. https://doi.org/10.1093/jn/134.8.1887

De Boeck, S., Rymen, T., & Stockx, J. (1975). Adenosine deaminase in chicken-egg yolk and its relation to homologous enzymes in liver and plasma of the adult hen. *European Journal of Biochemistry*, *52*(1), 191–195. https://doi.org/10.1111/j.1432-1033.1975.tb03986.x

Debruyne, I. (1982). Hen's egg yolk alkaline phosphatase: General characterization and kinetic study with inhibitors. *The International Journal of Biochemistry*, *14*(6), 519–528. https://doi.org/10.1016/0020-711x(82)90120-3

Debruyne, I., & Stockx, J. (1977). Some properties of the alkaline phosphatase from the hen's egg yolk. *Archives Internationales de Physiologie et de Biochimie*, *85*(5), 961–963. https://doi.org/10.3109/13813457709053304

Farzi-Khajeh, H., Safa, K. D., & Dastmalchi, S. (2017). Arsanilic acid modified superparamagnetic iron oxide nanoparticles for purification of alkaline phosphatase from hen's egg yolk. *Journal of Chromatography B*, *1061*, 26–33. https://doi.org/10.1016/j.jchromb.2017.06.048

Geng, F., Xie, Y., Wang, J., Majumder, K., Qiu, N., & Ma, M. (2018). N-glycoproteomic analysis of chicken egg yolk. *Journal of Agricultural and Food Chemistry*, *66*(43), 11510–11516. https://doi.org/10.1021/acs.jafc.8b04492

Ghasemi, H. A., Hajkhodadadi, I., Hafizi, M., Fakharzadeh, S., Abbasi, M., Kalanaky, S., & Nazaran, M. H. (2022). Effect of advanced chelate compounds-based mineral supplement in laying hen diet on the performance, egg quality, yolk mineral content, fatty acid composition, and oxidative status. *Food Chemistry*, *366*, 130636. https://doi.org/10.1016/j.foodchem.2021.130636

Gravena, R. A., Marques, R. H., Roccon, J., Picarelli, J., Hada, F. H., da Silva, J. D. T., de Queiroz, S. A., & de Moraes, V. M. B. (2011). Egg quality during storage and deposition of minerals in eggs from quails fed diets supplemented with organic selenium, zinc and manganese. *Revista Brasileira de Zootecnia*, *40*, 2767–2775. https://doi.org/10.1590/S1516-35982011001200022

Grobas, S., Mendez, J., Bote, B. L., De Blas, B., & Mateos, G. (2002). Effect of vitamin E and A supplementation on egg yolk α-tocopherol concentration. *Poultry Science*, *81*(3), 376–381. https://doi.org/10.1093/ps/81.3.376

Haaland, J., Etheredge, E., & Rosenberg, M. (1971). Isolation of an ATPase from the membrane complex of the hen's egg. *Biochimica et Biophysica Acta (BBA)-Biomembranes*, *233*(1), 137–144. https://doi.org/10.1016/0005-2736(71)90366-X

Hencken, H. (1992). Chemical and physiological behavior of feed carotenoids and their effects on pigmentation. *Poultry Science*, *71*(4), 711–717. https://doi.org/10.3382/ps.0710711

Hrabia, A. (2022). Reproduction in the female. In *Sturkie's avian physiology* (pp. 921–966). Elsevier. https://doi.org/10.1016/B978-0-12-819770-7.00002-5

Huopalahti, R., Anton, M., López-Fandiño, R., & Schade, R. (2007). *Bioactive egg compounds* (Vol. 5). Springer. https://link.springer.com/content/pdf/10.1007/978-3-540-37885-3.pdf

Huyghebaert, G., Lippens, M., & Nollet, L. (2006). Enrichment of eggs with trace minerals using inorganic or organic trace sources: a feasibility? 12th European Poultry Conference, Verona, Italy, https://www.cabi.org/Uploads/animal-science/worlds-poultry-science-association/WPSA-italy-2006/10133.pdf

Immerseel, F. v., Nys, Y., & Bain, M. (2011). Volume 2: Egg safety and nutritional quality. In *Improving the safety and quality of eggs and egg products* (1st ed.). Elsevier.

Jang, I., Ko, Y., Kang, S., Kim, S., Song, M., Cho, K., Ham, J., & Sohn, S. (2013). Effects of dietary lutein sources on lutein-enriched egg production and hepatic antioxidant system in laying hens. *The Journal of Poultry Science*, *51*(1), 58–65. https://doi.org/10.2141/jpsa.0130017

Kim, W., & Patterson, P. H. (2005). Effects of dietary zinc supplementation on hen performance, ammonia volatilization, and nitrogen retention in manure. *Journal of Environmental Science and Health Part B*, *40*(4), 675–686. https://doi.org/10.1081/PFC-200061598

Kljak, K., Duvnjak, M., Bedeković, D., Kiš, G., Janječić, Z., & Grbeša, D. (2021). Commercial corn hybrids as a single source of dietary carotenoids: Effect on egg yolk carotenoid profile and pigmentation. *Sustainability*, *13*(21), 12287. https://doi.org/10.3390/su132112287

Leeson, S., & Caston, L. (2003). Vitamin enrichment of eggs. *Journal of Applied Poultry Research*, *12*(1), 24–26. https://doi.org/10.1093/japr/12.1.24

Leeson, S., & Caston, L. (2004). Enrichment of eggs with lutein. *Poultry Science*, *83*(10), 1709–1712. https://doi.org/10.1093/ps/83.10.1709

Lopez, R., Cabre, F., Franco, R., Cascante, H., & Canela, E. (1990). Purification of adenosine deaminase from chicken-egg yolk by affinity column chromatography. *Preparative Biochemistry*, *20*(3–4), 199–204. https://doi.org/10.1080/00327489008050196

Mann, K. (2008). Proteomic analysis of the chicken egg vitelline membrane. *Proteomics*, *8*(11), 2322–2332. https://doi.org/10.1002/pmic.200800032

Marounek, M., & Pebriansyah, A. (2018). Use of carotenoids in feed mixtures for poultry: A review. *Agricultura Tropica et Subtropica*, *51*(3), 107–111. https://doi.org/10.2478/ats 2018 0011

Mattila, P. H., Valkonen, E., & Valaja, J. (2011). Effect of different vitamin D supplementations in poultry feed on vitamin D content of eggs and chicken meat. *Journal of Agricultural and Food Chemistry*, *59*(15), 8298–8303. https://doi.org/10.1021/jf2012634

Milbradt, B. G., Muller, A. L. H., da Silva, J. S., Lunardi, J. R., Milani, L. I. G., de Moraes Flores, E. M., Callegaro, M. da G. K., & Emanuelli, T. (2015). Eggshell as calcium source for humans: Mineral composition and microbiological analysis/Casca de ovo como fonte de calcio para humanos: Composicao mineral e analise microbiologica. *Ciencia Rural*, *45*(3), 560–567.

Nabi, F., Arain, M. A., Rajput, N., Alagawany, M., Soomro, J., Umer, M., Soomro, F., Wang, Z., Ye, R., & Liu, J. (2020). Health benefits of carotenoids and potential application in poultry industry: A review. *Journal of Animal Physiology and Animal Nutrition*, *104*(6), 1809–1818. https://doi.org/10.1111/jpn.13375

Noda, S., Horn, F., Linder, D., & Schoner, W. (1986). Purified pyruvate kinases type M2 from unfertilized hen's egg are substrates of protein kinase C. *European Journal of Biochemistry*, *155*(3), 643–651. https://doi.org/10.1111/j.1432-1033.1986.tb09535.x

Obeid, R., Heil, S. G., Verhoeven, M. M., Van den Heuvel, E. G., De Groot, L. C., & Eussen, S. J. (2019). Vitamin B12 intake from animal foods, biomarkers, and health aspects. *Frontiers in Nutrition*, *6*, 93. https://doi.org/10.3389/fnut.2019.00093

Ramalho, H. M., Santos, V. V., Medeiros, V. P., Silva, K. H., & Dimenstein, R. (2006). Effect of thermal processing on retinol levels of free-range and caged hen eggs. *International Journal of Food Sciences and Nutrition*, *57*(3–4), 244–248. https://doi.org/10.1080/02656730600836469

Réhault-Godbert, S., Gautron, J., Labas, V., Belghazi, M., & Nys, Y. (2008). Identification and characterization of the precursor of chicken matrix metalloprotease 2 (pro-MMP-2) in hen egg. *Journal of Agricultural and Food Chemistry*, *56*(15), 6294–6303. https://doi.org/10.1021/jf8003948

Réhault-Godbert, S., Guyot, N., & Nys, Y. (2019). The golden egg: Nutritional value, bioactivities, and emerging benefits for human health. *Nutrients*, *11*(3), 684. https://doi.org/10.3390/nu11030684

Samaraweera, H., Zhang, W. g., Lee, E. J., & Ahn, D. U. (2011). Egg yolk phosvitin and functional phosphopeptides. *Journal of Food Science*, *76*(7), R143–R150. https://doi.org/10.1111/j.1750-3841.2011.02291.x

Schiavone, A., & Barroeta, A. (2011). Egg enrichment with vitamins and trace minerals. In *Improving the safety and quality of eggs and egg products* (pp. 289–320). Elsevier. https://doi.org/10.1533/9780857093929.3.289

Schlatterer, J., & Breithaupt, D. E. (2006). Xanthophylls in commercial egg yolks: Quantification and identification by HPLC and LC-(APCI) MS using a C30 phase. *Journal of Agricultural and Food Chemistry*, *54*(6), 2267–2273. https://doi.org/10.1021/jf053204d

Schmid, A., & Walther, B. (2013). Natural vitamin D content in animal products. *Advances in Nutrition*, *4*(4), 453–462. https://doi.org/10.3945/an.113.003780

Seko, A., Juneja, L., & Yamamoto, T. (2018). Enzymes in unfertilized hen eggs. In *Hen eggs* (pp. 135–144). CRC Press.

Seuss-Baum, I. (2007). Nutritional evaluation of egg compounds. In *Bioactive egg compounds* (pp. 117–144). Springer. https://doi.org/10.1007/978-3-540-37885-3_18

Singh, A., Kelkar, N., Natarajan, K., & Selvaraj, S. (2021). Review on the extraction of calcium supplements from eggshells to combat waste generation and chronic calcium deficiency. *Environmental Science and Pollution Research*, *28*(34), 46985–46998. https://doi.org/10.1007/s11356-021-15158-w

Sirri, F., & Barroeta, A. (2007). Enrichment in vitamins. In *Bioactive egg compounds* (pp. 171–182). Springer. https://link.springer.com/content/pdf/10.1007/978-3-540-37885-3_21.pdf

Siwko, S., & Mochly-Rosen, D. (2007). Use of a novel method to find substrates of protein kinase C delta identifies M2 pyruvate kinase. *The International Journal of Biochemistry & Cell Biology*, *39*(5), 978–987. https://doi.org/10.1016/j.biocel.2007.01.018

Song, W. O., & Kerver, J. M. (2000). Nutritional contribution of eggs to American diets. *Journal of the American College of Nutrition*, *19*(sup5), 556S–562S. https://doi.org/10.1080/07315724.2000.10718980

Surai, P., & Sparks, N. (2001). Designer eggs: From improvement of egg composition to functional food. *Trends in Food Science & Technology*, *12*(1), 7–16. https://doi.org/10.1016/S0924-2244(01)00048-6

Tang, G. (2014). Vitamin A value of plant food provitamin A—evaluated by the stable isotope technologies. *International Journal for Vitamin and Nutrition Research*, *84*(Suppl 1), 25–29. https://doi.org/10.1024/0300-9831/a000183

Vadehra, D., Nath, K., & Forsythe, R. (1973). Eggs as a source of protein. *Critical Reviews in Food Science & Nutrition*, *4*(2), 193–309. https://doi.org/10.1080/10408397309527158

Warren, M. F., Pitman, P. M., Hodgson, D. D., & Livingston, K. A. (2021). Super-doses of dietary vitamin D3 intake in aged laying hens illustrates limitation of 24, 25-dihydroxycholecalciferol conversion. *bioRxiv*. https://doi.org/10.1101/2021.10.14.464328

Watanabe, F. (2007). Vitamin B12 sources and bioavailability. *Experimental Biology and Medicine*, *232*(10), 1266–1274. https://doi.org/10.3181/0703-MR-67

Yamamoto, T., Juneja, L. R., Hatta, H., & Kim, M. (1996). *Hen eggs: Basic and applied science*. CRC Press.

Yilmaz, B., & Ağagündüz, D. (2020). Bioactivities of hen's egg yolk phosvitin and its functional phosphopeptides in food industry and health. *Journal of Food Science*, *85*(10), 2969–2976. https://doi.org/10.1111/1750-3841.15447

Yonekura, L., & Nagao, A. (2007). Intestinal absorption of dietary carotenoids. *Molecular Nutrition & Food Research*, *51*(1), 107–115. https://doi.org/10.1002/mnfr.200600145

Zaheer, K. (2017). Hen egg carotenoids (lutein and zeaxanthin) and nutritional impacts on human health: A review. *CYTA-Journal of Food*, *15*(3), 474–487. https://doi.org/10.1080/19476337.2016.1266033

Part III

Functional Properties of Egg Components

8

Egg White Protein Functionality

Yujie Chi* **and Zihong Ma**

CONTENTS

8.1 Introduction to the Functional Properties of Egg Components

The food industry manufactures large quantities of proteins for use as nutritional and functional additives in many products. Functional properties are defined as any attribute—other than nutritional value—that affects the utilization of food ingredients, and are key to the importance of proteins in a food application (Pomeranz, 1991). Protein functionality includes solubility, emulsion, foaming, and gelation characteristics (Hettiarachchy et al., 2013). The utilization of any food component is largely dependent on their functional properties, which allows for the formulation of food products with desirable characteristics. In recent years, the food processing industry has increasingly relied on fabricated components, which has increased the demand for desirable functional properties in food formulation and product development (Lu et al., 2019).

Eggs and their components are extensively used in domestic cooking and industrial processing owing to their superb functional properties. Egg whites, for example, are known for their favorable gelling and foaming properties; egg yolks, on the other hand, can form emulsions and gels. The egg components are also rich in bioactive compounds (Xiao et al., 2020; Xiao et al., 2021). Egg white contains approximately 110 g/kg proteins, while egg yolk contains approximately 150–170 g/kg proteins (Chang et al., 2018). The lipids are also concentrated in the egg yolk. The unique composition of the egg white and yolk imparts excellent functional and nutritional values to egg-containing foods (Réhault-Godbert et al., 2019).

The use and function of egg proteins and lipids are closely related to our daily needs. Eggs are common in various food formulations (Figure 8.1). When eggs are cooked, egg proteins undergo unfolding, denaturation, coagulation, and gelation, as observed for boiled eggs and scrambled eggs. We can also adjust the cooking temperature or time to obtain different textures and flavors according to our own or consumer needs. Product development, quality control, and sensory evaluation require a proper understanding of the functional properties of egg components, as well as their behavior in the different conditions of industrial processing (Kumbar et al., 2015).

Gels, foams, and emulsions account for the majority of formulated foods (Yang & Baldwin, 1995). Understanding the gelling, foaming, and emulsifying characteristics of food components is vital for their use in industrial processing and is conducive to the maintenance of nutrients.

* Corresponding author.

DOI: 10.1201/9781003254430-11

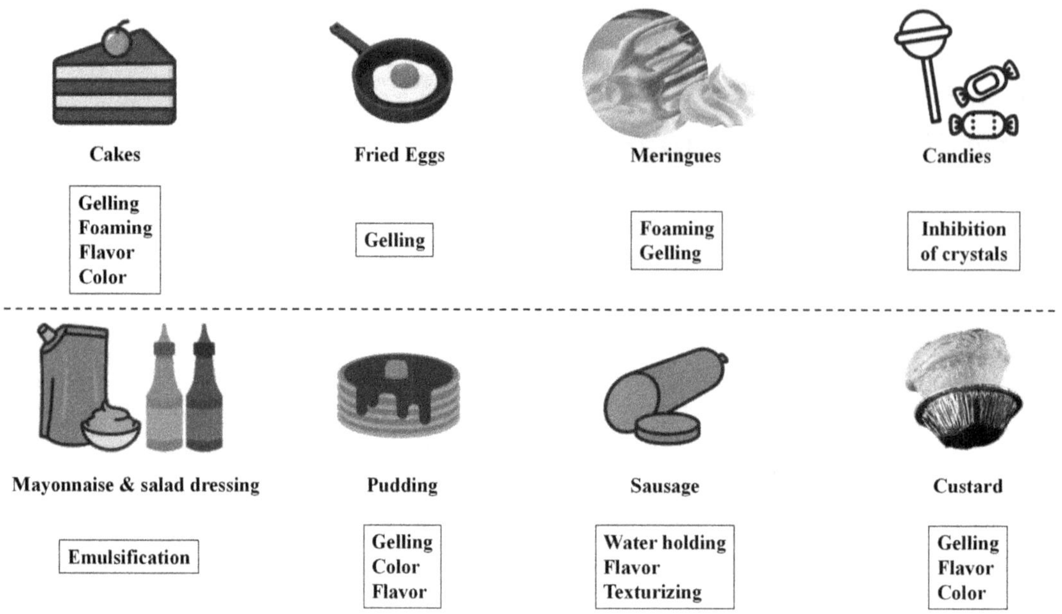

FIGURE 8.1 Functional properties of egg components involved in foods.

8.2 The Gelling Properties of Egg White Proteins

The gelling of proteins is an indispensable functional attribute in food processing and provides a structural matrix to bind other ingredients, nutrients, and bioactive compounds (Wang et al., 2021). The quality of many foods, especially texture and juiciness, is determined by gelling (Zayas, 1997). Generally, gelling refers to the sol–gel transformation of a protein—mostly induced by heat—where a critical equilibrium of attractive and repulsive forces is achieved (Ferry, 1948). To form a gel, proteins undergo a series of changes of denaturation, aggregation, and crosslinking (mainly noncovalent cross-linking, such as hydrophobic interactions) (Doi, 1993; Mine, 2002; Nicolai, 2019; Liu et al., 2022c).

There are two types of protein gels: turbid coagula and transparent gels (Damodaran, 1997a). The formation of a thermally turbid coagulum results from random protein interactions, thus forming soluble and insoluble aggregates; on the other hand, thermally transparent gels can be regarded as an ordered three-dimensional network (Hermansson, 1979). The type of gel is partly related to the amino acid composition of the proteins as well as the protein denaturation and aggregation rates during gel formation (Shimada & Matsushita, 1980; Hermansson, 1986; Weijers et al., 2002). In addition, the aqueous conditions, such as pH, ionic strength, and protein concentration, also play important roles (Renard & Lefebvre, 1992; Gao et al., 2021b). It has been proposed that a high proportion (> 30%) of hydrophobic residues (e.g., egg white albumin) promote the formation of coagulum-type gels, while a high proportion of hydrophilic amino acids (e.g., ovomucoid) contribute to the formation of transparent gels (Hermansson, 1979; Shimada & Matsushita, 1980). A higher rate of protein aggregation over protein denaturation favors the formation of a disordered gel network owing to random protein aggregation (Hermansson, 1979); in contrast, an ordered gel network with high transparency can be formed with highly mobile protein molecules (Ma et al., 2022).

Egg white contains various globular proteins with a dense well-defined secondary and tertiary structure, making it an excellent source of gel-forming ingredients (Nicolai, 2019). Ovalbumin is the most abundant protein in egg white, accounting for ~54% of the total proteins, and consequently dominates the gelling process (Mine, 2002). It is the only egg white protein with free sulfhydryl (SH) groups at the protein core (isoelectric point [pI] 4.5) (Mine, 2002; Gharbi & Labbafi, 2018). During storage, ovalbumin can change into more heat-stable forms, of which the most heat-stable is S-ovalbumin, and its content is negatively correlated with the Haugh unit—an important indicator for egg quality (Yang & Baldwin, 1995; Huang et al., 2012). Ovotransferrin and lysozyme are another two important proteins in egg white, and they are stabilized by 15 and 4 disulfide (SS) bonds, respectively. Notably, ovotransferrin is the most heat-liable egg white protein (Alleoni et al., 2006).

During processing, there are several factors that can directly affect the gelling of egg white proteins, including thermal treatment (Liu et al., 2020d; Ma et al., 2021a), salt/metal ions (Li et al., 2018a; Liu et al., 2022d; Tian et al., 2022a), saccharide treatment (Mu et al., 2020; He et al., 2021), pH adjustment (Khemakhem et al., 2019; Alavi et al., 2020c; Gao et al., 2020), mechanical treatment (Naderi et al., 2017; Hong et al., 2021; Xue et al., 2021; Joeres et al., 2022), and chemical modification (Alavi et al., 2019; Ma et al., 2019a; Liu et al., 2020b; Yang et al., 2021). The discrepancies in the physicochemical behaviors of egg white protein gels can impart unique texture, shape, and mouth feel to the end products, and can directly determine consumer acceptability of new food products.

8.2.1 Mechanism of Egg White Protein Gel Formation

8.2.1.1 Heat-Induced Egg White Protein Gels

The heat-induced gelling of egg white protein is common in domestic cooking and industrial production and is the most frequently used mode of gel preparations (Lv et al., 2022). Multiple food manufacturers have taken advantage of the variety and ease of thermal gel processing, which has paved the way for the development of unique food formulas (Wang et al., 2022a).

The mechanism of heat-induced gelling has been extensively studied and remains a research focus. Thermally induced egg white protein gels are formed owing to protein–water interactions and protein–protein cross-linking, where hydrophobic interactions are the dominant driving forces (Banerjee & Bhattacharya, 2012). Generally, heat-induced gelling can be divided into three stages (Figure 8.2): (i) proteins unfold and denature to form spherical aggregates via hydrophobic and electrostatic interactions; (ii) the preformed aggregates are strengthened and further aggregated due to sulfhydryl–disulfide (SH–SS) exchange and a three-dimensional network is formed, trapping water and other solvents in it; and (iii) cooling promotes the rapid formation of hydrogen bonds that fortify the gel texture (Mine, 1995; Lv et al., 2022; Zheng et al., 2022a).

Although the main aggregation process is dominated by the formation of SS bonds (Visschers & de Jongh, 2005), they are not essential in the gel forming process but can fortify gel integrity (Gharbi & Labbafi, 2018). A strong, direct correlation (r = 0.97) was found between egg white protein surface SH group concentration and the gel strength of its heat-induced gels (Margoshes, 1990). During heating, the gelling process is dominated by hydrophobic interactions, especially at higher temperatures, while during cooling, hydrogen bonds prevail, thus increasing the elasticity of the cooled gel; the electrostatic interactions, on the other hand, are susceptible to changes in pH, molecule charges, ionic strength, and divalent ions (Sikorski, 2006).

A "molten globule state" transition occurs during the heat-induced denaturation of ovalbumin (Hagolle et al., 1997). Other than heating, the molten globule state can also be initiated by extreme pH, high-pressure treatment, and exogenous mutagens (Zhu et al., 2021a). The molten globule state—found in various proteins—is a compact intermediate protein conformation different from either the native or totally denatured forms (Ohgushi & Wada, 1983; Zhu et al., 2021b; Rico-Pasto et al., 2022). The existence of this intermediate state may accelerate the aggregation of protein molecules, which is closely related to the thermally induced gelation mechanism (Mine, 1995). Apart from improving gelling properties, the molten globule state also improves the emulsifying properties of egg white proteins (Yu et al., 2021b; Yu et al., 2021c).

During heating, changes in the structure of egg white proteins can be detected. Large polymers of ovalbumin are formed during heating, as confirmed by native–PAGE analysis (Sun & Hayakawa, 2002). The formation of stable intermolecular β-sheet structures occurs during heat denaturation and aggregation of egg white or ovalbumin, as revealed by Raman and circular dichroism (CD) spectroscopy (Painter & Koenig, 1976; Kato & Takagi, 1988; Ma et al., 2022). Protein aggregation is typically linked to increased β-sheet content, which is produced in large antiparallel areas between ovalbumin molecules; β-sheets also interact with ovotransferrin and lysozyme (Nakamura & Doi, 2000; Joshi et al., 2014).

Increasing salt concentrations and molecular weight is conducive to increasing the β-sheet content of heat-induced ovalbumin aggregates. For extended protein structures, hydrophobic interactions play a major role in stabilizing β-sheet conformations (Narayanan & Dias, 2013; Alavi et al., 2019).

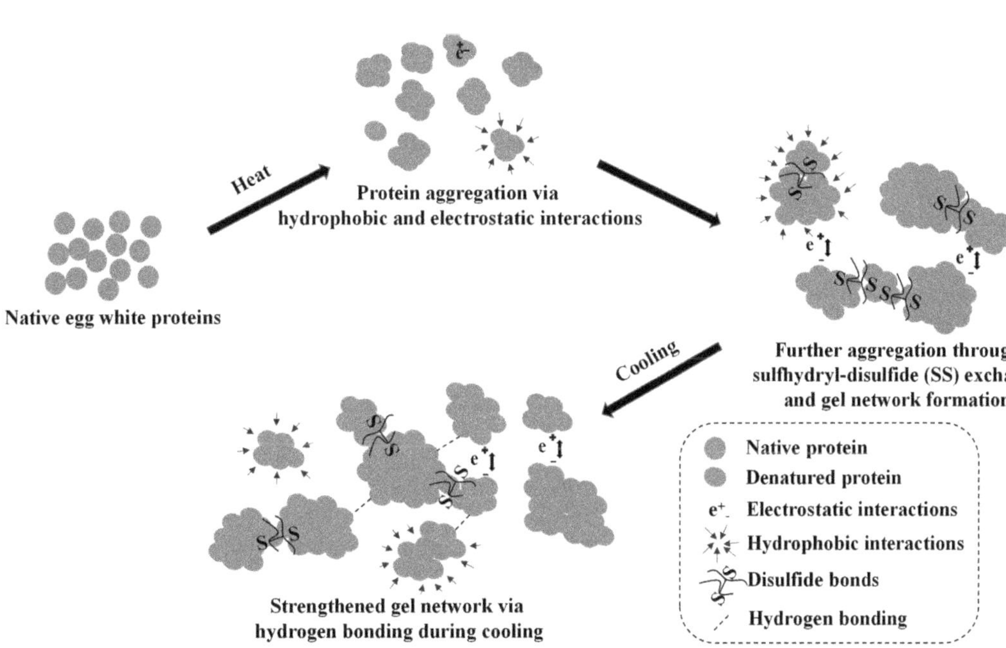

FIGURE 8.2 Schematic diagram of egg white protein gel formation.

Furthermore, β-sheets are critical components in stabilizing denatured protein aggregations, and protein molecules with a high fraction of β-sheets are more likely to form tight gels (Su et al., 2015).

Proteomics is frequently utilized to elucidate the mechanism of heat-induced gel formation by egg white proteins (Liu et al., 2020d; Liu et al., 2021; Wang et al., 2022a). Comparative proteomic analysis showed that the abundance of ovalbumin and ovomucoid increase, while the abundance of ovotransferrin, lysozyme, and ovomucin (mucin 5B and mucin 6) decrease in the early heating stage (72°C, 8 min) of heat-induced gelling (Wang et al., 2022a). Ovomucin is thought to play a crucial role in the formation and modulation of egg white thermal gel characteristics (Wang et al., 2022a).

Begam et al. (2021) evaluated the kinetics of heat-induced gelling and microscopic dynamics of egg white gels using X-ray photon correlation spectroscopy and ultrasmall-angle X-ray scattering, and found that the gel exhibits stress-activated dynamics with an exponential rise in the relaxation time (aging) and a subsequent steady-state ballistic motion displaying significant temporal heterogeneity. Their study provides an empirical basis for future experiments of protein gelling and aggregation at nano- and micrometer scales, and sheds light on processes highly relevant for the food industry and soft matter physics (Begam et al., 2021).

8.2.1.2 High Pressure-Induced Egg White Protein Gels

In the food industry, high hydrostatic pressure is increasingly being employed to improve the microbiological safety, functioning, and bioactive qualities of food items (Naderi et al., 2017). Pressure treatment could induce the denaturation and aggregation of egg white proteins, depending on pressure level, protein concentration, treatment time, pH, and temperature (Gharbi & Labbafi, 2018).

Egg white treated with high hydrostatic pressure (>600 MPa) produces a softer and more elastic gel than those produced with heat treatment. With increasing pressure, the gel hardness increases and adhesiveness decreases (Hayashi et al., 1989; Okamoto et al., 1990). The pressure level necessary for gel induction is affected by pH, presence of salts or sugars and depends also on the protein type and concentration (Strohalm et al., 2000). Pressure treatment does not alter the vitamin content or any taste of egg white gels (Hayashi et al., 1989). Okamoto et al. (1990) reported that, in the denatured state, the rearrangement of water molecules around amino acid residues may be important in increasing the glossiness and transparency of pressure-induced gels. It has also been suggested that alterations in protein SS links alter the egg white structure, resulting in gel formation (Naderi et al., 2017). The soft gel texture may be explained by the high level of SH group exposure, which leads to oxidation and SH–SS bond exchange that produces soluble aggregates (Van der Plancken et al., 2005b). Zhang et al. (2020) prepared an egg white–whey protein mixture under acidic conditions, which was subjected to high pressure processing to take advantage of high pressure-induced egg white protein gelling to mask the interaction sites of whey protein with salivary protein and reduce its astringency. They found that egg white did form gels after high pressure processing; however, the high pressure treated samples exhibit a higher astringency, possibly due to the exposure of more hydrophobic sites on the proteins.

8.2.1.3 Alkali-Induced Egg White Protein Gels

A highly elastic egg white protein gel can be gradually formed under alkali conditions without heat treatment (Cunningham & Cotterill, 1962; Zhao et al., 2016). The most traditional application of alkali-induced gelling is to preserve eggs, also known as "*Pidan*." Strong alkali leads to the denaturation, aggregation, cross-linking, and interaction of solvent molecules with egg white proteins; ionic bonds, SS bonds, and hydrophobic interactions are also involved (Zhao et al., 2016).

The gelling behavior of alkali-treated egg white in shell eggs varies inside and outside of the shell. The gels obtained from preserved eggs are harder and more stable than those of alkali-treated liquid (i.e., without the shell) egg white (Gharbi & Labbafi, 2018). However, the microstructural characteristics of ovalbumin gels treated with strong alkali are comparable between these conditions (Zhao et al., 2016). Ovalbumin is the dominant component involved in egg white gel formation under strong alkali conditions (Chen et al., 2015). Ionic bonds mainly contribute to gel formation, with SS bonds playing a lesser role (Zhao et al., 2014, 2016). Alkali treatment converts the α-helical structure of ovalbumin to a β-sheet structure, accompanied by the exposure of amino acid residues to a polar environment (Zhao et al., 2016).

Interestingly, a sol–gel–sol transition of egg white proteins with alkali treatment occurs during pickling (Gao et al., 2020; Gao et al., 2021b). The gelling process is accompanied by a deepening in color caused by Maillard reactions, and gel darkness generally corresponds with a softer gel texture (Chen et al., 2015; Gao et al., 2020; Tan et al., 2021). The gel–sol transition may be caused by the partial unfolding or degradation of proteins, which inhibits protein aggregation (Gao et al., 2020). Meanwhile, the content of β-sheets decreases, and the system gradually becomes disordered (Gao et al., 2021b). Generally, this liquefaction is undesirable for the production of preserved eggs, and, in traditional pickling, generally requires inhibition using heavy metal ions. However, due to health concerns, safer additives are needed. Jin et al. (2022b) reported that high concentrations of sodium ascorbate (2% and 3%) possess the potential to inhibit the "alkali injury liquefaction" of preserved egg whites without heavy metals.

8.2.1.4 Acid-Induced Egg White Protein Gels

Below pH 8.0, the viscosity of egg white decreases gradually, reaching a minimum at approximately pH 4.0. However, the viscosity increases dramatically when acid is added. At pH values below 2.2, egg white forms a cloudy, milk-like gel that remains stable until neutralized; acidic egg white gels can be reconstituted with water and solidify over time (Cunningham & Cotterill, 1964). Interestingly, solid gels can still form when egg white is diluted by three volumes of water (Cunningham & Cotterill, 1964).

Heat-induced soluble protein aggregates subjected to acid trigger the association of aggregates to form gel networks (i.e., acid-induced gelation) (Alavi et al., 2020a). Cold-set egg white protein gels have greater potential for application in the development of health foods and drug delivery systems (Alavi et al., 2020c). The formation of acid-induced cold-set gels involves two stages. At neutral pH and low ionic strength in the first stage, a heat-denatured protein solution of soluble aggregates is produced. Gelling is induced by adjusting the pH in the second stage to the protein's pI at room temperature. This encourages the development of a percolation network by reducing electrostatic repulsion between the aggregates; different gel network structures and rheological properties can be obtained by adjusting the type of aggregates produced in the first stage (Weijers et al., 2006). Pu et al. (2022) prepared acid-induced whey protein concentrate and egg white protein composite gels and found that the self-supporting gels prepared by mixing whey and egg white protein have better performance than single whey protein concentrate gels; their findings may provide opportunities to develop functional food formulations enriched with whey protein and egg white proteins.

8.2.1.5 Ethanol-Induced Egg White Protein Gels

"*Zaodan*"—another unique traditional food in China—is made by preserving fresh eggs (mainly duck eggs) in rice wine, which has a milky white or jam yellow jelly egg white and an orange semisolid egg yolk. Researchers ascribed this gelling phenomenon to ethanol—the main component in rice wine. By mixing egg white solution with ethanol at different concentrations, increased egg white viscosity was observed with increased ethanol concentration due to the formation of SS bonds promoted by high ethanol concentrations and gels were formed at ethanol concentrations above 25% (Yu et al., 2020). The hydroxyl groups of ethanol form hydrogen bonds with the protein molecules to enhance the gel hardness, and the dielectric constant of water is altered by ethanol, leading to the denaturation and aggregation of egg white protein and good compactness of the gel network (Yu et al., 2020). Yao et al. (2020) reported that egg white protein aggregated under the influence of high concentrations of ethanol due to the improvement of molecular forces reflected by the increase in the absolute value of zeta potential and the decrease in free sulfhydryl content and intrinsic fluorescence intensity.

Changes in the *zaodan* gel microstructure and molecular interactions under the action of *baijiu* have also been reported (Yu et al., 2021a). The three-dimensional network structure of the gel gradually changes from a loose and porous sponge to a compact sponge-like structure with prolonged *baijiu* incubation, which exhibits higher gel hardness and water holding capacity. In this process, egg white proteins are expanded with *baijiu* entering the eggs by osmotic pressure, decreasing ionic bonds, exposing internal free SH groups, forming more SS bonds, and increasing surface hydrophobicity. The authors found that all other proteins were destroyed except for ovomucoid during pickling, which plays an important role in the formation of gels. The ethanol-induced aggregation or gelling of egg white proteins has been utilized to encapsulate bioactive compounds such as curcumin and tea polyphenols (Chang et al., 2019; Yao et al., 2020).

8.2.2 Factors Affecting the Gelling Behaviors of Egg White Proteins

Protein concentration: A critical protein concentration is essential for gel formation. However, this concentration is not fixed, but is associated with the net charge of the protein, which can be affected by the pH and ionic strength of the solvent environment (Grinberg et al., 1985; Renard & Lefebvre, 1992). The protein concentration mainly affects the hydrophobic and electrostatic interactions between protein molecules (Lv et al., 2022). Strong electrostatic repulsions lead to clear and fine-stranded gel structures, while weak electrostatic repulsions result in coarse lumpy structures (Grinberg et al., 1985; Clark, 1991). Heat-set egg white protein gel strength is positively correlated with protein concentration (Ma et al., 2021a).

Temperature: Temperature is an important driver of protein gelling as it directly causes protein domains to unfold (Totosaus et al., 2002). Protein denaturation increases ~600 times with every 10°C increase in temperature in the range typical of denaturation (Bove et al., 1997). A high gelling temperature coefficient leads to a higher denaturation rate than the aggregation rate. Small attractive forces between denatured protein chains cause sluggish aggregation, resulting in a fine and transparent gel network for a given denaturation rate. (Ferry, 1948; Hermansson, 1979). Generally, increasing temperature promotes the formation of a fine network (Pomeranz, 1991), and texture profile analysis (TPA) regarding heated thick egg white gel revealed that hardness, springiness, cohesiveness, and chewiness decline and then increase with increasing temperature (Luo et al., 2022).

Temperature has an important impact on changes in the free SH groups caused by pressure treatment. At lower temperatures, a lower pressure is required to expose SH groups, while higher levels of exposed SH groups could be obtained with high pressure treatment (>550 MPa) (Van der Plancken et al., 2005a; Van der Plancken et al., 2005b). An inverse relationship exists between temperature and pressure: At ambient pressure, increasing temperature causes the rate constants of the kinetic models describing the changes in physicochemical attributes, whereas the converse is seen at elevated pressures (Van der Plancken et al., 2007b).

In the case of alkali-treated egg white protein gels, Zhao et al. (2014) and Tan et al. (2022) reported that temperature significantly affects the rate of NaOH permeation into the egg white: High temperatures lead to high permeation rates and increased gel darkness. However, a high pickling temperature and a considerably long curing time disrupt the egg white gel structure (liquefaction), reducing product quality (Yang et al., 2012). Similarly, a low pickling temperature also poses a negative effect on product quality: The processing cycle is prolonged, and defective yellow eggs are formed (Lu et al., 2005). After an alkaline-treated egg white gel is formed, an additional heating treatment (100–121°C for 20 min) decreases the mechanical and rheological properties of the gels by destroying intermolecular forces, including hydrogen bonds, hydrophobic interactions, and SS bonds (Fan et al., 2021). For acid-induced cold-set egg white protein gels, their microstructure compactness, mechanical and rheological properties, and water holding capacity are improved by increasing the preheating temperature and protein content during the formation of protein aggregates (Alavi et al., 2020c).

Salted eggs are traditional egg products in China and Southeast Asia and are conventionally made by marinating fresh duck eggs in saturated brine or mixing them with salt paste and soil for approximately 4 weeks (Kaewmanee et al., 2011). The high salt concentration makes egg white lose its viscosity, and it often requires a boiling step before salted egg white is consumed. With different salting times, the heat-induced egg white gels prepared at 100°C and 120°C exhibit different gelling properties (Gao et al., 2021a). A higher heating temperature accelerates the migration of water in the gel, damages more protein tertiary structures, and promotes more SS bonds or SH–SS exchange, leading to greater hardness, chewiness, and gumminess. However, the total protein digestibility at 100°C is higher than that at 121°C.

Protein composition: Egg whites consist of various proteins that exhibit different thermal behaviors. The structure of egg white proteins can be denatured by heat treatment over 60–65°C (Yamashita et al., 1998). Donovan et al. (1975) studied the thermal denaturation of egg white proteins using differential scanning calorimetry (DSC) and observed three responses in the thermogram at neutral pH, corresponding to the denaturation of ovotransferrin (61°C), lysozyme (75°C), and ovalbumin (84°C).

Liu et al. (2020d) investigated the DSC curves of lyophilized thick/thin egg white proteins and reported their average thermal denaturation temperatures (T_d) (72.51 and 67.01°C, respectively), indicating that thin egg white is more susceptible to high temperatures. This might be explained by the difference in protein composition: The abundance of ovotransferrin in thin egg white is 10.1% higher than that in thick egg white (Iwashita et al., 2019; Liu et al., 2020d). Ovotransferrin is the least heat-stable and is therefore likely a crucial factor in determining the characteristics of thin and thick heat-induced gels. The same researchers showed that thin egg white has a larger denaturation enthalpy (ΔH) than thick egg white, implying that the protein profiles of thin egg white are ordered, allowing for greater heat absorption (Liu et al., 2020d). Hammershøj et al. (2001) also reported that an increased concentration of ovotransferrin in albumen has a positive effect, whereas lysozyme has a negative effect on the albumen gel texture.

Mixed proteins in a system with varying pI values can attract each other through both hydrophobic and electrostatic interactions, which promotes protein aggregation (Gharbi & Labbafi, 2018). Each protein coexisting under the same conditions produces aggregates at a lower temperature in a mixed compared to a single protein solution, which affects the final characteristics of the gel (Matsudomi et al., 1991; Hong et al., 2017; Iwashita et al., 2017). A recent study showed that the gelling of an ovalbumin-lysozyme binary system benefits from a synergistic effect during co-gelation, which is dependent on the nucleation of heteroprotein co-aggregate units. Furthermore, these binary systems have extremely low critical gelling concentrations (Liu et al., 2022e).

pH and acid: Generally, gel network development is aided by small electrostatic repulsive protein–protein interactions (Campbell et al., 2003). Given sufficient protein concentrations, pH values near the pI facilitate protein aggregation. Protein solubility, which influences protein gelling, can also be affected by changes in pH values. Egg white protein solubility increases with increased pH, with higher solubility at pH 9.0 and lower solubility at pH 4.6 (Machado et al., 2007). The textural profiles, water holding capacity, color, and microstructure of heat-induced egg white gels are all pH dependent (Hatta et al., 1986; Handa et al., 1998). Heat-induced egg white gel prepared close to the pI of ovalbumin (4.5) becomes brittle and soft with the lowest gel hardness; with increasing pH (4.5–7.0), the gel hardness also exhibits an increasing trend; however, a slight decrease in gel hardness was observed at pH 8.0 (Khemakhem et al., 2019). A low linear relationship between heat-set egg white gel firmness and pH (R^2 0.59) was observed (Raikos et al. 2007b). Near the pI of most egg white proteins, a coagulum is formed after heat induction, and the resultant coagulum has a coarse appearance, low viscoelasticity and water holding capacity; at higher pH (7 and 9), egg white gels are more viscoelastic due to the higher protein charge and promoted SH–SS exchange (Croguennec et al., 2002). The heat-set egg white gel prepared at pH 5 exhibits a porous structure, while that prepared at pH 9 exhibits a compact microstructure, and the latter has a lower fluorescein isothiocyanate-pepsin diffusion coefficient (Somaratne et al., 2020). Hatta et al. (1986) also reported a pH-dependent ovalbumin gel hardness and turbidity (with the presence of 20 mM NaCl). However, the authors found that the maximum gel hardness was obtained at pH 3.5 and 6.5, with slight turbidity, and at more acidic and alkaline pH values, transparent gels were observed. The transparent gel prepared at approximately pH 3.5 and 7.0 is harder than the turbid gel prepared at approximately pH 5.0.

Interestingly, after egg white protein powders are heated in the dry state in alkaline pH region for several days, their subsequent hydrothermal gels also exhibit firmer and more elastic characteristics (Mine, 1996). Similar results were observed in radiofrequency-assisted-thermal-treated egg white powders: A higher pH (9.9) leads to increased gel firmness (Kar et al., 2020). An increased pH promotes the proteins to unfold and aggregate owing to the changes in protein net charge and the enhanced gel firmness is ascribed to accelerated SH–SS exchange.

The acid type can also impact egg white protein gel properties. Ascorbic acid, for example, is commonly used in food products and is notable for its involvement in both Maillard reactions and free radical cycles (Farahnaky et al., 2003). Ascorbic acid decreases the thermal denaturation temperature of egg white proteins, which also decreases the mechanical properties (failure strength, failure strain, and Young's modulus) (Nafchi et al., 2013). In the case of marinated eggs, Xue et al. (2022a) reported that the gel strength of heat-induced egg white is increased by shikimic acid braising, which could be used as a novel enhancer of egg white gel.

In the case of alkaline-induced egg white gels, high alkalinity (high NaOH concentrations) rapidly increases the strength of the duck egg white gel to the maximum value and then rapidly decreases; the strength of low alkalinity (low NaOH

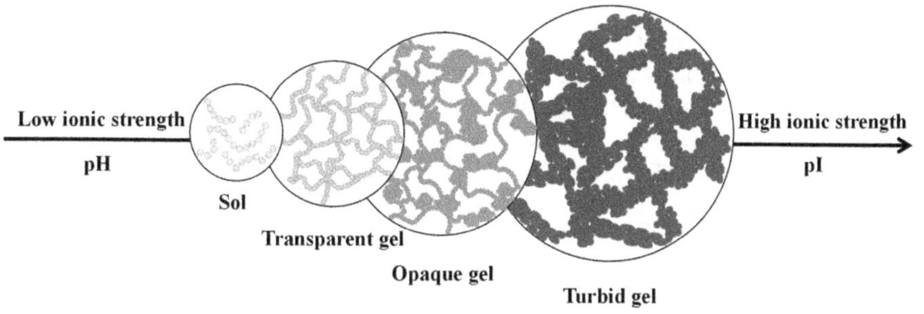

FIGURE 8.3 Effect of pH and ionic strength on the gelling behaviors of egg white protein.

concentrations)-treated gels is highly stable and exhibits gradual changes over time (Zhao et al., 2014). Generally, the pH environment of preserved eggs should be controlled to obtain the desired texture.

Ionic strength and salt: Water absorption, swelling, and solubility of proteins can be affected by ionic strength through the establishment of competitive linkages (Totosaus et al., 2002). At neutral pH, ovalbumin appears as a linear chain with ionic strength-dependent branching (Weijers et al., 2004). Figure 8.3 depicts the influence of ionic strength on the appearance of heat-induced ovalbumin gels (Doi, 1993).

At a pH near the pI or high ionic strength, the electrostatic repulsion between ovalbumin molecules is weak, and the attraction between protein molecules (i.e., hydrophobic interactions) induces the formation of disordered aggregates, thus leading to a turbid gel with decreased hardness. However, at pH values far below/above the pI or low ionic strength, little or no shielding exists for the surface charge of protein molecules and strong electrostatic repulsion hinders the formation of disordered aggregates, promotes the formation of linear aggregates, and finally forms transparent gels (Kitabatake et al., 1988a; Weijers et al., 2003; Weijers et al., 2005; Weijers et al., 2006; Khemakhem et al., 2019). Kitabatake et al. (1988b) proposed a two-step heating with salt method: Diluted egg white heated for 1 h at 80°C gives a transparent solution that becomes a translucent gel when adding 0.3 M NaCl during reheating. The authors also observed that a transparent gel made with 150 mM NaCl is more firm and less sticky than a turbid gel made with a one-step heating approach under the same conditions. A rapid decrease ($p < 0.05$) in the fracture strength/strain of egg white gels was found as the NaCl concentration increased (0–3%), and NaCl addition affects the water holding capacity of egg white gels by increasing the mobility and distribution of free water (Li et al., 2018a).

The type of salt ions also plays an important role. The viscoelasticity of egg white gel is affected by Ca^{2+} and Mg^{2+} at neutral pH because these ions are fixed to ovalbumin phosphoserine residue. In the presence of Ca^{2+}, the denaturation and aggregation temperatures decrease, thus leading to less homogeneous egg white gels (Hegg et al., 1979; Croguennec et al., 2002). Jin et al. (2021) showed that phosphates have a favorable effect on the heat-induced egg white gel (increased gel hardness, springiness, and water holding capacity and finer gel network). Compared with NaCl and KCl, egg white protein gelling with Na_2SO_4 showed improved mechanical properties, including texture profiles, elastic modulus, as well as the lowest gelation concentration endpoint, which is due to a firmer gel network structure with thicker chains (Liu et al., 2022d).

Regarding preserved eggs, the type and concentration of salt ions play important roles in the aggregation of egg white proteins. Multivalent ions (K^+, Ca^{2+}, Zn^{2+} and Fe^{3+}) can promote the particle size of alkali-induced egg white aggregates, and K^+-treated samples show better gel properties (Deng et al., 2020). Furthermore, a low Cu^{2+} level favors protein–protein interactions, but high ionic strength has a deleterious impact on alkaline-induced gels (Shao et al., 2017). Zhao et al. (2020) compared the effect of three different metal compounds (lead oxide, copper sulfate, mixed compound of zinc sulfate and copper sulfate) on the quality of preserved eggs. They found that egg whites pickled by lead oxide possess moderate hardness and good elasticity and that the mixture of copper and zinc sulfate possess the most compact structure; the content of SS bonds in preserved egg white gel pickled by lead oxide is the highest. Regarding the effect of NaCl, Li et al. (2018c) showed that the addition of NaCl (0.3–1.2%) during alkali pickling results in the formation of opaque egg white gel and an increase in surface hydrophobicity, particle turbidity and zeta potential; when subjected to additional heat treatment, NaCl addition can inhibit the unfolding of egg white proteins.

Furthermore, Kaewmanee et al. (2011) reported that NaCl plays a role in the formation of a coagulum-type gel of cooked salted egg white, which has a coarse texture with opaqueness: Larger clusters of protein aggregates were observed with increasing NaCl concentrations. To reduce the Na content in salted egg white, several other salts are used to partially replace NaCl during salting (Liu et al., 2022b; Tian et al., 2022a). Their findings showed that the gel strength of the $CaCl_2$-treated group is significantly higher, while the gel strength of the KCl-treated group was similar to that of the NaCl-treated group ($P > 0.05$).

Saccharide: Saccharides play important roles in many food formulations, and their presence can have a direct impact on the gelling properties of egg white proteins. Mu et al. (2020) reported that an unmodified egg white protein solution forms a more elastic gel, whereas the addition of a saccharide (glucose, maltodextrin, dextran T20, and dextran T40) results in a more viscous gel with reduced hardness; this is due to the addition of saccharides potentially inhibiting the denaturation and unfolding of the protein molecules. Indeed, saccharides have been confirmed to alter the thermal behavior and increase the thermal stability of egg white solutions (Nasabi & Labbafi,

2020). Studies have also reported that egg white protein gels prepared with gellan gum/xanthan gum exhibit harder texture, which also drastically increases their stability against in vitro gastrointestinal degradation and enhances their applications in site-specific bioactive delivery and expands the specific utilization of hybrid gels (Babaei et al., 2019; Zhang et al., 2019). Therefore, saccharide type plays an important role in gel formation.

However, the effect of saccharides on egg white protein functionality can sometimes be detrimental. Egg white contains a small number of sugars (mainly glucose) (Sunwoo & Gujral, 2015). During the spray-drying process, these sugars may interact with egg white proteins via the Maillard reaction and lend an undesirable appearance to the final product. Thus, a desugaring process is often conducted before egg whites are spray-dried (Wang et al., 2022b).

Under controlled conditions, the Maillard reaction can potentially improve protein functionality via the formation of protein–saccharide conjugates. The Maillard conjugation of egg white proteins with saccharides has received much attention in recent years (Gharbi & Labbafi, 2019). Depending on the desired result, the Maillard reaction can be conducted in different modes, such as dry heating, microwave heating, wet heating, and ultrasonic induction, among others (Ma et al., 2019a; Chang et al., 2021; Li et al., 2022a; Zheng et al., 2022b). Wang et al. (2020a) reported that egg white proteins glycated with isomalto-oligosaccharide via dry heating showed improved gelling properties, as revealed by increased proportions of intramolecular and intermolecular β-sheets and decreased α-helical structure content. The ovalbumin-xylose conjugates formed via high-intensity ultrasound also present increased gel hardness and better gel microappearance (Sheng et al., 2022). Therefore, glycosylation between proteins and saccharides can improve the gelling properties of egg white proteins.

Drying method: Egg white protein powders are used by many manufacturers owing to their extended shelf life, reduced product volume and weight, improved food safety, and lower production costs (Ma et al., 2022). Dried egg white proteins can be manufactured by spray drying and freeze drying. Spray drying, which is frequently used in the food industry, is the transformation of feed from a fluid to a dried state by spraying the feed into a flowing hot drying medium, usually air (Samborska et al., 2005). However, deterioration in the gelling properties of egg white has been observed when using the spray-drying process (Katekhong & Charoenrein, 2018; Preethi et al., 2021). This is a result of instantaneous high temperature as well as the partial denaturation of egg white proteins (Lechevalier et al., 2007). The conventional solution is to further heat the dried egg white under controlled temperature (> 60°C) and relative humidity, also known as dry heat (Kato et al., 1989). Dry heat treatment can not only destroy harmful microorganisms but can also offset the adverse effect of spray drying on egg white protein functionality (Wang et al., 2018). Improved protein gelling performance can be attributed to the formation of specific low-molecular-weight soluble aggregates (Ma et al., 2019b; Ma et al., 2021a; Ma et al., 2022). By preforming dry heating at 75°C and a relative humidity of 60% for 21 d, the protein surface charge and the levels of hydrophobic and functional groups can be increased. Some novel technologies have also been investigated, such as microwave heating, which has higher efficiency than dry heating (Wang et al., 2020b). Radiofrequency-assisted thermal processing is also applied to improve the gelling properties of egg white powders, which greatly shortens the total processing time (Boreddy et al., 2016). It significantly increases the gel firmness of egg white proteins at elevated treatment temperatures and pH values (Kar et al., 2020).

On the other hand, freeze drying avoids protein denaturation, but it is time-consuming and expensive (Monge-Morera et al., 2021). The hardness and water holding capacity of gels made from fresh egg white are comparable to those of freeze-dried eggs, indicating that the freeze-drying process does not affect the gelling properties (Katekhong & Charoenrein, 2016, 2017). Microwave freeze drying—an emerging technology—can enhance gel strength with the drying time and generate a denser gel network (Liu et al., 2022f). Liu et al. (2020a) compared the structural and functional changes of hydrolyzed/glycosylated ovalbumin under spray drying and microwave freeze drying, and found that spray-dried samples have increased gel hardness compared with microwave freeze-dried samples, while the latter exhibits better thermal stability.

Storage conditions: The functional properties of egg white are affected by several environmental conditions, such as storage time, temperature, humidity, and the composition of the gaseous environment (Qiu et al., 2012). Under increased storage temperature, ovalbumin degradation is accelerated, the formation of ovalbumin complexes is increased, and the levels of lipocalin family proteins are decreased (Qiu et al., 2012). With increased storage time (37 ± 1°C, 0–12 d), the egg white gels have a harder texture and the thick egg whites undergo thinning; this improvement in functional properties is related to the disordered structure of the egg white proteins (Huang et al., 2022; Liu et al., 2022g). Basso et al. (2021) reported that egg white subjected to hyperbaric storage at 200 MPa at room temperature has a slightly decreased gelling capacity during storage, which is due to partial conversion of ovalbumin into S-ovalbumin. "Fenghuang egg"—a delicacy of fertilized eggs cooked on the 12th day of storage (at 37°C and 85% relative humidity)—shows different gel properties compared with fresh egg white, as manifested by increased gel hardness, cohesiveness, and water holding capacity (Liu et al., 2022j).

Spray-dried egg white proteins subjected to additional dry heat storage have enhanced gelling and foaming properties (Kato et al., 1989; Ma et al., 2021a). Both spray- and freeze-dried egg white proteins stored at high temperature and relative humidity levels show changes in protein conformation, with increased gel hardness commonly observed (Katekhong & Charoenrein, 2016, 2017, 2018). However, Ma et al. (2019a) reported that the gel hardness of dry heated and Maillard glycated egg white proteins decreases linearly with increasing storage time (25 ± 2°C, 65 ± 3% relative humidity, 0–12 months), but the hardness is still greater than that of unmodified egg white proteins.

Feed additive: Nutritionally fortified eggs, also known as nutrition-enriched or functional eggs, are produced by enhancing the nutrients in poultry feed, which are now experiencing rapid growth in recent years worldwide (Tian et al., 2022b). The different feed compositions may influence the protein composition/content of egg whites, thus affecting their technofunctional properties. There are selenium-enriched eggs, folic acid-enriched eggs, unsaturated fatty acid-enriched eggs, and astaxanthin-enriched eggs. Liu et al. (2022i) reported that selenium enrichment improves the textural properties, microstructure, and T_d of heat-induced gels of egg white, which might be attributed to the increase in ovomucin content. After selenium enrichment, egg white proteins also exhibit better digestibility in the gastrointestinal tract. A general linear pattern was also observed between egg white gel hardness and supplemental Lys levels in the feed by Prochaska et al. (1996). However, Ding et al. (2020b) reported that albumen gel strength, hardness, gumminess, and chewiness decrease with increasing tea polyphenol enrichment, with minimal values obtained with the addition of 400 mg/kg tea polyphenols. No effects of increasing tea polyphenols on the adhesiveness, springiness, cohesiveness, and resilience value of albumen gel was found.

8.2.3 Modification of Egg White Protein Gelling Properties

During the production and processing of egg white or egg white proteins, their functional properties are more or less negatively affected, which is detrimental to their application. Improving the gel properties of egg white proteins is still a demanding issue in the food industry, and novel approaches and insights have been introduced, as shown in Table 8.1.

8.3 The Foaming Properties of Egg White Proteins

In many macro food systems, foaming plays a vital role in the production of aerated foods, such as fluffy aerated candies and baked products, and imparts desirable visual and textural characteristics. Liquid foam is generated in a two-phase system in which a discontinuous gas phase is distributed in a continuous liquid phase (Wouters et al., 2018). However, due to the high-energy state of air–liquid interfaces, foams are unstable and easily collapse.

Stable food foams rely on the surface activity and film-forming capabilities of small amounts of protein components (Yang & Baldwin, 1995). Proteins are extensively used in aerated systems owing to their amphiphilic chemical nature (Damodaran, 1997b). They adsorb and orient with their hydrophilic groups toward the water phase, while their hydrophobic groups are oriented toward the air phase (Gharbi & Labbafi, 2019). As a result of noncovalent intermolecular interactions and covalent disulfide cross-linking, proteins produce a viscoelastic film that provides some stability to foam. They may also form large-volume stable foams that coagulate with heating (Damodaran, 1997b; Zhao et al., 2021).

Compared to other proteins, egg white proteins are particularly effective at stabilizing foams (Duan et al., 2018). Foam is formed by air bubbles that are trapped in the egg white liquid during vigorous mixing/whipping (Figure 8.4). With increasing whipping duration, the size of the air bubbles decreases, the number of the air bubbles increases, and the transparent albumin becomes opaque yet remains wet (Yang & Baldwin, 1995). As more air is integrated into the foam, it stiffens and becomes more viscous. With continued whipping, the foam becomes friable and loses its wet and glossy appearance (Yang & Baldwin, 1995). Egg white proteins are the most prevalent industrial foaming additives, and impart a distinctive structure and feel (Gharbi & Labbafi, 2019; Alavi et al., 2020b). They are commonly used for the preparation of meringues, cakes, and leavened bakery products, among others (Figure 8.1) (Anton et al., 2016; Gharbi & Labbafi, 2019).

8.3.1 Mechanism of Egg White Protein Foam Formation

In general, pure liquid cannot produce stable foams. Similar to emulsions, foams are thermodynamically unstable due to drainage, disproportionation, and coalescence (Zhan et al., 2022); unlike emulsions, they have a considerably shorter lifespan (Lazidis et al., 2016; Alavi et al., 2020b). Due to the uniformity of the liquid surface and interior, it is impossible to form an elastic film. Bubbles are unstable owing to the high tension at the air–liquid interface. Consequently, the formation of an adsorbed molecular layer is required to produce stable foams (Wierenga & Gruppen, 2010). Surfactants, such as proteins, reduce the tension at the liquid–air interface and stabilize bubble formation (Damodaran, 2005). Proteins need to diffuse to the air–liquid interface and then shift to the "adsorbed" state to form layers (Gharbi & Labbafi, 2018). The hydrophobic regions of proteins enable adsorption at the interface during whipping, which is followed by partial unfolding (Poole, 1989). The partially unfolded proteins create a stabilizing film around the bubbles (Yang & Baldwin, 1995).

The two fundamental parameters influencing foam quality are foam capacity (foamability) and foam stability. Foam capacity is associated with an increase in foam volume, which is dependent on the quantity of air incorporated in the protein solution (Gharbi & Labbafi, 2019; Sun et al., 2022a). The protein adsorption and unfolding rate at the air–liquid interface ultimately determines the foaming capacity, and better foaming capacity is related to faster protein diffusion and adsorption at the interface (Wierenga & Gruppen, 2010; Patino & Pilosof, 2011; Sun et al., 2022a). Meanwhile, foam stability is critical for the appearance and functionality of aerated foods, as it reflects their capacity to maintain structural integrity during processing, including mixing, heating, and cutting (Foegeding et al., 2006; Raikos et al., 2007a; Sun et al., 2022a).

Proteins with exceptional foaming qualities have three features: (i) rapid adsorption at the interface during whipping, (ii) structural alteration and rearrangement at the interface, and (iii) formation of a viscoelastic film via intermolecular interactions (Jiang et al., 2020; Zhao et al., 2021). The amphiphilic characteristics of egg white proteins, which have a comparatively high surface hydrophobicity, are likely related to the exceptional foaming capabilities of egg white and contribute to the efficient adsorption rate at the air–liquid interface. The molecular flexibility of egg white proteins allows for conformational rearrangement at the interface and the construction

TABLE 8.1
Approaches to Improving the Gelling Properties of Egg White Proteins

Approaches	Characteristics	Improved Gelling Properties	References
Dry heating	Formation of linear soluble aggregates with low molecular weight Partial denaturation "Molten globule"	*In vitro* digestibility Gel strength Gel transparency Water holding capacity	Kato et al. (1989) Wang et al. (2018) Ma et al. (2019b) Ma et al. (2021a)
Microwave heating	High efficiency Less pronounced protein unfolding, denaturation, and cross-linking Lanthionine cross-links	Gel firmness Gels with more mobile water	Rombouts et al. (2020) Sun et al. (2020c)
Radiofrequency (RF) assisted heating	High efficiency SH-SS interchange	Gel firmness Water holding capacity	Boreddy et al. (2016) Kar et al. (2020)
High pressure (HP)	A high degree of exposure of SH group Reduced extreme aggregation Destruction of fibrilliar ovomucin-lysozyme complex	Gel transparency Promoted gel formation	Panozzo et al. (2014) Naderi et al. (2017) Zhang et al. (2020)
Pulsed electric field (PEF)	Formation of soluble aggregates Lower levels of denaturation and gelation	Gel firmness	Yogesh (2016) Joeres et al. (2022)
Ultrasound treatment	Faster thermal aggregation	Gel strength Water holding capacity	Xue et al. (2021) Xiong et al. (2016a)
Glycosylation	Exposure of active groups and intensified hydrophobic interactions	Gel hardness Water holding capacity	Ma et al. (2019a) Wang (2020a)
Phosphorylation	Increased the solubility and negative surface charge	Better structural integrity Gel hardness	Jin et al. (2021)
Succinylation	Increased electrostatic repulsion Decreased surface hydrophobicity	Thermal stability	He et al. (2019) Basak and Singhal (2022)
Acylation	Enhanced hydrophobic interactions	Light transmission Gel hardness Resilience Water holding capacity	Hu et al. (2022)
Oxidation	Loss of free SH groups, α-helix structure, and intrinsic fluorescence intensity	Gel strength Antioxidant activity after in vitro digestibility	Bao et al. (2017) Cheng et al. (2021)
Enzymatic treatment	Improved mechanical properties More exposed amino acid residues	Elasticity Gel strength Thermal stability	Alavi et al. (2020a) Liu et al. (2020b)
Alkali treatment	Exposure of hydrophobic, SH and SS groups	More stable Gel springiness	Gao et al. (2020) Huang et al. (2019)
Ethanol treatment	Increased free SH groups content and viscosity	Conducive to the formation of gels Water holding capacity	Yao et al. (2020) Yu et al. (2020)
Polyphenol treatment	Enhanced cross-linking	Gel hardness Digestibility	Xue et al. (2022b) Ai et al. (2019b)
Saccharide treatment	More uniformly distributed gel particles	Elasticity Heat stability	He et al. (2021) Nasabi and Labbafi (2020)
Salt treatment	Decreased immobile water	Gel harness Gel appearance	Tian et al. (2022a) Liu et al. (2022b)

of a continuous viscoelastic network (particularly when denaturation occurs) (Anton et al., 2016; Gharbi & Labbafi, 2019).

The synergistic adsorption of the component proteins is largely responsible for the foaming capabilities of egg white (Lechevalier et al., 2005; Zhao et al., 2021). Several studies have considered the molecular mechanisms of egg white foam formation. Globulin is a particularly important surfactant that reduces surface tension, which encourages the formation of tiny bubbles during the early phase of foam formation, thus generating a smooth texture (MacDonnell et al., 1955; Yang & Baldwin, 1995; Gharbi & Labbafi, 2019).

Proteomic analysis has identified over 140 egg white proteins (D'Ambrosio et al., 2008). Ovalbumin plays an important role in egg white functioning given its tendency to spontaneously aggregate upon adsorption at the air–liquid interface, which also provides heat-setting properties (Yang & Baldwin,

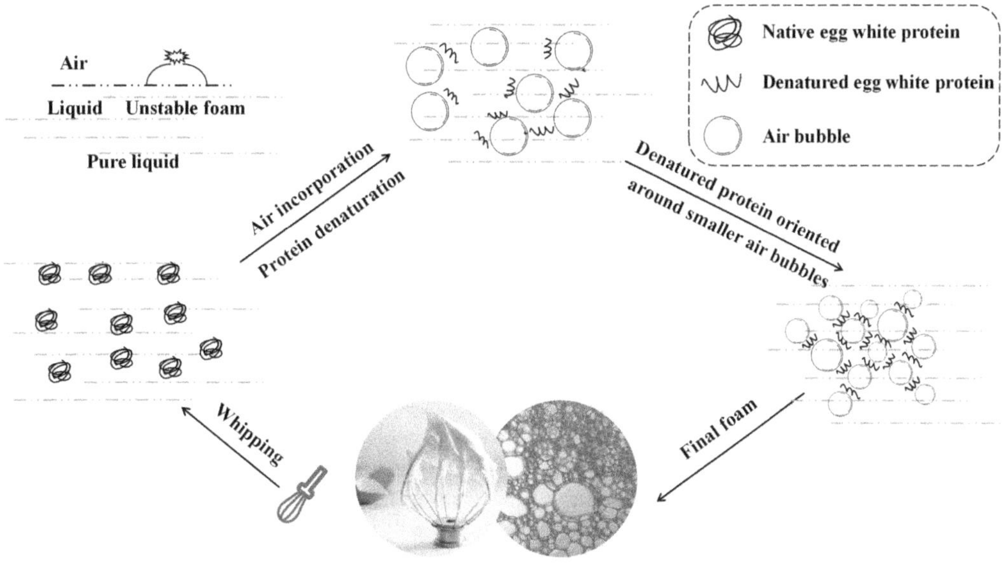

FIGURE 8.4 Schematic diagram of egg white foam formation.

1995; Lechevalier et al., 2005; Murray, 2011). Lysozyme has poor foaming properties because it has a rigid molecular structure and low surface hydrophobicity (Desfougères et al., 2008). However, in the absence of lysozyme, the foaming properties of egg white may decrease. Zhao et al. (2021) reported that the foaming capacity and stability of egg white without lysozyme drop by 45.1% and 35.3%, respectively. The surface tension of egg white increases with lysozyme depletion, which is detrimental to protein adsorption, stretching, and rearrangement at the interface. This might be attributed to the interactions between lysozyme and other proteins, which provide favorable conditions for stable foam formation (Johnson & Zabik, 1981; Zhao et al., 2021).

The co-adsorption behavior of various proteins at the air–liquid interface has been found to increase foaming capacity (Damodaran et al., 1998; Lechevalier et al., 2005; Sun et al., 2022a). For example, a complex consisting of lysozyme and ovomucin renders a high viscosity, which is one of the key reasons for the outstanding stability of egg white foam (Omana et al., 2010). Lysozyme has been shown to increase the strength and durability of foam films, and ovomucoid reduces foam release (Dickinson, 2011). Lysozyme is positively charged, and the adsorbed proteins (ovalbumin and ovotransferrin) are negatively charged. These proteins are electrostatically attached in the viscoelastic network at the interface (Le Floch-Fouéré et al., 2009; Le Floch-Fouéré et al., 2010). In addition, the protein conformation in foam formed by the mixture of ovalbumin and lysozyme changes dynamically in a time-dependent manner (Jin et al., 2022a): Proteins become relaxed at the air–liquid interface and are embedded as a hinge. Although ovomucin can impart high viscosity to egg white liquid, it becomes insoluble with excess whipping, which reduces the elasticity of the bubbles (Omana et al., 2010; Panozzo et al., 2014). However, if we extract egg white from foams, foam can still be generated from a second whipping, but only in the presence of considerable amounts of uncoagulated ovomucin and with initially unstable foam (MacDonnell et al., 1955).

Jin et al. (2023) employed label-free quantitative proteomics techniques to identify and analyze key proteins that affect the egg white foam capacity and stability. A total of 35 proteins are identified as key proteins affecting foam capacity, including 14 up-regulated and 21 down-regulated proteins; a total of 25 proteins are considered to be closely related to foam stability, including 13 up-regulated and 12 down-regulated proteins. The authors found that hydrophobic and hydrophilic fragments always alternate, and the hydrophobic domain is often buried in the interior of the molecule and moderate grand average of hydropathicity values for proteins might be more beneficial for favorable air–water interface properties. The contribution of ovostatin to foam formation cannot be ignored, while its abundance in liquefied interfacial proteins is significantly down-regulated due to the high molecular weight and the insoluble aggregates formed.

8.3.2 Factors Affecting the Foaming Properties of Egg White Proteins

Many factors influence egg white foaming behavior. Particle size, for example, is an important factor affecting protein diffusion and adsorption rates at the air–liquid interface, which is directly related to the foaming properties of protein solutions (Chang et al., 2020). Surface hydrophobicity and electrostatic repulsion are key factors influencing protein adsorption. On the one hand, surface hydrophobicity lowers the adsorption barrier at the air–liquid interface, thus increasing adsorption. On the other hand, lowering the surface charge (or electrostatic repulsion) lowers the electrostatic barrier to adsorption, thus improving protein adsorption at the interface (Delahaije et al., 2014).

Foaming properties can also be affected by pretreatment (either physical or chemical) and added ingredients. During processing, treatments and ingredients that produce undesirable characteristics should be avoided, while optimized modifications can improve foaming properties and end-product quality.

Whipping methods and parameters: In the baking industry, the foaming properties of egg white play an indispensable role in determining the quality of the final products. The difference in foaming performance between hand-whipped and electric-whipped egg white was confirmed by John and Flor (1930) and Henry and Barbour (1933). The rate, motion, and duration of whipping can affect the amount of air incorporated into the foam, and excessive whipping could lead to unstable and dry foams, which are undesirable (Nakamura, 1963).

Because of a reduction in ovomucin fiber length during whipping, whipping rate and product volume can be increased (Forsythe & Bergquist, 1951). Homogenization also affects ovomucin structure, with homogenized albumin resulting in shorter whipping times and decreased product volume (Yang & Baldwin, 1995).

Water: Before whipping, adding up to 40% more water to the egg white can enhance the volume without affecting the foam stability. However, above this amount, water begins to separate from the foam with time (Yang & Baldwin, 1995).

Heat treatment: Heat treatment is one of the most important treatments in the food industry. It can alter protein conformation and affect foaming performance by changing the surface charges and hydrophobic group content (Chang et al., 2020). Heat treatment of egg white at >55°C promotes rearrangement of the protein structure and the generation of numerous protein aggregates, thus leading to enhanced foaming capacity; heating at lower temperatures (<45°C) induces the partial hydrolysis of egg white proteins, which promotes the formation of polymers with lower molecular weights that quickly diffuse and adsorb to the air–liquid interface (Ding et al., 2022). Gharbi et al. (2017) also found that heating egg white at 60°C reduces foam volume; however, when heated to 65°C, foam volume increased considerably due to the exposure of SH groups in ovalbumin. This indicates that heating temperature directly determines the foaming properties of egg white proteins. Heat-induced egg white protein nanofibrils are reported to have better foaming properties, and fibril-based meringue batter has more desirable characteristics (Alavi et al., 2020b).

Pasteurization also affects the foaming properties of egg white: Foam formation requires a longer beating time when using pasteurized albumin with the same specific gravity as unpasteurized albumin foam (Yang & Baldwin, 1995). Therefore, the egg processing industry requires optimized technologies to offset the adverse effects of processing.

Subcritical water treatment is a novel heating method that uses hot water at temperatures between 100 and 374°C with high pressure to maintain the liquid state; Chang et al. (2022) compared the effects of general heat treatment (90°C) and subcritical water treatment (120°C) on the structure and foaming properties of egg white proteins, and found that subcritical water treatment enhanced foaming capacity by 3-fold by increasing the surface hydrophobicity and flexibility.

Heat treatment of egg white protein powders under regulated temperature and relative humidity could improve foam stability and performance (Kato et al., 1989; Croguennec et al., 2007; Desfougères et al., 2008; Talansier et al., 2009; Nicorescu et al., 2011; Chang et al., 2020). Dry heat can facilitate the formation of small soluble aggregates in aqueous solutions, which have greater surface hydrophobicity and surface activity, and can quickly and effectively adsorb to the air–liquid interface (Chang et al., 2020; Ma et al., 2022). According to Talansier et al. (2009), the beneficial impact of dry heat on the functional characteristics of egg white protein is accompanied by a decrease in the viscoelasticity of the interfacial layer (causing stiffness) and a reduction in bubble size. Microwave heating has recently been employed to improve the foaming properties of egg white proteins (Li et al., 2018d; Wang et al., 2020b). Wang et al. (2020b) reported that microwave heating could increase the surface hydrophobicity and flexibility of egg white proteins, thus enhancing adsorption at the interface and overall foaming capacity. Microwave-treated egg white proteins also show higher solubility, decreased surface tension, and increased hydrophobicity and viscosity.

Drying method: Spray-drying is a routine practice in the production of egg white powders; however, the partial denaturation of proteins could exert a negative effect on egg white foaming properties (Lechevalier et al., 2007).

Freeze–thaw: Frozen egg products offer obvious benefits and a wide range of applications (Ma et al., 2021b). Fernández-Martín et al. (2017) reported that egg white subjected to magnet-assisted freezing exhibits reduced foam ability, which might be caused by freeze-induced protein aggregation and denaturation. However, Duan et al. (2017) found that, except for ovalbumin, the foaming ability of egg white proteins improves after freezing and thawing; moreover, the foaming ability of egg white is higher than that of albumin proteins and improves after several freeze-thaw cycles. By controlling pH levels, ionic strength, and freezing time, the foaming properties of egg white aggregates can be modified by freezing (Fang & Nakagawa, 2021).

Pressure: The use of high hydrostatic pressure can improve microbiological safety without causing denaturation (Naderi et al., 2017). Its positive effect on the foaming properties of egg white proteins has been confirmed in many studies. Albumin that has been subjected to high hydrostatic pressure has a greater viscosity, yet it retains its foaming capabilities as a consequence of protein unfolding (Iametti et al., 1999). According to Van der Plancken et al. (2007a), foams made from pressure-treated egg white solutions have a moister and creamier appearance, with a smaller bubble size and lower susceptibility to bubble coalescence, as indicated by the reduced likelihood of these foams collapsing. However, excessive coagulation of proteins under high-pressure treatment could negatively influence their foam stability. Basso et al. (2021) reported that egg white subjected to hyperbaric storage at 200 MPa at room temperature exhibits a higher foaming capacity, which can be employed not only for egg white storage under optimal hygienic conditions, but also for its pasteurization.

Irradiation: Radiation technology has been used the egg industry, particularly for sterilization. Ball & Gardner (1968) found that gamma radiation treatments extend the whipping time of liquid egg white and impair its foam stability, whereas post radiation freezing storage restores the foaming capabilities. Egg white exposed to modest levels of UV radiation (10.6 kJ m^{-2}) creates stable foams due to protein clump clogging and/or the increased viscosity of the aqueous phase (de Souza et al., 2015; Gharbi & Labbafi, 2018).

Ultrasound: The foaming capacity of egg white protein is likewise affected by ultrasound frequency, temperature, and time, and foaming stability can be improved with ultrasound treatment (Sun et al., 2020b). Because of lower viscosity and surface tension as well as a higher solubility, proteins can more easily bind at the air–liquid interface and undergo molecular rearrangement (Sheng et al., 2018). These characteristics are associated with increased free SH concentrations and surface hydrophobicity that render the egg white proteins more hydrophobic and flexible (Chen et al., 2019). Stefanović et al. (2017) reported that the length of ultrasonication exhibits an important effect on the particle size, uniformity, and charge of egg white proteins. The improved foaming properties of ultrasound-treated egg white proteins are attributed to decreased particle size and increased zeta potential. To decrease the adverse effect of pasteurization on the functional properties of liquid egg white, integrated ultrasound and microwave treatment was proposed by Liu et al. (2022a). The authors found that the foaming ability of egg white increases after the integrated treatment, but the increase of pasteurized egg white is not significant compared with untreated egg white; the foaming stability of treated egg white is not significantly different from pasteurized and untreated egg white, while that of pasteurized egg white is decreased compared with untreated egg white.

Ball milling: Milling and shearing are widely utilized to produce homogeneous systems. The foaming ability and stability of ovalbumin is decreased after ball milling treatment (Li et al., 2019a). However, ball milling is reported to be an effective pretreatment of glycosylation to improve the foaming properties of egg white proteins (Liu et al., 2022h; Wu et al., 2022).

Supercritical carbon dioxide: Supercritical carbon dioxide is a green, novel, and practical technology in food processing. It was reported that the foaming capacity of egg white is significantly promoted by supercritical carbon dioxide treatment at 9 MPa; the exposure of hydrophobic groups through the synergy of high pressure and CO_2 solvent was also found (Ding et al., 2020a).

pH or acid: pH plays an important role in egg white foaming behavior. An increase in pH (4–7) in the binary system of egg white proteins and soy peptides promoted protein unfolding, exposed SH groups and hydrophobic amino acids, and increased flexibility and surface tension, facilitating the rapid adsorption, rearrangement, and protein network formation at the air–liquid interface (Tang et al., 2022a). It is common practice to add lemon juice or plain vinegar to egg white before whipping to achieve better foaming performance. This was confirmed by Rhodes et al. (1960), who made angel cakes using acidified duck egg white that compared favorably to those prepared with chicken egg white. The acid reportedly affected the ovomucin, thus reducing the time needed for whipping (Yang & Baldwin, 1995). Meanwhile, citric acid was reported to increase the average pore size (higher porosity) of meringues with softer texture (Licciardello et al., 2012). Hammershøj and Larsen (1999) reported that, after 30 min, the foam at pH 7.0 had the highest drainage resistance, while on a longer-term scale, the foam at pH 4.8 had the best drainage resistance. This is due to the stiffer surface behavior at pH 4.8 and the development of tiny bubbles, resulting in a sluggish draining of liquid from the foam and reduced dynamic surface tension, resulting in the high overrun. Bonilla et al. (2022) confirmed that the shape of the bubbles under acidic conditions makes the foams extend their linear response to deformation: Preparation of foams under acidic condition reshapes the bubble perimeter to more 'hexagonal-like' shapes, allowing bubbles to 'swap' positions during increasing oscillatory deformations and therefore conserving the structure of the foam.

Saccharide: Saccharides play a critical role in the formation and stability of egg white foam (Sun et al., 2022a). When sugar is added to egg white, it delays foam development (Raikos et al., 2007a), which is most noticeable during the first few minutes of whipping. Saccharides decrease water mobility and bubble size and increase foam viscosity, thus reducing the air-phase fraction, inhibiting drainage, and promoting foam stability (Ochi et al., 2000; Yang & Foegeding, 2011). The addition of sugar reduces the linear region by bringing smaller bubbles closer together, promoting bubble coalescence, which provokes structural breakdown when increasing oscillatory deformation is applied (Bonilla et al., 2022). Sun et al. (2022a) reported that egg white solutions supplemented with 3% saccharide (glucose, sucrose, and maltodextrin) exhibit increased surface hydrophobicity and reduced surface tension, leading to an increase in foam capacity and stability, and the best foam performance of the compound egg white solution was achieved with the presence of 3% maltodextrin, along with the smallest and most homogeneous bubbles. The addition of hydrocolloids (e.g., pectin, Persian gum, and Xanthan gum) to egg white solution was also reported to improve foam stability (Muthukumaran et al., 2008; Mohammadian & Alavi, 2016; Sadahira et al., 2016; Dabestani & Yeganehzad, 2019). The formation of soluble complexes between egg white and hydrocolloid possibly increases the continuous phase viscosity and enhances the foam stability by slowing liquid drainage (Sadahira et al., 2016). However, some negative effects of hydrocolloids on the overrun and foam density of egg white proteins are undeniable (Dabestani & Yeganehzad, 2019).

Salt: NaCl plays controversial roles in the foaming behavior of egg white. Salt can reduce the stability of albumin foams (Hanning, 1944). Other reports have shown that NaCl improves foaming ability and increases overrun (Raikos et al., 2007a). NaCl counter ions may bind to charged protein molecules, reducing electrostatic repulsion between adsorbed and non-adsorbed protein molecules and facilitating adsorption at the air-liquid interface. The extra charge and decreasing surface tension of egg white dispersions at pH > 7.0/NaCl ≥ 0.6% result in the destabilization of foams (Li et al., 2018b). Alavi et al. (2020d) evaluated foams and meringues produced from preheated egg white proteins in the presence of $CaCl_2$, and the authors found that the addition of $CaCl_2$ significantly decreases the drainage of the preheated egg white protein foams, probably by forming a viscoelastic network at the bubble surface.

Egg yolk or oil: Before the whipping of egg white, the complete absence of egg yolk or oil is often needed; even blending a small amount of egg yolk or oil in egg white can significantly reduce foamability (Forsythe, 1957; Wang & Wang, 2009a). Yolk plasma—largest proportion of egg yolk—contributes to the poor foam stability of contaminated egg white, whereas yolk granules preserve foam stability for dispersions (Li et al., 2019c). Li et al. (2022b) confirmed that a mixture of egg white and yolk, even at a low percentage, produced a cake batter with a greater density, larger bubble size, and less uniform bubble distribution, as well as a more compact crumb.

To address this issue, Liu et al. (2014) and Yao et al. (2014) developed a non-invasive method utilizing VIS spectroscopy for rapid and quantitative determination of yolk contamination in egg white, with a 0.001% detection limit. Raman spectroscopy has also been developed to detect yolk contamination in egg white based on its molecular optical properties, and is rapid and cost effective (Cluff et al., 2016).

Researchers have assessed several methods to improve the foaming properties of yolk-contaminated egg white, such as exogenous protein addition, enzymatic hydrolysis, and optimized heat treatment, among others. Modified soy proteins have been proven to offset the adverse effect of yolk contamination (Wang & Wang, 2009b; Wang et al., 2012). Li et al. (2021) found that the treatment of yolk-contaminated egg white with lipase and phospholipase A_2 improved the foamability and stability of the modified egg white.

Additional peptides: Peptides—products of protein hydrolysis—have better physicochemical properties than natural proteins and are indispensable compounds for the development of functional nutritional foods. The effects of different peptides (soy peptides, corn peptides, whey peptides, and fish skin peptides) on the foaming properties of egg white proteins were evaluated (Tang et al., 2021; Tang et al., 2022a). The authors found that the addition of soy peptides, corn peptides, and fish skin peptides enhances the foaming properties of egg white proteins, while whey peptides significantly reduce the foam stability, and the optimal foaming performance was obtained by adding 9% soy peptides to egg white proteins under neutral pH conditions. Furthermore, the cake made by soy peptide-incorporated egg white proteins shows a microstructure that is superior to that of an unmodified egg white protein cake and a quality that is closer to that of a cake prepared using liquid egg white (Tang et al., 2022b). Hydrolyzed egg white peptides with 1% addition are also reported to have a positive effect on the foaming properties of pasteurized egg products, which have great potential in the baking industry (Ho et al., 2021).

Grafting: Maillard glycosylation is a common way to modify protein functionality. O'Charoen et al. (2015) proposed that glycation with D-arolone sugar improved the foaming capabilities of egg white protein. The positive effect of glycosylation combined with other physical treatments on the foaming properties of egg white proteins has been confirmed by many studies (Chang et al., 2021; Li et al., 2022a; Wu et al., 2022). The egg white protein-pullulan conjugates prepared via the Maillard reaction show better foaming stability than non-glycosylated proteins. The conjugates display smaller and more homogeneous bubble sizes (Sheng et al., 2020). Liu et al. (2022h) and Wu et al. (2022) treated egg white proteins with ball-milling, followed by glycosylation. The authors found that glycosylated egg white protein powders after ball-milling pretreatment possess lower free amino acid content and higher solubility and foaming abilities.

Phosphorylated proteins have improved foaming, solubility, emulsification, calcium absorption, and gelling qualities, among others. Dry heat is a commonly used method for the phosphorylation of egg white protein (Li et al., 2003; Hayashi et al., 2009), but microwave-assisted heating is increasingly used owing to its high efficiency (Li et al., 2018d; Li et al., 2019b). Phosphorylation can improve the foaming properties of egg white protein due to its electronegativity.

Acylating reagents can be used to obtain a one or a two-unit shift in charge of modified protein functional groups (Ball & Winn, 1982). Acylation considerably improves the foaming capabilities of egg white, and greatly decreases the adverse effects of heat damage on foaming (Sato & Nakamura, 1977).

Succinylation is preferred over other chemical modifications due to the dual effect of electronegative charge and steric hindrance imparted to proteins (Basak & Singhal, 2022). Succinylation has been reported to increase the foaming capacity of proteins but decrease foam stability, which might be due to reduced protein-protein interactions due to an increase in negative charge (Basak & Singhal, 2022).

Oxidation: Oxidation can modify the molecular arrangement and interaction of egg white proteins via SH–SS interchange, surface hydrophobicity and secondary structures (Takahashi et al., 2016; Duan et al., 2018). Egg white proteins subjected to ascorbic acid and H_2O_2 treatment lead to the oxidation of free SH groups, increased surface hydrophobicity and zeta-potential, and protein aggregation via SS bonds (Alavi et al., 2019). A dramatically enhanced foam stability was obtained by the egg white proteins aggregated at pH 11. Duan et al. (2018) utilized 2,2′-azobis (2-amidinopropane) dihydrochloride (AAPH) to induce the oxidation of egg white proteins and found that moderate oxidation leads to exposure of hydrophobic groups, whereas excessive oxidation leads to a decrease in surface hydrophobicity. The authors reported that excessive oxidation increases the foaming ability of egg white proteins, while decreasing their foam stability.

Enzymatic hydrolysis: It is generally believed that the moderate hydrolysis of proteins is conducive to improved functional qualities (Behnke et al., 1986). Because egg white contains a large number of protease inhibitors, heat treatment is required to inactivate these inhibitors and improve the degree or speed of enzymatic hydrolysis. However, preheating may affect the structural and functional characteristics of egg white proteins (Hammershøj et al., 2008). Restriction enzyme hydrolysis can increase foaming performance by reducing the molecular weight of egg white proteins, exposing hydrophobic groups and increasing protein solubility and flexibility (Stefanović et al., 2018). These methods can improve the adsorption capacity of hydrolysate at the air–liquid interface. However, excessive hydrolysis may reduce surface hydrophobicity and impair foaming performance (Gharbi & Labbafi, 2019; Liu et al., 2020b).

Fermentation: Fermentation is one of the oldest natural methods in food manufacture that benefits many properties of products. Jiang et al. (2020) reported that lactic acid fermentation treatment could be used to enhance the foaming ability of egg white proteins, as revealed by decreased free SH groups and increased surface hydrophobicity. Jia et al. (2021) further investigated the individual contribution of fermented albumen proteins to the improved foaming properties and the related mechanisms. The authors found that the foaming ability of ovalbumin exhibits similar patterns as its surface hydrophobicity during fermentation (first increases and then decreases); fermented ovomucin and ovomucoid have improved surface characteristics and appropriate apparent viscosity and they both exhibit enhanced foaming ability and stability after fermentation. The cakes made with fermented egg white exhibit lower baking loss and improved flavor, suggesting that lactic acid fermentation has the potential to be applied in egg products (Jia et al., 2022).

Fermentation has also been utilized during the desugarization process of egg white, which is an important technique in industrial egg processing to maintain the color and shelf life of egg white powder (Asaithambi et al., 2022). Asaithambi et al. (2022) reported that yeast-fermented egg white has a higher foaming and emulsifying ability than non-desugared samples. Azizah et al. (2022) also found that the foaming capacity of *Lactobacillus bulgaricus*-fermented egg white powder increase as the percentage of *Lactobacillus bulgaricus* increase and an addition of 6% in egg white powder was optimized.

8.4 The Emulsifying Properties of Egg White Proteins

Emulsion is a non-equilibrium system with a limited lifespan but is nevertheless important in food production and storage (Nishinari et al., 2014). Proteins are excellent emulsifiers (Kim et al., 2020), with the egg yolk showing particularly good emulsifying properties.

Egg white proteins are typically thought to have rather weak emulsifying capacity compared to whey proteins or soybean proteins, which limits their use in emulsion systems (Chang et al., 2016). In neutral and alkaline environments, most egg white protein hydrophobic amino acid residues are concealed within the molecule and rarely exposed on the surface. Consequently, egg white proteins exhibit high hydrophilic qualities but weak emulsifying capabilities (Yu et al., 2021b). Therefore, this section covers the modification of egg white proteins to improve their emulsifying properties (the emulsifying properties of egg yolk are discussed in Chapter 9).

8.4.1 Modification of the Emulsifying Properties of Egg White Proteins

Dry heating: Dry heating has been extensively utilized in egg white protein processing for sterilization and protein modification. Several studies have confirmed the improved functional properties of egg white proteins as a result of dry heating (Kato et al., 1989; Kato et al., 1990). Egg white proteins are expanded by dry heating, thus improving their surface properties. In addition, a reduction in the enthalpy of egg white proteins by dry heating may enhance the rate of diffusion of the thermally altered molecules to the air–liquid interface, facilitating protein-protein interactions. The improved foaming and emulsifying properties of dry heated proteins reflect the faster rates of unfolding and greater levels of intermolecular interaction at the air-liquid interface (Kato et al., 1990). Dry heating has also been used for chemical modification of egg white proteins, including glycosylation and phosphorylation (Li et al., 2004; Ai et al., 2021a).

pH: pH-shifting treatments have been utilized to improve the emulsifying properties of egg white proteins. Adjusting the pH of egg white solutions to alkaline conditions combined with mild heating can induce transitions in their tertiary and secondary structures, during which higher surface hydrophobicity and free sulfhydryl groups were also observed. Egg white proteins subjected to extremely alkaline shifting treatment (pH = 13) or combined pH-shifting (pH = 12) and heating treatment have desirable emulsifying properties (Yu et al., 2021b; Yu et al., 2021c).

Acid or acid-heat treatments can also induce protein tertiary structure transitions; The emulsifying capacity of egg white proteins is enhanced after acid or acid-heat treatment, with more hydrophobic amino acids and charged groups exposed on the surface but few changes in secondary structure (Chang et al., 2016).

Ultrasound: High-intensity ultrasound and heating treatment can effectively improve the emulsifying properties of egg white protein peptides, mainly due to the exposure of hydrophobic groups, increased surface loading charge, and alterations to protein molecular structures (Arzeni et al., 2012; Stefanović et al., 2017; Ai et al., 2021b).

Irradiation: Irradiated egg white proteins can potentially serve as novel emulsifiers/stabilizers in various food systems (Kuan et al., 2011; Kuan et al., 2013). The improved stability of UV-irradiated egg white proteins can be explained by decreased droplet size as well as altered conformational and surface properties; protein denaturation and changes in the conformational structure would affect surface hydrophobicity and subsequently lead to better adsorption of the oil-in-water emulsion system (Jambrak et al., 2009). Gamma-irradiated egg white proteins also have better emulsification ability and stability than untreated egg white proteins, which is mainly due to the conversion of α-helix and β-sheet structures into β-turn and random coil structures (Liu et al., 2020c).

Grafting: Glycosylation is an effective method to improve the emulsifying property of proteins, and the grafting of egg white proteins with sugars has been extensively investigated. The enhancement of emulsifying properties is highly correlated with molecular structure and grafting degree, which is influenced by the molecular flexibility, surface hydrophobicity, tertiary structure changes, and free sulfhydryl content of egg white proteins (Wang et al., 2019; Ai et al., 2021a; Yao et al., 2022). Glycosylation is conventionally achieved via dry-heating or wet-heating (Kan et al., 2021; Zheng et al., 2022b).

To improve its emulsifying properties, ovalbumin can be phosphorylated via dry-heating in the presence of sodium tripolyphosphate, during which covalent interactions (C–O–P and C–N–P bonds) are formed between the phosphate groups bound to the -OH on serine and threonine and the $-NH_2$ on lysine and arginine (Xiong et al., 2016b). The phosphorylated protein may become increasingly unfolded and have an enhanced surface hydrophobicity due to the repelling effects of the added phosphate groups' negative charges, which improves the emulsifying properties by promoting an amphiphilic balance and preventing the coalescence of oil droplets (Li et al., 2004; Li et al., 2005; Li et al., 2010).

Egg white proteins can also be grafted with phenol, which provides a novel method for enhancing protein functionalities (Liu et al., 2019). The emulsifying activity of egg white proteins is significantly improved by grafting with tea polyphenol at pH 7.0, but is negatively correlated with tea polyphenol concentration (Sun et al., 2022b). Apart from tea polyphenol, the conjugation of egg white proteins with catechin and chlorogenic acid also improves their emulsifying properties (Jing et al., 2020; Sun et al., 2020a).

Aggregation induction: The formation of soluble aggregates can improve the functional properties of proteins. The aggregation of egg white peptides (< 1 kDa) was induced by proanthocyanidins via a two-step binding (Wen et al., 2022). Hydrogen bonding and hydrophobic interactions are involved in the first step, while hydrogen bonding prevails in the second step. The formed aggregates exhibit improved emulsifying properties by reducing surface tension and improving the contact angle.

Enzymatic hydrolysis: Enzymatic modification is an effective and safe way to improve the functional properties of proteins. Hydrolysis can expose hydrophobic residues especially tyrosine, and emulsions stabilized by hydrolyzed egg white proteins are less sensitive to heat or salt (Chang et al., 2017). Different proteases have been applied to regulate the emulsifying properties of protein hydrolysates (Ai et al., 2019a; Ling et al., 2020). Therefore, choosing the correct type of protease is as important as optimizing hydrolysis conditions.

8.5 Summary

Eggs are an important part of modern diets worldwide, and their consumption increases continuously. Eggs have been applied in a wide variety of food products owing to their excellent functional properties. Their gelling, foaming, and emulsifying properties have been extensively used in domestic cooking and industrial food production, and a proper understanding of these functional characteristics would promote the development and modernization of eggs and egg products.

REFERENCES

Ai, M., Tang, T., Zhou, L., Ling, Z., Guo, S., & Jiang, A. (2019a). Effects of different proteases on the emulsifying capacity, rheological and structure characteristics of preserved egg white hydrolysates. *Food Hydrocolloids*, *87*, 933–942. https://doi.org/10.1016/j.foodhyd.2018.09.023

Ai, M., Xiao, N., & Jiang, A. (2021a). Molecular structural modification of duck egg white protein conjugates with monosaccharides for improving emulsifying capacity. *Food Hydrocolloids*, *111*, 106271. https://doi.org/10.1016/j.foodhyd.2020.106271

Ai, M., Zhang, Z., Fan, H., Cao, Y., & Jiang, A. (2021b). High-intensity ultrasound together with heat treatment improves the oil-in-water emulsion stability of egg white protein peptides. *Food Hydrocolloids*, *111*, 106256. https://doi.org/10.1016/j.foodhyd.2020.106256

Ai, M., Zhou, Q., Guo, S., Ling, Z., Zhou, L., Fan, H., & Jiang, A. (2019b). Effects of tea polyphenol and $Ca(OH)_2$ on the intermolecular forces and mechanical, rheological, and microstructural characteristics of duck egg white gel. *Food Hydrocolloids*, *94*, 11–19. https://doi.org/10.1016/j.foodhyd.2019.03.002

Alavi, F., Emam-Djomeh, Z., & Chen, L. (2020a). Acid-induced gelation of thermal co-aggregates from egg white and hempseed protein: Impact of microbial transglutaminase on mechanical and microstructural properties of gels. *Food Hydrocolloids*, *107*, 105960. https://doi.org/10.1016/j.foodhyd.2020.105960

Alavi, F., Emam-Djomeh, Z., Mohammadian, M., Salami, M., & Moosavi-Movahedi, A. A. (2020b). Physico-chemical and foaming properties of nanofibrillated egg white protein and its functionality in meringue batter. *Food Hydrocolloids*, *101*, 105554. https://doi.org/10.1016/j.foodhyd.2019.105554

Alavi, F., Emam-Djomeh, Z., Momen, S., Hosseini, E., & Moosavi-Movahedi, A. A. (2020c). Fabrication and characterization of acid-induced gels from thermally-aggregated egg white protein formed at alkaline condition. *Food Hydrocolloids*, *99*, 105337. https://doi.org/10.1016/j.foodhyd.2019.105337

Alavi, F., Emam-Djomeh, Z., Momen, S., Mohammadian, M., Salami, M., & Moosavi-Movahedi, A. A. (2019). Effect of free radical-induced aggregation on physicochemical and interface-related functionality of egg white protein. *Food Hydrocolloids*, *87*, 734–746. https://doi.org/10.1016/j.foodhyd.2018.08.048

Alavi, F., Tian, Z., Chen, L., & Emam-Djomeh, Z. (2020d). Effect of $CaCl_2$ on the stability and rheological properties of foams and high-sugar aerated systems produced by preheated egg white protein. *Food Hydrocolloids*, *106*, 105887. https://doi.org/10.1016/j.foodhyd.2020.105887

Alleoni, A. C. C. (2006). Albumen protein and functional properties of gelation and foaming. *Scientia Agricola*, *63*(3), 291–298. https://doi.org/10.1590/S0103-90162006000300013

Anton, M., Lechevalier, V., & Nau, F. (2016). From eggs to egg products. In T. C. R. Jeantet, P. Schuck, & G. Brulé (Eds.), *Handbook of food science and technology 3: Food biochemistry and technology* (pp. 115–143). ISTE Ltd and John Wiley & Sons Inc. https://doi.org/10.1002/9781119296225.ch3

Arzeni, C., Pérez, O. E., & Pilosof, A. M. R. (2012). Functionality of egg white proteins as affected by high intensity ultrasound. *Food Hydrocolloids*, *29*(2), 308–316. https://doi.org/10.1016/j.foodhyd.2012.03.009

Asaithambi, N., Singha, P., & Singh, S. K. (2022). Comparison of the effect of different desugarization techniques on the functionality of egg white protein hydrolysates. *Applied Food Research*, *2*(2), 100152. https://doi.org/10.1016/j.afres.2022.100152

Azizah, N. F. N., Evanuarini, H., & Widyastuti, E. S. (2022). Physicochemical characteristics of egg white powder using lactobacillus bulgaricus. *IOP Conference Series: Earth and Environmental Science*, *1020*(1), 012028. https://doi.org/10.1088/1755-1315/1020/1/012028

Babaei, J., Khodaiyan, F., & Mohammadian, M. (2019). Effects of enriching with gellan gum on the structural, functional, and degradation properties of egg white heat-induced hydrogels. *International Journal of Biological Macromolecules*, *128*, 94–100. https://doi.org/10.1016/j.ijbiomac.2019.01.116

Ball, H. R., & Gardner, F. A. (1968). Physical and functional properties of gamma irradiated liquid egg white. *Poultry Science*, *47*(5), 1481–1487. https://doi.org/10.3382/ps.0471481

Ball, H. R., & Winn, S. E. (1982). Acylation of egg white proteins with acetic anhydride and succinic anhydride1. *Poultry Science*, *61*(6), 1041–1046. https://doi.org/10.3382/ps.0611041

Banerjee, S., & Bhattacharya, S. (2012). Food gels: Gelling process and new applications. *Critical Reviews in Food Science and Nutrition*, *52*(4), 334–346. https://doi.org/10.1080/10408398.2010.500234

Bao, Z., Wu, J., Cheng, Y., & Chi, Y. (2017). Effects of lipid peroxide on the structure and gel properties of ovalbumin. *Process Biochemistry*, *57*, 124–130. https://doi.org/10.1016/j.procbio.2017.03.009

Basak, S., & Singhal, R. S. (2022). Succinylation of food proteins- a concise review. *LWT*, *154*, 112866. https://doi.org/10.1016/j.lwt.2021.112866

Basso, F., Manzocco, L., Maifreni, M., & Nicoli, M. C. (2021). Hyperbaric storage of egg white at room temperature: Effects on hygienic properties, protein structure and technological functionality. *Innovative Food Science & Emerging Technologies*, *74*, 102847. https://doi.org/10.1016/j.ifset.2021.102847

Begam, N., Ragulskaya, A., Girelli, A., Rahmann, H., Chandran, S., Westermeier, F., & Schreiber, F. (2021). Kinetics of network formation and heterogeneous dynamics of an egg white gel revealed by coherent X-ray scattering. *Physical Review Letters*, *126*(9), 098001. https://doi.org/10.1103/PhysRevLett.126.098001

Behnke, U., Kiss, E., Nádudvari, V., & Rutiloff, H. (1986). Enzymatic modification of egg-white protein and some of its functional properties. *Food/Nahrung*, *30*(3–4), 319–326. https://doi.org/10.1002/food.19860300330

Bonilla, J. C., Sørensen, J. L., Warming, A. S., & Clausen, M. P. (2022). Quantitative image analysis of protein foam microstructure and its correlation with rheological properties: Egg white foam. *Food Hydrocolloids*, *133*, 108010. https://doi.org/10.1016/j.foodhyd.2022.108010

Boreddy, S. R., Thippareddi, H., Froning, G., & Subbiah, J. (2016). Novel radiofrequency-assisted thermal processing improves the gelling properties of standard egg white powder. *Journal of Food Science*, *81*(3), E665–E671. https://doi.org/10.1111/1750-3841.13239

Bove, J., Ma, C., & Harwalkar, V. (1997). Thermal denaturation and coagulation of proteins. In S. Damodaran & A. Paraf (Ed.), *Food proteins and their applications* (pp. 25–56). CRC Press.

Campbell, L., Raikos, V., & Euston, S. R. (2003). Modification of functional properties of egg-white proteins. *Nahrung*, *47*(6), 369–376. https://doi.org/10.1002/food.200390084

Chang, C., Lahti, T., Tanaka, T., & Nickerson, M. T. (2018). Egg proteins: Fractionation, bioactive peptides and allergenicity. *Journal of the Science of Food and Agriculture*, *98*(15), 5547–5558. https://doi.org/10.1002/jsfa.9150

Chang, C., Li, X., Li, J., Niu, F., Zhang, M., Zhou, B., & Yang, Y. (2017). Effect of enzymatic hydrolysis on characteristics and synergistic efficiency of pectin on emulsifying properties of egg white protein. *Food Hydrocolloids*, *65*, 87–95. https://doi.org/10.1016/j.foodhyd.2016.11.004

Chang, C., Meikle, T. G., Su, Y., Wang, X., Dekiwadia, C., Drummond, C. J., & Yang, Y. (2019). Encapsulation in egg white protein nanoparticles protects anti-oxidant activity of curcumin. *Food Chemistry*, *280*, 65–72. https://doi.org/10.1016/j.foodchem.2018.11.124

Chang, C., Niu, F., Su, Y., Qiu, Y., Gu, L., & Yang, Y. (2016). Characteristics and emulsifying properties of acid and acid-heat induced egg white protein. *Food Hydrocolloids*, *54*, 342–350. https://doi.org/10.1016/j.foodhyd.2015.09.026

Chang, C., Su, Y., Gu, L., Li, J., & Yang, Y. (2021). Microwave induced glycosylation of egg white protein: Study on physicochemical properties and baking performance. *Food Hydrocolloids*, *118*, 106569. https://doi.org/10.1016/j.foodhyd.2020.106569

Chang, C., Xu, Y., Shi, M., Su, Y., Li, X., Li, J., & Yang, Y. (2020). Effect of dry-heat and guar gum on properties of egg white powder: Analysis of forming capacity and baking performance. *Food Hydrocolloids*, *99*, 105333. https://doi.org/10.1016/j.foodhyd.2019.105333

Chang, K., Jiang, W., & Liu, J. (2022). Effect of subcritical water treatment on the structure and foaming properties of egg white protein. *Food Hydrocolloids*, *124*, 107241. https://doi.org/10.1016/j.foodhyd.2021.107241

Chen, Y., Sheng, L., Gouda, M., & Ma, M. (2019). Impact of ultrasound treatment on the foaming and physicochemical properties of egg white during cold storage. *LWT*, *113*, 108303. https://doi.org/10.1016/j.lwt.2019.108303

Chen, Z., Li, J., Tu, Y., Zhao, Y., Luo, X., Wang, J., & Wang, M. (2015). Changes in gel characteristics of egg white under strong alkali treatment. *Food Hydrocolloids*, *45*, 1–8. https://doi.org/10.1016/j.foodhyd.2014.10.026

Cheng, Y., Chi, Y., Geng, X., & Chi, Y. (2021). Effect of 2,2′-azobis(2-amidinopropane) dihydrochloride (AAPH) induced oxidation on the physicochemical properties, in vitro digestibility, and nutritional value of egg white protein. *LWT*, *143*, 111103. https://doi.org/10.1016/j.lwt.2021.111103

Clark, A. H. (1991). Structural and mechanical properties of biopolymer gels. In E. Dickinson (Ed.), *Food polymers, gels and colloids* (pp. 322–338). Woodhead Publishing. https://doi.org/10.1533/9781845698331.322

Cluff, K., Konda Naganathan, G., Jonnalagada, D., Mortensen, I., Wehling, R., & Subbiah, J. (2016). Determination of yolk contamination in liquid egg white using Raman spectroscopy. *Poultry Science*, *95*(7), 1702–1708. https://doi.org/10.3382/ps/pew095

Croguennec, T., Nau, F., & Brulé, G. (2002). Influence of pH and salts on egg white gelation. *Journal of Food Science*, *67*(2), 608–614. https://doi.org/10.1111/j.1365-2621.2002.tb10646.x

Croguennec, T., Renault, A., Beaufils, S., Dubois, J., & Pezennec, S. (2007). Interfacial properties of heat-treated ovalbumin. *Journal of Colloid and Interface Science*, *315*(2), 627–636. https://doi.org/10.1016/j.jcis.2007.07.041

Cunningham, F. E., & Cotterill, O. J. (1962). Factors affecting alkaline coagulation of egg white. *Poultry Science*, *41*(5), 1453–1461. https://doi.org/10.3382/ps.0411453

Cunningham, F. E., & Cotterill, O. J. (1964). Factors affecting acid coagulation of the egg white. *Poultry Science*, *43*(1), 53–59. https://doi.org/10.3382/ps.0430053

D'Ambrosio, C., Arena, S., Scaloni, A., Guerrier, L., Boschetti, E., Mendieta, M. E., & Righetti, P. G. (2008). Exploring the chicken egg white proteome with combinatorial peptide ligand libraries. *Journal of Proteome Research*, *7*(8), 3461–3474. https://doi.org/10.1021/pr800193y

Dabestani, M., & Yeganehzad, S. (2019). Effect of Persian gum and xanthan gum on foaming properties and stability of pasteurized fresh egg white foam. *Food Hydrocolloids*, *87*, 550–560. https://doi.org/10.1016/j.foodhyd.2018.08.030

Damodaran, S. (1997a). Food proteins: An overview. In S. Damodaran & Paraf, A. (Ed.), *Food proteins and their applications* (pp. 1–24). CRC Press. https://doi.org/10.1201/9780203755617

Damodaran, S. (1997b). Protein-stabilized foams and emulsions. In S. Damodaran & A. Paraf (Ed.), *Food proteins and their applications* (pp. 57–110). CRC Press.

Damodaran, S. (2005). Protein stabilization of emulsions and foams. *Journal of Food Science*, *70*(3), R54–R66. https://doi.org/10.1111/j.1365-2621.2005.tb07150.x

Damodaran, S., Anand, K., & Razumovsky, L. (1998). Competitive adsorption of egg white proteins at the air–water interface: Direct evidence for electrostatic complex formation between lysozyme and other egg proteins at the interface. *Journal of Agricultural and Food Chemistry*, *46*(3), 872–876. https://doi.org/10.1021/jf970722i

de Souza, P. M., Müller, A., Beniaich, A., Mayer-Miebach, E., Oehlke, K., & Fernández, A. (2015). Functional properties and nutritional composition of liquid egg products treated in a coiled tube UV-C reactor. *Innovative Food Science & Emerging Technologies*, *32*, 156–164. https://doi.org/10.1016/j.ifset.2015.09.004

Delahaije, R. J. B. M., Gruppen, H., Giuseppin, M. L. F., & Wierenga, P. A. (2014). Quantitative description of the parameters affecting the adsorption behaviour of globular proteins. *Colloids and Surfaces B: Biointerfaces*, *123*, 199–206. https://doi.org/10.1016/j.colsurfb.2014.09.015

Deng, C., Shao, Y., Xu, M., Yao, Y., Wu, N., Hu, H., & Tu, Y. (2020). Effects of metal ions on the physico-chemical, microstructural and digestion characteristics of alkali-induced egg white gel.*Food Hydrocolloids*, *107*, 105956. https://doi.org/10.1016/j.foodhyd.2020.105956

Desfougères, Y., Lechevalier, V., Pezennec, S., Artzner, F., & Nau, F. (2008). Dry-heating makes hen egg white lysozyme an efficient foaming agent and enables its bulk aggregation. *Journal of Agricultural and Food Chemistry*, *56*(13), 5120–5128. https://doi.org/10.1021/jf703715j

Dickinson, E. (2011). Mixed biopolymers at interfaces: Competitive adsorption and multilayer structures. *Food Hydrocolloids*, *25*(8), 1966–1983. https://doi.org/10.1016/j.foodhyd.2010.12.001

Ding, L., Xia, M., Zeng, Q., Zhao, Q., Cai, Z., & Zhu, Z. (2022). Foaming properties and aggregation mechanism of egg white protein with different physical treatments. *LWT*, *153*, 112505. https://doi.org/10.1016/j.lwt.2021.112505

Ding, L., Zhao, Q., Zhou, X., Tang, C., Chen, Y., & Cai, Z. (2020a). Changes in protein structure and physicochemical properties of egg white by super critical carbon dioxide treatment. *Journal of Food Engineering*, *284*, 110076. https://doi.org/10.1016/j.jfoodeng.2020.110076

Ding, X., Du, J., Zhang, K., Bai, S., Zeng, Q., Peng, H., & Wang, J. (2020b). Tandem mass tag-based quantitative proteomics analysis and gelling properties in egg albumen of laying hens feeding tea polyphenols. *Poultry Science*, *99*(1), 430–440. https://doi.org/10.3382/ps/pez523

Doi, E. (1993). Gels and gelling of globular proteins. *Trends in Food Science & Technology*, *4*(1), 1–5. https://doi.org/10.1016/S0924-2244(05)80003-2

Donovan, J. W., Mapes, C. J., Davis, J. G., & Garibaldi, J. A. (1975). A differential scanning calorimetric study of the stability of egg white to heat denaturation. *Journal of the Science of Food and Agriculture*, *26*(1), 73–83. https://doi.org/10.1002/jsfa.2740260109

Duan, X., Li, J., Zhang, Q., Zhao, T., Li, M., Xu, X., & Liu, X. (2017). Effect of a multiple freeze-thaw process on structural and foaming properties of individual egg white proteins. *Food Chemistry*, *228*, 243–248. https://doi.org/10.1016/j.foodchem.2017.02.005

Duan, X., Li, M., Shao, J., Chen, H., Xu, X., Jin, Z., & Liu, X. (2018). Effect of oxidative modification on structural and foaming properties of egg white protein. *Food Hydrocolloids*, *75*, 223–228. https://doi.org/10.1016/j.foodhyd.2017.08.008

Fan, H., Ai, M., Cao, Y., Long, J., Li, S., & Jiang, A. (2021). Understanding the hydration of alkali-induced duck egg white gel at high temperature. *LWT*, *142*, 110976. https://doi.org/10.1016/j.lwt.2021.110976

Fang, B., & Nakagawa, K. (2021). Effect of pH, ionic strength, and freezing treatment on a colloidal suspension of egg white aggregates. *Food Structure*, *27*, 100181. https://doi.org/10.1016/j.foostr.2021.100181

Farahnaky, A., Gray, D. A., Mitchell, J. R., & Hill, S. E. (2003). Ascorbic acid and hydrogen peroxide (Fenton's reagent) induced changes in gelatin systems. *Food Hydrocolloids*, *17*(3), 321–326. https://doi.org/10.1016/S0268-005X(02)00093-0

Fernández-Martín, F., Pérez-Mateos, M., Dadashi, S., Gómez-Guillén, C. M., & Sanz, P. D. (2017). Impact of magnetic assisted freezing in the physicochemical and functional properties of egg components. Part 1: Egg white. *Innovative Food Science & Emerging Technologies*, *44*, 131–138. https://doi.org/10.1016/j.ifset.2017.07.004

Ferry, J. D. (1948). Protein gels. In M. L. Anson & J. T. Edsall (Eds.), *Advances in protein chemistry* (pp. 1–78). Academic Press. https://doi.org/10.1016/S0065-3233(08)60004-2

Foegeding, E. A., Luck, P. J., & Davis, J. P. (2006). Factors determining the physical properties of protein foams. *Food Hydrocolloids*, *20*(2), 284–292. https://doi.org/10.1016/j.foodhyd.2005.03.014

Forsythe, R. (1957). Some factors affecting the use of eggs in the food industry. *Cereal Science Today*, *2*, 211–216.

Forsythe, R. H., & Bergquist, D. H. (1951). The effect of physical treatments on some properties of egg white*. *Poultry Science*, *30*(2), 302–311. https://doi.org/10.3382/ps.0300302

Gao, X., Guo, W., Wu, N., Yao, Y., Du, H., Xu, M., & Tu, Y. (2021a). Effects of salt and heat treatment on the physicochemical properties, microstructure, secondary structure, and simulated in vitro gastrointestinal digestion of duck egg white. *Journal of the Science of Food and Agriculture*, *101*(14), 6093–6103. https://doi.org/10.1002/jsfa.11266

Gao, X., Yao, Y., Wu, N., Xu, M., Zhao, Y., & Tu, Y. (2021b). The underlying mechanism of alkali-induced ovalbumin gel transforms to sol: Physicochemical properties, structure and quantitative protein degradation analysis. *Food Hydrocolloids*, *120*, 106954. https://doi.org/10.1016/j.foodhyd.2021.106954

Gao, X., Yao, Y., Wu, N., Xu, M., Zhao, Y., & Tu, Y. (2020). The sol-gel-sol transformation behavior of egg white proteins induced by alkali. *International Journal of Biological Macromolecules*, *155*, 588–597. https://doi.org/10.1016/j.ijbiomac.2020.03.209

Gharbi, N., & Labbafi, M. (2018). Effect of processing on aggregation mechanism of egg white proteins. *Food Chemistry*, *252*, 126–133. https://doi.org/10.1016/j.foodchem.2018.01.088

Gharbi, N., & Labbafi, M. (2019). Influence of treatment-induced modification of egg white proteins on foaming properties. *Food Hydrocolloids*, *90*, 72–81. https://doi.org/10.1016/j.foodhyd.2018.11.060

Gharbi, N., Labbafi, M., & Madadlou, A. (2017). Effect of heat treatment on foaming properties of ostrich (Struthio camelus) egg white proteins. *International Journal of Food Properties*, *20*(12), 3159–3169. https://doi.org/10.1080/10942912.2017.1280676

Grinberg, N., Bikbov, T., Grinberg, V., Chaika, T., Vaintraub, I., & Tolstoguzov, V. (1985). Thermotropic gelation of ovalbumin: 2. Boundary conditions on the formation and the concentration dependence of equilibrium elastic modulus for thermotropic ovalbumin gels. *International Journal of Biological Macromolecules*, *7*(3), 152–160. https://doi.org/10.1016/0141-8130(85)90018-2

Hagolle, N., Relkin, P., Dalgleish, D. G., & Launay, B. (1997). Transition temperatures of heat-induced structural changes in ovalbumin solutions at acid and neutral pH. *Food Hydrocolloids*, *11*(3), 311–317. https://doi.org/10.1016/S0268-005X(97)80061-6

Hammershøj, M., & Larsen, L. (1999). Foaming of ovalbumin and egg albumen fractions and the role of the disulfide bonds at various pH levels. Proceeding of VII European Symposium on the Quality of Eggs and Egg Products. Bologna.

Hammershøj, M., Larsen, L. B., Ipsen, R. H., & Qvist, K. B. (2001). Effect of hen egg production and protein composition on textural properties of egg albumen gels. *Journal of Texture Studies*, *32*(2), 105–129. https://doi.org/10.1111/j.1745-4603.2001.tb01037.x

Hammershøj, M., Nebel, C., & Carstens, J. H. (2008). Enzymatic hydrolysis of ovomucin and effect on foaming properties. *Food Research International*, *41*(5), 522–531. https://doi.org/10.1016/j.foodres.2008.03.004

Handa, A., Takahashi, K., Kuroda, N., & Froning, G. W. (1998). Heat-induced egg white gels as affected by pH. *Journal of Food Science*, *63*(3), 403–407. https://doi.org/10.1111/j.1365-2621.1998.tb15752.x

Hanning, F. M. (1944). *Effect of sugar or salt upon denaturation produced by beating and upon the ease of formation and the stability of egg white foams*. Iowa State University.

Hatta, H., Kitabatake, N., & Doi, E. (1986). Turbidity and hardness of a heat-induced gel of hen egg ovalbumin. *Agricultural and Biological Chemistry*, *50*(8), 2083–2089. https://doi.org/10.1271/bbb1961.50.2083

Hayashi, R., Kawamura, Y., Nakasa, T., & Okinaka, O. (1989). Application of high pressure to food processing: Pressurization of egg white and yolk, and properties of gels formed. *Agricultural and Biological Chemistry*, *53*(11), 2935–2939. https://doi.org/10.1080/00021369.1989.10869784

Hayashi, Y., Nagano, S., Enomoto, H., Li, C. P., Sugimoto, Y., Ibrahim, H., & Aoki, T. (2009). Improvement of foaming property of egg white protein by phosphorylation through dry-heating in the presence of pyrophosphate. *Journal of Food Science*, *74*(1), C68–C72. https://doi.org/10.1111/j.1750-3841.2008.01019.x

He, D., Lv, Y., & Tong, Q. (2019). Succinylation improves the thermal stability of egg white proteins. *Molecules*, *24*(20), 3783. https://doi.org/10.3390/molecules24203783.

He, W., Xiao, N., Zhao, Y., Yao, Y., Xu, M., Du, H., & Tu, Y. (2021). Effect of polysaccharides on the functional properties of egg white protein: A review. *Journal of Food Science*, *86*(3), 656–666. https://doi.org/10.1111/1750-3841.15651

Hegg, P.-O., Martens, H., & Löfqvist, B. (1979). Effects of pH and neutral salts on the formation and quality of thermal aggregates of ovalbumin. A study on thermal aggregation and denaturation. *Journal of the Science of Food and Agriculture*, *30*(10), 981–993. https://doi.org/10.1002/jsfa.2740301006

Henry, W. C., & Barbour, A. D. (1933). Beating properties of egg white. *Industrial & Engineering Chemistry*, *25*(9), 1054–1058. https://doi.org/10.1021/ie50285a035

Hermansson, A. M. (1979). Aggregation and denaturation involved in gel formation. In Pour-El, A. (Ed.), *Functionality and protein structure* (pp. 81–103). American Chemical Society. https://doi.org/10.1021/bk-1979-0092.ch005

Hermansson, A. M. (1986). Soy protein gelation. *Journal of the American Oil Chemists' Society*, *63*(5), 658–666. https://doi.org/10.1007/BF02638232

Hettiarachchy, N., Kannan, A., Schäfer, C., & Wagner, G. (2013). Gelling of plant based proteins. In U. Bröckel, W. Meier and G. Wagner (Ed.), *Product design and engineering* (pp. 221–245). Wiley. https://doi.org/10.1002/9783527654741.ch8

Ho, H., Ciou, J., Qiu, Y., Hsieh, S., Shih, M., Chen, M., & Hou, C. (2021). Improvement of foaming characteristics and stability of sterilized liquid egg with egg white hydrolysate (EWH). *Foods*, *10*(6), 1326. https://doi.org/10.3390/foods10061326

Hong, T., Iwashita, K., Han, J., Nishinami, S., Handa, A., & Shiraki, K. (2021). Aggregation of hen egg white proteins with additives during agitation. *LWT*, *146*, 111378. https://doi.org/10.1016/j.lwt.2021.111378

Hong, T., Iwashita, K., Handa, A., & Shiraki, K. (2017). Arginine prevents thermal aggregation of hen egg white proteins. *Food Research International*, *97*, 272–279. https://doi.org/10.1016/j.foodres.2017.04.013

Hu, G., Ma, M., Batool, Z., Sheng, L., Cai, Z., Liu, Y., & Jin, Y. (2022). Gel properties of heat-induced transparent hydrogels from ovalbumin by acylation modifications. *Food Chemistry*, *369*, 130912. https://doi.org/10.1016/j.foodchem.2021.130912

Huang, Q., Liu, L., Wu, Y., Huang, X., Wang, G., Song, H., & Luo, P. (2022). Mechanism of differences in characteristics of thick/thin egg whites during storage: Physicochemical, functional and molecular structure characteristics analysis. *Food Chemistry*, *369*, 130828. https://doi.org/10.1016/j.foodchem.2021.130828

Huang, Q., Qiu, N., Ma, M., Jin, Y., Yang, H., Geng, F., & Sun, S. (2012). Estimation of egg freshness using S-ovalbumin as an indicator. *Poultry Science*, *91*(3), 739–743. https://doi.org/10.3382/ps.2011-01639

Huang, X., Li, J., Chang, C., Gu, L., Su, Y., & Yang, Y. (2019). Effects of NaOH/NaCl pickling on heat-induced gelation behaviour of egg white. *Food Chemistry*, *297*, 124939. https://doi.org/10.1016/j.foodchem.2019.06.006

Iametti, S., Donnizzelli, E., Pittia, P., Rovere, P. P., Squarcina, N., & Bonomi, F. (1999). Characterization of high-pressure-treated egg albumen. *Journal of Agricultural and Food Chemistry*, *47*(9), 3611–3616. https://doi.org/10.1021/jf9808630

Iwashita, K., Handa, A., & Shiraki, K. (2017). Co-aggregation of ovalbumin and lysozyme. *Food Hydrocolloids*, *67*, 206–215. https://doi.org/10.1016/j.foodhyd.2017.01.014

Iwashita, K., Handa, A., & Shiraki, K. (2019). Co-aggregation of ovotransferrin and lysozyme. *Food Hydrocolloids*, *89*, 416–424. https://doi.org/10.1016/j.foodhyd.2018.11.022

Jambrak, A. R., Lelas, V., Mason, T. J., Krešić, G., & Badanjak, M. (2009). Physical properties of ultrasound treated soy proteins. *Journal of Food Engineering*, *93*(4), 386–393. https://doi.org/10.1016/j.jfoodeng.2009.02.001

Jia, J., Ji, B., Tian, L., Li, M., Lu, M., Ding, L., & Duan, X. (2021). Mechanism study on enhanced foaming properties of individual albumen proteins by lactobacillus fermentation. *Food Hydrocolloids*, *111*, 106218. https://doi.org/10.1016/j.foodhyd.2020.106218

Jia, J., Xiong, D., Bai, J., Yuan, Y., Song, Q., Lan, T., & Duan, X. (2022). Investigation on flavor and physicochemical properties of angel food cakes prepared by lactic acid fermented egg white. *LWT*, *164*, 113659. https://doi.org/10.1016/j.lwt.2022.113659

Jiang, Y., Jia, J., Xiong, D., Xu, X., Yang, Y., Liu, X., & Duan, X. (2020). Effects of short-term fermentation with lactic acid bacteria on egg white: Characterization, rheological and foaming activities. *Food Hydrocolloids*, *101*, 105507. https://doi.org/10.1016/j.foodhyd.2019.105507

Jin, H., Chen, J., Zhang, J., & Sheng, L. (2021). Impact of phosphates on heat-induced egg white gel properties: Texture, water state, micro-rheology and microstructure. *Food Hydrocolloids*, *110*, 106200. https://doi.org/10.1016/j.foodhyd.2020.106200

Jin, H., Jin, Y., Pan, J., Mi, S., Zeng, Q., Li, Z., & Sheng, L. (2023). Comprehensive identification and hydrophobic analysis of key proteins affecting foam capacity and stability during the evolution of egg white foam. *Food Hydrocolloids*, *134*, 108033. https://doi.org/10.1016/j.foodhyd.2022.108033

Jin, H., Sun, Y., Pan, J., Fang, Y., Jin, Y., & Sheng, L. (2022a). Adsorption kinetics of ovalbumin and lysozyme at the air-water interface and foam properties at neutral pH. *Food Hydrocolloids*, *124*, 107352. https://doi.org/10.1016/j.foodhyd.2021.107352

Jin, Y., Yao, Y., Wu, N., Du, H., Xu, M., Zhao, Y., & Tu, Y. (2022b). Inhibition of the liquefaction of alkali-induced egg white gel by sodium ascorbate. *Food Chemistry*, *381*, 132220. https://doi.org/10.1016/j.foodchem.2022.132220

Jing, H., Sun, J., Mu, Y., Obadi, M., McClements, D. J., & Xu, B. (2020). Sonochemical effects on the structure and antioxidant activity of egg white protein–tea polyphenol conjugates. *Food & Function*, *11*(8), 7084–7094. https://doi.org/10.1039/D0FO01636E

Joeres, E., Schölzel, H., Drusch, S., Töpfl, S., Heinz, V., & Terjung, N. (2022). Ohmic vs. conventional heating: Influence of moderate electric fields on properties of egg white protein gels. *Food Hydrocolloids*, *127*, 107519. https://doi.org/10.1016/j.foodhyd.2022.107519

John, J. S., & Flor, I. H. (1930). A study of whipping and coagulation of eggs of varying quality. *Poultry Science*, *10*(2), 71–82. https://doi.org/10.3382/ps.0100071

Johnson, T. M., & Zabik, M. E. (1981). Egg albumen proteins interactions in an angel food cake system. *Journal of Food Science*, *46*(4), 1231–1236. https://doi.org/10.1111/j.1365-2621.1981.tb03029.x

Joshi, V., Shivach, T., Yadav, N., & Rathore, A. S. (2014). Circular dichroism spectroscopy as a tool for monitoring aggregation in monoclonal antibody therapeutics. *Analytical Chemistry*, *86*(23), 11606–11613. https://doi.org/10.1021/ac503140j

Kaewmanee, T., Benjakul, S., & Visessanguan, W. (2011). Effect of NaCl on thermal aggregation of egg white proteins from duck egg. *Food Chemistry*, *125*(2), 706–712. https://doi.org/10.1016/j.foodchem.2010.09.072

Kan, X., Chen, G., Zhou, W., & Zeng, X. (2021). Application of protein-polysaccharide Maillard conjugates as emulsifiers: Source, preparation and functional properties. *Food Research International*, *150*, 110740. https://doi.org/10.1016/j.foodres.2021.110740

Kar, A., Wei, X., Majumder, K., Eskridge, K., Handa, A., & Subbiah, J. (2020). Effect of traditional and radiofrequency assisted thermal processing on the gel firmness of egg white powder. *LWT*, *133*, 110091. https://doi.org/10.1016/j.lwt.2020.110091

Katekhong, W., & Charoenrein, S. (2016). Changes in physical and gelling properties of freeze-dried egg white as a result of temperature and relative humidity. *Journal of the Science of Food and Agriculture*, *96*(13), 4423–4431. https://doi.org/10.1002/jsfa.7653

Katekhong, W., & Charoenrein, S. (2017). Color and gelling properties of dried egg white: Effect of drying methods and storage conditions. *International Journal of Food Properties*, *20*(9), 2157–2168. https://doi.org/10.1080/10942912.2016.1233429

Katekhong, W., & Charoenrein, S. (2018). Influence of spray drying temperatures and storage conditions on physical and functional properties of dried egg white. *Drying Technology*, *36*(2), 169–177. https://doi.org/10.1080/07373937.2017.1307218

Kato, A., Ibrahim, H. R., Watanabe, H., Honma, K., & Kobayashi, K. (1989). New approach to improve the gelling and surface functional properties of dried egg white by heating in dry state. *Journal of Agricultural and Food Chemistry*, *37*(2), 433–437. https://doi.org/10.1021/jf00086a036

Kato, A., Ibrahim, H. R., Watanabe, H., Honma, K., & Kobayashi, K. (1990). Enthalpy of denaturation and surface functional properties of heated egg white proteins in the dry state. *Journal of Food Science*, *55*(5), 1280–1283. https://doi.org/10.1111/j.1365-2621.1990.tb03916.x

Kato, A., & Takagi, T. (1988). Formation of intermolecular β-sheet structure during heat denaturation of ovalbumin. *Journal of Agricultural and Food Chemistry*, *36*. https://doi.org/10.1021/jf00084a007

Khemakhem, M., Attia, H., & Ayadi, M. A. (2019). The effect of pH, sucrose, salt and hydrocolloid gums on the gelling properties and water holding capacity of egg white gel. *Food Hydrocolloids*, *87*, 11–19. https://doi.org/10.1016/j.foodhyd.2018.07.041

Kim, W., Wang, Y., & Selomulya, C. (2020). Dairy and plant proteins as natural food emulsifiers. *Trends in Food Science & Technology*, *105*, 261–272. https://doi.org/10.1016/j.tifs.2020.09.012

Kitabatake, N., Shimizu, A., & Doi, E. (1988a). Preparation of heat-induced transparent gels from egg white by the control of pH and ionic strength of the medium. *Journal of Food Science*, *53*(4), 1091–1095. https://doi.org/10.1111/j.1365-2621.1988.tb13537.x

Kitabatake, N., Shimizu, A., & Doi, E. (1988b). Preparation of transparent egg white gel with salt by two-step heating method. *Journal of Food Science*, *53*(3), 735–738. https://doi.org/10.1111/j.1365-2621.1988.tb08943.x

Kuan, Y. H., Bhat, R., & Karim, A. A. (2011). Emulsifying and foaming properties of ultraviolet-irradiated egg white protein and sodium caseinate. *Journal of Agricultural and Food Chemistry*, *59*(8), 4111–4118. https://doi.org/10.1021/jf104050k

Kuan, Y. H., Bhat, R., Patras, A., & Karim, A. A. (2013). Radiation processing of food proteins–A review on the recent developments. *Trends in Food Science & Technology*, *30*(2), 105–120. https://doi.org/10.1016/j.tifs.2012.12.002

Kumbar, V., Nedomova, S., Strnkova, J., & Buchar, J. (2015). Effect of egg storage duration on the rheology of liquid egg products. *Journal of Food Engineering*, *156*, 45–54. https://doi.org/10.1016/j.jfoodeng.2015.02.011

Lazidis, A., Hancocks, R. D., Spyropoulos, F., Kreuß, M., Berrocal, R., & Norton, I. T. (2016). Whey protein fluid gels for the stabilization of foams. *Food Hydrocolloids*, *53*, 209–217. https://doi.org/10.1016/j.foodhyd.2015.02.022

Le Floch-Fouéré, C., Beaufils, S., Lechevalier, V., Nau, F., Pézolet, M., Renault, A., & Pezennec, S. (2010). Sequential adsorption of egg-white proteins at the air–water interface suggests a stratified organization of the interfacial film. *Food Hydrocolloids*, *24*(4), 275–284. https://doi.org/10.1016/j.foodhyd.2009.10.006

Le Floch-Fouéré, C., Pezennec, S., Lechevalier, V., Beaufils, S., Desbat, B., Pézolet, M., & Renault, A. (2009). Synergy between ovalbumin and lysozyme leads to non-additive interfacial and foaming properties of mixtures. *Food Hydrocolloids*, *23*(2), 352–365. https://doi.org/10.1016/j.foodhyd.2008.01.007

Lechevalier, V., Croguennec, T., Pezennec, S., Guérin-Dubiard, C., Pasco, M., & Nau, F. (2005). Evidence for synergy in the denaturation at the air–water interface of ovalbumin, ovotransferrin and lysozyme in ternary mixture. *Food Chemistry*, *92*(1), 79–87. https://doi.org/10.1016/j.foodchem.2004.07.006

Lechevalier, V., Jeantet, R., Arhaliass, A., Legrand, J., & Nau, F. (2007). Egg white drying: Influence of industrial processing steps on protein structure and functionalities. *Journal of Food Engineering*, *83*(3), 404–413. https://doi.org/10.1016/j.jfoodeng.2007.03.033

Li, C., Hayashi, Y., Shinohara, H., Ibrahim, H. R., Sugimoto, Y., Kurawaki, J., & Aoki, T. (2005). Phosphorylation of ovalbumin by dry-heating in the presence of pyrophosphate: Effect on protein structure and some properties. *Journal of Agricultural and Food Chemistry*, *53*(12), 4962–4967. https://doi.org/10.1021/jf047793j

Li, C., Enomoto, H., Hayashi, Y., Zhao, H., & Aoki, T. (2010). Recent advances in phosphorylation of food proteins: A review. *LWT - Food Science and Technology*, *43*(9), 1295–1300. https://doi.org/10.1016/j.lwt.2010.03.016

Li, C., Ibrahim, H. R., Sugimoto, Y., Hatta, H., & Aoki, T. (2004). Improvement of functional properties of egg white protein through phosphorylation by dry-heating in the presence of pyrophosphate. *Journal of Agricultural and Food Chemistry*, *52*(18), 5752–5758. https://doi.org/10.1021/jf0498259

Li, C., Salvador, A. S., Ibrahim, H. R., Sugimoto, Y., & Aoki, T. (2003). Phosphorylation of egg white proteins by dry-heating in the presence of phosphate. *Journal of Agricultural and Food Chemistry*, *51*(23), 6808–6815. https://doi.org/10.1021/jf030043+

Li, J., Li, X., Wang, C., Zhang, M., Xu, Y., Zhou, B., & Yang, Y. (2018a). Characteristics of gelling and water holding properties of hen egg white/yolk gel with NaCl addition. *Food Hydrocolloids*, *77*, 887–893. https://doi.org/10.1016/j.foodhyd.2017.11.034

Li, J., Wang, C., Li, X., Su, Y., Yang, Y., & Yu, X. (2018b). Effects of pH and NaCl on the physicochemical and interfacial properties of egg white/yolk. *Food Bioscience*, *23*, 115–120. https://doi.org/10.1016/j.fbio.2017.12.004

Li, J., Zhang, Y., Fan, Q., Teng, C., Xie, W., Shi, Y., & Yang, Y. (2018c). Combination effects of NaOH and NaCl on the rheology and gel characteristics of hen egg white proteins. *Food Chemistry*, *250*, 1–6. https://doi.org/10.1016/j.foodchem.2018.01.031

Li, M., Li, M., Tan, W., Liu, X., & Duan, X. (2019a). Effects of ball-milling treatment on physicochemical and foaming activities of egg ovalbumin. *Journal of Food Engineering*, *261*, 158–164. https://doi.org//10.1016/j.jfoodeng.2019.06.015

Li, P., Sheng, L., & Jin, Y. (2019b). Using microwave-assisted phosphorylation to improve foaming and solubility of egg white by response surface methodology. *Poultry Science*, *98*(12), 7110–7117. https://doi.org/10.3382/ps/pez424

Li, P., Sun, Z., Ma, M., Jin, Y., & Sheng, L. (2018d). Effect of microwave-assisted phosphorylation modification on the structural and foaming properties of egg white powder. *LWT*, *97*, 151–156. https://doi.org/10.1016/j.lwt.2018.06.055

Li, S., Zhang, S., Liu, Y., Fu, X., Xiang, X., & Gao, S. (2022a). Effects of ultrasound-assisted glycosylation on the interface and foaming characteristics of ovotransferrin. *Ultrasonics Sonochemistry*, *84*, 105958. https://doi.org/10.1016/j.ultsonch.2022.105958

Li, X., Li, J., Chang, C., Wang, C., Zhang, M., Su, Y., & Yang, Y. (2019c). Foaming characterization of fresh egg white proteins as a function of different proportions of egg yolk fractions. *Food Hydrocolloids*, *90*, 118–125. https://doi.org/10.1016/j.foodhyd.2018.12.014

Li, X., Wang, Y., Lv, J., & Yang, Y. (2022b). Investigations of foaming, interfacial and structural properties of dispersions, batters and cakes formed by industrial yolk-contaminated egg white protein. *LWT*, *154*, 112776. https://doi.org/10.1016/j.lwt.2021.112776

Li, X., Wang, Y., Sun, C., Lv, J., & Yang, Y. (2021). Comparative study on foaming properties of egg white with yolk fractions and their hydrolysates. *Foods*, *10*(9), 2238. https://doi.org/10.3390/foods10092238

Licciardello, F., Frisullo, P., Laverse, J., Muratore, G., & Del Nobile, M. A. (2012). Effect of sugar, citric acid and egg white type on the microstructural and mechanical properties of meringues. *Journal of Food Engineering*, *108*(3), 453–462. https://doi.org/10.1016/j.jfoodeng.2011.08.021

Ling, Z., Ai, M., Zhou, Q., Guo, S., Zhou, L., Fan, H., & Jiang, A. (2020). Fabrication egg white gel hydrolysates-stabilized oil-in-water emulsion and characterization of its stability and digestibility. *Food Hydrocolloids*, *102*, 105621. https://doi.org/10.1016/j.foodhyd.2019.105621

Liu, B., Jin, F., Li, Y., Wang, H., Chi, Y., Tian, B., & Feng, Z. (2022a). Pasteurization of egg white by integrating ultrasound and microwave: Effect on structure and functional properties.

Innovative Food Science & Emerging Technologies, *79*, 103063. https://doi.org/10.1016/j.ifset.2022.103063

Liu, H., Feng, F., Xue, H., Gao, B., Han, T., Li, R., & Zhao, Y. (2022b). Effects of partial replacement of NaCl by KCl and $CaCl_2$ on physicochemical properties, microstructure, and textural properties of salted eggs. *Journal of Food Science*, *87*(2), 795–807. https://doi.org/10.1111/1750-3841.16033

Liu, J., Chai, J., Yuan, Y., Zhang, T., Saini, R. K., Yang, M., & Shang, X. (2022c). Dextran sulfate facilitates egg white protein to form transparent hydrogel at neutral pH: Structural, functional, and degradation properties. *Food Hydrocolloids*, *122*, 107094. https://doi.org/10.1016/j.foodhyd.2021.107094

Liu, J., Jiang, H., Zhang, M., Gong, P., Yang, M., Zhang, T., & Liu, X. (2022d). Ions-regulated aggregation kinetics for egg white protein: A promising formulation with controlled gelation and rheological properties. *International Journal of Biological Macromolecules*, *200*, 263–272. https://doi.org/10.1016/j.ijbiomac.2021.12.185

Liu, J., Jiang, H., Zhang, M., Yang, M., Zhang, T., Du, Z., & Liu, X. (2022e). Relationship of co-gelation and co-aggregation on egg white ovalbumin-lysozyme heteroprotein complex: Formation and thermodynamics. *Food Chemistry*, *388*, 133030. https://doi.org/10.1016/j.foodchem.2022.133030

Liu, J., Yong, H., Yao, X., Hu, H., Yun, D., & Xiao, L. (2019). Recent advances in phenolic–protein conjugates: Synthesis, characterization, biological activities and potential applications. *RSC Advances*, *9*(61), 35825–35840. https://doi.org/10.1039/C9RA07808H

Liu, L., Dai, X., Kang, H., Xu, Y., & Hao, W. (2020a). Structural and functional properties of hydrolyzed/glycosylated ovalbumin under spray drying and microwave freeze drying. *Food Science and Human Wellness*, *9*(1), 80–87. https://doi.org/10.1016/j.fshw.2020.01.003

Liu, L., Li, Y., Dai, X., Zhu, Y., Hao, W., & Yang, X. (2020b). Effects of synergistic modification with enzymatic hydrolysis and phosphorylation on functional and structural properties of ovalbumin. *Journal of Food Processing and Preservation*, *44*(12), e14934. https://doi.org/10.1111/jfpp.14934

Liu, L., Shi, S., Zhang, Y., Li, Y., Guo, J., & Zhang, M. (2022f). Effect of microwave freeze drying on moisture migration and gel characteristics of egg white protein. *Journal of Food Processing and Preservation*, 46(9), e16899. https://doi.org/10.1111/jfpp.16899

Liu, L., Wang, J., Wang, G., Song, H., Geng, F., Zeng, Q., & Huang, Q. (2022g). Quantitative proteomics provides a new perspective on the mechanism of network structure depolymerization during egg white thinning. *Food Chemistry*, *392*, 133320. https://doi.org/10.1016/j.foodchem.2022.133320

Liu, L., Wang, Q., Wu, Y., Wang, G., Geng, F., Song, H., & Huang, Q. (2022h). Effect of ball milling-assisted glycosylation modification on the structure and foaming property of egg white protein. *Journal of Food Science*, *87*(7), 3117–3128. https://doi.org/10.1111/1750-3841.16218

Liu, M., Yao, L., Wang, T., Li, J., & Yu, C. (2014). Rapid determination of egg yolk contamination in egg white by VIS spectroscopy. *Journal of Food Engineering*, *124*, 117–121. https://doi.org/10.1016/j.jfoodeng.2013.10.004

Liu, Q., Wang, Q., He, P., Zhang, Y., Pan, L., Chen, Y., & Zhang, M. (2022i). Heat-induced gel properties and gastrointestinal digestive properties of egg white produced by hens fed with selenium-enriched yeast. *Food Chemistry*, *366*, 130712. https://doi.org/10.1016/j.foodchem.2021.130712

Liu, X., He, L., Zhang, M., Zhao, S., Yang, F., Liang, Y., & Jin, G. (2020c). Effect of different doses of Co-60 gamma-ray irradiation treatment on the micro-rheological and emulsifying properties of liquid egg white. *Colloids and Surfaces A: Physicochemical and Engineering Aspects*, *589*, 124461. https://doi.org/10.1016/j.colsurfa.2020.124461

Liu, X., Ma, L., Zhong, M., Wei, S., Liu, Y., Dong, S., & Li, S. (2022j). Formation mechanism of high-viscosity gelatinous egg white among "Fenghuang Egg": Phenomenon, structure, and substance composition. *International Journal of Biological Macromolecules*, *217*, 803–813. https://doi.org/10.1016/j.ijbiomac.2022.07.089

Liu, X., Wang, J., Huang, Q., Cheng, L., Gan, R., Liu, L., & Geng, F. (2020d). Underlying mechanism for the differences in heat-induced gel properties between thick egg whites and thin egg whites: Gel properties, structure and quantitative proteome analysis. *Food Hydrocolloids*, *106*, 105873. https://doi.org/10.1016/j.foodhyd.2020.105873

Liu, X., Wang, J., Liu, L., Cheng, L., Huang, Q., Wu, D., & Geng, F. (2021). Quantitative N-glycoproteomic analyses provide insights into the effects of thermal processes on egg white functional properties. *Food Chemistry*, *342*, 128252. https://doi.org/10.1016/j.foodchem.2020.128252

Lu, F., Zheng, M., & Chen, L. (2005). Process of lead-free preserved eggs by gradient temperature control. *Journal of Fujian Agriculture and Forestry University*, *34*, 527–530.

Lu, Y., Mao, L., Hou, Z., Miao, S., & Gao, Y. (2019). Development of emulsion gels for the delivery of functional food ingredients: From structure to functionality. *Food Engineering Reviews*, *11*(4), 245–258. https://doi.org/10.1007/s12393-019-09194-z

Luo, X., Wang, Q., Wu, Y., Duan, W., Zhang, Y., Geng, F., & An, F. (2022). Mechanism of effect of heating temperature on functional characteristics of thick egg white. *LWT*, *154*, 112807. https://doi.org/10.1016/j.lwt.2021.112807

Lv, X., Huang, X., Ma, B., Chen, Y., Batool, Z., Fu, X., & Jin, Y. (2022). Modification methods and applications of egg protein gel properties: A review. *Comprehensive Reviews in Food Science and Food Safety*. https://doi.org/10.1111/1541-4337.12907

Ma, Y., Shan, A., Wang, R., Zhao, Y., & Chi, Y. (2021a). Characterization of egg white powder gel structure and its relationship with gel properties influenced by pretreatment with dry heat. *Food Hydrocolloids*, *110*, 106149. https://doi.org/10.1016/j.foodhyd.2020.106149

Ma, Y., Zhao, Y., & Chi, Y. (2019a). Changes in the gel characteristics of two hen egg white powders modified by dry heating and the Maillard reaction during long-term storage. *LWT*, *109*, 123–129. https://doi.org/10.1016/j.lwt.2019.04.003

Ma, Y., Zhao, Y., Jiang, Y., & Chi, Y. (2019b). Effect of dry heating on the aggregation behaviour and aggregate morphologies of ovalbumin. *Food Chemistry*, *285*, 296–304. https://doi.org/10.1016/j.foodchem.2019.01.170

Ma, Z., Chi, Y., Zhang, H., Chi, Y., & Ma, Y. (2022). Inhibiting effect of dry heat on the heat-induced aggregation of egg white protein. *Food Chemistry*, *387*, 132850. https://doi.org/10.1016/j.foodchem.2022.132850

Ma, Z., Ma, Y., Wang, R., & Chi, Y. (2021b). Influence of antigelation agents on frozen egg yolk gelation. *Journal of*

Food Engineering, 302, 110585. https://doi.org/10.1016/j.jfoodeng.2021.110585

MacDonnell, L. R., Feeney, R. E., Hanson, H. L., Campbell, A., & Sugihara, T. F. (1955). The functional properties of the egg-white protein. *Food Technology*, *9*, 49–53.

Machado, F. F., Coimbra, J. S. R., Garcia Rojas, E. E., Minim, L. A., Oliveira, F. C., & Sousa, R. d. C. S. (2007). Solubility and density of egg white proteins: Effect of pH and saline concentration. *LWT - Food Science and Technology*, *40*(7), 1304–1307. https://doi.org/10.1016/j.lwt.2006.08.020

Margoshes, B. (1990). Correlation of protein sulfhydryls with the strength of heat-formed egg white gels. *Journal of Food Science*, *55*(6), 1753–1753. https://doi.org/10.1111/j.1365-2621.1990.tb03619.x

Matsudomi, N., Takasaki, M., & Kobayashi, K. (1991). Heat-induced aggregation of lysozyme with ovotransferrin. *Agricultural and Biological Chemistry*, *55*(6), 1651–1653. https://doi.org/10.1080/00021369.1991.10870828

Mine, Y. (1995). Recent advances in the understanding of egg white protein functionality. *Trends in Food Science & Technology*, *6*(7), 225–232. https://doi.org/10.1016/S0924-2244(00)89083-4

Mine, Y. (1996). Effect of pH during the dry heating on the gelling properties of egg white proteins. *Food Research International*, *29*(2), 155–161. https://doi.org/10.1016/0963-9969(96)00008-7

Mine, Y. (2002). Recent advances in egg protein functionality in the food system. *World's Poultry Science Journal*, *58*(1), 31–39. https://doi.org/10.1079/WPS20020005

Mohammadian, M., & Alavi, F. (2016). The effects of Iranian gum tragacanth on the foaming properties of egg white proteins in comparison with guar and xanthan gums. *Journal of Food Biosciences and Technology*, *6*(2), 59–68.

Monge-Morera, M., Lambrecht, M. A., Deleu, L. J., Godefroidt, T., Goos, P., Rousseau, F., & Delcour, J. A. (2021). Drying mode and hydrothermal treatment conditions govern the formation of amyloid-like protein fibrils in solutions of dried hen egg white. *Food Hydrocolloids*, *112*, 106276. https://doi.org/10.1016/j.foodhyd.2020.106276

Mu, Y., Sun, J., Obadi, M., Chen, Z., & Xu, B. (2020). Effects of saccharides on the rheological and gelling properties and water mobility of egg white protein. *Food Hydrocolloids*, *108*, 106038. https://doi.org/10.1016/j.foodhyd.2020.106038

Murray, B. S. (2011). Rheological properties of protein films. *Current Opinion in Colloid & Interface Science*, *16*(1), 27–35. https://doi.org/10.1016/j.cocis.2010.06.005

Muthukumaran, A., Ratti, C., & Raghavan, V. G. (2008). Foam-mat freeze drying of egg white and mathematical modeling Part I optimization of egg white foam stability. *Drying Technology*, *26*(4), 508–512. https://doi.org/10.1080/07373930801929581

Naderi, N., House, J. D., Pouliot, Y., & Doyen, A. (2017). Effects of high hydrostatic pressure processing on hen egg compounds and egg products. *Comprehensive Reviews in Food Science and Food Safety*, *16*(4), 707–720. https://doi.org/10.1111/1541-4337.12273

Nafchi, A. M., Tabatabaei, R. H., Pashania, B., Rajabi, H. Z., & Karim, A. A. (2013). Effects of ascorbic acid and sugars on solubility, thermal, and mechanical properties of egg white protein gels. *International Journal of Biological Macromolecules*, *62*, 397–404. https://doi.org/10.1016/j.ijbiomac.2013.09.050

Nakamura, R. (1963). Studies on the foaming property of the chicken egg white. *Agricultural and Biological Chemistry*, *27*(6), 427–432. https://doi.org/10.1080/00021369.1963.10858129

Nakamura, R., & Doi, E. (2000). Egg processing. In N. Shuryo & H. W. Modler (Eds.), *Food proteins: Processing applications* (pp. 171–207). Vch Pub.

Narayanan, C., & Dias, C. L. (2013). Hydrophobic interactions and hydrogen bonds in β-sheet formation. *Journal of Chemical Physics*, *139*(11), 115103. https://doi.org/10.1063/1.4821596

Nasabi, M., & Labbafi, M. (2020). Thermal aggregation of egg white proteins as affected by saccharides. *Journal of Food and Bioprocess Engineering*, *3*(2), 110–115. https://doi.org/10.22059/jfabe.2020.307261.1060

Nicolai, T. (2019). Gelation of food protein-protein mixtures. *Advances in Colloid and Interface Science*, *270*, 147–164. https://doi.org/10.1016/j.cis.2019.06.006

Nicorescu, I., Vial, C., Talansier, E., Lechevalier, V., Loisel, C., Valle, D., & Legrand, J. (2011). Comparative effect of thermal treatment on the physicochemical properties of whey and egg white protein foams. *Food Hydrocolloids*, *25*(4), 797–808. https://doi.org/10.1016/j.foodhyd.2010.09.020

Nishinari, K., Fang, Y., Guo, S., & Phillips, G. O. (2014). Soy proteins: A review on composition, aggregation and emulsification. *Food Hydrocolloids*, *39*, 301–318. https://doi.org/10.1016/j.foodhyd.2014.01.013

O'Charoen, S., Hayakawa, S., & Ogawa, M. (2015). Food properties of egg white protein modified by rare ketohexoses through Maillard reaction. *International Journal of Food Science & Technology*, *50*(1), 194–202. https://doi.org/10.1111/ijfs.12607

Ochi, A., Katsuta, K., Maruyama, E., Kubo, M., & Ueda, T. (2000). Effects of sugars on stability of egg foam and their rheological properties. In K. Nishinari (Ed.), *Hydrocolloids* (pp. 275–280). Elsevier Science. https://doi.org/10.1016/B978-044450178-3/50095-0

Ohgushi, M., & Wada, A. (1983). 'Molten-globule state': A compact form of globular proteins with mobile side-chains. *FEBS Letters*, *164*(1), 21–24. https://doi.org/10.1016/0014-5793(83)80010-6

Okamoto, M., Kawamura, Y., & Hayashi, R. (1990). Application of high pressure to food processing: Textural comparison of pressure- and heat-induced gels of food proteins. *Agricultural and Biological Chemistry*, *54*(1), 183–189. https://doi.org/10.1080/00021369.1990.10869911

Omana, D. A., Wang, J., & Wu, J. (2010). Ovomucin – A glycoprotein with promising potential. *Trends in Food Science & Technology*, *21*(9), 455–463. https://doi.org/10.1016/j.tifs.2010.07.001

Painter, P., & Koenig, J. (1976). Raman spectroscopic study of the proteins of egg white. *Biopolymers*, *15*, 2155–2166. https://doi.org/10.1002/bip.1976.360151105

Panozzo, A., Manzocco, L., Calligaris, S., Bartolomeoli, I., Maifreni, M., Lippe, G., & Nicoli, M. C. (2014). Effect of high pressure homogenisation on microbial inactivation, protein structure and functionality of egg white. *Food Research International*, *62*, 718–725. https://doi.org/10.1016/j.foodres.2014.04.051

Patino, J. M. R., & Pilosof, A. M. R. (2011). Protein–polysaccharide interactions at fluid interfaces. *Food Hydrocolloids*, *25*(8), 1925–1937. https://doi.org/10.1016/j.foodhyd.2011.02.023

Pomeranz, Y. (1991). Proteins: General. In Y. Pomeranz (Ed.), *Functional properties of food components* (pp. 147–192). Academic Press. https://doi.org/10.1016/B978-0-12-561281-4.50009-8

Poole, S. (1989). The foam-enhancing properties of basic biopolymers. *International Journal of Food Science & Technology*, *24*(2), 121–137. https://doi.org/10.1111/j.1365-2621.1989.tb00626.x

Preethi, R., Shweta, D., Moses, J. A., & Anandharamakrishnan, C. (2021). Conductive hydro drying as an alternative method for egg white powder production. *Drying Technology*, *39*(3), 324–336. https://doi.org/10.1080/07373937.2020.1788073

Prochaska, J. F., Carey, J. B., & Shafer, D. J. (1996). The effect of L-Lysine intake on egg component yield and composition in laying hens1. *Poultry Science*, *75*(10), 1268–1277. https://doi.org/10.3382/ps.0751268

Pu, X., Zhang, M., Lin, X., Tuo, Y., Murad, M. S., Mu, G., & Jiang, S. (2022). Development and characterization of acid-induced whey protein concentrate and egg white protein composite gel. *LWT*, *164*, 113624. https://doi.org/10.1016/j.lwt.2022.113624

Qiu, N., Ma, M., Zhao, L., Liu, W., Li, Y., & Mine, Y. (2012). Comparative proteomic analysis of egg white proteins under various storage temperatures. *Journal of Agricultural and Food Chemistry*, *60*(31), 7746–7753. https://doi.org/10.1021/jf302100m

Raikos, V., Campbell, L., & Euston, S. R. (2007a). Effects of sucrose and sodium chloride on foaming properties of egg white proteins. *Food Research International*, *40*(3), 347–355. https://doi.org/10.1016/j.foodres.2006.10.008

Raikos, V., Campbell, L., & Euston, S. R. (2007b). Rheology and texture of hen's egg protein heat-set gels as affected by pH and the addition of sugar and/or salt. *Food Hydrocolloids*, *21*(2), 237–244. https://doi.org/10.1016/j.foodhyd.2006.03.015

Réhault-Godbert, S., Guyot, N., & Nys, Y. (2019). The golden egg: Nutritional value, bioactivities, and emerging benefits for human health. *Nutrients*, *11*(3), 684. https://www.mdpi.com/2072-6643/11/3/684

Renard, D., & Lefebvre, J. (1992). Gelation of globular proteins: Effect of pH and ionic strength on the critical concentration for gel formation. A simple model and its application to β-lactoglobulin heat-induced gelation. *International Journal of Biological Macromolecules*, *14*(5), 287–291. https://doi.org/10.1016/S0141-8130(05)80042-X

Rhodes, M. B., Adams, J. L., Bennett, N., & Feeney, R. E. (1960). Properties and food uses of duck eggs1,2. *Poultry Science*, *39*(6), 1473–1478. https://doi.org/10.3382/ps.0391473

Rico-Pasto, M., Zaltron, A., Davis, S. J., Frutos, S., & Ritort, F. (2022). Molten globule-like transition state of protein barnase measured with calorimetric force spectroscopy. *Proceedings of the National Academy of Sciences*, *119*(11), e2112382119. https://doi.org/10.1073/pnas.2112382119

Rombouts, I., Wouters, A. G. B., Lambrecht, M. A., Uten, L., Van Den Bosch, W., Vercruysse, S. A. R., & Delcour, J. A. (2020). Food protein network formation and gelation induced by conductive or microwave heating: A focus on hen egg white. *Innovative Food Science & Emerging Technologies*, *66*, 102484. https://doi.org/10.1016/j.ifset.2020.102484

Sadahira, M. S., Rodrigues, M. I., Akhtar, M., Murray, B. S., & Netto, F. M. (2016). Effect of egg white protein-pectin electrostatic interactions in a high sugar content system on foaming and foam rheological properties. *Food Hydrocolloids*, *58*, 1–10. https://doi.org/10.1016/j.foodhyd.2016.02.007

Samborska, K., Witrowa-Rajchert, D., & Gonçalves, A. (2005). Spray-drying of α-amylase—The effect of process variables on the enzyme inactivation. *Drying Technology*, *23*(4), 941–953. https://doi.org/10.1081/DRT-200054243

Sato, Y., & Nakamura, R. (1977). Functional properties of acetylated and succinylated egg white. *Agricultural and Biological Chemistry*, *41*(11), 2163–2168. https://doi.org/10.1080/00021369.1977.10862830

Shao, Y., Zhao, Y., Xu, M., Chen, Z., Wang, S., & Tu, Y. (2017). Effects of copper ions on the characteristics of egg white gel induced by strong alkali. *Poultry Science*, *96*(11), 4116–4123. https://doi.org/10.3382/ps/pex213

Sheng, L., Liu, Q., Dong, W., & Cai, Z. (2022). Effect of high intensity ultrasound assisted glycosylation on the gel properties of ovalbumin: Texture, rheology, water state and microstructure. *Food Chemistry*, *372*, 131215. https://doi.org/10.1016/j.foodchem.2021.131215

Sheng, L., Tang, G., Wang, Q., Zou, J., Ma, M., & Huang, X. (2020). Molecular characteristics and foaming properties of ovalbumin-pullulan conjugates through the Maillard reaction. *Food Hydrocolloids*, *100*, 105384. https://doi.org/10.1016/j.foodhyd.2019.105384

Sheng, L., Wang, Y., Chen, J., Zou, J., Wang, Q., & Ma, M. (2018). Influence of high-intensity ultrasound on foaming and structural properties of egg white. *Food Research International*, *108*, 604–610. https://doi.org/10.1016/j.foodres.2018.04.007

Shimada, K., & Matsushita, S. (1980). Relationship between thermocoagulation of proteins and amino acid compositions. *Journal of Agricultural and Food Chemistry*, *28*(2), 413–417. https://doi.org/10.1021/jf60228a019

Sikorski, Z. E. (2006). The role of proteins in food. In Z. E. Sikorski (Ed.), *Chemical and functional properties of food components* (pp. 129–175). CRC Press. https://doi.org/10.1201/9781420009613

Somaratne, G., Nau, F., Ferrua, M. J., Singh, J., Ye, A., Dupont, D., & Floury, J. (2020). Characterization of egg white gel microstructure and its relationship with pepsin diffusivity. *Food Hydrocolloids*, *98*, 105258. https://doi.org/10.1016/j.foodhyd.2019.105258

Stefanović, A. B., Jovanović, J. R., Balanč, B. D., Šekuljica, N. Ž., Tanasković, S. M. J., Dojčinović, M. B., & Knežević-Jugović, Z. D. (2018). Influence of ultrasound probe treatment time and protease type on functional and physicochemical characteristics of egg white protein hydrolysates. *Poultry Science*, *97*(6), 2218–2229. https://doi.org/10.3382/ps/pey055

Stefanović, A. B., Jovanović, J. R., Dojčinović, M. B., Lević, S. M., Nedović, V. A., Bugarski, B. M., & Knežević-Jugović, Z. D. (2017). Effect of the controlled high-intensity ultrasound on improving functionality and structural changes of egg white proteins. *Food and Bioprocess Technology*, *10*(7), 1224–1239. https://doi.org/10.1007/s11947-017-1884-5

Strohalm, J., Novotna, P., Houska, M., Kyhos, K., Vavreinova, S., Gabrovska, D., & Rysova, J. (2000). Influence of high pressure treatment on rheological and other properties of egg white. *High Pressure Research*, *19*(1–6), 137–143. https://doi.org/10.1080/08957950008202547

Su, Y., Dong, Y., Niu, F., Wang, C., Liu, Y., & Yang, Y. (2015). Study on the gel properties and secondary structure of soybean protein isolate/egg white composite gels. *European Food Research and Technology*, *240*(2), 367–378. https://doi.org/10.1007/s00217-014-2336-3

Sun, J., Chang, C., Su, Y., Gu, L., Yang, Y., & Li, J. (2022a). Impact of saccharides on the foam properties of egg white: Correlation between rheological, interfacial properties and foam properties. *Food Hydrocolloids*, *122*, 107088. https://doi.org/10.1016/j.foodhyd.2021.107088

Sun, J., Jing, H., Mu, Y., McClements, D. J., Dong, S., & Xu, B. (2020a). Fabrication of antioxidant emulsifiers from natural ingredients: Conjugation of egg white proteins with catechin and chlorogenic acid. *Food Hydrocolloids*, *108*, 106019. https://doi.org/10.1016/j.foodhyd.2020.106019

Sun, J., Liu, T., Zhang, F., Huang, Y., Zhang, Y., & Xu, B. (2022b). Tea polyphenols on emulsifying and antioxidant properties of egg white protein at acidic and neutral pH conditions. *LWT*, *153*, 112537. https://doi.org/10.1016/j.lwt.2021.112537

Sun, J., Mu, Y., Jing, H., Mohammed, O., Chen, Z., & Xu, B. (2020b). Effects of single- and dual-frequency ultrasound on the functionality of egg white protein. *Journal of Food Engineering*, *277*, 109902. https://doi.org/10.1016/j.jfoodeng.2020.109902

Sun, J., Mu, Y., Mohammed, O., Dong, S., & Xu, B. (2020c). Effects of single-mode microwave heating and dextran conjugation on the structure and functionality of ovalbumin–dextran conjugates. *Food Research International*, *137*, 109468. https://doi.org/10.1016/j.foodres.2020.109468

Sun, Y., & Hayakawa, S. (2002). Heat-induced gels of egg white/ovalbumins from five avian species: Thermal aggregation, molecular forces involved, and rheological properties. *Journal of Agricultural and Food Chemistry*, *50*(6), 1636–1642. https://doi.org/10.1021/jf0109975

Sunwoo, H. H., & Gujral, N. (2015). Chemical composition of eggs and egg products. In P. C. K. Cheung & B. M. Mehta (Eds.), *Handbook of food chemistry* (pp. 331–363). Springer. https://doi.org/10.1007/978-3-642-36605-5_28

Takahashi, M., Handa, A., Yamaguchi, Y., Kodama, R., & Chiba, K. (2016). Anodic oxidative modification of egg white for heat treatment. *Journal of Agricultural and Food Chemistry*, *64*(34), 6503–6507. https://doi.org/10.1021/acs.jafc.6b02785

Talansier, E., Loisel, C., Dellavalle, D., Desrumaux, A., Lechevalier, V., & Legrand, J. (2009). Optimization of dry heat treatment of egg white in relation to foam and interfacial properties. *LWT - Food Science and Technology*, *42*(2), 496–503. https://doi.org/10.1016/j.lwt.2008.09.013

Tan, J., Liu, T., Yao, Y., Wu, N., Du, H., Xu, M., & Tu, Y. (2021). Changes in physicochemical and antioxidant properties of egg white during the Maillard reaction induced by alkali. *LWT*, *143*, 111151. https://doi.org/10.1016/j.lwt.2021.111151

Tan, J., Yao, Y., Wu, N., Du, H., Xu, M., Liao, M., & Tu, Y. (2022). Color, physicochemical characteristics and antioxidant activities of preserved egg white pickled at different temperatures. *LWT*, *164*, 113685. https://doi.org/10.1016/j.lwt.2022.113685

Tang, T., Du, H., Tang, S., Jiang, Y., Tu, Y., Hu, M., & Xu, M. (2021). Effects of incorporating different kinds of peptides on the foaming properties of egg white powder. *Innovative Food Science & Emerging Technologies*, *72*, 102742. https://doi.org/10.1016/j.ifset.2021.102742

Tang, T., Liu, J., Tang, S., Xiao, N., Jiang, Y., Tu, Y., & Xu, M. (2022a). Effects of soy peptides and pH on foaming and physicochemical properties of egg white powder. *LWT*, *153*, 112503. https://doi.org/10.1016/j.lwt.2021.112503

Tang, T., Wu, S., Tang, S., Xiao, N., Wu, L., Tu, Y., & Xu, M. (2022b). Effect of modified egg white powder on the properties of angel cakes. *Journal of Food Engineering*, *326*, 111012. https://doi.org/10.1016/j.jfoodeng.2022.111012

Tian, Y., Jin, H., Guo, S., Lin, S., & Bao, Z. (2022a). Effects of different metal ions on the physicochemical properties and microstructure of egg white gel. *Journal of the Science of Food and Agriculture*. https://doi.org/10.1002/jsfa.11677

Tian, Y., Zhu, H., Zhang, L., & Chen, H. (2022b). Consumer preference for nutritionally fortified eggs and impact of health benefit information. *Foods*, *11*(8), 1145. https://doi.org/10.3390/foods11081145

Totosaus, A., Montejano, J. G., Salazar, J. A., & Guerrero, I. (2002). A review of physical and chemical protein-gel induction. *International Journal of Food Science & Technology*, *37*(6), 589–601. https://doi.org/10.1046/j.1365-2621.2002.00623.x

Van der Plancken, I., Van Loey, A., & Hendrickx, M. E. (2005a). Combined effect of high pressure and temperature on selected properties of egg white proteins. *Innovative Food Science & Emerging Technologies*, *6*(1), 11–20. https://doi.org/10.1016/j.ifset.2004.10.002

Van der Plancken, I., Van Loey, A., & Hendrickx, M. E. (2007a). Foaming properties of egg white proteins affected by heat or high pressure treatment. *Journal of Food Engineering*, *78*(4), 1410–1426. https://doi.org/10.1016/j.jfoodeng.2006.01.013

Van der Plancken, I., Van Loey, A., & Hendrickx, M. E. (2007b). Kinetic study on the combined effect of high pressure and temperature on the physico-chemical properties of egg white proteins. *Journal of Food Engineering*, *78*(1), 206–216. https://doi.org/10.1016/j.jfoodeng.2005.09.018

Van der Plancken, I., Van Loey, A., & Hendrickx, M. E. G. (2005b). Changes in sulfhydryl content of egg white proteins due to heat and pressure treatment. *Journal of Agricultural and Food Chemistry*, *53*(14), 5726–5733. https://doi.org/10.1021/jf050289+

Visschers, R. W., & de Jongh, H. H. J. (2005). Disulphide bond formation in food protein aggregation and gelation. *Biotechnology Advances*, *23*(1), 75–80. https://doi.org/10.1016/j.biotechadv.2004.09.005

Wang, C., Li, J., Li, X., Chang, C., Zhang, M., Gu, L., & Yang, Y. (2019). Emulsifying properties of glycation or glycation-heat modified egg white protein. *Food Research International*, *119*, 227–235. https://doi.org/10.1016/j.foodres.2019.01.047

Wang, C., Li, J., Li, X., Zhang, M., Gu, L., Chang, C., & Yang, Y. (2020a). Molecular forces and gelling properties of heat-induced gel from egg white protein glycated with isomalto-oligosaccharide. *Food Hydrocolloids*, *99*, 105356. https://doi.org/10.1016/j.foodhyd.2019.105356

Wang, G., Troendle, M., Reitmeier, C. A., & Wang, T. (2012). Using modified soy protein to enhance foaming of egg white protein. *Journal of the Science of Food and Agriculture*, *92*(10), 2091–2097. https://doi.org/10.1002/jsfa.5588

Wang, G., & Wang, T. (2009a). Effects of yolk contamination, shearing, and heating on foaming properties of fresh egg white. *Journal of Food Science*, *74*(2), C147–C156. https://doi.org/10.1111/j.1750-3841.2009.01054.x

Wang, G., & Wang, T. (2009b). Improving foaming properties of yolk-contaminated egg albumen by basic soy protein. *Journal of Food Science*, *74*(8), C581–C587. https://doi.org/10.1111/j.1750-3841.2009.01306.x

Wang, J., Chi, Y., Cheng, Y., & Zhao, Y. (2018). Physicochemical properties, in vitro digestibility and antioxidant activity of dry-heated egg white protein. *Food Chemistry*, *246*, 18–25. https://doi.org/10.1016/j.foodchem.2017.10.128

Wang, J., Liu, X., Li, S., Ye, H., Luo, W., Huang, Q., & Geng, F. (2022a). Ovomucin may be the key protein involved in the early formation of egg-white thermal gel. *Food Chemistry*, *366*, 130596. https://doi.org/10.1016/j.foodchem.2021.130596

Wang, T., Chang, C., Gu, L., Su, Y., Zhang, M., Yang, Y., & Li, J. (2022b). Comparison of the functionality of egg white liquid with different desugaring treatments. *International Journal of Food Science & Technology*. https://doi.org/10.1111/ijfs.15872

Wang, X., Gu, L., Su, Y., Li, J., Yang, Y., & Chang, C. (2020b). Microwave technology as a new strategy to induce structural transition and foaming properties improvement of egg white powder. *Food Hydrocolloids*, *101*, 105530. https://doi.org/10.1016/j.foodhyd.2019.105530

Wang, Y., Tao, H., Wang, K., Li, M., Wen, X., Fu, R., & Ni, Y. (2021). Nanocapsules formed by interactions between chondroitin sulfate and egg white protein for encapsulating hydrophilic ingredients. *Green Chemistry*. https://doi.org/10.1039/D1GC01706C

Weijers, M., Barneveld, P. A., Cohen Stuart, M. A., & Visschers, R. W. (2003). Heat-induced denaturation and aggregation of ovalbumin at neutral pH described by irreversible first-order kinetics. *Protein Science*, *12*(12), 2693–2703. https://doi.org/10.1110/ps.03242803

Weijers, M., de Hoog, E. H. A., Cohen Stuart, M. A., Visschers, R. W., & Barneveld, P. A. (2005). Heat-induced formation of ordered structures of ovalbumin at low ionic strength studied by small angle X-ray scattering. *Colloids and Surfaces A: Physicochemical and Engineering Aspects*, *270–271*, 301–308. https://doi.org/10.1016/j.colsurfa.2005.07.001

Weijers, M., van de Velde, F., Stijnman, A., van de Pijpekamp, A., & Visschers, R. W. (2006). Structure and rheological properties of acid-induced egg white protein gels. *Food Hydrocolloids*, *20*(2), 146–159. https://doi.org/10.1016/j.foodhyd.2005.02.013

Weijers, M., Visschers, R. W., & Nicolai, T. (2002). Light scattering study of heat-induced aggregation and gelation of ovalbumin. *Macromolecules*, *35*(12), 4753–4762. https://doi.org/10.1021/ma0120198

Weijers, M., Visschers, R. W., & Nicolai, T. (2004). Influence of the ionic strength on the structure of heat-set globular protein gels at pH 7. Ovalbumin. *Macromolecules*, *37*(23), 8709–8714. https://doi.org/10.1021/ma048918o

Wen, H., Li, Z., Li, Y., Hao, Y., Du, Z., Liu, X., & Zhang, T. (2022). Aggregation of egg white peptides (EWP) induced by proanthocyanidins: A promising fabrication strategy for EWP emulsion. *Food Chemistry*, *400*, 134019. https://doi.org/10.1016/j.foodchem.2022.134019

Wierenga, P. A., & Gruppen, H. (2010). New views on foams from protein solutions. *Current Opinion in Colloid & Interface Science*, *15*(5), 365–373. https://doi.org/10.1016/j.cocis.2010.05.017

Wouters, A. G. B., Rombouts, I., Fierens, E., Brijs, K., Blecker, C., Delcour, J. A., & Murray, B. S. (2018). Foaming and air-water interfacial characteristics of solutions containing both gluten hydrolysate and egg white protein. *Food Hydrocolloids*, *77*, 176–186. https://doi.org/10.1016/j.foodhyd.2017.09.033

Wu, Y., Zhang, Y., Duan, W., Wang, Q., An, F., Luo, P., & Huang, Q. (2022). Ball-milling is an effective pretreatment of glycosylation modified the foaming and gel properties of egg white protein. *Journal of Food Engineering*, *319*, 110908. https://doi.org/10.1016/j.jfoodeng.2021.110908

Xiao, N., Huang, X., He, W., Yao, Y., Wu, N., Xu, M., & Tu, Y. (2021). A review on recent advances of egg byproducts: Preparation, functional properties, biological activities and food applications. *Food Research International*, *147*, 110563. https://doi.org/10.1016/j.foodres.2021.110563

Xiao, N., Zhao, Y., Yao, Y., Wu, N., Xu, M., Du, H., & Tu, Y. (2020). Biological activities of egg yolk lipids: A review. *Journal of Agricultural and Food Chemistry*, *68*(7), 1948–1957. https://doi.org/10.1021/acs.jafc.9b06616

Xiong, W., Wang, Y., Zhang, C., Wan, J., Shah, B. R., Pei, Y., & Li, B. (2016a). High intensity ultrasound modified ovalbumin: Structure, interface and gelation properties. *Ultrasonics Sonochemistry*, *31*, 302–309. https://doi.org/10.1016/j.ultsonch.2016.01.014

Xiong, Z., Zhang, M., & Ma, M. (2016b). Emulsifying properties of ovalbumin: Improvement and mechanism by phosphorylation in the presence of sodium tripolyphosphate. *Food Hydrocolloids*, *60*, 29–37. https://doi.org/10.1016/j.foodhyd.2016.03.007

Xue, H., Tu, Y., Zhang, G., Xin, X., Hu, H., Qiu, W., & Zhao, Y. (2021). Mechanism of ultrasound and tea polyphenol assisted ultrasound modification of egg white protein gel. *Ultrasonics Sonochemistry*, *81*, 105857. https://doi.org/10.1016/j.ultsonch.2021.105857

Xue, H., Xu, M., Zhang, G., Wang, P., Yu, L., Zhao, Y., & Zhao, Y. (2022a). Study on the mechanism of enhanced gel strength of heat-induced egg white by shikimic acid braising. *Poultry Science*, *101*(5), 101774. https://doi.org/10.1016/j.psj.2022.101774

Xue, H., Zhang, G., Han, T., Li, R., Liu, H., Gao, B., & Zhao, Y. (2022b). Improvement of gel properties and digestibility of the water-soluble polymer of tea polyphenol-egg white under thermal treatment. *Food Chemistry*, *372*, 131319. https://doi.org/10.1016/j.foodchem.2021.131319

Yamashita, H., Ishibashi, J., Hong, Y. H., & Hirose, M. (1998). Involvement of ovotransferrin in the thermally induced gelation of egg white at around 65°C. *Bioscience, Biotechnology, and Biochemistry*, *62*(3), 593–595. https://doi.org/10.1271/bbb.62.593

Yang, M., Liu, J., Yang, X., Li, S., Li, C., Liu, B., & Yu, Y. (2021). Effect of glycation degree on the in vitro simulated gastrointestinal digestion: A promising formulation for egg white gel with controlled digestibility. *Food Chemistry*, *349*, 129096. https://doi.org/10.1016/j.foodchem.2021.129096

Yang, S. C., & Baldwin, R. E. (1995). Functional properties of eggs in foods. In W. J. Stadelman & O. J. Cotterill (Eds.), *Egg science and technology* (pp. 405–463). CRC Press. https://doi.org/10.1201/9780203758878

Yang, X., & Foegeding, E. A. (2011). The stability and physical properties of egg white and whey protein foams explained based on microstructure and interfacial properties. *Food Hydrocolloids*, *25*(7), 1687–1701. https://doi.org/10.1016/j.foodhyd.2011.03.008

Yang, Y., Zhao, Y., Tu, Y., Huang, X., Li, J., Luo, X., & Wang, J. (2012). Change of alkalinity, pH and texture properties during pidan pickling. *Science and Technology of Food Industry*, *33*, 111–114.

Yao, L., Jiang, A., & Chen, L. (2020). Characterization of ethanol-induced egg white gel and transportation of active nutraceuticals. *LWT*, *130*, 109530. https://doi.org/10.1016/j.lwt.2020.109530

Yao, L., Zhou, W., Wang, T., Liu, M., & Yu, C. (2014). Quantification of egg yolk contamination in egg white using UV/Vis spectroscopy: Prediction model development and analysis. *Food Control*, *43*, 88–97. https://doi.org/10.1016/j.foodcont.2014.02.037

Yao, X., McClements, D. J., Su, Y., Li, J., Chang, C., Wang, J., & Gu, L. (2022). Fabrication, structural and emulsifying properties of egg white protein-dextran conjugates through Maillard reaction. *Food Biophysics*. https://doi.org/10.1007/s11483-022-09745-8

Yogesh, K. (2016). Pulsed electric field processing of egg products: A review. *Journal of Food Science and Technology*, *53*(2), 934–945. https://doi.org/10.1007/s13197-015-2061-3

Yu, L., Xiong, C., Li, J., Luo, W., Xue, H., Li, R., & Zhao, Y. (2020). Ethanol induced the gelation behavior of duck egg whites. *Food Hydrocolloids*, *105*, 105765. https://doi.org/10.1016/j.foodhyd.2020.105765

Yu, L., Xue, H., Xiong, C., Xin, X., Wang, P., Feng, F., & Zhao, Y. (2021a). Characterization of duck egg white gel under the action of baijiu (Chinese liquor). *LWT*, *147*, 111487. https://doi.org/10.1016/j.lwt.2021.111487

Yu, Y., Guan, Y., Liu, J., Hedi, W., Yu, Y., & Zhang, T. (2021b). Molecular structural modification of egg white protein by pH-shifting for improving emulsifying capacity and stability. *Food Hydrocolloids*, *121*, 107071. https://doi.org/10.1016/j.foodhyd.2021.107071

Yu, Y., Guan, Y., Wen, H., Zhang, Y., Liu, J., & Zhang, T. (2021c). Mild heating assisted alkaline pH shifting modify the egg white protein: The mechanism and the enhancement of emulsifying properties. *LWT*, *151*, 112094. https://doi.org/10.1016/j.lwt.2021.112094

Zayas, J. F. (1997). *Functionality of proteins in food*. Springer. https://doi.org/10.1007/978-3-642-59116-7_7

Zhan, F., Youssef, M., Shah, B. R., Li, J., & Li, B. (2022). Overview of foam system: Natural material-based foam, stabilization, characterization, and applications. *Food Hydrocolloids*, *125*, 107435. https://doi.org/10.1016/j.foodhyd.2021.107435

Zhang, H., Yang, L., Tu, Y., Wu, N., Jiang, Y., & Xu, M. (2019). Changes in texture and molecular forces of heated-induced egg white gel with adding xanthan gum. *Journal of Food Process Engineering*, *42*(4), e13071. https://doi.org/10.1111/jfpe.13071

Zhang, Z., Li, Y., Lee, M., Ravanfar, R., Padilla-Zakour, O., & Abbaspourrad, A. (2020). The impact of high-pressure processing on the structure and sensory properties of egg white-whey protein mixture at acidic conditions. *Food and Bioprocess Technology*, *13*. https://doi.org/10.1007/s11947-019-02397-6

Zhao, Q., Ding, L., Xia, M., Huang, X., Isobe, K., Handa, A., & Cai, Z. (2021). Role of lysozyme on liquid egg white foaming properties: Interface behavior, physicochemical characteristics and protein structure. *Food Hydrocolloids*, *120*, 106876. https://doi.org/10.1016/j.foodhyd.2021.106876

Zhao, Y., Cao, D., Shao, Y., Xiong, C., Li, J., & Tu, Y. (2020). Changes in physico-chemical properties, microstructures, molecular forces and gastric digestive properties of preserved egg white during pickling with the regulation of different metal compounds. *Food Hydrocolloids*, *98*, 105281. https://doi.org/10.1016/j.foodhyd.2019.105281

Zhao, Y., Chen, Z., Li, J., Xu, M., Shao, Y., & Tu, Y. (2016). Formation mechanism of ovalbumin gel induced by alkali. *Food Hydrocolloids*, *61*, 390–398. https://doi.org/10.1016/j.foodhyd.2016.04.041

Zhao, Y., Tu, Y., Li, J., Xu, M., Yang, Y., Nie, X., & Du, H. (2014). Effects of alkaline concentration, temperature, and additives on the strength of alkaline-induced egg white gel. *Poultry Science*, *93*(10), 2628–2635. https://doi.org/10.3382/ps.2013-03596

Zheng, L., Regenstein, J. M., Zhou, L., & Wang, Z. (2022a). Soy protein isolates: A review of their composition, aggregation, and gelation. *Comprehensive Reviews in Food Science and Food Safety*, 1–18. https://doi.org/10.1111/1541-4337.12925

Zheng, Y., Chang, Y., Luo, B., Teng, H., & Chen, L. (2022b). Molecular structure modification of ovalbumin through controlled glycosylation with dextran for its emulsibility improvement. *International Journal of Biological Macromolecules*, *194*, 1–8. https://doi.org/10.1016/j.ijbiomac.2021.11.130

Zhu, P., Bian, X., Yang, Y., Fu, Y., Wu, N., Yu, D., & Na, Z. (2021a). Preparation, processing characteristics, and stability analysis of molten globule protein. *ES Food & Agroforestry*, *5*, 4–13. https://doi.org/10.30919/esfaf511

Zhu, P., Yang, Y., Zou, L., Gao, J., Wang, B., Bian, X., & Zhang, N. (2021b). The effect of trehalose on the thermodynamic stability and emulsification of soybean 11S globulin in the molten globule state. *Food Hydrocolloids*, *118*, 106811. https://doi.org/10.1016/j.foodhyd.2021.106811

9

Egg Yolk Lipid and Protein Functionality

Yujie Chi* **and Zihong Ma**

CONTENTS

9.1 Introduction

Egg yolks are used in a variety of food products because of their color, flavor, and remarkable emulsifying and gelation qualities. Egg yolks are highly nutritious and contain easily digestible proteins, among other nutrients (Seuss-Baum & Nau, 2011). They stabilize emulsions by lowering the interfacial tension between oil and water (Anton et al., 2016) and add richness to baked products due to their high fat content and act as thickeners when heated, during which the proteins denature and form gels. Due to their outstanding nutritional value and techno-functionalities, including strong emulsifying activity and gelation characteristics, egg yolks have been a prominent ingredient in a range of food formulations (Li et al., 2021b). There is simply no replacement in formulated foods that can match the superior functional attributes of egg yolks (Laca et al., 2014).

Egg yolks can be fractionated into plasma and granules by mild centrifugation (10 000 × *g*, 30 min) (Mc Bee & Cotterill, 1979) (Figure 9.1). The plasma (supernatant) mainly consists of 85% low density lipoprotein (LDL) and 15% yolk globin (livetins); the granule (precipitated fraction) contains 70% high density lipoprotein (HDL), 16% yolk high phosphorus protein (phosvitin), and 12% LDL (Anton, 2013). The major functions of yolk fractions are listed in Table 9.1.

It is generally believed that yolk plasma contributes more to yolk functionality than yolk granules. Nearly 90% of yolk lipids are concentrated in the plasma, which is a rich source of lecithin. Egg yolk lecithin has many important physiological functions as an antioxidant, antibacterial, and anti-inflammatory agent, among others (Xiao et al., 2020); it also has excellent emulsification properties and is widely used in the food and cosmetic industry owing to its moisturizing effect (Alhajj et al., 2020). Lecithin is also incorporated in in situ drug-delivery to promote the bioavailability of drugs and other substances (Shen et al., 2020). Yolk plasma can also be used as a sperm cryoprotectant because the LDL in the yolk can adhere to and protect the sperm-cell membrane (Bajuk et al., 2018; Mehdipour et al., 2018).

Yolk granules have great potential in the development of low cholesterol mayonnaise and novel encapsulating agents (Puertas & Vázquez, 2019; Marcet et al., 2022), but their insoluble nature at low ionic strength limits further application. Nevertheless, there is a continuing need to optimize the functional properties of egg yolks, which requires a proper understanding of the current paradigm of egg yolk functionalities.

9.2 Egg Yolk Emulsion

Food emulsions are two-phase systems containing a continuous water phase and a disperse phase (Ferrari et al., 2022). Emulsifiers promote the formation of emulsions and improve their stability; furthermore, emulsifying activity and stability are common indicators used to characterize emulsifier performance (Derderian et al., 1979). Emulsifying stability refers

* Corresponding author.

DOI: 10.1201/9781003254430-12

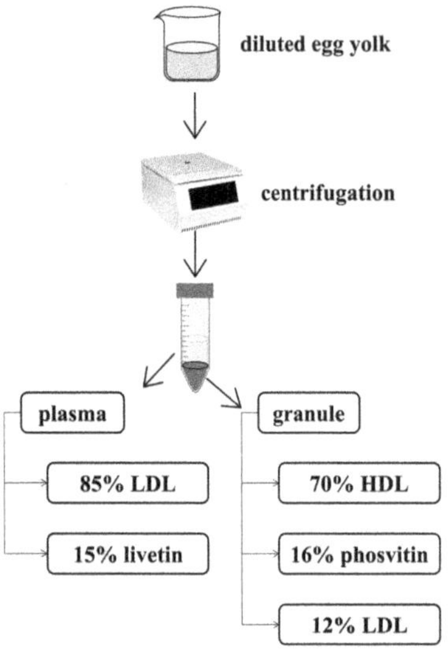

FIGURE 9.1 Composition of hen egg yolk (LDL, low-density lipoprotein; HDL, high-density lipoprotein).

to the ability of the emulsifying system to maintain stability under certain conditions, and emulsifying activity refers to the maximum expansion of the emulsifier surface area per unit mass in a specific emulsifying system (Tian et al., 2021). The abundant protein content of egg yolks imparts several functional properties. As a natural complex protein lipid emulsifier system, it has excellent emulsifying properties. Therefore, egg yolk is commonly used as a natural food emulsifier in the food industry.

9.2.1 Mechanism and Components of Egg Yolk Emulsion Formation

Emulsifiers with good emulsifying properties need to meet the following three conditions: (i) rapid interfacial adsorption; (ii) rapid expansion and reorientation on the interface; and (iii) interaction with adjacent molecules to form a film with strong adhesion and viscoelasticity, which can withstand heating and mechanical shear (Yang & Baldwin, 1995).

Egg yolk is a complex mixture, and different components have their own emulsifying behaviors. Egg yolk livetins exhibit poor adsorption at the oil–water interface (Le Denmat et al., 2000), while LDL is the most important determinant of emulsifying capacity (Speroni et al., 2005). Emulsions made from egg yolk show comparable properties to those made from egg yolk plasma, but not those made of yolk granules (Dyer-Hurdon & Nnanna, 1993; Anton & Gandemer, 1997). The excellent emulsifying property of LDL is related to the unique combination of proteins and lipids (Jolivet et al., 2006): The removal of lipids or the reduction of soluble protein is known to reduce emulsifying ability (Mizutani & Nakamura, 1987; Paraskevopoulou & Kiosseoglou, 1994). LDLs are disrupted at the interface to release neutral lipids from the lipoprotein core, enabling their spread. Each lipid class on the LDL film seems to behave independently at the interface (Martinet et al., 2003). Le Denmat et al. (2000) deduced that steric repulsion

TABLE 9.1

The Major Functions of Yolk Fractions

Fraction	Function	Application
Plasma (LDL)	Emulsifying properties	Emulsifying agent (Ali & Wu, 2021; Patil et al., 2022)
	Emulsion stabilization	Encapsulation and delivery (Li et al., 2022d; Ye et al., 2022)
	Gelation	Thermal yolk gel (Anton, 2013; Blume et al., 2015); salted egg (Xu et al., 2019a); preserved egg (Yang et al., 2019); 3D printing (Charbonneau et al., 2019; Anukiruthika et al., 2020; Xu et al., 2020)
	Cell membrane protection	Cryoprotection of spermatozoa (Bencharif & Dordas-Perpinya, 2020; da Silva et al., 2022)
	Film-forming properties	Preservation and packaging of food (Fuertes et al., 2017)
	Lecithin provider	Emulsifying agent (Yu et al., 2022a); antioxidant, antibacterial, and anti-inflammatory agent (Xiao et al., 2020; Abeyrathne et al., 2022); cosmetics production (Zhang et al., 2021)
Livetin (α, β, γ)	Immune response properties	Antibody production (Leiva et al., 2020; Jiang et al., 2022)
	Encapsulate and retain air	Foaming agent (Chalamaiah et al., 2017)
	Emulsifying properties	Emulsifying agent (Chalamaiah et al., 2017)
Granule	Emulsifying properties	Emulsifying agent (Li et al., 2020)
	Emulsion stabilization	Encapsulation and delivery (Sassi et al., 2020; Li et al., 2021a; Li et al., 2022a)
	Encapsulate and retain air	Foaming agent (Marcet et al., 2014)
	Film-forming properties	Preservation and packaging of food (Fuertes et al., 2017)
	Folate provider	Nutrition enhancer (Naderi et al., 2017)
Phosvitin	Emulsifying properties	Emulsifying agent (Zhang et al., 2019)
	Metal-binding properties	Metal chelator (Jung et al., 2012)
	Antimicrobial properties	Immune enhancer (Kim et al., 2020)
	Antioxidant properties	Antioxidant (Benedé & Molina, 2020)

is the main interaction opposing to droplet aggregation in food emulsions made with yolk and a driving contribution of the proteinaceous part of plasma, namely, apo-LDL, was hypothesized.

Lecithin is a nutritional surfactant with outstanding hypolipidemic, emulsifying, diffusing, and infiltrating characteristics (Blesso, 2015; Asomaning & Curtis, 2017). Highly pure lecithin, on the other hand, lacks the emulsifying characteristics of unpurified phosphatides, implying that the most effective yolk emulsifiers are combinations or complexes (Yeadon et al., 1958).

The contribution of egg yolk granules to egg yolk emulsification is relatively small. Its protein solubility is low with poor surface activity and cannot be effectively adsorbed when competing with soluble proteins. When the physical and chemical conditions allow the particles to dissolve, the dissolved HDL can provide additional emulsification, which is conducive to greater interface area coverage (Causeret et al., 1991). Phosvitins account for 16% of egg yolk granules and are often applied as stabilizers (Takeuchi et al., 2022). They are capable of easily capturing Ca^{2+} and Fe^{2+} ions and possess good preservative, antioxidant, and emulsifying properties (Marcet et al., 2022). The triblock-like structure of phosvitin (Fleer et al., 1998), with two relatively hydrophobic outer blocks and a large hydrophilic charged block, drives the surface activity of phosvitins at the oil–water interface. Despite its multiple negative charges, phosvitin can be anchored at the air–water interface. Despite the limited solubility of granules, the associated lipovitellins make up the majority of egg yolk surface components. Granule lipovitellins are thought to be hydrophobic proteins that can interact with oil surfaces in an insoluble and complexed form (Mine, 1998).

9.2.2 Factors Affecting the Emulsifying Properties of Egg Yolks and Their Components

Heat: Heat treatment is essential in the processing of egg yolks and their products. Pasteurization is used to promote the microbiological safety of liquid egg yolks and usually entails a heat treatment at temperatures ranging from 60 to 68°C for periods varying from a few seconds to approximately 10 min, depending on the temperature (Guilmineau & Kulozik, 2006a). However, due to the heat-sensitive nature of egg yolk itself, high-temperature heat treatments will degrade the functional characteristics of egg proteins, resulting in economically unappealing end products (Liu et al., 2018). Therefore, great efforts have been made to improve the heat stability of egg yolks for modified heat resistance and preserved emulsifying properties.

Heat treatment at 69°C has no effect on the emulsifying characteristics of egg yolk and plasma (Le Denmat et al., 1999); however, emulsifying activity diminishes at temperatures above 72°C, regardless of the granules maintaining their stability. Similarly for heated LDL, the emulsifying capacity and stability of LDL decrease above 75°C (Tsutsui, 1988). The emulsion stability of mayonnaise prepared from thermally processed egg yolk was significantly better than that prepared from unheated egg yolk; the emulsion stability of mayonnaise was high at 62°C to 68°C, and decreased above 69°C (Ho et al., 2022). Le Denmat et al. (1999) found that aggregated egg yolk proteins could not adsorb at the oil–water interface, thus reducing their emulsifying performance. However, the addition of salt and sugar in the heating process promoted their adsorption (Campbell et al., 2005). Guilmineau and Kulozik (2006a) freported that with increased protein denaturation, the tension at the oil–water interface diminishes, suggesting that proper heat treatment can improve the emulsifying behaviors of egg yolk (Thamjedsada et al., 2018).

pH and ionic strength: The use of egg yolk encompasses a wide pH and ionic strength range, depending on the product composition (Guilmineau & Kulozik, 2006b). Egg yolk proteins aggregate into larger emulsion particles at pH 3.0 (Mine, 1998), and the mean droplet size of the emulsions decreases as the pH increases. Egg yolk particles have various sizes under different pH and ionic strength values, and particle size influences the emulsifying stability index of egg yolk dispersions (Li et al., 2018a).

Yolk granule solubility approaches 80% at an ionic strength >0.3 mol L^{-1} NaCl (Cook & Martin, 1969; Anton & Gandemer, 1997). When the ionic strength exceeds 1.71 mol L^{-1} NaCl, the granules are completely disrupted (Anton, 2013). Granules can form stable dispersions at pH below 4.0 or above 7.0, but phase separation is found at pH 4.0–7.0: By disrupting calcium bridges, NaCl (0.3 M and 0.5 M) is able to dissociate the granules at pH 4.0–7.0 (Li et al., 2022b). Emulsions stabilized with yolk granules are associated with a higher emulsifying stability at an acidic pH higher than 2.0. Excellent emulsifying stability can be attained at pH 4.0 (Wang et al., 2020). Notably, although the emulsification properties of granules are improved when they are at least partially solubilized, the emulsifying properties of the native granules have been described as similar to those of whole egg yolk when they are used to prepare mayonnaise, which explains the good performance of the granules at low pH of this emulsion, usually around pH 4.0 (McWhorter et al., 2021).

In addition, Mi et al. (2022) reported that granules are able to form stable high internal phase emulsions at NaCl ionic strengths over 0.2 M and at pH over 5.0, with an NaCl ionic strength of 0.3 M. The granules easily adsorb and cover the oil–water interface to form emulsion droplets with small particle sizes.

Alkaline conditions or high ionic strength were able to disrupt the calcium-phosphate bridges, thus increasing the solubility of egg yolk HDL under certain conditions; the emulsifying ability of extracted HDL shows a strong correlation with pH and HDL concentration (Zhang et al., 2022b). In addition, adding NaCl to phosvitin lowers the emulsifying ability between pH 3.0 and 10.0 and decreases the emulsifying stability between pH 3.0 and 9.0. At 0.5 M NaCl, increased emulsion instability results in oil droplet coalescence (Chung & Ferrier, 1992).

Dilution and dehydration: Varadarajulu and Cunningham (1972) discovered that lower emulsion stability is likely due to a decrease in egg yolk viscosity caused by dilution, although the decrease in solid content is most likely the major cause. However, during the industrial separation of egg white and yolk, a small amount of egg white is mixed with the yolk. Therefore, dilution of yolk with egg white should be maintained below 20% to maintain emulsifying performance.

Dehydration is a routine practice in the production of egg yolk powders. According to Schultz et al. (1966), the extractability of "free lipids" increases rapidly during drying, which is deleterious to the yolk's emulsifying capacity. Different drying methods exert different impacts on the emulsifying properties of egg yolk lecithin. Egg yolk lecithin prepared through vacuum freeze-drying presents the highest emulsifying capacity, followed by samples processed through vacuum drying, hot-air drying, spray drying, and ultrasonic vacuum drying (Bai et al., 2020).

Freeze–thaw: Liquid egg yolks can be frozen to retain freshness, protect heat-sensitive nutrients, and increase shelf life. The effect of freezing and thawing on the emulsifying ability of egg yolk remains controversial. It is generally believed that gelation of egg yolk induced by freezing reduces emulsifying performance (Wakamatsu et al., 2018). However, frozen yolk was proven to be a more effective emulsifier in mayonnaise than fresh yolk, while it was also shown that mayonnaise required more frozen yolk than fresh yolk (Miller & Winter, 1951). Although the qualities of frozen yolk mayonnaise differ from those produced with fresh yolk, its consumer acceptability is unaffected (Huang et al., 2016).

A significant increase in emulsion stability was observed in magnetic-assisted frozen egg yolk, which could be related to an increase in free SH groups, leading to the formation of SS bonds or SS–SH exchanges (Fernández-Martín et al., 2018). Freeze–thaw treatments could also improve the emulsifying and biological activities of egg phosvitin (Li et al., 2018b).

Saccharide: Polysaccharides can act as thickeners or stabilizers in the emulsifying system to improve stability or modify protein structure and improve functionality through protein–polysaccharide interactions (Felix et al., 2017). Emulsion gels prepared with egg yolk and alginate complexes showed rheological properties similar to those of a full-fat mayonnaise product; these emulsion gels are also more tolerant to heating and freezing treatment (Yang et al., 2020a). The addition of cationic guar gum and oleic acid can also improve the stability of plasma emulsions prepared with enzymatically hydrolyzed egg yolk (Zou et al., 2020). Chitosan (2 g kg^{-1}[w/w]) is also reported to improve the emulsifying stability of egg yolk hydrolysates, and hydrolyzed egg yolk-prepared composite emulsions possess better thermal resistance (Chang et al., 2021). In fact, sugars have been applied to improve heat stability, thus improving the emulsifying properties of heated egg yolk. Furthermore, LDL-pectin complexes were also used to stabilize high internal phase Pickering emulsions, which exhibit superior emulsifying and stabilization capability than LDL alone, and pectin plays critical roles in the formation of stable gel-like structures in the prepared emulsions (Li et al., 2023).

Emulsifiers: The addition of 0.05% sodium stearoyl-2-lactylate greatly boosts egg yolk emulsifying capacity, and the reverse is observed with the addition of lecithin (Varadarajulu & Cunningham, 1972). However, an increase in the emulsion stability of yolk granules was observed with the addition of soybean lecithin. The appropriate addition of lecithin (≤0.25%) could be helpful in the formation of highly stable emulsions with low particle size by further dissociating the aggregate structure and slightly reducing the contact angle while increasing the surface net charge (Shen et al., 2020).

Pressure: Pressure can affect the molecular structure of a protein, thereby changing its functional characteristics. The degree of change at the molecular level depends on the structure of the protein and the conditions of the reaction environment, including pH and protein concentration. The solubility and emulsifying activity of egg yolk decreased gradually with high-pressure treatment at 100–500 MPa for 10 min; however, at 100 MPa, the emulsifying stability was slightly higher than that of fresh egg yolk, and at 200 MPa, the emulsifying stability was significantly reduced (Yan et al., 2010). This is mainly due to the gradual denaturation of protein after high-pressure treatment, which inhibits the expansion and adsorption capacity at the interface, resulting in the reduction of emulsifying activity (Yan et al., 2010). Microfluidization pressure can increase viscosity, decrease particle size, and enhance the emulsifying characteristics of egg yolk, demonstrating that steady emulsion formation may be achieved using microfluidized egg yolk (Suhag et al., 2021). Regarding yolk LDL, it seems that high-pressure treatment combined with an alkaline pH decreases droplet flocculation of LDL dispersions (Speroni et al., 2005).

Ultrasound: To increase the functional qualities of egg yolk proteins, high-intensity ultrasound might be an effective and environmentally friendly method. The egg yolk solution shows improved emulsifying, foaming, and gel characteristics after high-intensity ultrasound treatment, but with reduced foam stability. High-intensity ultrasound reduces the average particle size, enhances the absolute zeta potential value, and modifies the aggregation of yolk LDL and granules (Xie et al., 2020b), while their molecular interface becomes more flexible and stretched, improving the zeta potential and surface hydrophobicity of LDL, as well as its solubility and emulsifying characteristics (Xie et al., 2020a). The depolymerization of yolk granules induced by high-intensity ultrasound is thought to slightly improve their emulsifying activity index (Geng et al., 2021).

Ball-milling: Ball-milling treatment can modify the surface, electrical, and thermal properties of phosvitin, which enhances its emulsifying ability but decreases emulsifying stability (Zhang et al., 2019).

Enzymatic hydrolysis: Proteinase improves the solubility of proteins. Indeed, modulated enzymatic hydrolysis has been reported to improve the emulsifying properties and heat stability of egg yolk as well as its components (Bao et al., 2017; Gao et al., 2019; Tang et al., 2019; Fu et al., 2020; Li et al., 2020). The emulsion stability index of subtilisin-treated granules increased nearly 8-fold compared with untreated granules (Li et al., 2020). Hydrolyzed HDL with more soluble peptide fragments and exposed ionizable residues also showed better emulsifying properties, which correlated with hydrolysis time (Liu et al., 2022). Freeze–thawed egg yolk pellets, the by-products of IgY extraction, have poor functionalities; however, after limited hydrolysis, their solubility and emulsifying capacity are remarkably increased (Su et al., 2022). Enzymatic hydrolysis can also be combined with other novel techniques to achieve better modification. The emulsification activity and capacity of egg yolk processed with both enzymatic hydrolysis and high-intensity ultrasound were greater than those of yolk treated only by hydrolysis (Yu et al., 2022b).

Notably, hydrolysis of the lipid fraction of egg yolk improves emulsifying properties. After an enzymatic treatment with phospholipase A_1 and A_2, the solubility and emulsifying capacity of egg yolk increased, since phospholipids in egg yolk lipoproteins are changed into lysophospholipids, which are more hydrophilic due to the elimination of fatty acids (Li et al., 2021b). The interfacial tension of the phospholipase A_2-modified egg yolk solution is significantly higher, whereas the surface dilatational modulus is much lower, and the kinetics of interfacial adsorption are faster (Gazolu-Rusanova et al., 2020). Furthermore, phospholipase A_2-treated egg yolk plasma and granules show better surface activity even after being spray-dried (Strixner et al., 2013). Emulsions prepared with phospholipase A_2-treated egg yolk and granules had better freeze–thaw stability, while the stability of a plasma-stabilized emulsion decreased (Gmach et al., 2022).

Fermentation: Fermentation by lactic acid bacteria is widely used in the food industry to modify biologically active molecules, such as proteins and lipids (Pescuma et al., 2008). The emulsifying activity of egg yolk increased with fermentation, and emulsifying stability peaked after 3 h of incubation, which could be attributed to a reduction in free SH groups and surface hydrophobicity of egg yolk proteins (Tian et al., 2021). Fermentation of yolk phosvitin degrades the protein into short peptides and substantially improves surface hydrophobicity, thermal stability, and viscosity (Tian et al., 2022).

Oxidation: During egg yolk processing and storage, some lipid peroxidation products might be produced. Moderate oxidizing conditions can enhance the emulsification of yolk LDL, while high-intensity oxidation decreases emulsification (Bao et al., 2019). Lipid oxidation is also responsible for the development of unpleasant flavors and odors. Its effective suppression requires the addition of certain antioxidants, such as 0.1 g/100 g vanillin or thymol; thymol at 0.1 g/100 g also increases emulsion activity (0.63 + 0.01 Abs) (Gouda et al., 2017). Moderate ozonation treatment ($\leq$ 20 min) can expose the interior hydrophobic groups of egg yolk proteins and increase protein flexibility, which greatly improves the emulsifying properties of egg yolk (Li et al., 2022c).

Grafting: Chen et al. (2014) reported that conjugation of phosvitin with dextran improved the emulsion stability in acidic pH conditions. Improved emulsifying properties of phosvitin were also observed after grafting with pectin, which is due to increased surface hydrophobicity (Cui et al., 2019). Yang et al. (2021) prepared Na_2SO_3 treated wheat gluten, which was subsequently cross-linked with phosvitin via transglutaminase catalysis. A higher surface hydrophobicity, thermal stability, and a lower zeta potential and apparent viscosity were observed, thus affecting its emulsifying properties. The mixture has better emulsifying activity than individual phosvitin. Ren et al. (2020) reported that both phosphorylation and succinylation change the structure of egg yolk proteins, and improve the emulsion stability of egg yolk.

Extraction: With the rapid development of the egg processing industry, the functional and bioactive components of egg yolk are generally extracted to achieve efficient utilization (Huang & Ahn, 2019; Su et al., 2020). Some health risk factors, such as cholesterol, can also be removed to meet the needs of healthier diets, although the dietary cholesterol restriction of 300 mg/d has been removed from dietary guidelines in several countries (Zhu et al., 2018; Puertas & Vázquez, 2019; Zhong et al., 2019).

The extraction of egg yolk lipids requires an organic solvent, which could diminish the functional properties of the egg yolk, including protein solubility, emulsifying activity, and emulsifying stability (Chung & Ferrier, 1991; Paraskevopoulou & Kiosseoglou, 1994). Supercritical CO_2 treatment is another way of extracting egg yolk lipids. The solubility and emulsifying activity of supercritical CO_2-treated egg yolk powder is slightly lower than that of untreated egg yolk powder, but its emulsifying stability is significantly improved (Su et al., 2020). Partial defatting is conducive to maintaining the solubility of some protein subunits and avoiding more severe protein denaturation and sulfhydryl oxidation (Zhang et al., 2022a). Enzymatic hydrolysis is often used to improve the functional properties of defatted egg yolk (Bao et al., 2017; Zhang et al., 2022a). Enzymatic hydrolysis can also be performed before lipid extraction to preserve the functional properties of egg yolk proteins (Long et al., 2019).

The effect of egg yolk protein extraction is relatively mild. After phosvitin extraction, granules show better foaming and emulsifying properties than untreated granules because phosvitin removal results in the disruption of granule structure (Chalamaiah et al., 2018).

9.3 Egg Yolk Gelation

Gelation is an essential functional aspect of egg yolk, as it determines the rheological and textural qualities of popular foods, such as cakes, egg-based creams, omelets, and confectionaries, as well as their consumer acceptance. The molecular foundation of gel formation is the high concentration and diversity of proteins and lipids in the egg yolk. Heat, cold, salt, alkali, and other treatments alter the structure of proteins, resulting in protein–protein and protein–lipid interactions and gel formation (Zhao et al., 2021).

9.3.1 Mechanism of Egg Yolk Gel Formation

Egg yolk is abundant in proteins, which are essential to the formation of a gel network. Yolk gelation can be regarded as the destabilization of yolk particles caused by the denaturation of their surface-stabilizing protein molecules, which leads to attractive particle interactions and, eventually, the creation of interparticle networks, which is closely related to yolk LDL (Kiosseoglou, 2003a). Anton et al. (2001) showed that the yolk gelation process is driven by LDL apolipoprotein molecular interactions since yolk plasma and liquid yolk follow similar gelation patterns, whereas granules are less efficient in gel network formation. The difference in gelation behavior of yolk LDL and granules is attributed to the lower heat sensitivity of granular proteins owing to their globular form, as opposed to LDL apolipoproteins, which denature at a lower temperature (Le Denmat et al., 1999). In addition, the formation of egg yolk gel is influenced by many external factors, such as heating time, temperature, pH, ionic strength, and protein species. With variation of these external conditions, egg yolk gels can be produced with different textures (hardness, elasticity, and cohesiveness) and water holding capacities.

9.3.1.1 Heat-Induced Egg Yolk Gelation

Heat-induced egg yolk gel has been extensively studied. Plasma livetin and LDL are the main constituents involved in heat-induced egg yolk gelation, but granules also play an important role (Kiosseoglou & Paraskevopoulou, 2005). Protein denaturation, aggregation of partly denatured protein molecules, and random association of aggregates are thought to be involved in the development of egg yolk thermal gels (Cordobés et al., 2004). Hydrophobic interactions and covalent disulfide bonds are believed to play important roles in gel structure development (Kiosseoglou, 2003b; Kiosseoglou & Paraskevopoulou, 2005). The release of lipids is observed in heated egg yolk gel, which produces an ordered, stable, and flat gel network, suggesting that the migration of lipids contributes to the final characteristics of the gel (Xiang et al., 2020).

9.3.1.2 Freeze-Induced Egg Yolk Gelation

Freezing is an important method of extending the shelf life of egg yolks (Gorman et al., 1977). However, egg yolk subjected to freezing below −6°C loses its fluidity and forms a gel after thawing (Moran, 1925). As reported by Au et al. (2015), this gelation phenomenon is dynamic even at −20°C. It has been proposed that gelation is driven by the removal of water induced by freezing, with the involvement of both plasma and granule components, including LDL, HDL, and α-livetin proteins, where granules play a role in stabilizing the gel network; the aggregated components formed during long-term freezing storage facilitate interactions and self-association, which strengthens gel formation (Powrie et al., 1963; Wakamatu et al., 1982; Au et al., 2015; Primacella et al., 2018b; Wang et al., 2021). The freeze-induced gelation of egg yolk is influenced by a number of factors, with hydrogen bonds being the most important, followed by hydrophobic interactions and disulfide bonds, as reported by Wang et al. (2022). Great efforts have been made to inhibit the freeze-induced gelation of egg yolk, with sucrose and NaCl (10%, w/w) being promising antigelation candidates (Gorman et al., 1977). Furthermore, rapid freezing, enzymatic hydrolysis, and hydrolyzed egg white/yolk proteins can reportedly inhibit gelation (Lopez et al., 1954; Lopez et al., 1955; Primacella et al., 2018a; Primacella et al., 2020; Ma et al., 2021).

9.3.1.3 High Pressure-Induced Egg Yolk Gelation

Egg yolks subjected to high pressure can solidify and form a gel, similar to high pressure-induced egg white gels. Fresh egg yolk treated with pressurization at 4000 or higher MPa makes a stiff gel, which results from changes in noncovalent bonds, i.e., hydrogen bonds, ionic bonds, and hydrophobic bonds (Hayashi et al., 1989).

9.3.1.4 Salt-Induced Egg Yolk Gelation

Salting is a traditional method of food preservation and processing, and NaCl endosmosis during salting can induce gelation of egg yolk, plasma, and granules (Xu et al., 2019b). Studies have shown that egg weight and moisture content decrease with salting (Xu et al., 2019a; Liu et al., 2022). During salting, the egg yolk gradually solidifies due to protein aggregation, and salting treatment disrupts the egg yolk proteins and liberates the constituents (phospholipids, neutral lipids, and proteins), which interact to form large aggregates. Hydrophobic interactions or intermolecular disulfide bonds are involved in this process (Xu et al., 2019a; Xu et al., 2019b).

9.3.1.5 Alkali-Induced Egg Yolk Gelation

Preserved eggs are a popular traditional delicacy and present a gel-like appearance and unique mouth feel. Alkaline treatment promotes protein aggregation and the development of lipoprotein gels by boosting SH oxidation and SS–SH exchange (Yang et al., 2020b). Electrostatic interactions and hydrogen bonds dominate the intermolecular forces of the egg yolk plasma gel under alkaline conditions, while electrostatic interactions, hydrophobic interactions, and disulfide bonds dominate the granule gel, and electrostatic interactions and SS bonds dominate the whole egg yolk gel (Yang et al., 2020c).

9.3.2 Factors Affecting the Gelation Behaviors of Egg Yolk

Temperature: Yolk LDL has a relatively low denaturation temperature, as is evident from the thickening of egg yolk when heated above 64°C and coagulation at approximately 70°C (Rossi & Schiraldi, 1992; Le Denmat et al., 1999; Kiosseoglou & Paraskevopoulou, 2005). Granular proteins are less susceptible to heat treatment, and they start to unfold at temperatures above 75°C (Dixon & Cotterill, 1981). Due to the discrepancy in the thermal and structural properties of plasma and granules, higher temperatures are required for granules to form gel networks.

For freeze-induced gelation, the freezing temperature also plays a role in the gelation behaviors of egg yolk. As described above, egg yolk subjected to temperatures below −6°C undergoes gelation (Moran, 1925). Commercial or household freezing generally requires a central temperature of −18°C; however, the gelation of egg yolk is the most effective with slow freezing to this temperature. Therefore, rapid freezing is utilized to modulate the undesirable gelation of egg yolk. Lopez et al. (1954) found that yolk subjected to liquid nitrogen freezing (−196°C) shows acceptable fluidity after thawing.

Saccharide and salt: Campbell et al. (2005) reported that the presence of sucrose and NaCl can improve the thermal stability of egg yolk, suggesting that higher temperatures are required to form a gel. Similar results also revealed that the addition of sugar and salt led to delayed gel formation (Raikos et al., 2007). Indeed, sugars have been considered as protein stabilizers in both heating and freezing processes. Salt, on the other hand, can alter the ionic strength of the protein environment, thus affecting protein–protein interactions. However, Huang et al. (2021) reported that the gelling point of yolk is decreased with the presence of κ-carrageenan, which can be attributed to electrostatic interactions, and heat-induced gel is enhanced with increasing κ-carrageenan concentration.

Meanwhile, sucrose and NaCl at 10% (w/w) concentration can act as cryoprotective agents to suppress freeze-induced gelation of egg yolk (Moran, 1925; Gorman et al., 1977). Furthermore, the synergistic effect of low-level sucrose and NaCl has also been reported to inhibit gelation (Ma et al., 2021). Studies have shown that NaCl can increase the amount of unfreezable water, which exists in the form of an LDL–water–NaCl complex, to inhibit gelation (Wakamatu et al., 1983; Primacella et al., 2018a).

pH: The textural profiles of heat-set egg gels are all pH dependent (Raikos et al., 2007). The pH of the protein environment can alter the charge distribution among the amino acid side chains. In addition, heat processing at pH levels far from the isoelectric point (pI) reveals denser and more homogeneous microstructures compared with heat treatment near the pI (Guerrero et al., 2004). Aguilar et al. (2019) also found that both pH and the acid type can affect heat-induced gelation. Alkaline conditions can induce structural changes as well as SS and hydrophobic interactions, which lead to the formation of solidified egg yolk (Yang et al., 2020b).

Enzymatic treatment: Enzymatic hydrolysis has been extensively used to improve interfacial properties and thermal stability; however, it is detrimental to the gelling properties of egg yolks. The hydrolyzed peptides are unable to cross-link and form a stable gel network. On the other hand, enzymatic crosslinking can modify the gelling properties of proteins with glutamine transaminase, which can be applied to egg yolk or egg yolk powder for better gelling performance (Rodil et al., 2016).

Grafting: Ren et al. (2020) reported that egg yolk modified by phosphorylation and succinylation exhibits higher gel hardness, and fine and compact egg yolk gel network after grafting was observed.

9.4 Summary

Egg yolks are important food ingredients with excellent functional properties that play indispensable roles in food processing and storage. Despite their desirable attributes, egg yolks are susceptible to fouling under some processing conditions. Therefore, understanding the mechanism of egg yolk functionality during food processing and exploiting modification methods for improving functionality are important for the informed development of the egg industry.

REFERENCES

Abeyrathne, E. D. N. S., Nam, K. C., Huang, X., & Ahn, D. U. (2022). Egg yolk lipids: Separation, characterization, and utilization. *Food Science and Biotechnology*, *31*(10), 1243e1256. https://doi.org/10.1007/s10068-022-01138-4

Aguilar, J. M., Cordobés, F., Bengoechea, C., & Guerrero, A. (2019). Heat-induced gelation of egg yolk as a function of pH. Does the type of acid make any difference? *Food Hydrocolloids*, *87*, 142–148. https://doi.org/10.1016/j.foodhyd.2018.07.045

Alhajj, M. J., Montero, N., Yarce, C. J., & Salamanca, C. H. (2020). Lecithins from vegetable, land, and marine animal sources and their potential applications for cosmetic, food, and pharmaceutical sectors. *Cosmetics*, *7*(4), 87. https://doi.org/10.3390/cosmetics7040087

Ali, M., & Wu, J. (2021). Emulsifying properties of plasma fractionated from egg yolk using low centrifugal forces—Mayonnaise preparation. *Journal of Food Processing and Preservation*, *45*(6), e15553. https://doi.org/10.1111/jfpp.15553

Anton, M. (2013). Egg yolk: Structures, functionalities and processes. *Journal of the Science of Food and Agriculture*, *93*(12), 2871–2880. https://doi.org/10.1002/jsfa.6247

Anton, M., & Gandemer, G. (1997). Composition, solubility and emulsifying properties of granules and plasma of egg yolk. *Journal of Food Science*, *62*(3), 484–487. https://doi.org/10.1111/j.1365-2621.1997.tb04411.x

Anton, M., Le Denmat, M., Beaumal, V., & Pilet, P. (2001). Filler effects of oil droplets on the rheology of heat-set emulsion gels prepared with egg yolk and egg yolk fractions. *Colloids and Surfaces B: Biointerfaces*, *21*(1), 137–147. https://doi.org/10.1016/S0927-7765(01)00167-9

Anton, M., Lechevalier, V., & Nau, F. (2016). From eggs to egg products. In T. C. Romain Jeantet, Pierre Schuck, & G. Brulé (Eds.), *Handbook of food science and technology 3* (pp. 115–143). ISTE Ltd and John Wiley & Sons Inc. https://doi.org/10.1002/9781119296225.ch3

Anukiruthika, T., Moses, J., & Anandharamakrishnan, C. (2020). 3D printing of egg yolk and white with rice flour blends. *Journal of Food Engineering*, *265*, 109691. https://doi.org/10.1016/j.jfoodeng.2019.109691

Asomaning, J., & Curtis, J. M. (2017). Enzymatic modification of egg lecithin to improve properties. *Food Chemistry*, *220*, 385–392. https://doi.org/10.1016/j.foodchem.2016.09.155

Au, C., Acevedo, N. C., Horner, H. T., & Wang, T. (2015). Determination of the gelation mechanism of freeze-thawed hen egg yolk. *Journal of Agricultural and Food Chemistry*, *63*(46), 10170–10180. https://doi.org/10.1021/acs.jafc.5b04109

Bai, X., Gao, J., Yang, Y., Zhu, W., & Fan, J. (2020). Effects of drying methods on the structure and emulsifying capacity of egg yolk lecithin. *International Journal of Agricultural and Biological Engineering*, *13*(4), 238–244. https://doi.org/10.25165/j.ijabe.20201304.5648

Bajuk, B. P., Pihlar, T., Pogačnik, N., & Klinc, P. (2018). Dialysis of the goat semen and its effect on the quality of frozen/thawed spermatozoa processed in the presence of egg yolk. *Animal Reproduction Science*, *198*, 65–73. https://doi.org/10.1016/j.anireprosci.2018.09.001

Bao, Z., Kang, D., Xu, X., Sun, N., & Lin, S. (2019). Variation in the structure and emulsification of egg yolk high-density lipoprotein by lipid peroxide. *Journal of Food Biochemistry*, *43*(11), e13019. https://doi.org/10.1111/jfbc.13019

Bao, Z., Zhao, Y., Wang, X., & Chi, Y. (2017). Effects of degree of hydrolysis (DH) on the functional properties of egg yolk hydrolysate with alcalase. *Journal of Food Science & Technology*, *54*(3), 1–10. https://doi.org/10.1007/s13197-017-2504-0

Bencharif, D., & Dordas-Perpinya, M. (2020). Canine semen cryoconservation: Emerging data over the last 20 years. *Reproduction in Domestic Animals*, *55*(S2), 61–65. https://doi.org/10.1111/rda.13629

Benedé, S., & Molina, E. (2020). Chicken egg proteins and derived peptides with antioxidant properties. *Foods*, *9*(6), 735. https://doi.org/10.3390/foods9060735

Blesso, C. N. (2015). Egg phospholipids and cardiovascular health. *Nutrients*, *7*(4), 2731–2747. https://doi.org/10.3390/nu7042731

Blume, K., Dietrich, K., Lilienthal, S., Ternes, W., & Drotleff, A. M. (2015). Exploring the relationship between protein secondary structures, temperature-dependent viscosities, and technological treatments in egg yolk and LDL by FTIR and rheology. *Food Chemistry*, *173*, 584–593. https://doi.org/10.1016/j.foodchem.2014.10.084

Campbell, L., Raikos, V., & Euston, S. R. (2005). Heat stability and emulsifying ability of whole egg and egg yolk as related to heat treatment. *Food Hydrocolloids*, *19*(3), 533–539. https://doi.org/10.1016/j.foodhyd.2004.10.031

Causeret, D., Matringe, E., & Lorient, D. (1991). Ionic strength and pH effects on composition and microstructure of yolk granules. *Journal of Food Science*, *56*(6), 1532–1536. https://doi.org/10.1111/j.1365-2621.1991.tb08634.x

Chalamaiah, M., Esparza, Y., Hong, H., Temelli, F., & Wu, J. (2018). Physicochemical and functional properties of leftover egg yolk granules after phosvitin extraction. *Food Chemistry*, *268*, 369–377. https://doi.org/10.1016/j.foodchem.2018.06.033

Chalamaiah, M., Esparza, Y., Temelli, F., & Wu, J. (2017). Physicochemical and functional properties of livetins fraction from hen egg yolk. *Food Bioscience*, *18*, 38–45. https://doi.org/10.1016/j.fbio.2017.04.002

Chang, C., Gao, Y., Su, Y., Gu, L., Li, J., & Yang, Y. (2021). Influence of chitosan on the emulsifying properties of egg yolk hydrolysates: Study on creaming, thermal and oxidative stability. *Journal of the Science of Food and Agriculture*, *101*(11), 4691–4698. https://doi.org/10.1002/jsfa.11114

Charbonneau, A. M., Kinsella, J. M., & Tran, S. D. (2019). 3D cultures of salivary gland cells in native or gelled egg yolk plasma, combined with egg white and 3D-printing of gelled egg yolk plasma. *Materials*, *12*(21), 3480. https://doi.org/10.3390/MA12213480

Chen, H., Jin, Y., Ding, X., Wu, F., Bashari, M., Chen, F., & Xu, X. (2014). Improved the emulsion stability of phosvitin from hen egg yolk against different pH by the covalent attachment with dextran. *Food Hydrocolloids*, *39*, 104–112. https://doi.org/10.1016/j.foodhyd.2013.12.031

Chung, S. L., & Ferrier, L. K. (1991). Partial lipid extraction of egg yolk powder: Effects on emulsifying properties and soluble protein fraction. *Journal of Food Science*, *56*(5), 1255–1258. https://doi.org/10.1111/j.1365-2621.1991.tb04746.x

Chung, S. L., & Ferrier, L. K. (1992). pH and sodium chloride effects on emulsifying properties of egg yolk phosvitin. *Journal of Food Science*, *57*(1), 40–42. https://doi.org/10.1111/j.1365-2621.1992.tb05419.x

Cook, W. H., & Martin, W. G. (1969). Structural and functional aspects of lipoproteins in living systems. In E. Tria & A. Scanu (Eds.), *Egg lipoproteins* (pp. 579–615). Academic Press.

Cordobés, F., Partal, P., & Guerrero, A. (2004). Rheology and microstructure of heat-induced egg yolk gels. *Rheologica Acta*, *43*(2), 184–195. https://doi.org/10.1007/s00397-003-0338-3

Cui, Y., Li, X., Lu, M., Liu, X., & Duan, X. (2019). Role of polysaccharide conjugation in physicochemical and emulsifying properties of egg phosvitin and the calcium binding capacity of its phosphopeptides. *Food & Function*, *10*(4), 1808–1815. https://doi.org/10.1039/C8FO02464B

da Silva, E. A., Corcini, C. D., de Assis Araújo Camelo Junior, F., Martins, D., Meneghello Gheller, S. M., Hädrich, G., Dora, C. L., & Varela Junior, A. S. (2022). Probe ultrasonification of egg yolk plasma forms low-density lipoprotein nanoparticles that efficiently protect canine semen during cryofreezing. *Journal of Biological Chemistry*, *298*(7), 101975. https://doi.org/10.1016/j.jbc.2022.101975

Derderian, E. J., Glass, J. E., & Bryant, G. M. (1979). The relationship between microemulsion specific refraction and its interfacial tension behavior versus excess oil and brine. *Journal of Colloid and Interface Science*, *68*(1), 184–189. https://doi.org/10.1016/0021-9797(79)90270-4

Dixon, D. K., & Cotterill, O. J. (1981). Electrophoretic and chromatographic changes in egg yolk proteins due to heat. *Journal of Food Science*, *46*(4), 981–983. https://doi.org/10.1111/j.1365-2621.1981.tb02971.x

Dyer-Hurdon, J. N., & Nnanna, I. A. (1993). Cholesterol content and functionality of plasma and granules fractionated from egg yolk. *Journal of Food Science*, *58*(6), 1277–1281. https://doi.org/10.1111/j.1365-2621.1993.tb06165.x

Felix, M., Romero, A., & Guerrero, A. (2017). Viscoelastic properties, microstructure and stability of high-oleic O/W emulsions stabilised by crayfish protein concentrate and xanthan gum. *Food Hydrocolloids*, *64*, 9–17. https://doi.org/10.1016/j.foodhyd.2016.10.028

Fernández-Martín, F., Pérez-Mateos, M., Dadashi, S., Gómez-Guillén, C. M., & Sanz, P. D. (2018). Impact of magnetic assisted freezing in the physicochemical and functional properties of egg components. Part 2: Egg yolk. *Innovative Food Science & Emerging Technologies*, *49*, 176–183. https://doi.org/10.1016/j.ifset.2017.11.006

Ferrari, M., Handgraaf, J. W., Boccardo, G., Buffo, A., Vanni, M., & Marchisio, D. L. (2022). Molecular modeling of the interface of an egg yolk protein-based emulsion. *Physics of Fluids*, *34*(2), 021903. https://doi.org/10.1063/5.0079883

Fleer, G. J., Stuart, M. A. C., Scheutjens, J. M. H. M., Cosgrove, T., & Vincent, B. (1998). *Polymers at interfaces*. Springer. https://doi.org/10.1007/978-94-011-2130-9_6

Fu, X., Huang, X., Jin, Y., Zhang, S., & Ma, M. (2020). Characterization of enzymatically modified liquid egg yolk: Structural, interfacial and emulsifying properties. *Food Hydrocolloids*, *105*, 105763. https://doi.org/10.1016/j.foodhyd.2020.105763

Fuertes, S., Laca, A., Oulego, P., Paredes, B., Rendueles, M., & Díaz, M. (2017). Development and characterization of egg yolk and egg yolk fractions edible films. *Food Hydrocolloids*, *70*, 229–239. https://doi.org/10.1016/j.foodhyd.2017.04.007

Gao, Y., Li, J., Chang, C., Wang, C., Yang, Y., & Su, Y. (2019). Effect of enzymatic hydrolysis on heat stability and emulsifying properties of egg yolk. *Food Hydrocolloids*, *97*, 105224. https://doi.org/10.1016/j.foodhyd.2019.105224

Gazolu-Rusanova, D., Mustan, F., Vinarov, Z., Tcholakova, S., Denkov, N., Stoyanov, S., & de Folter, J. W. J. (2020). Role of lysophospholipids on the interfacial and liquid film properties of enzymatically modified egg yolk solutions. *Food Hydrocolloids*, *99*, 105319. https://doi.org/10.1016/j.foodhyd.2019.105319

Geng, F., Xie, Y., Wang, Y., & Wang, J. (2021). Depolymerization of chicken egg yolk granules induced by high-intensity ultrasound. *Food Chemistry*, *354*, 129580. https://doi.org/10.1016/j.foodchem.2021.129580

Gmach, O., Golda, J., & Kulozik, U. (2022). Freeze-thaw stability of emulsions made with native and enzymatically modified egg yolk fractions. *Food Hydrocolloids*, *123*, 107109. https://doi.org/10.1016/j.foodhyd.2021.107109

Gorman, J. M., Kohlenberg, O. J., & Powrie, W. D. (1977). Freezing of egg products. In W. D. Norman & K. T. Donald (Ed.), *Fundamentals of food freezing* (pp. 396–412). Springer. https://doi.org/10.1007/978-94-011-7726-9_10

Gouda, M., Zhang, S., Liu, Y., Sheng, L., & Ma, M. (2017). Effects of four natural antioxidant phenyl terpenes on emulsifying and rheological properties of egg yolk. *LWT - Food Science and Technology*, *83*, 59–67. https://doi.org/10.1016/j.lwt.2017.04.075

Guerrero, A., Carmona, J. A., Martínez, I., Cordobés, F., & Partal, P. (2004). Effect of pH and added electrolyte on the thermal-induced transitions of egg yolk. *Rheologica Acta*, *43*(5), 539–549. https://doi.org/10.1007/s00397-004-0395-2

Guilmineau, F., & Kulozik, U. (2006a). Impact of a thermal treatment on the emulsifying properties of egg yolk. Part 1: Effect of the heating time. *Food Hydrocolloids*, *20*(8), 1105–1113. https://doi.org/10.1016/j.foodhyd.2005.12.005

Guilmineau, F., & Kulozik, U. (2006b). Impact of a thermal treatment on the emulsifying properties of egg yolk. Part 2: Effect of the environmental conditions. *Food Hydrocolloids*, *20*(8), 1114–1123. https://doi.org/10.1016/j.foodhyd.2005.12.006

Hayashi, R., Kawamura, Y., Nakasa, T., & Okinaka, O. (1989). Application of high pressure to food processing: Pressurization of egg white and yolk, and properties of gels formed. *Agricultural and Biological Chemistry*, *53*(11), 2935–2939. https://doi.org/10.1080/00021369.1989.10869784

Ho, J., Lee, T., Namai, N., Sakai, S., Lou, S., Handa, A., & Lin, W. (2022). Thermal processing of liquid egg yolks modulates physio-chemical properties of mayonnaise. *Foods*, *11*(10), 1426. https://doi.org/10.3390/foods11101426

Huang, X., & Ahn, D. U. (2019). How can the value and use of egg yolk be increased? *Journal of Food Science*, *84*(2), 205–212. https://doi.org/10.1111/1750-3841.14430

Huang, M., Mao, Y., Li, H., & Yang, H. (2021). Kappa-carrageenan enhances the gelation and structural changes of egg yolk via electrostatic interactions with yolk protein. *Food Chemistry*, *360*, 129972. https://doi.org/10.1016/j.foodchem.2021.129972

Huang, L., Wang, T., Han, Z., Meng, Y., & Lu, X. (2016). Effect of egg yolk freezing on properties of mayonnaise. *Food Hydrocolloids*, *56*, 311–317. https://doi.org/10.1016/j.foodhyd.2015.12.027

Jiang, X., Mu, H., Hsieh, Y.-H. P., & Rao, Q. (2022). Isolation and characterization of chicken serum albumin (hen egg alpha-livetin, gal d 5). *Foods*, *11*(11), 1637. https://doi.org/10.3390/foods11111637

Jolivet, P., Boulard, C., Beaumal, V., Chardot, T., & Anton, M. (2006). Protein components of low-density lipoproteins purified from hen egg yolk. *Journal of Agricultural and Food Chemistry*, *54*(12), 4424–4429. https://doi.org/10.1021/jf0531398

Jung, S., Kim, D. H., Son, J. H., Nam, K., Ahn, D. U., & Jo, C. (2012). The functional property of egg yolk phosvitin as a melanogenesis inhibitor. *Food Chemistry*, *135*(3), 993–998. https://doi.org/10.1016/j.foodchem.2012.05.113

Kim, H. J., Lee, J. H., Ahn, D. U., & Paik, H.-D. (2020). Anti-biofilm effect of egg yolk phosvitin by inhibition of biomass production and adherence activity against streptococcus mutans. *Food Science of Animal Resources*, *40*(6), 1001. https://doi.org/10.5851/kosfa.2020.e71

Kiosseoglou, V. (2003a). Egg yolk protein gels and emulsions. *Current Opinion in Colloid & Interface Science*, *8*(4), 365–370. https://doi.org/10.1016/S1359-0294(03)00094-3

Kiosseoglou, V. (2003b). Functional properties of egg yolk. Proceedings of the Xth European Symposium on the quality of eggs and egg products, France.

Kiosseoglou, V., & Paraskevopoulou, A. (2005). Molecular interactions in gels prepared with egg yolk and its fractions. *Food Hydrocolloids*, *19*(3), 527–532. https://doi.org/10.1016/j.foodhyd.2004.10.027

Laca, A., Paredes, B., Rendueles, M., & Díaz, M. (2014). Egg yolk granules: Separation, characteristics and applications in food industry. *LWT - Food Science and Technology*, *59*(1), 1–5. https://doi.org/10.1016/j.lwt.2014.05.020

Le Denmat, M., Anton, M., & Beaumal, V. (2000). Characterisation of emulsion properties and of interface composition in O/W emulsions prepared with hen egg yolk, plasma and granules. *Food Hydrocolloids*, *14*(6), 539–549. https://doi.org/10.1016/S0268-005X(00)00034-5

Le Denmat, M., Anton, M., & Gandemer, G. (1999). Protein denaturation and emulsifying properties of plasma and granules of egg yolk as related to heat treatment. *Journal of Food Science*, *64*(2), 194–197. https://doi.org/10.1111/j.1365-2621.1999.tb15863.x

Leiva, C. L., Gallardo, M. J., Casanova, N., Terzolo, H., & Chacana, P. (2020). IgY-technology (egg yolk antibodies) in human medicine: A review of patents and clinical trials. *International Immunopharmacology*, *81*, 106269. https://doi.org/10.1016/j.intimp.2020.106269

Li, J., Chang, C., Gu, L., Su, Y., Yang, Y., & Zhai, J. (2022a). Improved retention ratio and bioaccessibility of lutein loaded in emulsions stabilized by egg yolk granules-lecithin complex. *Journal of the Science of Food and Agriculture*, *102*(12), 5153–5161. https://doi.org/10.1002/jsfa.11867

Li, J., Shen, Y., Zhai, J., Su, Y., Gu, L., Chang, C., & Yang, Y. (2021a). Enhancing the oxidative stability of algal oil powders stabilized by egg yolk granules/lecithin composites. *Food Chemistry*, *345*, 128782. https://doi.org/10.1016/j.foodchem.2020.128782

Li, J., Wang, C., Li, X., Su, Y., Yang, Y., & Yu, X. (2018a). Effects of pH and NaCl on the physicochemical and interfacial properties of egg white/yolk. *Food Bioscience*, *23*, 115–120. https://doi.org/10.1016/j.fbio.2017.12.004

Li, J., Zhai, J., Gu, L., Su, Y., Gong, L., Yang, Y., & Chang, C. (2021b). Hen egg yolk in food industry - A review of emerging functional modifications and applications. *Trends in Food Science & Technology*, *115*, 12–21. https://doi.org/10.1016/j.tifs.2021.06.031

Li, M., Li, X., Li, J., Lu, M., Liu, X., & Duan, X. (2018b). Effects of multiple freeze–thaw treatments on physicochemical and biological activities of egg phosvitin and its phosphopeptides. *Food & Function*, *9*(9), 4602–4610. https://doi.org/10.1039/C8FO01101J

Li, Q., Tang, S., Mourad, F. K., Zou, W., Lu, L., & Cai, Z. (2020). Emulsifying stability of enzymatically hydrolyzed egg yolk granules and structural analysis. *Food Hydrocolloids*, *101*, 105521. https://doi.org/10.1016/j.foodhyd.2019.105521

Li, T., Su, H., Zhu, J., & Fu, Y. (2022b). Janus effects of NaCl on structure of egg yolk granules. *Food Chemistry*, *371*, 131077. https://doi.org/10.1016/j.foodchem.2021.131077

Li, Z., Sun, Y., Jin, H., Wang, Q., Jin, Y., Huang, X., & Sheng, L. (2022c). Improvement and mechanism of emulsifying properties of liquid egg yolk by ozonation technology. *LWT*, *156*, 113038. https://doi.org/10.1016/j.lwt.2021.113038

Li, Z., Wang, Y., & Luo, Y. (2022d). High internal phase Pickering emulsions stabilized by egg yolk low density lipoprotein for delivery of curcumin. *Colloids and Surfaces B: Biointerfaces*, *211*, 112334. https://doi.org/10.1016/j.colsurfb.2022.112334

Li, Z., Xiong, Y., Wang, Y., Zhang, Y., & Luo, Y. (2023). Low density lipoprotein-pectin complexes stabilized high internal phase pickering emulsions: The effects of pH conditions and mass ratios. *Food Hydrocolloids*, *134*, 108004. https://doi.org/10.1016/j.foodhyd.2022.108004

Liu, W., Li, X., Li, S., Xu, B., & Zhang, B. (2022). Emulsifying and emulsion stabilizing properties of hydrolysates of high-density lipoprotein from egg yolk. *Food Chemistry*, *369*, 130891. https://doi.org/10.1016/j.foodchem.2021.130891

Liu, X., Guo, J., Wan, Z., Liu, Y., Ruan, Q., & Yang, X. (2018). Wheat gluten-stabilized high internal phase emulsions as mayonnaise replacers. *Food Hydrocolloids*, *77*, 168–175. https://doi.org/10.1016/j.foodhyd.2017.09.032

Liu, Y., Ma, Y., Chi, Y., & Chi, Y. (2022). Change in rapid salting kinetics and characteristics of hen egg yolks. *Journal of Food Engineering*, *329*, 111090. https://doi.org/10.1016/j.jfoodeng.2022.111090

Long, J., Wang, F., Jiao, A., Xu, X., Xie, Z., & Jin, Z. (2019). Preparation, characterization and physicochemical properties of novel low-phosphorus egg yolk protein. *Journal of the Science of Food and Agriculture*, *99*(4), 1740–1747. https://doi.org/10.1002/jsfa.9363

Lopez, A., Fellers, C. R., & Powie, W. D. (1954). Some factors affecting gelation of frozen egg yolk. *Journal of Milk and Food Technology*, *17*(11), 334–339. https://doi.org/10.4315/0022-2747-17.11.334

Lopez, A., Fellers, C., & Powrie, W. (1955). Enzymic inhibition of gelation in frozen egg yolk. *Journal of Milk and Food Technology*, *18*(3), 77–80. https://doi.org/10.4315/0022-2747-18.3.77

Ma, Z., Ma, Y., Wang, R., & Chi, Y. (2021). Influence of anti-gelation agents on frozen egg yolk gelation. *Journal of Food Engineering*, *302*, 110585. https://doi.org/10.1016/j.jfoodeng.2021.110585

Marcet, I., Alvarez, C., Paredes, B., & Diaz, M. (2014). Inert and oxidative subcritical water hydrolysis of insoluble egg yolk granular protein, functional properties, and comparison to enzymatic hydrolysis. *Journal of Agricultural and Food Chemistry*, *62*(32), 8179–8186.

Marcet, I., Sáez-Orviz, S., Rendueles, M., & Díaz, M. (2022). Egg yolk granules and phosvitin. Recent advances in food technology and applications. *LWT*, *153*, 112442. https://doi.org/10.1016/j.lwt.2021.112442

Martinet, V., Saulnier, P., Beaumal, V., Courthaudon, J.-L., & Anton, M. (2003). Surface properties of hen egg yolk low-density lipoproteins spread at the air–water interface. *Colloids and Surfaces B: Biointerfaces*, *31*(1), 185–194. https://doi.org/10.1016/S0927-7765(03)00139-5

Mc Bee, L. E., & Cotterill, O. J. (1979). Ion-exchange chromatography and electrophoresis of egg yolk proteins. *Journal of Food Science*, *44*(3), 656–667. https://doi.org/10.1111/j.1365-2621.1979.tb08469.x

McWhorter, A. R., Khan, S., Sexton, M., Moyle, T. S., & Chousalkar, K. K. (2021). Acidification and extended storage at room temperature of mayonnaise reduce salmonella typhimurium virulence and viability. *Food Research International*, *141*, 110117. https://doi.org/10.1016/j.foodres.2021.110117

Mehdipour, M., Daghigh Kia, H., Moghaddam, G., & Hamishehkar, H. (2018). Effect of egg yolk plasma and soybean lecithin on rooster frozen-thawed sperm quality and fertility. *Theriogenology*, *116*, 89–94. https://doi.org/10.1016/j.theriogenology.2018.05.013

Mi, S., Xia, M., Zhang, X., Liu, J., & Cai, Z. (2022). Formation of natural egg yolk granule stabilized pickering high internal phase emulsions by means of NaCl ionic strength and pH change. *Foods*, *11*(2), 229. https://doi.org/10.3390/foods11020229

Miller, C., & Winter, A. R. (1951). Pasteurized frozen whole egg and yolk for mayonnaise production. *Journal of Food Science*, *16*(1–6), 43–49. https://doi.org/10.1111/j.1365-2621.1951.tb17347.x

Mine, Y. (1998). Emulsifying characterization of hens egg yolk proteins in oil-in-water emulsions. *Food Hydrocolloids*, *12*(4), 409–415. https://doi.org/10.1016/S0268-005X(98)00054-X

Mizutani, R., & Nakamura, R. (1987). Emulsifying properties of a complex between apoprotein from hen's egg yolk low density lipoprotein and egg yolk lecithin. *Agricultural and Biological Chemistry*, *51*(4), 1115–1119. https://doi.org/10.1080/00021369.1987.10868169

Moran, T. (1925). The effect of low temperature on hens' eggs. *Proceedings of the Royal Society of London*, *98*(691), 436–456. https://doi.org/10.1098/rspb.1925.0046

Naderi, N., Doyen, A., House, J. D., & Pouliot, Y. (2017). The use of high hydrostatic pressure to generate folate-enriched extracts from the granule fraction of hen's egg yolk. *Food Chemistry*, *232*, 253–262. https://doi.org/10.1016/j.foodchem.2017.03.144

Paraskevopoulou, A., & Kiosseoglou, V. (1994). Cholesterol and other lipid extraction from egg yolk using organic solvents: Effects on functional properties of yolk. *Journal of Food Science*, *59*(4), 766–768. https://doi.org/10.1111/j.1365-2621.1994.tb08123.x

Patil, S., Rao, B., Matondkar, M., Bhushette, P., & Sonawane, S. K. (2022). A review on understanding of egg yolk as functional ingredients. *Journal of Microbiology, Biotechnology and Food Sciences*, *11*(4), e4627–e4627. https://doi.org/10.55251/jmbfs.4627

Pescuma, M., Hébert, E. M., Mozzi, F., & Font de Valdez, G. (2008). Whey fermentation by thermophilic lactic acid bacteria: Evolution of carbohydrates and protein content.

Food Microbiology, *25*(3), 442–451. https://doi.org/10.1016/j.fm.2008.01.007

Powrie, W., Little, H., & Lopez, A. (1963). Gelation of egg yolk. *Journal of Food Science*, *28*, 38–46. https://doi.org/10.1111/j.1365-2621.1963.tb00156.x

Primacella, M., Acevedo, N. C., & Wang, T. (2020). Effect of freezing and food additives on the rheological properties of egg yolk. *Food Hydrocolloids*, *98*, 105241. https://doi.org/10.1016/j.foodhyd.2019.105241

Primacella, M., Fei, T., Acevedo, N., & Wang, T. (2018a). Effect of food additives on egg yolk gelation induced by freezing. *Food Chemistry*, *263*, 142–150. https://doi.org/10.1016/j.foodchem.2018.04.071

Primacella, M., Wang, T., & Acevedo, N. C. (2018b). Use of reconstitued yolk systems to study the gelation mechanism of frozen-thawed hen egg yolk. *Journal of Agricultural and Food Chemistry*, *66*(2), 512–520. https://doi.org/10.1021/acs.jafc.7b04370

Puertas, G., & Vázquez, M. (2019). Advances in techniques for reducing cholesterol in egg yolk: A review. *Critical Reviews in Food Science and Nutrition*, *59*(14), 2276–2286. https://doi.org/10.1080/10408398.2018.1448357

Raikos, V., Campbell, L., & Euston, S. R. (2007). Rheology and texture of hen's egg protein heat-set gels as affected by pH and the addition of sugar and/or salt. *Food Hydrocolloids*, *21*(2), 237–244. https://doi.org/10.1016/j.foodhyd.2006.03.015

Ren, L., Liu, J., Zhang, X., Zhao, S., Lv, Y., & Guo, H. (2020). Emulsion, gelation, physicochemical properties and microstructure of phosphorylated and succinylated egg yolk. *LWT*, *131*, 109675. https://doi.org/10.1016/j.lwt.2020.109675

Rodil, A., Laca, A., Paredes, B., Rendueles, M., Meana, Á., & Díaz, M. (2016). Gels prepared from egg yolk and its fractions for tissue engineering. *Biotechnology Progress*, *32*(6), 1577–1583. https://doi.org/10.1002/btpr.2364

Rossi, M., & Schiraldi, A. (1992). Thermal denaturation and aggregation of egg proteins. *Thermochimica Acta*, *199*, 115–123. https://doi.org/10.1016/0040-6031(92)80255-U

Sassi, C. B., Marcet, I., Rendueles, M., Díaz, M., & Fattouch, S. (2020). Egg yolk protein as a novel wall material used together with gum Arabic to encapsulate polyphenols extracted from Phoenix dactylifera L pits. *LWT*, *131*, 109778. https://doi.org/10.1016/j.lwt.2020.109778

Schultz, J., Snyder, H., & Forsythe, R. (1966). Interactions of carbohydrates with egg lipoproteins. *Poultry Science*, *45*(5), 1122.

Seuss-Baum, I., & Nau, F. (2011). The nutritional quality of eggs. In F. Van Immerseel, Y. Nys, & M. Bain (Eds.), *Improving the safety and quality of eggs and egg products* (pp. 201–236). Woodhead Publishing. https://doi.org/10.1533/9780857093929.3.201

Shen, Y., Chang, C., Shi, M., Su, Y., Gu, L., Li, J., & Yang, Y. (2020). Interactions between lecithin and yolk granule and their influence on the emulsifying properties. *Food Hydrocolloids*, *101*, 105510. https://doi.org/10.1016/j.foodhyd.2019.105510

Speroni, F., Puppo, M. C., Chapleau, N., de Lamballerie, M., Castellani, O., Añón, M. C., & Anton, M. (2005). High-pressure induced physicochemical and functional modifications of low-density lipoproteins from hen egg yolk. *Journal of Agricultural and Food Chemistry*, *53*(14), 5719–5725. https://doi.org/10.1021/jf0502808

Strixner, T., Würth, R., & Kulozik, U. (2013). Combined effects of enzymatic treatment and spray drying on the functional properties of egg yolk main fractions granules and plasma. *Drying Technology*, *31*(13–14), 1485–1496. https://doi.org/10.1080/07373937.2013.790411

Su, Y., Gao, J., Chang, C., Li, J., Gong, L., Gu, L., & Yang, Y. (2022). Study on functional improvement of freeze–thaw egg yolk pellet by enzymatic hydrolysis. *Journal of Food Science and Technology*. https://doi.org/10.1007/s13197-022-05511-3

Su, Y., Ji, M., Li, J., Chang, C., Dong, S., Deng, Y., & Gu, L. (2020). Subcritical fluid extraction treatment on egg yolk: Product characterization. *Journal of Food Engineering*, *274*, 109805. https://doi.org/10.1016/j.jfoodeng.2019.109805

Suhag, R., Dhiman, A., Thakur, D., Kumar, A., & Upadhyay, A. (2021). Physico-chemical and functional properties of microfluidized egg yolk. *Journal of Food Engineering*, *294*, 110416. https://doi.org/10.1016/j.jfoodeng.2020.110416

Takeuchi, M., Mashima, T., Sztucki, M., Petukhov, A. V., Vis, M., Friedrich, H., & Tuinier, R. (2022). Characterization of hen phosvitin in aqueous salt solutions: Size, structure, and aggregation. *Food Hydrocolloids*, *129*, 107545. https://doi.org/10.1016/j.foodhyd.2022.107545

Tang, S., Zhou, X., Gouda, M., Cai, Z., & Jin, Y. (2019). Effect of enzymatic hydrolysis on the solubility of egg yolk powder from the changes in structure and functional properties. *LWT*, *110*, 214–222. https://doi.org/10.1016/j.lwt.2019.04.070

Thamjedsada, N., Pradipasena, P., & Pradistsuwana, C. (2018). Effect of thermal processing and ingredients on egg yolk quality. *Pakistan Journal of Agricultural Sciences*, *55*(1). https://doi.org/10.21162/PAKJAS/18.5249

Tian, L., Hu, S., Jia, J., Tan, W., Yang, L., Zhang, Q., & Duan, X. (2021). Effects of short-term fermentation with lactic acid bacteria on the characterization, rheological and emulsifying properties of egg yolk. *Food Chemistry*, *341*, 128163. https://doi.org/10.1016/j.foodchem.2020.128163

Tian, L., Xiong, D., Jia, J., Liu, X., Zhang, Y., & Duan, X. (2022). Mechanism study on enhanced emulsifying properties of phosvitin and calcium-binding capacity of its phosphopeptides by lactic acid bacteria fermentation. *LWT*, *155*, 113002. https://doi.org/10.1016/j.lwt.2021.113002

Tsutsui, T. (1988). Functional properties of heat-treated egg yolk low density lipoprotein. *Journal of Food Science*, *53*(4), 1103–1106. https://doi.org/10.1111/j.1365-2621.1988.tb13539.x

Varadarajulu, P., & Cunningham, F. E. (1972). A study of selected characteristics of hens' egg yolk.: I. Influence of albumen and selected additives1,2,3. *Poultry Science*, *51*(2), 542–546. https://doi.org/10.3382/ps.0510542

Wakamatsu, H., Handa, A., & Chiba, K. (2018). Observations using phosphorus-31 nuclear magnetic resonance (31P-NMR) of structural changes in freeze-thawed hen egg yolk. *Food Chemistry*, *244*, 169–176. https://doi.org/10.1016/j.foodchem.2017.10.032

Wakamatu, T., Sato, Y., & Saito, Y. (1982). Identification of the components responsible for the gelation of egg yolk during freezing. *Journal of the Agricultural Chemical Society of Japan*, *46*(6), 1495–1503. https://doi.org/10.1080/00021369.1982.10865281

Wakamatu, T., Sato, Y., & Saito, Y. (1983). On sodium chloride action in the gelation process of low density lipoprotein

(LDL) from hen egg yolk. *Journal of Food Science*, *48*, 507–512. https://doi.org/10.1111/j.1365-2621.1983.tb10777.x

Wang, A., Xiao, Z., Wang, J., Li, G., & Wang, L. (2020). Fabrication and characterization of emulsion stabilized by table egg-yolk granules at different pH levels. *Journal of the Science of Food and Agriculture*, *100*(4), 1470–1478. https://doi.org/10.1002/jsfa.10154

Wang, R., Ma, Y., Zhang, L., Zhang, Z., Chi, Y., & Chi, Y. (2021). Changes in egg yolk gelation behaviour and mechanisms during freezing. *LWT*, *151*, 112223. https://doi.org/10.1016/j.lwt.2021.112223

Wang, R., Zhang, L., Chi, Y., & Chi, Y. (2022). Forces involved in freeze-induced egg yolk gelation: Effects of various bond dissociation reagents on gel properties and protein structure changes. *Food Chemistry*, *371*, 131190. https://doi.org/10.1016/j.foodchem.2021.131190

Xiang, X., Liu, Y., Liu, Y., Wang, X., & Jin, Y. (2020). Changes in structure and flavor of egg yolk gel induced by lipid migration under heating. *Food Hydrocolloids*, *98*, 105257. https://doi.org/10.1016/j.foodhyd.2019.105257

Xiao, N., Zhao, Y., Yao, Y., Wu, N., Xu, M., Du, H., & Tu, Y. (2020). Biological activities of egg yolk lipids: A review. *Journal of Agricultural and Food Chemistry*, *68*(7), 1948–1957. https://doi.org/10.1021/acs.jafc.9b06616

Xie, Y., Wang, J., Shi, Y., Wang, Y., Cheng, L., Liu, L., & Geng, F. (2020a). Molecular aggregation and property changes of egg yolk low-density lipoprotein induced by ethanol and high-density ultrasound. *Ultrasonics Sonochemistry*, *63*, 104933. https://doi.org/10.1016/j.ultsonch.2019.104933

Xie, Y., Wang, J., Wang, Y., Wu, D., Liang, D., Ye, H., & Geng, F. (2020b). Effects of high-intensity ultrasonic (HIU) treatment on the functional properties and assemblage structure of egg yolk. *Ultrasonics Sonochemistry*, *60*, 104767. https://doi.org/10.1016/j.ultsonch.2019.104767

Xu, L., Gu, L., Su, Y., Chang, C., Wang, J., Dong, S., & Li, J. (2020). Impact of thermal treatment on the rheological, microstructural, protein structures and extrusion 3D printing characteristics of egg yolk. *Food Hydrocolloids*, *100*, 105399. https://doi.org/10.1016/j.foodhyd.2019.105399

Xu, L., Zhao, Y., Xu, M., Nie, X., Wu, N., & Tu, Y. (2019a). Formation mechanism of low-density lipoprotein gel induced by NaCl. *Poultry Science*, *98*(10), 5166–5176. https://doi.org/10.3382/ps/pez232

Xu, L., Zhao, Y., Xu, M., Yao, Y., Wu, N., Du, H., & Tu, Y. (2019b). Changes in physico-chemical properties, microstructure, protein structures and intermolecular force of egg yolk, plasma and granule gels during salting. *Food Chemistry*, *275*, 600–609. https://doi.org/10.1016/j.foodchem.2018.09.078

Yan, W., Qiao, L., Gu, X., Li, J., Xu, R., Wang, M., & Yang, Y. (2010). Effect of high pressure treatment on the physicochemical and functional properties of egg yolk. *European Food Research and Technology*, *231*(3), 371–377. https://doi.org/10.1007/s00217-010-1286-7

Yang, L., Jia, J., Zhou, X., Liu, M., Zhang, Q., Tian, L., & Duan, X. (2021). Phosvitin-wheat gluten complex catalyzed by transglutaminase in the presence of Na_2SO_3: Formation, cross-link behavior and emulsifying properties. *Food Chemistry*, *346*, 128903. https://doi.org/10.1016/j.foodchem.2020.128903

Yang, S., & Baldwin, R. E. (1995). Functional properties of eggs in foods. In W. J. Stadelman & O. J. Cotterill (Eds.), *Egg science and technology* (pp. 405–463). CRC Press. https://doi.org/10.1201/9780203758878

Yang, X., Li, A., Yu, W., Li, X., Sun, L., Xue, J., & Guo, Y. (2020a). Structuring oil-in-water emulsion by forming egg yolk/alginate complexes: Their potential application in fabricating low-fat mayonnaise-like emulsion gels and redispersible solid emulsions. *International Journal of Biological Macromolecules*, *147*, 595–606. https://doi.org/10.1016/j.ijbiomac.2020.01.106

Yang, Y., Zhao, Y., Xu, M., Wu, N., Yao, Y., Du, H., & Tu, Y. (2019). Changes in physico-chemical properties, microstructure and intermolecular force of preserved egg yolk gels during pickling. *Food Hydrocolloids*, *89*, 131–142. https://doi.org/10.1016/j.foodhyd.2018.10.016

Yang, Y., Zhao, Y., Xu, M., Yao, Y., Wu, N., Du, H., & Tu, Y. (2020b). Alkali induced gelation behavior of low-density lipoprotein and high-density lipoprotein isolated from duck eggs. *Food Chemistry*, *311*, 125952. https://doi.org/10.1016/j.foodchem.2019.125952

Yang, Y., Zhao, Y., Xu, M., Yao, Y., Wu, N., Du, H., & Tu, Y. (2020c). Effects of strong alkali treatment on the physicochemical properties, microstructure, protein structures, and intermolecular forces in egg yolks, plasma, and granules. *Food Chemistry*, *311*, 125998. https://doi.org/10.1016/j.foodchem.2019.125998

Ye, H., Wang, J., Wang, N., Wu, D., Li, H., & Geng, F. (2022). Ultrasound-assisted pH-shifting remodels egg-yolk low-density lipoprotein to enable construction of a stable aqueous solution of vitamin D3. *Current Research in Food Science*, *5*, 964–972. https://doi.org/10.1016/j.crfs.2022.05.013

Yeadon, D. A., Goldblatt, L. A., & Altschul, A. M. (1958). Lecithin in oil-in-water emulsions. *Journal of the American Oil Chemists Society*, *35*(8), 435–438. https://doi.org/10.1007/BF02632564

Yu, X., Zhao, Y., Sun, M., Liu, L., Li, X., Zhang, X., & Jiang, S. (2022a). Effects of egg yolk lecithin/milk fat globule membrane material ratio on the structure and stability of oil-in-water emulsions. *LWT*, 113891. https://doi.org/10.1016/j.lwt.2022.113891

Yu, Z., Hu, J., Gao, S., Han, R., Ma, L., & Chen, Y. (2022b). Joint effects of enzymatic and high-intensity ultrasonic on the emulsifying properties of egg yolk. *Food Research International*, *156*, 111146. https://doi.org/10.1016/j.foodres.2022.111146

Zhang, F., Li, J., Chang, C., Gu, L., Su, Y., & Yang, Y. (2022a). Selective removal effect of subcritical fluid extraction on egg yolk lipids and characterization and enzymatic improvement of defatted egg yolk powder. *Innovative Food Science & Emerging Technologies*, *80*, 103090. https://doi.org/10.1016/j.ifset.2022.103090

Zhang, Q., Yang, L., Hu, S., Liu, X., & Duan, X. (2019). Consequences of ball-milling treatment on the physicochemical, rheological and emulsifying properties of egg phosvitin. *Food Hydrocolloids*, *95*, 418–425. https://doi.org/10.1016/j.foodhyd.2019.04.064

Zhang, X., Chelliappan, B., & Antonysamy, R., M. (2021). Recent advances in applications of bioactive egg compounds in nonfood sectors. *Frontiers in Bioengineering and Biotechnology*, *9*, 738993. https://doi.org/10.3389/fbioe.2021.738993

Zhang, Y., Chen, Y., Xiong, Y., Ding, B., Li, Z., & Luo, Y. (2022b). Preparation of high internal phase Pickering emulsions stabilized by egg yolk high density lipoprotein: Stabilizing mechanism under different pH values and protein concentrations. *LWT, 157,* 113091. https://doi.org/10.1016/j.lwt.2022.113091

Zhao, Y., Feng, F., Yang, Y., Xiong, C., & Tu, Y. (2021). Gelation behavior of egg yolk under physical and chemical induction: A review. *Food Chemistry, 355*(5), 129569. https://doi.org/10.1016/j.foodchem.2021.129569

Zhong, V., Van Horn, L., Cornelis, M. C., Wilkins, J. T., Ning, H., Carnethon, M. R., Greenland, P., Mentz, R. J., Tucker, K. L., Zhao, L., Norwood, A. F., Lloyd-Jones, D. M., & Allen, N. B. (2019). Associations of dietary cholesterol or egg consumption with incident cardiovascular disease and mortality. *The Journal of the American Medical Association, 321*(11), 1081–1095. https://doi.org/10.1001/jama.2019.1572

Zhu, H., He, Z., Kwek, E., Liu, J., Hao, W., Liang, N., & Chen, Z. (2018). Dose-dependent increases in liver cholesterol but not plasma cholesterol from consumption of one to five whole eggs and no effects from egg whites on liver or plasma cholesterol in hamsters. *Journal of Agricultural and Food Chemistry, 66*(48), 12805–12814. https://doi.org/10.1021/acs.jafc.8b04730

Zou, W., Tang, S., Li, Q., Hu, G., Liu, L., Jin, Y., & Cai, Z. (2020). Addition of cationic guar-gum and oleic acid improved the stability of plasma emulsions prepared with enzymatically hydrolyzed egg yolk. *Food Hydrocolloids, 105,* 105827. https://doi.org/10.1016/j.foodhyd.2020.105827

Part IV

Processing of the Egg and Its Application in the Food System

10

Egg Processing Plant Design: From Shell Eggs to Liquid and Powder Egg Products

Hans Chr. D. Rasmussen, Roberto Colavitti, Giosuè Casagrande, Jan Høy Carstens*, and Paolo Bonato

CONTENTS

DOI: 10.1201/9781003254430-14

10.1 Introduction

While egg processing may have started in the 1950s as an outlet for excess inventory of shell eggs, the use of egg products is now common in the food service and food industry sector, bringing additional safety and uniformity to the use of eggs. Egg products are also reaching retail stores, providing consumers with more opportunities to eat eggs. While egg processing is rather complex with the presence of egg proteins bringing some challenges not seen with other processed foods, a series of well-defined principles and specifically-designed pieces of equipment allow to repeatedly produce quality and safe products. This chapter will review some key principles and steps involved for the production of liquid egg and egg powder products.

10.2 Egg Breaking Lines

Until the introduction of egg breaking machines in the 1950s, the breaking of shell eggs was performed manually (Figure 10.1). A very skilled person was able to break up to 2,000 eggs per hour, including the removal of eggs deemed not suitable for processing (Johansen, 1998). A system for the separation of the yolk from the egg white was developed, consisting of two slanted chutes along a table, where the employees were located. The upper chute was equipped with holes or a sleeve, allowing the egg white to drop onto the lower chute (Johansen, 1998).

Compared to the current breaking machines with capacities of over of 200,000 eggs per hour, the first breaking machines had a much smaller capacity, around 8,000 eggs per

FIGURE 10.1 Manual egg breaking.

hour (Figure 10.2). However, they represented considerable improvements in terms of labor costs.

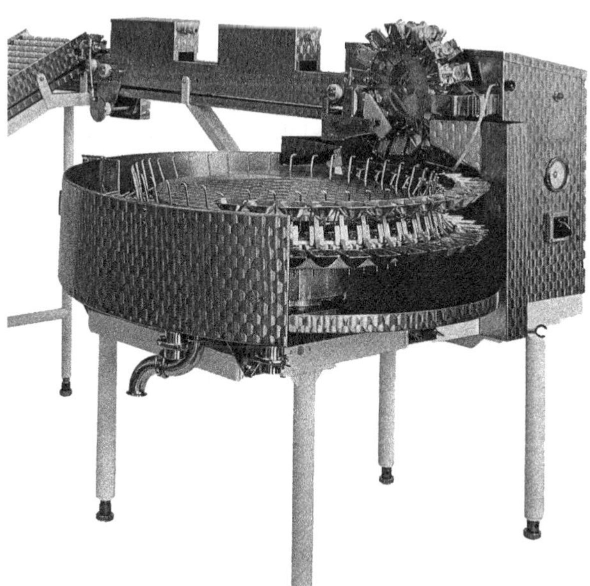

FIGURE 10.2 The first breaking machines.

10.2.1 Loading of Eggs

While the eggs were initially placed manually directly onto the infeed conveyors of the breaking machine, the automated loading of breaking machines was introduced to boost production capacity. Eggs used by breaking plants are usually delivered to the facilities in trays, each containing 30 eggs, and stacked in 6 layers. The stacks are placed onto pallets or trolleys. Single trays or stacks of six trays are placed on the loader infeed belt (Figure 10.3). Each single tray is picked up by the loader, supported at the bottom side by hooks. The eggs are then lifted off the tray, with a suction cup positioned on top of each egg. The empty trays, remaining on a conveyor belt, are usually automatically stacked while the eggs are positioned onto a special conveyor often called roller table, with rollers bringing the eggs to the breaking machine in a number of rows matching the breaking machine.

The manual loading of stacks of six trays at a time is heavy and labor-intensive work. Aids have been developed

FIGURE 10.3 Loading of eggs for breaking.

to facilitate such work, e.g., as pneumatic manipulators (Figure 10.4), making one operator able to place six stacks at a time onto the loader infeed conveyor using only limited strength.

Due to the increased capacity of breaking lines in recent years, some automation has been introduced, with robots used for depalletization and the placement of stacks of trays onto the loader infeed systems. While egg trays have traditionally been made of pulp material, automation has increased the use of plastic trays since plastic is a more rigid material more suitable for automation (Figure 10.5). A complete system of pallets, divider plates between every six stacks of trays, and trays, all made of plastic, were therefore developed for safer transportation and greater ease for the automatic loading of eggs onto breaking machines.

The plastic cargo system has also become popular for its cleanability. Unlike wooden pallets or pulp trays, the plastic cargo system is easy to clean and sanitize, thus reducing the risks of bacterial contamination in the breaking area or spreading diseases back to egg farms. On the other hand, pulp trays will weaken under pressure or within a humid environment. When stacks of six trays are placed on a pallet, up to six layers high, the load on the bottom trays is considerable. Pulp trays do not have sufficient strength to withstand such load. Consequently, the eggs will have to carry the load, resulting in several cracked eggs. Plastic trays can withstand a much higher load and the eggs will not crack. The automation of loading often includes the in-line washing of trays, dividers and pallets, as well as their robotic palletizing prior to being returned to the farms (Figure 10.6).

In Japan and other Eastern Asian countries, a special loading method is used, called Japan Washer or Aqua Loading (Figure 10.7). Eggs, from a single tray, are manually dropped into a water tank with a submerged roller conveyor bringing the eggs toward the breaking machine. The method has some hygienic issues since the water becomes quickly contaminated despite frequent replacement. This kind of loading is also very labor-intensive and not suitable for large capacity breaking lines. However, this is their preferred loading method since the dimensions of egg tray in Japan and some other Asian countries have not been standardized, thus making automated loading difficult.

10.2.2 In-Line Operations

Especially in the USA, companies integrating a breaking facility with a farm with multiples barns are common. The eggs are conveyed onto rod conveyers directly from the layer houses (most often using cage housing systems) to the breaking machines, thus eliminating the handling of eggs between farms and a breaking facility. Although very efficient, the system is vulnerable to outbreaks of avian influenza and other diseases,

FIGURE 10.4 Pneumatic manipulator.

FIGURE 10.5 Cargo System: Plastic pallets, divider plates, and trays allow for safer transportation and easier automatic loading of eggs.

as both farm and breaking facility would be part of the same quarantine perimeter and shut down in case of an outbreak.

As eggs are randomly spread onto the rod conveyors to bring them toward the breaking lines, they must be guided into separate rows on the conveyor leading to the breaking machine (Figure 10.8). This is done by using an accumulation table. Despite the name, its ability to accumulate eggs is rather limited.

There are plants combining in-line and off-line operations, with the possibility to load eggs from trays onto the accumulating table of an in-line feed system.

10.2.3 Dirt and Leak Detectors

In recent years, the automatic dirt and leak detection system, similar to that used in grading plants, has been integrated, in some cases, in egg breaking lines after the egg washing step. A dirt detection system consists of a camera, a light source and the appropriate software. Such system can detect spots as small as 1 mm^2. A leak detection system can detect holes in the eggshell also as small as 1 mm^2. The dirt and leak detector is positioned inside an enclosed cabinet to avoid disturbances from any light sources. Dirty eggs and eggs with leaks detected by this system are automatically removed from the roller conveyor through a drop gate in the conveyor.

10.2.4 Egg Washing

The washing of eggs prior to breaking is mandatory in the USA, unless the eggs are completely clean. Therefore, all breaking

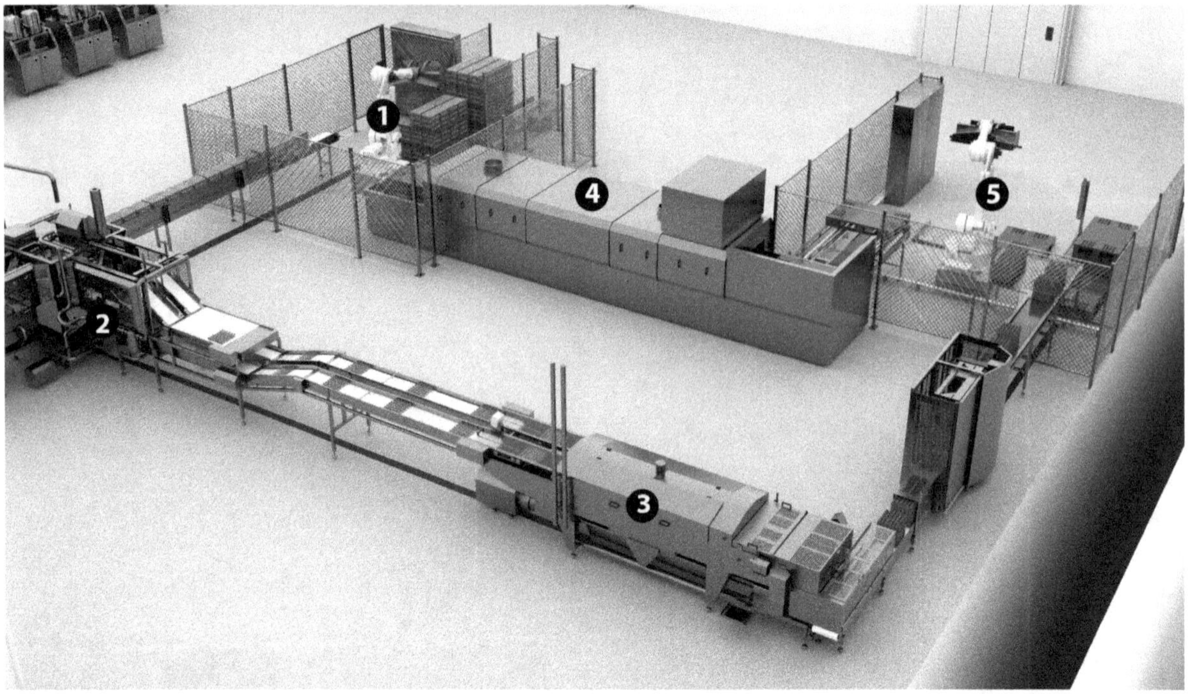

FIGURE 10.6 (1) Robotic depalletizer. The stacks of eggs are placed on the infeed conveyor of the loader (2) the divider plates and the pallets are placed on the infeed conveyor of the pallet washing machine (4). The empty trays are directed from the loader (2) to a tray washing machine (3). The clean trays and pallets and dividers from (4) are palletized at (5).

lines in the USA are equipped with washing machines. While egg washing is also performed prior to egg breaking in several countries in South America and Asia, it is not a common practice in Europe. The argument for not washing eggs relates to the fact that a protective coating, the cuticle, is laid on the outside of the egg by the hen at oviposition. By washing, the risk of contamination is increased as the cuticle may be removed and the water with impurities could be sucked through the pores in the shell into the egg (see Chapter 18).

When washing eggs, a convey called the roller table, bringing the eggs toward the breaking machine, is equipped with a washing machine (Figure 10.9). In the washing machine, the eggs on the roller table are passing through a tunnel. Washing

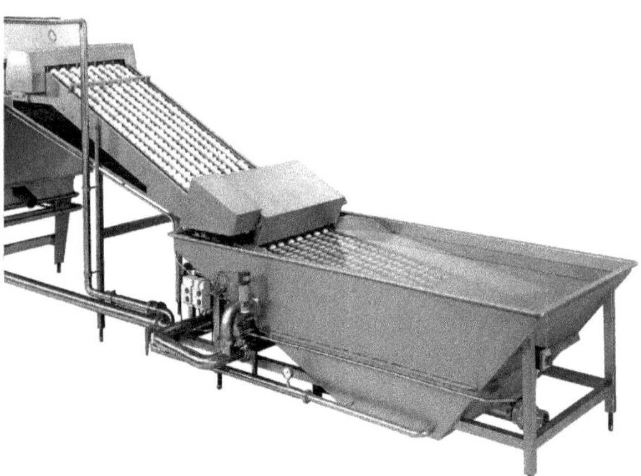

FIGURE 10.7 Japan washer.

FIGURE 10.8 Accumulating table.

FIGURE 10.9 Egg washing machine.

is normally performed by the combination of spraying warm water on the eggs and moving brushes rubbing against the eggs. Underneath the washer roller table, a tank is installed for the collection and circulation of the wash water. Since the rollers on the conveyor are rotating, the eggs are also rotating while going through the washing machine, thus increasing the washing efficiency. The brushes are designed to follow the shape of the eggs for a more even cleaning of the eggshells. The proper cleaning of the egg extremities may be a challenge when using brushes. Wash water is also sprayed over the brushes to keep them clean. The height of the brushes above the rollers can be adjusted, depending on the size of the eggs. A lower position of the brushes will increase the mechanical cleaning force, but also increase the number of broken eggs. When washing non-graded eggs, the height of the brushes is defined as a compromise between the number of small eggs not sufficiently cleaned and the number of larger broken eggs occurring during the washing process. Some egg washing machines using only nozzles have been developed. The advantage of this system is to be able to better wash both extremities of eggs. In this case, the washing efficiency does not depend on the egg size and there are fewer leaking eggs, with portions of the shell missing. Concurrently with the addition of a detergent in the wash water, an anti-foaming agent may be added to avoid the formation of foam in the wash water, created by proteins from broken eggs.

The following factors are important to achieve the satisfactory washing of eggs: wash time, temperature, mechanical force and use of a detergent. If one factor is suboptimal, another factor must compensate to achieve the same washing result. To increase the wash time, the speed of the roller table must be reduced, thereby reducing the capacity of the breaking line. Alternatively, the length of the wash tunnel may be increased. There are some limitations for the water temperature since the process should not cook the eggs. Finally, too much mechanical force will increase breakage during washing. Therefore, the best compromise must be found. Although the wash time is typically up to 60 seconds, the speed may be slowed down to extend the washing time of very dirty eggs. According to the Food Safety and Inspection Service (FSIS), the temperature of the wash water in the USA must be at a minimum of 32°C (90°F) and at least 11°C (20°F) higher than the temperature of the eggs (see Chapter 18). The water temperature in the wash tank must not exceed 49°C (120°F) to prevent thermal cracks in the eggshell. The level of the wash water in the tank is maintained by adding water to compensate for the water losses. The right concentration of detergent should be maintained automatically, by continuous dosing the detergent. Finally, to maintain good sanitary conditions, the wash water must be changed frequently.

10.2.5 Sanitizing

In the USA, sanitizing is mandatory after washing and is a common practice elsewhere. Sanitizing is performed by spraying water with a sanitizer such as chlorine, hydrogen peroxide or other approved sanitizers. The sanitizing liquid may be saved and returned to the washer tank as replacement water.

10.2.5.1 UV-Light

UV-light is used for disinfection of the eggshell surface prior to egg breaking. A cabinet with UV-lamps, reflectors and cooling fans is placed on the roller table in the breaking line, normally right before the eggs enter the breaking machine. UV-light will only sanitize the surface of the eggs and will not penetrate deeply into dirt on the eggs. With only a few seconds exposure, this method is not effective at fast processing speeds. Tests have shown a Total Viable Count (TVC) reduction of 0.8 log on a 6-row egg conveyor after exposure by 10 lamps, 55 W, for 2 seconds.

10.2.5.2 Other Sanitizing Methods

A new method, combining steam and ultrasound treatment, has shown very promising results with a TVC reduction of 3–5 log for a 1.5 second exposure. The catalyzing effect of ultrasound accelerates the heat transfer to the eggs, resulting in an instant increase in temperature at the surface. The disinfection

FIGURE 10.10 Candling booth.

treatment is halted before the heat penetrates the eggs and causes unwanted thermal damages.

10.2.6 Candling

When a candling booth is used, it is usually placed after the washing machine. During the candling process, the eggs are mechanically rotated several times by the roller table, over a bright light shining underneath, for visual inspection by operators (Figure 10.10). With the internal inspection of the eggs also possible, the candling process will allow the removal of dirty eggs, eggs with meat spots or blood spots, and other eggs deemed not suitable for breaking. Candling is mainly used in the USA and Canada.

10.2.7 Breaking Machines

All breaking machines contain at the minimum a cracker assembly consisting of grippers to receive the eggs from the roller table, a knife function to penetrate and open the eggshells, and a holding mechanism to retain the opened eggshell until discarded (Figure 10.11). The knife assembly consists of two thin knife blades, placed closely to each other, when they penetrate the egg. After penetrating, the two knife blades will spread apart, thus facilitating the release of the egg content.

Using this setup, the breaking machine will only be able to produce whole eggs. In order to separate the egg white and egg yolk, a receiving cup assembly must be placed right under the cracker assembly. The egg content is dropped into a cup, designed in such a way to retain the yolk, while the egg white flows over the edge and is collected by another cup underneath (Figure 10.12). The cup holding the yolk is vibrating for better separation of the egg white from the yolk.

After the separation and a possible inspection by operators, the liquid product is emptied into collecting trays, usually integrated at the bottom of the machine. If the vitelline membrane around the yolk is ruptured during the breaking process, the yolk will also flow from the yolk cup into the egg white cup. The presence of yolk particles in the egg white will increase its fat content and reduce the foamability of the egg white. For this reason, egg white cups containing some traces of yolk will be emptied into the tray collecting liquid whole egg. When emptied, the cups are washed, and air dried in a cup wash section.

FIGURE 10.11 Cracker unit.

For many years, breaking machines had cracker and cup assemblies mounted on a carousel or a chain conveyor moving in a circular or ellipse-shaped pattern (Figure 10.13). The eggs were received from the roller table by grippers on the cracker assembly. The next steps along the carousel were the breaking and opening of the eggs, separating and emptying of cups before returning to the position where new eggs were received, repeating the process all over again. Similarly, each cup was inspected by an operator and a pin on the cup assembly was activated when some egg white contaminated with yolk was observed, emptying the cup into the whole egg tray. Likewise, the observation of rotten eggs would lead to the cups being emptied into a waste pan.

FIGURE 10.12 With separation cup.

FIGURE 10.13 Carousel egg breaking machine.

In the late 1980s, a scanning system was developed for monitor the presence of yolk traces into egg white cups. The system was sending a light at a specific wavelength and measured the wavelength of the reflecting light from the content of the egg white cup. If yolk particles were detected, the separation cups would automatically be emptied into the whole egg tray. This feature allowed to eliminate the need of operator inspection in many countries. However, in some countries, including the USA, Canada and Japan, local regulations still demand an inspection with human operators. When rotten eggs are detected, the machine must be stopped and the cracker and separation cup assemblies must be detached for cleaning.

Even though double decker machines with two carousels were developed to double the breaking capacity for the same plant footprint, it became evident that the carousel system was no longer ideal to meet the demand for higher breaking capacity. In the 1990s, linear breaking machines were introduced with several parallel breaking lanes (Figure 10.14). This design gave a much higher capacity when compared to the footprint of the earlier machines. On the carousel-type breaking machines, the same cracker and separation cup assemblies follow each other all the way around and the movement took place on a horizontal plane. On the linear breaking machines, the cracking assembly and the separation assembly are mounted on two separate chain systems, and the movement takes place on a vertically plane. The cracker chain system is placed above the separation cup chain system, with the speed and position of these two systems synchronized for correct delivery of the liquid from the crackers to the separation cups. The cracker conveyor is shorter, thus making inspection of the separation cups possible, since inspection for rotten eggs is still required in the USA, Canada and Japan.

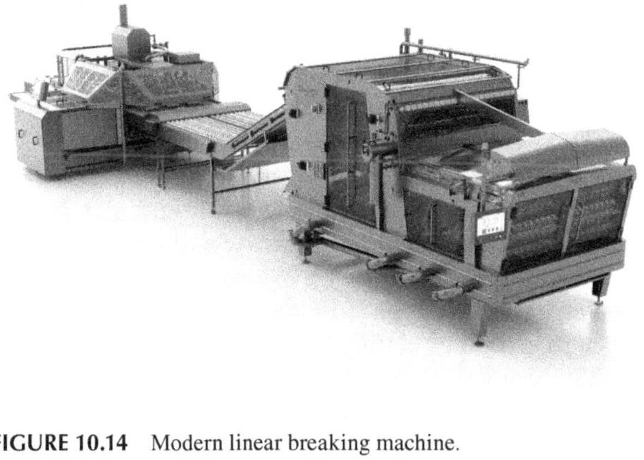

FIGURE 10.14 Modern linear breaking machine.

FIGURE 10.15 Plastic separation cups.

Over time, linear breaking machines have been improved with a new scanning technique developed based on a computer vision system. Separation cups made of plastic were also introduced, offering several benefits (Figure 10.15). The light source for scanning can be placed underneath the cups, giving a diffuse light for the camera, and unlike stainless cups, plastic cups do not generate light reflection, therefore ensuring optimum detection. Another advantage of plastic cups is less mass to be transported, reducing the overall power consumption and the wear and tear on the chain links.

The breaking line is the heart of any egg processing facility. The performance of the breaking machine affects the yield and the quality of finished products. The cost of shell eggs is by far the highest cost in an egg processing facility. Consequently, reducing product losses is essential for

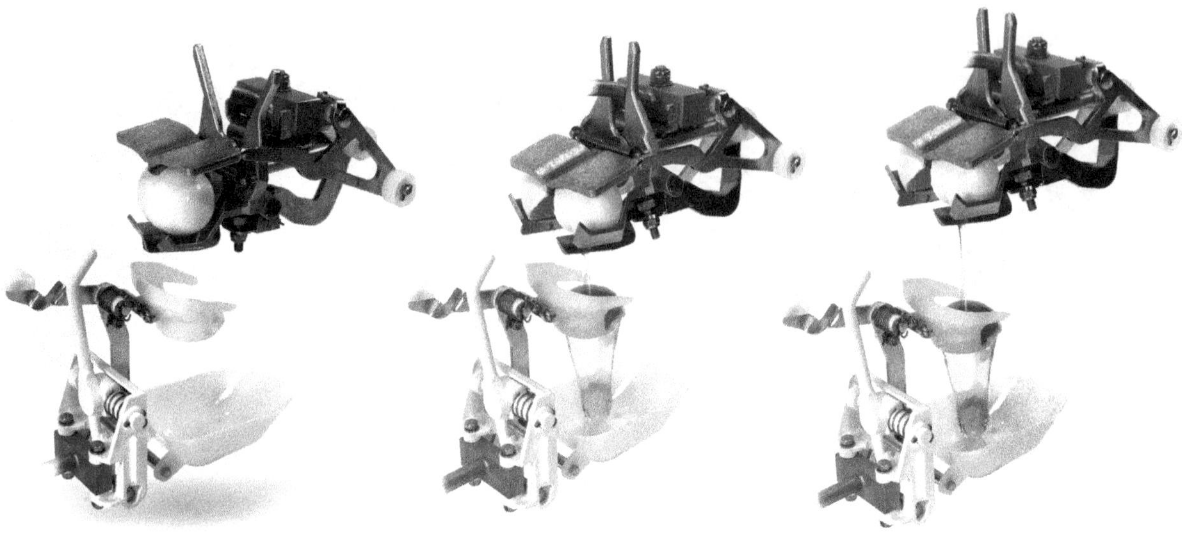

FIGURE 10.16 Increasing distance between egg shell and separation cup.

a profitable production. Losses can occur at all steps, from the transport of shell eggs to the cleaning of pipes and tanks parts of the processing equipment.

Regarding the breaking process, one very critical point is the transfer from rollers to cracker assemblies. The speed of the crackers is typically double the speed of eggs on the rollers. At the transfer point of eggs, the speed is increased, meaning some risk that the crackers will not catch the eggs and lose them. This risk can be minimized by tilting the cracker assembly in a backward movement when passing the transfer point, thereby slowing the speed of the cracker relative to the speed of the eggs on the roller conveyor. To optimally catch the eggs, the cracker must be in different positions when compared to the position of eggs on the roller at the time of transfer, depending on the speed of the line. This can be obtained by electronically adjusting the position of the crackers relative to the position of the rollers, based on the speed of the machine. After the eggs have arrived safely in the crackers, the yield can be maximized by ensuring a large opening angle of the eggshell and a long dripping time. Additional measures can be taken. For instance, by slowly increasing the distance between the cracker assembly and the separation cups after delivering the liquid eggs into the separation cups, the more viscous part of the egg white adhering to the shell will be affected by an increased gravity power and released from the shell (Figure 10.16).

Finally, the remaining egg white inside the eggshells can be sucked out, prior to discharging the shells. While the yield improvement obtained from the measures mentioned above will depend on the quality of the eggs, they do have an impact on the overall economic profitability of egg processing. For many egg processors, a goal is to minimize the total amount of whole egg produced and to obtain an egg yolk with a high solid content, meaning minimizing the amount of egg white in the yolk. To minimize the amount of whole egg, the knife design of the crackers is essential. When the egg is emptied from its shell, there is a risk of breaking the vitelline membrane around the yolk. This can be minimized by tilting the knives during the emptying of the eggs, thus avoiding the breaking of the yolk membrane when passing the corner edge of the knifes. A minimal drop distance between the eggshell and yolk cup will also minimize the risk of broken membranes. To ease the separation, the yolk cup is vibrating, but there is also the risk that vibrations too powerful will increase the risk of breaking the yolk membrane, thus resulting in producing more whole egg product. Finally, an optimal scanning system with no false positive for the detection of traces of yolk in the egg white cups is also important.

The design of the yolk cup, improved over the years, as well as the separation time have an influence on the dry matter of the yolk. The goal should be to leave as little space as possible for the egg white in the yolk cup, without dropping the yolk into the egg white cup. With small eggs, the amount of remaining egg white in the yolk cup will be higher than for larger sized eggs.

When the separation into yolk and egg white is not required, machines without separation cups can be used (Figure 10.17). In this case, the liquid will drop directly from the cracker assembly into a tray. A separation is still possible on this type of machine, by guiding the eggs from the tray into a chute with a sleeve used for the drainage of egg white (Figure 10.18). In these machines, the separation of egg white from yolk spots must be done manually by scooping with a spoon or similar device. Using this method, a satisfying separation into egg white and yolk can only be obtained with very high-quality, fresh eggs since the vitelline membrane will remain firm and unbroken.

A simpler way to break shell eggs into liquid whole egg is by using a centrifuge, normally used for the separation of the remaining egg white from shells. This method is only legal in a few countries as through this process, the egg liquid obtained is in direct contact with the exterior of the eggshells.

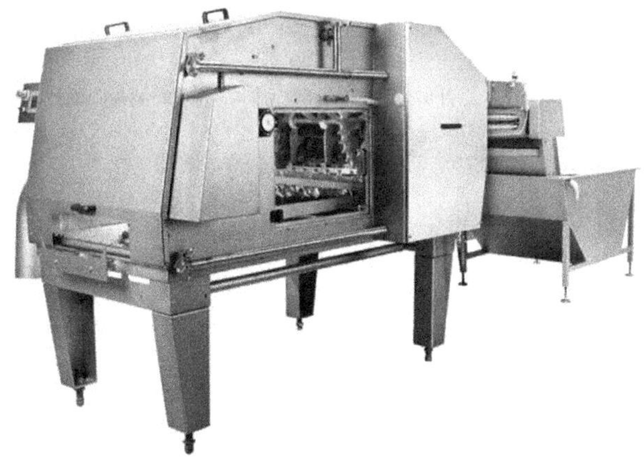

FIGURE 10.17 Breaking machine without separation cups.

FIGURE 10.18 Separation chute.

10.2.8 Super Whole Egg and Super Yolk

In Japan and other countries in East Asia, non-pasteurized egg products, known as Super whole egg and Super yolk, are popular. In these products, the membrane around the yolk must remain intact and they are packed, often in small volume jars, directly from the breaking machine (Figure 10.19). The integrity of the vitelline membrane is considered as a guarantee of the freshness of these products. The manufacture of these types of products required some additional efforts to ensure a very gentle handing of eggs, with extra precautions taken to minimize the drop distance of the egg yolk when emptying the egg cups.

FIGURE 10.19 Super whole egg.

FIGURE 10.20 Egg shell centrifuge.

10.3 Egg Shell Treatment

From the breaking machines, the eggshells are conveyed to a shell room for further treatment and processing. This conveyance is typically performed by augers, wire belts or pneumatic conveyance. The liquid egg remaining in the shells expelled from the breaking machine is separated from the shells and the shells are then crushed by a centrifuge (Figure 10.20) or a screw press (Figure 10.21) to minimize the volume. The centrifugal force creatıng by spinning or pressure from the screw allows the remaining liquid to be separated from the shells.

In a screw press, the eggshells are slowly moved forward toward a piston by the action of a screw inside a tubular screen, forcing the liquid through the screen. The intense pressure exercised by the screw cruses the eggshells. Compared to the centrifuge method, the remaining moisture in the shells

FIGURE 10.21 Screw press.

FIGURE 10.22 Kiln oven egg shell dryer.

will be slightly higher when using the screw press. While in most countries the liquid egg obtained is considered as inedible and not fit for human consumption, it may be used for the production of animal feeds. In many cases, the liquid will be spray dried to extend its shelf life. The crushed eggshells may be discarded or used as a feed supplement for animal feeds since it has a high content of highly bioavailable calcium (see Chapter 11). For the same reason, it can also be spread on fields as a calcium source.

Due to the relatively high moisture content and high bacterial count, the eggshells will rot very fast and consequently emit an unpleasant odor, making them unsuitable for animal feed production. This degradation of the shells can be avoided by drying them. Shells are usually dried in rotary kiln ovens (Figure 10.22), but fluid bed drying can also be used (Figure 10.23). In some cases, mostly in Asian countries, dried eggshells are also used for human consumption.

In the European Union, since the BSE crisis, it is forbidden to give animal byproducts from a species to the same species. Therefore, eggshells are not allowed as a source of calcium for the formulation of poultry rations.

10.4 Liquid Egg Products

After the breaking stage, the three liquid products (egg white, egg yolk and whole egg) potentially obtained must be delivered to raw storage tanks in the most hygienic and gentle way possible to reduce any risk of contamination and mechanical stress to a minimum. After breaking, liquid products have lost their protective shells and have been in contact with the equipment, the external environment as well as small particles of shells inevitably generated during the breaking process. It is therefore extremely important to keep the bacteria level in the raw products as low as possible by removing the shell fragments as much as possible. Since most bacteria found on eggshells have a limited growth at low temperatures, it is also essential to cool down these raw products to a temperature <4°C to limit the bacterial growth. While it is easy to understand the importance of removing eggshell pieces from a product quality standpoint, they also constitute a major source of potential contamination. For these reasons, the filtration of the raw product is an extremely important step to consider.

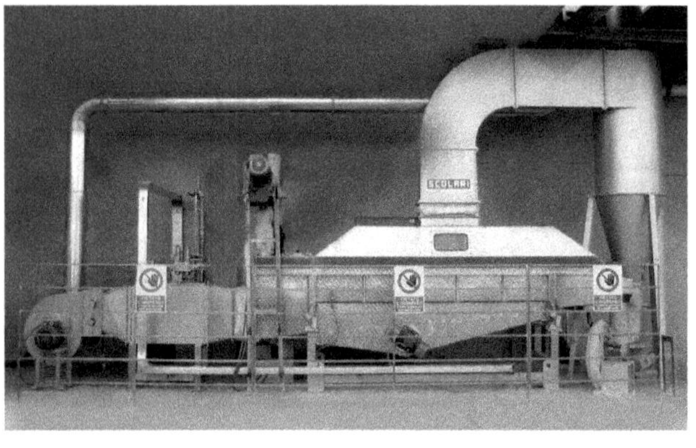

FIGURE 10.23 Fluid bed egg shell dryer.

10.4.1 Product Receiving

In general, the liquid products, flowing by gravity from the breaker, are collected in vats to create a good buffer of product to be pumped to the storage tanks. The collecting vats (1 for each type of product) are positioned close to the breakers in ensure a good flow of product by gravity. In each collecting vat, a sensor is installed to time, based on the level of product, the on-off feature of the unloading pump used to convey the raw product to its destination (Figure 10.24). During this phase, it is important to have a good control system to avoid having the pump working with not enough product in the inlet line. Otherwise, some air will also be pumped and mixed in the product, thus creating a lot of foam and a source of problems later on in the production lines.

A plant setup without collecting vats is also possible. In this case the product is flowing by gravity from the breaker directly to the pump and a pressure/level sensor is placed before the pump to control the level of the product in the pipes (Figure 10.25). The speed of the pump is adjustable, depending on the quantity of the product in the pipe, to adapt the flowrate with operation speed of the breaker. When compared with the on/off functionality for the collecting vats, this continuous level control system allows a constant flow of the product in the pipes and a more constant refrigeration of the product.

FIGURE 10.24 Breaker line collecting VAT-filter-cooler.

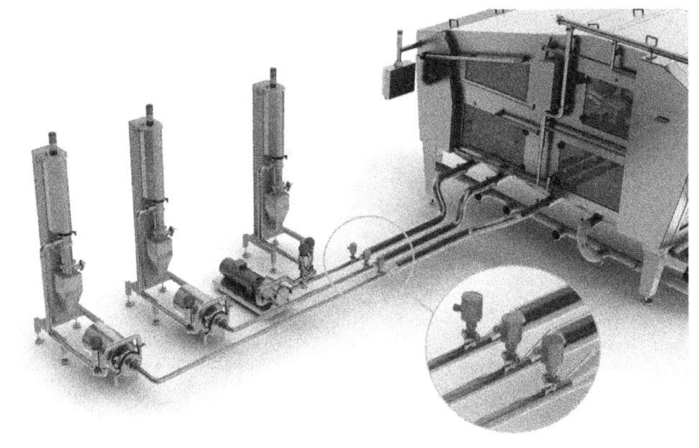

FIGURE 10.25 Breaker line continuous level control filter cooler.

10.4.2 Pumps

Each collecting vat is equipped with level sensor to control a pump to unload the vat and deliver the product to the next processing steps. While different type of pumps can be used for this purpose, two main types have been identified, Dynamic pumps and Positive Displacement pumps:

Dynamic pumps (i.e., centrifugal pumps): Through a high-speed impeller, the centrifugal pump creates a centrifugal effect that moves the liquid egg, transforming the mechanical energy into kinetic energy. Due to the high speed and shear stress these pumps apply to the product, they are mainly used in liquid egg processing plants for the whole egg and egg yolk lines.

Positive displacement (i.e., lobe Pumps): The liquid egg is pumped by means of the displacement or variation of volume. These pumps are gentler than the centrifugal pumps and, with a lower shear stress, and are usually installed in lines dedicated to Egg white since proteins are very sensitive to mechanical force that may damage irreversibly the functional properties of the product.

10.4.3 Filtering

To remove all the small parts of shells carried from the breakers along with the product, filters are inserted in the production lines after the collecting vats. The simplest way to filter the product for shell particles as well as other potential impurities is to use manual filters (Figure 10.26).

While the product is flowing from inside to the outside of the filter, all shell particles are blocked inside the cartridge. The filter must be opened regularly, and the internal cartridge cleaned manually by operators when dirty. Since production must be stopped during the manual cleaning of filters, this type of filter is, very often, mounted with two sets of filters to allow production to continue using one filter while the second one is being cleaned (Figure 10.27).

It is important to clean the filters often, and not only when the filter is blocked, since the pieces of shell present on the filter will be in contact with the product flowing from the inside

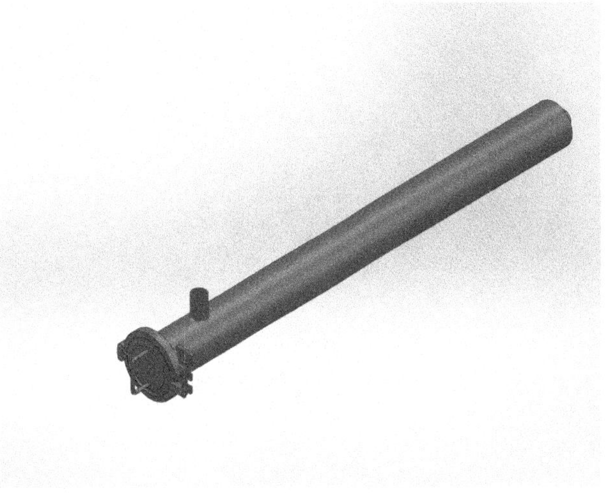

FIGURE 10.26 Illustration of manual filter.

FIGURE 10.27 Illustration of manual double filter.

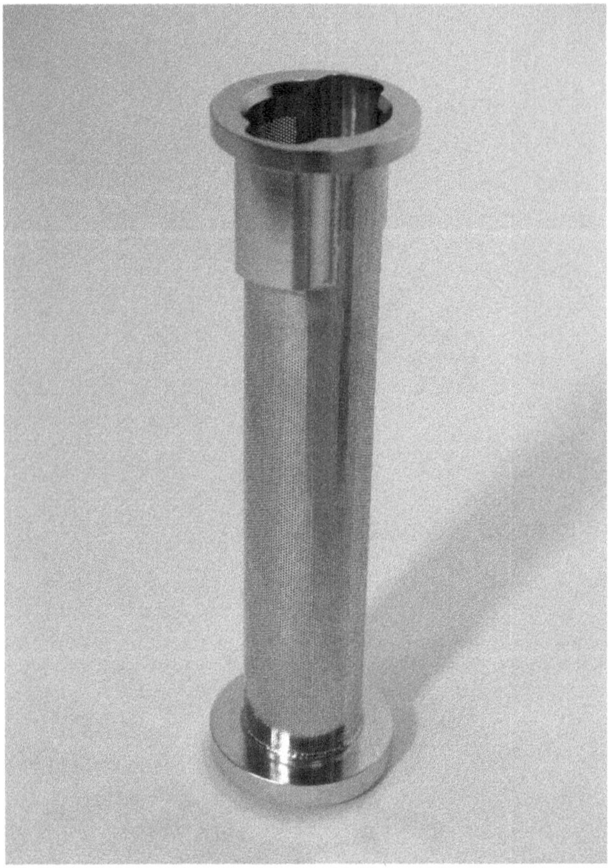

FIGURE 10.28 Illustration of a cartridge with perforated plate.

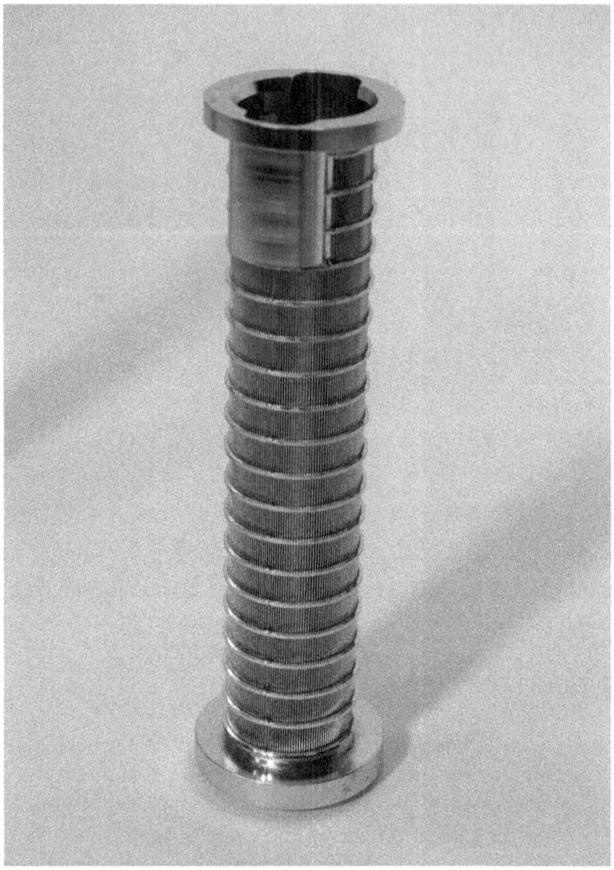

FIGURE 10.29 Illustration of a cartridge with wedge wire screen.

and therefore will contribute to further contamination of the raw product. The main feature of the filters is the size and shape of the holes in the filter cartridge. Although there are many designs for filtering cartridge available commercially, the most used commonly used for the processing of egg yolk and liquid whole egg is a cartridge with perforated filter inserts of the order of 1 mm (Figure 10.28).

During the processing of liquid egg white, the use of filters may cause some issues with the chalazae obstructing the filter very quickly. Therefore, filters with wedge wire screen with dimensions in the order of tenths of a millimeter are commonly used on egg white lines (Figure 10.29).

Instead of using manual filters in egg product plants, it is preferable to use automatic self-cleaning filters. Their basic concept is the same as that manual filters although they are equipped with a piston or a scraper that will continuously remove the shell particles from the cartridge, thus reducing to the minimum the time of contact between the shells and the product filtered and limiting potential contamination (Figure 10.30).

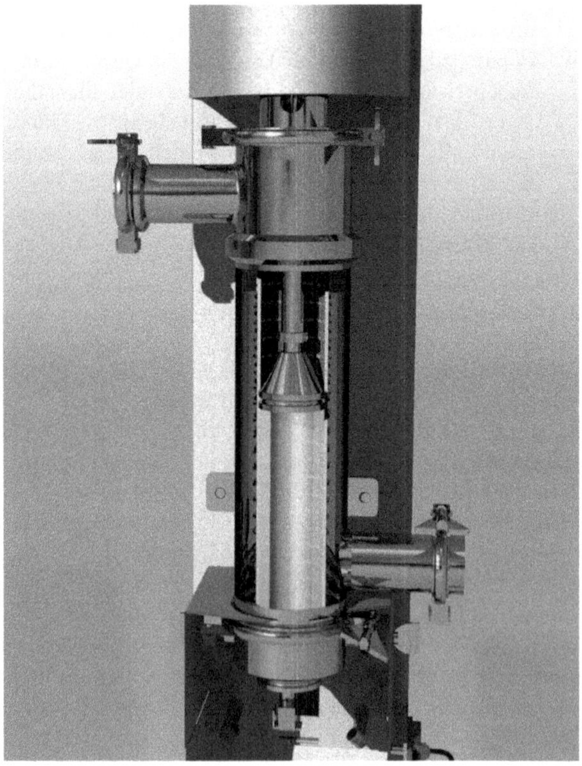

FIGURE 10.30 Illustration of an automatic filter where a piston is pushing the shell particles out from the bottom.

10.4.4 Cooling

Once the product has been filtered to remove as much as possible of the shell particles, it must be cooled down to a temperature <4°C to limit bacterial growth. Coolers, based on plate or tubulars heat exchangers, are used with glycol water recirculated as a cooling media (Figure 10.31).

FIGURE 10.31 After filtering, the cooling of product can be achieved with plates (right) or tubular heat exchangers (left).

The cooling of the product is extremely important, and it must be continuously controlled to ensure the right temperature of the products before reaching the storage tanks. The glycol water is regulated by valves, controlled automatically by a temperature transmitter installed in the product line, and will close when product pumps stop to avoid freezing the product. For various reasons, it is important that the filtration of product occurs before this cooling step. First reason is when using Plate Heat Exchangers, the shell particles may block the cooler due to small passages allowed within the plates. A second reason is that the cooling step will increase the viscosity of the product considerably, especially egg yolk, and would make filtration post-cooling very difficult. After passage through the coolers, the liquid products are delivered to their respective storage tanks.

10.4.5 Raw Storage Tanks

One of the important functions of the raw product storage section is to create a production buffer that will allow the continuous supply of product to the pasteurization step occurring downstream of these tanks. It is also important to let the product rest before pasteurization to reduce the amount of air contained (de-aeration) in the product and ensure an efficient pasteurization. However, the storage of the raw products before pasteurization has another important function: it allows to standardize the product to meet the desired characteristics. The characteristics of liquid egg products vary depending on the quality of the eggs used (e.g., age of hens, seasonality, temperature, type of feeds provided to the hens) and on the quality of the machine used for breaking (e.g., efficiency in breaking, separation of the products, drainage time of the shell). All these variables lead to the non-uniformity of the raw products obtained, potentially varying in dry matter content (DM), fat content, color and raw products may not comply with the desired specifications of the finished products. It is therefore essential to check products in the raw storage tanks and make the appropriate adjustments prior to sending a product for pasteurization since further adjustment will not possible after pasteurization.

In the raw product storage section, it is not only the chemical-physical characteristics of the plain products that will be checked and adjusted, but it is here that other ingredients (typically salt, sugar, enzymes, gums, milk and other ingredients) will be added to produce specially formulated egg products.

The operations of control and preparation of these mixtures require some time since products stored in the tanks must be refrigerated and kept in motion to avoid any separation effect. The tanks are equipped with a refrigeration jacket in which cold water recirculates and with an agitator, used to keep the product moving (stirring effect) or to mix any ingredient (mixing). Generally, in the storage area, there are tanks dedicated to the simple storage of products and adjustment of

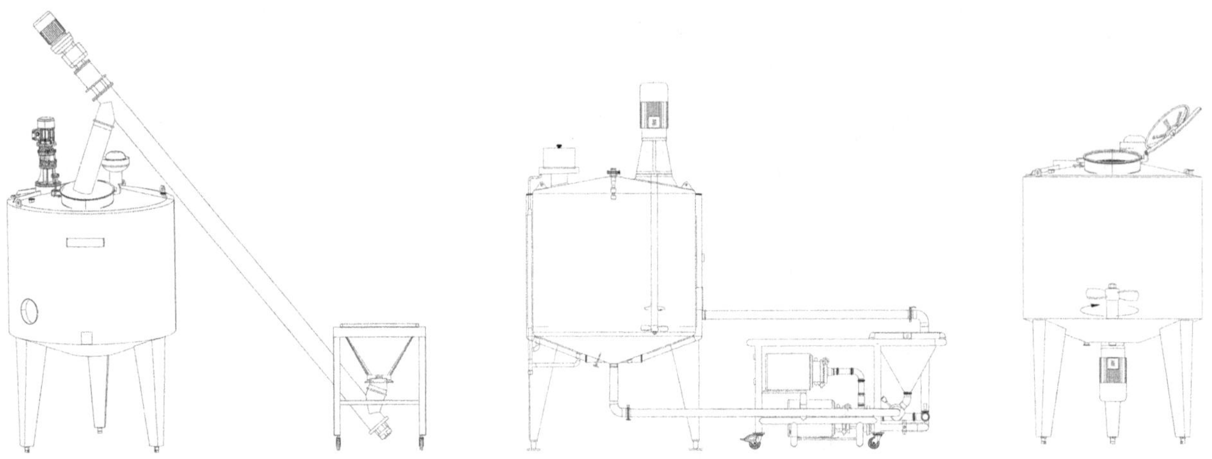

FIGURE 10.32 Schematic illustration of the different ways to incorporate and mix various ingredients into liquid egg products.

the chemical-physical characteristics (storage tank) and tanks dedicated to the mixing with other ingredients (mixing tank or blender). The parameters to control in the storage tanks are not numerous but they are extremely important to ensure a correct production, the legality of the product (some countries mandate a specific DM value for certain products) and to reduce to the minimum any risk of production failure. In particular, as minimum requirements, a temperature transmitter must be used to monitor the temperature of the stored product, a level sensor at the bottom to detect the presence or absence of product in the tank interlocked with any cleaning function to avoid cross contamination and another level sensor located at the top of the storage tank to stop any feeding function when the tank is full. These are the minimum required instruments installed in any storage tank to monitor the correct execution of the process. Very often, a weighting system is installed and will provide the continuous monitoring of the quantity of product stored. This instrument is particularly important for the mixing tanks in order to calculate the right amount of ingredients to mix in the product.

The incorporation of ingredients into the product is generally made through a manhole in the tank when small quantities are added or automatically through a blending system when the quantity added is more significant (typically salt and sugar with for instance the 50% sugared egg yolk product with an equal amount of sugar and egg yolk). A third possibility is to use a high shear mixing tank used for batches preparation. The product is loaded to the tank (typically around a 1,000 liters), then the ingredient is added, mixed for few minutes with a high-speed blade positioned at the bottom of the tank and then pumped into the larger storage tank (Figure 10.32).

Once the liquid product has been de-aerated, standardized, mixed, controlled and approved by the quality assurance personnel, it will be ready to be sent to the pasteurizer.

10.5 Egg Pasteurization

Pasteurization is a thermal process that destroys harmful bacterial before packaging products. The process consists of heating the food at a defined temperature and time to reduce or eliminate pathogens and extend its shelf life. Pasteurization is important in reducing the risk of foodborne diseases by killing organisms and enzymes that could lead to food spoilage. The pasteurization technique was developed in the mid-1800s by the French biologist Louis Pasteur and it is now applied to almost all foods that can contain pathogen bacteria. A good example are egg products which are used in restaurants, hospitals, hotels and by food processing companies as an ingredient in foods such as mayonnaise and ice cream and many others. Today, liquid egg products are also becoming common at retail, thus increasing the importance of pasteurization for the safety of all egg products. To understand the pasteurization process, it is best to explain the complete process based on the evolution of the pasteurization systems over the years, mainly evolving due to market demands and customer needs.

10.5.1 Batch Pasteurization

The first pasteurizers used the batch technology. A finite amount of product was heated in a double-walled (double jacketed) steel tank container with hot water or steam recirculated in the space between the inner and outer walls. The product was continuously stirred to ensure a uniform temperature distribution and the heat was transferred to the product through the walls of the vessel. Once the pasteurization temperature was reached, it was maintained for the desired time and then cooled within the tank by introducing a coolant liquid into the space of the double-jacketed tank (Figure 10.33)

Depending on the volume considered, it may take a long time to bring the temperature of liquid eggs to the desired temperature (typically 63°C). The temperature of the tank walls cannot be too high to prevent the coagulation of the product in contact with the walls. Furthermore, to be sure that the temperature of the walls will be transmitted thoroughly to the entire mass of product, the holding times were typically very long (e.g., at least 30 minutes). The process of batch pasteurization in a tank is summarized in Figure 10.34, with 3 phases: heating, holding and cooling.

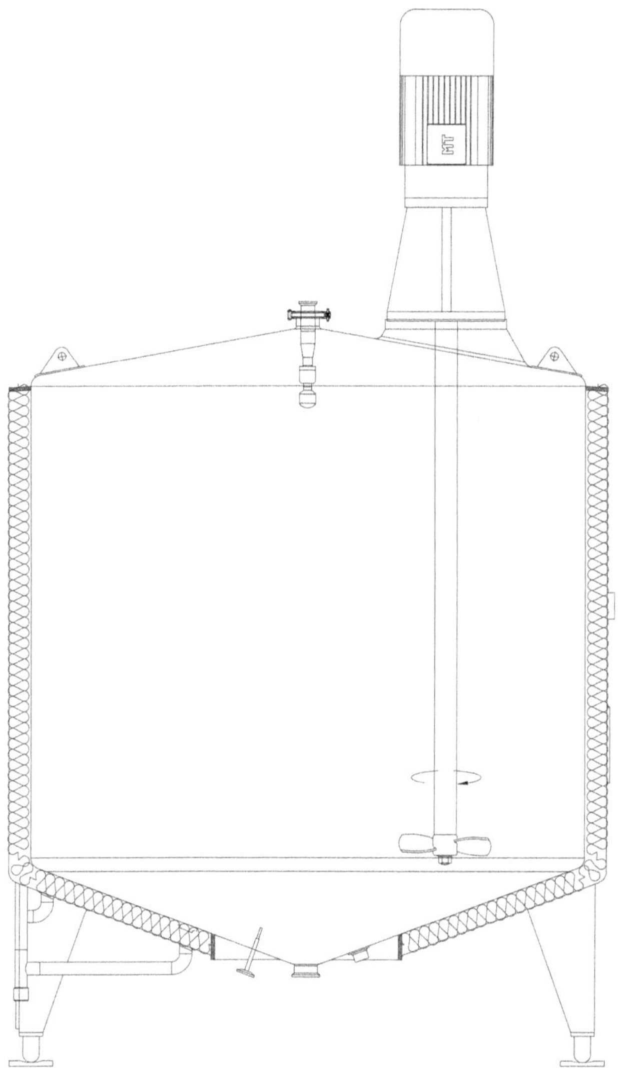

FIGURE 10.33 Double jacketed batch pasteurizer tank.

10.5.2 Continuous Pasteurization

In 1922, APV (Aluminium Plant & Vessel Company Limited), a company making processing equipment, developed the first High Temperature Short Time (HTST) pasteurizer that established the basis for modern pasteurization systems for all liquid food, including egg products. The HTST pasteurizer was developed for milk products and reduced the thermal cycle from 30 minutes to reach 63°C in the batch pasteurizer, (also called Low Temperature Long Time or LTLT), to 15 seconds to reach 72°C in the HTST pasteurizer. A further development in HTST pasteurization was the heat treatment system called Ultra High Temperature (UHT) sterilization where the products are heated up to 138–150°C for a few seconds. Packaged in sterile, hermetically sealed containers, these products can then be stored without refrigeration for several months. However, this system cannot be applied to egg products due to the detrimental effects of high temperatures on liquid eggs, causing complete coagulation of the products.

The continuous improvement of the HTST technology was driven by the increase in the types and volume of products treated and prompted the following performance needs: reduction of operating costs (lower energy consumption), better accuracy and uniformity for temperature control, ease and efficiency of cleaning with automatic control, reduction of manual interventions by operators and longer shelf life for products. Although some very small egg processing plants may still be using batch pasteurization technology, the majority of pasteurizers used today in the egg processing sector are of the continuous HTST type. HTST pasteurizers use Plate Heat Exchangers or Tubular Heat Exchangers to transfer the energy from the heating and cooling medias to the product.

As opposed to the batch pasteurization where the total amount of product is divided in batches, HTST pasteurizers are based on a continuous flow of product. Pasteurization consists in raising the temperature of the product, keeping it at a specific temperature for a well-defined time, and then cooling

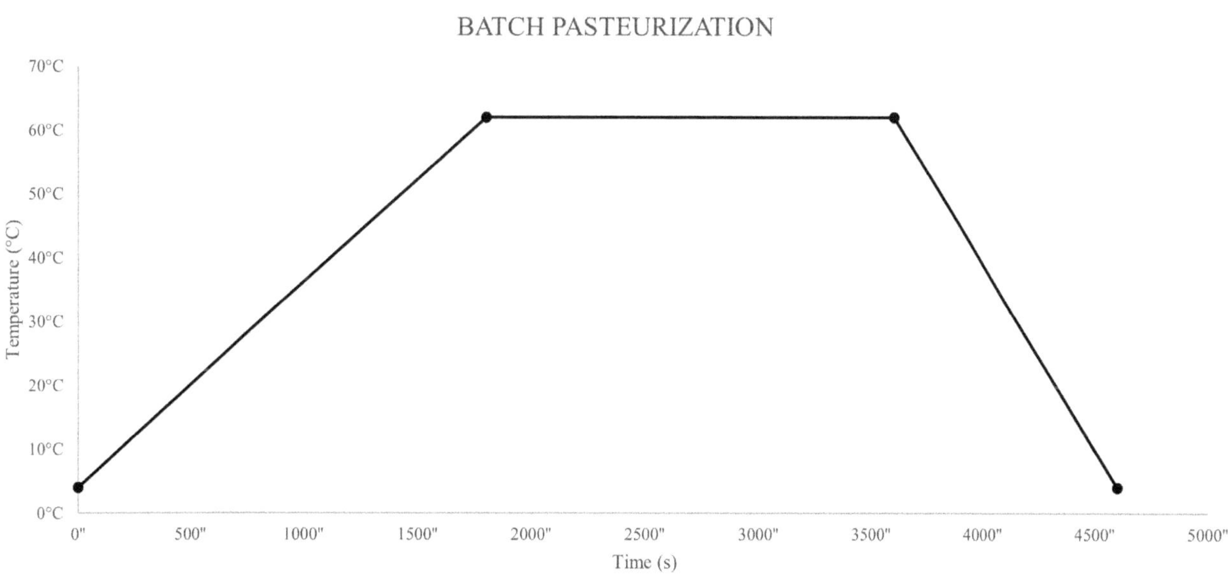

FIGURE 10.34 The Three phases of batch pasteurization.

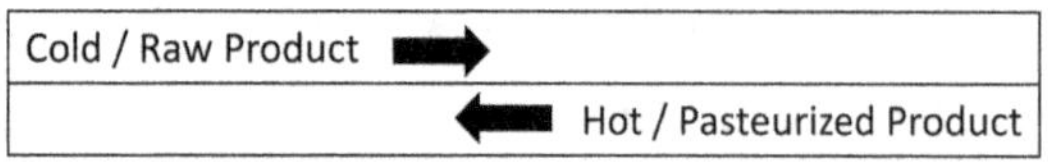

FIGURE 10.35 Schematic illustration of the concept of heat transfer to saving energy during pasteurization.

it again. In order to achieve some energy savings, the idea to use the hot pasteurized product to pre-heat the cold raw product and, conversely, to use the cold incoming raw product to pre-cool the pasteurized product emerged (Figure 10.35).

The thermal cycle (Temperature/Time Diagram) of a HTST pasteurizer, presented in Figure 10.36 will require at a minimum 6 items: balance tank, pump, heat regeneration system, heating section, holding section and a cooling section.

In Figure 10.37, the 6 essential components of a HTST pasteurizer is illustrated. Starting from the raw product holding tanks, the raw product is delivered to the Balance Tank (1) where a feeding product pump (2) is distributing the liquid through the entire system. The product enters the Balance Tank at a low temperature required to minimize bacterial proliferation during storage. Therefore, the product should be pre-heated to a temperature that will depend on the efficiency of the Heat Regenerator (3). Warmer product will then flow through the Heater (4) and reach and be maintained at the desired pasteurization temperature, within a coil of pipes called the Holder (5). The product will then be pre-cooled in the Heat Regenerator (3) by heat exchanges with the cold product coming from the Balance Tank. The pasteurized product will then be refrigerated in the Cooler (6) at temperature < 4°C (Figure 10.37).

10.5.3 Main Aspects of Liquid Egg Pasteurization

The pasteurization of eggs is a highly complex procedure; therefore, a good understanding of eggs is a fundamental requirement to design and operate an egg processing plant. Liquid egg is extremely delicate, as egg proteins are much more sensitive to heat than proteins from other liquid foods such as milk or juices. The main reason is that the egg is made of two fundamental components: the yolk and the egg white, which have very different compositions and behaviors, but will interact when mixed together.

For instance, the denaturation of egg white occurs below 58°C while for yolk, it will start at a temperature around 65°C. When the yolk and the egg white are mixed correctly, the

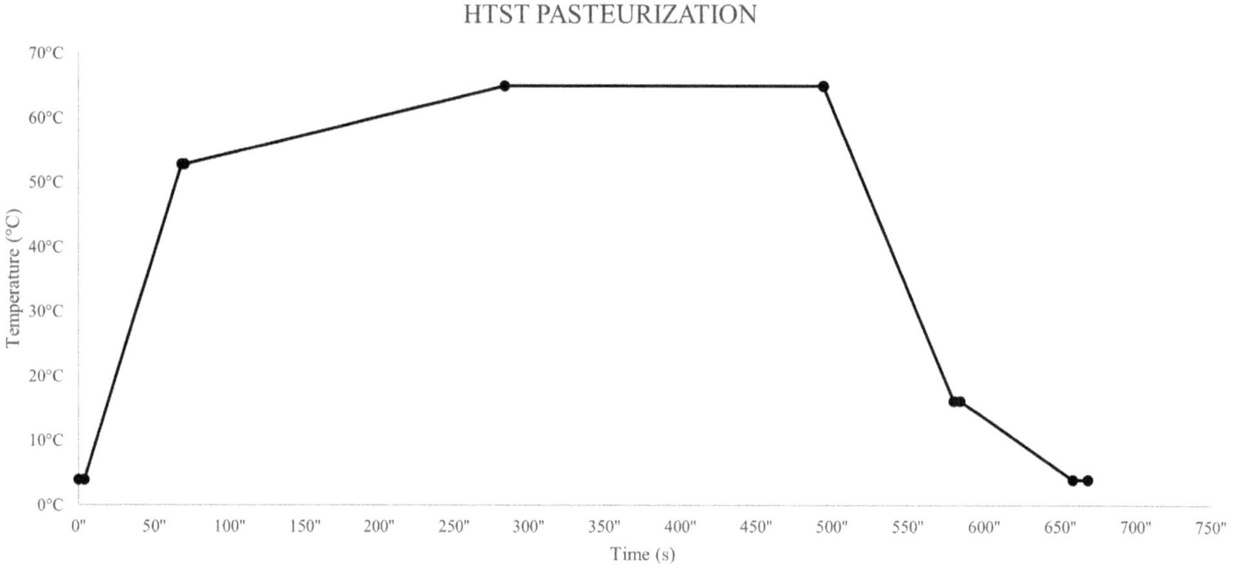

FIGURE 10.36 HTST pasteurization cycle.

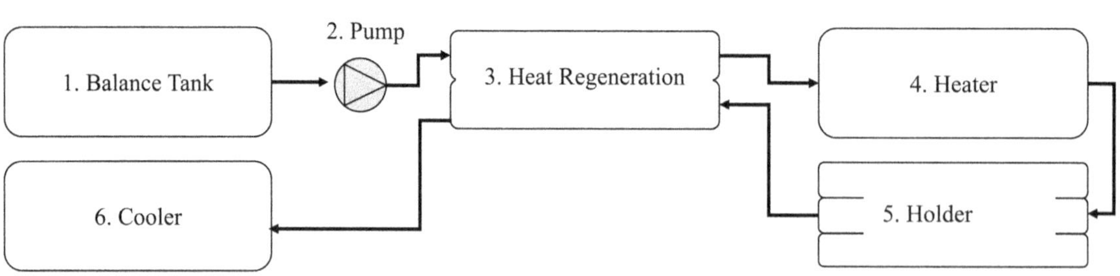

FIGURE 10.37 The Six essential components of a HTST pasteurizer.

pasteurization temperature can be increased to higher levels without experiencing the negative effects of denaturation. The better egg yolk and egg white are mixed, the higher the pasteurization temperature applied without any risk of curdling.

The type of eggs used will also have an impact on the response of liquid product to pasteurization since brown and white eggs have different dry matter and fat contents, resulting in different denaturing temperatures. For yolk production, the outcome depends a lot on the type of eggs, their freshness and temperature as well as on the efficiency of the breaking machine used.

10.5.3.1 Good Analysis Is Essential for Optimum Performance

In order to standardize the production, it is best to adjust the settings of the equipment to the type of products being produced. If the egg product is mixed with other ingredients, such as salt or sugar, additional behavioral aspects are introduced, including high viscosity, high resistance to pasteurization and more complex cooling requirements. During the design of a pasteurizer for egg products, all these aspects must be considered while doing an in-depth analysis of all the different types of products potentially treated. It is incorrect to design a pasteurizer only for standard liquid egg products; all types of products should be defined and considered. A good analysis of the types of products to pasteurize and the method of handling the eggs prior to pasteurization is fundamental to plan the most suitable pasteurizer for egg products. Apart from the different types of products, there are some important details to keep in mind while planning a pasteurization plant. These are very specific to the pasteurization of egg products and are rarely encountered during the pasteurization of other liquid food products such as milk and juices.

10.5.3.2 Foam Causes Malfunctioning

A particular feature of egg products is their whipping ability, that is their propensity to capture air and keep it inside their protein structure. This feature is important for certain food products using eggs as an ingredient and must therefore be preserved throughout the processing of eggs. If the egg product capture air prior to pasteurization, this will create a high quantity of foam causing the malfunctioning of the electronic instrumentation and negatively impacting the holding time. It also causes a considerable reduction of the efficiency of pasteurization and increases the coagulation of the product inside the plates or tubular heat exchangers. Therefore, attention should be placed on the equipment used to pump the product throughout the process, the tank storage, the pasteurizer infeed, etc., to reduce foaming to an absolute minimum.

10.5.3.3 Denaturing Process

Egg products are pasteurized at temperatures near the limit of their denaturing process in order to obtain the highest safety with bacteriological reduction or complete elimination. This often causes a denaturing effect of proteins and, even if not seen during the pasteurizing process, some changes in the functional properties will become evident when the pasteurized egg products are used in various food applications. In order to avoid this denaturation process, the heating process should be controlled accurately to maintain a steady temperature and set the hot water temperature (heating liquid) as low as possible. Moreover, temperature controls should be incorporated into the system to ensure high stability and avoid that the hot water temperature rise above the defined temperature limits. This can be obtained with the use of sequential controls made with highly developed automation systems.

10.5.3.4 Autonomy in Production

Within the food manufacturing sector, a prolonged production time is a very important component of cost-efficient production. The longer the production time before requiring even some minor cleaning phase, the lower the cost of production. Shorter production times result in more plant cleaning (even minor), increasing the costs for personnel, energy, water and cleaning chemicals. To obtain long production time, some very important technical solutions should be introduced. The use of a greater heat exchange surface in the pasteurizer is one solution. Using a milk or juice pasteurizer will result in under proportioning the setup for egg products and production time will be considerably reduced. For this reason, it is best to use a pasteurizer specifically designed to process liquid eggs. The use of a homogenizer will substantially increase the production time as a good mix of yolk and egg white will ensure better pasteurization and minimize product coagulation. In association with the use of the homogenizer, the installation of indirect heat recovery systems will significantly increase the production time. The use of a tubular heat exchanger instead of a plate heat exchanger will also contribute to increasing the production time.

10.5.3.5 Varying Viscosity

Egg product viscosity varies depending on the physical composition, the temperature treatments they undergo (both heating and cooling) and the presence of other ingredients mixed with the eggs (usually salt and sugar). This phenomenon does not occur with other liquid pasteurizing processes, such as milk where the viscosity stays almost the same during heating or cooling. Based on the treatment applied to egg products, either thermal or physical, an egg processing plant should be planned according to this. Sometimes these basic concepts are not taken under consideration during the planning stage, especially if the project manager does not have experience with egg processing. During pasteurization, the liquid egg product should not be exposed for a long period of time to heat or put in contact with surfaces that are too cold (in cooling sections when recycling energy and mixing systems are used), as this would result in a much higher viscosity and consequently a higher-pressure differential within the system.

10.5.3.6 Thorough Cleaning

Due to the formation of product build up on the surface of pipes, plates, and tubes, the cleaning of egg pasteurization plants represents a critical phase. The cleaning cycle must completely remove all proteins from steel surfaces since they constitute a source of bacterial contamination throughout the process line. Therefore, this cleaning phase must be completed in optimal fashion. Considering some basics about cleaning, it is essential to realize that when sodium hydroxide gets in contact with egg proteins, it will solidify and create a thick glue-like gel which makes cleaning difficult. Moreover, there are some areas where eggshells will typically accumulate and must be removed. Under these circumstances, a traditional cleaning will not be sufficient. Therefore, cleaning must be performed using very high flow rates to obtain high turbulence in the pipes and plates and consequently a high mechanical action. To increase the washing efficiency, several techniques may be used: the pasteurization plant can be divided into several circuits to reduce pressure drops and increase the speed of the washing solutions in each circuit; reverse the cleaning flow to make sure that all areas inside the pipes and plates are reached; the pasteurization plant can be left soaking with acid solutions overnight in order to remove shell residues.

10.5.3.7 Automation and Control

The automation and control systems are very important parts of the pasteurization process. Even the simpler and smaller pasteurizers should have a control system with Programmable Logic Controller (PLC), allowing a better handling of the process. The accuracy of the temperature and flow settings, the speed of response to alarm warnings occurring during pasteurization, the presence of software interlocks to avoid operator errors, and optimal and correct cleaning cycles are the different situations that a control system must be designed and able to handle. Therefore, it is very important to install automatic controls for these critical points in a pasteurizing plant. For example, what will happen when power runs out in the middle of the pasteurization? If the control system is not able to restart immediately and retrieve the initial conditions, it will cause a complete coagulation of the product inside the heating sections and in the holding tubes. While this problem is less critical for pasteurizers used to process for products less delicate than liquid eggs, this is a major concern for the egg processing plant.

10.5.3.8 Variability of Liquid Eggs Products

As addressed previously, the pasteurization of eggs is one of the most complex pasteurization processes because it depends on many variables and because we are not dealing with a constant and uniform product. In general, the purpose of pasteurization is to make the product safe from bacterial contamination and the killing of bacteria depends on the pasteurization temperature and the corresponding holding time. In addition, there are some temperature and time limits since pasteurization must be achieved while preserving the physical, chemical, and organoleptic characteristics of the product treated. In any case, pasteurization, even if minimal, always influences the characteristics of the product. Heat applied to the product affects the proteins and egg products are one of the richest in proteins.

Adhesion, aeration, binding, browning/color, and coating are just some of the functions of egg products linked to the activity of proteins in the formulation of final food products (see https://www.incredibleegg.org/professionals/manufacturers/real-egg-functionality). It is important to keep in mind that proteins have a different behavior during pasteurization depending on the type of pasteurizer used, on the type of product being pasteurized and on the level of mechanical stress to which they are subjected during the process. Therefore, the design of a pasteurizer is of fundamental importance for the preservation of the functional properties of proteins. For egg products, we must also consider the wide range of products produced, each differing from each other in terms of physical and chemical characteristics.

Liquid egg white: low dry matter (considered about 11–12% in engineering calculations), with a high protein content and a low denaturation temperature. Very sensitive to mechanical stress.

Yolk: high dry matter (considered from 42% to 46% in engineering calculations depending on the types of products it is intended for), medium resistance to temperature but very viscous at low temperatures.

Whole egg: a mixture of yolk and egg white with a dry matter between 21% and 25%, obtained with different proportions of yolk and egg white mixed together, depending on the intended use, requires a good mixing of the components (homogenization), pasteurized at different temperatures for different purposes.

The previous three products are often mixed with additional ingredients (typically salt and sugar) which will significantly change the characteristics of the final product. One of the most difficult products to process is 11–12% salted yolk, which, especially cold, can reach very high viscosities. The 50% sweetened yolk is another product that is widely used and delicate to treat since a heat treatment that is too intense (both in time and temperature) can cause the product to turn brown due to the Maillard reaction of the sugars and proteins contained in the product. Therefore, to correctly size a pasteurizer for egg products, it must be kept in mind that it can be used for all the types of products listed above as well as the following criteria: Very heat sensitive products with strong denaturation; Products that require milder heating; Products with increasing viscosity if heated too much; Products with increasing viscosity if cooled too much; Products that need to be well mixed to perform a good pasteurization––products with different viscosity that have a different behavior when pumped through the pipes.

Very often, it is not possible to design a pasteurization system dedicated to a single product since the processing plant will likely produce a wide range of products (Mixed whole egg, yolk, egg white, mixed and salted or sweetened yolks). The design of a pasteurizer that takes all the points listed above into consideration is not always easy and it will require finding the best compromising solution.

10.5.4 Design of a Liquid Egg Pasteurization Plant

During the planning stage of the design of a plant producing all types of egg products, the first choice the engineering firm has to make is about the selection of the right heat exchanger based on the heat exchange rate/efficiency and the pressure drops that the product will have to undergo during the various processing steps. The most critical sections are the heating and cooling section because this is where the treated product is subjected to the most extreme temperatures. In general, the viscosity of egg products, (especially the more viscous ones such as salted and sugared yolks), will significantly increase at low temperatures. For cooling, the plate exchangers are mainly used in a configuration where the plates allow the parallel passage of fluids with some significant losses in pressures due to pressure drops. The heat recovery section may use plate or tubular devices, but the use of plate sections is preferred, due to cost savings and more flexibility of use.

Different considerations are made for the pasteurization section whether plate or tubular sections are used. Pasteurization with plates is faster in terms of heating time (so it should be preferred, for example, for sugared products to minimize the browning effect). On the other hand, using tubular exchangers for the pasteurization section extends the production time since the passages in the tubes are faster than through plates. A fundamental point to consider while designing a heating section for a liquid egg pasteurizer is the high denaturation of the proteins. The pasteurization temperature is usually very close to the coagulation limit of the products. In the heating section, the hot water is heating the product, so it is fundamental to keep the temperature of the hot water as low as possible (just a few tenth of a degree higher than the pasteurization temperature) in order to avoid the phenomena of overheating at the proximity of the tubes and plates surfaces. In summary, the main elements considered by an engineering firm for the design of the correct pasteurizer are: 1) Heat Regeneration: as efficient as possible to minimize the consumption of energy and ensure a safe heat/cool exchange between hot and cold product without cross-contamination; 2) Cooling Section: able to cool down all egg products while limiting any pressure drops in the section to a minimum. 3) Heating Section: designed to allow the minimal temperature difference between the heating media and the product.

10.5.5 Pasteurization Temperatures and Holding Times

Once the heating, cooling and thermal recovery sections have been defined, the holder must be dimensioned, which varies from product to product. Generally, during the design of a pasteurizer, whole egg is used as the reference, requiring a holding time of 3.5 minutes (210 seconds). The holding time is obtained by dividing the total volume of the holding pipes by the flow rate of the pasteurizer (for example a 3,000 l/h pasteurizer with a holder volume of 200 liters will have a holding time equal to 200/3,000 = 0.066 h or 237.6 sec.). In order to have shorter or longer holding times, the pasteurizer capacity must be increased or decreased. The pasteurizer must therefore be designed to obtain the desired holding times, required for microbial inactivation, by regulating its flow rate.

10.5.5.1 Microbial Inactivation

All thermal processes involve three distinct steps: a heating time, a holding time, and a cooling time. In theory, all three periods may contribute to the reactions taking place although, in situations where heating and cooling are rapid, the holding time is the most significant one. Procedures are needed to evaluate each of these periods individually to determine the overall effect. By far, the holding period is the easiest to deal with since it takes place at constant temperature. We must then establish how reaction rates are affected by changes in temperature during heating and cooling. To simplify this analysis, microbial inactivation is first measured at a constant temperature. This is usually followed by observing how microbial inactivation changes with temperature.

10.5.5.1.1 Heat Resistance at Constant Temperature

When heat inactivation studies are carried out at constant temperature, it is often observed that microbial inactivation follows first order reaction kinetics. The rate of inactivation is directly proportional to the bacterial population. This can be illustrated by plotting the log of the population against time and finding that there is a straight-line relationship (Figure 10.38).

The heat resistance of bacteria is characterized by its decimal reduction time (D_T), which is defined as the time required to reduce the population by 90% or by one order of magnitude or one log cycle (i.e., from 10^3 to 10^2), at a constant temperature, T. Each type of microorganisms has its own characteristic heat resistance and the higher its D value, the greater its heat resistance. Heat resistance is also affected by a wide range of environmental factors such as pH, water activity and the presence of other solutes, such as sugar and salts. The extent of microbial inactivation is measured by the number of decimal reductions which can be achieved. This is given by $\log(N_0/N)$, where N_0 is the initial population and N is the final population; and it is determined from the following equation:

$$\log\left(\frac{N_0}{N}\right) = \frac{H_{time}}{D_T}$$

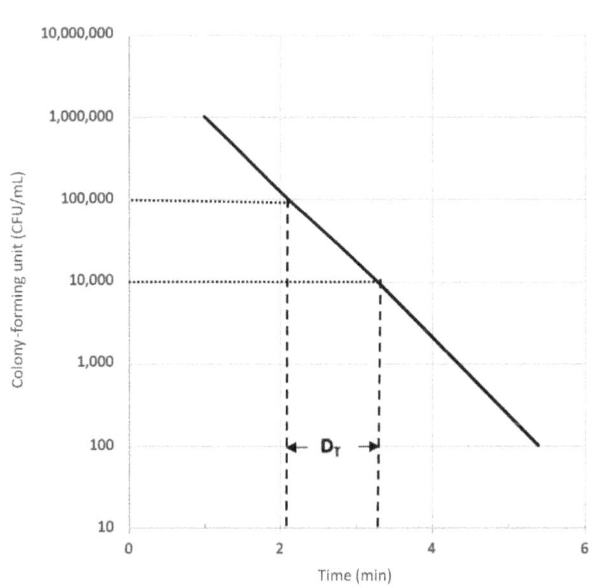

FIGURE 10.38 Evolution of bacterial population over time of treatment at a constant temperature.

There are two important aspects associated with the first order reaction kinetics: it is not theoretically possible to achieve 100% reduction, and, for a specific heat treatment, the final population increases as the initial population increases.

If the D_{T70} for a bacteria spp. is 5 sec., heating for 5 sec. will obtain a 1D (90%) reduction in the population, 10 sec. heating will achieve 2D (99%), 15 sec. will achieve 3D (99.9%) and 30 sec. achieve 6D (99.9999%). In summary, it is not theoretically possible to achieve 100% reduction.

Improving the heating process, by extending the holding time or increasing the temperature of the heat treatment, will reduce the chance of any type of surviving bacteria. In other words, if the D_{T70} for a bacteria spp. is 5 sec. and the initial population 10^5 ml^{-1}, after a 40 sec. heating the final population would be 10^{-3} ml^{-1}. It may be difficult to imagine a fraction of an organism, but another way of expressing this would be the presence of one organism per 1,000 ml. Thus, if the sample has been packed in 10 mL portions, it is therefore possible to find one surviving microorganism in every 100 samples analyzed or 99 packs are free from microorganisms.

The same heat treatment applied to a raw material with higher count (10^6 ml^{-1}) would give one surviving microorganism in 10 samples analyzed or 9 packs out of 10 are free from microorganisms.

Note that just one surviving bacterium in a product or package can create product spoilage. In practice, deviation from first order reaction kinetics is often encountered and the reasons previously discussed (Cerf et al., 1990).

10.5.5.1.2 Temperature Dependence

Most processes do not take place at constant temperature but involve a heating and cooling periods. Therefore, although it is easy to evaluate the effect of the holding period on the heat resistance and the lethality number of decimal reductions of a process, it is important to appreciate that heating and cooling may also contribute to the overall lethality. It is known that reaction rates increase as temperature increases and microbial inactivation is no exception. Food scientists use a parameter known as the z value, to describe the DT temperature dependence. This is based on the observation that, over a limited temperature range, there is a linear relationship between the log of the decimal reduction time and the temperature (Figure 10.39).

The z-value is defined as the temperature change, which results in a tenfold change in the decimal reduction time. The z value for most heat resistant spores is about 10°C, whereas the z value for vegetative bacteria is considerably lower; usually between 4°C and 8°C. A low z value implies that the reaction in question is very temperature sensitive. In general, microbial inactivation is very temperature- sensitive, with inactivation of vegetative bacteria being more temperature sensitive than heat-resistant spores.

In contrast to microbial inactivation, chemical reaction rates are much less temperature-sensitive than microbial inactivation, having higher z values (20–40°C). This is also the case for many heat-resistant enzymes, although heat-labile enzymes such as alkaline phosphatase or a-amylase are exceptions to this rule. This difference between chemical reactions and microbial inactivation has some important implications for quality improvement when using higher temperatures for shorter time.

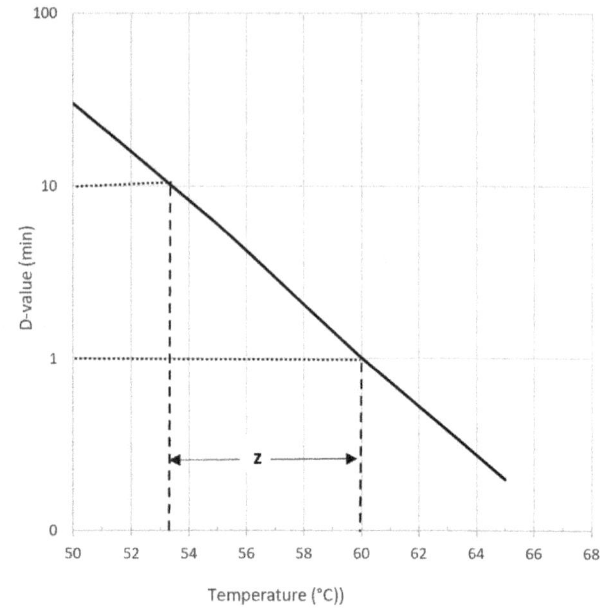

FIGURE 10.39 Defining the z value, based on the observation that, over a limited temperature range, there is a linear relationship between the log of the decimal reduction time and the temperature.

The relationship between D values at two different temperatures and the z value is given by the following formula:

$$\log\left(\frac{D_1}{D_2}\right)=\left(\frac{(\theta_2-\theta_1)}{z}\right)$$

An alternative way of using z values is for comparing processes. If it is known that a temperature of 65°C (q) is effective for 210 seconds (3.5 mins) and the z value is 7°C, then an equally effective process will be at 72°C for 21 seconds (0.35 mins) or 2100 seconds (35 mins) at 58°C. In another way, 1 min at q is equivalent to 0.1 min at (q + z) or 10 min at (q - z). The D_T & z-value of Salmonella spp., the main pathogenic bacteria of concern during the processing of liquid egg products, are presented in Table 10.1.

Tables 10.2 and Table 10.3 are listing the D_T & z-value of the most pathogenic and spoiling bacteria species in liquid medium comparable to egg products (Sörqvist, 2003).

Perhaps best known are the standard lethality table against Salmonella spp. in different liquid egg-products (Table 10.4) obtained in industrial processes and devices under turbulent flow (Froning et al., 2002).

The lethality at any experimental temperature (T) can be compared to that at the reference temperature (q), using the following equation:

$$\log L=\frac{(T-\theta)}{z}$$

Thus, for a standard reference temperature of 60°C and an experimental temperature of 57°C using z = 5°C. then L = 0.25. Therefore, 1 min at 57°C is equivalent to 0.25 min. at 60°C.

The proteins of egg white have different denaturation temperatures, starting at 61°C and, considering the holding times used in the egg pasteurization, the limit temperatures can

TABLE 10.1

Thermal Resistance of *Salmonella* in Liquid Egg Products

Food Product	Salmonella Inoculum	T (°C)	D-Value (min.)	Z-value (°C)
Homogenized whole egg	*Senftenberg 775 W*	55.0	34.00	8.2
		60.0	5.60	
		64.0	2.80	
Homogenized whole egg	*Enteritidis PT4*	55.0	6.40	6.0
		60.0	0.44	
		64.0	0.22	
Liquid egg white	Cocktail of 5 strains[e]	55.1	8.00	3.5
		58.3	1.00	
Liquid egg yolk	Cocktail of 5 strains[e]	60.0	0.28	4.3
		62.2	0.09	
Cocktail of 5 strains[e]	Typhimurium /ATCC 14028, type strain), Typhimurium (2564-2, egg layer house solate), Enteritidis (ME-15, shell egg isolate), Enteritidis (ME-18, live poultry isolate), Enteritidi (ATCC 4931, clinical isolate).			

be even lower (Li-Chen et al., 1995). The first objective of a pasteurization cycle is the elimination of pathogenic bacteria (e.g., *Salmonella spp., Listeria monocytogenes*) and the time/temperature configurations are therefore established to ensure the inactivation of these pathogenic bacteria. Generally, the temperatures and times recommended are far above the denaturation temperatures of certain proteins. It should also be kept in mind that sometimes pathogenic bacteria are not the only ones that will cause the deterioration of the product. The shelf life of egg products is often linked to the presence of other bacteria which, although not pathogenic, will cause some spoilage of the product.

Regulations for the pasteurization of egg products vary from country to country. In the USA for example, a reference is made to the International Egg Pasteurization Manual which reports the minimum time/temperature requirements for the most common egg products. However, in Europe there is no legislation that defines the minimum pasteurization temperature for egg products, and the rule is very general. It is the responsibility of each processing plant to define the appropriate parameters in order to ensure the absence of pathogenic bacteria in egg products. It seems that the USA will soon adopt the same approach, leaving egg processors with the responsibility to define their own process and procedures and to have a solid HACCP plan (FSIS, 2020).

10.5.5.2 Egg Pasteurization and Product Functionality

Many times, the minimum pasteurization temperature is sufficient to ensure a safe product but not sufficient to ensure a long shelf life, a product feature increasingly required by the market to further expand the use of these products. In order to produce Extended Shelf Life (ESL) products, the temperature and time mentioned in the International Egg Pasteurization Manual (Froning et al., 2002) are, sometimes, not sufficient, and higher temperatures (> 67°C) are used to ensure a better bacterial inactivation. Although the demand for ESL products is increasing rapidly, the extension of shelf life with higher temperature triggering some damage to the product, may

TABLE 10.2

z-Values Review for Pathogenic and Spoiling Bacteria Species Content in Liquid Medium Comparable to the Egg Products

	z-Values (°C)		
Bacterium/Bacterial Group	**Range**	**Average ± SD**	**N⁰ of Records[a]**
Enterococcus faecium	3.6–14.3	10.2 ± 3.3	24
Enterococcus faecalis	2.2–14.2	8.1 ± 3.2	36
Listeria monocytogenes	4.3–11.4	6.3 ± 1.3	103
Escherichia coli	3.2–9.1	5.4 ± 1.5	33
Yersinia enterocolitica	4.0–13.7	6.6 ± 2.7	20
Salmonella spp. (No Salm. senftenberg 775 W)	3.2–9.5	5.1 ± 1.6	63
Salmonella senftenberg 775 W	4.5–9.1	6.2 ± 1.1	16
Campylobacter jejuni/coli	2.8–8.0	5.5 ± 1.1	24

Source: Adapted from Sörqvist (2003).

[a] z-values where figured out from reported or calculated D-values.

TABLE 10.3

Values of D for Microbial Inactivation

		D-Values [s]	
Temperature (°C)	**Average**	**Min.**	**Max.**
	55	3813.0	3095.0–4697.0
	60	1150.0	1017.0–1300.0
	65	347.0	315.0–381.0
	72	65.0	53.0–79.0
	55	1393.0	1089.0–1783.0
	60	415.0	361.0–476.0
	65	123.0	108.0–141.0
	72	23.0	17.0–30.0
	55	643.0	577.0–715.0
	60	87.0	81.0–93.0
	65	12.0	11.0–13.0
	72	0.7	0.6–0.8
	55	266.0	239.0–297.0
	60	39.0	35.0–42.0
	65	5.6	5.1–6.2
	72	0.4	0.3–0.5
	55	168.0	124.0–227.0
	60	30.0	24.0–37.0
	65	5.4	4.0–7.4
	72	0.5	0.3–0.9
	55	222.0	208.0–237.0
	60	24.0	23.0–26.0
	65	2.6	2.3–2.9
	72	0.1	0.1–0.2
	55	50.0	44.0–57.0
	60	8.2	6.5–10.0
	65	1.3	0.9–2.0
	72	0.1	0.1–0.2

Source: Adapted from Sörqvist (2003).

TABLE 10.4

Lethality Values for Salmonella Cocktail Strains in Liquid Egg Products

	Minimum Temperature	Minimum Holding Time[a]
Liquid Egg Product	**[°C]**	**[s]**
Egg white with adjusted pH	56.7	4.3
Egg white without adjusted pH	57.7	6.3
Egg white (pH = 8.6) with H_2O_2	54.4	3.5
USDA scrambled egg mix (30% solids)	62.2	2.0
Scrambled egg mix (22% solids)	60.0	2.4
Imitation egg product (pH = 8.8)	56.7	4.6
Salted yolk (10%)	63.3	4.5
Salted whole Egg (10%) without storage	63.3	5.7
Salted whole Egg (10%) with 96-hours storage	63.3	3.5
Sugared yolk (10%)	63.3	3.5
Sugared whole Egg (10%)	61.1	3.5
Plain yolk	61.1	3.5
	60.0	6.2
Plain whole egg	60.0	3.5

Source: Adapted from Froning et al. (2002)

[a] Based on a 5-log reduction

come at the expense of some functional characteristics of the product, even to complete loss of them.

The behavior of the pasteurization of whole egg in relation to the loss of functionality of the pasteurized product (compared with the raw product) is presented in Figure 10.40. Two limits may be identified: 1) The limit for loss of functionality of about 20% (still acceptable in the egg product market); 2) The limit for the loss of functionality of 100% by identifying the product that has lost most of its functional characteristics and is no longer marketable.

The intersection of the time/temperature points (60°C for 210 sec.) suggested by the International Egg Pasteurization Manual is clearly visible in this diagram and is in an area where the functional damage is negligible. However, the pasteurization of ESL egg products is generally above the functional damage curve of 20%. Typically, ESL products use temperatures ranging from 67°–70°C for 210 seconds. Analyzing the graph, it is clear that this combination of time/temperature is above the functional damage curve by 20%. In order to produce ESL products at high temperatures (especially for whole egg) an accurate mixing of the yolk with the egg white is required so that the fat contained in the yolk can interact at the interface of the proteins, in some way protecting them from the damage by temperature. The pasteurizers used for ESL products are equipped with a homogenizer that is positioned after the thermal recovery for an optimal homogenization of the pre-heated product (and fat). The circuit of an ESL pasteurizer with the homogenizer is shown in Figure 10.41.

The various elements considered during the design of an egg product pasteurization plant for optimum product performance, both in terms of bacteria inactivation and functional properties, are presented in the following section.

10.5.5.2.1 Pumping System

Proteins in egg products are sensitive, not only to the pasteurization temperature, but also to the Shear Stress caused by the pumping of the product during the various phases of the process.

Generally, positive displacement pumps (e.g., lobe pumps) are used to pump products in a gentle manner while the centrifugal pumps are used for the proper washing cycles of the system. However, the presence of both lobe and centrifugal pumps requires the use of by-pass sections and shut-off valves, making the circuit more complex. The trend observed over the past few years is for the use of double screw pumps that can be

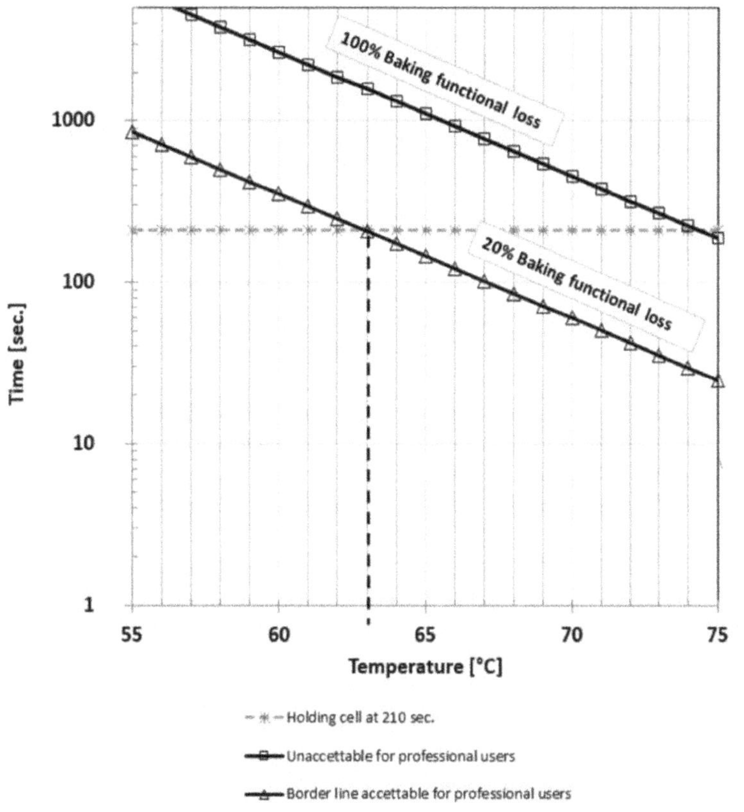

Diagram: t-T relationship for Microbial destruction (6-9 Log Red) and "Baking-Functional Loss (FL)".

FIGURE 10.40 The behavior of the pasteurization of whole egg in relation to the loss of functionality of the pasteurized product. Identification of two limits (20% and 100% of product functionality affected).

used for both the gentle pumping of the product and the washing phases of the system with higher flow rates. Pasteurizers designed to handle high viscosity products will require extra pumping systems to maintain the operating pressures low. Typically, a booster pump must be installed before the cooling section. The pumping system must be regulated automatically to keep a constant flow rate of product and ensure the correct holding time. Variation in flow rates cause instability within the system, with the risk of failing to maintain the correct holding time and consequentially resulting in insufficient pasteurization of the product.

10.5.5.2.2 Temperature Control and Heat Regeneration System

The heat regeneration is a very important part in the pasteurization of liquid eggs. As addressed previously, the pasteurization cycles are based on the continuous heating of products to the required temperature, keeping them at such temperature

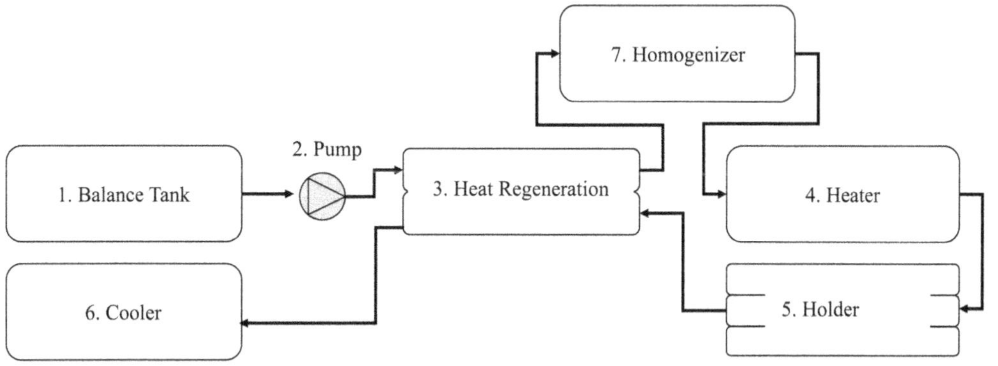

FIGURE 10.41 The seven essential components of an ESL pasteurizer.

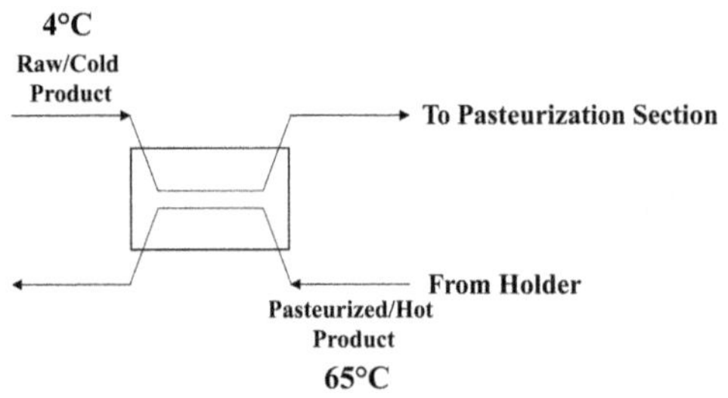

FIGURE 10.42 Schematic illustration of the concept of heat transfer for saving energy during pasteurization.

for the desired time (holding time) and then cooling them down to a temperature where the bacterial growth is minimized. Typical pasteurization cycle for liquid egg is (whole egg on traditional systems) will include: 1) Inlet temperature: <4°C; 2) Pasteurization temperature 60°C–68°C (depending on the type of treatment and final use of product); 3) Holding time 210 seconds and 4) Outlet Temperature <4°C. The heating of the product is generally made with hot water heated by steam (or electrical energy in some cases) and the cooling of the product is achieved with a recirculation of a cooling media (e.g., cold water, glycol). The continuous process of heating and cooling allows the possibility to save a large part of the energy needed for heating and cooling by using the heated product (pasteurized) to pre-heat the cold product (raw) and vice-versa (cooling the pasteurized product with the cold incoming raw product) (Figure 10.42).

10.5.5.2.3 Energy Regenerative Efficiency

All HTST pasteurizers are equipped with an energy recovery system that is fundamental in the pasteurization process as it will drastically reduce the production costs. When pasteurizing eggs, there are different aspect to consider for the design of the heat regeneration process. The energy saved during the pasteurization cycle reduces the production costs and therefore the main goal is to design a system with the highest possible "regeneration". This efficiency is strictly linked to the number of plates/tubulars (Heating/Cooling Surface) used. The design is determined by finding the best compromise among different factors: efficiency, pressure drops, production running time, cleaning flow rate and costs. When keeping under consideration these elements, it is possible to generate energy savings in a liquid egg pasteurizer ranging from 60 to 85% (newer systems developed recently allow an energy saving in the order of 90%).

To illustrate this point regarding potential energy savings, let's consider the following pasteurization parameters: 1) Inlet Temperature: 4°C, 2) Pasteurization Temperature: 65°C and 3) Outlet Temperature: 4°C. The product must be heated from 4°C to 65°C or an increase of 61°C. Installing a system to save 70% of energy means that only the 30% of the total thermal energy and cooling energy must be used in the process. Therefore, 70% of the temperature differential must be achieved by the regeneration unit or 61°C × 70% = 42.7°C. The Thermal cycle will be then modified as follow: 1) Inlet Temperature: 4°C; 2) Heat Regenerative System Outlet: 46.7°C (4°C + 42.7°C recovered by the system); 3) Pasteurization Temperature: 65°C; 4) Heat Regenerative System Outlet: 22.3°C (65°C – 42.7°C recovered by the system); 5) Outlet Temperature: 4°C (Figure 10.43).

The energy needed for heating and cooling and the savings realized may be calculated as follow (the calculations are made considering water to simplify the formulas):

Without Heating/Cooling regeneration:
  4°C → 65°C → 4°C or about 71 KW needed to heat up and 71 kW to cool down each ton of processed product.

With Heating/Cooling regeneration:
  4°C → 46.7°C → 65°C → 22.3°C → 4°C or about 21 kW needed to heat up and 21 kW to cool down each ton of processed product.

A second very important aspect to keep into consideration during the design of the Pasteurization Energy Regenerative Systems, is the possible cross contamination of product. Since a section in the pasteurizer will have the pasteurized product exchange energy with the raw product, any contact between the raw and pasteurized products must be avoided to prevent any contamination of the pasteurized product. This potential risk is prevented by using double wall plates, differential pressure control and indirect heat regeneration, three widely accepted systems present in all the pasteurization and sterilization processes.

10.5.5.2.3.1 Double wall plates heat exchangers Each plate is made up of a pair of identical plates, joined by leaving an intermediate free channel where the product may flow in case of a leak or damage to a plate. The direct contact between the raw and pasteurized products is avoided by the presence of this intermediate channel (Figure 10.44)

Since the efficiency of such system configuration ranges between 60% and 75%, the efficiency of a double wall plate

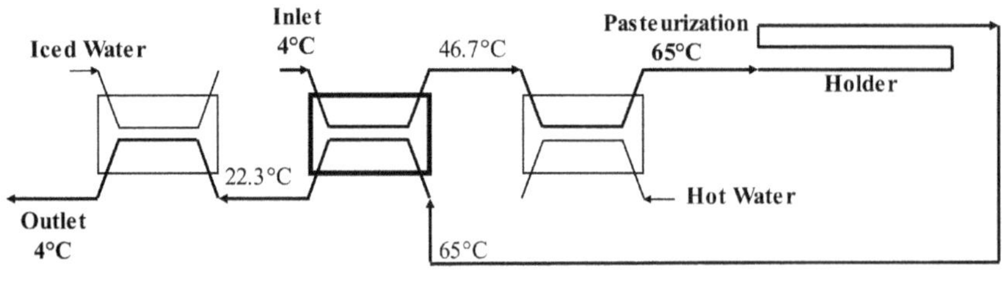

FIGURE 10.43 Schematic of heat regeneration system achieving significant energy savings.

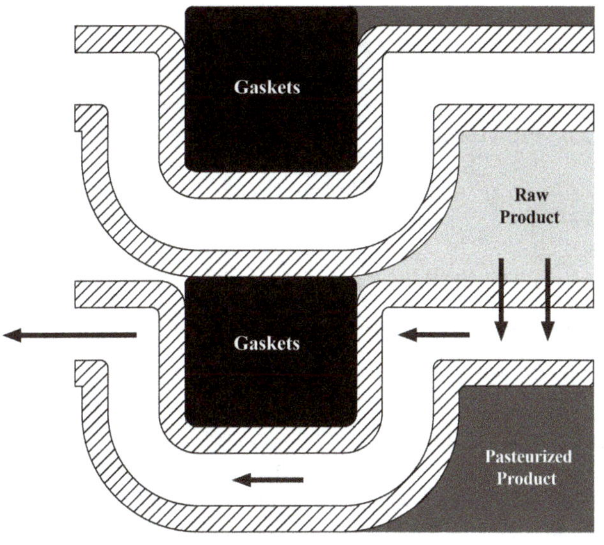

FIGURE 10.44 Schematic of the double wall plates design to prevent the risk of cross-contamination between the pasteurized and raw products in the energy regeneration section.

system is about 30% less than the efficiency of a single plate system.

10.5.5.2.3.2 Differential Pressure control system The pressure of the pasteurized product is controlled and always kept at a higher value than that of the raw product (Figure 10.45). In case of a plate damage, the pasteurized product will flow toward the raw product instead of the opposite. The pressure differential will therefore prevent the contamination of the pasteurized product.

The efficiency of such design is very high, in the order of 80–85%.

10.5.5.2.3.3 Indirect Heat Regenerative systems This system is achieved with a double plate heat exchanger, to keep the raw and the pasteurized products completely separated, and water is recirculating in the intermediate channel to transfer the energy from one section to the other, heating up the raw product and cooling down the pasteurized product (Figure 10.46).

The efficiency for this type of configuration is between 70% and 75%.

It is often underestimated, that there is a third important aspect to keep into consideration during the design of the energy regenerative system used for the pasteurization of liquid egg product, mainly related to liquid whole egg. Liquid whole egg is composed of egg white and egg yolk, both having very different behaviors during pasteurization. The egg white is rich in fragile proteins, easily damaged during pasteurization. For this reason, liquid egg white is typically pasteurized at lower temperatures (56–58°C) while the liquid egg yolk in traditional pasteurizers is pasteurized at temperature around 65°C. When the egg white and egg yolk are mixed to produce whole egg, there is a mutual interference of the 2 components and the fat present in the egg yolk protects, to some extent, the egg white proteins during the pasteurization process. This phenomena allow the pasteurization of whole egg at higher temperatures, as required for the production of ESL products. The pasteurization of whole egg can reach temperature

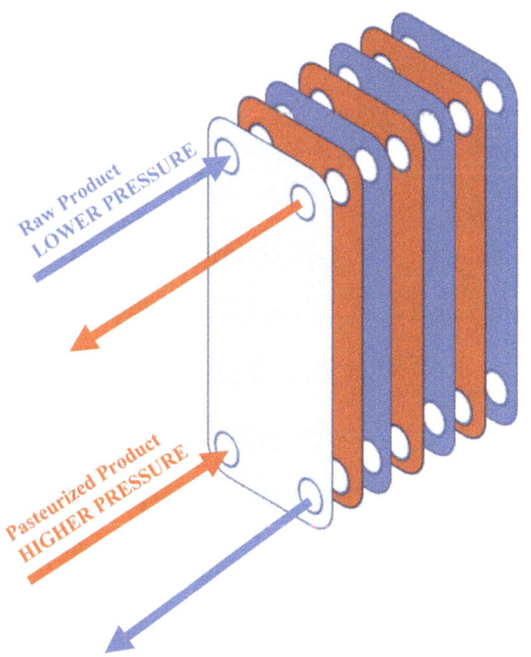

FIGURE 10.45 Pressure differential between the pasteurized and raw sides in the energy regeneration section will prevent risk of cross-contamination of pasteurized products.

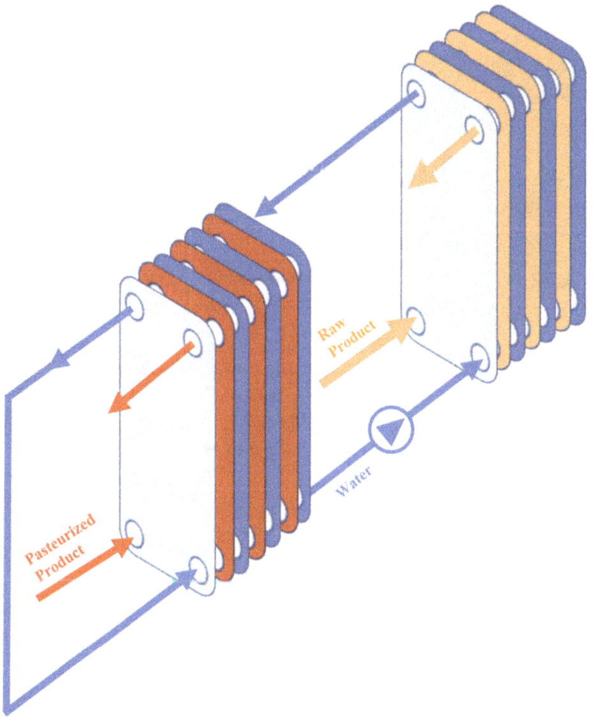

FIGURE 10.46 Schematic of the indirect heat regeneration system designed to prevent the risk of cross-contamination between the pasteurized and raw products in the energy regeneration section.

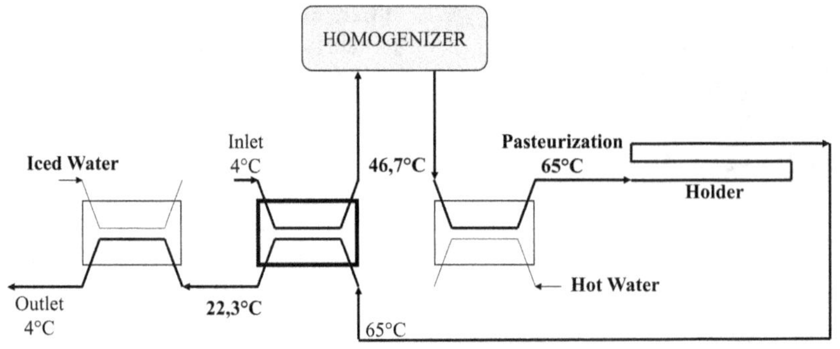

FIGURE 10.47 Optimum positioning of the homogenizer.

of 68°–70°C in traditional systems and even higher (around 75°C) with some new systems. Currently, to produce ESL products, increasing the pasteurization temperature is one of the main objectives. The focus is to pasteurize the product at the highest possible temperature without compromising its quality and without compromising the length of production runs, potentially limited by faster product fouling when the complete pasteurizer design is incorrect. The homogenization of the product will allow to operate a pasteurizer at higher temperatures.

10.5.5.2.4 Homogenization

A homogenizer is recommended to ensure an effective pasteurization of whole egg and avoid the subsequent separation of these 2 phases. The effect of high pressure for a very short time will create a very good, homogenous mixing of whole egg products. It has been demonstrated that the homogenization around 80–100 bar has the maximum effect if the product is pre-heated to a temperature ideal to melt the fat contained in the egg yolk. The homogenizer should therefore be installed just after the first regenerative section (Figure 10.47).

Homogenization and product quality: The cold raw product entering the regenerative section is not homogenized and therefore the egg white is free to flow within the egg yolk component without any protection against heating (Figure 10.48).

The pasteurized and homogenized whole egg, coming back from the holding section at pasteurization temperature, is heating the raw not-homogenized product. The point where

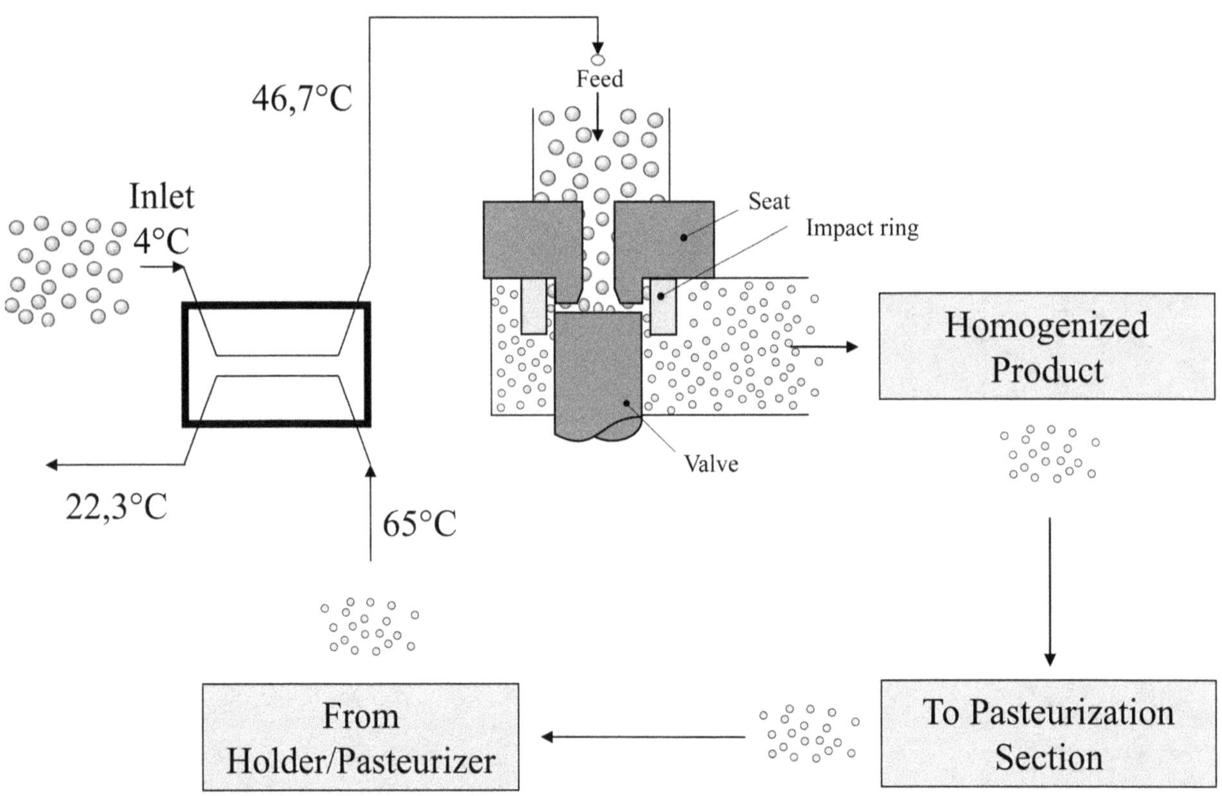

FIGURE 10.48 Effects of homogenization of the size of yolk fat particles.

the non-homogenized raw product encounters the hot pasteurized product at very high temperature in the first passage of the plate heat exchanger, (65°C in Figure 10.48 but potentially higher for ESL production), creates unavoidable damages to the proteins in the egg white, unprotected against these temperature-induced changes (Figure 10.49).

This is a very common problem found during the pasteurization of liquid eggs. The first section of the zone of heat regeneration will get dirty and blocked before the pasteurization section due to the fouling of the product. This phenomenon can reduce the quality of the pasteurized liquid whole egg and drastically reduces the operating time of the pasteurizer. It also requires frequent cleaning during production, thus drastically reducing the efficiency of the pasteurizer and increasing the cost of production. The only way to avoid this problem is to limit the maximum temperature that the raw (non-homogenized) product reaches during the process.

Cold homogenization: A solution may be to homogenize the raw product before the energy regeneration system in order to heat only homogenized product (Figure 10.50). However, this may compromise the effect of the homogenization since it has been demonstrated that the best homogenization is achieved when the product is pre-heated (Figure 10.51).

Product cooling: Another solution, sometimes adopted, is to cool down, after the holding section, the product to a temperature below the critical temperature. This will ensure that the temperature of the non-homogenized whole egg will never go higher than the temperature triggering the denaturation of the egg white proteins. However, this solution drastically reduces the efficiency of the energy regenerative system as the product is cooled down after the holder section without any possibility of recovering energy.

Indirect heat regeneration: The third and more efficient solution, in terms of energy and flow control, is the use of the

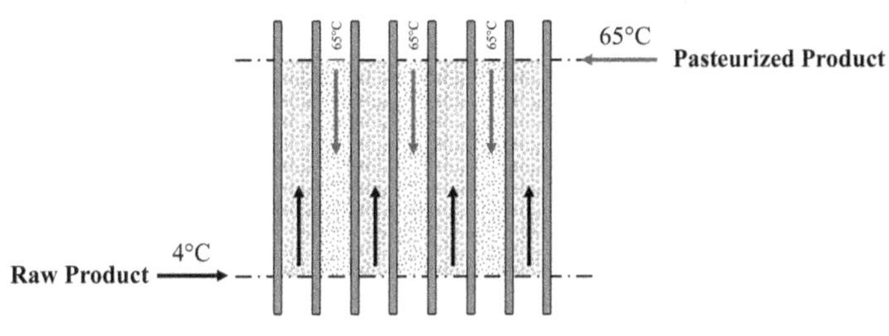

FIGURE 10.49 Damages potentially caused in the heat regeneration.

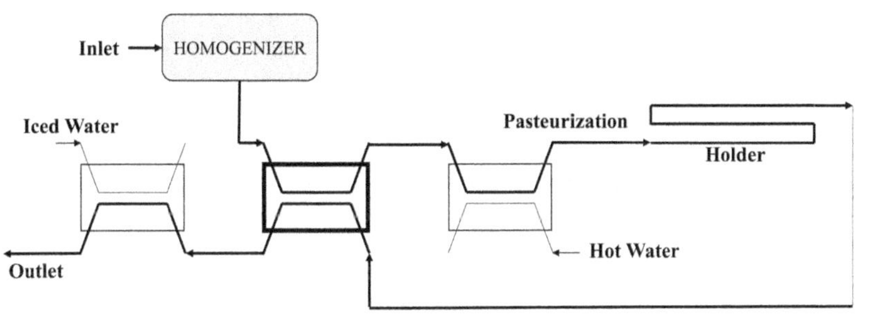

FIGURE 10.50 Position of the homogenizer in the "Cold Homogenization" design.

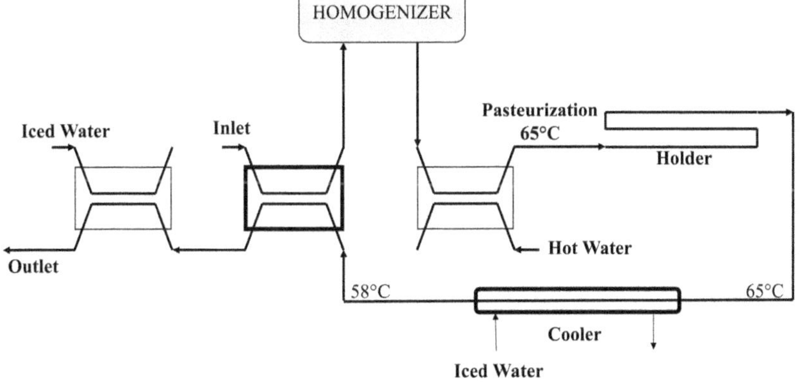

FIGURE 10.51 Position of the homogenizer in the "Product Cooling" design.

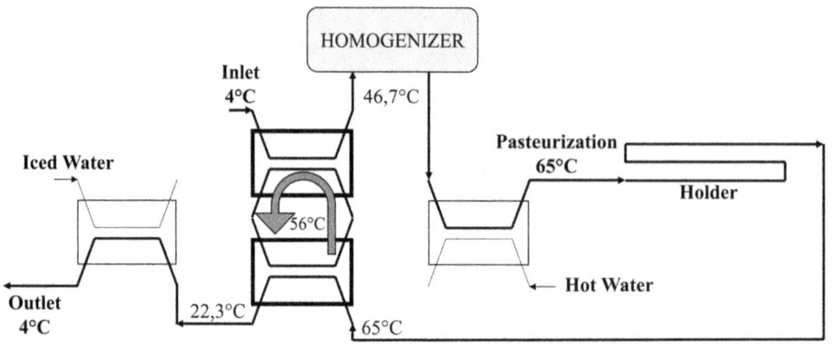

FIGURE 10.52 Position of the homogenizer in the "Indirect Heat Regeneration" design.

TABLE 10.5
Advantages of Different Energy Regeneration Systems

System Types	Efficiency (E)	Safety	Product Quality/Long Run
Double wall plates	60% < E < 75%, typically ~ 70%	Ensured by Free Separation Area	Short production time, unless to place a cooler reducing the efficiency to <70%
Differential pressure control	80% < E <85%, typically ~ 80%	Counter pressure valves and pumps required	Short production time, unless to place a cooler reducing the efficiency to <70%
Indirect regenerative system	70% < E <75%, typically ~ 73%	Ensured by 2 separate circuits	Ensured by Design
Mixing regeneration system	90%	Combination of indirect and direct system	Ensured by design

Indirect Heating (Figure 10.52). As explained previously, this system uses a water recirculation system to transfer energy from the hot to the cold product and vice-versa.

The indirect heat regeneration system is designed to keep the highest temperature of the recirculation water below the critical temperature and avoid any denaturation of the egg white proteins in the non-homogenized liquid whole egg. This system is the only one that ensure the safety of the system (no cross contamination) and, at the same time, avoid the coagulation of the egg white. The limit in this system is that due to the configurations of the plates/tubulars and the limitations for the maximum temperatures used, the efficiency achieved is only about 73%.

Some innovative systems, combining direct and indirect heat regeneration systems, can reach energy savings in the order of 90% and still avoid the problems of egg white protein denaturation (Table 10.5).

10.5.5.2.5 Heat/Cool Exchangers

There are many ways to transfer heat (or remove heat) from products during the pasteurization process. In most cases, the indirect heat exchange is considered, meaning that the heat exchange occurs indirectly through heat exchangers from a hot liquid to a cold liquid. Heat exchangers can be divided into plate heat exchangers and tubular heat exchangers.

The plate heat exchangers are composed of a series of plates that allow the passage of the hot fluid and the cold fluid, keeping them separated from each other by means of specially designed gaskets (Figure 10.53, left). The tubular exchangers, on the other hand, are composed of a main tube (jacket) in which 1 to N tubes are inserted and run through the entire length of the jacket (Figure 10.53, right).

10.5.5.3 Evaluation of Pasteurization Effects

It is common practice to measure the effect of the pasteurization process in terms of bacteria that the process is able to kill (killing rate or LOG reduction) and how long the product will last during storage/shelf life. Often, the commercialization of a product is based on the evaluation of these parameters to clearly demonstrate compliance with local regulations. Such

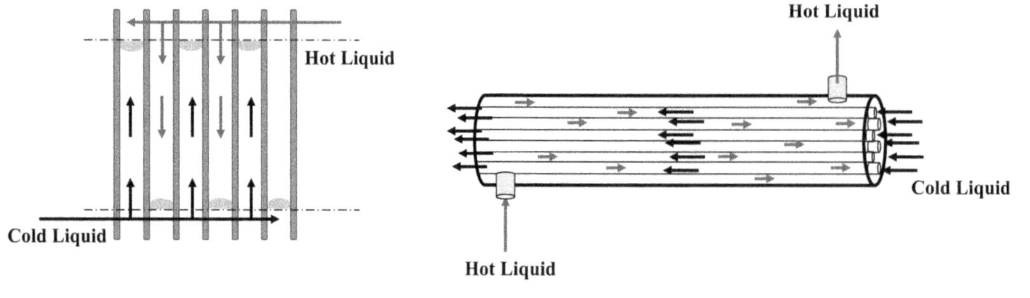

FIGURE 10.53 Schematic of a plate heat exchanger (left) and a tubular heat exchanger (right).

parameters are generally only linked to safety aspects, such as the total absence of pathogenic bacteria like *Salmonella spp.* and *Listeria monocytogenes.* During the shelf life of a product, the following degradation phenomena may occur: bad smell, coagulation, and/or loss of functionalities, such as heat setting properties, whipping, binding, and emulsifying properties. These phenomena are generally caused by non-pathogenic bacteria that survive the pasteurization.

During the standard pasteurization process, the way to increase the killing rate (and consequently the shelf life) is by increasing the pasteurization temperature or increasing the holding time (Humphrey et al., 1990). Experience has confirmed that the right way to go is by increasing the temperature of pasteurization, thus the preference for HTST-systems by trying to reach as high temperature as possible, and reducing, if possible, the holding time. This type of pasteurization concept will improve the shelf life of the product considerably. Since the early 2010s, several alternative pasteurization technologies at lower temperatures (e.g., infrared, high pressure, radio frequency, microwave, pulsed electric field, etc.) have been developed but until now, the only technology that ensures the highest killing rate is based on the combined effect of temperature and holding time.

Egg products are pasteurized at temperatures close to the limit of their denaturing process in order to obtain the highest safety and bacteriological reduction or complete elimination. This often causes a denaturing effect of proteins and, even if not seen during the pasteurizing process, it will be evident during the use of the pasteurized egg product with their changed functional properties. In order to avoid this denaturation process, the heating should be kept as accurate as possible, maintaining a steady temperature and maintaining the temperature of the heating liquid (hot water) as low as possible. Moreover, a precise regulation of temperature should be included into the design to provide a high stability to the system and prevent hot water from surpassing the admissible limits. This can be obtained by sequential regulations made with highly developed automation systems. Various analyzes were made to study the behavior of pasteurized liquid whole egg at different temperatures evaluating its performances on two bakery applications: genoise (sponge cake) and choux; the first one is a sweet bakery application and the second one is a salty bakery application and they are both influenced by almost all the functional properties of the liquid egg, including the heat setting properties. The limit curves for denaturation of the proteins in the liquid whole egg used for the preparation of the two above-mentioned pastries are presented in Figure 10.54. It shows two distinct areas, one with functional loss of 100%, and one with 20% functional loss, normally accepted for commercial liquid whole egg products.

The upper line is showing the limit of the 100% loss of properties and the lower line is showing the limit for the 20% property loss (20% PL). All the combinations of temperature (x-axis in Celsius) and time (ordinate in seconds) around the 20% PL can be considered acceptable for commercial use. Therefore, as shown on Figure 10.54, three areas can be identified: good (light gray), admissible (white) and not acceptable (dark gray).

If we consider the standard holding time of 210 seconds (3.5 minutes), the following diagrams (Figure 10.55) show that temperatures around 63°C has no significant influence on the properties of whole egg.

Figure 10.56 presents the Combination Temperature-Times to Get a Minimum Theoretical 6 LOG Reduction of *Salmonella Enteritidis*, *Pseudomonas spp.* and *Micrococcus spp.*

Salmonella Enteritidis: Thermosensitive, pathogenic bacteria that has a resistance to heat treatment comparable to the resistance of enterobacteriaceae (9 LOG reduction required).

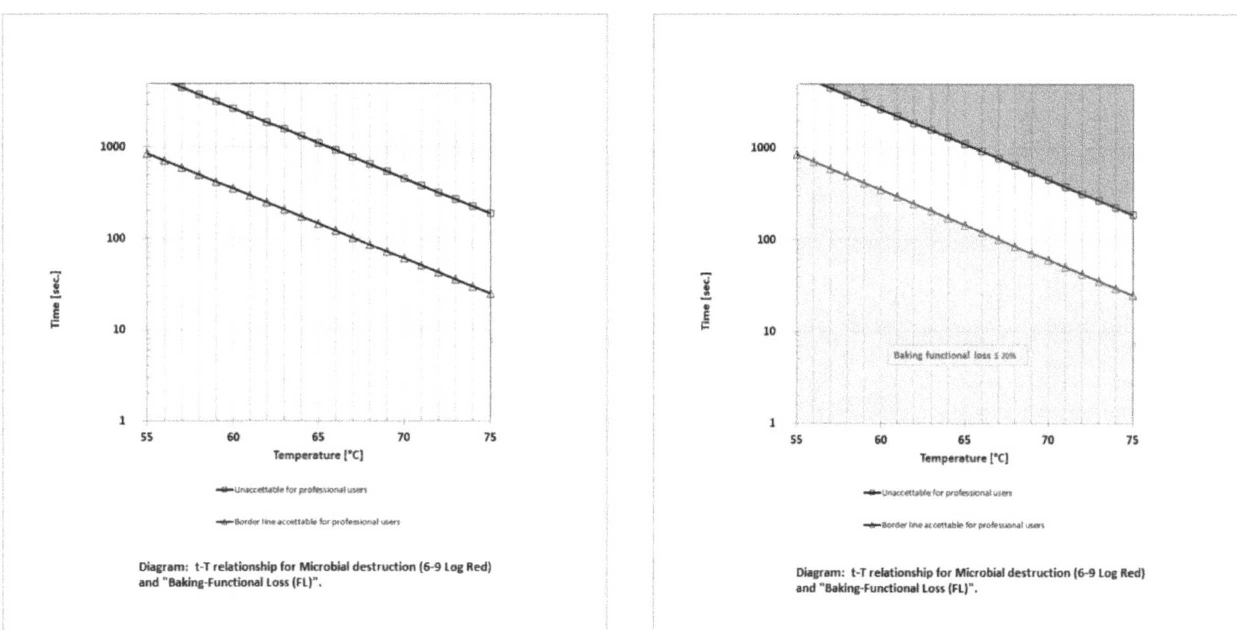

FIGURE 10.54 Limit curves for denaturation of the proteins in liquid whole egg used for the preparation of genoise (sponge cake) and choux.

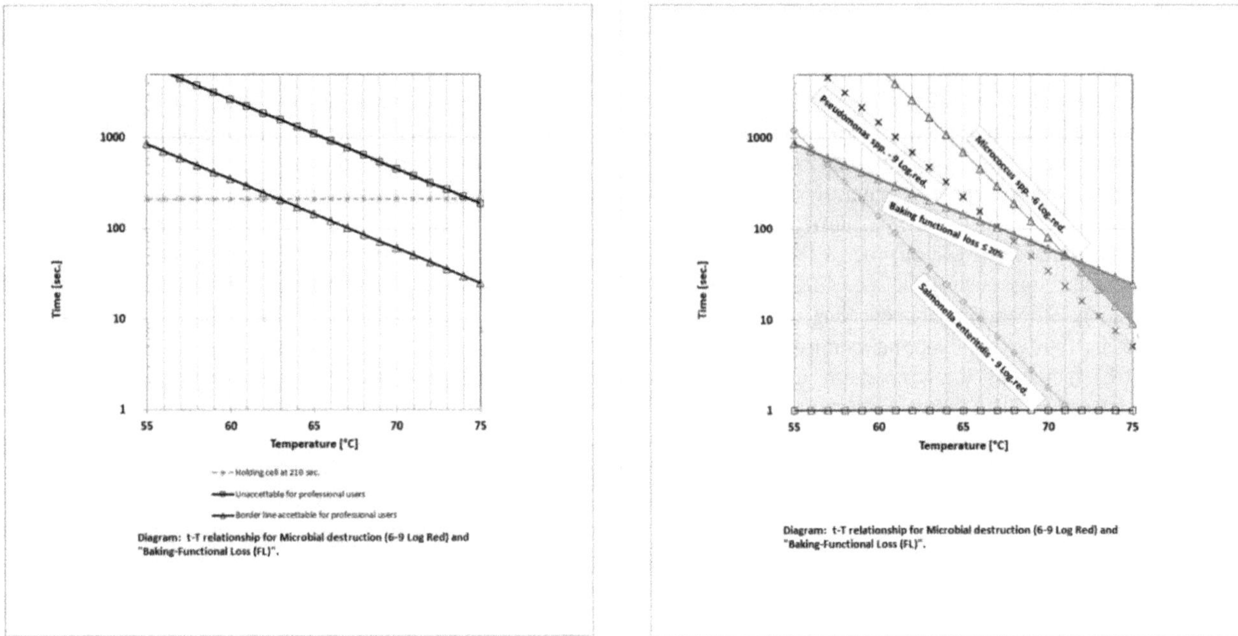

FIGURE 10.55 Optimization of thermal processing for a liquid egg pasteurizer.

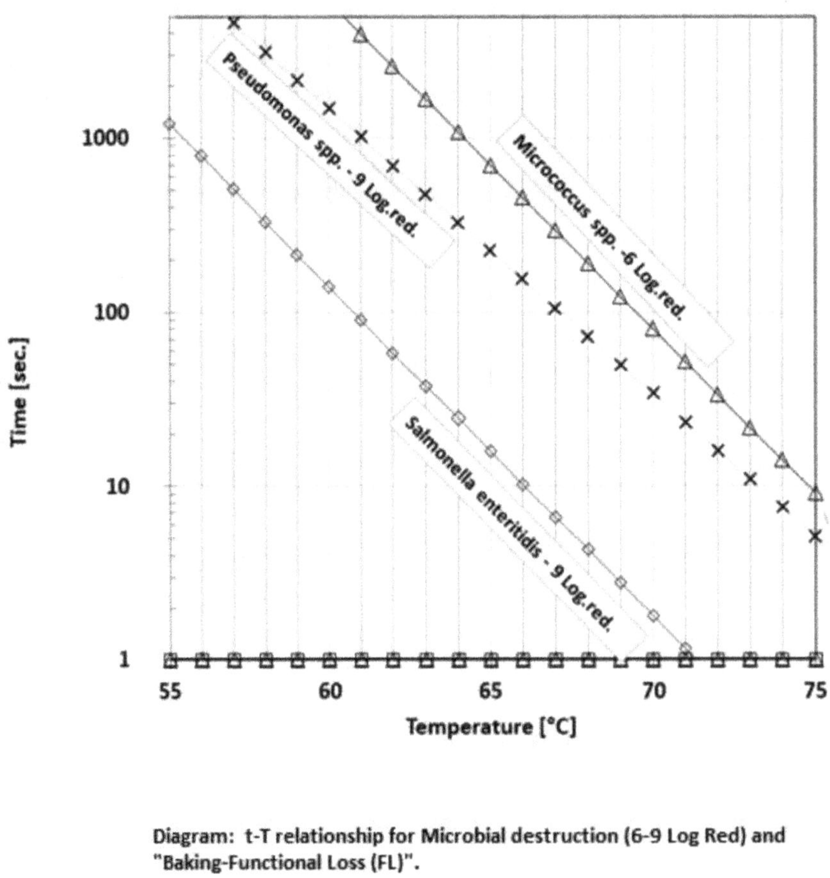

FIGURE 10.56 Combination temperature–time to get a minimum theoretical 6 LOG Reduction of *Salmonella enteritidis*, *Pseudomonas spp.* and *Micrococcus spp.*

Pseudomonas spp.: Thermoresistant, pathogenic bacteria that requires a higher temperature than the enterobacteriaceae (10–15°C higher) for a good inactivation (9 LOG reduction required).

Micrococcus spp.: Thermoresistant spoiling, non-pathogenic bacteria to humans but responsible for damages to the product quality even when refrigerating. The inactivation of these bacteria occurs at temperatures 15–30°C higher than the temperature required for the enterobacteriaceae (6 LOG reduction required).

Using Figure 10.56 as a baseline, two important areas can be identified: 1) the area above all bacteria inactivation curves; and 2) the area below the 20% Protein Loss curve. By combining the curves for the bacteria inactivation and the curve for the admissible denaturation, we will identify the acceptable time-temperature combinations (Figure 10.57).

The intersection between the "safe" area and the "good" area is identifying the region where the maximum killing rate will be obtained with the minimum denaturation of the product pasteurized (showed in red in Figure 10.57). The ideal system should be acting with "surgical precision" within the dark gray area. In addition, the system must be Quick, Precise and Uniform: Quick since the holding time in that area is very limited (fast heating of product), Precise as we are dealing with very high temperatures and a very low margin for errors (higher risk of product coagulation) and Uniform, with all product particles instantaneously reaching the same temperature.

Due to the mechanical configuration of the traditional heating systems (tubulars and plates), it is impossible to fulfill the above-mentioned characteristics. In fact, the heating with indirect systems like plates and tubulars requires minutes for heating time. The thermal inertia in the system is huge since the volumes of water and product present in the circuits are very large. And the uniformity is not guaranteed as the heat is transferred from stainless steel surfaces (heated by hot water) to the product and it is unavoidable that the product in contact with the walls of the heat exchangers will be hotter than the product flowing in the internal layers.

Since we cannot use indirect heating to fulfill the Quick, Precise, and Uniform factors, a direct heating system must be used. There are mainly two types of direct heating systems commonly used in the dairy and juice processing sectors: Steam Injection, where steam is introduced in the pipe in which the product is flowing and then removed by a vacuum chamber; Steam infusion, using the same concept of steam injection but, in this case, the product is injected into a chamber where the steam is keeping the temperature at < 130°C. It is easy to understand why it would be extremely difficult to apply such systems to the liquid egg sector, since the high

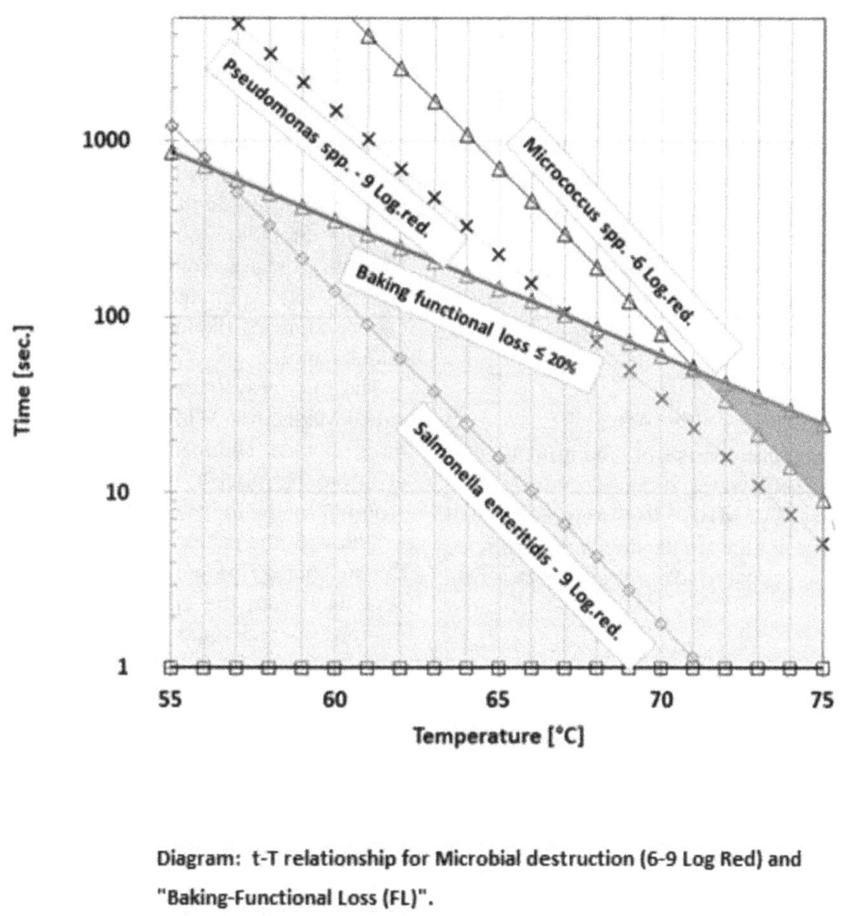

FIGURE 10.57 Definition of the optimum time–temperature combination for optimum killing efficiency and minimum protein losses.

temperature of the steam will immediately cause the coagulation of the product.

Other systems, seen more and more in pasteurization and sterilization processes are based on the joule effect, the principle whereby a body crossed by an electric field heat up according to the resistance of the body itself. The system most often used is ohmic (or joule) heating, based on electro-heating consisting in the direct passage of an electric current through the food by using special applicators containing a set of modules with several electrodes and insulators.

10.5.5.3.1 Ohmic Heating Technology

Each liquid can be thought as containing a set of positively and negatively charged ions. For simplicity, we refer to water but it is known that in pure water, there is a non-zero concentration of hydrogen ions (positively charged) and a non-zero charge of hydroxyl ions (OH-).

If a voltage is applied to water by means of an electric field, the negative ions move toward the positive electrode, while the positive ions migrate toward the negative one. Moving ions collide with other ions, molecules, and thus generating some heat. The greater the ion concentration, the greater the applied electric field and the greater the amount of heat generated. The liquid within the section where the electric field is applied may be compared to an electric resistance. When traversed by an electric current, it generates heat (Figure 10.58).

Figure 10.58 highlights the motion phenomenon of ions inside a liquid subjected to an electric field. Although a small amount of heat is generated with each collision, ions are in very high number and there are many collisions. According to the kinetic theory of fluids, the total heat generated is the sum of millions and millions of these small molecular collisions. Fortunately, liquid eggs have an excellent concentration in ions, and lends itself very well to this technology. For eggs, the typical conductivity values are about 5 mS/cm at 25°C and can double at temperatures considered during egg processing (approximately around 70°C). The typical values of the electrical conductivity of liquid egg products (yolk, white and mixed) are presented in Table 10.6.

10.5.5.3.2 Radio Frequency Technology

The radio frequency technique consists in submitting the flow of product to electromagnetic waves, of length and frequency typical of radio waves (27.12 MHZ). With this process, all water molecules tend to point toward the direction of the electric field, and in response, generate themselves an electric field.

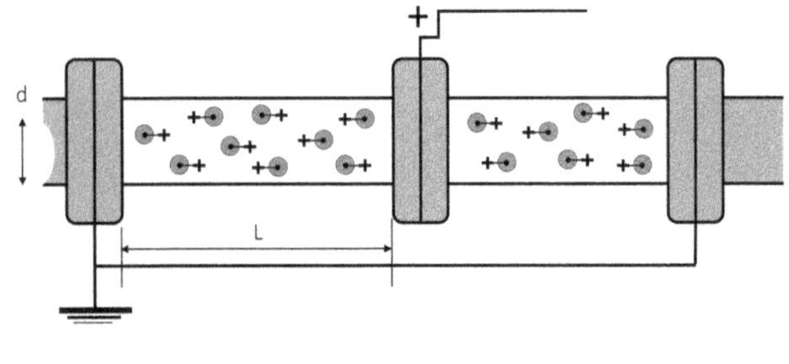

FIGURE 10.58 Liquid may be considered as resistance when an electric current circulates through it.

TABLE 10.6

Product Conductivity (mS/cm) at Temperatures Ranging from 5°C to 70°C (SANOVO Field Data)

Liquid Egg Product	Conductivity [mS/cm]			
	5°C	20°C	50°C	70°C
Egg white	4.88	7.83	13.73	17.66
Plain yolk	2.32	3.44	5.69	7.19
Plain whole egg	3.44	5.62	9.98	12.89

The continuous reversion of the electric field polarity generates collisions, that is kinetics energy and therefore heating (dielectric mechanism). At the same time, the ions also tend to orientate themselves, therefore generating heating (ionic mechanism). While the liquid heats up homogeneously with all liquid in transit subjected to the electromagnetic field, this technology has a very low efficiency in terms of energy consumption. To obtain an electric field of a defined intensity and oscillating at those frequencies (27.12 MHZ), a triode (composed of a cathode filament, an anode plate and a control grid installed in a evacuated glass container) must be used but it is very expensive and has a limited lifetime (typically 3 years). The triode must be polarized at its trigger voltage, so a minimum level of energy is always supplied to make it run, thus implying a significant drop in the overall performance of the device. For these reasons, the typical efficiency of this system is very low (below 60%).

10.5.5.3.3 Microwave Technology

The technology using microwaves to generate heat is well known and used in every home environment. This concept is completely analogous to the concept of radiofrequency. The difference resides in the frequency of the waves used: MHz is used for the radiofrequency system and GHz for microwaves. Currently, the usage of Microwave system within the egg processing sector is limited by its high cost and level of sophistication, requiring the intervention of expert technicians for its maintenance.

The microwaves are generated by a thermo-ionic device called Magnetron. While magnetrons used for domestic ovens have a wide commercial distribution (therefore available at affordable prices), magnetrons for the high-power usage required by the food sector s are very expensive and difficult to find with the paucity of manufacturers. The application of this technology present also some issues: the need to shield personnel from the electromagnetic field generated by the microwaves and some difficulties to concentrate the energy into the right point.

10.5.5.3.4 Pulsed Electric Field Technologies

The Pulsed Electric Field (PEF) technologies are the latest technology developed for food applications. It is an Electroheating type technology, similar to the Ohmic technology, but different in the way the voltage pulses are applied. This technology tends to carry out pasteurization not so much by thermal effect, but by electroporation. Compared with the Ohmic heating, higher voltage pulses are applied to the liquid egg in order to open pores within the cell membrane of bacteria, thus weakening their resistance. In liquid eggs, the PEF are commonly used in combination with thermal treatments since electric pulses are unable to completely inactivate *Salmonell spp.* (Bonfà & Bertin, 2016).

10.6 Spray Drying

A large number of foods can be dried with the primary purpose of preserving products (e.g., grapes for raisins, plums for prunes). The extension of the shelf life is secured by removing water from products to a level stopping bacterial activity. Liquids containing dry solids can also be dried by evaporating the liquid (mostly water) and the suspended dry solids remaining as powder. In addition, the spray drying of liquids reduces the volume considerably, resulting in less storage and transport costs. Spray dried products are soluble in water, enabling them to be easily reconstitutes and returned back into a liquid form.

In the egg industry, the price of eggs, the raw product for spray drying, has seasonal fluctuations. Around Christmas and Easter seasons, the demand for shell eggs is high and prices will rise. The occurrence of disease outbreaks (e.g., Avian influenza) in many flocks will also limit the availability of eggs, thus increasing prices. Therefore, the prices of processed eggs, whether liquid or powder products, tend to follow closely the prices of shell eggs. Spray drying may take advantage of these seasonal price adjustments by spray drying eggs when the price of shell eggs is low and selling egg powders when prices have increased.

The earliest mention of spray drying involved egg drying in 1865 (Ternes et al., 1994) and industrial applications of spray drying, of any significance, occurred in the milk industry in the 1920s. During the Second World War, the sudden need for dehydrated foods to reduce shipping volume, weight and transportation costs created a high demand for this technology and the use of spray dryers blossomed.

Although the spray drying process is highly energy intensive and entails considerable capital expenditure, it offers also a high degree of automation and hence, low labor costs. In order to maximize their return on investment, most plants will run the operations around-the-clock, six to seven days a week. Nowadays, spray drying is widely used by the food processing industry for the drying of various products, including eggs. The liquid to be dried is pumped into the drying chamber and atomized either at high pressure by means of nozzles, or by a centrifugal atomizer. The purpose of atomization is to provide the liquid with a large surface area in contact with heated ambient air. The energy of the heated air is used to evaporate the water content in the product.

10.6.1 Spray Drying of Egg Products

Liquid whole eggs as well as egg white and yolk are all suitable for drying. The solids content of liquid whole eggs is approximately 24%, whereas that of liquid yolk is approximately 44%. The water content of yolk and whole egg powder is normally 3%. Liquid whole egg and yolk can be mixed with salt and other ingredients prior to pasteurization and drying, thus providing some distinct properties to the final powder. The addition of ingredients will increase the solid content of the liquid. Mixing prior to pasteurization is preferred to prevent any risk of contamination of the liquid egg by the ingredients. The solids content of liquid egg white is approximately 11% and is dried to approximately 7% water content in a powder.

After breaking and before drying, both the egg yolk and whole eggs are pasteurized. Normally, yellow products are cooled after pasteurization and stored at approximately 3°C (37°F) in tanks prior to drying. It is possible to dry cooled products directly from the tank. However, higher capacity and improved atomization of the product is obtained by preheating to approximately 55°C (130°F) immediately before spray drying. It is possible, but not very common, to concentrate whole eggs, either by evaporation or UF-filtration.

The egg white contains glucose and the heat effect during the drying process will discolor the product, due to the Millard reaction. Consequently, the glucose is removed by fermentation prior to drying. The fermentation process is performed by using yeast, enzyme treatment, lactic acid bacteria or a mother culture of egg white. With a solid content as low as 11%, it is profitable to concentrate liquid egg white before drying. This concentration step is performed by membrane filtration, ultrafiltration or reverse osmosis, increasing the solids content to at least 22%. It is not common to pre-heat the egg white prior to spray drying. The fermented dried egg white powder is an unpasteurized product and must be heat treated to reduce bacterial counts and improve product properties (e.g., gel strength). Since egg white coagulates at temperatures around 55°C, a satisfactory pasteurization result is difficult to achieve in liquid egg white with the additional bacteria present as a result of the fermentation process. The problem is resolved by providing a heat treatment to the egg white in its powder form. This heat treatment also improves the gel strength of the egg white. The heat treatment usually takes place after packaging in heated chambers named hot rooms. Compared to whole egg and yolk having a high fat content, the egg white has a minimum of fat content, a characteristic essential for its good whip ability properties. Therefore, it is not possible to switch from yolk or whole egg to egg white drying without a thorough wet cleaning of all product contact areas inside a dryer. Likewise, since egg white is dried without pasteurization and liquid egg yolk and whole egg are pasteurized prior to drying, a thorough wet cleaning of the dryer is also required when switching from egg white to whole egg or yolk. Wet cleaning is time consuming step and involves a labor-intensive cleaning set-up. Consequently, the standard in the egg industry is to have one dryer dedicated to egg white and another dryer dedicated to yellow products.

10.6.2 Structure of a Spray Drying Plant

The actual conversion of liquid into solids takes place in the drying chamber, present in all spray drying systems. Heated air and liquid product are fed into the drying chamber. The drying chamber will discharge drying air and product. Since a certain amount of product is contained in the discharge air, it should be recovered before the air is removed from the system (Figure 10.59). Consequently, all dryers consist of at least the following components: 1) Drying chamber; 2) Equipment to

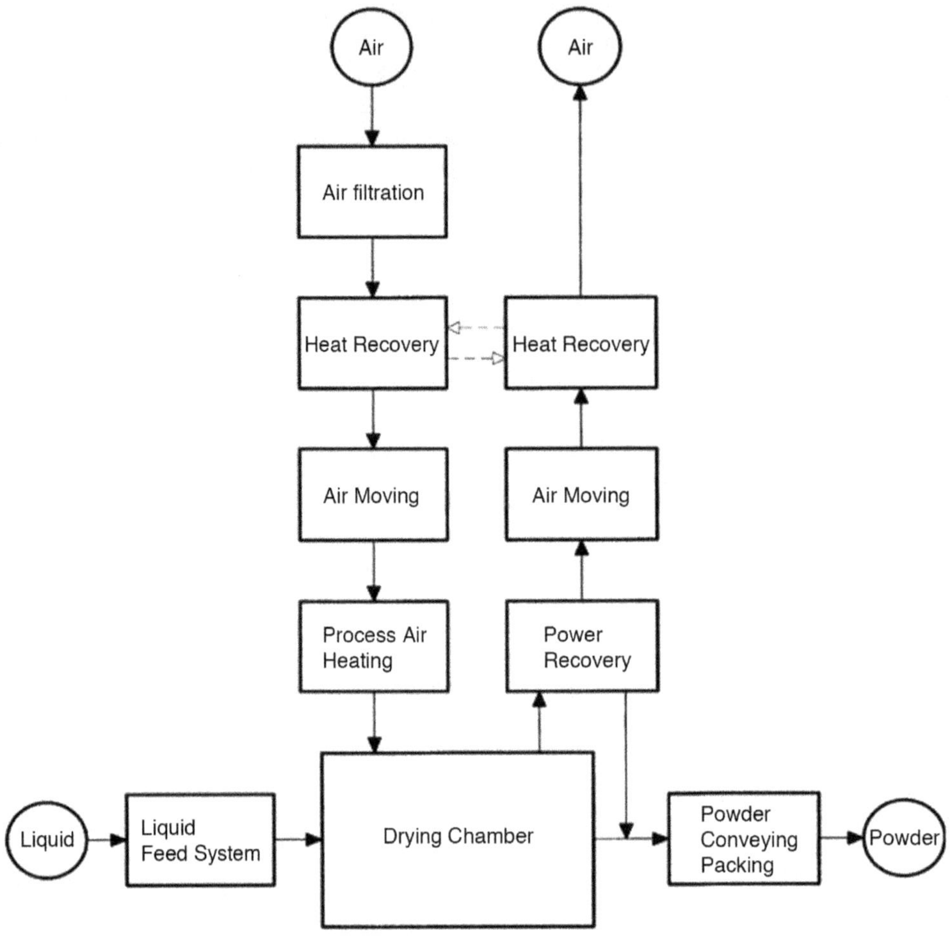

FIGURE 10.59 Structure of a spray drying plant.

feed the liquid product into the drying chamber and to atomize the product; 3) Equipment to recover the powder from the discharge air; 4) Equipment for removal of dried product from drying plant and for packing of product; 5) Equipment for heating and moving of drying air into drying chamber; 6) Equipment for removal of discharge air from drying chamber.

10.6.3 The Drying Process

The liquid to be dried is pumped into the drying chamber and atomized. The purpose of atomization is to provide the liquid with a large surface area. Heated ambient air flows around the atomized product. The energy of the heated air is used to evaporate the water content in the product. Nature will always seek balance. When the hot dry air meets the wet water drop (containing solids particles) in the drying chamber, water will evaporate into the dry air. However, the energy required for water evaporation (water transformed from a liquid to gas phase), will simultaneously cool down the drying air. In theory, this process will continue up to a point of balance between the humidity in the air and the humidity in the powder particle. Since powder and drying air are separated, this process is interrupted in the drying chamber before this balance is reached.

During spray drying, the remaining water content of the powder being discharged from the drying chamber is an important factor. The reason for this is that the product should be dried to the extent where all bacteria activity has stopped: typically, less than 8% for egg white powder and less than 4% for whole egg and yolk powders. In order to maintain a constant moisture content in the powder, both inlet air temperature and outlet air temperature must be constant. If the inlet temperature is increased, the outlet temperature must be increased, but not as much as the inlet temperature to maintain the water vapor content constant in the outlet air and thereby also in the moisture in the powder at the same constant level.

A description of what happens with the individual drop/powder particle, passing through the plant: when the drop is formed in the hot air, water will evaporate immediately from its surface. This water evaporation will occur at a constant speed. Simultaneously, the water content of the ambient air will be increased. In theory, this process, described in Figure 10.60, implies that the water during the entire process is able to evaporate freely from the liquid drop transformed into a powder particle.

In practice, the liquid at the surface of the drop will evaporate first, and the solids that remains will form a shell,

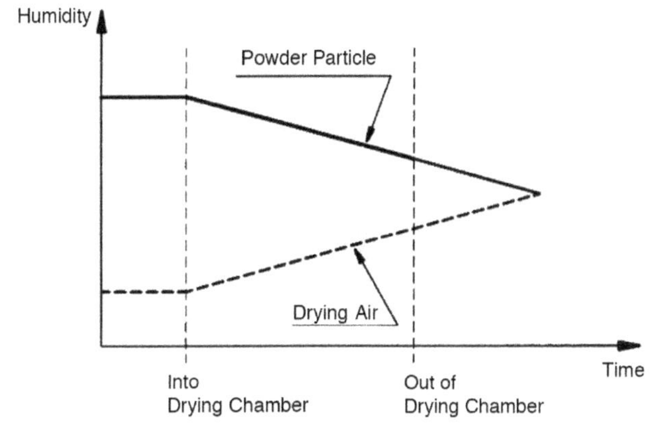

FIGURE 10.60 Theoretical humidity balance.

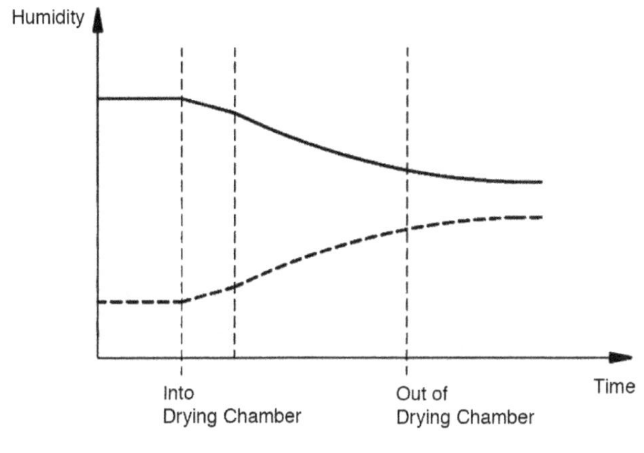

FIGURE 10.61 Practical humidity balance.

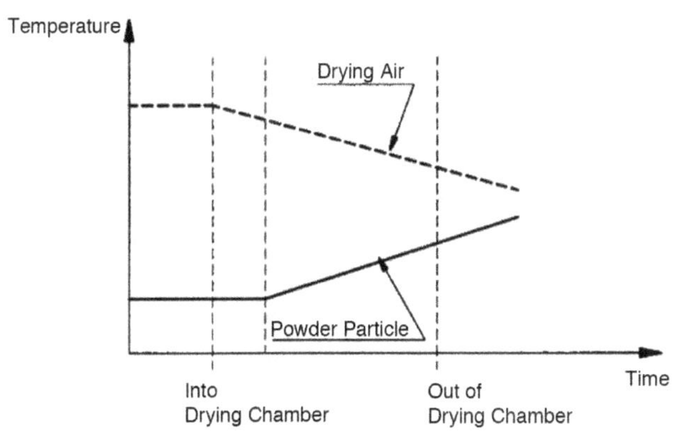

FIGURE 10.62 Temperature balance.

encircling water at the center of the particle. This will inhibit evaporation, and the drying process (Figure 10.61).

As long as water can evaporate freely from the drop, all energy contained in the hot air will be used for evaporation, and the water drop/powder particle will not be heated. When the shell has been formed, evaporation will be inhibited and part of the energy contained in the hot air will be used for heating the powder particle (Figure 10.62).

Consequently, the air inlet temperature has an influence on the exposure of the powder particle to heat. This is important for the drying of eggs since proteins will denature at too high heat exposure. Consequently, there are limits to how high an inlet temperature that can be used without causing damage to the product. During drying, the powder temperature will typical be 25 to 40°C lower than the temperature of drying air outlet. The key operating rules for egg drying are: 1) Raising the inlet air temperature and keeping the outlet air temperature constant = Higher capacity and higher moisture content in the powder; 2) Lowering the inlet air temperature and keeping the outlet air temperature constant = Lower capacity and lower moisture content in the powder; 3) Keeping the inlet air temperature constant and raising the outlet temperature = Lower capacity and lower moisture content in the powder; 4) Keeping the inlet air temperature constant and lowering the outlet temperature = Higher capacity and higher moisture content in the powder.

Normally the drying air temperature is kept constant and the moisture content in the powder is regulated by increasing or decreasing the dryer outlet temperature by controlling the feed rate to the dryer. When drying egg products, the inlet temperature will range between 150°C (300°F) and 200°C (395°F) and the outlet temperature between 60°C (140°F) and 90°C (195°F). Samples of the powder are collected frequently to measure the moisture content and the drying air outlet temperature is adjusted accordingly to maintain the moisture content in the powder.

10.6.4 Components of a Spray Drying Plant

10.6.4.1 Design of Drying Chambers

Historically, two types of drying chambers have been available on the market. The Vertical drying chambers or towers (Figure 10.63) have cylindrical chambers with conical

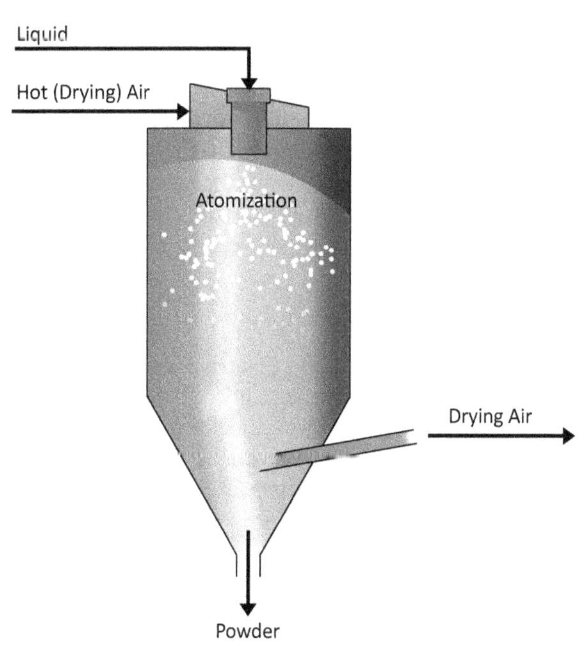

FIGURE 10.63 Vertical drying chamber.

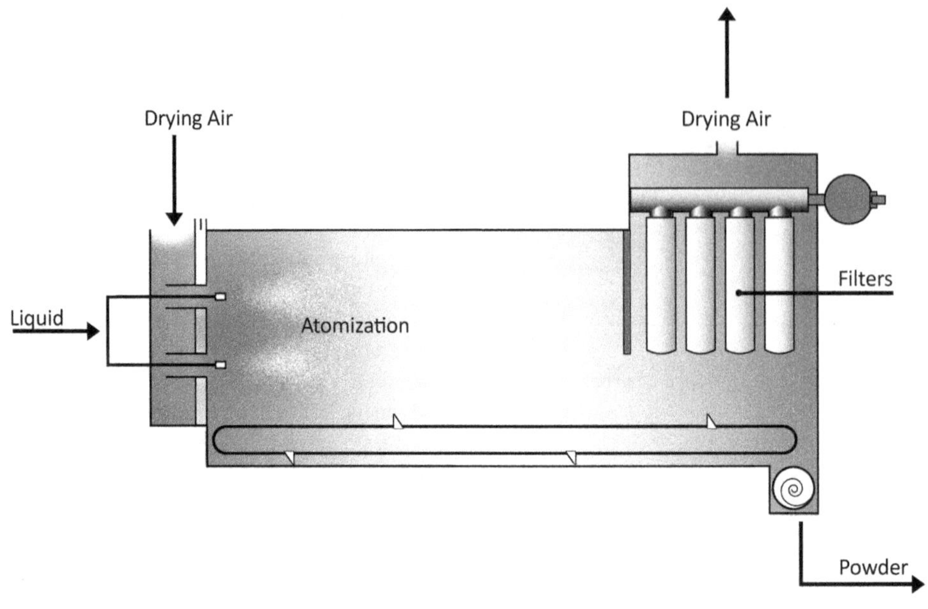

FIGURE 10.64 Horizontal drying chamber.

bottoms. Product and air are fed at the top and the drying air is rotated in a special screw-shaped air distributor. The powder is removed at the bottom by means of gravity, and the drying air is removed on the side of the conical bottom. Atomization is normally affected by means of a centrifugal atomizer. Atomization by means of nozzles is also applicable in tower systems.

Horizontal drying chambers or box dryers (Figure 10.64) are rectangular chambers, supplying air and product at one end and discharging at the opposite end. In box dryers, only atomization by means of nozzles is applicable. The powder recovery system, in the shape of a bag collector, is often integrated in the drying chamber. The powder falls on the floor of the drying chamber and is discharged mechanically.

Tower dryers were developed in Europe and have been prevailing in Europe, whereas box dryers were developed in America and are prevailing in the Americas. Over the years, both system designs have been adopted globally and the same powder quality is obtained for standard egg products. However, box driers have always been the preferred dryer type for egg products.

10.6.4.2 Feed System

Normally, the supply of a drying plant includes product outtake of the tank storing the product to be dried. The construction of a feed system depends on the product to be dried, degree of automation and use of nozzles or centrifugal atomization. A brief description of the main components for centrifugal atomization and nozzle atomization are described in the following section (Table 10.7).

10.6.4.2.1 Centrifugal Atomization

Centrifugal atomization is only applicable in spray towers. The atomizer is placed at the top of the drying chamber in the center of the air distributor (Figure 10.65).

TABLE 10.7
Comparison of Centrifugal Atomization and Nozzle Atomization

Centrifugal Atomizer	Nozzles
Higher investment	Improved dust collection of products is required
Particle size independent of capacity	Particle size depends on capacity/pressure
Large particles can only be produced on large plants	Large particles can also be produced on small plants
The powder is more agglomerated	The product is less agglomerated
Mean particles size up to 200 microns	Mean particles size up to 350 microns
Lower density of powder	Higher density of powder
More free fat in powder	Less free fat in powder
Reduced free flowing character	Improved free flowing character

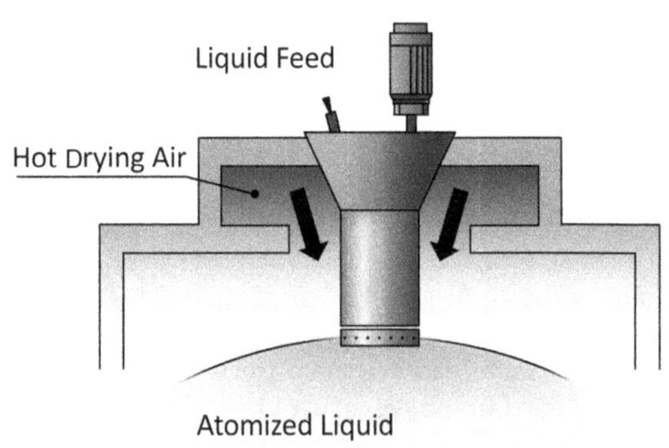

FIGURE 10.65 Centrifugal atomization.

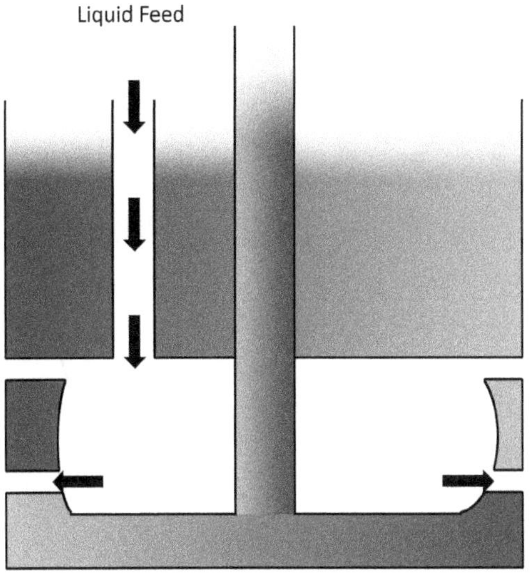

FIGURE 10.66 Atomizer wheel.

The most important part of the atomizer is the atomizer wheel rotating at several thousand rpm. The liquid is fed into the center of the atomizer wheel, accelerated and thrust out by centrifugal force through bored tubes, leaving the wheel in the form of atomized liquid (Figure 10.66). The liquid is pumped to the atomizer by a positive pump to allow the control of the supplied quantity.

10.6.4.2.2 Nozzle Atomization

During nozzle atomization, the product is atomized in a number of specially designed nozzles placed into the hot air supply to the spraying chamber (Figure 10.67).

For the spray drying of foods, a nozzle pressure between 100 bar (1450 psi) and 250 bar (3700 psi) is normally applied. To achieve such pressure, a high-pressure pump is used as a feed pump for the plant. A high-pressure pump is a piston pump that requires a pressure of 1 bar (15 psi) to 3 bar (45 psi) at product inlet. Consequently, a pump is required between feed tank and high-pressure pump. There are several suppliers of spray nozzles specially designed for spray dryers. A nozzle consists of the following parts: 1) Body affixed the nozzle stem; 2) Core

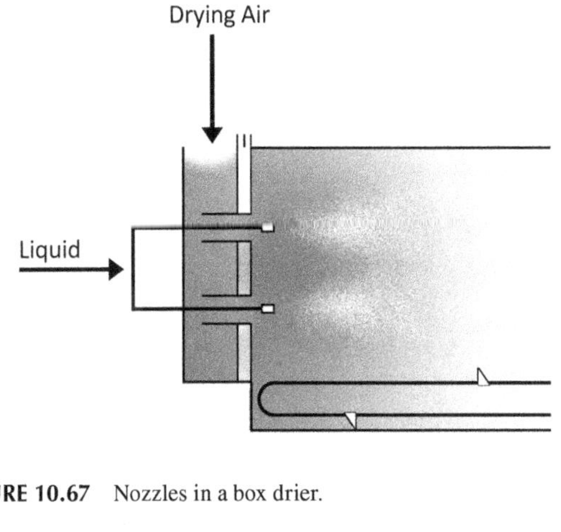

FIGURE 10.67 Nozzles in a box drier.

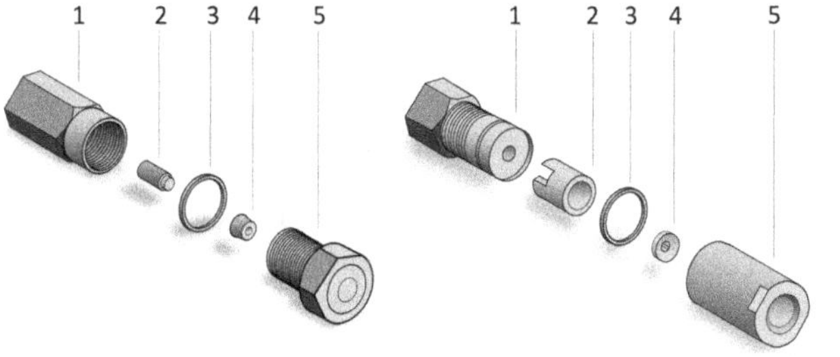

FIGURE 10.68 Nozzle assembly.

or Swirl Chamber, rotating the product; 3) Gasket; 4) Orifice, which is the constriction building the pressure; 5) Cap that assembles the nozzle (Figure 10.68).

The ratio of nozzle capacity to nozzle pressure is governed by the dimension of the Orifice constriction. The spray angle is governed by the design of the Core/Swirl chamber. Different capacities and spray angles can be obtained by a combination of Core and Orifice.

Each individual nozzle can be removed during operation for inspection, cleaning and replacement of the core or orifice. Some special factors to remember in connection with nozzle atomization are:

- Larger flow through identical nozzle size offers higher pressure, smaller particles, larger particle surface and, as a consequence, powder that is drier;
- Higher product viscosity offers lower capacity, larger particles and, as a consequence, powder that has more moisture;
- Higher capacity is achieved by using larger cores or by increasing the pressure

10.6.4.3 Powder Recovery

Cyclones and bag collectors are normally used for powder recovery from the drying air. In cyclones, powder separation is obtained by means of centrifugal force, also called dynamic separation (Figure 10.69). The efficiency of cyclones is lower than that of bag collectors and the egg white regulatory requirements of maximum powder rate in discharge air cannot be met. For this reason, bag collectors are mainly used by the egg industry (Figure 10.70).

The filtration of air is affected in bag collectors by retaining the powder particles. A bag collector consists of a number of cylindrical collector bags through which air passes from the outside separating the powder from the air. The powder sticks at the outside of the bag. At defined time intervals, the bag is cleaned with compressed-air briefly blown through the bag, shaking the bag and causing the separated powder to fall off. To avoid that the bags collapse during air sucking, they are kept extended by a wire cage. Collectors are designed for replacement from below, inside the drying chamber (in the collector department) or from above (in the upper plenum). Their replacement from the top is easiest but will require a distance from the top of collector to the ceiling of at least 2 m (7 feet), depending on the bag length.

Various materials are used for bag collectors, the most common being needle felt of polyester. Polyethylene and Gore-Tex are used as well. It is essential to use the correct bag collector material for the right powder product. Material properties, perfect for one powder product type may be disastrous for another. In recent years, bag collectors have been developed to be cleaned automatically through a Clean-In-Place (CIP) system. They are rather high-priced and mostly used in the pharmaceutical industry where frequent product changes occur. Finally, to avoid static electricity which may cause explosions, some bags are anti-static treated through interweaving carbon fibers or steel wire in the bag material.

10.6.4.4 Powder Conveying Systems

After powders are removed from the drier, they are conveyed to buffer store silos or directly to a sifter and packing place. In any case, the conveyance required may be achieved through various means: Screw conveyors; Belts; Vibration conveyors; Bucket conveyors; Pneumatic conveyance where the powder is conveyed in a tube and separated from the air in a cyclone or by a bag collector (Figure 10.71). When using cold dry air, the powder can be cooled during conveyance; Fluid lift plant, where conveyance is achieved by pressure differential obtained from a rotary blower (Figure 10.72).

Pneumatic conveyance and particularly Fluid Lift enable longer conveyance. Conveyance is obtained either by overpressure or vacuum, with the mounting of rotary valves required for feeding and discharge of product to and from the system.

Powder cooling: Typically, the product is discharged from the drying chamber at a temperature between 25°C (70°F) and 40°C (100°F) lower than the outlet temperature of the drying air. In many cases, such temperature will be too high for packing powders or storing them in silos. If the powder has a high fat content, part of the fat will remain fluent and the powder will clump together during cooling in the silo or in packing. Consequently, product cooling may be required. The following equipment may be used for this purpose:

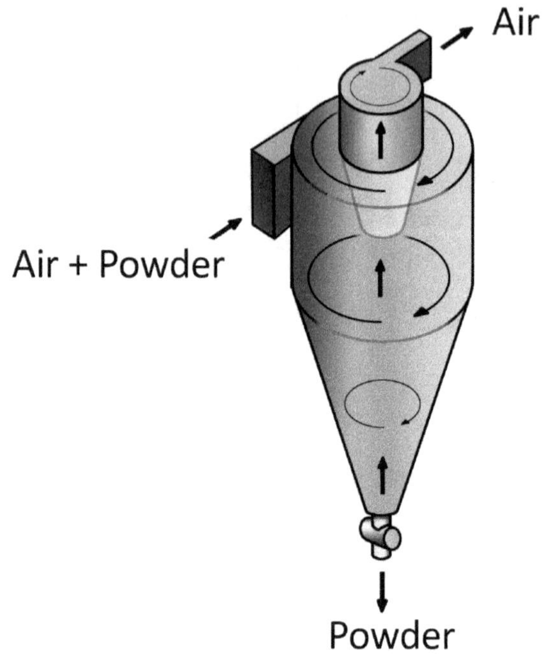

FIGURE 10.69 Cyclone.

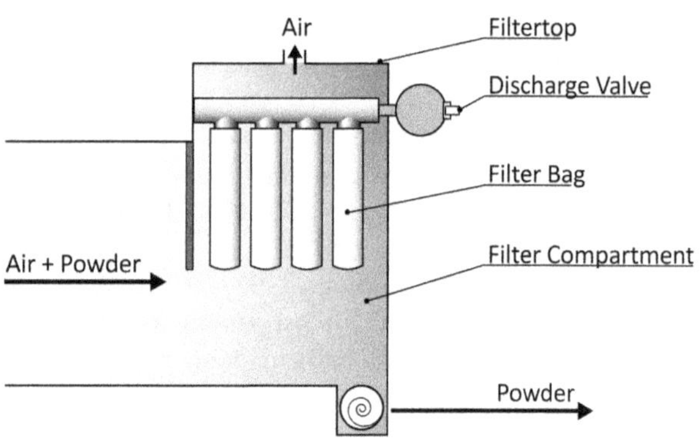

FIGURE 10.70 Bag collector.

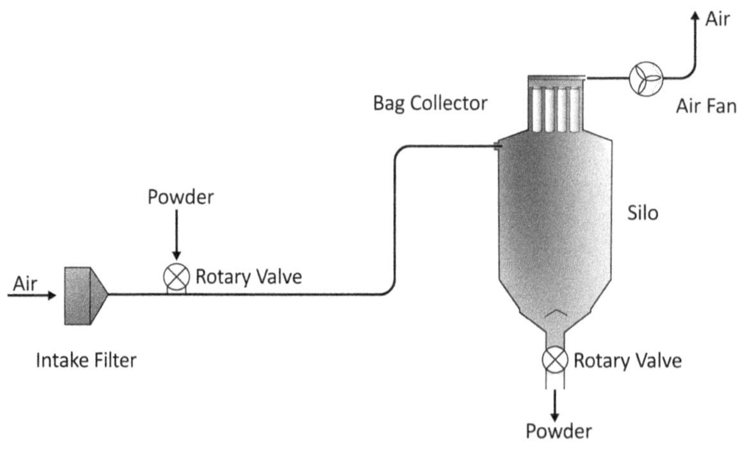

FIGURE 10.71 Pneumatic conveyance.

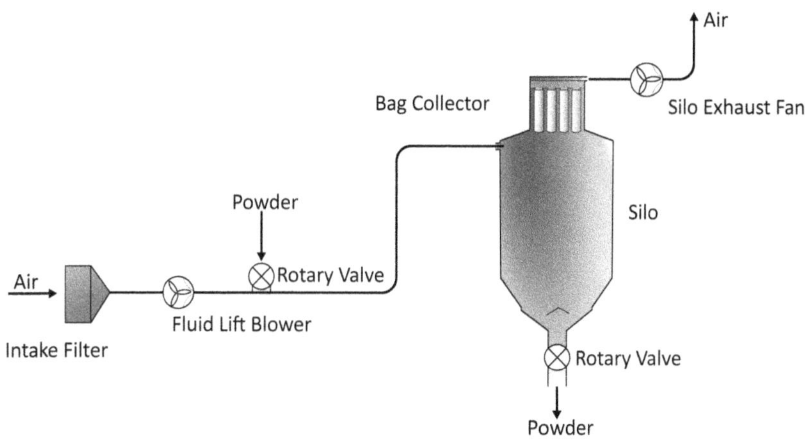

FIGURE 10.72 Fluid lift conveyance.

Fluid bed cooler (Figure 10.73): In this system the product is conveyed over a perforated plate through which cooling air is being blown. The product is fluidized and behaves as a liquid. The excellent contact between air and powder in a fluid bed cooler provides efficient cooling,

Pneumatic cooling/conveying plant: This is a pneumatic conveying plant as described previously. The temperature of the conveyed air is lower than that of the powder, thus, cooling the product during conveyance. Normal ambient air can be used in a pneumatic conveying system but, it may be advantageous to use cooled dry air.

Although a fluid bed cooling is a high investment, it allows the gentlest product treatment and offers lower operating expenses.

10.6.4.5 Powder Storage

Frequently, powders are stored in a buffer in silos, mobile containers or tote bins prior to packing. Even in a three-shift drying operation, the packing of products, a rather labor-intensive process, is possible during one single shift. Most silos are fitted at their bottom with a mechanical device (bin actuator or silo vibrator) to prevent powders from blocking. If a pneumatic or fluid lift system is used for powder conveyance to the silo, the conveying air should be removed from the silo (Figure 10.74). Therefore, a bag collector is installed at the top of the silo, retaining powder from the conveying air.

10.6.4.6 Blending

Various ingredients in powder form may be added to egg powders. This will be done by using a batch mixer or by dosing the ingredient into the drying chamber or the powder stream coming out of the drying chamber. When an ingredient is added into the drying chamber, the explosiveness of the ingredient and the resulting mixed powder must be evaluated and be within the explosion protection specifications for the drying chamber. The addition of ingredients (e.g., starch, extra proteins) to egg powders is mainly dictated by specific demands and needs of customers. A free-flowing agent is often added to yolk powder to improve its flow characteristics and avoid caking. Since whole egg and yolk powder have already been pasteurized, it is important to keep strict sanitary conditions and a high level of hygiene when incorporating ingredients at this stage.

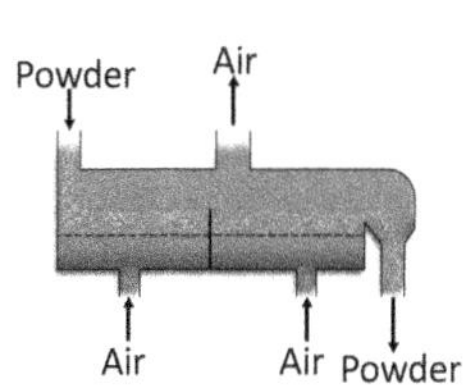

FIGURE 10.73 Fluid bed cooler.

10.6.4.7 Sifters

The powder will typically pass through a sifter immediately before packing. The purpose of sifting is to remove any potential foreign material or lumps from the powder. Another

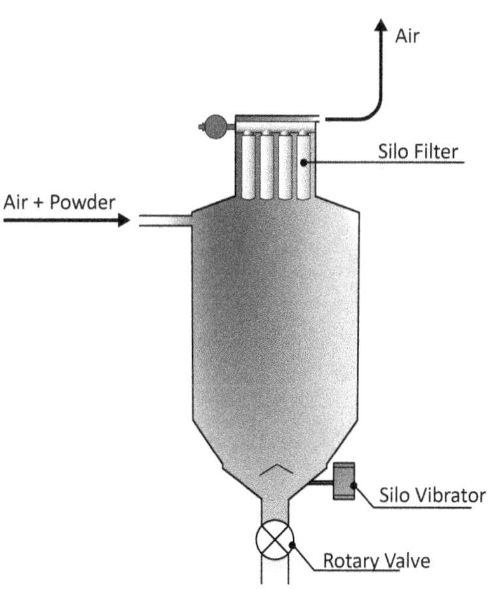

FIGURE 10.74 Powder silo.

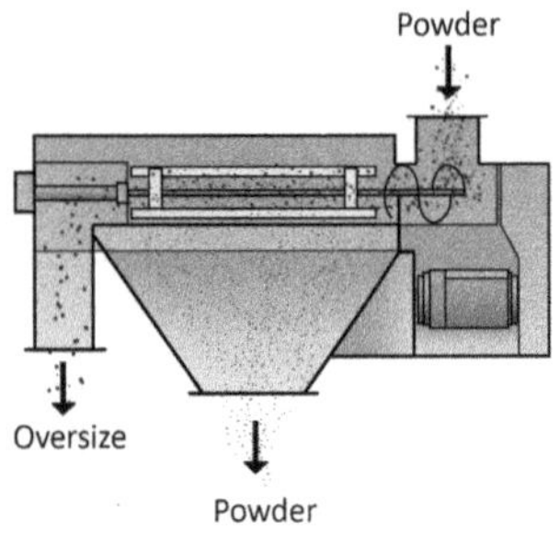

FIGURE 10.75 Centrifugal sifter.

purpose of sifting may be to separate the powder into a number of particle fractions.

Two types of sifters may be used: Centrifugal sifters, leading the product into a rotating cylindrical sifter (Figure 10.75) and Vibration sifters, the sifter screen vibrating in alternating linear and circular movements (Figure 10.76). The treatment of the product is rougher in centrifugal sifters than in vibration sifters.

Vibration sifters may be fitted with a deck of rubber balls or plastic rings under the sifter deck to keep the sifter screen clean during the sifting of products with high fat content. Also, a line of magnet rods can be mounted under the sifter screen for the detection of possible metal particles in the powder, and to catch metal threads from the sifter screen.

10.6.4.8 Metal Detectors and Magnets

Magnets and/or metal detectors are fitted in the plant. The optimum placement of magnets is in a descending powder flow, immediately before powder packing. Magnet rods are placed in a descending powder flow attracting magnetizable particles. The magnets are removable for cleaning. The efficiency depends on the magnetizing force and distance between magnets. Magnetic field strength of up to 11,000 Gauss may be

FIGURE 10.76 Vibration sifter.

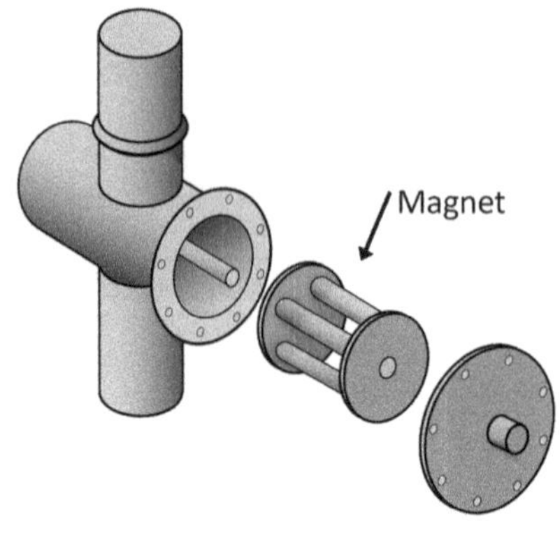

FIGURE 10.77 Rotating magnet.

applied. A shorter distance between magnets offers the most efficient catching of metal particles. On the other hand, powder build-up may occur between the magnet rods, causing the powder flow to stop. This can be remedied by making the magnets moveable (Figure 10.77). A magnet attracts far smaller particles than a metal detector is able to detect. Magnets are normally not applied to attract normal stainless steel AISI 304, which is non-magnetic. However, small particles which have been worn off the plant equipment should normally be attracted, as such particles become magnetic after detachment.

A metal detector does not perform the same function as magnet rods. It can be placed in a descending powder flow close to packing of the powders. A metal detector should not be placed near the magnets, as the magnetic field will interfere with the proper functioning of the metal detector. If both magnet and metal detector are installed for security optimization, the magnet should be installed prior to packing to capture all small metal particles in the powder while the metal detector will check for the presence of larger metal parts in the packed powder products.

In principle, a metal detector is similar to the device applied in airport security checks. The product is conveyed through a magnetic field, and any interference of this magnetic field will be detected. The larger the product passage, the less the sensitivity.

10.6.4.9 Powder Packing

Powder products are typically packed by weight into bags or cartons of 20 kg (40 lb) to 25 kg (50 lb), with polyethylene liners made of PE HD (Polyethylene High Density) with a thickness of 40 to 70 microns. It is preferable to choose a blue or green color for the bags to facilitate the detection of potential pieces of plastic in the product. In some cases, big bags over 100 kg may be used. The packing can be manually; the powder falling into the packing area, directly from the sifter, and the weight of the packed product adjusted by adding or removing powder. The entire packing process can also be automated, with an automatic feed of the product being packed, automatic weighing, and packing sealing. The level of automation depends on customer requirements and plant capacity. In many cases, fully automatic packing is not an economically viable option, when powder quantities are too small and supervisory staff still required. If the powder is dusty, a dust extractor should be installed above the packing spot to prevent dust nuisance in the room as well as for worker safety.

10.6.4.10 Packing and Heat Treatment of Egg White Powder

When packing egg white powder for hot room treatment, the material for cardboard boxes must not be made of recycle cardboards and should be stable at a temperature of 80°C in a humid atmosphere. The glue on the adhesive tape and stickers used on the boxes must be made of materials that will resist the environment in a hot room. The boxes of powder are normally placed on wooden pallets in three or four tiers. The pallets are placed on racks in the hot room. A distance of 2–3 cm between boxes must be respected during palletization to enable sufficient circulation of hot air between the boxes. When ready for heat treatment in the hot room, the loaded pallets are brought into the hot room and placed in racks. In order to achieve the pasteurization of the egg white powder and the desired gel strength, the powder must be stored at 60–80°C for two to three weeks.

10.6.4.10.1 Hot Room Design

Some factories have one single hot room where powders are stored and taken out daily. Other plants may have three or more individual hot rooms, each able to store the volume of one week of production. Once a week production is collected, it is placed in one of the hot rooms for the start of heat treatment. This way, it will be possible to heat treat the powder for a period of three weeks on a continuous basis without any traffic in and out of the hot room. When using one single large hot room, powder products will normally be taken in and out on a daily basis.

10.6.4.10.2 Building Materials

The floor should be made of concrete or tiles and be well insulated. The walls and the roof are normally made of sandwich panels with polyurethane foam and painted steel plates. The polyurethane foam should be made of a heat resistant material, not losing its adhesive properties after long term exposure to 80°C. The walls are typically 80 mm thick. The joints between the sandwich panels should be sealed with silicone in order to prevent condensation and corrosion.

10.6.4.10.3 Air Circulation System

The air in the hot room is circulated and heated by a separate unit for each hot room, placed outside the hot room, typically on the roof. The air enters at the bottom of one end of the hot room and is exhausted at the top on the opposite end to achieve good heat distribution. A control of the humidity of the air in the room will prevent further drying of powders. The air treatment unit consist of a filter unit, a fan, a heating unit and possibly a humidifier. The hot room can be heated by a steam heat exchanger or electrical heaters. The temperature must be controlled in the range of 60–80°C. The humidity of the circulated air is controlled by spraying steam into the heated air. Steam can be used from the steam boiler or generated from water by an electrical steam generator.

10.6.4.11 Filtration of Processing Air

The air used during the drying process should be filtered to prevent the transfer of impurities from the drying air to the powder. The filters are typically placed in the inlet duct system, as suction filters. The filters used are classified according to the global standard ISO 16890 for testing and clarification of air filters. The filters are available in standard dimensions and are fixed to frames placed in the air intake system. Filtration is normally done in two steps, with a coarser filter designed to extend the length of use of the second filter, more expensive due its higher degree of filtration.

10.6.4.12 Fans

Fans are applied for the movement of air throughout the drying plant, (Figure 10.78). The fans are dimensioned for the required air quantity and pressure, caused by system pressure drop. As it is desirable to adjust drying air quantity, the fan speed is adjustable by a frequency control.

In most cases, fans should operate at approximately 90–100% of maximum air quantity. The speed of fans is controlled by frequency inverters.

10.6.4.13 Heating of Drying Air

The energy consumption of spray drying equipment has a significant impact on production costs and profitability. The majority of the energy needs is for heating the drying air. With global warming and climate change, the focus should therefore be on energy preservation. The efficiency with which the drying air is heated is also crucial. Heating is usually done by steam, indirectly by a heat exchanger, and directly or indirectly with oil or gas. The efficiency of a steam boiler

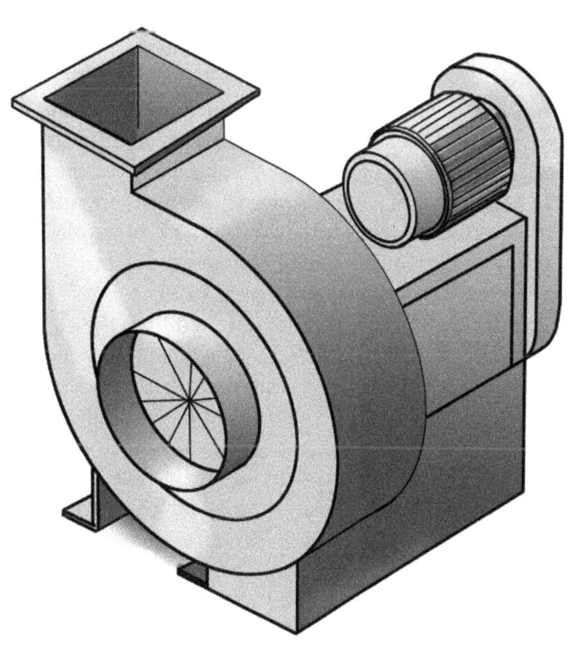

FIGURE 10.78 Centrifugal fan.

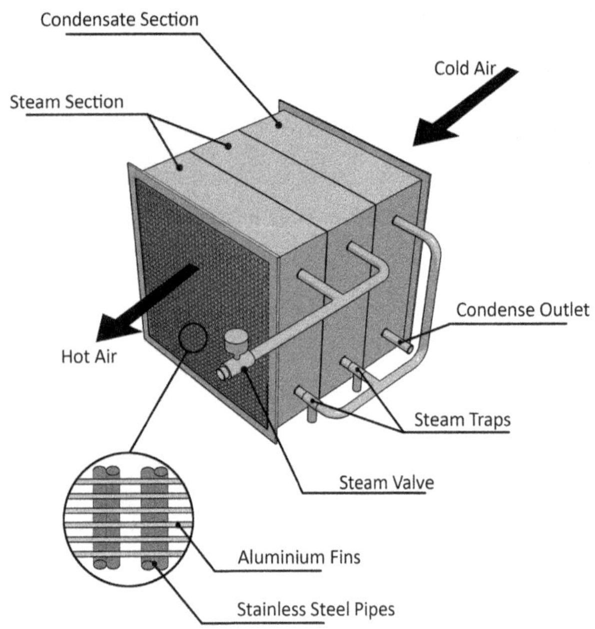

FIGURE 10.79 Steam air heater.

is typically between 85 and 90%. The same efficiencies can be achieved with indirect gas or oil heaters. The highest efficiency is obtained with direct gas heating. However, the disadvantage of using direct gas is that the combustion fumes are mixed with the air and comes into contact with the product to be dried. For health reasons this type of heating is prohibited in many countries. Highly efficient gas- or oil-fired heaters with an efficiency close to that of direct gas heating are available on the market. Some more sustainable heating sources such as biofuel and electricity produced by wind turbines or solar cells are now emerging as viable options.

10.6.4.13.1 Air Heating with Steam

Steam heating is always indirect. If steam was blown directly into the drying air, the air would humidify and be unfit for drying. Steam coils is placed along the air flow and steam is led through the steam coils. Condensation of steam is initiated by the colder drying air, thereby transferring energy to the drying air. To obtain an efficient heat transfer, the tubes are fitted with fins. Normally the tubes are manufactured of stainless steel and the fins are manufactured of aluminum. The principle of a steam air heater is presented in Figure 10.79.

To ensure that the steam is not removed from the air heater before condensation, a steam trap is mounted. Normally, the air heater is also fitted with a section where the condensate is further cooled before being returned to the steam boiler. The entire air heater and the subsequent ducts to the drying chamber are typically insulated. The attainable air temperature depends on the steam pressure available. The higher steam pressure, the higher drying air inlet temperature. For instance, at a steam pressure of 15 bar (225 psi), the maximum attainable air temperature will be 180°C (335°F), even though the evaporation temperature is 201.2°C (397°F). This is due to the pressure drop in the steam tube and at the steam adjusting valve as well as by a necessary temperature difference between steam and air. To obtain the drying air inlet temperatures suitable for spray drying, a steam boiler is required with a higher steam pressure than often available in a factory. The thermal efficiency in steam heating is determined by the efficiency of the steam boiler and the heat losses in the piping.

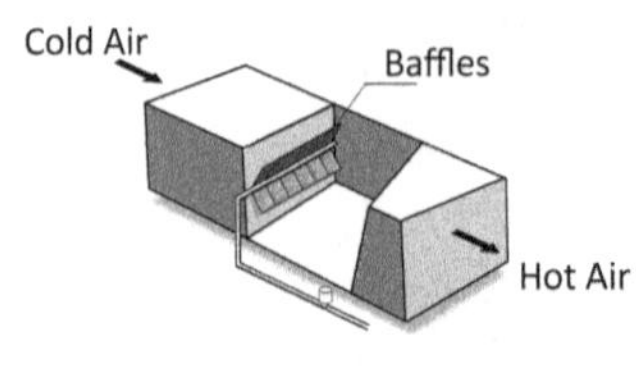

FIGURE 10.80 Direct air heating with gas.

10.6.4.13.2 Direct Air Heating with Gas

The gas is burned directly in the drying air flow by placement of a line of baffles in the drying air duct (Figure 10.80). Direct gas heating is used in a large number of eggs drying plants as well as plants drying other food products. The advantage of direct gas heating is its efficiency close to 100% since there is no heat loss to the surrounding environment. It should be noted, however, that water vapor will be released during combustion and added to the drying air, slightly reducing the plant evaporation capacity per added energy unit compared with indirect heating. The combustion of gas forms nitric oxides NOx. Such nitric oxides produce chemical compounds in the product, suspected to be carcinogenic. For this reason, direct heating is prohibited in certain countries while other countries have mandated a maximum acceptable content of NOx in final products.

10.6.4.13.3 Direct Low NOx Air Heating

To meet the legal requirements of a maximum product NOx content, special Low NOx air heaters have been developed to control the flame temperature by combustion (Figure 10.81). Such air heaters are primarily developed for processing air in the food processing industry and should not be mistaken for Low NOx burners, which are manufactured with a aim to reduce contamination of the external surroundings.

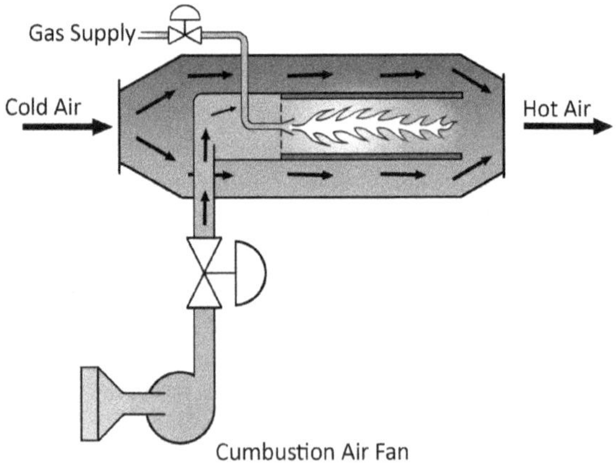

FIGURE 10.81 Direct low NOx air heating.

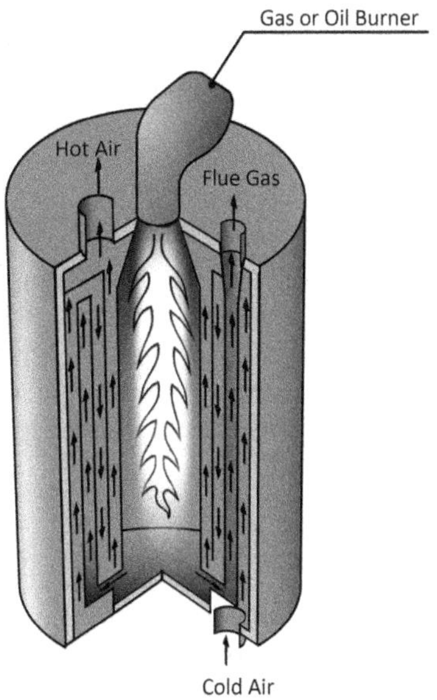

FIGURE 10.82 Indirect gas or oil air heater.

10.6.4.13.4 Indirect Air Heating with Gas or Oil

In indirect combustion drying, air and flue gas are kept apart. Combustion takes place in a cylindrical combustion chamber. Combustion air is heat exchanged with drying air by passing the air in counter flow at each side of cylindrical plate shells placed around the combustion chamber (Figure 10.82). The construction of the air heater is identical whether fitted with a gas burner or an oil burner. Heat transfer is partly through radiation heat from the combustion chamber and partly through convection heat. The temperature of the flue gas removed from the air heater is between 200°C (400°F) and 300°C (600°F), depending on the required inlet temperature to the drying chamber. This method will be the equivalent of an efficiency of 85%.

10.6.4.13.5 High Efficiency Indirect Gas Air Heating

Due to rising energy prices and increased authority concerns about direct combustion, high efficiency indirect air heaters have been developed. From the combustion chamber flue, gas is conveyed through a three-stage tube heat exchanger. Part of the flue gas is recirculated in the tube bundles to improve the heat transfer rate. By cooling the flue gas to approximately 100°C (210°F), an efficiency of 97–98% can be obtained by application of this design (Figure 10.83).

10.6.4.13.6 Thermal Oil Air Heating

The term 'thermal oil' does not refer to the fuel, but to a heat transfer fluid heated in a thermal oil heater, discharging heat energy in an air heater placed in the air flow to the drier. A thermal oil air heater is a tube heat exchanger with hot thermal oil flowing in the tubes and drying air passing around the tubes. The tubes are normally manufactured of stainless steel with aluminum fins (Figure 10.84).

The Thermal Oil heater has in principle the same design as an indirect gas or oil air heater. The Thermal oil is circulated in in tubes in the combustion chamber instead of the drying air (Figure 10.85).

Figure 10.86 illustrates the construction of a thermal oil system, certainly resembling a central heating plant for domestic space heating.

The boiling point of thermal oil is high and can be used to heat to higher temperatures than water. The system operates at far lower pressure than a steam plant. Multiple users can be connected to the same thermal oil system in the same way as a

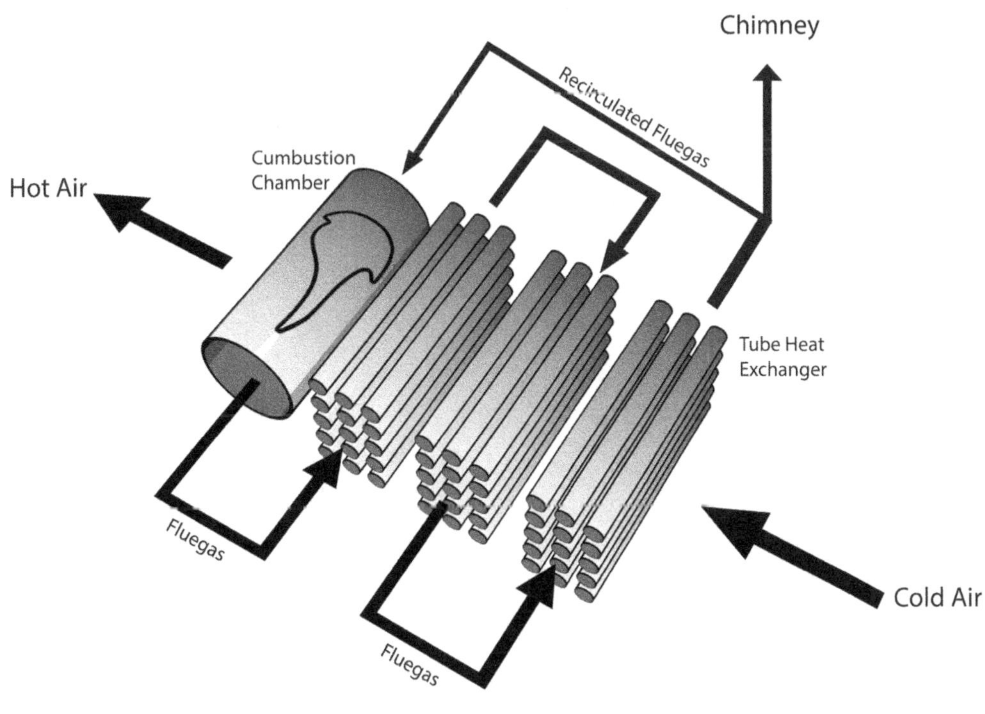

FIGURE 10.83 High efficiency indirect gas air heating.

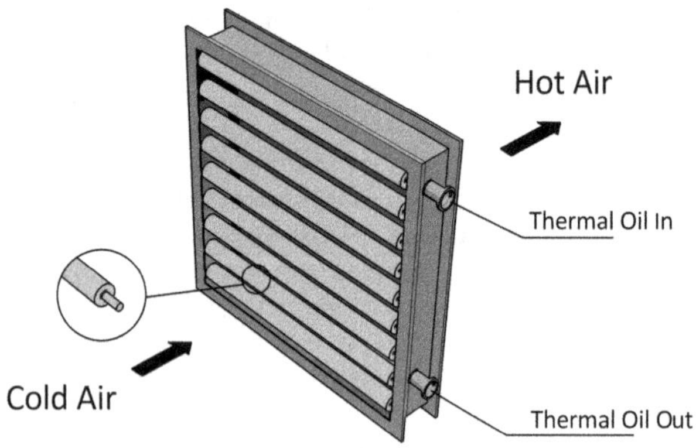

FIGURE 10.84 Thermal oil air heater.

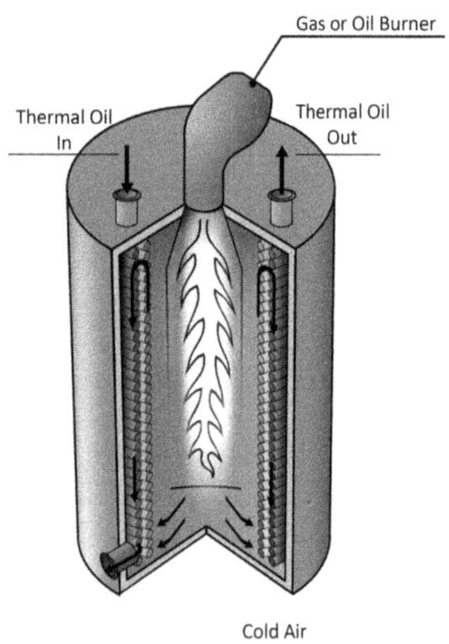

FIGURE 10.85 Thermal oil heater.

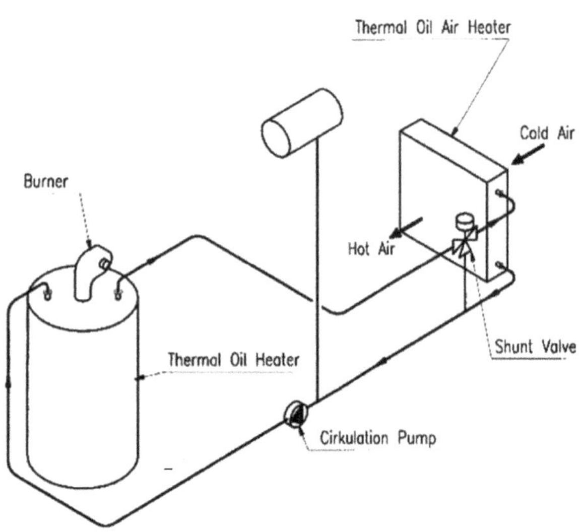

FIGURE 10.86 Thermal oil installation.

stream system. A wide range of thermal oils are available with different properties and temperature ranges. Although leakage oil would get in contact with the product, some oils are FDA-approved for food contact, which minimize the potential problems related to this.

10.6.4.13.7 Electrical Air Heating

The heating of drying air with electrical heating elements built into the drying air stream is used only in small laboratory and test plants. With a stronger concern about climate change and more incentive to electricity by renewable sources, electrical heating of dryers will likely increase in the future.

10.6.4.13.8 Biofuel

Biofuel may be wood chips or manure burning. In both cases, it will not be acceptable to bring the "fuel" inside the production area, like commonly done with oil or gas in a closed piping system. The burning of the biofuel must take place in a furnace away from the production area and the energy must be fed to the spray dryer using steam or thermal oil.

10.6.4.14 Heat Recovery

The discharged drying air has a temperature ranging between 50°C (120°F) and 100°C (210°F), depending on the drier and product. Therefore, it would make sense to attempt to use the discharged air to heat the inlet air. One technique is crossing the two air flows, for instance in a plate or a tube heat exchanger. Heat exchangers are typically manufactured in stainless steel or aluminum (Figure 10.87).

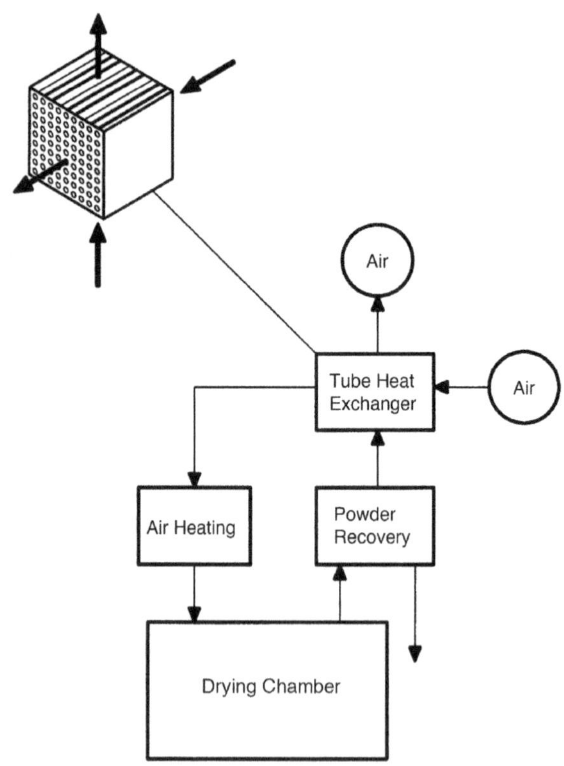

FIGURE 10.87 Air to air heat exchanger.

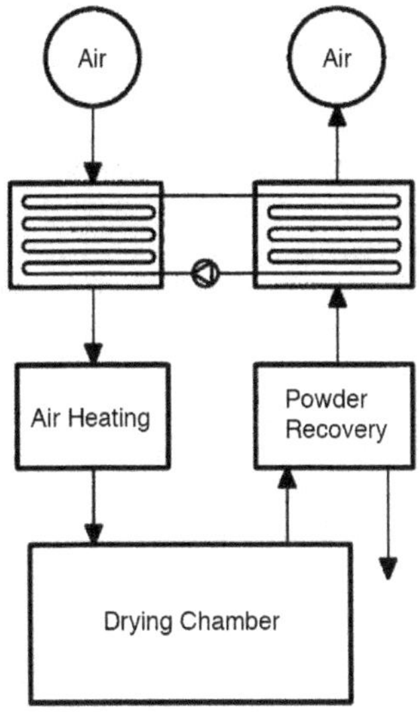

FIGURE 10.88 Air to liquid to air heat exchanger.

Another technique is the application of an 'indirect air-to-liquid-to-air, heat exchanging system (Figure 10.88), where the outlet air heats the liquid contained in the tubes in a tube heat exchanger. The liquid is water mixed with glycol to prevent the liquid from freezing in the winter time, The liquid is pumped to a similar heat exchanger in the intake air duct and the absorbed heat energy is discharged to the air flow. The tube heat exchanger is typically manufactured of stainless steel with aluminum fins.

Since heat recovery is 50–75% of the energy in the discharged air, energy savings of up to 28% may be realized, depending on operating parameters and climate. During wintertime, an even higher degree of heat recovery can be achieved with the cold inlet air cooling the discharged air to such an extent that it will condense.

10.6.4.15 Explosion Securing and Fire Extinguishing

The drying of most products, including egg products, creates a risk of powder explosion. Even distribution of powders through the air, the right mix of powder and oxygen as well as the presence of a sufficient amount of energy (spark) for ignition may cause a powder to explode. These conditions may occur mostly in fluid beds and in bag collectors. Furthermore, fire may break out in moist powder build-ups constantly exposed to hot drying air. This especially applies to the atomization area.

In most countries, explosion protection of drying and powder handling equipment is mandatory. In the EU, an explosion protection standard named ATEX must be followed when designing drying equipment. Many countries outside the European Union will follow these rules or have established rules similar to the ATEX regulation.

Explosion relief or Explosion suppression are measures designed to reduce the consequences of an explosion:

10.6.4.15.1 Explosion Relief

The installation of pressure relief membranes, breaking at a pressure lower than of the pressure in the rest of plant is one option. The explosion pressure is normally vented to the outdoor surroundings through ducts toward the roof. However, this solution limits the layout possibilities. As an alternative, boxes, including membrane, a flame arresting and a dust filter with integrated pressure wave absorber, can be installed directly on the equipment, and will facilitate the pressure relief for the room where the equipment is installed.

10.6.4.15.2 Explosion Suppression

The installation of pressure sensors or spark sensors will allow to spray an inert gas into the plant, as soon as an explosion is detected, thus displacing the air and consequently the explosion feed. The explosion is therefore suppressed.

10.6.4.15.3 Fire Extinguishing

Some spray drying plants are fitted with fire extinguishing equipment. The fire extinguishing equipment typically consists of a temperature detector in the outlet air that stops the fans and supplies water to the plant through a number of nozzles to suppress the conditions having created the unusual high outlet temperature.

10.6.4.16 Cleaning/CIP (Cleaning in Place)

Dry cleaning as well as wet cleaning are possible in spray drying plants. Wet cleaning can be done manually or automatically (CIP). Normally, wet cleaning of areas outside the dryer with powder deposits should be minimized, as such cleaning, must be a hundred percent perfect to prevent bacteriological problems (presence of humidity and heat).

Large tower dryers are normally fitted with CIP equipment in the drying chamber, fluid beds and powder conveying ducts. Box dryers are typically not fitted with CIP equipment since the drying chamber is easily accessible for manual cleaning and there are no powder conveying ducts. Manual cleaning is quite manageable in a box drier and needs to be performed less often than for towers.

CIP cleaning of a spray drying plant is done by lowering a rotating turbine into the drying chamber. A number of CIP cleaning balls are built into ducts and fluid beds. The drying chamber of a box drier can also be CIP-cleaned by lowering a cleaning turbine into the drying chamber. The bag collector can be CIP designed. If not, the bags should be removed before cleaning. Dry cleaning is the mechanical or pneumatic removal of powder from the internal surfaces of the plant. Such cleaning can be done by using a broom. Towers can also be fitted with an air sweep. Air is injected through a slot in the jacket of the drying chamber. The air sweeps over the walls of the drying chamber pulling the powder off the wall.

Powder deposits in drying plants may be removed during the drying process by installing a Sonic Horn (Figure 10.89) which makes powder fall off ceilings and walls through low-frequency vibrations generated by compressed air. Typically, the horn is activated a few seconds every 10th minute.

Another possibility for cleaning the drying chamber during operation is to install hammers on the drying chamber walls.

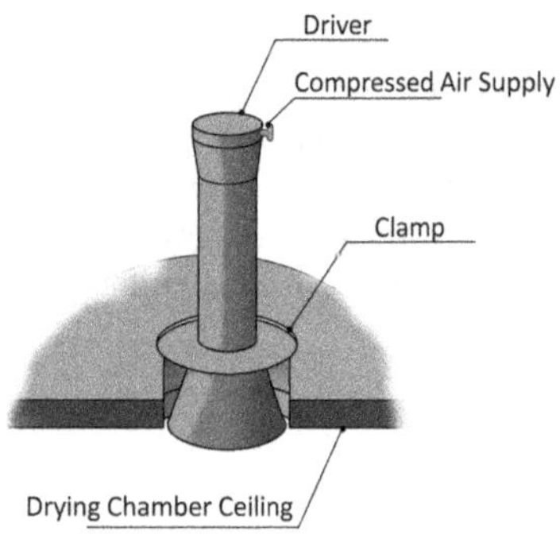

FIGURE 10.89 Sonic horn.

The hammer, typically pneumatically activated, will knock on the drying chamber wall at intervals and the vibrations will cause the powder to fall off. Hammers are especially used in towers, but a drawback is that they will eventually produce cracks on the inside of the drying chamber wall.

10.6.4.17 Control System

A spray drying plant is well suited for automation, and Programmable Logic Control (PLC) with a Human Machine Interface (HMI) is the standard system. The degree of automation can vary, and the PLC control system can be for the dryer only with a small touch panel or may be part of the entire plant control with a Supervisory Control and Data Acquisition (SCADA) system. An automation system reduces the need for operator supervision of the dryer, and more importantly, it can limit the possibility of human errors, improve the dryer safety and ensure a constant quality of products.

The maintenance of a constant outlet temperature of drying air from a drying plant is normally managed by a control loop controlling the quantity of liquid being pumped into the drying chamber. Thus, the stable water content in the powder during production is secured. In addition, the drying air temperature being fed into the drying chamber is kept constant by adjusting the steam valve or the burner position, based on the recorded inlet air temperature to the drying chamber. The control system includes a number of other functions guaranteeing product quality and securing both equipment and personnel.

10.6.5 Typical Laboratory Evaluations

A sample of powder is routinely collected every hour during production to measure the moisture content. Based on the result of the test, the drying air outlet temperature may be adjusted to maintain the desired moisture content in the powder.

The following evaluations or tests are done daily or for each batch: 1) Total bacterial count of the powder; 2) Solubility of the powder when redissolved in water; 3) Number of scorched particles; 4) Water content; and 5) pH.

For egg white powders, the following evaluations may also be added: 1) Foaming properties when redissolved in water; and 2) Gel strength of the heat-treated egg white powder.

10.7 Cleaning-in-Place

During operations in a food processing plant, fouling events will occur due to the deposition of a soil layer on the internal surfaces of piping, vessels or other equipment. Fouling may be generated by the contact of the product with the internal walls of the plant or by the deposition of various substances generated by product denaturation due to heating, cooling, or time-dependent phenomena during the process. From a mechanical point of view, fouling layers cause an increase in pressure drop and lower efficiency in heating or cooling devices. From a food safety standpoint, the presence of fouling may generate zones where microorganisms can grow, affecting the quality of products or even invalidating the effect of pasteurization.

For these reasons, cleaning-in-place (CIP) operation plays a critical role in the food processing industry. The following section will introduce the fundamentals of cleaning operations in processing plants and suggest some ways to best manage the CIP of egg processing plants.

Each CIP procedure must be validated, on a case-by-case basis. Each product may behave differently from others and generate different types of fouling due to differences in product composition or plant environmental conditions, such as the chemical and physical properties of the water used in the plant. A great number of variables can influence the results of the cleaning operations.

10.7.1 Egg Processing Fouling

Several kinds of fouling can be generated during liquid egg processing. Fouling can be caused by many elements with different behaviors. Fouling is typically defined into two major categories: Water soluble substances, that are generally easy to remove; and Water insoluble substances, that are generally harder to remove and composed of organic or inorganic substances.

10.7.1.1 Organic Substances

When organic substances like fats, proteins, carbohydrates oils, greases, proteins are heated during processing, different reactions may happen, and make them difficult to remove. For instance, egg proteins can denature in the heat exchangers and create a solid layer difficult to remove. Normally organic matters in a system like a pasteurizer or tanks can be removed by using alkaline cleaning solutions.

10.7.1.2 Inorganic Substances

Inorganic substances are composed of minerals and salts deposits. One of the most commonly inorganic soils is limestone (calcium carbonate), precipitating in hard water at high temperature. Inorganic substances are generally dissolved by acid detergents.

10.7.2 CIP Management

CIP is the term used to describe the systems necessary to make food processing equipment ready for further processing. This will be conducted at various frequencies such as when changing product or recipe, or simply before starting production or at the end of a day of production. As this method is applied without taking the equipment apart, it is widely used today within the food industry.

10.7.2.1 Cleaning Parameters

There are several strategies to remove fouling from the internal walls in an egg processing plant. Fouling can be removed through mechanical action, chemical action or thermal action applied for a certain period of time. In order to define the correct washing cycle for a plant, a right balance between the different forces must be taken into account, thus guaranteeing the best results at the minimum operating costs. The fundamental specifications to define for a CIP process are: the washing flowrate, the chemical concentration, the temperature of the washing solution and the operating time for each phase of the washing process.

10.7.2.1.1 Washing Flowrate

The greater the mechanical action that will be applied on fouled surfaces, the better the result of the cleaning will be. In the typical processing plants, the mechanical action on the internal surfaces is applied by the flow of the cleaning solution tangential to the internal walls. The strength applied by the washing solution to the fouling layer is a function of the velocity of the fluid in that portion of the plant. The fluid velocity is proportional to the shear force applied on the internal wall, necessary for the mechanical removing of soiling material. To evaluate the flow velocity in a certain section of the plant, it is sufficient to divide the volumetric flowrate of the washing solution by the cross section flowing area.

To perform an adequate CIP process, the correct washing flowrate needs to be sized, taking into account the flow velocity in the section of the plant with the largest area. The flow velocity can be correlated to the efficiency of the cleaning process and can be used as a parameter to size the correct CIP procedure. A general rule is to have a flow velocity not lower than 1.5 m/s.

10.7.2.1.2 Chemical Concentration

The action of chemicals on the fouling layer depends on the concentration of the chemicals in the solution and on the ability of the chemicals to react with the different kind of soiled surfaces. Usually alkaline detergents are used first, to dissolve the proteins, fats, and sugars. Alkaline detergents are based on sodium hydroxide (NaOH). Sodium hydroxide is commonly used at a concentration of 0.5–2.0%w/w although in some cases, a higher concentration may be necessary. Using too high a concentration of sodium hydroxide could be counterproductive since it may induce cross-linking phenomena in proteins, leading to more difficulties to efficient cleaning operations.

Acid cleaning is usually conducted after the caustic step and often relies on nitric acid (HNO3). Nitric acid is typically used at a concentration equal to 0.5–1.5%w/w and allows the removal of any inorganic compound present on food contact surfaces. Higher acid concentration may have negative consequences on the integrity of steel and polymeric material gaskets.

The decision regarding the chemical concentrations is always challenging. It is important to use the maximum allowed concentration of a chemical to ensure the best cleaning results but on the other hand, some financial and environmental considerations would suggest to use the minimum necessary amount of chemicals. A good decision for setting up the concentration of chemicals will allow a satisfactory CIP operation, at the minimum cost and the least impact on the environment.

10.7.2.1.3 Thermal Action

The effects of the CIP operation will be maximized by selecting the best process temperature, since the solubility of a substance in a solution is proportional to the temperature and most chemical reactions are faster at higher temperatures. This parameter needs to be evaluated, balancing its effect on cleaning with the incremental process cost. Furthermore, in highly thermal-sensitive fluids such as egg products, using too high a temperature during CIP operations may increase the denaturation phenomena of the fouling, with negative consequences on the cleaning results.

In summary, all the food processing plants should be equipped with automatic CIP to guarantee:

- An adequate flowrate during the CIP operations to have the necessary friction phenomena on each internal surface of the plant;
- An adequate concentration of chemicals during the CIP operations to guaranty the effective chemical action on the fouling layers;
- An adequate washing time to give the necessary time to the mechanical, thermal, and chemical actions to complete successfully the cleaning operation;
- An adequate rinse time, to ensure that all chemical residues have been removed from the plant, and prevent any contaminations with the food products

10.7.3 Validation of the Cleaning Procedure

The same cleaning procedure will often have different results when applied in different food processing plants or different environmental conditions. For this reason, it is essential to validate the cleaning procedure adopted onsite. Once a CIP protocol has been defined, it is necessary to establish a monitoring program and validate its performance during the initial CIP operation. Various samples will be taken and analyzed, performing some tests to verify the correct results of the CIP operation. Some adjustments may be made to the CIP protocol until the best washing conditions are achieved. A monitoring program is also useful to avoid to "over-clean" a plant, resulting in unnecessary use of chemicals, water, and energy and therefore additional costs. The tests commonly used to monitor the effectiveness of a CIP protocol are presented in the following section.

10.7.3.1 ATP Test

ATP test allows to rapidly evaluate the presence of actively growing microorganisms, through the detection of adenosine triphosphate (ATP). The ATP is a molecule present in living cells as a source of energy and is a clear indicator of the presence of microorganisms on the process plant internal walls. The analysis mechanism is based on the reaction between the ATP and a light-emitting enzyme. The amount of light emitted, measured by a luminometer, is proportional to the amount of ATP present, measured in relative light units (RLU).

10.7.3.2 Protein Residue Test

The protein residue test is an easy and accurate way to verify the cleanliness of food equipment surfaces. The most widespread methods involve the swabbing of a surface following by placing the swab into a container filled with a reagent. The result is based on color changes in the reagent, with color varying depending on the concentration of protein residues on the swab.

10.7.3.3 Microbiological Test

Microbiological tests require the swabbing of surfaces, plating on petri dishes or films, and encouraging the growth of specific bacteria by providing them the right conditions (e.g., temperature, nutrients). The results are verified measured and expressed by counting bacteria colonies at the end of the incubation period.

10.7.4 Cleaning-In-Place Procedure

The general procedure for a CIP protocol involves a series of washing steps with alkaline solutions, acid solutions and rinses.

10.7.4.1 Customer Water

All cleaning procedures must consider several parameters, including the types of products processed in a plant and certain environmental factors. One of these environmental factors is the chemical and physical properties of the water available in a plant.

The optimal water properties to guarantee the good performance of a CIP operation is presented in Table 10.8.

TABLE 10.8

Optimal Water Properties to Guarantee the Good Performance of a CIP Operation

Parameter	Range Values	Notes
pH	6.5–8.5	More acid water could increase the corrosion risk. More basic water could reduce the efficiency of some disinfectants
Hardness	<25°F	Higher hardness water involves the necessity of higher chemical concentrations.
Conductivity	<500 mS/cm	Water with higher values could generate more solid precipitations
Sulphates	<50 ppm	Higher concentrations involve more accurate disinfection procedures to prevent microbial proliferation
Iron	<0.3 ppm	Higher levels could generate rust deposition
Chlorides	<100 ppm	Higher concentration could promote the pitting phenomena
Silicates	<25 ppm	Higher concentration could have negative effects on the materials
Sulphides	< 250 ppm	Higher concentrations are responsible of bad smell

10.7.4.2 Typical Cleaning Procedure

The correct sequence of a general cleaning procedure for the internal surfaces of a food processing plant has these fundamental steps:

- **First rinse/plant preparation:** this step allows to remove all gross residues of product that can be removed without chemical action;
- **Alkaline washing:** this step will remove all the contaminants that are soluble in a basic environment like proteins, greases, and oils;
- **Second rinse:** this rinse is necessary to remove all the residues of lye in the plant and avoid any interference with the acid products used during the following step (potential waste of acid chemical and source of dangerous reactions);
- **Acid washing:** this washing step is necessary to remove any substances soluble in an acidic environment such as salts or limestone fouling;
- **Third rinse:** this rinse is necessary to remove all residues of acid in the plant and avoid any interference with products used during the following step;
- **Disinfection:** this step will disinfect the internal surfaces of the process plant. This step does not influence the washing performances and may be eliminate, depending on the overall plant cleaning management strategy;
- **Final rinse** this is the most important rinse since it must remove any contaminant substance from the plant internal surfaces to avoid any risk of contamination of the food products during process operation

Based on the plant design, the products processed and product flow, changes can be made to these washing steps. A CIP procedure without the acid washing step is commonly defined as "single stage cleaning". A CIP procedure, using both acid and alkaline washing, is commonly defined as "dual stage cleaning". In some cases where a thermal sterilization will be conducted prior to initiating production, the disinfection step may be omitted.

Protein and fat fouling is removed using strong alkaline cleaners, anyway it is preferable not to use simple sodium hydroxide solutions to perform a CIP.

Proteins which have been treated become more resistant to cleaning solutions, with protein, carbohydrate and mineral interactions, such as gel formation and viscosity increase, precipitation of salts, foaming and many other negative effects.

Due to these reasons, caustic soda solutions are avoided and it is preferable to use blended mixture of alkaline substances and other components to improve the CIP efficiency.

10.7.5 Biofilm Treatment

Bacteria can create biofilm on the steel surfaces, and it is a strong accumulation of bacteria, difficult to remove and it may be a source of contamination during food processing.

To prevent the biofilm it is necessary to use disinfectant agents during the rinsing phase of the CIP, one of the most effective agent is the chlorine dioxide.

Once the biofilm is present, only strong mechanical action, together with disinfectant agents can remove it. In case of biofilm occurrence in egg processing, peracetic acids can be circulated in concentrations higher than for normal disinfection. Peracetic acid penetrates the biofilm, kills the living organisms and, when over-neutralized, acts as an oxidizer to remove the starch-like structure of biofilm.

10.8 Enzymes in Egg Processing

Enzymes are part of a distinct group of proteins and are found in all living things. Without enzymes, there would be no life. In the human body, many processes involve enzymes, both at a macro level for the digestion of foods for instance and at the cellular level in various metabolic pathways.

Enzymes act as catalysts of chemical reactions. They perform an action, but do not become part of the action. Furthermore, they are very specific and perform only certain tasks and nothing else. Enzymes are like a set of keys, with each key fitting only one type of lock. Enzymes speed up the rate of reaction between substances without being consumed themselves in the chemical reaction. As such, they are vital to functions such as digestion, and they make processes possible that normally could not occur except at temperatures so high they would threaten the well-being of the body.

10.8.1 Temperatures and Enzymes

When the temperature of the environment is too high or too low, no enzymatic reaction will take place. If the temperature is too low, an enzyme will be immobilized (freezing stage), but can get active again once the temperature rises. Above a certain temperature, the denaturation of an enzyme will lead to the irreversible loss of its activity.

10.8.2 Production of Enzymes

Enzymes, extracted or manufactured at an industrial scale, are used within many industries and their applications are numerous. They are used within detergents and cleaning agents to remove fat, proteins and carbohydrates. They are used in a variety of food industries for example the bread industry to improve the volume and texture of bread, the oil industry to purify oils, the dairy industry to make cheese, or the starch industry to turn starch into sugars.

Enzymes work best under a defined set of conditions that vary from enzyme to enzyme. For instance, certain enzymes require high temperatures while others are active at cold temperatures. Other conditions such as the pH value, the presence of certain minerals, oxygen or other molecules can also be critical to get the desired outcome from the use of enzymes.

As an illustration of these specific conditions for enzymatic reaction, the industry has managed to select enzymes acting at lower temperatures to allow detergents to be effective at lower water temperature, thus maintaining the quality of clothing and saving energy.

10.8.3 Conditions That Can Influence the Action of Enzymes

The main conditions influencing the activity of enzymes are: the availability of substrate, Humidity, Temperature, pH, Concentration (activity) and the Reaction time.

In nature as well as during the industrial use of enzymes, the goal is to convert large molecules into smaller and usable fragments. Enzymes used by the industry (Industrial enzymes) cover a wide range of food and non-food applications.

Industrial enzymes have been produced for more than 100 years, typically extracted from animal and plant sources or produced by microorganisms through a fermentation process in steel tanks.

10.8.4 Enzymes in Egg Processing

A range of enzymes are used by the egg processing sector for various applications:

- Phospholipase A2: Improvement of the emulsifying properties of yolk and whole egg
- Glucose oxidase: Removal of glucose in egg products before spray drying
- Catalase: Removal of hydrogen peroxide, when used for cold pasteurization, or for making oxygen available for the desugaring process
- Lipases: Improvement of the foaming properties of egg white
- Proteases: For nutritional purposes and for higher pasteurization temperatures

10.8.4.1 Phospholipase A2

This enzyme may be obtained from different sources. While it may be extracted from porcine pancreas, its application is somewhat limited by its non-kosher and non-halal status. Phospholipase A2 can also be produced by microbial fermentation, using *Aspergillus Niger.* In addition to its Kosher and Halal status, the fermentation-produced enzyme has a better functionality than the animal-derived enzyme.

Phospholipase A2 is mainly used to produce what in the industry is called “heat stable egg yolk”. Egg yolks treated with phospholipase A2 have improved emulsification properties, and make emulsions heat stable. These characteristics allow the optimization of mayonnaise and dressings recipes. For instance, the improved emulsifying properties of the egg yolk allows the reduction of stabilizers in the recipe. Better emulsifying properties allow also to use less egg yolk, thus reducing the cost of ingredients in the recipe. Furthermore, the heat stable emulsions made with heat stable egg yolk can withstand pasteurization, cooking and heat stress tests which is very important test in they mayonnaise and dressing industry in Eastern Europe, and helps to prevent separation of the product during storage in warm condition

A mayonnaise produced with heat stable egg yolk will also have a higher viscosity, a characteristic sometimes required by manufacturers or essential to optimize a recipe. Although phospholipase A2 can also improve the emulsifying properties of whole egg, its use is not common for whole egg products.

Phospholipase A2 hydrolyzes the fatty acids on the second position of the phospholipids in the egg yolk and converts it into lysophospholipid. The speed of the reaction depends mainly on the temperature of the product, faster at 50°C and slower at 5°C. While the speed of the reaction is somewhat depending on the presence of calcium ions, the presence of sufficient calcium in egg products makes the addition of calcium not necessary. This enzyme works well at the natural pH of eggs.

10.8.4.2 Glucose Oxidase

Glucose oxidase is used to remove the glucose present in especially egg white before spray drying, but also in some egg yolk and whole egg powders.

The glucose must be removed from egg white to prevent the Maillard reaction causing a brown coloring when glucose and proteins are heated. This browning effect will happen during spray drying or powder pasteurization of egg white, resulting in a light brown egg white powder, making the powders less desirable by customers. In addition, this reaction will also have a negative effect on protein solubility of the egg white powder, thus reducing its functional properties.

Likewise, the removal of glucose in liquid whole egg and yolk products prior to spray drying will give a better protein solubility throughout the shelf life of the products. This will have a positive effect on baking properties, with proteins easier to solubilize when the egg powders are hydrated for use by food processors.

Glucose oxygen works in the presence of oxygen. To get this available for the enzyme, hydrogen peroxide is added to the product as well. Then the enzyme catalase is also added which hydrolyze the hydrogen peroxide into oxygen and water, so oxygen will be present in the egg product.

10.8.4.3 Catalase

Besides its use during the desugaring of egg products, catalase is also used for a process called "cold pasteurization". This is a process where hydrogen peroxide is added to egg white to kill bacteria or make them more sensitive to the pasteurization step. It is then possible to pasteurize products at lower temperatures, thus causing less protein denaturation during pasteurization. and enhancing the foaming properties in the end product. Catalase is then used to remove hydrogen peroxide from the product.

10.8.4.4 Lipases

Some lipases are used to minimize the negative effects that traces of egg yolk can have on the foaming properties of egg white. Lipases will hydrolyze the lipids from the egg yolk and minimize the negative effects of them. But the use of lipases can also generate potential problems if there are active lipases in final products. The lipases can be activated again when the egg white powder is used for the manufacturing of food products (e.g., chocolate bars), with the hydrolysis of lipids in the food products resulting in unwanted off flavors or altered consistency.

10.8.4.5 Proteases

Proteases are used in egg products for different reasons. Some food processors may use them for nutritional products, improving their digestibility, or allowing pasteurization especially of egg white products, at higher temperatures which can be at 70°C (158°F) or higher. Since some functional properties may be changed, proteases are mainly used for protein drinks and products with special nutritional purposes. Like all other enzymes, proteases will require a specific pH range to work, and certain temperatures of the products or presence of other catalysts to speed up the reaction.

10.9 Conclusion

While the design and operation of an egg processing plant may seem complex, a set of well-defined rules and protocols allow to produce a wide range of egg products, meeting repeatedly the quality and safety required for a sustainable egg processing business. Over the years, manufacturers have developed new pieces of equipment more efficient, more adapted to the gentle care required for egg proteins and more automated. The use of the latest technologies allows to produce more with the same input requirements, to deliver customers with better quality products and to ensure the safety of all products while minimizing the impact of the egg processing sector on the environment. It is anticipated that further development in the egg processing sector will further contribute to the overall sustainability of the egg sector.

REFERENCES

Bonfà, F., & Bertin, I. (2016). The innovative PEF technology for bacteria inactivation in foods (In Italian). https://www.enea.it/it/Ricerca_sviluppo/documenti/ricerca-di-sistema-elettrico/adp-mise-enea-2015-2017/processi-e-macchinari-industriali/rds_par2015-071.pdf

Cerf, O., Dousset, X., & Brossard, J. (1990). Pasteurization and sterilization by heat (In Italian), da *Microbiologia Alimentare* (pp. 295–30). Tecniche Nuove.

Froning, G. W., Peters, D., Muriana, P., Eskridge, K., Travnicek, D., & Summer, S. S. (2002). International Egg Pasteurization Manual. https://www.incredibleegg.org/wp-content/uploads/2020/07/Pasteurization_Manual.pdf

FSIS (2020). Egg product inspection regulations. Final rule. https://www.fsis.usda.gov/sites/default/files/media_file/2021-01/2020-%20Speech-Egg%20Products%20Inspection%20Regulations%20Final%20Rule.pdf

Humphrey, T. J., Chapman, P. A., & Gilbert, R. J. (1990). A comparative study of heat resistance of salmonellas in homogenized whole egg, egg yolk or albumen. *Epidemiology and Infection*, *104*, 237–241. https://doi.org/10.1017/s0950268800059409

Johansen, H. C. (1998). *A Danish niche production (In Danish)*. Odense University Press Odense, ISNB 87-7838-355-2.

Li-Chen, E. C. Y., Powrie, W. D., & Nakai, S. (1995). The chemistry of eggs and egg products. In W.J. Stadelman & O.J. Owen (Eds.), *Egg science and technology* (4th ed., pp. 105–175). Food Product Press.

Schuman, J. D., & Sheldon, B. W. (1997). Thermal resistance of *Salmonella* spp. and listeria monocytogenes in liquid yolk and egg white. *Journal of Food Protection*, *60*, 634–638.

Sörqvist, S. (2003). Heat resistance in liquids of *Enterococcus* spp., *Listeria* spp., *Escherichia* coli, *Yersinia* enterocolitica, *Salmonella* spp. and *Campylobacter* spp. *Acta Veterinaria Scandinavica*, *44*, 1–19. https://doi.org/10.1186/1751-0147-44-1

Ternes, W., Acker, L., & Scholtyssek, S. (1994). Egg and egg products (In German). ISBN 3-489-63114-5.

11

Eggshell Calcium

Hiroyuki Shidara

CONTENTS

11.1 Introduction

The global production of shell eggs in 2022 accounted for 86.7 million tons (see Chapter 2), meaning that about 8.6 million shell eggs were generated by the egg sector.

Combined with the number of eggs produced for broiler breeder flocks, the poultry sector generated more than 30 million tons of eggshells on an annual basis. The use of these shells, a great source of calcium, is mostly under-utilized and should receive more attention from the poultry and egg sector. This chapter will discuss in detail the benefits and principles of eggshell calcium intake, including some comparisons with other calcium sources, and will review another potential uses of eggshell calcium.

11.2 The Industrial Production Method for Eggshell Calcium

Eggshell account for about 10% of the mass of shell eggs used by egg processing plants. After collection, eggshells are ground into a powder to produce eggshell calcium. However, since eggshell membranes, egg whites, and egg yolks will remain attached to eggshells, a process to remove them is necessary. The egg white and yolk can be easily removed by washing eggshells with water. The eggshell membranes cannot be easily separated from eggshells as these membranes are entangled with the shell during its formation process (see Chapter 4). Several patents have been issued for methods of separating them, including the process described below (Miyazaki et al., 1995).

Eggshells are crushed to 2~3 mm in size by a wet-type crusher. The crushed eggshells are fed into a device that generates a swirling movement, thus twisting the crushed eggshells. Centrifugal force separates the eggshells from the eggshell membranes, as the lower specific gravity eggshell membranes are suspended while the eggshells with high specific gravity are collected from the bottom of the vessel.

After collection, the separated eggshells are further crushed in a dry crusher such as a jet mill to a predetermined size. Commercially available products generally have an average particle diameter of about 8 μm, but some may be ground down to 2~3 μm. It is possible to grind eggshells to the nm level for average particle size by using a wet grinder. However, due to complications during the drying process, eggshells ground down to the nm size are not commercially available.

11.3 Nutritional Function of Eggshell Calcium

Calcium is one of the minerals essential to the body, not only for maintaining the body's structure as a component of bones and teeth but also for muscle contraction, the normal functioning of many enzymes, blood coagulation, and the maintenance of a normal heart rhythm. The amount of calcium in cells and blood is strictly regulated in the body since the human body cannot maintain its normal functioning without the right amount of calcium available (Pravina et al., 2013). Calcium is ingested from foods and other sources and is eliminated through the urine, and the intake of at least 1,000–1,500 mg of calcium per day is needed to maintain normal blood calcium levels. When the calcium intake is low, and calcium levels drop, parathyroid hormones intervene to maintain the blood calcium level. Parathyroid hormones are produced by the four parathyroid glands located behind the thyroid gland. These hormones stimulate the bones to release calcium into the bloodstream, reduce the amount of calcium discharged into the urine by the kidneys, and increase calcium absorption either directly by stimulating the digestive tract or by stimulating the activation of vitamin D in the kidneys. When the calcium concentration in the blood decreases, the concentration of the parathyroid hormones in the blood increases, thus helping to maintain calcium levels in the blood. The calcium is transferred from the bones to the bloodstream, resulting in its depletion in bones. As a result, the bones become weaker, eventually leading to osteoporosis. But the consequences of low calcium levels in

DOI: 10.1201/9781003254430-15

the blood are not just osteoporosis. Calcium is essential in managing adequate blood pressure, with lower calcium intake associated with associated with hypertension (Langford & Watson, 1973). It has been suggested that the blood pressure-lowering effect of calcium is due to the suppression of blood calcium by the decreased secretion of parathyroid hormone and the subsequent decrease in calcium content of vascular smooth muscle cells (Simonetti & Mohaupt, 2007).

Heaney et al. (2002) have reported negative associations between calcium intake and body weight during adulthood and calcium and body fat accumulation during childhood. They stated that there is a reasonably consistent effect and that for every 300 mg increase in average calcium intake, there is about 1 kg less body fat in children and 2.5 to 3.0 kg less body weight in adults. People who do not have the habit of consuming regularly dairy products or who live in volcanic regions where the concentration of calcium in soils (and therefore in crops produced) is low in the soil are at risk of suffering from calcium deficiency.

In addition to whey calcium in milk, there are other natural sources of calcium such as cow bone meal, fishbone meal and eggshells. The eggshell has a calcium content of 38%, which is very high among the natural sources of calcium. In addition, it is characterized by a phosphorus content of 99 mg per 100 g, low in comparison to beef bone meal and fishbone meal. Methods such as grinding eggshells into powder with a pestle and mortar or coffee mill or dissolving eggshells in vinegar or other acids have been used to make them edible. In recent years, improvements in the grinding technology have made it possible to pulverize eggshell particles into microns. As a result, industrially produced eggshell powders are now used for calcium fortification purposes and improve the quality of various foods.

It is well recognized that the eggshell is not only designed to protect the developing embryo from the outside environment but also to be used as a source of calcium for bone development during the second half of chick embryonic development (Ahmed et al., 2022; Halgrain et al., 2022). Therefore, it is inferred that calcium has the property of dissolving and being easily absorbed by the body without significant changes. This is different from calcium carbonate, which exists as a mineral source of calcium. Electron micrographs of pulverized calcium eggshells and calcium carbonate are shown in Figure 11.1. Calcium carbonate is crystallized and has a smooth surface, while eggshell calcium has a layered and porous structure. Eggshell has a matrix consisting of small amounts of proteins, glycoproteins and proteoglycans that become occluded within the eggshell represents ~2% of its weight (Hincke et al., 1992). Kusumi et al. (1999) compared the solubility of calcium in the stomach of elderly patients with cerebrovascular disease after 2 hours of treatment with mineral calcium carbonate and calcium eggshell. They found that calcium carbonate dissolved less than 70% of the calcium in the stomach, while calcium eggshell dissolved about 95%. Tashiro et al. (1997) speculated that the pepsin in the gastric juice hydrolyzed the proteins in the eggshells, leading the eggshells to disintegrate into smaller pieces and making them more soluble in gastric acid. Ito et al. (2020) extracted phosphate compounds with calcification-alienating activity from the inner side of ostrich eggshells using a 10% acetic acid solution. They reported that

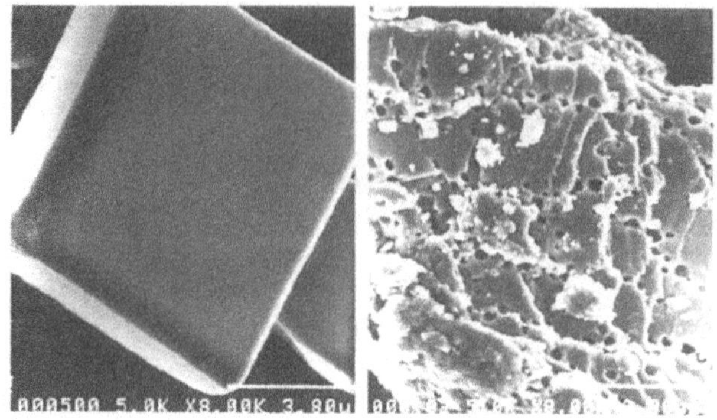

Mineral calcium carbonate x8000 Eggshell calcium x8000

FIGURE 11.1 Electron micrographs of mineral calcium carbonate (left) and egg shell calcium (right).

the extracted D-myo-inositol 4,5-diphosphate (Ins (4,5) P2) facilitates the dissolution of ostrich eggshell calcium during embryonic growth by partially forming amorphous calcium carbonate on the inside of the eggshell. The same compound has been described to be present in chicken eggshells and is thought to be one of the reasons for the readily dissolution of chicken eggshells in acid and its easy absorption into the body.

Regarding the absorption of eggshell calcium, a comparative study with calcium carbonate was conducted in male rats (Goto et al., 1981). According to this study, calcium carbonate, rough ground calcium eggshell, and finely ground calcium eggshell all showed high absorption rates of 97% during the growth period, while there was a difference in their absorption rates during the maturation period: 13.7%, 21.3%, and 34.8%, respectively. Mijan et al. (2014) studied the effect of finely ground eggshells and rough ground eggshell powder on postmenopausal osteoporosis in rats, comparing the results with ovariectomized rats. Rats fed with finely ground eggshell showed 6.6% higher bone mineral density than ovariectomized rats, while rats fed with loosely ground eggshells showed a 2.2% higher bone mineral density. These results suggest that the size of the pulverized eggshells may affect their utilization in the body.

In addition to the excellent absorption of eggshell calcium, studies have also evaluated its contribution to bone formation after absorption. Niiyama and Sakamoto (1984) measured the calcium content accumulated in the maternal femur of pregnant rats fed a diet containing milk casein as a protein source and 0.6% calcium from crystalline calcium carbonate or eggshell powder as a calcium source. Omi and Ezawa (1998) measured bone mineral density in an osteoporosis model using ovariectomized rats and eggshell calcium were added to provide 0.3% calcium. The results showed that the bone mineral density of the lumbar vertebrae and proximal tibia in the calcium-egg-shell group was statistically significantly higher than that in the calcium-carbonate group. They also reported that the bone mineral density of the tibia diaphysis and bone strength ash weight/dry weight in the calcium eggshell group tended to be higher than those in the calcium carbonate group. These results suggest that calcium eggshell is significantly absorbed and contributes to bone formation.

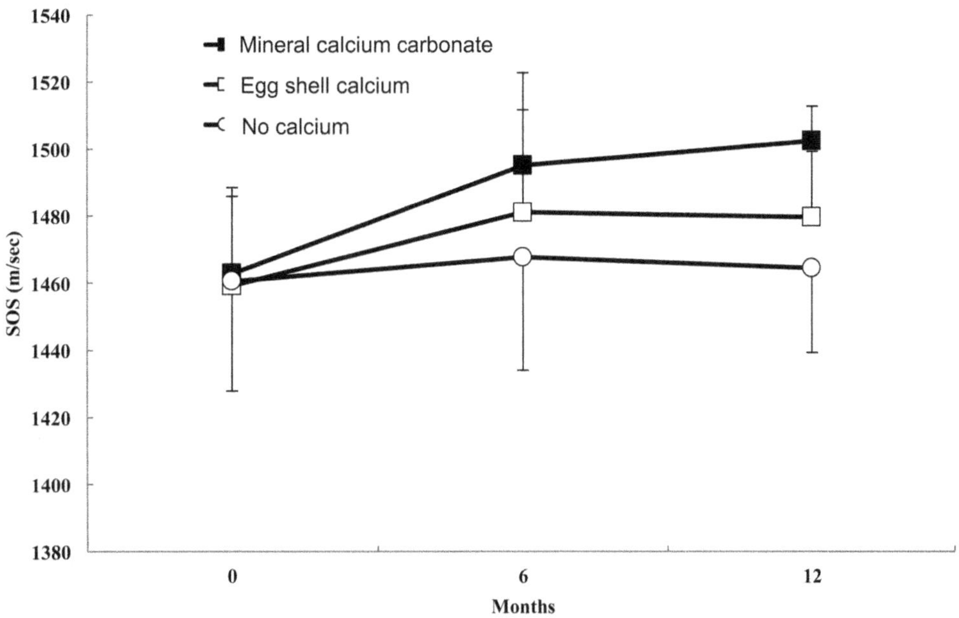

FIGURE 11.2 Changes in bone mass after calcium intake.

Sakai et al. (2017) studied the effect of calcium supplementation with eggshell on bone mass in 54 postmenopausal Vietnamese women living in the countryside. They were divided into 3 groups, receiving daily 300 g of calcium from eggshell, 300 mg of calcium carbonate, or no calcium, respectively. The bone mass in women of the eggshell group increased significantly at 12 months (p = 0.05). It greatly exceeded the speed of sound (SOS) in the placebo and calcium carbonate groups at 12 months, suggesting that calcium eggshell was more effective than calcium carbonate to increase bone mass in postmenopausal women (Figure 11.2).

The pathogenesis of age-related osteoporosis involves secondary hyperparathyroidism due to a decrease in blood calcium concentration caused by insufficient calcium intake or absorption. Therefore, calcium supplementation is thought to normalize blood PTH (parathyroid hormone) levels and inhibit increased bone resorption, thus preventing the progression of osteoporosis.

Masaki et al. (2000) examined PTH and bone resorption markers in older women, each given 2 g of calcium carbonate or calcium in eggshell for three months, containing the same amount of calcium. The results showed that calcium eggshell suppressed serum intact-PTH and the decreased urinary deoxypyridinoline excretion more than calcium carbonate. This suggests that the calcium from eggshells may have a more substantial inhibitory effect on bone resorption than calcium carbonate.

Recently, Kung et al. (2000) also reported that when osteoporosis patients were supplemented for one year with 500 mg of calcium, NTX (type I collagen N-terminal terror peptide) was suppressed by 14%.

Although additional well-controlled studies are needed, it is believed that calcium supplementation suppresses bone resorption in older women and the high bone turnover type of osteoporosis.

Removing excess phosphorus from the bloodstream is a vital issue for kidney disease patients undergoing hemodialysis. This is because phosphorus is one of the most challenging substances to pull, even with the current dialysis technology. In the past, the aluminum gel was used to adsorb phosphorus and expel it in the feces. Still, due to some concerns about potential brain damage and bone deposition of aluminum, calcium carbonate is now the most commonly used phosphorus adsorbent. However, the phosphorus adsorption effect of calcium carbonate is inferior to that of aluminum gel, and it also has the disadvantage of increasing the calcium concentration in the blood.

Ebisu et al. (1998) studied the phosphorus adsorption effect of calcium eggshell on six chronic renal failure dialysis patients with hyperphosphatemia. The results showed that the serum calcium concentration showed no increase with calcium eggshell (from 10.1 ± 1.3 mg/dL to 10.3 ± 2.0 mg/dL after treatment), and that the serum phosphorus level was reduced from 7.3 ± 1.1 mg/dL to 5.7 ± 0.8 mg/dL. Furthermore, when calcium eggshell was replaced by calcium carbonate, the serum phosphorus level increased from 5.7 ± 0.8 mg/dL to 6.8 ± 1.1 mg/dL, thus confirming the usefulness of calcium eggshell over calcium carbonate.

Gastric hyperacidity due to increased gastrin secretion is a complication in hemodialysis patients and the antacid He-blocker has been used to suppress gastric acid secretion in these patients. However, He-blocker inhibits the phosphorus adsorption effect of calcium carbonate.

Ogihara et al. (1996) compared the effects of He-blocker in dialysis patients with hyperphosphatemia. Their results showed that while calcium carbonate did not affect serum phosphorus levels, calcium eggshells significantly decreased these levels, thus confirming the phosphorus adsorption effect.

Active vitamin D inhibits bone resorption by promoting calcium absorption from the intestinal tract and inhibiting PTH secretion. It also acts directly on osteoblasts to stimulate bone formation. However, its side effects include hypercalciuria and hypercalcemia and kidney damage such as stone formation. Steroid-induced osteoporosis induces a hypo-turning metabolism, so calcium and active vitamin D should be prescribed together. In this case, however, special attention should be paid to the potential risk of hypercalciuria and hypercalcemia.

In an ovariectomized osteoporosis model animal, Hirasawa et al. (2001) compared the effects on bone mineral density and serum calcium concentration when calcium carbonate and eggshell calcium were used as calcium sources and combined with active vitamin D3. The results showed that while calcium carbonate increased serum calcium concentration, calcium eggshell did not. In addition, the bone mineral density in the proximal tibia increased in the case of eggshell calcium. It is thought that calcium eggshell does not quickly increase serum calcium while contributing to an increase in bone density.

In this chapter, we have discussed the usefulness of eggshell calcium from a nutritional perspective. Eggshell calcium is also believed to improve physical properties and is used in various foods such as snacks and fried foods. In Japan, eggshell is mainly used as a soil conditioner, but it is expected that eggshell will be used for both nutritional improvement and nourishment properties, and their consumption will increased.

11.4 Use of Eggshell Calcium in Foods and Industrial Products

Eggshell calcium, refined and micronized for easy use in food processing, is widely used as a calcium source (Figure 11.3). Its use as an ingredient in foods also improve their physical properties as demonstrated in Table 11.1 (Sasagawa, 2016).

Besides being an edible calcium source, eggshells are also utilized as soil conditioners. When lots of chemical fertilizers are used, the soil will become acidic, and the insufficient cations will make it difficult for crops to grow. Sprinkling eggshell calcium neutralizes the ground and makes crops' growth easier. Recent studies have also shown that sprinkling eggshells on paddy fields improves rice quality and reduces the effects of inclement weather to improve yields. It is also used in the production of chalk and studless tires.

11.5 Conclusion

Eggshell calcium has excellent functions such as good absorbability and is valuable as a calcium source. Its efficacy as a source of calcium has been demonstrated in animal models and human studies. However, the current use of eggshells as a source of calcium remains limited. Most eggshells generated by the Poultry sector are discarded at worst or used as acid soil improvers. One of the reasons for this lack of applications is the difficulty of separating the eggshells from the eggshell membranes. Eggshells still containing some eggshell membranes will burn through the action of heat generated during pulverization, thus producing a very unpleasant odor. Therefore, to obtain good-quality eggshell calcium, the separation of eggshell and eggshell membranes is an essential step. In the future, the development of a simple separation method will be essential to supply quality eggshell calcium and further boost its use in a wide range of applications in the market.

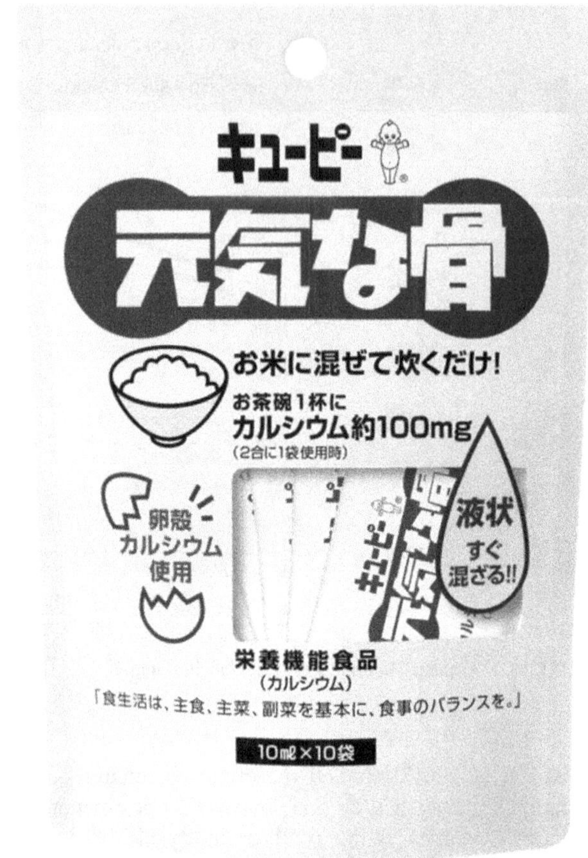

FIGURE 11.3 Products using eggshell calcium as a source of fortification.

TABLE 11.1
Use of Eggshell Calcium in Foods

Example of Use	Effect of Addition	Guideline for Amount to Add
Processed meat products	Reinforces bonding inhibits water separation	0.2–1% of meat
Kamaboko (fish cake)	Increased elasticity, increased whiteness	0.05–0.2 % for surimi
Noodles	Increased elasticity	0.5–1% for wheat flour
Bread	Improved retention	0.1–0.5% for flour
Cookies	Improved texture and retention	2–5% for flour
Swollen food	Improved swelling condition Improved texture	1–10% for flour
Fried food batter	Texture improvement and maintenance	1.5–3% for flour
Retort pouch foods	Prevents discoloration of white fish and poultry	2–3% for product

REFERENCES

Ahmed, T. A. E., Cordeiro, C. M. M., Elebute, O., & Hincke, M. T. (2022). Proteomic analysis of chicken chorioallantoic membrane (CAM) during embryonic development provides functional insight. 2022, ID 7813921. https://doi.org/10.1155/2022/7813921

Ebisu, N., Ebi, N., & Kita, Y. (1998). Study on the usefulness of eggshell calcium for hyperphosphatemia in chronic patients. *Journal of Wakayama Society for Dialysis Therapy*, *13*(2), 11–16 (in Japanese).

Goto, M., Suzuki, K., & Kanke, H. (1981). Use of Eggshell as a source of Ca. *35th Annual Meeting of the Japanese Society of Nutrition and Food Science*, p. 124.

Halgrain, M., Bernardet, N., Crepeau, M., Même, N., Narcy, A., Hincke, M., & Réhault-Godbert, S. (2022). Eggshell decalcification and skeletal mineralization during chicken embryonic development: Defining candidate genes in the chorioallantoic membrane. *Poultry Science*, *101*(2), 101622. http://doi.10.1016/j.psj.2021.101622

Heaney, R. P., Davies, K. M., & Barger-Lux, M. J. (2002). Calcium and weight: Clinical studies. *Journal of the American College of Nutrition*, *21*(2), 152–155. https://doi.org/10.1080/07315724.2002.10719213

Hincke, M. T., Bernard, A. M., Lee, E. R., Tsang, C. P. W., & Narbaitz, R. (1992). Soluble protein constituents of the domestic fowl eggshell. *British Poultry Science*, *33*(3), 505–516. https://doi.10.1080/00071669208417489

Hirasawa, T., Omi, N., & Ezawa, I. (2001). Effect of 1 alpha – Hydroxyvitamin D and egg-shell calcium on bone metabolism in ovariectomized osteoporotic model rats. *Journal of Bone and Mineral Metabolism*, *19*, 84–88. http://doi.10.1007/s007740170045

Ito, T., Kato, S., Kubo, A., Suematsu, M., Nakata, M., & Saikawa, Y. (2020). Small molecules assisting eggshell calcium dissolution for embryonic bone formation, *Tetrahedron*, *76*, 130854. https://doi.10.1016/j.tet.2019.130854

Kung, A. W., Yeung, S. S., & Chu, L. W. (2000). The efficacy and tolerability of alendronate in postmenopausal osteoporotic Chinese women: A randomized placebo-controlled study. *Calcified Tissue International*, *67*, 286–290. https://doi.10.1007/s0022330001142

Kusumi, N., Nakamura, M., Tando, Y., Suda, T.,. & Kudou, K. (1999). Gastric solubility of eggshell calcium preparation. Nutrition-Evaluation and Treatment Japanese *Journal of Nutritional Assessment*, *16*(2), 291–296.

Langford, H. G., & Watson, R. L. (1973), Electrolytes, environment and blood pressure. *Clinical Science*, *45*, 111–113. https://doi.org/10.1042/cs045111s

Masaki, H., Nakatsuka, Y., Miki, T., Takamoto, K., Ohnishi, T., Suguro, N., & Morii, H. (2000). Inhibitory effect of eggshell calcium on bone resorption related to parathyroid function in the elderly -comparison with calcium carbonated. *Osteoporosis Japan*, *8*(2), 245–247.

Mijan, M. A., Lee, Y. K., & Kwak, H. S. (2014). Effect of nanopowdered egg shell on post-menopausal osteoporosis: A rat study. *Food Science and Biotechnology*, *23*(5), 1667–1676. https://doi.10.1016/j.phanu.2013.11.109

Miyazaki, N., Mitani, T., Torisaki, A., & Inoue, M. (1995). Method and device for separating membrane from egg shell. *Japanese Patent JPH08173838A*

Niiyama, Y., & Sakamoto, T. (1984). Calcium utilization in pregnant rats treated with isolated soy protein. *Report of the Soy Protein Research Committee*, *5*, 53–58 (in Japanese).

Ogihara, M., Suzuki, T., & Umeda, H. (1996). Weakening of the action of phosphorus adsorbents when taking histamine H. antagonists: A comparison of precipitated calcium carbonate and powdered eggshell calcium I. *Kidney and Dialysis*, *41*(5), 695–698.

Omi, N., & Ezawa, I. (1998). Effect of egg-shell ca on preventing of bone loss after ovariectomy. *Journal Home Economics Japan*, *49*(3), 277–282. https://doi.org/10.11428/jhej1987.49.277

Pravina, P., Sayaji, D., & Avinash, M. (2013). Calcium and its role in human body. *International Journal of Research in Pharmaceutical and Biomedical Sciences*, *4*(2), 659–668.

Sakai, S., Thi, V., Hien, T., Tuyen, L. D., Duc, H. A., Masuda, Y., & Yamamoto, S. (2017). Effects of eggshell calcium supplementation on bone mass in postmenopausal Vietnamese women. *Journal of NutrItion Science and Vitaminolgy*, *63*, 120–124. https://doi.org/10.3177/jnsv.63.120

Sasagawa, N. (2016). Application of calcium: Application of CalHope™ to processed food. *Food Chemical*, *32*, 57–59 (in Japanese).

Simonetti, G., & Mohaupt, M. (2007). Calcium and blood pressure *therapeutische umsche*, *Revue Therapeutique*, *64*, 249–252. https://doi.org./10.1024/0040-5930.64.5.249

Tashiro, M. (1997). Study on Solubility of Eggshell-Derived Calcium, *Kieikai Research Foundation Summary of Research Reports*, p. 23 (in Japanese).

12

Application of Eggs in Food Product Development

Hiroyuki Shidara*, Kazuto Arimitsu, Takeya Miyashita, Gouta Uematsu, Satoshi Akiniwa, Noriaki Nishijima, Yoshifumi Oda, Hidekazu Takeuchi, and Takashi Kojima

CONTENTS

* Corresponding author.

DOI: 10.1201/9781003254430-16

12.1 Introduction

Eggs are delicious and have an appealing color. In addition, eggs possess some valuable physical properties such as coagulation, foaming, and emulsification. These properties allow eggs to be prepared, processed, and consumed in various ways. This chapter will introduce the processed egg products mainly consumed in Japan and present them according to their functions. Heat-coagulated egg products will first be introduced, along with some issues related to the large-scale commercial production of boiled eggs. Dried egg whites and their numerous applications will then be discussed. The foaming properties of eggs will be explained along with a review of foods taking advantage of the foaming ability of eggs. Mayonnaise will be taken as an example to illustrate emulsification and its applications in the preparation of foods. Finally, the importance of colors and nutrition will be discussed for various egg products.

12.2 Application of Eggs Using Heat Coagulation

Eggs have the property of coagulating when heated, and this function is due to the denaturation and coagulation of egg proteins upon heating. This characteristic is not exclusive to eggs as it is also observed with meat (beef, pork, chicken), seafood, and plant proteins. However, eggs are the only ones in a liquid state prior to heating. Eggs are utilized as raw materials for various egg dishes such as boiled eggs, omelets, fried eggs, confectionery such as puddings, sponge cakes, custard cream, and processed foods such as burger buns, sausages, fish cakes, and noodles due to this thermal coagulant property.

In addition to temperature, the thermal coagulability of eggs is also greatly affected by salts, sugar, and pH (Table 12.1). Adding salts or sugar or changing the pH changes the structure of the egg proteins, making them more stable or unstable toward heat. Also, depending on protein concentration, denaturation may result in soluble aggregates that do not gel.

Another feature of heat coagulation of eggs is seen in the differences in coagulability between the egg white and egg yolk. Egg whites are composed of multiple proteins, each having different sensitivity to heat. Egg whites start to become cloudy around 55–57°C, form a soft and non-flowing jelly at 60°C or higher, retain their shape at 75°C, and turn into a stiff gel at 80°C or higher. On the other hand, egg yolk gradually begins to solidify around 65°C, showing a

TABLE 12.1
Factors and Their Effects on Egg Protein Gel

Factor	Effect
Temperature	The higher the temperature, the harder the gel. The longer the heating time, the firmer the gel.
Concentration	The thicker the gel, the harder the gel. The thinner the gel, the less gel it may form.
Salt	Proteins gel even at low concentrations.
Sugar	Sugar raises the coagulating point of proteins. Heat-agglomerated gels of protein solutions containing sugar are softer than those without.
Acid	Acid lowers the coagulation point of proteins. No gelation occurs below pH 4.

TABLE 12.2
Condition of Eggs When Heated at Various Temperatures

Temperature	60°C	65°C	70°C	75°C	80°C	85°C
Yolk	Liquid	Liquid	Solid	Solid	Solid	Solid
Whole egg	Liquid	Liquid	Paste	Solid	Solid	Solid
Egg white	Liquid	Paste	Paste	Solid	Solid	Solid

Note: All samples were heated for 40 min.

sticky paste-like appearance around 70°C. Around 75°C, it completely solidifies into the hard gel. Since the whole egg is composed of egg white and yolk, it exhibits intermediate properties between those of the egg white and yolk or slightly similar to egg yolk. These properties are well utilized in egg recipes (Table 12.2).

12.2.1 Boiled Egg

The surface of the egg white may be damaged when boiled eggs are mechanically peeled after heating. It is essential to avoid this type of damage to produce good-quality boiled eggs. In this section, manufacturing points affecting the quality of boiled egg products are discussed in terms of processing points and how to deal with them. The following three points are necessary to maintain the quality of boiled egg products:

- Selection of raw eggs
- Preparation of raw eggs
- Control of the heating process

The best way to manage with each of them will be presented in the following section.

12.2.1.1 Selection of Raw Eggs

As hens age, the protein concentration in egg whites decreases, and with it the breaking strength of the egg white heating gel (Figure 12.1). As the breaking strength of the egg white gel decreases, the egg white portion is destroyed during peeling, and the quality of the boiled egg product declines. In addition, as laying hens grow, the weight per egg produced becomes heavier. When subjected to mechanical and physical shocks, the egg white portion is more likely to fracture due to the force of its weight (Isobe et al., 2013).

In the case of hard-boiled eggs, the heating time is long enough, and the eggs are sufficiently thermally coagulated so that eggs laid by old hens can be used as raw materials for boiled eggs. However, the mechanical peeling of large quantities of half-boiled eggs or half-boiled eggs with runny yolks may damage the egg white parts of boiled eggs during production, making it impossible to maintain a good quality. Therefore, it is necessary to select shell eggs most suitable for boiling. When producing half-boiled eggs, it is preferable to use small eggs laid by young hens. This is because the egg white of eggs laid by young hens will be a more rigid gel. However, a certain percentage of smaller eggs are also laid by older hens, so size alone is not a distinguishing factor. By controlling the age of a flock prior to boiling, it is easy to select eggs most suitable for the type of production desired.

12.2.1.2 Preparation of Raw Eggs (Aging)

Boiled eggs made from eggs stored at 10°C for a certain period are easier to peel off than boiled eggs made from freshly laid eggs. The taste (texture) of egg white is unique to boiled eggs (hydrogen sulfide). It gives off an odor and improves the taste of boiled eggs. It is thought that this is because CO_2 is released from the egg white and affects the texture and flavor of the egg white. For this reason, the eggs used to make boiled eggs should be stored properly, and not immediately after oviposition.

Similarly, when boiled eggs are produced in the food industry, they are produced after releasing CO_2 contained in chicken

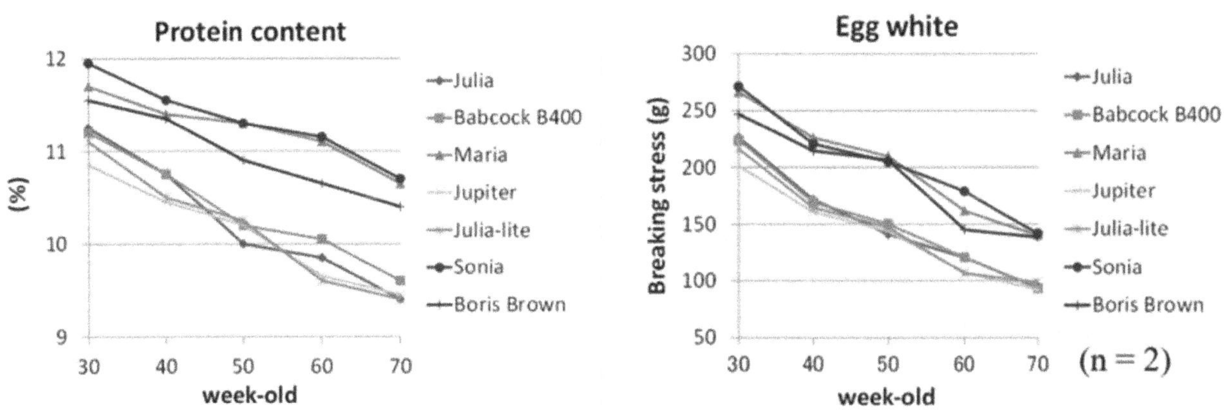

FIGURE 12.1 Relationship between the age of the hens and protein content in egg white and hardness of egg white gel.

eggs. Even if eggs can be collected in-line at a large-scale poultry farm, they should not be used immediately as a raw material for boiled eggs.

After being stored in constant temperature conditions for several days, eggs may be used. This storage process is called aging (tempering). The standard for aging is the pH of the egg white. The egg white pH of chicken eggs immediately after oviposition is pH 8–8.5. For boiling, a pH 8.8 or higher is desirable. The pH of egg white increases over time with the releasing of CO_2.

On the other hand, the Haugh unit decreases with aging. A decrease in Haugh unit means a decrease in the thick egg white fraction and an increase in watery egg white. This reduction in thick albumen will cause the yolk to no longer remain in a central position inside the egg. Therefore, with increasing storage time, shell eggs lose their commercial value for boiling. As a guide, it is best to use eggs with Haugh units of at least 70.

12.2.1.2.1 Example of a Test to Determine Aging Conditions

The definition of the USDA Haugh unit standards (HU 72 or more: AA, 60 to less than 72: A, 31 to less than 60: B, less than 31: C) is used as a criterion for judging the quality of eggs. Even eggs at HU93.2 (AA) on the day of lay will retain the AA grade of 79.3 (AA) after storage at 10°C for 10 days. The pH measured at the sharp end was 8.6 after 3 days, 8.8 after for 6 days, and 8.9 after 10 days (mean value of n = 10, Kewpie Co., Ltd. data). The ease of peeling the shell was improved when eggs were stored for 6 days or longer compared to eggs stored at 10°C for 3 days (Kojima et al., 2005). Figures 12.2–12.4 shows the condition of boiled eggs, the ratio of non-defective products to defective products, and the evaluation of taste.

The rate of CO2 release from eggs is affected by temperature. When aged at room temperature, the egg white pH quickly rises, and the HU also quickly decreases. While HU will decrease much less when eggs are aged at low temperature, storage at lower temperature has the disadvantage of delaying the rising of the pH of egg white. Therefore, it has been proposed to release CO2 using a carbon dioxide adsorption facility. By using this method, it is said that the pH of egg white can be raised, while the HU does not change, thus allowing in a short period of time to produce good quality boiled eggs (Wakamatsu, 2009).

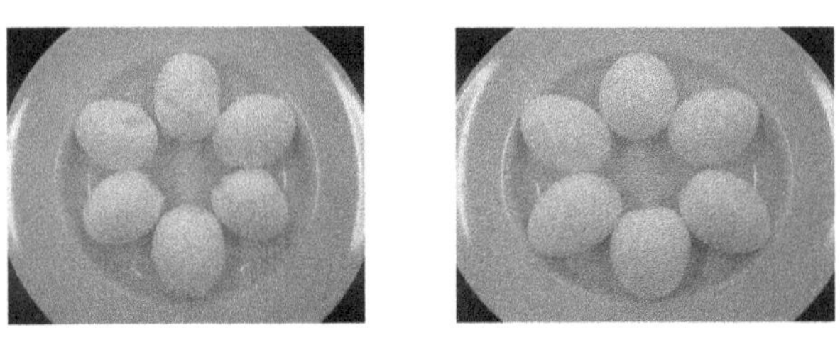

Boiled eggs made from eggs laid the day after laying and eggs stored at 10 °C for 10 days

FIGURE 12.2 Boiled eggs made from eggs laid the day before are difficult to peel and often have large chips.

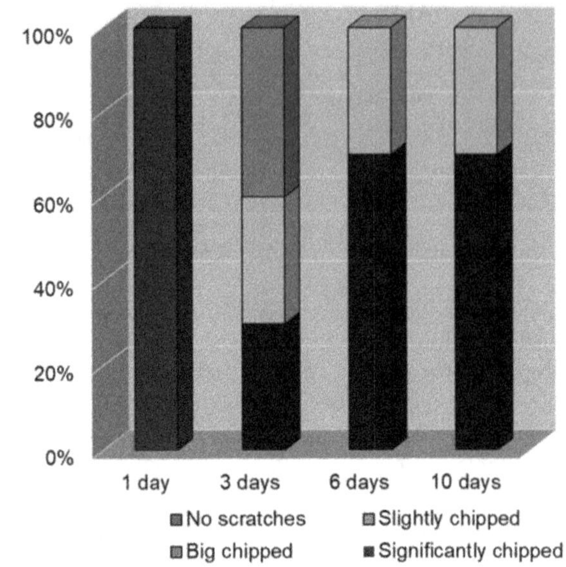

FIGURE 12.3 Relationship between the number of days after lay and the ease with which boiled eggs can be peeled off.

12.2.1.3 Control of the Heating Process

To produce large quantities of boiled eggs in the food industry, it is necessary to strictly control the temperature of the egg yolk part during the heating process, especially during the production of semi-boiled products. To produce high-quality semi-boiled eggs, temperature variations are minimized by controlling the size of the raw shell egg and the type of heating medium used; i.e., boiling water or steam. When boiling eggs at home, starting with cold water will prevent the shells from cracking. In the case of commercial production, it is not practical and time-consuming to start the process with cold water in order to control the yolk temperature. When the eggs are directly heated in boiling water maintained at a constant temperature for some predetermined period of time and then immediately cooled down, the yolk can meet the required level

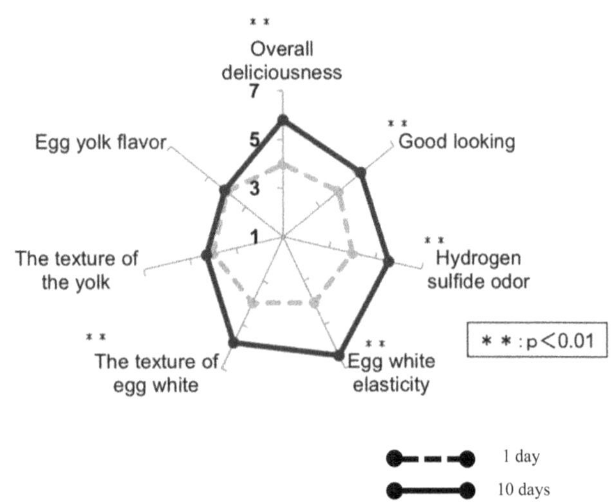

FIGURE 12.4 Evaluation of aged eggs and non-aged eggs.

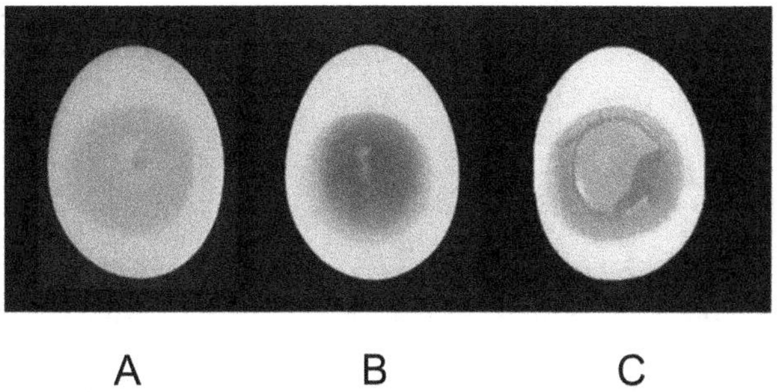

FIGURE 12.5 Condition of boiled egg yolk sold in the Japanese market.

of cooking. However, rapid heating of cold eggs can cause them to break easily. Raising the temperature of the eggs to 40–60°C before heating can prevent this breakage.

When making a boiled egg, it is necessary to peel the shell without damaging the egg white gel. One of the methods of peeling the shell without causing any damage is to puncture a hole in the blunt end with a needle-like device before heating. Home cookware is commercially available to proceed as described. Alternatively, it has been proposed to crack the shell in a way to respect the integrity of the eggshell membranes. When the shell is cracked, the CO_2 gas dissolved in the egg is vaporized by heating. As the inside of the egg expands, the cracked part expands, and the generated gas quickly escapes. After heating for a predetermined time according to the desired yolk aspect, the eggs are cooled.

Nevertheless, the inside of the egg will contract and become negatively pressurized, allowing the cooling water to penetrate the eggshell. This would cause the shell to flake off more easily. Figure 12.5 shows the various conditions of the yolk in boiled eggs sold in Japan.

12.2.1.3.1 Reducing Explosions Caused by Microwave Heating

Heating boiled eggs in a microwave oven can cause them to explode. Even boiled eggs cut in half or heated only with egg whites and yolks can explode. Freezing and denaturing boiled eggs can prevent them from exploding. However, it should be noted that if techniques such as quick freezing or quick thawing, having freeze-denature prevention effects, are used, the boiled eggs will not be sufficiently frozen and denatured and will still explode during microwave overheating.

The explosion of egg white gel upon microwave heating is due to the water vapor produced by partial overheating. When egg white gel is frozen, a portion of the tissue is destroyed. From the destroyed areas, water vapor can escape. This prevents the egg white gel from exploding. The texture of the egg white gel will deteriorate when frozen. If this does not cause any concern, freezing is a good option (Hanada, 2017).

Boiled eggs for commercial use are often used in lunch boxes and oden because of their ease of use and are usually heated by microwave oven heating. Boiled eggs are widely used in the market, and there are rare reports of boiled eggs bursting. Under which conditions are the boiled egg less likely to burst when heated in a microwave oven. When one boiled egg is heated in a microwave oven at 500W, it always bursts in about 40–50 seconds. If you cut the boiled egg in half, put it in a container, and add the seasoning liquid until the boiled egg is thoroughly pickled, it will not burst even if it is overheated for 4 minutes (50 seconds after the start of heating). When the seasoning liquid was poured to half the height, it burst on average in 2 minutes 51 seconds. It has been reported that the salt concentration of 1.0% rather than 0.5% of the fluid around the boiled egg is less likely to cause rupture (Hanada, 2010).

If these methods are carefully implemented, the rupture will not occur. Still, the physical properties are not changed fundamentally changed, so special attention should be paid. The boiled egg will overheat during heating in a microwave oven if the boiled egg is exposed (boiled egg protruding from the surrounding ingredients or floating). After heating, it was reported that an impact applied to grab a boiled egg with chopsticks or during chewing may cause the egg to burst, so it is necessary to refrain from overheating.

12.2.1.4 Boiled Egg Products on the Market

In Europe and the United States, hard-boiled egg products are sold in presentations ranging from one to six eggs sealed in deep-drawn or plastic containers (see Chapter 14). The distribution temperature is <4°C (33–40°F), and the shelf life is about 70 days. Some products may be offered in Modified Atmosphere Packages (MAP), e.g., a defined portion of CO_2 and oxygen, while others contain citric acid, sodium benzoate, potassium sorbate, etc. In Japan, the use of sodium benzoate in processed egg products is prohibited, and the distribution temperature must be below 10°. According to the current regulations in Japan, sodium acetate, glycine, nisin, etc., may be added to guarantee shelf life. Major boiled egg products for commercial use are sealed in a seasoning solution, which extends their shelf life and makes them easier to use in lunchboxes and prepared foods. Depending on the condition of the yolk, boiled eggs come in various variations, from solid orange to runny. These variations attract customers mainly in the restaurant industry, such as in ramen stores (Figure 12.6).

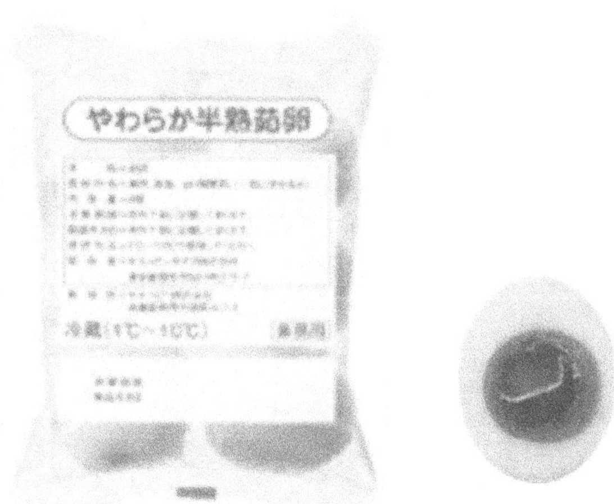

FIGURE 12.6 Boiled eggs sold all over the world.

12.2.1.5 Seasoned Boiled Egg with Shell

Salty seasoned boiled shell eggs are available on the market. After preheating shell eggs at 40–60°C, boiled eggs are prepared by heating at 90°C and higher, with air-pressurized to 2 kg/cm^2 and immersed in a saturated saline at 15°C for 30 minutes. During this process, salt will penetrate the eggshell and impart some flavor to the boiled eggs. Then, boil heating is performed again to remove any salt adhering to the surface of the shell.

Furthermore, this secondary boil heating can kill bacteria that may have passed through the shell and reduce the risk of secondary contamination. Finally, the cooling process will follow. Cooling by blowing air rather than cooling with chilled water will also be effective to prevent the entry of bacteria through the shell and improve the stability of shelf life (Ohno & Shinagawa, 1998).

Which product may be safer to consume, raw eggs, or heat-treated shell eggs such as boiled eggs or hot spring eggs? Lysozyme, one of the proteins in egg whites, has some antibacterial activity. Furthermore, egg whites are alkaline with a pH of about 9, preventing some degree of spoilage even if bacteria enter the shell. If stored in the refrigerator, the best-by date of commercially available eggs is about 2–3 weeks. However, the heat-treated egg white will lose its antibacterial effect while the protein has been denatured by heat. After heat treatment, bacterial growth (rot) could occur if bacteria penetrated through the pores of the eggshell. Secondarily, it is important not to contaminate the finished products during distribution. Another risk issue is *Salmonella enteritidis* (SE), which may be present inside the egg. When producing heat-treated shell eggs, it is necessary to destroy SE.

Eggs need to be heated at 63°C and higher, a temperature reached at the center of the egg, in order to kill SE bacteria, while preventing the coagulation of the egg components.

In the United States, pasteurized shell eggs are available in grocery stores. These pasteurized shell eggs are treated at a lower temperature than processed shell egg products. As shown in Figure 12.7, the egg white and yolk part of the pasteurized shell eggs have the same aspect as that of raw eggs. These pasteurized eggs are safe and preferable for fried eggs, poached eggs, and soft scrambled eggs (Vandepopuliere & Cotterill, 1999; Davidson, 2006).

FIGURE 12.7 Pasteurized shell eggs.

12.2.1.6 Egg Salad

One of the most popular menus using boiled eggs is egg salads. Boiled eggs are best used as egg filling for sandwiches, side dish bread, and salad mixed with vegetables offered in delicatessen. A basic recipe is diced boiled eggs, mayonnaise, and other ingredients such as seasonings, starches, and bacteriostatic agents to maintain the best-by date. Sterilization by heat can expand the best-by date to some extent. In the food industry, without using boiled eggs, some egg salad products have both the cooked egg white and yolk parts made from pasteurized liquid eggs.

12.2.2 Omelets

Omelets are one of the egg dishes easily prepared by placing beaten eggs on a frying pan. The origin of the omelet dates back to ancient Persia, and the present omelet style was first seen in France in the 16th century.

Omelets are easy to cook, and many variations and recipes are available worldwide. In this sense, omelets have no standard. Eggs are used as a wrapper for some omelets and confectionery. For seasoning, tomato ketchup, demiglace sauce, white sauce, and Chinese bean paste may be used.

12.2.2.1 Plain Omelet

A plain omelet is the basic one, made of beaten eggs seasoned with salt, pepper, milk, cream, etc., into a round, half-moon, and spindle (lemon-like) shape.

12.2.2.2 Souffle Omelet (France)

Souffle omelet is made by mixing egg yolk with whipped whole eggs or whipped egg whites and then baking them in a frying pan or oven. Its fluffy texture is the best feature obtained by whipping. The most famous souffle omelet is "La Mère Poulard" omelet created in the 19th century by a Chef originating from the Mont Saint Michel in the Gulf of Saint-Malo on the west coast of France.

12.2.2.3 Tortilla (Spain)

Tortilla is made of potatoes, bacon, spinach, etc., combined with beaten eggs and placed in an oiled frying pan. Tortilla shapes may be flat and round, like a pancake. No particular rules are applied regarding the ingredients used; thus, they vary based on the preferences of each household.

12.2.2.4 Hoppel Poppel (Germany)

"Hoppel Poppel" has a rustic flavor with chopped potatoes, bacon, and onions.

12.2.2.5 Frittata (Italy)

Frittata is made by mixing ingredients with beaten eggs and baking them in a frying pan or oven.

12.2.2.6 Tember Omelet (America)

Tember omelet is made with ham, paprika, onions, tomatoes, potatoes, cheese, mushrooms, spinach, etc., mixed with beaten eggs or fried eggs.

12.2.2.7 Egg Foo Young, Crab Ball (China)

Egg foo young is an egg dish made by mixing loosened crabmeat, shiitake mushrooms, wood ear mushrooms, ginger, and other vegetables with plenty of hot oil to make the texture soft.

12.2.2.8 Takuan (Taiwan)

Takuan, made by adding dried radish to beaten eggs, is a breakfast menu staple in Taiwan. Oyster omelet is also famous as street food, made by binding oysters and beaten eggs with water-soluble sweet potato starch.

12.2.2.9 Processed Product

Processed products have been quite common in the Japanese market, especially as frozen foods, for general household use, and as calcium-fortified products for school meals. Photos are shown in Figure 12.8.

12.2.2.10 Omelet Rice

Omelet rice is an egg dish of stir-fried rice wrapped in thinly fried eggs. Some records say it has more than a 100-year-old history, first being served around 1900. At that time, the current wrapping styles did not exist. All ingredients and seasonings with eggs were mixed and fried and it was a dish which was regarded as a quick meal easy to eat with a spoon. "Omelet rice" has changed considerably in response to customers' preferences, and now the most popular one is soft-boiled eggs or

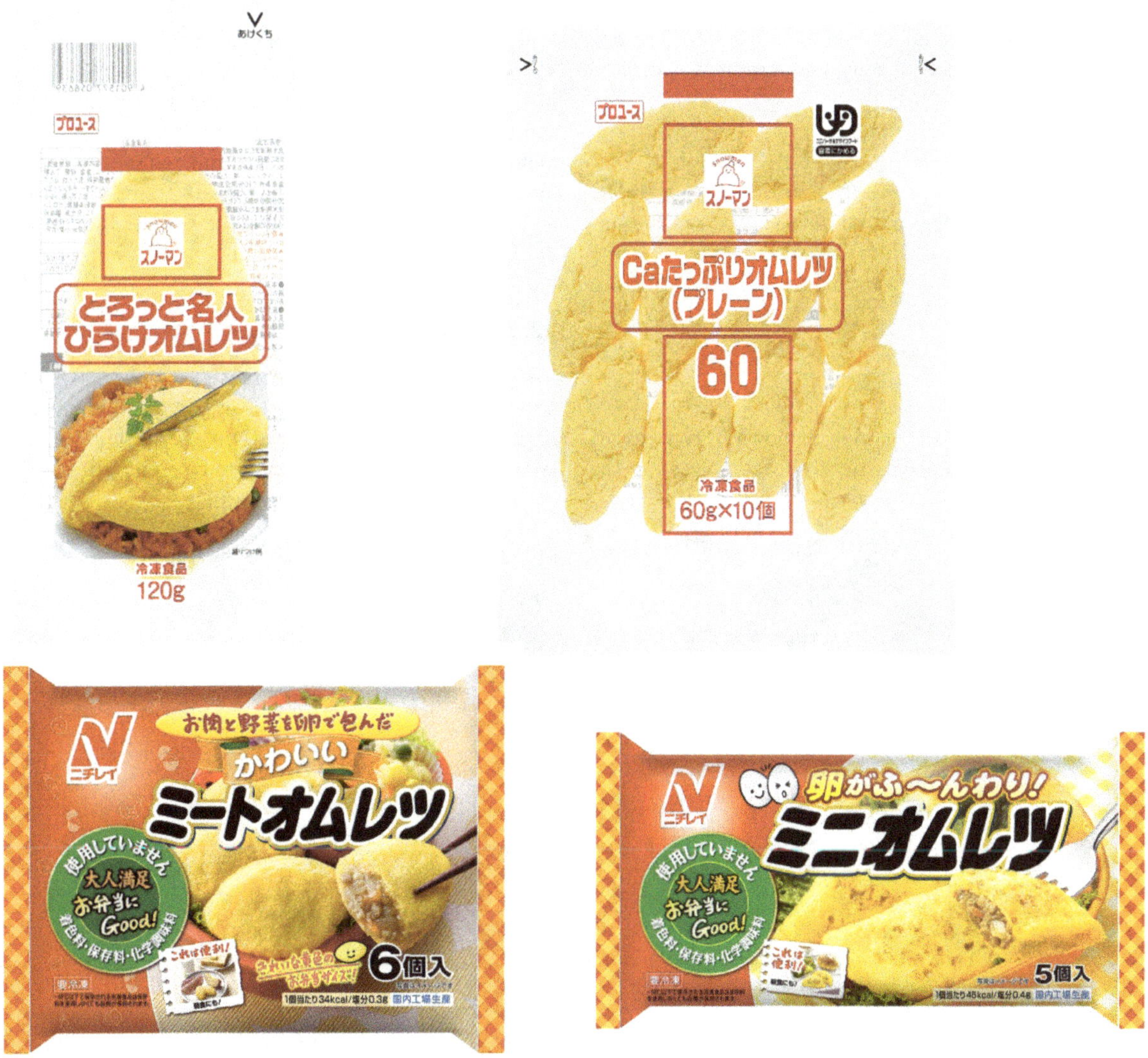

FIGURE 12.8 Processed egg products sold in the Japanese market.

half-cooked omelets put on top of stir-fried rice. This style is also unique to omelet rice created and developed in Japan.

12.2.3 Fried Eggs

One of the simplest of egg dishes, fried eggs, is used in various recipes worldwide. For example, it is served with bacon, ham, French fries, and vegetables, used as toppings for nasi goreng, loco moco, and mie goreng, and is best with sandwiches and hamburgers. In France, it is mainly fried on a plate, but in the United States, it is often fried in a frying pan.

In the United States, various names are used for fried eggs, depending on the cooking process and the egg's shape.

12.2.3.1 Sunny-Side up

Sunny-side up is baked on one side only. Since heat is not transferred to the top of the yolk, the yolk is not heated and remains fluid.

12.2.3.2 Turn over

Turn over is fried on both sides. When one side is cooked, the egg is turned over and cooked on the other side. The egg whites and yolks are cooked evenly, and burn marks may be seen on both sides.

12.2.3.3 Basted Egg

Like Sunny-side up, the eggs are not turned over, but the surface of the yolk will become white like a thin film by scooping oil on the frying pan, such as warm butter, until the yolks are cooked well.

12.2.3.4 Over Easy

Over easy used to be called "sunny-side down." Over easy is fried on both sides and appeals mostly to consumers in the United States. The egg white part is soft, and the yolk part is semi-cooked and remains fluid. When the egg whites part is cooked hard, turn it over without losing the yolk's fluidity.

12.2.3.5 Over Medium

Over medium has half-solidified yolk by grilling on both sides.

12.2.3.6 Over Hard

Over hard used to be called "over well." Eggs are cooked on both sides, and the yolks will become firm. The appearance is shown in Figure 12.9.

12.2.3.7 Fried Egg Analogs

Figure 12.10 shows processed egg products looking like fried eggs and sold as frozen products. The texture of the egg white part will remain the same even when frozen. Moreover, the yolk part will not coagulate even when reheated.

12.2.4 Egg Rolls

Egg roll is a Japanese dish of beaten eggs seasoned with salt, soy sauce, soup stock, sugar, or sweet sake (mirin) cooked in a square pan. It is often cooked with plain recipes, but sometimes vegetables such as spinach and carrots are added, and/or other ingredients such as minced meat or eel, to give more flavor. An egg roll (dashi maki tamago) cooking pan is shown in Figure 12.11, and the formulation is shown in Table 12.3.

12.2.4.1 Atsuyaki Tamago

Atsuyaki Tamago is an omelet seasoned with soup stock, sugar, soy sauce, etc., to a light salty or sweet type. It is made using an egg roll-cooking pan. Especially in the vicinity of Tokyo, the sweet type of thick omelet tends to be preferred.

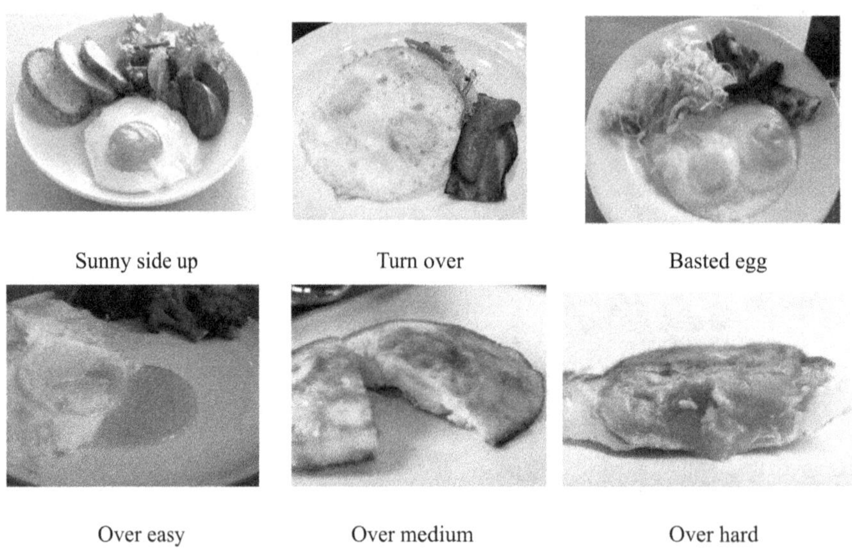

FIGURE 12.9 Fried eggs with different degrees of cooking.

FIGURE 12.10 Processed egg product mimicking a fried egg.

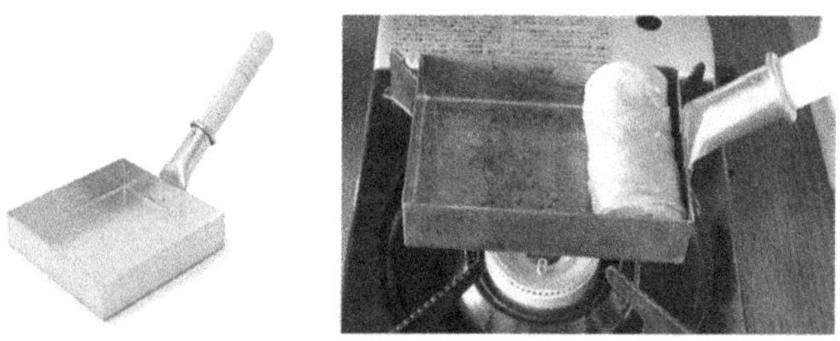

FIGURE 12.11 Egg rolls (Dashimaki Tamago) making pan.

TABLE 12.3

Japanese Style Fried Egg Formulation

	Atsuyaki	Dashimaki
Whole egg	430 g	250 g
Broth	100 g	100 g
Soy sauce	18 g	14 g
Mirin	16 g	18 g
Sugar	8 g	–
Salt	–	1g

12.2.4.2 Dashimaki Tamago

Dashimaki Tamago has a higher proportion of soup stock than a thick omelet and is a soft omelet seasoned with soy sauce, salt, and sweet sake. A thinly cooked egg will be rolled and layered. After cooking, it is shaped up while it is hot.

12.2.4.3 Usuyaki Tamago Kinshi Tamago

Beaten eggs are seasoned with salt and sugar and cooked thinly for sushi rolls. Shredded thinly cooked eggs are called "Kinshi tamago (brocade eggs)" and are used for "Chirashizushi (scattered sushi)" and toppings of chilled Chinese noodles.

12.2.4.4 Datemaki

Eggs and minced fish are well mixed and cooked in a square pan;, and after cooking, they are rolled using a "makisu (rolling mat)" to shape them up.

12.2.4.5 Processed Egg (Egg Rolls)

Like the above-processed egg products, egg rolls have been commonly cooked at home in Japan and have become available in supermarkets and retail stores. Figure 12.12 shows an egg roll machine.

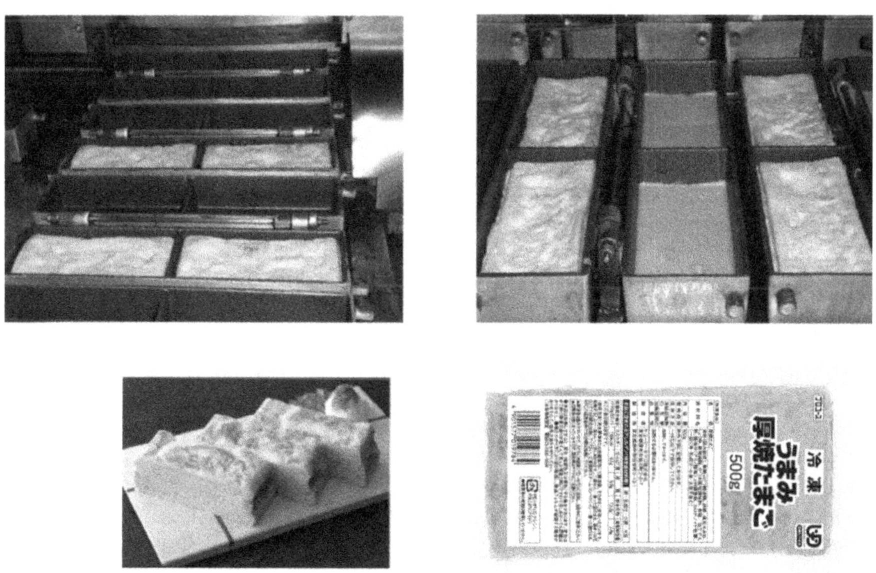

FIGURE 12.12 Egg-baking machine and products.

FIGURE 12.13 Oyakodon (chicken and egg bowl).

12.2.5 Other Products

12.2.5.1 Oyakodon

Oyakodon is an egg dish made of boiled chicken and onions, dressed with semi-cooked eggs. It is one of the familiar cooking menus in Japan's unique food culture, first recorded in the literature around 1887. In the beginning, the remaining broth from the hot chicken pot was used as a side dish for white rice with eggs. At that time, it was served with eggs fully cooked. The current style using rice with one half-boiled egg has become familiar over time. Figure 12.13 shows two examples of *Oyakodon* serving.

12.2.5.2 The Development of a Product Reproducing Egg Yolk

In Japan, many egg dishes are served with raw egg yolk on the top or used as a sauce. Typical recipes are "Yakiniku donburi (grilled meat bowl)," "Kaisen donburi (seafood bowl)," and "Tsukune (chicken meatball)." In these dishes, their colorful appearance with the yolk appeals to consumers, and the yolk deepens each dish's flavor. However, using raw egg yolks for institutional applications, mainly lunch boxes and lunches served at entertainment facilities, has been an issue because of the difficulty in separating and collecting many egg yolks and the risk of foodborne diseases due to SE. For this reason, processed egg analog products have been developed by adding vegetable oils, processed starches, and polysaccharide thickeners to egg yolks. These ingredients make egg yolk analog products safe, do not solidify when heated, and have excellent freeze-thaw stability (Figure 12.14). They can be served in a thawed state and are used in various applications, particularly in lunch boxes sold in convenience stores.

FIGURE 12.14 Processed product imitating the shape of egg yolk.

12.3 Applications Using the Heat Coagulation Ability of Egg Whites

12.3.1 Characteristics of Dried Eggs

Dried eggs are made by breaking chicken eggs and drying liquid egg whites, whole liquid eggs, and liquid egg yolks into powders. There are multiple methods for drying liquid eggs, and the one used in general production is spray drying (see Chapter 10). Spray drying will instantly dehydrate liquid eggs by spraying the liquids in mist and drying them with heat. Dried eggs have the following features compared to liquid eggs and frozen eggs:

1. Dried egg can be stored at room temperature.
2. Dried egg can be stored for a more extended period.
3. Dried egg can save storage space per solid content.
4. Dried egg can be reconstituted to any concentration when used.
5. Dried egg can be dissolved in other liquids as well as water.

In terms of the quality of dried whole eggs and yolks, some lipoproteins in the yolks will be broken during the high-temperature drying process when compared with liquid eggs. In frozen eggs, functions such as emulsifying and foaming properties are inferior. In addition, depending on the storage temperature and time after drying off, odor will become noticeable due to the oxidation of lipids and the Maillard reaction. Some additives are effective, and the manufacturing process is devised to prevent quality deterioration. Dried egg products have been developed based on customers' requirements and preferences.

In the case of dried egg white, quality improvement in gelling is possible by modifying the pasteurizing parameters. When reconstituted with water and heated, a higher-quality dried egg white produces a tough or supple gel. The quality improvement effect suitable for the physical characteristics of this gel has been studied.

12.3.2 Use for Meat Products

Many processed products using meat from livestock, such as hams, sausages, hamburgers, pork cutlets, fried chicken, and dumplings, are available worldwide.

Historically, hams and sausages were first made in Europe as a means to preserve foods and are now indispensable food items for home cooking and processed ready-to-eat dishes. They vary tremendously based on the preference of consumers for taste and flavors, and new products with various flavors,

Type	Standard	Hard	Very Hard	Elastic
Character	Hardness of boiled egg whites	Twice as hard as standard	Four times harder than standard	Supple and elastic A few separate water after frozen
	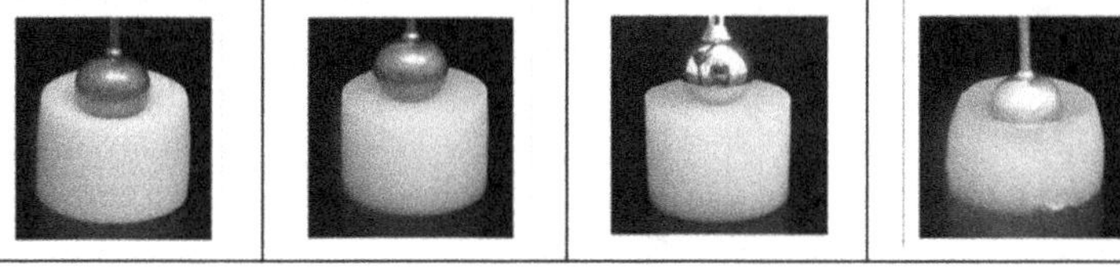			

FIGURE 12.15 Gel Quality of various dried egg whites.

textures, and price ranges are constantly being developed to respond to consumer demands.

Manufacturing equipment, methods, and formulations are the focus of much research to meet these diverse needs. In Japan, the Japanese Agricultural Standards (JAS) were introduced to categorize the types of bacon, ham, sausage, etc., and divide them into three classes, "standard," "advanced," and "special grade," based on the quality and type of raw materials used. Various types of proteins are widely used as indispensable raw materials, directly contributing to the yield of final products, texture (hardness, juiciness), and freeze-thaw stability.

Egg white proteins have excellent heat coagulating and emulsifying properties compared to other proteins such as milk and vegetable. Egg white proteins are indispensable for the production of current Japanese highly hydrous meat products. Some types of dried egg white will give better freeze-thaw stability and higher-water retention ability to meat products. As an example, below are the details of ham processing. The manufacturing process of loin hams will typically follow the following steps:

> Raw material processing → salting → filling → smoke → heating → cooling → slicing/packaging
>
> **(Okada, 2010)**

Traditional ham production takes weeks or months due to the time required for the salting process. Two types of salt pickling are used in the production. One is the "dry salt method" and the other, the "wet salt method." The former consists in applying salt and spices directly onto the meat, while the latter requires the soaking raw meats into a liquid seasoning (pickle liquid) containing salt, spices, and seasonings. For large-scale production, the pickle injection method is used where the pickling liquid is forcibly injected with needles into the meat.

Proteins used in the pickling liquid injected in meats play a key role to improve both yield and texture. Without additional proteins, the pickling liquid injected during the heating process will escape, resulting in a decrease in output. If too many proteins are added to the pickling liquid, the texture will negatively be affected. Dried egg white products have been developed to achieve higher water retention while maintaining the appropriate texture. A combination of dried egg white products will therefore meet the manufacturers' requirements of higher yields and good quality, translating also into satisfied consumers.

Test results of hardness and yield of representative ham recipes applying different types of dried egg white in the pickling liquid (Figure 12.15) are shown below (Table 12.4, Figures 12.16 and 12.17).

TABLE 12.4

Simple Ham Test Pickle Formulation

Formulation of Pickling Liquid	
Salt	4.0%
Seasoning	2.3%
Phosphate	0.9%
Nitrite and Vt. C	0.3%
Dextrin	10.4%
Egg white powder	10.0%
Food colors	0.1%
water	72.0%

12.3.3 Use for Fish Products

In Japan, eggs have long been used in surimi-seafood products. The egg white gives elasticity or good texture to kamaboko and improves their appearance, color, and luster. Whole

Production method

Minced pork lean and pickle liquid 100:80

↓

Mix in the absence of air

↓

Refrigerate overnight

↓

Filled in a casing with a folding diameter of 60 mm

↓

Heat at 85 ° C for 40 minutes

↓

Cool with ice water and leave overnight

↓

Cut to 3 cm width

↓

Evaluation of the physical properties

FIGURE 12.16 Simple ham test process.

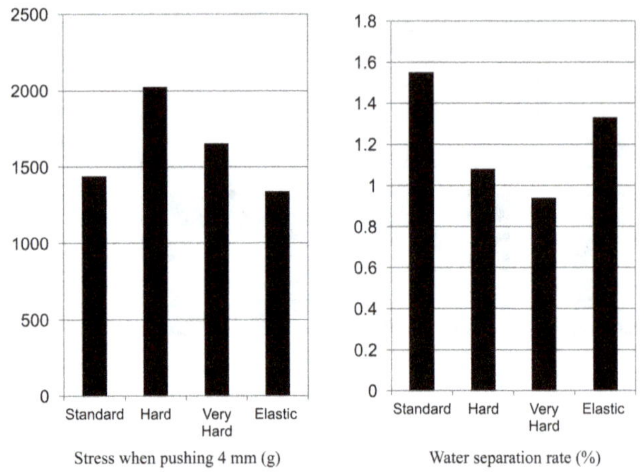

FIGURE 12.17 Hardness and water separation rate of simple Ham containing various dried egg whites.

eggs and egg yolks are used as raw materials for rolled omelets mixed with fish paste (*datemaki*). The eggs are added to hold air bubbles in the products and give them a soft texture and a vivid yellowish color (Okada, 2000).

Surimi is also popular overseas, and the Japanese word "*surimi*" is considered a universal term. Imitation crabmeat and crab analogs are world famous. In Spain, this Japanese-originated technology is used to produce an imitation of the traditional recipe for fried eel (Anguirus). In addition, surimi noodles, claiming a high content in protein, are also being marketed overseas due to the growing demand for protein intake, especially in the United States. In Japan, crabmeat imitations are once again attracting some attention due to their protein content and efficient role to increase muscle mass (Figure 12.18).

12.3.3.1 Effects of Egg Components on the Quality of Kamaboko

Egg white main function is to improve the gel formation of surimi-seafood products. When the surimi is ground with salt, the myosin filaments of filaments that make up the myofibrils break down to form myosin alone. The myosin is combined with actin filaments to create a highly viscous actomyosin. When heat is applied, the myosin portions of actomyosin form a network with each other, forming a gel with the elasticity characteristic of fish paste (Yamashita, 1995). Gelation of kamaboko has the unique characteristic of "sitting" and "returning," and interactions between surimi and egg components are also being studied. The effects of adding egg white can be roughly divided into elasticity reinforcement effect, sitting effect, and himodori effect.

FIGURE 12.18 Imitation crab leg.

12.3.3.2 Elasticity Reinforcement Effect

Some liquid egg white was added to surimi (0–20% of surimi weight) and cooked at 90°C for 30 minutes to make a fish paste. Gel strength was higher with the addition of liquid egg white than with kamaboko, to which no egg white was added. The addition of 5% liquid egg white gave the highest gel strength while it was reduced when 10% or more was added (Iso et al., 1985).

It has been reported that egg white does not bind to myogenic proteins forming the mesh structure of kamaboko but plays the role of filler for the mesh (Noda, 2018). When the egg white is used to reinforce elasticity, adding up to 10% to the surimi is considered appropriate. When egg white proteins (ovalbumin, ovotransferrin, ovomucoid, and lysozyme) were added separately to kamaboko, only the addition of ovalbumin and lysozyme resulted in improved gel strength in the kamaboko. Since lysozyme is present in very small amounts in liquid egg white, it was thought that the improvement in gel strength of kamaboko was due to the effect of ovalbumin (Iso et al., 1985).

12.3.3.3 Sitting Effect

Gels will form when the surimi with salt is kept at a relatively lower temperature between 10 and 40°C. This phenomenon is called the sitting effect. The sitting effect refers to the phenomenon in which myofibrillar proteins (myosin) dissolved by salt are cross-linked and polymerized by a transglutaminase (TGase) to form an insoluble network, including a gel. The efficiency of the sitting effect is affected by protein denaturation and aggregation, the amount of TGase activity and enzyme in surimi, pH, the structural position of the binding site to be cross-linked, the amount of Ca ions that activates TGase, and the presence or absence of inhibitors (Suzuki, 2008).

A study reported no difference related to the presence or absence of egg white at 10 and 40°C for Alaska pollack surimi (Iso et al., 1985). Another study showed that when Atka mackerel, walleye pollock, and mixed surimi were heated at 25°C or 40°C for 9 hours, the formation of the sitting warmed gel was enhanced with the addition of dried egg white (Watanabe, 2008). It was suggested that the egg white may reinforce the elasticity of surimi that may lack or have a lower sitting effect.

12.3.3.4 Modori Effect

When surimi is slowly heated from 50°C to 70°C, the gel will become weaker and then disintegrate. This phenomenon is called the Modori effect. The Modori effect is caused by the non-uniformity of protein network structure when heating surimi. Actin in the network structure, formed by heating at around 50°C, becomes denatured and dissociated, thus weakening the gel. Alternatively, myosin causes non-uniformity in the network structure and weakens the gel. The Modori effect is mostly observed around 60°C. This phenomenon occurs as some protease activity in muscles causes myosin to break down and the network structure becomes non-uniform or is disrupted (Suzuki, 2008).

It is well recognized that egg white has the effect of suppressing the Modori effect. The jelly strength of gels formed by heating meat paste containing 10% liquid egg white at 60°C for 0–3 hours and then at 90°C for 30 minutes was suppressed. Gels were also maintained when ovalbumin, ovotransferrin, and ovomucoid (egg white protein components) were added alone. However, the trypsin inhibitor ovomucoid was less effective than the other two egg white proteins. This result suggested that the effect of egg white does not occur because it suppresses trypsin activity (Iso et al., 1985).

12.3.4 Use for Noodle Making

Eggs are widely used in Chinese noodles, buckwheat noodles, udon noodles, and pasta, and egg white, whole egg, and egg yolk are blended for the following effects.

The effects of adding egg white are:

1. Strengthening of the hardness of noodles.
2. Maintaining a good texture while and after cooking.
3. Holding starches inside while cooking.

The effects of adding egg yolk are:

1. Strengthening the extensibility and stickiness of noodles.
2. Giving some brighter colors.
3. Adding some egg flavor.

The addition of whole egg will have effects intermediate to these presented for egg white and yolk.

Dried egg white can form a gel inside the noodle with higher water retention capacity. The gel holding water will prevent moving inside the noodle, and avoid the phenomenon of "noodles getting soggy" in the early stage.

In particular, alkaline type dried egg white shows flexible and water-retaining gel giving the supple texture to noodles, and delays uniformity of water distribution in the cooked noodles. When 3% of dried egg white is added to semolina flour, starch will stop flowing out into the cooking water, and the weight reduction of the noodles is reduced. In addition, frozen noodles have become popular and are easy to prepare in the microwave. With dried egg whites, the gel in the egg white prevents structural damage and reduces water loss during thawing, thus minimizing texture loss during microwave heating.

The physical characteristics of egg white gel (high water retention and freezing resistance) will effectively maintain or improve the quality of noodle products. In particular, alkaline dried egg white tends to have more SH groups on their molecular surface. After constant heating, fewer S-S bonds were seen when compared to standard dried egg white, due to the molecular structure of the egg white protein. It is hypothesized that alkaline type dried egg white has more free SH groups on the surface, can form S-S bonds with SH groups of gluten and strengthens the network among proteins (Xie et al., 2020).

12.4 Applications Using the Foaming Properties of Eggs

When chicken eggs are beaten, they become foamy while entrapping air. This function of eggs is called "foaming" and is used for the preparation of various foods, especially confectioneries.

When a liquid foams, the following three conditions must be satisfied.

1. Surface tension of solution is small.
2. Foam film is air denatured and durable.
3. Bubbles are not bound together.

The foaming characteristics of egg whites, whole eggs, and egg yolks are shown below (Figure 12.19).

12.4.1 Egg White Foaming

Egg whites exhibit much better foaming properties than whole eggs or egg yolks. The main component of egg whites, the proteins, mainly albumin, ovotransferrin, and globulin, come into contact with air and adsorb at the interface, thus reducing surface tension.

A structural diagram of the egg white foam is shown in the left figure of Figure 12.20. Egg white proteins have a

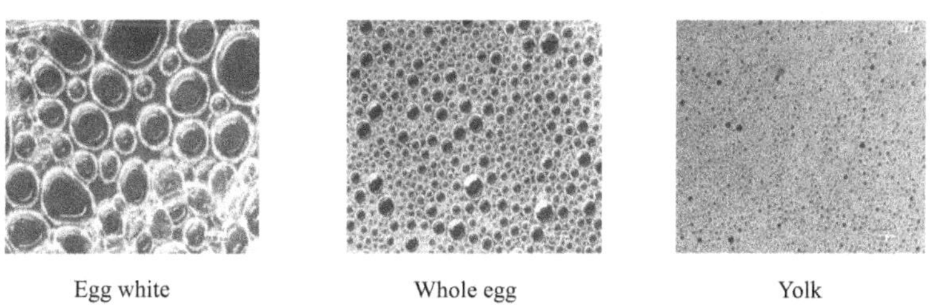

FIGURE 12.19 Comparison of the foam of egg white, whole egg and yolk.

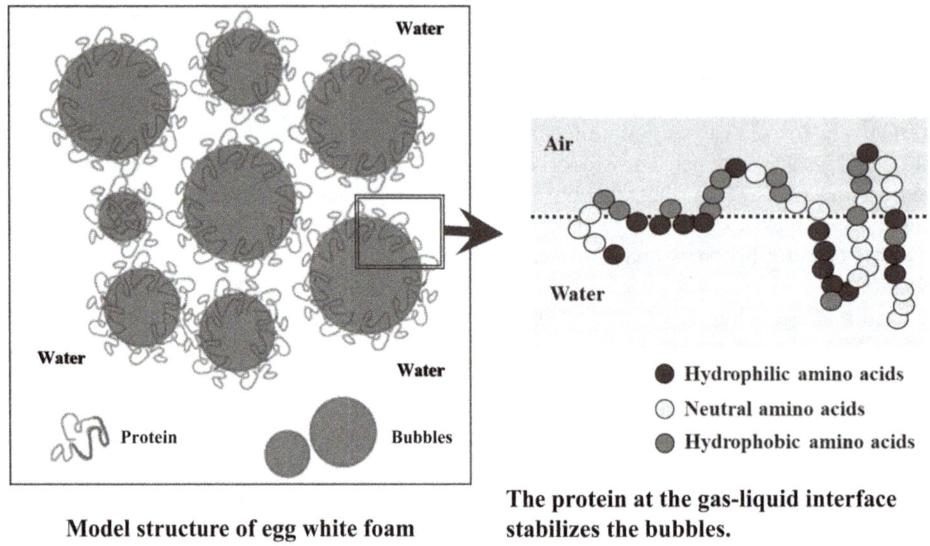

FIGURE 12.20 Model structure of egg white foam.

"hydrophilic part (shown in light gray)" that has a significant affinity for water and a "hydrophobic part (shown in dark gray)" for air. The hydrophilic and hydrophobic parts are lined up at the gas-liquid interface. Furthermore, the amino acid sequence of the protein on the right of Figure 12.20, shows the presence of hydrophilic amino acids (hydrophilic part), hydrophobic amino acids (hydrophobic region), and neutral amino acids, all part of the same protein, and contributing to the formation of a stable film along the gas-liquid interface, as shown in the figure (Ichimura, 2017). When chicken eggs are beaten, they become foamy while entrapping air. This function of eggs is called "foaming" and is used in various foods, especially confectioneries.

When liquid foams, the following three conditions must be satisfied.

1. Surface tension of solution is small.
2. Foam film is air denatured and durable.
3. Bubbles are not bound together.

Properties required for egg white foam (meringue) are fine and firm texture and good stability without breaking the foam. The properties and states of foam are affected by factors such as beating conditions (speed and time), egg white temperature, viscosity, and various materials added, as shown in Table 12.5.

TABLE 12.5
Effect of "Additives" on the Foaming Property of Egg White

Additives	Ease of Foaming	Hardness of Foam
Lemon juice, potassium hydrogen tartrate	++	++
Sugar	±	++
Yolk	±	±
Milk, cream	±	±
Butter, margarine	–	–
Salad oil, shortening	–	–

Note: ++: Good; ±: Not good; –: Bad.

12.4.2 Foaming of Whole Eggs

When mixed with fats and oils, egg whites destroy the membrane structure of the egg white foam and reduce its foaming ability. However, when whole eggs are whipped, they do not form bubbles as quickly as when egg whites are whipped. Whole eggs must be beaten thoroughly over time to obtain a fine, uniform foam.

Egg yolk contains more than 30% of fats and oils (lipids). The fats and oils do not exist in isolation, unlike in butter and salad oil. As shown in Figure 12.21, proteins and lipids have a minute oil droplet-like structure of about 4–150 μm called "lipid-protein." Therefore, the fat (yolk lipids) will not directly destroy the bubbles in the egg white, and the whole egg, regarded as a mixture of yolk and egg white, can show a specific foaming property (Ichimura, 2018).

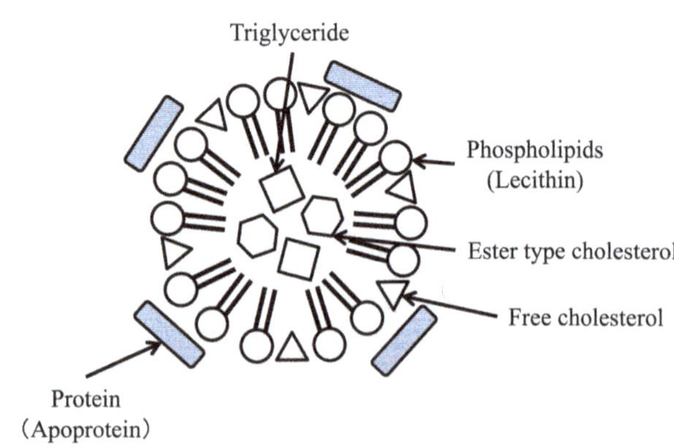

FIGURE 12.21 Structure of egg yolk lipoprotein (schematic diagram).

12.4.3 Foaming of Egg Yolk

Egg yolk is not often whipped for foaming applications but is only used after adding sugar and whipping until it becomes whitish. Egg yolk shows some foaming properties due to the presence "low-density lipoproteins" with different properties from to egg white proteins. Even in the case of egg yolk, the air is taken in, and bubbles are dispersed by beating. However, egg yolks have a significantly lower propensity to sustain bubbles than egg whites and cannot hold as much air. Therefore, the bubbles created are not large enough to be visible by the naked eye.

12.4.4 Sponge Cake (Sponge Dough)

Typical methods for making sponge dough are commonly used: the "separate method" and the "Tomodate method." Although the mixing ratios of eggs, sugar, flour, fats, and oils are the same, a significant difference can be seen between the two methods.

The "separate method" uses egg whites and yolks that are whipped separately and then mixed to make the dough. Adding sugar to the egg whites comes first, followed by whisking to make the mixture firm and stable. Then egg yolks, flour, etc., are added to make the dough to avoid breaking the bubbles. Sponge cake will have a soft and light texture and melt in the mouth.

The "Tomodate method" is the most popular dough for sponge cake, whipping whole eggs and sugar together. The "separation method" has a relatively coarse and firm texture (air bubbles) and is suitable for fluffy biscuits and chiffon cakes. In contrast, the "Tomodate method" has smaller air bubbles and uniform texture, described as a "sponge." The dough features are a lot of air bubbles inside and rich in fluidity like cream. Therefore, the "Tomodate method" dough is preferably used for rolling products that are poured into molds and baked moistly or thinly into a top plate and baked.

12.4.5 Meringue and Foods Using the Foaming Ability of Eggs

Whipped egg whites are called meringue and are used in various foods. Meringues can be broadly classified into three types according to their preparation procedures.

French Meringue (Cold meringue)

Method: Whisked egg whites with sugar several times without heating.

Characteristics: Produces light and flexible meringue (no heat denaturation).

Applications: Sponge dough, etc.

Swiss Meringue (Hot Meringue)

Method: Egg whites are heated to about 50°C (122°F) in a hot water bath with sugar before whisking.

Characteristics: Meringue with fine texture, viscosity, and firmness can be adjusted with the amount of sugar added once warm.

Applications: Meringue confectionery, etc.

Italian Meringue

Method: Add gradually syrup heated up to 118–120°C to the meringue and whisk until the mixture is heated to a temperature of 118–120°C.

Characteristics: Produces a firm, stable meringue.

Applications: Creams, mousses, squeezed out for decoration, etc.

Other preparations using the foaming ability of eggs are:

Pâte à bombe (*Cannonball dough* in French)

Egg yolks are beaten with heated syrup, or the yolks and syrup are mixed and then beaten in a hot water bath to incorporate air. Light and rich texture can be used as a base for buttercream, mousses, glasses, etc., and often mixed with chocolate, nuts, and coffee flavors. Bombe means "cannonball," used for bomb glacé dough, and ice cream pastries like cannonball, spherical or conical shape.

Angel Food Cake (Angel Cake)

Angel food cake is made by adding sugar and flour to beaten egg whites and baking, resulting in a very light batter and soft and fluffy cakes. To improve heat transfer during baking and to maintain its shape after baking, a unique mold with a cylindrical hole in the center is often used, and the appearance resembles an angel's ring, hence the name.

Macaroon

Macarons are made by adding sugar, almond powder, and other ingredients to egg whites, then extruding the dough into a round shape and baking it. A smooth surface characterizes its texture. A cracked bulge on the side is called "pied" (foot in English) and jam, cream, or ganache are sandwiched between two macarons. A necessary procedure during the preparation of macarons is mixing the whipped meringue by scraping it in a bowl using a card or other tool, which moderately reduces the number of air bubbles in the meringue and creates a smooth, flowing, and glossy dough.

Mousse

Mousse means "foam" in French. For sweets, mousse is generally made by adding fruit puree or chocolate to whipped egg whites, egg yolks, or whipped cream. The name "mousse" is thought to be given because meringue gives it a light, foamy texture.

Traditionally, mousse has been made by cooling and hardening meringue with whipped cream alone. In recent years, gelatin has been commonly used to improve shape retention.

Soufflé

Soufflé means "puffed up" in French. A soufflé is made by mixing whipped egg whites (meringue) with other ingredients and baking them. Heating will cause air bubbles inside to expand, resulting in a fluffy texture, and when cooled, it is no longer able to keep its shape and gradually becomes smaller. It can be used not only as sweet mixed with custard cream, etc. but also as a main dish with white sauce, lined vegetables, meat, fish, etc.

Derivative recipes of soufflé have become popular such as soufflé omelets baked by blending beaten egg whites (meringue) with egg yolks and soufflé pancakes baked by adding meringue to the batter.

Fritters

Fritto means "fried" in Italian. A dish with seafood, meat, and vegetables is battered with a mixture of flour, egg yolks, milk, water, meringue, and deep-fried in oil. Compared to other fried foods, its light and fluffy texture are favorable.

Cloud Bread

Cloud bread is made by baking meringue with egg yolks, cream cheese, sugar, baking powder, etc. "Cloud bread" is named for its appearance and texture associated with clouds. It is often eaten as a substitute for bread made with wheat flour, with ingredients sandwiched between or with toppings. Since it is gluten-free and low in calories and sugar content, it is often eaten for weight control and sugar restriction purposes and has become one of the hot food items in recent years (Aoyama, 2008).

12.4.6 Pasteurized Eggs Retaining Their Foaming Power

From a food safety standpoint, liquid whole eggs and frozen whole eggs need to be heat treated to kill Salmonella. However, heat treatment can reduce the egg's ability to foam, negatively impacting food products using this foaming ability. Lots of efforts have been made to make quality sponges from pasteurized eggs. Methods such as adding alum to protect the egg protein or emulsifiers to reinforce their foaming power have been used. Although these methods are excellent to maintain foaming power, they will not give good texture and are not clean label compatible (USDA, 1969). Recently, a technique has been patented for maintaining the foaming property of eggs without adding food additives by low-temperature heat treating the liquid egg through a strainer having a coarse mesh of about 2 mm after mixing (Ichimura et al., 2005).

Loss of the foaming force of egg white is caused by heat treatment and contamination of egg yolk. To prevent contamination of egg yolks, the breaking of fresh eggs is recommended. At the factory, eggs are cracked and inspected, and yolk contaminated in egg whites is removed (see Chapter 10).

12.5 Application Using the Emulsifying Properties of Eggs

Among the various food applications of eggs, the emulsifying properties are one of the most critical features of eggs along with coagulation, foaming, nutrition, color, and flavor. Typical examples of products using this feature are mayonnaise, dressings, and ice cream. In addition, emulsified mixtures of fats and oils with egg yolk and egg white are used by the livestock meat processing, fishery processing, and the confectionery sectors.

In this section, mayonnaise will be considered as the main example of applications using the emulsifying properties of eggs. Seasonings emulsified with eggs are often called mayonnaise. Various types of mayonnaise are available and have defined standards in their respective country. The U.S. standards for mayonnaise are often used as a global reference (U.S. Food and Drug, 2022). The Food and Drug Administration (FDA) has clearly defined the types of food preparation allowed to use the term "mayonnaise," and ingredients such as vegetable oil, acidic components (vinegar, acetic acid, lemon juice, lime juice), and egg yolk must be used. In addition, 65% or more of vegetable oil needs to be mixed, and salt, nutritional carbohydrate sweeteners, and spices (excluding the use of saffron and turmeric that imitate egg yolk color) can be used as optional ingredients. As food additives, natural fragrances, MSG, and sequestrants (Ethylenediamine-tetraacetic acid, calcium(II), disodium salt, dihydrate (EDTA-CaNa2), Ethylenediamine-tetraacetic acid disodium salt dihydrate(EDTA-Na2), and others) are also allowed in their recipe.

Mayonnaise is often classified depending on the amount of oil used, with various types of mayonnaise produced. The amount of oil incorporated ranges from 75% or even more to about 10% to 50%. As per the FDA standards, preparations with an oil content of 30% or more are called salad dressings, not mayonnaise. Starches or similar ingredients are often used to maintain the mayonnaise-like viscosity (Imai, 1993). Mayonnaise is sometimes classified differently, based on the type of eggs used: yolk and whole egg. The yolk type is often called egg yolk only to emphasize its strong points. Mayonnaise will vary in terms of flavor and vinegars used are responsible for this flavor variation. Wine vinegar, grain vinegar, fruit vinegar, etc., will characterize the flavor of the mayonnaise. Mayonnaise with cheese or mustard, with distinct flavor and taste are also commercially available.

12.5.1 History of Mayonnaise

Many stories about the origin of mayonnaise have circulated over the years. The most widely believed origin is the port city of Mahón (or Maó) on Menorca, one of the Balearic Islands in the Mediterranean, now part of Spain. A record mentioned that mayonnaise-like seasonings were served as a meal in Mahón around 1756. This sauce was then introduced into France. In France, the sauce was initially called "Salsa de Mahon," taking after the name of the city and later changed to mayonnaise (Glenn & Larsen, 2013). Menorca is also the origin of a breed of chickens with black wings named Menorca. Homemade mayonnaise has used these local eggs, along with olive oil, lemon juice, wine vinegar, salt, and garlic, all ingredients readily available on Menorca, blessed with the warm Mediterranean climate allowing the availability of vegetables and fruits even in the winter.

12.5.2 The Production Method of Mayonnaise

When making mayonnaise, seasonings, spices, vinegar, etc., are first added to the eggs, and then vegetable oil is added little by little while stirring. This emulsion is called oil-in-water emulsification, which means that oil droplets are scattered in the water phase. To make this emulsion at home, one can use a whip and stir by hand. The same method is used for small-scale production, although a motorized mixer is used instead of a whip.

Due to its oil content and the presence of air after mixing, mayonnaise is prone to oxidation and the shelf life of commercial products is greatly affected by oxidation. Removing air from the mixing vessel in which the mayonnaise is made and preventing air from entering during mixing will prevent the presence of air in the mayonnaise, thus preventing the oil from oxidation and extending the shelf life of the mayonnaise. Using an emulsification device such as a colloid mill will reduce the size of the oil droplets and stabilize its emulsified state.

The average oil particle size of hand-made mayonnaise at home is relatively large, about 15–50 µm. On the other hand, commercial products manufactured using emulsifier equipment have excellent oil particle sizes ranging from 2 to 10 µm with relatively higher uniformity. In industrial products, oil droplets can be stably maintained in emulsified states and remain emulsified even after long-term storage. Figure 12.22 shows micrographs of homemade mayonnaise and factory-produced mayonnaise particles.

No significant difference can be seen in the manufacturing process of mayonnaise containing starches and/or thickeners. In mayonnaise, the emulsifying properties of eggs maintain the emulsified state of the finished product. In this type of mayonnaise, the presence of additives will increase the viscosity of the aqueous phase portion and prevent coalescence among oil droplets. As a result, the stability of mayonnaise is increased (Nakae, 1986).

As mayonnaise has a certain amount of eggs and is distributed at ambient temperature, consumers may think that some preservatives are used. In reality, no preservatives need to be used since ingredients such as vinegar and salt have antiseptic effects, preventing bacteria from growing in case of product contamination. However, after mixing mayonnaise and vegetables to make salads, for example, the concentration of vinegar and salt will be diluted and no longer prevent bacterial growth, thus making the salads more prone to spoilage. A certain number of microorganisms will increase, and thorough attention will be required out of food safety concerns.

12.5.3 Functional Mayonnaise

Mayonnaise flavors are broad in range to meet the diverse needs of consumers. In addition to these flavor variations,

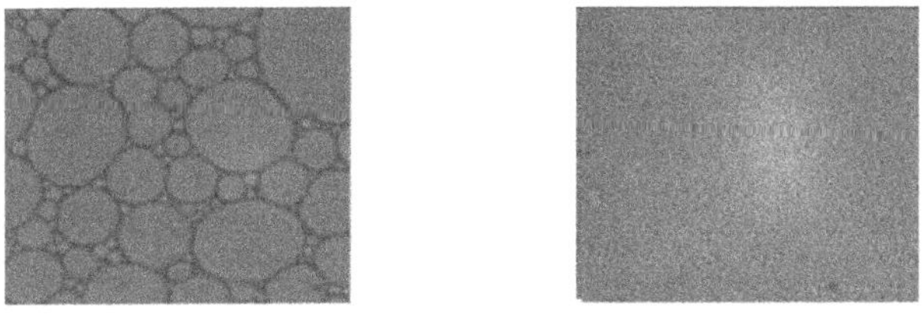

FIGURE 12.22 Difference in size of oil droplets between homemade mayonnaise and industrial products.

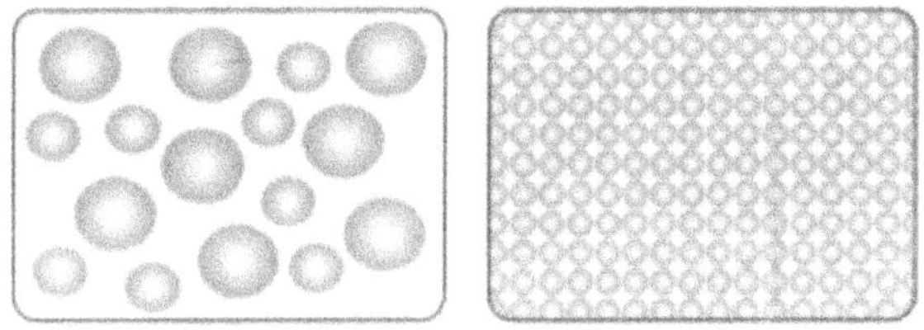

FIGURE 12.23 Difference in size of mayonnaise oil droplets.

healthy functional mayonnaises have also been introduced to the market. Some consumers do not eat mayonnaise because of its high-calorie content. For those who are concerned about calories, low-calorie products have been developed. Low-calorie mayonnaise can be made by reducing the percentage of oil and fat used in the recipes. However, reducing the oil and fat content will affect the unique physical properties of mayonnaise. Various methods have been proposed to prevent this loss of physical properties. One is to minimize the size of oil droplets in the aqueous phase and increase the surface area of oil droplets. The interaction with the aqueous phase becomes more extensive and vicious. Kewpie Corporation has named this method the "microemulsion method" and markets a product containing half the calories of regular mayonnaise (Figure 12.23).

Another method is as follows. Producing water-in-oil emulsion first and then dispersing it in water to make another emulsion state around the emulsion prepared first. This method is called the "double emulsification method" and enables to reduce the oil and fat content while maintaining the emulsified state identical to that of regular mayonnaise (Ajinomoto, 2022; Figure 12.24).

In addition to its calorie-reducing effect, double-emulsified mayonnaise has another function. When mayonnaise is mixed with vegetables, water is released from the vegetables, reducing the salad's appearance, retention, and taste. Double emulsification can absorb moisture from the vegetables, thus maintaining the quality of the salad (Takeuchi, 2020).

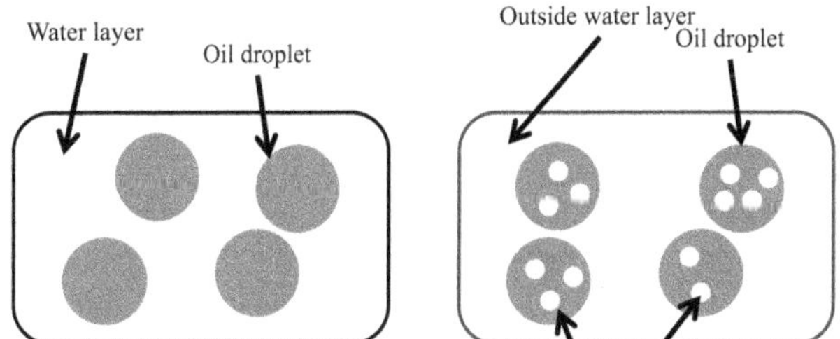

FIGURE 12.24 Schematic diagram of double emulsified mayonnaise.

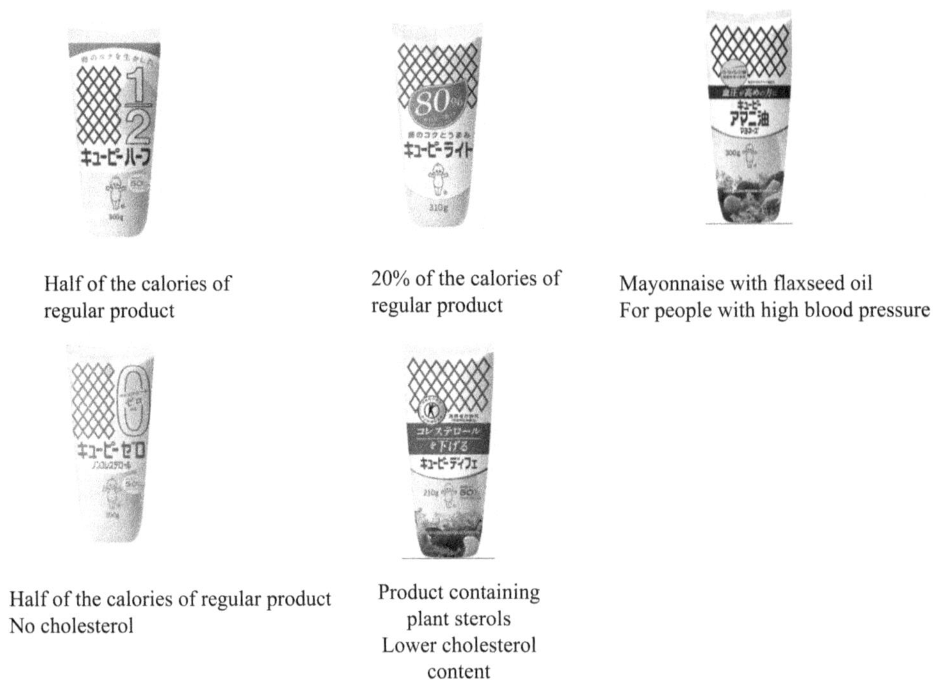

FIGURE 12.25 Health appeal mayonnaise.

The other types of functional mayonnaise are non-cholesterol, made with cholesterol-removed egg yolk, and mayonnaise containing plant sterols that inhibit cholesterol absorption in the intestine, all types available in the Japanese market (Figure 12.25).

The benefit of functional mayonnaise is not limited to health functions as they can also improve their suitability for cooking. When mayonnaise is heated, the egg yolk denatures and loses its emulsifying properties, causing oil and water to separate. Therefore, when cooking bread, mayonnaise is applied after the bread is baked. However, mayonnaise is made using egg yolks with improved heat resistance; as described below, mayonnaise-flavored cooked bread will not separate even when applied to the bread before baking. In addition, the combination of egg yolks with an emulsifier such as casein stabilizes the emulsifying power, resulting in mayonnaise that does not separate even when frozen. Since the mayonnaise does not separate even after freezing, it is possible to apply the mayonnaise to frozen bread and make it into a cooking bread produced by a combination of thawing and baking.

12.5.4 Emulsification Evaluation of Mayonnaise

There are typically two main methods to evaluate the emulsifying property of an ingredient, the emulsified volume method and the storage test method. Emulsification capacity is the method measuring the amount of oil entering the emulsion when adding oil to an aqueous phase containing an emulsifier and emulsifying it until it changes from an oil-in-water droplet form to a water-in-oil droplet form (so-called oil separation). The higher the amount of oil added, the higher the emulsification capacity. The storage test method evaluates the stability of emulsification, by assessing the state of an oil phase or an aqueous phase when the emulsion is left to stand for a certain period. The preservation test method evaluates the stability of emulsions by assessing the state of separation of the oil or water phase after the emulsion has been left for a certain period. Since mayonnaise must be an emulsion that does not separate during production and remains stable throughout its shelf life, the egg must have high emulsifying ability and stability.

Mayonnaise includes egg yolk, whole egg, egg yolk, and egg white mixed in a specific ratio. Generally, mayonnaise that uses only egg yolk is called egg yolk type, and mayonnaise that also contains egg white is called whole egg-type mayonnaise.

Mayonnaise emulsified with egg yolk or with whole egg were made and compared for emulsification capacity. Their evaluation showed that the emulsifying power of egg white is 1/4 that of the emulsifying power of egg yolk (Lowe, 1964), while that of whole egg is ranging between the emulsifying power of yolk and egg white. The effect of egg white on emulsion stability was also investigated. Industrially produced egg yolks may often contain minute traces of egg whites. For testing purposes, egg yolks with no egg white contamination were obtained. The mixture was then centrifuged to evaluate the amount of aqueous phase separation. The amount of separation increased as the mixing ratio of egg white increased (Varadarajulu & Cunningham, 1972). This study indicated that the excellent emulsification properties of chicken eggs are mainly due to the properties of the egg yolk.

12.5.5 Oxidation of Mayonnaise by Iron Ions in Egg Yolk

Oxidation of oil is one of the causes for deterioration of the mayonnaise flavor. We have tried to keep the mayonnaise from containing oxygen, which causes oxidation, but it is not easy to reduce it down to zero. Iron ions in the egg yolk bind to the phosphate group of phosbitin, an egg yolk protein in neutral pH, and form a soluble complex but is released in acidity. Since mayonnaise is acidic, iron ions are free and will promote oil oxidation by oxygen. Mayonnaise oil may be easily oxidized, and its flavor will deteriorate. It is essential to chelate iron ions to stabilize the taste of mayonnaise for an extended period. The US FDA permits the use of ethylenediaminetetraacetic acid for this purpose. Since it is not allowed in Japan, mayonnaise is manufactured by emulsifying fats and oils under a vacuum to remove dissolved oxygen in raw fats and oils and prevent air dissolution during stirring. After emulsification, it is filled in a multi-layer container making it difficult for oxygen to permeate. The headspace of the container is filled with nitrogen gas and covered by an aluminum seal (Kobayashi et al., 2005). Once opened, oxygen in the air may enter, and the mayonnaise oil may oxidize, although storage under refrigeration will keep oxidation to a minimum.

12.5.6 Heat Treatment of Egg Yolk for Emulsification

Egg yolks used for the production of mayonnaise are pasteurized for added safety. The pasteurization parameters vary from country to country, but they are usually performed at 60°C for 3–5 minutes. Mayonnaise was prepared from pure egg yolks obtained by treating egg yolks at 61, 63, and 65°C. for 4 minutes, respectively, and evaluated for the amount of aqueous phase separated by centrifugation. The results showed that the highest water separation was observed with the unheated treatment, followed by the 61°C treatment, while the separated aqueous phase is minimum when using egg yolks treated at 63 and 65°C (Varadarajulu & Cunningham, 1972). These results indicated that the heat treatment of egg yolks improves the emulsification stability of mayonnaise.

12.5.7 Processed Egg Yolk and Emulsifying Properties

As a source of egg yolks for the manufacture of mayonnaise, the FDA mentions the use of liquid egg yolk, frozen egg yolk, dried egg yolk, runny whole egg, frozen whole egg, and dried whole egg. If egg yolk is frozen as is, freeze denaturation will occur, the solubility decrease, and the emulsifying power decrease. Adding about 10% salt to yolks prevents freeze denaturation. In a study on the amount of salt added to egg yolk, freezing temperature, and viscosity of mayonnaise, 10% salted commercial egg yolk was stored at −18 to −23°C for 1–4 months prior to making mayonnaise. Almost no difference was observed (Palmer et al., 2013). As for the emulsion stability, the results show no difference in the emulsion stability even if the 10% salted commercial egg yolk is stored at −15°C for six months prior to making mayonnaise. Wakamatsu has studied in detail the changes caused by the freezing of egg yolk and the effects of salting and freezing temperature. By selecting the salt concentration and freezing temperature, the salted egg yolk does not freeze entirely and does not denature. Therefore, it was observed that mayonnaise using frozen salted yolks was equivalent to that produced with pasteurized egg yolks (Wakamatsu, 1974).

TABLE 12.6

Viscosity of Mayonnaise Prototype (Pa/s)

Ingredients	Immediately	1 Month After	4 Months After
Liquid yolk	200	160	140
Powdered yolk	235	180	155

Dried egg yolks are also used for the manufacture of mayonnaise. Dried eggs are made by spray drying or pan drying liquid eggs to reduce the water content to make them into powder or flakes (see Chapter 10).

Since liquid eggs quickly spoil at room temperature, they must be kept at chilled temperature or frozen during storage and distribution. In some parts of the world, cold chain distribution is not well developed, and it is considered advantageous to dry eggs for easier transportation and safe storage.

The authors made a prototype mayonnaise by replacing all the liquid egg yolk with dried egg yolk. A study confirmed that there was no oil separation even when dried eggs were used, and a good emulsified state was maintained for an extended period of time (Tables 12.6 and 12.7).

12.5.8 Enzyme-Treated Egg Yolk

When an enzyme acts on egg yolk, its function is altered. Phospholipase A2, converting phospholipids in the egg yolk into lysophospholipids and fatty acids, is known as an enzyme that modifies the emulsification-related functions of eggs. Egg yolks treated with phospholipase A2 acts will show increased emulsification and heat stability. Mayonnaise made with enzyme-treated egg yolks does not separate oil when heated and can be used as a topping for cooked bread (Figure 12.26).

The reason for the increased thermostability of phospholipase-treated egg yolks is not fully understood. When phospholipids are broken down into lysophospholipids and fatty acids, the hydrophilic-hydrophobic balance on the lipoprotein surface is altered. The increased adsorption of egg yolk at the oil droplet interface may improve the emulsion stability and thermostability of the emulsion.

12.5.9 Uses Other Than Seasonings

Mayonnaise is a condiment and can also be used as a food improver. Here are some examples. When making hamburgers at home, they sometimes become tough, and this phenomenon is caused by the meat protein being tightly bound by heating and the moisture evaporating. When mixing minced beef with mayonnaise, about 5% of the meat, the emulsified vegetable oil

TABLE 12.7

Average Particle Size D50% of Mayonnaise Prototype (μm)

Ingredients	Immediately	1 Month After	4 Months After
Liquid yolk	2.49	2.40	2.49
Powdered yolk	2.40	2.35	2.36

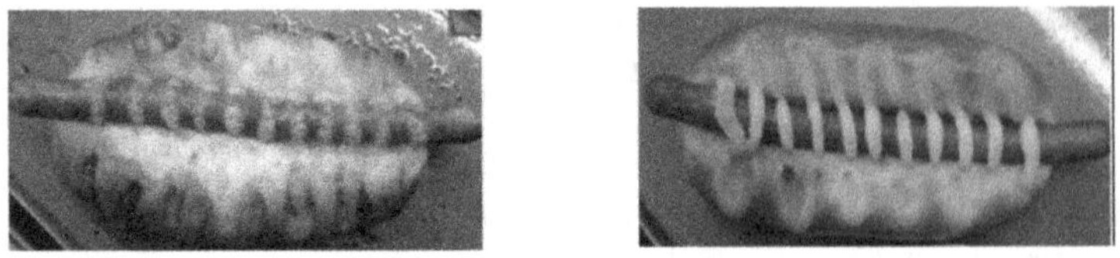

Mayonnaise with egg yolk
After baking, mayonnaise was separated and the oil removed

Mayonnaise with enzyme-treated egg yolk
Mayonnaise remains firmly after baking

FIGURE 12.26 Cooking example using heat-resistant mayonnaise.

prevents the meat protein from binding during heating, making the hamburger fluffy and juicy.

When added mayonnaise to pancakes, the texture of the pancake becomes softer. Mixing it with the pancake dough to keep the sponge-like texture makes something soft and crispy.

And when mayonnaise is mixed with fried rice, the lecithin in the egg yolk is adsorbed on the surface of the rice and prevents the rice from binding, making it delicious fried rice that is not sticky (Kewpie Corporation home pages, 2022).

Mayonnaise is now expected to be used as an essential seasoning and expanded into various menus.

12.6 Application of Eggs Using Color Tone and Nutritional Function

12.6.1 The Color Tones

The five senses (taste, hearing, smell, touch, and sight) are essential when tasting. Among the five senses listed above, information obtained from the eyes is the most critical for food tasting. Therefore, the inseparable relationship between food and color must be addressed. Eggs are used in a wide range of food categories such as cooking, sweets, and bakeries, but giving the yolk color to products is one of the essential purposes for using eggs. Egg-based dishes vary a lot. They include omelets, scrambled eggs, and sauces and seasonings such as mayonnaise, orange sauce, and carbonara, French toast, custard cream, and ice cream, and confectionery such as sponge cake. All these foods have beautiful egg color tones, one of the most critical factors related to the enjoyment of delicious meals using eggs. In most countries, yolk yellow color is recognized as an energizing color associated with "light" and "sun."

Yolk color varies a lot. Since pigments responsible for the yolk color cannot be synthesized by the laying hens, the yolk color is greatly affected by pigments present in the poultry feeds and transferred to the yolk (Ishikawa, 2008; see Chapter 15). Some chicken eggs contain many pigment components, while others may have almost no pigment. Some eggs may show unfavorable color due to the type of feed provided to the hens and egg producers may control the yolk color by modulation the composition of the ration provided to hens, with colors ranging from a light lemon to a deep reddish-orange color (Table 12.7). In Japan, many consumers prefer firm orange and yellow eggs to lighter ones. On the other hand, in some parts of Canada and Europe––where wheat is used as the main ingredient instead of corn––eggs with a light-colored yolk are sold and consumers tend not to like dark egg yolk (Table 12.8).

As mentioned previously, the yolk color intensity can easily be controlled, and it is possible to add value by producing eggs of a specific color, meaning local consumer needs.

We investigated the taste of boiled eggs, sponge cakes, puddings, and fried eggs using eggs adjusted to various yolk color. The preferred color tone was different depending on the product. Especially for boiled eggs, depending on the heating conditions, it is possible to make a hard-boiled egg in which the yolk is firmly solidified from the soft-boiled egg to the center. Soft-boiled eggs with soft yolks tended to have a solid reddish-orange color, and these were preferred, while hard-boiled eggs with completely set yolks with deep yellow color were preferred.

For the sponge cake, deep yellow colored egg yolks tend to be preferred to prepare the bright yellow sponge cake. On the other hand, if an egg with a deep yolk color is used for a cake using fruit such as a light-colored strawberry, the color of the fruit may not be expressed clearly. In this case, using eggs with a light yolk color is desirable. Furthermore, when a deep yellow food ingredient such as a pumpkin is used in preparing cakes, eggs with a deep orange or a deep yellow egg yolk are suitable to enhance the color of the cake.

Paprika is added to the feed to enhance the red color of the egg yolk. However, dyes derived from paprika may fade away when heated. When such eggs are used in boiled eggs, egg putties, sponge cakes, etc., the color may fade, and the color tone may differ from that prior to heating. It is important to consider this point, in order to safeguard the desired color of finished foods.

Egg yolks from chicken eggs can be used to enhance the visual aspect of food and stimulate appetite. Understanding

TABLE 12.8
Pigment Content in Yolk of Various Chicken Eggs (mg/100g)

	Egg A	Egg B	Egg C	Egg D
Lutein	1.0	6.3	1.1	0.1
Zeaxanthin	0.9	2.0	1.6	—
Capsanthin	0.3	0.3	1.9	<0.1
Cryptoxanthin	0.2	0.5	0.4	–
Total xanthiophyll	2.5	9.1	6.2	0.1

Note: Egg A: Ordinary chicken egg; Egg B: Egg with a yellow yolk; Egg C: Egg with red yolk; and Egg D: Egg yolk contains almost no pigment.

this, it is essential to use yolks having the color preferred in an area to give consumers the best impression about the food produced.

12.6.2 The Nutritional Function of Eggs

Eggs are nutritious and delicious; everyone loves eggs in Japan and the world (see Chapters 2 and 29). Egg menus are also vibrant and diverse. Egg prices remain less expensive and stable and is always a very familiar and essential food in Japan. Over the years, a false notion that "eating eggs raises blood cholesterol level" had been disseminated, and consumption declined as a result. However, since the 1990s, eggs have been steadily consumed in Japan, with an average of 2.5 million tons or more being consumed annually. Values of eggs can be immense: the WHO website (WHO 2022) says "adequate nutrition and hydration are essential" as nutritional advice during novel coronavirus outbreak, and eggs are mentioned along with meat, fish, and milk as excellent animal products (Table 12.9).

Proteins in eggs are excellent. Unlike other animal and plant proteins, eggs are used very efficiently in the human body (Lin & Huang, 1986) (Table 12.10).

Based on nutritional evidence, food products have been developed claiming the nutritional aspects of eggs, especially egg white proteins. In the United States, protein products focusing on the role of this high-quality egg protein form some specific markets. Similar effects have begun to be launched in Japan, and future trends are drawing attention (see Chapter 14).

It has been reported that lecithin contained in egg yolk has various functions. Lecithin, also called phosphatidylcholine, is a phospholipid (phospholipid containing phosphorus) containing about 13% choline and is the main component of cell membranes in the living body. Lecithin has an emulsifying effect that mixes water and oil, an antioxidant effect, and a water retention effect. Choline, which constitutes lecithin, functions as a neurotransmitter, "acetylcholine," that dilates blood vessels and lowers blood pressure (see Chapter 36). Various supplements are sold on the market for efficient intake of egg yolk lecithin. As a traditional Japanese food, supplements that combine egg yolk oil and garlic are also taken for fortification.

TABLE 12.9

Nutritional Advice during COVID-19 Pandemic Released by the World Health Organization

Nutritional Advice for Adults During the COVID-19 Pandemic	
Eat fresh and unprocessed foods every day	Fruits, vegetables, legumes, nuts and whole grains and foods from animal sources (e.g., meat, fish, eggs and milk).
Drink enough water every day	Drink 8–10 cups of water every day.
Eat moderate amounts of fat and oil	Consume unsaturated rather than saturated. Choose white meat (e.g., poultry) and fish
Eat less salt and sugar	Limit your daily salt intake to less than 5 g. Avoid foods (e.g., snacks) that are high in salt and sugar.
Avoid eating out	Eat at home to reduce your rate of contact with other people and lower your chance of being exposed to COVID-19. We recommend maintaining a distance of at least 1 m between yourself and anyone who is coughing or sneezing.

TABLE 12.10

Comparison of the Nutritional Value of Food Proteins

	Egg	Milk	Casein	Soybean	Beef
Amino acid score	100	100	100	100	92
Protein efficiency ratio	3.9	3.1	3.2	2.3	2.9
Biological value (%)	94	84	76	73	80
Net protein utilization	94	82	77	61	73

Source: Adapted from U.S. Dairy Export Council, and Animal Food Protein, Asakura Publish Co. Ltd. pp. 75.

While eggs are rich in nutrients, producing higher nutritious eggs has been attempted by raising nutritional components with health functions in the yolk, taking advantage of the fact that ingredients in feeds are transferred to the yolk (see Chapters 29 and 38).

Due to its porous structure, eggshell calcium (calcium carbonate derived from eggshell) is more easily dissolved in the stomach than calcium carbonate (calcium carbonate derived from limestone). Eggshell calcium is easily absorbed in the small intestine. Calcium-fortified nutritional foods containing eggshell calcium are also commercially available (see Chapter 11).

In the future, further research on eggs and functional ingredients will be conducted. New products are always expected, especially those incorporating available elements new to eggs, and innovation will further contribute to the promotion of good health and nutrition in Japan and around the world.

REFERENCES

Ajinomoto, H. P. (2022). https://www.ajinomoto.co.jp/pureselect/products/kokuuma/

Aoyama, S. (2008). Noodle in Egg Science and Function, 107–108. I & K Corporation (in Japanese).

Davidson, L. D. (2006). US Patent 5,843,505.

Glenn, J., & Larsen, E. (2013). The essential field guide to serious fun. Bloomsbury Publishing Plc. 158.

Hanada, T. (2010). Effect of salt on microwave heating of boiled eggs. Proceeding of Cookery Science of Japan Poster Session 2P-14 22(0), 171 (in Japanese).

Hanada, T. (2017). Japan Patent 6231426.

Ichimura, T. (2017). The world of eggs you do not know (4) Pain No.12, 12–15 Japan Institute of Baking (in Japanese).

Ichimura, T. (2018). The world of eggs you do not know (5) Pain No.10, 14–17, Japan Institute of Baking (in Japanese).

Ichimura, T., Arimitsu, K., & Tsukamoto, K. (2005). Japan Patent 3657078.

Imai, C. (1993). Knowledge of mayonnaise and dressing Saiwai shobo, 44–45 (in Japanese).

Ishikawa, S. (2008). Pigments and colors in egg science and function, 55–56, I & K Corporation (in Japanese).

Iso, N., Mizuno, H., Saito, T., Lin, C. Y., Fujita, T., & Nagahisa, E. (1985). The effects of additives (Egg white and soybean protein) on the rheological properties of kamaboko. *Bulletin of the Japanese Society of Scientific Fisheries*, *51*(3), 485–488.

Isobe, K., Irie, H., Oda, Y., Yoshida, M., Arimitsu, K., Goto, M., & Kawasaki, S. (2013). The relation between hen breed and the quality of shell eggs. IFT 2013, Poster Session 255-129.

Kewpie Corporation home pages (2022). (In Japanese). https://www.kewpie.co.jp/mayokitchen/urawaza/

Kobayashi, H., Takamiya, M., & Hasegawa, M. (2005). Prevention of mayonnaise oxidation by controlling dissolved oxygen. *OleoScience*, 473–479 (in Japanese). https://doi.org/10.5650/oleoscience.5.473

Kojima, K., Shimura, H., Sugie, A., Aoyama, S., & Yoshida, K. (2005). The effect of storage of chicken eggs at low temperatures on the taste characteristics of egg dishes. *Proceeding of Cookery Science of Japan Oral Session* P-16, 141 (in Japanese). https://doi.org/10.11402/ajscs.17.0.141.0

Lin, C. P., & Huang, P. C. (1986). Comparison of control diets containing various protein levels for determining net protein utilization by rats. *Journal Nutrition*, *116*(2), 216–222.

Lowe, B. (1964). *Experimental cookerys.* John Wiley & Son, 276.

Nakae, T. (1986). *Science of milk, meat and eggs characteristics and functions.* I & K Corporation, 98.

Noda, S. (2018). Factors affecting jelly strength of fish paste products. *Journal of Cookery Science Japan*, *51*(4), 23–240. https://doi.org/10.11402/cookeryscience.51.236

Ohno, T., & Shinagawa, S. (1998). Japan Patent 2851997.

Okada, K. (2010). Advanced and high-quality meat processing technology 2 Saiwai Shobo Co., Ltd. (in Japanese).

Okada, M. (2000). Science of "Kamaboko" 66 Seizando-Shoten Publishing co., Ltd. (in Japanese).

Palmer, H. H., Ijichi, K., Cimino, S. L., & Roff, H. (2013). Salted egg yolk 1. Viscosity and performance of pasteurized and frozen samples. *Food Technology*, 23, 148–153.

Suzuki, J. (2008). Effect of adding dried egg white on the quality of heated gel prepared from atka mackerel and walleye pollock and mixed meat paste. *Journal of the School of Marine Science and Technology*, *6*(2), 27–35. Tokai University. https://doi.org/10.1111/j.1365-2621.2004.00825.x

Takeuchi, H. (2020). Japan Patent 6682180.

U.S. Food and Drug. (2020). https://www.accessdata.fda.gov/scripts/cdrh/cfdocs/cfcfr/cfrsearch.cfm?fr=169.140

USDA (1969). Egg pasteurization manual, 14.

Vandepopuliere, J. M., & Cotterill, O. J. (1999). US Patent 6,004, 603.

Varadarajulu, P., & Cunningham, F. E. (1972). A study of selected characteristics of Hen's egg yolk. 1. Influence of albumen and selected additives. *Poultry Science*, 51, 542–546.

Wakamatsu, T. (1974). Factors and mechanisms affecting gelation of egg protein 2 gelation of egg yolk by cryopreservation. *New Food Industry*, *27*, 61–70.

Wakamatsu, T. (2009). Japan Patent 4780336.

Watanabe, K. (2008). *Noodle in egg science and function.* I & K Corporation, 107–108.

WHO (2022). Nutrition advice for adults during the COVID-19 outbreak. http://www.emro.who.int/nutrition/nutrition-infocus/nutrition-advice-for-adults-during-the-covid-19-outbreak.html

Xie, L., Nishijima, N., Oda, Y., Hand, A., Majumder, B., Xu, B., & Zhang, Y. (2020). Utilization of egg white solids to improve the texture and cooking quality of cooked and frozen pasta. LWT-Foods Science and Technology 122109031.

Yamashita, T. (1995). Effect of chicken egg component on the physical characteristics of Alaska pollack kamaboko gel. *Nippon Suisan Gakkaishi*, *61*(4), 580–587 (in Japanese).

13

Traditional Egg and Egg Products

Hajime Hatta*, **Yonggang Tu, Yan Zhao, Hui Xue, David Cassin, and Yoshinori Mine**

CONTENTS

* Corresponding author.

DOI: 10.1201/9781003254430-17

13.1 History of Chicken Egg Use

Eggs have a long history as a food source and are depicted in murals in caves inhabited by Cro-Magnon people (Toops, 2014). About 10,000 years ago, when our ancestors began to settle down, animals and birds that had previously been hunted were domesticated to support a sedentary lifestyle. Birds must have been the most helpful animal food because they provided both meat and eggs (West & Zhow, 1988).

About 9,000 years ago, wild red chickens were domesticated in Southeast Asia, and about 8,000 years ago, chickens were enlarged in China. It is then believed to have spread to West Asia via the Silk Road about 5,000 years ago, and the use of chicken eggs began in West Asia, India, China, and Egypt about 3,500–4,000 years ago. Chicken bones excavated from Mohenjodaro (near present-day Pakistan), the largest urban site of the Indus civilization, indicate that chickens were raised and used in Western India about 4,000–5,000 years ago (Toops, 2014).

13.1.1 Use in Western Europe

Around 1,000 BC, chickens were introduced to East Asia and Persia, and the Phoenician civilization flourished on the eastern coast of the Mediterranean. At that time, the Phoenicians enjoyed eating chicken and ostrich eggs. The Etruscans in ancient Italy also raised chickens, ducks, geese, and pigeons to obtain eggs. The earliest written mention of eggs as food is in cuneiform on ancient Assyrian clay tablets in Mesopotamia, where it is stated that in 879 BC, the king of Assyria held a great feast to celebrate the completion of his new city and served 10,000 eggs (Toops, 2014).

Next, in the Roman Empire, which lasted about 300 years, beginning in 27 BC, egg producers emerged, and the method of hatching eggs was passed down from parent to child as a secret tradition. At the height of the Roman Empire, the world's oldest cookbook, "De re Coquinaria (On the Subject of Cooking)" was written, in which egg dishes such as omelets, a pastry called "ribum," and pancakes were introduced. It is also reported that complete courses in that period began with eggs and ended with apples.

In the Middle Ages, around the 5th century, the Western Roman Empire was destroyed by the great migration of Germanic tribes, and the poultry industry, which produced and supplied eggs for food, declined. In the Middle Ages, which lasted for the next 1,000 years, poultry farming was carried out in the manors and monastery gardens of the nobility, and only various Roman egg dishes, such as omelets, were handed down. The nobility of this period was more interested in meat dishes that went well with wine than egg dishes, and of the 170 recipes in the 14th-century French cookbook, only four used eggs. The nobility drank wine with meat dishes, while monks drank "milk wine" (the prototype of eggnog) made with eggs and figs. Even amidst all this, in the 11th century, white bread puffed up with egg whites was created, leading to the birth of sponge cake (Toops, 2014).

The era was the Renaissance of the 14th century, and the 16th and 17th centuries saw the introduction of new ideas in egg dishes, especially the increased consumption and diversification of sugar in cooking. The prototype of custard was also born during this period: in 1544, the world's first sugar factory was established in London; in 1653, the meringue was developed in France by adding sugar to egg liquid and whisking it; the French Revolution of the 18th century and the Industrial Revolution that began in England further expanded the possibilities of using chicken eggs and led to the development of baked meringue, ice cream made by whisking eggs while freezing them, mayonnaise, and various sauces that took advantage of eggs' emulsifying properties were developed in France. Incidentally, there are various theories on the origin of the word "mayonnaise," but the most popular theory is that it is a sauce made in Maon on the Mediterranean island of Menorca (Maon's sauce).

13.1.2 Use in Japan

The oldest chicken bones in Japan were excavated from the historical ruins in Iki island, located between Kyu-shu island and the Korean peninsula, 2,100–2,200 years ago. Chickens were introduced from China via the Korean Peninsula during the Yayoi Era, about 2,000 years ago. At first, chickens were used in Japan not so much for food. They were kept as birds that tell the time with their chirps, for Shinto rituals, as a pet, and cock-fighting.

From around the end of the 6th century during the Asu-ka Period to the Nara Period (early to late 8th century), the prohibition against killing spread with the arrival of Buddhism (538), and Emperor Ten-mu issued a decree (675) stating "Do not eat the flesh of cattle, horses, dogs, monkeys, and chickens, and if you commit such an offense, you shall be punished. In 730, Emperor Shou-mu issued a decree forbidding the killing of animals, and the custom of eating the flesh of four-legged animals was discontinued. However, fish, chickens, and eggs were still eaten, boars were called "mountain whales," and rabbits (Usagi: in Japanese) were interpreted as a combination of two types of wild birds (cormorant and egret) in Japanese sounds.

In the early Heian period history book (797), there is a statement from a governor in the Kyushu area who encouraged farmers to raise boars and chickens. There was a record raising 30–50 chickens during the Heian period (from the late 8th to the end of the 12th century) and a description of a store selling chicken eggs. However, as Buddhism spread deeply into the ordinary people, people stopped eating chickens and eggs, as depicted in the "Picture of Hell," written around the 12th century, which describes a chicken hell. Japanese books did not reference cooking methods using eggs until the 16th century.

Later, in the Azuchi-Momoyama period (late 16th century), the value of eggs was reevaluated under the influence of Christian missionaries and Chinese traders. Eggs began to be used in Nan-ban (Western European influenced South-East Asian style) cuisines such as sponge cake, egg noodles (somen), and tempura dishes significantly promoted egg use in food. In "Ryouri-Monogatari (Cookery Tale)" (1643), the first book to introduce egg dishes in Japan, there are 5–6 recipes for egg dishes such as soft and fluffy egg and egg noodles (egg somen), egg sake (egg spirits), etc. In 1785, the first book on egg dishes was published. In " Manpou Ryori Himitsu Bako (A Treasure of Secret Egg Recipes)," published in 1785, 103 egg dishes were introduced. Most of the egg dishes eaten in Japan today are shown in these 103 egg dishes.

From the late Edo period (1603–1868), the use of eggs became increasingly popular among the public. In the "Higashi-honganji Gozensho" (Buddhist Temple Food Diary), which recorded the daily meals of the monks of the massive temple in Kyoto and their families from 1763 to 1869, egg dishes appeared on the menu every two to three days. Some of the egg dishes on the menu are: Tamago-Toji (softly hardened egg with dashi), Soft and Fluffy Egg, Boiled Egg, Browned Baked Egg, Scrambled Egg, Baked Egg with Wheat Gluten, Rolled Egg, and Fish Paste with Egg, Egg Miso (fermented soy paste with egg), and many others. In the Meiji period (1868–1912), commercial hen-raising began, and in "200 Kinds of Egg and Chicken Dishes and Home Cooking" (1904), the way of eating Sukiyaki with raw egg was introduced.

From the Taisho Era to the Showa Era, egg production and use spread further, especially after World War II, along with the period of rapid economic growth. As Japan's economy grew, so did egg consumption, significantly since around 1965, which has increased rapidly.

13.2 Japanese Eggs and Egg Products

13.2.1 Japanese Traditional Egg Dishes

Eggs have long been used as food and food ingredients. Humans have developed numerous egg dishes and egg-processed foods to enrich their diet. In particular, the Japanese are famous worldwide for their love of eggs and the many egg dishes they prepare. Currently, the number of eggs consumed per capita is 340 per year, ranking second in the world after Mexico's 380 eggs per capita (International Egg Commission Report, 2021). As of August 2022, "Cookpad," the world's largest recipe search site, introduced approximately 845,000 egg dishes, accounting for 23% of the total 3,720,000 recipes on the site (https://cookpad.com). The sheer number of recipes is a testament to the ingenuity of the Japanese people and their love of eggs.

13.2.2 "Manpoho Ryori Himitsu Bako (A Treasure of Secret Egg Recipes)"

The diversity of egg dishes in Japan has long been well known. Japan's oldest egg cookbook, "Manpoho Ryori Himitsu Bako," published in 1785 during the late Edo period (1603–1868), introduced a whopping 103 different egg dishes. The book introduces not only various Japanese dishes using boiled eggs but other egg dishes that can be found today, such as Sponge Cake, Egg Noodles, Fried Chicken with Sweet and Sour Sauce, Savory Egg Custard, and other table cuisines. The book's most unique and surprising egg dish is a "Reversed-boiled egg." It is a boiled egg in which the yolk and white are swapped. The recipe for the Reversed-boiled egg is introduced in this book, but for a long time, it was famous as a surprisingly boiled egg that has never been reproduced.

13.2.3 Preparing a Reversed-Boiled Egg (the Yolk Inside-Out)

A method for preparing a reversed boiled egg introduced in an old Japanese recipe book has been a mystery for a long time until we recently succeeded in reproducing it 220 years later. The recipe book was published in 1785, in which 103 recipes for cooking eggs were introduced. Although 102 recipes have been recreated, only the reversed boiled egg failed to be reproduced, despite numerous challenges having been carried out.

The recipe book reads, "Prepare a fresh egg and pierce the eggshell by about 3 cm depth with a needle. Leave it in a fermented, salted rice-bran paste for three days, and then boil the egg to ensure the yolk inside out." The key to solving the mystery lay in the idea that the eggs used for cooking in such an old period could be mostly fertilized eggs. We finally reproduced the reversed boiled egg using fertilized eggs incubated for three days at 38°C. For 3–4 days of incubation of fertilized eggs, water in the egg white is transferred to the egg yolk. As a result, the weight of the egg yolk doubled, and the viscosity of the egg yolk significantly decreased by one hundredth, becoming watery.

On the other hand, the egg white weight halved and got thick. During this process, the vitelline membrane became very weak and easily broken. Upon rupturing the vitellin membrane, the runny yolk diffused inside the shelled egg, and the runny yolk gelled around the thick albumen during heating. Thus, the reversed boiled egg (the Yolk Inside Out) was reproduced (Figure 13.1) (Hatta et al., 2021).

13.2.4 Hot Spring Boiled Egg (Onsen Tamago)

Hard-boiled and half-boiled eggs are well-known worldwide, while Onsen Tamago is another popular processed-boiled egg product in Japan. It is a kind of boiled egg in which the yolk is solidified, and the egg white is in a sol state and is made by taking advantage of the difference in heating coagulation temperature between the yolk and egg white. The protein in the egg white begins to denature at 60°C, becomes a cloudy sol, and hardens at 80°C or higher. On the other hand, egg yolks

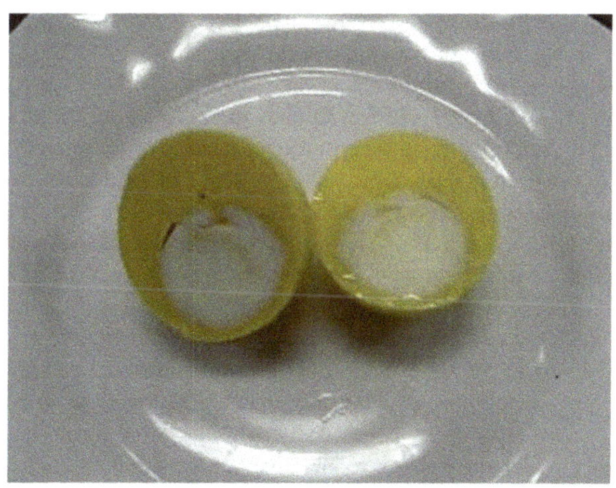

FIGURE 13.1 Reversed-boiled egg (the yolk inside-out) was recreated for the first time in 220 years.

FIGURE 13.2 F Hot spring boiled egg (*Onsen Tamago*).

begin to coagulate slowly at 65°C and harden at 75°C or higher. When eggs are soaked in hot water at approximately 70°C for 15–20 minutes, the yolks harden, and the egg whites are sol-like, producing Hot Spring Boiled Egg (Onsen Tamago).

Japan has many volcanoes and hot springs. Onsen Tamago began in various hot springs, where geothermally heated boiled eggs were served to hotel guests. In this process, where the temperature of the hot spring source is just around 70°C, the eggs are soaked to produce the Hot Spring Boiled Egg, in which the yolk hardens, and the egg white is remained soft sol state. The most common and tasty way to eat a Hot Spring Boiled Egg is to crack open the shell, serve the contents in a bowl, and pour Dashi soup over it (Figure 13.2).

In recent years, factories have been built to produce Hot Spring Boiled eggs in significant quantities by controlling the heating temperature. Hot Spring Boiled Egg with dashi soup has become a smash hit item in supermarkets and convenience stores throughout Japan.

13.2.5 Reversed Hot Spring Boiled Egg (Hyoutei Tamago)

A famous boiled egg in Kyoto can be described as the opposite of Onsen Tamago (hot spring boiled egg). It is a boiled egg in which the white hardens, and the yolk flows without hardening at all (Figure 13.3). It is famous in Kyoto as the specialty egg dish of "Hyotei," a luxury ryotei (Japanese-style restaurant) that has been in business for 400 years. The method of making this dish seems to be a secret, unexplored outside the gate as a dish handed down from one generation to the next. However, its preparation method focuses on the difference in heating gelation temperature between the egg white and egg yolk. If boiled from boiling water, it seems possible to reproduce it simply by adjusting the time. Then, through trial and error, we found heating conditions under which it could be made at home. First, fill a pot with about 2 liters of water, bring it to a boil, quickly add four eggs (at room temperature), and boil them for 5.5 minutes while rotating them in the boiling water. As shown in the photo, the eggs are removed and immediately placed in ice water, shelled, and cut with a knife into pieces to complete

FIGURE 13.3 Reversed hot spring boiled egg (*Hyoutei Tamago*).

the reversed Hot Spring Boiled Egg. Although it is not a real "Hyotei Tamago," I hope you will enjoy this weird taste, not found in conventional boiled eggs, as a home-cooked dish.

13.2.6 Dashi Maki Tamago (Japanese Style Omelet)

Dashi Maki Tamago is a kind of Japanese style omelet, and beaten eggs mixed with dashi soup are fried thin and rolled up thick. It has said to have originated at the "Ganso Kyo-Dashimaki Okamoto (specialty shop)" near Toji Temple in Kyoto (Fujishin et al., 2019). According to the fourth-generation owner of the store, it was initially started by her great-grandfather, an egg wholesaler, for effective use of eggs that were broken or cracked during transport and were no longer for sale.

By mixing broth of Japanese kelp, which has been commonly used in the Kansai region (around Osaka) since that time, with the egg liquid and baking it, "Dashi Maki Tamago" was born, which is different from the thickly Baked Egg in the Kanto region (around Tokyo) (Figure 13.4).

Dashi Maki Tamago contains more dashi broth than thickly Baked eggs. It is baked using only dashi broth and salt (without sugar) in an oblong omelet so that the egg is cooked thinly and rolled up quickly, leaving the egg's color intact. Depending on the rolling direction, there are two types of rolls: Kyoto-maki (roll), which is rolled from the front using a chopstick, and Osaka-maki (roll), which is rolled from the top to the front. After baking, the roll shape of the egg is formed using a bamboo mat.

FIGURE 13.4 Dashi Maki Tamago (Japanese style omelet).

Today, specialty stores in Kyoto specialize in Dashi Maki Tamago, and it is also served in catered lunches in Kyoto and at home as a side dish for lunch or rice. Recently, it is also sold at convenience stores and has become a famous dish throughout Japan (Fujishin et al., 2022).

13.2.7 Egg on Steamed Rice (Tamago Kake Gohan)

"Tamago Kake Gohan" is a unique Japanese egg dish consisting of freshly cooked rice topped with a raw egg and soy sauce (Figure 13.5). The history of this dish is surprisingly old. A cookbook entitled "Amateur Chef's Knife," written in 1805, around the middle of the Edo period, describes a dish like "Tamago Kake Gohan." The dish is called "Tamago-meshi (Egg Rice)" and consists of steamed rice topped with a beaten egg, covered with a lid, and steamed. The recipe notes that the egg is beaten, and a little soy sauce is added. Mix the cooked rice with the beaten egg. Serve in bowls and sprinkle with a small amount of chopped Japanese leek. The recipe uses one egg per two cups of rice (360g).

So, when was the first time that egg and cooked rice was eaten with a raw egg poured directly over the steamed rice?

FIGURE 13.5 Dropped egg on steamed rice (*Tamago Kake Gohan*).

Eggs eaten raw must be fresh. From this perspective, the closest place to obtain freshly laid eggs would be from farmers who raise chickens in their backyards. As for rice, it seems that in the mid-Edo period (18th century), rice production became so popular that even farmers could eat white rice. According to one theory, the origin of the "Tamago Kake Gohan" came from the farmers' breakfast of freshly cooked rice topped with a raw egg and seasoned with soy sauce to be called the "first food" for Japanese farmers (Kawae et al., 2015).

The first recorded instance of egg on rice was in the late 19th century, in the modern era, when Mr. Kishida Ginko (1833–1905), Japan's military reporter, was the first to eat egg on rice and recommend it to others. Later, during the food shortages after World War II, chicken eggs became a scarce commodity. Still, since 1955, chicken eggs have been attracting attention for their taste and nutritional qualities, and eggs and rice are now on the tables of the public.

Furthermore, in recent years, the eating habitat of egg on rice has become booming with the development of a special soy sauce for it and being called the "Tamagokake Gohan "as TKG in public. Its ease of preparation and the many variations of TKG with various toppings have led to the creation of specialty restaurants, the holding of TKG symposiums and research associations, and the publication of many TKG recipe books.

13.3 Chinese Eggs and Egg Products

The poultry egg is one of the most nutritious foods in nature, providing the body with rich protein, fat, minerals, and vitamins (Lunow et al., 2013). In addition, the protein of poultry eggs has a high biological valence, and their amino acids are appropriate, making them extremely easy to digest and absorb by the human body. Asia is the region with the highest production of poultry eggs, but most countries still consume mainly fresh eggs (Xue et al., 2022). At the same time, in the long-term life practice, Asian residents also developed some traditional and practical methods of processing egg products, which are simple to make and have a unique flavor and relatively long shelf life, mainly including pidan (century egg) or preserved egg (egg preserved in alkali solution), salted eggs (egg preserved in a salt solution), and zaodan (egg preserved in rice wine) (Xue et al., 2020). Some products even have a history of hundreds or even thousands of years. Each traditional egg product has an entirely different processing and formation mechanism, and each has its unique flavor.

Preserved egg white is a brown or greenish-brown gel, and preserved egg yolk is a dark green or green grass gel or cream-like substance. The preserved egg has a unique taste of delicious, cool, and refreshing flavor, and it can be eaten directly or made into preserved egg salad or congee with preserved egg and lean meat (Zhao et al., 2014). Cooked salted egg white is white colloid with a delicate texture and savory flavor, while salted egg yolk is a golden-yellow solid with oily surroundings. It is delicious and fragrant, with the freshness of crab yolk, and is usually eaten with porridge or made into the filling of dumplings (egg yolk dumplings) (Xu et al., 2018). Zaodan white is a soft, creamy white jelly with a semi-solid orange yolk. Zaodan is delicious, tender, rich in wine aroma, and generally eaten directly (Yu et al., 2020).

13.3.1 Preserved Egg

Preserved egg, also known as pidan and century egg, is a traditional egg product unique to Asia, with a delicious, flavorful, and refreshing taste. A preserved egg is rich in protein, essential amino acids, fatty acids, vitamins, and minerals such as calcium, magnesium, iron, zinc, copper, potassium, and sodium (Zhao et al., 2014). According to the statistics recorded in ancient Chinese books, there was a mature method of processing preserved eggs in China more than 700 years ago. Currently, China is the largest producer and consumer of preserved eggs in the world (Wang et al., 2021). In the early days, fresh poultry eggs were wrapped in a mixture of clay and plant ash and pickled at room temperature for dozens of days (Luo et al., 2020). During the pickling process of preserved egg, egg proteins are gradually degraded to produce small molecules of proteins, peptides, and amino acids (Meng et al., 2020), which not only facilitates the digestion of preserved eggs but also makes them have various functional activities, such as anti-inflammatory, hypolipidemic, and anti-tumor (Li et al., 2021). This makes preserved eggs a healthy, nutritious, and delicious egg product.

13.3.1.1 Processing Method of Preserved Egg

The production of preserved eggs has a long history, and there are various processing methods, but the main ones that have been used to date are mainly the mud coating method and the soaking method. The mud-coating method is the most traditional method, in which water, lime powder, soda ash, plant ash, and salt are first thoroughly mixed in a particular proportion, then the mixture is evenly wrapped around the surface of fresh duck eggs, and finally, the wrapped eggs are covered with rice hulls and pickled to form mature preserved eggs (Luo et al., 2020). In addition, the soaking method can be subdivided into the solution and turbid lipid methods. For the solution method, the fresh duck eggs are directly immersed in a pickling solution (usually containing sodium hydroxide, sodium chloride, and metal salt) and pickled directly with sodium hydroxide. At the same time, the turbid lipid method uses lime, water, and soda ash to produce sodium hydroxide. During this period, a large amount of calcium carbonate precipitation is also produced, which is not conducive to processing preserved eggs. It can be seen that the operation of the solution method is more straightforward than the turbid lipid method, and the quality of preserved eggs made by the solution method is good, so at present, the industrial production of preserved eggs is mainly made by the solution method (Li et al., 2021).

The main processing formula of the mud coating method is 1,000 duck eggs, 12–15 kg of plant ash, 15–20 kg of water, 1–1.5 kg of soda ash, 4–6 kg of quicklime, 0.5–1.2 kg of black tea and 1.2–1.5 kg of salt. The specific processing method of preserved eggs is as follows. The black tea is boiled with water and cooled overnight, then quicklime is added and stirred continuously, during which plant ash, soda ash, and salt are added, and these are mixed thoroughly to form the pickling mud. The mud is fully wrapped around the surface of the duck eggshells, 25–35 g of pickling mud per egg, with a thickness of about 0.2 cm. After wrapping, the eggs are rolled inside the rice husk so that the mud sticks to the husk and prevents the eggs from sticking together. Finally, the eggs are placed in a jar to pickle. Generally, the pickling time is 50–60 d in summer, 60–70 d in spring and autumn, and 70–80 d in winter.

The main processing formula of the solution method is 1,000 duck eggs, 3–4 kg of sodium hydroxide, 75–85 kg of water, 0.25–0.35 kg of copper sulfate (or other metal salts), and 2–3 kg of salt. The specific processing method for preserved eggs is as follows. The water is boiled and then cooled to room temperature. The copper sulfate is added and stirred thoroughly, then the sodium hydroxide and salt are added continuously and constantly stirred to mix the solution to form the pickling liquid. The duck eggs are then added to the pickling solution. The preserved egg can generally be matured after 30–45 d of pickling by solution.

The main processing formula of the turbid lipid method is as follows: 1,000 duck eggs, 45–55 kg of water, 15–18 kg of lime, 32–38 kg of soda ash, 0.1–0.2 kg of copper sulfate (or other metal salts), 1.5–2 kg of black tea, 1.2–1.5 kg of salt and 0.5–1.2 kg of plant ash. The specific processing method for preserved eggs is as follows. The black tea is boiled in water, then cooled overnight and the residue removed. Soda ash, copper sulfate, and salt are added and stirred continuously to form the primary pickling mixture. Then, lime is added to the primary pickling mixture, and the hot water is stirred and added until the mixture is entirely uniform, thus forming the ultimate pickling mixture. Finally, fresh eggs are added to the ultimate pickling mixture when the ultimate pickling mixture is cooled to room temperature. In general, the maturation time for duck eggs cured by the turbid lipid method is 25–30 d in summer, 30–35 d in spring and autumn, and 40–45 d in winter.

In recent years, different preserved egg processes have been developed to improve the production efficiency of preserved eggs, including gradient temperature control technology and vacuum pickling technology (Lv et al., 2005; Guo et al., 2013). For example, vacuum pickling technology can accelerate the formation of preserved egg white gel (Zhang et al., 2020). Sun et al. (2020) found that vacuum pickling technology can significantly shorten the pickling time compared with traditional processes. There were differences in the pickling time of different varieties of duck eggs, which might be related to the thickness and structure of the eggshell. However, the above processes are still at the research stage, and their industrial application needs further practical exploration.

13.3.1.2 Processing Principle of Preserved Egg

Egg products are processed using the gel-forming properties of proteins. It is a process whereby the protein solution of egg whites loses its fluidity under certain conditions, and the protein molecules aggregate to form a solid or colloid, during which the attractive and repulsive forces between the proteins reach a state of equilibrium, resulting in a highly ordered three-dimensional network structure that retains a large amount of water Li et al. (2018). For preserved eggs, alkali and heavy metal ions play a significant role in the pickling process.

FIGURE 13.6 Schematic diagram of preserved egg formation during processing.

A moderate concentration of alkali and heavy metal can synergistically induce the formation of preserved egg white gel, but a high concentration of alkali and heavy metal may liquefy the preserved egg gel. Therefore, appropriate processing technology is essential to produce preserved eggs.

Egg white and yolk can coagulate during the pickling process of preserved egg. According to the state of egg white, the formation of the preserved egg can be divided into four stages: hydration stage, solidification stage, color change stage, and maturation stage. During the hydration stage, the penetration of strong alkali into the egg causes the destruction of the tertiary and quaternary structures of proteins in egg white, the arrangement of protein molecules becomes looser from tight, and the bound water becomes free water. During the solidification stage, the protein secondary structure is disrupted, hydrogen bonds are broken, and hydrophilic groups increase, which leads to an increase in the hydrophilic of protein molecules and the interconnection of protein molecules to form a gel. During the color change stage, the primary structure of the proteins is disrupted, and reactions such as the Maillard reaction give preserved egg white its distinctive color. At the maturation stage, various intermediate products result from the Maillard reaction, eventually forming brown egg white gel (Ganesan & Benjakul, 2014; Ai et al., 2020). The formation of preserved egg yolk gel differs from that of egg white gel. Under the synergistic action of alkali and metal ions, the water and lipids in egg yolk are gradually bound, and a tight gel network structure is gradually formed between protein through ionic bonds and disulfide bonds (Zhao et al., 2016a). The preserved egg is formed by transforming egg white and yolk from sol to gel under the synergistic action of alkali and metal ions. In this process, preserved eggs' unique color, flavor, and taste are produced (Li et al., 2009). Figure 13.6 shows the formation process of preserved eggs prepared by the solution method.

13.3.1.3 Quality Formation and Control of Preserved Egg

- Quality formation of preserved egg

  The processing technology of preserved eggs is exceptional. Under the action of alkali, the preserved egg can form a typical non-thermally induced and highly elastic gel. During the pickling process, egg white and yolk gradually gel. Moreover, under the action of alkali and metal ions, a series of reactions take place in the egg, which eventually creates the unique color and flavor of preserved egg.

- Formation of preserved egg gel

  The protein content of poultry egg white is about 10%, and in the dry matter of poultry egg white, protein accounts for more than 90%. The egg white is rich in ovalbumin, ovomucin, ovotransferrin, and lysozyme, which provide the material conditions for forming preserved egg white gel (Mine, 2002). The formation of preserved egg white was mainly the result of the formation of protein gel under the synergistic induction of alkali and metal ions, as shown in Figure 13.7 (Zhao et al., 2016b). During the pickling process, egg white proteins gradually denature and unfold, and the hydrophobic and sulfhydryl groups buried inside the protein molecules are exposed (Zhao et al., 2016a). Subsequently, these unfolded proteins gradually gather to form macromolecular

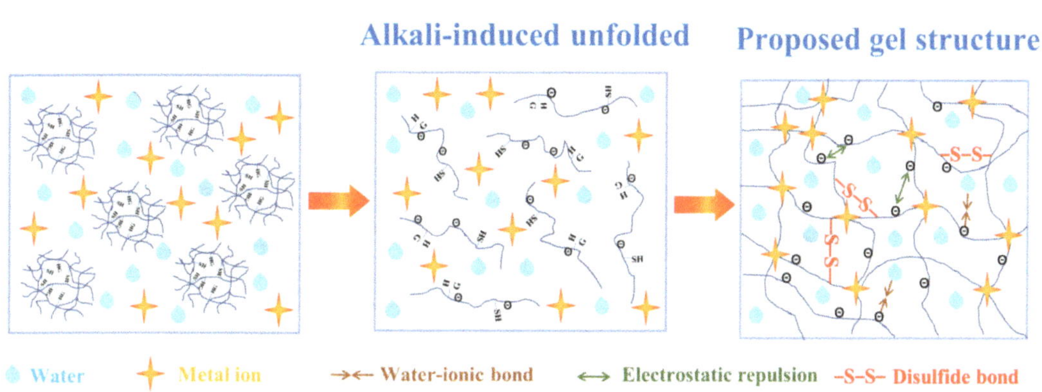

FIGURE 13.7 Formation mechanism of preserved egg white gel. (Adapted from Zhao et al., 2016b.)

aggregates through ionic bonds, disulfide bonds, and hydrophobic interactions. In this process, the metal ions also promote cross-linking between protein molecules. Alkali not only makes the surface of protein with many a net negative charge, which inhibits the aggregation between proteins, and promotes the formation of preserved egg transparent gel (Deng et al., 2020), but also ionizes some groups on the protein surface, accelerates the binding efficiency between water and protein, and improves the transparency of preserved egg white gels, and finally formed the unique appearance of preserved egg white.

Egg yolk contains 22.8%–32.6% lipid and 15.6%–16.8% protein, that is, about 65% lipid and 35% protein in the dry matter of the yolk, which is called a natural multi-scale protein-lipid supramolecular assembly, whose main components are phosvitin, low-density lipoprotein (LDL), high-density lipoprotein (HDL) and phospholipid (Anton et al., 2013). The formation of preserved egg yolk is mainly the result of the interaction between protein and lipids in the presence of alkali, as shown in Figure 13.8 (Yang et al., 2020a). LDL in egg yolk is a sphere-like molecule. The structure of LDL is disrupted, and lipids and proteins are released under the induction of strong alkali. Subsequently, proteins and lipids combine through hydrophobic interaction and hydrogen bonds, forming non-spherical LDL aggregates, but these aggregates cannot form a stable gel (due to the presence of strong electrostatic repulsion between proteins) (Yang et al., 2020a, 2020b). HDL is gradually unfolded under the action of alkali and then forms a dense and uniform network structure under the forces of hydrophobic interactions, disulfide bonds, and ionic bonds (Yang et al., 2020a, 2020b). Ultimately, the proteins and lipids migrate under the synergistic action of alkali and metal ions, culminating in forming a unique egg yolk gel.

- Formation of preserved egg color

Mature preserved egg white is a greenish-brown or dark brown gel, and preserved egg yolk is a dark green or green grass gel or cream-like substance. The color formation of a preserved egg is a very complicated biochemical process resulting from the interaction of multiple substances. There are different color formation mechanisms between preserved egg white and yolk. The color formation of preserved egg white is mainly caused by the Maillard reaction (Tan et al., 2021). In short, the proteins in egg white are decomposed into amino acids by the action of alkali, which reacts with reducing sugars (such as glucose) in the egg white to produce 5-methylfurfural. This compound combines easily with proteins to form melanin. Melanin is the primary substance responsible for the color development of preserved egg whites (Zhong et al., 2018; Tan et al., 2021). Moreover, when preserved eggs are pickled at a higher temperature, the rate of the Maillard reaction is higher, which accelerates the consumption of amino acids and the accumulation of melanin, thus promoting the deepening of the color of preserved egg white.

The color formation of preserved egg yolk is mainly due to the decomposition of proteins under the action of alkali, which produces cystine and cysteine, rich in sulfhydryl group (-SH) and disulfide group (-S-S). Cystine and cysteine easily combine with the metal ions in the egg to produce different chromogenic substances. The mixture of these chromogenic substances turns into green complexes under the action of hydrogen sulfide (Liang & Shi, 1992). In addition, cystine is further decomposed to produce sulfur ions under the action of alkali, which react with metal ions to form dark green complexes (Li et al., 1992). These complexes and the yellow color of the yolk form the distinctive color of preserved egg yolk.

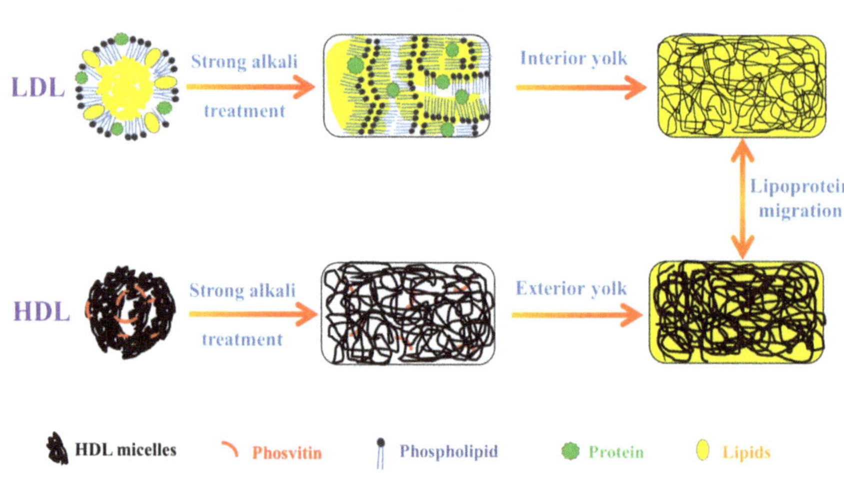

FIGURE 13.8 Formation mechanism of preserved egg yolk gel. (Based on Yang et al., 2020a.)

- Formation of preserved egg flavor

The flavor characteristics of preserved eggs are umami, salty, cool, and refreshing, which makes them widely spread. The flavor formation of a preserved egg is complicated. Firstly, the degradation of precursor substances in preserved eggs, mainly proteins and lipids. Under the action of alkali, some of the proteins are degraded into small peptides and even amino acids, which are essential flavor substances. In addition, amino acids contain flavor, including sour, sweet, bitter, salty, and umami. They are also a critical a precursor substance, which can produce volatile flavor substances through Strecker degradation or Maillard reaction or the interaction with lipids (Chen et al., 2022).

Moreover, amino acids are further degraded to produce thiophene compounds by alkali. The Maillard reaction between reducing sugars and amino acids also plays a role in the formation of the distinctive flavor of the egg. The oxidation of lipids is also an essential factor in the flavor formation of preserved eggs. Some unsaturated fatty acids are oxidized into peroxides under the action of alkali, and these peroxides are further degraded into flavor carbonyl compounds, including aliphatic hydrocarbons, aldehydes, ketones, alcohols, and hydroxy acids. Therefore, the proteins, reducing sugars, and lipids in the poultry egg undergo a series of reactions in the presence of alkali, producing the unique flavor of the preserved egg.

- Quality control of preserved egg

The pickling conditions are the key factors that affect the quality of a preserved egg. The formula of the pickling solution (including the alkalinity and metal ions) and the pickling temperature can significantly affect the quality of a preserved egg. If the pickling conditions are not properly controlled, alkali injury may occur (Li et al., 2009). Therefore, regulating the processing of preserved eggs is the key to controlling the quality of the preserved egg.

The alkalinity of the pickling solution is one of the critical factors directly affecting the quality of a preserved egg. When the pickling solution contains too much alkali and the pickling time is too long, the already gelatinized preserved egg white gel will be rehydrated into liquid. This phenomenon is called "alkali injury," which significantly influences the preserved egg yield. When the pickling solution contains too little alkali, on the one hand, the gel strength of the preserved egg is weak, and the quality is poor; on the other hand, it also needs a longer pickling time, which increases the production cost. Therefore, the amount of alkali used should be strictly controlled in the processing of preserved eggs (Luo et al., 2012). It was shown that under certain conditions, a concentration of about 4% NaOH in the pickling solution is more suitable alkalinity, and the prepared preserved egg has a higher quality (Li et al., 2009). Therefore, current studies on the production and processing of preserved eggs also tend to use around 4% NaOH (Luo et al., 2020).

Metal compounds (copper sulfate, zinc sulfate, iron sulfate) must be added to the pickling solution for a preserved egg to ensure that they will not be liquefied by alkali in the later pickling stage. If no heavy metal compounds are added to the pickling liquor, half of the preserved eggs will not reach the color change stage and will be liquefied at the maturation stage (Ma, 2007). These indicate that metal compounds play an essential regulatory role in forming preserved eggs. On the one hand, the protein gradually decomposes under the action of alkali to produce hydrogen sulfide. The hydrogen sulfide and the metal ions in the liquid produce precipitates and attach to the stomata of the eggshell. There are apparent differences in the form of precipitates of different metal compounds. The precipitates block the stomata to restrict the entry of alkali, which can effectively prevent the gel of the egg white from being liquefied (Zhao et al., 2020).

On the other hand, metal ions can also act in concert with strong alkali to promote or hinder the aggregation of protein molecules by influencing the intermolecular forces or by forming aggregates directly with protein molecules, and the effect of different metal compounds can vary depending on the type of metals (Zhao et al., 2020). For example, low concentrations of Ca^{2+}, Zn^{2+}, and Fe^{3+} can improve the gel strength and promote the formation of ordered microstructure; while at high concentrations, they have a negative effect. Compared with Ca^{2+}, Zn^{2+}, and Fe^{3+}, preserved eggs induced by K^{+} have better gel characteristics, such as higher gel strength, water holding capacity, and a denser microstructure (Deng et al., 2020). Thus, metal ions play an essential role in forming preserved eggs by directly and indirectly regulating the processing of preserved eggs in concert with a strong alkali.

The pickling temperature directly affects the penetration rate of the alkaline solution, thus regulating the quality of a preserved egg. When the pickling temperature is too high, on the one hand, it is easy to produce an unpleasant smell and destroy the gel structure of the protein. On the other hand, it is effortless to breed bacteria, resulting in decreased sensory quality of preserved eggs (Zhang, 2004). When the pickling temperature is too low, it is not conducive to the color change of preserved eggs, which can easily lead to rotten preserved eggs; and it also tends to lead to a longer processing cycle, which increases the processing cost. It was reported that the rated velocity of alkali penetration was stable at a pickling temperature of 25°C (compared with 13°C and 30°C), which accelerated the ripening rate of preserved egg and shortened the processing cycle, and better quality of the preserved egg produced at this temperature

(Wu, 2013). Li et al. (2009) found that the chewiness and cohesiveness of preserved egg gel decreased as the pickling temperature increased, while the adhesiveness would first decrease and then increase. When the pickling temperature was 20°C, the adhesiveness of preserved egg protein gel was the smallest, and preserved egg had the best quality. These results showed that the pickling temperature of preserved eggs is controlled at a relatively moderate temperature (20~30°C). Its primary purpose is to regulate the rate and amount of alkali and metal ions permeating into the egg by controlling the pickling temperature (Luo et al., 2020; Yang et al., 2020b).

13.3.2 Salted Eggs

Salted eggs, also known as pickled eggs, salt eggs, and flavored eggs, are a traditional Chinese characteristic egg processing product, and the processing of salted eggs has a history of more than 1600 years. Salted eggs are generally made of raw eggs pickled in brine or a mixture of straw ash and salt, and the pickling time varies from 20–50 days, depending on the processing method. A series of physicochemical changes occur in the egg due to salt penetration. Under the action of salt, the egg whites gradually hydrate, and the yolks solidify, forming unique qualities such as tender taste, vermillion oil, and rich and fluffy sand. In addition to direct consumption, salted egg yolks are also widely used as fillings for mooncakes, rice dumplings, and dishes influenced by traditional dietary habits. Salted eggs are famous among consumers because of their unique flavor, convenient consumption, rich nutrition, and easy preservation, and they have been exported to many countries and regions such as Malaysia, Singapore, Myanmar, and the United States.

13.3.2.1 Processing Methods

The traditional processing methods of salted eggs mainly include the straw ash method, the salt mud coating method, and the brine soaking method.

- straw ash method

  The straw ash method is subdivided into two kinds of methods: the pulp-lifted and ash-wrapped method and the ash material-wrapped egg method.

  The classic recipe for the pulp-lifted and ash-wrapped method is 1,000 duck eggs, 15–25 kg of straw ash, 5–7 kg of salt, and 13–15 kg of water. The standard of ingredients can be adjusted appropriately according to the processing season and different tastes. Before beating, the salt is dissolved in water, and then the straw ash is added. Subsequently, the brine and straw ash mixture is stirred by a beater to form a uniform mortar that does not flow, does not hold water, and does not lump, staying overnight for use. Lifting pulp is to flip the selected raw eggs over in the mortar so that a 2 mm thick layer of mortar can be evenly stuck to the surface of the eggshell. Wrapping ash is to roll the pulp-lifted eggs inside the dry straw ash as soon as possible to cover them with the 2 mm thick dry ash. After wrapping the ash, the ash materials are tightly pressed on the eggs, and it is noted that pressing the ash should be appropriately tight, and the layer of dry ash keeps even in thickness. After that, the eggs can be put into the jar. The salted eggs are pickled by this method for 20–30 days in summer and 40–50 days in spring and autumn (Chi et al., 2011).

  The main formula of the ash material-wrapped egg method is 1,000 duck eggs, 3.5–4 kg of salt, 50 kg of straw ash, and an appropriate amount of water. After the salt is dissolved in water, it is added to the straw ash and stirred thoroughly to make the ash material agglomerate. After washing and drying, the raw eggs are evenly wrapped with ash and put into the jar. The salted eggs are pickled by this method for about 15 days in summer, 30 days in spring and autumn, and 30–40 days in winter.

- The salt mud coating method

  The salt mud coating method is to mix salt and red soil with water to make mud, which is used to coat and wrap raw eggs to pickle salted eggs. The formula is 1,000 duck eggs, 6–7.5 kg of salt, 6.5 kg of red soil, and 4–4.5 kg of water. The salt is placed in a container and dissolved with water, and then added to the crushed red soil. After fully absorbing brine, the red soil is made into a paste-like salt mud. The selected duck eggs are placed in the prepared salt mud so that the surface of the eggs is covered with salt mud and then put in a tank. The salted eggs are pickled by this method for 25–30 days in summer, and 30–40 days in spring and autumn. Salted eggs coated with salt mud generally have a heavier salty taste, looser egg yolks, and more oil droplets.

- The brine soaking method

  The brine soaking method is based on the ratio of 1 kg of brine for pickling 1 kg of fresh eggs, and the brine with a mass fraction of 20% is prepared for later use. After being washed with cold boiled water and dried, the selected raw eggs are put into a jar, and then they are pressed with a fine-hole bamboo cover. Subsequently, the prepared brine is poured into the jar until the eggs can be fully immersed in the brine. The salted eggs are pickled by this method for 15–20 days in summer and 30 days in winter, and they should not be stored for a long time. Figure 13.9 shows salted eggs under different processing methods.

13.3.2.2 Processing Principles

During the pickling process, the high concentration of salt penetrates from the eggshell membrane to the inside of the egg, gradually forming a concentration difference and a pressure difference on both sides of the eggshell membrane, creating a high osmotic pressure inside and outside the egg,

FIGURE 13.9 Salted eggs under different processing methods.

thereby allowing the ions in salt solution to penetrate the egg. In contrast, the water in the egg seeps out from the stomata on the eggshells. When the salted eggs are mature, the osmotic pressure in the egg basically reaches a dynamic equilibrium. For egg yolks, as the pickling proceeds, salt penetration leads to the separation of hydrophobic groups from hydrophilic and lipophilic groups in egg yolk, and the bound water molecules in egg yolk are converted into free water molecules, and then the free water molecules gradually diffuse to the egg white, which ultimately causes the egg yolk to be concentrated. In addition, salt dehydration can also lead to the destruction of LDLs structure and the release of free lipids (Xu et al., 2019). Figure 13.10 shows the gel formation mechanism of egg yolk of salted eggs (Xu et al., 2019). At the same time, pickling leads to the gradual separation and exudation of delicate emulsified lipids that were initially uniformly dispersed, thus causing the aggregation of lipophilic groups to form oil droplets (Bougatef et al., 2010). As the salt concentration in the egg increases, the salt solution can promote the salting- out of proteins and accelerate the coagulation of the yolk proteins (Xu et al., 2018). For egg whites, the salt disrupts the structure that maintains the special conformation of the protein, turning rigid protein molecules into flexible molecules with loose structure, resulting in the decrease of egg white viscosity Li et al. (2018).

Salt can reduce the activity of protease in eggs and the ability of bacteria to produce protease, thereby slowing down the rate of spoilage and prolonging the shelf life of salted eggs. During the pickling process, the osmotic pressure of the microbial protoplasm is relatively small, and the cells lose water, which leads to bacteria being unable to grow and reproduce in higher salt concentrations and even die due to "plasma separation." The salt penetrates the egg, reducing the water activity in the egg and inhibiting the life activities of microorganisms. The high-concentration salt solution can also cause physiological toxicity to microorganisms. Therefore, pickling has a specific effect on the preservation of salted eggs (Wang & Pang, 2020).

13.3.2.3 Quality Formation and Control of Salted Eggs

- Quality formation of salted eggs

  The quality of salted eggs mainly includes the hydration of egg white, the solidification of egg yolk, the unique color and flavor of salted egg yolk, and the formation of white, delicate, and tender egg whites, loose, sandy, and oily egg yolks after pickling. The unique qualities of salted eggs are mainly due to the osmotic dehydration of salt, intermolecular interaction of proteins, and oxidation of lipids.

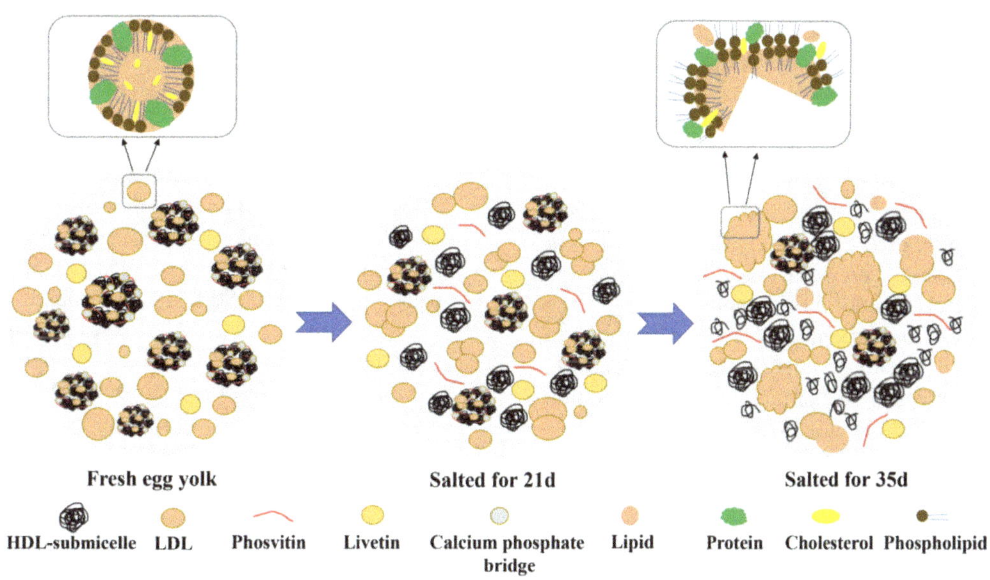

FIGURE 13.10 Diagram of the gel formation mechanism of egg yolk of salted eggs. (Modified from Xu et al., 2019.)

- Egg white hydration

The thick egg whites are mainly composed of ovomucin and lysozyme, the basis of gel formation. Thereinto, ovomucin is a fibrous protein intertwined to form the network structure through intermolecular interactions in egg white. Chemical changes during egg processing and storage will lead to the dissociation of the ovomucin-lysozyme complex structure, resulting in a decrease in thick egg whites, the weakening of protein molecular aggregation, and the unfolding of protein molecular structure, which ultimately leads to a decrease in egg white viscosity (Smith & Back, 1962). As the pickling with salt proceeds, the high salt concentration will dissociate the ovomucin structure, hydrate the egg white, and reduce the ovomucin content (Zuo, 2013). In addition, ovomucin contains many acidic amino acids, which contribute to many negative charges. The concentration of a salt ion in egg white may affect the charges of the protein, resulting in a decrease in the viscosity of egg white (Sun et al., 2018). During the heat-induced protein gel formation of salted egg white, it was found that salt can change the electrostatic interactions between proteins, and metal cations shield the negative charges of the protein, which reduces the repulsion between proteins. At the same time, metal ions also have a particular effect on hydrogen bonds. They can compete with proteins for water, inhibiting the interactions between water molecules and hydrophilic groups of proteins, promoting the gradual transformation of bound water into free water, and therefore protein-protein interactions are enhanced, and egg white protein molecules aggregate (Huang et al., 1999; Oliveira Dos Santos et al., 2008). Based on this, it can also be indirectly explained that the formation of a rough and uneven gel network of salted egg white is mainly caused by the hydration of egg white during pickling.

- Salted egg yolk gel curing

During the pickling process of salted eggs, with the increase of salt content in egg yolk, the native structures of protein molecules undergo changes, the consistency index gradually increases, and the flow activation energy decreases. Under the action of osmotic pressure, the solidified part of the egg yolk gradually extends to the inside of the egg yolk until the osmotic pressure inside and outside the egg yolk reaches equilibrium, and finally, the egg yolk changes from a semi-fluid and viscous state to a solid or semi-solid gel state (Ibarz, 1993; Xu et al., 2018). In the early stage of pickling, salt infiltration leads to a gradual increase in the soluble protein content of egg yolk, and sol is formed, which enhances the mobility of water, and the low salt concentration during the pickling period contributes to the exposure of sulfhydryl groups. In the later pickling stage, the higher salt concentration gradually dehydrates the egg yolk to form a gel network structure, and the water mobility gradually weakens. At the same time, the sulfhydryl-disulfide exchange reaction and the oxidation of sulfhydryl groups may also occur during this period to generate new disulfide bonds (Matsudomi et al., 2002), which play an essential role in the formation of egg yolk protein gel. In addition, with the increase of salt content in egg yolk, the egg yolk emulsion system is disrupted, and the value of zeta potential decreases, while the lower zeta potential leads to the aggregation of egg yolk proteins (Woodward & Cotterill, 1986). From the analysis of the secondary structure of egg yolk proteins, the content of β-sheet structures within the salted egg yolk molecules increases after pickling, probably because high salt content causes the rearrangement of hydrogen bonds and the hydrophobic interactions between protein molecules, thus resulting in the close combination of protein-water-lipid and the formation of a highly viscoelastic yolk gel (Xu et al., 2020).

- Yolk color and flavor formation

The orange transformation and hardening of egg yolk gradually developed to the center of egg yolk, and the thickness of the orange layer increased with the extension of pickling time. Pyrrole pigments are formed with the oxidation of fat and protein during pickling with salt, and the continuous accumulation of pyrrole pigment increases the concentration of pigments, resulting in the decrease of a* value of yolk (Yang & Chen, 2001; Zamora et al., 2005). In addition, another explanation is that the changes in the color of salted egg yolks are possibly due to the increase in pigment concentration caused by the dehydration of egg yolks during the pickling process (Chi & Tseng, 1998). Considering that different cooking methods significantly influence the color of salted egg yolks, some studies have found that after ultrasonic-assisted cooking, the a* and b* values of salted egg yolks increased, while L* value decreased significantly. The increase in a* and b* values may be due to the release of lutein and zeaxanthin, which are encapsulated in lipids, and the formation of Maillard products facilitated by ultrasonic treatment. The decrease in L* value was possibly due to the increased water loss and oil exudation caused by ultrasound (Li et al., 2021; Wang et al., 2021). Currently, commercially available salted eggs are usually vacuum-packed and then autoclaved. Still, vacuum-cooked salted egg yolks are prone to black circles, which may be formed by the joint action of sulfur ions, iron ions, and proteins (Li et al., 2021). The metal chelating agent can be added to the pickling solution to prevent the black circles of salted egg yolks.

Salt penetrates the egg and combines with the proteins to form a unique flavor substance, contributing to the salted eggs obtaining a unique flavor. Therefore, the flavor of salted eggs has a good correlation with the pickling process. Other studies have found that lipid oxidation and Strecker degradation are the primary way to produce the flavor of salted egg yolk. In addition, the content of esters, pyrimidines, and thiazoles in egg yolks increased after pickling with salt, which may be the main reason for the formation of the unique flavor of salted egg yolks (Zhou, 2012). Due to the long pickling cycle of the traditional processing method of salted eggs, to quickly obtain the flavoring base of salted egg yolks, it was found that fresh egg yolk can be induced to produce salted egg yolk flavor by the synergistic effect of enzymatic reaction and microwave. The formation of salted egg yolk flavor may be due to the disintegration of the egg yolk gel network by the enzymatic reaction, which is beneficial to the migration of moisture and lipids, and the microwave treatment further aggravates lipid oxidation. The volatile compounds such as hexanal, heptanal, benzaldehyde, and 2-amyl furan produced by the synergistic effect of enzymatic reaction and microwave may be related to the flavor of salted egg yolk (Su et al., 2021).

- Formation of salted egg yolk loose, sand, and oil quality

The loose, sandy, and oily characteristic of salted egg yolks is an essential criterion in evaluating the quality of salted eggs. During the pickling process of salted egg, the interaction between Na^+ and negatively charged hydrophobic groups leads to the structure change of LDL in the yolk, and the balance between oil and water is broken so that the emulsified fine lipid droplets originally evenly distributed in the egg yolk are gradually separated and oozed out. Finally, the free lipids aggregate in the form of large oil droplets due to similar compatibility (Bao et al., 2020). At the late pickling stage, the oil exudation of raw and cooked salted egg yolks increased to varying degrees. With the extension of pickling time, the dehydration of egg yolk is enhanced, resulting in the strengthening of the connection between egg yolk particles, and the shape of egg yolk particles changes from spherical to irregular polygonal. The gap between egg yolk proteins is reduced, and the structure becomes much more compact (Xu et al., 2017). In addition, yolk granules mainly consist of calcium-phosphorus bridge linked HDL-phosvitin complexes (Le Denmat et al., 1999). With the extension of pickling time, the increase of salt content in the egg yolk leads to the replacement of Ca^{2+} by Na^+ in the calcium-phosphorus bridge that maintains the structure of the yolk particles, thus resulting in the destruction of egg yolk granules and the dissolution of soluble phosvitin (Causeret et al., 1991).

- Quality control of salted eggs

The traditional processing methods of salted eggs have problems such as the long processing cycle, the unstable quality, and the unrecyclable pickling liquid, so the industrialization of the salted egg cannot further achieve a qualitative leap. Reduce the salt content of egg white, improve the quality of salted eggs and realize the industrial production of salted eggs to shorten the pickling period of salted eggs. Some salted egg processing methods combine traditional salted egg pickling technology with new salted egg pickling technology. For example, the vacuum decompression pickling technology utilizes the hydrodynamic mechanism and the deformation relaxation phenomenon caused by the pressure difference to increase the pickling rate. During the processing of salted eggs, the gas in the raw eggs can escape by vacuuming, which can make the pickling liquid fill the gas in the eggs and speed up the pickling rate of salted eggs (Xu et al., 2015). Ultrasonic cavitation was used to break macromolecules, expand mass transfer channels, reduce the viscosity of egg white, reduce the osmotic resistance of salt, increase the mass transfer rate, and improve the quality of salted eggs (Sun et al., 2018). Pulse pressure processing can shorten the pickling period and improve the quality of salted products through a reasonable pressure cycle process. Salted eggs with pulsating pressure technology can effectively promote the maturity of salted eggs and shorten the pickling time (Wang et al., 2013). There are problems such as varying salt concentration in the pickling solution and significant differences in the saltiness of egg whites in the static pickling of salted eggs by the brine immersion method. The salting egg with circulating water may improve the properties of some hydrophilic substances in the eggs, reduce the salt penetration resistance, and cause the salt in the pickling solution to enter the eggs constantly to achieve a better pickling effect (Pu et al., 2010).

With the improvement of living standards, there are increasing risk of high blood pressure, kidney disease, and cardiovascular disease among human beings. A low-sodium diet has gradually become a general trend because it can improve the basal metabolic rate and avoid a rise in blood pressure. Most of the traditional salted eggs are pickled with ordinary table salt, leading to a high sodium salt content in the finished product. Other salts have been utilized to replace sodium salt for pickling to obtain high-quality salted eggs. At present, it has been found that the use of potassium salt as a sodium salt substitute can accelerate the migration of water in salted eggs, promote the exudation of oil, and achieve the same effect as salted eggs with sodium salt (Liu et al., 2022). For shortening the pickling cycle of salted eggs, duck eggs are treated with 1.0 N HCl for 120 min or pretreated with 5% acetic acid for 30 min before pickling, then pretreated with 0.25% neutral enzyme for 90 min, and finally pickled in an aqueous salt solution, which can regulate the permeability of eggshell and effectively shorten the pickling cycle of salted eggs (Lai et al., 1997; Kaewmanee et al., 2012). Since the quality of salted eggs often decreases due to oxidation, the addition of Curcuma phaeocaulis Valeton concentrate, Alpinia officinarum Hance extract, and Aloe vera solution to the pickling solution can increase the antioxidant activity of salted eggs to a certain extent (Sholehah et al., 2017; Thohari & Rosyidi, 2017; Novia & Juliyarsi, 2019).

13.3.3 Zaodan (Egg Pickled in Rice Wine)

Zaodan (egg pickled in rice wine) enjoys the reputation of being "one of the best Chinese foods" (Wang & You, 2010). It is a traditional Chinese pickled food made of fresh duck eggs pickled in rice wine. The traditional production cycle of zaodan ranges from 5–12 months due to the different processing methods. During the pickling process of fresh duck eggs, substances such as alcohols, sugars, organic acids in rice wine, and the added salt gradually penetrate the eggs, leading to the physicochemical changes in egg white and yolk. Under the action of ethanol and organic acids, egg white and yolk progressively denature and solidify (Meng et al., 2010). Mature zaodan has soft eggshell, tender flesh, jelly-like egg whites with milky white or brown, semi-solidified egg yolks with orange-red color, rich and mellow wine aroma, and a lingering aftertaste after eating. Due to its bright color, long mellow fragrance, and rich nutrition, zaodan is well received by consumers; it is a unique special dish for families and banquets and has a high reputation in Chinese and Southeast Asian markets.

13.3.3.1 The Processing Methods of Zaodan

Zaodan is slightly produced by factories and folks around China, with different flavors due to different production techniques. The most famous zaodan are Zhejiang Pinghu zaodan and Sichuan Yibin Xufu zaodan, as shown in Figure 13.11.

Zhejiang Pinghu zaodan (Wang, 2001): A typical formula comprises 7.8~8.4 kg of fresh duck eggs (120 eggs), 10 kg of refined white glutinous rice, 1.88 kg of salt, 32 g of distiller's yeast, and 6.5 g of sweet Chinese yeast. Glutinous rice is added with water and steamed until it is done. After cooled and drained, the steamed glutinous rice is poured into a jar and sprinkled with the distiller's yeast and sweet Chinese yeast (the dosage of yeasts can adequately increase or decrease according to the temperature); then, the rice pieces are repeatedly rubbed with hands and turned up and down to form a trumpet-shaped circular nest of which the diameter of the bottom is about 20 cm, and the surface of the rice is flat. After 6~7 d, the fermented glutinous rice wine is mixed evenly and then standing in the jar for 15 d until the ethanol content in rice wine reaches 15% for the preparation of zaodan. Cleaned fresh duck eggs are knocked with bamboo slices until the eggshells are broken, and the egg membranes keep intact. Subsequently, a layer of fermented glutinous rice and a layer of eggs are, in turn, placed into a sterilized jar until all the eggs are installed. Finally, the top of the jar is covered with a layer of fermented glutinous rice where the salt is evenly sprinkled. The jar is sealed with kraft paper and stored in a warehouse. Generally, after about 5 months of pickling, Pinghu zaodan can be mature.

Pinhu *zaodan* Xufu *zaodan*

FIGURE 13.11 Zaodan under different processing methods.

Sichuan Yibin Xufu zaodan (Zhu, 1991): A representative formula consists of fresh duck eggs (150 eggs), 5 kg of fermented glutinous rice, 1 kg of brown sugar, 1 kg of Chinese baijiu, 1.5 kg of salt, 25 g of tangerine peel and 25 g of Sichuan Pepper. Fermented glutinous rice, brown sugar, Chinese baijiu, and salt are mixed. Then a layer of the mixture is placed in the bottom of a sterilized jar, where a layer of fresh duck eggs is successively put into until all the eggs are installed. Eventually, the top of the jar is covered by a layer of the mixture, and the jar is sealed. After pickling for 3–4 months in the jar, the eggshells are stripped, and egg membranes cannot be punctured. Subsequently, the peeled eggs are placed into Chinese baijiu pickling for 2 d, then taken out one by one and put into the original jar. In addition to the original ingredients, brown sugar, salt, tangerine peel, Sichuan Pepper, boiled syrup (brown sugar and the right amount of water are boiled until the sugar becomes silk), and then the jar is sealed and stored in a dry, cool and ventilated place. After 2–3 months, turn over the eggs, that is, the upper part on the bottom and the lower part on the top, so that all the eggs can be evenly pickled. After that, the jar is still sealed and stored. From the production of duck eggs into the jar to storing after more than a year, Xufu zaodan are mature.

13.3.3.2 The Processing Principles of Zaodan

Distiller's yeast and sweet Chinese yeast used for the fermentation of glutinous rice are saccharification bacteria and yeast products, and their microorganisms mainly include Rhizopus and yeast. Rhizopus decomposes the starch into sugars, and then a tiny amount of yeast will convert sugars into alcohols (mainly ethanol) through glycolysis, and part of the alcohols are oxidized to form acetic acid. In rice wine, ethanol, acetic acid, sugars, and other substances gradually penetrate the raw duck eggs through infiltration and diffusion, resulting in protein denaturation and gelation. Egg white and yolk are solidified, and zaodan obtains wine flavor and slight sweetness (Meng et al., 2010; Wang & You, 2010). At the same time, due to the action of acetic acid in rice wine, the calcium carbonate in the eggshells solves, and the eggshells become thinner and softer, as well as gradually separate from the inner membrane of the eggshells. Besides, the ethanol content in rice wine is not high, but in the pickling process for a long time, it can work with the added salt to kill the microorganisms in eggs, especially salmonella. Therefore, zaodan can be eaten raw.

13.3.3.3 The Quality Formation and Control of Zaodan

- The quality formation of zaodan

  The processing technologies of traditional zaodan are complicated. However, the formation of their quality is mainly due to the denaturation and aggregation of proteins under the action of ethanol and acetic acid in rice wine, as well as a series of physicochemical changes caused by ethanol, sugars, organic acid, salt, and other substances in pickling liquid penetrating the eggs.
- The gel formation of zaodan

  Ethanol is the main component of rice wine. After pickling for an extended period, the ethanol content in egg white increases, which promotes the interactions between proteins and solvent and leads to the decrease of free water and the increase of the viscosity of egg white (Yao et al., 2020). Egg white gradually transforms from liquid to semi-solid jelly. Ovomucoid plays a crucial role in the formation of egg white gel of zaodan (Yu et al., 2020; Yu et al., 2021). During the pickling process of zaodan, with the infiltration of ethanol, polypeptide chains of proteins in egg white are unfolded, the internal free sulfhydryl groups are exposed, more disulfide bonds are formed, the secondary structure of proteins undergoes changes, and denaturation and aggregation of proteins occur (Yu et al., 2021). Except for ovomucoid, all other protein molecules are degraded, forming a porous structure with dense structure and tiny pores, thus enhancing the water retention and hardness of egg white gel (Yu et al., 2021). In addition, during the process of ethanol-induced egg white gel formation, it was found that ethanol can reduce the dielectric constant of water medium and enhance protein intramolecular and intermolecular electrostatic interactions, thus resulting in the unfolding of polypeptide chains and the exposure of sulfhydryl groups (Yu et al., 2020). The exposed sulfhydryl groups were conducive to crosslinking new polypeptide chains, facilitating the closer combination between proteins and water molecules and promoting egg white protein aggregation.

  As the pickling proceeds, the ethanol content in egg yolk increases as well, the secondary structure of proteins in egg yolk changes, and the number of β-sheet structures increases, resulting in the dense packing of proteins and macroscopically forming the highly compact gel (Zeng et al., 2022). Moreover, ethanol can enhance electrostatic repulsion among protein molecules by altering the free energy of charged groups in egg yolk and interacting with solvent-solvent to enhance protein-protein interactions, thus promoting egg yolk protein aggregation (Yoshizawa et al., 2014). The protein content of egg yolk is about 15.6%~16.8%, of which low-density lipoprotein (LDL) accounts for two-thirds of the total weight of egg yolk dry material (Anton et al., 2003). The aggregation of LDL in egg yolk is generally thought to be the leading cause of egg yolk gelation (Kaewmanee et al., 2011). Ethanol treatment could result in the denaturation and unfolding of LDL. It exposes some polar residues and hydrophilic groups (such as Asp, Glu, and post-translational modified oligosaccharide chains (Geng et al., 2018)). It increases the amount of negative charge on the surface of LDL, facilitating the closer binding of water and protein molecules, thus promoting the aggregation of LDL molecules (Xie et al., 2020). Additionally, as the acetic acid in rice wine penetrates egg yolk, the pH of egg yolk is acidic. The advanced structure of the protein can be destructed under acidic conditions, leading to the unfolding of the regular secondary structure and the changes in the content of the secondary structure, thus promoting egg yolk protein denaturation and gelation (Wang et al., 2013).
- The formation of flavor and taste of zaodan

  During the fermentation and maturation of zaodan, the Rhizopus and yeast contained in rice wine can degrade proteins in raw eggs to form small peptides and free amino acids directly related to flavor, among which glutamate is an important umami substance. Its existence plays an essential role in the formation of the flavor of zaodan (Meng et al., 2010). In addition, the yeasts for the fermentation of glutinous rice contain saccharification bacteria, which can convert the starch in glutinous rice into sugars, and yeasts ferment the sugars to produce alcohol. Some alcohols can be oxidized to form acetic acid. In the pickling process of zaodan, ethanol, sugars, organic acids, and active enzymes in pickling liquid penetrate the interior of fresh duck eggs through the eggshell. Thus, egg membrane undergoes a long-term biochemical reaction, leading to the increase of free amino acids, sugars, alcohols, organic acids, aromatic esters, and other ingredients, which brings the wine taste and the characteristics of easy digestion of zaodan, and produces a unique flavor (Wang & You, 2010). This flavor combines the fragrance of wine and ester, thus forming a variety of complex flavors. Consequently, eating zaodan has an endless aftertaste and nourishing effect.
- The quality control of zaodan

  Zaodan is a semi-gel that fresh duck eggs gradually solidify under the action of ethanol and acetic acid in rice wine. Egg white proteins can rapidly aggregate to form a high elastic gel under ethanol treatment alone (Yao et al., 2020; Yu et al., 2020). Thus, ethanol concentration in the pickling liquid has a particular impact on the formation of zaodan. If the ethanol concentration is too low, it can lead to the insignificant protein aggregation degree of egg white, and the formed gel is easy to collapse and of poor quality under the action of external forces (Yao et al., 2020); meanwhile, it can result in longer pickling time, which will increase the production cost. However, when the ethanol concentration is too high, it will destroy the hydrogen bonds that maintain the rigid tertiary structure of the protein and further destroy the hydrophobic region of the protein to enhance the ethanol-water interaction, leading to the formation of insoluble precipitations (Yao et al., 2020), which will affect the quality of zaodan to some extent.

Moreover, the traditional pickling methods of zaodan are generally carried out at room temperature and constant pressure, which have problems such as complex processes, high production cost, uncontrollable production process, unstable quality, and long production cycle. These problems affect enterprises' development and economic benefits and limit the large-scale production of zaodan. Therefore, in recent years, some processing methods of zaodan combine the traditional pickling technology with new pickling technology. For example, vacuum technology is used to pickle zaodan (Liu et al., 2012). In the pickling process of zaodan, the container is decompressed and vacuumed so that the raw eggs are pickled under vacuum conditions. The rate at which the pickling liquid penetrates the raw eggs increases under vacuum and the negative pressure can quickly and evenly to the whole container, not limited by the shape and volume of raw eggs, which shortens the production cycle of zaodan (Liu et al., 2012; Wang et al., 2013). During the stress reduction process, due to diffusion and infiltration, the soluble ingredients inside zaodan are redistributed with the liquid and water, so the quality of all the eggs can be more uniform. Therefore, by regulating the processing technology of zaodan, the penetration rate of pickling liquid can be controlled indirectly, thereby influencing the formation of zaodan.

13.4 European Eggs and Egg Products

Egg products are diversified in Europe due to the culinary diversity of European countries. Some simple products are very commonly consumed, such as the hard-boiled eggs, as the basis of European culinary culture, but others have national specificities. These products are marketed in supermarkets, in collective catering, and as ingredients for the food industry, depending on the country.

13.4.1 Cooked Eggs

Here we refer to cooked products derived from the processing and direct cooking of shell eggs.

13.4.1.1 Hard-boiled Eggs

Hard-boiled eggs are cooked in their shells, then shelled and stored in brine or protective atmosphere packaging. The European market of hard-boiled eggs is overwhelmingly focused on "small" graded eggs, which are difficult to place on the market of shell eggs, and for reasons of practicality for the consumer because of the size homogeneity. Shell eggs are, therefore, pre-graded in the packing centers.

Egg storage before hard-boiled egg processing is quite specific, as egg freshness substantially impacts peeling quality. As we can all experience at home, it is sometimes difficult to remove the shell of some eggs after cooking, while for other eggs, the shell can be removed very easily. This results from the difference in freshness between eggs: the fresher the egg, the more difficult it is to remove the shell. This phenomenon is not fully explained, but it has been proven that the ease of peeling is directly related to the egg white pH, which should be above 8.5 (Britton & Fletcher, 1987).

At the industrial level, it is possible to control the freshness of the eggs by fine-tuning the management of supplies and storage conditions of eggs. A sampling control over receipt of shell eggs by the hard-boiled egg manufacturer makes it possible to know the eggs' average initial level of freshness. In the second phase, egg storage at the controlled temperature over a defined period, depending on the results of the initial freshness control, makes it possible to reach the optimum range for good peeling after cooking. Depending on the quality of eggs in stock and the production requirements, it is possible to vary the storage temperature to accelerate or slow down the aging process.

Besides its control and regularity, the main difficulty with cooking is the bursting of the eggs when heated, with economic (loss of material) and hygienic (organic contamination of cooking and cooling equipment) consequences. Indeed, the expansion of the egg content due to heating leads to a compression of the egg's interior. Therefore, a potential shell bursting can be avoided by weakening the shell in the air space to allow the pressure to escape during cooking. Several techniques can be used, such as piercing the shell with metal spikes or "dropping" the eggs a few millimeters above a flat surface. Whichever method is used, it must be adapted to the industrial context and, therefore, to the classic 30-egg tray format. In addition, weakening the shell will increase the efficiency of the shelling process.

Once shell weakening has been carried out, the eggs are traditionally cooked in a steam-saturated chamber or a water bath. The cooking scale varies according to the manufacturers and the cooking equipment; it is usually between 95 and 100°C for 10–15 min. The core temperature of the eggs must be around 85°C to obtain a wholly coagulated yolk while limiting the appearance, around the yolk, of a black border at the white-yellow interface when the eggs are returned to the user's premises. This border harms the appearance and presentation of the product. It is in no way a microbiological alteration of the product but oxidation induced by the heat treatment. It is also possible to observe the same border at home when hard-boiled eggs are cooked beyond 10 min. This is due to the formation of iron sulfide (FeS) because of the reaction between the iron present in the yolk and the hydrogen sulfide (H_2S) produced in the egg white during cooking (Germs, 1973).

After cooking, the eggs are immediately cooled by immersion in cold-water baths. This cooling aims to bring the eggs back to around 30°C before peeling. After cooling, the shelling operation consists of removing the shell from the cooked egg. This is done mechanically by rubbing the eggs (between two metal plates, for example), which allows the shell to fall off as if the eggs were rubbed between two hands. The most sensitive part of this step is both not to crush the eggs and to limit the penetration of shell pieces into the cooked egg white.

After the peeling stage, the cooked eggs can be left in an acidification bath for several hours (continuously or in batch). This particular stage prevents the appearance of the black border mentioned above. Finally, shelled hard-boiled eggs are packaged in two main ways: in brine or a protective atmosphere.

Pickling, i.e., storage in brine, has the advantage of being a simple and easy-to-use preservation method. The eggs are completely covered with an acidic (pH 3–4) and salty brine, which considerably limits bacterial growth. Anti-fungal agents (such as sorbate and sodium benzoate) can also be added to the brine. Moreover, this packaging is portable for the user. Its opening does not affect the conservation of the product, as long as the brine still covers the eggs. The few bacteria that can develop at these acidic pH levels are lactic acid, but they do not bring any organoleptic changes to the product and do not pose any microbiological risk for consumers. The main disadvantage of bringing is the lower taste quality of the product due to its acidity.

Protective atmosphere packaging comes in small packages of 10 or 30 eggs, up to bulk bags of over 100 eggs. The principle is to inject a gas mixture into the packaging, made up of carbon dioxide and nitrogen, to deprive the microorganisms of the oxygen present. The taste quality of the eggs is thus preserved because these gases are inert to the taste. On the other hand, these products can favor the development of anaerobic bacterial species during conservation. In addition, the opening of the packaging requires total consumption of the product since it causes oxygen ingress.

Culturally, hard-boiled eggs are the basis of culinary specialties such as "scotch eggs" in the United Kingdom, where hard-boiled eggs are surrounded by sausage meat, then breaded and fried. This recipe was invented in London by Fortnum & Mason in 1738, and it is not, contrary to popular belief, a Scottish dish.

13.4.1.2 Poached Eggs

Poached eggs are also cooked eggs, but unlike hard-boiled eggs, cooking occurs after the shell has been removed. The eggs are therefore broken raw, directly into the cooking system. There are two ways of doing this: eggs are either broken into individual molds, which are then immersed in a hot water bath or are broken and poured directly into a hot water bath. The cooking schedule is adapted in both situations to allow the egg white to coagulate while preserving the liquid and runny yolk. It is possible to obtain a viscous egg yolk depending on the cooking time. If a mold is used, baking can also take place directly in the packaging. Industrially, this is the least complex system, but the finished product takes the shape of a mold. This means that the visual appearance is no longer traditional and leaves little doubt as to the industrial nature of the product.

When cooking a broken egg directly in hot water, it is necessary to recreate a vortex in the water bath to allow the egg white to wrap around the yolk. This is the traditional method in a pan, and a cook has to generate a continuous vortex with his fork. Industrially, this phase is very complex to obtain, susceptible, and involves more losses during cooking. However, the visual rendering of the finished product is remarkable, approaching a "gastronomic" visual. In this process, as for hard-boiled eggs, poached eggs can be packaged in a protective atmosphere or immersed in a conservation brine.

Here too, poached eggs can be a basis for culinary specialties. As with hard-boiled eggs, British 'Scotch eggs' can be made from poached eggs, replacing the hard-boiled egg. In France, a gastronomic specialty combines a slice of bread (or English muffin in variant), bacon, poached eggs, and a hollandaise sauce (an emulsion of lemon juice, butter, and egg yolk); this specialty is called "les oeufs Bénédicte." The story goes that the recipe originated in the Waldorf Hotel in New York one morning in 1892, to satisfy a guest named Lemuel Benedict.

13.4.1.3 Fried Eggs

The need for sanitary production for collective catering (hospital catering in particular) has led to the development of ready-to-use products such as fried eggs. For this product, the eggs are broken in a flat mold. The real difficulty is mastering the cooking process, as the yolk must remain sufficiently liquid but still coagulate around the edges to avoid bursting during packaging and transport. The packaging is exclusively in trays, under a modified atmosphere, the most protective packaging for the product.

13.4.2 Cooked Egg Products

Here we refer to cooked products resulting from the processing and cooking of egg products, with or without previous separation of the egg yolk and egg white.

13.4.2.1 Omelets

There is currently no legal or normative definition of omelet, and there is a wide variety of recipes in Europe. Even if it is agreed that the basis of the omelet is a whole egg, the latter can be enriched with milk or egg white to improve the final taste. In addition, depending on the country to which the product is addressed, the omelet may be creamy.

The omelet maker can break the eggs himself, buy a pasteurized whole egg, or a "ready-made" mixture of his preparation for direct cooking. Sourcing a pasteurized mixture limits the risk of contamination from handling the ingredients in isolation.

Formulation work is essential to ensure product homogeneity and industrial baking. The use of texturizers is common. These texturizers have a double role: the first is to ensure an adapted texture of the preparation during pumping and passage on the cooking molds; the second is to limit the release of water (syneresis) observed during the defrosting of frozen omelets.

Cooking an omelet is relatively simple. The liquid preparation is pumped onto a hot plate, generally in molds. The cooking schedule depends on the desired result of the omelet and, in particular, on the desired level of creaminess. Unlike hard-boiled eggs, the cooking time is relatively short (2–5 min), which prevents the omelets from turning green during production. However, iron sulfide can be formed when the product is heated up again before eating, so citric acid should be added to the preparation before cooking.

More elaborate cooking techniques, such as baking ovens, steam cooking, and infrared cooking, can be used. Many parameters will influence the choice of cooking method: productivity, the final result (browned or white omelet, more or less creamy), whether or not to add oil during cooking, the subsequent processing of the omelet (e.g., omelet cubes for the industry), etc.

There are two methods of adding ingredients (filling) into an omelet. The first is to add the ingredients to the preparation before cooking. This technique's main advantage is limiting microbiological contamination from handling the ingredients. On the other hand, the manufacturing process must be adapted since the pumps supplying the preparation to the heating plates must be calibrated to transfer liquid food, including markers of a few millimeters. This technique of incorporating the ingredients into the heart of the omelet is also easily automated, thus saving labor. The second method, which is simpler to implement, consists of adding the filling to the cooked omelet, generally before folding the omelet. The main disadvantages of this technique are that it is much more challenging to automate, and there is a greater risk of microbiological contamination due to the handling of the ingredients by the operators. This hygienic feature requires precise management of operator hygiene, flow management, ingredients, etc.

After cooling, the omelets are quickly vacuum-packed or packed in a protective atmosphere at +4°C or frozen at −18°C.

13.4.2.2 Scrambled Eggs

This traditional "English breakfast" dish seems to have been described by a French author in 1664. The recipe consists of stirring a liquid whole egg with a fatty substance, generally a dairy fat (butter, cream), over low heat.

In this recipe, it is the creaminess that is desired. Industrially, the egg and fat mixture are cooked on a flat surface, with a system of stirring or scraping that prevents the egg from coagulating in a mass. Once cooked and the texture obtained, the scrambled eggs are packed in trays. Subsequent pasteurization in the packaging can be carried out, but at temperatures not exceeding the coagulation point of the egg to maintain the creamy texture.

Culturally, there are many variants of this recipe in Europe, such as in Belgium, where hop shoots accompany scrambled eggs and beer; in Switzerland, the recipe may incorporate Emmental cheese; in Poland, country sausage (podwawelska) can accompany scrambled eggs; and in Spain, scrambled eggs are served with fennel and ham (kinojos con jamón).

13.4.2.3 Hard-Boiled Egg Bars

Although it may seem the most elaborate and technical, the hard-boiled egg bar remains one of the first elaborate egg products made in Europe. However, few food companies have the necessary equipment to produce this product.

The manufacturing principle of this product is as ingenious as it is technical: liquid egg white and liquid egg yolk are cooked individually in cylindrical cooking molds. The egg white is first placed in a cylindrical ring (created by the space between two concentric cylinders) heated by a double jacket. Once the egg white has coagulated sufficiently, the central cylinder, around which the egg white has been heated, is removed. Then, the yolk is poured into the free space. The cooking process is then continued until the yolk has coagulated. The resulting egg bar is removed from the cylinder and immediately vacuum-packed, cooled, or frozen. The liquid white and yolk are generally heated beforehand to reduce the cooking time. The technicality of hard-boiled egg bar production is related to the excellent adhesion of the white and yolk for a good appearance of the egg slices during use. Therefore, the egg white coagulation level before the yolk is injected must be perfectly controlled.

The main advantage of this product is that homogeneous slices can be served over the whole egg bar; there is no loss of product as the ratio of white to the yolk is regular. As for omelets, using hydrocolloids prevents syneresis when the product is defrosted, and the addition of citric acid limits the formation of iron sulfide.

13.4.3 Beaten Egg Whites

Beaten egg whites are one of the few sweet egg products. The production process is straightforward as it is similar to household techniques. Sweetened liquid egg white (around 20% sugar) is foamed in an industrial mixer by injecting air or specific gas mixtures. In order to ensure an acceptable microbial level, the sweetened egg white used is generally previously pasteurized. The sweetened egg white foam is then dosed into a plastic tray with a touch of caramel. Lastly, the product is baked to stabilize the foam, and then it is quickly cooled.

The product density is determined by the manufacturer, accordingly to customer requirements. It is, on average, around 150 g/l. Beaten egg whites are sold chilled only, and the customers mainly use this product as a floating island, i.e., added to custard.

13.4.4 Egg-Based Alcohols

We will end with two European references to egg-based spirits.

13.4.4.1 Marsala with Egg

Marsala with egg or "cremovo" is an Italian aperitif and dessert wine made in Sicily. Marsala is a sweet wine with a registered designation of origin produced in the region of the Italian town of Marsala in Sicily. Cremovo is made by mixing Marsala wine with egg yolk (around 10%) and sugar. Its alcoholic strength is between 16 and 18%.

13.4.4.2 Advocaat

Advocaat is a smooth liqueur of Dutch origin, made from egg yolk, sugar, and alcohol. It has a slight taste reminiscent of almonds, with an alcohol content usually between

14 and 20%. The original advocaat was a liqueur created by the Dutch in Suriname and Recife, including avocados. Back in the Netherlands, where this fruit was not available, they managed to recreate an identical texture with egg yolk. A similar drink is also produced in Germany under the name "Eierlikoer."

13.5 Others

13.5.1 Eggnog

Eggnog, historically also known as a milk punch or an egg milk punch when alcoholic beverages are added. A traditional holiday drink dating back hundreds of years, eggnog is made with eggs (hence the name), milk, cream, spices like nutmeg and vanilla, and fortified with rum, whisky, and/or brandy and whipped egg white. The "eggnog" is an American term introduced in 1775, consisting of the words "egg" and "nog," with "nog" meaning "strong ale." Milk, eggs, and sherry were foods of the wealthy, so eggnog was often used in toasts to prosperity and good health.

13.5.2 Pisco Sour

A pisco sour is an alcoholic cocktail of Peruvian origin that is typical of the cuisines from Peru and Chile. The drink's name comes from pisco, which is its base liquor, and the cocktail term sour, in reference to sour citrus juice and sweetener components. There are many ways to make a pisco sour cocktail. Peruvian recipe is made with pisco brandy, simple syrup, lime juice, egg white, and a few dashes of bitters, the pisco sour's taste is a tantalizing mix of tart, sweet, silky, and herbaceous. The Chilean version is similar, but uses Chilean pisco and Pica lime, and excludes the bitters and egg white. Cocktail connoisseurs consider the pisco sour a South American classic. Chile and Peru both claim the pisco sour as their national drink, and each asserts ownership of the cocktail's base liquor—pisco Chile and Peru bicker fiercely over the pisco sour's origin (and that of pisco, too), but by most accounts, the drink's genesis is tied to a United States citizen.

13.6 Conclusion

Eggs have long been used as food and food ingredients. The history of the expansion of egg consumption in Western Europe, China and Japan, including the food cultures of each period was introduced in this chapter, including traditional egg preservation methods and their uses, as egg dishes of the regions and their origins, including Japan, China, the European Union, and Latin and North America. Japanese are famous worldwide for their love of eggs and the many varieties of traditional egg dishes. As traditional Asian egg products, preserved eggs, salted eggs, and zaodan have been consumed in east Asia and different processing techniques and processing principles are applied. Preserved eggs are formed under the synergistic induction of strong alkali and metal ions and have a unique taste of delicious, cool, and refreshing flavor. Salted eggs are formed under the synergistic induction of high salt and temperature, which are delicious and fragrant. The zaodan is mainly induced by ethanol and organic acids, which are delicious, tender, and rich in wine aroma. These traditional egg products have their characteristics and unique flavor and are deeply loved by consumers. In Europe and North and South America, egg products are diversified due to the culinary diversity of these countries. Some simple products are very commonly consumed, such as the hard-boiled eggs, as the basis of their culinary culture, but others have national specificities. Egg-based alcohols are also consumed in Western countries and Latin America.

REFERENCES

Ai, M., Zhou, Q., Guo, S., Fan, H., Cao, Y., Ling, Z., Zhou, L., & Jiang, A. (2020a). Characteristics of intermolecular forces, physicochemical, textural and microstructural properties of preserved egg white with Ca(OH)2 addition. *Food Chemistry*, *314*, 126206. https://doi.org/10.1016/j.foodchem.2020.126206

Anton, M., Martinet, V., Dalgalarrondo, M., Beaurnal, V., David-Briand, E., & Rabesona, H. (2003). Chemical and structural characterisation of low-density lipoproteins purified from hen egg yolk. *Food Chemistry*, *83*(2), 175–183. https://doi.org/10.1016/S0308-8146(03)00060-8

Bao, Z., Kang, D., Li, C., Zhang, F., & Lin, S. (2020). Effect of salting on the water migration, physicochemical and textural characteristics, and microstructure of quail eggs. *LWT - Food Science and Technology*, *132*, 109847. https://doi.org/10.1016/j.lwt.2020.109847

Bougatef, A., Nedjar-Arroume, N., Manni, L., Ravallec, R., Barkia, A., Guillochon, D., & Nasri, M. (2010). Purification and identification of novel antioxidant peptides from enzymatic hydrolysates of sardinelle (*Sardinella aurita*) by-products proteins. *Food Chemistry*, *118*(3), 559–565. https://doi.org/10.1016/j.foodchem.2009.05.021

Britton, W. M., & Fletcher, D. L. (1987). Influence of storage environment on ease of shell removal from hard-cooked eggs. *Poultry Science*, *66*(3), 453–457. https://doi.org/10.3382/ps.0660453

Causeret, D., Matringe, E., & Lorient, D. (1991). Ionic strength and pH effects on composition and microstructure of yolk granules. *Journal of Food Science*, *56*(6), 1532–1536. https://doi.org/10.1111/j.1365-2621.1991.tb08634.x

Chen, Z., Zhu, Y., Cao, W., Zhou, L., Zhang, C., Qin, X., Zheng, H., Lin, H., & Gao, J. (2022). Novel insight into the role of processing stages in nutritional components changes and characteristic flavors formation of noble scallop Chlamys nobilis adductors. *Food Chemistry*, *378*, 132049. https://doi.org/10.1016/j.foodchem.2022.132049

Chi, S. P., & Tseng, K. H. (1998). Physicochemical properties of salted pickled yolks from duck and chicken eggs. *Journal of Food Science*, *63*(1), 27–30. https://doi.org/10.1111/j.1365-2621.1998.tb15668.x

Chi, Y., Xu, W., & Liu, Z. (2011). Salted egg and salted egg yolk pickling technology. *Guide to Chinese Poultry* (9), 49–50 (in Chinese).

Deng, C., Shao, Y., Xu, M., Yao, Y., Wu, N., Hu, H., Zhao, Y., & Tu, Y. (2020). Effects of metal ions on the physicochemical, microstructural and digestion characteristics of alkali-induced egg white gel. *Food Hydrocolloids*, *107*, 105956. https://doi.org/10.1016/j.foodhyd.2020.105956

Fujishin, R., Kitagawa, H., Takada, C., & Hatta, H. (2019). Taste and evaluation of physical properties of Kyoto style Dashi egg roll. *Journal of Food Science, Kyoto Women's University*, *73*, 13–19. (in Japanese) http://hdl.handle.net/11173/2806

Fujishin, R., Kubo, N., & Hatta, H. (2022). Taste and evaluation of physical properties of Kyoto style Dashi egg roll part 2: A study of home cooking to reproduce the taste of specialty shops. *Journal of Food Science, Kyoto Women's University*, *76*, 35–40. (in Japanese) http://hdl.handle.net/11173/3389

Ganesan, P., & Benjakul, S. (2014). Effect of glucose treatment on texture and colour of pidan white during storage. *Journal of Food Science and Technology*, *51*(4), 729–735. https://doi.org/10.1007/s13197-011-0555-1

Geng, F., Xie, Y. X., Wang, J. Q., Majumder, K., Qiu, N., & Ma, M. H. (2018). N-glycoproteomic analysis of chicken egg yolk. *Journal of Agricultural and Food Chemistry*, *66*(43), 11510–11516. https://doi.org/10.1021/acs.jafc.8b04492

Germs, A. C. (1973). Hydrogen sulphide production in eggs and egg products as a result of heating. *Journal of the Science of Food and Agriculture*, *24*(1), 7–16. https://doi.org/10.1002/jsfa.2740240103

Guo, C., Gao, Z., & Wu, H. (2013). Optimization of pulse pressure technology for preserved-eggs by response surface methodology. *Food Science and Technology*, *38*(8), 118–122. (in Chinese).

Hatta, H., Nakamoto, A., & Horimoto, Y. (2021). Reproduction of a reversed boiled egg (the yolk inside out) listed in an old Japanese recipe book. *International Journal of Food Science and Agriculture*, *5*(2), 293–302. https://doi.org/10.26855/ijfsa.2021.06.013

Huang, J. J., Tsai, J. S., & Pan, B. S. (1999). Pickling time and electrodialysis affects functional properties of salted duck egg white. *Journal of Food Biochemistry*, *23*(6), 607–618. https://doi.org/10.1111/j.1745-4514.1999.tb00589.x

Ibarz, A. (1993). Rheology of salted egg yolk. *Journal of Texture Studies*, *24*(1), 63–71. https://doi.org/10.1111/j.1745-4603.1993.tb01277.x

International Egg Commission Report. (2021). https://www.internationalegg.com/resource/global-egg-production-continues-to-increase-at-an-average-of-3-per-year/. (accessed on 5th June 2023)

Kaewmanee, T., Benjakul, S., & Visessanguan, W. (2011). Effects of salting processes and time on the chemical composition, textural properties, and microstructure of cooked duck egg. *Journal of Food Science*, *76*(2), S139–S147. https://doi.org/10.1111/j.1750-3841.2010.01975.x

Kaewmanee, T., Benjakul, S., & Visessanguan, W. (2012). Effect of acetic acid and commercial protease pretreatment on salting and characteristics of salted duck egg. *Food and Bioprocess Technology*, *5*(5), 1502–1510. https://doi.org/10.1007/s11947-011-0510-1

Kawae, M., Yamashita, M., & Hatta, H. (2015). Sensory evaluation of rice with raw egg prepared by Rosanjin's recipe. *Journal of Food Science, Kyoto Women's University*, *70*, 15–19. (in Japanese) http://hdl.handle.net/11173/2276

Lai, K., Ko, W., & Lai, T. (1997). Effect of NaCl penetration rate on the granulation and oil-off of the yolk of salted duck egg. *Food Science and Technology International, Tokyo*, *3*(3), 269–273. https://doi.org/10.3136/fsti9596t9798.3.269

Le Denmat, M., Anton, M., & Gandemer, G. (1999). Protein denaturation and emulsifying properties of plasma and granules of egg yolk as related to heat treatment. *Journal of Food Science*, *64*(2), 194–197. https://doi.org/10.1111/j.1365-2621.1999.tb15863.x

Li, C.-H., Hsieh, C.-H., Hung, C.-C., & Cheng, C.-W. (2021). Nondestructive detection of the gel state of preserved eggs based on dielectric impedance. *Foods*, *10*(2), 394. https://doi.org/10.3390/foods10020394

Li, J., Hou, C., Xiong, S., & Zhao, S. (2009). Effects of preserved conditions on quality of preserved egg. *Food Research and Development*, *8*, 101–105. (in Chinese).

Li, J., Li, X., Wang, C., Zhang, M., Xu, Y., Zhou, B., Su, Y., & Yang, Y. (2018). Characteristics of gelling and water holding properties of hen egg white/yolk gel with NaCl addition. *Food Hydrocolloids*, *77*, 887–893. https://doi.org/10.1016/j.foodhyd.2017.11.034

Li, J., Zhang, M., Chang, C., Gu, L., Peng, N., Su, Y., & Yang, Y. (2018). Molecular forces and gelling properties of heat-set whole chicken egg protein gel as affected by NaCl or pH. *Food Chemistry*, *261*, 36–41. https://doi.org/10.1016/j.foodchem.2018.03.079

Li, Q., liu, H., Li, Y., Dai, Y., Li, H., & Liu, Y. (2021). Analysis of the causes of the "black circles" in the vacuum cooked salted duck eggs. *Modern Food Science and Technology*, *37*(9), 234–241. (in Chinese).

Li, S., Huang, P., & Wang, Q. (1992). Study on the color change mechanism of egg yolk during the processing and storage of preserved eggs. *Food Science*, *2*, 18–22. (in Chinese).

Li, Z., Zheng, Y., Sun, Q., Wang, J., Zheng, B., & Guo, Z. (2021). Structural characteristics and emulsifying properties of myofibrillar protein-dextran conjugates induced by ultrasound Maillard reaction. *Ultrasonics Sonochemistry*, *72*, 105458. https://doi.org/10.1016/j.ultsonch.2020.105458

Liang, Q., & Shi, Q. (1992). Study on the stage and color during the processing of preserved eggs. *Journal of Gansu Agricultural University*, *2*, 171–176. (in Chinese)

Liu, H., Feng, F., Xue, H., Gao, B., Han, T., Li, R., Hu, X., Tu, Y., & Zhao, Y. (2022). Effects of partial replacement of NaCl by KCl and $CaCl_2$ on physicochemical properties, microstructure, and textural properties of salted eggs. *Journal of Food Science*, *87*(2), 795–807. https://doi.org/10.1111/1750-3841.16033

Liu, H. P., Liu, P. W., Zhang, X. Y., Ji, L., & Han, Z. F. (2012). A Method for Egg Pickled in Rice Wine under Low Pressure-vacuum Processing. CN102068010B. China. (in Chinese)

Lunow, D., Kaiser, S., Brückner, S., Gotsch, A., & Henle, T. (2013). Selective release of ACE-inhibiting tryptophan-containing dipeptides from food proteins by enzymatic hydrolysis. *European Food Research and Technology*, *237*(1), 27–37. https://doi.org/10.1007/s00217-013-2014-x

Luo, W., Xue, H., Xiong, C., Li, J., Tu, Y., & Zhao, Y. (2020). Effects of temperature on quality of preserved eggs during storage. *Poultry Science*, *99*(6), 3144–3157. https://doi.org/10.1016/j.psj.2020.01.020

Luo, X., Zhao, Y., TU, Y., Wang, J., Li, J., Yang, Y., & Deng, W. (2012). Research progress in protein gel formation and regulation of preserved eggs. *Science and Technology of Food Industry*, *33*(23), 383–387. (in Chinese)

Lv, F., Zheng, M., & Chen, L. (2005). Process of lead-free preserved eggs by gradient temperature control. *Journal of Fujian Agriculture and Forestry University (Natural Science Edition)*, *34*(4), 527–530. (in Chinese)

Ma, M. (2007). *Egg and egg product processing*. China Agricultural Press. (in Chinese)

Matsudomi, N., Oka, H., & Sonoda, M. (2002). Inhibition against heat coagulation of ovotransferrin by ovalbumin in relation to its molecular structure. *Food Research International*, *35*(9), 821–827. https://doi.org/10.1016/S0963-9969(02)00085-6

Meng, Y. C., Xu, K., & Chen, J. (2010). Preliminary study on physicochemical properties and microbial safety of Pinghu Zao-egg. *China Poultry*, *32*(08), 36–39. (in Chinese)

Meng, Y., Chen, C., Qiu, N., & Keast, R. (2020). Modulation of gut microbiota in rats fed whole egg diets by processing duck egg to preserved egg. *Journal of Bioscience and Bioengineering*, *130*(1), 54–62. https://doi.org/10.1016/j.jbiosc.2020.02.015

Mine, Y. (2002). Recent advances in egg protein functionality in the food system. *World's Poultry Science Journal*, *58*(1), 31–39. https://doi.org/10.1079/WPS20020005

Novia, D., & Juliyarsi, I. (2019). Quality characteristics of salted egg soaked with Aloe vera solution. *International Journal on Advanced Science, Engineering and Information Technology*, *9*, 434–440. https://doi.org/10.18517/ijaseit.9.2.1805

Oliveira Dos Santos, D., Selia Dos Reis Coimbra, J., De Sousa, R. D. C. S., Da Silva, C. A. S., Barreto, S. L. D. T., & Minim, L. A. (2008). Solubility of quail (Coturnix japonica) egg white protein. *Journal of Food Process Engineering*, *31*(5), 684–693. https://doi.org/10.1111/j.1745-4530.2007.00182.x

Pu, Y., Du, J., Liang, Z., Pi, J., Pan, A., Wu, Y., Shen, J., Li, Q., Xiang, J., & Xu, X. (2010). Study on processing of salted egg with circulating water. *China Poultry*, *16*, 25–27. (in Chinese)

Sholehah, F., Thohari, I., & Jaya, F. (2017). Pengaruh penambahan Sari lengkuas merah (alpinia purpurata k. Schum) dan lama simpan telur asin terhadap total mikroorganisme, aktivitas antioksidan, aktivitas air dan tekstur. *Jurnal Ilmu dan Teknologi Hasil Ternak*, *10*(2), 18–27. https://doi.org/10.21776/ub.jitek.2015.010.02.2

Smith, M., & Back, J. F. (1962). Modification of ovalbumin in stored eggs detected by heat denaturation. *Nature*, *193*(4818), 878–879. https://doi.org/10.1038/193878a0

Su, Y., Chen, Z., Li, J., Chang, C., Gu, L., & Yang, Y. (2021). Characterization of salted egg yolk flavoring prepared by enzyme hydrolysis and microwave irradiation. *Food Chemistry*, *338*, 127913. https://doi.org/10.1016/j.foodchem.2020.127913

Sun, N., Liu, H., Wen, Y., Yuan, W., Wu, Y., Gao, J., & Li, C. (2020). Comparative study on Tianjin and Baiyangdian preserved eggs pickled by vacuum technology. *Journal of Food Processing and Preservation*, *44*(4), e14405. https://doi.org/10.1111/jfpp.14405

Sun, X., He, L., Yang, H., Wu, W., Yue, L., Peng, W., Jin, G., & Jin, Y. (2018). Intermittent ultrasound assisted in speeding up the pickling speed of salted eggs. *Science and Technology of Food Industry*, *39*(22), 204–211. (in Chinese)

Sun, X., Huang, J., Zeng, H., & Wu, J. (2018). Protein-resistant property of egg white ovomucin under different pHs and ionic strengths. *Journal of Agricultural and Food Chemistry*, *66*(42), 11034–11042. https://doi.org/10.1021/acs.jafc.8b03905

Tan, J., Liu, T., Yao, Y., Wu, N., Du, H., Xu, M., Liao, M., Zhao, Y., & Tu, Y. (2021). Changes in physicochemical and antioxidant properties of egg white during the Maillard reaction induced by alkali. *LWT-Food Science and Technology*, *143*, 111151. https://doi.org/10.1016/j.lwt.2021.111151

Thohari, I., & Rosyidi, D. (2017). Pengaruh kosentrasi sari kunyit putih (Curcuma zediaria) terhadap kualitas telur asin ditinjau dari aktivitas antioksidan, total fenol, kadar protein dan kadar garam. *Jurnal Ilmu dan Teknologi Hasil Ternak*, *10*(1), 46–53. https://doi.org/10.21776/ub.jitek.2015.010.01.6

Toops, D. (2014). *Eggs: A global history, Reaktion books* (New ed.). (April 15, 2014) ISBN-13 : 978-1780232645.

Wang, H. N., Liu, H. P., Liu, P. W., & Yuan, W. (2013). Changes in yolk protein secondary structure of the preserved egg in wine during low pressure-vacuum processing. *Modern Food Science and Technology*, *29*(06), 1262–1265. (in Chinese)

Wang, J. G. (2001). Zhejiang famous specialty: Pinghu Zao-egg. *Jiangsu Condiment and Subsidiary Food*, *1*, 31. (in Chinese).

Wang, J. G., & You, M. T. (2010). Process techniques and features of Chinese wine cured egg. *China Brewing*, *9*, 138–141. (in Chinese).

Wang, S., Wang, S., Zhang, Y., & Zhang, R. (2013). Parameter optimization for quickly salted egg by using ultrasonic-pulsed pressure technology. *Transactions of the Chinese Society of Agricultural Engineering*, *29*(23), 286–292. (in Chinese)

Wang, X., Huang, Y., Zhou, B., Xu, W., Xiang, X., Huang, Q., & Li, S. (2021). Improvement of quality and flavor of salted egg yolks by ultrasonic assisted cooking. *Ultrasonics Sonochemistry*, *75*, 105579. https://doi.org/10.1016/j.ultsonch.2021.105579

Wang, Y., & Pang, M. (2020). Study on salting mechanism and method of salted duck eggs. *Modern Food*, *22*, 52–54. (in Chinese).

Wang, Y., Xiong, C., Luo, W., Li, J., Tu, Y., & Zhao, Y. (2021). Effects of packaging methods on the quality of heavy metals–free preserved duck eggs during storage. *Poultry Science*, *100*(5), 101051. https://doi.org/10.1016/j.psj.2021.101051

West, B., & Zhow, B.-X. (1988). Did chickens go North? New evidence for domestication. *Journal of Archaeological Science*, *15*, 515–533. https://doi.org/10.1016/0305-4403(88)90080-5

Woodward, S., & Cotterill, O. (1986). Texture and microstructure of heat-formed egg white gels. *Journal of Food Science*, *51*(2), 333–339. https://doi.org/10.1111/j.1365-2621.1986.tb11123.x

Wu, H. (2013). Effects of different temperatures on quality of pidan. *The Food Industry*, *1*, 36–38. (in Chinese).

Xie, Y. X., Wang, J. Q., Shi, Y. N., Wang, Y., Cheng, L., Liu, L. L., Wang, N., Li, H. M., Wu, D., & Geng, F. (2020). Molecular aggregation and property changes of egg yolk low-density lipoprotein induced by ethanol and high-density ultrasound. *Ultrasonics Sonochemistry*, *63*, 104933. https://doi.org/10.1016/j.ultsonch.2019.104933

Xu, J., Liu, H., Yang, X., Zhang, C., Wang, Y., Zhao, F., & Liu, X. (2015). Speed up the process of salted eggs by vacuum technique and the changes of fatty acid in yolk. *Food Science and Technology*, 40, (6), 102–107. (in Chinese).

Xu, L., Gu, L., Su, Y., Chang, C., Wang, J., Dong, S., Liu, Y., Yang, Y., & Li, J. (2020). Impact of thermal treatment on the rheological, microstructural, protein structures and extrusion 3D printing characteristics of egg yolk. *Food Hydrocolloids*, *100*, 105399. https://doi.org/10.1016/j.foodhyd.2019.105399

Xu, L., Zhao, Y., Xu, M., Yao, Y., Nie, X., Du, H., & Tu, Y. (2018). Changes in aggregation behavior of raw and cooked salted egg yolks during pickling. *Food Hydrocolloids*, *80*, 68–77. https://doi.org/10.1016/j.foodhyd.2018.01.026

Xu, L., Zhao, Y., Xu, M., Yao, Y., Nie, X., Du, H., & Tu, Y.-g. (2017). Effects of salting treatment on the physicochemical properties, textural properties, and microstructures of duck eggs. *PLoS One*, *12*(8), e0182912. https://doi.org/10.1371/journal.pone.0182912

Xu, L., Zhao, Y., Xu, M., Yao, Y., Wu, N., Du, H., & Tu, Y. (2019). Changes in physico-chemical properties, microstructure, protein structures and intermolecular force of egg yolk, plasma and granule gels during salting. *Food Chemistry*, *275*, 600–609. https://doi.org/10.1016/j.foodchem.2018.09.078

Xue, H., Tu, Y., Xu, M., Liao, M., Luo, W., Guo, W., Zhang, G., & Zhao, Y. (2020). Changes in physicochemical properties, gel structure and in vitro digestion of marinated egg white gel during braising. *Food Chemistry*, *330*, 127321. https://doi.org/10.1016/j.foodchem.2020.127321

Xue, H., Zhang, G., Han, T., Li, R., Liu, H., Gao, B., Tu, Y., & Zhao, Y. (2022). Improvement of gel properties and digestibility of the water-soluble polymer of tea polyphenol-egg white under thermal treatment. *Food Chemistry*, *372*, 131319. https://doi.org/10.1016/j.foodchem.2021.131319

Yang, S., & Chen, K. (2001). The oxidation of cholesterol in the yolk of selective traditional Chinese egg products. *Poultry Science*, *80*(3), 370–375. https://doi.org/10.1093/ps/80.3.370

Yang, Y., Zhao, Y., Xu, M., Yao, Y., Wu, N., Du, H., & Tu, Y. (2020a). Alkali induced gelation behavior of low-density lipoprotein and high-density lipoprotein isolated from duck eggs. *Food Chemistry*, *311*, 125952. https://doi.org/10.1016/j.foodchem.2019.125952

Yang, Y., Zhao, Y., Xu, M., Yao, Y., Wu, N., Du, H., & Tu, Y. (2020b). Effects of strong alkali treatment on the physicochemical properties, microstructure, protein structures, and intermolecular forces in egg yolks, plasma, and granules. *Food Chemistry*, *311*, 125998. https://doi.org/10.1016/j.foodchem.2019.125998

Yao, L., Jiang, A. M., & Chen, L. (2020). Characterization of ethanol-induced egg white gel and transportation of active nutraceuticals. *LWT-Food Science and Technology*, *130*, 109530. https://doi.org/10.1016/j.lwt.2020.109530

Yoshizawa, S., Arakawa, T., & Shiraki, K. (2014). Dependence of ethanol effects on protein charges. *International Journal of Biological Macromolecules*, *68*, 169–172. https://doi.org/10.1016/j.ijbiomac.2014.04.041

Yu, L., Xiong, C. H., Li, J. K., Luo, W. X., Xue, H., Li, R., Tu, Y. G., & Zhao, Y. (2020). Ethanol induced the gelation behavior of duck egg whites. *Food Hydrocolloids*, *105*, 105765. https://doi.org/10.1016/j.foodhyd.2020.105765

Yu, L., Xue, H., Xiong, C. H., Xin, X. J., Wang, P., Feng, F., Cao, D. H., Tu, Y. G., & Zhao, Y. (2021). Characterization of duck egg white gel under the action of baijiu (Chinese liquor). *LWT-Food Science and Technology*, *147*, 111487. https://doi.org/10.1016/j.lwt.2021.111487

Zamora, R., Nogales, F., & Hidalgo, F. J. (2005). Phospholipid oxidation and nonenzymatic browning development in phosphatidylethanolamine/ribose/lysine model systems. *European Food Research and Technology*, *220*(5), 459–465. https://doi.org/10.1007/s00217-004-1114-z

Zeng, H. H., Liang, X. P., Li, S. Q., & Liu, F. G. (2022). Effects of combined treatments with ethanol and transglutaminase on the gel properties of whey protein isolates. *Journal of Shaanxi Normal University (Natural Sciences Edition)*, *50*(2), 1–11. (in Chinese).

Zhang, F. (2004). Changes of alkalinity during preserved chicken egg processing. *Food and Fermentation Industries*, *30*(10), 81–83. (in Chinese).

Zhang, X., Liu, X., Gao, J., Liu, H., Xu, X., & Tian, W. (2020). Mechanism of accelerating the formation of preserved egg white gel by vacuum decompression. *Modern Food Science and Technology*, *36*(6), 172–180. (in Chinese).

Zhao, Y., Cao, D., Shao, Y., Xiong, C., Li, J., & Tu, Y. (2020). Changes in physico-chemical properties, microstructures, molecular forces and gastric digestive properties of preserved egg white during pickling with the regulation of different metal compounds. *Food Hydrocolloids*, *98*, 105281. https://doi.org/10.1016/j.foodhyd.2019.105281

Zhao, Y., Chen, Z., Li, J., Xu, M., Shao, Y., & Tu, Y. (2016a). Changes of microstructure characteristics and intermolecular interactions of preserved egg white gel during pickling. *Food Chemistry*, *203*, 323–330. https://doi.org/10.1016/j.foodchem.2016.02.044

Zhao, Y., Chen, Z., Li, J., Xu, M., Shao, Y., & Tu, Y. (2016b). Formation mechanism of ovalbumin gel induced by alkali. *Food Hydrocolloids*, *61*, 390–398. https://doi.org/10.1016/j.foodhyd.2016.04.041

Zhao, Y., Tu, Y., Xu, M., Li, J., & Du, H. (2014). Physicochemical and nutritional characteristics of preserved duck egg white. *Poultry Science*, *93*(12), 3130–3137. https://doi.org/10.3382/ps.2013-03823

Zhong, L., Ma, N., Wu, Y., Zhao, L., Ma, G., Pei, F., & Hu, Q. (2018). Characterization and functional evaluation of oat protein isolate-*Pleurotus ostreatus* β-glucan conjugates formed via Maillard reaction. *Food Hydrocolloids*, *87*, 459–469. https://doi.org/10.1016/j.foodhyd.2018.08.034

Zhou, Q. (2012). *Studies on texture and flavor of salty yolk.* [Master dissertation, Tianjin University of Science and Technology]. Tianjing, China.

Zhu, Y. (1991). The processing technique of egg preserved in rice wine. *Meat Research*, *1*, 27–28. (in Chinese).

Zuo, S. (2013). *The gel property of ovomucin and its effect on egg white gel.* [Master dissertation, Huazhong Agricultural University]. Wuhan, China.

14

Innovative Strategies for New Egg Product Development

Hajime Hatta, Vincent Guyonnet, Yoshie Ueno, Hiroyuki Shidara, Tetsuya Miyamoto, Akihiro Handa, and Yoshinori Mine*

CONTENTS

14.1 Introduction

The global growth in egg consumption is linked to providing consumers more opportunities to eat eggs at any time of the day. Egg products, in the form of snack foods, quick ready-to-eat meals or drinks, have therefore the potential to provide consumers a wide range of options for more meal opportunities with eggs. The egg product segment is well developed in both the USA and Japan, servicing the food processing industry as well as the retail sector. In fact, Japan is the only country in the

* Corresponding author.

DOI: 10.1201/9781003254430-18

world where consumers typically eat more eggs in the form of egg products than shell eggs (Guyonnet, 2020a). This chapter will review the current market situation for egg products in the USA and Japan, highlighting the different egg products developed in these countries. The future development of the egg product segment will depend on its ability to address the evolving demand of consumers, especially as it relates to price, convenience, taste, authenticity and sustainability.

14.2 The Development of the Egg Product Business in North America

While the egg product business started in North America as a niche market, offering an outlet for excess inventory of shell eggs (see Chapter 2), it quickly emerged as a mass market, servicing initially the food manufacturing and food service sectors followed by a broad distribution into the retail sector.

14.2.1 Food Service Segment

14.2.1.1 Traditional Liquid and Frozen Egg Products

From an outlet for eggs in time of oversupply, the processing of eggs through investment and innovations offers now many options to add value to the egg sector. As a proof that egg processing is no longer a sector built on surplus eggs, in several countries, the entire production of farms is solely dedicated to egg processing, where the eggs are automatically conveyed from the layer houses to the processing plant (see Chapter 2). For many years, the processing of eggs has grown along with the increasing demand from the food service sector (Quick Service Restaurants or QSR) and food manufacturers. Combining convenience with food safety, fresh and frozen liquid products, ranging in presentation size from 250 grams to 20,000 kilograms, have been adopted by many and represent today about 60% of the total egg usage by the food service and food manufacturing sectors (Guyonnet, 2021a).

14.2.1.2 Hard-Boiled Eggs

The production of hard-boiled eggs has allowed egg producers and processors to find a profitable use for small and medium size shell eggs not desired by retail consumers. The production of hard-boiled eggs is quite simple and standardized. Over the past 20 years, manufacturers of egg processing equipment have automated the manufacturing processes for boiling, cooling and peeling the eggs. It is possible to produce up to 20,000 peeled hard-boiled eggs per hour without much extra labor, thus minimizing the cost of production. Initially, the demand for hard-boiled eggs was driven by the food service sector, with eggs used for salad bars and as an ingredient for the preparation of sandwiches. Chefs and restaurants dealing with a large number of customers welcomed this new alternative which saved them both time and labor and also reduced the probability of food safety issues (Guyonnet, 2021a).

14.2.1.3 Frozen Omelets and Egg Sandwiches

Catering for the travel and hotel industries has also created more opportunities for the egg processing sector. Omelets served at breakfast in North American hotel chains are usually processed by egg companies, offering a range of flavors, shapes and weights per serving portion. Egg sandwiches are recent items added to many menus of QSR. The global growth of the QSR, often linked to longer commuting time and a busier lifestyle, will likely stimulate innovations in the egg processing sector over the next decade.

14.2.2 Retail Segment

14.2.2.1 Cholesterol-Free Egg Substitutes

The first widely distributed egg product, Egg Beaters®, was launched in the USA in 1972, at a time when consumers were concerned about the level of cholesterol in eggs. Through processing, the yolk was removed after breaking and some flavoring and coloring agents, along with some gums for texture, were blended with the egg white. The final product looked and somewhat tasted like an egg but did not contain any cholesterol. Since the yolk was removed from the final product, the term used under the US legislation for this new product was "egg substitute". Unlike other egg products, this type of product has been under the jurisdiction of the US Food and Drug Administration (FDA). However, as of October 30, 2023, egg substitutes will be recognized as egg products and will be under the jurisdiction of the US Food Safety and Inspection Services (Federal Register, 2021). These products were initially used by consumers with dietary restrictions due to high blood cholesterol and related cardiovascular conditions and were sold frozen, in small presentation sizes (less than 200 ml). Later on, egg substitutes were adopted by more and more consumers worried about cholesterol in eggs and many US retailers launched their own brand of refrigerated egg substitutes (called private labels in the USA), with increasing presentation sizes (up to 1 liter) (Figure 14.1).

14.2.2.2 Liquid Egg White Products

Liquid egg white products appeared on retail shelves in the 1990s, initially to circumvent the patent obtained by ConAgra

FIGURE 14.1 Cholesterol-free egg substitute products were the first egg products widely available in US retail stores.

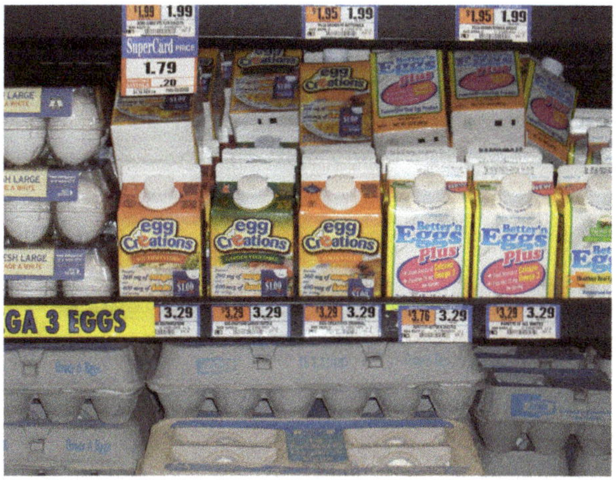

FIGURE 14.2 Ready-to-cook liquid omelets have recently gained popularity in US retail stores.

and offered an alternative to consumers concerned about cholesterol. Consumers baking at home quickly adopted these products for convenience when recipes required only egg whites. Later on, fitness aficionados, convinced of the nutritional benefits of egg white proteins (see Chapter 34), drove the need for bigger presentation sizes (1 liter or pack of three 1-liter containers) (Figure 14.1).

FIGURE 14.3 Functional liquid retail egg products (enriched in omega 3 fatty acids and lutein) on US retail shelves.

14.2.2.3 *Liquid Egg Omelets*

Following the dispelling of the cholesterol myth in the mid 2000s, several egg processors introduced new retail liquid egg products containing less yolk than in whole eggs products (one yolk for eight egg whites) or even based on the composition of whole eggs. In addition, ready-to-cook liquid omelets were developed, making it easier for consumers to cook an omelet without cracking an egg, Particulates such as pieces of cheese, red peppers, green peppers or small cubes of meat were added to egg substitutes or whole egg products (Figure 14.2).

14.2.2.4 *Functional Liquid Egg Products*

Initially in Canada, and later on in the USA, several liquid egg products were enriched with omega-3 polyunsaturated fatty acids (omega-3 PUFA), lutein, vitamin E or calcium. While some egg products are prepared with the use of shell eggs naturally enriched in omega-3 PUFA or other nutrients, most functional egg products are enriched through the addition of extra nutrients during the blending process of these products. In retail stores assigning a nutritional ranking to products on shelves, these products are often rated higher than regular liquid egg products (Figure 14.3).

14.2.2.5 *Hard-Boiled Eggs*

In many countries, consumers have a strong preference toward shell eggs of large or extra-large caliber, often leaving producers with inventories of medium or small eggs difficult to sell at a reasonable profit. Therefore, the development of the hard-boiled egg segment has been a blessing for egg producers and marketers. Since the mid-2010s, this market segment has been growing tremendously in North America. And while the demand for hard-boiled eggs was initially driven by the food service sector, the "meal on the go!" wave in North America, where consumers grab some foods and drinks conveniently located at the entrance of retail stores, has offered the opportunity to sell hard-boiled eggs at retail (Figure 14.4). With the use of a wide range of shell eggs, including organic, free range, omega-3 PUFA enriched eggs, or eggs stored in brine solutions imparting a specific flavor to the eggs, a complete line of hard-boiled eggs may be offered at retail. Overall, hard-boiled eggs have allowed egg processors to enter the snack food market, one of the fastest-growing segments of the retail food sector (Guyonnet, 2019).

14.2.2.6 *Egg Wraps, Egg Bites and Snacks*

With the expansion of the snack food retail segment, more egg processing companies are using eggs as the main ingredient to prepare convenient snack foods such as egg white-based wraps, quiches and bite-size egg-based foods. These products are usually sold as frozen items, allowing for a long shelf-life and in a ready-to-eat format, requiring only quick warming in microwave ovens.

It is expected that the egg product retail segment will attract more players over the next decade with product innovation likely contributing to the major growth of the egg category (combined sales of shell eggs and egg products) at retail.

FIGURE 14.4 Hard-boiled Eggs, available in retail stores and service stations, have become extremely popular in North America for snacking and quick meal options at retail stores and service stations.

14.3 New Egg Product Development in Japan

As the only country in the world where consumers typically eat more eggs in the form of egg products than shell eggs, the development of new egg products has been one of the main drivers behind the growth of the Japanese egg sector. Japan is likely the country with the widest variety of egg products sold at retail and available to food processors for the manufacture of food products.

14.3.1 Processed Egg Products Using Fermentation Technology

Fermentation is a technique using mold and yeast for processing food. This technique has been used since ancient times and applied to produce foods such as yogurt, cheese and natto from milk and soybeans, respectively. This technology improves the flavor and enables the long-term storage of perishable foods. There has never been a traditional egg product using fermentation technology, likely due to the naturally long-storage shelf-life of shell eggs. In recent years, this technology was tested with eggs in an attempt to further enhance their taste.

14.3.1.1 Lactic Acid Fermented Egg White

Since the egg white is rich in protein and fat-free, it constitutes an excellent food for athletes who want to strengthen their muscles but do not want to gain weight (see Chapter 34). However, raw egg whites are alkaline with a slimy texture, making them not very tasty. After heat treatment, egg whites coagulate, thus improving their texture. However, the denaturation of proteins and decomposition of amino acids leads to an unpleasant smell due to released sulfur combining with hydrogen to form hydrogen sulfide (rotten egg smell). Fermentation with lactic acid bacteria can resolve this concern and make egg whites easier to eat. An egg white product (sold under the Lacty Egg trademark), produced by fermentation with lactic acid bacteria, improves greatly the taste and flavor of egg whites, making the product easy to drink by itself, similar to a yogurt drink. Fermentation further improved the physical properties of egg whites (Arimitsu et al., 2015).

Mixing (Egg white, Water, Sugar, Yeast extract, etc.)
⇩
Heating
⇩
Inoculation of lactic acid bacteria
⇩
Fermented (42℃×24h)
⇩
Pasteurize
⇩
Homogenized
⇩
Packed

FIGURE 14.5 Manufacturing process for lactic fermented egg whites.

Egg white is a material that does not quickly ferment lactic acid since it contains little sugar, a nutrient source for lactic acid bacteria, and is also rich in proteins with antibacterial properties such as lysozyme. A manufacturing process (Figure 14.5) was developed to ferment lactic acid in egg white, combining a heating step to inactivate lysozyme (inhibiting the growth of lactic acid) with the addition of a fermentation aid to promote the growth of lactic acid bacteria (Yoshimi, 2011).

The analysis of the volatile components produced in lactic acid fermented egg white revealed the production of aroma components such as acetaldehyde and diacetyl, and the reduction of hydrogen sulfide causing the foul odor of egg white. The amount of hydrogen sulfide generated from egg white will change depending on the pH value, decreasing in acidic pH values. The amount of hydrogen sulfide from lactic acid fermented egg white was significantly lower than that of egg white, even when the pH was adjusted to 4.3 with hydrochloric acid (Figure 14.6). Therefore, it was demonstrated that lactic acid fermented egg white will have a better flavor and can easily be consumed alone, with no need to combine with other ingredients (Arimitsu et al., 2015).

Adding lactic acid fermented egg white to foods will confer them the moist feeling of sponge cakes and suppress the peculiar odor of foods packed in retort pouches. Figure 14.7 shows the results of a sensory evaluation and aroma analysis after mixing 2% and 5% of lactic acid-fermented egg white with the ingredients of a retortable packaged beef bowl and

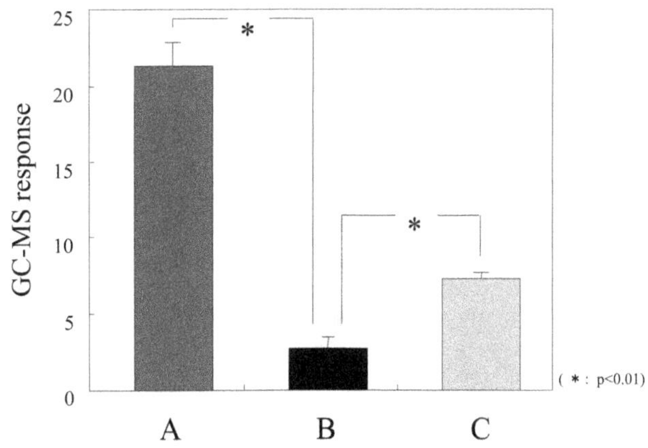

FIGURE 14.6 Hydrogen sulfide generated from egg white products: A: Egg white at pH 7.0; B: fermented egg white at pH 4.3; C: egg white at pH 4.3. (Adapted from Arimitsu et al, 2015.)

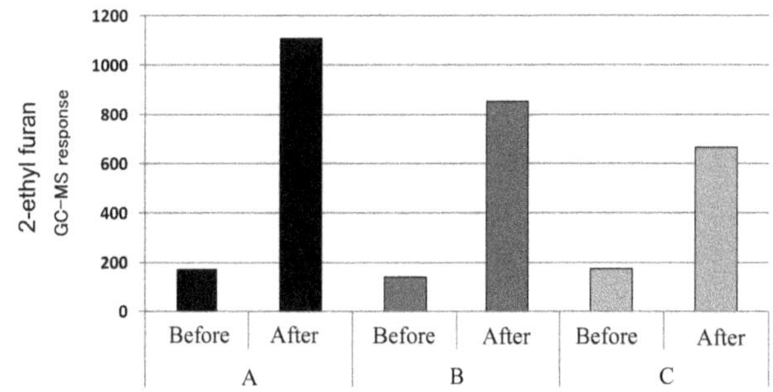

FIGURE 14.8 Thermally derived off-flavor compounds of a retortable packaged beef bowl preparation before and after heat treatment: A: Beef bowl filing; B: beef bowl filling with 2% fermented egg white; C: beef bowl with 5% fermented egg white. (Adapted from Arimitsu et al, 2015.)

heating at 121°C for 40 minutes. The study confirmed that the odor generated by heating as well as the peculiar odor of beef meat could be suppressed by blending lactic acid fermented egg white. The amount of 2-ethyl furan, generated by heating the retort pouch, was reduced with the incremental addition of fermented egg whites (Figure 14.8) (Arimitsu et al., 2015).

Lacto-fermented egg whites have also shown some excellent health benefits. In a study with 88 adult males with total blood cholesterol levels of 200–260 mg/dL, the consumption of 8 g of Lacto-fermented egg whites for 8 weeks resulted in a significant reduction ($p < 0.05$) of blood cholesterol levels by 11.0 ± 3.7 mg/dL Furthermore, CT examination suggested also a reduction in visceral fat (Usuda et al., 2011).

Lacto-fermented egg whites, combining the ease of consumption in a drink, the improved taste when combined to other foods, and the great nutritional values of egg proteins, represent an innovative way to consume more egg whites.

14.3.1.2 Aged Egg Yolk

Koji is an essential element in Japanese cuisine and other East Asian cuisines. It is generally made by covering steamed grains

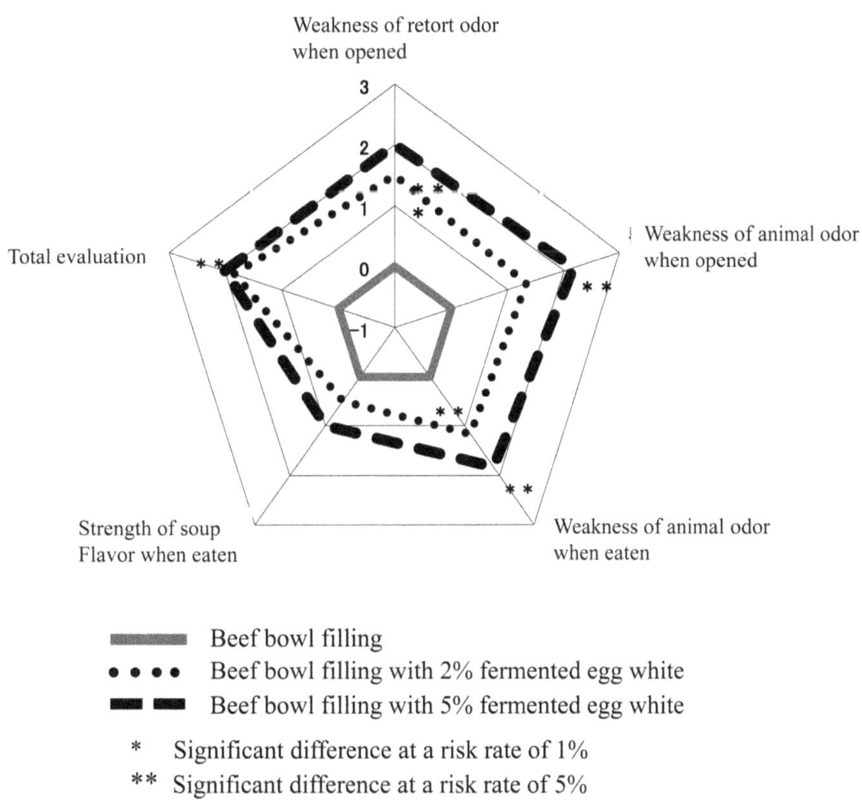

FIGURE 14.7 Sensory evaluation of a retortable packaged beef bowl preparation. (Adapted from Arimitsu et al., 2015.)

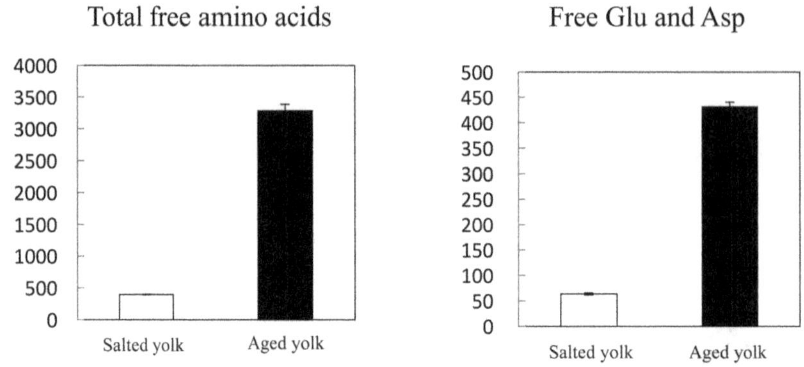

FIGURE 14.9 Differences in total free amino acids (left) and free glutamine (Glu) and aspartic acid (Asp) (Right) between salted yolk and aged yolk. (Adapted from Miyamoto et al., 2019.)

such as rice, wheat or soybeans with koji mold derived from koji spores and letting them ferment. Koji is a fermentation starter for various Japanese staple foods, including sake, miso and soy sauce. The production of aged egg yolk applies this traditional processing method to egg yolks. Aged egg yolk is produced by mixing egg koji with egg yolk and leave the mixture to age for a certain period of time. Malted Egg is made by fermenting eggs with koji mold. During the fermentation process, an enzyme added to the egg yolk will break down the fat and proteins. This aging process further enhances the flavor.

Compared to salted egg yolk, aged egg yolk has more umami and sweetness components, with an enhanced specific egg yolk flavor (Figure 14.9). 1-octen-3-ol, an aromatic component enriching the richness of the food, increased by more than 150 times in aged egg yolk. The aging process generated isovaleraldehyde, phenylacetaldehyde and methional, aromatic components enhancing umami flavor, richness, and cheese flavor, respectively (Miyamoto et al., 2019).

Carbonara is an Italian pasta dish made with eggs, hard cheese, cured pork and black pepper. Carbonara dishes were prepared using either regular egg yolks or aged egg yolks and were compared for egg yolk and cheese flavor intensity and overall richness. Thirty-one panelists, recruited for the taste test, rated the carbonara prepared with aged egg yolks as having a stronger and richer egg and cheese taste than the carbonara prepared with regular egg yolks (Figure 14.10).

In addition, when eight expert panelists tested the change in flavor when eating the carbonara, they evaluated the flavor of the carbonara prepared with aged egg yolks as stronger in intensity and lasting longer in the mouth (Figure 14.11) (Miyamoto et al., 2020). Although the use of aged yolks by food processors is currently limited to the production of a few commercial preparations, it is anticipated that such seasoning has the potential to improve the taste of foods cooked at home.

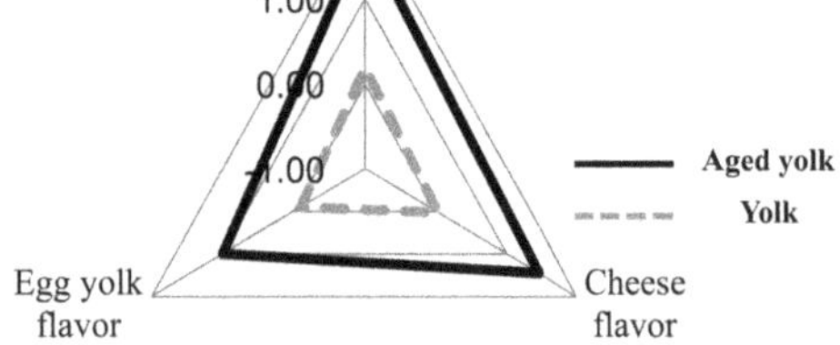

FIGURE 14.10 Differences in flavor of carbonara prepared with regular egg yolk and aged yolk. (Adapted Miyamoto et al., 2020.)

14.3.2 Egg Koji and Egg White Fermented Seasoning (Tamago Shoyu)

Soy sauce made from soybeans and wheat is widely used as the essential seasoning in Japanese cuisine, including for home cooking, although traditional fish seasoning (fermented fish with salt) is also used in certain regions. The soy sauce is a fermented seasoning obtained by the breakdown of proteins in soybeans and wheat with a protease produced by Koji mold (Kotaro & Asahi, 2021). In addition, Japan has the food culture of eating raw eggs, exemplified by "*Tamago Kake Gohan*", a dish of steamed rice mixed with a raw egg and seasoned with soy sauce. A new egg white fermented seasoning made from egg whites was developed to use for the traditional "*Tamago Kake Gohan*" (egg on top of rice).

Typical soy sauce is made by growing koji mold (*Aspergillus oryzae* or *Aspergillus sojae*) in a medium of soybeans and

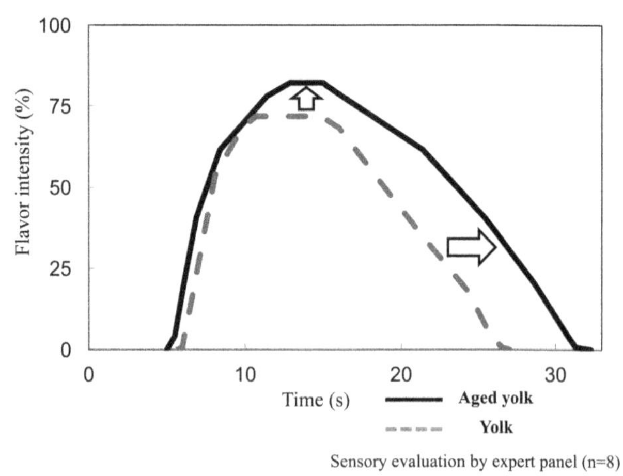

FIGURE 14.11 Time-intensity curve of richness of carbonara prepared with regular and aged yolks. (Adapted from Miyamoto et al., 2020.)

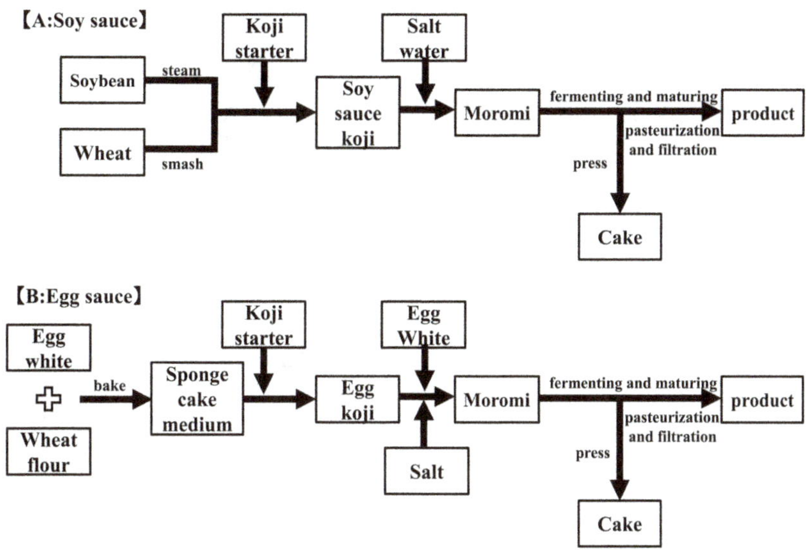

FIGURE 14.12 Manufacturing process of soy sauce and egg sauce.

FIGURE 14.13 Growth of koji mold on various media (cross-sections).

wheat, mixing the resulting soy sauce malt with salted water and fermenting the mixture at room temperature for about six months. For the preparation of egg sauce, flour and egg whites were used (Figure 14.12). Generally, in soy sauce malt, koji mold grows only at the surface of soybeans and wheat grains. For the egg sauce koji, koji malt was cultivated in a sponge cake with a broad porous surface, allowing koji malt to extend its mycelium growth well into the sponge cake.

Koji mold culture media was prepared with a mixture of egg white and flour, with boiled eggs (Boiled egg culture media), cooked eggs (omelet culture media) and sponge cake (sponge cake culture media). When koji mold was cultured in these media, it grew only at the surface of the boiled egg and omelet egg media. On the other hand, the koji mold was able to grow inside the sponge cake medium (Figure 14.13).

Protease activity, essential for the production of egg sauce, was the highest in sponge cake medium (Figure 14.14). The presence of air inside the sponge cake medium allowed koji mold to grow inside the sponge cake while the absence of air space in the boiled egg and omelet media limited its growth.

In addition, compared to the traditional soy sauce koji (soybeans and wheat), the activity of enzymes such as protease and α-amylase were approximately 1.4 to 3.5 times higher in egg koji (Table 14.1) (Sakiko et al., 2014). Sponge cake medium appeared to be the most suitable for the culture of koji mold and enzyme activity and may be used in various applications.

Egg malt, on which koji mold was propagated in a sponge cake medium, was added with salt and egg whites and allowed to ferment and mature for six months. This process resulted in the decomposition of proteins and the production of egg soy sauce containing high levels of glutamic acid and other flavor-enhancing amino acids. This egg soy sauce contains almost the same proportion of amino acids as regular soy sauces (Table 14.2).

Generally, soybeans and non-fat soybeans are rich in reducing sugars, and the color of soy sauces made from regular soybeans will become darker over time due to the Maillard

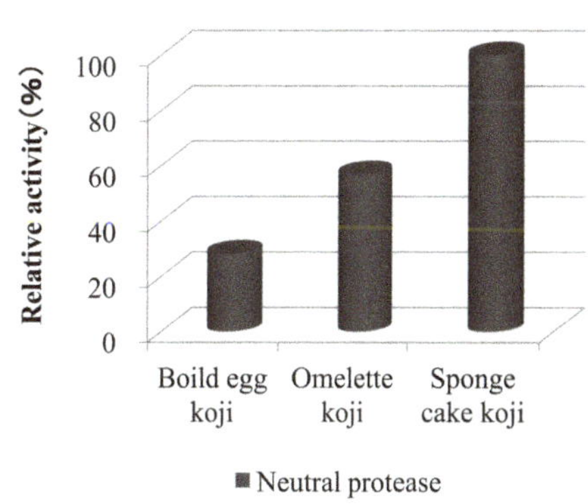

FIGURE 14.14 Neutral protease activity on various koji.

TABLE 14.1

Various Enzyme Activity of the Koji, Expressed in U/g

Enzyme	Egg Koji	Soy Sauce Koji
Neutral protease	90,800 ± 5,290	49,500 ± 6,170
Acid protease	48,800 ± 2,760	27,100 ± 700
Acid carboxypeptidase	55,900 ± 5,450	30,600 ± 609
Glutaminase	144 ± 24	104 ± 10
α-Amylase	4,470 ± 420	2,260 ± 159
Glucoamylase	1,810 ± 124	511 ± 13

Source: Adapted from Sakiko et al. (2014).

reaction. Instead of soybeans, egg white soy sauce is made from egg whites, containing very few reducing sugars, and thus less affected by the Maillard reaction and its effect on coloring (Figure 14.15). As such, the color tone of fermented egg white sauce is much lighter than that of regular soy sauces and its use in cooking will not affect the color of the ingredients used. The use of egg soy sauce is well suited for Japanese cuisine, allowing it to make the most of the natural color of all ingredients. It is also a good match for the many egg dishes appreciated in Japan.

FIGURE 14.15 Fermented egg white sauce products.

14.4 Production of Cholesterol-Removed Egg Yolk for Cholesterol-Free Foods

In recent years, lifestyle-related diseases have been increasing due to diversified diets, unhealthy lifestyles and lack of exercise. There is also a growing demand for foods with added functionality or reduced content in specific components and various technologies can be used to meet these objectives. One of them is the extraction technology using supercritical carbon dioxide (SCD). Carbon dioxide is a suitable extraction solvent used in the food industry, meeting its safety and efficacy requirements as a non-toxic, non-flammable and inexpensive reagent (Essien et al., 2020). SCD extraction technology has two objectives: the extraction of useful target substances and the removal of undesirable target substances. This technology has been used for the removal of cholesterol from egg yolk and the development of cholesterol-free mayonnaise-type seasoning products and other food products.

TABLE 14.2

Free Amino Acids of Egg Sauce and Soy Sauce (Expressed in g/100 ml)

Amino Acid	Egg Sauce	Soy Sauce A	Soy Sauce B
Asp	0.25	0.52	0.15
Thr	0.39	0.24	0.30
Ser	0.45	0.35	0.42
Glu	0.97	1.05	1.32
Pro	0.21	0.19	0.21
Gly	0.17	0.18	0.21
Ala	0.59	0.29	0.80
Cys	0.00	0.03	0.00
Val	0.52	0.32	0.44
Met	0.31	0.09	0.11
Ile	0.41	0.32	0.42
Leu	0.76	0.53	0.69
Tyr	0.08	0.05	0.06
Phe	0.51	0.31	0.41
His	0.15	0.13	0.15
Lys	0.41	0.28	0.34
Arg	0.14	0.38	0.52
Total	**6.32**	**5.25**	**6.55**

14.4.1 Application of SCD to Food Products

In Europe and the USA, SCD extraction is used in the decaffeination of green coffee beans (Vandeponseele et al., 2020) and flavor extraction from hops (Sievers & Eggers, 1996). Decaffeinated coffee is extremely popular and sold commonly along with regular coffee in most retail stores or restaurants. In Germany, there is a large facility for SCD extraction producing a large amount of hop extracts at low cost and distributed to beer manufacturers around the world. On the other hand, the scale of SCD extraction is small in Japan and limited to high value-added fields in the food sector such as the extraction of flavoring ingredients, spice oils, dyes, functional food ingredients and the separation of oils. For instance, Octacosanol, a functional food ingredient, is produced by a high-pressure two-step separation method, including SCD extraction.

Research focusing on the extraction of functional food ingredients using SCD is rather active around the world (Azmir et al., 2013; Wang et al., 2021). For example, laboratory research is focusing on the extraction of antioxidant and anti-inflammatory substances from seaweed (Gan & Baroutian, 2022), crustacean (Ahmadkelayeh & Hawboldt, 2020), plant seeds (Daud et al., 2022), plant peels (Romano et al., 2022) and legumes (Moreno-García et al., 2022).

14.4.2 Chicken Eggs and Cholesterol

14.4.2.1 Effects of Dietary Cholesterol on Health

While eggs contain most nutrients required for good nutrition, they are also a cholesterol-rich food with 100 g of yolk containing 1,400 mg of cholesterol (see Chapter 29). With the average weight of one egg estimated at 64 g, the yolk weighs about 17 g and will provide about 238 mg of dietary cholesterol. Cholesterol is an essential component in the body, as a constituent of cell membranes and as a molecule required for the synthesis of bile acids and steroid hormones. A healthy person may synthesize *de novo* up to 800 mg of cholesterol per day, but this amount is controlled by the amount of cholesterol ingested. Since the absorption rate of cholesterol from the intestinal tract is about 50% of the intake, and since eggs contain phospholipids and proteins that maintain normal serum cholesterol levels, there is absolutely no need for healthy individuals to reduce their consumption of shell eggs out of concern for cholesterol (see Chapter 30). In spite of this scientific evidence, a consumer demand remains for cholesterol-reduced or cholesterol-free foods.

14.4.2.2 Various Methods to Remove Cholesterol from Egg Yolk

Methods for removing cholesterol from egg yolk include extraction with organic solvents (Paraskevopoulou & Kiosseoglou, 2006; Puertas & Vázquez, 2019), extraction with vegetable oils, inclusion with cyclic oligosaccharides (Sun et al., 2018; Güven et al., 2019), conversion with microbial enzymes (Liu et al., 2013), fractionation of egg yolk (Lamas et al., 2016) and extraction with SCD. Each method has its own set of challenges: it is difficult to completely remove the organic solvents used, the method using vegetable oils generates a large amount of cholesterol-containing oils as by-products, the method using cyclic oligosaccharides requires some consideration regarding the utilization of cyclic oligosaccharides encapsulated with cholesterol, while the methods using microbial enzymes and the fractionation of egg yolk both cannot reach a sufficient cholesterol removal rate. For all these reasons, the method using SCD has been adopted commercially in Japan.

14.4.3 Extraction with SCD

14.4.3.1 Definition of a Supercritical State

In the case of carbon dioxide, it becomes supercritical at temperatures above 31.1°C and pressures above 7.3 MPa (Figure 14.16). The supercritical state is a fluid intermediate state between the liquid and gas phases. While this supercritical state has a high density similar to that of a liquid, it has the advantage of a viscosity closer to that of a gas with excellent permeation and diffusion into materials. Its diffusion coefficient is much higher than that of liquids, even when pressurized. In the supercritical state, the density of the fluid can be significantly changed by slight changes in temperature and pressure, so that the target component can be selectively extracted by setting the appropriate temperature and pressure.

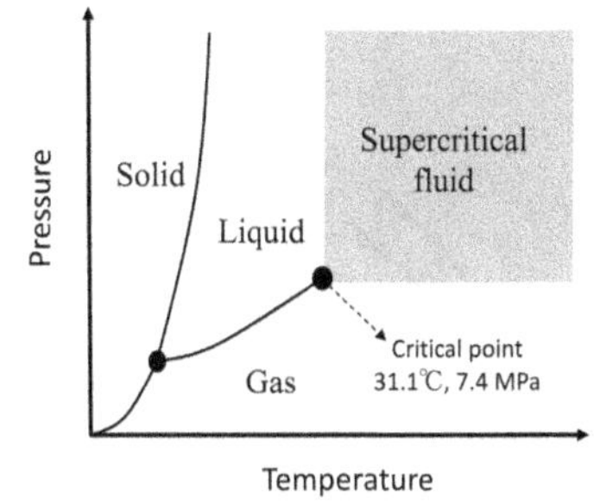

FIGURE 14.16 Carbon dioxide pressure – temperature phase diagram.

14.4.3.2 Extraction of Egg Components by SCD

The results of the analysis of the components extracted from egg yolk by SCD are presented in Table 14.3. The extracted components were lipids consisting of neutral lipids and cholesterol, which were selectively extracted. In addition to neutral lipids and cholesterol, the egg yolk contains phospholipids within its lipid fraction. Considering their valuable emulsifying ability, it was important to retain them within the yolk. As shown in Table 14.3, the presence of phospholipids in the extracts was not detected at any time, confirming the suitability of the SCD extraction. Since the solubility of neutral lipids in SCD is higher than that of cholesterol, the percentage of cholesterol in the extract is relatively higher in the late stage of extraction than in the early stage of extraction.

The changes in the percentage of extracted lipids over time at three different extraction pressures (25 MPa, 30 MPa and 35 MPa) at a constant SCD feed rate indicated that increasing the extraction pressure resulted in a higher lipid extraction rate (Figure 14.17). While the percentage of extracted lipids in this test was about 43%, the processing time may be shortened by increasing the applied pressure. As the density of SCD increases with increasing pressure, the amount of neutral lipids and cholesterol dissolved in SCD will increase at higher pressures. Therefore, for an efficient extraction of cholesterol, it is necessary to extract with SCD at the highest possible pressure.

TABLE 14.3

Composition of the Yolk Extract (%) at Different Extraction Stages

Components	Initial Stage	Middle Stage	Late Stage
Neutral lipids	96.2	95.6	90.5
Cholesterol	3.8	4.4	9.5
Phospholipids	N.D.	N.D.	N.D.

Note: N.D. = not detected.

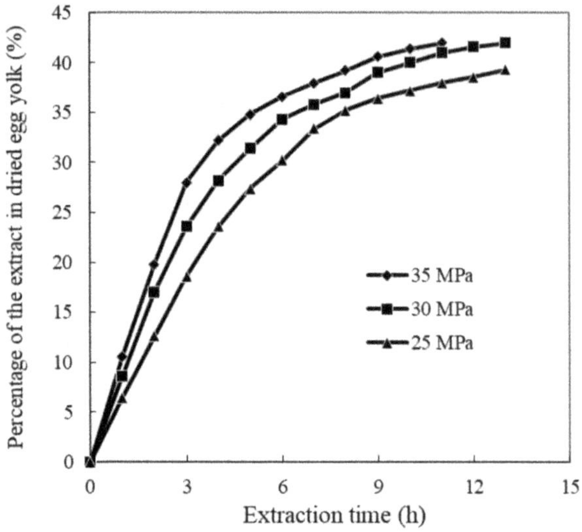

FIGURE 14.17 Evolution of the percentage extraction from dried egg yolk at 25 MPa, 30 MPa and 35 MPa based on the extraction time (expressed in hours).

The overall composition of the extract was not influenced by the differences in processing pressure.

14.4.3.3 SCD Extraction Plant

A schematic diagram of the SCD extraction process under commercial conditions is presented in Figure 14.18. The equipment required consists of a liquefied carbon dioxide tank, a pressurizing pump, a pressure valve, a heating unit, an extraction tank, a separation tank and a condenser.

Egg yolks are placed in the extraction tank, and liquefied carbon dioxide is pumped into the tank. After being pressurized to a defined pressure set by a pressure valve, the liquid goes through a heating unit and is then sent to the extraction tank as SCD. In the extraction tank, neutral lipids and cholesterol are dissolved in the SCD when they come into contact with the egg yolk due to the osmotic and diffusion properties of SCD. The dissolved neutral lipids and cholesterol are sent from the extraction tank to the separation tank. A pressure valve is installed between those tanks to reduce the pressure to less than 7.3 MPa, making both neutral lipids and cholesterol insoluble and ready for collection in the separation tank. The state of the solvent is also changed from SCD to carbon dioxide gas by passing through the pressure valve. Afterward, the carbon dioxide gas is cooled down by a condenser and will be recovered as liquefied carbon dioxide in a tank, ready to be used again as an extraction solvent. For an efficient extraction, it is important to increase the capacity of the liquefied carbon dioxide pump to have a higher amount of liquefied carbon dioxide pumped per unit of time. Although carbon dioxide has some negative aspects including its global warming potential, most of the carbon dioxide is recycled during the SCD extraction, therefore making this extraction method extremely environmentally friendly and sustainable.

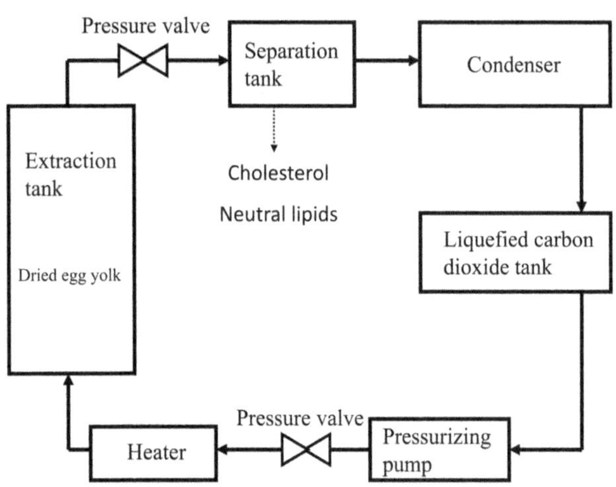

FIGURE 14.18 Schematic diagram of the SCD extraction process under commercial conditions.

14.4.3.4 SCD Processing of Egg Yolk on a Commercial Scale

The egg yolk is a highly nutritious fraction of shell eggs, consisting of 50% water, 33% lipid and 17% protein (see Chapters 6 and 29). However, egg yolks are prone to microbial spoilage in the liquid state. Therefore, dehydrated egg yolk (dried egg yolk) is generally used as the raw material in SCD extraction. There are several methods for drying egg yolk, including spray-drying, freeze-drying and vacuum drum drying (Bergquist, 1990), and the spray-drying method is the most suitable for its consistent quality and cost (see Chapter 10). The average diameter of spray-dried egg yolk particles is less than 100 μm, and their particle size seems to make them excellent candidates for SCD extraction. However, under commercial extraction conditions, the speed of SCD moving through the extraction tank is so high that fine particles may clog the extraction tank. Therefore, spray-dried egg yolk should be granulated prior to SCD extraction.

The higher the processing pressure and the faster the transfer rate of SCD, the higher the extraction efficiency. The largest commercial supercritical extraction facility for foods in Japan is operated under a pressure of 35 MPa and a flow rate of 3,000 L/h. The heating temperature is set at 35–40°C to avoid the denaturation and flavor degradation of egg yolk due to thermal treatment. By processing egg yolk under these conditions, it is possible to extract more than 95% of the cholesterol contained in egg yolks. The composition of dried egg yolks before and after SCD extraction (Figure 14.19) shows that neutral lipids and cholesterol are almost completely removed, and that the relative content of proteins and phospholipids will be significantly increased.

14.4.4 Applications of Cholesterol-Removed Egg Yolks in Food Products

14.4.4.1 Applications as a Food Ingredient

In addition to their basic food functions of nutrition and taste, egg yolks have very important physical properties such as emulsification, thermal coagulation and foaming. The application

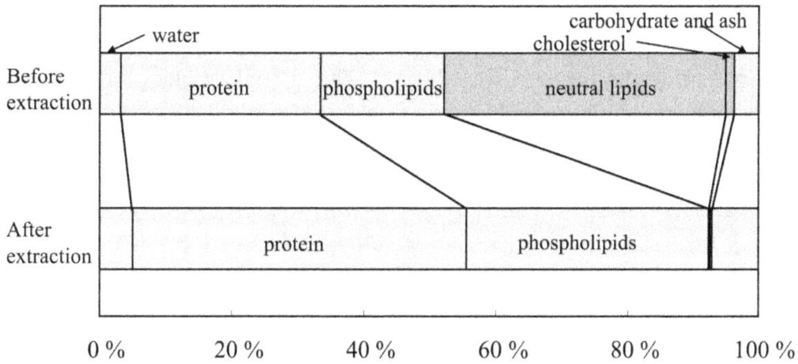

FIGURE 14.19 Composition of dried egg yolk before and after extraction (expressed in %).

of the SCD extraction method must therefore preserve these key properties of egg yolks. The emulsification capacity of dried egg yolks was measured before and after SCD extraction by evaluating the amount of vegetable oil emulsified by a certain amount of egg yolks under acidic conditions (pH 4.0). The emulsification capacity of dried egg yolks after SCD extraction was 96% of that of dried egg yolks before SCD extraction, indicating that the emulsification capacity of the dried egg yolks is maintained even after SCD extraction.

In order to evaluate the thermal coagulation function, the oil extracted by the SCD treatment was replaced with vegetable oil, then water was added to equalize the egg yolk composition and mixed to reconstitute the egg yolk. The reconstituted egg yolk was filled into a casing tube, heated at 80°C for 40 minutes and then cooled down. The breaking strength of the resulting gel was evaluated and compared to a control gel made of dried egg yolks before SCD treatment, prepared in a similar fashion. The results showed a decrease in breaking strength of about 15%, indicating the possibility of producing processed egg products with cholesterol removed by adding water and vegetable oil to SCD-treated dried egg yolk.

As for the foaming property, a comparison of the volume of sponge cakes made from dried egg yolk before and after SCD treatment showed that the volume increased with SCD treatment under practical extraction conditions, indicating that SCD treatment enhances the foaming function of dried egg yolk (Froning et al., 1990).

14.4.4.2 Cholesterol Labelling in Food Products

In response to the increased number of consumers unsubstantially concerned about cholesterol in food products due to the rise in lifestyle-related diseases, a standard for labeling cholesterol was established in 1999 in Japan. In order to label a food product as containing no cholesterol (e.g., zero, cholesterol-free), the cholesterol content in 100 g of food must be less than 5 mg, and the saturated fatty acid content must be less than 1.5 g (see Chapter 18).

14.4.4.3 Designing a Cholesterol-Free Mayonnaise Type Product

Mayonnaise is a processed food consisting of vegetable oil, eggs, vinegar, spices and seasonings. The cholesterol content per 100 g of mayonnaise made with egg yolk is 150 mg. According to the Japanese labeling standard, it is not possible to label this product as cholesterol-free. Since the majority of cholesterol in mayonnaise is derived from egg yolk, a mayonnaise-type product without egg yolk can contain less than 5 mg cholesterol. However, the flavor and texture of a mayonnaise without egg yolk are rather different from those of regular mayonnaise. Many people concerned about cholesterol are also concerned about calorie intake. Therefore, a product was developed to address the needs to be both cholesterol-free and low-calorie, while aiming for the same egg flavor and texture as real mayonnaise.

The egg yolks used in the non-cholesterol-free mayonnaise were SCD-treated egg yolks. Their use makes it possible to reduce the cholesterol content while imparting the characteristic egg flavor to the finished product. However, the texture of mayonnaise depends largely on vegetable oils. When the amount of vegetable oil is reduced, the unique texture of mayonnaise cannot be obtained. In order to solve this problem, the technology used did focus on reducing the size of oil droplets in the product. After numerous tests, a technique controlling the average size of oil droplets to 1–2 μm was developed by blending vegetable oils, egg whites and a small amount of thickening agent in a well-balanced manner. Based on the combined SCD extraction technology and technique to control the average particle size of oil droplets, a cholesterol-free mayonnaise-type product is now produced and commercialized in Japan. This product contains less than 5 mg of cholesterol per 100 g and less than half the calories of conventional mayonnaise.

14.4.4.4 Development of Nutritional Supplements

While the ethanol extraction of dried egg yolks will yield total egg yolk lipids including cholesterol, the ethanol extraction of SCD-treated dried egg yolks results in highly pure egg yolk phospholipids in the absence of cholesterol. Egg yolk phospholipids are used in nutritional supplements as a source of choline, with demonstrated health functions such as brain health maintenance and hepatoprotection (see Chapter 36). Since a study demonstrated that the combined administration of egg yolk phosphatidylcholine, the main component of egg yolk phospholipids and vitamin B_{12} improved learning ability in rats (Masuda et al., 1998), several commercial products containing egg yolk phosphatidylcholine and vitamin B_{12} have also been developed in Japan. It is anticipated that more nutritional supplements will be developed in the future, using the full potential of the SCD extraction to obtain valuable nutrients present in the egg yolk (see Chapters 6, 23 and 25).

14.5 Vegan Eggs

Due to their excellent functionalities such as gelling, foaming and emulsification, eggs are widely used in the preparation of foods (see Chapters 8 and 9). Eggs are also a great nutritional food as they contain numerous proteins, vitamins, minerals, essential fatty acids and other micronutrients, distributed between the egg yolk and egg white (see Chapter 29). The egg white is rich in proteins, namely ovomucins, ovalbumin, ovotransferrin, antibacterial lysozyme and peptides while the egg yolk is rich in lipids (65–70% on dry basis), proteins (30% on dry basis) and represents also a good source of lutein, zeaxanthin and vitamins (see Chapters 5 and 6). Eggs are versatile products available in the market in various forms, such as shell eggs, liquid and dry powders of whole eggs, egg whites and egg yolks. Eggs are consumed worldwide regardless of religious beliefs and represent one of the most nutritious and affordable staple foods (Guyonnet, 2012). However, eggs are one of the most common causes of food allergies in young children, often persisting into adulthood. The allergic reaction is triggered by ovomucoid (Gal d 1), ovalbumin (Gal d 2), ovotransferrin/conalbumin (Gal d 3) and lysozyme (Gal d 4), mainly located in egg whites and, to a lesser extent, in egg yolks. This allergy may cause serious skin reactions, nasal congestion, and gastrointestinal and respiratory symptoms (Chapter 31).

Consumers are turning towards plant-based food substitutes as a response to rising concerns related to health, the economy, food safety, ethics, religion and environmental issues. There is also a trend in rich countries of reducing the consumption of meat and other animal products (Boukid & Gagaoua, 2022). Conversely, plant-based diets are gaining popularity with an increasing number of consumers identified as vegetarians or vegans. The global meat substitute manufacturing sector uses several plant-based protein sources, including soy-based, grain-based, vegetable/plant-based and single-cell protein-based ingredients. Asia is the second-largest market for meat substitutes after North America, where consumption is growing at an annual rate of 4.2% (Shim et al., 2021).

Creating egg substitute products meeting both the functionality and nutritional properties of shell eggs is a challenging task (Guyonnet, 2021b). Initially, animal proteins such as milk powder, casein, whey and bovine plasma protein were mainly used, due to their excellent functional properties including solubility, emulsification, foaming and heat-induced gelation (Mohamed et al., 1995; Myhara & Kruger, 1998). While the presence of animal-sourced ingredients is still predominant in the human diet, plant-based foods are gaining popularity around the world due to their health benefits, environmental sustainability and ethical merit (Fehér et al., 2020). To address this growing vegan and health-conscious market segment, manufacturers have developed vegan eggs using different types of plant-based ingredients (e.g., proteins, polysaccharide hydrocolloids, or emulsifiers), alone or combined, to replace regular eggs in the manufacture of food products.

The global market for vegan eggs is expanding rapidly and is projected to have a compound annual growth rate of 5.8% over the next 8 years, with revenue expected to reach USD 1.5 billion by 2026 (Boukid & Gagaoua, 2022). More and more consumers are opting for vegan protein sources and allergen-reduced plant-based dietary ingredients. Moreover, the demand for high-protein, high-fiber and low-saturated-fat ingredients such as legume seed products, especially chickpeas, lentils and soybean, is increasing steadily. Soybean, which has been widely used as an ingredient in meat substitutes, contains not only high-quality protein but is also rich in oleic and linoleic acid, which can mitigate the effects of vascular diseases (Shim et al., 2021).

Based on the current global market, vegan eggs are marketed in different forms (powder, liquid and egg-shaped) (Table 14.4) with the powder form as the most dominant. The most sold vegan eggs are made from starches, plant-based proteins, soy products (lecithin, tofu and tahini), algae flours and other ingredients (e.g., fruit purees and vinegar). While North America is estimated to account for 47.8% of the global

TABLE 14.4

Attributes of the Global Market Segment for Vegan Eggs

Segment Attributes	Details
Presentation form	Powder, liquid, and egg-shaped
Main ingredient	Starch Soy products lecithin, tofu and tahini Plant proteins (e.g., pea, chickpea) Algal flour Other (fruit purees and vinegar)
Applications	Mayonnaise Biscuits and cookies Cakes / Pastries / Muffins / Bread Chocolates Noodles & Pasta
Main players	Glambia plc Ingredion Incorporated Cargill Bob's Red Mill Natural Foods, Inc. House Foods America Corporation EVO Foods Mantiqeira (n.ovo) JUST Inc. Orgran Foods Terra Vegane Free and Easy Follow Your Heart The Vegg Vezlay Foods Private Limited Now Foods The Neat Egg ConAgra brands, Inc. Ener-G
Region	North America Latin America Europe, Middle-East and Africa Asia - Pacific

Source: Adapted from Fact.MR (2022).

sales of vegan eggs, this segment is now trending in Europe due to its applications in reformulating snacks and meat alternatives. The most important producers of vegan eggs are Corbion NV (Amsterdam, The Netherlands), Glanbia Plc (Kilkenny, Ireland), Tate & Lyle Plc (London, UK), Ingredion Incorporated (Westchester, IL, USA), Ener-G Foods, Inc. (Seattle, WA, USA), Natural Products, Inc. (Grinnell, IA, USA), Orchard Valley Foods Limited (Tenbury Wells, UK), Puratos Group (Dilbeck, Belgium), TerraVia Holdings, Inc. (San Francisco, CA, USA) and Archer Daniels Midland Company (Chicago, IL, USA). Vegan eggs are mostly used as substitutes for eggs in bakery products, desserts and confectionaries. Claims such as natural, organic, clean label, Halal or Kosher certified, dairy-free, GMO-free and gluten-free are also boosting the market of vegan eggs (Fact.MR, 2021). Since sustainability is now playing an essential role within the food sector (Alcorta et al., 2021), 83.3% of vegan eggs also carry environmental or ethical claims, including recycling food waste and the use of sustainable packaging.

14.6 New Consumers Demands to Further Guide Innovation

The attitudes, perceptions and behaviors of consumers towards shell egg and egg products have been previously reviewed (Hansstein, 2011; Guyonnet, 2012; Bertechini, 2017; Rondoni et al., 2020) and were also presented in Chapter 2. By 2027, the Gen Z, people born between 1998 and 2016 will be the largest demographic segment, representing 28% of the global population. Combined with Millennials, those born between 1981 and 1997, these 2 generations will represent 52% of the global population and represent the largest consumer segment for food purchases (Donnan, 2018). In addition to price, taste and convenience – the 3 most commonly buying factors reported – the growing influence of younger generations of consumers has also brought food authenticity and sustainability as new key buying factors for foods.

With the rising level of disposable income for foods in many countries and younger generations of consumers less concerned about price, most consumers no longer mention price as their top buying factor, offering more options for innovation and the development of premium egg products. In addition, a global survey conducted on almost 19,000 consumers in 28 countries showed that 44% of food consumers were purpose-driven and therefore considering buying factors other than price in their buying decisions (IBM, 2020). On the other hand, the development of new processes which may reduce the cost of production of egg products will convince more companies to enter the egg processing business and further expand the overall market size for egg products.

Taste, the second most important buying factor, is a key attribute for the development of any food new product. Consumers are less likely to compromise on taste for additional benefits provided by foods. For instance, studies have clearly demonstrated that consumers were unlikely to purchase functional foods, by definition better for their health, if they did not taste as well as regular foods (Verbeke, 2006; Guyonnet, 2011; Baker et al., 2022).

Convenience is probably best expressed by the rise of hard-boiled eggs, ready to be eaten by consumers at any time of the day. With the hard-boiled eggs, egg processors have been able to enter the snack food market, one of the fastest segments of the retail food business. The transformation of shell eggs into hard-boiled eggs has contributed to some incremental profit for egg producers and egg processors, while also delivering great nutrition (Guyonnet, 2019). In many countries, the vast majority of eggs are consumed for breakfast, but mostly only on weekends. People are often too busy during the week to take time to cook for breakfast. Therefore, the recent focus of R&D has been on finding products easy to prepare for breakfast during the weekdays. While some new convenient products may involve very limited research – just combine several foods, like cheese and nuts, to the hard-cooked egg, package them together and you have a new product – others will require more investment to develop new recipes and to design automated manufacturing processes to deliver great tasting products at the best price. Convenience will not only relate to the way the egg products are consumed but also to the way the products have to be transported and stored. A shelf-stable, ready-to-eat egg product should certainly have the favor of consumers (Guyonnet, 2021a).

Consumers, especially younger generations of consumers, have expressed a desire for authenticity in the foods they purchase. In their mind, authenticity relates to food composition and the overall healthiness of foods. The healthiness of egg products can easily be enhanced by the addition of natural ingredients with functional properties such as the promotion of a healthier digestive system, better eyesight or brain development in young children (Guyonnet, 2015). And the egg processing sector is in an excellent position to develop functional products since the functional ingredient can either be added during the manufacture of the egg product or via the diet fed to the laying hens. The first approach is certainly more efficient while the second approach is truly natural and answers well to the desire of consumers for authenticity (Guyonnet, 2020b). The authenticity of foods is also defined in the mind of consumers as foods with a short list of ingredients, with ingredients recognized for their healthiness, with natural ingredients as opposed to synthesized molecules and without artificial color or preservatives. Retail food products like liquid egg white or liquid whole egg fit perfectly this description – simple, natural food without any ingredients added. However, the new generation of consumers is also unpredictable. With some negative perceptions about modern egg production, with birds housed in cages, younger generations of consumers have shown interest in products claiming to be an egg while only made of plant-sourced ingredients, even though these products have typically a long list of ingredients in order to mimic an egg (Guyonnet, 2021b).

The last key buying decision factor relates to sustainability. Younger generations of consumers do care more about the environment and the sustainability of the foods they consume (Guyonnet, 2020b; IBM, 2020). This interest in sustainability will offer the egg processing sector a number of areas on which to focus and further demonstrate to consumers its commitment to sustainability, ranging from the type of ingredients used and where there are sourced, innovation in processing that will reduce the use of chemicals for the cleaning of equipment or the use of energy for egg processing, and to a careful evaluation of equipment and tools that are disposable (that is single usage) compared to those that can easily be washed and safely re-used.

Overall, the changing demands of consumers combined with the rising level of incomes in many countries offer the egg processing sector a number of strategic directions for the development of many new egg products over the coming years.

14.7 Conclusion

While egg processing accounts for less than 10% of global egg production, innovations within this sector are providing customers and consumers more opportunities to consume eggs. Over the past 40 years, the development of new egg products has been one of the main drivers for the growth of the global egg sector. Applications using the different forms of egg products are numerous, especially in Japan and North America, and new technologies are able to further improve the functionality and nutritional value of eggs. Although it is impossible to replicate shell eggs, one of the most affordable and nutritious foods provided by nature, alternatives using plant-based ingredients have been developed and commercialized. In the future, innovative strategies addressing the five main consumer buying factors – price, taste, convenience, authenticity and sustainability – should guide the development of new egg products and further contribute to the sustainability of the egg processing sector.

REFERENCES

Ahmadkelayeh, S., & Hawboldt, K. (2020). Extraction of lipids and astaxanthin from crustacean by-products: A review on supercritical CO_2 extraction. *Trends in Food Science and Technology*, *103*, 94–108. https://doi.org/10.1016/j.tifs.2020.07.016

Alcorta, A., Porta, A., Tárrega, A., Alvarez, M. D., & Vaquero, M. P. (2021). Foods for plant-based diets: Challenges and innovations. *Foods (Basel, Switzerland)*, *10*(2), 293. https://doi.org/10.3390/foods10020293

Arimitsu, K., Kojima, T., Matsuoka, R., Narita, K., & Idei, A. (2015). The world where the new material "Lacty egg" made by lactic acid fermentation of egg white spreads. *Japan Journal of Food Engineering*, *16*(1), 79–82. (in Japanese) https://doi.org/10.11301/jsfe.16.79

Azmir, J., Zaidul, I. S. M., Rahman, M. M., Sharif, K. M., Mohamed, A., Sahena, F., & Omar, A. K. M. (2013). Techniques for extraction of bioactive compounds from plant materials: A review. *Journal of Food Engineering*, *117*(4), 426–436. https://doi.org/10.1016/j.jfoodeng.2013.01.014

Baker, M. T., Lu, P., Parrella, J. A., & Leggette, H. R. (2022). Consumer acceptance toward functional foods: A scoping review. *International Journal of Environmental Research and Public Health*, *19*, 1217. https://doi.org/10.3390/ijerph19031217

Bergquist D. H. (1990). Egg dehydration. In W. J. Stadelman & O. J. Cotterill (Eds.), *Egg science and technology* (4th ed., pp. 335–376). Food Products Press, Inc.

Bertechini, A. G. (2017). Economic and cultural aspects of the table egg as an edible commodity. In P. Y. Hester (Ed.), *Egg innovations and strategies for improvement* (pp. 223–232). Academic Press. http://dx.doi.org/10.1016/B978-0-12-800879-9.00021-4

Boukid, F., & Gagaoua, M. (2022). Vegan egg: A future-proof food ingredient? *Foods*, *11*, 161. https://doi.org/10.3390/foods11020161

Daud, N. M., Putra, N. R., Jamaludin, R., Norodin, N. S. M., Hamzah, M. H. S., Nasir, H. M., Zaidel, D. N. A., Yunus, M. A. C., & Salleh, L. M. (2022). Valorisation of plant seed as natural bioactive compounds by various extraction methods: A review. *Trends in Food Science and Technology*, *119*, 201–214. https://doi.org/10.1016/j.tifs.2021.12.010

Donnan, D. (2018). Influence vs. affluence: The changing menu of food choices. Retrieved April 28, 2022), from https://www.kearney.com/documents/20152/4955955/Influence+vs.+Affluence%E2%80%94The+Changing+Menu+of+Food+Choices.pdf/02a54bd9-9e84-b5e7-ea6d-ea50af038d22?t=1516394321000

Essien, S. O., Young, B., & Baroutian, S. (2020). Recent advances in subcritical water and supercritical carbon dioxide extraction of bioactive compounds from plant materials. *Trends in Food Science and Technology*, *97*, 156–169. https://doi.org/10.1016/j.tifs.2020.01.014

Fact.MR (2022, September 14). Vegan Egg Market Outlook. Available online: https://www.factmr.com/report/vegan-eggs-market

Federal Register (2021). Mandatory inspection of egg substitutes and freeze-dried egg products imported into the United States. *Federal Register*, *86*(170), 50013–50015. https://www.govinfo.gov/content/pkg/FR-2021-09-07/pdf/2021-19264.pdf

Fehér, A., Gazdecki, M., Véha, M., Szakály, M., & Szakály, Z. A. (2020). Comprehensive review of the benefits of and the barriers to the switch to a plant-based diet. *Sustainability*, *12*, 4136. https://doi.org/10.3390/su12104136

Froning, G. W., Wehling, R. L., Cuppet, S. L., Pierce, M. M., Niemann, L., & Siekman, D. K. (1990). Extraction of cholesterol and other lipids from dried egg yolk using supercritical carbon dioxide. *Journal of Food Science*, *55*(1), 95–98. https://doi.org/10.1111/j.1365-2621.1990.tb06025.x

Gan, A., & Baroutian, S. (2022). Current status and trends in extraction of bioactives from brown macroalgae using supercritical CO_2 and subcritical water. *Journal of Chemical Technology and Biotechnology*, *97*(8), 1929–1940. https://doi.org/10.1002/jctb.7063

Güven, G., Evli, S., Uygun, M., & Uygun, D. A. (2019). Cholesterol removal by β-cyclodextrin modified cryogel column. *Journal of Liquid Chromatography and Related Technologies*, *42*(17–18), 537–545. https://doi.org/10.1080/10826076.2019.1632894

Guyonnet, V. (2011). Strategic planning for the development of the egg nutraceutical industry. In Y. Nys, M. Bain, & F. Van Immerseel (Eds.), *Improving the safety and quality of eggs and egg products. Vol. 2: Egg safety and nutritional quality chemistry* (pp. 374–399). Woodhead Publishing.

Guyonnet, V. (2012, August 5–9). *Eggs and egg products: Consumers' attitudes, perceptions and behaviours [conference invited lecture]*. World's Poultry Science Congress. http://www.facta.org.br/wpc2012-cd/pdfs/plenary/Vincent_Guyonnet.pdf

Guyonnet, V. (2015). Strengths and weaknesses of functional eggs for human health. In R. R. Watson & F. De Meester (Eds.), *Handbook of eggs in human function* (pp. 153–165). Wageningen Academic Publishers. https://doi.org/10.3920/978-90-8686-804-9

Guyonnet, V. (2019). Positioning hard-boiled eggs in the snacking market. *Poultry International*, *58*(6), 30–31. https://www.wattagnet.com/articles/37290-positioning-hard-boiled-eggs-in-the-snacking-market

Guyonnet, V. (2020a). Japan consuming more eggs, despite population stagnation. *Poultry International*, *59*(11), 20–21. https://www.wattagnet.com/articles/41113-japan-consuming-more-eggs-despite-population-stagnation

Guyonnet, V. (2020b). How the egg industry can grow its appeal to Gen Z. *Poultry International*, *59*(7), 30–33. https://www.wattagnet.com/articles/40211-how-the-egg-industry-can-grow-its-appeal-to-gen-z

Guyonnet, V. (2021a). Value through processing – As simple as boiling an egg! *Poultry International*, *60*(1), 18–19. https://www.wattagnet.com/articles/41640-value-through-processing—as-simple-as-boiling-an-egg

Guyonnet, V. (2021b). Eggs: Often imitated, never equated. *Poultry International*, *60*(11), 18–19. https://www.wattagnet.com/articles/43453-eggs-often-imitated-never-equaled

Hansstein, F. (2011). Profiling the egg consumer: Attitudes, perceptions and behaviours. In Y. Nys, M. Bain, & F. Van Immerseel (Eds.), *Improving the safety and quality of eggs and egg products. Vol. 1: Egg chemistry, production and consumption* (pp. 39–61). Woodhead Publishing.

IBM (2020). Research Insights - Meet the 2020 consumers driving changes. Retrieved on March 3, 2022 from https://www.ibm.com/downloads/cas/EXK4XKX8

Kotaro, I., & Asahi, M. (2021). Koji molds for Japanese soy Sauce brewing: Characteristics and key enzymes. *Journal of Fungi (Basel)*, *7*(8), 658. https://doi.org/10.3390/jof7080658

Lamas, A., Anton, X., Miranda, J. M., Roca-Saavedra, P., Cardelle-Cobas, A., Ibarra, I. S., & Cepeda, A. (2016). Technological strategies for the development of egg-derived products with reduced content of cholesterol. *Food and Bioprocess Technology*, *9*(1), 81–90. https://doi.org/10.1007/s11947-015-1599-4

Liu, H., Yang, C., Jing, Y., Li, Z., Zhong, W., & Li, G. (2013). Ability of lactic acid bacteria isolated from mink to remove cholesterol: In vitro and in vivo studies. *Canadian Journal of Microbiology*, *59*(8), 563–569. https://doi.org/10.1139/cjm-2013-0200

Masuda, Y., Kokubu, T., Yamashita, M., Ikeda, H., & Inoue, S. (1998). Egg phosphatidylcholine combined with vitamin B_{12} improved memory impairment following lesioning of nucleus basalis in rats. *Life Sciences*, *62*(9), 813–822. https://doi.org/10.1016/S0024-3205(97)01183-1

Miyamoto, T., Kihara, A., & Shidara, H. (2019). Development of "aged egg yolk", an egg yolk that brings out the deliciousness using Koji The Society for Biotechnology, Japan The 71th annual meeting

Miyamoto, T., Kihara, A., & Shidara, H. (2020). Aged egg yolk as an enhancer of umami flavor and richness SHIFT20 Virtual Experience

Mohamed, S., Lajis, S. M. M., & Hamid, N. A. (1995). Effects of protein from different sources on the characteristics of sponge cakes, rice cakes (apam), doughnuts and frying batters. *Journal of the Science of Food* and *Agriculture*, *68*, 271–277. https://doi.org/10.1002/jsfa.2740680303

Moreno-García, K. L., Antunes-Ricardo, M., Martínez-Ávila, M., Milán-Carrillo, J., & Guajardo-Flores, D. (2022). Evaluation of the antioxidant, anti-inflammatory and anti-hyperglycemic activities of black bean (Phaseolus vulgaris L.) by-product extracts obtained by supercritical CO_2. *Journal of Supercritical Fluids*, *183*, 105560. https://doi.org/10.1016/j.supflu.2022.105560

Myhara, R. M., & Kruger, G. (1998). The performance of decolorized bovine plasma protein as a replacement for egg white in high ratio white cakes. *Food Quality and Preference*, *9*, 135–138. https://doi.org/10.1016/S0950-3293(97)00046-3

Paraskevopoulou, A., & Kiosseoglou, V. (2006). Cholesterol and other lipid extraction from egg yolk using organic solvents: Effects on functional properties of yolk. *Journal of Food Science*, *59*(4), 766–768. https://doi.org/10.1111/j.1365-2621.1994.tb08123.x

Puertas, G., & Vázquez, M. (2019). Advances in techniques for reducing cholesterol in egg yolk: A review. *Critical Reviews in Food Science and Nutrition*, *59*(14), 2276–2286. https://doi.org/10.1080/10408398.2018.1448357

Romano, R., De Luca, L., Aiello, A., Rossi, D., Pizzolongo, F., & Masi, P. (2022). Bioactive compounds extracted by liquid and supercritical carbon dioxide from citrus peels. International *Journal of Food Science and Technology*, *57*(6), 3826–3837. https://doi.org/10.1111/ijfs.15712

Rondoni, A., Asiolia, D., & Millana, E. (2020). Consumer behaviour, perceptions, and preferences towards eggs: A review of the literature and discussion of industry implications. *Trends in Food Science & Technology*, *106*, 391–401. https://doi.org/10.1016/j.tifs.2020.10.038

Sakiko, S., Yosie, U., Hajime, H., & Hiroshi, N. (2014). High-density cultivation of koji using sponge cake made from egg white and development of fermented egg white sauce. *Nippon Shokuhin Kagaku Kogaku Kaishi*, *61*(2), 77–84. (in Japanese) https://doi.org/10.3136/nskkk.61.77

Shim, Y. Y., He, Y., Kim, J. H., Cho, J. Y., Meda, V., Hong, W. S., & Reaney, M. J. T. (2021). Aquafaba from Korean soybean: A functional vegan food additive. *Foods*, *10*, 2433. https://doi.org/10.3390/foods10102433

Sievers, U., & Eggers, R. (1996). Heat recovery in supercritical fluid extraction process with separation at subcritical pressure. *Chemical Engineering and Processing: Process Intensification*, *35*(4), 239–246. https://doi.org/10.1016/0255-2701(95)04124-9

Sun, J., Xu, B., Mu, Y., Ma, H., & Qu, W. (2018). Functional magnetic nanoparticles for highly efficient cholesterol removal. *Journal of Food Science*, *83*(1), 122–128. https://doi.org/10.1111/1750-3841.13999

Usuda, M., Ogino, Y., Kimura, M., Fujiwara, S., Okuyama, Y., Kunika, H., & Utsunomiya, K. (2011). Lactic acid fermented egg white lowers blood cholesterol levels in humans The Japanese Society for Food Science and Technology, The 58th Annual Meeting. (in Japanese)

Vandeponseele, A., Draye, M., Piot, C., & Chatel, G. (2020). Subcritical water and supercritical carbon dioxide: Efficient and selective eco-compatible solvents for coffee and coffee by-products valorization. *Green Chemistry*, *22*(24), 8544–8571. https://doi.org/10.1039/D0GC03146A

Verbeke, W. (2006). Functional foods: Consumer willingness to compromise on taste for health? *Food Quality and Preference*, *17*(1–2), 126–131. https://doi.org/10.1016/j.foodqual.2005.03.003

Wang, W., Rao, L., Wu, X., Wang, Y., Zhao, L., & Liao, X. (2021). Supercritical carbon dioxide applications in food processing. *Food Engineering Reviews*, *13*, 570–591. https://doi.org/10.1007/s12393-020-09270-9

Yoshimi, K. (2011). Development of new egg white material lactic acid fermented egg white. *Food Style 21*, *15*(1), 44–46. (in Japanese)

Part V

Shell Egg Quality

15

Eggshell and Yolk Color

Takehiko Hayakawa

CONTENTS

15.1 Appearance of Eggs and Indicator of Egg Freshness

Egg quality is one of the most important purchasing factors when buying eggs, but eggshell quality (strength), cleanliness, eggshell color, and egg yolk color are especially important, next to the price and the freshness of the eggs, because these factors are visually distinguishable (Research International, 2003; Metrixlab, 2018). The first thing that catches the eye when first purchasing a chicken egg is the appearance of the egg, i.e., the eggshell.

Whether or not eggs are washed before placing on the market depends on the regulations and market trends in each country. In countries where eggs are distributed and sold under refrigeration, they are often washed and sterilized; in countries where eggs are distributed and sold under room temperature conditions, they are often not washed. Washing and sterilization removes most of the dirt from the eggshell, but it also removes the cuticle layer that is naturally present on the outside of the eggshell to prevent bacteria from penetrating into the egg. For this reason, eggs that have been washed and sterilized must be stored and controlled under refrigeration.

Regardless of whether the eggs are washed or unwashed, it is necessary to be careful because a large temperature difference in the environment from refrigerated storage to room temperature storage tends to cause condensation on the eggshells, which creates an environment in which bacteria can easily multiply. On the other hand, although an unwashed egg can be stored at room temperature for a certain period of time, it should be noted that the Yolk index and Haugh unit, which are indicators of the freshness of the yolk and albumen as internal egg quality, are significantly lower than those observed under refrigerated storage. Yolk index does not show a clear difference under cold storage, but the decrease is more significant at room temperature as the days go by (Figure 15.1).

15.2 Eggshell

15.2.1 Eggshell Color

The differences in eggshell colors, such as white, brown, light pink, and light blue, are due to the genetic traits of the hens: white-feathered hens lay white eggs and colored-feathered hens lay colored eggs, but the tones of the feathers do not correspond to the tones of the eggshells. There are various arguments as to why birds inherently lay colored eggs, for example, as a protective color to blend in with the surrounding environment or as an appeal to the male.

Colored hens differ from white hens in that colored hens clearly synthesize higher amounts of protoporphyrin and biliverdin; brown eggs are affected by protoporphyrin, blue-green eggs are due to biliverdin, and the shade of eggshell color is dependent on the synthesis of these substances. The eggshell coloration depends on the amount of these substances synthesized. It has been reported that the shade of eggshell color is also affected by factors such as hen's age, the environment affecting health, and stress related to the health of the oviducts that synthesize and secrete these eggshell pigments (Lu et al., 2021). Tsushima et al. (2013) also reported a positive correlation between protoporphyrin and eggshell strength in a study with Hy-Line Brown hens. This may be due to a decrease in the synthesis of protoporphyrin.

DOI: 10.1201/9781003254430-20

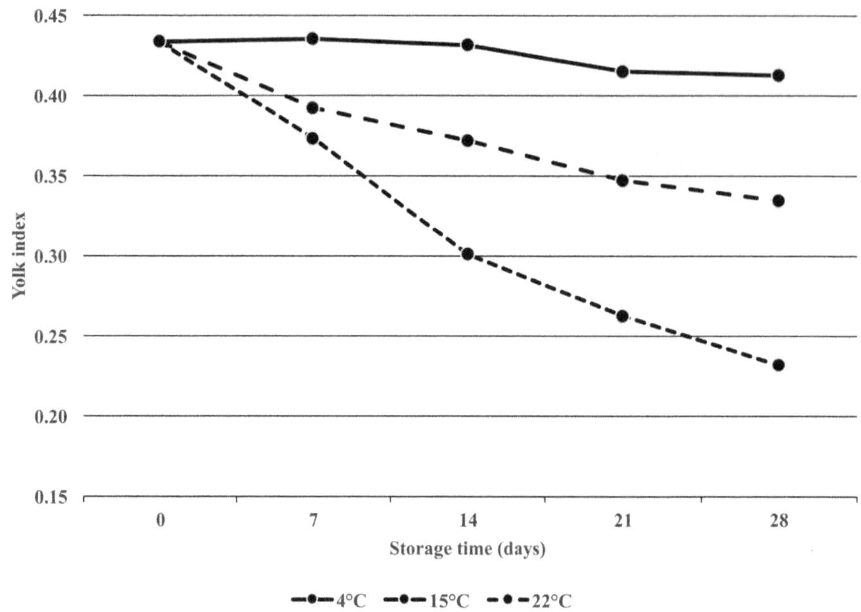

FIGURE 15.1 Change in yolk index at different storage temperatures and days.

15.2.2 Market Preference for Eggshell Color

Consumer preference for eggshell color is different by countries, but according to a report by the International Egg Commission (2020), brown eggs are preferred in Europe and Oceania, while white eggs are preferred in Scandinavian countries, and North and South America also have a high percentage of white eggs (Figure 15.2). Asian countries show a wide variety, and brown eggs are used in some cases to differentiate hen eggs from duck eggs.

In countries where white eggs dominate, brown eggs are often used for branded eggs, enriched eggs, and other high value-added eggs to differentiate them in the market. Shell color fan is available in the Japanese market and can be scored from 1 to 10 as a tool for measuring, evaluating, and comparing the eggshell color of brown eggs (Figure 15.3).

15.2.3 Eggshell Quality

Although eggshell strength is a particularly important indicator for maintaining and controlling egg quality, no numerical guideline indicator has yet been established. There are two types of strength display units in use, namely kgf and N, but in

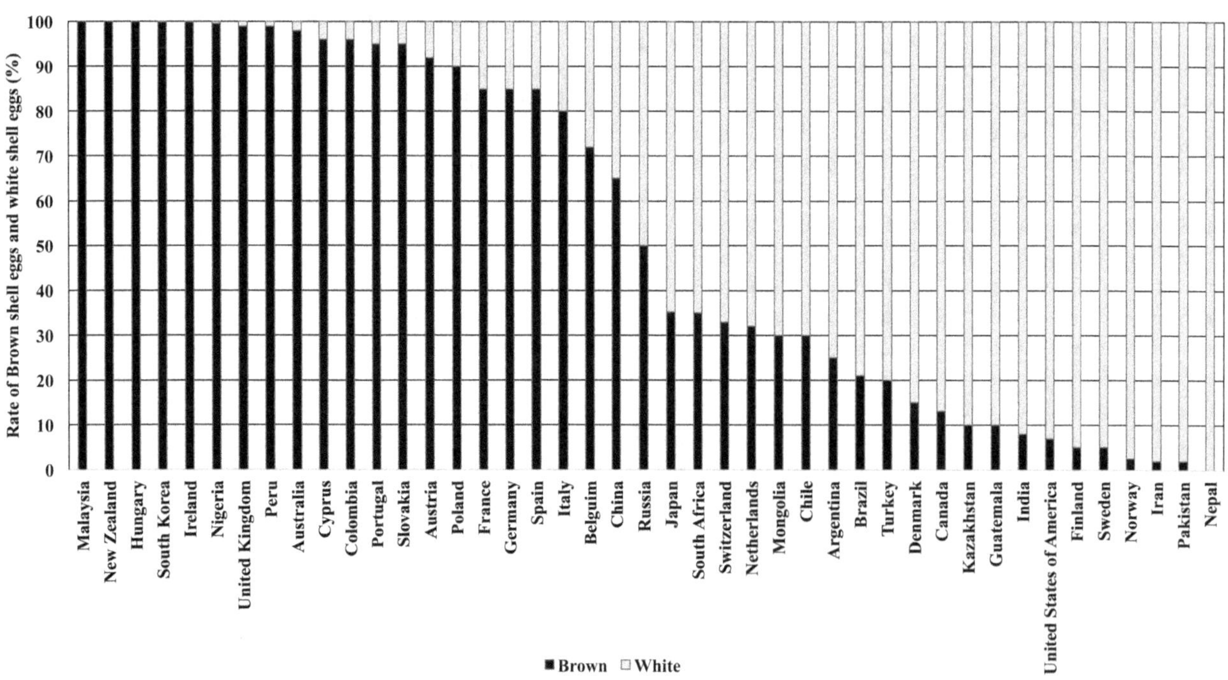

FIGURE 15.2 Percentage of white and brown eggs in different countries. (Adapted from IEC, 2020.)

FIGURE 15.3 Shell color fan. The photo on the right shows the back of the shell color fan. (Courtesy of GHEN Corporation, Japan.)

many cases kgf is preferred. The eggshell strength is measured by mechanically applying a load to the eggshell, either vertically (small end and large end) or horizontally (equator). Which direction is used depends on the country or market. In most cases where packaged eggs are sold, pressure is applied vertically. On the other hand, some say that it is better to measure from the side because eggs roll horizontally. However, since the strength from the vertical direction is generally stronger than that measured in the horizontal direction, a comparison cannot be made between the vertical and horizontal methods, so the decision should be based on which is generally adopted in the country and the market. The reason why it is difficult to set an index of eggshell strength is, besides the aforementioned differences in measurement methods, due to trends that vary with the age of the hen, differences in eggshell size, etc. Miranda et al. (2016) reported a wide range of 1–7.5 kgf, even when measured vertically. In comparison with the results of previous measurements, it was found that eggshell strength below 3 kgf was often associated with hairline cracks and egg breakage. Eggs of 3 kgf or more without eggshell problem are often found marketed as table eggs. It is known that the size of a hen's egg increases as the hen grows, while eggshell strength tends to weaken (Figure 15.4); but eggshell thickness does not change as significantly as eggshell strength. Kemps et al. (2006) reported

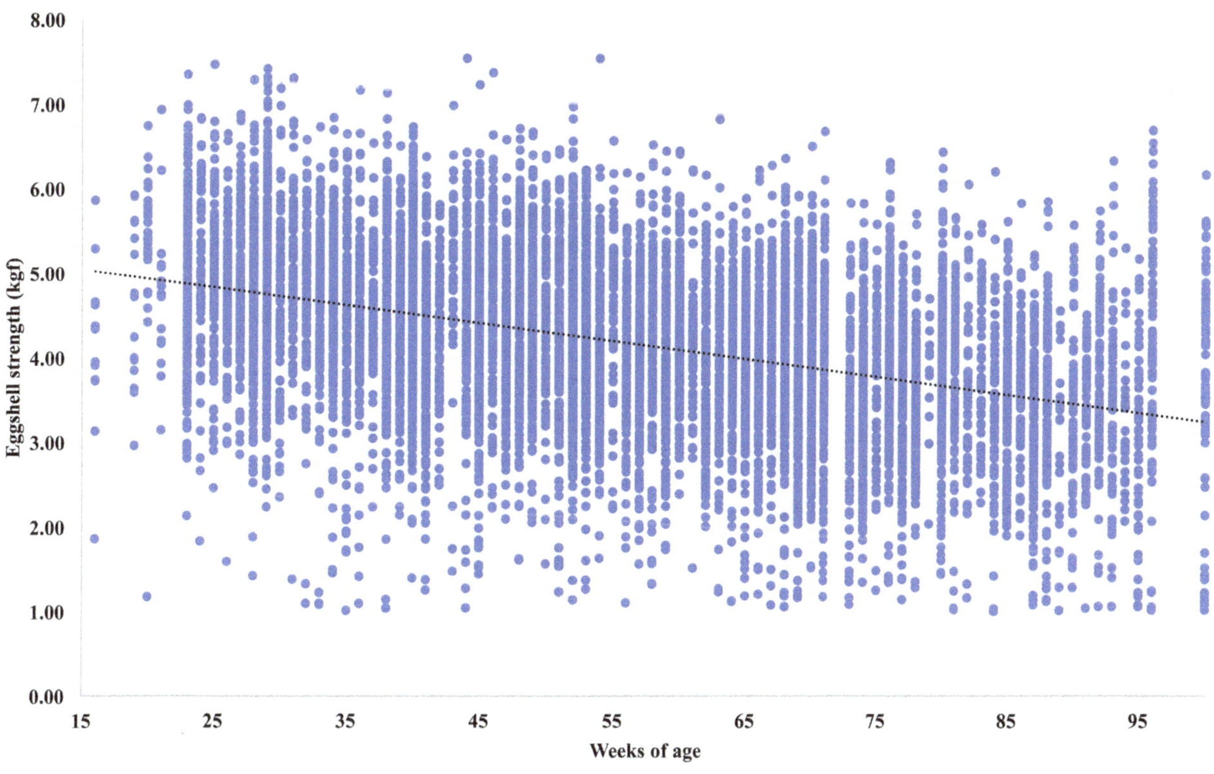

FIGURE 15.4 Relationship between eggshell strength and week age of laying hens. (Courtesy of Miranda et al., 2016.)

that when eggshell quality was compared between early and late egg-laying stages, eggshell weight itself did not change significantly, but there was a significant decrease in eggshell membrane weight, suggesting that the eggshell membrane may play an important role in eggshell strength and egg quality.

15.2.4 Factors Influencing Eggshell Quality

Eggshell quality is influenced by the health and age of the hen, but egg breakage is also largely caused by differences in henhouse systems and distribution systems (Mertens et al., 2006). The component nutrients that significantly affect eggshell quality itself are mainly calcium, phosphorus, and trace minerals, but the absorption of calcium in particular decreases with age (Al-Batshan et al., 1994).

The trigger for this calcium uptake is 25-hydroxyvitamin D_3 (25-OH-D_3), a metabolite of vitamin D3. This 25-OH-D_3 is linked to blood calcium concentrations, which are regulated primarily by two hormones, parathyroid hormone (PTH) and calcitonin (CT) (Figure 15.5). When blood calcium levels decrease, parathyroid hormone stimulates the metabolic conversion of 25-OH-D_3, which is detached from circulating vitamin D-binding protein (VDP) in the blood, to the active form, 1α-25-$(OH)_2$-D_3, in the kidney. 1α-25-$(OH)_2$-D_3 promotes the uptake of calcium in the intestinal tract and stimulates osteoclasts and osteoblasts to incorporate calcium from cortical and medullary bone into the blood to help form eggshells. Conversely, when the calcium concentration in the blood rises, calcitonin acts to prevent the uptake of excess calcium without metabolic conversion to 1α-25-$(OH)_2$-D_3 by keeping 25-OH-D_3 in the blood circulation with VDP. Although 1α-25-$(OH)_2$-D_3 is the direct substance that takes in calcium necessary for eggshell formation, it is a hormonal substance and thus cannot be metabolically controlled. Therefore, 25-OH-D_3 is the substance that regulates the balance of calcium uptake, which is important for eggshell formation.

After 40 weeks of age, when the peak egg-laying period has passed, eggshell strength tends to weaken, and in many cases, high calcium supplementation is used. However, it is important to note that if calcium alone is added to the diet, without enough vitamin D intake, especially 25-OH-D_3, the necessary calcium cannot be taken up.

The calcium content of eggshells is almost constant at about 2.2 g per egg regardless total egg weight, with about 60% of the calcium source coming from the feed and 40% from the bones. Consequently, not only do eggshells become weaker in old age during the egg-laying period, but also osteoporosis and leg fractures become more frequent. Modern laying hen breeds are high producing, and the production target for these hens is 500 eggs at 100 weeks of age. Thus, it is very important to take care of not only calcium but also its absorbing factor, vitamin D, since bone growth during the growing period greatly influences subsequent growth in order to maintain eggshell strength over a long period of time.

Although nutritional status during the growing period may seem to be unconnected to eggshell quality issues during the laying period, modern hens begin laying eggs at approximately 17 weeks of age, but skeletal growth is 95% developed at 13 weeks of age (Hy-Line, 2016), so maximum growth during the growing period is necessary to continue sustainable production. Cortical bone, which makes up the skeleton, develops significantly during the growing period, as mentioned above, and bone growth almost stops with sexual maturity, while the medullary bone is formed in hens as estrogen hormones elevate. This medullary bone is primarily used as a source of calcium for eggshell formation. During the laying period, medullary bone is not only utilized, but is also continuously formed and accumulated. However, the bone mass of the cortical bone continues to decrease during the egg-laying period because osteoclasts, which take up calcium, use not only the medullary bone but also the cortical bone, which forms the skeleton, as a source of calcium for the eggshell. This means

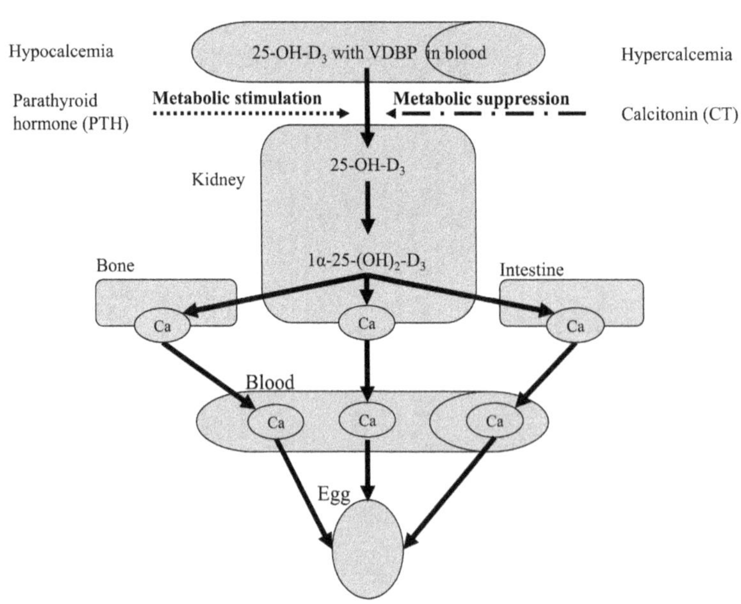

FIGURE 15.5 Relationship between calcium uptake and vitamin D required for eggshell formation.

that bone growth during the growing period is critical for the maintenance of high-quality eggshell formation.

In addition, environmental factors, especially reduced feed intake due to heat and poor absorption due to a poor intestinal environment, have a negative impact on eggshell quality. Since the only way for the chickens to regulate their body temperature in a hot environment is to dissipate heat through respiration, they will hyperventilate and produce excessive amounts of carbon dioxide, resulting in a decrease in blood carbon dioxide levels and consequently affecting the synthesis of calcium carbonate, the component of eggshells. Therefore, it is necessary to mitigate the effects on eggshell quality by adding sodium in order to balance electrolytes and administering drinking water with vitamins.

FIGURE 15.6 DSM YolkFan™. (Courtesy of DSM Nutritional Products AG.)

15.3 Egg Yolk Color

15.3.1 Egg Yolk Color Preference

The color of egg yolk depends on the feed consumed by the hen, and it can produce a bright yellow-orange hue. The origin of the word "egg yolk" in the Chinese and Japanese languages means "yellow egg," and it can be inferred from these words that hen egg yolks are traditionally brightly colored. Egg yolk color is one of the qualities of chicken eggs that can be easily distinguished by the human eye, and thus can easily lead to differentiation, but it is also subject to complaints in the same way as the eggshell quality. In recent years, the egg industry has been using digital equipment to objectively measure and quantify egg quality, but DSM YolkFan™ (formerly known as Roche Yolk Color Fan) remains a worldwide standard for egg yolk color evaluation that is classified from No. 1 to No. 16 (Figure 15.6).

As shown in Figure 15.7, egg yolk color, as well as eggshell color, varies from country to country, but it does not necessarily correspond to consumer preferences. The results of this study were obtained by measuring the egg yolk color of eggs actually sold in the market. Then, preparing four kinds of eggs with different egg yolk from light yellow to dark orange, these eggs were shown to consumers, and they were asked which one looked better. Many consumers in countries prefer eggs with a darker egg yolk color than light egg yolk color. This suggests that many consumers desire a darker egg yolk color against the usually available eggs with light egg yolk color. These differences in egg yolk color preferences and the eggs sold can be attributed to differences in food culture and national characteristics. In recent years, eggs with a darker egg yolk color have generally been value-added eggs, such as specialty eggs and enriched eggs. Consumers believe that there is a relationship between the

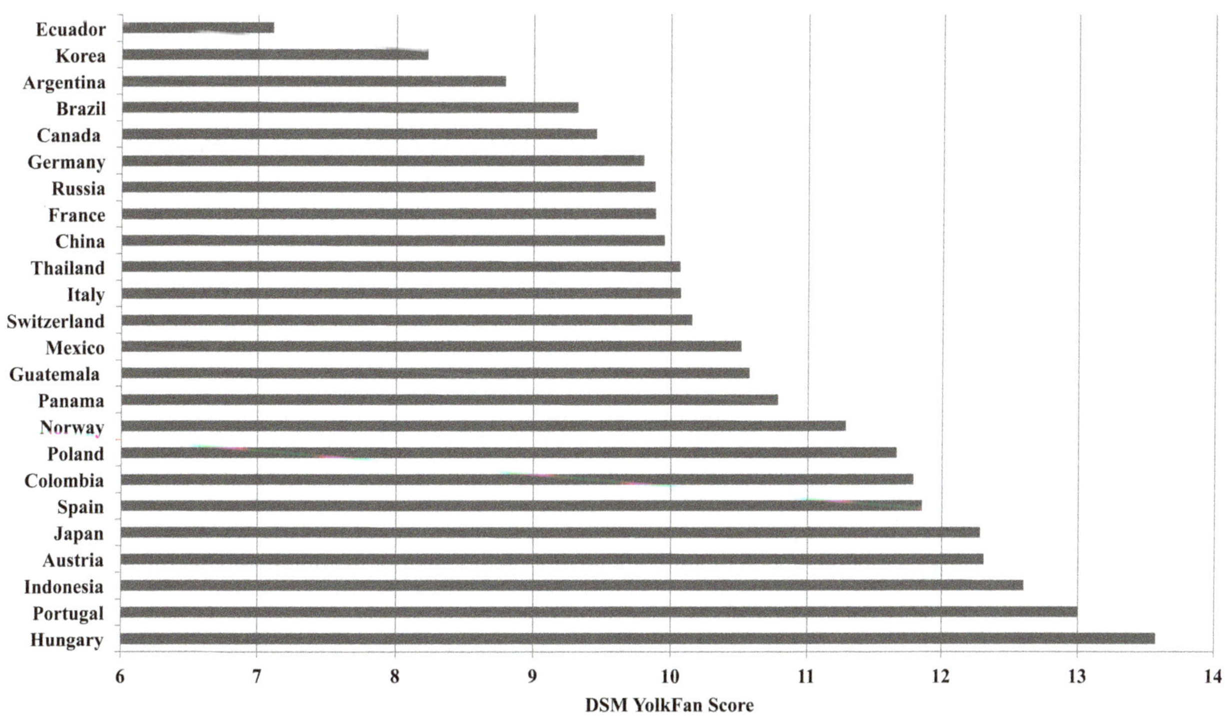

FIGURE 15.7 Egg yolk color difference in various countries. (Courtesy of DSM Nutritional Products AG.)

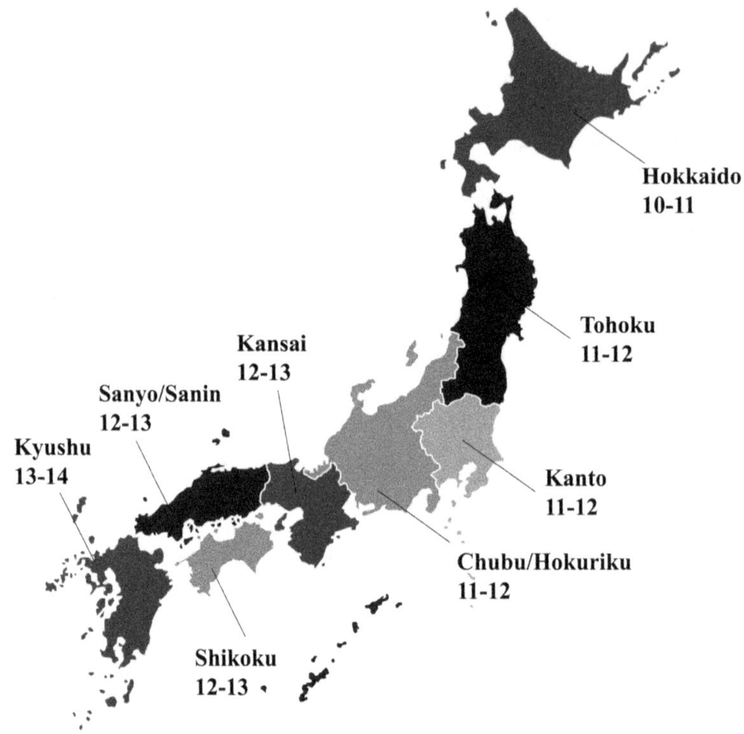

FIGURE 15.8 Trends in egg yolk color by region in Japan. Values are DSM YolkFan score.

nutritional value of eggs and the shade of egg yolk color, since regular eggs tend to be lighter, and egg yolk color tends to be darker in countries where the egg market is highly differentiated and where consumers have a strong desire to buy eggs.

Also, the Japanese market, which is said to have the largest variety of branded eggs in the world with 1,481 brands (Takahashi, 2022), the average egg yolk color in 1994 was YolkFan number 8, while in 2019 it was number 12. The egg yolk color score has thus increased by 4 points over the past 25 years (Central Egg Standards Trading Council, 2020). This can be attributed to the Japanese food culture of enjoying the color of food by eating eggs raw with various dishes, and to the competition for branded eggs using egg yolk color, which is easy to differentiate in egg yolk. The average egg yolk color varies by region of the Japanese archipelago, which stretches from east to west, and although it is not clear whether this is due to climate or regional food culture, a market survey found that egg yolk color was higher in the southern region than in the northern region (Figure 15.8).

15.3.2 Source of Egg Yolk Color

The shade of egg yolk color depends on the feed the hens eat. Egg yolk color ranges from light yellow to dark orange, depending on the type and content of naturally occurring compounds called carotenoids. Carotenoids cannot be synthesized in the hen's body, so they must be obtained from the feed. The main carotenoid components used in poultry feed are shown in Figure 15.9. They can be divided into yellow carotenoids

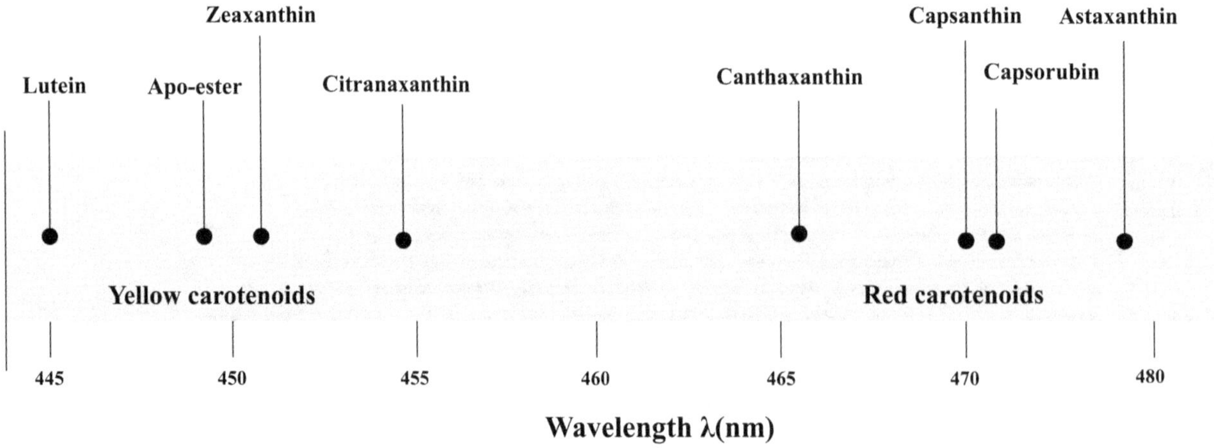

FIGURE 15.9 Wavelength of carotenoids in poultry feed.

and red carotenoids, and each carotenoid has a different hue depending on its wavelength.

Common laying hen diets can be divided into corn- or wheat-based diets. The predominant feed depends on the market: Europe and Oceania regions tend to use wheat-based diets, while the rest of the world is corn-based. Yet, carotenoids are present in corn and its by-products, but not in wheat. The YolkFan score increases by one point for every 10% of corn added to wheat-based diets, and while corn alone can change egg yolk color up to number 6, the addition of red carotenoids is necessary for numbers 7 and above. Neither yellow carotenoids alone nor red carotenoids alone can produce a bright egg yolk color, and an optimal balance of both types of carotenoids is necessary. In addition, the carotenoid content in feed ingredients such as corn is affected by the place of origin, conditions at harvest, and storage conditions. Therefore, there is a need to add yellow carotenoids depending on the condition of the feed ingredients.

The absorption efficiency of carotenoids also varies by animal species, with the following order of carotenoid deposition from feed to egg yolk in laying hens.

In yellow carotenoids
Apo-ester > Zeaxanthin > Lutein > Citranaxanthin

In red carotenoids
Canthaxanthin > Capsanthin/Capsorbin > Astaxanthin

15.3.3 Development of Egg Yolk and Change in Egg Yolk Color

The egg yolk takes approximately 7–12 days to develop from the primary follicle to the tertiary follicle in the ovary. Nutrients such as carotenoids and vitamins accumulated in the liver during this formation period are deposited daily via the bloodstream into the developing follicles in concentric circles. If red carotenoids are added to laying hen diets that were fed only yellow carotenoids, the color changes from the outside of the circle toward the inside one day at a time, as shown in Figure 15.10. Practically, it takes 7–11 days to change the egg yolk color, but commercially, it will take about 2 weeks to see the egg yolk color change, since there are individual differences among hens.

If egg yolk color is not uniform, it may be due to variations in carotenoid content in the feed, low titer, or poor nutrient absorption and metabolism due to heat or poor intestinal environment. So uniform egg yolk color is a barometer of laying hen health.

15.3.4 Egg Yolk Color and Its Function

Surai et al. (2001) reported that carotenoid levels were higher in wild birds than in commercial hens. Carotenoids not only bring about bright yolk coloration, but also have physiological implications. For example, when comparing fertilized eggs with high carotenoid content and those with low carotenoid content, many studies have reported that fertility and hatchability are clearly higher for eggs with a brighter egg yolk color, which contains more carotenoids, and this is practiced in actual breeder and hatchery farms. Although the shade of egg yolk color itself is not important for wild birds, the darker egg yolk color means higher carotenoid content, which is used for reproductive purposes according to the theory of species conservation.

Among the carotenoids, canthaxanthin, which has a high deposition rate in egg yolk, has a high antioxidant capacity compared to other carotenoids, as well as astaxanthin (Terao, 1989; Camera, 2008; Zhang et al., 2011). In fertilized eggs with darker egg yolk color that contain carotenoids such as canthaxanthin, the antioxidant properties help to strengthen the vitelline membrane, preventing direct exposure of the early embryo to the high pH albumen and increasing embryo survival. Damaziak et al. (2018) also indicated that the addition of canthaxanthin and iodine to the diet influenced vitelline membrane strength and Salmonella growth inhibition in eggs laid.

The results shown in Figure 15.11 come from a study of the yolk breakage rate of eggs under room temperature storage conditions in hens fed a normal corn-soybean meal diet with the addition of canthaxanthin and vitamin E. The trial results showed that the yolk breakage rate was significantly lower in the eggs laid by hens fed with canthaxanthin and vitamin E in comparison with the control group. It is thought that eggs with darker egg yolk color may be more valuable not only for preservation but also for increasing the commercialization rate of processed eggs, especially since it is effective when yolk and egg white need to be mechanically separated in food factories.

Consequently, eggs with a darker egg yolk color are not only a value-added egg in terms of color, but also in terms of nutritional and physiological functions compared to eggs with a lighter egg yolk color.

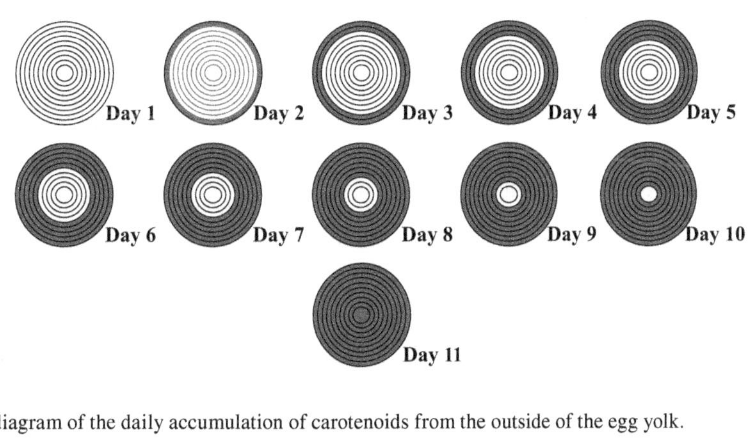

FIGURE 15.10 Schematic diagram of the daily accumulation of carotenoids from the outside of the egg yolk.

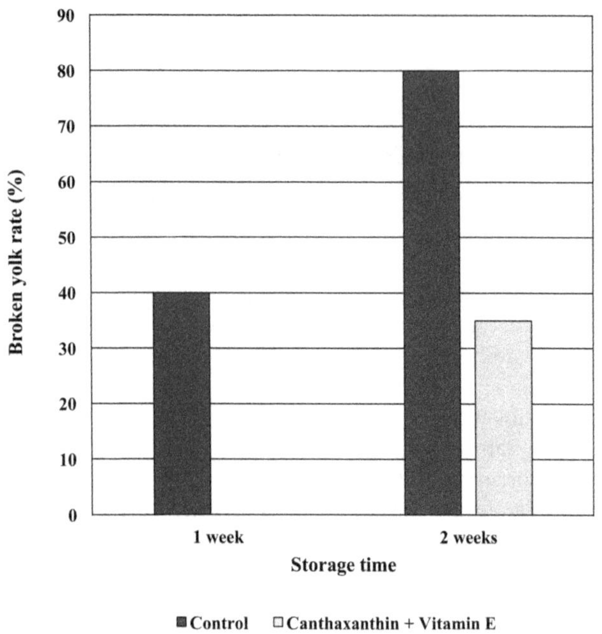

FIGURE 15.11 Comparison between regular layer diet and canthaxanthin + vitamin E supplementation on broken yolk rate of eggs.

15.4 Conclusion

Eggshell and egg yolk color are the most important elements of egg quality in differentiating eggs, and they are also a barometer of the health of the hens that lay them. If the hens are not in good health, they will not be able to absorb nutrients from the feed, which will weaken the eggshells and cause the egg yolk color to pale. Therefore, only healthy hens can produce eggs with strong eggshells and bright egg yolk color, and these egg qualities can be improved by observing the hens and paying attention to the nutrition of their feed and the environment in which they live.

A strong eggshell prevents bacteria from invading the egg and allows as many saleable eggs as possible to be distributed, thus maintaining their value as table eggs. This will contribute to sustainable food from the perspective of food loss and waste.

Consumers often associate these factors with the freshness and nutritional value of eggs. On the other hand, egg yolk color has been said to be not important in terms of nutritional value, but it is a carotenoid that has been reported to suppress the generation of active oxygen, protect cell membranes, have an antitumor effect, and stimulate immunity. Therefore, chicken eggs can be expected to have nutritional value as a carotenoid-rich food.

REFERENCES

Al-Batshan, H. A., Scheideler, S. E., Black, B. L., Garlich, J. D., & Anderson, K. E. (1994). Duodenal calcium uptake, femur ash, and eggshell quality decline with age and increase following molt. *Poultry Science*, *73*, 1590–1596. https://doi.org/10.3382/ps.0731590

Camera, E., Mastrofrancesco, A., Fabbril, C., Daubrawa, F., Picardo, M., Sies, H., & Stahl, W. (2008). Astaxanthin, canthaxanthin and b-carotene differently affect UVA-induced oxidative damage and expression of oxidative stress-responsive enzymes. *Experimental Dermatology*, *18*, 222–231. https://doi.org/10.1111/j.1600-0625.2008.00790.x

Central Egg Standards Trading Council. (2020). Outline of standards and quality inspection for packaged retail eggs in 2019. Central Egg Standards Trading Council. https://www.jpa.or.jp/chuo_root/kensa/pdf/results_2019.pdf

Damaziak, K., Marzec, A., Riedel, J., Szeliga, J., Koczywas, E., Cisneros, F., Michalczuk, M., Lukasiewicz, M., Gozdowski, D., Siennicka, A., Kowalska, H., Niemiec, J., & Lenart, A. (2018). Effect of dietary canthaxanthin and iodine on the production performance and egg quality of laying hens. *Poultry Science*, *97*, 4008–4019. https://doi.org/10.3382/ps/pey264

Hy-Line. (2016). *Technical update: Growing management of commercial pullets*. Hy-Line International. https://www.hyline.com/Upload/Resources/TU%20PULLET%20MGMT%20ENG.pdf

International Egg Commission. (2020). *Country data 2015–2020*. International Egg Commission. https://marketdata.internationalegg.com/graphs/

Kemps, B. J., Govaerts, T., De Ketelaere, B., Mertens, K., Bamelis, F. R., Bain, M. M., Decuypere, E. M., & De Baerdemaeker, J. G. (2006). The influence of line and laying period on the relationship between different eggshell and membrane strength parameters. *Poultry Science*, *85*, 1309–1317. https://doi.org/10.1093/ps/85.7.1309

Lu, M.-Y., Xu, L., Qi, G.-H., Zhang, H.-J., Qui, K., Wang, J., & Wu, S.-G. (2021). Mechanisms associated with the depigmentation of brown eggshells: A review. *Poultry Science*, *100*, 101273. https://doi.org/10.1016/j.psj.2021.101273

Mertens, K., Bamelis, F., Kemps, B., Kamers, B., Verhoelst, E., De Ketelaere, B., Bain, M., Decuypere, E., & De Baerdemaeker, J. (2006). Monitoring of eggshell breakage and eggshell strength in different production chains of consumption eggs. *Poultry Science*, *85*, 1670–1677. https://doi.org/10.1093/ps/85.9.1670

Metrixlab. (2018) Egg study report. Metrixlab. Contact the author for referencing this reference, as this is personal communication.

Miranda, J. F., Moretto, S., Roos, F., Adami, A., Bittencourt, L., & Sechinato, A. (2016). Quality of eggs in Brazil: Correlation between parameters evaluated. *Poultry Science*, *95*(E-Suppl. 2), 41. https://poultryscience.org/files/galleries/2016_LATAM_Abstracts.pdf

Research International. (2003). Report for evaluation of eggs and understanding market perceptions. Contact the author for referencing this reference, as this is personal communication.

Surai, P. F., Speake, B. K., & Sparks, N. H. C. (2001). Carotenoids in avian nutrition and embryonic development. 1. Absorption, availability and levels in plasma and egg yolk. *Journal of Poultry Science*, *38*, 1–27. https://doi.org/10.2141/jpsa.38.1

Takahashi, S. (2022). Introduction of branded eggs. Tamago museum. http://takakis.la.coocan.jp/tokusyu-new.htm

Terao, J. (1989). Antioxidant activity of beta carotene related carotenoids in solution. *LIPIDS*, *24*(7), 659–661. https://doi.org/10.1007/bf02535085

Tsushima, Y., Kikuchi, M., Ohmori, S., Kurita, A., Suzuki, N., Maeda, R., Kougo, K., Sakamoto, M., Ohta, Y., Yoshida, T., Nakano, N., & Tanaka, M. (2013). Research on eggshell color in poultry 4. *Animal Husbandry*, *67*(12), 1177–1183. https://agriknowledge.affrc.go.jp/RN/2010870911.pdf

Zhang, W., Zhang, K. Y., Ding, X. M., Bai, S. P., Hernandez, J. M., Yao, B., & Zhu, Q. (2011). Influence of canthaxanthin on broiler breeder reproduction, chick quality, and performance. *Poultry Science*, *90*, 1516–1522. https://doi.org/10.3382/ps.2010-01126

16

Measurement of Shell Egg Quality

Nanase Kubo and Hajime Hatta*

CONTENTS

16.1 Quality Evaluation Method for Shell Eggs

Since hen's eggs are perishable foods, if they are stored for an extended period in an inappropriate environment, bacteria can proliferate and cause food poisoning. Especially in countries with a culture of eating eggs raw, the quality preservation of shell eggs is essential from a food sanitation perspective. There are two methods for evaluating the quality and freshness of shell eggs: one is a nondestructive method that does not require cracking the egg (nondestructive method), and the other is a method that cracks the egg and examines the state of the egg's white and yolk (cracked egg inspection method).

16.1.1 Non-Destructive Testing Methods

16.1.1.1 Transmitted Light Inspection (Candling Test)

This method involves shining a 60 W light bulb from a 3 cm-diameter hole through the egg and observing the transmitted light inside the egg. Although dedicated equipment exists, a simple measurement can be made by shining light through the blunt end of a shelled egg using a household flashlight. When the light is shined on the shelled egg, the white portion appears bright, and the yolk portion appears dark, allowing visual observation of the yolk position (Miura & Kimura, 2006) (Figure 16.1). The egg yolk is considered fresh when it is centered and has a clear outline. It is more difficult to visually confirm the freshness of a brown egg than a white egg because light does not penetrate as well, and the entire egg looks dark. Although it is not suitable for strict freshness determination, it helps inspect cracked eggshells, rotten eggs, and eggs contaminated with foreign substances (eggs containing pieces of meat or blood in the egg white) and is currently used mainly at GP centers for quality inspection of shell eggs. In addition, the Japanese egg trading standards (Japanese Ministry of Agriculture, Forestry and Fisheries, 2000) and the U.S. Egg Grading Manual (United States Department of Agriculture, 2000) require the Transmitted light inspection in addition to visual observation (Tables 16.1 and 16.2).

16.1.1.2 Measurement of the Air Chamber Height

Similar to the transmitted light inspection, this method estimates freshness by shining light on shelled eggs and examining the height of the air chambers at the blunt end (Jacob et al., 2000). Since water evaporates from the pores of the eggshell, the longer the eggshell is stored, the larger the air chamber volume becomes. At the same time, the height (the distance from the top of the blunt end to the bottom of the air chamber) becomes higher. The height of the air chamber of a fresh egg is about 2 mm, while that of an egg stored at 30°C and

* Corresponding author.

DOI: 10.1201/9781003254430-21

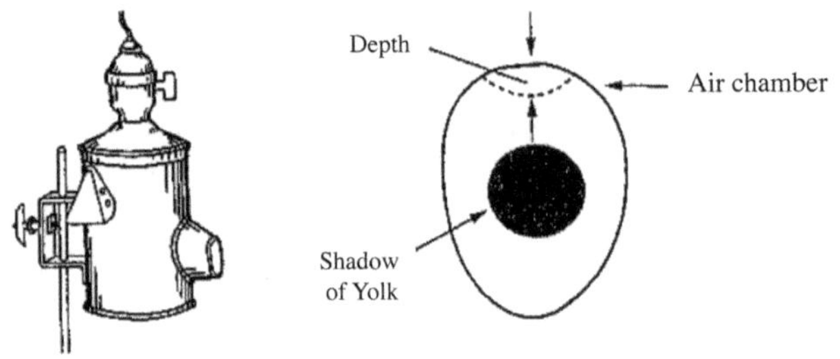

FIGURE 16.1 Device for transmitted light inspection (candling test) and observed image of shell eggs.

50% humidity for 12 days is about 4 to 5 mm, and then the height of the air chamber expands further. Since the amount of water evaporation varies with humidity, it is difficult for the air chamber to become larger under a humid storage environment, and the correlation with other freshness indices may become poor. In addition, if the bottom of the air chamber is sloped, it is difficult to measure the appropriate height, and this causes significant individual differences in measurement, especially in eggs that have lost freshness.

16.1.1.3 Specific Gravity Measurement

Contrary to the air chamber height, the specific gravity of the whole shelled egg decreases as the water in the eggshell evaporates with prolonged storage. The specific gravity of fresh eggs is around 1.08, but it decreases to around 1.02 for eggs stored at 30°C and 50% humidity for 12 days. A simple method to determine the freshness of shelled eggs by floating them in a solution with a specific gravity of about 1.08 (e.g., salt water) is generally known (Voisey & Hamilton, 1977).

16.1.2 Cracked Egg Method

16.1.2.1 Haugh Unit (HU)

Developed by Raymond Haugh in 1937, this method is currently the most common method for evaluating the freshness of shell eggs worldwide (Williams, 1992; Samli et al., 2005). It is calculated using the height (mm) of the egg's thick albumen when broken on a flat plate, taking advantage of the fact that the thick albumen height becomes watery as the freshness decreases. The egg weight (W g) is measured in advance, and then the height (H mm) of the thick albumen is measured by breaking the egg gently on a horizontal glass plate. Haugh units are calculated from the egg weight and the height of thick albumen using the following formula.

$$\text{Haugh Unit (HU)} = 100 \times \log(\text{H} - 1.7\,\text{W}^{0.37} + 7.6)$$

The height of thick albumen is generally determined using an egg quality device and a flat glass table (Figure 16.2a), but a simplified egg quality scale (Figure 16.3) can also be used at farm sites. In recent years, specialized measuring digital devices have also been developed, which can be operated more easily (Figure 16.2b). The HU of fresh eggs ranges from 80–90, and the value decreases with decreasing freshness. In the USDA Egg Grading Manual (2000), AA (freshest and best quality eggs) has HU 72 or higher, A (best quality eggs) has HU 71–60, and B (eggs for liquid eggs) has less than 59.

TABLE 16.1

Egg Trading Standards in Japan

Factors \ Grade		Special Grade (For Raw Consumption)	1st Grade (For Raw Consumption)	2nd Grade (For Heat Processing)	Others (Inedible)
Visual and transmitted light inspection	Shell	Oval, dense, fine-grained, normal in color, clean, intact, and in good condition	Slightly abnormal, such as snaggle, roughness, faded color, etc. Mildly stained eggs, slightly stained eggs Intact eggs	Malformed eggs Severely coarse Soft eggs Severely stained eggs Broken eggs without leakage	Moldy eggs Broken eggs with leaks Stinky eggs
Transmitted light inspection	Egg yolk	Located in the center, the outline is slightly and not flattened	Slightly off-center Outline is clear Slightly flattened	Fairly off-center Flat and enlarged Disarray for physical reasons	Rotten eggs Hatch aborted Eggs Eggs with blood spots Disarray eggs Foreign-contaminated egg
	Egg white	Transparent and not soft	Transparent, but slightly soft	Soft and liquid form	
	Air chamber	Depth less than 4 mm and almost constant	Depth is less than 8 mm	Depth exceeds 8 mm	
Cracked egg inspection	Diffusion area	Small	Usual	Quite wide	
	Yolk	Circularly raised	Slightly flattened	Flat, and yolk membrane is soft and weak	
	Thick albumen	Occupying a large quantity, raised, and well enclosing the yolk	Small quantities, flattened, and not fully enclosing the yolk	Little or nothing	
	Thin albumen	Small quantity	Normal quantity	Occupying a large quantity	

Source: Attachment to the Egg Standard Transaction Guidelines, Egg Transaction Standard, dated December 1, 2000.

TABLE 16.2

Summary of U.S. Standards for Quality of Individual Shell Eggs

	Specifications for Each Quality Factor		
Quality Factor	**AA Quality**	**A Quality**	**B Quality**
Shell	Clean Unbroken Practically normal	Clean Unbroken Practically normal	Clean to slightly stained[a] Unbroken Abnormal
Air cell	1/8 inch or less in depth Unlimited movement and free or bubbly	3/16 inch or less in depth Unlimited movement and free or bubbly	Over 3/16 inch in depth Unlimited movement and free or bubbly
White	Clear Firm	Clear Reasonably firm	Weak and watery Small blood and meat spots present[b]
Yolk	Outline slightly defined Practically free from defects	Outline fairly well defined Practically free from defects	Outline plainly visible Enlarged and flattened Clearly visible germ development but no blood Other serious defects
For Eggs with Dirty or Broken Shells, the Standards of Quality Provide Two Additional Qualities:			
Dirty Unbroken, adhering dirt or foreign material, prominent stains, moderate stains in excess of B quality		Check Broken or cracked shell, but membranes intact, not leaking[c]	

Source: Egg Grading Manual (United States Department of Agriculture, 2000).

[a] Moderately stained areas permitted (1/32 of surface if localized, or 1/16 if scattered).

[b] If they are small (aggregation not more than 1/8 inch in diameter).

[c] Leaker has broken or cracked shell membranes, and contents leaking or free to leak.

16.1.2.2 Yolk Index (YI)

The yolk height (H mm) and diameter (D mm) are measured on a horizontal glass plate. The yolk index is calculated as height/diameter (H/D) (Sauter et al., 1951; Wolk et al., 1952). Generally, it can be measured using an egg quality device or a specially designed digital device, but as described later, the YI can also be measured using a vernier caliper (Figure 16.2c). The yolk index of fresh eggs is 0.44~0.36, and the value decreases with decreasing freshness. This decrease is because the yolk membrane becomes weak with decreasing freshness, resulting in a larger yolk diameter and a lower yolk height (Feeney et al., 1956).

16.1.2.3 Measurement of Yolk Membrane Strength

A method was developed to measure the strength of the yolk membrane with relative reproducibility. In this method, egg yolks are placed in 20 ml beakers to hold, and the yolk membranes are broken by pressing on them from above with the cylindrical plunger of a compression measuring device. The strength of the yolk membrane directly correlates with the decline in the egg's freshness, and the yolk membrane breaks at the slightest pressure.

Another classic but more straightforward method for evaluating yolk membrane strength is to pinch the yolk with the thumb and forefinger, hold it for 5 seconds, and observe whether the yolk membrane breaks or not (Mochiage method). The yolk of an old chicken egg cannot be pinched and lifted.

16.1.2.4 Measurement of Egg White pH

The pH of fresh egg white is around 7.5, but as the storage period is prolonged, the pH gradually increases as carbon dioxide in egg whites dissipates outside eggshells, eventually reaching a pH of about 10.0.

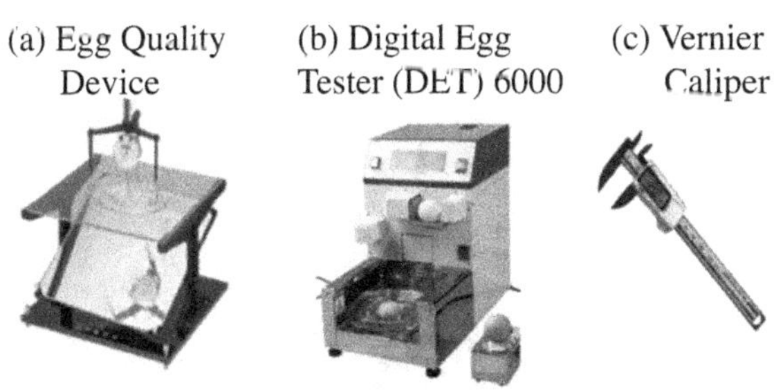

FIGURE 16.2 Devices for measuring the freshness of shell eggs. (a) Egg quality device, (b) digital egg tester (DET) 6000, and (c) vernier caliper.

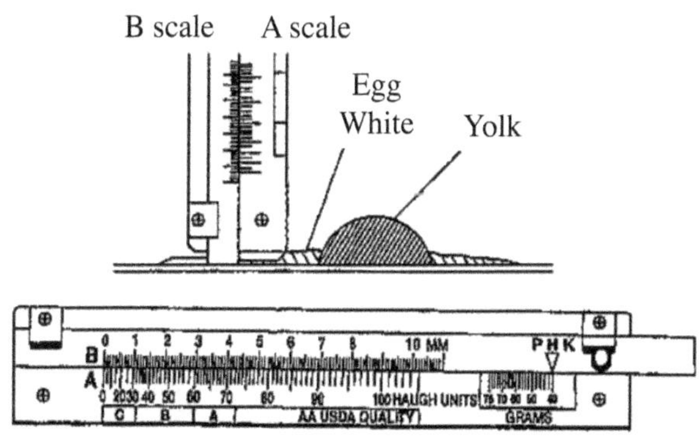

FIGURE 16.3 Egg quality scale for measuring Haugh unit (HU).

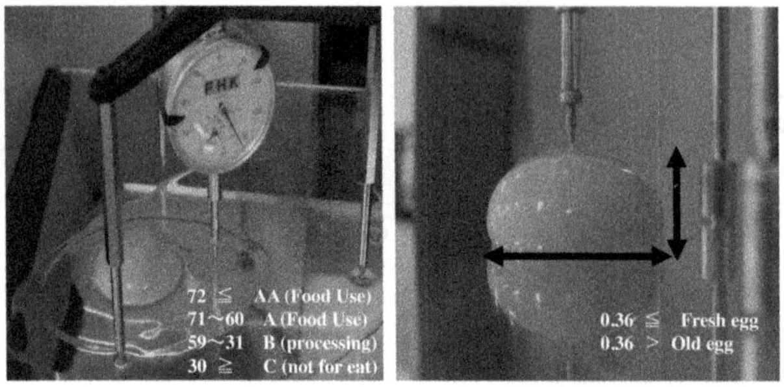

$HU = 100 \times \log(H - 1.7W^{0.37} + 7.6)$

H : Height of thick albumen (㎜)

W: Egg weight (g)

Yolk Index

= Height (㎜)/ Diameter (㎜) of EY

FIGURE 16.4 Haugh unit and Yolk index measurements and formulas.

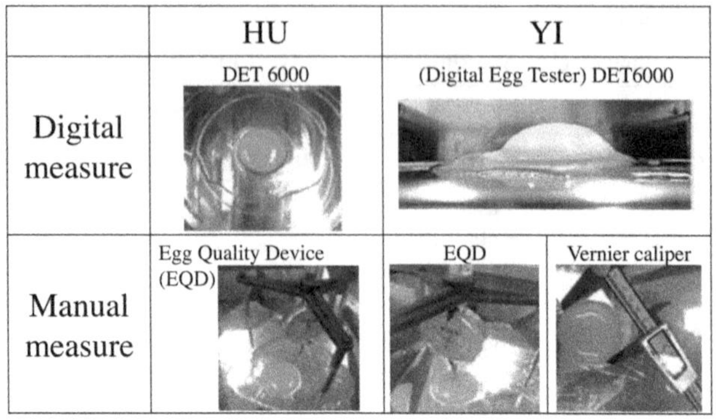

FIGURE 16.5 Measurement details for each freshness index.

16.2 Comparison of HU and YI

16.2.1 Manual vs. Digital Measurement

In the conventional measurement method, HU requires an egg quality device and a scale, while in YI, an egg quality device and a ruler are all required for measurement (manual measurement) (Figure 16.4). In recent years, digital measuring machines have been developed for measuring egg freshness and are used at various GP centers. We used an egg quality measuring machine (digital egg tester; DET6000) manufactured by Nabel Corporation that can measure HU and YI, and investigated whether there is any difference in the freshness indices of HU and YI between the conventional manual measurements using an egg quality device and the digital measurements using the DET6000.

Freshly laid eggs (Boris Brown, first day of laying) were stored at 30°C/75% humidity, 20°C/65% humidity, and 5°C (in a refrigerator) at each storage-temperature until expiration and sampled over time for freshness measurements. In addition, various commercially available packaged eggs were purchased, and 402 eggs were used for freshness measurements. First, as a digital measurement method, eggs were cracked on a DET6000 sample dish, and the HU and yolk index were measured on whole eggs without separating egg whites and yolks. Next, the same egg samples were transferred on a horizontal glass plate, and the HU and yolk index were measured manually using an egg quality device. For the manual measurement of the yolk index, the yolk and egg white were separated, and the yolk was placed on a measuring table. Finally, as the most straightforward measurement method, the yolk width and height were manually measured in millimeters using a vernier caliper, and the yolk index was calculated. (Figure 16.5)

Digital and corresponding manual measurements of all eggs measured for each freshness index (HU and YI) were plotted in XY coordinates (Figure 16.6). Regression lines and coefficient of determination (R^2 values) were calculated by the least squares method. The correlation between digital and manual measurements was better for YI than HU. Especially at low freshness (HU < 60), the HU's variability of the digital and manual measurements was more remarkable, resulting in lower R^2 values than for YI. The coefficient of determination (R^2 values) for each measurement method used in this study is summarized in Table 16.3. The correlation between manual and digital measurements was higher for the YI determined by DET6000 and either an egg quality device or a vernier caliper than for that of HU, the most used index in the world. YI values measured by an egg quality device and by a vernier caliper

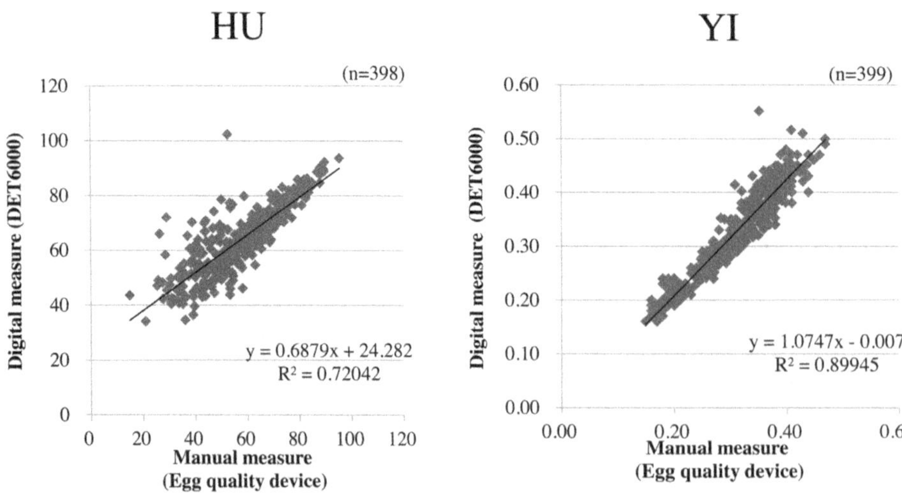

FIGURE 16.6 Comparison of digital and manual data.

TABLE 16.3

Coefficient of Determination (R^2) between Digital and Manual Data in the Least-Square Method

	Digital Measure		Manual Measure	Coefficient of Determination (R^2)
HU	DET6000	vs.	Egg quality device	0.72
YI ①	DET6000	vs.	Egg quality device	**0.90**
YI ②	DET6000	vs.	Vernier caliper	0.89

were comparable to the digitally measured YI values, and the R^2 Values were 0.90 and 0.89, respectively. It should be noted that although the YI measurement with the vernier caliper cannot process many eggs, the data obtained yielded a freshness index value (YI) comparable to the digital measurement, which can apply to egg-laying poultry in developing countries.

16.2.2 Evaluation of Variation in Measurements

Figure 16.7 shows the changes in the freshness of eggs stored at 30°C, 20°C, and 5°C in terms of HU and yolk index (digitally measured values), respectively, it was shown that the HU and yolk index decreased with the lengthening of the storage period at each temperature. On the other hand, if we look at each value's standard deviation (variation), we see that it is wide for the HU and tiny for the yolk index. The student t-test results showed a significant difference in CV (Coefficient of Variation) values between HU and YI in the digital measurement at a 1% significance level. These results indicate that the yolk index has little variability than the HU and that the yolk index value is much more stable.

16.2.3 Advantages of Using YI

Based on the above results, the YI is a freshness index with the slightest fluctuation of value and the most accurate value in multiple measurement methods. As shown in Figure 16.8, the coefficient of determination (R^2) of the yolk index between the vernier caliper and the egg quality device was as high as approximately 0.98. Therefore, it can be said that if YI is used as a freshness index, it is possible to evaluate freshness using a vernier caliper or an egg quality device, which is equivalent to digital measurement (Table 16.3). These results indicate that YI, which measures changes in yolk height and diameter due to the weakening of the yolk membrane, is more accurate than HU, which is a conventionally used freshness index that measures the thickness of thickened albumen, in evaluating freshness. The advantages of using YI as a freshness index are listed below.

First, the easiest and quickest measurement is the YI in digital measurement. The digital determination is ideal for mass measurements at GP centers, as they can be done at the touch of a button without separating the yolks from egg white. In addition, YI has less variation and is a more objective indicator of freshness than HU. Therefore, accurate and reliable values can be obtained even in small-volume measurements at poultry farms using conventionally used egg quality devices. The yolk index can be used as a uniform freshness index regardless of the poultry farm's location (country/region) or business size.

Currently, methods for evaluating the freshness of shell eggs at the time of shipment are not standardized in each country. For example, as mentioned above, the United States has a transmitted light inspection, and Japan has standards in the transmitted light inspection and broken egg inspection. However,

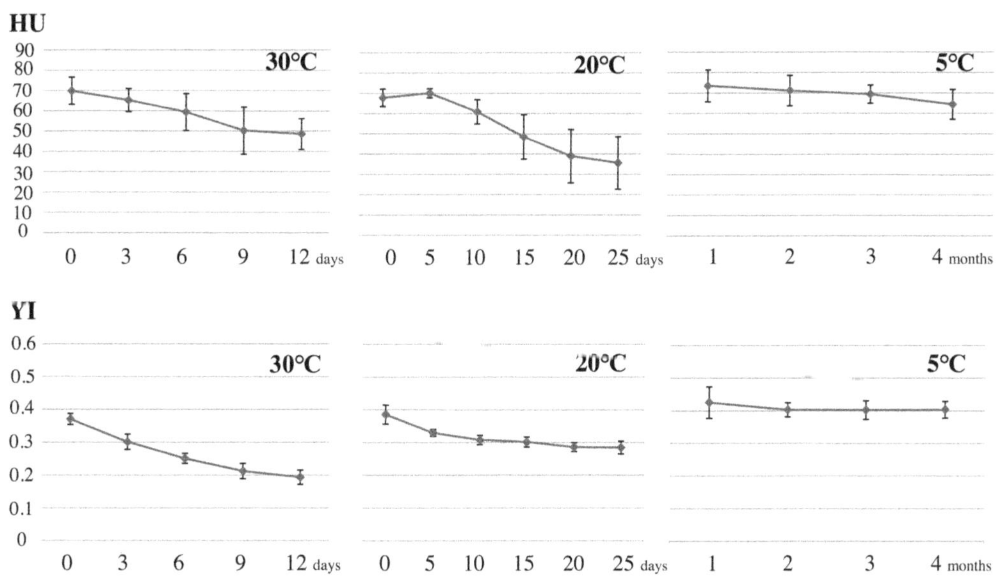

FIGURE 16.7 Changing of HU and YI in fresh eggs.

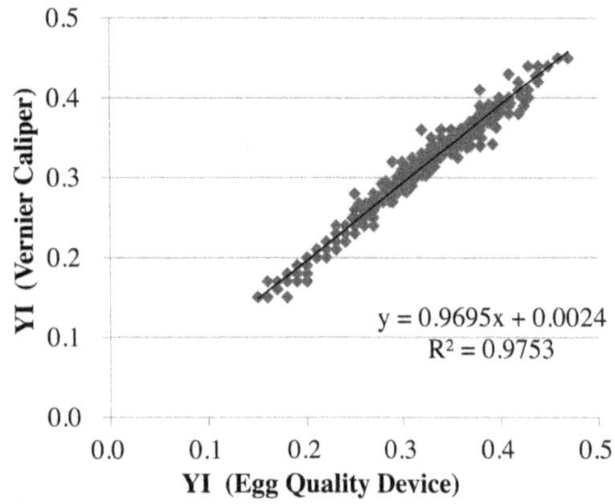

FIGURE 16.8 Relationship between YI measured by egg quality device and vernier caliper.

the transmitted light inspection is an indicator that remains ambiguous, and there is no specific measurement method or absolute value for the egg freshness test. International import and export of eggs are expected to increase in the future, and it is important to unify accurate freshness evaluation standards for shell eggs on a global scale, from a food sanitation perspective. Under such circumstances, it would be beneficial to use YI as a freshness index because it is more reliable and straightforward and can be used for both large- and small-egg businesses.

16.3 Conclusion

In this chapter, we have outlined the measurement methods and principles of nondestructive and cracked egg inspection methods used to evaluate the quality and freshness of shell eggs. In addition, a comparison was made between digital measurements (using the DET6000; egg quality tester) and manual measurements (using an egg quality device and vernier calipers) of HU and YI, the most well-known evaluation indices by the cracked egg inspection. The results showed that YI had the highest Coefficient of Determination (R^2 value) between digital and manual measurements and the lower coefficient of variation (CV value) for eggs of different freshness levels. In addition, YI, unlike HU, showed a high R^2 value of 0.9 between digital measurements (using the DET6000; egg quality tester) and manual measurements (using an egg quality device). The R^2 showed almost the exact value of 0.89 (Table 16.3) even when a vernier caliper was used as a more straightforward measurement method. It is predicted that the number of shell eggs imported and exported worldwide will increase, and we propose using the yolk index as a new, unified indicator of egg freshness.

REFERENCES

Feeney, R. E., Weaver, J. M., Jones, J. R., & Rhodes, M. B. (1956). Studies of the kinetics and mechanisms of yolk deterioration in shell eggs. *Poultry Science, 35*(5), 1061–1066. https://doi.org/10.3382/ps.0351061

Jacob, J. P., Miles, R. D., & Mather, F. B. (2000). Egg quality. *Cooperative extension service, institute of food and agricultural sciences (IFAS)* (p. 7). https://ufdcimages.uflib.ufl.edu/ir/00/00/42/62/00001/ps02000.pdf

Japanese Ministry of Agriculture, Forestry and Fisheries. (2000). *Egg Transaction Standard, Attachment to the Egg Standard Transaction Guidelines*. (in Japanese). https://www.jpa.or.jp/chuo_root/gaiyo/gaiyo_20201124_01.pdf

Miura, H., & Kimura, S. (2006). *Handbook of food preservation and distribution technology*. Kenpakusha. (in Japanese). https://www.kenpakusha.co.jp/np/isbn/9784767961040/

Samli, H. E., Agma, A., & Senkoylu, N. (2005). Effects of storage time and temperature on egg quality in old laying hens. *Journal of Applied Poultry Research, 14*(3), 548–553. https://doi.org/10.1093/japr/14.3.548

Sauter, E. A., Stadelman, W. J., Harns, V., & Mclaren, B. A. (1951). Methods for measuring yolk index. *Poultry Science, 30*(4), 629–632. https://doi.org/10.3382/ps.0300629

United States Department of Agriculture. (2000). *Egg grading manual*. https://www.ams.usda.gov/sites/default/files/media/Egg%20Grading%20Manual.pdf

Voisey, P. W., & Hamilton, R. M. G. (1977). Sources of error in egg specific gravity measurements by the flotation method. *Poultry Science, 56*(5), 1457–1462. https://doi.org/10.3382/ps.0561457

Williams, K. C. (1992). Some factors affecting albumen quality with particular reference to Haugh unit score. *World's Poultry Science Journal, 48*(1), 5–16. https://doi.org/10.1079/WPS19920002

Wolk, J., McNally, E. H., & Brant, A. W. (1952). Yolk measurements use d as an indication of temperature deterioration of eggs. *Poultry Science, 31*(4), 586–588. https://doi.org/10.3382/ps.0310586

17

Shelf Life of Eggs

Hajime Hatta* **and Nanase Kubo**

CONTENTS

17.1 Introduction

Chicken eggs are an excellent source of protein and fat, as evidenced by the fact that fertilized eggs produce chicks, and they are also an incredibly nutritious fresh food with a good balance of vitamins and minerals (Hatta et al., 2008; Yuceer & Caner, 2014). Therefore, depending on storage conditions, it is prone to spoilage and is consistently listed as one of the tops causes of foodborne illnesses worldwide (Cardoso et al., 2021). From the perspective of preventing food poisoning, commercial eggs and processed egg products have an expiration date set to ensure safe and tasty consumption. However, due to differences in food culture and the handling of shell eggs in countries around the world, the method of setting expiration dates have not been standardized and differ significantly from country to country. Japan, in particular, is a country that has long eaten eggs raw, and its method of setting expiration dates and the scientific basis for doing so is fascinating.

In some countries, eggs are not labeled with expiration dates. In such cases, consumers must rely on self-inspection of the egg to protect themselves. Is the surface of the eggshell slimy? Are the egg whites and yolks watery when the eggs are cracked? Is there a rancid smell to determine whether or not the eggs are edible? Humans are generally sensitive to rancid odors, but a specific type of food poisoning bacteria, *Salmonella enterica subsp. enterica serovar* Enteritidis (SE) may multiply in eggs and not produce a rancid odor, so from the standpoint of preventing food poisoning, it is essential to label eggs with an expiration date.

Furthermore, the rationale for setting an expiration date must be explained correctly to consumers. Japan has long had a culture of eating eggs raw, and the best-before labeling is the only example in the world that indicates when eggs can be eaten raw. In this chapter, we will show the best-before dates for eggs in major egg-consuming countries as much as possible and explain the basis for setting the best-before dates in Japan.

17.2 Labeling of Expiration Dates of Chicken Eggs in Various Countries

The expiration date of chicken eggs is labeled to ensure their safety as food based on its freshness and quality over time. There are countries where eggs are eaten raw, where eggs are always heated, where shell eggs are washed, where washing is prohibited, where eggs are distributed refrigerated, and where eggs are distributed at room temperature. Each of these countries has different laws regarding producing and selling shell eggs, so naturally, labeling different expiration dates (Yimenu et al., 2017). For

* Corresponding author.

DOI: 10.1201/9781003254430-22

example, EU law prohibits shell eggs from being washed in any way and must not be refrigerated before being sold to the final consumer. Thus, the generation of water droplets due to temperature changes on the eggshell surface is prevented. In addition, the shelf life in the EU should not exceed 28 days after laying eggs (EUR-Lex, 2008). The shelf life in the United Kingdom is also 28 days after the egg-laying date (National Health Service, 2015).

In the United States, the U.S. Department of Agriculture (USDA) requires that shell eggs be washed and disinfected before reaching the consumer and stored at 7.2°C or lower to prevent the growth of pathogens. The shelf life label of packed eggs is optional, but if indicated, it should be printed in month/day format, preceded by the appropriate prefix "EXP," "Sell By," etc., and 30 days after packing. When indicating the maximum recommended period that the consumer can maintain the quality of the eggs when stored under ideal conditions, terms such as "Use by," "Use before," or "Best before" are used to indicate how long the eggs should be consumed before their overall quality deteriorates. In this case, the period cannot exceed 45 days, including the day the eggs were packed (USDA-FSIS, 2019).

The Australian and New Zealand governments have stated that eggs do not need to be refrigerated for distribution (Food Standards Australia New Zealand, 2011). However, the Consumer Site of Australian Egg Farmers (2014) recommends that raw eggs be refrigerated and have a shelf life of six weeks.

In China, all packed eggs must be washed and sterilized. Their shelf life is set at 30 days after packing at room temperature and 45 days at refrigerated. Packaged eggs are usually sold at supermarkets at room temperature. However, in some traditional markets, piles of eggs are sold by weight with no best-before labeling. The current egg grading regulations in Korea have existed since 2007 and stipulate that shell eggs must be stored and distributed at 0–15°C, in case of processed eggs at 10°C or lower. In addition, the shelf life of packed eggs is not specified (Lee et al., 2016).

Japan has long had a food culture of eating raw eggs, and the best-before-date labeling for eggs is the most unique in the world. The date on eggs in Japan indicates when eggs can be eaten raw. Shell eggs are cleaned and sterilized at a Grading and Packaging Center (GP Center) and inspected for cracks in the shell. A transmitted light test should be conducted to produce packed eggs. The expiration time is labeled for 21 days from the laying date whenever stored at 25°C or less. Packaged eggs sold in the Japanese market can be eaten raw within this expiration date. This is the number of days that *Salmonella* Enteritidis (SE), a well-known cause of food poisoning in eggs, will not multiply even if it is present in the eggs. The basis for setting the number of days that SE cannot multiply in eggs will be discussed in more detail later, along with the history of the legalization of Japan's special best-before-date labeling.

17.3 Pathways of Bacterial Contamination of Shelled Eggs

17.3.1 On-Egg Contamination

Generally, the inside of a chicken egg is sterile immediately after laying (Qi et al., 2020), but the surface of the eggshell is contaminated with a thousand to one million bacteria. This is because the cloaca (gross excretory cavity), where feces and urine are excreted, is connected to the exit of the oviduct of the laying hen. Therefore, the intestinal bacteria of the hens adheres to the eggshell surface. In addition, soil bacteria from the poultry farm environment adhere mainly to the egg's surface after laying, and with time the desiccation-sensitive intestinal bacteria decrease, and the desiccation-resistant soil bacteria become predominant. The eggshell surface bacteria are physically protected from penetration into the egg by the cuticle of the eggshell surface cover, the firm eggshell structure, and the eggshell membrane. However, when the cuticle is detached by egg washing or the eggshell surface becomes wet, the bacteria easily pass through the approximately 10,000 pores in the eggshell and reach the eggshell membrane. Nevertheless, fresh egg whites are highly viscous and contain antibacterial components (lysozyme, ovotransferrin, avidin, etc.), making it difficult for bacteria to multiply.

On the other hand, egg yolk is rich in nutrients and is an excellent medium for bacteria. When bacteria are inoculated in egg white, they rarely multiply, but when egg yolk is mixed in, the bacteria multiply explosively. Generally, the yolk of fresh eggs is retained in the center of the egg by the thick egg white and chalaza layer, protecting it from contact with bacteria on the eggshell surface. However, as egg freshness declines, especially when the thick egg white changes to watery egg white, the yolk, with its low specific gravity, float up in the egg and adheres to the eggshell membrane. In this case, bacteria that have passed through the eggshell and reached the eggshell membrane come into direct contact with the yolk, resulting in explosive growth and spoilage. This spoilage of chicken eggs caused by bacteria on the eggshell surface entering the egg is called On-egg contamination.

17.3.2 In-Egg Contamination

In recent years, when *Salmonella* Enteritidis (SE), a foodborne pathogen, infects laying hens and settles in the ovaries or oviducts, it has become problematic for them to lay eggs with SE inside the shell eggs (yolk and albumen), although this is rare. This is called In-egg contamination, which occurs at a rate of 2–3 eggs per 100,000 eggs in Japan. Usually, the number of SE in In-egg contaminated eggs are less than 10, and since these bacteria are iron-demanding, their growth in the egg white has been stopped. However, when egg yolk components and iron ions leak from the yolk as the freshness of the egg declines, the SE on the yolk membrane multiplies explosively, causing salmonella food poisoning.

In any case, it is essential to understand quality preservation techniques for shell eggs because of changes in shell egg quality will eventually lead to the internal spoilage of eggs by bacteria, which can cause food poisoning. Hygienic management and cold storage during each production stage, distribution, and marketing are critical when handling hen's eggs.

17.4 Quality Changes of Shell Eggs

17.4.1 Change in Egg Quality over Time

Egg quality changes in shelled eggs begin with a detachment of the cuticle from the eggshell surface layer, changes in the pH and viscosity of the egg white, and changes in the strength of the eggshell and yolk membranes (chemical and physical quality changes). Depending on the storage temperature and duration, bacteria on the eggshell surface may invade the interior and ultimately spoil the contents (microbial quality changes).

Usually, shell eggs undergo significant environmental changes before and after laying. The partial pressure of CO_2 in the oviduct of a laying hen is 0.1 atm; however, the partial pressure of CO_2 in the air outside is 0.0003 atm. Therefore, immediately after egg laying, CO_2 dissolved in the egg white begins to dissipate outward through the pores of the eggshell. Accordingly, the pH of the egg white rises from 7.5 (immediately after egg laying) to about 9.5 in a few days.

In addition, the water solubilization of thick egg white and the reduction in the strength of the eggshell and yolk membranes progress over time due to the degradation of the β-ovomucin subunit and O-glycosidic bonds, the disruption of the ovomucin-lysozyme complex, and the reduction of albumen-protein interactions (Obianwuna et al., 2022). The higher the environmental temperature, the faster these chemical and physical qualities changes occur. The above chemical and physical quality changes proceed faster at higher environmental temperatures.

17.4.2 Changes in Freshness over Time

Currently, the Haugh unit (HU), Yolk Index (YI), egg white pH, egg yolk pH, thickened egg white percentage (%), yolk membrane strength, air chamber diameter, air chamber height, egg weight, and specific gravity are known as indicators to evaluate egg freshness (Dong et al., 2017; Jiang et al., 2018) and in Chapter 16. These freshness indices are known to be affected by the storage temperature (Fikiin et al., 2020; Jin et al., 2011), storage humidity (Yimenu et al., 2018), and storage duration (Yimenu et al., 2017) of shelled eggs, respectively. Sami et al. found that the freshness indices most affected by storage period and temperature were HU, egg white pH, and air chamber diameter (Samli et al., 2005).

HU is the best-known indicator of egg freshness over time worldwide and is used to grade shell eggs in several countries (Li, 2013). However, the HU range of its grading standard varies slightly among countries; in the EU, the three quality classifications are Class A, Extra fresh, Class A, category I, and Class B, with Class A being fresh eggs (UNECE, 2017). Furthermore, in the United States, eggs are classified into three categories, AA (HU>72), A (72–60), and B (<60) (United States Department of Agriculture, 2000), while in China, eggs are classified into four grades of freshness, AA (HU>72), A (72–60), B (59–31), and C (31>) (Qi et al., 2020).

HU showed a decreasing trend as hens aged, with HU 90.9 in fresh eggs at 180 days of age decreasing to 77.8 at 450 days of age, a decrease of about 16%, and then recovering to HU 85.5, a decrease of about 10%, after subsequently forced feather replacement (Williams, 2019). Breeding hens can also improve HU, and breeding hens that lay shell eggs with high HU has been conducted. Although HU is regarded as the world standard for the freshness index of chicken eggs, there is room for improvement in its measurements, etc., and a more accurate freshness index to replace HU is underway (Narushin et al., 2021).

In the authors' study, fresh chicken eggs were also stored at 30°C and 75% humidity for 12 days, and the freshness indices, HY, YI, yolk pH, egg white pH, yolk membrane strength, and thick egg white ratio, were examined every three days over time. The results showed that HU, YI, and yolk membrane strength had a linear relationship with the number of days of storage. However, yolk membrane strength could not be detected after 9 days of storage because the yolk membrane was ruptured before measurement. Therefore, HU and YI were the only freshness indices that could estimate the number of days of storage. Furthermore, YI, which has a smaller variation in measured values than HU, was found to be a more accurate indicator for estimating the number of storage days (Kubo & Hatta, Chapter 16, this book) (Figure 17.1).

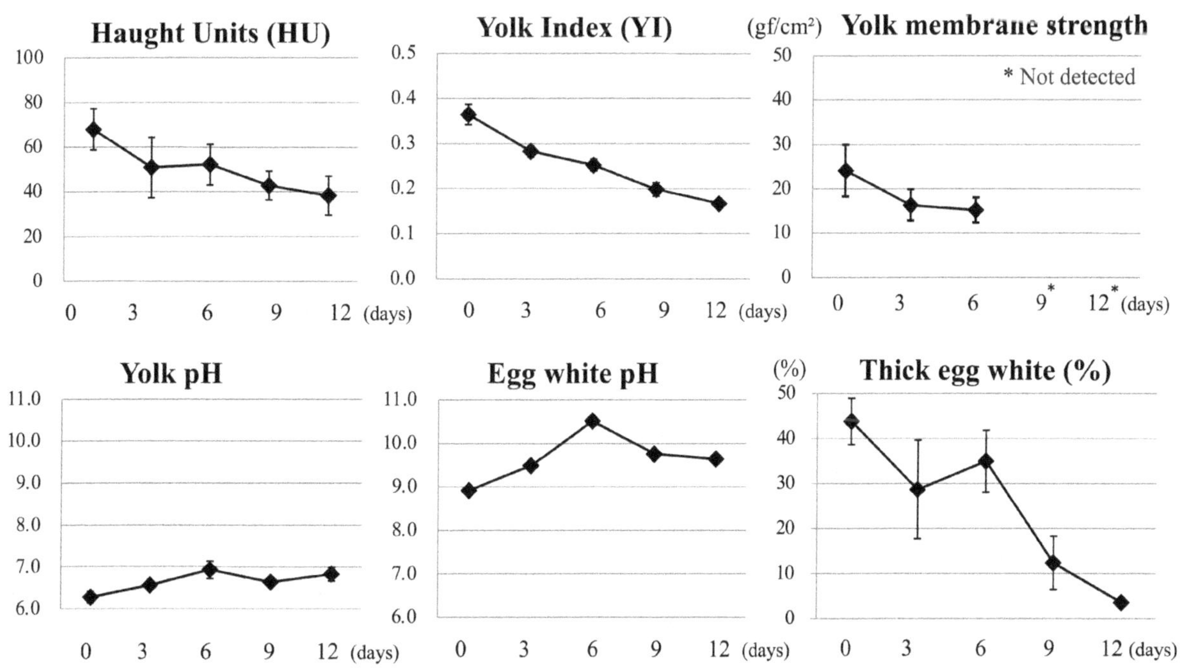

FIGURE 17.1 Changes in shelled egg quality during storage (at 30°C, 75% humidity).

Thus, research is actively being conducted to investigate the changes over time of various freshness indices in the storage of chicken eggs, to examine the most accurate method for evaluating the decline in the freshness of chicken eggs starting immediately after laying, and to establish techniques for evaluating freshness parameters (Akter et al., 2014). The use of the latest technologies in the evaluation of egg freshness includes dielectric property analysis (Ragni et al., 2006), the combination of visible light near-infrared transmission and low-resolution proton nuclear magnetic resonance spectroscopy (Kemps et al., 2007), the use of FT-NIR spectroscopy (Giunchi et al., 2008), and electronic nose system (Yongwei et al., 2009), and others have received attention.

17.5 Quality Maintenance of Shelled Eggs

17.5.1 Cleaning and Drying of Eggshell Surfaces

Shelled eggs are delivered from the poultry farm to the nearest GP (Grading and Packaging) Center, where the eggshell surface is washed and dried depending on countries. The eggs are then inspected by transmitted light (to remove cracked eggs) to ensure these qualities. Then the eggs weigh for grading and are packed for shipment. Some countries argue that washing shell eggs is undesirable because it removes the cuticle from the eggshell surface, but eggs with poultry manure have no commercial value, and packaged eggs in Japan are washed. Although egg washing reduces the number of bacteria on eggshells to 1/10 to 1/100, there is a risk that wet eggshells will allow bacteria to pass through the pores in the shell easily, so eggs must be dried immediately after washing. Drying the eggshells makes it more difficult for bacteria to penetrate the pores. In other words, dry eggshells are essential for maintaining the quality of shelled eggs.

17.5.2 On-Egg Contamination Prevention

It should be recognized that the surface of eggshells is not sterile even after washing and sterilization. Therefore, when shelled eggs are distributed for sale or stored at home, care must be taken in handling shelled eggs to prevent contamination of other foods with eggshell-associated bacteria. It is natural to discard eggs with cracked eggshells.

The eggshell surface has 10,000 pores with an average diameter of 10–30 μm, and bacteria are generally less than 1 μm in size. When the eggshell surface is dry, the bacteria cannot move, but when it is wet, bacteria move by Brownian motion and fall deep into the pores (Okubo et al., 1997). Therefore, eggshell-associated bacteria are more likely to enter the eggshell through the pores if the eggshell is wet.

In an experiment where chicken eggs were washed, bacteria were observed to pass through the pores and invade the eggshell and eggshell membrane (Hutchison et al., 2004). Therefore, care should be taken to prevent water droplets on the eggshell surface while selling packed eggs. Furthermore, being dried shell egg surface is critical during storage at home, especially when placing the eggs in and out of the refrigerator.

EU law prohibits the refrigerated storage of Class A shell eggs. The EU recognizes that preventing water droplets from forming on the surface of the eggshell due to temperature differences at the time of purchase is very important for the hygiene of shell eggs.

17.5.3 In-Egg Contamination Prevention

To prevent In-egg contamination, it is necessary to strengthen the quarantine of imported chicken chicks to eliminate chickens infected with SE bacilli. One effective measure against SE infection is vaccination. Inactivated SE are injected into chickens to produce IgY antibodies in their bodies and prevent SE infection. According to the Japan Poultry Farmers Association, approximately 40–50% of the eggs currently distributed in Japan are laid by hens vaccinated against SE.

The United Kingdom implemented a national Lion Code initiative and was the first to spread the SE vaccine. As a result, Salmonella food poisoning has been dramatically reduced, and now more than 90% of shell eggs laid by vaccinated hens are sold under the Lion Egg brand (ACMSF-Ad Hoc Group on Eggs, 2016).

In addition, since Japanese people have a habit of eating raw eggs, the Ministry of Health and Welfare has taken the lead in considering measures to prevent In Egg-contamination, and legislation was enacted to label "best-before" dates for shell eggs. As a result, "best-before" labeling of packed shell eggs began in October 1998. The best-before date was calculated based on the growth rate of SE in eggs in Japan.

17.5.4 Low-Temperature Storage of Shell Eggs

Chemical and physical quality changes, such as decreased Haugh units (HU) and Yolk Index (YI), which indicate the freshness of shell eggs, are accelerated at higher storage temperatures. Figure 17.2 shows how units when shell eggs are stored at various temperatures. Storage in a refrigerator (4–6°C) and in a winter room (4.5–15°C), suppresses the decline of the HU. Shell eggs should be distributed and sold under storage conditions of at least 15°C or lower and stored in refrigerators at home to maintain egg freshness.

Regarding the relationship between bacterial contamination of shell eggs and storage temperature, it was reported that in the case of On-egg contamination, no bacteria were found to invade the eggs for 14 days under natural temperatures in August after egg washing. However, bacteria of order 10^7 were found in the eggs after 21 days.

On the other hand, in the case of In-egg contamination, SE does not grow in shell eggs at 4°C, and their growth is reported to be very slow at 10°C. Usually, the percentage of eggs contaminated with SE (In-egg contamination) in shell eggs is about 0.03%, and the number of contaminating bacteria is only a few per egg, so it is not necessary to consider the growth rate of SE in the "best-before" date.

The number of contaminated eggs is only a few per egg, and there is no risk of food poisoning within the "best-before" date, which considers the speed of growth of SE. Implementing a cold chain to maintain the freshness and quality of shell eggs at temperatures lower than 15°C throughout distribution, sale, and consumption is preferred.

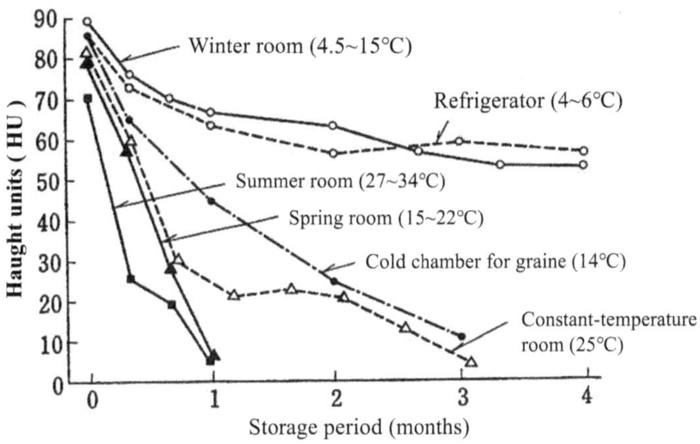

FIGURE 17.2 Changes in HU of shelled egg at each storage condition. (From Sato et al. (1989). *The Science of Egg Cooking and Health* (in Japanese).)

17.6 Japan's Unique System of Labeling Best-Before-Date for Chicken Eggs

Around 1989, food poisoning caused by *Salmonella* Enteritidis (SE bacteria) increased rapidly; hence, hen eggs and their processed products were suspected of being the causative food. Although On-egg contamination of shell eggs can be prevented by the thorough implementation of hygiene control measures issued by the Ministry of Health, Labor and Welfare and the Ministry of Agriculture, Forestry and Fisheries, the rapid increase in food poisoning caused by SE bacteria have been attributed to In-egg contaminated eggs, which already contain SE bacteria in the shell at the time of laying.

The expiration date of a food product usually indicates the time it can be enjoyed without deterioration in quality, while the expiration date of shell eggs in Japan indicates the time they can be eaten raw. In 1990, In-Egg contaminated eggs were found in several out of every 10,000 eggs sold in the egg market, and Salmonella food poisoning was the leading cause of bacterial infection, making food safety measures against In-Egg contaminated eggs an urgent necessity. On the other hand, in the United Kingdom, the country where In-egg-contaminated eggs originated by SE-infected hens developed, Dr. Humphrey and his colleagues had been researching the infection of SE bacteria in eggs and their ability to multiply in eggs and had achieved tremendous results (Humphrey & Whitehead, 1993).

For example, levels of Salmonella in intact eggs are usually as low as 10 CFU/egg or less (Humphrey et al., 1989, 1991). SE growth is limited in egg whites, which contain antimicrobial components even at room temperature (Kang et al., 2006; Schoeni et al., 1989). The development time of SE in egg yolks incubated at 5–15.5°C and 37°C was 3.5 h and 35 min, respectively (Bradshaw et al., 1990). SE bacterial growth was not observed after 7 weeks of storage at 10°C (Chen et al., 2005). SE growth in egg yolks is inhibited when temperatures are below 8° C (Bradshaw et al., 1990; Lublin & Sela, 2008), and so on. For a variety of papers examining the relationship between SE proliferation in eggs and storage temperature, is shown to refer to the review by Dr. Humphrey and colleagues (Gantois et al., 2009).

The Japanese egg organizations established a committee to study the labeling of egg dates, etc. They developed a formula for calculating the shelf life of eggs that can be consumed raw; a relationship between storage temperature and storage period suppresses the growth of SE even contaminated within an egg. The formula has been called Dr. Humphrey's formula in Japan. In June 1998, the "Date Labeling Manual for Poultry Eggs" was formulated, and a system for labeling the best-before dates for eggs that can be consumed raw was started.

Normally, healthy people can eat even In-egg contaminated eggs raw without any problem as long as the number of bacteria in the egg is low. However, as the egg's freshness decreases, the strength of the yolk membrane decreases, and its permeability increases. When nutrients such as iron and yolk nutrient leak from the yolk into the egg white, SE bacteria multiply rapidly (Clay & Board, 1991), and such contaminated eggs can cause food poisoning (Figure 17.3). The number of days until the yolk membrane strength decreases inside SE-contaminated eggs, and yolk components leak out. SE bacteria rapidly multiply related

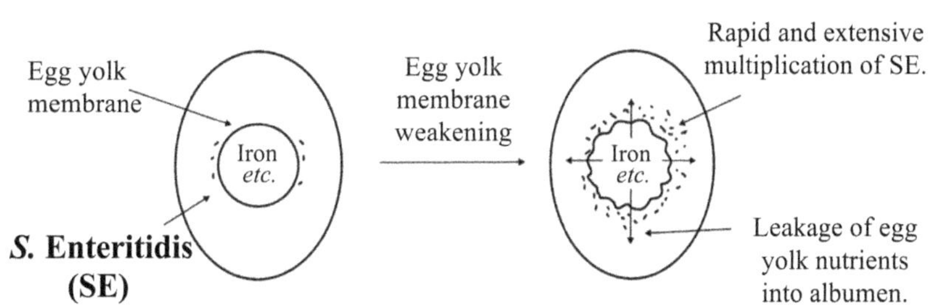

FIGURE 17.3 Images of SE multiplication in egg. (From Committee for the Revision of Egg Date Labeling [in Japan Poultry Association]. (2010). *Manual for Date Labeling of Poultry Eggs, Revised Edition* (in Japanese).)

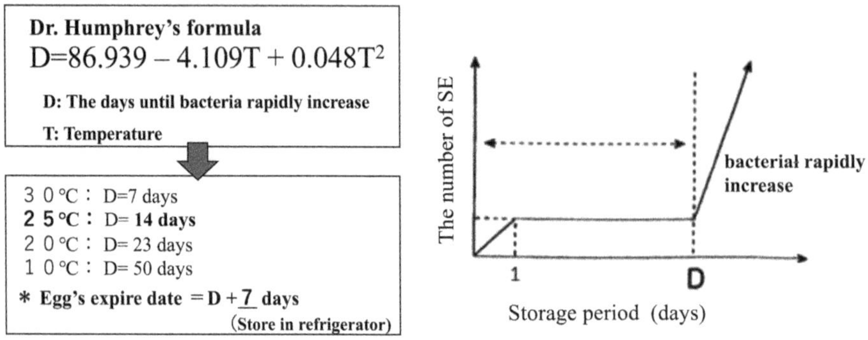

FIGURE 17.4 Dr. Humphrey's formula for calculation of egg's expiry date in Japan. (From Committee for the Revision of Egg Date Labeling [in Japan Poultry Association]. (2010). *Manual for Date Labeling of Poultry Eggs, Revised Edition* (in Japanese).)

to storage temperature: 50 days from the egg laying date when stored at 10°C, 23 days at 20°C, and 7 days at 30°C (Figure 17.4).

The shelf life of shell eggs was determined by adding to this the period that each household keeps the eggs in the refrigerator (below 10°C), which is set at 7 days. Therefore, the best-before date had been set at a particular date after packing eggs, depending on the storage temperature. However, since April 2010, in response to growing demand from consumers for information on the actual date of egg laying, the egg industry has voluntarily established a "best-before date of no more than 21 days from the time of laying (distribution at 25°C or less)" under its own rules. Since April 2010, the egg industry has standardized the best-before date to be within 21 days from the date of laying eggs (distributed at 25°C or below).

17.7 Verification of the Shelf Life of Japanese Eggs

The authors experimentally verified the validity of Dr. Humphrey's formula, which has become the standard for the shelf life of eggs in Japan. Fresh raw eggs (stored at 5°C after laying) and eggs stored at 30°C (75% humidity) for 6 and 12 days were prepared. Substituting 30°C into Dr. Humphrey's formula, the day (D) of the explosive growth of SE on the yolk membrane is 7 days. First, each egg's HU, YI, and yolk membrane strength were measured to check for freshness loss (Figure 17.5).

Next, the eggs were cracked, the yolk was separated to avoid damaging the yolk membrane, and the yolk membrane surface

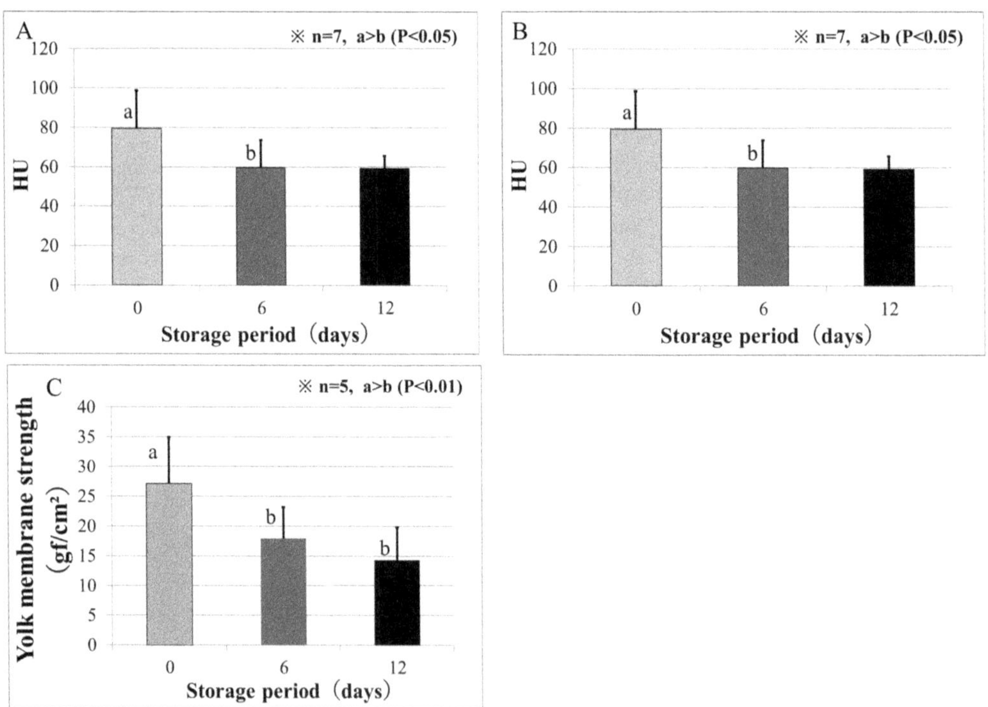

FIGURE 17.5 Changes in HU (A), YI (B), and yolk membrane strength (C) during storage at 30°C. (From Kubo et al. (2019). *Weakening of Egg Yolk Membrane and Growth of E. coli in Eggs Stored for Long Periods Beyond Their Expire Dates* (in Japanese).)

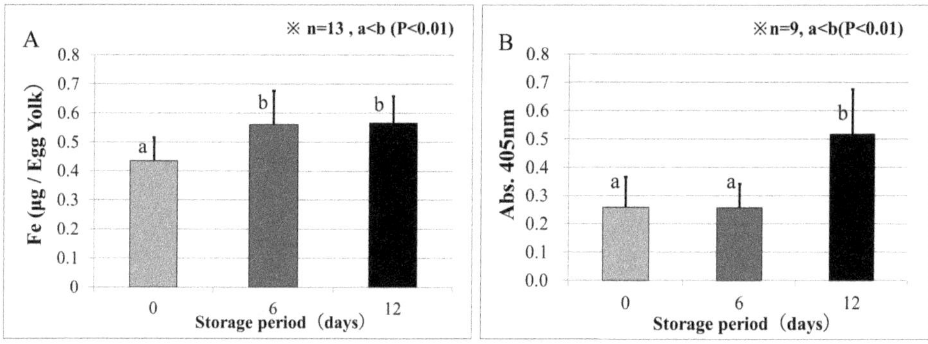

FIGURE 17.6 Changes in the amount of Fe (A) and IgY (B) on yolk membranes during storage at 30°C. (From Kubo et al. (2019). *Weakening of Egg Yolk Membrane and Growth of E. coli in Eggs Stored for Long Periods Beyond Their Expire Dates* (in Japanese).)

was washed with saline solution. This yolk membrane surface wash was used as a sample to measure the amount of iron (Fe) that leaked onto the yolk membrane surface using an ICP optical emission spectrometer. In addition, the egg yolk IgY antibody, which is found only in egg yolk, was measured by ELISA. The results of each are shown in Figure 17.6. Leakage of yolk components to the egg white side was observed to increase with decreasing freshness. In addition, a significant increase in Fe and IgY was observed on days 6 and 12, respectively, when stored at 30°C. We believe these differences in the leakage of egg yolk components are related to the difference in molecular weight of Fe and IgY.

The yolk surface was washed with the medium, and the yolk material leached was collected in the medium. Then, *E. coli* JM109 was cultured in the medium, and changes in the turbidity of the bacterial solution were examined. In another experiment, the eggs were broken, and the yolk was carefully separated to avoid damaging the yolk membrane, soaked in the bacterial solution (saline solution containing 1×10^7 CFU/ml of *E. coli* JM109). The growth of *E. coli* was compared by measuring the amount of bacterial ATP using chemiluminescence. The results showed that eggs stored at 30°C for extended periods were more prone to bacterial growth (Figure 17.7). Overall, these trends were extreme for eggs stored at 30°C for 12 days beyond the D value (approximately 7 days), confirming the validity of Dr. Humphrey's formula as a standard for setting expiration dates in Japan.

17.8 Conclusion

In Japan, where eggs are eaten raw, a system has been established to ensure that unheated raw eggs are safe and secure. The main reason for this is the introduction of the "best-before" date labeling system for eggs in 1998, although it is due to the improved sanitary conditions of poultry farms, the spread of SE vaccines, and the increased hygienic awareness of consumers (individuals and companies). Dr. Humphrey's formula, which is based on the estimated time until the explosive growth of SE from the storage temperature and storage period of shell eggs, has been effectively used.

It has already been about 30 years since Japan's egg expiration date system started, and the number of cases of salmonella food poisoning has decreased dramatically, and the detection rate of In-egg contaminated eggs has dropped to 3 per 100,000 eggs (0.03%). According to calculations, even if a person continues to eat one egg every day, he or she will not be exposed to In-egg contaminated eggs for 91 years. Thus, Japan's egg sanitation management system is probably the safest in the world and is in line with the domestic raw food culture.

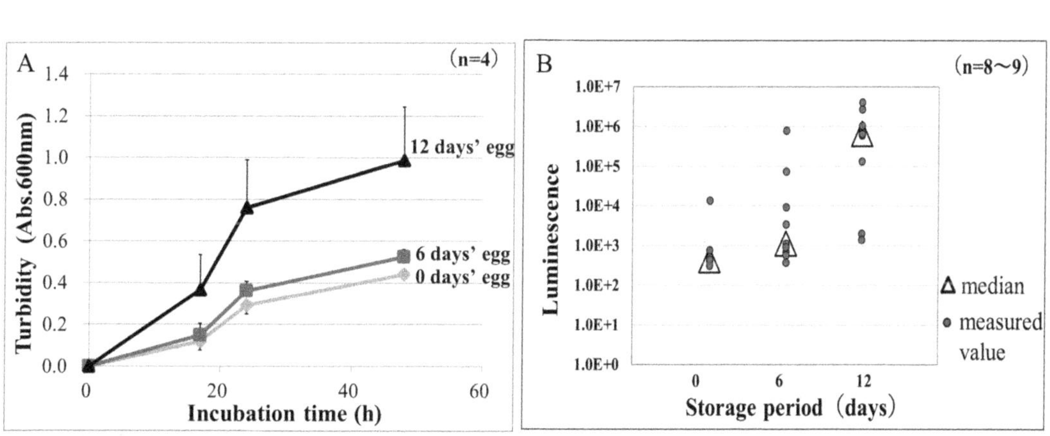

FIGURE 17.7 Increase of *E. coli* in each storage period (at 30°C) of eggs. Growth of *E. coli* in yolk surface washed medium (A) and in saline solution soaked with egg yolk (B) is shown. (From Kubo et al. (2019). *Weakening of Egg Yolk Membrane and Growth of E. coli in Eggs Stored for Long Periods Beyond Their Expire Dates* (in Japanese).)

REFERENCES

ACMSF-Ad Hoc Group on Eggs. (2016). An update on the microbiological risk from shell eggs and their products. https://webarchive.nationalarchives.gov.uk/20171207164948/https://www.food.gov.uk/sites/default/files/acmsf-egg-reportv1.pdf

Akter, Y., Kasim, A., Omar, H., & Sazili, A. Q. (2014). Effect of storage time and temperature on the quality characteristics of chicken eggs. *Journal of Food, Agriculture and Environment*, *12*, 87–92.

Bradshaw, J. G., Shah, D. B., Forney, E., & Madden, J. M. (1990). Growth of Salmonella enteritidis in yolk of shell eggs from normal and seropositive hens. *Journal of Food Protection*, *53*(12), 1033–1036. https://doi.org/10.4315/0362-028x-53.12.1033

Cardoso, M. J., Nicolau, A. I., Borda, D., Nielsen, L., Maia, R. L., Møretrø, T., Ferreira, Y., Knøchel, S., Langsrud, S., & Teixeira, P. (2021). Salmonella in eggs: From shopping to consumption—A review providing an evidence-based analysis of risk factors. *Comprehensive Reviews in Food Science and Food Safety*, *20*, 2716–2741. https://doi.org/10.1111/1541-4337.12753

Chen, J., Shallo Thesmar, H., & Kerr, W. L. (2005). Outgrowth of salmonellae and the physical property of albumen and vitelline membrane as influenced by egg storage conditions. *Journal of Food Protection*, *68*(12), 2553–2558. https://doi.org/10.4315/0362-028x-68.12.2553

Clay, C. E., & Board, R. G. (1991). Growth of salmonella enteritidis in artificially contaminated hens' shell eggs. *Epidemiol Infect*, *106*, 271–81. https://doi.org/10.1017/S095026880004841X

Committee for the Revision of Egg Date Labeling [in Japan Poultry Association]. (2010). Manual for date labeling of poultry eggs. Revised Edition (in Japanese) https://www.jz-tamago.co.jp/wp/wp-content/uploads/2020/03/E05_3_m_1.pdf

Consumer Site of Australian Egg Farmers. (2014). Storing eggs. Accessed October, 2022. http://www.eggs.org.au/facts-and-tips/storing-eggs/

Dong, X., Dong, J., Peng, Y., & Tang, X. (2017). Comparative study of albumen pH and whole egg pH for the evaluation of egg freshness. *Spectroscopy Letters*, *50*, 463–469.

Dong, X., Li, Z., Shen, Z., & Tang, X. (2018). Nondestructive egg freshness assessment from the equatorial and blunt region based on visible near infrared spectroscopy. *Spectroscopy Letters*, *51*, 540–546. https://doi.org/10.1080/00387010.2018.1525409

United States Department of Agriculture (Revised 2000), Egg-grading manual. In Agricultural handbook (Vol. 75, p. 56).

EUR-Lex. (2008). EU egg marketing regulations. Accessed October 20, 2022. http://data.europa.eu/eli/reg/2008/589/oj

Fikiin, K., Akterian, S., & Stankov, B. (2020). Do raw eggs need to be refrigerated along the food chain? Is the current EU regulation ensuring high-quality shell eggs for the European consumers? *Trends in Food Science & Technology*, *100*, 359–362. https://doi.org/10.1016/j.tifs.2020.04.003

Food Standards Australia New Zealand. (2011). Primary production and processing standards for eggs and egg products. Accessed October, 2022. http://www.foodstandards.gov.au/code/proposals/Documents/P301%20Eggs%20PPPS%20IAR%20FINAL.doc

Gantois, I., Ducatelle, R., Pasmans, F., Haesebrouck, F., Gast, R., Humphrey, T. J., & Immerseel, F. V. (2009). Mechanisms of egg contamination by salmonella enteritidis. *FEMS Microbiology Reviews*, *33*(4), 718–738. https://doi.org/10.1111/j.1574-6976.2008.00161.x

Giunchi, A., Berardinelli, A., Ragni, L., Fabbri, A., & Silaghi, F. (2008). Non-destructive freshness assessment of shell eggs using FTNIR spectroscopy. *Journal of Food Engineering*, *89*, 142–148. https://doi.org/10.1016/j.jfoodeng.2008.04.013

Hatta, H., Kapoor, M. P., & Juneja, L. R. (2008). Bioactive components in egg yolk. In Y. Mine (Ed.), *Egg bioscience and biotechnology* (pp. 185–237). Wiley & Sons, Inc.

Humphrey, T. J., Baskerville, A., Mawer, S., Rowe, B., & Hopper, S. (1989). Salmonella enteritidis phage type 4 from the contents of intact eggs: A study involving naturally infected hens. *Epidemiology and Infection*, *103*(3), 415–423. https://www.ncbi.nlm.nih.gov/pubmed/2691262

Humphrey, T. J., & Whitehead, A. (1993). Egg age and the growth of salmonella enteritidis PT4 in egg contents. *Epidemiology & Infection*, *111*(2), 209–219. https://doi.org/10.1017/S0950268800056910

Humphrey, T. J., Whitehead, A., Gawler, A. H., Henley, A., & Rowe, B. (1991). Numbers of salmonella enteritidis in the contents of naturally contaminated hens' eggs. *Epidemiology and Infection*, *106*(3), 489–496. https://doi.org/10.1017/s0950268800067546

Hutchison, M. L., Gittune, J., Walker, A., Sparks, N., Humphrey, T. J., Burton, C., & Moore, A. (2004). An assessment of the microbiological risks involved with egg washing under commercial conditions. *Journal of Food Protection*, *67*, 4–11. https://doi.org/10.4315/0362-028X-67.1.4

Jiang, S., Zhu, T., Jia, D., Yao, J., & Jiang, Y. (2018). Effect of egg freshness on their automatic orientation. *Journal of the Science of Food and Agriculture*, *98*, 2642–2650.

Jin, Y. H., Lee, K. T., Lee, W. I., & Han, Y. K. (2011). Effects of storage temperature and time on the quality of eggs of laying hens at peak production. *Asian-Australasian Journal of Animal Sciences*, *24Äi0*(2), 279–284. https://doi.org/10.5713/ajas.2011.10210

Kang, H., Loui, C., Clavijo, R. I., Riley, L. W., & Lu, S. (2006). Survival characteristics of *Salmonella enterica* serovar Enteritidis in chicken egg albumen. *Epidemiology and Infection*, *134*, 967–976. https://doi.org/10.1017/S0950268806006054

Kemps, B. J., De Ketelaere, B., Bamelis, F. R., Mertens, K., Decuypere, E. M., De Baerdemaeker, J. G., & Schwagele, F. (2007). Albumen freshness assessment by combining visible near-infrared transmission and low-resolution proton nuclear magnetic resonance spectroscopy. *Poultry Science*, *86*, 752–759. https://doi.org/10.1093/ps/86.4.752

Kubo, N., Yamashita, M., & Hatta, H. (2019). Weakening of egg yolk membrane and growth of *E. coli* in eggs stored for long periods beyond their expire dates. *Kyoto Women's University Food Society Journal*, *73*, 7–12. (in Japanese) http://hdl.handle.net/11173/2802

Lee, M. H., Cho, E. J., Choi, E. S., & Sohn, S. H. (2016). The effect of storage period and temperature on egg quality in commercial eggs. *Korean Journal of Poultry Science*, *43*, 31–38 http://dx.doi.org/10.5536/KJPS.2016.43.1.31

Li, X. (2013). The Haugh unit is an important indicator of the egg quality inspection. *Journal of Inspection & Quarantine, 23*(2), 48–49.

Lublin, A., & Sela, S. (2008). The impact of temperature during the storage of table eggs on the viability of salmonella enterica serovars enteritidis and Virchow in the eggs. *Poultry Science, 87*, 2208–2214. https://doi.org/10.3382/ps.2008-00153

Narushin, V. G., Romanov, N., Griffin, M. N., & K, D. (2021). A novel egg quality index as an alternative to Haugh unit score. *Journal of Food Engineering, 289*, 110176. https://doi.org/10.1016/j.jfoodeng.2020.110176

National Health Service. (2015). The healthy way to eat eggs. Accessed October, 2022. https://familyserviceshub.havering.gov.uk/kb5/havering/directory/advice.page?id=NhsMiz6zsts

Obianwuna, U. E., Oleforuh-Okoleh, V. U., Wang, J., Zhang, H.-J., Qi, G.-H., Qiu, K., & Wu, S.-G. (2022). Natural products of plants and animal origin improve albumen quality of chicken eggs. *Frontiers in Nutrition, 9*. Article 875270. https://doi.org/10.3389/fnut.2022.875270

Okubo, T., Akachi, S., & Hatta, H. (1997). Chapter 1: Structure of the hen eggs and physiology of egg laying. In T. Yamamoto, L. R. Juneja, H. Hatta, & M. Kim (Eds.), *Hen eggs; Their basic and applied science* (pp. 1–12). CRC Press. https://doi.org/10.1201/9780203752081

Qi, L., Zhao, M., Li, Z., Shen, D., & Lu, J. (2020). Non-destructive testing technology for raw eggs freshness: A review. *SN Applied Sciences, 2*, 1113. https://doi.org/10.1007/s42452-020-2906-x

Ragni, L., Gradari, P., Berardinelli, A., Giunchi, A., & Guarnieri, A. (2006). Predicting quality parameters of shell eggs using a simple technique based on the dielectric properties. *Biosystems Engineering, 94*, 255–262. https://doi.org/10.1016/j.biosystemseng.2006.03.003

Samli, H. E., Agma, A., Senkoylu, N., & Person, E. (2005). Effects of storage time and temperature on egg quality in old laying hens. *Journal of Applied Poultry Research, 14*, 548–553. https://doi.org/10.1093/japr/14.3.548

Sato, Y., Tanabe, N., Nakamura, R., & Watanabe, K. (1989). The science of egg cooking and health. *Kogaku Shuppan*. (in Japanese) https://iss.ndl.go.jp/books/R100000002-I000002015744-00

Schoeni, J. L., Glass, K. A., NcDermott, J. L., & Wong, A. C. L. (1989). Growth and penetration of Salmonella enteritidis, Salmonella heidelberg and Salmonella typhimurium in eggs. *International Journal of Food Microbiology*, 24, 385–396. https://doi.org/10.1016/0168-1605(94)00042-5

UNECE. (2017). *Egg-1 UNECE standard eggs-in-shell* (p. 16). United Nations

USDA-FSIS, Last Updated: Nov. 04, (2019). Dating of Cartons. https://www.fsis.usda.gov/food-safety/safe-food-handling-and-preparation/eggs/shell-eggs-farm-table#18

Williams, K. C. (2019). Some factors affecting albumen quality with particular reference to Haugh unit score. *World's Poultry Science Journal, 48*, 5–16. https://doi.org/10.1079/WPS19920002

Yimenu, S. M., Kim, J. Y., & Kim, B. S. (2017). Prediction of egg freshness during storage using electronic nose. *Poultry Science, 96*, 3733–3746. https://doi.org/10.3382/ps/pex193

Yimenu, S. M., Kim, J. Y., Koo, J., & Kim, B. S. (2017). Predictive modeling for monitoring egg freshness during variable temperature storage conditions. *Poultry Science, 96*, 2811–2819. http://dx.doi.org/10.3382/ps/pex038

Yimenu, S. M., Koo, J., Kim, J. Y., Kim, J. H., & Kim, B. S. (2018). Kinetic modeling impacts of relative humidity, storage temperature, and air flow velocity on various indices of hen egg freshness. *Poultry Science, 97*, 4384–4391. https://doi.org/10.3382/ps/pey334

Yongwei, W., Wang, J., Zhou, B., & Lu, Q. (2009). Monitoring storage time and quality attribute of egg based on electronic nose. *Analytica Chimica Acta, 650*, 183–188. https://doi.org/10.1016/j.aca.2009.07.049

Yuceer, M., & Caner, C. (2014). Antimicrobial lysozyme-chitosan coatings affect functional properties and shelf life of chicken eggs during storage. *Journal of the Science of Food and Agriculture, 94*, 153–162. https://doi.org/10.1002/jsfa.6322

18

Regulations for Shell Eggs in Different Countries

Vincent Guyonnet

CONTENTS

DOI: 10.1201/9781003254430-23

18.1 Introduction

Shell eggs are one of the most nutritious and affordable animal-sourced foods consumed on all continents. Consumers expect the provision of safe and nutritious foods and most often rely on governments to monitor and control the safety of their food supply chains. The principle of preventing food safety risks is widely accepted as the most effective and economic way to provide safe foods to consumers. After a review of the main hazards putting at risk the safety and quality of shell eggs, this chapter will review some general principles for the hygienic production and commercialization of shell eggs. The regulations in key markets, either global leaders or leaders on their respective continents, will then be discussed while the results of an extensive survey will present the regulatory situation in an additional 24 countries on 5 continents. Finally, the rise of global private food safety standards will also be discussed as they very often exceed national regulations and will play a bigger role in the future.

18.2 Hazard Analysis for Shell Eggs

18.2.1 HACCP Principles

The concept of Hazard Analysis and Critical Control Point (HACCP) was initially developed by the private sector in collaboration with the United States National Aeronautics and Space Administration (NASA) to produce safe foods for space travel. The HACCP system was designed to ensure the production of the safest food products while reducing the number of tests of finished products, emphasizing the control of the production process and quality as far upstream as possible. By identifying the potential risks and their origin, they were able to implement a series of measures to eliminate food safety risks or at least reduce them below an acceptable level. The Codex Alimentarius Commission, the joint group of experts from the World Health Organization (WHO) and the Food and Agriculture Organization of the United Nations (FAO) dealing with food production and food safety standards, recognized the value of HACCP for food safety. The Codex Alimentarius Commission adopted in 1993 the Guidelines for the application of the HACCP system and incorporated them within the General Principles of Food Hygiene, initially published in 1969 (Codex Alimentarius, 2011). The design and implementation of the HACCP system for food safety relies on 7 principles and a series of 12 tasks (Table 18.1).

In many countries, the concept of HACCP has been adopted by the food sector, including the egg sector (Ropkins & Beck, 2000). Given the recognized value of HACCP for the safety of shell eggs and regardless of the regulatory need to do so, many egg companies are implementing HACCP plans, both at primary production level (egg farms) and in grading/packing

TABLE 18.1
HACCP Principles and Tasks for Implementation

HACCP Principles	HACCP Key Tasks
1. Conduct a hazard analysis	1. Assemble HACCP team
2. Determine the critical control points (CCPs)	2. Describe product
	3. Identify the intended use
3. Establish the critical limits for each CCP	4. Construct process flow diagram
	5. On-site confirmation of flow diagram
4. Establish process to monitor the CCPs	6. List all potential hazards / Conduct hazard analysis and consider control measures
5. Define the corrective actions when the CCPs are not controlled	7. Determine CCPs
6. Establish the verification procedures	8. Establish validated critical limits for each CCP
7. Establish the documentation for procedures and implementation verification	9. Establish a monitoring system for each CCP
	10. Establish corrective actions
	11. Validate the HACCP plan and establish verification procedures
	12. Establish documentation and record keeping

Source: Adapted from Codex Alimentarius (2011).

plants, to minimize the potential health and food safety risks related to the consumption of shell eggs (Viator et al., 2007).

The identification of potential hazards relies on the experience of farm and plant operators and on publications by international organizations (FAO, 2002) and scientists, as well as a wide range of published or internet-searchable databases (CFIA, 2008, 2022; EFSA, 2022; Ministry of Primary Industries, New Zealand, 2022; USDA, 2022). A review of the potential biological, chemical and physical hazards related to the production, grading and distribution of shell eggs is presented below.

18.2.2 Biological Hazards

The microbiology of shell eggs is presented in Chapter 20 and was previously reviewed (Musgrove, 2011; Chemaly & Salvat, 2011; Raspoet et al., 2011). Among all potential biological hazards, *Salmonella spp.* plays an important role, having been linked to numerous food-borne illnesses (Whiley & Ross, 2015; Cardoso et al., 2021). The main two serotypes incriminated are *Salmonella enterica* serotype *enteritidis* and *typhimurium*, both able to contaminate eggs either vertically (colonization of the reproductive track and deposition into the different components during egg formation) or horizontally (deposition at the surface of the shell followed by penetration through the shell and shell membranes). Other Salmonella serovars (e.g., *S. Mbandaka, S. Livingstone, S. Heibelderg, S. Hadar, S. Infanis* and *S. Virchow*) have also been isolated from eggshells although at a lower frequency (Martelli & Davies, 2012). While a number of bacteria are able to penetrate through the eggshell and shell membranes, gram-negative, motile and non-clustering bacteria (*Pseudomonas, Alcaligenes sp., Carnobacterium sp. and Acinetobacter Baumannii*) have been reported more frequently for their ability to penetrate the eggshell (Gantois et al., 2009). The bacterial contamination at the surface of the shell may be the result of the infection of the vagina at the time of oviposition or the presence of contaminated fecal materials. While the contamination of the shell is unlikely once the eggs are packed in cartons, eggs are still distributed and sold in bulk in a number of countries. This mode of distribution may bring new biological hazards, depending on the environmental conditions of transportation.

Although a few viruses can also be vertically transmitted, none have been implicated in food-borne diseases in humans. A recent assessment of the transmission of low pathogenic avian influenza virus via the consumption of raw shell eggs established a negligible risk for humans (Gonzales et al., 2018).

Microscopic fungi may also be identified at the surface of the eggshell. A study of eggs produced in different commercial housing systems (cage, barn, free-range and organic) in Poland showed that the main fungal species identified were *Alternaria sp., Penicillium sp., Chaetomium sp.* and *Fusarium sp.* (Tomczyk et al., 2018).

As detailed in Chapter 20, the storage conditions of shell eggs (duration, temperature and relative humidity) may influence the level of risks caused by bacteriological hazards. Storage at lower temperature under low relative humidity will typically reduce the food safety risks due to the presence of biological hazards.

18.2.3 Chemical Hazards

The presence of chemical residues in eggs, caused by intentional or unintentional events, was previously reviewed (Jondreville et al., 2011). The broader use of the outdoors by laying hens increases the potential exposure to environmental contaminants (e.g., dioxins, polychlorinated biphenyls or PCBs, pesticides and heavy metals) (Piskorska-Pliszczynska et al., 2019). Drug residues in eggs (e.g., antibiotics, anticoccidials, antiparasitic drugs) can be the results of accidental administration at feed mills via carry-over from medicated feeds intended for other animal species, overdosing with approved medicines or illegal use of products (Goetting et al., 2011; Reich & Triacchini, 2018). The contamination of poultry feeds with mycotoxins (e.g., aflatoxins, ochratoxin and zearalenone) may also be a source of contamination although mycotoxins are rarely identified in eggs (Lee et al., 2016; Tomczyk et al., 2018; Wang et al., 2018). The use of non-food sanitizers and disinfectants on farms or in a grading plant may also result in the presence of chemical residues in the egg.

Depending on the structure of these chemical hazards, they may be present preferentially in the egg white (water-soluble), the yolk (lipo-soluble) or the eggshell (heavy metals). Due to the nature of egg formation with the simultaneous presence of several follicles at various stages of development, chemicals depositing preferentially in the yolk will accumulate in greater amount and represent a greater risk to food safety (Goetting et al., 2011).

18.2.4 Physical Hazards

Due to the protective nature of the eggshell, the presence of physical hazards inside the eggs will always be related to a loss of integrity of the shell. Therefore, physical hazards such as the presence of pieces of plastic, wood, metal or dirt are usually prevented by the removal of broken or leaking eggs, either at farm level or at receiving by the grading/packing plant.

The presence inside the egg of pieces of tissues sloughed off the reproductive tract (meat spot) and of blood spots may be considered as physical hazards. Therefore, in some countries, the regulations will call for the removal of eggs containing meat and blood spots during the candling process.

18.3 General Principles – Code of Hygiene

In countries where there are no specific regulations for shell eggs, the Code of hygienic practice for eggs and egg products (CAC/RCP 15-1976), published initially in 1976 by the Codex Alimentarius and regularly reviewed, provides some guidance to the egg sector regarding the suitability of shell eggs and egg products for human consumption (Codex Alimentarius, 2007). This code applied specifically to the egg sector, supplementing the previously published General principles of Food Hygiene (CAC/RCP 1-1969) (Codex Alimentarius, 2011). The Code of hygienic practice for eggs and egg products addresses food hygiene and safety related to the production of eggs (primary production) and to the processing of shell eggs, including practices and measures that can be implemented by smallholder farmers and operators. This code relies on 3 main principles: 1) from the production site to the point of retail, control measures should be in place to achieve the appropriate level of public health protection; 2) these measures should be implemented within the context of HACCP systems and 3) the control measures should be effective and validated. Since the Code of hygienic practice for shell eggs and egg products may constitute the sole legal document available in some countries, a brief summary is presented below (Codex Alimentarius, 2007).

18.3.1 Primary Production

The egg production stage represents the main source of food safety risks for shell eggs with the possibility to contaminate eggs directly from the laying hens, from poor environmental conditions and management practices, from contaminated feeds and water and from poor handling and storage of eggs at the farm. Most biosecurity measures already implemented on farms to guarantee the good health and productivity of laying hens ensure also the safety of eggs.

18.3.1.1 Environmental Hygiene

With the rise of free-range production systems in Western countries and the prevalence of cage-free systems in developing countries (see Chapter 2), the potential contamination of eggs from the surrounding environment must be addressed carefully. The contamination of the environment may come from the previous uses of the land, the presence of contaminants, the presence of polluted surface water and the presence of microbiological or chemical hazards from manure and other animal production in the vicinity of a farm. The presence of wild birds and other wildlife should be monitored, both from a biosecurity standpoint and from a food safety standpoint, as they may be a source of contamination of the land and water supply used by laying hens. The quality of the air, both outside and inside the layer houses can also impact the level of contamination of the eggshell. The building materials for layer houses and the equipment used inside these barns should be selected to minimize the risks of harboring pathogens and contaminants and should be easy to clean and disinfect between flocks. In addition, it is recommended to implement a monitoring program for *Salmonella enteritidis* with regular sampling of feces, litter, dust and other elements indicative of a potential contamination of the environment.

18.3.1.2 Flock Management and Animal Health

The provision of day-old pullets from breeder flocks and hatcheries free of specific pathogens will significantly reduce the risk of contamination of shell eggs by a number of pathogens, present either inside the egg or on the shell surface.

Water and feeds constitute also other sources of biological, chemical and physical contaminants that may affect the safety of eggs. Municipal water should be preferred, although well water is appropriate when adequately treated (e.g., filtration, UV or ozone treatment) and tested regularly. The use of recycled water as drinking water given to laying hens should be discouraged due the level of risk brought by this practice.

Feeds should be purchased from feed mills implementing good manufacturing practices to minimize the risks of chemical and biological contamination. When feeds are produced at farm level, feed ingredients should be sourced from suppliers able to guarantee their composition and quality, and finished feeds should be part of a routine testing program. Since feeds are a potential source of pathogen contamination, heat (pelletization) or other treatments should be applied to reduce or eliminate pathogens. Feeds should be stored under conditions limiting their deterioration and contamination, including spillage attracting pest and wild birds.

Good management practices will reduce the risk of diseases in a flock, thus limiting the need to medicate birds and reducing the risk of residues in shell eggs. Dead birds should be removed promptly from the houses and disposed of using methods preventing the potential transmission of pathogens to other animals or farm workers. A health management program, using a wide range of vaccines often administered during the pullet stage, is effective to maintain a healthy flock and to reduce the need for medical treatment. When required, veterinary drugs and medicated premixes must be used under the supervision of a veterinarian, in compliance with local regulations in terms of usage level and withholding period. Proper records should be kept for all vaccines and medicines administered to a flock. Agricultural chemicals should be transported, stored, used and disposed of according to the instructions provided by manufacturers and local regulations.

Pest such as rodents and insects are known vectors for the introduction of various pathogens inside layer houses. A well-designed and implemented pest control program will reduce the risk of pathogen contamination of shell eggs while preventing their contamination by the chemicals used to control pests.

18.3.1.3 Egg Collection, Handling, Storage and Distribution

Whether collected manually or automatically, the process should be managed in a way that minimizes any damage to the eggshell, a major risk of contamination. Cracked eggs and leaking eggs should be excluded from direct commercialization. Dirty eggs should also be excluded from direct consumption unless regulations allow for the cleaning of shell eggs. Tools and equipment used for the collection and storage of shell eggs should be made of materials designed for food contact and easy to wash and sanitize. Eggs should be transported from the farms under sanitary conditions (e.g., clean truck, absence of chemicals or other sources of contamination) with the time and temperature of distribution defined to minimize the risks of contamination.

18.3.2 Grading Establishment

Although the vast majority of the global egg production is still sold as non-graded eggs, the grading of shell eggs is a mandatory step before commercialization in all Western and rich nations (see Chapter 2). Grading establishments play a key role in ensuring the commercial availability of safe eggs. Throughout the grading process, shell eggs should be handled in a manner that avoids damage, minimizes the deposition of moisture on the eggshell (duration of processing, relative humidity and temperature fluctuations) and prevents contamination.

18.3.2.1 Receiving and Handling of Shell Eggs

At receiving, a visual inspection should segregate shell eggs deemed unsafe or unsuitable for human consumption (broken and leaking eggs, eggs with bacterial or fungal rots, aged eggs). Where egg cleaning is allowed, the receiving of dirty eggs (within a certain percentage of the lot) is permitted and these eggs must be processed according to local regulations. The cleaning of eggs by the grading establishment must be done using approved dry or wet methods, taking special care to minimize damage to the shell surface. The dry method should minimize damage caused to the protective cuticle and may be followed by the spraying of food-grade oil to coat the eggshell. Eggs may also be washed with water, using detergents and sanitizers designed for food applications. Eggs should never be immersed in water and must be dried after washing to minimize the presence of moisture on the surface of the shell, as it will promote the growth of molds. The application of a sanitizer or food grade oil may also be allowed after washing and drying the eggs.

Other methods such as in-shell pasteurization treatment used to eliminate pathogens may also be allowed as long as they respect the safety and suitability of eggs for human consumption.

18.3.2.2 Storage and Shelf-Life

During storage, temperature fluctuations should be kept to a minimum as they will contribute to the deposition of moisture at the surface of the eggs. The length of storage, the temperature and relative humidity conditions during storage are important factors that will impact the safety of shell eggs. A lower temperature of storage will minimize microbial growth, including *Salmonella enteritidis*.

The shelf-life of shell eggs depends on various factors: storage conditions (temperature, temperature variations and relative humidity), method for handling the eggs (segregation of cracked, leaking or dirty eggs at receiving and cleaning of eggs) and the type of packaging used for the eggs. Unless defined by the regulatory authorities, a grading establishment should define the shelf-life of the eggs based on factors mentioned previously, as well as on information related to the conditions of the eggs prior to receiving (length between production and receiving, storage and transportation conditions prior to receiving).

18.3.2.3 Maintenance and Sanitation

Proper maintenance and sanitation of the grading establishment will also contribute to the availability of safe shell eggs. A grading establishment should implement a validated cleaning program for the equipment and premises, establish a pest control program and manage waste removal in a manner to limit the risk of contamination. The effectiveness of the programs and measures in place should be periodically monitored to ensure the safe grading of shell eggs.

18.3.3 Distribution and Commercialization

When leaving the grading establishment, eggs should be transported in a way that will minimize breakage, damage and contamination. Vehicles used for transportation should be visibly clean and easy to clean thoroughly. Likewise, any reusable container used during distribution should be properly cleaned and disinfected prior to being reused.

Information should be provided on the packaging materials to identify production lot (traceability) as well as relevant information for consumers (storage and handling conditions, advice about the safe handling use and consumption of shell eggs).

18.4 Shell Egg Regulations in Different Countries

The regulations in a few of the main egg-producing countries or regions will be reviewed, outlining the key elements directly related to shell eggs as well as some related regulations (e.g., labeling rules, general food safety). A survey was designed to gather specific information about shell egg regulations in 24 additional countries and the results are presented at the end of this section.

18.4.1 China

The regulations for food products are contained in a series of National Standards (*guo biao* in Chinese), listed by the letters GB followed by the document number and its date of publication. Technical specifications, identified by the letter T, may be related to a National Standard (GB/T) or to documents addressing the Agricultural sector (*nong ye* in Chinese), listed as NY/T. The basic regulatory document for the egg sector, GB 2749-2015, is entitled National Food Safety Standard Eggs and Egg Products (China, 2015). Published in 2015, it replaced GB 2748-2003 (Hygiene standard for fresh eggs) and GB 2749-2003 (Hygiene standard for egg products). With the rising level of income in China and higher expectations of consumers in terms of food safety and quality, there is a growing demand for graded and nicely packaged shell eggs.

TABLE 18.2

Quality Grading Requirements for Shell Eggs in China

Indicators		Premium	Class 1	Class 2
Appearance		Inherent color of eggshell, uniform, without any sign of defects (sandpaper or rough shell, spots or deformed)		
	Eggshell cleanliness	No visible stain on the surface of the shell		Visible stains on the surface, each spot less than 4 mm² and in total less than 8 mm²
Egg Contents	Egg yolk	Intact, not ruptured		
	Haugh units	> 72	> 60	> 55
	Egg white	Viscous and transparent. Clear distinction between thick and thin albumen	Less viscous and transparent. Clear distinction between thick and thin albumen	Not so viscous but still transparent
	Germinal disc	No obvious sign of development		
	Foreign bodies	Blood spots and meat spots (diameter inferior to 2 mm) are allowed. No other foreign bodies allowed		

Source: Adapted from China (2020).

The national standard, GB/T- 39438-2020, recently published, covers specifically the requirements for shell eggs packaged in egg cartons (China, 2020), while the Agricultural standard NY/T 1551-2007 addresses aspects related to the washing and grading of eggs (China, 2007).

18.4.1.1 Grading Characteristics

At receiving, eggs should be inspected and visibly damaged (broken, leaking) or deformed eggs removed to limit the further risk of contamination of the packing plant. While GB 2749-2015 covers the basic sensory requirements for shell eggs (color, odor and general conditions), the detailed quality characteristics are listed in Appendix A of GB/T 39438-2020 (China, 2020) and presented in Table 18.2. Based on these characteristics, shell eggs are categorized as Premium, Class I and Class 2.

Shell eggs are packed under 3 grades, AA, A and B, according to the following specifications for a given production lot:

- AA: at least 90% Premium eggs, less than 10% Class 1, no Class 2 and no more than 1% broken eggs
- A: at least 90% Premium or Class 1 eggs, less than 10% Class 2 eggs and no more than 2% broken eggs
- B: other combinations of Premium, Class 1 and Class 2 and no more than 3% broken eggs

18.4.1.2 Egg Cleaning/Washing

The cleaning of eggs is allowed but no mandatory. The water used for washing must meet the requirement of GB 5749-2022 (Sanitary standards for drinking water) and its temperature must be between 10°C and 32°C. Detergents used should be designed for the purpose, both in terms of efficacy and safety for use on food products. After washing, the eggs may be dried with warm air (less than 45°C). Disinfectants (GB 149030.2-2012) or film coating (e.g., edible vegetable oil, silicone oil, beeswax) may be applied after drying (China, 2007).

18.4.1.3 Storage Conditions

Prior to grading, shell eggs may be stored for up to 24 hours, at temperatures ranging from 4°C to 25°C (China, 2007). After grading, the use of refrigeration (rather uncommon in China) is at the discretion of the packing plant, and this decision is often made by the retail chains. Packed shell eggs should be kept at a temperature lower than 25°C with a relative humidity between 70% and 88% (China, 2020).

18.4.1.4 Sorting by Weight

Based on their weight, shell eggs are allocated into 6 levels, ranging from level 1 (more than 65 grams) to level 6 (less than 45 grams), as shown in Table 18.3 (China, 2007).

TABLE 18.3

Sorting of Shell Eggs by Weight in China, USA, India, EU27 Countries, Japan, Mexico and South Africa

China		USA		EU 27		Japan		Mexico		South Africa	
1	> 65	Jumbo	≥ 70.9			LL	≥ 70			S-Jumbo	> 72
2	≥ 60	X-large	≥ 63.8	X-Large	≥ 73	L	≥ 64	X-Grande	≥ 64	Jumbo	> 66
3	≥ 55	Large	≥ 56.7	Large	≥ 63	M	≥ 58	Grande	60–64	X-Large	> 59
4	≥ 50	Medium	≥ 49.6	Medium	≥ 53	MS	≥ 52	Mediano	55–60	Large	> 51
5	≥ 45	Small	≥ 42.5	Small	< 53	S	≥ 46	Chico	50–55	Medium	> 43
6	< 45	Peewee	< 42.5			SS	≥ 40	Canica	< 50	Small	≤ 43

Source: Adapted from USDA (2000); China (2007); EU (2008); Mexico (2016b); South Africa (2020).

18.4.1.5 Packing Centers

The requirements related to grading establishments are outlined in NY/T 1551-2007, with elements related to their location (e.g., isolated, accessible via a paved road), design (e.g., building materials, ventilation, welfare areas for workers), equipment (design, cleaning and maintenance) and worker hygiene (China, 2007).

18.4.1.6 Shelf-Life

There is currently no mandatory shelf-life for shell eggs. When indicated on egg packaging, the statement relates to a "Eat before" date, defined by the grading station. Given the wide range of conditions used for the storage and distribution of shell eggs, there is also no accepted industry standard and a "eat before" date is rarely printed on egg cartons (China, 2020).

18.4.1.7 Egg Marking and Traceability

There is no mandatory requirement for any identification on shell eggs. Marking on eggs may include some information to identify the company (code or logo) and the date of packing, using a safe, food grade ink. Other information may be applied at the discretion of companies. In practice, egg marking is mostly applied for marketing purposes and rarely from a traceability standpoint (China, 2020).

18.4.1.8 Packaging Information

The information required on egg cartons is outlined in Appendix D of GB/T 39438-2020. Any marking should be clearly visible, easily understandable, science-based and not misleading consumers. The net content may be indicated by quantity (number of eggs), by weight or both (e.g., 12 eggs × 60 grams or 720 grams (60 grams × 12)) and the date of packing listed as YYYY-MM-DD. The packaging should provide some information related to storage conditions after purchase and potentially provide different eat-before date whether stored at room temperature or under refrigeration (China, 2020).

NY/T 2798.11-2016 (Quality and safety control – Technical production of non-pollution agricultural products: Part 11: Fresh eggs) addresses the specific requirements related to the production and grading of shell eggs within the Green Food quality scheme (China, 2016). Other regulations related to the safe production, processing and distribution of shell eggs are:

- GB 9685 – 2016: Use of additives for food contact materials and products
- GB 4806.1 – 2016: General safety requirements for food contact materials and products
- GB 2763 – 2021: Maximum Residue Limits of Pesticides in Food
- GB 31650 – 2019: Maximum residue limits of veterinary drugs in food
- GB 14930.1 – 2015: National standard for food safety – Detergents
- NY/T 1167 – 2006: Environmental Quality and Sanitary Control Code for Livestock and Poultry Farms
- NY/T 5032 – 2006: Guidelines for the use of pollution-free food animal feed and feed additives

18.4.2 USA

The regulations for shell eggs are governed by two separate Acts: The Egg Products Inspection Act (EPIA, 1970) and the Agricultural Marketing Act (AMS, 1946). The inspection of shell eggs is mandatory for all producers and producer-packers with more than 3,000 laying hens selling through retail channels. Designed to ensure the safety of shell eggs, the United State Department of Agriculture (USDA) food safety program consists in a random inspection of a facility once per quarter (EPIA, 1970). The safety of shell eggs was further enhanced with the Egg Safety Rule issued by the Food and Drug Administration (FDA) in 2009, requiring the testing for *Salmonella enteritidis* in barns housing pullets (14–16 weeks of age) and laying hens (40–45 weeks of age) along with the testing of shell eggs in case of a positive environmental sample result (FDA, 2009). The grading of shell eggs for quality is voluntary and designed to ensure the integrity, quality and consistency of the business transaction between farmers/packers and consumers. Any egg packing station or company choosing to carry the USDA grade shield on its egg cartons must pay for the grading service provided by over 400 USDA employees (called USDA graders) continuously present in about 220 packing plants across the USA. Local State agencies monitor the compliance to the grading process of egg packing plants that do not use the USDA grading service.

18.4.2.1 Grading Characteristics

The quality of an egg is determined by its grade and not its size. Shell eggs, graded based on the U.S. Standards for interior and exterior quality factors, are allocated to 3 different US consumer grades: AA, A or B (Table 18.4). Some tolerances have also been defined for each grade, both at the origin (packing center) and at destination (retail stores). For instance, US consumer Grade AA at origin must consist of at least 87% AA quality eggs and a maximum of 13% A and B quality eggs (but no more than 1% B eggs due to air cell depth, blood spot dimension or yolk defects); at destination, at least 72% of the shell eggs must be AA quality grade, at least 10% A quality grade and the remaining B quality grade (but no more than 1% due to excessive air cell depth, blood spot dimension or yolk defects). The standards evaluating the quality of eggs apply to all shell eggs sold at retail, whether or not they carry the USDA shield on the egg cartons (USDA, 2000).

18.4.2.2 Egg Cleaning/Washing

The washing of shell eggs is mandatory using equipment that is sanitarily designed, maintained in a clean and sanitary manner and thoroughly cleaned at the end of each day of operation. The water used to wash shell eggs must be maintained at a temperature of at least 32.2°C (90°F) and always be at least 6.7°C (20°F) warmer than the internal temperature of the eggs being washed.

TABLE 18.4
Quality Grading Requirements for Shell Eggs in the USA

	AA Quality	A Quality	B Quality
Shell	Clean, intact and practically normal		Clean to slightly stained (0.80 mm if localized, 1.6 mm if scattered), unbroken, abnormal
Air Cell	Depth ≤ 3.2 mm,	Depth ≤ 4.8 mm	Depth > 4.8 mm,
		Unlimited movement and free or bubbly	
Egg White	Transparent, Firm	Transparent, reasonable firm	Weak and watery, some blood or meat spots (totaling ≤ 3.2 mm in diameter)
Yolk	Practically free of defects, slightly defined during candling	Practically free of defects, fairly well defined during candling	Visible germ development, outline visible during candling, dark, enlarged or flattened, some blood or meat spots (totaling ≤ 3.2 mm in diameter)

Source: Adapted from USDA (2000).

Wash water should be changed approximately every 4 hours or more frequently if needed to ensure good sanitary conditions. An approved detergent or detergent sanitizer must be added to the wash water while rinse water, slightly warmer than the wash water, may contain an approved sanitizer. Other approved chemicals (e.g., defoamers) may also be added in the wash water.

After washing, eggs must be dried, using drying equipment sanitarily designed and properly maintained. Filters must be cleaned and regularly changed to ensure the proper filtration of the air used for drying. The washing, rinsing and drying of shell eggs must be continuous and completed rapidly to prevent the presence of moisture (sweating) on shell eggs.

The use of oil, filtered prior to its application, may be used to protect shell eggs and maximize the conservation of egg quality. However, eggs with excess moisture on the shell should not be oil-treated (AMS, 1946).

As long as the process for the destruction of *Salmonella spp.* was approved by the FDA (at least a 5-log reduction in total *Salmonella* count)), shell eggs may be pasteurized and will carry the statement "USDA certified Pasteurized".

18.4.2.3 Storage Conditions

As recently mandated by the FDA, shell eggs must be stored and transported at the maximum ambient temperature of 7.2°C (45°F), beginning 36 hours after the time of lay (FDA, 2009). In the USA, most farms holding a large number of laying hens are directly connected via an in-line (conveyor) system to a packing center. In packing centers, cooler rooms containing shell eggs to be sold at retail must also be maintained at the maximum ambient temperature of 7.2°C and with a relative humidity (70–85%) minimizing shrinkage (AMS, 1946). Accurate thermometers and hygrometers must be available to properly monitor the storage conditions. Shell eggs sold at retail, including pasteurized shell eggs, must also be distributed and maintained under refrigeration (ambient temperature ≤7.2°C or 45°F) until the final sale.

18.4.2.4 Sorting by Weight

Based on their weight, shell eggs are assigned to 6 weight classes, ranging from Peewee (< 42.5 grams or 1.25 ounces per egg) to Jumbo (≥70.9 grams or 2.5 ounces per egg), as shown in Table 18.3. There are also some minimum net weights established for a case (30 dozen) and an egg carton holding a dozen eggs, since a certain tolerance (margin of error) is allowed for the weight of individual eggs. For instance, a production lot may contain 3.3% of shell eggs below their weight class limit (but within the next weight class) as long as there is no individual case in the lot that exceeds 5% (USDA, 2000).

18.4.2.5 Packing Centers

The requirements for equipment (e.g., digital-display scales, adequate candling, washing and grading equipment) and dedicated rooms for key activities (e.g., storage, candling, washing) are listed in the 7 CFR Part 56. Operational procedures, including general hygiene for employees are also described in the regulations. In grading centers under the supervision of the USDA, the Pre-Operational inspection of the plant is conducted by USDA staff to validate the good sanitary conditions of the grading facility (AMS, 1946).

18.4.2.6 Shelf-Life

The expiry date ("Sell by" or "Not to be sold after") is calculated from the date shell eggs are packed into retail cartons and may not exceed 30 days, including the date of pack. Other terms, "Use before", "Use by" or "Best before", indicate the maximum period for the expected quality and may not exceed 45 days (USDA, 2000).

18.4.2.7 Egg Marking and Traceability

There is currently no regulatory mandate to mark shell eggs individually for traceability purpose. Any marking on the shell surface is applied on a voluntary basis and often performed for marketing reasons or requested by retailers. The FDA has included shell eggs in the Food Traceability list currently under evaluation as part of the Food Safety Modernization Act (FSMA) designed to further protect public health. However, the inclusion of shell eggs in this list will require, once finalized, only additional traceability recordkeeping requirements and will not mandate individual egg marking (FDA, 2021).

18.4.2.8 Packaging Information

Only shell eggs, processed under the continuous monitoring of USDA graders, may be packed in USDA-approved packaging materials bearing the USDA grade shield. Egg cartons must carry at the minimum the following information: USDA grade (AA, A or B), weight class, date of grading expressed as Julian date, lot code and plant identifier. The placement of the expiration date (Exp. or Sell by) on egg cartons is optional.

All shell egg cartons, including pasteurized shell eggs, must carry the following safe handling statement: "SAFE HANDLING INSTRUCTIONS: To prevent illness from bacteria: Keep eggs refrigerated, cook eggs until yolks are firm, and cook foods containing eggs thoroughly".

The USDA also certifies claims related to production system (e.g., organic, cage-free) or specific nutritional content.

18.4.3 India

Although egg production in India has increased by 209.2% between 2000 and 2020, contributing more needed proteins to the Indian diet, there are currently no specific regulations for shell eggs. However, shell eggs are one of the food products addressed by the Food safety and Standards (Food products standards and Food additives) (India, 2011a).

Shell eggs should be collected at least on a weekly basis from farms and packed and graded within 24 hours after receiving them. The packing plants should handle only eggs not visibly broken, leaking, stained with fecal contamination or previously incubated. Shell eggs should not be washed or dry-cleaned and may be sorted by weight, although there are no government rules for egg sizes (India, 2011a). A guidance note from the India government mentions that shell eggs can maintain their freshness for up to 10–12 days after production when stored at 28°C and the shelf-life will be reduced with increasing temperature of storage. Refrigeration will keep shell eggs fresh longer and also minimize the risk of bacterial growth (India, 2018).

The labeling on egg cartons must comply with the general food labeling regulations, with the main requirement related to the printing of a brown color symbol on egg cartons to notify the "non-vegetarian" status of shell eggs (India, 2011b).

18.4.4 European Union (EU27)

In the 27 countries of the European Union, shell eggs are one of the agricultural products covered by EU Regulation 1308/2013, requiring that only Class A (fresh) eggs be sold to consumers after grading by weight and proper marking for traceability (EU, 2013). The marketing standards for shell eggs are outlined in the EU Regulation 589/2008 and are summarized below (EU, 2008).

18.4.4.1 Grading Characteristic

Class A shell eggs must meet the following criteria:

- Shell and cuticle: normal shape, clean and undamaged;
- Air space (dimension of the air cell) does not exceed 6 mm and 4 mm for eggs labeled as "extra fresh";
- The yolk is visible during candling as a shadow only, without a discernible outline and is slightly mobile when rotating the eggs but returning to a central position;
- The egg white is clear, translucent and the germinal disc shows no sign of development;
- The eggs exhibit no sign of foreign matter or unexpected smell.

Class B shell eggs do not meet the quality characteristics specified for Class A eggs, either immediately at receiving or overtime (downgrading from Class A to Class B) and received a different marking to clearly segregate them throughout the value chain.

18.4.4.2 Egg Cleaning/Washing

Class A eggs should not be washed or cleaned since EU regulators have determined that this practice can cause damage to the egg shell and cuticle. However, in Member States where packing plants were authorized to wash shell eggs on 1 June 2003, this practice may be continued, provided it is in accordance with national guidelines for egg washing and washed eggs are sold only in EU countries authorizing egg washing. However, certain practices such as the treatment of eggs with ultra-violet rays, is allowed.

18.4.4.3 Storage Conditions

As a measure to prevent the presence of condensation at the surface of the eggs, the storage and transportation of eggs will preferably be done at a constant temperature, with no refrigeration allowed before the sale to final consumers. The transportation of eggs at a temperature below 5°C for not more than 24 hours or the storage at retail for not more than 72 hours is permissible. An exemption to this rule is made for the storage and distribution of shell eggs to the French overseas departments.

18.4.4.4 Sorting by Weight

The grading of eggs, based on four categories, very large (XL), large (L), medium (M) and small (S), is presented in Table 18.3. This grading is indicated by the corresponding letters or terms, with the optional mention of the corresponding weight ranges. Derogations are given to pack eggs of different sizes within the same carton, providing the minimum net weight of the eggs and the statement "Eggs of different sizes" appear on the outer surface of the pack.

18.4.4.5 Packing Centers

The grading and packing establishments are regulated by national authorities and are allocated an identification number, indicating their suitability to perform properly the various tasks related to the inspection, grading and packing of shell eggs.

18.4.4.6 Shelf-Life (Minimum Durability)

Eggs must be graded, marked and packed within 4 days for eggs sold as "extra fresh" and within 10 days for other eggs. The "best before" date or date of minimum durability must be indicated at packing with a maximum of 28 days from the laying day. Packed shell eggs must be delivered to consumers within a maximum time limit of 21 days of laying. The date printed on an egg carton is determined by the laying date of the oldest egg contained in the carton.

18.4.4.7 Egg Marking and Traceability

During the grading and packing process (within 10 days of laying), shell eggs must be marked for identification and traceability. The first number for this marking indicates the type of production system used: 0 = Organic; 1 = Free-range; 2 = Barn; 3 = Cage (enriched) while the next 2 letters indicate the country of production (e.g., FR = France, ES = Spain, …). The rest of the marking clearly identifies a specific layer house within a farm, using different methods in the UE27 countries (e.g., France: XXX (Farm identifier) - XX (Layer house identifier); Spain: XX (Province) – XXX (Municipality) – XXX (layer house identifier)).

18.4.4.8 Packaging Information

Egg cartons (packs) containing Class A eggs must contain on their outer surface the following information, easily visible and clearly legible:1) the packing center code; 2) Quality grading ("Class A" or the letter "A", alone or in combination with the word "fresh"; 3) the weight grading (XL, L, M or S); 4) the date of minimum durability ("best before date"); 5) the wording "washed eggs" if applicable; 6) Some indications advising consumers to keep eggs refrigerated (chilled) after purchase. In addition, the pack must also indicate the type of production system (organic, free-range, barn and cage) and the producer code must be explained on or inside the pack.

The term "extra" or "extra fresh" may be used on egg cartons only until the ninth day after laying of the eggs. The laying date and the time limit of nine days must be easily visible and clearly legible on egg cartons.

When indicating the type of rations fed to the laying hens, the reference may be made to cereals as a feed ingredient only when accounting for at least 60% in weight of the feed provided, with no more than 15% of cereal by-products.

EU 27 countries have developed national regulations and enforcement mechanisms for the EU Regulation 589/2008. In addition, various EU regulations and national regulations ensure the quality, safety and legality of egg production on farms, including detailed rules about housing systems, on-farm monitoring for *Salmonella sp.* and Avian influenza and nutritional practices to reduce the environmental impact of egg production.

When compared to other countries, EU 27 nations must comply with the most diverse and strictest set of rules. However, these regulations are not without criticism as noted by the absence of specific requirements for relative humidity and fluctuations in temperature during the storage and distribution of shell eggs and the inconsistent message about refrigeration (no refrigeration allowed from farm to retail but strongly recommended after purchase) (Fikiin et al., 2020).

18.4.5 Japan

The regulations for shell eggs in Japan are found in 2 key documents: the Hygiene management guidelines for egg grading and packaging facilities (MHW, 1998) and the Egg standard transaction guidelines (Bulletin 12-Livestock A-3266, Ministry of Agriculture, Forestry and Fisheries (MAFF) issued in 2000), recently translated in English (Kashimori, 2017). In addition, MAFF issued in 2005 Notice No. 8441 on Egg Salmonella comprehensive measures guidelines to further reduce the risk of *Salmonella* contamination of shell eggs since many dishes in Japan are using raw eggs (MAFF, 2005).

18.4.5.1 Grading Characteristics

Shell eggs obviously dirty, rotten, broken or moldy should not be transported to egg grading and packaging centers (GP). After an initial inspection at the GP, spoiled, moldy and broken eggs should also be removed and all relevant information for each lot received should be collected (e.g., quantity, farm location and laying house identification, date of production, date of collections). Shell eggs are graded for quality by evaluating, through visual inspection, candling and observations after breaking, the key components (shell, air cell, egg white and yolk) as presented in Table 18.5. Shell eggs assigned to Special Grade and Grade 1 may be used for raw consumption while Grade 2 shell eggs should always be cooked prior to consumption. Moldy, leaking, smelly, rotten, partially incubated or eggs with ruptured yolk or blood spots are considered Off-Grade and unfit for human consumption (Kashimori, 2017).

18.4.5.2 Egg Cleaning/Washing

After removing shell eggs visibly broken (Off-Grade and some Grade 2), egg washing must be performed using warm water

TABLE 18.5
Quality Grading Requirements for Shell Eggs in Japan

	Special Grade	Grade 1	Grade 2
Shell	Oval shape, clean, intact, normal color, thick, normal texture	Minor defects (shape, texture or color), intact,	Abnormal shape, rough texture, soft shell, checks (broken but not leaking)
Air Cell	Depth ≤ 4.0 mm, mostly stable during rotation	Depth ≤ 8 mm, moving slightly during rotation	Depth > 8 mm, moving significantly during rotation
Egg White	Transparent, firm. mainly thick albumen, good height, surrounded the yolk	Transparent, reasonable firm, less thick albumen, less height and not surrounding the yolk, normal thin albumen	Weak and watery, little thick albumen, mostly thin albumen
Yolk	Centered, outline slightly visible (candling), round and exhibiting good height	Slightly off-centered, clear outline (candling), slight flattening	Significantly off-centered, flattened and enlarged, weak vitelline membrane

Source: Adapted from Kashimori (2017).

(at less 30°C and no less than 5°C warmer than the internal temperature of the eggs). Shell eggs must be washed and rinsed with sodium hypochlorite solution (>150 ppm) or a similar type of sanitizers. The rinse water temperature must be at least 5°C warmer than that of the wash water. After washing, shell eggs must be dried quickly using clean and sanitary equipment (MHW, 1998). The pasteurization of shell eggs is allowed although not commonly practiced.

18.4.5.3 Storage Conditions

Prior to grading, shell eggs should be stored in a cool place, not exposed directly to sunlight and be clearly identified (origin and date of production) to ensure the "First-In, First-Out" (FIFO) management of inventories. While it is considered desirable to keep graded shell eggs under refrigeration (temperature ≤10°C) during storage and distribution, the law requires only to keep them at a temperature as low as possible. However, attention should be made to limit large temperature fluctuations throughout the distribution channels and refrigeration, when applied, must be provided consistently until the point of sale (MWH, 1998).

18.4.5.4 Sorting by Weight

Based on weight, shell eggs are assigned to 6 different categories, ranging from SS (40–46 grams) to LL (70–76 grams), as shown in Table 18.3 (Kashimori, 2017).

18.4.5.5 Packing Centers

GP facilities must comply with the standards for management and operation specified in the Food Sanitation Act. For example, at the end of each day of operation, the different rooms and equipment must be cleaned and disinfected (MWH, 1998). HACCP certification is not mandatory in Japan although some large egg companies have obtained the HACCP certification or other types of certifications, including FSSC22000, ISO22000 and SQF. Since June 2021, almost all food businesses involved in production, processing, cooking and sales, including GP plants, are required to introduce hygiene practices in line with HACCP. The Japanese egg industry association issued in 2019 a comprehensive HACCP handbook to facilitate the transition and compliance to these new requirements (JEIA, 2019).

18.4.5.6 Shelf-Life

There is no mandated shelf-life date for shell eggs. The best-before (BB) date is defined as the date after which consumers must cook the eggs prior to consumption. The BB date printed on egg cartons must be scientifically and rationally defined, based on microbiological, physiochemical and sensory tests (Kashimori, 2017).

18.4.5.7 Egg Marking and Traceability

Although a traceability coding on eggs, similar to the EU system is under consideration, there is currently no regulatory requirement to mark eggs individually. Most marking (traceability number or sticker on individual egg) is performed on a voluntary basis by GP plants and egg companies, using their own systems (Kashimori, 2017).

18.4.5.8 Packaging Information

Due to frequent use of raw eggs in the Japanese cuisine, packaging requirements are different for shell eggs intended for raw consumption and those to be cooked prior to consumption. In the first case, printed information, clearly visible on the outside of packaging, must indicate name and address of GP plant, BB date, storage conditions (refrigeration after purchase at a temperature ≤10°C) and some statements indicating that these eggs may be consumed raw prior to the BB date but must be cooked after the BB date. For egg cartons containing shell eggs not intended for raw consumption, the following information must be provided to consumers on the outside of the packaging: name and address of GP plant, date of lay, date of packing and BB date and a statement indicating that these eggs must be cooked prior to consumption (MHW, 1998; Kashimori, 2017).

18.4.6 Mexico

Regulations for shell eggs are included in two key documents: NOM-159-SSA1-2016 (Egg and its products. Sanitary measures and specifications. Testing procedures) and NMX-FF-127-SCFI-2016 (Poultry products – Fresh egg from laying hen – Specifications and testing procedures) (Mexico, 2016a, 2016b).

18.4.6.1 Grading Characteristics

At receiving, shell eggs visibly dirty (fecal or blood stains) or broken must be removed before proceeding with the allocation of fresh eggs into four grades, Extra, Category 1, Category 2 and no grade. The grading is based on the aspect of the eggshell, the position and height of the air cell, the aspect and consistence of the egg white and yolk (Table 18.6.). Interestingly, the evaluation of the yolk color, using the DSM Yolkfan™, with yolk color of increasing intensity ranging from 1 (pale yellow) to 16 (bright, intense orange), is one of the criteria considered for the quality grading of shell eggs, emphasizing the importance of yolk color in the Mexican market place. Some tolerances for all grading criteria are also outlined in section 8 of the NMX-FF-127-SCFI-2016 document (Mexico, 2016b).

18.4.6.2 Egg Cleaning/Washing

Although the washing of eggs is required for egg processing, shell eggs sold at retail must not be washed (Mexico, 2016a).

18.4.6.3 Storage Conditions

Apart from the basic mention that eggs should always be handled in a way minimizing any risk of damage and breakage, the Mexican regulations provide few requirements in terms of storage conditions. However, if shell eggs are refrigerated (defined as storage at a temperature ≤7°C) at any stage of the process, they must be maintained under refrigeration all the way until the point of sale (Mexico, 2016a).

TABLE 18.6
Quality Grading Requirements for Shell Eggs in Mexico

	Extra	Category 1	Category 2	No grade
Shell	Normal aspect, intact and clean		Intact but defects allowed except for stains and foreign material	Dirty, stained, broken or eggs have been washed
Air Cell	Normal, ≤ 3.2 mm	Normal to slightly mobile, ≤ 5.0 mm	Limited motion, ≤ 6.0 mm	Mobile, foamy, or > 6.0 mm
Egg White	Firm, transparent, height > 5.5 mm or Haugh units > 70	Firm, transparent, height > 5.0 mm or Haugh units 61–70	Weak and watery, some blood or meat spots (≤ 3.1mm), height > 2.2 mm or Haugh units 31–60	Cloudy or blood and meat spots (> 3.1 mm)
Yolk	Round, centered, free of defects, absence of blood and meat spots, germinal disc no visible. Color between 9 and 13 (DSM YolkFan™)		Slightly flattened or elongated, not centered, slightly visible germinal disc. Absence of blood spots. Color between 9 and 13 (DSM YolkFan™)	Dark, not centered, abnormal shape, germinal disc development and/or signs of contamination

Source: Adapted from Mexico (2016b).

18.4.6.4 Sorting by Weight

After weighing, shell eggs are allocated to five categories, ranging from Peewee (*Canica*, <50 grams) to Extra Large (*Extra grande*, ≥64 grams). The breakdown of weight ranges is presented in Table 18.3.

18.4.6.5 Packing Centers

Although NOM-159-SSA1-2016 clearly identifies in its introduction that *Salmonella spp.* represents the greatest food safety risk and mandates the implementation of HACCP for egg processing plants, there is no similar mention for packing centers (Mexico, 2016a). The general requirements for packing centers are outlined in NOM-251-SSA1-2009 (Hygienic practices for the processing of foods, drinks or dietary supplements). Some additional requirements covering worker hygiene and documentation, applicable to both shell eggs and egg products, are provided in NOM-159-SSA1-2016 (Mexico, 2016a). The packing of shell eggs may be done in dedicated packing centers or in dedicated rooms located on farms (Mexico, 2016b).

18.4.6.6 Shelf-Life

Defined as the date of expiration, it establishes the length of time after which the quality and safety of shell eggs, stored under the conditions recommended by the packing centers, may be affected and should no longer be marketed or consumed. There is no mandated shelf-life for graded shell eggs (Mexico, 2016a).

18.4.6.7 Egg Marking and Traceability

There are no mandated requirements to mark graded shell eggs for quality or traceability purposes. The marking of eggs is allowed and performed by some egg graders for marketing purposes, as a way to demonstrate the freshness of their products.

18.4.6.8 Packaging Information

Some general information is provided for the selection of packaging materials used for shell eggs (e.g., innocuous material, design to prevent breakage) along with the requirement to have an inviolability seal applied to the egg cartons. Using clear, legible fonts, the following information must be printed on egg cartons: name of product, brand and designation of company, country of origin, grade and size, date of production, expiry date, lot and net content. In addition, egg cartons must provide some general information regarding storage conditions, use and preparation. For instance, egg cartons must carry the statement "Must be stored in a cool and dry location" (Mexico, 2016b).

18.4.7 South Africa

The document entitled "Regulations regarding the grading, packing and marking of eggs intended for sale in the Republic of South Africa" is part of the Agricultural Product Standards Act (1990) and was recently revised (South Africa, 2020).

18.4.7.1 Grading Characteristics

Based on the criteria outlined in Table 3 of these regulations, shell eggs are divided into 3 grades: Gradel, Grade 2 and Grade 3. The main differences in quality factors between these 3 grades are: Haugh units (>55 for Grade 1, >35 for Grade 2 and no value for Grade 3), air cell depth (maximum 6 mm for Grade 1, 9 mm for Grade 2 and no value for Grade 3). For each grade, the permissible occurrence of certain defects (e.g., irregular shell shape, internal defects) are defined in Tables 4 and 5 of the regulations. Any egg not meeting the criteria for Grade 1–3 is deemed unsuitable for sale.

18.4.7.2 Egg Cleaning/Washing

Egg washing is not allowed, and only clean eggs may be considered for Grade 1 or 2. Some stains, nesting materials or even fecal spots at the surface of the shell are permissible, within some specified limits. Although the pasteurization of shell eggs is allowed, it is not commonly practiced by egg packers in South Africa.

18.4.7.3 Storage Conditions

There are no mandatory conditions for the storage, distribution and retail of shell eggs. Shell eggs are typically stored at room temperature. Eggs may be oiled using food-grade mineral oil, providing that it is properly mentioned on the egg cartons.

18.4.7.4 Sorting by Weight

The sorting of eggs by weight is not mandatory and eggs (at least 33 grams) may be packaged and sold in "mixed sizes" cartons. A large percentage of the national production is sold in the informal market and sometimes one egg at a time. When eggs are sorted by size, they are assigned to 6 different designations, ranging from Small (less than 43 grams) to Super Jumbo (more than 72 grams), as indicated in Table 18.3.

18.4.7.5 Packing Centers

Most farmers grade and pack their own eggs. The regulations for shell eggs do not specify any requirements for the packing centers. Egg grading and packing centers are regulated at the national, provincial and local levels through the Foodstuffs, Disinfectants and Cosmetics Act (Act 54 of 1972), the National Health Act (Act 61 of 2003) and the Standards Act (Act 8 of 2008).

18.4.7.6 Shelf-Life

The "best before" or "best quality before date" is defined as the date after which an egg carton should no longer be sold as the eggs may not retain their grading quality (e.g., Haugh unit). However, these eggs may still be acceptable for human consumption. The best before date was defined as 40 days from the date of lay for regular shell eggs and 56 days for pasteurized shell eggs. Current practices call for shorter Best-before periods with seasonal variation since summer and winter ambient temperatures differ markedly.

18.4.7.7 Egg Marking and Traceability

The presence of a code on shell eggs is not required but may be applied at the discretion of the grading establishment.

18.4.7.8 Packaging Information

The information on egg cartons must be clearly legible and the minimum vertical height of letters for a number of elements printed on the package (e.g., size and grade of the eggs, the word "Eggs", an indication regarding the packing plant, best before date…) is provided in Table 1 of the regulations.

The indication "fresh" on egg cartons is allowed for Grade 1 eggs (Haugh units > 55) but not for Grade 2, 3 and pasteurized eggs. Any claim made on egg cartons, which implies that these eggs are healthier than other eggs and free of additives or nutritionally enhanced, must be substantiated by scientific evidence.

Shell eggs sold as "free-range" must have been produced on a farm meeting some specific criteria (e.g., cage-free indoor housing with no more than 10 laying hens/m^2, daily outdoor access with no more than 5 birds/m^2). Due to a number of recent Avian Influenza outbreaks in South Africa, the regulations allow to maintain this free-range status during indoor confinement up to 24 weeks. For eggs produced in a cage system, it is not required to indicate the production system on the packaging. Current regulations for production systems are not well aligned with other jurisdictions and a long process to modernize the production system regulations is underway.

Pasteurized eggs must be clearly identified as pasteurized and a traceability system must be in place along with regular testing of layer flocks for *Salmonella spp.*

18.4.8 Other Countries

A survey was sent to egg industry experts to gather information related to shell egg regulations in 57 countries, across Asia/Oceania, Europe, North America, Latin America and Africa. The collected information was related to the availability of national regulations specific to shell eggs, the mandate or possibility to wash shell eggs, the sorting by weight, any regulated storage conditions and the mandate for on-farm *Salmonella* testing. The results are presented in Table 18.7. More than 80% of the countries surveyed have some specific regulations for shell eggs, often supplemented by regulations related to food labeling, food safety or hygiene practices in food production. The importance of on-farm *Salmonella* testing to ensure the safety of shell eggs is widely accepted, with over 80% of the countries mandating testing plans. Apart from the USA and Japan previously reviewed, Canada is the only country mandating that shell eggs be washed prior to packing. The washing of shell eggs is allowed in 15 countries surveyed, including Australia, Brazil, Egypt, Korea and Ukraine, while it is specifically prohibited in 39 countries, including, in addition to the EU27 countries, countries like Chile, Thailand or the UK. Apart from the USA and Canada, very few countries (Algeria, Libya and Tunisia) mandate a specific temperature for the storage and distribution of shell eggs while Australia and Korea require refrigeration only if shell eggs have been washed by packing centers.

In the vast majority of the countries surveyed, shell eggs are sorted by weight prior to commercialization, either required by law or as a well-accepted industry practice. The data previously reported (Table 18.3) showed that weight classes ranged from 4 (EU 27 countries) to 6 (China, USA, Japan and South Africa). In the remaining 24 countries surveyed, there are between 3 (Egypt and Libya) and 6 weight classes (e.g., Brazil, Canada, Colombia and Thailand), with some slight variations in the weight range assigned to each class (Table 18.8).

On a global basis, it is evident that the safety and quality of shell eggs is guaranteed in different ways (Guyonnet, 2018). Overall, government regulations and sector-driven practices are widely applied and contribute to the provisions of safe and nutritious shell eggs to local populations.

TABLE 18.7

Shell Eggs Regulations and Industry Practices in Various Countries

Country	Egg Washing		Mandatory Refrigeration (Graded Eggs)	Sorting by Weightor Grade	Published Regulations	On-Farm *Salmonella* Testing	Other Regulations
	Mandatory	Allowed					
Asia / Oceania (12)							
Australia	No	Yes	Yes if washed	Yes	Yes	Yes	Various
Bangladesh	No	Yes	No	No	No	No	No
China, PR	No	Yes	No	Yes	Yes	No	Various
China, Taiwan	No	No	No	Yes	Yes	No	Feed regulations
India	No	No	No	No	No	No	Food safety & standards
Indonesia	No	Yes	No	Yes	No	No	No
Japan	Yes	–	No	Yes	Yes	Yes	Various
Korea	No	Yes	Yes if washed	Yes	Yes	Yes	Various
Malaysia	No	Yes	No	Yes	No	Yes	No
Philippines	No	Yes	No	Yes	No	Yes	No
Thailand	No	No	No	Yes	Yes	No	No
Turkey	No	No	Yes	Yes	Yes	No	No
Europe (31)							
EU 27	No	No	No	Yes	Yes	Yes	Various
United Kingdom	No	No	No	Yes	Yes	Yes	Various
Ukraine	No	Yes	No	Yes	Yes	Yes	No
North America (2)							
Canada	Yes	–	$< 10°C$	Yes	Yes	Yes	Various
USA	Yes	–	$< 7°C$	Yes	Yes	Yes	Various
Latin America (7)							
Argentina	No	Yes	No	Yes	Yes	Yes	No
Brazil	No	Yes	No	Yes	Yes	Yes	No
Chile	No	No	No	Yes	Yes	Yes	*Salmonella* testing in plants
Colombia	No	No	No	Yes	Yes	Yes	*Salmonella pullorum* and *S. gallinarum* plans
Guatemala	No	Yes	No	No	No	Yes	No
Mexico	No	No	No	Yes	Yes	Yes	No
Peru	No	Yes	No	Yes	Yes	No	No
Africa (7)							
Algeria	No	No	2–8°C	Yes	No	Yes	No
Benin	No	No	No	No	No	No	No
Burkina Faso	No	Yes	No	No	No	No	No
Egypt	No	Yes	No	Yes	No	No	Mycoplasma in feed ingredients
Libya	No	Yes	15°C	Yes	Yes	Yes	No
South Africa	No	No	No	Yes	Yes	Yes	No
Tunisia	No	No	4–8°C	No	Yes	Yes	No

TABLE 18.8

Classification of Shell Eggs by Weight (Size) in Some Selected Countries Surveyed

	Egg Class / Size / Caliber Listed by Weight (in grams)					
Country	**Pee-Wee**	**Small**	**Medium**	**Large**	**Extra-Large**	**Jumbo**
Asia / Oceania						
Korea	< 44	44–51	52–59	60–67	> 67	
Philippines	< 50	50–55	56–60	61–65	66–70	> 70
Thailand[a]	< 50	50–54	55–59	60–64	65–69	> 69
Turkey		< 53	53–62	63–72	> 73	
Latin America						
Brazil	< 45	45–49	50–54	55–59	60–65	> 65
Chile	< 40	40–47	48–54	55–61	62–68	> 68
Colombia[b]	< 46	46–53	54–60	61–67	67–78	> 78
Peru	< 50	50–55	56–62.5	62.6–68.9	69–72.2	> 72.2
North America						
Canada	< 42	43–48	49–55	56–63	64–69	> 69
Africa						
Algeria		< 57	< 60	< 64	> 64	
Egypt[c]		< 53	53–62	> 62		
Libya		< 50	50–60	> 60		

[a] Size numbered from O (≥ 70 grams to 5 (< 50 grams).
[b] Size expressed with letters: Extra-large = AAA, Large = A, Medium = A, Small = B and Pee-Wee = C.
[c] Size not regulated but industry standards.

18.5 Private Standards

Government regulations have often been established for a long time and are rather unfrequently revised through lengthy processes. As a response to several major food safety events in the late 1980s–early 1990s, a number of private food standards (e.g., Safe Quality Foods (SQF), British Retail Council (BRC)) were developed and adopted by retailers as a guarantee for a safe food supply chain. Since the different retailers a company was dealing with could never agree on a common private standard, food plants, including egg grading and packing stations, were often subjected to several annual audits, each with a different set of requirements (Weinroth et al., 2018). As a way to enhance food safety while providing some consistency for food processors, major stakeholders within the global food industry established in 2000 the Global Food Safety Initiative (GFSI) with the aim to develop global food safety standards. Mainly made up of global retailers and manufacturers members of the Consumer Goods Forum (CGF), GFSI developed global food safety standards in partnership with food safety experts and in consultation with the private section (GFSI, 2022). While the stated goal of GFSI is to ensure the delivery of safe foods to consumers worldwide and simplify the auditing process for food plants, it is also an initiative designed to protect global retailers and brand owners from negative publicity. GFSI defines the minimum requirements for food safety evaluation and certification programs, and most often requires more than complying with local regulations. GFSI implements also a comprehensive benchmarking process to assess the applicability of current food safety private standards to the GFSI global standards. Since 2000, GFSI has recognized various private standards including BRC, SQF, Food Safety System Certification (FSSC 22000), The International Features Standard (IFS) and Global G.A.P. (GFSI, 2022). Most often, retailers requesting an annual third-party audit of a GP plant will accept any GFSI-recognized private food safety standard, although some retailers still request specific items to be covered during the audit. Each private food safety standard covers a wide range of points to ensure that the plant is in full control of its processes and the quality, safety and legality of its products. For instance, the BRC Global Standard for Food Safety contains 7 sections applicable to the grading and packing of shell eggs and covers a total of 268 items (Table 18.9). The section related

TABLE 18.9

Key Elements for Shell Egg Grading/Packing Centers Addressed by the BRC Global Standards for Food Safety

Section Topic	Number of Items Evaluated
1. Senior management commitment	15
2. The food safety plan – HACCP	20
3. Food Safety & Quality management system	44
4. Site standards	113
5. Product control	38
6. Process control	17
7. Personnel	21

Source: Adapted from GSFI (2018).

to Site standards contains the most items (113) while the commitment of Senior Management is evaluated through 15 items. Some items are considered as fundamental and the failure to address any of them adequately during an audit will result in the immediate failure of the food plant to meet the GFSI standards (BRC, 2018).

As a business-driven initiative, GFSI thrives for the continuous improvement of food safety management systems. As such, private standards recognized by GFSI are evaluated regularly, deleting or, more often, adding new items. For instance, since the first version of the BRC Global Standard Food Safety published in 1998, this standard has been revised several times, with version 9 issued in 2022. The speed and flexibility to adapt to new situations is certainly the strength of private standards and they are becoming more and more relevant to the egg sector, especially in countries with weaker oversight of government agencies in terms of plant hygiene and food safety.

18.6 Conclusion

While shell eggs are a highly nutritious and very affordable animal-sourced food, their safety and quality play also a critical role in their global appeal to consumers. The highest risk of a food safety outbreak, the presence of *Salmonella* inside the egg, is well controlled by programs implemented both on farms and in egg grading and packing plants. The wide application of grading and sorting by weight of shell eggs allows consumers to purchase products with a consistent quality and appeal. Although governments have often taken different approaches toward food safety and issued regulations that may seem at times in opposition, it is also evident that eggs have naturally some excellent mechanisms to prevent contamination. Regardless, the global egg sector should expect a demand for higher food safety standards, both on farms and in egg grading and packing plants, mostly driven by the increasing demand of retailers to comply with private food safety standards.

Acknowledgment

The information presented in this chapter would not have been possible without the appreciated contribution of the following colleagues: Algeria: Radhouane Kara; Argentina: Javier Prida; Bangladesh: Bishiwajit Roy; Benin: Geoffroy Djossa; Brazil: Marcelo Checo; Burkina Faso: Souleymane Guiro; Chile: Patricio Kurte; China PR: Lihai Song, Qi Su and William Li; China, Taiwan: Jay Chen; Colombia: Sergio Gómez; Egypt: Ramy Alazzouny; Europe Union: María del Mar Fernández Poza; Guatemala: Peggy Contreras; India: Mrigen Dutta and D.J. Kalita; Indonesia: Tony Unandar, Agnes Festi and Vin; Japan: Takehiko Hayakawa; Korea: James Lee and Juyoung Yoon; Libya: Abdulwahad Kammon; Malaysia: Keat Fu and Leandro Ferreira Bezerra;; Mexico: Rodolfo Valadez; Peru: Raúl Zegarra; Philippines: Warlito Mingrajal; South Africa: Charles Costello and Kevin Lovell; Thailand: Sarawut Tanasatian; Tunisia: Moncef Bouzouaia; Turkey: Serdar Ertas; Ukraine: Wojciech Hodorowicz

REFERENCES

AMS. (1946). Agriculture marketing act. 7CFR Part 56, voluntary grading of shell eggs. Code of Federal Regulations. https://www.ecfr.gov/current/title-7/subtitle-B/chapter-I/subchapter-C/part-56

BRC. (2018). *Global standard food safety, version 8*. BRC Global Standards

Cardoso, M. J., Nicolau, A. I., Borda, D., Nielsen, L., Maia, R. L., Møretrø, T., & Teixeira, P. (2021). Salmonella in eggs : From shopping to consumption—A review providing an evidence-based analysis of risk factors. *Comprehensive Reviews in Food Science and Food Safety*, *20*(3), 2716–2741. https://doi.org/10.1111/1541-4337.12753

CFIA. (2008). Reference database for hazard identification. https://ag.umass.edu/sites/ag.umass.edu/files/projects/related-files/reference_database_for_hazard_identification.pdf

CFIA. (2022, July 22). Reference database for hazard identification (RDHI). https://active.inspection.gc.ca/rdhi-bdrid/english/rdhi-bdrid/introe.aspx?i=1

Chemaly, M., & Salvat, G. (2011). Foodborne disease associated with eggs: Microbial hazrads and *salmonella enteritidis* risk assessment. In Y. Nys, M. Bain, & F. Van Immerseel (Eds.), *Improving the safety and quality of eggs and egg products. Vol. 2: Egg safety and nutritional quality*, (pp. 34–61). Woodhead Publishing.

China. (2007). Agriculture sector standard of the People's Republic of China, Technical specification for sorting, cleaning and grading chicken eggs, NY/T 1551-2007 (In Chinese). http://down.foodmate.net/wap/index.php?itemid=48415&moduleid=23

China. (2015). National standard of the People's Republic of China. Food safety national standard for Egg and egg products, GB 2749-2015 (In Chinese). http://down.foodmate.net/wap/index.php?itemid=48415&moduleid=23

China. (2016). Quality and safety control – Technical production of non-pollution agricultural products: Part 11: Fresh eggs, NY/T 2798.11-2016 (In Chinese). http://down.foodmate.net/wap/index.php?itemid=48415&moduleid=23

China. (2020). National standard of the People's Republic of China. Pack for shell eggs, GB/T 39438-2020 (In Chinese). http://down.foodmate.net/wap/index.php?itemid=48415&moduleid=23

Codex Alimentarius. (2007). Code of hygienic practice for eggs and egg products, CAC/RCP 15-1976. https://www.fao.org/fao-who-codexalimentarius/codex-texts/codes-of-practice/en/

Codex Alimentarius. (2011). General principles of food hygiene, CXC 1-1969. Rome, Italy. https://www.fao.org/fao-who-codexalimentarius/codex-texts/codes-of-practice/en/

EFSA. (2022, July 22). Chemical hazards database (OpenFoodTox). https://www.efsa.europa.eu/en/data-report/chemical-hazards-database-openfoodtox

EPIA. (1970). Egg products inspection act. 7 CFR Part 57, Inspection of eggs. Code of Federal Regulations. https://www.ecfr.gov/current/title-7/subtitle-B/chapter-I/subchapter-C/part-57

EU. (2008). Commission regulation (EC) No 589/2008 of 23 June 2008 laying down detailed rules for implementing Council Regulation (EC) No 1234/2007 as regards marketing standards for eggs. https://eur-lex.europa.eu/legal-content/EN/TXT/?uri=celex%3A32008R0589

EU. (2013). Part VI – Eggs of hens of the *Gallus gallus* species. In Regulation (EU) No 1308/2013 of the European Parliament and of the Council of 17 December 2013 establishing a common organisation of the markets in agricultural products and repealing Council Regulations (EEC) No 922/72, (EEC) No 234/79, (EC) No 1037/2001 and (EC) No1234/2007 (pp. 818–819). https://eur-lex.europa.eu/legal-content/EN/TXT/?uri=CELEX:32013R1308

FAO. (2002). Risk assessments of *Salmonella* in eggs and broiler chickens: Interpretative summary. In *Microbiological risk assessment series 1.* FAO. https://www.fao.org/3/Y4393E/y4393e00.htm

FDA. (2009). 21 CFR parts 16 and 118 prevention of salmonella enteritidis in shell eggs during production, storage, and transportation; Final rule. *Federal Register, 74*(130), 33030–33101. https://www.govinfo.gov/content/pkg/FR-2009-07-09/pdf/E9-16119.pdf

FDA. (2021, January 12). Food traceability list. https://www.fda.gov/food/food-safety-modernization-act-fsma/food-traceability-list?utm_medium=email&utm_source=govdelivery#List

Fikiin, K., Akterian, S., & Stankov, B. (2020). Do raw eggs need to be refrigerated along the food chain?: Is the current EU regulation ensuring high-quality shell eggs for the European consumers? *Trends in Food Science & Technology, 100*, 359–362. https://doi.org/10.1016/j.tifs.2020.04.003

Gantois, I., Ducatelle, R., Pasmans, F., Haesebrouck, F., Gast, R., Humphrey, T. J., & Van Immerseel, F. (2009). Mechanisms of egg contamination by salmonella enteritidis. *FEMS Microbiology Reviews, 33*(4), 718–738. https://doi.org/10.1111/j.1574-6976.2008.00161.x

GFSI. (2022, July 21). Overview – Safe food for people everywhere. Retrieved July 21, 2022, from https://mygfsi.com/who-we-are/overview/

Goetting, V., Lee, K. A., & Tell, L. A. (2011). Pharmacokinetics of veterinary drugs in laying hens and residues in eggs: A review of the literature. *Journal of Veterinary Pharmacology and Therapeutics, 34*(6), 521 556. https://doi.org/10.1111/j.1365-2885.2011.01287.x

Gonzales, J. L., Roberts, H., Smietanka, K., Baldinelli, F., Ortiz-Pelaez, A., & Verdonck, F. (2018). Assessment of low pathogenic avian influenza virus transmission via raw poultry meat and raw table eggs. *EFSA Journal, 16*(10), e05431. https://doi.org/10.2903/j.efsa.2018.5431

Guyonnet, V. (2018). Refrigerated eggs versus shelves: Is one method safer? *Poultry International, 57*(2), 14–15. https://www.wattagnet.com/articles/32905-refrigerated-eggs-versus-shelves-is-one-method-safer

India. (2011a). Food safety and standards (Food products standards and Food additives) Regulations, version XII (08.09.2020).. https://www.fssai.gov.in/upload/uploadfiles/files/Food_Additives_Regulations.pdf

India. (2011b). Food Safety and Standards (Packaging and Labelling) Regulation, version III (27.08.2020). https://www.fssai.gov.in/upload/uploadfiles/files/Packaging_Labelling_Regulations.pdf

India. (2018). Egg quality and safety. Guidance Note No. 03/2018. https://www.fssai.gov.in/upload/uploadfiles/files/Guidance_Note_Plastic_Eggs_08_08_2018.pdf

JEIA. (2019). HACCP Implementation guide. Egg sorting and packaging facility/liquid egg manufacturing facility (In Japanese), Japan Egg Industry Association (*Nihon Rangyo Kyokai* Association). http://www.nichirankyo.or.jp/haccp-dounyu/haccp-guide.pdf

Jondreville, C., Fournier, A., & Feidt, C. (2011). Chemical residues and contaminants of eggs. In Y. Nys, M. Bain, & F. Van Immerseel (Eds.), *Improving the safety and quality of eggs and egg products. Vol. 2: Egg safety and nutritional quality* (pp. 62–80). Woodhead Publishing.

Kashimori, A. (2017). *The illustrated egg handbook.* Appendix (pp. 141–149). Context Publications. https://www.contextbookshop.com/user_uploads/files/booksamples/The-Illustrated-Egg-Handbook.pdf

Lee, M., Seo, D. J., Jeon, S. B., Ok, H. E., Jung, H., Choi, C., & Chun, H. S. (2016). Detection of foodborne pathogens and mycotoxins in eggs and chicken feeds from farms to retail markets. *Korean Journal for Food Science of Animal Resources, 36*(4), 463–468. https://doi.org/10.5851/kosfa.2016.36.4.463

MAFF. (2005). Egg salmonella comprehensive measures guidelines, Notice No. 8441 (In Japanese). https://www.maff.go.jp/j/syouan/douei/eisei/e_kanri_kizyun/sal/pdf/keiran_sogo.pdf

Martelli, F., & Davies, R. H. (2012). Salmonella serovars isolated from table eggs: An overview. *Food Research International, 45*(2), 745–754. https://doi.org/10.1016/j.foodres.2011.03.054

Mexico. (2016a). Egg and its products. Sanitary measures and specifications. Testing procedures, Official Mexican Norm, NOM-159-SSA1- 2016 (In Spanish). https://www.snice.gob.mx/~oracle/SNICE_DOCS/NOM-159-SSA1-2016-NOM-159-DOF_20180719-20180116.pdf

Mexico. (2016b). Poultry products – Fresh egg from laying hen – Specifications and testing procedures, Official Mexican Norm, NMX-FF-127-SCFI-2016 (In Spanish). http://sitiosl.dif.gob.mx/alimentacion/docs/NMX-FF-127-SCFI-2016_Huevo_fresco.pdf

MHW. (1998). Hygiene management guidelines for egg sorting and packaging facilities (Ministry of Health and Welfare Notice No. 1674 (In Japanese). https://www.jz-tamago.co.jp/wp/wp-content/uploads/2020/03/E05_3_3.pdf

Ministry of Primary Industries, New Zealand. (2022, July 22). Food safety. Hazard database. https://www.foodsafety.govt.nz/registers-lists/hazards/

Musgrove, M. T. (2011). Microbiology and safety of table eggs. In Y. Nys, M. Bain, & F. Van Immerseel (Eds.), *Improving the safety and quality of eggs and egg products. Vol. 2: Egg safety and nutritional quality* (pp. 3–33). Woodhead Publishing.

Piskorska-Pliszczynska, J., Struciński, P., & Kan, C. A. (2019). Contaminants in eggs: dioxins/PCBs and other toxic substances and their possible health implications. In F. J. M. Smulders, I. M. C. M. Rietjens, & M. Rose (Eds.), *ECVPH food safety assurance* (Vol. 7, pp. 415–441). https://doi.org/10.3920/978-90-8686-877-3_17

Raspoet, R., Gantois, I., Devloo, R., Pasmans, F., Haesebrouck, F., Ducatelle, R., & Van Immerseel, F. (2011). Internal contamination of eggs by *Salmonella enteritidis.* In Y. Nys, M. Bain, & F. Van Immerseel (Eds.), *Improving the safety and quality of eggs and egg products. Vol. 2: Egg safety and nutritional quality* (pp. 46–61). Woodhead Publishing.

Reich, H., & Triacchini, G. A. (2018). Occurrence of residues of fipronil and other acaricides in chicken eggs and poultry muscle/fat. *EFSA Journal, 16*(5), e05164. https://doi.org/10.2903/j.efsa.2018.5164

Ropkins, K., & Beck, A. J. (2000). Evaluation of worldwide approaches to the use of HACCP to control food safety. *Trends in Food Science & Technology, 11*(1), 10–21. https://doi.org/10.1016/S0924-2244(00)00036-4.

South Africa. (2020). Regulations regarding the grading, packing and marking of eggs intended for sales in the Republic of South Africa, No. R. 345, Government Gazette. https://www.gov.za/sites/default/files/gcis_document/202003/43108rg11055gon345.pdf

Tomczyk, Ł, Stępień, Ł, Urbaniak, M., Szablewski, T., Cegielska-Radziejewska, R., & Stuper-Szablewska, K. (2018). Characterisation of the mycobiota on the shell surface of table eggs acquired from different egg-laying hen breeding systems. *Toxins*, *10*(7), 293. https://doi.org/10.3390/toxins10070293

USDA. (2000). Egg grading manual https://www.ams.usda.gov/sites/default/files/media/Egg%20Grading%20Manual.pdf

USDA. (2022, July 22). Food Safety Research Information Office (FSRIO). https://www.nal.usda.gov/programs/fsrio

Viator, C. L., Cates, S. C., Karns, S. A., Muth, M. K., & Meekho, R. L. (2007). Food safety practices used at egg packing and egg products establishments. *Food Protection Trends*, *27*(4), 238–245.

Wang, L., Zhang, Q., Yan, Z., Tan, Y., Zhu, R., Yu, D., & Wu, A. (2018). Occurrence and quantitative risk assessment of twelve mycotoxins in eggs and chicken tissues in China. *Toxins*, *10*(11), 477. http://dx.doi.org/10.3390/toxins10110477

Weinroth, M. D., Belk, A. D., & Belk, K. E. (2018). History, development, and current status of food safety systems worldwide. *Animal Frontiers*, *8*(4), 9–15. https://doi.org/10.1093/af/vfy016

Whiley, H., & Ross, K. (2015). Salmonella and eggs: From production to plate. *International Journal of Environmental Research and Public Health*, *12*(3), 2543–2556. https://doi.org/10.3390/ijerph120302543

Part VI

Microbiology, Quality, and Safety

19

Egg Proteomics

Yaqi Meng, Fang Geng, and Ning Qiu*

CONTENTS

19.1 Overview of the Egg Proteome

19.1.1 Technical Development of Egg Proteomics

The last two decades have delivered significant progress in the field of egg proteomics, especially with the shift from gel-based proteomics to mass spectrometry-based (MS) high-throughput proteomics. For protein separation, the gel-based methods, particularly, two-dimensional electrophoresis (2DE), are versatile to visualize protein information, such as protein mass, isoelectric point, and posttranslational modifications (PTMs), and have been widely used in the early stage. Moreover, the combination of 2DE and MS facilitated the visualization and identification of egg proteins that were differentially modulated under certain conditions (Guérin-Dubiard et al., 2006). However, owing to the limitations of 2DE-based proteomics, including the weak ability to detect proteins in minute amounts, narrow linear dynamic detection range, time-consuming process, copious sample handling, and low reproducibility, chromatographic techniques have been developed for proteomics to achieve direct, automated sample introduction and protein (or peptide) complex separation (Fournier et al., 2007).

The liquid chromatography (LC)-based separation approach then facilitated the development of shotgun proteomics, allowing the indirect measurement of proteins by comparing the tandem mass spectra of peptide fragmentation to a database with theoretical tandem mass spectra of peptides derived from numerous in-silico digested proteins (Zhang et al., 2013). The advances in MS and data analysis software for shotgun proteomics have facilitated high-throughput proteome analysis, such as proteomics based on label-free methods, tandem mass tags (TMT), and isobaric tags for relative and absolute quantification (iTRAQ) (Rauniyar & Yates, 2014). For the PTMs of egg proteins, the focus has been on egg phosphoproteome and egg N-glycoproteome for the past five years. The analytical workflow of the phosphoproteome requires enrichment of phosphopeptides before LC-MS/MS identification using immobilized metal affinity chromatography (Sun et al., 2020). The omics strategy for the identification of N-glycoproteins and N-glycosites usually relies on enzyme digestion of proteins, glycopeptides enrichment using hydrophilic interaction liquid chromatography (HILIC), deglycosylation by PNGase F in $H_2{}^{18}O$ water, and finally LC-MS/MS identification (Figure 19.1) (Geng et al., 2017). In recent years, with the deepening of bioscience research, tissue heterogeneity and spatial differences have gradually been found in biosamples. The research trend of proteomics has produced nanoproteomics from the traditional whole tissue to the study of a microcell group (less than 5000 cells) or even a single cell, to obtain more accurate biological system information (Zhu et al., 2018). Moreover, with the new generation of timsTOF Pro mass spectrometry platform (based on 3D separation (retention time, m/z, and ion intensity)), the fourth dimension of ion mobility separation has been added to reduce the complexity of samples, improve detection sensitivity, and comprehensively improve detection performance in many aspects, making proteomics enter a new 4D era (Meire et al., 2018).

* Corresponding author.

DOI: 10.1201/9781003254430-25

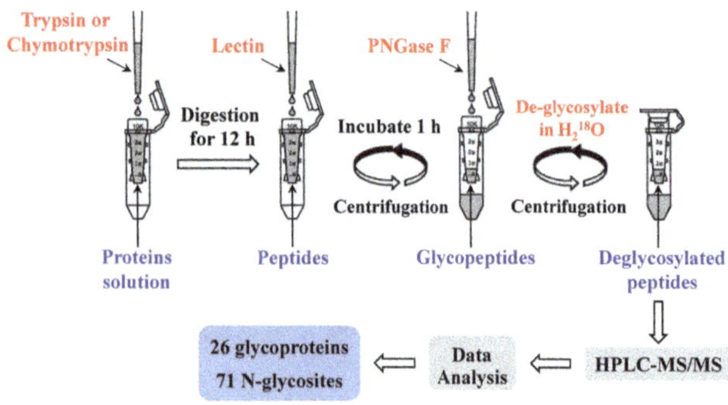

FIGURE 19.1 The workflow of chicken egg withe N-glycoproteomic analysis.

19.1.2 Egg White Proteome

Egg white does not only function as the shock-absorber, antimicrobial barrier, and source of nutrition for the developing embryo but also as an inexpensive ingredient for the food industry with gelling, emulsifying, and foaming properties. In addition, some proteins, including ovalbumin and ovotransferrin, are widely used in biomedical research and protein chemistry. The first proteome of chicken egg white was analyzed 20 years ago when comparatively investigating protein composition using different electrophoretic separations. The isoforms of ovalbumin, ovotransferrin, and ovomucoid were visualized using isoelectric separation. Although numerous spots were unknown, a large number of protein spots were separated using 2DE, which was then the most effective separation method for the subsequent comprehensive analysis of the proteome (Desert et al., 2001). Shortly after, Guearin-Dubiard et al. (Guérin-Dubiard et al., 2006) reported 16 new proteins and improved the identification of protein polymorphism and functional protein families from egg white through 2DE-based analyses. These results suggested that there were more new proteins in the egg white that need to be identified. In 2007, 78 chicken egg white proteins were identified, 54 of which were found through the combination of LC-MS/MS and MS^3 (Mann, 2007). As the most comprehensive egg white proteome at that time, these results benefited from the improved quantitate methods of calculating the exponentially modified protein abundance index. By limiting the hindering effect of proteins in high abundance on the identification of proteins in minute amounts, the use of two types of peptide ligand libraries, which contained modified hexapeptides terminating with a primary amine or a terminal carboxyl group, allowed the identification of 148 egg white proteins, with a great number not previously reported (D'Ambrosio et al., 2008). Subsequently, Mann et al. reported a novel identification approach using the LTQ Orbitrap Velos, which is fast enough to simultaneously isolate and fragment multipeaks with the acquisition of high-resolution mass full-scan spectrum, and found an additional 79 new proteins (Mann & Mann, 2011).

In the early 2DE-based studies of the egg white proteome, a large number of protein spots with different molecular weights or different isoelectric point values of the same protein were found to be homologous (Desert et al., 2001; Guérin-Dubiard et al., 2006); mainly modified through glycosylation and phosphorylation, which increased research interest. Geng et al. reported 71 N-glycosites in 26 chicken egg white glycoproteins, of which ovomucin, the most heavily glycosylated protein, may possess some antiviral activity (Geng et al., 2017). N-glycoproteome analysis of duck egg white identified 231 N-glycosites from 68 N-glycoproteins. The biological functions of duck egg white N-glycoproteins were compared with chicken orthologs, and the results showed different allergenicities and antibacterial abilities (Meng et al., 2019). These differences might be initiated by specific alterations in glycosylation, which can enhance proteolysis resistance and protein steric hindrance. Phosphorylation of proteins is another important and pleiotropic PTM. Phosphoproteomic analysis of chicken egg white identified 25 phosphoproteins carrying 41 phosphosites (Yang et al., 2019). The following year, 189 phosphosites from 96 phosphoproteins were identified using trypsin and trypsin/glutamyl endoproteinase digestion combined with immobilized metal affinity chromatography, thereby suggesting that the trypsin/Glu-C digestion approach could greatly increase the identification of proteins (Sun et al., 2020). Subsequently, the phosphoproteome of duck egg white was reported. A total of 92 phosphosites representing 36 phosphoproteins were identified, and the results provided fundamental insight into the biological functions of duck egg white phosphoproteins (Arshad et al., 2020).

Recently, various high-throughput peptidomic studies have identified some native functional peptides in the egg white which may give a boost to the development of functional foods and medicinal compounds (Arena et al., 2020; Zhang et al., 2020). Affinity chromatography enrichment can specifically screen the egg white proteins with a specific biological activity. A total of 20 heparin-binding proteins from chicken egg white were identified, including known antimicrobial proteins (such as avidin, lysozyme, and TENP). The antibacterial activity of three new egg candidates was tested against *Listeria monocytogenes* and/or *Salmonella enterica* Enteritidis (Guyot et al., 2016). Another egg white-derived peptide QIGLF was studied to clarify its underlying antihypertensive mechanisms in spontaneously hypertensive rats using the tandem mass tag quantitative proteomic method. QIGLF might exert antihypertensive effects by inhibiting Na+ reabsorption and oxidative stress, restoring gap and tight junctions (Zhou et al., 2022). Through the use of egg white proteomics, peptidomics, and PTM omics, more proteins and their related functions will be identified in future studies. The application of these proteins in food production and nutritional supplements represents an important direction for future research.

19.1.3 Egg Yolk Proteome

The chicken egg yolk, surrounded by the vitelline membrane, consists of approximately 17% protein, 33% lipid, and 48% water. The major egg yolk proteins are distributed within the plasma and granule compartments of the yolk. The precursors of the egg yolk proteins are synthesized mainly by the liver of the hen, secreted into the blood, and then transported through the circulatory system to the ovary, where they are transferred into the egg yolk. The precursors of egg yolk low-density lipoprotein (LDL) are the hen serum very low-density lipoproteins (VLDL), which comprise two main apoproteins: apolipoprotein B (Apo B) and apovitellenin-1 (Jolivet et al., 2006). Hen serum Apo B contains only one unit (523 kDa), and is enzymatically cleaved into seven fragments during its transfer into the egg yolk. This fragmentation and recombination are conducive to the stability of egg yolk LDL (Jolivet et al., 2008). For the main proteins in the yolk granule, vitellogenins (VTGs), precursors of several main egg yolk proteins, are synthesized in the liver and then transported to the ovary, where they enter the follicle via receptor-mediated endocytosis in the follicle membrane (Schneider, 2016). During transport, VTGs are cleaved to phosvitin, lipovitellin-1, lipovitellin-2, and yolk glycoproteins by the action of cathepsin. Then, lipovitellin-1 and -2 bind to phospholipids to form yolk high-density lipoprotein (HDL), which subsequently assembles with phosvitin and yolk low-density lipoprotein (LDL) to form yolk granules.

To systematically recognize and comprehensively identify the proteins in the yolk, Mann et al. used multistage mass spectrometry identification methods based on LTQ-FT mass spectrometry and reported 119 proteins, including the discovery of 86 new proteins. Most of the proteins are present in both the plasma and granule, but their quantities vary greatly (Mann & Mann, 2008). Similar to the solution adopted for the study of the egg white proteome, Farinazzo et al. increased the number of identified egg yolk proteins to 255 with the combinatorial effect of peptide ligand libraries to avoid the interference from proteins in high quantity (Farinazzo et al., 2009). Using proteomics, bioinformatics analyses of the egg yolk proteins revealed their role in the function of cell-to-cell interaction, cell development, and proliferation, as well as the development of the hematological system (D'Alessandro et al., 2010). In addition to the above egg yolk proteomic studies intended to identify the most abundant proteins, a comprehensive peptidomic research was conducted. A total of 628 native peptides were identified and may have the potential to become novel bioactive compounds (e.g., having antimicrobial, antihypertensive, antiviral, and anti-inflammatory activities), functional foods, and cosmetics (Arena & Scaloni, 2018).

The above studies also predicted the occurrence of PTMs of egg yolk proteins, which have important effects on protein structure, physical, and chemical properties. Thereafter, Geng et al. reported the first comprehensive N-glycoproteomic analysis of the chicken egg yolk, indicating valuable structural information for food processing and insights on embryo development from the analysis of 217 N-glycosylation sites and 86 glycoproteins (Geng et al., 2018). Subsequently, an N-glycoproteomic study of duck egg yolk identified 359 N-glycosites corresponding to 113 N-glycoproteins, suggesting a higher N-glycosylation level in duck egg yolk proteins (Meng et al., 2020). Functional annotation of the duck egg yolk N-glycoproteins revealed that biological adaption meets the requirements for embryo development by modulating protein structure and binding ability, including immune response and ligand transport. For the phosphoproteome of egg yolk, 253 phosphorylation sites assigned to 66 phosphoproteins were identified in duck egg yolk, of which VTG-1, VTG-2, and fibrinogen alpha chain were the high phosphoproteins (Arshad et al., 2022). Biological function annotation showed that duck egg yolk phosphoproteins were mainly involved in binding, catalytic, immune response, and metabolic activities.

19.1.4 Vitelline Membrane Proteome

The vitelline membrane is an important structure for the transmembrane transport of egg yolk components, protection of the blastodisc, and separation of egg white and egg yolk. Using high-throughput and high-end LC-MS^n methods, 137 proteins were identified from the chicken egg vitelline membrane, a barrier to microbial infection, as well as the structure preventing the mixing of egg yolk with egg white. The main proteins in the vitelline membrane, for instance, lysozyme C, ovomucin, ovalbumin, and ovotransferrin, are also present in other egg compartments while other proteins, including eight ZP proteins, two ATPases, and one oviductin protease, are only found in the vitelline membrane (Mann, 2008). These results revealed the diversity of protein sources in the vitelline membrane, although our understanding of these proteins remains incomplete, including the possible PTMs. Recently, an N-glycoproteome study of the chicken egg vitelline membrane reported 435 N-glycosites of 208 N-glycoproteins, some of which may regulate proteinases/inhibitors and play a role in the transmembrane transport of lipids (Xiao et al., 2020).

Interestingly, the N-glycosite locations of mucin-5B (the alpha subunit of ovomucin) from different egg parts are quite different (Figure 19.2). There may be several reasons for this difference:

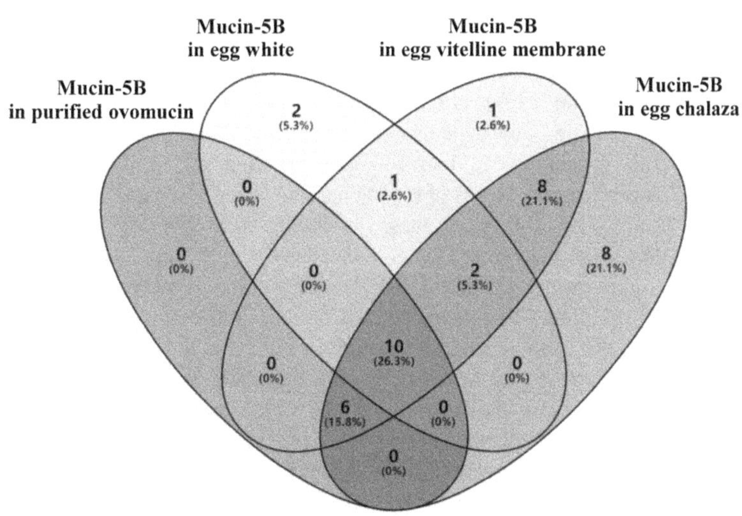

FIGURE 19.2 Comparison of the N-glycosites of mucin-5B from different egg parts.

1) The analysis method differed among studies; 2) The relative content of mucin-5B differs among egg parts, and a higher relative content would generally be beneficial for improving the depth of identification; 3) The N-glycosylation of mucin-5B is heterogeneous. The tissue specificity of the phosphorylation/N-glycosylation was also observed in other egg proteins, and is an "economical and efficient" mode acquired by living organisms in the long-term evolutionary process: a protein template, yielding multiple structures, with various activities.

During the chick embryo development process, the vitelline membrane is gradually integrated into the chorioallantoic membrane, a placenta-like structure produced by the embryo for acid-base balance, respiration, and calcium solubilization from the eggshell for bone mineralization. The chorioallantoic membrane proteome at two stages of the embryo development was characterized and the contribution of the embryonic blood serum proteome to identifying unique proteins was assessed. The study identified 1796 unique proteins, which contribute to functions such as calcium ion transport, gas exchange, vasculature development, and chemical protection against invading pathogens, therefore providing a crucial basis to expand its biomedical applications for pharmaceutical and cancer research (Ahmed et al., 2022).

19.1.5 Eggshell and Eggshell Membrane

Chicken eggshell is a calcitic porous container with the function of preventing the egg from physical damage and microbial contamination, as well as being a source of calcium and regulating gas exchange during embryogenesis (Nys et al., 2001). The eggshell consists of three parts, the outermost cuticle, the calcified layer, and the innermost eggshell membranes. Approximately half of the organic matrix in the eggshell calcified layer is soluble in acid. Using LC-MS/MS, the number of proteins identified increased to 520, all of which were newly discovered except 10 proteins previously reported (Mann et al., 2006). The eggshell-specific proteins, ovocleidins, and ovocalyxins, along with other proteins produced by the eggshell gland cells, may play a specific role in eggshell formation. However, this study was designed only to identify the proteins in the eggshell calcified layer, irrespective of their possible functions. Subsequently, Rose and Hincke reviewed the omics analyses of the eggshell matrix, which contributed to our understanding of the impact of these proteins on eggshell formation and function (Rose & Hincke, 2009). The two eggshell-specific matrix proteins (ovocleidins and ovocalyxins) were suggested to play roles in the regulation of mineralization and antimicrobial defense. A proteomic analysis of the eggshell cuticle, the outermost cuticle of avian eggshell, was subsequently conducted by Rose-Martel et al. (Rose-Martel et al., 2012). They identified 47 proteins using LC-MS/MS. Ovocalyxin-32 and a protein similar to Kunitz-like protease inhibitor were the most abundant proteins in the cuticle. The researchers reported the first comprehensive proteome of the eggshell cuticle and laid the molecular basis for understanding its antimicrobial properties. The understanding of eggshell biomineralization has significantly improved recently with the advent of multi-omics, including proteomics, in which the evolution of selected matrix proteins and their underlining role in the biological toolkit required for eggshell calcification in avian species was analyzed (Roy et al., 2021).

During eggshell proteomics, one important objective is to verify the impact of PTMs on eggshell components since they greatly alter the properties and interactions of proteins. In 2007, Mann et al. identified 39 phosphoproteins in the eggshell matrix using LC-MS^n, 22 of which were newly reported. PTMs such as phosphorylation often affect the assembly of eggshell substrates through protein–protein interactions (Mann et al., 2007). Yang et al. performed another systematic analysis involving the proteome, phosphoproteome, and glycoproteome of the chicken eggshell matrix and reported 297 N-glycosylation sites on 182 N-glycoproteins and 112 phosphorylation sites on 69 phosphoproteins. The N-glycoproteins mainly contributed to the immune response regulation while phosphoproteins mainly regulated enzyme activity (Yang et al., 2020).

The eggshell membrane, another important component of the eggshell, protects the egg from pathogen infection and provides the platform for eggshell calcitic nucleation. With the extensive cross-linkages and stabilized characteristics of the eggshell membrane, the identification of proteins is technically challenging, requiring various extraction/solubilization conditions to conduct proteomic analysis. These methods allowed the identification of 472 proteins, essential for fibrous structure formation and the prevention of pathogen invasion (Ahmed et al., 2017). The advances in proteomics have further facilitated the functional discovery of eggshell membrane, which can modulate the gut microbiota to prevent murine pre-cachexia through suppression of T helper cell differentiation (Jia et al., 2022).

19.2 Factors Affecting the Initial Egg Proteome

When shelled eggs are laid by hens, their proteomes are different. Many factors determine the composition of egg proteins and can further alter their bioavailability, functionality, and bioactivity by affecting protein migration, hydrolysis, aggregation, and PTM, which brings about differences in raw material quality.

19.2.1 Interspecies Variability

It is well known that chicken eggs dominate the table-egg market worldwide. However, duck and quail eggs are also widely consumed in many countries in Asia, Latin America, and Europe. Owing to the long-term environmental adaptation and evolution, different avian species present unique genetic characteristics and their eggs are different in many respects, including size, composition, physicochemical properties, and nutrient composition (Song et al., 2000; Damaziak et al., 2020). Hence, a comparative proteomic analysis of egg proteins from different poultry species offers the possibility to decipher the species-specific biological unknowns associated with proteins and could help us understand the evolutionary relationship as well as indicate the level of adaptation to the environment.

A previous study reported that the protein content of egg white is higher in duck and quail eggs than in chicken eggs (Miguel et al., 2005). Using 2DE separation and MALDI-TOF MS/MS identification with egg white as a reference, 10 and

9 unique proteins were found in quail and duck egg whites, respectively. The distribution of duck egg white proteins differed on the 2D gel, while the quail and chicken egg proteins were distributed similarly. This finding is consistent with the evolutionary distance of the three species: chickens and quails belong to *Galliformes*, while ducks belong to *Anseriformes*. From those egg white proteome patterns, the interspecies diversity can be seen to correlate with the specific functional needs of the proteins, such as microbial defense and embryogenesis regulation (Hu et al., 2016).

Another comprehensive proteomic study was performed on the egg white of six major poultry species using tandem mass tags quantification technology. The number of proteins identified varied across species: 148 (chicken), 138 (duck), 183 (quail), 162 (turkey), 179 (pigeon), and 150 (goose). A total of 52 conserved proteins were common across these six species and their plausible functions seemed to prevent the early death of embryonic cells and provide a secure environment for the developing embryo. Conversely, the species-specific proteins, including serpin-like protein HMSD in duck, haptoglobin in pigeon, and ovodefensin in chicken, were screened and found to be associated with some species-dependent biological processes. The study was instructive for our understanding of the species-specific biological challenges and provided—from a proteomics point of view—some insights into the poultry evolution–domestication processes (Sun et al., 2017).

The amino acid sequences of some specific ovalbumin peptides differ among eggs from different avian species, constituting a good model to understand the amino acid diversity of homologous proteins caused by the evolution process and the adaptation to diverse environments (Ren et al., 2017.). Furthermore, duck lysozyme has a stronger antimicrobial activity against *Salmonella enterica* Enteritidis than that of chicken lysozyme, which may be due to its evolution under the pressure of humidity and an environment with a higher bacterial load (Naknukool et al., 2009). Thus, the interspecies variability of different egg proteins may have diverse effects on their composition, functionality, and bioactivity. Differences in egg proteomes between breeds of domestic chickens were also observed. The quantity, but not the composition, of 189 egg white proteins is different among five domestic fowl breeds (Araucana, Czech golden pencilled, Minorca, Booted bantam, and Rosecomb bantam) (Bílková et al., 2018). This suggests that the interbreed differences in egg white proteome are much smaller than the interspecies differences.

Interspecies comparisons of the egg yolk proteome have also received attention. Egg yolk proteins identified from three poultry species (chicken, quail, and duck) share a high level of homology in protein sequence and structure, which are supposed to relate to the adaptability of avian species to different environmental stresses. In addition, some protein-specific regions on 2D gels can be used as biomarkers of the authenticity of origin (Liu et al., 2018). To verify the species-specific diversity of egg proteins, a comparison of egg yolk proteins between Tibetan and lowland chickens using label-free proteomics revealed a higher antioxidant capacity in Tibetan chicken egg yolk, which may be associated with the PIT54 protein and glutathione peroxidase 3 (Liu et al., 2020). The proteomic profiles of Silky fowl egg yolk and Leghorn egg yolk were compared using label-free LC-MS/MS analysis, and the results showed that there were 57 unique proteins in Silky fowl egg yolk. Among the commonly quantified yolk proteins, 26 of them were differentially abundant, especially, the abundance of vitelline membrane outer layer protein 1, transthyretin, and ovoinhibitor were more than 10 times higher in Silky fowl egg yolk. Bioinformatics analysis revealed that proteins with higher abundance in Silky fowl egg yolk were more involved in protease inhibitory activity, coagulation-related activity, and neuroactive ligand–receptor interaction (Wu et al., 2022). These cross-species and breed comparisons give a comprehensive understanding of the biological activities of proteins and provide a new direction for investigating the effect of phosphorylation on the nutritional characteristics of egg proteins.

Because different PTMs between proteins can further reflect their functional differences, the interspecific comparison of egg PTMs was performed. Meng et al. conducted the N-glycoproteome analyses of duck egg white and yolk and compared them with orthologs of the chicken egg. They demonstrated that species-specific allergenicity, antibacterial ability, and antivirus properties were related to the differences in glycosylation (Meng et al., 2019; Meng et al., 2020). An integrated proteomic, phosphoproteomic, and N-glycoproteomic analyses of quail egg identified 175 proteins, 109 N-glycoproteins (293 N-glycosites), and 23 phosphoproteins (84 phosphorylation sites). Compared with chicken and duck egg proteins, the type and modification site of the quail egg proteins differ considerably. Functional analysis indicated that quail egg N-glycoproteins were mainly involved in protein hydrolysis and enzyme activity regulation, while phosphoproteins were mainly related to lipid transport (Liu et al., 2020). This type of study facilitates the application of egg proteins from various sources to benefit the development of functional foods by the food industry.

19.2.2 Feeding Management and the Stress of a Laying Hen

A 2DE-based proteomic analysis among six different specialty eggs (such as omega-3-enriched eggs, organic eggs, and free-range eggs) reported 23 proteins, of which 19 were observed in significantly different quantities. These results suggested that the egg proteome could be influenced by the feed composition and breeding management (Wang et al., 2012). The difference in egg white proteomes between free-range and barn-raised chicken was successfully differentiated using a combination of optimization of ultrasound-assisted protein extraction method and MALDI-TOF-MS identification; 23 and 29 proteins were identified in free-range chicken and barn-raised chicken egg white, respectively (He et al., 2014).

A stressed environment can affect the protein composition of eggs by altering gene expression and protein translation in the oviduct of laying hens. The 2DE patterns of eggs changed significantly after five days of stress application induced by corticosterone, suggesting that the egg proteome was significantly altered. Further MALDI-TOF/TOF MS/MS identification revealed that the levels of TENP, hemopexin, IgY-Fc upsilon 3-4, and extracellular fatty acid-binding protein decreased while the levels of ovoinhibitor and ovalbumin-related protein X increased under stress (Kim & Choi, 2014). Vanadium-induced stress in laying hens can be alleviated with epigallocatechin-3-gallate (EGCG). iTRAQ-based quantitative proteome analysis of egg white showed that 20 proteins responded to the vanadium stress, and significant changes in the abundance of 28 proteins were accompanied with stress relief after EGCG application. The detoxification mechanism of EGCG under vanadium stress may act through regulating metal-binding mediation and immune function-related proteins (Wang et al., 2017).

To fully understand the different mechanisms of the egg proteome and explore the nutritional properties of the egg, we need to pay attention to the origin of the egg: the formation of egg yolk (formation of the follicle) and the secretion of egg white. Currently, there is a gap between food science and animal science: research in the field of food science is focused on the egg proteome differences between species and breeds, while research in the field of animal science is focused on the production performance of laying hens. The egg formation process has become a blank area between the two research fields. Areas, such as the transport patterns of nutrients, the mechanisms of assembly of major nutrients (such as HDL and yolk granules), and the secretion sequence of thick and thin egg white during egg formation remain unclear. Therefore, proteomic research on egg formation is urgently required. Recently, a quantitative chicken egg yolk proteomic analysis during its formation showed that the abundance of apoproteins (Apo B, apovitellenin-1, and VTGs) in egg yolk LDL changed significantly during yolk formation, and was associated with varied protein hydrolases (cathepsin E-A-like protein and carboxypeptidase) and isomerases (peptidyl-prolyl cis-trans isomerase) related to yolk proteins processing (Wang et al., 2022). Proteins related to the transport of vitamins (retinol, riboflavin, and vitamin D) and cholesterol were most abundant during the final period of egg yolk formation. The overall abundance of immunoglobulins decreased while the complement proteins increased with yolk formation, indicating that the immune system was adjusted dynamically. These new findings provide an important contribution to the details of nutrient transportation during egg yolk formation, and are valuable for the quality, preservation, and processing of egg yolk.

19.3 Proteome Changes during Storage

Storage conditions and time of egg highly affect its quality. In many countries, there is no regulatory requirement to store eggs under refrigeration (temperature below 10°C) and eggs lose quality with longer storage times. The environmental conditions, temperature, and length of storage time modify the quality of raw eggs and their functional properties mainly by affecting the inner egg compartments. Over the last few years, as a high-throughput analysis method, proteomics has been a very effective tool for screening egg quality-related biomarkers and has revealed the mechanisms of egg quality deterioration during storage.

Proteomic analysis of egg white has shown that for a large number of proteins, such as ovalbumin, ovoinhibitor, and ovotransferrin, the number of proteins greatly changed during the initial storage period (22 ± 2°C), while further storage caused minor changes in the proteome. This study suggested that the degradation of ovalbumin and clusterin during long-term storage may be responsible for the thinning of the egg white (Omana et al., 2011). A recent quantitative proteomic analysis indicated that the reduced abundance of β-ovomucin may be the direct cause of egg white thinning (Liu et al., 2022). Another 2DE-based proteomic study focusing on the effect of storage at various temperatures (4, 20, and 37°C for 15 days) on egg white indicated that the degradation of ovalbumin was possibly due to the reduction in antiprotease, further providing some insights into the thermally-induced biochemical variations during the deterioration process of egg quality. Clusterin, a protein identified in the 2DE-based proteome decreased during high-temperature storage conditions and has been proposed as a valid biomarker for the egg industry to evaluate egg quality (Qiu et al., 2012). A comparative 2DE proteomic method followed by MALDI-TOF-MS/MS was used to verify the alterations in egg yolk proteins during 40 days of storage. Most egg yolk proteins presented some degradation, accelerating over time (Gao et al., 2017). The impact of carbon dioxide on chicken egg white was assessed using an iTRAQ-based quantitative proteomic analysis. In addition, the potential predictors (β-hexosaminidase, trypsin inhibitor ClTI-1, and apolipoprotein D) of egg freshness were screened using correlation analysis between protein abundance and freshness indexes (Xu et al., 2018). Furthermore, an integrated peptidomics procedure conducted to identify the biomarkers of egg freshness revealed 1974 and 836 unique variable molecules from the egg yolk plasma and white, respectively. Using stringent criteria for filtering data and validating through label-free based proteomics, 25 (egg yolk) and 19 peptides (egg white) were selected as egg aging markers (Renzone et al., 2021).

In addition to the types and abundance, the PTMs of the egg proteins also change during storage. The phosphorylation status of egg yolk proteins at different storage temperatures (4 and 37°C for 15 days) was further studied. Most egg yolk proteins with an enhanced degree of phosphorylation may be involved in the catalysis, binding, and transport functions that correlate with the deterioration of egg quality (Wang et al., 2020). Studying the alterations in the chicken egg vitelline membrane caused by storage at 30°C for 10 days, a comparative proteomic and N-glycoproteomic analysis revealed that changes in the number of structural proteins and related glycosites were possibly the result of the disintegration of the complexes, producing a loss in the fibrous structure that led to membrane deterioration during high-temperature storage (Zhou et al., 2020; Zhou et al., 2021). The changes in the egg protein PTMS during the storage process may be dominated by endogenous enzymes, and the underlying mechanisms need further research.

19.4 Proteome Changes during Processing

The processing mechanisms of natural food are challenging due to the complexity of a natural food system. High-throughput omics technology provides a very effective tool for exploring the processing mechanisms of complex food systems. Protein is the dominant nutrient for eggs, and in recent years, reports on egg processing mechanisms based on proteome analysis have gradually increased.

Heat-induced gel formation by egg proteins is one of the most important processes in the cooking and production of eggs, especially egg white. Fresh egg white can be easily categorized into two types, namely the colloid-like egg white and the solution-like egg white, which are called thick egg white (TKEW) and thin egg white (TNEW), respectively. The heat-induced gel properties of TKEW and TNEW were systematically analyzed and compared. The results showed that TKEW (72.51°C) had a higher temperature of heat denaturation than TNEW (67.01°C). The texture profile analysis demonstrated that TKEW gel appeared soft and tough (lower hardness and higher cohesiveness), while TNEW gel appeared hard and brittle (higher hardness and springiness, but lower cohesiveness). Scanning electron microscope images showed that the TKEW gel exhibited a "mesh structure" with "mountain-like" protrusions and micropores, while the TNEW gel exhibited a "block structure" with a dense section and "rope-head" protrusions. Quantitative proteomic analysis revealed that a higher content of ovotransferrin in TNEW might be the main reason for its lower temperature of thermal denaturation, and a higher content of ovomucin in TKEW might regulate the microstructure of the gel during heating and cause the differences in gel texture properties (Liu et al., 2020). Subsequently, quantitative N-glycoproteomic analyses between the fresh egg white and spray-dried egg white showed that 3 N-glycoproteins (TENP, clusterin, and BPI fold containing family C) were significantly upregulated, and 18 N-glycoproteins (mainly mucin-5B, macroglobulin family members, and carboxypeptidase) were significantly downregulated after spray-drying (Liu et al., 2021). HILIC and subsequent identification using LC-MS/MS revealed that the downregulated N-glycosylation may be because the terminal ends of the N-glycans were cross-linked with the protein side chains during spray-drying, preventing enrichment. Those results showed that the N-glycans of egg white proteins (mainly ovomucin) were likely to be covalently cross-linked during spray-drying, which enriched the thermal gel with a "skeleton" structure and stronger cohesiveness. These two proteomic analyses related to egg white gel formation showed that among the main proteins in egg white, ovomucin may be the main factor influencing the properties of egg white gel: the content and N-glycosylation of ovomucin were closely related to the differences/changes in gel properties.

Further studies reconfirmed the critical role of ovomucin in the formation of egg white gel: quantitative proteomic analysis revealed that ovotransferrin, lysozyme, and ovomucin (mucin 5B and mucin 6) participated in the formation of the egg white gel insoluble skeleton, and their abundance decreased in the gel supernatant after centrifugation; ovalbumin and ovomucoid were not involved in the early formation of the egg white gel skeleton, and their abundance increased in the gel supernatant after centrifugation. Interestingly, if the initial interaction of egg white proteins is changed through ultrasonic pretreatment, the participation of ovomucin in the formation of the gel skeleton can be inhibited, resulting in a change in the type of egg white gel (Figure 19.3) (Wang et al., 2022).

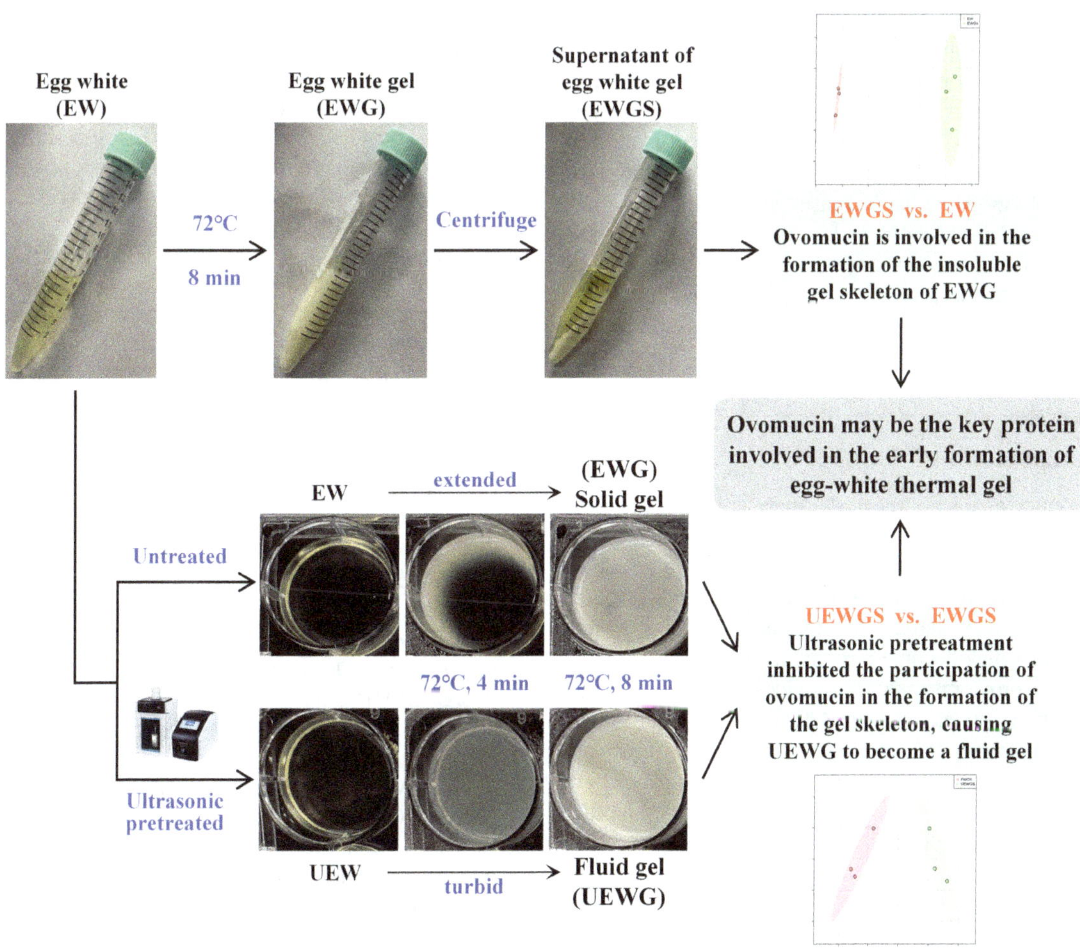

FIGURE 19.3 Ovomucin may be the key protein involved in the early formation of egg-white thermal gel.

Based on the research mentioned above, the "skeleton effect" of ovomucin in the formation of egg white gel was hypothesized. Because of its structure and interaction with other egg white proteins, ovomucin participates in the initiation and leads the formation of the egg white gel skeleton, and accordingly, plays a key role in the formation of egg white gel. To verify this hypothesis, the molecular structure of ovomucin should be analyzed further (especially, its heterogeneous molecular structure due to high glycosylation modification). In addition, the interaction between ovomucin and the other main proteins in egg white should be characterized because this could be a key approach for ovomucin to exert its "skeletal effect". Such findings will provide accurate quantitative information on egg white proteins involved in the early formation of a gel. Similar strategies can be applied in future studies on the functional properties of eggs, such as the foaming properties of egg white and the emulsifying properties of egg yolk.

Proteomic-based techniques enable a more detailed understanding of egg digestibility after processing. The influences of high-pressure treatment on the digestibility of egg white were evaluated, and the results showed that high-pressure treatment (400–800 MPa and 9°C) of egg white could promote the hydrolysis of egg white proteins by pepsin. In addition, a greater digestibility was obtained after 800 MPa treatment compared with thermal treatment at 95°C. Further 2DE results showed that high-pressure treatment promotes pepsin digestion of the band at 48 kDa, and LC-MS/MS results suggested that the majority of the pepsin-digested peptides were derived from ovalbumin (Hoppe et al., 2013). The digestibility of egg white protein under common food processing conditions (pH, salt concentration, and heat) was studied, and the amino acid modification of egg white proteins was examined using proteomic analysis. The results showed that heating improved the digestibility of egg white but increased amino acid oxidation, nitration, deamidation, and dehydration. The in vitro digestibility of egg white was not altered by pH and salt denaturation. In addition, the Maillard reaction could attenuate the digestion of egg white by inducing cross-linking and protein glycation (Lassé et al., 2015).

19.5 Conclusion and Perspectives

Egg proteomics enables large-scale analysis of the complete set or the dynamic set of proteins, especially their location, abundance, and PTMs, by bridging the molecular information to phenotypic attributes that are related to egg composition, authentication, the bioactivity of proteins, and general quality traits, which may become a routine analysis tool in future research. First, proteomic analysis provides a more comprehensive view and revisits egg characteristics in a big-data fashion. Furthermore, an integrated proteomic, metabolomic, and lipidomic analyses could provide a more comprehensive perspective for understanding the nutritional properties of eggs, and also would be an effective technical tool for identifying egg species, origin, and feeding methods. Moreover, using proteomic analysis to revisit important questions in egg science will provide new insights. Similar to the application of proteomic analysis to the study of the formation mechanisms of egg white thermogels, which suggested that ovomucin may be the key protein, the same research strategy can also be applied to the study of the foaming property of egg white and the emulsifying property of egg yolk, thereby confirming the key egg white/yolk proteins involved in the water–air and oil–water interfaces. In addition, artificial eggs have emerged, and recombinantly expressed ovalbumin has achieved low-cost large-scale production. However, recombinant egg proteins for use in food need to be carefully scrutinized in terms of structure, processing properties, and nutritional properties. In particular, the similarities and differences in the PTMs of recombinant and natural egg proteins need to be given more attention. Because protein sequences can be guaranteed to be consistent, but phosphorylation and glycosylation are variable, comparative phosphoproteomic and glycoproteomic analyses are essential.

REFERENCES

Ahmed, T. A. E., Suso, H.-P., & Hincke, M. T. (2017). In-depth comparative analysis of the chicken eggshell membrane proteome. *Journal of Proteomics*, *155*, 49–62. https://doi.org/10.1016/j.jprot.2017.01.002

Ahmed, T. A. E., Cordeiro, C. M. M., Elebute, O., & Hincke, M. T. (2022). Proteomic analysis of chicken chorioallantoic membrane (CAM) during embryonic development provides functional insight. *BioMed Research International*, e7813921. https://doi.org/10.1155/2022/7813921

Arena, S., Renzone, G., & Scaloni, A. (2020). A multi-approach peptidomic analysis of hen egg white reveals novel putative bioactive molecules. *Journal of Proteomics*, *215*, 103646. https://doi.org/10.1016/j.jprot.2020.103646

Arena, S., & Scaloni, A. (2018). An extensive description of the peptidomic repertoire of the hen egg yolk plasma. *Journal of Agricultural and Food Chemistry*, *66*(12), 3239–3255. https://doi.org/10.1021/acs.jafc.8b01183

Arshad, R., Meng, Y. Q., Qiu, N., Geng, F., Mine, Y., Keast, R., & Zhu, C. X. (2022). Phosphoproteomic analysis of duck egg yolk provides novel insights into its characteristics and biofunctions. *Journal of the Science of Food and Agriculture*, *102*(3), 1165–1173. https://doi.org/10.1002/jsfa.11453

Arshad, R., Meng, Y. Q., Qiu, N., Sun, H. H., Keast, R., & Rehman, A. (2020). Phosphoproteomic analysis of duck egg white and insight into the biological functions of identified phosphoproteins. *Journal of Food Biochemistry*, *44*(10). https://doi.org/10.1111/jfbc.13367

Bílková, B., Świderská, Z., Zita, L., Laloë, D., Charles, M., Beneš, V., Stopka, P., & Vinkler, M. (2018). Domestic fowl breed variation in egg white protein expression: Application of proteomics and transcriptomics. *Journal of Agricultural and Food Chemistry*, *66*(44), 11854–11863. https://doi.org/10.1021/acs.jafc.8b03099

D'Alessandro, A., Righetti, P. G., Fasoli, E., & Zolla, L. (2010). The egg white and yolk interactomes as gleaned from extensive proteomic data. *Journal of Proteomics*, *73*(5), 1028–1042. https://doi.org/10.1016/j.jprot.2010.01.002

Damaziak, K., Kieliszek, M., & Gozdowski, D. (2020). Structural and proteomic analyses of vitelline membrane proteins of blackbird (Turdus merula) and song thrush (Turdus philomelos). *Scientific Reports*, *10*(1), 19344. https://doi.org/10.1038/s41598-020-76559-4.

D'Ambrosio, C., Arena, S., Scaloni, A., Guerrier, L., Boschetti, E., Mendieta, M. E., Citterio, A., & Righetti, P. G. (2008). Exploring the chicken egg white proteome with combinatorial peptide ligand libraries. *Journal of Proteome Research*, *7*(8), 3461–3474. https://doi.org/10.1021/pr800193y

Desert, C., Guérin-Dubiard, C., Nau, F., Jan, G., Val, F., & Mallard, J. (2001). Comparison of different electrophoretic separations of hen egg white proteins. *Journal of Agricultural and Food Chemistry*, *49*(10), 4553–4561. https://doi.org/10.1021/jf001423n

Farinazzo, A., Restuccia, U., Bachi, A., Guerrier, L., Fortis, F., Boschetti, E., Fasoli, E., Citterio, A., & Righetti, P. G. (2009). Chicken egg yolk cytoplasmic proteome, mined via combinatorial peptide ligand libraries. *Journal of Chromatography A*, *1216*(8), 1241–1252. https://doi.org/10.1016/j.chroma.2008.11.051

Fournier, M. L., Gilmore, J. M., Martin-Brown, S. A., & Washburn, M. P. (2007). Multidimensional separations-based shotgun proteomics. *Chemical Reviews*, *38*(47), 3654–3686. https://doi.org/10.1021/cr068279a

Gao, D., Qiu, N., Liu, Y. P., & Ma, M. H. (2017). Comparative proteome analysis of egg yolk plasma proteins during storage: Proteomic analysis of egg yolk plasma proteins during storage. *Journal of the Science of Food and Agriculture*, *97*(8), 2392–2400. https://doi.org/10.1002/jsfa.8052

Geng, F., Wang, J. Q., Liu, D. Y., Jin, Y. G., & Ma, M. H. (2017). Identification of *N* -glycosites in chicken egg white proteins using an omics strategy. *Journal of Agricultural and Food Chemistry*, *65*(26), 5357–5364. https://doi.org/10.1021/acs.jafc.7b01706

Geng, F., Xie, Y. X., Wang, J. Q., Majumder, K., Qiu, N., & Ma, M. H. (2018). *N* -glycoproteomic analysis of chicken egg yolk. *Journal of Agricultural and Food Chemistry*, *66*(43), 11510–11516. https://doi.org/10.1021/acs.jafc.8b04492

Guérin-Dubiard, C., Pasco, M., Mollé, D., Désert, C., Croguennec, T., & Nau, F. (2006). Proteomic analysis of hen egg white. *Journal of Agricultural & Food Chemistry*, *54*(11), 3901–3910. https://doi.org/10.1021/jf0529969

Guyot, N., Labas, V., Harichaux, G., Chessé, M., Poirier, J.-C., Nys, Y., & Réhault-Godbert, S. (2016). Proteomic analysis of egg white heparin-binding proteins: Towards the identification of natural antibacterial molecules. *Scientific Reports*, *6*(1), 27974. https://doi.org/10.1038/srep27974

He, J. N., Yan, H. T., & Fan, C. L. (2014). Optimization of ultrasound-assisted extraction of protein from egg white using response surface methodology (RSM) and its proteomic study by MALDI-TOF-MS. *RSC Advances*, *4*(80), 42608–42616. https://doi.org/10.1039/C4RA07272C

Hoppe, A., Jung, S., Patnaik, A., & Zeece, M. G. (2013). Effect of high pressure treatment on egg white protein digestibility and peptide products. *Innovative Food Science & Emerging Technologies*, *17*, 54–62. https://doi.org/10.1016/j.ifset.2012.11.003

Hu, S., Qiu, N., Liu, Y. P., Zhao, H. Y., Gao, D., Song, R., & Ma, M. H. (2016). Identification and comparative proteomic study of quail and duck egg white protein using 2-dimensional gel electrophoresis and matrix-assisted laser desorption/ionization time-of-flight tandem mass spectrometry analysis. *Poultry Science*, *95*(5), 1137–1144. https://doi.org/10.3382/ps/pew033

Jia, H., Lyu, W., Hirota, K., Saito, E., Miyoshi, M., Hohjoh, H., Furukawa, K., Saito, K., Haritani, M., Taguchi, A., Hasebe, Y., & Kato, H. (2022). Eggshell membrane modulates gut microbiota to prevent murine pre-cachexia through suppression of T helper cell differentiation. *Journal of Cachexia, Sarcopenia and Muscle*. https://doi.org/10.1002/jcsm.13019

Jolivet, P., Boulard, C., Beaumal, V., Chardot, T., & Anton, M. (2006). Protein components of low-density lipoproteins purified from hen egg yolk. *Journal of Agricultural and Food Chemistry*, *54*(12), 4424–4429. https://doi.org/10.1021/jf0531398

Jolivet, P., Boulard, C., Chardot, T., & Anton, M. (2008). New insights into the structure of apolipoprotein B from low-density lipoproteins and identification of a novel YGP-like protein in hen egg yolk. *Journal of Agricultural and Food Chemistry*, *56*(14), 5871–5879. https://doi.org/10.1021/jf800321m

Kim, J., & Choi, Y.-H. (2014). Differential abundance of egg white proteins in laying hens treated with corticosterone. *Journal of Agricultural and Food Chemistry*, *62*(51), 12346–12359. https://doi.org/10.1021/jf504469t

Lassé, M., Deb-Choudhury, S., Haines, S., Larsen, N., Gerrard, J. A., & Dyer, J. M. (2015). The impact of pH, salt concentration and heat on digestibility and amino acid modification in egg white protein. *Journal of Food Composition and Analysis*, *38*, 42–48. https://doi.org/10.1016/j.jfca.2014.08.007

Liu, L., Wang, J., Wang, G., Song, H., Geng, F., Zeng, Q., & Huang, Q. (2022). Quantitative proteomics provides a new perspective on the mechanism of network structure depolymerization during egg white thinning. *Food Chemistry*, *392*, 133320. https://doi.org/10.1016/j.foodchem.2022.133320

Liu, L., Yang, R., Luo, X., Dong, K., Huang, X., Song, H., Gao, H., Li, S., & Huang, Q. (2020). Omics analysis of holoproteins and modified proteins of quail egg. *Food Chemistry*, *326*, 126983. https://doi.org/10.1016/j.foodchem.2020.126983

Liu, X., Wang, J., Liu, L., Cheng, L., Huang, Q., Wu, D., Peng, L., Shi, X., Li, S., & Geng, F. (2021). Quantitative N-glycoproteomic analyses provide insights into the effects of thermal processes on egg white functional properties. *Food Chemistry*, *342*, 128252. https://doi.org/10.1016/j.foodchem.2020.128252

Liu, X., Wang, J. Q., Huang, Q., Cheng, L., Gan, R., Liu, L. L., Wu, D., Li, H. M., Peng, L. X., & Geng, F. (2020). Underlying mechanism for the differences in heat-induced gel properties between thick egg whites and thin egg whites: Gel properties, structure and quantitative proteome analysis. *Food Hydrocolloids*, *106*, 105873. https://doi.org/10.1016/j.foodhyd.2020.105873

Liu, Y. P., Qiu, N., Gao, D., & Ma, M. H. (2018). Comparative proteomic analysis of chicken, duck, and quail egg yolks. *International Journal of Food Properties*, *21*(1), 1311–1321. https://doi.org/10.1080/10942912.2018.1489836

Liu, Y. P., Sheng, L., Ma, M. H., & Jin, Y. G. (2020). Proteome-based identification of chicken egg yolk proteins associated with antioxidant activity on the Qinghai-Tibetan plateau. *International Journal of Biological Macromolecules*, *150*, 1093–1103. https://doi.org/10.1016/j.ijbiomac.2019.10.115

Mann, K. (2007). The chicken egg white proteome. *Proteomics*, *7*(19), 3558–3568. https://doi.org/10.1002/pmic.200700397

Mann, K. (2008). Proteomic analysis of the chicken egg vitelline membrane. *Proteomics*, *8*(11), 2322–2332. https://doi.org/10.1002/pmic.200800032

Mann, K., Maček, B., & Olsen, J. V. (2006). Proteomic analysis of the acid-soluble organic matrix of the chicken calcified eggshell layer. *Proteomics*, *6*(13), 3801–3810. https://doi.org/10.1002/pmic.200600120

Mann, K., & Mann, M. (2008). The chicken egg yolk plasma and granule proteomes. *Proteomics*, *8*(1), 178–191. https://doi.org/10.1002/pmic.200700790

Mann, K., & Mann, M. (2011). In-depth analysis of the chicken egg white proteome using an LTQ orbitrap velos. *Proteome Science*, *9*(1), 7. https://doi.org/10.1186/1477-5956-9-7

Mann, K., Olsen, J. V., Maček, B., Gnad, F., & Mann, M. (2007). Phosphoproteins of the chicken eggshell calcified layer. *Proteomics*, *7*(1), 106–115. https://doi.org/10.1002/pmic.200600635

Meire, F., Brunner, A., Koch, S., Koch, H., Lubeck, M., Krause, M., Goedecke, N., Decker, J., Kosinski, T., Park, M. A., Bache, N., Hoerning, O., Cox, J., Räther, O., & Mann, M. (2018). Online parallel accumulation-serial fragmentation (PASEF) with a novel trapped ion mobility mass spectrometer. *Molecular Cellular Proteomics*, *17*(12), 2534–2545. https://doi.org/10.1074/mcp.TIR118.000900

Meng, Y. Q., Qiu, N., Geng, F., Huo, Y. Q., Sun, H. H., & Keast, R. (2019). Identification of the duck egg white n-glycoproteome and insight into the course of biological evolution. *Journal of Agricultural and Food Chemistry*, *67*(35), 9950–9957. https://doi.org/10.1021/acs.jafc.9b03059

Meng, Y. Q., Qiu, N., Geng, F., Keast, R., Li, B., & Zheng, X. X. (2020). N-glycoproteomic analysis of duck egg yolk proteins: Implications for biofunctions and evolution. *International Journal of Biological Macromolecules*, *151*, 19–26. https://doi.org/10.1016/j.ijbiomac.2020.02.038

Miguel, M., Manso, M. A., López-Fandiño, R., & Ramos, M. (2005). Comparative study of egg white proteins from different species by chromatographic and electrophoretic methods. *European Food Research and Technology*, *221*(3–4), 542–546. https://doi.org/10.1007/s00217-005-1182-8

Naknukool, S., Hayakawa, S., Uno, T., & Ogawa, M. (2009). Antimicrobial activity of duck egg lysozyme against salmonella enteritidis. In *Global issues in food science and technology* (pp. 293–307). Elsevier. https://doi.org/10.1016/B978-0-12-374124-0.00018-1

Nys, Y., Gautron, J., McKee, M. D., Garcia-Ruiz, J. M., & Hincke, M. T. (2001). Biochemical and functional characterisation of eggshell matrix proteins in hens. *World's Poultry Science Journal*, *57*(4), 401–413. https://doi.org/10.1079/WPS20010029

Omana, D. A., Liang, Y., Kav, N. N. V., & Wu, J. (2011). Proteomic analysis of egg white proteins during storage. *Proteomics*, *11*(1), 144–153. https://doi.org/10.1002/pmic.201000168

Qiu, N., Ma, M. H., Zhao, L., Liu, W., Li, Y. Q., & Mine, Y. (2012). Comparative proteomic analysis of egg white proteins under various storage temperatures. *Journal of Agricultural and Food Chemistry*, *60*(31), 7746–7753. https://doi.org/10.1021/jf302100m

Rauniyar, N., & Yates, J. R. (2014). Isobaric labeling-based relative quantification in shotgun proteomics. *Journal of Proteome Research*, *13*(12), 5293–5309. https://doi.org/10.1021/pr500880b

Ren, J. D., Hu, J. H., Chen, L., Liu, Y. L., Xu, X. Q., He, J., Shen, J. L., & Lu, L. Z. (2017). Identification of egg white proteins and divergence in the regulatory region of the ovalbumin Gene in avians. *Protein & Peptide Letters*, *24*(1), 12–25. https://doi.org/10.2174/0929866523666161025120953

Renzone, G., Novi, G., Scaloni, A., & Arena, S. (2021). Monitoring aging of hen egg by integrated quantitative peptidomic procedures. *Food Research International*, *140*, 110010. https://doi.org/10.1016/j.foodres.2020.110010

Rose, M. L. H., & Hincke, M. T. (2009). Protein constituents of the eggshell: Eggshell-specific matrix proteins. *Cellular and Molecular Life Sciences*, *66*(16), 2707–2719. https://doi.org/10.1007/s00018-009-0046-y

Rose-Martel, M., Du, J., & Hincke, M. T. (2012). Proteomic analysis provides new insight into the chicken eggshell cuticle. *Journal of Proteomics*, *75*(9), 2697–2706. https://doi.org/10.1016/j.jprot.2012.03.019

Roy, N. L., Stapane, L., Gautron, J., & Hincke, M. T. (2021). Evolution of the avian eggshell biomineralization protein toolkit – New insights from multi-omics. *Frontiers in Genetics*, *12*, 672433. https://doi.org/10.3389/fgene.2021.672433.

Schneider, W. J. (2016). Chapter 17—Lipoprotein receptors. In N. D. Ridgway & R. S. McLeod (Eds.), *Biochemistry of lipids, lipoproteins and membranes* (6th ed., pp. 489–518). Elsevier. https://doi.org/10.1016/B978-0-444-63438-2.00017-1

Song, K. T., Choi, S. H., & Oh, H. R. (2000). A comparison of egg quality of pheasant, chukar, quail and Guinea fowl. *Asian-Australasian Journal of Animal Sciences*, *13*(7), 986–990. https://doi.org/10.5713/ajas.2000.986

Sun, C. J., Liu, J. N., Li, W. B., Xu, G. Y., & Yang, N. (2017). Divergent proteome patterns of egg albumen from domestic chicken, duck, goose, Turkey, quail and pigeon. *Proteomics*, *17*(17–18), 1700145. https://doi.org/10.1002/pmic.201700145

Sun, Y., Jin, H. B., Sun, H. H., & Sheng, L. (2020). A comprehensive identification of chicken egg white phosphoproteomics based on a novel digestion approach. *Journal of Agricultural and Food Chemistry*, *68*(34), 9213–9222. https://doi.org/10.1021/acs.jafc.0c03174

Wang, H., Qiu, N., Mine, Y., Sun, H. H., Meng, Y. Q., Bin, L., & Keast, R. (2020). Quantitative comparative integrated proteomic and phosphoproteomic analysis of chicken egg yolk proteins under diverse storage temperatures. *Journal of Agricultural and Food Chemistry*, *68*(4), 1157–1167. https://doi.org/10.1021/acs.jafc.9b06912

Wang, J. P., Bai, X., Ding, X. M., Bai, S. P., Zeng, Q. F., Mao, X. B., & Zhang, K. Y. (2017). Quantitative proteomic analysis reveals the role of tea polyphenol EGCG in egg whites in response to vanadium stress.*Nutrition*, *39–40*, 20–29. https://doi.org/10.1016/j.nut.2017.02.007

Wang, J. P., Liang, Y., Omana, D. A., Kav, N. N. V., & Wu, J. P. (2012). Proteomics analysis of egg white proteins from different egg varieties. *Journal of Agricultural and Food Chemistry*, *60*(1), 272–282. https://doi.org/10.1021/jf2033973

Wang, J. Q., Liu, X., Li, S. G., Ye, H. L., Luo, W., Huang, Q., & Geng, F. (2022). Ovomucin may be the key protein involved in the early formation of egg-white thermal gel. *Food Chemistry*, *366*, 130596. https://doi.org/10.1016/j.foodchem.2021.130596

Wang, Y., Wang, J. Q., Shi, Y. Q., Ye, H. L., Luo, W., & Geng, F. (2022). Quantitative proteomic analyses during formation of chicken egg yolk. *Food Chemistry*, *374*, 131828. https://doi.org/10.1016/j.foodchem.2021.131828

Wu, R., Chen, C., & Zhang, X. Y. (2022). Label-free LC-MS/MS analysis reveals different proteomic profiles between egg yolks of silky fowl and ordinary chickens. *Foods*, *11*(7), 1035. https://doi.org/10.3390/foods11071035

Xiao, J., Wang, J. Q., Cheng, L., Gao, S. H., Li, S. G., Qiu, N., Li, H. M., Peng, L. X., & Geng, F. (2020). A puzzle piece of protein N-glycosylation in chicken egg: N-glycoproteome of chicken egg vitelline membrane. *International Journal of Biological Macromolecules*, *164*, 3125–3132. https://doi.org/10.1016/j.ijbiomac.2020.08.193

Xu, L., Jia, F., Luo, C. Y., Yu, Q. Q., Dai, R. T., & Li, X. M. (2018). Unravelling proteome changes of chicken egg whites under carbon dioxide modified atmosphere packaging. *Food Chemistry*, *239*, 657–663. https://doi.org/10.1016/j.foodchem.2017.06.128

Yang, R., Geng, F., Huang, X., Qiu, N., Li, S., Teng, H., Chen, L., Song, H., & Huang, Q. (2020). Integrated proteomic, phosphoproteomic and N-glycoproteomic analyses of chicken eggshell matrix.*Food Chemistry*, *330*, 127167. https://doi.org/10.1016/j.foodchem.2020.127167

Yang, R., Song, H. B., Huang, Q., Liu, L. L., Qiu, N., & Ma, M. H. (2019). Identification and analysis of phosphoproteome in chicken egg white. *Food Science*, *40*(11), 30. https://doi.org/10.7506/spkx1002-6630-20181217-195

Zhang, Y.-H., Bai, J., Jiang, W.-N., Zhao, C.-R., Ji, J.-J., Wang, J.-Z., & Liu, Y.-W. (2020). Promising hen egg-derived proteins/peptides (EDPs) for food engineering, natural products and precision medicines. *Research in Veterinary Science*, *128*, 153–161. https://doi.org/10.1016/j.rvsc.2019.11.011

Zhang, Y. Y., Fonslow, B. R., Shan, B., Baek, M.-C., & Yates, J. R. (2013). Protein analysis by Shotgun/Bottom-up proteomics. *Chemical Reviews*, *113*(4), 2343–2394. https://doi.org/10.1021/cr3003533

Zhou, M., Zhao, W., Xue, W., Liu, J., & Yu, Z. (2022). Potential antihypertensive mechanism of egg white-derived peptide QIGLF revealed by proteomic analysis. *International Journal of Biological Macromolecules*, *218*, 439–446. https://doi.org/10.1016/j.ijbiomac.2022.07.149

Zhou, Y., Qiu, N., Mine, Y., Keast, R., & Meng, Y. Q. (2021). Comparative N-glycoproteomic analysis provides novel insights into the deterioration mechanisms in chicken egg vitelline membrane during high-temperature storage. *Journal of Agricultural and Food Chemistry*, *69*(7), 2354–2363. https://doi.org/10.1021/acs.jafc.0c07557

Zhou, Y., Qiu, N., Mine, Y., Meng, Y. Q., Keast, R., & Zhu, C. X. (2020). Quantitative comparative proteomic analysis of chicken egg vitelline membrane proteins during high-temperature storage. *Journal of Agricultural and Food Chemistry*, *68*(36), 9816–9825. https://doi.org/10.1021/acs.jafc.0c03538

Zhu, Y., Piehowski, P. D., Kelly, R. T., & Qian, W. (2018). Nanoproteomics comes of age. *Expert Review of Proteomics*, *15*(11), 865–871. https://doi.org/10.1080/14789450.2018.1537787

20

Microbiology of Egg and Egg Products

Florence Baron*, **Sophie Jan, and Clarisse Techer**

CONTENTS

20.1 Introduction

Eggs constitute a readily available and a low-cost source of proteins, fats, and micronutrients with an important nutritional role in the context of growing human population. Therefore, eggs are an important part of the human diet and are consumed all over the world as shell eggs (table eggs) or as egg products (e.g. egg yolk, egg white, and whole egg in liquid, dried, or frozen forms). They are used in many food products (sauces, meat, seafood, dairy products, etc.). Eggs represent a

* Corresponding author.

DOI: 10.1201/9781003254430-26

major economic market and the safety of eggs is a major concern for the egg industry and consumers.

Shell egg is considered safe because it is intended to meet the nutritional needs of the developing embryo and to protect it from any microbial invasion during its development. It thus has a remarkable arsenal of antimicrobial defenses, which can however be affected by the conditions during egg production, collection, packaging, storage, and manufacturing in the case of egg products. Scientific studies have provided significant advances in egg safety research and have improved our understanding of the link between egg quality and safety and the risk of foodborne pathogens as well as mitigation strategies. However, eggs and egg products continue to be implicated in food poisoning resulting from bacterial contamination. In the European Union, 45.6% of reported *Salmonella* outbreaks in 2018 were caused by the consumption of eggs or egg products (EFSA and ECDC, 2019). This *Salmonella*-egg vehicle combination is also responsible for the most widespread foodborne illnesses in the United States (Dewey-Mattia et al., 2018). Events of eggs or egg products spoilage also can lead to high economic losses for the industry. Preventing microbial contamination thus represents a challenge for the egg industry.

This chapter focuses on microbial contamination of shell eggs and egg products and reviews the origin and the type of microorganisms involved, depending on the type of eggs and egg products. We will focus on the antimicrobial defense mechanisms of eggs and the adaptations put in place by microorganisms to bypass them. Finally, measures and strategies to limit the presence and multiplication of microorganisms in the eggs and egg products will be discussed.

20.2 Shell Egg Microbiology

20.2.1 Microorganisms Involved and Routes of Contamination

Under normal layer management, the egg content is generally sterile just after oviposition. However, it can be contaminated by a diversified microbiota composed of food spoilage microorganisms and sometimes, pathogenic bacteria. Two possible routes of egg contamination have been described, either (i) vertical transmission where eggs are contaminated during their formation in the ovary or in the oviduct of infected hens, or (ii) horizontal transmission corresponding to the penetration of the hen's fecal microorganisms or other microorganisms present in the environment, through the eggshell.

20.2.1.1 Salmonella Enteritidis *and the Egg*

Salmonella Enteritidis, in an outstanding way, epidemiologically associated with eggs (EFSA and ECDC, 2019) and has been implicated in recent egg-related outbreaks in different countries (Cardoso et al., 2021).

Salmonellae are facultative-anaerobic bacteria belonging to the *Enterobacteriaceae* family. *Salmonella* spp. are Gram-negative, rod-shaped, non-spore-forming bacteria. They are intestinal parasites and intracellular pathogens in many mammalian hosts and are also found in birds, reptiles, amphibians, and plants (McQuiston et al., 2008).

The genus *Salmonella* contains two species, *Salmonella enterica* and *Salmonella bongori. S. enterica* is divided into six subspecies (*enterica, salamae, arizonae, diarizonae, houtenae,* and *indica*) comprising more than 2,500 serotypes. Most pathogenic strains, with a few exceptions, belong to the subspecies *Salmonella enterica* subsp. *enterica* (Parry, 2006). Worldwide, *Salmonella enterica* Enteritidis is recognized as the most implicated serotype in the outbreaks linked to the consumption of eggs and egg products (Cardoso et al., 2021). *S.* Enteritidis often causes gastroenteritis, the main clinical signs being diarrhea (87% of cases), abdominal pain (84%), fever (75%), nausea and muscle pain (65%), headaches, and vomiting (25%) (Stevens et al., 1989). These symptoms start usually between 12 and 48 h after consuming the contaminated food. In addition to gastroenteritis, certain strains can cause localized or generalized sepsis which can lead to death (Humphrey, 2006). The amount of *Salmonella* that must be ingested to develop the clinical disease varies depending on the strain and on the health of the host. The infectious threshold is usually 6 to 8 log CFU (colony forming unit)/mL or per g, but it can be much lower in some cases.

Unlike in humans, few symptoms are observed in adult hens infected by *S.* Enteritidis. There are no clinical signs, excessive mortality, or noticeable decrease in egg production. This "healthy" carrier state is involved in the dissemination of bacteria between hens within the same flock, in particular through feces, and in the contamination of eggs (Suzuki, 1994).

The eradication, in the 1970s, of *S.* Gallinarum and *S.* Pullorum serotypes (pathogenic serotypes for hens, but rarely for humans) probably favored the development of the *S.* Enteritidis serotype in laying hen farms (Humphrey, 2006). In addition, *S.* Enteritidis has a particular tropism for the reproductive organs (Gantois et al., 2008) and can therefore contaminate the egg during its formation (vertical contamination).

20.2.1.2 Vertical Contamination

The vertical contamination, also called endogenic contamination, has been shown by different teams, which carried out artificial inoculation with *S.* Enteritidis and observed subsequent contamination of eggs. However, this transmission in eggs occurred only at a low frequency, intermittently, and for high inoculation levels (in the order of 10^8–10^9 cells per hen inoculated orally, intravenously, or peritoneally). Depending on the authors, the levels of contaminated eggs varied from 0 to 8.1% (Shivaprasad et al., 1990; Gast & Beard, 1992; Keller et al., 1995; Gast et al., 2002), with a low level of *Salmonella*, i.e. often less than 10 CFU/egg (Humphrey et al., 1991). In the case of eggs from farms naturally contaminated with *Salmonella*, the analysis revealed even lower rates of egg contamination: 0.06% according to Poppe et al. (1991) and 0.6% according to Lock and Board (1992).

The prevalence of *Salmonella* in commercial table eggs is generally low in most developed countries, although the values reported vary depending on the country. For example, a value of 0.005% contaminated eggs is estimated for the United States by Ebel and Schlosser (2000), and 0.37% is reported in

Europe by EFSA and ECDC (2019), where positive eggs are only reported for 13 member states. *Salmonella* contamination in layer farms is tending to decrease, but there are still differences between countries depending on the control programs put in place.

When contaminated eggs are analyzed at the time of lay, *Salmonella* is more frequently isolated from the egg contents than from the eggshell, suggesting a contamination of the ovary or the upper part of the oviduct (magnum), and contamination of the developing egg before the shell deposition (Gast & Beard, 1990; Suzuki, 1994). In addition, there is no direct relationship between fecal carrier status and the presence of *Salmonella* in egg contents (Humphrey et al., 1991; Gast & Beard, 1992). *Salmonella* can indeed be isolated from the reproductive tract in the absence of intestinal contamination (De Buck et al., 2004).

Most of the authors conclude that the egg white is more often contaminated than the yolk, and they found that the magnum, where the albumen is deposited, is more contaminated than the other oviduct parts (Gast & Beard, 1990; Shivaprasad et al., 1990; Keller et al., 1995). On the contrary, other authors implicate the ovary, knowing that they observe contamination of the yolk or of the vitelline membrane (Gast & Holt, 2000a, 2001). Hopper and Mawer (1988) suggest that the low frequency of contaminated eggs laid by infected hens is related to the fact that infected and healthy follicles can coexist within the same ovary. Other authors assume that the high frequency of contamination of shell membranes could be linked to the contamination of the distal part of the oviduct (Okamura et al., 2001; De Buck et al., 2004). Thus, vertical contamination is possible and *S.* Enteritidis could contaminate the egg yolk, the vitelline membrane, the egg white, or the shell membranes, depending on the regions of the hen's reproductive tract colonized by these bacteria.

Keller et al. (1995) noted that, after inoculation via oral route, the level of contaminated eggs collected in the oviduct is higher (28.25%) than the rate of contaminated eggs collected at the time of lay (0 to 0.6%). These authors suggested that the inhibitory factors present in the egg content, such as antibodies, anti-bacterial proteins, protease inhibitors, and chelating proteins of essential elements, might control *S.* Enteritidis in the egg before it is laid. This hypothesis is supported by the gap between the large number of contaminated ovarian follicles in infected hens and the low frequency of infection of freshly laid eggs (Okamura et al., 2001).

Vertical contamination was also shown for two other serotypes, namely *Salmonella* Heidelberg (Gast et al., 2007) and *Salmonella* Pullorum (Wigley et al., 2001), with a low number of infected eggs after oral contamination for both. For other serotypes, such as *Salmonella* Typhimurium (Keller et al., 1995), *Salmonella* Infantis, and *Salmonella* Hadar (Okamura et al., 2001), although colonization of reproductive tissues has been shown, infected eggs were never found after oral or intravenous inoculation.

With regard to other pathogens, Sahin et al. (2003) showed vertical contamination by *Campylobacter jejuni*. This species, frequently isolated from hens, is responsible for a large number of childhood diarrhea in humans. After a high dose oral inoculation (10^7 cells/mL), these authors observed approximately 4–6% of infected eggs immediately after oviposition, while no contaminated eggs were detected one week after oviposition. Likewise, they did not find any contaminated eggs in a positive flock (i.e. *Campylobacter* found in hen's feces). These authors concluded that vertical contamination by this microorganism is possible, but that this event is rare and that it does not play a very important role in the contamination of eggs.

Vertical contamination is also possible for the avian influenza (AI) virus (influenza type A virus), which causes virosis in several species of birds, including chickens. Most strains are found primarily in the respiratory tract and digestive system, but some highly pathogenic strains are able to rapidly spread to all tissues of infected birds. Although egg production ceases in infected birds, eggs laid at the onset of the disease can be contaminated in all compartments (egg yolk, egg white, or eggshell). Through experimental infections, Swayne and Beck (2004) isolated the virus from eggs with frequencies of 85–100% of positive eggs. However, the same authors did not find the virus in the 9,930 eggs analyzed during the 1996–1998 AI episode, although it was present in the oviduct of hens (Swayne & Beck, 2004). For the 2001–2003 AI episode, the virus was not found in the eggs analyzed by Lu et al. (2004), while its presence was shown in cloacal contents. Moreover, a risk assessment conducted by EFSA concluded that the likelihood of viral transmission to consumers is very low (EFSA et al., 2018).

To conclude, all studies show that, if vertical contamination is possible, the contamination of the egg by this route mainly concerns *S.* Enteritidis, and it remains low and intermittent, given the low rate of contaminated eggs.

20.2.1.3 Horizontal Contamination

Horizontal contamination, also called exogenic contamination, corresponds to the contamination of the eggshell surface and the possible microbial penetration throughout the eggshell after oviposition. This route of contamination concerns a wide range of microorganisms, including *S.* Enteritidis but also other pathogenic bacteria or spoilage microorganisms. The horizontal contamination occurs through contact between eggs and hen's feces, or microorganisms from the environment (air, dust, soil).

The contamination of the shell egg varies depending on the contamination of the surface of the eggshell, which itself depends on the environment of the egg, at the time and after oviposition. The egg production practices, the geographic area, the hen's contact with eggs within the nest, the housing, management, and hygienic practices in the farm, in the packaging centers and during egg transport and storage; all of these factors potentially impacting the eggshell microbiota. The aerobic mesophilic flora on the egg surface varies from 10^2 to up to 10^7 CFU/egg, depending on the study (De Reu et al., 2008).

Regarding the types of contamination, there is a lot of old studies (before 1990), some focusing on hatching eggs, others aiming to compare the egg production systems, such housing systems or egg washing practices (De Reu et al., 2008). It is generally accepted that the dominant flora is Gram-positive bacteria due to their ability to resist the dry conditions on the eggshell surface. The genera commonly isolated are

Micrococcus, Streptococcus, Aerococcus, Staphylococcus, Enterococcus, and *Bacillus*. Gram-negative bacteria may also be present on the eggshell, such as *Alcaligenes*, *Pseudomonas*, and bacteria belonging to the *Enterobacteriaceae* family (De Reu et al., 2008). For this family, Musgrove et al. (2004) were able to isolate *E. coli* from eggshell, and different species of the genera *Yersinia, Enterobacter, Klebsiella, Serratia, Citrobacter,* and *Salmonella*.

Eggs may also be contaminated by yeast and fungi (De Reu et al., 2008). A recent study showed that the *Alternaria, Penicillium, Chaetomium,* and *Fusarium* genera were the most frequently fungi genera isolated on the eggshell surface (Tomczyk et al., 2018).

All the microorganisms present on the eggshell surface could potentially penetrate into the egg but, fortunately, the egg possesses an arsenal of defenses to fight bacterial invasion.

20.2.2 Egg Defense Against Microorganism Penetration and Invasion

Even if the eggshell is contaminated, the egg has different barriers that prevent the penetration and multiplication of microorganisms, from the surface of the eggshell to the egg white, then to the egg yolk, a rich compartment that allows bacterial growth. The barriers to penetration are, from the outside to the inside, the cuticle, the shell, the shell membranes, the egg white, and finally the vitelline membrane.

20.2.2.1 The Cuticle, Shell, and Shell Membranes

The cuticle is a mucoprotein envelope that represents the first line of egg defense as a physical barrier to microorganism penetration. The effectiveness of this protective layer is limited in time, due to its cracking under the effect of desiccation, or treatments that can impair it, such as washing (Vadehra et al., 1970; Tung et al., 1979; De Reu et al., 2006). The cuticle plugs the pore of the eggshell and reduces the microbial penetration into internal contents. Eggs with a thicker cuticle are more able to resist to microbial penetration (Bain et al., 2013). Moreover, several antimicrobial proteins have been identified in the cuticle (lysozyme, ovotransferrin, ovocalyxin-32, and ovocleidin-17) (Rose-Martel et al., 2012). De Reu et al. (2006) and Messens et al. (2007) showed, in experimental penetration studies, a correlation between a good cuticle deposition, i.e. a cuticle covering the entire surface of the shell, and the absence of *Salmonella* penetration through the eggshell.

The mineral portion of the eggshell, thereafter referred to as the shell, constitutes a calcified framework, which represents an inefficient physical barrier due to the possible passage of microorganisms through its pores. However, no link is observed between the number of pores and the penetration rate through the shell (De Reu et al., 2006; Messens et al., 2007; Chen et al., 2019). In addition, the shell contains lysozyme and ovotransferrin, which may play an important physiological role in protecting against penetration (Hincke et al., 2000; Gautron et al., 2001). Concerning the shell thickness, De Reu et al. (2006) and Messens et al. (2005) reported that it was not a factor influencing bacterial invasion, while Chen et al. (2019) showed that eggs with thin shells are more likely to be contaminated with *E. coli*. However, in this latter study, the shells studied were much thinner than in the previous two.

Shell membranes, i.e. the outer membrane, which adheres to the shell, and the inner membrane, which adheres to the albumen, are more efficient to prevent microbial penetration. They consist of glycoprotein fibers organized in networks functioning as anti-bacterial filters (Tung et al., 1979). They contain lysozyme, β-N-acetylglucosaminidase, and ovotransferrin, which antimicrobial activities increase the mechanical protection against the penetration of bacteria already trapped in the protein network (Hincke et al., 2000; Gautron et al., 2001). The inner membrane, although thinner, is more efficient than the outer membrane because of its greater richness in lysozyme (Vadehra et al., 1970; Hincke et al., 2000).

Most bacteria, yeast and fungi are able to pass through the shell and shell membranes, but De Reu et al. (2006) showed that motile Gram-negative bacteria penetrate most frequently the eggshell.

20.2.2.2 The Egg White

In addition to the eggshell (cuticle, shell, and shell membranes), the egg white is another line of the natural egg defenses. Various bacterial species may penetrate into the egg, but few will survive in the harmful egg white environment. Egg white is a saline aqueous solution of proteins (10.6%) containing carbohydrates and minerals at concentrations above the needs for bacterial growth (Cochet et al., 2021). However, egg white is very effective to deal against invading microorganisms. Hence, this biological fluid exhibits a wide range of antimicrobial activities supported by an arsenal of antimicrobial molecules, and physicochemical factors as well, in particular its heterogeneous structure, viscosity, and alkaline pH.

Depending on the type of microorganisms and environmental conditions, especially temperature, egg white could be bactericidal or bacteriostatic, or could allow weak growth. Gram-positive bacteria are more sensitive to egg white than Gram-negative species. Tranter and Board (1984) demonstrated the bactericidal activity of egg white against many Gram-positive bacterial species: *B cereus, B. megaterium, S. aureus, S. epidermidis, Brochothrix thersphacta, Kurthia zopfii, Micrococcus luteus,* and *Streptococcus faecalis*. Regarding Gram-negative bacteria, the same authors demonstrated that *Pseudomonas fluorescens* and *P. aeruginosa, Enterobacter aerogenes, Serratia marcescens, E. coli,* and *Salmonella* are able to survive in egg white. Notably, *S.* Enteritidis exhibits an exceptional ability to survive in egg white (Lock & Board, 1992; Clavijo et al., 2006; Gantois et al., 2009; Baron et al., 2016). All the studies highlighted the key role of temperature on the survival of the bacteria tested. For example, for *S.* Enteritidis, a bacteriostatic effect is observed at refrigerated temperature, while ambient temperatures are slightly growth permissive and, above 37°C and particularly at 42°C, which matches the laying hen's body temperature, a bactericidal activity is observed (Baron et al., 2016).

Egg white is a heterogeneous medium consisting of four physically separated layers (see Chapter 5). The chalazae layer, composed of spiral filaments extending from the egg yolk to both ends of the egg, contributes to maintain the yolk in a central position. The external liquid (thin) egg white is in direct contact with the shell membranes. The thick egg white, which is fixed to the two extremities of the egg, has the appearance of a gel. Lastly, the internal liquid (thin) egg white is located between the thick egg white and egg yolk. However, egg white is globally considered as a highly viscous medium (Lang & Rha, 1982), which together with the specific distribution of these layers of different rheological characteristics, may be considered as a physical barrier affecting the motility of flagellated bacteria, and thus limiting their dissemination throughout the egg, and their progression to the egg yolk that contains many nutrients. Moreover, proteins and minerals are probably not distributed uniformly within the egg white. For example, calcium and magnesium, partly bound to proteins, are heterogeneously distributed between the thick and thin egg white layers (Sauveur, 1988).

The pH of the egg white increases from 7.6 to 9.3 in a few days after oviposition, due to the loss of carbon dioxide through the pores of the eggshell. The increase in the pH value depends on the egg storage temperature conditions: the higher the storage temperature, the quicker the pH increase, and the higher the final pH value (Sauveur, 1988). This alkaline pH value is unusual for a biological medium and has potential consequences on bacteria that must maintain their intracellular pH around its optimum value (7.6–7.8) to preserve their proton motive force (pmf). The pmf is essential for the production of energy and for the maintenance of the biological functions of the bacterial cytoplasmic membrane. At alkaline pH, to preserve their pmf, bacteria accelerate the uptake of protons and block their export; they also suppress the systems responsible for high energy expenditure, such as those associated with flagella biosynthesis and chemotaxis (Maurer et al., 2005). Jan et al. (2013) have shown that *E. coli* lacks flagella in contact with egg white, and Baron et al. (2017) observed that *S.* Enteritidis down-regulated their flagella/motility-related genes. Yet, the lack of flagella could impair the migration of microorganisms throughout egg white and limit their access to nutrients.

Kang et al. (2006) showed that at 37°C, egg white can allow the growth of *S.* Enteritidis at pH 8, but not at pH 9. Tranter and Board (1984) showed that at 39.5°C, egg white has a bactericidal effect against *E. coli* at pH above 8.8, but not at pH 7.5. At 20°C, Messens et al. (2004) observed a one-$\log_{10}$ increase in *S.* Enteritidis growth at pH 8.2 compared to 9.3. Alabdeh et al. (2011) reported that the increase of pH from 7.8 to 9.3 potentiates the antibacterial effect of egg white on *E. coli* and *S.* Enteritidis at different temperatures, ranging from 37 to 48°C. Huang et al. (2019) hypothesized that the alkaline pH of egg white increases the *Salmonella* sensitivity to egg white antimicrobial molecules by inhibiting the bacterial systems phoP/phoQ and pmrA/pmrB, both involved in antimicrobial peptide resistance.

Numerous proteins and peptides have been identified in egg white (Mann & Mann, 2011), a number of them with putative or confirmed antimicrobial properties. Some have been commonly cited to explain the antimicrobial activity of egg white. The mechanisms of action of these egg white molecules include perturbation or damage of the bacterial envelopes, chelation of key elements to limit their availability for bacterial growth, inhibiting bacterial proteases, thus limiting bacterial spread. These proteins and peptides are more or less effective depending on the environmental conditions and the type of microorganisms. Some of the latter have developed resistance mechanisms to deal with these molecules.

Lysozyme is probably the most well-known egg white protein, also present in many other biological fluids and tissues. The best-known activity of lysozyme is its muramidase activity involved in the degradation of the peptidoglycan of Gram-positive bacteria, which could lead to cell death in low osmotic strength media. Gram-negative bacteria are generally resistant to lysozyme, because of the protection of the outer membrane that prevents the entrance into the cell of large molecules (Nikaido, 2003). In addition, some bacteria such as *Enterobacteriaceae*, the family to which *Salmonella* belongs, express lysozyme inhibitors in the periplasm. The latter is part of the resistance mechanism developed by bacteria to counteract the effect of the widespread lysozyme, and to protect bacteria in case of outer membrane disturbance and possible entrance of lysozyme into the periplasm (Callewaert et al., 2008).

Like other molecules, lysozyme is also able to disrupt the outer membrane of bacteria: it is the second mechanism of lysozyme activity, which is independent of its muramidase activity. The permeabilization of the outer membrane of *E. coli* has been showed by Pellegrini et al. (2000) and Derde et al. (2013). These last authors showed that lysozyme was able to cause reorganization of the lipopolysaccharides (LPS) of the outer membrane, leading to pore formation in this membrane. The pore formation was not observed in the inner membrane, but the membrane permeability was modified by the presence of lysozyme (Derde et al., 2013).

Finally, the induction of autolysins against *B. subtilis* (Ibrahim et al., 2001) and streptococci (Laible & Germaine, 1985) has also been attributed to lysozyme. The autolysins are peptidoglycan hydrolases whose induction may lead to the degradation of the peptidoglycan.

One of these mechanisms may prevail over the others depending on the microorganisms and the conditions. Under the alkaline conditions of the egg white, the muramidase activity of lysozyme does presumably not play a major role, because lysozyme is enzymatically inactive at high pH (Ibrahim et al., 1996). Recent studies showed that, in the egg white context, the outer membrane permeabilization of Gram-negative bacteria is probable. Baron et al. (2020) observed an induction of the gene encoding the lysozyme inhibitor protein in presence of egg white fractions (>10 kDa, containing lysozyme) and Wang et al. (2018) showed that *S.* Enteritidis survival in egg white is reduced by deletion of the lysozyme inhibitor gene.

Ovotransferrin, also called conalbumin, is another well-known egg white protein, widely distributed in physiological fluids and, like all transferrins, it is a metal-binding transport glycoprotein. Ovotransferrin is able to bind iron with extremely high affinity and can also chelate other divalent cations such as chromium, copper, manganese, zinc, nickel, cobalt, and cadmium, although with lower affinities (Tan & Woodworth, 1969). It is largely recognized that iron chelation by ovotransferrin is a key factor limiting bacterial growth in egg white (Legros et al., 2021). Considering the high concentration of ovotransferrin in egg white and the low amounts of iron (3.6–18 μM), it is generally concluded that all the iron present in egg white is bound to ovotransferrin (Legros et al., 2021). Thus, the iron-binding capacity of ovotransferrin creates an iron-deficient environment, which restricts bacterial growth. This conclusion is supported by numerous studies that showed that egg white bacteriostatic activity is entirely reversed by the addition of iron in sufficient amount for ovotransferrin oversaturation (Garibaldi, 1970; Lock & Board, 1992; Baron et al., 1997; Kang et al., 2006).

However, many bacteria are able to synthetize extracellular iron-chelating molecules, called siderophores, that can efficiently remove iron from transferrin for delivery to bacteria (Andrews et al., 2003), and therefore circumvent the iron restriction conditions encountered in biological fluids such as the egg white. Actually, studies with the egg white showed that the production of siderophores is important for bacterial survival, but is not sufficient to support growth without additional iron, even though the egg white contains sufficient iron (3.6–18 μM) to support bacterial growth (Legros et al., 2021). Besides, even if bacteria produce siderophores, their production in sufficient amount, or their import and use might be altered under egg white conditions (i.e. viscosity, pH, alteration of siderophore synthesis and uptake due to membrane disturbance). Moreover, egg white contains a siderophore-sequestering lipocalin protein, namely Ex-FABP (Guérin-Dubiard et al., 2006), which can inhibit the growth of both *E. coli* and *Bacillus subtilis* (Correnti et al., 2011), and also a strain of *S.* Enteritidis that produces only the enterobactin siderophore (Julien et al., 2020) under iron-limited *in vitro* conditions. Actually, Ex-FABP is able to bind some siderophores such as enterobactin and bacillibactin produced by *Enterobacteriacae* (*E. coli, S.* Enteritidis) and *B. subtilis,* respectively, but neither pyochelin and pyoverdine produced by *Pseudomonas aeruginosa*, nor salmochelin produced by some strains of *Salmonella* (Correnti et al., 2011). To date, the exact role of siderophores and Ex-FABP in egg white remains unclear and further experimentations are needed. However, the major role of ovotransferrin in preventing bacterial growth in egg white by creating iron-restricted conditions is unquestionable, even for *S.* Enteritidis, the most resistant bacterium dealing with egg white. Thereby, recent studies showed that the siderophore synthesis and iron-related genes are among the most highly induced mechanisms when *S.* Enteritidis is exposed to egg white (Baron et al., 2017; Huang et al., 2019; Qin et al., 2019).

Other antimicrobial mechanisms, independent of iron deprivation, are also described for ovotransferrin. The ability of transferrin to chelate divalent cations, which stabilize the outer membrane through neutralization of the strong negative charges of the lipopolysaccharide molecules, is suggested to have an effect on outer membrane permeabilization (Ellison et al., 1988). The effect on the bacterial membranes requires a direct interaction with the protein (Valenti et al., 1986). However, it remains unclear whether ovotransferrin has the capacity for direct access to the bacterial membrane within the egg, due to the egg white viscosity, or to remove cations from bacterial outer membranes because of the high level of divalent cations within the egg white, which may support the integrity of the bacterial outer membrane (Legros et al., 2021). However, Aguilera et al. (2003) showed that ovotransferrin can permeabilize the outer membrane of *E. coli*, access the inner membrane and cause dissipation of the pmf involved in the cytoplasmic membrane function. Likewise, Baron et al. (2014), by using an egg white model medium, showed that ovotransferrin provokes a disruption of the membrane potential in *B. cereus*. This mechanism is believed to be involved in *B. cereus* cell death in egg white. Baron et al. (1997) reported a bacteriostatic effect of ovotransferrin in the same egg white model medium against *S.* Enteritidis. Environmental conditions can modulate the antimicrobial effect of ovotransferrin. Thus, Baron et al. (2014) observed that the depolarization of the membrane of *B. cereus* by ovotransferrin in egg white model medium is stronger at pH 9.3 than at pH 7.8. The viscosity of egg white could also affect the direct interaction of ovotransferrin with bacterial membranes. Moreover, ovotransferrin (or iron-ovotransferrin complexes) are probably not distributed uniformly within egg white due to the heterogeneous structure of the latter (Li-Chan & Nakai, 1989), which can influence both the contact with the bacterial membrane and the bacterial ability to remove iron from ovotransferrin via the use of siderophores.

In conclusion, the well-known iron-deprivation effect of ovotransferrin largely explains the bacteriostatic effect of egg white. However, several studies suggest a direct interaction of ovotransferrin with microorganisms, leading to the permeation of the outer membrane of Gram-negative bacteria, and to the disturbance of the inner membrane via the dissipation of the pmf, thus enhancing bacteriostasis or provoking the cell destruction, depending on the bacterium. The role and the precise mechanism of this altered membrane activity need to be clarified and further investigated under egg white conditions.

Egg white contains several vitamin-chelating proteins, namely riboflavin-binding protein, thiamine-binding protein, and avidin (binding biotin). By chelating compounds essential for life, these proteins are considered part of the defense system of egg white. However, a bacteriostatic effect has been shown only for avidin (Banks et al., 1986). Since many bacteria are able to synthesize biotin when needed (Beckett, 2007), the biotin-deprivation activity of avidin is limited. Recent studies in the egg white context showed that biotin-synthesis-related genes (Baron et al., 2017, 2020; Huang et al., 2019) and proteins (Qin et al., 2019) are strongly induced by *S.* Enteritidis after an exposure to egg white. These results indicate that *S.* Enteritidis synthetizes biotin in response to the poor biotin availability in the egg white, resulting from its chelation by avidin. Moreover, Raspoet et al. (2014) and Wang et al. (2018) observed a reduced survival ability in egg white for biotin-synthesis-genes knock-out mutants of *S.* Enteritidis. Taking together, these studies suggest that the presence of avidin in egg white submits bacteria to biotin deficiency, which could have an antibacterial effect if bacteria are not able to synthetize biotin.

The presence of anti-proteases in the egg white is believed to have a role of inhibiting the spread of bacteria. By degrading the host's proteins, microbial proteases facilitate the assimilation of nutrients by pathogens, limit the effectiveness of the host immune defense, and induce tissue damage, which facilitates the pathogen dissemination (Réhault-Godbert et al., 2007). Ovostatin, also called ovomacroglobulin, is a minor protein of the egg white, which is known to inhibit various proteases produced by *Serratia marcescens*, *Pseudomonas aeruginosa*, and *Bacillus stearothermophilus* (Molla et al., 1987). Cystatin, another minor egg white protein, targets cysteine proteases and exhibits antimicrobial activity toward *S.* Typhimurium (Nakai, 2000). Ovomucoid and ovoinhibitor, both major egg white proteins, exhibit a wide spectrum of anti-protease activities, but to date, their antimicrobial role is not well documented. Although the egg white protease inhibitors have anti-protease activities with some value for therapeutic applications, their involvement in the natural defense system of egg white against bacterial contamination has not yet been demonstrated, and will probably be minimal (Baron et al., 2016).

Ovalbumin-related protein X (OVAX), belonging to the serine protease inhibitor family (serpin family), was detected in egg white (Guérin-Dubiard et al., 2006). Surprisingly, OVAX does not exhibit any anti-protease activity, but its antimicrobial activity has been described against *Listeria monocytogenes* and *S.* Enteritidis (Réhault-Godbert et al., 2013). The same research team identified novel antibacterial proteins, present at low abundance in the egg white: VMO-1 (Vitelline Membrane Outer layer protein 1), pleiotrophin, and beta-microseminoprotein-like (Guyot et al., 2016). VMO-1, initially identified in the outer vitelline membrane of the chicken egg, exhibits antibacterial activity against *Listeria monocytogenes*, but not against *S.* Enteritidis. Pleiotrophin and beta-microseminoprotein-like possess antibacterial activities against these two foodborne pathogens (Guyot et al., 2016). Antibacterial activities of these four proteins were revealed using radial diffusion assays with high protein concentration (around 100 g/L). This strategy is interesting for discovering new therapeutic agents, but the involvement of these proteins in the egg white context (i.e. alkaline pH, low concentration) remains undefined.

In addition to proteins, the egg white contains natural antimicrobial peptides. The avian β- defensin 11 (AvBD11) and gallin, both belonging to the defensin family, have been identified in the chicken egg white (Mann, 2007). Defensins are small polypeptides (< 10 kDa) able to bind to the microbial cell membrane, and to provoke efflux of essential ions. The AvBD11 inhibits the growth of *S.* Enteritidis, *S.* Typhimurium, *E. coli*, *Listeria monocytogenes*, and *S. aureus* by radial diffusion assays (Hervé-Grépinet et al., 2010). Gallin inhibits the growth of *E. coli* in phosphate buffered salt (PBS) (Gong et al., 2010).

Recent studies have highlighted the role of egg white peptide fractions. Baron et al. (2020) showed that the <10 kDa egg white filtrate exerts a bactericidal effect toward *S.* Enteritidis at 45°C. Huang et al. (2020) showed that a proteinase K treatment cancels the antibacterial activity of a <3 kDa egg white filtrate on *S.* Enteritidis at 37°C. Cochet et al. (2021) identified novel antimicrobial peptides from the <10 kDa egg white fraction. Taking together, these results suggest that peptides smaller than 3 or 10 kDa have an active role in egg white defenses, probably enhanced by the pH and temperature conditions.

Moreover, the egg white probably contains antimicrobial proteins or peptides that have not yet been identified. For the past decade, powerful analytical methods have revealed a great number of proteins in egg white, with putative antimicrobial roles based on functional homologies. However, most of these proteins are not yet been investigated experimentally. For example, the protein TENP, revealed for the first time in hen egg white in 2006, has strong homology with the Bactericidal Permeability Increasing protein/Lipopolysaccharide Binding Protein (BPI/LBP) family (Guérin-Dubiard et al., 2006). As this family is known to inhibit bacterial growth due to the ability to disrupt bacterial membranes (Elsbach & Weiss, 1998), TENP is believed to have a similar role. However, to date, this assumption has never been confirmed for this protein. Purification from egg white or overexpression in recombinant microorganisms is needed before testing its predicted function and clarifying its role in egg white defense.

It is important to keep in mind that antimicrobial activity of molecules, as demonstrated *in vitro*, may differ in the egg white context (viscosity, alkaline pH), and depending on the temperature and the potential interactions that probably occur between different egg white molecules. For example, the already described cooperation between lysozyme and lactoferrin (Facon & Skura, 1996) suggests that the same type of cooperation could occur between lysozyme and ovotransferrin. By chelating divalent cations, ovotransferrin could destabilize the outer membrane of bacteria and allow lysozyme to access the peptidoglycan. Park et al. (2006) have shown that animal mucins inhibit the muramidase activity of lysozyme. Hence, it is reasonable to imagine that such an effect may occur in egg white, where ovomucin and lysozyme are closely associated via electrostatic interactions. Moreover, the potential interactions between antimicrobial egg white molecules could be modulated by alkaline pH and the temperature encountered.

In conclusion, egg white is a complex harmful biological fluid with multicomponent and multifactorial antimicrobial effects, adapted to microbial diversity. Concerning *S.* Enteritidis, the ablest to survive in the egg white, the inhibitory mechanism involved is iron restriction imposed by ovotransferrin, and disturbance of the bacterial envelopes through the action of some antimicrobial molecules and peptides. This mechanism matches with the results of numerous studies focused on the *S.* Enteritidis genes essential to its survival in egg white (Clavijo et al., 2006; Kang et al., 2006; Gantois et al., 2008; Raspoet et al., 2014), or genes up-regulated after egg white exposure (Baron et al., 2017; Huang et al., 2019). Indeed, the genes highlighted by these studies are mainly iron-related genes and genes implicated in cell envelop repair.

20.2.2.3 Vitelline Membrane

The vitelline membrane, surrounding the yolk, separates this latter from the egg white, preventing the transfer of bacteria from the egg white to the egg yolk. It is composed of two layers, namely the inner and the outer vitelline membranes. The latter is connected to the chalazae that contribute to maintain the yolk in a central position. The inner membrane is composed of glycoproteins, and the outer membrane contains numerous proteins that it shares with egg white (ovalbumin, ovotransferrin, lysozyme, and ovomucin), together with some others that are more specific, such as the vitelline membrane outer layer proteins I (VMOI) and II (VMOII), and several ATPases and proteases (Mann, 2007). A more detailed description of the vitelline membrane is available Chapter 3.

Most of the identified proteins in the vitelline membrane have unknown functions, but the VMOI and VMOII have been shown by Guyot et al. (2016) and by Hervé-Grépinet et al. (2010), respectively, to have antibacterial activities. The vitelline membrane represents the last obstacle to the migration of the microorganisms to the egg yolk (Chen et al., 2005), where bacterial multiplication is possible and could lead within a few hours to a high level of contamination at permissive temperatures (Růzicková, 1994; Schoeni et al., 1995; Gast & Holt, 2000a, 2001; Gumudavelli et al., 2007).

20.2.3 Egg Content Contamination

In spite of protective barriers such as the cuticle, eggshell, and shell membranes, microorganisms are able to move through the surface of the egg, especially at the large end, where the shell membranes separate to form an air cell (Berrang et al., 1999). After oviposition, temperature changes between oviduct (42°C) and ambient temperature during storage and transport, combined with high humidity, could favor water condensation on the eggshell surface, and thus penetration by retraction of the egg contents with the absorption of water and microorganisms present on the shell (Miyamoto et al., 1998; De Reu et al., 2008). However, even if penetration is possible, the survival, multiplication and migration into the egg contents is not possible for all microorganisms, and the arsenal of antimicrobial properties of the egg white exerts a high selection pressure on microorganisms. Whereas the dominating flora present at the surface of the eggshell is composed of Gram-positive bacteria, the contamination of the egg contents involves mainly Gram-negative bacteria.

Among them, *S.* Enteritidis is the ablest to survive and multiply into the egg contents, but other *Salmonella* serotypes, such as *S.* Typhimurium have the ability to penetrate and survive as well (Miyamoto et al., 1998). *Salmonella* contamination could lead to toxi-infection risks for consumers (salmonellosis), and the *S.* Enteritidis serotype is the most frequently associated with egg-borne outbreaks. Nevertheless, other serovars may also be involved (Threlfall et al., 2014). Furthermore, other bacteria could be found in egg contents such as *Pseudomonas, Proteus, Alcaligenes, Escherichia, Aeromonas* and, to a lesser extent, *Flavobacterium, Serratia, Bacillus, Arthrobacter, Micrococcus* and *Streptococcus* (Techer et al., 2014). These bacteria are involved in egg spoilage rather than egg-borne diseases but could be responsible for high economic losses for the industry. These bacteria could provoke off-odors, mainly rotten odors, and color changes with fluorescent green, red, pink, blue, and black rot depending on the microorganism involved (Techer et al., 2014). Yeast and fungi, especially the genera *Penicillium, Alternaria, Fusarium,* and *Mucor,* have also been identified in egg contents (Tomczyk et al., 2018). These species are associated with egg spoilage, but attention should be paid with the presence of *Fusarium* spp. due to the potential risk of mycotoxin production if egg storage conditions allow their growth (Tomczyk et al., 2018).

It is important to note that, following contamination, the survival or multiplication in egg contents varies depending on the microorganisms, the egg compartment infected (shell membranes, egg white, egg yolk), the integrity of the egg internal structures and contents, and the temperature. Bacterial load of eggshells is important in terms of risk of penetration into the egg contents. A limit of 5 log CFU/egg surface is considered as an acceptable hygienic quality criterion (De Reu et al., 2008).

20.2.4 Factors Impacting Egg Contamination

20.2.4.1 Factors Related to the Hens

The leading potential sources of introduction of microorganisms, and especially *S.* Enteritidis, into laying flocks are the replacement of the hens themselves. Therefore, the breeder flocks may constitute a primary source of contamination, and a strict control of the microbial status of these flocks is necessary.

Some authors have studied the feasibility of genetic selection to improve the resistance of hens, or eggs, to *Salmonella* colonization, supporting by the fact that *S.* Enteritidis presence has been shown to differ between genetic lines of egg-laying hens (Sadeyen et al., 2006; Gast et al., 2019). Genes associated with high resistance to *S.* Enteritidis presence have been identified (Calenge et al., 2011) but, to date, none of these genes have been demonstrated to have a major effect. Concerning the eggs, selection to improve egg quality is an important strategy to satisfy consumer demands, but few studies have focused on the resistance of eggs to bacterial contamination. One study demonstrated that the antibacterial potential of egg white is moderately inheritable (Sellier et al., 2007), but the genetic component of the egg white antimicrobial properties remains to be explored.

The laying cycle of the hen begins at 16–20 weeks of age in commercial flocks and ends commonly at 70–75 weeks, due to expensive costs to maintain older hens with lower egg laying rates. Although the hen age affects shell characteristics, it seems to have no significant influence on the eggshell penetration and on the egg contamination (Messens et al., 2005; De Reu et al., 2006). The Haugh unit score, which is a logarithmic expression of the height of the thick egg white surrounding the yolk, and the strength of the vitelline membrane, significantly decrease in older hens when compared to younger ones (Perić et al., 2017; Marzec et al., 2019). Moreover, egg white pH was found to be 7.8 for fresh eggs from younger hens *versus* 8.2 for those from older hens (Perić et al., 2017). However, although egg white quality is influenced by hen's age, to date, there is no literature available on the influence of the hen's age on egg white antimicrobial or anti-*Salmonella* properties. In the future, the typical laying cycle is expected to extend to 100 weeks, thanks to improved genetics, hen management, nutrition, and disease prevention. The benefits of this extension would improve the efficacy and sustainability of the egg production system, necessary to cope with the human population growth and the increasing demand for eggs. Nevertheless, the impact of this evolution should be clarified with regard to antimicrobial egg defense and, consequently, to egg safety.

Vaccination is also a strategy available worldwide, and several vaccines have shown efficacy against *Salmonella* colonization of hens and to successfully reduce egg contamination (Methner, 2018). Since 2006, the Commission Regulation 2006/1177/EC mandates the use of vaccination programs against *S.* Enteritidis in the European member states if its prevalence exceeds 10% in poultry flocks.

Factors causing stress to laying hens are known to play a role in the susceptibility of hens to contract salmonellosis. The level of stress observed at the egg-laying peak, appears to be responsible for a decrease in immune defenses, and on the influx of macrophages to the reproductive organs. Wigley et al. (2005) observed heavy contamination of the ovaries and high release rates of *Salmonella* in the feces of hens artificially contaminated with *S.* Enteritidis or *S.* Pullorum serotypes during this peak. The molting is also a stress for the hen: the egg production stops, the genital tract regresses, the feathers fall. The molting, which allows the regeneration of the tissues of the oviduct after a period of intense laying, is often caused by insufficient nutrition, in the wild or in commercial flocks, where molting is forced by using feed with low energy content (the past practice of feed restriction is now illegal in most countries). These limiting feeding conditions provide, in the digestive system of the hen, a favorable environment for the development of *Salmonella* (Ricke, 2003).

20.2.4.2 Housing Environment

For animal welfare reasons, a transformation of the housing system is progressively initiated worldwide, with differences between countries and regulations. Conventional cages are prohibited in the European Union (EU) since 2012, in favor of furnished cages or alternative houses systems. Among them, floor housing or barn, aviaries, free-range systems, and organic production have been developed and spread to member states. Around the world, a number of countries are considering to ban battery cages, and some are in a transition phase, from conventional cages to enriched cages or alternative housing systems, or have local bans on battery cages, such as Canada, the USA, Australia, and New Zealand (see Chapter 2). Furnished cages provide at least 750 cm^2 per hen (in comparison to 550 cm^2 for conventional cages), a nest, a dust bath, and 15 cm perch per bird. Non-cage systems provide 1,111 cm^2 per hen, nest, perch, a litter area, and platforms with floors at different heights (aviary) or at one height (floor or barn housing). When hens have access to outdoor runs, the systems are called free-range systems (De Reu et al., 2008).

Literature is abundant on the effect of the type of housing system on egg quality, but results are conflicting. This is probably due to the difficulty of isolating the impact of the system tested from other related effects, such as age of the building and equipment, feeding, cleaning practices, laying behavior, geographical area, farm management, and egg collection. Beyond all of these systems, the hen house environment plays an important role. Thus, there is a correlation between the contamination of the eggshell and the contamination of the air, itself related to the dustiness. The latter varies according to the activity and behavior of the hens, the use of cages and their type (conventional or furnished), the use of aviaries, the equipment (nest, perch, scratching area), litter and feed used (De Reu et al., 2005).

Concerning the cage system, some authors found no difference in eggshell contamination between eggs from conventional *versus* furnished cages. However, some others observed a slight difference due probably to the higher bacterial load on the eggs collected outside the nest in the furnished cages, compared to the conventional cages (De Reu et al., 2005, 2008; Mallet et al., 2006; Huneau-Salaün et al., 2010; Kone et al., 2013).

Numerous authors found lower bacterial contamination on eggshell surfaces from cage systems compared to alternative housing systems, where the probability of egg contact with feces or bedding material is increased (De Reu et al., 2005; Mallet et al., 2006; Huneau-Salaün et al., 2010). De Reu et al. (2005) showed much higher total counts of aerobic bacteria in cage-free systems compared to cage systems, especially on eggs laid on the floor of the aviaries, with differences less pronounced in experimental units than in commercial farms. This was not confirmed by Huneau-Salaün et al. (2010) who found no difference in the load of aerobic bacteria on the eggshells in furnished cages compared with cage-free systems. With regard to Gram-negative bacteria, lower contamination was found on eggs from cage-free systems, in experimental and in commercial houses (De Reu et al., 2005). Concerning cage-free systems, no difference was found for eggshell contamination between aviary and floor systems, but differences were found between farms, mainly attributed to differences in farm design and management (De Reu et al., 2008).

The outdoor access in the free-range systems increases the risk of hen's interaction with wild animals and raises the question of *S.* Enteritidis contamination of the flocks and the probability of production of *Salmonella*-contaminated eggs. The transition from confined housing to free-range is a recent situation. The studies examining *Salmonella* recovery from free-range *versus* confined housing are scarce, due to the limited numbers of commercial free-range flocks in the world. Investigation is very difficult because free-range systems can vary from housing hens with access to extensive pasture areas or less outdoor space, and with the different behaviors of hens within the same farm, with high- and low-range user hens. Among these studies, all conducted in the EU where free-range systems are more prevalent, most found little difference between free-range *versus* no free-range systems (Holt, 2021).

Beyond the housing systems, a seasonal influence on the eggshell contamination is highlighted. The bacterial contamination is lower in the winter period, probably due to the lower temperature that restricts the bacterial growth in comparison to the summer period (De Reu et al., 2005; Mallet et al., 2006).

Some studies have investigated the influence of the housing system on eggshell quality, to better measure the impact of the transition to alternative systems. Depending on the authors (Rossi & De Reu, 2011), no or some slight differences were reported for eggshell thickness and strength of eggs between cage and cage-free systems, with thicker shells observed in free-range and organic eggs. More interestingly, some authors (Rossi & De Reu, 2011) have studied the influence of the housing systems on the occurrence of cracked eggshells, increasing the possibility of bacterial penetration inside the egg contents. Higher levels of cracks have been observed in furnished cages, with great variations between farms, mainly attributed to the design of the nest, the egg accumulation in the egg saver line, or the low frequency of egg collection. However, at the market and retail levels, no significant differences were found between eggs from cage and alternative systems, due to the egg candling and sorting carried out in the packaging centers.

Data on the influence of the housing system on egg contents and structure mainly concern the height (Haugh unit) and pH of the egg white. These data are very conflicting (Rossi & De Reu, 2011) and, in our opinion, they do probably reflect more on differences in egg collection, in egg storage time and temperature, and in retail turnover in the case of market studies (for organic eggs, for example).

The cleanliness of layer houses is important because microbial load on the surface of the shells is linked to the level of contamination of the housing environment of the hens (De Reu et al., 2005). As the hens use the house throughout their laying cycle, a high level of soiling (manure, feed residues, dust, microorganisms) can be expected. Moreover, the possibility to thoroughly clean and disinfect the layer houses only comes during the all-out phase between flocks. *Salmonella* can survive in the rearing environment and promote contamination between two successive flocks. Davies and Breslin (2003) showed the persistence of *S.* Enteritidis in the environment of a farm, 26 months after the removal of Salmonella-positive birds, while the buildings were empty. The cleaning and disinfection of the buildings between flocks of laying hens is now almost systematic, but farmers should take the advantage of this hen-free phase to create a *Salmonella*–free environment before introducing new hens. The cleanability of building and equipment could be influenced by the housing system. However, regardless of housing system, an effective disinfection requires to be preceded by effective cleaning. The cleaning phase usually consists in a dry cleaning for dust and organic matter removal, followed by a wet cleaning. The building and the equipment such as feeders and cages are then disinfected, once or twice. The disinfectant may be applied by a steam treatment, i.e. thermo-nebulization, or by a spray treatment, i.e. pulverization. However, it must consider the difficulty to eradicate bacteria under field conditions, especially for *Salmonella* that can still be found in layer houses after cleaning and disinfecting (Davies & Breslin, 2003). Some bacteria are able to adhere to the building, industrial or equipment surfaces, and to form biofilms constituted by a strongly adherent cell community attached to a surface and embedded in an extracellular polymeric matrix. The sessile life-style is a survival strategy allowing bacteria to deal with fluctuating conditions encountered in a natural or industrial environment, such as cleaning and disinfection. The bacteria present in a biofilm are difficult to eradicate due to their resistant phenotypes, and they are a source of persistent contaminations in the food industry. Biofilms may play a crucial role in the survival of *Salmonella* in poultry houses. The best strategy to eradicate biofilms is to prevent their formation, using well-cleaning equipment associated with effective cleaning and disinfection measures. Once the biofilm is established, the most effective measures are the use of mechanical action to disrupt the biofilm matrix, and to make the microorganisms accessible to chemicals. Numerous strategies to combat *Salmonella* biofilms in the poultry industry have been tested and continue to be studied, such as the use of enzymes, surfactants, lactic acid bacteria spread on building surfaces, the use of phages, natural antimicrobial compounds, and nanomaterials (Merino et al., 2019). Controlling the effectiveness of cleaning and disinfection is also crucial, and it is essential to monitor the surface flora and, more specifically *Salmonella*, by microbiological analysis after the disinfection procedures and throughout the life of the flocks.

The contribution of flies (Holt et al., 2007), rodents (Lapuz et al., 2007), red mite (Valiente Moro et al., 2007) as a vector of *S.* Enteritidis in farms is well-documented. Preventing the introduction and multiplication of pests by effective pest management measures is therefore critical.

20.2.4.3 Feeding

With the end of the routine use of antibiotics in poultry production, feed supplements have been developed to eliminate or prevent the colonization of the hen's gut by pathogens. A number of different approaches for reducing the *Salmonella* colonization in poultry have been explored thus far, but with various degrees of success and with many drawbacks. These approaches include feeding chickens with organic acids, prebiotic, competitive exclusion bacteria, phytogenic additives, or bacteriophages.

Organic acids, often referred to as fatty acids (volatile or not), carboxylic acids or other weak acids, are naturally-occurring (hence organic) compounds with acidic properties (Broom, 2015). Due to their antimicrobial properties, organic acids have been used for a long time as an effective means of reducing or controlling microbial contamination of feed and raw materials, and to reduce pathogens like *Salmonella* in the food chain. Various solutions based on short-chain organic acids (including propionic, butyric, and formic acids), and/or mixed with medium-chain organic acids (caproic, caprylic, capric, and lauric acids) or derivatives (esterified forms) are commercially available for their antibacterial properties, and are frequently used in poultry production (Parry, 2006). Efficacy of these compounds against pathogens can vary according to their nature, their mode of administration (water/feed), their concentration in the feed, and other parameters such as their bioavailability in the digestive tract of the animals. As it is generally the case for organic acids, Gram-negative bacteria (i.e. *E. coli*, *Salmonella*) are much more inhibited than Gram-positive bacteria. However, some acids, such as lauric acid or esterified form (monolaurate), appear as very effective on *Clostridia*. Benzoic acid, frequently used in poultry, has been reported as having an important role in lowering numbers of many pathogenic bacteria such as *Campylobacter jejuni*, *Escherichia coli*, *Listeria monocytogenes*, and *Salmonella enterica* (Friedman et al., 2003).

Another strategy is to prevent the development of *Salmonella* in the digestive system of the hen by adding pre- or probiotics to the feed. Prebiotics are non-digestible food components supposed to improve host health and to serve as a barrier to pathogen colonization, by selective stimulation of certain bacteria already resident in the digestive tract. Most of the prebiotics traditionally used are carbohydrates, including mono-, di-, oligo-, and polysaccharides. In the last decades, development of gut microbiome sequencing, metabolomics, and transcriptomics studies have allowed to identify the microbial populations responding to prebiotic administration and better understand their impact on *Salmonella* colonization. These approaches led to the exploration of novel sources of prebiotics to limit *Salmonella* in poultry production, and their use and efficacy by the poultry sector is always a subject of intensive research (Ricke, 2018). The use of probiotics is also possible, and search of living microorganisms able to confer health host benefits and to inhibit pathogens in poultry has increased due to the higher demand for antibiotic-free poultry farming. A variety of microorganisms has been studied, including *Bacillus, Bifidobacterium, Enterococcus, Lactobacillus, Lactococcus, Streptococcus,* and yeast species (especially *Saccharomyces*). Since 1990, commercially available probiotic mixtures have shown anti-*Salmonella* effects *in vitro* and *in vivo* (Carter et al., 2009), and are used in poultry farming to reduce *Salmonella* colonization. The selection of probiotic strains, the design of the mixture supplements, and their practical application in poultry farming are still the challenge. A better scientific understanding of probiotic mechanisms should improve the effectiveness of probiotics against colonization by pathogens. The combination of pre- and probiotic administration is also possible, and feed supplements appear now as an integral part of *Salmonella* control strategy.

Plant-derived substances, or phytogenic additives, have been used for centuries for the purposes of flavoring, as well as for their beneficial effects on feed intake stimulation (appetizing effect), digestion through digestive secretion stimulation, and gut protection through antioxidant, anti-inflammatory, antimicrobial, anthelmintic, and antiviral effects. One class of phytogenic additives that has recently received a large amount of media attention is essential oils, as many of them present these different properties (Dey et al., 2017; Huang & Lee, 2018; Ruvalcaba-Gómez et al., 2022). Natural essential oils are generally obtained by distillation and are not simple compounds, but complex mixtures of various volatile, aromatic molecules. Factors such as plant species, geographic origin, ecological factors and climatic conditions, harvesting time, part of the plant used, and method of extraction, all affect the chemical composition of essential oils, and also their properties. This variability complicates their assessment and application (Zeng et al., 2015). For instance, it is difficult to get a constant concentration in active compounds, or to check their stability through a feed production process. For these reasons, it is usual to supplement poultry feeds with pure phytogenic molecules, also called nature-identical molecules, meaning they are produced synthetically, but are chemically identical to their natural counterparts. Generally, high antimicrobial activities are obtained from phenolic compounds like carvacrol and thymol (from oregano and thyme), and cinnamaldehyde (from cinnamon). Essential oils and aromatic plants are well known to exert antibacterial, antifungal, and antiviral activity using *in vitro* experiments. In general, Gram-positive bacteria appear to be more susceptible than Gram-negative bacteria, but many studies have shown the antimicrobial activity of certain essential oils toward different strains and serotypes of *Salmonella* or *E. coli.* Kim et al. (1995) reported that carvacrol showed the most potent antibacterial activity, compared to citral and geraniol, against *Salmonella* Typhimurium, including a rifampicin-resistant strain. They reported with carvacrol minimum inhibitory and bactericidal concentrations (MIC and MBC) of 250 μg/mL for the tested strains. Trans-cinnamaldehyde, another compound of interest, supplemented for 66 days at 1 and 1.5% to 40-wk and 25-wk old layer hens, has shown to reduce *Salmonella* Enteritidis on the eggshell and in the yolk, without causing any harmful effect on the growth, egg production, and consumer acceptability of eggs (Galiş et al., 2013). Vicente et al. (2007) tested two levels (18 and 36 ppm) of capsaicin supplementation on hens challenged with *S.* Enteritidis on day 25. They reported that *Salmonella* liver and spleen invasion was reduced when hens were supplemented with the higher level (36 ppm). Capsaicin also increased the deposition of red pigment in egg yolk (Vicente et al., 2007).

In brief, the use of essential oils to reduce the colonization of the digestive tract and the fecal excretion of *Salmonella* in poultry farms have been largely documented in different studies (Andoni et al., 2020; Ruvalcaba-Gómez et al., 2022). However, even presented as promising compounds *in vitro*, comparing the effectiveness of the essential oils, especially *in vivo*, is a difficult task due to the variability of chemical properties of the essential oils, various environmental and experimental conditions, and variability of responses of the animals. Knowledge of their use in poultry nutrition is still insufficient and requires further research to clarify their mode of action, as well as the appropriate supplementation level, and their interaction with feed ingredients for desired effects. Strategies to combine organic acids and essential oils for their synergistic or additive activities, or to use encapsulation technology that allows a protection of active compounds from degradation or rapid absorption before reaching the target area within the gastro intestinal tract of animal, also represent promising avenues that have to be investigated further.

Bacteriophages (phages) are viruses that specifically target and infect prokaryotes (bacteria) or archaea. They are the most abundant organisms on earth, with an estimated total number of phage particles in the biosphere approximating 10^{31}, which is 10 times more than the estimated number of bacterial cells (Żbikowska et al., 2020). They are ubiquitous and can be found in all habitats colonized by bacteria, such as water, plants, and food. They are recognized as important components of the natural microbiome of humans and animals. According to their interactions with bacteria and their life cycle, they are classified as lytic (virulent, productive) and/or lysogenic (temperate, dormant). Only the lytic bacteriophages are suitable for the application such as phage therapy in poultry. Phage therapy includes their use as anti-bacterial agents, as feed additives, or their use in poultry prior to slaughter for prevention, treatment, and control of bacterial infections. Even if literature is abundant on the search, identification, and characterization of lytic bacteriophages against *Salmonella* spp., few products, i.e. three at the time of writing this book, are commercially available in some countries for feed application (Żbikowska et al.,

2020; Ruvalcaba-Gómez et al., 2022). This can be explained by regulatory limitations, especially in the EU, or other aspects such as stability, resistance emergence, or consumer acceptance that have to be addressed before the common use of this great potential technology.

20.2.4.4 Collection, Sorting, and Packaging

Concerning egg collection on-farm, sorting, and packaging, the practices and regulations are quite different between countries (see Chapter 18). For example, the processing and handling of eggs after oviposition are quite contradictory between on the one hand, the US practices that promote immediate chilling, washing, and in-shell egg pasteurization, and on the other, the EU practices and regulations, based on minimal intervention principles.

Rapid cooling of shell eggs, in the early hours after oviposition, has demonstrated benefits in reducing *Salmonella* risk. Thus, if *S.* Enteritidis is present on the eggshell or in the egg contents, the rapid cooling and maintenance of the cold chain until delivery allow to inhibit bacteria penetration, migration, and proliferation. Indeed, Gast et al. (2007) showed that a rapid cooling at 4°C within 36 h of oviposition avoids the penetration of *Salmonella* through the vitelline membrane. However, rapid cooling process for shell eggs is not currently used worldwide. For example, the USA requires shell eggs to be placed at/or below 7°C after egg collection, but no temperature is defined for the products themselves (eggs), and no time is fixed to reach this refrigeration temperature. Thus, depending on the starting temperature and the cooling system, the time for the eggs to reach 7°C could be several days. However, the technology for a rapid cooling of shell eggs using cryogenic CO_2 is patented (Curtis et al., 2000). In Europe, regulations stipulate that eggs should not be refrigerated before sale to the final consumers, in order to limit the risk of temperature changes throughout the supply chain, and to avoid water condensation on the eggshell surface. These differences of practices and regulations between countries all over the world, although they may cause some confusion, have probably their origin in the climatic conditions (temperature, humidity) of each country, but also in the supply chain organization.

Nevertheless, it is generally accepted that everything should be done to avoid water condensation on the eggshell surface, in order to avoid the risk of penetration and multiplication of microorganisms. Water condensation generally occurs when chilled eggs are transferred to a higher temperature environment and depends on the relative humidity of that environment as well. Thus, it is important to optimize the control of stable relative humidity and temperature, from egg collection on the farm, to transport, sorting, storage, distribution, and delivery to the consumers. A visual check for the absence of water condensation on the eggshell surface, at each step of egg processing, might be a good idea. In 2009, The EFSA opinion stated that cooling of table eggs below 7°C has a beneficial effect to limit the growth of pathogens, but disruption of the cold chain is one factor that increases the risk of condensation, and this could favor bacterial penetration into the egg. Maintaining a constant temperature throughout the supply chain is the key point and the challenge in terms of logistics.

During egg collection and packaging, it is also essential to preserve the eggshell integrity. In packaging plant, damage to the shell must be avoided during loading and unloading of the eggs.

Sorting and candling should remove eggs with external and internal quality defects, i.e. leaky, soiled, or cracked eggs. The elimination of cracked eggs, the contents of which may spill, and cleaning and disinfection of machines are important to avoid the contamination of intact eggs. However, a slight shock can cause a micro-crack in the shell structure (less than 5 mm in length), difficult to detect. Yet, these micro-cracks can compromise the effectiveness of the shell in protecting against contamination by pathogens such as *Salmonella* (Bain et al., 2006). The analysis of micro-cracks and the fragility of the shell remain a weak point because many eggs are weakened downstream of the sorting process. Several technologies for measuring eggshell integrity have already been tested (Ketelaere et al., 2004), but the need for improvement in detecting micro-cracks is always highly relevant to reduce the potential risks of bacterial contamination of eggs.

Before packaging, the eggs may be washed and/or pasteurized. These practices, detailed in the next paragraph, are not used in all countries. The field of active packaging, and in particular antibacterial films, is promising. Some studies propose the transfer of these technologies to the coating of the eggshell. Chitosan, a natural polysaccharide, known for its antimicrobial activity and film-forming properties, has been used for eggshell coating. The chitosan coating has been shown to reduce eggshell contamination and to prevent microbial shell penetration (Kim et al., 2007).

20.2.4.5 Egg Washing and Sanitation

In the EU, washing of table eggs is prohibited (except in Sweden), while it is mandatory in the USA, Canada, and Japan, and allowed in many other countries (see Chapter 18). When it is practiced, egg-washing processes generally involve four steps: (1) pre-washing with a spray of warm water; (2) washing with a spray containing detergents (often sodium hypochlorite) and rubbing with brushes; (3) rinsing with clean hot water to remove chemical residues; and (4) drying with air jet.

The subject of egg washing is very controversial and still debated. The arguments in favor of this practice are the reduction in the level of bacterial contamination on the eggshell and, consequently, in the level of internal contamination. On the opposite, critics cite the weakening of the egg's external barriers, such as the cuticle, promoting penetration and, consequently, increasing the level of internal contamination. In addition, this practice does not provide any benefit for eggs contaminated vertically during egg formation (EFSA, 2005). The EFSA report (2005) concluded that egg washing has clear advantages if it is well done, especially in countries with a low prevalence of *Salmonella*, but that poor practices increase the risk. This EFSA opinion gave several requirements for well-washing procedures:

1. Egg storage time: less than seven days after oviposition at room temperature and in a dry place, sorting of eggs with the elimination of dirty and cracked eggs.

2. Pre-washing and washing with pH and temperature control: the water temperature must be at least 12°C higher than the temperature of the egg to avoid the phenomenon of depression, leading to the retraction of the egg contents and the penetration of microorganisms, but not exceeding 45°C to avoid damaging the cuticle and inducing cracking of the eggshell.
3. Control of the water quality: low calcium and iron concentrations to avoid calcification and to allow the maintenance of the natural antimicrobial defenses of the egg, respectively.
4. Chemical cleaning and/or disinfection agents: low iron content, and which do not damage the cuticle and leave no residue after drying.
5. Rinsing and quick drying, followed or not by waxing of the eggshell surface to seal the pores, in order to increase the storage time.

Concerning the "pasteurization" of shell eggs, it is a commercial available practice in the USA, but less than 1% of all shell eggs are actually pasteurized (Keener, 2017). The technologies, which aim to pasteurize the whole shell eggs or the eggshell surface, are well protected by international patents, and some are still under study and development. The first studies started with immersing the shell eggs in hot water, which is the basis of Davidson's patent (Davidson, 2001); this process allows shell egg pasteurization (>5 log reduction of bacterial load) without significant impact on egg quality. Michael Food patented a hot air steaming system (Ball et al., 2002), while EggBert has developed a system combining hot air and microwave (Erasmus & Rossouw, 2005). Currently, there are a number of studies evaluating egg microwave pasteurization, alone or in combination with other technologies. Egg radiofrequency pasteurization is also investigated since radiofrequencies have greater penetration in eggs than microwaves. This technology appears promising, but studies must be carried out at the industrial scale.

In the USA, shell egg treatment by irradiation (gamma rays, electron beams, and X-rays) is authorized for a maximum dose of 3 kGy. Nevertheless, this treatment is not used commercially because of the poor acceptability of this preservation process by consumers, and because it alters organoleptic and physicochemical properties of the egg (Min et al., 2005; Keener, 2017).

Other alternative technologies have been studied to reduce the bacterial load of eggshell surface (Keener, 2017):

1. Electrolyzed oxidative water: an electric current is applied to water (or a salt solution), just before its application for washing the eggs, to obtain reactive ionized oxygen species that are very harmful to bacteria.
2. Ultraviolet (UV) light: the effectiveness of UV technology as a disinfectant has been well documented, applicable at an industrial scale for egg sanitation. However, it is limited by the low penetration of UV, which requires the use of a high UV dose and that the treated eggs are already clean and free of soiling.
3. Ozone: ozone is classified as "generally recognized as safe" (GRAS) in the USA, and can be used by the food industry. Ozone is generated by a high electrical field applied to pure oxygen gas. The bactericidal effect of ozone, alone or combined with another treatment, is well documented and has been studied for egg sanitation applications. However, its application at an industrial scale is difficult because ozone needs to be generated in enclosure, upon request.
4. Gas plasma: The use of atmospheric cold plasma for surface and food decontamination is a promising technology due to its low energetic cost. An electric current produced by a high voltage generator is applied to a gas, and changes it from a gaseous state to a plasma state. This generates chemical species (radicals, atoms, excited states, and charged species) and UV that are active on microorganisms.

While these technologies have demonstrated effective microbial reductions and may be promising, they have often been studied with artificially contaminated eggs. It is therefore necessary to validate them under natural conditions of contamination, including adherent bacteria present inside biofilms, which are known to offer greater resistance to cleaning and disinfection treatments. In addition, the development of these technologies must be studied at an industrial scale, and it is necessary to be sure that the risk-benefit balance is positive.

20.2.4.6 Storage

Most countries have regulations and requirements for temperature and humidity control during egg storage, and countries lacking regulations often follow guidelines recommended by the others ones (see Chapter 18 and Fikiin et al., 2020 for details of regulations and recommendations worldwide). However, the temperature of egg storage varies from 2 to 25°C depending on the countries. In countries that impose chilling after oviposition, such as in the USA, a temperature below 7°C is required throughout the supply chain. In countries where eggs are not chilled after oviposition, such as in France, temperatures between 5 and 25°C are recommended. Here again, the most important is to maintain a constant temperature throughout the supply chain and during storage, to avoid any water condensation phenomenon on the shell surface. The relative humidity recommended varies from 60 to 88% depending on the storage temperature. The shelf life of table eggs ranges from 21 to 35 days, depending on the country and the practices (Fikiin et al., 2020).

The conditions of egg storage induce internal changes that could have an impact on bacterial ability to survive and invade the egg contents. These changes are essentially due to water and carbon dioxide losses through the eggshell, and water and component exchange through the vitelline membrane between the egg white and the egg yolk (Thapon, 1994).

The water loss is related to the degradation of the cuticle after oviposition, and is influenced by the storage temperature, relative humidity, and storage time. The loss of carbon dioxide induces a pH increase and, consequently, the decrease in egg white viscosity, described as egg white thinning (see Chapter 5). The thinning of egg white and the breakdown of the chalazae promote the migration of the yolk from its central position to the sides of the egg, in contact with the shell membranes. The difference in osmotic pressure between the two sides of the vitelline membrane leads to a transfer of water and minerals (essentially calcium and magnesium) from the albumen to the yolk. This phenomenon leads to an increase in egg yolk volume, a decrease in its viscosity, and the weakening of the vitelline membrane. The loss of integrity of the vitelline membrane favors the entrance of nutrients from the egg yolk into the egg white, and allows the migration and invasion of bacteria to the nutrient-rich egg yolk. To summarize, it must be pointed out that storage conditions may have an impact on the natural structures of the egg, the main contributor to its defense against invasion by microorganisms (cuticle, egg white viscosity, vitelline membrane, yolk position). The time and temperature of storage are thus important: the higher the storage temperature, the faster these changes are occurring over storage time.

Many authors have studied the effect of storage time and temperature on the thinning of egg white by measuring the decrease of Haugh units, and on the integrity of the vitelline membrane by measuring its mechanical strength and the yolk index (measurement of the spherical shape of the yolk). Generally, Haugh units, yolk index, and the mechanical strength of the vitelline membrane remain stable when eggs are stored at 4°C, but decrease in the first two weeks of storage at ambient temperature (EFSA, 2014). In other words, refrigeration delays the aging process of eggs and maintains their natural structure. However, if refrigeration contributes to avoid bacterial contamination, egg white antimicrobial molecules play a more important role and are also affected by storage conditions.

Kato et al. (1988) suggested that the lysozyme activity is reduced during storage. Rehault-Godbert et al. (2010) analyzed changes in egg white proteins after storage at three temperatures, namely 4°, 20°C, and 37°C. They showed that long storage at 37°C reduces the anti-tryptic and anti-chymotryptic activity of egg white, and increases the degradation of ovalbumin and ovotransferrin. With the same temperatures tested, Qiu et al. (2012) revealed that storage at 37°C increases the degradation of ovalbumin and the formation of ovalbumin-lysozyme complexes, and decreases the lipocalin proteins and clusterin contents. At ambient temperature (22°C), Omana et al. (2011) observed a significant degradation of ovalbumin, clusterin, ovoinhibitor, ovotransferrin and prostaglandin D2 synthase after 20 days of storage. Liu et al. (2018) showed an increase in the egg white peptide content after prolonged storage at 22°C, resulting from the degradation of ovotransferrin, ovomucin, ovomucoid, and ovoinhibitor. These authors observed also a rapid degradation of ovodefensin. They concluded that the degradation of the major antimicrobial proteins could lead to a higher susceptibility of the egg white to bacterial contamination. Depending on the authors and on the temperature tested, the effects on proteins may vary, but it cannot be denied that the storage, especially over time and at high temperatures, leads to protein degradation. Nevertheless, it is important to note that protein hydrolysis can also lead to the release of peptides with antimicrobial activity. For instance, the study of Cochet et al. (2021) suggested that antimicrobial peptides released from non-antimicrobial proteins, such as clusterin, is also possible.

If storage affects the egg natural defenses, it also has an impact on the microorganism metabolism and their ability to survive, migrate, and multiply. These abilities mainly depend on the environmental temperature, and on their location on/in the eggs.

During egg storage, the survival of *Salmonella* is possible on the surface of the eggshell. This survival lasts longer under low temperature and high humidity conditions (Messens et al., 2006), but the probability of penetration into the egg and multiplication is lower at refrigerated temperature. In egg white, refrigeration temperatures below 8°C prevent the growth of *Salmonella* (Růzicková, 1994; Schoeni et al., 1995) and may have a bactericidal effect below 4°C (Lock & Board, 1992; Schoeni et al., 1995; Chen et al., 2005). At 10°C, some authors observed the same bacteriostatic effect than at 8°C (Clay & Board, 1991; Chen et al., 2005), while others observed a weak growth (Schoeni et al., 1995). *Salmonella* penetration through the vitelline membrane has never been observed at refrigeration temperatures (Gast & Holt, 2001; Gast et al., 2006). In egg yolk, a weak growth is possible at 10°C (Gast & Holt, 2000b; Gumudavelli et al., 2007), but it would no longer be possible at 7°C (Gumudavelli et al., 2007). To sum up, the refrigerated storage of eggs helps controlling *Salmonella* penetration and multiplication. However, it should be kept in mind that *Salmonella* can survive at these temperatures and can therefore multiply again if temperature rises, especially if bacteria are present in the egg yolk (in case of transovarian contamination, for example), which is very rich in nutrients. Maintaining refrigeration throughout the supply chain until consumption is therefore crucial.

Since egg storage at ambient temperature is common in the entire world except the USA, Canada, and Japan, its impact on *Salmonella* growth has been widely studied. In the egg white, at temperatures between 20 and 30°C, the number of *Salmonella* may increase from 1 to 4 log units, depending on the *Salmonella* strain, the temperature, and the storage time (Clay & Board, 1991; Schoeni et al., 1995; Cogan et al., 2001; Duboccage et al., 2001; Gast & Holt, 2001; Chen et al., 2005; Messens et al., 2005; Murase et al., 2005). Moreover, at these temperatures, the penetration through the vitelline membrane is possible. Several authors using similar methods have demonstrated this: intact egg yolks (with intact vitelline membranes) were separated from egg whites, and *Salmonella* were inoculated onto the vitelline membrane. After the inoculation, the egg whites were used to cover the egg yolks. The *Salmonella* detection/enumeration into the egg yolk after one and/or several day(s) of storage allowed to know if there had been penetration through the vitelline membrane. The rate of penetration into the vitelline membrane varies between authors, due to the disparity of the strains used, the inoculum size and medium, and the incubation temperatures. Nevertheless, whether they are present in the egg yolk after passing through the vitelline membrane, or due to deposition in the yolk during egg formation, *Salmonella* can reach a high population number at ambient temperatures (20–30°C), increasing by 8 log units within less than two days (Růzicková, 1994; Schoeni et al., 1995; Gast & Holt, 2001; Gumudavelli et al., 2007).

Since *Salmonella* are more often found in the egg white in the case of vertical contamination, and since they must deal with this harmful medium (either in the case of vertical or horizontal contamination), numerous studies have focused on bacterial growth in egg white under ambient temperature. As already noticed, between 20 and 30°C, *Salmonella* may exhibit some weak growth, but this growth appears favored in eggs stored for less than one week after oviposition, compared to eggs stored at the same temperature for two to four weeks (Duboccage et al., 2001; Rehault-Godbert et al., 2010). This difference is probably due to the alkalinization of the egg white during egg storage, since high pH is known to enhance antimicrobial properties of egg white proteins (Tranter & Board, 1984). Huang et al. (2019) suggested that the alkaline pH environment enhances the bacterial sensitivity to egg white molecules by inhibiting the expression of two systems in *Salmonella*, namely phoP/phoQ and pmrA/pmrB that regulate the genes related to antimicrobial peptide resistance.

Overall, there does not appear to be any difference in the antibacterial properties of egg white in the presence or absence of egg yolk, provided that the egg yolk is surrounded by an intact vitelline membrane. However, Gast and Holt (2000b) and Murase et al. (2005) observed better growth of *Salmonella* inoculated in the egg white near the vitelline membrane, compared to cells inoculated away from this membrane. Therefore, it seems that some exchanges occur between the egg white and the egg yolk through the vitelline membrane, which may allow local bacterial growth. These exchanges are undoubtedly more important during the aging of the egg, which is accompanied by a decrease in the integrity of the vitelline membrane.

At temperatures above 37°C, no growth is observed in the egg white, even becoming bactericidal toward *Salmonella*, once the temperature reaches ≥42°C (Tranter & Board, 1984; Růzicková, 1994; Clavijo et al., 2006; Kang et al., 2006; Alabdeh et al., 2011; Baron et al., 2017). It is noticeable that this temperature matches the natural conditions encountered during egg formation in the hen's genital tract (42°C). Hence, it is believed that the temperature of the hen's body (around 40°C–42°C) enhances the antimicrobial activity of egg white proteins during the egg formation in the oviduct (Baron et al., 2016). Keller et al. (1995) and Okamura et al. (2001) noted that the ovarian follicles and/or the eggs collected in the oviduct are more contaminated than the eggs collected at the time of oviposition. These observations suggest that the control of *Salmonella* during egg formation may be exerted by the egg white antimicrobial activity at the hen's body temperature. The impact of high temperature on the antimicrobial activity of egg white was used to develop a patent, whereby liquid egg white is incubated at 42–45°C for 1–5 days, resulting in stability of the egg white at room temperature for several months (Liot & Anza, 1996). However, this patent is only effective in liquid egg white alone, but not in the context of table eggs or whole liquid egg products, due to the presence of egg yolk canceling this bactericidal effect.

At 42°C, Guan et al. (2006) showed that the vitelline membrane is impermeable to *Salmonella*, avoiding its penetration into the yolk during the formation of the egg in the oviduct before oviposition. These authors specified that penetration into the egg yolk through the vitelline membrane might occur following oviposition, if eggs are stored under conditions allowing *Salmonella* multiplication.

While high temperatures (around 42°C) are effective in controlling *Salmonella* in egg white and in the egg during its formation in the oviduct (less than 24 h), they cannot be used to store table eggs. Indeed, long-time storage at temperatures higher than 37°C is usually associated with rapid breakdown of the vitelline membrane and denaturation of egg white proteins (Rehault-Godbert et al., 2010), resulting in the loss of the egg natural defenses against bacteria.

The effect of storage conditions on other pathogenic bacteria or spoilage microorganisms has not been studied as much as on *Salmonella*. However, similar trends can be observed depending on the bacteria (EFSA, 2014). For example, refrigerated temperatures may induce the selection of psychrotrophic bacteria such as *Pseudomonas*, but the rate of multiplication will be in any event slower at refrigerated temperatures than at ambient temperatures.

To avoid water condensation on the eggshell surface and to limit the bacterial penetration during storage, it is possible to cover the eggshell surface with protective hydrophobic films, such as oils, waxes or silicates. The obstruction of the pores limits also the CO_2 leakage and, consequently, the pH increase of the egg white and its thinning. The use of modified atmosphere packaging (under CO_2, O_2, or N_2) was also proposed to limit the egg aging process during storage. Rocculi et al. (2009) showed that packaging with 100% O_2 or 100% CO_2 stabilizes the egg white pH and Haugh units of eggs stored for 28 days at 25°C. However, further experiments must be carried out to precise to what extent this type of packaging has an impact on bacteria/*Salmonella* survival/growth.

At the consumer's level, the temperature used to store eggs varies depending on the age and on the vulnerability of the consumer, and above all, the country and the tradition. Some consumers store eggs in the refrigerator, and others at ambient temperature, with no correlation between the prevalence of *Salmonella* in eggs in the country where the consumers live (Cardoso et al., 2021). When eggs are stored in the refrigerator, they are often placed in the door, although it is at door level where the highest and most fluctuating temperatures are observed, due to frequent door opening. According to EFSA (2014), the refrigerated storage is recommended to reduce the risk of *Salmonella* infections due to extended storage. We can add that storage of the eggs in the refrigerator door should be avoided to prevent water condensation on the eggshells.

The consumers use and eat eggs in general within the best-before date (3–4 weeks after production), but attention should be paid to a suggestion widely found on the web to consume eggs after the best-before date. Recommendations for consumers must be clear. It is important to keep in mind that the fresher is the egg, the more its natural defenses are preserved and, therefore, the less the risk of penetration and/or multiplication of bacteria such as *Salmonella*. Consumers should favor the use of eggs less than one-week old for the preparation of foods or recipes containing raw or undercooked eggs: soft-boiled eggs, casserole eggs, fried eggs, poached eggs, and tartare steak, but also sauces such as mayonnaise, hollandaise sauce, and desserts such as ice cream or chocolate mousse. Eggs can be used between 10 and 21 days after oviposition for the cooking of omelets and scrambled eggs. They may be used up to 28 days for hard-boiled eggs, cooked sauces, cooked cakes and pastries. Whether the egg is used in a cooked or raw preparation, it is important to check that the shell is intact. Even a newly laid egg should not be used if its shell is cracked. Raw egg dishes should be stored immediately in the refrigerator, for a maximum of 24–48 h. Preparations based on cooked eggs can be kept longer in refrigerator, but it is important to keep in mind that once the egg is cooked, it loses its natural antimicrobial defenses.

To conclude, shell eggs have remarkable natural defenses against microorganisms to protect the embryo. However, the microbial contamination of table eggs, especially by *Salmonella*, remains the major preoccupation of egg producers. The contamination of eggs by *Salmonella* still represents a serious public health issue, since it can lead to human *Salmonella* outbreaks. This contamination has also a significant socio-economic cost, due to many *Salmonella* detection analyses, preventing measures, and egg recalls. The reduction of hen's contamination by *Salmonella* and maintaining the antimicrobial defenses of shell eggs to prevent microbial invasion and multiplication are therefore crucial issues to manage throughout the supply chain.

20.3 Egg Products Microbiology

An increasing proportion of eggs are processed by the egg processing industry, also known as the egg-breaking industry. Egg products intended for human consumption are produced, after removal of the shell and the shell membranes, from the egg's inner components, which are either separated or mixed together. Several forms of egg products are available in the market. The predominant products of the egg processing industry are egg products known as "first-stage processing" such as the egg white, the yolk and the whole egg. They are mainly in the form of refrigerated liquid eggs or powders and, more marginally as frozen products. In addition, "specialties" or "further processed egg products" are also manufactured. These products result from the cooking of either table eggs (hard-boiled, poached, and fried eggs), or formulated egg products (omelets, scrambled eggs, whipped egg whites, or pickled eggs). They are sold either directly to consumers, or through the food service sector.

Because of their various functional properties (e.g. foaming, binding, gelling, emulsifying or thickening properties), "first-stage processing" egg products are widely used in the food industry in pastries, meat and fish products, sauces, ice cream, cakes, biscuits, desserts, etc. Controlling their microbiological quality is imperative, especially when they are used as ingredients in raw or undercooked foods. The microbiota of egg products depends essentially on that of the raw material (egg), and is strongly influenced by the processes of transformation and stabilization that are employed in the plants. Egg quality, hygiene practices, and the use of the appropriate processing and preservation technologies are therefore critical.

20.3.1 Origin, Reception, and Storage of Eggs

Nowadays, eggs can come from a wide variety of housing and production systems, and the percentage of eggs from nonconventional production is growing, and responds to consumer desires for greater choice in egg buying options. This is the case for table eggs, but also for egg-based products. Then egg product users often require that egg products be processed from eggs produced in alternative housing systems. The specifications required by customers may be extremely diverse depending on the final product in which the egg product will be incorporated, and the type of market/consumers targeted: cage, aviary, free range, or organic systems of egg production, but also various nutritional claims (eggs enriched with omega-3 fatty acids or some vitamins and minerals). Recent studies comparing the microbiological status of table eggs of various hens' housing and production systems, hen's strain and feeding regimen, were often conflicting, so it is difficult to extrapolate these results for an overall evaluation around the world (see Section 20.2.4). Moreover, there is a lack of studies investigating the effects of these various egg production conditions on the overall microbiology of egg products. It is essential to investigate further this aspect, since egg products' contamination is highly dependent on the eggshell contamination.

The reception is the step allowing for checking the quality of the shell eggs. Freshness and percentage of dirty or cracked eggs are important parameters to control the bacterial load of the final products. However, it is very difficult to have requirements in terms of quality with suppliers, knowing that some eggs used in the breaking process are downgraded eggs, i.e. eggs that have been discarded from the table-egg market by the packaging centers for various reasons, or even eggs coming from *Salmonella* positive farms, or from imports.

For the manufacture of egg products, the European regulation states that the following eggs only are allowed (Regulation (EC) No 853/2004):

1. Non-incubated eggs suitable for human consumption and with a fully developed shell and no defects.
2. Cracked eggs provided that they are delivered directly to an approved egg-breaking plant preparing pasteurized egg products.
3. Eggs from hens, turkeys, guinea fowl, ducks, geese, and quails that meet the above criteria, the mixing of eggs from different species being prohibited.

The same rules apply in the USA, Canada, and Japan. Controls at reception focus on the external appearance of the packaging and eggs. At each delivery of eggs, a freshness check should be carried out by measuring the Haugh units (Sauveur, 1988).

The delay between the time of lay and the receiving of egg by the egg-breaking facility is highly variable (from one to three days up to 21 days). Good storage conditions (i.e. low and constant temperature, controlled relative humidity) are also essential to limit the penetration, migration, and multiplication of microorganisms, especially *Salmonella*, in the egg. The eggs are often stored for an average period of three to four days at room temperature, or in air-conditioned premises at a temperature of 15°C. For longer storage periods, eggs are placed in cold storage at 4°C before breaking. In the USA, the results of a national census survey of egg product plants (n = 57) conducted in 2014 showed that 71% of all eggs received at plants were stored three days or less before breaking; about one-third of all eggs were broken in less than one day. About one-quarter of the plants did not process restricted eggs (i.e. dirty, check, inedible, or loss eggs); about two-thirds of the annual production using restricted eggs was from eggs that were seven to ten days old, or even younger when received by the plant. Less than one-half (44%) of plants reported that they stored eggs at temperatures at or below 7°C, and more than one-half of the plants (55%) tempered eggs at room temperature before breaking (Viator et al., 2016).

Because eggs can also be contaminated during storage, good cleaning, disinfection, pest control, and stock management are crucial to maintain microbiological quality. The use of inventory methods such as FIFO principle (First In, First Out) appears also as a helpful way to manage egg stock, to limit their storage period. This last criterion is key because the flora of the eggs determines the level of contamination of the egg products, and storage conditions may act as selection pressure, and favor the maintenance or the growth of some bacteria. It is therefore crucial to receive eggs with the lowest possible microbial load, and to maintain the storage conditions (temperature, relative humidity, and duration) that limit the development of bacteria.

20.3.2 Egg Breaking and Raw Egg Product Contamination

While the washing of table eggs is not allowed in the EU (see Chapter 18), it is allowed for eggs intended for the manufacture of egg products (Regulation (EC) No 853/2004, Regulation (EC) No 589/2008). In practice, washing is rarely carried out in the EU because it represents an additional cost for the industry, especially as the European regulation requires the shell to be dried immediately after washing. However, the European situation is rather unique, since practically all the countries in the world wash the eggs before breaking. It is even mandatory in many countries, such as the USA, Canada, Japan, and Australia.

Egg-breaking machines are designed to break eggs, remove the eggshells, and separate the egg white from the yolk. Contact of the egg's inner contents with eggshells and the surfaces of the machines leads to systematic contamination. Moreover, once broken, eggs lose their natural antimicrobial defenses. Whole egg and egg yolk are hence ideal environments for the development of microorganisms, whereas egg white limits microbial growth, providing that its defense mechanisms are not compromised by the presence of egg yolk (even traces have an impact), along with iron and nutrients (Baron et al., 1997). The eggshells are quickly removed in order to minimize their contact with the egg products. They are then crushed and centrifuged, an action that should be performed in a separate room from egg product manufacturing, in order to prevent cross-contamination. The egg products are filtered to make them more homogenous and to remove eggshell debris and chalazae. The machine and filter cleaning and disinfecting procedures must be effective in order to avoid contamination and biofilm formation.

Concerning the levels and type of microorganisms found in raw egg products, Protais et al. (2006) reported total flora levels of around $10^{5.7}$ CFU/mL in 144 collected samples of whole liquid. The dominant genus or bacterial families were *Enterobacteriacea*, *Streptococcus* sp., *Enterococcus* sp., *E. coli*, *Pseudomonas* sp. and lactic acid bacteria. To a lesser occurrence, they identified also *Brochothrix thermosphacta*, *Bacillus cereus* group bacteria, and *Clostridium* sp. (Protais et al., 2006). These results were comparable with those by Vieira et al. (2019), who used a 16S rRNA sequencing approach, and reported predominantly *Lactobacillales*, *Pseudomonales*, *Enterobacteriales* in whole liquid samples from Brazil. These bacteria were present in the three samples that were studied, and represented above 60–80%, 5–20%, and 5–15% of total flora, respectively. Less frequent bacteria were *Xanthomonadales*, *Alteromonadales*, and *Bacillales* (Vieira et al., 2019).

In liquid raw egg white, Techer et al. (2015), based on samples from 30 batches tested, reported that 93% were contaminated by bacteria (positive samples), at levels ranging from $10^{0.3}$ to $10^{8.3}$ CFU/mL, and with an average value of $10^{1.7}$ CFU/mL. All positive samples (100%) were contaminated with *Enterobacteriaceae*, mainly with *Enterobacter* sp., *Escherichia* sp., *Serratia* sp., *Shigella* sp., and *Pantoea* sp., which were detected in 78%, 50%, 33%, 28%, and 11% of the samples, respectively. Other *Enterobacteriaceae* such as *Klebsiella* sp., *Yersinia* sp., *Rahnella* sp., and *Salmonella* were rarely detected (in a single batch, corresponding to 6% of the samples analyzed). Other bacteria such as *Enterococcus* sp., *Pseudomonas* sp. and the *Bacillus cereus* group were identified in 39%, 6%, and 6% of the positive samples, respectively (Techer et al., 2015).

Concerning the occurrence of the main pathogens of interest in raw egg products, Ohtsuka et al. (2005) detected *Salmonella*, by different cultural or molecular biology-based methods, in 100% of 24 samples of whole egg from egg-breaking plants in Japan. The level of contamination with *Salmonellae* ranged from <1 CFU/g to 2.4·10^2 CFU/g. From another series of 110 raw whole egg samples from four egg-breaking plants, these authors reported that 90% were contaminated with *Salmonella* (Ohtsuka et al., 2005). Ebel et al. (1993) reported, from 937 unpasteurized liquid egg samples collected in breaking plants in the USA, that 19% of the samples (179 samples out of 937) were *S.* Enteritidis positive. In France, Protais et al. (2007b) investigated the contamination of raw whole egg by *Campylobacter* and *Listeria* in 144 samples from three egg-breaking plants. They reported few contaminations by *Campylobacter*, i.e. 6%, and 17% of *Listeria monocytogenes*-positive samples. In raw whole egg as well, Leasor and Foegeding (1989) found 36% *Listeria*-positive samples in frozen samples collected from 11 processing establishments in the USA, at levels ranging from 1 to 8 CFU/mL. These authors estimate that contamination of fresh products is probably less than 100 CFU/mL when samples are positive.

After filtration, the egg products undergo various processing and/or stabilization steps, depending on their destination. These might include pasteurization, sugaring, salting, concentration, refrigeration, freezing, or drying.

20.3.3 Impact of Processing and Stabilization Steps

20.3.3.1 Pasteurization

20.3.3.1.1 Traditional Liquid Egg Product Pasteurization

In the egg product industry, the destruction of microorganisms is mainly achieved by heat treatment. Heat treatment can affect the structure of proteins, DNA, and RNA of the microorganisms. It can alter membrane permeability by destroying surface structures. Microbial proteins are certainly the most sensitive, with denaturation possible at a temperature as low as 42°C (Lindquist & Craig, 1988). However, due to the heat sensitivity of egg proteins, the pasteurization parameters applied to egg products do not allow the destruction of sporulated microorganisms.

The destruction of microorganisms at constant temperature follows a first-order kinetics:

$$\log \frac{N_0}{N} = \frac{t}{D_\theta}$$

where N_0 and N are the initial and final (after treatment) viable population, respectively, t is the treatment time at a given temperature θ, and D_θ is the decimal reduction time at that temperature, i.e. the time required to reduce the population to one-tenth of its initial value. This equation indicates that complete destruction is never possible, and that the number of surviving microorganisms depends directly on the initial number of microorganisms.

Increasing the temperature decreases the decimal reduction time (D_θ). The relationship between D_θ and temperature is given by the following equation:

$$\log \frac{D_\theta}{D_{\theta ref}} = \frac{\theta_{ref} - \theta}{z}$$

where $D_{\theta ref}$ is the value of D_θ at the reference temperature θ_{ref}, and z is the temperature increase required to divide the decimal reduction time by ten. The thermal resistance characteristics of a microorganism are fully described if the two parameters D at a given temperature θ and z are known. These values may vary depending on the type (species, strain) and the physiological state of the microorganism (spore or vegetative form, potential adaptation to a previous stress), and the nature and composition, the pH and the water activity (Aw) of the egg product.

Current performance levels require that pasteurization treatments aim for a five to sixdecimal reductions to be considered effective. With regard to pathogenic microorganisms, many studies on thermal resistance have been carried out with *Salmonella* Enteritidis, given its importance in the egg product industry. However, the studies are very difficult to compare, as thermal resistance depends on many factors: the strain; the inoculum culture conditions (temperature, composition of the medium); the inoculation rate; the matrices tested, with or without additives; the physiological conditions of the cells (cells in stationary or exponential phase, planktonic or sessile, etc.) ; the equipment used for the heat treatment (capillary systems, tubes, plate exchangers, etc.); and, finally, the conditions for counting the surviving bacteria (revival of injured cells).

Whole egg pasteurization requirements can vary according to the legislation in a country. In Australia, New Zealand, United Kingdom, or China, authorities require that liquid whole egg be heated to a temperature not lower than 64°C for at least 2.5 min. Although a lower temperature is required in the USA, namely 60°C for at least 3.5 min, this time-temperature appeared sufficient to ensure satisfactory pasteurization of liquid whole egg and yolk. In France, heat treatments generally applied to egg products varied from 65 to 68°C for 5–6 min for whole egg and yolk, and from 55 to 57°C for 2–5 min for liquid egg white. Based on these ranges of time-temperature values applied worldwide, the reduction levels can be calculated by taking into account the D and z-values found in the literature for *S.* Enteritidis in egg products (Table 20.1). It can be concluded that the pasteurization parameters allow satisfactory levels of destruction for the whole egg and yolk, but are insufficient for the egg white. US authorities also concluded that the regulatory pasteurization time and temperature for liquid egg white (56–57°C for 3.5 min in the USA) did not achieve a 5-log reduction of Salmonella. Moreover, these time-temperature combinations were based on pH of about 9, while current practice allows for the egg white to reach pH close to 7.8, which can significantly affect the efficiency of pasteurization for *Salmonella* Enteritidis. However, it should be highlighted once more that egg white is not a favorable environment for bacterial development (see Section 20.2.2.2).

For many microorganisms, the heat treatment that does not result in destruction (sublethal treatment) may result in an increase in tolerance to subsequent heat stress. Insufficient heat treatment can therefore lead to an increase in the thermal

TABLE 20.1
Efficacy of the Treatments Used Worldwide for Pasteurization of Liquid Egg Products

	Whole Egg	Egg Yolk	Egg White
D-values (min)	D 60°C = 0.22–0.44	D 63°C = 0.27–0.35	D 57°C = 1.7–2.1
z-values (°C)	3.3	4.6–6.6	4.3–5
Treatments	60–68°C for 3.5–6 min	60–68°C for 3.5–6 min	55–57°C for 2–5 min
Number of decimal reductions	8–151	45–500	2.4–2.9
Treatment efficacy	Sufficient	Sufficient	**Insufficient**

Source: Denis et al. (1995); Doyle and Mazzotta (2000); Humphrey (1990); Palumbo et al., (1996, 1995); and Shah et al. (1991).
Note: The number of decimal reductions was calculated using the D and z-values found in literature for *Salmonella* enteritidis in egg products.

resistance of the target population, the acquisition and maintenance of thermal resistance depending on the temperature applied during the sublethal stress period, as well as its duration. The increased thermal resistance results from the induction of a set of specific proteins called "heat shock proteins", the majority of which are in low quantities in the unstressed cells, where they play an important role in cellular metabolism. Their induction is thought to protect proteins from the toxic effects of further stress, by binding proteins to prevent them from clotting. Anyway, it should be kept in mind that a low-intensity treatment may cause a resistance-inducing stress beneficial for bacteria, which will make therefore a subsequent pasteurization treatment ineffective. The time-temperature applied must then be rigorously monitored.

Moreover, certain proteins induced by the deprivation of nutrients confer to the microorganisms a resistance to stresses to which they can be exposed in the food industry, such as thermal stress, or oxidative stress due to the use of cleaning and disinfecting products. Bacterial stress and induction of resistance are therefore essential phenomena to be taken into account when processing a food product, but also when establishing predictive models of the evolution of flora in food.

In the specific case of *Salmonella* in egg products, an increase in the thermal resistance of the bacteria is observed when (i) a sublethal treatment is applied beforehand, allowing the induction of heat shock proteins (Shah et al., 1991); (ii) the bacteria cells are pre-cultured in egg product rather than in culture medium (Muriana, 1996); (iii) the pH of egg white is adjusted to 8 (instead of 9), or the eggs used are fresh (pH close to 8) (Palumbo et al., 1996); (iv) or salt or sugar is added (Palumbo et al., 1995).

Generally, the thermal resistance of microorganisms increases when the Aw decreases. A study performed by Denis et al. (1995) showed that the heat resistance of *Salmonella* Enteritidis is increased by a factor of seven at 60°C in the whole egg with 11% added salt (Aw=0.91) compared to the regular whole egg (Aw=0.98).

On the contrary, some treatments increase the sensitivity of microorganisms to heat treatments. This is the case with irradiation and the addition of EDTA, polyphosphates, lactic acid, hydrogen peroxide, or nisin (Doyle & Mazzotta, 2000). However, for economic, legal, or consumer confidence reasons, only the addition of salt or sugar is widely used by egg processors around the world.

In the case of the avian influenza virus, Swayne and Beck (2004) reported effective parameters to inactivate the virus: 56.7°C for 3.5 min, and 57.7°C for 6.3 min for liquid egg white; 60°C for 3.5 min, and 63.3°C for 3.5 min for liquid whole egg; 63.3°C for 3.5 min for yolk with 10% salt added. The World Health Organization (WHO) also concluded that the pasteurization protocols typically used by the industry for liquid egg products are effective to inactivate the virus (WHO, 2005).

After pasteurization, the number of microorganisms in egg products decreases and, in particular, a significant decrease is obtained for a number of enterobacteria, *E. coli* and bacteria of the genera *Brochothrix* and *Pseudomonas*, on which pasteurization seems to be effective (Protais et al., 2006). The effectiveness of pasteurization also appears to be good for *Campylobacter*, *Salmonella* (Protais et al., 2007b) and *Listeria* (Moore & Madden, 1993). For the latter, Moore and Madden (1993) did not find *Listeria* after the enrichment of 500 samples of whole egg and then pasteurization. The contamination of pasteurized egg products by this bacterium has been rarely mentioned in the literature, except by Protais et al. (2006). The characteristics of *Listeria* (pathogenicity, ability to create biofilms, and growth at low temperatures) require manufacturers to be very vigilant in respecting good heat treatment, cleaning and disinfection practices, and maintaining cold chain temperatures. In the USA, to enhance public health protection associated with pasteurized egg products for consumers, the Food safety and inspection service (FSIS) modernized its egg products sampling programs to mirror FSIS' other ready-to-eat (RTE) testing programs. Since 2016, all domestic and imported pasteurized egg products that FSIS analyzes for *Salmonella* are also analyzed for *Listeria monocytogenes.*

It is also important to highlight that the incidence of *Salmonella* and *Listeria* continues to be low in egg products. European food safety authority indicated that only 0.16% of the 1,246 egg product samples analyzed in 2019 in the EU were *Salmonella*-positive and, on 26 samples analyzed, none were positive for *Listeria* (EFSA and ECDC, 2021). A comparable level, i.e. 0.14% of *Salmonella*-positive samples, has been reported in the USA by FSIS/USDA between 2008 and 2017 on 15,258 samples of pasteurized egg products analyzed, including liquid and dried egg whites, whole eggs and yolks, with or without added ingredients (FSIS, 2021). During 2016–2017, they also reported 0.09% *Listeria monocytogenes*-positive pasteurized egg product samples (FSIS, 2021).

Concerning *Streptococci, Enterococci*, and *Bacillus* spores, they are less affected by pasteurization than other bacteria (Protais et al., 2006). Even if the importance of contamination by *Streptococci* and their involvement in sanitary and/or spoilage issues must not be underestimated, particular attention has been paid by the egg products field to the presence and potential issues linked to *Enteroccus* and *Bacillus.*

Bacteria of the genus *Enterococcus* are among the most thermotolerant of the non-spore-forming bacteria. They are able to adapt to many substrates and growth conditions over a temperature range of 5–50°C, with an optimum temperature of 35°C, and over a pH range of 4.6– 9.9, with an optimum around 7.5. Due to their ubiquitous nature and thermal resistance, these bacteria are frequently found as contaminants in certain heat-treated products (Foulquié Moreno et al., 2006). Knowing that this genus is part of the commensal flora of the chicken digestive tract (Bortolaia et al., 2016), it is not surprising to find it on the eggshell and subsequently in the egg product. Protais et al. (2007a) showed that, after a 2-day storage at 2°C, the proportion of samples with *Enterococcus* was 92.4% for raw whole egg, and 20.1% for pasteurized whole egg (Table 20.2). In both raw and pasteurized whole egg, *E. faecalis* species was found as predominant (64.4%), followed by *E. durans* (12.1%), *E. faecium* (8.8%), *E casseliflavus* (7.4%), *E. hirae* (5.5%), *E. gallinarium* (1.4%), and *E. avium* (0.5%). The distribution of these species is not the same for the raw and pasteurized whole egg. The heat treatment seems to be more effective on *E. faecalis*, while some species such as *E. faecium* and *E. durans*, which are in minority in the raw whole egg, seem more resistant. In the study of Miller et al. (2010), lactic acid bacteria,

TABLE 20.2

Prevalence of *Enterococcus* in Raw and Pasteurized Liquid Whole Egg (Protais et al., 2007a) and Egg White (Techer et al., 2015), and Main Species Identified

	Whole Egg	Egg White
Raw	92.4%	38.8%
Pasteurized	20.1%	93.4%
Total number of samples	144 (raw) 144 (pasteurized)	18 (raw) 15 (pasteurized)
Incidence of species in positive samples	*E. faecalis* (64.6%) *E. durans* (12.1%) *E. faecium* (8.8%) *E. casseliflavus* (7.4%) *E. hirae* (5.5%) *E. gallinarum* (1.4%) *E. avium* (0.5%)	*E. faecalis* (33%) *E. rivorum* (6.1%) *E. avium* (6.1%) *E. hirae* (3.1%) *E. faecium* (3.1%) *E.pseudoavium* (3.1%)

represented by the species *E. faecalis* and *E. faecium*, constituted the dominant flora of pasteurized whole eggs. Of the 45 strains isolated, 67% belonged to the *E. faecium* species, and 27% to the *E. faecalis* species. All strains were resistant to a pasteurization treatment at 60°C for 30 min, and even 80°C for 3 min for certain strains of *E. faecium*. Finally, in the study by Techer *et al.* (2015), 39% of raw egg white samples were found to be contaminated with *Enterococcus*, whereas contaminated pasteurized samples showed a higher prevalence (93%) (Table 20.2). This illustrates that pasteurization exerts a selection pressure. By eliminating the heat-sensitive microflora, pasteurization allows to *Enterococcus* to become the dominant flora. However, the levels of microbial contamination published in this study are very low, with less than 2 and 1 log CFU/mL for aero-anaerobic mesophilic flora, in raw and pasteurized egg white, respectively, bearing in mind that egg white is a very poor medium for harboring microorganisms, and very unlikely to support their growth.

Protais et al. (2006) showed low contamination rates by *Bacillus* spores for both raw and pasteurized whole egg ($10^{0.63}$ and $10^{0.48}$ CFU/mL, respectively). This illustrates the near ineffectiveness of pasteurization parameters on spore forms. The proportion of samples with *Bacillus* spores in this study was 58% (28 out of 48) for the raw whole egg, and 54% (13 out of 24) after pasteurization.

Among *Bacillus spp*., the *B. cereus* group is the main concern. This group, also known as *Bacillus cereus sensu lato* (*s.l.*), are Gram-positive, rod-shaped, motile, spore-forming opportunistic pathogen that are commonly found in soil, air, grains, rice (raw and cooked), vegetables, meat, milk, and egg products, due to the bacterial capability to grow at temperatures from 4 to 50°C, and to resist heat and chemical treatments. Even if the taxonomy of this group is complex and plagued by inconsistencies, it comprises actually nine closely related species: *B. anthracis*, *B. cereus sensu stricto* (*s.s.*), *B. thuringiensis*, *B. mycoides*, *B. pseudomycoides*, *B. weihenstephanensis*, *B. cytotoxicus*, *B. toyonensis*, and *B. wiedmannii* (Liu et al., 2017). In order to better differentiate the strains in this group, *B. cereus s.l.* has been subdivided into seven genetic groups, with the former species each falling into one or more groups (Guinebretière et al., 2008). The seven genetic groups have different thermal adaptation capacities, from the most psychrotrophic to the most thermotolerant, and different degrees of association with foodborne illness. Group VI is the most psychrotrophic and has not yet been associated with foodborne illness. Groups II, IV, V are psychrotrophic to mesophilic, and groups III and VII are mesophilic to thermotolerant. Thermal resistance of these bacteria varies according to their affiliation into phylogenetic groups, with $D_{90°C}$ values for *B. cereus* spores ranging from 1 to 7 min for group VI, 20–30 min for groups II, IV, V, and 40–90 min for groups III and VII. To obtain six decimal reductions, it has been shown for example that heat treatments of 70°C for 12 s, and 105°C for 36 min, for vegetative and sporulated forms, respectively, were needed for pork meat.

Due to the ubiquitous nature and the ability of *Bacillus* to sporulate, it is also difficult to avoid contamination of the raw and pasteurized products. The problem of elimination is particularly acute in the egg products, which can only be subjected to moderate heat treatment (pasteurization) due to the heat sensitivity of their constituent proteins. Byrne et al. (2006) reported D-values for *B. cereus* spores ranging from 2.0 min (95°C) to 32.1 min (85°C), and a calculated z-value of 8.5°C. Compared to the treatments generally applied to egg products, i.e. 65–68°C for 5–6 min for the whole egg and yolk, and 55–57°C for 2–5 min for the liquid egg white, it can be concluded that treatments are ineffective on spore-forming microorganisms. In addition, these heat treatments can trigger the germination of *Bacillus* spores and lead to the destruction of competing microorganisms potentially present in a vegetative form.

20.3.3.1.2 Other Pasteurization Treatments

As mentioned above, the effective parameters against *Salmonella* are higher than that allowed for liquid egg white due to the high heat sensitivity of egg white proteins, thus explaining the interest to investigate other treatments and applications by the egg-processing sector.

- French patent Liot

  In 1996, a patent for the bacterial decontamination of egg white according to an original process was filed by Liot and Anza (1996). This process consists of treating liquid egg white, previously conditioned, at moderate temperatures (40 to 48°C) for a period between one and five days. The best results are obtained at temperatures between 42 and 45°C. This patent differs from usual heat treatments for the type of parameters applied (low temperature - long time), but above all for its effectiveness. Indeed, it allows such a level of microbial destruction that the liquid egg white thus treated is stable at room temperature for several months. Moreover, the original physicochemical and techno-functional characteristics of the egg white are preserved (Liot & Anza, 1996). According to the authors of the patent, this process takes advantage of the "natural self-defense power of egg white" (see Section 20.2.2.2).

  Although the only criteria of time-temperature pairing of the "Liotisation treatment" is not sufficient to explain the destruction of the microorganisms, it is no less important, as it is particularly adapted to the natural physicochemical conditions imposed by the egg white. Indeed, it has been shown that egg white has a bactericidal effect on *Salmonella* Enteritidis at this temperature range. At 42°C, i.e. the body temperature of the hen, studies agree on a bactericidal effect of egg white. The destruction varies from less than 2–3.5 decimal reductions, for treatment times between 24 and 96 h (Guan et al., 2006; Kang et al., 2006; Gantois et al., 2008). For *E. coli*, Tranter and Board (1984) observed the destruction of 6.5 log CFU/mL after 16 h and 8 h at 39.5°C and 44°C respectively, for the same inoculum level of $10^{6.5}$ CFU/mL.

Alabdeh et al. (2011), based on results from 180 different experimental conditions, have shown that the bactericidal activity of egg white is the result of interactions between temperature, protein fraction, and pH. The most influential factor is the incubation temperature. Bactericidal activity starts at 42°C for *Salmonella* Enteritidis, and 40°C for *E. coli*. Alkaline pH is also a key element in the bactericidal effect of egg white: pH interacts with temperature to increase the bactericidal effect of egg white against *Salmonella* Enteritidis (Alabdeh et al., 2011). It should be noted that some authors have already mentioned the influence of the hen's body temperature (42°C) on the destruction of *Salmonella* Enteritidis in the egg white during egg formation in the oviduct, which lasts for more than 20 h (Keller et al., 1995; Guan et al., 2006). These conditions (temperature close to 42°C, over a long period) are also found during incubation, and undoubtedly contribute to the protection of the chick embryo from microbial invasion.

Jan et al. (2013) were able to observe, using atomic force microscopy, transient steps in the process of *E. coli* cell destruction. It includes the permeabilization of the outer and cytoplasmic membranes, and the destruction of the peptidoglycan (wall), the formation of spheroplasts (bacteria without a wall), the lysis of the latter, and the leakage of the cell content, lastly leading to the death of the bacteria (Jan et al., 2013).

Baron et al. (2017, 2020) and Cochet et al. (2021) have been interested in the drastic bactericidal activity observed during incubation of egg white at 45°C. The hypotheses that are currently emerging point toward complex molecular systems, proteins and peptides that operate optimally together at 42–45°C to generate a highly effective antibacterial cocktail leading to *Salmonella* cells death under ionic and pH conditions of egg white.

In conclusion, it can be pointed out that, if the molecular mechanisms involved are not yet known, the effectiveness of this patent is definitely based on the ability of the egg white medium to develop, under certain conditions (notably temperature around 40°C–42°C), a strong bactericidal activity. However, slight growth is allowed at lower temperatures, in the range from 20 to 37°C.

- Dry-heating of the egg white powder

In the case of egg white powder, extensive heat treatments can be applied without damaging the techno-functional qualities of the product. This specific pasteurization process, carried out in hot chambers, is called "dry-heating". Since bacteria are generally more heat resistant in dry products (low Aw), D-values are measured in days rather than minutes. Two types of dry-heating programs are used: a standard treatment of about 10–15 days at 67°C, and another of about 10–15 days at 75–80°C. The latter allows, in addition to the destruction of microorganisms, a huge increase of the functional properties of the egg white powder (Kato et al., 1989).

Dry-heating at 67°C and 75°C, even for a very short time (one day), allows the total elimination of endogenous flora and *Salmonella* Enteritidis inoculated at a rate of 10^6 CFU/mL (Baron et al., 2003). However, this treatment can affect, to a greater or lesser extent depending on the parameters used, the ability of egg white, once reconstituted with water, to limit bacterial growth and in particular that of *Salmonella*. In dry-heated egg white reconstituted in water, the number of *Salmonella* increased from 10^3 to 10^6 cells/mL in 24 h at 30°C when dry-heating lasted one day, and up to 10^8 cells/mL when dry-heating lasted 15 days. In comparison, an increase of 10^3 to 10^4 cells/mL was observed in egg white collected after breaking, or in egg white reconstituted from powder that has not been dry-heated (Baron et al., 2003).

This loss of bacteriostatic capacity seems to be due to the thermal denaturation of ovotransferrin, which plays a very important role in the inhibition of bacteria in the egg white (see Section 20.2.2.2). Ovotransferrin denaturation, and the subsequent loss of bacteriostatic properties, are greater at 75°C than at 67°C. Dry-heating at 75°C is, however, widely used in industry for techno-functional reasons. By testing both the techno-functional properties and the bacteriostatic properties, Baron et al. (2003) concluded that dry-heating at 67°C improves the foaming properties, while maintaining the bacteriostatic potential, but does not improve the gelling properties. Dry-heating at 75°C, which is very effective in improving the gelling properties, does not allow any additional gain compared to 67°C for the foaming properties, but it strongly decreases the bacteriostatic properties.

- Non-thermal treatments

Irradiation is used by the food industry to reduce the bacterial population. The principle is similar to pasteurization. The fundamental difference between food irradiation and pasteurization is the source of the energy used to destroy the microorganisms. While conventional pasteurization relies on heat, irradiation relies on the energy of ionizing radiation. The FDA first approved the use of irradiation in 1963 to kill pests in wheat and flour. To date, the FDA and the USDA have approved food irradiation for use on fruits, vegetables, spices, raw poultry, and red meats. The use of irradiation for egg products appeared limited in the USA. Although the French regulation authorizes the treatment of liquid, dried and frozen egg white with doses of 3 kGy, it currently has little or no application in the egg processing industry. Actually, the poor consumers' acceptance of food irradiated by radioactive sources, and safe handling of radioactive materials have become the bottlenecks for food irradiation and sterilization. Moreover, the investment and maintenance costs of the irradiation sterilization equipment have been another bottleneck restricting its application by the egg product industry.

Processes such as pulsed electric fields have been developed recently, and they seemed to be one of the most interesting alternatives to heat treatments for industrial use. They consist of applying electric fields with an amplitude between 5 and 50 kV.cm^{-1} over durations of 2 µs to 1 ms, with the aim to inactivate microorganisms at low temperatures. Their use for the stabilization of egg products has been the subject of a limited number of studies. They can achieve six decimal reductions of *E. coli* at 37°C for whole egg (Martin-Belloso et al., 2007), and 3.5 decimal reductions of *Salmonella* Enteritidis at 30°C in batch-treated egg white, without apparent protein denaturation (Jeantet et al., 1999). This level of decontamination, although higher than that obtained by the heat treatment of liquid egg white, is still insufficient to guarantee the sanitary quality of industrial products. Furthermore, Jeantet et al. (2004) showed that the destruction of *Salmonella* was limited in the presence of egg white proteins. The destruction observed in the previous studies seems to correspond to the dual effect of the electric fields and the ohmic heating of the product induced during the treatment (Jeantet et al., 2004).

High-hydrostatic pressure (HHP) is a technology that immerses a product under water, and exposes it to a hydrostatic pressure of several hundred megapascal in a HP vessel. The product is commonly packed in a high-pressure suitable packaging, and it is held under pressure for specific times, until decompression. HHP treatment can take place between 100 and 1,000 MPa, and at temperatures between 0 and 120°C. This technology appeared as a technology with a great potential since it reduces efficiently the microbial load of egg products, without impacting their nutritional and sensorial characteristics when it is used at a level below 400 MPa (Tóth et al., 2017). Depending on the type of egg products (whole egg, yolk, and egg white), higher levels can be used to satisfy an accepted level of decontamination. Indeed, Tóth et al. (2016) studied the effects of HHP at 400 MPa for 600 s on liquid egg yolk and whole egg, and showed that whole egg's proteins were not denatured by the treatment; however, 40% of the egg yolk's proteins were denatured due to processing. Other effects have been observed, in particular texture modifications due to protein aggregation and hydrolysis for both products. Using lower level (from 200 to 350 MPa), these authors reported that significant denaturation was not detected for liquid egg (egg white, yolk, and whole egg), with a decrease of viable cell count even at 250 MPa. Differences were observed for microbial reduction, and the most HHP-sensitive liquid egg type was the egg white, with a 3-log reduction at 300–350 MPa. Bacterial reduction was lower for whole egg (less than one log reduction for higher treatment level used), suggesting also that higher pressure values are appropriate for this product.

Other strategies consisting in the combination of HHP and antimicrobial agents such as nisin, or successive use of HHP with heat or other non-thermal methods such as pulsed electric field or ultrasound, have been tried. The objective was to reach a maximum microbial inactivation, with minimum changes in physical, chemical, and functional properties of liquid whole egg (Buzrul, 2021).

Although HHP technology appeared promising for various applications, it appeared also limited for use in egg products, especially due to its cost. Aganovic et al. (2021) recently estimated the treatment cost between 0.1 and 0.3 EUR/kg, depending on the size of the HP vessel, filling ratio, and treatment conditions (pressure and treatment time). For comparison, the cost of a thermal treatment (e.g. high temperature short time, HTST) is usually around 0.005–0.01 EUR/kg, depending on the level of energy recovery, while the cost of a treatment by pulsed electric fields (PEF) is around 0.01–0.2 EUR/kg, depending on the throughput and annual operating hours.

20.3.3.2 Refrigeration

Refrigeration, widely used in the food industry, simultaneously slows down the growth kinetics of microorganisms, and the production of toxins by pathogenic bacteria. The growth rate of a microorganism is maximum when it is placed at a temperature described as the optimal growth temperature. On either side of this optimum temperature, growth, and therefore metabolic reactions, are slowed down, or even inhibited. Depending on their optimum growth temperature, a distinction is made between thermophilic microorganisms (optimum temperature between 40 and 55°C), mesophilic microorganisms (optimum temperature between 20 and 40°C), and psychrophilic microorganisms (optimum temperature between 5 and 20°C). Beyond this terminology, microorganisms that retain significant activity at temperatures of 7°C or less, but have an optimal growth temperature between 20 and 30°C, are referred to as psychrotrophic. These bacteria are of particular interest to the food industry.

The food industry has artificially recreated the thermal environment of the ecological niches of the Arctic and Antarctic zones by using cold rooms, cold stores, or household refrigerators. The extension of refrigeration has certainly improved the quality and shelf life of products, but it is also responsible for the selection of a psychrotrophic microbial flora that can colonize foods. Psychrotrophic microorganisms belong to a large number of genera and species, which are able to survive and multiply at temperatures closed to refrigeration temperature.

Their slow growth rate allows them to invade food environments within one to three weeks, with a generation time of about 24 h (Cousin, 1982). Their metabolism remains very active at refrigeration temperatures, particularly for the production of hydrolytic enzymes such as lipases or proteases, which pose technological (protein coagulation, color change) and organoleptic issues (rancidity, appearance of off-flavors and off-tastes). Colonization by these microorganisms is the cause of unsanitary conditions and/or shortened storage times for foodstuffs, and even production losses, which can have a considerable economic impact.

Psychrotrophic microorganisms that cause foodborne illness include *Listeria monocytogenes*, *Yersinia enterolitica*, *Bacillus cereus*, and *Clostridium botulinum* type E (Palumbo, 1986). Some strains of *Salmonella* and *E. coli* are also likely to grow between 5 and 7°C, but they remain atypical and these microorganisms are not considered psychrotrophic (Catteau, 1999). Among the psychrotrophic bacteria responsible for food spoilage, *Pseudomonas* are predominant, but also the genera *Shewanella*, *Alcaligenes*, *Acinetobacter*, *Alteromonas*, *Flavobacterium*, *Enterobacter*, *Serratia*, *Hafnia*, *Micrococcus*, and *Brochothrix* (Bornert, 2000). To control this flora, a temperature of 2°C provides a very effective inhibition, but very small variations in temperature can be risky, since significant metabolic activity can be observed at temperatures as low as 4°C, although at very low growth rates. For instance, *Pseudomonas* at 4°C shows slow growth, but high production of enzymes responsible for the hydrolysis of food substrates (Bourgeois & Leveau, 1996). Moreover, some psychrotrophic bacteria are very strong producers of lipases and proteases, which can show a thermal resistance much higher than that of the bacteria cell itself. Toxin production is stopped in most microorganisms at refrigeration temperatures, except for some psychrotrophic bacteria. Some strains of *Bacillus cereus*, able to grow at 4°C, can produce toxins starting at 6°C (Catteau, 1999). Baron et al. (2007) observed enterotoxin production by a strain of *Bacillus weihenstephanensis* (belonging to the *Bacillus cereus* group) after 21 days of incubation at 6°C, and seven days at 8°C in liquid whole egg.

Protais et al. (2006) observed that the occurrence of mesophilic and psychrotrophic flora, counted on Plate Count Agar (PCA) after respectively three days at 30°C, and eight days at 8°C, is very similar ($10^{5.7}$ and $10^{5.1}$ CFU/mL), showing that the majority of species contaminating egg products are able to grow at low temperatures. Control of psychrotrophic flora in egg products requires control of the sources of the contamination, and the stability of refrigeration temperatures. Good cleaning and disinfection practices can control the establishment of bacteria such as *Pseudomonas* and *Listeria*, which are able to colonize refrigerated and with high relative humidity rooms, such as cold rooms (Salvat et al., 1997). For refrigeration, care should be taken to ensure that intense and continuous cold is applied. Knowing that the psychrotrophic flora is more inhibited as the temperature approaches 0°C, it seems essential to better control the cold chain and to have high-performance refrigeration facilities at all stages of the production and distribution chain of refrigerated products. These practices must be accompanied by adequate means of monitoring the compliance with the cold chain (temperature recorders, probes, etc.).

20.3.3.3 Stabilization by Water Activity Reduction

The water activity (Aw) of a product is indirectly a measurement of the amount of free water, which is an essential factor for the growth of microorganisms. Water is essential for the hydrolysis reactions involved in central metabolism, for the transfer of nutrients, and for the excretion of waste materials. When a bacterium is in a water-free environment, plasmolysis occurs: the water escapes from the cell compartment. This loss of water limits bacterial growth and can, in some cases, lead to the death of the cell.

The amount of free water available to microorganisms varies greatly from one food to another. Some of the water is unavailable because it is strongly bound to certain molecules in the food, such as sugars, salt, and proteins. The lower the Aw of a food, the more difficult it is for microbial growth and survival. However, microorganisms are unequal in their ability to grow or survive at low Aw. Generally, bacteria cannot grow below an Aw of 0.95 (0.91 for Gram-positive bacteria); yeasts and molds can grow at low Aw (down to 0.65 for osmophilic yeasts); and no microorganism can grow below an Aw of 0.6. It should be noted that microorganisms, even if inhibited, are not necessarily destroyed, especially spore-forming microorganisms.

Liquid egg products (egg white, yolk, and whole egg) have an Aw close to 1. The aim of technological processes such as freezing, concentration, drying, or the addition of sugar or salt, is to lower this value by eliminating or mobilizing water.

- Freezing

  By involving water in a crystal lattice, freezing stabilizes egg products by reducing Aw. The forming ice crystals can furthermore alter the cell structures and lead to the death of the bacteria. There is a better bacterial resistance to rapid freezing treatment resulting in the formation of intracellular microcrystals, compared to large crystals formed during slow freezing. In general, Gram-positive bacteria are more resistant than Gram-negative bacteria (Imai et al., 1986). Before the use of frozen egg products, thawing has to be correctly achieved; otherwise, an increase in the level of flora could occur.
- Concentration

  The concentration step, prior to drying or in order to limit transport costs, is done either by ultrafiltration or by reverse osmosis. It is applied to whole egg and egg white, allowing the dry matter content of these products to be increased up to 44% and 33%, respectively. The naturally high dry matter content of egg yolk makes it economically unfeasible to concentrate this product by membrane technology. However, the concentration levels achieved only slightly decrease Aw of whole egg and egg white (about 0.9 after filtration). These techniques therefore have limited influence on the development of microorganisms, unless they are combined with the addition of sugar and/or salt. The ultrafiltration equipment, or any other equipment allowing concentration, must be cleaned, disinfected, and replaced (membranes) regularly, in order to prevent contamination.

- Drying

  Egg products are dried by spraying the product in a co-current of hot air. Egg powders obtained have a moisture content of approximately 5% for the whole and yolk, and 8% for the egg white. Aw of egg powders ranges from 0.2 to 0.3, meaning that no bacterial growth is possible under these conditions. However, spray drying does not lead to the death of bacteria, especially spore-formers.

  Egg white is usually concentrated twice by ultrafiltration before drying. It must then be desugared of the free glucose it contains to avoid non-enzymatic browning (Maillard reaction) during spray drying and dry-heating. Desugaring can be carried out by fermentation, either with bacteria belonging to the genera *Streptococcus* and *Lactobacillus,* or with *Saccharomyces cerevisiae,* the baker's yeast. The use of yeast allows a rapid fermentation (2 to 4 h) and is easy to implement. However, it can cause off-odors if fermentation is not stopped as soon as no more free glucose is available. After acidification to pH 6–7, the yeast is added to the liquid egg white at 0.2–0.5% (w:w), and the mixture is incubated at 30°C. After fermentation, the mixture is refrigerated, and the yeast can be removed by centrifugation.
- Sugar or salt addition

  The addition of salt and/or sugar allows to apply higher pasteurization temperatures, to preserve the functional properties of the proteins during heat or freezing treatments, or to improve the preservation of the products by lowering the Aw below 0.85. In the latter case, addition levels can be as high as 50% for sugar, and 10% for salt. This is achieved by simple addition in a blender before pasteurization (Galet et al., 2010); it can be combined with concentration in the case of egg white and whole egg.

20.3.3.4 Addition of Microbial Inhibitors

Microbial inhibitors such as sodium benzoate, sodium and calcium propionate, sorbic acid, ethyl formate, and sulfur dioxide are examples of commercial food additives that can be used to preserve the microbiological quality of liquid egg products. These chemical compounds remain largely used by the food industry, due to their effectiveness against a broad spectrum of microorganisms and their low cost. Nonetheless, with increasing concerns about toxicological effects of synthetic molecules (e.g. sorbate, benzoate, or phosphates), and market trends that show increased popularity of foods with no artificial substances, egg processors have recently focused on the application of naturally occurring preservatives, such as microbial products with antimicrobial activity, organic acids, hydrogen peroxide, bacteriocins or phages (Gálvez et al., 2014; Xu, 2021). In recent years, significant advances have been documented in the literature, highlighting the potential for these natural molecules to be used for preservation (Miller et al., 2010; Niamah, 2018; Necidová et al., 2019; Wang et al., 2021). However, the commercial use of certain of these products remains limited, especially in the EU, for various reasons such as regulatory acceptance, effectiveness of pathogen elimination, high price, or the impact they may have on sensory characteristics of products. In the EU, approved food additives, including preservative materials and conditions for their use, are listed in the Commission Regulation (EU) No 1129/2011 (European Commission, 2011).

Among authorized products, Laktocid (a blend of lactic acid, acetic acid, and sodium lactate), Galimax Flavor V50 (low sodium buffered vinegar), and Defence JB (natural product of fermentation by LAB and yeast starter culture) are registered and commercially available preparations in the EU countries. These products have been evaluated by Necidová et al. (2019) with other alternative preservative materials on the microbiological quality and sensory characteristics of pasteurized liquid whole egg. These authors have demonstrated that a mixture of organic acids (acetic acid, lactic acid, and sodium lactate) has a great potential for extending the shelf life of pasteurized liquid whole egg up to 45 days when stored at 4°C. However, it has also a significant effect on the pH value and sensory characteristics (mainly color) of liquid whole egg, when compared with the control sample.

Other alternative preservatives tested, i.e. bacteriocin (nisin), natural product of fermentation by starter cultures, or vinegar, have shown to be effective in controlling the undesirable development of microorganisms in pasteurized liquid whole egg for a limited period of three weeks, while having a minimum impact on the sensory characteristics. However, none of the tested traditional/alternative preservatives have been able to substantially extend the shelf life of liquid whole egg, while preserving its optimum properties (Necidová et al., 2019).

Bacteriocin is defined as small antimicrobial proteins or peptides produced by bacteria, and that inhibit or kill other related and unrelated microorganisms (Necidová et al., 2019). The best-known and used bacteriocin, namely nisin, is produced by *Lactococcus lactis* and belongs to the family of bacteriocins called lantibiotics. Nisin is available as a commercial product that is regulated as a food additive, and is approved for use in most major food-producing countries (Davidson et al., 2005). In EU, it can be used in liquid egg products at inclusion rate up to 6.25 mg/kg (Annex II of Regulation (EC) No 1333/2008).

Nisin inhibits the growth of a wide range of Gram-positive bacteria, including *Listeria monocytogenes*, *Staphylococcus aureus*, *Bacillus cereus*, and LAB species (Delves-Broughton et al., 1992; Miller et al., 2010; Shrestha & Hariram, 2022). Delves-Broughton et al. (1992) also reported that nisin, when used at a level of 5 mg/L, protects the liquid egg from the growth of *Bacillus cereus*, resulting in a significant increase in the refrigerated shelf life of pasteurized liquid whole egg from 6–11 days up to 17–20 days. In another study with liquid whole egg, where lactic acid bacteria were identified as the main spoilage agents, Miller et al. (2010) reported that the shelf life was extended by a minimum of five weeks with the addition of nisin at 6.25 mg/L, and storage under refrigerated temperature (Miller et al., 2010).

Other bacteriocin-based products have been investigated or developed such as Pediocin, produced by *Pediococcus spp,* or Micocin X, a mixed of three bacteriocins (piscicolin 126, carnobacteriocin BM1, and carnocyclin A) produced by a single strain of *Carnobacterium maltaromaticum*. The former was shown to reduce the survival of *L. monocytogenes* in whole egg (Muriana, 1996). The latter, authorized in some countries such as the USA, Canada, and a few South American countries, has shown to be effective against spoilage lactic acid bacteria isolated from pasteurized liquid whole egg (Miller et al., 2010). Miller et al. (2010) reported a shelf life of liquid whole egg of eight weeks when Micocin X was added at 500 mg/mL, compared to three weeks without the inclusion of a microbial inhibitor (Miller et al., 2010).

Synergistic effects between bacteriocin and physical treatments, such as heat (Boziaris et al., 1998; Knight et al., 1999) or pulsed electric fields (Calderón-Miranda et al., 1999), have also been reported. Knight et al. (1999) determined D and z-values for *Listeria monocytogenes* in salted (10% NaCl) or unsalted liquid whole egg, with or without nisin (0 or 10 μg/mL). These authors reported that the addition of NaCl increased D-values at all temperatures tested, i.e. from 50 to 67°C. The addition of nisin to unsalted liquid whole egg resulted in a rapid 4-log reduction in viable counts within the first hour. Nisin significantly decreased D-values at lower (<58°C) temperatures in both unsalted and salted liquid whole egg, but had little effect on the D-values at current minimum U.S. and Canadian pasteurization temperatures (60°C without NaCl; 63°C with NaCl). However, when nisin was added 2 h prior to heat treatment, D-values were significantly reduced, resulting in a reduction at pasteurization conditions of 3.2 log and 1.3 log in the presence and absence of salt, respectively (Calderón-Miranda et al., 1999). Although nisin is not active on Gram-negative bacteria, Boziaris et al. (1998) have shown that nisin addition also increased the heat sensitivity of *Salmonella* Enteritidis PT4 in liquid whole egg and in egg white during pasteurization. Calderón-Miranda et al. (1999) reported that the highest extent of microbial inactivation with pulsed electric fields (50 kV/cm - 32 pulses) was 3.5 decimal reductions, and that the same parameters of treatment combined with nisin addition at 37 IU/mL in liquid whole egg, resulted in a decrease of 4.4 log in the population of *L. innocua*.

Over the past 20 years, literature has shown an increasing interest for the use of bacteriophages for food safety or as food preservatives. The use of phage-based products is particularly advanced in the USA, and has received approval in several other countries such as Israel, Canada, Switzerland, Australia, and New Zealand, but it is not yet approved in the EU. In 2020, fourteen phage biocontrol products have been approved for food safety applications under various regulatory frameworks in the USA. Out of these fourteen products, eleven products were approved as GRAS by the Center for Food Safety and Applied Nutrition (CFSAN) of the FDA (Vikram et al., 2021). The first approval was for the *L. monocytogenes* specific phage cocktail called ListShield™, based on an amendment of the FDA's "food additive regulations" in 2006 (Food and Drug Administration, 2006). This phage cocktail is also listed as a suitable antimicrobial intervention for ready-to-eat meat and poultry products. Other commercialized products, mainly targeting *E. coli* (Secure Shield E1, EcoShield PX™, Finalise), *Salmonella spp.* (SalmoFresh™, PhageGuard S™, SalmoPro®), or *Shigella* spp. (ShigaShield™) are also used by the food industry. Their commercial uses are linked to a specific type of food according to the main bacteriological issue of the food. In egg products, bacteriophage technology has been particularly studied for *Salmonella* (Guenther et al., 2012; Hong et al., 2016; Zhang et al., 2021). For instance, Zhang et al. (2021) recently described the efficiency of a new Pu20 phage exhibiting broad host range, high pH and heat resistance, and showing significant inhibition (0.87 to 4.60 decimal reductions) of *Salmonella spp* (*S.* Typhimurium and *S.* Enteritidis) in liquid egg white and yolk at 4°C and 25°C. Similar results have been reported by Hong et al. (2016), with a monophage preparation consisting of SJ2, which was able to significantly reduce *Salmonella* in liquid egg during storage at 4°C and 21°C. Guenther et al. (2012) reported that *Salmonella* populations were initially reduced by about 0.5–2 log in egg yolk after phage (FO1-E2) addition, but bacterial recovery matched controls in egg yolks after two days. As bacteriophages are targeting specific strains of bacteria, it is logical to consider that results are highly dependent on the bacteriophage/bacterial strains and conditions of use.

The industrial use of bacteriophages in egg products appeared nowadays limited to some countries, where legislation allowed their use. However, the commercialization of phage-based products has started and these technologies should be exploited further to their full potential for successful use in the food chain. More than their effectiveness in killing specific targeted pathogenic bacteria, bacteriophages offer various advantages such as their natural origin, narrow potential for resistance and lack of cross-resistance with antibiotics, low inherent toxicity, and no adverse impact on organoleptic, nutritional, and rheological properties of the food (Vikram et al., 2021). The actual limitations for their regular use, such as phage stability, inactivation in various food matrices, broad host range, resistance emergence, have to be addressed with many new analytical technologies and innovative industrial strategies. Moreover, for a wider market success and acceptance by consumers, the next step will be to work on communication and raising the awareness about the use, benefits, need, and safety of these edible viruses.

20.3.3.5 Contamination in Production Lines

The adherence of microorganisms to industrial surfaces, and the subsequent biofilm formation, are frequently observed in the food industry. Biofilms are defined as bacterial aggregates attached to various biotic and abiotic surfaces, which interact with each other to adapt to environmental stressors compared to planktonic existence. Bacterial biofilm provides a protected mode of growth that allows microorganisms to resist in hostile and fluctuating environmental conditions encountered in the food industry production lines, during food processing or cleaning and disinfection procedure (variation of pH, temperature, nutrients, shear forces, and presence of disinfectants). Biofilms are difficult to eliminate due to their resistant phenotype, and adhering bacteria can be the source of recalcitrant contaminations, leading to transmission of spoilage and pathogenic bacteria in food.

Techer et al. (2019) identified by 16S DNA sequencing the adhering microorganisms on stainless steel chips disposed two months inside the pipes of seven French egg-breaking plants, at two locations (before and after the pasteurizer), and at two consecutive seasons (summer and winter). They showed that bacteria present in production lines were mainly facultative aerobic-anaerobic Gram-positive bacteria, such as *Streptococcus*, *Staphylococcus*, *Bacillus*, and *Kocuria*, and that most of the bacteria identified from chips belonged to several genera. Hence, this study highlighted the multi-species organization in biofilm ecosystem, which interact in different ways to constitute a complex and dynamic network. Main adhering bacteria were identified at both locations, and during both seasons, suggesting that the pasteurization process and season do not select for specific bacteria among those present in raw egg products (Techer et al., 2019).

Even if few data are available on flora present in egg-breaking plant pipelines, biofilm formation abilities have been particularly studied using stainless-steel coupons under conditions that mimic egg-processing industry, in the aim to evaluate the effectiveness of sanitation methods. Yang et al. (2017) investigated the biofilm formation by three *S.* Enteritidis strains in a simulated liquid egg processing environment using 0.1% peptone water (control), 10% whole egg, 10% egg yolk, and 10% egg white as growth media, and evaluated the effectiveness of chlorine (200 ppm, 5 min) and hot water (71°C, 30 s) treatments against *S.* Enteritidis biofilms. They showed that *S.* Enteritidis formed significantly denser biofilms in peptone water and egg white, compared to those in whole egg and egg yolk. However, biofilms formed in peptone water were less resistant to chlorine treatment than that formed in whole egg, egg yolk, and egg white, with average decimal reductions of 6.34, 2.28, 0.67, and 0.95 CFU/cm^2, respectively. Microscopic observation showed that biofilm morphology was greatly affected by the growth medium, and the egg matrices might act as protective barriers, contributing to the greater chlorine resistance.

Iñiguez-Moreno et al. (2019) investigated the biofilm development by pathogenic (*Bacillus cereus*, *Escherichia coli*, *Listeria monocytogenes*, *S.* Enteritidis, and *S.* Typhimurium) and spoilage (*Bacillus cereus* and *Pseudomonas aeruginosa*) microorganisms onto stainless steel and polypropylene B chips, under conditions that mimic different food processing industry (dairy, meat, and egg products). They reported that pathogens and spoilage microorganisms developed mono-species biofilms with higher cellular densities than multi-species biofilms. In multi-species biofilms, the time to reach each biofilm phase varied depending on environmental factors. Cell count decrease of 1.12–2.44 log CFU/cm^2 occurred at 48 h and 240 h, and was most notable in the biofilms developed at 9°C. Moreover, cell density reached by each microorganism was different; *P. aeruginosa* and *B. cereus* were the dominant microorganisms in the biofilms, whereas *B. cereus* showed the lower densities until undetectable levels (Iñiguez-Moreno et al., 2019). These results highlight the complex behavior of multi-species biofilms, that should be taken into consideration for better risk evaluation, or for cleaning and disinfection procedures optimization.

20.3.3.6 Packaging

Packaging varies from small containers to large tankers, depending on the final destination of the products (i.e. consumer, food service, or industry), and the product category. The shelf life depends on the type of product and on its packaging. It ranges from one or two weeks at 4°C for liquid egg products conditioned in bulk packages and intended for the food industry, and from several days and up to 60 days for the small packaging (1 or 2 kg) intended for food service, catering, or for consumers in some countries. A shelf life of several months is obtained for frozen egg products (12 months) and concentrated products with high sugar or salt content (up to 24 months at room temperature). Pre-cooked egg products such as egg patties and curds or omelets are often available in frozen state with shelf life of 12 months. Hard-boiled eggs in vacuum bag, or in brine, have shelf life of 70 days below 4°C.

The type of packaging and storage conditions influence the microbial shelf life of products, but it represent also an important parameter for the selection of some bacteria. For example, O_2-free packaging limit the growth of aerobic microorganisms, and minimize oxidative quality changes. Packaging materials with a poor gas barrier will promote microbial growth of aerobes and facultative anaerobes.

20.3.4 Contamination of Egg Products

20.3.4.1 Residual Contamination

The flora of the pasteurized egg products is significantly less diverse than that of the raw egg products, confirming the efficiency of the pasteurization. Generally, the remaining bacteria include Gram-positive bacteria, especially *Enterococcus* and other lactic acid bacteria, and bacteria belonging to *Bacillus* group (Hidalgo et al., 2008; Miller et al., 2010; Techer et al., 2020; Wang et al., 2020). Moreover, even if pasteurization treatment appeared as efficient toward some Gram-negative bacteria, especially *Pseudomonas*, storage and transport conditions are also key points explaining that these bacteria are frequently reported as dominant bacteria in chilled egg products (Vieira et al., 2019; Wang et al., 2020). The risk of post-pasteurization contamination should not be underestimated and could involve all types of bacteria, including Gram-negative bacteria and *Salmonella*.

Although whole egg or yolk products are more likely to be contaminated than egg white because they constitute optimal media, as opposed to egg white, their pasteurization is possible under stronger treatment than those used for egg white. The resulting level of contamination reported in the literature is nearly similar for all of liquid egg products, and is generally less than 10^2 CFU/mL (Protais et al., 2006; Correa et al., 2008; Miller et al., 2010; Techer et al., 2015). Residual contamination levels have clearly improved due to preventive measures put in place by the manufacturers to meet the regulations, but also to meet the increasingly demanding requirements of customers of egg products.

20.3.4.2 Evolution of Contamination

Overall, egg yolk and whole egg can be considered as ideal media for the development of microorganisms, whereas egg white has a bacteriostatic or bactericidal effect depending on the type of microorganisms and the environmental conditions. However, the behavior of microorganisms in an egg product depends on the nature of the egg product (liquid, dried, concentrated, with added salt or sugar), its initial level of contamination and, above all, the temperature. Indeed, only by keeping temperatures below 4°C can bacterial growth in liquid egg products be limited.

Studies on the behavior of microorganisms in egg products focus mainly on *Salmonella*, *Listeria*, and *Bacillus*. As previously mentioned (see Section 20.3.3.2), *Salmonella* growth at temperatures between 4 and 10°C does not seem to be possible, or is minimal in liquid egg products. Therefore, refrigeration can be considered as an efficient way of control to limit the development of this pathogen in egg products. Between 20 and 30°C, *Salmonella* growth is possible in the different categories of liquid egg products, with a high growth rate in yolk, where the population can increase very rapidly by seven to eight log units within two days of incubation. In contrast, growth is weaker in egg white, with the population increasing by one to four log units depending on temperature and incubation time.

Listeria growth, although slow, is possible at refrigeration temperatures in all three types of egg products. For yolk and whole egg stored at 4°C, Notermans et al. (1991) observed an increase of two log units in 22 days, while other authors have observed higher growth, i.e. about six log units after 21 days (Foegeding & Leasor, 1990). The differences can be explained by the use of more or less psychrotrophic strains. At temperatures above 20°C, *Listeria* growth is faster in egg yolk than in whole egg. Notermans et al. (1991) showed an increase of three log units in one day in egg yolk at 22°C, compared to two log units after two to nine days in whole egg. Under these conditions, a bactericidal effect is observed in the egg white. Indeed, Notermans et al. (1991) observed the disappearance of a population of 10^5 CFU/mL in nine days in egg white incubated at 22°C, and Wang and Shelef (1991) observed the same phenomenon after 24 h incubation of the egg white at 35°C.

Baron et al. (2007) studied the growth of a psychrotrophic strain of *Bacillus cereus* group in liquid whole egg. The bacterial population increased by two log units after seven days at 8°C, and 14 days at 6°C, compared to 4.5 log units at 10°C within seven days. In contrast, no development was observed at 4°C. This work also focused on the characterization of this strain in terms of toxin expression. They showed that the strain was able to produce enterotoxins and express cytotoxic activity in whole egg after a seven-day incubation at 8°C, corresponding to a population of $10^{5.1}$ CFU/mL. The authors concluded that refrigeration temperatures should be strictly respected and, if possible, kept below 4°C in order to control these psychrotrophic strains (Baron et al., 2007).

Compliance with refrigeration temperatures therefore appears to be one of the most effective means of controlling microbial growth in egg products, including psychrotrophic strains. It is also important to consider that according to countries and companies, practices for stabilization and supply chain organization (conditions of transport, temperature…) can differ, impacting microbiota of raw materials and further liquid processed egg products. Vieira et al. (2019) also highlighted that the practices employed by suppliers and transportation distances are key in modulating the microbiota of chilled raw materials and further liquid-processed egg products. They studied, using a 16S rRNA-based amplicon sequencing, the microbiota of Brazilian samples of liquid eggs (raw materials) produced by different suppliers, and final products (chilled pasteurized liquid whole egg and chilled pasteurized liquid egg yolk). Raw material suppliers were located at different distances (100–1,500 km) from the liquid egg processing facility, and used varying practices after egg breaking (pasteurization and/or addition of hydrogen peroxide), followed by chilled transportation. In this study, it is important to note that raw materials consisted of samples without pre-treatment, or were pre-treated by pasteurization and/or hydrogen peroxide. Results showed that final products presented lower diversity indexes than raw materials, suggesting that processing and storage act toward shaping their microbiota. *Proteobacteria* and Firmicutes dominated the sequences of 16S rRNA in raw materials and final product samples, while the abundance of classes was found to vary considerably among initial and final products. *Lactobacillales* and *Pseudomonadales* were the most abundant order in almost all raw material samples, whereas *Enterobacteriales* were abundant in few raw material samples. The principal genera found in final products were *Pseudomonas*, *Carnobacterium*, and less frequently, *Clostridium* sp. *Carnobacterium spp.* was detected in four pasteurized whole egg products out of six, with abundance levels from 34 to 46% of total flora, and at a lower level in raw material from suppliers. Although these bacteria have never been clearly involved in liquid egg products spoilage events, they have been identified in spoiled desserts-based egg products (Techer et al., 2020).

20.3.4.3 Impact of Contamination on the Egg Products

The main risk associated with the development of bacteria in egg products remains sanitary, mainly due to *Salmonella*. However, the implementation of farm surveillance programs and good manufacturing practices (pasteurization process, storage conditions…) allow to reduce significantly the presence of these pathogenic bacteria in egg products, and to limit their development. In addition to the control of pathogens, manufacturers have to deal with the development of spoilage bacteria.

The growth of spoilage flora can lead to technological or sensorial problems (coagulation, changes in color or in flavor) that often involve the production of hydrolytic enzymes, such as lipases, phospholipases, or proteases. Spoilage manifestations vary according to the type of products, i.e. depending on whether egg products are ingredients such as whole egg, yolk, or egg white, or more processed egg products (mixed with other ingredients and/or cooked). Indeed, the consequences of spoilage are highly dependent on the nutritional composition of the egg products, and the process, packaging and storage conditions they have been submitted to.

Main manifestation of whole and egg yolk spoilage is visual defects, such as color changes or coagulation characterized by the appearance of white slurry cream. The main enzymatic activities involved in these defaults are lipolytic and proteolytic activities, expressed for example by *Enterococcus* spp. or *Pseudomonas* spp., and the lecithinase activity of the *B. cereus* group. Other spoilage characteristics can be observed through physicochemical, rheological and technological changes.

Wang et al. (2020) reported in pasteurized liquid whole egg stored at 4°C, a decrease of pH and an increase in total volatile basic nitrogen with the growth of microorganisms, mainly dominated by *Pseudomonas* and lactic acid bacteria. Among dominant bacteria, strains identified were *Pseudomonas* spp., *Microbacterium* spp., *Bacillus* spp., *Enterococcus faecium*, *Stenotrophomonas* spp., *Streptococcus* spp., *Acinetobacter* spp., *Brevundimonas*, and *Rhodococcus* spp. These authors also reported an increase of particle size of liquid whole egg, and a decrease of their functional, mainly gelling, foaming and emulsifying properties after a week at 4°C. Another study from this research team highlighted the ability of most of these strains to express high hydrolytic enzymatic activities in solid egg-based medium at low temperature, and their involvement in coagulation of liquid whole egg at 4°C (Wang et al., 2021). Other spoilage manifestations, characterized by off-flavor, gas production leading to packaging distension, and protein hydrolysis, have also been observed for liquid whole eggs. Vieira et al. (2019) reported that this type of deterioration matches with complaints received by the egg processing company, and can be caused by *Clostridium* since they identified *Clostriales* as abundant, up to 30% of microbiota, in some chilled pasteurized liquid whole egg samples (Vieira et al., 2019).

In egg yolk, the *B. cereus* group lecithinase may lead to the destabilization of the yolk binding properties, hence causing a marked modification in the colloidal state of the egg components. Nevertheless, the involvement of other enzymatic activity and bacteria in the spoilage of egg products has not received sufficient attention for establishing a clear relationship between a type of bacterial enzyme and a specific characteristic of spoilage. Moreover, the putative involvement of heat-resistant enzymes in spoilage events, well known in the dairy industry, has never been investigated in the egg product environment.

The visual spoilage of egg white is difficult to evaluate. Liquefaction occurs throughout storage and, for the most part, is due to protein denaturation accompanied by the growth of microorganisms. Nevertheless, considering the antimicrobial properties of egg white described previously (see Section 20.2.2.2), this medium does not favor bacterial growth, contrary to egg yolk and whole egg. Events of egg white spoilage mainly imply defects in the breaking process, providing nutrients from the egg yolk in the case of poor separation of the egg white and the yolk. However, when the egg white enters in the composition of finished products, such as certain desserts, these defensive properties are no longer effective. Bacteria that have persisted in the egg white are found in finished products, sometimes undercooked or not cooked, generally very rich in metabolizable substrates, allowing them to grow and alter the product. Thus, the persistence and the development of certain bacteria can involve spoilage issues of finished products. Thus, low-cooked and rich (high nutritional) products, which contain egg white in their formula, are often a concern for food manufacturers.

Cooked egg product spoilage depends on numerous factors such as storage temperature, number, and types of microorganisms, ingredients used in the formulation, and type of finished product packaging. Under aerobic packaging, spoilage is caused by *Pseudomonas*, *Serratia* species, yeasts, molds and other microorganisms found in egg processing plants. Yeasts and molds are also capable of spoiling hard-cooked eggs packed in highly acidic brines.

Boiled eggs spoilage mainly involves psychrotrophic lactic acid bacteria. Therefore, from recall spoiled boiled eggs, preserved in brine and stored in vacuum packaging in Belgium, Pothakos et al. (2014) identified *Leuconostoc gelidum* as the dominant flora. Spoilage manifestation was a pungent and sour off-odor and the development of a thick and viscous texture of the brine due to extreme formation of ropy slime.

Noonim and Venkatachalam (2021) studied physicochemical and microbial qualities of salted duck egg white meringues stored in different alternative packaging and at two alternative storage temperatures (30°C and 40°C) over a period of 120 days. They reported that meringues kept at 40°C experienced more loss of quality than those kept at 30°C. During storage, the lightness and yellowness of the meringues diminished, while its redness increased, and the sample volume shrunk. The sample's weight, moisture, Aw, and pH were highest in the control samples when stored at 40°C. Texture features such as hardness, chewiness, cohesiveness, fracturability, gumminess, and springiness decreased in all cases with storage time, whereas adhesiveness and resilience increased. The control samples had significant levels of pathogenic bacteria (*Escherichia coli* and *Salmonella*) and spoilage microorganisms (yeasts and molds), reducing the shelf life to 90 days. In contrast, alternative packaging tested allowed to extend the shelf life of meringues by maintaining the microbial and textural qualities throughout the testing period of 120 days (Noonim & Venkatachalam, 2021).

In vanilla cream pudding containing egg yolk, Moschonas et al. (2021) identified *Bacillus subtilis* subsp. *subtilis* as the dominant spoilage microorganism. From cream stored aerobically at 4, 8, 12, and 15°C for a maximum period of 40 days, spoilage occurred only at 12°C and 15°C. They reported notable biochemical changes with an increase of lactic acid content, and to a lesser extent of formic and acetic acids, and an increase in the concentration of glucose and fructose at the expense of sucrose and lactose. The increase of lactic acid has equally been reported by Techer et al. (2020). From 64 spoiled desserts composed of custard cream, these authors reported that total aerobic microbial populations varied from $10^{4.5}$ to $10^{10.5}$ CFU/mL, with an average value of $10^{7.5 \pm 1.3}$ CFU/mL. The spoilage microorganisms mainly consisted of bacteria, and only three desserts were co-contaminated by yeasts and bacteria. Moreover, 81% of the spoiled desserts were contaminated with members of the *B. cereus* group. Coagulase negative *Staphylococcus, Enterococcus spp.* and *Leuconostoc spp.* were present in 25%, 19%, and 9% of the spoiled desserts, respectively. To a lower extent (less than 5%), 13 other genera were also detected, belonging to Firmicutes, *Actinobacteria,* and *Proteobacteria.* Further challenge testing with main spoilage species were investigated in custard cream stored at 10°C to check the spoilage capabilities of bacteria, and to evaluate more precisely biochemical changes. *B. cereus* group was associated with a high proteolysis, phospholipolysis, and the production of a range of compounds, including esters, compounds derived from sugar metabolism such as succinic acid, and compounds derived from amino acid catabolism, such as 2- and 3-methylbutanoic acid, 2-methylpropanol, and sulfur-containing compounds (dimethyl disulfide, dimethyl trisulfide, and mercaptoacetone). Some *B. cereus* group strains were particularly associated with the production of branched-chain volatile compounds, and other by their inability to use sucrose. *Staphylococcus* spp. isolates were mainly associated with the production of phenolic compounds, and low or no decrease of custard cream pH. *Enterococcus* spp. and *Leuconostoc* spp. isolates were mainly associated with high levels of compounds derived from sugar metabolism, such as diacetyl, acetoin, 2, 3-pentanedione, acetic and lactic acids, high acidification, and sensory defects such as the occurrence of yoghurt-odors and custard cream phase shift (Techer et al., 2020).

Some of the examples of egg-based food spoilage presented above are summarized in Table 20.3.

20.3.4.4 The Risk Due to Bacillus Cereus in Egg Products: an Illustration of the Selection Imposed by Preservation Process

Specific attention has to be considered for bacteria such as *Bacillus cereus* group, whose presence and development are mainly linked to the selection pressure due to the processes implemented by the egg-processing sector. As an illustration, Baron and collaborators isolated, in 2007, a psychrotrophic bacteria belonging to the *Bacillus cereus* group from a spoiled sample of liquid whole egg. They have shown that this bacterium was able to grow in liquid whole egg and to produce toxins, even at refrigerated temperatures. Moreover, it expressed adhesion capabilities that allowed it to form biofilms (Baron et al., 2007). These results highlighted both sanitary and spoilage issues that these bacteria may cause to the egg product sector. Further investigations were also performed at different levels of the field.

In 50 conventional layer farms in Western France, eggs were collected in warm and cold seasons. Egg surfaces were analyzed for the presence of *B. cereus* group bacteria, environmental measurements were recorded, and production practices were identified through a questionnaire filled out by the farmers. A total of 44% of the farms were contaminated by mesophilic bacteria, and 10% by psychrotrophic *B. cereus* group bacteria. No significant effect of the season was observed. Several procedures were associated with reduced eggshell contamination by mesophilic bacteria, including the limitation of dust formation from manure and feeding, and efficient disinfection of the feed silo, houses and the sanitary wall between houses (Kone et al., 2013).

At the egg products production stage, a study performed in six egg-breaking factories have shown contamination by

TABLE 20.3

Examples of Egg-Based Food Spoilage: Microorganisms Involved and Consequences

Products	Storage Conditions	Microbial Evolution/ Spoilage Microorganisms	Sensory and/or Chemical Changes
Boiled eggs (Pothakos et al., 2014)	Refrigerated (from local supermarket) Storage in brine or in vacuum packaging	*Leuconostoc gelidum* as the dominant spoilage microorganism	Pungent and sour off-odor Development of thick and viscous texture of the brine due to extreme formation of ropy slime
Salted duck egg white meringues (Noonim & Venkatachalam, 2021)	Package: paper control Storage at 30°C and 40°C for 120 days	Significant levels of pathogenic bacteria (*Escherichia coli* and *Salmonella*) and spoilage microorganisms (yeasts and molds)	Decrease of lightness and yellowness; increase of redness Decrease of hardness, chewiness, cohesiveness, fracturability, gumminess; increase of adhesiveness and resilience
Vanilla cream pudding containing egg yolk (Moschonas et al., 2021)	Aerobic storage at 4°C, 8°C, 12°C, or 15°C for maximum 40 days	*Bacillus subtilis* subsp. *subtilis* as the dominant spoilage microorganism. Spoilage occurred only at 12°C and 15°C	Increase of lactic acid, and to a lesser extent formic and acetic acids Increase of the glucose and fructose concentration, at the expense of sucrose and lactose
"Ile flottante" dessert : egg white foam + vanilla cream pudding containing whole egg (Techer et al., 2020)	Accelerated shelf life tests at 10°C and 30°C	Total aerobic microbial populations: $10^{4.5}$ and $10^{5.7}$ CFU/mL *B. cereus* group, coagulase negative Staphylococcus, *Enterococcus* spp., *Leuconostoc* spp.: in 81%, 25%, 19%, and 9% of the spoiled desserts, respectively; yeasts and bacteria in 5%.	pH decrease Sensorial changes (mainly off-odors)
Vanilla cream pudding containing whole egg (Techer et al., 2020)	Storage at 10°C for 21 days	*B. cereus* group	Low acidification High proteolysis, phospholipolysis, Increase of esters, succinic acid, 2- and 3-methylbutanoic acid, 2-methylpropanol, dimethyl disulfide, dimethyl trisulfide, and mercaptoacetone
		Staphylococcus spp.	Low or no acidification Production of phenolic compounds
		Enterococcus spp. and *Leuconostoc* spp.	Significant acidification Yoghurt off-odors Custard cream phase shift Increased levels of diacetyl, acetoin, 2, 3-pentanedione, acetic and lactic acids

Bacillus cereus in all investigated plants. Bacteria occurrence has been studied after the enrichment of samples at two temperatures, 37°C and 10°C, to favor mesophilic or psychrotrophic bacteria growth, respectively. Higher *Bacillus cereus* incidence was observed in samples enriched at 37°C (76.1% positive samples) compared to 10°C (43.7% positive samples) (Jan et al., 2011). A significant effect of the season was observed for enrichments performed at 10°C, with a higher occurrence in the cold season than in the warm season.

These two studies showed few *Bacillus cereus* at the farm level, but high occurrence in egg products, probably linked to the bacteria ability to adhere and form biofilms in the egg-breaking environment, as reported by Techer et al. (2019). These latter authors were then interested in spoilage flora of finished products considered as sensitive, because of low or no cooked egg products that compose them. A study performed on 64 spoiled custard cream desserts manufactured by two French companies has confirmed *B. cereus* group bacteria as the main bacteria involved in spoilage. These bacteria were identified as dominating flora in more than 80% of the desserts (Techer et al., 2020). These works highlight that *B. cereus* group bacteria are among the most persistent of heat-resistant psychrotrophic bacteria in pasteurized liquid egg products, and they can express spoilage activities, in the egg product itself and when used as an ingredient in different finished products, even at low temperature.

20.3.5 Control of the Microbiological Quality of Egg Products

The production of safe food requires the application of good hygiene practices, good agricultural practices, as well as the principles of Hazard Analysis Critical Control Point (HACCP). The HACCP system is defined as the food safety system, including the HACCP plan, to prevent, eliminate, or reduce food safety hazards. The HACCP plan includes the hazard analysis, supporting documentation, including prerequisite programs and scientific documentation used to support decisions, and all HACCP records. These approaches enable the development and implementation of a total food-safety management system that will control most reliably the significant hazards in the food produced. Some hazards are better addressed through good agricultural practices or good hygiene practices (e.g. controlling the initial levels of a hazard through good cleaning and disinfection practices). Others are clearly better addressed through HACCP, with a defined Critical Control Point (CCP) that has been validated to control the hazard of concern (e.g. reducing the level of a hazard by pasteurization).

The use of testing, including microbiological analysis, to validate and check the good hygiene practices and CCP is essential. Microbiological criteria may be established for quality, as well as safety concerns and can relate to raw materials, ingredients, semi-processed, and finished products.

A microbiological criterion should include an indication of the microorganisms and/or toxins/metabolites whose presence is undesirable; the analytical methods for their detection and/or quantification; the plan for the number of units to be sampled; the microbiological limits considered appropriate for the food; and the number of units that must comply with these limits. A microbiological criterion should also specify the food and the stage(s) in the food chain to which it applies, and any action to be taken when it is not met. The microbiological criterion thus indicates a maximum tolerable number of specified microorganisms enumerated by a specified method, and in a specified food at a specified stage of manufacture. In general, microbiological criteria are used to distinguish between raw materials, ingredients, products, and batches that are acceptable or unacceptable to regulatory agencies and/or food establishment management. They can also be used to verify and/or validate the effectiveness of good hygienic practices and the HACCP system.

Distinctions can be made between:

1. Microbiological standards: standards are contained in international, national, and regional laws and regulations. Exceeding a standard for a pathogen, such as *Salmonella* or *Listeria*, may lead to a product recall and potentially punitive action.
2. Microbiological specifications: Purchase specifications are agreements between the supplier and the client of a product as a basis for sale. These criteria can be considered mandatory, and failure of the vendor to meet specifications can be used as a basis for product rejection.
3. Microbiological guidelines: guidelines are internal, advisory criteria established by a processor, a trade association, or sometimes a government. Failure to meet guidelines serves as an alert to the processor, indicating that remedial action should be taken. A wide variety of criteria fit into this category, such as results on pre-operational swabs from equipment, in-process samples from product, or equipment and environmental samples tested for pathogens.

The sampling plans classically used in food microbiology, initially defined by the ICMSF (International Commission on Microbiological Specifications for Foods), are designed as two or three class plans (attribute-based control plans).

The following symbols and terms are used in sampling plans:

i. n represents the number of sample units to be taken at random from a lot. This number may vary depending on the risk, lot size, and sometimes the number of units available; "n = 5" is often used.
ii. m represents the limit of microorganism concentration corresponding to satisfactory hygiene of the processes considered, concentration being usually expressed in number of CFU per g or mL or cm^2; m can also be expressed as "detected or not detected in x g of product" for two-class plans only.
iii. M (three-class plans only) represents the limit of unsatisfactory hygiene concentration, usually expressed as number of CFU per g or mL or cm^2.
iv. c represents the maximum allowable number of sample units, either of acceptable quality for a three-class design (i.e. the maximum number of values between m and M), or of unsatisfactory quality for a two-class design (i.e. the maximum number of values above m). If the number of sample units of acceptable or

unsatisfactory quality, as the case may be, is greater than c, the lot from which the sample is taken is considered as unacceptable. Three-class plans are typically with c = 2 and are less stringent than two-class plans; they are appropriate where health risk is relatively low. Two-class plans with c = 0 are usually used for situations where the health risk is significant, and more stringent control is needed.

The ideal situation for a manufacturer wishing to know the microbiological quality of its foodstuffs from the time of production is to be able, on one hand, to analyze all its production and, on the other hand, to search for and count all the pathogenic and spoilage microorganisms. However, the diversity of microorganisms, the heavy and cost nature of the experiments, and the destructive effect of the analyses force the food processor, who aims for efficiency, to select a reduced number of analyses, sufficiently informative to play the role of quality indicator. Consequently, the strategy currently followed consists, on one hand, in analyzing the main pathogens that may be encountered in the food and, on the other hand, in evaluating the potential of the food to be preserved without being altered by the search for indicator microorganisms.

The ICMSF proposed aerobic colony count, coliform, and *Salmonella* criteria for liquid and frozen egg products. At the point of manufacture, aerobic colony count may provide some information to verify the adequacy of the pasteurization process, as well as the general quality of the product. The numeration of *Enterobacteriaceae* is recommended because they represent a broader group of microorganisms, which should be inactivated during pasteurization, and including coliforms and some pathogenic bacteria such as *Salmonella*.

In Europe, egg products and ready-to-eat (RTE) foods containing raw egg are subject to a food safety criterion for *Salmonella* of n = 5, c = 0, m = absence in 25 g, and a process hygiene criterion for *Enterobacteriaceae* of n = 5, c = 2, m = 10/g and M = 10^2/g (Regulation (EC) 2073/2005). The number of sample units of the sampling plan may be reduced if the food business operator can demonstrate with historical documentation that effective procedures are in place.

In the USA, all egg products plants must comply with different requirements described under the 9 Code of Federal Regulations (CFR). Egg product manufacturers are mainly concerned by part 590 (Inspection of eggs and egg products), part 591 (Sanitation Requirements and Hazard Analysis and CCP), and part 592 (Voluntary Inspection of egg products). Pasteurized egg products must be free of detectable pathogens (*Salmonella* and *Listeria monocytogenes*), have no undesired levels of residues (pesticides, heavy metals, and persistent organic pollutants) and detectable toxins or levels of toxin-producing organisms that would be a public health concern. The U.S. Department of Agriculture's (USDA) Food Safety and Inspection Service (FSIS) published in 2020 a revised guideline giving information to assist plants producing egg products that undergo pasteurization, heat treatment, cooling, and freezing in complying with regulatory requirements associated with the safe production of egg products; HACCP; options to achieve lethality and/or safe cooling and freezing; and recommendations for meeting the sampling requirements for the detection of *Salmonella* spp. in egg products (FSIS, 2020). Four types of sampling programs are also conducted with egg products to fulfill regulatory requirements.

Salmonella surveillance sampling Program is required by regulation to be conducted by each Official Egg Products Plant. Each plant selects samples from the lots of pasteurized egg products that they produced, and determine the lots negative for the presence of *Salmonella* as product eligible for sale. Plants that can demonstrate a history of compliance by producing *Salmonella*-negative product can reduce the number of lots sampled for liquid and frozen egg products. FSIS has established a program that sets forth the level, frequency and number of samples that have to be analyzed for liquid and frozen egg products. The frequency of sampling moves from each lot (for new plants with no history of prior production, or plants that introduce a new product category (new type)) to every other sample or a ratio of 1:2. After establishing a history of compliance at the 1:2 ratio, the plant can then move to a 1:4 ratio, and eventually a 1:8 ratio. The 1:8 ratio is the lowest level of sampling permitted. All dried egg products must be tested at 100% and no reduced sampling is permitted for these products.

The second program, i.e. FSIS microbiological testing program for pasteurized egg products, aims to determine if egg product plants are adequately pasteurizing egg products as required by the 9 CFR § 590, 570, and 575, and the Egg Products Inspection Act (EPIA). FSIS collects and analyzes egg product samples for the presence of *Salmonella* and *Listeria monocytogenes*. By testing for the presence of these two bacteria, this program verifies the efficacy of each egg product plant's *Salmonella* surveillance program, and enhances the public health protection against pathogens associated with pasteurized egg products. All amenable egg products (pasteurized liquid, frozen, and dried egg products) are tested under the FSIS Microbiological Testing Program.

The third program is dedicated for certification (e.g. school lunch program) or export. Sampling is realized by the plant under FSIS inspection program personnel supervision, in accordance with the requirements of the certification program, or the Export Library for the importing country. All inspection activities that are required outside the scope of mandatory inspection are to be billed to the plant under the Voluntary Egg Products Inspection Program.

The fourth program concerns the analysis for chemical residues of veterinary drugs (drugs including sulfonamides), pesticides, and environmental contaminants (such as arsenics). Plant employee under the monitoring of inspection personal performs it. A violation occurs when a chemical residue is detected, and the residue is in excess of an established tolerance or action level.

Other sampling programs can be realized for egg products, for example in case of food-borne illness outbreak, or because FSIS is conducting a nationwide raw liquid egg microbiological baseline survey. Moreover, because pasteurized egg products are used in institutional settings (e.g. hospitals, long-term care), stringent sampling plans should be considered for products targeted for that market. Assessment and control of microbial loads in food processing environments is also important, because there is plenty of evidence that post-processing contamination can affect product quality and safety. Environmental testing is undertaken to ensure that good hygiene procedures are effective in minimizing product contamination from the processing environment. Swab or sponge samples can be collected before the start of operations to verify the effectiveness of cleaning and sanitation. Analysis for aerobic colony count is a common method. Typical aerobic colony counts on thoroughly cleaned and sanitized food contact surfaces are $<10^2$ CFU/cm^2. Higher numbers will be encountered during production.

Although not mandatory, some microbiological tests are performed in egg products factory in order to satisfy requirement specifications expected by clients for specific applications, and/or to anticipate eventual complaints. To evaluate hygienic quality of raw material or potential spoilage of egg products or finished products, methods targeting mainly bacterial growth or markers can also be applied. Conventional microbiological analyses focus on the search for psychrotrophic flora, or a bacterial genus known to cause spoilage problems, such as the *Bacillus* genus. Nevertheless, there may be a delay in obtaining results, especially when psychrotrophic flora are searched for.

Alternative methods such as PCR testing are commonly used for a rapid detection of pathogenic (*Salmonella* and *Listeria*), and spoilage bacteria. Thus, Postollec et al. (2010) developed a method to identify spore-forming bacteria among the residual flora of many foods, including egg products, using molecular biology tools. This sensitive method allowed the identification of certain species that were not detected by classical microbiology methods, or to identify and discriminate certain species within the *B. cereus* group. However, it does not allow to discriminate the spoilage bacteria from the no-spoiling ones. Indeed, the search for a specific species does not necessarily imply that this species is involved in spoilage problems because the spoilage potential can vary from one strain to another, as it has already been shown with strains from milk (Dogan & Boor, 2003). Other approaches using optical or calorimetric methods have been used to study the growth of bacteria involved in egg product spoilage (Riva et al., 2001; Correa et al., 2008). Although convenient to provide rapid results, the lack of sensitivity and the use of equipment requiring significant technical skills limit their routine use by the egg processing industry.

Egg manufacturers can also use methods based on the detection and quantification of biochemical markers using sensorial analysis, convenient tools such as enzymatic/immunological kits, or more advanced technology such as chromatographic methods. Among these markers, volatile compounds (e.g. dimethyl disulfide), organic acids (lactic, succinic acids), and uracil were the most studied (Brown et al., 1986; Morris et al., 1989; Alamprese et al., 2004; Hidalgo et al., 2008). The use of an "electronic nose" was proposed by Suman et al. (2007) to replace the panels of sensory analysts, and/or the traditional methods of quality control (organic acids, microbiological analyses) to evaluate the freshness of egg products. These authors have shown that this method has a strong discriminating power, and that the results obtained are actually correlated to chemical (organic acids production) and microbiological (bacterial growth) changes during the degradation process. Moreover, the "electronic nose" provides rapid responses, in 30 to 40 min for three repetitions for the same sample, compared to the hours required to complete traditional chemical and microbiological analyses. Regarding the economic aspect, even if the cost of the equipment is high, the use of this device is advantageous compared to the use of enzymatic kits (Suman et al., 2007).

It is also important to highlight that there are, to date, no universal markers that can be used for both evaluating the quality of all types of egg products, and anticipating spoilage issue. The search for specific markers depends on the context and has to be adapted to the aim of the analysis, e.g. to evaluate the freshness of the raw material or finished product quality that has been stored in specific conditions, to anticipate potential spoilage issue for egg-based sensitive products, or as criteria in specific tests such as aging tests. The latest or "Accelerated Shelf Life Testing" (ASLT) is defined by the study of the behavior of microorganisms naturally present in a food matrix (NF V01-002 standard). These tests consist of storing products under conditions of time and temperature that accelerate the development of microorganisms and, consequently, the spoilage. Among manufacturers of egg products, these tests are generally not used routinely, except for egg products packaged in small volumes (e.g. 1 L). In the case of dessert manufacturers, producing sensitive products using egg products as ingredients, these tests can be an integral part of the verification or validation of the control, with regard to the microbiological quality of the products.

20.4 Conclusion

Egg content from healthy flocks is generally sterile; however, microorganisms can contaminate it. The risk of contamination, and in particular by *Salmonella*, is a major concern of egg sector and egg industry. However, the shell egg has remarkable self-defense properties, originally intended to preserve the embryo and limit or reduce bacterial penetration and growth of bacteria from endogenous or exogenous origin. The housing systems, the egg collection and storage conditions, and the packaging must prevent the multiplication of microorganisms and the damage to the natural antimicrobial egg defenses. A good knowledge of the microorganisms and egg antimicrobial properties, and compliance with good hygiene practices, from farm to egg commercialization, should ensure a satisfactory microbiological egg quality.

In egg industry, egg breaking results in an unavoidable contamination of the egg products, and in the loss of much of the natural egg defenses. Even though the flora of egg products is highly related to the flora of the incoming shell eggs, it is mainly influenced by transformation and stabilization processes as well. Thus, the most represented bacteria in egg products are not necessarily the most represented in the flora of the eggs before heat treatment, but rather the most resistant to the processes, especially pasteurization. Additional selection pressure is applied during egg product storage at refrigeration temperatures; species able to grow at low temperatures will then develop. If sanitary risk, especially contamination by *Salmonella,* which remains the major safety preoccupation of the sector, is now under the control of surveillance programs on farms and of effective pasteurization processes, manufacturers have to deal with microorganisms selected by the preservation processes used.

To limit the presence of pathogens and spoilage bacteria, and considering different aspects such as economic and evolution of consumers' demand, various promising research works dealing with new food processing technologies, natural antibacterial technology, or the development of analytical tools have to be explored further.

REFERENCES

Aganovic, K., Hertel, C., Vogel, R. F., Johne, R., Schlüter, O., Schwarzenbolz, U., & Heinz, V. (2021). Aspects of high hydrostatic pressure food processing : Perspectives on technology and food safety. *Comprehensive Reviews in Food Science and Food Safety*, *20*(4), 3225–3266. https://doi.org/10.1111/1541-4337.12763

Aguilera, O., Quiros, L. M., & Fierro, J. F. (2003). Transferrins selectively cause ion efflux through bacterial and artificial membranes. *FEBS Letters*, *548*(1), 5–10. https://doi.org/10.1016/S0014-5793(03)00719-1

Alabdeh, M., Lechevalier, V., Nau, F., Gautier, M., Cochet, M. F., Gonnet, F., & Baron, F. (2011). Role of incubation conditions and protein fraction on the antimicrobial activity of egg white against *Salmonella* enteritidis and *Escherichia coli*. *Journal of Food Protection*, *74*(1), 24–31. https://doi.org/10.4315/0362-028X.JFP-10-157

Alamprese, C., Rossi, M., Casiraghi, E., Hidalgo, A., & Rauzzino, F. (2004). Hygienic quality evaluation of the egg product used as ingredient in fresh egg pasta. *Food Chemistry*, *87*(2), 313–319. https://doi.org/10.1016/j.foodchem.2003.12.030

Andoni, E., Ozuni, E., Bijo, B., Shehu, F., Maçi, R., & Ranucci, D. (2020, December 19–20). Utilize of essential oils as an alternative method to reduce *Salmonella* in poultry production. A review. *Online International Conference on Life Sciences* (ICLS20). Turkey

Andrews, S. C., Robinson, A. K., & Rodríguez-Quiñones, F. (2003). Bacterial iron homeostasis. *FEMS Microbiology Reviews*, *27*(2–3), 215–237. https://doi.org/10.1016/S0168-6445(03)00055-X

Bain, M. M., Dunn, I. C., Wilson, P. W., Joseph, N., De Ketelaere, B., De Baerdemaeker, J., & Waddington, D. (2006). Probability of an egg cracking during packing can be predicted using a simple non-destructive acoustic test. *British Poultry Science*, *47*(4), 462–469. https://doi.org/10.1080/00071660600829233

Bain, M. M., Mcdade, K., Burchmore, R., Law, A., Wilson, P. W., Schmutz, M., & Dunn, I. C. (2013). Enhancing the egg's natural defence against bacterial penetration by increasing cuticle deposition. *Animal Genetics*, *44*(6), 661–668. https://doi.org/10.1111/age.12071

Ball, H. R., Meester, F. D., & Schuman, J. D. (2002). Treatment of food products using humidity controlled air (United States Patent No US6455094B1). https://patents.google.com/patent/US6455094B1/en

Banks, J. G., Board, R. G., & Sparks, N. H. (1986). Natural antimicrobial systems and their potential in food preservation of the future. *Biotechnology and Applied Biochemistry*, *8*(2–3), 103–147.

Baron, F., Bonnassie, S., Alabdeh, M., Cochet, M.-F., Nau, F., Guérin-Dubiard, C., & Jan, S. (2017). Global Gene-expression analysis of the response of *Salmonella* enteritidis to egg white exposure reveals multiple egg white-imposed stress responses. *Frontiers in Microbiology*, *8*. https://www.frontiersin.org/article/10.3389/fmicb.2017.00829

Baron, F., Cochet, M.-F., Alabdeh, M., Guérin-Dubiard, C., Gautier, M., Nau, F., & Jan, S. (2020). Egg-white proteins have a minor impact on the bactericidal action of egg white toward *Salmonella* enteritidis at 45°C. *Frontiers in Microbiology*, *11*. https://www.frontiersin.org/article/10.3389/fmicb.2020.584986

Baron, F., Cochet, M.-F., Grosset, N., Madec, M.-N., Briandet, R., Dessaigne, S., & Jan, S. (2007). Isolation and characterization of a psychrotolerant toxin producer, *Bacillus* weihenstephanensis, in liquid egg products. *Journal of Food Protection*, *70*(12), 2782–2791. https://doi.org/10.4315/0362-028X-70.12.2782

Baron, F., Gautier, M., & Brule, G. (1997). Factors involved in the inhibition of growth of *Salmonella* enteritidis in liquid egg white. *Journal of Food Protection*, *60*(11), 1318–1323. https://doi.org/10.4315/0362-028X-60.11.1318

Baron, F., Jan, S., Gonnet, F., Pasco, M., Jardin, J., Giudici, B., & Nau, F. (2014). Ovotransferrin plays a major role in the strong bactericidal effect of egg white against the *Bacillus* cereus group. *Journal of Food Protection*, *77*(6), 955–962. https://doi.org/10.4315/0362-028X.JFP-13-473

Baron, F., Nau, F., Guérin-Dubiard, C., Bonnassie, S., Gautier, M., Andrews, S. C., & Jan, S. (2016). Egg white versus *Salmonella* enteritidis ! A harsh medium meets a resilient pathogen. *Food Microbiology*, *53*, 82–93. https://doi.org/10.1016/j.fm.2015.09.009

Baron, F., Nau, F., Guérin-Dubiard, C., Gonnet, F., Dubois, J. J., & Gautier, M. (2003). Effect of dry heating on the microbiological quality, functional properties, and natural bacteriostatic ability of egg white after reconstitution. *Journal of Food Protection*, *66*(5), 825–832. https://doi.org/10.4315/0362-028x-66.5.825

Beckett, D. (2007). biotin sensing : Universal influence of biotin status on transcription. *Annual Review of Genetics*, *41*(1), 443–464. https://doi.org/10.1146/annurev.genet.41.042007.170450

Berrang, M. E., Cox, N. A., Frank, J. F., & Buhr, R. J. (1999). Bacterial penetration of the eggshell and shell membranes of the chicken hatching egg : A review. *Journal of Applied Poultry Research*, *8*(4), 499–504. https://doi.org/10.1093/japr/8.4.499

Bornert, G. (2000). Importance des bactéries psychrotrophes en hygiène des denrées alimentaires. *Revue Médicale Vétérinaire*, *151*, 4–13.

Bortolaia, V., Espinosa-Gongora, C., & Guardabassi, L. (2016). Human health risks associated with antimicrobial-resistant enterococci and Staphylococcus aureus on poultry meat. *Clinical Microbiology and Infection*, *22*(2), 130–140. https://doi.org/10.1016/j.cmi.2015.12.003

Bourgeois, C. M., & Leveau, J. Y. (1996). *Microbiologie alimentaire. Vol. 1: Aspect microbiologique de la sécurité et de la qualité alimentaire*. Lavoisier.

Boziaris, I. S., Humpheson, L., & Adams, M. R. (1998). Effect of nisin on heat injury and inactivation of *Salmonella* enteritidis PT4. *International Journal of Food Microbiology*, *43*(1–2), 7–13. https://doi.org/10.1016/s0168-1605(98)00083-x

Broom, L. J. (2015). Organic acids for improving intestinal health of poultry. *World's Poultry Science Journal*, *71*(4), 630–642. https://doi.org/10.1017/S0043933915002391

Brown, M. L., Holbrook, D. M., Hoerning, E. F., Legendre, M. G., & St. angelo, A. J. (1986). Volatile indicators of deterioration in liquid egg products. *Poultry Science*, *65*(10), 1925–1933. https://doi.org/10.3382/ps.0651925

Buzrul, S. (2021). High hydrostatic pressure applications on liquid whole egg. *World's Poultry Science Journal*, *77*(1), 71–90. https://doi.org/10.1080/00439339.2020.1866963

Byrne, B., Dunne, G., & Bolton, D. J. (2006). Thermal inactivation of *Bacillus* cereus and clostridium perfringens vegetative cells and spores in pork luncheon roll. *Food Microbiology*, *23*(8), 803–808. https://doi.org/10.1016/j.fm.2006.02.002

Calderón-Miranda, M. L., Barbosa-Cánovas, G. V., & Swanson, B. G. (1999). Inactivation of *Listeria* innocua in liquid whole egg by pulsed electric fields and nisin. *International Journal of Food Microbiology*, *51*(1), 7–17. https://doi.org/10.1016/S0168-1605(99)00070-7

Calenge, F., Vignal, A., Demars, J., Fève, K., Menanteau, P., Velge, P., & Beaumont, C. (2011). New QTL for resistance to *Salmonella* carrier-state identified on fowl microchromosomes. *Molecular Genetics and Genomics*, *285*(3), 237–243. https://doi.org/10.1007/s00438-011-0600-9

Callewaert, L., Aertsen, A., Deckers, D., Vanoirbeek, K. G. A., Vanderkelen, L., Herreweghe, J. M. V., & Michiels, C. W. (2008). A new family of lysozyme inhibitors contributing to lysozyme tolerance in Gram-negative bacteria. *PLoS Pathogens*, *4*(3), e1000019. https://doi.org/10.1371/journal.ppat.1000019

Cardoso, M. J., Nicolau, A. I., Borda, D., Nielsen, L., Maia, R. L., Møretrø, T., & Teixeira, P. (2021). *Salmonella* in eggs : From shopping to consumption—A review providing an evidence-based analysis of risk factors. *Comprehensive Reviews in Food Science and Food Safety*, *20*(3), 2716–2741. https://doi.org/10.1111/1541-4337.12753

Carter, A. J., Adams, M. R., Woodward, M. J., & La Ragione, R. M. (2009). Control strategies for *Salmonella* colonisation of poultry : The probiotic perspective. *Food Science & Technology Bulletin: Functional Foods*, *5*(9), 103–115. https://doi.org/10.1616/1476-2137.15682

Catteau, M. (1999). Pathogènes rencontrés lors de la conservation par le froid. In Office des publications officielles des Communautés européennes, *La microbiologie prévisionnelle appliquée à la conservation des aliments réfrigérés* (pp. 333).

Chen, X., Li, X., Guo, Y., Li, W., Song, J., Xu, G., & Zheng, J. (2019). Impact of cuticle quality and eggshell thickness on egg antibacterial efficiency. *Poultry Science*, *98*(2), 940–948. https://doi.org/10.3382/ps/pey369

Chen, J., Thesmar, H. S., & Kerr, W. L. (2005). Outgrowth of *Salmonellae* and the physical property of albumen and vitelline membrane as influenced by egg storage conditions. *Journal of Food Protection*, *68*(12), 2553–2558. https://doi.org/10.4315/0362-028X-68.12.2553

Clavijo, R. I., Loui, C., Andersen, G. L., Riley, L. W., & Lu, S. (2006). Identification of genes associated with survival of *Salmonella* enterica serovar enteritidis in chicken egg albumen. *Applied and Environmental Microbiology*, *72*(2), 1055–1064. https://doi.org/10.1128/AEM.72.2.1055-1064.2006

Clay, C. E., & Board, R. G. (1991). Growth of *Salmonella* enteritidis in artificially contaminated hens' shell eggs. *Epidemiology & Infection*, *106*(2), 271–281. https://doi.org/10.1017/S095026880004841X

Cochet, M.-F., Baron, F., Bonnassie, S., Jan, S., Leconte, N., Jardin, J., & Nau, F. (2021). Identification of new antimicrobial peptides that contribute to the bactericidal activity of egg white against *Salmonella* enterica serovar enteritidis at 45°C. *Journal of Agricultural and Food Chemistry*, *69*(7), 2118–2128. https://doi.org/10.1021/acs.jafc.0c06677

Cogan, T. A., Domingue, G., Lappin-Scott, H. M., Benson, C. E., Woodward, M. J., & Humphrey, T. J. (2001). Growth of *Salmonella* enteritidis in artificially contaminated eggs : The effects of inoculum size and suspending media. *International Journal of Food Microbiology*, *70*(1), 131–141. https://doi.org/10.1016/S0168-1605(01)00540-2

Correa, E. C., Diaz-Barcos, V., Fuentes-Pila, J., Barreiro, P., & Gonzalez, M. C. (2008). Modeling ovoproduct spoilage with red LED light. *Acta Horticulturae*, *802*, 265–272. https://doi.org/10.17660/ActaHortic.2008.802.34

Correnti, C., Clifton, M. C., Abergel, R. J., Allred, B., Hoette, T. M., Ruiz, M., & Strong, R. K. (2011). Galline ex-FABP is an antibacterial siderocalin and a lysophosphatidic acid sensor functioning through dual ligand specificities. *Structure*, *19*(12), 1796–1806. https://doi.org/10.1016/j.str.2011.09.019

Cousin, M. A. (1982). Presence and activity of psychrotrophic microorganisms in milk and dairy products : A Review. *Journal of Food Protection*, *45*(2), 172–207. https://doi.org/10.4315/0362-028X-45.2.172

Curtis, P. A. M., Anderson, K. E., Conner, D. E., Hughes, L. A., & Keener, K. M. (2000). Use of co2 cooling in treatment of poultry eggs (World Intellectual Property Organization Patent No WO2000007453A1).

Davidson, L. J. (2001). Pasteurized in-shell chicken eggs and method for production thereof (United States Patent No US6322833B1). https://patents.google.com/patent/US6322833B1/en

Davidson, P. M., Sofos, J. N., & Branen, A. L. (2005). *Antimicrobials in food* (3rd ed.). CRC Press. https://doi.org/10.1201/9781420028737

Davies, R. H., & Breslin, M. (2003). Investigation of *Salmonella* contamination and disinfection in farm egg-packing plants. *Journal of Applied Microbiology*, *94*(2), 191–196. https://doi.org/10.1046/j.1365-2672.2003.01817.x

De Buck, J., Van Immerseel, F., Haesebrouck, F., & Ducatelle, R. (2004). Colonization of the chicken reproductive tract and egg contamination by *Salmonella*. *Journal of Applied Microbiology*, *97*(2), 233–245. https://doi.org/10.1111/j.1365-2672.2004.02294.x

De Reu, K., Grijspeerdt, K., Heyndrickx, M., Zoons, J., De Baere, K., Uyttendaele, M., Debevere, J., & Herman, L. (2005). Bacterial eggshell contamination in conventional cages, furnished cages and aviary housing systems for laying hens. *British Poultry Science*, *46*(2), 149–155. https://doi.org/10.1080/00071660500065359

De Reu, K., Grijspeerdt, K., Messens, W., Heyndrickx, M., Uyttendaele, M., Debevere, J., & Herman, L. (2006). Eggshell factors influencing eggshell penetration and whole egg contamination by different bacteria, including *Salmonella* enteritidis. *International Journal of Food Microbiology*, *112*(3), 253–260. https://doi.org/10.1016/j.ijfoodmicro.2006.04.011

De Reu, K., Messens, W., Heyndrickx, M., Rodenburg, T. B., Uyttendaele, M., & Herman, L. (2008). Bacterial contamination of table eggs and the influence of housing systems. *World's Poultry Science Journal*, *64*(1), 5–19. https://doi.org/10.1017/S0043933907001687

Delves-Broughton, J., Williams, G. C., & Wilkinson, S. (1992). The use of the bacteriocin, nisin, as a preservative in pasteurized liquid whole egg. *Letters in Applied Microbiology*, *15*(4), 133–136. https://doi.org/10.1111/j.1472-765X.1992.tb00746.x

Denis, C., Protais, J., Corbion, B., Picoche, B., Colin, P., & Lahellec, C. (1995). Thermorésistance of *Salmonella* enteritidis inoculated in egg products. Influence of water activity on thermoresistance. *Industries Alimentaires et Agricoles*, *112*, 659–663.

Derde, M., Lechevalier, V., Guérin-Dubiard, C., Cochet, M.-F., Jan, S., Baron, F., & Nau, F. (2013). Hen egg white lysozyme permeabilizes *Escherichia coli* outer and inner membranes. *Journal of Agricultural and Food Chemistry*, *61*(41), 9922–9929. https://doi.org/10.1021/jf4029199

Dewey-Mattia, D., Manikonda, K., Hall, A. J., Wise, M. E., & Crowe, S. J. (2018). Surveillance for foodborne disease outbreaks—United States, 2009–2015. *MMWR Surveillance Summaries*, *67*(10), 1–11. https://doi.org/10.15585/mmwr.ss6710a1

Dey, A., Misra, S. S., Dahiya, S. S., Balhara, A. K., Kumar, K., Das, A. K., & Imaz, J. A. (2017). Essential oils as phytogenic feed additive: Potential benefits on environment, livestock health and production. *International Journal of Economic Plants*, *4*(1), 177–181.

Dogan, B., & Boor, K. J. (2003). Genetic diversity and spoilage potentials among *Pseudomonas spp.* Isolated from fluid milk products and dairy processing plants. *Applied and Environmental Microbiology*, *69*(1), 130–138. https://doi.org/10.1128/AEM.69.1.130-138.2003

Doyle, M. E., & Mazzotta, A. S. (2000). Review of studies on the thermal resistance of *Salmonellae*. *Journal of Food Protection*, *63*(6), 779–795. https://doi.org/10.4315/0362-028X-63.6.779

Duboccage, L., Heyndrickx, M., Grijspeerdt, K., & Herman, L. (2001). Growth of *Salmonella* in egg white. *Mededelingen (Rijksuniversiteit Te Gent Fakulteit van de Landbouwkundige En Toegepaste Biologische Wetenschappen*, *66*(3b), 531–534.

Ebel, E. D., Mason, J., Thomas, L. A., Ferris, K. E., Beckman, M. G., Cummins, D. R., & Smithhisler, N. M. (1993). Occurrence of *Salmonella* enteritidis in unpasteurized liquid egg in the United States. *Avian Diseases*, *37*(1), 135–142. https://doi.org/10.2307/1591466

Ebel, E., & Schlosser, W. (2000). Estimating the annual fraction of eggs contaminated with *Salmonella* enteritidis in the United States. *International Journal of Food Microbiology*, *61*(1), 51–62. https://doi.org/10.1016/S0168-1605(00)00375-5

EFSA and ECDC (European Food Safety Authority and European Centre for Disease Prevention and Control). (2019). The European union one health 2018 zoonoses report. *EFSA Journal*, *17*, 276. https://doi.org/10.2903/j.efsa.2019.5926

EFSA and ECDC (European Food Safety Authority and European Centre for Disease Prevention and Control). (2021). The European union one health 2019 zoonoses report. *EFSA Journal*, *19*(2), 6406. https://doi.org/10.2903/j.efsa.2021.6406

EFSA, E. F. S. (2005). Opinion of the scientific panel on biological hazards (BIOHAZ) related to the microbiological risks on washing of table eggs. *EFSA Journal*, *3*(10), 269. https://doi.org/10.2903/j.efsa.2005.269

EFSA, E. F. S. (2014). Scientific opinion on the public health risks of table eggs due to deterioration and development of pathogens. *EFSA Journal*, *12*(7), 3782. https://doi.org/10.2903/j.efsa.2014.3782

EFSA, E. F. S., Gonzales, J. L., Roberts, H., Smietanka, K., Baldinelli, F., Ortiz-Pelaez, A., & Verdonck, F. (2018). Assessment of low pathogenic avian influenza virus transmission via raw poultry meat and raw table eggs. *EFSA Journal*, *16*(10), e05431. https://doi.org/10.2903/j.efsa.2018.5431

Ellison, R. T., Giehl, T. J., & LaForce, F. M. (1988). Damage of the outer membrane of enteric gram-negative bacteria by lactoferrin and transferrin. *Infection and Immunity*, *56*(11), 2774–2781. https://doi.org/10.1128/iai.56.11.2774-2781.1988

Elsbach, P., & Weiss, J. (1998). Role of the bactericidal/permeability-increasing protein in host defence. *Current Opinion in Immunology*, *10*(1), 45–49. https://doi.org/10.1016/S0952-7915(98)80030-7

Erasmus, C., & Rossouw, M. (2005). In-shell pasteurization of eggs. https://patentscope.wipo.int/search/en/detail.jsf?docId=WO2005102064

European Commission. (2011). Directive (EC) No. 1129/2011, Office of the Journal of the European Union L295, 33–176.

Facon, M. J., & Skura, B. J. (1996). Antibacterial activity of lactoferricin, lysozyme and EDTA against *Salmonella* enteritidis. *International Dairy Journal*, *6*(3), 303–313. https://doi.org/10.1016/0958-6946(95)00004-6

Fikiin, K., Akterian, S., & Stankov, B. (2020). Do raw eggs need to be refrigerated along the food chain? : Is the current EU regulation ensuring high-quality shell eggs for the European consumers? *Trends in Food Science & Technology*, *100*, 359–362. https://doi.org/10.1016/j.tifs.2020.04.003

Foegeding, P. M., & Leasor, S. B. (1990). Heat resistance and growth of *Listeria* monocytogenes in liquid whole egg. *Journal of Food Protection*, *53*(1), 9–14. https://doi.org/10.4315/0362-028X-53.1.9

Food and Drug Administration. (2006). Food additives permitted for direct addition to food for human consumption; Bacteriophage Preparation. 21 CFR 172, 71 FR 47729.

Foulquié Moreno, M. R., Sarantinopoulos, P., Tsakalidou, E., De Vuyst, L. (2006). The role and application of enterococci in food and health. *International Journal of Food Microbiology*, *106*(1), 1–24. https://doi.org/10.1016/j.ijfoodmicro.2005.06.026

Friedman, M., Henika, P. R., & Mandrell, R. E. (2003). Antibacterial activities of phenolic benzaldehydes and benzoic acids against campylobacter jejuni, *Escherichia coli*, *Listeria* monocytogenes, and *Salmonella* enterica. *Journal of Food Protection*, *66*(10), 1811–1821. https://doi.org/10.4315/0362-028x-66.10.1811

FSIS. (2020). FSIS food safety guideline for egg products. 42.

FSIS. (2021). Food safety and inspection service. U.S. Department of Agriculture. Microbiological Testing Program for Pasteurized Egg Products, Calendar Years 1995–2017. Retrieved on April 23, 2022, from https://www.fsis.usda.gov.

Galet, O., Cassin, D., & Jeantet, R. (2010). Les ovoproduits. In F. Nau, C. Guérin-Dubiard, F. Baron, & J. L. Thapon (Eds.), *Science et technologie de l'œuf. Vol. 2. De l'œuf aux ovoproduits* (pp. 223–277). Lavoisier.

Galiş, A. M., Marcq, C., Marlier, D., Portetelle, D., Van, I., Beckers, Y., & Théwis, A. (2013). Control of *Salmonella* contamination of shell eggs—Preharvest and postharvest methods : A review. *Comprehensive Reviews in Food Science and Food Safety*, *12*(2), 155–182. https://doi.org/10.1111/1541-4337.12007

Gálvez, A., López, R. L., Pulido, R. P., & Burgos, M. J. G. (2014). Natural antimicrobials for food biopreservation. In A. Galvez, M. J. Grande Burgos, R. Lucas López, & R. Pérez Pulido (Éds.), *Food biopreservation* (pp. 3–14). Springer. https://doi.org/10.1007/978-1-4939-2029-7_2

Gantois, I., Ducatelle, R., Pasmans, F., Haesebrouck, F., Gast, R., Humphrey, T. J., & Van Immerseel, F. (2009). Mechanisms of egg contamination by *Salmonella* enteritidis. *FEMS Microbiology Reviews*, *33*(4), 718–738. https://doi.org/10.1111/j.1574-6976.2008.00161.x

Gantois, I., Eeckhaut, V., Pasmans, F., Haesebrouck, F., Ducatelle, R., & Van Immerseel, F. (2008). A comparative study on the pathogenesis of egg contamination by different serotypes of *Salmonella*. *Avian Pathology: Journal of the W.V.P.A*, *37*(4), 399–406. https://doi.org/10.1080/03079450802216611

Garibaldi, J. A. (1970). Role of microbial iron transport compounds in bacterial spoilage of eggs. *Applied Microbiology*, *20*(4), 558–560. https://doi.org/10.1128/am.20.4.558-560.1970

Gast, R. K., & Beard, C. W. (1990). Isolation of *Salmonella* enteritidis from internal organs of experimentally infected hens. *Avian Diseases*, *34*(4), 991–993. https://doi.org/10.2307/1591394

Gast, R. K., & Beard, C. W. (1992). Detection and enumeration of *Salmonella* enteritidis in fresh and stored eggs laid by experimentally infected hens. *Journal of Food Protection*, *55*(3), 152–156. https://doi.org/10.4315/0362-028X-55.3.152

Gast, R. K., Guard-Petter, J., & Holt, P. S. (2002). Characteristics of *Salmonella* enteritidis contamination in eggs after Oral, aerosol, and intravenous inoculation of laying hens. *Avian Diseases*, *46*(3), 629–635. https://doi.org/10.1637/0005-2086(2002)046[0629:COSECI]2.0.CO;2

Gast, R. K., Guraya, R., Guard-Bouldin, J., & Holt, P. S. (2007). In vitro penetration of egg yolks by *Salmonella* enteritidis and *Salmonella* Heidelberg strains during thirty-six-hour ambient temperature storage. *Poultry Science*, *86*(7), 1431–1435. https://doi.org/10.1093/ps/86.7.1431

Gast, R. K., & Holt, P. S. (2000a). Deposition of phage type 4 and 13a *Salmonella* enteritidis strains in the yolk and albumen of eggs laid by experimentally infected hens. *Avian Diseases*, *44*(3), 706–710. https://doi.org/10.2307/1593116

Gast, R. K., & Holt, P. S. (2000b). Influence of the level and location of contamination on the multiplication of *Salmonella* enteritidis at different storage temperatures in experimentally inoculated eggs. *Poultry Science*, *79*(4), 559–563. https://doi.org/10.1093/ps/79.4.559

Gast, R. K., & Holt, P. S. (2001). Assessing the frequency and consequences of *Salmonella* enteritidis deposition on the egg yolk membrane. *Poultry Science*, *80*(7), 997–1002. https://doi.org/10.1093/ps/80.7.997

Gast, R. K., Holt, P. S., & Guraya, R. (2006). Effect of refrigeration on in vitro penetration of *Salmonella* enteritidis through the egg yolk membrane. *Journal of Food Protection*, *69*(6), 1426–1429. https://doi.org/10.4315/0362-028X-69.6.1426

Gast, R. K., Regmi, P., Guraya, R., Jones, D. R., Anderson, K. E., & Karcher, D. M. (2019). Colonization of internal organs by *Salmonella* enteritidis in experimentally infected laying hens of four commercial genetic lines in conventional cages and enriched colony housing. *Poultry Science*, *98*(4), 1785–1790. https://doi.org/10.3382/ps/pey541

Gautron, J., Hincke, M. T., Panheleux, M., Garcia-Ruiz, J. M., Boldicke, T., & Nys, Y. (2001). Ovotransferrin is a matrix protein of the hen eggshell membranes and basal calcified layer. *Connective Tissue Research*, *42*(4), 255–267. https://doi.org/10.3109/03008200109016840

Gong, D., Wilson, P. W., Bain, M. M., McDade, K., Kalina, J., Hervé-Grépinet, V., & Dunn, I. C. (2010). Gallin; An antimicrobial peptide member of a new avian defensin family, the ovodefensins, has been subject to recent gene duplication. *BMC Immunology*, *11*(1), 12. https://doi.org/10.1186/1471-2172-11-12

Guan, J., Grenier, C., & Brooks, B. W. (2006). In vitro study of *Salmonella* enteritidis and *Salmonella* typhimurium definitive type 104 : Survival in egg albumen and penetration through the vitelline membrane. *Poultry Science*, *85*(9), 1678–1681. https://doi.org/10.1093/ps/85.9.1678

Guenther, S., Herzig, O., Fieseler, L., Klumpp, J., & Loessner, M. J. (2012). Biocontrol of *Salmonella* typhimurium in RTE foods with the virulent bacteriophage FO1-E2. *International Journal of Food Microbiology*, *154*(1), 66–72. https://doi.org/10.1016/j.ijfoodmicro.2011.12.023

Guérin-Dubiard, C., Pasco, M., Mollé, D., Désert, C., Croguennec, T., & Nau, F. (2006). Proteomic analysis of hen egg white. *Journal of Agricultural and Food Chemistry*, *54*(11), 3901–3910. https://doi.org/10.1021/jf0529969

Guinebretière, M.-H., Thompson, F. L., Sorokin, A., Normand, P., Dawyndt, P., Ehling-Schulz, M., & … De Vos, P. (2008). Ecological diversification in the *Bacillus* cereus group. *Environmental Microbiology*, *10*(4), 851–865. https://doi.org/10.1111/j.1462-2920.2007.01495.x

Gumudavelli, V., Subbiah, J., Thippareddi, H., Velugoti, P. R., & Froning, G. (2007). Dynamic predictive model for growth of *Salmonella* enteritidis in egg yolk. *Journal of Food Science*, *72*(7), M254–M262. https://doi.org/10.1111/j.1750-3841.2007.00444.x

Guyot, N., Labas, V., Harichaux, G., Chessé, M., Poirier, J.-C., Nys, Y., & Réhault-Godbert, S. (2016). Proteomic analysis of egg white heparin-binding proteins : Towards the identification of natural antibacterial molecules. *Scientific Reports*, *6*(1), 27974. https://doi.org/10.1038/srep27974

Hervé-Grépinet, V., Réhault-Godbert, S., Labas, V., Magallon, T., Derache, C., Lavergne, M., & Nys, Y. (2010). Purification and characterization of avian beta-defensin 11, an antimicrobial peptide of the hen egg. *Antimicrobial Agents and Chemotherapy*, *54*(10), 4401–4409. https://doi.org/10.1128/AAC.00204-10

Hidalgo, A., Franzetti, L., Rossi, M., & Pompei, C. (2008). Chemical markers for the evaluation of raw material hygienic quality in egg products. *Journal of Agricultural and Food Chemistry*, *56*(4), 1289–1297. https://doi.org/10.1021/jf072201t

Hincke, M. T., Gautron, J., Panheleux, M., Garcia-Ruiz, J., McKee, M. D., & Nys, Y. (2000). Identification and localization of lysozyme as a component of eggshell membranes and eggshell matrix. *Matrix Biology*, *19*(5), 443–453. https://doi.org/10.1016/S0945-053X(00)00095-0

Holt, P. S. (2021). Centennial review : A revisiting of hen welfare and egg safety consequences of mandatory outdoor access for organic egg production. *Poultry Science*, *100*(12), 101436. https://doi.org/10.1016/j.psj.2021.101436

Holt, P. S., Geden, C. J., Moore, R. W., & Gast, R. K. (2007). Isolation of *Salmonella* enterica serovar enteritidis from houseflies (Musca domestica) found in rooms containing *Salmonella* serovar enteritidis-challenged hens. *Applied and Environmental Microbiology*, *73*(19), 6030–6035. https://doi.org/10.1128/AEM.00803-07

Hong, Y., Schmidt, K., Marks, D., Hatter, S., Marshall, A., Albino, L., & Ebner, P. (2016). Treatment of *Salmonella*-contaminated eggs and pork with a broad-spectrum, single bacteriophage : Assessment of efficacy and resistance development. *Foodborne Pathogens and Disease*, *13*(12), 679–688. https://doi.org/10.1089/fpd.2016.2172

Hopper, S. A., & Mawer, S. (1988). *Salmonella* enteritidis in a commercial layer flock. *The Veterinary Record*, *123*(13), 351. https://doi.org/10.1136/vr.123.13.351

Huang, X., Hu, M., Zhou, X., Liu, Y., Shi, C., & Shi, X. (2020). Role of yoaE gene regulated by CpxR in the survival of *Salmonella* enterica serovar enteritidis in antibacterial egg white. *mSphere*, *5*(1), e00638–19. https://doi.org/10.1128/mSphere.00638-19

Huang, C. M., & Lee, T. T. (2018). Immunomodulatory effects of phytogenics in chickens and pigs—A review. *Asian-Australasian Journal of Animal Sciences*, *31*(5), 617–627. https://doi.org/10.5713/ajas.17.0657

Huang, X., Zhou, X., Jia, B., Li, N., Jia, J., He, M., He, Y., Qin, X., Cui, Y., Shi, C., Liu, Y., & Shi, X. (2019). Transcriptional sequencing uncovers survival mechanisms of *Salmonella* enterica serovar enteritidis in antibacterial egg white. *mSphere*, *4*(1), e00700–18. https://doi.org/10.1128/mSphere.00700-18

Humphrey, T. (2006). Public health aspects of *Salmonella* enterica in food production. In P. Mastroeni & D. Maskell (Éds.), *Clinical, immunological and molecular aspects* (Vol. 4, pp. 89–116). Cambridge University Press.

Humphrey, T. J. (1990). Heat resistance in *Salmonella* enteritidis phage type 4 : The influence of storage temperatures Before heating. *Journal of Applied Bacteriology*, *69*(4), 493–497. https://doi.org/10.1111/j.1365-2672.1990.tb01540.x

Humphrey, T. J., Whitehead, A., Gawler, A. H. L., Henley, A., & Rowe, B. (1991). Numbers of *Salmonella* enteritidis in the contents of naturally contaminated hens' eggs. *Epidemiology & Infection*, *106*(3), 489–496. https://doi.org/10.1017/S0950268800067546

Huneau-Salaün, A., Michel, V., Huonnic, D., Balaine, L., & Le Bouquin, S. (2010). Factors influencing bacterial eggshell contamination in conventional cages, furnished cages and free-range systems for laying hens under commercial conditions. *British Poultry Science*, *51*(2), 163–169. https://doi.org/10.1080/00071668.2010.482462

Ibrahim, H.R., Higashiguchi, S., Juneja, L.R., Kim, M., & Yamamoto, T. (1996). A structural phase of heat-denaturated lysozyme with novel antimicrobial action. *Journal of Agricultural and Food Chemistry*, *44*(6), 1416–1423. https://doi.org/10.1021/jf9507147

Ibrahim, H. R., Thomas, U., & Pellegrini, A. (2001). A helix-loop-helix peptide at the upper lip of the active site cleft of lysozyme confers potent antimicrobial activity with membrane permeabilization action. *The Journal of Biological Chemistry*, *276*(47), 43767–43774. https://doi.org/10.1074/jbc.M106317200

Imai, C., Saito, J., & Ishikawa, M. (1986). Storage stability of liquid egg products below 0 C. *Poultry Science*, *65*(9), 1679–1686. https://doi.org/10.3382/ps.0651679

Iñiguez-Moreno, M., Gutiérrez-Lomelí, M., & Avila-Novoa, M. G. (2019). Kinetics of biofilm formation by pathogenic and spoilage microorganisms under conditions that mimic the poultry, meat, and egg processing industries. *International Journal of Food Microbiology*, *303*, 32–41. https://doi.org/10.1016/j.ijfoodmicro.2019.04.012

Jan, S., Baron, F., Alabdeh, M., Chaari, W., Grosset, N., Cochet, M. F., & Nau, F. (2013). Biochemical and micrographic evidence of Escherichia coli membrane damage during incubation in egg white under bactericidal conditions. *Journal of Food Protection*, *76*(9), 1523–1529. https://doi.org/10.4315/0362-028X.JFP-12-418

Jan, S., Brunet, N., Techer, C., Le Maréchal, C., Koné, A. Z., Grosset, N., & Baron, F. (2011). Biodiversity of psychrotrophic bacteria of the *Bacillus* cereus group collected on farm and in egg product industry. *Food Microbiology*, *28*(2), 261–265. https://doi.org/10.1016/j.fm.2010.05.029

Jeantet, R., Baron, F., Nau, F., Roignant, M., & Brulé, G. (1999). High intensity pulsed electric fields applied to egg white : Effect on *Salmonella* enteritidis inactivation and protein denaturation. *Journal of Food Protection*, *62*(12), 1381–1386. https://doi.org/10.4315/0362-028X-62.12.1381

Jeantet, R., Fernández, J., Keag, J., Grosset, N., Baron, F., & Korolczuk, J. (2004). Traitement d'ovoproduits en continu par champs électriques pulsés. *Sciences des Aliments*, *24*(2) 137–158. https://doi.org/10.3166/sda.24.137-158.

Julien, L. A., Fau, C., Baron, F., Bonnassie, S., Guérin-Dubiard, C., Nau, F., Gautier, M., Karatzas, K. A., Jan, S., & Andrews, S.C. (2020). The three lipocalins of egg-white: Only Ex-FABP inhibits siderophore-dependent iron sequestration by *Salmonella* Enteritidis. *Frontiers in Microbiology*, *11*, 913. https://doi.org/10.3389/fmicb.2020.00913

Kang, H., Loui, C., Clavijo, R. I., Riley, L. W., & Lu, S. (2006). Survival characteristics of *Salmonella* enterica serovar enteritidis in chicken egg albumen. *Epidemiology & Infection*, *134*(5), 967–976. https://doi.org/10.1017/S0950268806006054

Kato, A., Ibrahim, H. R., Watanabe, H., Honma, K., & Kobayashi, K. (1989). New approach to improve the gelling and surface functional properties of dried egg white by heating in dry state. *Journal of Agricultural and Food Chemistry*, *37*(2), 433–437. https://doi.org/10.1021/jf00086a036

Kato, A., Shibata, M., Yamaoka, H., & Kobayashi, K. (1988). Deamidation of lysozyme during the storage of egg white. *Agricultural and Biological Chemistry*, *52*(8), 1973–1978. https://doi.org/10.1080/00021369.1988.10868982

Keener, K. M. (2017). Chapter 16 - shell egg pasteurization. In P. Y. Hester (Éd.), *Egg innovations and strategies for improvements* (pp. 165–175). Academic Press. https://doi.org/10.1016/B978-0-12-800879-9.00016-0

Keller, L. H., Benson, C. E., Krotec, K., & Eckroade, R. J. (1995). *Salmonella* enteritidis colonization of the reproductive tract and forming and freshly laid eggs of chickens. *Infection and Immunity*, *63*(7), 2443–2449. https://doi.org/10.1128/iai.63.7.2443-2449.1995

Ketelaere, B. D., Bamelis, F., Kemps, B., Decuypere, E., & Baerdemaeker, J. D. (2004). Non-destructive measurements of the egg quality. *World's Poultry Science Journal*, *60*(3), 289–302. https://doi.org/10.1079/WPS200417

Kim, J. M., Marshall, M., Cornell, J. A., Iii, J. F. P., & Wei, C. I. (1995). Antibacterial activity of carvacrol, citral, and geraniol against *Salmonella* typhimurium in culture medium and on fish cubes. *Journal of Food Science*, *60*(6), 1364–1368. https://doi.org/10.1111/j.1365-2621.1995.tb04592.x

Kim, S. H., No, H. K., & Prinyawiwatkul, W. (2007). Effect of molecular weight, type of chitosan, and chitosan solution pH on the shelf-life and quality of coated eggs. *Journal of Food Science*, *72*(1), S044–S048. https://doi.org/10.1111/j.1750-3841.2006.00233.x

Knight, K. P., Bartlett, F. M., McKellar, R. C., & Harris, L. J. (1999). Nisin reduces the thermal resistance of *Listeria* monocytogenes Scott A in liquid whole egg. *Journal of Food Protection*, *62*(9), 999–1003. https://doi.org/10.4315/0362-028x-62.9.999

Kone, A. Z., Jan, S., Le Marechal, C., Grosset, N., Gautier, M., Puterflam, J., & Baron, F. (2013). Identifying risk factors for eggshell contamination by *Bacillus* cereus group bacteria in French laying farms. *British Poultry Science*, *54*(3), 298–305. https://doi.org/10.1080/00071668.2013.783900

Laible, N. J., & Germaine, G. R. (1985). Bactericidal activity of human lysozyme, muramidase-inactive lysozyme, and cationic polypeptides against streptococcus sanguis and streptococcus faecalis : Inhibition by chitin oligosaccharides. *Infection and Immunity*, *48*(3), 720–728. https://doi.org/10.1128/iai.48.3.720-728.1985

Lang, E. R., & Rha, C. (1982). Apparent shear viscosity of native egg white. International *Journal of Food Science & Technology*, *17*(5), 595–606. https://doi.org/10.1111/j.1365-2621.1982.tb00219.x

Lapuz, R., Tani, H., Sasai, K., Shirota, K., Katoh, H., & Baba, E. (2007). An epidemiological analysis of *Salmonella* enteritidis contamination in a rat-infested chicken layer farm, an egg processing facility, and liquid egg samples by pulsed-field gel electrophoresis. *The Journal of Veterinary Medical Science*, *69*(6), 649–652. https://doi.org/10.1292/jvms.69.649

Leasor, S. B., & Foegeding, P. M. (1989). *Listeria* species in commercially broken raw liquid whole egg. *Journal of Food Protection*, *52*(11), 777–780. https://doi.org/10.4315/0362-028X-52.11.777

Legros, J., Jan, S., Bonnassie, S., Gautier, M., Croguennec, T., Pezennec, S., & Baron, F. (2021). The role of ovotransferrin in egg-white antimicrobial activity : A review. *Foods*, *10*(4), 823. https://doi.org/10.3390/foods10040823

Li-Chan, E., & Nakai, S. (1989). Biochemical basis for the properties of egg white. *Critical Reviews in Poultry Biology (USA)*, *2*, 21–57.

Lindquist, S., & Craig, E. A. (1988). The heat-shock proteins. *Annual Review of Genetics*, *22*(1), 631–677. https://doi.org/10.1146/annurev.ge.22.120188.003215

Liot, R., & Anza, L. (1996). Procede de traitement de blanc d'oeuf liquide (European Union Patent No EP0917429B1). https://patents.google.com/patent/EP0917429B1/fr

Liu, Y., Du, J., Lai, Q., Zeng, R., Ye, D., Xu, J., & Shao, Z. (2017). Proposal of nine novel species of the *Bacillus* cereus group. *International Journal of Systematic and Evolutionary Microbiology*, *67*(8), 2499–2508. https://doi.org/10.1099/ijsem.0.001821

Liu, T., Lv, B., Zhao, W., Wang, Y., Piao, C., Dai, W., & Sun, F. (2020). Effects of ultrahigh temperature pasteurization on the liquid components and functional properties of stored liquid whole eggs. *BioMed Research International*, *2020*, e3465465. https://doi.org/10.1155/2020/3465465

Liu, M., Yu, W., Ren, F., & Wu, J. (2018). Formation and characterization of peptides in egg white during storage at ambient temperature. *Food Chemistry*, *263*, 135–141. https://doi.org/10.1016/j.foodchem.2018.04.042

Lock, J. L., & Board, R. G. (1992). Persistence of contamination of hens' egg albumen in vitro with *Salmonella* serotypes. *Epidemiology & Infection*, *108*(3), 389–396. https://doi.org/10.1017/S095026880004989X

Lu, H., Dunn, P. A., Wallner-Pendleton, E. A., Henzler, D. J., Kradel, D. C., Liu, J., & Miller, P. (2004). Investigation of H7N2 avian influenza outbreaks in two broiler breeder flocks in Pennsylvania, 2001-02. *Avian Diseases*, *48*(1), 26–33. https://doi.org/10.1637/6063

Mallet, S., Guesdon, V., Ahmed, A. M. H., & Nys, Y. (2006). Comparison of eggshell hygiene in two housing systems : Standard and furnished cages. *British Poultry Science*, *47*(1), 30–35. https://doi.org/10.1080/00071660500468132

Mann, K. (2007). The chicken egg white proteome. *Proteomics*, *7*(19), 3558–3568. https://doi.org/10.1002/pmic.200700397

Mann, K., & Mann, M. (2011). In-depth analysis of the chicken egg white proteome using an LTQ orbitrap velos. *Proteome Science*, *9*(1), 7. https://doi.org/10.1186/1477-5956-9-7

Martin-Belloso, O., Vega-Mercado, H., Qin, B., Chang, F., Barbosa-Canovas, G., & Swanson, B. (2007). Inactivation of Escherichia coli suspended in liquid egg using pulsed electric fields. *Journal of Food Processing and Preservation*, *21*, 193–208. https://doi.org/10.1111/j.1745-4549.1997.tb00776.x

Marzec, A., Damaziak, K., Kowalska, H., Riedel, J., Michalczuk, M., Koczywąs, E., & Niemiec, J. (2019). Effect of hens age and storage time on functional and physiochemical properties of eggs. *Journal of Applied Poultry Research*, *28*(2), 290–300. https://doi.org/10.3382/japr/pfy069

Maurer, L. M., Yohannes, E., Bondurant, S. S., Radmacher, M., & Slonczewski, J. L. (2005). PH regulates genes for flagellar motility, catabolism, and oxidative stress in *Escherichia coli* K-12. *Journal of Bacteriology*, *187*(1), 304–319. https://doi.org/10.1128/JB.187.1.304-319.2005

McQuiston, J. R., Herrera-Leon, S., Wertheim, B. C., Doyle, J., Fields, P. I., Tauxe, R. V., & Logsdon, J. M. (2008). Molecular phylogeny of the *Salmonellae* : Relationships among *Salmonella* species and subspecies determined from four housekeeping genes and evidence of lateral gene transfer events. *Journal of Bacteriology*, *190*(21), 7060–7067. https://doi.org/10.1128/JB.01552-07

Merino, L., Procura, F., Trejo, F. M., Bueno, D. J., & Golowczyc, M. A. (2019). Biofilm formation by *Salmonella* sp. in the poultry industry: Detection, control and eradication strategies. *Food Research International*, *119*, 530–540.. https://doi.org/10.1016/j.foodres.2017.11.024. Epub 2017 Nov 21. PMID: 30884686.

Messens, W., Duboccage, L., Grijspeerdt, K., Heyndrickx, M., & Herman, L. (2004). Growth of *Salmonella* serovars in hens' egg albumen as affected by storage prior to inoculation. *Food Microbiology*, *21*(1), 25–32. https://doi.org/10.1016/S0740-0020(03)00045-5

Messens, W., Grijspeerdt, K., De Reu, K., De Ketelaere, B., Mertens, K., Bamelis, F., & Herman, L. (2007). Eggshell penetration of various types of hens' eggs by *Salmonella* enterica serovar enteritidis. *Journal of Food Protection*, *70*(3), 623–628. https://doi.org/10.4315/0362-028X-70.3.623

Messens, W., Grijspeerdt, K., & Herman, L. (2005). Eggshell characteristics and penetration by *Salmonella* enterica serovar enteritidis through the production period of a layer flock. *British Poultry Science*, *46*(6), 694–700. https://doi.org/10.1080/00071660500395582

Messens, W., Grijspeerdt, K., & Herman, L. (2006). Eggshell penetration of hen's eggs by *Salmonella* enterica serovar enteritidis upon various storage conditions. *British Poultry Science*, *47*(5), 554–560. https://doi.org/10.1080/00071660600954601

Methner, U. (2018). Immunisation of chickens with live *Salmonella* vaccines – Role of booster vaccination. *Vaccine*, *36*(21), 2973–2977. https://doi.org/10.1016/j.vaccine.2018.04.041

Miller, P., Haveroen, M. E., Solichova, K., Merkl, R., McMullen, L. M., Mikova, K., & Chumchalova, J. (2010). Shelf life extension of liquid whole eggs by heat and bacteriocin treatment. *Czech Journal of Food Sciences*, *28*(4), 280–289. http://www.agriculturejournals.cz/web/cjfs.htm

Min, B. R., Nam, K. C., Lee, E. J., Ko, G. Y., Trampel, D. W., & Ahn, D. U. (2005). Effect of irradiating shell eggs on quality attributes and functional properties of yolk and white. *Poultry Science*, *84*(11), 1791–1796. https://doi.org/10.1093/ps/84.11.1791

Miyamoto, T., Horie, T., Baba, E., Sasai, K., Fukata, T., & Arakawa, A. (1998). *Salmonella* penetration through eggshell associated with freshness of laid eggs and refrigeration. *Journal of Food Protection*, *61*(3), 350–353. https://doi.org/10.4315/0362-028X-61.3.350

Molla, A., Matsumura, Y., Yamamoto, T., Okamura, R., & Maeda, H. (1987). Pathogenic capacity of proteases from *Serratia* marcescens and *Pseudomonas* aeruginosa and their suppression by chicken egg white ovomacroglobulin. *Infection and Immunity*, *55*(10), 2509–2517. https://doi.org/10.1128/iai.55.10.2509-2517.1987

Moore, J., & Madden, R. H. (1993). Detection and incidence of *Listeria* species in blended raw egg. *Journal of Food Protection*, *56*(8), 652–654. https://doi.org/10.4315/0362-028X-56.8.652

Morris, C. E., Hoerning, E. F., Angelo, A. J., & St. (1989). Uracil as a potentially useful indicator of spoilage in egg products. *Journal of Food Science*, *54*(3), 581–583. https://doi.org/10.1111/j.1365-2621.1989.tb04657.x

Moschonas, G., Lianou, A., Nychas, G.-J. E., & Panagou, E. Z. (2021). Spoilage potential of *Bacillus* subtilis in a neutral-pH dairy dessert. *Food Microbiology*, *95*, 103715. https://doi.org/10.1016/j.fm.2020.103715

Murase, T., Holt, P. S., & Gast, R. K. (2005). Growth of *Salmonella* enterica serovar enteritidis in albumen and yolk contents of eggs inoculated with this organism onto the vitelline membrane. *Journal of Food Protection*, *68*(4), 718–721. https://doi.org/10.4315/0362-028X-68.4.718

Muriana, P. M. (1996). Bacteriocins for control of *Listeria* spp. In food. *Journal of Food Protection*, *59*(13), 54–63. https://doi.org/10.4315/0362-028X-59.13.54

Musgrove, M. T., Jones, D. R., Northcutt, J. K., Cox, N. A., & Harrison, M. A. (2004) Identification of Enterobacteriaceae from washed and unwashed commercial shell eggs. *Journal of Food Protection*, *67*(11), 2613–2616. https://doi.org/10.4315/0362-028x-67.11.2613

Nakai, S. (2000). Molecular modifications of egg proteins for functional improvement. In J. S. Sim, S. Nakai, & W. Guenter (Eds.), *Egg nutrition and biotechnology* (pp. 205–217). CAB International.

Necidová, L., Bursová, S., Ježek, F., Haruštiaková, D., Vorlová, L., & Golian, J. (2019). Effect of preservatives on the shelf-life and sensory characteristics of pasteurized liquid whole egg stored at 4°C. *Poultry Science*, *98*(11), 5940–5948. https://doi.org/10.3382/ps/pez378

Niamah, A. K. (2018). Structure, mode of action and application of pediocin natural antimicrobial food preservative : A review. *Basrah Journal of Agricultural Sciences*, *31*(1), 59–69. https://doi.org/10.37077/25200860.2018.76

Nikaido, H. (2003). Molecular basis of bacterial outer membrane permeability revisited. *Microbiology and Molecular Biology Reviews: MMBR*, *67*(4), 593–656. https://doi.org/10.1128/MMBR.67.4.593-656.2003

Noonim, P., & Venkatachalam, K. (2021). Quality changes and shelf life of salted duck egg white meringue stored in alternative packages at two temperatures. *Quality Assurance and Safety of Crops & Foods*, *13*, 62–73. https://doi.org/10.15586/qas.v13i3.947

Notermans, S., Soentoro, P. S. S., Bolder, N. M., & Mulder, R. W. A. W. (1991). Adaptation of *Listeria* in liquid egg containing sucrose resulting in survival and outgrowth. *International Journal of Food Microbiology*, *13*(1), 55–61. https://doi.org/10.1016/0168-1605(91)90136-D

Ohtsuka, K., Yanagawa, K., Takatori, K., & Hara-Kudo, Y. (2005). Detection of *Salmonella* enterica in naturally contaminated liquid eggs by loop-mediated isothermal amplification, and characterization of *Salmonella* isolates. *Applied and Environmental Microbiology*, *71*(11), 6730–6735. https://doi.org/10.1128/AEM.71.11.6730-6735.2005

Okamura, M., Miyamoto, T., Kamijima, Y., Tani, H., Sasai, K., & Baba, E. (2001). Differences in abilities to colonize reproductive organs and to contaminate eggs in intravaginally inoculated hens and in vitro adherences to vaginal explants between *Salmonella* enteritidis and other *Salmonella* serovars. *Avian Diseases*, *45*(4), 962–971. https://doi.org/10.2307/1592875

Omana, D. A., Liang, Y., Kav, N. N. V., & Wu, J. (2011). Proteomic analysis of egg white proteins during storage. *Proteomics*, *11*(1), 144–153. https://doi.org/10.1002/pmic.201000168

Palumbo, S. A. (1986). Is refrigeration enough to restrain foodborne pathogens? *Journal of Food Protection*, *49*(12), 1003–1009. https://doi.org/10.4315/0362-028X-49.12.1003

Palumbo, M. S., Beers, S. M., Bhaduri, S., & Palumbo, S. A. (1995). Thermal resistance of *Salmonella* spp. And *Listeria* monocytogenes in liquid egg yolk and egg yolk products†. *Journal of Food Protection*, *58*(9), 960–966. https://doi.org/10.4315/0362-028X-58.9.960

Palumbo, M. S., Beers, S. M., Bhaduri, S., & Palumbo, S. A. (1996). Thermal resistance of *Listeria* monocytogenes and *Salmonella* spp. In liquid egg white. *Journal of Food Protection*, *59*(11), 1182–1186. https://doi.org/10.4315/0362-028X-59.11.1182

Park, W.-K., Chung, J.-W., Kim, Y.-K., Chung, S.-C., & Kho, H.-S. (2006). Influences of animal mucins on lysozyme activity in solution and on hydroxyapatite surfaces. *Archives of Oral Biology*, *51*(10), 861–869. https://doi.org/10.1016/j.archoralbio.2006.04.002

Parry, C. (2006). Epidemiological and clinical aspects of human typhoid fever. In D. Maskell & P. Mastroeni (Eds.), *Salmonella infections: Clinical, immunological and molecular aspects (Advances in molecular and cellular microbiology)* (pp. 1–24). Cambridge University Press. https://doi.org/10.1017/CBO9780511525360.002

Pellegrini, A., Thomas, U., Wild, P., Schraner, E., & Von Fellenberg, R. (2000). Effect of lysozyme or modified lysozyme fragments on DNA and RNA synthesis and membrane permeability of Escherichia coli. *Microbiological Research*, *155*(2), 69–77. https://doi.org/10.1016/S0944-5013(00)80040-3

Perić, L., Stojčić, M. Đ., & Bjedov, S. (2017). The effect of storage and age of hens on the quality of table eggs. *Advanced Research in Life Sciences*, *1*(1), 64–67. https://doi.org/10.1515/arls-2017-0011

Poppe, C., Irwin, R. J., Forsberg, C. M., Clarke, R. C., & Oggel, J. (1991). The prevalence of *Salmonella* enteritidis and other *Salmonella* spp. Among Canadian registered commercial layer flocks. *Epidemiology & Infection*, *106*(2), 259–270. https://doi.org/10.1017/S0950268800048408

Postollec, F., Bonilla, S., Baron, F., Jan, S., Gautier, M., Mathot, A. G., & Sohier, D. (2010). A multiparametric PCR-based tool for fast detection and identification of spore-forming bacteria in food. *International Journal of Food Microbiology*, *142*(1), 78–88. https://doi.org/10.1016/j.ijfoodmicro.2010.06.005

Pothakos, V., Taminiau, B., Huys, G., Nezer, C., Daube, G., & Devlieghere, F. (2014). Psychrotrophic lactic acid bacteria associated with production batch recalls and sporadic cases of early spoilage in Belgium between 2010 and 2014. *International Journal of Food Microbiology*, *191*, 157–163. https://doi.org/10.1016/j.ijfoodmicro.2014.09.013

Protais, J., Boscher, E., Quéguiner, S., Chidaine, B., & Fravalo, P. (2007a, Mars 28–29). *Identification of the enterococcus genus in the matrix of liquid whole eggs* (pp. 528–531). 7èmes Journées de la Recherche Avicole.

Protais, J., Gérault, P., Queguiner, S., Bosher, E., Chidaine, B., Ermel, G., & Lorthioir, P. (2006). Identification et comportement des bactéries d'altération dans la matrice oeuf entier liquide. *Sciences et Techniques Avicoles*, *57*, 4–13.

Protais, J., Quéguiner, S., Boscher, E., Chidaine, B., Ermel, G., Gérault, P., & Jugiau, F. (2007b, Mars 28–29). *Campylobacter sp. et Listeria monocytogenes dans l'œuf entier liquid* (pp. 532–535). 7èmes Journées de la Recherche Avicole.

Qin, X., He, S., Zhou, X., Cheng, X., Huang, X., Wang, Y., & Shi, X. (2019). Quantitative proteomics reveals the crucial role of YbgC for *Salmonella* enterica serovar enteritidis survival in egg white. *International Journal of Food Microbiology*, *289*, 115–126. https://doi.org/10.1016/j.ijfoodmicro.2018.08.010

Qiu, N., Ma, M., Zhao, L., Liu, W., Li, Y., & Mine, Y. (2012). Comparative proteomic analysis of egg white proteins under various storage temperatures. *Journal of Agricultural and Food Chemistry*, *60*(31), 7746–7753. https://doi.org/10.1021/jf302100m

Raspoet, R., Shearer, N., Appia-Ayme, C., Haesebrouck, F., Ducatelle, R., Thompson, A., & Van Immerseel, F. (2014). A genome-wide screen identifies *Salmonella* enteritidis lipopolysaccharide biosynthesis and the HtrA heat shock protein as crucial factors involved in egg white persistence at chicken body temperature. *Poultry Science*, *93*(5), 1263–1269. https://doi.org/10.3382/ps.2013-03711

Réhault-Godbert, S., Anton, M., Nau, F., Gautron, J., & Nys, Y. (2007). Les activités biologiques de l'oeuf. *Productions Animales*, *20*(4), 337–348.

Réhault-Godbert, S., Labas, V., Helloin, E., Hervé-Grépinet, V., Slugocki, C., Berges, M., & Nys, Y. (2013). Ovalbumin-related protein X is a heparin-binding ov-serpin exhibiting antimicrobial activities. *The Journal of Biological Chemistry*, *288*(24), 17285–17295. https://doi.org/10.1074/jbc.M113.469759

Rehault-Godbert, S., Baron, F., Mignon-Grasteau, S., Labas, V., Gautier, M., Hincke, M. T., & Nys, Y. (2010). Effect of temperature and time of storage on protein stability and anti-*Salmonella* activity of egg white. *Journal of Food Protection*, *73*(9), 1604–1612. https://doi.org/10.4315/0362-028X-73.9.1604

Ricke, S. (2003). The gastrointestinal tract ecology of *Salmonella* enteritidis colonization in molting hens. *Poultry Science*, *82*(6), 1003–1007. https://doi.org/10.1093/ps/82.6.1003

Ricke, S. C. (2018). Impact of prebiotics on poultry production and food safety. *The Yale Journal of Biology and Medicine*, *91*(2), 151–159.

Rocculi, P., Tylewicz, U., Pękosławska, A., Romani, S., Sirri, F., Siracusa, V., & Rosa, M. D. (2009). MAP storage of shell hen eggs, part 1 : Effect on physico-chemical characteristics of the fresh product. *LWT - Food Science and Technology*, *42*(3), 758–762. https://doi.org/10.1016/j.lwt.2008.09.017

Rose-Martel, M., Du, J., & Hincke, M. T. (2012). Proteomic analysis provides new insight into the chicken eggshell cuticle. *Journal of Proteomics*, *75*(9), 2697–2706. https://doi.org/10.1016/j.jprot.2012.03.019

Rossi, M., & De Reu, K. (2011). Alternative hen housing systems and egg quality. In Y. Nys, M. Bain, & F. Van Immerseel (Éds.), *Improving the safety and quality of eggs and egg products* (pp. 351–375). Woodhead Publishing. https://doi.org/10.1533/9780857093912.3.351

Ruvalcaba-Gómez, J. M., Villagrán, Z., Valdez-Alarcón, J. J., Martínez-Núñez, M., Gomez-Godínez, L. J., Ruesga-Gutiérrez, E., & Villarruel-López, A. (2022). Non-antibiotics strategies to control *Salmonella* infection in poultry. *Animals*, *12*(1), 102. https://doi.org/10.3390/ani12010102

Růzicková, V. (1994). Growth and survival of *Salmonella* enteritidis in selected egg foods. *Veterinarni Medicina*, *39*(4), 187–195.

Sadeyen, J.-R., Trotereau, J., Protais, J., Beaumont, C., Sellier, N., Salvat, G., & Lalmanach, A.-C. (2006). *Salmonella* carrier-state in hens : Study of host resistance by a gene expression approach. *Microbes and Infection*, *8*(5), 1308–1314. https://doi.org/10.1016/j.micinf.2005.12.014

Sahin, O., Kobalka, P., & Zhang, Q. (2003). Detection and survival of campylobacter in chicken eggs. *Journal of Applied Microbiology*, *95*(5), 1070–1079. https://doi.org/10.1046/j.1365-2672.2003.02083.x

Salvat, G., Toquin, M. T., & Ermel, G. (1997). Epidémiologie de *Listeria* monocytogenes dans la filière porcine. *Viandes et Produits Carnés*, *18*(6), 264–268.

Sauveur, B. (1988). Structure, composition et valeur nutritionelle de l'oeuf. In B. Sauveur (Ed.), *Reproduction des volailles et production d'œufs*. INRA.

Schoeni, J. L., Glass, K. A., McDermott, J. L., & Wong, A. C. L. (1995). Growth and penetration of *Salmonella* enteritidis, *Salmonella* Heidelberg and *Salmonella* typhimurium in eggs. *International Journal of Food Microbiology*, *24*(3), 385–396. https://doi.org/10.1016/0168-1605(94)00042-5

Sellier, N., Vidal, M. L., Baron, F., Michel, J., Gautron, J., Protais, M., & Nys, Y. (2007). Estimations of repeatability and heritability of egg albumen antimicrobial activity and of lysozyme and ovotransferrin concentrations. *British Poultry Science*, *48*(5), 559–566. https://doi.org/10.1080/00071660701592367

Shah, D. B., Bradshaw, J. G., & Peeler, J. T. (1991). Thermal resistance of egg-associated epidemic strains of *Salmonella* enteritidis. *Journal of Food Science*, *56*(2), 391–393. https://doi.org/10.1111/j.1365-2621.1991.tb05287.x

Shivaprasad, H. L., Timoney, J. F., Morales, S., Lucio, B., & Baker, R. C. (1990). Pathogenesis of *Salmonella* enteritidis infection in laying chickens. I. Studies on egg transmission, clinical signs, fecal shedding, and serologic responses. *Avian Diseases*, *34*(3), 548–557. https://doi.org/10.2307/1591243

Shrestha, S., & Hariram, U. (2022). Control of *Bacillus* weihenstephanensis in pasteurized liquid whole eggs formulated with nisin. *Journal of Food Protection*, *85*(4), 647–652. https://doi.org/10.4315/JFP-21-316

Stevens, A., Joseph, C., Bruce, J., Fenton, D., O'Mahony, M., Cunningham, D., & Rowe, B. (1989). A large outbreak of *Salmonella* enteritidis phage type 4 associated with eggs from overseas. *Epidemiology & Infection*, *103*(3), 425–433. https://doi.org/10.1017/S095026880003082X

Suman, M., Riani, G., & Dalcanale, E. (2007). MOS-based artificial olfactory system for the assessment of egg products freshness. *Sensors and Actuators B: Chemical*, *125*(1), 40–47. https://doi.org/10.1016/j.snb.2007.01.031

Suzuki, S. (1994). Pathogenicity of *Salmonella* enteritidis in poultry. *International Journal of Food Microbiology*, *21*(1–2), 89–105. https://doi.org/10.1016/0168-1605(94)90203-8

Swayne, D. E., & Beck, J. R. (2004). Heat inactivation of avian influenza and Newcastle disease viruses in egg products. *Avian Pathology*, *33*(5), 512–518. https://doi.org/10.1080/03079450400003692

Tan, A. T., & Woodworth, R. C. (1969). Ultraviolet difference spectral studies of conalbumin complexes with transition metal ions. *Biochemistry*, *8*(9), 3711–3716. https://doi.org/10.1021/bi00837a033

Techer, C., Baron, F., & Jan, S. (2014). Microbial spoilage of eggs and egg products. In C. Batt & P. Patel (Eds.), *Encyclopedia of food microbiology* (2nd ed., p. 3248). Academic Press, Elsevier. https://doi.org/10.1016/B9780123847300.00371-2

Techer, C., Daoud, A., Madec, M.-N., Gautier, M., Jan, S., & Baron, F. (2015). Microbial quality of industrial liquid egg white : Assumptions on spoiling issues in egg-based chilled desserts. *Journal of Food Science*, *80*(2), M389–M398. https://doi.org/10.1111/1750-3841.12764

Techer, C., Jan, S., Gonnet, F., Grosset, N., Gautier, M., & Baron, F. (2019). Bacterial diversity on stainless steel surfaces of egg processing companies and potential of selected isolates to spoil liquid whole egg products. *Journal of Applied Microbiology*, *127*(5), 1501–1510. https://doi.org/10.1111/jam.14403

Techer, C., Jan, S., Thierry, A., Maillard, M.-B., Grosset, N., Galet, O., & Baron, F. (2020). Identification of the bacteria and their metabolic activities associated with the microbial spoilage of custard cream desserts. *Food Microbiology*, *86*, 103317. https://doi.org/10.1016/j.fm.2019.103317

Thapon, J. L. (1994). Evolution de l'œuf au cours de sa conservation. In J. L. Thapon & C. M. Bourgeois (Eds.), *L'œuf et les ovoproduits* (pp. 84–94). Lavoisier.

Threlfall, E. J., Wain, J., Peters, T., Lane, C., Pinna, E., Little, C. L., & Davies, R. H. (2014). Egg - borne infections of humans with *Salmonella*: Not only an s. Enteritidis problem. *World's Poultry Science Journal*, *70*(1), 15–26. https://doi.org/10.1017/S0043933914000026

Tomczyk, Ł., Stępień, Ł., Urbaniak, M., Szablewski, T., Cegielska-Radziejewska, R., & Stuper-Szablewska, K. (2018). Characterisation of the mycobiota on the shell surface of table eggs acquired from different egg-laying hen breeding systems. *Toxins*, *10*(7), 293. https://doi.org/10.3390/toxins10070293

Tóth, A., Németh, C., Juhász, R., Zeke, I., Csehi, B., Bényi, D., & Friedrich, L. (2016). Effects of HPH processing at 400 MPa on proteins of liquid egg products. *Review on Agriculture and Rural Development*, *5*(1–2), 153–157. https://doi.org/10.14232/rard.2016.1-2.153-157

Tóth, A., Németh, C., Palotás, P., Surányi, J., Nyulas-Zeke, I. C., Csehi, B., & Balla, C. (2017). HHP treatment of liquid egg at 200-350 MPa. *Journal of Physics: Conference Series*, *950*, 042008. https://doi.org/10.1088/1742-6596/950/4/042008

Tranter, H. S., & Board, R. G. (1984). The influence of incubation temperature and pH on the antimicrobial properties of hen egg albumen. *Journal of Applied Bacteriology*, *56*(1), 53–61. https://doi.org/10.1111/j.1365-2672.1984.tb04696.x

Tung, M. A., Garland, M. R., & Gill, P. K. (1979). A scanning electron microscope study of bacterial invasion in Hen's egg shell. *Canadian Institute of Food Science and Technology Journal*, *12*(1), 16–22. https://doi.org/10.1016/S0315-5463(79)73030-6

Vadehra, D. V., Baker, R. C., & Naylor, H. B. (1970). Role of cuticle in spoilage of chicken eggs. *Journal of Food Science*, *35*(1), 5–6. https://doi.org/10.1111/j.1365-2621.1970.tb12354.x

Valenti, P., Visca, P., Antonini G., & Orsi, N. (1986). Interaction between lactoferrin and ovotransferrin and Candida cells. *FEMS Microbiology Letters*, *33*(2–3), 271–275. https://doi.org/10.1111/j.1574-6968.1986.tb01285.x

Valiente Moro, C., Fravalo, P., Amelot, M., Chauve, C., Zenner, L., & Salvat, G. (2007). Colonization and organ invasion in chicks experimentally infected with dermanyssus gallinae contaminated by *Salmonella* enteritidis. *Avian Pathology*, *36*(4), 307–311. https://doi.org/10.1080/03079450701460484

Van Immerseel, F., Russell, J. B., Flythe, M. D., Gantois, I., Timbermont, L., Pasmans, F., & Ducatelle, R. (2006). The use of organic acids to combat *Salmonella* in poultry: A mechanistic explanation of the efficacy. *Avian Pathology*, *35*(3), 182–1888. doi: 10.1080/03079450600711045. PMID: 16753609.

Viator, C. L., Noyes, G., Muth, M. K., Karns, S. A., & Cates, S. C. (2016). Food safety practices in the egg products industry. *Journal of Food Protection*, *79*(7), 1210–1215. https://doi.org/10.4315/0362-028X.JFP-15-520

Vicente, J. L., Lopez, C., Avila, E., Morales, E., Hargis, B. M., & Tellez, G. (2007). Effect of dietary natural capsaicin on experimental *Salmonella* enteritidis infection and yolk pigmentation in laying hens. *International Journal of Poultry Science*, *6*, 393–396.

Vieira, D. A. P., Cabral, L., Noronha, M. F., Júnior, G. V. L., & Sant'Ana, A. S. (2019). Microbiota of eggs revealed by 16S rRNA-based sequencing : From raw materials produced by different suppliers to chilled pasteurized liquid products. *Food Control*, *96*, 194–204. https://doi.org/10.1016/j.foodcont.2018.09.016

Vikram, A., Woolston, J., & Sulakvelidze, A. (2021). Phage biocontrol applications in food production and processing. *Current Issues in Molecular Biology*, *40*, 267–302. https://doi.org/10.21775/cimb.040.267

Wang, J., Gu, L., Su, Y., Chang, C., Xu, L., Yang, Y., & Li, J. (2020). Changes in microbial, physiochemical, and functional properties of pasteurized liquid whole egg during refrigerated storage. *Journal of the Science of Food and Agriculture*, *100*(7), 2873–2879. https://doi.org/10.1002/jsfa.10311

Wang, Y., Jia, B., Xu, X., Zhang, L., Wei, C., Ou, H., & Shi, X. (2018). Comparative genomic analysis and characterization of two *Salmonella* enterica serovar enteritidis isolates from poultry with notably different survival abilities in egg whites. *Frontiers in Microbiology*, *9*. https://www.frontiersin.org/article/10.3389/fmicb.2018.02111

Wang, C., & Shelef, L. A. (1991). Factors contributing to antilisterial effects of raw egg albumen. *Journal of Food Science*, *56*(5), 1251–1254. https://doi.org/10.1111/j.1365-2621.1991.tb04745.x

Wang, J., Su, Y., Gu, L., Chang, C., Xu, L., Yang, Y., & Li, J. (2021). The inhibition of cell-free supernatants of several lactic acid bacteria on the selected psychrophilic spoilage bacteria in liquid whole egg. *Food Control*, *123*, 107753. https://doi.org/10.1016/j.foodcont.2020.107753

WHO. (2005). Word Health Organization. No bird flu risk for consumers from properly cooked poultry and eggs. Retrieved on 26 April 2022 from https://apps.who.int.

Wigley, P., Berchieri, A., Page, K. L., Smith, A. L., & Barrow, P. A. (2001). *Salmonella* enterica serovar Pullorum persists in splenic macrophages and in the reproductive tract during persistent, disease-free carriage in chickens. *Infection and Immunity*, *69*(12), 7873–7879. https://doi.org/10.1128/IAI.69.12.7873-7879.2001

Wigley, P., Hulme, S. D., Powers, C., Beal, R. K., Berchieri, A., Smith, A., & Barrow, P. (2005). Infection of the reproductive tract and eggs with *Salmonella* enterica serovar pullorum in the chicken is associated with suppression of cellular immunity at sexual maturity. *Infection and Immunity*, *73*(5), 2986–2990. https://doi.org/10.1128/IAI.73.5.2986-2990.2005

Xu, Y. (2021). Phage and phage lysins : New era of bio-preservatives and food safety agents. *Journal of Food Science*, *86*(8), 3349–3373. https://doi.org/10.1111/1750-3841.15843

Yang, Y., Hoe, Y. W., Zheng, Q., Chung, H.-J., & Yuk, H.-G. (2017). Biofilm formation by *Salmonella* enteritidis in a simulated liquid egg processing environment and its sensitivity to chlorine and hot water treatment. *Food Control*, *73*, 595–600. https://doi.org/10.1016/j.foodcont.2016.09.008

Żbikowska, K., Michalczuk, M., & Dolka, B. (2020). The use of bacteriophages in the poultry industry. *Animals*, *10*(5), 872. https://doi.org/10.3390/ani10050872

Zeng, Z., Zhang, S., Wang, H., & Piao, X. (2015). Essential oil and aromatic plants as feed additives in non-ruminant nutrition : A review. *Journal of Animal Science and Biotechnology*, *6*(1), 7. https://doi.org/10.1186/s40104-015-0004-5

Zhang, Y., Ding, Y., Li, W., Zhu, W., Wang, J., & Wang, X. (2021). Application of a novel lytic podoviridae phage Pu20 for biological control of drug-resistant *Salmonella* in liquid eggs. *Pathogens*, *10*(1), 34. https://doi.org/10.3390/pathogens10010034

Part VII

Bioactives of Egg Components

21

Eggshell Membrane

Lingjiao Zhu, Limei Wang, Yuting Zheng, and Xi Huang*

CONTENTS

21.1 Introduction

Eggshell membrane (ESM) is a natural fibrous biopolymer that surrounds the egg white, providing physical and biological protection to the embryo (Makkar et al., 2015; Shi et al., 2021). The ESM is composed mainly of proteins (~90%), lipids (~3%), and carbohydrates (~2%) (Chen et al., 2014; Kulshreshtha et al., 2022). As shown in Figure 21.1, numerous studies indicate that the ESM is an outstanding source of biologically active substances; it contains more than 500 proteins, some bioactive peptides, and various glycosaminoglycans (GAGs), mainly composed of hyaluronic acid (81%) (Jia et al., 2017), all of which possess many biological activities (Sah & Rath, 2016; Ahmed et al., 2017; Shi et al., 2021).

It is estimated that more than 8 million tons of eggshell waste are produced worldwide yearly (Xiao et al., 2021). As one of the main by-products of poultry farming and egg processing, ESM accounts for about 10–11% of eggshell waste by weight, generating a large amount of waste ESM (Waheed et al., 2020; Xiao et al., 2021). The ESM and their derived products showed various biological activities, including antioxidant, anti-inflammatory, antibacterial, joint health promotion, cosmetic skincare, and other health-improving bioactivities (Huang et al., 2010; Gil-Quintana et al., 2018; Zhao et al., 2019; Kulshreshtha et al., 2022; Zhu et al., 2022). Studying the bioactive fractions of ESM can provide a better understanding of their roles in embryo development and protection and the key components and mechanisms involved in the bioactivities of ESM and its derivatives, which are important for their future applications. This chapter provides a comprehensive review of the bioactive compounds in ESM and provides a theoretical basis for its role in improving egg quality and the value-added utilization of ESM later.

* Corresponding author.

DOI: 10.1201/9781003254430-28

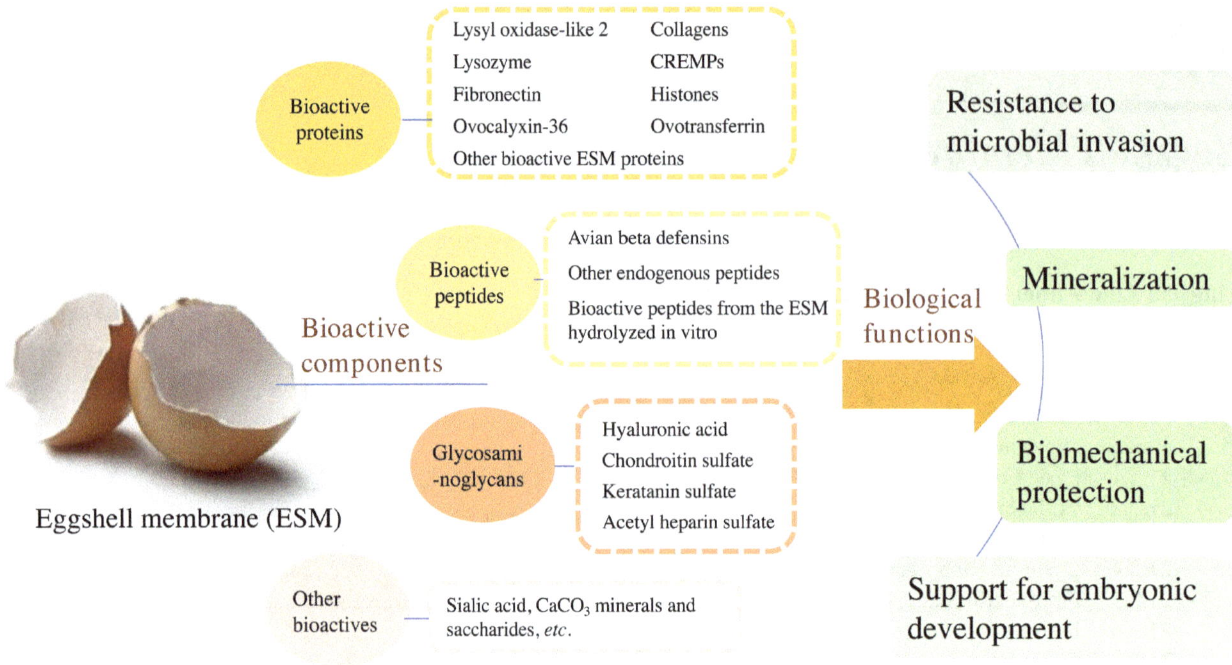

FIGURE 21.1 Schematic illustration of bioactive components in the eggshell membrane and their biological function.

21.2 Bioactive Proteins and Peptides in Eggshell Membrane

As shown in Table 21.1, ESM is rich in biologically active proteins, including collagen, lysyl oxidases, lysozyme, cysteine-rich eggshell membrane proteins (CREMPs), fibronectin, histones, ovocalyxin-36 (OCX-36), and ovotransferrin (Sah & Rath, 2016; Kulshreshtha et al., 2022). These bioactive proteins are the major contributors to ESM's structure and various biological functions, including the resistance to pathogen invasion, support of eggshell calcification, and stabilization of the encapsulated egg whites (Ahmed et al., 2017; Xiao et al., 2021).

21.2.1 Lysyl Oxidase-Like 2 (LOXL2)

Lysyl oxidases are a family of enzymes that oxidize tertiary lysine residues on collagen and elastin in the extracellular matrix (ECM) with copper ions, resulting in the covalent linkages between collagen and elastin (Du et al., 2015; Ahmed et al., 2017). Lysyl oxidases play a critical role in forming functional connective tissues, which has been demonstrated experimentally and clinically by inhibiting lysyl oxidases and Menkes' disease and X-linked keratosis relaxants, respectively (Wu et al., 1992).

There are five genes encoding lysine oxidase activity in chickens: lysyl oxidase (LOX), 4 lysyl oxidase-like proteins (LOXL1, LOXL2, LOXL3, and LOXL4), of which LOXL2 expression is the most abundant in ESM (Du et al., 2015; Ahmed et al., 2017). LOXL2 catalyzes the cross-linking of collagen and elastin in ESM, which gives ESM high insolubility and structural flexibility, which can enclose the egg white and provide firm support for eggshell calcifications (Ahmed et al., 2017).

The deficiency of copper in hen's feed destroys ESM tissue, suggesting a key role of lysyl oxidase activity during the ESM formation (Chowdhury, 1990). Besides, LOXL2 is one of the most abundant proteins during the early calcification stages (Du et al., 2015), suggesting that LOXL2 may be involved in eggshell mineralization.

Moreover, LOXL2 is proven to modulate angiogenesis in both *in vitro* and *in vivo* models (Bignon et al., 2011) and may have a possible role in the improvement of angiogenesis (Ahmed et al., 2019; Li et al., 2019a, 2019b), the important potential for application of ESM in promoting wound healing.

21.2.2 Collagens

Collagens are the main component of ESM, where each ESM fiber consists of a collagen-rich core and a glycoprotein-rich outer membrane (Osuoji, 1971; Sah & Rath, 2016). To date, collagens I, III, IV, V, VII, and X have been identified in ESM (Wong et al., 1984; Ahmed et al., 2017). The collagen molecule consists of three α chains ($\alpha1$, $\alpha2$, and $\alpha3$). The collagen triple helix consists of a repetitive $(Gly\text{-}X\text{-}Y)_n$ sequence with the X and Y positions occupied by proline (Pro) and hydroxyproline (Hyp), respectively (Bhattacharjee & Bansal, 2005).

As a major component of the ECM, collagen plays an important role in various developmental processes and tissue dynamic homeostasis. Collagen can provide tissues with optimal mechanical strength, thermal stability, and interactions with other biomolecules to meet basic tissue requirements (Ahmed et al., 2017). Various collagens and extensive lysine-mediated cross-linking (i.e., desmosine and isodesmosine) can confer specific mechanical properties to ESM (Kulshreshtha et al., 2022). In addition, collagen has various biological activities that promote wound healing, histocompatibility, and biomineralization (Sah & Rath, 2016; Ahmed et al., 2019; Shi et al., 2021).

TABLE 21.1

Bioactive Proteins and Endogenous Peptides of Eggshell Membrane

Constituents	Chemical Compounds	Function	References
Proteins (>500 species)	Lysyl oxidase-like 2	Promotion of functional connective tissues formation, and catalysis of the cross-linking of collagen and elastin; modulation of angiogenesis	Bignon et al. (2011); Du et al. (2015); Ahmed et al. (2017)
	Collagens	Optimal mechanical strength, thermal stability, wound healing, histocompatibility, and biomineralization	Sah and Rath (2016); Shi et al. (2021)
	Lysozyme	Antimicrobial activity	Ahlborn and Sheldon (2006); Kozuka et al. (2015)
	Cysteine-rich eggshell membrane proteins (CREMPs)	Mechanical stability, promotion of wound healing	Kodali et al. (2011); Cordeiro and Hincke (2016)
	Fibronectin	Promotion of cell anchoring, proliferation, migration, and differentiation, and wound healing	Grinnell (1984); Lenselink (2015)
	Histones	Antimicrobial activity	Marino-Ramirez et al. (2005); Kawasaki and Iwamuro (2008)
	Ovocalyxin-36	Antimicrobial properties and positive immunomodulatory effects	Cordeiro et al. (2013); Kovacs-Nolan et al. (2014)
	Ovotransferrin	Iron binding, antimicrobial effects, antioxidant properties, and immunomodulatory functions	Gautron et al. (2001); Wang et al. (2021b)
	Other bioactive proteins: Keratins, laminins, agrin, titin, dynein, obscurin, myosin, nebulin, protease inhibitors, antiproteases, mucoid substances etc.	Biomechanical function, cytoskeletal organization and anchoring functions, adhesion and differentiation, biomineralization, antimicrobial activity, and molecular coordination for embryonic development	Huopalahti et al. (2007); Lieleg et al. (2012); Makkar et al. (2015); Cordeiro and Hincke (2016); Ahmed et al. (2017)
Endogenous peptides	Avian beta defensins	Antimicrobial and immunomodulatory	Ahmed et al. (2017)
	Arginine-Glycine-Aspartic acid	Cell adhesion, reduction in the risk of immune reactions or pathogen transfer	Sah and Rath (2016)
	Other endogenous peptides: (Peptides derived from clusterin, actin regulator protein, etc.)	Antimicrobial activities and cytoprotective effect	Mann et al. (2004); Makkar et al. (2015)

Collagen X, short-chain collagen expressed in hypertrophic cartilage, is the major collagen in ESM and is regarded as a mineralization inhibitor that prevents the mineralization of ESM itself (Ahmed et al., 2017). The removal of the non-helical domain of collagen X by protein hydrolysis promotes the cell-mediated ESM mineralization *in vitro*, whereas an intact collagen X does not promote cell-mediated mineralization; therefore, the non-helical domain of collagen X has an inhibitory effect on mineralization *in vivo*, thereby establish a boundary against mineral deposition (Arias et al., 1997; Ahmed et al., 2017).

Due to its good bioactivity, biocompatibility, and biodegradability, collagen is widely used in food, cosmetic, and pharmaceutical industries (Kulshreshtha et al., 2022). The ESM-derived collagen has the advantages of low allergic reactions and high biosafety, especially free from bovine zoonotic diseases or potential autoimmune/allergic reactions, making the ESM a preferred choice for collagen extraction (King'Ori, 2011; Xiao et al., 2021).

21.2.3 Lysozyme

Lysozyme, also known as muramidase or N-acetylmuramic acid hydrolase, hydrolyzes the β-1,4-glycosidic bonds between N-acetylglucosamine and N-acetylmuramic acid in the peptidoglycans of the bacterial cell wall to exert antibacterial activity against microorganisms, especially Gram-positive bacteria (Ahlborn et al., 2006; Khorshidian et al., 2022). Due to the presence of a protective lipopolysaccharide layer on the outer membrane, lysozyme is significantly weaker against Gram-negative bacteria, but the modification of its physical or chemical can improve the antibacterial activity of lysozyme against Gram-negative bacteria (Ferraboschi et al., 2021; Khorshidian et al., 2022).

Lysozyme, the most abundant antimicrobial protein (~10%) in ESM, is mainly responsible for antimicrobial activity (Ahlborn & Sheldon, 2006; Ahmed et al., 2019). In ESM, the combination of other antimicrobial proteins, transferrin, and β-N-acetylglucosaminidase interfere with the interactions between bacterial lipopolysaccharides, which play an important role in protecting the embryo from microbial contamination (Hincke et al., 2000; Kulshreshtha et al., 2022). In addition to the well-known antimicrobial properties during embryonic development, lysozyme from ESM may also function as a structural protein because it affects calcium carbonate deposition in a soluble form during calcification. Calcite crystals grown in the presence of purified hen lysozyme not only altered crystal morphology but also accelerated the precipitation of amorphous calcium phosphate from sub-stable solutions (Hincke et al., 2000).

A previous study showed that lysozyme was readily purified from the ESM, and the activity of lysozyme purified was comparable to that from the egg whites at pH 4–7 or 25–50°C. However, the ESM lysozyme had lower activity at alkaline pH (8.0) or extreme temperatures than the egg whites lysozyme, indicating that the ESM lysozyme is more susceptible to denaturation than the egg white lysozyme (Kozuka et al., 2015). In addition, the activity of ESM lysozyme varies with the breed and age of the hen as well as the treatment and storage time of the ESM, with the younger hens producing higher ESM lysozyme activity than the older hens (Ahlborn & Sheldon, 2005).

As a natural enzyme with antibacterial activity, lysozyme is widely used in medical, pharmaceutical, and especially food industries as a natural food additive to control bacterial growth. The problem of antibiotic resistance has stimulated the search for new molecules or applications of known compounds, and the use of lysozyme as an alternative to antibiotics has gained great interest. Lysozyme from eggs is considered a major industrial source for practical use because of its availability and cost-effectiveness (Ferraboschi et al., 2021; Khorshidian et al., 2022).

21.2.4 Cysteine-Rich Eggshell Membrane Proteins (CREMPs)

CREMPs, a prominent constituent of ESM (~30%), are a family of cysteine-rich ESM proteins similar to spore coat protein SP75 of cellular slime molds (Cordeiro & Hincke, 2016; Ahmed et al., 2019; Kulshreshtha et al., 2022). CREMPs are widely coupled via irreversible lysine-derived cross-links and are important structural proteins in the ESM (Kodali et al., 2011; Ahmed et al., 2019). These proteins can only be identified under the most intense extraction/solubilization conditions, suggesting that CREMPs lead to the extreme stability of ESM and are the main structural components of mechanical stability (Ahmed et al., 2017). Moreover, CREMPs are considered one of the main wound-healing promoters besides collagen, with a high potential to promote wound healing (Ahmed et al., 2019).

21.2.5 Fibronectin

Fibronectin is a dimeric form of glycoprotein that functions to activate or bind proteins in the ESM (Shi et al., 2021). As a ubiquitous cell adhesion ECM protein, fibronectin can promote cell anchoring, proliferation, migration, and differentiation, making it crucial in the overall wound healing process (Grinnell, 1984; Valenick et al., 2005), especially in the formation and re-epithelialization of the ECM (Lenselink, 2015).

Fibronectin plays several different roles in wound healing due to the specific functional domains and binding sites in its structure. During blood coagulation, plasma FN is deposited together with fibronectin, a major component of the provisional matrix. As repair and remodeling proceed, fibronectin expression is upregulated. The locally produced cellular fibronectin is a major granulation component that replaces the temporary matrix (Valenick et al., 2005; Lenselink, 2015). In one study, fragmented fibronectin promoted $\alpha4\beta1$ integrin-mediated contraction of the temporary fibronectin-fibronectin matrix, an important step in the healing process (Valenick et al., 2005). In addition, it was reported that fibronectin regulated the inflammatory response (Ahmed et al., 2019). This biological activity of fibronectin present in ESM should contribute to therapeutic wound healing, which has been reported in numerous studies (Jun et al., 2013; Ahmed et al., 2019; Li et al., 2019b).

21.2.6 Histones

Histones are superfamily proteins with strong antimicrobial activity and play a central role in chromatin folding and compacting (Marino-Ramirez et al., 2005; Kawasaki & Iwamuro, 2008). Different members of histones have been found in the ESM, including histones H1.11L, H1.11R, H2A, H2B, H3 family 3C, and H5 (Ahmed et al., 2017). Histones H1, H2A, and H2B from chickens have a wide-spectrum antimicrobial activity, suggesting that they may be involved in immune defense mechanisms (Li et al., 2010). Histones H1, H2B, H2A, H3, H4, and H5 isolated from chicken erythrocytes exhibited potent antimicrobial activity against Gram-negative and Gram-positive bacteria and played a role in pathogen recognition by the vertebrate immune system. In addition, a mixture of histones H1, H2A, and H2B isolated from the hen ovaries and oviducts also possessed this antimicrobial activity. Besides, the histones-derived molecules can potentially develop novel antibiotics (Rose-Martel & Hincke, 2014). Thus, histones in the ESM are another potential source of antimicrobial protection for embryos (Ahmed et al., 2017).

21.2.7 Ovocalyxin-36 (OCX-36)

OCX-36 is a bioactive protein with antimicrobial properties and positive immunomodulatory effects found in eggshells and ESM (Cordeiro et al., 2013; Kovacs-Nolan et al., 2014). The OCX-36 protein is unique to avian species, and no corresponding OCX-36 gene has been detected in either mammalian or boney fish genomes (Ahmed et al., 2017). To date, OCX36 has been detected as one of the most abundant proteins in the eggshell proteome of chickens, quail, turkeys, and zebra finches, suggesting a fundamental role of OCX36 in eggshell mineralization and/or function (Mann & Mann, 2015; Ahmed et al., 2017).

The amino acid sequence comparison revealed that OCX-36 (also known as OVO-36, BPIFB3, or BPIFB8) shares protein sequence homology to bactericidal permeability-increasing protein (BPI) and lipopolysaccharide-binding protein (LBP) and palate, lung, and nasal epithelium clone (PLUNC) proteins, which play an important role in innate immune protection (Gautron et al., 2007). The OCX-36 extracted from the ESM exhibited inhibitory activity against the growth of *Staphylococcus aureus* and bound to lipopolysaccharide (LPS) and lipoteichoic acid (LTA). These are important activities in neutralizing endotoxic and non-endotoxic pyrogens during the inflammatory response (Cordeiro et al., 2013). In addition, OCX-36 has immunomodulatory effects *in vitro* by reducing TNF-a secretion from the LPS-induced Raw264.7 cells and *in vivo* by significantly reducing MPO activity in a mouse model of endotoxemia, an indicator of neutrophil infiltration in the intestinal mucosa (Kovacs-Nolan et al., 2014).

Therefore, OCX-36 may be involved in the innate defense of eggs from pathogens, which may play important roles in protection against bacterial invasion of both embryonated eggs (relevant to avian reproductive success) and unfertilized table eggs (relevant to food safety).

21.2.8 Ovotransferrin

Ovotransferrin, also known as conalbumin, is an iron-binding glycoprotein present in ESM, as well as in egg whites and eggshells (Cordeiro & Hincke, 2016). The ESM ovotransferrin has been reported to play dual roles: it influences the nucleation and growth of calcite crystals and is a bacteriostatic filter that inhibits the growth of *Salmonella* in egg proteins (Gautron et al., 2001).

It is well known that ovotransferrin possesses antibacterial activity against Gram-negative and Gram-positive bacteria, including *Escherichia coli*, *Listeria monocytogenes*, and *Staphylococcus aureus* (Legros et al., 2021; Wang et al., 2021b). Notably, ovotransferrin is the only protein known to inhibit the growth of *Salmonella enterica* (Baron et al., 1997). The antibacterial activity of ovotransferrin is mainly ascribed to its ability to deprive iron and other minerals (e.g., zinc) of microorganisms, thereby inhibiting the formation of biofilms (Cordeiro & Hincke, 2016; Legros et al., 2021).

In addition, ovotransferrin shares many similar bioactivities with lactoferrin, another member of the transferrin family, such as iron binding, antimicrobial effects, antioxidant properties, and immunomodulatory functions. Due to the various biological activities of ovotransferrin, it is considered a promising functional food ingredient, a high-potential substitute for lactoferrin, for enhancing iron and calcium absorption, preventing osteoporosis and oxidative damage (Brand et al., 2016; Wang et al., 2021b).

21.2.9 Other Bioactive ESM Proteins

In addition to the previous proteins, there are many other biologically active proteins in the ESM, including a range of structural proteins, mineralization-related proteins, proteins associated with the biomechanical functions of muscle, antimicrobial proteins, proteins for cytoskeletal organization and anchoring functions, and proteins associated with adhesion and differentiation functions (Makkar et al., 2015; Cordeiro & Hincke, 2016; Ahmed et al., 2017). Keratins, laminins, and agrin are associated with the structural components of the membrane, which forms the fibrous meshwork of the ESM together with collagens and CREMPs. Some proteins associated with the biomechanical functions of muscle, such as titin, dynein, obscurin, myosin, and nebulin, may provide biomechanical support and elasticity to the ESM to protect the egg from fall injury. In addition, protease inhibitors and antiproteases, such as ovalbumin Y, ovomacroglobulin, ovoglycan, serine proteases, and β-N-acetylglucosaminidase also possess antimicrobial activity (Huopalahti et al., 2007; Ahmed et al., 2017). The mucoid substances, including ovomucin and muin similarly, can protect against viruses (Lieleg et al., 2012). Besides, ovocleidin-116 and ovocalyxin are associated with the mineralization process, providing support for eggshell mineralization on the ESM. ESM is a matrix rich in various bioactive proteins that provide antimicrobial and mechanical protection to the egg and offer molecular coordination for embryonic development (Makkar et al., 2015).

21.2.10 Bioactive Peptides

Although proteins have both a structural and functional foundation in tissues, peptides also play essential roles in many biological processes, such as signaling, transport, and host defense (Makkar et al., 2015). ESM is rich in a diverse range of bioactive proteins and peptides, many of which have different protective and supportive functions for the embryo. Avian beta-defensins (AvBDs) are currently ESM's most investigated endogenous peptides. In addition, bioactive peptides from ESM proteins can be released by *in vitro* hydrolysis, and these bioactive peptides exhibit excellent antioxidant and anti-inflammatory activities.

21.2.10.1 Avian Beta Defensins (AvBDs)

AvBDs, small cationic nonglycosylated peptides (1 to 9 kDa), are host defense peptides (HDPs) present in birds (Herve-Grepinet et al., 2010). The HDPs are a large class of small peptides found in almost all life forms and are an important first-line defense with antimicrobial and immunomodulatory properties (Zhang & Sunkara, 2014; Ahmed et al., 2017). HDPs have broad activity against Gram-negative and Gram-positive bacteria, mycobacteria, fungi, viruses, and cancer cells. They are a low-risk mechanism triggering microbial resistance due to their actions on microbial membranes or the host immune system (Zhang & Sunkara, 2014).

Defensins are cysteine-rich HDPs classified into three major groups, i. e., α, β, and θ defensins, based on the locations of disulfide bridge pairings and the positions of the conserved six-cysteine residues (Lynn et al., 2007; Zhang & Sunkara, 2014). The chicken genome encodes 14 different β-defensin genes, and AvBDs-9, -10, and -11 were identified in the ESM (Rose-Martel et al., 2015; Ahmed et al., 2017; Ahmed et al., 2019). AvBDs-9 and AvBDs-11 have been successfully tested for their antimicrobial activity. They have a large antimicrobial spectrum, exhibiting antimicrobial activity toward the Gram-positive and Gram-negative bacteria (Herve-Grepinet et al., 2010; Zhang & Sunkara, 2014). The presence of AvBDs in ESM is likely to reinforce the antimicrobial defenses (Cordeiro & Hincke, 2016; Ahmed et al., 2017), which can contribute to the protection of the embryo during the development and production of pathogen-free eggs (Herve-Grepinet et al., 2010).

21.2.10.2 Other Endogenous Peptides

In addition, several extractable endogenous bioactive peptides of ESM, such as peptides derived from clusterin, phosphatase and actin regulator (PHAR) proteins, and a sporozoite surface protein 2-like protein, were identified (Makkar et al., 2015). Clusterin is a secreted glycoprotein with cytoprotective functions that protect against various environmental microbes and oxidative stress (Mann et al., 2004). Clusterin peptides may also exhibit similar effects to clusterin but need further investigation. Some keratin peptides detected from the ESM also showed antimicrobial activities (Makkar et al., 2015).

Arginine-Glycine-Aspartic acid (RGD) is a tripeptide glycoprotein motif that is a cell adhesion peptide sequence in the ESM (Sah & Rath, 2016). RGD can minimize the risk of immune reactions or pathogen transfer that often occurs in xenografts (Bellis, 2011; Sah & Rath, 2016). In addition, the RGD sequence induces intracellular Ca^{2+} release, and the surface modification of RGD-containing peptides can promote osteoblast differentiation (Yang et al., 2005)

21.3 Bioactive Peptides from the ESM Proteins Hydrolyzed In Vitro

The hydrolysis of bioactive proteins *in vitro* is one way of obtaining ESM bioactive peptides. A variety of peptides prepared by *in vitro* hydrolysis of active proteins identified in ESM exhibit biological activities, which are even higher than those of the source protein. For example, as shown in Table 21.2, the OCX-36 peptide obtained by enzymatic digestion of OCX-36 showed better immunomodulatory effects than the OCX-36 *in vitro* and a mouse model, suggesting that OCX-36-derived peptides have potential therapeutic applications in sepsis and could be used to develop a new endotoxin-neutralizing therapeutic agent (Kovacs-Nolan et al., 2014). In addition, the derived peptides obtained by the hydrolysis of fibronectin replaced the intact fibronectin in a fibrin-fibronectin provisional matrix when cell adhesion and matrix contraction of CHOalpha4 cells were significantly enhanced, which facilitated wound healing (Valenick et al., 2005). Ovotransferrin hydrolysis and their derived peptides have also shown stronger antioxidant activity than ovotransferrin (Lee et al., 2017; Wang et al., 2021b). Additionally, as one of the main bioactive proteins found in ESM, collagens hydrolyzed *in vitro* have been demonstrated to produce bioactive peptides with a wide range of biological activities, such as antioxidant, antihypertensive, and antidiabetic effects (Fu et al., 2019; Devita et al., 2021).

The protease hydrolysis of ESM can also yield a variety of bioactive peptides, and these hydrolysates have been reported to exhibit antioxidant (Huang et al., 2010; Shi et al., 2014a, 2014b; Zhu et al., 2022), anti-inflammatory activities (Shi et al., 2014c) and beneficial effects on joint and skin health. (Hewlings et al., 2019; Furukawa et al., 2021; Xiao et al., 2021; Kulshreshtha et al., 2022), as displayed in Table 21.2. Among these effects, antioxidant activity associated with multiple health-protective benefits has gained increasing attention from researchers. The specific sequences and biological activities of antioxidant peptides derived from ESM hydrolysates were studied recently: Zhao et al. (2019) identified five antioxidant peptides (ESYHLPR, NVIDPPIYAR, MFAEWQPR, LLFAMTKPK, MLKMLPFK) from enzymatic hydrolysates of ESM with a significant *in vitro* antioxidant activity. Zhu et al. (2022) characterized six antioxidant peptides from the ESM hydrolysates by a combined *in vitro* and *in silico* study, confirmed their antioxidant activity at *in vitro* chemical and cellular levels, and found that they were associated with activating the Keapl (Kelch-like ECH-associated protein 1)-Nrf2 (nuclear factor erythroid 2-related factor 2) anti-oxidation signaling pathway, a key cellular defense system against oxidative stress (Boyenle et al., 2021).

21.4 Glycosaminoglycans (GAGs)

GAGs (5–10,000 kDa) are linear and negatively charged polysaccharides composed of a variable number of repeating disaccharide units (Fraser et al., 1997; Sodhi & Panitch, 2020) in the

TABLE 21.2

Representative Studies of Bioactive Peptides Prepared by Hydrolysis of Proteins Identified in ESM or Hydrolysis of ESM Directly

Protein Source	Hydrolysis Method	Bioactivity	References
Ovocalyxin-36	Hydrolysis by pepsin or thermolysin	Immunomodulatory effects	Kovacs-Nolan et al. (2014)
Fibronectin	Hydrolysis by chymotrypsin	Promotion of wound healing: regulation cell morphology and cell-mediated matrix contraction	Valenick et al. (2005)
Ovotransferrin	Hydrolysis by promod 278P, thermolysin and a combination of the two enzymes	Antioxidant and anticancer effects	Lee et al. (2017)
Collagen	Hydrolysis by alcalase, neutrase, and savinase	Antioxidant, antihypertensive, and antidiabetic effects	Devita et al. (2021)
ESM	Hydrolysis by alcalase	Antioxidant activity: free radical scavenging activity, protective effect on DNA damage, chelating power, and reducing ability	Huang et al. (2010); Zhao et al. (2019)
ESM	Hydrolysis by alcalase and protease S	Antioxidant activity: free radical scavenging activity, reducing ability, chelating power and protection against oxidative damage in Caco-2 cells Amelioration of intestinal inflammation	Shi et al. (2014a); Shi et al. (2014b); Shi et al. (2014c)
ESM	Hydrolysis by basic solution	Improvement of joint health	Hewlings et al. (2019)
ESM	Hydrolysis by papain	Improvement of skin health	Furukawa et al. (2021)
ESM	Hydrolysis by alcalase	Antioxidant activity: regulation of the cellular anti-oxidation Keap1-Nrf2 signaling pathway and protection against oxidative damage in RAW264.7 cells	Zhu et al. (2022)

ESM (Kulshreshtha et al., 2022). According to their repeating disaccharide structure, GAGs can be classified into non-sulfated hyaluronic acid (HA), chondroitin sulfate (CS), keratan sulfate (KS), and heparan sulfate (HS) (Neves et al., 2020; Puri et al., 2020; Wang et al., 2021a). Previous investigations have demonstrated that four kinds of GAGs are present in the ESM, with HA being the most abundant, followed by CS, KS, and HS (Liu et al., 2014; Liu et al., 2016; Vuong et al., 2017). Numerous studies suggested that GAGs possessed various biological properties, including antioxidant, anti-inflammatory, wound healing, anti-aging, immunomodulatory, prebiotic, and mineral absorption properties, and prevented cardiovascular and other diseases. Therefore, the GAGs are widely used in the pharmaceutical, healthcare, cosmetic, and food industries (Handel et al., 2005; Neves et al., 2020; Puri et al., 2020; Sodhi & Panitch, 2020; Sharma et al., 2021). The studies reveal that the GAGs of ESM can improve the physiology and overall appearance of human skin and the appearance of aging skin (Yoo et al., 2014; Kalman & Hewlings, 2020). The ESM-derived carbohydrate fractions containing sulfated and non-sulfated GAGs possess NF-κB-mediated immunomodulatory and anti-inflammatory activity in the LPS-triggered human immune cells by significantly reducing NF-κB activity and the secretion of pro-inflammatory IL-1β and IL-6 (Vuong et al., 2017). In addition, the GAGs content in the ESM has a significant correlation with eggshell strength (Ha et al., 2007). The dietary Mn supplementation can improve eggshell quality by increasing the content of GAGs and uronic acids in the ESM (Xiao et al., 2014). Moreover, ESM contains natural GAGs and proteins that are essential for maintaining the health of articular cartilage and surrounding synovial membranes in previous studies (Ruff et al., 2009a, 2009b, 2012).

21.4.1 Hyaluronic Acid (HA)

Recently, GAGs with HA as a major component have been found in the ESM (Liu et al., 2014, 2016). HA is a biodegradable linear polysaccharide found mainly in vertebrate connective tissues, such as synovial fluid of joints and vitreous humor of eyes. Compared with CS, KS, and HS, HA is the simplest and the highest molecular-weight GAGs, which neither contains sulfhydryl groups nor is linked to core proteins (Liu et al., 2016; Sodhi & Panitch, 2020). Several studies show that high molecular-weight HA ($6–7 \times 10^6$ Da) exerts anti-inflammatory and immunosuppressive effects, while low molecular weight HA ($0.5–1 \times 10^6$ Da) contributes to the stimulation of gene expression and the synthesis of pro-inflammatory proteins (Wang et al., 2021a; Kulshreshtha et al., 2022). Another study also found that HA in the ESM plays a dual role in wound healing, with both pro-inflammatory and anti-inflammatory effects, which further confirms that HA is a wound-healing mediator and its activity is related to the size of the molecule (Vuong et al., 2017). Furthermore, HA has a great water-holding capacity, which contributes to the shock absorption and lubrication of synovial joints, and increases matrix biomineralization of the bone cells (Liu et al., 2016; Sodhi & Panitch, 2020). As a result of its attractive bioactivity, HA is used in the treatment of osteoarthritis, tendon diseases, cancer nanomedicine conjugates, plastic surgery, and anti-wrinkle moisturizing cosmetics (Liu et al., 2014; Puri et al., 2020; Sodhi & Panitch, 2020; Wang et al., 2021a).

21.4.2 Chondroitin Sulfate (CS)

CS is a component of solid connective tissues such as cartilage, bone, skin, ligaments, and tendons. Like HA, CS plays a key role in retaining water when combined with proteoglycans due to its high density of anionic groups, and its anti-compression properties make it a key player in cushioning and lubrication of joints (Sodhi & Panitch, 2020). In addition, CS is widely used in cosmetics and the healing of traumatic wounds because of its antioxidant, anti-inflammatory, and wound-healing effects. Several studies showed that CS could stimulate chondrocyte, proteoglycan, and cartilage matrix synthesis, inhibit cartilage degradation, repair cartilage tissue, and relieve osteoarthritis pain (Martel-Pelletier et al., 2010; Yang et al., 2020; Mishra & Ganguli, 2021). A recent study on ESM demonstrates that carbohydrates containing CS can exert anti-inflammatory effects by reducing the activity of NF-κB and decreasing the secretion of pro-inflammatory IL-1β and IL-6 (Vuong et al., 2017).

21.4.3 Keratan Sulfate (KS)

KS is a complex molecule with multiple structures, consisting of repeated disaccharides 3Galβ1-4GlcNAcβ1- (N-acetyl lactosamine, also known as L1), and Gal, GlcNAc, and natural GAGs are present in the hydrolyzed ESM (Caterson & Melrose, 2018; Kalman & Hewlings, 2020). The previous study reveals that shell strength depends on the KS content in the ESM with an increase in the shell weight (Ha et al., 2007). KS is also a functional electrosensory and neuro-instructive molecule related to brain development, cognitive learning, neuroregulation, and neural repairing (Melrose, 2019). Furthermore, KS is also associated with certain specific disease processes, involving macular corneal dystrophy (Hasegawa et al., 2000) and amyotrophic lateral sclerosis (Hirano et al., 2013), and Alzheimer's disease (Lindahl et al., 1996). KS fragments could trigger anti-inflammatory innate immunomodulatory responses via specific receptors, such as PAMP lectin receptors, Siglec, and Langerin (Gonzalez-Gil et al., 2018; Kizuka et al., 2018). It has been reported that the highly sulfated KS disaccharide could block inflammation *in vivo* and be an effective candidate for treating chronic obstructive pulmonary disease (COPD) and its progression (Xu et al., 2005; Shirato et al., 2013).

21.4.4 Heparan Sulfate (HS)

HS is the least abundant GAGs in the ESM and is a linear and sulfated polysaccharide composed of glucuronic acid attached to N-acetylglucosamine and is considered the most prevalent and complex GAGs (Liu et al., 2014; Liu et al., 2016). HS is widely present in animal tissues, and its structure varies due to changes in the sulfation pattern of its chains (Handel et al., 2005; Sodhi & Panitch, 2020). HS has cell adhesion, cell growth and proliferation regulation, blood coagulation, cell surface binding of lipoprotein lipase and other proteins, viral invasion, and tumor metastasis functions (Neves et al., 2020; Puri et al., 2020; Wang et al., 2021a).

21.5 Other Bioactives

Sialic acid is another important bioactive found in the ESM (Nakano et al., 1994; Nakano et al., 2003). It is a nine-carbon carbohydrate molecule involved in the structure of complex carbohydrates, including glycoproteins, glycolipids, and keratan sulfate proteoglycans. There are more than 50 types of salivary acids. Among them, the most widely distributed salivary acids are N-acetylneuraminic acid (Neu5Ac), deaminoneuraminic acid (2-keto-3-deoxy-nonulononic acid or Kdn), and N-glycolylneuraminic acid (Neu5Gc). Neu5Ac is considered the biosynthetic precursor of all other types of salivary acids (Nakano et al., 1994; Angata & Varki, 2002).

Sialic acid has various biological activities, including promoting brain and cognitive development, immune enhancement, anti-hypertension, anti-cancer, and skin whitening, which are essential for biological functions and pharmacological applications in human nutrients (Pongracz et al., 2019). Therefore, salivary acids can be used as functional ingredients in various industries (Ling et al., 2022). The potential value of sialic acid as a pharmaceutical and food ingredient requires large-scale economic production of this compound to meet the growing demands, and sialic acid from egg fractions has been identified as a potential commercial source. Obtaining sialic acid from relatively inexpensive sources such as ESM is economically important in the poultry industry and has a large potential for development (Nakano et al., 1994). In addition to the various bioactive compounds discussed, $CaCO_3$ and small amounts of saccharides are also present in the ESM (Sah & Rath, 2016; Shi et al., 2021).

21.6 Conclusion

The ESM is a complex bioactive library with various active proteins such as collagens and lysozyme, some functional peptides, GAGs (e.g., HA and CS), sialic acid, and other bioactive components. These bioactive compounds collectively contribute to the biological functions of the ESM that provide resistance to microbial invasion, biomechanical protection, biomineralization of eggshells, and support for embryonic development. Knowledge of the bioactive components of ESM allows a better understanding of its role in embryo development and protection, egg quality improvement, and the development of functional ESM-derived products to promote the use of this agricultural waste. The functional studies of ESM bioactive components and the exploitation of their potential applications will constitute the next frontier of egg science.

REFERENCES

Ahlborn, G. J., Clare, D. A., Sheldon, B. W., & Kelly, R. W. (2006). Identification of eggshell membrane proteins and purification of ovotransferrin and beta-NAGase from hen egg white. *Protein Journal*, *25*(1), 71–81.

Ahlborn, G., & Sheldon, B. W. (2005). Enzymatic and microbiological inhibitory activity in eggshell membranes as influenced by layer strains and age and storage variables. *Poultry Science*, *84*(12), 1935–1941.

Ahlborn, G., & Sheldon, B. W. (2006). Identifying the components in eggshell membrane responsible for reducing the heat resistance of bacterial pathogens. *Journal of Food Protection*, *69*(4), 729–738.

Ahmed, T. A. E., Suso, H. P., & Hincke, M. T. (2017). In-depth comparative analysis of the chicken eggshell membrane proteome. *Journal of Proteomics*, *155*(23), 49–62.

Ahmed, T. A. E., Suso, H. P., Maqbool, A., & Hincke, M. T. (2019). Processed eggshell membrane powder: Bioinspiration for an innovative wound healing product. *Materials Science & Engineering C-Materials for Biological Applications*, *95*, 192–203.

Angata, T., & Varki, A. (2002). Chemical diversity in the sialic acids and related α-keto acids: An evolutionary perspective. *Chemical Reviews*, *102*(2), 439–470.

Arias, J. L., Nakamura, O., Fernandez, M. S., Wu, J. J., Knigge, P., Eyre, D. R., & Caplan, A. I. (1997). Role of type X collagen on experimental mineralization of eggshell membranes. *Connective Tissue Research*, *36*(1), 21–33.

Baron, F., Gautier, M., & Brule, G. (1997). Factors involved in the inhibition of growth of salmonella enteritidis in liquid egg white. *Journal of Food Protection*, *60*(11), 1318–1323.

Bellis, S. L. (2011). Advantages of RGD peptides for directing cell association with biomaterials. *Biomaterials*, *32*(18), 4205–4210.

Bhattacharjee, A., & Bansal, M. (2005). Collagen structure: The Madras triple helix and the current scenario. *IUBMB Life*, *57*(3), 161–172.

Bignon, M., Pichol-Thievend, C., Hardouin, J., Malbouyres, M., Brechot, N., Nasciutti, L., & Germain, S. (2011). Lysyl oxidase-like protein-2 regulates sprouting angiogenesis and type IV collagen assembly in the endothelial basement membrane. *Blood*, *118*(14), 3979–3989.

Boyenle, I. D., Divine, U. C., Adeyemi, R., Ayinde, K. S., Olaoba, O. T., Apu, C., & Adelusi, T. I. (2021). Direct Keapl-kelch inhibitors as potential drug candidates for oxidative stress-orchestrated diseases: A review on in silico perspective. *Pharmacological Research*, *167*, 105577.

Brand, J., Dachmann, E., Pichler, M., Lotz, S., & Kulozik, U. (2016). A novel approach for lysozyme and ovotransferrin fractionation from egg white by radial flow membrane adsorption chromatography: Impact of product and process variables. *Separation & Purification Technology*, *161*, 44–52.

Caterson, B., & Melrose, J. (2018). Keratan sulfate, a complex glycosaminoglycan with unique functional capability. *Glycobiology*, *28*(4), 182–206.

Chen, L., Kang, J., & Sukigara, S. (2014). Preparation and characterization of polyurethane/soluble eggshell membrane nanofibers. *Bio-Medical Materials and Engineering*, *24*(6), 1979–1989.

Chowdhury, S. D. (1990). Shell membrane protein system in relation to lathyrogen toxicity and copper deficiency. *World's Poultry Science Journal*, *46*(2), 153–169.

Cordeiro, C. M. M., Esmaili, H., Ansah, G., & Hincke, M. T. (2013). Ovocalyxin-36 is a pattern recognition protein in chicken eggshell membranes. *PLoS One*, *8*(12), e84112.

Cordeiro, C. M. M., & Hincke, M. T. (2016). Quantitative proteomics analysis of eggshell membrane proteins during chick embryonic development. *Journal of Proteomics*, *130*, 11–25.

Devita, L., Lioe, H. N., Nurilmala, M., & Suhartono, M. T. (2021). The bioactivity prediction of peptides from tuna skin collagen using integrated method combining in vitro and in silico. *Foods*, *10*(11), 2739.

Du, J. W., Hincke, M. T., Rose-Martel, M., Hennequet-Antier, C., Brionne, A., Cogburn, L. A., & Gautron, J. (2015). Identifying specific proteins involved in eggshell membrane formation using gene expression analysis and bioinformatics. *BMC Genomics*, *16*, 792.

Ferraboschi, P., Ciceri, S., & Grisenti, P. (2021). Applications of lysozyme, an innate immune defense factor, as an alternative antibiotic. *Antibiotics (Basel)*, *10*(12), 1534.

Fraser, J. R. E., Laurent, T. C., & Laurent, U. B. G. (1997). Hyaluronan: Its nature, distribution, functions and turnover. *Journal of Internal Medicine*, *242*(1), 27–33.

Fu, Y., Therkildsen, M., Aluko, R., & Lametsch, R. (2019). Exploration of collagen recovered from animal by-products as a precursor of bioactive peptides: Successes and challenges. *Critical Reviews in Food Science and Nutrition*, *59*(13), 2011–2027.

Furukawa, K., Kono, M., Kataoka, T., Hasebe, Y., Jia, H., & Kato, H. (2021). Effects of eggshell membrane on keratinocyte differentiation and skin aging in vitro and in vivo. *Nutrients*, *13*(7), 2144.

Gautron, J., Hincke, M. T., Panheleux, M., Garcia-Ruiz, J. M., Boldicke, T., & Nys, Y. (2001). Ovotransferrin is a matrix protein of the hen eggshell membranes and basal calcified layer. *Connective Tissue Research*, *42*(4), 255–267.

Gautron, J., Murayama, E., Vignal, A., Morisson, M., McKee, M. D., Rehault, S., & Hincke, M. T. (2007). Cloning of ovocalyxin-36, a novel chicken eggshell protein related to lipopolysaccharide-binding proteins, bactericidal permeability-increasing proteins, and plunc family proteins. *Journal of Biological Chemistry*, *282*(8), 5273–5286.

Gil-Quintana, E., Fenaux, M., Nuez, M. L., Molero, A., & Aguirre, A. (2018). Short-term effects of ovomet®, eggshell membrane, in joint pain. A double-blind and placebo study. *Journal of Osteoporosis and Physical Activity*, *6*(1), 1000211.

Gonzalez-Gil, A., Porell, R. N., Fernandes, S. M., Wei, Y., Yu, H., Carroll, D. J., & Schnaar, R. L. (2018). Sialylated keratan sulfate proteoglycans are siglec-8 ligands in human airways. *Glycobiology*, *28*(10), 786–801.

Grinnell, F. (1984). Fibronectin and wound healing. *Journal of Cellular Biochemistry*, *26*(2), 107–116.

Ha, Y. W., Son, M. J., Yun, K. S., & Kim, Y. S. (2007). Relationship between eggshell strength and keratan sulfate of eggshell membranes. *Comparative Biochemistry and Physiology A-Molecular & Integrative Physiology*, *147*(4), 1109–1115.

Handel, T. M., Johnson, Z., Crown, S. E., Lau, E. K., Sweeney, M., & Proudfoot, A. E. (2005). Regulation of protein function by glycosaminoglycans–as exemplified by chemokines. *Annual Review of Biochemistry*, *74*, 385–410.

Hasegawa, N., Torii, T., Kato, T., Miyajima, H., Furuhata, A., Nakayasu, K., & Habuchi, O. (2000). Decreased GlcNAc 6-O-sulfotransferase activity in the cornea with macular corneal dystrophy. *Investigative Ophthalmology & Visual Science*, *41*(12), 3670–3677.

Herve-Grepinet, V., Rehault-Godbert, S., Labas, V., Magallon, T., Derache, C., Lavergne, M., & Nys, Y. (2010). Purification and characterization of avian beta-defensin 11, an antimicrobial peptide of the hen egg. *Antimicrobial Agents & Chemotherapy*, *54*(10), 4401–4408.

Hewlings, S., Kalman, D., & Schneider, L. V. (2019). A randomized, double-blind, placebo-controlled, prospective clinical trial evaluating water-soluble chicken eggshell membrane for improvement in joint health in adults with knee osteoarthritis. *Journal of Medicinal Food*, *22*(9), 875–884.

Hincke, M. T., Gautron, J., Panheleux, M., Garcia-Ruiz, J., Mckee, M. D., & Nys, Y. (2000). Identification and localization of lysozyme as a component of eggshell membranes and eggshell matrix. *Matrix Biology*, *19*(5), 443–453.

Hirano, K., Ohgomori, T., Kobayashi, K., Tanaka, F., Matsumoto, T., Natori, T., & Kadomatsu, K. (2013). Ablation of keratan sulfate accelerates early phase pathogenesis of ALS. *PLoS One*, *8*(6), e66969.

Huang, X., Zhou, Y. H., Ma, M. H., Cai, Z. X., & Li, T. (2010). Chemiluminescence evaluation of antioxidant activity and prevention of DNA damage effect of peptides isolated from soluble eggshell membrane protein hydrolysate. *Journal of Agricultural and Food Chemistry*, *58*(23), 12137–12142.

Huopalahti, R., López-Fandiño, R., Anton, M., & Schade, R. (2007). *Bioactive egg compounds*. Springer.

Jia, H., Hanate, M., Aw, W., Itoh, H., Saito, K., Kobayashi, S., & Kato, H. (2017). Eggshell membrane powder ameliorates intestinal inflammation by facilitating the restitution of epithelial injury and alleviating microbial dysbiosis. *Scientific Reports*, *7*, 43993.

Jun, H. J., Oh, K. H., Yoo, J., Han, W. G., Chang, J. W., Jung, H. H., & Choi, J. (2013). A new patch material for tympanic membrane perforation by trauma: The membrane of a hen egg shell. *Acta Oto Laryngologica*, *134*(3), 250–254.

Kalman, D. S., & Hewlings, S. (2020). The effect of oral hydrolyzed eggshell membrane on the appearance of hair, skin, and nails in healthy middle-aged adults: A randomized double-blind placebo-controlled clinical trial. *Journal of Cosmetic Dermatology*, *19*(6), 1463–1472.

Kawasaki, H., & Iwamuro, S. (2008). Potential roles of histones in host defense as antimicrobial agents. *Infectious Disorders - Drug Targets*, *8*(3), 195–205.

Khorshidian, N., Khanniri, E., Koushki, M. R., Sohrabvandi, S., & Yousefi, M. (2022). An overview of antimicrobial activity of lysozyme and its functionality in cheese. *Frontiers in Nutrition*, *9*, 833618.

King'Ori, A. M. (2011). A review of the uses of poultry eggshells and shell membranes. *International Journal of Poultry Science*, *10*(11), 908–912.

Kizuka, Y., Mishra, S., Yamaguchi, Y., & Taniguchi, N. (2018). Implication of c-type lectin receptor langerin and keratan sulfate disaccharide in emphysema. *Cellular Immunology*, *333*, 80–84.

Kodali, V. K., Gannon, S. A., Paramasivam, S., Raje, S., Polenova, T., & Thorpe, C. (2011). A novel disulfide-rich protein motif from avian eggshell membranes. *PLoS One*, *6*(3), e18187.

Kovacs-Nolan, J., Cordeiro, C., Young, D., Mine, Y., & Hincke, M. (2014). Ovocalyxin-36 is an effector protein modulating the production of proinflammatory mediators. *Veterinary Immunology and Immunopathology*, *160*(1–2), 1–11.

Kozuka, M., Murao, S., Yamane, T., Inoue, T., Ohkubo, I., & Ariga, H. (2015). Rapid and simple purification of lysozyme from the egg shell membrane. *Journal of Nutritional Science and Vitaminology*, *61*(1), 101–103.

Kulshreshtha, G., Diep, T., Hudson, H. A., & Hincke, M. T. (2022). High value applications and current commercial market for eggshell membranes and derived bioactives. *Food Chemistry*, *382*, 132270.

Lee, J., Moon, S., Kim, H., Park, E., Ahn, D., & Paik, H. (2017). Antioxidant and anticancer effects of functional peptides from ovotransferrin hydrolysates. *Journal of the Science of Food and Agriculture*, *97*(14), 4857–4864.

Legros, J., Jan, S., Bonnassie, S., Gautier, M., Croguennec, T., Pezennec, S., & Baron, F. (2021). The role of ovotransferrin in egg-white antimicrobial activity: A Review. *Foods*, *10*(4), 823.

Lenselink, E. A. (2015). Role of fibronectin in normal wound healing. *International Wound Journal*, *12*(3), 313–316.

Li, X., Cai, Z., Ahn, D. U., & Huang, X. (2019a). Development of an antibacterial nanobiomaterial for wound-care based on the absorption of AgNPs on the eggshell membrane. *Colloids and Surfaces B: Biointerfaces*, *183*, 110449.

Lieleg, O., Lieleg, C., Bloom, J., Buck, C. B., & Ribbeck, K. (2012). Mucin biopolymers as broad-spectrum antiviral agents. *Biomacromolecules*, *13*(6), 1724–1732.

Li, X., Ma, M., Ahn, D. U., & Huang, X. (2019b). Preparation and characterization of novel eggshell membrane-chitosan blend films for potential wound-care dressing: From waste to medicinal products. *International Journal of Biological Macromolecules*, *123*, 477–484.

Li, G. H., Mine, Y., Hincke, M. T., & Nys, Y. (2010). Isolation and characterization of antimicrobial proteins and peptide from chicken liver. *Journal of Peptide Science*, *13*(6), 368–378.

Lindahl, B., Eriksson, L., Spillmann, D., Caterson, B., & Lindahl, U. (1996). Selective loss of cerebral keratan sulfate in Alzheimer's disease. *Journal of Biological Chemistry*, *271*(29), 16991–16994.

Ling, A. J. W., Chang, L. S., Babji, A. S., Latip, J., Koketsu, M., & Lim, S. J. (2022). Review of sialic acid's biochemistry, sources, extraction and functions with special reference to edible bird's nest. *Food Chemistry*, *367*, 130755.

Liu, Z., Sun, X., Cai, C., He, W., Zhang, F., & Linhardt, R. J. (2016). Characteristics of glycosaminoglycans in chicken eggshells and the influence of disaccharide composition on eggshell properties. *Poultry Science*, *95*(12), 2879–2888.

Liu, Z., Zhang, F., Li, L., Li, G., He, W., & Linhardt, R. J. (2014). Compositional analysis and structural elucidation of glycosaminoglycans in chicken eggs. *Glycoconjugate Journal*, *31*, 593–602.

Lynn, D. J., Higgs, R., Lloyd, A. T., O'Farrelly, C., Hervé-Grépinet, V., Nys, Y., & Lehrer, R. I. (2007). Avian beta-defensin nomenclature: A community proposed update. *Immunology Letters*, *110*(1), 86–89.

Makkar, S. K., Liyanage, R., Kannan, L., Packialakshmi, B., Lay, J., & Rath, N. C. (2015). Chicken egg shell membrane associated proteins and peptides. *Journal of Agricultural and Food Chemistry*, *63*(44), 9888–9898.

Mann, K., Gautron, J., Nys, Y., Mckee, M. D., BaJari, T., Schneider, W. J., & Hincke, M. T. (2004). Disulfide-linked heterodimeric clusterin is a component of the chicken eggshell matrix and egg white. *Matrix Biology*, *22*(5), 397–407.

Mann, K., & Mann, M. (2015). Proteomic analysis of quail calcified eggshell matrix: A comparison to chicken and Turkey eggshell proteomes. *Proteome Science*, *13*, 22.

Marino-Ramirez, L., Kann, M. G., Shoemaker, B. A., & Landsman, D. (2005). Histone structure and nucleosome stability. *Expert Review of Proteomics*, *2*(5), 719–729.

Martel-Pelletier, J., Kwan Tat, S., & Pelletier, J. P. (2010). Effects of chondroitin sulfate in the pathophysiology of the osteoarthritic joint: A narrative review. *Osteoarthritis Cartilage*, *18*, S7–S11.

Melrose, J. (2019). Functional consequences of keratan sulfate sulfation in electrosensory tissues and in neuronal regulation. *Advanced Biosystems*, *3*(4), 1800327.

Mishra, S., & Ganguli, M. (2021). Functions of, and replenishment strategies for, chondroitin sulfate in the human body. *Drug Discovery Today*, *26*(5), 1185–1199.

Nakano, T., Ikawa, N. I., & Ozimek, L. (2003). Chemical composition of chicken eggshell and shell membranes. *Poultry Science*, *82*(3), 510–514.

Nakano, K., Nakano, T., Ahn, D. U., & Sim, J. S. (1994). Sialic acid contents in chicken eggs and tissues. *Canadian Journal of Animal Science*, *74*(4), 601–606.

Neves, M. I., Araujo, M., Moroni, L., da Silva, R. M. P., & Barrias, C. C. (2020). Glycosaminoglycan-inspired biomaterials for the development of bioactive hydrogel networks. *Molecules*, *25*(4), 978.

Osuoji, C. I. (1971). Acid glycosaminoglycan of eggshell membranes. *Biochimica et Biophysica Acta-General Subjects*, *244*(2), 481–483.

Pongracz, T., Wuhrer, M., & de Haan, N. (2019). Expanding the reaction space of linkage-specific sialic acid derivatization. *Molecules*, *24*(19), 3617.

Puri, S., Coulson-Thomas, Y. M., Gesteira, T. F., & Coulson-Thomas, V. J. (2020). Distribution and function of glycosaminoglycans and proteoglycans in the development, homeostasis and pathology of the ocular surface. *Frontiers in Cell and Developmental Biology*, *8*, 731.

Rose-Martel, M., & Hincke, M. T. (2014). Antimicrobial histones from chicken erythrocytes bind bacterial cell wall lipopolysaccharides and lipoteichoic acids. *International Journal of Antimicrobial Agents*, *44*(5), 470–472.

Rose-Martel, M., Smiley, S., & Hincke, M. T. (2015). Novel identification of matrix proteins involved in calcitic biomineralization. *Journal of Proteomics*, *116*, 81–96.

Ruff, K. J., Devore, D. P., Leu, M. D., & Robinson, M. A. (2009a). Eggshell membrane: A possible new natural therapeutic for joint and connective tissue disorders. Results from two open-label human clinical studies. *Clinical Interventions in Aging*, *4*, 235–240.

Ruff, K. J., Endres, J. R., Clewell, A. E., Szabo, J. R., & Schauss, A. G. (2012). Safety evaluation of a natural eggshell membrane-derived product. *Food and Chemical Toxicology*, *50*(3–4), 604–611.

Ruff, K. J., Winkler, A., Jackson, R. W., DeVore, D. P., & Ritz, B. W. (2009b). Eggshell membrane in the treatment of pain and stiffness from osteoarthritis of the knee: A randomized, multicenter, double-blind, placebo-controlled clinical study. *Clinical Rheumatology*, *28*, 907–914.

Sah, M. K., & Rath, S. N. (2016). Soluble eggshell membrane: A natural protein to improve the properties of biomaterials used for tissue engineering applications. *Materials Science & Engineering C-Materials for Biological Applications*, *67*, 807–821.

Sharma, R., Kataria, A., Sharma, S., & Singh, B. (2021). Structural characterisation, biological activities and pharmacological potential of glycosaminoglycans and oligosaccharides: A review. *International Journal of Food Science & Technology*, *57*(1), 4–15.

Shi, Y., Kovacs-Nolan, J., Jiang, B., Tsao, R., & Mine, Y. (2014a). Antioxidant activity of enzymatic hydrolysates from eggshell membrane proteins and its protective capacity in human intestinal epithelial Caco-2 cells. *Journal of Functional Foods*, *10*, 35–45.

Shi, Y., Kovacs-Nolan, J., Jiang, B., Tsao, R., & Mine, Y. (2014b). Peptides derived from eggshell membrane improve antioxidant enzyme activity and glutathione synthesis against oxidative damage in Caco-2 cells. *Journal of Functional Foods*, *11*, 571–580.

Shirato, K., Gao, C., Ota, F., Angata, T., Shogomori, H., Ohtsubo, K., & Taniguchi, N. (2013). Flagellin/Toll-like receptor 5 response was specifically attenuated by keratan sulfate disaccharide via decreased EGFR phosphorylation in normal human bronchial epithelial cells. *Biochemical and Biophysical Research Communications*, *435*(3), 460–465.

Shi, Y., Rupa, P., Jiang, B., & Mine, Y. (2014c). Hydrolysate from eggshell membrane ameliorates intestinal inflammation in mice. *International Journal of Molecular Sciences*, *15*(12), 22728–22742.

Shi, Y., Zhou, K., Li, D., Guyonnet, V., Hincke, M. T., & Mine, Y. (2021). Avian eggshell membrane as a novel biomaterial: A review. *Foods*, *10*(9), 2178.

Sodhi, H., & Panitch, A. (2020). Glycosaminoglycans in tissue engineering: A review. *Biomolecules*, *11*(1), 29.

Valenick, L. V., Hsia, H. C., & Schwarzbauer, J. E. (2005). Fibronectin fragmentation promotes alpha4beta1 integrin-mediated contraction of a fibrin-fibronectin provisional matrix. *Experimental Cell Research*, *309*(1), 48–55.

Vuong, T. T., Rønning, S. B., Suso, H. P., Schmidt, R., Prydz, K., Lundström, M., & Pedersen, M. E. (2017). The extracellular matrix of eggshell displays anti-inflammatory activities through NF-κ B in LPS-triggered human immune cells. *Journal of Inflammation Research*, *10*, 83–96.

Waheed, M., Yousaf, M., Shehzad, A., Inam-Ur-Raheem, M., Khan, M. K. I., Khan, M. R., & Aadil, R. M. (2020). Channelling eggshell waste to valuable and utilizable products: A comprehensive review. *Trends in Food Science & Technology*, *106*, 78–90.

Wang, S. T., Neo, B. H., & Betts, R. J. (2021a). Glycosaminoglycans: Sweet as sugar targets for topical skin anti-aging. *Clinical Cosmetic and Investigational Dermatology*, *14*, 1227–1246.

Wang, X., Wei, Z., & Xue, C. (2021b). The past and future of ovotransferrin: Physicochemical properties, assembly and applications. *Trends in Food Science & Technology*, *116*, 47–62.

Wong, M., Hendrix, M. J. C., Mark, K. V. D., Little, C., & Stern, R. (1984). Collagen in the egg shell membranes of the hen. *Developmental Biology*, *104*(1), 28–36.

Wu, Y., Rich, C., Lincecum, J., Trackman, P., Kagan, H., & Foster, J. (1992). Characterization and developmental expression of chick aortic lysyl oxidase. *The Journal of Biological Chemistry*, *267*(34), 24199–24206.

Xiao, N., Huang, X., He, W., Yao, Y., Wu, N., Xu, M., & Tu, Y. (2021). A review on recent advances of egg byproducts: Preparation, functional properties, biological activities and food applications. *Food Research International*, *147*, 110563.

Xiao, J. F., Zhang, Y. N., Wu, S. G., Zhang, H. J., Yue, H. Y., & Qi, G. H. (2014). Manganese supplementation enhances the synthesis of glycosaminoglycan in eggshell membrane: A strategy to improve eggshell quality in laying hens. *Poultry Science*, *93*(2), 380–388.

Xu, H., Kurihara, H., Ito, T., Kikuchi, H., Yoshida, K., Yamanokuchi, H., & Asari, A. (2005). The keratan sulfate disaccharide Gal(6S03) beta1,4-GlcNAc(6S03) modulates interleukin 12 production by macrophages in murine Thy-1 type autoimmune disease. *Journal of Biological Chemistry*, *280*(21), 20879–20886.

Yang, J., Shen, M., Wen, H., Luo, Y., Huang, R., Rong, L., & Xie, J. (2020). Recent advance in delivery system and tissue engineering applications of chondroitin sulfate. *Carbohydrate Polymers*, *230*, 115650.

Yang, F., Williams, C. G., Wang, D. A., Lee, H., Manson, P. N., & Elisseeff, J. (2005). The effect of incorporating RGD adhesive peptide in polyethylene glycol diacrylate hydrogel on osteogenesis of bone marrow stromal cells. *Biomaterials*, *26*(30), 5991–5998.

Yoo, J., Park, K., Yoo, Y., Kim, J., Yang, H., & Shin, Y. (2014). Effects of egg shell membrane hydrolysates on anti-inflammatory, anti-wrinkle, anti-microbial activity and moisture-protection. *Korean Journal for Food Science of Animal Resources*, *34*(1), 26–32.

Zhang, G., & Sunkara, L. T. (2014). Avian antimicrobial host defense peptides: From biology to therapeutic applications. *Pharmaceuticals*, *7*(3), 220–247.

Zhao, Q. C., Zhao, J. Y., Ahn, D. U., Jin, Y. G., & Huang, X. (2019). Separation and identification of highly efficient antioxidant peptides from eggshell membrane. *Antioxidants*, *8*(10), 495.

Zhu, L., Xiong, H., Huang, X., Guyonnet, V., Ma, M., Chen, X., & Hu, G. (2022). Identification and molecular mechanisms of novel antioxidant peptides from two sources of eggshell membrane hydrolysates showing cytoprotection against oxidative stress: A combined in silico and in vitro study. *Food Research International*, *157*, 111266.

List of Acronyms and Abbreviations

Full Name	Abbreviations
Eggshell membrane	ESM
Glycosaminoglycans	GAGs
Cysteine-rich eggshell membrane proteins	CREMPs
Lysyl oxidase	LOX
Ovocalyxin-36	OCX-36
Lysyl oxidase-like 2	LOXL2
Extracellular matrix	ECM
Proline	Pro
Hydroxyproline	Hyp
Bactericidal permeability-increasing protein	BPI
Lipopolysaccharide-binding protein	LBP
Palate, lung, and nasal epithelium clone	PLUNC
Lipopolysaccharide	LPS

Lipoteichoic acid	LTA	Keratan sulfate	KS
Avian beta-defensins	AvBDs	Heparan sulfate	HS
Host defense peptides	HDPs	Chronic obstructive pulmonary disease	COPD
Phosphatase and actin regulator	PHAR	N-acetylneuraminic acid	Neu5Ac
Arginine-Glycine-Aspartic acid	RGD	N-glycolylneuraminic acid	Neu5Gc
Hyaluronic acid	HA	Kelch-like ECH-associated protein 1	Keap1
Chondroitin sulfate	CS	Nuclear factor erythroid 2-related factor 2	Nrf2

22

Bioactives of Egg White Proteins and Peptides

Xiaoyu Bao*, **Zihan Wang***, **and Jianping Wu**†

CONTENTS

* Equal contribution to this chapter.
† Corresponding author.

DOI: 10.1201/9781003254430-29

22.1 Introduction

Chicken egg white accounts for around 63% of the total egg weight and contains approximately 10% protein (albumen) and 90% water. Ovalbumin (54%), ovotransferrin (12%), ovomucoid (11%), ovomucin (3.5%), and lysozyme (3.4%) are the major components, while avidin (0.05%), cystatin (0.05%), ovomacroglobulin (0.5%), and ovoinhibitor (1.5%) are the minor components (summarized by Kovacs-Nolan et al., 2005). As the most affordable high-quality proteins that are easily digestible and contain balanced essential amino acids, egg white proteins are also the most versatile food ingredient. Egg white contains a reservoir of bioactive proteins with potential applications as antimicrobials, antioxidants, nutraceuticals, and pharmaceuticals (Wu, 2019). An increasing number of studies also reveal the value of bioactive peptides derived from egg white proteins; bioactive peptides are mostly composed of ~2–50 amino acids and exert their biological activities after being released during food processing and/or digestion (Liao et al., 2018). Biological activity is the capacity of a molecule to exert a defined biological effect on a target and is measured by biological assays – procedures conducted for determining a biological indicator in living organisms or tissues (Jackson et al., 2007). Hence, studies based on chemical tests are excluded except for findings reporting the antioxidative activity of proteins and peptides by *in vitro* test-tube assays. This chapter covers our current knowledge of the bioactivities of egg white proteins and their derived peptides, aiming to consolidate the potential of egg white proteins and their derived peptides for alleviating chronic diseases and improving human health.

22.2 Ovalbumin

Ovalbumin is the major protein (~54%) in avian egg white with a molecular weight of 45 kDa (Abeyrathne et al., 2013). Based on mRNA and protein structure (McReynolds et al., 1978; Nisbet et al., 1981), ovalbumin is comprised of 386 amino acids, among which six are cysteines and one disulfide bond is formed between Cys_{74} and Cys_{121} (Takahashi et al., 1996). Albeit as a member of serpin superfamily, ovalbumin does not possess any protease inhibitory activity like others (Huntington & Stein, 2001). Ovalbumin is extensively used in the food industry due to its multiple functional properties (He et al., 2019). Furthermore, ovalbumin is also an excellent source of bioactive peptides/hydrolysates with antioxidative activity, angiotensin-converting enzyme-inhibitory (ACEi) activity, antiviral activity, and among others.

22.2.1 Antioxidative Activity

Enzymatic hydrolysis has been considered an effective way to modify the antioxidative activity of proteins (Venuste et al., 2013; Aguilar et al., 2019; Gomes & Kurozawa, 2020) by which antioxidant peptides are released (Cai & Corke, 2000; Lacou et al., 2016). Apart from the proven antioxidative activity in various *in vitro* assays (Bueno-Gavilá et al., 2021), ovalbumin hydrolysates provide the antioxidant defense of aged mice model by the elevation of the antioxidant enzymes such as superoxide dismutase (SOD), catalase (CAT) and glutathione peroxidase (GSH-Px), thus decreasing oxidant production and preventing age-related oxidative damage (Xu et al., 2007).

Conjugation between ovalbumin and other compounds exhibits improved antioxidant activity. When combined with some antioxidant compounds like curcumin and catechin, the antioxidative effect of ovalbumin synergistically increased (Almajano et al., 2007; Y. Liu et al., 2017; Feng et al., 2018). In other cases, glycosylation through Maillard reaction using different saccharides was also reported to improve ovalbumin's antioxidative activities and functional properties (Sun et al., 2006; Huang et al., 2012; Tu et al., 2015; Hu et al., 2019). The changed antioxidative activities might be ascribed to the formation of the browning compounds, mainly composed of melanoidins, during the Maillard reaction, which contributes to the radical scavenging capacity and metal chelating capacity (Wang et al., 2011). Likewise, selenization markedly improved the antioxidative activity of ovalbumin by enhancing its radical scavenging activity (Li et al., 2014).

22.2.2 Angiotensin-Converting Enzyme-Inhibitory (ACEi) and Antihypertensive Activity

Central to the renin-angiotensin system (RAS) necessary for blood pressure regulation, ACE is responsible for the formation of angiotensin II (Ang II), a vasoconstrictor that raises blood pressure (Ruiz-Ortega et al., 2001). Thus, inhibition of ACE activity is one of the strategies for treating hypertension. ACEi drugs are widely used; however, they often generate side effects over the long-term use. Natural compounds, including food-derived ACEi peptides, have been identified as safer alternatives (Gu et al., 2011; Rawendra et al., 2013; Aluko, 2015; Ling et al., 2018; Fan & Wu, 2021). Furthermore, as an abundant protein in eggs, ovalbumin is an excellent source of bioactive peptides, including ACEi peptides (Liao et al., 2018).

The octapeptide FRADHPFL, also known as ovokinin, is the first ACEi peptide identified from ovalbumin (IC_{50} = 3.2 μM) and possesses relaxing activity for the canine mesenteric artery (Fujita et al., 1995; Miguel et al., 2004). Oral administration of ovokinin could lower blood pressure in spontaneously hypertensive rats (SHRs) at a dose of 100 mg/kg as a peptide solution (dissolved in saline). This hypotensive effect was potentiated by preparing the peptide in an emulsion containing 30% egg yolk, in which only 25 mg/kg of peptides was needed (Fujita et al., 1995). Since the report of ovokinin, many ACEi peptides derived from ovalbumin have been reported. Among them, RADHPFL, a fragment of ovokinin, with comparable ACEi activity (IC_{50} value = 6.2 μM) also showed a blood-lowering effect in SHRs (Miguel et al., 2004; Miguel et al., 2005). Similarly, RADHPF exerted a vasodilatory effect in an isolated mesenteric artery from SHRs and significantly reduced the systolic blood pressure of SHRs at a dose of 10 mg/kg (Matoba et al., 1999). RADHP, the end product of the gastrointestinal digestion of RADHPFL, is also likely to be the active component of ovokinin (Miguel et al., 2006). Interestingly, Quirós et al. suggested that ovalbumin hydrolysis prepared under the pressure of 200-400 MPa could accelerate and promote the release of ovokinin and RADHPFL (Quirós et al., 2007). Additionally, several gastrointestinal (GI)-resistant ACEi peptides, like KLP, LKISQ, SQAVH, and FDKLPGFGD were identified from thermolysin-digested egg white hydrolysate (T-EWH) (Fan et al., 2019). Identification of four peptides with ACEi activity (IAAEVYEHTEGSTTSY, HLFGPPGKKDPV, PIAAEVYEHTEGSTTSY, and YAEERYPI from ovalbumin hydrolysates prepared by artichoke protease also confirm the feasibility to identify ACEi peptides from ovalbumin (Bueno-Gavilá et al., 2021). Using an integrated bioinformatics approach, ten ACEi peptides (CF, KM, ELPF, AM, ADHPF, LPR, PR, FR, PRM and GR) were identified from ovalbumin; these ACEi peptides also possessed dipeptidyl peptidase-4 (DPP-4) inhibitory activity as well, which might activate insulin and consequently lower the blood sugar level (Salim & Gan, 2020). Likewise, several peptides isolated from ovalbumin digested by gastrointestinal proteases were predicted to have ACEi activity, among which dipeptide VY also possesses antioxidative activity (Silva et al., 2019). However, a double-blind, placebo-controlled randomized crossover trial involving 65 participants with mildly elevated blood pressure failed to support the blood pressure-lowering effect of ovalbumin-derived hydrolysates (Lucey et al., 2019).

22.2.3 Antiviral Activity

3-hydroxyphthalic anhydride-modified ovalbumin (HP-OVA) has been widely reported to have antiviral activity against human immunodeficiency virus (HIV) (Li et al., 2010), human papillomavirus (HPV) (Hua et al., 2019) and Middle East respiratory syndrome-related coronavirus (MERS-CoV) (Zhao et al., 2013). Recent evidence suggests that HP-OVA was able to block the viral attachment and S protein-mediated cell-cell fusion, which could serve as a potential therapeutic agent for the treatment of COVID-19 (Liang et al., 2021). The elevated positively charged side chains of lysine and arginine might account for the antiviral activity of HP-OVA since the native OVA had no inhibitory effect on SARS-CoV-2 (Li et al., 2010).

22.3 Ovotransferrin

Ovotransferrin, formerly known as conalbumin (Schade & Caroline, 1944), is the second most abundant protein in egg white proteins (about 12%), which is important in the growth and development of chicken embryo. Structurally speaking, ovotransferrin is made of a 686-residue single chain containing fifteen disulfide bridges but no free sulfhydryl groups (Williams et al., 1982). Like other members of the transferrin family, ovotransferrin is comprised of two globular lobes with an iron-binding site each, while the N- and C-terminal half molecules endow ovotransferrin with great antimicrobial effect against a wide range of microorganisms (Kurokawa et al., 1995; Wu & Acero-Lopez, 2012; Legros et al., 2021). Ovotransferrin was reported to have antiviral activity (Zhou et al., 2018; Mann & Ndung'u, 2020), antioxidative activity (Kim et al., 2012; Hou et al., 2015), immunomodulatory activity (Ru et al., 2021), and osteogenic activity (Shang & Wu, 2018, 2019); some of these activities are shared by its hydrolysates and peptides (Wu & Acero-Lopez, 2012). Based on the abovementioned bioactivities, ovotransferrin exhibits immense application potential as a natural food preservative or a functional food ingredient or a nutraceutical (Wang et al., 2021).

22.3.1 Antimicrobial Activity

Antimicrobial activity is one of the well-studied biological activities of ovotransferrin (Wu & Acero-Lopez, 2012; Legros et al., 2021). Since it contains two inner iron-binding sites, the iron-sequestering capacity enables ovotransferrin to deplete the irons essential for the growth of microorganisms (Bullen et al., 1978). As reported, ovotransferrin is effective in inhibiting most foodborne pathogens, including *Staphylococcus aureus* (Ibrahim et al., 1998), *Bacillus cereus* (Baron et al., 2014), *Listeria monocytogenes* (Moon et al., 2012) and *Escherichia coli* O157:H7 (Ibrahim et al., 1998). Gram-positive bacteria are more sensitive than gram-negative bacteria to ovotransferrin due to the different compositions of cell membranes (Craig & Reeves, 2002). It was reported that native chicken ovotransferrin exerted minor bacteriostatic activity against *S. aureus*, whereas it had no effect on the growth of *E. coli* O157:H7 (Varon et al., 2013). Bicarbonate could enhance the bacteriostatic effectiveness against *E. coli* O157:H7 and *L. monocytogenes* of ovotransferrin as carbonate (CO_3^{2-}) anion is involved in the formation of "transferrin-ferric-bicarbonate" complex (Phelps & Antonini, 1975; Ko et al., 2008b), though sodium bicarbonate alone has no antibacterial effect against *E. coli* O157:H7. However, additional involvement of citric acid or citrate would antagonize the antibacterial effect of this complex (Ko et al., 2008b). Interestingly, citric acid itself shows some inhibiting actions on *E. coli* O157:H7 and has a synergistic effect with ovotransferrin (Valenti et al., 1981); the similar bacteriostatic effect of EDTA against different strains of L. monocytogenes was also reported (Ko et al., 2008a).

It is interesting to note that the iron-depriving mechanism is not essential to the antibacterial activity of ovotransferrin as Zn^{2+}-saturated ovotransferrin appeared to have a more potent antimicrobial effect against *E. coli* 803 than apo-ovotransferrin, in which Zn-ovotransferrin complex contact the bacterial surface as a whole (Valenti et al., 1987). Also, the surface interaction between ovotransferrin/lactoferrin and *C. krusei*, but not iron deprivation, involves this antifungal activity (Valenti et al., 1986).

E. coli is one of the most common foodborne pathogens and often causes food recalls of meat products (Moghadam et al., 2013; Currie et al., 2019). Although ovotransferrin alone or with $NaHCO_3$/EDTA possess antibacterial activity *in vitro*, their bactericidal effect against *E. coli* was not observed in pork chops and commercial hams (Ko et al., 2009), this difference is thought to be due to the distribution or dilution effect of ovotransferrin on these products or the inhomogeneous distribution within the system (Ellison, 1991). Moon et al. (2011) also pointed out that meat additives like salt (2%) or polyphosphate (0.05%) could lessen the antibacterial effect of ovotransferrin against *L. monocytogenes*. In addition to being possible food additives, ovotransferrin was suggested to act as the antimicrobial agent in food packing for extending the shelf life of meat products; κ-carrageenan films containing ovotransferrin and EDTA showed strong antimicrobial activity against *E. coli, S. aureus, S. typhimurium and C. albicans* (Seol et al., 2009).

22.3.2 Antiviral Activity

Marek's disease virus (MDV), an avian alpha herpes virus, causes Marek's disease in chickens (Bertzbach et al., 2018). The antiviral activity of ovotransferrin-derived peptides DQKDEYELL and KDLLFK against MDV was reported (Giansanti et al., 2005); two bovine lactoferrin peptides (ADRDQYELL and EDLIWK) with sequence homology showed similar activity against herpes simplex virus (Siciliano et al., 1999), suggesting the importance of these peptides in the antiviral activity of intact transferrin proteins against herpes viruses (Giansanti et al., 2002). Importantly, ovotransferrin inhibits the replication of vesicular stomatitis virus (VSV) via the upregulation of type-1 interferon expression in virus-infected macrophages (Zhou et al., 2018). Binding to the essential cell-surface components provides another strategy to inhibit the virus cell entry. Lactoferrin shows the potential antiviral activity in host cell against SARS-CoV infection through binding with heparan sulfate proteoglycans which provides the first anchoring sites on the cell surface for virus entry (Belting, 2003; Lang et al., 2011); its antiviral activity is also effective in the newly emerging virus SARS-CoV-2 through direct recognition with the spike glycoprotein, consequently hindering the spike binding to the virus receptor ACE2 (Campione et al., 2021). Ovotransferrin is speculated to be the potential antiviral agent against SARS-CoV-2 as it shares many same activities as human/bovine lactoferrin (Giansanti et al., 2012; Hou et al., 2015; Mann & Ndung'u, 2020).

22.3.3 Antioxidative Activity

Ovotransferrin was reported to have SOD-like superoxide anion scavenging activity due to its ability of binding copper and manganese (Ibrahim et al., 2007). The antioxidative activity was enhanced in ovotransferrin-catechin conjugates based on the oxygen radical absorbance capacity (ORAC) assay (You et al., 2014). Its hydrolysates showed comparable protective effects against H_2O_2-induced DNA damage in human leukocytes (Kim et al., 2012), suggesting the possible role of ovotransferrin in protecting the developing embryo from iron-dependent oxidative damage *in ovo* (Ishikawa et al., 2004).

Peptides with antioxidative activity were also reported from ovotransferrin. Using ORAC assay, the antioxidative activity of 14 peptides was compared, identified from ovotransferrin hydrolyzation by thermolysin. Among 14 peptides, a tetrapeptide, WNIP (f144-147), showed the highest ORAC value (15.47 μmol/μmol); peptide GWNI (f143-146) had the similar ORAC value (13.90 μmol/μmol), which reveals the necessary motif WNI (f144-146) responsible for their high antioxidative activity (Shen et al., 2010). However, their antioxidative activity was not supported by the further cell study (Jahandideh et al., 2016). IRW, initially identified as an ACEi peptide (Majumder & Wu, 2011), was also shown to have antioxidative activity. The antioxidative activity of peptide IRW with a high ORAC value (6.63 or 13.98 μmol TE/mg) is contributed by di-peptide RW. Combination with some phytochemicals like vitamin C, caffeic acid, and epigallocatechin gallate (EGCG) synergistically improved the oxygen radical-scavenging effects of IRW (Huang et al., 2010a). The antioxidative activity of IRW was also validated in human umbilical vein endothelial cells, in which IRW inhibited TNF-α-induced oxidative stress along with the decreased levels of superoxide ions (Huang et al., 2010b).

Application of ovotransferrin regarding antioxidative activity is still limited. However, lactoferrin, also recognized as the antioxidant, has been suggested as an infant formula ingredient to prevent several diseases related to free radical oxidation happened in low-birth-weight premature infants (Satué-Gracia et al., 2000; Raghuveer et al., 2002). Interestingly, Pastori et al. reported that lactoferrin-loaded contact lenses protected eyes from many ocular diseases characterized by high levels of oxidative stress (Pastori et al., 2015).

22.3.4 Immunomodulatory Activity

Ovotransferrin was characterized as a positive acute-phase protein (APP) in chickens. It increases during an acute phase response in various induced infections and inflammation (Hallquist & Klasing, 1994; Xie et al., 2002). However, transferrin in mammals seems to act as the negative APP with down-regulated expression during inflammation (Safi, 2012). Immunomodulatory activity of ovotransferrin and its hydrolysates in RAW264.7 macrophages via activated mitogen-activated protein kinase (MAPK) signaling pathway were confirmed in current studies (Lee et al., 2018; Ru et al., 2021). This immune-enhancing activity also shows an intestinal protective role in the amelioration of the dysbiosis of immunomodulatory function and intestinal microbiota possibly due to the regulated expression of cytokines (TNF-α and IL-10) and immunoglobulins (Zhu et al., 2018; Zhu et al., 2019). In addition, ovotransferrin hydrolysates prepared by pepsin inhibited the mutation of bone marrow-derived dendritic cells via suppressing LPS-induced activation of the mitogen-activated protein kinase (MAPK) p38, c-Jun N-terminal kinases (JNK), and extracellular signal-regulated kinases (ERK), suggesting its critical role in the modulation of immune responses and several diseases related to immune dysfunction (Liu et al., 2017). Taken together, these findings provide the possibilities for applying ovotransferrin as a functional food ingredient with immune-enhancing activity.

22.3.5 Angiotensin-Converting Enzyme-Inhibitory (ACEi) Activity

Because of the essential role of ACE in generating vasoconstrictor Ang II, ACE inhibitor is the most common target for mitigating hypertension (Aluko, 2015). Until now, numerous ACEi peptides from ovotransferrin have been identified as safe alternatives of antihypertensive drugs. Ovotransferrin-derived peptide EWL exhibited *in vitro* ACEi activity (IC_{50} value: $38 \pm 10 \mu M$) when interacting with S1 and S2' pockets of ACE through pi interaction and hydrogen bond (Yu et al., 2019). Lee et al. reported the ACEi activity of another ovotransferrin-derived peptide KVRQGT and its parent peptide (KVRQGTTY) (Lee et al., 2006). Although the hydrolyzed product KVRQGT has about 11-fold higher inhibitory activity *in vitro*, a similar hypotensive ability of these two peptides was observed in spontaneously hypertensive rats (SHRs). Furthermore, peptide RVPSL isolated from Alcalase ovotransferrin hydrolysate showed an IC_{50} value of 20 μM, whereas adding one amino acid Met (M) at C terminal of RVPSL negated its ACEi activity, suggesting the critical role of the terminal amino acid (Liu et al., 2010). In addition, ACEi peptide IRW, IQW, and LKP were identified from thermolysin/pepsin digested-ovotransferrin through an integrated *in silico* digestion and quantitative structure and activity relationship (QSAR) study (Majumder & Wu, 2010); their antihypertensive activity in SHRs was also reported despite the blood pressure reduction induced by IRW is contributed by other mechanisms rather than ACE inhibition (Majumder et al., 2013; Majumder et al., 2015)

22.3.6 Osteogenic Activity

As transferrin stimulates cell proliferation by supplying iron for DNA synthesis (Laskey et al., 1988), the osteogenic activity of ovotransferrin by stimulating proliferation and differentiation of osteoblasts was confirmed (Shang & Wu, 2018) and lipoprotein receptor-related protein (LRP1) was partially responsible for the promotion of bone formation (Shang et al., 2020); another important member of transferrin family lactoferrin could preserve bone mass and microarchitecture in ovariectomized rats, indicating its potential function in the treatment of estrogen-dependent bone loss (Guo et al., 2009) or other bone disorders (Cornish et al., 2004).

22.4 Ovomucin

Ovomucin is a glycoprotein of high molecular weight in egg white. The molecular weight of different forms of ovomucin varies dramatically as summarized in the previous review (Omana et al., 2010). Ovomucin consists of a carbohydrate-rich β-subunit (~400 kDa) and a carbohydrate poor α-subunit (~220 kDa) with roughly 60% and 15% carbohydrate contents, respectively (Kato & Sato, 1971; Robinson & Monsey, 1971; Itoh et al., 1987). On average, it consists of 33% of carbohydrate content, and the unfractionated form contains 10–12% hexosamine, 15% hexose, and 2.6–8% sialic acid (Mine, 1995). The carbohydrate structures are mainly composed of 3–5 units of N-acetylgalactosamine, galactose, N-acetylglucosamine, and sialic acid (Kato et al., 1978; Strecker et al., 1987, 1989; Strecker et al., 1992). Ovomucin has little α-helix and appears to contain about 70% β-structure and 30% random coil at neutral pH in 1 M KCl (Donovan et al., 1970). It is inherently heterogeneous and connected by disulfide bonds, the reduction of which contributes to depolymerization of ovomucin, decrease in foaming property, and egg white thinning (Donovan et al., 1970; Donovan et al., 1972; Kato et al., 1985). However, as dissociation proceeds, the emulsifying property of ovomucin increases due to greater surface hydrophobicity (Kato et al., 1985). Previous research has well summarized its physiochemical properties and extraction methods (Omana et al., 2010). To date, studies have demonstrated that ovomucin and its hydrolysates show various bioactivities, including anti-inflammatory, anti-adhesive, antioxidative, anti-tumor, bifidogenic, and ACEi activities (summarized in Tables 22.1 and 22.2).

22.4.1 Anti-Inflammatory Activity

The latest study showed that intact ovomucin could attenuate dextran sulfate sodium (DSS)-induced colitis in mice and reduce mRNA expression and protein levels of inflammatory cytokines, including TNF-α, IL-6, IL-1 β, IFN-γ, IL-8, and IL-10 in colon tissues, as well as TNF-α levels in serum (Tu et al., 2021). Ovomucin hydrolysate prepared by Alcalase could inhibit tumor necrosis factor-mediated nuclear factor-κB (NF-κB) activation and decrease the expression of the proinflammatory protein intercellular cell adhesion molecule-1 (ICAM-1) in human dermal fibroblasts (Sun et al., 2016). NF-κB activation is induced by various stimuli like cytokines, endotoxins, pathogens, and stress thus modulating the transcription of genes encoding inflammatory cytokines, inflammatory enzymes, adhesion molecules, etc. (Ahn & Aggarwal, 2005). Mitogen-activated protein kinase (MAPK) pathway is another stress signaling paradigm that functions with NF-κB in response to different stimuli (Kyriakis & Avruch, 2001). Enzyme Protex 26L prepared ovomucin hydrolysate inhibited LPS-activated NF-κB and MAPK signaling pathways and showed anti-inflammatory activity in LPS-treated Caco-2 cells (Bao & Wu, 2021). These studies indicate the ability of intact ovomucin and its derived peptides to prevent inflammatory response and further potential application in managing chronic and/or acute inflammation that is associated with various diseases.

22.4.2 Anti-Adhesive Activity

Sugar chains on glycoproteins have been widely recognized as receptors of divergent pathogens like bacteria, viruses, and mycoplasma (Paulson, 1989). Thus, glycoconjugates are explored to serve as decoy receptors to inhibit the adhesion and colonization of pathogens (Lane et al., 2010). Ovomucin was initially reported to be a competitive inhibitor for hemagglutination induced by the influenza virus (Gottschalk & Lind, 1949). Later investigations also demonstrated its binding activity to bovine rotavirus, hen Newcastle disease virus, and human influenza virus (Tsuge et al., 1996a, 1996b, 1997a; Wang et al., 2018). The α-subunit and β-subunit showed binding differences and the sialic acid residues in the β-subunit greatly contributed to the binding of ovomucin to Newcastle disease virus (Tsuge et al., 1996b, 1997a, 1997b). Ovomucin could bind to avian influenza viruses H_5N_1 and H_1N_1, and sialic acid was involved (Xu et al., 2018). This is consistent with the acknowledgment that the sialic acid moiety of the sugar chains is the most frequent recognition ligand of microorganisms (Wasik et al., 2016). Besides, interaction with Mg^{2+} can affect the adhesion capacity of ovomucin to Newcastle disease virus (Shan et al., 2014).

TABLE 22.1
Egg White Proteins and Their Hydrolysates with Bioactivities

Egg White Protein	Hydrolysate/Modifications	Bioactivity	Effects	References
Ovalbumin	Pepsin	Antioxidative	↓ oxidant production; ↓ age-related oxidative damage by ↑ antioxidant enzymes	Xu et al. (2007)
	/	Antiviral	↓ SARS-CoV-2 infection by ↓ the viral attachment & S protein-mediated cell-cell fusion	Liang et al. (2021)
Ovotransferrin	/	Antimicrobial	Deplete the irons essential for the growth of microorganisms	Bullen, Roger, & Griffiths (1978)
			Surface interaction involves the antifungal activity	Valenti et al. (1986)
			N-terminal half molecule responsible in the bactericidal actions	Ibrahim et al. (1998)
		Antiviral	↓ the replication of virus via ↑ type-1 interferon expression in virus-infected macrophages	Zhou et al. (2018)
		Antioxidative	Possess DPPH radical scavenging activity	Kim et al. (2012)
			↑ antioxidative activity in ORAC assay after the conjugation with catechin	You et al. (2014)
		Immunomodulatory	↑MAPK signaling pathway	Ru et al. (2021)
			Regulation on the expression of TNF-α, IL-10, and IgA	Zhu et al. (2018, 2019)
	Pepsin/trypsin		↓ mutation of bone marrow-derived dendritic cell via activation of ERK and JNK	Liu et al. (2017)
	Alcalase	ACEi	ACEi peptides might contribute to the ACEi activity of hydrolysates	Liu et al. (2010)
		Osteogenic	↑ proliferation and differentiation of osteoblasts	Shang & Wu (2018)
Ovomucin	/	Anti-inflammatory & gut microbiota modulating	↓ DSS-induced colitis in mice; ↓ levels of seral TNF-α and colonic TNF-α, IL-6, IL-1 β, IFN-γ, IL-8, and IL-10; ↓ the activation of NF-κB pathway; ↑ the relative abundance of intestinal beneficial bacteria like *Lactobacilli*, *Faecalibaculum*, *Ruminococcus*, etc. ↑ the production of short chain fatty acids; ↑ intestinal morphology and barrier function	Tu et al. (2021)
		Microbe modulating	*Bifidobacterium bifidum* and *Akkermansia muciniphila* degrade *O*-glycans on ovomucin; *O*-glycans serve as a carbon source to support the *in vitro* growth of *A. muciniphila*	Takada et al. (2020)
	Pepsin/pancreatin	Microbe modulating	Support growth of *Bifidobacterium infantis*; ↑ production of lactate after 24 h incubation as a sole carbon source	Sun, Gänzle et al. (2016)
	/	Immunomodulatory	↑ proliferation of mouse spleen lymphocytes stimulated by LPS	Otani & Maenishi (1994)
	Pronase E /papain		*O*-linked sulfated carbohydrate chains contribute to its activity; ↑ the differentiation of J774.1 cells; ↑ H_2O_2 generation and IL-1 production	Tanizaki et al. (1997)
	Alcalase	Anti-inflammatory	↓ TNF-induced activation of NF-κB pathway; ↓ ICAM-1 expression in human dermal fibroblasts	Sun & Chakrabarti et al. (2016)
	Protex 26L		↓ LPS-induced activation of NF-κB & MAPK pathways in Caco-2 cells	Bao & Wu (2021)

(Continued)

TABLE 22.1 ***(Continued)***

Egg White Proteins and Their Hydrolysates with Bioactivities

Egg White Protein	Hydrolysate/Modifications	Bioactivity	Effects	References
	/	Anti-adhesive	↓ hemagglutination caused by virus; sialic acid residues in the β-subunit contribute to ovomucin binding to newcastle disease virus	Gottschalk & Lind (1949); Tsuge et al. (1996a, 1996b, 1997a, 1997b); Wang et al. (2018)
			Bind to avian influenza viruses H5N1 and H1N1	Xu et al. (2018)
			↓ adhesion of urease (the surface adhesin of *Helicobacter pylori*) to porcine gastric mucin; ↓ *H. pylori* colonization in mouse stomach	Kodama & Kimura (2001)
	Pronase		Attach to Escherichia coli O157:H7 cells	Kobayashi et al. (2004)
	Ovomucin-protease P, protease N, acid protease II, Pronase, Protex 26L, and pepsin/pancreatin		↓ agglutination of porcine erythrocytes induced by porcine K88 ETEC	Sun et al. (2017)
	Protease P, Protex 26L, pepsin/pancreatin, and Pronase		↓ adhesion of K88 ETEC strains to porcine small intestinal epithelial (IPEC-J2) cells	Sun et al. (2019)
	Glycosylated fragments (220 kDa & 120 kDa) from ovomucin-Pronase hydrolysate	Immunomodulatory & Anti-tumor	Modulate immune system of mice with Meth-A fibrosarcoma cell-xenografted tumor, cure the treated tumor and inhibit the distant xenograft; sialic acid residues in the 120 kDa fragment might contribute to inhibiting the distant tumor	Watanabe et al. (1998)
	Ovomucin β-subunit		↑ neutrophils, macrophages, and lymphocytes on the margins of the degenerated and necrotic tumors in sarcoma-180 cell-xenografted mice	Yokota et al. (1999a)
	A glycosylated fragment of 70 kDa from ovomucin α-subunit-Pronase hydrolysate		↑ neutrophils, macrophages, and lymphocytes to the degenerated and necrotic tumor tissues in Meth-A fibrosarcoma cell-xenografted mice inhibit angiogenesis	Oguro et al. (2000)
	Ovomucin β-subunit	Anti-tumor	Exert cytotoxicity on SEKI and 3LL cells; ↓ proliferation of sarcoma-180 cells, induce swelling and bleb formation of microvilli and chromatin clumping, induce cell necrosis	Ohami (1993); Yokota et al. (1999b)
	A glycosylated fragment of 120 kDa from ovomucin β-subunit-Pronase hydrolysate		Interact with the basic fibroblast growth factor receptor on sarcoma-180 cell surface, ↓ cell growth and induce cell necrosis	Yokota et al. (2000)
	Protamex, Flavourzyme, and Alcalase hydrolysates	Antioxidative	Free radical scavenging activity by ATBS assay	Chang et al. (2013)
	Heat treatment		TBARS value of oil emulsion	Abeyrathne et al. (2016)
	/	Hypocholesterolemic	↓ cholesterol uptake by Caco-2 cells; interact with cholesterol mixed micelles in the jejunum, ↓ cholesterol absorption into serum, ↓ total lipid level in liver, ↓ bile acid reabsorption in ileum, ↑ the fecal excretion of bile acids	Nagaoka et al. (2002)
Lysozyme	/	Antiviral	Herpes simplex virus	Cisani et al. (1984)
			Poliovirus	Oderinde et al. (2017)
			Measles	Artym & Zimecki (2013)

(Continued)

TABLE 22.1 ***(Continued)***

Egg White Proteins and Their Hydrolysates with Bioactivities

Egg White Protein	Hydrolysate/Modifications	Bioactivity	Effects	References
			↓ human immunodeficiency virus (HIV-1) by degrading viral polysaccharides and RNA transcripts, or genomic RNA	Lee-Huang et al. (1999)
		Antioxidative	↓ redox balance at baseline and oxidant stress against paraquat, advanced glycation end products (AGE), and H_2O_2 in HepG2 cells by ↓ ROS generation and oxidative stress response genes i.e. AGE-induced $p66^{shc}$ expression and c-Jun phosphorylation; transgenic mice expressing hen egg white lysozyme could mitigate chronic exogenous oxidant stress	Liu et al. (2006)
			↓ betel quid chewing-aggravated gastric oxidative stress, restore mucosal levels of GSH, ↓ mucosal lipid peroxide generation, attenuate mucus liberation and acid-back diffusion in diabetic and atherosclerotic rats	Hung (2005); Chen & Wang (2007)
		Anti-tumor	↓ metastasis formation, prolong mouse life-span; ↑ the number of peroxidase-positive cells in lung tissues due to neutrophils and monocytes; retain morphological structures of the spleen that is responsible for host defenses; liberate bacterial peptidoglycans for immunopotentiation against tumor metastases	Sava, Perissin et al. (1988); Sava et al. (1986); Ceschia et al. (1988); Sava et al. (1991); Sava, Ceschia, et al. (1988); Sava et al. (1995); Sava (1989); Sava et al. (1989)
		Anti-inflammatory	Interact with LPS and *E. coli*, influence the immunostimulatory activity of LPS and *E. coli* on lymphocytes and macrophages; ↓ release of inflammatory cytokine TNF and IL-6 in RAW264.7 cells and septic shock mice during infections	Ohno & Morrison (1989a, 1989b, 1989c); Takada et al. (1994a, 1994b, 1994c)
			↓ LPS- and lipid A analogue-induced activation of MAPK JNK pathway in mouse peritoneal macrophages and RAW264.7 cells; ↓ levels of inflammatory cytokines in serum (TNF-α and IL-6), liver (IL-6), and spleen (IL-6, *IL-6*, *IL-1*β, and *IL-12*), and ↓ LPS-induced systemic inflammation in mice	Tanida et al. (1992); Tagashira et al. (2018); Tagashira et al. (2019)
			Maintain vascular barrier permeability, ↓ inflammatory biomarkers, including TNF-α, IL-6, ICAM-1, VCAM-1, and E-selectin in polyphosphate-stressed HUVECs and mice; ↓ hepatic and renal injury	Chung et al. (2016)
			↑ morphological changes and mucin expression, mitigate DSS-induced intestinal inflammation, ↓ levels of proinflammatory cytokines TNF-α, IL-6, IFN-γ, IL-8, and IL-17 and ↑ anti-inflammatory cytokines IL-4 and TGF-β, ↑ regulatory T cells	Lee et al. (2009b)
			↓ ETEC proliferation and colonization in piglet ileum, protect intestinal mucosa, and prevent inflammation	Nyachoti et al. (2012); May et al. (2012); Oliver & Wells (2013, 2015); Oliver et al. (2014); Wells et al. (2015)

(Continued)

TABLE 22.1 *(Continued)*
Egg White Proteins and Their Hydrolysates with Bioactivities

Egg White Protein	Hydrolysate/Modifications	Bioactivity	Effects	References
		Antibacterial	Lyse broad spectrum of Gram-positive bacteria, including *Bacillus stearothermophilus*, *Clostridium thermosaccharolyticum*, *Clostridium tyrobutyricum*, etc.	Thompson (1952); Losso et al. (2000)
			Permeabilize the outer membrane of *E. coli* by inducing the formation of large size pore	Derde et al. (2013)
	Lysozyme/lysozyme polymers by thermochemical methods	Antibacterial	↓ the viability of *Proteus mirabilis*, *Pseudomonas Xuorescens*, *E. coli*, *Staphylococcus epidermidis*, *Staphylococcus aureus*	Cegielska-Radziejewska et al. (2009); Kummer et al. (2021); Wei et al. (2021)
	Heat/DTT denatured lysozyme		*Micrococcus lysodeikticus*, *E. coli*, *Staphylococcus carnosus* through destructing bacterial cell walls	Vilcacundo et al. (2018)
	Disulfide bond-reduced lysozyme		Bind to LPS structures, permeabilize the bacterial outer membrane, and display significantly higher antimicrobial activity than native and heat-denatured lysozyme due to ↑ flexibility and surface hydrophobicity	Touch et al. (2003)
	Pepsin		↓ *Bacillus* spoilage organisms as a commercial preservative	Abdou et al. (2007)
	Native/heated lysozyme-pepsin hydrolysate		Some acetic acid bacteria, including *Acetobacter aceti* CIAL-106, *Gluconobacter oxydans* CIAL-107, *Lactobacillus casei* CIAL-52, *Oenococcus oeni* CIAL-91, *O. oeni* CIAL-96, *E. coli*, *Pseudomonas fluorescens*, *Shigella sonnei*, and *Shigella flexneri*	Masschalck et al. (2001); Carrillo et al. (2014)
	Native/heated lysozyme-pepsin Hydrolysates, with 65 peptides identified		*E. coli* and *Staphylococcus carnosus*; cationic peptides rich in Lys and Arg and containing Trp in the sequence need further validation in bioassays	Carrillo et al. (2018); Carrillo & Ramos (2018)
	Nanostructures (folic acid-tagged) of self-assembled lysozyme	Anti-tumor	Damage cell membrane, induce the formation of cytoplasmic granules, and ↑ cell death in MCF-7 breast cancer cells	Mahanta et al. (2015)
	Lysozyme coupled with polyoxyethylenglycol		Induce cell cycle arrest, ↓ cells expressing adhesion molecules, modulate host immune system, ↓ lung metastases	Pacor et al. (1999)
	Native lysozyme/denatured lysozyme by heat or DTT/lysozyme-pepsin hydrolysate	Anti-inflammatory & anti-nociceptive	↓ paw edema of carrageenan-stressed mice, ↓ acetic acid-induced abdominal writhing in mice	Carrillo et al. (2016)
	Pepsin/trypsin	Anti-inflammatory	↓ IL-8 production in Caco-2 cells	Lee et al. (2009c)
	Aminopeptidases from *Aspergillus sp.*		↓ DSS-induced colitis in piglets, ↓ colonic levels of TNF-α, IL-6, IL-1b, IFN-γ, IL-8, and IL-17, maintain mucin expression and colonic morphology and barrier integrity	Lee et al. (2009a)
	Lysozyme/heat-denatured lysozyme	Immunomodulatory	↑ immunoglobulin production in human-human hybridoma HB4C5 cells and human peripheral blood lymphocytes possibly through ↑ the translation process and incorporation into cells	Murakami et al. (1997); Sugahara et al. (2000); Sugahara et al. (2002)

(Continued)

TABLE 22.1 ***(Continued)***

Egg White Proteins and Their Hydrolysates with Bioactivities

Egg White Protein	Hydrolysate/Modifications	Bioactivity	Effects	References
	Alcalase	Antioxidative	ORAC assay, DPPH, and ATBS scavenging	You et al. (2010)
	Simulated gastrointestinal digestion (pepsin, α-chymotrypsin, and trypsin)		DPPH scavenging, reducing power assay, and lipid peroxidation inhibition assay	Rao et al. (2012)
	Trypsin, papain, and a combination of trypsin with papain		DPPH and ATBS scavenging	Memarpoor-Yazdi et al. (2012a)
		ACEi		Asoodeh et al. (2012); Memarpoor-Yazdi et al. (2012b)
Ovomucoid	/	Trypsin-inhibitory	↑ the trypsin-like activity and the serum levels of the trypsinogen activation peptide in rats with experimental pancreatitis, ↓ the primary pancreas destruction	Valueva et al. (1992)
			↓ the activity of proteolytic enzymes to stabilize the bioactivity of insulin, ↑ insulin diffusion through the intestinal epithelia into bloodstream	Agarwal et al. (2001); Plate et al. (2002)
			Benefit drug delivery to protect salmon calcitonin for intestinal digestion by ↓ the activity of trypsin, α-chymotrypsin, and elastase	Shah & Khan (2004)
		Immunomodulatory	Exhibit T and/or B cell epitopes and IgE binding epitopes along the molecule, modulate activity of T cells to produce IL-4, IL-6, IL-10, IL-13, TNF-α, and IFN-γ	Holen et al. (2001)
Avidin	/	Antibacterial	↓ biotin-requiring microorganisms due to its high affinity with biotin	Campbell et al. (1972); Board & Fuller (1974); Green (1975); Pugliese et al. (1994)
		Antibacterial & immunomodulatory	Bind to bacterial outer membrane porin, ↓ bacterial uptake and utilization of biotin, ↓ the growth and proliferation of pathogens, potentially affect the interaction of pathogens with the host and mediate host defense	Campbell et al. (1972); Korpela et al. (1984)
		Immunomodulatory	↓ concanavalin A and pokeweed mitogen-induced proliferation of lymphocytes and ↑ antibody production through direct binding to mitogen and participation of accessory cells	Otani & Maenishi (1994)
		Embryo benefiting	Maintain embryo viability during the late embryonic developmental phase possibly by changing the proportion of carbohydrates to amino acids for embryogenesis; ↓ residual tarsus length due to binding to biotin	Krkavcová et al. (2018)
	Avidin-biotin complex	Drug delivery	Couple drugs to the vectors with high efficiency, contribute to delivery and liberation of the therapeutics, enable the slow release of TNF and target biotinylated TNF to tumors	Corti et al. (1998); Gasparri et al. (1999); Bickel et al. (2001)
			↑ pharmacokinetics and ↓ immunogenicity; ↑ incorporation of the antimicrobial peptide KYE28 to the system and suppress *E. coli*	Wilchek & Bayer (1984); Caliceti et al. (2002); Nystrom et al. (2018)

(Continued)

TABLE 22.1 *(Continued)*
Egg White Proteins and Their Hydrolysates with Bioactivities

Egg White Protein	Hydrolysate/Modifications	Bioactivity	Effects	References
Cystatin	/	Antibacterial	↓ the growth of *Porphyromonas gingivalis* probably by interfering with the bacterial peptide transport system	Blankenvoorde et al. (1996); Blankenvoorde et al. (1998)
			↓ the growth of the Gram-positive strains *Staphylococcus aureus*, *S. xylosus*, and *S. gallinarum* and exhibit bactericidal activity against *E. coli*, *Pseudomonas aeruginosa*, *Acinetobacter lwoffii*, and *Oligella sp.* strains	Wesierska et al. (2005)
		Antifungal	The most frequent human fungal pathogens of the genus *Candida*, including *Candida parapsilosis*, *Candida tropicalis*, and *Candida glabrata*; combat efflux-mediated multidrug resistance	Kolaczkowska et al. (2010)
		Antiviral	Alter the intracellular proteolytic processing of poliovirus proteins and ↓ virus production in human Hela cells	Korant et al. (1985)
			↓ the cysteine protease activity of human rhinovirus 1A to ↓ viral replication	Korant et al. (1986)
			Interfere the replication of human rotavirus, ↓ virus-induced diarrhea in suckling mice, and recover histopathological changes in the mouse intestine	Ebina & Tsukada (1991)
		Immunomodulatory	↑ the synthesis of TNF-α and IL-10 to activate nitric oxide (NO) synthase and ↑ NO release from IFN-γ-activated mouse peritoneal macrophages	Verdot et al. (1996); Verdot et al. (1999)
			↑ the IL-6 and IL-8 production by human gingival fibroblasts and IL-6 by murine splenocytes	Kato et al. (2000)
			Synergize with suboptimal IFN-γ to ↓ spleen burden of *Leishmania donovani* that causes visceral leishmaniasis, ↓ the growth of amastigotes in macrophages and the production of IL-14 but ↑ the levels of IL-12 and NO, ↑ Th2-Th1 conversion as long-term resistance against the parasite infection	Das et al. (2001)
		Bone health benefiting	↓ calcium release from calvarial bone explants into the medium, induce more osteoclasts associated with demineralized non-degraded bone matrix, reversibly ↓ bone matrix degradation in the resorption lacunae	Brand et al. (2004)
		Anti-tumor	↓ invasion of human ras-transformed breast epithelial cells and ↓ the activity of cathepsin B and L in gastric cancer tissues, ↓ tumoral invasion, and metastasis as an inhibitor against most cysteine proteinases by ↓ various proteases crucial for tumor development	Koblinski et al. (2000); Premzl et al. (2001); Saleh et al. (2003)
	Hybrids of a cystatin loop with human uPA fragment	Anti-tumor	↓ the activity of cathepsins B and L	Muehlenweg et al. (2000)

(Continued)

TABLE 22.1 *(Continued)*

Egg White Proteins and Their Hydrolysates with Bioactivities

Egg White Protein	Hydrolysate/Modifications	Bioactivity	Effects	References
	Bifunctional inhibitors composed of cystatin and N-terminal domain of human matrix metalloproteinase inhibitor TIMP-1 or TIMP-3; trifunctional inhibitors composed of cystatin, TIMP-1/3, and uPA		↓ the invasion of human ovarian cancer cells and ↓ tumor burden in nude mice by regulating tumor metastasis-associated proteolytic systems	Krol, Kopitz et al., (2003); Krol, Sato et al. (2003)
	Cystatin with chitosan in hydroxypropyl methylcellulose-formed films	Antibacterial	↓ *Micrococcus flavus*, *Bacillus cereus*, *E. coli*, *Pseudomonas fluorescens*, and *Candida famata*	Zimoch-Korzycka et al. (2015)
Ovomacroglobulin (ovostatin)	/	Protease-inhibitory & antibacterial	↓ activity of proteases from *S. marcescens* (56K, 60K, and 73K proteases), *P. aeruginosa* elastase and alkaline protease, and thermolysin from *B. stearothermophilus* and ↓ bacterial growth, vascular permeability, and corneal destruction in guinea pigs, and ↑ wound healing in the serratial and pseudomonal keratitis in rabbits	Molla et al. (1986); Molla, Oda et al. (1987); Miyagawa, Nishino et al. (1991); Molla, Matsumura et al. (1987); Miyagawa, Kamata, et al. (1991)
		Antibacterial	Synergize with ofloxacin and ↓ stromal degradation in pseudomonal keratitis	Miyagawa et al. (1994)
		Antibacterial & immunomodulatory	↓ ring abscess formation induced by living *Pseudomonas aeruginosa* and *pseudomonal elastase* through ↑ the migration of polymorphonuclear leukocytes toward the pathogenic bacteria	Ijiri et al. (1993)
		Protease-inhibitory	↓ the metalloprotease produced by *Vibrio vulnificus* and ↓ septicemia by ↓ intravenous bacterial dissemination and lethality in mice	Maruo et al. (1998)
		Anti-inflammatory	↓ the activity of protease medullasin from invading granulocytes in the inflammatory tissues	Ikai et al. (1989)
		Fibroblast migrating & wound healing	↑ fibroblast growth, collagen deposition, and capillary formation in granulation tissues, ↑ surgical wound healing	Ofuji et al. (1992)
			↑ the migration of 3T6 mouse embryonic fibroblasts and human skin fibroblasts, ↑ cell adhesion to extracellular matrix, ↓ intercellular aggregation, strengthen cytoskeleton	Geng et al. (2016)

TABLE 22.2
Egg White Protein-Derived Peptides with Bioactivities

Parent Protein	Preparation	Peptide Sequence	Bioactivity	Mechanisms	References
Ovalbumin	Pepsin	FRADHPFL, RADHPFL	ACEi & antihypertensive	Vasorelaxation in spontaneously hypertensive rats (SHRs)	Fujita et al. (1995); Fujita et al. (1995); Scruggs et al. (2004); Yuko et al. (2002)
		LW	Antihypertensive		Fujita et al. (2000)
		FGRCVSP, ERKIKVYL, FFGRCVSP, FCF, NIFYCP, LW, FFGRCVSP	ACEi		
	Chymotrypsin	RADHPF	Antihypertensive	Vasorelaxation in SHRs, ↑ stimulating the nitric oxide (NO)	Matoba et al. (1999)
	Pepsin/corolase PP (trypsin & chymotrypsin)	RADHP (end gastrointestinal hydrolysis product of RADHPFL)	ACEi & antihypertensive	Other mechanisms other than ACE inhibition in SHRs	Miguel et al. (2006)
	Pepsin/pancreatin	KLP, LKISQ, SQAVH, and FDKLPGFGD	ACEi		Fan et al. (2019)
	Artichoke protease	IAAEVYEHTEGSTTSY, HLFGPPGKKDPV, PIAAEVYEHTEGSTTSY & YAEERYPIL			Bueno-Gavilá et al. (2021)
	Integrative bioinformatics approach	CF, KM, ELPF, AM, ADHPF, LPR, PR, FR, PRM & GR			Salim & Gan (2020)
Ovotransferrin	Acid	OTAP-92	Antimicrobial	Destroy the outer membrane of Gram-negative bacteria	Ibrahim et al. (2000)
	Sequence alignment with bovine lactoferrin and chemical	DQKDEYELL and KDLLFK	Antiviral	Both amino acid sequence and the conformation of peptides in the intact protein essential for their antiviral activity	Giansanti et al. (2005, 2012)
	Thermolysin	WNIP and GWNI	Antioxidative	Motif WNI necessary in antioxidative activity	Shen et al. (2010)
	Thermolysin/pepsin	IRW		Di-peptide RW with strong antioxidative activity	Huang et al. (2010a)
	In silico method with virtual enzymolysis and pharmacophore study and flexible docking	EWL	ACEi	Pi interaction and hydrogen bond important in the ACE-EWL complex	Yu et al. (2019)
	Alcalase	RVPSL			Liu et al., (2010)
	Chymotrypsin	KVRQGT and KVRQGTTY	ACE-inhibitory & antihypertensive	ACE inhibition in SHRs	Lee et al. (2006)
Ovomucin	Pepsin	WNWAD	Antioxidative	↓ROS generation, ↓mitochondria-mediated cell apoptosis in H_2O_2-stressed HEK-293 cells	Liu et al. (2014)
	Protamex digestion of ovomucin α-subunit	LDEPDPL (f686–692) and NIQTDDFRT (f539–547)		DPPH scavenging acitivity	Chang et al. (2013)

(Continued)

TABLE 22.2 *(Continued)*

Egg White Protein-Derived Peptides with Bioactivities

Parent Protein	Preparation	Peptide Sequence	Bioactivity	Mechanisms	References
Lysozyme	*Chemical synthesis	DITASVNCAKKIVSDGNGMNAWVAWRNR (f87-114)*, DITASVNCAKKIVS (f87-100)*, DGNGMN (f101-106)*, AWVAWRNR (f107-114)*	Antibacterial	Cross the outer membrane of Gram-negative bacteria via self-promoted uptake and damage the inner membrane through channel formation	Ibrahim et al. (2001)
	Clostripain *Chemical synthesis	IVSDGNGMNAWVAWR (f98–112), IVSDGNGM (f98-105)*, NAWVAWR (f106-112)*, **R**AWVAWR (Arg replaces Asn), Denantiomer of **R**AWVAWR, **R**WVAWR (Arg replaces Ala)		Block DNA and RNA synthesis; ↑outer membrane permeabilization causing cell death of *E. coli*	Pellegrini et al. (1997); Pellegrini et al. (2000); Hunter et al. (2005)
	Pepsin	KVFGRCEL (f1-8), KVFGRCELAAAMKRHGL (f1-17), DNYRGYSL (f18-25), f1-38, f18-56, f18-62, f114-118, f125-129		Disrupt bacterial cytoplasmic membrane of both Gram-positive and Gram-negative bacteria by forming pores into the membrane	Ibrahim et al. (2005)
	Pepsin/trypsin	IVSDGNGMNAW (f98-108), HGLDNYR (f15-21)			Mine et al. (2004)
	Papain/trypsin	NTDGSTDYGILQINSR (f46-61)			Memarpoor-Yazdi et al. (2012a)
	Pepsin	f(23–28) YSLGNW, f(123–129) WIRGCRL, and f(122–129) AWIRGCRL, f(124–129) IRGCRL		Positive charge and Trp in their sequence essential for the antibacterial activity	Carrillo et al. (2018); Carrillo & Ramos (2018)
	Trypsin	TAKPEGLSY (f20-28) from goose egg white lysozyme		The specific peptide structure carrying center Pro residue and terminal Lys residue as well as the formation of a helix structure	Pukcothanung et al., (2007); Thammasirirak et al. (2010)
	Pepsin	VAWRNRCKGTD (f109–119), WRNRCKGTD (f111–119), AWIRGCRL (f122–129), WIRGCRL (f123–129), and IRGCRL (f124–129)	Antioxidative	ORAC assay; inhibit lipid peroxidation in Zebrafish larvae by TBARS test	Carrillo, Gómez-Ruiz et al. (2016)
	Alcalase	KRHGLDNY (f13–20), RHGLDNYRGY (f14–23), and RHGLDNYRGY (f13–23)			You et al. (2010)
	Trypsin and papain	NTDGSTDYGILQINSR (f46-61)		DPPH radical scavenging	Memarpoor-Yazdi et al. (2012a)
	Simulated gastrointestinal digestion (pepsin, α-chymotrypsin, and trypsin)	RGY (f21-23), WIR (f123-125), and VAW (f109-111) among the 38 identified peptides		Reducing power assay, and lipid peroxidation inhibition assay	Rao et al. (2012)
		RGY, WIR, and VAW, KVF (f1-3), MKR (f12-14), AMK (11-13), AKF (32-34), and GIL (f54-56)	ACEi		
	Trypsin and papain	FESNFNTQATNR (f34-45), NTDGSTDYGILQINSR (f46-61) & VFGR (f2-5)			Asoodeh et al. (2012); Memarpoor-Yazdi et al. (2012b)

Note: * means "chemical synthesis."

In addition, ovomucin also showed anti-adhesive activity toward bacteria. Urease is a surface adhesin of *Helicobacter pylori* that causes gastric infection and peptic ulcers. The study revealed that ovomucin could inhibit urease adhesion to porcine gastric mucin and eliminate pathogen colonization from the stomach of *H. pylori*-infected mice (Kodama & Kimura, 2001). Ovomucin-derived glycopeptide prepared by Pronase digestion was illustrated to attach to *Escherichia coli* O157:H7 (Kobayashi et al., 2004). Later research unraveled the anti-adhesive property of ovomucin hydrolysates digested by different enzymes, including protease P, protease N, acid protease II, Pronase, Protex 26L, and pepsin/pancreatin, against porcine K88 enterotoxigenic *E. coli* (ETEC) by a hemagglutination assay (Sun et al., 2017). Based on this assay, glycopeptides from ovomucin by Protease P, Protex 26L, pepsin/pancreatin, and Pronase digestion inhibited ETEC strains adhesion to porcine small intestinal epithelial cells (Sun et al., 2019). And consistent with that reported from intact ovomucin, the carbohydrate moieties in glycopeptides contributed to its anti-adhesive ability (Sun et al., 2019).

22.4.3 Antioxidative Activity

Ovomucin hydrolysates prepared by Protamex, Flavourzyme, and Alcalase showed free radical scavenging activity at 90.4, 89.4, and 88.6%, respectively, by the 2,2′-azinobis (3-ethylbenzothiazoline-6-sulfonic acid) (ABTS) assay. Also, peptides LDEPDPL (f686–692) and NIQTDDFRT (f539–547) from the ovomucin α-subunit were identified from the hydrolysate by enzyme Protamex and exhibited antioxidative activity of 51.8% and 24.7%, respectively, as determined by the 2,2-diphenyl-1-picrylhydrazyl (DPPH) assay (Chang et al., 2013). Ovomucin hydrolysate obtained by heating at 100°C for 15 min at pH 12.0 also showed a strong antioxidative activity (Abeyrathne et al., 2016). Further biological assays are needed to validate their antioxidative activity as these chemical assays are not based on living organisms. A pentapeptide (WNWAD) derived from egg white ovomucin pepsin hydrolysate was reported to protect HEK-293 cells against H_2O_2-induced oxidative stress by reducing intracellular ROS accumulation and inhibiting the mitochondria-mediated cell apoptosis pathways (Liu et al., 2014).

22.4.4 Anti-Tumor Activity

β-subunit of ovomucin was firstly proved to exert cytotoxicity on cultured human melanoma (SEKI) and Lewis lung cancer cells (3LL) (Ohami, 1993). Later, Yokota et al. (1999b) reported that β-subunit of ovomucin significantly reduced the proliferation of sarcoma-180 cells and promoted the swelling and bleb formation of microvilli and abnormal chromatin clumping. The shape and structure of the nuclear and organelles also showed obvious changes associated with cell necrosis. And then they reported that a highly glycosylated fragment (120 kDa) in β-subunit from the Pronase-digested ovomucin could interact with the basic fibroblast growth factor receptor on the surface of sarcoma-180 cells to inhibit cell growth and induce cell necrosis (Yokota et al., 2000). *In vivo* studies also demonstrated its anti-tumor activity. By inoculating Meth-A fibrosarcoma cells to establish a double-grafted tumor system in mice, Watanabe et al. (1998) reported that highly glycosylated fragments (220 kDa and 120 kDa) from Pronase-treated ovomucin could modulate the immune system and cure the right (treated) tumor and inhibit the growth of the left (distant) one. In the same study, they found that the sialic acid residues in the 120 kDa fragment might play a crucial role in regressing the distant tumors (Watanabe et al., 1998). This was further evidenced that the injection of ovomucin β-subunit recruited neutrophils, macrophages, and lymphocytes on the margin of the degenerated and necrotic tumor tissue in the sarcoma-180 cell-xenografted mice (Yokota et al., 1999a). A 70 kDa highly glycosylated fragment in the α-subunit separated from Pronase-treated ovomucin was also illustrated to exhibit similar anti-tumor effects as the β-subunit and inhibit angiogenesis (Oguro et al., 2000).

22.4.5 Other Bioactivities

The latest *in vivo* study demonstrated that ovomucin could increase the relative abundance of intestinal beneficial bacteria like *Lactobacilli*, *Faecalibaculum*, *Ruminococcus*, etc. and elevate the production of short-chain fatty acids (Tu et al., 2021). In this study, ovomucin was shown to mitigate colitis severity, improve intestinal morphology, boost intestinal barrier function by restoring the expression of *MUC2* and tight junction proteins, and regulate the release of inflammatory cytokines (Tu et al., 2021). Ovomucin hydrolysate by pepsin digestion for 3 h followed by pancreatin for 3 h showed bifidogenic activity as it could promote the *in vitro* growth of *Bifidobacterium infantis* and increase the production of lactate after 24 h incubation as a sole carbon source (Sun et al., 2016). *O*-glycans on ovomucin could be degraded by *B. bifidum* and *Akkermansia muciniphila* and serve as a carbon source to support the *in vitro* growth of *A. muciniphila*, indicating the potential of ovomucin to modulate gut-beneficial microbes and improve gut health (Takada et al., 2020).

In addition to those mentioned above that ovomucin/fragments showed immunomodulatory activity in tumor cell xenografted mice and colitis mice, ovomucin could enhance the proliferation of mouse spleen lymphocytes stimulated by lipopolysaccharide (Otani & Maenishi, 1994). Pronase E/papain-digested ovomucin hydrolysate could promote the differentiation of cultured macrophage-like cells J774.1 and increase H_2O_2 generation and IL-1 production (Tanizaki et al., 1997). The activation of macrophage cells was attributed to the *O*-linked sulfated carbohydrate chains, among which N-acetylneuraminyl-(2,3)-galactosyl-(1,3)-N-acetylgalactosamine-6-sulfate was identified as the main sulfated oligosaccharide structure in ovomucin (Tanizaki et al., 1997).

Nagaoka et al. reported that ovomucin inhibited cholesterol uptake by Caco-2 cells and increased the fecal excretion of cholesterol and bile acids in rats (Nagaoka et al., 2002). Ovomucin could directly interact with cholesterol-mixed micelles in the jejunal epithelia and inhibit reabsorption of bile acids in the ileum to suppress cholesterol absorption into serum and total lipids level in liver and attenuate hypercholesterolemia in rats. Ovomucin hydrolysates prepared by Alcalase, papain, and trypsin showed ACEi activity by test-tube assays (Abeyrathne et al., 2016), which needs to be verified by bioassays. These studies indicate the potential of ovomucin and its derived bioactive peptides in promoting metabolic syndrome, but further research is warranted. In summary, ovomucin is a crucial component of egg white proteins and possesses diverse bioactivities. It can be acknowledged as a promising source of bioactive peptides for developing novel functional foods.

22.5 Lysozyme

Lysozyme accounts for around 3.4% of total hen egg white proteins and is a classic member of the lysozyme family called *c* type (chicken- or conventional-type) lysozymes. It comprises 129 amino acid residues with a molecular weight of 14.3 kDa and an isoelectric point of 10.7 (Canfield, 1963; Yang et al., 1998). Lysozyme molecule consists of two domains linked by a long α-helix; domain 1 (residues 40 to 88) is composed primarily of β-sheet and domain 2 (residues 1 to 39 and 89 to 129) is chiefly a-helix (Young et al., 1994). Chicken hen egg white lysozyme is well demonstrated with antimicrobial activity due to its ability to catalyze the hydrolysis of β-(1-4) glycosidic linkages between amino sugar residues in polysaccharide components of bacterial cell walls (Salton, 1957; Phillips, 1967). It has been commercialized as a natural food preservative to control spoilage and developed for pharmaceutical application to prevent infection (Proctor et al., 1988; Hughey et al., 1989; Losso et al., 2000; Tenovuo, 2002). With an increasing number of studies investigating its biological activities, it has been revealed with anti-virus, anti-inflammatory, antioxidative, and ACEi activities (Tables 22.1 and 22.2).

22.5.1 Antimicrobial Activity

Lysozyme can degrade peptidoglycan, the structural component of the bacterial cell wall, by cleaving the β(1–4) linkage between N-acetlymuraminic acid and N-acetylglucosamine (Salton, 1957). In early 1950s, egg white lysozyme was reported to lyse Gram-positive bacteria (Thompson, 1952). Lysozyme was summarized to be effectively bacteriolytic to a broad spectrum of Gram-positive bacteria, including *Bacillus stearothermophilus*, *Clostridium thermosaccharolyticum*, *Clostridium tyrobutyricum*, etc. (Losso et al., 2000). However, egg white lysozyme was also shown to promote the growth of cheese starter organisms, *Streptococcus cremoris.* AM (slow fermentative strain) and HP (fast fermentative strain) (Akashi, 1972). Generally, the limited effect of egg white lysozyme on Gram-negative bacteria is attributed to the enveloped cell walls that can block the access of lysozyme. However, studies reported the effect of egg white lysozyme against Gram-negative bacteria and indicated its catalytically independent antimicrobial activity to induce membrane permeabilization and DNA/RNA synthesis perturbation (Pellegrini et al., 2000; Ibrahim et al., 2001, 2005; Derde et al., 2013).

Modification of egg white lysozyme by conjugation with various molecules i.e. saturated fatty acids and polysaccharides have been well demonstrated to broaden its antimicrobial effect to Gram-negative bacteria without reducing its enzymatic activity (Aminlari et al., 2014). Heat-denatured and pepsin-digested egg white lysozyme could increase its antibacterial spectrum against some acetic acid bacteria, including *Acetobacter aceti* CIAL-106, *Gluconobacter oxydans* CIAL-107, *Lactobacillus casei* CIAL-52, *Oenococcus oeni* CIAL-91, *O. oeni* CIAL-96, *Proteus mirabilis*, *Pseudomonas Xuorescens*, *E. coli*, and *Staphylococcus* epidermidis (Cegielska-Radziejewska et al., 2009; Carrillo et al., 2014). Also, thermal and DL-dithiothreitol (DTT) treatment increased the antimicrobial activity of lysozyme against *Escherichia coli*, independent of its lytic activity (Vilcacundo et al., 2018). Amyloid fibrils and nanoparticles formed from hen egg white lysozyme showed improved antimicrobial activity against *Staphylococcus aureus* and *E. coli* compared to native lysozyme (Kummer et al., 2021; Wei et al., 2021). The reduction of disulfide bonds in egg white lysozyme by thiol-disulfide exchange reaction under heat denaturing condition could introduce new half-cystine residues and render the lysozyme molecule more flexible with dramatically increased surface hydrophobicity. Thus, the reduced lysozyme could be more likely to bind to LPS structures and permeabilize the bacterial outer membrane, displaying significantly higher antimicrobial activity than native and heat-denatured lysozyme (Touch et al., 2003). Under high hydrostatic pressure, lysozyme and heat-denatured lysozyme had a bactericidal or bacteriostatic effect on *E. coli*, *Pseudomonas fluorescens*, *Shigella sonnei*, and *Shigella flexneri*; and the enzymatic activity was shown indispensable for its antimicrobial activity (Masschalck et al., 2001)

Enzymatic hydrolysis is another method of modification of egg white lysozyme to induce the release of antimicrobial peptides. Digestion of chicken egg white lysozyme by clostripain released a pentadecapeptide, IVSDGNGMNAWVAWR (f98-112) with broad-spectrum antimicrobial action *in vitro* (Ibrahim et al., 2001). The peptide NAWVAWR at the C-terminal of the pentadecapeptide retained some bactericidal activity and replacement of asparagine 106 with arginine (RAWVAWR) significantly increased the bactericidal activity (Pellegrini et al., 1997). These modified peptides could block DNA and RNA synthesis and increase the outer membrane permeabilization, causing cell death of *E. coli* strain ML-35p (Pellegrini et al., 2000; Hunter et al., 2005). Pepsin and trypsin digestion of chicken egg white lysozyme showed bactericidal activity against Gram-positive bacteria (*Staphylococcus aureus* 23-394) and Gram-negative bacteria (*E. coli* K-12); peptide IVSDGNGMNAW (resides 98–108), located in the middle part of the helix-loop-helix, inhibited *E. coli* K-12 and another peptide HGLDNYR (residues 15–21) inhibited *S. aureus* 23-394 (Mine et al., 2004). Pepsin A digestion yielded bactericidal peptides from an N-terminal helix-loop-helix domain (Lys^1-Phe^{38}) with the first two β-strands (Asn^{39}-Leu^{56}) of the β-domain of lysozyme; the bactericidal peptides were shown predominantly α-helical and helix-sheet motifs (Ibrahim et al., 2005). Pepsin-digested egg white lysozyme commercially available as a food preservative also inhibited *Bacillus* spoilage organisms (Abdou et al., 2007). Goose egg white lysozyme digested by trypsin also showed antimicrobial activity against broad spectrum of Gram-positive and Gram-negative bacteria. And a peptide with the sequence of TAKPEGLSY, residues 20–28 at the N-terminal of goose lysozyme, was identified to form a helical structure by the centered

Pro residue and the terminal Lys residue after binding to the cell membrane and exhibited antimicrobial activity with a broad spectrum (Pukcothanung et al., 2007; Thammasirirak et al., 2010). A potent peptide NTDGSTDYGILQINSR (residues 46-61) was identified from papain- and trypsin-digested lysozyme hydrolysate to exert inhibitory effects on *E. coli* and *Leuconostoc mesenteroides* (Memarpoor-Yazdi et al., 2012a). Pepsin-digested lysozyme (both native and heated) hydrolysates also showed antimicrobial activity against *E. coli* and *Staphylococcus carnosus* (Carrillo et al., 2018; Carrillo & Ramos, 2018). More than sixty peptides, cationic sequences containing Trp and rich in Lys and Arg, were identified from the lysozyme-pepsin hydrolysate. Four peptides YSLGNW f(23–28), WIRGCRL f(123–129), AWIRGCRL f(122–129), and IRGCRL f(124–129) were demonstrated to inhibit the growth of *E. coli* and *S. carnosus*; the rest screened sequences awaited further validation in biological assays (Carrillo et al., 2018; Carrillo & Ramos, 2018).

22.5.2 Anti-Viral Activity

Cisani et al. reported that heat and enzymatically modified hen egg white lysozyme could inhibit herpes simplex virus independent of the enzymatic activity of the lysozyme compounds (Cisani et al., 1984). Egg white lysozyme was also reported to have shown activity against human immunodeficiency virus (HIV-1) by degrading viral polysaccharides and RNA transcripts, or genomic RNA (Lee-Huang et al., 1999). The latest research showed that egg white lysozyme exerted antiviral effects on poliovirus (Oderinde et al., 2017) and measles (Artym & Zimecki, 2013).

22.5.3 Antioxidative Activity

Liu et al. (2006) demonstrated that hen egg white lysozyme could inhibit the generation of reactive oxygen species (Liu et al., 2006). Transgenic mice expressing hen egg lysozyme were resistant to chronic exogenous oxidant stress and hepatocytes preincubated with lysozyme showed reduced redox balance at baseline and suppressed oxidant stress (Liu et al., 2006). Intragastric administration of lysozyme chloride suppressed betel quid chewing-aggravated gastric oxidative stress and restored the decreased mucosal GSH, mucus liberation, elevated mucosal lipid peroxide generations, and acid-back diffusion in diabetic (Hung, 2005) and atherosclerotic rats (Chen & Wang, 2007). Alcalase-digested lysozyme also showed antioxidative activity using oxygen radical absorbance capacity-fluorescein (ORAC-FL), 2,20-Azino-bis(3-ethylbenzothiazoline-6-sulfonic acid) diammonium salt radical cation ($ABTS\cdot^{+}$) and 2,2-di(4-tert-octylphenyl)-1-picrylhydrazyl (DPPH) radical scavenging methods. And positively charged peptides from two antioxidant fractions were identified with the sequences of KRHGLDNY (f13–20), RHGLDNYRGY (f14–23), and RHGLDNYRGY (f13–23) (You et al., 2010). Trypsin, papain, and trypsin–papain hydrolysates of hen egg white lysozyme exhibited an antioxidative activity as indicated by quenching DPPH and ATBS radicals; the peptide NTDGSTDYGILQINSR (f46-61) was identified from the most active fraction of the hydrolysate prepared by a combination of the two enzymes and reported with DPPH radical scavenging activity (Memarpoor-Yazdi et al., 2012a). Lysozyme hydrolysate prepared by a simulated gastrointestinal digestion showed an antioxidative activity. More than 30 peptides were identified from the most active fraction and the peptides RGY (f21-23), WIR (f123-125), and VAW (f109-111) showed strong antioxidative effects with RGY and VAW close to the reducing power of glutathione at the same concentration and WIR exhibiting the highest activity (Rao et al., 2012). Five positively charged peptides, VAWRNRCKGTD (f109–119), WRNRCKGTD (f111–119), AWIRGCRL (f122–129), WIRGCRL (f123–129), and IRGCRL (f124–129) were isolated from pepsin digested hen egg lysozyme and showed antioxidative activity using ORAC-FL method. All five synthetic peptides also efficiently inhibited lipid peroxidation in Zebrafish larvae by TBARS test (Carrillo, Gómez-Ruiz et al., 2016). To our knowledge, chemical test-tube assays are commonly conducted in studies exploring the antioxidative activity of food-derived bioactive peptides, thus requiring biological assays for further validation.

22.5.4 Anti-Tumor Activity

The anti-tumor activity of hen egg white lysozyme was reported against Lewis lung carcinoma and MCa mammary carcinoma in mice (Sava et al., 1986; Sava, Perissin et al., 1988). Lysozyme could suppress metastasis formation and prolong the life-span of the animals. A later study showed that lysozyme increased the number of peroxidas-positive cells in the lung tissues, which was postulated by the presence of neutrophils and neo-arrived monocytes. Also, lysozyme retained the tissue structure of the spleens. These indicated that oral administration of hen egg white lysozyme could cause significant modification in organs and tissues responsible for host defenses against tumor metastasis (Sava, Ceschia et al., 1988; Ceschia et al., 1988; Sava et al., 1991; Sava et al., 1995). The serum concentration of lysozyme remained undetectable in the orally treated mice even at the dosage up to 400 mg/kg·d, suggesting the anti-tumor and anti-metastatic roles of lysozyme might be mediated by host defenses (Sava et al., 1988; Sava, 1989). Liberation of bacterial substances i.e peptidoglycans for immunopotentiation by lysozyme was regarded as another possible mechanism ascribed to its antitumor activity (Sava et al., 1989). Coupling hen egg white lysozyme with polyoxyethylenglycol was reported to decrease lung metastases by causing tumor cell cycle arrest, reducing cells expressing adhesion molecules, and modulating the host immune system (Pacor et al., 1999). Self-assemble nanostructured hen egg white lysozyme induced cell membrane damage and cytoplasmic granules formation in MCF-7 breast cancer cells and initiated cell-killing process. This lysozyme nanostructure tagged with folic acid showed enhanced cytotoxicity against the tumor cells (Mahanta et al., 2015).

22.5.5 Anti-Inflammatory Activity

The outer membrane of gram-negative bacteria comprises complex lipopolysaccharide (Phelps & Antonini, 1975) structures, which are among the most potent immunostimulants of the bacteria and have been demonstrated to interact with egg white lysozyme to inhibit its enzymatic activity. Meanwhile, lysozyme could affect their immunostimulatory activity on B-lymphocyte proliferation and differentiation and on macrophage production of lymphocyte-activating factor activity (Ohno & Morrison, 1989a, 1989b, 1989c). Lysozyme-LPS complex could completely abrogate the production of tumor necrosis factor (TNF) and the mortality caused by LPS-triggered septic shock in specific-pathogen-free male ICR mice (Takada et al., 1994a). Lysozyme could also bind to heat-killed *E. coli* O111:B4 cells and inhibit TNF synthesis (Takada et al., 1994a). The complexes formed by lysozyme and *E. coli* or LPS could restore levels of TNF and IL-6 in serum of septic shock mice and in mouse macrophage-like RAW264.7 cells, indicating the role of lysozyme in neutralizing endotoxin and regulating the systemic inflammation during pathogenic infections (Takada et al., 1994b, 1994c). Lysozyme could penetrate macrophages and regulate the immunostimulatory properties of LPS and a lipid-A analogue in macrophages from mouse peritoneal macrophages and RAW264.7 cells through inhibiting the phosphorylation of the element c-Jun N-terminal kinases (JNK) in the mitogen-activated protein kinase (MAPK) signaling pathway (Tanida et al., 1992; Tagashira et al., 2018). Oral administration of lysozyme could suppress levels of inflammatory cytokines in serum, liver, and spleen and ameliorate LPS-induced systemic inflammation in mice (Tagashira et al., 2019). The function of lysozyme was also consistently reported in polyphosphate-induced inflammation in human umbilical vein endothelial cells (HUVECs) and mice (Chung et al., 2016). Lysozyme could suppress severe vascular inflammatory responses by recovering vascular barrier permeability and modulating the expression of inflammatory cytokines such as TNF-α and IL-6 and activation of NF-κB and MAPK p38 molecules. Inflammatory biomarkers like vascular cell adhesion molecule-1 (VCAM-1), intercellular adhesion molecule-1 (ICAM-1), and E-selectin correlated with the increased binding and migration of leukocytes were also inhibited by lysozyme. Polyphosphate-elevated serum levels of the hepatic injury markers aspartate transaminase (AST) and alanine transaminase (ALT), renal injury markers creatinine and bun urea nitrogen (BUN), and tissue injury marker lactate dehydrogenase (LDH) were also effectively reduced by lysozyme (Chung et al., 2016). Lysozyme also mitigated DSS-induced inflammation, morphological changes, and mucin expression in piglets (Lee et al., 2009b). It functioned as an immunomodulator to reduce the production of pro-inflammatory cytokines TNF-α, IL-6, IFN-γ, IL-8, and IL-17 and increase the expression of the anti-inflammatory mediators IL-4 and TGF-β as well as activating regulatory T cells, thus restoring gut homeostasis (Lee et al., 2009b). Lysozyme was well demonstrated to benefit piglets during ETEC challenge by minimizing the proliferation of pathogenic ETEC in the ileum, protecting the intestinal mucosa, modulating the immunological response in piglets, and preventing inflammation (Nyachoti et al., 2012). Lysozyme is used as an alternative to growth antibiotics in swine production as it improves growth performance and intestinal morphology, effectively controls zoonotic pathogens, and ameliorates the effects of a chronic indirect immune challenge (May et al., 2012; Oliver & Wells, 2013, 2015; Oliver et al., 2014; Wells et al., 2015).

Lysozyme hydrolysate prepared by pepsi-trypsin reduced IL-8 production in Caco-2 cells (Lee et al., 2009c). Lysozyme hydrolysate prepared by aminopeptidases of *Aspergillus* sp. origin attenuated the clinical severity of DSS-induced colitis in piglets by reducing the expression of pro-inflammatory cytokines TNF-α, IL-6, IL-1b, IFN-γ, IL-8, and IL-17 in colon tissues and retaining the functional morphology of the colon, mucin expression, and colonic barrier integrity (Lee et al., 2009a). Lysozyme and modified forms of lysozyme by heat (at 90°C, pH 6.0 for 20 min), chemical (by DTT) or pepsin digestion presented anti-inflammatory and anti-nociceptive activity against carrageenan-induced paw edema in mice, indicating the independence of their natural enzymatic activity (Carrillo et al., 2016). Intriguingly, denatured lysozyme showed the highest anti-inflammatory activity against paw edema, suggesting that the hydrophobic and positively charged unfolding protein could likely interact with inflammatory mediators and modulate the paw edema formation (Carrillo et al., 2016).

Hen egg white lysozyme could also stimulate immunoglobulin production in human-human hybridoma HB4C5 cells and human peripheral blood lymphocytes likely through accelerating the translation process (Murakami et al., 1997; Sugahara et al., 2000). Denature of lysozyme by heating at 83°C for 30 min could dramatically increase its immunoglobulin production stimulating activity in HB4C5 cells. And heating with 4 M urea solution and/or 0.5 mM 2-mercaptoethanol could facilitate the incorporation of lysozyme into the cells and further enhance its stimulating activity (Sugahara et al., 2002).

22.5.6 Angiotensin-Converting Enzyme-Inhibitory (ACEi) Activity

Together with the three peptides RGY, WIR, and VAW discussed above with strong antioxidant effects, peptides KVF, MKR, AMK, AKF, RGY, WIR, VAW, and GIL from the lysozyme hydrolysate prepared by the simulated gastrointestinal digestion showed strong ACEi activity (Rao et al., 2012). Lysozyme hydrolysates prepared by trypsin, papain, and a combination of the two also exhibited ACEi activity and peptides FESNFNTQATNR (Asoodeh et al., 2012), NTDGSTDYGILQINSR and VFGR (Memarpoor-Yazdi et al., 2012b) were identified from the papain-trypsin hydrolysate and shown to act as uncompetitive inhibitors against ACE. Similar to studying the antioxidative activity of peptides, these peptides with potential ACEi activity need to be further studied in living organisms and/or tissues.

22.6 Ovomucoid

Ovomucoid, comprising around 11% of egg white protein, is a glycosylated protein comprising 186 amino acid residues organized as three tandem domains with 22–29% of carbohydrate content and the molecular weight of approximately 28.0 kDa (Davis et al., 1971; Kato et al., 1987; Yousif & Kan, 2002). It is well known as a trypsin inhibitor and a main food allergen in egg white (Kato et al., 1987; Cooke & Sampson, 1997). Duck egg white ovomucoid has been reported to stabilize the bioactivity of insulin immobilized in the polymer hydrogel particles ascribed to the neutralization of proteolytic enzymes by ovomucoid. Also, duck egg ovomucoid inclusion increased the rate of insulin diffusion from the particles through the intestinal epithelia by the chemical binding of ovomucoid to lectin in the gastrointestinal tract, which further elevated insulin diffusion into bloodstream (Agarwal et al., 2001; Plate et al., 2002). Duck ovomucoid and turkey ovomucoid have also been shown to inhibit the effects of trypsin, α-chymotrypsin, and elastase and serve as promising components of drug delivery systems to protect salmon calcitonin from intestinal digestion (Shah & Khan, 2004). Intravenous injection of duck ovomucoid to rats with experimental pancreatitis was demonstrated to decrease the trypsin-like activity and the serum levels of the trypsinogen activation peptide to the levels in intact rats and reduce the primary pancreas destruction (Valueva et al., 1992).

Ovomucoid and its derived peptides exhibit immunomodulating activity because the most allergen-reactive epitopes along the molecule have been well identified to be T and/or B cell epitopes and IgE binding epitopes, with the ability to modulate T cell lines secreting cytokines, including IL-4, IL-6, IL-10, IL-13, TNF-α, and IFN-γ (Holen et al., 2001). Peptides with T cell epitopes can be systematically modified for the development of synthetic vaccines, T cell receptor antagonists, etc. (Holen et al., 2001). Ovomucoid hydrolysates prepared by pepsin, Alcalase, Alcalase with papain, and Alcalase with trypsin showed strong antioxidative activity by testing the amounts of 2-thiobarbituric acid reactive substances (TBARS) value of oil emulsion (Abeyrathne et al., 2015). Pepsin-digested ovomucoid showed >80% ACEi activity, with the other three hydrolysates exhibiting 60-70% ACEi activity (Abeyrathne et al., 2015). However, these analyses were conducted by chemical methods. Further *in vitro* cell lines as well as *in vivo* animal models are needed to identify responsible peptides present in the hydrolysates and certify their bioactivities.

22.7 Avidin

Avidin is a glycoprotein with its isoelectric point at pH 10 (Woolley & Longsworth, 1942) and only accounts for 0.05% of egg white protein (Green, 1975). It consists of four essentially identical polypeptide chains with each comprising 128 amino acid residues and bearing a single biotin-binding site and a single oligosaccharide chain that mainly includes mannose and N-acetyl glucosamine residues. Its molecular weight is around 62.4 kDa and the carbohydrate chains accounts for roughly 10% (DeLange & Huang, 1971). Egg white avidin has high affinity to bind biotin, partially accounting for its antimicrobial activity against biotin-requiring microorganisms (Campbell et al., 1972; Board & Fuller, 1974; Green, 1975; Pugliese et al., 1994). Avidin can also bind bacterial cells, especially to porins to effectively trap biotin and prevent bacterial uptake and utilization of biotin (Korpela et al., 1984). However, very few data reported the possible inhibition of avidin on the growth and proliferation of pathogens not relying on exogenous biotin supplementation (Campbell et al., 1972). These suggested its potential role in host defense mechanisms, independent of biotin utilization, as analogous to affect the interaction of pathogens with the host (Korpela et al., 1984). Avidin was demonstrated to markedly inhibit concanavalin A and pokeweed mitogen-induced proliferation of lymphocytes and enhance antibody production at the cell level (Otani & Maenishi, 1994). This modulatory property was associated with direct binding between avidin and mitogen as well as the participation of accessory cells. And both carbohydrate and polypeptide portions of avidin make contributes as digestion by α-mannosidase or pepsin could lower its activity (Otani & Maenishi, 1994). Tissue injury and inflammation can induce avidin expression in the oviduct and non-oviductal tissues (Elo et al., 1975; Elo et al., 1979a, 1979b), indicating the participation of avidin in a nonspecific defense system against infection (Elo et al., 1980). Avidin induction in tumorous tissue and viral infection-induced cell damage (Korpela et al., 1982, 1983) and its secretion in large amounts by macrophages also suggest its role in host defense (Korpela, 1984).

The avidin-biotin complex has also been shown to be advantageous in coupling drugs to the vector with high efficiency and liberating the therapeutics (i.e. bioactive peptides) after cleaving bonds that link the therapeutics and the vector (Bickel et al., 2001). Corti et al. (1998) and Gasparri et al. (1999) demonstrated the feasibility for targeting biotinylated tumor necrosis factor (TNF) to tumors by pretargeting with biotinylated antibodies and avidin. This targeting complex enabled the slow release of bioactive TNF in the microenvironment of the targeted cells and improved the anti-tumor activity of TNF. Avidin-biotin complex and other modified avidin complex has been widely studied for optimizing drug delivery, for example via tumor-specific antibodies, due to improved pharmacokinetics and reduced immunogenicity (Wilchek & Bayer, 1984; Caliceti et al., 2002). Microgel multilayer formation based on avidin-biotin complex improved incorporation of the antimicrobial peptide KYE28 (KYEITTIHNLFRKLTHRLFRRNFGYTLR), showed more pronounced suppression to *E. coli*, and promoted disruption of *E. coli* cells attached to the multilayers (Nystrom et al., 2018).

Avidin also plays an important role in maintaining embryo viability, avian hatching morphology, and immune phenotype. Experimental enrichment of avidin in egg white improved embryo survival during the late embryonic developmental phase probably by changing the proportion of carbohydrates to amino acids for embryogenesis. However, as it binds biotin, enrichment of avidin exhibited a negative effect on structural body size i.e. residual tarsus length (Krkavcová et al., 2018).

22.8 Cystatin

Egg white cystatin has a molecular weight approximately 13 kDa and consists mainly of a straight five-turn α-helix, a five-stranded antiparallel β-pleated sheet, and an appending segment of partially α-helical geometry (Bode et al., 1988). It can bind tightly to papain-like cysteine proteinases (Nicklin & Barrett, 1984). Egg white cystatin has been revealed with antimicrobial, antiviral, antitumor, immunomodulatory activities, etc. (Patryk et al., 2020). Studies on cystatins from human and animals, including mouse, rat, dog, and cow expand our understanding of the bioactivities of structurally and functionally related proteins (Patryk et al., 2020). In this section, present knowledge about the bioactivities of egg white cystatin is described and summarized (Table 22.1).

22.8.1 Antimicrobial Activity

Blankenvoorde et al. reported that egg white cystatin could inhibit the growth of *Porphyromonas gingivalis* probably by interfering with the bacterial peptide transport system (Blankenvoorde et al., 1996, 1998). Egg white cystatin also inhibited the growth of the Gram-positive strains *Staphylococcus aureus*, *S. xylosus*, and *S. gallinarum* and exhibited bactericidal activity against *E. coli*, *Pseudomonas aeruginosa*, *Acinetobacter lwoffii*, and *Oligella* sp. strains (Wesierska et al., 2005). In addition, cystatin showed antifungal activity against the most frequent human fungal pathogens of the genus *Candida*, including *Candida parapsilosis*, *Candida tropicalis*, and *Candida glabrata*, and had great potential to combat efflux-mediated multidrug resistance (Kolaczkowska et al., 2010). Incorporation of egg white cystatin/lysozyme with chitosan in the protective films formed from hydroxypropyl methylcellulose could inhibit yeast growth and exhibit antimicrobial activity against *Micrococcus flavus*, *Bacillus cereus*, *E. coli*, *Pseudomonas fluorescens*, and *Candida famata* (Zimoch-Korzycka et al., 2015).

22.8.2 Antiviral Activity

Chicken egg white cystatin could alter the intracellular proteolytic processing of poliovirus proteins and reduced virus production in human Hela cells (Korant et al., 1985). It also showed antiviral activity against human rhinovirus 1A encoding its own cysteine protease for replication like poliovirus (Korant et al., 1986). Ebina et al. reported that egg white cystatin prevented human rotavirus-induced diarrhea in suckling mice and recovered histopathological changes in the small intestine, indicating the ability of cystatin to inhibit human rotavirus infection by interfering viral replication (Ebina & Tsukada, 1991).

22.8.3 Anti-Tumor Activity

Tumoral invasion and metastasis involve the function of various proteases, including urokinase-type plasminogen activator (uPA)/plasmin, cathepsins B and L, and matrix metalloproteases (Koblinski et al., 2000). As an inhibitor against most cysteine proteinases, egg white cystatin could inhibit invasion of human ras-transformed breast epithelial cells (Premzl et al., 2001), and decrease the activity of cathepsin B and L in gastric cancer tissues to non-cancerous values (Saleh et al., 2003). Hybrids of egg white cystatin by substituting a loop of cystatin with uPA-derived peptides covering the uPA receptor binding site retained efficient inhibitory effects on cathepsins B and L (Muehlenweg et al., 2000); further modification of the variants with N-terminal domain of the human matrix metalloproteinase inhibitors or uPA fragment significantly decreased the invasion of human ovarian cancer cells and reduced tumor burden in mice, representing promising agents for prevention of tumor migration and metastasis by mediating different tumor-associated proteolytic systems (Krol, Kopitz et al., 2003; Krol, Sato et al., 2003).

22.8.4 Immunomodulatory Activity

Egg white cystatin was reported to induce the synthesis of TNF-α and IL-10 to activate nitric oxide (NO) synthase and upregulate NO release from IFN-γ-activated mouse peritoneal macrophages (Verdot et al., 1996, 1999). It also upregulated the IL-6 and IL-8 production in human gingival fibroblasts and IL-6 in murine splenocytes, independent on its papain-inhibitory site (Kato et al., 2000). Egg white cystatin could synergize with a suboptimal concentration of IFN-γ to eliminate the spleen burden of *Leishmania donovani*, the causative agent of visceral leishmaniasis. Cystatin and IFN-γ inhibited the growth of amastigotes in macrophages and the production of IL-14 but increased the levels of IL-12 and NO, demonstrating the cure and protective effects of cystatin on NO generation that kills the parasite and on the induction of Th2-Th1 conversion that functions as long-term resistance against the parasite infection (Das et al., 2001).

22.8.5 Other Bioactivities

Egg white cystatin showed inhibitory effect on osteoclast-mediated bone resorption (Brand et al., 2004), indicating its potential in mediating bone health. As more research advances our knowledge on egg white cystatin and its related proteins of different origins, further investigations are warranted to unravel its multiple bioactivities and potentiate its application in food and pharmaceutical formulations for managing diseases and human health (Kupaj et al., 2021).

22.9 Ovomacroglobulin

Chicken egg white ovomacroglobulin, also termed as ovostatin, is a glycoprotein with an isoelectric point of 4.9 and an estimated molecular weight of approximately 780 kDa, containing four subunits connected in pairs by disulfide bonds (Miller & Feeney, 1966; Kitamoto et al., 1982; Nagase et al., 1983; Geng et al., 2015). These four cylindrical subunits associate midway from both ends to form a tetramer with the proteinase-susceptible bait regions located near the molecular center (Ruben et al., 1988). When a proteinase reaches the bait region, the subunit arms curl and entrap the enzyme within the ovomacroglobulin molecule to inhibit the activity of proteinases (Nagase & Harris, 1983; Ruben et al., 1988). Egg white ovomacroglobulin showed a protease inhibitory activity toward trypsin, papain, and thermolysin (Kitamoto et al., 1982; Nagase & Harris, 1983), Serratia 56K protease and cytotoxicity (Molla et al., 1986; Molla, Oda et al., 1987), *Serratia marcescens* proteases (56K, 60K, and 73K proteases), *Pseudomonas aeruginosa* elastase and alkaline protease, and *Bacillus stearothermophilus* thermolysin (Molla, Matsumura et al., 1987). Ovomacroglobulin inhibited the enzymatic activity of bacterial proteases, suppressed the bacterial growth (Miyagawa, Nishino et al., 1991), and prevented corneal destruction and vascular permeability elevation (Molla, Matsumura et al., 1987). Consistent findings were also reported in the serratial and pseudomonal keratitis in rabbits with the acceleration of wound healing (Miyagawa, Kamata et al., 1991). Ovomacroglobulin effectively mitigated living *Pseudomonas aeruginosa*- and pseudomonal elastase-induced ring abscess formation, likely through facilitating the migration of polymorphonuclear leukocytes toward the pathogenic bacteria (Ijiri et al., 1993). Chicken egg white ovomacroglobulin also showed a synergistic therapeutic effect with ofloxacin in preventing stromal degradation in pseudomonal keratitis (Miyagawa et al., 1994). Ovomacroglobulin could strongly inhibit the metalloprotease produced by *Vibrio vulnificus* without appreciable antibacterial effect *in vitro*. It dramatically ameliorated septicemia as evidenced by significant reduction of intravenous bacterial dissemination and lethality in mice (Maruo et al., 1998). Intriguingly, different from ovomacroglobulin from chicken egg white, duck egg white ovomacroglobulin has a conserved reactive thiol ester that can form covalent bond and is capable of inhibiting both metallo- and serine proteinases stoichiometrically (Nagase et al., 1986; Nagase & Brew, 1987).

Egg white ovomacroglobulin also inhibited the protease medullasin, a proinflammatory protein released from invading granulocytes in the inflammatory tissues, thus possessing anti-inflammatory property (Ikai et al., 1989). Ofuji et al. found that ovomacroglobulin could promote fibroblast growth, collagen deposition, and capillary formation in granulation tissues and enhance surgical wound healing (Ofuji et al., 1992). Ovomacroglobulin promoted the migration of mouse embryonic fibroblasts and human skin fibroblasts and cell adhesion to extracellular matrix, demonstrating its role in accelerating wound healing and mediating cancer pathological processes (Geng et al., 2016).

22.10 Ovoinhibitor

Ovoinhibitor accounts for 1.5% of egg white protein and has a molecular weight around 49.0 kDa with an isoelectric point of 5.1 (Davis et al., 1969; Stevens, 1991; Tomimatsu et al., 1966). It is an inhibitor of serine proteinases such as trypsin, chymotrypsin (Rhodes et al., 1960; Feeney et al., 1963; Ryan & Clary, 1964; Ryan et al., 1965), proteinase K and F (Birk et al., 1983), a bacterial proteinase from *Bacillus subtilis* var. *biotecus* (Matsushima, 1958), and a fungal proteinase from *Aspergillus oryzae* (Feeney et al., 1963). Egg white ovoinhibitor is expressed by the oviduct, while the precursor of egg yolk ovoinhibitor is expressed by liver and secreted into the blood to finally form ovoinhibitor in the follicle. Ovoinhibitor is also present in all other egg compartments. The latest research reported that egg yolk ovoinhibitor showed antimicrobial activity against *Bacillus thuringiensis*, indicating its significant role in protecting eggs and chick embryo from bacterial contamination (Bourin et al., 2011). Ovoinhibitors from egg yolk and white share identical amino-acid composition, N-terminal amino acid sequence, and reactivity to antisera. However, they show different molecular weights, carbohydrate contents and moieties, motility on native PAGE, and isoelectric points. (Sugimoto et al., 1996). These characteristics suggest that ovoinhibitors from egg yolk and white possess similar but also probably distinct bioactivities. Further studies are needed to broaden our knowledge about their novel functions and promising applications.

22.11 Conclusions and Future Perspectives

Egg whites are a versatile food ingredient, mostly in the liquid and dried forms, due to its nutritional qualities and foaming and gelling properties. It is now well acknowledged that egg whites contain diverse proteins and peptides imparting various biological functions. Chemical modifications and enzymatic digestion have also been reported to increase their uses. Among the egg white proteins, 1) ovalbumin and avidin have great potential for use in the drug delivery systems, 2) ovotransferrin can be used as an antimicrobial agent and iron carrier to promote bone health, 3) lysozyme is used as an antimicrobial agent and a preservative in the food industry, 4) ovomucin, ovomucoid and cystatin can function as antimicrobial and immunomodulating agents against infections, 5) ovomacroglobulin (ovostatin) might promote wound healing promoter as an antimicrobial agent, and 6) lysozyme and ovomucin can function as elements of biocompatible and functional biomaterials in biomedical applications for tissue engineering (i.e., implant scaffolds and treatment of osteoporosis) (Martins et al., 2009; Carpena et al., 2017; Jalili-Firoozinezhad et al., 2020; Li et al., 2020; Liu et al., 2020). Egg white proteins have also been researched as promising nutraceuticals for alleviating metabolic syndrome and benefiting human health (Matsuoka et al., 2008; Abeyrathne et al., 2013; Matsuoka et al., 2017; Shirouchi & Matsuoka, 2019). However, the translation of egg white proteins and their peptides for human health applications is in the infancy due largely to a lack of funding support for well-designed human trials, and the concern over this well-known source of allergens, despite an increasing number of publications on the bioactivities of egg proteins/peptides. Understanding the biological activities of egg white proteins and their peptides provide the egg industry new opportunities to create novel products with benefits to animals and humans in the future, which in turn will add value to the egg industry.

REFERENCES

Abdou, A. M., Higashiguchi, S., Aboueleinin, A., Kim, M., & Ibrahim, H. R. (2007). Antimicrobial peptides derived from hen egg lysozyme with inhibitory effect against *Bacillus* species. *Food Control*, *18*(2), 173–178.

Abeyrathne, E. N. S., Lee, H., & Ahn, D. U. (2013). Egg white proteins and their potential use in food processing or as nutraceutical and pharmaceutical agents—A review. *Poultry Science*, *92*(12), 3292–3299.

Abeyrathne, E., Lee, H., Jo, C., Suh, J., & Ahn, D. (2016). Enzymatic hydrolysis of ovomucin and the functional and structural characteristics of peptides in the hydrolysates. *Food Chemistry*, *192*, 107–113.

Abeyrathne, E., Lee, H., Jo, C., Suh, J., & Ahn, D. U. (2015). Enzymatic hydrolysis of ovomucoid and the functional properties of its hydrolysates. *Poultry Science*, *94*(9), 2280–2287.

Agarwal, V., Nazzal, S., Reddy, I. K., & Khan, M. A. (2001). Transport studies of insulin across rat jejunum in the presence of chicken and duck ovomucoids. *Journal of Pharmacy and Pharmacology*, *53*(8), 1131–1138.

Aguilar, J. G. d. S., Granato Cason, V., & de Castro, R. J. S. (2019). Improving antioxidant activity of black bean protein by hydrolysis with protease combinations. *International Journal of Food Science & Technology*, *54*(1), 34–41.

Ahn, K. S., & Aggarwal, B. B. (2005). Transcription factor NF-κB: A sensor for smoke and stress signals. *Annals of the New York Academy of Sciences*, *1056*(1), 218–233.

Akashi, A. (1972). The effect of egg-white lysozyme on the growth of cheese starter organisms. I.

Almajano, M. P., Delgado, M. E., & Gordon, M. H. (2007). Albumin causes a synergistic increase in the antioxidant activity of green tea catechins in oil-in-water emulsions. *Food Chemistry*, *102*(4), 1375–1382.

Aluko, R. E. (2015). Antihypertensive peptides from food proteins. *Annual Review of Food Science and Technology*, *6*, 235–262.

Aminlari, L., Mohammadi Hashemi, M., & Aminlari, M. (2014). Modified lysozymes as novel broad spectrum natural antimicrobial agents in foods. *Journal of Food Science*, *79*(6), R1077–R1090.

Artym, J., & Zimecki, M. (2013). Milk-derived proteins and peptides in clinical trials. *Postepy Hig Med Dosw (Online)*, *67*, 800–816.

Asoodeh, A., Yazdi, M. M., & Chamani, J. (2012). Purification and characterisation of angiotensin I converting enzyme inhibitory peptides from lysozyme hydrolysates. *Food Chemistry*, *131*(1), 291–295.

Bao, X., & Wu, J. (2021). Egg white ovomucin hydrolysate inhibits intestinal integrity damage in LPS-treated Caco-2 cells. *Journal of Functional Foods*, *87*, 104822.

Baron, F., Jan, S., Gonnet, F., Pasco, M., Jardin, J., Giudici, B., & Nau, F. (2014). Ovotransferrin plays a major role in the strong bactericidal effect of egg white against the bacillus cereus group. *Journal of Food Protection*, *77*(6), 955–962.

Belting, M. (2003). Heparan sulfate proteoglycan as a plasma membrane carrier. *Trends in Biochemical Sciences*, *28*(3), 145–151.

Bertzbach, L. D., Kheimar, A., Ali, F. A. Z., & Kaufer, B. B. (2018). Viral factors involved in Marek's disease virus (MDV) pathogenesis. *Current Clinical Microbiology Reports*, *5*(4), 238–244.

Bickel, U., Yoshikawa, T., & Pardridge, W. M. (2001). Delivery of peptides and proteins through the blood–brain barrier. *Advanced Drug Delivery Reviews*, *46*(1–3), 247–279.

Birk, Y., Khalef, S., & Jibson, M. D. (1983). Purification and properties of protease F, a bacterial enzyme with chymotrypsin and elastase specificities. *Archives of Biochemistry and Biophysics*, *225*(2), 451–457.

Blankenvoorde, M. F., Henskens, Y., Van't Hof, W., Veerman, E. C., & Nieuw Amerongen, A. (1996). Inhibition of the growth and cysteine proteinase activity of *Porphyromonas gingivalis* by human salivary cystatin S and chicken cystatin. *Biological Chemistry Hoppe-Seyler*, *377*, 847–850.

Blankenvoorde, M. F., Van't Hof, W., Walgreen-Weterings, E., Van Steenbergen, T., Brand, H., Veerman, E., & Nieuw Amerongen, A. (1998). Cystatin and cystatin-derived peptides have antibacterial activity against the pathogen *Porphyromonas gingivalis*. *Biological Chemistry*, *379*(11), 1371–1376.

Board, R., & Fuller, R. (1974). Non-specific antimicrobial defences of the avian egg, embryo and neonate. *Biological Reviews*, *49*(1), 15–49.

Bode, W., Engh, R., Musil, D., Thiele, U., Huber, R., Karshikov, A., & Turk, V. (1988). The 2.0 AX-ray crystal structure of chicken egg white cystatin and its possible mode of interaction with cysteine proteinases. *The EMBO Journal*, *7*(8), 2593–2599.

Bourin, M., Gautron, J., Berges, M., Attucci, S., Le Blay, G., Labas, V., & Rehault-Godbert, S. (2011). Antimicrobial potential of egg yolk ovoinhibitor, a multidomain Kazal-like inhibitor of chicken egg. *Journal of Agricultural and Food Chemistry*, *59*(23), 12368–12374.

Brand, H., Lerner, U., Grubb, A., Beertsen, W., Amerongen, A. N., & Everts, V. (2004). Family 2 cystatins inhibit osteoclast-mediated bone resorption in calvarial bone explants. *Bone*, *35*(3), 689–696.

Bueno-Gavilá, E., Abellán, A., Girón-Rodríguez, F., Cayuela, J. M., & Tejada, L. (2021). Bioactivity of hydrolysates obtained from chicken egg ovalbumin using artichoke (Cynara scolymus L.) Proteases. *Foods*, *10*(2), 246.

Bullen, J., Rogers, H. J., & Griffiths, E. (1978). Role of iron in bacterial infection. *Current Topics in Microbiology and Immunology*, *80*. Springer. https://doi.org/10.1007/978-3-642-66956-9_1

Cai, Y. Z., & Corke, H. (2000). Production and properties of spray-dried Amaranthus betacyanin pigments. *Journal of Food Science*, *65*(7), 1248–1252.

Caliceti, P., Chinol, M., Roldo, M., Veronese, F. M., Semenzato, A., Salmaso, S., & Paganelli, G. (2002). Poly (ethylene glycol)–avidin bioconjugates: Suitable candidates for tumor pretargeting. *Journal of Controlled Release*, *83*(1), 97–108.

Campbell, A., del Campillo-Campbell, A., & Chang, R. (1972). A mutant of Escherichia coli that requires high concentrations of biotin. *Proceedings of the National Academy of Sciences*, *69*(3), 676–680.

Campione, E., Lanna, C., Cosio, T., Rosa, L., Conte, M. P., Iacovelli, F., & Franchin, E. (2021). Lactoferrin against SARS-CoV-2: In vitro and in silico evidences. *Frontiers in Pharmacology*, *12*, 666600.

Canfield, R. E. (1963). The amino acid sequence of egg white lysozyme. *Journal of Biological Chemistry*, *238*(8), 2698–2707.

Carpena, N. T., Abueva, C. D., Padalhin, A. R., & Lee, B. T. (2017). Evaluation of egg white ovomucin-based porous scaffold as an implantable biomaterial for tissue engineering. *Journal of Biomedical Materials Research Part B: Applied Biomaterials*, *105*(7), 2107–2117.

Carrillo, W., García-Ruiz, A., Recio, I., & Moreno-Arribas, M. (2014). Antibacterial activity of hen egg white lysozyme modified by heat and enzymatic treatments against oenological lactic acid bacteria and acetic acid bacteria. *Journal of Food Protection*, *77*(10), 1732–1739.

Carrillo, W., Gómez-Ruiz, J., Miralles, B., Ramos, M., Barrio, D., & Recio, I. (2016). Identification of antioxidant peptides of hen egg-white lysozyme and evaluation of inhibition of lipid peroxidation and cytotoxicity in the zebrafish model. *European Food Research and Technology*, *242*(10), 1777–1785.

Carrillo, W., Lucio, A., Gaibor, J., Morales, D., & Vásquez, G. (2018). Isolation of antibacterial hydrolysates from hen egg white lysozyme and identification of antibacterial peptides. *Journal of Medicinal Food*, *21*(8), 808–818.

Carrillo, W., & Ramos, M. (2018). Identification of antimicrobial peptides of native and heated hydrolysates from hen egg white lysozyme. *Journal of Medicinal Food*, *21*(9), 915–926.

Carrillo, W., Spindola, H., Ramos, M., Recio, I., & Carvalho, J. E. (2016). Anti-inflammatory and anti-nociceptive activities of native and modified hen egg white lysozyme. *Journal of Medicinal Food*, *19*(10), 978–982.

Cegielska-Radziejewska, R., Lesnierowski, G., & Kijowski, J. (2009). Antibacterial activity of hen egg white lysozyme modified by thermochemical technique. *European Food Research and Technology*, *228*(5), 841–845.

Ceschia, V., Sava, G., Gagliardi, R., & Zabucchi, G. (1988). Effects of lysozyme on spleen and lungs in mice with Lewis lung carcinoma. *Pharmacological Research Communications*, *20*, 615.

Chang, O. K., Ha, G. E., Han, G. S., Seol, K.-H., Kim, H. W., Jeong, S.-G., & Ham, J. S. (2013). Novel antioxidant peptide derived from the ultrafiltrate of ovomucin hydrolysate. *Journal of Agricultural and Food Chemistry*, *61*(30), 7294–7300.

Chen, W. H., & Wang, P. S. (2007). Protective effect of lysozyme chloride on gastric oxidative stress and hemorrhagic ulcers in severe atherosclerotic rats. *Medical Science Monitor*, *13*(12), BR271–BR279.

Chung, J., Ku, S. K., Lee, S., & Bae, J. S. (2016). Suppressive effects of lysozyme on polyphosphate-mediated vascular inflammatory responses. *Biochemical and Biophysical Research Communications*, *474*(4), 715–721.

Cisani, G., Varaldo, P. E., Ingianni, A., Pompei, R., & Satta, G. (1984). Inhibition of herpes simplex virus-induced cytopathic effect by modified hen egg-white lysozymes. *Current Microbiology*, *10*(1), 35–40.

Cooke, S. K., & Sampson, H. A. (1997). Allergenic properties of ovomucoid in man. *The Journal of Immunology*, *159*(4), 2026–2032.

Cornish, J., Callon, K. E., Naot, D., Palmano, K. P., Banovic, T., Bava, U., & Chen, Q. (2004). Lactoferrin is a potent regulator of bone cell activity and increases bone formation in vivo. *Endocrinology*, *145*(9), 4366–4374.

Corti, A., Gasparri, A., Sacchi, A., Curnis, F., Sangregorio, R., Colombo, B., & Magni, F. (1998). Tumor targeting with biotinylated tumor necrosis factor α: Structure-activity relationships and mechanism of action on avidin pretargeted tumor cells. *Cancer Research*, *58*(17), 3866–3872.

Craig, D., & Reeves, R. (2002). *High value opportunities from the chicken egg*. RIRDC Publication.

Currie, A., Honish, L., Cutler, J., Locas, A., Lavoie, M.-C., Gaulin, C., & Taylor, M. (2019). Outbreak of Escherichia coli O157: H7 infections linked to mechanically tenderized beef and the largest beef recall in Canada, 2012. *Journal of Food Protection*, *82*(9), 1532–1538.

Das, L., Datta, N., Bandyopadhyay, S., & Das, P. K. (2001). Successful therapy of lethal murine visceral leishmaniasis with cystatin involves up-regulation of nitric oxide and a favorable T cell response. *The Journal of Immunology*, *166*(6), 4020–4028.

Davis, J. G., Mapes, C. J., & Donovan, J. W. (1971). Batch purification of ovomucoid and characterization of the purified product. *Biochemistry*, *10*(1), 39–42.

Davis, J. G., Zahnley, J. C., & Donovan, J. W. (1969). Separation and characterization of the ovoinhibitors from chicken egg white. *Biochemistry*, *8*(5), 2044–2053.

DeLange, R. J., & Huang, T. S. (1971). Egg white avidin: III. Sequence of the 78-residue middle cyanogen bromide peptide. Complete amino acid sequence of the protein subunit. *Journal of Biological Chemistry*, *246*(3), 698–709.

Derde, M., Lechevalier, V., Guérin-Dubiard, C., Cochet, M. F., Jan, S., Baron, F., Gautier, M., Vié, V., & Nau, F. (2013). Hen egg white lysozyme permeabilizes Escherichia coli outer and inner membranes. *Journal of Agricultural and Food Chemistry*, *61*(41), 9922–9929.

Donovan, J., Davis, J., & White, L. (1970). Chemical and physical characterization of ovomucin, a sulfated glycoprotein complex from chicken eggs. *Biochimica et Biophysica Acta (BBA)-Protein Structure*, *207*(1), 190–201.

Donovan, J. W., Davis, J. G., & Wiele, M. B. (1972). Viscosimetric studies of alkaline degradation of ovomucin. *Journal of Agricultural and Food Chemistry*, *20*(2), 223–228.

Ebina, T., & Tsukada, K. (1991). Protease inhibitors prevent the development of human rotavirus-induced diarrhea in suckling mice. *Microbiology and Immunology*, *35*(7), 583–588.

Ellison, R. (1991). 3rd and TJ Giehl, killing of Gram-negative bacteria by lactoferrin and lysozyme. *J. Clin. Invest*, *88*, 1080–1091.

Elo, H. A., Kulomaa, M. S., & Tuohimaa, P. J. (1979a). Avidin induction by tissue injury and inflammation in male and female chickens. *Comparative Biochemistry and Physiology. B, Comparative Biochemistry*, *62*(3), 237–240.

Elo, H. A., Kulomaa, M. S., & Tuohimaa, P. J. (1979b). Progesterone-independent avidin induction in chick tissues caused by tissue injury and inflammation. *European Journal of Endocrinology*, *90*(4), 743–752.

Elo, H., Räisänen, S., & Tuohimaa, P. (1980). Induction of an antimicrobial biotin-binding egg white protein (avidin) in chick tissues in septic *Escherichia coli* infection. *Experientia*, *36*(3), 312–313.

Elo, H., Tuohimaa, P., & Jänne, O. (1975). Cumulative superinduction of avidin in the chick oviduct by tissue damage and actinomycin D. *Molecular and Cellular Endocrinology*, *2*(3), 203–211.

Fan, H., Wang, J., Liao, W., Jiang, X., & Wu, J. (2019). Identification and characterization of gastrointestinal-resistant angiotensin-converting enzyme inhibitory peptides from egg white proteins. *Journal of Agricultural and Food Chemistry*, *67*(25), 7147–7156.

Fan, H., & Wu, J. (2021). Purification and identification of novel ACE inhibitory and ACE2 upregulating peptides from spent hen muscle proteins. *Food Chemistry*, *345*, 128867.

Feeney, R. E., Stevens, F. C., & Osuga, D. T. (1963). The specificities of chicken ovomucoid and ovoinhibitor. *Journal of Biological Chemistry*, *238*(4), 1415–1418.

Feng, J., Xu, H., Zhang, L., Wang, H., Liu, S., Liu, Y., & Li, C. (2018). Development of nanocomplexes for curcumin vehiculization using ovalbumin and sodium alginate as building blocks: Improved stability, bioaccessibility, and antioxidant activity. *Journal of Agricultural and Food Chemistry*, *67*(1), 379–390.

Fujita, H., Sasaki, R., & Yoshikawa, M. (1995). Potentiation of the antihypertensive activity of orally administered ovokinin, a vasorelaxing peptide derived from ovalbumin, by emulsification in egg phosphatidylcholine. *Bioscience, Biotechnology, and Biochemistry, 59*(12), 2344–2345.

Fujita, H., Usui, H., Kurahashi, K., & Yoshikawa, M. (1995). Isolation and characterization of ovokinin, a bradykinin B1 agonist peptide derived from ovalbumin. *Peptides, 16*(5), 785–790.

Fujita, I., Yokoyama, K., & Yoshikawa, M. (2000). Classification and antihypertensive activity of angiotensin I-converting enzyme inhibitory peptides derived from food proteins. *Journal of Food Science, 65*(4), 564–569.

Gasparri, A., Moro, M., Curnis, F., Sacchi, A., Pagano, S., Veglia, F., & Corti, A. (1999). Tumor pretargeting with avidin improves the therapeutic index of biotinylated tumor necrosis factor α in mouse models. *Cancer Research, 59*(12), 2917–2923.

Geng, F., Huang, X., & Ma, M. (2016). Hen egg white ovomacroglobulin promotes fibroblast migration via mediating cell adhesion and cytoskeleton. *Journal of the Science of Food and Agriculture, 96*(9), 3188–3194.

Geng, F., Huang, X., Majumder, K., Zhu, Z., Cai, Z., & Ma, M. (2015). Mass spectrometry and two-dimensional electrophoresis to characterize the glycosylation of hen egg white ovomacroglobulin. *Journal of Agricultural and Food Chemistry, 63*(37), 8209–8215.

Giansanti, F., Leboffe, L., & Antonini, G. (2012). Antiviral activity of lactoferrin and ovotransferrin derived peptides towards herpesviridae. *Herpesviridae—A Look into This Unique Family of Viruses, 2012*, 295–320.

Giansanti, F., Massucci, M. T., Giardi, M. F., Nozza, F., Pulsinelli, E., Nicolini, C., & Antonini, G. (2005). Antiviral activity of ovotransferrin derived peptides. *Biochemical and Biophysical Research Communications, 331*(1), 69–73.

Giansanti, F., Rossi, P., Massucci, M. T., Botti, D., Antonini, G., Valenti, P., & Seganti, L. (2002). Antiviral activity of ovotransferrin discloses an evolutionary strategy for the defensive activities of lactoferrin. *Biochemistry and Cell Biology, 80*(1), 125–130.

Gomes, M. H. G., & Kurozawa, L. E. (2020). Improvement of the functional and antioxidant properties of rice protein by enzymatic hydrolysis for the microencapsulation of linseed oil. *Journal of Food Engineering, 267*, 109761.

Gottschalk, A., & Lind, P. E. (1949). Product of interaction between influenza virus enzyme and ovomucin. *Nature, 164*(4162), 232–233.

Green, N. M. (1975). Avidin. In *Advances in protein chemistry* (Vol. 29, pp. 85–133). Elsevier.

Gu, Y., Majumder, K., & Wu, J. (2011). QSAR-aided in silico approach in evaluation of food proteins as precursors of ACE inhibitory peptides. *Food Research International, 44*(8), 2465–2474.

Guo, H. Y., Jiang, L., Ibrahim, S. A., Zhang, L., Zhang, H., Zhang, M., & Ren, F. Z. (2009). Orally administered lactoferrin preserves bone mass and microarchitecture in ovariectomized rats. *The Journal of Nutrition, 139*(5), 958–964.

Hallquist, N. A., & Klasing, K. C. (1994). Serotransferrin, ovotransferrin and metallothionein levels during an immune response in chickens. *Comparative Biochemistry and Physiology Part B: Comparative Biochemistry, 108*(3), 375–384.

He, W., Xu, H., Lu, Y., Zhang, T., Li, S., Lin, X., & Wu, X. (2019). Function, digestibility and allergenicity assessment of ovalbumin–EGCG conjugates. *Journal of Functional Foods, 61*, 103490.

Holen, E., Bolann, B., & Elsayed, S. (2001). Novel B and T cell epitopes of chicken ovomucoid (Gal d 1) induce T cell secretion of IL-6, IL-13, and IFN-γ. *Clinical & Experimental Allergy, 31*(6), 952–964.

Hou, Z., Imam, M. U., Ismail, M., Azmi, N. H., Ismail, N., Ideris, A., & Mahmud, R. (2015). Lactoferrin and ovotransferrin contribute toward antioxidative effects of edible bird's nest against hydrogen peroxide-induced oxidative stress in human SH-SY5Y cells. *Bioscience, Biotechnology, and Biochemistry, 79*(10), 1570–1578.

Hu, Y., Chen, H., Xiao, L., Chu, L., Wang, S., & Wang, H. (2019). Comparison of ovalbumin glycation by microwave irradiation and conventional heating. *LWT, 116*, 108560.

Hua, C., Zhu, Y., Wu, C., Si, L., Wang, Q., Sui, L., & Jiang, S. (2019). The underlying mechanism of 3-hydroxyphthalic anhydride-modified bovine beta-lactoglobulin to block human papillomavirus entry into the host cell. *Frontiers in Microbiology*, 2188.

Huang, W., Chakrabarti, S., Majumder, K., Jiang, Y., Davidge, S. T., & Wu, J. (2010b). Egg-derived peptide IRW inhibits TNF-α-induced inflammatory response and oxidative stress in endothelial cells. *Journal of Agricultural and Food Chemistry, 58*(20), 10840–10846.

Huang, W. Y., Majumder, K., & Wu, J. (2010a). Oxygen radical absorbance capacity of peptides from egg white protein ovotransferrin and their interaction with phytochemicals. *Food Chemistry, 123*(3), 635–641.

Huang, X., Tu, Z., Xiao, H., Wang, H., Zhang, L., Hu, Y., & Niu, P. (2012). Characteristics and antioxidant activities of ovalbumin glycated with different saccharides under heat moisture treatment. *Food Research International, 48*(2), 866–872.

Hughey, V. L., Wilger, P. A., & Johnson, E. A. (1989). Antibacterial activity of hen egg white lysozyme against *Listeria* monocytogenes Scott A in foods. *Applied and Environmental Microbiology, 55*(3), 631–638.

Hung, C. R. (2005). Effect of lysozyme chloride on betel quid chewing aggravated gastric oxidative stress and hemorrhagic ulcer in diabetic rats. *World Journal of Gastroenterology: WJG, 11*(37), 5853.

Hunter, H. N., Jing, W., Schibli, D. J., Trinh, T., Park, I. Y., Kim, S. C., & Vogel, H. J. (2005). The interactions of antimicrobial peptides derived from lysozyme with model membrane systems. *Biochimica et Biophysica Acta (BBA)-Biomembranes, 1668*(2), 175–189.

Huntington, J. A., & Stein, P. E. (2001). Structure and properties of ovalbumin. *Journal of Chromatography B: Biomedical Sciences and Applications, 756*(1–2), 189–198.

Ibrahim, H. R., Hoq, M. I., & Aoki, T. (2007). Ovotransferrin possesses SOD-like superoxide anion scavenging activity that is promoted by copper and manganese binding. *International Journal of Biological Macromolecules, 41*(5), 631–640.

Ibrahim, H. R., Inazaki, D., Abdou, A., Aoki, T., & Kim, M. (2005). Processing of lysozyme at distinct loops by pepsin: A novel action for generating multiple antimicrobial peptide motifs in the newborn stomach. *Biochimica et Biophysica Acta (BBA)-General Subjects, 1726*(1), 102–114.

Ibrahim, H. R., Iwamori, E., Sugimoto, Y., & Aoki, T. (1998). Identification of a distinct antibacterial domain within the N-lobe of ovotransferrin. *Biochimica et Biophysica Acta (BBA)-Molecular Cell Research*, *1401*(3), 289–303.

Ibrahim, H. R., Sugimoto, Y., & Aoki, T. (2000). Ovotransferrin antimicrobial peptide (OTAP-92) kills bacteria through a membrane damage mechanism. *Biochimica et Biophysica Acta -General Subjects*, 1523(2–3), 196–205.

Ibrahim, H. R., Thomas, U., & Pellegrini, A. (2001). A helix-loop-helix peptide at the upper lip of the active site cleft of lysozyme confers potent antimicrobial activity with membrane permeabilization action. *Journal of Biological Chemistry*, *276*(47), 43767–43774.

Ijiri, Y., Yamamoto, T., Kamata, R., Aoki, H., Matsumoto, K., Okamura, R., & Kambara, T. (1993). The role of *Pseudomonas* aeruginosa elastase in corneal ring abscess formation in pseudomonal keratitis. *Graefe's Archive for Clinical and Experimental Ophthalmology*, *231*(9), 521–528.

Ikai, A., Nakashima, M., & Aoki, Y. (1989). Inhibition of inflammatory proteinase, medullasin, by α2-macroglobulin and ovomacroglobulin. *Biochemical and Biophysical Research Communications*, *158*(3), 831–836.

Ishikawa, S., Yano, Y., Arihara, K., & Itoh, M. (2004). Egg yolk phosvitin inhibits hydroxyl radical formation from the Fenton reaction. *Bioscience, Biotechnology, and Biochemistry*, *68*(6), 1324–1331.

Itoh, T., Miyazaki, J., Sugawara, H., & Adachi, S. (1987). Studies on the characterization of ovomucin and chalaza of the hen's egg. *Journal of Food Science*, *52*(6), 1518–1521.

Jackson, C. M., Esnouf, M. P., Winzor, D. J., & Duewer, D. L. (2007). Defining and measuring biological activity: Applying the principles of metrology. *Accreditation and Quality Assurance*, *12*(6), 283–294.

Jahandideh, F., Chakrabarti, S., Davidge, S. T., & Wu, J. (2016). Antioxidant peptides identified from ovotransferrin by the ORAC method did not show anti-inflammatory and antioxidant activities in endothelial cells. *Journal of Agricultural and Food Chemistry*, *64*(1), 113–119.

Jalili-Firoozinezhad, S., Filippi, M., Mohabatpour, F., Letourneur, D., & Scherberich, A. (2020). Chicken egg white: Hatching of a new old biomaterial. *Materials Today*, *40*, 193–214.

Kato, A., Hirata, S., & Kobayashi, K. (1978). Structure of the sulfated oligosaccharide chain of ovomucin. *Agricultural and Biological Chemistry*, *42*(5), 1025–1029.

Kato, T., Imatani, T., Miura, T., Minaguchi, K., Saitoh, E., & Okuda, K. (2000). Cytokine-inducing activity of family 2 cystatins. *Biological Chemistry*, *381*(11), 1143–1147.

Kato, A., Oda, S., Yamanaka, Y., Matsudomi, N., & Kobayashi, K. (1985). Functional and structural properties of ovomucin. *Agricultural and Biological Chemistry*, *49*(12), 3501–3504.

Kato, A., & Sato, Y. (1971). The separation and characterization of carbohydrate rich component from ovomucin in chicken eggs. *Agricultural and Biological Chemistry*, *35*(3), 439–440.

Kato, I., Schrode, J., Kohr, W. J., & Laskowski Jr, M. (1987). Chicken ovomucoid: Determination of its amino acid sequence, determination of the trypsin reactive site, and preparation of all three of its domains. *Biochemistry*, *26*(1), 193–201.

Kim, J., Moon, S., Ahn, D. U., Paik, H., & Park, E. (2012). Antioxidant effects of ovotransferrin and its hydrolysates. *Poultry Science*, *91*(11), 2747–2754.

Kitamoto, T., Nakashima, M., & Ikai, A. (1982). Hen egg white ovomacroglobulin has a protease inhibitory activity. *The Journal of Biochemistry*, *92*(5), 1679–1682.

Kobayashi, K., Hattori, M., Hara-Kudo, Y., Okubo, T., Yamamoto, S., Takita, T., & Sugita-Konishi, Y. (2004). Glycopeptide derived from hen egg ovomucin has the ability to bind enterohemorrhagic Escherichia coli O157: H7. *Journal of Agricultural and Food Chemistry*, *52*(18), 5740–5746.

Koblinski, J. E., Ahram, M., & Sloane, B. F. (2000). Unraveling the role of proteases in cancer. *Clinica Chimica Acta*, *291*(2), 113–135.

Kodama, Y., & Kimura, N. (2001). *Inhibitor of helicobacter pylori colonization*. United States patent US 6,235,709. U.S. Patent and Trademark Office.

Kolaczkowska, A., Kolaczkowski, M., Sokolowska, A., Miecznikowska, H., Kubiak, A., Rolka, K., & Polanowski, A. (2010). The antifungal properties of chicken egg cystatin against *Candida* yeast isolates showing different levels of azole resistance. *Mycoses*, *53*(4), 314–320.

Ko, K., Mendonca, A. F., & Ahn, D. U. (2008a). Effect of ethylenediaminetetraacetate and lysozyme on the antimicrobial activity of ovotransferrin against *Listeria* monocytogenes. *Poultry Science*, *87*(8), 1649–1658.

Ko, K., Mendonca, A. F., & Ahn, D. U. (2008b). Influence of zinc, sodium bicarbonate, and citric acid on the antibacterial activity of ovotransferrin against *Escherichia coli* O157: H7 and listeria monocytogenes in model systems and ham. *Poultry Science*, *87*(12), 2660–2670.

Ko, K., Mendoncam, A., Ismail, H., & Ahn, D. U. (2009). Ethylenediaminetetraacetate and lysozyme improves antimicrobial activities of ovotransferrin against Escherichia coli O157: H7. *Poultry Science*, *88*(2), 406–414.

Korant, B. D., Brzin, J., & Turk, V. (1985). Cystatin, a protein inhibitor of cysteine proteases alters viral protein cleavages in infected human cells. *Biochemical and Biophysical Research Communications*, *127*(3), 1072–1076.

Korant, B. D., Towatari, T., Ivanoff, L., Pettewa, Jr., S. Y., Brzin, J., Lenarcic, B., & Turk, V. (1986). Viral therapy: Prospects for protease inhibitors. *Journal of Cellular Biochemistry*, *32*(2), 91–95.

Korpela, J. (1984). Chicken macrophages synthesize and secrete avidin in culture. *European Journal of Cell Biology*, *33*(1), 105–111.

Korpela, J., Kulomaa, M., Tuohimaa, P., & Vaheri, A. (1982). Induction of avidin in chickens infected with the acute leukemia virus OK 10. *International Journal of Cancer*, *30*(4), 461–464.

Korpela, J., Kulomaa, M., Tuohimaa, P., & Vaheri, A. (1983). Avidin is induced in chicken embryo fibroblasts by viral transformation and cell damage. *The EMBO Journal*, *2*(10), 1715–1719.

Korpela, J., Salonen, E.-M., Kuusela, P., Sarvas, M., & Vaheri, A. (1984). Binding of avidin to bacteria and to the outer membrane porin of Escherichia coli. *FEMS Microbiology Letters*, *22*(1), 3–10.

Kovacs-Nolan, J., Phillips, M., & Mine, Y. (2005). Advances in the value of eggs and egg components for human health. *Journal of Agricultural and Food Chemistry*, *53*(22), 8421–8431.

Krkavcová, E., Kreisinger, J., Hyánková, L., Hyršl, P., & Javůrková, V. (2018). The hidden function of egg white antimicrobials: Egg weight-dependent effects of avidin on avian embryo survival and hatchling phenotype. *Biology Open*, *7*(4), bio031518.

Krol, J., Kopitz, C., Kirschenhofer, A., Schmitt, M., Magdolen, U., Krüger, A., & Magdolen, V. (2003). Inhibition of intraperitoneal tumor growth of human ovarian cancer cells by bi-and trifunctional inhibitors of tumor-associated proteolytic systems. *Biological Chemistry*, *384*(7), 1097–1102.

Krol, J., Sato, S., Rettenberger, P., Assfalg-Machleidt, I., Schmitt, M., Magdolen, V., & Magdolen, U. (2003). Novel bi-and trifunctional inhibitors of tumor-associated proteolytic systems. *Biological Chemistry*, *384*, 1085–1096.

Kummer, N., Wu, T., De France, K. J., Zuber, F., Ren, Q., Fischer, P., & Nyström, G. (2021). Self-assembly pathways and antimicrobial properties of lysozyme in different aggregation States. *Biomacromolecules*, *22*(10), 4327–4336.

Kupaj, P., Korus, M., Malgorzata, K., & Kopec, W. (2021). Egg white cystatin - A review. *Acta Veterinaria-Beograd*, *70*(4), 401–422.

Kurokawa, H., Mikami, B., & Hirose, M. (1995). Crystal structure of diferric hen ovotransferrin at 2.4 Å resolution. *Journal of Molecular Biology*, *254*(2), 196–207.

Kyriakis, J. M., & Avruch, J. (2001). Mammalian mitogen-activated protein kinase signal transduction pathways activated by stress and inflammation. *Physiological Reviews*, *81*(2), 807–869.

Lacou, L., Léonil, J., & Gagnaire, V. (2016). Functional properties of peptides: From single peptide solutions to a mixture of peptides in food products. *Food Hydrocolloids*, *57*, 187–199.

Lane, J. A., Mehra, R. K., Carrington, S. D., & Hickey, R. M. (2010). The food glycome: A source of protection against pathogen colonization in the gastrointestinal tract. *International Journal of Food Microbiology*, *142*(1–2), 1–13.

Lang, J., Yang, N., Deng, J., Liu, K., Yang, P., Zhang, G., & Jiang, C. (2011). Inhibition of SARS pseudovirus cell entry by lactoferrin binding to heparan sulfate proteoglycans. *PLoS One*, *6*(8), e23710.

Laskey, J., Webb, I., Schulman, H. M., & Ponka, P. (1988). Evidence that transferrin supports cell proliferation by supplying iron for DNA synthesis. *Experimental Cell Research*, *176*(1), 87–95.

Lee, J. H., Ahn, D. U., & Paik, H. D. (2018). In vitro immune-enhancing activity of ovotransferrin from egg white via MAPK signaling pathways in RAW 264.7 macrophages. *Korean Journal for Food Science of Animal Resources*, *38*(6), 1226.

Lee, N., Cheng, J., Enomoto, T., & Nakano, Y. (2006). One peptide derived from hen ovotransferrin as pro-drug to inhibit angiotensin converting enzyme. *Journal of Food and Drug Analysis*, *14*(1), 31.

Lee-Huang, S., Huang, P. L., Sun, Y., Huang, P. L., Kung, H. F., Blithe, D. L., & Chen, H. C. (1999). Lysozyme and RNases as anti-HIV components in β-core preparations of human chorionic gonadotropin. *Proceedings of the National Academy of Sciences*, *96*(6), 2678–2681.

Lee, M., Kovacs-Nolan, J., Archbold, T., Fan, M. Z., Juneja, L. R., Okubo, T., & Mine, Y. (2009a). Therapeutic potential of hen egg white peptides for the treatment of intestinal inflammation. *Journal of Functional Foods*, *1*(2), 161–169.

Lee, M., Kovacs-Nolan, J., Yang, C., Archbold, T., Fan, M. Z., & Mine, Y. (2009b). Hen egg lysozyme attenuates inflammation and modulates local gene expression in a porcine model of dextran sodium sulfate (DSS)-induced colitis. *Journal of Agricultural and Food Chemistry*, *57*(6), 2233–2240.

Lee, M., Young, D., Mine, Y., & Jo, C. (2009c). Reduction of interlukin-8 by peptides from digestive enzyme hydrolysis of hen egg lysozyme. *Food Science and Biotechnology*, *18*(3), 706–711.

Legros, J., Jan, S., Bonnassie, S., Gautier, M., Croguennec, T., Pezennec, S., & Baron, F. (2021). The role of ovotransferrin in egg-white antimicrobial activity. A Review. *Foods*, *10*(4), 823.

Liang, T., Qiu, J., Niu, X., Ma, Q., Zhou, C., Chen, P., & Liu, S. (2021). 3-hydroxyphthalic anhydride-modified chicken ovalbumin as a potential candidate inhibits SARS-CoV-2 infection by disrupting the interaction of spike protein with host ACE2 receptor. *Frontiers in Pharmacology*, *11*, 2263.

Liao, W., Jahandideh, F., Fan, H., Son, M., & Wu, J. (2018). Egg protein-derived bioactive peptides: Preparation, efficacy, and absorption. *Advances in Food and Nutrition Research*, *85*, 1–58.

Li, L., He, L., Tan, S., Guo, X., Lu, H., Qi, Z., & Liu, S. (2010). 3-hydroxyphthalic anhydride-modified chicken ovalbumin exhibits potent and broad anti-HIV-1 activity: A potential microbicide for preventing sexual transmission of HIV-1. *Antimicrobial Agents and Chemotherapy*, *54*(5), 1700–1711.

Li, C.-P., He, Z., Wang, X., Yang, L., Yin, C., Zhang, N., & Zhao, H. (2014). Selenization of ovalbumin by dry-heating in the presence of selenite: Effect on protein structure and antioxidant activity. *Food Chemistry*, *148*, 209–217.

Ling, Y., Liping, S., & Yongliang, Z. (2018). Preparation and identification of novel inhibitory angiotensin-I-converting enzyme peptides from tilapia skin gelatin hydrolysates: Inhibition kinetics and molecular docking. *Food & Function*, *9*(10), 5251–5259.

Liu, J., Chen, Z., He, J., Zhang, Y., Zhang, T., & Jiang, Y. (2014). Anti-oxidative and anti-apoptosis effects of egg white peptide, Trp-Asn-Trp-Ala-Asp, against H_2O_2-induced oxidative stress in human embryonic kidney 293 cells. *Food & Function*, *5*(12), 3179–3188.

Liu, L., Xu, M., Tu, Y., Du, H., Zhou, Y., & Zhu, G. (2017). Immunomodulatory effect of protease hydrolysates from ovotransferrin. *Food & Function*, *8*(4), 1452–1459.

Liu, Y., Ying, D., Cai, Y., & Le, X. (2017). Improved antioxidant activity and physicochemical properties of curcumin by adding ovalbumin and its structural characterization. *Food Hydrocolloids*, *72*, 304–311.

Liu, J., Yu, Z., Zhao, W., Lin, S., Wang, E., Zhang, Y., & Chen, F. (2010). Isolation and identification of angiotensin-converting enzyme inhibitory peptides from egg white protein hydrolysates. *Food Chemistry*, *122*(4), 1159–1163.

Liu, S., Zhang, H., Hu, Q., Shen, Z., Rana, D., & Ramalingam, M. (2020). Designing vascular supportive albumen-rich composite bioink for organ 3D printing. *Journal of the Mechanical Behavior of Biomedical Materials*, *104*, 103642.

Liu, H., Zheng, F., Cao, Q., Ren, B., Zhu, L., Striker, G., & Vlassara, H. (2006). Amelioration of oxidant stress by the defensin lysozyme. *American Journal of Physiology-Endocrinology and Metabolism*, *290*(5), E824–E832.

Li, K., Zhuang, P., Tao, B., Li, D., Xing, X., & Mei, X. (2020). Ultra-small lysozyme-protected gold nanoclusters as nanomedicines inducing osteogenic differentiation. *International Journal of Nanomedicine*, *15*, 4705.

Losso, J., Nakai, S., & Charter, E. (2000). Lysozyme. *Natural Food Antimicrobial Systems*, *185*, 210.

Lucey, A. J., Heneghan, C., Manning, E., Kroon, P. A., & Kiely, M. E. (2019). Effect of an egg ovalbumin-derived protein hydrolysate on blood pressure and cardiovascular risk in adults with a mildly elevated blood pressure: A randomized placebo-controlled crossover trial. *European Journal of Nutrition*, *58*(7), 2823–2833.

Mahanta, S., Paul, S., Srivastava, A., Pastor, A., Kundu, B., & Chaudhuri, T. K. (2015). Stable self-assembled nanostructured hen egg white lysozyme exhibits strong antiproliferative activity against breast cancer cells. *Colloids and Surfaces B: Biointerfaces*, *130*, 237–245.

Majumder, K., Chakrabarti, S., Morton, J. S., Panahi, S., Kaufman, S., Davidge, S. T., & Wu, J. (2013). Egg-derived tri-peptide IRW exerts antihypertensive effects in spontaneously hypertensive rats. *PLoS One*, *8*(11), e82829.

Majumder, K., Liang, G., Chen, Y., Guan, L., Davidge, S. T., & Wu, J. (2015). Egg ovotransferrin-derived ACE inhibitory peptide IRW increases ACE2 but decreases proinflammatory genes expression in mesenteric artery of spontaneously hypertensive rats. *Molecular Nutrition & Food Research*, *59*(9), 1735–1744.

Majumder, K., & Wu, J. (2010). A new approach for identification of novel antihypertensive peptides from egg proteins by QSAR and bioinformatics. *Food Research International*, *43*(5), 1371–1378.

Majumder, K., & Wu, J. (2011). Purification and characterisation of angiotensin I converting enzyme (ACE) inhibitory peptides derived from enzymatic hydrolysate of ovotransferrin. *Food Chemistry*, *126*(4), 1614–1619.

Mann, J. K., & Ndung'u, T. (2020). The potential of lactoferrin, ovotransferrin and lysozyme as antiviral and immune-modulating agents in COVID-19. *Future Virology*, *15*(9), 609–624.

Martins, A. M., Pham, Q. P., Malafaya, P. B., Raphael, R. M., Kasper, F. K., Reis, R. L., & Mikos, A. G. (2009). Natural stimulus responsive scaffolds/cells for bone tissue engineering: Influence of lysozyme upon scaffold degradation and osteogenic differentiation of cultured marrow stromal cells induced by CaP coatings. *Tissue Engineering Part A*, *15*(8), 1953–1963.

Maruo, K., Akaike, T., Ono, T., & Maeda, H. (1998). Involvement of bradykinin generation in intravascular dissemination of Vibrio vulnificus and prevention of invasion by a bradykinin antagonist. *Infection and Immunity*, *66*(2), 866–869.

Masschalck, B., Van Houdt, R., Van Haver, E. G., & Michiels, C. W. (2001). Inactivation of gram-negative bacteria by lysozyme, denatured lysozyme, and lysozyme-derived peptides under high hydrostatic pressure. *Applied and Environmental Microbiology*, *67*(1), 339–344.

Matoba, N., Usui, H., Fujita, H., & Yoshikawa, M. (1999). A novel anti-hypertensive peptide derived from ovalbumin induces nitric oxide-mediated vasorelaxation in an isolated SHR mesenteric artery. *FEBS Letters*, *452*(3), 181–184.

Matsuoka, R., Kimura, M., Muto, A., Masuda, Y., Sato, M., & Imaizumi, K. (2008). Mechanism for the cholesterol-lowering action of egg white protein in rats. *Bioscience, Biotechnology, and Biochemistry*, *72*(6), 1506–1512.

Matsuoka, R., Shirouchi, B., Umegatani, M., Fukuda, M., Muto, A., Masuda, Y., & Sato, M. (2017). Dietary egg-white protein increases body protein mass and reduces body fat mass through an acceleration of hepatic β-oxidation in rats. *British Journal of Nutrition*, *118*(6), 423–430.

Matsushima, K. (1958). An undescribed trypsin inhibitor in egg white. *Science*, *127*(3307), 1178–1179.

May, K., Wells, J., Maxwell, C., & Oliver, W. (2012). Granulated lysozyme as an alternative to antibiotics improves growth performance and small intestinal morphology of 10-day-old pigs. *Journal of Animal Science*, *90*(4), 1118–1125.

McReynolds, L., O'Malley, B. W., Nisbet, A. D., Fothergill, J. E., Givol, D., Fields, S., & Brownlee, G. G. (1978). Sequence of chicken ovalbumin mRNA. *Nature*, *273*(5665), 723–728.

Memarpoor-Yazdi, M., Asoodeh, A., & Chamani, J. (2012a). A novel antioxidant and antimicrobial peptide from hen egg white lysozyme hydrolysates. *Journal of Functional Foods*, *4*(1), 278–286.

Memarpoor-Yazdi, M., Asoodeh, A., & Chamani, J. (2012b). Structure and ACE-inhibitory activity of peptides derived from hen egg white lysozyme. *International Journal of Peptide Research and Therapeutics*, *18*(4), 353–360.

Miguel, M., Aleixandre, M., Ramos, M., & López-Fandiño, R. (2006). Effect of simulated gastrointestinal digestion on the antihypertensive properties of ACE-inhibitory peptides derived from ovalbumin. *Journal of Agricultural and Food Chemistry*, *54*(3), 726–731.

Miguel, M., López-Fandino, R., Ramos, M., & Aleixandre, A. (2005). Short-term effect of egg-white hydrolysate products on the arterial blood pressure of hypertensive rats. *British Journal of Nutrition*, *94*(5), 731–737.

Miguel, M., Recio, I., Gomez-Ruiz, J., Ramos, M., & López-Fandiño, R. (2004). Angiotensin I–converting enzyme inhibitory activity of peptides derived from egg white proteins by enzymatic hydrolysis. *Journal of Food Protection*, *67*(9), 1914–1920.

Miller, H. T., & Feeney, R. E. (1966). The physical and chemical properties of an immunologically cross-reacting protein from avian egg whites. *Biochemistry*, *5*(3), 952–958.

Mine, Y. (1995). Recent advances in the understanding of egg white protein functionality. *Trends in Food Science & Technology*, *6*(7), 225–232.

Mine, Y., Ma, F., & Lauriau, S. (2004). Antimicrobial peptides released by enzymatic hydrolysis of hen egg white lysozyme. *Journal of Agricultural and Food Chemistry*, *52*(5), 1088–1094.

Miyagawa, S., Kamata, R., Matsumoto, K., Okamura, R., & Maeda, H. (1991). Inhibitory effects of ovomacroglobulin on bacterial keratitis in rabbits. *Graefe's Archive for Clinical and Experimental Ophthalmology*, *229*(3), 281–286.

Miyagawa, S., Kamata, R., Matsumoto, K., Okamura, R., & Maeda, H. (1994). Therapeutic intervention with chicken egg white ovomacroglobulin and a new quinolone on experimental *Pseudomonas keratitis*. *Graefe's Archive for Clinical and Experimental Ophthalmology*, *232*(8), 488–493.

Miyagawa, S., Nishino, N., Kamata, R., Okamura, R., & Maeda, H. (1991). Effects of protease inhibitors on growth of *Serratia* marcescens and *Pseudomonas aeruginosa*. *Microbial Pathogenesis*, *11*(2), 137–141.

Moghadam, A. K., Schmidt, C., & Grier, K. (2013). The impact of *E. Coli* O157: H7 recalls on live cattle futures prices: Revisited. *Food Policy*, *42*, 81–87.

Molla, A., Matsumoto, K., Oyamada, I., Katsuki, T., & Maeda, H. (1986). Degradation of protease inhibitors, immunoglobulins, and other serum proteins by *Serratia* protease and its toxicity to fibroblast in culture. *Infection and Immunity*, *53*(3), 522–529.

Molla, A., Matsumura, Y., Yamamoto, T., Okamura, R., & Maeda, H. (1987). Pathogenic capacity of proteases from *Serratia* marcescens and *Pseudomonas* aeruginosa and their suppression by chicken egg white ovomacroglobulin. *Infection and Immunity*, *55*(10), 2509–2517.

Molla, A., Oda, T., & Maeda, H. (1987). Different binding kinetics of *Serratia* 56 K protease with plasma α2-macroglobulin and chicken egg white ovomacroglobulin. *The Journal of Biochemistry*, *101*(1), 199–205.

Moon, S. H., Lee, J. H., Lee, Y. J., Paik, J. Y., Ahn, D. U., & Paik, H. D. (2012). Antioxidant, antimicrobial, and cytotoxic activities of ovotransferrin from egg white. *Food Science of Animal Resources*, *32*(5), 612–617.

Moon, S., Paik, H. D., White, S., Daraba, A., Mendonca, A. F., & Ahn, D. U. (2011). Influence of nisin and selected meat additives on the antimicrobial effect of ovotransferrin against *Listeria* monocytogenes. *Poultry Science*, *90*(11), 2584–2591.

Muehlenweg, B., Assfalg-Machleidt, I., Parrado, S. G., Bürgle, M., Creutzburg, S., Schmitt, M., & Magdolen, V. (2000). A novel type of bifunctional inhibitor directed against proteolytic activity and receptor/ligand interaction: Cystatin with a urokinase receptor binding site. *Journal of Biological Chemistry*, *275*(43), 33562–33566.

Murakami, F., Sasaki, T., & Sugahara, T. (1997). Lysozyme stimulates immunoglobulin production by human-human hybridoma and human peripheral blood lymphocytes. *Cytotechnology*, *24*(2), 177–182.

Nagaoka, S., Masaoka, M., Zhang, Q., Hasegawa, M., & Watanabe, K. (2002). Egg ovomucin attenuates hypercholesterolemia in rats and inhibits cholesterol absorption in Caco-2 cells. *Lipids*, *37*(3), 267–272.

Nagase, H., & Brew, K. (1987). Amino acid sequence of a 32-residue region around the thiol ester site in duck ovostatin. *FEBS Letters*, *222*(1), 83–88.

Nagase, H., Harris, E. Jr, & Brew, K. (1986). Evidence for a thiol ester in duck ovostatin (ovomacroglobulin). *Journal of Biological Chemistry*, *261*(3), 1421–1426.

Nagase, H., Harris, E. Jr, Woessner, J. Jr, & Brew, K. (1983). Ovostatin: A novel proteinase inhibitor from chicken egg white. I. Purification, physicochemical properties, and tissue distribution of ovostatin. *Journal of Biological Chemistry*, *258*(12), 7481–7489.

Nagase, H., & Harris, E. (1983). Ovostatin: A novel proteinase inhibitor from chicken egg white. II. Mechanism of inhibition studied with collagenase and thermolysin. *Journal of Biological Chemistry*, *258*(12), 7490–7498.

Nicklin, M. J., & Barrett, A. J. (1984). Inhibition of cysteine proteinases and dipeptidyl peptidase I by egg-white cystatin. *Biochemical Journal*, *223*(1), 245–253.

Nisbet, A. D., SAUNDRY, R. H., Moir, A. J., Fothergill, L. A., & Fothergill, J. E. (1981). The complete amino-acid sequence of hen ovalbumin. *European Journal of Biochemistry*, *115*(2), 335–345.

Nyachoti, C., Kiarie, E., Bhandari, S., Zhang, G., & Krause, D. (2012). Weaned pig responses to Escherichia coli K88 oral challenge when receiving a lysozyme supplement. *Journal of Animal Science*, *90*(1), 252–260.

Nystrom, L., Al-Rammahi, N., Malekkhaiat Häffner, S., Stromstedt, A. A., Browning, K. L., & Malmsten, M. (2018). Avidin–biotin cross-linked microgel multilayers as carriers for antimicrobial peptides. *Biomacromolecules*, *19*(12), 4691–4702.

Oderinde, B., Agbede, O., Iheukwumere, I., Ghamba, P., Medugu, J., & Oku, E. (2017). Antiviral activity of hen egg-white lysozyme on polio virus. *Sokoto Journal of Medical Laboratory Science*, *2*(3), 109–120.

Ofuji, Y., Suzuki, T., Yoshie, H., Hara, K., & Adachi, M. (1992). The effects of ovomacroglobulin on gingival wound healing in rats. *Periodontal Clinical Investigations: Official Publication of the Northeastern Society of Periodontists*, *14*(2), 13–22.

Oguro, T., Watanabe, K., Tani, H., Ohishi, H., & Ebina, T. (2000). Morphological observations on antitumor activities of 70 kDa fragment in α-subunit from pronase-treated ovomucin in a double grafted tumor system. *Food Science and Technology Research*, *6*(3), 179–185.

Ohami, H. (1993). Cytotoxic effect of sialoglycoprotein derived from avian egg white ovomucin on the cultured tumor cell. *Medicine & Biology*, *126*, 19–23.

Ohno, N., & Morrison, D. C. (1989a). Effects of lipopolysaccharide chemotype structure on binding and inactivation of hen egg lysozyme. *European Journal of Biochemistry*, *186*(3), 621–627.

Ohno, N., & Morrison, D. C. (1989b). Lipopolysaccharide interaction with lysozyme: Binding of lipopolysaccharide to lysozyme and inhibition of lysozyme enzymatic activity. *Journal of Biological Chemistry*, *264*(8), 4434–4441.

Ohno, N., & Morrison, D. C. (1989c). Lipopolysaccharide interactions with lysozyme differentially affect lipopolysaccharide immunostimulatory activity. *European Journal of Biochemistry*, *186*(3), 629–636.

Oliver, W., & Wells, J. (2013). Lysozyme as an alternative to antibiotics improves growth performance and small intestinal morphology in nursery pigs. *Journal of Animal Science*, *91*(7), 3129–3136.

Oliver, W., & Wells, J. (2015). Lysozyme as an alternative to growth promoting antibiotics in swine production. *Journal of Animal Science and Biotechnology*, *6*(1), 1–7.

Oliver, W., Wells, J., & Maxwell, C. (2014). Lysozyme as an alternative to antibiotics improves performance in nursery pigs during an indirect immune challenge. *Journal of Animal Science*, *92*(11), 4927–4934.

Omana, D. A., Wang, J., & Wu, J. (2010). Ovomucin – A glycoprotein with promising potential. *Trends in Food Science & Technology*, *21*(9), 455–463.

Otani, H., & Maenishi, K. (1994). Effects of hen egg proteins on proliferative responses of mouse spleen lymphocytes. *LWT-Food Science and Technology*, *27*(1), 42–47.

Pacor, S., Gagliardi, R., Di Daniel, E., Vadori, M., & Sava, G. (1999). In vitro down regulation of ICAM-1 and E-cadherin and in vivo reduction of lung metastases of TS/A adenocarcinoma by a lysozyme derivative. *International Journal of Molecular Medicine*, *4*(4), 369–444.

Pastori, V., Tavazzi, S., & Lecchi, M. (2015). Lactoferrin-loaded contact lenses: Eye protection against oxidative stress. *Cornea*, *34*(6), 693–697.

Patryk, K., Maciej, K., Malgorzata, K., & Wieslaw, K. (2020). Egg white Cystatin–A review. *Acta Veterinaria*, *70*(4), 401–422.

Paulson, J. C. (1989). Glycoproteins: What are the sugar chains for? *Trends in Biochemical Sciences*, *14*(7), 272–276.

Pellegrini, A., Thomas, U., Bramaz, N., Klauser, S., Hunziker, P., & Von Fellenberg, R. (1997). Identification and isolation of a bactericidal domain in chicken egg white lysozyme. *Journal of Applied Microbiology*, *82*(3), 372–378.

Pellegrini, A., Thomas, U., Wild, P., Schraner, E., & Von Fellenberg, R. (2000). Effect of lysozyme or modified lysozyme fragments on DNA and RNA synthesis and membrane permeability of Escherichia coli. *Microbiological Research*, *155*(2), 69–77.

Phelps, C. F., & Antonini, E. (1975). A study of the kinetics of iron and copper binding to hen ovotransferrin. *Biochemical Journal*, *147*(3), 385–391.

Phillips, D. C. (1967). The hen egg-white lysozyme molecule. *Proceeding of the National Academy Sciences*, *57*(3), 483–495.

Plate, N., Valuev, I., Sytov, G., & Valuev, L. (2002). Mucoadhesive polymers with immobilized proteinase inhibitors for oral administration of protein drugs. *Biomaterials*, *23*(7), 1673–1677.

Premzl, A., Puizdar, V., Zavaénik-Bergant, V., Kopitar-Jerala, N., Lah, T. T., Katunuma, N., & Kos, J. (2001). Invasion of ras-transformed breast epithelial cells depends on the proteolytic activity of cysteine and aspartic proteinases. *Biological Chemistry*, *382*(5), 853–858.

Proctor, V. A., Cunningham, F., & Fung, D. Y. (1988). The chemistry of lysozyme and its use as a food preservative and a pharmaceutical. *Critical Reviews in Food Science & Nutrition*, *26*(4), 359–395.

Pugliese, L., Malcovati, M., Coda, A., & Bolognesi, M. (1994). Crystal structure of apo-avidin from hen egg-white. *Journal of Molecular Biology*, *235*(1), 42–46.

Pukcothanung, Y., Booncharoen, S., & Thammasirirak, S. (2007). Antimicrobial peptides derived from lysozymes. *Asia-Pacific Journal of Science and Technology*, *12*(3), 265–276.

Quirós, A., Chichón, R., Recio, I., & López-Fandiño, R. (2007). The use of high hydrostatic pressure to promote the proteolysis and release of bioactive peptides from ovalbumin. *Food Chemistry*, *104*(4), 1734–1739.

Raghuveer, T. S., McGuire, E. M., Martin, S. M., Wagner, B. A., Rebouché, C. J., Buettner, G. R., & Widness, J. A. (2002). Lactoferrin in the preterm infants' diet attenuates iron-induced oxidation products. *Pediatric Research*, *52*(6), 964–972.

Rao, S., Sun, J., Liu, Y., Zeng, H., Su, Y., & Yang, Y. (2012). ACE inhibitory peptides and antioxidant peptides derived from in vitro digestion hydrolysate of hen egg white lysozyme. *Food Chemistry*, *135*(3), 1245–1252.

Rawendra, R. D., Chang, C. I., Chen, H. H., Huang, T. C., & Hsu, J. L. (2013). A novel angiotensin converting enzyme inhibitory peptide derived from proteolytic digest of Chinese soft-shelled turtle egg white proteins. *Journal of Proteomics*, *94*, 359–369.

Rhodes, M. B., Bennett, N., & Feeney, R. E. (1960). The trypsin and chymotrypsin inhibitors from avian egg whites. *Journal of Biological Chemistry*, *235*(6), 1686–1693.

Robinson, D., & Monsey, J. (1971). Studies on the composition of egg-white ovomucin. *Biochemical Journal*, *121*(3), 537–547.

Ru, Z., Xu, M., Zhu, G., Tu, Y., Jiang, Y., & Du, H. (2021). Ovotransferrin exerts bidirectional immunomodulatory activities via TLR4-mediated signal transduction pathways in RAW264. 7 cells. *Food Science & Nutrition*, *9*(11), 6162–6175.

Ruben, G., Harris, E., & Nagase, H. (1988). Electron microscopic studies of free and proteinase-bound duck ovostatins (ovomacroglobulins). Model of ovostatin structure and its transformation upon proteolysis. *Journal of Biological Chemistry*, *263*(6), 2861–2869.

Ruiz-Ortega, M., Lorenzo, O., Ruperez, M., Esteban, V., Suzuki, Y., Mezzano, S., & Egido, J. (2001). Role of the renin-angiotensin system in vascular diseases: Expanding the field. *Hypertension*, *38*(6), 1382–1387.

Ryan, C., & Clary, J. J. (1964). Some reactions of chicken chymotrypsin and turkey trypsin with substrates and naturally occurring protease inhibitors. *Archives of Biochemistry and Biophysics*, *108*, 169–171.

Ryan, C., Clary, J. J., & Tomimatsu, Y. (1965). Chicken chymotrypsin and turkey trypsin: Part II: Physical and enzymic properties. *Archives of Biochemistry and Biophysics*, *110*(1), 175–183.

Safi, S. (2012). Acute phase proteins-analysis, clinical applications and potentials. *Inflammatory Diseases-Immunopathology, Clinical and Pharmacological Bases*, 351–381.

Saleh, Y., Siewiński, M., Kielan, W., Ziółkowski, P., Gryboś, M., & Rybka, J. (2003). Regulation of cathepsin B and L expression in vitro in gastric cancer tissues by egg cystatin. *Journal of Experimental Therapeutics and Oncology*, *3*(6), 319–324.

Salim, M. A. S. M., & Gan, C.-Y. (2020). Dual-function peptides derived from egg white ovalbumin: Bioinformatics identification with validation using in vitro assay. *Journal of Functional Foods*, *64*, 103618.

Salton, M. (1957). The properties of lysozyme and its action on microorganisms. *Bacteriological Reviews*, *21*(2), 82–100.

Satué-Gracia, M. T., Frankel, E. N., Rangavajhyala, N., & German, J. B. (2000). Lactoferrin in infant formulas: Effect on oxidation. *Journal of Agricultural and Food Chemistry*, *48*(10), 4984–4990.

Sava, G. (1989). Reduction of B16 melanoma metastases by oral administration of egg-white lysozyme. *Cancer Chemotherapy and Pharmacology*, *25*(3), 221–222.

Sava, G., Benetti, A., Ceschia, V., & Pacor, S. (1989). Lysozyme and cancer: Role of exogenous lysozyme as anticancer agent. *Anticancer Research*, *9*(3), 583–591.

Sava, G., Ceschia, V., Pacor, S., & Zabucchi, G. (1991). Observations on the antimetastatic action of lysozyme in mice bearing Lewis lung carcinoma. *Anticancer Research*, *11*(3), 1109–1113.

Sava, G., Ceschia, V., & Zabucchi, G. (1988). Evidence for host-mediated antitumor effects of lysozyme in mice bearing the MCa mammary carcinoma. *European Journal of Cancer and Clinical Oncology*, *24*(11), 1737–1743.

Sava, G., Pacor, S., Dasić, G., & Bergamo, A. (1995). Lysozyme stimulates lymphocyte response to ConA and IL-2 and potentiates 5-fluorouracil action on advanced carcinomas. *Anticancer Research*, *15*(5B), 1883–1888.

Sava, G., Perissin, L., & Zorzet, S. (1988). Antimetastatic action of orally administered lysozyme in mice bearing Lewis lung carcinoma. *Clinical & Experimental Metastasis*, *6*(3), 245–253.

Sava, G., Perissin, L., Zorzet, S., & Callerio, C. (1986). Antineoplastic effects of egg-white lysozyme in mice bearing solid metastasizing tumors. *Anticancer Research*, *6*(2), 183–186.

Schade, A. L., & Caroline, L. (1944). Raw hen egg white and the role of iron in growth inhibition of *Shigella* dysenteriae, *Staphylococcus aureus*, *Escherichia coli* and *Saccharomyces cerevisiae*. *Science*, *100*(2584), 14–15.

Scruggs, P., Filipeanu, C. M., Yang, J., Chang, J.K., Dun, N.J. (2004). Interaction of ovokinin (2–7) with vascular bradykinin 2 receptors. *Regulatory peptides*, *120*(1–3), 85–91.

Seol, K. H., Lim, D. G., Jang, A., Jo, C., & Lee, M. (2009). Antimicrobial effect of κ-carrageenan-based edible film containing ovotransferrin in fresh chicken breast stored at 5 C. *Meat Science*, *83*(3), 479–483.

Shah, R. B., & Khan, M. A. (2004). Protection of salmon calcitonin breakdown with serine proteases by various ovomucoid species for oral drug delivery. *Journal of Pharmaceutical Sciences*, *93*(2), 392–406.

Shan, Y., Xu, Q., & Ma, M. (2014). Mg2+ binding affects the structure and activity of ovomucin. *International Journal of Biological Macromolecules*, *70*, 230–235.

Shang, N., Bhullar, K. S., & Wu, J. (2020). Ovotransferrin exhibits osteogenic activity partially via low-density lipoprotein receptor-related protein 1 (LRP1) activation in MC3T3-El cells. *Journal of Agricultural and Food Chemistry*, *68*(35), 9427–9435.

Shang, N., & Wu, J. (2018). Egg white ovotransferrin shows osteogenic activity in osteoblast cells. *Journal of Agricultural and Food Chemistry*, *66*(11), 2775–2782.

Shang, N., & Wu, J. (2019). Egg white ovotransferrin attenuates RANKL-induced osteoclastogenesis and bone resorption. *Nutrients*, *11*(9), 2254.

Shen, S., Chahal, B., Majumder, K., You, S. J., & Wu, J. (2010). Identification of novel antioxidative peptides derived from a thermolytic hydrolysate of ovotransferrin by LC-MS/MS. *Journal of Agricultural and Food Chemistry*, *58*(13), 7664–7672.

Shirouchi, B., & Matsuoka, R. (2019). Alleviation of metabolic syndrome with dietary egg white protein. *Journal of Oleo Science*, *68*(6), 517–524.

Siciliano, R., Rega, B., Marchetti, M., Seganti, L., Antonini, G., & Valenti, P. (1999). Bovine lactoferrin peptidic fragments involved in inhibition of herpes simplex virus type 1 infection. *Biochemical and Biophysical Research Communications*, *264*(1), 19–23.

Silva, M. B. R., Falcão, H. G., Kurozawa, L. E., Prudencio, S. H., de Camargo, A. C., Shahidi, F., & Ida, E. I. (2019). Ultrasound-and hemicellulase-assisted extraction increase β-glucosidase activity, the content of isoflavone aglycones and antioxidant potential of soymilk. *Journal of Food Bioactives*, *6*, 140–147.

Stevens, L. (1991). Egg white proteins. *Comparative Biochemistry and Physiology Part B: Comparative Biochemistry*, *100*(1), 1–9.

Strecker, G., Wieruszeski, J., Cuvillier, O., Michalski, J., & Montreuil, J. (1992). 1H and 13C-NMR assignments for sialylated oligosaccharide-alditols related to mucins. Study of thirteen components from hen ovomucin and swallow nest mucin. *Biochimie*, *74*(1), 39–51.

Strecker, G., Wieruszeski, J. M., Martel, C., & Montreuil, J. (1989). Complete 1H-and 13C-NMR assignments for two sulphated oligosaccharide alditols of hen ovomucin. *Carbohydrate Research*, *185*(1), 1–13.

Strecker, G., Wieruszeski, J.-M., Martel, C., & Montreuil, J. (1987). Determination of the structure of sulfated tetra- and pentasaccharides obtained by alkaline borohydride degradation of hen ovomucin. A fast atom bombardment-mass spectrometric and1H-NMR spectroscopic study. *Glycoconjugate Journal*, *4*(4), 329–337.

Sugahara, T., Murakami, F., Yamada, Y., & Sasaki, T. (2000). The mode of actions of lysozyme as an immunoglobulin production stimulating factor. *Biochimica et Biophysica Acta (BBA)-General Subjects*, *1475*(1), 27–34.

Sugahara, T., Yamada, Y., Yano, S., & Sasaki, T. (2002). Heat denaturation enhanced immunoglobulin production stimulating activity of lysozyme from hen egg white. *Biochimica et Biophysica Acta (BBA)-General Subjects*, *1572*(1), 19–24.

Sugimoto, Y., Kusakabe, T., Nagaoka, S., Nirasawa, T., Tatsuguchi, K., Fujii, M., & Koga, K. (1996). A proteinase inhibitor from egg yolk of hen is an ovoinhibitor analog. *Biochimica et Biophysica Acta (BBA)-Protein Structure and Molecular Enzymology*, *1295*(1), 96–102.

Sun, X., Chakrabarti, S., Fang, J., Yin, Y., & Wu, J. (2016). Low-molecular-weight fractions of Alcalase hydrolyzed egg ovomucin extract exert anti-inflammatory activity in human dermal fibroblasts through the inhibition of tumor necrosis factor–mediated nuclear factor κB pathway. *Nutrition Research*, *36*(7), 648–657.

Sun, X., Gänzle, M., Field, C. J., & Wu, J. (2016). Effect of proteolysis on the sialic acid content and bifidogenic activity of ovomucin hydrolysates. *Food Chemistry*, *212*, 78–86.

Sun, X., Gänzle, M., & Wu, J. (2017). Identification and characterization of glycopeptides from egg protein ovomucin with anti-agglutinating activity against porcine K88 enterotoxigenic Escherichia coli strains. *Journal of Agricultural and Food Chemistry*, *65*(4), 777–783.

Sun, X., Gänzle, M., & Wu, J. (2019). Glycopeptides from egg white ovomucin inhibit K88ac enterotoxigenic Escherichia coli adhesion to porcine small intestinal epithelial cell-line. *Journal of Functional Foods*, *54*, 320–328.

Sun, Y., Hayakawa, S., Chuamanochan, M., Fujimoto, M., Innun, A., & Izumori, K. (2006). Antioxidant effects of Maillard reaction products obtained from ovalbumin and different D-aldohexoses. *Bioscience, Biotechnology, and Biochemistry*, *70*(3), 598–605.

Tagashira, A., Nishi, K., Matsumoto, S., & Sugahara, T. (2018). Anti-inflammatory effect of lysozyme from hen egg white on mouse peritoneal macrophages. *Cytotechnology*, *70*(3), 929–938.

Tagashira, A., Nishi, K., & Sugahara, T. (2019). Lysozyme from hen egg white ameliorates lipopolysaccharide-induced systemic inflammation in mice. *Cytotechnology*, *71*(2), 497–506.

Takada, H., Katoh, T., & Katayama, T. (2020). Sialylated O-glycans from hen egg white ovomucin are decomposed by mucin-degrading gut microbes. *Journal of Applied Glycoscience*, *67*(2), 31–39.

Takada, K., Ohno, N., & Yadomae, T. (1994a). Binding of lysozyme to lipopolysaccharide suppresses tumor necrosis factor production in vivo. *Infection and Immunity*, *62*(4), 1171–1175.

Takada, K., Ohno, N., & Yadomae, T. (1994b). Detoxification of lipopolysaccharide (LPS) by egg white lysozyme. *FEMS Immunology & Medical Microbiology*, *9*(4), 255–263.

Takada, K., Ohno, N., & Yadomae, T. (1994c). Lysozyme regulates LPS-induced interleukin-6 release in mice. *Circulatory Shock*, *44*(4), 169–174.

Takahashi, N., Tatsumi, E., Orita, T., & Hirose, M. (1996). Role of the intrachain disulfide bond of ovalbumin during conversion into S-ovalbumin. *Bioscience, Biotechnology, and Biochemistry*, *60*(9), 1464–1468.

Tanida, N., Ohno, N., Adachi, Y., Matsuura, M., Nakano, M., Kiso, M., & Yadomae, T. (1992). Modification of immunopharmacological activities of synthetic monosaccharide lipid A analogue, GLA60, by lysozyme. *The Journal of Biochemistry*, *112*(5), 616–623.

Tanizaki, H., Tanaka, H., Iwata, H., & Kato, A. (1997). Activation of macrophages by sulfated glycopeptides in ovomucin, yolk membrane, and chalazae in chicken eggs. *Bioscience, Biotechnology, and Biochemistry*, *61*(11), 1883–1889.

Tenovuo, J. (2002). Clinical applications of antimicrobial host proteins lactoperoxidase, lysozyme and lactoferrin in xerostomia: Efficacy and safety. *Oral Diseases*, *8*(1), 23–29.

Thammasirirak, S., Pukcothanung, Y., Preecharram, S., Daduang, S., Patramanon, R., Fukamizo, T., & Araki, T. (2010). Antimicrobial peptides derived from goose egg white lysozyme. *Comparative Biochemistry and Physiology Part C: Toxicology & Pharmacology*, *151*(1), 84–91.

Thompson, A. R. (1952). The c-terminal residue of lysozyme. *Nature*, *169*(4299), 495–496.

Tomimatsu, Y., Clary, J. J., & Bartulovich, J. J. (1966). Physical characterization of ovoinhibitor, a trypsin and chymotrypsin inhibitor from chicken egg white. *Archives of Biochemistry and Biophysics*, *115*(3), 536–544.

Touch, V., Hayakawa, S., Fukada, K., Aratani, Y., & Sun, Y. (2003). Preparation of antimicrobial reduced lysozyme compatible in food applications. *Journal of Agricultural and Food Chemistry*, *51*(17), 5154–5161.

Tsuge, Y., Shimoyamada, M., & Watanabe, K. (1996a). Binding of egg white proteins to viruses. *Bioscience, Biotechnology, and Biochemistry*, *60*(9), 1503–1504.

Tsuge, Y., Shimoyamada, M., & Watanabe, K. (1996b). Differences in hemagglutination inhibition activity against bovine rotavirus and hen Newcastle disease virus based on the subunits in hen egg white ovomucin. *Bioscience, Biotechnology, and Biochemistry*, *60*(9), 1505–1506.

Tsuge, Y., Shimoyamada, M., & Watanabe, K. (1997a). Bindings of ovomucin to Newcastle disease virus and anti-ovomucin antibodies and its heat stability based on binding abilities. *Journal of Agricultural and Food Chemistry*, *45*(12), 4629–4634.

Tsuge, Y., Shimoyamada, M., & Watanabe, K. (1997b). Structural features of Newcastle disease virus-and anti-ovomucin antibody-binding glycopeptides from pronase-treated ovomucin. *Journal of Agricultural and Food Chemistry*, *45*(7), 2393–2398.

Tu, Z., Hu, Y., Wang, H., Huang, X., Xia, S., & Niu, P. (2015). Microwave heating enhances antioxidant and emulsifying activities of ovalbumin glycated with glucose in solid-state. *Journal of Food Science and Technology*, *52*(3), 1453–1461.

Tu, A., Wang, X. C., Chen, H., Jia, X., Wang, T., Yi, Y., & Shan, Y. (2021). Ovomucin ameliorates intestinal barrier and intestinal bacteria to attenuate DSS-induced colitis in mice. *Journal of Agricultural and Food Chemistry*, *69*(21), 5887–5896.

Valenti, P., De Stasio, A., Mastromerino, P., Seganti, L., Sinibaldi, L., & Orsi, N. (1981). Influence of bicarbonate and citrate on the bacteriostatic action of ovotransferrin towards staphylococci. *FEMS Microbiology Letters*, *10*(1), 77–79.

Valenti, P., Visca, P., Antonini, G., & Orsi, N. (1986). Interaction between lactoferrin and ovotransferrin and *Candida* cells. *FEMS Microbiology Letters*, *33*(2-3), 271–275.

Valenti, P., Visca, P., Antonini, G., Orsi, N., & Antonini, E. (1987). The effect of saturation with Zn2+ and other metal ions on the antibacterial activity of ovotransferrin. *Medical Microbiology and Immunology*, *176*(3), 123–130.

Valueva, T., Matveev, N., Mosolov, V., & Penin, V. (1992). Protein proteinase inhibitor therapy in experimental pancreatitis: Pharmacological characterization of the inhibitor. *Agents and Actions. Supplements*, *38*, 203–210.

Varon, O., Allen, K. J., Bennett, D. C., Mesak, L. R., & Scaman, C. H. (2013). Purification and characterization of tinamou egg white ovotransferrin as an antimicrobial agent against foodborne pathogenic bacteria. *Food Research International*, *54*(2), 1836–1842.

Venuste, M., Zhang, X., Shoemaker, C. F., Karangwa, E., Abbas, S., & Kamdem, P. E. (2013). Influence of enzymatic hydrolysis and enzyme type on the nutritional and antioxidant properties of pumpkin meal hydrolysates. *Food & Function*, *4*(5), 811–820.

Verdot, L., Lalmanach, G., Vercruysse, V., Hartmann, S., Lucius, R., Hoebeke, J., & Vray, B. (1996). Cystatins up-regulate nitric oxide release from interferon-γ-activated mouse peritoneal macrophages. *Journal of Biological Chemistry*, *271*(45), 28077–28081.

Verdot, L., Lalmanach, G., Vercruysse, V., Hoebeke, J., Gauthier, F., & Vray, B. (1999). Chicken cystatin stimulates nitric oxide release from interferon-g-activated mouse peritoneal macrophages via cytokine synthesis. *European Journal of Biochemistry*, *266*, 1111–1117.

Vilcacundo, R., Méndez, P., Reyes, W., Romero, H., Pinto, A., & Carrillo, W. (2018). Antibacterial activity of hen egg white lysozyme denatured by thermal and chemical treatments. *Scientia Pharmaceutica*, *86*(4), 48.

Wang, H. Y., Qian, H., & Yao, W. R. (2011). Melanoidins produced by the Maillard reaction: Structure and biological activity. *Food Chemistry*, *128*(3), 573–584.

Wang, Z., Tu, A., Tang, D., & Shan, Y. (2018). Effectively preparing soluble ovomucin with high antiviral activity from egg white. *International Journal of Biological Macromolecules*, *118*, 504–510.

Wang, X., Wei, Z., & Xue, C. (2021). The past and future of ovotransferrin: Physicochemical properties, assembly and applications. *Trends in Food Science & Technology*, *116*, 47–62.

Wasik, B. R., Barnard, K. N., & Parrish, C. R. (2016). Effects of sialic acid modifications on virus binding and infection. *Trends in Microbiology*, *24*(12), 991–1001.

Watanabe, K., Tsuge, Y., Shimoyamada, M., Ogama, N., & Ebina, T. (1998). Antitumor effects of pronase-treated fragments, glycopeptides, from ovomucin in hen egg white in a double grafted tumor system. *Journal of Agricultural and Food Chemistry*, *46*(8), 3033–3038.

Wei, Z., Wu, S., Xia, J., Shao, P., Sun, P., & Xiang, N. (2021). Enhanced antibacterial activity of hen egg-white lysozyme against *Staphylococcus aureus* and *Escherichia coli* due to protein fibrillation. *Biomacromolecules*, *22*(2), 890–897.

Wells, J., Berry, E., Kalchayanand, N., Rempel, L., Kim, M., & Oliver, W. (2015). Effect of lysozyme or antibiotics on faecal zoonotic pathogens in nursery pigs. *Journal of Applied Microbiology*, *118*(6), 1489–1497.

Wesierska, E., Saleh, Y., Trziszka, T., Kopec, W., Siewinski, M., & Korzekwa, K. (2005). Antimicrobial activity of chicken egg white cystatin. *World Journal of Microbiology and Biotechnology*, *21*(1), 59–64.

Wilchek, M., & Bayer, E. A. (1984). The avidin-biotin complex in immunology. *Immunology Today*, *5*(2), 39–43.

Williams, J., Elleman, T. C., Kingston, I. B., Wilkins, A. G., & Kuhn, K. A. (1982). The primary structure of hen ovotransferrin. *European Journal of Biochemistry*, *122*(2), 297–303.

Woolley, D., & Longsworth, L. (1942). Isolation of an antibiotin factor from egg white. *Journal of Biological Chemistry*, *142*(1), 285–290.

Wu, J. (2019). *Eggs as functional foods and nutraceuticals for human health.* Royal Society of Chemistry.

Wu, J., & Acero-Lopez, A. (2012). Ovotransferrin: Structure, bioactivities, and preparation. *Food Research International*, *46*(2), 480–487.

Xie, H., Huff, G., Huff, W., Balog, J., Holt, P., & Rath, N. (2002). Identification of ovotransferrin as an acute phase protein in chickens. *Poultry Science*, *81*(1), 112–120.

Xu, M., Shangguan, X., Wang, W., & Chen, J. (2007). Antioxidative activity of hen egg ovalbumin hydrolysates. *Asia Pac. J. Clin. Nutr*, *16*, 178–182.

Xu, Q., Shan, Y., Wang, N., Liu, Y., Zhang, M., & Ma, M. (2018). Sialic acid involves in the interaction between ovomucin and hemagglutinin and influences the antiviral activity of ovomucin. *International Journal of Biological Macromolecules*, *119*, 533–539.

Yang, H. H., Li, X. C., Amft, M., & Grotemeyer, J. (1998). Protein conformational changes determined by matrix-assisted laser desorption mass spectrometry. *Analytical Biochemistry*, *258*(1), 118–126.

Yokota, T., Ohishi, H., & Watanabe, K. (1999a). Antitumor effects of β-subunit from egg white ovomucin on xenografted sarcoma-180 cells in mice. *Food Science and Technology Research*, *5*(3), 279–283.

Yokota, T., Ohishi, H., & Watanabe, K. (1999b). In vitro studies of cytotoxic effect on sarcoma-180 cells of β-subunit from egg white ovomucin. *Food Science and Technology Research*, *5*(3), 273–278.

Yokota, T., Tani, H., Ohishi, H., Oguro, T., & Watanabe, K. (2000). Interaction between 120 kDa fragment in β-subunit from egg white ovomucin and sarcoma-180 cells through basic fibroblast growth factor receptor. *Food Science and Technology Research*, *6*(4), 275–279.

You, J., Luo, Y., & Wu, J. (2014). Conjugation of ovotransferrin with catechin shows improved antioxidant activity. *Journal of Agricultural and Food Chemistry*, *62*(12), 2581–2587.

Young, A. C., Tilton, R. F., & Dewan, J. C. (1994). Thermal expansion of hen egg-white lysozyme: Comparison of the 1·9 Å Resolution structures of the tetragonal form of the enzyme at 100 K and 298 K. *Journal of Molecular Biology*, *235*(1), 302–317.

Yousif, A. N., & Kan, J. W. (2002). Visualization of chicken ovomucoid in polyacrylamide gels. *Analytical Biochemistry*, *311*(1), 93–97.

You, S. J., Udenigwe, C. C., Aluko, R. E., & Wu, J. (2010). Multifunctional peptides from egg white lysozyme. *Food Research International*, *43*(3), 848–855.

Yu, Z., Chen, Y., Zhao, W., Zheng, F., Ding, L., & Liu, J. (2019). Novel ACE inhibitory tripeptides from ovotransferrin using bioinformatics and peptidomics approaches. *Scientific Reports*, *9*(1), 1–10.

Yuko, Y., Nobuyuki, M., Hachiro, U., Kunihiko, O., & Masaaki, Y. (2002). Design of a Highly Potent Anti-hypertensive Peptide Based on Ovokinin. *Biotechnology, and Biochemistry*, *66*(7), 1213–1217.

Zhao, G., Du, L., Ma, C., Li, Y., Li, L., Poon, V. K., & Jiang, S. (2013). A safe and convenient pseudovirus-based inhibition assay to detect neutralizing antibodies and screen for viral entry inhibitors against the novel human coronavirus MERS-CoV. *Virology Journal*, *10*(1), 1–8.

Zhou, Y., Tang, Q., Du, H., Tu, Y., Wu, S., Wang, W., & Xu, M. (2018). Antiviral effect of ovotransferrin in mouse peritoneal macrophages by up-regulating type I interferon expression. *Food and Agricultural Immunology*, *29*(1), 600–614.

Zhu, G., Jiang, Y., Yao, Y., Wu, N., Luo, J., Hu, M., & Xu, M. (2019). Ovotransferrin ameliorates the dysbiosis of immunomodulatory function and intestinal microbiota induced by cyclophosphamide. *Food & Function*, *10*(2), 1109–1122.

Zhu, G., Luo, J., Du, H., Jiang, Y., Tu, Y., Yao, Y., & Xu, M. (2018). Ovotransferrin enhances intestinal immune response in cyclophosphamide-induced immunosuppressed mice. *International Journal of Biological Macromolecules*, *120*, 1–9.

Zimoch-Korzycka, A., Rouilly, A., Bobak, Ł, Jarmoluk, A., & Korzycki, M. (2015). Chitosan and cystatin/lysozyme preparation as protective edible films components. *International Journal of Polymer Science*, *2015*, Article ID 139617.

23

Egg Yolk Proteins and Peptides

Xi Feng, Xi Huang, and Dong U. Ahn*

CONTENTS

23.1 Introduction

Lipovitellins (36%, α- and β-forms), livetins (38%, α-, β-, and γ-forms), low-density lipoproteins (LDL, 17%), and phosvitin (8.0%) are the major components of the egg yolk protein (Stadelman & Cotterill, 1995). However, γ-livetin (γ-globulin or immunoglobulin Y) and phosvitin are the most important value-added egg yolk proteins, and much research has been focused on these two proteins. Chickens produce immunoglobulin Y (IgY) from the B-lymphocytes, release it into the serum, and transfer it to the yolk, where it is stored (Patterson et al., 1962; Alberts et al., 2015). Specific antibodies can be produced by immunizing chickens with antigens. It is also widely used as an antimicrobial agent to cure animal and human diseases or prevent the colonization of bacteria in animal organs (Hatta et al., 1993; Landon et al., 1995). Phosvitin has 217 amino acids, 123 of which are serine residues. Most serine residues are phosphorylated (Mecham & Olcott, 2002). As a result, phosvitin can be used as a metal-chelating agent, which can also be an excellent candidate as a natural antioxidant or antibacterial agent (Sattar Khan et al., 2000; Ishikawa et al., 2004). Phosvitin also plays an important role in embryo development: phosvitin stimulates osteoblast differentiation (Liu et al., 2017), plays a critical role in the immunity of embryos (Wang et al., 2011), and protects the developing embryo from oxidative damages (Ishikawa et al., 2004), and phosvitin phosphorous is involved in chicken embryo bone formation through dephosphorylation (Li et al., 2014). The rest of the major egg yolk proteins include lipovitellins, livetins (α- and β-form), and low-density lipoproteins, but their separation methods, functions, and utilization are largely unknown. This chapter will cover the most recent progress in separating egg yolk proteins and their functionalities. Also, their potential application in food and the non-food system will be discussed.

23.2 Separation of Functional Proteins from Egg Yolk

IgY and phosvitin are the egg yolk's two most important functional proteins. IgY is a water-soluble protein in the plasma fraction of egg yolk, while phosvitin is also a soluble protein but embedded in the granules structure of egg yolk. IgY comes from chicken serum IgG (Kovacs-Nolan & Mine, 2004) and is stored in the egg yolk, and the amount in the chicken egg is considerably higher than other resources. The amount of IgY that one laying hen can produce in one month equals that of a large mammal, such as sheep or goats. Thus, producing immunoglobulin from chicken eggs is much more efficient and cost-effective than using mammals (Fayer & Jenkins, 1992; Shimizu et al., 1994; Svendsen et al., 1995; Michael et al., 2010), and invasive blood collection from animals could be avoided, important for animal welfare. In addition, chicken antibodies cannot recognize the corresponding human substrate, preventing false-positive responses (Larsson et al., 1992).

Several methods have been applied to separate IgY; however, most of them are for laboratory scales with complicated procedures (Jensenius et al., 1981; Polson et al., 1985; Akita & Nakai, 1992; Dong et al., 2008). All the protocols' current challenges are due to the high lipid content in the egg yolk. In order to remove lipids from the egg yolk and improve extraction and separation efficiencies of IgY, xanthan (Hatta et al., 1988) or carrageenan gum (Hatta et al., 1990) and acidic water dilution (Akita & Nakai, 1993) were used. Dilution with 9 volumes of acidic water with additional precipitation agents is commonly used to precipitate most water-insoluble components from the egg yolk. The treated solution has soluble IgY, which can be further concentrated and precipitated with $(NH_4)_2SO_4$ or sodium sulfate (Akita & Nakai, 1992, 1993). Polyethylene glycol (Polson et al., 1985), pectin (Hernández-Campos et al., 2009), dextran sulfate (Schade et al., 2005), and

* Corresponding author.

DOI: 10.1201/9781003254430-30

ammonium sulfate (Chang et al., 2000) have also been used to precipitate of IgY from the egg yolk. The dilution of egg yolk with large volumes of water decreases the IgY separation efficacy, but ultrafiltration to concentrate the diluted solution can significantly improve the separation efficiency. The IgY produced using the precipitation methods is low in yield and purity, and affinity chromatography and ion-exchange chromatography were used to increase the purity of separated IgY (Ko & Ahn, 2007; Dong et al., 2008). Recently, Redwan et al. (2021) used the caprylic acid and ultrafiltration combination method to separate IgY from the 5x diluted egg yolk at pH 5.0. The method used 3 consecutive precipitations of diluted egg yolk with 2% caprylic acid, centrifugation, and then concentration with ultrafiltration. The authors claimed that the proposed protocol is simple, fast, direct, cheap, and has great potential for scale-up processing, and the produced IgY had 97% purity and 99% activity. However, the 3 consecutive precipitations and centrifugation involve too many steps and consume a large amount of caprylic acid. Another method by Lee et al. (2014) diluted the egg yolk with 2 volumes of water (3x dilution) and then centrifugated it to separate the plasma fraction. The collected plasma was further diluted with 2 volumes of water (3x dilution) and then centrifuged to collect the supernatant that contained water-soluble proteins. The solution was concentrated using ultrafiltration, and IgY was separated using the ammonium sulfate-NaCl precipitation. The purity of the separated IgY was 98.7%, and the yield was 80.9%. The method was simple and efficient and showed a high potential for scale-up production of IgY from the egg yolk.

The separation of phosvitin from the egg yolk generally involves three steps: the removal of lipids from the egg yolk or yolk granules using organic solvents, extraction of phosvitin from the lipid-free yolk granules or egg yolk with a salt solution, and then precipitation of phosvitin with $MgSO_4$ or $(NH_4)_2SO_4$ (Joubert & Cook, 1958; Losso & Nakai, 1994; Castellani et al., 2003; Ko et al., 2011). However, the use of non-food grade organic solvents in the lipid removal step limits the final application of phosvitin in humans, and the conformation or configuration of phosvitin could be changed by contact with organic solvents, which could cause the loss of its functionality. Castellani et al. (2003) developed a solvent-free method to separate phosvitin from yolk granules: they dissolved egg yolk granules in a 1.74 M NaCl solution at pH 7.25 dialyzed against distilled water and centrifuged at 10 000 *g* for 30 min. The supernatant was diluted to 0.2 M Mg^{2+} using 0.9 M $MgSO_4$ solution to precipitate phosvitin. Anion exchange and affinity chromatography are frequently applied to increase the purity of phosvitin (Connelly & Taborsky, 1961; Castellani et al., 2003; Lei & Wu, 2012; Zhang et al., 2012). Zhang et al. (2012) and Lei and Wu (2012) also developed solvent-free methods that used polyethylene glycol, heat treatment, and carbonate-bicarbonate buffer to separate phosvitin. However, those protocols were not practical for industrial application as anion exchange chromatography and dialysis were used in the purification procedure. Lee et al. (2014) developed a sequential separation method for phosvitin and IgY from the egg yolk without using solvents. They have successfully used alkaline NaCl solution (1.74 M or 10% NaCl in 0.05 N NaOH solution, pH > 10.5–11.0) to break the phosphocalcic bonds between phosvitin and the phosphate groups in the submicelles of the egg yolk granules (Shade & Chacana, 2007; Strixner et al., 2014). The purity of separated phosvitin was 97.2%, and the yield was 98.7%. The method was simple, efficient, solvent-free, and showed a high potential for scale-up production of phosvitin from the egg yolk.

Other major proteins found in egg yolk include lipovitellins (α- and β-form), livetins (α- and β-form), low-density lipoproteins, and yolkin. Low-density lipoprotein is the major component in the egg yolk, which accounts for 2/3 on the wet basis (Martin et al., 1964). LDL can be used as an emulsifier in the food system. The saturated ammonium precipitation method is used to acquire LDLs. Lipovitellins is a high-density lipoprotein composed of 75–80% proteins and 20–25% lipids (65% phospholipids, 30% triglycerides, and 5% cholesterol) (Cook & Martin, 1962). However, the separation and specific use of lipovitellin proteins from the egg yolk are hardly studied. Livetins, which include α, β, and γ-forms, represent about 9.3% of hen egg yolk proteins (Chalamaiah et al., 2017). The livetins in egg yolk were separated after 10 times dilution with acidified water to final pH 5.0 (Ahn et al., 2006). Gums, dextran, and cellulose are added to remove lipoproteins in the supernatant (Jensenius et al., 1981; Hatta et al., 1988; Yokoyama et al., 1993). Ultrafiltration cartridges were used to collect and concentrate livetins. The separation and use of γ-form of livetin (IgY) are discussed above. However, no specific separation methods and uses of α- and β-forms of livetin are reported.

Yolkin is a peptide with a molecular weight of 1–35 kDa. A sequential separation method for IgY and yolkin has been developed: nine volumes of distilled water were added to the egg yolk to precipitate lipoproteins and lipids from the water-soluble components. The unprecipitated lipoproteins in the supernatant were removed using 0.01% charcoal and pH adjustment (pH 4.0), and then IgY- and yolkin-rich fragments were precipitated using 40% saturation of $(NH_4)_2SO_4$. The precipitant was dialyzed with distilled water and 0.1M phosphate buffer at pH 7.2. IgY and yolkin were finally separated with the Sephacryl S-100 gel filtration column. The molecular weight range of the final yolkin product was 15–35 kDa (Polanowski et al., 2013). Another study used 50% ethanol to replace the previous size exclusion chromatography (Zambrowicz et al., 2017). The detailed composition of egg yolk plasma and granules is shown in Table 23.1.

23.3 Functional Characteristics and the Potential Use of Egg Yolk Proteins

IgY has received considerable attention because it can be an alternative to antibiotics for growth improvement (Cook, 2004; Cook & Trott, 2010; Huang & Ahn, 2019; Jahangiri et al., 2021). As it can neutralize antigenic molecules, it plays a vital role in body defense, which can also be used in diagnosis and immunotherapy (da Silva et al., 2021). Besides as a specific antibody, it has major applications in the control of enteric bacteria or viruses and foodborne pathogens, as anti-venomous or immunodiagnostic agents, or to treat various human diseases, such as Chagas disease, tetanus, celiac disease, rheumatism, dental caries, and wheat allergy (Vega et al., 2011; Wen

TABLE 23.1

Composition of Yolk Plasma and Granules

Granule (22%)	Plasma (78%)
56% solids (60% Protein, 34% lipids, and 6% ash)	50% solids (24% protein, 74% lipids, and 2% ash)
Proteins (60%)	**Proteins (24%)**
Lipovitellins: 70% (as HDL)	Livetins (water-soluble proteins): 22%
- α-Lipovitellins: 58%	- α-livetin (serum albumin): 4.4%
- β-Lipovitellins: 12%	- β- ivetin (glycoprotein): 11.0%
Phosvitin: 18% (226 mg/egg: 7.9% of yolk protein)	- γ-livetin (g-globulin): 6.6% (IgY: 105 mg/egg, 3.67% of yolk protein)
Low-density lipoproteins: 12%	Low-density lipoproteins (water-insoluble proteins): 78%
Lipids (34%)	**Lipids (74%)**
40% phospholipids	26% phospholipids
- 82% PC	- 74% PC
- 15% PE	- 18% PE
- 3% sphingomyelin and lysophospholipids	- 8% sphingomyelin and lysophospholipids
58% TG	70% TG
2% cholesterol	4% cholesterol
Ash (6%)	**Ash (2%)**

et al., 2012; Alvarez et al., 2013; Yang et al., 2014; Amirijavid & Hashemi, 2015; Selim et al., 2015; da Rocha et al., 2017; Grando et al., 2017; Thibodeau et al., 2017; Kumaran et al., 2018; Karachaliou et al., 2021). The specific chicken egg yolk IgY prepared against *S. Typhimurium* modulated the intestinal mucosal immune response in a mouse model of Salmonella typhimurium infection by mobilizing $CD8^+$ and TCRγδ T cells in the epithelium and $CD4^+$ and $CD8^+$ T cells in the lamina propria (Li et al., 2016). The Fab fragments of IgY reduced the mRNA expression of the Toll-like receptor and αVβ3 integrin and inhibited the inflammatory effect in the LPS-induced RAW 264.7 macrophages by blocking the NF-κB and MAPKs signaling pathways (Zhou et al., 2021). Meram and Wu (2017) also reported that egg yolk livetins hydrolysates (peptides) prepared with pepsin and alcalase showed anti-inflammatory activities by inhibiting the production of nitric oxide (NO), TNF-α, interleukin-1β (IL-1β), and interleukin-6 (IL-6). They also indicated that livetins and their hydrolysates enhanced the phagocytic capacity of the RAW 264.7 macrophages.

IgY can be taken as oral administration to treat human diseases. However, IgY has lower stability during processing, storage, and digestion (Kovacs-Nolan & Mine, 2012; Li et al., 2015). Purified IgY can be stored frozen without losing its functionality for several years (Larsson et al., 1993; Schade et al., 2005). IgY is unstable at low and high pH. When pH < 3.5 or > 12.), it loses its functions dramatically. The pH changes modify the conformation in the antigen-binding site. Also, it is not stable at high-temperature treatment (> 70°C) (Hatta et al., 1993; Lee et al., 2002). IgY is resistant to trypsin and chymotrypsin hydrolysis, but it is easy to be hydrolyzed by pepsin when pH < 4.5 (Schmidt et al., 1989; Shimizu et al., 1993). As all the factors discussed above, IgY stability in the gut depends on enzyme levels, transition time, and diets. Encapsulation can improve the biological activity retention rates in the gut (Kovacs-Nolan et al., 2005; Li et al., 2007), as the encapsulated IgY was mainly released in the small intestine rather than in the stomach (Gu et al., 2021). With stabilizers (non-reducing sugars, sorbitol, etc.), the stability of IgY can increase its stability to heat, acid, and other treatments (Shimizu et al., 1994; Jaradat & Marquardt, 2010). Encapsulation of IgY with alginate/carrageenan protected IgY from simulated gastric fluid conditions and slowed its release, helping carry the IgY to the small intestine, the main action site (Gu et al., 2021). The sugar coating (encapsulating) of *Salmonella Infantis*-specific IgY produced by hyper-immunizing hens also protected the IgY in the digestive systems and reduced the colonization of *Salmonella* in the liver and ceca (Isfahani et al., 2020).

Phosvitin accounts for 8.8% of yolk protein, and 57% of the amino acid residues are serine (Taborsky & Mok, 1967; Cook, 1968). Most serine residues are phosphorylated and have a metal-binding capacity (Byrne et al., 1984; Sattar Khan et al., 2000). During embryo development, phosvitin plays a role in delivering metal ions. However, the bioactivity is low when the metal ions are bound to phosvitin for food or supplements (Albright et al., 1984; Ishikawa et al., 2007). Phosvitin can stimulate bone formation by activating the proliferation and differentiation of osteoblastic MC3T3-E1 cells and inhibiting bone resorption (Ji et al., 2007). Phosvitin is also involved in bone formation by dephosphorylation (Li et al., 2014), which is proved by enhancing mineralization both *in vivo* and *in vitro* (Liu et al., 2013; Zhang et al., 2015, 2016; Liu et al., 2017; Zhang et al., 2017). Other research has reported that phosvitin can improve immune activity by stimulating the production of pro-inflammatory mediators. Phosvitin increases NO production in RAW 264.7 macrophages and mRNA expression for iNOS, TNF-α, and IL-1β in the RAW 264.7 macrophages, respectively (Lee et al., 2017). Phosvitin has antimicrobial activity against *Escherichia coli* and *Staphylococcus aureus*. It can bind to lipopolysaccharide (LPS), an endotoxin responsible for initiating host responses to inflammation. Phosvitin can also prevent LPS-induced tumor necrosis factor (TNF)-release from murine RAW264.7 cells, reduce serum TNF levels in mice and promote the survival rate of the endotoxemia mice (Ma et al., 2013). Recent discoveries indicated that phosvitin could be used in cosmetics to decrease melanogenesis and inhibit tyrosinase activity and melanin synthesis. Another study showed that phosvitin could be used as an anticancer agent. In *in vitro* test, phosvitin can inhibit the growth of cancer cells (MCF-7, SK-MES-1, HeLa, AGS, and HepG2) but Protected leucocytes (Moon et al., 2014).

In the food industry, phosvitin can be used as a preservative for its strong metal binding capacity to prevent metal-catalyzed lipid oxidation (Lu & Baker, 1986; Ishikawa et al., 2004; Shade & Chacana, 2007). It can also be used as an emulsifier (Chung & Ferrier, 1992; Khan et al., 1998). Recent studies found that lyophilized phosvitin subjected to a ball-milling treatment (Zhang et al., 2019) or conjugating phosvitin with pectin improved its emulsifying properties (Cui et al., 2019).

The antioxidant activity of natural phosvitin is well documented, and the peptides derived from phosvitin (phosvitin phosphopeptides, PPPs) also exhibited strong antioxidant activity (Samaraweera et al., 2011). Xu et al. (2007) reported that PPPs have stronger antioxidant capability than the natural phosvitin in a linoleic acid system. The PPPs showed more efficient radical-scavenging activity, but their iron-chelating capability was weaker than the intact phosvitin. They have attributed the strong antioxidant activity of PPPs to the amino acid composition (histidine, methionine, and tyrosine) rather than to the content of phosphoserine ligands. However, the contribution of histidine, methionine, and tyrosine to the antioxidant effect of the PPPs or phosvitin is questionable because their content in the phosvitin is very low (Byrne et al., 1984).

Yolkin in the hen egg yolk plays an important role in developing an embryo due to cytokine-inducing activity. It can induce TNF-α production, cytokine IL-1β, Il-6, and IL-10 release (Polanowski et al., 2013). Another research also reported that yolkin could stimulate nitric oxide release from the macrophage, mitigate aging behaviors and improve cognitive learning and memory in rats (Polanowski et al., 2012; Lemieszewska et al., 2016). All yolkins isolated from different bird species showed comparable immunomodulatory activity to that obtained from the hen egg yolk, and their peptide composition was very similar, indicating that yolkin might have common biological functions (Zabłocka et al., 2021).

Egg yolk powder has excellent physicochemical and nutritional properties and thus can be used as a functional food ingredient. However, its utilization in food is limited because of its high lipid content and low water solubility. Efforts were made to improve the functionality and utilization of egg yolk or yolk components such as HDL in food processing: Gu et al. (2021) combined subcritical fluid extraction and enzymatic hydrolysis to produce egg yolk powder with low lipid content and high water-solubility. Liu et al. (2022) separated HDL from the egg yolk and hydrolyzed it using neutral protease, trypsin, and alkaline protease. The solubility of hydrolysates was significantly higher than that of HDL, and the emulsifying properties of the hydrolysates also increased because of the generation of soluble peptides fragments and the exposure of ionizable residues from the HDL. Also, the stability of emulsions prepared with the hydrolysates was higher than that of the HDL. Also, a few attempts have been made to produce bioactive peptides from the egg yolk or the by-products after extracting lipids and phospholipids from the egg yolk. Eckert et al. (2019) used serine protease and aspartyl protease from yeast to hydrolyze the by-product after extracting phospholipids from the egg yolk and showed that the peptides have strong antioxidants ACE-inhibitory activities.

These research results indicated that several approaches could improve the functionality and utilization of egg yolk components, especially egg yolk proteins, in food processing. Especially, enzymatic hydrolysis can effectively widen the applications and functionality of egg yolk proteins. The combinations of various enzymes, pH, temperature, and ionic strength conditions may enable egg yolk protein products with different functional characteristics such as solubility, interactions with other food components, bioactivities, and stability, which will broaden the base of their final applications in the food industry.

23.4 Summary and Future Research

There are many functional proteins in the egg. However, the application of these proteins on an industrial scale is still limited. The study to separate and purify IgY and phosvitin has provided solid evidence on the lab scale, but more research should be conducted on the industrial scale. Also, limited toxic compounds should be used in the separation and purification as the final products will be consumed by humans. Besides the egg yolk proteins, the peptides from phosvitin and IgY have cytotoxic, anticancer, immune-modulating, antimicrobial, anti-inflammatory, and antioxidant activities, which can be used in nutraceuticals, cosmeceutical, and pharmaceutical industries. For example, phosvitin has unusual amino acid composition and structural characteristics, and thus the production of functional phosphopeptides from phosvitin using enzymes was extremely difficult (Xu et al., 2012; Samaraweera et al., 2013; Samaraweera et al., 2014). Huang et al. (2019) developed an innovative high-temperature mild pressure (HTMP) pretreatment to hydrolyze phosvitin. The results showed that the pretreatment effectively cut peptide bonds at random sites, and the combination of enzymes to the HTMP pretreatment helped further hydrolysis of the peptides. The HTMP-enzyme combinations could produce hundreds of phosphopeptides (PPPs) from phosvitin with varying sizes and amino acid sequences with similar structural characteristics to the casein phosphopeptides (CPPs), well-known functional phosphopeptides used as calcium or iron-supplementing agents. However, little information on the functionality and the utilization of the PPPs are available, and studies on the functionality of PPPs using animals are very important for the future applications of PPPs in nutraceutical, pharmaceutical, and cosmeceutical areas.

Meanwhile, there is almost no information on the characteristics and utilization of lipovitellins, livetins (α- and β-forms), high-density lipoproteins, and low-density lipoproteins. It is also important to develop simple, industry-scale separation and purification methods for various egg yolk proteins and their enzyme hydrolysates. State-of-the-art instruments should further analyze the structure and conformation of the newly developed peptides. Adding value-added egg yolk protein to other food systems or developing more innovative products should be further studied.

REFERENCES

Ahn, D., Lee, S., Singam, H., Lee, E. J., & Kim, J. C. (2006). Sequential separation of main components from chicken egg yolk. *Food Science and Biotechnology, 15*(2), 189–195.

Akita, E. M., & Nakai, S. (1992). Immunoglobulins from egg yolk: Isolation and purification. *Journal of Food Science, 57*(3), 629–634.

Akita, E. M., & Nakai, S. (1993). Comparison of four purification methods for the production of immunoglobulins from eggs laid by hens immunized with an enterotoxigenic *E. coli* strain. *Journal of Immunological Methods, 160*(2), 207–214.

Alberts, B., Johnson, A., Lewis, J., Morgan, D., Raff, M., Roberts, K., & Walter, P. (2015). Chapter 24. The innate and adaptive immune systems. In *Molecular biology of the cell* (6th ed., pp. 1297–1342). Garland Science.

Albright, K. J., Gordon, D. T., & Cotterill, O. J. (1984). Release of iron from phosvitin by heat and food additives. *Journal of Food Science, 49*(1), 78–81.

Alvarez, A., Montero, Y., Jimenez, E., Zerpa, N., Parrilla, P., & Malave, C. (2013). IgY antibodies anti-*Tityus caripitensis* venom: Purification and neutralization efficacy. *Toxicon, 74*, 208–214.

Amirijavid, S., & Hashemi, M. (2015). Detection of anticancer and apoptotic effect of the produced IgYs against the three extracellular domains of human DR5 protein. *Iranian Journal of Cancer Prevention, 2*, 109–115.

Byrne, B. M., van het Schip, A. D., van deKlundert, J. A. M., Arnberg, A. C., Gruber, M., & Ab, G. (1984). Amino acid sequence of phosvitin derived from the nucleotide sequence of part of the chicken vitellogenin gene. *Biochemistry, 23*(19), 4275–4279.

Castellani, O., Martinet, V., David-Briand, E., Guérin-Dubiard, C., & Anton, M. (2003). Egg yolk phosvitin: Preparation of metal-free purified protein by fast protein liquid chromatography using aqueous solvents. *Journal of Chromatography B, Analytical Technologies in the Biomedical and Life Sciences, 791*(1–2), 273–284.

Chalamaiah, M., Esparza, Y., Temelli, F., & Wu, J. (2017). Physicochemical and functional properties of livetins fraction from hen egg yolk. *Food Bioscience, 18*, 38–45.

Chang, H. M., Yang, C. C., & Chang, Y. C. (2000). Rapid separation of lysozyme from chicken egg white by reductants and thermal treatment. *Journal of Agricultural and Food Chemistry, 48*(2), 161–164.

Chung, S. L., & Ferrier, L. K. (1992). pH and sodium chloride effects on emulsifying properties of egg yolk phosvitin. *Journal of Food Science, 57*(1), 40–42.

Connelly, C., & Taborsky, G. (1961). Chromatographic fractionation of phosvitin. *Journal of Biological Chemistry, 236*(5), 1364–1368.

Cook, G. M. (1968). Glycoproteins in membranes. *Biological Reviews, 43*(3), 363–391.

Cook, M. E. (2004). Antibodies: Alternatives to antibiotics in improving growth and feed efficiency. *Journal of Applied Poultry Research, 13*(1), 106–119.

Cook, W. H., & Martin, W. G. (1962). Composition and properties of soluble lipoproteins in relation to structure. *Canadian Journal of Biochemistry and Physiology, 40*, 1273–1285.

Cook, M. E., & Trott, D. L. (2010). IgY – Immune component of eggs as a source of passive immunity for animals and humans. *World's Poultry Science Journal, 66*(2), 215–225.

Cui, Y., Li, X., Lu, M., Liu, X., & Duan, X. (2019). Role of polysaccharide conjugation in physicochemical and emulsifying properties of egg phosvitin and the calcium-binding capacity of its phosphopeptides. *Food & Function, 10*(4), 1808–1815.

da Rocha, D. G., Fernandez, J. H., de Almeida, C. M. C., da Silva, C. L., Magnoli, F. C., da Silva, O. E., & da Silva, W. D. (2017). Development of IgY antibodies against anti-snake toxins endowed with highly lethal neutralizing activity. *European Journal of Pharmaceutical Science, 106*, 404–412.

da Silva, M. T. L., Deodato, R. M., & Villar, L. M. (2021). Exploring the potential usefulness of IgY for antiviral therapy: A current review. *International Journal of Biological Macromolecules, 189*, 785–791.

Dong, D., Liu, H., Xiao, Q., & Li, R. (2008). Affinity purification of egg yolk immunoglobulins (IgY) with a stable synthetic ligand. *Journal of Chromatography. B, Analytical Technologies in the Biomedical and Life Sciences, 870*(1), 51–54.

Eckert, E., Zambrowicz, A., Bobak, L., Zabłocka, A., Chrzanowska, J., & Trziszka, T. (2019). Production and identification of biologically active peptides derived from a by-product of hen egg-yolk phospholipid extraction. *International Journal of Peptide Research and Therapeutics, 25*, 669–680.

Fayer, R., & Jenkins, M. C. (1992). Colostrum from cows immunized with Eimeria acervulina antigens reduces parasite development in vivo and in vitro. *Poultry Science, 71*(10), 1637–1645.

Grando, T. H., Baldissera, M. D., de Sa, M. F., do Carmo, G. M., Porto, B. C. Z., Aguirre, G. S. V., Azevedo, M. I., de Jesus, F. P. K., Santurio, J. M., Sagrillo, M. R., Stefani, L. M., & Monteiro, S. G. (2017). Avian antibodies (IgY) against *Trypanosoma cruzi*: Purification and characterization studies. *Journal of Immunological Methods, 449*, 56–61.

Gu, L., Jiao, H., McClements, D. J., Ji, M., Li, J., Chang, C., Dong, S., Su, Y., & Yang, Y. (2021). Improvement of egg yolk powder properties through enzymatic hydrolysis and subcritical fluid extraction. *LWT-Food Science and Technology, 150*, 112075.

Gu, L., McClements, D. J., Li, J., Su, Y., Yang, Y., & Li, J. (2021). Formulation of alginate/carrageenan microgels to encapsulate, protect and release immunoglobulins: Egg yolk IgY. *Food Hydrocolloids, 112*, 106349.

Hatta, H., Kim, M., & Yamamoto, T. (1990). A novel isolation method for hen egg yolk antibody, "IgY.". *Agricultural and Biological Chemistry, 54*(10), 2531–2535.

Hatta, H., Sim, J. S., & Nakai, S. (1988). Separation of phospholipids from egg yolk and recovery of water-soluble proteins. *Journal of Food Science, 53*(2), 425–427.

Hatta, H., Tsuda, K., Akachi, S., Kim, M., & Yamamoto, T. (1993). Productivity and some properties of egg yolk antibody (IgY) against human rotavirus compared with rabbit IgG. *Bioscience, Biotechnology, and Biochemistry, 57*(3), 450–454.

Hernández-Campos, F. J., Brito-De La Fuente, E., & Torrestiana-Sánchez, B. (2009). Purification of egg yolk immunoglobulin (IgY) by ultrafiltration: Effect of pH, ionic strength, and membrane properties. *Journal of Agricultural and Food Chemistry, 58*(1), 187–193.

Huang, X., & Ahn, D. U. (2019). How can the value and use of egg yolk be increased? *Journal of Food Science, 84(2),* 205–212.

Huang, X., Moon, S. H., Lee, J. H., Paik, H. D., Lee, E. J., & Ahn, D. U. (2019). Effective preparation method of phosphopeptides from phosvitin and the analysis of peptide profiles using tandem mass spectrometry. *Journal of Agricultural and Food Chemistry, 67,* 14086–14101.

Isfahani, N. H., Rahimi, S., Rasaee, M. J., Torshizi, M. A. K., Salehi, T. Z., & Grimes, J. L. (2020). The effect of capsulated and noncapsulated egg-yolk–specific antibodies to reduce colonization in the intestine of *Salmonella enterica ssp. enterica* serovar infantis–challenged broiler chickens. *Poultry Science, 99,* 1387–1394.

Ishikawa, S. I., Tamaki, S., Arihara, K., & Itoh, M. (2007). Egg yolk protein and egg yolk phosvitin inhibit calcium, magnesium, and iron absorptions in rats. *Journal of Food Science, 72*(6), S412–S419.

Ishikawa, S. I., Yano, Y., Arihara, K., & Itoh, M. (2004). Egg yolk phosvitin inhibits hydroxyl radical formation from the Fenton reaction. *Bioscience, Biotechnology, and Biochemistry, 68*(6), 1324–1331.

Jahangiri, A., Owlia, P., Rasooli, I., Salimian, J., Derakhshanifar, E., Aghajani, Z., Abdollahi, S., Khalili, S., Talei, D., & Eslam, E. D. (2021). Specific egg yolk immunoglobulin as a promising non-antibiotic biotherapeutic product against *Acinetobacter baumannii* pneumonia infection. *Scientific Reports, 11*(1), 1–11.

Jaradat, Z. W., & Marquardt, R. R. (2010). Studies on the stability of chicken IgY in different sugars, complex carbohydrates and food materials. *Food and Agricultural Immunology, 12*(4), 263–272.

Jensenius, J. C., Andersen, I., Hau, J., Crone, M., & Koch, C. (1981). Eggs: Conveniently packaged antibodies. Methods for purification of yolk IgG. *Journal of Immunological Methods, 46*(1), 63–68.

Ji, M. Y., Leem, K. H., Kim, M., & Kim, H. K. (2007). Egg yolk soluble protein stimulates the proliferation and differentiation of osteoblastic MC3T3-E1 cells. *Bioscience, Biotechnology and Biochemistry, 71*(5), 1327–1329.

Joubert, F. J., & Cook, W. H. (1958). Preparation and characterization of phosvitin from hen egg yolk. *Canadian Journal of Biochemistry and Physiology, 36*(1), 399–408.

Karachaliou, C. E., Vassilakopoulou, V., & Livaniou, E. (2021). IgY technology: Methods for developing and evaluating avian immunoglobulins for the in vitro detection of biomolecules. *World Journal of Methodology, 11(5),* 243–262.

Khan, M. A. S., Babiker, E. E., Azakami, H., & Kato, A. (1998). Effect of protease digestion and dephosphorylation on high emulsifying properties of hen egg yolk phosvitin. *Journal of Agricultural and Food Chemistry, 46*(12), 4977–4981.

Ko, K. Y., & Ahn, D. U. (2007). Preparation of immunoglobulin Y from egg yolk using ammonium sulfate precipitation and ion-exchange chromatography. *Poultry Science, 86*(2), 400–407.

Ko, K. Y., Nam, K. C., Jo, C., Lee, E. J., & Ahn, D. U. (2011). A simple and efficient method for preparing partially purified phosvitin from egg yolk using ethanol and salts. *Poultry Science, 90*(5), 1096–1104.

Kovacs-Nolan, J., & Mine, Y. (2004). Avian egg antibodies: Basic and potential applications. *Avian and Poultry Biolohy Review, 15*(1), 25–46.

Kovacs-Nolan, J., & Mine, Y. (2012). Egg yolk antibodies for passive immunity. *Annual Review of Food Science and Technology, 3*(1), 163–182.

Kovacs-Nolan, J., Phillips, M., & Mine, Y. (2005). Advances in the value of eggs and egg components for human health. *Journal of Agricultural and Food Chemistry, 53*(22), 8421–8431.

Kumaran, T., Thirumalaikumar, E., Lelin, C., Palanikumar, P., Michaelbabu, M., & Citarasu, T. (2018). Physicochemical properties of anti-vibrio harveyi egg yolk antibody (IgY) and its immunological influence in Indian white shrimp *Fenneropenaeus indicus. Fish and Shellfish Immunology, 74,* 349–362.

Landon, J., Woolley, J. A., & McLean, C. (1995). Antibody production in the hen. In J. Landon & T. Chard (Eds.), *Therapeutic antibodies.* Springer.

Larsson, A., Bålöw, R. M., Lindahl, T. L., & Forsberg, P. O. (1993). Chicken antibodies: Taking advantage of evolution - A review. *Poultry Science, 72*(10), 1807–1812.

Larsson, A., Wejåker, P. E., Forsberg, P. O., & Lindahl, T. (1992). Chicken antibodies: A tool to avoid interference by complement activation in ELISA. *Journal of Immunological Methods, 156*(1), 79–83.

Lee, H. Y., Abeyrathne, E. D. N., Choi, S., Suh, I. W., & Ahn, J. W. (2014). Sequential separation of immunoglobulin Y and phosvitin from chicken egg yolk. *Poultry Science, 93,* 2668–2677.

Lee, S. K., Han, J. H., & Decker, E. A. (2002). Antioxidant activity of phosvitin in phosphatidylcholine liposomes and meat model systems. *Journal of Food Science, 67*(1), 37–41.

Lee, J. H., Moon, S. H., Kim, H. S., Park, E., Ahn, D. U., & Paik, H. D. (2017). The immune-enhancing activity of phosvitin by stimulating the production of a pro-inflammatory mediator. *Poultry Science, 96*(11), 3872–3878.

Lei, B., & Wu, J. (2012). Purification of egg yolk phosvitin by anion exchange chromatography. *Journal of Chromatography A, 1223,* 41–46.

Lemieszewska, M., Jakubik-Witkowska, M., Stańczykiewicz, B., Zambrowicz, A., Zabłocka, A., Polanowski, A., Triziszke, T., & Rymaszewska, J. (2016). Pro-cognitive properties of the immunomodulatory polypeptide complex, yolkin, from chicken egg yolk and colostrum-derived substances: Analyses based on animal model of age-related cognitive deficits. *Archivum Immunologiae et Therapiae Experimentalis, 64*(5), 425–434.

Li, C., Geng, F., Huang, X., Ma, M., & Zhang, X. (2014). Phosvitin phosphorus is involved in chicken embryo bone formation through dephosphorylation. *Poultry Science, 93*(12), 3065–3072.

Li, X. Y., Jin, L. J., McAllister, T. A., Stanford, K., Xu, J. Y., Lu, Y. N., Zhen, Y. H., Sun, Y. X., & Xu, Y. P. (2007). Chitosan-alginate microcapsules for oral delivery of egg yolk immunoglobulin (IgY). *Journal of Agricultural and Food Chemistry, 55*(8), 2911–2917.

Li, X., Wang, L., Zhen, Y., Li, S., & Xu, Y. (2015). Chicken egg yolk antibodies (IgY) as non-antibiotic production enhancers for use in swine production: A review. *Journal of Animal Science and Biotechnology, 6*(1), 1–10.

Li, X., Yao, Y., Wang, X., Zhen, Y., Thacker, P. A., Wang, L., Shi, M., Zhao, J., Zong, Y., Wang, N., & Xu, Y. (2016). Chicken egg yolk antibodies (IgY) modulate the intestinal mucosal immune response in a mouse model of *Salmonella* typhimurium infection. *International Immunopharmacology*, *36*, 305–314.

Liu, J., Czernick, D., Lin, S. C., Alasmari, A., Serge, D., & Salih, E. (2013). Novel bioactivity of phosvitin in connective tissue and bone organogenesis revealed by live calvarial bone organ culture models. *Developmental Biology*, *381*(1), 256–275.

Liu, Q., Li, C., Geng, F., Huang, X., & Ma, M. (2017). Hen egg yolk phosvitin stimulates osteoblast differentiation in the absence of ascorbic acid. *Journal of the Science of Food and Agriculture*, *97*(13), 4532–4538.

Liu, W. J., Li, X. L., Li, S. G., Xu, B. C., & Zhang, B. (2022). Emulsifying and emulsion stabilizing properties of hydrolysates of high-density lipoprotein from egg yolk. *Food Chemistry*, *369*, 130891.

Losso, J. N., & Nakai, S. A. (1994). A simple procedure for the isolation of phosvitin from chicken egg yolk. In J. S. Sim & S. Nakai (Eds.), *Egg uses and processing technologies - New developments*, (pp. 150–157). CAB International.

Lu, C. L., & Baker, R. C. (1986). Characteristics of egg yolk phosvitin as an antioxidant for inhibiting metal-catalyzed phospholipid oxidations. *Poultry Science*, *65*(11), 2065–2070.

Ma, J., Wang, H., Wang, Y., & Zhang, S. (2013). Endotoxin-neutralizing activity of hen egg phosvitin. *Molecular Immunology*, *53*(4), 355–362.

Martin, W. G., Augustyniak, J., & Cook, W. H. (1964). Fractionation and characterization of the low-density lipoproteins of hen's egg yolk. *Biochimica et Biophysica Acta*, *84*(6), 714–720.

Mecham, D. K., & Olcott, H. S. (2002). Phosvitin, the principal phosphoprotein of egg yolk. *Journal of the American Chemical Society*, *71*(11), 3670–3679.

Meram, C., & Wu, J. (2017). Anti-inflammatory effects of egg yolk livetins (α, β, and γ-livetin) fraction and its enzymatic hydrolysates in lipopolysaccharide-induced RAW 264.7 macrophages. *Food Research International*, *100*, 449–457.

Michael, A., Meenatchisundaram, S., Parameswari, G., Subbraj, T., Selvakumaran, R., & Ramalingam, R. S. (2010). Chicken egg yolk antibodies (IgY) as an alternative to mammalian antibodies. *Indian Journal of Science and Technology*, *3*(4), 468–474.

Moon, S. H., Lee, J. H., Lee, M., Park, E., Ahn, D. U., & Paik, H. D. (2014). Cytotoxic and antigenotoxic activities of phosvitin from egg yolk. *Poultry Science*, *93*(8), 2103–2107.

Patterson, R., Youngner, J. S., Weigle, W. O., & Dixon, F. J. (1962). Antibody production and transfer to egg yolk in chickens. *Journal of Immunology*, *89*, 272–278.

Polanowski, A., Sosnowska, A., Zabłocka, A., Janusz, M., & Trziszka, T. (2013). Immunologically active peptides that accompany hen egg yolk immunoglobulin Y: Separation and identification. *Biological Chemistry*, *394*(7), 879–887.

Polanowski, A., ZabŁocka, A., Sosnowska, A., Janusz, M., & Trziszka, T. (2012). Immunomodulatory activity accompanying chicken egg yolk immunoglobulin Y. *Poultry Science*, *91*(12), 3091–3096.

Polson, A., Coetzer, T., Kruger, J., von Maltzahn, E., & van der Merwe, K. J. (1985). Improvements in the isolation of IgY from the yolks of eggs laid by immunized hens. *Immunological Investigations*, *14*(4), 323–327.

Redwan, E. M., Aljadawi, A. A., & Uversky, V. N. (2021). Simple and efficient protocol for immunoglobulin Y purification from chicken egg yolk. *Poultry Science*, *100*, 100956.

Samaraweera, H., Moon, S. H., Lee, E. J., Grant, J., Fouks, J., Choi, I. W., Suh, J. W., & Ahn, D. U. (2014). Characterization of phosvitin phosphopeptides using MALDI-TOF mass spectrometry. *Food Chemistry*, *165*, 98–103.

Samaraweera, H., Moon, S. H., Lee, E. J., Yang, S. H., Suh, J. W., & Ahn, D. U. (2013). Chemical hydrolysis of phosvitin and the functional properties of its hydrolysates. *International Journal of Research in Engineering and Technology*, *2*, 3373–3386.

Samaraweera, H., Zhang, W., Lee, E. J., & Ahn, D. U. (2011). Egg yolk phosvitin and functional phosphopeptides - Review. *Journal of Food Science*, *76(7)*, R143–R150.

Sattar Khan, M. A., Nakamura, S., Ogawa, M., Akita, E., Azakami, H., & Kato, A. (2000). Bactericidal action of egg yolk phosvitin against *Escherichia coli* under thermal stress. *Journal of Agricultural and Food Chemistry*, *48*(5), 1503–1506.

Schade, R., Calzado, E. G., Sarmiento, R., Chacana, P. A., Porankiewicz-Asplund, J., & Terzolo, H. R. (2005). Chicken egg yolk antibodies (IgY-technology): A review of progress in production and use in research and human and veterinary medicine. *Alternatives to Laboratory Animals: ATLA*, *33*(2), 129–154.

Schmidt, P., Hafner, A., Reubel, G. H., Wanke, R., Franke, V., Lösch, U., & Dahme, E. (1989). Production of antibodies to canine distemper virus in chicken eggs for immunohistochemistry. *Journal of Veterinary Medicine. Series B*, *36*(9), 661–668.

Selim, A. M., Ibrahim, E. M., El Meshad, A. H., & Hamouda, F. K. (2015). Development of IgY antibodies for control of tetanus. *Biotechnology in Animal Husbandry*, *31*, 109–122.

Shade, R., & Chacana, P. A. (2007). Livetin fractions (IgY). In R. Huopalahti, R. López-Fandiño, M. Anton, & R. Schade (Eds.), *Bioactive egg compounds*, (pp. 25–32). Springer.

Shimizu, M., Miwa, Y., Hashimoto, K., & Goto, A. (1993). Encapsulation of chicken egg yolk immunoglobulin G (IgY) by liposomes. *Bioscience, Biotechnology, and Biochemistry*, *57*(9), 1445–1449.

Shimizu, M., Nagashima, H., Hashimoto, K., & Suzuki, T. (1994). Egg yolk antibody (Ig Y) stability in aqueous solution with high sugar concentrations. *Journal of Food Science*, *59*(4), 763–765.

Stadelman, W. J., & Cotterill, O. J. (1995). *Egg science and technology*. AVI Publishing Co.

Strixner, T., Sterr, J., Kulozik, U., & Gebhardt, R. (2014). Structural study on hen-egg yolk high-density lipoprotein (HDL) granules. *Food Biophysics*, *9*, 314–321.

Svendsen, L., Crowley, A., Ostergaard, L. H., Stodulski, G., & Hau, J. (1995). Development and comparison of purification strategies for chicken antibodies from egg yolk. *Laboratory Animal Science*, *45*(1), 89–93.

Taborsky, G., & Mok, C. C. (1967). Phosvitin: Homogeneity and molecular weight. *Journal of Biological Chemistry*, *242*(7), 1495–1501.

Thibodeau, A., Fravalo, P., Perron, A., Laurent-Lewandowski, S., & Letellier, A. (2017). Production and characterization of anti-*campylobacter jeju*ni IgY from egg yolk. *Acta Veterinaria Scandinavica*, *59*, 80. https://doi.org/10.1186/s13028-017-0346-4.

Vega, C., Bok, M., Chacana, P., Saif, L., Fernandez, F., & Parreno, V. (2011). Egg yolk IgY: Protection against rotavirus-induced diarrhea and modulatory effect on the systemic and mucosal antibody responses in newborn calves. *Veterinary Immunology and Immunopathology*, *142*, 156–169.

Wang, S., Wang, Y., Ma, J., Ding, Y., & Zhang, S. (2011). Phosvitin plays a critical role in the immunity of zebrafish embryos via acting as a pattern recognition receptor and an antimicrobial effector. *Journal of Biological Chemistry*, *286*(25), 22653–22664.

Wen, J., Zhao, S., He, D., Yang, Y., Li, Y., & Zhu, S. (2012). Preparation and characterization of egg yolk immunoglobulin Y specific to influenza B virus. *Antiviral Research*, *93*, 154–159.

Xu, X., Katayama, S., & Mine, Y. (2007). Antioxidant activity of tryptic digests of hen egg yolk phosvitin. *Journal of the Science of Food Agriculture*, *87*, 2604–2608.

Xu, C., Yang, C., Yin, Y., Liu, J., & Mine, Y. (2012). Phosphopeptides (PPPs) from hen egg yolk phosvitin exert anti-inflammatory activity via modulation of cytokine expression. *Journal of Functional Foods*, *4*, 718–726.

Yang, Y., Wen, J., Zhao, S., Zhang, K., & Zhou, Y. (2014). Prophylaxis and therapy of pandemic H1N1 virus infection using egg yolk antibody. *Journal of Virological Methods*, *206*, 19–26.

Yokoyama, H., Peralta, R. C., Horikoshi, T., Hiraoka, J., Ikemori, Y., Kuroki, M., & Kodama, Y. (1993). A two-step procedure for purification of hen egg yolk immunoglobulin G: Utilization of hydroxypropylmethylcellulose phthalate and synthetic affinity ligand gel (Avid AL®). *Poultry Science*, *72*(2), 275–281.

Zabłocka, A., Bobak, T., Macała, J., Rymaszewska, J., Kazana, W., & Zambrowicz, A. (2021). Comparative studies of yolkin preparations isolated from egg yolks of selected bird species. *Chemistry & Biodiversity*, *18*, e2100178.

Zambrowicz, A., Zabłocka, A., Bobak, Macała, J., Janusz, M., Polanowski, A., & Trziszka, T. (2017). A simple and rapid method of isolation of active polypeptide complex, yolkin, from chicken egg yolk. *Food Chemistry*, *230*, 705–711.

Zhang, X., Geng, F., Huang, X., & Ma, M. (2015). Acceleration of the initial phase transformation of mineralization by phosvitin. *Journal of Crystal Growth*, *409*, 44–50.

Zhang, X., Geng, F., Huang, X., & Ma, M. (2016). Calcium-binding characteristics and structural changes of phosvitin. *Journal of Inorganic Biochemistry*, *159*, 76–81.

Zhang, X., Huang, X., & Ma, M. (2017). Role of phosphorylation of phosvitin in the phase transformation of mineralization. *International Journal of Biological Macromolecules*, *101*, 712–718.

Zhang, J., Sun, J., Liu, Y., Li, J., Su, Y., Xia, W., & Yang, Y. (2012). Separation and purification of phosvitin phosphopeptides using immobilized metal affinity nanoparticles. *Journal of Chromatography. B, Analytical Technologies in the Biomedical and Life Sciences*, *893–894*, 121–126.

Zhang, Q., Yang, L., Hu, S., Liu, X., & Duan, X. (2019). Consequences of ball-milling treatment on the physicochemical, rheological and emulsifying properties of egg phosvitin. *Food Hydrocolloids*, *95*, 418–425.

Zhou, X., Ahn, D. U., Xia, M., Zeng, Q., Li, X., & Cai, Z. (2021). Fab fragment of immunoglobulin Y modulates NF-κB and MAPK signaling through TLR4 and αVβ3 integrin and inhibits inflammatory effect in R264.7 macrophage. *Journal of Agricultural and Food Chemistry*, *69*, 8747–8757.

24

Egg Bioactives and Chronic Degenerative Diseases

Jack N. Losso*, Manish Thapaliya, and Achyut Adhikari

CONTENTS

24.1 Introduction

Avian egg contains a variety of bioactive compounds that are essential for the growth of the chick. These bioactives have also been identified as important for human health (Board et al., 1995). Proteins, phospholipids, fat-soluble vitamins, water-soluble vitamins, carotenoids, and minerals make the group of bioactives of interest. The review will first address major bioactive compounds in decreasing order in egg white. The egg yolk is a good source of both water-and lipid-soluble bioactives and the section on egg yolk will address these bioactives accordingly.

24.2 Egg White Bioactives

24.2.1 Introduction

Egg white is a good source of more than 40 proteins most of which are rich in essential amino acids (Lewis et al., 1950), very little carbohydrates, lipids, and vitamins or minerals. Proteins with antimicrobial, anti-inflammatory, immunoprotective, anti-viral, anti-tumor, mineral-binding, vitamin-binding, and protease inhibitory activities are present in egg white. The major proteins in egg white are ovalbumin (54%), ovotransferrin (12%), ovomucoid (11%), ovoglobulin (G2 and

* Corresponding author.

DOI: 10.1201/9781003254430-31

G3, 8%), ovomucin (3.5%), and lysozyme (3.5%). The minor proteins are ovoinhibitor (1.5%), ovoglycoprotein (1.0%), ovoflavoprotein (0.8%), ovomacroglobulin (0.5%), avidin (0.05%), and cystatin (0.05%). However, most of these proteins can induce unfavorable pro-inflammatory responses in individuals who are allergic to egg proteins. Limited *in vitro* studies and animal studies will be reported in this chapter for the simple reason that there is not always a direct extrapolation of animal study results to humans as many *in vitro* and animal studies have failed to be validated *in vivo* in clinical trials. Attempts were made to discuss only translational human/clinical studies.

24.2.2 Egg White Proteins

24.2.2.1 Ovalbumin

Ovalbumin is the major protein in egg white and makes 54% of total egg white proteins. Ovalbumin is a 45,000 daltons water-soluble molecular weight protein. The protein regulates the redox status through its ability to chelate metal through the free thiol (SH) group (Nakamura et al., 1992). While ovalbumin imparts excellent functionalities among proteins as far as developing food products, it is the major allergen in eggs and limits ovalbumin and egg white usage in many chronic disease prevention or treatment applications. The protein is rather used as a disease trigger in several biochemical experiments (Langeland, 1985; Holen & Elsayed, 1990; Bernhisel-Broadbent et al., 1994). Ovalbumin allergenicity is characterized by an increased production of immunoglobulin E (IgE) antibodies and Th2 cytokines and IgE-mediated type I hypersensitivity (Savage et al., 2007). Ovalbumin allergenicity can be reduced by heating or irradiation followed by heating (Kim et al., 2002). These investigators showed that protein hydrophobic amino acids such as the epitope Asp95-Ala102 are very critical to binding to IgE, followed by polar and charged amino acid residues.

Recent studies have attempted to determine whether *in vitro* and animal studies that demonstrate vasodilatory effects of ovalbumin-derived peptides can be reproduced *in vivo* in humans. A double-blind, placebo-controlled randomized crossover trial investigated the efficacy of ovalbumin-derived peptides (3g/day) or placebo (3g/day) in 75 adults aged 50–70 years with systolic blood pressure ($130 - \leq 150$ mmHg). The study involved two 6-week periods during which participants were either on ovalbumin-peptides or a placebo separated by a 3-week washout period (Lucey et al., 2019). Results from 65 participants with a mean systolic blood pressure (135.1 ± 11 mmHg) indicated that mean office and central blood pressure and arterial stiffness or pulse wave analysis, blood lipids, and glucose concentrations did not change over time and no significant differences were observed between the ovalbumin peptides and placebo groups ($P > 0.05$). The health benefits of hen egg white ovalbumin or albumin peptides need to be discovered.

24.2.2.2 Ovotransferrin

Ovotransferrin, also known as conalbumin, is the second major egg white protein. It accounts for about 12% of the total egg white proteins. The neutral protein is a monomer of 686 amino acids with a molecular weight of 77.9 Kilodaltons. Ovotransferrin primary amino acid sequence shows 51% homology to human transferrin. The protein has numerous potential *in vivo* applications in chronic disease prevention. Ovotransferrin is well-known for its iron and zinc chelating ability. The protein can inhibit Gram-negative and Gram-positive bacteria. *In vitro* studies have shown ovotransferrin to possess strong iron-related antibacterial activities against *Pseudomonas spp*., *E. coli*, *Streptococcus mutans* in the presence of bicarbonate ions (alkaline pH) and weak action in citrate (acidic) environment while *Staphylococcus aureus*, *Proteus spp*., and *Klebsiella* are resistant (Valenti et al., 1983). Ovotransferrin was similar to milk lactoferrin when it was investigated to protect guinea pigs *in vivo* against *E. coli* by gastrointestinal route (Valenti et al., 1983). Iron-saturated ovotransferrin can still kill bacteria. The bactericidal activity of *apo* or *holo* ovotransferrin is also related to the protein cationic domain 109-200 ability to cross the bacterial outer membrane, which leads to killing of Gram-negative bacteria (Superti et al., 2007). Ovotransferrin chelates iron at two binding sites, binds directly to the surface of the bacteria, penetrates the outer membrane of the bacteria, and enters the inner membrane of the bacteria leading to iron leakage and subsequent uncoupling of the respiratory-dependent energy production. The efficacy of ovotransferrin as an iron chelator has been shown against Gram-negative bacteria, including *E. coli* O157:H7, *Pseudomonas sp*., and *Salmonella enteritidis* as an iron chelator (Valenti et al., 1983; Baron et al., 1997). *In vivo*, ovotransferrin has been used as an antibacterial in the guinea pig model of bacterial infection (Sellier et al., 2007). The activities of ovotransferrin against Gram-negative bacteria open the possibility of using this naturally occurring bioactive protein in human diseases and reestablish eubiosis. Dysbiosis is an imbalance of gut microbiota, in favor of mostly Gram-negative bacteria, which is associated with the progression and sustainability of several chronic degenerative diseases, including obesity, diabetes, hypertension, cardiovascular disease, immunosuppressive diseases, neurological diseases, cancer, and COVID-19. The inhibitory activity against Gram-positive bacteria such as *Staphylococcus aureus* KCCM 32,395, *Bacillus cereus* KCCM 40,935, *Listeria monocytogenes* ATCC 15,313 indicate the wide broad range of antibacterial activity ascribed to ovotransferrin.

The overlapping antimicrobial, fungal, and viral properties of ovotransferrin compared to lactoferrin, which has demonstrated activities against SARS-CoV, offer possibilities for investigating ovotransferrin against SARS-CoV-2 and potential opportunistic fungi and viruses that come along with COVID-19. The rationale is that SARS-CoV-2 is infrequently found in tissues or secretions such as tears wherein lactoferrin or lysozyme is found in the human body (Wu et al., 2020; Xia et al., 2020). A clinical trial was carried out in Italy in which 200 mg of ovotransferrin was orally administered as adjuvant therapy in cases of mild and moderate COVID-19 and awaiting results publication (Hilpert, 2021).

Ovotransferrin has shown in young (3 weeks old) and adult (8 weeks old) mice the ability to modulate the gut microbiota and morphology by increasing the *in vivo* population of the beneficial microbiota population such as *Akkermansia* and the expression of tight junction protein occludin and crypt depth (Zhang et al., 2020).

Iron-bound ovotransferrin is more resistant to gastrointestinal digestion and well absorbed making it more valuable for human disease (Azari & Feeney, 1958; Brock et al., 1976). Ovotransferrin has immunomodulatory, antioxidant, anti-inflammatory, and angiotensin-converting enzyme-inhibitory activities (Mann & Ndung'u, 2020). Ovotransferrin peptides similar to human transferrin peptides are readily transported into human intestinal cells and have antiviral properties that should be helpful for human and animal diseases (Ding et al., 2015; Mann & Ndung'u, 2020).

Ovotransferrin also enhances phagocytic activity as well as cytokine production of macrophages, which enhances intestinal immune responses through dendritic cell maturation, Th1/Th2 balance restoration, and humoral immunity promotion (Mann & Ndung'u, 2020). Ovotransferrin peptides downregulate lipopolysaccharide-induced inflammatory interleukin-6 (IL-6), tumor necrosis factor-alpha (TNF-α), and myeloperoxidase activity in peritonitis (Jiao et al., 2019). Ovotransferrin peptides display ACE inhibitory activity (antihypertensive) as they easily bind to angiotensin II receptor type 1 to inhibit angiotensin II pro-inflammatory activity (Liao et al., 2018). Ovotransferrin can protect against chronic diseases by inhibiting bacteria and reducing the concentrations of lipopolysaccharides that induce inflammation, because in the worst-case scenario, excessive lipopolysaccharide concentration in the blood causes sepsis. Ovotransferrin peptides are bioavailable and viable treatment agents for improving endothelial dysfunction and preventing cardiovascular disease development and progression (Siciliano et al., 1999). The protein and its peptides scavenge superoxide radicals. Zn^{2+}-, Mn^{2+}- and Cu^{2+}-ovotransferrin possess potent superoxide dismutase (SOD) anion (**O_2–**) scavenging protein with higher scavenging activity than ascorbate or serum albumin (Ibrahim et al., 2007).

Ovotransferrin and its peptides have antiviral properties. Ovotransferrin peptides DQKDEYELL and KDLLFK are efficient against *Herpes simplex* virus (Siciliano et al., 1999). The anti-herpes viral activity of ovotransferrin peptides 222-230 (ADRDQYELL) and 264-269 (EDLIWK) has also been demonstrated (Siciliano et al., 1999).

Ovotransferrin has anti-fungal activity. Ovotransferrin and iron-saturated ovotransferrin sensitize more than 100 *Candida* than bacteria. The protein may interact with the fungi through *Candida* cells' adsorption of the protein rather through iron chelation-mediated inhibition (Valenti et al., 1985).

In vitro HCT-116 colon and MCF-7 breast cancer cell inhibitory activity of ovotransferrin has been reported (Ibrahim & Kiyono, 2009). The inhibition of colon cancer is very interesting since colon cancer is a leading cause of cancer-related death in the world.

Ovotransferrin is an immunomodulator. In mice, the protein ameliorated cyclophosphamide-induced dysbiosis, enhanced the secretion of immunoglobulin A (IgA) and secretory immunoglobulin A (slgA), and promoted intestinal humoral immune response (Zhu et al., 2019).

Should ovotransferrin be effective *in vivo* in eliminating or inhibiting Gram-negative bacteria, fungi, and viruses, the protein will/can provide significant health benefits to human and animal health. However, the enzymatic activity of pancreatin on ovotransferrin needs to be investigated.

Despite all the functional and health benefits ascribed to ovotransferrin, the protein is not commercially available in large quantities from vendors. Similarly, the protein is less investigated by food scientists or those who devote time and efforts to egg proteins. Ovotransferrin allergenicity cannot be ignored while investigating this protein (Langeland, 1985).

24.2.2.3 Ovomucoid

Ovomucoid is a 28 Kilodaltons glycoprotein with a pI of 4.1 that makes about 12% of the egg white proteins(Kovacs-Nolan et al., 2005). The protein has good functionality in food product development. Ovomucoid is a trypsin inhibitor and binds 1:1 with trypsin (De Oliveira et al., 2009). However, ovomucoid remains one the major allergenic proteins in egg white (Langeland, 1985). A double-blind, placebo-controlled food challenge in 38 subjects with high levels of IgE antibodies for egg white was performed in order to compare the allergenicities of heated and ovomucoid-depleted egg white, freeze-dried egg white, and heated egg white (Urisu et al., 1997). Radioallergosorbent test (RAST) was used to determine the levels of IgE antibodies for egg white, ovomucoid, ovalbumin, ovotransferrin, and lysozyme. Fifty-five percent of the participants who developed high levels of IgE antibodies to freeze-dried egg white showed negative response when challenged with heated egg white. And 94.1% of participants who had high levels of IgE to heated egg whites had insignificant levels of IgE to heated egg white or ovomucoid-depleted egg white. Participants with positive challenge responses to freeze-dried egg white tended to have higher IgE antibody values to ovomucoid than those with negative responses. Participants with positive response to heated egg white preparations had higher levels of IgE than those with no response. Participants showed no significant differences in the levels of IgE antibodies to ovomucin, ovalbumin, ovotransferrin, and lysozyme in the negative-response and positive-response groups in challenge tests with freeze-dried and heated egg white. Freeze-dried egg white preparations were more allergenic than heated and ovomucoid-depleted egg white preparation. It was concluded that ovomucoid is one of the most allergenic proteins in egg white. Overcoming ovomucoid allergenicity is the major step toward developing health benefits against chronic disease using this protein.

24.2.2.4 Ovomucin

Ovomucin is a glycoprotein mixture of 3 subunits, including the α1-, α2-, and β subunit linked together by disulfide bonds for a total of about 770,000 daltons. The α1-subunit is a 150,000 daltons, the α2- subunit is a 220,000 daltons, and the β-subunit is a 400,000 daltons protein (Itoh et al., 1987). The glycoprotein represents 3.5% of egg white proteins.

Antibacterial, antiviral, antitumor, and cholesterol absorption inhibition activities have been ascribed to ovomucin. Dysbiosis is caused by the overabundance of Gram-negative bacteria over Gram-positive ones. Any compound that can reduce or inhibit Gram-negative bacteria population has or will have a significant effect on chronic disease. Ovomucin through its sialic moiety has the ability to bind to *Escherichia coli* O157: H7, *Streptococcus*, and *Helicobacter pylori* (Tu et al., 2020). *H. pylori* affects over 50% of the world population and the bacteria has very high infectivity rates in human populations (GonzÁlez et al., 2018). *Helicobacter pylori* is associated with gastrointestinal disease and can lead to gastric, pancreatic, and lung cancer development (Venerito et al., 2017; GonzÁlez et al., 2018; Wu et al., 2019; Engstrand & Graham, 2020). Ovomucin binds urease on the surface of *Helicobacter pylori* cells and inhibits *Helicobacter pylori* adhesion to gastric mucosa receptors (Kodama & Kimura, 2001). The anti-tumor activity of the β-subunit of ovomucin has been explained by the potential of the protein to activate the immune system that can lead to tumor regression (Tanizaki et al., 1997).

The antiviral activity of ovomucin has also been ascribed to the ability of the sialic acid moiety of the glycoprotein to bind to the virus (Xu et al., 2018). The sialic acid has been identified as the recognition site for the glycoprotein (Xu et al., 2018).

In vitro assays using Caco-2 cells and *in vivo* study in rats have shown that ovomucin has a cholesterol-lowering effect compared to casein. Ovomucin lowers serum cholesterol by binding to cholesterol-mixed micelles in the jejunal epithelium and decreasing cholesterol solubility (Schauer, 2004). By inhibiting cholesterol absorption ovomucin can significantly prevent the development of cardiovascular disease.

Ovomucin peptides may be metabolized by different *Bifidobacterium* or other members of the intestinal microbiota and form short-chain fatty acids which in term protect the gastrointestinal tract barrier against lipopolysaccharides translocation to circulation (Jukes et al., 2020). The protein and its peptides have good anti-inflammatory, antioxidant, immunity-regulating, and other biological functions that are good for immune and intestinal health. Ovomucin peptides obtained from alkaline hydrolysis of heat-denatured protein showed inhibitory activities against the angiotensin-converting enzyme (ACE) through metal-chelating abilities (Abeyrathne et al., 2016). Ovomucin health-enhancing properties against chronic diseases in individuals who are not allergic to egg proteins should be investigated and promoted.

24.2.2.5 Lysozyme

Lysozyme represents 3.5% of egg white proteins. It is a 14.3 Kilodalton protein containing four disulfide bonds that increase its rigidity. The protein is an inhibitor of Gram-positive bacteria and can also after modification inhibit the growth of Gram-negative bacteria (see Chapter 26). Lysozyme is an anti-inflammatory protein. Lysozyme attenuated inflammation in a porcine model of dextran sodium sulfate (DSS)-induced colitis (Lee et al., 2009). Whereas DSS treatment-sinduced weight loss, severe mucosal and submucosal inflammation, colonic crypt distortion, muscle wall thickening, down-regulation of mucin gene expression, and increased gastric permeability, egg lysozyme significantly reduced the local expression of pro-inflammatory cytokines, including TNF-α, IL-6, IFN-γ, IL-8, and IL-17. Egg lysozyme also increased the expression of the anti-inflammatory mediators IL-4 and TGF-β, suggesting that egg lysozyme has anti-inflammatory and immunomodulating properties. Egg lysozyme may also restore gut homeostasis by increasing TGF-β and Foxp3 levels and activating regulatory T cells, which are important in the regulation of the mucosal immune system (Lee et al., 2009). Egg lysozyme is bioavailable and may restore health *in vivo*. However, egg lysozyme may play a role in the pathogenesis of pulmonary emphysema since it impairs the ability of hyaluronic acid to prevent elastase injury to elastic fibers which undergo degradation in this disease (Cantor et al., 2002).

A cross-over clinical trial design investigated the effects of 1, 2 or 5 g of lysozyme peptides (NWT-03) on daytime, 36-h, and night-time systolic and diastolic blood pressure (SBP and DBP) in 92 generally healthy subjects with normal BP (n = 29), high-normal BP (n = 34) or mild hypertension (n = 29) (Plat et al., 2017). The study had six treatment arms (1 g NWT-03 or placebo in period 1 and placebo or 1 g NWT-03 in period 2, 2 g NTW-03 or placebo in period 1 and placebo or 2 g NWT-03 in period 2, or 5 g NTW-03 or placebo in period 1 and placebo or 5 g NTW-03 in period 2). A comparable number of subjects from each BP class were included in each study arm. Participants completed a 7-day period for both treatments in each arm, separated by 5-day washout periods. In mild-hypertensive subjects, 2 g NWT-03 significantly decreased daytime SBP (7.9 mmHg; P = 0.006), daytime DBP (4.2 mmHg; P = 0.009), 36-h SBP (6.9 mmHg; P = 0.015) and 36-h DBP (3.5 mmHg; P= 0.035) compared with placebo subjects. Also, in mild-hypertensive subjects, 5 g NWT-03 significantly decreased night-time SBP (14.8 mmHg; P = 0.008) and night-time DBP (8.4 mmHg; P = 0.020) compared to blood pressure data in placebo subjects. While additional studies are needed, data from this trial should be interpreted with care because no dose–response relationship was evident.

24.2.2.6 Protease Inhibitors

The egg white contains several protease inhibitors, including ovostatin, ovoinhibitors, and cystatin. Ovostatin also known as ovomacroglobulin is a 780,000 daltons α2-macroglobulin with an isoelectric point of 4.9 (Nagase et al., 1983). Ovomacroglobulin is a glycoprotein containing four subunits of 175,000 KDa as determined by SDS-PAGE and carbohydrate making 11.5% of the protein molecule (Kitamoto et al., 1982; Geng et al., 2015). MALDI-TOF-MS which is more accurate than DSD-PAGE showed that the OVMG subunit [M + H](+) ion has a peak at m/z 183297 daltons (Geng et al., 2015). It is homologous to human α2-macroglobulin (Kitamoto et al., 1982). The protein is a trypsin inhibitor and has strong anti-collagenase activity (Nagase et al., 1983). Collagenases such as the matrix metalloproteinase -1 (MMP-1), MMP-13, and MMP-18 hydrolyze the peptide bonds of fibrillar collagen type I, II, and III into fragments that become susceptible to more degradation by gelatinases such as MMP-2 and MMP-9 (Chung et al., 2004; Nagase et al., 2006). MMP-13 is thought to be a major contributor to the degenerative changes that occur in the cartilage during osteoarthritis pathogenesis (Mehana et al., 2019). Humans, horses, and dogs can suffer from osteoarthritis and solution found for humans will also benefit horses and dogs. Ovostatin also inhibits papain and thermolysin (Kitamoto et al., 1982). Glycosylation of ovomacroglobulin increases the protein binding to IgE and therefore its allergenicity. *In vivo* studies are awaiting to ascertain the potential health effects of this glycoprotein.

Ovoinhibitors are heat-stable glycoproteins and protease inhibitors that inhibit serine proteases, including trypsin and chymotrypsin and a variety of bacterial and fungal proteases (Li-Chan & Nakai, 1989; Huopalahti et al., 2007)). Serine proteases are involved in colon cancer angiogenesis and progression and potential dietary inhibitors of proteases would contribute to colon cancer prevention and inhibition (Clemente et al., 2012).

Cystatin has a molecular weight of 13 kDa with two major isoelectric forms at pH 6.5 and 5.6 that differ in the occurrence of phosphorylation on serine (Anastasi et al., 1983; Huopalahti et al., 2007). Cystatin inhibits the activity of papain and related cysteine proteinases (cathepsin B, H, and L; ficin; bromelain). Hen cystatins have the potential to be used in pharmaceutical formulations as inhibitors of enzymes associated with the onset and/or progression of a wide range of pathological processes, including poliomyelitis caused by poliovirus, inflammations, infections, osteoporosis, gingivitis, periodontitis, and cancer may also be prevented by cystatins (Blankenvoorde et al., 2000).

Ovoglycoprotein is a glycoprotein of three subunits. The molecular mass of the deglycosylated subunits is 30,000 Daltons; 28,400 Daltons; and 21,400 Daltons; respectively (Li-Chan & Kim, 2008). The pI of the different subunits are 4.37–4.51; 4.34–4.44; and 4.17–4.4.3 (Haginaka et al., 2000). The glycoprotein is heat stable in boiling water and stable in trichloroacetic acid. (Li-Chan & Kim, 2008). Purified glycosylated ovoglycoprotein inhibits Helicobacter pylori colonization and has the potential for use in gastric ulcers (Kodama & Kimura, 2001)

Ovoflavoprotein is a riboflavin-binding protein. The protein is a single polypeptide of 219 amino acids and several α- and β-helices. The riboflavin-binding site is made of the 4 α-helix (A-D) and 4 β-helix (Stadelman & Cotterill, 1977).

Cystatin is a 13 kDa composite of two subunits with pI of 6.5 and 5.6, respectively (Huopalahti et al., 2007). Phosphorylation affects the isoelectric point of the protein subunits (Anastasi et al., 1983). Cystatin inhibits papain and ficin (Huopalahti et al., 2007).

24.2.2.7 Avidin

Avidin is a basic tetrameric glycoprotein of 14.3 KDa size made of 128 amino acids. Each monomer of avidin binds one molecule of biotin and the interaction avidin-biotin has a dissociation constant of 10^{-15} M which has been reported as the strongest known interaction between a protein and ligand (Wilchek & Bayer, 1988). The potential health benefits of avidin are associated with its ability to inhibit microorganisms (Stadelman & Cotterill, 1977; Huopalahti et al., 2007). The strong interaction between avidin and biotin finds applications *in vivo* to localize and image tumor cells and also pretarget drugs to tumors (Hytönen et al., 2003). In addition, avidin is also used in drug delivery through the blood–brain barrier (Bickel et al., 2001). Avidin can also be used as an antidote of biotinylated drugs (Petronzelli et al., 2010).

24.2.3 Egg White Lipids, Carbohydrates, Minerals, and Vitamins

Lipids make 0.03% w/w of the egg albumen (Watkins et al., 2003). The major lipids in the group include palmitic (C16:0; 18.30% of total lipids), stearic (C18:0; 7.76% of total lipids), linoleic (C18:2 *n-6*; 20.92% of total lipids), and arachidonic (C20:4 *n-6*; 9.92% of total lipids) acids.

Glucose is the main free sugar of the albumen (Watkins et al., 2003). It represents about 0.8–1% by weight of the albumen. Glucose is also found in most of the albumen glycoproteins. The sugar is important for the growth of the chick before and after hatching. Glucose level increases in dry egg white. Technologically, it is removed from many egg white products to prevent potential nonenzymatic browning through the Maillard reaction during high-temperature processing. Besides its potential interaction through the nonenzymatic browning, egg white glucose has no health-enhancing benefits in humans worth mentioning.

The major minerals in egg albumen include chlorine (66.1 mg), sodium (63 mg), sulfur (62 mg), and potassium (57 mg) (Li-Chan & Kim, 2008). The minor and trace minerals in egg white are phosphorus (8 mg), magnesium (4.15 mg), calcium (3.8 mg), iron (0.053 mg), zinc (0.05 mg), cupper (0.009 mg), manganese (0.002 mg), and iodine (0.001mg) (Li-Chan & Kim, 2008). Egg white is not a major mineral contributor to human health.

The vitamins in egg white include niacin (1.52 mg), folic acid (0.46 mg), riboflavin (0.11 mg), pyridoxine (0.09 mg), pantothenic acid (0.035 mg), riboflavin (0.008 mg), and thiamine (0.004 mg). Since vitamins are effective at lower levels, egg white can contribute to vitamin intake.

While these bioactives are essential for the egg biochemistry and chick growth, they are not major contributors to cause or prevent chronic diseases.

Table 24.1 summarizes clinical trials performed on egg white bioactives. The allergenicity associated with most of these bioactives renders testing difficult because of ethical issues. However, clinical trials can be carried out with non-allergenic very low molecular weight peptides derived from these bioactives.

24.3 Egg Yolk Bioactives

24.3.1 Introduction

Egg yolk contains 32.0% lipids, 16.0% proteins, 1.0% carbohydrates, 1.0% minerals on wet weight basis. On dry weight basis, the yolk is 50% solids of which 65–70% is lipids and proteins make about 30% (Li-Chan & Kim, 2008). Egg yolk contains granules suspended in yellow liquid also known as plasma.

Egg yolk proteins include apovitellinin I–IV (37.3%), lipovitellin apoproteins (α-lipovitellin (26.7%) and β-lipovitellin (13.3%), livetins (α-livetin or serum albumin (2.7%), β-livetin or α-2-glycoprotein (4.0%), γ-livetin or γ-globulin (2.7%), phosvitin (13.3%), and biotin-binding protein (trace). The lipovitellinins or low-density lipoproteins (LDL) contain 12% proteins on a dry weight basis and 86.7% total lipids.

TABLE 24.1

Clinical Trials with Egg White Bioactives

Egg White Bioactive	Participant's Health Condition	Physiological Activity Tested	Outcome	References
Ovalbumin peptides (3g/d; 6 weeks)	High blood pressure (130 – ≤150 mmHg), age 50–70; $n = 65$	Vasodilatory effect	No effect and no difference vs placebo	Lucey et al. (2019)
Ovotransferrin (200 mg/d; 10 days)	Moderate COVID-19	Systolic blood pressure	Results awaiting publication	Hilpert (2021)
Ovomucoid	Allergy to egg white, $n = 38$ (ovomucoid-rich vs ovomucoid-depleted)	Antiviral and Immunomodulatory IgE	Ovomucoid is very allergenic	Urisu et al. (1997)
Lysozyme (1,2, or 5 g peptides; 7 days)	High blood pressure, $n = 92$	Daytime, 36-h, and night-time systolic and diastolic blood pressure	2–5 g decreased systolic and diastolic blood pressure. But no dose-response relationship was evident	Plat et al. (2017)

The high-density lipoprotein (HDL) or lipovitellin includes the α-and β-lipovitellin, both of which are glycoconjugates with mannose, galactose, glucosamine, and sialic acid groups (Li-Chan et al., 1995).

The bioactive lipids of the yolk include the triglycerides (65% of yolk lipids on dry weight basis), phospholipids (28–30% on dry weight basis), cholesterol (4–5%), and glycolipids (trace amounts).

The carbohydrates represent about 0.7–1.0% of the yolk and include 0.3% free glucose and the remainder is made of glycoconjugates. Pigments are very low and less than 1000 μg per 100 g. Each group will be described and potentials to chronic disease will be discussed.

24.3.2 Egg Yolk Lipids

24.3.2.1 Introduction

The main components of egg yolk solids are lipids which make 32–36% of the total solids. Egg yolk lipids' main components are triglycerides (65%), 28–30% phospholipids and 4–5% cholesterol. The fatty acid composition of the egg yolk lipids varies as a function of genotype, diet, and age.

24.3.2.2 Triglycerides

In most cases, palmitic acid and stearic acid make about 30–38% of the total fatty acids of the yolk. The polyunsaturated (linoleic and arachidonic acids) and the monounsaturated (mostly oleic) make also one-third of the yolk triglycerides. Diet can affect yolk lipids which have implications on yolk quality for disease prevention or development and progression. Red mold rice added to hen's diet significantly reduced cholesterol, LDL, and triglycerides concentration compared to control group (Wang & Pan, 2003).

24.3.2.3 Phospholipids

Phosphatidylcholine and phosphatidylethanolamine are the major phospholipids in egg yolk. Phosphatidylcholine makes 81% of yolk phospholipids and phosphatidylethanolamine makes 12% of the yolk lipids. The remaining phospholipids include lysophosphaditylcholine, lysophosphatidylethanolamine, and sphingomyelins. The *sn-1* position of the phospholipids is filled with saturated fatty acids and the *sn-2* position is filled with polyunsaturated fatty acids.

Diarrhea is a common and serious complication in patients receiving enteral nutrition and affects 2–95% of patients on enteral nutrition. Recent reports indicate that egg yolk lecithin or phosphatidylcholine can improve diarrhea in enteral formulations. Enteral formula emulsified with egg yolk lecithin improved diarrhea and promoted lipid absorption by preventing the destruction of emulsified substances by gastric acid (Akashi et al., 2017). It was also shown that egg yolk lecithin was better than soybean lecithin in improving diarrhea.

Egg yolk is a good source of choline. Dietary choline is an essential nutrient required for fetal development and normal human liver and muscle functions. However, studies have shown that under dysbiosis, choline may also be used by Gram-negative bacteria in the gut and be converted into trimethylamine-N-oxide (TMAO), which is purported to be a risk factor for developing heart disease. These toxic metabolites are eliminated through the kidney and may also be damaging to the kidney. These compounds are referred to as gut-derived uremic toxins (GDUT).

A healthy gut microbiome has the ability to process and digest complex carbohydrates into short-chain fatty acids (SCFA). These SCFA function as signaling molecules, immune-modulating molecules, and energy sources. Gut dysbiosis is the imbalance in the gut microbiome whereby unhealthy pathogenic microbiota dominate the microflora. These bacteria metabolize egg yolk phosphatidylcholine, red meat carnitine, choline, or betaine into an overproduction of trimethylamine-N-oxide (TMAO). In 4007 patients referred to Cleveland Clinic for coronary angiograms, those with plasma levels in the top quartile of TMAO levels after a challenge of two hard-boiled eggs had a 2.5-fold increase in the 3-year risk of stroke, myocardial infarction or vascular death (Tang et al., 2013; Spence et al., 2021; Manolis et al., 2022). Gut dysbiosis also leads to increased intestinal permeability allowing the microbiome-derived lipopolysaccharide (LPS), a bacterial endotoxin, to enter the blood circulation and trigger inflammatory responses. Excessive LPS in the blood can lead to septic shock and death. Elevated TMAO and LPS levels alter bile acid metabolism pathways and increase the risk of cardiovascular diseases including atherosclerosis, myocardial infarction, thrombosis, arrhythmias, and stroke.

In clinical trial registration ClinicalTrials.gov NCT01906554, 6 volunteers previously on a low-choline diet participated in a longitudinal, double-blind, randomized dietary intervention, during which the participants were fed 0, 1, 2, 4, or 6 egg yolks (Miller et al., 2014). Each dose of egg yolk was separated by a 2-week washout period. Results indicate that consumption of ≥ 2 egg yolks was associated with a 14% conversion of egg yolk choline into TMAO, an increase in plasma TMAO although there was considerable variation between individuals in the TMAO (from near 0 μM to 30 μM) response. Plasma levels of high-sensitivity C-reactive protein (hsCRP) or oxidized LDL concentrations showed no difference after egg doses. The study had a low number of participants and an additional study was recommended to confirm the association between egg yolk and TMAO and cardiovascular disease and make any medical recommendations on choline. Gut microbiome belonging to the phyla *Firmicutes* and *Proteobacteria* and dysbiosis has been identified as contributing to the differences in TMAO production (Brown & Hazen, 2014; Romano et al., 2015; Arias et al., 2020).

In another trial registration, Clinicaltrials.gov (Protocol NCT03877003), 23 participants with metabolic syndrome were randomly assigned to consume 3 eggs per day and/or a choline supplement for 4 weeks in a controlled crossover study. The eggs and the supplement had a choline equivalent of 400 mg/day (Thomas et al., 2022). Plasma TMAO, was not different between treatments compared to the baseline ($p > 0.05$).

Among 316 patients attending stroke prevention clinics (age 66.74 ± 10.42 years; 59.7% men) in a cohort study, results of nutrient intake assessed by the 131-item Harvard Food Frequency Questionnaire and TMAO assessed by UPLC-TOF-MS indicated that higher plasma levels of GDUT was induced by even a moderate impairment of renal function (Pignanelli et al., 2019). The investigators suggested that patients at risk of atherosclerosis, particularly patients with impaired renal function should limit their intake of choline rich food such as egg yolk.

Kang and Zivkovic concluded recently that it would be beneficial to check in clinical studies if eggs may be beneficial to be incorporated or kept into the diet of individuals whose plasma cholesterol and TMAO are not affected by egg cholesterol and choline intake (Kang & Zivkovic, 2022). For patients at risk of cardiovascular disease and given that egg is an excellent source of both macro- and micronutrients, patient's genetic background, absorption, bioavailability differences, the integrity of the patient's gastrointestinal tract and associated digestive organs, and the individual's microbiome should all be considered before recommending egg yolk intake to prevent the potential of plasma cholesterol and TMAO increase. More clinical studies are needed in order to develop Precision Nutrition to help identify patients whose egg intake is not at risk of developing cardiovascular disease and those whose egg intake is limited because of their susceptibility to increase plasma TMAO or cholesterol. Since TMAO is high in fish but has no effect on consumer's cardiovascular disease, more research is warranted to examine the association of chronic egg yolk consumption and serum TMAO responses in individuals with chronic conditions such as diabetics.

24.3.2.4 Cholesterol

One egg yolk contains about 200–240 mg cholesterol. As a result, it was decided to remove the dietary cholesterol restriction of 300 mg per day from the US Dietary Guidelines (Millen et al., 2016). Therefore, consumption of one egg per day is not associated with risk factors for cardiovascular disease, because observational and interventional studies have shown that dietary cholesterol when not accompanied by high levels of saturated fats, as with egg yolk, does not raise plasma cholesterol to alarming levels. (Millen et al., 2016; Soliman, 2018; Drouin-Chartier et al., 2020).

The association of egg yolk consumption and increase in plasma cholesterol seems to depend on the individual because, at the current level of knowledge, only individuals with genetic polymorphism in the ABCG5 and ABCG8 genes are susceptible to cholesterol hyperabsorption at the level of the intestinal tract. Dysbiosis may also increase cholesterol absorption by forming coprostanol and showing lower levels of fecal and serum cholesterol (Kenny et al., 2020). Thus, whether eggs raise plasma cholesterol depends on the individual. In most cases, dietary cholesterol intake in the form of eggs since it is not accompanied also by an increased saturated fat intake may even improve high-density lipoprotein (HDL) cholesterol efflux capacity. Other studies have shown that consumption of 2–3 whole eggs per day had no effect on total cholesterol or LDL-cholesterol but increased the cholesterol efflux capacity of HDL particles in overweight postmenopausal women and paraoxonase-1 activity in healthy individuals (DiMarco et al., 2017).

More research at the clinical trial levels is needed to determine how to identify individuals at risk for CVD for whom egg yolk intake would be contraindicated vs. those for whom egg yolk intake is not associated with gut microbiome-derived TMAO metabolites, uremic toxic compounds, and negative effects on plasma lipid profiles.

24.3.3 Egg Yolk Proteins

24.3.3.1 Low-Density Lipoproteins

Low-density lipoproteins are found in the plasma where they represent 85% of the plasma proteins and in the granules (LDLg) where they represent 12% of the proteins (Anton, 2013). Low-density lipoprotein (LDL) is an endogenous cholesterol carrier, which makes it undesirable for human consumption.

Male participants (four men per treatment with age range 50–62 years old, not in cholesterol-lowering medication, the average weight of 79 ± 6.5 kg, and average waist circumference of 87 ± 6.8 cm) consumed chicken egg, goose egg, ostrich egg, or turkey egg for 30 days (Golzar Adabi et al., 2013). Plasma low-density lipoprotein (LDL) increased ($P < 0.05$) in all male subjects. The LDL-C:HDL-C ratio (which is a strong predictor of coronary heart disease risk) increased, although non-significantly, by consuming chicken, quail, and ostrich eggs.

24.3.3.2 High-Density Lipoprotein

Egg yolk high-density lipoprotein (HDL) represents 70% of yolk granule proteins and is poorly soluble and dispersible in water and entrapped in the undissolved granules of egg yolk (Anton, 2013). HDL is complexed to phosvitin in a non-soluble form and the complex is linked by phosphocalcic bridges made possible by the phosphate groups that bind calcium. Human HDL can eliminate cholesterol from blood vessels by promoting cholesterol efflux from macrophages as the first step in reversing cholesterol transport and protecting against inflammation and oxidative stress (Barter et al., 2004; Soran et al., 2009; Rosenson et al., 2012). It has been hypothesized that hen yolk HDL will elevate human serum HDL. Cholesterol efflux capacity is negatively correlated with incident of cardiovascular disease (Rohatgi et al., 2014; Shea et al., 2019). A clinical trial was performed to evaluate how diet affects the functionality of HDL in postmenopausal women who often are at higher risk for cardiovascular disease (CVD) than their younger counterparts (Sawrey-Kubicek et al., 2019). Also, it is not known whether higher HDL cholesterol correlates with changes in HDL function (Rosenson et al., 2012; El Khoudary et al., 2016). Clinical trial registered at clinicaltrials.gov as NCT92445638 enrolled 20 overweight, postmenopausal women (mean ± SD age of 57.7 ± 5.3 years and mean ± SD BMI of 28.3 ± 2.9) in a 14-week, single-blind, randomized crossover dietary trial. "The crossover treatments were frozen breakfast meals containing 100 g of liquid whole eggs compared with 100 g of yolk-free eggs per day, separated by a 4-week washout. Fasting blood samples were taken at the beginning and end of each treatment period to determine the effects on HDL composition and function." Results indicate that the consumption of 2 whole eggs/day by overweight, postmenopausal women showed a significant increase in cholesterol efflux capacity (mean ± SD percentage change: +5.69% ± 9.9%) compared with the yolk-free egg treatment (−3.69% ± 5.3%) ($P < 0.01$). No significant changes were observed in HDL functions or antioxidant or inflammatory markers. Similarly, egg treatment did not change ApoA-I, total cholesterol (TC), LDL cholesterol, and HDL cholesterol. The investigators suggested that HDL function rather than HDL cholesterol should be addressed in this population.

Metabolic syndrome (MetS) is associated with reductions in plasma lipoprotein. In a study that enrolled participants with metabolic syndrome to investigate the effects of whole egg feeding on plasma and lipoprotein carotenoids, participants consumed 3 whole eggs per day (EGG, $n = 20$) or the equivalent amount of yolk-free egg substitute (SUB, $n = 17$), as part of a carbohydrate-restricted diet (CRD) for 12 weeks (Blesso et al., 2013). Significant enrichment of isolated HDL and LDL fractions in lutein (+20% HDL, +9% LDL) and zeaxanthin (+57% HDL, +46% LDL) was observed in the EGG group after 12 weeks. HDL increased lutein and zeaxanthin transport ($P < 0.01$), β-cryptoxanthin transport ($P < 0.05$), and lycopene transport ($P < 0.05$) after 12 weeks intervention. The trial also showed that daily intake of 3 whole eggs, as part of a CRD, increased HDL which carried along lutein and zeaxanthin. Egg yolk may represent an important food source to improve HDL intake. The potentials of HDL in population at risk of cardiovascular disease such as diabetes warrant further studies.

In a randomized cross-over design, 33 men and women, > 60 years of age, participated in an 18-week study, which included one run-in and one washout period of no eggs prior to and between two 5-week interventions of either consuming 1 egg or egg substitute/day (Goodrow et al., 2006). Serum concentrations of total cholesterol, LDL cholesterol, HDL cholesterol, and triglycerides were not affected. One egg a day in older adults had no effect on HDL

24.3.3.3 Livetins

Livetins are globular glycoproteins that account for 30% of plasma proteins, correspond to chicken blood serum proteins and are present in egg yolk plasma as α (serum albumin), β (α_2-glycoprotein), and γ-forms, in which γ-livetin is primarily dominated by immunoglobulin Y (IgY) to be an ideal alternative to mammalian IgG antibodies (Kovacs-Nolan & Mine, 2012).

There are several advantages of using IgY in human health. The use of chickens for the production of polyclonal antibodies provides several advantages over the traditional method of producing antibodies in mammals. In contrast to mammalian serum, egg yolk, in contrast to mammalian serum, contains only a single class of antibody, IgY, which can be easily purified from the yolk by simple precipitation techniques (Gassmann et al., 1990). IgY recognizes different epitopes than mammalian antibodies IgG, giving access to a different antibody repertoire than with mammalian antibodies (Carlander et al., 1999). The method of producing antibodies in hens is much less invasive and requires only egg collection and is therefore less stressful on the bird (Schade et al., 1991). Sustained high titers in chickens reduce the need for frequent injections (Gassmann et al., 1990). Birds cost less than most animals. The animal care and IgY production costs are also lower for the chicken compared to that for mammals. Akita and Nakai employed water dilution at pH 5.0–5.2 to separate water-soluble proteins from egg yolk granules and by the combination of salt precipitation, ultrafiltration, and gel filtration, recovered 100 mg of electrophoretically pure egg yolk immunoglobulin IgY from one egg. and ion exchange chromatography (Akita & Nakai, 1992). IgY is stable in the egg for months, and purified IgY can be stored for years in the cold (Larsson et al., 1993). IgY elicits low or no allergenicity. Akita and Nakai studied the allergenicity of using passive cutaneous anaphylaxis and observed that IgY or its antigen binding (Fab′) and crystallizable (pFc′) fragments produced by pepsin digestion, elicited low IgE antibody response in mice when compared to egg white, a known allergen (Akita et al., 1999). The investigators used ELISA and determined that the pFc′ fragment of IgY was more antigenic than the Fab′ fragment. It was concluded that the low IgE antibody response to the egg yolk proteins was associated with a higher suppressor T-lymphocytes or lower helper T-lymphocytes, or both, in comparison to egg white-derived lymphocytes.

IgY has been used in several studies for oral passive immunization against various bacteria and viruses. has several reported *in vitro* and *in vivo* bioactivities. The FAb' fragment produced by peptic digestion of IgY was found to be effective in neutralizing heat-labile toxin produced by *enterotoxigenic* Escherichia coli (ETEC) *strain H10407* (Akita et al., 1998).

Kovacs and Mine summarized very well the immunotherapeutic applications of IgY in animal husbandry, aquaculture, and human disease (Kovacs-Nolan & Mine, 2004). Rotavirus, coronavirus, *E.coli*, *Salmonella*, *Yersinia*, *Edwardsiella*, *Staphylococcus*, *Streptococcus*, *Vibrio anguillarum*, shrimp white spot virus syndrome, and *Pseudomonas* in human, fish, birds, or animals can be eliminated using IgY. IgY-specific antibodies to Streptococcus mutants added in a mouthwash were investigated for passive immunization against dental plaque (Hatta et al., 1997; Smith et al., 2001).

Helicobacter pylori is an important etiologic agent of human gastroduodenal ulcers and malignancies, which infects over 50% of the world population. *Helicobacter pylori* urease catalyzes the organism's colonization in the gastroduodenal mucosa. Hens immunized with the purified urease produce a highly specific anti-*H. pylori* urease immunoglobulin (IgY-urease) in their egg yolks (Horie et al., 2004). A functional drinking yogurt containing 1% of egg yolk IgY-urease, *Lactobacillus acidophilus* and *Bifidobacterium* spp was given to 42 volunteers who tested positive for *H. pylori* using a ^{13}C-urea breath test. Participants consumed 150 mL of yogurt three times a day for 4 weeks. After 2 and 4 weeks, participants who tested positive showed a significant decrease in the urea breath test compared to the control group.

IgY has also been used in the prevention of *Candida albicans*-induced oral candidiasis in chemotherapy-treated immunocompromised children (Wilhelmson et al., 2005). *In vitro* study on IgY against acne was pursued by showing that IgY produced had strong specific antibody reactivity (Selvan et al., 2012). Rotavirus in cows can be treated with IgY (Vega et al., 2015). Clinical efficacy of IgY-against rotavirus has been reported in children where the yolk immunoglobulin significantly reduced the duration of intravenous fluid administration (Rahman et al., 2012).

Cystic fibrosis (CF) pulmonary disease is characterized by progressive pulmonary decline governed by a persistent, exaggerated inflammatory response dominated by polymorphonuclear neutrophils (PMNs). Chronic *Pseudomonas aeruginosa* (PA) biofilm infection is the principal contributor which attracts, activates PMNs and perpetuate the inflammation in all patients. Strategies to prevent initial airway colonization with *P. aeruginosa* by augmenting the phagocytic competence of PMNs may postpone the deteriorating chronic biofilm infection. While it is impossible to eradicate PA when a patient has been chronically colonized, immunotherapy-specific egg-yolk antibodies (IgY) may be an alternative to antibiotics for the prevention of PA infections. CF patients gargled daily with an IgY-antibody preparation obtained from eggs of hens immunized with PA bacteria and compared to a group of CF patients who did not gargle with the preparation (Kollberg et al., 2003). Both groups had their first colonization with PA eradicated by antibiotics. In CF patients on IgY preparation, the period between the first and second colonization with PA was significantly prolonged compared to the control group (Kaplan-Meier $P = 0.015$, Breslow test). In the prolonged study, the treated group had only 2.5 sputum cultures positive for PA per 100 months of observation, and none of these patients became chronically colonized with PA. No adverse events were reported. In the control group, 13.7 cultures per 100 months of observation were positive for PA, and 24% patients became chronically colonized with PA. The study shows that antipseudomonal IgY has the potential to effectively prevent PA colonization without any severe adverse effects. A phase III study was awaiting.

24.3.3.4 Phosvitin

Phosvitin represents 16% of yolk proteins and 4% of yolk dry matter (Anton et al., 2007). Phosvitin is the most phosphorylated protein known in nature. The protein contains 10% phosphorus and 54% of the amino acids are serine residues of which 90% of phosphorylated (Clark, 1985). The phosphoserines can carry up to 15 consecutive residues which makes it challenging to hydrolyze this protein. Phosvitin is a mixture of two major subunits, including the α-phosvitin (160 kDa) and β-phosvitin (190 kDa) (Itoh et al., 1983). α-phosvitin is an aggregate of three or four sub-units of 35 to 40 kDa, and β-phosvitin contains four or five sub-units of 45 kDa (Anton et al., 2007). Phosvitin is a polyanionic polymer and therefore has the ability to chelate divalent cations, including Fe (II), Fe (III), Co (II), Mn (II), Ca (II), and Mg (II). As a result, phosvitin is one of the strongest metal-chelating, naturally occurring compound (Hegenauer et al., 1979). Phosvitin also has strong antioxidative properties. The protein is resistant to proteolysis because the negative charge of the protein phosphate groups prevents enzymes to access the peptide bonds. Protocols to isolate phosvitin have been developed and continue to be developed (Losso et al., 1993; Losso & Nakai, 1994; Zhang et al., 2011).

The potential health benefits of phosvitin against chronic diseases are numerous. LPS in the blood at very high levels can cause septic shock. Over 50% of mortality in critically ill patients has been associated with septic shock for which effective treatments are still being sought after. Phosvitin or its peptides can neutralize Gram-negative bacteria-generated lipopolysaccharides (LPS) or endotoxins. Phosvitin can inhibit the growth of Gram-negative bacteria *E. coli* and Gram-positive bacteria *Staphylococcus*, bind to Gram-negative bacterial LPS-induced tumor necrosis factor-alpha (TNF-α), lipoteichoic acid, and peptidoglycan (Ma et al., 2013). Phosvitin was not cytotoxic to murine macrophage RAW264.7 and showed no hemolytic activity toward human red blood cells. Similar endotoxin-neutralizing activities have been observed with zebrafish phosvitin peptides (Hu et al., 2013).

Phosvitin at 50μg/ml, added to B16F10 melanoma cells inhibited tyrosinase activity by approximately 42% and melanin synthesis by 17% compared to control cells without phosvitin (Jung et al., 2012). In the same work, phosvitin also inhibited the expression of tyrosinase-related protein 1 (TRP-1), TRP-2, and microphthalmia-associated transcription factor (MITF) in the B16F10 melanoma cells.

Given the properties of phosvitin or its phosphopeptides, it is hypothesized that this protein or its phosphopeptides can enhance health by way of its antioxidative activities or its ability to inhibit metalloproteinases, enzymes associated with angiogenesis and the progression of many degenerative diseases.

24.3.4 Egg Yolk Xanthophylls

Lutein and zeaxanthin are carotenoids with very minimal structural differences between them. Each xanthophyll contains a hydrophilic ionone ring structure with hydroxyl groups on each end, connected by conjugated C = C bonds (Nimalaratne et al., 2012). Isomers including all-E-lutein, all-E-zeaxanthin, 131 -Z-lutein, 13-Z-zeaxanthin are predominant in the egg yolk and are also diet-dependent. Lutein and zeaxanthin represent microgram quantities in the egg yolk. While their levels are low, these bioactives are easily bioavailable because of the presence of the phospholipids that they are soluble in and facilitate their absorption. The health benefits associated with lutein and zeaxanthin include reducing macular degeneration and improving eye health and vision.

Egg yolk is the best source of bioavailable polar xanthophylls lutein and zeaxanthin. The food matrix and processing have a significant effect on bioavailability and bioaccessibility of egg yolk xanthophylls. Boiled egg provides the most bioaccessible and absorbable xanthophylls (Nimalaratne et al., 2015). Lipoproteins in egg, specially the high-density lipoproteins, facilitate xanthophylls transport and human *in vivo* studies have shown that xanthophylls are better transported in the blood by egg lipoproteins than when consumed alone as a dietary supplement or with spinach, which is another good of lutein (Chung et al., 2004). Eleven moderately hypercholesterolemic men and women consumed two separate baseline diets, one diet contained tallow supplemented with 1.3 cooked egg yolk per day and the other contained corn oil supplemented with 1.3 egg yolk per day (Handelman et al., 1999). Plasma lutein concentration in the beef tallow diet and corn oil diet increased by 28% ($P < 0.05$) and 50%, ($P < 0.05$), respectively. Plasma zeaxanthin in tallow diet and corn oil diet increased by 142% ($P < 0.001$) and 114% ($P < 0.001$), respectively. It was concluded that egg yolk is an excellent carrier of lutein and zeaxanthin and these bioactives can counteract the effect of LDL-cholesterol in plasma.

In a study that looked at the potentials of lutein and zeaxanthin on macular degeneration, older adults (60 years +) consumed 2 followed by 4 egg yolks per day for 5 weeks each with a washout period of 4 weeks at baseline and between the 2 interventions (Vishwanathan et al., 2009). In 52 participants, serum lutein increased by only 16% after consumption of 2 egg yolks and by 24% after consumption of 4 egg yolks ($P <$ 0.05). Serum zeaxanthin increased by 36% after consumption of 2 egg yolks and by 82% after consumption of 4 egg yolks ($P < 0.001$) in serum zeaxanthin ($n = 52$) after consumption of 2 and 4 egg yolks, respectively. In 37 participants with low MPOD, macular pigment optical density (MPOD) increased more (≤ 50%) in individuals when they consumed 4 egg yolks than when individuals consumed 2 egg yolks (31%). Serum HDL cholesterol increased by 5% ($P < 0.05$) while serum LDL did not change after consumption of 2 and 4 egg yolks, most participants being on cholesterol-lowering statins.

In another study, when women aged 24–59 years old consumed 6 eggs per week for 12 weeks, serum zeaxanthin and macular pigment optical density increased significantly (Wenzel et al., 2006).

Clinical trial NCT00527553 investigated the effect of lutein or zeaxanthin-enriched eggs, or a lutein enriched egg-yolk based buttermilk beverage on serum lutein and zeaxanthin concentrations and macular pigment levels in 100 healthy men (Kelly et al., 2014). Participants consumed the carotenoid-rich egg for 90 days. Lutein- and zeaxanthin-enriched egg were developed by feeding excess amounts of the carotenoids. Group 1 consumed one normal egg in their daily diet. Group 2 received a lutein-enriched egg yolk in a beverage. Group 3 consumed one lutein-enriched egg in their diet. Group 4 added one zeaxanthin-enriched egg to their diet. Group 5 was the control and individuals in this group stay the course with their daily diet. Serum lutein and zeaxanthin concentrations and macular pigment densities were determined by heterochromatic flicker photometry at baseline, day 45, and day 90.

Serum lutein and zeaxanthin concentration increased in the lutein-enriched egg and egg yolk-based beverage groups ($P < 0.001$, 76%, and 77%) and in individuals receiving zeaxanthin-enriched eggs ($P < 0.001$, 430%), respectively. Macular densities of lutein or zeaxanthin in various groups did not change. It was concluded that egg yolk rich in lutein or zeaxanthin can increase the serum concentration of these bioactives to levels comparable with the daily intake of 5 mg supplements.

Lutein transport and concentration in the plasma parallels high-density lipoprotein transport (Blesso et al., 2013). Lutein accumulates in the retina where it increases macular pigment density thereby protecting the retina against the development of macular degeneration (Abdel-Aal et al., 2013).

In a clinical trial registration, Clinicaltrials.gov (Protocol NCT03877003), 23 participants with metabolic syndrome were randomly assigned to consume 3 eggs per day and/or a choline supplement for 4 weeks (Thomas et al., 2022). The eggs and the supplement had a choline equivalent of 400 mg/day. Plasma lutein and astaxanthin increased suggesting that xanthophylls in egg yolk can protect against oxidative stress, a common condition in metabolic syndrome.

In a randomized cross-over design, 33 men and women, > 60 years of age, participated in a 18-week study, which included one run-in and one washout period of no eggs prior to and between two 5-week interventions of either consuming 1 egg or egg substitute/day (Goodrow et al., 2006). Serum lutein increased by 26% ($P < 0.001$) and zeaxanthin increased by 38% ($P < 0.001$) after 5-week of 1 egg/day compared with the phase prior to consuming eggs. One egg a day increased carotenoids in older adults.

24.3.5 Egg Yolk Carbohydrates, Minerals, and Vitamins

Egg yolk contains about 0.7–1.0% carbohydrate of which glucose is present as free carbohydrate at the level of 0.3% and the remainder is either bound to proteins as glycoproteins or bound to lipids as glycolipids (Li-Chan & Kim, 2008). Sialic

TABLE 24.2
Clinical Trials with Egg Yolk Bioactives

Egg Yolk Bioactive	Participant's Health Condition	Physiological Activity Tested	Outcome	References
Egg yolk, *n* = 6	Healthy volunteers	Plasma trimethylamine-N-oxide (TMAO)	Considerable variation in TMAO among individuals	Miller et al. (2014)
Egg (3/d) containing 400 mg choline equivalent or Choline supplement (400 mg), *n* = 23; 4 weeks	Participants with metabolic syndrome	TMAO	TMAO not different among groups	Thomas et al. (2022)
Low-density lipoproteins (*n* = 4), 30 days	Participants not on low-cholesterol medication	Plasma LDL	Plasma LDL increased	Golzar et al. (2013)
Whole eggs (2 eggs/day); *n* = 20; 14-weeks	Postmenopausal women (57.7 ± 5.3 years) and BMI (28.3 ± 2.9)	HDL, inflammatory markers	Increase in cholesterol efflux capacity, No effect on HDL	Sawrey-Kubicek et al. (2019)
Whole egg (3 whole eggs/day); *n* = 20; 12 weeks	Metabolic syndrome	Carotenoids	Increase of HDL, lutein, zeaxanthin, β-cryptoxanthin, lycopene	Blesso et al. (2013)
Whole egg (1 egg/day); *n* = 33 (men and women); > 60 years; 18 week	Healthy	Cholesterol, HDL, LDL	No effect	Goodrow et al. (2006)
IgY against urease, *n* = 42; 2–4 weeks	*Helicobacter pylori*	Urea breath test	Significant decrease in the urea breath test compared to the control group.	Horie et al. (2004)
IgY against Pseudomonas aeruginosa	Cystic fibrosis	*Pseudomonas* colonization	Prevented colonization without severe effects	Kollberg et al. (2003)
Egg yolk *n* = 11, 1.3 egg/day	Hypercholesterolemic	Plasma lutein and zeaxanthin	Significant increase over 100%	Handelman et al. (1999)
Egg yolk; *n* = 52; 2 then 4 egg yolks/day; 15 weeks	Old people (60 years+)	Macular pigment	Macular pigment density increased with 4 eggs	Vishwanathan et al. (2009)
Whole egg, 6 eggs/week; 12 weeks	Pre and post menopause women (24–59 years old)	Macular pigment	Macular pigment density increased	Wenzel et al. (2006)
Xanthophyll-enriched eggs; *n* = 100; 90 days	Healthy men	Serum lutein, zeaxanthin and macular pigment	Lutein and zeaxanthin increased. Macular pigment did not	Kelly et al. (2014)
Whole egg (3 eggs/day)or choline (400 mg); *n* = 23; 4 weeks	Metabolic syndrome	Plasma lutein or zeaxanthin	Plasma lutein and zeaxanthin increased	Thomas et al. (2022)
Egg or egg substitute (1 per day) *n* = 33; 18 weeks	Healthy volunteers (60 years+)	Plasma lutein or zeaxanthin	Serum lutein and zeaxanthin increased	Goodrow et al. (2006)

acid (N-acetylneuraminic acid) is common in egg yolk glycoproteins or glycolipids at a total level of 5.2 ± 0.4 μg/mg yolk (Nakano et al., 1994).

Egg yolk contains 1% minerals of which phosphorus (102 mg) is the major one because of its presence in phospholipids and phosvitin (Watkins, 1995). The yolk also contains chloride (29.9 mg), sulfur (28 mg), calcium (25.2), potassium (17 mg), sodium (9 mg), magnesium (2.15 mg), iron (1.02 mg), manganese (0.019 mg), cupper (0.024 mg), and iodine (0.024 mg).

The egg yolk is a very good source of vitamin A (260 IU, 12% of recommended daily allowance), D (27 IU, 6% of recommended daily allowance), riboflavin (0.057 mg, 9% of recommended daily allowance), and pantothenic acid (0.01 mg, 8% of recommended daily allowance) (Watkins, 1995). From the same reference, other vitamins in yolk included folic acid (238 mg), niacin (4.35 mg), vitamin E (0.87 mg), pyridoxine (0.73 mg), and vitamin B_{12} (0.48 mg).

Table 24.2 Summarizes clinical trials carried out with egg yolk bioactives. While there are more trials performed with egg yolk bioactives than egg white bioactives, there is still a need for more investigations for bioactives with the potential for enhancing human health.

24.4 Conclusion and Future Perspectives

Egg white is a good source of water-soluble proteins, most of which have pro-inflammatory properties while others have a combination of pro-allergenicity and anti-inflammatory activities. Peptides from some of these proteins such as lysozyme peptides, ovotransferrin peptides, and ovomucin peptides are bioactives with health-enhancing activities. Food scientists and nutritionists are encouraged to further investigate the *in vivo* clinical activities of peptides from egg white peptides.

In the yolk, phosvitin has been significantly investigated but mostly if not all *in vitro* only. Egg yolk is a good source and carrier of xanthophylls lutein and zeaxanthin. Xanthophyll-rich diets need to be promoted for hens in an effort to increase yolk xanthophylls for human health. More human trials have been performed with egg yolk bioactives than egg white bioactives. However, clinical trials are needed for phosvitin. High-density lipoproteins also need human trials as a potential substitute for human HDL. Hen egg LDL consumption should be reduced or limited. More work is needed on the interaction of egg yolk phospholipids and human gut microbiome during eubiosis and dysbiosis, respectively, to draw a conclusion on the effect of egg yolk phospholipids on human health.

REFERENCES

Abdel-Aal, E.-S. M., Akhtar, H., Zaheer, K., & Ali, R. (2013). Dietary sources of lutein and zeaxanthin carotenoids and their role in eye health. *Nutrients*, *5*(4), 1169–1185.

Abeyrathne, E., Lee, H., Jo, C., Suh, J., & Ahn, D. (2016). Enzymatic hydrolysis of ovomucin and the functional and structural characteristics of peptides in the hydrolysates. *Food Chemistry*, *192*, 107–113.

Akashi, T., Muto, A., Takahashi, Y., & Nishiyama, H. (2017). Enteral formula containing egg yolk lecithin improves diarrhea. *Journal of Oleo Science*, *66*(9), 1017–1027. https://doi.org/10.5650/jos.ess17007

Akita, E. M., & Nakai, S. (1992). Immunoglobulins from egg yolk: Isolation and purification. *Journal of Food Science*, *57*(3), 629–634.

Akita, E., Jang, C., Kitts, D., & Nakai, S. (1999). Evaluation of allergenicity of egg yolk immunoglobulin Y and other egg proteins by passive cutaneous anaphylaxis. *Food and Agricultural Immunology*, *11*(2), 191–201.

Akita, E., Li-Chan, E., & Nakai, S. (1998). Neutralization of enterotoxigenic *Escherichia coli* heat-labile toxin by chicken egg yolk immunoglobulin Y and its antigen-binding fragments. *Food and Agricultural Immunology*, *10*(2), 161–172.

Anastasi, A., Brown, M., Kembhavi, A., Nicklin, M., Sayers, C. A., Sunter, D., & Barrett, A. (1983). Cystatin, a protein inhibitor of cysteine proteinases. Improved purification from egg white, characterization, and detection in chicken serum. *Biochemical Journal*, *211*(1), 129–138.

Anton, M. (2013). Egg yolk: Structures, functionalities and processes. *Journal of the Science of Food and Agriculture*, *93*(12), 2871–2880.

Anton, M., Castellani, O., & Guérin-Dubiard, C. (2007). Phosvitin. In *Bioactive egg compounds* (pp. 17–24). Springer.

Arias, N., Arboleya, S., Allison, J., Kaliszewska, A., Higarza, S. G., Gueimonde, M., & Arias, J. L. (2020). The relationship between choline bioavailability from diet, intestinal microbiota composition, and its modulation of human diseases. *Nutrients*, *12*(8), 2340.

Azari, P. R., & Feeney, R. E. (1958). Resistance of metal complexes of conalbumin and transferrin to proteolysis and to thermal denaturation. *Journal of Biological Chemistry*, 293–302.

Baron, F., Gautier, M., & Brule, G. (1997). Factors involved in the inhibition of growth of salmonella enteritidis in liquid egg white. *Journal of Food Protection*, *60*(11), 1318–1323. https://doi.org/10.4315/0362-028x-60.11.1318

Barter, P. J., Nicholls, S., Rye, K. A., Anantharamaiah, G. M., Navab, M., & Fogelman, A. M. (2004). Antiinflammatory properties of HDL. *Circulation Research*, *95*(8), 764–772. https://doi.org/10.1161/01.Res.0000146094.59640.13

Bernhisel-Broadbent, J., Dintzis, H. M., Dintzis, R. Z., & Sampson, H. A. (1994). Allergenicity and antigenicity of chicken egg ovomucoid (Gal d III) compared with ovalbumin (Gal d I) in children with egg allergy and in mice. *Journal of Allergy and Clinical Immunology*, *93*(6), 1047–1059.

Bickel, U., Yoshikawa, T., & Pardridge, W. M. (2001). Delivery of peptides and proteins through the blood-brain barrier. *Advanced Drug Delivery Reviews*, *46*(1–3), 247–279. https://doi.org/10.1016/s0169-409x(00)00139-3

Blankenvoorde, M. F., Brand, H. S., Henskens, Y. M., Veerman, E. C., & Amerongen, A. V. N. (2000). Protease inhibitors in health and disease control—Medical and industrial aspects. In *Recombinant protease inhibitors in plants* (pp. 203–214). Landes Bioscience.

Blesso, C. N., Andersen, C. J., Bolling, B. W., & Fernandez, M. L. (2013). Egg intake improves carotenoid status by increasing plasma HDL cholesterol in adults with metabolic syndrome. *Food & Function*, *4*(2), 213–221. https://doi.org/10.1039/c2fo30154g

Board, R., Tranter, H., Stadelman, W., & Cotterill, O. (1995). *Egg science and technology* (W. J. Stadelman & O. J. Cotterill, Eds.). Haworth Press, Inc.

Brock, J., Arzabe, F., Lampreave, F., & Pineiro, A. (1976). The effect of trypsin on bovine transferrin and lactoferrin. *Biochimica et Biophysica Acta (BBA)-Protein Structure*, *446*(1), 214–225.

Brown, J. M., & Hazen, S. L. (2014). Meta-organismal nutrient metabolism as a basis of cardiovascular disease. *Current Opinion in Lipidology*, *25*(1), 48.

Cantor, J. O., Shteyngart, B., Cerreta, J. M., & Turino, G. M. (2002). The effect of lysozyme on elastase-mediated injury. *Experimental Biology and Medicine (Maywood)*, *227*(2), 108–113. https://doi.org/10.1177/153537020222700205

Carlander, D., Stålberg, J., & Larsson, A. (1999). Chicken antibodies: A clinical chemistry perspective. *Upsala Journal of Medical Sciences*, *104*(3), 179–189.

Chung, H.-Y., Rasmussen, H. M., & Johnson, E. J. (2004). Lutein bioavailability is higher from lutein-enriched eggs than from supplements and spinach in men. *Journal of Nutrition*, *134*(8), 1887–1893.

Chung, L., Dinakarpandian, D., Yoshida, N., Lauer-Fields, J. L., Fields, G. B., Visse, R., & Nagase, H. (2004). Collagenase unwinds triple-helical collagen prior to peptide bond hydrolysis. *The EMBO Journal*, *23*(15), 3020–3030. https://doi.org/10.1038/sj.emboj.7600318

Clark, R. C. (1985). The primary structure of avian phosvitins. Contributions through the Edman degradation of methylmercaptovitins prepared from the constituent phosphoproteins. *The International Journal of Biochemistry*, *17*(9), 983–988.

Clemente, A., Carmen Marín-Manzano, M., Jiménez, E., Carmen Arques, M., & Domoney, C. (2012). The anti-proliferative effect of TI1B, a major bowman-birk isoinhibitor from pea (Pisum sativum L.), on HT29 colon cancer cells is mediated through protease inhibition. *British Journal of Nutrition*, *108*(Suppl 1), S135–S144. https://doi.org/10.1017/s000711451200075x

De Oliveira, F. C., dos Reis Coimbra, J. S., Mendes da Silva, L. H., Garcia Rojas, E. E., & do Carmo Hespanhol da Silva, M. (2009). Ovomucoid partitioning in aqueous two-phase systems. *Biochemical Engineering Journal*, *47*(1–3), 55–60.

DiMarco, D. M., Norris, G. H., Millar, C. L., Blesso, C. N., & Fernandez, M. L. (2017). Intake of up to 3 eggs per day is associated with changes in HDL function and increased plasma antioxidants in healthy, young adults. *Journal of Nutrition*, *147*(3), 323–329.

Ding, L., Wang, L., Zhang, Y., & Liu, J. (2015). Transport of antihypertensive peptide RVPSL, ovotransferrin 328–332, in human intestinal Caco-2 cell monolayers. *Journal of Agricultural and Food Chemistry*, *63*(37), 8143–8150.

Drouin-Chartier, J.-P., Chen, S., Li, Y., Schwab, A. L., Stampfer, M. J., Sacks, F. M., & Bhupathiraju, S. N. (2020). Egg consumption and risk of cardiovascular disease: Three large prospective US cohort studies, systematic review, and updated meta-analysis. *BMJ*, *368*. https://doi.org/10.1136/bmj.m513

El Khoudary, S. R., Wang, L., Brooks, M. M., Thurston, R. C., Derby, C. A., & Matthews, K. A. (2016). Increase HDL-C level over the menopausal transition is associated with greater atherosclerotic progression. *Journal of Clinical Lipidology*, *10*(4), 962–969. https://doi.org/10.1016/j.jacl.2016.04.008

Engstrand, L., & Graham, D. Y. (2020). Microbiome and gastric cancer. *Digestive Diseases and Sciences*, *65*(3), 865–873. https://doi.org/10.1007/s10620-020-06101-z

Gassmann, M., Thömmes, P., Weiser, T., & Hübscher, U. (1990). Efficient production of chicken egg yolk antibodies against a conserved mammalian protein. *The FASEB Journal*, *4*(8), 2528–2532.

Geng, F., Huang, X., Majumder, K., Zhu, Z., Cai, Z., & Ma, M. (2015). Mass spectrometry and two-dimensional electrophoresis to characterize the glycosylation of hen egg white ovomacroglobulin. *Journal of Agricultural and Food Chemistry*, *63*(37), 8209–8215. https://doi.org/10.1021/acs.jafc.5b02618

Golzar Adabi, S. H., Ahbab, M., Fani, A. R., Hajbabaei, A., Ceylan, N., & Cooper, R. G. (2013). Egg yolk fatty acid profile of avian species–influence on human nutrition. *Journal of Animal Physiology and Animal Nutrition (Berl)*, *97*(1), 27–38. https://doi.org/10.1111/j.1439-0396.2011.01239.x

GonzÁlez, I., Araya, P., & Rojas, A. (2018). Helicobacter pylori infection and lung cancer: New insights and future challenges. *Zhongguo Fei Ai Za Zhi*, *21*(9), 658–662. https://doi.org/10.3779/j.issn.1009-3419.2018.09.03

Goodrow, E. F., Wilson, T. A., Houde, S. C., Vishwanathan, R., Scollin, P. A., Handelman, G., & Nicolosi, R. J. (2006). Consumption of one egg per day increases serum lutein and zeaxanthin concentrations in older adults without altering serum lipid and lipoprotein cholesterol concentrations. *Journal of Nutrition*, *136*(10), 2519–2524. https://doi.org/10.1093/jn/136.10.2519

Haginaka, J., Matsunaga, H., & Kakehi, K. (2000). Separation of enantiomers on a chiral stationary phase based on ovoglycoprotein: VIII. Chiral recognition ability of partially and completely deglycosylated ovoglycoprotein. *Journal of Chromatography B: Biomedical Sciences and Applications*, *745*(1), 149–157.

Handelman, G. J., Nightingale, Z. D., Lichtenstein, A. H., Schaefer, E. J., & Blumberg, J. B. (1999). Lutein and zeaxanthin concentrations in plasma after dietary supplementation with egg yolk. *The American Journal of Clinical Nutrition*, *70*(2), 247–251.

Hatta, H., Tsuda, K., Ozeki, M., Kim, M., Yamamoto, T., Otake, S., & Michalek, S. (1997). Passive immunization against dental plaque formation in humans: Effect of a mouth rinse containing egg yolk antibodies (IgY) specific to streptococcus mutans. *Caries Research*, *31*(4), 268–274.

Hegenauer, J., Saltman, P., & Nace, G. (1979). Iron (III)-phosphoprotein chelates: Stoichiometric equilibrium constant for interaction of iron (III) and phosphorylserine residues of phosvitin and casein. *Biochemistry*, *18*(18), 3865–3879.

Hilpert, K. (2021). Peptides in COVID-19 clinical trials—A snapshot. *Biologics*, *1*(3), 300–311. https://doi.org/10.3390/biologics1030018

Holen, E., & Elsayed, S. (1990). Characterization of four major allergens of hen egg-white by IEF/SDS-PAGE combined with electrophoretic transfer and IgE-immunoautoradiography. *International Archives of Allergy and Immunology*, *91*(2), 136–141.

Horie, K., Horie, N., Abdou, A., Yang, J.-O., Yun, S.-S., Chun, H.-N., & Hatta, H. (2004). Suppressive effect of functional drinking yogurt containing specific egg yolk immunoglobulin on helicobacter pylori in humans. *Journal of Dairy Science*, *87*(12), 4073–4079.

Hu, L., Sun, C., Wang, S., Su, F., & Zhang, S. (2013). Lipopolysaccharide neutralization by a novel peptide derived from phosvitin. *The International Journal of Biochemistry & Cell Biology*, *45*(11), 2622–2631. https://doi.org/10.1016/j.biocel.2013.09.002

Huopalahti, R., Anton, M., López-Fandiño, R., & Schade, R. (2007). *Bioactive egg compounds* (Vol. 5). Springer.

Hytönen, V. P., Laitinen, O. H., Grapputo, A., Kettunen, A., Savolainen, J., Kalkkinen, N., & Kulomaa, M. S. (2003). Characterization of poultry egg-white avidins and their potential as a tool in pretargeting cancer treatment. *Biochemical Journal*, *372*(Pt 1), 219–225. https://doi.org/10.1042/bj20021531

Ibrahim, H. R., & Kiyono, T. (2009). Novel anticancer activity of the autocleaved ovotransferrin against human colon and breast cancer cells. *Journal of Agricultural and Food Chemistry*, *57*(23), 11383–11390.

Ibrahim, H. R., Hoq, M. I., & Aoki, T. (2007). Ovotransferrin possesses SOD-like superoxide anion scavenging activity that is promoted by copper and manganese binding. *International Journal of Biological Macromolecules*, *41*(5), 631–640.

Itoh, T., Abe, Y., & Adachi, S. (1983). Comparative studies on the α and α-Phosvitin from Hen's egg yolk. *Journal of Food Science*, *48*(6), 1755–1757.

Itoh, T., Miyazaki, J., Sugawara, H., & Adachi, S. (1987). Studies on the characterization of ovomucin and chalaza of the hen's egg. *Journal of Food Science*, *52*(6), 1518–1521.

Jiao, H., Zhang, Q., Lin, Y., Gao, Y., & Zhang, P. (2019). The ovotransferrin-derived peptide IRW attenuates lipopolysaccharide-induced inflammatory responses. *BioMed Research International*, *2019*, Article ID 8676410, 7. https://doi.org/10.1155/2019/8676410

Jukes, C., Ricci, L., von Coburg, E., Nasr, M., Walker, A., Duncan, S., & Hegazy, A. (2020). P021 the influence of short-chain fatty acids produced by commensal bacteria on macrophage phenotype. *Journal of Crohn's and Colitis*, *14*(Supplement_1), S141–S142.

Jung, S., Kim, D. H., Son, J. H., Nam, K., Ahn, D. U., & Jo, C. (2012). The functional property of egg yolk phosvitin as a melanogenesis inhibitor. *Food Chemistry*, *135*(3), 993–998. https://doi.org/10.1016/j.foodchem.2012.05.113

Kang, J. W., & Zivkovic, A. M. (2022). Are eggs good again? A precision nutrition perspective on the effects of eggs on cardiovascular risk, taking into account plasma lipid profiles and TMAO. *Journal of Nutritional Biochemistry*, *100*, 108906. https://doi.org/10.1016/j.jnutbio.2021.108906

Kelly, E. R., Plat, J., Haenen, G. R., Kijlstra, A., & Berendschot, T. T. (2014). The effect of modified eggs and an egg-yolk based beverage on serum lutein and zeaxanthin concentrations and macular pigment optical density: Results from a randomized trial. *PLoS One*, *9*(3), e92659.

Kenny, D. J., Plichta, D. R., Shungin, D., Koppel, N., Hall, A. B., Fu, B., & Balskus, E. P. (2020). Cholesterol metabolism by uncultured human gut bacteria influences host cholesterol level. *Cell Host & Microbe*, *28*(2), 245–257. https://doi.org/10.1016/j.chom.2020.05.013

Kim, M.-J., Lee, J.-W., Yook, H.-S., Lee, S.-Y., Kim, M.-C., & Byun, M.-W. (2002). Changes in the antigenic and immunoglobulin E–binding properties of hen's egg albumin with the combination of heat and gamma irradiation treatment. *Journal of Food Protection*, *65*(7), 1192–1195.

Kitamoto, T., Nakashima, M., & Ikai, A. (1982). Hen egg white ovomacroglobulin has a protease inhibitory activity. *Journal of Biochemistry*, *92*(5), 1679–1682. https://doi.org/10.1093/oxfordjournals.jbchem.a134097

Kodama, Y., & Kimura, N. (2001). Inhibitor of helicobacter pylori colonization. Google Patents. https://patents.google.com/patent/CA2292012C/en

Kollberg, H., Carlander, D., Olesen, H., Wejåker, P. E., Johannesson, M., & Larsson, A. (2003). Oral administration of specific yolk antibodies (IgY) may prevent *Pseudomonas aeruginosa* infections in patients with cystic fibrosis: A phase I feasibility study. *Pediatric Pulmonology*, *35*(6), 433–440.

Kovacs-Nolan, J., & Mine, Y. (2004). Avian egg antibodies: Basic and potential applications. *Avian and Poultry Biology Reviews*, *15*(1), 25–46.

Kovacs-Nolan, J., & Mine, Y. (2012). Egg yolk antibodies for passive immunity. *Annual Review of Food Science and Technology*, *3*, 163–182.

Kovacs-Nolan, J., Phillips, M., & Mine, Y. (2005). Advances in the value of eggs and egg components for human health. *Journal of Agricultural and Food Chemistry*, *53*(22), 8421–8431.

Langeland, T. (1985). Allergy to hen's egg white in atopic dermatitis. *Acta Dermato-Venereologica Supply (Stockh)*, *114*, 109–112. https://doi.org/10.2340/00015555114109112

Larsson, A., Balow, R.-M., Lindahl, T. L., & Forsberg, P.-O. (1993). Chicken antibodies: Taking advantage of evolution—A review. *Poultry Science*, *72*(10), 1807–1812.

Lee, M., Kovacs-Nolan, J., Yang, C., Archbold, T., Fan, M. Z., & Mine, Y. (2009). Hen egg lysozyme attenuates inflammation and modulates local gene expression in a porcine model of dextran sodium sulfate (DSS)-induced colitis. *Journal of Agricultural and Food Chemistry*, *57*(6), 2233–2240. https://doi.org/10.1021/jf803133b

Lewis, J., Snell, N., Hirschmann, D., & Fraenkel-Conrat, H. (1950). Amino-acid composition of egg proteins. *Journal of Biological Chemistry*, *186*, 23–35.

Liao, W., Fan, H., & Wu, J. (2018). Egg white-derived antihypertensive peptide IRW (Ile-Arg-Trp) inhibits angiotensin II-stimulated migration of vascular smooth muscle cells via angiotensin type I receptor. *Journal of Agricultural and Food Chemistry*, *66*(20), 5133–5138.

Li-Chan, E., & Nakai, S. (1989). Biochemical basis for the properties of egg white. *CRC Critical Reviews in Poultry Biology*, *2*, 21–58.

Li-Chan, E., Powrie, W., & Nakai, S. (1995). *Egg science and technology* (W. J. Stadelman & O. J. Cotterill, Eds.). The Haworth Press.

Li-Chan, E. C., & Kim, H. O. (2008). Structure and chemical composition of eggs. *Egg Bioscience and Biotechnology*, *10*, 9780470181249.

Losso, J. N., Bogumil, R., & Nakai, S. (1993). Comparative studies of phosvitin from chicken and salmon egg yolk. *Comparative Biochemistry and Physiology. B, Comparative Biochemistry*, *106*(4), 919–923.

Losso, J. N., & Nakai, S. (1994). A simple procedure for the isolation of phosvitin from chicken egg yolk. In J. S. Sim & S. Nakai (Eds.), *Egg uses and processing technologies: New developments* (pp. 150–157). CAB International.

Lucey, A. J., Heneghan, C., Manning, E., Kroon, P. A., & Kiely, M. E. (2019). Effect of an egg ovalbumin-derived protein hydrolysate on blood pressure and cardiovascular risk in adults with a mildly elevated blood pressure: A randomized placebo-controlled crossover trial. *European Journal of Nutrition*, *58*(7), 2823–2833.

Ma, J., Wang, H., Wang, Y., & Zhang, S. (2013). Endotoxin-neutralizing activity of hen egg phosvitin. *Molecular Immunology*, *53*(4), 355–362.

Mann, J. K., & Ndung'u, T. (2020). The potential of lactoferrin, ovotransferrin and lysozyme as antiviral and immune-modulating agents in COVID-19. *Future Virology*, *15*(9), 609–624.

Manolis, A. A., Manolis, T. A., Melita, H., & Manolis, A. S. (2022). Gut Microbiota and cardiovascular disease: Symbiosis versus dysbiosis. *Current Medicinal Chemistry*, *29*(23), 4050–4077. https://doi.org/10.2174/0929867328666211213112949

Mehana, E.-S. E., Khafaga, A. F., & El-Blehi, S. S. (2019). The role of matrix metalloproteinases in osteoarthritis pathogenesis: An updated review. *Life Sciences*, *234*, 116786.

Millen, B. E., Abrams, S., Adams-Campbell, L., Anderson, C. A., Brenna, J. T., Campbell, W. W., & Neuhouser, M. L. (2016). The 2015 dietary guidelines advisory committee scientific report: Development and major conclusions. *Advances in Nutrition*, *7*(3), 438–444.

Miller, C. A., Corbin, K. D., da Costa, K. A., Zhang, S., Zhao, X., Galanko, J. A., & Zeisel, S. H. (2014). Effect of egg ingestion on trimethylamine-N-oxide production in humans: A randomized, controlled, dose-response study. *The American Journal of Clinical Nutrition*, *100*(3), 778–786. https://doi.org/10.3945/ajcn.114.087692

Nagase, H., Harris, E. Jr, Woessner, J. Jr, & Brew, K. (1983). Ovostatin: A novel proteinase inhibitor from chicken egg white. I. Purification, physicochemical properties, and tissue distribution of ovostatin. *Journal of Biological Chemistry*, *258*(12), 7481–7489.

Nagase, H., Visse, R., & Murphy, G. (2006). Structure and function of matrix metalloproteinases and TIMPs. *Cardiovascular Research*, *69*(3), 562–573. https://doi.org/10.1016/j.cardiores.2005.12.002

Nakamura, S., Kato, A., & Kobayashi, K. (1992). Enhanced antioxidative effect of ovalbumin due to covalent binding of polysaccharides. *Journal of Agricultural and Food Chemistry*, *40*(11), 2033–2037.

Nakano, K., Nakano, T., Ahn, D., & Sim, J. (1994). Sialic acid contents in chicken eggs and tissues. *Canadian Journal of Animal Science*, *74*(4), 601–606.

Nimalaratne, C., Lopes-Lutz, D., Schieber, A., & Wu, J. (2012). Effect of domestic cooking methods on egg yolk xanthophylls. *Journal of Agricultural and Food Chemistry*, *60*(51), 12547–12552.

Nimalaratne, C., Savard, P., Gauthier, S. F., Schieber, A., & Wu, J. (2015). Bioaccessibility and digestive stability of carotenoids in cooked eggs studied using a dynamic in vitro gastrointestinal model. *Journal of Agricultural and Food Chemistry*, *63*(11), 2956–2962.

Petronzelli, F., Pelliccia, A., Anastasi, A. M., Lindstedt, R., Manganello, S., Ferrari, L. E., & De Santis, R. (2010). Therapeutic use of avidin is not hampered by antiavidin antibodies in humans. *Cancer Biotherapy and Radiopharmaceuticals*, *25*(5), 563–570. https://doi.org/10.1089/cbr.2010.0797

Pignanelli, M., Bogiatzi, C., Gloor, G., Allen-Vercoe, E., Reid, G., Urquhart, B. L., & Spence, J. D. (2019). Moderate renal impairment and toxic metabolites produced by the intestinal microbiome: Dietary implications. *Journal of Renal Nutrition*, *29*(1), 55–64. https://doi.org/10.1053/j.jrn.2018.05.007

Plat, J., Severins, N., Morrison, S., & Mensink, R. P. (2017). Effects of NWT-03, an egg-protein hydrolysate, on blood pressure in normotensive, high-normotensive and mild-hypertensive men and women: A dose-finding study. *British Journal of Nutrition*, *117*(7), 942–950.

Rahman, S., Higo-Moriguchi, K., Htun, K. W., Taniguchi, K., Icatlo, F. C. Jr, Tsuji, T., & Oo, H. N. (2012). Randomized placebo-controlled clinical trial of immunoglobulin Y as adjunct to standard supportive therapy for rotavirus-associated diarrhea among pediatric patients. *Vaccine*, *30*(31), 4661–4669.

Rohatgi, A., Khera, A., Berry, J. D., Givens, E. G., Ayers, C. R., Wedin, K. E., & Shaul, P. W. (2014). HDL cholesterol efflux capacity and incident cardiovascular events. *The New England Journal of Medicine*, *371*(25), 2383–2393. https://doi.org/10.1056/NEJMoa1409065

Romano, K. A., Vivas, E. I., Amador-Noguez, D., & Rey, F. E. (2015). Intestinal microbiota composition modulates choline bioavailability from diet and accumulation of the proatherogenic metabolite trimethylamine-N-oxide. *MBio*, *6*(2), e02481–02414.

Rosenson, R. S., Brewer, H. B. Jr., Davidson, W. S., Fayad, Z. A., Fuster, V., Goldstein, J., & Yvan-Charvet, L. (2012). Cholesterol efflux and atheroprotection: Advancing the concept of reverse cholesterol transport. *Circulation*, *125*(15), 1905–1919. https://doi.org/10.1161/circulationaha.111.066589

Savage, J. H., Matsui, E. C., Skripak, J. M., & Wood, R. A. (2007). The natural history of egg allergy. *Journal of Allergy and Clinical Immunology*, *120*(6), 1413–1417.

Sawrey-Kubicek, L., Zhu, C., Bardagjy, A. S., Rhodes, C. H., Sacchi, R., Randolph, J. M., & Zivkovic, A. M. (2019). Whole egg consumption compared with yolk-free egg increases the cholesterol efflux capacity of high-density lipoproteins in overweight, postmenopausal women. *The American Journal of Clinical Nutrition*, *110*(3), 617–627. https://doi.org/10.1093/ajcn/nqz088

Schade, R., Pfister, C., Halatsch, R., & Henklein, P. (1991). Polyclonal IgY antibodies from chicken eggyolk: An alternative to the production of mam-malian IgG type antibodies in rabbit. *ATLA*, *19*, 403–419.

Schauer, R. (2004). Sialic acids: Fascinating sugars in higher animals and man. *Zoology*, *107*(1), 49–64.

Sellier, N., Vidal, M. L., Baron, F., Michel, J., Gautron, J., Protais, M., & Nys, Y. (2007). Estimations of repeatability and heritability of egg albumen antimicrobial activity and of lysozyme and ovotransferrin concentrations. *British Poultry Science*, *48*(5), 559–566. https://doi.org/10.1080/00071660701592367

Selvan, K., Sentila, R., & Michael, A. (2012). Generation and characterization of chicken egg yolk antibodies against propionibacterium acnes for the prevention of acne vulgaris. *Indian Journal of Dermatology*, *57*(1), 15.

Shea, S., Stein, J. H., Jorgensen, N. W., McClelland, R. L., Tascau, L., Shrager, S., & Tall, A. R. (2019). Cholesterol mass efflux capacity, incident cardiovascular disease, and progression of carotid plaque. *Arteriosclerosis, Thrombosis, and Vascular Biology*, *39*(1), 89–96. https://doi.org/10.1161/atvbaha.118.311366

Siciliano, R., Rega, B., Marchetti, M., Seganti, L., Antonini, G., & Valenti, P. (1999). Bovine lactoferrin peptidic fragments involved in inhibition of herpes simplex virus type 1 infection. *Biochemical and Biophysical Research Communications*, *264*(1), 19–23. https://doi.org/10.1006/bbrc.1999.1318

Smith, D. J., King, W. F., & Godiska, R. (2001). Passive transfer of immunoglobulin Y antibody to streptococcus mutans glucan binding protein B can confer protection against experimental dental caries. *Infection and Immunity*, *69*(5), 3135–3142.

Soliman, G. A. (2018). Dietary cholesterol and the lack of evidence in cardiovascular disease. *Nutrients*, *10*(6), 780.

Soran, H., Younis, N. N., Charlton-Menys, V., & Durrington, P. (2009). Variation in paraoxonase-1 activity and atherosclerosis. *Current Opinion in Lipidology*, *20*(4), 265–274. https://doi.org/10.1097/MOL.0b013e32832ec141

Spence, J. D., Srichaikul, K. K., & Jenkins, D. J. A. (2021). Cardiovascular harm from egg yolk and meat: More than just cholesterol and saturated fat. *Journal of the American Heart Association*, *10*(7), e017066. https://doi.org/10.1161/jaha.120.017066

Stadelman, W. J., & Cotterill, O. J. (1977). *Egg science and technology*. Avi Pub. Co.

Superti, F., Ammendolia, M. G., Berlutti, F., & Valenti, P. (2007). Ovotransferrin. In *Bioactive egg compounds* (pp. 43–50). Springer.

Tang, W. H., Wang, Z., Levison, B. S., Koeth, R. A., Britt, E. B., Fu, X., & Hazen, S. L. (2013). Intestinal microbial metabolism of phosphatidylcholine and cardiovascular risk. *The New England Journal of Medicine*, *368*(17), 1575–1584. https://doi.org/10.1056/NEJMoa1109400

Tanizaki, H., Tanaka, H., Iwata, H., & Kato, A. (1997). Activation of macrophages by sulfated glycopeptides in ovomucin, yolk membrane, and chalazae in chicken eggs. *Bioscience, Biotechnology, and Biochemistry*, *61*(11), 1883–1889.

Thomas, M. S., DiBella, M., Blesso, C. N., Malysheva, O., Caudill, M., Sholola, M., & Fernandez, M. L. (2022). Comparison between egg intake versus choline supplementation on gut microbiota and plasma carotenoids in subjects with metabolic syndrome. *Nutrients*, *14*(6). https://doi.org/10.3390/nu14061179

Tu, A., Zhao, X., Shan, Y., & Lü, X. (2020). Potential role of ovomucin and its peptides in modulation of intestinal health: A review. *International Journal of Biological Macromolecules*, *162*, 385–393.

Urisu, A., Ando, H., Morita, Y., Wada, E., Yasaki, T., Yamada, K., & Wakamatsu, T. (1997). Allergenic activity of heated and ovomucoid-depleted egg white. *The Journal of Allergy and Clinical Immunology*, *100*(2), 171–176. https://doi.org/10.1016/s0091-6749(97)70220-3

Valenti, P., Antonini, G., Von Hunolstein, C., Visca, P., Orsi, N., & Antonini, E. (1983). Studies of the antimicrobial activity of ovotransferrin. *International Journal of Tissue Reactions*, *5*(1), 97–105.

Valenti, P., Antonini, G., Von Hunolstein, C., Visca, P., Orsi, N., & Antonini, E. (1983). Studies of the antimicrobial activity of ovotransferrin. *International Journal of Tissue Reactions*, *5*(1), 97–105.

Valenti, P., Visca, P., Antonini, G., & Orsi, N. (1985). Antifungal activity of ovotransferrin towards genus candida. *Mycopathologia*, *89*(3), 169–175.

Vega, C., Bok, M., Saif, L., Fernandez, F., & Parreño, V. (2015). Egg yolk IgY antibodies: A therapeutic intervention against group A rotavirus in calves. *Research in Veterinary Science*, *103*, 1–10.

Venerito, M., Vasapolli, R., Rokkas, T., Delchier, J. C., & Malfertheiner, P. (2017). Helicobacter pylori, gastric cancer and other gastrointestinal malignancies. *Helicobacter*, *22*(Suppl 1). https://doi.org/10.1111/hel.12413

Vishwanathan, R., Goodrow-Kotyla, E. F., Wooten, B. R., Wilson, T. A., & Nicolosi, R. J. (2009). Consumption of 2 and 4 egg yolks/d for 5 wk increases macular pigment concentrations in older adults with low macular pigment taking cholesterol-lowering statins. *The American Journal of Clinical Nutrition*, *90*(5), 1272–1279.

Wang, J.-J., & Pan, T.-M. (2003). Effect of red mold rice supplements on serum and egg yolk cholesterol levels of laying hens. *Journal of Agricultural and Food Chemistry*, *51*(16), 4824–4829.

Watkins, B. (1995). The nutritive value of the egg. In W. J. Stadelman & O. J. Cotten'll (Eds.), *Egg science and technology*. Routledge. ISBN-IS: 9781560228554.

Watkins, B. A., Feng, S., Strom, A. K., DeVitt, A. A., Yu, L., & Li, Y. (2003). Conjugated linoleic acids alter the fatty acid composition and physical properties of egg yolk and albumen. *Journal of Agricultural and Food Chemistry*, *51*(23), 6870–6876.

Wenzel, A. J., Gerweck, C., Barbato, D., Nicolosi, R. J., Handelman, G. J., & Curran-Celentano, J. (2006). A 12-wk egg intervention increases serum zeaxanthin and macular pigment optical density in women. *Journal of Nutrition*, *136*(10), 2568–2573.

Wilchek, M., & Bayer, E. A. (1988). The avidin-biotin complex in bioanalytical applications. *Analytical Biochemistry*, *171*(1), 1–32.

Wilhelmson, M., Carlander, D., Kreuger, A., Kollberg, H., & Larsson, A. (2005). Oral Treatment with yolk antibodies for the prevention of C. albicans infections in chemotherapy treated children. A feasibility study. *Food and Agricultural Immunology*, *16*(1), 41–45.

Wu, J. Y., Lee, Y. C., & Graham, D. Y. (2019). The eradication of helicobacter pylori to prevent gastric cancer: A critical appraisal. *Expert Rev Gastroenterol Hepatol*, *13*(1), 17–24. https://doi.org/10.1080/17474124.2019.1542299

Wu, P., Duan, F., Luo, C., Liu, Q., Qu, X., Liang, L., & Wu, K. (2020). Characteristics of ocular findings of patients with coronavirus disease 2019 (COVID-19) in Hubei province, China. *JAMA Ophthalmology*, *138*(5), 575–578.

Xia, J., Tong, J., Liu, M., Shen, Y., & Guo, D. (2020). Evaluation of coronavirus in tears and conjunctival secretions of patients with SARS-CoV-2 infection. *Journal of Medical Virology*, *92*(6), 589–594.

Xu, Q., Shan, Y., Wang, N., Liu, Y., Zhang, M., & Ma, M. (2018). Sialic acid involves in the interaction between ovomucin and hemagglutinin and influences the antiviral activity of ovomucin. *International Journal of Biological Macromolecules*, *119*, 533–539. https://doi.org/10.1016/j.ijbiomac.2018.07.186

Zhang, Q., Tan, W., Yang, L., Lu, M., Dong, S., Liu, X., & Duan, X. (2020). Multi-omics analysis of the effects of egg ovotransferrin on the gut environment in mice: Mucosal Gene expression, Microbiota composition, and intestinal structural homeostasis. *Molecular Nutrition & Food Research*, *64*(7), e1901024. https://doi.org/10.1002/mnfr.201901024

Zhang, X., Qiu, N., Geng, F., & Ma, M. (2011). Simply and effectively preparing high-purity phosvitin using polyethylene glycol and anion-exchange chromatography. *Journal of Separation Science*, *34*(22), 3295–3301.

Zhu, G., Jiang, Y., Yao, Y., Wu, N., Luo, J., Hu, M., & Xu, M. (2019). Ovotransferrin ameliorates the dysbiosis of immunomodulatory function and intestinal microbiota induced by cyclophosphamide. *Food & Function*, *10*(2), 1109–1122.

25

Egg Yolk Lipids

Fang Yang* **and Luyan Yu**

CONTENTS

25.1 Introduction

The edible portions of egg yolk include protein, lipids, and other substances. The egg yolk contributes around 36% of the total weight of a fresh egg. As shown in Table 25.1, most lipids in egg yolks exist as lipoproteins, which comprise 60~65% of the dry egg yolk (Huopalahti et al., 2007; Xiao et al., 2020). Approximately 62~65% of egg yolk lipids are triglycerides, 30~33% phospholipids, 4~5% cholesterol, and less than 1% carotenoids (Mine, 2008). Lutein and zeaxanthin, which give egg yolks their color and also have important biological functions, are the two primary carotenoids. In addition, 68% low-density lipoproteins (LDL), 16% high-density lipoproteins (HDL), 10% globular proteins (livetins), 4% phosphoprotein (phosvitin), and 2% minor proteins constitute the dry matter of egg yolk (Vallejo-Cordoba et al., 1986).

25.2 Egg Yolk Oil

Egg yolk oil (EYO) is traditionally the fat-soluble liquid mixture extracted from egg yolk, containing triacylglycerols, fatty acids, phospholipids, soluble vitamins, cholesterol, and carotenoids (Salari et al., 2016). In the well-known Chinese medical encyclopedia Flora Sinensis, EYO is listed as a cure for multiple diseases. It can remove heat and moisten dryness, treat skin and head ulcers, moderate fever in children with diarrhea, and heal myogenic and painful wounds (Hu et al., 2012; Xie & Pang, 2012). Historically, direct heat, banking, and organic solvent methods have been used for extraction. However, modern procedures utilize enzyme or supercritical CO_2 fluid extractions (Wu et al., 2016). Each procedure has its benefits and drawbacks (Wu et al., 2016). Dry distillation and solvent extraction are currently the most used industrial processes. Studies have revealed that EYO has bioactivities that reduce inflammation and promote wound healing (Rastegar et al., 2011; Shen et al., 2015).

25.2.1 Anti-Inflammatory

Inflammation is a natural biological process that occurs when injured, or stimulated tissues are present (Zhao et al., 2017). Egg components have been found to possess anti-inflammatory properties. At a dose of 2 mL, EYO appeared to reduce oxidative stress and body dysfunction in burnt mice while

* Corresponding author.

DOI: 10.1201/9781003254430-32

TABLE 25.1

Composition of Dry Hen Egg Yolk

Constituent	Main Compositions	Relative %, w/w
Lipids		62.5%
	Triglycerides	38.7%
	Phospholipids	20.6%
	Cholesterol	<3.1%
Carotenoids		<0.6%
Proteins		33.0%
	Proteins in LDL	7.6%
	Proteins in HDL	11.6%
	Livetin	9.9%
	Phosphoprotein	3.6%
	Other proteins	0.3%
Carbohydrates		1.2%
Minerals		3.5%

Source: Data from Xiao et al. (2020).

increasing vascular permeability (Xie & Pang, 2012). It has been shown that EYO possesses a unique antioxidant property that promotes the development of reversible inflammation in the body. However, the abovementioned research evaluated the efficacy of two extraction methods without considering the differences among egg species. According to Mahmoudi et al., EYO obtained from three types of egg yolks (organic hens, duck, and machine made) by direct heat demonstrated anti-inflammatory activity at 100~300 mg/kg compared to olive oil ($p < 0.001$) in hot plate test (Mahmoudi et al., 2013). Nuclear transcription factor-κB (NF-κB), a crucial transcription factor, is involved in and plays an essential role in the signaling pathway associated with inflammation (Lawrence, 2009). Shen et al. reported that only the partial delipidating fraction of EYO possessed an anti-inflammatory effect, which was mediated by a reduction in NF-κB signaling transduction and expression (Shen et al., 2015). However, it is unclear how adequate delipidation affects the beneficial properties of EYO and how this affects EYO's actual action. Due to the EYO combination's complexity and the conventional technique's high temperature, yields, and bioactive matrix may be compromised (Wu et al., 2016). There is less evidence regarding the specific issues and molecular mechanisms of inflammation clinical treatment. To make therapeutic recommendations, more evidence of its anti-inflammatory efficacy at different levels and using a variety of animals must be investigated.

25.2.2 Wound Healing

A wound is the rupture or defect of the skin and other tissues caused by an external force on the body (Lazarus et al., 1994). Recently, reducing the high frequency of wounds and mitigating the possibility of repercussions has been a hot topic. Natural products and dietary components have been identified as readily available treatment sources with no adverse effects distinct from conventional medications (Xu et al., 2012). EYO's efficacy in conventional burn and dermatitis treatments has been demonstrated, and current research has confirmed these properties and expanded its application (Wu et al., 2016). EYO was discovered to improve wound-damaged surfaces, stimulate re-epithelialization in third-degree burn wounds, and treat acute conditions (Rastegar et al., 2011; Salari et al., 2016). Due to their hydrophobic nature and anti-inflammatory effect, egg yolk lipids may play a significant role in therapy and foster a favorable recovery environment (Ehehalt et al., 2004). According to reports, 100 ug/ml EYO in chitosan gel formulation significantly reduced inflammation and aided the healing process (Yenilmez et al., 2015). In conclusion, EYO positively affects wound healing and is anticipated to be utilized in clinical settings. However, the equilibrium between EYO and medications must be coordinated to ensure efficacy and the protection of human health.

25.3 Triacylglycerols and Fatty Acids

Fatty acids can be divided into three groups based on their degree of unsaturation: saturated fatty acids (SFA), monounsaturated fatty acids (MUFA), and polyunsaturated fatty acids (PUFA). n-3 and n-6 fatty acids are the two types of essential unsaturated fatty acids in nutrition. Fatty acids are essential components of lipid molecules, which are the foundation for comprehending the detailed information about lipid molecules in egg yolk (Brady et al., 2003; Park et al., 2015; Aguillón-Páez et al., 2020). Table 25.2 displays the typical structures and compositions of fatty acids and triglycerides in egg yolk. The fatty acid profile of lipids is approximately 30~35% SFA, 40~45% MUFA, and 20~25% PUFA based on a standard hen diet. There are three important fatty acids in egg yolks—oleic acid, palmitic acid, and linoleic acid—which make up 40~45%, 20~30%, and 15~20%, respectively, of the total fatty acid profile (Kuksis, 1992; Golzar Adabi et al., 2013). Compared with the commercial egg, the abundance of oleic acid, palmitic acid, and linoleic acid was higher in all native egg yolks on the fatty acid composition, and the sum of these three fatty acids exceeds 80% of total fatty acids (Lordelo et al., 2020).

Eggs are an inexpensive, nutrient-dense food that is an excellent source of n-3 and n-6 fatty acids. Unsaturated fatty acids have multiple functions, including enhancing antioxidant capacity, regulating membrane dependence, and reducing inflammation, as well as having anticancer properties (Xiao et al., 2020). It has been shown that the egg-derived phospholipids and HDL increased their antioxidant potential (Chen et al., 2019), which is discussed in the following statement. Zhang et al. examine the effects of n-3 PUFAs and n-6 FAs on the survival of some cancer cells such as the undifferentiated human colon cancer cell LoVo and the semi-differentiated human colon adenocarcinoma cell RKO. It was discovered that n-3 and n-6 fatty acids caused the death of colon cancer cells by modifying the mitochondrial-mediated pathway at concentrations greater than 120 μM (Zhang et al., 2015). Yang et al. assessed the impact of six fatty acids (palmitic acid, oleic acid, linoleic acid, arachidonic acid, eicosapentanoic acid (EPA), and docosahexaenoic acid (DHA)) on cholesterol uptake and transport in Caco-2 cells. The results demonstrated that EPA and DHA inhibited cholesterol uptake and transport in Caco-2 cells in a dose-dependent manner (Yang et al., 2018b). In addition to lowering cholesterol and serum triglyceride

TABLE 25.2

The Typical Structure and Composition of Fatty Acids and Triglycerides in Egg Yolk

Classification	Common Name	Systematic Name	Abbreviation	Formula	Structure	Content %
Fatty acids	Palmitic acid (C16:0)	hexadecanoic acid	P	C16H32O2		24.5±0.48[a]
	Palmitoleic acid (C16:1 n-7)	9Z-hexadecenoic acid	Po	C16H30O2		2.98±0.21[a]
	Stearic acid (C18:0)	octadecanoic acid	St	C18H36O2		7.95±0.33
	Oleic acid (C18:1 n-9)	9Z-octadecenoic acid	O	C18H34O2		43.2±1.10
	Linoleic acid (C18:2 n-6)	9Z,12Z-octadecadienoic acid	LA	C18H32O2		14.9±1.92[b]
	α-Linolenic acid (C18:3 n-3)	9Z,12Z,15Z-octadecatrienoic acid	α-Ln	C18H30O2		0.59±0.03[b]
	Arachidonic acid (20:4 n-6)	5Z,8Z,11Z,14Z-eicosatetraenoic acid	An	C20H32O2		1.82±0.06[b]
	Eicosapentanoic acid (C20:5 n-3)	11,18-dihydroxy -5Z,8Z,12E,14Z,16E -eicosapentaenoic acid	EPA	C20H30O4	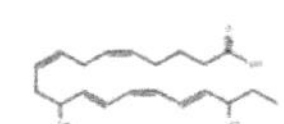	nd
	Docosahexaenoic acid (C22:6 n-3)	4Z,7Z,10Z,13Z,16Z,19Z -docosahexaenoic acid	DHA	C22H32O2		0.89±0.04[b]
Triglycerides	16:0/18:1(9Z)/18:1(9Z)	1-hexadecanoyl-2,3-di-(9Z-octadecenoyl)-sn-glycerol	TG	C55H102O6	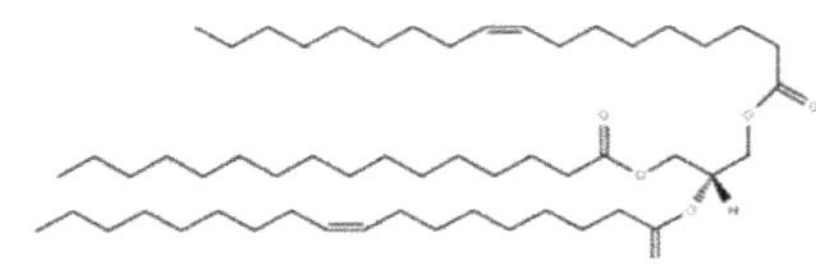	

Source: Data from Aguillón-Páez et al. (2020).

Note: nd = not detected.

[a-b] Within a row, means with different superscripts are significantly different ($P < 0.05$).

levels, these fatty acids have additional health benefits, such as relieving skin problems, improving female sexual premenstrual syndrome, compressive ability, reducing allergic reactions, reducing asthma, enhancing organ tissue inflammation, and promoting the brain development of children (Xiao et al., 2020).

25.4 Sterols and Sterol Esters

Sterols are hydroxyl-containing steroids with cyclopentane polyhydrophenanthrene as their fundamental structure and include cholesterol in animals, phytosterols such as stigmasterol and sitosterol in plants, and ergosterol in fungi. More than 40 phytosterols have been identified to date, with campesterol, β-sitosterol, and trace amounts of stanols constituting the majority. Cholesterol and phytosterols are triterpenoids with tetracyclic and carbon-linked side chains. Phytosterols are structurally distinct from cholesterol at the C24 side chain (Makhmudova et al., 2021). Cholesterol in egg yolk derives partly from hen feed and partly from the synthesis in the liver during lipoprotein synthesis. Almost all cholesterol (150~200 mg/egg) is found in the egg yolk (Kuang et al., 2018), which accounts for ~5% of total lipids in free (85~90%) or esterified (10~15%) form (Bitman & Wood, 1980). Cholesterol esters are present in the lipid core of LDL, composed of 35% oleic acids, 33% palmitic acids, 12% linoleic acids, and 11% stearic acid (Kuksis, 1992). In addition to its role in forming cell membranes, cholesterol provides steroids for embryonic development. The cholesterol levels of fertilized and unfertilized eggs decreased gradually with increasing incubation time, with fertilized eggs exhibiting lower cholesterol levels. After 15 days of incubation, the cholesterol level in fertilized eggs decreased from 0.33 to 0.21 g/egg, with the greatest decrease occurring during days 9 and 12 (Duan et al., 2013).

25.5 Phospholipids

Phospholipids are the primary constituents of biological membranes, which are divided into two categories. Glycerophospholipids with a glycerol backbone include phosphatidyl choline (PC), phosphatidyl ethanolamine (PE), phosphatidyl serine (PS), phosphatidyl inositol (PI), phosphatidyl glycerol (PG), cardiolipin (CL), lysophospholipids, and plasmalogen (Lagace & Ridgway, 2013). Sphingolipids, the second largest group of membrane lipids, consist of a polar head group and two nonpolar tails (a long hydrocarbon chain of fatty acid and a sphingoid base backbone), including sphingoid bases, ceramides, sphingomyelin (SM), glycosphingolipids, and other sphingolipids (Fahy et al., 2005). Egg yolk phospholipids consist of 84% PC, 12% PE, 2% SM, and 2% lysoPC and other minor compounds (Hatta et al., 2008). Using UPLC-Q-TOF-MS, Ali et al. have identified and quantified 57 molecular species of phospholipids in egg yolk. Among the various classes of egg yolk phospholipids, PC (16:0–18:1), PE (18:0-20:4), PI (18:0-18:2), PS (18:0-18:2), SM (d18:1/16:0), and lysoPC (16:0) were the most prevalent. Meanwhile, C16:0, C18:0, C18:1, C18:2, and C20:4 were the major fatty acids in egg yolk phospholipids (Ali et al., 2017). Table 25.3 displays

TABLE 25.3

The Typical Structure and Composition of Phospholipids in Egg Yolk

Common Name	Systematic Name	Structure
PC(18:1(11Z)/16:0)	1-(11Z-octadecenoyl)-2-hexadecanoyl-sn-glycero-3-phosphocholine	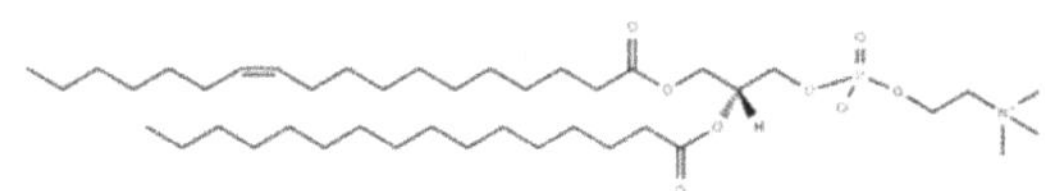
PE(20:4(5Z,8Z,11Z,14Z)/18:0)	1-(5Z,8Z,11Z,14Z-eicosatetraenoyl)-2-octadecanoyl-glycero-3-phosphoethanolamine	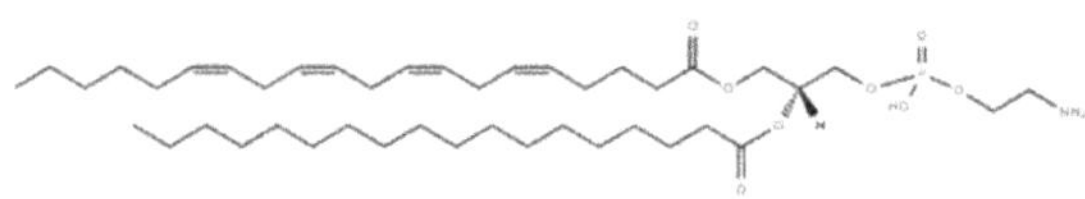
PI(18:2(9Z,12Z)/18:0)	1-(9Z,12Z-octadecadienoyl)-2-octadecanoyl-glycero-3-phospho-(1′-myo-inositol)	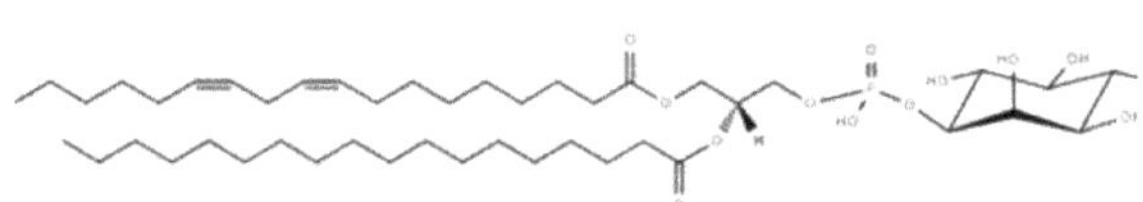
PS(18:2(9Z,12Z)/18:0)	1-(9Z,12Z-octadecadienoyl)-2-octadecanoyl-glycero-3-phosphoserine	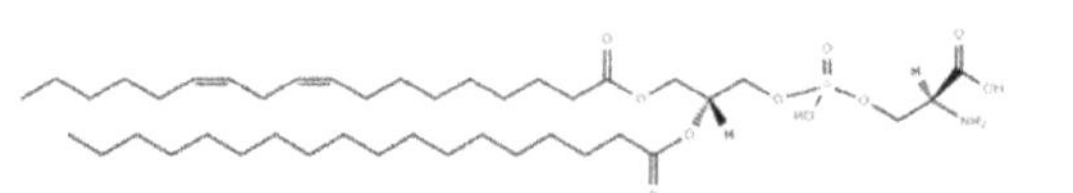
SM (d18:1/16:0)	N-(hexadecanoyl)-sphing-4-enine-1-phosphocholine	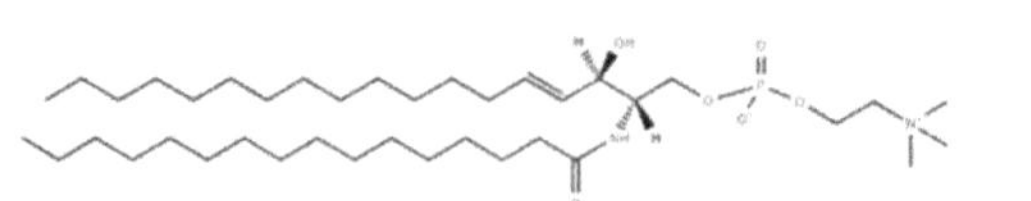
PC(18:2(9Z,12Z)/0:0)	1-(9Z,12Z-octadecadienoyl)-sn-glycero-3-phosphocholine	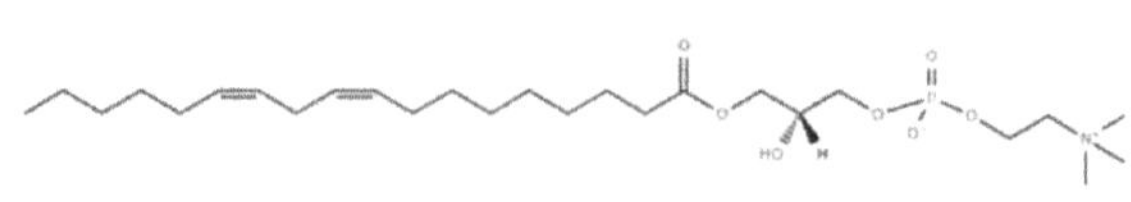

the typical structure and composition of phospholipids in egg yolk. PC and PE are abundant in yolk phospholipids, comprising 76% and 22% of total phospholipids, respectively, with PC being three times higher than soy phospholipids. PI, PS, SM, CL, lysoPC, and lysoPE levels were relatively low. PUFA accounts for 30~40% of total fatty acids in phospholipids, SFA accounts for 45%, and MUFA accounts for 20~25% (Kuksis, 1992). Phospholipids are a possible source of the bioactive lipids found in egg yolks, and they have multiple effects on antioxidants, anti-inflammatory properties, neuroprotective effects, and cholesterol absorption inhibition.

25.5.1 Antioxidant

According to King et al., egg yolk phospholipids have antioxidant activity in a refined salmon oil model system, and the presence of nitrogen enhances the phospholipids' antioxidant activity (King et al., 1992). The auto-oxidation of lipids and unsaturated fatty acids is a chain reaction of free radicals. The addition of antioxidants can delay lipids' oxidation and prolong food's shelf life. By chelating pro-oxidation metals as antioxidative products of Maillard reactions or in combination with other antioxidants, antioxidative phospholipids in food are frequently utilized (Cui & Decker, 2016). Egg yolk phospholipids' antioxidant activity is strongly correlated with the degree of fatty acid chain saturation (Young et al., 2011). Egg yolk lipids can be divided into polar and neutral fractions (Shen et al., 2015). Polar phospholipids have a greater antioxidant capacity than neutral lipids, positively correlated with the unsaturated fatty acids. In contrast, higher saturation of fatty acid side chains has the opposite effect (Sugino et al., 1997). In addition to being the second-most flexible glycerophospholipid, PE can scavenge free radicals at a concentration of 6 mg/ml, making it a crucial post-translational modification matrix (Calzada et al., 2016; Sun et al., 2019). Thus, unsaturated fatty acids are essential to the biological activities of phospholipids (Calzada et al., 2016). Even though they are considered excellent antioxidants, phospholipids may also prevent oxidative stress *in vivo* in alcohol-induced oxidative stress prevention (Jayaraman et al., 2008).

25.5.2 Anti-Inflammatory

Crohn's disease and ulcerative colitis are forms of inflammatory bowel disease (IBD), characterized by persistent colonic damage (Xu et al., 2012). The PC or lysoPC levels in patients with ulcerative colitis intestinal mucus are lower. However, it is unknown whether PC affects the intestinal lumen (Ehehalt et al., 2004). C-reactive protein (CRP) levels decreased, and adiponectin levels increased in the plasma of obese men who consumed three eggs per day for 12 weeks. In contrast, these improvements were not observed when using a yolk-free egg substitute (Ratliff et al., 2008). The anti-inflammatory compound may inhibit the activation of the NF-κB and mitogen-activated protein kinase (MAPK) inflammatory signaling pathways (Bengmark & Jeppsson, 1995; Lichtenberger, 1995; Ehehalt et al., 2004; Treede et al., 2007). It has been demonstrated that PC molecules inhibit tumor necrosis factor-α (TNF-α)-induced overexpression of pro-inflammatory genes. Moreover, the anti-inflammatory efficacy of PC containing unsaturated fatty acids was greater than that of PC containing saturated fatty acids indicating that unsaturated fatty acids play a significant role in regulating inflammatory responses in the body (Treede et al., 2009).

25.5.3 Neuroprotective Effect

Regarding molecular characteristics and side chain groups, phospholipids may play a role in preventing and treating neurodegenerative diseases (Chen et al., 2019). Alzheimer's is a neurodegenerative disorder that causes progressive memory and cognitive impairment (Che et al., 2018). Choline, the molecule's polar component, is essential for brain development, liver function, and cancer prevention. It is commonly added to commercial baby formula as an essential nutrient to ensure that infants receive the required nutrients (Gutierrez et al., 2018). Researchers have discovered that consuming yolk phospholipids may reduce Alzheimer's disease symptoms in certain individuals (Yamamoto et al., 1996). In rats, egg yolk PC increased acetylcholine levels and improved memory while, to some extent preserving cerebral function (Chung et al., 1995). There is a strong correlation between advancing age and cognitive decline. The consumption of egg PC (5 g/100 g diet) improved the cognitive abilities and brain function of aged mice in the maze (Lim & Suzuki, 2000).

The phospholipids' unsaturated fatty acid side chain may provide indirect protection (Sun et al., 2019). Due to their exceptional amphiphilicity phospholipid molecules readily cross the blood-brain barrier, enabling their use for memory and cognitive function enhancement and preventing and delaying the onset of neurodegenerative diseases (Kato-Kataoka et al., 2010; Che et al., 2018). According to Che et al (Che et al., 2018), phospholipids may reduce inflammation, oxidative stress, and age-related disorders. Egg yolk PC (0.2 mg/ml) protects PC12 cells from PUFA side-chain damage by inhibiting the formation of acetylcholinesterase and lowering oxidized product levels (Chen et al., 2019). These findings suggest that the phospholipid's unsaturated fatty acid side chain is more effective at preserving the proper function of neural networks. In addition, it possesses considerable regulatory capacities, particularly for the EPA and DHA (Pandey et al., 2009; Che et al., 2018; Chen et al., 2019).

25.5.4 Inhibition of Cholesterol Absorption

Due to the presence of phospholipids in the egg yolk, eggs have the additional benefit of preventing cholesterol absorption in the small intestine (Blesso, 2015). Small intestinal cell lines and tissue were used *in vitro* to demonstrate that phospholipids in mixed micelles or lipid emulsions can inhibit the hydrolysis of triglycerides by altering the rate of micelle formation and diffusion, thereby preventing small intestinal cholesterol absorption (Eckhardt et al., 2002; Feng et al., 2010). Experiments with animal feeding demonstrated that natural phospholipids such as PC, PE, and SM inhibit cholesterol absorption in a dose-dependent manner (Imaizumi et al., 1983; Imaizumi et al., 1991; Jiang et al., 2001; Noh & Koo, 2003, 2004; Duivenvoorden et al., 2006). Jiang et al. demonstrated, using rats with cannulated lymph ducts, that duodenal infusion of a lipid emulsion containing egg PC decreased cholesterol absorption significantly. In the same study, egg PC inhibited cholesterol absorption into lymphatics more effectively than more unsaturated soy PC (Jiang et al., 2001). Egg SM inhibits cholesterol absorption and transport in Caco-2 cells significantly more effectively than PC, according to Yang et al. (Yang et al., 2018a). They are lipids with a similar structure in comparison to proteins, and their absorption is achieved by assembling into micelles with cholesterol and other substances before being transported to the enterocyte. Phospholipids and fatty acids have the potential to modulate the absorption and transport of cholesterol in concert.

Recent research has demonstrated that SM and its hydrolyzed metabolites, such as sphingosines, ceramide, and sphingosine-1-phosphate, play a crucial role in regulating lipid metabolism, obesity, and chronic metabolic diseases (Ohlsson et al., 2010; Norris & Blesso, 2017; Norris et al., 2019; Yamashita et al., 2021). Slotte found that saturated SM has a larger interface region than other phospholipids, and its ability to form strong hydrogen bonds with cholesterol is essential to form a liquid-ordered phase in the cell membrane (Slotte, 2016). Egg SM did not increase atherosclerosis in apoE$^{-/-}$ mice fed a high-fat diet, but it had an anti-atherosclerosis effect in apoE$^{-/-}$ mice fed a standard diet. Through LXR-SREBP-1c pathway, it can reduce liver lipid, intestinal cholesterol absorption, and liver lipids of ApoE$^{-/-}$ mice fed high-fat diet, and increase fecal lipid and cholesterol excretion (Chung et al., 2013). Two other studies also showed that egg SM did not affect SM levels in the blood of ApoE$^{-/-}$ mice fed a high-fat diet and reduced lipid accumulation in the aortic root to alleviate atherosclerosis (Chung et al., 2017; Millar et al., 2019). In Caco-2 and HT-29-D4 cells, Garmy et al. discovered that sphingosines significantly reduced cholesterol absorption (Garmy et al., 2005). The critical target involved in cholesterol cellar absorption is Niemann-Pick C1-Like 1 (NPC1L1), which can be inhibited by Ezetimibe (Zhang et al., 2022). Yang Fang et al. discovered that egg SM inhibits cholesterol transport in the Caco-2 monolayer by downregulating the expression levels of key protein for sterol absorption NPC1L1 and cell transcription factor SREBP-2 in Caco-2 cells (Yang et al., 2018a).

In addition, egg SM can improve the distribution and abundance of intestinal microbiota in mice fed a high-fat diet; however, it remains to be determined whether its effect on intestinal flora inhibits cholesterol absorption. In the colon, SM may regulate intestinal permeability and influence the expression of intestinal tight junctions (Milard et al., 2019), influence the type and abundance of intestinal flora (Norris et al., 2016; Chung et al., 2017; Millar et al., 2019; Vors et al., 2020), and attenuate lipopolysaccharide-induced intestinal inflammation (Jiang et al., 2001). Egg SM significantly decreased the relative abundance of *Ruminococcus* in the gut of apoE$^{-/-}$ mice fed a high-fat diet, but no significant differences in microbiota at the phylum level were observed (Millar et al., 2019). A multicenter, double-blind, crossover study of 58 overweight postmenopausal women (56–62 years old) revealed that a diet rich in milk phospholipids might reduce cardiometabolic risk-related lipid markers, increase cholesterol-ileal excretion and fecal prostanol loss. It improved cardiometabolic health without affecting the predominant gut microbiota species or short-chain fatty acids (SCFA) in feces (Vors et al., 2020). Overall, eggs are an appropriate dietary recommendation as a phospholipid-rich and inexpensive food, and the PUFA found in egg yolk is more advantageous for maximizing their efficacy.

25.6 Egg Yolk Lipoproteins

Egg yolk is primarily composed of 68% LDL, 16% HDL, 10% livetins, and 4% phosvitins (Anton et al., 2003). LDLs are spherical particles (17~60 nm in diameter, with a mean of approximately 35 nm) with a liquid lipid core (triglycerides and cholesterol esters) encased in a phospholipid and protein monofilm (Tria & Scanu, 1969; Evans et al., 1973). LDL is comprised of 11~17% protein and 83~89% lipid, out of which the latter 74% is neutral lipid and 26% phospholipids (Martin et al., 1964). Their population is composed of two sub-groups: LDL_1 (10×10^6 Da) and LDL_2 (3×10^6 Da). LDL_1 represents 20% of the total LDL and contains twice the amount of proteins as LDL_2 (Martin et al., 1964). Due to the correlation between high LDL levels and consumption and cardiovascular diseases (CVDs), it has been assumed that LDL is a harmful substance (Yu et al., 2019). Under certain conditions, LDL can become oxidized (oxLDL), leading to a chronic condition known as atherosclerosis (Trpkovic et al., 2015). Mature HDL consists of a hydrophobic core made up of cholesterol esters and TG, proteins encased in a lipid monolayer that contains phospholipids and free cholesterol as its primary constituents (Eftekhar et al., 2015). In contrast to LDLs, HDLs have a pseudo-molecular structure like that of globular proteins instead of a micelle-like spherical shape. HDLs exist as dimers consisting of two 200 kDa monomers, which comprise 75~80% proteins and 20~25% lipids, of which 65% phospholipids and 5% cholesterol make up the lipids (Tria & Scanu, 1969).

25.6.1 Antioxidant

Granule is the primary antioxidant substance isolated from egg yolk, which contains ~80% HDL (Yamamoto et al., 1990). HDL can inhibit the oxidative transformation of LDL (Van Lenten et al., 1995), reducing their atherogenicity and promoting cardiovascular health (Barter et al., 2004). The anti-inflammatory and antioxidant properties of HDL have been linked to Apolipoprotein A-I (apoA-I), apoA-II, and antioxidant enzymes such as paraoxnase (Barter et al., 2004; Ferretti & Bacchetti, 2012). ApoA-I is the principal apolipoprotein of HDL by removing oxidized lipids to delay the oxidative degree and reduce the production of oxLDL (Barter et al., 2004). HDL is believed to prevent the formation of atherosclerosis by inhibiting the production of adhesion molecules in endothelial cells and the recruitment of blood monocytes into the artery wall; both functions are deemed crucial in avoiding atherosclerosis formation (Barter et al., 2004). *In vitro*, HDL also inhibits the formation of intercellular adhesion molecule-1 (ICAM-1), vascular cell adhesion molecule-1 (ICAM-1), and E-selectin in human umbilical vein endothelial cells induced by cytokines (Rye & Barter, 2008). Indeed, vitamin E and carotenoids carried by lipoproteins may also exert underlying antioxidant effects (Blesso et al., 2013).

25.6.2 Anti-Atherosclerosis

Due to the high HDL content of egg yolk, the second component of egg yolk may have important functional properties in lipid metabolism, which are closely related to its composition of more PC, PI, and less triglyceride (Yu et al., 2019). Atherosclerosis is caused by increased total cholesterol, triglycerides, and excessive accumulation of LDL or oxLDL (Trpkovic et al., 2015). HDL exhibits significant anti-atherosclerosis activity through reverse cholesterol transport, antioxidant, and anti-inflammatory actions (Eftekhar et al., 2015). Reverse cholesterol transport occurs in three phases: extravascular, intravascular, and intrahepatic. Purified HDL from egg yolk has an anti-atherosclerosis effect in an experimental model of atherosclerosis (Eftekhar et al., 2015). HDL is the primary carrier for the transfer of cholesterol from peripheral cells to the liver for excretion and catabolism (Miller, 1990). HDL may reduce cholesterol build-up on vessel walls by forming a complex. It suggested that egg yolks are good dietary supplements to increase serum HDL levels for cardiovascular protection, and the consumption of egg yolk HDL at dosages of 100 and 400 mg/dL significantly reduced atherosclerosis in rabbits (Eftekhar et al., 2015). In obese mice induced by a high-fat diet, 0.6 mg/g of egg yolk HDL per day significantly reduced obesity and downregulated the expression of target genes (Yu et al., 2019). In addition, HDL is also recognized for its antithrombotic capability. By stimulating the production of prostacyclin, it protects and inhibits the formation of thrombosis in blood vessels. HDL's ability to inhibit tissue factor and selectin production may explain its antithrombotic properties (Mineo et al., 2006).

25.7 Carotenoids

Carotenoids are pigments in a wide variety of plants, fungi, and green leafy vegetables (Miranda et al., 2015). There are 30~40 different types of carotenoids in the human body, with lutein and zeaxanthin being the most abundant (Zimmer & Hammond, 2007). Carotenoids are isoprenoids with a long polyene chain that contains 3~15 conjugated double bonds, which determines their absorption spectrum. Carotenoids are effective singlet oxygen quenchers and free radical scavengers due to their core structure of conjugated carbon-carbon bonds (Fiedor & Burda, 2014). In addition, carotenoids have high lipophilicity due to their structure, which allows them to delay lipid peroxidation and maintain lipid-protein complexes such as cell membranes. Zeaxanthin is the stereoisomer of lutein, which is structurally identical to lutein except for the position of a single C=C double bond; however, their biological activities are distinct. Carotenoids, especially lutein, and zeaxanthin are essential components and pigments of egg yolk except proteins, phospholipids, and fatty acids (Zaheer, 2017). Additionally, egg yolk contains several additional carotenoids, including beta-cryptoxanthin, cis isomers of lutein and zeaxanthin, and synthetic xanthophylls (Nimalaratne et al., 2012).

Figure 25.1 depicts the chemical structure of egg yolk xanthophylls, including all-trans lutein, all-trans zeaxanthin, all-trans canthaxanthin (synthetic xanthophyll), all-trans β-apo-8′-carotenoic acid ethyl ester (synthetic xanthophyll), and their cis isomers. The diet of the hen influences the number of carotenoids deposited in the yolk (Nimalaratne et al., 2013). Carotenoids are highly lipophilic and hydrophobic substances that prefer to be dissolved in a lipid matrix. They can only carry out their physiological functions when absorbed and transported to the bloodstream (Gopinath et al., 2020). its lipophilic nature makes it easier to flow through the cell membrane or adhere to the surface of the membrane, thereby increasing its bio-accessibility (Zaheer, 2017). Unfortunately, these carotenoids undergo structural changes due to alterations in function and bioavailability caused by exposure to light, heat, and oxygen. Although the digestive stability of carotenoids in egg yolk is unaffected by cooking methods, their bioavailability varies considerably (Nimalaratne et al., 2012, 2015). The lipid matrix of the yolk enables efficient absorption of these carotenoids.

25.7.1 Antioxidant

It is estimated that the yolk contains approximately 200~300 μg of lutein and zeaxanthin on average. The bioavailability of lutein and zeaxanthin in egg yolks is greater than that of green leafy vegetables, indicating that egg yolks are an excellent source of these antioxidants (Zaheer, 2017). Under normal conditions, at least 20 different carotenoids can be identified in human serum (Fiedor & Burda, 2014). However, lutein and zeaxanthin are the only pigments that can aggregate in the retinal cell membrane (Widomska & Subczynski, 2014). It may also be found in the central nervous system, but the highest concentration of the compound is in the retina (Zaheer, 2017). Lutein and zeaxanthin offer a variety of potential health benefits, the most important of which are antioxidation and the prevention of macular degeneration (age-related macular degeneration [AMD]) (Fiedor & Burda, 2014). Carotenoids are effective at quenching singlet oxygen and reactive oxygen species, delaying lipid peroxidation, preserving cell membrane structure, and reducing oxidative stress in cells and tissues due to double bonds (Zaheer, 2017). An essential function of this characteristic is to protect the retina from light-induced free radical damage caused by superoxide anion (O_2–), hydroxyl radical (OH–), and peroxy radical (ROO–), which react with vital biomolecules (e.g., lipids, proteins) (Fiedor & Burda, 2014). Humans may consider eggs a good source of physiologically active carotenoids.

25.7.2 Prevention of AMD

AMD is a multifactorial disease that affects the macular portion of the retina, with age, heredity, and environment all contributing to the disease's development; severe AMD results in significant vision loss and lasting visual impairment (Mitchell et al., 2018). There is no effective treatment for advanced AMD, although zinc and vitamin supplements may prevent and delay the condition (Mitchell et al., 2018). Carotenoids can only be obtained through diet, as the human body cannot produce them

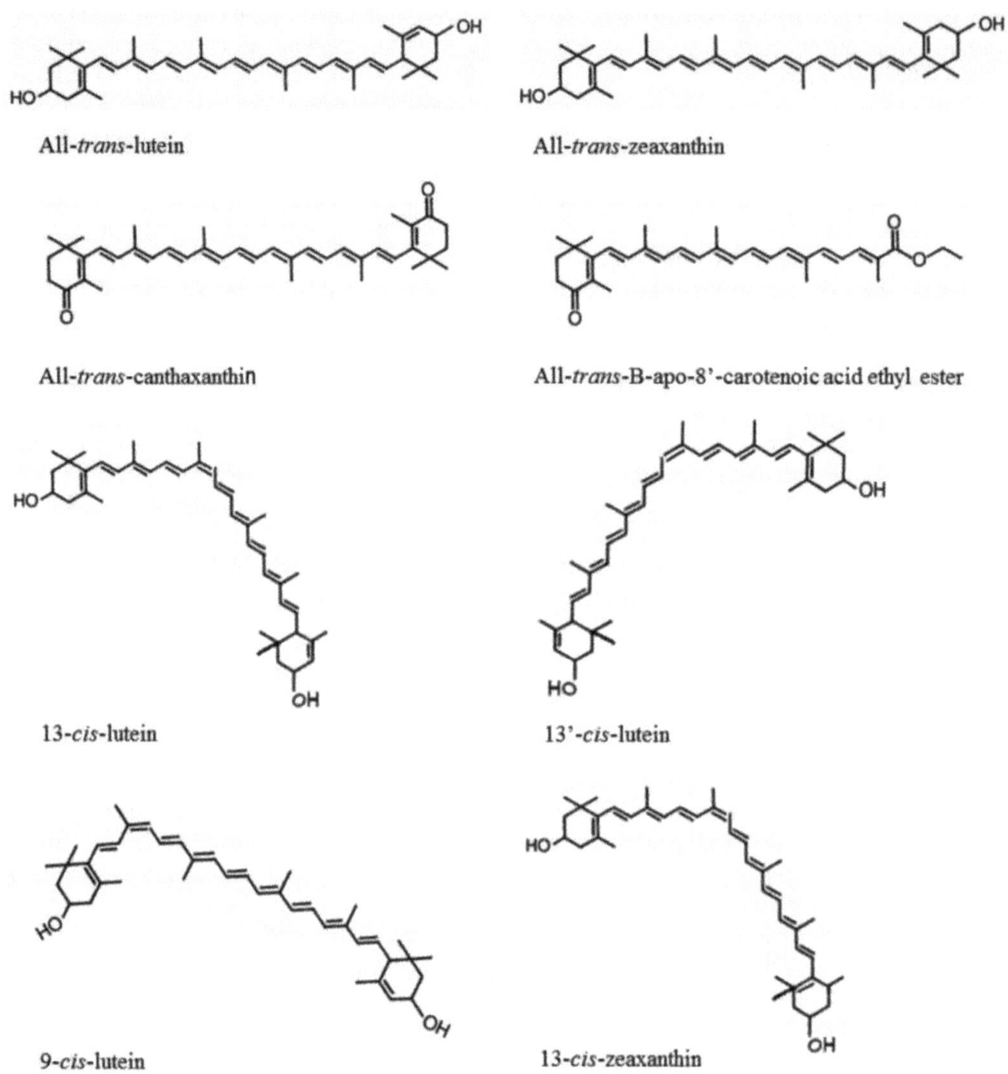

FIGURE 25.1 Structures of carotenoids in egg yolk. (From Nimalaratne et al., 2013.)

(Miranda et al., 2015). Egg yolk is rich in carotenoids and HDL, which is also a carotenoid transporter (Gopinath et al., 2020). According to epidemiological studies, increasing moderate egg consumption significantly reduces the incidence of advanced AMD (Goldberg et al., 1988; Gopinath et al., 2020). Egg consumption increases the amount of HDL in the blood, which has a synergistic effect on the levels of lutein and zeaxanthin in the blood (Blesso et al., 2013). The ocular effects of oxidative stress and blue light stimulation are significant risk factors for vision loss. Lutein and zeaxanthin owe may its antioxidant properties to absorb or filter blue light (Miranda et al., 2015).

25.7.3 Prevention of Metabolic Syndrome

Apart from playing critical roles in eye health, lutein and zeaxanthin are crucial in the prevention or reduction of the severity of CVDs, stroke, cancer (Mares-Perlman et al., 2002; Shin et al., 2013), oxidative stress (American Chemical Society, 2011; Fiedor & Burda, 2014), and neurodegenerative illnesses (Nataraj et al., 2016). Egg consumption may improve carotenoid status in individuals with metabolic syndrome by elevating plasma HDL concentrations (Blesso et al., 2013). Egg yolk may be an essential source of carotenoid in a population at high risk for cardiovascular disease and type 2 diabetes (Shin et al., 2013). Higher lutein and zeaxanthin intake from egg yolks increases lutein and zeaxanthin concentrations in serum and retina but does not affect other carotenoids, tocopherol, and retinol levels (Vishwanathan et al., 2010). Egg yolks' lipid matrix increases the bioavailability of carotenoid pigments such as lutein and zeaxanthin without affecting lipoprotein or glucose metabolism (Ballesteros et al., 2015). Moreover, they may help treat skin conditions caused by excessive exposure to UV radiation, such as sunburn (Blesso et al., 2013).

25.8 Fat-Soluble Vitamins

The yolk of a fresh egg contains most of the egg's vitamins, including all the B vitamins, A, D, E, and K, and trace amounts of vitamin C. Egg yolks contain fat-soluble vitamins

A, D, and E that are bioavailable to humans. The composition of the egg yolk varies based on genetics, egg production rate, season, and diet (Huopalahti et al., 2007). One egg contains approximately 13% of a woman's recommended daily allowance of vitamin B_2 and 22% of her recommended daily allowance of vitamin A. These vitamins have various physiological functions, including promoting growth, reproduction, and maintaining normal bone, epithelial tissue, vision, and mucosal epithelium secretions. Around 30% of the recommended daily allowance of vitamin D is provided by egg yolk in the human diet (Huopalahti et al., 2007). This deficiency has been associated with a higher risk of osteoporosis, diabetes, multiple sclerosis, cancer, and rheumatoid arthritis (Huopalahti et al., 2007). The average egg contains approximately 1.1 mg of vitamin E (Anton, 2007), equivalent to 8.5% of the daily value. The most active form of vitamin E, α-tocopherol, is a well-known lipophilic chain-breaking antioxidant that is known to protect against LDL oxidation to reduce the incidence of CVDs (Burton & Traber, 1990; Traber & Atkinson, 2007). Furthermore, vitamin E-enhanced eggs can provide up to 150% of the recommended daily intake without developing an off-flavor and protect the yolk from oxidation of long-chain fatty acids (Surai et al., 2000).

25.9 Conclusion and Prospects

As a nutrient-dense food source, eggs contain proteins and lipids with superior functional properties and biological activities. Approximately 30% of the weight of egg yolk is composed of lipids, which are highly concentrated in triglycerides and phospholipids with 60~70% unsaturated fatty acids. In recent years, the significant nutritional value of egg yolk lipids has garnered widespread attention from the domestic and international medical, nutrition, and food industries. This section describes and emphasizes the biological functions of egg yolk lipids, such as egg yolk oil, phospholipids, and fatty acids, as well as their potential health benefits. Numerous *in vitro* and *in vivo* studies have demonstrated that these lipid components play a crucial role in maintaining and improving the body's nutrition and health, particularly in terms of their antioxidant, anti-inflammatory, retinal protection, neuroprotective, antimicrobial, lipid metabolism, and cardiovascular protective properties. Eggs are a natural repository of bioactive chemicals. Therefore, there are numerous development opportunities in the deep processing of eggs and product development. Natural active compounds in egg yolk that have a high added value should be given special attention and exploited to their fullest potential.

REFERENCES

Aguillón-Páez, Y. J., Romero, L. A., & Diaz, G. J. (2020). Effect of full-fat sunflower or flaxseed seeds dietary inclusion on performance, egg yolk fatty acid profile and egg quality in laying hens. *Animal Nutrition*, *6*(2), 179–184. https://doi.org/10.1016/j.aninu.2019.12.005

Ali, A. H., Zou, X., Lu, J., Abed, S. M., Yao, Y., Tao, G., Jin, Q., & Wang, X. (2017) Identification of phospholipids classes and molecular species in different types of egg yolk by using UPLC-Q-TOF-MS. *Food Chemistry*, *221*, 58–66. https://doi.org/10.1016/j.foodchem.2016.10.043

American Chemical Society. (2011). *Oxidative stress: Diagnostics, prevention, and therapy*. ACS Publications. https://doi.org/10.1021/bk-2011-1083.ot001

Anton, M. (2007). Composition and structure of hen egg yolk. In *Bioactive egg compounds* (pp. 1–6). Springer. https://doi.org/10.1007/978-3-540-37885-3_1

Anton, M., Martinet, V., Dalgalarrondo, M., Beaumal, V., David-Briand, E., & Rabesona, H. (2003). Chemical and structural characterisation of low-density lipoproteins purified from hen egg yolk. *Food Chemistry*, *83*(2), 175–183. https://doi.org/10.1016/S0308-8146(03)00060-8

Ballesteros, M. N., Valenzuela, F., Robles, A. E., Artalejo, E., Aguilar, D., Andersen, C. J., Valdez, H., & Fernandez, M. L. (2015). One egg per day improves inflammation when compared to an oatmeal-based breakfast without increasing other cardiometabolic risk factors in diabetic patients. *Nutrients*, *7*(5), 3449–3463. https://doi.org/10.3390/nu7053449

Barter, P. J., Nicholls, S., Rye, K.-A., Anantharamaiah, G., Navab, M., & Fogelman, A. M. (2004). Anti-inflammatory properties of HDL. *Circulation Research*, *95*(8), 764–772. http://doi.org/10.1161/01.RES.0000146094.59640.13

Bengmark, S., & Jeppsson, B. (1995). Gastrointestinal surface protection and mucosa reconditioning. *Journal of Parenteral and Enteral Nutrition*, *19*(5), 410–415. https://doi.org/10.1177/0148607195019005410

Bitman, J., & Wood, D. (1980). Cholesterol and cholesteryl esters of eggs from various avian species. *Poultry Science*, *59*(9), 2014–2023. https://doi.org/10.3382/ps.0592014

Blesso, C. N. (2015). Egg phospholipids and cardiovascular health. *Nutrients*, *7*(4), 2731–2747. https://doi.org/10.3390/nu7042731

Blesso, C. N., Andersen, C. J., Bolling, B. W., & Fernandez, M. L. (2013). Egg intake improves carotenoid status by increasing plasma HDL cholesterol in adults with metabolic syndrome. *Food & Function*, *4*(2), 213–221. http://doi.org/10.1039/C2FO30154G

Brady, D., Lowe, N., Gaines, S., Fenelon, L., Mc Partlin, J., & O'Farrelly, C. (2003). Inhibition of *Streptococcus mutans* growth by hen egg-derived fatty acids. *Journal of Food Science*, *68*(4), 1433–1437. https://doi.org/10.1111/j.1365-2621.2003.tb09662.x

Burton, G. W., & Traber, M. G. (1990). Vitamin E: Antioxidant activity, biokinetics, and bioavailability. *Annual Review of Nutrition*, *10*(1), 357–382. https://doi.org/10.1146/annurev.nu.10.070190.002041

Calzada, E., Onguka, O., & Claypool, S. M. (2016). Phosphatidylethanolamine metabolism in health and disease. *International Review of Cell and Molecular Biology*, *321*, 29–88. https://doi.org/10.1016/bs.ircmb.2015.10.001

Che, H., Fu, X., Zhang, L., Gao, X., Wen, M., Du, L., Xue, C., Xu, J., & Wang, Y. (2018). Neuroprotective effects of n-3 polyunsaturated fatty acid-enriched phosphatidylserine against oxidative damage in PC12 cells. *Cellular and Molecular Neurobiology*, *38*(3), 657–668. https://doi.org/10.1007/s10571-017-0516-y

Chen, J., Lin, S., Sun, N., Bao, Z., Shen, J., & Lu, X. (2019). Egg yolk phosphatidylcholine: Extraction, purification and its potential neuroprotective effect on PC12 cells. *Journal of*

Functional Foods, *56*, 372–383. https://doi.org/10.1016/j.jff.2019.03.037

Chung, R. W., Kamili, A., Tandy, S., Weir, J. M., Gaire, R., Wong, G., Meikle, P. J., Cohn, J. S., & Rye, K.-A. (2013). Dietary sphingomyelin lowers hepatic lipid levels and inhibits intestinal cholesterol absorption in high-fat-fed mice. *PLoS One*, *8*(2), e55949. https://doi.org.10.1371/journal.pone.0055949

Chung, S.-Y., Moriyama, T., Uezu, E., Uezu, K., Hirata, R., Yohena, N., Masuda, Y., Kokubu, Y., & Yamamoto, S. (1995). Administration of phosphatidylcholine increases brain acetylcholine concentration and improves memory in mice with dementia. *The Journal of Nutrition*, *125*(6), 1484–1489. https://doi.org/10.1093/jn/125.6.1484

Chung, R. W., Wang, Z., Bursill, C. A., Wu, B. J., Barter, P. J., & Rye, K.-A. (2017). Effect of long-term dietary sphingomyelin supplementation on atherosclerosis in mice. *PLoS One*, *12*(12), e0189523. https://doi.org/10.1371/journal.pone.0189523

Cui, L., & Decker, E. A. (2016). Phospholipids in foods: Prooxidants or antioxidants? *Journal of the Science of Food and Agriculture*, *96*(1), 18–31. https://doi.org/10.1002/jsfa.7320

Duan, X., Li, M., Wu, F., Yang, N., Nikoo, M., Jin, Z., & Xu, X. (2013). Postfertilization changes in nutritional composition and protein conformation of hen egg. *Journal of Agricultural and Food Chemistry*, *61*(49), 12092–12100. https://doi.org/10.1021/jf403099q

Duivenvoorden, I., Voshol, P. J., Rensen, P. C., van Duyvenvoorde, W., Romijn, J. A., Emeis, J. J., Havekes, L. M., & Nieuwenhuizen, W. F. (2006). Dietary sphingolipids lower plasma cholesterol and triacylglycerol and prevent liver steatosis in APOE* 3 Leiden mice–. *The American Journal of Clinical Nutrition*, *84*(2), 312–321. https://doi.org/10.1093/ajcn/84.2.312

Eckhardt, E. R., Wang, D. Q. H., Donovan, J. M., & Carey, M. C. (2002). Dietary sphingomyelin suppresses intestinal cholesterol absorption by decreasing thermodynamic activity of cholesterol monomers. *Gastroenterology*, *122*(4), 948–956 https://doi-org.tue.80599.net/10.1053/gast.2002.32539

Eftekhar, S., Parsaei, H., Keshavarzi, Z., Yazdi, A. T., Hadjzadeh, M.-A.-R., Rajabzadeh, A., & Malayeri, S. O. (2015). The prevention and treatment effects of egg yolk high density lipoprotein on the formation of atherosclerosis plaque in rabbits. *Iranian Journal of Basic Medical Sciences*, *18*(4), 343–349.

Ehehalt, R., Wagenblast, J., Erben, G., Lehmann, W. D., Hinz, U., Merle, U., & Stremmel, W. (2004). Phosphatidylcholine and lysophosphatidylcholine in intestinal mucus of ulcerative colitis patients. A quantitative approach by nanoelectrospray-tandem mass spectrometry. *Scandinavian Journal of Gastroenterology*, *39*(8), 737–742. https://doi.org/10.1080/00365520410006233

Evans, R. J., Bauer, D. H., Bandemer, S. L., Vaghefi, S. B., & Flegal, C. J. (1973). Structure of egg yolk very low density lipoprotein. Polydispersity of the very low density lipoprotein and the role of lipovitellenin in the structure. *Archives of Biochemistry and Biophysics*, *154*(2), 493–500. https://doi.org/10.1016/0003-9861(73)90001-5

Fahy, E., Subramaniam, S., Brown, H. A., Glass, C. K., Merrill, A. H., Murphy, R. C., Raetz, C. R., Russell, D. W., Seyama, Y., & Shaw, W. (2005). A comprehensive classification system for lipids1. *Journal of Lipid Research*, *46*(5), 839–861. https://doi.org/10.1194/jlr.E400004-JLR200

Feng, D., Ohlsson, L., Ling, W., Nilsson, Å., & Duan, R.-D. (2010). Generating ceramide from sphingomyelin by alkaline sphingomyelinase in the gut enhances sphingomyelin-induced inhibition of cholesterol uptake in Caco-2 cells. *Digestive Diseases and Sciences*, *55*(12), 3377–3383. https://doi.org/10.1007/s10620-010-1202-9

Ferretti, G., & Bacchetti, T. (2012). Effect of dietary lipids on paraoxonase-1 activity and gene expression. *Nutrition, Metabolism and Cardiovascular Diseases*, *22*(2), 88–94. https://doi.org/10.1016/j.numecd.2011.08.011

Fiedor, J., & Burda, K. (2014). Potential role of carotenoids as antioxidants in human health and disease. *Nutrients*, *6*(2), 466–488. https://doi.org/10.3390/nu6020466

Garmy, N., Taïeb, N., Yahi, N., & Fantini, J. (2005). Interaction of cholesterol with sphingosine: Physicochemical characterization and impact on intestinal absorption. *Journal of Lipid Research*, *46*(1), 36–45. https://doi.org/10.1194/jlr.M400199-JLR200

Goldberg, J., Flowerdew, G., Smith, E., Brody, J. A., & Tso, M. O. (1988). Factors associated with age-related macular degeneration: An Analysis of Data from the fi1r8t National health and nutrition Examination survey. *American Journal of Epidemiology*, *128*(4), 700–710. https://doi.org/10.1093/oxfordjournals.aje.a115023

Golzar Adabi, S., Ahbab, M., Fani, A., Hajbabaei, A., Ceylan, N., & Cooper, R. (2013). Egg yolk fatty acid profile of avian species–influence on human nutrition. *Journal of Animal Physiology and Animal Nutrition*, *97*(1), 27–38. https://doi.org/10.1111/j.1439-0396.2011.01239.x

Gopinath, B., Liew, G., Tang, D., Burlutsky, G., Flood, V. M., & Mitchell, P. (2020). Consumption of eggs and the 15-year incidence of age-related macular degeneration. *Clinical Nutrition*, *39*(2), 580–584. https://doi.org.tue.80599.net/10.1016/j.clnu.2019.03.009

Gutierrez, M., Takahashi, H., & Juneja, L. (2018). Nutritive evaluation of hen eggs. In *Hen eggs* (pp. 25–35). CRC Press

Hatta, H., Kapoor, M. P., & Juneja, L. R. (2008). Bioactive components in egg yolk. In *Egg bioscience and biotechnology* (pp. 185–237). John Wiley & Sons Inc. https://doi.org/10.1002/9780470181249

Hu, T.-M., Lan, T.-H., & Loh, E.-W. (2012). An emerging complimentary medicine-yolk oil made from heating method. *Journal of Traditional and Complementary Medicine*, *2*(3), 202–203. https://doi.org/10.1016/S2225-4110(16)30100-6

Huopalahti, R., Anton, M., López-Fandiño, R., & Schade, R. (2007). *Bioactive egg compounds*. Springer. https://doi.org/10.1007/978-3-540-37885-3_1

Imaizumi, K., Mawatari, K., Murata, M., Ikeda, I., & Sugano, M. (1983). The contrasting effect of dietary phosphatidylethanolamine and phosphatidylcholine on serum lipoproteins and liver lipids in rats. *The Journal of Nutrition*, *113*(12), 2403–2411. https://doi.org/10.1093/jn/113.12.2403

Imaizumi, K., Sekihara, K., & Sugano, M. (1991). Hypocholesterolemic action of dietary phosphatidylethanolamine in rats sensitive to exogenous cholesterol. *The Journal of Nutritional Biochemistry*, *2*(5), 251–254. https://doi.org/10.1016/0955-2863(91)90083-H

Jayaraman, T., Kannappan, S., Ravichandran, M., & Anuradha, C. (2008). Impact of Essentiale L on ethanol-induced changes in rat brain and erythrocytes.*Singapore Medical Journal*, *49*(4), 320.

Jiang, Y., Noh, S. K., & Koo, S. I. (2001). Egg phosphatidylcholine decreases the lymphatic absorption of cholesterol in rats. *The Journal of Nutrition*, *131*(9), 2358–2363. https://doi.org/10.1093/jn/131.9.2358

Kato-Kataoka, A., Sakai, M., Ebina, R., Nonaka, C., Asano, T., & Miyamori, T. (2010). Soybean-derived phosphatidylserine improves memory function of the elderly Japanese subjects with memory complaints. *Journal of Clinical Biochemistry and Nutrition*, 1009240057–1009240057. https://doi.org/10.3164/jcbn.10-62

King, M., Boyd, L., & Sheldon, B. (1992). Antioxidant properties of individual phospholipids in a salmon oil model system. *Journal of the American Oil Chemists Society*, *69*(6), 545–551. https://doi.org/10.1007/BF02636106

Kuang, H., Yang, F., Zhang, Y., Wang, T., & Chen, G. (2018). The impact of egg nutrient composition and its consumption on cholesterol homeostasis. *Cholesterol*, *2018*. https://doi.org/10.1155/2018/6303810

Kuksis, A. (1992). Yolk lipids. *Biochim Biophys Acta*, *1124*(3), 205–222. https://doi.org/10.1016/0005-2760(92)90132-f

Lagace, T. A., & Ridgway, N. D. (2013). The role of phospholipids in the biological activity and structure of the endoplasmic reticulum. *Biochimica et Biophysica Acta (BBA)-Molecular Cell Research*, *1833*(11), 2499–2510. https://doi.org/10.1016/j.bbamcr.2013.05.018

Lawrence, T. (2009). The nuclear factor NF-kappaB pathway in inflammation. *Cold Spring Harbor Perspectives in Biology*, *1*(6), a001651–a001651. https://doi.org/10.1101/cshperspect.a001651

Lazarus, G. S., Cooper, D. M., Knighton, D. R., Margolis, D. J., Percoraro, R. E., Rodeheaver, G., & Robson, M. C. (1994). Definitions and guidelines for assessment of wounds and evaluation of healing. *Wound Repair and Regeneration*, *2*(3), 165–170. https://doi.org/10.1046/j.1524-475X.1994.20305.x

Lichtenberger, L. (1995). The hydrophobic barrier properties of gastrointestinal mucus. *Annual Review of Physiology*, *57*(1), 565–583. https://doi.org/10.1146/annurev.ph.57.030195.003025

Lim, S.-Y., & Suzuki, H. (2000). Intakes of dietary docosahexaenoic acid ethyl ester and egg phosphatidylcholine improve maze-learning ability in young and old mice. *The Journal of Nutrition*, *130*(6), 1629–1632. https://doi.org/10.1093/jn/130.6.1629

Lordelo, M., Cid, J., Cordovil, C. M., Alves, S. P., Bessa, R. J., & Carolino, I. (2020). A comparison between the quality of eggs from indigenous chicken breeds and that from commercial layers. *Poultry Science*, *99*(3), 1768–1776. https://doi.org/10.1016/j.psj.2019.11.023

Mahmoudi, M., Ebrahimzadeh, M., Pourmorad, F., Rezaie, N., & Mahmoudi, M. (2013). Anti-inflammatory and analgesic effects of egg yolk: A comparison between organic and machine made. *European Review for Medical and Pharmacological Sciences*, *17*(4), 472–476. https://doi.org/10.1016/j.peptides.2012.10.009

Makhmudova, U., Schulze, P. C., Lütjohann, D., & Weingärtner, O. (2021). Phytosterols and cardiovascular disease. *Current Atherosclerosis Reports*, *23*(11), 1–8. https://doi.org/10.1007/s11883-021-00964-x

Mares-Perlman, J. A., Millen, A. E., Ficek, T. L., & Hankinson, S. E. (2002). The body of evidence to support a protective role for lutein and zeaxanthin in delaying chronic disease. Overview. *The Journal of Nutrition*, *132*(3), 518S–524S. https://doi.org/10.1093/jn/132.3.518S

Martin, W., Augustyniak, J., & Cook, W. (1964). Fractionation and characterization of the low-density lipoproteins of hen's egg yolk. *Biochimica et Biophysica Acta (BBA)-Specialized Section on Lipids and Related Subjects*, *84*(6), 714–720. https://doi.org/10.1016/0926-6542(64)90029-0

Milard, M., Penhoat, A., Durand, A., Buisson, C., Loizon, E., Meugnier, E., Bertrand, K., Joffre, F., Cheillan, D., & Garnier, L. (2019). Acute effects of milk polar lipids on intestinal tight junction expression: Towards an impact of sphingomyelin through the regulation of IL-8 secretion? *The Journal of Nutritional Biochemistry*, *65*, 128–138. https://doi.org/10.1016/j.jnutbio.2018.12.007

Millar, C. L., Norris, G. H., Vitols, A., Garcia, C., Seibel, S., Anto, L., & Blesso, C. N. (2019). Dietary egg sphingomyelin prevents aortic root plaque accumulation in apolipoprotein-E knockout mice. *Nutrients*, *11*(5), 1124. https://doi.org/10.3390/nu11051124

Miller, N. E. (1990). HDL metabolism and its role in lipid transport. *European Heart Journal*, *11*(suppl_H), 1–3. https://doi.org/10.1093/eurheartj/11.suppl_H.1

Mine, Y. (2008). *Egg bioscience and biotechnology*. John Wiley & Sons. https://doi.org/10.1002/9780470181249

Mineo, C., Deguchi, H., Griffin, J. H., & Shaul, P. W. (2006). Endothelial and antithrombotic actions of HDL. *Circulation Research*, *98*(11), 1352–1364. https://doi.org/10.1161/01.RES.0000225982.01988.93

Miranda, J. M., Anton, X., Redondo-Valbuena, C., Roca-Saavedra, P., Rodriguez, J. A., Lamas, A., Franco, C. M., & Cepeda, A. (2015). Egg and egg-derived foods: Effects on human health and use as functional foods. *Nutrients*, *7*(1), 706–729. https://doi.org/10.3390/nu7010706

Mitchell, P., Liew, G., Gopinath, B., & Wong, T. Y. (2018). Age-related macular degeneration. *The Lancet*, *392*(10153), 1147–1159. https://doi.org/10.1016/S0140-6736(18)31550-2

Nataraj, J., Manivasagam, T., Thenmozhi, A. J., & Essa, M. M. (2016). Lutein protects dopaminergic neurons against MPTP-induced apoptotic death and motor dysfunction by ameliorating mitochondrial disruption and oxidative stress. *Nutritional Neuroscience*, *19*(6), 237–246. https://doi.org/10.1179/1476830515Y.0000000010

Nimalaratne, C., Lopes-Lutz, D., Schieber, A., & Wu, J. (2012). Effect of domestic cooking methods on egg yolk xanthophylls. *Journal of Agricultural and Food Chemistry*, *60*(51), 12547–12552. https://doi.org/10.1021/jf303828n

Nimalaratne, C., Savard, P., Gauthier, S. F., Schieber, A., & Wu, J. (2015). Bioaccessibility and digestive stability of carotenoids in cooked eggs studied using a dynamic in vitro gastrointestinal model. *Journal of Agricultural and Food Chemistry*, *63*(11), 2956–2962. https://doi.org/10.1021/jf505615w

Nimalaratne, C., Wu, J., & Schieber, A. (2013). Egg yolk carotenoids: Composition, analysis, and effects of processing on their stability. In *Carotenoid cleavage products* (pp. 219–225). ACS Publications. https://doi.org/10.1021/bk-2013-1134.ch018

Noh, S. K., & Koo, S. I. (2003). Egg sphingomyelin lowers the lymphatic absorption of cholesterol and α-tocopherol in rats. *The Journal of Nutrition*, *133*(11), 3571–3576. https://doi.org/10.1093/jn/133.11.3571

Noh, S. K., & Koo, S. I. (2004). Milk sphingomyelin is more effective than egg sphingomyelin in inhibiting intestinal absorption of cholesterol and fat in rats. *The Journal of Nutrition*, *134*(10), 2611–2616. https://doi.org/10.1093/jn/134.10.2611

Norris, G. H., & Blesso, C. N. (2017). Dietary sphingolipids: Potential for management of dyslipidemia and nonalcoholic fatty liver disease. *Nutrition Reviews*, *75*(4), 274–285. https://doi.org/10.1093/nutrit/nux004

Norris, G. H., Jiang, C., Ryan, J., Porter, C. M., & Blesso, C. N. (2016). Milk sphingomyelin improves lipid metabolism and alters gut microbiota in high fat diet-fed mice. *The Journal of Nutritional Biochemistry*, *30*, 93–101. https://doi.org/10.1016/j.jnutbio.2015.12.003

Norris, G. H., Milard, M., Michalski, M.-C., & Blesso, C. N. (2019). Protective properties of milk sphingomyelin against dysfunctional lipid metabolism, gut dysbiosis, and inflammation. *The Journal of Nutritional Biochemistry*, *73*, 108224. https://doi.org/10.1016/j.jnutbio.2019.108224

Ohlsson, L., Hertervig, E., Jönsson, B. A., Duan, R.-D., Nyberg, L., Svernlöv, R., & Nilsson, Å. (2010). Sphingolipids in human ileostomy content after meals containing milk sphingomyelin. *The American Journal of Clinical Nutrition*, *91*(3), 672–678. https://doi.org/10.3945/ajcn.2009.28311

Pandey, N., Sultan, K., Twomey, E., & Sparks, D. (2009). Phospholipids block nuclear factor-kappa B and tau phosphorylation and inhibit amyloid-beta secretion in human neuroblastoma cells. *Neuroscience*, *164*(4), 1744–1753. https://doi.org/10.1016/j.neuroscience.2009.09.062

Park, J., Upadhaya, S., & Kim, I. (2015). Effect of dietary marine microalgae (Schizochytrium) powder on egg production, blood lipid profiles, egg quality, and fatty acid composition of egg yolk in layers. *Asian-Australasian Journal of Animal Sciences*, *28*(3), 391. https://doi.org/10.5713/ajas.14.0463

Rastegar, F., Azarpira, N., Amiri, M., & Azarpira, A. (2011). The effect of egg yolk oil in the healing of third degree burn wound in rats. *Iranian Red Crescent Medical Journal*, *13*(10), 739–743.

Ratliff, J. C., Mutungi, G., Puglisi, M. J., Volek, J. S., & Fernandez, M. L. (2008). Eggs modulate the inflammatory response to carbohydrate restricted diets in overweight men. *Nutrition & Metabolism*, *5*(1), 1–9. https://doi.org/10.1186/1743-7075-5-6

Rye, K.-A., & Barter, P. J. (2008). *Anti-inflammatory actions of HDL: A new insight* (Vol. 28, pp. 1890–1891). Am Heart Assoc. https://doi.org/10.1161/ATVBAHA.108.173575

Salari, M., Salari, R., Dadgarmoghadam, M., Khadem-Rezaiyan, M., & Hosseini, M. (2016). Efficacy of egg yolk and nitroglycerin ointment as treatments for acute anal fissures: A randomized clinical trial study. *Electronic Physician*, *8*(10), 3035–3041. https://doi.org/10.19082/3035

Shen, Q., Riedl, K. M., Cole, R. M., Lehman, C., Xu, L., Alder, H., Belury, M. A., Schwartz, S. J., & Ziouzenkova, O. (2015). Egg yolks inhibit activation of NF-κB and expression of its target genes in adipocytes after partial delipidation. *Journal of Agricultural and Food Chemistry*, *63*(7), 2013–2025. https://doi.org/10.1021/jf5056584

Shin, J. Y., Xun, P., Nakamura, Y., & He, K. (2013). Egg consumption in relation to risk of cardiovascular disease and diabetes: A systematic review and meta-analysis. *The American Journal of Clinical Nutrition*, *98*(1), 146–159. https://doi.org/10.3945/ajcn.112.051318

Slotte, J. P. (2016). The importance of hydrogen bonding in sphingomyelin's membrane interactions with co-lipids. *Biochimica Et Biophysica Acta (BBA)-Biomembranes*, *1858*(2), 304–310. https://doi.org/10.1016/j.bbamem.2015.12.008

Sugino, H., Ishikawa, M., Nitoda, T., Koketsu, M., Juneja, L. R., Kim, M., & Yamamoto, T. (1997). Antioxidative activity of egg yolk phospholipids. *Journal of Agricultural and Food Chemistry*, *45*(3), 551–554. https://doi.org/10.1021/jf960416p

Sun, N., Chen, J., Bao, Z., Wang, D., An, B., & Lin, S. (2019). Egg yolk phosphatidylethanolamine: Extraction optimization, antioxidative activity, and molecular structure profiling. *Journal of Food Science*, *84*(5), 1002–1011. https://doi.org/10.1111/1750-3841.14512

Surai, P., MacPherson, A., Speake, B., & Sparks, N. (2000). Designer egg evaluation in a controlled trial. *European Journal of Clinical Nutrition*, *54*(4), 298–305. https://doi.org/10.1038/sj.ejcn.1600939

Traber, M. G., & Atkinson, J. (2007). Vitamin E, antioxidant and nothing more. *Free Radical Biology and Medicine*, *43*(1), 4–15. https://doi.org/10.1016/j.freeradbiomed.2007.03.024

Treede, I., Braun, A., Jeliaskova, P., Giese, T., Füllekrug, J., Griffiths, G., Stremmel, W., & Ehehalt, R. (2009). TNF-α-induced up-regulation of pro-inflammatory cytokines is reduced by phosphatidylcholine in intestinal epithelial cells. *BMC Gastroenterology*, *9*(1), 1–11. https://doi.org/10.1186/1471-230X-9-53

Treede, I., Braun, A., Sparla, R., Kühnel, M., Giese, T., Turner, J. R., Anes, E., Kulaksiz, H., Füllekrug, J., & Stremmel, W. (2007). Anti-inflammatory effects of phosphatidylcholine. *Journal of Biological Chemistry*, *282*(37), 27155–27164. https://doi.org/10.1074/jbc.M704408200

Tria, E., & Scanu, A. M. (1969). Structural and functional aspects of lipoproteins in living systems. *Structural and functional aspects of lipoproteins in living systems.*

Trpkovic, A., Resanovic, I., Stanimirovic, J., Radak, D., Mousa, S. A., Cenic-Milosevic, D., Jevremovic, D., & Isenovic, E. R. (2015). Oxidized low-density lipoprotein as a biomarker of cardiovascular diseases. *Critical Reviews in Clinical Laboratory Sciences*, *52*(2), 70–85. https://doi.org/10.3109/10408363.2014.992063

Vallejo-Cordoba, B., Nakai, S., Powrie, W., & Beveridge, T. (1986). Protein hydrolysates for reducing water activity in meat products.

Van Lenten, B. J., Hama, S. Y., de Beer, F. C., Stafforini, D. M., McIntyre, T. M., Prescott, S. M., La Du, B. N., Fogelman, A. M., & Navab, M. (1995). Anti-inflammatory HDL becomes pro-inflammatory during the acute phase response. Loss of protective effect of HDL against LDL oxidation in aortic wall cell cocultures. *The Journal of Clinical Investigation*, *96*(6), 2758–2767. https://doi.org/10.1172/JCI118345

Vishwanathan, R., Gendron, C. M., Goodrow-Kotyla, E. F., Wilson, T. A., & Nicolosi, R. J. (2010). Increased consumption of dietary cholesterol, lutein, and zeaxanthin as egg yolks does not decrease serum concentrations and lipoprotein distribution of other carotenoids, retinol, and tocopherols. *Nutrition Research*, *30*(11), 747–755. https://doi.org/10.1016/j.nutres.2010.10.007

Vors, C., Joumard-Cubizolles, L., Lecomte, M., Combe, E., Ouchchane, L., Drai, J., Raynal, K., Joffre, F., Meiller, L., & Le Barz, M. (2020). Milk polar lipids reduce lipid cardiovascular risk factors in overweight postmenopausal women: Towards a gut sphingomyelin-cholesterol interplay. *Gut*, *69*(3), 487–501. http://dx.doi.org/10.1136/gutjnl-2018-318155

Widomska, J., & Subczynski, W. K. (2014). Why has nature chosen lutein and zeaxanthin to protect the retina? *Journal of Clinical & Experimental Ophthalmology*, *5*(1), 326. https://doi.org/10.4172/2155-9570.1000326

Wu, P., Pan, Y., Yan, J., Huang, D., & Li, S. (2016). Assessment of egg yolk oil extraction methods of for ShiZhenKang oil by pharmacodynamic index evaluation. *Molecules*, *21*(1). https://doi.org/10.3390/molecules21010106

Xiao, N., Zhao, Y., Yao, Y., Wu, N., Xu, M., Du, H., & Tu, Y. (2020). Biological activities of egg yolk lipids: A review. *Journal of Agricultural and Food Chemistry*, *68*(7), 1948–1957. https://doi.org/10.1021/acs.jafc.9b06616

Xie, X.-m, & Pang, X.-b (2012). Pharmacodynamic study on scald wound of egg yolk oil by ethanol extract. *Journal of Henan University (Medical Science)*. https://doi.org/10.3969/j.issn.1672-7606.2012.02.006

Xu, C., Yang, C., Yin, Y., Liu, J., & Mine, Y. (2012). Phosphopeptides (PPPs) from hen egg yolk phosvitin exert anti-inflammatory activity via modulation of cytokine expression. *Journal of functional Foods*, *4*(4), 718–726. https://doi.org/10.1016/j.jff.2012.04.011

Yamamoto, T., Juneja, L. R., Hatta, H., & Kim, M. (1996). *Hen eggs: Basic and applied science*. CRC Press

Yamamoto, Y., Sogo, N., Iwao, R., & Miyamoto, T. (1990). Antioxidant effect of egg yolk on linoleate in emulsions. *Agricultural and Biological Chemistry*, *54*(12), 3099–3104. https://doi.org/10.1080/00021369.1990.10870480

Yamashita, S., Kinoshita, M., & Miyazawa, T. (2021). Dietary sphingolipids contribute to health via intestinal maintenance. *International Journal of Molecular Sciences*, *22*(13), 7052. https://doi.org/10.3390/ijms22137052

Yang, F., Chen, G., Ma, M., Qiu, N., Zhu, L., & Li, J. (2018a). Egg-Yolk sphingomyelin and phosphatidylcholine attenuate cholesterol absorption in Caco-2 cells. *Lipids*, *53*(2), 217–233. https://doi.org/10.1002/lipd.12018

Yang, F., Chen, G., Ma, M., Qiu, N., Zhu, L., & Li, J. (2018b). Fatty acids modulate the expression levels of key proteins for cholesterol absorption in Caco-2 monolayer. *Lipids in Health and Disease*, *17*(1), 1–14. https://doi.org/10.1186/s12944-018-0675-y

Yenilmez, E., Başaran, E., Arslan, R., Berkman, M., Güven, U., Bayçu, C., & Yazan, Y. (2015). Chitosan gel formulations containing egg yolk oil and epidermal growth factor for dermal burn treatment. *Die Pharmazie-An International Journal of Pharmaceutical Sciences*, *70*(2), 67–73. https://doi.org/10.1691/ph.2015.4126

Young, D., Nau, F., Pasco, M., & Mine, Y. (2011). Identification of hen egg yolk-derived phosvitin phosphopeptides and their effects on gene expression profiling against oxidative stress-induced Caco-2 cells. *Journal of Agricultural and Food Chemistry*, *59*(17), 9207–9218. https:/doi.org/10.1021/jf202092d

Yu, Z., Mao, C., Fu, X., & Ma, M. (2019). High density lipoprotein from egg yolk (EYHDL) improves dyslipidemia by mediating fatty acids metabolism in high fat diet-induced obese mice. *Food Science of Animal Resources*, *39*(2), 179–196. https://doi.org/10.5851/kosfa.2018.e38

Yu, Z., Xiang, X., Jin, Y., Wang, N., & Ma, M. (2019). High-throughput lipidomic profiling of high-density lipoprotein from egg yolk (EYHDL): Comparison based on UPLC-MS/MS and GC–MS. *European Food Research and Technology*, *245*(8), 1665–1675. https://doi.org/10.1007/s00217-019-03275-3

Zaheer, K. (2017). Hen egg carotenoids (lutein and zeaxanthin) and nutritional impacts on human health: A review. *CYTA-Journal of Food*, *15*(3), 474–487. https://doi.org/10.1007/s00217-019-03275-3

Zhang, R., Liu, W., Zeng, J., Meng, J., Jiang, H., Wang, J., & Xing, D. (2022). Niemann-pick C1-like 1 inhibitors for reducing cholesterol absorption. *European Journal of Medicinal Chemistry*, 114111. https://doi.org.tue.80599.net/10.1016/j.ejmech.2022.11411

Zhang, C., Yu, H., Shen, Y., Ni, X., Shen, S., & Das, U. N. (2015). Polyunsaturated fatty acids trigger apoptosis of colon cancer cells through a mitochondrial pathway. *Archives of Medical Science: AMS*, *11*(5), 1081–1094. https://doi.org/10.5114/aoms.2015.54865

Zhao, Y., Yao, Y., Xu, M., Wang, S., Wang, X., & Tu, Y. (2017). Simulated gastrointestinal digest from preserved egg white exerts anti-inflammatory effects on Caco-2 cells and a mouse model of DSS-induced colitis. *Journal of Functional Foods*, *35*, 655–665. https://doi.org/10.1016/j.jff.2017.06.028

Zimmer, J. P., & Hammond, B. R. Jr. (2007). Possible influences of lutein and zeaxanthin on the developing retina. *Clinical Ophthalmology (Auckland, NZ)*, *1*(1), 25.

26

Lysozyme-Antimicrobial Agent and Applications

Jack N. Losso*, Arissara Phosanam, and Achyut Adhikari

CONTENTS

26.1 Introduction

Lysozyme, also known as N-acetylmuramic hydrolase or muramidase (EC 3.2.1.17), was discovered by Alexander Fleming in 1922 from his own nasal mucus as a bacteriolytic protein (Fleming, 1922). In 1945, Alderon et al. identified lysozyme in hen's egg albumen and found it to be similar to egg white protein known as globulin G1 that was reported by Longworth in 1940 using moving-boundary electrophoresis (Alderton et al., 1945).

Lysozyme belongs to the class of enzymes that lyse the cell walls of Gram-positive bacteria by hydrolyzing the β-1,4 linkage between *N*-acetylmuramic acid (NAM) and *N*-acetyl-glucosamine (NAG) of gigantic polymers $(NAM\text{-}NAG)_n$ in the peptidoglycan (murein). Gram-negative bacteria, fungal, and viral inhibition by lysozyme have recently been reported and will be discussed in detail in the subsequent sections.

* Corresponding author.

DOI: 10.1201/9781003254430-33

26.2 Lysozyme Occurrence

26.2.1 Lysozyme Sources

Sources of lysozyme include animals, avians, plants, and invertebrates (Table 26.1). In avian sources, lysozyme is found in egg albumen, shell, and vitelline membrane. Lysozyme has also been recently produced through recombinant technology in transgenic animals. Recombinant expression of lysozyme in *E.coli* and different strains of *Aspergillus niger* have been attempted (Salazar & Asenjo, 2007).

In the animal kingdom, three dominant types of lysozymes, including the chicken type (*c*-type), goose type (*g*-type), and invertebrate type (*i*-type) have been identified based on characteristics such as structure, catalysis, and immunization. The c-type and g-type lysozyme have alkaline pI values and therefore are very basic proteins. There are three other types of lysozymes, including the bacterial-type, phage-type, and plant-type (Callewaert & Michiels, 2010). The enzyme is also found in large quantities in human organs, tissues, and human and other species secretions, including spleen, placenta, skin milk, tear, saliva, and serum, bacteria secretions, and papaya secretions (papaya latex)(Ogawa et al., 1971).

Egg white lysozyme is the classic representative of the lysozyme family, and the related enzymes are called *c*-type (chicken- or conventional-type) lysozymes (Losso et al., 2000). Lysozyme content of avian egg is 0.3–0.4 g per egg and 3.4% of egg white (Jalili-Firoozinezhad et al., 2020). Lysozyme can also be isolated from hen eggshell membrane but has weaker antimicrobial properties compared to egg white lysozyme (Losso et al., 2000). The *g*-type is typical for goose egg white. However, lysozyme occurs simultaneously in both the "*c*-type" and "*g*-type" in the goose and swan (Hindenburg et al., 1974). Emu egg white has a very low level of *g*-type lysozyme with a very low level of activity (Maehashi et al., 2012). The *i*-type represents the invertebrate type of lysozyme (Nilsen & Myrnes, 2001; Datta et al., 2008; Callewaert & Michiels, 2010). It is much smaller (12, 985.6 Da) in size by MALDI-TOF-MS than the *g*-type (Q.-G. Xue et al., 2007). The presence of *i*-type lysozyme has been confirmed in oyster *Crassostrea virginica* and other bivalves such as the *Tapes Japonica* (Q. Xue et al., 2010) and sea cucumber *Stichopus japonicus* (Cong et al., 2009).

TABLE 26.1

Activities of Lysozyme of Various Origins[a]

Lysozyme	Lytic Activity (%)[b]
Chicken egg	100
Japanese pheasant egg	123
Turkey egg	176
Rabbit milk	204
Pig milk	245
Rat milk	255
Human milk	396

[a] Jolles (1996).

[b] The lytic activity values was calculated compared with chicken egg lysozyme (Lytic activity (%) = 100 × Lytic activity of sample/Lytic activity of chicken egg).

There is no evidence for the occurrence of lysozyme *g* or its gene in biological systems other than birds (Prager & Jolles, 1996). A viral type (υ) family has also been defined for phage lysozymes. Phage lambda (λ) is a transglycosidase and active on *E. coli* cell walls but not on *Micrococcus luteus* or on chitin (β-1,4-polyNAG). Two types of lysozymes have been identified in plants, including the *h* type (hevamine type) and the *b* type (barley type) (Beintema & Terwisscha van Scheltinga, 1996).

26.2.2 Lysozyme Screening/Identification

Several techniques can be used to identify lysozyme. Electrophoresis and specifically the matrix-assisted laser desorption-time of flight-mass spectrometry (MALDI-TOF-MS) can provide the molecular size of lysozyme at 14,309 Daltons with great accuracy. The pI of the protein at value of 10.5 is also selective for the protein. Other techniques, including amino acid sequence, affinity, ion exchange, gel filtration chromatography, circular dichroism, differential scanning calorimetry, NMR, Raman, and FTIR can be used for lysozyme identification.

26.2.3 Quantitative and Activity Assays for Lysozyme

As a pure protein, lysozyme purity can be confirmed by the methods in Section 2.2. above. The protein has a high content of aromatic amino acids and should have a good absorbance reading at 280 nm. The methods for determining lysozyme activity include the turbidimetric and agar plate method that uses the Gram-positive *Micrococcus luteus*, immunoassays, fluorometric and colorimetric assays, and histochemical and cytochemical location. The turbidimetric assay provides a simple rapid accurate and specific method for lysozyme activity determination; however, the method has relatively low sensitivity. *Leuconostoc eonos* and specially *Leuconostoc eonos* L294 was proposed to be used as a suitable substrate for lysozyme activity at acidic pH because lysozyme does not lyse the standard substrate *Micrococcus luteus* that is commonly used to evaluate lysozyme activity (Pitotti & Boschelle, 1991). The proposed method is applicable to acidic pH where lysozyme has great solubility, very rapid, sensitive, reproducible, and with good precision. Helal and Melzig proposed a highly sensitive fluorescence method to determine lysozyme activity at a low level such as cell cultures (Helal & Melzig, 2008). For unknown reasons, until now, the *Micrococcus luteus* method remains the most used method for lysozyme activity determination (Beaussart et al., 2021; Naveed et al., 2022).

26.3 Molecular Properties

26.3.1 Isolation/Purification

Lysozyme purification methods have included direct crystallization with 5% NaCl at pH 9.5 and 40–90% recovery and

100% purity (Alderton & Fevold, 1946); cation exchange with 90–95% lysozyme recovery (Li-Chan et al., 1986; Guérin-Dubiard et al., 2005) selective precipitation using anionic surfactant, heat-induced denaturation, and gelation, or 30% ethanol (Chang et al., 2000; Jiang et al., 2001; Shin et al., 2003); aqueous two-phase system such as polyethylene glycol/sodium sulfate providing 70-85% pure lysozyme (Su & Chiang, 2006); affinity chromatography with 79–95% lysozyme recovery (Altıntaş & Denizli, 2006; Li et al., 2014); expanded bed adsorption chromatography with lysozyme recovery of 84–87% (Tong et al., 2002). An extended summary can be found in (Shahmohammadi, 2018).

26.3.2 Physico-Chemistry

The hen *c*-type and other avian egg *g*-type lysozyme have been the most investigated. The *c*-type remains the most studied, easiest, and most economical source for lysozyme recovery. The protein is 14.3 Kilodaltons in size with 129 amino acid residues maintained by four disulfide bridges, and a pI of 10.5. The molecular weight of lysozyme as determined by MALDI-TOF-MS is 14,309 (Yang et al., 1998). The accuracy of MALDI-MS is 0.05%. Hen egg white lysozyme has a higher extinction coefficient at 280 nm (26.4 vs 4–15) than lysozyme from other sources or many other proteins because of the high content of the aromatic amino acids in the lysozyme molecule. There is no sequence homology between hen egg white *c* type lysozyme (129 amino acids) and *g* type lysozyme (185 amino acids). The *g* type lysozyme is three times more active than the *c* type (Canfield & McMurry, 1967). The amino acid sequence of hen egg white, human, cow, and oyster lysozyme is shown in Figure 26.1. Human and cow lysozyme are more homologous to each other than to hen egg white or oyster.

26.3.3 Structure

Hen egg white lysozyme contains four disulfide bonds at positions 6-12, 30-115, 64-80, and 76-94 and has no free sulfhydryl (SH-) group (Figure 26.2). The enzyme molecule has two lobes, i.e. α- and β-lobes linked by a long α-helix Asp87-Arg114 between which lies the active site. The substrate binding site of lysozyme contains six subsites (A-F) that can bind six saccharide residues (Blake et al., 1967; Imoto, 1996). The active center of the enzyme is located in the region of the two potential catalytic sites Glu-35 and Asp-52 between subsites E and F (Kawaguchi et al., 2015). As a result, the glycosidic bond disintegrates.

Lysozyme is stable at acidic pH even when heated at 100 °C. The enzyme loses some of it hydrolytic activity when heated at a high temperature in alkaline solutions. The stability of lysozyme at alkaline pH under a high temperature can be preserved by the presence of polyols and sugars (Leśnierowski & Yang, 2021). Lysozyme can upon heating at 80 °C for various times and form oligomers with increased antimicrobial activities against Gram-negative bacteria *E. coli K12* in a dose-dependent manner, while having anti-bactericidal activity against Gram-positive *Staphylococcus aureus* that was almost similar to that of the native lysozyme (Ibrahim et al., 1996). Lysozyme can also form dimers during prolonged storage as a result of changes in the disulfide bonds of the amino acids in the enzyme. However, in a natural environment, the dimer is unstable and tends to revert to the stable monomeric state (Leśnierowski & Yang, 2021). Lysozyme trimer has also been reported (Cegielska-Radziejewska & Szablewski, 2013). Both dimeric and trimeric lysozymes can be obtained by heating and cooling the monomeric lysozyme and have higher hydrophobicity than the monomeric lysozyme (Cegielska-Radziejewska & Szablewski, 2013). Lysozyme is stable in a carbonated environment and acidic pH (Pilevar et al., 2022). Recently, it was shown that lysozyme activity is higher in eggs from older hens than younger ones (Banaszewska et al., 2019). Lysozyme can also form oligomers with anti-parallel β-sheet conformation at room temperature and form amyloid fibril with parallel β-sheet conformation at acidic and elevated temperature that are endowed with enhanced antibacterial activities against Gram-positive and Gram-negative bacteria (Zou et al., 2013).

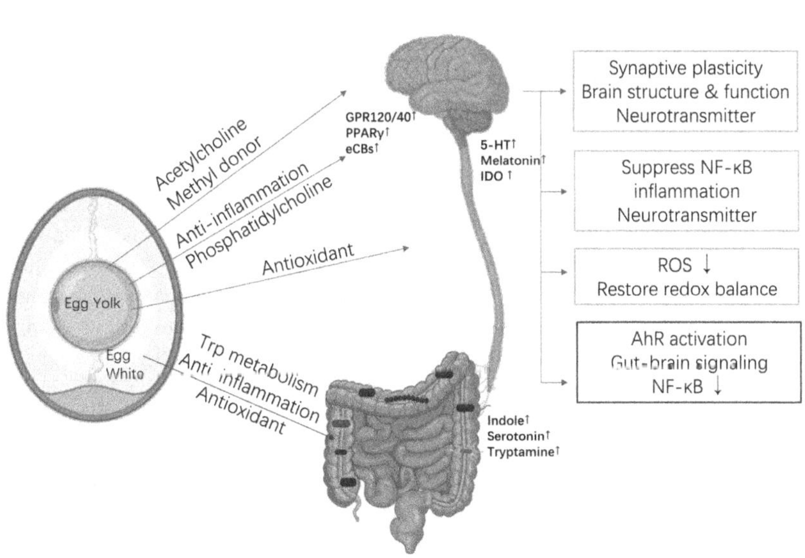

FIGURE 26.1 The egg-gut-brain axis in health and disease.

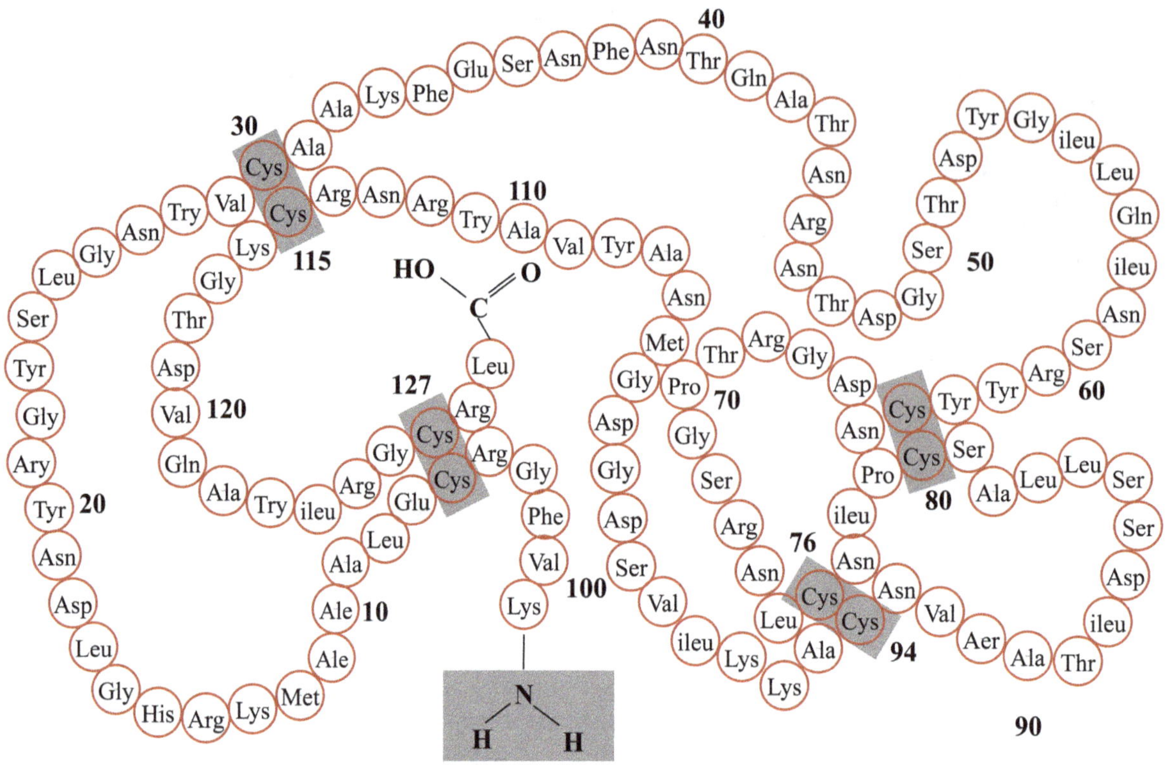

FIGURE 26.2 Amino acid sequence of hen egg white, human, cow, and oyster lysozyme.

26.4 Antimicrobial Activity

26.4.1 Introduction

The anti-microbial/catalytic activity of lysozyme depends on several factors, including the enzyme solubility in aqueous environment, pH, ionic strength, substrate concentration, temperature, water activity, presence of lysozyme inhibitors, and the concentrations of sodium and potassium in the reaction mixture (Davies et al., 1969). Since the effects of these factors are interdependent, it is difficult to evaluate the effect of individual parameters alone.

Lysozyme inhibits pathogenic and nonpathogenic bacteria. Lysozyme antimicrobial activity is mostly directed against pathogenic and spore-forming Gram-positive bacteria including *L. monocytogenes* and *Clostridium botulinum*. The enzyme can also inhibit the vegetative forms of the bacteria and the outgrowth of their spores (Bester & Lombard, 1990). Lysozyme at high concentration can also inhibit coliforms (Bester & Lombard, 1990). Nonspore-forming *Leuconostoc* and *Micrococcus* are very sensitive to the action of lysozyme. While most Gram-positive bacteria are sensitive to lysozyme, some nonspore-forming bacteria, including *Staphylococcus aureus*, are very resistant to the action of the enzyme. *Micrococcus luteus* is resistant at acidic pH.

26.4.2 Mechanism of Action

The uniqueness of lysozyme as an antimicrobial protein lies in its ability to use either its lytic or non-lytic properties to inhibit bacterial, fungal, or viral growth or its ability to bind to nucleic acids in microorganisms and cause bacteria genetic material to mutate or disintegrate. Lysozyme can act as a lytic muramidase to hydrolyze the β-1,4 glycosidic bonds between N-acetylmuramic acid (NAM) and N-acetylglucosamine (NAG) of cell wall peptidoglycan (PG) in Gram-positive. As a result, the cytoplasmic content leaks and cell death ensues. Lysozyme exhibits antimicrobial activity against multiple foodborne Gram-positive bacterial organisms, including *Micrococcus, Clostridium, Listeria, Streptococcus, and Staphylococcus spp.,* amongst others. Lysozyme has some or no effect on Gram-negative bacteria, fungi, and viruses.

26.4.2.1 *Lytic Mechanism for the Antibacterial Activity of Lysozyme*

During the lytic action, Glu-35 of lysozyme donates a proton to the glycosidic ether linkage between NAG and NAM, protonates the glycosidic oxygen, creates an oxonium ion which is followed by nucleophilic displacement of the hydroxy NAG and concurrent formation of an ester linkage of NAM to Asp-52 (Anastas et al., 2021). The ester is then hydrolyzed to provide a terminal hydroxy NAM, completing the scission of the glycosidic bond, leaving the enzyme unchanged (Vocadlo et al., 2001)

Lysozyme is more effective against Gram-positive bacteria than Gram-negative bacteria, fungi, mold, or viruses. In Gram-positive bacteria, the cell wall is composed of several layers of peptidoglycan, a polymer of *N*-acetyl-D-glucosamine units β(1→4)-linked to *N*-acetylmuramic acid, responsible for the strength of the wall, which are associated by a small group of amino acids and amino acid derivatives, forming the glycan-tetrapeptide which are repeated many times through the wall (Kollár et al., 1997). The peptidoglycan is connected to negatively charged teichoic acids and polysaccharides toward the outside of the bacterial cell wall. Teichoic acid is negatively charged which may bind the lysozyme and prevent its motion and diminish its hydrolytic action on the peptidoglycan (Meyer et al., 2003). At least two of the disulfide bonds in the lysozyme molecule must remain unbroken in order for the enzyme to maintain its catalytic activity. The spectrum of *in vitro* lysozyme inhibitory activity against Gram-positive bacteria is shown in Table 26.2

In Gram-negative bacteria, the main component on the outer surface is a protective lipopolysaccharide (LPS) layer along with proteins and phospholipids, where the access of lysozyme to the peptidoglycan layer of the cell wall is retarded by the outer LPS layer (Ibrahim et al., 1991). The spectrum of in vitro lysozyme inhibitory activity against Gram-negative bacteria is shown in Table 26.3.

Lysozyme can be combined with potentiating chemicals such as lactic acid or EDTA which allow disruption of the cell wall components of the outer membrane of Gram-negative bacteria and provide lysozyme better access to the peptidoglycan for enhanced lytic activity (Hughey & Johnson, 1987).

TABLE 26.2

Antimicrobial Spectrum of Lysozyme against Gram-Positive Bacteria

Family	Species	Sensitivity
MICROCOCCACEAE	*Micrococcus lysodeikticus*	+
	Staphylococcus aureus	-/+
	Sacina lutea	+
LACTOBACILLACEAE	*Lactobacillus bulgaricus*	nds/-
	Lactobacillus helveticus	+
	Lactobacillus acidophilus	-
	Lactobacillus lactis	-/+
	Lactobacillus casei	-
	Lactobacillus plantarum	-
	Lactobacillus fermentym	-
STREPOCOCCACEAE	*Streptococcus thermophilus*	+
	Streptococcus lactis	-/+
	Streptococcus cremoris	-/+
	Streptococcus diacetilactis	-
	Streptococcus faecalis	-
	Streptococcu faecium	+
	Streptococcus faecalis var. liquefaciens	-/nds
	Streptococcus faecalis var. zymogenes	-
	Streptococcus bovis	-
LEUCONOSTOC	*Leuconostoc citrovorum*	-
PEDIOCOCCUS	*Pediococcus cerevisiae*	ndr
BIFIDOBACTERIUM	*Bifidobacterium spp.*	+
BACILLACEAE	*Bacillus spp.*	+
	Bacillus subtilis	+/nds
	Bacilus cereus	-/+
	Bacilus megaterium	+
	Bacillus stearothermophilus var. calidolactis	+
CLOSTRIDIUM	*Clostridium welchii*	+
	Clostridium tyrobytyricum	+
	Clostridium butyricum	nds
	Clostridium sporogenes	+
	Clostridium perfringens	-
	Clostridium botulinum A, B, E	m+
PRPPIONOBACTERIACEAE	*Propionobacterium spp.*	-/ndr
CORYNEBACTERIUM	*Corynebacterium betae*	+
	Corynebacterium flacumfaciens	++
	Corynebacterium poinsettae	-
	Corynebacterium tritici	

Source: Modified from Cisani et al. (1984) and Johnson (1994).

Note: + = sensitive, - = resistance, m+ = moderately sensitive, nds = not demonstrated sensitivity, ndr = not demonstrated resistance.

TABLE 26.3
Antimicrobial Spectrum of Lysozyme against Gram-Negative Bacteria

Family	Species	Sensitivity
ENTEROBACTERIACEAE	*Escherichia coli*	-/ndr/+
	Salmonella	-/+
	typhymurium	+
	Shigella	-
	Proteus vulgaris	-
	Serratia marcescens	+
	Erwinia spp.	m+
	Yersinia enterocolitica	
PSEUDOMONADIACEAE	*P. aeruginosa*	+
	P. fluorescence	ndr
	Achromobacter spp.	-
PASTEURELLA	*Pasteurella spp.*	+
NEISSERIACEAE	*Neisseria spp.*	+
VIBRIONACEAE	*Vibrio cholerea*	-
CAMPYLOBACTERIACEAE	*Campylobacter jejuni*	m+
KLEBSIELLEAE	*Klebsiella pneumoniae*	-

Source: Modified from Cinani et al. (1984) and Johnson (1994).
Note: + = sensitive, - = resistance, m+ = moderately sensitive, nds = not demonstrated sensitivity, ndr = not demonstrated resistance.

26.4.2.2 Non-Lytic Mechanism for Antibacterial Activity of Lysozymes

The spectrum of lysozyme activity is broadened when lysozyme uses its structural factors including its cationic and hydrophobic properties to inhibit Gram-positive and Gram-negative bacteria. The structural changes involve loss of muramidase activity, changes in the secondary and tertiary structure, heat denaturation, and/or breakage of disulfide bonds by dithiothreitol (DTT). The lytic activity of the enzyme is reduced as a result of the loss of disulfide bonds, but this is compensated by the increase in the enzyme ability to inhibit Gram-negative bacteria. Lysozyme treated with 2 mM DTT at pH 8, 30°C for 30 min to 240 min had enhanced surface hydrophobicity and increased binding affinity and greater bactericidal activity against *S. enteritidis* than *Escherichia coli*, *Proteus mirabilis* or *Serratia marcescens* (Touch et al., 2004). Heating and high pH do not completely abolish the neuramidase activity; rather they increase protein hydrophobicity and the ability to bind the LPS of the Gram-negative bacteria and lysozyme fungistatic activities (Düring et al., 1999).

Lysozyme is endowed with the ability to bind to lipopolysaccharides (LPS). Lysozyme binds electrostatically to charged groups of the endotoxin molecules with stoichiometry of endotoxin and lysozyme of 3: 1 molar ratio for both LPS Re and lipid A, suggesting that lysozyme predominately binds to lipid A-phosphates (Brandenburg et al., 1998).

Lysozyme peptides have FDA GRAS status and are approved for use as an ingredient in supplements; the European Union (UE) approved the use of lysozyme hydrolysates (with a new food regulation UE 2018/991) in food supplements for the adult population (Vilcacundo et al., 2018). Lysozyme peptides have shown enhanced inhibitory activity against Gram-negative bacteria. Mine et al. developed lysozyme peptides with broadened antimicrobial activity against Gram-negative bacteria (Mine et al., 2004). It was shown that a subsequent combination of pepsin and trypsin hydrolysis of lysozyme generates peptides of less than 1,000 Da highly effective against Gram-negative bacteria. A lysozyme peptide of the following amino acid sequence NTDGSTDYGILQINSR and MW: 1753.98 ± 0.5 Da, obtained by enzymatic hydrolysis with a combination of papain and trypsin, was inhibitory against both Gram-negative and Gram-positive bacteria with a minimum inhibition concentration (MIC) values of 355.64 (±2.2) μg/ml against *Escherichia coli* and 442.25 (±2.8) μg/ml against *Leuconostoc mesenteroides*, respectively (Memarpoor-Yazdi et al., 2012). In these studies, and many others in the literature, the mechanism(s) of bacterial inhibition have not been reported.

26.4.2.3 Binding of Lysozyme to Viruses

Lysozyme binds to nucleic acids DNA and RNA and has activities against HIV-1 virus (Lee-Huang et al., 1999; Steinrauf et al., 1999). Lysozyme-DNA association is predominantly electrostatic, salt-sensitive, and entropy-driven (Potty et al., 2011). Lysozyme uses its cationic nature to bind to viruses (Naidu, 2000). Lysozyme has antiviral activity (Pitotti et al., 1991). The antiviral activity can be potentiated by combination with other bioactive compounds such as lactoferrin and nisin (Małaczewska et al., 2019).

26.4.2.4 Lysozyme and Fungi

The yeast cell wall is mainly composed of mannoprotein and fibrous β (1→3) glucans with some branches of β(1→6) glucans, the latter are responsible for the mechanical strength, rigidity, shape, and elasticity of the cell wall. The inner wall is made of β (1→3) glucan–chitin complex which forms the fibrous scaffold of the wall. The outer surface of the yeast cell is made of densely packed mannoproteins, which limit yeast cell wall permeability to solutes. As a result of thick cell wall membrane, lysozyme muramidase activity is not effective against the yeast cell wall. However, lysozymes in combination with enzymes, including glucanase, protease, mannanase, and chitinase that can hydrolyze the yeast cell wall can synergistically degrade yeast cell walls (Salazar & Asenjo, 2007).

26.4.2.5 Lysozyme and Coliforms

The high-water content of certain cheese such as the burrata, mozzarella, and Stracciatella cheeses makes these products very susceptible to coliform spoilage. Lysozyme inhibitory activity against coliforms in mozzarella cheese was enhanced during 7 days of storage when lysozyme (0.25 mg/ml) was combined with Na_2-EDTA (10, 20, or 50 mmol/L) to the combined brine (Sinigaglia et al., 2008). Lysozyme (500mg/Kg)/ Na_2-EDTA (50 mM) and modified-atmosphere packaging prolonged the shelf life of burrata cheese against coliform growth (Conte et al., 2011).

26.4.3 Factors that Affect Lysozyme Antimicrobial Activity

Several factors affect lysozyme stability and lytic activity. Water, ionic strength, pH, water activity, temperature, protein size, hydrophobicity, carbohydrates, carboxylic acids, sulfonic acids, surface active agents, gums, metal ions, potentiating chemicals, chelators (phosphates, EDTA), and polymerization affect the lytic activity of lysozyme. Lysozyme is inactive in water but increases activity as the ionic strength of the solution increases. Low ionic strength around 0.1 and potassium or sodium salts enhance lysozyme activity (Salton, 1958). High ionic strength (0.2 M salt) inhibits the enzyme lytic activity. pH has a significant effect on lysozyme solubility and activity. Lysozyme exhibits enhanced solubility at reduced pH. Lysozyme activity is maximal at pH 5 and low at acidic (< pH 3.8) and alkaline pH (the protein is basic with a pI around 10.5).

Lysozyme can be stored dry at a refrigeration temperature of 4 °C for many years or at 30 °C for several months without losing its lytic activity. A temperature between 25 °C and 60°C has no effect on the enzyme as its denaturation temperature is 75°C. A temperature as high as 100°C for 2 min and low pH enhances the enzyme activity because the presence of four disulfide bonds confers stability to the protein molecule. High temperature and alkaline pH inactivate the enzyme. Heat denaturation, including microwave heating, increases lysozyme hydrophobicity and antimicrobial activity against Gram-negative bacteria (Yang & Leśnierowski, 2019). Lysozyme has four disulfide bonds which confer heat stability to the enzyme. Polyols and sugars can protect lysozyme against heat destabilization and increase its heat stability (Losso et al., 2000). Alcohol at 25% stabilizes lysozyme (Losso et al., 2000). Lysozyme activity is inhibited by polysaccharides with carboxylic acid and sulphonic acid groups but adding salt can offset these inhibitory effects (Cunningham et al., 1991). However, Amir et al. indicated that lysozyme-dextran conjugate was effective against *E. coli* in cheese curd as it reduced the bacterial count by 3 log after 40 days of storage (Amiri et al., 2008). Glycated lysozyme maintained its structural integrity and enzymatic after milling better than lysozyme alone (Xing & Yaylayan, 2019).

Metal ions bind to the enzyme active site at Glu 35 and -Asp 52 and inactivate the enzyme (Cunningham et al., 1991). Divalent cations such as calcium and magnesium contribute to the stability of the outer membrane of Gram-negative by providing electrostatic interactions between proteins and polysaccharides (Delves-Broughton, 1993).

Lysozyme dimerization, oligomerization, and polymerization increase and expand lysozyme spectrum of stronger inhibitory activity against both Gram-positive and Gram-negative bacteria than monomeric lysozyme (Cegielska-Radziejewska et al., 2010).

26.4.4 Synergism or the Hurdle Effect

Lysozyme is effective against the limited number of Gram-positive bacteria and is mostly ineffective against Gram-negative bacteria. The hurdle effect allows a combination of preservative or facilitating agents which in concert control microbial spoilage and make it unlikely for any microorganism to survive. The hurdle may be a combination of temperature, water activity, pH, redox potential, enzyme antimicrobials, salt, sugar, pasteurization, bacteriocin, chelators, ultra-high pressure, hydrostatic pressure, complexation, thermochemical processes, glycation, edible coatings, gas packaging, and bioconservation. The preservative factors may disturb one or several of the homeostatic mechanisms of microorganisms such as pH habituation of resistant cells and, as a result, the microorganism will not multiply, but instead will remain inactive or even die. Hurdle technology as applied to lysozyme has enhanced and expanded the spectrum of lysozyme antimicrobial activity to include Gram-negative, fungi, viruses, and coliforms. The combination of lysozyme with natural preservatives that accommodate clean labels such as nisin, lactoferrin, glycine, organic acids, carbohydrates, detergents, egg white proteins, trypsin, aprotinin, gelatin, ultra-high pressure, electroporation, or lysozyme hydrolysis has in most cases performed better against a wide spectrum of microorganisms than lysozyme used singly. Typical cases of hurdle effects will be provided in different applications in real food systems as described below in the sections on dairy products, meat products, fruits and vegetables, seafood products, alcoholic beverages and juices, and packaging.

26.5 Biological Advantage

26.5.1 Additive Advantages to Foods

Lysozyme is a protein endogenous to foods of animal or plant origins. It is safe and stable in conditions encountered in most foods because most foods are at a pH below 6. Lysozyme can be used as liquid or powder. The lyophilized form is granulated to reduce the formation of aerosols in food processing. Lysozyme does not induce adverse chemical or organoleptic changes that would be of concern under normal processing conditions. In most cases, lysozyme does not interfere with the viability of microorganisms considered beneficial for human health.

26.5.2 Physiological Advantages to Host

Lysozyme added to foods is destroyed in the human stomach. Lysozyme binds to lipopolysaccharides with a high binding affinity to produce lysozyme-LPS complex which leads to LPS detoxification and lysozyme inactivation. The binding of lysozyme to LPS reduces the ability of LPS to stimulate the production of inflammatory cytokines, including the TNF-α *in vitro* and *in vivo* (Takada et al., 1994). IL-6 production in serum by LPS in carrageenan (CAR)-primed mice was reduced when a low dose (1 μg of LPS per mouse) was used but remained high at a lethal dose (100 μg of LPS per mouse) was used. Lysozyme may also display anti-endotoxic effect in periodontal therapy (Ito et al., 1997). Lipopolysaccharide activates the immune system through TLR-4 pathway. High mobility group box 1 (HMGB1) is an important extracellular mediator of sepsis. Lysozyme posttreatment suppressed LPS-mediated release of HMGB1, inhibited HMGB1-mediated hyperpermeability and leukocyte migration, inhibited the HMGB1-mediated activation of Akt, nuclear factor-κB (NF-κB), extracellular regulated kinases (ERK) 1/2 and production of interleukin (IL)-1β, IL-6, tumor necrosis factor-α (TNF-α), and chemoattractant protein-1 (MCP-1) in HUVECs (Lee et al., 2015). Lysozyme is an immunostimulant, antiviral and anti-inflammatory in individuals with herpes lesions and individuals with cancer (Anton et al., 2006).

26.6 Suitability/Safety/Tolerance

Lysozyme has been recognized as safe (GRAS) by the U.S. Food and Drug Administration (FDA) for use in cheese. However, because lysozyme is derived from eggs and most egg proteins are allergenic, the FDA has concluded that foods containing egg white lysozyme are required to mention that the food contains lysozyme (Khorshidian et al., 2022).

The Joint FAO/WHO Expert Committee on Food Additives (JECFA) considers egg white lysozyme as a class I enzyme and food. For JECFA, intake of lysozyme in cheese made from treated milk is not a risk to consumer health because the amount is very low although and does induce an increase of IgE RAST without clinical reaction (Joint FAO/WHO Expert Committee on Food Additives, World Health Organization, & Food and Agriculture Organization of the United Nations, 1993; Iaconelli et al., 2008). Lysozyme has been approved and is being used in many countries to control the growth of spoilage organisms in foods (Wang & Shelef, 1992). Lysozyme is approved by the European Additives Directive to be used as preservative (E1105) in ripened cheese to prevent late blowing caused by *Clostridium tyrobutyricum, Clostridium* sporogenes, *Clostridium butyricum, Clostridium beijerinckii, Clostridium cochlearium, Clostridium prrfingens,* and/or *Clostrodium septicum* (Bassi et al., 2015).

26.7 Applications

The literature is rich in *in vitro* studies on the efficacy of lysozyme alone, lysozyme peptides, or lysozyme or in combination with other potentiating agents to inhibit the activity of undesirable microorganisms in foods. The sections below will mostly address studies performed on real food systems and studies involved with natural preservatives that are not being banned or removed from usage because of health concerns.

26.7.1 Dairy Products

The occurrence of lysozyme in human milk and the similarity to hen egg white lysozyme has been a push for the use of hen egg white lysozyme as a natural preservative in dairy products. Lysozyme at 300–600 μg/mL can be added to liquid milk, especially pasteurized milk or UHT milk because lysozyme is heat resistant at the pH of milk, usually adding before packaging and effectively prolong the shelf life of milk products. Lysozyme addition does not affect the organoleptic attributes of milk. Lysozyme and nisin can synergistically prolong the shelf life of milk without affecting the taste and sensory qualities of milk. Lysozyme addition to infant formula promotes milk digestion (Cunningham et al., 1991).

Cheese is one the major dairy products, and applications of lysozyme as a preservative in cheese have been investigated extensively. Lysozyme is preferred as natural preservative in dairy products for the inhibition of spoilage bacteria growth, prevention of cheese texture deterioration, late cheese blowing caused by *Clostridium tyrobutyricum*, and unpleasant cheese flavor and taste.

Lysozyme has been used as an antimicrobial agent in Italian cheeses Edam or Gouda. In milk, butyric acid bacteria can convert lactic acid into butyric acid, carbon dioxide, and hydrogen which causes cheese to blow and become unpalatable (Losso et al., 2000). Lysozyme is added to milk and binds electrostatically to casein and 90% of the lysozyme activity resides with the curd. Lysozyme is preferred over other inhibitors, including formaldehyde, nitrate, or nisin in milk because these compounds may be inhibitory to the starter cultures or secondary cultures during cheese ripening. Lysozyme can inhibit or stimulate the growth of coliforms (Bester & Lombard, 1990). During the aging process, the lysozyme will prevent the growth of *C. tyrobutyricum*. At 21 mg/L lysozyme in milk can kill 99% of 5×10^5 resting vegetative cells of *C. tyrobutyricum* within 24 h at 25 °C.

Listeria monocytogenes is of greatest concern in milk and soft cheese. Milk and dairy products have contributed the most to the outbreak of listeriosis. Soft-ripened cheeses such as Camembert are good sources of listeria transfer to humans. The use of lysozyme to prevent the growth of listeria in camembert span over four decades (Goudedranche et al., 1986). These investigators showed that direct addition of lysozyme (0.5–1.0 g kg^{-1}) to ultra-filtration Saint-Paulin and Camembert pre-cheeses in the retentate led to pH change, change in mesophilic bacteria, and increased proteolysis during ripening. As a result, the taste, texture, and the flavor of UF-St Paulin and Camembert cheeses were improved. The potential mechanism associated with the improvement in organoleptic properties of the Camembert cheeses is that lysozyme which binds to casein micelles was dissociated from the milk protein as ripening progressed.

In Gouda cheese, lysozyme at 250 U/ml inhibited the growth of vegetative cells of *C. tyrobutyricum* BZ15 and C611; lysozyme at 500 to 1000 units/ml inhibited the growth of three cultures of lactobacillus and lysozyme at a concentration up to 2,500 U/ml did not affect the activities of the starters cultures (Bester & Lombard, 1990). Spore germination of C. *tyrobutyricum* was not affected up to lysozyme concentration of 2,500 U/mL (Bester & Lombard, 1990).

Lysozyme alone does not inhibit biofilm formation by *S. aureus* but in combination with nisin the inhibition of biofilm by 13 strains in raw milk or 12 strains in cheese can be achieved (Sudagidan & Yemenicioğlu, 2012). Lysozyme can also be used as anti-spoilage edible film in a variety of cheese packaging. The composite lysozyme-chitosan, lysozyme-dextran, lysozyme-xantham gum, lysozyme-whey protein can significantly reduce the growth of *Penicillium chrysogenum*,

E. coli O157:H7, *P. fluorescens*, *L. monocytogenes*, and *S. aureus* with no adverse effect on the sensory attributes of the dairy products (Khorshidian et al., 2022). The hurdle effect enhances lysozyme bioactivity.

Lysozyme-Na_2-EDTA or lysozyme- Na_2-EDTA- modified atmosphere packaging (MAP) combination can be effective against coliforms, *Pseudomonas* spp, or total bacterial counts (Khorshidian et al., 2022). EDTA (21 CFR 172.135) is permitted in US by the Food and Drug Association (FDA) and the World Health Organization (WHO) for use at 2.5 mg/kg/person/day or roughly 150 mg/day for the calcium disodium salt. Current European Union legislation (European Parliament and Council Directive No. 95/2/EC) and the Codex Alimentarius (Codex Standard 283-1978) allow the use of egg white lysozyme in the range from 50 mg to 350 mg per kilo of cheese as a preservative (Bärtschi et al., 2006). However, since lysozyme derives from egg, and egg proteins are associated with 90% of food allergies, lysozyme in a food product must be declared on the label as required by FDA and EU. However, the use of EDTA in food products is being disapproved by regulatory agencies around the world.

26.7.2 Meats and Meat Products

Raw meat, because of this composition and attributes, is an excellent environment for the growth of numerous bacteria, including food poisoning causing pathogenic bacteria. The decontamination of meat and poultry carcasses can help to reduce bacterial load on meat and subsequent contamination (Bolder, 1997). Lysozyme alone or in combination with natural preservatives, including nisin, lactate, lactoferrin, ovotransferrin, hydrostatic pressure, and modified atmosphere packaging, can be better alternatives to chemical preservatives, including nitrates, EDTA, and others and extend the application of lysozyme to the inhibition of resistant Gram-positive and Gram-negative bacteria. Lysozyme at 0.5% (5000 ppm) in a post-chill immersion tank was effective against *Salmonella* and *Campylobacter* without affecting the sensory attributes of the chicken breast meat prepared thereafter (Nagel et al., 2013). Lysozyme at 48,000 U/mL caused a 20-fold reduction in the initial bacterial count of skin-on legs (thighs and drumsticks) chicken meat after 5 days at 4 °C (Kijowski et al., 2013). The combination of lysozyme (1.5% w/w), vacuum packaging and rosemary oil or lysozyme, and oregano oil and vacuum packaging was very effective against the growth of Gram-negative, Gram-positive bacteria, and to a lesser extent on yeasts in semi-cooked coated chicken fillets (Ntzimani et al., 2010). Lysozyme at 250 µg of antimicrobial/ml was sufficient to inhibit *B. thermosphacta* B2 for 27 days at 2°C in pork juice and in combination with nisin was also effective (Nattress et al., 2001).

The synergistic effect of lysozyme at 0.5% and EDTA at 2% can inhibit the growth of a wide range of pathogenic bacteria, including *Brochotrix thermosphacta* in buffalo meat during 8 days of storage (Cannarsi et al., 2008). Ham and bologna sausages were prepared with or without the addition of 500 mg kg^{-1} lysozyme:nisin, 1:3, and 500 mg. kg^{-1} EDTA. Sausages were inoculated with *Brochothrix thermosphacta*, *E.coli* O157:H7, *Lactobacillus sakei*, *Serratia grimesii Lactobacillus curvatus*, *Leuconostoc mesenteroides*, *Listeria monocytogenes*, *Salmonella typhimurium*, or *Shewanella putrefaciens*, vacuum packed, and stored for 4 weeks at 8°C (Gill & Holley, 2000). The synergistic effect of the inhibitors prevented the growth of *Escherichia coli* O157:H7 and *Brochothrix thermosphacta* for 4 weeks, reduced growth of *Lb. curvatus* on ham and bologna up to Week 3, reduced the growth of *Lc. mesenteroides* and *L. monocytogenes* on bologna for 2 weeks, reduced the growth of *E. coli* O157:H7 on ham for 4 weeks, but increased the growth of S. *typhimurium* from week 3. Mangalassary et al. showed that in-package pasteurization (65°C for 32 s) + lysozyme and nisin (10 mg lysozyme + 2 mg/ml) reduced *L. monocytogenes* population below detectable levels during 2-3 week refrigerated storage (Mangalassary et al., 2008). Lysozyme (400 µg/g) in combination with *Ziziphora clinopodioides* essential oil (0.3%) showed good antimicrobial against *Listeria monocytogenes* and antioxidant activities without affecting the organoleptic properties ($p > 0.05$) of Balkan type fresh sausage under modified atmosphere packaging (MAP) during 13-day refrigerated storage condition (Ajourloo et al., 2021). The essential oil alone at 0.3% was better than lysozyme alone at 400 µg/g.

Lysozyme in combination with chitosan increased the shelf life of minced meat by inhibiting the growth of *B. cereus*, *E. coli* and *Pseudomonas fluorescens* and reduced the growth of *S. aureus* (Rao et al., 2008). The synergistic antimicrobial activity of lysozyme (250 ppm), nisin (250 ppm), and disodium ethylenediamine tetraacetic acid (EDTA) (20 mM) against *Listeria monocytogenes* and meat-borne spoilage bacteria in ostrich patties packaged in air and vacuum decreased the *L. monocytogenes* population below the official limit of the European Union (<2 log CFU/g) and did not affect *Enterobacteriaceae* and *Pseudomonas* spp (Mastromatteo et al., 2010). However, the total viable counts for the air or vacuum-treated samples showed a reduction of 1 log cycle only for 2 days of storage.

26.7.3 Fruits and Vegetables

Lysozyme has been experimented in the storage of climacteric and non-climacteric fruits.

Lysozyme coating alone or in combination with 1 µL/L of 1-Methylcyclopropene (1-MCP) (lysozyme at 0.08%, 2 min and 1-MCP at 0.9 µl/L, 22 h) can reduce ethylene production and respiratory rate and increase ascorbate peroxidase, superoxide dismutase, and catalase activity in treated kiwi fruits (Xu et al., 2019). Because of allergenicity, commercial operations may not be interested in applying lysozyme on fruits and vegetables.

26.7.4 Seafoods

Spoilage of fish after harvest begins right after fish dead because fish is high in moisture, nutrient content, has weak connective tissue and a pH close to pH 7, and is very susceptible to autocatalytic activity and microbial growth. As a result, fresh fish and fish products require efficient methods of preservation; and natural preservatives are better candidates for seafood products' preservation.

Lysozyme at 3 mg/mL in H_2O suppressed the growth of *Listeria monocytogenes* on raw cod fish fillets stored at 20°C for 3 days or at 5°C for 17 days; lysozyme (3 mg/mL) and Na_2EDTA (25 mM) suppressed the growth of *Listeria monocytogenes* and the natural microflora on the fillets and also prevented slime formation (Wang & Shelef, 1992). Lysozyme (0.5%) and collagen (4%) as the coating agent were effective in preserving fresh salmon-fillets against total viable counts over 15 days at refrigerated temperature (4 ± 1°C) (Wang et al., 2017). Lysozyme at 0.5 mg/mL was effective in preventing the growth of *Shewanella* and *Acinetobacter* in spoiled carp fillets while the combination of lysozyme (0.25 mg/mL) and phytic (0.25 mg/ml) was very effective in reducing the proportion of *Shewanella*, *Acinetobacter*, and *Pseudomonas* in spoiled grass carp fillets (Sun et al., 2020). Lysozyme-pomegranate peel extract combination addition on gelatin–polycaprolactone composite film was effective in increasing the shelf life by 9 days and preserving the quality of mackerel *Scomberomorus commerson* filets (Khodanazary, 2019). Chitosan film containing 0.6 mg/mL lysozyme inhibited the growth of *S. aureus* and maintained the sensory quality of refrigerated large yellow croaker (Wu et al., 2018). Lysozyme-galactomannan or lysozyme-palmitic acid was effective against Gram-negative *Edwardsiella tarda* infection in carp, *Cyprinus carpio* L. fish (Nakamura et al., 1996). The conjugate improved fish survival compared to control lysozyme.

26.7.5 Alcoholic Beverages and Juices

Lysozyme has been used in wine to control malolactic fermentation during wine making and aging. Lysozyme has been tried as a potential substitute for sulfite in alcoholic fermentation because it can control the growth of Gram-positive and spoilage bacteria with no effect on yeast growth and sulfites pose health issues to asthmatics. Additionally, lysozyme in wine and other alcoholic beverages can eliminate the growth of acidogenic bacteria. Lysozyme at 250 mg/L added to red wines after malolactic fermentation promoted microbiological (*Oenococcus oeni, Lactobacillus spp., Pediococcus damnosus,* and *Pediococcus parvulus*) stabilization and showed no increase in acetic acid or biogenic amines for six-month storage at 18°C (Gerbaux et al., 1997). Other work indicated that lysozyme in combination with enological tannins in wine must results in a beverage with distinct sensory impact when compared to conventional musts fermented with sulfites (Sonni et al., 2009). Recent studies have shown that lysozyme added to Italian Riesling ice wine improved protein stability and stabilized the wine shelf life for 6 months with no effect on wine flavor (Chen et al., 2017). Lysozyme at 125 mg/L or 250 mg/L did not affect yeast growth and sugar reduction in lab scale Chardonnay wine inoculated with *Lactobacillus kunkeei, Lactobacillus brevis, Pediococcus parvulus*, and *Pediococcus damnosus* at pH 3.8 and significantly reduced acetic acid production (Gao et al., 2002). It was found that lysozyme may be better for high-pH wines in which sulfites are less effective.

In beer, lysozyme at 100 ppm extended the stability and shelf life of unpasteurized beer by exerting strong inhibitory activity against spoilage lactic acid bacteria (Silvetti et al., 2010). Sensory tests revealed that lysozyme had no unfavorable influence on beer flavor; and even months after the expiry date panelists still approved the lysozyme-treated beer.

In fruit juices, lysozyme alone showed weak reduction of *S. typhimurium* in non-pulpy orange juice (0.05 log CFU/mL) and in pulpy orange juice (0.08 log CFU/mL), whereas the combination of lysozyme and nisin or pulsed electric field showed greater bactericidal effect (Liang et al., 2002). Similarly, egg lysozyme was a weak inhibitor of *S. typhimurium, Shigella flexneri, E. coliO157:H7, Yersinia enterocolitica*, and spoilage microorganisms but was a stronger inhibitor when combined with high hydrostatic pressure, nisin, or heat in red grape and banana juice or apple cider (Liang et al., 2006; Raybaudi-Massilia et al., 2009).

Because of the allergenicity issue associated with lysozyme, wines, beers, or juices with lysozyme must be labeled as such and people who are looking for an egg-free diet may not be interested in beverages with allergens. Italian ciders treated with lysozyme without label declaration are on the market and may constitute a danger for asthmatic wine consumers (Mainente et al., 2017). To overcome lysozyme sensitization, immobilized lysozyme has been proposed to control malolactic fermentation during wine making; however, addition of lysozyme to alcoholic beverages or any other beverage must be declared on the label (Liburdi et al., 2014).

26.7.6 Active Packaging

Alternative packaging from natural sources including biopolymers is gaining traction over synthetic petroleum-based packaging materials because naturally occurring biopolymers including polysaccharides and proteins are biodegradable, available, low cost, and have excellent film-forming properties. Edible coatings containing lysozyme + gelatin, lysozyme + chitosan, lysozyme + whey protein, lysozyme + collagen, lysozyme + lactoperoxidase, and lysozyme + zein have been developed and applied on seafoods, eggs, vegetables, meats, fruits, and vegetables, and dairy products (Zhang & Rhim, 2022). Lysozyme has good stability and solubility in water and above all serves as an antimicrobial agent.

26.8 Trends and Future Perspectives

Egg white lysozyme has demonstrated efficacy against Gram-positive bacteria. Through physicochemical modification or hurdle technology, the efficacy of lysozyme can be enhanced and expanded to include both Gram-positive and Gram-negative bacteria, coliforms, fungi, and viruses. The protein has demonstrated allergenicity and, as a result, it has to be mentioned on the label when present in the food or in any ingredient. Despite the FDA's GRAS status granted to lysozyme, it remains one of the least investigated antimicrobial egg proteins in US Food Science programs. Consumers want a natural preservative and food with a "clean label" that does not promote disease. Developing cost-effective purification methods and reducing egg lysozyme allergenicity while maintaining or improving functionality will be fundamental to increased research on the application and the acceptability of the protein in many foods.

REFERENCES

Ajourloo, M., Khanjari, A., Misaghi, A., Akhondzadeh Basti, A., Kamkar, A., Yadegar, F., & Fallah, F. (2021). Combined effects of *Ziziphora clinopodioides* essential oil and lysozyme to extend shelf life and control *Listeria* monocytogenes in Balkan-style fresh sausage. *Food Science & Nutrition*, *9*(3), 1665–1675.

Alderton, G., & Fevold, H. (1946). Direct crystallization of lysozyme from egg white and some crystalline salts of lysozyme. *Journal of Biological Chemistry*, *164*(1), 1–5.

Alderton, G., Wabd, W., & Fevold, H. (1945). Isolation of lysozyme from egg white. *Journal of Biological Chemistry*, *157*, 43–58.

Altıntaş, E. B., & Denizli, A. (2006). Monosize poly (glycidyl methacrylate) beads for dye-affinity purification of lysozyme. *International Journal of Biological Macromolecules*, *38*(2), 99–106.

Amiri, S., Ramezani, R., & Aminlari, M. (2008). Antibacterial activity of dextran-conjugated lysozyme against *Escherichia coli* and *Staphylococcus aureus* in cheese curd. *Journal of Food Protection*, *71*(2), 411–415.

Anastas, P. T., Rodriguez, A., de Winter, T. M., Coish, P., & Zimmerman, J. B. (2021). A review of immobilization techniques to improve the stability and bioactivity of lysozyme. *Green Chemistry Letters and Reviews*, *14*(2), 302–338.

Anton, M., Nau, F., & Nys, Y. (2006). Bioactive egg components and their potential uses. *World's Poultry Science Journal*, *62*(3), 429–438.

Banaszewska, D., Biesiada-Drzazga, B., Ostrowski, D., Drabik, K., & Batkowska, J. (2019). The impact of breeder age on egg quality and lysozyme activity. *Turkish Journal of Veterinary & Animal Sciences*, *43*(5), 583–589.

Bärtschi, F., Muralt, L., Rieder, K., Kämpfer, U., & Schaller, J. (2006). Bestimmung von lysozym in Käse mittels LC-MS. *Mitteilungen aus Lebensmitteluntersuchung und Hygiene*, *97*, 478–488.

Bassi, D., Puglisi, E., & Cocconcelli, P. S. (2015). Understanding the bacterial communities of hard cheese with blowing defect. *Food Microbiology*, *52*, 106–118.

Beaussart, A., Retourney, C., Quilès, F., Morais, R. D. S., Gaiani, C., Fiérobe, H.-P., & El-Kirat-Chatel, S. (2021). Supported lysozyme for improved antimicrobial surface protection. *Journal of Colloid and Interface Science*, *582*, 764–772.

Beintema, J., & Terwisscha van Scheltinga, A. (1996). Plant lysozymes. *EXS*, *75*, 75–86.

Bester, B. H., & Lombard, S. H. (1990). Influence of lysozyme on selected bacteria associated with Gouda cheese. *Journal of Food Protection*, *53*(4), 306–311.

Blake, C., Johnson, L. N., Mair, G., North, A., Phillips, D. C., & Sarma, V. (1967). Crystallographic studies of the activity of hen egg-white lysozyme. *Proceedings of the Royal Society of London. Series B. Biological Sciences*, *167*(1009), 378–388.

Bolder, N. (1997). Decontamination of meat and poultry carcasses. *Trends in Food Science & Technology*, *8*(7), 221–227.

Brandenburg, K., Koch, M. H., & Seydel, U. (1998). Biophysical characterisation of lysozyme binding to LPS Re and lipid A. *European Journal of Biochemistry*, *258*(2), 686–695.

Callewaert, L., & Michiels, C. W. (2010). Lysozymes in the animal kingdom. *Journal of Biosciences*, *35*(1), 127–160.

Canfield, R. E., & McMurry, S. (1967). Purification and characterization of a lysozyme from goose egg white. *Biochemical and Biophysical Research Communications*, *26*(1), 38–42.

Cannarsi, M., Baiano, A., Sinigaglia, M., Ferrara, L., Baculo, R., & Del Nobile, M. A. (2008). Use of nisin, lysozyme and EDTA for inhibiting microbial growth in chilled buffalo meat. *International Journal of Food Science & Technology*, *43*(4), 573–578.

Cegielska-Radziejewska, R., & Szablewski, T. (2013). Effect of modified lysozyme on the microflora and sensory attributes of ground pork. *Journal of Food Protection*, *76*(2), 338–342.

Cegielska-Radziejewska, R., Lesnierowski, G., Szablewski, T., & Kijowski, J. (2010). Physico-chemical properties and antibacterial activity of modified egg white—Lysozyme. *European Food Research and Technology*, *231*(6), 959–964.

Chang, H.-M., Yang, C.-C., & Chang, Y.-C. (2000). Rapid separation of lysozyme from chicken egg white by reductants and thermal treatment. *Journal of Agricultural and Food Chemistry*, *48*(2), 161–164.

Chen, K., Han, S. Y., Li, M., & Sheng, W. J. (2017). Use of lysozyme and oligomeric proanthocyanidin to reduce sulfur dioxide and the evolution of volatile compounds in Italian Riesling ice wine during aging process. *Journal of Food Processing and Preservation*, *41*(1), e12755.

Cisani, G., Varaldo, P. E., Ingianni, E., Pompo, R., & Satta, G. (1984). Inhibition of herpes simplex virus-induced cytopathic effect by modified hen egg-white lysozymes. *Current Microbiology*, *10*, 35–40.

Cong, L., Yang, X., Wang, X., Tada, M., Lu, M., Liu, H., & Zhu, B. (2009). Characterization of an i-type lysozyme gene from the sea cucumber *Stichopus* japonicus, and enzymatic and nonenzymatic antimicrobial activities of its recombinant protein. *Journal of Bioscience and Bioengineering*, *107*(6), 583–588.

Conte, A., Brescia, I., & Del Nobile, M. A. (2011). Lysozyme/EDTA disodium salt and modified-atmosphere packaging to prolong the shelf life of burrata cheese. *Journal of Dairy Science*, *94*(11), 5289–5297.

Cunningham, F., Proctor, V., & Goetsch, S. (1991). Egg-white lysozyme as a food preservative: An overview. *World's Poultry Science Journal*, *47*(2), 141–163.

Datta, S., Janes, M., Xue, Q. G., Losso, J., & La Peyre, J. (2008). Control of *Listeria* monocytogenes and *Salmonella anatum* on the surface of smoked salmon coated with calcium alginate coating containing oyster lysozyme and nisin. *Journal of Food Science*, *73*(2), M67–M71.

Davies, R., Neuberger, A., & Wilson, B. (1969). The dependence of lysozyme activity on pH and ionic strength. *Biochimica et Biophysica Acta (BBA)-Enzymology*, *178*(2), 294–305.

Delves-Broughton, J. (1993). The use of EDTA to enhance the efficacy of nisin towards Gram-negative bacteria. *International Biodeterioration & Biodegradation*, *32*(1–3), 87–97.

Düring, K., Porsch, P., Mahn, A., Brinkmann, O., & Gieffers, W. (1999). The non-enzymatic microbicidal activity of lysozymes. *FEBS Letters*, *449*(2–3), 93–100.

Fleming, A. (1922). On a remarkable bacteriolytic element found in tissues and secretions. *Proceedings of the Royal Society of London. Series B, Containing Papers of a Biological Character, 93*(653), 306–317.

Gao, Y. C., Zhang, G., Krentz, S., Darius, S., Power, J., & Lagarde, G. (2002). Inhibition of spoilage lactic acid bacteria by lysozyme during wine alcoholic fermentation. *Australian Journal of Grape and Wine Research, 8*(1), 76–83.

Gerbaux, V., Villa, A., Monamy, C., & Bertrand, A. (1997). Use of lysozyme to inhibit malolactic fermentation and to stabilize wine after malolactic fermentation. *American Journal of Enology and Viticulture, 48*(1), 49–54.

Gill, A. O., & Holley, R. A. (2000). Inhibition of bacterial growth on ham and bologna by lysozyme, nisin and EDTA. *Food Research International, 33*(2), 83–90.

Goudedranche, H., Ducruet, P., Vachot, J., Pannetier, R., & Maubois, J. (1986). *Use of lysozyme for managing ripening of cheeses made through MMV process.* Saint-Paulin, Camembert; Ultrafiltration.

Guérin-Dubiard, C., Pasco, M., Hietanen, A., del Bosque, A. Q., Nau, F., & Croguennec, T. (2005). Hen egg white fractionation by ion-exchange chromatography. *Journal of Chromatography A, 1090*(1–2), 58–67.

Helal, R., & Melzig, M. (2008). Determination of lysozyme activity by a fluorescence technique in comparison with the classical turbidity assay. *Die Pharmazie-An International Journal of Pharmaceutical Sciences, 63*(6), 415–419.

Hindenburg, A., Spitznagel, J., & Arnheim, N. (1974). Isozymes of lysozyme in leukocytes and egg white: Evidence for the species-specific control of egg-white lysozyme synthesis. *Proceedings of the National Academy of Sciences, 71*(5), 1653–1657.

Hughey, V., & Johnson, E. (1987). Antimicrobial activity of lysozyme against bacteria involved in food spoilage and food-borne disease. *Applied and Environmental Microbiology, 53*(9), 2165–2170.

Iaconelli, A., Fiorentini, L., Bruschi, S., Rossi, F., Mingrone, G., & Piva, G. (2008). Absence of allergic reactions to egg white lysozyme additive in Grana Padano cheese. *Journal of the American College of Nutrition, 27*(2), 326–331.

Ibrahim, H. R., Higashiguchi, S., Juneja, L. R., Kim, M., & Yamamoto, T. (1996). A structural phase of heat-denatured lysozyme with novel antimicrobial action. *Journal of Agricultural and Food Chemistry, 44*(6), 1416–1423.

Ibrahim, H. R., Kato, A., & Kobayashi, K. (1991). Antimicrobial effects of lysozyme against gram-negative bacteria due to covalent binding of palmitic acid. *Journal of Agricultural and Food Chemistry, 39*(11), 2077–2082. https://doi.org/10.1021/jf00011a039

Imoto, T. (1996). Engineering of lysozyme. *EXS, 75*, 163–181.

Ito, H. O., Hirata, M., & Koga, T. (1997). Hen egg-white lysozyme inhibits biological activities of lipopolysaccharides from periodontopathic bacteria. *Journal of Periodontal Research, 32*(3), 295–299.

Jalili-Firoozinezhad, S., Filippi, M., Mohabatpour, F., Letourneur, D., & Scherberich, A. (2020). Chicken egg white: Hatching of a new old biomaterial. *Materials Today, 40*, 193–214.

Jiang, C. M., Wang, M. C., Chang, W. H., & Chang, H. M. (2001). Isolation of lysozyme from hen egg albumen by alcohol-insoluble cross-linked pea pod solid ion-exchange chromatography. *Journal of Food Science, 66*(8), 1089–1093.

Johnson E. (1994). Egg white lysozyme as a preservative for use in foods. In J. S. Sim & S. Nakai (Eds.), *Egg uses and processing technologies: New developments* (pp. 177–191). CAB International.

Joint FAO/WHO Expert Committee on Food Additives, World Health Organization, & Food and Agriculture Organization of the United Nations. (1993). Evaluation of certain food additives and contaminants: Forty-first report of the Joint FAO/WHO Expert Committee on Food Additives [Meeting held in Geneva from 9 to 18 February 1993]. World Health Organization.

Jolles, P. (1996). *Lysozymes: Model enzymes in biochemistry and biology.* Birkhäuser Basel.

Kawaguchi, Y., Yoneda, K., Torikata, T., & Araki, T. (2015). Asp48 function in the hydrogen-bonding network involving Asp52 of hen egg-white lysozyme. *Bioscience, Biotechnology, and Biochemistry, 79*(2), 196–204.

Khodanazary, A. (2019). Quality characteristics of refrigerated mackerel *Scomberomorus commerson* using gelatin-polycaprolactone composite film incorporated with lysozyme and pomegranate peel extract. *International Journal of Food Properties, 22*(1), 2057–2071.

Khorshidian, N., Khanniri, E., Koushki, M. R., Sohrabvandi, S., & Yousefi, M. (2022). An overview of antimicrobial activity of lysozyme and its functionality in cheese. *Frontiers in Nutrition, 9*, 325. https://doi.org/10.3389/fnut.2022.833618

Kijowski, J., Marciszewska, C., Cegielska-Radziejewska, R., & Popiół, A. (2013). Effect of lysozyme treatment on quality and bacterial contamination of chilled chicken legs. *Journal of Veterinary Research, 57*(1), 79–84.

Kollár, R., Reinhold, B. B., Petráková, E., Yeh, H. J., Ashwell, G., Drgonová, J., & Cabib, E. (1997). Architecture of the yeast cell wall: β (1→ 6)-glucan interconnects mannoprotein, β (1→ 3)-glucan, and chitin. *Journal of Biological Chemistry, 272*(28), 17762–17775.

Lee, W., Ku, S.-K., Na, D. H., & Bae, J.-S. (2015). Anti-inflammatory effects of lysozyme against HMGB1 in human endothelial cells and in mice. *Inflammation, 38*(5), 1911–1924.

Lee-Huang, S., Huang, P. L., Sun, Y., Huang, P. L., Kung, H.-F., Blithe, D. L., & Chen, H.-C. (1999). Lysozyme and RNases as anti-HIV components in β-core preparations of human chorionic gonadotropin. *Proceedings of the National Academy of Sciences, 96*(6), 2678–2681.

Leśnierowski, G., & Yang, T. (2021). Lysozyme and its modified forms: A critical appraisal of selected properties and potential. *Trends in Food Science & Technology, 107*, 333–342.

Li, Z., Cao, M., Zhang, W., Liu, L., Wang, J., Ge, W., & William, W. Y. (2014). Affinity adsorption of lysozyme with reactive red 120 modified magnetic chitosan microspheres. *Food Chemistry, 145*, 749–755.

Liang, Z., Cheng, Z., & Mittal, G. (2006). Inactivation of spoilage microorganisms in apple cider using a continuous flow pulsed electric field system. *LWT-Food Science and Technology, 39*(4), 351–357.

Liang, Z., Mittal, G. S., & Griffiths, M. W. (2002). Inactivation of *Salmonella* typhimurium in orange juice containing antimicrobial agents by pulsed electric field. *Journal of Food Protection, 65*(7), 1081–1087.

Liburdi, K., Benucci, I., & Esti, M. (2014). Lysozyme in wine: An overview of current and future applications. *Comprehensive Reviews in Food Science and Food Safety, 13*(5), 1062–1073.

Li-Chan, E., Nakai, S., Sim, J., Bragg, D., & Lo, K. (1986). Lysozyme separation from egg white by cation exchange column chromatography. *Journal of Food Science, 51*(4), 1032–1036.

Losso, J., Nakai, S., & Charter, E. (2000). Lysozyme. *Natural Food Antimicrobial Systems, 185*, 210.

Maehashi, K., Matano, M., Irisawa, T., Uchino, M., Kashiwagi, Y., & Watanabe, T. (2012). Molecular characterization of goose-and chicken-type lysozymes in emu (Dromaius novaehollandiae): Evidence for extremely low lysozyme levels in emu egg white. *Gene, 492*(1), 244–249.

Mainente, F., Simonato, B., Pasini, G., Franchin, C., Arrigoni, G., & Rizzi, C. (2017). Hen egg white lysozyme is a hidden allergen in Italian commercial ciders. *Food Additives & Contaminants: Part A, 34*(2), 145–151.

Małaczewska, J., Kaczorek-Łukowska, E., Wójcik, R., & Siwicki, A. K. (2019). Antiviral effects of nisin, lysozyme, lactoferrin and their mixtures against bovine viral diarrhoea virus. *BMC Veterinary Research, 15*(1), 1–12.

Mangalassary, S., Han, I., Rieck, J., Acton, J., & Dawson, P. (2008). Effect of combining nisin and/or lysozyme with in-package pasteurization for control of *Listeria* monocytogenes in ready-to-eat Turkey Bologna during refrigerated storage. *Food Microbiology, 25*(7), 866–870.

Mastromatteo, M., Lucera, A., Sinigaglia, M., & Corbo, M. R. (2010). Synergic antimicrobial activity of lysozyme, nisin, and EDTA against *Listeria* monocytogenes in ostrich meat patties. *Journal of Food Science, 75*(7), M422–M429.

Memarpoor-Yazdi, M., Asoodeh, A., & Chamani, J. (2012). A novel antioxidant and antimicrobial peptide from hen egg white lysozyme hydrolysates. *Journal of Functional Foods, 4*(1), 278–286.

Meyer, W., Seegers, U., Herrmann, J., & Schnapper, A. (2003). Further aspects of the general antimicrobial properties of pinniped skin secretions. *Diseases of Aquatic Organisms, 53*, 177–179.

Mine, Y., Ma, F., & Lauriau, S. (2004). Antimicrobial peptides released by enzymatic hydrolysis of hen egg white lysozyme. *Journal of Agricultural and Food Chemistry, 52*(5), 1088–1094. https://doi.org/10.1021/jf0345752

Nagel, G. M., Bauermeister, L., Bratcher, C., Singh, M., & McKee, S. (2013). Salmonella and campylobacter reduction and quality characteristics of poultry carcasses treated with various antimicrobials in a post-chill immersion tank. *International Journal of Food Microbiology, 165*(3), 281–286.

Naidu, A. (2000). *Natural food antimicrobial systems*. CRC Press.

Nakamura, S., Gohya, Y., Losso, J. N., Nakai, S., & Kato, A. (1996). Protective effect of lysozyme-galactomannan or lysozyme-palmitic acid conjugates against Edwardsville tarda infection in carp, Cyprinus carpio L. *FEBS Letters, 383*(3), 251–254.

Nattress, F., Yost, C., & Baker, L. (2001). Evaluation of the ability of lysozyme and nisin to control meat spoilage bacteria. *International Journal of Food Microbiology, 70*(1–2), 111–119.

Naveed, M., Tianying, H., Wang, F., Yin, X., Chan, M. W. H., Ullah, A., & Abbas, Q. (2022). Isolation of lysozyme producing *Bacillus subtilis* strains, identification of the new strain *Bacillus subtilis* BSN314 with the highest enzyme production capacity and optimization of culture conditions for maximum lysozyme production. *Current Research in Biotechnology, 4*, 290–301. https://doi.org/10.1016/j.crbiot.2022.06.002

Nilsen, I. W., & Myrnes, B. (2001). The gene of chlamysin, a marine invertebrate-type lysozyme, is organized similar to vertebrate but different from invertebrate chicken-type lysozyme genes. *Gene, 269*(1–2), 27–32.

Ntzimani, A. G., Giatrakou, V. I., & Savvaidis, I. N. (2010). Combined natural antimicrobial treatments (EDTA, lysozyme, rosemary and oregano oil) on semi cooked coated chicken meat stored in vacuum packages at 4 C: Microbiological and sensory evaluation. *Innovative Food Science & Emerging Technologies, 11*(1), 187–196.

Ogawa, H., Miyazaki, H., & Kimura, M. (1971). Isolation and characterization of human skin lysozyme. *Journal of Investigative Dermatology, 57*(2), 111–116.

Pilevar, Z., Abhari, K., Tahmasebi, H., Beikzadeh, S., Afshari, R., Eskandari, S., & Hosseini, H. (2022). Antimicrobial properties of lysozyme in meat and meat products: Possibilities and challenges. *Acta Scientiarum, Animal Sciences, 44*. https://doi.org/10.4025/actascianimsci.v44i1.5526

Pitotti, A., & Boschelle, A. D. (1991). Assay of lysozyme by its lytic action on *Leuconostoc oenos*: A suitable substrate at acidic PH 1. *Journal of Food Biochemistry, 15*(6), 393–403.

Potty, A. S., Kourentzi, K., Fang, H., Schuck, P., & Willson, R. C. (2011). Biophysical characterization of DNA and RNA aptamer interactions with hen egg lysozyme. *International Journal of Biological Macromolecules, 48*(3), 392–397.

Prager, E., & Jolles, P. (1996). Animal lysozymes c and g: An overview. *EXS, 75*, 9–31.

Rao, M., Chander, R., & Sharma, A. (2008). Synergistic effect of chitooligosaccharides and lysozyme for meat preservation. *LWT-Food Science and Technology, 41*(10), 1995–2001.

Raybaudi-Massilia, R. M., Mosqueda-Melgar, J., Soliva-Fortuny, R., & Martín-Belloso, O. (2009). Control of pathogenic and spoilage microorganisms in fresh-cut fruits and fruit juices by traditional and alternative natural antimicrobials. *Comprehensive Reviews in Food Science and Food Safety, 8*(3), 157–180,

Salazar, O., & Asenjo, J. A. (2007). Enzymatic lysis of microbial cells. *Biotechnology Letters, 29*(7), 985–994.

Salton, M. (1958). The lysis of micro-organisms by lysozyme and related enzymes. *Microbiology, 18*(2), 481–490.

Shahmohammadi, A. (2018). Lysozyme separation from chicken egg white: A review. *European Food Research and Technology, 244*(4), 577–593.

Shin, Y. O., Rodil, E., & Vera, J. (2003). Selective precipitation of lysozyme from egg white using AOT. *Journal of Food Science, 68*(2), 595–598.

Silvetti, T., Brasca, M., Lodi, R., Vanoni, L., Chiolerio, F., De Groot, M., & Bravi, A. (2010). Effects of lysozyme on the microbiological stability and organoleptic properties of unpasteurized beer. *Journal of the Institute of Brewing, 116*(1), 33–40.

Sinigaglia, M., Bevilacqua, A., Corbo, M. R., Pati, S., & Del Nobile, M. A. (2008). Use of active compounds for prolonging the shelf life of mozzarella cheese. *International Dairy Journal, 18*(6), 624–630.

Sonni, F., Cejudo Bastante, M. J., Chinnici, F., Natali, N., & Riponi, C. (2009). Replacement of sulfur dioxide by lysozyme and oenological tannins during fermentation:

Influence on volatile composition of white wines. *Journal of the Science of Food and Agriculture*, *89*(4), 688–696.

Steinrauf, L., Shiuan, D., Yang, W. -J., & Chiang, M. Y. (1999). Lysozyme association with nucleic acids. *Biochemical and Biophysical Research Communications*, *266*(2), 366–370.

Su, C.-K., & Chiang, B. H. (2006). Partitioning and purification of lysozyme from chicken egg white using aqueous two-phase system. *Process Biochemistry*, *41*(2), 257–263.

Sudagidan, M., & Yemenicioğlu, A. (2012). Effects of nisin and lysozyme on growth inhibition and biofilm formation capacity of *Staphylococcus aureus* strains isolated from raw milk and cheese samples. *Journal of Food Protection*, *75*(9), 1627–1633.

Sun, X., Hong, H., Jia, S., Liu, Y., & Luo, Y. (2020). Effects of phytic acid and lysozyme on microbial composition and quality of grass carp (*Ctenopharyngodon idellus*) fillets stored at 4 C. *Food Microbiology*, *86*, 103313.

Takada, K., Ohno, N., & Yadomae, T. (1994). Lysozyme regulates LPS-induced interleukin-6 release in mice. *Circulatory Shock*, *44*(4), 169–174.

Tong, X.-D., Dong, X.-Y., & Sun, Y. (2002). Lysozyme adsorption and purification by expanded bed chromatography with a small-sized dense adsorbent. *Biochemical Engineering Journal*, *12*(2), 117–124.

Touch, V., Hayakawa, S., & Saitoh, K. (2004). Relationships between conformational changes and antimicrobial activity of lysozyme upon reduction of its disulfide bonds. *Food Chemistry*, *84*(3), 421–428.

Vilcacundo, R., Méndez, P., Reyes, W., Romero, H., Pinto, A., & Carrillo, W. (2018). Antibacterial activity of hen egg white lysozyme denatured by thermal and chemical treatments. *Scientia Pharmaceutica*, *86*(4), 48.

Vocadlo, D. J., Davies, G. J., Laine, R., & Withers, S. G. (2001). Catalysis by hen egg-white lysozyme proceeds via a covalent intermediate. *Nature*, *412*(6849), 835–838.

Wang, C., & Shelef, L. A. (1992). Behavior of *Listeria* monocytogenes and the spoilage microflora in fresh cod fish treated with lysozyme and EDTA. *Food Microbiology*, *9*(3), 207–213.

Wang, Z., Hu, S., Gao, Y., Ye, C., & Wang, H. (2017). Effect of collagen-lysozyme coating on fresh-salmon fillets preservation. *LWT*, *75*, 59–64.

Wu, T., Ge, Y., Li, Y., Xiang, Y., Jiang, Y., & Hu, Y. (2018). Quality enhancement of large yellow croaker treated with edible coatings based on chitosan and lysozyme. *International Journal of Biological Macromolecules*, *120*, 1072–1079.

Xing, H., & Yaylayan, V. (2019). Mechanochemically induced controlled glycation of lysozyme and its effect on enzymatic activity and conformational changes. *Journal of Agricultural and Food Chemistry*, *67*(11), 3249–3255.

Xu, F., Liu, S., Liu, Y., Xu, J., Liu, T., & Dong, S. (2019). Effectiveness of lysozyme coatings and 1-MCP treatments on storage and preservation of kiwifruit. *Food Chemistry*, *288*, 201–207.

Xue, Q., Hellberg, M. E., Schey, K. L., Itoh, N., Eytan, R. I., Cooper, R. K., & La Peyre, J. F. (2010). A new lysozyme from the eastern oyster, *Crassostrea virginica*, and a possible evolutionary pathway for i-type lysozymes in bivalves from host defense to digestion. *BMC Evolutionary Biology*, *10*(1), 1–17.

Xue, Q.-G., Itoh, N., Schey, K., Li, Y.-L., Cooper, R., & La Peyre, J. (2007). A new lysozyme from the eastern oyster (*Crassostrea virginica*) indicates adaptive evolution of i-type lysozymes. *Cellular and Molecular Life Sciences*, *64*(1), 82–95.

Yang, H. H., Li, X. C., Amft, M., & Grotemeyer, J. (1998). Protein conformational changes determined by matrix-assisted laser desorption mass spectrometry. *Analytical Biochemistry*, *258*(1), 118–126.

Yang, T., & Leśnierowski, G. (2019). Changes in selected physicochemical properties of lysozyme modified with a new method using microwave field and oxidation. *PLoS One*, *14*(2), e0213021.

Zhang, W., & Rhim, J.-W. (2022). Functional edible films/coatings integrated with lactoperoxidase and lysozyme and their application in food preservation. *Food Control*, *133*, 108670.

Zou, Y., Li, Y., Hao, W., Hu, X., & Ma, G. (2013). Parallel β-sheet fibril and antiparallel β-sheet oligomer: New insights into amyloid formation of hen egg white lysozyme under heat and acidic condition from FTIR spectroscopy. *The Journal of Physical Chemistry B*, *117*(15), 4003–4013.

27

Egg Yolk Antibody-IgY

Shofiqur Rahman

CONTENTS

DOI: 10.1201/9781003254430-34

27.1 Introduction

27.1.1 History of Immunoglobulins in Yolk (IgY)

In 1969, G.A. Leslie and L.W. Clem suggested the name Immunoglobulin Y. Other synonymous names are Chicken IgG, Egg Yolk IgG, and 7S-IgG. In 1893, the first scientific report of IgY was published by Felix Klemperer. In 1959, Russell and Burch proliferated animal welfare research extensively. Since the 1980s, IgY has been frequently studied due to the revolution of overall technology. In the 1990s, the term "IgY technology" was introduced to describe a procedure to produce polyclonal antibodies of the Y class (IgYs).

In 1995 Warr et al., reported that the IgY is a key isotype in antibody evolution. IgY is thought to have diverged from an ancestral IgM, and it is a widely held belief that an IgM gene duplication event led to the formation of IgY (Warr et al., 1995). IgY is also thought to be the precursor of IgG and IgE.

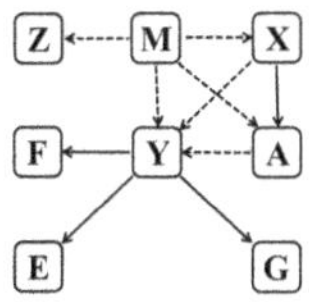

FIGURE 27.1 The central position of IgY in immunoglobulin evolution. Solid arrows indicate an orthologous relationship between isotypes. Broken arrows connect isotypes that have a putative orthologous relationship, not yet verified. The broken arrow that relates IgA to IgY refers to an ancestral form of IgA. (Adapted from Zhang et al., 2017.)

At some point during the evolution of the mammalian lineage, IgY underwent a gene duplication event and diversified into IgE and IgG. Thus, the proposal that IgM gave rise to the mucosal antibody IgX and then to IgA, which gave rise to IgY and the serum antibodies IgG and IgE. These relationships are depicted in Figure 27.1, which highlights the central role of IgY.

Physicochemical and antigenic evidence obtained during the past three decades has indicated that IgY occurs throughout the vertebrate classes Amphibia, Reptilia, and Aves. It is clear from the diverse capabilities of IgY in so many species that molecular genetic studies of this molecule will broadly contribute to our understanding of Ig evolution. Perhaps looking backward in evolution will take us forwards in our knowledge of mammalian antibody function.

In 1996, the European Centre for the Validation of Alternative Methods to animal testing (ECVAM) strongly recommended avian antibodies as alternatives to mammalian ones (Schade et al., 1996). In parallel, in 1999, the IgY technology was approved as an alternative method for supporting animal welfare by the Veterinary Office of the Swiss Government. The field is more than 120 years old. However, in the last decades, major advances in research and development areas such as Genetics, Biochemistry, Bioengineering, and Bioprocessing, have prompted new approaches to this old technology.

In 2001, the world's first standardized laboratory practice of IgY technology, known as the "IgY laboratory manual" was reported (Schade et al., 2001). In 2002–2006, the project *Multidisciplinary Hen Egg Research* was started through a Cooperative Organization Science and Technology action (COST 923) in the European Union framework for the versatile utilization of eggs. In 2007, Huopalahti et al. summarized one project where the biomedical use of IgY became the focus of the action plan. In 2011, a Chinese version of IgY monograph was published. NCBI database with different search terms, namely "IgY Technology", "IgY Antibodies" and "IgY" surveyed for timeline 1893–2022. NCBI database surveys for the three time periods, (e.g., 1893~1955, 1956~1987, 1988~2022) of avian IgY publications, show a progressive increase in the number of published works since the 1988s (Figure 27.2).

27.2 Chicken: Natural Source of IgY

27.2.1 Chicken Immunity and Immune System

The avian immune system is the system of biological structures and cellular processes that protects *birds* from disease. Like other avian immune systems, the immune system of chickens is made up of two types of mechanisms—nonspecific and specific. Nonspecific immune mechanisms include the inherent ways (e.g., genetic factors, body temperature, anatomic features, normal microflora, respiratory tract-cilia) in which a chicken resists disease. Specific immune mechanisms, which make up the acquired immune system, comprise noncellular (humoral) and cellular components.

The noncellular component includes immunoglobulins (or antibodies) and the cells that produce them. The cellular component of the specific immune mechanisms includes all the cells that react with specificity to antigens except those associated with antibody production. The cells associated with this

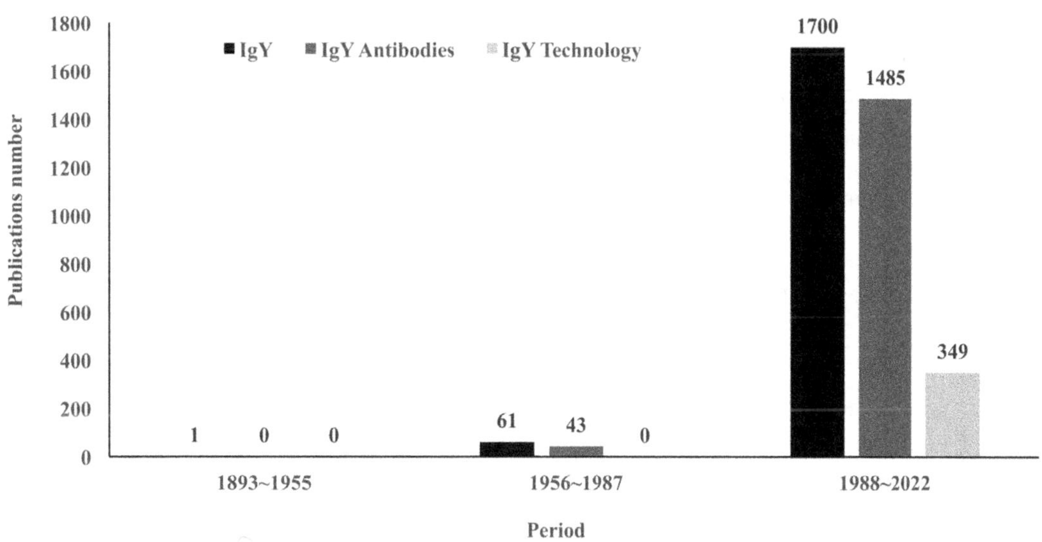

FIGURE 27.2 Evolution of publications on avian IgY antibodies. A search of publications on avian IgY antibodies and IgY Technology was performed on the NCBI database with different search terms over a time window from 1893 to 2022. The search terms were "IgY Technology", "IgY Antibodies" and "IgY" and the total number of publications for each term were 301, 1368, and 1583, respectively (June 4, 2022).

TABLE 27.1

The Avian Immune System

Organs/Tissues	Cellular Elements	Humoral Elements
Primary Lymphoid Organs: Bursa of fabricius and thymus Secondary Lymphoid Organs: Spleen, bone marrow, Harderian gland, pineal gland, mucosa-associated lymphoid tissue (MALT), and lymphoid nodules	Lymphocytes, T-cells, B-cells, and macrophages	Immunoglobulins (IgY, IgA, IgM), complement, and cytokines

system, T-lymphocytes (T-cells), begin as the same stem cells as B-lymphocytes (B-cells). Birds are the first vertebrates in which a clear dichotomy of the lymphoid system has been established: (1) Thymus-derived (T) lymphocytes, the effector cells in cell-mediated immunity, and (2) Bursa-derived (B) lymphocytes are the precursor cells of the antibody-synthesizing plasma cell. Table 27.1 summarizes the chicken immune system.

27.2.2 Immunoglobulin Classes in Chickens

Avians and mammals' specific or adaptive immune system is based on immunoglobulins. Birds produce three types of immunoglobulins (IgM, IgY, and IgA), and mammals five (IgM, IgD, IgG, IgE, and IgA) (Benedict et al., 1963; Leslie & Chem, 1969). A basic comparison of the Immunoglobulin classes or isotypes between avian and mammals is shown in Table 27.2. Immunoglobulin classes in other avian species (Ostrich, Quail, Turkey, Duck, and Goose) are also the same (Härtle et al., 2014) like chicken, e.g., IgM, IgA, and IgY.

27.3 Properties of IgY

27.3.1 Maternal Antibody (MA) Transfer from the Hen to the Progeny

Maternal Antibodies (MA), also known as passive immunity, are the natural transfer of immunoglobulins from one individual to another. Evidence for maternal IgY transfer from chicken to egg yolk for embryo protection was first reported more than 100 years ago (Klemperer, 1893). In birds, maternal antibodies are passed from hyper-immunized or naturally infected breeder hens to the progeny through the egg. This passive immunity has a relatively short duration, commonly 1–2 weeks and generally less than 4 weeks, and its function is to protect young chicks during a period (first few weeks) when their immune system is not fully developed to properly react to an early challenge. The chicken transfer MA to the egg by depositing the antibodies [IgY, IgA, and IgM)] in the egg yolk and albumin. Transport of IgY from maternal serum to the offspring (Ferreira Júnior et al., 2018) is unique process. This process comprises two steps. First step: MA Transfer from the Hen to the Egg; the first is the transfer from circulating maternal blood to the yolks of maturing oocytes in ovarian follicles, (which is in analogy to the cross placental transfer of antibodies in mammals). The passage of IgY into the ova is regulated by the follicular epithelium, which goes through morphologic changes as the ova grows. This epithelium becomes flatter and thinner in the larger ovum allowing the passage of a large amount of IgY. The transfer of IgY through the ovarian follicular epithelium reaches its maximum 3–4 days before ovulation. It starts to decrease due to the development of the vitelline membrane between the ovum and the follicular epithelium of the ovary in preparation for ovulation. Therefore, as a single hen has several ovas in different stages of development and the amount of IgY transferred to each one is not the same. The IgA and IgM are mainly found in the albumen (Rose et al., 1974) and they are transferred to the albumen due to mucosal secretion in the oviduct, specifically in the Magnum. Second step: MA Transfer from the Egg to Embryo. The second involves IgY transfer from the egg yolks to the embryonic circulation (to the developing embryo) through the yolk sac membrane (Linden & Roth, 1978; Tressler & Roth, 1987) (Figure 27.3). The second step of MA transfer relies on the IgY Fc receptor, FcRY (West et al., 2004); the relevant receptor involved in IgY transport is unknown. The IgY is transferred from the egg yolk to the offspring via embryonic circulation. The transfer starts from day 7 of embryonic development and reaches its maximum rate 3–4 days before hatch.

TABLE 27.2

Comparison of the Immunoglobulin Classes between Avian and Mammals

	Vertebrate Class	
Immunoglobulin	Avian	Mammals
Isotype	IgM (10%)	IgM (10%)
	Homologus proposed by Chen et al. (1982)	IgD (1%)
	IgY (75%)	IgG (70–75%)
	IgA (15%)	IgA (10–15%)
	Homologus proposed by Burns and Maxwell (1981)	IgE (0.001%)

27.4 Characteristics of IgY

27.4.1 Molecular Structural Characterization of IgY

Phylogenetic studies have shown that the IgY antibody, a homolog of mammalian IgG, has similarities with both mammalian IgG and IgE antibodies. Although chicken IgY is the functional equivalent of mammalian IgG, there are some profound differences in their molecular structure. The general structure of the IgY molecule is the same as the IgG molecule, with two heavy (H) chains and two light (L) chains, but IgY has a molecular mass of 180 kDa, which is larger than that of mammalian IgG (150 kDa) (Figure 27.4). The molecular mass (67–70 kDa) of the H chain in IgY is larger than the H chain in mammals (50 kDa). The greater molecular mass of IgY is due to an increased number of heavy-chain constant domains and carbohydrate chains (Warr et al., 1995). IgG has 3 C regions ($C_\gamma1$–$C_\gamma3$/CH1–CH3),

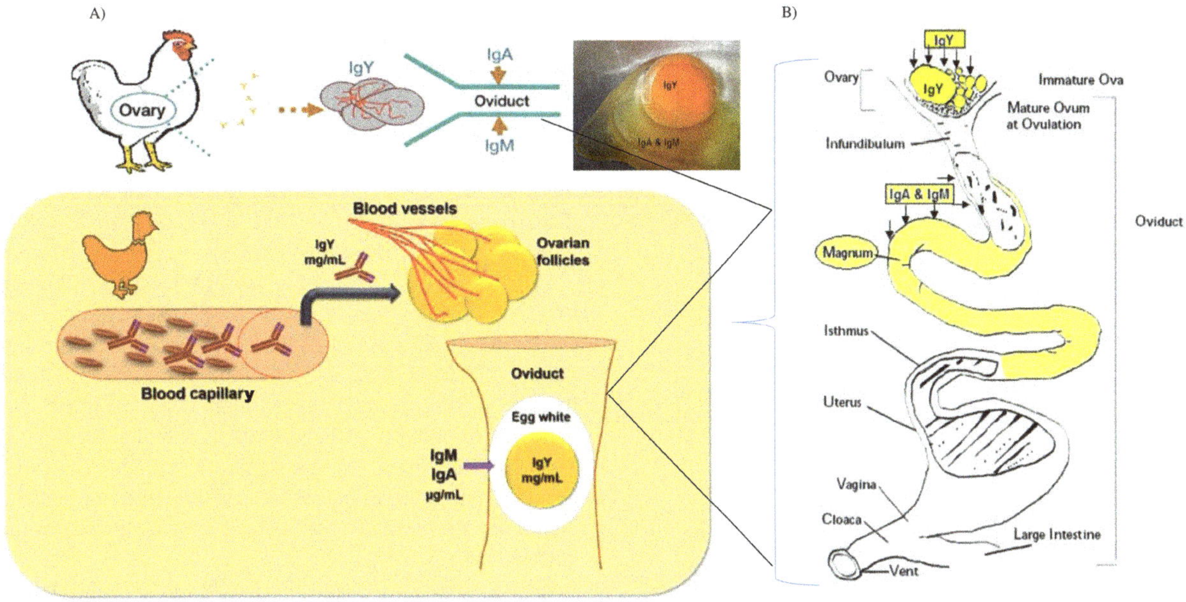

FIGURE 27.3 (A) Maternal antibody (MA) transfer from the hen to the egg. (B) Schematic representation of the IgY antibody translocation from the hens' blood to egg yolk in the ovarian follicle, and IgM and IgA deposition into egg whites through the oviduct epithelium. (Modified and adapted from Ferreira Júnior et al., 2018.)

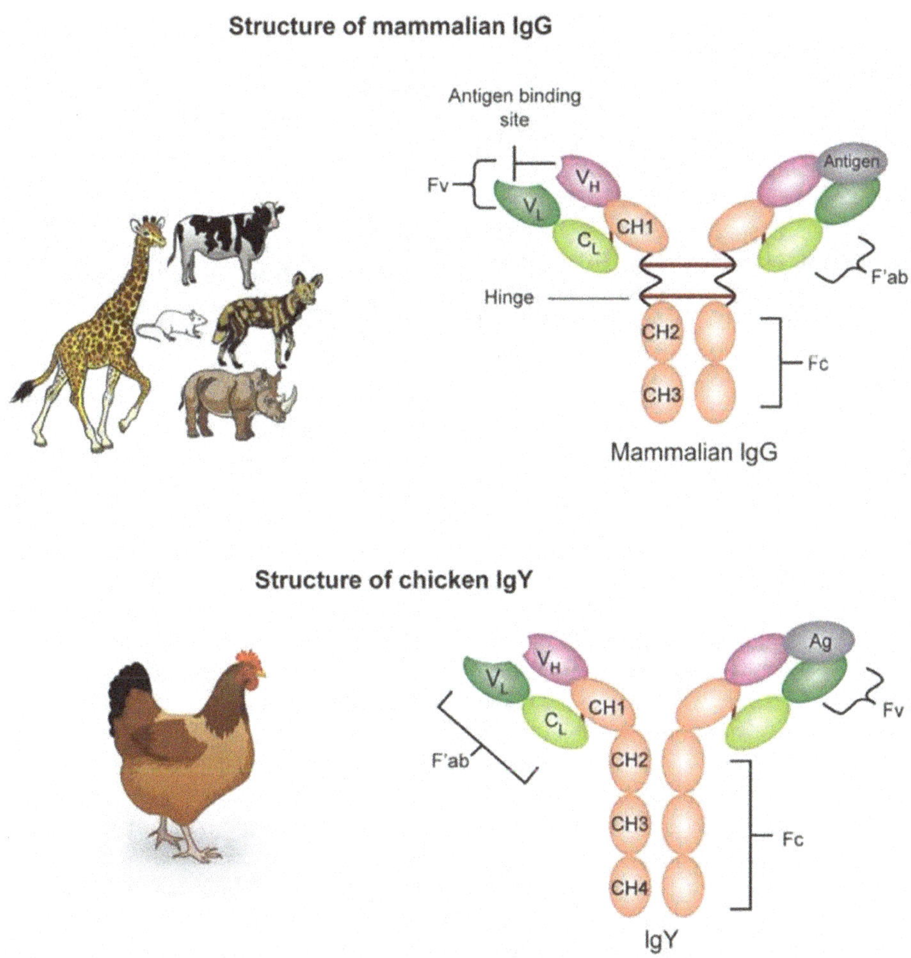

FIGURE 27.4 Structural comparison between mammalian IgG and avian IgY. (Adapted from Abbas et al., 2019.)

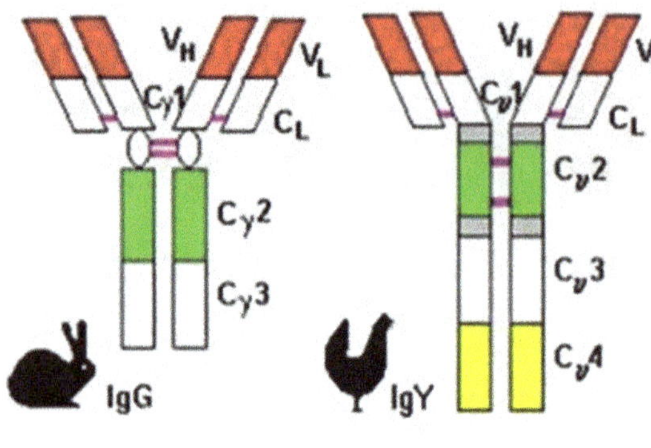

FIGURE 27.5 Comparison of the characteristics of mammalian IgG (rabbit) and chicken IgY. (Modified and adapted from Schade et al., 2005.)

while IgY has 4 C regions ($C_{\upsilon}1$–$C_{\upsilon}4$/CH1–CH4), and the presence of one additional C region with its two corresponding carbohydrate chains logically results in a greater molecular mass of IgY compared with IgG (Figure 27.5).

Other differences in the structure include the fact that the hinged region of IgY is much less flexible than mammalian IgG. It has also been suggested that IgY is a more hydrophobic molecule than IgG (Dávalos-Pantoja et al., 2000). Finally, IgY has an isoelectric point of pH 5.7–7.6, whereas IgG lies between 6.1 and 8.5 (Sun et al., 2001). Unlike mammalian IgG, IgY does not fix mammalian complement and does not interact with mammalian Fc and complement receptors (Carlander et al., 1999; Carlander et al., 2000). As well, IgY does not bind to protein A, protein G, or rheumatoid factor, so no false positives are obtained on immunoassay, which is a problem with IgG-based mammalian assays. These differences provide significant advantages to the application of IgY technology in many areas of research, such as diagnostics (Erhard et al., 2000), antibiotic-alternative therapy (Carlander et al., 2000), and xenotransplantation (Fryer et al., 1999). Figure 27.5 and Table 27.3 summarize the overall comparison of the characteristics of mammalian IgG and chicken IgY.

27.4.2 Immuno-Functional Characterization of IgY and Passive Immunity

Passive immunity is the transfer of active humoral immunity in the form of ready-made antibodies from one individual to another. Passive immunization was first introduced more than 100 years ago by Albert Calmette, and others (Calmette, 1896). As such, passive immunotherapy by antigen specific IgY acquires a special value as a tool for infection control and immunologic research with global commercial application as raw material for nutraceutical and pharmaceutical products and for applications in numerous medical and research fields since the 1980s. Specific IgY antibodies are obtained by immunizing the hen with the antigen of interest. The antibody fragment (Fab) domain containing a structure of the IgY with no hinge region, (Figure 27.4) gives IgY less flexibility to antigen binding with a broad array of antigenic epitopes (on, e.g., proteins, carbohydrates, and nucleic acids). A small amount of antigen in the milligram or microgram range usually elicits enough IgY response, and the antibody titers persist over several weeks to several months. It is reported that antigen-antibody interactions are characterized as noncovalent interactions (like the "lock and key" fit of enzyme-substrate). This interaction does not lead to irreversible alteration of antigen (Ag) or antibody (Ab) (Figure 27.6).

Compared to mammalian IgG, chicken IgY has 3–5 times more affinity and reacts more rapidly to the same antigens (Stuart et al., 1988; Ikemori et al., 1993; Lemamy et al., 1999) when tested in competition assays.

The advantages of using chicken IgY have been recognized by many authors (Schade et al., 2005). Since antibiotics are commonly used or misused for the treatment of gastrointestinal infections, the frequency of antibiotic-resistance organisms has been on the rise at an alarming rate against a backdrop of decreasing numbers of new antibiotics being developed and added to the market. We are therefore compelled to fall back to simple and yet effective natural remedies of which IgY comprises the most potent and easily generated substitute to antibiotics.

IgY can have different qualities because its production is "tailor-made", production conditions can be optimized to produce the standardized IgY antibody by producing a specific antibody, where the animals must be challenged with the respective specific antigen. With standardized antibody titers per batch, the product can deliver consistent results. IgY is more effective than IgG because (1) customized production: IgY is tailor-made and is specific against gut/infected area pathogens (compared to nonspecific IgG); (2) genetic selection theory: repeated hyperimmunization creates stronger antibody molecules; (3) molecular structure: IgY is much bigger than IgG; size is bigger; (4) protease resistance (against pancreatic enzymes: trypsin and chymotrypsin; sensitive to pepsin and papain; (5) maternal antibody transfer mechanism: God-gifted, genetically strong. Because milk-based IgG is provided daily baby till certain periods (few days to years). But Yolk-based IgY was provided only once to the chick before hatching; (6) repeated vaccination (hyperimmunization) with the same antigen creates stronger IgY (antibody) molecules as trained soldiers, protect the country from imminent enemies with well pre-prepared know-hows; and (7) IgY enhances colostrum IgG uptake. (Erhard et al., 1999; Quezada-Tristán et al., 2014). Moreover, IgG in colostrum (1) using bovine colostrum for swine or other animals will be less effective due to different pathogens; (2) high-quality colostrum is an excellent source of Ig, but the quality of colostrum is very variable.

IgY immunotherapy has several attractive features, including: (1) lack of reactivity with the human complement system and human Fc-receptors, thereby preventing nonspecific inflammation (Larsson et al., 1993); (2) excludes the use of toxic compounds or additives for their preparation from egg yolks; (3) egg cholesterols and triglycerides can be controlled to infinitesimally low levels (Nilsson et al., 2008); (4) IgY exerts beneficial antimicrobial and immuno-stimulatory effects in conjunction with other egg proteins (Kovacs-Nolan et al., 2005); and (5) high content of sialic acid (Gilgunn et al., 2016), which is reported to increase the half-life of the drug (Liu, 2015) compared with those with lower sialic acid content. It indicates that IgY-based therapy could have a longer circulating half-life, increasing its efficacy against infections. Egg allergies usually involve egg albumin components which

TABLE 27.3

Comparison of the Characteristics of IgG and IgY

Characteristics	IgG	IgY
Species	Mammals	Birds, reptiles, amphibians and lungfish
Sites of generation	Lymph nodes, spleen and bone marrow	Bursa of fabricius, spleen, and bone marrow
Antibody subclasses	IgG_1, IgG_2, IgG_3 and IgG_4	Not for chicken
Source of antibodies	Serum	Serum and Egg
Antibody collection	Invasive, painful	Meets 3R principle of animal welfare
Average antibody levels per animal	5 mg/mL of blood, blood collection up to 40 mL/month	50–100 mg/egg yolk
Monthly antibody yield per animal	200 mg/rabbit/month	1,000–2,800 mg/chicken/ month
Concentration (mg/mL)	10–15 (Serum)	8–10[a] (Serum), 10–20[b] (egg yolk)
Amount of antigen-specific antibodies	50–200 µg/mL serum	0.5% –10% total IgY
Phylogenetic distance to mammals	Near	310 Mya
Immune response to mammalian conserved antigens	Weak	Strong
Affinity	$1 \times 10^{-8} - 1 \times 10^{-10}$ mol/L	Comparable to rabbit IgG
Antibody avidity	High	3–4 times higher in compare to IgG
Molecular weight (kDa)	150	180
Isoelectric point (pI)	6.4–9.0[c]	5.7–7.6[d]
pH stability	2.0–11.0	3.5–11.0
Extinction coefficient (mL mg^1 cm^1 at 280 nm)	1.4	1.094, 1.33, 1.35
Trypsin stability	Weaker than IgY	Generally stable
Heat stability	Generally higher than IgY, Up to 75°C–80°C	Generally Up to 70 °C, specifically at 65°C > 2 months; 100°C > 6 min; 4°C > 6 months,
Proteolytic degradation	Pepsin, papain, trypsin and chymotrypsin	Pepsin and papain
Pepsin stability	Higher than IgY,	91% IgY activity at pH 4 for 1 hour, 63% after 10 hour
Fc chain (type of chain/number of domains/hinge region between Fab and Fc chains)	γ chain/2 constant domains/yes	υ chain/3 constant domains/ no
Hinge region	Yes	No
Number of constant domains	4 (3H and 1L)	5 (4H and 1L)
Fab chain (type of chain/number of domains)	κ or λ chains/1 variable domain and 1 constant domain	λ chain/1 variable domain and 1 constant domain
Mammalian complement binding	Yes	No
Rheumatoid factor binding	Yes	No
Fc receptor binding	Yes	No
Mediates anaphylaxis	No	Yes
Binding to protein A and G	Yes	No
Cross-reactivity with rheumatoid factor	Yes	No
Cross-reactivity with human anti- mouse antibody (HAMA)	Yes	No
Reaction to hetero-agglutinins(Coombs, blood group classification)	Yes	No
Non-affinity separation	Remove plasma mixture from blood	Remove lipid mixture from egg yolk
Purification by salting-out	Yes	Yes
Purification by affinity columns	Yes (protein A and G)	Protein M[e]
Conjugation with enzymes, fluorophores and colloidal gold	Yes	Yes
Immunoprecipitation	Good	Relatively inefficient
Immunosuppression applications	Several products are under development	Can be used for xenotransplantation
Diagnostic applications	Widely used, especially monoclonal antibodies	Studied and applied for detection/ diagnosis
Therapeutic applications	Well developed	Under development
Generation of full-length monoclonal antibody by hybridoma technology	Routinely and commercially applied	Difficult, and lower yield of monoclonal IgY obtained
Generation of scFv antibody fragment by phage display	Easily produced	Easily produced

Source: Adapted and Modified from Ferreira Júnior et al. (2021).

[a] Wang et al., 2000

[b] Carlander, D., 2002

[c] Li et al., 2002

[d] Dávalos-Pantoja et al., 2000

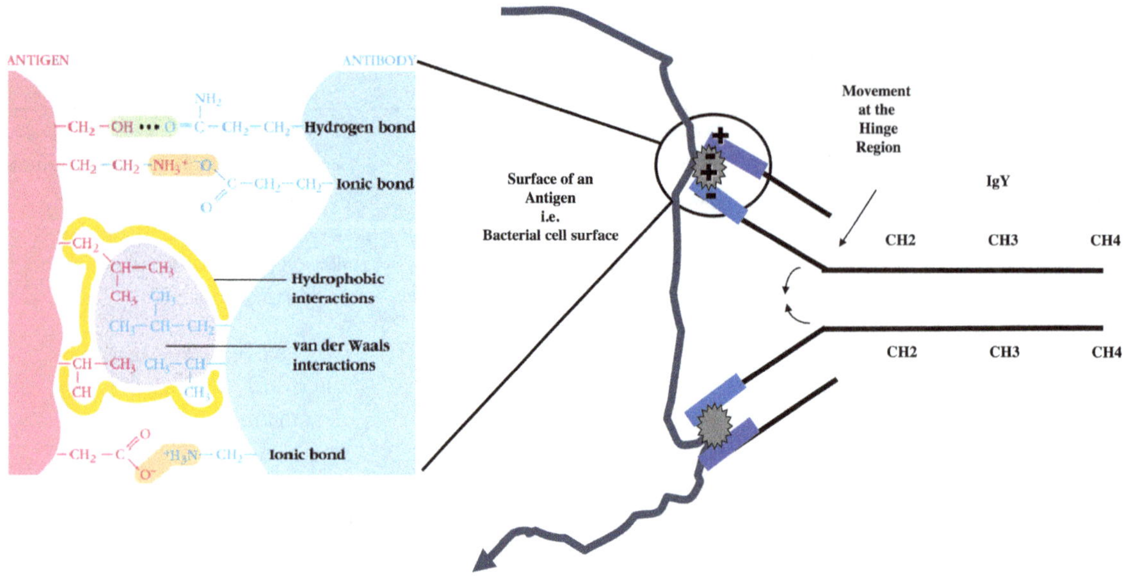

FIGURE 27.6 The noncovalent interactions that form the basis of antigen-antibody (Ag-Ab) binding. (Adapted from Goldsby et al., 2000.)

may explain why no reactivity issues have been encountered in consumer use of several products now in the market containing purified IgY.

Compared to vaccination, passive immunotherapy using IgY has distinct advantages such as (1) rapid and local onset of action, (2) highly specific activity, and (3) applicability to a broader age range of patients from infants to adults, including immunodeficient patients, and (4) it is nontoxic being a normal part of the human diet. While immunity derived from passive immunization lasts for only a short period coterminous with antibodies in the recipient, it nonetheless provides immediate and efficient host protection when properly concentrated onto the target organ.

27.4.3 Physicochemical Stability of IgY

Effects of heat, atmospheric pressure, pH, pepsin, and gut passage on IgY stability were studied extensively. IgY is the predominant immunoglobulin isotype in chicken eggs and acts as a major immunoglobulin fraction that confers passive gut immunity. IgY is proteinaceous and is therefore sensitive to heat, pH, and pepsin, properties that pose real challenges to its oral application for various digestive disorders. Within the past decade, several studies have been conducted to overcome these problems with various degrees of success. Shin et al. evaluated the heat, pH, and pepsin stabilities of anti-*Helicobacter pylori* IgY (IgY-Hp) (Figure 27.7) (Shin et al., 2002).

The binding activity of IgY with antigen decreased with increasing temperature and heating time. IgY is stable at temperatures ranging between 30°C and 70°C. The activity of IgY decreased by heating for 15 min at 70°C or higher, and IgY was denatured significantly when treated at temperatures higher than 75°C. IgY is relatively stable to pressure up to 4,000 kg per cm². The addition of high levels of sucrose, maltose, glycerol, or glycine conferred additional protection against pressure and thermal denaturation of IgY.

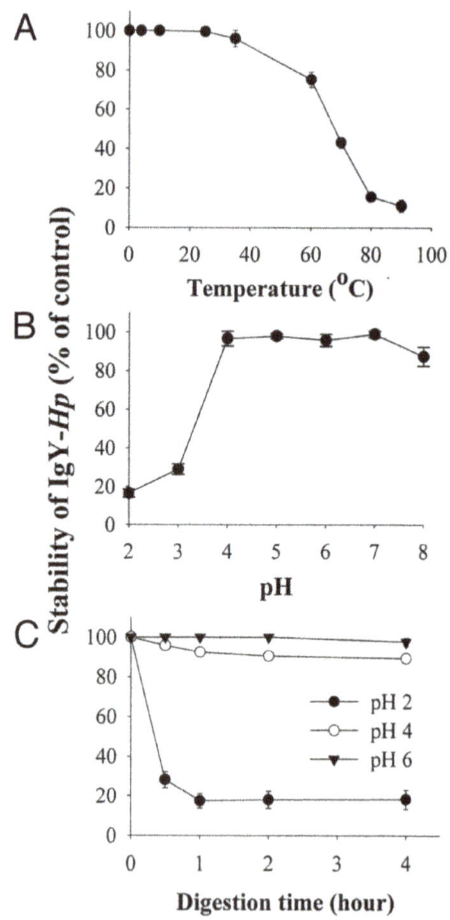

FIGURE 27.7 Effect of heat, pH, and pepsin on IgY-Hp. IgY-Hp was treated at various temperatures for 10 min (A), at various pHs for 4 hour (B), and with pepsin (15 mg/mL) (C) at pH 2, 4, and 6 for 0.5, 1, 2, and 4 hours. Remaining activities after the treatments were measured using ELISA and are expressed as a percentage of the initial activity. (Adapted with permission from Shin et al., 2002.)

The isoelectric point (pI) of IgY is in the range of 5.7–7.6, whereas that of IgG lies between 6.1 and 8.5 (Dávalos-Pantoja et al., 2000; Sun et al., 2001). The most hydrophobic moiety of the Ab molecule is the Fc fragment. Since the Fc fragment of the IgY is bigger than the IgG's, the IgY molecule is more hydrophobic than IgG. It has been demonstrated that this property might be useful for achieving stable adsorption of IgY onto latex particles (Dávalos-Pantoja et al., 2000). It could be demonstrated that preferentially at pH 8, the Fc part of IgY is firmly bound to the latex particles (Dávalos-Pantoja et al., 2001), hence IgY coated to latex microspheres retains its specific binding activity. The stability of IgY to acid and alkali has been studied under various conditions. It was found that the activity range of IgY for pH was pH 3.5~11. The stability of IgY at pH 3 was increased in the presence of sorbitol. IgY is quite resistant to trypsin and chymotrypsin inactivation but is degraded by pepsin (Hatta et al., 1993). The stability of IgY against pepsin appears to be highly dependent on pH and the enzyme/substrate ratio. At pH 5 or higher, IgY was fairly resistant to pepsin and retained its antigen-binding and cell-agglutinating activities. However, at pH 4.5 or below, both activities were lost. IgY digested with pepsin at pH 4 retained 91% and 63% of its activity after 1 hour and 4 hour incubation, respectively.

Several strategies to protect IgY from hydrolysis by gastric enzymes and acidic conditions have been investigated like dissolving in sodium carbonate buffer, encapsulation with liposomes, egg lecithin/cholesterol liposomes, and chitosan-alginate. Encapsulated IgY was released smoothly in in-vitro studies and was found to cure enteric colibacillosis in pigs more rapidly than non-coated IgY (Li et al., 2009). Encapsulated IgY was more resistant to pepsin and gastric conditions (Shimizu et al., 1993), but the uncoated IgY showed a better effect than the commonly used antibiotic. Another report showed that IgY and freeze-dried IgY coated with gum arabic were protected against hydrolysis by trypsin, chymotrypsin, and pepsin (Chang et al., 1999).

Researchers have investigated the in vivo passage and the efficacy of IgY in the gastrointestinal tract of piglets (Yokoyama et al., 1993) and calves (Ikemori et al., 1996). Results indicated that IgY powder was transported as immunologically functional molecules from the stomach down to the small intestine of calves while retaining much of their original biological activity (Figure 27.8).

27.4.4 Storage Stability of IgY

IgY is naturally protected by the yolk granules. The stability of IgY during storage is reasonably good under specified conditions. Dried IgY preparations were stored for 5–10 years at 4°C without significant loss of antibody activity and the preparations also retained activity for 6 months at room temperature and for 1 month at 37°C (Larsson et al., 1993). While

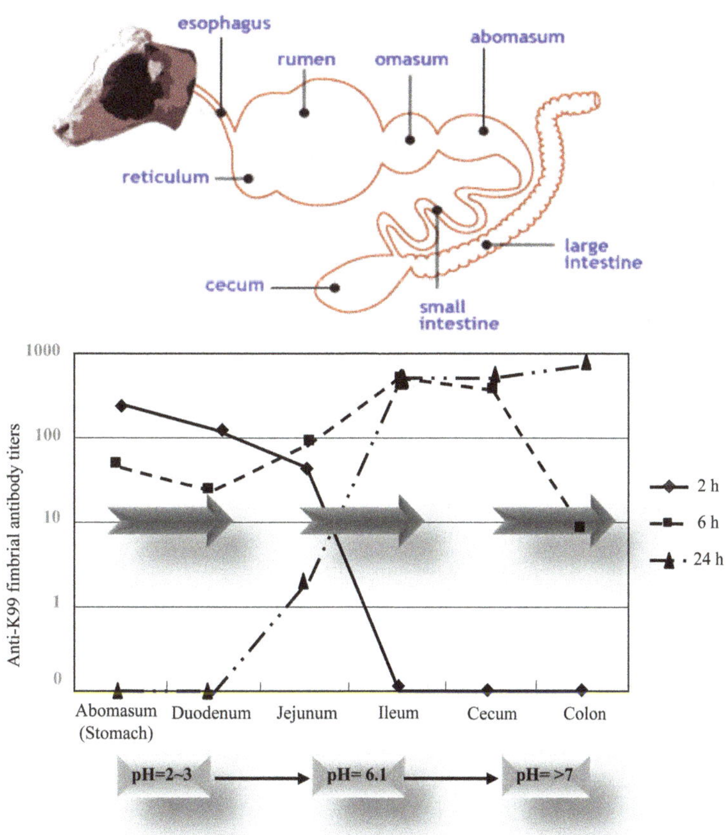

FIGURE 27.8 In vivo passage of IgY in the gastrointestinal tract of calves. Anti-K99 fimbriae antibody titers of IgY in the gastrointestinal tract of calves after 2, 6, and 24 hours post-administration. (Modified and adapted from Ikemori et al., 1996.)

lyophilization minimizes bacterial growth and maintains the better structural integrity of purified IgY, spray-drying is more economical (Yokoyama et al., 1992). The activity of IgY was well preserved after freeze-drying. Freezing at –70°C can cause the loss of up to 50% of IgY activity (Staak et al., 2000), so storage at –20°C is preferable. The addition of high levels of sucrose, maltose, glycerol or glycine displayed effective additional protection against thermal denaturation of IgY. If encapsulated, they are particularly resistant to pH and digestive enzymes. Encapsulation of IgY with egg lecithin/cholesterol liposomes reduced the activity loss of IgY under gastric conditions. IgY may be stable in 0.9% NaCl, 0.02% NaN_3 (sodium azide) at 4°C for periods ranging from months to a few years without any significant loss of antibody titer. Adding 0.03% w/v thimerosal or 50 μg/mL gentamicin in IgY solution helped to retard microbial growth (Schade et al., 2005). Two important points should be considered when storing IgY solutions with NaN_3 (sodium azide): (1) sodium azide can inhibit horseradish peroxidase (HRP) activity, especially when primary antibody is used in dilutions lower than 1:1,000; (2) IgY solutions stored with sodium azide should be dialyzed before the labeling of IgY antibodies with, for example, HRP. To avoid the effects of repetitive thawing/freezing of IgY, 10–50% (v/v) glycerol can be added. Collectively, these unique biological attributes make IgY an effective natural food antimicrobial system and immunotherapeutic agent.

27.4.5 Biological Safety of Oral IgY

Several properties make IgY attractive for oral immunotherapy (Schade et al., 2007). While the mouth is the portal of entry for many infectious agents, it is, therefore, logical to use this as the route for IgY to target specific infectious entities within the alimentary tract. IgY does not pass as intact molecules from the intestines to the blood circulation, thus precluding any systemic effect (Losonsky et al., 1985). IgY use is associated with much lower risk of inducing specific resistance among pathogenic microorganisms since it is directed to multiple antigenic targets that require multiple genes for their synthesis. IgY is superior to mammalian IgG in terms of safety issue. IgYs are safer than IgGs because they do not bind to human Fc receptors or fix mammalian complement components, hence they do not trigger potentially dangerous immune responses (Carlander et al., 1999). Kubickova et al exposed immortalized human lung epithelial cells to IgY, using lipopolysaccharide as a positive control and phosphate-buffered saline as a negative control. Treatments also included exposure to human and goat IgG, and exposure lasted 24 hours for all treatments. The researchers found that the levels of pro-inflammatory cytokines tumor necrosis factor (TNF)-α, interleukin (IL)-1β, IL-6, and granulocyte-macrophage colony-stimulating factor (GM-CSF) were very low in cell cultures treated with IgY compared with the high levels of TNF-α and GM-CSF in cells treated with lipopolysaccharides, indicating that IgYs do not cause inflammatory responses in lung cells and can thus be safely used for prevention of airway infections (Kubickova et al., 2014). Additionally, oral IgY antibodies have been applied to treat rotavirus infections in humans (Sarker et al., 2001; Rahman et al., 2012; Thu et al., 2017) and to treat pulmonary Pseudomonas aeruginosa infections, and no negative side effects of IgY treatment have been observed in up to 10 years of use (Nilsson et al., 2007b).

Being an ingredient in our regular diet, poultry eggs are considered generally safe. As natural components of eggs, the IgY antibodies tend to be nontoxic. IgY antibodies are not deposited in meat, hence avoiding potential violations of regulations in countries that forbid the use of antibiotics for poultry and livestock (Li et al., 2015). Allergic reactions may occur upon ingestion of egg-derived components particularly those that contain appreciable amount of egg white. However, the water-soluble IgY materials purified from egg yolk (delipidated) are not usually associated with allergic reactions. The risk of allergy is lower when administering the antibodies orally than by other routes (Russo et al., 2001). Moreover, oral administration of egg protein (mainly ovalbumin) has been shown to induce systemic tolerance (Matsunaga et al., 2000).

Generally Recognized as Safe (GRAS) status from both the U.S. Department of Agriculture (USDA) and the Food and Drug Administration (FDA) for IgY is being tried to obtain. Approval of individual products by the FDA for using egg antibodies in human patients is relatively easy. Since the activity of IgY was well preserved and easy to apply for human patients, some companies (e.g., EW Nutrition Japan) (Table 27.4) have started to develop various food products with this IgY like tablets, lozenge, candy, pastilles, sachets, yoghurt (regular, drinking) and baby milk etc. (Table 27.5) (Thu et al., 2017). This would be easier to handle, both for the patients and for the pharmacy or hospital, due to the ease of storing IgY at room temperature.

27.5 Production of IgY

The major choice of animal for IgY production is the avian species chicken. IgY production steps include (1) antigen of interest; (2) antigen + adjuvant mixture, emulsion; (3) immunization; (4) immunization routes; (4) hyperimmunized egg collection; (5) egg breaking; (6) pasteurization of egg liquid; (7) spray-drying; and (8) IgY powder production. IgY production results from immunization, but its production and immunization are not very predictable. Parameters that may influence the immune response of the immunized chicken include the antigen nature and dose, the type of adjuvant used, the route of administration, characteristics of chicken (such as keeping conditions, age, breed, effect on egg laying capacity), and overall immunization schedule. Different type of antigens is being used to produce IgY, such as nucleic acid (Bachtiar et al., 2016), protein (Lee et al., 2016), lipid, and carbohydrates (Zhen et al., 2011). In addition, an immune humoral response must be elicited by immunization with recombinant proteins (Nasiri et al., 2016) or peptides (Hodek et al., 2015). Both, complex antigens, e.g., whole viruses, bacteria, and parasites (Grando et al., 2017; Amro et al., 2018; Lopes et al., 2019; de Faria et al., 2019; Silva et al., 2020) and individual biomolecules, e.g., large proteins (Skottrup et al., 2019; Lu et al., 2020), or small peptides conjugated to a suitable carrier protein, such as keyhole limpet hemocyanin (KLH) (Grzywa et al., 2014; Łupicka-Słowik et al., 2014), have been used to stimulate the

TABLE 27.4
Summary of IgY Purification

Step	Method	Reference
1. De-lipidation		Schade et al. (2005); Pereira et al. (2019)
	a) Freeze and thaw at neutral pH	Jensenius et al. (1981)
	b) Water dilution method	Akita and Nakai (1993); Losonczy et al. (1999); IGY Life Sciences (2021)
	c) Organic solvents (chloroform, acetone, isopropanol)	Bade and Stegemann (1984); Chung and Ferrier (1991); Bizhanov and Vyshniauskis (2000)
	d) Organic acids (caprylic acid, trichloroacetic acid)	Araújo et al. (2010)
	e) Natural gums (polyanionic polysaccharides, e.g., xanthans)	Hatta et al. (1990)
	f) Supercritical carbon dioxide extraction (SCE) delipidation.	Hiidenhovi et al. (2005); Aro et al. (2009)
	g) Polysaccharides (e.g., pectin, λ-carrageenan, carboxymethylcellulose, methylcellulose, dextran sulfate)	Chang et.al. (2000); Tong et al. (2015)
2. Precipitation	a) 3.5% polyethylene glycol (PEG)	Polson et al. (1980, 1985); Svendsen et al. (1995); Meulenaer et al. (2001)
	b) Caprylic acid followed by ultrafiltration at pH 9.0.	Redwan et al. (2021)
	c) Dextran sulphate and calcium chloride	Jensenius et al. (1981)
	d) Phosphotungstic acid and magnesium chloride	Vieira et al. (1984)
	e) Saturated sodium sulphate	Jensenius et al. (1981)
	f) Saturated ammonium sulphate	Ambrosius and Hadge (1987); Garvey et al. (1977)
	g) Precipitation using 12% PEG	Polson et al. (1980); Polson et al. (1985)
3. Purification		
A. Chromatographic Method	a) Size-exclusion chromatography (Sephadex 100G)	Pour Amir & Rasaei (2001)
	b) Low-pressure chromatography	Amro et al. (2018)
	c) Ion exchange (Anion, DEAE-Sephacel) and Cation (CM-Cellulose) chromatography	Akita and Nakai (1993); Ko and Ahn (2007)
	d) Higher solution chromatography through multicolumn systems	Constantin et al. (2020)
	e) Hydrophobic interaction chromatography	Hassl and Aspöck (1988)
	f) Thiophylic interaction chromatography	Hansen et al. (1998)
	g) Gel-filtration chromatography	Pour Amir & Rasaei (2001)
	h) Affinity chromatography	Verdoliva et al. (2000); Behn et al. (2001); Chen et al. (2002); Constantinoiu et al. (2007); Dong et al. (2008); Sui et al. (2011); Jiang et al. (2016a)
B. Filtration	Gel filtration, funnel filtration, and ultrafiltration	Akita & Nakai (1993); Kim & Nakai (1998); Meulenaer et al. (2001); Hernández-Campos et al. (2010)

development of specific IgYs in hens. The antigen dose may be critical since too much or too little antigen can lead to an undesirable immune response (Schade et al., 2001).

27.5.1 Processing of Hyperimmunized Eggs into Egg (IgY) Products

Eggs processing can start with egg washing. Eggs are then individually broken with machines making also white and yolk separation, as well as eggshell removal. Immediately after breaking and separation, yolk and whole egg are filtered and cooled before transfer into pasteurization tanks. After that, these two fractions can be pasteurized before packaging, to obtain liquid egg products, but they can also be spray-dried for marketing as powders. In that case, yolk and whole egg are pasteurized before spray-drying described. Snapshots of different processing steps of hyperimmunized eggs into egg (IgY) products are shown in Figure 27.9.

Different drying technologies have been applicable for producing a whole egg, egg yolk, and egg white powder products. Some of them like foam-mat drying (Kudra & Ratti, 2006; Thirupathi et al., 2008), pan drying, belt drying, and spray drying (Patel et al., 2009; Nandiyanto & Okuyama, 2011; Anandharamakrishnan & Ishwarya, 2015), freeze-drying (Jaekel et al., 2008), and hot air (oven) (Ratti, 2001) was applied within their specific criteria and their limitations.

Spray drying is a unit operation technology widely used in foods (Loh et al., 2005), pharmaceuticals, encapsulation,

TABLE 27.5
IgY Administration Route and Dosage Form for Animal and Human Health Application

Category	Administration Route	
Animal:	**Enteral (Oral) Dosage form**	**Topical (Local) Dosage form**
Feed	Whole egg powder	
Additive	Top dressing	
	Pellet	
Premix	Oil-coated pellet	Lotions
	Tablet	Eye drops
	Enteric tablet	Nasal drops
	Kibble	Spray
	Dental biscuit	Toothpaste
	Dental rope	Wet tissue/paper
	Dental gum	Dental soluble film
	Paste in bottle	Inhalers
	Paste in syringe	
Human:	Egg yolk powder	
	Sachet	Lotions
Food	Troche	Eye drops
	Tablet/coated tablet	
	Candy	
	Lozenges	
	Lollipop	Nasal drops
	Capsule	
Supplement	Chewing gum	Spray
	Oral saline sachet	Toothpaste
	Baby milk powder	Mouth wash
	Drinking yoghurt	Wet tissue/paper
	Regular yoghurt	Dental soluble film
	Shell egg	Face mask
		Air filter
		Inhalers
		Bath salt (nyokuzai)

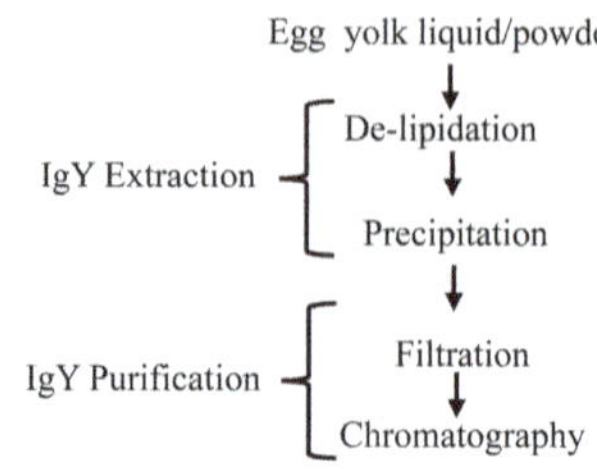

FIGURE 27.10 Flow diagram of IgY purification steps.

plastic resins, paint pigments, ceramic materials, and catalyst supports, for use with microalgae and the chemical industry. The various flow types manufactured in spray dryers include (1) co-current flow dryer, (2) counter-current flow dryer, (3) mixed flow dryer, (4) open cycle dryer, (5) closed cycle dryer, (6) semi-closed cycle dryer, (7) single stage dryer, (8) two-stage dryer, (9) vertical dryer, and (10) horizontal dryer. Most vertical chambers are cylindrical, ending with an inverted cone on their base. (Cal & Sollohub, 2010).

27.6 Purification of IgY

Purification of IgY (Figure 27.10) mainly consists of 2 steps: IgY extraction and purification. The major components of egg yolk are lipids (32%), proteins (16%), and water (50%). About one-third of the lipids are phospholipids, mainly phosphatidylcholine (about 80%) (Figure 27.11). The high concentration of lipids and lipoproteins is one of the major barriers to IgY extraction from egg yolk (Verdoliva et al., 2000, Schade et al., 2005). Therefore, in most cases, IgY isolation involves the removal of lipids ("de-lipidation" step). Various IgY extraction methods were reviewed in detail by De Meulenaer & Huyghebaert (Hatta et al., 1990; Akita & Nakai, 1992; Akita & Nakai, 1994; Fischer et al., 1996; De Meulenaer & Huyghebaert 2001). These methods can be divided into two steps: de-lipidation and precipitation, followed by purification.

IgY purification is summarized in Table 27.4.

It is reported that a combination of the methods mentioned above, e.g., a combination of PEG precipitation with affinity

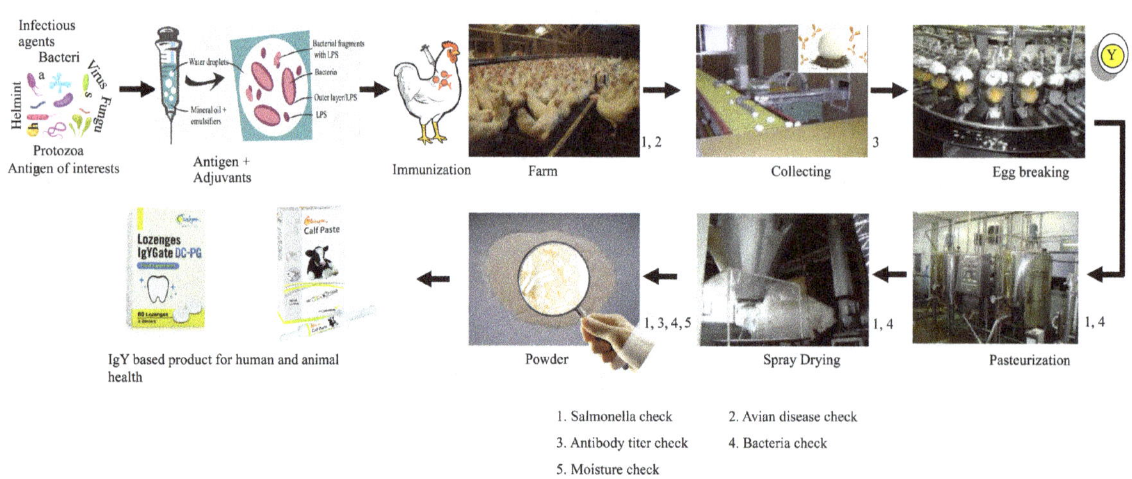

FIGURE 27.9 Snapshots of processing steps of hyperimmunized eggs into IgY products.

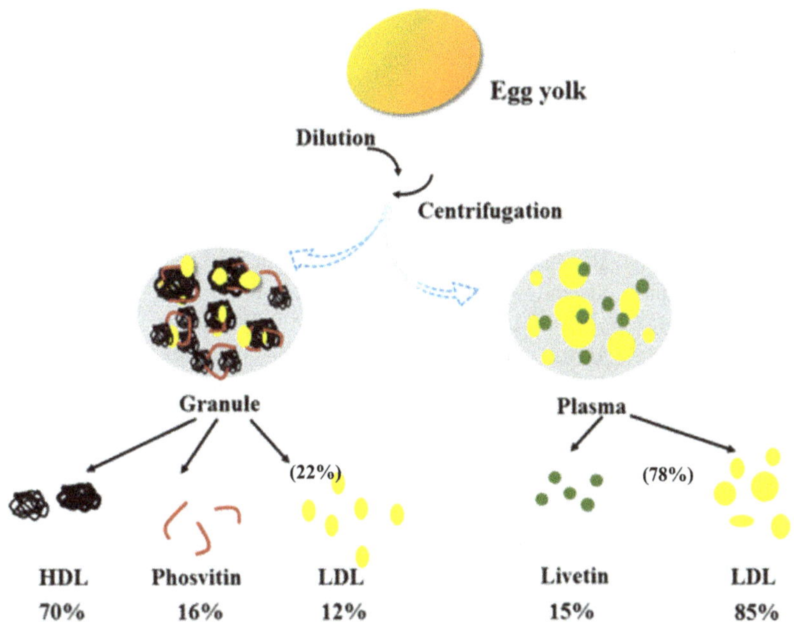

FIGURE 27.11 Constituents of plasma and granules from hen egg yolk (LDL, low-density lipoprotein; HDL, high-density lipoprotein). (Adapted from Xiao et al., 2020.)

chromatography (Pauly et al., 2011) or ammonium sulfate precipitation with ion exchange chromatography (Ko & Ahn, 2007), can further increase the purity of the IgY preparation. Moreover, sequential precipitation with 31% ammonium sulfate and 12% PEG resulted in IgY antibodies of more than 95% purity without any loss in immunoreactivity (Constantin et al., 2020). Despite the numerous protocols described in the literature, the most popular isolation strategy of IgYs from immune eggs involves a de-lipidation step, in which IgY is extracted in the supernatant after treating the egg yolk with 10 volumes of acidic water and a subsequent precipitation step, in which IgY precipitates with ammonium sulfate or PEG, at suitable concentrations (Amro et al., 2018). Purity level is checked by SDS-PAGE and Western blotting (Redwan et al., 2021). Various types of commercial IgY extraction kits are available in the market. Trouble-shooting advice and protocols for other IgY purification methods can be found in the IgY manual (Staak et al., 2000).

In conclusion, IgY extraction methods require an initial delipidation step before extraction of insoluble lipids and lipoproteins. The water dilution method was most efficient for extracting IgY from the water-soluble protein fraction. Further purification of IgY after crude extraction can be achieved by selective precipitation. The precipitation with sodium or ammonium sulfate, sodium chloride as well as polyethylene glycol offers a cheap and easy methodology and can be used in laboratory practice. Column chromatography and ultrafiltration are expensive and impractical for the large-scale production of antibodies.

Large-scale production is continuously improving to achieve a sustainable turnover for industries. Bioactive peptides of egg components like lysozyme, avidin, and IgY antibodies are currently under industrial production using standardized purification protocols, is feasible with different techniques involving supercritical fluid technology, magnetic particles for separation of specific egg compounds by immune magnetic separation (Huopalahti et al., 2007), and enzymatic hydrolysis of egg white albumin and yolk followed by ultrafiltration to obtain low-molecular-weight peptides combined with chromatographic techniques (Nimalaratne & Wu, 2015). The choice of a suitable method depends on the yield and purity desired, final application of the IgY as well as material cost, technology, labor skills, and scale of extraction.

27.7 IgY Administration Route and Dosage Form

IgY has been extensively used in human and veterinary health (Rahman et al., 2013). But the IgY activities are in question when used in different forms. IgY is a protein in nature, may be structural alterations happened due to the digestive enzyme activity and gastrointestinal acidity. At the same time, they pass through the gastrointestinal tract and their biological activity is at risk. To overcome these situations of IgY activity and to improve IgY delivery as well, different types of coating materials (e.g., chitosan-alginate microcapsules, hydrogels containing acrylamide and acrylic, acid methacrylic acid copolymers, liposomes, polymeric microspheres and multiple emulsification) and conventional dosage forms have been developed for different gastrointestinal and topical administrations (Table 27.5). These above coating materials are commonly used to encapsulate and thus protect IgY from digestive degradation (Holser, 2013; Paul et al., 2014; Soumaila Garba et al., 2019). The study of IgY delivery and dosage form design is still at an early stage of development. Some delivery methods reported such as microneedle-based transdermal, transdermal patch, inhalation, aerosol, film-based buccal targeting on oral mucosal surface, nasal sprays, and self-administration allowed subcutaneous delivery, such as NanoPass MicronJet technology and iontophoresis technology (Anselmo et al., 2019).

27.8 Chicken Monoclonal IgY Antibodies

Monoclonal antibodies are clones of just one antibody, and they bind to one antigen only. Polyclonal antibodies come from several different types of immune cells and will bind to more than one antigen. Monoclonal IgY (mIgY) is a new antibody development that combines the advantages of IgY and Monoclonal antibodies (mAbs). Table 27.6 shows the difference between monoclonal and polyclonal antibodies.

Different types of functional chicken antibodies are reported with variable degrees of immunogenicity. As chicken antibodies are immunogenic in mammals, their clinical applications are limited (Tateishi et al., 2008). In particular, the constant region contributes a significant component to immunogenicity (Morrison et al., 1984). To overcome this problem, genetic engineering techniques have been applied to produce chimeric (Andris-Widhopf et al., 2000; Nishibori et al., 2004), humanized (Jones et al., 1986; Foote & Winter, 1992; Studnicka et al., 1994), and recombinant antibody (Bird et al., 1988; Nakamura et al., 2000) formats of chicken antibodies.

Methods for monoclonal IgY production include (1) Hybridoma Technology (Pink, 1986; Nishinaka et al., 1989; Nishinaka et al., 1991; Nishinaka et al., 1996; Matsushita et al., 1998; Matsuda et al., 1999; Nakamura et al., 2000); (2) DT40 Cell Line, a chicken lymphoma cell line (Cumbers et al., 2002; Seo et al., 2005; Seo et al., 2006); and (3) Antibody Display Technologies: Phage Display (Smith, 1985; Andris-Widhopf et al., 2000), Yeast Surface Display (Boder & Wittrup, 1997; Bogen et al., 2020), and Ribosomal Display (Hanes & Plückthun, 1997; Hanes et al., 1998; He & Taussig, 2007; Qi et al., 2009; Plückthun, 2012).

The hybridoma technology is difficult in the chicken system due to the lack of a robust fusion partner. The simplest, easier, and more efficient way of mIgY production by using phage display library selection for scFv gene constructs with higher affinity against targeted pathogens (Kyung et al., 2005; Finlay et al., 2006). mIgY can be developed into small molecular weight antibody fragments (particularly scFv), chimeric, and humanized antibodies to address the many challenges ahead to developing mAb for various biomedical applications, especially those for which conventional antibodies are ineffective. Table 27.7 shows the overview of mIgY preparations using different antibody engineering technology and selection methods. mIgY has potential applications for immunological detection and diagnosis, for screening and validating biomarkers and drug targets, and to produce antibodies against conserved

TABLE 27.6

Difference between Polyclonal and Monoclonal Antibodies

Parameter	Monoclonal Antibodies	Polyclonal Antibodies
Animal	Rat, mouse, chicken, rabbit, human, etc.	Rabbit, guinea pig, goat, sheep, rat, mouse, chicken, etc.
Form	Hybridoma	Antiserum
Production:		
- Time	Long (up to a year)	Short (3–4 months)
- Cost	Moderate	Low
- Ease	Difficult	Very Easy
Class, subclass	Single Class	Mixed Classes
Epitope	React to a single epitope	React to multiple epitopes
Targets	Immunogenic targets	Immunogenic targets
Binding site	Single specific site	Different areas of target molcules
Specificity	High if good quality antibodies are selected	Moderate. Lower than monoclonal antibodies because multiple types of antibodies are present
Sensitivity	Moderate to high	Variable
Engineering	Only after converting to rAb	Not possible
Antibody nature	Single antibody species	Mixed population of antibodies
Reproducibility	The same antibodies are produced indefinitely	Variable among lots
Stability	Moderate. Binding ability may be lost if the epitope is lost by fixation/denaturation of the antigen, because monoclonal antibodies are homogeneous.	High. Binding ability tends to be unaffected by fixation/ denaturation of the antigen, because multiple different antibody molecules are present.
	Tend to be sensitive to modifications, such as labeling and removal of the Fc region	Tolerate modifications, such as labeling and removal of the Fc region
Tolerance	May only recognize a particular protein form (phosphorylation, dimersied) Infinitely renewable	Tolerant of small changes in protein structure (denaturation, dimerization, phospholyration)
Reproducibility	Virtually reproducible	Limited
Availability	Commercially available	Commercially available

Source: Modified and adapted from Frenzel et al. (2013).

TABLE 27.7

Overview of mIgY Preparations Using Different Antibody Engineering Technology and/or Selection Methods

mIgY Type	Antigen	Use	Eng./Sel.Method	Reference
Chicken/ human chimeric fab	Recombinant hemagglutinin protein	H5N1 diagnostic testing	Phage display	Pitaksajjakul et al. (2010)
scFv	Gentamicin	Antibiotic detection	Phage display	Li et al. (2016a)
scFv	Salbutamol	Sensitive assay (detection)	Phage display	Lee et al. (2018a)
scFv	Inactivated cobra venom proteins	Diagnosis of snakebites and antibody affinity test	Phage display	Lee et al. (2018b)
scFv	Epidermal growth factor receptor (EGFR)	Detection	Yeast surface display	Bogen et al. (2020)

Source: Adapted from Leow et al. (2021).

mammalian proteins. In addition, the production of antibody fragments and chimeric antibodies make mIgY more widely clinical applicable. We believe that mIgY have enormous prospects in the development of new antibody platforms. New technologies and new types of antibodies are likely to broaden the application of antibodies. As a molecular-targeted drug, monoclonal antibodies that could not be obtained in mammals rather it will be obtained by utilizing the antibody production of chickens of different species. We hope humanized antibody drugs based on monoclonal IgY antibodies will be developed. In the future, chicken mIgY's are likely to play a more important role in disease diagnosis and treatment, and in basic and applied antibody research (Table 27.8).

27.9 Production of IgY in Transgenic Chickens

Transgenic animal refers to an animal in which there has been a deliberate modification of the genome - the material responsible for inherited characteristics - in contrast to spontaneous mutation. Genetically modified animals have significantly contributed to our understanding of different aspects related to immunity, infectious diseases, neurology, behavior, and developmental biology (Yeh et al., 2002; Lyall et al., 2011; Lalonde et al., 2012; Pinkert, 2014; Park et al., 2017). While mice were the first animals to be genetically modified (Costantini & Lacy, 1981; Gordon & Ruddle, 1981). Significant progress was made in generating recombinant proteins, including mAbs for therapeutic applications, in genetically modified chickens over the last decades (Bahrami et al., 2020; Park et al., 2020). Schematic representation of the original immunoglobulin Y technology, Monoclonal IgY and IgY in Transgenic chickens and are also shown (periphery, left and right) in Figure 27.12 (Karachaliou et al., 2021).

TABLE 27.8

Examples of Monoclonal Antibody Treatments Approved by the Food and Drug Administration

Name	Target	mAb Type	Approval Date
Butuximab	CD30	Chimeric	2011.08.19
Dituximab	GD2	Chimeric	2015.03.10
Basiliximab	CD25	Chimeric	1999.03.09
Patoizumab	HER-2	Humanized	2012.06.08
Otozumab	CD20	Humanized	2013.11.01
Perizumab	PD-1	Humanized	2014.09.04
Daremumab	CD38	Humanized	2015.11.06
Elotuzumab	SLAMF7	Humanized	2015.11.30
Nesimumab	EGFR	Recombinant human	2015.11.24
Olarezumab	PDGFR-α	Human	2016.10.19
Avelumab	PD-L1	Human	2017.03.23

Source: Adapted from Leow et al. (2021).

To successfully generate genetically modified chickens, the following methods have been used such as (a) development of in vitro culture system for chickens, (b) utilization of primordial germ cells and genome editing in chickens, (c) creation of chicken embryonic stem (ES) cells and their utilization, and (d) search for leukemia inhibitory factor (LIF) that maintains totipotency of chicken ES cells.

The first genetically modified chicken was generated by the insertion of retroviral foreign DNA delivered by the avian leukosis virus that was successfully integrated to the germline (Salter et al., 1987). Since then, various viral vectors have been used to generate transgenic chickens to produce recombinant proteins (Salter et al., 1987; Harvey et al., 2002; Kamihira et al., 2005), including mAbs (Kamihira et al., 2009).

One of the most effective approaches to producing transgenic chickens is the in vitro transfection of avian cell lines, such as primordial germ cells (PGCs) and embryonic stem cells (ES), the clonal selection and reinsertion into the embryo leading to fully transgenic progeny in the next generation (Zhu et al., 2005; van de Lavoir et al., 2006; Kim et al., 2018).

The OmniChicken by Ligand Pharmaceuticals Inc. is a worldwide unique platform to produce human monoclonal antibodies from chickens using the phylogenetic difference between mammals and birds. A study conducted by Oishi and colleagues demonstrated the ability to integrate human interferon beta (hIFN-β) into the chicken ovalbumin locus in order to produce hIFN-β in egg white (Oishi et al., 2018). Authors demonstrated the ability to produce foreign proteins in eggs, which would have industrial and therapeutic applications.

Several advantages are provided by newly invented gene editing technologies, including the simplicity of design and application combined with high efficiency (Chira et al., 2017). Understanding the host cell behavior during host-pathogen interactions may help target pathogen-specific receptors and viral cellular transport (Heaton et al., 2016). Determining new target genes associated with disease susceptibility should fill the research gap and open the door for

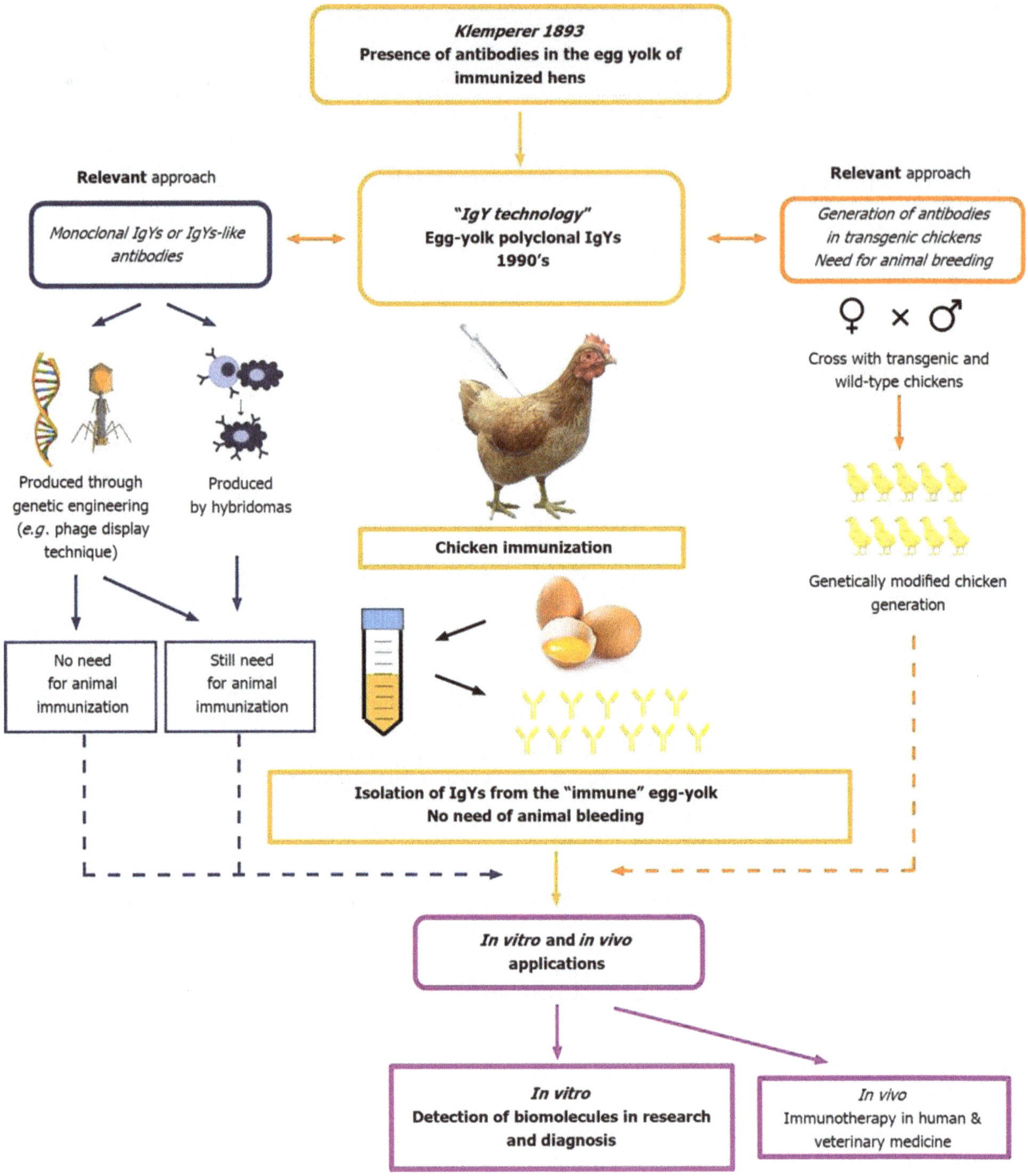

FIGURE 27.12 Schematic representation of the main parts comprising the original immunoglobulin Y technology (central axis); promising relevant approaches are also shown (periphery, left and right). (Adapted from Karachaliou et al., 2021.)

new therapeutical approaches. Although the debate about using genetically modified animals in food production will continue to be stimulated, we may obtain new breeds of chickens in the future that are resistant for specific pathogens. We speculate that spending more efforts connecting gene editing technologies with the prevention of infectious diseases will change the way we use to fight pathogens and will probably improve the animal welfare.

27.10 Use of IgY in Immunoassay and Diagnosis

IgY-based immunoassay is a widespread field to diagnosis biomolecules of interest in biological specimen and infectious diseases pathogen in humans, animals, aquaculture and plants. Enzyme-linked immunosorbent assay (ELISA), western blotting (WB), immunofluorescence (IF), and immunohistochemistry

(IHC) methods are the most common applications for IgY (Behn et al., 2001). Relevant to consider that antigenic target recognition by using antibodies can show some differences according to the method used. Regarding the conformational epitope structure, the performance of antibodies recognizing a given epitope by WB can represent nothing about the same antibody performance in ELISA with the same antigen (Uhlen et al., 2016). IgY antibody shows advantages compared to the mammal IgG use in immunoassay. In this context, the rheumatoid factor (RF) is a major source of interference in many immunoassays using mammalian polyclonal or monoclonal antibodies by interacting with IgG and causing false-positive results. IgY antibodies do not react with RF avoiding false-positive results in ELISA (Larsson et al., 1991). Because of this advantage conferred by IgY's structure, the use of IgY antibodies in immunoassays may result in less background noise, fewer false positives, and decreased aggregation of antibodies, which are common issues observed with both monoclonal and polyclonal mammalian antibodies. Furthermore, IgY antibodies have also shown high binding specificity and low cross-reactivity with other antigens comparable to current industry standards and may have value in a variety of applications to detect pathogens.

27.10.1 IgY in Immunoassay

Use of IgY has been extensively studied in various immunoassay such as:

a. ELSIA (Enzyme-linked immunosorbent assay): IgY-based ELISA does not demand expressive modifications from the most common procedures and reagents used to carry out this protocol (Ferreira Júnior et al., 2012; da Silva et al., 2016). Egg yolk IgY can be labeled with horseradish peroxidase for use in immunoenzymatic assays (Ruan et al., 2007). Some possibilities for application of yolk antibodies-based ELISA are: screening molecules related to drug-drug interactions (Jiang et al., 2016b); sandwich ELISA to capture bacterial toxins (Nagaraj et al., 2016); screening animal diseases by using recombinant antigens (Zhang et al., 2016b); characterization of maternal IgY transferred to egg yolk (Murai et al., 2016); coproantigen to capture ELISA for intestine human parasite (Teimoori et al., 2016); screening antibiotic residues in food samples (He et al., 2016); detection of blood circulating helminth antigens by immunomagnetic bead ELISA (Nie et al., 2014); potential for clinical application diagnosing cancer antigen (Grzywa et al., 2014);
b. Immunofluorescence assay: Specific IgY was used in an indirect immunofluorescence assay to detect microbial antigens in biological specimens (Camenisch et al., 1999; Cipoll al., 2001; Sesarman et al., 2008; Shin et al., 2009; Bentes et al., 2015);
c. Flow cytometry assay: (Santoro et al., 2004). IgY antibodies are suitable reagents for flow cytometry and as well as certain monoclonal antibodies, for example, to study human and rabbit platelet physiology (Santoro et al., 2004). When using phage display-based single chain variable fragments (scFvs), polyclonal IgY production and further the flow cytometry assays it is possible to develop immunoreagents for the isolation and characterization of stem cells, molecular diagnostics and therapeutics of lung cancers (Leu et al., 2010; Bowes et al., 2011);
d. Immunochromatographic assay: The immunochromatographic assay (ICA) requires no instruments and has a detection time of less than 10 min and it is portable and easy to perform in the field. The development of IgY-based strip could be a promising on-site tool for screening infection or disease outbreaks. IgY-gold complexes depositing onto the conjugate pad as detector reagents showed high specificity (He et al., 2015; Zhang et al., 2015a);
e. In diagnostic and analytical applications: IgY antibodies serve as essential components in a variety of diagnostic assays used for the qualitative and quantitative determination of a wide range of substances (Cakir-Koc et al., 2020; Kota et al., 2020; Porte et al., 2020); and
f. As an immunoaffinity ligand and Proteomic: Separation of complex protein mixtures that have a wide dynamic range of concentration, such as plasma or serum, is a challenge for proteomic analysis. Sample preparation to remove high-abundant proteins is essential for proteomics analysis. Immunoglobulin yolk (IgY) antibodies have unique and advantageous features that enable specific protein removal to aid in the detection of low-abundant proteins and biomarker discovery. (Huang et al., 2005; Fang & Zhang, 2008; Nilsson et al., 2008; Qian et al., 2008; Rajic et al., 2009; Polaskova et al., 2010; Magagnotti et al., 2010; Manhani et al., 2011; Kovacs-Nolan & Mine, 2011).

27.10.2 IgY in Diagnosis

a. Viral infections diagnosis: In the detection of canine parvovirus (He et al., 2015), bovine viral diarrhea virus (BVDV) (Zhang et al., 2016), *rhesus* monkeys hepatitis A virus (HAV) (Silva et al., 2012), soft-shelled turtle systemic septicemia spherical virus (STSSSV) (Zhang et al., 2015b), acute respiratory syndrome associated to coronavirus (SARS-CoV) (Palaniyappan et al., 2012), dengue virus 2 (DENV2) (Figueiredo et al., 2015).
b. Bacterial infections diagnosis: In the diagnosis of *Staphylococcus aureus* (Richman et al., 1982; Walczak et al., 2015), staphylococcal enterotoxin A (SEA), B (SEB) C (SEC), toxic shock syndrome toxin (TSST) and α-hemolysin of *S. aureus* (Reddy et al., 2013; Mudili et al., 2015) methicillin resistant *S. aureus* (MRSA) strains (Yamada et al., 2013).
c. Parasitic infections diagnosis: in detecting the protozoan *Toxoplasma gondii* (Cakir-Koc, 2016; Sert et al., 2017), the helminth *Opisthorchirs viverrini, Taenia* spp., *Echinostoma* spp. and *Minute Intestinal Fluke* (MIF) (Teimoori et al., 2017). Miura et al.

produced IgY against the GP60 protein, from the *Cryptosporidium hominis* protozoan. These IgY bound to GP60 in Western blot and to *C. parvum* sporozoites in fecal samples by indirect immunofluorescence, suggesting that anti-GP60 IgY could be used in the diagnosis of cryptosporidiosis caused by both *C. hominis* and *C. parvum* (Miura et al., 2017).

d. Tumors diagnosis: IgY against the peptide antigen CA 15-3, a commonly used breast cancer marker, was used as secondary antibody in a sandwich ELISA aiming to detect CA 15-3, showing potential for clinical use (Grzywa et al., 2014). Sun et al. produced IgY against two portions of the transmembrane glycoprotein HER2, is promising for use in breast cancer diagnosis (Sun et al., 2015). In another study, Łupicka-Słowik et al. developed a direct ELISA test using a lysate of human malignant tumor cells and IgY against bovine adenosine deaminase (c-ADA) for the detection and quantification of ADA for pleural tuberculosis diagnosis (Łupicka-Słowik et al., 2018). IgY developed to detect prostatic specific antigen (PSA) and two peptide fragments of this protein demonstrated specificity (Łupicka-Słowik et al., 2014).

e. Hematological tests: Hens immunized with umbilical cord sera produced specific IgY against IgG and the complement fractions C3b and C3d. These antibodies did not react with the C4b fraction — which configures a higher specificity, since anti-C4b antibodies often cross-reacts with the antigens of Chido/Rodgers RBC group — nor with erythrocyte antigens from ABO group. These antibodies are, therefore, promising as a reagent for Coombs test (Calzado et al., 2017).

f. Enzyme detection: IgY immunoassays were used to detect the hepatic expression of Cytochrome P450 2E1 (CYP2E1) in mice treated with medicinal herbs and products derived from plants rich in flavonoids. Anti-CYP2E1 IgY was specific, without reacting with other cytochromes, and was able to detect the reduction of hepatic CYPE2E1 expression due to the ingestion of natural products with hepatoprotective effects (Jiang et al., 2016b).

g. Harmful substances identification: The ability of IgY in identifying harmful substances in consumer products has been evaluated. An ELISA test was developed to detect the staphylococcal enterotoxin G (SEG), using specific IgY, and showed satisfactory sensitivity and specificity, reducing the interference of protein A that occurs in IgG tests. This test was successfully used to detect SEG in milk and dairy products samples and could therefore be used to identify the toxin in food (Nagaraj et al., 2016). Bittner et al. used IgY in ELISA to detect potentially allergenic proteins in commercially available latex gloves. This assay presented similar results to that of the gold standard test, which uses mammalian IgG (Bittner et al., 2016).

h. Antibiotic residues identification: IgY can also be used in assays to identify antibiotic residues in food products of animal origin, as demonstrated in a study by He et al., in which produced anti-gentamycin IgY specifically detected the target antibiotic present in animal origin products (He et al., 2016). Following this rationale, Li et al. used specific IgY to detect kanamycin and gentamycin residues in milk and meat samples by means of competition ELISA and FPIA (fluorescence polarization immunoassay) (Li et al., 2017).

i. Toxin substance identification: The potential of IgY in identifying substances has also been used to evaluate the toxicity of a natural product employed in the alternative medicine. IgY labeled with Quantum dot were successfully used in a lateral flow assay for the detection of rhein; a toxic substance found in the plant *Rheum officinale*, widely used in Chinese traditional medicine; in plant samples and serum from users (Zhang et al., 2018).

In general, these findings suggest that the phylogenetic distance between birds and mammals, that ensures a stronger immune response against mammalian antigens by birds than by other mammals (Gassmann et al., 1990), makes the production of IgY against various substances, advantageous for usage in several types of immunoassays and diagnoses. Thus development of polyclonal or monoclonal IgY based immunoassays could be a promising alternative to mammalian antibodies. Use of IgYs in Immuno-assay and diagnosis has been summarized in Table 27.9.

27.11 Use of IgY in Animal Health

Oral administration of chicken egg yolk antibody (IgY) is a promising nutritional strategy to control pathogen infections. Oral passive immunization using IgY has been focused on as an alternative to antibiotics for the treatment and control of diseases in animals. This section was focused to determine the effect of IgY in controlling and preventing various diseases in domesticated animals, including Bovine, Swine, Poultry, and Pets, (Wani et al., 2022) which have been studied during the last three decades (1980~2022) and are summarized in Table 27.10.

27.11.1 Use of IgY in Bovine

27.11.1.1 Bovine Mastitis

Worldwide bovine mastitis continues to be a costly disease for the dairy industry (Zhen et al., 2008a; Zhen et al., 2008b; Zhen et al., 2009). Numerous pathogens can cause mastitis and these can be classified into contagious pathogens (primary Staphylococcus aureus and Streptococcus agalactiae) or environmental pathogens (primary *E. coli*) (Riffon et al., 2001). *S. aureus* is implicated in ruminant mastitis, which affects the quality and quantity of milk produced, as well as the health of infected animals (Baselga et al., 1994). Although there are antibiotic treatments and vaccines available, they are not fully effective and *S. aureus*-associated

TABLE 27.9

Use of IgYs in Immuno-Assay and Diagnosis

Category of Use	IgY Using Area Covered	Reference
A. Immuno-assay		
ELISA	IgY-based ELISA modifications from the most common procedures and reagents	da Silva et al. (2016)
	Labeled IgY with horseradish peroxidase for use in immunoenzymatic assays	Ruan et al. (2007)
	Screening molecules related to drug-drug interactions	Jiang et al. (2016b)
	Sandwich ELISA to capture bacterial toxins	Nagaraj et al. (2016)
	Screening animal diseases by using recombinant antigens	Zhang et al. (2016b)
	Characterization of maternal IgY transferred to egg yolk	Murai et al. (2016)
	Coproantigen to capture ELISA for intestine human parasite	Teimoori et al. (2016, 2017)
	Screening antibiotic residues in food samples	He et al. (2016)
	Detection of blood circulating helminth antigens by immunomagnetic bead ELISA	Nie et al. (2014)
	Clinical diagnosing cancer antigen	Grzywa et al. (2014)
	Evaluation of ADA as a cancer biomarker	Łupicka-Słowik et al. (2018)
	Diagnosis of human strongyloidiasis	de Faria et al. (2019)
	Detection of neurocysticercosis	Silva et al. (2020)
	Diagnosis of Hookworm infection	Souza et al. (2020)
	Diagnosis of human ascariasis	Lopes et al. (2019)
	Diagnosis of human prostate cancer	Łupicka-Słowik et al. (2019)
	Detection of *Fusarium verticillioides* (and prediction of fumonisin contamination) in poultry feed	Omori et al. (2019)
	Evaluation of karilysin (i.e., an enzyme secreted by the periontopathogen *Tannerella forsythia*) as a biomarker for the diagnosis of periodontitis	Skottrup et al. (2019)
ELISA, FPIA	Detection of veterinary drug residues (SMZ) in milk	Liang et al. (2018)
Immunofluorescence	Detection of antigenic targets on cells and tissues samples	Camenisch et al. (1999)
	To clarify the pathogeny of mammalian autoimmune diseases	Sesarman et al. (2008)
	To detect mycobacterium avium subsp. paratuberculosis pathogen inside the cytoplasm of infected macrophages	Shin et al. (2009)
	Detection of hepatitis A virus in frozen liver sections	Bentes et al. (2015)
Fluorescence sensor assay	Diagnosis of hand-foot-and-mouth disease caused by enterovirus 71 infection	Nie et al. (2019)
Fluorescence immunochromatographic rapid-antigen test	Diagnosis of COVID-19	Porte et al. (2020)
In vitro immunochemicaltechniques	Diagnosis of infection with *Salmonella typhimurium* and *Salmonella enteritidis*	Esmailnejad et al. (2019)
Tissue indirect immunofluorescence assay	Diagnosis of human ascariasis	Lopes et al. (2019)
Immunocapture PCR assay	Detection of *Staphylococcus aureus* in food samples, skin and nasal swabs	Kota et al. (2020)
Flow cytometry	To study human and rabbit platelet physiology	Santoro et al. (2004)
	For the isolation and characterization of stem cell	Leu et al. (2010)
	Molecular diagnostics and therapeutics of lung cancers	Bowes et al. (2011)
Immunochromatographic assay (ICA)	IC strip to detect swimming crab Portunus trituberculatus reovirus	Zhang et al. (2015a)
	For canine parvovirus detection	He et al. (2015)
Immunocytochemistry, Immunohistochemistry	Diagnosis of infection with influenza A virus	da Silva et al. (2018)
Lateral flow immunoassay	Detection of fumonisin B1 and fumonisin B2 in maize	Tran et al. (2019)
Immuno-dot blot assay (with the use of IgY-colloidal gold nanoparticles conjugates)	Detection of indoor dust mite allergens	Egea et al. (2019)

(Continued)

TABLE 27.9 *(Continued)*

Use of IgYs in Immuno-Assay and Diagnosis

Category of Use	IgY Using Area Covered	Reference
Paper-based microfluidic immunochromatographic test	Differential diagnosis of Russell's viper envenomation	Lin et al. (2020)
Latex agglutination assay	Diagnosis of Toxoplasmosis	Cakir-Koc et al. (2020)
Latex agglutination assay	Diagnosis of infection with influenza A virus	Budama-Kilinc et al. (2018)
B. Diagnosis		
Viral infections diagnosis	Canine parvovirus detection	He et al. (2015)
	Bovine viral diarrhea virus (BVDV)	Zhang et al. (2016b)
	Hepatitis A virus (HAV)	Silva et al. (2012)
	Detection of hepatitis A virus in frozen liver sections	Bentes et al. (2015)
	Soft-shelled turtle systemic septicemia spherical virus (STSSSV) in serum and feces samples of infected turtles	Zhang et al. (2015a)
	Nucleocapsid protein (NP) of coronavirus (CoV)	Palaniyappan et al. (2012)
	Severe acute respiratory syndrome-associated coronavirus (SARS-CoV)	
	Non-structural protein 1 (NS1) of dengue virus 2 (DENV2)	Figueiredo et al. (2015)
Bacterial infections diagnosis	Diagnosis of *Staphylococcus aureus* infection	Walczak et al. (2015)
	Detection of α-hemolysin of *S. aureus* on food and clinical samples	Mudili et al. (2015)
	Detection of α-hemolysin of *S. aureus*	Reddy et al. (2013)
	Detection of methicillin-resistant *Staphylococcus aureus*	Yamada et al. (2013)
Parasitic infections diagnosis	Detection of the protozoan *Toxoplasma gondii*	Cakir-Koc (2016)
	FITC-labeled IgY antibody: fluorescence imaging *Toxoplasma gondii*	Sert et al. (2017)
	IgY-based coproantigen capture ELISA for diagnosis of human opisthorchiasis (*Opisthorchirs viverrine*)	Teimoori et al. (2017)
	Diagnosis of cryptosporidiosis caused by both *C. hominis* and *C. parvum*	Miura et al. (2017)
Diagnosis of tumors	Cancer antigen 15-3	Grzywa et al. (2014)
	Detection of cancer antigen human epidermal growth factor receptor 2 (HER2) in breast cancer diagnosis	Sun et al. (2015)
	IgY against bovine adenosine deaminase (c-ADA) in detecting human adenosine deaminase (h-ADA) present in the tumor cells lysate	Łupicka-Słowik et al. (2018)
	Anti-prostate-specific antigen (PSA) IgY for prostate cancer diagnostics.	Łupicka-Słowik et al. (2014)
In proteomics	Efficacy in removing high-abundant proteins (HAP) from plasma, serum, CSF, urine, and other body fluid or cellular sources	Huang et al. (2005)
	Specific phage-display peptides discriminate diferente forms of neurocysticercosis by antibody detection in the serum samples	Manhani et al. (2011)
	Detection of low abundance human plasma proteins using a tandem IgY12-SuperMix immunoaffinity separation strategy	Qian et al. (2008)
	Human protein depletion	Rajic et al. (2009)
	Human plasma biomarker discovery	Polaskova et al. (2010)
Hematological tests	As a polyspecific reagent for coombs test	Calzado et al. (2017)
Enzyme detection	To detect the hepatic expression of Cytochrome P450 2E1 (CYP2E1)	Jiang et al. (2016b)
Identification of substances	Detection of staphylococcal enterotoxin G (SEG) in milk and dairy products	Nagaraj et al. (2016)
	Detection allergenic proteins in commercially available latex gloves	Bittner et al. (2016)
	Identification of antibiotic gentamicin residues in food products of animal origin	He et al. (2016)
	Detection of kanamycin and gentamicin residues in animal-derived food	Li et al. (2017)
	Evaluation of the toxicity of a natural product employed in the alternative medicine	Zhang et al. (2018)

Source: Modified and adapted from Karachaliou et al. (2021).

TABLE 27.10
Use of IgY to Control Diseases in Animals

Animal Spp	Pathogen	IgY Outcome	Reference
A. Bovine			
Mastitis	*Staphylococcus aureus* and *Streptococcus agalactiae*, *E. coli*	Prevented bacterial infection	Riffon et al. (2001)
			Zhen et al. (2008a, 2008b, 2009)
			Wang et al. (2011)
Diarrhea	Rotavrus A	Inactivate virus/bacteria in vivo	Kuroki et al. (1994)
	Enterotoxigenic *E. coli* (ETEC)		Ikemori et al. (1992)
	Corona virus		Ikemori et al. (1997)
Pneumonia	Respiratory syncytial virus, pneumovirus	Neutralize virus in vitro	Ferella et al. (2012)
Enzootic bovine leukosis	Leukemia virus	Bound to BLV particles	Martínez et al. (2014)
B. Swine			
Diarrhea			
	Enterotoxigenic *E. coli* (ETEC)	Prevented K88+, K99+, 987P+ ETEC infection	Yokoyama et al. (1992)
		Inhibits the adhesion of K88+ ETEC to piglet intestinal mucosa	Jin et al. (1998)
		Prevention of K88+ ETEC infection in neonatal and early weaned piglets	Marquardt et al. (1999)
		Fast protection from diarrhea in piglets orally treated with anti-K88+ ETEC IgY encapsulated on chitosan-alginate microparticles	Li et al. (2007)
		Protection of ETEC infected piglets from diarrhea with IgY delivered in hydrogel-carbon nanotubes composites	Alustiza et al. (2016)
		Neutralize the activity of heat-stable enterotoxins (ST) and heat-labile enterotoxin (LT)	You et al. (2011)
Diarrhea	Porcine Epidemic Diarrhea virus (PEDV)	Partial protection of piglets against PEDV associated mortality	Kweon et al. (2000)
		Protected of neonatal piglets against PEDV	Lee et al. (2015)
Diarrhea	Porcine transmissible gastroenteritis virus (TGEV)	Prophylactic administration: Increase in piglets' survival rate after challenge	Zuo et al. (2009)
		Therapeutic administration: Reduction in mortality	
Diarrhea	Rotavirus group A (RVA)	Protection of gnotobiotic piglets from human RVA-associated diarrhoea	Vega et al. (2012)
C. Poultry			
Lesion	*E. coli*	Protection from homologous challenge by *E. coli* 078	Kariyawasam et al. (2004)
		Reduced symptoms, lesions	Tamilzarasan et al. (2009)
		Decreased ileal *E. coli* counts and the circulating	Mahdavi et al. (2010a, 2010b)
Campylobacteriosis	*Campylobacter jejuni*	Prophylactic (99%) and therapeutic treatments (80%–95%) reduction in bacteria	Tsubokura et al. (1997)
	C. Jejuni	Significantly reduced bacterial cell counts	Hermans et al. (2014)
	C. Jejuni	Significantly reduced bacterial cell counts	Vandeputte et al. (2019)
Salmonellosis	*Salmonella enteritidis* or *S. typhimurium*		Lee et al. (2002)
	S. enteritidis		Rahimi et al. (2007); Tellez et al. (2001)
	S. enteritidis	*Salmonella* contamination of eggs	Gürtler et al. (2004)
	S. enteritidis and S. typhimurium		Chalghoumi et al. (2009)

(Continued)

TABLE 27.10 ***(Continued)***

Use of IgY to Control Diseases in Animals

Animal Spp	Pathogen	IgY Outcome	Reference
	S. enteritidis		Fulton et al. (2002)
	Gallibacterium anatis		Zhang et al. (2019)
Infectious bursal disease (IBD)	IBD virus	Protection of chickens against IBD virud	Muhammad et al. (2001); Malik et al. (2006); El-Ghany (2011) ; Farooq et al. (2012)
Newcastle disease		Protection of chickens against Newcastle disease	Wills and Luginbuhl (1963); Box et al. (1969)
Duck hepatitis	Duck hepatitis virus	Protection of ducklings against duck hepatitis	Gui-rong & Yun-ying (2011)
Arthritis	Reovirus	Diagnostic with high consistency	Zhang et al. (2015a)
Coccidiosis	Genus Eimeria (coccidia)	Diagnostic and therapeutic treatment	Thirumalai et al. (2019)
	E. acervuline, E. maxima, E. tenella, E. praecox and *E. necatrix*		
D. Pet			
Canin (Dog) Periodontitis	*Porphyromonas gingivalis* and *P. gulae*	Reduced bacterial cell counts and protection against periodontitis dose dependently	Shofiqur et al. (2011)
Canin Diarrhea	Canine parvovirus type 2 (CPV-2)	Dose dependently minimized the excretion of virus from stool samples	Van Nguyen et al. (2006)
	Canine parvovirus field isolates	Intravenous IgY immunotherapy was effective in dogs after oral challenge with a highly pathogenic CPV isolate Kinetics of anti CPV IgY antibodies was studied by regression analysis to determine the level of IgY in serum from the time of injection	Suartini et al. (2014) Suartini et al. (2016) Guimaraes et al. (2008)
		To detect canine parvovirus in feces by indirect ELISA.	
Canin cold, bronchitis, pneumonia, and gastroenteritis	Canine morbillivirus (formerly termed Canine distemper virus, CDV)	Protection against Canin cold, bronchitis, pneumonia, and gastroenteritis in pets	Guimaraes et al. (2009a)
Allergies to cats (Human allergy)	Cat allergen protein Fel d 1	Neutralized salival Fel d 1 protein in cat	Satyaraj et al. (2019)
Rabbit hemorrhagic disease (RHD)	Rabbit hemorrhagic virus	Significant reduction in onset, duration and severity of RHD infection	Li et al. (2014)

infections often recur (Hwang et al., 2000). The control of this disease is necessary to ensure the sustainability of dairy farming and the production of milk which meets high quality global standards (Ruegg, 2017). At present, antibiotics are primarily used for therapy of mastitis, although milk-containing antibiotics must be discarded (Nair et al., 2005). Staphylococcus aureus, the primary pathogen causing bovine mastitis, is increasingly resistant to antibiotic treatment and has a propensity to recur chronically (Gill et al., 2006). As a result, alternative therapies for mastitis are needed. IgY as an inexpensive and easily produced antibody has received much attention and was found to efficiently prevent or control pathogen infections in animals (Peralta et al., 1994; Erhard et al., 1996; Kuroki et al., 1997; Jin et al., 1998; Carlander et al., 2000; Sugita-Konishi et al., 2000; Gürtler et al., 2004). IgY produced against *S. aureus* was highly specific to mastitis-causing strains, enhanced the phagocytic activity of milk macrophages (Zhen et al., 2008a), and reduced bacterial growth in culture (Mahenthiran et al., 2013). In addition, specific IgY blocked the internalization and infection of bovine mammary epithelial cells by *S.aureus in vitro* (Wang et al., 2011). Wang et al. reported that specific IgY against encapsulated type 5 (IgY-T5) and type 8 (IgY-T8) and non-encapsulated type 336 (IgY-T336) S. aureus strains (at 5 mg/mL) significantly blocked the internalization of their homologous strains by bovine mammary epithelial cells (MAC-T cells) within 6 hour (Wang et al., 2011). At a concentration of 20 mg/mL, IgY anti-*S.aureus* infused by insertion into the teat canal decreased somatic cell and bacterial counts, while curing most experimentally challenged lactating cows (Zhen et al., 2009). Similar cure rates were also observed in challenged buffaloes with mastitis, and specific IgY administered through intramammary infusions at a dose rate of 20 mg/mL improved milk yield (Iqbal et al., 2013). It was reported that polyclonal IgY against Streptococcus agalactiae and *S. aureus* was effective in reducing somatic cell count (SCC) in dairy cows (Coleman, 1996). These results suggest that IgY acts to control mastitis by prevention of uptake rather than by impacting on growth of the pathogen.

27.11.1.2 Bovine Diarrhea

Neonatal calf diarrhea is a common disease affecting newborn calves and is the leading cause of calf mortality before weaning in both beef and dairy calves (Cho & Yoon, 2014). Diarrhea is more common between 1 and 21 days of age, with a peak incidence at 2 weeks, but it can extend up to 30–45 days of age (Bartels et al., 2010). Diarrhea is caused by a combination of agents which may result in more severe disease than a single agent alone (Hoet et al., 2003; Izzo et al., 2011; Cho et al., 2013; Ferragut et al., 2016). Historically, calf diarrhoea has been attributed to bovine rotavirus group A (RVA), bovine coronavirus (BCoV), Salmonella spp., *Escherichia coli* strains with virulent factors and Cryptosporidium parvum as the main aetiological agents (Foster & Smith, 2009; Heller & Chigerwe, 2018; Brunauer et al., 2021)

Antibodies from bovine colostrum protect against RVA-associated diarrhea in calves (Parreño et al., 2004; Parreño et al., 2010). However, this strategy is industrially unfeasible at a large scale and several bovine infectious agents could be spread as adventitious virus/bacteria, like bovine leukemia virus. It has been shown that IgY antibodies can resist digestion in the gastrointestinal tract of calves, remaining biologically active (Vega et al., 2011). The therapeutic value and safety of using oral anti-rotavirus IgY in animals is now well-established after extensive studies over the past decades (Vega et al., 2015). In a review of the effect of IgY in the treatment of rotavirus infection (Thu et al., 2017), it is speculated that its mode of action involves either blocking the entry of the rotavirus into the host cells and/or minimizing the cell to cell spread of the virus. The oral administration of egg-derived preparations to new born calves has not only shown protection against diarrhea but also an increase in weight gain and an improvement in growth performance (DiLorenzo et al., 2006; Vega et al., 2015).

The passive protective effects of using anti-ETEC IgY administration on fatal enteric colibacillosis in neonatal calves have been studied (Ikemori et al., 1992). Calves fed milk containing IgY had only transient diarrhea, 100% survival and exhibited good body weight gains during the course of the study. In contrast, the controls which received no antibody had severe diarrhea and all died within 6 days of infection. The effect of IgY in controlling and preventing diarrhea (including rotaviral diarrhea) in domesticated animals from 1994 to 2015 has been a subject of critical review by Diraviyam T et al. (Diraviyam et al., 2014). The authors pooled accumulated data from 49 studies of 4 different animal species (piglets, mice, poultry and calves) that revealed that IgY significantly reduced the risk of diarrhea in treatment groups when compared to corresponding placebo groups. This general observation based on data from 49 studies supports the view that IgY is a useful tool for prophylaxis and treatment of diarrhea in animals. A promising, economically feasible and practical strategy which has been explored is the supplementation of the milk diet of calves with specific IgY antibodies from egg yolk (Kuroki et al., 1994, 1997; Mine & Kovacs-Nolan, 2002; Vega et al., 2011, 2015).

27.11.1.3 Bovine Pneumonia

Bovine respiratory syncytial virus is a pneumovirus in the Paramyxovirus family that afflicts young calves and is difficult to diagnose and treat due to its lability and poor growth in cell culture (Larsen, 2000). However, IgY against bovine respiratory syncytial virus specifically recognized and neutralized the virus *in vitro* when analyzed in dot blot and virus neutralization assays (Ferella et al., 2012). Because of the success of IgY production and activity *in vitro*, IgY may be a novel prophylactic treatment to combat bovine respiratory disease in infected calves.

27.11.1.4 Bovine Leukemia

Bovine leukemia virus (BLV) is a retrovirus that causes enzootic bovine leukosis, a chronic and slow-developing disease in cattle (Ghysdael et al., 1984). BLV can easily be transmitted through birth and contaminated colostrum, milk, blood, exudates, and tissue (Hopkins & DiGiacomo, 1997). Currently, methods to minimize BLV transmission include careful herd management given the unavailability of effective antiviral drugs and vaccines (Martínez et al., 2014). IgY antibodies generated against the whole virus or the p24 core protein specifically bound to BLV particles (in an infected cell line), purified p24, and supernatants from *ex vivo* cultures of peripheral blood mononuclear cells from naturally infected animals. IgY against BLV may warrant evaluation as a passive immunization against this virus for enzootic bovine leukosis.

27.11.2 Swine

27.11.2.1 Swine Diarrhea

Swine diarrhea, especially neonatal porcine diarrhea (NPD), was recognized as a serious problem in the late 1950s and 1960s with the emergence of the modern pig industry. Over the years, various aetiological agents have been described. Going forward, neonatal diarrhea not related to enterotoxigenic *E. coli* (ETEC), *C. perfringens* type C, TGE, or coccidiosis, was reported. Several potentially pathogenic, causative agents have been suggested, including C. perfringens type A, *Clostridioides (C.) difficile*, previously largely overlooked *E. coli* strains such as enteropathogenic *E. coli* (EPEC), rotavirus, and members of the Enterococcus (E.) faecium species group (*E. durans, E. hirae, E. villorum*). NPD has been given various names: neonatal diarrhea neonatal colibacillosis, baby pig scours, enterotoxemia in baby pigs, infectious gastroenteritis of suckling pigs, neonatal hemorrhagic and necrotic enteritis, necrotizing enteritis, neonatal necrotic enterotyphlocolitis (Uzal & Songer, 2019), epidemic diarrhoea type II, transmissible gastroenteritis, or new neonatal porcine diarrhoea (NNPD) (Jacobson, 2022).

The most common cause of enteric colibacillosis is Enterotoxigenic *E. coli* (ETEC) in neonatal (Alexander, 1994) and post-weaned pigs (Yokoyama et al., 1992). The strains of *E. coli* associated with intestinal colonization which cause severe diarrhea are the K88, K99 and 987P fimbrial adhesins. Among the ETEC, those expressing the K88+ fimbrial antigen are the most prevalent forms causing *E. coli* infection worldwide (Rapacz & Hasler-Rapacz, 1986). It has been estimated that K88+ ETEC are responsible for more than half of the piglet mortality which occurs each year (Waters & Sellwood, 1982), causing significant economic loss for the pig industry.

IgY is recognized as an alternative source of antibodies for the prevention of ETEC coli infection because it has been found to inhibit binding of *E. coli* to the intestinal mucosa (Jin et al., 1998). IgY has been orally administered to piglets and offers a potential prophylactic and therapeutic approach for controlling *E. coli*-induced diarrhea. Yokoyama et al. showed that orally administered IgY generated against *E. coli* K88, K99, or 987P fimbriae was protective against infection from each of the three homologous strains of *E. coli* in a dose-dependent manner. *E. coli* K88, K99, and 987P strains adhered equally to porcine epithelial cells from the duodenum and ileum but failed to so in the presence of homologous anti-fimbrial IgY (Yokoyama et al., 1992). A group of researchers at the University of Manitoba (Winnipeg, Canada) have carried out some excellent studies on the passive protective effect of IgY against ETEC K88 fimbriae in the control of neonatal and early-weaned piglets in vitro and in vivo (Jin et al., 1998; Marquardt et al., 1999).

Porcine epidemic diarrhea virus (PEDV) is a highly contagious enteric pathogen of swine causing high mortality rates in piglets. PEDV outbreaks have occurred in most swine-producing countries globally and leading to large economic losses for pig industries (Kikuti et al., 2022). Oral administration of anti-PEDV IgY efficiently protects neonatal piglets against PEDV, suggesting its potential as a prophylactic or therapeutic agent against acute PEDV infection (Kweon et al., 2000; Lee et al., 2015; Umeda et al., 2019).

27.11.3 Avian (Poultry)

27.11.3.1 Salmonellosis

Salmonella infections are thought to be responsible for a variety of acute and chronic diseases of poultry. It has been shown that specific IgY against *Salmonella enteritidis* or *Salmonella typhimurium* inhibits bacterial growth in vitro (Lee et al., 2002) and reduces the colonization of *Salmonella* in market-aged broilers (Rahimi et al., 2007b, Tellez et al., 2001). The use of whole egg powder (containing antibody) as a feed additive may be an alternative way to reduce the rate of *Salmonella* contamination of eggs (Gürtler et al., 2004).

27.11.3.2 Campylobacteriosis

Campylobacter jejuni has become a major concern to the commercial broiler, turkey and commercial egg-producing flocks in all countries. Tsubokura et al. used egg yolk antibodies for prophylactic and therapeutic applications in *Campylobacter*-infected chickens. In a prophylaxis experiment, it was found that these antibodies caused a 99% decrease in the number of *Campylobacter* observed, whereas in a therapy trial (antibodies were given after establishment of the infection), the number of bacteria in the feces was 80–95% lower (Tsubokura et al., 1997).

27.11.3.3 Infectious Bursal Disease

Infectious bursal disease (IBD) is an acute, highly contagious immunosuppressive disease of young chickens caused by IBD virus (Chettle et al., 1989; Qin & Zheng, 2017; Dey et al., 2019). Antibiotic therapy is the most readily available approach for controlling IBD-induced secondary bacterial infections in affected flocks. It has been shown that specific IgY has a great potential as an alternative to antibiotics for IBD. Muhammad et al. demonstrated that yolks from hyper-immunized hens can be used to control IBD in commercial laying hens (Muhammad et al., 2001).

27.11.3.4 Newcastle Disease

Newcastle disease is a severe viral infection causing a respiratory nervous disorder in several species of poultry. This disease is endemic in commercial poultry from many countries and can cause great economic loss due to high mortality rates (Lancaster, 1976). Vaccination has been used to prevent this disease in endemic areas, but vaccines are not always effective and vaccinated flocks may still be infected. It has been shown that egg yolk antibodies conferred protection in chickens against Newcastle disease (Wills & Luginbuhl, 1963; Box et al., 1969). Wills and Luginbuhl have found that subcutaneous administration of egg yolk containing high levels of IgY antibody specific to Newcastle disease protected 80% of the birds during a four-week study period (Wills & Luginbuhl, 1963).

27.11.3.5 Hepatitis Virus

Numerous studies have been carried out to develop IgY preparations against duck hepatitis virus. A related veterinary drug has been licenced in China for protection of ducklings against duck hepatitis by prophylactic subcutaneous or intramuscular injection (Gui-rong & Yun-ying, 2011).

27.11.3.6 Coccidiosis

Avian coccidiosis is an intestinal disease caused by infection with a protozoan parasite of the genus Eimeria (coccidia). There are several distinct species that infect chickens, including *E. acervuline, E. maxima, E. tenella, E. praecox*, and *E. necatrix*. The reader is referred to a review of the use of IgY as both a diagnostic and therapeutic treatment of parasitic infections (Thirumalai et al., 2019).

27.11.3.7 Arthritis Reovirus

Most frequent reovirus-associated disease in poultry is arthritis with malabsorption syndrome, immunosuppression, pericarditis, myocarditis, and osteoporosis as other common features (Fahey & Crawley, 1954; Jones, 2013). Although several commercial vaccines have been developed against avian reovirus, it is difficult to detect and treat infected flocks with multiple or novel strains of circulating reovirus (Sellers, 2017). Specific IgY against avian reovirus in infected birds displayed a high sensitivity to the virus, detected the presence of the virus in contaminated tissue, and neutralized the virus in BHK-21 cells without binding to heterologous viral strains (Jung et al., 2014).

27.11.4 Pet

27.11.4.1 Canin Periodontitis

Porphyromonas gulae is one of the key microorganisms in biofilm dysbiosis that leads to periodontal disease, a prevalent disease in pet (e.g., dog and cat). Gingipains are proteases secreted that promote the disruption of cell adhesion and the differentiation of osteoclasts. The use of anti-gingipain immunoglobulin Y (IgY-GP) has emerged as a promising alternative to conventional prevention and treatment methods. The consumption of IgY-GP reduces plaque accumulation, which may lead to an improvement in the oral health of adult dogs (Rahman et al., 2011) and cats (Oba et al., 2018).

27.11.4.2 Canine Parvovirus

Canine parvovirus (CPV) infection is a fatal disease of dogs, believed to have originated from cats and parvoviruses of carnivores, including bush dogs, cats, coyotes, bears and wolves, causes hemorrhagic diarrhea and myocarditis resulting from severe gastroenteritis. The aetiological agent is Canine parvovirus type 2 (CPV-2). IgY antibodies have been used for the diagnosis, prophylaxis and treatment of diseases of dogs. Oral passive immunization with IgY antibodies specific to CPV-2 virus, administered as an egg yolk powder to dogs challenged orally with a CPV-2 viral strain for 7 days controlled the infection to a great extent (Van Nguyen et al., 2006). Animal groups treated with 2 g of IgY antibodies in powdered form for 16 days post challenge showed significant weight gain and shorter duration of virus shedding than the control group. Apart from oral administration of antibodies, an intravenous route of anti CPV-2 IgY treatment has been used to protect dogs from CPV infection. Experimental infection induced by an oral challenge of dogs with CPV, after treatment with 1,000 and 10,000 PD (Protective Doses) of Anti CPV IgY antibodies showed recovery rates of 25% and 100%, respectively. Higher doses produce increased antibody titres which suppressed the viral load and minimized the excretion of virus from stool samples (Suartini et al., 2014). Therefore, in addition to oral therapy, intravenous immunization with antibodies also proves to be an efficient passive therapy method. The kinetics of anti CPV IgY antibodies was studied by regression analysis to determine the level of IgY in serum from the time of injection (Suartini et al., 2016). The use of monoclonal IgY antibodies and scFv to treat disease has become more prominent, including the treatment of CPV. Oral passive immunotherapy can be extended to the development of supplementary oral products to protect dogs from parvoviral infections

27.11.4.3 Canine Morbillivirus

Canine morbillivirus (formerly termed Canine distemper virus, CDV) is an extremely contagious immunosuppressive disease that affects dogs. Morbilliviruses belong to the family Paramyxoviridae and the genus Morbillivirus and cause moderate to severe immunosuppressive, respiratory, gastrointestinal and neurological diseases in a variety of hosts from humans (measles) to canines; and present an interesting model to research inter-species jumping (Quintero-Gil et al., 2019). The early symptoms are similar to a cold, followed by bronchitis, catarrhal pneumonia, and gastroenteritis. In the later stages of the disease, there are neurological symptoms such as convulsions. In some cases, there may be a high degree of keratosis of the nose and hardening of the skin of the footpads of the paws (hard pad disease). Dogs are vaccinated to protect from the disease, but there are geographical genetic variations in the virus (Pratelli, 2011).

Due to the broad clinical symptoms, laboratory tests are required to confirm the disease and a range of biological samples have been used to measure the virus, mainly by PCR but also by immunochromatography, immunofluorescence and ELISA (Costa et al., 2019). Currently, there is no specific drug for the treatment of this disease. Antibody based therapy has been considered as a relatively efficient intervention. IgY antibodies have been generated by immunizing laying hens with CDV (Guimaraes et al., 2008; Guimaraes et al., 2009a).

27.11.4.4 Rabbit Hemorrhagic Virus

Global rabbit production has been estimated at more than one million tonnes per year, according to the FAO (FAOSTAT 2012 ; Alexandratos & Bruinsma, 2012) and therefore infections of rabbits need to be controlled. Rabbit hemorrhagic disease (RHD) is a contagious and highly lethal viral disease of rabbits. Hens were immunized with the N-terminal of the viral VP-60 capsid protein and IgY treated rabbits showed a significant reduction in the onset, duration and severity of RHD infection (Li et al., 2014).

27.11.4.5 Cat Allergen Fel d 1

Allergies to cats are the most common human allergy of animal origin; the major cat allergen is Fel d 1 (Satyaraj et al., 2019). Hens were immunized with this allergen and allergen specific IgY fed to cats was shown to neutralize Fel d 1 after its production and lessen the release into the environment thus decreasing the allergic response of the owners.

27.12 Fish Health (Aquaculture)

Globally fish production has been estimated to be about 179 million tonnes in 2018, of which aquaculture accounted for 46% (82 million tonnes); aquaculture production is projected to reach 109 million tonnes in 2030 (FAO, 2020). The leading producers of farmed fish are China, Egypt, Chile, India, Indonesia, Viet Nam, Bangladesh, and Norway. Of the major finfish species produced in 2018, carp accounted for 33%, salmon for 4.5%, and rainbow trout for 1.6% of the total production; of the world production of crustaceans, white shrimp accounted for 53% (FAO, 2020). One of the major factors impacting the productivity of the aquaculture sector is the outbreak of diseases;

the widespread use of antimicrobials is problematic and can lead to the development of antibiotic resistance. While finfish have innated and adaptive immune systems, crustaceans depend completely on the innate immune system for defense and thus vaccination is ineffective (Vazquez et al., 2009). Oral administration of specific IgY antibodies is effective against various intestinal pathogen (Mine & Kovacs-Nolan, 2002). This section presents an overview of the potential to use immunotherapy with specific IgY for the prevention and treatment of fish diseases. Oral intubation and feeding of anti-fish pathogens IgY resulted in different levels of protection against various fish diseases are summarized in Table 27.11.

The IgY-coated diets helped to reduce the pathogen load and boosted the immune system in fish against both pathogens challenge. The oral administration of specific IgY against fish pathogens could provide an alternative method to antibiotics and chemotherapy for the prevention of microbial diseases of fish in a fish farm.

The use of passive immunization in aquaculture poses a number of specific challenges which need to be considered. For routine application of antibodies oral administration would be preferred but may be species specific. Administration of IgY through oral and anal routes was not detectable in gastric rainbow trout probably impacted by pepsin activity in the gut (Winkelbach et al., 2015a, 2015b); in contrast uptake in the gastric carp (lacking acidic stomach enzymes) demonstrated an efficient uptake of IgY (Winkelbach et al., 2015a, 2015b) and probably increased IgY transcytosis in carp compared to

TABLE 27.11

IgY on Fish Health

Pathogen	Fish Species	IgY Form	IgY Outcome	Reference
Edwardsiella tarda	Japanese eels (*Anguilla japonica*	Oral	Non-treated infected eels died within 15 days whereas the eels receiving IgY survived and no disease symptoms were observed	Gutierrez et al. (1993); Hatta et al. (1994)
White spot syndrome virus (WSSV)	Shrimp	Injected IM	0.01 mg/10 μL, 0.1 and 0.5 mg/μL exhibited a survival rate of 50%, 85%, and 83%	Kim et al. (2004)
White spot syndrome virus (WSSV)	Shrimp	Injected IM	When challenged with virus, 73% survival when IgY prepared with inactivated virus; 33% with IgY prepared from WSSV-DNA	Lu et al. (2008)
WSSV	Crayfish	Injected IM, oral, immersion	IgY from inactivated WSSV and DNA vaccine were, respectively, 20% and 80% mortality. Diet added 10% egg yolk powder and 1% IgY power showed 53.3% and 67.7% mortality, respectively, and the immersion showed 46.7% mortality	Lu et al. (2009)
White spot syndrome virus	Shrimp	IgY in feed	Highly resistanttTo WSSV challenge	Kumaran et al. (2010)
White spot syndrome virus (WSSV)	Crayfish	Injected IM, feed or immersion in IgY solution	All methods provided some protection of crayfish against infection	Lu et al. (2009)
V. harveyi and V. parahaemolyticus	Shrimp	β-Cyclodextrin encapsulated egg yolk powder	Protected from infection	Gao et al. (2016a); Gao et al. (2016b)
Vibrio parahaemolyticus	Shrimp	Whole egg powder with specific IgY fed at 20%	Survival of 86% when treated with recombinant PirA- like toxin induced IgY	Nakamura et al. (2019)
Yersinia ruckeri	Rainbow trout	Microencapsulated IgY	Lower mortalities than control fish	Lee et al. (2000)
Piscirickettsia salmonis	Salmon	Chitosan-alginate micro encapsulated IgY	Encapsulated IgY can resist degradation and is absorbed into the bloodstream	Oliver et al. (2015)
Reovirus	Swimming crabs	Rapid immunochromatographic test strip	Anti-swimming crab reovirus IgY effectively detected the virus in contaminated samples with high consistency	Zhang et al. (2015a)
Vibrio anguillarum	Rainbow trout	Intraperitoneal(IP) injection, oral intubation, or feeding	Injection conferred protection for 14 days. Oral intubation and feeding gave comparable results in some cases	Arasteh et al. (2004)
Edwardsiella tarda	Turbot	Chitosan-alginate encapsulated IgY at 1%, 3%, and 5% in feed	Survival rates of the1%, 3% and 5% micro-encapsulated specific IgY groups were 20%, 56.7% 63.3%, on the tenth day post infection with 10^7 CFU *E. tarda*	Xu et al. (2020)

(Continued)

TABLE 27.11 *(Continued)*

IgY on Fish Health

Pathogen	Fish Species	IgY Form	IgY Outcome	Reference
V. alginolyticus	Abalone	5% or 10% alginate encapsulated IgY in feed	Survival rates of fed with 5 or 10% anti-vibrio IgY egg powders ranged from 65–70% 14 days post- *V. alginolyticus* challenge (1×10^{6}CFU)	Elston & Lockwood (1983); Anguiano-Beltrán et al. (1998); Liu et al. (2000); Wu et al. (2011)
Vibrio	Ayu (*Plecoglossus altivelis*)		Anti-Vibrion IgY supplemented feed, significantly increased the survival of Ayu fish, challenged by *V. alginolyticus.*	Kanno et al. (1989)
Vibrio splendidus	Sea Cucumber (*Apostichopus japonicus*)		Increased the survival of Sea Cucumber, challenged by *V. splendidus*	Li et al. (2016b)
V. anguillarum	Half-smooth tongue sole (*Cynoglossus semilaevi*)	Oral feed	Increased the survival of Half-smooth tongue sole, challenged by *V. anguillarum*	Gao et al. (2016b)
Shewanella marisflavi	Sea cucumber	Oral yolk powder at 25, 5 and 1 mg/mL	25, 5, and 1 mg/mL anti-S. marisflavi AP629 IgY gave77.5%, 50%, and 22.5% survival rates at day 12, respectively, when challenged with 4.2 10^6 CFU *S. marisflavi*	Xu et al. (2019)
Non-O1 Vibrio Cholerae	Carp	Feed additive	Survival rate of carp fed with IgY significantly elevated, when challenged with non-O1 *V. cholerae.*	Gao et al. (2017)
Vibrio harveyi IgY	Indian white shrimp	Feed supplement	Anti-Vibrio harveyi IgY effective against V. harveyi	Kumaran et al. (2018)
Red-spotted grouper nervous necrosis virus (RGNNV)	Marine fish larvae	10% lyophilized egg yolk powder in feed	Protection rates of groupers under treatment after RGNNV infection and continuous feeding during RGNNV infection were 68% and 70%, respectively,	Liu et al. (2021)
Vibrio parahaemolyticus	Pacific white shrimp	IgY fed at a dose of 0, 0.2%, 0.3%, 0.4%, and 0.5%	Reduced mortality rate (25–40%) in contrast to the control group (60%)	Keetanon et al. (2021)

trout. Other considerations are the period of immunization which is most effective for uptake and the duration for which passively transmitted IgY are present in the different fish species; continuous administration may be necessary in some cases (Rajan et al., 2017). Despite these challenges, advances have been made in the use of IgY antibodies for the treatment of fish diseases and these are summarized in Table 27.11.

Some references IgY development for fish health are being discussed here.

Vibrio parahaemolyticus is a pathogen associated with acute hepatopancreatic necrosis disease (AHPND) and can cause up to 100% mortality in post-larvae shrimp. IgY has been shown to confer protection to this infection (Gao et al., 2016a).

Piscirickettsia salmonis causes the disease piscirickettsiosis in salmon, resulting in significant mortality and economic losses in fish (Oliver et al., 2015). Specific IgY was produced by immunizing hens with *P. salmonis* proteins and was effective against the infection in SHK-1 cells; further studies are necessary to show its efficacy in fish.

Vibrio anguillarum is an important pathogen of marine and fresh water fish and alternates to vaccination which can be stressful for fish were sought (Arasteh et al., 2004). Oral treatment is preferable and was shown to be as effective in some cases as intraperitoneal injection.

Edwardsiella tarda, a Gram-negative bacterium belonging to the family Enterobacteriaceae, is considered a common pathogen infecting mainly economically important fish species such as Japanese eel (*Anguilla japonica*), red sea bream (*Pagrus major*), yellowtail (*Seriola quinqueradiata*), channel catfish (*Ictalurus punctatus*), and turbot (*Scophthalmus maximus*) (Park et al., 2012). The use of encapsulated IgY shows promise for the treatment of turbot (Xu et al., 2020).

IgY has been tested against *Vibrio alginolyticus*, which causes infections and high rates of mortality in abalone (marine snails) (Elston & Lockwood, 1983; Casandra et al., 1998; Liu et al., 2000). When incorporated into the feed, IgY anti-*V. alginolyticus* significantly increased the survival of small abalones challenged by *V. alginolyticus* (Wu et al., 2011). Similar results were obtained using IgY against *Vibrio anguillarum* in rainbow trout (Arasteh et al., 2004), ayu fish (Kanno et al., 1989; Li et al., 2016b), half-smooth tongue sole (Gao et al., 2016b), shrimp (Keetanon et al., 2021), and carp (Gao et al., 2017).

Reovirus is a growing problem in aquaculture, causing high mortality rates of swimming crabs. Use of anti-swimming crab reovirus IgY effectively detected the virus in contaminated samples with high consistency, suggesting potential benefit in reovirus-associated disease outbreaks in aquaculture (Zhang et al., 2015a).

Aeromonas is commonly found in aquatic environments. IgY-specific antibodies have been investigated as an alternative method to diagnose *Aeromonas*-diseased aquatic animals. IgY produced against *Aeromonas hydrophila* detected *A. hydrophila* in tissues and phagocytes in infected Nile tilapia (Fernandes et al., 2019a), neutralized bacterial adhesins and toxins released by the bacteria, promoted agglutination, inhibited bacterial growth, and enhanced the phagocytic activity of infected Nile tilapia, blunt snout bream (Qin et al., 2018; Fernandes et al., 2019b), and polyploid gibel carp (Li et al., 2006). When added to rearing water, specific IgY eliminated the development of skin ulcers, as well as transmission of the infection between different fish (Gan et al., 2015).

27.13 Use of IgY in Plant Health

Viruses are one of the major plant pathogens, drastically influencing crop yield and productivity in the whole world (Fraser, 1990; Kang et al., 2005). Although there is no statistical data on crop yield losses due to virus attacks, the damage varies from 50 to 80% and sometimes reaches 100% depending upon the severity of virus spread (Roger, 2009). The history of plant virus infections starts in 1898. There are no natural effective control measures discovered yet except plants' own internal wide range of defense mechanisms to protect themselves against virus infection (Sanseverino et al., 2010). In these circumstances, the production of polyclonal IgY against various etiological agents or viruses of plants may have an important role in immuno-diagnosis in epidemiological studies with the prevention and treatment of plant diseases.

27.13.1 IgY in Immuno-Diagnosis of Plant Viruses

With rapid international expansion of trade, horticultural production of ornamental plants has become a global industry with commercial production of flowers and bulbs in many countries. Various plant viruses infect plants with variable infection frequencies (3% ~100%), followed by great economical loss. There was no efficient method discovered yet to detect the viruses. Detection of viruses in field samples in epidemiological studies can be performed routinely using the IgY. Based on a literature survey, it is reported that IgY has already developed for the diagnosis of fique Furcraea Necrotic Streak Virus-FNSV (Toloza-Moreno et al., 2022), Potato viruses X and Y (PVX, PVY) (Weilbach & Sander, 2000), Lily symptomless virus (LSV), Lily mottle virus (LMoV), Cucumber mosaic virus (CMV) (Yoo & Jung, 2014), orchid Cymbidium mosaic virus (CyMV) (Vejaratpimol et al., 1999), Citrus yellow mosaic virus (CYMV) (Kumar et al., 2018). In conclusion, using this IgY was established for detecting plant viruses. This method can be used for virus surveillance to help reduce economic losses in the natural, agricultural, and ornamental plant industries.

27.13.2 IgY Against Plant Viruses

Chinese sacbrood virus (CSBV) infects Apis cerana larvae, resulting in the inability of the larvae to pupate and their consequent death, which may pose a serious threat to entire colonies (Sun et al., 2018). Apis mellifera Colony Collapse Disorder by Deformed Wing Virus (DWV) is also reported (Nordseth, 2020). As there is no effective medical treatment for CSBV and DWV infections. IgY against CSBV (Sun et al., 2018) and DWV (Nordseth, 2020) has been developed. Experimental results revealed that anti-CSBV IgY and anti-DWV IgY has protection ability against the infection of CSBV infections and DWV, which would significantly reduce the diseases. This technology could have a profound impact on the future of beekeeping – an industry that is key to avoiding a global food sustainability crisis.

27.14 Use of IgY in Environmental Health (Historical World Pandemic)

Cholera, influenza, and SARS-CoV-2 are some of the most brutal killers in human history. Outbreaks of these diseases across international borders are properly defined as a pandemic. IgY may be useful to control such pandemic diseases. The use of IgY in environmental health is summarized in Table 27.12.

TABLE 27.12

Use of IgY in Environmental Health/World Pandemic

World Pandemic Disease	Pathogen	IgY Studied Based on	Reference
Cholera	*Vibrio cholerae* (Cholera Toxin B Subunit)	In vitro and in vivo mouse	Hirai et al. (2010); Barati et al. (2018)
	Vibrio cholerae Lipopolysaccharide (LPS)	In vitro and in vivo mouse	Akbari et al. (2018)
Influenza	Influenza virus A/H1N1 2009	In vitro	Adachi et al. (2011); Tsukamoto et al. (2011); Yang et al. (2014)
	Influenza Virus H5N1 and H1N1	In vitro and in vivo Mice	Nguyen et al. (2010); Wallach et al. (2011)
	Influenza B Virus	In vitro and in vivo mice	Wen et al. (2012)
COVID-19	Severe acute respiratorysyndrome coronavirus 2 (SARS-CoV-2) infection	Review	Constantin et al. (2020)
		In vitro	Lu et al. (2020); Shen et al. (2021); Wei et al. (2021)

Based on these data, IgY may be inexpensive and rapid prophylaxis that can be given *via* orally after mixing them into dairy products, powdered milk, foods, water, or ORS to prevent or treat cholera in humans, because they can supply water and nutrition in addition to antibody activity. IgY-supplemented tablet, lozenge, face mask intranasal spray, or drops are also applied to prevent or treat influenza and SARS-CoV-2 infection to bind to numerous epitopes on the virus and thus prevent viral entry into the body (Yang et al., 2014). This approach may be practical for those where vaccination is unavailable or as a temporary measure (such as during travel) before or to supplement vaccination. IgY can be easily given to people of all ages, from babies to the elderly, even under serious or miserable conditions such as those occurring after a natural disaster.

27.15 Use of IgY in Human Health

Acute microbial gastrointestinal infections destroy the body's first line of defense in the alimentary tract, which includes the innate component of the immune response, particularly the epithelial barriers, as well as the adaptive mucosal immunity resulting in the development of severe and complicated forms of the infectious disease. Gastrointestinal infections with various pathogens are mainly treated with antibiotics or antimicrobials. However, a dramatic increase in antibiotic resistance among common bacterial pathogens has impacted negatively impacted the efficacy of antimicrobial chemotherapy (Carlander et al., 2000). It is re-shaping the topography of research for novel and alternative infection control modalities. The past decade or so has seen increasing numbers of studies on and use of IgY in the treatment and prevention of infectious diseases in a variety of animal species (Xu et al., 2011) as well as in the development of functional food for human application (Schade et al., 2007). It has been established that oral administration of antimicrobial immunoglobulins derived from bovine milk (Korhonen et al., 2000) and poultry egg is an effective way to provide protective immunity against a variety of viral or bacterial pathogens (Narat, 2003) which might reduce the clinical use of antibiotics and thereby minimize the risk of bacteria developing antibiotic resistance.

Here we provide a comprehensive overview of previous applications of IgY in humans, including viral, bacterial, fungal, parasitic infections and others.

27.15.1 Viral Infection

27.15.1.1 Hepatitis B Virus

Hepatitis B viruses have a narrow range of host specificity (Maenz et al., 2007). Similar to the human hepatitis B virus (HBV), duck HBV infection can lead to chronic hepatitis, fibrosis, cirrhosis, and hepatocellular carcinoma (Omata et al., 1983). In eggs laid by ducks that were immunized against duck HBV, significant titers of specific IgY were found in the yolk (Abouzid et al., 2006; Rollier et al., 2000). Also, uninfected treated ducks were protected against the virus, and treated carriers had decreased serum and liver levels of duck HBV (Thermet et al., 2004). Despite the availability of an effective vaccine, rates of vaccination in humans are poor and immunotherapies may be unavailable in resource-limited settings, or not tolerated. IgY may also warrant evaluation as a therapeutic for the treatment of postexposure HBV infection in humans where standard treatments are not readily available.

27.15.1.2 Rotavirus

Among the etiologic agents of diarrhea, rotavirus is the most important being responsible for over two million diarrhea episodes among infants with 600,000 deaths annually, mainly in developing countries (Parashar et al., 2003). Although mortality rate from diarrhea have decreased, morbidity rates remain high. Several independent guidelines based on systematic reviews of the best available evidence related to rotavirus vaccination of infants and to the management of acute gastroenteritis among infants and young children were published (Guarino et al., 2012; Szajewska and Dziechciarz, 2010). There is agreement in the scientific community that antimicrobials should not be routinely administered to children with gastroenteritis. The uptrend in the frequency of antibiotic-resistant bacteria, the widespread treatment of diarrhea with antimicrobials that sometimes do not respond to antibiotics, and the increasing number of immuno-compromised individuals has prompted much research into alternative approaches to management of diarrhea. The oral administration of IgY specific for any of the causative agents of diarrhea has proved successful for treatment of a variety of gastrointestinal infections (Rahman et al., 2013).

Rotavirus is the most common cause of severe diarrhea in children and outbreaks can occur in both vaccinated and unvaccinated children (Burke et al., 2018). When administered to pediatric patients who tested positive for rotavirus infection, orally administered treatments of 20% IgY sachets (Rahman et al., 2012) or 4 doses of 10 g of IgY powder (Sarker et al., 2001) significantly reduced the time for viral clearance in the feces, volume and duration of oral rehydration, intravenous fluid administration, duration of diarrhea, and recovery time. Wang et al. (Wang et al., 2019b) conducted a meta-analysis involving 2626 infants with rotavirus diarrhea from 17 randomized clinical trials. Among these infants, 1347 received anti-rotavirus IgY taken orally and 1279 received conventional treatment. Anti-rotavirus IgY treatment was significantly more effective than conventional treatment. These preliminary data demonstrate the utility of IgY as an antiviral therapy for infantile rotavirus enteritis (Wang et al., 2019b).

27.15.1.3 Zika Virus

In 2016, a global health emergency was declared by the World Health Organization after the observation of hundreds of thousands of infections by the Zika virus. Zika virus is transmitted by Aedes mosquitoes (Dick, 1952). The illness is characterized by fever, rash, arthritis (Duffy et al., 2009), and less commonly, Guillain-Barré syndrome (Cao-Lormeau et al., 2016). In pregnant women, Zika virus infection may cause severe birth defects, including microcephaly (Brasil et al., 2016; Rasmussen et al., 2016). There is no effective drug or vaccine for Zika virus infection in part because of antibody-dependent enhancement, a phenomenon in which prior infection results in virus-specific antibodies enhancing replication of virus into monocytes/macrophages and granulocytic cells through interaction with Fc or complement receptors (Tirado & Yoon, 2003). IgY against the Zika virus was able to neutralize the virus *in vitro* at a concentration of 25 μg/mL (O'Donnell et al., 2019). Furthermore, intraperitoneal injection with 1 mg of specific IgY protected 3-week-old IFNAR–/– mice that received a lethal challenge of Zika virus without inducing antibody-dependent enhancement. Zika virus-specific IgY may warrant further evaluation as passive immunotherapy, with caution for the potential to elicit an allergic response.

27.15.1.4 Dengue Virus

Dengue fever, dengue hemorrhagic fever, and dengue shock syndrome are tropical, mosquito-borne diseases that are caused by infection with one of four dengue virus serotypes (Gubler & Clark, 1995). In the last 50 years, dengue virus-associated diseases have re-emerged, causing millions of infected cases and tens of thousands of deaths annually (Wilder-Smith et al., 2020). Although there is a commercially available vaccine to treat dengue fever, the World Health Organization recommends only vaccinating seropositive individuals who have a history of dengue virus infection, or by age 9 years in areas where the infection is prevalent (WHO, 2018). Currently, IgG-coupled enzyme-linked immunosorbent assays (ELISAs) are used for serological testing for dengue virus infection, suggesting that IgY can also be used in a diagnostic test for this disease.

Specific IgY has been developed against nonstructural protein 1, which is secreted by the dengue virus during infection and is detectable for up to 9 days after infection (Dussart et al., 2006; Lapphra et al., 2008). IgY anti-nonstructural protein 1 that is highly specific can neutralize the virus in immunoassays (Figueiredo et al., 2015; O'Donnell et al., 2017; O'Donnell et al., 2020). Specific IgY generated against dengue virus serotype 2 also demonstrated a similar ability to neutralize the virus (Fink et al., 2017). In vivo, IgY antibodies given at a dose of 150 μg (Figueiredo et al., 2015) or 1 mg through intraperitoneal injection (Fink et al., 2017) protected mice lethally challenged with dengue virus, suggesting that specific IgY may be used as a treatment of dengue virus-associated diseases in humans provided that a severe immune response (e.g., anaphylactic shock) to the IgY will not be generated (Yao et al., 2015). IgY antibodies are not likely to lead to immune amplification of dengue virus infection, unlike IgG where enhanced viral uptake *via* IgG-bound dengue virus enters into monocytes/macrophages *via* Fc receptors (Goncalvez et al., 2007).

27.15.1.5 Hantavirus

Hantavirus Pulmonary Syndrome (HPS) is a rare, severe, and potentially fatal respiratory disease caused by infection with hantaviruses (most commonly, Andes virus) (Zaki et al., 1995; Padula et al., 1998; Lee et al., 2021). Infection is believed to occur primarily through inhalation or ingestion of rodent feces, urine, and saliva, or by rodent bites. Person-to-person transmission is also recognized. The mortality rate is estimated at 38% (CDC, 2021). There are no current immunotherapies, antiviral treatments, or vaccines to treat HPS (Custer et al., 2003). Specific IgY developed after immunization of ducks with Andes virus neutralized the virus *in vitro* (Brocato et al., 2012; Haese et al., 2015) and protected Syrian hamsters administered a dose of 12,000 neutralizing antibody units/kg through intranasal delivery (Brocato et al., 2012) or 20,000 neutralizing antibody units/kg through subcutaneous injection (Haese et al., 2015) after receiving intramuscular and intranasal challenge. This suggests that IgY may warrant evaluation as a treatment following infection with the Andes virus to prevent the onset of HPS, especially in settings of clustering of cases (Lee et al., 2021).

27.15.1.6 Ebola Virus

Ebola virus infection is rare but results in high mortality rates in humans and nonhuman primates. Ebola virus also can persist in survivors and relapse has been documented. In addition to supported care, two monoclonal antibody treatments and a vaccine have been approved (WHO, 2021).

Anti-Ebola virus IgY was harvested from hens immunized with a recombinant vesicular stomatitis virus vector encoding Ebola virus glycoproteins. Anti-Ebola virus IgY was then evaluated in newborn Balb/c mice challenged with a lethal dose of Ebola pseudovirus 2 or 24 hours after infection (Zhang et al., 2021). Animals receiving a high dose of anti-Ebola virus IgY showed complete protection, while the low dose group showed partial protection. All mice receiving naïve IgY (i.e., from hens not immunized with Ebola glycoproteins) died. Zhang et al. note that because Ebola epidemics typically occur in impoverished hot African areas where electricity and cold-chain storage may be limited, advantages of low-cost mass production and avoidance of antibody-dependent enhancement may make anti-Ebola virus IgY an especially novel approach that warrants further investigation (Zhang et al., 2021).

27.15.2 Bacterial Infection

27.15.2.1 Dental Caries

Dental caries is one of the most common infectious diseases among children and adolescents affecting up to 90% of the world's inhabitants. The economic burden of the disease is therefore quite high with dental caries costs alone exceeding the total healthcare budget for children in many low-income countries. *Streptococcus mutans* (*S. mutans*) is the main odontopathogen implicated in the development of dental carries in humans (Loesche, 1986). Although most antibiotic treatments of *S. mutans* are effective, resistance to penicillin, erythromycin, amoxicillin, clindamycin, and lincomycin is common (Al-Shami et al., 2019). IgY developed against *S. mutans* is highly stable and cross-reacts with other serotypes, including *S. salivarum* (Hatta et al., 1997; Chang et al., 1999; Moreno et al., 2011). IgY against *S. mutans* inhibited *in vitro* bacterial growth, biofilm development, and binding to bacterial adhesion proteins, and inducing agglutination of *S. mutans* (Nguyen et al., 2011; Bachtiar et al., 2015). Rats exposed to

S. mutans and fed a caries-inducing diet had reduced dental caries when lyophilized anti-*S. mutans* IgY was incorporated into their feed (Otake et al., 1991; Fan et al., 2003), topical gel (Bachitar et al., 2016b), or chitosan-enriched soy milk (Bachtiar et al., 2015).

The efficacy of anti-*S. mutans* IgY in humans has also been evaluated. IgY anti-*S. mutans* incorporated into a mouth rinse significantly reduced levels of *S. mutans* in saliva within 4 hours, and in plaque within 7 days (Hatta et al., 1997). In comparison to commercial toothpaste, anti-S. *mutans* IgY toothpaste reduced levels of S. *mutans* more quickly and colonization was suppressed as long as 2 weeks after discontinuation (Chi et al., 2004).

In a randomized, double-blind, placebo-controlled clinical trial, Nguyen et al. reported that lozenges containing anti-CA-GTase IgY can significantly and selectively suppress oral colonization by salivary mutans after 5 days (Nguyen et al., 2011. Prior to this, effective local protection against dental caries was achieved with anti-*Streptococcus mutans* IgY in an animal model (Krüger et al., 2004). In this latter trial, a direct correlation was found between a given IgY dose and a reduction in the incidence of dental caries. Hatta et al. evaluated the efficacy of oral IgY anti-*S. mutans* rinses in human volunteers. This IgY inhibited *S. mutans* adherence to saliva-coated hydroxyapatite discs by 59%, while the control IgY from non-immunized hens only gave an 8% inhibition (Hatta et al., 1997). All these results strongly support the efficacy of oral treatments with anti-*S. mutans* IgY as a valid alternative for preventing dental plaque in humans.

27.15.2.2 Periodontitis

Periodontitis is a bacterial biofilm-induced oral disease, mostly caused by *Aggregatibacter actinomycetemcomitans* (*A. actinomycetemcomitans*) (Li et al., 2020), *Porphyromonas gingivalis* (*P. gingivalis*), *Prevotella intermedia, Solobacterium moorei, Fusobacterium nucleatum*. Oral administration of chicken egg yolk antibody (IgY) is a promising nutritional strategy to control pathogen infections.

A review on the effectiveness of mechanical, chemical, and antibiotic plaque removal in subjects with periodontal disease has outlined their various degrees of limitations. While oral care products containing chlorhexidine exert anti-plaque effects as indicated by meta-analyses (Sugano, 2012), IgE antibodies against chlorhexidine have been detected in the majority of sera from a small group of predominantly Japanese individuals showing anaphylactic-type adverse reactions directed against chlorhexidine. The use of chlorhexidine at a concentration effective for oral care has thus been banned in Japan.

Meanwhile, the emergence of antimicrobial resistance is currently posing a major challenge globally, with an increasing number of strains, including commensal and pathogenic oral bacteria, becoming resistant to commonly used antibiotics. Due to these current limitations, new therapeutic approaches for the control of biofilm are clearly required. The search for adjuncts in biofilm control has led to the exploration of oral passive immunotherapy by IgY as a biological plaque controller.

27.15.2.2.1 *Porphyromonas gingivalis*

Porphyromonas gingivalis (*P. gingivalis*), which causes biofilm on teeth (Costalonga & Herzberg, 2014), is quite resistant to the host immune response and leads to inflammation and bone loss associated with periodontitis (Hajishengallis & Lamont, 2012). Anti-*P. gingivalis* IgY against gingipains, a protein family released by *P. gingivalis*, inhibited attachment of the bacteria in cultured human epithelial cells (Yokoyama et al., 2007a). In addition, IgY against a 40 kDa outer membrane protein prevented the aggregation of *P. gingivalis* with *Streptococcus gordonii*, another bacterial strain implicated in periodontitis (Hamajima et al., 2007). Furthermore, IgY-anti-*P. gingivalis* in dry feed or administered as an ointment in a dog model reduced biofilm formation and inflammation (Rahman et al., 2011). Sublingual application of IgY anti-*P. gingivalis* in five periodontitis patients reduced the levels of the bacteria and gum bleeding (Yokoyama et al., 2007b). A clinical trial using egg yolk antibody against gingipains (IgY-GP) was performed in five patients with chronic periodontitis. IgY-GP containing ointment was administered directly into the periodontal pocket. Scaling and root planning (SRP) combined with the use of IgY-GP reduced the probing depth, bleeding on probing, and levels of *P. gingivalis* at 4 weeks as compared with SRP only (Yokoyama et al., 2007b). Sugano N also investigated the effect of IgY-GP on periodontitis by IgY-GP supplemented tablets in 42 patients after scaling and root planning employing a double-blind placebo-controlled approach (Sugano, 2009). A significant improvement in mean probing depth was noted in the IgY-GP group at 12 weeks after therapy. Parallel to the clinical changes, the number of *P. gingivalis* cells in subgingival plaque from the deepest pocket was significantly reduced. These results indicated that daily administration of tablet containing IgY-GP, in conjunction with scaling and root planning, in patients produced significantly better clinical and microbiological results. Near-infrared (NIR) photo-antimicrobial targeting therapy (PAT) for periodontitis is reported by Maruyama et al. The antimicrobial effect of PG-IgY-PAT is dependent on the NIR-light dose. PG-IgY-PAT significantly reduces the area of ulcers in a mouse model of PG-infected cutaneous ulcers, indicating that PG-IgY-PAT is a new promising therapeutic method for PG infection (Maruyama et al., 2022).

27.15.2.2.2 *Prevotella intermedia*

Like *Porphyromonas gingivalis*, *Prevotella intermedia* also causes gingivitis and other periodontal diseases (Maeda et al., 1998) and can be associated in humans with systemic diseases such as diabetes mellitus (Schara et al., 2013), respiratory illnesses (Widmer, 2010), cardiovascular disease (Beck et al., 1996, ischemic stroke (Joshipura et al., 2003), osteoporosis (von Wowern et al., 1994), and risk of low birthweight preterm pregnancies (Offenbacher et al., 1996). In rat models challenged with *P. intermedia* on gingivae, IgY anti-*P. intermedia* treatment protected against gingivitis by decreasing gingival index, plaque index, bleeding on probing, white blood cell counts, and local inflammation typically associated with periodontal disease (Hou et al., 2014). Because of the success of anti-*P. intermedia* IgY in rat models, as well as the general challenge due to increased resistance to antibiotics, IgY treatment may provide an alternative in humans.

27.15.2.2.3 Solobacterium moorei

Solobacterium moorei causes oral halitosis, periodontitis, and gingivitis (Haraszthy et al., 2008; Vancauwenberghe et al., 2013). *S. moorei* is susceptible to common antibiotics such as penicillin, vancomycin, and moxifloxacin (Pedersen et al., 2011). Specific IgY inhibited bacterial growth in liquid media and biofilm formation *in vitro* (Li et al., 2012). In a mouse model challenged with *S. moorei*, 20–40 mg/mL of specific IgY decreased bacterial counts in the oral cavities of treated animals (Li et al., 2012). Benefit in humans has not been determined.

27.15.2.2.4 Fusobacterium nucleatum

Fusobacterium nucleatum is one of many pathogenic bacterial strains that contributes to halitosis and periodontitis (Pianotti et al., 1986). Although available treatments include chemical antiseptics, antimicrobials, and mechanical therapy (Wang et al., 2019a), efficacy is limited by poor compliance and the development of antibiotic resistance. IgY anti-*F. nucleatum* has been suggested as an immunotherapeutic alternative to mediating the development of *F. nucleatum* and other bacteria in the oral cavity. In a periodontitis rat model, IgY anti-*F. nucleatum* inhibited the development of volatile sulfurous and odorous compounds and decreased the malodor index, levels of anti-inflammatory cytokines, and alveolar bone loss, while aiding periodontal restoration (Wang et al., 2019a).

27.15.2.3 Gastritis (*Helicobacter pylori*)

Helicobacter pylori infection may lead to gastric cancer which is the fourth most common cancer and second leading cause of cancer-related deaths worldwide (Herszenyi & Tulassay, 2010). *H. pylori* infects approximately 50% of the world's inhabitants and the number of newly diagnosed cases was calculated as 750,000 persons per year. *H. pylori* is the first bacterium to be classified as a class 1 carcinogen by the World Health Organization. Eradication of *H. pylori* infection both in animal models and in human subjects invariably fails when using an antibiotic as a monotherapeutic regimen even when the organism is susceptible to said antibiotic in vitro. Current first-line treatment regimens generally employ a potent acid-suppressing agent plus two antibiotics (such as amoxicillin, metronidazole, or tetracycline) but this approach is also associated with a variety of problems, including induction of antimicrobial resistance and high cost of treatment. Because of the development of resistance to antibiotics, treatment of *H. pylori* requires the incorporation of multiple antibiotics. *H. pylori* exhibits resistance to metronidazole (78%), levofloxacin (56%), multidrug treatments (53%), and clarithromycin (31%) (Wang et al., 2019d). Thus, orally administered immunoglobulins, and particularly IgY, have been suggested as an alternative approach to treat *H. pylori*-related infections because they limit the development of antibiotic resistance. IgY has been developed against *H. pylori* urease (HPU), a protein likely required for bacterial adhesion to mucin or the surface of epithelial cells in the gastric mucosa (Marcus & Scott, 2001).

IgY have been used successfully to reduce *H. pylori* colonization and diminish the severity of mucosal inflammation in the stomach in a mouse model of infection (Ameri Shah Reza et al., 2012). In animal models, a chow diet containing 25 mg anti-HPU IgY/g and 0.16 mg famotidine/g reduced *H. pylori* activity in infected Mongolian gerbils and prevented colonization of *H. pylori* in the gastrointestinal tracts of uninfected controls (Nomura et al., 2005). In the Mongolian gerbil model, anti- *H. pylori* IgY reduced inflammation, neutrophil and leukocyte infiltration, and gastric mucosal injury by interfering with the adhesion of *H. pylori via* its urease (Shin et al., 2002). A similar anti-inflammatory effect and reduction of *H. pylori* in the gastric mucosa were also observed in C57BL6/j mice treated with 60 mg of anti-*H. pylori* urease C IgY in either powder form or dissolved in phosphate-buffered saline (Malekshahi et al., 2011). Importantly, orally administered IgY (100–500 mg) in male C57BL/6 mice were more efficacious in eliminating *H. pylori* compared to treatment with the commonly used proton pump inhibitor pantoprazole (Yang et al., 2012). IgY against other *H. Pylori* antigens with good efficacy includes IgY against the outer inflammatory protein (Dossumbekova et al., 2006; Franco et al., 2008), the neutrophil-activating protein (Evans et al., 1995), and the native or recombinant VacA protein 1 (Xun et al., 2010; Xia et al., 2011).

In a clinical trial performed in 16 volunteers and designed to evaluate the protective effect of a yogurt drink fortified with anti-*H. pylori* urease IgY, values of urea breath and *H. pylori* stool antigen (HpSA) among the treatment group decreased significantly (Yamane et al., 2003). Interestingly, the number of volunteers with complaint of gastric pain decreased over the three-month treatment period. Although reduction and not elimination of *H. pylori* load upon oral treatment with IgY against *H. pylori* urease was observed, such reduction may have been enough to improve the quality of life of *H. pylori* infected patients. An egg yolk powder dietary supplement containing IgY anti-HPU administered to a cohort of asymptomatic *H. pylori*-positive patients reduced the levels of *H. pylori* and aided the treatment of *H. pylori*-associated gastritis, with no side effects reported (Suzuki et al., 2004). Another clinical study (Horie et al., 2004) on anti-urease IgY involving 42 volunteers revealed significant reduction in urea breath values among patients in the treated group. Protection by anti-*H. pylori* IgY has also been investigated in animals (Nomura et al., 2005) and humans (Suzuki et al., 2004). In this study of 42 *H. pylori*-positive subjects, drinkable yogurt fortified with anti-HPU IgY significantly suppressed *H. pylori* infection and was well tolerated with no adverse effects (Horie et al., 2004). Some studies on anti-*H. pylori* IgY in animals also demonstrated a prophylactic effect. While IgY does not bring about a total eradication of *H. pylori*, it may serve as an adjunct to standard treatment of *H. pylori* infection.

27.15.2.4 Salmonellosis

The *Salmonella* species, particularly *S. typhimurium* and *S. enteritidis*, are human and chicken pathogens (Haselbeck et al., 2017; Broom & Kogut, 2019). *Salmonella*-specific IgY inhibited bacterial cell growth (Lee et al., 2002) by binding to and structurally altering antigens on the surface of the bacterium or by causing bacterial agglutination (Mine, 1997). In a human epithelial Caco2 cell model, IgY anti-*Salmonella* antibodies also prevented the adhesion of the bacterium to cells (Chalghoumi et al., 2009).

IgY anti-*S. typhimurium* and IgY anti-*S. enteritidis* exhibited significant cross-reactivity and agglutination (Mine, 1997; Terzolo et al., 1998), which indicates that IgY against a specific *Salmonella* serovar may be useful in treating a broad range of different *Salmonella* strains. IgY anti-*S. typhimurium* reduced immune cell recruitment and cytokine release in a mouse model infected with these bacteria (Li et al., 2016c). A combination treatment of a probiotic with IgY anti-*S. enteritidis* also decreased colonization and fecal shedding in young, market-aged broiler chicks challenged with *S. enteritidis* (Rahimi et al., 2007), indicating the additional potential benefit of using IgY anti-*Salmonella* antibodies in animals consumed as food.

27.15.2.5 Clostridiosis (*Clostridium difficile*)

Clostridium difficile (CD) is a cause of morbidity due to diarrhea and mortality due to inflammation of the colon, especially in the elderly, immunosuppressed, and after chronic antibiotic use. This serious condition has been increasing in incidence. IgY anti-CD has shown promise for patients based on animal studies (Khanna & Pardi, 2016). For example, 0.5 mg of IgY against the CD's FliD colonization-associated factor administered by gavage prevented the adhesion of CD and significantly enhanced survival rates in CD-challenged Syrian hamsters (Mulvey et al., 2011). Oral gavage treatments of 0.6 mg of anti-CD spore IgY delayed diarrhea onset and reduced spore adhesion to intestinal cells in mouse models, especially when coupled with an existing antibiotics treatment such as vancomycin (Pizarro-Guajardo et al., 2017). IgY anti-CD toxin A and B neutralized toxins and prevented recurrent infections in a hamster model (Kink & Williams, 1998) (209). Delivery of IgY anti-CD toxins to the colon instead of the upper gastrointestinal tract was enhanced when IgY was coated on microbeads (Zhang et al., 2016a) or encapsulated in chitosan-Ca pectinate microbeads (Xing et al., 2017) in a rat model.

27.15.2.6 Tetanus (*Clostridium tetani*)

Tetanus neurotoxin (TeNT), the product of Clostridium tetani, is the causative agent of the fatal disease tetanus. It is estimated 58,000 neonates and an unknown number of mothers die every year from tetanus (Thwaites et al., 2015). The tetanus bacteria typically enter the body through a cut or puncture wound and release the toxin in increasing amounts. When the neurotoxin reaches the nervous system it triggers increased rigidity of voluntary muscles, mainly those of the face, body, legs, neck, and tail (in animals). The steady and prolonged rigidity of the affected muscles ultimately leads to spasms and death. The neurotoxin is composed by two chains, one of 100 kDa (heavy chain) and another of 50 kDa (light chain), interacting through an interchain disulphide bond. The heavy chain is responsible for binding to polysialogangliosides and nidogen of nerve cell membranes and the light chain has a catalytic function, cleaving VAMP/synaptobrevin and blocking inhibitory synaptic vesicle release. Finally, an imbalance occurs between the inhibitory and excitatory synaptic vesicles on the motor neurons, leading to interruptions in muscle contraction and spastic paralysis (Surana et al., 2018). In a study, Selim et al., investigated the application of IgY both prophylactically and therapeutically for tetanus treatment (Selim et al., 2015). Results showed that all mice in the therapeutic groups as well as a prophylactic groups survived after a challenge with 2 minimum lethal dose (MLD) of *C. tetani*; similar results were obtained in donkeys. These data confirmed that the IgY approach was as effective as the equine IgG approach in tetanus therapy (Selim et al., 2015).

27.15.2.7 Botulism (*Clostridium botulinum*)

Botulinum neurotoxin (BoNT) is produced by *Clostridium botulinum* under anaerobic conditions and is known as one of the most poisonous substances in the world (Peck, 2009). To date a total of seven different BoNT toxin types are described (A to G). Botulism usually occurs as a food poisoning caused by botulinum neurotoxin produced by *C. botulinum*. BoNT consists of a heavy chain (100 KDa) and a light chain (50 KDa) that interact via a disulphide bond. Just like TeNT, the heavy chain binds to polysialogangliosides on the nerve plasma membrane, to G1b gangliosides, with high specificity and affinity and whole toxin is internalized to the cytoplasm via endocytic vesicles. The light chain has an endopeptidase activity and cleaves the SNARE proteins VAMP/synaptobrevin 1–3 and syntaxin. This phenomena inhibits neurotransmitter (acetylcholine) vesicle release which results in muscle fiber paralysis (Rossetto et al., 2011). In a number of studies, BoNT/A or BoNTB or a combination were used to generate IgY for treatment of mice or birds (Pauly et al., 2009; Li et al., 2013; You et al., 2014a). In all cases IgY showed a protective effect. Another important fact to significantly reduce botulism mortality, is the need for a fast diagnosis of BoNT toxin. Indeed a rapid and accurate test for the botulinum neurotoxin is essential for BoNT prevention and therapy. IgY was developed against either a linear peptide substrate (SNAP25) (Li et al., 2013) or BoNT A/B and D (Doellgast et al., 1997) and both showed robust results for toxin detection applications, namely in food or clinical conditions. Notably, Doellgast et al., develop an enzyme linked immunosorbent assay and an enzyme linked coagulation assay (ELISA-ELCA) for high sensitivity detection of anti-neurotoxin in human sera (Doellgast et al., 1997).

27.15.2.8 Necrotic Enteritis (*Clostridium perfringens*)

Clostridium perfringens type A is the most common bacterial infection associated with necrotic enteritis (NE). The main toxin produced by *Clostridium perfringens* type A is alpha toxin (CPA), which can have critical roles in pathogenesis of NE. This toxin can induce mucosal damage in chicken intestinal loops and has been associated with NE lesions in germ-free chickens. Anti-CPA serum applications have been shown to effectively neutralize the effects of the toxin (Doellgast et al., 1997). Specific IgY has been shown to be effective in protecting birds lowering the presentation of NE lesions.

27.15.2.9 Gastroenteritis (*Campylobacter jejuni*)

The *Campylobacter* species, in particular *Campylobacter jejuni*, is the most common cause of gastroenteritis in humans worldwide (WHO, 2021a). Most human *C. jejuni* infections are caused by the consumption of contaminated poultry. However, *C. jejuni* seems to have a commensal relationship with chickens while acting as a pathogen in humans (Young et al., 2007). Because of the use of growth-promoting antibiotics in the meat-producing industry, there is a rise in antibiotic-resistant *Campylobacter* strains (Alfredson & Korolik, 2007). IgY against *C. jejuni* may provide an alternative to antibiotic use (Al-Adwani et al., 2013; Thibodeau et al., 2017). Because of its high specificity and limited cross-reactivity, IgY anti-*C. jejuni* also can provide a highly accurate method to detect food contamination with *C. jejuni* (Hochel et al., 2004; Horak & Hochel, 2005).

To reduce the spread of *C. jejuni* from poultry to humans, IgY has been administered as a passive vaccine to chickens. IgY against *C. jejuni* has prophylactic and therapeutic effects through its ability to decrease overall and fecal bacterial levels in *C. jejuni*-challenged chickens (Tsubokura et al., 1997; Hermans et al., 2014; Nothaft et al., 2016). Similar results were also observed using IgY against *C. jejuni* adhesins and flagellins, which significantly reduced caecal colonization by *C. jejuni* and initiated the production of *C. jejuni* specific antibodies, although these may not play a role in protection (Neal-McKinney et al., 2014; Chintoan-Uta et al., 2016). Administration of anti-*C. jejuni* IgY also resulted in a significant reduction in transmission of *C. jejuni* to non-inoculated birds (Hermans et al., 2014), without altering the microflora of the intestinal tract (Nothaft et al., 2016).

27.15.2.10 Sepsis

27.15.2.10.1 *Escherichia coli*

Escherichia coli is an integral constituent of the mammalian microflora, with several pathotypes of *E. coli* implicated in the development of enteric and extraintestinal diseases such as diarrhea, sepsis, meningitis, and urinary tract infections (Kaper et al., 2004). Among the enteric *E. coli*-associated diseases, there are at least six different categories: enterotoxigenic *E.coli*, enteropathogenic *E. coli*, enterohemorrhagic *E. coli*, enteroaggregative *E. coli*, enteroinvasive *E. coli*, and diffusely adherent *E. coli* (Nataro & Kaper, 1998). Due to widespread antibiotic resistance of *E.coli* (Sáenz et al., 2004), IgY may serve as an alternative method to neutralize virulent *E. coli* in food, animals, and humans.

The major factor in the pathogenicity of *E. coli* is production of Shiga-like toxin. Numerous assays using IgY have been developed to detect the presence of Shiga-toxin-producing enterohemorrhagic *E.coli*. For instance, an ELISA assay involving IgY against *E.coli* O157:H7 was able to detect as little as 40 CFU/mL of *E. coli* O157:H7, suggesting that such assays can be used for detecting foodborne pathogens (Sunwoo et al., 2006). Furthermore, because Shiga-toxin is uniformly expressed by all enterohemorrhagic *E. coli*, IgY can be used to detect different serotypes and variants of Shiga-toxin-producing *E. coli* (Parma et al., 2012). Similarly, toxin-specific IgY has also been used to detect and neutralize heat-labile toxin produced by enterotoxigenic *E. coli* (Akita et al., 1998). Chicken or ostrich IgY against *E. coli* O157:H7 and O78:K80 inhibited bacterial growth in liquid medium (Sunwoo et al., 2006; Mahdavi et al., 2010a; Meenatchisundaram et al., 2011; Tobias et al., 2012).

A randomized, double-blind, placebo-controlled trial evaluated 301 Guatemalan children (154 intervention and 147 placebo) with acute non-bloody diarrhea who received PTM202 (combined IgY specifically targets rotavirus, enterotoxigenic *E. coli*, Shiga toxin-positive *E. coli*, and salmonella) or placebo for 3 days (Gaensbauer et al., 2017). PTM202 led to a reduction in duration of diarrhea among children whose diarrheal stool at enrollment contained one or more PTM202-targeted organisms. No adverse events were reported.

27.15.2.10.2 Enterohemorrhagic *Escherichia coli* Infection

Enterohemorrhagic *Escherichia coli* (EHEC) causes a spectrum of human diseases, including diarrhea, hemorrhagic colitis, disordered consciousness, renal failure and hemolytic uremic syndrome (HUS). The bacterium colonizes the large intestine and produces shiga-toxins (Stxs) as a major virulence factors. EHEC produces two types of Stxs, Stx-1, and Stx-2. Stxs released into the intestinal lumen enter the systemic circulation and reach target organs. Stxs are multimeric proteins that consist of an A subunit and a pentamer of B subunits. The B pentamer is responsible for the toxin binding to target cell-surface glycolipid receptors (Linderberg et al., 1987; Lingwood et al., 1987). After binding, the toxin is internalized into the cells by endocytosis, and then a fragment cleaved from A subunit exerts its toxic activity by inhibiting protein synthesis.

Monoclonal antibodies against Stxs have been investigated for the prevention of Stx-mediated diseases (Dowling et al., 2005; López et al., 2010). The antibodies have been designed mostly for parenteral administration to neutralize Stx in the systemic circulation. However, EHEC produces Stx in the intestine, from where the toxin enters the bloodstream to reach the target organ. While we believe that neutralization of Stxs in the intestine by oral treatment is more desirable than neutralization in the vessel for preventing Stx-mediated diseases, oral administration requires greater quantities of antibodies compared with parenteral administration. Recently, oral administration of chicken IgY antibody has been demonstrated to prevent or control several intestinal infections. Chicken IgY is thought to be an economical source to obtain large amounts of antibody that can be given orally (Mine & Kovacs-Nolan, 2002). We focused IgY antibodies against Stxs for oral use to prevent Stx-mediated diseases (Neri et al., 2011; Neri et al., 2012).

In conclusion, immunization with Stx-1 and Stx-2 holotoxins induced IgY antibodies in chickens. Their neutralizing activity against Stxs was exerted through binding activity to polymeric form of Stx B subunit. The oral administration of anti-Stx-2 IgY prevented the death of mice infected intestinally with Stx-2-producing EHEC O157:H7. Immunization with recombinant Stx B subunit also induced neutralizing IgY antibody in chickens. Oral administration of the IgY is thought to be promising to prevent Stx-mediated diseases. The IgY may have the potential to reduce the risk of HUS due to the use of antibiotics by the combined treatment.

27.15.2.10.3 *Staphylococcus aureus*

Staphylococcus aureus is a pathogenic bacterial strain that causes food poisoning, toxic shock syndrome, endocarditis, sepsis, soft tissue infections, and in-hospital infections (Lowy, 1998; Henderson and Nimmo, 2018). Although *S. aureus* is normally found in mammals and 30%–50% of humans, it remains a dangerous pathogen due to its endotoxicity, virulence, invasiveness, and antibiotic resistance (Le et al., 2003). *S. aureus* has increasingly displayed resistance to common antibiotic treatments such as methicillin and vancomycin (Smith et al., 1999; Klevens et al., 2007) and antimicrobial-resistant strains are now detected in the community, not just in healthcare settings (Chambers, 2001; Saveli et al., 2011). Stronger measures to contain *S. aureus* infections, as well as alternative and combination treatments, are now promoted to combat increasing antibiotic resistance. In vitro, specific IgY had high binding specificity and inhibited bacterial growth in culture, possibly by interrupting interactions with surface antigens (Guimarães et al., 2009b). IgY generated against *S. aureus* also caused agglutination of the bacterium, and did not show cross-reactivity with other bacterial strains such as *Streptococcus epidermidis*, *Escherichia coli*, and *Pseudomonas aeruginosa* (Jagadeeswari et al., 2015). The anterior nare of the nose is the most frequent carriage site for *S. aureus* (Kluytmans et al., 1997). When the nares are treated topically to eliminate nasal carriage, in most cases the organism also disappears from other body areas. Anti-*S. aureus* IgY given intranasally may be of special interest in the treatment of this pathogen, including for the growing threat of methicillin-resistant *S. aureus* (Jernigan et al., 2020).

27.15.2.10.4 *Aeromonas*

Aeromonas is found in aquatic environments and the micro flora of animals and humans. However, certain strains of *Aeromonas* have been implicated in the development of sepsis and gastroenteritis in humans (Merino et al., 1995), as well as fish, other animals, and environmental reservoirs (Parker & Shaw, 2011). Notably, almost all subspecies of *Aeromonas* express strong resistance to beta-lactam antibiotics such as penicillin, ampicillin, and carbenicillin, which has led to the pursuit of alternative and combination antimicrobial therapies (Altwegg & Geiss, 1989). IgY-specific antibodies have been investigated as an alternative method to diagnose, prevent and treat the *Aeromonas*-diseased aquatic animals (Li et al., 2006; Gan et al., 2015; Qin et al., 2018; Fernandes et al., 2019a).

27.15.2.11 Lung Infections

27.15.2.11.1 Treatment of *Pseudomonas aeruginosa* in Cystic Fibrosis

Cystic fibrosis is a common, hereditary, and life-threatening disease associated with damage to the lungs, pancreas, and male sex organs (Collins, 1992; Govan & Deretic, 1996). Patients with cystic fibrosis are especially prone to debilitating chronic lung infections caused by bacteria such as *Pseudomonas aeruginosa* (Ulrich et al., 2010). Due to a fear of developing antibiotic-resistant strains, alternative treatments to chronic antibiotic therapy have been studied, including the use of IgY against *P. aeruginosa* as a method of passive immunization. Anti-*P. aeruginosa* IgY significantly increased the neutrophil-mediated respiratory burst and subsequent bacterial killing of *P. aeruginosa in vitro* (Thomsen et al., 2015; Thomsen et al., 2016). Anti-*P. aeruginosa* IgY also inhibited murine pneumonia when administered intranasally as evidenced by reduced bacterial burden, inflammatory cytokines, inflammation of the lung tissue, and clinical symptoms, an effect enhanced by pretreatment with azithromycin 8 (Thomsen et al., 2021). The benefit of specific IgY anti-*P. aeruginosa* is believed to be against the flagellin protein implicated in the motility, adhesion, and inflammation of *P. aeruginosa* (Nilsson et al., 2007a).

In 17 patients with cystic fibrosis, prophylactic continuous oral treatment (solution gargled for 2 min and swallowed in the evening) with specific IgY against *P. aeruginosa* to prevent pulmonary infections for up to 12 years (114 patient-years) showed significant reduction in *P. aeruginosa* infections compared with 23 cystic fibrosis control patients, with no adverse events (Nilsson et al., 2008).

A randomized, double-blind, placebo-controlled Phase 3 trial of 164 patients age 5 and older with cystic fibrosis was conducted at 47 European sites from 2011 to 2015 to evaluate treatment with oral anti-*P. aeruginosa IgY* gargling solution ($n = 83$) vs. placebo ($n = 81$). Study drug was to be gargled and then swallowed once daily in the evening. Study duration (and primary outcome measure) was until the next *P. aeruginosa* infection was diagnosed or two years, whichever came first. There was no significant difference between treatment groups in time to first recurrence of *P. aeruginosa* infection (median, 26.3 months for IgY-treated group) or secondary endpoints of number of exacerbations, number of days of illness, and use of antibiotics (EU Clinical Register, 2021). Despite findings of no efficacy, this study provided an excellent safety database. A total of 1972 adverse events (AEs), mostly mild in severity, were reported, 989 of which were in the placebo group and 980 in the IgY group. The incidence of AEs was also similar between the two groups. The most commonly reported AEs were abdominal pain, vomiting, pyrexia, nasopharyngitis, and upper respiratory tract infection. No deaths occurred. Only 5 AEs in the IgY-treated group and 20 AEs in the placebo group (none serious) were judged to be related to the study drug.

27.15.2.12 Tuberculosis

Mycobacterium tuberculosis (MBTC) is responsible for the development of tuberculosis, a potentially fatal respiratory disease that also can cause extrapulmonary disease (e.g., of the urinary system). MBTC is increasingly becoming more difficult to treat due to antibiotic resistance (Chauhan et al., 2021). In a rat peripheral blood mononuclear cell model, administration of high concentrations of IgY anti-MBTC increased interleukin-2 and interferon expression (Sudjarwo et al., 2017a). IgY against MBTC may warrant evaluation for use in combination with other immunotherapeutic treatments of tuberculosis (Sudjarwo et al., 2017b). In addition, IgY may also be of interest as a novel treatment of pulmonary nontuberculous mycobacteria, regarded as more challenging to treat because of frequent antimicrobial intolerance, toxicities, resistance, and drug-drug interactions (Shulha et al., 2019).

27.15.2.13 Nosocomial Infections

Acinetobacter baumannii, a gram-negative bacterium, is the cause of nosocomial infections and outbreaks in hospitals worldwide, such as sepsis, urinary tract infections, pneumonia, or surgical wound infections. Due to its resistance to desiccation and antimicrobial agents, *A. baumannii* is associated with significant mortality, costs, and hospital stays, particularly in vulnerable patients (Fournier & Richet, 2006). Shi et al. produced specific IgY against multi-drug resistant strains of *Acinetobacter baumannii*. In a mouse model of *A. baumannii*-associated pneumonia, intraperitoneal anti-*A. baumannii* IgY specific to pan-drug-resistant strains reduced levels of inflammatory cytokines, lung inflammation, and mortality (Shi et al., 2017). Similar results were also seen with intraperitoneal injections of 40 μg of IgY developed against the inactivated whole-cell or outer membrane proteins of *A. baumannii*, which protected against nasally challenged mice, possibly by inhibiting bacterial adhesion (Jahangiri et al., 2019).

27.15.2.14 Skin-Related Infections

27.15.2.14.1 Propionibacterium

Acne vulgaris is a skin condition that affects most humans at some time and is thought to be caused by multiple factors, including increased sebaceous gland sebum production, hormones, cytokines, nutrition, and bacteria such as *Propionibacterium acnes* (Kurokawa et al., 2009). Because of rising antibiotic resistance, IgY has also been proposed as a cost-effective alternative to antimicrobial treatments of acne. IgY anti-*P. acnes* inhibited growth of *P. acnes* colonies as well as biofilm development by preventing bacterial adhesion (Revathy et al., 2014).

27.15.3 Fungal Infection

27.15.3.1 Candidiasis

Candidiasis is one of the most common oral fungal infections. From over 8.7 million eukaryotic species identified to the current date, the kingdom of fungi has approximately 611,000 species, making up 7% of all eukaryotic species (Mora et al., 2011). At the cellular level, fungi are more related to humans than bacteria and belong to the Eumycota group, as chemoheterotrophic organisms (Khan et al., 2010). However, only 600 species of fungi are able to cause infections in humans (Mora et al., 2011). The genus *Candida* includes about 150 species, but many species are endosymbionts of humans, causing infections mainly in immunosuppressed hosts. Around 80% of infections are caused by *Candida albicans,* although *Candida non-albicans* infections (*Candida glabrata*, *Candida tropicalis, Candida krusei, Candida dubliniensis*) are becoming more and more frequent (Mark & Roberts, 2016; Ciurea et al., 2020). Candidiasis is one of the most common oral fungal infections in patients with impaired immune system and has a high morbidity with approximately 85% of patients being infected at some point during their illness. As we know that HIV and AIDS patients are susceptible to opportunistic infections. Oral candidiasis or thrush is the primary manifestation of fungal infection in these patients. (Noël de Tilly & Tharmalingam, 2022). In this section, author highlighted the novel treatment options of candidiasis with anti C. albicans IgY added in gel (Takeuchi et al., 2016); denture base (Kamikawa et al., 2016), tablet (Ibrahim et al., 2007), photosensitizing phthalocyanine dye (IRDye700Dx; IR700) (CA-IgY-IR700) (Yasui et al., 2021), including intravenous immunoglobulin antibodies (Pedraza-Sanchez et al., 2018).

A gel preparation for oral use containing IgY against *Candida albicans* was tested by Tekeuchi et al. and caused a reduction in the number of colony-forming units (CFU) on the oral cavity of elderly people, showing promise for prophylactic use against *C. albicans* oral infection (Takeuchi et al., 2016). In other research, specific IgY inhibited the adhesion of *Candida albicans* and *Candida glabrata* to denture base material. Anti-*C. albicans* IgY was more effective against *C. albicans* than anti-*C. glabrata* IgY, while both antibodies were equally effective in preventing the adhesion of *C. glabrata* (Kamikawa et al., 2016).

Treatment of oral candidiasis is relatively simple and effective for the healthy patient. Typically, topical medications are adequate usually involving the use of a commonly prescribed anti-fungal agent, nystatin oral suspension. To be effective, topical medications must be in contact with the organism to eliminate it. Since patients are usually unable to hold liquids in their mouth cavity, antibiotic-supplemented lozenge tablets are used wherein the tablet dissolves slowly allowing the drug to be present for a longer length of time in the oral cavity. From another perspective, development of cross resistance has primarily been a problem with fluconazole in AIDS treatment. Inasmuch as antibiotic resistance correlates with clinical failure, oral passive immunization with IgY for the control of oral candidiasis acquires a special relevance. Toward this end, a clinical trial was performed by our group in two healthy elderly volunteer subjects. A tablet supplemented with egg yolk antibody against *Candida albicans* (CA-IgY) was prescribed for 4 weeks daily and treatment was stopped for 4 weeks. This treatment cycle was repeated three more times with both subjects being examined every week. The patient who received tablets containing anti-CA IgY revealed reduced the number of salivary Candida CFU. With each 4-week pause in treatment, *Candida albicans* count gradually increased to previous level (Ibrahim et al., 2007) indicating that the mode of action was specific for the anti-CA IgY and reduction of Candida CFU is feasible with regular treatment.

Near-Infrared Photo-Antimicrobial Targeting Therapy for Candidiasis is reported by Yasui et al. (Yasui et al., 2021). Near-infrared photoimmunotherapy (NIR-PIT), originally developed as a cancer treatment, specifically kills cancer cells via a photosensitizing phthalocyanine dye (IRDye700Dx; IR700)-conjugated monoclonal antibody, and irradiating NIR light. IgY-photo-antimicrobial targeting therapy (IgY-PAT), exploiting NIR-PIT, is investigated to destroy only microorganisms. IR700 is conjugated with anti-*Candida albicans* IgY

(CA-IgY) to generate CA-IgY-IR700, which specifically binds various *Candida* spp. (and not human skin cells). The antimicrobial effect of CA-IgY-PAT is dependent on the NIR-light dose. CA-IgY-PAT significantly reduces the area of ulcers in a mouse model of CA-infected cutaneous ulcers ($p < 0.0001$), indicating that CA-IgY-PAT is a new promising therapeutic method for CA infection. (Yasui et al., 2021).

These novel forms of therapy are important in consideration of the arms race in successfully treating fungal infections while preventing the evolution of resistant strains. Like bacteria, fungi are constantly evolving and developing resistance to existing treatments and prophylaxis application (Noel de Tilly and Tharmalingam, 2022).

27.15.4 Protozoal Infection

27.15.4.1 Trypanosoma

The protozoan parasite, *Trypanosoma cruzi,* found in mammals and triatomine bugs in the Americas, causes Chagas disease, a zoonotic disease that can be transmitted to humans by insect vectors, blood-sucking triatomine bugs. Chronic infection can lead to heart and gastrointestinal disease that can be life-threatening (Bern et al., 2011; Elliot et al., 2015). Vaccines developed against *T. cruzi* are only partially protective since defined antigens must be used to prevent the occurrence of cryptic infections (Brener, 1973). Furthermore, drugs approved for the treatment of Chagas disease can have toxic, mutagenic, and other adverse side effects (Castro et al., 2006). Since no cytotoxic or proliferative effects were observed on mononuclear and VERO cells *in vitro* when treated with IgY against *T. cruzi*, specific IgY has been considered as a possible therapeutic for Chagas disease (Grando et al., 2017). In a mouse model, anti-*T. cruzi* IgY administered prophylactically at 50 mg/kg reduced parasitemia post-challenge and prevented the development of cardiac lesions by amastigotes. These same effects were also observed with the therapeutic administration of 50 mg/kg of IgY, which also improved the immune response by preventing an increase in activity of E-NTPDase and E- ADA activities in the splenic lymphocytes of the animals (Grando et al., 2018).

Another member of the *Trypanosoma* family, *T. evansi*, infects a wide range of domesticated livestock worldwide (Luckins, 1988), causing anemia (Trail et al., 1990). One case of *T. evansi* infection in humans has also been documented, possibly by blood transmission from an infected animal (Joshi et al., 2005). Current strategies for controlling the dissemination of *T. evansi* include herd culling (Herrera et al., 2004) and chemical therapy of infected animals, although this latter approach has limited use due to high toxicity and the development of drug-resistant strains (Silva et al., 2002). In contrast, no significant cytotoxic or genotypic toxicity was observed when IgY anti-*T. evansi* was used to treat peripheral blood samples, although an increase in cell viability and lymphocyte proliferation was observed when a concentration of 10 mg/mL of specific IgY was used (Sampaio et al., 2014a). In vivo, specific IgY against *T. evansi* administered intraperitoneally at a dose of 10 mg/kg increased the longevity and survival of infected animals (Sampaio et al., 2014b).

27.15.4.2 Cryptosporidiosis

Several members of the *Cryptosporidium* family are implicated in the development of cryptosporidiosis, an intestinal infection that causes diarrhea and, less commonly, pneumonia in humans (Fayer, 1997). There are no therapies to fully treat cryptosporidiosis or prevent the infection in humans and animals (Pinto & Vinayak, 2021) although hydration and passive immunization through the administration of monoclonal antibodies (Perryman et al., 1993), nitazoxanide, or hyperimmune bovine colostrum have limited efficacy (Nord et al., 1990; Ungar et al., 1990). IgY has been explored as a treatment against *Cryptosporidium* infection. In vitro, IgY antibodies generated against *C. parvum* oocyst antigens were highly specific (Hashemzadeh and Shahbazi, 2016), decreased binding of the parasite to Caco-2 cells, and blocked the vitality of *C. parvum* (Kobayashi et al., 2004). However, in a severe combined immunodeficiency mouse model, treatment using feed containing 25% specific IgY powder and a 20% specific IgY solution was only capable of partially reducing oocyst shedding in challenged animals, and was unable to eliminate infection. Similarly, IgY against the P23 protein in *C. parvum* also has high specificity for the parasite (Shahbazi et al., 2009; Omidian et al., 2014). Using a mouse model, the anti-P23 IgY reduced oocyst shedding by 70%. Specific IgY against the GP60 glycoprotein in *Cryptosporidium hominis*, another strain implicated in cryptosporidiosis, was also found to specifically bind to the antigen, as well as the parasite (Miura et al., 2017). The high specificity and protective effectiveness of these antibodies suggest that IgY warrants evaluation as a novel diagnostic test for cryptosporidiosis and a passive immunization treatment in immunocompromised individuals.

27.15.5 Other Important Applications of IgY

27.15.5.1 Antiobesity Activity

IgY raised against porcine pancreatic lipase was used against the enzyme *in vitro* and *in vivo*. Later, mice with obesity induced by high fat diet were orally treated with the antibody, which was given concomitantly with food, and a reduction of adipose tissue and liver fat level was observed, as well as an increase of fecal excretion of triglycerides and their decrease in blood plasma. Anti-lipase IgY inhibited the hydrolysis of diet fat and reduced its intestinal absorption, showing anti-obesity activity (Hirose et al., 2013; Tarigan et al., 2016).

27.15.5.2 Antiallergic Activity

Wei-xu et al. evaluated the antiallergic effect of specific IgY against the pro-inflammatory cytokines IL-β1 and TNF-α in guinea pigs with induced allergic rhinitis. A reduction of the eosinophils number in the blood and in the nasal and bronchial lavages was found, as well as a decrease of eosinophils, neutrophils and lymphocytes infiltration into the nasal mucosa and the lungs of animals treated with anti-IL-β1 and anti-TNF-α IgY, alone or jointly (Wei-xu et al., 2016).

27.15.5.3 Antitumor Activity

The phylogenetic distance between birds and mammals ensures a stronger immune response against mammalian antigens by birds (Gassmann et al., 1990). Such a feature may be advantageous to produce IgY against human tumor antigens. Following this rationale, Amirijavid et al. produced highly specific IgY against a sequence of 21 amino acids present on the ectodomain of the TRAIL (TNF-related apoptosis-inducing ligand) receptor TRAIL-R2 (DR5). The antibodies bound to the amino acid sequence and activated the DR5 receptors in human breast cancer cells MCF7, acting as a TRAIL agonist and inducing apoptosis (Amirijavid et al., 2016). IgY against other receptors, such as the HER2 receptor, was tested coupled to single walled carbon nanotubes (SWNTs) and specifically detected the HER2 receptors on the surface of SK-BR-3 cells. The binding of the complex to the receptors was measured by Raman signals emitted by the nanotubes. SWNT has a near infrared absorption (NIR), which can be used for tumor ablation, and, coupled to anti-HER2 IgY, was able to kill SK-BR-3 cells without needing internalization of the complex by the cell (Xiao et al., 2009). These findings show that IgY produced against tumor antigens is an attractive alternative for a more selective treatment of cancers and its use could, therefore, minimize the side effects of traditional chemotherapy.

27.15.5.4 Anti-Venom Activity

One of the side effects that occur in individuals receiving anti-venom serum produced in goats, sheep and horses is due to the presence of serum proteins on the anti-venom serum derived from these animals, in which IgG is not sufficiently purified (Araújo et al., 2010; Sjostrom et al., 1994). One advantage of using IgY in anti-venom serotherapy is that it is easily purified, which would minimize the occurrence of side effects due to nonspecific proteins. Arqo et al. demonstrated this property when specific IgY was produced as anti-venom of the snake genus *Bothrops* sp. These antibodies neutralized a pool of venoms from five *Bothrops* species, with an ED50 of 150 μL/2LD50, showing little to no side effects in mice (Araújo et al., 2010).

Mendoza et al. also produced IgY capable of neutralizing the venom of the peruvian snake *Bothrops atox*. The anti-venom IgY showed considerable cross reaction with the venom of *Bothrops brazili* and could be used not only as *B. atox* anti-venom, but also as a tool for the research of cross reaction with venoms from different species (Mendoza et al., 2012).

In another elegant work, Andrade et al. produced IgY against a pool of venoms from snakes of the genus *Bothrops* and against the venom from the species *Crotalus durissus terrificus*. Anti-venom IgY extracted from eggs was compared to the horse anti-venom IgG in Western blot. The results showed that specific egg's IgY recognized the same antigens as the equine anti-venom (de Andrade et al., 2013).

IgY against coral snake venom was first produced in response to a pool of venoms from different species of *Micrurus*. These antibodies recognized, by Western blot, venom proteins from several snakes: *M. isozonus*, *M. surinamensis*, *M. f. fulvius*, *Naja kaouthia*, *N. pallida*, *Bothrops colombiensis*, *Crotalus durissus cumanensis*, and *C. vegrandis* and could, therefore, be used as a broad-spectrum snake anti-venom (Aguilar et al., 2014).

Zolfagharian and Dounighi produced IgY by inoculating the *Vipera lebetina* snake venom, inactivated by γ radiation, in hens (Zolfagharian & Dounighi, 2015). These antibodies were effective in neutralizing the crude venom of *Vipera lebetina* in mice.

Anti-venom IgY were also obtained from eggs of hens immunized with the venom of the snake *Trimeresurus albolabris*. These IgY recognized, by Western blot, most of the proteins present in the *T. albolabris* venom and neutralized it in mice in a dose-dependent manner (Duan et al., 2016).

More recently, Liu et al. extracted and purified IgY from eggs of hens inoculated with the venom of the *Deinagkistrodon actus* snake. These antibodies were able to neutralize the lethal effects of the venom, such as bleeding, edema formation and myotoxicity in a dose-dependent manner (Liu et al., 2017).

da Rocha et al. produced IgY against ophidian toxins of *Crotalus durissus terrificus*, *Bothrops jararaca* and *Bitis arietans*. The antibodies were able to bind to specific components of the venoms in Western blot and protected 100% of the intoxicated mice when obtained after the ninth inoculation. The authors recommended the use of a small antigen dose (20 μL) applied in successive inoculations for IgY production, since this dose was enough to genetically alter the V(D)J segments on the naïve cells and to generate immunological memory (da Rocha et al., 2017).

However, IgY raised against the venom of the snake *Oxyuranus scutellatus* was less effective than equine IgG, being unable to neutralize the neurotoxic and coagulant effects of the venom (Navarro et al., 2016). Nonetheless, this result cannot be extended to IgY produced against other venoms.

IgY against the *Tityus caripitensis* scorpion venom, produced by Alvarez et al., neutralized not only the venom of *T. caripitensis*, but also that of other *Tityus* species (*T. quirogae*, *T. discrepans*, and *Tityus gonzalespongai*), and inactivated the hyaluronidase, an enzyme that facilitates the toxin spread in the tissues, present in the *T. serrulatus* venom (Alvarez et al., 2013). Thus, IgY raised against *T. caripitensis* venom could be used as a broad- spectrum anti-scorpionic serum.

27.15.5.5 Prophylaxis of Celiac Disease

Another application of IgY technology, the prophylaxis of Celiac disease, was demonstrated by Gujral et al., who developed powdered egg yolk formula with protective sugars containing anti-gliadin IgY, among which, the formula with mannitol (EYP-M) retained its activity after being submitted, *in vitro*, to chemical conditions analogous to those of the stomach and small intestine. The formula IgY-EPY-M neutralized *in vitro* both the isolated gliadin and that present in food matrix and inhibited its intestinal absorption in mice, showing promising for the prevention of Celiac disease (Gujral et al., 2012).

27.15.5.6 Prophylaxis of Toxicity

Bobeck et al. used IgY against the human intestinal alkaline phosphatase (hIAP) to assess the influence of IAP on increased bioavailability of phytate phosphate in the presence of 1α-dihydroxycholecalciferol (vitamin D3) in chickens. Anti-hIAP IgY was ingested by chickens and reduced the absorption of phytate phosphate, which suggests that although it performed less adequately than sevelamer chorhydrate, already used for the same purpose, anti-hIAP IgY can be optimized for the prevention of phytate phosphate toxicity induced by the consumption of the active form of vitamin D (Bobeck et al., 2016).

27.15.5.7 Food Preservation

IgY raised against the bacterium *Listeria monocytogenes* showed a significant inhibitory effect of bacterial growth in liquid medium and in fish samples stored between 0°C and 6°C in a dose-dependent manner, which indicates that anti-*L. monocytogenes* IgY is a potential antimicrobial for use in the food industry (Sui et al., 2011). Taking into consideration the versatility and the range of IgY already tested against several bacteria, this result could be easily apply to other food poisoning bacteria and viruses.

27.15.5.8 Bioterrorism Circumstance

Among the importance of this technology, LeClair et al. demonstrated that IgY produced against the staphylococcal enterotoxin B (SEB), a potential biological weapon, can save individuals exposed to this material. The results with *Rhesus* monkeys showed that animals that received anti-SEB IgY 30 min before or 4 hour after exposure to a le- thal SEB aerosol survived (LeClaire et al., 2002), which indicates that anti-SEB IgY could be used to protect populations in a hypothetical context of bioterrorism involving SEB (Shade & Terzolo, 2006).

Previous applications of IgY in humans, including viral, bacterial, fungal, parasitic infections, and others are summarized in Table 27.13.

It may be mentioned here again the advantages of IgY in controlling animal, plant, fish, human and other diseases are (1) highly effective, (2) cost-effective, (3) egg collection is non-invasive, (4) treatment is safe and live organisms are not used, (5) procedure is environmentally friendly, (6) no toxic residues are produced and there is no development of resistance, and (7) treatment can be used to control many different types of pathogens.

27.16 IgY-Technology in Human Medicine: Patents and Clinical Trials

IgY technology has been successfully used for the development of many potential therapeutics (Pereira et al., 2019) and many of them are being tested in human clinical trials (Leiva et al., 2020). Leiva et al. reviewed the patents and clinical trials of IgY (egg yolk antibodies) in human medicine. This review provided a comprehensive analysis of IgY-based biologicalls for human medicine, including patent applications and clinical trials during the period 2010–2018, and addressed how IgY-technology can lead to innovation in the production of biologicals for the treatment and prophylaxis of a wide range of infectious and non-communicable diseases. Authors also listed IgY patent applications and clinical trials in tabular form (Leiva et al., 2020).

27.17 Utilization of IgY in Feeds and Foods

United States Code of Federal Regulations has granted egg powder, including IgY, as "GRAS generally recognized as safe" for use as food or food ingredients. Avian immunoglobulin is a super-drug (used in more than 250,000 doses of anti-*Pseudomonas* IgY in phase I, II, and III studies on protection for Pseudomonas infection in the lungs of CF patients.) that immediately must be considered in the world's fight against antibiotic resistance (Kollberg, 2017). GI Immune DF is a self-affirmed GRAS supplement that provides IgY immunoglobulins and immunoregulating molecules from a hyperimmune chicken egg. GI-Immune-DFHUGHPA.pdf (naplescfm.com) distributed by Hughes Center for Functional Medicine 800 Goodlette Road North # 270 Naples, FL 34102 239-649-7400.

IgY is gradually moving from feed to functional food to active pharmaceutical ingrédients. The results of all clinical trials (Rahman et al., 2012; Thu et al., 2017) indicate that the use of oral dosage form of anti-rotavirus IgY in a child-friendly carrier (such as infant milk formula or maltitol) is a promising, safe, and effective adjunct to the management of acute diarrhea in pediatric patients. The infant milk formula is particularly attractive as it addresses the nutritional and therapeutic needs of rotavirus-infected children in a user-friendly manner. For diarrhea patients below 5 years of age and who are lactose tolerant, the infant milk formula made available in powder form is an ideal vehicle for oral IgY delivery. It must be noted however that lozenges, tablets, or capsules for older children may also be used as oral delivery systems as had been done for some IgY products with various target infections now being sold in the Japanese and East Asian markets (Table 27.14).

While all these products are currently classified as food supplements, the forward trend is toward the pharmaceuticalization of IgY, which may only be a few years away in the future. The re-classification of IgY as a drug would provide a fresh breath of air in the field of therapeutics, where the dearth in novel synthetic antimicrobials has cast a pall of uncertainty to the healthcare industry and medicine end-users, especially now that we are facing increasingly resistant populations of pathogenic microorganisms.

27.18 IgY for Life: Holistic Approach

The health of humans, animals, and the environment are inextricably linked. The last decade of the twentieth century and the first decade of the twenty-first century saw the emergence of a plethora of public health challenges at the convergence of human, animal, and environmental health, including bovine spongiform encephalopathy (BSE) and variant Creutzfeldt–Jakob Disease, H5N1 influenza, Nipah virus, West Nile Virus, 9/11 and the threat of bioterrorism, SARS, and the impact of climate change on global food systems. Potential outcomes from the one health approach: (1) more interdisciplinary programs in education, training, research, and established policy;

TABLE 27.13

Use of IgY in Human Health

Pathogen	Disease	IgY Study Based on	Reference
A. Viral origin			
Rotavirus	Diarrhea	Randomized clinical trials	Rahman et al. (2012)
		Randomized clinical trials	Thu et al. (2017)
		Meta-analysis of 17 randomized clinical trials	Wang et al. (2019b)
Hepatitis B viruses (HBV)	Hepatitis	Duck HBV model	Thermet et al. (2004)
Zika virus	Fever, rash, arthritis, and microcephaly	In vitro and in vivo mice model	O'Donnell et al. (2019)
Dengue virus	Dengue fever	In vitro and mice model	Fink et al. (2017)
Hantavirus	Hantavirus pulmonary syndrome (HPS)	In vitro and in vivo Syrian hamsters	Brocato et al. (2012); Haese et al. (2015)
Ebola virus	Ebola virus infection	In vivo mice model	Zhang et al. (2021)
B. Bacterial origin			
Streptococcus mutans	Dental caries	Randomized clinical trials	Nguyen et al. (2011)
	Dental plaque	Human trial	Hatta et al. (1997)
Porphyromonas gingivalis	Periodontitis	Periodontitis patients	Yokoyama et al. (2007b)
	Periodontitis	Randomized clinical trials	Sugano (2009)
	Porphyromonas gingivalis infection	Near-infrared photo-antimicrobial targeting therapy	Maruyama et al. (2022)
Prevotella intermedia	Periodontitis	In vivo rat	Hou et al. (2014)
Solobacterium moorei	Halitosis, periodontitis, and gingivitis	In vitro and in vivo mouse	Li et al. (2012)
Fusobacterium nucleatum	Halitosis and periodontitis	In vivo rat model	Wang et al. (2019a)
Candida albicans	Candidiasis	In vitro, in vivo mouse and human	Ibrahim et al. (2007) (2008); Takeuchi et al. (2016)
	Candida infection	In vitro and Near-Infrared Photo-Antimicrobial Targeting Therapy	Yasui et al. (2021)
Helicobacter pylori	Gastritis		Marcus & Scott (2001)
		In vivo mouse	Malekshahi et al. (2011); Ameri Shah Reza et al. (2012); Yang et al. (2012)
		In vivo Mongolian gerbils	Nomura et al. (2005); Shin et al. (2002)
		In vivo humans	Yamane et al. (2003); Horie et al. (2004); Suzuki et al. (2004)
Salmonella typhimurium and *S. enteritidis*	Salmonellosis	In vitro	Lee et al. (2002); Chalghoumi et al. (2009)
	S. typhimurium infection	In vivo mouse model	Li et al. (2016c)
		In vivo young broiler chicks	Rahimi et al. (2007)
Clostridium difficile	Clostridiosis	In vivo Syrian hamsters	Mulvey et al. (2011)
		in mouse models	Pizarro-Guajardo et al. (2017)
		In vivo hamster model	Kink and Williams (1998)
Clostridium tetani, Tetanus neurotoxin (TeNT)	Tetanus	In vivo mouse models	Selim et al. (2015)
Clostridium botulinum Botulinum neurotoxin (BoNT)	Botulism	In vivo mice or birds	Pauly et al. (2009); Li et al. (2013); You et al. (2014a)
Campylobacter jejuni,	Gastroenteritis	In vivo birds	Hermans et al. (2014)
E. coli O157:H7 and O78:K80	Extraintestinal: sepsis, meningitis, and urinary tract infections	In vitro	Sunwoo et al. (2006); Mahdavi et al. (2010a)
E. coli	Respiratory, enteric, and septicemic diseases	In vivo birds	Kariyawasam et al. (2004)

(Continued)

TABLE 27.13 ***(Continued)***

Use of IgY in Human Health

Pathogen	Disease	IgY Study Based on	Reference
Enterotoxigenic *E. coli K88*	Gastrointestinal infection, Diarrhea	In vivo piglet	Wang et al. (2019c)
Enterohemorrhagic Stx-2-producing *Escherichia coli* (EHEC) O157:H7	Diarrhea, hemorrhagic colitis	In vivo mouse	Neri et al. (2011, 2012)
Enterotoxigenic Shiga toxin-positive *E. coli*	Non-bloody diarrhea	Randomized clinical trials	Gaensbauer et al. (2017)
Staphylococcus aureus	Food poisoning, toxic shock syndrome, endocarditis, and sepsis	In vitro	Guimarães et al. (2009b)
Aeromonas hydrophila	Sepsis and gastroenteritis in humans	In vivo fish	Merino et al. (1995); Fernandes et al. (2019a); Fernandes et al. (2019b)
Aeromonas Salmonicida	Gastroenteritis in humans; skin ulcer in fish	In vivo fish	Gan et al. (2015)
Propionibacterium acnes	Skin infection in human	In vitro	Revathy et al. (2014)
Pseudomonas aeruginosa	Cystic fibrosis, pneumonia	In vitro and In vivo mouse	Thomsen et al. (2016); Thomsen et al. (2021)
		Randomized clinical trials	Nilsson et al. (2008); EU Clinical Trials Register (2021)
Mycobacterium tuberculosis	Tuberculosis	In vivo rat	Sudjarwo et al. (2017a, 2017b)
Acinetobacter baumannii	Nosocomial infections in hospitals	In vivo mouse	Shi et al. (2017); Jahangiri et al. (2019)
Listeria monocytogenes	Food contaminants	In vitro	Sui et al. (2011)
(Food preservation)			
Staphylococcal enterotoxin B (SEB)	Bioterrorism	In vitro and in vivo monkey	LeClaire et al. (2002); Shade and Terzolo (2006)
C. Protozoal origin			
Trypanosoma cruzi	Chagas disease	In vitro and in vivo mouse	Grando et al. (2017, 2018)
Trypanosoma evansi	Anemia	In vitro and in vivo animal	Joshi et al. (2005); Sampaio et al. (2014a, 2014b)
Cryptosporidium parvum	Cryptosporidiosis	In vitro and in vivo mouse	Kobayashi et al. (2004); Hashemzadeh & Shahbazi (2016)
Cryptosporidium hominis	Cryptosporidiosis	In vitro	Miura et al. (2017)
D. Others			
Lipase	Obesity	In vivo mouse	Hirose et al. (2013)
Pro-inflammatory cytokines IL-β1 and TNF-α	Allergic rhinitis	In vivo guinea pigs	Wei-xu et al. (2016)
Human tumor antigens	Tumor, Cancer	In vitro	Xiao et al. (2009);
			Amirijavid et al. (2016)
Bothrops sp	Sanke Venom toxigenosis	In vitro and in vivo mouse	Araújo et al. (2010); Mendoza et al. (2012)
Crotalus durissus terrificus		In vitro	de Andrade et al. (2013)
Micrurus spp		In vitro	Aguilar et al. (2014)
Vipera lebetina		In vitro and in vivo mouse	Zolfagharian & Dounighi (2015)
Trimeresurus albolabris		In vitro and in vivo mouse	Duan et al. (2016)
Deinagkistrodon actus		In vitro and in vivo mouse	Liu et al. (2017)
Crotalus durissus terrificus,Bothrops jararaca, Bitis arietans		In vitro and in vivo mouse	da Rocha et al. (2017)
Tityus caripitensis, can cross neutralize other *Tityus* species (*T. quirogae, T. discrepans* and *T. gonzale- spongai*)			Alvarez et al. (2013)
Gliadin in food matrix	Celiac disease	In vitro and in vivo mouse	Gujral et al. (2012)
Intestinal alkaline phosphatase hIAP)	Phytate phosphate toxicity	In vitro and in vivo chick	Bobeck et al. (2016)

TABLE 27.14

IgY Supplemented Food Products Available Commercially in Global Biohealth Market

Product Name	Product Type	Country	Sales Start
Ovalgen HP Anti-*Helicobacter pylori* IgY for gastritis	Drinking yoghurt	Korea	2001.5
	Capsule	Japan	2001.10
	Shell egg	Japan	2003.4
	Tablet	Japan	2003.10
	Regular yoghurt	Japan	2004.7
	Drinking yoghurt	Taiwan	2004.8
	Tablet	Japan	2005.6
	Tablet	Japan	2005.6
	Drinking yoghurt	Japan	2010.7
	Sachet	Viet Nam	2015.2
Ovalgen DC Anti-*Streptococcus mutans* IgY for dental caries	Lozenge	Japan	2005.9
	Drinking yoghurt	Korea	2006.9
	Tablet	Japan	2007.5
	Lozenge	America	2009.8
	Lozenge	Japan	2010.5
	Lozenge	Viet Nam	2013.11
Ovalgen PG Anti-*Porphyromonas gingivalis* IgY for periodontitis/gingivitis	Lozenge	Japan	2005.9
	Lozenge	America	2009.8
	Lozenge	Viet Nam	2013.11
Ovalgen CA Anti-*Candida albicans* IgY for oral thrush or candidiasis	Dental gel	Japan	2012.5
Ovalgen FL Anti-Influenza IgY for seasonal flu	AC filter	Japan	2003.10
	Mask	Japan	2005.10
	Mask	Japan	2006.10
	Tablet	Japan	2008.12
	Lozenge	Viet Nam	2013.10
Ovalgen RV Anti-human rotavirus IgY for rotaviral diarrhea	Baby milk	Korea	2009.6
	Baby milk	Korea	2014.9
Ovalgen CS Anti- *Cronobacter sakazakii* IgY for infant milk formula	Baby milk	Korea	2014.9

Source: Adapted from Thu et al. (2017).

(2) more information sharing related to disease detection, diagnosis, education, and research; (3) more prevention of diseases, both infectious and chronic; and (4) development of new therapies and approaches to treatments

IgY antibodies are a sustainable and efficacious therapeutics for life and Global One Health. Application of egg-derived, supernatural, tailor-made IgY is the God-gifted miracle holistic approach for health. Diseases can be treated with holistic agent biological/medicine with a holistic approach. Many biohealth companies (like EW Nutrition Japan, IGY Life Sciences, NABAS, GenWay Biotech, Creative Biolabs, Aves lab, Creative diagnostics, IFY Inc.Mitacs, Bioinnovo IgY-DNT, Thermo Fisher, Sigma-Aldrich, Charles River laboratories, and so on (Table 27.15)) has optimized the sustainable generation of highly specific and efficacious egg-derived IgY antibodies, for human and animal health. These companies' pipeline includes marketed immune and sports health products, and lead compounds against COVID-19 and African swine fever and are ready upfront to prepare suitable solutions for the upcoming emergence of novel pathogens.

27.19 IgY and Egg Protein Industries and Markets

Some IgY companies in the global market are shown in Table 27.15. Many companies retail products and services based on IgY-derived technologies for research purposes, including R&D antibodies, antibody-based solutions, and diagnostic kits. For reference, these include companies such as Abcore Inc. (US), Agrisera AB (Sweden), Covalab (France), Creative Biolabs Inc. (US), Immuno Reagents Inc. (US), Innovagen AB (Sweden) and OriGene Technologies Inc. (US). To date, there is no systematic report on the IgY antibody markets, so one

TABLE 27.15

Summary of IgY Companies and Industries in the World

Name	Website	Country
AD Biotech Co.	http://adbiotech.com	South Korea
Aves Labs, Inc.	http://www.aveslab.com/	United States
Avianax LLC	https://www.avianax.com/	United States
Bioinnovo	http://bioinnovo.com.ar/	Argentina
Crystal Bioscience[a]	http://www.crystalbioscience.com/	United States
DAN Biotech, Inc.	http://www.danbio.com	South Korea
Davids Biotechnologie GmbH	https://www.davids-bio.de/	Germany
Eggcellent Proteins	https://www.eggcellentproteins.com/	Scotland
EW Nutrition	https://ew-nutrition.com/	Germany
Gallus Immunotech, Inc.[b]	https://www.exalpha.com/	United States
Gentian AS	https://www.gentian.com/	Norway
Good Biotech Corp.	http://www.good-biotech.com/	China
HenBiotech	http://www.henbiotech.com/	Portugal
IgY Immunologix	http://www.igylx.com/	India
IGY Life Sciences, Inc.	http://www.igylifesciences.com/	Canada
IgY Nutrition	https://www.igynutrition.com/	United States
IgYTechnology.com	https://www.igytechnology.com/	Portugal
Immune Biosolutions	https://immunebiosolutions.com/	Canada
NABAS	http://www.nabas.no/	Norway
Ovagen Group Limited	http://www.ovagen.ie/	Ireland

[a] Acquired by Ligand Pharmaceuticals Inc. in 2017 (CA, USA).
[b] Acquired by Exalpha Biologicals, Inc. in 2018 (MA, USA). Working webpages as of September 2022.

can only extrapolate the market value of particular IgY applications in specific market niches, and it is rather challenging and maybe misleading, to combine all of them in a single number reflecting the overall IgY businesses market value.

27.20 Perspective of IgY Technology Study

IgY field is more than 120 years old. Interests in IgY technology span many areas of scientific research, from basic biology to applications for both human and animal welfare (Zhang et al., 2021). Until now, the number of articles, patents, clinical studies, and commercial products on IgY technology has increased significantly. Wu et al. (2022) surveyed 1,029 IgY-related papers, including 981 journal articles and 48 reviews, and reported scientometric analysis of IgY. These Web of Science (WoS) database analyses showed an increasing trend in IgY-related publications over the 4 decades, especially from 2008 to 2021.

IgY is central to our understanding of immunoglobulin evolution across species. A cross-disciplinary approach may enhance our understanding of IgY and promote IgY study and application. IgY is an effective immunologic tool to fight infection involving microbes colonizing the alimentary tract of humans. It is relatively safe being a functional foodstuff found in the daily human diet and can exert its activity within the entire length of the alimentary tract in a predictable fashion. While it (IgY) may not exert total microbial eradication, it may significantly reduce infectious pathogen load to a point where the patient's immunity can finish the job of host protection. A major force that is drawing more and more attention to IgY's reliable and customizable antimicrobial mechanism is the gloomy prospect in the long-term fight against pathogenic microbes whose resistance to many antimicrobials is thwarting current treatment efforts. Together with other developments in recent antimicrobials and chemotherapeutic research, IgY has the potential to play a contributory role in delaying the advent of the dreaded post-antibiotic era. Non- parenterally administered IgY products, such as oral ingestible, nasal sprays, and nasal drops may provide widespread protection against pathogens that can colonize, infect, or damage the gastrointestinal and respiratory tracts. Neutralizing IgY antibodies may also warrant evaluation to target specific pathogens circulating in the bloodstream or localized in a specific area. Because IgY administration before the infection has been demonstrated to have significant protective effects *in vivo* in animal models, evaluating its use as prophylactic therapy in humans may be of special interest. Besides passive immunotherapy, IgY antibodies have also shown promise as a potential therapeutic agent for a wide spectrum of clinical applications. Specific IgY has detected and neutralized both surface and internal pathogenic antigens when administered after infection or consumption in multiple preclinical models. In most of these applications, IgY is not curative and has a greater therapeutic benefit with greater protection when used as a prophylactic treatment or in conjunction to supplement existing standard treatments.

In conclusion, IgY technology has made significant progress in the last 2 decades (during 1988 ~ 2022) (Figure 27.2) and has proven that it can be applied for diagnostic, prophylactic, or treatment purposes. IgY technology is expected to continue developing with the rapid advancements of modern biotechnology and biomedicine. There remains a requirement to standardize the technology components to make it more accessible to new researchers in the field and to broaden its potential value-added applications. In addition, developing laboratory methods and applying the other technologies to this field has led to more maturity of IgY technology toward commercialization.

27.21 Future Directions of IgY Technology (Feed to Food to Pharma)

Future directions of IgY inclined from feed to pharma. Within the biopharmaceutical disciplines, IgY technology constitutes a relatively novel field that started to draw serious attention more than two decades ago (1988 ~ 2022), with pioneering efforts largely coming from Japan. With the availability of IgY products as functional foodstuffs delivered via lozenge tablets or food carriers such as fermented milk products in the past decade in the Japanese and other East Asian markets, the practical impact of IgY for human application has created ripples in the biomedical sphere, with its broad potential only now starting to unfold. The relatively low cost of producing

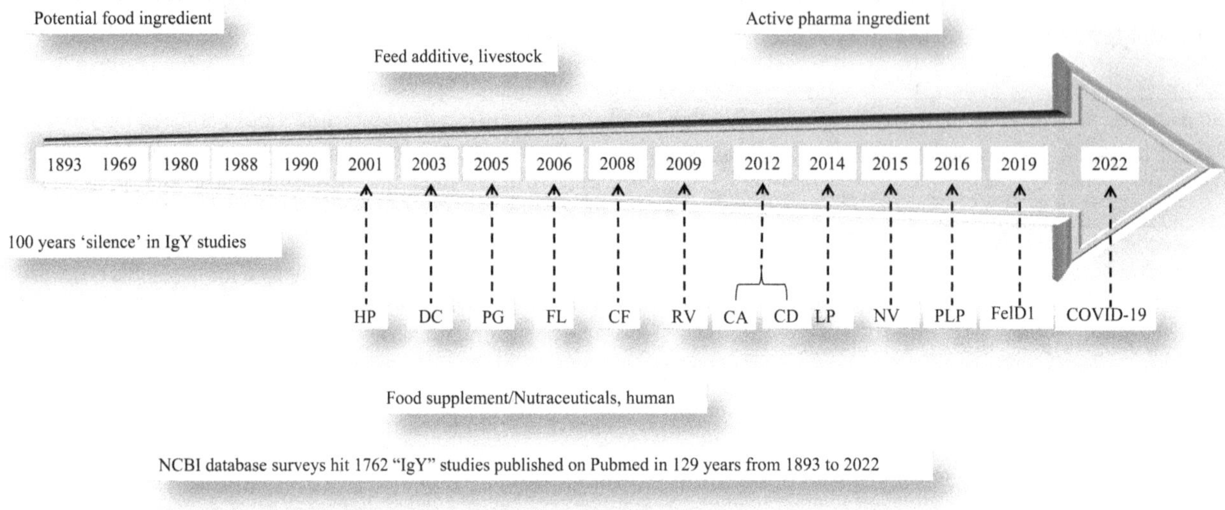

FIGURE 27.13 Future direction of IgY research: The timeline shows 100 years of progression from feed/food additive or health supplement toward nutraceutical and active pharmaceutical ingredient status. HP = Anti-*Helicobacter pylori* IgY for gastritis (2001); DC = Anti-*Streptococcus mutans* IgY for dental caries (2003); PG = Anti-*Porphyromonas gingivalis* IgY for periodontitis (2005); CF = Anti-*Pseudomonas aeruginosa* IgY for cystis fibrosis (2008); RV = Anti-Rotavirus IgY for rotaviral diarrhea (2009); CA = Anti-*Candida albicans* IgY for candidiasis (2012); CD = Anti-Gliadin IgY for celiac diseases (2012); LP = Anti-Lipase IgY for metabolic syndrome (2014); NV = Anti-Norovirus IgY for norovirus infection (2015); PLP = Anti-Phospholipase IgY for pain relieve (2016); FelD1 = Anti-Feld1 IgY for cat allergen (2019); and COVID-19 = Anti-SARS COV-2 IgY for severe acute respiratory syndrome coronavirus 2 (SARS-CoV-2) infection (2022).

antibodies from poultry eggs is an attractive side of IgY technology. The need for an inexpensive alternative to anti-infectious regimens has become even more urgent with sharply escalating healthcare costs, the prospect of an aging population in many industrialized and newly-developed countries, and scarcity of financial resources among Third World economies. Likewise, the current trend among consumers shifting from synthetics to natural materials to alleviate medical concerns has provided further impetus to the growth of the IgY market. As a functional foodstuff, IgY is well-positioned to expand its niche in both pharmaceutical and dietary supplement areas. With the expected application of advances in drug delivery systems to IgY delivery, IgYs are destined for pharmaceuticalization. They are expected to devolve toward other important clinical targets, including microbial toxins or other high-value targets such as metabolic syndrome (Hirose et al., 2013). Timeline shows 100 years of progression of IgY from functional food ingredient status, feed/food additive, or health supplement toward nutraceutical and active pharmaceutical ingredient status (Figure 27.13).

REFERENCES

Abbas, A. T., El-Kafrawy, S. A., Sohrab, S. S., & Azhar, E. (2019). IgY antibodies for the immunoprophylaxis and therapy of respiratory infections. *Human Vaccines & Immunotherapeutics*, *15*(1), 264–275. https://doi.org/10.1080/21645515.2018.1514224

Abouzid, K., Ndeboko, B., Durantel, S., Jamard, C., Zoulim, F., Buronfosse, T., & Cova, L. (2006). Genetic vaccination for production of DNA-designed antibodies specific to Hepadnavirus envelope proteins. *Vaccine*, *24*(21), 4615–4617. https://doi.org/10.1016/j.vaccine.2005.08.085

Adachi, K., Takama, K., Tsukamoto, M., Inai, M., Handharyani, E., Hiroi, S., & Tsukamoto, Y. (2011). Ostrich produce cross-reactive neutralization antibodies against pandemic influenza virus A/H1N1 following immunization with a seasonal influenza vaccine. *Experimental and Therapeutic Medicine*, *2*(1), 41–45. https://doi.org/10.3892/etm.2010.180

Aguilar, I., Sánchez, E. E., Girón, M. E., Estrella, A., Guerrero, B., & Rodriguez-Acosta, F. A. (2014). Coral snake antivenom produced in chickens (*Gallus domesticus*). *Revista do Instituto de Medicina Tropical de Sao Paulo*, *56*(1), 61–66. https://doi.org/10.1590/S0036-46652014000100009

Akbari, M. R., Ahmadi, A., Mirkalantari, S., & Salimian, J. (2018). Anti-vibriocholerae IgY antibody inhibits mortality in suckling mice model. *Journal of the National Medical Association*, *110*(1), 84–87. https://doi.org/10.1016/j.jnma.2017.04.001

Akita, E. M., Li-Chan, E. C. Y., & Nakai, S. (1998). Neutralization of enterotoxigenic *Escherichia coli* heat-labile toxin by chicken egg yolk immunoglobulin Y and its antigen-binding fragments. *Food Agriculture Immunology* 10, 161–72. https://doi.org/10.1080/09540109809354979

Akita, E. M., & Nakai, S. (1993). Comparison of four purification methods for the production of immunoglobulins from eggs laid by hens immunized with an enterotoxigenic *E. coli* strain. *Journal of Immunological Methods*, *160*(2), 207–214. https://doi.org/10.1016/0022-1759(93)90179-b

Akita, E. M., & Nakai, S. (1992). Immunoglobulins from egg yolks: Isolation and purification. *Journal of Food Science*, *57*, 629–634.

Akita, E. M., & Nakai, S. (1994). Preparation and purification of Fab' immunoreactive fragments from chicken egg immunoglobulin using pepsin and *Aspergillus saitoi* protease. In J. S. Sim & S. Nakai (Eds.). *Egg uses and processing technologies* (pp. 228–240), Wallingford, UK: CAB International.

Al-Adwani, S. R., Crespo, R., & Shah, D. H. (2013). Production and evaluation of chicken egg-yolk-derived antibodies against *Campylobacter jejuni* colonization-associated proteins. *Foodborne Pathogens and Disease*, *10*(7), 624–631. https://doi.org/10.1089/fpd.2012.1313

Alexander, T. J. L. (1994). Neonatal diarrhoea in pigs. In C. L. Gyles (Ed.). *Escherichia coli* in domestic animals and humans (pp. 151–170), Wallingford Oxon, UK: CAB International.

Alexandratos, N., & Bruinsma, J. (2012). World agriculture towards 2030/2050: The 2012 revision. ESA Working paper No. 12-03. Rome, FAO. https://www.fao.org/3/ap106e/ap106e.pdf

Alfredson, D. A., & Korolik, V. (2007). Antibiotic resistance and resistance mechanisms in *Campylobacter jejuni* and *Campylobacter coli*. *FEMS Microbiology Letters*, *277*(2), 123–132. https://doi.org/10.1111/j.1574-6968.2007.00935.x

Al-Shami, I., Al-Shamahy, H., & Lutf, A. (2019). Efficacy of some antibiotics against Streptococcus mutans associated with tooth decay in children and their mothers. *Online J Dentistry Oral Health*, *2*, 1–4. https://doi.org/10.33552/OJDOH.2019.02.000530

Altwegg, M., & Geiss, H. K. (1989). Aeromonas as a human pathogen. *Critical Reviews in Microbiology*, *16*(4), 253–286. https://doi.org/10.3109/10408418909105478

Alustiza, F., Bellingeri, R., Picco, N., Motta, C., Grosso, M. C., Barbero, C. A., Acevedo, D. F., & Vivas, A. (2016). IgY against enterotoxigenic *Escherichia coli* administered by hydrogel-carbon nanotubes composites to prevent neonatal diarrhoea in experimentally challenged piglets. *Vaccine*, *34*(28), 3291–3297. https://doi.org/10.1016/j.vaccine.2016.05.004

Alvarez, A., Montero, Y., Jimenez, E., Zerpa, N., Parrilla, P., & Malavé, C. (2013). IgY antibodies anti-Tityus caripitensis venom: Purification and neutralization efficacy. *Toxicon: Official Journal of the International Society on Toxinology*, *74*, 208–214. https://doi.org/10.1016/j.toxicon.2013.08.058

Ambrosius, H., & Hadge, D. (1987). Chicken immunoglobulins. *Veterinary Immunology and Immunopathology*, *17*(1–4), 57–67. https://doi.org/10.1016/0165-2427(87)90127-9

Ameri Shah Reza, M., Mousavi Gargari, S. L., Rasooli, I., Jalali Nadoushan, M., & Ebrahimizadeh, W. (2012). Inhibition of *H. pylori* colonization and prevention of gastritis in murine model. *World Journal of Microbiology & Biotechnology*, *28*(7), 2513–2519. https://doi.org/10.1007/s11274-012-1059-5

Amirijavid, S., Entezari, M., Movafagh, A., Hashemi, M., Mosavi-Jarahi, A., & Dehghani, H. (2016). Apoptotic killing of breast cancer cells by IgYs produced against a small 21 aminoacid epitope of the human TRAIL-2 receptor. *Asian Pacific Journal of cancer Prevention: APJCP*, *17*(S3), 293–297. https://doi.org/10.7314/apjcp.2016.17.s3.293

Amro, W. A., Al-Qaisi, W., & Al-Razem, F. (2018). Production and purification of IgY antibodies from chicken egg yolk. *Journal, Genetic Engineering & Biotechnology*, *16*(1), 99–103. https://doi.org/10.1016/j.jgeb.2017.10.003

Anandharamakrishnan, C., & Ishwarya, S. P. (2015). *Spray drying techniques for food ingredient encapsulation*. John Wiley & Sons, Ltd., Hoboken. https://doi.org/10.1002/9781118863985

Andris-Widhopf, J., Rader, C., Steinberger, P., Fuller, R., & Barbas, C. F. 3rd (2000). Methods for the generation of chicken monoclonal antibody fragments by phage display. *Journal of Immunological Methods*, *242*(1–2), 159–181. https://doi.org/10.1016/s0022-1759(00)00221-0

Anguiano-Beltrán, C., Searcy-Bernal, R., & Lizárraga-Partida, M. L. (1998). Pathogenic effects of *Vibrio alginolyticus* on larvae and postlarvae of the red abalone *Haliotis rufescens*. *Disease of Aquatic Organisms*, *33* (2), 119–122. https://doi.org/10.3354/dao033119

Anselmo, A. C., Gokarn, Y., & Mitragotri, S. (2019). Non-invasive delivery strategies for biologics. *Nature Reviews. Drug Discovery*, *18*(1), 19–40. https://doi.org/10.1038/nrd.2018.183

Arasteh, N., Aminirissehei, A. H., Yousif, A. N., Albright, L. J., & Durance, T. D. (2004). Passive immunization of rainbow trout (*Oncorhynchus mykiss*) with chicken egg yolk immunoglobulins (IgY). *Aquaculture*, *231*(1), 23–36. https://doi.org/10.1016/j.aquaculture.2003.11.004

Araújo, A. S., Lobato, Z. I., Chávez-Olórtegui, C., & Velarde, D. T. (2010). Brazilian IgY-*Bothrops* antivenom: Studies on the development of a process in chicken egg yolk. *Toxicon*, *55*(4), 739–744. https://doi.org/10.1016/j.toxicon.2009.11.004

Aro, H., Järvenpää, E. P., Könkö, K., Shivonen, M., & Hietaniemi, V. (2009). Isolation and purification of egg yolk phospholipids using liquid extraction and pilot-scale supercritical fluid techniques. *European Food Research Techno*logy, 228(6), 857–863. https://doi.org/10.1007/s00217-008-0998-4

Bachtiar, E. W., Afdhal, A., Meidyawati, R., Soejoedono, R. D., & Poerwaningsih, E. (2016b). Effect of topical anti-Streptococcus mutans IgY gel on quantity of *S. mutans* on rats' tooth surface. *Acta microbiologica et immunologica Hungarica*, *63*(2), 159–169. https://doi.org/10.1556/030.63.2016.2.2

Bachtiar, E. W., Bachtiar, B. M., Soejoedono, R. D., Wibawan, I. W., & Afdhal, A. (2016a). Biological and immunogenicity property of IgY anti S. mutans ComD. *The Open Dentistry Journal*, *10*, 308–314. https://doi.org/10.2174/1874210601610010308

Bachtiar, E. W., Soejoedono, R. D., Bachtiar, B. M., Henrietta, A., Farhana, N., & Yuniastuti, M. (2015). Effects of soybean milk, chitosan, and anti-Streptococcus mutans IgY in malnourished rats' dental biofilm and the IgY persistency in saliva. *Interventional Medicine & Applied Science*, *7*(3), 118–123. https://doi.org/10.1556/1646.7.2015.3.6

Bade, H., & Stegemann, H. (1984). Rapid method of extraction of antibodies from hen egg yolk. *Journal of Immunological Methods*, *72*(2), 421–426. https://doi.org/10.1016/0022-1759(84)90010-3

Bahrami, S., Amiri-Yekta, A., Daneshipour, A., Jazayeri, S. H., Mozdziak, P. E., Sanati, M. H., & Gourabi, H. (2020). Designing a transgenic chicken: Applying new approaches toward a promising bioreactor. *Cell Journal*, *22*(2), 133–139. https://doi.org/10.22074/cellj.2020.6738

Barati, B., Ebrahimi, F., & Nazarian, S. (2018). Production of chicken egg yolk antibody (IgY) against recombinant cholera toxin B subunit and evaluation of its prophylaxis potency in mice. *Iranian Journal of Immunology: IJI*, *15*(1), 47–58.

Bartels, C. J., Holzhauer, M., Jorritsma, R., Swart, W. A., & Lam, T. J. (2010). Prevalence, prediction and risk factors of enteropathogens in normal and non-normal faeces of young Dutch dairy calves. *Preventive Veterinary Medicine*, *93*(2-3), 162–169. https://doi.org/10.1016/j.prevetmed.2009.09.020

Baselga, R., Albizu, I., & Amorena, B. (1994). *Staphylococcus aureus* capsule and slime as virulence factors in ruminant mastitis. A review. *Veterinary Microbiology, 39*(3–4), 195–204. https://doi.org/10.1016/0378-1135(94)90157-0

Beck, J., Garcia, R., Heiss, G., Vokonas, P. S., & Offenbacher, S. (1996). Periodontal disease and cardiovascular disease. *Journal of Periodontology, 67*(10 Suppl), 1123–1137. https://doi.org/10.1902/jop.1996.67.10s.1123

Behn, I., Erhard, M., Hlinak, A., & Staak, C. (2001). Chicken egg yolk antibodies, production and application. In R. Schade, I. Behn, M. Erhard, A. Hlinak & C. Staak (Eds.). *IgY-Technology* (p. 255). Berlin Heidelberg: Springer-Verlag. https://doi.org/10.1007/978-3-662-04488-9

Benedict, A. A., Brown, R. J., & Hersh, R. T. (1963). The temporal synthesis and some chromatographic and ultracentrifugal characteristics of chicken antibodies. *Journal of Immunology (Baltimore, Md.: 1950), 90*, 399–411.

Bentes, G. A., Lanzarini, N. M., Lima, L. R., Manso, P. P., da Silva, A., Mouta Junior, S., Guimarães, J. R., de Moraes, M. T., Pelajo-Machado, M., & Pinto, M. A. (2015). Using immunoglobulin Y as an alternative antibody for the detection of hepatitis A virus in frozen liver sections. *Memorias do Instituto Oswaldo Cruz, 110*(4), 577–579. https://doi.org/10.1590/0074-02760140457

Bern, C., Kjos, S., Yabsley, M. J., & Montgomery, S. P. (2011). Trypanosoma cruzi and Chagas' disease in the United States. *Clinical Microbiology Reviews, 24*(4), 655–681. https://doi.org/10.1128/CMR.00005-11

Bird, R. E., Hardman, K. D., Jacobson, J. W., Johnson, S., Kaufman, B. M., Lee, S. M., Lee, T., Pope, S. H., Riordan, G. S., & Whitlow, M. (1988). Single-chain antigen-binding proteins. *Science (New York, N.Y.), 242*(4877), 423–426. https://doi.org/10.1126/science.3140379

Bittner, C., Garrido, M. V., Krach, L. H., & Harth, V. (2016). Content of asthmagen natural rubber latex allergens in commercial disposable gloves. *Advances in Experimental Medicine and Biology, 921*, 37–44. https://doi.org/10.1007/5584_2016_227

Bizhanov, G., & Vyshniauskis, G. (2000). A comparison of three methods for extracting IgY from the egg yolk of hens immunized with Sendai virus. *Veterinary Research Communications, 24*(2), 103–113. https://doi.org/10.1023/a:1006460506303

Bobeck, E. A., Hellestad, E. M., Helvig, C. F., Petkovich, P. M., & Cook, M. E. (2016). Oral antibodies to human intestinal alkaline phosphatase reduce dietary phytate phosphate bioavailability in the presence of dietary 1α-hydroxycholecalciferol. *Poultry Science, 95*(3), 570–580. https://doi.org/10.3382/ps/pev341

Boder, E. T., & Wittrup, K. D. (1997). Yeast surface display for screening combinatorial polypeptide libraries. *Nature Biotechnology, 15*(6), 553–557. https://doi.org/10.1038/nbt0697-553

Bogen, J. P., Grzeschik, J., Krah, S., Zielonka, S., & Kolmar, H. (2020). Rapid generation of chicken immune libraries for yeast surface display. *Methods in molecular Biology (Clifton, N.J.), 2070*, 289–302. https://doi.org/10.1007/978-1-4939-9853-1_16

Bowes, T., Hanley, S. A., Liew, A., Eglon, M., Mashayekhi, K., O'Kennedy, R., Barry, F., Taylor, W. R., O'Brien, T., Griffin, M. D., Finlay, W. J., & Greiser, U. (2011). Developing cell-specific antibodies to endothelial progenitor cells using avian immune phage display technology. *Journal of Biomolecular Screening, 16*(7), 744–754. https://doi.org/10.1177/1087057111407067

Box, P. G., Stedman, R. A., & Singleton, L. (1969). Newcastle disease. I. The use of egg yolk derived antibody for passive immunisation of chickens. *Journal of Comparative Pathology, 79*(4), 495–506. https://doi.org/10.1016/0021-9975(69)90070-x

Brasil, P., Pereira, J. P. Jr, Moreira, M. E., Ribeiro Nogueira, R. M., Damasceno, L., Wakimoto, M., Rabello, R. S., Valderramos, S. G., Halai, U. A., Salles, T. S., Zin, A. A., Horovitz, D., Daltro, P., Boechat, M., Raja Gabaglia, C., Carvalho de Sequeira, P., Pilotto, J. H., Medialdea-Carrera, R., Cotrim da Cunha, D., Abreu de Carvalho, L. M., … Nielsen-Saines, K. (2016). Zika virus infection in pregnant women in Rio de Janeiro. *The New England Journal of Medicine, 375*(24), 2321–2334. https://doi.org/10.1056/NEJMoa1602412

Brener, Z. (1973). Biology of *Trypanosoma cruzi*. *Annual Review of Microbiology, 27*, 347–382. https://doi.org/10.1146/annurev.mi.27.100173.002023

Brocato, R., Josleyn, M., Ballantyne, J., Vial, P., & Hooper, J. W. (2012). DNA vaccine-generated duck polyclonal antibodies as a postexposure prophylactic to prevent hantavirus pulmonary syndrome (HPS). *PloS One, 7*(4), e35996. https://doi.org/10.1371/journal.pone.0035996

Broom, L. J., & Kogut, M. H. (2019). Deciphering desirable immune responses from disease models with resistant and susceptible chickens. *Poultry Science, 98*(4), 1634–1642. https://doi.org/10.3382/ps/pey535

Brunauer, M., Roch, F. F., & Conrady, B. (2021). Prevalence of worldwide neonatal calf diarrhoea caused by bovine rotavirus in combination with bovine coronavirus, *Escherichia coli* K99 and *Cryptosporidium* spp.: A meta-analysis. *Animals: an Open Access Journal from MDPI, 11*(4), 1014. https://doi.org/10.3390/ani11041014

Budama-Kilinc, Y., Cakir-Koc, R., Ozdemir, B., Kaya, Z., & Badur, S. (2018). Production and characterization of a conserved M2e peptide-based specific IgY antibody: Evaluation of the diagnostic potential via conjugation with latex nanoparticles. *Preparative Biochemistry & Biotechnology, 48*(10), 930–939. https://doi.org/10.1080/10826068.2018.1525564

Burke, R. M., Tate, J. E., Barin, N., Bock, C., Bowen, M. D., Chang, D., Gautam, R., Han, G., Holguin, J., Huynh, T., Pan, C. Y., Quenelle, R., Sallenave, C., Torres, C., Wadford, D., & Parashar, U. (2018). Three rotavirus outbreaks in the postvaccine era - California, 2017. *MMWR. Morbidity and Mortality Weekly Report, 67*(16), 470–472. https://doi.org/10.15585/mmwr.mm6716a3

Burns, R. B., & Maxwell, M. H. (1981). Probable occurrence of IgE in the adult domestic fowl (*Gallus domesticus*) after horse serum stimulation. *Veterinary Research Communications, 5*(1), 67–72. https://doi.org/10.1007/BF02214970

Cakir-Koc, R. (2016). Production of anti-SAG1 IgY antibody against *Toxoplasma gondii* parasites and evaluation of antibody activity by ELISA method. *Parasitology Research, 115*(8), 2947–2952. https://doi.org/10.1007/s00436-016-5047-9

Cakir-Koc, R., Budama-Kilinc, Y., Ustun, E., & Babur, C. (2020). Conjugation and characterization of latex particles with *Toxoplasma gondii*-specific immunoglobulin Y antibodies for diagnostic aim and evaluation efficiency in in vitro culture. *Journal of Equine Veterinary Science, 92*, 103145. https://doi.org/10.1016/j.jevs.2020.103145

Cal, K., & Sollohub, K. (2010). Spray drying technique. I: Hardware and process parameters. *Journal of Pharmaceutical Sciences, 99*(2), 575–586. https://doi.org/10.1002/jps.21886

Calmette, A. (1896). The treatment of animals poisoned with snake venom by the injection of antivenomous serum. *British Medical Journal, 2*(1859), 399–400. https://doi.org/10.1136/bmj.2.1859.399

Calzado, E., Heredia, M. T., Duharte, J. F., & Zarnani, A. H. (2017). Evaluation of IgY antibody as a polyspecific coombs-reagent. *Avicenna Journal of Medical Biotechnology, 9*(2), 87–93.

Camenisch, G., Tini, M., Chilov, D., Kvietikova, I., Srinivas, V., Caro, J., Spielmann, P., Wenger, R. H., & Gassmann, M. (1999). General applicability of chicken egg yolk antibodies: The performance of IgY immunoglobulins raised against the hypoxia-inducible factor 1alpha. *FASEB Journal: Official Publication of the Federation of American Societies for Experimental Biology, 13*(1), 81–88. https://doi.org/10.1096/fasebj.13.1.81

Cao-Lormeau, V. M., Blake, A., Mons, S., Lastère, S., Roche, C., Vanhomwegen, J., Dub, T., Baudouin, L., Teissier, A., Larre, P., Vial, A. L., Decam, C., Choumet, V., Halstead, S. K., Willison, H. J., Musset, L., Manuguerra, J. C., Despres, P., Fournier, E., Mallet, H. P., ... Ghawché, F. (2016). Guillain-Barré Syndrome outbreak associated with Zika virus infection in French Polynesia: A case-control study. *Lancet (London, England), 387*(10027), 1531–1539. https://doi.org/10.1016/S0140-6736(16)00562-6

Carlander, D. (2002). Avian IgY antibody. In vitro and in vivo. Acta Universitatis Upsaliensis. Comprehensive Summaries of Uppsala Dissertations from the Faculty of Medicine 1119. 53 pp. Uppsala. ISBN 91-554-5227-2.

Carlander, D., Kollberg, H., Wejåker, P. E., & Larsson, A. (2000). Peroral immunotherapy with yolk antibodies for the prevention and treatment of enteric infections. *Immunologic Research, 21*(1), 1–6. https://doi.org/10.1385/ir:21:1:1

Carlander, D., Stålberg, J., & Larsson, A. (1999). Chicken antibodies: A clinical chemistry perspective. *Upsala Journal of Medical Sciences, 104*(3), 179–189. https://doi.org/10.3109/03009739909178961

Casandra, A. B., Searcy-Bernal, R., & Lizárraga-Partida, M. L. (1998). Pathogenic effects of *Vibrio alginolyticus* on larvae and postlarvae of the red abalone *Haliotis rufescens*. *Diseases Aquatic Organism, 33*, 119–122. https://doi.org/10.3354/dao033119

Castro, J. A., de Mecca, M. M., & Bartel, L. C. (2006). Toxic side effects of drugs used to treat Chagas' disease (American trypanosomiasis). *Human & Experimental Toxicology, 25*(8), 471–479. https://doi.org/10.1191/0960327106het653oa

CDC (2021). Centers for Disease Control and Prevention. (2021). Available at: https://www.cdc.gov/hantavirus/hps/symptoms.html (Accessed April 15, 2021).

Chalghoumi, R., Théwis, A., Beckers, Y., Marcq, C., Portetelle, D., & Schneider, Y. J. (2009). Adhesion and growth inhibitory effect of chicken egg yolk antibody (IgY) on *Salmonella enterica* serovars Enteritidis and Typhimurium in vitro. *Foodborne Pathogens and Disease, 6*(5), 593–604. https://doi.org/10.1089/fpd.2008.0258

Chambers, H. F. (2001). The changing epidemiology of Staphylococcus aureus? *Emerging Infectious Diseases, 7*(2), 178–182. https://doi.org/10.3201/eid0702.010204

Chang, H. M., Lu, T. C., Chen, C. C., Tu, Y. Y., & Hwang, J. Y. (2000). Isolation of immunoglobulin from egg yolk by anionic polysaccharides. *Journal of Agricultural and Food Chemistry, 48*(4), 995–999. https://doi.org/10.1021/jf990539k

Chang, H. M., Ou-Yang, R. F., Chen, Y. T., & Chen, C. C. (1999). Productivity and some properties of immunoglobulin specific against Streptococcus mutans serotype c in chicken egg yolk (IgY). *Journal of Agricultural and Food Chemistry, 47*(1), 61–66. https://doi.org/10.1021/jf980153u.

Chauhan, A., Kumar, M., Kumar, A., & Kanchan, K. (2021). Comprehensive review on mechanism of action, resistance and evolution of antimycobacterial drugs. *Life Sciences, 274*, 119301. https://doi.org/10.1016/j.lfs.2021.119301

Chen, C. C., Tu, Y. Y., Chen, T. L., & Chang, H. M. (2002). Isolation and characterization of immunoglobulin in yolk (IgY) specific against hen egg white lysozyme by immunoaffinity chromatography. *Journal of Agricultural and Food Chemistry, 50*(19), 5424–5428. https://doi.org/10.1021/jf011567h

Chen, C. L., Lehmeyer, J. E., & Cooper, M. D. (1982). Evidence for an IgD homologue on chicken lymphocytes. *Journal of Immunology (Baltimore, Md.: 1950), 129*(6), 2580–2585.

Chettle, N., Stuart, J. C., & Wyeth, P. J. (1989). Outbreak of virulent infectious bursal disease in East Anglia. *The Veterinary Record, 125*(10), 271–272. https://doi.org/10.1136/vr.125.10.271

Chi, Z. B., Gao, Y. X., Pan, Y., Zhang, B., & Feng, X. P. (2004). *Shanghai kou qiang yi xue = Shanghai journal of stomatology*, 13(4), 256–258.

Chintoan-Uta, C., Cassady-Cain, R. L., & Stevens, M. P. (2016). Evaluation of flagellum-related proteins FliD and FspA as subunit vaccines against Campylobacter jejuni colonisation in chickens. *Vaccine, 34*(15), 1739–1743. https://doi.org/10.1016/j.vaccine.2016.02.052

Chira, S., Gulei, D., Hajitou, A., Zimta, A. A., Cordelier, P., & Berindan-Neagoe, I. (2017). CRISPR/Cas9: Transcending the reality of genome editing. *Molecular Therapy. Nucleic Acids, 7*, 211–222. https://doi.org/10.1016/j.omtn.2017.04.001

Cho, Y. I., & Yoon, K. J. (2014). An overview of calf diarrhea - infectious etiology, diagnosis, and intervention. *Journal of Veterinary Science, 15*(1), 1–17. https://doi.org/10.4142/jvs.2014.15.1.1

Cho, Y. I., Han, J. I., Wang, C., Cooper, V., Schwartz, K., Engelken, T., & Yoon, K. J. (2013). Case-control study of microbiological etiology associated with calf diarrhea. *Veterinary Microbiology, 166*(3-4), 375–385. https://doi.org/10.1016/j.vetmic.2013.07.001

Chung, S. L., & Ferrier, L. K. (1991). Conditions affecting emulsifying properties of egg yolk phosvitin. *Journal of Food Science, 56*(5), 1259–1262. https://doi.org/10.1111/j.1365-2621.1991.tb04747.x

Cipolla, A., Cordeviola, J., Terzolo, H., Combessies, G., Bardón, J., Noseda, R., Martínez, A., Medina, D., Morsella, C., & Malena, R. (2001). "Campylobacter fetus diagnosis: Direct immunofluorescence comparing chicken IgY and rabbit IgG conjugates." *ALTEX - Alternatives to Animal Experimentation, 18*(3), 165–170. Retrieved May 3, 2023, from https://altex.org/index.php/altex/article/view/1180

Ciurea, C. N., Kosovski, I. B., Mare, A. D., Toma, F., Pintea-Simon, I. A., & Man, A. (2020). *Candida* and Candidiasis—

Opportunism versus pathogenicity: A review of the virulence traits. *Microorganisms*, *8*(6), 857. https://doi.org/10.3390/microorganisms8060857

Coleman, Marilyn A. (1996). Oral administration of chicken yolk immunoglobulins to lower somatic cell count in the milk of lactating ruminants. US Patent 5585098.

Collins, F. S. (1992). Cystic fibrosis: Molecular biology and therapeutic implications. *Science (New York, N.Y.)*, *256*(5058), 774–779. https://doi.org/10.1126/science.1375392

Constantin, C., Neagu, M., Diana Supeanu, T., Chiurciu, V., & Spandidos, A., D. (2020). IgY - turning the page toward passive immunization in COVID-19 infection (review). *Experimental and Therapeutic Medicine*, *20*(1), 151–158. https://doi.org/10.3892/etm.2020.8704

Constantinoiu, C. C., Molloy, J. B., Jorgensen, W. K., & Coleman, G. T. (2007). Purification of immunoglobulins from chicken sera by thiophilic gel chromatography. *Poultry Science*, *86*(9), 1910–1914. https://doi.org/10.1093/ps/86.9.1910

Costalonga, M., & Herzberg, M. C. (2014). The oral microbiome and the immunobiology of periodontal disease and caries. *Immunology Letters*, *162*(2 Pt A), 22–38. https://doi.org/10.1016/j.imlet.2014.08.017

Costa, V. G. D., Saivish, M. V., Rodrigues, R. L., Lima Silva, R. F., Moreli, M. L., & Krüger, R. H. (2019). Molecular and serological surveys of canine distemper virus: A meta-analysis of cross-sectional studies. *PLoS One*, *14*(5), e0217594. https://doi.org/10.1371/journal.pone.0217594

Costantini, F., & Lacy, E. (1981). Introduction of a rabbit beta-globin gene into the mouse germ line. *Nature*, *294*(5836), 92–94. https://doi.org/10.1038/294092a0

Cumbers, S. J., Williams, G. T., Davies, S. L., Grenfell, R. L., Takeda, S., Batista, F. D., Sale, J. E., & Neuberger, M. S. (2002). Generation and iterative affinity maturation of antibodies in vitro using hypermutating B-cell lines. *Nature Biotechnology*, *20*(11), 1129–1134. https://doi.org/10.1038/nbt752

Custer, D. M., Thompson, E., Schmaljohn, C. S., Ksiazek, T. G., & Hooper, J. W. (2003). Active and passive vaccination against hantavirus pulmonary syndrome with Andes virus M genome segment-based DNA vaccine. *Journal of Virology*, *77*(18), 9894–9905. https://doi.org/10.1128/jvi.77.18.9894-9905.2003

da Rocha, D. G., Fernandez, J. H., de Almeida, C., da Silva, C. L., Magnoli, F. C., da Silva, O. É., & da Silva, W. D. (2017). Development of IgY antibodies against anti-snake toxins endowed with highly lethal neutralizing activity. *European Journal of Pharmaceutical Sciences: Official Journal of the European Federation for Pharmaceutical Sciences*, *106*, 404–412. https://doi.org/10.1016/j.ejps.2017.05.069

da Silva Raposo, R., Santarém, V. A., Merigueti, Y. F., Rubinsky-Elefant, G., de Lima Cerazo, L. M., Pereira, L., Zampieri, B. P., Silva, da, & Laposy, A. V., C. B. (2016). Kinetic and avidity of IgY anti-Toxocara antibodies in experimentally infected chickens. *Experimental Parasitology*, *171*, 33–41. https://doi.org/10.1016/j.exppara.2016.09.009

Silva, da, Schaefer, M. C., Gava, R., Souza, D., da Silva Vaz, C. K., Bastos, I. Jr, & Venancio, A. P., E. J. (2018). Production and application of anti-nucleoprotein IgY antibodies for influenza A virus detection in swine. *Journal of Immunological Methods*, *461*, 100–105. https://doi.org/10.1016/j.jim.2018.06.023

Dávalos-Pantoja, L., Ortega-Vinuesa, J. L., Bastos-González, D., & Hidalgo-Álvarez, R. (2001). Colloidal stability of IgG- and IgY-coated latex microspheres. *Colloids and Surfaces. B, Biointerfaces*, *20*(2), 165–175. https://doi.org/10.1016/s0927-7765(00)00189-2

Dávalos-Pantoja, L., Ortega-Vinuesa, J. L., Bastos-González, D., & Hidalgo-Alvarez, R. (2000). A comparative study between the adsorption of IgY and IgG on latex particles. *Journal of Biomaterials Science. Polymer Edition*, *11*(6), 657–673. https://doi.org/10.1163/156856200743931

de Andrade, F. G., Eto, S. F., Santos Ferraro, N., Gonzales Marioto, A. C., Vieira, D. T., Cheirubim, N. J., de Paula Ramos, A. P., & Venâncio, S., E. J. (2013). The production and characterization of anti-bothropic and anti-crotalic IgY antibodies in laying hens: A long term experiment. *Toxicon: Official Journal of the International Society on Toxinology*, *66*, 18–24. https://doi.org/10.1016/j.toxicon.2013.01.018

de Faria, L. S., de Souza, D., Ribeiro, R. P., de Sousa, J., Borges, I. P., Ávila, V., Ferreira-Júnior, Á., Goulart, L. R., & Costa-Cruz, J. M. (2019). Highly specific and sensitive anti-*Strongyloides venezuelensis* IgY antibodies applied to the human strongyloidiasis immunodiagnosis. *Parasitology International*, *72*, 101933. https://doi.org/10.1016/j.parint.2019.101933

De Meulenaer, B., & Huyghebaert, A. (2001). Isolation and purification of chicken egg yolk immunoglobulins: A review. *Food and Agricultural Immunology 13*(4), 275–288. https://doi.org/10.1080/09540100120094537

Dey, S., Pathak, D. C., Ramamurthy, N., Maity, H. K., & Chellappa, M. M. (2019). Infectious bursal disease virus in chickens: Prevalence, impact, and management strategies. *Veterinary Medicine (Auckland, N.Z.)*, *10*, 85–97. https://doi.org/10.2147/VMRR.S185159

Dick, G. W. (1952). Zika virus. II. Pathogenicity and physical properties. *Transactions of the Royal Society of Tropical Medicine and Hygiene*, *46*(5), 521–534. https://doi.org/10.1016/0035-9203(52)90043-6

DiLorenzo, N., Diez-Gonzalez, F., & DiCostanzo, A. (2006). Effects of feeding polyclonal antibody preparations on ruminal bacterial populations and ruminal pH of steers fed high-grain diets. *Journal of Animal Science*, *84*(8), 2178–2185. https://doi.org/10.2527/jas.2005-489

Diraviyam, T., Zhao, B., Wang, Y., Schade, R., Michael, A., & Zhang, X. (2014). Effect of chicken egg yolk antibodies (IgY) against diarrhea in domesticated animals: A systematic review and meta-analysis. *PloS One*, *9*(5), e97716. https://doi.org/10.1371/journal.pone.0097716

Doellgast, G. J., Brown, J. E., Koufman, J. A., & Hatheway, C. L. (1997). Sensitive assay for measurement of antibodies to Clostridium botulinum neurotoxins A, B, and E: Use of hapten-labeled-antibody elution to isolate specific complexes. *Journal of Clinical Microbiology*, *35*(3), 578–583. https://doi.org/10.1128/jcm.35.3.578-583.1997

Dong, D., Liu, H., Xiao, Q., & Li, R. (2008). Affinity purification of egg yolk immunoglobulins (IgY) with a stable synthetic ligand. *Journal of Chromatography. B, Analytical Technologies in the Biomedical and Life Sciences*, *870*(1), 51–54. https://doi.org/10.1016/j.jchromb.2008.05.036

Dossumbekova, A., Prinz, C., Mages, J., Lang, R., Kusters, J. G., Van Vliet, A. H., Reindl, W., Backert, S., Saur, D., Schmid, R. M., & Rad, R. (2006). *Helicobacter pylori* HopH

(OipA) and bacterial pathogenicity: Genetic and functional genomic analysis of *hopH* gene polymorphisms. *The Journal of Infectious Diseases*, *194*(10), 1346–1355. https://doi.org/10.1086/508426

Dowling, T. C., Chavaillaz, P. A., Young, D. G., Melton-Celsa, A., O'Brien, A., Thuning-Roberson, C., Edelman, R., & Tacket, C. O. (2005). Phase 1 safety and pharmacokinetic study of chimeric murine-human monoclonal antibody c alpha Stx2 administered intravenously to healthy adult volunteers. *Antimicrobial Agents and Chemotherapy*, *49*(5), 1808–1812. https://doi.org/10.1128/AAC.49.5.1808-1812.2005

Duan, H. L., He, Q. Y., Zhou, B., Wang, W. W., Li, B., Zhang, Y. Z., Deng, Q. P., Zhang, Y. F., & Yu, X. D. (2016). Anti-*Trimeresurus albolabris* venom IgY antibodies: Preparation, purification and neutralization efficacy. *The Journal of Venomous Animals and Toxins Including Tropical Diseases*, *22*(1), 23. https://doi.org/10.1186/s40409-016-0078-3

Duffy, M. R., Chen, T. H., Hancock, W. T., Powers, A. M., Kool, J. L., Lanciotti, R. S., Pretrick, M., Marfel, M., Holzbauer, S., Dubray, C., Guillaumot, L., Griggs, A., Bel, M., Lambert, A. J., Laven, J., Kosoy, O., Panella, A., Biggerstaff, B. J., Fischer, M., & Hayes, E. B. (2009). Zika virus outbreak on Yap Island, Federated States of Micronesia. *The New England Journal of Medicine*, *360*(24), 2536–2543. https://doi.org/10.1056/NEJMoa0805715

Dussart, P., Labeau, B., Lagathu, G., Louis, P., Nunes, M. R., Rodrigues, S. G., Storck-Herrmann, C., Cesaire, R., Morvan, J., Flamand, M., & Baril, L. (2006). Evaluation of an enzyme immunoassay for detection of dengue virus NS1 antigen in human serum. *Clinical and Vaccine Immunology: CVI*, *13*(11), 1185–1189. https://doi.org/10.1128/CVI.00229-06

Egea, E., Mendoza, D., Garavito, G., Saavedra, S., Gómez, H., & Sanjuan, M. (2019). Nanogold - IgY antibodies. An immunoconjugated for the detection of house dust mite (Dermatophagoides) allergens. *Journal of Immunological Methods*, *464*, 15–21. https://doi.org/10.1016/j.jim.2018.08.013

El-Ghany, W. A. A. (2011). Comparison between immunoglobulins IgY and the vaccine for prevention of infectious bursal disease in chickens. *Global Veterinaria*, *6*(1), 16–24. http://www.idosi.org/gv/gv6(1)11/3.pdf

Elliot, S. L., Rodrigues, J., Lorenzo, M. G., Martins-Filho, O. A., & Guarneri, A. A. (2015). Trypanosoma cruzi, etiological agent of Chagas disease, is virulent to its triatomine vector Rhodnius prolixus in a temperature-dependent manner. *PLoS Neglected Tropical Diseases*, *9*(3), e0003646. https://doi.org/10.1371/journal.pntd.0003646

Elston, R., & Lockwood, G. S. (1983). Pathogenesis of vibriosis in cultured juvenile red abalone, *Haliotis rufescens* Swainson. *Journal of Fish Disease 6*, 111–128. https://doi.org/10.1111/j.1365-2761.1983.tb00059.x

Erhard, M. H., Amon, P., Younan, M., Ali, Z., & Stangassinger, M. (1999). Absorption and synthesis of immunoglobulins G in newborn calves. *Reproduction in Domestic Animals*, *34*(3–4), 173–175. https://doi.org/10.1111/j.1439-0531.1999.tb01237.x

Erhard, M. H., Bergmann, J., Renner, M., Hofmann, A., & Heinritzi, K. (1996). Prophylaktische Wirkung von spezifischen Dotterantikörpern bei *Escherichia coli* K88 (F4)-bedingten Durchfallerkrankungen von Absatzferkeln [Prophylactic effect of specific egg yolk antibodies in diarrhea caused by *Escherichia coli* K88 (F4) in weaned piglets]. *Zentralblatt fur Veterinarmedizin. Reihe A*, *43*(4), 217–223.

Erhard, M. H., Schmidt, P., Zinsmeister, P., Hofmann, A., Münster, U., Kaspers, B., Wiesmüller, K. H., Bessler, W. G., & Stangassinger, M. (2000). Adjuvant effects of various lipopeptides and interferon-gamma on the humoral immune response of chickens. *Poultry Science*, *79*(9), 1264–1270. https://doi.org/10.1093/ps/79.9.1264

Esmailnejad, A., Abdi-Hachesoo, B., Hosseini Nasab, E., & Shakoori, M. (2019). Production, purification, and evaluation of quail immunoglobulin Y against *Salmonella typhimurium* and *Salmonella enteritidis*. *Molecular Immunology*, *107*, 79–83. https://doi.org/10.1016/j.molimm.2019.01.012

EU Clinical Trials Register. (2021). Clinical trial results: Prospective randomized, placebo-controlled, double blind, multicenter study (Phase III) to evaluate clinical efficacy and safety of avian polyclonal anti-pseudomonas antibodies (IgY), in: *Prevention of recurrence of Pseudomonas Aeruginosa infection in cystic fibrosis patients*. Available at: https://www.clinicaltrialsregister.eu/ctr-search/trial/2011-000801-39/results (Accessed April 15, 2021).

Evans, D. J. Jr, Evans, D. G., Takemura, T., Nakano, H., Lampert, H. C., Graham, D. Y., Granger, D. N., & Kvietys, P. R. (1995). Characterization of a *Helicobacter pylori* neutrophil-activating protein. *Infection and Immunity*, *63*(6), 2213–2220. https://doi.org/10.1128/iai.63.6.2213-2220.1995

Fahey, J. E., & Crawley, J. F. (1954). Studies on chronic respiratory disease of chickens II. Isolation of a virus. *Canadian Journal of Comparative Medicine and Veterinary Science*, *18*(1), 13–21.

Fan, M., Jiang, Q., & Bian, Z. (2003). *Hua xi kou qiang yi xue za zhi = Huaxi kouqiang yixue zazhi = West China journal of stomatology*, *21*(5), 339–341.

Fang, X., & Zhang, W. W. (2008). Affinity separation and enrichment methods in proteomic analysis. *Journal of Proteomics*, *71*(3), 284–303. https://doi.org/10.1016/j.jprot.2008.06.011

FAO. (2020). *The state of world fishery and aquaculture 2020 (SOFIA)*. Rome, Italy: Food and Agriculture Organization of the United Nations. http://www.fao.org/publications/sofia/2020/en/

FAOSTAT. (2012). Food and agricultural organization statistical database. http://faostat3.fao.org/faostatgateway/go/to/download/Q/QL/E

Farooq, A., Rabbani, M., Khushi, M., Akram, Z., Ahad, A., Fatima, Z., Khurram, T., & Anwar, Z. (2012). Passive immunization in infectious bursal disease virus infected birds using chemically purified immune yolk immunoglobulins (IgY). *African Journal of Microbiology Research*, *6*(12), 2993–2998. https://doi.org/10.5897/AJMR12.049

Fayer, R. (Ed.). (1997) *The general biology of Cryptosporidium*. Boca Raton, FL: CRC Press.

Ferella, A., Bellido, D., Chacana, P., Wigdorovitz, A., Santos, M. J. D., & Mozgovoj, M. V. (2012). Chicken egg yolk antibodies against bovine respiratory syncytial virus neutralize the virus in vitro. *Procedia in Vaccinology*, *6*, 33–38. https://doi.org/10.1016/j.provac.2012.04.006

Fernandes, D. C., Eto, S. F., Funnicelli, M. I. G., Fernandes, C. C., Charlie-Silva, I., Belo, M.A.A., & Pizauro, J. M.

(2019a). Immunoglobulin Y in the diagnosis of *Aeromonas hydrophila* infection in Nile tilapia (*Oreochromis niloticus*). *Aquaculture*, *500*, 576–585. https://doi.org/10.1016/j.aquaculture.2018.10.045.

Fernandes, D. C., Eto, S. F., Moraes, A. C., Prado, E., Medeiros, A., Belo, M., Samara, S. I., Costa, P. I., & Pizauro, J. M. (2019b). Phagolysosomal activity of macrophages in Nile tilapia (*Oreochromis niloticus*) infected in vitro by *Aeromonas hydrophila*: Infection and immunotherapy. *Fish & Shellfish Immunology*, *87*, 51–61. https://doi.org/10.1016/j.fsi.2018.12.074

Ferragut, F., Vega, C. G., Mauroy, A., Conceição-Neto, N., Zeller, M., Heylen, E., Uriarte, E. L., Bilbao, G., Bok, M., Matthijnssens, J., Thiry, E., Badaracco, A., & Parreño, V. (2016). Molecular detection of bovine noroviruses in Argentinean dairy calves: Circulation of a tentative new genotype. *Infection, Genetics and Evolution: Journal of Molecular Epidemiology and Evolutionary Genetics in Infectious Diseases*, *40*, 144–150. https://doi.org/10.1016/j.meegid.2016.02.034

Ferreira Júnior, Á., Morgan, P. M., Zhang, X., & Schade, R. (2021). Biology and molecular structure of avian IgY antibody. In X.-Y. Zhang et al. (Eds.), *IgY-technology: Production and application of egg yolk antibodies*. Nature Switzerland AG: Springer. https://doi.org/10.1007/978-3-030-72688-1_5

Ferreira Júnior, Á., Santiago, F. M., Silva, M. V., Ferreira, F. B., Macêdo Júnior, A. G., Mota, C. M., Faria, M. S., Silva Filho, H. H., Silva, D. A., Cunha-Júnior, J. P., Mineo, J. R., & Mineo, T. W. (2012). Production, characterization and applications for *Toxoplasma gondii*-specific polyclonal chicken egg yolk immunoglobulins. *PloS One*, *7*(7), e40391. https://doi.org/10.1371/journal.pone.0040391

Ferreira Júnior, Á., Santos, J. P. D., Sousa, I. D. O., Martin, I., Alves, E. G. L., & Rosado, I. R. (2018). *Gallus gallus domesticus*: Immune system and its potential for generation of immunobiologics. *Ciência Rural*, *48*(8), e20180250. https://doi.org/10.1590/0103-8478cr20180250

Figueiredo, A., Vieira, N. C., dos Santos, J. F., Janegitz, B. C., Aoki, S. M., Junior, P. P., Lovato, R. L., Nogueira, M. L., Zucolotto, V., & Guimarães, F. E. (2015). Electrical detection of dengue biomarker using egg yolk immunoglobulin as the biological recognition element. *Scientific Reports*, *5*, 7865. https://doi.org/10.1038/srep07865

Fink, A. L., Williams, K. L., Harris, E., Alvine, T. D., Henderson, T., Schiltz, J., Nilles, M. L., & Bradley, D. S. (2017). Dengue virus specific IgY provides protection following lethal dengue virus challenge and is neutralizing in the absence of inducing antibody dependent enhancement. *PLoS Neglected Tropical Diseases*, *11*(7), e0005721. https://doi.org/10.1371/journal.pntd.0005721

Finlay, W. J., Shaw, I., Reilly, J. P., & Kane, M. (2006). Generation of high-affinity chicken single-chain Fv antibody fragments for measurement of the *Pseudonitzschia pungens* toxin domoic acid. *Applied and Environmental Microbiology*, *72*(5), 3343–3349. https://doi.org/10.1128/AEM.72.5.3343-3349.2006

Fischer, M., Hlinak, A., Montag, T., Claros, M., Schade, R., & Ebner, D. (1996). Vergleich von Standardmethoden zur Präparation von Dotterantikörpern. *Tierärztliche Praxis*, *24*, 411–418.

Foote, J., & Winter, G. (1992). Antibody framework residues affecting the conformation of the hypervariable loops. *Journal of Molecular Biology*, *224*(2), 487–499. https://doi.org/10.1016/0022-2836(92)91010-m

Foster, D. M., & Smith, G. W. (2009). Pathophysiology of diarrhea in calves. *The Veterinary Clinics of North America. Food Animal Practice*, *25*(1), 13–xi. https://doi.org/10.1016/j.cvfa.2008.10.013

Fournier, P. E., & Richet, H. (2006). The epidemiology and control of *Acinetobacter baumannii* in health care facilities. *Clinical Infectious Diseases: an Official Publication of the Infectious Diseases Society of America*, *42*(5), 692–699. https://doi.org/10.1086/500202

Franco, A. T., Johnston, E., Krishna, U., Yamaoka, Y., Israel, D. A., Nagy, T. A., Wroblewski, L. E., Piazuelo, M. B., Correa, P., & Peek, R. M. Jr (2008). Regulation of gastric carcinogenesis by *Helicobacter pylori* virulence factors. *Cancer Research*, *68*(2), 379–387. https://doi.org/10.1158/0008-5472.CAN-07-0824

Fraser, R. S. S. (1990). The genetics of plant-virus interactions: Mechanisms controlling host range, resistance and virulence. In: Fraser, R.S.S. (eds) *Recognition and response in plant-virus interactions*. NATO ASI Series, vol 41. Springer, Berlin, Heidelberg.

Frenzel, A., Hust, M., & Schirrmann, T. (2013). Expression of recombinant antibodies. *Frontiers in Immunology*, *4*, 217. https://doi.org/10.3389/fimmu.2013.00217

Fryer, J., Firca, J., Leventhal, J., Blondin, B., Malcolm, A., Ivancic, D., Gandhi, R., Shah, A., Pao, W., Abecassis, M., Kaufman, D., Stuart, F., & Anderson, B. (1999). IgY antiporcine endothelial cell antibodies effectively block human antiporcine xenoantibody binding. *Xenotransplantation*, *6*(2), 98–109. https://doi.org/10.1034/j.1399-3089.1999.00015.x

Fulton, R. M., Nersessian, B. N., & Reed, W. M. (2002). Prevention of *Salmonella enteritidis* infection in commercial ducklings by oral chicken egg-derived antibody alone or in combination with probiotics. *Poultry Science*, *81*(1), 34–40. https://doi.org/10.1093/ps/81.1.34

Gaensbauer, J. T., Melgar, M. A., Calvimontes, D. M., Lamb, M. M., Asturias, E. J., Contreras-Roldan, I. L., Dominguez, S. R., Robinson, C. C., & Berman, S. (2017). Efficacy of a bovine colostrum and egg-based intervention in acute childhood diarrhoea in Guatemala: A randomised, double-blind, placebo-controlled trial. *BMJ Global Health*, *2*(4), e000452. https://doi.org/10.1136/bmjgh-2017-000452

Gan, H., He, H., Sato, A., Hatta, H., Nakao, M., & Somamoto, T. (2015). Ulcer disease prophylaxis in koi carp by bath immersion with chicken egg yolk containing anti-*Aeromonas salmonicida* IgY. *Research in Veterinary Science*, *99*, 82–86. https://doi.org/10.1016/j.rvsc.2015.01.016

Gao, X., Chen, N., Zhang, Y., Zhang, X., & Bing, X. (2017). Non-O1 *Vibrio cholerae* pathogen from *Cyprinus carpio* and control with anti-non-O1 *V. cholerae* egg yolk powder (IgY). *Aquaculture*, *479*, 69–74. https://doi.org/10.1016/j.aquaculture.2017.05.015.

Gao, X., Zhang, X., Lin, L., Yao, D., Sun, J., Du, X., Li, X., & Zhang, Y. (2016a). Passive immune-protection of *Litopenaeus vannamei* against *Vibrio harveyi* and *Vibrio parahaemolyticus* infections with anti-*Vibrio* egg yolk (IgY)-encapsulated feed. *International Journal of Molecular Sciences*, *17*(5), 723. https://doi.org/10.3390/ijms17050723

Gao, X., Zhang, X., Sun, J., Du, X., Li, X., Zhang, Y., & Lin, L. (2016b). Passive protection effect of anti-*Vibrio anguillarum* IgY-encapsulated feed on half-smooth tongue sole (*Cynoglossus semilaevi*) against *V. anguillarum. Fish Shellfish Immunol* 56:483–8. https://doi.org/10.1016/j.fsi.2016.07.041

Garvey, J. S., Cremer, N. E., & Sussdorf, D. H. (1977). Ammonium sulphate precipitation, in *Methods in immunology*, 3rd edn, Addison-Wesley, Reading, MA, pp. 218–219.

Gassmann, M., Thömmes, P., Weiser, T., & Hübscher, U. (1990). Efficient production of chicken egg yolk antibodies against a conserved mammalian protein. *FASEB Journal: Official Publication of the Federation of American Societies for Experimental Biology*, *4*(8), 2528–2532. https://doi.org/10.1096/fasebj.4.8.1970792

Ghysdael, J., Bruck, C., Kettmann, R., & Burny, A. (1984). Bovine leukemia virus. *Current Topics in Microbiology and Immunology*, *112*, 1–19. https://doi.org/10.1007/978-3-642-69677-0_1

GI-Immune-DFHUGHPA.pdf (naplescfm.com) distributed by Hughes Center for Functional Medicine 800 Goodlette Road North # 270 Naples, FL 34102 239-649-7400.

Gilgunn, S., Millan Martin, S., Wormald, M. R., Zapatero-Rodríguez, J., Conroy, P. J., O'Kennedy, R. J., Rudd, P. M., Saldova, R., & Mondelli, M. U. (2016). Comprehensive N-Glycan profiling of avian immunoglobulin Y. *PLoS One*, *11*(7), e0159859. https://doi.org/10.1371/journal.pone.0159859

Gilgunn, S., Millán Martín, S., Wormald, M. R., Zapatero-Rodríguez, J., Conroy, P. J., O'Kennedy, R. J., Rudd, P. M., & Saldova, R. (2016).

Gill, J. J., Pacan, J. C., Carson, M. E., Leslie, K. E., Griffiths, M. W., & Sabour, P. M. (2006). Efficacy and pharmacokinetics of bacteriophage therapy in treatment of subclinical *Staphylococcus aureus* mastitis in lactating dairy cattle. *Antimicrobial Agents and Chemotherapy*, *50*(9), 2912–2918. https://doi.org/10.1128/AAC.01630-05

Goldsby, R. A., Kindt, T. J., Osborne, B. A., & Kuby, J. (2000). *Kuby immunology*. New York: W.H. Freeman. http://www.worldcat.org/oclc/41528664

Goncalvez, A. P., Engle, R. E., St Claire, M., Purcell, R. H., & Lai, C. J. (2007). Monoclonal antibody-mediated enhancement of dengue virus infection in vitro and in vivo and strategies for prevention. *Proceedings of the National Academy of Sciences of the United States of America*, *104*(22), 9422–9427. https://doi.org/10.1073/pnas.0703498104

Gordon, J. W., & Ruddle, F. H. (1981). Integration and stable germ line transmission of genes injected into mouse pronuclei. *Science (New York, N.Y.)*, *214*(4526), 1244–1246. https://doi.org/10.1126/science.6272397

Govan, J. R., & Deretic, V. (1996). Microbial pathogenesis in cystic fibrosis: Mucoid *Pseudomonas aeruginosa* and *Burkholderia cepacia*. *Microbiological Reviews*, *60*(3), 539–574. https://doi.org/10.1128/mr.60.3.539-574.1996

Grando, T. H., Baldissera, M. D., de Sá, M. F., do Carmo, G. M., Porto, B., Aguirre, G., Azevedo, M. I., de Jesus, F., Santurio, J. M., Sagrillo, M. R., Stefani, L. M., & Monteiro, S. G. (2017). Avian antibodies (IgY) against *Trypanosoma cruzi*: Purification and characterization studies. *Journal of Immunological Methods*, *449*, 56–61. https://doi.org/10.1016/j.jim.2017.07.002

Grando, T. H., Baldissera, M. D., Do Carmo, G., Oliveira, C. B., Santi, E. T., Doleski, P. H., Leal, D., Stefani, L. M., Mendes, R. E., Da Silva, A. S., & Monteiro, S. G. (2018). Ecto-enzymes activities in splenic lymphocytes of mice experimentally infected by *Trypanosoma cruzi* and treated with specific avian immunoglobulins: An attempt to improve the immune response. *Molecular and Cellular Biochemistry*, *448*(1–2), 9–15. https://doi.org/10.1007/s11010-018-3308-x

Grzywa, R., Łupicka-Słowik, A., Walczak, M., Idzi, M., Bobrek, K., Boivin, S., Gaweł, A., Stefaniak, T., Oleksyszyn, J., & Sieńczyk, M. (2014). Highly sensitive detection of cancer antigen 15-3 using novel avian IgY antibodies. *ALTEX*, *31*(1), 43–52. https://doi.org/10.14573/altex.1309181

Guarino, A., Dupont, C., Gorelov, A. V., Gottrand, F., Lee, J. K., Lin, Z., Lo Vecchio, A., Nguyen, T. D., & Salazar-Lindo, E. (2012). The management of acute diarrhea in children in developed and developing areas: From evidence base to clinical practice. *Expert Opinion on Pharmacotherapy*, *13*(1), 17–26. https://doi.org/10.1517/14656566.2011.634800

Gubler, D. J., & Clark, G. G. (1995). Dengue/dengue hemorrhagic fever: The emergence of a global health problem. *Emerging Infectious Diseases*, *1*(2), 55–57. https://doi.org/10.3201/eid0102.952004

Guimaraes, M. C., Amaral, C., Borges, L. G., Vieira, F. V., Matta, H. P. L., & Matta, C. G. F. (2009a). Characterization of an IgY polyclonal antibodies directed against the canine distemper virus. *Journal of Medical and Biological Science*, *8*(1), 18–25. http://repositorio.ufba.br/ri/handle/ri/20562

Guimaraes, M. C. C., Amaral, L. G., Borges, F. V., Vieira, H. P. L., Shimoya, A., Gomes, C., Matta, F., & Matta, R. (2008). Production and use of egg-yolk antibody for detection of canine parvovirus in feces. *Journal of Medical and Biological Science*, *7*(3), 241–248. http://repositorio.ufba.br/ri/handle/ri/20548

Guimarães, M. C., Amaral, L. G., Rangel, L. B., Silva, I. V., Matta, C. G., & Matta, M. F. (2009b). Growth inhibition of *Staphylococcus aureus* by chicken egg yolk antibodies. *Archivum immunologiae et therapiae experimentalis*, *57*(5), 377–382. https://doi.org/10.1007/s00005-009-0041-x

Gui-rong, Z., & Yun-ying, Z. (2011). Development of yolk antibodies against young duck viral hepatitis. *China Animal Husbandry Veterinary Medicine*, *38*(10), 142–144. http://www.chvm.net/EN/Y2011/V38/I10/142

Gujral, N., Löbenberg, R., Suresh, M., & Sunwoo, H. (2012). In-vitro and in-vivo binding activity of chicken egg yolk immunoglobulin Y (IgY) against gliadin in food matrix. *Journal of Agricultural and Food Chemistry*, *60*(12), 3166–3172. https://doi.org/10.1021/jf205319s

Gürtler, M., Methner, U., Kobilke, H., & Fehlhaber, K. (2004). Effect of orally administered egg yolk antibodies on *Salmonella enteritidis* contamination of hen's eggs. *Journal of Veterinary Medicine. B, Infectious Diseases and Veterinary Public Health*, *51*(3), 129–134. https://doi.org/10.1111/j.1439-0450.2004.00739.x

Gutierrez, M. A., Miyazaki, T., Hatta, H., & Kim, M. (1993). Protective properties of egg yolk IgY containing anti-*Edwardsiella tarda* antibody against paracolo disease in the Japanese eel, *Anguilla japonica* Temminck & Schlegel. *Journal of Fish Disease*, 16 (2), 113–122. https://doi.org/10.1111/j.1365-2761.1993.tb00854.x

Haese, N., Brocato, R. L., Henderson, T., Nilles, M. L., Kwilas, S. A., Josleyn, M. D., Hammerbeck, C. D., Schiltz, J., Royals, M., Ballantyne, J., Hooper, J. W., & Bradley, D. S. (2015).

Antiviral biologic produced in DNA vaccine/goose platform protects hamsters against hantavirus pulmonary syndrome when administered post-exposure. *PLoS Neglected Tropical Diseases*, *9*(6), e0003803. https://doi.org/10.1371/journal.pntd.0003803

Hajishengallis, G., & Lamont, R. J. (2012). Beyond the red complex and into more complexity: The polymicrobial synergy and dysbiosis (PSD) model of periodontal disease etiology. *Molecular oral Microbiology*, *27*(6), 409–419. https://doi.org/10.1111/j.2041-1014.2012.00663.x

Hamajima, S., Maruyama, M., Hijiya, T., Hatta, H., & Abiko, Y. (2007). Egg yolk-derived immunoglobulin (IgY) against *Porphyromonas gingivalis* 40-kDa outer membrane protein inhibits coaggregation activity. *Archives of oral Biology*, *52*(7), 697–704. https://doi.org/10.1016/j.archoralbio.2006.12.013

Hamal, K. R., Burgess, S. C., Pevzner, I. Y., & Erf, G. F. (2006). Maternal antibody transfer from dams to their egg yolks, egg whites, and chicks in meat lines of chickens. *Poultry science*, *85*(8), 1364–1372. https://doi.org/10.1093/ps/85.8.1364

Hanes, J., Jermutus, L., Weber-Bornhauser, S., Bosshard, H. R., & Plückthun, A. (1998) Ribosome display efficiently selects and evolves high-affinity antibodies in vitro from immune libraries. *Proceedings of the National Academy of Sciences U S A*, *95*(24), 14130–14135. https://doi.org/10.1073/pnas.95.24.14130

Hanes, J., & Plückthun, A. (1997). In vitro selection and evolution of functional proteins by using ribosome display. *Proceedings of the National Academy of Sciences of the United States of America*, *94*(10), 4937–4942. https://doi.org/10.1073/pnas.94.10.4937

Hansen, P., Scoble, J. A., Hanson, B., & Hoogenraad, N. J. (1998). Isolation and purification of immunoglobulins from chicken eggs using thiophilic interaction chromatography. *Journal of Immunological Methods*, *215*(1–2), 1–7. https://doi.org/10.1016/s0022-1759(98)00050-7

Haraszthy, V. I., Gerber, D., Clark, B., Moses, P., Parker, C., Sreenivasan, P. K., & Zambon, J. J. (2008). Characterization and prevalence of *Solobacterium moorei* associated with oral halitosis. *Journal of Breath Research*, *2*(1), 017002. https://doi.org/10.1088/1752-7155/2/1/017002

Härtle, S., Magor, K. E., Gobel, T. W., Davidson, F., & Kaspers, B. (2014). Chapter 6- Structure and evolution of avian immunoglobulins. In: *Avian immunology* (pp. 103–120), Academic Press. https://doi.org/10.1016/B978-0-12-396965-1.00006-6.

Harvey, A. J., Speksnijder, G., Baugh, L. R., Morris, J. A., & Ivarie, R. (2002). Consistent production of transgenic chickens using replication-deficient retroviral vectors and high-throughput screening procedures. *Poultry Science*, *81*(2), 202–212. https://doi.org/10.1093/ps/81.2.202

Haselbeck, A. H., Panzner, U., Im, J., Baker, S., Meyer, C. G., & Marks, F. (2017). Current perspectives on invasive nontyphoidal *Salmonella* disease. *Current Opinion in Infectious Diseases*, *30*(5), 498–503. https://doi.org/10.1097/QCO.0000000000000398

Hashemzadeh, F., & Shahbazi, P. (2016). Production of specific egg yolk antibody (IgY) against Cryptosporidium parvum oocysts. *Medical Laboratory Journal*, *10*(6), 38–42. http://mlj.goums.ac.ir/article-1-920-en.html

Hassl, A., & Aspöck, H. (1988). Purification of egg yolk immunoglobulins. A two-step procedure using hydrophobic interaction chromatography and gel filtration. *Journal of Immunological Methods*, *110*(2), 225–228. https://doi.org/10.1016/0022-1759(88)90107-x

Hatta, H., Kim, M., & Yamamoto, T. (1990) A novel isolation method for hen egg yolk antibody, "IgY". *Agricultural Biological Chemistry*, *54* (10), 2531–2535. https://doi.org/10.1271/bbb1961.54.2531

Hatta, H., Mabe, K., Kim, M., Yamamoto, T., Gutierrez, M. A., & Miyazaki, T. (1994). Prevention of fish disease using egg yolk antibody. In J. S. Sim & S. Nakai (Eds.). *Egg uses and processing technologies: New developments* (pp. 241–249), Wallingford: CAB International.

Hatta, H., Tsuda, K., Akachi, S., Kim, M., Yamamoto, T., & Ebina, T. (1993). Oral Passive immunization effect of anti-human rotavirus IgY and its behavior against proteolytic enzymes. *Bioscience, Biotechnology, and Biochemistry*, *57*(7), 1077–1081. https://doi.org/10.1271/bbb.57.1077

Hatta, H., Tsuda, K., Ozeki, M., Kim, M., Yamamoto, T., Otake, S., Hirasawa, M., Katz, J., Childers, N. K., & Michalek, S. M. (1997). Passive immunization against dental plaque formation in humans: Effect of a mouth rinse containing egg yolk antibodies (IgY) specific to *Streptococcus* mutans. *Caries Research*, *31*(4), 268–274. https://doi.org/10.1159/000262410

He, J., Hu, J., Thirumalai, D., Schade, R., Du, E., & Zhang, X. (2016). Development of indirect competitive ELISA using egg yolk-derived immunoglobulin (IgY) for the detection of Gentamicin residues. *Journal of Environmental Science and Health. Part. B, Pesticides, Food Contaminants, and Agricultural Wastes*, *51*(1), 8–13. https://doi.org/10.1080/03601234.2015.1080479

He, J., Wang, Y., Sun, S., & Zhang, X. (2015). Evaluation of chicken IgY generated against canine parvovirus viral-like particles and development of enzyme-linked immunosorbent assay and immunochromatographic assay for canine parvovirus detection. *Viral Immunology*, *28*(9), 489–494. https://doi.org/10.1089/vim.2015.0030

He, M., & Taussig, M. J. (2007). Eukaryotic ribosome display with in situ DNA recovery. *Nature Methods*, *4*(3), 281–288. https://doi.org/10.1038/nmeth1001

Heaton, N. S., Moshkina, N., Fenouil, R., Gardner, T. J., Aguirre, S., Shah, P. S., Zhao, N., Manganaro, L., Hultquist, J. F., Noel, J., Sachs, D., Hamilton, J., Leon, P. E., Chawdury, A., Tripathi, S., Melegari, C., Campisi, L., Hai, R., Metreveli, G., Gamarnik, A. V., … Marazzi, I. (2016). Targeting viral proteostasis limits influenza virus, HIV, and dengue virus infection. *Immunity*, *44* (1), 46–58. https://doi.org/10.1016/j.immuni.2015.12.017

Heller, M. C., & Chigerwe, M. (2018). Diagnosis and treatment of infectious enteritis in neonatal and juvenile ruminants. *The Veterinary Clinics of North America. Food Animal Practice*, *34*(1), 101–117. https://doi.org/10.1016/j.cvfa.2017.08.001

Henderson, A., & Nimmo, G. R. (2018). Control of healthcare- and community-associated MRSA: Recent progress and persisting challenges. *British Medical Bulletin*, *125*(1), 25–41. https://doi.org/10.1093/bmb/ldx046

Hermans, D., Van Steendam, K., Verbrugghe, E., Verlinden, M., Martel, A., Seliwiorstow, T., Heyndrickx, M., Haesebrouck, F., De Zutter, L., Deforce, D., & Pasmans, F. (2014). Passive immunization to reduce *Campylobacter jejuni* colonization and transmission in broiler chickens. *Veterinary Research, 45*, 27. https://doi.org/10.1186/1297-9716-45-27

Hernández-Campos, F. J., Brito-De la Fuente, E., & Torrestiana-Sánchez, B. (2010). Purification of egg yolk immunoglobulin (IgY) by ultrafiltration: Effect of pH, ionic strength, and membrane properties. *Journal of Agricultural and Food Chemistry, 58*(1), 187–193. https://doi.org/10.1021/jf902964s

Herrera, H. M., Dávila, A. M., Norek, A., Abreu, U. G., Souza, S. S., D'Andrea, P. S., & Jansen, A. M. (2004). Enzootiology of *Trypanosoma evansi* in Pantanal, Brazil. *Veterinary Parasitology, 125*(3–4), 263–275. https://doi.org/10.1016/j.vetpar.2004.07.013

Herszényi, L., & Tulassay, Z. (2010). Epidemiology of gastrointestinal and liver tumors. *European Review for Medical and Pharmacological Sciences, 14*(4), 249–258.

Hiidenhovi, J., Hietanen, A., Mäkinen, J., Huopalahti, R., & Ryhänen, E.-L. (2005) Hydrolysis of ovomucin by different enzymes. *Proceedings of Eleventh European Symposium on the Quality of Eggs and Egg Product*, 23–26 May, Doorwerth, Netherlands, CD-ROM 251–256.

Hirai, K., Arimitsu, H., Umeda, K., Yokota, K., Shen, L., Ayada, K., Kodama, Y., Tsuji, T., Hirai, Y., & Oguma, K. (2010). Passive oral immunization by egg yolk immunoglobulin (IgY) to *Vibrio cholerae* effectively prevents cholera. *Acta Medica Okayama, 64*(3), 163–170. https://doi.org/10.18926/AMO/40008

Hirose, M., Ando, T., Shofiqur, R., Umeda, K., Kodama, Y., Nguyen, S. V., Goto, T., Shimada, M., & Nagaoka, S. (2013). Anti-obesity activity of hen egg anti-lipase immunoglobulin yolk, a novel pancreatic lipase inhibitor. *Nutrition & Metabolism, 10*(1), 70. https://doi.org/10.1186/1743-7075-10-70

Hochel, I., Viochna, D., Skvor, J., & Musil, M. (2004). Development of an indirect competitive ELISA for detection of *Campylobacter jejuni* subsp.jejuni O:23 in foods. *Folia microbiologica, 49*(5), 579–586. https://doi.org/10.1007/BF02931537

Hodek, P., Hrdinova, J., Macova, I., Soucek, P., Mrizova, I., Burdova, K., Kizek, R., Hudecek, J., & Stiborova, M. (2015). Preparation and application of anti-peptide antibodies for detection of orphan cytochromes P450. *Neuro Endocrinology Letters, 36*(Suppl 1), 38–45. PubMed PMID: 26757124

Hoet, A. E., Nielsen, P. R., Hasoksuz, M., Thomas, C., Wittum, T. E., & Saif, L. J. (2003). Detection of bovine torovirus and other enteric pathogens in feces from diarrhea cases in cattle. *Journal of Veterinary Diagnostic Investigation: Official Publication of the American Association of Veterinary Laboratory Diagnosticians, Inc, 15*(3), 205–212. https://doi.org/10.1177/104063870301500301

Holser, R. (2013). Lipid encapsulated phenolic compounds by fluidization. *Journal of Encapsulation Adsorption Science 3*(1), 13–15. https://doi.org/10.4236/jeas.2013.31002.

Hopkins, S. G., & DiGiacomo, R. F. (1997). Natural transmission of bovine leukemia virus in dairy and beef cattle. *The Veterinary Clinics of North America. Food Animal Practice, 13*(1), 107–128. https://doi.org/10.1016/s0749-0720(15)30367-4

Horák, D., & Hochel, I. (2005). Magnetic poly(glycidyl methacrylate) microspheres for *Campylobacter jejuni* detection in food. *e-Polymers* 5, 1–12. https://doi.org/10.1515/epoly.2005.5.1.64

Horie, K., Horie, N., Abdou, A. M., Yang, J. O., Yun, S. S., Chun, H. N., Park, C. K., Kim, M., & Hatta, H. (2004). Suppressive effect of functional drinking yogurt containing specific egg yolk immunoglobulin on *Helicobacter pylori* in humans. *Journal of Dairy Science, 87*(12), 4073–4079. https://doi.org/10.3168/jds.S0022-0302(04)73549-3

Hou, Y. Y., Zhen, Y. H., Wang, D., Zhu, J., Sun, D. X., Liu, X. T., Wang, H. X., Liu, Y., Long, Y. Y., & Shu, X. H. (2014). Protective effect of an egg yolk-derived immunoglobulin (IgY) against *Prevotella intermedia*-mediated gingivitis. *Journal of Applied Microbiology, 116*(4), 1020–1027. https://doi.org/10.1111/jam.12419

Huang, L., Harvie, G., Feitelson, J. S., Gramatikoff, K., Herold, D. A., Allen, D. L., Amunngama, R., Hagler, R. A., Pisano, M. R., Zhang, W. W., & Fang, X. (2005). Immunoaffinity separation of plasma proteins by IgY microbeads: Meeting the needs of proteomic sample preparation and analysis. *Proteomics, 5*(13), 3314–3328. https://doi.org/10.1002/pmic.200401277

Huopalahti, R., López-Fandiño, R., Anton, M., & Schade, R. (Eds.). (2007). *Bioactive egg compounds.* Berlin, Germany: Springer Publication.

Hwang, C. Y., Pak, S. I., & Han, H. R. (2000). Effects of autogenous toxoid-bacterin in lactating cows with *Staphylococcus aureus* subclinical mastitis. *The Journal of Veterinary Medical Science, 62*(8), 875–880. https://doi.org/10.1292/jvms.62.875

Ibrahim, E. M., Rahman, A. K. M. S., Isoda, R., Umeda, K., Kodama, Y., & Maeda, N. (2007). Anti-Candida Albicans Egg Yolk Immunoglobulin: Cross Activity and Pilot Study. The 55th Annual Meeting of Japanese Association for Dental Research (JADR), 2007, November 17–18, Yokohama, Japan.

Ibrahim, El-Sayed Moustafa, Rahman, A. K., Isoda, R., Umeda, K., Van Sa, N., & Kodama, Y. (2008). In vitro and in vivo effectiveness of egg yolk antibody against *Candida albicans* (anti-CA IgY). *Vaccine, 26*(17), 2073–2080. https://doi.org/10.1016/j.vaccine.2008.02.046

IGY Life Sciences. (2021). IgY antibodies—Sustainable and efficacious therapeutics for human and animal health. [cited 4 February 2021]. https://www.nature.com/articles/d43747-020-01049-5.

Ikemori, Y., Kuroki, M., Peralta, R. C., Yokoyama, H., & Kodama, Y. (1992). Protection of neonatal calves against fatal enteric colibacillosis by administration of egg yolk powder from hens immunized with K99-piliated enterotoxigenic Escherichia coli. *American Journal of Veterinary Research, 53*(11), 2005–2008.

Ikemori, Y., Ohta, M., Umeda, K., Icatlo, F. C. Jr, Kuroki, M., Yokoyama, H., & Kodama, Y. (1997). Passive protection of neonatal calves against bovine coronavirus-induced diarrhea by administration of egg yolk or colostrum antibody powder. *Veterinary Microbiology, 58*(2–4), 105–111. https://doi.org/10.1016/s0378-1135(97)00144-2

Ikemori, Y., Ohta, M., Umeda, K., Peralta, R. C., Kuroki, M., Yokoyama, H., & Kodama, Y. (1996). Passage of chicken egg yolk antibody treated with hydroxypropyl methylcellulose phthalate in the gastrointestinal tract of calves. *The Journal of Veterinary Medical Science, 58*(4), 365–367. https://doi.org/10.1292/jvms.58.365

Ikemori, Y., Peralta, R. C., Kuroki, M., Yokoyama, H., & Kodama, Y. (1993). Research note: Avidity of chicken yolk antibodies to enterotoxigenic *Escherichia coli* fimbriae. *Poultry Science*, *72* (12), 2361–2365. https://doi.org/10.3382/ps.0722361

Iqbal, M., Ahmad, T., Yousaf, A., Sajjad-ur, R., Saqib, M., Nadeem, M., & Muhammad, G. (2013). In vivo comparison of specific activity of egg yolk immunoglobulins (IgY) and antibiotic against *Staphylococcus aureus* causing mastitis in buffaloes (*Bubalus bubalis*). *Buffalo Bulletin*, *32*, 1017–20.

Izzo, M. M., Kirkland, P. D., Mohler, V. L., Perkins, N. R., Gunn, A.A., & House, J. K. (2011). Prevalence of major enteric pathogens in Australian dairy calves with diarrhoea. *Australian Veterinary Journal*, *89*(5), 167–173. https://doi.org/10.1111/j.1751-0813.2011.00692.x

Jacobson, M. (2022). On the infectious causes of neonatal piglet diarrhoea—A review. *Veterinary Science*, *9*, 422. https://doi.org/10.3390/vetsci9080422

Jaekel, T., Dautel, K., & Ternes, W. (2008). Preserving functional properties of hen's egg yolk during freeze-drying. *Journal of Food Engineering*, *87*(4), 522–526. https://doi.org/10.1016/j.jfoodeng.2008.01.006

Jagadeeswari, M., Dhanabalan, R., & Saranya, G. (2015). Detection of *Staphylococcus aureus* from clinical and sub clinical Bovine Mastitis milk samples using chicken egg yolk antibodies (IgY). *International Journal of Applied Pure Science Agriculture*, *7*, 1–6.

Jahangiri, A., Owlia, P., Rasooli, I., Salimian, J., Derakhshanifar, E., Naghipour Erami, A., Darzi Eslam, E., & Darvish Alipour Astaneh, S. (2019). Specific egg yolk antibodies (IgY) confer protection against *Acinetobacter baumannii* in a murine pneumonia model. *Journal of Applied Microbiology*, *126*(2), 624–632. https://doi.org/10.1111/jam.14135

Jensenius, J. C., Andersen, I., Hau, J., Crone, M., & Koch, C. (1981). Eggs: Conveniently packaged antibodies. Methods for purification of yolk IgG. *Journal of Immunological Methods*, *46*(1), 63–68. https://doi.org/10.1016/0022-1759(81)90333-1

Jernigan, J. A., Hatfield, K. M., Wolford, H., Nelson, R. E., Olubajo, B., Reddy, S. C., McCarthy, N., Paul, P., McDonald, L. C., Kallen, A., Fiore, A., Craig, M., & Baggs, J. (2020). Multidrug-resistant bacterial infections in U.S. hospitalized patients, 2012-2017. *The New England Journal of Medicine*, *382*(14), 1309–1319. https://doi.org/10.1056/NEJMoa1914433

Jiang, X., Diraviyam, T., & Zhang, X. (2016a). Affinity purification of egg yolk immunoglobulins (IgY) using a human mycoplasma protein. *Journal of Chromatography. B, Analytical Technologies in the Biomedical and Life Sciences*, *1012-1013*, 37–41. https://doi.org/10.1016/j.jchromb.2016.01.012

Jiang, Z., Jiang, X., Li, C., Xue, H., & Zhang, X. (2016b). Development of an IgY antibody-based immunoassay for the screening of the CYP2E1 Inhibitor/Enhancer from herbal medicines. *Frontiers in Pharmacology*, *7*, 502. https://doi.org/10.3389/fphar.2016.00502

Jin, L. Z., Baidoo, S. K., Marquardt, R. R., & Frohlich, A. A. (1998). In vitro inhibition of adhesion of enterotoxigenic *Escherichia coli* K88 to piglet intestinal mucus by egg-yolk antibodies. *FEMS Immunology and Medical Microbiology*, *21*(4), 313–321. https://doi.org/10.1111/j.1574-695X.1998.tb01179.x

Jones, P. T., Dear, P. H., Foote, J., Neuberger, M. S., & Winter, G. (1986). Replacing the complementarity-determining regions in a human antibody with those from a mouse. *Nature*, *321*(6069), 522–525. https://doi.org/10.1038/321522a0

Jones, R. C. (2013). "Reovirus infections". In: D.E. Swayne (Ed.), *Diseases of poultry* (pp. 351–373), 13th edition. Hoboken: Wiley-Blackwell.

Joshi, P. P., Shegokar, V. R., Powar, R. M., Herder, S., Katti, R., Salkar, H. R., Dani, V. S., Bhargava, A., Jannin, J., & Truc, P. (2005). Human trypanosomiasis caused by *Trypanosoma evansi* in India: The first case report. *The American Journal of Tropical Medicine and Hygiene*, *73*(3), 491–495.

Joshipura, K. J., Hung, H. C., Rimm, E. B., Willett, W. C., & Ascherio, A. (2003). Periodontal disease, tooth loss, and incidence of ischemic stroke. *Stroke*, *34*(1), 47–52. https://doi.org/10.1161/01.str.0000052974.79428.0c

Jung, K. M., Bae, E. H., Jung, Y. T., & Kim, J. W. (2014). Use of IgY antibody to recombinant avian reovirus σC protein in the virus diagnostics. *Acta Virologica*, *58*(2), 108–113. https://doi.org/10.4149/av_2014_02_108

Kamihira, M., Kawabe, Y., Shindo, T., Ono, K., Esaka, K., Yamashita, T., Nishijima, K., & Iijima, S. (2009). Production of chimeric monoclonal antibodies by genetically manipulated chickens. *Journal of Biotechnology*, *141*(1–2), 18–25. https://doi.org/10.1016/j.jbiotec.2009.02.022.

Kamihira, M., Ono, K., Esaka, K., Nishijima, K., Kigaku, R., Komatsu, H., Yamashita, T., Kyogoku, K., & Iijima, S. (2005). High-level expression of single-chain Fv-Fc fusion protein in serum and egg white of genetically manipulated chickens by using a retroviral vector. *Journal of Virology*, *79*(17), 10864–10874. https://doi.org/10.1128/JVI.79.17.10864-10874.2005

Kamikawa, Y., Fujisaki, J., Nagayama, T., Kawasaki, K., Hirabayashi, D., Hamada, T., Sakamoto, R., Mukai, H., & Sugihara, K. (2016). Use of Candida-specific chicken egg yolk antibodies to inhibit the adhering of *Candida* to denture base materials: Prevention of denture stomatitis. *Gerodontology*, *33*(3), 342–347. https://doi.org/10.1111/ger.12163

Kang, B. C., Yeam, I., & Jahn, M. M. (2005). Genetics of plant virus resistance. *Annual Review of Phytopathology*, *43*, 581–621. https://doi.org/10.1146/annurev.phyto.43.011205.141140

Kanno, T., Nakai, T., & Muroga, K. (1989). Mode of transmission of Vibriosis among Ayu *Plecoglossus altivelis*. *Journal of Aquatic Animals Health*, *1*(1), 2–6. https://doi.org/10.1577/1548-8667(1989)001<0002:MOTOVA>2.3.CO;2

Kaper, J. B., Nataro, J. P., & Mobley, H. L. (2004). Pathogenic Escherichia coli. *Nature Reviews. Microbiology*, *2*(2), 123–140. https://doi.org/10.1038/nrmicro818

Karachaliou, C. E., Vassilakopoulou, V., & Livaniou, E. (2021). IgY technology: Methods for developing and evaluating avian immunoglobulins for the *in vitro* detection of biomolecules. *World Journal of Methodology*, *11*(5), 243–262. https://doi.org/10.5662/wjm.v11.i5.243

Kariyawasam, S., Wilkie, B. N., & Gyles, C. L. (2004). Resistance of broiler chickens to *Escherichia coli* respiratory tract infection induced by passively transferred egg-yolk antibodies. *Veterinary Microbiology*, *98*(3–4), 273–284. https://doi.org/10.1016/j.vetmic.2003.10.022

Keetanon, A., Chuchird, N., Bae, H.-D., Won, M.-K., Kim, S.-Y., & Elahi, F. (2021). Effects of IgY antibody on growth, survival, immune responses and protection against *Vibrio parahaemolyticus* in Pacific white shrimp. *Journal of Fisheries and Environment*, *45*(1), 1–6.

Khan, M. S. A., Ahmad, I., Aqil, F., Owais, M., Shahid, M., & Musarrat, J. (2010). Virulence and pathogenicity of fungal pathogens with special reference to *Candida albicans*. In I. Ahmad, M. Owais, M. Shahid & F. Aqil (Eds.). *Combating fungal infections: Problems and remedy* (pp. 21–45). Berlin/Heidelberg, Germany: Springer.

Khanna, S., & Pardi, D. S. (2016). Clinical implications of antibiotic impact on gastrointestinal microbiota and *Clostridium difficile* infection. *Expert Review of Gastroenterology & Hepatology*, *10*(10), 1145–1152. https://doi.org/10.1586/17474124.2016.1158097

Kikuti, M., Drebes, D., Robbins, R., Dufresne, L., Sanhueza, J. M., & Corzo, C. A. (2022). Growing pig incidence rate, control and prevention of porcine epidemic diarrhea virus in a large pig production system in the United States. *Porcine Health Management*, *8*, 23. https://doi.org/10.1186/s40813-022-00268-9

Kim, D. K., Jang, I. K., Seo, H. C., Shin, S. O., Yang, S. Y., & Kim, J. W. (2004). Shrimp protected from WSSV disease by treatment with egg yolk antibodies (IgY) against a truncated fusion protein derived from WSSV. *Aquaculture*, *237*(1–4), 21–30. https://doi.org/10.1016/j.aquaculture.2004.03.015

Kim, H., & Nakai, S. (1998). Simple separation of immunoglobulin from egg yolk by ultrafiltration. *Journal of Food Science*, *63*(3), 485–490. https://doi.org/10.1111/j.1365-2621.1998.tb15769.x

Kim, Y. M., Park, J. S., Kim, S. K., Jung, K. M., Hwang, Y. S., Han, M., Lee, H. J., Seo, H. W., Suh, J. Y., Han, B. K., & Han, J. Y. (2018). The transgenic chicken derived anti-CD20 monoclonal antibodies exhibits greater anti-cancer therapeutic potential with enhanced Fc effector functions. *Biomaterials*, *167*, 58–68. https://doi.org/10.1016/j.biomaterials.2018.03.021

Kink, J. A., & Williams, J. A. (1998). Antibodies to recombinant *Clostridium difficile* toxins A and B are an effective treatment and prevent relapse of *C. difficile*-associated disease in a hamster model of infection. *Infection and Immunity*, *66*(5), 2018–2025. https://doi.org/10.1128/IAI.66.5.2018-2025.1998

Klemperer, F. (1893). Ïber natürliche Immunität und ihre Verwerthung für die Immunisirungstherapie. *Archive of Experimental Pathology Pharmakology*, *31*, 356–382.

Klevens, R. M., Morrison, M. A., Nadle, J., Petit, S., Gershman, K., Ray, S., Harrison, L. H., Lynfield, R., Dumyati, G., Townes, J. M., Craig, A. S., Zell, E. R., Fosheim, G. E., McDougal, L. K., Carey, R. B., & Fridkin, S. K., & Active Bacterial Core surveillance (ABCs) MRSA Investigators (2007). Invasive methicillin-resistant *Staphylococcus aureus* infections in the United States. *JAMA*, *298*(15), 1763–1771. https://doi.org/10.1001/jama.298.15.1763

Kluytmans, J., van Belkum, A., & Verbrugh, H. (1997). Nasal carriage of Staphylococcus aureus: Epidemiology, underlying mechanisms, and associated risks. *Clinical Microbiology Reviews*, *10*(3), 505–520. https://doi.org/10.1128/CMR.10.3.505

Ko, K. Y., & Ahn, D. U. (2007). Preparation of immunoglobulin Y from egg yolk using ammonium sulfate precipitation and ion exchange chromatography. *Poultry Science*, *86*(2), 400–407. https://doi.org/10.1093/ps/86.2.400

Kobayashi, C., Yokoyama, H., Nguyen, S. V., Kodama, Y., Kimata, T., & Izeki, M. (2004). Effect of egg yolk antibody on experimental *Cryptosporidium parvum* infection in *scid* mice. *Vaccine*, *23*(2), 232–235. https://doi.org/10.1016/j.vaccine.2004.05.034

Kollberg, H. (2017). Avian Antibodies (IgY)-A New Weapon against Antibiotic Resistance. Clinical Microbiology: Open Access, *6*(6), 304–305. https://doi.org/ 10.4172/2327-5073.1000304

Korhonen, H., Marnila, P., & Gill, H. S. (2000). Bovine milk antibodies for health. *The British Journal of Nutrition*, *84 Suppl 1*, S135–S146. https://doi.org/10.1017/s0007114500002361

Kota, R. K., Reddy, P. N., & Sreerama, K. (2020). Application of IgY antibodies against staphylococcal protein A (SpA) of *Staphylococcus aureus* for detection and prophylactic functions. *Applied Microbiology and Biotechnology*, *104*(21), 9387–9398. https://doi.org/10.1007/s00253-020-10912-5

Kovacs-Nolan, J., & Mine, Y. (2011). Using egg IgY antibodies for health, diagnostic and other industrial applications. *Improving the Safety and Quality of Eggs and Egg Products*, 346–373. https://doi.org/10.1533/9780857093929.3.346

Kovacs-Nolan, J., Phillips, M., & Mine, Y. (2005). Advances in the value of eggs and egg components for human health. *Journal of Agricultural and Food Chemistry*, *53*(22), 8421–8431. https://doi.org/10.1021/jf050964f

Krüger, C., Pearson, S. K., Kodama, Y., Vacca Smith, A., Bowen, W. H., & Hammarström, L. (2004). The effects of egg-derived antibodies to glucosyltransferases on dental caries in rats. *Caries Research*, *38*(1), 9–14. https://doi.org/10.1159/000073914

Kubickova, B., Majerova, B., Hadrabova, J., Noskova, L., Stiborova, M., & Hodek, P. (2014). Effect of chicken antibodies on inflammation in human lung epithelial cell lines. *Neuro Endocrinology Letters*, *35*(*Suppl 2*), 99–104.

Kudra, T., & Ratti, C. (2006). Foam-mat drying: Energy and cost analyses. *Canadian Biosystematic Engineering*, *48*, 3.27–3.32.

Kumar, P. V., Sharma, S. K., Rishi, N., Ghosh, D. K., & Baranwal, V. K. (2018). An isothermal based recombinase polymerase amplification assay for rapid, sensitive and robust indexing of citrus yellow mosaic virus. *Acta virologica*, *62*(1), 104–108. https://doi.org/10.4149/av_2018_113

Kumaran, T., Michaelbab, M., Thangaswamy, S., Albindhas, S., & Citarasu, T. (2010). Production of anti WSSV IgY edible antibody using herbal immunoadjuvant *Asparagus racemosus* and its immunological influence against WSSV infection in *Penaeus monodon*. *Journal of Aquaculture Feed Science and Nutrition*, *2*(1), 1–5. https://doi.org/10.3923/joafsnu.2010.1.5

Kumaran, T., Thirumalaikumar, E., Lelin, C., Palanikumar, P., Michaelbabu, M., & Citarasu, T. (2018). Physicochemical properties of anti *Vibrio harveyi* egg yolk antibody (IgY) and its immunological influence in Indian white shrimp *Fenneropenaeus indicus*. *Fish & Shellfish Immunology*, *74*, 349–362. https://doi.org/10.1016/j.fsi.2017.12.062

Kurokawa, I., Danby, F. W., Ju, Q., Wang, X., Xiang, L. F., Xia, L., Chen, W., Nagy, I., Picardo, M., Suh, D. H., Ganceviciene, R., Schagen, S., Tsatsou, F., & Zouboulis, C. C. (2009). New developments in our understanding of acne pathogenesis and treatment. *Experimental Dermatology*, *18*(10), 821–832. https://doi.org/10.1111/j.1600-0625.2009.00890.x

Kuroki, M., Ohta, M., Ikemori, Y., Icatlo, F. C. Jr, Kobayashi, C., Yokoyama, H., & Kodama, Y. (1997). Field evaluation of chicken egg yolk immunoglobulins specific for bovine rotavirus in neonatal calves. *Archives of Virology*, *142*(4), 843–851. https://doi.org/10.1007/s007050050123

Kuroki, M., Ohta, M., Ikemori, Y., Peralta, R. C., Yokoyama, H., & Kodama, Y. (1994). Passive protection against bovine rotavirus in calves by specific immunoglobulins from chicken egg yolk. *Archives of. Virology*, 138, 143–148. https://doi.org/10.1007/BF01310045

Kweon, C. H., Kwon, B. J., Woo, S. R., Kim, J. M., Woo, G. H., Son, D. H., Hur, W., & Lee, Y. S. (2000). Immunoprophylactic effect of chicken egg yolk immunoglobulin (Ig Y) against porcine epidemic diarrhea virus (PEDV) in piglets. *The Journal of Veterinary Medical Science*, *62*(9), 961–964. https://doi.org/10.1292/jvms.62.961

Kyung, J. P., Dong, W. P., Chun, H. K., Beom, K. H., Tae, S. P., Jae, Y. H., Hyun, S. L., & Jin-Kyoo, K. (2005). Development and characterization of a recombinant chicken single-chain Fv antibody detecting Eimeria acervulina sporozoite antigen. *Biotechnol. Letter*, *27*, 289–295. https://doi.org/10.1007/s10529-005-0682-8

Lalonde, R., Fukuchi, K., & Strazielle, C. (2012). Neurologic and motor dysfunctions in APP transgenic mice. *Reviews in the Neurosciences*, *23*(4), 363–379. https://doi.org/10.1515/revneuro-2012-0041

Lancaster, J. (1976). A history of Newcastle disease with comments on its economic effects. *World's Poultry Science Journal*, *32*(2), 167–175. https://doi.org/10.1079/WPS19760001

Lapphra, K., Sangcharaswichai, A., Chokephaibulkit, K., Tiengrim, S., Piriyakarnsakul, W., Chakorn, T., Yoksan, S., Wattanamongkolsil, L., & Thamlikitkul, V. (2008). Evaluation of an NS1 antigen detection for diagnosis of acute dengue infection in patients with acute febrile illness. *Diagnostic Microbiology and Infectious Disease*, *60*(4), 387–391. https://doi.org/10.1016/j.diagmicrobio.2007.11.010

Larsen, L. E. (2000). Bovine respiratory syncytial virus (BRSV): A review. *Acta veterinaria Scandinavica*, *41*(1), 1–24. https://doi.org/10.1186/BF03549652

Larsson, A., Bålöw, R. M., Lindahl, T. L., & Forsberg, P. O. (1993). Chicken antibodies: Taking advantage of evolution–a review. *Poultry Science*, *72*(10), 1807–1812. https://doi.org/10.3382/ps.0721807

Larsson, A., Karlsson-Parra, A., & Sjöquist, J. (1991). Use of chicken antibodies in enzyme immunoassays to avoid interference by rheumatoid factors. *Clinical Chemistry*, *37*(3), 411–414.

Le Loir, L. Y., Baron, F., & Gautier, M. (2003). *Staphylococcus aureus* and food poisoning. *Genetics and Molecular Research: GMR*, *2*(1), 63–76.

LeClaire, R. D., Hunt, R. E., & Bavari, S. (2002). Protection against bacterial superantigen staphylococcal enterotoxin B by passive vaccination. *Infection and Immunity*, *70*(5), 2278–2281. https://doi.org/10.1128/IAI.70.5.2278-2281.2002

Lee, C. H., Lee, Y. C., Leu, S. J., Lin, L. T., Chiang, J. R., Hsu, W. J., & Yang, Y. Y. (2016). Production and characterization of neutralizing antibodies against *Bungarus multicinctus* snake venom. *Applied and Environmental Microbiology*, *82*(23), 6973–6982. https://doi.org/10.1128/AEM.01876-16

Lee, C. H., Leu, S. J., Lee, Y. C., Liu, C. I., Lin, L. T., Mwale, P. F., Chiang, J. R., Tsai, B. Y., Chen, C. C., Hung, C. S., & Yang, Y. Y. (2018b). Characterization of chicken-derived single chain antibody fragments against venom of *Naja Naja Atra*. *Toxins*, *10*(10), 383. https://doi.org/10.3390/toxins10100383

Lee, D. H., Jeon, Y. S., Park, C. K., Kim, S., Lee, D. S., & Lee, C. (2015). Immunoprophylactic effect of chicken egg yolk antibody (IgY) against a recombinant S1 domain of the porcine epidemic diarrhea virus spike protein in piglets. *Archives of Virology*, *160*, 2197–2207. https://doi.org/10.1007/s00705-015-2494-z

Lee, E. N., Sunwoo, H. H., Menninen, K., & Sim, J. S. (2002). In vitro studies of chicken egg yolk antibody (IgY) against *Salmonella enteritidis* and *Salmonella typhimurium*. *Poultry Science*, *81*(5), 632–641. https://doi.org/10.1093/ps/81.5.632

Lee, L., Samardzic, K., Wallach, M., Frumkin, L. R., & Mochly-Rosen, D. (2021). Immunoglobulin Y for potential diagnostic and therapeutic applications in infectious diseases. *Frontiers in Immunology*, *12*, 696003. https://doi.org/10.3389/fimmu.2021.696003

Lee, S. B., Mine, Y., & Stevenson, R. M. (2000). Effects of hen egg yolk immunoglobulin in passive protection of rainbow trout against *Yersinia ruckeri*. *Journal of Agricultural and Food Chemistry*, *48*(1), 110–115. https://doi.org/10.1021/jf9906073

Lee, W., Syed, A., Leow, A., Tan, C. Y., & Leow, S. C., C. H. (2018a). Isolation and characterization of a novel anti-salbutamol chicken scFv for human doping urinalysis. *Analytical Biochemistry*, *555*, 81–93. https://doi.org/10.1016/j.ab.2018.05.009

Leiva, C. L., Gallardo, M. J., Casanova, N., Terzolo, H., & Chacana, P. (2020). IgY-technology (egg yolk antibodies) in human medicine: A review of patents and clinical trials. *International Immunopharmacology*, *81*, 106269. https://doi.org/10.1016/j.intimp.2020.106269

Lemamy, G. J., Roger, P., Mani, J. C., Robert, M., Rochefort, H., & Brouillet, J. P. (1999). High-affinity antibodies from hen's egg yolks against human mannose-6-phosphate/insulin-like growth-factor-II receptor (M6P/IGFII-R): Characterization and potential use in clinical cancer studies. *International Journal of Cancer*, *80*(6), 896–902. https://doi.org/10.1002/(SICI)1097-0215(19990315)80:6<896::AID-IJC16>3.0.CO;2-J

Leow, C. H., Xu, L., Harley, C. A., Vieira-Pires, R. S., & Zhang, X. (2021). Monoclonal IgY antibodies. In Zhang, X.-Y. et al. (Eds.). *IgY-technology: Production and application of egg yolk antibodies*. Nature Switzerland AG: Springer. https://doi.org/10.1007/978-3-030-72688-1_13

Leslie, G. A., & Clem, L. W. (1969). Phylogeny of immunoglobulin structure and function. *Journal of Experimental Medicine*, *130* (6), 1337–1352. https://doi.org/10.1084/jem.130.6.1337

Leu, S. J., Lee, Y. C., Shih, N. Y., Huang, I. J., Liu, K. J., Lu, H. F., Huang, S. Y., & Yang, Y. Y. (2010). Generation and characterization of anti-alpha-enolase single-chain antibodies in chicken. *Veterinary Immunology and Immunopathology*, *137*(3–4), 251–260. https://doi.org/10.1016/j.vetimm.2010.06.001

Li, C., He, J., Ren, H., Zhang, X., Du, E., & Li, X. (2016a). Preparation of a chicken scFv to analyze gentamicin residue in animal derived food products. *Analytical Chemistry*, *88*(7), 4092–4098. https://doi.org/10.1021/acs.analchem.6b00426

Li, C., Zhang, Y., Eremin, S. A., Yakup, O., Yao, G., & Zhang, X. (2017). Detection of kanamycin and gentamicin residues in animal-derived food using IgY antibody based ic-ELISA and FPIA. *Food Chemistry*, *227*, 48–54. https://doi.org/10.1016/j.foodchem.2017.01.058

Li, G., Stewart, R., Conlan, B., Gilbert, A., Roeth, P., & Nair, H. (2002). Purification of human immunoglobulin G: a new approach to plasma fractionation. *Vox sanguinis*, *83*(4), 332–338. https://doi.org/10.1046/j.1423-0410.2002.00241.x

Li, T., Liu, H., Cai, K., Tian, M., Wang, Q., Shi, J., Gao, X., & Wang, H. (2013). Hypersensitive detection and quantitation of BoNT/A by IgY antibody against substrate linear-peptide. *PLoS One 8* (3), e58908. https://doi.org/10.1371/journal.pone.0058908

Li, X., He, P., Yu, L., He, Q., Jia, C., Yang, H., Lu, M., Wei, X., & Zhao, S. (2020). Production and characteristics of a novel chicken egg yolk antibody (IgY) against periodontitis-associated pathogens. *Journal of Oral Microbiology*, *12*(1), 1831374. https://doi.org/10.1080/20002297.2020.1831374

Li, X., Jing, K., Wang, X., Li, Y., Zhang, M., Li, Z., Xu, L., Wang, L., & Xu, Y. (2016b). Protective effects of chicken egg yolk antibody (IgY) against experimental *Vibrio splendidus* infection in the sea cucumber (*Apostichopus japonicus*). *Fish & Shellfish Immunology*, *48*, 105–111. https://doi.org/10.1016/j.fsi.2015.11.024

Li, X., Liu, H., Xu, Y., Xu, F., Wang, L., You, J., Li, S., & Jin, L. (2012). Chicken egg yolk antibody (IgY) controls *Solobacterium moorei* under in vitro and in vivo conditions. *Applied Biochemistry and Biotechnology*, *168*(6), 1448–1458. https://doi.org/10.1007/s12010-012-9869-3

Li, X., Wang, L., Zhen, Y., Li, S., & Xu, Y. (2015). Chicken egg yolk antibodies (IgY) as non-antibiotic production enhancers for use in swine production: A review. *Journal of Animal Science and Biotechnology*, *6*(1), 40. https://doi.org/10.1186/s40104-015-0038-8

Li, X., Yao, Y., Wang, X., Zhen, Y., Thacker, P. A., Wang, L., Shi, M., Zhao, J., Zong, Y., Wang, N., & Xu, Y. (2016c). Chicken egg yolk antibodies (IgY) modulate the intestinal mucosal immune response in a mouse model of *Salmonella typhimurium* infection. *International Immunopharmacology*, *36*, 305–314. https://doi.org/10.1016/j.intimp.2016.04.036

Li, X. L., Shuai, J. B., & Fang, W. H. (2006). Protection of *Carassius auratus Gibelio* against infection by *Aeromonas hydrophila* using specific immunoglobulins from hen egg yolk. *Journal of Zhejiang University. Science. B*, *7*(11), 922–928. https://doi.org/10.1631/jzus.2006.B0922

Li, X. Y., Jin, L. J., McAllister, T. A., Stanford, K., Xu, J. Y., Lu, Y. N., Zhen, Y. H., Sun, Y. X., & Xu, Y. P. (2007). Chitosan-alginate microcapsules for oral delivery of egg yolk immunoglobulin (IgY). *Journal of Agricultural and Food Chemistry*, *55*(8), 2911–2917. https://doi.org/10.1021/jf062900q

Li, X. Y., Jin, L. J., Uzonna, J. E., Li, S. Y., Liu, J. J., Li, H. Q., Lu, Y. N., Zhen, Y. H., & Xu, Y. P. (2009). Chitosan-alginate microcapsules for oral delivery of egg yolk immunoglobulin (IgY): In vivo evaluation in a pig model of enteric colibacillosis. *Veterinary Immunology and Immunopathology*, *129*(1–2), 132–136. https://doi.org/10.1016/j.vetimm.2008.12.016

Li, Z. X., Hu, W. D., Li, B. C., Li, T. Y., Zhou, X. Y., & Zhang, Z. (2014). Egg yolk IgY against RHDV capsid protein VP60 promotes rabbit defense against RHDV infection. *Veterinary Immunology and Immunopathology*, *157*(1–2), 97–104. https://doi.org/10.1016/j.vetimm.2013.10.002

Liang, X., Sheng, Y., Yu, W., Zhao, S., Shan, H., Zhang, Q., & Wang, Z. (2018). Comparison of chicken IgY and mammalian IgG in three immunoassays for detection of sulfamethazine in milk. *Food Analytical Methods*, *11*, 3452–3463. https://doi.org/10.1007/s12161-018-1316-9

Lin, J. H., Lo, C. M., Chuang, S. H., Chiang, C. H., Wang, S. D., Lin, T. Y., Liao, J. W., & Hung, D. Z. (2020). Collocation of avian and mammal antibodies to develop a rapid and sensitive diagnostic tool for Russell's vipers snakebite. *PLoS Neglected Tropical Diseases*, *14*(9), e0008701. https://doi.org/10.1371/journal.pntd.0008701

Lindberg, A. A., Brown, J. E., Strömberg, N., Westling-Ryd, M., Schultz, J. E., & Karlsson, K. A. (1987). Identification of the carbohydrate receptor for Shiga toxin produced by Shigella dysenteriae type 1. *The Journal of Biological Chemistry*, *262*(4), 1779–1785.

Linden, C. D., & Roth, T. F. (1978). IgG receptors on foetal chick yolk sac. *Journal of Cell Science*, *33*, 317–328. https://doi.org/10.1242/jcs.33.1.317

Lingwood, C. A., Law, H., Richardson, S., Petric, M., Brunton, J. L., De Grandis, S., & Karmali, M. (1987). Glycolipid binding of purified and recombinant *Escherichia coli* produced verotoxin in vitro. *The Journal of Biological Chemistry*, *262*(18), 8834–8839.

Liu, L. (2015). Antibody glycosylation and its impact on the pharmacokinetics and pharmacodynamics of monoclonal antibodies and Fc-fusion proteins. *Journal of Pharmaceutical Sciences*, *104*(6), 1866–1884. https://doi.org/10.1002/jps.24444

Liu, J., He, Q., Wang, W., Zhou, B., Li, B., Zhang, Y., Luo, C., Chen, D., Tang, J., & Yu, X. (2017). Preparation and neutralization efficacy of IgY antibodies raised against *Deinagkistrodon acutus* venom. *The Journal of Venomous Animals and Toxins Including Tropical Diseases*, *23*, 22. https://doi.org/10.1186/s40409-017-0112-0

Liu, J., Qin, Y., Yan, L., Liu, W., Shi, H., Lu, Y., & Liu, X. (2021). Protective effects of egg yolk immunoglobulins (IgY) on juvenile groupers (*Epinephelus fuscoguttatus* × *Epinephelus lanceolatus*) with red spotted grouper nervous necrosis virus infection. *Aquaculture*, *545*, 737218. https://doi.org/10.1016/j.aquaculture.2021.737218.

Liu, P. C., Chen, Y. C., Huang, C. Y., & Lee, K. K. (2000). Virulence of Vibrio parahaemolyticus isolated from cultured small abalone, Haliotis diversicolor supertexta, with withering syndrome. *Letters in Applied Microbiology*, *31*(6), 433–437. https://doi.org/10.1046/j.1365-2672.2000.00843.x

Loesche, W. J. (1986). Role of Streptococcus mutans in human dental decay. *Microbiological Reviews*, *50*(4), 353–380. https://doi.org/10.1128/mr.50.4.353-380.1986

Loh, S. K., Chan Man, Y. B., Tan, C. P., Osman, A., & Hamid, N. S. A. (2005). Process optimisation of encapsulated pandan (Pandanus amaryllifolius) powder using spray-drying method. *Journal of the Science of Food and Agriculture*, *85*(12), 1999–2004. https://doi.org/10.1002/jsfa.2169

Lopes, C. A., de Faria, L. S., de Sousa, J., Borges, I. P., Ribeiro, R. P., Bueno, L. L., Rodrigues Ávila, V. M., Ferreira Júnior, Á., & Costa-Cruz, J. M. (2019). Anti-*Ascaris suum* immunoglobulin Y as a novel biotechnological tool for the diagnosis of human ascariasis. *Journal of Helminthology*, *94*, e71. https://doi.org/10.1017/S0022149X19000701

López, E. L., Contrini, M. M., Glatstein, E., González Ayala, S., Santoro, R., Allende, D., Ezcurra, G., Teplitz, E., Koyama, T., Matsumoto, Y., Sato, H., Sakai, K., Hoshide, S., Komoriya, K., Morita, T., Harning, R., & Brookman, S. (2010). Safety and pharmacokinetics of urtoxazumab, a humanized monoclonal antibody, against Shiga-like toxin 2 in healthy adults and in pediatric patients infected with Shiga-like toxin-producing Escherichia coli. *Antimicrobial Agents and Chemotherapy*, *54*(1), 239–243. https://doi.org/10.1128/AAC.00343-09

Losonczy, S., Szabó, C., Kiss, Z., & Bárdos, L. (1999). Application of an anti-HQIgY antibody for the measurement of IgY concentrations of hen's and quail's serum and yolk. *Acta Physiologica Hungarica*, *86*(3–4), 253–258.

Losonsky, G. A., Johnson, J. P., Winkelstein, J. A., & Yolken, R. H. (1985). Oral Administration of human serum immunoglobulin in immunodeficient patients with viral gastroenteritis. A pharmacokinetic and functional analysis. *The Journal of Clinical Investigation*, *76*(6), 2362–2367. https://doi.org/10.1172/JCI112248

Lowy, F. D. (1998). *Staphylococcus aureus* infections. *The New England Journal of Medicine*, *339*(8), 520–532. https://doi.org/10.1056/NEJM199808203390806

Lu, Y., Liu, J., Jin, L., Li, X., Zhen, Y., Xue, H., Lin, Q., & Xu, Y. (2009). Passive immunization of crayfish (*Procambius clarkiaii*) with chicken egg yolk immunoglobulin (IgY) against white spot syndrome virus (WSSV). *Applied Biochemistry and Biotechnology*, *159*(3), 750–758. https://doi.org/10.1007/s12010-009-8555-6

Lu, Y., Liu, J., Jin, L., Li, X., Zhen, Y., Xue, H., You, J., & Xu, Y. (2008). Passive protection of shrimp against white spot syndrome virus (WSSV) using specific antibody from egg yolk of chickens immunized with inactivated virus or a WSSV-DNA vaccine. *Fish & Shellfish Immunology*, *25*(5), 604–610. https://doi.org/10.1016/j.fsi.2008.08.010

Lu, Y., Wang, Y., Zhang, Z., Huang, J., Yao, M., Huang, G., Ge, Y., Zhang, P., Huang, H., Wang, Y., Li, H., & Wang, W. (2020). Generation of chicken IgY against SARS-COV-2 spike protein and epitope mapping. *Journal of Immunology Research*, *2020*, 9465398. https://doi.org/10.1155/2020/9465398

Luckins A. G. (1988). Trypanosoma evansi in Asia. *Parasitology Today (Personal ed.)*, *4*(5), 137–142. https://doi.org/10.1016/0169-4758(88)90188-3

Łupicka-Słowik, A., Grzywa, R., Leporowska, E., Procyk, D., Oleksyszyn, J., & Sieńczyk, M. (2019). Development and evaluation of an immunoglobulin Y-based ELISA for measuring prostate specific antigen in human serum. *Annals of Laboratory Medicine*, *39*(4), 373–380. https://doi.org/10.3343/alm.2019.39.4.373

Łupicka-Słowik, A., Psurski, M., Grzywa, R., Bobrek, K., Smok, P., Walczak, M., Gaweł, A., Stefaniak, T., Oleksyszyn, J., & Sieńczyk, M. (2018). Development of adenosine deaminase-specific IgY antibodies: Diagnostic and inhibitory application. *Applied Biochemistry and Biotechnology*, *184*(4), 1358–1374. https://doi.org/10.1007/s12010-017-2626-x

Łupicka-Słowik, A., Walczak, M., Grzywa, R., Bobrek, K., Łęcka, M., Boivin, S., Gaweł, A., Stefaniak, T., Oleksyszyn, J., & Sieńczyk, M. (2014). Generation and application of polyclonal IgY antibodies specific for full-length and nicked prostate-specific antigen. *Bioanalysis*, *6*(23), 3197–3213. https://doi.org/10.4155/bio.14.172

Lyall, J., Irvine, R. M., Sherman, A., McKinley, T. J., Núñez, A., Purdie, A., Outtrim, L., Brown, I. H., Rolleston-Smith, G., Sang, H., & Tiley, L. (2011). Suppression of avian influenza transmission in genetically modified chickens. *Science (New York, N.Y.)*, *331*(6014), 223–226. https://doi.org/10.1126/science.1198020

Maeda, N., Okamoto, M., Kondo, K., Ishikawa, H., Osada, R., Tsurumoto, A., & Fujita, H. (1998). Incidence of *Prevotella intermedia* and *Prevotella nigrescens* in periodontal health and disease. *Microbiology and Immunology*, *42*(9), 583–589. https://doi.org/10.1111/j.1348-0421.1998.tb02328.x

Maenz, C., Chang, S. F., Iwanski, A., & Bruns, M. (2007). Entry of duck hepatitis B virus into primary duck liver and kidney cells after discovery of a fusogenic region within the large surface protein. *Journal of Virology*, *81*(10), 5014–5023. https://doi.org/10.1128/JVI.02290-06

Magagnotti, C., Fermo, I., Carletti, R., Ferrari, M. & Bachi, A. (2010). Comparison of different depletion strategies for improving resolution of the human urine proteome. *Clinical Chemistry and Laboratory Medicine*, *48*(4), 531–535. https://doi.org/10.1515/CCLM.2010.109

Mahdavi, A. H., Rahmani, H. R., Nili, N., Samie, A. H., Soleimanian-Zad, S., & Jahanian, R. (2010b). Effects of dietary egg yolk antibody powder on growth performance, intestinal *Escherichia coli* colonization, and immunocompetence of challenged broiler chicks. *Poultry Science*, *89*(3), 484–494. https://doi.org/10.3382/ps.2009-00541

Mahdavi, A., Rahmani, H., Nili, N., Samie, A., & Soleimania, S. (2010a). Chicken egg yolk antibody (IgY) powder against *Escherichia coli* O78:K80. *Journal of Animal Veterinary Advances*, *9*(2), 366–373. https://doi.org/10.3923/javaa.2010.366.373

Mahenthiran, R., Naveen, S., Diraviyam, T., Kumar, K. R., & Michael, A. (2013). Generation purification and neutralization potential of chicken egg yolk antibodies IgY against mastitis causing *Escherichia coli* and *Staphylococcus aureus*. *International Journal of Pharma and Bio Sciences*, *4*, 687–96.

Malekshahi, Z. V., Gargari, S. L., Rasooli, I., & Ebrahimizadeh, W. (2011). Treatment of *Helicobacter pylori* infection in mice with oral administration of egg yolk-driven anti-UreC immunoglobulin. *Microbial Pathogenesis*, *51*(5), 366–372. https://doi.org/10.1016/j.micpath.2011.06.002

Malik, M. W., Ayub, N., & Qureshi, I. Z. (2006). Passive immunization using purified IgYs against infectious bursal disease of chickens in Pakistan. *Journal of Veterinary Science*, *7*(1), 43–46. https://doi.org/10.4142/jvs.2006.7.1.43

Manhani, M. N., Ribeiro, V. S., Cardoso, R., Ueira-Vieira, C., Goulart, L. R., & Costa-Cruz, J. M. (2011). Specific phage-displayed peptides discriminate different forms of neurocysticercosis by antibody detection in the serum samples. *Parasite Immunology*, *33*(6), 322–329. https://doi.org/10.1111/j.1365-3024.2011.01283.x

Marcus, E. A., & Scott, D. R. (2001). Cell lysis is responsible for the appearance of extracellular urease in *Helicobacter pylori*. *Helicobacter*, *6*(2), 93–99. https://doi.org/10.1046/j.1523-5378.2001.00014.x

Mark, H. B., & Roberts, S. P. (2016). *The Merck manual of diagnosis and therapy*, 18th Edition. Kenilworth, NJ, USA: Merck Sharp & Dohme Corp.

Marquardt, R. R., Jin, L. Z., Kim, J. W., Fang, L., Frohlich, A. A., & Baidoo, S. K. (1999). Passive protective effect of egg-yolk antibodies against enterotoxigenic *Escherichia coli* K88+ infection in neonatal and early-weaned piglets. *FEMS Immunology and Medical Microbiology, 23*(4), 283–288. https://doi.org/10.1111/j.1574-695X.1999.tb01249.x

Martínez, C., Gutiérrez, G., Alvarez, I., Porta, N., Lomónaco, M., Wigdorovitz, A., Chacana, P., & Trono, K. (2014). Egg yolk antibodies (IgY) against Bovine leukemia virus. *Retrovirology, 11*(Suppl 1), 46. https://doi.org/10.1186/1742-4690-11-S1-P46

Maruyama, H., Sakai, K., Sato, K., Sakaguchi, K., Koma, Y., Watanabe, J., & …Hibi, H. (2022). Near infrared photo-antimicrobial targeting therapy for periodontal disease. International Association for Dental Research, IADR/AADR/CADR General Session (Virtual). Presented in July 23rd, 2022.

Matsuda, H., Mitsuda, H., Nakamura, N., Furusawa, S., Mohri, S., & Kitamoto, T. (1999). A chicken monoclonal antibody with specificity for the N-terminal of human prion protein. *FEMS Immunology and Medical Microbiology, 23*(3), 189–194. https://doi.org/10.1111/j.1574-695X.1999.tb01238.x

Matsunaga, Y., Wakatsuki, Y., Tabata, Y., Kawasaki, H., Usui, T., Yoshida, M., Itoh, T., Habu, S., & Kita, T. (2000). Oral immunization with size-purified microsphere beads as a vehicle selectively induces systemic tolerance and sensitization. *Vaccine, 19*(4–5), 579–588. https://doi.org/10.1016/s0264-410x(00)00120-1

Matsushita, K., Horiuchi, H., Furusawa, S., Horiuchi, M., Shinagawa, M., & Matsuda, H. (1998). Chicken monoclonal antibodies against synthetic bovine prion protein peptide. *The Journal of Veterinary Medical Science, 60*(6), 777–779. https://doi.org/10.1292/jvms.60.777

Meenatchisundaram, S., Shanmugam, V., & Anjali, V. M. (2011). Development of chicken egg yolk antibodies against *Streptococcus mitis* - purification and neutralizing efficacy. *Journal of Basic and Clinical Pharmacy, 2*(2), 109–114.

Mendoza, J. C., Vivas, D., Rodríguez, E., Inga, R., Sandoval, G., & Lazo, F. (2012). Eficacia Experimental de Anticuerpos IgY Producidos en Huevos, Contra el Veneno de la Serpiente Peruana *Bothrops atrox*. *Revista Peruana de Medicina Experimental y Salud Pública (RPMESP), 29*(1), 69–75. https://doi.org/10.17843/rpmesp.2012.291.310

Merino, S., Rubires, X., Knochel, S., & Tomas, J. M. (1995). Emerging pathogens: Aeromonas spp. *International Journal of Food Microbiology, 28*(2), 157–168. https://doi.org/10.1016/0168-1605(95)00054-2

Meulenaer, B., De, & Huyghebaert, A. (2001). Isolation and purification of chicken egg yolk immunoglobulins: A review. *Food and Agricultural Immunology, 13*(4), 275–288. https://doi.org/10.1080/09540100120094537

Mine, Y. (1997). Separation of *Salmonella enteritidis* from experimentally contaminated liquid eggs using a hen IgY immobilized immunomagnetic separation system. *Journal of Agricultural Food Chemistry 45*, 3723–37. https://doi.org/10.1021/jf9701998

Mine, Y., & Kovacs-Nolan, J. (2002). Chicken egg yolk antibodies as therapeutics in enteric infectious disease: A review. *Journal of Medicinal Food, 5*(3), 159–169. https://doi.org/10.1089/10966200260398198

Miura, V. C., Aoki, S. M., Junior, P. P., Pires, L. C., Dalmagro, P., & Nakamura, A. A. (2017). Evaluation of recombinant *Cryptosporidium hominis* GP60 protein and anti-GP60 chicken polyclonal IgY for research and diagnostic purposes. *Revista Brasileira de Parasitologia Veterinaria, 26*(2), 205–210. https://doi.org/10.1590/S1984-29612017032

Mora, C., Tittensor, D. P., Adl, S., Simpson, A. G., & Worm, B. (2011). How many species are there on earth and in the ocean?. *PLoS Biology, 9*(8), e1001127. https://doi.org/10.1371/journal.pbio.1001127

Moreno, L., Moreno, C., Bilbao, V., Acevedo, A., Felizzola, O., Zerpa, N., & Malave, C. (2011), Producción Y Evaluación *In Vitro* De IgY Contra Streptococcus Mutans. *Rev la Sociedad Venezolana Microbiología 31.* 118–23. (article in Spanish). http://www.redalyc.org/articulo.oa?id=199422818007

Morrison, S. L., Johnson, M. J., Herzenberg, L. A., & Oi, V. T. (1984). Chimeric human antibody molecules: Mouse antigen-binding domains with human constant region domains. *Proceedings of the National Academy of Sciences of the United States of America, 81*(21), 6851–6855. https://doi.org/10.1073/pnas.81.21.6851

Mudili, V., Makam, S.S., Sundararaj, N., Siddaiah, C., Gupta, V. K., & Rao, P. (2015). A novel IgY-Aptamer hybrid system for cost-effective detection of SEB and its evaluation on food and clinical samples. *Scientific Reports, 5*, 15151. https://doi.org/10.1038/srep15151 (Retraction published Sci Rep. 2022 Jun 29;12(1):10939)

Muhammad, K., Rabbi, F., Hussain, I., Riaz, A., Saeed, K., & Hussain, T. (2001). Passive immunization against infectious bursal disease in chicks. *International Journal of Agriculture and Biology, 3*(4), 413–415. https://eurekamag.com/research/003/876/003876367.php

Mulvey, G. L., Dingle, T. C., Fang, L., Strecker, J., & Armstrong, G. D. (2011). Therapeutic potential of egg yolk antibodies for treating *Clostridium difficile* infection. *Journal of Medical Microbiology, 60*(Pt 8), 1181–1187. https://doi.org/10.1099/jmm.0.029835-0

Murai, A., Kakiuchi, M., Hamano, T., Kobayashi, M., Tsudzuki, M., Nakano, M., Matsuda, Y., & Horio, F. (2016). An ELISA for quantifying quail IgY and characterizing maternal IgY transfer to egg yolk in several quail strains. *Veterinary Immunology and Immunopathology, 175*, 16–23. https://doi.org/10.1016/j.vetimm.2016.04.013

Nagaraj, S., Ramlal, S., Kingston, J., & Batra, H. V. (2016). Development of IgY based sandwich ELISA for the detection of staphylococcal enterotoxin G (SEG), an egc toxin. *International Journal of Food Microbiology, 237*, 136–141. https://doi.org/10.1016/j.ijfoodmicro.2016.08.009

Nair, M. K., Joy, J., Vasudevan, P., Hinckley, L., Hoagland, T. A., & Venkitanarayanan, K. S. (2005). Antibacterial effect of caprylic acid and monocaprylin on major bacterial mastitis pathogens. *Journal of Dairy Science, 88*(10), 3488–3495. https://doi.org/10.3168/jds.S0022-0302(05)73033-2

Nakamura, N., Aoki, Y., Horiuchi, H., Furusawa, S., Yamanaka, H. I., Kitamoto, T., & Matsuda, H. (2000). Construction of recombinant monoclonal antibodies from a chicken hybridoma line secreting specific antibody. *Cytotechnology, 32*(3), 191–198. https://doi.org/10.1023/A:1008149815908

Nakamura, R., Pedrosa-Gerasmio, I. R., Alenton, R., Nozaki, R., Kondo, H., & Hirono, I. (2019). Anti-PirA-like toxin immunoglobulin (IgY) in feeds passively immunizes shrimp against acute hepatopancreatic necrosis disease. *Journal of Fish Diseases*, *42*(8), 1125–1132. https://doi.org/10.1111/jfd.13024

Nandiyanto, A. B. D., & Okuyama, K. (2011). Progress in developing spray-drying methods for the production of controlled morphology particles: From the nanometer to submicrometer size ranges. *Advanced Powder Technology*, *22*(1), 1–19. https://doi.org/10.1016/j.apt.2010.09.011

Narat, M. (2003). Production of antibodies in chickens. *Food Technology Biotechnology*, *41*(3), 259–267. https://hrcak.srce.hr/118637

Nasiri, K., Zibaee, S., Nassiri, M., Tahmoorespur, M., & Haghparast, A. (2016). Production of specific IgY antibody to the recombinant FanC protein produced in *Escherichia coli*. *Iranian Journal of Basic Medical Sciences*, *19*(8), 883–889.

Nataro, J. P., & Kaper, J. B. (1998). Diarrheagenic *Escherichia coli*. *Clinical Microbiology Reviews*, *11*(1), 142–201. https://doi.org/10.1128/CMR.11.1.142

Navarro, D., Vargas, M., Herrera, M., Segura, Á., Gómez, A., Villalta, M., Ramírez, N., Williams, D., Gutiérrez, J. M., & León, G. (2016). Development of a chicken-derived antivenom against the taipan snake (*Oxyuranus scutellatus*) venom and comparison with an equine antivenom. *Toxicon: Official Journal of the International Society on Toxinology*, *120*, 1–8. https://doi.org/10.1016/j.toxicon.2016.06.018

Neal-McKinney, J. M., Samuelson, D. R., Eucker, T. P., Nissen, M. S., Crespo, R., & Konkel, M. E. (2014). Reducing *Campylobacter jejuni* colonization of poultry via vaccination. *PloS One*, *9*(12), e114254. https://doi.org/10.1371/journal.pone.0114254

Neri, P., Tokoro, S., Kobayashi, R., Sugiyama, T., Umeda, K., Shimizu, T., Tsuji, T., Kodama, Y., Oguma, K., & Mori, H. (2011). Specific egg yolk immunoglobulin as a new preventive approach for Shiga-toxin-mediated diseases. *PloS One*, *6*(10), e26526. https://doi.org/10.1371/journal.pone.0026526

Neri, P., Tokoro, S., Sugiyama, T., Umeda, K., Shimizu, T., Tsuji, T., Kodama, Y., & Mori, H. (2012). Recombinant Shiga toxin B subunit can induce neutralizing immunoglobulin Y antibody. *Biological & Pharmaceutical Bulletin*, *35*(6), 917–923. https://doi.org/10.1248/bpb.35.917

Nguyen, H. H., Tumpey, T. M., Park, H. J., Byun, Y. H., Tran, L. D., Nguyen, V. D., Kilgore, P. E., Czerkinsky, C., Katz, J. M., Seong, B. L., Song, J. M., Kim, Y. B., Do, H. T., Nguyen, T., & Nguyen, C. V. (2010). Prophylactic and therapeutic efficacy of avian antibodies against influenza virus H5N1 and H1N1 in mice. *PLoS One*, *5*(4), e10152. https://doi.org/10.1371/journal.pone.0010152

Nguyen, S. V., Icatlo, F. C. Jr, Nakano, T., Isogai, E., Hirose, K., Mizugai, H., Kobayashi-Sakamoto, M., Isogai, H., & Chiba, I. (2011). Anti-cell-associated glucosyltransferase immunoglobulin Y suppression of salivary mutans streptococci in healthy young adults. *Journal of the American Dental Association (1939)*, *142*(8), 943–949. https://doi.org/10.14219/jada.archive.2011.0301

Nie, G., Wang, T., Lu, S., Liu, W., Li, Y., & Lei, J. (2014). Detection of *Clonorchis sinensis* circulating antigen in sera from Chinese patients by immunomagnetic bead ELISA based on IgY. *PloS One*, *9*(12), e113208. https://doi.org/10.1371/journal.pone.0113208

Nie, W., Zhao, C., Guo, X., Sun, L., Meng, T., Liu, Y., Song, X., Xu, K., Wang, J., & Li, J. (2019). Preparation and identification of chicken egg yolk immunoglobulins against human enterovirus 71 for diagnosis of hand-foot-and-mouth disease. *Analytical Biochemistry*, *573*, 44–50. https://doi.org/10.1016/j.ab.2019.02.029

Nilsson, E., Amini, A., Wretlind, B., & Larsson, A. (2007a). *Pseudomonas aeruginosa* infections are prevented in cystic fibrosis patients by avian antibodies binding *Pseudomonas aeruginosa* flagellin. *Journal of Chromatography. B, Analytical Technologies in the Biomedical and Life Sciences*, *856*(1–2), 75–80. https://doi.org/10.1016/j.jchromb.2007.05.029

Nilsson, E., Kollberg, H., Johannesson, M., Wejåker, P. E., Carlander, D., & Larsson, A. (2007b). More than 10 years' continuous oral treatment with specific immunoglobulin Y for the prevention of *Pseudomonas aeruginosa* infections: A case report. *Journal of Medicinal Food*, *10*(2), 375–378. https://doi.org/10.1089/jmf.2006.214

Nilsson, E., Larsson, A., Olesen, H. V., Wejåker, P. E., & Kollberg, H. (2008). Good effect of IgY against *Pseudomonas aeruginosa* infections in cystic fibrosis patients. *Pediatric Pulmonology*, *43*(9), 892–899. https://doi.org/10.1002/ppul.20875

Nimalaratne, C., & Wu, J. (2015). Hen egg as an antioxidant food commodity: A review. *Nutrients*, *7*(10), 8274–8293. https://doi.org/10.3390/nu7105394

Nishibori, N., Shimamoto, T., Nakamura, N., Shimokawa, M., Horiuchi, H., Furusawa, S., & Matsuda, H. (2004). Expression vectors for chicken-human chimeric antibodies. *Biologicals: Journal of the International Association of Biological Standardization*, *32*(4), 213–218. https://doi.org/10.1016/j.biologicals.2004.09.002

Nishinaka, S., Akiba, H., Nakamura, M., Suzuki, K., Suzuki, T., Tsubokura, K., Horiuchi, H., Furusawa, S., & Matsuda, H. (1996). Two chicken B cell lines resistant to ouabain for the production of chicken monoclonal antibodies. *The Journal of Veterinary Medical Science*, *58*(11), 1053–1056. https://doi.org/10.1292/jvms.58.11_1053

Nishinaka, S., Matsuda, H., & Murata, M. (1989). Establishment of a chicken X chicken hybridoma secreting specific antibody. *International Archives of Allergy and Applied Immunology*, *89*(4), 416–419. https://doi.org/10.1159/000234985

Nishinaka, S., Suzuki, T., Matsuda, H., & Murata, M. (1991). A new cell line for the production of chicken monoclonal antibody by hybridoma technology. *Journal of Immunological Methods*, *139*(2), 217–222. https://doi.org/10.1016/0022-1759(91)90191-h

Noël de Tilly, A., & Tharmalingam, S. (2022). Review of treatments for oropharyngeal fungal infections in HIV/AIDS patients. *Microbiology Research*, *13*, 219–234. https://doi.org/10.3390/microbiolres13020019

Nomura, S., Suzuki, H., Masaoka, T., Kurabayashi, K., Ishii, H., Kitajima, M., Nomoto, K., & Hibi, T. (2005). Effect of dietary anti-urease immunoglobulin Y on *Helicobacter pylori* infection in Mongolian gerbils. *Helicobacter*, *10*(1), 43–52. https://doi.org/10.1111/j.1523-5378.2005.00290.x

Nord, J., Ma, P., DiJohn, D., Tzipori, S., & Tacket, C. O. (1990). Treatment with bovine hyperimmune colostrum of cryptosporidial diarrhea in AIDS patients. *AIDS (London, England)*, *4*(6), 581–584. https://doi.org/10.1097/00002030-199006000-00015

Nordseth, T. (2020). "IgY Antibody immunotherapy for deformed wing virus in Western honeybees (Apis Mellifera)" *IdeaFest*. 6. https://red.library.usd.edu/idea/6

Nothaft, H., Davis, B., Lock, Y. Y., Perez-Munoz, M. E., Vinogradov, E., Walter, J., Coros, C., & Szymanski, C. M. (2016). Engineering the *Campylobacter jejuni* N-glycan to create an effective chicken vaccine. *Scientific Reports*, *6*, 26511. https://doi.org/10.1038/srep26511

O'Donnell, K. L., Espinosa, D. A., Puerta-Guardo, H., Biering, S. B., Warnes, C. M., Schiltz, J., Nilles, M. L., Li, J., Harris, E., & Bradley, D. S. (2020). Avian anti-NS1 IgY antibodies neutralize dengue virus infection and protect against lethal dengue virus challenge. *Antiviral Research*, *183*, 104923. https://doi.org/10.1016/j.antiviral.2020.104923

O'Donnell, K. L., Fink, A., Nilles, M. L., & Bradley, D. S. (2017). Dengue NS1-specific IgY antibodies neutralizes dengue infection without inducing antibody dependent enhancement. *J Immunol*, *198* (1 Supplement).

O'Donnell, K. L., Meberg, B., Schiltz, J., Nilles, M. L., & Bradley, D. S. (2019). Zika virus-specific IgY results are therapeutic following a lethal Zika virus challenge without inducing antibody-dependent enhancement. *Viruses*, *11*(3), 301. https://doi.org/10.3390/v11030301

Oba, P. M., Devito, F. C., Santos, J. P. F., Stipp, R. N., Gomes, M., de, O. S., Carciofi, A. C., & Brunetto, M. A. (2018). Effects of passive immunization by anti-Gingipain IgY on the Oral health of cats fed Kibble diets. *Journal of Veterinary Dentistry*, *35*(4), 275–280. https://doi.org/10.1177/0898756418814010

Offenbacher, S., Katz, V., Fertik, G., Collins, J., Boyd, D., Maynor, G., McKaig, R., & Beck, J. (1996). Periodontal infection as a possible risk factor for preterm low birth weight. *Journal of Periodontology*, *67*(10 Suppl), 1103–1113. https://doi.org/10.1902/jop.1996.67.10s.1103

Oishi, I., Yoshii, K., Miyahara, D., & Tagami, T. (2018). Efficient production of human interferon beta in the white of eggs from ovalbumin gene-targeted hens. *Scientific Reports*, *8*(1), 10203. https://doi.org/10.1038/s41598-018-28438-2

Oliver, C., Valenzuela, K., Silva, H., Haro, R. E., Cortés, M., Sandoval, R., Pontigo, J. P., Álvarez, C., Figueroa, J. E., Avendaño-Herrera, R., Troncoso, J. M., & Yáñez, A. J. (2015). Effectiveness of egg yolk immunoglobulin against the intracellular salmonid pathogen *Piscirickettsia salmonis*. *Journal of Applied Microbiology*, *119*(2), 365–376. https://doi.org/10.1111/jam.12857

Omata, M., Uchiumi, K., Ito, Y., Yokosuka, O., Mori, J., Terao, K., Wei-Fa, Y., O'Connell, A. P., London, W. T., & Okuda, K. (1983). Duck hepatitis B virus and liver diseases. *Gastroenterology*, *85*(2), 260–267.

Omidian, Z., Ebrahimzadeh, E., Shahbazi, P., Asghari, Z., & Shayan, P. (2014). Application of recombinant *Cryptosporidium parvum* P23 for isolation and prevention. *Parasitology Research*, *113*(1), 229–237. https://doi.org/10.1007/s00436-013-3648-0

Omori, A. M., Ono, E. Y. S., Hirozawa, M. T., de Souza Suguiura, I. M., Hirooka, E. Y., Pelegrinelli Fungaro, M. H., & Ono, M. A. (2019). Development of indirect competitive enzyme-linked immunosorbent assay to detect *Fusarium verticillioides* in poultry feed samples. *Toxins (Basel)*, *11*(1), 48. https://doi.org/10.3390/toxins11010048

Otake, S., Nishihara, Y., Makimura, M., Hatta, H., Kim, M., Yamamoto, T., & Hirasawa, M. (1991). Protection of rats against dental caries by passive immunization with hen-egg-yolk antibody (IgY). *Journal of Dental Research*, *70*(3), 162–166. https://doi.org/10.1177/00220345910700030101

Padula, P. J., Edelstein, A., Miguel, S. D., López, N. M., Rossi, C. M., & Rabinovich, R. D. (1998). Hantavirus pulmonary syndrome outbreak in Argentina: Molecular evidence for person-to-person transmission of Andes virus. *Virology*, *241*(2), 323–330. https://doi.org/10.1006/viro.1997.8976

Palaniyappan, A., Das, D., Kammila, S., Suresh, M. R., & Sunwoo, H. H. (2012). Diagnostics of severe acute respiratory syndrome-associated coronavirus (SARS-CoV) nucleocapsid antigen using chicken immunoglobulin Y. *Poultry Science*, *91*(3), 636–642. https://doi.org/10.3382/ps.2011-01916

Parashar, U. D., Hummelman, E. G., Bresee, J. S., Miller, M. A., & Glass, R. I. (2003). Global illness and deaths caused by rotavirus disease in children. *Emerging Infectious Diseases*, *9*(5), 565–572. https://doi.org/10.3201/eid0905.020562

Park, J. S., Lee, K. Y., & Han, J. Y. (2020). Precise genome editing in poultry and its application to industries. *Genes*, *11*(10), 1182. https://doi.org/10.3390/genes11101182

Park, K. J., Park, D. W., Kim, C. H., Han, B. K., Park, T. S., Han, J. Y., Lillehoj, H. S., & Kim, J. K. (2005). Development and characterization of a recombinant chicken single-chain Fv antibody detecting *Eimeria acervulina* sporozoite antigen. *Biotechnology Letters*, *27*(5), 289–295. https://doi.org/10.1007/s10529-005-0682-8

Park, M. H., Lee, H. J., Lee, H. L., Son, D. J., Ju, J. H., Hyun, B. K., Jung, S. H., Song, J. K., Lee, D. H., Hwang, C. J., Han, S. B., Kim, S., & Hong, J. T. (2017). Parkin knockout inhibits neuronal development via regulation of proteasomal degradation of p21. *Theranostics*, *7*(7), 2033–2045. https://doi.org/10.7150/thno.19824

Park, S. B., Aoki, T., & Jung, T. S. (2012). Pathogenesis of and strategies for preventing Edwardsiella tarda infection in fish. *Veterinary Research*, *43*(1), 67. https://doi.org/10.1186/1297-9716-43-67

Parker, J. L., & Shaw, J. G. (2011). Aeromonas spp. clinical microbiology and disease. *The Journal of Infection*, *62*(2), 109–118. https://doi.org/10.1016/j.jinf.2010.12.003

Parma, Y. R., Chacana, P. A., Lucchesi, P. M., Rogé, A., Granobles Velandia, C. V., Krüger, A., Parma, A. E., & Fernández-Miyakawa, M. E. (2012). Detection of Shiga toxin-producing *Escherichia coli* by sandwich enzyme-linked immunosorbent assay using chicken egg yolk IgY antibodies. *Frontiers in Cellular and Infection Microbiology*, *2*, 84. https://doi.org/10.3389/fcimb.2012.00084

Parreño, V., Béjar, C., Vagnozzi, A., Barrandeguy, M., Costantini, V., Craig, M. I., Yuan, L., Hodgins, D., Saif, L., & Fernández, F. (2004). Modulation by colostrum-acquired maternal antibodies of systemic and mucosal antibody responses to rotavirus in calves experimentally challenged with bovine rotavirus. *Veterinary Immunology and Immunopathology*, *100*(1–2), 7–24. https://doi.org/10.1016/j.vetimm.2004.02.007

Parreño, V., Marcoppido, G., Vega, C., Garaicoechea, L., Rodriguez, D., Saif, L., & Fernández, F. (2010). Milk supplemented with immune colostrum: Protection against rotavirus diarrhea and modulatory effect on the systemic and mucosal antibody responses in calves experimentally challenged with bovine rotavirus. *Veterinary Immunology and Immunopathology*, *136*(1–2), 12–27. https://doi.org/10.1016/j.vetimm.2010.01.003

Patel, R., Patel, M., & Suthar, A. (2009). Spray drying technology: An overview. *Indian Journal of Science and Technology*, *2*(10), 44–47. https://doi.org/10.17485/ijst/2009/v2i10.3

Paul, N. C., Al-Adwani, S., Crespo, R., & Shah, D. H. (2014). Evaluation of passive immunotherapeutic efficacy of hyperimmunized egg yolk powder against intestinal colonization of *Campylobacter jejuni* in chickens. *Poultry Science*, *93*(11), 2779–2787. https://doi.org/10.3382/ps.2014-04234

Pauly, D., Chacana, P. A., Calzado, E. G., Brembs, B., & Schade, R. (2011). IgY technology: Extraction of chicken antibodies from egg yolk by polyethylene glycol (PEG) precipitation. *Journal of Visualized Experiments: JoVE*, (51), 3084. https://doi.org/10.3791/3084

Pauly, D., Dorner, M., Zhang, X., Hlinak, A., Dorner, B., & Schade, R. (2009). Monitoring of laying capacity, immunoglobulin Y concentration, and antibody titer development in chickens immunized with ricin and botulinum toxins over a two-year period. *Poultry Science*, *88*(2), 281–290. https://doi.org/10.3382/ps.2008-00323

Peck, M. W. (2009). Biology and genomic analysis of *Clostridium botulinum*. *Advances in Microbial Physiology*, *55*, 183–320. https://doi.org/10.1016/S0065-2911(09)05503-9

Pedersen, R. M., Holt, H. M., & Justesen, U. S. (2011). Solobacterium moorei bacteremia: Identification, antimicrobial susceptibility, and clinical characteristics. *Journal of Clinical Microbiology*, *49*(7), 2766–2768. https://doi.org/10.1128/JCM.02525-10

Pedraza-Sánchez, S., Méndez-León, J. I., Gonzalez, Y., Ventura-Ayala, M. L., Herrera, M. T., Lezana-Fernández, J. L., Bellanti, J. A., & Torres, M. (2018). Oral Administration of human polyvalent IgG by mouthwash as an adjunctive treatment of chronic oral Candidiasis. *Frontiers in Immunology*, *9*, 2956. https://doi.org/10.3389/fimmu.2018.02956

Peralta, R. C., Yokoyama, H., Ikemori, Y., Kuroki, M., & Kodama, Y. (1994). Passive immunisation against experimental salmonellosis in mice by orally administered hen egg-yolk antibodies specific for 14-kDa fimbriae of *Salmonella enteritidis*. *Journal of Medical Microbiology*, *41*(1), 29–35. https://doi.org/10.1099/00222615-41-1-29

Pereira, E., van Tilburg, M. F., Florean, E., & Guedes, M. (2019). Egg yolk antibodies (IgY) and their applications in human and veterinary health: A review. *International Immunopharmacology*, *73*, 293–303. https://doi.org/10.1016/j.intimp.2019.05.015

Perryman, L. E., Kegerris, K. A., & Mason, P. H. (1993). Effect of orally administered monoclonal antibody on persistent *Cryptosporidium parvum* infection in *scid* mice. *Infection and Immunity*, *61*(11), 4906–4908. https://doi.org/10.1128/iai.61.11.4906-4908.1993

Pianotti, R., Lachette, S., & Dills, S. (1986). Desulfuration of cysteine and methionine by *Fusobacterium nucleatum*. *Journal of Dental Research*, *65*(6), 913–917. https://doi.org/10.1177/00220345860650061101

Pink, J. R. (1986). Counting components of the chicken's B cell system. *Immunological Reviews*, *91*, 115–128. https://doi.org/10.1111/j.1600-065x.1986.tb01486.x

Pinkert, C. A. (Ed.). (2014). Introduction to transgenic animal technology. In *Transgenic animal technology*, 3rd edition (pp. 3–13), London: Elsevier.

Pinto, D. J., & Vinayak, S. (2021). *Cryptosporidium*: Host-parasite interactions and pathogenesis. *Current Clinical Microbiology Reports*, *8*(2), 62–67. https://doi.org/10.1007/s40588-021-00159-7

Pitaksajjakul, P., Lekcharoensuk, P., Upragarin, N., Barbas, C. F., III., Ibrahim, M. S., Ikuta, K., & Ramasoota, P. (2010). Fab MAbs specific to HA of influenza virus with H5N1 neutralizing activity selected from immunized chicken phage library. *Biochemistry Biophysics Research Communication*, *395*(4), 496–501. https://doi.org/10.1016/j.bbrc.2010.04.040

Pizarro-Guajardo, M., Díaz-González, F., Álvarez-Lobos, M., & Paredes-Sabja, D. (2017). Characterization of chicken IgY specific to *Clostridium difficile* R20291 spores and the effect of oral administration in mouse models of initiation and recurrent disease. *Frontiers in Cellular and Infection Microbiology*, *7*, 365. https://doi.org/10.3389/fcimb.2017.00365

Plückthun, A. (2012). Ribosome display: A perspective. *Methods in Molecular Biology (Clifton, N.J.)*, *805*, 3–28. https://doi.org/10.1007/978-1-61779-379-0_1

Polaskova, V., Kapur, A., Khan, A., Molloy, M. P., & Baker, M. S. (2010). High-abundance protein depletion: Comparison of methods for human plasma biomarker discovery. *Electrophoresis*, *31*(3), 471–482. https://doi.org/10.1002/elps.200900286

Polson, A., Coetzer, T., Kruger, J., von Maltzahn, E., & van der Merwe, K. J. (1985). Improvements in the isolation of IgY from the yolks of eggs laid by immunized hens. *Immunological Investigations*, *14*(4), 323–327. https://doi.org/10.3109/08820138509022667

Polson, A., von Wechmar, M. B., & van Regenmortel, M. H. (1980). Isolation of viral IgY antibodies from yolks of immunized hens. *Immunological Communications*, *9*(5), 475–493. https://doi.org/10.3109/08820138009066010

Porte, L., Legarraga, P., Vollrath, V., Aguilera, X., Munita, J. M., Araos, R., Pizarro, G., Vial, P., Iruretagoyena, M., Dittrich, S., & Weitzel, T. (2020). Evaluation of a novel antigen-based rapid detection test for the diagnosis of SARS-CoV-2 in respiratory samples. *International Journal of Infectious Diseases: IJID: Official Publication of the International Society for Infectious Diseases*, *99*, 328–333. https://doi.org/10.1016/j.ijid.2020.05.098

Amir, P., & Rasaei, M. (2001). Purification of immunoglobulin Y (IgY) from egg yolk by gel filtration chromatography. *Journal of Babol University of Medical Sciences*, *3*(2), 7–11. http://jbums.org/article-1-2868-en.html

Pratelli, A. (2011). The evolutionary processes of canine coronaviruses. *Advances in Virology*, *2011*, 562831. https://doi.org/10.1155/2011/562831

Qi, Y., Wu, C., Zhang, S., Wang, Z., Huang, S., Dai, L., Wang, S., Xia, L., Wen, K., Cao, X., Wu, Y., & Shen, J. (2009). Selection of anti-sulfadimidine specific ScFvs from a hybridoma cell by eukaryotic ribosome display. *PloS One*, *4*(7), e6427. https://doi.org/10.1371/journal.pone.0006427

Qian, W. J., Kaleta, D. T., Petritis, B. O., Jiang, H., Liu, T., Zhang, X., Mottaz, H. M., Varnum, S. M., Camp, D. G., 2nd, Huang, L., Fang, X., Zhang, W. W., & Smith, R. D. (2008). Enhanced detection of low abundance human plasma proteins using a tandem IgY12-SuperMix immunoaffinity separation strategy. *Molecular & Cellular Proteomics: MCP*, *7*(10), 1963–1973. https://doi.org/10.1074/mcp.M800008-MCP200

Qin, Y., & Zheng, S. J. (2017). Infectious bursal disease virus-host interactions: Multifunctional viral proteins that perform multiple and differing jobs. *International Journal of Molecular Sciences*, *18*(1), 161. https://doi.org/10.3390/ijms18010161

Qin, Z., Babu, V. S., Li, N., Fu, T., Li, J., Yi, L., Zhao, L., Li, J., Zhou, Y., & Lin, L. (2018). Protective effects of chicken egg yolk immunoglobulins (IgY) against experimental *Aeromonas hydrophila* infection in blunt snout bream (*Megalobrama amblycephala*). *Fish & Shellfish Immunology*, *78*, 26–34. https://doi.org/10.1016/j.fsi.2018.04.001

Quezada-Tristán, T., García-Flor, V. L., Ortiz-Martínez, R., Arredondo-Figueroa, J. L., Medina-Esparza, L. E., Valdivia-Flores, A. G., & Montoya-Navarrete, A. L. (2014). Biochemical parameters in the blood of Holstein calves given immunoglobulin Y-supplemented colostrums. *BMC Veterinary Research 10*, 159. https://doi.org/10.1186/1746-6148-10-159

Quintero-Gil, C., Rendon-Marin, S., Martinez-Gutierrez, M., & Ruiz-Saenz, J. (2019). Origin of canine distemper virus: Consolidating evidence to understand potential Zoonoses. *Frontiers in Microbiology*, *10*, 1982. https://doi.org/10.3389/fmicb.2019.01982

Rahimi, S., Shiraz, Z., Salehi, T., Karimi Torshizi, M. A., & Grimes, J. (2007). Prevention of *Salmonella* infection in poultry by specific egg-derived antibody. *International Journal of Poultry Science*, *6*(4), 230–235. https://doi.org/10.3923/ijps.2007.230.235

Rahman, S., Galila, E., Isoda, R., Umeda, K., Nguyen, V., & Kodama, Y. (2011), Effect of passive immunization by anti-Gingipain Igy on periodontal health of dogs. *Veterinary Science Development*, *1*, 35–39. https://doi.org/10.4081/vsd.2011.2204

Rahman, S., Higo-Moriguchi, K., Htun, K. W., Taniguchi, K., Icatlo, F. C. Jr., Tsuji, T., Kodama, Y., Van Nguyen, S., Umeda, K., Oo, H. N., Myint, Y. Y., Htut, T., Myint, S. S., Thura, K., Thu, H. M., Fatmawati, N. N., & Oguma, K. (2012). Randomized placebo-controlled clinical trial of immunoglobulin Y as adjunct to standard supportive therapy for rotavirus-associated diarrhea among pediatric patients. *Vaccine*, *30*(31), 4661–4669. https://doi.org/10.1016/j.vaccine.2012.04.091

Rahman, S., Van Nguyen, S., Icatlo, F. C. Jr, Umeda, K., & Kodama, Y. (2013). Oral passive IgY-based immunotherapeutics: A novel solution for prevention and treatment of alimentary tract diseases. *Human Vaccines & Immunotherapeutics*, *9*(5), 1039–1048. https://doi.org/10.4161/hv.23383

Rajan, B., Løkka, G., Koppang, E. O., & Austbø, L. (2017). Passive immunization of farmed fish. *Journal of Immunology*, *198*(11), 4195–4202. https://doi.org/10.4049/jimmunol.1700154

Rajic, A., Stehmann, C., Autelitano, D. J., Vrkic, A. K., Hosking, C. G., Rice, G. E., & Ilag, L. L. (2009). Protein depletion using IgY from chickens immunised with human protein cocktails. *Preparative Biochemistry & Biotechnology*, *39*(3), 221–247. https://doi.org/10.1080/10826060902952915

Rapacz, J., & Hasler-Rapacz, J. (1986). Polymorphism and inheritance of swine small intestinal receptors mediating adhesion of three serological variants of Escherichia coli-producing K88 pilus antigen. *Animal Genetics*, *17*(4), 305–321. https://doi.org/10.1111/j.1365-2052.1986.tb00724.x

Rasmussen, S. A., Jamieson, D. J., Honein, M. A., & Petersen, L. R. (2016). Zika virus and birth defects–reviewing the evidence for causality. *The New England Journal of Medicine*, *374*(20), 1981–1987. https://doi.org/10.1056/NEJMsr1604338

Ratti, C. (2001). Hot air and freeze-drying of high value food: A review. *Journal of Food Engineering*, *49*(4), 311–319. https://doi.org/10.1016/S0260-8774(00)00228-4

Reddy, P. K., Shekar, A., Kingston, J. J., Sripathy, M. H., & Batra, H. (2013). Evaluation of IgY capture ELISA for sensitive detection of alpha hemolysin of *Staphylococcus aureus* without staphylococcal protein A interference. *Journal of Immunological Methods*, *391*(1–2), 31–38. https://doi.org/10.1016/j.jim.2013.02.004

Redwan, E. M., Aljadawi, A. A., & Uversky, V. N. (2021). Simple and efficient protocol for immunoglobulin Y purification from chicken egg yolk. *Poultry Science*, *100*(3), 100956. https://doi.org/10.1016/j.psj.2020.12.053

Revathy, J., Karthika, S., Sentila, R., & Michael, A. (2014). In vitro evaluation of the efficacy of chicken egg yolk antibodies (IgY) generated against *Propionibacterium acnes*. *International Journal of Cosmetic Science*, *36*(1), 68–73. https://doi.org/10.1111/ics.12097

Richman, D. D., Cleveland, P. H., Oxman, M. N., & Johnson, K. M. (1982). The binding of staphylococcal protein A by the sera of different animal species. *Journal of Immunology (Baltimore, Md.: 1950)*, *128*(5), 2300–2305. https://www.jimmunol.org/content/128/5/2300.long

Riffon, R., Sayasith, K., Khalil, H., Dubreuil, P., Drolet, M., & Lagacé, J. (2001). Development of a rapid and sensitive test for identification of major pathogens in bovine mastitis by PCR. *Journal of Clinical Microbiology*, *39*(7), 2584–2589. https://doi.org/10.1128/JCM.39.7.2584-2589.2001

Robbe, D., & Imogen, F. (2021). Chapter 1 - Traditional and novel sources of long-chain omega-3 fatty acids. In Pedro J. García-Moreno, Charlotte Jacobsen, Ann-Dorit Moltke Sørensen, Betül Yesiltas, (Eds.). *Omega-3 delivery systems* (pp. 3–23). Academic Press.

Roger, H. (2009). *Comparative plant virology*. Second edition. Elsevier Academic Press, USA. ISBN 13: 978-0-12-374154-7 http://www.elsevierdirect.com/companions/9780123741547

Rollier, C., Charollois, C., Jamard, C., Trepo, C., & Cova, L. (2000). Early life humoral response of ducks to DNA immunization against hepadnavirus large envelope protein. *Vaccine*, *18*(27), 3091–3096. https://doi.org/10.1016/s0264-410x(00)00130-4

Rose, M. E., Orlans, E., & Buttress, N. (1974). Immunoglobulin classes in the hen's egg: Their segregation in yolk and white. *European Journal of Immunology*, *4*(7), 521–523. https://doi.org/10.1002/eji.1830040715

Rossetto, O., Pirazzini, M., Bolognese, P., Rigoni, M., & Montecucco, C. (2011). An update on the mechanism of action of tetanus and botulinum neurotoxins. *Acta chimica Slovenica*, *58*(4), 702–707.

Ruan, G. P., Ma, L., Meng, X. J., Meng, M. J., Wang, X. N., Lin, Y., Wu, Z. Q., He, X. W., Wang, J. F., & Zhu, Y. (2007). Quantification of antibody (IgY) titers in hen eggs following immunization and their use in detecting cell surface molecules on nitrocellulose membranes. *Journal of Immunoassay & Immunochemistry*, *28*(1), 35–45. https://doi.org/10.1080/15321810601026083

Ruegg, P. L. (2017). A 100-year review: Mastitis detection, management, and prevention. *Journal of Dairy Science*, *100* (12), 10381–10397. https://doi.org/10.3168/jds.2017-13023.

Russell, W. M. S., & Burch, R. L. (1959). *The principles of humane experimental technique*. Methuen & Co Ltd.: London, UK.

Russo, M., Nahori, M. A., Lefort, J., Gomes, E., de Castro Keller, A., Rodriguez, D., Ribeiro, O. G., Adriouch, S., Gallois, V., de Faria, A. M., & Vargaftig, B. B. (2001). Suppression of asthma-like responses in different mouse strains by oral tolerance. *American Journal of Respiratory Cell and Molecular Biology*, *24*(5), 518–526. https://doi.org/10.1165/ajrcmb.24.5.4320

Sáenz, Y., Briñas, L., Domínguez, E., Ruiz, J., Zarazaga, M., Vila, J., & Torres, C. (2004). Mechanisms of resistance in multiple-antibiotic-resistant *Escherichia coli* strains of human, animal, and food origins. *Antimicrobial Agents and Chemotherapy*, *48*(10), 3996–4001. https://doi.org/10.1128/AAC.48.10.3996-4001.2004

Salter, D. W., Smith, E. J., Hughes, S. H., Wright, S. E., & Crittenden, L. B. (1987). Transgenic chickens: Insertion of retroviral genes into the chicken germ line. *Virology*, *157*(1), 236–240. https://doi.org/10.1016/0042-6822(87)90334-5

Sampaio, L. C. L., Baldissera, M. D., Grando, T. H., Gressler, L. T., Capeleto, D., de Sa, M. F., de Jesus, F.P., dos Santos, A. G. Jr, Anciuti, A. N., Colonetti, K., Stainki, D. R., & Monteiro, S. G. (2014a). Production, purification and therapeutic potential of egg yolk antibodies for treating *Trypanosoma evansi* infection. *Veterinary Parasitology*, *204*(3–4), 96–103. https://doi.org/10.1016/j.vetpar.2014.05.032

Sampaio, L. C. L., Baldissera, M. D., Sagrilo, M. R., Heres, T. S., Oliveira, C. B., Stainki, D. R., & Monteiro, S. G. (2014b). *In vitro* cytotoxicity and genotoxicity of chicken egg yolk antibodies (IgY) against *trypanosoma evansi* in human lymphocytes. *International Journal of Pharmacy and Pharmaceutical Science*, *6*(3), 167–170.

Sanseverino, W., Roma, G., De Simone, M., Faino, L., Melito, S., Stupka, E., Frusciante, L., & Ercolano, M. R. (2010). PRGdb: A bioinformatics platform for plant resistance gene analysis. *Nucleic Acids Research*, *38*(Database issue), D814–D821. https://doi.org/10.1093/nar/gkp978

Santoro, M. L., Barbaro, K. C., Flores da Rocha, T. R., Soares Torquato, R. J., Hirata, I. Y., & Sano-Martins, I. S. (2004). Simultaneous isolation of platelet factor 4 and glycoprotein IIb-IIIa complex from rabbit platelets, and characterization of specific chicken antibodies to assay them. *Journal of Immunological Methods*, *284*(1–2), 55–72. https://doi.org/10.1016/j.jim.2003.10.005

Sarker, S. A., Casswall, T. H., Juneja, L. R., Hoq, E., Hossain, I., Fuchs, G. J., & Hammarström, L. (2001). Randomized, placebo-controlled, clinical trial of hyperimmunized chicken egg yolk immunoglobulin in children with rotavirus diarrhea. *Journal of Pediatric Gastroenterology and Nutrition*, *32*(1), 19–25. https://doi.org/10.1097/00005176-200101000-00009

Satyaraj, E., Wedner, H. J., & Bousquet, J. (2019). Keep the cat, change the care pathway: A transformational approach to managing Fel d 1, the major cat allergen. *Allergy*, *74 Suppl 107*(Suppl 107), 5–17. https://doi.org/10.1111/all.14013

Saveli, C. C., Belknap, R. W., Morgan, S. J., & Price, C. S. (2011). The role of prophylactic antibiotics in open fractures in an era of community-acquired methicillin-resistant Staphylococcus aureus. *Orthopedics*, *34*(8), 611–617. https://doi.org/10.3928/01477447-20110627-25

Schade, R., Behn, I., Erhard, M., Hlinak, A., & Staak, C. (2001). Chicken egg yolk antibodies, production and application – *IgY technology*. Berlin: Springer Verlag.

Schade, R., Calzado, E. G., Sarmiento, R., Chacana, P. A., Porankiewicz-Asplund, J., & Terzolo, H. R. (2005). Chicken egg yolk antibodies (IgY-technology): A review of progress in production and use in research and human and veterinary medicine. *Alternatives to Laboratory Animals: ATLA*, *33*(2), 129–154. https://doi.org/10.1177/026119290503300208

Schade, R., Staak, C., Hendriksen, C., Erhard, M., Hugl, H., Koch, G., Larsson, A., Pollmann, W., van Regenmortel, M., Rijke, E., Spielmann, H., Steinbusch, H., & Straughan, D. (1996). The production of avian (egg yolk) antibodies: IgY. The report and recommendations of ECVAM workshop 21. *Alterntatives to Laboratory Animals*, *24*(6), 925–934. https://doi.org/10.1177/026119299602400607

Schade, R., Zhang, X. Y., & Terzolo, H. R. (2007). Use of IgY antibodies in human and veterinary medicine. In R. Huopalahti, R. López-Fandiño, M. Anton & R. Schade (Eds.). *Bioactive egg compounds* (pp. 213–222), Berlin, Heidelberg: Springer.

Schara, R., Skaleric, E., Seme, K., & Skaleric, U. (2013). Prevalence of periodontal pathogens and metabolic control of type 1 diabetes patients. *Journal of the International Academy of Periodontology*, *15*(1), 29–34.

Selim, A. M., Ibrahim, E. M., El, M., & Hamouda, F. K. (2015). Development of IgY antibodies for control of tetanus. *Biotechnology Animal Husbandry*, *31(1)*, 109–122. https://doi.org/10.2298/BAH1501109S

Sellers H. S. (2017). Current limitations in control of viral arthritis and tenosynovitis caused by avian reoviruses in commercial poultry. *Veterinary Microbiology*, *206*, 152–156. https://doi.org/10.1016/j.vetmic.2016.12.014

Seo, H., Hashimoto, S., Tsuchiya, K., Lin, W., Shibata, T., & Ohta, K. (2006). An ex vivo method for rapid generation of monoclonal antibodies (ADLib system). *Nature Protocols*, *1*(3), 1502–1506. https://doi.org/10.1038/nprot.2006.248

Seo, H., Masuoka, M., Murofushi, H., Takeda, S., Shibata, T., & Ohta, K. (2005). Rapid generation of specific antibodies by enhanced homologous recombination. *Nature Biotechnology*, *23*(6), 731–735. https://doi.org/10.1038/nbt1092

Sert, M., Cakir Koc, R., & Budama Kilinc, Y. (2017). Novel FITC-labeled IgY antibody: Fluorescence imaging *Toxoplasma gondii in vitro*. *Scientific Reports*, *7*(1), 852. https://doi.org/10.1038/s41598-017-00930-1

Sesarman, A., Mihai, S., Chiriac, M. T., Olaru, F., Sitaru, A. G., Thurman, J. M., Zillikens, D., & Sitaru, C. (2008). Binding of avian IgY to type VII collagen does not activate complement and leucocytes and fails to induce subepidermal blistering in mice. *The British Journal of Dermatology*, *158*(3), 463–471. https://doi.org/10.1111/j.1365-2133.2007.08388.x

Shade, R., & Terzolo, H. (2006). *EPC 2006-12th European Poultry Conference, Verona, Italy, 10–14 September.* IgY-technology: Application and trends.

Shahbazi, P., Shayan, P., Ebrahimzadeh, E., & Rahbari, S. (2009). Specific egg yolk antibody against recombinant *Cryptosporidium parvum* P23 protein. *Iran Journal Parasitology*, *4*, 15–24. https://doi.org/10.1007/S00436-013-3648-0

Shen, H., Cai, Y., Zhang, H., Wu, J., Ye, L., Yang, P., Lin, X., Jiang, S., & Liao, M. (2021). Anti-SARS-CoV-2 IgY isolated from egg yolks of hens immunized with inactivated SARS-CoV-2 for immunoprophylaxis of COVID-19. *Virologica Sinica*, *36*(5), 1080–1082. https://doi.org/10.1007/s12250-021-00371-1

Shi, H., Zhu, J., Zou, B., Shi, L., Du, L., Long, Y., Wang, H., Xu, H., Zhen, Y., & Sun, L. (2017). Effects of specific egg yolk immunoglobulin on pan-drug-resistant acinetobacter baumannii. *Biomedicine & Pharmacotherapy = Biomedecine & Pharmacotherapie*, *95*, 1734–1742. https://doi.org/10.1016/j.biopha.2017.09.112

Shimizu, M., Miwa, Y., Hashimoto, K., & Goto, A. (1993). Encapsulation of chicken egg yolk immunoglobulin G (IgY) by liposomes. *Bioscience, Biotechnology, and Biochemistry*, *57*(9), 1445–1449. https://doi.org/10.1271/bbb.57.1445

Shin, J. H., Yang, M., Nam, S. W., Kim, J. T., Myung, N. H., Bang, W. G., & Roe, I. H. (2002). Use of egg yolk-derived immunoglobulin as an alternative to antibiotic treatment for control of *Helicobacter pylori* infection. *Clinical and Diagnostic Laboratory Immunology*, *9*(5), 1061–1066. https://doi.org/10.1128/cdli.9.5.1061-1066.2002

Shin, Nr, N., Kim, J., & Yoo, H. (2000). Control of swine respiratory disease using egg yolk antibodies I. Analysis of immunogenes of Bordetella bronchiseptica, Pasteurella multocida and Actinobacillus pleuropneumoniae and production of IgY. *Korean Journal of Veterinary Research*, *40*(3), 551–561.

Shin, S. J., Lee, S. S., Manning, E. J., & Collins, M. T. (2009). Production of and applications for a polyclonal IgY diagnostic reagent specific for *Mycobacterium avium* subsp. paratuberculosis. *Journal of Microbiology*, *47*, 600–609. https://doi.org/10.1007/s12275-009-0052-7

Shofiqur, R. A. K. M., Ibrahim, S. M., Isoda, R., Umeda, K., Nguyen, V. S., & Kodama, Y. (2011). Effect of passive immunization by anti-gingipain IgY on periodontal health of dogs. *Veterinary Science Development*, *1(8)*, 35–39. https://doi.org/10.4081/vsd.2011.e8.

Shulha, J. A., Escalante, P., & Wilson, J. W. (2019). Pharmacotherapy approaches in nontuberculous mycobacteria infections. *Mayo Clinic Proceedings*, *94*(8), 1567–1581. https://doi.org/10.1016/j.mayocp.2018.12.011

Silva, A., Vasconcelos, G. A., Kappel, L. A., Pinto, M. A., & Paula, V. S. (2012). An immunoenzymatic assay for the diagnosis of hepatitis A utilising immunoglobulin Y. *Memorias do Instituto Oswaldo Cruz*, *107*(7), 960–963. https://doi.org/10.1590/s0074-02762012000700022

Silva, G., Faria, L., Lopes, C. A., Nunes, D. S., Ribeiro, V. S., de Sousa, J., Paiva, G., Gonçalves-Pires, M., Borges, I. P., Santos, M. M., Ávila, V., Júnior, Á. F., & Costa-Cruz, J. M. (2020). Egg yolk immunoglobulin Y as a promising tool to detect immune complexes in neurocysticercosis serum samples. *Transactions of the Royal Society of Tropical Medicine and Hygiene*, *114*(8), 585–592. https://doi.org/10.1093/trstmh/traa028

Silva, R., Rivera Davila, A. M., Seidl, A., & Ramirez, L. (2002) Trypanosoma evansi E Trypanosoma Vivax: Biologia, Diagnóstico E Controle. *Corumbá: Embrapa Pantanal-Livro Científico (ALICE)*. (book in Portuguese).

Sjostrom, L., al-Abdulla, I. H., Rawat, S., Smith, D. C., & Landon, J. (1994). A comparison of ovine and equine antivenoms. *Toxicon: Official Journal of the International Society on Toxinology*, *32*(4), 427–433. https://doi.org/10.1016/0041-0101(94)90294-1

Skottrup, P. D., López, R., Ksiazek, M., Højrup, P., Baelum, V., Potempa, J., & Kaczmarek, J. Z. (2019). An IgY-based immunoassay to evaluate the biomarker potential of the *Tannerella forsythia* virulence factor karilysin in human saliva. *Journal of Immunological Methods*, *469*, 26–32. https://doi.org/10.1016/j.jim.2019.03.003

Smith, G. P. (1985). Filamentous fusion phage: Novel expression vectors that display cloned antigens on the virion surface. *Science (New York, N.Y.)*, *228*(4705), 1315–1317. https://doi.org/10.1126/science.4001944

Smith, T. L., Pearson, M. L., Wilcox, K. R., Cruz, C., Lancaster, M. V., Robinson-Dunn, B., Tenover, F. C., Zervos, M. J., Band, J. D., White, E., & Jarvis, W. R. (1999). Emergence of vancomycin resistance in Staphylococcus aureus. Glycopeptide-intermediate *Staphylococcus aureus* working group. *The New England Journal of Medicine*, *340*(7), 493–501. https://doi.org/10.1056/NEJM199902183400701

Soumaila Garba, A., Thibodeau, A., Perron, A., Laurent-Lewandowski, S., Letellier, A., & Fravalo, P. (2019). In vitro efficacy of potentiated egg yolk powder against *Campylobacter jejuni* does not correlate with in vitro efficacy. *PloS One*, *14*(3), e0212946. https://doi.org/10.1371/journal.pone.0212946

Souza, D. C., de Faria, L. S., Sousa, J., Lopes, C. A., Ribeiro, V., da Silva, V. J., Ribeiro, R. P., Rabelo, É., Rodrigues Ávila, V. M., Ferreira Júnior, Á., & Costa-Cruz, J. M. (2020). Use of polyclonal IgY antibodies to detect serum immune complexes in patients with active hookworm infection. *Parasitology*, *147*(6), 715–720. https://doi.org/10.1017/S0031182020000220

Staak, C., Schwarzkopf, C., Behn, I., Hommel, U., Hlinak, A., Schade, R., & Erhard, M. (2000). Isolation of IgY from yolk. In R. Schade, I. Behn, M. Erhard, A. Hlinak & C. Staak (Eds.). *Chicken egg yolk antibodies, production and application: IgY technology* (pp. 65–107), Berlin, Heidelberg, Germany and New York, USA: Springer Laboratory Manuals.

Stuart, C. A., Pietrzyk, R. A., Furlanetto, R. W., & Green, A. (1988). High affinity antibody from hen's eggs directed against the human insulin receptor and the human IGF-I receptor. *Analytical Biochemistry*, *173*(1), 142–150. https://doi.org/10.1016/0003-2697(88)90171-6

Studnicka, G. M., Soares, S., Better, M., Williams, R. E., Nadell, R., & Horwitz, A. H. (1994). Human-engineered monoclonal antibodies retain full specific binding activity by preserving non-CDR complementarity-modulating residues. *Protein Engineering*, *7*(6), 805–814. https://doi.org/10.1093/protein/7.6.805

Suartini, G. A. A., Suprayogi, A., Wibawan, W. T., Sendow, I., & Mahardika, G. (2014). Intravenous administration of chicken immunoglobulin has a curative effect in experimental infection of canine parvovirus. *Global Veterinaria*, *13*(5), 801–808. https://doi.org/10.5829/idosi.gv.2014.13.05.86180

Suartini, I. G. A. A., Sendow, I., Agustini, N. L. P., Suprayogi, A., Wibawan, I. W. T., & Mahardika, I. G. N. K. (2016). Kinetika Immunoglobulin Kuning Telur Antiparvovirus Anjing Pada Anjing (Kinetics of anti-canine parvovirus yolk iimmunoglobulin in dogs). *Jurnal Veteriner*, *17*(2), 292–299. https://ojs.unud.ac.id/index.php/jvet/article/view/22133

Sudjarwo, S. A., Eraiko, K., Sudjarwo, G. W., & Koerniasari (2017a). The activity of immunoglobulin Y anti-*Mycobacterium tuberculosis* on proliferation and cytokine expression of rat peripheral blood mononuclear cells. *Pharmacognosy Research*, *9*(Suppl 1), S5–S8. https://doi.org/10.4103/pr.pr_66_17

Sudjarwo, S. A., Eraiko, K., Sudjarwo, G. W., & Koerniasari (2017b). The potency of chicken egg yolk immunoglobulin (IgY) specific as immunotherapy to *Mycobacterium tuberculosis* infection. *Journal of Advanced Pharmaceutical Technology & Research*, *8*(3), 91–96. https://doi.org/10.4103/japtr.JAPTR_167_16

Sugano, N. (2009). Adjunctive effects of anti-*Porphyromonas gingivalis* egg yolk antibody with scaling and root planning: A randomized, placebo-controlled clinical trial. *Journal of Periodontology*, *80*, 1901–1903.

Sugano, N. (2012). Biological plaque control: Novel therapeutic approach to periodontal disease. *Journal of oral Science*, *54*(1), 1–5. https://doi.org/10.2334/josnusd.54.1

Sugita-Konishi, Y., Ogawa, M., Arai, S., Kumagai, S., Igimi, S., & Shimizu, M. (2000). Blockade of *Salmonella enteritidis* passage across the basolateral barriers of human intestinal epithelial cells by specific antibody. *Microbiology and Immunology*, *44*(6), 473–479. https://doi.org/10.1111/j.1348-0421.2000.tb02522.x

Sui, J., Cao, L., & Lin, H. (2011). Antibacterial activity of egg yolk antibody (IgY) against Listeria monocytogenes and preliminary evaluation of its potential for food preservation. *Journal of the Science of Food and Agriculture*, *91*(11), 1946–1950. https://doi.org/10.1002/jsfa.4381

Sun, L., Li, M., Fei, D., Diao, Q., Wang, J., Li, L., & Ma, M. (2018). Preparation and application of egg yolk antibodies against Chinese Sacbrood virus infection. *Frontiers in Microbiology*, *9*, 1814. https://doi.org/10.3389/fmicb.2018.01814

Sun, S., Mo, W., Ji, Y., & Liu, S. (2001). Preparation and mass spectrometric study of egg yolk antibody (IgY) against rabies virus. *Rapid Communication of Mass Spectrometry*, *15*(9), 708–712. https://doi.org/10.1002/rcm.271

Sun, Y., Yang, Y., Wang, L., Lv, L., Zhu, J., Han, W., Wang, E., Guo, X., & Zhen, Y. (2015). Highly sensitive detection of cancer antigen human epidermal growth factor receptor 2 using novel chicken egg yolk immunoglobulin. *Biologicals: Journal of the International Association of Biological Standardization*, *43*(3), 165–170. https://doi.org/10.1016/j.biologicals.2015.03.002

Sunwoo, H. H., Wang, W. W., & Sim, J. S. (2006). Detection of *Escherichia coli* O157:H7 using chicken immunoglobulin Y. *Immunology Letters*, *106*(2), 191–193. https://doi.org/10.1016/j.imlet.2006.05.005

Surana, S., Tosolini, A. P., Meyer, I., Fellows, A. D., Novoselov, S. S., & Schiavo, G. (2018). The travel diaries of tetanus and botulinum neurotoxins. *Toxicon: Official Journal of the International Society on Toxinology*, *147*, 58–67. https://doi.org/10.1016/j.toxicon.2017.10.008

Suzuki, H., Nomura, S., Masaoka, T., Goshima, H., Kamata, N., Kodama, Y., Ishii, H., Kitajima, M., Nomoto, K., & Hibi, T. (2004). Effect of dietary anti-*Helicobacter pylori*-urease immunoglobulin Y on *Helicobacter pylori* infection. *Alimentary Pharmacology & Therapeutics*, *20*(Suppl 1), 185–192. https://doi.org/10.1111/j.1365-2036.2004.02027.x

Svendsen, L., Crowley, A., Ostergaard, L. H., Stodulski, G., & Hau, J. (1995). Development and comparison of purification strategies for chicken antibodies from egg yolk. *Laboratory Animal Science*, *45*(1), 89–93.

Szajewska, H., & Dziechciarz, P. (2010). Gastrointestinal infections in the pediatric population. *Current Opinion in Gastroenterology*, *26*(1), 36–44. https://doi.org/10.1097/MOG.0b013e328333d799

Takeuchi, S., Motohashi, J., Kimori, H., Nakagawa, Y., & Tsurumoto, A. (2016). Effects of oral moisturising gel containing egg yolk antibodies against *Candida albicans* in older people. *Gerodontology*, *33*(1), 128–134. https://doi.org/10.1111/ger.12139

Tamilzarasan, K. B., Dinakaran, A. M., Selvaraju, G., & Dorairajan, N. (2009). Efficacy of egg yolk immunoglobulins (IgY) against enteric pathogens in poultry. *Tamilnadu Journal of Veterinary Science*, *5*(6), 264–268.

Tarigan, R., Kusumorini, N., & Manalu, W. (2016). Effectivity of immunoglobulin Y anti lipase as a pancreatic lipase inhibitor for prevention of obesity. *Pakistan Journal of Nutrition*, *15*, 752–759. https://doi.org/10.3923/pjn.2016.752.759

Tateishi, Y., Nishimichi, N., Horiuchi, H., Furusawa, S., & Matsuda, H. (2008). Construction of chicken-mouse chimeric antibody and immunogenicity in mice. *The Journal of Veterinary Medical Science*, *70*(4), 397–400. https://doi.org/10.1292/jvms.70.397

Teimoori, S., Arimatsu, Y., Laha, T., Kaewkes, S., Sereerak, P., Sripa, M., Tangkawattana, S., Brindley, P. J., & Sripa, B. (2016). Chicken IgY-based coproantigen capture ELISA for diagnosis of human pisthorchiasis. *Parasitology International*, *pii:S1383-5769*(16), 30093–30099.

Teimoori, S., Arimatsu, Y., Laha, T., Kaewkes, S., Sereerak, P., Sripa, M., Tangkawattana, S., Brindley, P. J., & Sripa, B. (2017). Chicken IgY-based coproantigen capture ELISA for diagnosis of human opisthorchiasis. *Parasitology International*, *66*(4), 443–447. https://doi.org/10.1016/j.parint.2015.10.011

Tellez, G., Petrone, V. M., Escorcia, M., Morishita, T. Y., Cobb, C. W., Villaseñor, L., & Promsopone, B. (2001). Evaluation of avian-specific probiotic and *Salmonella* Enteritidis-, *Salmonella* Typhimurium-, and *Salmonella* Heidelberg- specific antibodies on cecal colonization and organ invasion of *Salmonella* Enteritidis in broilers. *Journal of Food Protection*, *64*(3), 287–291. https://doi.org/10.4315/0362-028x-64.3.287

Terzolo, H. R., Sandoval, V. E., Caffer, M. I., Terragno, R., & Alcain, A. (1998). Aglutinación de inmunoglobulinas de yema de huevo de gallina (IgY) contra Salmonella enterica serovariedad enteritidis [Agglutination of hen egg-yolk immunoglobulins (IgY) against Salmonella enterica, serovar enteritidis]. *Revista Argentina de microbiologia*, *30*(2), 84–92.

Thermet, A., Robaczewska, M., Rollier, C., Hantz, O., Trepo, C., Deleage, G., & Cova, L. (2004). Identification of antigenic regions of duck hepatitis B virus core protein with antibodies elicited by DNA immunization and chronic infection. *Journal of Virology*, *78*(4), 1945–1953. https://doi.org/10.1128/jvi.78.4.1945-1953.2004

Thibodeau, A., Fravalo, P., Perron, A., Lewandowski, S. L., & Letellier, A. (2017). Production and characterization of anti-*Campylobacter jejuni* IgY derived from egg yolks. *Acta veterinaria Scandinavica*, *59*(1), 80. https://doi.org/10.1186/s13028-017-0346-4

Thirumalai, D., Visaga Ambi, S., Vieira-Pires, R. S., Xiaoying, Z., Sekaran, S., & Krishnan, U. (2019). Chicken egg yolk antibody (IgY) as diagnostics and therapeutics in parasitic infections - a review. *International Journal of Biological Macromolecules*, *136*, 755–763. https://doi.org/10.1016/j.ijbiomac.2019.06.118

Thirupathi, V., Sasikala, S., & Rajkumar, P. (2008). Studies on foam mat drying of whole egg liquid in a cabinet dryer. *Madras Agricultural Journal*, *95(1–6)*, 141–150.

Thomsen, K., Christophersen, L., Bjarnsholt, T., Jensen, P. Ø., Moser, C., & Høiby, N. (2016). Anti-*Pseudomonas aeruginosa* IgY antibodies augment bacterial clearance in a murine pneumonia model. *Journal of Cystic Fibrosis: Official Journal of the European Cystic Fibrosis Society*, *15*(2), 171–178. https://doi.org/10.1016/j.jcf.2015.08.002

Thomsen, K., Christophersen, L., Bjarnsholt, T., Jensen, P. Ø., Moser, C., & Høiby, N. (2015). Anti-*Pseudomonas aeruginosa* IgY antibodies induce specific bacterial aggregation and internalization in human polymorphonuclear neutrophils. *Infection and immunity*, *83*(7), 2686–2693. https://doi.org/10.1128/IAI.02970-14

Thomsen, K., Christophersen, L., Lerche, C. J., Holmgaard, D. B., Calum, H., Høiby, N., & Moser, C. (2021). Azithromycin potentiates avian IgY effect against *Pseudomonas aeruginosa* in a murine pulmonary infection model. *International Journal of Antimicrobial Agents*, *57*(1), 106213. https://doi.org/10.1016/j.ijantimicag.2020.106213

Thu, H. M., Myat, T. W., Win, M. M., Thant, K. Z., Rahman, S., Umeda, K., Nguyen, S. V., Icatlo, F. C. Jr, Higo-Moriguchi, K., Taniguchi, K., Tsuji, T., Oguma, K., Kim, S. J., Bae, H. S., & Choi, H. J. (2017). Chicken egg yolk antibodies (IgY) for prophylaxis and treatment of rotavirus diarrhea in human and animal neonates: A concise review. *Korean Journal for Food Science of Animal Resources*, *37*(1), 1–9. https://doi.org/10.5851/kosfa.2017.37.1.1

Thwaites, C. L., Beeching, N. J., & Newton, C. R. (2015). Maternal and neonatal tetanus. *Lancet (London, England)*, *385*(9965), 362–370. https://doi.org/10.1016/S0140-6736(14)60236-1

Tirado, S. M., & Yoon, K. J. (2003). Antibody-dependent enhancement of virus infection and disease. *Viral Immunology*, *16*(1), 69–86. https://doi.org/10.1089/088282403763635465

Tobias, F. L., Garcia, L. N., Kanashiro, M. M., Medina-Acosta, E., Brom-de-Luna, J. G., de Almeida, C. M., Azevedo Junior, R. R., Lemos, M., & Vieira-da-Motta, O. (2012). Growth inhibition of *Staphylococcus aureus* and *Escherichia coli* strains by neutralizing IgY antibodies from ostrich egg yolk. *Brazilian Journal of Microbiology: [publication of the Brazilian Society for Microbiology]*, *43*(2), 544–551. https://doi.org/10.1590/S1517-83822012000200015

Toloza-Moreno, D. L., Villamizar-Rivero, L. F., Cuartas-Otalora, P. E., & Barrera-Cubillos, G. P. (2022). Immunodetection of *Furcraea Necrotic Streak Virus*-FNSV in fique plants (*Furcraea macrophylla* Baker) using a polyclonal antibody IgY produced in chicken egg yolk. *Journal of Immunological Methods*, *503*, 113232. https://doi.org/10.1016/j.jim.2022.113232

Tong, C., Geng, F., He, Z., Cai, Z., & Ma, M. (2015). A simple method for isolating chicken egg yolk immunoglobulin using effective delipidation solution and ammonium sulfate. *Poultry Science*, *94*(1), 104–110. https://doi.org/10.3382/ps/peu005

Trail, J. C., d'Ieteren, G. D., Feron, A., Kakiese, O., Mulungo, M., & Pelo, M. (1990). Effect of trypanosome infection, control of parasitaemia and control of anaemia development on productivity of N'Dama cattle. *Acta tropica*, *48*(1), 37–45. https://doi.org/10.1016/0001-706x(90)90063-6

Tran, T. V., Do, B. N., Nguyen, T., Tran, T. T., Tran, S. C., Nguyen, B. V., Nguyen, C. V., & Le, H. Q. (2019). Development of an IgY-based lateral flow immunoassay for detection of fumonisin B in maize. *F1000Research*, *8*, 1042. https://doi.org/10.12688/f1000research.19643.2

Tressler, R. L., & Roth, T. F. (1987). IgG receptors on the embryonic chick yolk sac. *The Journal of Biological Chemistry*, *262*(32), 15406–15412. https://doi.org/10.1016/S0021-9258(18)47740-X

Tsubokura, K., Berndtson, E., Bogstedt, A., Kaijser, B., Kim, M., Ozeki, M., & Hammarström, L. (1997). Oral administration of antibodies as prophylaxis and therapy in *Campylobacter jejuni*-infected chickens. *Clinical and Experimental Immunology*, *108*(3), 451–455. https://doi.org/10.1046/j.1365-2249.1997.3901288.x

Tsukamoto, M., Hiroi, S., Adachi, K., Kato, H., Inai, M., Konishi, I., Tanaka, M., Yamamoto, R., Sawa, M., Handharyani, E., & Tsukamoto, Y. (2011). Antibodies against swine influenza virus neutralize the pandemic influenza virus A/H1N1. *Molecular Medicine Reports*, *4*(2), 209–214. https://doi.org/10.3892/mmr.2011.410

Uhlen, M., Bandrowski, A., Carr, S., Edwards, A., Ellenberg, J., Lundberg, E., Rimm, D. L., Rodriguez, H., Hiltke, T., Snyder, M., & Yamamoto, T. (2016). A proposal for validation of antibodies. *Nature Methods*, *13*(10), 823–827. https://doi.org/10.1038/nmeth.3995

Ulrich, M., Worlitzsch, D., Viglio, S., Siegmann, N., Iadarola, P., Shute, J. K., Geiser, M., Pier, G. B., Friedel, G., Barr, M. L., Schuster, A., Meyer, K. C., Ratjen, F., Bjarnsholt, T., Gulbins, E., & Döring, G. (2010). Alveolar inflammation in cystic fibrosis. *Journal of Cystic Fibrosis: Official Journal of the European Cystic Fibrosis Society*, *9*(3), 217–227. https://doi.org/10.1016/j.jcf.2010.03.001

Umeda, K., Suzuki, T., Ohashi, S., Moon, S. C., Rahman, S., & Sa, V. N. (2019). Passive immunization by specific chicken egg yolk immunoglobulin antibodies against porcine epidemic diarrhea (in Japanese). *Proceedings of Japan Pig Veterinary Society*, *73*, 25–31. https://tonbyo.com/proceedings/1365.html

Ungar, B. L., Ward, D. J., Fayer, R., & Quinn, C. A. (1990). Cessation of Cryptosporidium-associated diarrhea in an acquired immunodeficiency syndrome patient after treatment with hyperimmune bovine colostrum. *Gastroenterology*, *98*(2), 486–489. https://doi.org/10.1016/0016-5085(90)90842-o

Uzal, F. A., Songer, J. G. (2019). Clostridial diseases. In Zimmermann, J. J., Karriker, L.A., Ramirez, A., Schwartz, K.J., Stevenson, G.W., & Zhang, J. (Eds.). *Diseases of swine* (pp. 792–806), 11th Edition. John Wiley & Sons, Inc.: Hoboken, NJ, USA.

van de Lavoir, M. C., Diamond, J. H., Leighton, P. A., Mather-Love, C., Heyer, B. S., Bradshaw, R., Kerchner, A., Hooi, L. T., Gessaro, T. M., Swanberg, S. E., Delany, M. E., & Etches, R. J. (2006). Germline transmission of genetically modified primordial germ cells. *Nature*, *441*(7094), 766–769. https://doi.org/10.1038/nature04831

Van Nguyen, S., Umeda, K., Yokoyama, H., Tohya, Y., & Kodama, Y. (2006). Passive protection of dogs against clinical disease due to canine parvovirus-2 by specific antibody from chicken egg yolk. *Canadian Journal of Veterinary Research = Revue canadienne de recherche veterinaire*, *70*(1), 62–64.

Vancauwenberghe, F., Dadamio, J., Laleman, I., Van Tornout, M., Teughels, W., Coucke, W., & Quirynen, M. (2013). The role of *Solobacterium moorei* in oral malodour. *Journal of Breath Research*, *7*(4), 046006. https://doi.org/10.1088/1752-7155/7/4/046006

Vandeputte, J., Martel, A., Canessa, S., Van Rysselberghe, N., De Zutter, L., Heyndrickx, M., Haesebrouck, F., Pasmans, F., & Garmyn, A. (2019). Reducing *Campylobacter jejuni* colonization in broiler chickens by in-feed supplementation with hyperimmune egg yolk antibodies. *Scientific Reports*, *9*(1), 8931. https://doi.org/10.1038/s41598-019-45380-z

Vazquez, L., Alpuche, J., Maldonado, G., Agundis, C., Pereyra-Morales, A., & Zenteno, E. (2009). Review: Immunity mechanisms in crustaceans. *Innate Immunity*, *15*(3), 179–188. https://doi.org/10.1177/1753425909102876

Vega, C. G., Bok, M., Vlasova, A. N., Chattha, K. S., Fernández, F. M., Wigdorovitz, A., Parreño, V. G., & Saif, L. J. (2012). IgY antibodies protect against human rotavirus induced diarrhea in the neonatal gnotobiotic piglet disease model. *PloS One*, *7*(8), e42788. https://doi.org/10.1371/journal.pone.0042788

Vega, C., Bok, M., Chacana, P., Saif, L., Fernandez, F., & Parreño, V. (2011). Egg yolk IgY: Protection against rotavirus induced diarrhea and modulatory effect on the systemic and mucosal antibody responses in newborn calves. *Veterinary Immunology and Immunopathology*, *142*(3–4), 156–169. https://doi.org/10.1016/j.vetimm.2011.05.003

Vega, C., Bok, M., Saif, L., Fernandez, F., & Parreño, V. (2015). Egg yolk IgY antibodies: A therapeutic intervention against group A rotavirus in calves. *Research in Veterinary Science*, *103*, 1–10. https://doi.org/10.1016/j.rvsc.2015.09.005

Vejaratpimol, R., Channuntapipat, C., Pewnim, T., Ito, K., Iizuka, M., & Minamiura, N. (1999). Detection and serological relationships of cymbidium mosaic potexvirus isolates. *Journal of Bioscience and Bioengineering*, *87*(2), 161–168. https://doi.org/10.1016/s1389-1723(99)89006-9

Verdoliva, A., Basile, G., & Fassina, G. (2000). Affinity purification of immunoglobulins from chicken egg yolk using a new synthetic ligand. *Journal of Chromatography. B, Biomedical Sciences and Applications*, *749*(2), 233–242. https://doi.org/10.1016/s0378-4347(00)00426-6

Vieira, J. G., Oliveria, M. A., Russo, E. M., Maciel, R. M., & Pereira, A. B. (1984). Egg yolk as a source of antibodies for human parathyroid hormone (hPTH) radioimmunoassay. *Journal of Immunoassay*, *5*(1–2), 121–129. https://doi.org/10.1080/01971528408063002

von Wowern, N., Klausen, B., & Kollerup, G. (1994). Osteoporosis: A risk factor in periodontal disease. *Journal of Periodontology*, *65*(12), 1134–1138. https://doi.org/10.1902/jop.1994.65.12.1134

Walczak, M., Grzywa, R., Łupicka-Słowik, A., Skoreński, M., Bobrek, K., Nowak, D., Boivin, S., Brown, E. L., Oleksyszyn, J., & Sieńczyk, M. (2015). Method for generation of peptide-specific IgY antibodies directed to *Staphylococcus aureus* extracellular fibrinogen binding protein epitope. *Biopolymers*, *104*(5), 552–559. https://doi.org/10.1002/bip.22695

Wallach, M. G., Webby, R. J., Islam, F., Walkden-Brown, S., Emmoth, E., Feinstein, R., & Gronvik, K. O. (2011). Cross-protection of chicken immunoglobulin Y antibodies against H5N1 and H1N1 viruses passively administered in mice. *Clinical and Vaccine Immunology: CVI*, *18*(7), 1083–1090. https://doi.org/10.1128/CVI.05075-11

Wang, D., Guo, Q., Yuan, Y., & Gong, Y. (2019d). The antibiotic resistance of *Helicobacter pylori* to five antibiotics and influencing factors in an area of China with a high risk of gastric cancer. *BMC Microbiology*, *19*(1), 152. https://doi.org/10.1186/s12866-019-1517-4

Wang, F., Qiao, W., Bao, B., Wang, S., Regenstein, J., Shi, Y., Wu, W., & Ma, M. (2019a). Effect of IgY on periodontitis and halitosis induced by *Fusobacterium nucleatum*. *Journal of Microbiology and Biotechnology*, *29*(2), 311–320. https://doi.org/10.4014/jmb.1810.10044

Wang, L. H., Li, X. Y., Jin, L. J., You, J. S., Zhou, Y., Li, S. Y., & Xu, Y. P. (2011). Characterization of chicken egg yolk immunoglobulins (IgYs) specific for the most prevalent capsular serotypes of mastitis-causing Staphylococcus aureus. *Veterinary Microbiology*, *149*(3–4), 415–421. https://doi.org/10.1016/j.vetmic.2010.11.029

Wang, X., Song, L., Tan, W., & Zhao, W. (2019b). Clinical efficacy of oral immunoglobulin Y in infant rotavirus enteritis: Systematic review and meta-analysis. *Medicine*, *98*(27), e16100. https://doi.org/10.1097/MD.0000000000016100

Wang, Y.W., Cherian, G., Sunwoo, H.H., & Sim, J.S. (2000). Dietary polyunsaturated fatty acids significantly affect laying hen lymphocyte proliferation and immunoglobulin G concentration in serum and egg yolk. *Canadian Journal of Animal Science*, *80*, 597–604.

Wang, Z., Li, J., Li, J., Li, Y., Wang, L., Wang, Q., Fang, L., Ding, X., Huang, P., Yin, J., Yin, Y., & Yang, H. (2019c). Protective effect of chicken egg yolk immunoglobulins (IgY) against enterotoxigenic *Escherichia coli* K88 adhesion in weaned piglets. *BMC Veterinary Research*, *15*(1), 234. https://doi.org/10.1186/s12917-019-1958-x

Wani, M. Y., Pandit, A. A., Begum, J., Ahmed, N., Mir, M. S., & Makhdoomi, D. M. (2022). Egg yolk antibodies: Production and applications in the diagnosis and treatment of animal diseases - A review: Applications of egg yolk antibodies. *Letters In Animal Biology*, *2*(1), 32–40. Retrieved from https://liabjournal.com/index.php/liab/article/view/77

Warr, G. W., Magor, K. E., & Higgins, D. A. (1995). IgY: Clues to the origins of modern antibodies. *Immunology Today*, *16*(8), 392–398. https://doi.org/10.1016/0167-5699(95)80008-5

Waters, J. R., & Sellwood, R. (1982). Aspects of genetic resistance to K88 *E. coli* in pigs. In: Proceedings of the 2nd World Congress on Genetics Applied to Livestock Production. Madrid: International committee for World Congress on Genetics Applied to Livestock Production; 4-8 October 1982. p. 362.

Wei, S., Duan, S., Liu, X., Wang, H., Ding, S., Chen, Y., Xie, J., Tian, J., Yu, N., Ge, P., Zhang, X., Chen, X., Li, Y., & Meng, Q. (2021). Chicken egg yolk antibodies (IgYs) block the binding of multiple SARS-CoV-2 spike protein variants to human ACE2. *International Immunopharmacology*, *90*, 107172. https://doi.org/10.1016/j.intimp.2020.107172

Weilbach, A. D., & Sander, E. (2000). Quantitative detection of potato viruses X and Y (PVX, PVY) with antibodies raised in chicken egg yolk (IgY) by ELISA variants. *Journal of Plant Diseases and Protection*, *107*(3), 318–328. https://www.jstor.org/stable/43387000

Wei-xu, H., Wen-yun, Z., Xi-ling, Z., Zhu, W., Li-hua, W., Xiao-mu, W.,...Guo-zhu, H. (2016). Anti-interleukin-1 beta/tumor necrosis factor-alpha IgY antibodies reduce pathological allergic responses in guinea pigs with allergic rhinitis. *Mediators of Inflammation, 2016*, 1–11. https://doi.org/10.1155/2016/3128182

Wen, J., Zhao, S., He, D., Yang, Y., Li, Y., & Zhu, S. (2012). Preparation and characterization of egg yolk immunoglobulin Y specific to influenza B virus. *Antiviral Research, 93*(1), 154–159. https://doi.org/10.1016/j.antiviral.2011.11.005

West, A. P. Jr, Herr, A. B., & Bjorkman, P. J. (2004). The chicken yolk sac IgY receptor, a functional equivalent of the mammalian MHC-related Fc receptor, is a phospholipase A2 receptor homolog. *Immunity, 20*(5), 601–610. https://doi.org/10.1016/s1074-7613(04)00113-x

WHO (2018). Dengue vaccine: WHO position paper, September 2018 - recommendations. (2019). *Vaccine, 37*(35), 4848–4849. https://doi.org/10.1016/j.vaccine.2018.09.063

WHO (2021). World Health Organization. *Ebola* (2021). Available at: https://www.who.int/health-topics/ebola (Accessed April 15, 2021).

WHO (2021a). World Health Organization. *Campylobacter* (2021). Available at: https://www.who.int/news-room/fact-sheets/detail/campylobacter (Accessed April 15, 2021).

Widmer, R. P. (2010). Oral health of children with respiratory diseases. *Paediatric Respiratory Reviews, 11*(4), 226–232. https://doi.org/10.1016/j.prrv.2010.07.006

Wilder-Smith, A., Lindsay, S. W., Scott, T. W., Ooi, E. E., Gubler, D. J., & Das, P. (2020). The Lancet Commission on dengue and other Aedes-transmitted viral diseases. *Lancet (London, England), 395*(10241), 1890–1891. https://doi.org/10.1016/S0140-6736(20)31375-1

Wills, F. K., & Luginbuhl, R. E. (1963). The use of egg yolk for passive immunization of chickens against Newcastle disease. *Avian Disease, 7*(1), 5–12.

Winkelbach, A., Günzel, D., Schulz, C., & Wuertz, S. (2015a). Differences in IgY gut absorption in gastric rainbow trout (*Oncorhynchus mykiss*) and agastric common carp (*Cyprinus carpio*) assessed in vivo and in vitro. Comparative biochemistry and physiology. *Toxicology & Pharmacology: CBP, 167*, 58–64. https://doi.org/10.1016/j.cbpc.2014.09.001

Winkelbach, A., Schade, R., Schulz, C. & Wuertz S. (2015b). Comparison of oral, rectal and intraperitoneal administration of IgY antibodies in passive immunization of rainbow trout (*Oncorhynchus mykiss*). *Aquaculture International, 23*, 427–438. https://doi.org/10.1007/s10499-014-9823-1

Wu, C. J., Wang, H., Chan, Y. L., & Li, T. L. (2011). Passive immune-protection of small abalone against *Vibrio alginolyticus* infection by anti-*Vibrio* IgY-encapsulated feed. *Fish & Shellfish Immunology, 30*(4–5), 1042–1048. https://doi.org/10.1016/j.fsi.2011.01.026

Wu, R., Yakhkeshi, S., & Zhang, X. (2022). Scientometric analysis and perspective of IgY technology study. *Poultry Science, 101*(4), 101713. https://doi.org/10.1016/j.psj.2022.101713

Xia, L.-J., Yang, Z.-B., Huang, W., Zhang, R.-F., Jiang, R.-J., & Jiang, Y. (2011). Preventive effect of Hpaa- Vaca Igy on intragastric infection with *Helicobacter pylori* in mice. *Chinese Journal of Biology, 24*, 34–36. +40. (article in Chinese).

Xiao, N., Zhao, Y., Yao, Y., Wu, N., Xu, M., Du, H., & Tu, Y. (2020). Biological activities of egg yolk lipids: A Review. *Journal of Agricultural and Food Chemistry, 68*(7), 1948–1957. https://doi.org/10.1021/acs.jafc.9b06616

Xiao, Y., Gao, X., Taratula, O., Treado, S., Urbas, A., Holbrook, R. D., Cavicchi, R. E., Avedisian, C. T., Mitra, S., Savla, R., Wagner, P. D., Srivastava, S., & He, H. (2009). Anti-HER2 IgY antibody-functionalized single-walled carbon nanotubes for detection and selective destruction of breast cancer cells. *BMC Cancer, 9*, 351. https://doi.org/10.1186/1471-2407-9-351

Xing, P., Shi, Y., Dong, C., Liu, H., Cheng, Y., Sun, J., Li, D., Li, M., Sun, K., & Feng, D. (2017). Colon-targeted delivery of IgY against *Clostridium difficile* toxin A and B by encapsulation in Chitosan-Ca pectinate microbeads. *AAPS PharmSciTech, 18*(4), 1095–1103. https://doi.org/10.1208/s12249-016-0656-2

Xu, L., Che, J., Xu, Y., Chen, Y., Li, Y., Murtaza, B., Wang, L., Zhang, M., & Li, X. (2020). Oral administration of microencapsulated egg yolk immunoglobulin (IgY) in turbot (*Scophthalmus maximus*) to combat against *Edwardsiella tarda* 2CDM001 infections. *Fish & Shellfish Immunology, 106*, 609–620. https://doi.org/10.1016/j.fsi.2020.08.024

Xu, L., Xu, Y., He, L., Zhang, M., Wang, L., Li, Z., & Li, X. (2019). Immunomodulatory effects of chicken egg yolk antibodies (IgY) against experimental Shewanella marisflavi AP629 infections in sea cucumbers (Apostichopus japonicus). *Fish & Shellfish Immunology, 84*, 108–119. https://doi.org/10.1016/j.fsi.2018.09.073

Xu, Y., Li, X., Jin, L., Zhen, Y., Lu, Y., Li, S., You, J., & Wang, L. (2011). Application of chicken egg yolk immunoglobulins in the control of terrestrial and aquatic animal diseases: A review. *Biotechnology Advances, 29*(6), 860–868. https://doi.org/10.1016/j.biotechadv.2011.07.003

Xun, Z., Li-yuan, G., Zhibang, Y., & Xiaoping, C. (2010). Protective effects of sucralfate on anti-*H. pylori* Vaca Igy *in vivo* and *in vitro*. *African Journal of Microbiology Research, 4*, 1091–1099.

Yamada, K., Wanchun, J., Ohkura, T., Murai, A., Hayakawa, R., Kinoshita, K., & ...Ohta, M. (2013). Detection of methicillin-resistant *Staphylococcus aureus* using a specific anti-PBP2a chicken IgY antibody. *Japan Journal of Infectious Disease, 66*(2), 103–108. https://doi.org/10.7883/yoken.66.103

Yamane, T., Saito, Y., Takizawa, S., Goshima, H., Kodama, Y., Horie, N., & Kim, M. (2003). Development of anti-*Helicobacter pylori* urease IgY and its application for food product. *Food and Development, 38*, 70.

Yang, Y. H., Park, D., Yang, G., Lee, S. H., Bae, D. K., Kyung, J., Kim, D., Choi, E. K., Son, J. C., Hwang, S. Y., & Kim, Y. B. (2012). Anti-*Helicobacter pylori* effects of IgY from egg york of immunized hens. *Laboratory Animal Research, 28*(1), 55–60. https://doi.org/10.5625/lar.2012.28.1.55

Yang, Y.-e, Wen, J., Zhao, S., Zhang, K., & Zhou, Y. (2014). Prophylaxis and therapy of pandemic H1n1 virus infection using egg yolk antibody. *Journal of Virology Methods, 206*, 19–26. https://doi.org/10.1016/j.jviromet.2014.05.016

Yao, L., Zhao, H., Tang, H., Song, J., Dong, H., Zou, F., & Cai, S. (2015). Chicken IgY facilitates allergic airway inflammation in a chemical-induced murine asthma model by potentiating IL-4 release. *Toxicology Letters, 239*(1), 22–31. https://doi.org/10.1016/j.toxlet.2015.08.1108

Yasui, H., Takahashi, K., Taki, S., Shimizu, M., Koike, C., Umeda, K., Rahman, S., Akashi, T., Nguyen, V. S., Nakagawa, Y., &

Sato, K. (2021). Near infrared photo-antimicrobial targeting therapy for *Candida albicans*. *Advanced Therapeutics*, *4*(2), 2000221. https://doi.org/10.1002/adtp.202000221

Yeh, S., Tsai, M. Y., Xu, Q., Mu, X. M., Lardy, H., Huang, K. E., Lin, H., Yeh, S. D., Altuwaijri, S., Zhou, X., Xing, L., Boyce, B. F., Hung, M. C., Zhang, S., Gan, L., & Chang, C. (2002). Generation and characterization of androgen receptor knockout (ARKO) mice: An in vivo model for the study of androgen functions in selective tissues. *Proceedings of the National Academy of Sciences of the United States of America*, *99*(21), 13498–13503. https://doi.org/10.1073/pnas.212474399

Yokoyama, H., Peralta, R. C., Diaz, R., Sendo, S., Ikemori, Y., & Kodama, Y. (1992). Passive protective effect of chicken egg yolk immunoglobulins against experimental enterotoxigenic *Escherichia coli* infection in neonatal piglets. *Infection and Immunity*, *60*(3), 998–1007. https://doi.org/10.1128/iai.60.3.998-1007.1992

Yokoyama, H., Peralta, R. C., Sendo, S., Ikemori, Y., & Kodama, Y. (1993). Detection of passage and absorption of chicken egg yolk immunoglobulins in the gastrointestinal tract of pigs by use of enzyme-linked immunosorbent assay and fluorescent antibody testing. *American Journal of Veterinary Research*, *54*(6), 867–872.

Yokoyama, K., Sugano, N., Rahman, A. K. M. S., Oshikawa, M., & Ito, K., (2007a). Activity of anti-*Porphyromonas gingivalis* egg yolk antibody against gingipains in vitro. *Oral Microbiology Immunology*, *22*(5), 352–355. https://doi.org/10.1111/j.1399-302X.2007.00358.x

Yokoyama, K., Sugano, N., Shimada, T., Shofiqur, R. A., Ibrahim, e., Isoda, R., Umeda, K., Sa, N. V., Kodama, Y., & Ito, K. (2007b). Effects of egg yolk antibody against Porphyromonas gingivalis gingipains in periodontitis patients. *Journal of oral Science*, *49*(3), 201–206. https://doi.org/10.2334/josnusd.49.201

Yoo, H. N., & Jung, Y. T. (2014). Expression of Lily mottle virus coat protein and preparation of IgY antibody against the recombinant coat protein. *Korean Journal Horticultural Science & Technology*, *32* (4), 544–549. https://doi.org/10.7235/hort.2014.13167

You, J., Xu, Y., He, M., McAllister, T. A., Thacker, P. A., Li, X., Wang, T., & Jin, L. (2011). Protection of mice against enterotoxigenic *E. coli* by immunization with a polyvalent enterotoxin comprising a combination of LTB, STa, and STb. *Applied Microbiology and Biotechnology*, *89*(6), 1885–1893. https://doi.org/10.1007/s00253-010-2991-7

You, Z., Yang, H., Xin, W., Kang, L., Gao, S., Wang, J., Zhang, T., & Wang, J. (2014a). Preparation of egg yolk antibodies against BoNT/B and their passive protection in mouse models. *Human Vaccines & Immunotherapeutics*, *10*(8), 2321–2327. https://doi.org/10.4161/hv.29433

Young, K. T., Davis, L. M., & Dirita, V. J. (2007). *Campylobacter jejuni*: Molecular biology and pathogenesis. *Nature Reviews. Microbiology*, *5*(9), 665–679. https://doi.org/10.1038/nrmicro1718

Zaki, S. R., Greer, P. W., Coffield, L. M., Goldsmith, C. S., Nolte, K. B., Foucar, K., Feddersen, R. M., Zumwalt, R. E., Miller, G. L., & Khan, A. S. (1995). Hantavirus pulmonary syndrome. Pathogenesis of an emerging infectious disease. *The American Journal of Pathology*, *146*(3), 552–579.

Zhang, J. J., Kang, T. Y., Kwon, T., Koh, H., Chandimali, N., Huynh, D. L., Wang, X. Z., Kim, N., & Jeong, D. K. (2019). Specific chicken egg yolk antibody improves the protective response against *Gallibacterium anatis* infection. *Infection and Immunity*, *87*(3), e00619–18. https://doi.org/10.1128/IAI.00619-18

Zhang, L., Li, D., Liu, L., & Zhang, G. (2015a). Rapid immunochromatographic test strip to detect swimming crab *Portunus trituberculatus* reovirus. *Diseases of Aquatic Organisms*, *117*(1), 21–29. https://doi.org/10.3354/dao02921

Zhang, L., Li, D., Liu, L., Fang, J., Xu, R., & Zhang, G. (2015b). Development of a colloidal gold immunochromatographic strip for the rapid detection of soft-shelled turtle systemic septicemia spherical virus. *Journal of Virological Methods*, *221*, 39–45. https://doi.org/10.1016/j.jviromet.2015.04.016

Zhang, S., Xing, P., Guo, G., Liu, H., Lin, D., Dong, C., Li, M., & Feng, D. (2016a). Development of microbeads of chicken yolk antibodies against *Clostridium difficile* toxin A for colonic-specific delivery. *Drug Delivery*, *23*(6), 1940–1947. https://doi.org/10.3109/10717544.2015.1022836

Zhang, X., Calvert, R. A., Sutton, B. J., & Doré, K. A. (2017). IgY: A key isotype in antibody evolution. *Biological Reviews of the Cambridge Philosophical Society*, *92*(4), 2144–2156. https://doi.org/10.1111/brv.12325

Zhang, X., Chelliappan, B., & Antonysamy, R., M. (2021). Recent advances in applications of bioactive egg compounds in nonfood sectors. *Frontiers in Bioengineering and Biotechnology*, *9*, 738993. https://doi.org/10.3389/fbioe.2021.738993

Zhang, X., Diraviyam, T., Li, X., Yao, G., & Michael, A. (2016b). Preparation of chicken IgY against recombinant E2 protein of bovine viral diarrhea virus (BVDV) and development of ELISA and ICA for BVDV detection. *Bioscience, Biotechnology, and Biochemistry*, *80*(12), 2467–2472. https://doi.org/10.1080/09168451.2016.1217144

Zhang, Y., Kong, H., Liu, X., Cheng, J., Zhang, M., Wang, Y., Lu, F., Qu, H., & Zhao, Y. (2018). Quantum dot-based lateral-flow immunoassay for rapid detection of rhein using specific egg yolk antibodies. *Artificial Cells, Nanomedicine, and Biotechnology*, *46*(8), 1685–1693. https://doi.org/10.1080/21691401.2017.1389749

Zhang, X., Morgan, P. M., & Vieira-Pires, R. S. (2021). Perspectives on IgY technology monoclonal IgY antibodies. In X.-Y. Zhang et al. (Eds.), *IgY-technology: Production and application of egg yolk antibodies*. Nature Switzerland AG, Springer. https://doi.org/10.1007/978-3-030-72688-1_18

Zhang, Y., Wei, Y., Li, Y., Wang, X., Liu, Y., Tian, D., Jia, X., Gong, R., Liu, W., & Yang, L. (2021). IgY antibodies against Ebola virus possess post-exposure protection in a murine pseudovirus challenge model and excellent thermostability. *PLoS Neglected Tropical Diseases*, *15*(3), e0008403. https://doi.org/10.1371/journal.pntd.0008403

Zhen, Y. H., Fang, R., Ding, C., Jin, L. J., Li, X. Y., Diao, Y. P., Shu, X. H., Ma, X. C., & Xu, Y. P. (2011). Efficacy of specific IgY for treatment of lipopolysaccharide-induced endotoxemia using a mouse model. *Journal of Applied Microbiology*, *111*(6), 1524–1532. https://doi.org/10.1111/j.1365-2672.2011.05155.x

Zhen, Y. H., Jin, L. J., Guo, J., Li, X. Y., Li, Z., & Fang, R. (2008a). Characterization of specific egg yolk immunoglobulin (IgY) against mastitis-causing *Staphylococcus aureus*. *Journal Applied Microbiology*, *105*(5), 1529–1535. https://doi.org/10.1111/j.1365-2672.2008.03920.x

Zhen, Y. H., Jin, L. J., Guo, J., Li, X. Y., Lu, Y. N., Chen, J., & Xu, Y. P. (2008b). Characterization of specific egg yolk immunoglobulin (IgY) against mastitis-causing *Escherichia coli*. *Veterinary Microbiology*, *130*(1–2), 126–133. https://doi.org/10.1016/j.vetmic.2007.12.014

Zhen, Y. H., Jin, L. J., Li, X. Y., Guo, J., Li, Z., Zhang, B. J., Fang, R., & Xu, Y. P. (2009). Efficacy of specific egg yolk immunoglobulin (IgY) to bovine mastitis caused by *Staphylococcus aureus*. *Veterinary Microbiology*, *133*(4), 317–322. https://doi.org/10.1016/j.vetmic.2008.07.016

Zhu, L., van de Lavoir, M. C., Albanese, J., Beenhouwer, D. O., Cardarelli, P. M., Cuison, S., Deng, D. F., Deshpande, S., Diamond, J. H., Green, L., Halk, E. L., Heyer, B. S., Kay, R. M., Kerchner, A., Leighton, P. A., Mather, C. M., Morrison, S. L., Nikolov, Z. L., Passmore, D. B., Pradas-Monne, A., … Etches, R. J. (2005). Production of human monoclonal antibody in eggs of chimeric chickens. *Nature Biotechnology*, *23*(9), 1159–1169. https://doi.org/10.1038/nbt1132

Zolfagharian, H., & Dounighi, N. M. (2015). Study on development of *Vipera lebetina* snake anti-venom in chicken egg yolk for passive immunization. *Human Vaccines & Immunotherapeutics*, *11*(11), 2734–2739. https://doi.org/10.4161/21645515.2014.985492

Zuo, Y., Fan, J., Fan, H., Li, T., & Zhang, X. (2009). Prophylactic and therapeutic effects of egg yolk immunoglobulin against porcine transmissible gastroenteritis virus in piglets. *Frontiers of Agriculture in China*, *3*(1), 104–108. https://doi.org/10.1007/s11703-008-0080-9

28

Minor Components of Egg

Yaqi Meng and Ning Qiu*

CONTENTS

28.1 Methods Used in the Identification of Novel Minor Proteins

Egg proteins were previously studied using classical biochemical techniques such as chromatographic and electrophoretic separation and Edman sequence analysis. The development of molecular biology (cDNA and ESTs libraries, bioinformatic analysis of the chicken genome, transcriptomics) and proteomics, all major scientific advances, led to the identification and characterization of a number of minor egg components not previously known. The number of minor egg proteins has dramatically increased in the last two decades. The recent development of high-throughput methods used in combination with the available chicken genome sequence and bioinformatics tools to predict functions have generated new insights for the characterization of minor proteins in different egg compartments.

The publication of the chicken genome sequence was one of the major advances in identifying novel chicken genes and proteins. By using the next-generation sequencing technologies (NGS), tens of millions of chicken transcript fragments are now available online (http://www.ncbi.nlm.nih.gov/sra/?term=chicken). The availability in the database of sequences of genes and their corresponding proteins are fundamental for the application of high-throughput technologies (proteomics and transcriptomics) and the identification of new proteins from various egg compartments (the organic matrix of eggshell, egg white, yolk and vitelline membrane). In proteomic studies, the proteins are described by using identifiers provided by various databases, including GeneBank (http://www.ncbi.nlm.nih.gov/genbank/), ENSEMBL (http://www.ensembl.org) or UniProt (Universal Protein Resource, http://www.uniprot.org/). A total of 1,261 non-redundant protein sequences (gene products) obtained by egg proteomic studies were reported by Gautron (Gautron et al., 2007).

For these egg proteomic analyses, different strategies and methods have been used to bypass some technical problems due to specific properties related to the nature and composition of the egg. For instance, in the egg white, the abundance differs widely from one protein to another, and ovalbumin represents more than 50% of total proteins, thus making the detection of minor proteins extremely challenging. The use of combinatorial peptide ligand libraries (CPLLs) allowed the discovery and identification of a large number of previously unreported proteins in egg white and yolk (Ambrosio et al., 2008; Farinazzo et al., 2009). One of the main and most visible advantages of CPLL protein extract treatments is the concomitant decrease in concentration of high-abundance proteins and increase concentration of low and very low-abundance proteins, making them more amenable to further analysis via various MS tools (Farinazzo et al., 2009). By exploiting this CPLL technology, comparative proteomic analyses of low-abundance egg white proteins were conducted during the rapid embryonic growth period and allowed to better understand their biological activities (Liu et al., 2015).

Other minor egg proteins were identified in the analyses of protein complexes formed during the storage of eggs or embryonic development. Some of these minor components were identified as protein fragments with domains having some specific biological functions. In the study investigating candidate proteins interacting with ovalbumin, a minor peptide similar to RYR-1 was identified under thermal-induced condition (Pan et al., 2017). This protein is thought to be associated with meat quality. In another comparative proteomic study of egg white proteins during the early phase of embryonic development, a complex of ovalbumin and an uncharacterized protein similar to poly (rC) binding protein 3 (PCBP3) was identified. PCBP3 is a widely conserved protein family containing RNA-binding domains and playing some roles in mRNA stabilization, translational activation, and translational silencing. This interaction

* Corresponding author.

DOI: 10.1201/9781003254430-35

between ovalbumin and PCBP3 is thought to play a key role in pre-mRNA processing, as well as in important pathways regulating embryonic development (Qiu et al., 2012). Other minor egg proteins, including those derived from post-translational modifications (PTMs), have been detected after specific enrichment methods (Wang et al., 2020; Zhu et al., 2019).

28.2 Minor Proteins in the Egg White

Chicken egg albumen participates in the embryonic development by providing essential nutrients as well as antimicrobial protection. According to Gautron's analysis, 273 different proteins have been determined in egg white, of which most are minor components with very low or trace abundance. The biological significance of these minor proteins is yet to be explained. Some of them were also found in chicken plasma or cerebrospinal fluid, suggesting their systemic presence in body fluids, including oviduct fluid, and that their presence in eggs might be derived from decaying epithelial cells lining the oviduct. However, a wide range of minor components have shown potential bioactive functions in transporting nutrients and protecting eggs from microbial infection. These minor compounds can be classified into different groups according to their structural or functional properties.

Some minor egg proteins, including LBP-BPI-PLUNC family proteins and defensins, exhibit antimicrobial properties and contribute to the defense mechanisms protecting the embryo against invading bacteria (Table 28.1). The LBP (LPS-binding protein)/BPI (bactericidal permeability increasing protein)/PLUNC (palate, lung and nasal epithelium clone protein) family consists of proteins involved in host defense against bacteria. Members of this family are able to bind bacterial lipopolysaccharide (LPS), a surface component of Gram-negative bacteria, and mediate alterations of the outer membrane as well as cause damages to the inner membrane of bacteria, and therefore, lead to the inhibition of bacterial growth. Several proteins of this family, including OCX-36 and TENP (Transiently Expressed in Neural Precursor), have been detected in egg white. Defensins are cysteine-rich cationic peptides of about 2–6 kDa that are involved in the host innate defense. Most of these peptides possess a broad-spectrum activity against Gram-positive and Gram-negative bacteria, and also against fungi and viruses. AvBD11, ovodefensins gallin/OvoDA1 and OvoDB1 are typical defensins identified in egg white. Some of these minor egg proteins are also able to interact with heparin, indicative of antibacterial properties, since the heparin structure shows similarities to the negatively charged molecules present at the surface of bacteria.

Another set of minor egg white proteins are related to the transport of essential nutrients to the developing embryo, including extracellular fatty acid-binding protein, Selenium-binding protein 1, avidin and riboflavin-binding protein. Avidin and riboflavin-binding protein are the main egg vitamin-binding proteins widely present within all egg compartments and are involved in the transport of biotin and riboflavin, respectively. It is also hypothesized that they are involved in antimicrobial defenses by preventing the uptake of these vitamins by some specific microorganisms. Other minor egg white proteins, such as Dickkopf-related protein 3, are supposed to be involved in nutrient anabolism or organ development during the chicken embryo development.

28.3 Minor Proteins in the Egg Yolk and Vitelline Membrane

While the egg yolk contains all the fat and cholesterol present in eggs, it also hold almost half of its protein content, as well as various vitamins and minerals. Based on prior work on non-redundant protein in the egg (Gautron et al., 2007), 230 and 144 different proteins have been detected in egg yolk and vitelline membrane, respectively. Proteins in the egg yolk, mainly in the form of lipoproteins, contain high-density lipoproteins (HDLs), low-density lipoproteins (LDLs), and very low-density lipoproteins (VLDLs). It is speculated that

TABLE 28.1

Minor Egg Components and Their Related Functions

Egg Compartment	Minor Components	Functions	Reference
Egg white	LBP-BPI-PLUNC family proteins (e.g., OCX-36 and TENP), Defensins (Cysteine-rich cationic peptides, mainly including AvBD11, OvoDA1 and OvoDB1)	Contains antimicrobial properties that protect the embryo against bacteria	Gautron et al. (2007)
	Extracellular fatty acid-binding protein, Selenium-binding protein 1, Avidin and Riboflavin-binding protein	Transports essential nutrients to the developing embryo	Qiu et al. (2012)
Egg yolk and vitelline membrane	Superoxide dismutase, Glutathione peroxidase, Thioredoxin, and PIT 54	Protects the egg yolk from oxidation during its storage	Zhou et al. (2020)
	Coagulation factor X, Fibrinogen C-terminal domain-containing protein, Fibrinogen alpha chain, VWFD domain-containing protein, and Vascular endothelial growth factor A	Aids in wound healing	Zhou et al. (2020)
	Sialic acid and related derivatives	Promotes the memory and intellectual development in infants	Nakamura (2003)
Egg shell	Ovocleidin-17	Induces the formation of calcite crystals from amorphous calcium carbonate nanoparticles	Rose & Hincke (2009)

egg yolk lipids play a critical role in exerting their bioactive functionality, and its auto-oxidation process was considered to bring negative effect to human health and needs to be controlled (Xiao et al., 2020; Meng et al., 2021). Some minor proteins or peptides in the egg yolk and vitelline membrane, such as superoxide dismutase and glutathione peroxidase, have been suggested to play an important role in protecting the yolk from oxidation and maintain the quality of the chicken egg during storage (Zhou et al., 2020). Some antioxidant proteins, including glutathione peroxidase (GPX), thioredoxin (Thr), and PIT 54, have also been suggested to be involved in the anti-oxidation process. GPX is an essential protein of the protective system against lipid peroxidation through the glutathione metabolism pathway. Thr participates in peroxide removal by peroxiredoxin and serves as a scavenger of reactive oxygen species (ROS). Meanwhile, the Thr system is able to reduce GSSG to GSH to retain sufficiently high GSH concentrations. During the embryonic development, hemoglobin (Hb) released from erythrocytes probably caused the oxidation of proteins and lipids, thus, PIT54, as an Hb-binding glycoprotein, may act as a potent antioxidant.

Different minor proteins in egg yolk and vitelline membrane are involved in wound healing. These proteins include coagulation factor X, fibrinogen C-terminal domain-containing protein, fibrinogen alpha chain, VWFD-domain-containing protein, and vascular endothelial growth factor A (VEGF A). Blood coagulation and platelet-mediated primary hemostasis are important physiological defense mechanisms occurring in all vertebrates to block blood loss following vascular injury. Coagulation factor Xa is a vitamin K-dependent glycoprotein that turns prothrombin into thrombin during blood clotting in the presence of factor Va, calcium, and phospholipid. Thrombin cleaves fibrinogen to produce fibrins that polymerize to yield the insoluble fibrin clot that blocks sites of vascular injury and prevent further blood loss. The Von Willebrand factor is a large multimeric glycoprotein that performs two important functions in hemostasis, including participating in the adhesion of platelets to sub-endothelial connective tissue and binding the blood clotting factor VIII. In addition, angiogenesis is a very crucial process during wound healing, and VEGF can promote wound healing through angiogenesis.

In a quantitative proteomic analysis of chicken egg yolk under diverse storage temperatures, a total of 6 proteases (lysosomal protective protein precursor, carboxypeptidase N catalytic chain precursor, hepatocyte growth factor-like protein precursor, cathepsin E-A-like, plasminogen, hyaluronan-binding protein 2 isoform) and 8 protease inhibitors (interalpha-trypsin inhibitor heavy chain H2 precursor, interalpha-trypsin inhibitor heavy chain H3, antithrombin-III precursor, serpin peptidase inhibitor, alpha-1-antiproteinase, kininogen-1, alpha-2-HS-glycoprotein, coagulation factor V isoform X2) were detected. The cathepsin E-A-like, also known as "similar to nothepsin", is a new member of the aspartic protease family, and similarly to cathepsin D, has been suggested to be involved in the proteolytic maturation of yolk proteins. Most proteases retain their activities at higher temperatures (the physiological incubation temperature of chicken embryo being 37.8°C) while they are inactive at temperature of refrigeration. Therefore, it is expected that protease activity is increased during high-temperature storage and accelerate the degradation of egg yolk proteins. Decreases in antiprotease activity during the storage of eggs at high temperature might correlate with the deterioration of some pivotal proteins. Furthermore, another study has found that the increased degradation of protease inhibitors during prolonged storage led to an imbalance between proteases and antiproteases in egg yolk, likely playing a key role in the degradation of CEY proteins (Gao et al., 2017). The thermally-induced decrease in the abundance of proteases and their inhibitors was believed to be one of the major reasons for the accelerated degradation of egg yolk proteins over storage time.

28.4 Minor Proteins in the Egg Shell

The chicken egg is a reproductive structure that has been shaped to resist physical, microbial and thermal challenges from the external environment during the embryonic development, with the eggshell playing the main protective role against external aggressions. The egg shell represents about 10% of the total egg mass, or 5–6 g per egg. Its complex structure gifts the egg shell with remarkable mechanical properties and abilities to regulate the exchange of metabolic gases and provide calcium to the growing embryo. The shell is mainly composed of calcium carbonate (95%), associated with an organic matrix composed of proteins, glycoproteins and proteoglycans, termed "eggshell matrix proteins". Although the organic shell matrix, made up of proteoglycans and proteins, represents only 3.5% (175 mg) of the total egg mass, more than 900 different proteins have been identified. While these shell proteins are at very low concentrations, they do play some key roles in influencing the construction of this biomaterial and/or in participating in the antimicrobial protection of the developing embryo (Rose & Hincke, 2009).

In general, these soluble egg shell matrix proteins modify calcium carbonate crystal growth, and therefore regulate the macroscopic properties of the resulting bio-ceramic. Investigations have demonstrated that eggshell organic fractions exhibit calcium binding properties linked to the presence of proteins or keratan and dermatan sulfate proteoglycans. For example, Ovocleidin-17 (OC-17) was the first eggshell-specific matrix protein isolated and characterized through chromatographic purification after eggshell decalcification. This protein contains 142 amino acids, including 2 phosphorylated serines. Metadynamics simulations revealed that OC-17 induces the formation of calcite crystals from amorphous calcium carbonate nanoparticles. With an abundance of about 40 μg/g of shell, OC-17 is not considered as a minor component among eggshell-matrix proteins. Nevertheless, its trace existence was also detected in the chicken egg yolk plasma, egg white and vitelline membrane. Moreover, it also occurs in the eggshell soluble organic matrix as a minor form glycosylated at Asn-59 (23 kDa, OC-23) (Rose & Hincke, 2009).

Since shell strength is a key selection criteria for poultry genetics companies to improve the sustainability of the egg sector, a large number of proteins identified in the egg shell may be used as biological markers during genetic selection. In a proteomic study, Sun et al. reported that 15 proteins, mainly relating to eggshell matrix specific proteins, calcium binding and transportation, protein folding and sorting, bone development, and thyroid hormone activity. These proteins were considered closely associated with the formation of strong eggshell (Sun et al., 2013). The eggshell matrix proteins undergo posttranslational modifications, including phosphorylation and glycosylation, suggested to have important regulatory effects on the functional properties of proteins. In a recent systematic study, a total of 112 phosphorylation sites from 69 phosphoproteins and 297 N-glycosylation sites from 182 N-glycoproteins were identified in the chicken egg shell matrix (Zhou et al., 2020). Functional analyses indicated that the phosphoproteins were enriched in brain development and cell adhesion, and glycoproteins were enriched in the immune process and hydrolase activity of glycosidic bonds. These studies on egg shell matrix proteins and their posttranslational modifications provide comprehensive protein structure information to further explore some critical issues related to eggshells.

28.5 Other Minor Components of Egg

Sialic acid, a class of natural derivatives of nine-carbon carboxy sugars, usually exists in the form of oligosaccharides, glycolipids, or glycoproteins (Nakano et al., 2003), which exists widely in a variety of organisms, and is generally distributed at the ends of the carbohydrate chains of glycoproteins and glycolipids on the cell surface. In addition, sialic acid is a natural brain nutrient that promotes the memory and intellectual development in infants (Nakamura, 2003). Eggs are rich in sialic acid, distributed in eggshell (0.028%), eggshell membrane (0.02%), egg white (0.01%), egg yolk (0.095%), egg yolk membrane (0.153%), and frenulum (0.18%), but the part with the highest quantities is the yolk. The total content of sialic acid in egg yolk accounted for 78.8% of the total sialic acid content of the whole egg. In 2003, Koketsu et al. investigated the content and distribution of sialic acid in eggs, and found that the sialic acid content of egg yolk, albumen, and frenulum were 205.2, 11.96, and 0.83 mg, respectively (Koketsu, 2003). Sialic acid in egg yolk mainly exists in glycoproteins in the form of complex oligosaccharides, which are macromolecules and usually be prepared by acid or enzymatic hydrolysis. Sialic acid in eggshell membrane mainly exists in glycoproteins in the form of complex oligosaccharides. Sialic acid oligosaccharide in egg yolk is beneficial to improve the learning ability of infant rats. In addition, sialic acid compounds also have physiological functions such as controlling cell concentration, anti-recognition, and anti-tumor. There is still a lot of space for the development of research on the physiological activities of avian egg sialic acid and its derivatives.

28.6 Conclusion

In the last 20 years, developments in the biochemistry (proteomics) and molecular biology of the egg have allowed the identification of hundreds of minor egg protein components with potential nutritive values to benefit human health and further applications by the industry. These proteins are involved in the protection and development of the chicken embryo and also contribute to the nutritive value of this basic food ingredient. The bioinformatics analyses and experimental investigations have suggested that these minor components have a broad range of biological activities with potential interests for many sectors, including the pharmaceutical, cosmetic, food and bio-materials industries. Among the minor proteins in the shell matrix, their key functions are related to the calcification process, critical to control shell quality. Minor proteins with antimicrobial ability, as a part of innate immunity, are mostly present in egg white and vitelline membranes. Other studies have revealed some antioxidant activities or specific biochemical properties, especially among the egg yolk minor components, that are currently used to develop diagnostic tools and strategies for targeted therapy. The next frontier in egg science will likely be the identification of the biological activities of these novel egg protein components and their use to develop innovative products for improved human nutrition and health.

REFERENCES

Ambrosio, C. D., Arena, S., Scaloni, A., Guerrier, L., Boschetti, E., Mendieta, M. E., Citterio, A., & Righetti, P. G. (2008). Exploring the chicken egg white proteome with combinatorial peptide ligand libraries. *Journal of Proteome Research*, *7*(8), 3461–3474. https://doi.org/10.1021/pr800193y

Farinazzo, A., Restuccia, U., Bachi, A., Guerrier, L., Fortis, F., Boschetti, E., Fasoli, E., Citterio, A., & Righetti, P. G. (2009). Chicken egg yolk cytoplasmic proteome, mined via combinatorial peptide ligand libraries. *Journal of Chromatography A*, *1216*(8), 1241–1252. https://doi.org/10.1016/j.chroma.2008.11.051

Gao, D., Qiu, N., Liu, Y., & Ma, M. (2017). Comparative proteome analysis of egg yolk plasma proteins during storage. *Journal of the Science of Food and Agriculture*, *97*(8), 2392–2400. https://doi.org/10.1002/jsfa.8052

Gautron, J., Nau, F., Mann, K., Guerin-Dubiard, C., Rehault, S., Hincke, M. T., & Nys, Y. (2007). Molecular approaches for the identification of novel egg components. *World's Poultry Science Journal*, *63*(1), 82–90. https://doi.org/10.1017/S0043933907001298

Koketsu, M. (2003). Distribution of N-acetylneuraminic acid and sialylglycan in eggs of the silky fowl. *British Poultry Science*,*44*,145–148. https://doi.org/10.1080/000716603100 0085328

Liu, Y., Qiu, N., & Ma, M. (2015). Comparative proteomic analysis of egg white proteins during the rapid embryonic growth period by combinatorial peptide ligand libraries. *Poultry Science*, *94*(10), 2495–2505. https://doi.org/10.3382/ps/pev176

Meng, Y., Qiu, N., Mine, Y., & Keast, R. (2021). Comparative lipidomics of chick yolk sac during the embryogenesis provides insight into understanding the development-related lipid supply. *Journal of Agricultural and Food Chemistry*, *69*(26), 7467–7477. https://doi.org/10.1021/acs.jafc.1c01728

Nakamura, T. (2003) Concentrations of sialyloligosaccharides in bovine colostrum and milk during the prepartum and early lactation. *Journal of Dairy Science*, *86*(4),1315–1320. https://doi.org/10.3168/jds.s0022-0302(03)73715-1

Nakano, T., Ikawa., N. I., & Ozimek, L. (2003) Chemical composition of chicken eggshell and shell membranes. *Poultry Science*, *82*(3), 510–514. https://doi.org/10.1093/ps/82.3.510

Pan, P., Qiu, N., Zhao, H., Liu, Y., & Gao, D. (2017). Identification of candidate proteins interacted with ovalbumin during the early phase of embryonic development. *International Journal of Food Properties*, *20*(Sup3), S2305–S2312. https://doi.org/10.1080/10942912.2017.1326056

Qiu, N., Ma, M., Cai, Z., Jin, Y., Huang, X., Huang, Q., & Sun, S. (2012). Proteomic analysis of egg white proteins during the early phase of embryonic development. *Journal of Proteomics*, *75*(6), 1895–1905. https://doi.org/10.1016/j.jprot.2011.12.037

Rose, M. L. H., & Hincke, M. T. (2009). Protein constituents of the eggshell: Eggshell-specific matrix proteins. *Cellular and Molecular Life Sciences*, *66*, 2707–2719. https://doi.org/10.1007/s00018-009-0046-y

Sun, C., Xu, G., & Yang, N. (2013). Differential label-free quantitative proteomic analysis of avian eggshell matrix and uterine fluid proteins associated with eggshell mechanical property. *Proteomics*, *13*(23–24), 3523–3536. https://doi.org/10.1002/pmic.201300286

Wang, H., Qiu, N., Mine, Y., Sun, H., Meng, Y., Li, B., & Keast, R. (2020). Quantitative comparative integrated proteomic and phosphoproteomic analysis of chicken egg yolk proteins under diverse storage temperatures. *Journal of Agricultural and Food Chemistry*, *68*(4), 1157–1167. https://doi.org/10.1021/acs.jafc.9b06912

Xiao, N., Zhao, Y., Yao, Y., Wu, N., Xu, M., Du, H., & Tu, Y. (2020). Biological activities of egg yolk lipids: A review. *Journal of Agricultural and Food Chemistry*, *68*(7), 1948–1957. https://doi.org/10.1021/acs.jafc.9b06616

Zhou, Y., Qiu, N., Mine, Y., Meng, Y., Keast, R., & Zhu, C. (2020). Quantitative comparative proteomic analysis of chicken egg vitelline membrane proteins during high-temperature storage. *Journal of Agricultural and Food Chemistry*, *68*(36), 9816–9825. https://doi.org/10.1021/acs.jafc.0c03538

Zhu, F., Qiu, N., Sun, H., Meng, Y., & Zhou, Y. (2019). Integrated proteomic and N-glycoproteomic analyses of chicken egg during embryonic development. *Journal of Agricultural and Food Chemistry*, *67*(42), 11675–11683. https://doi.org/10.1021/acs.jafc.9b05133

Part VIII

Egg Nutrition and Human Health

29

Nutritional Facts about Eggs

Vincent Guyonnet

CONTENTS

29.1 Introduction

Eggs are consumed across all continents, appreciated for their great taste, convenience and versatility and used in many dishes and food products. While their nutritional value is often unknown to consumers, the eggs represent one of the most nutritious and affordable animal-sourced foods, making some important contribution to nutrition and food security in many countries. After being decried for many years for its cholesterol content, the egg is now viewed as a great nutrient-dense food, rich in high quality proteins, choline, and a range of vitamins and minerals. While consumers have in general a limited knowledge about nutrition, the efforts of many national organizations of egg producers are convincing more and more consumers of the great nutritional value of eggs (see Chapter 2). After reviewing the composition of shell eggs, this chapter will present the various contribution of egg nutrition to meeting the recommended daily intake for different nutrients. Some key nutrients will be discussed, outlining specific key features making the egg an ideal source. The contribution of shell eggs to human nutrition, at all stages of life, will conclude this chapter serving as an introduction to the following nine chapters on Egg Nutrition and Human Health.

29.2 Egg Composition

Depending on the nutrients considered, their concentration in eggs may vary greatly based on the composition of the diet provided to laying hens. The term "regular shell eggs" is often used to describe eggs that are produced by birds fed the standard diet available in a given country. The term "functional eggs" is used to describe shell eggs produced by laying hens provided with a ration enhanced in specific nutrients such as vitamins, minerals or polyunsaturated fatty acids (PUFA). In addition, the concentration of nutrients is also affected by the analytical methods used, the frequency of assays performed and the number of samples considered to establish average values. For these reasons, the use of national nutrient databases is recommended to report on the composition of shell eggs as they are periodically revised and serve often as the basis for any nutritional information provided on shell egg cartons (Table 29.1).

29.2.1 Regular Eggs

The composition of shell eggs is usually expressed in serving size or portion, excluding the eggshell which is considered as inedible (although it is a great source of highly bioavailable

DOI: 10.1201/9781003254430-37

TABLE 29.1

Regional and National Nutrient Databases

Country/Region	URL for Retrieving the Nutrition Database
Africa (West)	https://www.fao.org/3/ca7779b/CA7779B.PDF
Argentina	http://www.unlu.edu.ar/~argenfoods/Tablas/Tabla.htm
ASEAN	http://www.inmu.mahidol.ac.th/aseanfoods/composition_data.html
Australia	https://www.foodstandards.gov.au/science/monitoringnutrients/ausnut/foodnutrient/Pages/default.aspx
Brazil	http://www.tbca.net.br/
Canada	https://www.canada.ca/en/health-canada/services/food-nutrition/healthy-eating/nutrient-data.html
China	https://www.chinanutri.cn/yyzyxxpt/sjzx/swcfsjk/
FAO	https://www.fao.org/infoods/infoods/tables-and-databases/en/
France	https://ciqual.anses.fr/#/aliments/22000/egg-raw
Japan	https://www.mext.go.jp/en/policy/science_technology/policy/title01/detail01/sdetail01/sdetail01/1385122.htm
Korea	http://koreanfood.rda.go.kr/eng/fctFoodSrchEng/engMain
Mexico	https://www.incmnsz.mx/2019/TABLAS_ALIMENTOS.pdf
South Africa	https://safoods.mrc.ac.za/database.html
Spain	https://www.bedca.net/bdpub/index.php
UK	https://www.gov.uk/government/publications/composition-of-foods-integrated-dataset-cofid
USA	https://fdc.nal.usda.gov/fdc-app.html#/food-details/748967/nutrients

calcium). For instance, in the USA, large shell eggs have a mass equal or superior to 56.7 grams but the portion size declared on egg cartons is 50 grams. Since serving sizes are based on the amount of food people typically consume rather than what should be consumed, there are differences among countries around the world. In addition, some countries also regularly update the serving size to reflect the amount people will typically eat. For ease of comparison, the nutritional value of eggs in Australia, Canada, France, Japan and the USA are reported in Table 29.2, based on an edible portion of 100 grams (AUSNUT, 2011; MEXT, 2015; CIQAL, 2020; CNF, 2022; USDA, 2022).

The results for the proximate analysis are usually comparable between national databases, varying from 4.8% for total fats to 7.6% for proteins, and reflect little variations in composition related to diets, breed of birds or management practices. Energy values of a 100-gram serving size varies from 140 kcal (France) to 151 kcal (Japan) or by 9% and is related to the difference reported for fat content, ranging from 9.83 grams in France to 10.3 grams in Japan. The cholesterol content of eggs ranged from 366 mg (Canada) to 398 mg (France) or a variation of 8.7% among databases. As previously reported, hens have little dietary cholesterol intake and therefore must synthesize in the liver the cholesterol present in eggs. Potential genetic, nutritional and pharmacological interventions have often resulted in less than 5% in reduction of the cholesterol content of eggs (Guyonnet & Jing, 2015).

Greater variations are observed for the mineral and vitamin content of shell eggs since a number of them are easily affected by the composition of the diets given to laying hens. For instance, the concentration in iron ranged from 1.39 mg (Canada) to 1.88 mg (France) or a 35% variation. The content of eggs in phosphorus and iodine showed even greater variation, with 62% and 135%, respectively, among the values reported in these 5 countries. The variations observed for the content in vitamins are even more significant: 33.8% for riboflavin (Vit B2), 158.8% for folate (Vit B9), 361.5% for biotin (Vit B7) and 464.3% for vitamin D. The reasons for these variations were previously reviewed and poultry diets were identified as a major source of variation (Seuss-Baum & Nau, 2011). The importance of the formulation of the diet fed to laying hens is clearly demonstrated by the low content of eggs in vitamin D in Canada (1.4 μg/100 grams) where the regulatory authorities have limited the incorporation of vitamin D at 162.5 mg per tonne of layer feed. In Australia where there is no such restriction, regular eggs contain 7.9 μg/100 grams or more than 5 times the content in Canada.

Although Choline is recognized as an important nutrient in the egg (see Chapter 36) and included in the national database of Canada and the USA, it was not reported in the databases from Australia, France and Japan. It is therefore important to review regularly these databases as new nutrients are incorporated or values are modified to reflect new results available.

Most national databases will also report separately the nutritional values of egg white and yolk. In addition, some databases present the nutritional values of shell eggs after cooking (soft boiled, hard boiled, fried, poached or scrambled) and report some minimum losses during the heating process (AUSNUT, 2011; CIQAL, 2020; UK, 2021). A recent review, using the USDA nutrient database (now part of FoodData Central) reported no clear evidence of the effect of cooking on the denaturation of minerals and vitamins. While proteins are subjected to some tertiary structure during cooking, their relative amount is not affected by the cooking process and will in fact increase the digestibility of egg proteins (Réhault-Godbert et al., 2019).

In various countries, consumers are able to purchase at retail eggs from quails, ducks, turkey, geese, ostrich or even pigeon. While the focus of this chapter is the egg produced by laying hens, it is interesting to compare the composition of eggs in these different species. The Composition of foods integrated dataset (CoFID) from the UK was used to provide the data reported in Table 29.3 for eggs from laying hens, ducks, quails and turkeys (UK, 2021).

Proteins are the nutrient varying the least among eggs from chicken, duck, quail and turkey, ranging from 12.6 grams in chicken to 14.3 grams in duck. The total fat content is the highest in turkey eggs (12.2 grams) and the lowest in chicken eggs (9 grams). Chicken eggs contain the lowest amount of saturated, monounsaturated and polyunsaturated fats among the four species.

In line with the fat content, the energy value in 100 grams ranges from 131 kcal in chicken eggs to 165 kcal in turkey eggs. According to the UK database, cholesterol is the nutrient with the widest variation in concentration among species: 350 mg in

TABLE 29.2

Nutritional Values for Raw Shell Eggs (Based on 100 Grams Edible Portion), as Reported on Governmental Databases in Australia, Canada, France, Japan and the USA

Nutrient Name	Unit	Australia	Canada	France	Japan	USA
Proteins	g	12.2	11.8	12.7	12.2	12.4
Total Fat	g	9.9	10.01	9.83	10.3	9.96
Saturated	g	3.07	3.04	2.64	2.84	3.20
Monounsaturated	g	3.55	4.18	3.66	3.69	3.63
Polyunsaturated	g	1.60	1.45	1.65	1.66	1.82
Carbohydrates	g	1.4	0.98	0.27	0.4	0.96
Sugars, total	g	0.3	0.2	0.27	NR	0.2
Energy	kcal	142	145	140	151	143
Cholesterol	mg	382	366	398	370	370
Minerals						
Copper	mg	NR	0.055	0.055	0.05	ND
Iodine	µg	47.0	NR	21.0	33.0	49.1
Iron	mg	1.8	1.39	1.88	1.5	1.67
Phosphorus	mg	171	126	204	180	184
Selenium	µg	28	34.2	ND	24	31.1
Zinc	mg	1	1.16	1.01	1.1	1.24
Vitamins						
Vit A (Retinol)	µg	98	191	182	210	180
Vit B1 (Thiamin)	mg	0.05	0.071	0.055	0.06	0.077
Vit B2 (Riboflavin)	mg	0.4	0.495	0.45	0.37	0.419
Vit B3 (Niacin)	mg	NR	0.117	0.063	0.1	ND
Vit B5 (Pantothenic acid)	mg	1.07	2.44	1.57	1.16	1.6
Vit B6 (Pyridoxine)	mg	0.02	0.072	0.15	0.09	0.063
Vit B7 (Biotin)	µg	5.2	NR	NR	24.0	22.0
Vit B9 (Folate)	µg	88	72	34	49	71
Vit B12 (Cobalamin)	µg	0.33	1.90	1.45	1.10	1.02
Vit D	µg	7.9	1.4	1.88	3.8	2.46
Vit E (alpha-tocopherol)	mg	1.9	2.45	1.43	1.3	1.0
Vit K	µg	NR	0.3	0.3	1.2	NR
Other nutrients						
Choline	mg	NR	301.7	NR	NR	335
Lutein & Zeaxanthin	µg	NR	332	NR	NR	459

Source: Adapted from AUSNUT (2011), MEXT (2020), CIQAL (2020), CNF (2022), USDA (2022).
Note: ND = Not detected (below detection level); NR = Not reported in government database.

chicken eggs, 680 mg in duck and turkey eggs and 900 mg in quail eggs or a 157% variation. However, a detailed analysis of the lipids and fatty acids profile in eggs from various bird species reported lower cholesterol levels per 100 grams in eggs from quail (496 mg), duck (377 mg) and turkey (511 mg) (Sinanoglou et al., 2011). The differences in vitamin concentrations among eggs from these four species ranged from 68% (Riboflavin or vitamin B2) to 100% (Thiamin or vitamin B1, Niacin or vitamin B3 and Cobalamin or vitamin B12), with chicken eggs often having the lowest concentration and the duck eggs the highest concentration for the vitamins listed in Table 29.3. The data from the UK nutrient database seemed to indicate some similar observations for the content in minerals of these eggs although only two minerals could be used for comparison. The same observations were made among eggs of chickens, quails, ducks, geese and turkeys, using the USDA nutrient file (Réhault-Godbert et al., 2019). The differences in the composition of rations given to these different species account certainly for most of the variations reported among species.

Overall, the use of nutrient database is extremely helpful and serve very often as the legal base for any nutritional information printed on egg cartons (see Chapter 18). However, due

TABLE 29.3

Comparison of Egg Composition (Raw, Whole) from Chicken, Duck, Quail and Turkey

Nutrient Name	Unit	Chicken	Duck	Quail	Turkey
Proteins	g	12.6	14.3	12.9	13.7
Total Fat	g	9.0	11.8	11.1	12.2
Saturated	g	2.52	2.90	3.10	3.70
Monounsaturated	g	3.44	4.90	4.90	4.70
Polyunsaturated	g	1.44	2.00	1.30	1.40
Trans fat	g	0.01	0.10	0.02	NR
Energy	kcal	131	163	151	165
Cholesterol	mg	350	680	900	680
Minerals					
Iron	mg	1.72	2.90	3.70	4.10
Phosphorus	mg	179	200	230	170
Vitamins					
Vit B1 (Thiamin)	mg	0.08	0.16	0.13	0.11
Vit B2 (Riboflavin)	mg	0.50	0.47	0.79	0.47
Vit B3 (Niacin)	mg	0.1	0.2	0.2	ND
Vit B6 (Pyridoxine)	mg	0.13	0.25	0.15	NR
Vit B9 (Folate)	µg	47	80	NR	NR
Vit B12 (Cobalamin)	µg	2.7	5.4	NR	NR

Source: Adapted from UK (2021).

Note: Results expressed per 100 g edible portion. ND = Not detected (below detection level); NR = Not reported in government database

to the small number of samples analyzed and reported, there are some limitations in their use to report the nutrient content of eggs from species other than chickens.

29.2.2 Functional Eggs

As described in Chapter 2, the attitudes, perceptions and behaviors of consumers toward eggs have changed and authenticity and sustainability, in addition to price, taste and convenience, are now considered as key buying factors. Over the past decade, the importance of these two new buying factors is at the foundation of the growth of the organic food segment, considered by consumers as safer, healthier and more sustainable. Functional eggs are also considered by consumers as healthier since they do offer higher levels of vitamins, minerals, omega 3 PUFA or lutein.

Since the early 1990s, numerous studies have demonstrated that variations in the diet composition provided to laying hens can greatly influence the level of nutrients such as vitamin A, vitamin D3, vitamin K, vitamin B1, B2 and B6, biotin, niacin, vitamin E, folic acid, pantothenic acid, choline, lutein iron and selenium, total omega 3 and long chain omega 3 PUFA such as docosahexaenoic acid or DHA (Caston & Leeson, 1990; Naber, 1993; House et al., 2002; Leeson & Caston, 2003, 2004; Skřivan et al., 2005; Finisin et al., 2008).

In some early work, the efficiency at which laying hens are able to transfer additional vitamins in their diet into the eggs was defined and allowed to identified 4 groups of vitamins categories, presented in Table 29.4 (Naber, 1993).

Poultry nutritionists have developed well established diet formula allowing easy, predictable and repeatable ways to increase the level of certain nutrients in the eggs. The repeatability of this process has allowed the commercialization of functional eggs, as both consumers and regulatory agencies except and require truthfulness in labeling (Guyonnet, 2011). Functional eggs can also be enriched simultaneously in various nutrients. In a study in France, researchers obtained functional eggs with simultaneous enrichment for 2 minerals, 5 vitamins, omega 3 PUFA and lutein (Table 29.5).

TABLE 29.4

Classification of Vitamins Based on Their Relative Transfer Efficiency from the Laying Hen Diet to the Egg

Evaluation of Efficiency Transfer	Vitamin
Low efficiency (~ 5–10%)	Vitamin B1 (Thiamin)
	Vitamin B9 (Folate)
	Vitamin K_1
Medium efficiency (15–25%)	Vitamin D_3
	Vitamin E
High efficiency (40–50%)	Vitamin B2 (Riboflavin)
	Vitamin B5 (Pantothenic acid)
	Vitamin B7 (Biotin)
	Vitamin B12 (Cobalamin)
Very high efficiency (60% to 80%)	Vitamin A

Source: Adapted from Naber (1993).

TABLE 29.5

Composition of Functional Eggs (Multi-Enrichment) Compared to Regular Shell Eggs

Nutrient Name	Unit	Regular Egg	Functional Egg	Enhancement Factor
Iodine	µg	60	150	× 2.5
Selenium	µg	7	28	× 4
Lutein + Zeaxanthin	µg	250	1,500	× 6
Vitamin A	µg	175	180	× 1
Vitamin B9 (Folate)	µg	60	240	× 4
Vitamin B12	µg	1.0	1.4	× 1.4
Vitamin D	µg	0.5	1.5	× 3
Vitamin E	µg	1,300	8,000	× 6
Omega 3 ALA	mg	50	300	× 6
Omega 3 DHA	mg	40	120	× 3

Source: Adapted from Bourre (2005).
Note: Results expressed per 100 g edible portion.

The enhancement ranged from ×1.4 for vitamin B12 to ×6 for Vitamin E, Selenium and Omega 3 alpha-linoleic acid (ALA), confirming the potential value of the multi-enrichment of eggs to deliver needed micronutrients to the human diet (Bourre, 2005). Interestingly, their work did not report any increase in vitamin A although its status as a nutrient with very high efficiency transfer is widely accepted (Schiavone & Barroeta, 2011). The relation between the diet fed to laying hens and the concentration of nutrients in eggs was further established for 12 vitamins (Leeson & Caston, 2003) and is presented in Table 29.6. The work confirmed the excellent transfer for Vitamin B5 with a doubling of the level of vitamin B5 in the feed resulting in a 60% increase concentration in the egg (79% efficiency for the transfer of additional vitamin B5). For vitamin D_3, increasing its level by 2.4 times in the ration translated into a 192% increased concentration in the egg or a 122% efficiency transfer for the incremental dietary vitamin D_3. With the exception of vitamin B3 (only 15% of the incremental amount transferred to the egg), most other vitamins were in the medium to high efficiency transfer described previously.

Additionally, laying hens also act as natural "bioconverters and condensers". For example, hens convert the ALA from flax into DHA (Caston & Leeson, 1990) at a much higher efficiency than the ~ 2-10% typically reported in humans (Swanson et al., 2012). From a human nutrition standpoint, it is therefore more efficient to consume eggs enriched in DHA rather than foods rich in ALA. Overall, the retail market for functional eggs has grown over the past 20 years and addressed some important nutritional needs.

29.3 Eggs and Human Nutrition

In 2020, the annual global consumption of eggs was estimated at 185 eggs per person although some wide variations existed across continents, ranging from 284 in North America

TABLE 29.6

Comparison between Layer Ration Enhancements and the Multi-Enrichment of Shell Eggs in Various Vitamins

Nutrient Name	Regular Egg (µg)	Level of Enhancement in Layer Diet	Functional Egg (µg)	Enhancement Factor in Eggs	Estimated Transfer Efficiency
Vitamin A	98	× 4	125	× 1.3	32%
Vit B1 Thiamin	82	× 6	112	× 1.4	23%
Vit B2 Riboflavin	365	× 5.2	408	× 1.1	22%
Vit B3 Niacin	78	× 11	128	× 1.6	15%
Vit B5 Pantothenic acid	1,272	× 2	2,008	× 1.6	79%
Vit B6 Pyridoxine	45	× 3	55	× 1.2	41%
Vit B7 Biotin	28	× 6	30	× 1.1	18%
Vit B9 Folate	15	× 3	17	× 1.1	37%
Vit B12 Cobalamin	1.45	× 11	5.58	× 3.8	35%
Vitamin D3	0.65	× 2.4	1.90	× 2.9	122%
Vitamin E	2,200	× 7.7	6,267	× 2.8	37%
Vitamin K	217	× 2.7	217	× 1	37%

Source: Adapted from Leeson & Caston (2003).
Note: Results expressed in µg per 100 g edible portion.

to 44 eggs in Africa (see Chapter 2). An egg, by nature, is produced by laying hens to hatch, after a 21-day incubation period, a chick able to walk, see, eat regular poultry feed, fully feathered and immunocompetent. In order to achieve this amazing achievement, it is evident that the egg contains the right balance of nutrients, as detailed in previous chapters (see Part II of this book). The nutrients contained in the egg have also an important role in human nutrition as presented in the following sections.

29.3.1 Nutrition Reference Values (NRVs)

Nutrient Reference Values (NRVs), established for the general population (individuals older than 36 months), are a set of values designed to help consumers estimate the relative contribution of individual products to the overall healthiness of their dietary intake, and compare the nutrient content between food items. At a national level, the NVRs are established for the general population, by weighting science-based reference values for daily intakes for age-sex groups using census data and the proportions of each age sex group. NRVs are defined for each nutrient to meet specific criteria for adequacy in healthy individuals. Over the years, various terms have been used, often to express similar concept. The definitions presented in the latest revision of the Codex Alimentarius, the joint group of experts from the World Health Organization (WHO) and the Food and Agriculture Organization of the United Nations (FAO), will be provided along with some alternative terms often used in national regulations (Lewis, 2019).

Daily Intake Reference Values (DIRVs): The term refers to reference nutrient intake values that are established by the FAO/WHO or other recognized authoritative scientific bodies (RASBs). These DIRVs may be expressed as single values or a range, and are applicable to the general population or to a specific segment of the population (e.g., specified age group, pregnant women).

Individual Nutrient Level 98 (INL98): The value represents the daily intake reference value for a given nutrient, estimated to meet the requirement of 98% of apparently healthy individuals in a specific life stage and sex group. In this definition, the term "apparently healthy" refers to the absence of disease based on clinical signs and symptoms of micronutrient deficiency and normal function assessed by laboratory analysis and physical evaluation. INL98 will vary by age, gender and whether a woman is pregnant or breastfeeding. The same concept is expressed with other terms in different countries: Recommended Dietary Allowance or Recommended Daily Allowance (e.g., Canada, USA), Reference Nutrient Intake (e.g., UK) or Population Reference Intake (e.g., EU27).

Adequate Intake (AI): The term refers to the observed or experimentally derived intake level by a defined population group that appears to sustain health.

Daily Values (DV): The term is used for instance in the USA and Canada to define the recommended amounts of nutrients to consume or not to exceed each day. Defined by the competent regulatory authorities, a single DV for each nutrient is established for all people aged 4 years and older. Therefore, DV do not account for specific circumstance such as higher nutritional needs during pregnancy or lactation. In Canada, for instance, the DV for zinc is currently defined as 11 mg and that of vitamin B2 (riboflavin) as 1.3 mg.

% Daily Value (% DV): This percentage indicates the contribution of one serving of a food to meet the daily value for a given nutrient. For instance, in a serving of 100 grams of eggs, the intake of zing and riboflavin will be 1.16 mg and 0.495 mg, respectively. Therefore, the % DV of one 100-gram serving of eggs will be 10.5% for zinc and 38.1% for vitamin B2

Estimated average requirement (EAR): This value estimates the daily intake reference value for a given nutrient to meet the requirement of 50% of apparently healthy individuals in a specific life stage and sex group. The term *Average Requirement* (AR) is used in the EU27 countries to express this median value. Therefore, the terms EAR or AR have little value from a nutritional standpoint and should not be confused for an average value being recommended when consulting NRVs tables (EFSA, 2017).

As NRVs are reviewed by governments or the WHO/FAO when new scientific data become available, it is important to consult the appropriate websites regularly. Typically, governments will inform the food sector prior to releasing new NRVs and will allow a grace period for the implementation of new nutritional information on food packaging (Australia, 2006; EFSA, 2017).

29.3.2 Nutritional Content Claims

Recognizing the need to encourage the adoption of healthy eating habits, a number of countries, following the recommendations of the Codex Alimentarius, have allowed a number of nutrition-related claims. A Nutrition content claim (NCC) means "any representation which states, suggests or implies that a food has particular nutritional properties including, but not limited to the energy value and to the content of protein, fat and carbohydrates, as well as the content of vitamins and minerals" (CODEX Alimentarius, 2013). The NCC will describe the level of a nutrient present in a food in a manner easily understood by consumers (e.g., "source of zinc", "high in proteins", "low in sugars", "fat free"), based on the % DV or a specific amount provided in one serving size.

Among the countries mentioned in Table 29.7, Australia, the EU27 countries, Japan and the USA have defined two levels of NCC ("source", "contains" or "good source" and "high" or "good source") while Canada has introduced three NCC levels ("source", "good source" and "excellent source"). The EU27 countries have the most restrictive rules with the first level of NCC requiring a food to deliver at least 15% of the DV for a nutrient (reported as Recommended Allowance in the EU legislation) while Australia and the USA require the food to provide at least 10% DV and Canada 5%. The highest NCC level requires each serving size of a food to deliver 20% DV in the USA, 25% DV in Australia and Canada and 30% DV in the EU27 countries (Australia, 2016; Australia, 2018; CFIA, 2022a; EU, 2022; FDA, 2022). Japan has a different method to define NCC, providing some specific values for each nutrient eligible for a NCC. For instance, the NCC "contain protein" may be applied to a food providing 7.5 grams per 100 grams while 15 grams are required for the NCC "high in protein". In total, the Japanese authorities have established NCC threshold

TABLE 29.7

Nutritional Content and Other Claims Allowed for Regular Shell eggs in the USA, Canada, Australia, the EU and Japan

Country	Nutritional Content Claim Rules	Nutritional Content Claims for Regular Egg per 100 g Edible Portion (1)	Other Claim(s) Allowed
USA	"Good source", "contains" or "provides" if 10%–19% DV "High", "rich in" or "excellent source of" if > 20% DV	"Good source" for phosphorus, zinc, vitamin D and folate (B9) "High" for Protein, iodine, selenium, vitamin A, Riboflavin (B2), Pantothenic acid (B5), Biotin (B7), Cobalamin (B12) and Choline	Various Structure/Function Claims
Canada	"Source" if 5-14% DV "Good source" if 15–24% DV "Excellent source" if >25%	"Source" for copper, iron, phosphorus, zinc, Thiamin (B1) and vitamin D "Good source" for vitamin A, Folate (B9) and vitamin E "Excellent source" for Protein, selenium, Riboflavin (B2), Pantothenic acid (B5), Cobalamin (B12) and Choline	"Complete protein" claim can be used in some cases to describe a protein containing all essential amino acids Various nutrient function claim (e.g., Proteins: helps build and repair body tissues; helps build antibodies; helps build strong muscles)
Australia	"Source" if 10%–24% DV "Good source" if >25% DV	"Source" for Protein, iron, phosphorus, potassium, vitamin A, Pantothenic acid (B5), Biotin (B7), Folate (B9), Cobalamin (B12), and vitamin E "Good source" for iodine, selenium, Riboflavin (B2), vitamin D and Choline	Several general level health claims allowed (See Schedule 4 – Nutrition, health and related claims) (Australia, 2016)
Europe	"Source" if >15% RDA "High" if >30% RDA	"Source" for phosphorus, vitamin A, pantothenic acid (B5) and Folate (B9) "High" for Protein, Riboflavin (B2), and Cobalamin (B12)	Sugars-free
Japan	"Contains" or "enhanced" – specific amount for each nutrient; "High" – specific amount for each nutrient	"Contains" for Protein, iron, zinc, pantothenic acid (B5) and Folate (B9) "High" for Biotin (B7), vitamin A, Riboflavin (B2), Cobalamin (B12) and vitamin D	Various nutrient function claims: for example, Biotin – helps to maintain skin and mucosa healthy. (Mandatory warning indication: increased intake of this product will not result in curing diseases nor promoting health.) Please comply with the advisable daily intake. Vitamin A – helps to maintain vision in the dark, and helps to maintain skin and mucosa healthy. (Mandatory warning indication: increase intake of this product will not result in curing diseases nor promoting health.) Please comply with the advisable daily intake. Women within the third months of pregnancy or women considering to be pregnant should be careful of over consumption Vitamin D – promotes to absorb calcium in gut intestine and aids in the growth of bone. (Mandatory Warning indication: increase intake of this product will not result in curing diseases nor promoting health.) Please comply with the advisable daily intake.

Source: Adapted from Guyonnet (2011).

values for proteins, dietary fibers, five minerals and 12 vitamins (MHLW, 2022). In order to calculate the different NCC permissible per 100 grams of eggs, the different national databases were used to obtain local values for egg composition: Australia (AUSNUT, 2011), Canada (CNF, 2022), France as a representative of the EU27 countries (CIQAL, 2020), Japan (MEXT, 2020) and the USA (USDA, 2022). For each country, the published recommendations for daily intake of nutrients were consulted: Australia (Australia, 2006), Canada (Health Canada, 2016), France (EU, 2008) and the USA (NIH, 2022). For each nutrient reported in these databases, the % DV was calculated and compared to the threshold values for NCC as defined by each country. For Japan, the nutritional values of shell eggs were directly compared to the concentration thresholds defined for each nutrient (MHLW, 2022). Depending on the countries, proteins must meet different set of criteria to qualify for NCC. For instance, in the EU27 countries, the NCC "high in proteins" requires that at least 20% of the energy value of one serving of food is provided by proteins while in Canada the NNC for proteins is based on the protein rating, calculated as grams of proteins intake in a reasonable amount × 3.1 (protein efficiency ratio for eggs).

Overall, the total number of NCC per 100-gram serving of eggs range from 7 (EU27 countries) to 15 (Australia and Canada). Australia offers the most opportunities to make a "source" claim for shell eggs (10) while the regulations in the USA allow the highest number of "high" NCC (9). For minerals, phosphorus, iron and zinc are most often associated with a "source" NCC while iodine and selenium met the higher NCC level in both Australia and the USA. In total, 7 different minerals (Phosphorus, iron, selenium, zinc, iodine, copper and potassium) have the potential for a NCC although the highest number of NCC for minerals in a given country is five (Australia and Canada). In the EU27 countries, phosphorus is the only mineral eligible for a "source" claim.

For vitamins, vitamin A, vitamin B2 (Riboflavin), vitamin B5 (pantothenic acid), vitamin B9 (folic acid) and vitamin B12 (Cobalamin) are most often candidates for a NCC while vitamin B1 (Thiamin) is eligible for a "source" NCC only in Canada (lower threshold for NCC). Canada and Australia allowed the most NCC for vitamins (8) while only 5 NCC can be made in the EU27 countries for vitamins in shell eggs. The most "high" NCC for vitamins (5) can be made on shell egg cartons in Japan and the USA. A NCC for choline is allowed in Australia ("good source"), Canada ("excellent source") and the USA ("high") while the nutrient is not recognized for a NCC in Europe and Japan.

The NCC allowed for proteins ranged from "excellent source" in Canada, "high" in the UE27 countries and the USA, to "source" or "contains" in Australia and Japan.

In addition to NCC, some structural or functional claims may be applied to "regular" shell eggs. These nutrient functional claims (NFC) are defined as stating some well accepted effects derived from the nutritive value of foods. In some countries, regulators have established some standard wording that must be used for each nutrient. For instance, in Canada, a food identified as a "source of proteins" can also carry the following NFC related to proteins: "helps build and repair body tissues", "helps build antibodies", or "helps build strong muscles". The NFC allowed in Canada cover also 13 vitamins, 10 minerals, DHA, carbohydrates and fat (CFIA, 2022b). In Australia, these NFC are referred to as "general level health claims" and an exhaustive list of permissible wording covers 13 minerals, 12 vitamins as well as other nutrients such choline or DHA. For choline, the allowed claim must state "contributes to normal homocysteine metabolism", "contributes to normal fat metabolism" or "contributes to the maintenance of normal liver function" (Australia, 2016). In Japan, NFC have also been defined for 12 vitamins and 5 minerals. For instance, a serving of food containing between 135 and 600 µg of vitamin A (210 µg per a 100-gram serving of eggs) may claim: "helps to maintain vision in the dark, and helps to maintain skin and mucosa healthy". However, the following mandatory warning indication must also follow the claim made for vitamin A: "increase intake of this product will not result in curing diseases nor promoting health. Please comply with the advisable daily intake. Women within the third months of pregnancy or women considering to be pregnant should be careful of over consumption" (MHLW, 2022). In many cases, these warning indications will preclude the use of such NFC on Japanese egg cartons (Guyonnet, 2011).

There are some reservations about the real impact of NCC on the potential to improve the healthiness of food choices as the food industry may be able to influence consumers with perceived benefits (Kaur et al., 2017; Iles et al., 2018). However, as a naturally nutrient-rich food, shell eggs have not carried such suspicion and NCC are often used and help consumers appreciate the nutritional value of shell eggs. In addition, the growth of the functional egg segment (see Chapter 2) has also allowed more NCC to be made on egg packaging materials and further reinforce in the mind of consumers the great nutritional value of eggs.

29.3.3 Nutritional Information on Shell Egg Packaging

The regulations for shell eggs, including labeling information, were reviewed in Chapter 18. Many countries have well defined rules related to the nutritional information provided on shell egg packaging. While this section will take the Canadian regulations as an example, there are some differences between countries and local labeling rules should be consulted. There is often a minimum of information required on egg packaging (see Chapter 18). In Canada, for instance, the minimum required information to be provided to consumes will include calories, fat content (total, saturated and trans fats), carbohydrates (total, fibers and sugars), proteins and cholesterol. In addition, the label must also provide the content of one serving of eggs (two eggs or 105 grams) in sodium, potassium, calcium and iron. At the discretion of an egg packing company, more nutritional information can be provided, as long as the information is factual (Figure 29.1). For all additional vitamins and minerals mentioned in the extended version of the Nutrition Facts panel, the actual amount in one serving size and its significance in terms of meeting the daily value must be indicated on the label. The NCC "Excellent source of protein", applicable to all regular shell eggs, is also included on this label to remind consumers of the great nutritional value for eggs.

Additional NCC can be made for functional eggs enriched in various nutrients. For instance, in Canada, the "source of" claim for omega 3 polyunsaturated fatty acids (PUFA) can be made when one serving provides at least 0.3 g of omega-3 PUFA per serving size. When this NCC is made, the nutritional facts must indicate the total of omega 3 and omega 6 PUFA provided in one serving size. While there is no nutritional daily requirement established for lutein and docosahexaenoic acid (DHA), the actual content of one serving size is allowed on the label as factual information provided to consumers (Figure 29.2).

Overall, few countries, including the USA and Canada, require the provision of nutritional information for shell eggs packaging including both the content in a specific set of nutrients as well as their significance in terms of meeting daily nutritional requirements. Most countries will require only information about the nutritional composition of one serving size while some countries still have no specific regulations to provide some mandatory nutritional information on shell egg cartons (see Chapter 18).

29.4 Key Egg Nutrients

Eggs are typically enjoyed by consumers for their taste, convenience and affordable price. The functionality of shell eggs, such as their emulsifying, gelling and foaming characteristics, is another reason for consumers to appreciate the versatility of eggs. Since consumers have typically little knowledge about nutrition and the importance of different nutrient groups, the

Nutrition Facts Valeur nutritive Per 2 large eggs (105 g) pour 2 gros œufs (105 g)	
Calories 160	% Daily Value* % valeur quotidienne*
Fat / Lipides 11 g	15 %
Saturated / saturés 3.5 g + Trans / trans 0	18 %
Carbohydrate / Glucides 1 g	
Fibre / Fibres 0 g	0 %
Sugars / Sucres 0 g	0 %
Protein / Protéines 13 g	
Cholesterol / Cholestérol 400 mg	
Sodium 130 mg	6 %
Potassium 125 mg	3 %
Calcium 50 mg	4 %
Iron / Fer 1.75 mg	10 %

*5 % or less is **a little**, 15 % or more is **a lot**
*5 % ou moins c'est **peu**, 15 % ou plus c'est **beaucoup**

FIGURE 29.1 Nutrition Facts information provided on regular shell eggs in Canada, with the mandatory short version (left) and voluntary extended version (right).

following section will provide an overview of the nutritional value of some key nutrients present in shell eggs. More details are provided in the following nine chapters dealing with Egg Nutrition and Human Health. Through this section, it will become evident why the consumption of eggs plays an important role at each stage of our lives and can make a real contribution to food security.

29.4.1 Proteins

While the food processing sector and consumers baking at home appreciate proteins for their foaming and gelling abilities, their nutritional quality is extremely important at all stages of our lives. Proteins play a critical role in the growth, maintenance and functioning of the human body. For instance, athletes are frequently consuming large quantities of animal-sourced foods, including egg white, as a way to build up muscle mass (Hoffman & Falvo, 2004; Layman & Rodriguez, 2009).

With about 12.6 grams of proteins per 100-gram serving, eggs contain typically less proteins than meat products from terrestrial or aquatic animals (in the range of 20 – 21.5 grams) with the exception of milk (Table 29.8). Conversely, eggs contain more proteins than most plant-sourced foods excepted pulses such as lentils.

But the real value of egg proteins is less about their quantity and more about their high quality.

Nutrition Facts Valeur nutritive Per 2 large eggs (105 g) pour 2 gros œufs (105 g)	
Calories 160	% Daily Value* % valeur quotidienne*
Fat / Lipides 11 g	15 %
Saturated / saturés 3.5 g + Trans / trans 0 g	18 %
Polyunsaturated / polyinsaturés 2 g	
Omega-6 / oméga-6 1.5 g	
Omega-3 / oméga-3 0.8 g	
Monounsaturated / monoinsaturés 3.5 g	
Carbohydrate / Glucides 1 g	
Fibre / Fibres 0 g	0 %
Sugars / Sucres 0 g	0 %
Protein / Protéines 13 g	
Cholesterol / Cholestérol 400 mg	
Sodium 130 mg	6 %
Potassium 125 mg	3 %
Calcium 50 mg	4 %
Iron / Fer 1.75 mg	10 %
Vitamin A / Vitamine A 200 µg	22 %
Vitamin D / Vitamine D 1.5 µg	8 %
Vitamin E / Vitamine E 10.5 mg	70 %
Folate 70 µg DFE / ÉFA	18 %
Vitamin B_{12} / Vitamine B_{12} 1.55 µg	65 %
Zinc 1.25 mg	11 %

*5 % or less is **a little**, 15 % or more is **a lot**
*5 % ou moins c'est **peu**, 15 % ou plus c'est **beaucoup**

FIGURE 29.2 Nutrition Facts information and NCC for shell eggs enriched in omega 3 polyunsaturated fatty acids, lutein, vitamin D and vitamin B12.

TABLE 29.8

Main Nutrients in Shell Eggs Compared to Various Animal-Sourced and Plant-Based Foods

Food Name	NDB Number[a]	Energy (kcal)	Proteins (g)	Total Lipid (g)	Carbohydrates (g)	Dietary Fibers (g)	Cholesterol (mg)
Shell eggs, raw	1123	143	12.6	9.51	0.72	0	411
Pork tenderloin, lean, raw	10060	109	21.0	2.17	0	0	65
Beef tenderloin, lean, raw	23337	153	21.5	7.4	0	0	61
Chicken breast, raw	5057	172	20.8	9.25	0	0	64
Milk, 3.25% fat, raw	1077	61	3.15	3.25	4.8	0	10
Salmon, pink, raw	15083	127	20.5	4.4	0	0	46
Lentils, raw	16069	352	24.6	1.06	63.4	10.7	0
Yellow corn, raw	20014	365	9.42	4.74	74.3	7.3	0
Rice, white, raw	20444	365	7.13	0.66	80	1.3	0
Carrots, raw	11124	41	0.93	0.24	9.58	2.8	0
Peas, green, raw	11304	81	5.42	0.40	14.4	5.7	0
Potato with skin, raw	11352	77	2.05	0.09	17.5	2.1	0
Cassava, raw	11134	160	1.36	0.28	38.1	1.8	0

Source: Adapted from USDA (2018).

Note: Expressed per 100 g serving size.

[a] NDB = National Database food identification number (USDA, 2018).

Amino acids, the small building blocks of peptides and proteins, are considered as dispensable amino acids and indispensable amino acids (IAA) depending on the possibility for the human body to synthesize *de novo* a particular amino acid. By definition, IAA cannot be synthesized by the human body and must be found in our diets (WHO, 2002). Over the years, various methods have been used to evaluate the quality of dietary proteins, each with some level of variability and applicability to human nutrition (FAO, 2013):

Protein efficiency ratio (PER): This ratio establishes the efficiency of a protein source to contribute to growth in an animal model study. Rats are fed a protein source and their weight gain is related to their protein intake. The method measured the retention of proteins by body tissues and a protein source with a ratio exceeding 2.7 is usually considered as excellent. However, this animal model method does not correlate well with human nutritional needs and it is widely accepted that other methods should be considered for the evaluation of protein quality in the human diet.

Biological value (BV): The quality of proteins is measured by calculating the amount of nitrogen used for the formation of tissues in comparison to the amount of nitrogen ingested from a food and excreted. Reported as a percentage, it measures how efficiently food protein, once absorbed from the intestine, can be converted into body tissues.

Net protein utilization (NPU): This method evaluates the nitrogen retention by calculating the difference between the body nitrogen content of animals fed no protein and those fed a test protein. This value divided by the amount of protein consumed is the NPU which is defined as the "percentage of the dietary protein retained". While both NPU and BV are estimating the retained nitrogen, the calculation of NPU uses the total protein intake whereas that of BV considers the amount absorbed, accounting for digestibility. Therefore, NPU will be always lower than BV.

Protein Digestibility Corrected Amino Acid Score (PDCAAS): This method is based on comparing the concentration of the first limiting indispensable amino acid in the food protein tested with the concentration of that amino acid in a reference pattern defined for preschool-age children. For the calculation of PDCAAS, the limiting amino acid score (expressed as the ratio of the first-limiting amino acid in one gram of target food protein to that in a reference protein or requirement value) is multiplied by the true protein digestibility. Since this method assesses how well dietary protein can match the demand for amino acids, and allows the prediction of dietary protein utilization, PDCAAS was considered after a joint WHO/FAO expert meeting in 1989 as the preferred method for measuring the quality of proteins in human nutrition (Schaafsma, 2000). However, it was later recognized that the PDCAAS method did not credit extra nutritional value to high quality proteins, overestimated the protein quality of foods containing antinutritional factors and did not adequately take into account the bioavailability of individual amino acids (FAO, 2013).

Digestible Indispensable Amino Acid Score (DIAAS): Recognizing that protein digestibility does not always reflect the digestibility of individual dietary indispensable amino acids, a new method was recommended using a score based on individual dietary indispensable amino acid digestibility. The DIAAS is expressed with the following formula:

DIAAS % = 100 × [(mg of digestible dietary indispensable amino acid in 1 g of the dietary protein)/(mg of the same dietary indispensable amino acid in 1 g of the reference protein)].

For this method, the digestibility is based on the true ileal digestibility (i.e., determined at the end of the small intestine) of each amino acid preferably determined in humans, or if

TABLE 29.9

Evaluation of Protein Quality from Various Animal-Sourced and Plant-Based Foods

Food	PER	BV (%)	NPU (%)	PDCAAS	DIAAS
Eggs	3.8-3.9	100	94	100	101
Milk, Cow	2.5–3.1	91	82	100	114
Beef	2.9	80	73	92–100	112
Soy	2.1–2.2	74	61	91	90–91
Wheat	0.8–1.5	64	67	25–51	40–48
Corn	1.0	62	41	47	36–42

Source: Adapted from Schaafsma (2000), Hoffman & Falvo (2004), MacRae et al. (2005), FAO (2013), Ertl et al. (2016), Mathai et al. (2017), Phillips (2017), and Herreman et al. (2020).

Note: Range values are given for parameters reported by different authors. Maximum value for PDCAAS defined as 100, as per FAO recommendations.

not possible, in growing pigs or rats. The amino acid pattern used for calculating the protein quality should be breast milk for infants (less than 6 months of age) and the recommended amino acid scoring patterns for preschool children for all other ages (FAO, 2013).

With the exception of DIAAS, proteins from shell eggs had the best results for the different methods of quality evaluation, when compared to other animal-sourced foods (milk from cows and beef) or plant-based foods (soy, wheat and corn) (Table 29.9). Although the PDCAAS for eggs and milk have been reported as 118 and 121, respectively (Schaafsma, 2000), the FAO has recommended that the maximum value of 100 be used for PDCAAS, arguing that any or all indispensable amino acids present in excess may affect the absorption of dispensable amino acids (FAO, 2013).

Regardless of the methodology used, it is evident that egg proteins are of excellent quality for human nutrition. As exemplified by the wide acceptance of DIAAS as the best methodology to evaluate proteins from a human nutrition standpoint, the composition of foods in IAA is extremely important. The indispensable amino acids are leucine, isoleucine, valine, lysine, threonine, tryptophan, methionine, phenylalanine and histidine. In addition, the availability of tyrosine and cysteine, obtained during the metabolism of phenylalanine and methionine, respectively, is also dependent of the sufficient intake of phenylalanine and methionine in our diet. For this reason, data on IAA will typically report the combined requirements for methionine + cysteine and phenylalanine + tyrosine (WHO, 2002). The composition of shell eggs in IAA is excellent with 100 grams of eggs providing on average 51.8% of the recommended daily intake (RDI) for IAA (Table 29.10). Leucine is the limiting IAA, with 1.05 g per 100 grams or providing 38.5% of the RDI. Conversely, the highest RDI for IAA (76.5%) is providing with the sulfur amino acids methionine + cysteine, with 803 mg/100 grams of eggs. Although Tryptophan is the IAA with the lowest concentration in eggs (166 mg per 100 grams), one 100-gram serving will still contribute 59.3% of its RDI.

The excellent nutritional value of shell eggs is demonstrated by the fact that a 100-gram serving of eggs meets at least 25% of the recommended daily intake (RDI) for all 9 +2 IAA (Table 29.11). While the same 100-gram serving of chicken breast meat or pink salmon also provided for at least 25% RDI for all IAA, lentils were the only animal-sourced food that met this threshold for all IAA. Yellow corn and white rice met this requirement for 6 and 3 IAA, respectively,

TABLE 29.10

Indispensable Amino Acids (IAA) Requirements Met by a 100-Gram Serving of Shell Eggs

	Daily Requirements in mg Expressed per:					
Indispensable Amino Acid (IAA)	**kg bw**	**70 kg Adult**	**g Protein Intake[a]**	**Egg Composition (mg / 100 g)**	**% RDI in Adults**	**mg IAA per g of Egg Protein**
Histidine	10	700	15.2	283	40.4%	22.8
Isoleucine	20	1,400	30.3	616	44.0%	49.7
Leucine	39	2,730	59.1	1,050	38.5%	84.7
Lysine	30	2,100	45.5	832	39.6%	67.1
Methionine + *cysteine*	15	1,050	22.7	803	76.5%	64.8
Phenylalanine + *tyrosine*	25	1,750	37.9	1,172	67.0%	94.5
Threonine	15	1,050	22.7	594	56.6%	47.9
Tryptophan	4	280	6.1	166	59.3%	13.4
Valine	26	1,820	39.4	734	40.3%	59.2
Total IAA	**184**	**12,880**	**278.8**	**6,668**	**51.8%**	**504.0**

Source: Adapted from WHO (2002), USDA (2018).

[a] Recommendations for 0.66 g of protein per kg of body weight.

TABLE 29.11

Indispensable Amino Acids (IAA) Requirements Met by a 100-Gram Serving of Various Animal-Sourced and Plant-Based Foods

Food name	NDB Number[a]	Number of IAA Provided at > 25% RDI	Lowest % RDI Provided	Highest % RDI Provided
Shell eggs, raw	1123	9	Leu (38%)	Met + Cyst (76%)
Chicken breast raw	5057	9	Val, Leu (56%)	His (89%)
Salmon, pink, raw	15083	9	Leu (57%)	Thr (102%)
Lentils, raw	16069	9	Met + Cys (51%)	Phe + Tyr (107%)
Yellow corn, raw	20014	6	Lys (13%)	Phe + Tyr (48%)
Rice, white, raw	20444	3	Lys (12%)	Phe + Tyr (35%)
Milk, 3.25% fat, raw	1077	0	Met + Cys (10%)	Phe + Tyr (18%)
Carrots, raw	11124	0	Leu (4%)	Thr (18%)
Peas, green, raw	11304	0	Met + Cys (11%)	Thr (19%)
Lettuce, green leaf	11253	0	Leu (3%)	Iso (6%)
Broccoli, raw	11090	0	Leu (5%)	Try (12%)
Potato, skin, raw	11352	0	Leu (4%)	Try (8%)
Cassava, raw	11134	0	Leu (1%)	Try (7%)

Source: Adapted from WHO (2002), USDA (2018).
[a] NDB = National Database food identification number.

while milk, carrots, lettuce, broccoli, potato and cassava did not deliver at least 25% RDI for any IAA. While leucine is the most abundant amino acid in body tissues, it was often the limiting IAA in the foods listed in Table 29.11, especially plant-based foods.

Although shell eggs contain significantly less proteins and less total IAA per 100-gram serving than chicken breast meat, pink salmon or lentils, their proteins have more mg of IAA per gram of protein. Interestingly, although both yellow corn and white rice contain much less proteins than lentils, their ratio IAA/gram protein, 0.442 and 0.425, respectively, are higher than that of lentils (0.407). Overall, the nutritional quality of eggs is exemplified by the presence of 0.504 g of IAA per gram of proteins (Table 29.12).

The importance of amino acids composition is recognized by some national regulatory agencies. In Canada, although nutrient content claims such as "source of amino acids" or "source of essential amino acids" are not permitted, other types of statements such as quantitative declarations of amino acid content of a food outside the mandatory Nutrition Facts table are allowed when meeting a set of criteria, including the protein source claim (CFIA, 2022a).

Since proteins are important for development and growth, their adequate provision at an early age is essential to avoid retarded growth or stunting in children, with both protein quality and quantity as important elements to consider. The nutritional value of eggs at an early age to improve infant nutrition was clearly demonstrated and showed some promises to alleviate global food insecurity (Iannotti et al., 2014). For instance, a randomized controlled study in Ecuador showed that when young infants, 6–9 months of age, were given one egg per day for 6 months, the prevalence of stunting and underweight were decreased significantly, by 47% and 74%, respectively, when compared to infants received a standard diet without eggs (Iannotti et al., 2017). When comparing the profile of IAA from human milk and eggs, we can see some high similarities, ranging from 78.7% for tryptophan (i.e., 1 gram of egg protein provides 78.7% of the amount of tryptophan present in 1 gram of milk protein) to 196.7% for methionine + cysteine. For histidine, lysine, phenylalanine + tyrosine, threonine

TABLE 29.12

Protein and Indispensable Amino Acids (IAA) Contents in a 100-Gram Serving of Animal-Sourced and Plant-Based Foods

	NDB Number[a]	Proteins (g)	Total IAA (g)	IAA / Proteins
Shell eggs, raw	1123	12.4	6.25	0.504
Salmon, pink, raw	15083	20.5	9.53	0.465
Chicken breast, raw	5057	20.8	9.40	0.452
Yellow corn, raw	20014	9.42	4.16	0.442
White rice, raw	20444	7.13	3.03	0.425
Lentils, raw	16069	24.6	10.0	0.407

Source: Adapted from USDA (2018).
[a] NDB = National Database food identification number.

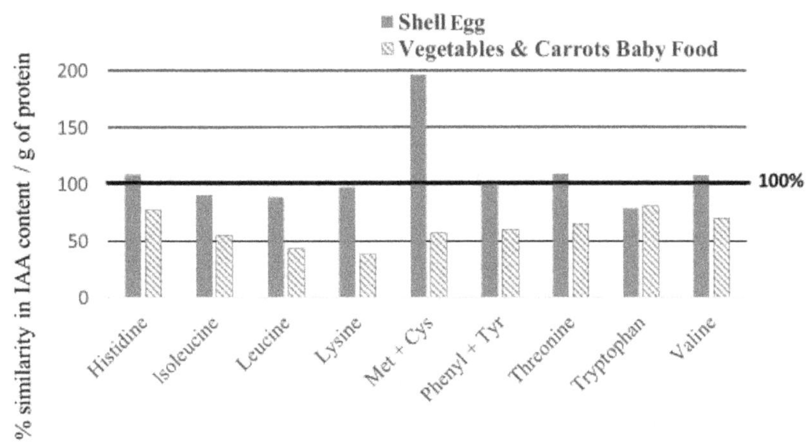

FIGURE 29.3 Similarities of shell eggs and a commercial vegetables and carrots baby food (NBD 173507) to human breast milk for the provision of essential amino acids. (Adapted from FAO, 2013; USDA, 2018.)

and valine, the similarities ranged between 97.2% and 108.9% (Figure 29.3). Conversely, a commercial baby food product for infants aged 8 months and up, made of carrots and vegetables, had lower similarities for IAA when compared to maternal milk. The highest similarity was for tryptophan (81%) and the lowest similarity for lysine (38%). Tryptophan was the only IAA for which the commercial baby food product exceeded the level provided per gram of egg protein.

29.4.2 Lipids

Lipids are predominantly present in the yolk and, to some extent, in the vitelline membrane surrounding the yolk. The content of eggs in lipids vary little between national databases, ranging from 9.83 g (per 100 g of edible portion) in France to 10.3 g in Japan (Table 29.2). In fact, it is recognized that the total lipid content of eggs does not vary much while the fatty acid composition is greatly affected by the diet provided to laying hens (Seuss-Baum & Nau, 2011; Réhault-Godbert et al., 2019). The lipid composition in regular eggs (i.e., eggs produced by hens given the standard local diet) is made up of triacylglycerols (63.1%), phospholipids (29.7%), free fatty acids (0.9%) along with free cholesterol (4.9%) and cholesterol ester (1.3%). The digestibility of both triglycerides (98%) and phospholipids (90%) is excellent, making egg yolk an excellent source (Seuss-Baum & Nau, 2011). Almost all lipids are present as lipoprotein complexes with an overall lipid to protein ration of 2. A detailed review of the lipid composition of the yolk is presented in Chapter 6. Fatty acids, the most abundant component of triaglycerols and phospholipids differ by chain length, the number and position of double bonds and the presence of isomers. Monounsaturated fatty acids represent between 36.9% and 41.8% of the total lipid fraction and are mostly made up of oleic acid (C18:1). Saturated fatty acids represent the second largest lipid fraction, between 26.9% and 32.1% of the total lipids. Palmitic acid (C16;0 and stearic acid (C18:0) are the main saturated fatty acids in shell eggs. The polyunsaturated fatty acids represent the smallest lipid fraction (between 14.5% and 18.3%) and are mainly linoleic acid (C18:2, 80%) and arachidonic acid (C20:4, 10%), both in the family of omega 6 polyunsaturated (first double bond in 6 position). The omega 3 PUFA (~ 5.6%) are mostly made up of docosahexaenoic acid (DHA, C22:6, ~3%), alpha-linolenic acid (ALA, C18:3, ~ 1.8%) and eicosapentaenoic acid (EPA, C20:5, ~ 0.5%) (Seuss-Baum & Nau, 2011). The PUFA lipid fraction is extremely interesting from a human nutrition standpoint as dietary modulations of the diets given to laying hens can greatly modify its composition. As described previously (see Section 29.2.2), the content of shell eggs in ALA and DHA can easily and consistently be increased 6-fold and 3-fold, respectively (Table 29.5). When laying hens are fed diets enriched in flaxseed or fish oil, the percentage of monounsaturated fatty acids in eggs will decreased while that of polyunsaturated fatty acids will increase (Mazalli et al., 2004). The nutritional and health benefits of omega 3 PUFA functional eggs have been demonstrated in numerous studies in healthy volunteers, patients with high cholesterol levels, during pregnancy and lactation as well as for infant nutrition (Guyonnet & Jing, 2015). A recent global survey demonstrated that the blood levels in EPA and DHA were extremely variable across regions and critically low in many countries, increasing the global risk for chronic diseases (Stark et al., 2016). Since laying hens convert more efficiently than humans short chain omega 3 fatty acids (ALA) into the long chain EPA and DHA, responsible for many of the benefits observed with omega 3 nutrition, omega 3 shell eggs are a better nutritional alternative to boost EPA and DHA blood levels than foods enriched in short chain PUFA. The contribution of the lipid fraction of chicken eggs to human nutrition will be further discussed in subsequent chapters. In other avian species (ostrich, turkey, quail, duck and goose), some significant differences have been identified in the lipid composition of their eggs, pointing to quail eggs as the best candidate to replace chicken eggs from a nutritional standpoint (Sinanoglou et al., 2011).

29.4.3 Cholesterol

The cholesterol content of shell eggs has often constituted an objection to more consumption (Guyonnet, 2012). Starting in the 1970s, medical doctors were taught that checking serum cholesterol and initiating early treatment if high was the best action to reduce the risk of Coronary Vascular Disease (CVD) (Constance, 2009). The public was advised to consume daily no more than 300 mg of cholesterol and therefore limit their consumption of eggs. However, the current knowledge about dietary cholesterol and eggs certainly brings some opportunities for a greater contribution of eggs to human nutrition.

The cholesterol content of eggs varies with avian species (Table 29.3). Eggs from chickens contain less cholesterol per 100 g or per gram of yolk compared to eggs produced by ducks, quails and other commercially produced eggs. In chickens, the cholesterol content varies also depending on breeds, nutrition and to a lesser extent housing system (Simčič et al., 2009; Naviglio et al., 2012). Some differences are reflected also by the different nutritional values for cholesterol reported by various national nutrition databases.

The role of eggs, dietary cholesterol and CVD have been frequently reviewed (Griffin, 2011; Blesso & Fernandez, 2018; Sugano & Matsuoka, 2021) and will be discussed in Chapter 30. Early studies did not take into consideration other diet components such as saturated fats and reinforced the message that "consuming dietary cholesterol increased blood cholesterol". Studies in the mid 90's or re-analysis of data then took into consideration the confounding influences of background diet. A recent meta-analysis of all papers published between 1966 and 2012 showed that consumption of eggs, up to one per day, was not associated with increased risk of CVD or stroke (Rong et al., 2013). With the difficulty to reach both physicians and patients, dispelling the cholesterol myth remained a challenge for many years (Gray & Griffin, 2009). Currently, the positive message about cholesterol and eggs has led a number of medical associations and heart foundations to remove any cautionary statement about egg consumption and risk of CVD. The USA was one of the last countries to remove such statement. In 2015, the Dietary Guidelines Advisory Committee (DGAC) concluded that the available evidence showed no appreciable relationship between the consumption of dietary cholesterol and serum cholesterol (USDA, 2015). The DGAC also stated that cholesterol was not a nutrient of concern for overconsumption in the USA. Most recent studies have demonstrated that dietary cholesterol related to egg consumption does not affect negatively serum lipids and may in fact modify the profile of lipoproteins and improve the functionality of high-density lipoprotein-cholesterol (HDL-C) (Blesso & Fernandez, 2018). This change in perspective related to dietary cholesterol and egg consumption does further enhance the value of eggs, a nutrient-dense food, for human nutrition.

29.4.4 Choline

Choline is an important nutrient that is present naturally in certain foods, including shell eggs (Zeisel et al., 2003). The recognition of the nutritional importance of choline is relatively recent, with an adequate intake of choline in adult men established in 1998 by the US Institute of Medicine (550 mg/day) and in 2016 by the EU (400 mg/day) (Zeisel & da Costa, 2009; EFSA, 2016). However, choline is still not part of the nutrients eligible for NCC in EU27 countries or Japan (EU, 2008; MHLW, 2022) and its nutritional value still relatively unrecognized by the medical profession (Zeisel & da Costa, 2009; Wallace et al., 2018; Derbyshire, 2019). Although choline can be synthesized in the liver through the methylation of phosphatidylethanolamine to phosphatidylcholine, this *de novo* synthesis is not sufficient to meet the human nutritional requirements (Zeisel & da Costa, 2009). Therefore, choline is an essential nutrient obtained from our diet. Among animal-sourced foods, shell eggs are one of the best sources of choline (335 mg/100 g), along with beef liver (333 mg/100 g), chicken liver (194 mg/100 g) while salmon (95 mg/100 g) and bacon (45 mg/100 g) contain relatively less choline. Shell eggs contain more choline than wheat germ cereals (179 mg/100 g), one of the best plant-based sources of choline, and also a favorite item for breakfast (USDA, 2018).

As a component of phosphatidylcholine, lysophosphatidylcholine and sphingomyelin, choline is an important molecule in all cell membranes while choline also plays a role in neurotransmission as a precursor of acetylcholine. Choline is now known to have many structural, biochemical, metabolic and neurological roles (see Chapter 36). With pregnancy depleting the maternal supply of choline and lactation requiring to meet a higher choline requirement for infant development, the daily adequate intake for women (425 mg in the USA and 400 mg in Europe) is increased for pregnant (450 mg in the USA and 480 mg in Europe) and lactating women (550 mg in the USA and 520 mg in Europe). However, in most cases, the mean intake falls short of these recommendations, especially in Europe when the medical profession does not routinely recommend choline supplementation (Wallace et al., 2018; Caudill et al., 2020). Since choline from shell eggs seems to be more bioavailable than from nutritional supplements (Lemos et al., 2018; Smolders et al., 2019; DiBella et al., 2020), the egg sector has recently increased its activities to raise the awareness within the health community and general population about choline and egg nutrition. While shell eggs are already one of the main dietary sources of choline in many countries, a greater consumption of eggs should further reduce the gap between adequate and mean intakes of choline and contribute further to human nutrition (Wallace & Fulgoni, 2017; Wallace et al., 2018; Wiedeman et al., 2018).

29.4.5 Vitamins and Minerals

As presented in Section 29.2.1 (see Table 29.2), the egg is rich in a wide range of vitamins and minerals. From the vitamin side, all vitamins are found in the egg with the exception of vitamin C. In addition to the fact that birds can satisfy their needs of vitamin C with a *de novo* synthesis from glucose, the absence of vitamin C may be due its lack of role during embryogenesis (Seuss-Baum & Nau, 2011; Réhault-Godbert et al., 2019). Fat-soluble vitamins (vitamin A, D, E and K) are found in the yolk while water-soluble vitamins (B vitamins) are present in both the yolk and the egg white. Vitamin B1 (Thiamin), vitamin B5 (pantothenic acid), vitamin B 6 (Pyridoxine), vitamin B7 (Biotin), vitamin B9 (Folic acid) and vitamin B12 (Cobalamin) are preferentially found in the yolk while only vitamin B3 (niacin) is found in greater amount in the egg white. As presented in Section 29.3.2 (see Table 29.7), many vitamins present in a 100-g serving of egg are eligible for a nutrient content claim, indicating their presence in a significant amount. From a human nutrition standpoint, it is evident that the yolk is an extremely rich source of vitamins and people avoiding the consumption of egg yolks will miss on this significant source of vitamins. In addition, as outlined in Section 29.2.2, the production of functional eggs has allowed to provide consumers with shell eggs further enriched in various vitamins. For instance, eggs enriched in vitamin D may assist in alleviating the potential health risks related to its current insufficient intake in various parts of the world (Cashman, 2022).

A summary of the different minerals present in eggs are presented in Table 29.2. While sodium, chlorine and potassium are preferentially found in the egg white, most other minerals will be found in greater quantity in the yolk (see Chapter 5 and Chapter 6). Like with vitamins, the content of certain minerals can easily be increased through the modulation of the diet fed to laying hens (e.g., iodine, selenium) and mineral-enriched shell eggs are widely available commercially. For instance, iodine-enriched shell eggs are extremely popular in Japan (Guyonnet, 2020a) while functional omega 3 PUFA shell eggs ae routinely enriched in selenium and vitamin E.

Since vitamins and minerals play a wide range of roles in numerous important metabolic pathways, their adequate provision is an essential part of good nutrition. Overall, shell eggs, especially with vitamins, constitute a valuable source for these micronutrients, often deficient in the diets of people living in the developing world and potentially exacerbated by the recent COVID-19 pandemic (Tulchinsky, 2010; Madzorera et al., 2021).

29.4.6 Lutein and Zeaxanthin

Lutein and its isomer, zeaxanthin, are natural compounds belonging to the class of xanthophylls present in plants and animal tissues Xanthophylls, along with carotenes, are part of the family of carotenoids, a group of about 700 fat-soluble compounds which provide color as we can see it in nature, responsible for the red in tomatoes and orange in carrots. Both lutein and zeaxanthin (L+Z), non-provitamin A carotenoids (i.e., not converted upon ingestion into vitamin A), cannot be synthesized by humans and must therefore be obtained from our diets (Thurnham, 2007; Eisenhauer et al., 2017; Demmig-Adams et al., 2020). However, there is currently no adequate intake established for lutein and zeaxanthin (Ranard et al., 2017) and these two nutrients are still not fully recognized by many national regulatory bodies nor included in their nutrient databases (Tables 29.2 and 29.7).

The recent interest in L+Z is mostly related to their presence in high concentration in the retina as part of the macular pigment and their potential role in filtering blue lights, reducing the oxidative stress on the retina and potentially preventing age-related macular degeneration (Krinsky et al., 2003; Thurnham, 2007; Eisenhauer et al., 2017; Li et al., 2020).

While lutein is the predominant carotenoid found in most foods, shell eggs, along with corn and corn products, contain about the same amount of lutein (230 μg/100 g) and zeaxanthin (229 μg/100 g) (USDA, 2018). Although shell eggs (about 332 to 459 μg/100 g) contain much less L+Z than cooked spinach (12,640 μg/100g), cooked kale (8,884 μg/100 g) or cilantro (7,703 μg/100 g), L+Z from shell eggs is highly bioavailable (Perry et al., 2009). In healthy men volunteers, the daily intake of 6 mg of lutein over a 10-day period resulted in significantly higher blood level when lutein was provided from lutein-enriched eggs when compared to spinach and lutein supplement products. The food matrix provided by the egg seems to constitute an excellent vehicle to increase the availability of lutein in tissues (Chung et al., 2004).

The enrichment of shell eggs in L+Z is well documented and demonstrated a clear dose response, reaching a plateau at about 3 mg per 100 g (Leeson & Caston, 2004). Although few L+Z enriched shell eggs are commercially available, some studies have established that their consumption will contribute to higher L+Z blood levels (Surai et al., 2000; Wenzel et al., 2006). Other studies, investigating the effects of a greater consumption of regular shell eggs (up to 12 eggs per week), also demonstrated some significant increases in L+Z blood levels (Vishwanathan et al., 2010; Aljohi et al., 2019).

The role of egg carotenoids in eye health will be reviewed in Chapter 37. Although global and national surveys on L+Z blood levels are few, it is fair to mention that most people do not meet the dietary recommendations of at least five serving of vegetable per day. Therefore, it is anticipated that the nutritional intake of L+Z will be less than the current suggested intake of 6 mg per day (Eisenhauer et al., 2017). Once a recommended dietary reference intake has been defined for L+Z, it is likely that shell eggs, both regular and functional eggs, will play a greater role in human nutrition by offering a highly bioavailable, nutritious and appetizing source of L+Z to all ages of consumers (Ranard et al., 2017; Wallace, 2018).

29.5 Shell Eggs to Contribute Further to Human Nutrition

Between 2000 and 2020, the global egg sector increased the production of shell eggs by 69.5%, leading to an increase in annual consumption from 139 eggs to 185 eggs per person. But there are huge differences among continents and countries. While Mexicans consume on a daily basis slightly more than one egg, there are also 55 countries in the world where people have access to less than 1 egg per week (see Chapter 2). Shell eggs have demonstrated some valuable nutritional properties at all stages of our lives, including some potential benefits to alleviate conditions such as obesity or sarcopenia (Gray & Griffin, 2009). It is therefore worth addressing how shell eggs may contribute further to human nutrition.

In many countries where the egg sector is well developed, the challenge is not to produce more eggs but to convince more consumers to eat more eggs. National association of egg producers are playing a key role to educate health professionals about the value of egg nutrition and raise the awareness of consumers about the nutritional and health benefits derived from eggs (Guyonnet, 2020b). For many years, the growth in egg consumption was limited by the flawed concern about the level of cholesterol in eggs (see Chapter 30). In many cases, the food-based dietary guidelines (FBDGs), issued by various governments to provide guidance about proper nutrition, were specifically restricting to 300 mg the dietary intake of cholesterol (Gray & Griffin, 2009). As new studies were published, the true relationship between dietary cholesterol intake and cardiovascular disease risk became evident and called for the removal of any restrictions about egg consumption (Kanter et al., 2012). Sine the early 2000s, a number of FBDGs have removed restrictions about the consumption of eggs and in fact have added eggs in their recommendations to the general public. In 2015, the USA were one of the last countries to update their FBDGs and remove any restrictive statement about egg consumption. In the 2020–2025 dietary guidelines for Americans, eggs are now mentioned 57 times, encouraging more consumption of eggs as a source of great proteins, vitamins, minerals and other much needed nutrients in local diets (USDA, 2020). However, these FBDGs are not available in all countries. For instance, only two out of 31 low-income countries and 12 out of 51 low to middle-income countries have FBDGs while 43 out of 80 high-income countries published their own FDDGs. And when FBDG are available, only in 30% of the cases are eggs mentioned as part of the recommendations made to consumers (Global Panel on Agriculture and Food Systems for Nutrition, 2016; Herforth et al., 2019). Globally, there is a need for further cooperation between the egg sector, the health community and governments to ensure that eggs are part of daily recommendations for good nutrition.

Food taboos are still prevalent in many countries and eggs are often cited for ill-outcomes after consumption. For instance, in the FBDG from Sierra Leone, spontaneous abortion, stillbirth, premature birth, or even "child will steal" are all taboos related to the consumption of local eggs while "disruption of the blood system" is mentioned for imported shell eggs (Sierra Leone, 2016). Overall, the persistence of these food taboos deprives pregnant women, young women and children of consuming eggs (Meyer-Rochow, 2009). Shell eggs have a great potential to alleviate food insecurity in many parts of the world and a greater emphasis should be placed by governments, the global egg sector and NGOs on increasing local production and consumption (Drewnowski, 2010; Morris et al., 2018; Walker & Baum, 2021; Guyonnet, 2022).

29.6 Conclusion

As a food enjoyed on all continents, shell eggs are able to contribute to human nutrition by providing high-quality proteins, omega 3 PUFA, choline, lutein + zeaxanthin as well a range of vitamins and minerals. In the following 9 chapters, we will see that the consumption of shell eggs may offer some opportunities to improve human nutrition and health. An interesting approach evaluated the health and financial benefits of increasing egg consumption in the USA. By assuming that the composition of shell eggs would have a positive impact on conditions such as Coronary Heart Disease, Eye health, Sarcopenia (muscle mass loss) and Fetal health (neural tube defects), it was estimated that about US$2.8 billion in health care expenditures could be saved annually with an increased egg consumption (Schmier et al., 2009). Therefore, the versatility of shell eggs, their ease of enrichment and their overall safety should make them one of the best options to improve the nutritional status in local populations, both in rich and developing countries.

REFERENCES

Aljohi, H., Dopler-Nelson, M., Cifuentes, M., & Wilson, T. A. (2019). The consumption of 12 eggs per week for 1 year does not alter fasting serum markers of cardiovascular disease in older adults with early macular degeneration. *Journal of Nutrition & Intermediary Metabolism*, *15*, 35–41. https://doi.org/10.1016/j.jnim.2018.11.004

AUSNUT (2011). Food nutrient database. https://www.foodstandards.gov.au/science/monitoringnutrients/ausnut/foodnutrient/Pages/default.aspx

Australia (2006). Nutrient reference values for Australia and New Zealand including recommended dietary intakes. https://www.nhmrc.gov.au/about-us/publications/nutrient-reference-values-australia-and-new-zealand-including-recommended-dietary-intakes#block-views-block-file-attachments-content-block-1

Australia (2016). Schedule 4 – Nutrition, health and related claims. https://www.legislation.gov.au/Details/F2017C00711

Australia (2018). Food standards of Australia and New Zealand, standard 1.2.7 nutrition, health and related claims. https://www.legislation.gov.au/Details/F2018C00942

Blesso, C., & Fernandez, M. (2018). Dietary cholesterol, serum lipids, and heart disease: Are eggs working for or against you? *Nutrients*, *10*(4), 426. https://doi.org/10.3390/nu10040426

Bourre, J.-M. (2005). Multi-enrichment of regular egg: Delivering higher nutritional levels in nutrients, notably polyunsaturated omega 3 fatty acids, vitamins, minerals and carotenoids (In French). *Médecine et Nutrition*, *41*(3), 116–134. https://doi.org/10.1051/mnut/2005413116

Cashman, K. D. (2022). 100 years of vitamin D: Global differences in vitamin D status and dietary intake: A review of the data. *Endocrine Connections*, *11*(1), e210282. https://doi.org/10.1530/EC-21-0282

Caston, L., & Leeson, S. (1990). Research note: Dietary flax and egg composition. *Poultry Science*, *69*(9), 1617–1620. https://doi.org/10.3382/ps.0691617

Caudill, M. A., Obeid, R., Derbyshire, E. J., Bernhard, W., Lapid, K., Walker, S. J., & Zeisel, S. H. (2020). Building better babies: Should choline supplementation be recommended for pregnant and lactating mothers? Literature overview and expert panel consensus. *European Gynecology and Obstetrics*, *2*(3), 149–161. https://www.egojournal.eu/wp-content/uploads/journal/2020/2.3/149/149-161.pdf

CFIA (2022a, July 31). Specific nutrient content claim requirements. https://inspection.canada.ca/food-labels/labelling/industry/nutrient-content/specific-claim-requirements/eng/1627085614476/1627085788924#a13

CFIA (2022b, July 31). Nutrient function claims. https://inspection.canada.ca/food-labels/labelling/industry/health-claims-on-food-labels/eng/1392834838383/1392834887794?chap=9#s16c9

Chung, H. Y., Rasmussen, H. M., & Johnson, E. J. (2004). Lutein bioavailability is higher from lutein-enriched eggs than from supplements and spinach in men. *The Journal of Nutrition*, *134*(8), 1887–1893. https://doi.org/10.1093/jn/134.8.1887

CIQAL (2020). Shell eggs, French food composition table. https://ciqual.anses.fr/#/aliments/22000/egg-raw

CNF (2022, July 25). Egg, chicken, whole, fresh or frozen, raw, food code 125, Canadian nutrient file. https://food-nutrition.canada.ca/cnf-fce/serving-portion.do?id=125

CODEX Alimentarius (2013). Guidelines for use of nutrition and health claims (CA/GL 23-1997). Joint FAO/WHO Codex Alimentarius Commission, Rome, Italy. https://www.fao.org/ag/humannutrition/32444-09f5545b8abe9a0c3baf01a4502ac36e4.pdf

Constance, C. (2009). The good and the bad: What researchers have learned about dietary cholesterol, lipid management and cardiovascular disease risk since the Harvard egg study. *International Journal of Clinical Practice*, *63*(Suppl. 163), 9–14. https://doi.org/10.1111/j.1742-1241.2009.02134.x

Demmig-Adams, B., López-Pozo, M., Stewart, J. J., & Adams, W. W. (2020). Zeaxanthin and lutein: Photoprotectors, anti-inflammatories, and brain food. *Molecules*, *25*(16), 3607. https://doi.org/10.3390/molecules25163607

Derbyshire, E. (2019). Could we be overlooking a potential choline crisis in the United Kingdom? *BMJ Nutrition Prevention & Health*, *2*(2), 86–89. https://doi,org/10.1136/bmjnph-2019-000037

DiBella, M., Thomas, M. S., Alyousef, H., Millar, C., Blesso, C., Malysheva, O., & Fernandez, M. L. (2020). Choline intake as supplement or as a component of eggs increases plasma choline and reduces interleukin-6 without modifying plasma cholesterol in participants with metabolic syndrome. *Nutrients*, *12*(10), 3120. https://doi.org/10.3390/nu12103120

Drewnowski, A. (2010). The nutrient rich foods index helps to identify healthy, affordable foods. *The American Journal of Clinical Nutrition*, *91*(4), 1095S–1101S. https://doi.org/10.3945/ajcn.2010.28450D

EFSA (2016). Dietary reference values for choline. *EFSA Journal*, *14*(8), 4484. https://doi.org/10.2903/j.efsa.2016.4484

EFSA (2017). Summary of dietary reference value. https://www.efsa.europa.eu/sites/default/files/assets/DRV_Summary_tables_jan_17.pdf

Eisenhauer, B., Natoli, S., Liew, G., & Flood, V. M. (2017). Lutein and zeaxanthin-food sources, bioavailability and dietary variety in age-related macular degeneration protection. *Nutrients*, *9*(2), 120. https://doi.org/10.3390/nu9020120

Ertl, P., Knaus, W., & Zollitsch, W. (2016). An approach to including protein quality when assessing the net contribution of livestock to human food supply. *Animal*, *10*(11), 1883–1889. https://doi.org/10.1017/S1751731116000902

EU (2008). Consolidated text: Council directive of 24 September 1990 on nutrition labelling for foodstuffs (90/496/EEC). https://eur-lex.europa.eu/legal-content/EN/TXT/PDF/?uri=CELEX:01990L0496-20081211&from=EN

EU (2022, July 25). Nutrition claims, European commission. https://food.ec.europa.eu/safety/labelling-and-nutrition/nutrition-and-health-claims/nutrition-claims_en#permitted-nutrition-claims

FAO (2013). Dietary protein quality evaluation in human nutrition. Report of an FAO expert consultation. FAO, food and nutrition paper 92. https://www.fao.org/ag/humannutrition/35978-02317b979a686a57aa4593304ffc17f06.pdf

FDA (2022, August 5). Label claims for conventional foods and dietary supplements. https://www.fda.gov/food/food-labeling-nutrition/label-claims-conventional-foods-and-dietary-supplements

Finisin, V., Papazyan, T., & Surai, P. (2008). Producing specialist poultry products to meet human nutrition requirements: Selenium enriched eggs. *World's Poultry Science Journal*, *64*(1), 85–98. https://doi.org/10.1017/S0043933907001742

Global Panel on Agriculture and Food Systems for Nutrition (2016). Food systems and diets: Facing the challenges of the 21st century. London, UK. http://glopan.org/sites/default/files/ForesightReport.pdf

Gray, J., & Griffin, B. (2009). Eggs and dietary cholesterol – Dispelling the myth. *Nutrition Bulletin*, *34*, 66–70. https://doi.org/10.1111/j.1467-3010.2008.01735.x

Griffin, B. A. (2011). Eggs, dietary cholesterol and disease: Facts and folklore. In Y. Nys, M. Bain, & F. Van Immerseel (Eds.), *Improving the safety and quality of eggs and egg products. Vol. 2: Egg safety and nutritional quality chemistry* (pp. 237–253). Woodhead Publishing.

Guyonnet, V., & Jing, H. (2015). The eggs and cholesterol: A liability or an opportunity? In R. R. Watson & F. De Meester (Eds.), *Handbook of cholesterol, biology, function and role in health and diseases* (pp.413–426). Wageningen Academic Publishers https://doi.org/10.3920/978-90-8686-821-6

Guyonnet, V. (2011). Strategic planning for the development of the egg nutraceutical industry. In Y. Nys, M. Bain, & F. Van Immerseel (Eds.), *Improving the safety and quality of eggs and egg products. Vol. 2: Egg safety and nutritional quality chemistry* (pp. 289–320). Woodhead Publishing.

Guyonnet, V. (2012). Eggs and egg products: Consumers' attitudes, perceptions and behaviours. World's Poultry Science Congress, Salvador, Bahia, Brazil. August 5–9, http://www.facta.org.br/wpc2012-cd/pdfs/plenary/Vincent_Guyonnet.pdf

Guyonnet, V. (2020a). Japan consuming more eggs, despite population stagnation. *Poultry International*, *59*(11), 20–21. https://www.wattagnet.com/articles/41113-japan-consuming-more-eggs-despite-population-stagnation

Guyonnet, V. (2020b). Egg associations key for sector's sustainable growth. *Poultry International*, *59*(1), 30–31. https://www.wattagnet.com/articles/39096-egg-associations-key-for-sectors-sustainable-growth

Guyonnet, V. (2022). Development of egg production and consumption in low-income countries in Africa (invited lecture). 7th Mediterranean poultry summit. Córdoba, Spain. World's Poultry Science Journal, Book of abstracts, 4.

Health Canada (2016). Nutrition labelling. Table of daily values. https://www.canada.ca/en/health-canada/services/technical-documents-labelling-requirements/table-daily-values.html

Herforth, A., Arimond, M., Álvarez-Sánchez, C., Coates, J., Christianson, K., & Muehlhoff, E. (2019). A global review of food-based dietary guidelines. *Advances in Nutrition*, *10*(4), 590–605. https://doi.org/10.1093/advances/nmy130

Herreman, L., Nommensen, P., Pennings, B., & Laus, M. C. (2020). Comprehensive overview of the quality of plant- and animal-sourced proteins based on the digestible indispensable amino acid score. *Food Science & Nutrition*, *8*(10), 5379–5391. https://doi.org/10.1002/fsn3.1809

Hoffman, J. R., & Falvo, M. J. (2004). Protein - which is best? *Journal of Sports Science & Medicine*, *3*(3), 118–130.

House, J., Braun, K., Ballance, D., O'Connor, C., & Guenter, W. (2002). The enrichment of eggs with folic acid through supplementation of the laying hen diet. *Poultry Science*, *81*(9), 1332–1337. https://doi.org/10.1093/ps/81.9.1332

Iannotti, L. L., Lutter, C. K., Bunn, D. A., & Stewart, C. P. (2014). Eggs: The uncracked potential for improving maternal and young child nutrition among the world's poor. *Nutrition Reviews*, *72*(6), 355–368. https://doi.org/10.1111/nure.12107

Iannotti, L. L., Lutter, C. K., Stewart, C. P., Gallegos Riofrío, C. A., Malo, C., Reinhart, G., & Waters, W. F. (2017). Eggs in early complementary feeding and child growth: A randomized controlled trial. *Pediatrics*, *140*(1), e20163459. https://doi.org/10.1542/peds.2016-3459

Iles, I. A., Nan, X., & Verrill, L. (2018). Nutrient content claims: How they impact perceived healthfulness of fortified snack foods and the moderating effects of nutrition facts labels. *Health Communication*, *33*(10), 1308–1316. https://doi.org/10.1080/10410236.2017.1351277

Kanter, M. M., Kris-Etherton, P. M., Fernandez, M. L., Vickers, K. C., & Katz, D. L. (2012). Exploring the factors that affect blood cholesterol and heart disease risk: Is dietary cholesterol as bad for you as history leads us to believe? *Advances in Nutrition*, *3*(5), 711–717. https://doi.org/10.3945/an.111.001321

Kaur, A., Scarborough, P., & Rayner, M. (2017). A systematic review, and meta-analyses, of the impact of health-related claims on dietary choices. *International Journal of Behavioral Nutrition and Physical Activity*, *14*(1), 93. https://doi.org/10.1186/s12966-017-0548-1

Krinsky, N. I., Landrum, J. T., & Bone, R. A. (2003). Biologic mechanisms of the protective role of lutein and zeaxanthin in the eye. *Annual Review of Nutrition*, *23*, 171–201. https://doi.org/10.1146/annurev.nutr.23.011702.073307

Layman, D. K., & Rodriguez, N. R. (2009). Egg protein as a source of power, strength, and energy. *Nutrition Today*, *44*(1), 43–48. https://doi.org/10.1097/NT.0b013e3181959cb2

Leeson, S., & Caston, L. (2003). Vitamin enrichment of eggs. *Journal of Applied Poultry Research*, *12*(1), 24–26. https://doi.org/10.1093/japr/12.1.24

Leeson, S., & Caston, L. (2004). Enrichment of eggs with lutein, *Poultry Science*, *83*(10), 1709–1712. https://doi.org/10.1093/ps/83.10.1709

Lemos, B. S., Medina-Vera, I., Malysheva, O. V., Caudill, M. A., & Fernandez, M. L. (2018). Effects of egg consumption and choline supplementation on plasma choline and trimethylamine-N-oxide in a young population. *Journal of the American College of Nutrition*, *37*(8), 716–723. https://doi.org/10.1080/07315724.2018.1466213

Leone, S. (2016). Sierra Leone food-based dietary guidelines for healthy eating. A guide for use for the general public. https://www.afro.who.int/publications/sierra-leone-food-based-dietary-guidelines-healthy-eating-2016

Lewis, J. (2019). *Codex nutrient reference values*. FAO and WHO. https://www.fao.org/3/ca6969en/CA6969EN.pdf

Li, L. H., Lee, J. C., Leung, H. H., Lam, W. C., Fu, Z., & Lo, A. (2020). Lutein supplementation for eye diseases. *Nutrients*, *12*(6), 1721. https://doi.org/10.3390/nu12061721

MacRae, J., O'Reilly, L., & Morgan, P. (2005). Desirable characteristics of animal products from a human health perspective. *Livestock Production Science*, *94*(1–2), 95–103. https://doi.org/10.1016/j.livprodsci.2004.11.030

Madzorera, I., Ismail, A., Hemler, E., Korte, M., Olufemi, A., Wang, D., & Fawzi, W. (2021). Impact of COVID-19 on nutrition, food security and dietary diversity and quality in Burkina Faso, Ethiopia and Nigeria. *Current Developments in Nutrition*, *5*(S.2), 234. https://doi.org/10.1093/cdn/nzab029_035

Mathai, J., Liu, Y., & Stein, H. (2017). Values for digestible indispensable amino acid scores (DIAAS) for some dairy and plant proteins may better describe protein quality than values calculated using the concept for protein digestibility-corrected amino acid scores (PDCAAS). *British Journal of Nutrition*, *117*(4), 490–499. https://doi.org/doi:10.1017/S0007114517000125

Mazalli, M. R., Faria, D. E., Salvador, D., & Ito, D. T. (2004). A comparison of the feeding value of different sources of fat for laying hens: 2. Lipid, cholesterol, and vitamin E profiles of egg yolk. *Journal of Applied Poultry Research*, *13*(2), 280–290. https://doi.org/10.1093/japr/13.2.280

MEXT (2020). *Standard tables of food composition in Japan, eight revised version. Ministry of education, culture, sports.* Science and Technology (MEXT). https://www.mext.go.jp/en/policy/science_technology/policy/title01/detail01/sdetail01/sdetail01/1385122.htm

Meyer-Rochow, V. B. (2009). Food taboos: Their origins and purposes. *Journal of Ethnobiology and Ethnomedicine*, *5*, 18. https://doi.org/10.1186/1746-4269-5-18

MHLW (2022, July 25). Food with health claims, food for special dietary uses, and nutrition labeling. https://www.mhlw.go.jp/english/topics/foodsafety/fhc/index.html

Morris, S. S., Beesabathuni, K., & Headey, D. (2018). An egg for everyone: Pathways to universal access to one of nature's most nutritious foods. *Maternal & Child Nutrition*, *14*(Suppl 3), e12679. https://doi.org/10.1111/mcn.12679

Naber, E. (1993). Modifying vitamin composition of eggs: A review. *Journal of Applied Poultry Research*, *2*(4), 385–393. https://doi.org/10.1093/japr/2.4.385

Naviglio, D., Gallo, M., Le Grottaglie, L., Scala, C., Ferrara, L., & Santini, A. (2012). Determination of cholesterol in Italian chicken eggs. *Food Chemistry*, *132*(2), 701–708. https://doi.org/10.1016/j.foodchem.2011.11.002

NIH (2022, 25 July 25). Nutrient recommendations and databases. https://ods.od.nih.gov/HealthInformation/nutrientrecommendations.aspx#dri

Perry, A., Rasmussen, H., & Johnson, E. J. (2009). Xanthophyll (lutein, zeaxanthin) content in fruits, vegetables and corn and egg products. *Journal of Food Composition and Analysis*, *22*(1), 9–15. https://doi.org/10.1016/j.jfca.2008.07.006

Phillips, S. M. (2017). Current concepts and unresolved questions in dietary protein requirements and supplements in adults. *Frontiers in Nutrition*, *4*(13). https://doi.org/10.3389/fnut.2017.00013

Ranard, K. M., Jeon, S., Mohn, E. S., Griffiths, J. C., Johnson, E. J., & Erdman, J. W. Jr (2017). Dietary guidance for lutein: Consideration for intake recommendations is scientifically supported. *European Journal of Nutrition*, *56*(Suppl 3), 37–42. https://doi.org/10.1007/s00394-017-1580-2

Réhault-Godbert, S., Guyot, N., & Nys, Y. (2019). The golden egg: Nutritional value, bioactivities, and emerging benefits for human health. *Nutrients*, *11*(3), 684. https://doi.org/10.3390/nu11030684

Rong, Y., Chen, L., Zhu, T., Song, Y., Yu, M., Shan, Z., Sands, A., Hu, F. B., & Liu, L. (2013). Egg consumption and risk of coronary heart disease and stroke: Dosed-response meta-analysis of prospective cohort studies. *British Medical Journal*, *346*, e8539. https://doi.org/10.1136/bmj.e8539

Schaafsma, G. (2000). The protein digestibility–corrected amino acid score. *The Journal of Nutrition, 130*(7), 1865S–1867S. https://doi.org/10.1093/jn/130.7.1865S

Schiavone, A., & Barroeta, A. (2011). Egg enrichment with vitamins and trace minerals. In Y. Nys, M. Bain, & F. Van Immerseel (Eds.), *Improving the safety and quality of eggs and egg products. Vol. 2: Egg safety and nutritional quality chemistry* (pp. 289–320). Woodhead Publishing.

Schmier, J. K., Barraj, L. M., & Tran, N. L. (2009). Single food focus dietary guidance: Lessons learned from an economic analysis of egg consumption. *Cost Effectiveness and Resource Allocation, 7*, 7. https://doi.org/10.1186/1478-7547-7-7

Seuss-Baum, I., & Nau, F. (2011). The nutritional quality of eggs. In Y. Nys, M. Bain & F. Van Immerseel (Eds.), *Improving the safety and quality of eggs and egg products. Vol. 2: Egg safety and nutritional quality chemistry* (pp. 201–236). Woodhead Publishing.

Sierra Leone (2016). Sierra Leone food-based dietary guidelines for healthy eating. A guide for use for the general public. https://www.afro.who.int/publications/sierra-leone-food-based-dietary-guidelines-healthy-eating-2016

Simčič, M., Stibilj, V., & Holcman, A. (2009). The cholesterol content of eggs produced by the Slovenian autochthonous Styrian hen. *Food Chemistry, 114*(1), 1–4. https://doi.org/10.1016/j.foodchem.2008.09.110

Sinanoglou, V., Strati, I., & Miniadis-Meimaroglou, S. (2011). Lipid, fatty acid and carotenoid content of edible egg yolks from avian species. *Food Chemistry, 124*(3), 971–977. https://doi.org/10.1016/j.foodchem.2010.07.037

Skřivan, M., Skřvanová, V., & Marounek, M. (2005). Effects of dietary zinc, iron, and copper in layer feed on distribution of these elements in eggs, liver, excreta, soil, and herbage. *Poultry Science, 84*(10), 1570–1575. https://doi.org/10.1093/ps/84.10.1570

Smolders, L., de Wit, N. J. W., Balvers, M. G. J., Obeid, R., Vissers, M. M. M., & Esser, D. (2019). Natural choline from egg yolk phospholipids is more efficiently absorbed compared with choline bitartrate; Outcomes of a randomized trial in healthy adults. *Nutrients, 11*(11), 2758. https://doi.org/10.3390/nu11112758

Stark, D. K., Van Elswyk, M. E., Higgins, E. R., Weatherford, C. H., & Salem, N. (2016). Global survey of the omega-3 fatty acids, docosahexaenoic acid and eicosapentaenoic acid in the blood stream of healthy adults. *Progress in Lipid Research, 63*, 132–152. https://doi.org/10.1016/j.plipres.2016.05.001

Sugano, M., & Matsuoka, R. (2021). Nutritional viewpoints on eggs and cholesterol. *Foods, 10*, 494. https://doi.org/10.3390/foods10030494

Surai, P. F., MacPherson, A., Speake, B. K., & Sparks, N. H. (2000). Designer egg evaluation in a controlled trial. *European Journal of Clinical Nutrition, 54*(4), 298–305. https://doi.org/10.1038/sj.ejcn.1600939

Swanson, D., Block, R., & Mousa, S. A. (2012). Omega-3 fatty acids EPA and DHA: Health benefits throughout life. *Advances in Nutrition, 3*(1), 1–7. https://doi.org/10.3945/an.111.000893

Thurnham, D. I. (2007). Macular zeaxanthins and lutein – A review of dietary sources and bioavailability and some relationships with macular pigment optical density and age-related macular disease. *Nutrition Research Reviews, 20*(2), 163–179. https://doi.org/10.1017/S0954422407842235

Tulchinsky, T. H. (2010). Micronutrient deficiency conditions: Global health issues. *Public Health Reviews, 32*, 243–255. https://doi.org/10.1007/BF03391600

UK (2021). Composition of foods integrated dataset (CoFID). McCance and Widdowson's 'composition of foods integrated dataset' on the nutrient content of the UK food supply. https://www.gov.uk/government/publications/composition-of-foods-integrated-dataset-cofid

USDA (2015). Scientific report of the 2015 dietary guidelines advisory committee. http://www.health.gov/dietaryguidelines/2015-scientific-report/

USDA (2018). USDA national nutrient database for standard reference, legacy release. https://data.nal.usda.gov/dataset/usda-national-nutrient-database-standard-reference-legacy-release/resource/c597fff5-f6ed-4db7-a473-7446c75af8c9

USDA (2020). *Dietary guidelines for Americans 2020–2025, Make Every Bite.* 9th edition. https://www.dietaryguidelines.gov/sites/default/files/2020-12/Dietary_Guidelines_for_Americans_2020-2025.pdf

USDA (2022, July 25). Food data central, eggs, Grade A, large, egg whole (NDB number 1123). https://fdc.nal.usda.gov/fdc-app.html#/food-details/748967/nutrients

Vishwanathan, R., Gendron, C. M., Goodrow-Kotyla, E. F., Wilson, T. A., & Nicolosi, R. J. (2010). Increased consumption of dietary cholesterol, lutein, and zeaxanthin as egg yolks does not decrease serum concentrations and lipoprotein distribution of other carotenoids, retinol, and tocopherols. *Nutrition Research, 30*(11), 747–755. https://doi.org/10.1016/j.nutres.2010.10.007

Walker, S., & Baum, J. I. (2021). Eggs as an affordable source of nutrients for adults and children living in food-insecure environments. *Nutrition Reviews, 80*(2), 178–186. https://doi.org/10.1093/nutrit/nuab019

Wallace, T. C. (2018). A comprehensive review of eggs, choline, and lutein on cognition across the life-span. *Journal of the American College of Nutrition, 37*(4), 269–285. https://doi.org/10.1080/07315724.2017.1423248

Wallace, T. C., Blusztajn, J. K., Caudill, M. A., Klatt, K. C., Natker, E., Zeisel, S. H., & Zelman, K. M. (2018). Choline: The under consumed and underappreciated essential nutrient. *Nutrition Today, 53*(6), 240–253. https://doi.org/10.1097/NT.0000000000000302

Wallace, T. C., & Fulgoni, V. (2017). Usual choline intakes are associated with egg and protein food consumption in the United States. *Nutrients, 9*(8), 839. https://doi.org/10.3390/nu9080839

Wenzel, A. J., Gerweck, C., Barbato, D., Nicolosi, R. J., Handelman, G. J., & Curran-Celentano, J. (2006). A 12-wk egg intervention increases serum zeaxanthin and macular pigment optical density in women. *The Journal of Nutrition, 136*(10), 2568–2573. https://doi.org/10.1093/jn/136.10.256

WHO (2002). Protein and amino acid requirement in human nutrition. WHO Technical Report Series 935. https://apps.who.int/iris/handle/10665/43411

Wiedeman, A. M., Barr, S. I., Green, T. J., Xu, Z., Innis, S. M., & Kitts, D. D. (2018). Dietary choline intake: Current state of knowledge across the life cycle. *Nutrients, 10*(10), 1513. https://doi.org/10.3390/nu10101513

Zeisel, S. H., & da Costa, K. A. (2009). Choline: An essential nutrient for public health. *Nutrition Reviews, 67*(11), 615–623. https://doi.org/10.1111/j.1753-4887.2009.00246.x

Zeisel, S. H., Mar, M. H., Howe, J. C., & Holden, J. M. (2003). Concentrations of choline-containing compounds and betaine in common foods. *The Journal of Nutrition, 133*(5), 1302–1307. https://doi.org/10.1093/jn/133.5.1302

30

Egg Consumption and Cardiovascular Disease

Tasnim F. Imran*, Trevor Toavs, Mahnoor Khalid, and Luc Djoussé

CONTENTS

30.1 Introduction

Humans have consumed eggs for thousands of years. Around the world, eggs constitute an important component of various diets with countless culinary applications. Eggs contain vitamins, minerals, and are a great source of protein (Kovacs-Nolan et al., 2005). A boiled egg contains 6.3 g of protein, 0.6 g of carbohydrate, 1.6 g of saturated fat, 2 g of monounsaturated fat, and 0.7 g of polyunsaturated fat. Eggs also contain calcium, iron, magnesium, phosphorus, and many vitamins (Kuang et al., 2018). In addition, they are an easily available and inexpensive food. However, dietary egg consumption and risk of cardiovascular disease is still debated by researchers today. Although many studies, including prospective cohort studies, case control studies, and meta-analyses have found no adverse cardiovascular effects of egg consumption (Scrafford et al., 2011; Diez-Espino et al., 2017; Guo et al., 2018; Qin et al., 2018), other studies reported adverse effects (Zhong et al., 2019; Xia et al., 2020). Subsequent studies found that moderate egg consumption (1 egg/day) was not associated with increased incidence of cardiovascular disease. (One large egg contains approximately 187 mg of cholesterol). Thus, the Scientific Dietary Advisory Committee of the American College of Cardiology/American Heart Association National Guidelines relaxed the daily cholesterol intake to 300 mg/day (Stone et al., 2014), which further propelled the debate. Similarly, the 2015–2020 Dietary Guidelines for Americans removed restriction on egg consumption (US Department of Health and Human Services & US Department of Agriculture, 2015).

In this chapter, we review and discuss current evidence of the association of egg consumption with cardiovascular disease, cardiovascular mortality, and serum lipids.

30.2 Egg Consumption and Risk of Cardiovascular Disease

30.2.1 Risk of All-Cause Mortality, Cardiovascular Disease, and Cardiovascular Mortality

In multiple large cohort studies, such as the Guangzhou Biobank Cohort Study, PURE (Prospective Urban Rural Epidemiology), ONTARGET (Ongoing Telmisartan Alone and in Combination with Ramipril Global End Point Trial), and TRANSCEND (Telmisartan Randomized Assessment Study in ACEI Intolerant Subjects with Cardiovascular Disease) cohort studies, and the China Health and Nutrition Survey (CHNS) study, moderate egg consumption was not associated with a higher risk of all-cause mortality (Xu et al., 2019; Dehghan et al., 2020; Zhuang et al., 2020). As shown in Table 30.1, there was no statistically significant association between egg consumption and all-cause mortality in these cohorts. Similarly, a single, large meta-analysis totaling over 850,000 participants from 4 studies reported that people who consumed 7 or more eggs per week were not at higher risk of mortality from all causes compared to people who consumed less than 1 egg per week (HR 1.09; 95% CI: 0.997–1.20) (Xu et al., 2019). In fact, a study performed by Zhuang et al. (2020), which included 18,914 Chinese adults ages 20 and older, found that people who consumed 1–2 eggs per day showed significantly lower rates of total mortality than those who consumed less than 1 egg per week, after multivariable adjustment (HR: 0.64; 95% CI: 0.52–0.78). Furthermore, researchers studying this Chinese population found cholesterol from eggs was protective while cholesterol from non-egg sources was associated with a higher risk of all-cause mortality (HR: 0.73 95% CI: 0.62–0.86, HR: 1.39 95% CI: 1.10–1.75, respectively).

* Corresponding author.

DOI: 10.1201/9781003254430-38

TABLE 30.1

All-Cause Mortality, Incident CVD, and CVD Mortality

Guangzhou Biobank Cohort Study	*N* = 28,024	CVD mortality and all-cause mortality	Adjusted HR for those who ate less than 1 egg/ week compared to those who ate 7 or more: CVD Mortality: 1.08 (95% CI: 0.93–1.24) All-cause mortality: 0.99 (95% CI: 0.76–1.27)	Xu et al. 2018
Updated meta-analysis	*N* = 363,565 individuals 9 studies	All-cause mortality	HR for those who ate 7+ eggs/week compared to those who ate <1 egg/week: 1.09 (95% CI: 0.97–1.2)	Xu et al. 2018
Cohort Study **Pure, Transcend, and Ontarget Cohorts**	PURE cohort *N* = 146,011 individuals 21 countries ONTARGET and TRANSCEND cohorts *N* = 31,544 individuals **Total = approx. 177,000 individuals** **50 countries** **6 continents**	Total mortality and major CVD	HR for those who consumed 7 + eggs per week compared to those who consume ≤ 1: PURE study Total mortality: 1.04 (95% CI: 0.94–1.15) Major CVD: 0.92 (95% CI: 0.83–1.01) ONTARGET/TRANSCEND Major CVD: HR: 0.97 (95% CI: 0.73–1.29) Total mortality: HR: 0.88 (95% CI: 0.62–1.24) Composite HR: 0.97 (95% CI: 0.73–1.29)	Dehghan et al. 2020
Prospective Cohort from Women's Health Initiative	*N* = 96,831 US Postmenopausal women	Incident CVD, CVD mortality, All-cause mortality	HR for those consuming ≥7 eggs/week compared to those ≤1 per week after multivariable adjustment: Incident CVD:1.14 (95% CI: 1.04–1.25) CVD mortality: 1.23 (95% CI: 1.10–1.39) All-cause mortality: 1.14 (95% CI: 1.07–1.22)	Chen et al. 2021
Dose-response meta-analysis	*N* = 1,117,033 Individuals	Incident CVD and CVD mortality	SRR (summary relative risks) for consuming 1 egg per day compared to no consumption: CVD Incidence: 0.94 (95% CI: 0.89–0.99) CVD Mortality: 0.95 (95% CI: 0.88–1.03)	Godos et al. 2021
NIH-AARP Prospective Cohort	*N* = 521,120	CVD mortality All-cause mortality	HR each additional half of an egg/day: All-cause mortality: 1.07 (95% CI: 1.06–1.08) CVD mortality: 1.07 (95% CI: 1.06–1.09)	Zhuang et al. 2021
Meta-analysis	*N* = 29,615 adults 6 prospective cohorts	Incident CVD and all-cause mortality	Adjusted HR for each additional ½ egg consumed per after multivariable adjustment: Incident CVD: 1.06 (95% CI: 1.03–1.10) Adjusted risk difference 1.11% All-cause mortality: 1.08 (95% CI: 1.04–1.11) Adjusted risk difference 1.93%	Zhong et al. 2019

(Continued)

TABLE 30.1 ***(Continued)***

All-Cause Mortality, Incident CVD, and CVD Mortality

NHS, NHS II, HPFS cohort studies with an updated meta-analysis	NHS *N* = 83,349 women NHS II *N* = 90,214 women HPFS *N* = 42,055 men Updated meta-analysis *N* = 1,720,108 individuals	Incident CVD	Pooled HR for consuming ≥1 egg/day compared to ≤1 egg/month after multivariable adjustment across cohorts: 0.93 (95% CI: 0.82–1.05) Pooled HR for consuming ≥2 eggs/day compared to ≤1 egg/month: 0.91 (95% CI: 0.72–1.15) HR for coming ≥1 egg/day compared to ≤1 egg/day in individuals with type 2 diabetes: 1.06 (95% CI: 0.81–1.39) Updated meta-analysis Pooled relative risk ratio for an increase in 1 egg/day: Total: 0.98 (95% CI: 0.93–1.03) Asia: 0.92 (95% CI: 0.85–0.99) Europe: 1.05 (95% CI: 0.92–1.19) US: 1.01 (95% CI: 0.96–1.06) Pooled relative risk comparing high vs low intake: 0.99 (95% CI: 0.93–1.06) Pooled relative risk for increase in 1 egg/day in patients with diabetes: 1.25 (95% CI: 0.99–1.59) RR for high vs low consumers with diabetes: 1.4 (95% CI: 1.00–1.97)	Drouin-Chartier et al. 2020
China Kadoorie Biobank Prospective Cohort Study (CKB)	*N* = 461,213 individuals	Incident CVD and CVD mortality	HR for egg consumers compared to non-consumers (average intake for consumers = 0.47 eggs/day or 5.32 eggs/week compared to non-consumers or rare consumers 1.03 eggs/week): 0.89 (95% CI: 0.87–0.92) CVD Mortality after multivariable adjustment: 0.82 (95% CI: 0.75–0.89)	Qin et al. 2018
Meta-analysis	*N* = 8 cohorts 348,420 participants *N* = 7,549 patients with diabetes	Incident CVD, and all-cause mortality Risk of diabetes CVD risk in patients with diabetes	Pooled HR for consuming 1 egg/day compared to never or ≤1 time/week: Incident CVD: 0.96 (95% CI: 0.88–1.05) All-cause mortality: 1.13 (95% CI: 0.95–1.33) Risk of diabetes: 1.42 (95% CI:1.09–1.86) CVD risk in patients with diabetes: 1.69 (95% CI: 1.09–2.62) Not associated with increased all-cause mortality	Shin et al. 2013
China Health and Nutrition Survey (CHNS) Prospective cohort	*N* = 18,914 Adults ≥ 20 yrs old	All-cause mortality	After multivariable adjustment, HR for consuming 1 egg/week, 3–7 eggs/week, 1–2 eggs/day, and ≥2 eggs/day compared to ≤1 egg per week: 1 egg/week: 0.69 (95% CI: 0.59–0.81) 3–7 eggs/week: 0.68 (95% CI: 0.59–0.79) 1–2 eggs/day: 0.78 (95% CI: 0.66–0.93) ≥2 eggs/day: 0.64 (95% CI: 0.52–0.78)	Zhuang et al. 2020
Meta-analysis	*N* = 23 prospective studies 1,415,839 individuals	CVD	Pooled HR for consumption of >1 egg/day compared to zero or 1 egg/day: 0.99 (95% CI: 0.93–1.06)	Krittanawong et al. 2021
Caerphilly Prospective Cohort Study (CAPS)	*N* = 1,781 men	CVD and all-cause mortality	HR for consumption of ≥5 eggs/week compared to ≤1 after multivariable adjustment: Incident CVD: 1.25 (95% CI: 0.94–1.66) All-cause mortality: 1.08 (95% CI: 0.84–1.38)	Guo et al. 2018

(Continued)

TABLE 30.1 ***(Continued)***

All-Cause Mortality, Incident CVD, and CVD Mortality

Physicians Health Study Cohort	$N = 21,327$ men	All-cause mortality and all-cause mortality in patients with diabetes	HR for consuming ≥1 eggs/day compared to ≤1/week after multivariable adjustment: Total all-cause mortality: 1.23 (95% CI: 1.11–1.36) All-cause mortality in patients with diabetes: 2.01 (95% CI: 1.26–3.20) Patients with diabetes who consume 5–6 eggs/week compared to less than 1: 2.27 (95% CI:1.28–4.03) Patients without diabetes: 1.22 (95% CI:1.09–1.35)	Djoussé & Gaziano 2008
Egg Consumption and Risk of CHD/CAD				
Dose-response Meta-analysis	$N = 11$ studies 14,370 cases	CHD	RR for comparing the highest to lowest levels of egg intake (0–75 g/day range): 0.99 (95% CI: 0.94–1.05) RR for dose response Each 50 g/day increase: 1.00 (95% CI: 0.95–1.06)	Bechthold et al. 2017
Dose-response Meta-analysis	$N = 6$ articles 9 reports	CHD / CHD in diabetic patients	Relative risk of CHD increasing consumption by 1 egg/day: 0.99 (95% CI: 0.85–1.15) Summary relative risk in patients with diabetes comparing the highest and lowest egg consumers: 1.54 (95% CI: 1.14–2.09)	Rong et al. 2013
NHS, NHS II, HPFS Cohort Studies with an Updated Meta-analysis	NHS: $N = 83,349$ women NHS II: $N = 90,214$ women HPFS: $N = 42,055$ men Updated meta-analysis $N = 1,411,261$ individuals	CHD	Pooled HR for consuming ≥1 egg/day compared to ≤1 egg/month after variable adjustment across cohorts: 0.90 (95% CI: 0.77–1.05)	Drouin-Chartier et al. 2020
Dose-response Meta-analysis	$N = 1,831,038$ individuals	CHD incidence and mortality	SRR (summary relative risks) for consuming 1 egg/day compared to no consumption: Incidence: 0.98 (95% CI: 0.89–1.08) Mortality: 0.95 (95% CI: 0.76–1.18)	Godos et al. 2020
Meta-analysis	$N = 23$ prospective studies 1,415,839 individuals	CAD	Pooled HR for consumption of >1 egg/day compared to no or 1 egg/day: 0.89 (95% CI: 0.86–0.93)	Krittanawong et al. 2021
Kuopio Ischemic Heart Disease Risk Factor Cohort Study	$N = 1,032$ men	CHD	HR for each additional egg consumed after multivariable adjustment: ApoE4 negative individuals: 1.17 (95% CI: 0.85–1.61) ApoE4 positive individuals: 0.93 (95% CI: 0.50–1.72)	Virtanen et al. 2016
Health Professions Follow-up Study and Nurses' Health Study Prospective Cohort Study	$N = 37,851$ men 80,082 women	CHD in patients with diabetes	Relative risk of consuming ≥1 eggs per day compared to ≤1 per week after multivariable adjustment: In men with diabetes: 2.02 (95% CI: 1.05–3.87) In women with diabetes: 1.49 (95% CI: 0.88–2.52)	Hu et al. 1999
Egg Consumption and Risk of Heart Failure				
Swedish Prospective Cohort (×2) **Cohort of Swedish Men** **Swedish Mammography Cohort**	$N = 37,766$ Males 72 cases $N = 32,805$ Females 37 Cases	Heart Failure	RR for consuming greater than or equal to 1 egg/day compared to 0–3/month: 1.30 (95% CI: 1.01–1.67) in Men 1.06 (95% CI: 0.74–1.52) in Women ≤6 eggs/week was not significant for men or women	Larsson et al. 2015

(Continued)

TABLE 30.1 ***(Continued)***

All-Cause Mortality, Incident CVD, and CVD Mortality

Dose-response Meta-analysis	N = 254,588 individuals	Heart Failure	Summary relative risks for consuming 1 egg per day compared to no consumption: 1.11 (95% CI: 0.99–1.25) 6 eggs/week 1.15 (95% CI: 1.02–1.30) 7 eggs/week 1.19 (95% CI: 1.04–1.36) 8 eggs/week 1.23 (95% CI:1.06–1.44) 9 eggs/week	Godos et al. 2020
Meta-analysis	N = 4 cohorts from 3 studies 106,989 individuals	Heart Failure	Pooled RR for highest vs lowest egg consumption 1.25 (95% CI: 1.12–1.39)	Khawaja et al. 2017
Atherosclerosis Risk in Communities (ARIC) Study	N = 14,153 Individuals	Heart Failure	RR per 1 serving/day 1.23 (95% CI: 1.08–1.41) after multivariable adjustment	Nettleton et al. 2008
Physicians' Health Study Prospective Cohort	N = 21,275 Individuals	Heart Failure	HR for consuming ≥7 eggs/week compared to <1 1.28 (95% CI: 1.02–1.61) 1/day 1.64 (95% CI: 1.08–2.49) 2+ /day	Djoussé & Gaziano. 2008
Dose-response Meta-analysis	N = 4 studies 5,059 cases	Heart Failure	RR for comparing the highest to lowest levels of egg intake (0–140 g/day range): 1.25 (95% CI: 1.12–1.39) RR for dose response Each 50 g/day increase:1.16 (95% CI: 1.03–1.31)	Bechthold et al. 2021
Caerphilly Prospective Cohort Study (CAPS)	N = 1,781 men	Heart Failure	HR for consumption of ≥5 eggs/week compared to ≤1 after multivariable adjustment: 0.89 (95% CI: 0.51–1.53) In people with type 2 diabetes/impaired glucose tolerance: 2.87 (95% CI: 1.13–7.27)	Guo et al. 2018
Egg Consumption and Risk of Stroke				
Swedish Prospective cohort (×2) **Cohort of Swedish men** **Swedish Mammography Cohort**	N = 37,766 M N = 32,805 F	Ischemic stroke Cases 57 M and 41 F Hemorrhagic stroke Cases = 23 M and 7 F	RR for consuming greater than or equal to 1 egg/day compared to 0–3/month: Ischemic stroke 0.87 (95% CI: 0.66–1.14) for men 1.06 (95% CI: 0.76–1.47) for women Hemorrhagic stroke 1.05 (95% CI: 0.59–1.88) for men 0.96 (95% CI: 0.44–2.12) for women Men and women: 1.03 (95% CI: 0.65–1.64)	Larsson et al. 2015
Dose-response Meta-analysis	N = 761,926 Individuals	Stroke incidence, stroke mortality, ischemic stroke and hemorrhagic stroke	SRR (summary relative risks) for consuming 1 egg/day compared to no consumption: Incidence: 0.98 (95% CI: 0.91–1.04) Mortality: 0.86 (95% CI: 0.69–1.06) Ischemic stroke: 0.98 (95% CI: 0.93–1.03) Hemorrhagic stroke: 0.98 (95% CI: 0.80–1.19)	Godos et al. 2020

(Continued)

TABLE 30.1 ***(Continued)***

All-Cause Mortality, Incident CVD, and CVD Mortality

Dose-response Meta-analysis	N = 24 prospective cohorts	Stroke	RR for highest quintile vs lowest quintile of egg consumption: 0.92 (95% CI: 0.84–1.01) Subgroup analysis: Asia: 0.83 (95% CI: 0.73–0.94) Europe: 1.02 (95% CI: 0.91–1.16) North America: 0.95 (95% CI: 0.77–1.16)	Tang et al. 2020
Dose-response Meta-analysis	N = 10 studies 12,735 cases	Stroke	RR for comparing the highest to lowest levels of egg intake (0–75 g/day range): 0.99 (95% CI: 0.93–1.05) RR for dose response Each 50 g/day increase 0.99 (95% CI: 0.93–1.05)	Bechthold et al. 2021
EPIC Prospective Cohort	N = 418,329 men and women from 9 European countries	Ischemic and Hemorrhagic stroke	HR per 20 g/day of egg consumed after multivariable adjustment Ischemic: 1.05 (95% CI: 0.97–1.14) Hemorrhagic: 1.25 (95% CI: 1.09–1.42)	Tong et al. 2020
Dose-response Meta-analysis	N = 6 articles with 6 reports 4 articles for stroke subgroup 3 articles for stoke in patients with diabetes	Stroke Hemorrhagic and Ischemic Stroke	Relative risk of consuming an additional egg/day: 0.91 (95% CI: 0.81–1.02) Hemorrhagic: 0.75 (95% CI: 0.57–0.99) Ischemic: 0.91 (95% CI: 0.82–1.01) Diabetic: 0.80 (95% CI: 0.29–2.15)	Rong et al. 2013
NHS, NHS II, HPFS Cohort Studies with an Updated Meta-analysis	NHS: N = 83,349 women NHS II: N = 90,214 women HPFS: N = 42,055 men	Stroke	Pooled HR for consuming ≥1 egg/day compared to ≤1 egg/month after variable adjustment across cohorts: 0.99 (95% CI: 0.81–1.22)	Drouin-Chartier et al. 2020
China Kadoorie Biobank Prospective Cohort Study (CKB)	N = 461,213 individuals	Stroke, stroke mortality, hemorrhagic stroke, and ischemic stroke	HR for egg consumers compared to non-consumers (average intake for consumers = 5.32 eggs/week compared to non-consumers/rare consumers 1.03 eggs/week): After multivariable adjustment: Hemorrhagic stroke: 0.74 (95% CI: 0.67–0.82) Hemorrhagic stroke mortality: 0.72 (95% CI: 0.62–0.84) Ischemic stroke: 0.90 (95% CI: 0.85–0.95)	Qin et al. 2018
Meta-analysis	N = 5 cohorts 241,900 individuals	Stroke and stroke mortality	Pooled HR for consuming 1 egg/day compared to never or ≤1 time/week: 0.93 (95% CI: 0.81–1.07) Stroke mortality: 0.92 (95% CI: 0.56–1.50)	Shin et al. 2013
Kuopio Ischemic Heart Disease Risk Factor Study (KIHD) Cohort	N = 2,682 men from eastern Finland	Total stroke, ischemic stroke and hemorrhagic stroke	HR for each additional egg consumed/day: Total stroke: 0.89 (95% CI: 0.64–1.22) Ischemic stroke: 0.90 (95% CI: 0.62–1.29) Hemorrhagic stroke: 0.73 (95% CI: 0.37–1.44)	Abdollahi et al. 2019

(Continued)

TABLE 30.1 *(Continued)*

All-Cause Mortality, Incident CVD, and CVD Mortality

Guangzhou Biobank Cohort Study with Updated Meta-analysis	N = 28,024 individuals Meta-analysis N = 363,565 individuals 9 studies	Stroke	Adjusted HR for those who ate < 1 egg/week compared to those who ate 7 or more: 0.88 (95% CI: 0.57–1.35) Meta-analysis HR for those who ate 7+ eggs/week compared to those who ate <1 egg/week: 0.91 (95% CI: 0.85–0.98)	Xu et al. 2018
Prospective Cohort from Women's Health Initiative	N = 96,831 US postmenopausal women	Ischemic stroke and hemorrhagic stroke	HR for those consuming ≥7 eggs/week compared to those ≤1 per week after multivariable adjustment: Ischemic stroke: 1.40 (95% CI: 1.18–1.66) Hemorrhagic stroke: 0.98 (95% CI: 0.66–1.46)	Chen et al. 2021
Caerphilly Prospective Cohort Study (CAPS)	N = 1,781 men	Stroke	HR for consumption of ≥5 eggs/week compared to ≤1 after multivariable adjustment: 1.60 (95% CI: 1.00–2.57)	Guo et al. 2018
Health Professions Follow-up Study and Nurses' Health Study Prospective Cohort Study	N = 37,851 men 80,082 women	Total stroke, ischemic stroke and hemorrhagic stroke	Relative risk of consuming ≥1 eggs/day compared to ≤1/week after multivariable adjustment: In men: 1.07 (95% CI: 0.66–1.75) Ischemic stroke: 0.93 (95% CI: 0.46–1.87) Hemorrhagic stroke consuming 5–6 eggs per day: 0.67 (95% CI: 0.22–2.07) In women: 0.89 (95% CI: 0.60–1.31) Ischemic stroke: 0.81 (95% CI: 0.46–1.42) Hemorrhagic stroke: 1.07 (95% CI: 0.56–2.03)	Hu et al. 1999
Physicians Health Study Cohort	N = 21,327 men	Stroke, ischemic stroke and hemorrhagic stroke	HR for consuming ≥1 eggs/day compared to ≤1/week after multivariable adjustment: Total stroke: 0.99 (95% CI: 0.80–1.23) Ischemic stroke: 0.99 (95% CI: 0.78–1.26) Hemorrhagic stroke: 1.07 (95% CI: 0.65–1.78)	Djoussé & Gaziano 2008

Additional studies also found no significant association between egg consumption and cardiovascular disease or death from cardiovascular events (Shin et al., 2013; Xu et al., 2019; Dehghan et al., 2020; Godos et al., 2021). A recent large meta-analysis conducted by Chartier et al. containing 1,720,108 individuals, found that increasing egg consumption by 1 egg per day did not lead to an increased risk of developing cardiovascular disease (pooled RR 0.98 95% CI: 0.93–1.03). This meta-analysis included cohorts from the Nurse's Health Study (NHS), NHS II, and the Health Professionals' Follow-Up Study (HPFS) and noted, after multivariable adjustment, no significant relationship between consuming 1 or more eggs per day compared to consuming less than 1 per month (pooled RR: 0.93 95% CI: 0.82–1.05). Moreover, when stratified by geographic region, there was no significant relationship between egg consumption and risk of CVD in US or European populations (pooled RRs for consuming 1 egg per day versus less than 1 egg per month: 1.01 (95% CI: 0.96–1.06 $I^2 = 30.8\%$) and 1.05 (95% CI: 0.92–1.19 $I^2 = 64.7\%$), respectively). However, in East Asian populations, egg consumption was associated with a lower risk of CVD (pooled RR: 0.92 (95% CI: 0.85–0.99 $I^2 = 44.8\%$) (Drouin-Chartier et al., 2020).

While this may be reassuring evidence for those who are avid egg consumers, some contradictory evidence also exists. The NIH-AARP cohort study, which consisted of 521,120 participants, found that people who consumed each additional half of a whole egg per day were at a 7% higher risk of both all-cause mortality and death from cardiovascular disease (HR: 1.07 95% CI: 1.06–1.09). Researchers from the NIH-AARP study also performed a hypothetical substitution analysis and found, when each half of a whole egg consumed was replaced with egg whites, there was a hypothetical 6% decrease in all-cause mortality and a 3% reduction in mortality from cardiovascular disease [7–10]. From the Physicians' Health Study cohort, male physicians were at an increased risk of mortality if they consumed seven or more eggs per week compared to those consuming 6 or less, after multivariable adjustment (HR: 1.05 95% CI: 0.93–1.19, HR: 1.23 95% CI: 1.11–1.36 for consumption of 5–6 and 7 or more eggs per week, respectively. Similarly, a study conducted by Zhong et al. involving 29,615 US adults found that people who consume an additional one half of an egg per day have a 6% higher risk of experiencing an adverse cardiovascular event and are at an 8% higher risk of death from all causes (HR 1.06 95% CI: 1.03–1.10 and 1.08 95% CI: 1.04–1.11, respectively). The associations of egg consumption with CVD, CVD mortality, and all-cause mortality were not significant after adjusting for dietary cholesterol, suggesting that dietary cholesterol was linked to these negative health outcomes. In addition, when examining incident CVD in subgroup populations, Zhong et al. found a significant relationship between each additional one-half egg consumed per day and adverse cardiovascular events in women but not in men (HR: 1.13 95% CI: 1.07–1.20) and only in participants over the age of 65 (HR: 1.07 95% CI: 1.03–1.11).

Notably, the risk of egg consumption and CVD, death from CVD, and all-cause mortality may vary in high-risk populations. One study involving postmenopausal women who consumed 7 or more eggs per week reported a 14% greater risk of all-cause mortality compared to women who consumed less than 1 egg per week (Chen et al., 2021). This study also noted a statistically significant increase in risk of incident CVD as well as mortality from CVD in postmenopausal women. In the Physicians' Health Study, male physicians with diabetes who consumed either 5–6 eggs per week or 7 or more eggs per week were at an increased risk of all-cause mortality compared to those who consumed less than 1 egg per week, (HR: 2.27 95% CI: 1.28–4.03, HR: 2.01 95% CI:1.26–3.20, respectively). Moreover, in a meta-analysis conducted by Shin et al., researchers found that people with diabetes who consumed more than 1 egg per day are more likely to develop CVD than those with diabetes who consume less than 1 egg per week (HR: 1.69 95% CI: 1.09–2.62). This evidence suggests that postmenopausal women and individuals who have diabetes may benefit by substituting whole eggs for other options such as egg whites; which is also consistent with researchers' recommendations for the general population from the NIH-AARP cohort study.

Evidence also suggests that patients with diabetes who consume more than 1 egg per day are at a greater risk for developing coronary heart disease (CHD). Compared to diabetic males who consume less than 1 egg per week, data from HPFS shows that male patients with diabetes who consume 1 egg per day are twice as likely to develop CHD (RR: 2.02 95% CI: 1.05–3.87 *P* for trend = 0.04). Contrary to this finding, data from the NHS suggest that diabetic women who consume similar amounts of eggs are not at a greater risk for developing CHD (RR: 1.49 95% CI: 0.88–2.52) (Hu et al., 1999).

While there is some evidence to suggest that egg consumption is linked to increased rates of all-cause mortality, incident CVD, and death from CVD, when reported, the evidence shows only a very small increase in absolute risk and at higher levels of egg consumption (Zhong et al., 2019). Overall, consumption of 6 or fewer eggs per week, as part of a healthy diet, is not likely to be associated with adverse health effects and in fact may be beneficial. There may be some health benefits to substituting whole eggs for egg whites when possible; especially for post-menopausal women or patients with diabetes. All together, these studies, which contain participants from more than 50 countries spanning 6 continents, provide supportive evidence that, in healthy individuals, moderate egg consumption can be included as part of a healthy diet (Shin et al., 2013; Xu et al., 2019; Dehghan et al., 2020; Drouin-Chartier et al., 2020; Zhuang et al., 2020; Godos et al., 2021). Interestingly, data from studies linking egg consumption to higher risk of CVD, death from CVD, and death from all-cause mortality come from cohorts in the United States. Alternatively, in studies including populations from various countries from around the world, there was no association between egg consumption and adverse health outcomes. There was even a benefit with egg consumption in East Asian populations. We stipulate that this noted benefit may be due to other healthy lifestyle factors/behaviors or food items with cardiovascular benefits in the East Asian populations.

30.2.2 Effect of Egg Consumption on Serum Lipids

On average, 1 large egg contains 200 mg of cholesterol, according to the Food and Nutrition database from US Department of Agriculture. The effect of egg consumption on blood lipid levels has been a topic of interest for many years. Hyperlipidemia, particularly higher LDL-c concentration is known to increase the risk of cardiovascular disease. The global prevalence of hyperlipidemia (elevated total cholesterol) is 39% (World Health Organization, 2014). Although cholesterol intake is associated with higher LDL-c levels to a degree, the relationship is more complex. Most of the cholesterol produced in the human body is endogenous via the process of biosynthesis (Blesso & Fernandez, 2018). In light of this controversy, several studies have examined the effects of egg consumption on serum cholesterol concentration (Table 30.2).

A meta-analysis including 17 randomized controlled trials in a healthy population showed a higher LDL-c/HDL-c ratio in individuals with higher egg consumption as compared to the control group (MD = 0.14, $p = 0.001$, $I^2 = 25\%$). The higher egg consumption group was defined as 3 experimental subgroups consuming 1, 2 and >3 eggs per day respectively, showing higher LDL cholesterol than the control group (MD = 8.14, $p < 0.0001$, $I^2 = 18\%$) (Li et al., 2020). Another meta-analysis from 2020, which included 66 randomized controlled trials with 3,185 participants, demonstrated a significant increase in total cholesterol, LDL cholesterol and TC/LDL-C ratio with egg consumption; there was no effect on ratio of LDL-C/HDL-C (Khalighi Sikaroudi et al., 2020). Similar results were seen in another study, where egg consumption increased total cholesterol by 5.60 mg/dL (95% CI: 3.11, 8.09; P < 0.0001), LDL cholesterol by 5.55 mg/dL (95% CI: 3.14, 7.69; P < 0.0001) and HDL cholesterol by 2.13 mg/dL (95% CI: 1.10, 3.16; P < 0.0001) compared to the control group; however, there was no effect on the ratios of LDL-C/HDL-C and TC/HDL-C (Rouhani et al., 2018). A meta-analysis showed that intake of >4 whole eggs/week in comparison to equal amounts of egg substitutes increased serum total cholesterol (WMD: 0.198 mmol/L; 95% CIs: 0.056, 0.339), LDL cholesterol (WMD: 0.171 mmol/L; 95% CIs: 0.028, 0.315), as well as HDL cholesterol (WMD: 0.068 mmol/L; 95% Cis: 0.006, 0.130) (Wang et al., 2019).

Certain individuals can be classified as 'hyper-responders' to dietary cholesterol intake, based on genetic variations. These individuals showed an increase in LDL-C concentrations in response to egg intake, as opposed to the other two-third of individuals who showed no significant change especially during weight loss conditions. This study also showed an increase in HDL-C with maintenance of LDL-C/HDL-C ratio, particularly when combined with either weight maintenance or weight loss (Blesso & Fernandez, 2018). Another review that combined data from 17 cohorts with 361,923 subjects and 19 trials with 632 subjects showed heterogenous results; dietary cholesterol showed statistically significant increase in both serum total cholesterol (net change: 11.2 mg/dL; 95% CI: 6.4, 15.9) and LDL-C (net change: 6.7 mg/dL; 95% CI: 1.7, 11.7 mg/dL) but increases in LDL cholesterol were no longer statistically significant when intervention doses exceeded 900 mg/d (Berger et al., 2015).

While most studies showed a strong association between egg consumption and increased blood cholesterol levels, other studies demonstrated neutral effects. A recent study published in February 2022, involving 728 participants, showed no statistical significance between egg consumption and lipid levels in participants with chronic metabolic diseases; in fact, it showed a better lipid profile with egg consumption in participants without any chronic conditions, with lower LDL-C levels (coefficient −7.01; 95% CI −13.39, −0.62) and a lower LDL-C/HDL-C ratio (coefficient −0.24, 95% CI −0.41, −0.06) (Mesas et al., 2022). A large study that combined results of PURE study, TRANSCEND and ONTARGET, showed no statistically significant relation between egg intake and blood lipid levels. The PURE study included 146,011 from 21 countries and showed that higher intake of egg (>7 egg/wk as opposed to <1 egg/wk) was not associated with higher lipid levels (HR: 0.96; 95% CI: 0.89, 1.04; P-trend = 0.74). The TRANSCEND/ONTARGET studies including 31,544 individuals showed similar outcomes (HR 0.97; 95% CI 0.76, 1.25; P-trend = 0.09) (Dehghan et al., 2020). In light of this emerging data, several dietary guidelines were revised; Dietary Guidelines for Americans 2020–2025 and Dietary Reference Intakes for Japanese, 2020, removed the cholesterol restriction from the diet. Another randomized controlled trial done in December 2014, showed that intake of dietary cholesterol only has a modest effect on LDL-C levels (Griffin & Lichtenstein, 2013). Some studies have found that egg consumption was associated with modestly increased HDL levels, which may be a protective factor [21,23].

The effect of egg consumption on the lipid profile among people with diabetes mellitus has not been studied extensively. One study evaluating the impact of egg consumption in people with diabetes demonstrated that high egg intake (>3–4/day) increased LDL-C in Type 1 Diabetes, but not Type 2 (Griffin, 2016). A randomized controlled trial including 140 participants [the Diabetes and Egg (DIABEGG) study], found that high egg intake (2 eggs/day for 6 d/wk) did not adversely affect the lipid profile in those with Type 2 diabetes over a 3 month follow up period. The mean difference between high and low egg consumption group (2 eggs/wk) was 0.02 mmol/L (−0.03, 0.08 mmol/L; P = 0.38), concluding that high-egg diet can be safely included as part of dietary regimen in people with diabetes (Fuller et al., 2015). A meta-analysis in 2012, including 22 cohorts with participants ranging from 1600 to 90,735, reviewed the risk of CVD in those with diabetes. Although there was no significant association between egg consumption and CVD risk in the general population [HR (95% CI): 0.96 (0.88, 1.05)], intake of >1 egg/day was associated with a 42% higher risk of CVD in people with diabetes [HR: 1.42 (1.09, 1.86)] (Shin et al., 2013).

The effect of egg intake on blood cholesterol level seems to have a neutral impact overall, with greater egg consumption associated with a modest increase in LDL. Studies over a large sample size show neutral effect of egg consumption on serum lipids. However, there seems to be an increase in LDL cholesterol with egg consumption among people with hyperlipidemia or diabetes. While eggs should be considered a part of a regular dietary pattern in healthy adults, greater egg consumption should be avoided, especially in people with dyslipidemia or diabetes.

30.2.3 Egg Consumption and Risk of Stroke or Transient Ischemic Attack

The effects of eggs on incidence of total stroke and stroke subtypes has been previously studied in large cohort studies

TABLE 30.2
Egg Consumption and Serum Lipids

Study Type	LDL-C	HDL-C	LDL HDL Ratio	Author
Systematic review and meta-analysis of Randomized controlled trials	Increased (p = 0.001) (MD = 8.14; 95% CI: 4.46 to 11.82)	NA	Increased (p = 0.0001) (MD = 1.27; 95% CI: –0.28 to 2.83)	Li et al. (2020)
Systematic review and meta-analysis of randomized clinical trials (*n* = 3185)	Increased LDL-c WMD: (7.389 mg/dL; 95% CI: 5.824, 8.955)	Increased (WMD: +1.411 mg/dL; 95% CI: 0.768, 2.054;	No change (WMD: 0.083 mg/dL; 95% CI: –0.011, 0.178)	Sikaroudi et al. (2020)
Systematic review and meta-analysis of randomized clinical trials	Increased (5.55 mg/dL 3.14 – 7.69; P < 0.0001)	Increased (2.13 mg/dL, 1.10– 3.16; P < 0.0001)	No change	Rouhani et al. (2018)
Systematic review and meta-analysis of randomized controlled trials	Increased (WMD: 0.171 mmol/L; 95% CIs: 0.028, 0.315)	Increased (WMD: 0.068 mmol/L; 95% CIs: 0.006, 0.130)	NA	Wang et al. (2019)
Meta-analysis	NA	Increased (0.005, 0.010 mmol/L; 0.2, 0.4 mg/dL)	NA	Weggemans et al. (2001)
Cohort study	Increased in hyper-responders (6.7 mg/dL net change)	Increased (3.2 mg/dL net change)	Increased in hyper-responders No change in hypo-responders or when combined with weight loss/weight maintenance	Blesso et al. (2018)
Randomized controlled trial	No change in people with chronic conditions Decreased in people with no chronic conditions (–7.01; 95% C1 –13.39, –0.62)	NA	No change in people with chronic conditions Decreased in people with no chronic conditions (–0.24, 95% CI –0.41, –0.06)	Mesas et al. (2022)
PURE (*N* = 146,011) TRANSCEND/ ONTARGET (*N* = 31,544)	No change (HR: 0.96; 95% CI: 0.89, 1.04; P-trend = 0.74) (HR 0.97; 95% CI 0.76, 1.25; P-trend = 0.09)	No change	-	Dehghan et al. (2020)
DIADEGG study, randomized controlled trial	No change in Type 2 Diabetes	-	-	Fuller et al. (2015)
Randomized controlled trial	No change (MD (95% CI) –0.02 (–0.22, 0.17) [0.81])	No change (0.024 (–0.030, 0.077) [0.38])	No change	Fuller et al. (2015)

(Rong et al., 2013; Shin et al., 2013; Bechthold et al., 2019; Xu et al., 2019; Drouin-Chartier et al., 2020; Godos et al., 2021; Tang et al., 2020; Tong et al., 2020). In a meta-analysis involving 761,926 individuals, Godos et al. found no increased risk of stroke incidence, mortality from stroke, or stroke subtype in individuals who consume 7 eggs/week compared to those who do not consume eggs (incidence 0.98 (95% CI: 0.91–1.04), mortality: 0.86 (95% CI: 0.69–1.06), ischemic stroke: 0.98 (95% CI: 0.93–1.03), hemorrhagic stroke: 0.98 (95% CI: 0.80–1.19)). Similarly, Shin et al. found no increased risk of stroke or stroke mortality in individuals who consumed 1 egg per day compared to those who ate less than 1 egg or no eggs per week [0.93 (95% CI: 0.81–1.07) or 0.92 (95% CI: 0.56–1.50)], respectively). Finally, in a very large, updated meta-analysis with 1,720,108 people, consumption of 1 or more eggs per day was not associated with an increased risk of stroke compared to intake of less than 1 egg per month (pooled HR: 0.99 (95% CI: 0.81–1.22) (Drouin-Chartier et al., 2020). In the Physicians' Health Study (21,327 male physicians), consumption of 7 or more eggs per week was not associated with risk of total stroke (HR 0.99 (95% CI: 0.80–1.23), ischemic stroke (HR: 0.99 (95% CI: 0.78–1.26), or hemorrhagic stroke (HR: 1.07 (95% CI: 0.65–1.78) compared to intake of less than 1 per week (Djousse & Gaziano, 2008b). When examining stroke risk between men and women in the Health Professions Follow-up Study and Nurses' Health Study respectively, Hu et al. found no significant relationship between egg consumption and stroke sub-type in either men or women (Table 30.1) (Hu et al., 1999).

In a meta-analysis involving 363,565 individuals from 9 studies, consumption of 7 or more eggs per week was not associated with the risk of stroke compared to intake of less than 1 egg per week (HR: 0.88 (95% CI: 0.57–1.35) (Xu et al., 2019). After multivariable adjustment, a meta-analysis involving 418,329 men and women from 9 European countries found that consuming an additional 20 grams of eggs per day was not associated with the risk of ischemic stroke (HR: 1.05 (95% CI: 0.97–1.14); however, there was a 25% higher risk of hemorrhagic stroke [HR: 1.25 (95% CI:1.09–1.42)] (Tong et al., 2020). In 2 Swedish prospective cohorts, the Cohort of Swedish Men and the Swedish Mammography Cohort, neither men or nor women were at a greater risk of hemorrhagic or ischemic stroke if they consumed 1 or more eggs per day compared to if they consumed 0–3 per month (Larsson et al., 2015). Tang et al. also reported no association between egg consumption and risk for stroke in a meta-analysis stratified for North American and European cohorts [RR: 0.95 (95% CI: 0.77–1.16) and 1.02 (95% CI: 0.91–1.16), respectively]. However, the incidence of stroke was significantly lower with egg consumption in East Asian cohorts (RR: 0.83 (95% CI: 0.73–0.94). The China Kadoorie Biobank Prospective Cohort Study (CKB), consisting of 461,213 individuals, found that people who consumed eggs daily were less likely to develop hemorrhagic stroke (HR 0.74 (95% CI: 0.67–0.82)) and ischemic stroke (HR: 0.90 (95% CI: 0.85–0.95)), and were less likely to experience mortality from hemorrhagic stroke (HR: 0.72 (95% CI: 0.62–0.84)) (Qin et al., 2018). On the other hand, Xu et al. found no relationship between egg consumption and stroke risk in an East Asian population in the Guangzhou Biobank Cohort Study (Adjusted HR: 0.88 (95% CI: 0.57–1.35) comparing consumption of 7+ eggs/week to less than 1 per egg/week) (Xu et al., 2019).

Patients with type 2 diabetes/impaired glucose tolerance (IGT) who consume large amounts of eggs may also be particularly susceptible to stroke (Guo et al., 2018). The Caerphilly prospective cohort study (CACS) found that patients with diabetes or IGT who consumed 5 or more eggs per week were at a significantly greater risk of stroke compared to patients with diabetes who consumed 0–1 eggs per week (HR: 2.87 95% CI: 1.13–7.27) (Guo et al., 2018). Additionally, postmenopausal women who consumed 7 or more eggs per week were at an increased risk for ischemic stroke (HR: 1.40 (95% CI: 1.18–1.66)) but not hemorrhagic stroke (HR: 0.98 (95% CI: 0.66–1.46)) (Chen et al., 2021). However, others have found that egg consumption may not be associated with stroke risk in patients with diabetes. In a dose-response meta-analysis conducted by Rong et al., the summary relative risk for patients with diabetes in the highest egg consuming group compared to the lowest was non-significant [SRR: 0.80 (95% CI: 0.29–2.15)] (Rong et al., 2013).

Taken together, current evidence suggests that egg consumption is not associated with the risk for total, ischemic, and hemorrhagic stroke in general. However, there is evidence in support of a lower risk of stroke with daily egg consumption in East Asian. Furthermore, egg consumption appears to confer a higher risk of stroke among post-menopausal women and individuals with diabetes for whom it will be advisable to limit egg consumption.

30.2.4 Risk of Heart Failure

Current studies demonstrate less heterogeneity in the effect sizes of egg consumption with risk of developing congestive heart failure (CHF)/heart failure (HF) contrary to CVD. In a large meta-analysis published by Godos et al. consisting of 6 data sets drawn from 4 studies, including 7,536 cases of HF, consumption of 1 egg per day was associated with an increased risk of developing heart failure (1.15 95% HR: 1.02–1.30) (Godos et al., 2021). However, when stratified by sex, women who consumed 4 eggs per week were less likely to develop heart failure compared to women who did not consume eggs (0.84 95% CI: 0.71–1.00). On the other hand, there was no significant association between egg consumption and CHF in women who consumed 1 egg per day compared to those who did not consume eggs (0.94 95% CI: 0.75–1.18) (Godos et al., 2021). Interestingly, Godos et al. found a significant increase in HF risk in men who consume on average 1 egg per day compared to none at all (1.21 95% CI: 1.01–1.45), while finding no increased risk in men who consumed only 4 eggs per week (1.05 95% CI: 0.94–1.18). Similarly, another meta-analysis including 105,999 people from 4 prospective cohorts also found a significant association between daily egg consumption and an increased risk of developing HF (pooled relative risk: 1.25 (95% CI: 1.12–1.39)) (Khawaja et al., 2017). Consistent with the results from this recent meta-analysis, a study involving 2 Swedish cohorts, the Cohort of Swedish Men and the Swedish Mammography Cohort totaling 37,766 men ages 45–79 and 32,805 women ages 49–83 respectively, found a significant association between consuming 7 eggs per week and an increased rate of CHF in men, but not in women (RR: 1.3 95% CI: 1.01–1.67, RR: 1.06 95% CI: 0.74–1.52, respectively: hazard ratios calculated after multivariable adjustment). Furthermore, the risk of HF was higher in men who consumed 2 or more eggs per day compared to those who almost never consumed eggs (RR: 1.99 (95% CI: 1.12–3.53); in contrast, egg consumption was not associated with HF risk in women (RR: 0.95 (95% CI: 0.38–2.35). After multivariable adjustment, the Atherosclerotic Risk in Communities (ARIC) cohort showed that people who consume 1 serving of egg per day are 1.23 times more likely to develop HF compared to those who consume eggs less than 1 time per week (RR: 1.23 95% CI: 1.01–1.16). The Physicians' Health Study found a positive and graded association of egg consumption with HF risk: multivariable adjusted HRs of 1.28 (95% CI: 1.02–1.61) and 1.64(95% CI: 1.08–2.44) for consumption of 1 and 2+ eggs per day, respectively)(Djousse & Gaziano, 2008a).

While more studies examining the causal relationship between egg consumption and the development of CHF are needed, the current evidence suggests that consuming 7 eggs or more per week, especially for men, is associated with a higher risk of developing heart failure. Furthermore, there is a lack of a significant association between egg consumption and heart failure in women. More studies are needed to elucidate these findings and explore potential biologic pathways by which egg consumption may modulate the pathogenesis of heart failure.

30.3 Conclusion

Overall, moderate egg consumption, as part of a healthy diet, does not appear to be associated with adverse health effects. There may be some health benefits to substituting whole eggs for egg whites when possible; especially for people with a high risk of CVD, including post-menopausal women, those with diabetes, established cardiovascular disease or dyslipidemia. Randomized trials suggest that egg consumption results in a modest increase in LDL-c cholesterol. Furthermore, egg consumption does not appear to raise the risk for total, ischemic, or hemorrhagic stroke in healthy individuals. However, most findings on egg consumption and cardiovascular disease have been derived from observational studies, where residual and/or unmeasured confounding makes it difficult to draw causal inference. Hence, additional intervention studies are needed to fill existing gaps and elucidate the pathophysiologic mechanisms underlying the association between egg consumption and cardiovascular disease.

REFERENCES

Abdollahi, A. M., Virtanen, H. E. K., Voutilainen, S., Kurl, S., Tuomainen, T. P., Salonen, J. T., & Virtanen, J. K. (2019). Egg consumption, cholesterol intake, and risk of incident stroke in men: The Kuopio ischaemic heart disease risk factor study. *The American Journal of Clinical Nutrition*, *110*(1), 169–176. https://doi.org/10.1093/ajcn/nqz066

Bechthold, A., Boeing, H., Schwedhelm, C., Hoffmann, G., Knuppel, S., Iqbal, K., De Henauw, S., Michels, N., Devleesschauwer, B., Schlesinger, S., & Schwingshackl, L. (2019). Food groups and risk of coronary heart disease, stroke and heart failure: A systematic review and dose-response meta-analysis of prospective studies. *Critical Reviews in Food Science and Nutrition*, *59*(7), 1071–1090. https://doi.org/10.1080/10408398.2017.1392288

Berger, S., Raman, G., Vishwanathan, R., Jacques, P. F., & Johnson, E. J. (2015). Dietary cholesterol and cardiovascular disease: A systematic review and meta-analysis. *The American Journal of Clinical Nutrition*, *102*(2), 276–294. https://doi.org/10.3945/ajcn.114.100305

Blesso, CN, & Fernandez, ML. (2018). Dietary Cholesterol, Serum Lipids, and Heart Disease: Are Eggs Working for or Against You? *Nutrients*. Mar 29; *10*(4): 426.

Chen, G. C., Chen, L. H., Mossavar-Rahmani, Y., Kamensky, V., Shadyab, A. H., Haring, B., Wild, R. A., Silver, B., Kuller, L. H., Sun, Y., Saquib, N., Howard, B., Snetselaar, L. G., Neuhouser, M. L., Allison, M. A., Van Horn, L., Manson, J. E., Wassertheil-Smoller, S., & Qi, Q. (2021). Dietary cholesterol and egg intake in relation to incident cardiovascular disease and all-cause and cause-specific mortality in postmenopausal women. *The American Journal of Clinical Nutrition*, *113*(4), 948–959. https://doi.org/10.1093/ajcn/nqaa353

Dehghan, M., Mente, A., Rangarajan, S., Mohan, V., Lear, S., Swaminathan, S., Wielgosz, A., Seron, P., Avezum, A., Lopez-Jaramillo, P., Turbide, G., Chifamba, J., AlHabib, K. F., Mohammadifard, N., Szuba, A., Khatib, R., Altuntas, Y., Liu, X., Iqbal, R., & Yusuf, S. (2020). Association of egg intake with blood lipids, cardiovascular disease, and mortality in 177,000 people in 50 countries. *The American Journal of Clinical Nutrition*, *111*(4), 795–803. https://doi.org/10.1093/ajcn/nqz348

Diez-Espino, J., Basterra-Gortari, F. J., Salas-Salvado, J., Buil-Cosiales, P., Corella, D., Schroder, H., Estruch, R., Ros, E., Gomez-Gracia, E., Aros, F., Fiol, M., Lapetra, J., Serra-Majem, L., Pinto, X., Babio, N., Quiles, L., Fito, M., Marti, A., Toledo, E., & Investigators, P. (2017). Egg consumption and cardiovascular disease according to diabetic status: The PREDIMED study. *Clinical Nutrition*, *36*(4), 1015–1021. https://doi.org/10.1016/j.clnu.2016.06.009

Djousse, L., & Gaziano, J. M. (2008a). Egg consumption and risk of heart failure in the Physicians' health study. *Circulation*, *117*(4), 512–516. https://doi.org/10.1161/CIRCULATIONAHA.107.734210

Djousse, L., & Gaziano, J. M. (2008b). Egg consumption in relation to cardiovascular disease and mortality: The Physicians' health study. *The American Journal of Clinical Nutrition*, *87*(4), 964–969. https://doi.org/10.1093/ajcn/87.4.964

Drouin-Chartier, J. P., Chen, S., Li, Y., Schwab, A. L., Stampfer, M. J., Sacks, F. M., Rosner, B., Willett, W. C., Hu, F. B., & Bhupathiraju, S. N. (2020). Egg consumption and risk of cardiovascular disease: Three large prospective US cohort studies, systematic review, and updated meta-analysis. *BMJ*, *368*, m513. https://doi.org/10.1136/bmj.m513

Fuller, N. R., Caterson, I. D., Sainsbury, A., Denyer, G., Fong, M., Gerofi, J., Baqleh, K., Williams, K. H., Lau, N. S., & Markovic, T. P. (2015). The effect of a high-egg diet on cardiovascular risk factors in people with type 2 diabetes:

The diabetes and egg (DIABEGG) study-a 3-mo randomized controlled trial. *The American Journal of Clinical Nutrition*, *101*(4), 705–713. https://doi.org/10.3945/ajcn.114.096925

Godos, J., Micek, A., Brzostek, T., Toledo, E., Iacoviello, L., Astrup, A., Franco, O. H., Galvano, F., Martinez-Gonzalez, M. A., & Grosso, G. (2021). Egg consumption and cardiovascular risk: A dose-response meta-analysis of prospective cohort studies. *European Journal of Nutrition*, *60*(4), 1833–1862. https://doi.org/10.1007/s00394-020-02345-7

Griffin, B. A. (2016). Eggs: Good or bad? *Proceedings of the Nutrition Society*, *75*(3), 259–264. https://doi.org/10.1017/S0029665116000215

Griffin, J. D., & Lichtenstein, A. H. (2013). Dietary cholesterol and plasma lipoprotein profiles: Randomized-controlled trials. *Current Nutrition Reports*, *2*(4), 274–282. https://doi.org/10.1007/s13668-013-0064-0

Guo, J., Hobbs, D. A., Cockcroft, J. R., Elwood, P. C., Pickering, J. E., Lovegrove, J. A., & Givens, D. I. (2018). Association between egg consumption and cardiovascular disease events, diabetes and all-cause mortality. *European Journal of Nutrition*, *57*(8), 2943–2952. https://doi.org/10.1007/s00394-017-1566-0

Hu, F. B., Stampfer, M. J., Rimm, E. B., Manson, J. E., Ascherio, A., Colditz, G. A., Rosner, B. A., Spiegelman, D., Speizer, F. E., Sacks, F. M., Hennekens, C. H., & Willett, W. C. (1999). A prospective study of egg consumption and risk of cardiovascular disease in men and women. *JAMA*, *281*(15), 1387–1394. https://doi.org/10.1001/jama.281.15.1387

Khalighi Sikaroudi, M., Soltani, S., Kolahdouz-Mohammadi, R., Clayton, Z. S., Fernandez, M. L., Varse, F., & Shidfar, F. (2020). The responses of different dosages of egg consumption on blood lipid profile: An updated systematic review and meta-analysis of randomized clinical trials. *Journal of Food Biochemistry*, *44*(8), e13263. https://doi.org/10.1111/jfbc.13263

Khawaja, O., Singh, H., Luni, F., Kabour, A., Ali, S. S., Taleb, M., Ahmed, H., Gaziano, J. M., & Djousse, L. (2017). Egg consumption and incidence of heart failure: A meta-analysis of prospective cohort studies. *Frontiers in Nutrition*, *4*, 10. https://doi.org/10.3389/fnut.2017.00010

Kovacs-Nolan, J., Phillips, M., & Mine, Y. (2005). Advances in the value of eggs and egg components for human health. *Journal of Agricultural and Food Chemistry*, *53*(22), 8421–8431. https://doi.org/10.1021/jf050964f

Krittanawong, C., Narasimhan, B., Wang, Z., Virk, H. U. H., Farrell, A. M., Zhang, H., & Tang, W. H. W. (2021). Association between egg consumption and risk of cardiovascular outcomes: A systematic review and meta-analysis. *The American Journal of Medicine*, *134*(1), 76–83.e72. https://doi.org/10.1016/j.amjmed.2020.05.046

Kuang, H., Yang, F., Zhang, Y., Wang, T., & Chen, G. (2018). The impact of egg nutrient composition and its consumption on cholesterol homeostasis. *Cholesterol*, *2018*, 6303810. https://doi.org/10.1155/2018/6303810

Larsson, S. C., Akesson, A., & Wolk, A. (2015). Egg consumption and risk of heart failure, myocardial infarction, and stroke: Results from 2 prospective cohorts. *The American Journal of Clinical Nutrition*, *102*(5), 1007–1013. https://doi.org/10.3945/ajcn.115.119263

Li, M. Y., Chen, J. H., Chen, C., & Kang, Y. N. (2020). Association between egg consumption and cholesterol concentration: A systematic review and meta-analysis of randomized controlled trials. *Nutrients*, *12*(7). https://doi.org/10.3390/nu12071995

Mesas, A. E., Garrido-Miguel, M., Fernandez-Rodriguez, R., Fernandez Franco, S., Lugones-Sanchez, C., Garcia-Ortiz, L., & Martinez-Vizcaino, V. (2022). Egg consumption and blood lipid parameters according to the presence of chronic metabolic disorders: The EVIDENT II study. *The Journal of Clinical Endocrinology and Metabolism*, *107*(3), e963–e972. https://doi.org/10.1210/clinem/dgab802

Nettleton, J. A., Steffen, L. M., Loehr, L. R., Rosamond, W. D., & Folsom, A. R. (2008). Incident heart failure is associated with lower whole-grain intake and greater high-fat dairy and egg intake in the atherosclerosis risk in communities (ARIC) study. *Journal of the American Dietetic Association*, *108*(11), 1881–1887. https://doi.org/10.1016/j.jada.2008.08.015

Qin, C., Lv, J., Guo, Y., Bian, Z., Si, J., Yang, L., Chen, Y., Zhou, Y., Zhang, H., Liu, J., Chen, J., Chen, Z., Yu, C., Li, L., & China Kadoorie Biobank Collaborative, G. (2018). Associations of egg consumption with cardiovascular disease in a cohort study of 0.5 million Chinese adults. *Heart*, *104*(21), 1756–1763. https://doi.org/10.1136/heartjnl-2017-312651

Rong, Y., Chen, L., Zhu, T., Song, Y., Yu, M., Shan, Z., Sands, A., Hu, F. B., & Liu, L. (2013). Egg consumption and risk of coronary heart disease and stroke: Dose-response meta-analysis of prospective cohort studies. *BMJ*, *346*, e8539. https://doi.org/10.1136/bmj.e8539

Rouhani, M. H., Rashidi-Pourfard, N., Salehi-Abargouei, A., Karimi, M., & Haghighatdoost, F. (2018). Effects of egg consumption on blood lipids: A systematic review and meta-analysis of randomized clinical trials. *Journal of the American College of Nutrition*, *37*(2), 99–110. https://doi.org/10.1080/07315724.2017.1366878

Scrafford, C. G., Tran, N. L., Barraj, L. M., & Mink, P. J. (2011). Egg consumption and CHD and stroke mortality: A prospective study of US adults. *Public Health Nutrition*, *14*(2), 261–270. https://doi.org/10.1017/S1368980010001874

Shin, J. Y., Xun, P., Nakamura, Y., & He, K. (2013). Egg consumption in relation to risk of cardiovascular disease and diabetes: A systematic review and meta-analysis. *The American Journal of Clinical Nutrition*, *98*(1), 146–159. https://doi.org/10.3945/ajcn.112.051318

Stone, N. J., Robinson, J. G., Lichtenstein, A. H., Bairey Merz, C. N., Blum, C. B., Eckel, R. H., Goldberg, A. C., Gordon, D., Levy, D., Lloyd-Jones, D. M., McBride, P., Schwartz, J. S., Shero, S. T., Smith, S. C. Jr., Watson, K., & Wilson, P. W., & American College of Cardiology/American Heart Association Task Force on Practice Guidelines. (2014). 2013 ACC/AHA guideline on the treatment of blood cholesterol to reduce atherosclerotic cardiovascular risk in adults: A report of the American college of Cardiology/American heart association task force on practice guidelines. *Journal of the American College of Cardiology*, *63*(25 Pt B), 2889–2934.

Tang, H., Cao, Y., Yang, X., & Zhang, Y. (2020). Egg consumption and stroke risk: A systematic review and dose-response meta-analysis of prospective studies. *Frontiers in Nutrition*, *7*, 153. https://doi.org/10.3389/fnut.2020.00153

Tong, T. Y. N., Appleby, P. N., Key, T. J., Dahm, C. C., Overvad, K., Olsen, A., Tjonneland, A., Katzke, V., Kuhn, T., Boeing, H.,

Karakatsani, A., Peppa, E., Trichopoulou, A., Weiderpass, E., Masala, G., Grioni, S., Panico, S., Tumino, R., Boer, J. M. A., & Perez-Cornago, A. (2020). The associations of major foods and fibre with risks of ischaemic and haemorrhagic stroke: A prospective study of 418 329 participants in the EPIC cohort across nine European countries. *European Heart Journal*, *41*(28), 2632–2640. https://doi.org/10.1093/eurheartj/ehaa007

US Department of Health and Human Services & US Department of Agriculture. (2015, December). *2015 – 2020 Dietary Guidelines for Americans,* 8th ed. https://health.gov/our-work/food-nutrition/previous-dietary-guidelines/2015

Virtanen, J. K., Mursu, J., Virtanen, H. E., Fogelholm, M., Salonen, J. T., Koskinen, T. T., Voutilainen, S., & Tuomainen, T. P. (2016). Associations of egg and cholesterol intakes with carotid intima-media thickness and risk of incident coronary artery disease according to apolipoprotein E phenotype in men: The Kuopio ischaemic heart disease risk factor study. *The American Journal of Clinical Nutrition*, *103*(3), 895–901. https://doi.org/10.3945/ajcn.115.122317

Wang, M. X., Wong, C. H., & Kim, J. E. (2019). Impact of whole egg intake on blood pressure, lipids and lipoproteins in middle-aged and older population: A systematic review and meta-analysis of randomized controlled trials. *Nutrition, Metabolism and Cardiovascular Diseases*, *29*(7), 653–664. https://doi.org/10.1016/j.numecd.2019.04.004

Weggemans, R. M., Zock, P. L., & Katan, M. B. (2001). Dietary cholesterol from eggs increases the ratio of total cholesterol to high-density lipoprotein cholesterol in humans: A meta-analysis. *The American Journal of Clinical Nutrition*, *73*(5), 885–891. https://doi.org/10.1093/ajcn/73.5.885

World Health Organization. (2014). Global health observatory (GHO) data-raised cholesterol. In *Global status report on noncommunicable diseases 2014*. WHO.

Xia, X., Liu, F., Yang, X., Li, J., Chen, J., Liu, X., Cao, J., Shen, C., Yu, L., Zhao, Y., Wu, X., Zhao, L., Li, Y., Huang, J., Lu, X., & Gu, D. (2020). Associations of egg consumption with incident cardiovascular disease and all-cause mortality. *Science China Life Sciences*, *63*(9), 1317–1327. https://doi.org/10.1007/s11427-020-1656-8

Xu, L., Lam, T. H., Jiang, C. Q., Zhang, W. S., Zhu, F., Jin, Y. L., Woo, J., Cheng, K. K., & Thomas, G. N. (2019). Egg consumption and the risk of cardiovascular disease and all-cause mortality: Guangzhou Biobank cohort study and meta-analyses. *European Journal of Nutrition*, *58*(2), 785–796. https://doi.org/10.1007/s00394-018-1692-3

Zhong, V. W., Van Horn, L., Cornelis, M. C., Wilkins, J. T., Ning, H., Carnethon, M. R., Greenland, P., Mentz, R. J., Tucker, K. L., Zhao, L., Norwood, A. F., Lloyd-Jones, D. M., & Allen, N. B. (2019). Associations of dietary cholesterol or egg consumption with incident cardiovascular disease and mortality. *JAMA*, *321*(11), 1081–1095. https://doi.org/10.1001/jama.2019.1572

Zhuang, P., Jiao, J., Wu, F., Mao, L., & Zhang, Y. (2020). Egg and egg-sourced cholesterol consumption in relation to mortality: Findings from population-based nationwide cohort. *Clinical Nutrition*, *39*(11), 3520–3527. https://doi.org/10.1016/j.clnu.2020.03.019

Zhuang, P., Wu, F., Mao, L., Zhu, F., Zhang, Y., Chen, X., Jiao, J., & Zhang, Y. (2021). Egg and cholesterol consumption and mortality from cardiovascular and different causes in the United States: A population-based cohort study. *PLoS Medicine*, *18*(2), e1003508. https://doi.org/10.1371/journal.pmed.1003508

31

Egg Allergy

Yoshinori Mine

CONTENTS

31.1 Introduction

Food allergy is defined as an immune-mediated adverse reaction un foods. Food allergy has been becoming an important global public health concern. The prevalence of food allergy is estimated to be around 5–10% among infants and 4.5% among school-age children (Sicherer & Sampson, 2014). Hen's egg is one of the most common causes of food allergy around the world. The estimated prevalence of egg allergy varies between 1.6% and 3.2% and represents the second most common cause of food allergies in children (Sampson, 2004; Urisu et al., 2015). In several industrialized countries, egg allergy has in fact been reported as the most prevalent food hypersensitivities in the pediatric population after cow's milk allergy (Mathew & Pfleghaar, 2022). Currently, the most efficient approach for egg allergy is total avoidance of the offending compounds. However, omnipresence of egg-derived components in cooked or manufactured food products renders the approach difficult (Allen et al., 2007) and inadvertent exposure may lead to life-threatening anaphylactic responses. Furthermore, when egg allergy occurs in association with other food allergies, such as cow's milk, the risks of nutritional deficiency represent an additional hurdle for the patients (Grimshaw, 2006; García & Lizaso, 2011). Therefore, the development of egg products safe for human consumption complemented with efficient forms of immunotherapy are highly warranted. This chapter provides the comprehensive information on egg protein allergenicity, and the effects of food processing on the allergenicity of egg proteins.

31.2 Natural History of Egg Allergy

Egg allergy usually develops within the first two years of life, and resolves by the time the child reaches school age (Heine et al., 2006; Teuber et al., 2006). Clinical adverse reactions sometimes emerge prior to any egg ingestion (Tan & Joshi, 2014). In these cases, sensitization may have occurred *in utero* or may result from the transfer of small doses of antigen into breast milk (Vance et al., 2005). However, there is currently no evidence that strict egg avoidance during pregnancy or lactation can decrease the risk of sensitization in infants. A few studies suggested that early childhood sensitization to hen's egg may favor the subsequent development of respiratory allergies (Kim & Burks, 2020).

Earlier studies documented that about two-thirds of children outgrew their allergy by the age of five (Urisu et al., 2015). A study, involving a cohort of more than 850 egg allergic patients aged 2–18 years old, reported that outgrowth of egg allergy may in fact occur at a later stage than previously reported (Savage et al., 2007). The study highlighted that the median time to develop tolerance was significantly increased in children suffering from other atopic diseases such as asthma or allergic rhinitis, and that specific immunoglobulin (Ig) E level ≥50kU/L was a good marker of persistent egg allergy. It was also observed that concomitant sensitization with other food allergens (e.g. peanut or milk) tends to delay the outgrowth of egg allergy (Savage et al., 2007; Hasan et al., 2013).

DOI: 10.1201/9781003254430-39

31.3 Mucosal Immune Response to Egg Allergens

Immunoglobulin (Ig)E-mediated food allergy, also known as type I food allergy, accounts for the majority of food-induced allergic responses and is characterized by the presence of elevated titers of antigen-specific serum IgE antibodies. Currently, it remains unclear how the mucosal immune system is oriented toward sensitization vs. immune tolerance when exposed to innocuous dietary antigens. However, it is suggested that upon ingestion, food proteins are capable to cross the intestinal epithelial barrier and be captured by the underlying immune system. Food proteins are then processed into peptidic fragments by a class of specialized immune cells, known as antigen presenting cells (APC). The peptidic fragments are displayed on their surface in association with major histocompatibility (MHC) class II molecules. The peptide-MHC II complexes can in turn be recognized by specific T cell receptors or TCR (van Wijk & Knippels, 2007).

An allergic immune response is believed to be orchestrated by a class of CD4+ T lymphocytes or helper T cells. Indeed cytokines produced by CD4+ T lymphocytes mediate a wide range of pro-inflammatory and anti-inflammatory responses. Most CD4+ T cells belong to either a Th1 or a Th2 subgroup, producing type-1 or type-2 cytokines respectively. Interferon (IFN)-γ is the archetypal type-1 cytokine, while type-2 cells typically produce a range of cytokines, including IL-4, IL-5, IL-9 and IL-13, contributing to the differentiation of B cells into IgE-producing plasma cells, and the recruitment of effector cells such as eosinophils, basophils and mast cells (Figure 31.1) (Prioult & Nagler-Anderson, 2005; Taams et al., 2006; Larché, 2007).

The mechanism of allergic sensitization is initiated by differentiation of naïve antigen-specific T helper cells into effector Thelper type-2 cells, leading to the differentiation of B cells into antibody-secreting plasma cells and the binding of allergen-specific IgE to high affinity Fc receptors (FcεR I) present on the surface of mast cells and basophils. Upon subsequent exposure to the allergen, the cross-linking of IgE-receptors leads to their degranulation. The release of vasoactive amines, cytokines and synthesis of a variety of arachidonic acid-derived inflammatory mediators ensues (Tordesillas et al., 2017) and leads ultimately to the manifestations of clinical symptoms commonly associated with egg allergy (e.g. atopic dermatitis, urticaria, angiodema) (Figure 31.2).

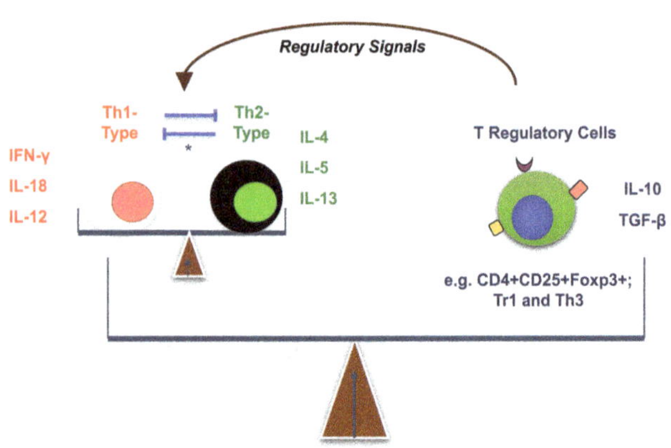

FIGURE 31.1 Model of immune homeostasis. (Adpated from Larché, 2007; Taams et al., 2006.)

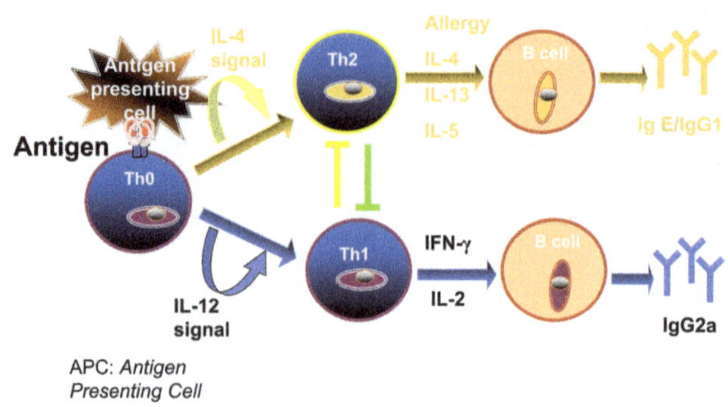

FIGURE 31.2 Naive T-cells (Th0) may differentiate into two distinct subsets. (Adapted from: http://www.immunocorp.com/beta-glucan/bio_research.cfm.)

A distinct population of CD4+ T cells, known as T regulatory (Treg) cells, is believed to control the balance between Thelper type-1 and type-2 responses. Regulatory T cells encompass "natural" T regulatory cells (e.g. CD4+CD25+ T cells) and "induced" or adaptive regulatory T cells such as IL-10-producing Tr1 cells and TGF-β- producing Th3 cells (Akdis et al., 2006). The manipulation of Treg cells functions in the prevention and treatment of (food) allergic diseases represent an attractive approach (Taams et al., 2006) (Figure 31.1). A detailed discussion of the role of T lymphocytes in the immunopathology of food allergy is beyond the scope of this review, however excellent reviews can be found (Chinthrajah et al., 2016; Yu et al., 2016).

31.4 Allergenic Components of the Egg

The chemical composition of hen eggs has been extensively investigated. Clinically relevant egg allergens have been identified in both the albumen (egg white) and the egg yolk fraction, however; reports have documented that the major egg allergens were mainly contained in the egg white (Anet et al., 1985).

Early studies involving a cohort of 342 patients, reported that the major egg allergens were, in increasing order; lysozyme>ovomucin>ovalbumin>ovomucoid, based on skin tests (Miller & Campbell, 1950). Using 13 egg allergic patients, further studies documented ovomucoid (OVM) as the dominant egg allergen (Bleumink & Young, 1971). Based on radioallergosorbent test (RAST) and crossed radio-immunoelectrophoresis (CRIE), studies completed in the 1980s established that ovalbumin (OVA), ovomucoid (OVM) and ovotransferrin (OVT) were the major egg allergens (Langeland, 1982b; Hoffman, 1983) and later, lysozyme (LYS) was also demonstrated to be a significant egg allergen (Holen & Elsayed, 1990). Allergy to lysozyme has been described in sensitization cases after exposure via inhalation. Frequency of recognition in egg allergic patients was reported to range from 6–67% for lysozyme (Frémont et al., 1997) and reach up to 22% for ovotransferrin (Djurtoft et al., 1991).

The two major allergens ovomucoid and ovalbumin constitute about 11% and 54% of egg white proteins, respectively (Kovacs-Nolan et al., 2005). Both proteins are glycosylated, with as high as 25% of the mass of ovomucoid comprising carbohydrates. Using egg allergic patients' sera, further studies confirmed that OVM was the dominant allergenic components in egg white (Cooke & Sampson, 1997; Urisu et al., 1997; Zhang & Mine, 1998; Zhang & Mine, 1999).

In an effort to establish which egg proteins were major allergens, a very recent study investigated binding of specific IgE to eight purified egg white and yolk proteins by (RAST), using sera from forty egg-sensitive children (Walsh et al., 2005). The study confirmed that the major egg allergens originated primarily from egg white, and include ovomucoid (Gal d 1), ovalbumin (Gal d 2), ovotransferrin (Gal d 3), and lysozyme (Gal d 4). The same report confirmed previous results that egg yolk also contains allergenic proteins, and were identified as apovitellenins I and VI and phosvitin (Walsh et al., 1987). In previous studies, other proteins such as ovoflavoprotein and ovoinhibitor were identified as antigenic, but lacked allergenic activity (Langeland, 1982a; Hoffman, 1983; Anet et al., 1985).

31.5 Stability of Egg Proteins and Their Allergenicity

The allergenicity of egg proteins depends a great deal, but not exclusively, on their resistance to heat and digestive enzymes (Astwood et al., 1996), reflecting their capacity to stimulate a specific immune response i.e. presence of B- and T-cell epitopes (Heine et al., 2006).

A common approach to evaluating the stability of egg allergens is to examine alterations of their IgE/IgG-binding capacities upon physical, chemical or genetic manipulations. In one study, the four major egg allergens were either physically or chemically modified and their binding activity to egg allergic patients specific IgG/IgE and rabbit specific-IgG were examined (Mine & Zhang, 2002). Thermal treatment was conducted for 15min at 95°C, and chemical modification with either urea 6M or carboxymethylation. Urea-treated OVA, LYS and OVT led to an increase in human IgG-binding activity, whilst carboxymethylation and thermal treatment of OVM and OVT led to a significant drop. Treatment with urea also enhanced the binding activity of rabbit IgG antibodies to OVT. While carboxymethylation of OVM, OVT and LYS decreased their allergenicity, IgE binding activity to OVA was not affected. These findings suggest that OVA contains sequential IgE epitopes, whilst OVM and LYS may contain both sequential and conformational IgE epitopes (Mine & Zhang, 2002).

Ovomucoid (OVM) is characterized by its high heat stability and can also be resistant to other forms of denaturation (e.g. urea) (Gremmel & Paschke, 2007), possibly related to the presence of its strong disulfide bonds (Djurtoft et al., 1991). In fact, *in vitro* stability of ovomucoid was reported in simulated intestinal fluid for at least 60 min (Takagi et al., 2003). A separate study reported on the other hand, that OVM can be easily denatured by trichloric acid-acetase or 8M urea procedures (Bernhisel-Broadbent et al., 1994). The chemical stability of OVM has been accounted for its dominance as a major egg allergen and its role in the prognosis of egg allergy. In a study involving 38 patients with allergic reactions to freeze-dried egg white, it was shown that most of them also developed a positive response to heated egg white (n = 21). On the other hand, 17 of these 21 subjects were tolerant to a challenge with heated-egg white depleted of OVM (Urisu et al., 1997). Experiments pertaining to the allergenic stability of OVM led to contradictory results. While some studies reported that reduction of OVM did not affect its IgE-binding capacity (Djurtoft et al., 1991), other studies reported a decrease (Cooke & Sampson, 1997) or an increase (Zhang & Mine, 1998), using human patients' sera. It was documented that heating of ovomucoid at 100°C for duration of 30 min could lead to irreversible conformational changes, as detected by monoclonal antibodies (Hirose et al., 2004). However, the protein is non-coagulable upon heat application and human specific-IgE were shown to bind to both native and denatured forms (Hirose et al., 2004). Using egg allergic patients sera, an interesting study reported that the IC50 concentration for specific IgE inhibition was 1700-fold higher for denatured ovomucoid compared to intact or oxidized ovomucoid (Holen & Elsayedl., 2001), suggesting that the allergenicity of OVM can significantly drop upon denaturation.

Interestingly, the allergenic and antigenic properties of OVM were shown to be maintained after peptic digestion (Kovacs-Nolan et al., 2000; Takagi et al., 2005). In *vitro* analyses showed that pepsin-digestion of OVM did not alter its IgE-binding capacity, as measured by ELISA using human egg allergic patients' sera (Kovacs-Nolan et al., 2000). The study also documented that pepsin-digestion of OVM did not significantly affect its trypsin inhibitory activity. On the other hand, reduction of OVM enhanced its digestibility, indicating the importance of its disulfide bonds. These findings were supported by previous studies investigating IgE binding activity of OVM after pepsin, chymotrypsin and trypsin digestion using RAST inhibition and Western blot tests, with serum samples obtained from egg allergic patients (Urisu et al., 1997; Besler et al., 1999b). Finally, five genetic mutants of OVM Gal d 1.3 were analyzed and compared *in vitro* to their native counterpart. It was demonstrated that the genetic mutation of phenylalanine (F) at position 37 with alanine (A) residue caused a drastic loss of binding activity to human allergic patients IgE, attributed to a disruption of the α-helix. Furthermore, substitution of glycine with methionine at position 32 led to synergistic effect on decreasing its binding activity (Mine et al., 2003).

Ovalbumin (OVA) is easily denatured by urea and guanidinium salts (Bernhisel-Broadbent et al., 1994). Furthermore, the protein is relatively heat labile, compared to OVM. Using egg allergic patients sera, reports have shown that the antigenicity of OVA could however resist heat treatment under certain conditions (Elsayed et al., 1986). As mentioned above, studies found that OVA can retain its human IgE binding capacity after chemical treatment such as reduction and carboxymethylation, heat or urea, indicating that IgE epitopes are linear and thermostable (Mine & Zhang, 2002). On the other hand, earlier report indicated that exposure of OVA to a temperature of 80°C for 3 min only, had the effect of decreasing its IgE-binding activity by 90% (Honma et al., 1994). The disparities observed from one study to another may be explained by variations inherent to patient cohort characteristics, an idea which was supported by subsequent reports (Walsh et al., 2005). Assessing resistance to enzymatic digestion represents an alternative manner to characterize egg allergen stability. It was documented that, in its native form, OVA is trypsin resistant but susceptible to pepsin digestion (Elsayed et al., 1986). Ovalbumin, along with phosvitin, was shown to be resistant to peptic digestion at pH 1.2 for more than 60 min (Astwood et al., 1996). Finally, it was reported that ovalbumin was relatively stable in simulated gastric and intestinal fluid, however preheating could lead to a decrease of its resistance to proteolysis (Takagi et al., 2003).

Lysozyme (LYS) stability has been the object of a number of studies. A recent report reviews the data obtained on LYS thermal stability (Petersen et al., 2004) and demonstrates that polyols such as sorbitol exert a significant stabilizing effect on lysozyme in its native form. In adduct-free solutions, hen's egg lysozyme was reported most stable in the pH range 3.5–5.0, with a denaturation temperature (Tm) between 75–80°C, while at lower pH conditions, its stability decreases rapidly and its Tm ranges within 52–56°C at pH 2 (Petersen et al., 2004). The mechanism of denaturation and unfolding of lysozyme has been finely characterized using the disulfide scrambling method (Chang & Li, 2002). LYS has also been shown to be resistant to pepsin digestion and proteinase K after incubation at 37° for 60 min (Polverino de Laureto et al., 2002). In order to examine the relationship between the conformational stability of LYS and its allergenicity, BALB/c mice were sensitized to variants of lysozyme with different conformational stability, but similar three-dimensional structures (So et al., 2001). Surprisingly, the least stable forms were associated with the strongest Th2 associated responses, as measured by IL-4 and specific IgE production.

Ovotransferrin (OVT) is a heat-labile allergen but it was reported that when coupled to bi- or trivalent metal ions, it could form heat-stable complexes (Gremmel & Paschke, 2007).

Egg yolk protein allergenicity has not been extensively studied compared to egg white proteins. Ovomucin (egg white), riboflavin and other egg yolk proteins seems to be less reactive with human patient sera IgE and they remain to be fully characterized. Ovomucin represents a distinct egg allergen as it has been associated with an increased susceptibility to baker's asthma, especially in central European population (Quirce et al., 2001). One study documented that heating of α-livetin can reduce, but does not eliminate its allergenicity (Quirce et al., 2001).

31.6 Cross-Reactivity of Egg Proteins

Reports have determined that cross-reactivity could occur between egg white proteins and egg yolk proteins. Earlier studies showed cross-reactivity between ovotransferrin and ovalbumin, and between apovitellenin I and ovalbumin (Walsh et al., 1987).

Although in Western countries, hen's eggs are the most commonly consumed, eggs from duck, quail and goose are sometimes consumed in other countries (Laundry, 2006). To this regard, serological cross-reactivity has been described between hen egg proteins and those of other bird eggs such as turkey, duck, goose and seagull (Langeland, 1982b; Anet et al., 1985). Mouse monoclonal IgG raised against Japanese quail ovomucoid were capable to bind to both hen and duck ovomucoid (Takahashi et al., 1999). A case of duck and goose egg allergy has also been reported, but the patient was tolerant to hen's egg (Anibarro et al., 2000). A recent study also reported a case of quail egg anaphylaxis in a child with hen egg allergy (Lee et al., 2021). The bird-egg syndrome describes the cross-reactivity that can occur between egg and other components of chicken serum or chicken meat. Earlier studies provided evidence that allergic reactions to egg yolk were induced upon respiratory sensitization to bird serum proteins (de Maat-Bleeker et al., 1985; Hoffman & Guenther, 1988). Sequence data are today available for egg proteins from many avian species and computational methods could further help predict IgE cross-reactivity (Aalberse & Stadler, 2006; Breiteneder & Mills, 2006).

31.7 Egg Yolk Allergenicity

It was initially believed that egg yolk was void of allergenic components (Ankier, 1969). However, as previously mentioned, subsequent studies reported the existence of significant IgE-binding activity against egg yolk components (Walsh et al., 2005). Clinical report of human specific IgE to egg yolk was first reported in the late 1980's (Carrillo et al., 1986). For example, α-livetin, also known as chicken serum albumin, has been identified as a minor allergen and is believed to be significantly involved in the pathogenesis of bird-egg syndrome (Quirce et al., 1998; Quirce et al., 2001; Martorell Aragonés et al., 2001). Similarly, a study determined that apovitellenins, a class of apoproteins representing up to 37% of total egg yolk proteins (Kovacs-Nolan et al., 2005; Urisu et al., 2015), also showed IgE-binding activity, as determined by RAST assays, and more specifically apovitellenins I and VI. Surprisingly, it has also been reported that egg-derived phospholipids (or lecithins) may have residual allergenic properties. However, their presence in manufactured food products was reported insufficient to elicit any adverse effects (Heine et al., 2006).

31.8 Effects of Processing on the Allergenicity of Egg Proteins

The emergence of egg allergy has had various implications for the food industry, affecting their manufacturing and labeling practices. While no efficient therapy currently exists for egg allergy, strict avoidance of the offending food remains the treatment of choice. However, the omnipresence of egg-derived components in prepackaged food products represents a significant hurdle for allergic consumers. Therefore regulatory bodies, food industrials, and research scientists have made joint efforts to implement preventive measures encompassing the production of hypoallergenic egg-containing products using chemical or physical processing methods.

Food processing methods may alter the allergenicity of food proteins in various ways. In general, food proteins present in a processed food will be in a denatured state, aggregated in protein networks, or interacting with carbohydrates (e.g. Maillard reactions) and lipids, leading either to the reduction of allergen content – and therefore potentially reducing its sensitizing potential – or to the formation of new epitopes (or neo-epitopes) (Thomas et al., 2007). With the intention to develop approaches for reducing egg allergenicity in food products, a number of studies have investigated the use of food processing methods such as heat application, enzymatic fragmentation, irradiation or high-pressure treatments.

31.8.1 Heat Application

Thermal processing is usually carried out to enhance texture and flavors, or to ensure microbiological safety, but is not primarily used to reduce allergenicity. However, the basis for this approach was demonstrated in a clinical study during which patients allergic to freeze-dried egg white did not react to cooked egg white (Urisu et al., 1997). Another case report described two patients who developed anaphylactic reactions upon raw egg ingestion, while no reaction was observed after cooked egg consumption (Eigenmann, 2000). Similarly, earlier *in vitro* studies showed that thermal treatment of egg white at 90°C for 10 min led to more than 50% decrease of RAST binding intensity using 16 egg allergic patients sera (Anet et al., 1985). It was documented that heating of α-livetin at 90°C for 30 min led to an 88% reduction of its IgE-reactivity (Quirce et al., 2001). On the other hand, other studies reported that residual IgG-binding activity against ovalbumin and ovomucoid could still be detected in both soft-boiled (100°C, 3 min) and hard-boiled (100°C, 20 min) eggs, as tested by radio-immunoelectrophoresis (Hoffman, 1983).

31.8.2 Enzymatic Fragmentation

Enzymatic processing represents a more specific approach than thermal processing and was reported efficient in the production of milk-based hypoallergenic formulas (Coombs & McLaughlan, 1984; Lee, 1992). Stability of the major egg allergens to enzymatic digestion was discussed in section 5. However, epitopes present on most food allergens are known to be sequential, so that reduction in egg allergenicity would only be expected when egg allergen epitopes are eliminated by the enzymatic fragmentation (Wal, 2003). Enzymatic hydrolysis has not been documented for reduction of egg allergenicity in food products. Indeed, it may not represent a viable alternative since egg proteins are often incorporated into food products for their unique functional properties e.g. foaming and gelling. Less denaturing approaches such as high-pressure or irradiation should be examined instead.

31.8.3 Gamma-Irradiation

Radiation technology has been explored in a number of studies for the modification of egg and other food allergens, such as shrimp and milk allergens (Byun et al., 2000; Lee et al., 2021). In the food industry, treatment doses up to 3kGy ensure a bacteriological quality of acceptable standard for liquid, frozen or dehydrated egg white preparations (Gremmel & Paschke, 2007). One study has investigated the use of γ-irradiation in the confection of hypoallergenic egg-based cakes (Seo et al., 2007). ELISA assays using egg-allergic patients' sera and OVA-specific rabbit IgG revealed a significant decrease in OVA content after γ-irradiation treatment. The same authors reported that γ-irradiation of OVA led to a hypoallergenic form of the protein (Seo et al., 2007). Results showed that γ-irradiation superior to 10 kGy altered the structure of OVA and led to a significantly decreased immunogenicity. Immunization trials in BALB/c mouse revealed that the titers of OVA-specific immunoglobulins and OVA-specific T cell-mediated responses were significantly lowered in groups of mice sensitized to irradiated-OVA. Furthermore, combination of γ-irradiation and heat treatment has been shown efficient in reducing the IgE-binding properties of OVM, the immunodominant egg allergen (Kim et al., 2002; Lee et al., 2002). These findings suggest that structural alteration of protein allergens may offer great opportunities for development of desensitization approaches in the context of egg allergy.

31.8.4 Glycation of Egg Proteins

The presence of N-glycosylation usually has a rigidifying or a stabilizing effect on protein structure, thereby conferring an enhanced resistance to denaturation (Pedrosa et al., 2000). It was reported that human IgE-binding to OVM domain III was greatly reduced by deglycosylation, but that mouse anti-OVM IgG could still bind to deglycosylated OVM domain III (Matsuda et al., 1985).

The contribution of glycosylation to egg allergenicity has therefore been explored and yielded mixed results. On the one hand, some studies suggested that the carbohydrate moieties did not contribute to the IgE or IgG binding activity of human patients sera (Cooke & Sampson, 1997; Besler, 1999a; Besler et al., 1999b). Furthermore, carbohydrate moieties were found not to be important for antigen presentation of OVM to T lymphocytes (Mizumachi & Kurisaki, 2003), as assessed in murine models. On the other hand, subsequent reports provided evidence that OVM carbohydrate chains may in fact have inhibitory effects on specific IgE binding activity (Zhang & Mine, 1998). The *in vitro* studies using recombinant glycosylated variants of OVM domain III, indicate that a masking effect exerted by N-glycans could contribute to a reduced allergenicity of OVM (Rupa et al., 2007). OVA-glycated mannose reduced maturation and uptake by dendritic cells. Less activation of T cells and type-2 cytokine response in DC-T-cell cocultures were observed with OVA-glycated mannose stimulation. (Rupa et al., 2014). A present study was to compare various glycated OVA-monosaccharides, including OVA-mannose (Man), -glucose, -ribose, and -fructose, in the attenuation of OVA-induced allergic response in a BALB/C mouse model. OVA-Man exhibited minimal DC internalization, DC processing, MHC antigen presentation, and antigen-specific T cell activation, resulting in an attenuated allergic response and validating its efficacy as a potential immunotherapeutic candidate to treat egg allergy (Rupa & Mine, 2019).

31.8.5 Other Food Processing Methods

A main obstacle to the use of food processing in reducing egg allergenicity is the risk to alter the unique functional attributes of egg proteins. Therefore, other processing methods such as high-pressure have been proposed as an alternative and milder approach. A recent review suggests that novel food processing methods, such as high-pressure and pulse-electric field, combined to physical and biochemical treatments, hold great promises for the development of hypoallergenic food products (Wichers et al., 2003). No such treatment has yet been reported for reduction of egg allergenicity. On the other hand, the production of egg food products with low OVM-content by solvent extraction has been explored. Ethanol (20%) precipitation successfully removed 70% of the OVM content from egg white powder, without affecting the whipping ability and foam stability of the remaining egg white proteins (Tanabe et al., 2000).

31.9 Conclusion

Hen's egg is one of the most common causes of food allergy around the world. The major egg allergens are proteins and are mainly present in the egg white. The two major allergens OVM and OVA constitute about 11% and 54% of egg white proteins, respectively. Egg yolk also contains allergenic proteins and were identified as apovitellenins I and VI and phosvitin. Egg allergy has various implications for the food industry, affecting their manufacturing and labeling practices. With the intention of developing approaches for reducing egg allergenicity in food products, a number of studies have investigated the use of food processing methods such as heat application, enzymatic fragmentation, irradiation, high-pressure treatments or glycation of egg allergens.

REFERENCES

Aalberse, R. C., & Stadler, B. M. (2006). In silico predictability of allergenicity: From amino acid sequence via 3-D structure to allergenicity. *Molecular Nutrition & Food Research*, *50*, 625–627. https://doi.org/10.1002/mnfr.200500270

Akdis, M., Blaser, K., & Akdis, C. A. (2006). T regulatory cells in allergy. *Chemical Immunology and Allergy*, *91*, 159–173. https://doi.org/10.1159/000090279

Allen, C. W., Campbell, D. E., & Kemp, A. S. (2007). Egg allergy: Are all childhood food allergies the same? *Journal of Paediatrics and Child Health*, *43*, 214–218. https://doi.org/10.1111/j.1440-1754.2007.00996.x

Anet, J., Back, J. F., Baker, R. S., Barnett, D., Burley, R. W., & Howden, M. E. (1985). Allergens in the white and yolk of hen's egg. A study of IgE binding by egg proteins. *International Archives of Allergy and Immunology*, *77*, 364–371. https://doi.org/10.1159/000233846

Anibarro, B., Seoane, F. J., Vila, C., & Lombardero, M. (2000). Allergy to eggs from duck and goose without sensitization to hen egg proteins. *The Journal of Allergy and Clinical Immunology*, *105*, 834–836. https://doi.org/10.1067/mai.2000.104547

Ankier, S.I. (1969). Reactivity of rat and man to egg-white. *Progress Drug Research*, *13*, 340–394. https://doi.org/10.1007/978-3-0348-7068-9_7

Astwood, J. D., Leach, J. N., & Fuchs, R. L. (1996). Stability of food allergens to digestion in vitro. *Nature Biotechnology*, *14*, 1269–1273. https://doi.org/10.1038/nbt1096-1269

Bernhisel-Broadbent, J., Dintzis, H. M., Dintzis, R. Z., & Sampson, H. A. (1994). Allergenicity and antigenicity of chicken egg ovomucoid (Gal d III) compared with ovalbumin (Gal d I) in children with egg allergy and in mice. *The Journal of Allergy and Clinical Immunology*, *93*, 1047–1059. https://doi.org/10.1016/s0091-6749(94)70054-0

Besler, M. (1999a). Hen's egg white. *Internet Symposium on Food Allergens*, *1*, 13–33.

Besler, M., Petersen, A., Steinhart, H., & Paschke, A. (1999b). Identification of IgE-binding peptides derived from chemical and enzymacti cleavage of ovomucoid (Gal d 1). *Internet Symposium on Food Allergens*, *1*, 1–12.

Bleumink, E., & Young, E. (1971). Studies on the atopic allergen in hen's egg. II. Further characterization of the skin-reactive fraction in egg-white; Immuno-electrophoretic studies. *International Archives of Allergy and Immunology*, *40*, 72–88. https://doi.org/10.1159/000230396

Breiteneder, H., & Mills, C. (2006). Structural bioinformatic approaches to understand cross-reactivity. *Molecular Nutrition & Food Research*, *50*, 628–632. https://doi.org/10.1002/mnfr.200500274

Byun, M. W., Kim, J. H., Lee, J. W., Park, J. W., Hong, C. S., & Kang, I. J. (2000). Effects of gamma radiation on the conformational and antigenic properties of a heat-stable major allergen in brown shrimp. *Journal of Food Protection*, *63*(940), 944. https://doi.org/10.4315/0362-028x-63.7.940

Carrillo, D. T., Cuevas, A. M., Moneo, G. I., Ibáñez Sandín, M. D., & Ureña, V. V. (1986). Detection of IgE specific for egg yolk by enzyme immunoassay, description of a case. *Allergologia et Immunopathologia* (Madr.), *14*, 9–14.

Chang, J. Y., & Li, L. (2002). The unfolding mechanism and the disulfide structures of denatured lysozyme. *FEBS Letters*, *511*, 73–78. https://doi.org/10.1016/s0014-5793(01)03284-7

Chinthrajah, R. S., Hernandez, J. D., Boyd, S. D., Galli, S. J., & Nadeau, K. C. (2016). Molecular and cellular mechanisms of food allergy and food tolerance. *Allergy and Clinical Immunology*, *137*(4),984–997. https://doi.org/10.1016/j.jaci.2016.02.004.

Cooke, S. K., & Sampson, H. A. (1997). Allergenic properties of ovomucoid in man. *Journal of Immunology*, *159*, 2026–2032.

Coombs, R. R. A., & McLaughlan, P. (1984). Allergenicity of food proteins and its possible modification. *Annals of Allergy*, *53*, 592–596.

de Maat-Bleeker, F., van Dijk, A. G., & Berrens, L. (1985). Allergy to egg yolk possibly induced by sensitization to bird serum antigens. *Annals of Allergy*, *54*, 245–248.

Djurtoft, R., Redersen, H. S., Aabin, B., & Barkholt, V. (1991). Studies of food allergens: Soybean and egg proteins. *Advances in Experimental Medicine and Biology*, *289*, 281–293. https://doi.org/10.1007/978-1-4899-2626-5_21

Eigenmann, P. A. (2000). Anaphylactic reactions to raw eggs after negative challenges with cooked eggs. *The Journal of Allergy and Clinical Immunology*, *105*, 587–588. https://doi.org/10.1067/mai.2000.104255

Elsayed, S., Hammer, A. S., Kalvenes, M. B., Florvaag, E., Apold, J., & Vik, H. (1986). Antigenic and allergenic determinants of ovalbumin. I. Peptide mapping, cleavage at the methionyl peptide bonds and enzymic hydrolysis of native and carboxymethyl OA. *International Archives of Allergy and Immunology*, *79*, 101–107.

Frémont, S., Kanny, G., Moneret-Vautrin, D. A., & Nicolas, J. P. (1997). Prevalence of lysozyme sensitization in an egg-allergic population. *Allergy*, *52*, 224–228. https://doi.org/10.1111/j.1398-9995.1997.tb00980.x

García, B. E., & Lizaso, M. T. (2011). Cross-reactivity syndromes in food allergy. *The Journal of Investigational Allergology and Clinical Immunology*, *21*(3), 162–170. https://doi.org/10.1159/00037541

Gremmel, S., & Paschke, A. (2007). Reducing allergens in egg and egg products (Chap. 11). In C. Mills, H. Wichers, & K. Hoffmann-Sommergruber (Eds.), *Managing allergens in food*. Woodhead Publishing Ltd.

Grimshaw, K. E. (2006). Dietary management of food allergy in children. *Proceedings of the Nutrition Society*, *65*, 412–417. https://doi.org/10.1017/s0029665106005167

Hasan, S. A., Wells, R. D., & Davis, C. M. (2013). Egg hypersensitivity in review. *Allergy & Asthma Proceedings*, *34*(1), 26–32. https://doi.org/10.2500/aap.2013.34.3621.

Heine, R. G., Laske, N., & Hill, D. J. (2006). The diagnosis and management of egg allergy. *Current Allergy and Asthma Reports*, *6*, 145–152. https://doi.org/10.1007/s11882-006-0053-0

Hirose, J., Kitabatake, N., Kimura, A., & Narita, H. (2004). Recognition of native and/or thermally induced denatured forms of the major food allergen, ovomucoid, by human IgE and mouse monoclonal IgG antibodies. *Bioscience, Biotechnology, and Biochemistry*, *68*, 2490–2497. https://doi.org/10.1271/bbb.68.2490

Hoffman, D. R. (1983). Immunochemical identification of the allergens in egg white. *The Journal of Allergy and Clinical Immunology*, *71*, 481–486. https://doi.org/10.1016/0091-6749(83)90465-7

Hoffman, D. R., & Guenther, D. M. (1988). Occupational allergy to avian proteins presenting as allergy to ingestion of egg yolk. *The Journal of Allergy and Clinical Immunology*, *81*, 484–488. https://doi.org/10.1016/0091-6749(88)90921-9

Holen, E., & Elsayed, S. (1990.) Characterization of four major allergens of hen egg-white by IEF/SDS-PAGE combined with electrophoretic transfer and IgE-immunoautoradiography. *International Archives of Allergy and Immunology*, *91*, 136–141. https://doi.org/10.1159/000235104

Honma, K., Kohno, Y., Saito, K., Shimojo, N., Niimi, H., & Tsunoo, H. (1994). Specificities of IgE, IgG and IgA antibodies to ovalbumin. Comparison of binding activities to denatured ovalbumin or ovalbumin fragments of IgE antibodies with those of IgG or IgA antibodies. *International Archives of Allergy and Immunology*, *103*, 28–35. https://doi.org/10.1159/000236602

Kim, M. J., Lee, J. W., Yook, H. S., Lee, S. Y., Kim, M. C., & Byun, M. W. (2002). Changes in the antigenic and immunoglobulin E-binding properties of hen's egg albumin with the combination of heat and gamma irradiation treatment. *Journal of Food Protection*, *65*, 1192–1195. https://doi.org/10.4315/0362-028x-65.7.1192

Kim, E. H., & Burks, A. W. (2020). Food allergy Immunotherapy: Oral Immunotherapy and epicutaneous Immunotherapy. *Allergy*, *75*(6), 1337–1346. https://doi.org/10.1111/all.14220.

Kovacs-Nolan, J., Phillips, M., & Mine, Y. (2005). Advances in the value of eggs and egg components for human health. *Journal of Agricultural and Food Chemistry*, *53*, 8421–8431. https://doi.org/10.1021/jf050964f

Kovacs-Nolan, J., Zhang, J. W., Hayakawa, S., & Mine, Y. (2000). Immunochemical and structural analysis of pepsin-digested egg white ovomucoid. *Journal of Agricultural and Food Chemistry*, *48*, 6261–6266. https://doi.org/10.1021/jf000358e

Langeland, T. (1982a). A clinical and immunological study of allergy to hen's egg white. II. Antigens in hen's egg white studied by crossed immunoelectrophoresis (CIE). *Allergy*, *37*, 323–333. https://doi.org/10.1111/j.1398-9995.1982.tb01918.x

Langeland, T. (1982b). A clinical and immunological study of allergy to hen's egg white. III. Allergens in hen's egg white studied by crossed radio-immunoelectrophoresis (CRIE). *Allergy*, *37*, 521–530. https://doi.org/10.1111/j.1398-9995.1982.tb02335.x

Larché, M. (2007). Update on the current status of peptide immunotherapy. *The Journal of Allergy and Clinical Immunology*, *119*(4), 906–909. https://doi.org/10.1016/j.jaci.2007.02.015.

Laundry, S. J. (2006). The relationship of T-cell epitope and allergen structure. In S. J. Maleki, A. W. Burks, & R. Helm (Eds.), *Food allergy* (pp. 123–259). ASM Press.

Lee, J., Gantulga, P., Lee, C., Jeong, K., Lee, E., & Lee, S. (2021). A preliminary study on cross-reactivity of heat-treated quail and hen's egg white proteins in young children. *Nutrients*, *13*(7), 2172. https://doi.org/10.3390/nu13072172

Lee, J. W., Lee, K. Y., Yook, H. S., Lee, S. Y., Kim, H. Y., Jo, C., & Byun, M. W. (2002). Allergenicity of hen's egg ovomucoid gamma irradiated and heated under different pH conditions. *Journal of Food Protection*, *65*, 1196–1199. https://doi.org/10.4315/0362-028x-65.7.1196

Lee, Y. H. (1992). Food-processing approaches to altering allergenic potential of milk-based formula. *The Journal of Pediatrics*, *121*, 47–50. https://doi.org/10.1016/s0022-3476(05)81406-4

Martorell Aragones, A., Bone Calvo, J., Garcia Ara, M. C., Nevot Falco, S., & Plaza Martin, A. M. (2001). Allergy to egg proteins. Food allergy committee of the Spanish society of pediatric clinical immunology and allergy. *Allergologia et Immunopathologia* (Madr), *29*, 72–95. https://doi.org/10.1016/s0301-0546(01)79022-2

Mathew, P., & Pfleghaar, J. L. (2022). Egg allergy. In *StatPearls [Internet]*. StatPearls Publishing.

Matsuda, T., Nakamura, R., Nakashima, I., Hasegawa, Y., & Shimokata, K. (1985). Human IgE antibody to the carbohydrate-containing third domain of chicken ovomucoid. *Biochemical and Biophysical Research Communications*, *129*, 505–510. https://doi.org/10.1016/0006-291x(85)90180-9

Miller, H., & Campbell, D. H. (1950). Skin test reactions to various chemical fractions of egg white and their possible clinical significance. *The Journal of Allergy and Clinical Immunology*, *21*, 522–524. https://doi.org/10.1016/0021-8707(50)90103-1

Mine, Y., Sasaki, E., & Zhang, J. W. (2003). Reduction of antigenicity and allergenicity of genetically modified egg white allergen, ovomucoid third domain. *Biochemical and Biophysical Research Communications*, *302*, 133–137. https://doi.org/10.1016/s0006-291x(03)00117-7

Mine, Y., & Zhang, J. W. (2002). Comparative studies on antigenicity and allergenicity of native and denatured egg white proteins. *Journal of Agricultural and Food Chemistry*, *50*, 2679–2683. https://doi.org/10.1021/jf0112264

Mizumachi, K., & Kurisaki, J. (2003). Localization of T cell epitope regions of chicken ovomucoid recognized by mice. *Bioscience, Biotechnology, and Biochemistry*, *67*, 712–719. https://doi.org/10.1271/bbb.67.712

Pedrosa, C., De Felice, F. G., Trisciuzzi, C., & Ferreira, S. T. (2000). Selective neoglycosylation increases the structural stability of vicilin, the 7S storage globulin from pea seeds. *Archives of Biochemistry and Biophysics*, *382*, 203–210. https://doi.org/10.1006/abbi.2000.2024

Petersen, S. B., Jonson, V., Fojan, P., Wimmer, R., & Pedersen, S. (2004). Sorbitol prevents the self-aggregation of unfolded lysozyme leading to and up to 13 degrees C stabilisation of the folded form. *Journal of Biotechnology*, *114*, 269–278. https://doi.org/10.1016/j.jbiotec.2004.07.004

Polverino de Laureto, P., Frare, E., Gottardo, R., Van Dael, H., & Fontana, A. (2002). Partly folded states of members of the lysozyme/lactalbumin superfamily: A comparative study by circular dichroism spectroscopy and limited proteolysis. *Protein Science*, *11*, 2932–2946. https://doi.org/10.1110/ps.0205802

Prioult, G., & Nagler-Anderson, C. (2005). Mucosal immunity and allergic responses: Lack of regulation and/or lack of microbial stimulation? *Immunological Reviews*, *206*, 204–218. https://doi.org/10.1111/j.0105-2896.2005.00277.x

Quirce, S., Díez-Gómez, M. L., Eiras, P., Cuevas, M., Baz, G., & Losada, E. (1998). Inhalant allergy to egg yolk and egg white proteins. *Clinical & Experimental Allergy*, *28*, 478–485. https://doi.org/10.1046/j.1365-2222.1998.00253.x

Quirce, S., Maranon, F., Umpierrez, A., de las Heras, M., Fernandez-Caldas, E., & Sastre, J. (2001). Chicken serum albumin (Gal d 5) is a partially heat-labile inhalant and food allergen implicated in the bird-egg syndrome. *Allergy*, *56*, 754–762. https://doi.org/10.1034/j.1398-9995.2001.056008754.x

Rupa, P., & Mine, Y. (2019). Comparison of glycated ovalbumin–monosaccharides in the attenuation of ovalbumin-induced allergic response in a BALB/C mouse model. *Journal of Agricultural and Food Chemistry, 67*(29), 8138–8148. https://doi.org/10.1021/acs.jafc.9b02132

Rupa, P., Nakamura, S., Katayama, S., & Mine, Y. (2014) Effects of ovalbumin glycoconjugates on alleviation of orally induced egg allergy in mice via dendritic-cell maturation and T cell activation. *Molecular Nutrition & Food Research*, *58*(2), 405–417. https://doi.org/10.1002/mnfr.201300067

Rupa, P., Nakamura, S., & Mine, Y. (2007). Genetically glycosylated ovomucoid third domain can modulate immunoglobulin E antibody production and cytokine response in BALB/c mice. *Clinical & Experimental Allergy*, *37*, 918–928. https://doi.org/10.1111/j.1365-2222.2007.02720.x

Sampson, H. A. (2004). Update on food allergy. *The Journal of Allergy and Clinical Immunology*, *113*, 805–819. https://doi.org/10.1016/j.jaci.2004.03.014

Savage, J. H., Matsui, E. C., Skripak, J. M., & Wood, R. A. (2007). The natural history of egg allergy. *The Journal of Allergy and Clinical Immunology*, *120*, 1413–1417. https://doi.org/10.1016/j.jaci.2007.09.040

Seo, J. H., Kim, J. H., Lee, J. W., Yoo, Y. C., Kim, M. R., Park, K. S., & Byun, M. W. (2007). Ovalbumin modified by gamma irradiation alters its immunological functions and allergic responses. *International Immunopharmacology*, *7*, 464–472. https://doi.org/10.1016/j.intimp.2006.11.012

Sicherer, S. H., & Sampson, H. A. (2014). Food allergy: Epidemiology, pathogenesis, diagnosis, and treatment. *The Journal of Allergy and Clinical Immunology*, *133*(2), 291–307. https://doi.org/10.1016/j.jaci.2013.11.020

So, T., Ito, H., Hirata, M., Ueda, T., & Imoto, T. (2001). Contribution of conformational stability of hen lysozyme to induction of type 2 T-helper immune responses. *Immunology*, *104*, 259–268. https://doi.org/10.1046/j.1365-2567.2001.01314.x

Taams, L. S., Palmer, D. B., Akbar, A. N., Robinson, D. S., Brown, Z., & Hawrylowicz, C. M. (2006). Regulatory T cells in human disease and their potential for therapeutic manipulation. *Immunology*, *118*, 1–9. https://doi.org/10.1111/j.1365-2567.2006.02348.x

Takagi, K., Teshima, R., Okunuki, H., Itoh, S., Kawasaki, N., Kawanishi, T., Hayakawa, T., Kohno, Y., Urisu, A., & Sawada, J. (2005). Kinetic analysis of pepsin digestion of chicken egg white ovomucoid and allergenic potential of pepsin fragments. *International Archives of Allergy and Immunology*, *136*, 23–32. https://doi.org/10.1159/000082581

Takagi, K., Teshima, R., Okunuki, H., & Sawada, J. (2003). Comparative study of in vitro digestibility of food proteins and effect of preheating on the digestion. *Biological and Pharmaceutical Bulletin*, *26*, 969–973. https://doi.org/10.1248/bpb.26.969

Takahashi, K., Horiguchi, M., Bando, N., Tsuji, H., Ogawa, T., & Asao, T. (1999). Immunochemical characterization of ovomucoid from Japanese quail egg white using monoclonal antibodies. *Journal of Nutritional Science and Vitaminology* (Tokyo), *45*, 491–500. https://doi.org/10.3177/jnsv.45.491

Tan, J. W., & Joshi, P. (2014). Egg allergy: An update. *Journal of Paediatrics and Child Health*, *50*(1), 11–5. https://doi.org/10.1111/jpc.12408.

Tanabe, S., Tesaki, S., & Watanabe, M. (2000). Producing a low ovomucoid egg white preparation by precipitation with aqueous ethanol. *Bioscience, Biotechnology, and Biochemistry*, *64*, 2005–2007. https://doi.org/10.1271/bbb.64.2005

Teuber, S. S., Beyer, K., Comstock, S., & Wallowitz, M. (2006). The big eight foods: Clinical and epidemiological overview. In S. J. Maleki, A. W. Burks, & R. M. Helm (Eds.), *Food allergy* (pp. 49–79). AMS Press.

Thomas, K., Herouet-Guicheney, C., Ladics, G., Bannon, G., Cockburn, A., Crevel, R., Fitzpatrick, J., Mills, C., Privalle, L., & Vieths, S. (2007). Evaluating the effect of food processing on the potential human allergenicity of novel proteins: International workshop report. *Food and Chemical Toxicology*, *45*, 1116–1122. https://doi.org/10.1016/j.fct.2006.12.016

Tordesillas, L., Berin, M. C., & Sampson, H. A. (2017). Immunology of food allergy. *Immunity*, *47*(1), 32–50. https://doi.org/10.1016/j.immuni.2017.07.004.

Urisu, A., Ando, H., Morita, Y., Wada, E., Yasaki, T., Yamada, K., Komada, K., Torii, S., Goto, M., & Wakamatsu, T. (1997). Allergenic activity of heated and ovomucoid-depleted egg white. *The Journal of Allergy and Clinical Immunology*, *100*, 171–176. https://doi.org/10.1016/s0091-6749(97)70220-3

Urisu, A., Kondo, Y., & Tsuge, I. (2015). Hen's egg allergy. *Chemical Immunology and Allergy*, *101*, 124–130. https://doi.org/10.1159/000375416

van Wijk, F., & Knippels, L. (2007). Initiating mechanisms of food allergy: Oral Tolerance versus allergic sensitization.

Biomedicine & Pharmacotherapy, *61*, 8–20. https://doi.org/10.1016/j.biopha.2006.11.003

Vance, G. H., Lewis, S., Grimshaw, K. E., Wood, P. J., Briggs, R. A., Thornton, C. A., & Warner, J. O. (2005). Exposure of the fetus and infant to hens' egg ovalbumin via the placenta and breast milk in relation to maternal intake of dietary egg. *Clinical & Experimental Allergy*, *35*, 1318–1326. https://doi.org/10.1111/j.1365-2222.2005.02346.x

Wal, J. M. (2003). Thermal processing and allergenicity of foods. *Allergy*, *58*, 727–729. https://doi.org/10.1034/j.1398-9995.2003.00225.x

Walsh, B. J., Elliott, C., Baker, R. S., Barnett, D., Burley, R. W., Hill, D. J., & Howden, M. E. (1987). Allergenic cross-reactivity of egg-white and egg-yolk proteins. An in vitro study. *International Archives of Allergy and Immunology*, *84*, 228–232. https://doi.org/10.1159/000234428

Walsh, B. J., Hill, D. J., Macoun, P., Cairns, D., & Howden, M. E. (2005). Detection of four distinct groups of hen egg allergens binding IgE in the sera of children with egg allergy. *Allergologia et Immunopathologia (Madr)*, *33*, 183–191. https://doi.org/10.1157/13077741

Wichers, H., Matser, A., Van Amerongen, A., & Soler-Rivas, C., Wickers, J. (2003). Monitoring of and technological effects on allergenicity of proteins in the food industry. In E. N. C. Mills, & P. R. Shewry (Eds.), *Plant food allergens* (pp. 196–212). Oxford.

Yu, W., Freeland, D. M. H., & Nadeau, K. C. (2016). Food allergy: Immune mechanisms, diagnosis and immunotherapy. *Nature Reviews Immunology*, *16*(12), 751–765. https://doi.org/10.1038/nri.2016.111

Zhang, J. W., & Mine, Y. (1998). Characterization of IgE and IgG epitopes on ovomucoid using egg-white-allergic patients' sera. *Biochemical and Biophysical Research Communications*, *253*, 124–127. https://doi.org/10.1006/bbrc.1998.9761

Zhang, J. W., & Mine, Y. (1999). Characterization of residues in human IgE and IgG binding site by chemical modification of ovomucoid third domain. *Biochemical and Biophysical Research Communications*, *261*, 610–613. https://doi.org/10.1006/bbrc.1999.1078

32

Dietary Egg and Chronic Diseases—Prevention of Metabolic Disorders

Yoshimi Kishimoto

CONTENTS

32.1 Dyslipidemia

32.1.1 Egg Consumption and Blood Lipids

Dyslipidemia is defined as elevated levels of total cholesterol (TC), low-density lipoprotein cholesterol (LDL-C), or triglycerides, or a low level of high-density lipoprotein cholesterol (HDL-C). Dyslipidemia is one of the most important risk factors of cardiovascular disease (CVD) (Tietge, 2014), which is the leading cause of mortality globally, responsible for ~18 million deaths in 2017, an increase from 12.3 million in 1990 (GBD 2017 Causes of Death Collaborators, 2018). Diet is a factor that influences the blood lipid profile, but since cholesterol is synthesized in the body, a reduction of the body's synthesis of cholesterol compensates for the absorption of additional dietary cholesterol. Given the necessity for cholesterol in human body, very effective intestinal uptake mechanisms and enterohepatic bile acid and cholesterol reabsorption cycles exist (Lecerf & de Lorgeril, 2011).

Eggs are a major source of dietary cholesterol, and many clinical trials and observational studies have thus investigated the relationship between egg consumption and blood lipid profiles. However, clinical studies and meta-analyses of those studies have come to very different conclusions about the effects of eggs on serum cholesterol levels and cardiovascular health. A meta-analysis of 17 trials (1974–1999) reported that dietary cholesterol increased total cholesterol and the ratio of TC/HDL-C concentrations. For each 100 mg/day increase in dietary cholesterol, the estimated change in TC and the ratio of TC/HDL-C were 2.2 mg/dL (0.056 mmol/L) and 0.020 units, respectively, (Weggemans et al., 2001). Another meta-analysis of 28 randomized controlled trials (RCTs) published from 2000 onward reported that compared to a control group, egg consumption increased TC by 5.60 mg/dL (95%CI: 3.11–8.09), LDL-C by 5.55 mg/dL (95%CI: 3.14–7.69), and HDL-C by 2.13 mg/dL (95%CI: 1.10–3.16), although increased egg consumption did not affect the ratios of LDL-C/HDL-C or TC/HDL-C, or the concentration of triglycerides (TG) (Rouhani et al., 2018).

The authors of a more recent meta-analysis of 66 RCTs reported that the consumption of more than one egg per day increased the levels of TC, LDL-C, HDL-C, TC/HDL-C, apolipoprotein (Apo) A1 and B100, but there were no significant changes in the ratio of LDL-C/HDL-C (Khalighi Sikaroudi et al., 2020). Another 2020 meta-analysis of 17 RCTs reported that egg consumption increased LDL-C and LDL-C/HDL-C (Li et al., 2020). The contrast in findings among all of the above-cited meta-analyses could be explained by the variation in criteria for study inclusion. According to the data obtained from large prospective studies, there appears to be a 2%–3% decrease in CVD risk for each 1 mg/dL increase in HDL-C (Gordon et al., 1989). Although that data are not yet conclusive, numerous studies have reported that egg consumption often

DOI: 10.1201/9781003254430-40

increases serum LDL-C and HDL-C levels but does not alter the LDL-C/HDL-C ratio much, suggesting that the negative effect of LDL-C on CVD may be compensated for by the positive effect of HDL-C.

Summarizing the results of several analyses, the estimated change in TC has been reported to be 2.2–2.5 mg/dL for each 100 mg/day increase in dietary cholesterol, equivalent to an approx. 2%–3% change in serum cholesterol per egg (McNamara, 2000; Blesso & Fernandez, 2018). This effect is relatively mild compared to those of modifications of other dietary components, such as saturated fatty acids (Mensink et al., 2003). Such small changes in dietary cholesterol intake are likely related to feedback control mechanisms that can limit our ability to absorb and synthesize cholesterol, as well as to an increase in the amount of cholesterol that we excrete from our bodies (Katan & Beynen, 1987; McNamara et al., 1987).

In a meta-regression analysis of results from 55 intervention trials (2,652 subjects), after adjusting the data for the effects of dietary fatty acids, the mean predicted changes in LDL-C for an increase of 100 mg dietary cholesterol/day were 1.90, 4.46, and 4.58 mg/dL for the linear model, nonlinear Michaelis-Menten model, and Hill model, respectively (Vincent et al., 2019). The nonlinear models predicted larger increases in LDL cholesterol from increases in dietary cholesterol of ≤400 mg/day compared with the linear model (Vincent et al., 2019).

32.1.2 Predictive Formulas

Predictive formulas for the response of serum cholesterol to dietary cholesterol have been proposed based on the results of clinical intervention trials. Formulas for estimating the effect of cholesterol and saturated fatty acid intake on blood cholesterol levels have been proposed for nearly 50 years, with Keys' and Hegsted's formulas being representative (Hegsted et al., 1965; Keys et al., 1965; Keys & Parlin, 1966):

$$\text{Key's formula}: \Delta\text{serum cholesterol (mg/dL)} = 2.7\Delta S - 1.35\Delta P + 1.5\sqrt{\Delta C}$$

$$\text{Hegsted's formula}: \Delta\text{serum cholesterol (mg/dL)} = 2.16\Delta S - 1.65\Delta P + 0.068\Delta C$$

where ΔS represents the change in saturated fatty acid intake (%energy), ΔP is the change in poly-unsaturated fatty acid intake (%energy), and ΔC is the change in cholesterol intake (mg of dietary cholesterol per 1,000 kcal).

Keys' and Hegsted's formulas indicate that a dietary intake of cholesterol increases the serum total cholesterol concentration. However, a clear association between dietary cholesterol and serum total or LDL cholesterol concentrations may be found only when the cholesterol intake is within a certain range. Connor et al. reported that cholesterol intake in the range of approx. 100–350 mg/day was strongly associated with serum cholesterol concentrations, but no clear association was found below or above that level (Connor & Connor, 2002). However, another study indicated that the association between cholesterol intake and the serum total cholesterol concentration is nearly linear in the range of cholesterol intake up to 400 mg/1,000 kcal (Hegsted, 1986).

32.1.3 Impacts of Saturated Fatty Acids

Eggs are rich in cholesterol but also contain less saturated and trans-fatty acids than other cholesterol-rich foods. In addition, eggs contain components that are capable of interfering with cholesterol absorption from the intestine and preventing elevated blood cholesterol (Matsuoka et al., 2014, 2017; Lee et al., 2019). This is in contrast to the case of beef, which contains relatively high levels of both cholesterol and saturated fatty acids (SFAs). Several lines of evidence indicate that the principal factor affecting serum cholesterol is not dietary cholesterol but saturated fats (Hegsted et al., 1965; Keys & Parlin, 1966; Nakamura et al., 2010). A meta-analysis showed that the response of LDL-C concentrations to changes in dietary cholesterol was weaker in the studies of a background diet low in saturated fat (ratio of polyunsaturated to saturated fat [P/S] >0.7) than in the studies of a background diet high in saturated fat (P/S ≤0.7) (Weggemans et al., 2001).

In Japan, eggs account for about half of the population's dietary cholesterol intake, while in the United States, meat is the main source, and eggs account for about one-quarter (Sugano & Matsuoka, 2021). These differences are of major interest from a nutritional point of view, because an increase in serum cholesterol in response to increased egg consumption is 1.7 times greater when the background diet is high in saturated fat compared to a low-saturated-fat background diet (Natoli et al., 2007). Japanese typically consume diets with low fat (daily energy intake from fat, <30%) and a high P/S ratio, which is a dietary style recommended for lowering serum LDL-C levels in Western countries. Recent observations have revealed an increasing trend of serum TC levels in Japan since 1980, and this trend has been reflected by an increase in HDL-C and a decreased ratio of TC/HDL-C (NCD Risk Factor Collaboration, 2020). The serum TC level is somewhat higher in Japanese than in the U.S. population, but CVD mortality is lower in Japanese (the CVD-related death rate is 21.5% in Japan and 31.7% in the United States; United Nations-World Population Prospects). Therefore, systematic evaluations of the Japanese diet are of interest in relation to egg consumption, the blood cholesterol level, and CVD risk.

32.1.4 Individual Differences

The evidence accumulated to date indicates that the impact of dietary cholesterol loading from eggs on serum lipids is highly variable. Approximately two-thirds of people have no increase or only a mild increase in serum cholesterol when they consume a large amount of dietary cholesterol. These individuals are classified as hypo-responders or compensators in that they can compensate for a dietary intake of cholesterol by reducing their biosynthesis, absorption, and excretion of cholesterol (Katan & Beynen, 1987; McNamara et al., 1987; Blesso & Fernandez, 2018). A smaller proportion of the population are hyper-responders; their serum cholesterol levels are increased by ≥2.3 mg/dL in response to 100 mg of dietary cholesterol intake.

Some of the genetic differences between people with and without a strong increase in their blood cholesterol concentration after increasing their cholesterol intake (hyper-responders) are being clarified. Variations in several genes have been shown to be associated with altered responsiveness to dietary cholesterol. This may be caused by polymorphisms in genes encoding proteins involved in the absorption or metabolic process of cholesterol; genes with well-known associations include MTP, APOE, and ABCG5.

Microsomal triglyceride transfer protein (MTP) is a lipid transfer protein that is present mainly in hepatocytes and enterocytes. MTP plays an essential role in the assembly process of very low-density lipoproteins (VLDLs) and chylomicrons and their secretion into the plasma (Sharp et al., 1993). Loss of activity due to mutations in the coding region of the MTP gene results in the development of a rare genetic disease, i.e., abetalipoproteinemia, with a suppression of chylomicron secretion and hypocholesterolemia (Wetterau et al., 1992; Sharp et al., 1993). The common polymorphism −493G/T (rs1800591) in MTP gene has been shown to alter the efficiency of sterol absorption from the small intestine (Wolff et al., 2011). Apolipoprotein E (APOE) can contribute to the variable response. ApoE is a multifunctional protein that transports and delivers cholesterol and other lipids in the plasma via binding to cell surface ApoE receptors (Zhao et al., 2018). The relatively common E4 alleles (rs429358 and rs7412) induce stronger elevations in blood cholesterol than other alleles because the encoded ApoE isoform is a more effective ligand for the LDL receptor (Egert et al., 2012). ATP-binding cassette protein G5 (ABCG5) and G8 (ABCG8) form heterodimers and are involved in the intestinal absorption and biliary excretion of neutral sterols, including cholesterol and plant sterols. Various polymorphisms (A632V, T400K, D19H, M429V, and C54Y) in the ABCG8 and ABCG5 (Q604E) genes have been found to be associated with several facets of cholesterol metabolism, including the baseline cholesterol level, the individual responsiveness of plasma cholesterol to dietary interventions for hypercholesterolemia, and an increased risk of gallstones (Rudkowska & Jones, 2008).

32.1.5 Lipoprotein Particle Composition

Recent results of egg intervention trials indicate that additional cholesterol intake does not adversely affect serum lipid concentrations; rather, it improves the LDL particle composition and HDL function in some cases. Herron et al. reported an increase in large LDL particles and a decrease in small LDL particles after increased dietary cholesterol intake from eggs (Herron et al., 2004). The lipoprotein particle composition has been associated with differences in the atherogenicity of lipoproteins. Small-density LDL particles are more easily oxidized and taken up by macrophages in the vessel wall compared to other lipoprotein particles (de Graaf et al., 1991; Brunzell, 2005). Several prospective cohort studies have reported that the number of small high-density LDL particles is a major predictor of CVD risk (Hoogeveen et al., 2014; Liou & Kaptoge, 2020).

When the lipoprotein particle size and subclasses were analyzed by nuclear magnetic resonance (NMR), significant increases in both LDL and HDL particles as well as an increase in lecithin cholesterol acyltransferase (LCAT) activity were observed after the consumption of whole egg (Mutungi et al., 2010). Although higher HDL-C is associated with a lower CVD risk, HDL function may be more important than the HDL-C concentration in deciphering the disease risk. The consumption of 2 whole eggs per day by overweight and postmenopausal women resulted in a significant increase in their cholesterol efflux capacity (Sawrey-Kubicek et al., 2019).

32.2 Cardiovascular Outcomes

32.2.1 Egg Consumption and CVD

Several observational studies and intervention trials concerning the relationship between dietary cholesterol intake and cardiovascular outcomes have been conducted. Some of those studies reported that dietary cholesterol or egg consumption increased the CVD risk (Keys & Parlin, 1966; Howell et al., 1997; Nissinen et al., 2008; Houston et al., 2011; Larsson et al., 2012; Zhong et al., 2019), whereas others reported a decreased risk or no change with higher cholesterol intake (Iso et al., 2001; Herron et al., 2004; Harman et al., 2008; Qin et al., 2018; Mohammadifard et al., 2022). Although many observational and intervention studies were conducted in developed countries, 3 large international prospective studies involving approx. 177,000 individuals in 50 countries on 6 continents recently reported no significant association between moderate egg intake and blood lipids, mortality, or major CVD events (Dehghan et al., 2020). Even a meta-analysis of prospective studies on egg consumption and the risk of cardiovascular disease has yielded inconsistent results, leading to confusion (Li et al., 2013; Berger et al., 2015; Drouin-Chartier et al., 2020a).

A 2015 systematic review by Berger et al., including 17 cohorts and 19 trials demonstrated that various sources of dietary cholesterol were not significantly associated with the CVD risk despite a mild increase in total and LDL cholesterol (Berger et al., 2015). A 2020 meta-analysis showed that moderate egg consumption (up to one egg per day) was not associated with the CVD risk overall and was associated with a potentially lower CVD risk in Asian populations (Drouin-Chartier et al., 2020a). An umbrella review of observational studies showed no significant association between egg consumption and health outcomes, including cancer, cardiovascular and metabolic disorders; evidence of possible beneficial effects toward the risk of stroke was also described (Marventano et al., 2020). Another umbrella review concluded that increased egg consumption was not associated with CVD risk in the general population (Mah et al., 2020). The results of the meta-analyses at this time do not allow definitive conclusions to be drawn due to heterogeneity among the studies and methodological issues; only the finding that there is no adverse effect of egg consumption on the risk of stroke seems conclusive.

32.2.2 Dietary Guidelines

The American Heart Association (AHA) and American College of Cardiology (ACC) guidelines on lifestyle management to reduce cardiovascular risk included a recommendation regarding dietary cholesterol, and the Dietary Guidelines for Americans (DGA) also recommended that cholesterol intake be limited to no more than 300 milligrams per day. However, the 2013 AHA/ACC guidelines did not include a recommendation for dietary cholesterol and concluded, "There is insufficient evidence to determine whether lowering dietary cholesterol reduces LDL-C" (Eckel et al., 2014). Similarly, in the current DGA (2020–), the recommendations for healthy individuals to limit dietary cholesterol intake has been removed (U.S. Department of Agriculture & U.S. Department of Health and Human Services, 2020).

The Dietary Reference Intakes for Japanese also set an upper limit for cholesterol intake as a tentative dietary goal (DG) for preventing lifestyle-related diseases, but this limit was eliminated in the 2015 edition. The current edition (2020) states, "DG of cholesterol is not developed, but this does not mean that there is no limit in allowable intake. In terms of prevention of the dyslipidemia progression, intake of cholesterol is recommended to be less than 200 mg/day" (Ministry of Health, Labour and Welfare of Japan, 2020). The restriction of cholesterol intake still exists in many countries' dietary guidelines for lipid management and the primary prevention of CVD (Ferraro et al., 2020).

32.2.3 Ethnic Differences

Ethnic differences in the effects of cholesterol intake have also been examined. Pan et al. investigated the associations of dietary cholesterol and egg consumption with cardiometabolic and all-cause mortality among Chinese and low-income Black and white Americans (Pan et al., 2021). Cholesterol intake showed a nonlinear association with increased all-cause mortality and a linear association with increased cardiometabolic mortality among Black Americans: the hazard ratios (HRs) (95%CIs) associated with 300 and 600 mg/day vs. 150 mg/day were 1.07 (1.03–1.11) and 1.13 (1.05–1.21) for all-cause mortality, and 1.10 (1.03–1.16) and 1.21 (1.08–1.36) for cardiometabolic mortality, respectively. Null associations with all-cause or cardiometabolic mortality were observed in white Americans. In contrast, nonlinear inverse associations were observed among Chinese: the HR (95%CI) for 300 vs. 150 mg/day was 0.94 (0.92–0.97) for all-cause mortality and 0.91 (0.87–0.95) for cardiometabolic mortality, but the inverse associations disappeared with a cholesterol intake >500 mg/day. Similarly, a positive association of egg intake with all-cause mortality was observed in Black Americans, but a null association in white Americans and a nonlinear inverse association in Chinese were observed. This study indicated that the associations of cholesterol and egg intake with cardiometabolic and all-cause mortality might differ among ethnic groups with different dietary patterns and cardiometabolic risk profiles.

32.3 Diabetes Mellitus

The relationship between egg consumption and the risk of type 2 diabetes mellitus (T2DM) is also complicated. Many observational studies from several countries found no association between egg consumption and T2DM in healthy subjects, or no adverse effect in individuals with T2DM (Djousse et al., 2016b; Tamez et al., 2016; Richard et al., 2017; Fuller et al., 2018; Geiker et al., 2018; Sabate et al., 2018; Drouin-Chartier et al., 2020b; Ni et al., 2020). In some reports, egg consumption was associated with a reduced risk of T2DM or with improved glycemic control (Kurotani et al., 2014; Lee & Kim, 2018; Pourafshar et al., 2018; Noerman et al., 2019; Wang et al., 2019). In contrast, several studies indicated a positive association between egg consumption and T2DM risk (Tamez et al., 2016; Jang et al., 2018; Y. Wang et al., 2021). Recent meta-analyses of prospective cohort studies were inconclusive regarding the association between egg consumption and T2DM risk. They reported positive associations for studies conducted among U.S. populations, but no association among studies conducted in Asia or Europe (Djousse et al., 2016a; Tamez et al., 2016; Wallin et al., 2016; Drouin-Chartier et al., 2020b). These discrepancies suggest that previous prospective analyses may have been unable to separate the effects of egg consumption per se from the background diet.

In the United States, egg consumption reflects Western dietary patterns, as eggs are often consumed with red and processed meats, refined grains, and sweetened beverages. In Europe, egg consumption is also mainly reflective of adherence to a Western dietary pattern, but this varies between Mediterranean and Northern European countries. In Asian countries, eggs are typically incorporated into various cuisines, which differ from Western egg consumption habits (Drouin-Chartier et al., 2020b). Dietary patterns, physical activity, and genetics may affect the predisposition of CVD and T2DM more than a single food item such as eggs (Geiker et al., 2018).

32.4 Hypertension

Hypertension is an important factor associated with CVD. Lower blood pressure reduces the risk of coronary heart disease, heart failure, stroke, and all-cause mortality (Ettehad et al., 2016), while higher blood pressure is associated with increased risks of stroke and myocardial infarction (Verdecchia et al., 2015). In the field of diet and hypertension, there has been a focus on the blood pressure-lowering effect of diet. Dietary approaches to stopping hypertension emphasize increased intakes of fruits, vegetables, whole grains, and low-fat dairy products and decreased intakes of fat, red meat, sodium, and added sugars, both of which have been shown to lower blood pressure (Saneei et al., 2014). Although the dietary pattern of fruit and vegetable intake was correlated with lower hypertension in China, eggs were also a contributor (egg intake ≥150 g/week, adjusted odds ratio, OR = 0.86, p = 0.029) (Yu et al., 2018).

In a meta-analysis of prospective cohort studies, egg consumption was associated with a lower risk of hypertension (adjusted relative risk [RR] = 0.79, p = 0.001) (Zhang & Zhang, 2018). In contrast, egg and cholesterol intakes were associated with a higher risk of hypertension in French women (MacDonald et al., 2020). A recent meta-analysis demonstrated that egg consumption had no significant effects on systolic and diastolic blood pressure in adults (Kolahdouz-Mohammadi et al., 2020). The inconsistency in these findings may depend on the amounts of eggs consumed.

The possible beneficial effects of egg consumption on blood pressure in humans can be attributed to the presence of components with antioxidant and antihypertensive effects in whole eggs. Bioactive peptides derived from eggs showed angiotensin-converting enzyme (ACE) inhibition and antihypertensive effects (Rao et al., 2012; Pokora et al., 2014; Yu et al., 2021).

32.5 Metabolic Syndrome (MetS)

Metabolic syndrome (MetS) is a set of symptoms that include abdominal obesity, hypertension, hyperglycemia, and dyslipidemia. Due to its associations with an increased risk of developing CVD and early all-cause mortality, MetS has become an important public health problem with a high economic burden worldwide (Lakka et al., 2002; Zimmet et al., 2005). Several studies have examined the association between egg consumption and MetS, and their findings have been inconsistent. Shin et al. conducted a cross-sectional study of a Korean population and reported that higher egg consumption was associated with a lower odds ratio for MetS (Shin et al., 2017). This association was confirmed only in women, not in men. Woo et al. studied 1,663 Korean subjects and found that higher egg consumption was correlated with a reduced risk of MetS in both men and women (Woo et al., 2016).

A recent review by Park et al. examined the association between egg consumption and MetS in Korean adults and revealed that a moderate level of egg consumption (4–6 times per week or once per day) was significantly associated with a reduced risk of MetS compared to a low level of egg consumption (less than once per month), whereas a high level of egg consumption (2 times or more per day) was not significantly associated with MetS (Park et al., 2018). Similarly, in a Chinese population, Wang et al. reported that higher egg consumption was associated with a lower odds ratio for MetS prevalence (Wang et al., 2021). A recent meta-analysis of 19 articles with 20 observational studies (331,667 participants) reported that higher egg consumption by adults was associated with a lower probability of having MetS (RR=0.92, 95%CI: 0.88–0.96; $p < 0.001$) (Ding & Zhang, 2022).

The inverse association between egg consumption and MetS may be attributed to some components of eggs that have favorable effects on the development of MetS. Matsuoka et al. investigated the effects of consuming egg white protein (EWP) or lactic-fermented egg white (LE), an easy-to-consume form of egg white, on the development of MetS in animal models and humans. In an in vitro experiment, EWP pepsin hydrolysate decreased the cholesterol micellar solubility and cholesterol transfer rate from micelles to the oil phase, and it increased the water-holding capacity, settling volume in water, and relative viscosity compared to casein pepsin hydrolysate (Matsuoka et al., 2014). In addition, a daily consumption of LE reduced serum total cholesterol and LDL cholesterol levels and the visceral fat area and improved the ratio of the visceral to subcutaneous fat area in men with mild hypercholesterolemia (Matsuoka et al., 2017).

A randomized control study comparing a whole-egg group with a yolk-free substitute group in terms of MetS reported that the whole egg had positive effects on the plasma carotenoid status and enrichment of HDL and LDL with lutein and zeaxanthin; these factors may influence the development of MetS-related degenerative conditions (Blesso et al., 2013c). Similar results were obtained in other randomized control studies (Andersen et al., 2013; Blesso et al., 2013a; Thomas et al., 2022). The severity of MetS is likely to be correlated with the degree of metabolic imbalance, an influential factor that affects lipoprotein metabolism (including low-grade systemic inflammation and insulin resistance) to a certain extent (Cornier et al., 2008). Blesso et al. administered a 12-week dietary intervention and weight loss program in patients with MetS and reported that egg yolk did not exacerbate the inflammatory status but instead improved inflammation when combined with carbohydrate restriction and weight loss (Blesso et al., 2013c). Thus, components in whole egg may be beneficial for the inflammation associated with MetS.

32.6 Oxidative Stress and Inflammation

Oxidative stress and chronic inflammation are known to play significant roles in the pathophysiology of numerous chronic diseases such as cardiovascular disease, T2DM, and hypertension. Lutein and zeaxanthin, 2 carotenoids present in egg yolk, are considered highly bioactive components because they protect against oxidative stress and inflammation. These carotenoids are major carotenoids in the eye and are known to have optical functions (such as light absorption and blue filtering) as well as antioxidant and anti-inflammatory functions. They are thus suspected to play a role in reducing the formation of macular degeneration and age-related cataracts (Bovier et al., 2014; Kelly et al., 2014). In older adults (≥60 years), the consumption or 2 or 4 eggs per day for 5 weeks increased serum lutein and zeaxanthin levels and the macular pigment optical density (Vishwanathan et al., 2009). A recent study of older Caucasian adults also demonstrated that the consumption of 2 lutein- and DHA-enriched eggs daily improved electrophysiological retina function without adversely altering plasma lipids (Walchuk et al., 2022).

The lutein and zeaxanthin content in eggs depends on the feed given to the laying hens. Variable contents of these carotenoids in non-enriched eggs have been observed: approx. 167–216 µg/yolk for lutein and approx. 85–185 µg/yolk for zeaxanthin (Kelly et al., 2014; Skřivan & Englmaierová, 2014; Miranda et al., 2015). Fukushima et al. recently established a carotenoid content database, including α-carotene, β-carotene, β-cryptoxanthin, lutein, zeaxanthin, lycopene, and fucoxanthin, and they reported the intakes of these carotenoids in Japanese women. In this database, eggs contain 270 µg of lutein and 230 µg of zeaxanthin per 100 g. Eggs were identified as the third-highest source of lutein in Japanese women (8.1% of its intake) and the main source of zeaxanthin (52% of its intake) (Fukushima et al., 2021). Lutein and zeaxanthin in eggs are known to be highly bioavailable compared to the lutein and zeaxanthin in other sources such as dietary supplements or spinach (Chung et al., 2004). This could be related to the fact that carotenoids depend on a lipophilic environment for optimal gastrointestinal uptake (Kelly et al., 2014). Carotenoids and dietary cholesterol share intestinal absorptive pathways, and lipoproteins are the major carriers of carotenoids in blood circulation (Clevidence & Bieri, 1993).

In subjects with MetS, a daily intake of 3 whole eggs as part of a carbohydrate-restricted diet for 12 weeks increased both plasma and lipoprotein lutein and zeaxanthin (Blesso et al., 2013c). A 5-week intervention study demonstrated that elderly subjects' consumption of only one egg per day significantly increased both serum lutein and zeaxanthin concentrations without elevating serum TC or LDL-C concentrations (Goodrow et al., 2006). Protection against LDL oxidation is an important strategy to prevent atherosclerosis, and many studies have investigated the relationship between serum carotenoid levels and the risk of subsequent CVD events (Morris et al., 1994; Iribarren et al., 19997). Kishimoto et al. evaluated the influence of daily egg consumption on the parameters of LDL oxidation in addition to the effects on serum lipid concentrations; a significant decrease in the serum malondialdehyde modified low-density lipoprotein (MDA-LDL) concentration and a prolonged lag time for LDL oxidation by the egg intervention was observed compared to the baseline values in healthy Japanese adults and those with moderately high blood cholesterol (Kishimoto et al., 2016, 2017; Taguchi et al., 2018). The subjects' serum lutein plus zeaxanthin concentrations were significantly higher than their baseline values and showed both an inverse relationship with MDA-LDL and a positive relationship with the LDL oxidation lag time (Kishimoto et al., 2017).

Phospholipids in egg may also have anti-inflammatory properties (Blesso, 2015). In overweight men, the consumption of

TABLE 32.1

Meta-Analyses and Umbrella Reviews of Egg Intake and Metabolic Disorders

Reference	Study Type	No. of Studies	Outcome	Result
Dyslipidemia				
Weggemans et al. (2001)	Meta-analysis	17 trials (1974–1999)	Serum lipid concentrations	↑ TC, TC/HDL-C
Rouhani et al. (2018)	Meta-analysis	28 RCTs (2000-)	Serum lipid concentrations	↑ TC, LDL-C, HDL-C ↔ TC/HDL-C, LDL-C/HDL-C, TG
Khalighi Sikaroudi et al. (2020)	Meta-analysis	66 RCTs (1977–2019)	Serum lipid concentrations	↑ TC, LDL-C, HDL-C, TC/HDL-C, Apo-A1/Apo-B100 ↔ LDL-C/HDL-C
Li et al. (2020)	Meta-analysis	17 RCTs (1984–2017)	Serum lipid concentrations	↑ LDL-C, LDL-C/HDL-C
CVD				
Li et al. (2013)	Meta-analysis	14 studies	Incident CVD	↑
Berger et al. (2015)	Meta-analysis	17 cohorts and 19 trials	Incident CVD	↔
Drouin-Chartier et al. (2020a)	Meta-analysis	28 studies	Incident CVD	↔ (overall) ↓ (Asian cohorts)
Mah et al. (2020)	Umbrella review	17 systematic reviews and 15 meta-analyses	Incident CVD	↔
Marventano et al. (2020)	Umbrella review		Incident CVD, CVD mortality	↔
T2DM				
Djousse et al. (2016a)	Meta-analysis	12 cohorts	T2DM risk	↑ (US) ↔ (European and Asian cohorts)
Tamez et al. (2016)	Meta-analysis	10 cohorts	T2DM risk	↑ (US) ↔ (European and Asian cohorts)
Wallin et al. (2016)	Meta-analysis	12 cohorts	T2DM risk	↑ (US) ↔ (European and Asian cohorts)
Drouin-Chartier et al. (2020b)	Meta-analysis	16 cohorts	T2DM risk	↑ (US) ↔ (European and Asian cohorts)
Hypertension				
Zhang & Zhang (2018)	Meta-analysis	10 cohorts	Hypertension risk	↓
Kolahdouz-Mohammadi et al. (2020)	Meta-analysis	15 RCTs	systolic and diastolic blood pressure	↔
MetS				
Ding & Zhang (2022)	Meta-analysis	20 observational studies	MetS risk	↓

3 eggs per day with a carbohydrate-restricted diet for 12 weeks decreased plasma C-reactive protein (CRP) and increased plasma adiponectin (Ratliff et al., 2008), and it decreased plasma TNFα and serum amyloid A in adults with MetS (Blesso et al., 2013b), whereas no changes were observed in subjects consuming a yolk-free egg substitute. Egg yolk contains a significant amount of choline. Dietary choline intake was shown to be inversely associated with serum inflammation markers in a Greek epidemiological study (Detopoulou et al., 2008). DiBella et al. reported that in patients with MetS, the consumption of 3 eggs per day lowered CRP, insulin, and insulin resistance compared to the baseline values (DiBella et al., 2020).

32.7 Discussion and Future Directions

Eggs have traditionally been associated with adverse factors in human health, due mainly to their cholesterol content. Nowadays, however, it is known that the response of cholesterol in the blood to dietary cholesterol consumption depends on several factors such as ethnicity, genetic makeup, and the dietary pattern and nutritional status of the consumer. Table 32.1 shows the summary of above-mentioned meta-analyses and umbrella reviews of egg intake and metabolic disorders. The influences of the various nutrients in eggs and of dietary patterns, including their interaction with gut microbes, merit further investigation. In addition, special consideration at the personalized level is needed for decisions regarding dietary cholesterol, not only for hypercholesterolemic patients but also for hyper-responsive healthy individuals. There is also evidence that eggs contain other biologically active compounds that may have a role in the treatment and prevention of chronic and infectious diseases. Although randomized controlled trials with long-term follow-up are needed to assess the associations between egg consumption and chronic diseases, the available information suggests that eggs are a healthy and cost-effective food worldwide.

REFERENCES

Andersen, C. J., Blesso, C. N., Lee, J., Barona, J., Shah, D., Thomas, M. J., & Fernandez, M. L. (2013). Egg consumption modulates HDL lipid composition and increases the cholesterol-accepting capacity of serum in metabolic syndrome. *Lipids, 48*(6), 557–567. https://doi.org/10.1007/s11745-013-3780-8

Berger, S., Raman, G., Vishwanathan, R., Jacques, P. F., & Johnson, E. J. (2015). Dietary cholesterol and cardiovascular disease: A systematic review and meta-analysis. *American Journal of Clinical Nutrition, 102*(2), 276–294. https://doi.org/10.3945/ajcn.114.100305

Blesso, C. N. (2015). Egg phospholipids and cardiovascular health. *Nutrients, 7*(4), 2731–2747. https://doi.org/10.3390/nu7042731

Blesso, C. N., Andersen, C. J., Barona, J., Volek, J. S., & Fernandez, M. L. (2013a). Whole egg consumption improves lipoprotein profiles and insulin sensitivity to a greater extent than yolk-free egg substitute in individuals with metabolic syndrome. *Metabolism: Clinical and Experimental, 62*(3), 400–410. https://doi.org/10.1016/j.metabol.2012.08.014

Blesso, C. N., Andersen, C. J., Barona, J., Volk, B., Volek, J. S., & Fernandez, M. L. (2013b). Effects of carbohydrate restriction and dietary cholesterol provided by eggs on clinical risk factors in metabolic syndrome. *Journal of Clinical Lipidology, 7*(5), 463–471. https://doi.org/10.1016/j.jacl.2013.03.008

Blesso, C. N., Andersen, C. J., Bolling, B. W., & Fernandez, M. L. (2013c). Egg intake improves carotenoid status by increasing plasma HDL cholesterol in adults with metabolic syndrome. *Food & Function, 4*(2), 213–221. https://doi.org/10.1039/c2fo30154g

Blesso, C. N., & Fernandez, M. L. (2018). Dietary cholesterol, serum lipids, and heart disease: Are eggs working for or against you? *Nutrients, 10*(4). https://doi.org/10.3390/nu10040426

Bovier, E. R., Renzi, L. M., & Hammond, B. R. (2014). A double-blind, placebo-controlled study on the effects of lutein and zeaxanthin on neural processing speed and efficiency. *PLoS One, 9*(9), e108178. https://doi.org/10.1371/journal.pone.0108178

Brunzell, J. D. (2005). Increased ApoB in small dense LDL particles predicts premature coronary artery disease. *Arteriosclerosis, Thrombosis, and Vascular Biology, 25*(3), 474–475. https://doi.org/10.1161/01.ATV.0000156537.78366.1d

Chung, H. Y., Rasmussen, H. M., & Johnson, E. J. (2004). Lutein bioavailability is higher from lutein-enriched eggs than from supplements and spinach in men. *Journal of Nutrition, 134*(8), 1887–1893. https://doi.org/10.1093/jn/134.8.1887

Clevidence, B. A., & Bieri, J. G. (1993). Association of carotenoids with human plasma lipoproteins. *Methods in Enzymology, 214*, 33–46. http://www.ncbi.nlm.nih.gov/pubmed/8469147

Connor, W. E., & Connor, S. L. (2002). Dietary cholesterol and coronary heart disease. *Current Atherosclerosis Reports, 4*(6), 425–432. https://doi.org/10.1007/s11883-002-0046-y

Cornier, M. A., Dabelea, D., Hernandez, T. L., Lindstrom, R. C., Steig, A. J., Stob, N. R., & Eckel, R. H. (2008). The metabolic syndrome. *Endocrine Reviews, 29*(7), 777–822. https://doi.org/10.1210/er.2008-0024

de Graaf, J., Hak-Lemmers, H. L., Hectors, M. P., Demacker, P. N., Hendriks, J. C., & Stalenhoef, A. F. (1991). Enhanced susceptibility to in vitro oxidation of the dense low density lipoprotein subfraction in healthy subjects. *Arteriosclerosis and Thrombosis, 11*(2), 298–306. https://doi.org/10.1161/01.atv.11.2.298

Dehghan, M., Mente, A., Rangarajan, S., Mohan, V., Lear, S., Swaminathan, S., Wielgosz, A., Seron, P., Avezum, A., Lopez-Jaramillo, P., Turbide, G., Chifamba, J., AlHabib, K. F., Mohammadifard, N., Szuba, A., Khatib, R., Altuntas, Y., Liu, X., Iqbal, R., Rosengren, A.,..., Yusuf, S. (2020). Association of egg intake with blood lipids, cardiovascular disease, and mortality in 177,000 people in 50 countries. *American Journal of Clinical Nutrition, 111*(4), 795–803. https://doi.org/10.1093/ajcn/nqz348

Detopoulou, P., Panagiotakos, D. B., Antonopoulou, S., Pitsavos, C., & Stefanadis, C. (2008). Dietary choline and betaine intakes in relation to concentrations of inflammatory markers in healthy adults: The ATTICA study. *American Journal of Clinical Nutrition, 87*(2), 424–430. https://doi.org/10.1093/ajcn/87.2.424

DiBella, M., Thomas, M. S., Alyousef, H., Millar, C., Blesso, C., Malysheva, O., Caudill, M. A., & Fernandez, M. L. (2020). Choline intake as supplement or as a component of eggs increases plasma choline and reduces interleukin-6 without modifying plasma cholesterol in participants with metabolic syndrome. *Nutrients*, *12*(10). https://doi.org/10.3390/nu12103120

Ding, J., & Zhang, Y. (2022). Relationship between egg consumption and metabolic syndrome. A meta-analysis of observational studies. *The Journal of Nutrition, Health & Aging*, *26*(4), 373–382. https://doi.org/10.1007/s12603-022-1765-0

Djousse, L., Khawaja, O. A., & Gaziano, J. M. (2016a). Egg consumption and risk of type 2 diabetes: A meta-analysis of prospective studies. *American Journal of Clinical Nutrition*, *103*(2), 474–480. https://doi.org/10.3945/ajcn.115.119933

Djousse, L., Petrone, A. B., Hickson, D. A., Talegawkar, S. A., Dubbert, P. M., Taylor, H., & Tucker, K. L. (2016b). Egg consumption and risk of type 2 diabetes among African Americans: The Jackson heart study. *Clinical Nutrition*, *35*(3), 679–684. https://doi.org/10.1016/j.clnu.2015.04.016

Drouin-Chartier, J. P., Chen, S., Li, Y., Schwab, A. L., Stampfer, M. J., Sacks, F. M., Rosner, B., Willett, W. C., Hu, F. B., & Bhupathiraju, S. N. (2020a). Egg consumption and risk of cardiovascular disease: Three large prospective US cohort studies, systematic review, and updated meta-analysis. *BMJ*, *368*, m513. https://doi.org/10.1136/bmj.m513

Drouin-Chartier, J. P., Schwab, A. L., Chen, S., Li, Y., Sacks, F. M., Rosner, B., Manson, J. E., Willett, W. C., Stampfer, M. J., Hu, F. B, & Bhupathiraju, S. N. (2020b). Egg consumption and risk of type 2 diabetes: Findings from 3 large US cohort studies of men and women and a systematic review and meta-analysis of prospective cohort studies. *American Journal of Clinical Nutrition*, *112*(3), 619–630. https://doi.org/10.1093/ajcn/nqaa115

Eckel, R. H., Jakicic, J. M., Ard, J. D., de Jesus, J. M., Houston Miller, N., Hubbard, V. S., Lee, I. M., Lichtenstein, A. H., Loria, C. M., Millen, B. E., Nonas, C. A., Sacks, F. M., Smith, S. C., Jr., & American College of Cardiology/American Heart Association Task Force on Practice Guidelines. (2014). 2013 AHA/ACC guideline on lifestyle management to reduce cardiovascular risk: A report of the American college of cardiology/American heart association task force on practice guidelines. *Circulation*, *129*(25 Suppl 2), S76–99. https://doi.org/10.1161/01.cir.0000437740.48606.d1

Egert, S., Rimbach, G., & Huebbe, P. (2012). ApoE genotype: From geographic distribution to function and responsiveness to dietary factors. *Proceedings of the Nutrition Society*, *71*(3), 410–424. https://doi.org/10.1017/S0029665112000249

Ettehad, D., Emdin, C. A., Kiran, A., Anderson, S. G., Callender, T., Emberson, J., Chalmers, J., Rodgers, A., & Rahimi, K. (2016). Blood pressure lowering for prevention of cardiovascular disease and death: A systematic review and meta-analysis. *Lancet*, *387*(10022), 957–967. https://doi.org/10.1016/S0140-6736(15)01225-8

Ferraro, R. A., Fischer, N. M., Xun, H., & Michos, E. D. (2020). Nutrition and physical activity recommendations from the United States and European cardiovascular guidelines: A comparative review. *Current Opinion in Cardiology*, *35*(5), 508–516. https://doi.org/10.1097/HCO.0000000000000763

Fukushima, Y., Taguchi, C., Kishimoto, Y., & Kondo, K. (2021). Japanese carotenoid database with alpha- and beta-carotene, beta-cryptoxanthin, lutein, zeaxanthin, lycopene, and fucoxanthin and intake in adult women. *International Journal for Vitamin and Nutrition Research*, 1–12. https://doi.org/10.1024/0300-9831/a000707

Fuller, N. R., Sainsbury, A., Caterson, I. D., Denyer, G., Fong, M., Gerofi, J., Leung, C., Lau, N. S., Williams, K. H., Januszewski, A. S., Jenkins, A. J., & Markovic, T. P. (2018). Effect of a high-egg diet on cardiometabolic risk factors in people with type 2 diabetes: The diabetes and egg (DIABEGG) study-randomized weight-loss and follow-up phase. *American Journal of Clinical Nutrition*, *107*(6), 921–931. https://doi.org/10.1093/ajcn/nqy048

GBD 2017 Causes of Death Collaborators. (2018). Global, regional, and national age-sex-specific mortality for 282 causes of death in 195 countries and territories, 1980-2017: A systematic analysis for the global burden of disease study 2017. *Lancet*, *392*(10159), 1736–1788. https://doi.org/10.1016/S0140-6736(18)32203-7

Geiker, N. R. W., Larsen, M. L., Dyerberg, J., Stender, S., & Astrup, A. (2018). Egg consumption, cardiovascular diseases and type 2 diabetes. *European Journal of Clinical Nutrition*, *72*(1), 44–56. https://doi.org/10.1038/ejcn.2017.153

Goodrow, E. F., Wilson, T. A., Houde, S. C., Vishwanathan, R., Scollin, P. A., Handelman, G., & Nicolosi, R. J. (2006). Consumption of one egg per day increases serum lutein and zeaxanthin concentrations in older adults without altering serum lipid and lipoprotein cholesterol concentrations. *Journal of Nutrition*, *136*(10), 2519–2524. https://doi.org/10.1093/jn/136.10.2519.

Gordon, D. J., Probstfield, J. L., Garrison, R. J., Neaton, J. D., Castelli, W. P., Knoke, J. D., Jacobs, D. R., Jr., Bangdiwala, S., & Tyroler, H. A. (1989). High-density lipoprotein cholesterol and cardiovascular disease. Four prospective American studies. *Circulation*, *79*(1), 8–15. https://doi.org/10.1161/01.cir.79.1.8

Harman, N. L., Leeds, A. R., & Griffin, B. A. (2008). Increased dietary cholesterol does not increase plasma low density lipoprotein when accompanied by an energy-restricted diet and weight loss. *European Journal of Nutrition*, *47*(6), 287–293. https://doi.org/10.1007/s00394-008-0730-y

Hegsted, D. M. (1986). Serum-cholesterol response to dietary cholesterol: A re-evaluation. *American Journal of Clinical Nutrition*, *44*(2), 299–305. https://doi.org/10.1093/ajcn/44.2.299

Hegsted, D. M., McGandy, R. B., Myers, M. L., & Stare, F. J. (1965). Quantitative effects of dietary fat on serum cholesterol in man. *American Journal of Clinical Nutrition*, *17*(5), 281–295. https://doi.org/10.1093/ajcn/17.5.281

Herron, K. L., Lofgren, I. E., Sharman, M., Volek, J. S., & Fernandez, M. L. (2004). High intake of cholesterol results in less atherogenic low-density lipoprotein particles in men and women independent of response classification. *Metabolism: Clinical and Experimental*, *53*(6), 823–830. https://doi.org/10.1016/j.metabol.2003.12.030

Hoogeveen, R. C., Gaubatz, J. W., Sun, W., Dodge, R. C., Crosby, J. R., Jiang, J., Couper, D., Virani, S. S., Kathiresan, S., Boerwinkle, E., & Ballantyne, C. M. (2014). Small dense low-density lipoprotein-cholesterol concentrations predict risk for coronary heart disease: The atherosclerosis risk in communities (ARIC) study. *Arteriosclerosis, Thrombosis, and Vascular Biology*, *34*(5), 1069–1077. https://doi.org/10.1161/ATVBAHA.114.303284

Houston, D. K., Ding, J., Lee, J. S., Garcia, M., Kanaya, A. M., & Tylavsky, F. A., Newman, A. B., Visser, M., Kritchevsky, S. B., & Health ABC Study. (2011). Dietary fat and cholesterol and risk of cardiovascular disease in older adults: The health ABC study. *Nutrition, Metabolism, and Cardiovascular Diseases*, *21*(6), 430–437. https://doi.org/10.1016/j.numecd.2009.11.007

Howell, W. H., McNamara, D. J., Tosca, M. A., Smith, B. T., & Gaines, J. A. (1997). Plasma lipid and lipoprotein responses to dietary fat and cholesterol: A meta-analysis. *American Journal of Clinical Nutrition*, *65*(6), 1747–1764. https://doi.org/10.1093/ajcn/65.6.1747

Iribarren, C., Folsom, A. R., Jacobs, D. R. Jr., Gross, M. D., Belcher, J. D., & Eckfeldt, J. H. (1997). Association of serum vitamin levels, LDL susceptibility to oxidation, and autoantibodies against MDA-LDL with carotid atherosclerosis. A case-control study. The ARIC study investigators. Atherosclerosis risk in communities. *Arteriosclerosis, Thrombosis, and Vascular Biology*, *17*(6), 1171–1177. https://doi.org/10.1161/01.atv.17.6.1171

Iso, H., Stampfer, M. J., Manson, J. E., Rexrode, K., Hu, F., Hennekens, C. H., Colditz, G. A., Speizer, F. E., & Willett, W. C. (2001). Prospective study of fat and protein intake and risk of intraparenchymal hemorrhage in women. *Circulation*, *103*(6), 856–863. https://doi.org/10.1161/01.cir.103.6.856

Jang, J., Shin, M. J., Kim, O. Y., & Park, K. (2018). Longitudinal association between egg consumption and the risk of cardiovascular disease: Interaction with type 2 diabetes mellitus. *Nutrition & Diabetes*, *8*(1), 20. https://doi.org/10.1038/s41387-018-0033-1

Katan, M. B., & Beynen, A. C. (1987). Characteristics of human hypo- and hyperresponders to dietary cholesterol. *American Journal of Epidemiology*, *125*(3), 387–399. https://doi.org/10.1093/oxfordjournals.aje.a114545

Kelly, E. R., Plat, J., Haenen, G. R., Kijlstra, A., & Berendschot, T. T. (2014). The effect of modified eggs and an egg-yolk based beverage on serum lutein and zeaxanthin concentrations and macular pigment optical density: Results from a randomized trial. *PLoS One*, *9*(3), e92659. https://doi.org/10.1371/journal.pone.0092659

Keys, A., Anderson, J. T., & Grande, F. (1965). Serum cholesterol response to changes in the diet: IV. Particular saturated fatty acids in the diet. *Metabolism: Clinical and Experimental*, *14*(7), 776–787. https://doi.org/10.1016/0026-0495(65)90004-1

Keys, A., & Parlin, R. W. (1966). Serum cholesterol response to changes in dietary lipids. *American Journal of Clinical Nutrition*, *19*(3), 175–181. https://doi.org/10.1093/ajcn/19.3.175

Khalighi Sikaroudi, M., Soltani, S., Kolahdouz-Mohammadi, R., Clayton, Z. S., Fernandez, M. L., Varse, F., & Shidfar, F. (2020). The responses of different dosages of egg consumption on blood lipid profile: An updated systematic review and meta-analysis of randomized clinical trials. *Journal of Food Biochemistry*, *44*(8), e13263. https://doi.org/10.1111/jfbc.13263

Kishimoto, Y., Taguchi, C., Saita, E., Suzuki-Sugihara, N., Nishiyama, H., Wang, W., Masuda, Y., & Kondo, K. (2017). Additional consumption of one egg per day increases serum lutein plus zeaxanthin concentration and lowers oxidized low-density lipoprotein in moderately hypercholesterolemic males. *Food Research International*, *99*(Pt 2), 944–949. https://doi.org/10.1016/j.foodres.2017.03.003

Kishimoto, Y., Taguchi, C., Suzuki-Sugihara, N., Saita, E., Usuda, M., Wang, W., Masuda, Y., & Kondo, K. (2016). The effect of the consumption of egg on serum lipids and antioxidant status in healthy subjects. *Journal of Nutritional Science and Vitaminology*, *62*(5), 361–365. https://doi.org/10.3177/jnsv.62.361

Kolahdouz-Mohammadi, R., Malekahmadi, M., Clayton, Z. S., Sadat, S. Z., Pahlavani, N., Sikaroudi, M. K., & Soltani, S. (2020). Effect of egg consumption on blood pressure: A systematic review and meta-analysis of randomized clinical trials. *Current Hypertension Reports*, *22*(3), 24. https://doi.org/10.1007/s11906-020-1029-5

Kurotani, K., Nanri, A., Goto, A., Mizoue, T., Noda, M., & Oba, S., Sawada, N., Tsugane, S, & Japan Public Health Center-based Prospective Study Group. (2014). Cholesterol and egg intakes and the risk of type 2 diabetes: The Japan public health center-based prospective study. *British Journal of Nutrition*, *112*(10), 1636–1643. https://doi.org/10.1017/S000711451400258X

Lakka, H. M., Laaksonen, D. E., Lakka, T. A., Niskanen, L. K., Kumpusalo, E., Tuomilehto, J., & Salonen, J. T. (2002). The metabolic syndrome and total and cardiovascular disease mortality in middle-aged men. *JAMA*, *288*(21), 2709–2716. https://doi.org/10.1001/jama.288.21.2709

Larsson, S. C., Virtamo, J., & Wolk, A. (2012). Dietary fats and dietary cholesterol and risk of stroke in women. *Atherosclerosis*, *221*(1), 282–286. https://doi.org/10.1016/j.atherosclerosis.2011.12.043

Lecerf, J. M., & de Lorgeril, M. (2011). Dietary cholesterol: From physiology to cardiovascular risk. *British Journal of Nutrition*, *106*(1), 6–14. https://doi.org/10.1017/S0007114511000237

Lee, J., & Kim, J. (2018). Egg consumption is associated with a lower risk of type 2 diabetes in middle-aged and older men. *Nutrition Research and Practice*, *12*(5), 396–405. https://doi.org/10.4162/nrp.2018.12.5.396

Lee, Y., Han, C. Y., Bae, M., Park, Y. K., & Lee, J. Y. (2019). Egg phospholipids exert an inhibitory effect on intestinal cholesterol absorption in mice. *Nutrition Research and Practice*, *13*(4), 295–301. https://doi.org/10.4162/nrp.2019.13.4.295

Li, M. Y., Chen, J. H., Chen, C., & Kang, Y. N. (2020). Association between egg consumption and cholesterol concentration: A systematic review and meta-analysis of randomized controlled trials. *Nutrients*, *12*(7). https://doi.org/10.3390/nu12071995

Li, Y., Zhou, C., Zhou, X., & Li, L. (2013). Egg consumption and risk of cardiovascular diseases and diabetes: A meta-analysis. *Atherosclerosis*, *229*(2), 524–530. https://doi.org/10.1016/j.atherosclerosis.2013.04.003

Liou, L., & Kaptoge, S. (2020). Association of small, dense LDL-cholesterol concentration and lipoprotein particle characteristics with coronary heart disease: A systematic review and meta analysis. *PLoS One*, *15*(11), e0241993. https://doi.org/10.1371/journal.pone.0241993

MacDonald, C. J., Madika, A. L., Bonnet, F., Fagherazzi, G., Lajous, M., & Boutron-Ruault, M. C. (2020). Cholesterol and egg intakes, and risk of hypertension in a large prospective cohort of French women. *Nutrients*, *12*(5). https://doi.org/10.3390/nu12051350

Mah, E., Chen, C. O., & Liska, D. J. (2020). The effect of egg consumption on cardiometabolic health outcomes: An umbrella review. *Public Health Nutrition*, *23*(5), 935–955. https://doi.org/10.1017/S1368980019002441

Marventano, S., Godos, J., Tieri, M., Ghelfi, F., Titta, L., Lafranconi, A., Gambera, A., Alonzo, E., Sciacca, S., Buscemi, S., Ray, S., Del Rio, D., Galvano, F. & Grosso, G. (2020). Egg consumption and human health: An umbrella review of observational studies. *International Journal of Food Sciences and Nutrition*, *71*(3), 325–331. https://doi.org/10.1080/09637486.2019.1648388

Matsuoka, R., Shirouchi, B., Kawamura, S., Baba, S., Shiratake, S., Nagata, K., Matsuoka, R., Shirouchi, B., Kawamura, S., Baba, S., Shiratake, S., Nagata, K., Imaizumi, K., & Sato, M. (2014). Dietary egg white protein inhibits lymphatic lipid transport in thoracic lymph duct-cannulated rats. *Journal of Agricultural and Food Chemistry*, *62*(44), 10694–10700. https://doi.org/10.1021/jf502741b

Matsuoka, R., Usuda, M., Masuda, Y., Kunou, M., & Utsunomiya, K. (2017). Lactic-fermented egg white reduced serum cholesterol concentrations in mildly hypercholesterolemic Japanese men: A double-blind, parallel-arm design. *Lipids in Health and Disease*, *16*(1), 101. https://doi.org/10.1186/s12944-017-0499-1

McNamara, D. J. (2000). The impact of egg limitations on coronary heart disease risk: Do the numbers add up? *Journal of the American College of Nutrition*, *19*(5 Suppl), 540S–548S. https://doi.org/10.1080/07315724.2000.10718978

McNamara, D. J., Kolb, R., Parker, T. S., Batwin, H., Samuel, P., Brown, C. D., Ahrens, E. H. Jr. (1987). Heterogeneity of cholesterol homeostasis in man. Response to changes in dietary fat quality and cholesterol quantity. *Journal of Clinical Investigation*, *79*(6), 1729–1739. https://doi.org/10.1172/JCI113013

Mensink, R. P., Zock, P. L., Kester, A. D., & Katan, M. B. (2003). Effects of dietary fatty acids and carbohydrates on the ratio of serum total to HDL cholesterol and on serum lipids and apolipoproteins: A meta-analysis of 60 controlled trials. *American Journal of Clinical Nutrition*, *77*(5), 1146–1155. https://doi.org/10.1093/ajcn/77.5.1146

Ministry of Health, Labour and Welfare of Japan. (2020). The dietary reference intakes for Japanese (2020). Retrieved July 5, 2022, from https://www.mhlw.go.jp/content/10900000/000862500.pdf

Miranda, J. M., Anton, X., Redondo-Valbuena, C., Roca-Saavedra, P., Rodriguez, J. A., Lamas, A., Franco, C. M., Cepeda, A. (2015). Egg and egg-derived foods: Effects on human health and use as functional foods. *Nutrients*, *7*(1), 706–729. https://doi.org/10.3390/nu7010706

Mohammadifard, N., Taheri, M., Haghighatdoost, F., Grau, N., Najafian, J., Sadeghi, M., Talaei, M., & Sarrafzadegan, N. (2022). Egg consumption and risk of cardiovascular events among Iranians: Results from Isfahan cohort study (ICS). *European Journal of Clinical Nutrition*. https://doi.org/10.1038/s41430-022-01118-1

Morris, D. L., Kritchevsky, S. B., & Davis, C. E. (1994). Serum carotenoids and coronary heart disease. The lipid research clinics coronary primary prevention trial and follow-up study. *JAMA*, *272*(18), 1439–1441. https://doi.org/10.1001/jama.272.18.1439

Mutungi, G., Waters, D., Ratliff, J., Puglisi, M., Clark, R. M., Volek, J. S., & Fernandez, M. L. (2010). Eggs distinctly modulate plasma carotenoid and lipoprotein subclasses in adult men following a carbohydrate-restricted diet. *The Journal of Nutritional Biochemistry*, *21*(4), 261–267. https://doi.org/10.1016/j.jnutbio.2008.12.011

Nakamura, Y., Okuda, N., Turin, T. C., Fujiyoshi, A., Okamura, T., & Hayakawa, T., Yoshita, K., Miura, K., Ueshima, H., & NIPPON DATA80/90 Research Group. (2010). Fatty acids intakes and serum lipid profiles: NIPPON DATA90 and the national nutrition monitoring. *Journal of Epidemiology*, *20(Suppl 3)*, S544–S548. https://doi.org/10.2188/jea.JE20090223

Natoli, S., Markovic, T., Lim, D., Noakes, M., & Kostner, K. (2007). Unscrambling the research: Eggs, serum cholesterol and coronary heart disease. *Nutrition & Dietetics*, *64*(2), 105–111. https://doi.org/10.1111/j.1747-0080.2007.00093.x

NCD Risk Factor Collaboration. (2020). National trends in total cholesterol obscure heterogeneous changes in HDL and non-HDL cholesterol and total-to-HDL cholesterol ratio: A pooled analysis of 458 population-based studies in Asian and Western countries. *International Journal of Epidemiology*, *49*(1), 173–192. https://doi.org/10.1093/ije/dyz099

Ni, L. P., Du, L. Y., Huang, Y. Q., & Zhou, J. Y. (2020). Egg consumption and risk of type 2 diabetes mellitus in middle and elderly Chinese population: An observational study. *Medicine (Baltimore)*, *99*(16), e19752. https://doi.org/10.1097/MD.0000000000019752

Nissinen, M. J., Gylling, H., & Miettinen, T. A. (2008). Responses of surrogate markers of cholesterol absorption and synthesis to changes in cholesterol metabolism during various amounts of fat and cholesterol feeding among healthy men. *British Journal of Nutrition*, *99*(2), 370–378. https://doi.org/10.1017/S0007114507811998

Noerman, S., Karkkainen, O., Mattsson, A., Paananen, J., Lehtonen, M., Nurmi, T., Tuomainen, T. P., Voutilainen, S., Hanhineva, K., & Virtanen, J. K. (2019). Metabolic profiling of high egg consumption and the associated lower risk of type 2 diabetes in middle-aged Finnish men. *Molecular Nutrition & Food Research*, *63*(5), e1800605. https://doi.org/10.1002/mnfr.201800605

Pan, X. F., Yang, J. J., Lipworth, L. P., Shu, X. O., Cai, H., Steinwandel, M. D., Blot, W. J., Zheng, W., Yu, D. (2021). Cholesterol and egg intakes with cardiometabolic and all-cause mortality among Chinese and low-income black and white Americans. *Nutrients*, *13*(6). https://doi.org/10.3390/nu13062094

Park, S. J., Jung, J. H., Choi, S. W., & Lee, H. J. (2018). Association between egg consumption and metabolic disease. *Korean Journal for Food Science and Animal Resources*, *38*(2), 209–223. https://doi.org/10.5851/kosfa.2018.38.2.209

Pokora, M., Zambrowicz, A., Dabrowska, A., Eckert, E., Setner, B., Szoltysik, M., Szewczuk, Z., Zablocka, A., Polanowski, A., Trziszka, T., & Chrzanowska, J. (2014). An attractive way of egg white protein by-product use for producing of novel anti-hypertensive peptides. *Food Chemistry*, *151*, 500–505. https://doi.org/10.1016/j.foodchem.2013.11.111

Pourafshar, S., Akhavan, N. S., George, K. S., Foley, E. M., Johnson, S. A., Keshavarz, B., Navaei, N., Davoudi, A., Clark, E. A., & Arjmandi, B. H. (2018). Egg consumption may improve factors associated with glycemic control and insulin sensitivity in adults with pre- and type II diabetes. *Food & Function*, *9*(8), 4469–4479. https://doi.org/10.1039/c8fo00194d

Qin, C., Lv, J., Guo, Y., Bian, Z., Si, J., Yang, L., Chen, Y., Zhou, Y., Zhang, H., Liu, J., Chen, J., Chen, Z., Yu, C., Li, L., & China Kadoorie Biobank Collaborative Group. (2018). Associations of egg consumption with cardiovascular disease in a cohort study of 0.5 million Chinese adults. *Heart*, *104*(21), 1756–1763. https://doi.org/10.1136/heartjnl-2017-312651

Rao, S.-Q., Ju, T., Sun, J., Su, Y.-J., Xu, R.-R., & Yang, Y.-J. (2012). Purification and characterization of angiotensin I-converting enzyme inhibitory peptides from enzymatic hydrolysate of hen egg white lysozyme. *Food Research International*, *46*(1), 127–134. https://doi.org/10.1016/j.foodres.2011.12.005

Ratliff, J. C., Mutungi, G., Puglisi, M. J., Volek, J. S., & Fernandez, M. L. (2008). Eggs modulate the inflammatory response to carbohydrate restricted diets in overweight men. *Nutrition & Metabolism*, *5*, 6. https://doi.org/10.1186/1743-7075-5-6

Richard, C., Cristall, L., Fleming, E., Lewis, E. D., Ricupero, M., Jacobs, R. L., & Field, C. J. (2017). Impact of egg consumption on cardiovascular risk factors in individuals with type 2 diabetes and at risk for developing diabetes: A systematic review of randomized nutritional intervention studies. *Canadian Journal of Diabetes*, *41*(4), 453–463. https://doi.org/10.1016/j.jcjd.2016.12.002

Rouhani, M. H., Rashidi-Pourfard, N., Salehi-Abargouei, A., Karimi, M., & Haghighatdoost, F. (2018). Effects of egg consumption on blood lipids: A systematic review and meta-analysis of randomized clinical trials. *Journal of the American College of Nutrition*, *37*(2), 99–110. https://doi.org/10.1080/07315724.2017.1366878

Rudkowska, I., & Jones, P. J. (2008). Polymorphisms in ABCG5/G8 transporters linked to hypercholesterolemia and gallstone disease. *Nutrition Reviews*, *66*(6), 343–348. https://doi.org/10.1111/j.1753-4887.2008.00042.x

Sabate, J., Burkholder-Cooley, N. M., Segovia-Siapco, G., Oda, K., Wells, B., Orlich, M. J., & Fraser, G. E. (2018). Unscrambling the relations of egg and meat consumption with type 2 diabetes risk. *American Journal of Clinical Nutrition*, *108*(5), 1121–1128. https://doi.org/10.1093/ajcn/nqy213

Saneei, P., Salehi-Abargouei, A., Esmaillzadeh, A., & Azadbakht, L. (2014). Influence of dietary approaches to stop hypertension (DASH) diet on blood pressure: A systematic review and meta-analysis on randomized controlled trials. *Nutrition, Metabolism, and Cardiovascular Diseases*, *24*(12), 1253–1261. https://doi.org/10.1016/j.numecd.2014.06.008

Sawrey-Kubicek, L., Zhu, C., Bardagjy, A. S., Rhodes, C. H., Sacchi, R., Randolph, J. M., Steinberg, F. M., & Zivkovic, A. M. (2019). Whole egg consumption compared with yolk-free egg increases the cholesterol efflux capacity of high-density lipoproteins in overweight, postmenopausal women. *American Journal of Clinical Nutrition*, *110*(3), 617–627. https://doi.org/10.1093/ajcn/nqz088

Sharp, D., Blinderman, L., Combs, K. A., Kienzle, B., Ricci, B., Wager-Smith, K., Gil, C. M., Turck, C. W., Bouma, M. E., Rader, D. J., Aggerbeck, L. P, Gregg, R. E., Gordon, D. A., & Wetterau, J. R. (1993). Cloning and gene defects in microsomal triglyceride transfer protein associated with abetalipoproteinaemia. *Nature*, *365*(6441), 65–69. https://doi.org/10.1038/365065a0

Shin, S., Lee, H. W., Kim, C. E., Lim, J., Lee, J. K., Lee, S. A., & Kang, D. (2017). Egg consumption and risk of metabolic syndrome in Korean adults: Results from the health examinees study. *Nutrients*, *9*(7). https://doi.org/10.3390/nu9070687

Skřivan, M., & Englmaierová, M. (2014). The deposition of carotenoids and α-tocopherol in hen eggs produced under a combination of sequential feeding and grazing. *Animal Feed Science and Technology*, *190*, 79–86. https://doi.org/10.1016/j.anifeedsci.2014.01.009

Sugano, M., & Matsuoka, R. (2021). Nutritional viewpoints on eggs and cholesterol. *Foods*, *10*(3). https://doi.org/10.3390/foods10030494

Taguchi, C., Kishimoto, Y., Suzuki-Sugihara, N., Saita, E., Usuda, M., Wang, W., Masuda, Y., & Kondo, K. (2018). Regular egg consumption at breakfast by Japanese woman university students improves daily nutrient intakes: Open-labeled observations. *Asia Pacific Journal of Clinical Nutrition*, *27*(2), 359–365. https://doi.org/10.6133/apjcn.042017.17

Tamez, M., Virtanen, J. K., & Lajous, M. (2016). Egg consumption and risk of incident type 2 diabetes: A dose-response meta-analysis of prospective cohort studies. *British Journal of Nutrition*, *115*(12), 2212–2218. https://doi.org/10.1017/S000711451600146X

Thomas, M. S., Puglisi, M., Malysheva, O., Caudill, M. A., Sholola, M., Cooperstone, J. L., & Fernandez, M. L. (2022). Eggs improve plasma biomarkers in patients with metabolic syndrome following a plant-based diet-a randomized crossover study. *Nutrients*, *14*(10). https://doi.org/10.3390/nu14102138

Tietge, U. J. (2014). Hyperlipidemia and cardiovascular disease: Inflammation, dyslipidemia, and atherosclerosis. *Current Opinion in Lipidology*, *25*(1), 94–95. https://doi.org/10.1097/MOL.0000000000000051

U.S. Department of Agriculture and U.S. Department of Health and Human Services. (2020). Dietary guidelines for Americans, 2020–2025. Retrieved July 5, 2022, from https://www.dietaryguidelines.gov/

Verdecchia, P., Reboldi, G., Angeli, F., Trimarco, B., Mancia, G., Pogue, J., J. Gao, P., Sleight, P., Teo, K., & Yusuf, S. (2015). Systolic and diastolic blood pressure changes in relation with myocardial infarction and stroke in patients with coronary artery disease. *Hypertension*, *65*(1), 108–114. https://doi.org/10.1161/HYPERTENSIONAHA.114.04310

Vincent, M. J., Allen, B., Palacios, O. M., Haber, L. T., & Maki, K. C. (2019). Meta-regression analysis of the effects of dietary cholesterol intake on LDL and HDL cholesterol. *American Journal of Clinical Nutrition*, *109*(1), 7–16. https://doi.org/10.1093/ajcn/nqy273

Vishwanathan, R., Goodrow-Kotyla, E. F., Wooten, B. R., Wilson, T. A., & Nicolosi, R. J. (2009). Consumption of 2 and 4 egg yolks/d for 5 wk increases macular pigment concentrations in older adults with low macular pigment taking cholesterol-lowering statins. *American Journal of Clinical Nutrition*, *90*(5), 1272–1279. https://doi.org/10.3945/ajcn.2009.28013

Walchuk, C., House, J. D., Aliani, M., & Suh, M. (2022). Lutein and docosahexaenoic acid enriched egg consumption improves retina function in healthy Caucasian older adults. *Journal of Functional Foods*, *89*, 104913. https://doi.org/10.1016/j.jff.2021.104913

Wallin, A., Forouhi, N. G., Wolk, A., & Larsson, S. C. (2016). Egg consumption and risk of type 2 diabetes: A prospective study and dose-response meta-analysis. *Diabetologia*, *59*(6), 1204–1213. https://doi.org/10.1007/s00125-016-3923-6

Wang, H., Wang, W., Shen, M., Yang, Z., Wang, N., Zhu, Z., Wu, Z., & Xie, D. (2021). Association between egg consumption and metabolic syndrome in Chinese population: A cross-sectional study. *BMJ Open*, *11*(12), e050317. https://doi.org/10.1136/bmjopen-2021-050317

Wang, X., Son, M., Meram, C., & Wu, J. (2019). Mechanism and potential of egg consumption and egg bioactive components on type-2 diabetes. *Nutrients*, *11*(2), 357. https://doi.org/10.3390/nu11020357

Wang, Y., Li, M., & Shi, Z. (2021). Higher egg consumption associated with increased risk of diabetes in Chinese adults - China health and nutrition survey. *British Journal of Nutrition*, *126*(1), 110–117. https://doi.org/10.1017/S0007114520003955

Weggemans, R. M., Zock, P. L., & Katan, M. B. (2001). Dietary cholesterol from eggs increases the ratio of total cholesterol to high-density lipoprotein cholesterol in humans: A meta-analysis. *American Journal of Clinical Nutrition*, *73*(5), 885–891. https://doi.org/10.1093/ajcn/73.5.885

Wetterau, J. R., Aggerbeck, L. P., Bouma, M. E., Eisenberg, C., Munck, A., Hermier, M., Schmitz, J., Gay, G., Rader, D. J., & Gregg, R. E. (1992). Absence of microsomal triglyceride transfer protein in individuals with abetalipoproteinemia. *Science*, *258*(5084), 999–1001. https://doi.org/10.1126/science.1439810

Wolff, E., Vergnes, M. F., Defoort, C., Planells, R., Portugal, H., Nicolay, A., & Lairon, D. (2011). Cholesterol absorption status and fasting plasma cholesterol are modulated by the microsomal triacylglycerol transfer protein -493 G/T polymorphism and the usual diet in women. *Genes & Nutrition*, *6*(1), 71–79. https://doi.org/10.1007/s12263-010-0174-x

Woo, H. W., Choi, B. Y., & Kim, M. K. (2016). Cross-sectional and longitudinal associations between egg consumption and metabolic syndrome in adults >/= 40 years old: The Yangpyeong cohort of the Korean genome and epidemiology study (KoGES_Yangpyeong). *PLoS One*, *11*(1), e0147729. https://doi.org/10.1371/journal.pone.0147729

Yu, G., Fu, H., Huang, W., Zhang, N., Deng, D., Li, G., & Lei, H. (2018). A dietary pattern of higher fish, egg, milk, nut, vegetable and fruit, and lower salt intake correlates with the prevalence and control of hypertension. *American Journal of Hypertension*, *31*(6), 679–686. https://doi.org/10.1093/ajh/hpy005

Yu, Z., Xue, W., Zhou, M., Wang, L., Wu, S., Zhao, W., & Ding, L. (2021). Potential antihypertensive mechanisms of the egg white-derived peptide QIGLF in spontaneously hypertensive rats revealed using untargeted serum metabolomics. *Journal of Agricultural and Food Chemistry*, *69*(40), 12063–12071. https://doi.org/10.1021/acs.jafc.1c05599

Zhang, Y., & Zhang, D. Z. (2018). Red meat, poultry, and egg consumption with the risk of hypertension: A meta-analysis of prospective cohort studies. *Journal of Human Hypertension*, 32(7), 507–517. https://doi.org/10.1038/s41371-018-0068-8

Zhao, N., Liu, C. C., Qiao, W.,, & Bu, G. (2018). Apolipoprotein E, receptors, and modulation of Alzheimer's disease. *Biological Psychiatry*, *83*(4), 347–357. https://doi.org/10.1016/j.biopsych.2017.03.003

Zhong, V. W., Van Horn, L., Cornelis, M. C., Wilkins, J. T., Ning, H., Carnethon, M. R., Greenland, P., Mentz, R. J., Tucker, K. L., Zhao, L., Norwood, A. F.,Lloyd-Jones, D. M., & Allen, N. B. (2019). Associations of dietary cholesterol or egg consumption with incident cardiovascular disease and mortality. *JAMA*, *321*(11), 1081–1095. https://doi.org/10.1001/jama.2019.1572

Zimmet, P., Magliano, D., Matsuzawa, Y., Alberti, G., & Shaw, J. (2005). The metabolic syndrome: A global public health problem and a new definition. *Journal of Atherosclerosis and Thrombosis*, *12*(6), 295–300. https://doi.org/10.5551/jat.12.295

33

Effect of Dietary Eggs on Microbiota

Xiang Duan* **and Ning Qiu**

CONTENTS

33.1 Introduction

The avian egg is an important source of nutrients, and contains all the proteins, lipids, vitamins, minerals, and growth factors required by the developing embryo, as well as some defense factors to protect it against bacterial and viral infection. Moreover, eggs are now understood to contain substances with biological functions beyond basic nutrition. This chapter mainly focuses on the effect of dietary eggs on gut microbiota.

The gut microbiota is the community of microorganisms that live in the digestive tracts of living organisms. In humans, the total number of bacteria that colonize the adult intestinal tract is approximately 10^{14}, which outnumbers the human cells in the body by tenfold (Ley et al., 2006). Generally, about 5 major divisions (phyla) of bacteria are present in the gut: Firmicutes, Actinobacteria, Fusobacteria, Proteobacteria, and Bacteroidetes. Among the microbiota, Firmicutes and Bacteroidetes are more numerous than the remaining 3 phyla, accounting for more than 90% of the gut microbiota. The proximal gastrointestinal tract is dominated by Firmicutes, *Lactobacilli*, and Proteobacteria, whereas the distal gastrointestinal tract contains mainly Bacteroidetes, Firmicutes, and *Akkermansia muciniphila*. All the types of microorganisms living in the intestine constitute gut microbiota, which has a complex structure and is in a dynamic balance.

Gut microbiota plays an essential role in human health, and increasing knowledge about the effect of diet on the composition of the gut microbial population is emerging, which consequently affects human health and disease. Increasing evidence suggests that the homeostasis of the gut microbiota is among the most important factors for maintaining the physical and mental health of the host. The disruption of the intestinal ecological balance is associated with multiple diseases, including obesity, allergy, diabetes, inflammatory bowel disease, and even cancer (Backhed et al., 2012; Lozupone et al., 2012).

The composition of the intestinal bacteria could be affected by a series of factors, such as the physiology, pathology, immune system, and lifestyle of the host, and diet is regarded as a key lifestyle factor in modulating gut microbiota (Brown et al., 2012). After the food has been digested and absorbed in the stomach and small intestine, there is still a substantial quantity of food entering the large intestine with the potential to modulate the gut bacteria (van Hylckama Vlieg et al., 2011; Rist et al., 2013). The gut microbiota metabolizes these dietary components and releases metabolites that influence host physiology either positively or negatively. The complex relationship between diet, gut microbiota, and human health has gained tremendous interest as a natural system to improve health and wellbeing (Wu et al., 2022). Previous studies have investigated the effects of different dietary proteins (derived from beef, pork, chicken, fish, casein, and soy) on gut bacteria, exhibiting significant differences in the profiles of gut bacteria among these groups (Zhu et al., 2015). Moreover, some studies have revealed that egg consumption improves the homeostasis of the intestinal flora (Meng et al., 2020; Yu et al., 2020). This chapter aims to delineate the effect of various egg components (proteins and lipids) on the relative abundances of the gut microbial populations, and to provide insights into their potential roles in metabolism and health. The effects of egg compositions on gut microbiota mentioned in this chapter have been summarized in Table 33.1.

* Corresponding author.

DOI: 10.1201/9781003254430-41

TABLE 33.1

The Effects of Some Egg Compositions on Gut Microbiota and Host

Egg Composition	Effects on Gut Microbiota	Functional Effects on Host	Reference
Eggshell			
Eggshell membrane	Regulating commensal microbiota	Regulating mucosal immune system to maintain intestinal homeostasis	Jia et al. (2017a)
	Alteration of gut microbiota composition	Reducing plasma triglycerides and liver total cholesterol in HFD-fed mice	Ramli et al. (2020)
	increasing *L. reuteri*	increasing bile acids level in blood	Il-Dong et al. (2016)
Egg white			
Egg white proteins	Increasing *Lactobacillus* and *Candidatus_ Saccharimonas*, and decreasing *Ruminiclostridium* and *Akkermansia*	Decreasing the levels of pro-inflammatory cytokines and the extent of crypt damage	Ge et al. (2021)
Egg white hydrolysate	Modulation of the gut microbiota	Exerting positive effects on intestinal dysbiosis	Requena et al. (2017)
Ovomucin	Inhibiting *E. coli, Streptococcus,* and *Helicobacter pylori*		Kelm & Schauer (1997); Schauer (2004)
	Suppressing the colonization of the intestinal epithelial cells by *E. coli, Salmonella,* and *Staphylococcus aureus*		Shan et al. (2014); Sun et al. (2017)
	Increasing *Lactobacilli, Faecalibaculum*, and *Ruminococcus*		Gotoh et al. (2018)
Ovomucin hydrolysates	Promoting the growth of *B. infantis*		Sun et al. (2016)
	Preventing *E. coli* adhesion to porcine small intestinal epithelial cells		Sun et al. (2017, 2019)
Ovotransferrin		Reducing inflammatory cytokines	Kobayashi et al. (2015)
	Increasing *Actinobacteria and Akkermansia,* and decreasing *TM 7, Ruminococcus* and *Anaerotruncus*		Zhang et al. (2020)
Ovotransferrin-derived peptides IRW and IQW	Decreasing Firmicutes and Actinobacteria species, and increasing Bacteroidetes and Proteobacteria species in *C. rodentium* group	Attenuating LPS-induced inflammatory responses	Ma et al., (2019)
Lysozyme	Exerting antimicrobial activity against gram-positive and gram-negative bacteria	Reducing intestinal gene expression of pro-inflammatory cytokines, and increasing the levels of anti-inflammatory IL-4 and transforming growth factor β	Lee et al. (2009)
Egg yolk			
Phosvtin	Bactericidal activity against E. coli		Sattar Khan et al. (2000)
	Increasing the number of *Bifidobacterium*, and decreasing *Helicobacter* and *Mucispirillum*		Li et al. (2019)
phospholipids	Decreasing the Firmicutes/Bacteroidetes ratio and the relative abundances of *Streptomyces, Arthrobacter,* and *Desulfovibrio*, and increasing the relative abundances of *Mucilaginibacter* and *Cellulomonas*	Having a potential modulating effect on intestinal disorders caused by HFD	Yin et al. (2021)

33.2 The Effect of Egg Proteins on Microbiota

A recent study showed that high protein diets containing egg or whey protein, together with inulin fiber and exercise, improved the energy balance in obese rats by decreasing food intake, enhancing energy expenditure, decreasing fat mass, improving glycemic control, and modulating metabolic hormones, gut microbiota composition and function, and select metabolites in the blood circulation (Avirineni et al., 2022). The following section summarizes the known effects of different egg proteins on gut microbiota and health.

33.2.1 The Effect of Eggshell Membrane Proteins on Microbiota

Eggshell membrane (ESM) is a source of nondigestible protein, which may have a physiological function similar to that of dietary fiber; it stimulates cecal fermentation and alters intestinal bacterial composition. ESM is associated with a restored ability to detoxify bacterial LPS, thereby regulating dysbiosis in inflammatory bowel disease (IBD) pathogenesis. Previous research has reported the effects and microbe-host interactions underlying an intervention using the fine powder of ESM against IBD. In a murine model of dextran sodium sulphate-induced colitis, ESM significantly suppressed the disease activity index and colon shortening. These effects were associated with considerable amelioration of the expression of genes related to inflammatory mediators, intestinal epithelial cell proliferation, restitution-related factors, and antimicrobial peptides. Concomitantly, cecal metagenomic information established the essential role of ESM in improving dysbiosis characterized by increasing the diversity of bacteria and decreasing absolute numbers of pathogenic bacteria, such as *Enterobacteriaceae* and *Escherichia coli*, as well as in the regulation of the expansion of Th17 cells by suppressing the overgrowth of segmented filamentous bacteria. Such modulations have functional effects on the host; i.e., repairing the epithelium, regulating energy requirements, and eventually alleviating mucosal inflammation. These findings suggest that ESM might cross-regulate commensal microbiota and the mucosal immune system to maintain intestinal homeostasis (Jia et al., 2017a). Notably, taking into consideration the minimal side effects of ESM as a by-product in the manufacturing of egg products, the use of this natural, low-priced waste product appears to be a promising candidate for the prevention and treatment of IBD, leading to a reduction in healthcare costs and a lessened environmental load.

The low digestibility of ESM, with physiological properties similar to that of dietary fiber, may enhance intestinal bacteria fermentation (Jia et al., 2017b) by providing a nutrient source, and hence might contribute to changes in the gut microbiota composition. A previous study reported that ESM supplementation in high-fat diet (HFD)-fed mice reduced plasma triglycerides and liver total cholesterol, possibly through alteration of the lipid metabolism gene expression and gut microbiota composition (Ramli et al., 2020). Interestingly, ESM supplementation increased *L. reuteri*, an anti-obesity bacterium associated with high levels of bile acids in the blood (Il-Dong et al., 2016). The abundance of another member of the phylum Firmicutes, *Roseburia faecis*, decreased in the ESM group, which is consistent with a prospective cohort study in obese postmenopausal women receiving a very low-calorie diet. Therefore, the improvement in lipid profile following ESM supplementation in HFD-fed mice may be at least partially due to the action of "resistant proteins" in ESM altering the microbial composition. Further studies, particularly regarding the mechanisms responsible for restoring a balanced community structure, are required to define potential targets for intervention strategies to improve colitis.

33.2.2 The Effect of Egg White Consumption on Microbiota

Hen egg white (HEW) is a complex system containing hundreds of proteins. These proteins have shown various bioactive properties, including antioxidation, anti-inflammation, antimicrobial, and anticancer activities. Ge et al. (2021) reported that the egg white proteins (EWPs) decreased the levels of pro-inflammatory cytokines and the extent of crypt damage in a dose-dependent manner (Ge et al., 2021). EWPs increased the relative abundances of beneficial bacteria *Lactobacillus* and *Candidatus_Saccharimonas*, and reduced the relative abundances of pathogenic bacteria *Ruminiclostridium* and *Akkermansia*.

Yu et al. (2020) reported a comparative study of the modulation of the gut microbiota in rats by dietary intervention with different sources of egg white (Yu et al., 2020). The relative abundance of *Akkermansia* in the group of rats fed with HEW was highest among all the groups, and the rats fed with duck egg white showed significantly higher relative abundances of *Proteobacteria* and *Peptostreptococcaceae* and lower relative abundance of *Lachnospiraceae*. The relative abundance of *Verrucomicrobia* in the HEW group was significantly higher than that of the other groups, mostly owing to the highest relative abundance of *Akkermansia* in the HEW group (Yu et al., 2020). *Akkermansia muciniphila* is a strictly anaerobic, gram-negative, degradable mucus bacterium that contributes to intestinal health (Derrien et al., 2004; Derrien et al., 2010). Despite being a beneficial probiotic, it has been reported that the relative abundance of *A. muciniphila* increased significantly in the feces of dextran sodium sulfate (DSS)-induced colitis mice (Zhang et al., 2016), and a higher abundance of *Akkermansia* spp in the intestine may aggravate the extent of acute colitis (Li et al., 2018). It is likely that protein, which is the major component of HEW, was the primary factor modifying the microbial population. The overall structure of the gut microbiota in rats fed egg white was more similar than that of rats fed casein. This might be because the amino acid composition and sequence of egg-white proteins in the egg groups were more similar than that of casein (milk protein).

A study of the gut microbiota revealed that pepsin-prepared egg white hydrolysate (EWH) treatment improved dysbiosis in obese rats; furthermore, short-chain fatty acid (SCFA) and lactate concentrations in feces were lower in the obese group (Requena et al., 2017). Mechanisms that could explain the phenomenon include maintenance of intestinal microbiota homeostasis or prevention of absorptive dysfunction by EWH. Nevertheless, Requena et al. (2017) hypothesized that the change in microbiota occurred secondary to peptide absorption, with their actions on target tissues serving as antioxidant and anti-inflammatory agents and leading to modulation of the gut microbiota (Requena et al., 2017). Indeed, the intestinal microbiota is a major factor in the development of innate and acquired immune responses and susceptibility to food allergy (Huang et al., 2017). Some studies have indicated that EWH can exert positive effects on intestinal dysbiosis (Requena et al., 2017), suggesting an interrelation between regulatory functions in the small and large intestine, provided by both bioactive peptides and commensal microorganisms, which might be involved in desensitization to food allergens.

In summary, EWPs and EWH contributed to the alleviation of colitis symptoms and intestinal injury through anti-inflammatory effects, repair of the intestinal mucosa, and modulation of the gut microbiota. This supports EWPs and EWH as alternative adjuvant therapeutic agents for treating intestinal diseases, particularly colitis.

33.2.3 Ovomucin

Ovomucin is characterized by its hydrogel-forming properties, high molecular weight, and extensive *O*-glycosylation with a high degree of sialylation. Many studies have reported the anti-infective activity of sialic acid against bacteria, such as *E. coli*, *Streptococcus*, and *Helicobacter pylori* (Kelm & Schauer, 1997; Schauer, 2004), and ovomucin has also shown similar characteristics. The mucins could bind to the adhesin of *H. pylori* and thus prevent the colonization of the mouse gastric mucosa by the bacterial cells (Kodama & Kimura, 2001). The sialic acid residues in ovomucin showed good adhesion to *E. coli* in vitro, which might act as a receptor analog and potentially compete for bacterial adhesion (Sun et al., 2017). Kobayashi et al. (2004) also proved that the sialic acid groups of pronase-treated ovomucin can bind to *E. coli O157: H7*, and this activity is lost after the ovomucin glycopeptide is treated with sialidase (Kobayashi et al., 2004). Ovomucin has good adhesion activity to *E. coli*, *Salmonella*, and *Staphylococcus aureus* (Shan et al., 2014), which can suppress the colonization of the intestinal epithelial cells by these bacteria (Sun et al., 2017). Ovomucin can also cause cell membrane damage or apoptosis in *S. aureus*, *S. saprophyticus*, and *Streptococcus mutans*, revealing that the bacteriostatic mechanism of ovomucin derivatives is by affecting the permeability of the cell membrane, leading to the eventual rupture of the cell wall and membrane (Tu et al., 2020).

Ovomucin can also inhibit the adhesion of urease, produced by *H. pylori*, to the gastric mucosa. Urease is located on the surface of *H. pylori* cells, and ovomucin inhibits the adhesion of *H. pylori* to the gastric mucosa receptors by predominantly binding to urease (Kodama & Kimura, 2001). The sialic acid at the end of the carbohydrate chain is not only used as sites in "molecule-cell" and "cell-cell" recognition but can also mask recognition sites.

In addition to its antiviral and anti-adhesive activities, the *O*-glycans of ovomucin can be a quality carbohydrate source for some bacteria, including specific gut microbes. Such mucin-degrading gut microbes in humans include *A. muciniphila*, which is notable for its association with the reduction of obesity and type 2 diabetes (Everard et al., 2013; Depommier et al., 2019) and *Bifidobacterium bifidum*, a probiotic species. These species possess sets of extracellular glycosidases involved in mucin glycan degradation (Turroni et al., 2010; Kiyohara et al., 2012; Shin et al., 2019). Moreover, the sugars released by their glycosidases are not only assimilated by these species but are also consumed by specific microbes or shared within the bacterial community (Gotoh et al., 2018), thus potentially influencing the structure of the gut microbiota. In a colitis C57BL/6J mice model, ovomucin administration significantly increased the relative abundance of intestinal beneficial bacteria, including *Lactobacilli*, *Faecalibaculum*, and *Ruminococcus*, and further upregulated the production of bacterial metabolites, such as SCFAs, which are a direct source of energy for the proliferation of epithelial and goblet cells.

In addition to the sialic acid, the ovomucin hydrolysates could also serve as an active substance. The *O*-glycan structure of the ovomucin degraded by mucus-degrading microorganisms can promote the growth of other beneficial microorganisms in the intestines (Takada et al., 2020). Pepsin–pancreatin ovomucin hydrolysate could promote the growth of *B. infantis* (Sun et al., 2016). Peptides released from ovomucin served as decoy receptors for the fimbriae of enterotoxigenic *E. coli* and prevented bacterial adhesion to porcine small intestinal epithelial cells (Sun et al., 2017, 2019). This anti-adhesive strategy is regarded as a promising approach to prevent pathogenic bacteria from colonizing the gastrointestinal tract and its associated enteric infection.

Therefore, ovomucin or its metabolites are likely to prevent intestinal mucus barrier damage by interfering with the structure of the intestinal microbiota, affecting colonization by intestinal *E. coli*, and then affecting the NF-κB signaling pathway and secretion of inflammatory factors. Therefore, when ovomucin and intestinal mucin coexist in the intestinal tissues, microorganisms using mucin as a carbohydrate source may decompose ovomucin at the same time, thus reducing the amount of intestinal mucin decomposed, thereby helping to maintain the integrity of the mucus layer of the intestine (Tu et al., 2020).

33.2.4 Ovotransferrin

Ovotransferrin, an iron-binding glycoprotein with antibacterial activity, has additionally been shown to reduce inflammatory colitis pathology in a DSS-induced mouse model of colitis (Giansanti et al., 2012; Kobayashi et al., 2015). Oral administration of ovotransferrin reduced inflammatory cytokines, while additionally mitigating the clinical markers of colitis, including weight loss and histological scores of the colon (Kobayashi et al., 2015).

Ovotransferrin increased the diversity of intestinal microflora and ameliorated gut microbiota disorders (Zhu et al., 2019), which indicates that ovotransferrin could play a biological function in the intestine by regulating the gut environment. Zhang et al. explored the potential biological activity of ovotransferrin on gut health with young (3-week-old) and adult (8-week-old) mouse models (Zhang et al., 2020). For the adult group, an increase in the abundance of *Actinobacteria* and a significant decrease in *TM 7* were observed at the phylum level after ovotransferrin supplementation. At the genus level, the proportion of *Akkermansia* significantly increased in both the young and adult groups. In addition, ovotransferrin suppressed several harmful bacteria, including *Ruminococcus* and *Anaerotruncus*, which were believed to be associated with inflammatory bowel disease (Hall et al., 2017) and obesity (Togo et al., 2016), although *Ruminococcus* also possesses beneficial effects, such as producing antimicrobial peptides active against gut pathogenic bacteria (Chiumento et al., 2019) and modulating the production and glycosylation of intestinal mucus (Graziani et al., 2016).

Citrobacter rodentium is a natural mouse pathogen that can be used to cause intestinal inflammation and replace enteropathogenic *E. coli* and enteropathogenic *E. coli*. Ma et al. evaluated the abilities of 2 peptides derived from ovotransferrin, IRW and IQW, to treat enteritis in a mouse model of *C. rodentium*-induced colitis by evaluating serum metabolomics and gut microbes (Ma et al., 2019). Compared with the basal diet (CTRL) group, the proportions of Firmicutes and

Actinobacteria species decreased significantly and the proportions of Bacteroidetes and Proteobacteria species increased in the *C. rodentium* group. Supplementation of IRW and IQW improves the morphological damage of the colon caused by *C. rodentium* while regulating the microbial composition of the colon and altering the serum metabolites through changes in microbial composition. In addition, IRW could attenuate LPS-induced inflammatory responses and regulate gut bacteria in rats (Jiao et al., 2019).

33.2.5 Lysozyme

Egg white-derived lysozyme naturally exerts antimicrobial activity against gram-positive and gram-negative bacteria through hydrolysis of the structural peptidoglycans in the bacterial cell walls, in addition to producing antibacterial peptides from within its complete protein structure through enzymatic hydrolysis (Pellegrini et al., 1997; Ibrahim et al., 2002). In a porcine model of DSS-induced colitis, hen egg lysozyme supplementation reduced intestinal gene expression of pro-inflammatory cytokines (TNF-α, IL-6, IFN-γ, IL-8, and IL-17) while increasing the levels of anti-inflammatory IL-4 and transforming growth factor β (TGF-β). Furthermore, lysozyme attenuated weight loss, colonic crypt distortion, muscle wall thickening, and gastric wall permeability observed in the control DSS-treated animals (Lee et al., 2009).

33.2.6 Phosvitin

Characterization of phosvitin as nutritionally negative is associated with reducing its absorption and bioavailability by forming complexes with minerals and being resistant to proteolytic enzymes (Ishikawa et al., 2007). However, phosvitin has significant bactericidal activity against *E. coli*, which is attributable to its high metal-chelating properties, in addition to its high surface activity under thermal stress (Sattar Khan et al., 2000). Li et al. (2019) reported the effects of phosvitin on the microbiota composition in young and adult mouse models, and indicated that phosvitin increased the number of *Bifidobacterium* in young individuals and decreased some pathogenic microbes in adult individuals, including *Helicobacter* and *Mucispirillum* (Li et al., 2019). *Helicobacter* is a commensal colonizer in the mucus of the gastrointestinal tract and is related to intestinal disease. *Mucispirillum* is associated with both inflammatory markers and active colitis in some mouse models. In this sense, phosvitin plays a positive role in the host intestine by inhibiting harmful bacteria.

33.3 The Effect of Egg Lipid on Microbiota

Eggs are a major source of phospholipids (PL) in Western diets. Dietary PL has emerged as a potential source of bioactive lipids that may have widespread effects on pathways related to inflammation, cholesterol metabolism, and high-density lipoprotein function.

The intestinal microbiota plays a vital role in life activities by influencing the absorption of substances in the intestinal tract, energy metabolism, immune organ development, and defense against pathogens. In a murine obesity model, PLs increased the reduction in SCFAs caused by HFD, and increased acetic acid, propionic acid, isobutyric acid, and total SCFAs, which effectively regulate the environment of intestinal microbial metabolism.

Egg consumption improves vascular function, which may be related to the alterations seen in the gut microbiota (Liu et al., 2022). Bacteria from the phyla Firmicutes and Bacteroidetes account for most of the gut microbiota (Stojanov et al., 2020), and the increased Firmicutes/Bacteroidetes ratio plays a role in obesity (Stojanov et al., 2020; Wu et al., 2020). PLs decrease the Firmicutes/Bacteroidetes ratio and the relative abundances of *Streptomyces*, *Arthrobacter*, and *Desulfovibrio*, and increase the relative abundances of *Mucilaginibacter* and *Cellulomonas* (Yin et al., 2021). These results suggest that PLs have a potential modulating effect on intestinal disorders caused by HFD. Egg consumption led to less tryptophan degradation enrichment. Enhanced degradation of tryptophan has been linked to inflammation and maintaining the homeostasis of tryptophan degradation may be a therapeutic target in patients with cardiovascular disease (CVD) (Liu et al., 2017). It is possible that this variation is attributable to differences in intestinal absorption of dietary cholesterol, which is high in healthy, non-insulin resistant individuals (Simonen et al., 2002; Pihlajamaki et al., 2004; Tannock et al., 2005; Paramsothy et al., 2011). However, it is possible that other factors, such as the composition of the microbiome or genetic variation, affect the dietary response to eggs (Tang et al., 2013; Tang & Hazen, 2014).

Egg phosphatidylcholine (PC) is a substrate for the generation of trimethylamine N-oxide (TMAO), a gut microbe-dependent metabolite associated with increased CVD risk (Blesso, 2015). Dietary choline can undergo catabolism by the gut microbiota in the gastrointestinal tract to form the gas trimethylamine, which is absorbed and rapidly oxidized to TMAO in the liver by the hepatic enzyme flavin-containing monooxygenase 3 (FMO3) (Al-Waiz et al., 1992; Lang et al., 1998). Hazen et al. showed that feeding apoE−/− mice diets rich in either choline (0.5% or 1% wt/wt) or TMAO (0.12% wt/wt) for 20 weeks resulted in increased aortic root lesion size despite the absence of changes in traditional CVD risk biomarkers (Wang et al., 2011). Highlighting the role of gut microflora in this process, TMAO formation from dietary PC and the atherogenic effect of choline feeding were abolished when the mice were germ-free or coadministered broad-spectrum antibiotics to deplete gut bacteria. Thus, the formation of blood TMAO subsequent to dietary PC intake requires intact gut microflora and may be influenced by hepatic FMO3 expression. Interestingly, TMAO levels do not appear to be strongly influenced by genetic factors in humans (Hartiala et al., 2014), suggesting that diet and gut microflora primarily determine the differences in plasma TMAO levels between individuals. Notably, healthy participants who consumed a dietary PC challenge showed acute increases in plasma TMAO in a gut microflora-dependent manner (Tang et al., 2013). Furthermore, Zeisel et al. performed a small, double-blind, randomized dose-response study where healthy subjects (n = 6) consumed a breakfast with 0, 1, 2, 4, or 6 egg yolks and the levels of plasma TMAO were measured at pre- and post-breakfast time intervals (up to 24 h). There were more than four-fold differences in plasma TMAO levels between subjects at similar dosages, which was potentially due to the differences in the gut microflora or hepatic FMO3 activity (Miller et al., 2014). Chronic egg intake could influence gut microflora composition and this would presumably impact TMAO formation from choline, similar to that observed with chronic exposure to carnitine (Koeth et al., 2013).

33.4 Conclusion

The differences in the species of food ingredients may result in greater differences in the intestinal flora. This chapter provides a reference for the effect of dietary eggs on microbiota. It is advantageous to examine the effect of dietary egg-induced changes on gut microbiota and to understand the corresponding metabolic functions. Owing to the complex relationship among dietary nutrition, intestinal microbes, and the host, further research is required to uncover the relative mechanisms related to the effect of dietary eggs on the host by altering the composition of the gut microbiota.

REFERENCES

Al-Waiz, M., Mikov, M., Mitchell, S. C., & Smith, R. L. (1992). The exogenous origin of trimethylamine in the mouse. *Metabolism*, *41*(2), 135–136. https://doi.org/10.1016/0026-0495(92)90140-6

Avirineni, B. S., Singh, A., Zapata, R. C., Phillips, C. D., & Chelikani, P. K. (2022). Dietary whey and egg proteins interact with inulin fiber to modulate energy balance and gut microbiota in obese rats. *Journal of Nutritional Biochemistry*, *99*, 108860. https://doi.org/10.1016/j.jnutbio.2021.108860

Backhed, F., Fraser, C. M., Ringel, Y., Sanders, M. E., Sartor, R. B., Sherman, P. M., Versalovic, J., Young, V., & Finlay, B. B. (2012). Defining a healthy human gut microbiome: Current concepts, future directions, and clinical applications. *Cell Host & Microbe*, *12*(5), 611–622. https://doi.org/10.1016/j.chom.2012.10.012

Blesso, C. N. (2015). Egg phospholipids and cardiovascular health. *Nutrients*, *7*(4), 2731–2747. https://doi.org/10.3390/nu7042731

Brown, K., DeCoffe, D., Molcan, E., & Gibson, D. L. (2012). Diet-induced dysbiosis of the intestinal Microbiota and the effects on immunity and disease. *Nutrients*, *4*(8), 1095–1119. https://doi.org/10.3390/nu4081095

Chiumento, S., Roblin, C., Kieffer-Jaquinod, S., Tachon, S., Leprètre, C., Basset, C., Aditiyarini, D., Olleik, H., Nicoletti, C., Bornet, O., Iranzo, O., Maresca, M., Hardré, R., Fons, M., Giardina, T., Devillard, E., Guerlesquin, F., Couté, Y., Atta, M., & Duarte, V. (2019). Ruminococcin C, a promising antibiotic produced by a human gut symbiont. *Science Advances*, *5*(9), eaaw9969. https://doi.org/10.1126/sciadv.aaw9969

Depommier, C., Everard, A., Druart, C., Plovier, H., Van Hul, M., Vieira-Silva, S., Falony, G., Raes, J., Maiter, D., Delzenne, N. M., Barsy, M., Loumaye, A., Hermans, M. P., Thissen, J. P., de Vos, W., & Cani, P. D. (2019). Supplementation with *Akkermansia muciniphila* in overweight and obese human volunteers: A proof-of-concept exploratory study. *Nature Medicine*, *25*(7), 1096–1103. https://doi.org/10.1038/s41591-019-0495-2

Derrien, M., van Passel, M. W. J., van de Bovenkamp, J. H. B., Schipper, R., de Vos, W., & Dekker, J. (2010). Mucin-bacterial interactions in the human oral cavity and digestive tract. *Gut Microbes*, *1*(4), 254–268. https://doi.org/10.4161/gmic.1.4.12778

Derrien, M., Vaughan, E. E., Plugge, C. M., & De Vos, W. M. (2004). *Akkermansia muciniphila* gen. nov., sp. nov., a human intestinal mucin-degrading bacterium. *International Journal of Systematic and Evolutionary Microbiology*, *54*(5), 1469–1476. https://doi.org/10.1099/ijs.0.02873-0

Everard, A., Belzer, C., Geurts, L., Ouwerkerk Janneke, P., Druart, C., Bindels Laure, B., Guiot, Y., Derrien, M., Muccioli, G. G., Delzenne, N. M., de Vos, W. M., Cani, P. D., & Cani Patrice, D. (2013). Cross-talk between *Akkermansia muciniphila* and intestinal epithelium controls diet-induced obesity. *Proceedings of the National Academy of Sciences*, *110*(22), 9066–9071. https://doi.org/10.1073/pnas.1219451110

Ge, H., Cai, Z., Chai, J., Liu, J., Liu, B., Yu, Y., Liu, J., & Zhang, T. (2021). Egg white peptides ameliorate dextran sulfate sodium-induced acute colitis symptoms by inhibiting the production of pro-inflammatory cytokines and modulation of gut microbiota composition. *Food Chemistry*, *360*, 129981–129991. https://doi.org/10.1016/j.foodchem.2021.129981

Giansanti, F., Leboffe, L., Pitari, G., Ippoliti, R., & Antonini, G. (2012). Physiological roles of ovotransferrin. *Biochimica et Biophysica Acta*, *1820*(3), 218–225. https://doi.org/10.1016/j.bbagen.2011.08.004

Gotoh, A., Katoh, T., Sakanaka, M., Ling, Y., Yamada, C., Asakuma, S., Urashima, T., Tomabechi, Y., Katayama-Ikegami, A., Kurihara, S., Yamamoto, K., Harata, G., He, F., Hirose, J., Kitaoka, M., Okuda, S., & Katayama, T. (2018). Sharing of human milk oligosaccharides degradants within bifidobacterial communities in faecal cultures supplemented with *Bifidobacterium* bifidum. *Scientific Reports*, *8*(1), 13958. https://doi.org/10.1038/s41598-018-32080-3

Graziani, F., Pujol, A., Nicoletti, C., Dou, S., Maresca, M., Giardina, T., Fons, M., & Perrier, J. (2016). *Ruminococcus* gnavus E1 modulates mucin expression and intestinal glycosylation. *Journal of Applied Microbiology*, *120*(5), 1403–1417. https://doi.org/10.1111/jam.13095

Hall, A. B., Yassour, M., Sauk, J., Garner, A., Jiang, X., Arthur, T., Lagoudas, G. K., Vatanen, T., Fornelos, N., Wilson, R., Bertha, M., Cohen. M., Garber, J., Khalili, H., Gevers, D., Ananthakrishnan, A. N., Kugathasan, S., Lander, E. S., Blainey, P., & Huttenhower, C. (2017). A novel *Ruminococcus* gnavus clade enriched in inflammatory bowel disease patients. *Genome Medicine*, *9*(1), 103. https://doi.org/10.1186/s13073-017-0490-5

Hartiala, J., Bennett, B. J., Tang, W. H., Wang, Z., Stewart, A. F., Roberts, R., McPherson, R., Lusis, A. J., Hazen, S. L., & Consortium, C. A. (2014). Comparative genome-wide association studies in mice and humans for trimethylamine N-oxide, a proatherogenic metabolite of choline and L-carnitine. *Arteriosclerosis, Thrombosis, and Vascular Biology*, *34*(6), 1307–1313. https://doi.org/10.1161/ATVBAHA.114.303252

Huang, Y. J., Marsland, B. J., Bunyavanich, S., O'Mahony, L., Leung, D. Y., Muraro, A., & Fleisher, T. A. (2017). The microbiome in allergic disease: Current understanding and future opportunities-2017 PRACTALL document of the American academy of allergy, asthma & immunology and the European academy of allergy and clinical immunology. *The Journal of Allergy and Clinical Immunology*, *139*(4), 1099–1110. https://doi.org/10.1016/j.jaci.2017.02.007

Ibrahim, R. H., Aoki, T., & Pellegrini, A. (2002). Strategies for new antimicrobial proteins and peptides: Lysozyme and aprotinin as model molecules. *Current Pharmaceutical Design*, *8*(9), 671–693. https://doi.org/10.2174/1381612023395349

Il-Dong, C., Sung-Hwan, K., Ji-Woong, J., Dong Eun, L., Chul-Sung, H., Seong Soo, H., & Jae-Hun, S. (2016). Triglyceride-lowering effects of two probiotics, lactobacillus plantarum KY1032 and lactobacillus curvatus HY7601, in a rat model of high-fat diet-induced hypertriglyceridemia. *Journal of Microbiology and Biotechnology*, *26*(3), 483–487. https://doi.org/10.4014/jmb.1512.12018

Ishikawa, S. I., Tamaki, S., Arihara, K., & Itoh, M. (2007). egg yolk protein and egg yolk phosvitin inhibit calcium, magnesium, and iron absorptions in rats. *Journal of Food Science*, *72*(6), S412–S419. https://doi.org/10.1111/j.1750-3841.2007.00417.x

Jia, H., Hanate, M., Aw, W., Itoh, H., Saito, K., Kobayashi, S., Hachimura, S., Fukuda, S., Tomita, M., Hasebe, Y., & Kato, H. (2017a). Eggshell membrane powder ameliorates intestinal inflammation by facilitating the restitution of epithelial injury and alleviating microbial dysbiosis. *Scientific Reports*, *7*, 43993. https://doi.org/10.1038/srep43993

Jia, H., Hanate, M., Aw, W., Itoh, H., Saito, K., Kobayashi, S., Hachimura, S., Fukuda, S., Tomita, M., Hasebe, Y., & Kato, H. (2017b). Eggshell membrane powder ameliorates intestinal inflammation by facilitating the restitution of epithelial injury and alleviating microbial dysbiosis. *Scientific Reports*, *7*, 43993. https://doi.org/10.1038/srep43993

Jiao, H., Zhang, Q., Lin, Y., Gao, Y., & Zhang, P. (2019). The ovotransferrin-derived peptide IRW attenuates lipopolysaccharide-induced inflammatory responses. *BioMed Research International*, *2019*, 8676410. https://doi.org/10.1155/2019/8676410

Kelm, S., & Schauer, R. (1997). Sialic acids in molecular and cellular interactions. In K. W. Jeon (Ed.), *International review of cytology* (Vol. 175, pp. 137–240). Academic Press. https://doi.org/10.1016/S0074-7696(08)62127-0

Kiyohara, M., Nakatomi, T., Kurihara, S., Fushinobu, S., Suzuki, H., Tanaka, T., Shoda, S.-I., Kitaoka, M., Katayama, T., Yamamoto, K., & Ashida, H. (2012). α-N-acetylgalactosaminidase from infant-associated *Bifidobacteria* belonging to novel glycoside hydrolase family 129 is implicated in alternative mucin degradation pathway. *Journal of Biological Chemistry*, *287*(1), 693–700. https://doi.org/10.1074/jbc.M111.277384

Kobayashi, K., Hattori, M., Hara-Kudo, Y., Okubo, T., Yamamoto, S., Takita, T., & Sugita-Konishi, Y. (2004). Glycopeptide derived from hen egg ovomucin has the ability to bind enterohemorrhagic *Escherichia coli* O157:H7. *Journal of Agricultural and Food Chemistry*, *52*(18), 5740–5746. https://doi.org/10.1021/jf0353335

Kobayashi, Y., Rupa, P., Kovacs-Nolan, J., Turner, P. V., Matsui, T., & Mine, Y. (2015). Oral Administration of hen egg white ovotransferrin attenuates the development of colitis induced by dextran sodium sulfate in mice. *Journal of Agricultural and Food Chemistry*, *63*(5), 1532–1539. https://doi.org/10.1021/jf505248n

Kodama, Y., & Kimura, N. (2001). *Inhibitor of helicobacter pylori colonization*. Google Patents.

Koeth, R. A., Wang, Z., Levison, B. S., Buffa, J. A., Org, E., Sheehy, B. T., Britt, E. B., Fu, X., Wu, Y., Li, L., Smith, J. D., Di Donato, J. A., Chen, J., Li, H., Wu, G. D., Lewis, J. D., Warrier, M., Brown, J. M., Krauss, R. M., & Hazen, S. L. (2013). Intestinal microbiota metabolism of l-carnitine, a nutrient in red meat, promotes atherosclerosis. *Nature Medicine*, *19*(5), 576–585. https://doi.org/10.1038/nm.3145

Lang, D., Yeung, C., Peter, R., Ibarra, C., Gasser, R., Itagaki, K., Philpot, R. M., & Rettie, A. (1998). Isoform specificity of trimethylamine N-oxygenation by human flavin-containing monooxygenase (FMO) and P450 enzymes: Selective catalysis by fmo3. *Biochemical Pharmacology*, *56*(8), 1005–1012. https://doi.org/10.1016/S0006-2952(98)00218-4

Lee, M., Kovacs-Nolan, J., Yang, C., Archbold, T., Fan, M. Z., & Mine, Y. (2009). Hen egg lysozyme attenuates inflammation and modulates local Gene expression in a porcine model of dextran sodium sulfate (DSS)-induced colitis. *Journal of Agricultural and Food Chemistry*, *57*(6), 2233–2240. https://doi.org/10.1021/jf803133b

Ley, R. E., Turnbaugh, P. J., Klein, S., & Gordon, J. I. (2006). Human gut microbes associated with obesity. *Nature*, *444*(7122), 1022–1023. https://doi.org/10.1038/4441022a

Liu, G., Chen, S., Zhong, J., Teng, K., & Yin, Y. (2017). Crosstalk between tryptophan metabolism and cardiovascular disease, mechanisms, and therapeutic implications. *Oxidative Med Cell Longev*, *2017*, 1602074. https://doi.org/10.1155/2017/1602074

Liu, X., Shao, Y., Sun, J., Tu, J., Wang, Z., Tao, J., & Chen, J. (2022). Egg consumption improves vascular and gut microbiota function without increasing inflammatory, metabolic, and oxidative stress markers. *Food Science & Nutrition*, *10*(1), 295–304. https://doi.org/10.1002/fsn3.2671

Li, M., Wu, Y., Hu, Y., Zhao, L., & Zhang, C. (2018). Initial gut microbiota structure affects sensitivity to DSS-induced colitis in a mouse model. *Science China Life Sciences*, *61*(7), 762–769. https://doi.org/10.1007/s11427-017-9097-0

Li, M., Zhang, Q., Wang, R., Lu, M., Xu, X., Liu, X., & Duan, X. (2019). Effects of egg phosvitin on mucosal transcriptional profiles and luminal microbiota composition in murine colon. *Food & Function*, *10*(5), 2805–2816. https://doi.org/10.1039/C9FO00074G

Lozupone, C. A., Stombaugh, J. I., Gordon, J. I., Jansson, J. K., & Knight, R. (2012). Diversity, stability and resilience of the human gut microbiota. *Nature*, *489*(7415), 220–230. https://doi.org/10.1038/nature11550

Ma, Y., Ding, S., Liu, G., Fang, J., Yan, W., Duraipandiyan, V., Al-Dhabi, N. A., Esmail, G. A., & Jiang, H. (2019). Egg protein transferrin-derived peptides IRW and IQW regulate citrobacter rodentium-induced, inflammation-related microbial and metabolomic profiles. *Frontiers in Microbiology*, *10*, 643. https://doi.org/10.3389/fmicb.2019.00643

Meng, Y., Chen, C., Qiu, N., & Keast, R. (2020). Modulation of gut microbiota in rats fed whole egg diets by processing duck egg to preserved egg. *Journal of Bioscience and Bioengineering*, *130*(1), 54–62. https://doi.org/10.1016/j.jbiosc.2020.02.015

Miller, C. A., Corbin, K. D., da Costa, K. A., Zhang, S., Zhao, X., Galanko, J. A., Blevins, T., Bennett, B. J., O'Connor, A., & Zeisel, S. H. (2014). Effect of egg ingestion on trimethylamine-N-oxide production in humans: A randomized, controlled, dose-response study. *The American Journal of Clinical Nutrition*, *100*(3), 778–786. https://doi.org/10.3945/ajcn.114.087692

Paramsothy, P., Knopp, R. H., Kahn, S. E., Retzlaff, B. M., Fish, B., Ma, L., & Ostlund, R. E. Jr. (2011). Plasma sterol evidence for decreased absorption and increased synthesis of cholesterol in insulin resistance and obesity. *The American Journal of Clinical Nutrition*, *94*(5), 1182–1188. https://doi.org/10.3945/ajcn.110.006668

Pellegrini, A., Thomas, U., Bramaz, N., Klauser, S., Hunziker, P., & Von Fellenberg, R. (1997). Identification and isolation of a bactericidal domain in chicken egg white lysozyme. *Journal of Applied Microbiology*, *82*(3), 372–378. https://doi.org/10.1046/j.1365-2672.1997.00372.x

Pihlajamaki, J., Gylling, H., Miettinen, T. A., & Laakso, M. (2004). Insulin resistance is associated with increased cholesterol synthesis and decreased cholesterol absorption in normoglycemic men. *Journal of Lipid Research*, *45*(3), 507–512. https://doi.org/10.1194/jlr.M300368-JLR200

Ramli, N. S., Jia, H., Sekine, A., Lyu, W., Furukawa, K., Saito, K., Hasebe, Y., & Kato, H. (2020). Eggshell membrane powder lowers plasma triglyceride and liver total cholesterol by modulating gut microbiota and accelerating lipid metabolism in high-fat diet-fed mice. *Food Science & Nutrition*, *8*(5), 2512–2523. https://doi.org/10.1002/fsn3.1545

Requena, T., Miguel, M., Garcés-Rimón, M., Martínez-Cuesta, M. C., López-Fandiño, R., & Peláez, C. (2017). Pepsin egg white hydrolysate modulates gut microbiota in Zucker obese rats. *Food & Function*, *8*(1), 437–443. https://doi.org/10.1039/C6FO01571A

Rist, V. T. S., Weiss, E., Eklund, M., & Mosenthin, R. (2013). Impact of dietary protein on microbiota composition and activity in the gastrointestinal tract of piglets in relation to gut health: A review. *Animal*, *7*(7), 1067–1078. https://doi.org/10.1017/S1751731113000062

Sattar Khan, M. A., Nakamura, S., Ogawa, M., Akita, E., Azakami, H., & Kato, A. (2000). Bactericidal action of egg yolk phosvitin against *Escherichia coli* under thermal stress. *Journal of Agricultural and Food Chemistry*, *48*(5), 1503–1506. https://doi.org/10.1021/jf990700r

Schauer, R. (2004). Victor Ginsburg's influence on my research of the role of sialic acids in biological recognition. *Archives of Biochemistry and Biophysics*, *426*(2), 132–141. https://doi.org/10.1016/j.abb.2004.03.008

Shan, Y., Xu, Q., & Ma, M. (2014). Mg2+ binding affects the structure and activity of ovomucin. *International Journal of Biological Macromolecules*, *70*, 230–235. https://doi.org/10.1016/j.ijbiomac.2014.06.056

Shin, J., Noh, J. R., Chang, D. H., Kim, Y. H., Kim, M. H., Lee, E. S., Cho, S., Ku, B. J., Rhee, M. S., Kim, B. C., Lee, C. H., & Cho, B. K. (2019). Elucidation of *Akkermansia muciniphila* probiotic traits driven by mucin depletion. *Frontiers in Microbiology*, *10*, 1137. https://doi.org/10.3389/fmicb.2019.01137

Simonen, P. P., Gylling, H. K., & Miettinen, T. A. (2002). Diabetes contributes to cholesterol metabolism regardless of obesity. *Diabetes Care*, *25*(9), 1511–1515. https://doi.org/10.2337/diacare.25.9.1511

Stojanov, S., Berlec, A., & Strukelj, B. (2020). The influence of probiotics on the Firmicutes/Bacteroidetes ratio in the treatment of obesity and inflammatory bowel disease. *Microorganisms*, *8*(11). https://doi.org/10.3390/microorganisms8111715

Sun, X., Gänzle, M., Field, C. J., & Wu, J. (2016). Effect of proteolysis on the sialic acid content and bifidogenic activity of ovomucin hydrolysates. *Food Chemistry*, *212*, 78–86. https://doi.org/10.1016/j.foodchem.2016.05.153

Sun, X., Gänzle, M., & Wu, J. (2017). Identification and characterization of glycopeptides from egg protein ovomucin with anti-agglutinating activity against porcine K88 enterotoxigenic Escherichia coli strains. *Journal of Agricultural and Food Chemistry*, *65*(4), 777–783. https://doi.org/10.1021/acs.jafc.6b04299

Sun, X., Gänzle, M., & Wu, J. (2019). Glycopeptides from egg white ovomucin inhibit K88ac enterotoxigenic Escherichia coli adhesion to porcine small intestinal epithelial cell-line. *Journal of Functional Foods*, *54*, 320–328. https://doi.org/10.1016/j.jff.2019.01.033

Takada, H., Katoh, T., & Katayama, T. (2020). Sialylated *O*-glycans from hen egg white ovomucin are decomposed by mucin-degrading gut microbes. *Journal of Applied Glycoscience*, *67*(2), 31–39. https://doi.org/10.5458/jag.jag.JAG-2019_0020

Tang, W. H., & Hazen, S. L. (2014). The contributory role of gut microbiota in cardiovascular disease. *Journal of Clinical Investigation*, *124*(10), 4204–4211. https://doi.org/10.1172/JCI72331

Tang, W. H., Wang, Z., Levison, B. S., Koeth, R. A., Britt, E. B., Fu, X., Wu, Y., & Hazen, S. L. (2013). Intestinal microbial metabolism of phosphatidylcholine and cardiovascular risk. *The New England Journal of Medicine*, *368*(17), 1575–1584. https://doi.org/10.1056/NEJMoa1109400

Tannock, L. R., O'Brien, K. D., Knopp, R. H., Retzlaff, B., Fish, B., Wener, M. H., Kahn, S. E., & Chait, A. (2005). Cholesterol feeding increases C-reactive protein and serum amyloid A levels in lean insulin-sensitive subjects. *Circulation*, *111*(23), 3058–3062. https://doi.org/10.1161/CIRCULATIONAHA.104.506188

Togo, A. H., Khelaifia, S., Valero, R., Cadoret, F., Raoult, D., & Million, M. (2016). 'Negativicoccus massiliensis', a new species identified from human stool from an obese patient after bariatric surgery. *New Microbes and New Infections*, *13*, 43–44. https://doi.org/10.1016/j.nmni.2016.05.018

Tu, A., Zhao, X., Shan, Y., & Lu, X. (2020). Potential role of ovomucin and its peptides in modulation of intestinal health: A review. *International Journal of Biological Macromolecules*, *162*, 385–393. https://doi.org/10.1016/j.ijbiomac.2020.06.148

Turroni, F., Bottacini, F., Foroni, E., Mulder, I., Kim, J.-H., Zomer, A., Sánchez, B., Bidossi, A., Ferrarini, A., Giubellini, V., Delledonne, M., Henrissat, B., Coutinho, P., Oggioni, M., Fitzgerald, G. F., Mills, D., Margolles, A., Kelly, D., van Sinderen, D., & Ventura, M. (2010). Genome analysis of *Bifidobacterium* bifidum PRL2010 reveals metabolic pathways for host-derived glycan foraging. *Proceedings of the National Academy of Sciences*, *107*(45), 19514–19519. https://doi.org/10.1073/pnas.1011100107

van Hylckama Vlieg, J. E. T., Veiga, P., Zhang, C., Derrien, M., & Zhao, L. (2011). Impact of microbial transformation of food on health—From fermented foods to fermentation in the gastro-intestinal tract. *Current Opinion in Biotechnology*, *22*(2), 211–219. https://doi.org/10.1016/j.copbio.2010.12.004

Wang, Z., Klipfell, E., Bennett, B. J., Koeth, R., Levison, B. S., Dugar, B., Feldstein, A. E., Britt, E. B., Fu, X., Chung, Y.-M., Wu, Y., Schauer, P., Smith, J. D., Allayee, H., Wilson Tang, W. H., DiDonato, J. A., Lusis, A. J., Hazen, S. L. (2011). Gut flora metabolism of phosphatidylcholine promotes cardiovascular disease. *Nature*, *472*(7341), 57–63. https://doi.org/10.1038/nature09922

Wu, S., Bhat, Z. F., Gounder, R. S., Ahmed Mohamed, I. A., Al-Juhaimi, F. Y., Ding, Y., & Bekhit, A. E. A. (2022). Effect of dietary protein and processing on gut Microbiota-a systematic review. *Nutrients*, *14*(3). https://doi.org/10.3390/nu14030453

Wu, T., Gao, Y., Hao, J., Geng, J., Zhang, J., Yin, J., Liu, R., Sui, W., Gong, L., & Zhang, M. (2020). Capsanthin extract prevents obesity, reduces serum TMAO levels and modulates the gut microbiota composition in high-fat-diet induced obese C57BL/6J mice. *Food Research International*, *128*, 108774. https://doi.org/10.1016/j.foodres.2019.108774

Yin, M., Zhang, L., Tang, S., Matsuoka, R., Xi, Y., Tao, N., & Wang, X. (2021). Egg yolk phospholipids modulate microbial imbalance in the intestinal tract of rats on a high-fructose diet. *European Journal of Lipid Science and Technology*, *124*(3). https://doi.org/10.1002/ejlt.202100131

Yu, H., Qiu, N., Meng, Y., & Keast, R. (2020). A comparative study of the modulation of the gut microbiota in rats by dietary intervention with different sources of egg-white proteins. *Journal of the Science of Food and Agriculture*, *100*(9), 3622–3629. https://doi.org/10.1002/jsfa.10387

Zhang, Q., Tan, W., Yang, L., Lu, M., Dong, S., Liu, X., & Duan, X. (2020). Multi-omics analysis of the effects of egg ovotransferrin on the gut environment in mice: Mucosal Gene expression, Microbiota composition, and intestinal structural homeostasis. *Molecular Nutrition & Food Research*, *64*(7), e1901024. https://doi.org/10.1002/mnfr.201901024

Zhang, Q., Wu, Y., Wang, J., Wu, G., Long, W., Xue, Z., Wang, L., Zhang, X., Pang, X., Zhao, Y., Zhao, L., & Zhang, C. (2016). Accelerated dysbiosis of gut microbiota during aggravation of DSS-induced colitis by a butyrate-producing bacterium. *Scientific Reports*, *6*(1), 27572. https://doi.org/10.1038/srep27572

Zhu, G., Jiang, Y., Yao, Y., Wu, N., Luo, J., Hu, M., Tu, Y., & Xu, M. (2019). Ovotransferrin ameliorates the dysbiosis of immunomodulatory function and intestinal microbiota induced by cyclophosphamide. *Food & Function*, *10*(2), 1109–1122. https://doi.org/10.1039/C8FO02312C

Zhu, Y., Lin, X., Zhao, F., Shi, X., Li, H., Li, Y., Zhou, G. (2015). Erratum: Meat, dairy and plant proteins alter bacterial composition of rat gut bacteria. *Scientific Reports*, *5*(1), 16546. https://doi.org/10.1038/srep16546

34

Egg and Sport Nutrition

Ning Qiu

CONTENTS

34.1 Introduction

With recent changes shaping the public attitude toward health, there is an increasing awareness of the importance of sports in the pursuit of high-quality lifestyles. Thus, more and more people are adopting an active lifestyle, which is leading to the conversion of many professional competitive sports into activities practiced by the general public. The growing health consciousness and the desire for fast results with bodybuilding has helped sports nutrition to become more appealing to mainstream consumers and has stimulated the development of the sports food industry. Sport foods provide nutrition to meet the metabolic and energy needs of active people and sports aficionados. The nutritional components of sports foods can be divided into basic nutrients and functional factors. Basic nutrients refer to the nutrients or metabolites required by the human body in daily life. Functional factors are bioactive ingredients that have the potential effects to enhance performance during workouts, such as the protection of articular cartilage, the improvement of muscle growth and/or the speed of recovery from exercise. Currently, there are various forms of sports foods on the market, including sports drinks, solid or semi-solid sports foods, and sports nutrition supplements. Whey protein, sports gels, or electrolyte replacement drinks are well-known examples of sports foods (Arenas-Jal et al., 2020).

Sports foods are typically consumed as dietary supplements. Their nutritional content and composition are significantly different from those of ordinary foods. Decades of research on sports nutrition have provided many insights into ideal nutrients for athletes and resulted in the huge increase of highly purified food supplements marketed to athletes (Cui et al., 2022). Nevertheless, these supplements are far from complete in terms of nutrition. Priority foods that are rich in various nutrients should be first considered for greater sport performance or recovery from training. Protein-rich and nutrient-dense wholesome foods, such as eggs, meet the standard of a so-called athletes' 'superfood'. With the robust nutritional profile, low cost and culinary versatility of eggs, sports enthusiasts would greatly benefit from including them in their usual diet and/or as a targeted postexercise nutritional strategy.

34.2 Use of Egg Protein Supplementation for Exercising

Adequate protein intake is imperative for athletes because proteins play various roles in supporting physical performance and recovery. The protein recommendations for athletes (1.2–2.0 g/kg body weight per day) are higher than the recommendations for the general public, since the ingestion of proteins influences a series of physiological factors including muscle remodeling, glycogen resynthesis, energy production, and maintenance of non-muscle structural tissues. Most of the scientific research investigating the effects of protein intake on exercise performance has focused on supplemental protein intake, with higher daily protein intakes recommended to ensure an anabolic environment for the athlete. The effect of different sources of proteins is also an important consideration for athletes. Among the dietary sources, whey protein is the preferred ingredient of sports products for athletes or fitness people in order to quickly replenish protein level. However, some studies have suggested the potential adverse effect of a diet during indiscriminating use of this type of supplement (Vasconcelos et al., 2021). Current literature provides some biologically plausible mechanisms by which eggs may provide some specific benefit over other protein sources (Kieffer & Walzem, 2015).

Avian egg is often thought of as an excellent source of protein for athletes due to its excellent digestibility and amino acid profile (see Chapter 29). For people involved in intense sports training, egg proteins may have a profound effect on their training results, as studies have shown an enhancement in skeletal muscles synthesis with the inclusion of egg protein in the diet (Herron & Fernandez, 2004). One large egg contains 75 kcal and 6 g of protein, but only 1.5 g of saturated fat, while the egg white compartment, fat-free, contains 16 kcal and 3.5 g of protein. At both 20 and 40 g doses, egg proteins have shown to significantly increase after resistance exercise both plasma

DOI: 10.1201/9781003254430-42

protein concentrations and protein synthesis in skeletal muscles. It is also well established that essential amino acids stimulate skeletal muscle protein synthesis in animal and human models, and that proteins in egg have the highest biological value (Glynn et al., 2010) (also see Chapter 29). Fifteen grams of egg white protein contain about 1,300 mg of leucine (the third most common amino acid in eggs after glutamic and aspartic acids), and is also an abundant source of branched amino acids and aromatic amino acids. Recent work showed that leucine induces the maximal skeletal muscle protein anabolic response in young people, which suggests that egg white protein intake might have an important effect on body mass accretion (Hida et al., 2012). More specifically, in animal models, leucine stimulates skeletal muscle synthesis independently of all other amino acids and is a potent stimulator of the cell hypertrophy mammalian target of rapamycin complex pathway. Additionally, leucine decreases muscle protein breakdown and breakdown-associated cellular signaling and mRNA expression (Miranda et al., 2015).

However, the supplementation with 15 g of egg white protein (raising the daily protein intake to 1.23 g of protein/kg body weight/day) over an 8-week period did not improve the maximal upper and lower body strengths in female subjects when compared to a carbohydrate placebo diet containing one gram of protein/kg body weight/day (Hida et al., 2012). In this study, egg white protein supplementation caused a significant increase in the resistance muscle strength as well as carbohydrate supplementation in female athletes. Although this protein supplementation was associated with some changes in protein metabolites when compared to the carbohydrate diet, there were no effects of the protein supplement on the body composition or strength measurements. It is also mentioned by the researchers that the most likely explanation for the improved fat free mass, one repetition maximum muscle strength and some of the free amino acids was that these well-trained female athletes continued training throughout the study period, rather than the dietary supplementation.

Additionally, Egg White Peptides (EWPs), produced by hydrolyzing egg white with a neutral protease, has strong anti-oxidative effects, and is expected to have some value as a functional ingredient for improving endurance training and decreasing post-training fatigue. The results of three human studies suggested that EWP is effective and may decrease muscle damage following physically demanding training sessions. Moreover, it is expected that the administration of EWP would be highly effective for the treatment of muscle fatigue during prolonged strenuous exercising (e.g. marathon running) and/or overtraining (Sugiyama et al., 2020).

34.3 Whole Egg as a Superfood for Performance Training

In addition to providing a cost effective, high-quality source of protein, eggs have also been identified as a functional food, as it is essential for athletes to select foods that not only meet protein requirements but also optimize health and prevent decrements in the immune function following intense training (Jäger et al., 2017). Some important nutrients provided by eggs include riboflavin (15% recommended daily allowance, RDA), selenium (17% RDA) and vitamin K (31% RDA) (See Chapter 29). Eggs are also rich in choline, a nutrient which may have positive effects on the cognitive function. Moreover, eggs provide an excellent source of the carotenoid-based antioxidants lutein and zeaxanthin. These non-protein nutrients also support the optimization of sports performance and the recovery after training. Thus, the consumption of whole eggs may promote a greater stimulation of postexercise muscle protein synthesis than the supplementation with egg white proteins (van Vliet et al., 2017).

A recent study with young men showed that the ingestion of whole eggs immediately after resistance exercise resulted in a greater stimulation of myofibrillar protein synthesis than did the ingestion of egg whites, despite being matched for protein content. The results of this study indicated that the ingestion of nutrient- and protein-dense foods differentially stimulates muscle anabolism compared with protein-dense foods (van Vliet et al., 2017). Further investigation has demonstrated that whole egg and egg white ingestion differentially influenced the co-localization of key regulators of mechanistic target of rapamycin complex 1 (mTORC1), with the greater postexercise muscle protein synthetic response with whole egg ingestion being related in part to an enhanced recruitment of mTORC1-Rheb complexes to the lysosome during recovery. These data suggested that non-protein dietary factors, presumably due to non-protein nutrients found within the egg yolk, influenced the postexercise regulation of mRNA translation in human skeletal muscle (Abou Sawan et al., 2018). Another study in resistance-trained young males found that the consumption of whole egg postexercise increased knee extension and handgrip strength, testosterone level, and reduced body fat percentage when compared the ingestion postexercise of egg white ingestion (Bagheri et al., 2021). Thus, whole eggs consumption are considered to be preferable, especially during resistance training programs geared toward the improvement of muscular strength and body fat percentage.

The removal of the yolk is often promoted for improved health when several eggs are consumed at once. This is an unsubstantiated belief related to the cholesterol and fat content of egg yolk (see Chapters 29 and 30). However, despite the failure of most recent large scale prospective studies to find an association between daily egg consumption and heart disease risk, eggs continue to be thought to promote heart disease by some health professionals. The yolk is nutrient dense and contains around 40% of the total protein contained in the egg, and its removal seems counterproductive for meeting the protein recommendations. Even though the current evidence suggests that egg yolk intake seems to enhance significant increases of myofibrillar protein synthesis, more studies are needed to better evaluate the effects of egg yolk (or whole eggs) intake on muscle mass, not only in young men, but also in other population segments such as women, older adults, and individuals with muscle wasting diseases (Table 34.1).

34.4 Conclusion

The practice of sports and intensive exercises increases the needs for energy and nutrients. The provision of a diet with the appropriate nutritional profile is critically important to reach

TABLE 34.1
Egg Components and Their Effects in Sport Performance

Egg Components	Findings of the Effects in Sport Performance	Participants	Reference
Egg white protein (15.0 g egg white protein; 75 kcal)	The protein supplement was associated with some changes in protein metabolites and caused a significant increase in the resistance muscle strength as well as carbohydrate supplementation.	Healthy female collegiate athletes	Hida et al. (2012)
Egg white peptides	It is expected to have some value as a functional ingredient for improving endurance training and decreasing post-training fatigue.	Strenuous exercise (marathon running) and/or overtraining	Sugiyama et al. (2020)
Whole egg	Ingestion of whole eggs immediately after resistance exercise resulted in greater stimulation of myofibrillar protein synthesis than did the ingestion of egg whites.	Resistance-trained men	van Vliet et al. (2017)
Whole egg	Whole egg ingestion, but not egg whites, induces mTOR colocalization to the lysosome during recovery from resistance exercise.	Resistance-trained men	Abou Sawan et al. (2018)
Whole egg	Postexercise whole egg ingestion increases knee extension, handgrip strength, testosterone, and reduces body fat percentage compared with postexercise egg white ingestion, despite no group differences in muscle mass.	Resistance-trained men	Bagheri et al. (2021)

maximum performance, reduce the risk of injury, and ensure the best recovery. There is a growing recognition that supporting optimum sport achievement through optimum nutrition can prevent health problems at various stages of life. Eggs contain proteins of high quality and bioavailability, a wide range of fatty acids having a positive impact from a cardiovascular point of view, as well as containing vitamins and minerals involved in energy and protein metabolism, cell growth, tissue repair, and defense against oxidative stress and inflammation. However, the analyses using eggs as the protein source for athletic performance and body composition is still limited, perhaps due to less funding opportunities relative to the support provided by the dairy industry. In addition to the provision of proteins, non-protein nutrients in the egg (e.g. choline, cholesterol, fatty acids, vitamins, minerals, and carotenoids) may also have a significant impact on training, recovery, and performance. While the egg is recognized as the gold standard for dietary proteins, the complex blend of micronutrients in eggs should entice athletes to include this "superfood" in their daily diets. Future work is required to identify the potential role of non-protein egg components in contributing to the stimulation of postprandial muscle protein synthesis rates and recovery.

REFERENCES

Abou Sawan, S., van Vliet, S., West, D. W., Beals, J. W., Paluska, S. A., Burd, N. A., & Moore, D. R. (2018).Whole egg, but not egg white, ingestion induces mTOR colocalization with the lysosome after resistance exercise. *American Journal of Physiology-Cell Physiology*, *315*(4), C537–C543. https://doi.org/10.1152/ajpcell.00225.2018

Arenas-Jal, M., J. Suñé-Negre, M., Pérez-Lozano, P., & García-Montoya, E. (2020). Trends in the food and sports nutrition industry: A review. *Critical Reviews in Food Science and Nutrition*, *60*(14), 2405–2421. https://doi.org/10.1080/10408398.2019.1643287

Bagheri, R., Moghadam, B. H., Ashtary-Larky, D., Forbes, S., Candow, C., Galpin, D. G., Kreider, A. J., & Wong, R. B., A. (2021). Whole egg vs. egg white ingestion during 12 weeks of resistance training in trained young males: A randomized controlled trial. *Journal of Strength and Conditioning Research*, *35*(2), 411–419. https://doi.org/doi:10.1519/JSC.0000000000003922

Cui, P., Li, M., Yu, M., Liu, Y., Ding, Y., Liu, W., & Liu, J. (2022). Advances in sports food: Sports nutrition, food manufacture, opportunities and challenges. *Food Research International*, *157*, 111258. https://doi.org/10.1016/j.foodres.2022.111258

Glynn, E. L., Fry, C. S., Drummond, M. J., Timmerman, K. L., Dhanani, S., Volpi, E., & Rasmussen, B. R. (2010). Excess leucine intake enhances muscle anabolic signaling but not net protein anabolism in young men and women. *The Journal of Nutrition*, *140* (11), 1970–1976. https://doi.org/10.3945/jn.110.127647

Herron, K. L., & Fernandez, M. L. (2004). Are the current dietary guidelines regarding egg consumption appropriate? *The Journal of Nutrition*, *134*(1), 187–190. https://doi.org/10.1093/jn/134.1.187

Hida, A., Hasegawa, Y., Mekata, Y., Usuda, M., Masuda, Y., Kawano, H., & Kawano, Y. (2012). Effects of egg white protein supplementation on muscle strength and serum free amino acid concentrations. *Nutrients*, *4*(10), 1504–1517. https://doi.org/10.3390/nu4101504

Jäger, R., Kerksick, C. M., Campbell, B., Cribb, I., Wells, P. J., Skwiat, S. D., & Antonio, T. M., J. (2017). International society of sports nutrition position stand: Protein and exercise. *Journal of the International Society of Sports Nutrition*, *14*(1), 1–25. https://doi.org/10.1186/s12970-017-0177-8

Kieffer, A. J., & Walzem, R. L. (2015). Modulation of lipids, lipoproteins, and other biomarkers with egg consumption and exercise. In *Handbook of eggs in human function* (pp. 233–251). https://doi.org/10.3920/978-90-8686-804-9_1

Miranda, J. M., Anton, X., Redondo-Valbuena, C., Roca-Saavedra, P., Rodriguez, J. A., Lamas, A., Franco, C. M., & Cepeda, A. (2015). Egg and egg-derived foods: Effects on human health and use as functional foods. *Nutrients* *7*(1), 706–729. https://doi.org/10.3390/nu7010706.

Sugiyama, K., Kuriki, D., Matsuoka, R., Masuda, Y., Kunou, M., & Daigo, E. (2020). Evaluation of the functionality of "Egg white peptide" as a dietary supplement for sports. *Journal of the Japanese Society for Food Science and Technology-Nippon Shokuhin Kagaku Kogaku Kaishi*, *67*(8), 245–256. https://doi.org/10.3136/nskkk.67.245

van Vliet, S., Shy, E. L., Abou Sawan, S., Beals, J. W., West, D. W., Skinner, S. K., Ulanov, A. V., Li, Z., Paluska, S. A., Parsons, C. M., Moore, D. R., Burd, N. A. (2017). Consumption of whole eggs promotes greater stimulation of postexercise muscle protein synthesis than consumption of isonitrogenous amounts of egg whites in young men. *American Journal of Clinical Nutrition*, *106*(6), 1401–1412. https://doi.org/10.3945/ajcn.117.159855

Vasconcelos, Q. D. J. S., Bachur, T. P. R., & Aragão, G. F. (2021). Whey protein supplementation and its potentially adverse effects on health: A systematic review. *Applied Physiology, Nutrition, and Metabolism*, *46*(1), 27–33. https://doi.org/10.1139/apnm-2020-0370

35

Egg and Bone Health

Pauline Duffuler, Khushwant S. Bhullar, and Jianping Wu*

CONTENTS

35.1 Introduction to Bone Health

Bones are living and dynamic tissue that plays essential mechanical, structural, and metabolic functions in the body (Morgan et al., 2018). In addition to providing the organizational framework of the body and protecting vital organs (Morgan et al., 2018), bones participate in the synthesis of red and white blood cells. Moreover, bones act as a storage pool for calcium and phosphorus, the vital structural and signaling minerals in the body (Morgan et al., 2013).

35.1.1 Bone Composition

Bones are composed of two main compounds: osteoid and hydroxyapatite (Boskey & Robey, 2013, 2018). Around 70% of the calcified bone is composed of inorganic mineral salts deposited within the matrix, called hydroxyapatite ($[Ca_3(PO_4)^2]_3Ca(OH)_2$) (Boskey & Robey, 2013, 2018). The presence of hydroxyapatite in the osteoid matrix is responsible for the hardness, strength, and rigidity of the bone (Boskey & Robey, 2018). Then, the organic bone matrix, known as osteoid, represents ~25% of the bone matrix and is composed of 98% of protein (Boskey & Robey, 2013, 2018). Type 1 collagen is the most abundant bone protein in the osteoid (~98%) and is the basic building block of the bone matrix, contributing to the flexibility and resilience of the bones (Boskey & Robey, 2018). About 30 other non-collagenous proteins are present in small quantities in the osteoid (bone sialoprotein, proteoglycan, gla protein, osteocalcin, etc.) (Al-Qtaitat & Aldalaen, 2014; Boskey & Robey, 2018). These non-collagenous proteins strengthen the collagen matrix and mainly contribute to the bone's biological functions (Heaney & Layman, 2008; Boskey & Robey, 2018).

* Corresponding author.

DOI: 10.1201/9781003254430-43

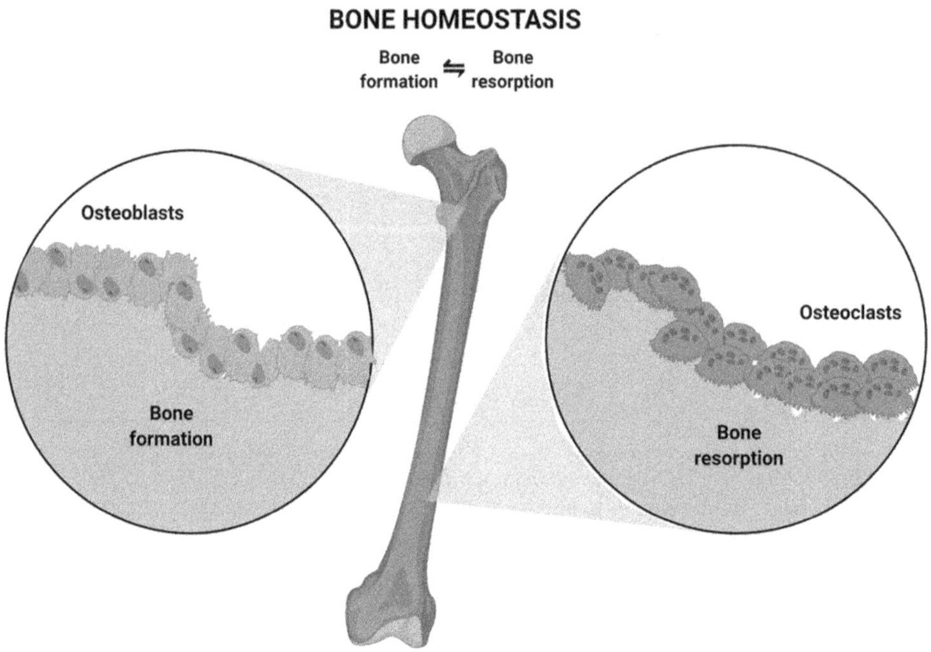

FIGURE 35.1 Bone homeostasis.

35.1.2 Bone Remodeling

From birth to adulthood, bones grow and fuse to form the 206 bones of the human skeleton. Once the bones reach their peak bone mass, they are continuously replaced by a new bone matrix, throughout life, in order to maintain bone structural homeostasis (Kini & Nandeesh, 2012; Siddiqui & Partridge, 2016). This crucial process, known as bone remodeling, is composed of two main steps: bone resorption and bone formation, as illustrated in Figure 35.1 (Siddiqui & Partridge, 2016). First, the bone-resorbing cells, osteoclasts, mature and resorb the old and damaged bone matrix (Siddiqui & Partridge, 2016). Secondly, osteoblasts, the bone-forming cells synthesize and deposit the (new) osteoid matrix (Siddiqui & Partridge, 2016). The newly deposited osteoid is mineralized to assume its final form (Siddiqui & Partridge, 2016). When a balance or physiological homeostasis is maintained between bone resorption and bone formation, bone health is preserved (Hadjidakis & Androulakis, 2006; Siddiqui & Partridge, 2016). However, an imbalance between bone formation and bone resorption can lead to the development of bone disorders such as osteoporosis, which is characterized by an excess of bone resorption compared to its formation (Hadjidakis & Androulakis, 2006).

35.2 Osteoporosis, a Worldwide Public Health Issue

35.2.1 Introduction to Osteoporosis

Osteoporosis is a major worldwide public health issue (Ensrud & Crandall, 2017), affecting 200 million people worldwide (World Health Organization, 2007) including 10 million Americans (Wright et al., 2014) and 32 million Europeans (Kanis et al., 2021). According to World Health Organization (WHO), osteoporosis is defined as "a multifactorial systemic skeletal disease characterized by low bone density and micro-architectural deterioration of bone tissue, with a consequent increase in bone fragility and susceptibility to fracture" (World Health Organization, 2007). In 2019, 178 million bone fractures occurred globally, indicating a high incidence of bone deterioration and disease (Wu et al., 2021). The most common fractures were related to the spine, hips, forearms, and vertebral fractures (Johnell & Kanis, 2006). Fractures impact deeply the life quality of patients as they can often lead to severe pains, loss of height, deformity, hunched posture, pulmonary dysfunction, immobility, morbidity, and even mortality (Magaziner et al., 1990; Graham et al., 1994; Sözen et al., 2017).

Osteoporosis, owing to its slow and silent progression, is also known as a "silent killer" or an "iceberg disease"(Sözen et al., 2017). This bone disorder is painless and appears progressively with no clinical manifestation until the first fracture occurs (Sözen et al., 2017). However, osteoporosis can easily be diagnosed by measuring bone mineral density (BMD) using a dual-energy x-ray absorptiometry; as a low BMD is correlated with a higher risk of fracture (Lorentzon & Cummings, 2015).

Age and gender are the two major risk factors associated with the development of osteoporosis. Osteoporosis affects mainly postmenopausal women (postmenopausal osteoporosis) as the reduction of estrogen leads to accelerated bone loss (Ji & Yu, 2015). According to Osteoporosis Canada, one in three women and one in five men over 50 years old, will experience osteoporosis in their lifetime (Curtis et al., 2016; Kanis et al., 2021). Although this bone disorder is often associated with women, the risk of osteoporosis undeniably increases with aging in men as well (Hawkins et al., 2021). Indeed, after reaching the peak bone mass, all individuals gradually experience a bone loss with increasing age, due to an excess in bone resorption compared to bone formation (Hawkins et al., 2021). After 50 years of age, osteoporosis affects 6.3% of men and 21.2% of women; and is an important risk for about 55% of the population (Kanis et al., 2008). Due to the burden of osteoporosis, it is necessary to develop strategies to manage osteoporosis and prevent bone loss.

35.2.2 Bone Loss Prevention and Osteoporosis Treatments

Despite the gravity of this bone disorder, bone loss and osteoporosis can be preventable by maintaining a healthy lifestyle. Indeed, adequate and healthy nutrition plays an important role in the development and preservation of bone health (Alekel et al., 2013). To prevent bone loss and fractures incidence, a diet should provide an adequate daily intake of protein (<0.8 g/kg body weight/day), calcium (700–1,200 mg/day), and vitamin D (1,000 IU/day) (Cranney et al., 2007). Indeed, supplementation of vitamin D and calcium are currently the key strategies used to prevent bone loss (Capozzi et al., 2020). Other micronutrients, such as vitamin K_2, magnesium, selenium, zinc, iron, and vitamin B, also contribute to the preventive of bone loss and must be supplied in an adequate amount in our diet (Capozzi et al., 2020). In addition, current preventive strategies include exercise and physical activity, maintaining a healthy body weight (BMI: 20–25 kg/m^2), and avoiding smoking and drinking (Eastell, 2017; Shanks et al., 2019).

In addition to prevention strategies, synthetic drugs are also used to counteract osteoporosis. The two main pharmaceutical drugs currently available include anabolic and antiresorptive agents (Salari Sharif et al., 2011). Antiresorptive agents target the inhibition of bone resorption by promoting osteoclast apoptosis, thus enhancing bone strength (Salari Sharif et al., 2011; Chen & Sambrook, 2012; Seeman & Martin, 2019). Thus, antiresorptive agents slow bone remodeling but do not abolish it (Seeman & Martin, 2019). The most common antiresorptive agents already used for osteoporotic patients are bisphosphonates (alendronate and risedronate) but other antiresorptive agents exist such as estrogen replacement therapy (ERT), hormone replacement therapies (HRT), raloxifene, calcitonin, and denosumab (Chen & Sambrook, 2012; Seeman & Martin, 2019). On the other hand, reconstruction of the bone matrix requires anabolic agents (Seeman & Martin, 2019). Indeed, anabolic agents stimulate bone formation by increasing the number of osteoblast precursors (Salari Sharif et al., 2011; Seeman & Martin, 2019). Only a few limited anabolic agents are currently available on the market, including parathyroid hormone analogs (teriparatide and abaloparatide) (Seeman & Martin, 2019).

Despite the remarkable aspects of the current pharmaceutical treatments, concerns are raised regarding their side effects, which lower the life quality of osteoporotic patients. Indeed, numerous mild and severe side effects were reported including nausea, dizziness, hot flushes, irregular heartbeat, loose bowel movement, atypical femur fractures, osteonecrosis of the jaw, development of cancer, and increased risk of developing cardiovascular diseases (Aki et al., 2003; Siris et al., 2005; Khan et al., 2009; Shane et al., 2013; Khosla & Lorenz, 2017). Additionally, their long-term efficacy remains contentious (Khosla & Lorenz, 2017; Nogués & Martinez-Laguna, 2018; Ukon et al., 2019; Langdahl, 2021). For example, the efficacy of bisphosphonates is unproven after five years of treatment (Whitaker et al., 2012). Due to these reasons, the compliance and the acceptance of the patients with the current drugs are challenging and became a major issue in osteoporosis treatment. Thus, it appears paramount to look for efficient, natural, and evidence-based safer alternatives to traditional therapies for the prevention and treatment of osteoporosis. Among all these alternatives, numerous food-derived components have shown positive effects on bone health, including eggs and their bioactive constituents.

35.3 Egg Consumption and Bone Health

One of the main strategies to prevent osteoporosis is to maintain a healthy diet and to include food rich in micronutrients and macronutrients that could benefit bone health (Seuss-baum, 2007). Eggs are a cheap and widely available food product; rich in proteins, lipids, vitamins, and minerals (Seuss-baum, 2007). Coheley et al. (2018) studied the impact of whole egg intake on bone health in healthy children. Their results demonstrated that whole egg intake was positively correlated with cortical bone strength, BMD, and with bone formation markers (Coheley et al., 2018). Moreover, egg consumption mitigates any possible increase in the risk of hip fractures in elderly Chinese subjects (Zeng et al., 2013). Unfortunately, no studies are currently available on the impact of whole egg consumption on the bone health of patients with osteoporosis, postmenopausal women, or among the elderly population. However, egg proteins and numerous egg-derived micronutrients demonstrate a positive influence on bone health (summarized in Table 35.1).

35.3.1 Egg Proteins and Bone Health

Protein(s) constitute ~50% of bone volume and a third of the total bone mass (Boskey & Robey, 2018). Proteins are quintessential nutrients to maintain bone homeostasis. Type 1 collagen provides the structure to the organic bone matrix whereas non-collagenous proteins play indispensable biological functions in the bone (Heaney & Layman, 2008). For example, alkaline phosphatase is involved in phosphate degradation whereas matrix GLA proteins (MGP) such as osteocalcin, regulate extracellular matrix calcification (Al-Qtaitat & Aldalaen, 2014). Therefore, adequate protein intake is essential to building and maintaining bone health (Bonjour, 2013; Shams-White et al., 2018). Dietary protein intake has been positively correlated with BMD in premenopausal women and older patients with osteoporosis (Feskanich et al., 1996; Promislow et al., 2002; Bonjour, 2013; Rizzoli et al., 2018). According to the European Safety Authority, the current recommended daily intake of protein is about 0.83 g/kg/day (European Food Safety Authority, 2012). However, regular dietary protein intake above the current RDA has a protective effect on bone health by enhancing BMD, reducing bone

TABLE 35.1

Egg Micronutrients and Macronutrients and Their Role in Bone Health

Compounds	One Egg (53 g) Expressed as % of RDI	Role in Bone Health	Association between Egg Compounds and Bone Health	References
Proteins	12.4 %	Building blocks of bone matrix Muscle strength ↑ ➔ prevention of fractures	Adequate protein intake: – positively associated with BMD ➔↓ bone loss	(Rizzoli et al., 2018; Promislow et al., 2002; Bonjour, 2013; Feskanich et al., 1996)
Lipids				
Ω-3 fatty acids	12.3 %	Anti-inflammatory	Positively associated with BMD ➔ ↓ risk of fracture – ↑ Calcium absorption – ↑ Bone regeneration – ↑ Structural strength – ↓ Calcium excretion in urine and feces	(Abdelhamid et al., 2019; Baggio et al., 2000; Ensrud & Taylor, 2013; Högström et al., 2007; Kruger et al., 2010; Kuroda et al., 2017; Orchard et al., 2013)
Ω-6 fatty acids	N/A	Pro-inflammatory	Negatively associated with BMD ➔↑ Bone loss ➔↑ risk of fracture	(Casado-Díaz et al., 2013; Orchard et al., 2013)
Minerals				
Phosphorus	16.4 %	Mineral compounds of the inorganic matrix (hydroxyapatite)	Phosphorus deficiency: – Calcium excretion ➔ impairment of bone mineralization Excess of phosphorus intake: – Disruption of calcium, vitamin D & phosphorus homeostasis ➔ impairment of bone mineralization – ↑ Bone resorption ➔ ↓ BMD	(Takeda et al., 2012; Calvo & Tucker, 2013)
Selenium	28.8 %	Cell proliferation/differentiation Prevention of bone resorption through osteoclast inactivation	Positively associated with BMD Inversely associated with bone turnover Selenium deficiency ➔↑ risk of growth retardation & osteopenia	(Moreno-Reyes et al., 2001; Hoeg et al., 2012)
Iron	8.8 %	Enzyme co-factor for the synthesis of collagen Transform vitamin D in its active form ➔ participate in calcium absorption	Positively associated with BMD	(Zheng et al., 2014)
Zinc	8.7 %	Involve in bone formation – Osteoblast activity – Collagen synthesis Involve in bone resorption (ALP activity)	Zinc deficiency ➔ risk factor of osteoporosis	(Zheng et al., 2014; Yamaguchi, 2015)
Vitamins	10 %			
Vitamin D	30.8 %	Essential vitamin for calcium absorption ↑ muscle strength ➔ prevent fractures Anti-inflammatory ➔ ↓ pro-inflammatory cytokine production	Vitamin D deficiency – ↑ bone loss & ↑ risk of fracture ➔↑ osteoporosis development – ↑ rickets development in children – ↑ osteomalacia development in adults	(Laird et al., 2010; Turner et al., 2012)
Vitamin K	38.5 %	Involved in osteoblast formation: – Coenzyme involved in the production of bone protein – Regulation of the transcription of osteoblast marker, Involved in bone resorption: – Supression of bone resorption – Regulation of osteoclast formation	Vitamin K deficiency ➔ ↓ BMD & ↑ risk of fracture	(Booth et al., 2003; Rodríguez-Olleros Rodríguez & Díaz Curiel, 2019)
Vitamin B12	35.3 %	Involved in osteoblast function Involved in iron metabolism	Vitamin B12 deficiency ➔ ↑ osteoporosis development	(Dhonukshe-Rutten et al., 2003)

Source: Based on Seuss-baum (2007).

Note: BMD: bone mineral density; RDI: recommended daily intake.

loss, and preventing hip fractures (Wallace & Frankenfeld, 2017; Rizzoli et al., 2018). Current studies did not show any adverse effects of high protein intake on bone health (Shams-White et al., 2018). Knowing the essential role of proteins in bone homeostasis, an adequate protein intake (0.83 g/kg/day) is essential to preserve bone homeostasis and could be supplied through adequate egg consumption. An egg constitutes ~12% of proteins, mainly contained in the albumen and the yolk, containing about 6.5 grams of protein, and thus providing ~12.4% of our protein daily intake (Seuss-baum, 2007). Additionally, egg proteins possess an impressive amino acids profile. Indeed, egg proteins are easily digestible and bioavailable compared to other sources, thus, making eggs a major source of proteins and other nutrients in our diet, that are beneficial for bone health.

35.3.2 Egg Vitamins and Bone Health

In addition to dietary proteins, eggs are also an important source of vitamins (Table 35.1). One egg provides more than 10% of the recommended vitamin daily intake (Seuss-baum, 2007). Among those vitamins, vitamins D, K, and B12 are essential micronutrients for bone health (Dhonukshe-Rutten et al., 2003; Lanham-New, 2008; Laird et al., 2010; Turner et al., 2012; Hamidi et al., 2013; Rodríguez-Olleros Rodríguez & Díaz Curiel, 2019).

As discussed above, calcium and vitamin D are the two vital nutrients to develop and preserve bone health (Capozzi et al., 2020). Provided ever through sun exposure and by the diet, vitamin D influences bone health through several cellular mechanisms. Together with parathyroid hormone, vitamin D modulates calcium and phosphorus metabolism as well as calcium absorption in the gut and in the kidneys tubules (Laird et al., 2010; Christakos et al., 2013). Additionally, vitamin D possesses immunoregulatory activity (Laird et al., 2010; Christakos et al., 2013). It modulates the production of pro-inflammatory cytokines and thus reduces inflammation and bone loss induced by osteoporosis (Laird et al., 2010). Therefore, vitamin D has a beneficial effect on BMD and fracture incidence by enhancing muscle strength and thus, preventing fractures induced by falls (Laird et al., 2010). On the contrary, insufficient vitamin D intake is associated with bone loss, an increased fracture risk, and osteoporosis development (Lanham-New, 2008; Laird et al., 2010; Hill & Aspray, 2017). Thus, to preserve bone health, a daily diet should include a dose of vitamin D of 800 IU/day in combination with a sufficient calcium intake (1,000 mg per day) (Cranney et al., 2007). Eggs are one of the dietary sources of vitamin D: one egg (53 g) provides 30.8% of our vitamin D daily intake requirement (Seuss-baum, 2007). Thus, incorporating eggs into our diet could supply daily intake of vitamin D and benefit bone health.

Further, vitamin K (including menaquinone and phylloquinone) also has a protective effect on bone health. Firstly, vitamin K is an essential coenzyme for the production of vitamin K-dependent bone proteins (MGP; osteocalcin), by converting protein-bound glutamate residues into gamma-carboxyglutamate residues (Hamidi et al., 2013). Once converted by vitamin K, these two proteins play an essential role in bone health. On one hand, matrix Gla Protein promotes bone formation by upregulating Wnt/ß-catenin signaling pathways (Zhang et al., 2019) Additionally, matrix Gla protein inhibits bone mineralization (Hamidi et al., 2013). On the other hand, osteocalcin, produced by osteoblasts, regulates the formation of hydroxyapatite crystals, by linking the calcium-binding Gla residues to the calcium ions of the hydroxyapatite (Fusaro et al., 2020). Osteocalcin can also inhibit bone mineralization as this protein inhibits the precipitation of calcium salts from saturated solution (Zoch et al., 2016; Fusaro et al., 2020). Vitamin K is a transcriptional regulator of the bone marker genes in the osteoblasts (Ichikawa et al., 2006). By binding to the nuclear receptor steroid and xenobiotic receptor (SXR), vitamin K promotes the accumulation of type 1 collagen in osteoblasts *in vivo* and thus, promotes bone formation (Ichikawa et al., 2006). Additionally, vitamin K stimulates bone formation by stimulating osteoblast differentiation *in vivo* through autophagy induction (Li et al., 2019). Finally, vitamin K can directly regulate the activity of bone cells by inhibiting the activation of the nuclear factor (NF-κB) signaling pathway (Yamaguchi & Weitzmann, 2011). In this way, vitamin K inhibits osteoclastogenesis and reduces bone resorption while promoting osteoblast differentiation and bone formation (Yamaguchi & Weitzmann, 2011). Thus, vitamin K mainly influences bone health through the gamma-carboxylation of matrix Gla protein and osteocalcin and by the inactivation of the NF-κB pathway. Deficiency in vitamin K has been associated with low BMD and a higher risk of fracture (Booth et al., 2003). Knowing the essential role of this vitamin in bone health, vitamin K should be supplied by 200 μg/day in our diet (Lanham-New, 2008). Eggs are one of the main sources of vitamin K: one egg provides 38.5% of vitamin K's daily intake requirement (Seuss-baum, 2007).

Finally, eggs are an important dietary source of vitamin B_{12} (Seuss-baum, 2007). Vitamin B12 (also called cobalamin) is also an important regulator of bone health (Clemens, 2014). Vitamin B12 circulating levels have been positively associated with BMD (Clemens, 2014). On the contrary, depletion in vitamin B_{12} levels has been associated with a higher risk of osteoporosis and with lower BMD in post-menauposal women and elderly women (Dhonukshe-Rutten et al., 2003; Tucker et al., 2005; Ouzzif et al., 2012; Tariq et al., 2018; Kalimeri et al., 2020; Saiedullah et al., 2021). The mechanisms are still not clearly understood. However, vitamin B_{12} significantly promotes bone formation, in a dose-dependent manner, by stimulating the proliferation of osteoblast and alkaline phosphatase (ALP) activity *in vivo* (Kim et al., 1996). Additionally, vitamin B_{12} deficiency increases the levels of homocysteine and methylmalonic acid, leading to an increase in osteoclastogenesis and a decrease in bone mass (Vaes et al., 2009). Based on the function of vitamin B_{12} in bone metabolism, the role of eggs in bone health is further supported (Seuss-baum, 2007).

35.3.3 Egg Minerals and Bone Health

Eggs also provide essential minerals for bone health including phosphorus, selenium, iron, and magnesium (Table 35.1; Seuss-baum, 2007). Along with calcium, phosphorus plays an essential role in the growth and development of bones

(Fukumoto, 2014). Around 85% of phosphorus exists in the form of hydroxyapatite $[Ca_{10}(PO_4)_6(OH)_2]$ that is deposited during the mineralization of bones and teeth (Fukumoto, 2014). Phosphorus directly impacts both bone resorption and bone formation (Serna & Bergwitz, 2020). On the one hand, phosphorus stimulates the differentiation of osteoblast and the maturation of osteocytes in the lacuna via the Extracellular Signal-Regulated Kinase 1/2 (ERK1/2)- Fra-1 signaling pathways (Julien et al., 2009; Zhang et al., 2011; Serna & Bergwitz, 2020). On the other hand, phosphorus decreases the expression of the receptor activator of nuclear factor-κB ligand (RANKL), which inhibits osteoclastogenesis and bone resorption (Hayashibara et al., 2007; Serna & Bergwitz, 2020). In order to accomplish its functions, phosphorus pool levels are needed to be maintained. In case of severe and prolonged phosphorus deficiency, bone health is altered to a hormonal disruption: an increased production of calcitriol, the active form of vitamin D and a decreased production of parathyroid hormone and fibroblast growth factor 23 (FGF23). These hormonal changes will stimulate bone resorption through calcitriol-mediated osteoclastogenesis via the receptor activator NF-κB (nuclear factor kappa light chain enhancer of activated B cells) signaling pathways (Serna & Bergwitz, 2020). Thus, prolonged phosphorus deficiency causes the demineralization of the skeleton, leading to a higher risk of fractures. Phosphorus deprivation can also cause rickets in children and osteomalacia in adults (Serna & Bergwitz, 2020). However, phosphorus is usually consumed in excess in western societies due to the high consumption of protein-rich foods (meat, milk, eggs) and processed foods (cereals, soda) (Takeda et al., 2012; Calvo & Tucker, 2013; Serna & Bergwitz, 2020). The recommended daily allowance for phosphorus is 700 mg/day. However, the phosphorus intake average is around 1,000–1,767 mg/day in western diets (Calvo & Tucker, 2013). Excess consumption of phosphorus leads to several adverse effects on bone health. Excess phosphorus intake leads to the formation of insoluble calcium-phosphate complexes, which decrease calcium absorption (Calvo & Tucker, 2013; Vorland et al., 2017; Serna & Bergwitz, 2020). Phosphorus can also directly influence bone cells by upregulating the expression of osteopontin and FGF23, which impairs bone mineralization and increases the risk of fractures (Vorland et al., 2017; Serna & Bergwitz, 2020). According to Pinhiero et al. (2009), every 100 mg of phosphorus excess increase by 9% the risk of fracture (Pinheiro et al., 2009). Knowing that one egg (53 g) provides 16.4% of phosphorus recommended daily intake, the consumption of eggs should not exceed 6 per day (Seuss-baum, 2007).

Besides phosphorus, several other egg-derived minerals are essential and beneficial for the development and maintenance of bone health. Among these, selenium plays a critical role in the prevention of bone resorption through osteoclast inactivation by suppressing the production of reactive oxygen species and NFκB signaling pathways (Moon et al., 2012). Selenium also promotes osteoblast differentiation by suppressing oxidative stress and ERK 1/2 signaling pathways (Liu et al., 2012). Low selenium intake has been associated with lower BMD, bone fractures, and bone disease development (Moreno-Reyes et al., 2001; Zheng et al., 2014; Reid, 2021; Wu et al., 2021). Additionally, iron also plays a critical role in bone health (Balogh et al., 2018). On one hand, iron acts as a co-factor of enzymes for collagen bone matrix synthesis (Katsumata et al., 2009). On the other hand, iron is involved in activating vitamin D to promote calcium absorption (Katsumata et al., 2009). Iron deficiency has been associated with a lower BMD, a lower bone turnover and with osteoporosis (Zheng et al., 2014). Finally, zinc, another egg micronutrient, also exerts a critical role in bone health. Indeed, zinc is an essential micronutrient osteoblast activity in the synthesis of collagen and alkaline phosphatase activity (Zheng et al., 2014; Yamaguchi, 2015). These physiological activities make zinc deficiency an important risk factor for osteoporosis (Zheng et al., 2014; Yamaguchi, 2015). Thus; iron, zinc, and selenium are essential nutrients that prevent bone loss and promote bone formation. Based on these key facts, eggs are categorized as an excellent dietary source of these minerals essential for bone health as one egg provides 28% of selenium RDA, 8.8% of iron RDA, and 8.7% of zinc RDA (Seuss-baum, 2007).

35.3.4 Polyunsaturated Fatty Acids Benefit Bone Health

Polyunsaturated fatty acids (PUFA) are composed of omega-3 fatty acids (ω-3) omega-6 fatty acids (ω-6) and omega 9 fatty acids (Saini & Keum, 2018). Both ω-3 and ω-6 are essential fatty acids that can significantly impact our bone health. Clinical evidence shows that higher consumption of PUFAs was negatively correlated with the prevalence of osteoporosis in Korean women aged over 50 years old (Lee et al., 2020). Additionally, BMD of elderly women was positively influenced by the total dietary PUFAs intake as well as ω-3 and ω-6 intake (Järvinen et al., 2012).

Numerous studies claim that ω-3 intake also has protective effects on bone health (Lavado-García et al., 2018). The ω-3 intake (DHA, EPA) was significantly associated with higher BMD, enhanced bone regeneration, and improved structural strength (Baggio et al., 2000; Högström et al., 2007; Kruger et al., 2010; Kuroda et al., 2017; Lavado-García et al., 2018; Abdelhamid et al., 2019). The mechanisms underlying the action of ω-3 on bone health are complex and not completely understood. However, studies demonstrate that ω-3 increases calcium absorption and decreases its urinary and fecal excretion (Baggio et al., 2000). Additionally, ω-3 fatty acids exhibit anti-inflammatory properties by reducing the production of proinflammatory cytokines, which decreases osteoclastogenesis and promotes osteoblastogenesis, and thus promotes bone homeostasis (Kruger et al., 2010).

Whereas ω-3 benefits bone health, the influence of ω-6 on bone metabolism remains controversial. Some studies have shown that ω-6 also exert a positive influence on bone health by promoting the production of eicosanoids and cytokines that are capable of increasing bone mineral feeding (Casado-Díaz et al., 2013). On the contrary, ω-6 inhibits osteoblastogenesis by inhibiting osteoblast differentiation, osteoblast proliferation and therefore bone formation, *in vitro* (Maurin et al., 2005; Casado-Díaz et al., 2013). At the same time, eicosanoids derived from ω-6 exert pro-inflammatory bioactivity (Schmitz & Ecker, 2008): ω-6 triggers the production of pro-inflammatory cytokines such as interleukin-1 (Il-1), tumor necrosis factor-α (TNF- α) (Priante et al., 2002). Additionally, ω-6 fatty acids supplementation in a high-fat diet stimulates bone resorption and reduces BMD *in vivo* (Mak et al., 2019).

Despite the different effects of these two essential fatty acids on bone metabolism, only the ratio ω-6/ω-3 is essential for bone health. The currently recommended ratio of ω-6/ω-3 is approximately 1:1 (Saini & Keum, 2018). However, this ratio is much higher in western society, between 15:1 and 16.7:1 which supports a high incidence of osteoporosis (Simopoulos, 2008). An increase in the ω-6/ω-3 ratio decreases bone formation *in vivo* and was associated with lower BMD in women (Li et al., 1999; Watkins et al., 2000). Thus, a lower ratio of ω-6/ω-3 is more desirable to prevent bone loss and reduce the prevalence of osteoporosis (Simopoulos, 2008; Casado-Díaz et al., 2013). As ω-3 and ω-6 are essential fatty acids that cannot be synthesized by the human body, they must be supplied to our diet. Egg yolk is a great source of ω-3 and ω-6: about 20% of egg fat is PUFA, of which 90% are ω-6 fatty acids and 10% are ω-3 (Seuss-baum, 2007). Consuming one egg provides 10.8% of essential fatty acid daily intake, including 12.3% of your ω-3 daily requirements (Seuss-baum, 2007). Thus, consuming eggs could contribute to our bone health by providing essential fatty acids.

Knowing the impact of whole eggs consumption and egg-derived micronutrients on bone health, eggs appear to be an excellent dietary source that provides all the micronutrients and proteins necessary for the development and maintenance of bone health (Table 35.1). As eggs are a cheap and widely available dietary source, incorporating eggs into our diet might be a viable strategy to preserve bone health. In addition to egg consumption, its constituent bioactive compounds could also be used as a food ingredient in nutraceuticals or functional foods to promote bone health.

35.4 Egg Bioactive Compounds and Bone Health

Due to the increasing concern regarding the efficacy and the safety of pharmaceutical drugs, research regarding the use of functional foods and nutraceuticals for the promotion of bone health, prevention of bone loss, and the management of osteoporosis have gained interest in the past decades. Numerous bioactive egg compounds have been reported for their bone health-promoting properties (Table 35.2).

35.4.1 Eggshells and Bone Health

Eggshells are a well-known biowaste of the egg industry. In 2018, 2.3 million tons of eggshells are being discarded, around the world, which leads to a detrimental impact on the environment (Ahmed et al., 2019, 2021). Thus, food scientists focused on developing sustainable waste managing strategies to add value to eggshells (Ahmed et al., 2019). Several studies demonstrate the potential of eggshells as a high bioavailable source of calcium and, for bone regeneration.

35.4.2 Calcium Carbonate, a Dietary Source of Calcium

Eggshells are composed of 95.1% minerals followed by 3.3% of proteins and 1.6% of water (Schaafsma et al., 2000). Calcium is the main mineral component of eggshells, representing 37.3% of eggshells' total weight. Eggshells' calcium is mainly present in the form of calcium carbonate ($CaCO_3$; 93.6%) (Schaafsma et al., 2000; Brun et al., 2013). Half an eggshell provides the daily calcium requirement of an adult (~1000 mg) (Brun et al., 2013). Knowing the essential role of calcium in bone health, eggshell supplementation is a cost-effective and environmentally-friendly strategy to promote bone health and prevent bone loss (Arnold et al., 2021).

Several studies investigate the impact of eggshell supplements on bone health. A diet containing eggshell powder was reported to enhance calcium absorption *in vivo* (Brun et al., 2013). To supplement calcium in our diet, eggshells powder was added to various food products such as bread (Brun et al., 2013; Alsuhaibani, 2018; Chilek et al., 2018; Platon et al., 2020), biscuits (Hassan, 2015), yogourt (Al Mijan et al., 2014; El-Shibiny et al., 2018), cakes (Ray et al., 2017), juice (Lachowicz et al., 2020), pizza (Brun et al., 2013), and spaghetti (Brun et al., 2013). Schaafsma and Pakan (1999) demonstrated that the consumption of dairy supplemented with eggshell powder has a positive influence, in short term, on the BMD of patients with osteoporosis (Schaafsma & Pakan, 1999). Additionally, 12 months of eggshells supplementation increase BMD in healthy late post-menopausal women (Schaafsma et al., 2002).

Thanks to these promising results, several eggshells-based nutraceuticals have been developed to promote bone health (Table 35.3). These eggshell-derived nutraceuticals include Bone Health Original™ capsules (Redd Remedies, n.d.), and OVOCET® (Eggnovo, n.d.) for human nutrition as well as Eggshell Calcium Supplement™ for dogs and cats (IrRAWsistible Pet Foods, n.d.). Additionally, eggshell powder has been used as a calcium supplement for laying hens (Cordeiro & Hincke, 2011). Overall, eggshell is an inexpensive source of calcium that can be used in addition to our diet, to increase calcium absorption and promote bone health.

35.4.2.1 Eggshells as a Biocompatible Material for Bone Tissue Regeneration

The natural coral skeleton has been widely used as a bone substitute in treating osseous defects (Shors, 1999). As hen eggshells possess a similar composition to the natural coral skeleton, food scientists start also investigating the potential of eggshells as a biocompatible alternative material for a bone substitute material. Hydroxyapatite ($Ca_{10}(PO_4)_6(OH)_2$) is one of the major mineral compounds found in the bone. Due to its vital role in bone health, hydroxyapatite is usually used for repairing bone defects (Sadat-Shojai et al., 2010). The majority of hydroxyapatite is usually chemically synthesized. However, hydroxyapatite can also be produced using natural by-products or biowastes as a starting material. Several studies reported the strong potential of eggshells biowaste for the production of hydroxyapatite (Naga et al., 2015; Wu et al., 2015; Dadhich

TABLE 35.2

Egg Bioactive Compounds and Bone Health

Egg Matrix	Egg Compounds	Outcomes on Bone Health	Potential Applications	References
Eggshells	Calcium carbonate ($CaCO_3$)	↑ Calcium absorption in human Synthesis of hydroxyapatite ↑ Osteogenesis *in vitro* Bone regeneration & bone healing *in vivo*	Calcium supplement Bone tissue healing & regeneration	(Park et al., 2008; Park et al., 2009; Kweon et al., 2011; Lee et al., 2012; Uraz et al., 2013; Balázsi et al., 2013; Lee et al., 2014; Arnold et al., 2021)
Egg Yolk	Carotenoids (lutein, zeaxanthin)	↑ BMD & ↓ bone loss in human ↓ Osteoclast formation & differentiation by inhibiting RANKL *in vitro*	Prevention of bone loss & Osteoporosis treatment	(Tominari et al., 2017; Takeda et al., 2017)
	Egg yolk hydrolysate	↑ Osteoblast differentiation, proliferation & bone mineralization *in vitro* ↓ Osteoclastogenesis *in vitro* ↑ Bone growth & BMD *in vivo* Prevention of BMD decline in menopausal women		(Ji et al. 2007; Leem et al., 2004; Kim et al., 2011)
	Phosvitin	↑ Osteoblast proliferation, differentiation & bone mineralization *in vitro* ↑ Production of type & collagen *in vitro* ↓ Production of pro-inflammatory markers *in vitro* ↓ Recruitment of osteoclasts by osteoblasts *in vitro* ↑ Calcium absorption *in vivo* ↓ Bone loss *in vivo*		(Choi et al., 2005; Zhang et al., 2015; Liu et al., 2017; Jie et al., 2017; Zhang et al., 2017; Jie et al., 2018; Zhang et al., 2018; Ren et al., 2019; Ren, 2019; Chakrabarti et al., 2020; Zhao et al., 2021)
Egg White	Ovotransferrin	↓ RANKL-induced osteoclastogenesis *in vitro* ↑ Osteogenic activity via LPR1 activation *in vitro* ↓ Bone loss & preserve bone microarchitecture *in vivo*	Prevention of bone loss & Osteoporosis treatment	(Shang & Wu, 2018; Shang, 2019; Shang & Wu, 2019)
	IRW	↓ LPS-induced osteoclastogenesis *in vitro* ↑ Osteoblasts differentiation and mineralization via RUNX2 pathway *in vitro*		(Shang et al., 2019; Shang & Wu, 2020)
	Cystatin	↓ Osteoclasts-mediated bone resorption *in vitro* ↓ Bone degradation *in vitro*		(Abrahamson et al., 2003; Brand et al., 2004)
	Ovalbumin	↑ Proliferation & differentiation of osteoblasts *in vitro* ↑ Expression of osteocalcin and ALP *in vitro*	Bone tissue regeneration	(Farrar et al., 2010)
	Ovomucin	↑ Cell adhesion & cell growth		(Carpena et al., 2017)
	Lysozyme	↑ Osteogenic differentiation, bone reformation, & bone mineralization *in vitro* Regulation of collagen degradation during bone remodeling		(Martins et al., 2009; Li et al., 2020)

Note: BMD: Bone mineral density; ALP: alkaline phosphatase; LRP1: low-density lipoprotein receptor-related protein 1, RANKL: receptor activator of nuclear factor- κB ligand; RUNX2: runt-related transcription factor 2.

et al., 2016; Huang et al., 2020). Indeed, eggshells $CaCO_3$ can be converted into hydroxyapatite, using mostly hydrothermal methods.

Eggshells hydroxyapatite can mimic the structure and the bioactivity of natural bone Numerous *in vitro* studies reported that eggshell-derived compounds guide bone regeneration and induce bone formation compared to an untreated empty defect. Indeed, eggshells particles enhance osteogenic differentiation and bone mineralization through the crosstalk of multiple pathways. However, the eggshells scaffold promotes osteogenesis *in vitro* by activating the phosphorylation of the ERK1/2 and protein kinase B (Akt) signaling pathways (Huang et al., 2020). Eggshells' bioactivity is improved when it is combined with other materials such as hyaluronan (Neunzehn et al., 2015) or, carboxymethyl chitosan (Huang et al., 2020). In addition to promising results *in vitro,* eggshells demonstrate promising results in rodents (Park et al., 2008, 2009; Uraz et al., 2013; Lee et al., 2014) and rabbits calvaria bone defect models (Kweon et al., 2011; Lee et al., 2012; Balázsi et al., 2013). Eggshells hydroxyapatite significantly promotes bone regeneration and bone healing *in vivo* (Park et al., 2008). Its efficiency was comparable with commercially available bone substitutes (Bio-OSS, DMBM, BioCoral, N/S) (Park et al., 2008).

In summary, eggshells are a safe, inexpensive, eco-friendly, and bioavailable starting material that has a strong potential to be used as an alternative for bone healing and bone regeneration. To our best knowledge, only one eggshell-derived complement for bone grafts is currently available on the market: Membrell's ® BONEhealthTM Plus D3 & K2. Salama et al. (2019) demonstrated that Membrell's ® BONEhealthTM Plus D3 & K2 enhance the formation of new bone and support BMD *in vivo* (Salama et al., 2019).

TABLE 35.3
Egg Bioactive Compounds Commercially Available for Bone Health

Egg Compounds	Value-added Application	Product Name	Product Functions	References
Eggshells calcium carbonate	Human nutrition	Membrell's BONEhealth™ Plus D3 & K2	Promotion of new bone formation BMD support Promotion of calcium absorption in healthy individuals	(Salama et al., 2019)
		OVOCET® (Eggnovo)	Calcium supplement to support calcium absorption	(Eggnovo, n.d.)
		Bone Health Original™ (Redd Remedies)	Calcium & vitamin D3 supplement to support calcium absorption	(Redd Remedies, n.d.)
	Animal nutrition	Eggshell Calcium Supplement™ (IrRAWsistible Pet Foods)	Calcium supplement to support calcium absorption in dogs and cats	(IrRAWsistible Pet Foods, n.d.)
Egg yolk hydrolysate powder	Human nutrition	Bonepep® (Pharma Foods)	Prevention of BMD decline Promotion of bone formation	(Pharma Foods International Co., n.d.; Kobayashi et al., 2015)

Note: BMD: bone mineral density.

35.4.3 Egg Yolk Bioactive Compounds and Bone Health

35.4.3.1 Egg Yolk Carotenoids: An Antioxidant for the Prevention of Bone Loss

Egg yolk is an interesting source of carotenoids, including lutein and zeaxanthin (Zia-ul-haq, 2021). Carotenoids are the natural yellow pigments in egg yolk, representing less than 1% of egg yolk lipids. One yolk contains approximately 167–216 μg of lutein and 85–185 μg of zeaxanthin (Skřivan & Englmaierová, 2014). Well-known for their anti-inflammatory and antioxidant activities, egg yolk carotenoids also exert a positive influence on bone health by stimulating bone formation and suppressing bone resorption.

In vitro, lutein acts on macrophages and suppresses osteoclast formation and differentiation by inhibiting the expression of the osteoclastogenesis factor RANKL, and therefore, suppressing bone resorption (Takeda et al., 2017; Tominari et al., 2017). Additionally, lutein also stimulates bone formation by promoting bone mineralization, by enhancing the expression of bone morphogenetic protein 2 (BMP-2) and by inhibiting the expression of RANKL expression, and sclerostin (Takeda et al., 2017; Tominari et al., 2017).

These previous findings have been confirmed *in vivo*. Four weeks of administration of lutein (1%) increase the BMD and the cortical bone mass of the femur in growing mice (Takeda et al., 2017). Additionally, lutein supplementation also protects ovariectomized (OVX) mice and rats against bone loss due to estrogen deficiency (Takeda et al., 2017; Li et al., 2018). Bovier et al. (2017) were the first to demonstrate the antiosteoporosis effects of lutein on humans. Indeed, healthy individuals with a diet rich in lutein/zeaxanthin, exhibit lower bone loss and higher BMD (Bovier & Hammond, 2017). This bone health-promoting bioactivity of lutein and zeaxanthin is explained in part by their antioxidant properties (Altindag et al., 2008; Bovier & Hammond, 2017). Indeed, lutein and zeaxanthin promote bone health by lowering oxidative stress and thus, reducing bone resorption (Altindag et al., 2008; Bovier & Hammond, 2017).

In conclusion, lutein and zeaxanthin possess some strong bone health-promoting properties thanks to their antioxidant properties. Despite promising results in lab conditions, no commercialized applications are currently available yet.

35.4.3.2 Bonepep®, a Commercialized Egg Yolk Hydrolysate for the Prevention of Bone Loss

Several egg yolk-derived proteins, hydrolysates, and peptides demonstrate bone health-promoting properties *in vitro* and *in vivo*. Ji et al. (2007) demonstrated that egg yolk water-soluble protein (50–5,000 μg/mL) stimulates bone formation by promoting osteoblast differentiation, proliferation, and mineralization *in vitro* (Ji et al., 2007; Ji et al. 2014). Additionally, egg water-soluble protein (100 μg/mL) also prevents bone resorption by inhibiting osteoclast formation *in vitro* (Kim et al., 2011). A few *in vivo* studies confirm the bone health-promoting properties of egg yolk proteins. Delipidated egg yolk protein promotes significantly longitudinal bone growth (Leem et al., 2004), BMD and, cortical thickness of rats tibia (Kim et al., 2011).

Among all egg yolk-derived compounds, Bonepep® is the only egg yolk compound currently available on the market to prevent bone loss. Bonepep® is a hydrolyzed egg yolk protein powder produced by enzymatic treatment of the water-soluble fraction of defatted chicken eggs. Oral administration of Bonepep® prevents the decline of BMD and promote the longitudinal growth of tibia in ovariectomized mouse (Song et al., 2011). Kobayashi et al. (2015) investigated the bone health-promoting properties of Bonepep® on orchiectomized dogs, an osteoporosis canine model. Oral administration of Bonepep® (2 months) significantly promotes bone formation *in vivo* (Kobayashi et al., 2015). Finally, oral administration of Bonepep® (100 mg/day; 6 months) prevents the decline of BMD in menopausal women (Pharma Foods International Co., n.d.). Due to its successful results, Bonepep® has been commercialized as a nutraceutical for the promotion of bone health in Europe, Asia, and North America (Pharma Foods International Co., n.d.; Song et al., 2011).

Overall, egg yolk proteins have a strong potential to promote bone formation and prevent bone resorption. Due to its successful results, food scientists started to investigate the bioactivity of individual egg yolk compounds on bone health, including the phosphoprotein phosvitin.

35.5 Phosvitin and Its Derived Phosphopeptides

The egg yolk protein, phosvitin is the most phosphorylated protein found in nature. It represents 4% of the egg proteins. Due to its high phosphorylation degree, phosvitin possesses a strong chelating capacity with minerals, including iron and calcium (Anton et al., 2007). Due to its structure and composition, phosvitin and its derived phosphopeptides exhibit numerous outstanding bioactivities: antimicrobial, antioxidant, anti-inflammatory, antioxidant, anti-cancer, etc (Lu & Baker, 1986; Sattar Khan et al., 2000; Anton et al., 2007; Xu et al., 2012; Jung et al., 2013). Compounds with antioxidant and anti-inflammatory properties promote bone health by increasing bone mass and preventing bone loss *in vivo* (Shen et al., 2009). As phosvitin and its phosphopeptides exhibit both anti-inflammatory and antioxidant properties, these molecules have strong potential to exhibit bone health-promoting properties.

First, phosvitin has an essential role in the bone formation of chicken embryos (Li et al., 2014). The phosphate content of phosvitin was inversely correlated with the body length growth and the activity of ALP, an osteoblast marker. Therefore, Li et al. (2014) hypothesized that phosvitin promotes bone formation in chicken embryos through protein dephosphorylation (Li et al., 2014). Given the essential role of phosvitin in chicken bone formation, the research community started to widely investigate the potential of phosvitin and phosvitin hydrolysates for the prevention and management of bone health.

The impacts of phosvitin and phosvitin hydrolysate were investigated on osteoblast bone cell lines (particularly in MC3T3-E1 cells). Several studies reported that phosvitin and phosvitin hydrolysates both significantly promote bone formation *in vitro* (Figure 35.2). First, phosvitin and phosvitin hydrolysate significantly inhibit the apoptosis of osteoblasts and thus, promote bone formation (Chakrabarti et al., 2020). Then, phosvitin and phosvitin hydrolysate significantly enhance osteoblasts proliferation and differentiation, in a dose-dependent manner, by stimulating the expression of osteoblast function markers including osteoprotegerin (OPG), osteocalcin, BMP-2, runt-related transcription factor 2 (RUNX2) and ALP (Jie et al., 2018; Ren et al., 2019; Chakrabarti et al., 2020; Zhao et al., 2021). Phosvitin and phosvitin hydrolysates also stimulate the production of type 1 collagen, one of the key bone components (Ren, 2019; Ren et al., 2019). Researchers have also demonstrated the beneficial contribution of phosvitin as an effective promoter of bone mineralization. Indeed, phosvitin mimics the role of ascorbic acid in osteoblastic cells and thus, significantly enhances bone mineralization (Liu et al., 2017). Indeed, phosvitin accelerates the initial phase transformation of mineralization (Zhang et al., 2015; Jie et al., 2017). The phosphorylated serine clusters of the protein are the active region for promoting mineralization (Zhang et al., 2017, 2018). Additionally, Jie et al. (2018) reported that the phosphate groups of phosvitin play an essential role in its osteogenic bioactivity. Indeed, phosvitin and its derived peptides, with a higher degree of phosphorylation, will better improve the proliferation and differentiation of osteoblasts (Jie et al., 2018). Moreover, phosvitin-derived phosphopeptides show stronger osteogenic bioactivity than the native protein (Chakrabarti et al., 2020; Zhao et al., 2021).

Moreover, osteoblasts produce numerous inflammatory biomarkers (normal T cell expressed and secreted (RANTES), monocyte chemoattractant protein-1 (MCP-1), cyclooxygenase-2 (COX-2) to recruit osteoclasts and maintain bone resorption (Ren, 2019). However, the addition of phosvitin and phosvitin hydrolysates inhibits the osteoblast's production of inflammatory factors via the activation of ERK 1/2 and AKT signaling pathways (Ren, 2019; Ren et al., 2019) and via the inhibition of RANKL expression *in vitro* (Chakrabarti et al., 2020). Thus, phosvitin-derived molecules also inhibit the recruitment of osteoclasts by osteoblasts and prevent bone resorption *in vitro* (Chakrabarti et al., 2020). However, the mechanisms of action are still mainly unknown.

As illustrated in Figure 35.2, phosvitin and its hydrolysates also inhibit the bone resorption activity of osteoclasts *in vitro* (Ren, 2019; Ren et al., 2021). Indeed, the expression of osteoclastogenesis transcription markers was decreased by phosvitin and phosvitin hydrolysates; including c-Fos, and nuclear factor of activated T-cells 1 (NFATc1) (Ren, 2019). Additionally, phosvitin and phosvitin hydrolysates both downregulate inflammatory biomarkers produced by osteoclast cell lines (TNF-α, MCP-1, RANTES, and iNOS) (Ren, 2019). Both phosvitin and phosvitin hydrolysates reduce osteoclastogenesis through the inhibition of RANKL via c-Jun N-terminal kinase (JNK), p38, NF-κB and Mitogen-activated protein kinases (MAPK) signaling pathways (Ren, 2019; Ren et al., 2021)

Choi et al. (2005) demonstrate that feeding rats with a diet enriched in phosvitin-derived phosphopeptides (<2.5% of total protein fed) increases the absorption and the accumulation of calcium in the tibia and femur, leading to higher BMD and bone mineral content. This animal study demonstrates, for the first time, the potential of phosvitin-derived phosphopeptides as an anabolic agent for the promotion of bone formation (Choi et al., 2005). Finally, Ren (2019) investigated the bone health-promoting properties of phosvitin and phosvitin hydrolysates *in vivo.* For this study, OVX rats (an osteoporosis animal model) were fed with 2% phosvitin hydrolysates or 2% phosvitin for 12 weeks. The results suggest that 2% phosvitin hydrolysate significantly improves bone microstructure and prevents bone loss in ovariectomized rats due to a potential combination of antiresorptive and anabolic activities (Ren, 2019).

Taken together, current literature suggests that phosvitin and its derived fractions could be used to prevent bone loss due to a combination of anabolic and antiresorptive activities (Figure 35.2). Thus, phosvitin could be a viable approach for the prevention and management of osteoporosis. Although the mechanisms of action are still unknown, the preliminary results demonstrated the strong potential of phosvitin in the prevention and management of bone health.

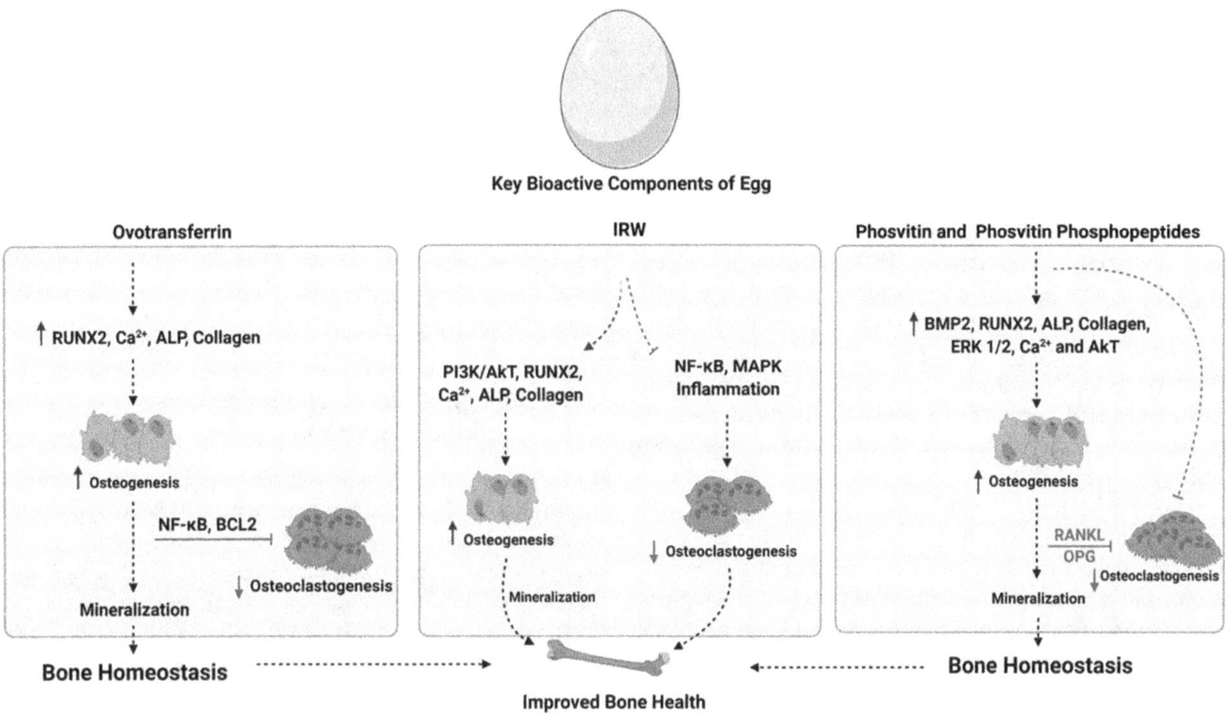

FIGURE 35.2 The impact of egg-derived bioactive components on bone health.

35.5.1 Bioactive Compounds of Egg White

35.5.1.1 Ovotransferrin and Its Derived Peptide IRW for the Prevention of Bone Loss

Ovotransferrin (78 kDa) is an iron-binding glycoprotein that accounts for 12% of the egg white protein (Giansanti et al., 2012). As a member of the transferrin family, ovotransferrin possesses antimicrobial, anti-inflammatory, immunomodulatory, and antioxidant bioactivities (Giansanti et al., 2012; Rathnapala et al., 2021).

Shang et al. (2019) were the first to investigate the bone health-promoting properties of ovotransferrin *in vitro* (Figure 35.2). In a dose-dependent manner, ovotransferrin (100; 1,000 μg/mL) significantly promotes bone formation *in vitro* by stimulating the proliferation and the differentiation of mouse preosteoblasts MC3T3-E1 cells as well as bone mineralization (Shang & Wu, 2018; Shang, 2019). Ovotransferrin interacts with low-density lipoprotein receptor-related protein 1 (LRP1) to enhance bone formation (Shang, 2019; Shang et al., 2020). Ovotransferrin stimulates the production of RUNX2, improves the level of Ca^{2}+, collagen synthesis as well as ALP activity via LRP1 activation (Shang, 2019; Shang et al., 2020). On the opposite, osteoblast cells MC3T3-E1 cells, with LRP1 deletion, display lower osteogenic activity (Shang, 2019; Shang et al., 2020).

Additionally, Shang et al. (2019) also investigated the impact of ovotransferrin on bone resorption *in vitro* (Figure 35.2). In preosteoblasts MC3T3-E1 cells, ovotransferrin prevents osteoclastogenesis by inhibiting the production of RANKL, while enhancing OPG production (Shang, 2019; Shang & Wu, 2019). A decrease in RANKL inhibits the activation of NF-κB and MAPK signaling pathways (Shang, 2019; Shang & Wu, 2019). Additionally, ovotransferrin downregulates the production of inflammatory markers associated with osteoclastogenesis such as TNF receptor-associated factor 6 (TRAF6), c-Fos, nuclear factor of activated T-cells cytoplasmic 1 (NFATc1), and cathepsin K (CathK). Then, ovotransferrin also promotes osteoclast apoptosis by regulating the Bcl-2 family (Shang, 2019; Shang & Wu, 2019).

Finally, Shang et al. (2019) investigate the impact of ovotransferrin in OVX rats. After 12 weeks of treatment, oral administration of ovotransferrin at 1% (w/w) slows bone loss and promotes the maintenance of BMD and bone microarchitecture (Shang, 2019). *In vivo,* supplementation with ovotransferrin also decreases the production of inflammatory markers TNF-α and IL-6 involved in bone resorption (Shang, 2019). Therefore, this study highlights the potential of ovotransferrin for the promotion of bone health and the prevention of bone loss (Shang, 2019). Despite these advances, most mechanisms remain uncleared and additional studies are needed before a future translation to clinical trials.

Additionally, IRW (Ile-Arg-Trp), a tripeptide derived from ovotransferrin hydrolysates also demonstrates bone health-promoting properties. This peptide demonstrates antihypertensive (Majumder et al., 2013; Liao et al., 2016; Wu, 2020), anti-inflammatory (Huang et al., 2010), and antioxidant bioactivities (Huang et al., 2010). Shang et al. (2019) were the first to demonstrate the bone health-promoting properties of IRW (Shang et al., 2019; Shang & Wu, 2020). The authors demonstrated that IRW significantly stimulated osteogenesis and increased osteoblast proliferation and mineralization *in vitro* (Shang et al., 2019). On the other hand, IRW upregulates the expression of RUNX2 and OPG and the production of ALP. Similarly, IRW decreases the expression of RANKL (Shang et al., 2019). Overall, egg-derived peptide IRW promotes bone formation via the activation of the PI3K-Akt-RUNX2 signaling pathway and its downstream effectors in MC3T3-E1 cells.

In addition to promoting bone formation, IRW also inhibits bone resorption. IRW decreases osteoclastogenesis induced by lipopolysaccharides (LPS) *in vitro* (Shang & Wu, 2020). Indeed, lipopolysaccharides (LPS) stimulate osteoclast activity through the production of inflammatory factors, leading to bone resorption. However, IRW (25–50 μM) inhibits the production of inflammatory factors (TNF-α, IL-6, NOS, COXII, NO, and PGE2, etc.), and inactivates the bone resorption signaling pathways NF-κB and MAPK, which prevent LPS-induced bone resorption activity (Shang & Wu, 2020). Additionally, IRW regulates the production of osteoclastogenesis factors, including TRAF6, c-Fos, NFATc1, and cathepsin K (Shang & Wu, 2020).

In summary, ovotransferrin and its derived peptide IRW regulate bone cell activity by promoting osteoblastic activity, enhancing bone formation as well as inhibiting osteoclastogenesis, and preventing bone loss *in vitro*. Moreover, ovotransferrin prevents bone loss and protects bone microarchitecture *in vivo*. These results demonstrate the potential applications of ovotransferrin and IRW, as functional foods or nutraceuticals, for the prevention of bone loss and the management of osteoporosis.

35.5.1.2 Egg White Protein-Based Scaffold for Bone Tissue Regeneration

Engineered scaffolds are widely used for the restoration of bone tissue to counteract skeletal disorders such as osteoporosis. Scaffolds contain bioactive agents to successfully promote the bone healing process, enhance osteoregeneration and reduce inflammation (Ramesh et al., 2018). They also stabilize the bone matrix until new bones are formed. In addition to eggshells calcium carbonate (4.1.2 section), several egg white proteins have been studied as bioactive molecules in scaffolds for bone regeneration including ovomucin; ovalbumin, and lysozyme.

Ovomucin is an egg white glycoprotein (2–4% of egg white proteins), mainly contributing to egg white gelling and foaming properties (Carpena et al., 2017). Carpena et al. (2017) demonstrate that ovomucin combined with gelatin was used to produce a porous scaffold with the desired pore size and porosity. Ovomucin combined with gelatin promotes bone formation by supporting cell adhesion and growth (Carpena et al., 2017).

Farrar et al. (2010) also study the use of egg white ovalbumin for bone tissue regeneration. The results demonstrate that the ovalbumin scaffold enhances the proliferation and the differentiation of osteoblast cells *in vitro* (Farrar et al., 2010).

Additionally, egg white lysozyme is also a potential implantable biomaterial for bone tissue. First, Martins et al. (2009) have developed a new material for bone regeneration: calcium-phosphate coated chitosan-based scaffold with lysozyme incorporated (Martins et al., 2009). In this *in vitro* study, this emerging material promotes osteogenic differentiation, bone reformation, and bone mineralization (Martins et al., 2009). Another study developed lysozyme-protected ultrasmall gold nanoclusters to promote bone formation (Li et al., 2020). This novel nanomaterial promotes osteogenic differentiation and decreases osteoclast activity *in vitro* (Li et al., 2020). According to the authors, this material could be used as a nanomedicine to rebuild bone in patients with osteoporosis (Li et al., 2020).

Thus, egg white ovomucin, lysozyme, and ovalbumin are natural, biocompatible, and functional biomaterials, that have strong potential in the biomedical industry for bone tissue engineering. As eggs are cheap and widely available, these proteins have strong potential to replace other synthetic and natural biopolymers.

35.5.1.3 Egg White Cystatin Prevents Bone Resorption

Lysosomal cysteine proteinases play an important role in bone resorption by digesting the bone matrix. The action of these proteinases can be inactivated by protease inhibitors such as egg white cystatin (Abrahamson et al., 2003). Egg white cystatin inhibits the action of several cysteine proteinases, including cathepsin B, H, and L (Abrahamson et al., 2003). Brand et al. (2004) have reported the impact of egg white cystatin on bone resorption (Brand et al., 2004). Their findings suggest that egg white cystatin inhibits osteoclasts-mediated bone resorption by regulating the activity of cysteine proteinases involved in bone matrix digestion. Moreover, egg white cystatin also decreases calcium release *in vitro* (Brand et al., 2004). Thus, egg white cystatin has a strong potential to be used as a therapeutical strategy for the prevention of bone loss.

35.6 Discussion and Future Directions

Osteoporosis is a major bone disorder that affects more than 200 million people worldwide (World Health Organization, 2007). This bone disorder is characterized by microarchitectural deterioration of the bone tissue accompanied by a low bone mineral density, which increases bone fragility and therefore augments the risk of fractures and related orthopedic complications (World Health Organization, 2007; Wu et al., 2021). Knowing the gravity and incidence of this bone disorder, osteoporosis has been characterized as a major public health issue worldwide (Curtis et al., 2018; Sietsema, 2020).

Despite the pathological severity of osteoporosis, this bone disorder is preventable. One of the key risks for osteoporosis is an undernourished imbalanced diet, particularly a diet poor in proteins and in essential micronutrients, the vital factors for the maintenance of bone health (Alekel et al., 2013). Indeed, a diet poor in proteins, calcium and/or vitamin D slowdowns bone formation, increases bone loss and enhances the risk of fractures (Hill & Aspray, 2017; Rizzoli et al., 2018). Therefore, nutrition remains the most effective strategy for the prevention of osteoporosis incidence (Chen et al., 2019; Hejazi et al., 2020). Knowing the essential role of proteins and other micronutrients in bone health, they need to be supplied in an adequate daily amount in our diet (Chen et al., 2019; Hejazi et al., 2020). Among foods rich in proteins and micronutrients, eggs are one of the dietary sources that provide most of the micronutrients and proteins essential for the development and the maintenance of bone health (Table 35.1; Seuss-baum, 2007).

Eggs are an inexpensive, widely available, and highly nutritious dietary source. Therefore, incorporating eggs into the diet seems to be a rational and viable strategy to preserve bone health and prevent bone loss. However, no studies are currently available on the impact of whole egg consumption on bone health, indicating a vital gap in knowledge. It would be interesting to study if incorporating eggs into the diet could prevent or slow bone loss in postmenopausal women or in osteoporotic patients. Moreover, phosphorus and ω-6 fatty acids are detrimental to bone health if consumed in excess (Simopoulos, 2008; Pinheiro et al., 2009). Indeed, every 100 mg of phosphorus excess increase by 9% the risk of fracture (Pinheiro et al., 2009). Likewise, a ratio of ω-6/ω-3 fatty acids higher than 15:1 has been correlated with osteoporosis (Simopoulos, 2008). As these two egg components are already consumed in excess in western diets, it would be essential to determine the recommended consumption of eggs per day, to benefit bone health without being impacted by the detrimental effects of phosphorus and ω-6 fatty acids. Therefore, additional clinical studies are needed to demonstrate and clarify the real benefits of whole egg consumption on bone health and osteoporosis.

In addition to preventing bone loss (antiresorptive agents), pharmaceutical treatments are also available to promote bone health (anabolic agents) (Seeman & Martin, 2019). Despite beneficial effects, their side effects and their efficacy remain controversial, which limits the compliance of patients with these treatments (Whitaker et al., 2012; Khosla & Lorenz, 2017). Therefore, to overcome these challenges, food scientists started looking for efficient, natural, and evidence-based safer alternatives to traditional therapies. Among these, a wide range of egg-derived bioactive compounds demonstrates numerous beneficial impacts on bone health (Table 35.2); for example, eggshells derived calcium carbonate promotes calcium absorption (Neunzehn et al., 2015; Salama et al., 2019), several egg white proteins (lysozyme, ovalbumin, ovomucin) demonstrate promising applications in bone scaffolds (Martins et al., 2009; Carpena et al., 2017; Li et al., 2020), and, several egg-derived proteins such as phosvitin, ovotransferrin, and their derived peptides, promote bone health and prevent bone loss *in vivo* (Ren, 2019; Shang, 2019). Therefore, these egg-derived bioactive compounds have great potential to promote bone health in healthy and osteoporotic individuals. These bioactive compounds could be either incorporated into the diet as functional foods to promote bone health in healthy individuals, or used to prevent or slow bone loss in individuals at risk of osteoporosis (e.g., elderly, postmenopausal women), in individuals with a low BMD or in osteoporotic patients.

However, despite promising results and a burgeoning body of research on bioactive egg components, only a few egg-derived products have been launched on the market (Table 35.3). Indeed, eggshell-derived calcium supplements have been commercialized to support calcium absorption in humans, laying hens, cats, and dogs (Cordeiro & Hincke, 2011; Eggnovo, n.d.; IrRAWsistible Pet Foods, n.d.; Redd Remedies, n.d.). Eggshells have also been commercialized as a biocompatible material for bone tissue regeneration (Salama et al., 2019) whereas only Bonepep®, an egg yolk hydrolysate, has been marketed in Eurasia and North America as a nutraceutical for the prevention of bone loss (Pharma Foods International Co., n.d.; Song et al., 2011). Despite these few exceptions, most of the egg-derived bioactive compounds did not translate yet into clinical trials (Duffuler et al., 2022). Even though there is a growing interest in our research community for egg-derived bioactive, most of these studies remain preliminary and clinical data are still far behind *in vitro* and *in vivo* data (Brown et al., 2018; Duffuler et al., 2022; Wang et al., 2022). Clinical trials are essential to demonstrate the efficacy of bioactive compounds in humans. However, clinical exploration of these egg bioactive is currently hindered due to the inadequate understanding of their mechanisms of action, underlying signaling pathways, and the lack of data on the bioavailability and the transportation of bioactive compounds to their targeted organs, the bones (Brown et al., 2018; Xu et al., 2019; Wang et al., 2022).

Therefore, future research should focus on filling this critical knowledge gap before any further applications. Since the pharmacokinetics and pharmacodynamics of phosvitin and ovotransferrin remains unknown, future *in vivo* studies are necessary to establish their pharmacokinetics profile, toxicity, and efficacy. Previous studies have demonstrated that phosvitin, ovotransferrin and their derived peptides can directly modulate bone health in cells. However, when orally administered *in vivo*, the bioactive proteins/peptides will go through the gastrointestinal tract where they will be digested enzymatically, absorbed in the intestinal and enter the systemic circulation where they will be able to reach their targeted organs (Udenigwe & Fogliano, 2017; Xu et al., 2019). Therefore, it could be interesting to study which of the intact or digested proteins/peptides are responsible for their bioactivity *in vivo*. Additionally, some proteins such as phosvitin are resistant to enzymatic hydrolysis and thus, can be fermented by the gut microbiota (Li et al., 2019). Therefore, it would be essential to study if phosvitin could regulate bone health *in vivo* through the modulation of the gut microbiota. Additional studies could also focus on the estimation of the most effective dosage necessary to minimize bone loss *in vivo*. In previous *in vivo* studies, ovariectomized (OVX) rats are usually used as an animal model for postmenopausal osteoporosis. Despite promising results in OVX rats, it remains to be established if phosvitin or ovotransferrin could show similar effects in other animal models, which would mimic other types of osteoporosis (glucocorticoid-induced osteoporosis, senile osteoporosis, or aging models). These potential *in vivo* studies would broaden the wide scope of applications of these egg-derived proteins and peptides. Finally, clinical trials could also be implemented to study the efficacy of phosvitin and ovotransferrin in the prevention of bone loss and osteoporosis, in postmenopausal women or in the elderly population. However, pharmacokinetics and pharmacodynamics must be established in advance.

In summary, eggs are an excellent dietary source that provides protein and essential micronutrients that benefit bone health. Additionally, given their impressive bioactivities, demonstrated both *in vitro* and *in vivo*, egg-derived bioactive components have great potentials to be used to promote bone health. These bioactives can be used in the biomedical industry for bone tissue engineering or in the food industry as nutraceuticals to prevent bone loss and contribute to the fight against osteoporosis. Despite promising initial results, additional studies are warranted to understand their in-depth mechanism of action and to confirm their efficacy on bone health before further clinical exploration and commercialization.

REFERENCES

Abdelhamid, A., Hooper, L., Sivakaran, R., Hayhoe, R. P. G., Welch, A., Ajabnoor, S., Alabdulghafoor, F., Alkhudairy, L., Biswas, P., Brainard, J., Bridges, C., Brown, T. J., Deane, K., Donaldson, D., Hanson, S., Jimoh, O. F., Hooper, L., Martin, N., O'Brien, A.,..., Winstanley, L. (2019). The relationship between omega-3, omega-6 and total polyunsaturated fat and musculoskeletal health and functional status in adults: A systematic review and meta-analysis of RCTs. *Calcified Tissue International*, *105*(4), 353–372. https://doi.org/10.1007/s00223-019-00584-3

Abrahamson, M., Alvarez-Fernandez, M., & Nathanson, C.-M. (2003). Cystatins. *Biochemical Society Symposia*, *70*(70), 179–199. https://doi.org/10.1042/bss0700179

Ahmed, T. A. E., Kulshreshtha, G., & Hincke, M. T. (2019). CHAPTER 19. Value-added uses of eggshell and eggshell membranes. In *Food chemistry, function and analysis* (Vols. 2019-Janua, Issue 14, pp. 359–397). The Royal Society of Chemistry. https://doi.org/10.1039/9781788013833-00359

Ahmed, T. A. E., Wu, L., Younes, M., & Hincke, M. (2021). Biotechnological applications of eggshell: Recent advances. *Frontiers in Bioengineering and Biotechnology*, *9*(July), 1–19. https://doi.org/10.3389/fbioe.2021.675364

Aki, S., Eskiyurt, N., Akarirmak, Ü, Tüzün, F., Eryavuz, M., Alper, S., Arpacioğlu, O., Atalay, F., Kavuncu, V., Kokino, S., Kuru, Ö, Nas, K., Özerbil, Ö, Savaş, G., Sendur, Ö. F., Soy, D., & Akyüz, G. (2003). Gastrointestinal side effect profile due to the use of alendronate in the treatment of osteoporosis. In *Yonsei Medical Journal* (Vol. 44, Issue 6, pp. 961–967). https://doi.org/10.3349/ymj.2003.44.6.961

Al Mijan, M., Choi, K. H., & Kwak, H. S. (2014). Physicochemical, microbial, and sensory properties of nanopowdered eggshell-supplemented yogurt during storage. *Journal of Dairy Science*, *97*(6), 3273–3280. https://doi.org/10.3168/jds.2013-7367

Alekel, D. L., Weaver, C. M., Ronis, M. J. J., & Ward, W. E. (2013). Nutritional influences on bone health and overview of methods. In *Bioactive food as dietary interventions for the aging population* (pp. 357–370). Elsevier. https://doi.org/10.1016/B978-0-12-397155-5.00096-9

Al-Qtaitat, A. I., & Aldalaen, S. M. (2014). A review of non-collagenous proteins; their role in bone. *American Journal of Life Sciences*, *2*(6), 351–355. https://doi.org/10.11648/j.ajls.20140206.14

Alsuhaibani, A. M. A. (2018). Rheological and nutritional properties and sensory evaluation of bread fortified with natural sources of calcium. *Journal of Food Quality*, *2018*. https://doi.org/10.1155/2018/8308361

Altindag, O., Erel, O., Soran, N., Celik, H., & Selek, S. (2008). Total oxidative/anti-oxidative status and relation to bone mineral density in osteoporosis. *Rheumatol Int*, *28*, 317–321. https://doi.org/10.1007/s00296-007-0452-0

Anton, M., Castellani, O., & Guérin-Dubiard, C. (2007). Phosvitin. *Bioactive Egg Compounds*, *1900*, 17–24. https://doi.org/10.1007/978-3-540-37885-3_4

Arnold, M., Rajagukguk, Y. V., & Gramza-Michałowska, A. (2021). Functional food for elderly high in antioxidant and chicken eggshell calcium to reduce the risk of osteoporosis—a narrative review. *Foods*, *10*(3). https://doi.org/10.3390/foods10030656

Baggio, B., Budakovic, A., Nassuato, M. A., Vezzoli, G., Manzato, E., Luisetto, G., & Zaninotto, M. (2000). Plasma phospholipid arachidonic acid content and calcium metabolism in idiopathic calcium nephrolithiasis. *Kidney International*, *58*(3), 1278–1284. https://doi.org/10.1046/j.1523-1755.2000.00283.x

Balázsi, C., Gergely, G., Balázsi, K., Chae, C. H., Sim, H. Y., Choi, J. Y., & Kim, S. G. (2013). Bone formation with nanohydroxyapatite from eggshell. *Materials Science Forum*, *729*, 25–30. https://doi.org/10.4028/www.scientific.net/MSF.729.25

Balogh, E., Paragh, G., & Jeney, V. (2018). Influence of iron on bone homeostasis. *Pharmaceuticals*, *11*(4), 107. https://doi.org/10.3390/PH11040107

Bonjour, J. P. (2013). Protein intake and bone health. *International Journal for Vitamin and Nutrition Research*, *81*(2–3), 134–142. https://doi.org/10.1024/0300-9831/a000063

Booth, S. L., Broe, K. E., Gagnon, D. R., Tucker, K. L., Hannan, M. T., McLean, R. R., Dawson-Hughes, B., Wilson, P. W. F., Cupples, L. A., & Kiel, D. P. (2003). Vitamin K intake and bone mineral density in women and men. *American Journal of Clinical Nutrition*, *77*(2), 512–516. https://doi.org/10.1093/ajcn/77.2.512

Boskey, A. L., & Robey, P. G. (2013). The regulatory role of matrix proteins in mineralization of bone. In *Osteoporosis* (4th ed.). Elsevier. https://doi.org/10.1016/B978-0-12-415853-5.00011-X

Boskey, A. L., & Robey, P. G. (2018). The composition of bone. *Primer on the Metabolic Bone Diseases and Disorders of Mineral Metabolism*, 84–92. https://doi.org/10.1002/9781119266594.ch11

Bovier, E., & Hammond, B. (2017). The macular carotenoids lutein and zeaxanthin are related to increased bone density in young healthy adults. *Foods*, *6*(9), 78. https://doi.org/10.3390/foods6090078

Brand, H. S., Lerner, U. H., Grubb, A., Beertsen, W., Nieuw Amerongen, A. V., & Everts, V. (2004). Family 2 cystatins inhibit osteoclast-mediated bone resorption in calvarial bone explants. *Bone*, *35*(3), 689–696. https://doi.org/10.1016/j.bone.2004.05.015

Brown, L., Caligiuri, S. P. B., Brown, D., & Pierce, G. N. (2018). Clinical trials using functional foods provide unique challenges. *Journal of Functional Foods*, *45*, 233–238. https://doi.org/10.1016/J.JFF.2018.01.024

Brun, L. R., Lupo, M., Delorenzi, D. A., di Loreto, V. E., & Rigalli, A. (2013). Chicken eggshell as suitable calcium source at home. *International Journal of Food Sciences and Nutrition*, *64*(6), 740–743. https://doi.org/10.3109/09637486.2013.787399

Calvo, M. S., & Tucker, K. L. (2013). Is phosphorus intake that exceeds dietary requirements a risk factor in bone health? *Annals of the New York Academy of Sciences*, *1301*(1), 29–35. https://doi.org/10.1111/nyas.12300

Capozzi, A., Scambia, G., & Lello, S. (2020). Calcium, vitamin D, vitamin K2, and magnesium supplementation and skeletal health. *Maturitas*, *140*(May), 55–63. https://doi.org/10.1016/j.maturitas.2020.05.020

Carpena, N. T., Abueva, C. D. G., Padalhin, A. R., & Lee, B. T. (2017). Evaluation of egg white ovomucin-based porous scaffold as an implantable biomaterial for tissue engineering. *Journal of Biomedical Materials Research - Part B Applied Biomaterials*, *105*(7), 2107–2117. https://doi.org/10.1002/jbm.b.33750

Casado-Díaz, A., Santiago-Mora, R., Dorado, G., & Quesada-Gómez, J. M. (2013). The omega-6 arachidonic fatty acid, but not the omega-3 fatty acids, inhibits osteoblastogenesis and induces adipogenesis of human mesenchymal stem cells: Potential implication in osteoporosis. *Osteoporosis International*, *24*(5), 1647–1661. https://doi.org/10.1007/s00198-012-2138-z

Chakrabarti, S., Ren, J., & Wu, J. (2020). phosvitin derived phospho-peptides show better osteogenic potential than intact phosvitin in MC3T3-E1 osteoblastic cells. *Nutrients*, *12*(10), 2998. https://doi.org/10.3390/nu12102998

Chen, J. S., & Sambrook, P. N. (2012). Antiresorptive therapies for osteoporosis: A clinical overview. *Nature Reviews Endocrinology*, *8*(2), 81–91. https://doi.org/10.1038/nrendo.2011.146

Chen, L. R., Hou, P. H., & Chen, K. H. (2019). Nutritional support and physical modalities for people with osteoporosis: Current opinion. *Nutrients, 11*(12), 2848. https://doi.org/10.3390/NU11122848

Chilek, T. Z. T., Kairuaman, N. A., Ahmad, F., Wahab, R. A., Zamri, A. I., & Mahmood, A. (2018). Development of white bread fortified with calcium derived from eggshell powder. *Malaysian Applied Biology*, *47*(6), 29–39.

Choi, I., Jung, C., Choi, H., Kim, C., & Ha, H. (2005). Effectiveness of phosvitin peptides on enhancing bioavailability of calcium and its accumulation in bones. *Food Chemistry*, *93*(4), 577–583. https://doi.org/10.1016/j.foodchem.2004.10.028

Christakos, S., Hewison, M., Gardner, D. G., Wagner, C. L., Sergeev, I. N., Rutten, E., Pittas, A. G., Boland, R., Ferrucci, L., & Bikle, D. D. (2013). Vitamin D: Beyond bone. *Annals of the New York Academy of Sciences*, *1287*(1), 45. https://doi.org/10.1111/NYAS.12129

Clemens, T. L. (2014). Vitamin B12 deficiency and bone health. *New England Journal of Medicine*, *371*(10), 963–964. https://doi.org/10.1056/NEJMcibr1407247

Coheley, L. M., Kindler, J. M., Laing, E. M., Oshri, A., Hill Gallant, K. M., Warden, S. J., Peacock, M., Weaver, C. M., & Lewis, R. D. (2018). Whole egg consumption and cortical bone in healthy children. *Osteoporosis International*, *29*(8), 1783–1791. https://doi.org/10.1007/s00198-018-4538-1

Cordeiro, C., & Hincke, M. (2011). Recent patents on eggshell: Shell and membrane applications. *Recent Patents on Food, Nutrition & Agriculture*, *3*(1), 1–8. https://doi.org/10.2174/2212798411103010001

Cranney, A., Horsley, T., O'Donnell, S., Weiler, H., Puil, L., Ooi, D., Atkinson, S., Ward, L., Moher, D., Hanley, D., Fang, M., Yazdi, F., Garritty, C., Sampson, M., Barrowman, N., Tsertsvadze, A., & Mamaladze, V. (2007). Effectiveness and safety of vitamin D in relation to bone health. In *Evidence report/technology assessment* (Issue 158, pp. 1–235).

Curtis, E. M., Harvey, N. C., & Cooper, C. (2018). The burden of osteoporosis. In *Osteoporosis: A Life course epidemiology approach to skeletal health* (pp. 1–20). https://doi.org/10.1201/9781351234627-1

Curtis, E. M., van der Velde, R., Moon, R. J., van den Bergh, J. P. W., Geusens, P., de Vries, F., van Staa, T. P., Cooper, C., & Harvey, N. C. (2016). Epidemiology of fractures in the United Kingdom 1988-2012: Variation with age, sex, geography, ethnicity and socioeconomic status. *Bone*, *87*, 19–26. https://doi.org/10.1016/j.bone.2016.03.006

Dadhich, P., Das, B., Pal, P., Srivas, P. K., Dutta, J., Ray, S., & Dhara, S. (2016). A simple approach for an eggshell-based 3D-printed osteoinductive multiphasic calcium phosphate scaffold. *ACS Applied Materials and Interfaces*, *8*(19), 11910–11924. https://doi.org/10.1021/acsami.5b11981

Dhonukshe-Rutten, R. A. M., Lips, M., de Jong, N., Marijke, J. M. C. A. P., Hiddink, G. J., van Dusseldorp, M., de Groot, L. C. P. G. M., & van Staveren, W. A. (2003). Vitamin B-12 status is associated with bone mineral content and bone mineral density in frail elderly women but not in men. *Journal of Nutrition*, *133*(3), 801–807. https://doi.org/10.1093/jn/133.3.801

Duffuler, P., Bhullar, K. S., Zani, S. C., de, C., & Wu, J. (2022). Bioactive peptides: From basic research to clinical trials and commercialization. *Journal of Agricultural and Food Chemistry*. https://doi.org/10.1021/ACS.JAFC.1C06289

Eastell, R. (2017). Prevention and management of osteoporosis. *Medicine (United Kingdom)*, *45*(9), 565–569. https://doi.org/10.1016/j.mpmed.2017.06.004

Eggnovo. (n.d.). *Ovocet: Natural calcium carbonate*. Retrieved December 6, 2021, from https://eggnovo.com/ingredients/ovocet/

El-Shibiny, S., El-Gawad, A., Assem, M. A. E. K. M., & El-Sayed, F. M., S. M. (2018). The use of nano-sized eggshell powder for calcium fortification of cow's and buffalo's milk yogurts. *Acta Scientiarum Polonorum, Technologia Alimentaria*, *17*(1), 37–49. https://doi.org/10.17306/J.AFS.2018.0541

Ensrud, K. E., & Crandall, C. J. (2017). Osteoporosis. *Annals of Internal Medicine*, *167*(3), ITC17. https://doi.org/10.7326/AITC201708010

Ensrud, K. E., & Taylor, B. C. (2013). Epidemiologic methods in studies of osteoporosis. In *Osteoporosis* (4th ed.). Elsevier. https://doi.org/10.1016/B978-0-12-415853-5.00024-8

European Food Safety Authority. (2012). Scientific opinion on dietary reference values for protein. *EFSA Journal*, *10*(2), 2557. https://doi.org/10.2903/j.efsa.2012.2557

Farrar, G., Barone, J., & Morgan, A. (2010). Ovalbumin-based porous scaffolds for bone tissue regeneration. *Journal of Tissue Engineering*, *1*(1), 1–6. https://doi.org/10.4061/2010/209860

Feskanich, D., Willett, W. C., Stampfer, M. J., & Colditz, G. A. (1996). Protein consumption and bone fractures in women. *American Journal of Epidemiology*, *143*(5), 472–479. https://doi.org/10.1093/oxfordjournals.aje.a008767

Fukumoto, S. (2014). Phosphate metabolism and vitamin D. *Bone Key Reports*, *3*, 497. https://doi.org/10.1038/bonekey.2013.231

Fusaro, M., Cianciolo, G., Brandi, M. L., Ferrari, S., Nickolas, T. L., Tripepi, G., Plebani, M., Zaninotto, M., Iervasi, G., la Manna, G., Gallieni, M., Vettor, R., Aghi, A., Gasperoni, L., Giannini, S., Sella, S., & Cheung, M., A. (2020). Vitamin K and osteoporosis. *Nutrients*, *12*(12), 3625. https://doi.org/10.3390/nu12123625

Giansanti, F., Leboffe, L., Pitari, G., Ippoliti, R., & Antonini, G. (2012). Physiological roles of ovotransferrin. *Biochimica et Biophysica Acta - General Subjects*, *1820*(3), 218–225. https://doi.org/10.1016/j.bbagen.2011.08.004

Graham, S. K., Parker, M. J., & Glyn, A. P. (1994). Mortality and morbidity after hip fractures. *BMJ*, *308*(6924), 343. https://doi.org/10.1136/bmj.308.6924.343b

Hadjidakis, D. J., & Androulakis, I. I. (2006). Bone remodeling. *Annals of the New York Academy of Sciences*, *1092*, 385–396. https://doi.org/10.1196/annals.1365.035

Hamidi, M. S., Gajic-Veljanoski, O., & Cheung, A. M. (2013). Vitamin K and bone health. *Journal of Clinical Densitometry*, *16*(4), 409–413. https://doi.org/10.1016/j.jocd.2013.08.017

Hassan, N. M. M. (2015). Chicken eggshell powder as dietary calcium source in biscuits. *World Journal of Dairy & Food Sciences*, *10*(2), 199–206. https://doi.org/10.5829/idosi.wjdfs.2015.10.2.1152

Hawkins, F., Garla, V., Allo, G., Males, D., Mola, L., & Corpas, E. (2021). Senile and postmenopausal osteoporosis: Pathophysiology, diagnosis, and treatment. In *Endocrinology of aging*. Elsevier Inc. https://doi.org/10.1016/b978-0-12-819667-0.00005-6

Hayashibara, T., Hiraga, T., Sugita, A., Wang, L., Hata, K., Ooshima, T., & Yoneda, T. (2007). Regulation of osteoclast differentiation and function by phosphate: Potential role of osteoclasts in the skeletal abnormalities in hypophosphatemic conditions. *The Journal of Bone and Mineral Research*, *22*, 1743–1751. https://doi.org/10.1359/JBMR.070709

Heaney, R. P., & Layman, D. K. (2008). Amount and type of protein influences bone health. *The American Journal of Clinical Nutrition*, *87*(5), 1567S–1570S. https://doi.org/10.1093/ajcn/87.5.1567S

Hejazi, J., Davoodi, A., Khosravi, M., Sedaghat, M., Abedi, V., Hosseinverdi, S., Ehrampoush, E., Homayounfar, R., & Shojaie, L. (2020). Nutrition and osteoporosis prevention and treatment. *Biomedical Research and Therapy*, *7*(4), 3709–3720. https://doi.org/10.15419/bmrat.v7i4.598

Hill, T. R., & Aspray, T. J. (2017). The role of vitamin D in maintaining bone health in older people. *Therapeutic Advances in Musculoskeletal Disease*, *9*(4), 89–95. https://doi.org/10.1177/1759720X17692502

Hoeg, A., Gogakos, A., Murphy, E., Mueller, S., Köhrle, J., Reid, D. M., Glüer, C. C., Felsenberg, D., Roux, C., Eastell, R., Schomburg, L., & Williams, G. R. (2012). Bone turnover and bone mineral density are independently related to selenium status in healthy euthyroid postmenopausal women. *Journal of Clinical Endocrinology and Metabolism*, *97*(11), 4061–4070. https://doi.org/10.1210/jc.2012-2121

Högström, M., Nordström, P., & Nordström, A. (2007). n-3 fatty acids are positively associated with peak bone mineral density and bone accrual in healthy men: The NO2 study. *American Journal of Clinical Nutrition*, *85*(3), 803–807. https://doi.org/10.1093/ajcn/85.3.803

Huang, W., Chakrabarti, S., Majumder, K., Jiang, Y., Davidge, S. T., & Wu, J. (2010). Egg-derived peptide IRW inhibits TNF-α-induced inflammatory response and oxidative stress in endothelial cells. *Journal of Agricultural and Food Chemistry*, *58*(20), 10840–10846. https://doi.org/10.1021/jf102120c

Huang, Y., Ji, Y., Kang, Z., Li, F., Ge, S., Yang, D.-P., Ruan, J., & Fan, X. (2020). Integrating eggshell-derived CaCO3/MgO nanocomposites and chitosan into a biomimetic scaffold for bone regeneration. *Chemical Engineering Journal*, *395*(November 2019), 125098. https://doi.org/10.1016/j.cej.2020.125098

Ichikawa, T., Horie-Inoue, K., Ikeda, K., Blumberg, B., & Inoue, S. (2006). Steroid and xenobiotic receptor SXR mediates vitamin K2-activated transcription of extracellular matrix-related genes and collagen accumulation in osteoblastic cells*. *Journal of Biological Chemistry*, *281*(25), 16927–16934. https://doi.org/10.1074/jbc.M600896200

IrRAWsistible Pet Foods. (n.d.). *Eggshell calcium supplement*. Retrieved December 6, 2021, from https://irrawsistiblepetfoods.com/products/irrawsistible-eggshell-calcium-supplement

Järvinen, R., Tuppurainen, M., Erkkilä, A. T., Penttinen, P., Kärkkäinen, M., Salovaara, K., Jurvelin, J. S., & Kröger, H. (2012). Associations of dietary polyunsaturated fatty acids with bone mineral density in elderly women. *European Journal of Clinical Nutrition*, *66*(4), 496–503. https://doi.org/10.1038/ejcn.2011.188

Ji, M., Leem, K.-H., Kim, M., & Kim, H. K. (2007). Egg yolk soluble protein stimulates the proliferation and differentiation of osteoblastic MC3T3-E1 cells. *Bioscience, Biotechnology, and Biochemistry*, *71*(5), 1327–1329. https://doi.org/10.1271/bbb.60217

Ji, M., Leem, K.-H., Kim, M., & Kim, H. K. (2014). Egg yolk soluble protein stimulates the proliferation and differentiation of osteoblastic MC3T3-E1 cells. *Bioscience, Biotechnology, and Biochemistry*, *71*(5), 1327–1329. https://doi.org/10.1271/bbb.60217

Ji, M.-X., & Yu, Q. (2015). Primary osteoporosis in postmenopausal women. *Chronic Diseases and Translational Medicine*, *1*(1), 9. https://doi.org/10.1016/J.CDTM.2015.02.006

Jie, Y., Cai, Z., Li, S., Xie, Z., Ma, M., & Huang, X. (2017). Hydroxyapatite nucleation and growth on collagen electrospun fibers controlled with different mineralization conditions and phosvitin. *Macromolecular Research*, *25*(9), 905–912. https://doi.org/10.1007/s13233-017-5091-z

Jie, Y., Li, X., Cai, Z., Ma, M., Jin, Y., Ahn, D. U., & Huang, X. (2018). Phosphorylation of phosvitin plays a crucial effects on the protein-induced differentiation and mineralization of osteoblastic MC3T3-E1 cells. *International Journal of Biological Macromolecules*, *118*, 1848–1854. https://doi.org/10.1016/j.ijbiomac.2018.07.029

Johnell, O., & Kanis, J. A. (2006). An estimate of the worldwide prevalence and disability associated with osteoporotic fractures. *Osteoporosis International*, *17*(12), 1726–1733. https://doi.org/10.1007/s00198-006-0172-4

Julien, M., Khoshniat, S., Lacreusette, A., Gatius, M., Bozec, A., Wagner, E. F., Wittrant, Y., Masson, M., Weiss, P., Beck, L., Magne, D., & Guicheux, J. (2009). Phosphate-dependent regulation of MGP in osteoblasts: Role of ERK1/2 and fra-1. *Journal of Bone and Mineral Research*, *24*(11), 1856–1868. https://doi.org/10.1359/JBMR.090508

Jung, S., Nam, K. C., Ahn, D. U., Kim, H. J., & Jo, C. (2013). Effect of phosvitin on lipid and protein oxidation in ground beef treated with high hydrostatic pressure. *Meat Science*, *95*(1), 8–13. https://doi.org/10.1016/j.meatsci.2013.04.005

Kalimeri, M., Leek, F., Wang, N. X., Koh, H. R., Roy, N. C., Cameron-Smith, D., Kruger, M. C., Henry, C. J., & Totman, J. J. (2020). Folate and vitamin B-12 status is associated with bone mineral density and hip strength of postmenopausal Chinese-Singaporean women. https://doi.org/10.1002/jbm4.10399

Kanis, J. A., McCloskey, E. V., Johansson, H., Oden, A., Melton, L. J., & Khaltaev, N. (2008). A reference standard for the description of osteoporosis. *Bone*, *42*(3), 467–475. https://doi.org/10.1016/j.bone.2007.11.001

Kanis, J. A., Norton, N., Harvey, N. C., Jacobson, T., Johansson, H., Lorentzon, M., McCloskey, E. V., Willers, C., & Borgström, F. (2021). SCOPE 2021: A new scorecard for osteoporosis in Europe. *Archives of Osteoporosis*, *16*(1). https://doi.org/10.1007/s11657-020-00871-9

Katsumata, S., Katsumata-Tsuboi, R., Uehara, M., & Suzuki, K. (2009). Severe iron deficiency decreases both bone formation and bone resorption in rats. *The Journal of Nutrition*, *139*(2), 238–243. https://doi.org/10.3945/jn.108.093757

Khan, A. A., Sándor, G. K. B., Dore, E., Morrison, A. D., Alsahli, M., Amin, F., Peters, E., Hanley, D. A., Chaudry, S. R., Lentle, B., Dempster, D. W., Glorieux, F. H., Neville, A. J., Talwar, R. M., Clokie, C. M., al Mardini, M., Paul, T., Khosla, S., Josse, R. G.,…, Compston, J. E. (2009). Bisphosphonate associated osteonecrosis of the jaw. *Journal of Rheumatology*, *36*(3), 478–490. https://doi.org/10.3899/jrheum.080759

Khosla, S., & Lorenz, H. (2017). Osteoporosis treatment: Recent developments and ongoing challenges. *Physiology & Behavior*, *176*(3), 139–148. https://doi.org/10.1016/S2213-8587(17)30188-2.Osteoporosis

Kim, G. S., Kim, C.-H., Park, J. Y., Lee, K.-U., & Park, C. S. (1996). Effects of vitamin B12 on cell proliferation and cellular alkaline phosphatase activity in human bone marrow stromal osteoprogenitor cells and UMR106 osteoblastic cells. *Metabolism*, *45*(12), 1443–1446. https://doi.org/10.1016/S0026-0495(96)90171-7

Kim, H. K., Lee, S., & Leem, K. H. (2011). Protective effect of egg yolk peptide on bone metabolism. *Menopause*, *18*(3), 307–313. https://doi.org/10.1097/gme.0b013e3181f31b1f

Kini, U., & Nandeesh, B. N. (2012). Physiology of bone formation, remodeling, and metabolism. In Radionuclide and *h*ybrid *b*one *i*maging (Vol. 9783642024, pp. 1–1046). https://doi.org/10.1007/978-3-642-02400-9

Kobayashi, T., Koie, H., Watanabe, A., Ino, A., Watabe, K., Kim, M., Kanayama, K., & Otsuji, K. (2015). Effects of food enriched with egg yolk hydrolysate (bone peptide) on bone metabolism in orchidectomized dogs. *Journal of Veterinary Medical Science*, *77*(4), 503–506. https://doi.org/10.1292/jvms.13-0421

Kobayashi, Y., Rupa, P., Kovacs-Nolan, J., Turner, P., Matsui, T., & Mine, Y. (2015). Oral Administration of hen egg white ovotransferrin attenuates the development of colitis induced by dextran sodium sulfate in mice. *Journal of Agricultural and Food Chemistry*, *63*(5), 1532–1539. https://doi.org/10.1021/jf505248n

Kruger, M. C., Coetzee, M., Haag, M., & Weiler, H. (2010). Long-chain polyunsaturated fatty acids: Selected mechanisms of action on bone. *Progress in Lipid Research*, *49*(4), 438–449. https://doi.org/10.1016/j.plipres.2010.06.002

Kuroda, T., Ohta, H., Onoe, Y., Tsugawa, N., & Shiraki, M. (2017). Intake of omega-3 fatty acids contributes to bone mineral density at the hip in a younger Japanese female population. *Osteoporosis International*, *28*(10), 2887–2891. https://doi.org/10.1007/s00198-017-4128-7

Kweon, H. Y., Lee, K. G., Chae, C. H., Balázsi, C., Min, S. K., Kim, J. Y., Choi, J. Y., & Kim, S. G. (2011). Development of nano-hydroxyapatite graft with silk fibroin scaffold as a new bone substitute. *Journal of Oral and Maxillofacial Surgery*, *69*(6), 1578–1586. https://doi.org/10.1016/j.joms.2010.07.062

Lachowicz, S., Oszmiański, J., Wilczyńska, M., Zaguła, G., Saletnik, B., & Puchalski, C. (2020). Impact mineralization of chokeberry and cranberry fruit juices using a new functional additive on the protection of bioactive compounds and antioxidative properties. *Molecules*, *25*(3), 659. https://doi.org/10.3390/molecules25030659

Laird, E., Ward, M., McSorley, E., Strain, J. J., & Wallace, J. (2010). Vitamin D and bone health; Potential mechanisms. *Nutrients*, *2*(7), 693–724. https://doi.org/10.3390/nu2070693

Langdahl, B. L. (2021). Overview of treatment approaches to osteoporosis. *British Journal of Pharmacology*, *178*(9), 1891–1906. https://doi.org/10.1111/BPH.15024

Lanham-New, S. A. (2008). Importance of calcium, vitamin D and vitamin K for osteoporosis prevention and treatment. *Proceedings of the Nutrition Society*, *67*(2), 163–176. https://doi.org/10.1017/S0029665108007003

Lavado-García, J., Roncero-Martin, R., Moran, J. M., Pedrera-Canal, M., Aliaga, I., Leal-Hernandez, O., Rico-Martin, S., & Canal-Macias, M. L. (2018). Long-chain omega-3 polyunsaturated fatty acid dietary intake is positively associated with bone mineral density in normal and osteopenic Spanish women. *PLoS One*, *13*(1), e0190539. https://doi.org/10.1371/JOURNAL.PONE.0190539

Lee, S. W., Balázsi, C., Balázsi, K., Seo, D., hyun, Kim, H. S., Kim, C. H., & Kim, S. G. (2014). Comparative study of hydroxyapatite prepared from seashells and eggshells as a bone graft material. *Tissue Engineering and Regenerative Medicine*, *11*(2), 113–120. https://doi.org/10.1007/s13770-014-0056-1

Lee, S.-W., Kim, S.-G., Balázsi, C., Chae, W.-S., & Lee, H.-O. (2012). Comparative study of hydroxyapatite from eggshells and synthetic hydroxyapatite for bone regeneration. *Oral Surgery, Oral Medicine, Oral Pathology and Oral Radiology*, *113*(3), 348–355. https://doi.org/10.1016/j.tripleo.2011.03.033

Lee, K., Kim, K. R., Lee, H., Choi, S., Jang, I. T., & Kim, H. (2020). Dietary intake of polyunsaturated fatty acids lowers the risk of osteoporosis in older Korean females. *Central Asian Journal of Medical Sciences*, *6*(3), 174–181. https://doi.org/10.24079/CAJMS.2020.09.009

Leem, K. H., Kim, M. G., Kim, H. M., Kim, M., Lee, Y. J., & Kim, H. K. (2004). Effects of egg yolk proteins on the longitudinal bone growth of adolescent male rats. *Bioscience, Biotechnology and Biochemistry*, *68*(11), 2388–2390. https://doi.org/10.1271/bbb.68.2388

Liao, W., Chakrabarti, S., Davidge, S. T., & Wu, J. (2016). Modulatory effects of egg white ovotransferrin-derived tripeptide IRW (Ile-Arg-Trp) on vascular smooth muscle cells against angiotensin II stimulation. *Journal of Agricultural and Food Chemistry*, *64*(39), 7342–7347. https://doi.org/10.1021/acs.jafc.6b03513

Li, C., Geng, F., Huang, X., Ma, M., & Zhang, X. (2014). Phosvitin phosphorus is involved in chicken embryo bone formation through dephosphorylation. *Poultry Science*, *93*(12), 3065–3072. https://doi.org/10.3382/ps.2014-04098

Li, H., Huang, C., Zhu, J., Gao, K., Fang, J., & Li, H. (2018). Lutein suppresses oxidative stress and inflammation by Nrf2 activation in an osteoporosis rat model. *Medical Science Monitor, 24*, 5071–5075. https://doi.org/10.12659/MSM.908699

Li, K., Zhuang, P., Tao, B., Li, D., Xing, X., & Mei, X. (2020). Ultra-small lysozyme-protected gold nanoclusters as nanomedicines inducing osteogenic differentiation. *International Journal of Nanomedicine, 15*, 4705–4716. https://doi.org/10.2147/IJN.S241163

Li, M., Zhang, Q., Wang, R., Lu, M., Xu, X., Liu, X., & Duan, X. (2019). Effects of egg phosvitin on mucosal transcriptional profiles and luminal microbiota composition in murine colon. *Food and Function, 10*(5), 2805–2816. https://doi.org/10.1039/c9fo00074g

Li, W., Zhang, S., Liu, J., Liu, Y., & Liang, Q. (2019). Vitamin K2 stimulates MC3T3-E1 osteoblast differentiation and mineralization through autophagy induction. *Molecular Medicine Reports, 49*(5), 3676–3684. https://doi.org/10.3892/MMR.2019.10040/HTML

Li, Y., Seifert, M. F., Ney, D. M., Grahn, M., Grant, A. L., Allen, K. G. D., & Watkins, B. A. (1999). Dietary conjugated linoleic acids alter serum IGF-I and IGF binding protein concentrations and reduce bone formation in rats fed (n-6) or (n-3) fatty acids. *Journal of Bone and Mineral Research, 14*(7), 1153–1162. https://doi.org/10.1359/jbmr.1999.14.7.1153

Liu, H., Bian, W., Liu, S., & Huang, K. (2012). Selenium protects bone marrow stromal cells against hydrogen peroxide-induced inhibition of osteoblastic differentiation by suppressing oxidative stress and ERK signaling pathway. *Biological Trace Element Research, 150*(1–3), 441–450. https://doi.org/10.1007/S12011-012-9488-4/FIGURES/5

Liu, Q., Li, C., Geng, F., Huang, X., & Ma, M. (2017). Hen egg yolk phosvitin stimulates osteoblast differentiation in the absence of ascorbic acid. *Journal of the Science of Food and Agriculture, 97*(13), 4532–4538. https://doi.org/10.1002/jsfa.8320

Lorentzon, M., & Cummings, S. R. (2015). Osteoporosis: The evolution of a diagnosis. *Journal of Internal Medicine, 277*(6), 650–661. https://doi.org/10.1111/joim.12369

Lu, C. L., & Baker, R. C. (1986). Characteristics of egg yolk phosvitin as an antioxidant for inhibiting metal-catalyzed phospholipid oxidations. *Poultry Science, 65*(11), 2065–2070. https://doi.org/10.3382/ps.0652065

Magaziner, J., Simonsick, E. M., Kashner, T. M., Hebel, J. R., & Kenzora, J. E. (1990). Predictors of functional recovery one year following hospital discharge for hip fracture: A prospective study. *Journals of Gerontology, 45*(3), 101–107. https://doi.org/10.1093/geronj/45.3.M101

Majumder, K., Chakrabarti, S., Davidge, S. T., & Wu, J. (2013). Structure and activity study of egg protein ovotransferrin derived peptides (IRW and IQW) on endothelial inflammatory response and oxidative stress. *Journal of Agricultural and Food Chemistry, 61*(9), 2120–2129. https://doi.org/10.1021/jf3046076

Mak, I. L., Lavery, P., Agellon, S., Rauch, F., Murshed, M., & Weiler, H. A. (2019). Arachidonic acid exacerbates diet-induced obesity and reduces bone mineral content without impacting bone strength in growing male rats. *The Journal of Nutritional Biochemistry, 73*, 108226. https://doi.org/10.1016/J.JNUTBIO.2019.108226

Martins, A. M., Pham, Q. P., Malafaya, P. B., Raphael, R. M., Kasper, F. K., Reis, R. L., & Mikos, A. G. (2009). Natural stimulus responsive scaffolds/cells for bone tissue engineering: Influence of lysozyme upon scaffold degradation and osteogenic differentiation of cultured marrow stromal cells induced by CaP coatings. *Tissue Engineering - Part A, 15*(8), 1953–1963. https://doi.org/10.1089/ten.tea.2008.0023

Maurin, A. C., Chavassieux, P. M., & Meunier, P. J. (2005). Expression of PPARγ and β/δ in human primary osteoblastic cells: Influence of polyunsaturated fatty acids. *Calcified Tissue International, 76*(5), 385–392. https://doi.org/10.1007/S00223-004-0108-Y/FIGURES/6

Moon, H. J., Ko, W. K., Han, S. W., Kim, D. S., Hwang, Y. S., Park, H. K., & Kwon, I. K. (2012). Antioxidants, like coenzyme Q10, selenite, and curcumin, inhibited osteoclast differentiation by suppressing reactive oxygen species generation. *Biochemical and Biophysical Research Communications, 418*(2), 247–253. https://doi.org/10.1016/J.BBRC.2012.01.005

Moreno-Reyes, R., Egrise, D., Nève, J., Pasteels, J. L., & Schoutens, A. (2001). Selenium deficiency-induced growth retardation is associated with an impaired bone metabolism and osteopenia. *Journal of Bone and Mineral Research, 16*(8), 1556–1563. https://doi.org/10.1359/jbmr.2001.16.8.1556

Morgan, E. F., Barnes, G. L., & Einhorn, T. A. (2013). The bone organ system. Form and function. In *Osteoporosis* (4th ed.). Elsevier. https://doi.org/10.1016/B978-0-12-415853-5.00001-7

Morgan, E. F., Unnikrisnan, G. U., & Hussein, A. I. (2018). Bone mechanical properties in healthy and diseased States. *Annual Review of Biomedical Engineering, 20*, 119–143. https://doi.org/10.1146/ANNUREV-BIOENG-062117-121139

Naga, S. M., El-Maghraby, H. F., Sayed, M., & Saad, E. A. (2015). Highly porous scaffolds made of nanosized hydroxyapatite powder synthesized from eggshells. *Journal of Ceramic Science and Technology, 6*(3), 237–243. https://doi.org/10.4416/JCST2014-00058

Neunzehn, J., Szuwart, T., & Wiesmann, H. P. (2015). Eggshells as natural calcium carbonate source in combination with hyaluronan as beneficial additives for bone graft materials, an in vitro study. *Head and Face Medicine, 11*(1), 1–10. https://doi.org/10.1186/s13005-015-0070-0

Nogués, X., & Martinez-Laguna, D. (2018). Update on osteoporosis treatment. *Medicina Clinica, 150*(12), 479–486. https://doi.org/10.1016/J.MEDCLI.2017.10.019

Orchard, T. S., Ing, S. W., Lu, B., Belury, M. A., Johnson, K., Wactawski-Wende, J., & Jackson, R. D. (2013). The association of red blood cell n-3 and n-6 fatty acids with bone mineral density and hip fracture risk in the women's health initiative. *Journal of Bone and Mineral Research, 28*(3), 505–515. https://doi.org/10.1002/jbmr.1772

Ouzzif, Z., Oumghar, K., Sbai, K., Mounach, A., Derouiche, E., Abdellah, M., & Maghraoui, E. (2012). Relation of plasma total homocysteine, folate and vitamin B12 levels to bone mineral density in Moroccan healthy postmenopausal women. *Rheumatology International, 32*(1), 123–128. https://doi.org/10.1007/s00296-010-1551-x

Park, J. W., Bae, S. R., Suh, J. Y., Lee, D. H., Kim, S. H., Kim, H., & Lee, C. S. (2008). Evaluation of bone healing with eggshell-derived bone graft substitutes in rat calvaria: A pilot study. *Journal of Biomedical Materials Research - Part A, 87*(1), 203–214. https://doi.org/10.1002/jbm.a.31768

Park, J. W., Jang, J. H., Bae, S. R., An, C. H., & Suh, J. Y. (2009). Bone formation with various bone graft substitutes in critical-sized rat calvarial defect. *Clinical Oral Implants Research*, *20*(4), 372–378. https://doi.org/10.1111/j.1600-0501.2008.01602.x

Pharma Foods International Co. (n.d.). *Bonepep: Bone metabolism improving substance derived from egg yolk*. Retrieved November 30, 2021, from https://www.pharmafoods.co.jp/en/products/bonepep

Pinheiro, M. M., Schuch, N. J., Genaro, P. S., Ciconelli, R. M., Ferraz, M. B., & Martini, L. A. (2009). Nutrient intakes related to osteoporotic fractures in men and women-The Brazilian Osteoporosis Study (BRAZOS). *Nutrition Journal*, *8*(1), 6. https://doi.org/10.1186/1475-2891-8-6

Platon, N., Arus, V. A., Georgescu, A. M., Nistor, I. D., & Barsan, N. (2020). White bread fortified with calcium from eggshell powder. *Revista de Chimie*, *71*(7), 299–306. https://doi.org/10.37358/RC.20.7.8248

Priante, G., Bordin, L., Musacchio, E., Clari, G., & Baggio, B. (2002). Fatty acids and cytokine mRNA expression in human osteoblastic cells: A specific effect of arachidonic acid. *Clinical Science*, *102*(4), 403–409. https://doi.org/10.1042/cs1020403

Promislow, J. H. E., Goodman-Gruen, D., Slymen, D. J., & Barrett-Connor, E. (2002). Protein consumption and bone mineral density in the elderly: The Rancho Bernardo study. *American Journal of Epidemiology*, *155*(7), 636–644. https://doi.org/10.1093/aje/155.7.636

Ramesh, N., Moratti, S. C., & Dias, G. J. (2018). Hydroxyapatite-polymer biocomposites for bone regeneration: A review of current trends. *Journal of Biomedical Materials Research Part B: Applied Biomaterials*, *106*(5), 2046–2057. https://doi.org/10.1002/jbm.b.33950

Rathnapala, E. C. N., Ahn, D. U., & Abeyrathne, S. (2021). Functional properties of ovotransferrin from chicken egg white and its derived peptides: A review. *Food Science and Biotechnology*, *30*(5), 619–630. https://doi.org/10.1007/s10068-021-00901-3

Ray, S., Kumar Barman, A., Kumar Roy, P., & Kumar Singh, B. (2017). *The Pharma Innovation Journal 2017; 6(9): 01–04 Chicken eggshell powder as dietary calcium source in chocolate cakes*. *6*(9), 1–4. www.thepharmajournal.com

Redd Remedies. (n.d.). *Bone Health Original™*. Retrieved December 6, 2021, from https://reddremedies.com/products/bone-health-original?variant=34271747080330

Reid, I. R. (2021). Nutrition and bone health: The case of selenium. *The Lancet Healthy Longevity*, *2*(4), e185–e186. https://doi.org/10.1016/S2666-7568(21)00057-X

Ren, J. (2019). *Phosvitin extraction from egg yolk and its potential as a functional food ingredient for improving bone health*. University of Alberta.

Ren, J., Chakrabarti, S., & Wu, J. (2019). Phosvitin and its hydrolysate promote differentiation and inhibit TNF-α induced inflammation in MC3T3-E1 cells via ERK and AKT pathways. *Journal of Functional Foods*, *53*(November 2018), 259–265. https://doi.org/10.1016/j.jff.2018.12.025

Ren, J., Chakrabarti, S., & Wu, J. (2021). Inhibition of osteoclastogenesis and inflammation by phosvitin and phosvitin hydrolysate via NF-κB and MAPK pathways in RAW 264.7 cells. *Journal of Food Bioactives*, *13*, 74–81. https://doi.org/10.31665/jfb.2020.13261

Rizzoli, R., Biver, E., Bonjour, J. P., Coxam, V., Goltzman, D., Kanis, J. A., Lappe, J., Rejnmark, L., Sahni, S., Weaver, C., Weiler, H., & Reginster, J. Y. (2018). Benefits and safety of dietary protein for bone health—An expert consensus paper endorsed by the European society for clinical and economical aspects of osteopororosis, osteoarthritis, and musculoskeletal diseases and by the international osteoporosis Foundation. *Osteoporosis International*, *29*(9), 1933–1948. https://doi.org/10.1007/s00198-018-4534-5

Rodríguez-Olleros Rodríguez, C., & Díaz Curiel, M. (2019). vitamin K and bone health: A review on the effects of vitamin K deficiency and supplementation and the effect of non-vitamin K antagonist oral anticoagulants on different bone parameters. *Journal of Osteoporosis*, *2019*. https://doi.org/10.1155/2019/2069176

Sadat-Shojai, M., Atai, M., Nodehi, A., & Khanlar, L. N. (2010). Hydroxyapatite nanorods as novel fillers for improving the properties of dental adhesives: Synthesis and application. *Dental Materials*, *26*(5), 471–482. https://doi.org/10.1016/j.dental.2010.01.005

Saiedullah, M., Begum, S., & Muraduzzaman, S. M. (2021). Association of vitamin B12 with bone mineral density in postmenopausal women in Bangladesh. *Saudi Journal of Biomedical Research Abbreviated Key Title: Saudi J Biomed Res*, *6*(9), 226–232. https://doi.org/10.36348/sjbr.2021.v06i09.002

Saini, R. K., & Keum, Y. S. (2018). Omega-3 and omega-6 polyunsaturated fatty acids: Dietary sources, metabolism, and significance — A review. *Life Sciences*, *203*, 255–267. https://doi.org/10.1016/J.LFS.2018.04.049

Salama, R., Khashaba, M., & el Rouby, D. (2019). Histomorphometric evaluation of a nano-sized eggshell-containing supplement as a natural alloplastic: An animal study. *Saudi Dental Journal*, *31*(3), 375–381. https://doi.org/10.1016/j.sdentj.2019.03.011

Salari Sharif, P., Abdollahi, M., & Larijani, B. (2011). Current, new and future treatments of osteoporosis. *Rheumatology International*, *31*(3), 289–300. https://doi.org/10.1007/s00296-010-1586-z

Sattar Khan, M. A., Nakamura, S., Ogawa, M., Akita, E., Azakami, H., & Kato, A. (2000). Bactericidal action of egg yolk phosvitin against *Escherichia coli* under thermal stress. *Journal of Agricultural and Food Chemistry*, *48*(5), 1503–1506. https://doi.org/10.1021/jf990700r

Schaafsma, A., & Pakan, I. (1999). Short-term effects of a chicken egg shell powder enriched dairy-based products on bone mineral density in persons with osteoporosis or osteopenia. *Bratislavske Lekarske Listy*, *100*(12), 651–656. http://www.ncbi.nlm.nih.gov/pubmed/10758743

Schaafsma, A., Pakan, I., Hofstede, G. J. H., Muskiet, F. A., van der Veer, E., & de Vries, P. J. F. (2000). Mineral, amino acid, and hormonal composition of chicken eggshell powder and the evaluation of its use in human nutrition. *Poultry Science*, *79*(12), 1833–1838. https://doi.org/10.1093/ps/79.12.1833

Schaafsma, A., van Doormaal, J. J., Muskiet, F. A. J., Hofstede, G. J. H., Pakan, I., & van der Veer, E. (2002). Positive effects of a chicken eggshell powder-enriched vitamin–mineral supplement on femoral neck bone mineral density in healthy late post-menopausal Dutch women. *British Journal of Nutrition*, *87*(3), 267–275. https://doi.org/10.1079/bjn2001515

Schmitz, G., & Ecker, J. (2008). The opposing effects of n−3 and n−6 fatty acids. *Progress in Lipid Research*, *47*(2), 147–155. https://doi.org/10.1016/J.PLIPRES.2007.12.004

Seeman, E., & Martin, T. J. (2019). Antiresorptive and anabolic agents in the prevention and reversal of bone fragility. *Nature Reviews Rheumatology*, *15*(4), 225–236. https://doi.org/10.1038/s41584-019-0172-3

Serna, J., & Bergwitz, C. (2020). Importance of dietary phosphorus for bone metabolism and healthy aging. *Nutrients*, *12*(10), 3001. https://doi.org/10.3390/NU12103001

Seuss-baum, I. (2007). Nutritional evaluation of egg compounds. In R. Huopalahti, R. López-Fandiño, M. Anton, & R. Schade (Eds.), *Bioactive egg compounds* (pp. 117–144). Springer. https://doi.org/10.1007/978-3-540-37885-3_18

Shams-White, M. M., Chung, M., Fu, Z., Insogna, K. L., Karlsen, M. C., LeBoff, M. S., Shapses, S. A., Sackey, J., Shi, J., Wallace, T. C., & Weaver, C. M. (2018). Dietary protein and bone health: A systematic review and meta-analysis from the national osteoporosis foundation. *PLoS One*, *13*(2), 1528–1543. https://doi.org/10.1371/journal.pone.0192459

Shane, E., Khosla, S., & Burr, D. (2013). Osteonecrosis of the jaw and atypical femoral fractures. In *Osteoporosis* (4th ed.). Elsevier. https://doi.org/10.1016/B978-0-12-415853-5.00081-9

Shang, N. (2019). The potential of egg white as a *functional* food ingredient against osteoporosis (Vol. 53, Issue 9). https://doi.org/10.1017/CBO9781107415324.004

Shang, N., Bhullar, K. S., Hubbard, B. P., & Wu, J. (2019). Tripeptide IRW initiates differentiation in osteoblasts via the RUNX2 pathway. *Biochimica et Biophysica Acta (BBA) - General Subjects*, *1863*(6), 1138–1146. https://doi.org/10.1016/J.BBAGEN.2019.04.007

Shang, N., Bhullar, K. S., & Wu, J. (2020). Ovotransferrin exhibits osteogenic activity partially via low-density lipoprotein receptor-related protein 1 (LRP1) activation in MC3T3-E1 cells. *Journal of Agricultural and Food Chemistry*, *68*(35), 9427–9435. https://doi.org/10.1021/acs.jafc.0c04064

Shang, N., & Wu, J. (2018). Egg white ovotransferrin shows osteogenic activity in osteoblast cells. *Journal of Agricultural and Food Chemistry*, *66*(11), 2775–2782. https://doi.org/10.1021/acs.jafc.8b00069

Shang, N., & Wu, J. (2019). Egg white ovotransferrin attenuates RANKL-induced osteoclastogenesis and bone resorption. *Nutrients*, *11*(9), 2254. https://doi.org/10.3390/nu11092254

Shang, N., & Wu, J. (2020). Egg-derived tripeptide IRW attenuates LPS-induced osteoclastogenesis in RAW 264.7 macrophages via inhibition of inflammatory responses and NF-κB/MAPK activation. *Journal of Agricultural and Food Chemistry*, *68*(22), 6132–6141. https://doi.org/10.1021/acs.jafc.0c01159

Shanks, G., Sharma, D., & Mishra, V. (2019). Prevention and treatment of osteoporosis in women. *Obstetrics, Gynaecology and Reproductive Medicine*, *29*(7), 201–206. https://doi.org/10.1016/j.ogrm.2019.04.001

Shen, C. L., Yeh, J. K., Cao, J. J., & Wang, J. S. (2009). Green tea and bone metabolism. *Nutrition Research*, *29*(7), 437–456. https://doi.org/10.1016/j.nutres.2009.06.008

Shors, E. C. (1999). Coralline bone graft substitutes. *Orthopedic Clinics*, *30*(4), 599–613. https://doi.org/10.1016/S0030-5898(05)70113-9

Siddiqui, J. A., & Partridge, N. C. (2016). Physiological bone remodeling: Systemic regulation and growth factor involvement. *Physiology (Bethesda, Md.)*, *31*(3), 233–245. https://doi.org/10.1152/PHYSIOL.00061.2014

Sietsema, D. L. (2020). Fighting the epidemic: Bone health and osteoporosis. *Nursing Clinics*, *55*(2), 193–202. https://doi.org/10.1016/J.CNUR.2020.02.002

Simopoulos, A. P. (2008). The importance of the omega-6/omega-3 fatty acid ratio in cardiovascular disease and other chronic diseases. *Experimental Biology and Medicine*, *233*(6), 674–688. https://doi.org/10.3181/0711-MR-311

Siris, E. S., Harris, S. T., Eastell, R., Zanchetta, J. R., Goemaere, S., Diez-Perez, A., Stock, J. L., Song, J., Qu, Y., Kulkarni, P. M., Siddhanti, S. P. S. P. R., Wong, M., & Cummings, S. R. (2005). Skeletal effects of raloxifene after 8 years: Results from the continuing outcomes relevant to evista (CORE) study. *Journal of Bone and Mineral Research*, *20*(9), 1514–1524. https://doi.org/10.1359/JBMR.050509

Skřivan, M., & Englmaierová, M. (2014). The deposition of carotenoids and α-tocopherol in hen eggs produced under a combination of sequential feeding and grazing. *Animal Feed Science and Technology*, *190*, 79–86. https://doi.org/10.1016/j.anifeedsci.2014.01.009

Song, M., Sauchi, Y., Yun, S., & Kim, M. (2011). *Development of bonepep bone metabolism-improving ingredient derived from egg yolk*. China Food Additives. https://en.cnki.com.cn/Article_en/CJFDTotal-ZSTJ201101033.htm

Sözen, T., Özışık, L., & Başaran, N. Ç. (2017). An overview and management of osteoporosis. *European Journal of Rheumatology*, *4*(1), 46. https://doi.org/10.5152/EURJRHEUM.2016.048

Takeda, H., Tominari, T., Hirata, M., Watanabe, K., Matsumoto, C., Grundler, F. M. W., Inada, M., & Miyaura, C. (2017). Lutein enhances bone mass by stimulating bone formation and suppressing bone resorption in growing mice. *Biological & Pharmaceutical Bulletin*, *40*(5), 716–721. https://doi.org/10.1248/bpb.b16-00897

Takeda, E., Yamamoto, H., Yamanaka-Okumura, H., & Taketani, Y. (2012). Dietary phosphorus in bone health and quality of life. *Nutrition Reviews*, *70*(6), 311–321. https://doi.org/10.1111/j.1753-4887.2012.00473.x

Tariq, S., Tariq, S., & Lone, K. P. (2018). Interplay of vitamin D, vitamin B12, homocysteine and bone mineral density in postmenopausal females. *Health Care for Women International*, *39*(12), 1340–1349. https://doi.org/10.1080/07399332.2018.1444042

Tominari, T., Matsumoto, C., Watanabe, K., Hirata, M., Grundler, F. M. W., Inada, M., & Miyaura, C. (2017). Lutein, a carotenoid, suppresses osteoclastic bone resorption and stimulates bone formation in cultures. *Bioscience, Biotechnology and Biochemistry*, *81*(2), 302–306. https://doi.org/10.1080/09168451.2016.1243983

Tucker, K. L., Hannan, M. T., Qiao, N., Jacques, P. F., Selhub, J., Cupples, L. A., & Kiel, D. P. (2005). Low plasma vitamin B12 is associated with lower BMD: The Framingham osteoporosis study. *Journal of Bone and Mineral Research*, *20*(1), 152–158. https://doi.org/10.1359/JBMR.041018

Turner, A. G., Anderson, P. H., & Morris, H. A. (2012). Vitamin D and bone health. *Scandinavian Journal of Clinical and Laboratory Investigation*, *72*(SUPPL. 243), 65–72. https://doi.org/10.3109/00365513.2012.681963

Udenigwe, C. C., & Fogliano, V. (2017). Food matrix interaction and bioavailability of bioactive peptides: Two faces of the same coin? *Journal of Functional Foods*, *35*, 9–12. https://doi.org/10.1016/j.jff.2017.05.029

Ukon, Y., Makino, T., Kodama, J., Tsukazaki, H., Tateiwa, D., Yoshikawa, H., & Kaito, T. (2019). Molecular-based treatment strategies for osteoporosis: A literature review. *International Journal of Molecular Sciences*, *20*(10). https://doi.org/10.3390/IJMS20102557

Uraz, A., Gultekin, S. E., Senguven, B., Karaduman, B., Sofuoglu, I. P., Pehlivan, S., Capan, Y., & Eren, K. (2013). Histologic and histomorphometric assessment of eggshell-derived bone graft substitutes on bone healing in rats. *Journal of Clinical and Experimental Dentistry*, *5*(1), e23–e29. https://doi.org/10.4317/jced.50968

Vaes, B. L. T., Lute, C., Blom, H. J., Bravenboer, N., de Vries, T. J., Everts, V., Dhonukshe-Rutten, R. A., Müller, M., de Groot, L. C. P. G. M., & Steegenga, W. T. (2009). Vitamin B12 deficiency stimulates osteoclastogenesis via increased homocysteine and methylmalonic acid. *Calcified Tissue International*, *84*(5), 413–422. https://doi.org/10.1007/S00223-009-9244-8/FIGURES/4

Vorland, C. J., Stremke, E. R., Moorthi, R. N., & Hill Gallant, K. M. (2017). Effects of excessive dietary phosphorus intake on bone health. *Current Osteoporosis Reports*, *15*(5), 473–482. https://doi.org/10.1007/s11914-017-0398-4

Wallace, T. C., & Frankenfeld, C. L. (2017). Dietary protein intake above the current RDA and bone health: A systematic review and meta-analysis. *Journal of the American College of Nutrition*, *36*(6), 481–496. https://doi.org/10.1080/07315724.2017.1322924

Wang, L., Wang, N., Zhang, W., Cheng, X., Yan, Z., Shao, G., Wang, X., Wang, R., & Fu, C. (2022). Therapeutic peptides: Current applications and future directions. *Signal Transduction and Targeted Therapy*, *7*(1), 1–27. https://doi.org/10.1038/s41392-022-00904-4

Watkins, B. A., Li, Y., Allen, K. G. D., Hoffmann, W. E., & Seifert, M. F. (2000). Dietary ratio of (n-6)/(n-3) polyunsaturated fatty acids alters the fatty acid composition of bone compartments and biomarkers of bone formation in rats. *Journal of Nutrition*, *130*(9), 2274–2284. https://doi.org/10.1093/jn/130.9.2274

Whitaker, M., Guo, J., Kehoe, T., & Benson, G. (2012). Bisphosphonates for osteoporosis—Where do we go from here? *New England Journal of Medicine*, *366*(22), 2048–2051. https://doi.org/10.1056/NEJMp1202619

World Health Organization. (2007). Who scientific group on the assessment of osteoporosis at primary health. *World Health*, *May*(May 2004), 1–13. https://doi.org/10.1016/S0140-6736(02)08761-5

Wright, N. C., Looker, A. C., Saag, K. G., Curtis, J. R., Delzell, E. S., Randall, S., & Dawson-Hughes, B. (2014). The recent prevalence of osteoporosis and low bone mass in the United States based on bone mineral density at the femoral neck or lumbar spine. *Journal of Bone and Mineral Research*, *29*(11), 2520–2526. https://doi.org/10.1002/jbmr.2269

Wu, J. (2020). A novel angiotensin converting enzyme 2 (ACE2) activating peptide: A reflection of 10 years of research on a small peptide Ile-Arg-Trp (IRW). *Journal of Agricultural and Food Chemistry*, *68*(49), 14402–14408. https://doi.org/10.1021/acs.jafc.0c05544

Wu, A.-M., Bisignano, C., James, S. L., Abady, G. G., Abedi, A., Abu-Gharbieh, E., Alhassan, R. K., Alipour, V., Arabloo, J., Asaad, M., Asmare, W. N., Awedew, A. F., Banach, M., Banerjee, S. K., Bijani, A., Birhanu, T. T. M., Bolla, S. R., Cámera, L. A., Chang, J.-C.,…, Vos, T. (2021). Global, regional, and national burden of bone fractures in 204 countries and territories, 1990–2019: A systematic analysis from the global burden of disease study 2019. *The Lancet Healthy Longevity*, *2*(9), e580–e592. https://doi.org/10.1016/S2666-7568(21)00172-0

Wu, C. C., Wang, C. K., Yang, A. M., Lu, C. S., & Lin, C. Y. (2021). Selenium status is independently related to bone mineral density, FRAX score, and bone fracture history: NHANES, 2013 to 2014. *Bone*, *143*, 115631. https://doi.org/10.1016/J.BONE.2020.115631

Wu, S.-C., Hsu, H.-C., Hsu, S.-K., Chang, Y.-C., & Ho, W.-F. (2015). Effects of heat treatment on the synthesis of hydroxyapatite from eggshell powders. *Ceramics International*, *41*(9), 10718–10724. https://doi.org/10.1016/j.ceramint.2015.05.006

Xu, C., Yang, C., Yin, Y., Liu, J., & Mine, Y. (2012). Phosphopeptides (PPPs) from hen egg yolk phosvitin exert anti-inflammatory activity via modulation of cytokine expression. *Journal of Functional Foods*, *4*(4), 718–726. https://doi.org/10.1016/j.jff.2012.04.011

Xu, Q., Hong, H., Wu, J., & Yan, X. (2019). Bioavailability of bioactive peptides derived from food proteins across the intestinal epithelial membrane: A review. *Trends in Food Science & Technology*, *86*, 399–411. https://doi.org/10.1016/J.TIFS.2019.02.050

Yamaguchi, M. (2015). Nutritional zinc plays a pivotal role in bone health and osteoporosis prevention. *Edorium Journal of Nutrition and Dietetics*, *1*, 1–8.

Yamaguchi, M., & Weitzmann, M. N. (2011). Vitamin K2 stimulates osteoblastogenesis and suppresses osteoclastogenesis by suppressing NF-κB activation. *International Journal of Molecular Medicine*, *27*(1), 3–14. https://doi.org/10.3892/IJMM.2010.562/HTML

Zeng, F., Fan, F., Xue, W., Xie, H., Wu, B., Tu, S., Ouyang, W., & Chen, Y. (2013). The association of red meat, poultry, and egg consumption with risk of hip fractures in elderly Chinese: A case-control study. *Bone*, *56*(2), 242–248. https://doi.org/10.1016/j.bone.2013.06.023

Zhang, J., Ma, Z., Yan, K., Wang, Y., Yang, Y., & Wu, X. (2019). Matrix Gla protein promotes the bone formation by up-regulating Wnt/β-catenin signaling pathway. *Frontiers in Endocrinology*, *10*, 891. https://doi.org/10.3389/fendo.2019.00891

Zhang, R., Lu, Y., Ye, L., Yuan, B., Yu, S., Qin, C., Xie, Y., Gao, T., Drezner, M. K., Bonewald, L. F., & Feng, J. Q. (2011). Unique roles of phosphorus in endochondral bone formation and osteocyte maturation. *Journal of Bone and Mineral Research*, *26*(5), 1047–1056. https://doi.org/10.1002/JBMR.294

Zhang, X., Geng, F., Huang, X., & Ma, M. (2015). Acceleration of the initial phase transformation of mineralization by phosvitin. *Journal of Crystal Growth*, *409*, 44–50. https://doi.org/10.1016/j.jcrysgro.2014.10.001

Zhang, X., Huang, X., & Ma, M. (2017). Role of phosphorylation of phosvitin in the phase transformation of mineralization. *International Journal of Biological Macromolecules*, *101*, 712–718. https://doi.org/10.1016/j.ijbiomac.2017.03.158

Zhang, X., Huang, X., & Ma, M. (2018). Phosphorylated serine clusters of phosvitin plays a crucial role in the regulatory mineralization. *International Journal of Biological Macromolecules*, *115*, 1109–1115. https://doi.org/10.1016/j.ijbiomac.2018.04.130

Zhao, M., Li, S., Ahn, D. U., & Huang, X. (2021). Phosvitin phosphopeptides produced by pressurized hea-trypsin hydrolysis promote the differentiation and mineralization of MC3T3-E1 cells via the OPG/RANKL signaling pathways. *Poultry Science*, *100*(2), 527–536. https://doi.org/10.1016/j.psj.2020.10.053

Zheng, J., Mao, X., Ling, J., He, Q., & Quan, J. (2014). Low serum levels of zinc, copper, and iron as risk factors for osteoporosis: A meta-analysis. *Biological Trace Element Research*, *160*(1), 15–23. https://doi.org/10.1007/s12011-014-0031-7

Zia-ul-haq, M. (2021). Carotenoids: Structure and function in the human body. In *Carotenoids: Structure and function in the human body*. https://doi.org/10.1007/978-3-030-46459-2

Zoch, M. L., Clemens, T. L., & Riddle, R. C. (2016). New insights into the biology of osteocalcin. *Bone*, *82*, 42–49. https://doi.org/10.1016/J.BONE.2015.05.046

36

Egg and Brain Health

Hua Zhang and Yoshinori Mine*

CONTENTS

36.1 Introduction

The chicken egg contains primarily a single embryonic cell, the yolk before fertilization. The egg is made up of approximately 27.5% yolk and 63% egg white (Cotterill & Geiger, 1977). Both the yolk and egg white are not only highly rich in macro- and micro-nutrients, including proteins, fatty acids, vitamins, and minerals; but also, a variety of dietary bioactive molecules such as carotenoids and peptides, designed to support the development of the chick embryo. The proteins and dietary-derived peptides as well as amino acids are distributed throughout the whole egg, while lipids, in a conjugated form with fat-soluble vitamins, are exclusively distributed in the egg yolk (Kovacs-Nolan et al., 2005). Due to the high content of these essential nutrients, chicken eggs gained attention from researchers for their numerous health-promoting properties. For centuries, humans have consumed chicken eggs and considered them as a highly nutritional food (Zaheer, 2015). In modern days, with the development of food processing technology, egg-derived proteins, lipids, and other constituents have been widely used for their foaming, gelling, and emulsifying properties by the food and pharmaceutical industry (Zaheer, 2015). Furthermore, clinical and epidemiologic studies have shown the efficacy of egg-derived bioactive compounds in preventing diseases (Réhault-Godbert et al., 2019). These findings imply that egg-derived bioactive molecules have far more food attributes and processing value than in their original forms. Thus, it is necessary to study the health benefits not only from chicken eggs but also egg-derived bioactive constituents, such as omega-3 fatty acids (ω-3), conjugated linoleic acids, fat-soluble vitamins and pigments, phospholipids, tryptophan-rich egg protein hydrolysate, and bioactive peptides (Kovacs-Nolan et al., 2005; Anton et al., 2006; Markus et al., 2010).

The brain is the central organ for human beings, controlling neurotransmission, endocrine hormone responses, and cognitive and behavior performance. The lifelong brain functioning is modulated by nutritional status, health conditions, and environmental stimuli. Both bioactive lipids and high-quality proteins are essential for supporting brain development throughout the entire pregnancy period and growing-up stages after birth (Lutter et al., 2018). Moreover, aging-related neurodegenerative diseases characterized by impaired memory and cognitive decline are becoming serious health risks for the elderly. The underlying cause of neurodegeneration is chronic neuroinflammation leading to the dysregulation of microglia (Parker et al., 2020). The maintenance of the nutritional status is essential for brain health and extends the quality of life during aging. Chicken eggs are healthy a nutritious food resource, ideal for stimulating human brain functioning and maintaining mental health. Brain development and function during the fetal, growing, and aging stages depend on essential nutrients such as proteins, lipids and vitamins. Hence, chicken eggs are a promising dietary source of macro-and micronutrients that support brain health.

Egg yolk lipids rich in both saturated and unsaturated fat are essential for structure and neurotransmission in neurons and glial cells, implicated in sustaining normal brain functions. The egg yolk phospholipid choline is required for cell division, growth, and membrane signaling (Caudill, 2010; Zeisel & Niculescu, 2006). Choline can prevent neural tube defects in the fetus and improve infant cognitive ability during the perinatal period (Caudill et al., 2018). Choline is also a precursor and metabolite of acetylcholine (Ach). Cognitive dysfunctions

* Corresponding author.

DOI: 10.1201/9781003254430-44

are found to be associated with a decrease of Ach (Oddo & LaFerla, 2006). In a study modeling Alzheimer's disease, the consumption of eggs contributed to increasing the level of Ach and improving memory retention in rats (Llorente-Ovejero et al., 2021). Moreover, some micronutrients, peptides, and pigments in the egg have neuroprotective actions against neurodegeneration and improve cognitive performance (Zaheer, 2015; Wallace, 2018; Sukik et al., 2021). A deficit of cognitive capacity, including attention, memory, thinking, learning, and perception, is commonly associated with aging and has a significant negative impact on the quality of life of elderly people (Statistics, 2020). For a single cell, the egg is uniquely rich in macronutrients, micronutrients, hormones, and bioactive constituents. Therefore, it is noteworthy to determine the potential impact of eggs on brain health for human beings. This chapter aims to comprehensively elucidate the neuroprotective effects of egg-derived bioactive constituents on brain functions and possible some mechanism underlying their action.

36.2 Egg-Contained Lipid Constituents

The egg yolk mainly consists of proteins and lipids (Sugino et al., 2018). The main egg yolk lipids are triacylglycerol (62%), phospholipids (33%), and cholesterol (5%) (Xiao et al., 2020). In addition, the majority of proteins in the egg yolk are in the form of lipoproteins, either in granule or plasma fractions. The granule fraction is comprised of high-density lipoproteins (HDL), phosvitin, and low-density lipoproteins (LDL) while the plasma fraction of the yolk is made up of the LDL fraction and water-soluble constituents, such as lipid-free globular proteins (Burley & Cook, 1961). Pigments, oligosaccharides, and glycosylated peptides constituents are also present in the yolk. As a rich source of bioactive lipids (i.e. phospholipid, choline, and PUFAs) as well as carotenoids, the egg yolk is an important food source for pregnant and lactating mothers, supporting the brain development in the fetus, neonate and infant, and the elderly, as shown in Figure 36.1 (Lutter et al., 2018). Moreover, as revealed in a cohort study, a long-term administration of one egg per week was shown to moderated-the-cognitive performance and slower memory decline in later life (Lee et al., 2021). Accordingly, increasing the intake of eggs rich in bioactive lipids contributes to maintaining brain health throughout the life span.

36.2.1 Phospholipids

Phospholipids (PL) are highly bioavailable and widely present in a variety of foods, including egg yolk, soybean, and dairy products (Küllenberg et al., 2012). Dietary phospholipids show a significant impact on the regulation of cognitive processes throughout the lifespan; and thus, are potential candidates for nutritional interventions to sustain the health of the brain and nervous system via suppressing the stress and aging-caused impairments (Schverer et al., 2020). An imbalanced phospholipid metabolism has a deleterious effect on axonal transport as well as peripheral neural signaling and communication, leading to the development of stress-related disorders (e.g. dementia, Parkinson's disease) (Müller et al., 2015; Zhao et al., 2018). The brain and neural tissue contain a large quantity of sphingomyelin which consists of a phosphocholine head group, fatty acid, and sphingosine (Küllenberg et al., 2012). Therefore, diets enriched in phospholipids are beneficial for the brain and nervous system, improving behavior and cognitive processes (Küllenberg et al., 2012). Phosphocholine, making up 65–70% of the phospholipids in the yolk, is a key constituent of colonic mucus which is involved in mucosal defense against pathogen invasion (Stremmel et al., 2010). Nonetheless, numerous results from clinical trials in the last 2 decades suggested that increased consumption of dietary phospholipids was passively correlated with improvements in the cognitive function and memory activity in humans (Cheatham et al., 2012; Gurnida et al., 2012; Ross et al., 2016; Boyle et al., 2019).

Choline is a well-known essential nutrient for neurodevelopment, including neurogenesis, differentiation, and myelination and synaptogenesis (Ramel & Georgieff, 2014). In recent

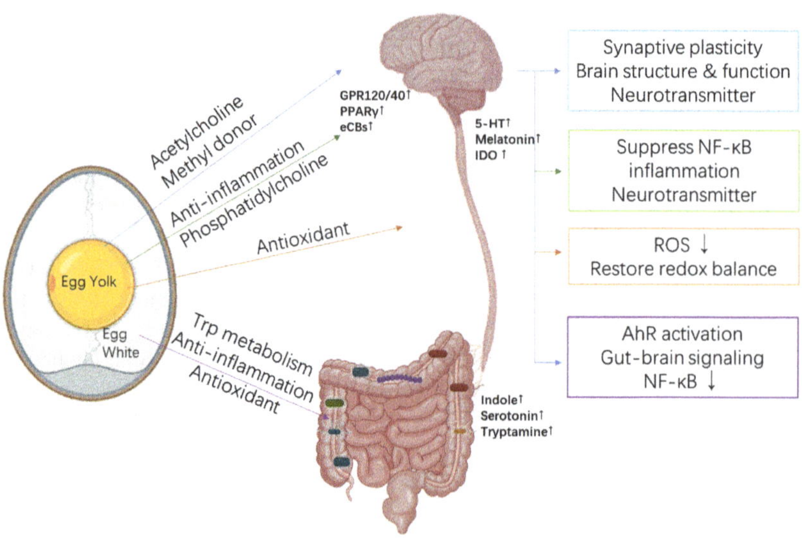

FIGURE 36.1 The modulatory effects of egg-derived bioactive components on brain health.

studies, it has been determined that blood choline levels are 6–7-fold higher in fetuses than in adults (Ilcol et al., 2002; Zeisel & Wurtman, 1981). Children, at the age of 7, showed a significant improvement in color-location memory performance when their mothers took 930 mg choline/day during pregnancy (Bahnfleth et al., 2019). A higher amount of choline is required by women during pregnancy to sustain the rapid division of fetal cells, due to the high demand for phosphatidylcholine during cell division and growth. Choline intake is not only necessary for brain development during the fetal stage, but also to maintain the cognitive health of elderly. During the aging process, the altered lipid composition of brain cells leads to the deterioration of memory functions. The phosphatidylcholines (PCs) and polyunsaturated fatty acids (PUFAs) are required to maintain functional neurotransmitters and efficient ion channels. However, a dramatic reduction in PCs and PUFAs in the brain has been found in the elderly, resulting in impaired learning and cognitive activities. Moreover, PC and PL from brain cell membranes can be disintegrated by chronic alcohol consumption since acetaldehyde, a metabolite of ethanol, can affect the antioxidant status and normal enzymic function of neurons and lead to neurodegeneration (Küllenberg et al., 2012). The increase in cholesterol/PL ratio strengthens cell membrane rigidity, suggesting that the intake of choline-rich foods may help improve the membrane structure of brain cells. PCs supplementation in rats was shown to maintain the antioxidative function against oxidative stress induced by ethanol on erythrocytes and brain cells (Jayaraman et al., 2008). In addition, Kable et al. (2015) demonstrated that for heavy drinkers, the supplementation of 750 mg/day choline during pregnancy could improve subsequent infant development tests (Kable et al., 2015). It is possible that the intake of choline prevents the alcohol-induced depletion of dimethylglycine and maintains glycine homeostasis during neurogenesis (Kable et al., 2015). Moreover, due to their status as a potential choline donor, PCs are also beneficial for the prevention or treatment of neurological diseases.

For physiologic development, choline is vital for supporting lipid metabolism and brain, liver, and muscle functions (Wallace et al., 2020). The physiological role of choline involves the modulation of the plasticity of neuronal cells, the support of the synthesis of membrane phosphatidylcholine in neuronal cells, the regulation of the synthesis of acetylcholine for neurotransmission, and the activation of lipid signaling cascades (Chin & Goh, 2019). In a previous study, the prenatal choline supplementation was shown to regulate gene networks that are implicated in psychological disorders. Among those, choline-suppressed genes were mainly involved in early-life iron deficiency, impacting negatively on the adult hippocampal transcriptome and resulting in long-term behavioral abnormalities (Tran et al., 2016). The neuron mitochondria have a relatively high energy demand to maintain cell functions and the intake of choline may contribute to maintain redox homeostasis in neurons. The findings from a recent *in vitro* study demonstrated that a prolong administration of choline suppressed neuroinflammation and restored synapse membrane formation by blocking nucleotide-binding oligomerization domain and leucine-rich repeat-containing receptor family pyrin domain containing (NLRP)-3 inflammasome activation (Wang et al., 2019). Owing to the health benefits of choline, dietary phospholipids rich in choline have some positive effects, improving the neuronal function and subsequently overall brain health. Whereas these health benefits result from lipid metabolites derived from dietary choline warrants further studies to comprehensively explore the altered lipid metabolic profile induced by choline supplementation to identify key metabolites responsible for sustaining the neuronal function.

Chicken eggs are an important food source for choline with 251 mg/100g (Zeisel et al., 2003). A previous clinal nutrition study demonstrated that pregnant women were able to reach the choline recommended level by consuming more than 1 egg per day, while women not consuming eggs achieved only 12.5% of the recommended level (Lewis et al., 2014). However, the intake of dietary choline promotes the biosynthesis of trimethylamine (TMA) by the gut microbiome, further converted into trimethylamine N-oxide (TMAO) in the liver, and will impair the hepatic and intestinal lipid metabolism. Nevertheless, a recent study with overweight postmenopausal women revealed that the consumption of 2 eggs per day led to elevated plasma choline and betaine levels, but not TMAO, suggesting that egg consumption may not increase the risk of developing cardiovascular diseases (Zhu et al., 2020). Moreover, the bioavailability of choline naturally derived from egg yolk phospholipids is relatively higher than that of choline bitartrate, indicating that the intake of choline through egg consumption has a positive impact on human health (Smolders et al., 2019). Donoso et al. (2020) demonstrated that egg yolk-derived phosphatidylethanolamine and phosphatidylglycerol, rather than choline, can protect cortical cells against corticosterone (CORT)-induced cytotoxicity and neuronal depletion, suggesting that supplementation with eggs has a protective effect on stress-related disorders (Donoso et al., 2020). In a case-control study, findings suggested that weekly consumption of 2 eggs effectively reduced the risk of glioma (Aminianfar et al., 2019). Furthermore, evidence from an omics study showed that egg consumption has a regulatory impact on transcriptomes of the prefrontal cortex (PFC), which is implicated in forming behavior, memory enhancement and emotional regulation (Bries et al., 2021). Based on these observations, increasing the consumption of eggs contributes to preventing deteriorations in cognitive performance thereby lowering the risk of developing neurological disorders.

36.2.2 Polyunsaturated Fatty Acids (PUFAs)

The fatty acids contained in egg yolk are composed of saturated fatty acids (35%), monounsaturated fatty acids (45%), and PUFAs (20%). PUFAs are divided into omega-3 fatty acids (ω-3), omega-6 fatty acids (ω-6), and omega-9 fatty acids (ω-9) (Anton & Gandemer, 1997). The ω-3 are eicosapentaenoic acid (EPA) and docosahexaenoic acid (DHA). Accumulating evidences suggest that a maternal diet supplemented with PUFAs contributes to the offspring's neuron and brain development (Smuts et al., 2003; Fleith & Clandinin, 2005; Martinat et al., 2021). ω-3 and ω-6 PUFAs accumulate in the brain of fetus during gestation and then increase approximately 30-fold by the age of 2 years. It was observed that insufficient intake of DHA during pregnancy has a negative impact on the brain development of preterm infants. Since only a limited level of DHA can be converted into EPA, dietary intake of EPA is essential for maintaining the balance of eicosanoids for a pregnant woman (Greenberg et al., 2008). According to the results of a cohort study, the group of preterm infants supplemented with EPA derived from fish oil showed higher brain weight gains compared to those without EPA supplementation (Woltil et al., 1999). Dietary intake of DHA by pregnant mothers sustained the pre and post-natal brain development, via the placenta during gestation and breast milk during lactation (Martinat et al., 2021). The daily intake recommended by WHO and FAO for DHA is 200 to 300 mg for pregnant women (Fao, 1995). Findings from a recent study suggested that a daily intake of 1000 mg DHA for women with low DHA status reduced the risk of early preterm birth (Carlson et al., 2021). In a rodent model for autism, scientific evidence demonstrated that daily supplementation of ω-3 at 200 mg/kg for 30 days effectively restored fatty acid metabolism and protected against the development of autistic-like manifestation (Alfawaz et al., 2018). According to the clinical evidence presented in a recent review, the higher intake of ω-3 is able to prevent signs in patients with autism (Martinat et al., 2021). Moreover, weak cognitive function and working memory performance were observed in children having a low level of DHA (Montgomery et al., 2013). Similarly, an inadequate dietary DHA intake showed a profound negative impact on cognitive function and psychomotor speed in middle-aged people (Kalmijn et al., 2004). Aging can alter the brain lipid composition, resulting in gradually deteriorating memory function and behavior performance of human beings (Küllenberg et al., 2012). The supplementation of PUFAs in elderly may increase the brain ω-3 content to improve the efficiency of neurotransmitter release, and consequently prevent memory decline (Favrelière et al., 2003). It is evident that PUFAs are not only implicated in the structure and function of cell membranes in the brain, but also in the control of its physiological function through the regulation of signaling pathways.

Emerging scientific evidence has indicated that ω-3 exerts a neuroprotective effect on neurons and glial cells by regulating the activity of glial cells, formation of syncytium in astrocytes, and preventing neuroinflammation in macroglia (Martinat et al., 2021). As a component of glial and microglia cell membranes, the bioavailability of PUFA and a molecular mechanism supporting their brain health-promoting action remain unknown. PUFAs uptake into the brain is primarily mediated by passive passage and specific membrane transporters distributed on the blood-brain barrier (BBB) (Bazinet & Layé, 2014). Several fatty acid transporters have been shown to facilitate this transfer. Firstly, the long-chain fatty acid CoA synthase (ACSL) is involved in forming a specific pool and increasing the esterification of DHA and arachidonic acid (ARA) into phospholipid membranes (Bazinet & Layé, 2014). Secondly, observations from a previous study indicated that the major facilitator superfamily domain-containing protein 2a (Mfsd2a) is regulated by lysophosphatidylcholine (LPC)-DHA to transport LPC-DHA into the fetus (Prieto-Sánchez et al., 2017). Thirdly, the fatty acid translocase (FAT/CD36) and fatty acid transport protein (FATP) widely distributed along the placental membrane also help to transfer DHA from the mother to the fetus (Layé et al., 2018). Finally, EPA, DHA, and ARA, all present in brain phospholipid membranes, can be de-esterified by specific phospholipase A2 (PLA2) and subsequently metabolized by cyclooxygenases (COXs), lipoxygenases (LOXs), and cytochrome P450 into bioactive oxylipins, including docosanoids from DHA or EPA, and eicosanoids from ARA (Gabbs et al., 2015). Oxylipins derived from ARA are involved in maintaining physiological homeostatic functions, such as inflammation, pain response, cell migration and proliferation, vascular tone and vasoconstriction (Gabbs et al., 2015). Accordingly, oxylipins exert some pro-inflammatory activity whereas DHA-derived resolvins and protectins have anti-inflammatory properties (Bazinet & Layé, 2014; Bosviel et al., 2017). It was demonstrated that a lower risk of inflammatory neurological disorders (i.e. Alzheimer's disease and Parkinson's disease) was associated with the consumption of a high amount of DHA (Laye, 2010). However, clinical investigations did not provide conclusive evidence due to the limited understanding of the mechanism underlying the neuroprotective action of PUFAs. At the molecular level, PUFAs and oxylipins can agonistically bind with the cell membrane and intracellular receptors, including oxysterols receptor (LXR), peroxisome proliferator-activated receptor (PPAR), hepatic nuclear factor 4A(HNF4A), chemokine like receptor 1(CHEMR23), and G-protein coupled receptor 32 (GPR32) to elicit the cellular signaling transduction. Subsequently, the protein kinase C (PKC) pathway is activated while the nuclear factor-κB (NFκB) pathway is inhibited in the cell (Bazinet & Layé, 2014). Inflammation disturbing microglial activity results in impairment of brain function. Among the cellular inflammatory signaling cascade, NFκB is a key transcriptional factor involved in the control of inflammatory responses. DHA and its derived mediator neuroprotection D (NPD)-1 were shown to suppress NFκB and COX2-mediated pro-inflammatory signaling events to protect against ischemia, and attenuate pro-inflammatory signaling events in microglia (Marcheselli et al., 2003; Lukiw & Bazan, 2010; Orr et al., 2013). Accordingly, these findings suggested that dietary PUFAs have the potential to promote brain health throughout life based on their neuroprotective effects.

It is well-known that ω-3 are deposited in the yolk of eggs when hens are fed with alga or flaxseeds that are rich in fish oil (Fraeye et al., 2012; Coorey et al., 2015). Nowadays, ω-3-enriched eggs are widely available in the marketplace, particularly for consumers who are looking for alternative foods rich in dietary DHA. Depending on the type of ω-3 source and fat content added to the ration given to laying hens, the amount of DHA deposited in egg yolk is varied from 1.4–146.6 mg per egg (Fraeye et al., 2012). The deposition efficiency of DHA in eggs is quite variable and affected by various factors such as hen's strain, age, and feed composition. For example, hens fed with heterotrophic microalgae can produce eggs containing up to 200 mg DHA per egg (Fraeye et al., 2012). Moreover, the recommended daily intakes of ω-3 are varied in different countries from 200 mg to 680 mg per day (Givens & Gibbs, 2008). Being an important dietary source, the enrichment of ω-3 foods can boost the intake of health-promoting fatty acids thereby eggs being a favorable dietary resource for brain health.

36.2.3 Egg-Derived Carotenoids

Lutein and zeaxanthin are the main lipophilic bioactive carotenoids in the egg yolk. These 2 carotenoids cannot be synthesized by humans, and thus must be obtained from dietary sources such as the egg. Carotenoids confer to the egg yolk its attractive red and golden appearance, leading to increasing egg sensory quality to meet consumer needs. Carotenoid-rich feeds can enhance the deposition of pigments in egg yolks. Supplementation of 3% carotenoid-rich plants in the diet leads to increasing carotenoid content over 20 μg/g and improving oxidative stability (Kljak et al., 2021). The lipophilic property of carotenoids improves their bio-accessibility and facilitates the uptake by cell membranes for structural stability (Britton, 1995). It is recognized that the bioavailability of carotenoids from fruit and vegetables is low, while it is relatively high from egg yolk (Handelman et al., 1999). Supplementation of lutein in feeds for laying hens was shown to significantly increase the lutein content in egg yolk (Leeson & Caston, 2004). The amount of lutein and zeaxanthin contained in an egg yolk is around 200–300 μg. After absorption, the efflux of carotenoids is transferred from enterocytes to the liver where they are used to synthesize lipoproteins subsequently distributed throughout the body or stored in adipose tissue (Mares, 2016; Murillo et al., 2019). In the human body, lutein and zeaxanthin are widely distributed in vision-related tissues, including the eye and brain (Johnson, 2014). Therefore, the egg is a promising dietary source of bioactive carotenoids for human consumption, and beneficial for maintaining brain and eye health (Johnson, 2014; Nimalaratne et al., 2015). Results from a clinical study validated that intake of an egg per day elevated serum level of lutein by 26% and zeaxanthin by 38% without detrimental effects on serum lipid profile, respectively (Goodrow et al., 2006). Adabi et al. (2010) identified that the plasma content of carotenoids in people around 60 years of age was significantly increased after consumption of 1 carotenoid-enriched egg per day (Adabi et al., 2010). Extensive evidence from epidemiological studies demonstrated that a high level of lutein and zeaxanthin was required for sustaining the fetus brain and neural development as well as for cognitive function in adults, suggesting that lutein and zeaxanthin have nutritional attributes for promoting human cognitive health through life (Johnson, 2014). Due to their high bioavailability, the consumption of egg yolk has a noticeable impact on promoting human brain and eye health. Both lutein and zeaxanthin are widely present in different brain regions and have bioactive properties to improve visual perception, cognition, decision-making, and motor coordination (Mewborn et al., 2018). The underlying mechanisms of the egg-derived carotenoids promoting brain health are summarized in the following section.

Carotenoids are lipophilic bioactive constituents and have strong antioxidant activities against oxidative stress. Upon pathogen invasion or tissue injury, the excess free radicals released from cell mitochondria elevate the oxidative stress status in a cell. At cellular level, endogenous antioxidants, including glutathione and antioxidant enzymes, are activated by oxidative stress and restore the redox balance. The dietary intake of natural antioxidants such as carotenoids can effectively quench free radicals and contribute to a healthy status (Murillo et al., 2019). In the eye, the conjugated double bonds of lutein and zeaxanthin exert photoprotective and antioxidant actions. As potent photoprotectants and antioxidants, lutein and zeaxanthin are able to reduce blue-light-caused chromatic aberration and absorb light in the visible light spectrum to protect lipid-dense retina from photochemical and oxidative damage (Sujak et al., 1999; Kijlstra et al., 2012). Results from a recent study indicated that lutein and zeaxanthin mitigated the release of pro-inflammatory mediators (i.e.5-hydroxyeicosatetraenoic acid and isoprostanes from the H_2O_2-induced retinal cells (Leung et al., 2020). The cellular oxidative stress elicits the production of inflammatory mediators, leading to the development of chronic inflammation diseases. The scientific evidences suggested that lutein directly modulates cellular oxidant-sensitive inflammatory signaling pathways, by which lutein exerts neuroprotective action of xanthophyll carotenoids (Ahn & Kim, 2021). It was shown that lutein can activate nuclear factor erythroid 2-related factor 2 (Nrf2), and subsequently modulate endogenous glutathione level, upregulate antioxidant genes of extracellular signal-regulated heme oxygenase (HO-1) and NAD(P)H quinone dehydrogenase (NQO1) expression, and enhance activities of various antioxidant enzymes (Ahn & Kim, 2021). In addition, findings from an animal model of traumatic brain injury suggested that lutein exerted strong anti-inflammatory action through the activation of the Nrf2 pathway, suppressing the expression of pro-inflammatory cytokines (i.e. IL-1β, IL-6, MCP-1) and reducing the level of serum reactive oxygen species (ROS) in an animal model of traumatic brain injury (Tan et al., 2017). Another recent evidence demonstrated that lutein can inhibit LPS-induced pro-inflammation signaling cascades, including phosphorylation of p38 kinase and activation of NF-κB in BV-2 microglia, validating that lutein has the neuroprotective potential to prevent inflammation-mediated neurodegenerative disorders (Wu et al., 2015). Moreover, a clinical nutrition study revealed that supplementation of dietary lutein and zeaxanthin were effectively suppressing IL-1β expression via interfering with inflammatory cascades, leading to enhanced cognitive performance (Stringham et al., 2019). Taking together, clinical evidence confirmed the promiosing neuroprotective activity of dietary lutein and zeaxanthin on eye and brain health in humans. The antioxidant and anti-inflammatory properties of lutein and zeaxanthin account for their regulatory role in the crosstalk between Nrf-2 and NF-κB at a cellular level, by which the cellular inflammatory signaling cascade and mediated inflammatory responses are suppressed as summarized in Table 36.1. Since eggs are an affordable dietary source of lutein and zeaxanthin, increasing egg consumption will be beneficial for the overall eye and brain health of human beings.

36.3 Egg-Contained Proteins, Peptides, and Amino Acids

The egg white accounts for 58% of the whole egg by weight and is constituted of approximately 90% water and 10% protein. Thus, the egg white is rich in proteins and widely recognized as a source of high-quality dietary proteins. Besides their nutritional value, egg white proteins and related hydrolysate

TABLE 36.1

Egg Yolk-Derived Lipophilic Composition and Their Role in Brain Health

Bioactive Egg Yolk Components	Role in Brain Health	Bioactive Constituents Derived from Egg on Brain Health	References
Choline	Neurodevelopment Neurotransmitter function Anti-inflammation	Membrane structure↑ Acetylcholine ↑ Neurotransmitter↑ Efficiency of ion channels↑ Plasticity of neural cells↑ NLRP-3↓	(Zeisel & Niculescu, 2006; Chin & Goh, 2019; Wang et al., 2019; Llorente-Ovejero et al., 2021)
ω-3	Sustain structure Neuroprotection Anti-inflammation	Memory performance↑ Neurotransmitter release↑ Neuroinflammation in macroglia↓ NFκB and COX-2↓	(Favrelière et al., 2003; Montgomery et al., 2013; Wu et al., 2015; Tan et al., 2017; Ahn & Kim, 2021)
Carotenoids	Antioxidant Vision improvement Anti-inflammation	Scavenging free radicals ↑ Nrf2↑ Glutathione↑ HO-1, NQO1↑ NFκB↓ Pro-inflammatory cytokines↓	(Kijlstra et al., 2012; Sujak et al., 1999)

made up of amino acids and peptides are involved in the modulation of physiological functions. Egg white-derived bioactive constituents (peptides or amino acids) released during the digestion process are subsequently implicated in the management of complex human health conditions, accounting for their various roles such as antioxidant, immunomodulatory, and anti-microbial functions (Moreno-Fernández et al., 2020). An egg white hydrolysate (EWH) was shown to mitigate the COX-2-mediated inflammation in microglia and prevent neurotoxicity in the hippocampus caused by chronic dietary exposure to aluminum in rats (Martinez et al., 2019). Moreover, it was reported that the intake of tryptophan (TRP)-rich egg protein hydrolysate (EPH) increased plasma TRP levels, leading to improving perceptual-motor and vigilance performance in human subjects (Markus et al., 2010). The following section will review the effects of bioactive peptides and amino acids derived from EPH on brain health status and associated cognitive function and behavioral performance.

36.3.1 Egg-Derived Bioactive Peptides

After the gastric and intestinal digestion of egg proteins, the released peptides and amino acids are up-taken and delivered through cell membrane transporters distributed in the intestine epithelium or through a paracellular gap into the circulation system. Their presence in the blood allows them to subsequently exert their regulatory effects on a wide range of physiological functions upon reaching a target site (Bruno et al., 2013). The digestibility and bioavailability of egg proteins are relatively high when compared to proteins from other food sources. However, the bioavailability of dietary peptides is low due to the epithelial barrier and efflux pumps. The short-chain peptides consisting of 2 or 3 amino acids are commonly delivered into cells of the gut by peptide transporter 1 (PepT1) (Miguel et al., 2008; Kovacs-Nolan et al., 2012; Xu et al., 2017). Non-absorbed egg-derived proteins or peptides can be further metabolized by the symbiotic microbiota of the lower gut (Liao et al., 2018). Subsequently, the uptake of released peptides and amino acids may also occur in the colon. Besides being an organ for digestion, the intestine is also engaged in controlling gut homeostasis by producing peptide hormones from enteroendocrine cells. These secreted peptide hormones modulate the central nervous system as well as the gastrointestinal tract. Owing to the wide variety of bioactive peptides released in the gut after the digestion of food proteins, some peptides may mimic neuropeptides or be directly involved in interactions with the endocrine system. It was recently shown that EWP released from gastrointestinal digestion promoted the secretion of glucagon-like peptide-1 (GLP-1), contributing to modulating gastrointestinal and homeostatic functions (Santos-Hernández et al., 2020). Therefore EWP may have some potential effect on ameliorating obesity and other related metabolic disorders.

Obesity elicits some alterations in brain functions by impairing high cerebral function and accelerating the aging-related cognitive decline (Nguyen et al., 2014). The systemic inflammation and related metabolic alterations were reported to play a vital role in the pathogenesis of cognitive decline in neurodegenerative diseases in the later stages of life (Nguyen et al., 2014; Wang et al., 2021). Recent studies identified that EWH supplementation has neuroprotective effects on the brain by mitigating the cognitive decline of rats exposed to low doses of mercury (Rizzetti et al., 2016a, 2016b). In a recent review, the neuroprotective peptides were identified as tripeptides and tetrapeptides containing hydrophobic amino acids, for their relatively high bioaccessibility and bioavailability, both *in vivo* and *in vitro* (Wang et al., 2021). In brain neurons, mitochondrial dysfunction augments the release of ROS to trigger neuronal cell apoptosis, leading to an increased risk of developing neurodegenerative disorders (Kopeikina et al., 2011). EWH demonstrated not only their neuroprotective effects against mercury-induced cognitive degradation but also some antioxidant properties to prevent mercury-induced vascular damage by neutralizing vascular ROS or nitric oxide (NO) (Escobar et al., 2020). Moreover, there is growing evidence demonstrating that

TABLE 36.2

Bioactive Egg White Components and Their Role in Brain Health

Bioactive Egg White Components	Role in Brain Health	Bioactive Constituents Derived from Egg White on Brain Health	References
EWH Peptides	Neuroprotection Neurohormonal regulation Mitigation of oxidative stress Anti-inflammation	MDA↑TNF-α↓ ROS and NO↓ BcL2↑ Regulation of microbial composition Pro-inflammatory cytokines↓ NFκB↓	(Rizzetti et al., 2016b; Liu et al., 2014; Escobar et al., 2020; Majumder et al., 2013; Wang et al., 2019)
Amino acids	Neuroprotection Neurohormonal regulation Immune regulation Maintaining emotional states	Memory performance↑ Sleep rhythm↑ Neurotransmitter release↑ Cognitive function↑ Memory↑ Stress↓	(Chyan et al., 1999; Agus et al., 2018; Chaves Filho et al., 2018; Colangeli et al., 2021)

EWPs have some promising antioxidative properties at a cellular level, including Trp-Asn-Trp-Ala-Asp, EWPH-III being a fraction of EWH-derived peptides (MW less than 1kDa), ovotransferrin-derived peptide IRW (Ile-Arg-Trp), VYLPR, and ovalbumin-derived peptides (Liu et al., 2014; Majumder et al., 2013; Liu et al., 2015; Wang et al., 2018; Zhang et al., 2019). These antioxidant EWPs exert some action against vascular inflammation and oxidative stress, thereby lowering the risk of developing cardiovascular diseases (Nguyen et al., 2014). Peptides derived from EWH, presenting antioxidant and anti-inflammation properties to promote brain health, are reviewed in Table 36.2. Future investigations on the mechanism underlying the neuroprotective effects of EWPs on brain function against cognitive impairments are required, a necessary step prior to conducting clinical trials with patients.

36.3.2 Egg-Derived Amino Acids

Chicken eggs are a rich food source of tryptophan (Trp), accounting for about 59% of the recommended daily intake in 100 grams (Moreno-Fernández et al., 2020). In humans, Trp is an essential amino acid, implying that it must be derived from foods such as milk, eggs, fish, and so on. L-Tryptophan is implicated in protein synthesis and is a precursor of multiple bioactive molecules (i.e. kynurenine, melatonin, and serotonin), playing a central role in gut-brain functions, immune responses, bone health, and mitochondrial function (Singh, 2019). An EPH rich in Trp, administered to healthy adults around 50 years of age, established the pharmacokinetics of Trp and demonstrated a dose-dependent increase in plasma Trp levels (Mitchell et al., 2011). Their results validated that EPH is an effective food ingredient to provide bioavailable Trp. In the small intestinal epithelium, EPH is absorbed directly by large neutral amino acid transporter (LAT1) or metabolized into kynurenine or serotonin and subsequently distributed into the submucosal layer (Kanai et al., 1998). Interestingly, LAT1 is also widely distributed on the luminal and abluminal membranes of brain capillary endothelial cells (BCECs) to transport neutral amino acids across the blood–brain barrier (BBB) to exert their regulatory actions on brain physiological functions (Campos-Bedolla et al., 2014).

Through main metabolism pathways, it is well accepted that Trp is an essential precursor for the synthesis of several active metabolites, key molecules involved in enzyme synthesis, brain function, and neurohormonal regulation (Nayak et al., 2022). Trp is composed of an indole ring with a side chain of alanine, giving Trp some ketogenic properties by which Trp metabolism undergoes the citric acid cycle (Nayak et al., 2022). After ingestion of Trp-rich food ingredients, Trp is first converted into $_{L}$-5-hydroxytryptamine (5-HT) by tryptophan hydrolase (TPH)-1 in enterochromaffin cells (ECs) within the gut or by TPH2 in the central and enteric nervous system. 5-HT is then metabolized into 5-hydroxytryptophan (5-HTP), a precursor of serotonin, and further converted into melatonin, an endogenous regulator of sleep rhythms (Roth et al., 2021). Serotonin and other TPH metabolites can be produced by specific gut microbiome strains such as *Lactococcus lactis cremoris*, *Streptococcus thermophilus*, *Escherichia coli K-12*, and *Morganella morganii* (Hata et al., 2017; Roth et al., 2021). Trp can also undergo the kynurenine degradation pathway to produce kynurenic acid (KA) and quinolinic acid (QA) via tryptophan 2,3-dioxygenase (TD) and indoleamine 2,3-dioxygenase (IDO) in the gastrointestinal tract, liver, and brain (Roth et al., 2021). Finally, the gut microbiome, primarily species such as *E.coli* and *Lactobacilli,* expresses Trp- and indole-active transporters, converting Trp into indole (Hubbard et al., 2015). Therefore, as Trp enters into the lower gut and interacts with gut microbiota, it undergoes the indole metabolism producing indole derivatives, such as tryptamine and 5-HT, implicated with gut barrier function via interplaying with gut chemosensors to regulate sensory neural pathways (Gheorghe et al., 2019). In addition, the indole-based molecules play an important role in controlling intestinal mucosal immune responses via the activation of aryl hydrocarbon receptor (AhR) signaling (Lamas et al., 2018). The growing evidence on the modulatory effects of Trp on microbiota-gut-brain axis highlights the key role of Trp metabolism on gut and neurologic health (Table 36.2).

Accordingly, it was reported that behavior and cognitive alternations are correlated with changes in gut microbiota distribution and dysfunctional Trp metabolism. In the gut, Trp metabolites activate the 5-HT receptor expressed by ECs to regulate gut motility, gut-brain signaling, secretion-absorption, and platelet functions through interplaying with submucosal neurons (Agus et al., 2018). On the one hand, the synthesis of serotonin occurs in both enteric and central neuron system (CNS) serotonergic neurons by decarboxylation from 5-HT. Accordingly, maintenance of 5-HTP level plays a vital role in the regulation of GI function controlled by the enteric nervous system (ENS); and therefore, against depression-suppressed GI motility (Israelyan et al., 2019). On the other hand, the crosstalk between serotonin and endocannabinoid (eCB) systems orchestrates the integrated neuronal activity, leading to the modulation of emotional states, stress homeostasis, cognitive function, and behavioral functions (Colangeli et al., 2021). Consequently, lower content of 5-HT in the brain is a factor contributing to the pathogenesis of depression. Serotonin is also a well-known neuromodulator to activate brain areas influenced by serotonergic neurons, such as the hippocampus and entorhinal cortex, important areas for learning and memory. Furthermore, Trp involved kynurenine degradation pathway has a significant impact on the metabolism of neurotransmitters (Agus et al., 2018). In blood, Trp is reduced through kynurenine degradation, resulting in limited neurotransmitter production, a condition linked to depression, anxiety, and stress-related disorders (Schwarcz et al., 2012). Findings from a previous study indicated that depression in obesity is characterized by indoleamine 2,3-dioxygenase (IDO)-1 activation and associated chronic inflammatory status (Chaves Filho et al., 2018). Therefore, a Trp deficiency due to insufficient Trp dietary intake is a factor involved in the depression observed in obese patients. Moreover, a bacterial metabolite of Trp, indole-3-propionic acid (IPA), may inhibit amyloid-β mediated neurotoxicity, indicating a neuroprotective role in Alzheimer's disease (Chyan et al., 1999). Therefore, the adequate dietary intake of Trp-rich proteins by people will contribute to preventing Trp pool depletion and elevating the level of serotonin in the brain; therefore, these are important factors for neural and brain health. Among all Trp-rich foods, eggs are the preferred food source due to the high Trp content in egg white proteins with high bioavailability and affordability. Through the evaluation of physiological parameters and neurophysiological responses, the supplementation of EPH in healthy adults has shown a positive impact on modulating stress and emotion (Zahar et al., 2022). Taking together, EWP is an important dietary source of Trp, which undergoes the eukaryotic and bacterial metabolism to produce derivatives (i.e. serotonin, melatonin, QA, and IDO) involved in a wide range of neurobiological functions. Findings from previous studies demonstrated some significant beneficial impact of EWH as well as Trp-rich bioactive peptides derived from EWP on cognitive function and behavioral performance, which was summarized in this chapter (Figure 36.1).

36.4 Conclusion and Future Directions

Chicken eggs are high in nutritional and sensory attributes, and therefore recognized as one of the most valuable nutritious food and source of ingredients for the manufacture of a variety of food products. Eggs are extensively consumed worldwide. Compared to meat and fish, eggs remain an affordable source of high-quality proteins for people. As a rich source of bioactive lipids (i.e. phospholipid, choline, and PUFAs) as well as carotenoids, the egg yolk is an important food source to maintain supporting brain development and cognitive performance. Additionally, egg white is composed primarily of a high level of proteins, subsequently hydrolyzed into functional peptides and amino acids that have modulatory effects on neurological disorders. However, there is still a limited understanding of the molecular mechanisms by which gut bacterial metabolism of EPH implicates the microbiome-gut-brain axis in neurological disorders. This lack of knowledge currently hinders further clinical exploration and translational applications. As a highly nutritious dietary source, chicken eggs are worthy of exploring the effects of egg-derived bioactive components on brain health status, particularly associated with cognitive function and behavioral performance throughout life. Although most of the mechanism remains unclear, future studies are necessary to understand the mechanisms of action by which underlying signaling pathways are regulated by egg-derived bioactive compounds, as well as the molecular basis of bioavailability and cellular uptake of these bioactives. Both *in vitro* and *in vivo* studies will be required before conducting human clinical trials to validate the efficacy of egg-derived bioactive constituents. Ultimately, egg-derived bioactive components may undergo industrial transformation and commercialized applications as functional foods or nutraceuticals for brain health.

REFERENCES

Adabi, S. G., Kamali, M., Davoudi, J., Cooper, R., & Hajbabaei, A. (2010). Quantification of lutein in egg following feeding hens with a lutein supplement and quantification of lutein in human plasma after consumption of lutein enriched eggs. *Archiv für Geflügelkunde*, *74*(3), 158–163.

Agus, A., Planchais, J., & Sokol, H. (2018). Gut microbiota regulation of tryptophan metabolism in health and disease. *Cell Host & Microbe*, *23*(6), 716–724.

Ahn, Y. J., & Kim, H. (2021). Lutein as a modulator of oxidative stress-mediated inflammatory diseases. *Antioxidants*, *10*(9), 1448.

Alfawaz, H., Al-Onazi, M., Bukhari, S. I., Binobead, M., Othman, N., Algahtani, N., … El-Ansary, A. (2018). The independent and combined effects of omega-3 and vitamin B12 in ameliorating propionic acid induced biochemical features in juvenile rats as rodent model of autism. *Journal of Molecular Neuroscience*, *66*(3), 403–413. https://doi.org/10.1007/s12031-018-1186-z

Aminianfar, A., Shayanfar, M., Mohammad-Shirazi, M., Sharifi, G., & Esmaillzadeh, A. (2019). Egg consumption in relation to glioma: A case–control study. *Nutrition and Cancer*, *71*(1), 41–49.

Anton, M., & Gandemer, G. (1997). Composition, solubility and emulsifying properties of granules and plasma of egg yolk. *Journal of Food Science*, *62*(3), 484–487.

Anton, M., Nau, F., & Nys, Y. (2006). Bioactive egg components and their potential uses. *World's Poultry Science Journal*, *62*(3), 429–438.

Bahnfleth, C., Canfield, R., Nevins, J., Caudill, M., & Strupp, B. (2019). Prenatal choline supplementation improves child color-location memory task performance at 7 y of age (FS05-

01-19). *Current Developments in Nutrition*, *3*(Supplement_1), nzz048. FS005-001-019.

Bazinet, R. P., & Layé, S. (2014). Polyunsaturated fatty acids and their metabolites in brain function and disease. *Nature Reviews Neuroscience*, *15*(12), 771–785.

Bosviel, R., Joumard-Cubizolles, L., Chinetti-Gbaguidi, G., Bayle, D., Copin, C., Hennuyer, N., & Porta, A. (2017). DHA-derived oxylipins, neuroprostanes and protectins, differentially and dose-dependently modulate the inflammatory response in human macrophages: Putative mechanisms through PPAR activation. *Free Radical Biology and Medicine*, *103*, 146–154.

Boyle, N. B., Dye, L., Arkbåge, K., Thorell, L., Frederiksen, P., Croden, F., & Lawton, C. (2019). Effects of milk-based phospholipids on cognitive performance and subjective responses to psychosocial stress: A randomized, double-blind, placebo-controlled trial in high-perfectionist men. *Nutrition*, *57*, 183–193.

Bries, A. E., Webb, J. L., Vogel, B., Carrillo, C., Day, T. A., Kimber, M. J., & McNeill, E. M. (2021). RNA sequencing reveals key metabolic pathways are modified by short-term whole egg consumption. *Frontiers in Nutrition*, *8*. https://doi.org/10.3389/fnut.2021.652192

Britton, G. (1995). Structure and properties of carotenoids in relation to function. *The FASEB Journal*, *9*(15), 1551–1558.

Bruno, B. J., Miller, G. D., & Lim, C. S. (2013). Basics and recent advances in peptide and protein drug delivery. *Therapeutic Delivery*, *4*(11), 1443–1467. https://doi.org/10.4155/tde.13.104

Burley, R. W., & Cook, W. (1961). Isolation and composition of avian egg yolk granules and their constituent α-and β-lipovitellins. *Canadian Journal of Biochemistry and Physiology*, *39*(8), 1295–1307.

Campos-Bedolla, P., Walter, F. R., Veszelka, S., & Deli, M. A. (2014). Role of the blood–brain barrier in the nutrition of the central nervous system. *Archives of Medical Research*, *45*(8), 610–638.

Carlson, S. E., Gajewski, B. J., Valentine, C. J., Kerling, E. H., Weiner, C. P., Cackovic, M., & Brown, A. R. (2021). Higher dose docosahexaenoic acid supplementation during pregnancy and early preterm birth: A randomised, double-blind, adaptive-design superiority trial. *EClinicalMedicine*, *36*, 100905.

Caudill, M. A. (2010). Pre-and postnatal health: Evidence of increased choline needs. *Journal of the American Dietetic Association*, *110*(8), 1198–1206.

Caudill, M. A., Strupp, B. J., Muscalu, L., Nevins, J. E., & Canfield, R. L. (2018). Maternal choline supplementation during the third trimester of pregnancy improves infant information processing speed: A randomized, double-blind, controlled feeding study. *The FASEB Journal*, *32*(4), 2172–2180.

Chaves Filho, A. J. M., Lima, C. N. C., Vasconcelos, S. M. M., de Lucena, D. F., Maes, M., & Macedo, D. (2018). IDO chronic immune activation and tryptophan metabolic pathway: A potential pathophysiological link between depression and obesity. *Progress in Neuro-Psychopharmacology and Biological Psychiatry*, *80*, 234–249. https://doi.org/10.1016/j.pnpbp.2017.04.035

Cheatham, C. L., Goldman, B. D., Fischer, L. M., da Costa, K.-A., Reznick, J. S., & Zeisel, S. H. (2012). Phosphatidylcholine supplementation in pregnant women consuming moderate-choline diets does not enhance infant cognitive function: A randomized, double-blind, placebo-controlled trial. *The American Journal of Clinical Nutrition*, *96*(6), 1465–1472.

Chin, E. W., & Goh, E. L. (2019). Modulating neuronal plasticity with choline. *Neural Regeneration Research*, *14*(10), 1697.

Chyan, Y.-J., Poeggeler, B., Omar, R. A., Chain, D. G., Frangione, B., Ghiso, J., & Pappolla, M. A. (1999). Potent neuroprotective properties against the alzheimer β-amyloid by an endogenous melatonin-related indole structure, indole-3-propionic acid. *Journal of Biological Chemistry*, *274*(31), 21937–21942.

Colangeli, R., Teskey, G. C., & Di Giovanni, G. (2021). Endocannabinoid-serotonin systems interaction in health and disease. *Progress in Brain Research*, *259*, 83–134.

Coorey, R., Novinda, A., Williams, H., & Jayasena, V. (2015). Omega-3 fatty acid profile of eggs from laying hens fed diets supplemented with chia, fish oil, and flaxseed. *Journal of Food Science*, *80*(1), S180–S187.

Cotterill, O., & Geiger, G. (1977). Egg product yield trends from shell eggs. *Poultry Science*, *56*(3), 1027–1031.

Donoso, F., Schverer, M., Rea, K., Pusceddu, M. M., Roy, B. L., Dinan, T. G., & Schellekens, H. (2020). Neurobiological effects of phospholipids in vitro: Relevance to stress-related disorders. *Neurobiology of Stress*, *13*, 100252.

Escobar, A. G., Rizzetti, D. A., Piagette, J. T., Peçanha, F. M., Vassallo, D. V., Miguel, M., & Wiggers, G. A. (2020). Antioxidant properties of egg white hydrolysate prevent mercury-induced vascular damage in resistance arteries. *Frontiers in Physiology*, *11*. https://doi.org/10.3389/fphys.2020.595767

Fao, W. (1995). WHO and FAO joint consultation: Fats and oils in human nutrition. *Nutrition Reviews*, *53*(7), 202–205.

Favrelière, S., Perault, M. C., Huguet, F., De Javel, D., Bertrand, N., Piriou, A., & Durand, G. (2003). DHA-enriched phospholipid diets modulate age-related alterations in rat hippocampus. *Neurobiology of Aging*, *24*(2), 233–243. https://doi.org/10.1016/s0197-4580(02)00064-7

Fleith, M., & Clandinin, M. (2005). Dietary PUFA for preterm and term infants: Review of clinical studies. *Critical Reviews in Food Science and Nutrition*, *45*(3), 205–229.

Fraeye, I., Bruneel, C., Lemahieu, C., Buyse, J., Muylaert, K., & Foubert, I. (2012). Dietary enrichment of eggs with omega-3 fatty acids: A review. *Food Research International*, *48*(2), 961–969.

Gabbs, M., Leng, S., Devassy, J. G., Monirujjaman, M., & Aukema, H. M. (2015). Advances in our understanding of oxylipins derived from dietary PUFAs. *Advances in Nutrition (Bethesda, Md.)*, *6*(5), 513–540. https://doi.org/10.3945/an.114.007732

Gheorghe, C. E., Martin, J. A., Manriquez, F. V., Dinan, T. G., Cryan, J. F., & Clarke, G. (2019). Focus on the essentials: Tryptophan metabolism and the microbiome-gut-brain axis. *Current Opinion in Pharmacology*, *48*, 137–145.

Givens, D. I., & Gibbs, R. A. (2008). Current intakes of EPA and DHA in European populations and the potential of animal-derived foods to increase them: Symposium on 'How can the n-3 content of the diet be improved?'. *Proceedings of the Nutrition Society*, *67*(3), 273–280.

Goodrow, E. F., Wilson, T. A., Houde, S. C., Vishwanathan, R., Scollin, P. A., Handelman, G., & Nicolosi, R. J. (2006). Consumption of one egg per day increases serum lutein and zeaxanthin concentrations in older adults without altering serum lipid and lipoprotein cholesterol concentrations. *Journal of Nutrition*, *136*(10), 2519–2524. https://doi.org/10.1093/jn/136.10.2519

Greenberg, J. A., Bell, S. J., & Van Ausdal, W. (2008). Omega-3 fatty acid supplementation during pregnancy. *Reviews in Obstetrics and Gynecology*, *1*(4), 162.

Gurnida, D. A., Rowan, A. M., Idjradinata, P., Muchtadi, D., & Sekarwana, N. (2012). Association of complex lipids containing gangliosides with cognitive development of 6-month-old infants. *Early Human Development*, *88*(8), 595–601.

Handelman, G. J., Nightingale, Z. D., Lichtenstein, A. H., Schaefer, E. J., & Blumberg, J. B. (1999). Lutein and zeaxanthin concentrations in plasma after dietary supplementation with egg yolk. *The American Journal of Clinical Nutrition*, *70*(2), 247–251.

Hata, T., Asano, Y., Yoshihara, K., Kimura-Todani, T., Miyata, N., Zhang, X.-T., & Sudo, N. (2017). Regulation of gut luminal serotonin by commensal microbiota in mice. *PLoS One*, *12*(7), e0180745.

Hubbard, T. D., Murray, I. A., & Perdew, G. H. (2015). Indole and tryptophan metabolism: Endogenous and dietary routes to Ah receptor activation. *Drug Metabolism and Disposition*, *43*(10), 1522–1535.

Ilcol, Y. O., Uncu, G., & Ulus, I. H. (2002). Free and phospholipid-bound choline concentrations in serum during pregnancy, after delivery and in newborns. *Archives of Physiology and Biochemistry*, *110*(5), 393–399. https://doi.org/10.1076/apab.110.5.393.11832

Israelyan, N., Del Colle, A., Li, Z., Park, Y., Xing, A., Jacobsen, J. P., & Saurman, V. (2019). Effects of serotonin and slow-release 5-hydroxytryptophan on gastrointestinal motility in a mouse model of depression. *Gastroenterology*, *157*(2), 507–521. e504.

Jayaraman, T., Kannappan, S., Ravichandran, M., & Anuradha, C. (2008). Impact of essential L on ethanol-induced changes in rat brain and erythrocytes. *Singapore Medical Journal*, *49*(4), 320.

Johnson, E. J. (2014). Role of lutein and zeaxanthin in visual and cognitive function throughout the lifespan. *Nutrition Reviews*, *72*(9), 605–612.

Kable, J., Coles, C., Keen, C. L., Uriu-Adams, J., Jones, K., Yevtushok, L., & Chambers, C. (2015). The impact of micronutrient supplementation in alcohol-exposed pregnancies on information processing skills in Ukrainian infants. *Alcohol*, *49*(7), 647–656.

Kalmijn, S., van Boxtel, M. P., Ocké, M., Verschuren, W. M., Kromhout, D., & Launer, L. J. (2004). Dietary intake of fatty acids and fish in relation to cognitive performance at middle age. *Neurology*, *62*(2), 275–280. https://doi.org/10.1212/01.wnl.0000103860.75218.a5

Kanai, Y., Segawa, H., Miyamoto, K., Uchino, H., Takeda, E., & Endou, H. (1998). Expression cloning and characterization of a transporter for large neutral amino acids activated by the heavy chain of 4F2 antigen (CD98). *Journal of Biological Chemistry*, *273*(37), 23629–23632. https://doi.org/10.1074/jbc.273.37.23629

Kijlstra, A., Tian, Y., Kelly, E. R., & Berendschot, T. T. (2012). Lutein: More than just a filter for blue light. *Progress in Retinal and Eye Research*, *31*(4), 303–315.

Kljak, K., Carović-Stanko, K., Kos, I., Janječić, Z., Kiš, G., Duvnjak, M., & Bedeković, D. (2021). Plant carotenoids as pigment sources in laying hen diets: Effect on yolk color, carotenoid content, oxidative stability and sensory properties of eggs. *Foods*, *10*(4), 721.

Kopeikina, K. J., Carlson, G. A., Pitstick, R., Ludvigson, A. E., Peters, A., Luebke, J. I., & Spires-Jones, T. L. (2011). Tau accumulation causes mitochondrial distribution deficits in neurons in a mouse model of tauopathy and in human Alzheimer's disease brain. *The American Journal of Pathology*, *179*(4), 2071–2082. https://doi.org/10.1016/j.ajpath.2011.07.004

Kovacs-Nolan, J., Phillips, M., & Mine, Y. (2005). Advances in the value of eggs and egg components for human health. *Journal of Agricultural and Food Chemistry*, *53*(22), 8421–8431.

Kovacs-Nolan, J., Zhang, H., Ibuki, M., Nakamori, T., Yoshiura, K., Turner, P. V., & Mine, Y. (2012). The PepT1-transportable soy tripeptide VPY reduces intestinal inflammation. *Biochimica et Biophysica Acta (BBA)-General Subjects*, *1820*(11), 1753–1763.

Küllenberg, D., Taylor, L. A., Schneider, M., & Massing, U. (2012). Health effects of dietary phospholipids. *Lipids in Health and Disease*, *11*(1), 1–16.

Lamas, B., Natividad, J. M., & Sokol, H. (2018). Aryl hydrocarbon receptor and intestinal immunity. *Mucosal Immunology*, *11*(4), 1024–1038. https://doi.org/10.1038/s41385-018-0019-2

Laye, S. (2010). Polyunsaturated fatty acids, neuroinflammation and well being. *Prostaglandins, Leukotrienes and Essential Fatty Acids (PLEFA)*, *82*(4–6), 295–303.

Layé, S., Nadjar, A., Joffre, C., & Bazinet, R. P. (2018). Anti-inflammatory effects of omega-3 fatty acids in the brain: Physiological mechanisms and relevance to pharmacology. *Pharmacological Reviews*, *70*(1), 12–38.

Lee, G. J., Oda, K., Morton, K. R., Orlich, M., & Sabate, J. (2021). Egg intake moderates the rate of memory decline in healthy older adults. *Journal of Nutritional Science*, *10*, e79.

Leeson, S., & Caston, L. (2004). Enrichment of eggs with lutein. *Poultry Science*, *83*(10), 1709–1712.

Leung, H. H., Galano, J.-M., Crauste, C., Durand, T., & Lee, J. C.-Y. (2020). Combination of lutein and zeaxanthin, and DHA regulated polyunsaturated fatty acid oxidation in H2O2-stressed retinal cells. *Neurochemical Research*, *45*(5), 1007–1019. https://doi.org/10.1007/s11064-020-02994-4

Lewis, E. D., Subhan, F. B., Bell, R. C., McCargar, L. J., Curtis, J. M., Jacobs, R. L., & Field, C. J. (2014). Estimation of choline intake from 24 h dietary intake recalls and contribution of egg and milk consumption to intake among pregnant and lactating women in Alberta. *British Journal of Nutrition*, *112*(1), 112–121. https://doi.org/10.1017/s0007114514000555

Liao, W., Jahandideh, F., Fan, H., Son, M., & Wu, J. (2018). Egg protein-derived bioactive peptides: Preparation, efficacy, and absorption. *Advances in Food and Nutrition Research*, *85*, 1–58.

Liu, J., Chen, Z., He, J., Zhang, Y., Zhang, T., & Jiang, Y. (2014). Anti-oxidative and anti-apoptosis effects of egg white peptide, Trp-Asn-Trp-Ala-Asp, against H_2O_2-induced oxidative stress in human embryonic kidney 293 cells. *Food & Function*, *5*(12), 3179–3188. https://doi.org/10.1039/c4fo00665h

Liu, J., Jin, Y., Lin, S., Jones, G. S., & Chen, F. (2015). Purification and identification of novel antioxidant peptides from egg white protein and their antioxidant activities. *Food Chemistry, 175*, 258–266. https://doi.org/10.1016/j.foodchem.2014.11.142

Llorente-Ovejero, A., Martínez-Gardeazabal, J., Moreno-Rodríguez, M., Lombardero, L., González de San Román, E., Manuel, I., & Rodríguez-Puertas, R. (2021). Specific phospholipid modulation by muscarinic signaling in a rat lesion model of Alzheimer's disease. *ACS Chemical Neuroscience, 12*(12), 2167–2181. https://doi.org/10.1021/acschemneuro.1c00169

Lukiw, W. J., & Bazan, N. G. (2010). Inflammatory, apoptotic, and survival gene signaling in Alzheimer's disease. *Molecular Neurobiology, 42*(1), 10–16.

Lutter, C. K., Iannotti, L. L., & Stewart, C. P. (2018). The potential of a simple egg to improve maternal and child nutrition. *Maternal & Child Nutrition, 14*, e12678.

Majumder, K., Chakrabarti, S., Davidge, S. T., & Wu, J. (2013). Structure and activity study of egg protein ovotransferrin derived peptides (IRW and IQW) on endothelial inflammatory response and oxidative stress. *Journal of Agricultural and Food Chemistry, 61*(9), 2120–2129. https://doi.org/10.1021/jf3046076

Marcheselli, V. L., Hong, S., Lukiw, W. J., Tian, X. H., Gronert, K., Musto, A., & Serhan, C. N. (2003). Novel docosanoids inhibit brain ischemia-reperfusion-mediated leukocyte infiltration and pro-inflammatory gene expression. *Journal of Biological Chemistry, 278*(44), 43807–43817.

Mares, J. (2016). Lutein and zeaxanthin isomers in eye health and disease. *Annual Review of Nutrition, 36*, 571.

Markus, C. R., Verschoor, E., Firk, C., Kloek, J., & Gerhardt, C. (2010). Effect of tryptophan-rich egg protein hydrolysate on brain tryptophan availability, stress and performance. *Clinical Nutrition, 29*(5), 610–616.

Martinat, M., Rossitto, M., Di Miceli, M., & Layé, S. (2021). Perinatal dietary polyunsaturated fatty acids in brain development, role in neurodevelopmental disorders. *Nutrients, 13*(4), 1185.

Martinez, C. S., Alterman, C. D., Vera, G., Márquez, A., Uranga, J.-A., Peçanha, F. M., & Miguel, M. (2019). Egg white hydrolysate as a functional food ingredient to prevent cognitive dysfunction in rats following long-term exposure to aluminum. *Scientific Reports, 9*(1), 1–13.

Mewborn, C. M., Terry, D. P., Renzi-Hammond, L. M., Hammond, B. R., & Miller, L. S. (2018). Relation of retinal and serum lutein and zeaxanthin to white matter integrity in older adults: A diffusion tensor imaging study. *Archives of Clinical Neuropsychology, 33*(7), 861–874. https://doi.org/10.1093/acn/acx109

Miguel, M., Davalos, A., Manso, M. A., de la Pena, G., Lasuncion, M. A., & López-Fandiño, R. (2008). Transepithelial transport across Caco-2 cell monolayers of antihypertensive egg-derived peptides. PepT1-mediated flux of Tyr-Pro-Ile. *Molecular Nutrition & Food Research, 52*(12), 1507–1513.

Mitchell, E. S., Slettenaar, M., Quadt, F., Giesbrecht, T., Kloek, J., Gerhardt, C., & Wiseman, S. (2011). Effect of hydrolysed egg protein on brain tryptophan availability. *British Journal of Nutrition, 105*(4), 611–617. https://doi.org/10.1017/S0007114510004150

Montgomery, P., Burton, J. R., Sewell, R. P., Spreckelsen, T. F., & Richardson, A. J. (2013). Low blood long chain omega-3 fatty acids in UK children are associated with poor cognitive performance and behavior: A cross-sectional analysis from the DOLAB study. *PLoS One, 8*(6), e66697.

Moreno-Fernández, S., Garcés-Rimón, M., & Miguel, M. (2020). Egg-derived peptides and hydrolysates: A new bioactive treasure for cardiometabolic diseases. *Trends in Food Science & Technology, 104*, 208–218.

Müller, C. P., Reichel, M., Mühle, C., Rhein, C., Gulbins, E., & Kornhuber, J. (2015). Brain membrane lipids in major depression and anxiety disorders. *Biochimica et Biophysica Acta (BBA)-Molecular and Cell Biology of Lipids, 1851*(8), 1052–1065.

Murillo, A. G., Hu, S., & Fernandez, M. L. (2019). Zeaxanthin: Metabolism, properties, and antioxidant protection of eyes, heart, liver, and skin. *Antioxidants (Basel, Switzerland), 8*(9), 390. https://doi.org/10.3390/antiox8090390

Nayak, B. N., Singh, R. B., & Buttar, H. S. (2022). Biochemical and dietary functions of tryptophan and its metabolites in human health. In *Functional foods and nutraceuticals in metabolic and non-communicable diseases* (pp. 783–798). Elsevier.

Nguyen, J. C., Killcross, A. S., & Jenkins, T. A. (2014). Obesity and cognitive decline: Role of inflammation and vascular changes. *Frontiers in Neuroscience, 8*, 375.

Nimalaratne, C., Savard, P., Gauthier, S. F., Schieber, A., & Wu, J. (2015). Bioaccessibility and digestive stability of carotenoids in cooked eggs studied using a dynamic in vitro gastrointestinal model. *Journal of Agricultural and Food Chemistry, 63*(11), 2956–2962.

Oddo, S., & LaFerla, F. M. (2006). The role of nicotinic acetylcholine receptors in Alzheimer's disease. *Journal of Physiology-Paris, 99*(2–3), 172–179.

Orr, S. K., Palumbo, S., Bosetti, F., Mount, H. T., Kang, J. X., Greenwood, C. E., & Bazinet, R. P. (2013). Unesterified docosahexaenoic acid is protective in neuroinflammation. *Journal of Neurochemistry, 127*(3), 378–393.

Parker, A., Fonseca, S., & Carding, S. R. (2020). Gut microbes and metabolites as modulators of blood-brain barrier integrity and brain health. *Gut Microbes, 11*(2), 135–157.

Prieto-Sánchez, M. T., Ruiz-Palacios, M., Blanco-Carnero, J. E., Pagan, A., Hellmuth, C., Uhl, O., & Koletzko, B. (2017). Placental MFSD2a transporter is related to decreased DHA in cord blood of women with treated gestational diabetes. *Clinical Nutrition, 36*(2), 513–521.

Ramel, S. E., & Georgieff, M. K. (2014). Preterm nutrition and the brain. *Nutritional Care of Preterm Infants, 110*, 190–200.

Réhault-Godbert, S., Guyot, N., & Nys, Y. (2019). The golden egg: Nutritional value, bioactivities, and emerging benefits for human health. *Nutrients, 11*(3), 684. https://doi.org/10.3390/nu11030684

Rizzetti, D. A., Altermann, C. D. C., Martinez, C. S., Peçanha, F. M., Vassallo, D. V., Uranga-Ocio, J. A., & Mello-Carpes, P. B. (2016a). Ameliorative effects of egg white hydrolysate on recognition memory impairments associated with chronic exposure to low mercury concentration. *Neurochemistry International, 101*, 30–37.

Rizzetti, D. A., Fernandez, F., Moreno, S., Ocio, J. A. U., Peçanha, F. M., Vera, G., & Wiggers, G. A. (2016b). Egg white hydrolysate promotes neuroprotection for neuropathic disorders induced by chronic exposure to low concentrations of mercury. *Brain Research, 1646*, 482–489.

Ross, R. G., Hunter, S. K., Hoffman, M. C., McCarthy, L., Chambers, B. M., Law, A. J., & Freedman, R. (2016). Perinatal phosphatidylcholine supplementation and early childhood behavior problems: Evidence for CHRNA7 moderation. *American Journal of Psychiatry*, *173*(5), 509–516.

Roth, W., Zadeh, K., Vekariya, R., Ge, Y., & Mohamadzadeh, M. (2021). Tryptophan metabolism and gut-brain homeostasis. *International Journal of Molecular Sciences*, *22*(6), 2973.

Santos-Hernández, M., Amigo, L., & Recio, I. (2020). Induction of CCK and GLP-1 release in enteroendocrine cells by egg white peptides generated during gastrointestinal digestion. *Food Chemistry*, *329*, 127188. https://doi.org/10.1016/j.foodchem.2020.127188

Schverer, M., O'Mahony, S. M., O'Riordan, K. J., Donoso, F., Roy, B. L., Stanton, C., & Cryan, J. F. (2020). Dietary phospholipids: Role in cognitive processes across the lifespan. *Neuroscience & Biobehavioral Reviews*, *111*, 183–193.

Schwarcz, R., Bruno, J. P., Muchowski, P. J., & Wu, H.-Q. (2012). Kynurenines in the mammalian brain: When physiology meets pathology. *Nature Reviews Neuroscience*, *13*(7), 465–477.

Singh, R. (2019). Role of tryptophan in health and disease: Systematic review of the anti-oxidant, anti-inflammation, and nutritional aspects of tryptophan and its metabolites. *World Heart Journal*, *11*(2), 161–178.

Smolders, L., de Wit, N. J. W., Balvers, M. G. J., Obeid, R., Vissers, M. M. M., & Esser, D. (2019). Natural choline from egg yolk phospholipids is more efficiently absorbed compared with choline bitartrate; Outcomes of a randomized trial in healthy adults. *Nutrients*, *11*(11), 2758.

Smuts, C. M., Huang, M., Mundy, D., Plasse, T., Major, S., & Carlson, S. E. (2003). A randomized trial of docosahexaenoic acid supplementation during the third trimester of pregnancy. *Obstetrics & Gynecology*, *101*(3), 469–479.

Statistics, F. I. F. o. A.-R. (2020, November 2). *Older Americans-key indicators of well-being*. National Institute on Aging. https://www.nia.nih.gov/news/new-report-indicators-well-being-among-older-americans

Stremmel, W., Hanemann, A., Ehehalt, R., Karner, M., & Braun, A. (2010). Phosphatidylcholine (lecithin) and the mucus layer: Evidence of therapeutic efficacy in ulcerative colitis? *Digestive Diseases*, *28*(3), 490–496.

Stringham, N. T., Holmes, P. V., & Stringham, J. M. (2019). Effects of macular xanthophyll supplementation on brain-derived neurotrophic factor, pro-inflammatory cytokines, and cognitive performance. *Physiology & Behavior*, *211*, 112650. https://doi.org/10.1016/j.physbeh.2019.112650

Sugino, H., Nitoda, T., & Juneja, L. (2018). General chemical composition of hen eggs. In *Hen eggs* (pp. 13–24). CRC press.

Sujak, A., Gabrielska, J., Grudziński, W., Borc, R., Mazurek, P., & Gruszecki, W. I. (1999). Lutein and zeaxanthin as protectors of lipid membranes against oxidative damage: The structural aspects. *Archives of Biochemistry and Biophysics*, *371*(2), 301–307.

Sukik, L., Liu, J., & Shi, Z. (2021). Association between egg consumption and cognitive function among Chinese adults: Long-term effect and interaction effect of iron intake. *British Journal of Nutrition*, 128(6), 1180–1189.

Tan, D., Yu, X., Chen, M., Chen, J., & Xu, J. (2017). Lutein protects against severe traumatic brain injury through anti-inflammation and antioxidative effects via ICAM-1/Nrf-2. *Molecular Medicine Reports*, *16*(4), 4235–4240.

Tran, P. V., Kennedy, B. C., Pisansky, M. T., Won, K.-J., Gewirtz, J. C., Simmons, R. A., & Georgieff, M. K. (2016). Prenatal choline supplementation diminishes early-life iron deficiency–induced reprogramming of molecular networks associated with behavioral abnormalities in the adult rat hippocampus. *Journal of Nutrition*, *146*(3), 484–493. https://doi.org/10.3945/jn.115.227561

Wallace, T. C. (2018). A comprehensive review of eggs, choline, and lutein on cognition across the life-span. *Journal of the American College of Nutrition*, *37*(4), 269–285.

Wallace, T. C., Blusztajn, J. K., Caudill, M. A., Klatt, K. C., & Zeisel, S. H. (2020). Choline: The neurocognitive essential nutrient of interest to obstetricians and gynecologists. *Journal of Dietary Supplements*, *17*(6), 733–752.

Wang, Y., Guan, X., Chen, X., Cai, Y., Ma, Y., Ma, J., & Bai, Y. (2019). Choline supplementation ameliorates behavioral deficits and Alzheimer's disease-like pathology in transgenic APP/PS1 mice. *Molecular Nutrition & Food Research*, *63*(18), 1801407.

Wang, J., Liao, W., Nimalaratne, C., Chakrabarti, S., & Wu, J. (2018). Purification and characterization of antioxidant peptides from cooked eggs using a dynamic in vitro gastrointestinal model in vascular smooth muscle A7r5 cells. *Npj Science of Food*, *2*(1), 7. https://doi.org/10.1038/s41538-018-0015-7

Wang, S., Sun-Waterhouse, D., Waterhouse, G. I. N., Zheng, L., Su, G., & Zhao, M. (2021). Effects of food-derived bioactive peptides on cognitive deficits and memory decline in neurodegenerative diseases: A review. *Trends in Food Science & Technology*, *116*, 712–732.

Woltil, H., Van Beusekom, C., Schaafsma, A., Okken, A., & Muskiet, F. (1999). Does supplementation of formula with evening primrose and fish oils augment long chain polyunsaturated fatty acid status of low birthweight infants to that of breast-fed counterparts? *Prostaglandins, Leukotrienes and Essential Fatty Acids (PLEFA)*, *60*(3), 199–208.

Wu, W., Li, Y., Wu, Y., Zhang, Y., Wang, Z., & Liu, X. (2015). Lutein suppresses inflammatory responses through Nrf2 activation and NF-κB inactivation in lipopolysaccharide-stimulated BV-2 microglia. *Molecular Nutrition & Food Research*, *59*(9), 1663–1673.

Xiao, N., Zhao, Y., Yao, Y., Wu, N., Xu, M., Du, H., & Tu, Y. (2020). Biological activities of egg yolk lipids: A review. *Journal of Agricultural and Food Chemistry*, *68*(7), 1948–1957.

Xu, Q., Fan, H., Yu, W., Hong, H., & Wu, J. (2017). Transport study of egg-derived antihypertensive peptides (LKP and IQW) using Caco-2 and HT29 coculture monolayers. *Journal of Agricultural and Food Chemistry*, *65*(34), 7406–7414.

Zahar, S., Schneider, N., Makwana, A., Chapman, S., Corthesy, J., Amico, M., & Hudry, J. (2022). Dietary tryptophan-rich protein hydrolysate can acutely impact physiological and psychological measures of mood and stress in healthy adults. *Nutritional Neuroscience*, *26*(4), 303–312.

Zaheer, K. (2015). An updated review on chicken eggs: Production, consumption, management aspects and nutritional benefits to human health. *Food and Nutrition Sciences*, *6*(13), 1208.

Zeisel, S. H., Mar, M. H., Howe, J. C., & Holden, J. M. (2003). Concentrations of choline-containing compounds and betaine in common foods. *Journal of Nutrition*, *133*(5), 1302–1307. https://doi.org/10.1093/jn/133.5.1302

Zeisel, S. H., & Niculescu, M. D. (2006). Perinatal choline influences brain structure and function. *Nutrition Reviews*, *64*(4), 197–203.

Zeisel, S. H., & Wurtman, R. J. (1981). Developmental changes in rat blood choline concentration. *Biochemical Journal*, *198*(3), 565–570. https://doi.org/10.1042/bj1980565

Zhang, B., Wang, H., Wang, Y., Yu, Y., Liu, J., Liu, B., & Zhang, T. (2019). Identification of antioxidant peptides derived from egg-white protein and its protective effects on H2O2-induced cell damage. *International Journal of Food Science & Technology*, *54*(6), 2219–2227. https://doi.org/10.1111/ijfs.14133

Zhao, H., Wang, C., Zhao, N., Li, W., Yang, Z., Liu, X., & Zhang, X. (2018). Potential biomarkers of Parkinson's disease revealed by plasma metabolic profiling. *Journal of Chromatography B*, *1081*, 101–108.

Zhu, C., Sawrey-Kubicek, L., Bardagjy, A. S., Houts, H., Tang, X., Sacchi, R., & Zivkovic, A. M. (2020). Whole egg consumption increases plasma choline and betaine without affecting TMAO levels or gut microbiome in overweight postmenopausal women. *Nutrition Research*, *78*, 36–41. https://doi.org/10.1016/j.nutres.2020.04.002

37

Egg Carotenoids and Eye Health

Katherine Alvarado, Claire Bordat, and Emmanuelle Reboul*

CONTENTS

37.1 Introduction: Background and Driving Forces

Carotenoids are pigments synthesized by photosynthetic bacteria, algae, plants, and fungi. These natural lipophilic compounds exhibit red, orange, and yellow colors. They are widely known for their antioxidant role, but some also have a pro-vitamin A activity. Carotenoids such as lutein and zeaxanthin were shown to play specific roles in human eye health. Indeed, these carotenoids can accumulate in the eye, where they display a protective role against both oxidation and cell degeneration.

As humans cannot synthesize carotenoids, therefore they must be obtained through the daily diet. A common source of carotenoids is egg yolk (Sirri et al., 2007). Indeed, egg yolk color is due to the presence of carotenoids such as lutein, zeaxanthin, canthaxanthin, capsanthin, capsorubin, citranaxanthin, or β-apo-8′-carotenoic acid ethyl esters (Wenzel et al., 2011; Çalışlar, 2019). Egg yolk can also contain astaxanthin (Honda et al., 2021) and lycopene (Kang et al., 2003). The amounts of carotenoids in egg yolk depend on hens' diet and the efficiency of carotenoid transfer into the egg. The main current sources of these carotenoids in poultry diet are marigold extract, paprika, corn, and synthetic carotenoids (Niu et al., 2008a). This chapter will review all these aspects, as well as carotenoid location, function, and importance in human eyes.

37.2 Chemical Structures of Carotenoids Found in the Human Eye (Lutein, Zeaxanthin, Canthaxanthin, Astaxanthin, and Lycopene)

Carotenoids can be divided into 2 subclasses according to their structure. While carotenes are pure hydrocarbons, xanthophylls contain one or more oxygens included in hydroxy, methoxy, carboxy, keto, or epoxy functional groups. Table 37.1 shows the chemical structure of different carotenoids found in the eye.

The human eyes mainly contain 3 xanthophylls, chemically identified by Bone et al.: lutein, *meso*-zeaxanthin and zeaxanthin, which share the same chemical formula "$C_{40}H_{56}O_2$" (Bone et al., 1993). Xanthophylls display a backbone containing 9 alternating conjugated double carbons and single carbon bonds, which enable their photoactivity and antioxidant properties (Arteni et al., 2015). The chain ends with an ionone ring to which oxygenated groups can be attached. Lutein and zeaxanthin present 4 methyl groups that are attached as side groups to the backbone. The difference between lutein and zeaxanthin is the location of one double bond in the cyclic end-group, which is shifted and no longer conjugated in lutein, and the location of the 3' carbon with opposite configuration (Ermakov et al., 2004). The extra-conjugated double-bond stabilizes zeaxanthin and improves its antioxidant properties compared to lutein (Mortensen & Skibsted,

* Corresponding author.

DOI: 10.1201/9781003254430-45

TABLE 37.1

Structure and Chemical Formula of Carotenoids in Both Eye and Eggs

Carotenoid	Structure	Molecular Formula
Lutein	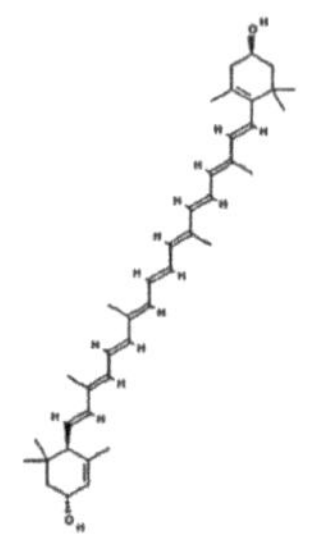	$C_{40}H_{56}O_2$
Zeaxanthin	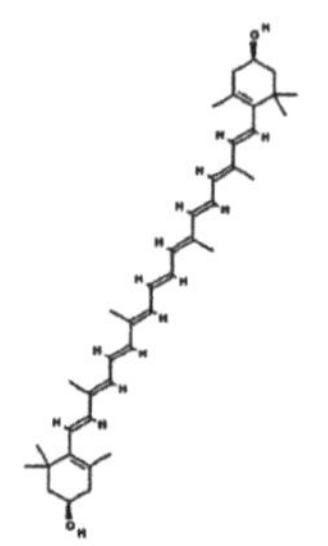	$C_{40}H_{56}O_2$
Meso-zeaxanthin	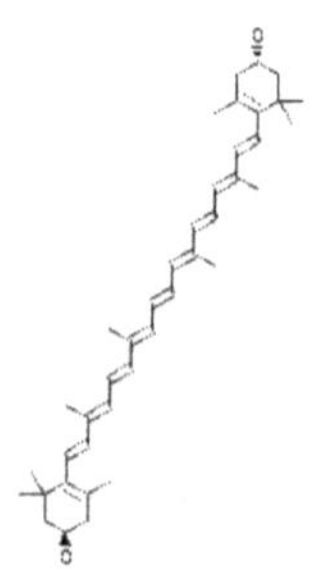	$C_{40}H_{56}O_2$
Canthaxanthin	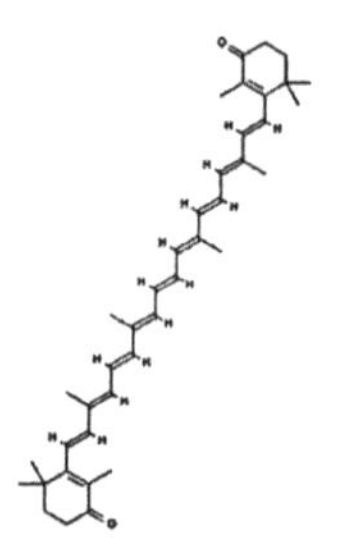	$C_{40}H_{52}O_2$
Astaxanthin	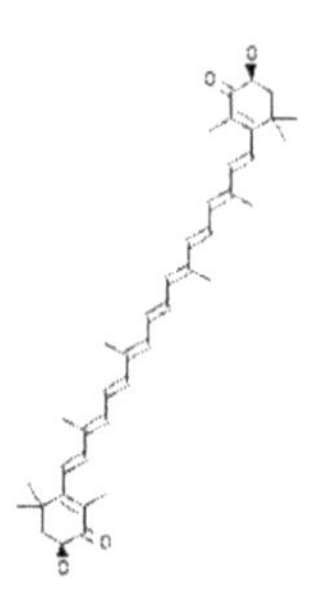	$C_{40}H_{52}O_4$
Lycopene	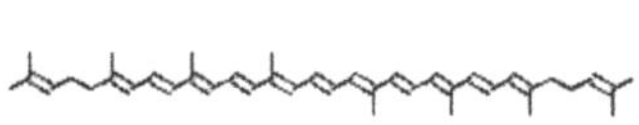	$C_{40}H_{56}$

Source: All the references were obtained from the database of the National Library of Medicine Pub Chem.

1997). The lutein present in the eye consists of a single stereoisomer [(3R,3′R,6′R)-β, ε-carotene-3,3′-diol], whereas zeaxanthin, a stereoisomer of lutein (Zaheer, 2017), consists in 3 stereoisomers: RR-zeaxanthin [(3R,3′R)-β, β-carotene-3,3′-diol], SS-zeaxanthin [(3S,3′S)-β, β-carotene-3,3′-diol] and RS-zeaxanthin or *meso*-zeaxanthin[(3R,3′ S)-β, β-carotene-3,3′-diol] (Bone et al., 1993). Usually, all carotenoids found in the eye display a *trans* configuration (Landrum & Bone, 2001).

The eye also contains 3 other xanthophylls in small amounts: β-cryptoxanthin, canthaxanthin, and astaxanthin. These xanthophylls are present only if they are highly consumed through the diet. β-cryptoxanthin ($C_{40}H_{56}O$) is a monohydroxylated derivative of β-carotene ((3R)-β, β-caroten-3-ol) with a pro-vitaminic A activity (Heying et al., 2018). Like β-carotene, it has 11 conjugated double bonds, and 2 β-rings, but contains a hydroxyl group at the number 3 carbon atom of one β-ring. Similarly, canthaxanthin, ($C_{40}H_{52}O_2$) is closely related to β-carotene (β,β-carotene-4,4′-dione) but contains 2 ketone groups, which makes it a diketone-carotenoid. The keto groups are substituted at the 4 and 4' positions of the β-ionone rings (Gupta et al., 1985). Astaxanthin ($C_{40}H_{52}O_4$) is a keto-carotenoid (β-carotene-4,4'-dione) with 2 asymmetric carbons located at the 3,3′ positions of the β-ionone ring with a hydroxyl group (-OH) on either end of the molecule, which provides both lipophilic and hydrophilic properties. Each double polyene bond can exist either as cis- or trans-geometric isomers. Further, due to the presence of 2 stereogenic carbon atoms at the 3,3′ positions, 3 different stereoisomers of astaxanthin, including a pair of enantiomers (3R,3′R- and 3S,3′S-astaxanthin) and an optically inactive mesoform (3R,3′S-astaxanthin) may be found (Dose et al., 2016).

Finally, lycopene can be detected in very small amounts if not at all, depending on dietary consumption. Lycopene is a carotene and has the exact same formula as β-carotene ($C_{40}H_{56}$), but lacks the β-ionic ring structure and pro-vitamin A activity. Lycopene is a symmetrical tetraterpene hydrocarbon polyene chain, assembled from 8 isoprene units, with eleven conjugated and 2 unconjugated double bonds that can be easily attacked by electrophilic reagents (Khachik et al., 2002a).

37.3 Location of Eye Carotenoids

Figure 37.1 presents the carotenoids found in the different eye structures. Most of the carotenoids in the human eye are concentrated in the *macula lutea*, a yellow oval-shaped region of the retina.

The macula is responsible for the central, high-acuity, color vision. It is essential for processing the fine details of an image and thus for facial recognition, reading, and driving (Li et al., 2020). Macula's yellow color is derived from its content in lutein, zeaxanthin, and *meso*-zeaxanthin in the photoreceptors (Beatty et al., 1999).

We can distinguish 4 parts on the macula. First, the concave depression at the center of the macula is called the *fovea centralis* (Snodderly, 1995). The tiny depression in the very center of fovea is known as the *umbo*. Around the fovea centralis is the *parafovea*. Finally, the *perifovea*, i.e. the outermost region of the macula, is considered as a transition zone between the

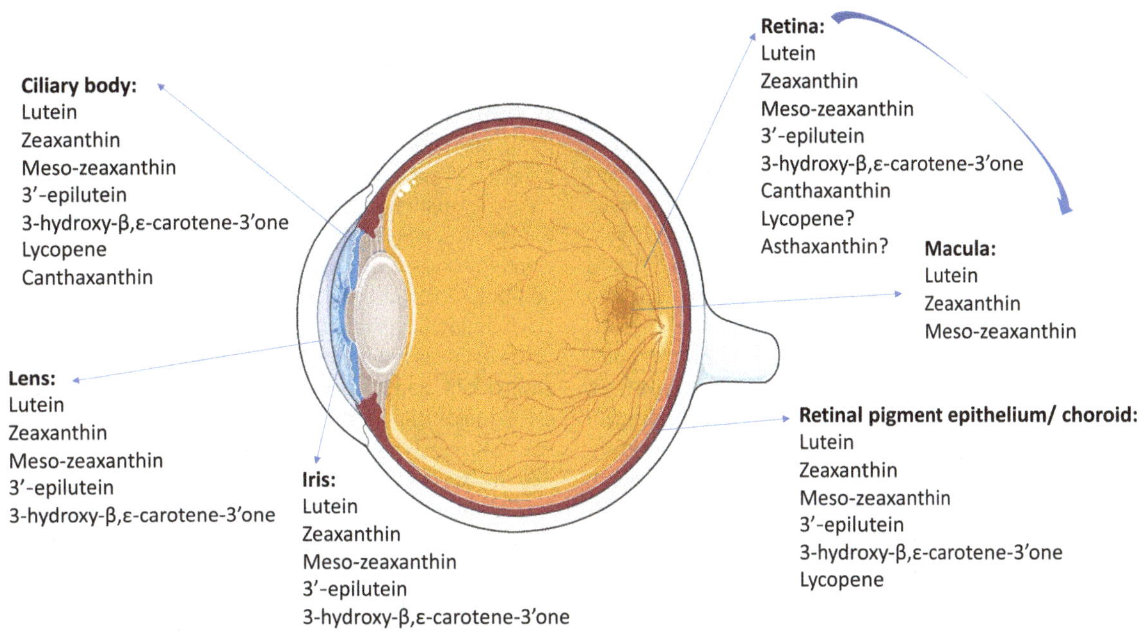

FIGURE 37.1 Scheme of the human eye, structures where carotenoids are present. Image created with Bernstein et al. (2001) and updated with literature. Meso-zeaxanthin, 3'-epilutein and 3-hydroxy-ß,ε-carotene-3'one are metabolites of lutein/zeaxanthin. The figure was partly generated using Servier Medical Art, provided by Servier, licensed under a Creative Commons Attribution 3.0 unported license.

highly specialized central regions and the periphery of the retina (Provis et al., 2005).

Mapping lutein and zeaxanthin is difficult due to their spatial distribution. The macular pigment density distribution within the retina was first described in 1984 in the primate eye (*Macaca fasciculatis, Macaca mulatta, Macaca nemetrina, Cebus apella*), using micro-spectrophotometry (Snodderly et al. 1984a; Snodderly et al. 1984b). The recent description of human macula is consistent with the data obtained in monkeys. A first analysis of the radial distribution of macular pigments showed a predominance of zeaxanthin in the very center of the *fovea centralis*, with radial decreasing concentrations of zeaxanthin and increasing concentrations of lutein in *parafovea* and *perifovea* (Widomska & Subczynski, 2014). This distribution creates an inverse ratio according to the degree of eccentricity, with a high zeaxanthin: lutein ratio (around 2.4:1) in the inner disk, and a high lutein: zeaxanthin ratio in the outer annulus (around 2:1) (Bernstein et al., 2001). Analysis of the vertical distribution pattern of macular pigments within the photoreceptor axons indicates high concentrations in the inner layers of the retina, and low concentrations at the level of the retinal pigment epithelium (RPE) and photoreceptor outer segments (Bernstein et al., 2001). In a recent study, mapping of human macular pigments distribution from the inner to outer limiting membranes in retinas from human donors was performed using confocal resonance Raman microscopy and was consistent with previous studies published by Snodderly et al. (1984b) or Bernstein et al. (2001). This study also confirmed that zeaxanthin is localized to the foveal center at high concentrations from the inner to the outer layers, while lutein is diffusely distributed across the inner retina at a much lower concentration. This suggests that zeaxanthin could be more important than lutein in the human macular health, because of its higher concentrations and lower diffusion compared to lutein (Li et al., 2020). Moreover, the concentration of macular pigments in the fovea is 100-fold higher than the peripherical retina.

Concerning the other carotenoids mentioned in this chapter, their location in the human eye is less studied because they are generally found in very small amounts or even absent. Lycopene, being highly present in eggs, tomatoes, and tomato-food-based products, is the most likely to be found in the human eye. A study by Khachik et al. found an average of 8 ng lycopene per 0.20 g of tissue in pooled extracts of human ciliary body and RPE/choroid, but the putative presence of lycopene in the retina is not documented (Khachik et al., 2002a). Canthaxanthin and astaxanthin, being mostly consumed as food additives, are less likely to be found in the human eye. However, they can accumulate in the eye in case of high consumption. Hence, canthaxanthin was shown to accumulate in the form of crystals in both human retina and ciliary body after a chronic overexposure (Daicker et al., 1987; Aguilar et al., 2010). Astaxanthin's precise location in the human eye is still unknown. However, an animal study demonstrated that astaxanthin can cross the blood–brain barrier and, similarly to lutein, can be found in the retina of mammals. Indeed, after 6 consecutive intraperitoneal injections of astaxanthin at 12-hour intervals, and at a concentration of about 37.5 mg/kg of body weight, the average concentration of astaxanthin in the rat retina was about 0.17 µg/mg of wet tissue 6 hours after the last injection (Tso & Lam, 1996).

37.4 Egg Yolk: A Source of Carotenoids

In nature, birds, including chickens, cannot produce carotenoids on their own and need to absorb them through their diet. Carotenoids in egg yolk are necessary for the incubation of the embryo (Çalışlar, 2019). Main carotenoids naturally found in most egg yolk analysis are xanthophylls, but other carotenoids can be found depending on the diet provided to the chicken (Moreno et al., 2016). The profile of carotenoids in egg yolk is highly dependent on the hens' diet. Indeed, chicken and hens can be specifically supplemented in carotenoids for commercial purpose, as meat and skin color are associated with hens' health and good rearing methods (Marounek & Pebriansyah, 2018). Carotenoid supplementation of poultry feed is allowed for all poultry categories, but for different reasons: either to optimize meat and skin pigmentation in broilers or to optimize yolk color in laying hens. Hence, by feeding the poultry with carotenoids, egg yolk with different tones (pink, pink orange, orange red, and yellow) can be produced depending on the targeted consumer populations, who tend to prefer more intense coloration of yolks (Marounek & Pebriansyah, 2018). Moreover, the supplementation of poultry diets with carotenoids increases the oxidative stability of poultry products.

Table 37.2 shows some of the carotenoids used for poultry pigmentation. In the European Union, carotenoids approved as feed additives for hens are lutein, zeaxanthin, β-cryptoxanthin, canthaxanthin, capsanthin, capsorubin, β–apo-8′-carotenoic acid ethyl ester, citranaxanthin, lycopene, and astaxanthin. In Canada, only β-apo-8′-carotenoic acid ethyl ester, lutein, and canthaxanthin are allowed. These regulatory differences explain the large discrepancy in carotenoid levels according to the origin of the eggs studied. Usually, lutein, zeaxanthin, and capsanthin are obtained from plant sources like paprika, marigold flower (*Tagetes erecta*), alfalfa (*Medicago sativa*), yellow corn, tomato, lucerne, algae, or commercial synthetic additives (Zaheer, 2017). Red pepper, more commonly known as paprika, is a natural source of capsanthin, but is less mentioned in literature for poultry feed (Schlatterer & Breithaupt, 2006). Other carotenoids are produced via chemical synthesis. When hens and eggs are organic, synthetic chemical products are not allowed. Organic egg yolk thus does not contain carotenoids such as canthaxanthin, which is mainly found in marine life (de Miguel et al., 2001). However, organic hens have access to dark green leafy vegetation when outside, which increases the amounts of lutein and zeaxanthin in egg yolk compared to caged laying hens (Schlatterer & Breithaupt, 2006). However, the amounts of carotenoids provided to these chickens may not be as precise as when given synthetic compounds (Zaheer, 2017). Thus, due to different legislations, different rearing methods and multiple possibilities of hen feed supplementations, egg yolk carotenoid profile can vary considerably from one brand to another, making it difficult to establish "normal" carotenoids composition (Mugnai et al., 2009; Nimalaratne et al., 2013). However, it is assumed that the amount of xanthophylls in eggs ranges from 0.3 to 0.5 mg per yolk with just over half present as lutein, but under the influence of the diet of laying hens, the lutein content can be 5–8 times higher (Steinberg et al., 2000). On average, egg yolk contains fairly high amounts of lutein and zeaxanthin at concentrations of 1,282 ± 182 (up to 2,000 in some studies) and 640 ± 140 μg/100 g of egg yolk, respectively (Nimalaratne et al., 2013).

The amount of carotenoid and color tone of egg yolk also depend on carotenoid digestibility, transfer, metabolism, and deposition in the target tissue, which in turn depend on carotenoid amount and composition in poultry diet. A study demonstrated that a 20-week supplementation with 40 ppm, 60 ppm, and 80 ppm of apo-carotenoid-esters from corn or synthetic sources gives a more intense pigmentation than 120 ppm, 180 ppm, or 240 ppm of marigold extract (Sirri et al., 2007). It is generally acknowledged that carotenoid polarity is a key parameter influencing egg yolk color. A study reported that polar carotenoids, such as xanthophylls, are more efficiently absorbed by chickens and transferred to the yolk than non-polar carotenoids, which increases pigmentation of muscle, skin, and egg yolk (Na et al., 2004).

Lycopene is not widely used in poultry feed, but it is a strong antioxidant, and it is well deposited in egg yolk when hens fed diets containing lycopene in a range from 65 to 650 mg/kg. An optimum was observed for a lycopene concentration of 420 mg/kg in the diet, which resulted in a yolk lycopene concentration of 4 mg/kg (Olson et al., 2008). Similarly, a study showed that adding tomato powder, a natural source of lycopene, at 5 or 10 g per kg of the basal diet of laying hens increased serum and yolk lycopene, and improved egg production persistency (Akdemir et al., 2012).

Canthaxanthin being mostly present in marine life (such as algae, shrimps, crabs, crayfishes, koi carp skin, wild salmon, and flamingo feathers), it is not present in hens' natural nutrition, and its presence in egg yolk is only due to a supplemented diet. As a food additive, its content in the feeding of laying hens and their eggs is determined mainly by legislative rules. Therefore, the Scientific Committee on Food (SCF) decided in 1997 to fix canthaxanthin acceptable daily intake level (ADI) to 0.03 mg/kg body weight for humans (Scientific Committee on Food, 1997) and assumed that the reliable residue level of canthaxanthin in eggs was 0.2 mg per egg (Scientific Committee on Food, 1997). In 2002, the European Union (EU) Scientific Committee for Animal Nutrition (SCAN) suggested a maximum supplementation of 8 mg canthaxanthin/kg diet for laying hens (SCAN (Scientific Committee on Animal Nutrition), 2002). A study estimated the deposition rate of canthaxanthin to be 40% independently of dietary supplementation level. Hence, the amount of canthaxanthin in yolk increased from 0.026 to 2.5 mg per egg when laying hens consume diet containing 0.5 to 64 mg canthaxanthin per kg of diet, respectively (Grashorn & Steinberg, 2002). More recently, a German study measured canthaxanthin concentration in egg yolks from different sources and estimated a concentration ranging from zero in the yolk from organic, free-ranged chicken to up to 6.92 ± 1.53 mg/kg of yolk from battery chickens, which corresponds to an average of 0.12 mg/egg yolk (Islam & Schweigert, 2015).

Similarly, astaxanthin is used commercially, mostly in the feed industry. Along with canthaxanthin, it is the most important and the most expensive pigment used in aquaculture for the pigmentation of salmon, trout, and shrimp meat, which affects consumers' preferences around the world. It is only recently that astaxanthin was found to have more effective antioxidant properties compared to β-carotene and lutein (Goto et al., 2001). Its use as a potential supplementation of laying-hen feed for egg yolk nutrient enrichment is even more recent, mostly because the first authorization of astaxanthin in animal and fish feeds at the maximum level of 8 mg/day was first granted in Sweden and by the FDA in 1995 (Fang & Chiou, 1996).

TABLE 37.2
Characteristics of the Main Carotenoids Used in Poultry Feeding

Coloration of Yolk	Name of the Molecule	European Community Number[a]	Chemical Abstracts Service Number[a]	Source Used for Poultry Feeding	References
Yellow Orange Orange-red	β-carotene	E160a	7235-40-7	Marigold (*Tagetes erecta*) Tomato powder Carrots[b]	(Sirri et al., 2007; Hammershøj et al., 2010)
Yellow Yellow-orange	Lutein	E161b	127-40-2	Marigold (*Tagetes erecta*) Tomato powder Paprika	(Niu et al., 2008b; Marounek & Pebriansyah, 2018)
Yellow Dark yellow-orange	Zeaxanthin	E161h	144-68-3	Marigold (*Tagetes erecta*) Corn Tangerine Paprika	(Schlatterer & Breithaupt, 2006; Lokaewmanee et al., 2009; Hammershøj et al., 2010; Marounek & Pebriansyah, 2018)
Light yellow-orange Dark orange-red	Canthaxanthin	E161g	514-78-3	Most used: synthetic from β -carotene Exists in natural form from algae and mushrooms, bacteria, crustaceans	(Marusich et al., 1960; Aguilar et al., 2010; Faruk et al., 2018; Marounek & Pebriansyah, 2018)
Blood-red	Astaxanthin	E161j	472-61-7	Microalgae (*Haematococcus pluvialis*) Yeast fungus (*Xanthophyllomyces dendrorhous*)	(Nimalaratne et al., 2012)
Light yellow-orange Dark orange Red	Capsanthin	E160c	465-42-9	Paprika	(Marusich et al., 1960; Marounek & Pebriansyah, 2018)
Red	Capsorubin	E160c	470-38-2	Paprika	(Marounek & Pebriansyah, 2018)
Yellow	Apo-Ester (β-apo-8'-carotenoic acid ethyl ester)	E160f	1109-11-1	Most used: chemical synthesis	(Sirri et al., 2007; Bories et al., 2009; Marounek & Pebriansyah, 2018)
Dark yellow Red-orange	β-apo-8'-carotenal	E160e	1107-26-2	Orange Tangerine Spinach Alfalfa Synthetic Endogenous formation	(Marusich et al., 1960; Bunnell et al., 1961; Fletcher et al., 1978; Rodriguez-Amaya, 2001; Bories et al., 2009)
Orange-red Red	Citranaxanthin	E161i	3604-90-8	Synthetic	Bories et al., 2009; Joint FAO/WHO Expert Committee on Food Additives & World Health Organization, 1988; Schlatterer & Breithaupt, 2006; Marounek & Pebriansyah, 2018)
Yellow-orange Orange-red	Lycopene	E160d	502-65-8	Tomato powder	(Suarez, 1969; Kang et al., 2003; Olson et al., 2008; Marounek & Pebriansyah, 2018)

[a] The numbers are taken from the pages of The European Community and Cas Common Chemistry for 2021. The existence of the number is not a guarantee that the additive is allowed. Laws can evolve in time and add or delete additives from allowed lists.

[b] Carrot can alter the egg mass production and feed intake but can lead to a higher color tone (Hammershøj et al., 2010), while carrot leaves seems to be a good option to feed hens (Titcomb et al., 2019).

However, recent scientific research has shown that the pigment has a positive influence on the color of egg yolk, as well as the skin and meat tissue (Walker et al., 2012). As for canthaxanthin, astaxanthin concentration in eggs depends directly on the amounts in the hen's diet. Indeed, in 2020, a study by Gao et al. reported that the astaxanthin egg yolk content gradually increased with increasing astaxanthin concentrations in hens' diet (from 25 to 100 mg/kg of diet). The highest level of astaxanthin enrichment was observed in the 100-mg/kg of diet group at the end of a 6-week supplementation, with a maximum concentration of 44.20 mg/kg of egg yolk (Gao et al., 2020). Likewise, in 2021, another study reported that a 28-day supplementation of laying hens with 8 mg/kg of diet, resulted in an astaxanthin content of 6.65 mg/kg of egg yolk when the Z-Isomer ratio was over 50% (Honda et al., 2021). Still in 2021, a study reported that after being fed for 24 weeks with a diet containing 42.6 mg/kg of astaxanthin, hens' eggs content was 20.23 mg/kg of egg yolk and enhanced egg fortification (Dansou et al., 2021). The deposition rate of astaxanthin into eggs has been reported to be between 3.6% (Akiba et al., 2000) and 14.4% (Moreno et al., 2020), and tends to be highly dependent on the isomers used for supplementation.

37.5 Effect of Egg Cooking and Processing

Eggs are usually cooked before consumption. The egg matrix can thus be affected by processing prior to consumption. Electron-rich carotenoids suffer from oxidation at high temperatures during long cooking procedures, during exposure to light or in the presence of oxygen, acids, and interaction with metals and radical species (Boon et al., 2010). Cooking and other procedures can also affect bioavailability of carotenoids in egg yolk (Nimalaratne et al., 2015). In general, carotenoids that change their natural form from *trans* to *cis* during cooking display modification in their bioavailability, crystallization capacity, and solubility, and lose their color (Aguilar et al., 2010).

Freeze-dried processing tends to decrease xanthophyll content in eggs, but carotenoids from fried-dried eggs are stable at room temperature if stored in the dark (Wenzel et al., 2011). Boiling can decrease the total carotenoid of egg yolk by 2.18% (Painsi et al., 2019). Compared to a raw egg yolk, cooking processes such as boiling, microwave cooking, and fried eggs can decrease total lutein content by 13–8%; total zeaxanthin by 5.9–11.6%; total canthaxanthin by 5.7–6.6%; and all trans-β-apo-8′-carotenoic acid ethyl ester by 6.9–9% (Nimalaratne et al., 2012). Losses are quantitative but not qualitative, as stereoisomers do not change upon heat treatment (Nimalaratne et al., 2013).

37.6 Egg Carotenoid Bioavailability and Metabolism from the Human Intestine to the Eye

37.6.1 Egg Carotenoid Digestion, Absorption, and Transport to Tissues

Human being, as other animals, cannot synthesize *de novo* carotenoids and need to obtain them from their diet (Abdel-Aal et al., 2013). Being lipophilic molecules, carotenoid absorption follows the same pathways than dietary lipids. It starts with carotenoid extraction from the food matrix, and their dissolution in the fat phase of the meal by chemical and mechanical disruption (Furr & Clark, 1997).

This fat phase is then converted into a lipidic emulsion with lipid droplets containing dissolved carotenoids. These droplets are first partially hydrolyzed by the gastric lipase in the stomach, then completely hydrolyzed by pancreatic lipases in the duodenum, which also allow the hydrolysis of carotenoid esters if any. Lipid digestion products and carotenoids are then incorporated into mixed micelles together with bile salts (Reboul, 2019). The degree of carotenoid solubilization and incorporation into mixed micelles is one of the crucial factors in determining the bioavailability of carotenoids. Dietary lipids, and especially phospholipids and other lipids from egg yolk, enhance carotenoid absorption by facilitating micelle formation. This explains why lutein and zeaxanthin from eggs have a high bioavailability compared to lutein and zeaxanthin from fruits and vegetables (Handelman et al., 1999).

Once incorporated in mixed micelles, carotenoids are absorbed by enterocytes in the small intestine through different mechanisms. The first mechanism described and demonstrated was simple diffusion (Hollander & Ruble, 1978). However, recent studies reported the existence of a receptor-mediated transport of xanthophylls at the apical membrane of enterocytes, with strong evidences for the involvement of SR-BI (Scavenger Receptor class B Type I), NPC1L1 (Niemann-Pick C1 Like 1), and CD36 (CD36 molecule) in this process (Reboul, 2019). Interestingly, SR-BI preferentially facilitates the absorption of xanthophylls (such as lutein) over carotenes (such as α- and β-carotene) (Kim et al., 2010). Lycopene absorption seems to be mediated by SR-BI, but not by NPC1L1 (Moussa et al., 2008). After carotenoids are taken up by the intestinal cells, they are packed into chylomicrons and secreted to the lymph for transport to the blood and liver (Reboul, 2019). The liver is one of the main organs that store lipophilic nutrients like carotenoids. If not metabolized in the liver, carotenoids are secreted back to the bloodstream associated with VLDLs (very low-density lipoproteins), further metabolized into LDLs (low-density lipoproteins) that transport the majority of plasma carotenoids (55% of total carotenoids) to be delivered to target tissues (Kotake-Nara & Nagao, 2011). Moreover, 33% of total carotenoids can also be found in HDLs (high-density lipoproteins) (Clevidence & Bieri, 1993). Interestingly, there is a plasmatic segregation of polar (xanthophylls) and nonpolar (carotenes) carotenoid molecules, with carotene being primarily associated with the LDLs, and xanthophylls being carried by the HDLs (Parker, 1996). Usually, lutein and zeaxanthin are incorporated into LDLs and HDLs, in ratios ranging from 2:1 to 4:1 (lutein: zeaxanthin) (Abdel-Aal et al., 2013). Conversely, lycopene was found to concentrate in the LDLs and VLDLs, but not in the HDL fraction of the serum (Stahl & Sies, 1996).

Uptake, transport, and stabilization of xanthophylls in the human retina are important components of a complex multistep process that is far from being understood. SR-BI likely facilitates the uptake of lutein, zeaxanthin, and other carotenoids into the retinal cells (During et al., 2008). Interphotoreceptor Retinoid Binding Protein (IRBP), known to be involved in the visual cycle, may also facilitate the transport of lutein and zeaxanthin to the retinal cells via CD36. However, the key binding proteins involved in carotenoid transport in the retina are GSTP1 (Glutathione S-Transferases P1) and StARD3 (StAR-related lipid transfer Domain 3). GSTP1, which plays an important role in detoxification and antioxidant system, was the first specific carotenoid binding protein to be identified in the human retina, with a high affinity for zeaxanthin and *meso*-zeaxanthin (Bhosale & Bernstein, 2005). Interestingly, GSTP1 can bind and stabilize zeaxanthin and form complexes with high antioxidant activities (Bhosale & Bernstein, 2005). StARD3, a sterol-binding protein involved in lipid trafficking, was also identified as a human retina's lutein-binding protein (Vachali et al., 2013). Regarding other xanthophylls found in the eye, no specific binding protein has been yet described and very little is known about their metabolism. Interestingly, unlike lutein and zeaxanthin, *meso*-zeaxanthin is not detectable in human plasma if not specifically added in the diet (Khachik et al., 2002b), which implies that isomerization from dietary precursors occurs in ocular tissues (Gorusupudi et al., 2016).

The number of publications concerning canthaxanthin metabolism is rather limited. Like other carotenoids, canthaxanthin follows the previously described metabolism, and is carried in the blood by LDL and HDL particles (Paetau et al., 1997). The localization of canthaxanthin in lipid membranes and its interaction with lipid molecules seem to play an important role in many of the biological functions of canthaxanthin (Sujak, 2009). Canthaxanthin is assumed to be stored in the liver and fat tissue. Retinal accumulation has been reported after high oral consumption (from 10 times to more than 1,000 times over the acceptable daily intake level) in the form of crystalline deposits in a reversible manner, but the transport mechanism involved is not well understood yet (Aguilar et al., 2010). Astaxanthin is also assumed to follow the carotenoid metabolism and transport processes described above.

Lastly, most carotenoids are stored in the liver and adipose tissue, but some carotenoids may specifically accumulate elsewhere. For example, lycopene is mostly found in the prostate and adrenal glands, while lutein and zeaxanthin are concentrated in the retina (Khachik et al., 1995).

37.6.2 Egg Carotenoid Bioavailability

The bioavailability of carotenoids in general depends on their bioaccessibility, which is affected by several factors such as fat content during or after processing or in the meal, food matrix characteristics, dietary fiber content, as well as processing. Usually, levels in plasma and human tissues reflect dietary intake. Human *in vivo* studies have assessed plasma and retina responses to egg lutein and zeaxanthin, to give some information on bioavailability. A study by Stringham & Stringham reported that an intake of 6.2 mg lutein + 0.7 mg zeaxanthin + 0.5 mg *meso*-zeaxanthin a day, was the most efficient dose regarding serum response, whereas the macular pigment optical density (MPOD) was efficiently increased after intakes of 10.9 mg lutein + 1.3 mg zeaxanthin + 0.9 mg *meso*-zeaxanthin (Stringham & Stringham, 2016). Several studies confirm that serum lutein levels in men were highest after eating eggs compared to supplements and spinach, because the bioavailability of these compounds from eggs is higher than from vegetable sources, most likely due to the fat content (Chung et al., 2004; Goodrow et al., 2006). Goodrow *et al.* estimated that the consumption of 1 egg per day for more than 5 weeks increases serum lutein levels by 26% and zeaxanthin levels by 38% (Goodrow et al., 2006). Other research reported that the daily intake of 3 eggs for 12 weeks increased serum lutein and zeaxanthin by 21% and 48%, respectively, in 20 adults (Blesso et al., 2013). However, these subjects suffered from metabolic syndrome meaning that results must be taken carefully, but these data still confirm the high lutein and zeaxanthin bioavailability in eggs. A study also assessed the bioavailability of xanthophylls from eggs enriched in lutein and zeaxanthin, compared to standard eggs. This study highlighted that the consumption of lutein- or zeaxanthin-enriched eggs, or a lutein-enriched egg-yolk-based buttermilk beverage, significantly increased serum lutein levels by 76% and 77%, respectively. A high increase in the serum zeaxanthin concentration was also observed in individuals receiving zeaxanthin-enriched eggs (Kelly et al., 2014).

Lycopene is the most predominant carotenoid in human plasma, with a plasma depletion half-life of about 2–3 days (Rock et al., 1992). Unlike other carotenoids, supplementation with lycopene is less certain to restore plasma concentrations (Carughi & Hooper, 1994), and a lycopene-free diet even for 2 weeks leads to a drastic drop in blood levels (Rao & Agarwal, 1998). In humans, lycopene absorption is in the range of 10–30% (Stahl & Sies, 1992).

Because canthaxanthin is not naturally found in fruits and vegetables unlike other carotenoids, its bioavailability is not extensively studied. Canthaxanthin is present in egg yolk only because of its application as a feed additive (E161g). Most of the studies concerning canthaxanthin bioavailability and metabolism thus focus on the evaluation of its safety as a food additive. An unpublished report submitted to the World Health Organization by F. Hoffmann-La Roche & Co reported a study conducted on 26 volunteers receiving respectively a single oral dose of 75 (n = 10) or 150 (n = 16) mg canthaxanthin, mixed with full milk cream. This study calculated a half-life of 4.5 days in both groups and estimated the absorbed dose to be 12% and 9% in each group (Joint FAO/WHO Expert Committee on Food Additives & World Health Organization, 1988). The report described a second study conducted on 10 volunteers who received doses of 1 mg for 5 days, or 8 mg for 2 days, 6 times a day, corresponding to a total dose of 30 mg or 96 mg, respectively. The average plasma concentrations of canthaxanthin after daily ingestion of 6 mg (6 times 1 mg) or 48 mg (6 times 8 mg) were 1.8 or 10.3 mg/L, respectively, with maximum serum concentrations 48 hours after ingestion and a half-life of 5.3 days. The absorption of canthaxanthin was no more than 34% of a 1 mg oral daily dose, and about 60% of the absorbed dose was transferred to fat tissue (Aguilar et al., 2010). Moreover, toxicokinetic studies in monkeys indicated that only 3–9% of orally administered radiolabeled canthaxanthin is absorbed, with fecal excretion being the major route of elimination (85 –89% of the dose).

Finally, astaxanthin absorption is assumed to be increased when consumed with dietary oils (Østerlie et al., 2000). Astaxanthin bioavailability in human plasma was studied in 3 middle-aged male volunteers (37–43 years). The oral administration of a single dose of 100 mg of a mixture of astaxanthin

TABLE 37.3
Half-Life of Carotenoids and % of Absorption in the Human Body

Carotenoid	Half-life	% of Absorption Bioaccessibility	References
Lutein	10–76 days	62–79	(Evans, 2006; Goñi et al., 2006; Bories et al., 2009; Juturu et al., 2016; Moran et al., 2018)
Zeaxanthin	12–76 days	1–111	(Thürmann et al., 2002; Thürmann et al., 2005; Moran et al., 2018; Tudor & Pintea, 2020)
Lycopene	2–3 days	10–30	(Stahl & Sies, 1992)
Canthaxanthin	5 days	9–34	(Aguilar et al., 2010)
Astaxanthin	11.4–32.1 hours	unknown	(Østerlie et al., 2000)

E/Z isomers successfully led to their absorption into plasma. However, this study being one of the few investigating astaxanthin plasma kinetics could not give definitive values regarding elimination half-life because of large variations (range 11.4–32.1 hours). Astaxanthin was present in all lipoprotein fractions in varying amounts at different times post dosing, although mainly in VLDLs/CM (36–64%) (Østerlie et al., 2000). A recent study reported that astaxanthin accumulation in rat plasma and liver was observed after feeding *Haematococcus* biomass as a source of astaxanthin (Ranga Rao et al., 2010).

Half-life of carotenoids and their percentage of absorption in the human body are reported in Table 37.3.

37.7 Egg Carotenoids and Visual Function

Lutein and zeaxanthin are the most studied carotenoids in the eye. These carotenoids are known to be interesting antioxidants for human health (Landrum & Bone, 2001). However, other functions have been discovered such as photoprotection against blue light that can directly reach the retina (390–540 nm), penetrating the cornea and the lens, only blocking ultraviolet (UV) below 300 nm and between 300 and 400 nm, respectively. This high-energy blue light can activate potent photosensitizers, and consequently generate reactive oxygen species (ROS) causing a severe photochemical injury to the retina (Widomska & Subczynski, 2014). However, lutein and zeaxanthin have a maximum absorption at 445 nm and 450 nm, respectively. Together, their absorption peak is at 460 nm, allowing them to protect the eye by absorbing 40–90% of the incident short-wavelength high-energy visible blue light (450-500 nm) prior to the formation of ROS, and by chemically and physically quenching ROS once they are formed (Cai et al., 2000).

The mechanism of protection from blue light depends on the oxidant species. Usually, carotenoids interfere with the reaction by reducing the chain propagation of oxidized organic molecules and radical species, by quenching or scavenging singlet oxygen (Widomska & Subczynski, 2019). In the macula, singlet oxygen is quenched by energy transfer to the carotenoids, generating triplet-state carotenoids that can dissipate harmful light through vibrational transitions and collisions, but without breaking bonds (Landrum & Bone, 2001). Such antioxidant properties have been demonstrated in a mouse model. Indeed, in a β5–/– mouse model that loose visual function with age because of primary deficiencies of RPE cells, an accumulation of pro-oxidant lipofuscin is extensively observed. Mice were then given 1.65 mg lutein and 66 μg zeaxanthin per mouse per day, or 52 mg lutein and 2 mg zeaxanthin per kg body weight for 9 months, which prevented RPE oxidation, cytoskeletal damage, and vision loss, providing a strong evidence of the importance of lutein's antioxidant properties for eye health (Yu et al., 2012).

Canthaxanthin is a particular case as this carotenoid can be found in the eye in the form of crystals, but its presence reflects a chronic overconsumption of this compound. When canthaxanthin consumption goes back to normal, crystals disappear. There is no evidence of any beneficial function of canthaxanthin in human eyes. Some authors even report that these crystals can cause retinopathy and crystalline disorders (Sujak., 2009; Beaulieu et al., 2013; Challiol et al., 2018).

Astaxanthin is also a special case, which seems to be a promising molecule for the future, although very little is known about this carotenoid. Indeed, astaxanthin gained interest when it was discovered that it has a higher antioxidant activity compared to other carotenoids such as α-carotene, lycopene, lutein, and β-carotene (Naguib, 2000). It was recently found that the nonpolar carotenoids lycopene and β-carotene disordered the membrane bilayer, and stimulated membrane lipid peroxidation on a model membrane enriched with polyunsaturated fatty acids, whereas astaxanthin preserved membrane structure and exhibited significant antioxidant activity (McNulty et al., 2008). Indeed, due to its molecular array, astaxanthin can bind and span the cell membrane, subsequently neutralizing free radicals both in the nonpolar and polar boundary zones. Thus, as astaxanthin efficiently prevents ROS formation with high scavenging activity, treatment with this xanthophyll micronutrient was rapidly expected to halt both the onset and the progression of ROS-related eye diseases. Soon after, it was demonstrated that astaxanthin has indeed a protective effect against retinal damage *in vitro* and *in vivo* (Nakajima et al., 2008; Shimokawa et al., 2019). Preclinical and clinical evidence support the potential use of astaxanthin in the prevention and treatment of several ocular diseases, such as cataract and age-related macular degeneration (AMD), in the same way as lutein and zeaxanthin, even though it is not a natural component of the retina.

37.8 Egg Carotenoids and Age-Related Cataracts

Although cataract is almost always a curable disease with surgery, it is still one of the most common causes of visual impairment and age-related eye diseases around the world. In

2020, among the 33.6 million adults aged 50 years and older worldwide who were blind, 15.2 million cases were affected by cataract ([95% UI: 12·7–18·0]) (Blindness, 2021). Cataract is a partial or total acquired opacification of the crystalline lens, which reduces the amount of light passing through to reach the retina, resulting in a blurred vision (Cabot et al., 2013).

Situated in the front of the eye, the lens is constantly subjected to oxidative stress by visual/UV light radiation exposure. Experimentally, the lens has been shown to be damaged by exposure to ultraviolet B radiation (Pitts et al., 1977). It is extensively recognized that oxidative stress is a significant factor in the genesis of senile cataract (the commonest cataract type), both in experimental animals and in cultured lens models (Gupta et al., 2003; Truscott, 2005). It is therefore not surprising that researchers have attempted to correlate lutein and zeaxanthin dietary intake and plasmatic levels with reduced risk for eye diseases, given their blue light filtering and antioxidant properties. Thus, since mid-1980s, several epidemiological studies have explored the possible role of known antioxidants in preventing the development of age-related lens opacities. In 1988, for the first time, results of a case-referent study (started in 1984) from the US Department of Agriculture (USDA) Human Nutrition Research Center at Tufts University were published, and showed an influence of antioxidants serum levels on the occurrence of cataract in humans (Jacques et al., 1988). Likewise, The Lens Opacities Case-Control Study, published in 1991, reported that high values of the index of antioxidant and vitamin intake decreased the risk of cortical, nuclear, and mixed cataract by more than half (Leske et al., 1991). The same results were reported in Finland in 1992 (Knekt et al., 1992).

To better characterize the antioxidants involved, Jacques and Chylack Jr conducted another epidemiological study to analyze the impact of plasma carotenoid levels, among others, on cataract risk. Interestingly, they reported for the first time no relationship between carotene intake and cataract risk, while analyses for fruit and vegetable consumption indicated significant decreases in the risk of cataract for persons who consume more than 3 1/2 servings of fruit and vegetables per day (Jacques & Chylack, 1991).

To support the results, a prospective analysis of The Nurses' Health Study published in 1992 (started in 1976) cohort, highlighted that consumption of spinach (rich in xanthophylls), but not carrots (rich in β-carotene), 5 times per week, was associated with lower risk for cataract extraction in women compared to less than once a month. Such diets being high in lutein and zeaxanthin, it was hypothesized that those xanthophylls, but not carotene, may play a role in reducing age-related eye disease risk (Hankinson et al., 1992). Similarly, a retrospective study by Mares-Perlman *et al.* reported that high consumption of spinach, but not carrots, was related to less severe nuclear sclerosis (Mares-Perlman et al., 1995). Later on, several other epidemiological studies strongly suggested a positive correlation between dietary intake and/or serum levels of carotenoids and decreased prevalence of age-related cataract (Jacques & Chylack Jr, 1991; Mares-Perlman et al., 1995; Tavani et al., 1996; Brown et al., 1999; Chasan-Taber et al., 1999).

However, only randomized clinical trials comparing the effectiveness of multivitamin supplementation in preventing the development of cataract would confirm this growing assumption. The first interventional clinical trial (randomized clinical trial) that suggested a relationship between vitamins and cataractogenesis was a trial started in 1985 in Linxian (China), designed to determine whether the vitamin/mineral supplements used in 2 cancer intervention trials affected the risk of developing age-related cataracts. The results showed that vitamin/mineral supplements decreased the risk of nuclear cataract after 5–6 years follow-up (Sperduto et al., 1993). The link with macular pigment was made in 1997, when Hammond *et al.* published a study showing that higher concentrations of xanthophylls in the retina correlate with higher concentrations in the lens, which impacts cataract development (Hammond et al., 1997). These results were already supported by transverse (Hammond et al., 1996; Ciulla et al., 2001; Curran-Celentano et al., 2001) and interventional studies (Hammond et al., 1997; Landrum et al., 1997), establishing a positive correlation between lutein and zeaxanthin intake and their serum and retinal concentrations.

To support the growing hypotheses that a high dose of lutein intake lowers age-related eye diseases, supplementation trials were initiated. In 1990, The Age-Related Eye Disease Study (AREDS) was the largest long-term multicenter, prospective, randomized study of persons aged from 55 to 80 years, which was designed by the National Eye Institute to assess the clinical course, prognosis, and risk factors of both AMD and cataract. The enrollment of the 4,757 volunteers began in 1992 and finished in 1998; they were then followed for an average of 6.3 years. The active treatment groups received antioxidants (vitamins C and E, and zinc), or 15 mg/day β-carotene, or a combination of both (later called "the AREDS formulation") (Age-Related Eye Disease Study Research, 1999). AMD results will be discussed below. Regarding cataract, results were less conclusive and gave rise to controversy about the protective effect of lutein on cataract development. Indeed, no effect of nutrient supplementation was found on the development of lens opacity. There was an equal proportion of subjects that underwent cataract surgery in treatments and control groups, meaning that none of the formulations decreased cataract risk (Age-Related Eye Disease Study Research, 2001a). Since AREDS study reports were published, demonstrating no beneficial effect of lutein on senile cataract risk, more attention has been paid to AMD, but still many studies indicate a positive effect of lutein on cataract risk, even though it remains more controversial. Hence, a pilot study reported by a group from Spain, showed that a high dose of lutein supplementation resulted in improved vision in cataract patients after a 2-year supplementation with 15 mg lutein daily (as lutein diesters) (Olmedilla et al., 2003).

Also, the population-based study POLA (Pathologies Oculaires Liées à l'Age study), started in 1995, showed that participants with the highest quintile of plasma zeaxanthin, but not with plasma lutein, had a significantly reduced risk of nuclear cataract (Delcourt et al., 2006). The same observation was made in a 10-year prospective study examining serum carotenoid levels in women from the Women's Health Study cohort, where high intake of lutein/zeaxanthin was associated to a lower risk of cataract (Christen et al., 2008). More recently, cross-sectional of dietary supplementation studies showed inverse associations between cataract and blood antioxidants, and the beneficial effects of lutein against age-related cataract (Dherani et al., 2008; Karppi et al., 2012). However, in accordance with AREDS results, several studies reported no or moderate beneficial effects of lutein on senile cataract risk (Brown et al., 1999; Chasan-Taber et al., 1999; Lyle et al., 1999; Taylor et al., 2002; Vu et al., 2006). Moreover, the AREDS2 study, launched in 2006 to test the value of adding omega-3 fatty acids or lutein/zeaxanthin in the AREDS study (known as the "AREDS formula"), confirmed that lutein supplementation was ineffective, both in preventing vision loss and in slowing progression toward cataract surgery (Chew et al., 2012). Yet, 2 meta-analyses reported a significant negative association between lutein serum levels and risk of developing nuclear cataract in a dose-linear response nuclear cataract (Liu et al., 2014; Ma et al., 2014).

Since the initiation of lens cataract is strongly associated with oxidative damage, the use of exogenous antioxidant still seems to be a strategy to protect visual functions. However, results of clinical studies vary from protective to no effect of carotenoid oral supplementation, while a positive correlation has been demonstrated for other ocular or oxidative stress-related pathologies. Interestingly, an analysis of the clinical studies discussed above tends to confirm that nutrients provided as a component of food are superior to mono-nutrient type supplements for the prevention and delay of cataract, meaning that mixed results might be explained by differences in xanthophyll bioaccessibility, bioavailability and by "food-matrix effect" (Braakhuis et al., 2019).

More recently, a new hope is rising regarding astaxanthin that tends to provide good pre-clinical results regarding protection against cataract formation. Indeed, in a model of steroid-induced cataract, astaxanthin administration efficiently prevented lens opacification by significantly recovering glutathione levels (Ishikawa et al., 2015). Moreover, in 2020, a study reported that astaxanthin supplementation delayed the development and progression of metabolic cataracts by inhibiting oxidative stress in diabetic mouse models (Yang et al., 2020). In summary, while the literature is mixed regarding the protective effect of xanthophylls, there appears to be a consensus that a diet high in fruits, vegetables, and especially eggs, which contain high levels of several xanthophylls, may be protective against cataract.

37.9 Egg Carotenoids and Age-related Macular Degeneration (AMD)

Age-related macular degeneration (AMD) is a chronic disease that affects the macular region of the retina, causing loss of central vision, characterized by progressive degeneration of the retinal photoreceptors, that occurs when the retinal pigment epithelium (RPE) and the Bruch's membrane (BM) cells are already severely altered (Coleman et al., 2008). AMD was first described almost 170 years ago, in 1855, by Donders, when he observed small drusen under the microscope in postmortem (Donders, 1855), and by Nagel who established in 1868 a link between white flecks around both macula and impaired visual acuity reported by a patient (Nagel, 1868). Since then, AMD has gained interest, especially in the last 70 years since the terminology "Age-related macular degeneration" was first used (Straub et al., 2008). In 1995, the grades of ARM were established and the terminology "Age-Related Maculopathy" was definitely accepted, which has facilitated epidemiological studies that were previously biased by a lack of clarity around the definition (Beatty et al., 1999).

Hence, in 2020, of the 33.6 million blind adults worldwide aged 50 years and over, ARM was responsible for 1.8 million cases ([1.3–2.4]), and was the fourth leading cause of global blindness (Steinmetz et al., 2021). Its prevalence is predicted to rise rapidly, particularly due to the increase in life expectancy (Wong et al., 2014).

There are 2 types of ARM identified. The atrophic form (non-neovascular), better known as the "dry form", accounts for 85–90% of cases, and is characterized by atrophy of the RPE. The neovascular form or "wet form", with a subretinal and sub-RPE proliferation of abnormal blood vessels that deviates the fluid, blood, and lipids, replaces the center of the retina with fibrous scarring.

ARM is a complex multifactorial disease largely influenced by a combination of environmental and genetic factors. Among the different etiologies and risk factors, we can mention demographic factors (age, sex, and ethnic origin), epidemiological risk factors (body mass index, smoking, diet, and genetic polymorphisms, e.g. mutations in complement cascade), and environmental risk factors (exposure to sun and chemicals) (Tikellis et al., 2007; Wang et al., 2009; Lechanteur et al., 2012; Klein et al., 2014; Wong et al., 2014; Hallak et al., 2019; Merle et al., 2019). At early stages of the disease, ARM is asymptomatic and patients initially report subtle changes, such as non-straight lines (metamorphopsia), blurred vision, and decreased night vision and contrasts sensitivity, with a characteristic scotoma, a blind spot in the middle of the image owing to retinal hemorrhage and fluid accumulation (Haute Autorité De Santé, 2022). The presence of neovascular ARM generally leads to more rapid and profound visual symptoms that may be acute or gradually worsen; and in 80% of the cases, this form leads to vision loss (Storey et al., 2014). The exact pathogenesis of ARM remains unknown, but one of the main components is thought to be age-related cumulative oxidative stimuli. With aging, RPE cells become less effective in maintaining the blood-retinal barrier participating in vision, and phagocytosing photoreceptor outer segments, which leads to pathological accumulation of cellular waste (Kannan et al., 2016; Kuo et al., 2020). With declining RPE function, the retina can no longer receive its proper nourishment and accumulates waste material (Chen et al., 2018). The addition of excessive oxidative stress ultimately leads to RPE cell degeneration, and subsequent death of photoreceptors because RPE cells lose the ability to provide them with oxygen and nutrients and remove waste materials (Abokyi et al., 2020). Macular degeneration results in a central relative or absolute scotoma, without affecting the peripheral visual field (Hawkins et al., 1999).

Unlike cataract, there is no consensus treatment for ARM, and given the evidence supporting the biological role of lutein and zeaxanthin in the eye, researchers gained interest in studying the effects of xanthophyll dietary intake and ARM prevalence, in the hope of preventing or slowing down the disease. One of the first encouraging results came from a study made on a sample from the National Health and Nutritional Examination Survey (NHANES) that showed that diets high in fruits and vegetables were inversely associated with ARM risk (Goldberg et al., 1988). These observations were confirmed after a 12-year follow-up on 77,466 women over 45 years (Chasan-Taber et al., 1999). In 1992, the Eye Disease Case Control Study, a multicenter observational study of risk factors for several retinal disorders, showed that total serum carotenoids were inversely related to ARM risk (Eye Disease Case Control Study, 1993). These results helped to provide a basis for the hypothesis that dietary carotenoids may have a protective effect against ARM. In 1994, an epidemiological study showed for the first time, a direct relationship between lutein intake and ARM (Seddon et al., 1994). Likewise, several epidemiological studies demonstrated a relationship between lutein and decreased risk of ARM (Eye Disease Case Control Study, 1993; Seddon et al., 1994; Snodderly, 1995; Lyle et al., 1999; Beatty et al., 2001; Mares-Perlman et al., 2001; Gale et al., 2003). Interestingly, observational case-control studies showed a relationship between lower lycopene serum level and the risk of ARM. In 1995, a study showed that among subjects with retinal pigment abnormalities (late ARM or neovascular and exudative ARM), very low levels of lycopene, but no other dietary carotenoids or tocopherols, were related to ARM (Mares-Perlman et al., 1995). Similarly, in an Italian study, Simonelli *et al.* reported that, of the serum carotenoids investigated, only lycopene was lower in the 2 ARM forms than in controls (Simonelli et al., 2002). In 2005, these observations were supported by a study that also reported lower lycopene serum levels in ARM patients (Cardinault et al., 2005). A study, conducted on a Chinese population, suggested that higher levels of serum carotenoids, in particular zeaxanthin and lycopene, are associated with a lower risk of exudative AMD (Zhou et al., 2011). Observational studies are listed in Table 37.4.

Results from observational studies being interesting, controlled intervention studies in humans were conducted to establish causality. In 1999, a case series pilot study showed improvements in various visual acuity tests in 14 ARM patients following a diet containing 5 ounces of spinach (around 14 mg lutein per day) and supplemented with purified lutein 4–7 times per week for up to a year. In 2001, AREDS (described above) results were published and reported a statistically significant decrease in the progression of advanced ARM compared to the placebo group (Age-Related Eye Disease Study Research, 2001a, 2001b). The same year, Bone *et al.* published a case-control comparative study of human retinas from 56 donors with ARM, and from 56 controls. They confirmed an inverse association between the risk of ARM and the amounts of lutein and zeaxanthin in the retina (Bone et al., 2001). At the same time, a study from a group in Milan, Italy, showed that supplementation of ARM patients with 15 mg of lutein per day for 18 months resulted in an improvement in visual acuity compared to the placebo group (Massacesi, 2001). Altogether, these studies established that lutein and zeaxanthin supplementation allow structural changes in the retina and positively affect visual acuity over time, so that the AREDS formulation rapidly became standard-of-care. However, since this formulation contained β-carotene instead of lutein and zeaxanthin, it became controversial, which leads to the design of AREDS2 study in 2006, a randomized, multi-centered, double-masked, placebo-controlled trial, including 4,203 subjects between 50 and 85 years with age-related eye disease, and followed for 6 years. In this study, participants who took antioxidants with lutein (10 mg/day) and zeaxanthin (2 mg/day) instead of β-carotene (AREDS2 formula) had significantly increased visual benefits and lower risk of advanced AMD, compared to those who took the AREDS formula (Age-Related Eye Disease Study 2 (AREDS2) Research, 2013). Since then, there has been a growing number of clinical studies demonstrating not only the preventive effects of a lutein-rich diet, but also the ameliorative effects of supplementation on visual functions, through measurement of macular pigment optical density (MPOD).

Among the most famous clinical studies reporting a significant beneficial effect of lutein supplementation on visual function in ARM patients, we can cite the Lutein Antioxidant Supplementation Trial (LAST) (Richer et al., 2004); the Taurine, Omega-3 fatty acids, Zinc, Antioxidants and Lutein (TOZAL) Study (Cangemi, 2007); the Carotenoids and Antioxidants in Age-Related Maculopathy Italian Study (CARMIS) (Parisi et al., 2008); the Carotenoids with Co-antioxidants in Age-Related Maculopathy (CARMA) study (Neelam et al., 2008); the LUTEGA study (Dawczynski et al., 2013); the Combination of Lutein Effects in the Aging Retina (CLEAR) study (Murray et al., 2013); the *Meso*-zeaxanthin Ocular Supplementation (MOST) study (Akuffo et al., 2015); and the Central Retinal Enrichment Supplementation Trials (CREST) study (Akuffo et al., 2017). Clinical studies that have established the positive effect of lutein on visual performance in patients with ARM are listed in Table 37.5.

All these clinical trials have shown an increased MPOD and improved macular function, visual acuity, and contrast sensitivity by long-term antioxidants and lutein/zeaxanthin supplementation, which was assumed to prevent ARM progression. The idea that lutein is essential for maintaining visual function and preventing age-related eye diseases is now widely accepted. Thus, in recent years, research tends to investigate the role of dietary habits in preventing the onset of ARM, considering the synergistic effects of food and nutrients consumed in combination with the diet. In 2009, the consumption of 2 or 4 egg yolks per day for 5 weeks was studied in 31 women and 21 men over 60 years. The results showed a convincing improvement in MPOD and an increase in serum lutein and zeaxanthin (Vishwanathan et al., 2009). Likewise, the Carotenoids in Age-Related Eye Disease Study (CAREDS) demonstrated a correlation between the decreased prevalence of early ARM and healthy behaviors diet in American women (Mares et al., 2011). In 2014, Kelly et al. studied the impact of adding enriched eggs (obtained from hens on a diet enriched with natural sources of lutein and zeaxanthin) in the diet on serum levels and MPOD, in 97 healthy subjects. As a result, there were no significant differences with the control group for MPOD, but serum samples showed that modified eggs had a significant positive impact on lutein and zeaxanthin levels in serum (Kelly et al., 2014).

TABLE 37.4

Observational Studies Indicating an Influence of Carotenoid Serum Levels on the Occurrence of Age-Related Eye Diseases

Study (Year)	Study Design	Observation	Results
(Jacques et al., 1988)	Observational. 112 participants aged 40–70 years with both lenses intact and not scheduled for lens removal.	Nutritional status determined by blood nutrient levels. Risk of any cataract.	Substantial decrease in risk of any cataract for subjects with moderate and high carotenoid levels.
the National Health and Nutritional Examination Survey (NHANES) (Goldberg et al., 1988)	3,082 persons aged ≥ 45 years.	Nutritional status. Prevalence of AMD.	Negative association of AMD degeneration with the frequency of consumption of vitamin A-rich fruits and vegetables.
(Jacques & Chylack, 1991)	Observational. 112 participants with cataracts (77) or clear lenses (35).	Determination of plasma carotenoid levels. Risk of any cataract.	The risk of developing cataract among persons with low (below the 20th percentile) and moderate (20th–80th percentiles) carotenoid plasma levels was over 5 times the risk than among persons with higher plasma carotenoid concentrations.
The multicenter Eye Disease Case-Control Study (EDCC) (Seddon et al., 1994)	Multicenter, case-control study. 876 participants with advanced AMD (356) or other ocular disease (520), aged between 55 and 80 years.	Carotenoid intake assessed by food-frequency questionnaire. Prevalence of AMD.	Subjects with the highest quintile of carotenoid intake had a 43% lower risk of AMD when compared with those in the lowest quintile. Furthermore, L and Z, from dark green leafy vegetables, were most strongly associated with a reduced risk of AMD.
Beaver Dam Eye Study (VandenLangenberg et al., 1998)	Population-based. 4,926 participants aged ≥ 43 years.	L, Z, lycopene, and α- and β-carotene intake through diet. Incidence of large drusen.	Significant inverse association between dietary intakes of pro-vitamin A carotenoids (and vitamin E) and the incidence of large drusen, and between zinc and the incidence of pigmentary abnormalities.
National Health and Nutrition Examination Survey (Klein et al., 1999)	Cross-sectional. 8,222 participants aged ≥ 40 years.	L, Z, and other carotenoids intake and serum levels. Incidence of pigmentary abnormalities and degeneration from early to late AMD.	Association between higher level of L and Z in the diet with lower risk of pigmentary abnormalities and of late AMD.
(Snellen et al., 2002)	Case-control. 138 participants with AMD (72) or no ocular conditions (66), aged ≥ 60 years.	L, Z, and other antioxidants intake through diet. Prevalence of neovascular AMD.	The prevalence rate of AMD in patients with low antioxidant intake and low L intake was about twice as high as that in patients with high intake.
(Gale et al., 2003)	Cross-sectional. 380 participants aged between 66 and 75 years.	L and Z serum levels. Risk of AMD assessed by the Wisconsin Age-Related Maculopathy Grading System.	Risk of AMD (early or late) was 2-fold increased in people with lower plasma concentrations of Z.
Pathologies Oculaires Liées à l'Age (POLA) Study (Delcourt et al., 2006)	Prospective study population-based. 899 participants aged ≥ 60 years.	Plasma carotenoid levels. Presence of late AMD (neovascular ARM, geographic atrophy) and/or soft indistinct drusen associated with pigmentary abnormalities. Determination of lens opacities.	The highest quintile of plasma Z level was significantly associated with reduced risk of AMD. AMD was also significantly associated with combined plasma L and Z. Cataract showed no such associations.
The Carotenoids in Age-Related Eye Disease Study (CARED) (Moeller et al., 2006)	Population-based. 1,787 women aged 50 to 79 years.	L, Z, β-carotene, ω-3 fatty acids and other antioxidants intake (food frequency questionnaire), and serum levels. Prevalence of AMD (presence of large drusen or extensive intermediate drusen).	There was no statistically significant difference in the prevalence of intermediate AMD among subjects with high and low intake of dietary L and Z. However, a reduced risk of AMD was observed for women younger than 75 years, with stable intake of L and Z (without a history of chronic diseases).
The Blue Mountains Eye Study (BMES) (Tan et al., 2008)	Prospective population–based cohort study. 2,454 Australian participants aged ≥ 49 years, reexamined after 5 years, 10 years, or both	Intake of L, Z, lycopene, α- and β-carotene, and other antioxidants (diet). 10-year incidence of AMD.	Higher dietary L and Z intake reduced the risk of long-term incident AMD.

Note: AMD: Age-related macular degeneration; ARM: Age-related maculopathy; L: lutein; Z: zeaxanthin.

TABLE 37.5

Interventional Studies Indicating an Influence of Lutein (L) Supplementation on Visual Function

Study (Year)	Design (Number of Participants)	Intervention and Supplementation (per Day)	Results
(Richer, 1999)	Case-series. 14 male participants with AMD; aged ≥ 60 years. 2 groups: G1 (11), G2 (3)	G1: 0.73 +/- 0.45 portions of dark-green, leafy vegetables/day base intake, with an additional portion of sautéed spinach (5 oz.) 4 to 7 times per week. G2: L-based antioxidant	Both patient groups demonstrated short-term positive effects in visual function, with improvements in the Amsler grid (87%), Snellen VA (71%), CS (92%), SKILL (65%), glare recovery (69%), and activities of daily vision subscale (60%). An obvious correlation between ophthalmoscopic appearance of the retina and visual outcome was lacking.
(Massacesi, 2001)	Randomized-control trial. 50 participants with early AMD.	L 15 mg	After 18 months, a 2-fold improvement in visual acuity.
(Olmedilla et al., 2001)	Case-series. 10 participants with cataract (G1, $n = 5$) or AMD (G2, $n = 55$); aged ≥ 55 years.	G1 + G2: L 15 mg	After 13 months, L supplementation significantly increased serum L levels and was associated with an improvement in visual acuity and glare sensitivity in both cataract and AMD patients, independently of the disease progression.
Age-Related Eye Disease Study (AREDS) **Cataract Report** (Age-Related Eye Disease Study Research, 2001a)	Randomized-control trial. 4,629 participants with intermediate AMD or cataract in one eye; aged from 55–80 years. 4 groups: G1 (853), G2 (869), G3 (1,451), G4 (1,456).	G1: 15 mg β-carotene + other antioxidants (no L or Z). G2: Zinc + Copper. G3: Antioxidants formulation. G4: Placebo.	No statistically significant effect of the antioxidant formulation was seen on the development or progression of age-related lens opacities. There was also no statistically significant effect of treatment in reducing the risk of progression for any of the 3 lens opacity types or for cataract surgery. AREDS formulation had no apparent effect on the 7-year risk of development or progression of age-related lens opacities or visual acuity loss.
Lutein and Antioxidant Supplementation Trial (LAST) (Richer et al., 2004)	Randomized-control trial. 90 participants with atrophic AMD; aged ≥ 70 years. 3 groups: G1 (29), G2 (30), G3 (31).	G1: L 10 mg. G2: L 10 mg + antioxidants + vitamins + minerals. G3: Placebo.	After 1 year, in G1 and G2, MPOD, Snellen equivalent, VA and CS were improved. G3 had no significant changes in any of the measured findings.
(Rosenthal et al., 2006)	Intervention trial. 45 participants with small or no drusen (15), large drusen (15), or advanced AMD (15); aged ≥ 60 years. 3 groups: G1 (15), G2 (15), G3 (15).	G1: L 2.5 mg G2: L 5 mg G3: L 10 mg	After 6 months, increase in L serum levels in the 3 groups, independently of eye disease severity, but no significant changes were detected in VA.
The Taurine, Omega-3 Fatty Acids, Zinc, Antioxidant, Lutein (TOZAL) study (Cangemi, 2007)	Intervention trial. 37 participants with at least one eye diagnosed with dry AMD; aged ≥ 70 years.	L 8 mg + Z 400 μg + Taurine 400 mg + Vitamin A 10,000 UI + β-carotene 18,640 IU	After 6 months, patients had a significant improvement in VA.
Carotenoids and co-antioxidants in Age-related Maculopathy (CARMA) Study (Neelam et al., 2008) (Beatty et al., 2013)	Randomized-control trial. 433 participants aged ≥ 55 years, with AMD patients. 2 groups: G1 (216), G2 (217).	G1: AREDS formulation. G2: Placebo.	By 24 months, a steadily increasing differential was observed in BCVA and L serum levels between both groups. MPOD declined steadily over time in G2, whereas a rise was observed in G1. But no change in CS was observed in both groups.
Carotenoids and Antioxidants in Age-Related Maculopathy Italian Study (CARMIS) (Parisi et al., 2008)	Randomized-control trial. 27 participants with non-advanced AMD; aged ≥ 60 years. 2 groups: G1 (15), G2 (12).	G1: L 10 mg + Z 1 mg + astaxanthin 4 mg. G2: Placebo.	After 12 months, a significant increase in response amplitude densities (assessed by Multifocal Electroretinogram) was observed in the central retina in G1, but not in G2. No significant effect was observed in peripheral retina in both groups.
Lutein Intervention Study Austria (LISA) (Weigert et al., 2011)	Randomized-control trial. 126 participants with AMD; aged ≥ 60 years. 2 groups: G1 (84), G2 (42).	G1: L 20 mg for 3 months, then L 10 mg for 3 months. G2: Placebo.	After 6 months, L supplementation significantly increased MPOD. No significant effect was observed on macular function (MDLT, as assessed by microperimetry) or VA.
The Zeaxanthin and Visual Function Study (ZVF) (Richer et al., 2011)	Randomized-control trial. 60 participants with early and moderate AMD; aged ≥ 60 years. 3 groups: G1 (10), G2 (25), G3 (25).	G1: L 9 mg ("Faux Placebo"). G2: Z 8 mg. G3: L 9 mg + Z 8 mg.	After 1 year, MPOD increased in all three groups. High contrast sensitivity increased in all three groups, but was greater in G2, whereas low-contrast sensitivity improved in G1 and G3, but not in G2.

(Continued)

TABLE 37.5 ***(Continued)***

Interventional Studies Indicating an Influence of Lutein (L) Supplementation on Visual Function

Study (Year)	Design (Number of Participants)	Intervention and Supplementation (per Day)	Results
(Sasamoto et al., 2011)	Intervention trial. 43 Japanese participants with AMD (33), or central serous chorioretinopathy (5), or no ocular condition (5); aged ≥ 40 years.	L 6 mg	After 1 year, no effect in MPOD was observed, but subjects had a significant improvement in CS and retinal sensitivity.
(Ma et al., 2012)	Randomized-control trial. 108 participants with early AMD; aged between 50 and 79 years. 4 groups: G1 (27), G2 (27), G3 (27), G4 (27) + Group of healthy controls (36).	G1: Placebo G2: L10 mg G3: L20 mg G4: L10 mg + Z 10 mg	After 48 weeks, MPOD increased in G2, G3 and G4. At week 48, significant increases of the CS at 3 cycles/degree from baseline were observed for all active treatment groups, and there was a significant group difference between G3 and the placebo group.
Combination of L Effects in the Aging Retina (CLEAR) study (Murray et al., 2013)	Randomized-control trial. 72 participants with early AMD; aged ≥ 60 years. 2 groups: G1 (36), G2 (36).	G1: L 10 mg G2: Placebo	After 12 months, significant increase of MPOD was observed in G1. No improvement in VA was observed in G1, but VA decreased in G2.
Long Term Effects of L/Z and ω-3 Supplementation on Optical Density of AMD Patients (LUTEGA) (Dawczynski et al., 2013)	Randomized-control trial. 145 participants with non-exudative AMD. 3 groups: G1 (50), G2 (55), G3 (40).	G1: L 10 mg + Z 1 mg + DHA 100 mg + EPA 30 mg. G2: L 20 mg + Z 2 mg + DHA 200 mg + EPA 60 mg. G3: Placebo.	After 12 months, all MPOD parameters and BCVA improved significantly in both G1 and G2.
Age-Related Eye Disease Study 2 (AREDS 2) (Age-Related Eye Disease Study 2 (AREDS2) Research, 2013)	Randomized-control trial. 4,203 participants with intermediate AMD in both eyes or intermediate AMD in one eye and advanced AMD in the other eye. 4 groups: G1 (1,007), G2 (1,038), G3 (1,062), G4 (1,069).	G1: AREDS formulation. G2: AREDS + L 10 mg + Z 2 mg. G3: AREDS + DHA 350 mg + EPA 650 mg. G4: AREDS + L 10 mg + Z 2 mg + DHA 350 mg + EPA 650 mg.	Adding omega-3 fatty acids or lutein + zeaxanthin to the AREDS formula had no additional overall effect on the risk of advanced AMD. However, AREDS2 participants who took L + Z instead of β-carotene (AREDS 2 formula) had an incremental increase in benefit, compared to those who took the AREDS formula.
(Berrow et al., 2013)	Randomized-control trial. 14 participants with AMD. 2 groups: G1 (8), G2 (6).	G1: L 12 mg G2: Placebo	After 40 weeks, no clinical effects; only minimal improvement in mfEGR.
(Sabour-Pickett et al., 2014)	Intervention trial. 52 participants with AMD. 3 groups: G1 (17), G2 (21), G3 (14).	G1: L 20 mg + Z 2 mg. G2: L 10 mg + Z 2 mg + MZ 10 mg. G3: L 3 mg + Z 2 mg + MZ 17 mg.	After 1 year, MPOD increased in all groups; the significant improvement in CS was much more effective in G3.
(Akuffo et al., 2015)	Intervention trial. 67 participants with AMD. 3 groups.	G1: L 20 mg + Z 2 mg. G2: L 10 mg + Z 2 mg + MZ 10 mg. G3: L 3 mg + Z 2 mg + MZ 17 mg.	After 3 years, all the groups showed a significant increase in MPOD, but no effects in reducing progression to advanced AMD or improving VA.
(Wolf-Schnurrbusch et al., 2015)	Intervention trial. 79 participants with AMD. 2 groups: G1 (40), G2 (39).	G1: L 10 mg + Z 1 mg. G2: L 10 mg + Z 1 mg + DHA / EPA 130 mg.	After 6 months and 1 year, MPOD and CS significantly increased in G1, but not VA. No significant effect in G2.
(Huang et al., 2015)	Randomized-control trial. 108 participants with early AMD. 4 groups: G1 (28), G2 (26), G3 (27), G4 (27).	G1: Placebo G2: L 10 mg G3: L 20 mg G4: L 10 mg + Z 10 mg	After 2 years, significant increase in MPOD and CS, no effect in VA, and flash recovery time. Same efficacy in all treatment groups.
(Fujimura et al., 2016)	Intervention trial. 20 participants with at least one eye with wet AMD or early AMD.	L 20 mg + Z 1 mg + DHA 200 mg	After 6 months, significant increase in MPOD and CS. Linear correlation between MPOD and serum concentrations of L.

Note: AMD: Age-related Macular Degeneration; CS: Contrast Sensitivity; DHA: Docosahexaenoic Acid; EPA: Eicosapentaenoic Acid; G: group; L: Lutein; MPOD: Macular Pigment Optical Density; BCVA: Best Corrected Visual Acuity; MDLT: Mean Differential Light Threshold; mfEGR: Multifocal Electroretinography; MZ: meso-Z; VA: Visual Acuity; Z: Zeaxanthin.

Similarly, the Regional Kale as Source of Lutein and Zeaxanthin to Improve Vision of AMD-patients (KALESIGHT) study reported that MPOD increased significantly in the kale group after 4 weeks of intervention, but it was not associated with an improved visual function after 4 weeks of wash-out (Arnold et al., 2013). More recently, the Coimbra study (Nunes et al., 2018) and the European Eye Study (Hogg et al., 2017) have shown that adhesion to a Mediterranean diet was associated with significantly reduced prevalence of any type of ARM and neovascular ARM, respectively. In 2020, the results from a retrospective study performed on The Blue Mountains Eye Study cohort were published. This study, which was conducted between 1992 and 1994 on 2,900 Australian subjects aged 49 to 97 years to investigate the consequences of protein, vitamin A, niacin, thiamin, and riboflavin intakes on nuclear cataract prevalence, showed interesting results on egg consumption over 15 years on the incidence of ARM. Participants who consumed 2 to 4 eggs/week had a reduced risk of incident late-stage ARM after 15 years, but no significant associations were observed between egg consumption and incident of early ARM (Gopinath et al., 2020). In 2022, clinical trials are still being launched to investigate the beneficial role of certain foods on eye health: for example, Tufts University started a study on "Pistachios and Neural Macular Pigment" (NCT05283941). The second aim of current research is to use pharmacological forms, allowing a better bioavailability of carotenoids to improve supplementation efficiency. For example, a study from Davey et al. was designed to compare a micronized lipid-based carotenoid liquid supplementation to AREDS-2 Formula, and reported a statistically significant improvement in contrast sensitivity function (CSF) in both eyes after 6 months (Davey et al., 2020). Thus, many studies support lutein's favorable effects on eye health, and high lutein intake with a balanced diet rich in fruit, vegetables, and eggs might be encouraged, particularly in the elderly.

Eggs are particularly interesting because, as discussed above, they might contain other xanthophylls that are not present in other dietary sources. Notably, astaxanthin arouses a growing interest because recent studies demonstrated its ability to protect the eye against oxidative damage. Hence, Parisi *et al.* conducted a 2-year comparative study that demonstrated that oral lutein/zeaxanthin (10 mg/1 mg per day) combined with an additional dose of astaxanthin (4 mg), daily for 12 months, can clinically improve and stabilize visual acuity, contrast sensitivity, and visual function in ARM patients (Parisi et al., 2008). However, astaxanthin was not used alone, and to date there has been no direct study that shows its effectiveness clinically. Yet, numerous animal studies have further confirmed the protective effects of astaxanthin on the eye, and its effectiveness in the prevention of ocular diseases, ranging from the ocular surface to the fundus, including phototoxic keratitis, cataract, diabetic retinopathy, ARM, and retinal ischemia injury (Lennikov et al., 2012; Otsuka et al., 2013, 2016; Yang et al., 2020; Janani et al., 2021). Finally, although low levels of lycopene were associated with a higher risk of ARM, no interventional studies have been conducted yet to study and characterize its precise role in human eye health.

37.10 Other Clinical Trials and Visual Function

As discussed above, the role of carotenoids in ARM has been greatly studied. The encouraging results have led to subsequent investigations into the role of carotenoids in other eye diseases, including diabetic retinopathy (Hu et al., 2011; Chous et al., 2016), glaucoma (Loughman et al., 2021), retinitis pigmentosa (Bahrami et al., 2006; Berson et al., 2010), macular telangiectasia type II (MacTel) (Zeimer et al., 2010), and retinopathy of prematurity (Gong & Rubin, 2015). Besides, after 2010, most active studies with lutein supplementation were related to a pathology other than ARM or cataract, with a greater interest toward the role of lutein and zeaxanthin in cognitive function in infants and in preventing the onset of dementia in elderly. We can cite the following studies: Enriched Eggs for Retina Health in Type 2 Diabetes (NCT04496817), Effect of Enriched QUARTET® Hen Eggs on Cardiovascular Function in Cardiovascular Patients and Healthy Individuals (NCT04564690), Effects of Lutein Supplementation on Cognition and MPOD in Multiple Sclerosis Patients (LuMES, NCT04843813). Similarly to ARM, low levels of lycopene tend to be associated with a higher risk of retinopathy, such as diabetic retinopathy (Brazionis et al., 2009; Shalini et al., 2021).

37.11 Conclusion

Carotenoids cannot be synthesized by animals and human beings. They are consumed by hens, either via vegetal products or via supplements, and accumulate in egg yolk. Among the carotenoids found in egg yolk, 2 main carotenoids originating from vegetal sources can be found in the human eye: lutein and zeaxanthin. Lycopene can also be found if animals are supplemented with specific vegetal products. Two others can be found only if hens received supplements: canthaxanthin and astaxanthin. These carotenoids are mainly located in the retinal region of the eye and play a crucial role in maintaining eye health. As they are powerful antioxidants, the first demonstrated role of xanthophylls is protecting the eye against light-induced oxidative stress and damages. More precisely, lutein and zeaxanthin are essential, as they are major components of the macula lutea. During the past 50 years, it has been extensively demonstrated that supplementing a population with such carotenoids helped in preventing the development of age-related eye diseases such as ARM or senile cataract. More recently, a growing interest toward astaxanthin arose because of its higher antioxidant properties compared to other carotenes or xanthophylls. Astaxanthin may be protective against both hens' egg degradation by oxidation and oxidative-related eye disease occurrence, as proven in pre-clinical studies. Interestingly, low serum levels of lycopene tend to be strongly associated with a higher risk of AMD, but no interventional studies have yet been conducted to rigorously determine its role in eye health. The most plausible explanation relies on its powerful antioxidant activities that protect organs, including eyes, from oxidative damages.

Overall, eggs represent a good and inexpensive source of carotenoids, and especially of lutein and zeaxanthin. Moreover, they are one of the only dietary sources of astaxanthin, a promising molecule. Egg yolk fat content also facilitates carotenoid absorption. Consuming eggs regularly would thus provide a good daily intake of bioavailable lutein and zeaxanthin and other active carotenoids.

REFERENCES

Abdel-Aal, E.-S., Akhtar, H., Zaheer, K., & Ali, R. (2013). Dietary sources of lutein and zeaxanthin carotenoids and their role in eye health. *Nutrients*, *5*(4), 1169–1185. https://doi.org/10.3390/nu5041169

Abokyi, S., To, C.-H., Lam, T. T., & Tse, D. Y. (2020). Central Role of oxidative stress in age-related macular degeneration : Evidence from a review of the molecular mechanisms and animal models. *Oxidative Medicine and Cellular Longevity*, *2020*. https://doi.org/10.1155/2020/7901270

Age-Related Eye Disease Study 2 (AREDS2) Research (2013). Lutein + zeaxanthin and omega-3 fatty acids for age-related macular degeneration : The age-related eye disease study 2 (AREDS2) randomized clinical trial. *JAMA*, *309*(19), 2005–2015.

Age-Related Eye Disease Study Research (1999). The age-related eye disease study (AREDS): Design implications AREDS report no. 1. *Controlled Clinical Trials*, *20*(6), 573.

Age-Related Eye Disease Study Research (2001a). A randomized, placebo-controlled, clinical trial of high-dose supplementation with vitamins C and E and beta carotene for age-related cataract and vision loss: AREDS report no. 9. *Archives of Ophthalmology*, *119*(10), 1439–1452.

Age-Related Eye Disease Study Research (2001b). A randomized, placebo-controlled, clinical trial of high-dose supplementation with vitamins C and E, beta carotene, and zinc for age-related macular degeneration and vision loss: AREDS report no. 8. *Archives of Ophthalmology*, *119*(10), 1417–1436.

Aguilar, F., Dusemund, B., Galtier, P., Gilbert, J., Gott, D., Grilli, R., & Woutersen, R. A. (2010). Scientific opinion on the re-evaluation of canthaxanthin (E161g) as a food additive. *EFSA Journal*, *8*(10), 1852. https://doi.org/10.2903/j.efsa.2010.1852

Akdemir, F., Orhan, C., Sahin, N., Sahin, K., & Hayirli, A. (2012). Tomato powder in laying hen diets: Effects on concentrations of yolk carotenoids and lipid peroxidation. *British Poultry Science*, *53*(5), 675–680. https://doi.org/10.1080/00071668.2012.729142

Akiba, Y., Sato, K., Takahashi, K., Takahashi, Y., Furuki, A., Konashi, S., & Nagao, H. (2000). Pigmentation of egg yolk with yeast phaffia rhodozyma containing high concentration of astaxanthin in laying hens fed on a low-carotenoid diet. *Japanese Poultry Science*, *37*(2), 77–85.

Akuffo, K. O., Beatty, S., Peto, T., Stack, J., Stringham, J., Kelly, D., & Nolan, J. M. (2017). The impact of supplemental antioxidants on visual function in nonadvanced age-related macular degeneration: A head-to-head randomized clinical trial. *Investigative Ophthalmology & Visual Science*, *58*(12), 5347–5360.

Akuffo, K. O., Nolan, J. M., Howard, A. N., Moran, R., Stack, J., Klein, R., & Thurnham, D. I. (2015). Sustained supplementation and monitored response with differing carotenoid formulations in early age-related macular degeneration. *Eye*, *29*(7), 902–912.

Arnold, C., Jentsch, S., Dawczynski, J., & Böhm, V. (2013). Age-related macular degeneration: Effects of a short-term intervention with an oleaginous kale extract—A pilot study. *Nutrition*, *29*(11–12), 1412–1417.

Arteni, A.-A., Fradot, M., Galzerano, D., Mendes-Pinto, M. M., Sahel, J.-A., Picaud, S., & Pascal, A. A. (2015). Structure and conformation of the carotenoids in human retinal macular pigment. *PLoS One*, *10*(8), e0135779.

Bahrami, H., Melia, M., & Dagnelie, G. (2006). Lutein supplementation in retinitis pigmentosa: PC-based vision assessment in a randomized double-masked placebo-controlled clinical trial [NCT00029289]. *BMC Ophthalmology*, *6*(1), 1–12.

Beatty, S., Boulton, M., Henson, D., Koh, H. H., & Murray, I. J. (1999). Macular pigment and age related macular degeneration. *British Journal of Ophthalmology*, *83*(7), 867–877. https://doi.org/10.1136/bjo.83.7.867

Beatty, S., Chakravarthy, U., Nolan, J. M., Muldrew, K. A., Woodside, J. V., Denny, F., & Stevenson, M. R. (2013). Secondary outcomes in a clinical trial of carotenoids with coantioxidants versus placebo in early age-related macular degeneration. *Ophthalmology*, *120*(3), 600–606.

Beatty, S., Murray, I. J., Henson, D. B., Carden, D., Koh, H.-H., & Boulton, M. E. (2001). Macular pigment and risk for age-related macular degeneration in subjects from a Northern European population. *Investigative Ophthalmology & Visual Science*, *42*(2), 439–446.

Beaulieu, R. A., Warwar, R. E., & Buerk, B. M. (2013). Canthaxanthin retinopathy with visual loss: A case report and review. *Case Reports in Ophthalmological Medicine*, *2013*, 1–4. https://doi.org/10.1155/2013/140901

Bernstein, P. S., Khachik, F., Carvalho, L. S., Muir, G. J., Zhao, D.-Y., & Katz, N. B. (2001). Identification and quantitation of carotenoids and their metabolites in the tissues of the human eye. *Experimental Eye Research*, *72*(3), 215–223.

Berrow, E. J., Bartlett, H. E., Eperjesi, F., & Gibson, J. M. (2013). The effects of a lutein-based supplement on objective and subjective measures of retinal and visual function in eyes with age-related maculopathy – A randomised controlled trial. *British Journal of Nutrition*, *109*(11), 2008–2014.

Berson, E. L., Rosner, B., Sandberg, M. A., Weigel-DiFranco, C., Brockhurst, R. J., Hayes, K. C., & Gaudio, A. R. (2010). Clinical trial of lutein in patients with retinitis pigmentosa receiving vitamin A. *Archives of Ophthalmology*, *128*(4), 403–411.

Bhosale, P., & Bernstein, P. S. (2005). Synergistic effects of zeaxanthin and its binding protein in the prevention of lipid membrane oxidation. *Biochimica et Biophysica Acta (BBA)-Molecular Basis of Disease*, *1740*(2), 116–121.

Blesso, C. N., Andersen, C. J., Bolling, B. W., & Fernandez, M. L. (2013). Egg intake improves carotenoid status by increasing plasma HDL cholesterol in adults with metabolic syndrome. *Food & Function*, *4*(2), 213–221.

Blindness, G. B. D. (2021). Vision impairment collaborators on behalf of the vision loss expert group of the global burden of disease study. Causes of blindness and vision impairment in 2020 and trends over 30 years, and prevalence of avoidable blindness in relation to vision 2020. The right to sight: An analysis for the global burden of disease study. *Lancet Glob Health*, *9*(2), e144–e160.

Bone, R. A., Landrum, J. T., Hime, G. W., Cains, A., & Zamor, J. (1993). Stereochemistry of the human macular carotenoids. *Investigative Ophthalmology and Visual Science*, *34*(6), 2033–2040.

Bone, R. A., Landrum, J. T., Mayne, S. T., Gomez, C. M., Tibor, S. E., & Twaroska, E. E. (2001). Macular pigment in donor eyes with and without AMD: A case-control study. *Investigative Ophthalmology & Visual Science*, *42*(1), 235–240.

Bone, R. A., Landrum, J. T., & Tarsis, S. L. (1985). Preliminary identification of the human macular pigment. *Vision Research*, *25*(11), 1531–1535. https://doi.org/10.1016/0042-6989(85)90123-3

Boon, C. S., McClements, D. J., Weiss, J., & Decker, E. A. (2010). Factors influencing the chemical stability of carotenoids in foods. *Critical Reviews in Food Science and Nutrition*, *50*(6), 515–532. https://doi.org/10.1080/10408390802565889

Bories, G., Brantom, P., Brufau De Barberà, J., Chesson, A., Cocconcelli, P. S., Debski, B., & Wester, P. (2009). Safety of use of colouring agents in animal nutrition—Part III: ß-apo-8'-carotenal, ethyl ester of ß-apo-8'-carotenoic acid, lutein, zeaxanthin and concluding remarks. *EFSA Journal*, *7*(5), 1098. https://doi.org/10.2903/J.EFSA.2009.1098

Braakhuis, A. J., Donaldson, C. I., Lim, J. C., & Donaldson, P. J. (2019). Nutritional strategies to prevent lens cataract: Current status and future strategies. *Nutrients*, *11*(5), 1186.

Brazionis, L., Rowley, K., Itsiopoulos, C., & O'Dea, K. (2009). Plasma carotenoids and diabetic retinopathy. *The British Journal of Nutrition*, *101*(2), 270–277. https://doi.org/10.1017/S0007114508006545

Brown, L., Rimm, E. B., Seddon, J. M., Giovannucci, E. L., Chasan-Taber, L., Spiegelman, D., & Hankinson, S. E. (1999). A prospective study of carotenoid intake and risk of cataract extraction in US men. *The American Journal of Clinical Nutrition*, *70*(4), 517–524.

Bunnell, R. H., Marusich, W. L., & Bauernfeind, J. C. (1961). β-Apo-8′-carotenal as an egg yolk pigmenter. *Poultry Science*, *41*(4), 1109–1115. https://doi.org/10.3382/ps.0411109

Cabot, F., Saad, A., McAlinden, C., Haddad, N. M., Grise-Dulac, A., & Gatinel, D. (2013). Objective assessment of crystalline lens opacity level by measuring ocular light scattering with a double-pass system. *American Journal of Ophthalmology*, *155*(4), 629–635.

Cai, J., Nelson, K. C., Wu, M., Sternberg, P. Jr, & Jones, D. P. (2000). Oxidative damage and protection of the RPE. *Progress in Retinal and Eye Research*, *19*(2), 205–221.

Çalışlar, S. (2019). The important of beta carotene on poultry nutrition. *Selcuk Journal of Agricultural and Food Sciences*, *33*(3), 252–259.

Cangemi, F. E. (2007). TOZAL study : An open case control study of an oral antioxidant and omega-3 supplement for dry AMD. *BMC Ophthalmology*, *7*(1), 1–10.

Cardinault, N., Abalain, J.-H., Sairafi, B., Coudray, C., Grolier, P., Rambeau, M., & Rock, E. (2005). Lycopene but not lutein nor zeaxanthin decreases in serum and lipoproteins in age-related macular degeneration patients. *Clinica Chimica Acta; International Journal of Clinical Chemistry*, *357*(1), 34–42. https://doi.org/10.1016/j.cccn.2005.01.030

Carughi, A., & Hooper, F. G. (1994). Plasma carotenoid concentrations before and after supplementation with a carotenoid mixture. *The American Journal of Clinical Nutrition*, *59*(4), 896–899. https://doi.org/10.1093/ajcn/59.4.896

Challiol, C. F., Bastien, A., Giambruni, J. M., & Fernandez Meijide, N. (2018). Retinal toxicity due to canthaxanthin. Case series. *Archivos De La Sociedad Espanola De Oftalmologia*, *93*(8), 411–415. https://doi.org/10.1016/j.oftal.2018.01.011

Chasan-Taber, L., Willett, W. C., Seddon, J. M., Stampfer, M. J., Rosner, B., Colditz, G. A., & Hankinson, S. E. (1999). A prospective study of carotenoid and vitamin A intakes and risk of cataract extraction in US women. *The American Journal of Clinical Nutrition*, *70*(4), 509–516.

Chen, C.-L., Chen, Y.-H., Liang, C.-M., Tai, M.-C., Lu, D.-W., & Chen, J.-T. (2018). Glucosamine-induced autophagy through AMPK–mTOR pathway attenuates lipofuscin-like autofluorescence in human retinal pigment epithelial cells in vitro. *International Journal of Molecular Sciences*, *19*(5), 1416.

Chew, E. Y., Clemons, T., SanGiovanni, J. P., Danis, R., Domalpally, A., McBee, W., & Group, A. R. (2012). The age-related eye disease study 2 (AREDS2): Study design and baseline characteristics (AREDS2 report number 1). *Ophthalmology*, *119*(11), 2282–2289.

Chous, A. P., Richer, S. P., Gerson, J. D., & Kowluru, R. A. (2016). The diabetes visual function supplement study (DiVFuSS). *British Journal of Ophthalmology*, *100*(2), 227–234.

Christen, W. G., Liu, S., Glynn, R. J., Gaziano, J. M., & Buring, J. E. (2008). Dietary carotenoids, vitamins C and E, and risk of cataract in women: A prospective study. *Archives of Ophthalmology*, *126*(1), 102–109.

Chung, H.-Y., Rasmussen, H. M., & Johnson, E. J. (2004). Lutein bioavailability is higher from lutein-enriched eggs than from supplements and spinach in men. *The Journal of Nutrition*, *134*(8), 1887–1893.

Ciulla, T. A., Curran-Celantano, J., Cooper, D. A., Hammond, B. R. Jr, Danis, R. P., Pratt, L. M., & Filloon, T. G. (2001). Macular pigment optical density in a midwestern sample. *Ophthalmology*, *108*(4), 730–737.

Clevidence, B. A., & Bieri, J. G. (1993). Association of carotenoids with human plasma lipoproteins. In L. Packer (Ed.), *Carotenoids part B: Metabolism, genetics, and biosynthesis*: *Vol. 214. Methods in enzymology* (pp. 33–46). Academic Press Inc. https://doi.org/10.1016/0076-6879(93)14051-J

Coleman, H. R., Chan, C.-C., Ferris, F. L. 3rd, & Chew, Y. E. (2008). Age-related macular degeneration. *The Lancet*, *372*(9652), 1835–1845.

Curran-Celentano, J., Hammond, B. R. Jr, Ciulla, T. A., Cooper, D. A., Pratt, L. M., & Danis, R. B. (2001). Relation between dietary intake, serum concentrations, and retinal concentrations of lutein and zeaxanthin in adults in a Midwest population. *The American Journal of Clinical Nutrition*, *74*(6), 796–802.

Daicker, B., Schiedt, K., Adnett, J.J., & Bermond, P. (1987). Canthaxanthine retinopathy. *Graefe's Archive for Clinical and Experimental Ophthalmology*, *225*, 189–197. https://doi.org/10.1007/BF02175448

Dansou, D. M., Wang, H., Nugroho, R. D., He, W., Zhao, Q., Tang, C., & Zhang, J. (2021). Effects of duration and supplementation dose with astaxanthin on egg fortification. *Poultry Science*, *100*(9), 101304.

Davey, P. G., Henderson, T., Lem, D. W., Weis, R., Amonoo-Monney, S., & Evans, D. W. (2020). Visual function and macular carotenoid changes in eyes with retinal drusen—An open label randomized controlled trial to compare a micronized lipid-based carotenoid liquid supplementation and AREDS-2 formula. *Nutrients, 12*(11), 3271.

Dawczynski, J., Jentsch, S., Schweitzer, D., Hammer, M., Lang, G. E., & Strobel, J. (2013). Long-term effects of lutein, zeaxanthin and omega-3-LCPUFAs supplementation on optical density of macular pigment in AMD patients : The LUTEGA study. *Graefe's Archive for Clinical and Experimental Ophthalmology = Albrecht Von Graefes Archiv Fur Klinische Und Experimentelle Ophthalmologie, 251*(12), 2711–2723. https://doi.org/10.1007/s00417-013-2376-6

de Miguel, T., Sieiro, C., Poza, M., & Villa, T. G. (2001). Analysis of canthaxanthin and related pigments from *Gordonia jacobaea* mutants. *Journal of Agricultural and Food Chemistry, 49*(3), 1200–1202.

Delcourt, C., Carriere, I., Delage, M., Barberger-Gateau, P., Schalch, W., & Group, P. S. (2006). Plasma lutein and zeaxanthin and other carotenoids as modifiable risk factors for age-related maculopathy and cataract: The POLA study. *Investigative Ophthalmology & Visual Science, 47*(6), 2329–2335.

Dherani, M., Murthy, G. V., Gupta, S. K., Young, I. S., Maraini, G., Camparini, M., & Fletcher, A. E. (2008). Blood levels of vitamin C, carotenoids and retinol are inversely associated with cataract in a North Indian population. *Investigative Ophthalmology & Visual Science, 49*(8), 3328–3335.

Donders, F. C. (1855). Beiträge zur pathologischen anatomie des auges. *Archiv Für Ophthalmologie*, 1(2), 106–118.

Dose, J., Matsugo, S., Yokokawa, H., Koshida, Y., Okazaki, S., Seidel, U., & Esatbeyoglu, T. (2016). Free radical scavenging and cellular antioxidant properties of astaxanthin. *International Journal of Molecular Sciences, 17*(1), 103.

During, A., Doraiswamy, S., & Harrison, E. H. (2008). Xanthophylls are preferentially taken up compared with β-carotene by retinal cells via a SRBI-dependent mechanism. *Journal of Lipid Research, 49*(8), 1715–1724.

Ermakov, I. V., Ermakova, M. R., Gellermann, W., & Bernstein, P. S. (2004). Noninvasive selective detection of lycopene and beta-carotene in human skin using Raman spectroscopy. *Journal of Biomedical Optics, 9*(2), 139–148. https://doi.org/10.1117/1.1646172

Evans, J. R. (2006). Antioxidant vitamin and mineral supplements for slowing the progression of age-related macular degeneration. *Cochrane Database of Systematic Reviews*, (2). https://doi.org/10.1002/14651858.CD000254.pub2

Eye Disease Case Control Study (1993). Antioxidant status and neovascular age-related macular degeneration. *Archives of Ophthalmology, 111*, 104–109.

Fang, T. J., & Chiou, T. Y. (1996). Batch cultivation and astaxanthin production by a mutant of the red yeast, phaffia rhodozyma NCHU-FS501. *Journal of Industrial Microbiology and Biotechnology, 16*(3), 175–181.

Faruk, M. U., Roos, F. F., & Cisneros-Gonzalez, F. (2018). A meta-analysis on the effect of canthaxanthin on egg production in brown egg layers. *Poultry Science, 97*(1), 84–87. https://doi.org/10.3382/PS/PEX236

Fletcher, D. L., Harms, R. H., & Janky, D. M. (1978). Yolk color characteristics, xanthophyll availability, and a model system for predicting egg yolk color using beta-apo-8′-carotenal and canthaxanthin. *Poultry Science, 57*, 624–629. https://doi.org/10.3382/ps.0570624

Fujimura, S., Ueda, K., Nomura, Y., & Yanagi, Y. (2016). Preliminary analysis of the relationship between serum lutein and zeaxanthin levels and macular pigment optical density. *Clinical Ophthalmology (Auckland, NZ), 10*, 2149.

Furr, H. C., & Clark, R. M. (1997). Intestinal absorption and tissue distribution of carotenoids. *The Journal of Nutritional Biochemistry, 8*(7), 364–377.

Gale, C. R., Hall, N. F., Phillips, D. I., & Martyn, C. N. (2003). Lutein and zeaxanthin status and risk of age-related macular degeneration. *Investigative Ophthalmology & Visual Science, 44*(6), 2461–2465.

Gao, S., Li, R., Heng, N., Chen, Y., Wang, L., Li, Z., & Xing, K. (2020). Effects of dietary supplementation of natural astaxanthin from *Haematococcus pluvialis* on antioxidant capacity, lipid metabolism, and accumulation in the egg yolk of laying hens. *Poultry Science, 99*(11), 5874–5882.

Goldberg, J., Flowerdew, G., Smith, E., Brody, J. A., & Tso, M. O. (1988). Factors associated with age-related macular degeneration: An analysis of data from the fi1r8t national health and nutrition examination survey. *American Journal of Epidemiology, 128*(4), 700–710.

Gong, X., & Rubin, L. P. (2015). Role of macular xanthophylls in prevention of common neovascular retinopathies: Retinopathy of prematurity and diabetic retinopathy. *Archives of Biochemistry and Biophysics, 572*, 40–48.

Goñi, I., Serrano, J., & Saura-Calixto, F. (2006). Bioaccessibility of β-carotene, lutein, and lycopene from fruits and vegetables. *Journal of Agricultural and Food Chemistry, 54*(15), 5382–5387.

Goodrow, E. F., Wilson, T. A., Houde, S. C., Vishwanathan, R., Scollin, P. A., Handelman, G., & Nicolosi, R. J. (2006). Consumption of one egg per day increases serum lutein and zeaxanthin concentrations in older adults without altering serum lipid and lipoprotein cholesterol concentrations. *The Journal of Nutrition, 136*(10), 2519–2524.

Gopinath, B., Liew, G., Tang, D., Burlutsky, G., Flood, V. M., & Mitchell, P. (2020). Consumption of eggs and the 15-year incidence of age-related macular degeneration. *Clinical Nutrition, 39*, 580–584. https://doi.org/10.1016/j.clnu.2019.03.009

Gorusupudi, A., Shyam, R., Li, B., Vachali, P., Subhani, Y. K., Nelson, K., & Bernstein, P. S. (2016). Developmentally regulated production of *meso*-zeaxanthin in chicken retinal pigment epithelium/choroid and retina. *Investigative Ophthalmology & Visual Science, 57*(4), 1853–1861.

Goto, S., Kogure, K., Abe, K., Kimata, Y., Kitahama, K., Yamashita, E., & Terada, H. (2001). Efficient radical trapping at the surface and inside the phospholipid membrane is responsible for highly potent antiperoxidative activity of the carotenoid astaxanthin. *Biochimica et Biophysica Acta (BBA)-Biomembranes, 1512*(2), 251–258.

Grashorn, M. A., & Steinberg, W. (2002). Deposition rates of canthaxanthin in egg yolks. *Archiv für Geflügelkunde, 66*(6), 258–262.

Gupta, A. K., Haberman, H. F., Pawlowski, D., Shulman, G., & Menon, I. A. (1985). Canthaxanthin. *International Journal of Dermatology, 24*(1), 528–532.

Gupta, S. K., Trivedi, D., Srivastava, S., Joshi, S., Halder, N., & Verma, S. D. (2003). Lycopene attenuates oxidative stress induced experimental cataract development: An in vitro and in vivo study. *Nutrition, 19*(9), 794–799.

Hallak, J. A., de Sisternes, L., Osborne, A., Yaspan, B., Rubin, D. L., & Leng, T. (2019). Imaging, genetic, and demographic factors associated with conversion to neovascular age-related macular degeneration: Secondary analysis of a randomized clinical trial. *JAMA Ophthalmology, 137*(7), 738–744.

Hammershøj, M., Kidmose, U., & Steenfeldt, S. (2010). Deposition of carotenoids in egg yolk by short-term supplement of coloured carrot (*Daucus carota*) varieties as forage material for egg-laying hens. *Journal of the Science of Food and Agriculture, 90*, 1163–1171. https://doi.org/10.1002/jsfa.3937

Hammond, B. R. Jr, Curran-Celentano, J., Judd, S., Fuld, K., Krinsky, N. I., Wooten, B. R., & Snodderly, D. M. (1996). Sex differences in macular pigment optical density: Relation to plasma carotenoid concentrations and dietary patterns. *Vision Research, 36*(13), 2001–2012.

Hammond, B. R. Jr, Wooten, B. R., & Snodderly, D. M. (1997). Density of the human crystalline lens is related to the macular pigment carotenoids, lutein and zeaxanthin. *Optometry and Vision Science: Official Publication of the American Academy of Optometry, 74*(7), 499–504.

Hammond, B. R., Johnson, E. J., Russell, R. M., Krinsky, N. I., Yeum, K.-J., Edwards, R. B., & Snodderly, D. M. (1997). Dietary modification of human macular pigment density. *Investigative Ophthalmology & Visual Science, 38*(9), 1795–1801.

Handelman, G. J., Nightingale, Z. D., Lichtenstein, A. H., Schaefer, E. J., & Blumberg, J. B. (1999). Lutein and zeaxanthin concentrations in plasma after dietary supplementation with egg yolk. *American Journal of Clinical Nutrition, 70*, 247–251. https://doi.org/10.1093/ajcn.70.2.247

Hankinson, S. E., Stampfer, M. J., Seddon, J. M., Colditz, G. A., Rosner, B., Speizer, F. E., & Willett, W. C. (1992). Nutrient intake and cataract extraction in women: A prospective study. *British Medical Journal, 305*(6849), 335–339.

Haute Autorité De Santé. (2022). Dégénérescence maculaire liée à l'âge : Outil d'amélioration des pratiques professionnelles. https://www.has-sante.fr/jcms/p_3383781/fr/degenerescence-maculaire-liee-a-l-age

Hawkins, B. S., Bird, A., Klein, R., & West, S. K. (1999). Epidemiology of age-related macular degeneration. *Molecular Vision, 5*, 26.

Heying, E. K., Ziemer, K. L., Tanumihardjo, J. P., Palacios-Rojas, N., & Tanumihardjo, S. A. (2018). β-cryptoxanthin–biofortified hen eggs enhance vitamin A status when fed to male Mongolian gerbils. *The Journal of Nutrition, 148*(8), 1236–1243. https://doi.org/10.1093/JN/NXY117

Hogg, R. E., Woodside, J. V., McGrath, A., Young, I. S., Vioque, J. L., Chakravarthy, U., & Soubrane, G. (2017). Mediterranean diet score and its association with age-related macular degeneration: The European eye study. *Ophthalmology, 124*(1), 82–89.

Hollander, D., & Ruble, P. E. (1978). Beta-carotene intestinal absorption: Bile, fatty acid, pH, and flow rate effects on transport. *The American Journal of Physiology, 235*(6), E686–E691. https://doi.org/10.1152/ajpendo.1978.235.6.E686

Honda, M., Kawashima, Y., Hirasawa, K., Uemura, T., Sun, J., & Hayashi, Y. (2021). Astaxanthin Z-isomer-rich diets enhance egg yolk pigmentation in laying hens compared to that in all-E-isomer-rich diets. *Animal Science Journal, 92*(1), e13512.

Hu, B.-J., Hu, Y.-N., Lin, S., Ma, W.-J., & Li, X.-R. (2011). Application of lutein and zeaxanthin in nonproliferative diabetic retinopathy. *International Journal of Ophthalmology, 4*(3), 303–306. https://doi.org/10.3980/j.issn.2222-3959.2011.03.19

Huang, Y.-M., Dou, H.-L., Huang, F.-F., Xu, X.-R., Zou, Z.-Y., Lu, X.-R., & Lin, X.-M. (2015). Changes following supplementation with lutein and zeaxanthin in retinal function in eyes with early age-related macular degeneration: A randomised, double-blind, placebo-controlled trial. *British Journal of Ophthalmology, 99*(3), 371–375.

Ishikawa, S., Hashizume, K., Nishigori, H., Tezuka, Y., Sanbe, A., & Kurosaka, D. (2015). Effect of astaxanthin on cataract formation induced by glucocorticoids in the chick embryo. *Current Eye Research, 40*(5), 535–540.

Islam, K. M. S., & Schweigert, F. J. (2015). Comparison of three spectrophotometric methods for analysis of egg yolk carotenoids. *Food Chemistry, 172*, 233–237.

Jacques, P. F., & Chylack, L. T. Jr (1991). Epidemiologic evidence of a role for the antioxidant vitamins and carotenoids in cataract prevention. *The American Journal of Clinical Nutrition, 53*(1), 352S–355S.

Jacques, P. F., Hartz, S. C., Chylack, L. T. Jr, McGandy, R. B., & Sadowski, J. A. (1988). Nutritional status in persons with and without senile cataract: Blood vitamin and mineral levels. *The American Journal of Clinical Nutrition, 48*(1), 152–158.

Janani, R., Anitha, R. E., Perumal, M. K., Divya, P., & Baskaran, V. (2021). Astaxanthin mediated regulation of VEGF through HIF1α and XBP1 signaling pathway: An insight from ARPE-19 cell and streptozotocin mediated diabetic rat model. *Experimental Eye Research, 206*, 108555.

Joint FAO/WHO Expert Committee on Food Additives, & World Health Organization. (1988). *Toxicological evaluation of certain food additives* (Vol. 31). Cambridge University Press.

Juturu, V., Bowman, J. P., Stringham, N. T., & Stringhan, J. (2016). Bioavailability of lutein/zeaxanthin isomers and macular pigment optical density response to macular carotenoid supplementation: A randomized double blind placebo controlled study. *New Frontiers in Ophthalmology, 2*, 140–145.

Kang, D. K., Kim, S. I., Cho, C. H., Yim, Y. H., & Kim, H. S. (2003). Use of lycopene, an antioxidant carotenoid, in laying hens for egg yolk pigmentation. *Asian-Australasian Journal of Animal Sciences, 16*(12), 1799–1803. https://doi.org/10.5713/ajas.2003.1799

Kannan, R., Sreekumar, P. G., & Hinton, D. R. (2016). Alpha crystallins in the retinal pigment epithelium and implications for the pathogenesis and treatment of age-related macular degeneration. *Biochimica et Biophysica Acta (BBA)-General Subjects, 1860*(1), 258–268.

Karppi, J., Laukkanen, J. A., & Kurl, S. (2012). Plasma lutein and zeaxanthin and the risk of age-related nuclear cataract among the elderly Finnish population. *British Journal of Nutrition, 108*(1), 148–154.

Kelly, E. R., Plat, J., Haenen, G. R. M. M., Kijlstra, A., & Berendschot, T. T. J. M. (2014). The effect of modified eggs and an egg-yolk based beverage on serum lutein and zeaxanthin concentrations and macular pigment optical density: Results from a randomized trial. *PLoS One, 9*(3), e92659. https://doi.org/10.1371/journal.pone.0092659

Khachik, F., Beecher, G. R., & Smith, J. C. (1995). Lutein, lycopene, and their oxidative metabolites in chemoprevention of cancer. *Journal of Cellular Biochemistry*, 22(Supplement), 236–246. https://doi.org/10.1002/jcb.240590830

Khachik, F., Carvalho, L., Bernstein, P. S., Muir, G. J., Zhao, D.-Y., & Katz, N. B. (2002a). Chemistry, distribution, and metabolism of tomato carotenoids and their impact on human health. *Experimental Biology and Medicine (Maywood, N.J.)*, *227*(10), 845–851. https://doi.org/10.1177/153537020222701002

Khachik, F., De Moura, F. F., Zhao, D.-Y., Aebischer, C.-P., & Bernstein, P. S. (2002b). Transformations of selected carotenoids in plasma, liver, and ocular tissues of humans and in nonprimate animal models. *Investigative Ophthalmology & Visual Science*, *43*(11), 3383–3392. PMID: 12407147.

Kim, Y.-S., Park, C.-S., & Oh, D.-K. (2010). Hydrophobicity of residue 108 specifically affects the affinity of human β-carotene 15, 15′-monooxygenase for substrates with two ionone rings. *Biotechnology Letters*, *32*(6), 847–853.

Klein, B. E., Howard, K. P., Iyengar, S. K., Sivakumaran, T. A., Meyers, K. J., Cruickshanks, K. J., & Klein, R. (2014). Sunlight exposure, pigmentation, and incident age-related macular degeneration. *Investigative Ophthalmology & Visual Science*, *55*(9), 5855–5861.

Klein, R., Klein, B. E., Jensen, S. C., Mares-Perlman, J. A., Cruickshanks, K. J., & Palta, M. (1999). Age-related maculopathy in a multiracial United States population: The national health and nutrition examination survey III. *Ophthalmology*, *106*(6), 1056–1065.

Knekt, P., Heliövaara, M., Rissanen, A., Aromaa, A., & Aaran, R.-K. (1992). Serum antioxidant vitamins and risk of cataract. *British Medical Journal*, *305*(6866), 1392–1394.

Kotake-Nara, E., & Nagao, A. (2011). Absorption and metabolism of xanthophylls. *Marine Drugs*, *9*, 1024–1037. https://doi.org/10.3390/md9061024

Kuo, C., Green, C. R., Rupenthal, I. D., & Mugisho, O. O. (2020). Connexin43 hemichannel block protects against retinal pigment epithelial cell barrier breakdown. *Acta Diabetologica*, *57*(1), 13–22.

Landrum, J. T., & Bone, R. A. (2001). Lutein, zeaxanthin, and the macular pigment. *Archives of Biochemistry and Biophysics*, *385*(1), 28–40. https://doi.org/10.1006/abbi.2000.2171

Landrum, J. T., Bone, R. A., Joa, H., Kilburn, M. D., Moore, L. L., & Sprague, K. E. (1997). A one-year study of the macular pigment: The effect of 140 days of a lutein supplement. *Experimental Eye Research*, *65*(1), 57–62. https://doi.org/10.1006/exer.1997.0309

Lechanteur, Y. T., van de Ven, J. P., Smailhodzic, D., Boon, C. J., Klevering, B. J., Fauser, S., & Hoyng, C. B. (2012). Genetic, behavioral, and sociodemographic risk factors for second eye progression in age-related macular degeneration. *Investigative Ophthalmology & Visual Science*, *53*(9), 5846–5852.

Lennikov, A., Kitaichi, N., Fukase, R., Murata, M., Noda, K., Ando, R., & Ishida, S. (2012). Amelioration of ultraviolet-induced photokeratitis in mice treated with astaxanthin eye drops. *Molecular Vision*, *18*, 455–464.

Leske, M. C., Chylack, L. T., & Wu, S.-Y. (1991). The lens opacities case-control study: Risk factors for cataract. *Archives of Ophthalmology*, *109*(2), 244–251.

Li, B., George, E. W., Rognon, G. T., Gorusupudi, A., Ranganathan, A., Chang, F. Y., & Bernstein, P. S. (2020). Imaging lutein and zeaxanthin in the human retina with confocal resonance Raman microscopy. *Proceedings of the National Academy of Sciences of the United States of America*, *117*(22), 12352–12358. https://doi.org/10.1073/pnas.1922793117

Liu, X.-H., Yu, R.-B., Liu, R., Hao, Z.-X., Han, C.-C., Zhu, Z.-H., & Ma, L. (2014). Association between lutein and zeaxanthin status and the risk of cataract: A meta-analysis. *Nutrients*, *6*(1), 452–465.

Lokaewmanee, K., Yamauchi, K. E., Komori, T., & Saito, K. (2009). Effects on egg yolk colour of paprika or paprika combined with marigold flower extracts. *Italian Journal of Animal Science*, *9*(4), 356–359. https://doi.org/10.4081/ijas.2010.e67

Loughman, J., Loskutova, E., Butler, J. S., Siah, W. F., & O'Brien, C. (2021). Macular pigment response to lutein, zeaxanthin, and *meso*-zeaxanthin supplementation in open-angle glaucoma: A randomized controlled trial. *Ophthalmology Science*, *1*(3), 100039.

Lyle, B. J., Mares-Perlman, J. A., Klein, B. E., Klein, R., & Greger, J. L. (1999). Antioxidant intake and risk of incident age-related nuclear cataracts in the beaver dam eye study. *American Journal of Epidemiology*, *149*(9), 801–809.

Ma, L., Hao, Z., Liu, R., Yu, R., Shi, Q., & Pan, J. (2014). A dose–response meta-analysis of dietary lutein and zeaxanthin intake in relation to risk of age-related cataract. *Graefe's Archive for Clinical and Experimental Ophthalmology*, *252*(1), 63–70.

Mares-Perlman, J. A., Brady, W. E., Klein, R., Klein, B. E., Bowen, P., Stacewicz-Sapuntzakis, M., & Palta, M. (1995). Serum antioxidants and age-related macular degeneration in a population-based case-control study. *Archives of Ophthalmology*, *113*(12), 1518–1523. https://doi.org/10.1001/archopht.1995.01100120048007

Mares-Perlman, J. A., Brady, W. E., Klein, B. E., Klein, R., Haus, G. J., Palta, M., & Shoff, S. M. (1995). Diet and nuclear lens opacities. *American Journal of Epidemiology*, *141*(4), 322–334. https://doi.org/10.1093/aje/141.4.322

Mares-Perlman, J. A., Fisher, A. I., Klein, R., Palta, M., Block, G., Millen, A. E., & Wright, J. D. (2001). Lutein and zeaxanthin in the diet and serum and their relation to age-related maculopathy in the third national health and nutrition examination survey. *American Journal of Epidemiology*, *153*(5), 424–432.

Mares, J. A., Voland, R. P., Sondel, S. A., Millen, A. E., LaRowe, T., Moeller, S. M., & Tinker, L. (2011). Healthy lifestyles related to subsequent prevalence of age-related macular degeneration. *Archives of Ophthalmology*, *129*(4), 470–480.

Marounek, M., & Pebriansyah, A. (2018). Use of carotenoids in feed mixtures for poultry: A review. *Agricultura Tropica et Subtropica*, *51*(3), 107–111. https://doi.org/10.2478/ats-2018-0011

Marusich, W., De Ritter, E., & Bauernfeind, J. C. (1960). Evaluation of carotenoid pigments for coloring egg yolks. *Poultry Science*, *39*(6), 1338–1345. https://doi.org/10.3382/ps.0391338

Massacesi, A. L. (2001). The effect of oral supplementation of macular carotenoids (lutein and zeaxanthin) on the prevention of age-related macular degeneration: 18 months of follow up study. *Association of Research in Vision and Ophthalmology*, *42*, 128–135.

Ma, L., Yan, S.-F., Huang, Y.-M., Lu, X.-R., Qian, F., Pang, H.-L., & Xiao, X. (2012). Effect of lutein and zeaxanthin on macular pigment and visual function in patients with early age-related macular degeneration. *Ophthalmology*, *119*(11), 2290–2297.

McNulty, H., Jacob, R. F., & Mason, R. P. (2008). Biologic activity of carotenoids related to distinct membrane physicochemical interactions. *The American Journal of Cardiology, 101*(10), S20–S29.

Merle, B. M., Colijn, J. M., Cougnard-Grégoire, A., Koning-Backus, de, Delyfer, A. P., Kiefte-de Jong, M.-N., & Samieri, J. C. (2019). Mediterranean diet and incidence of advanced age-related macular degeneration: The EYE-RISK consortium. *Ophthalmology, 126*(3), 381–390.

Moeller, S. M., Parekh, N., Tinker, L., Ritenbaugh, C., Blodi, B., Wallace, R. B., & Group, C. R. S. (2006). Associations between intermediate age-related macular degeneration and lutein and zeaxanthin in the carotenoids in age-related eye disease study (CAREDS): Ancillary study of the Women's health initiative. *Archives of Ophthalmology, 124*(8), 1151–1162.

Moran, N. E., Mohn, E. S., Hason, N., Erdman, J. W. Jr, & Johnson, E. J. (2018). Intrinsic and extrinsic factors impacting absorption, metabolism, and health effects of dietary carotenoids. *Advances in Nutrition, 9*(4), 465–492.

Moreno, J. A., Díaz-Gómez, J., Fuentes-Font, L., Angulo, E., Gosálvez, L. F., Sandmann, G., & Christou, P. (2020). Poultry diets containing (keto) carotenoid-enriched maize improve egg yolk color and maintain quality. *Animal Feed Science and Technology, 260*, 114334.

Moreno, J. A., Díaz-Gómez, J., Nogareda, C., Angulo, E., Sandmann, G., Portero-Otin, M., & Zhu, C. (2016). The distribution of carotenoids in hens fed on biofortified maize is influenced by feed composition, absorption, resource allocation and storage. *Scientific Reports, 6*(1), 1–11.

Mortensen, A., & Skibsted, L. H. (1997). Free radical transients in photobleaching of xanthophylls and carotenes. *Free Radical Research, 26*(6), 549–563.

Moussa, M., Landrier, J.-F., Reboul, E., Ghiringhelli, O., Coméra, C., Collet, X., & Borel, P. (2008). Lycopene absorption in human intestinal cells and in mice involves scavenger receptor class B type I but not Niemann-pick C1-like 1. *The Journal of Nutrition, 138*(8), 1432–1436. https://doi.org/10.1093/jn/138.8.1432

Mugnai, C., Dal Bosco, A., & Castellini, C. (2009). Effect of rearing system and season on the performance and egg characteristics of Ancona laying hens. *Italian Journal of Animal Science, 8*(2), 175–188.

Murray, I. J., Makridaki, M., van der Veen, R. L. P., Carden, D., Parry, N. R. A., & Berendschot, T. T. J. M. (2013). Lutein supplementation over a one-year period in early AMD might have a mild beneficial effect on visual acuity: The CLEAR study. *Investigative Ophthalmology & Visual Science, 54*(3), 1781–1788. https://doi.org/10.1167/iovs.12-10715

Na, J.-C., Song, J.-Y., Lee, B.-D., Lee, S.-J., Lee, C.-Y., & An, G.-H. (2004). Effect of polarity on absorption and accumulation of carotenoids by laying hens. *Animal Feed Science and Technology, 117*(3–4), 305–315.

Nagel, A. (1868). Ueber chorioiditis areolaris und über krystalle im augenhintergrund. On areolar choroiditis and crystalls in the eye fundus. *Klinische Monatsblätter für Augenheilkunde, 4*, 417–420.

Naguib, Y. M. (2000). Antioxidant activities of astaxanthin and related carotenoids. *Journal of Agricultural and Food Chemistry, 48*(4), 1150–1154. https://doi.org/10.1021/jf991106k

Nakajima, Y., Inokuchi, Y., Shimazawa, M., Otsubo, K., Ishibashi, T., & Hara, H. (2008). Astaxanthin, a dietary carotenoid, protects retinal cells against oxidative stress in-vitro and in mice in-vivo. *Journal of Pharmacy and Pharmacology, 60*(10), 1365–1374.

Neelam, K., Hogg, R. E., Stevenson, M. R., Johnston, E., Anderson, R., Beatty, S., & Chakravarthy, U. (2008). Carotenoids and co-antioxidants in age-related maculopathy: Design and methods. *Ophthalmic Epidemiology, 15*(6), 389–401.

Nimalaratne, C., Lopes-Lutz, D., Schieber, A., & Wu, J. (2012). Effect of domestic cooking methods on egg yolk xanthophylls. *Journal of Agricultural and Food Chemistry, 60*(51), 12547–12552. https://doi.org/10.1021/jf303828n

Nimalaratne, C., Savard, P., Gauthier, S. F., Schieber, A., & Wu, J. (2015). Bioaccessibility and digestive stability of carotenoids in cooked eggs studied using a dynamic in vitro gastrointestinal model. *Journal of Agricultural and Food Chemistry, 63*(11), 2956–2962. https://doi.org/10.1021/jf505615w

Nimalaratne, C., Wu, J., & Schieber, A. (2013). Egg yolk carotenoids: Composition, analysis, and effects of processing on their stability. *American Chemical Society, 1134*, 219–225. https://doi.org/10.1021/bk-2013-1134.ch018

Niu, Z., Fu, J., Gao, Y., & Liu, F. (2008a). Influence of paprika extract supplement on egg quality of laying hens fed wheat-based diet. *International Journal of Poultry Science, 7*(9), 887–889.

Niu, Z., Fu, J., Gao, Y., & Liu, F. (2008b). Influence of paprika extract supplement on egg quality of laying hens fed wheat-based diet. *International Journal of Poultry Science, 7*, 887–889. https://doi.org/10.3923/ijps.2008.887.889

Nunes, S., Alves, D., Barreto, P., Raimundo, M., Luz Cachulo, da, Farinha, M., & Ribeiro, C. (2018). Adherence to a Mediterranean diet and its association with age-related macular degeneration. The Coimbra. *Nutrition, 51*, 6–12.

Olmedilla, B., Granado, F., Blanco, I., & Vaquero, M. (2003). Lutein, but not α-tocopherol, supplementation improves visual function in patients with age-related cataracts: A 2-y double-blind, placebo-controlled pilot study. *Nutrition, 19*(1), 21–24.

Olmedilla, B., Granado, F., Blanco, I., Vaquero, M., & Cajigal, C. (2001). Lutein in patients with cataracts and age-related macular degeneration: A long-term supplementation study. *Journal of the Science of Food and Agriculture, 81*(9), 904–909.

Olson, J. B., Ward, N. E., & Koutsos, E. A. (2008). Lycopene incorporation into egg yolk and effects on laying hen immune function. *Poultry Science, 87*(12), 2573–2580. https://doi.org/10.3382/PS.2008-00072

Østerlie, M., Bjerkeng, B., & Liaaen-Jensen, S. (2000). Plasma appearance and distribution of astaxanthin E/Z and R/S isomers in plasma lipoproteins of men after single dose administration of astaxanthin. *The Journal of Nutritional Biochemistry, 11*(10), 482–490.

Otsuka, T., Shimazawa, M., Inoue, Y., Nakano, Y., Ojino, K., Izawa, H., & Hara, H. (2016). Astaxanthin protects against retinal damage : Evidence from in vivo and in vitro retinal ischemia and reperfusion models. *Current Eye Research, 41*(11), 1465–1472.

Otsuka, T., Shimazawa, M., Nakanishi, T., Ohno, Y., Inoue, Y., Tsuruma, K., & Hara, H. (2013). Protective effects of a dietary carotenoid, astaxanthin, against light-induced retinal damage. *Journal of Pharmacological Sciences*, *123*(3), 209–218. https://doi.org/10.1254/jphs.13066fp

Paetau, I., Chen, H., Goh, N. M., & White, W. S. (1997). Interactions in the postprandial appearance of beta-carotene and canthaxanthin in plasma triacylglycerol-rich lipoproteins in humans. *The American Journal of Clinical Nutrition*, *66*(5), 1133–1143. https://doi.org/10.1093/ajcn/66.5.1133

Painsi, C., Hesterberg, K., Lademann, J., Knorr, D., Patzelt, A., Vandersee, S., & Lange-Asschenfeldt, B. (2019). Influence of storage and preservation techniques on egg-derived carotenoids: A substantial source for cutaneous antioxidants. *Skin Pharmacology and Physiology*, *32*(2), 65–71. https://doi.org/10.1159/000493147

Parisi, V., Tedeschi, M., Gallinaro, G., Varano, M., Saviano, S., Piermarocchi, S., & Group, C. S. (2008). Carotenoids and antioxidants in age-related maculopathy Italian study: Multifocal electroretinogram modifications after 1 year. *Ophthalmology*, *115*(2), 324–333.

Parker, R. S. (1996). Absorption, metabolism, and transport of carotenoids. *The FASEB Journal*, *10*(5), 542–551. https://doi.org/10.1096/fasebj.10.5.8621054

Pitts, D. G., Cullen, A. P., & Hacker, P. D. (1977). Ocular effects of ultraviolet radiation from 295 to 365 nm. *Investigative Ophthalmology & Visual Science*, *16*(10), 932–939.

Provis, J. M., Penfold, P. L., Cornish, E. E., Sandercoe, T. M., & Madigan, M. C. (2005). Anatomy and development of the macula: Specialisation and the vulnerability to macular degeneration. *Clinical and Experimental Optometry*, *88*(5), 269–281.

Ranga Rao, A., Raghunath Reddy, R. L., Baskaran, V., Sarada, R., & Ravishankar, G. A. (2010). Characterization of microalgal carotenoids by mass spectrometry and their bioavailability and antioxidant properties elucidated in rat model. *Journal of Agricultural and Food Chemistry*, *58*(15), 8553–8559. https://doi.org/10.1021/jf101187k

Rao, A. V., & Agarwal, S. (1998). Bioavailability and in vivo antioxidant properties of lycopene from tomato products and their possible role in the prevention of cancer. *Nutrition and Cancer*, *31*(3), 199–203. https://doi.org/10.1080/01635589809514703

Reboul, E. (2019). Mechanisms of carotenoid intestinal absorption: Where do we stand? *Nutrients*, *11*(4), 838. https://doi.org/10.3390/nu11040838

Richer, S. (1999). ARMD–pilot (case series) environmental intervention data. *Journal of the American Optometric Association*, *70*(1), 24–36.

Richer, S. P., Stiles, W., Graham-Hoffman, K., Levin, M., Ruskin, D., Wrobel, J., & Thomas, C. (2011). Randomized, double-blind, placebo-controlled study of zeaxanthin and visual function in patients with atrophic age-related macular degeneration: The zeaxanthin and visual function study (ZVF) FDA IND #78, 973. *Optometry*, *82*(11), 667–680. https://doi.org/10.1016/j.optm.2011.08.008

Richer, S., Stiles, W., Statkute, L., Pulido, J., Frankowski, J., Rudy, D., & Nyland, J. (2004). Double-masked, placebo-controlled, randomized trial of lutein and antioxidant supplementation in the intervention of atrophic age-related macular degeneration: The veterans LAST study (lutein antioxidant supplementation trial). *Optometry-Journal of the American Optometric Association*, *75*(4), 216–229.

Rock, C. L., Swendseid, M. E., Jacob, R. A., & McKee, R. W. (1992). Plasma carotenoid levels in human subjects fed a low carotenoid diet. *The Journal of Nutrition*, *122*(1), 96–100. https://doi.org/10.1093/jn/122.1.96

Rodriguez-Amaya, D. B. (2001). *A guide to carotenoid analysis in foods*. ILSI PRESS, OMNI Research. ISBN 1-57881-072-8

Rosenthal, J. M., Kim, J., de Monastario, F., Thompson, D. J., Bone, R. A., Landrum, J. T., & Schleicher, R. L. (2006). Dose-ranging study of lutein supplementation in persons aged 60 years or older. *Investigative Ophthalmology & Visual Science*, *47*(12), 5227–5233.

Sabour-Pickett, S., Beatty, S., Connolly, E., Loughman, J., Stack, J., Howard, A., & Myers, C. E. (2014). Supplementation with three different macular carotenoid formulations in patients with early age-related macular degeneration. *Retina*, *34*(9), 1757–1766.

Sasamoto, Y., Gomi, F., Sawa, M., Tsujikawa, M., & Nishida, K. (2011). Effect of 1-year lutein supplementation on macular pigment optical density and visual function. Graefe's *Clinical and Experimental Ophthalmology*, *249*(12), 1847–1854.

SCAN (Scientific Committee on Animal Nutrition). (2002). Opinion of the Scientific Committee on Animal Nutrition on the use of canthaxanthin in feeding stuffs for salmon and trout, laying hens, and other poultry, opinion expressed 2002.

SCF (Scientific Committee on Food). (1997). Minutes of the 107th Meeting of the Scientific Committee for Food, held on 12–13 June 1997 in Brussels.

Schlatterer, J., & Breithaupt, D. E. (2006). Xanthophylls in commercial egg yolks : Quantification and identification by HPLC and LC-(APCI)MS using a C30 phase. *Journal of Agricultural and Food Chemistry*, *54*(6), 2267–2273. https://doi.org/10.1021/jf053204d

Seddon, J. M., Ajani, U. A., Sperduto, R. D., Hiller, R., Blair, N., Burton, T. C., & Miller, D. T. (1994). Dietary carotenoids, vitamins A, C, and E, and advanced age-related macular degeneration. *JAMA*, *272*(18), 1413–1420.

Shalini, T., Jose, S. S., Prasanthi, P. S., Balakrishna, N., Viswanath, K., & Reddy, G. B. (2021). Carotenoid status in type 2 diabetes patients with and without retinopathy. *Food & Function*, *12*(10), 4402–4410. https://doi.org/10.1039/d0fo03321a

Shimokawa, T., Yoshida, M., Fukuta, T., Tanaka, T., Inagi, T., & Kogure, K. (2019). Efficacy of high-affinity liposomal astaxanthin on up-regulation of age-related markers induced by oxidative stress in human corneal epithelial cells. *Journal of Clinical Biochemistry and Nutrition*, *64*(1), 27–35. https://doi.org/10.3164/jcbn.18-27

Simonelli, F., Zarrilli, F., Mazzeo, S., Verde, V., Romano, N., Savoia, M., & Sacchetti, L. (2002). Serum oxidative and antioxidant parameters in a group of Italian patients with age-related maculopathy. *Clinica Chimica Acta; International Journal of Clinical Chemistry*, *320*(1–2), 111–115. https://doi.org/10.1016/s0009-8981(02)00056-6

Sirri, F., Iaffaldano, N., Minelli, G., Meluzzi, A., Rosato, M. P., & Franchini, A. (2007). Comparative pigmentation efficiency of high dietary levels of apo-ester and marigold extract on quality traits of whole liquid egg of two strains of laying hens. *Journal of Applied Poultry Research*, *16*(3), 429–437. https://doi.org/10.1093/JAPR/16.3.429

Snellen, E. L. M., Verbeek, A. L. M., Van Den Hoogen, G. W. P., Cruysberg, J. R. M., & Hoyng, C. B. (2002). Neovascular age-related macular degeneration and its relationship to antioxidant intake. *Acta Ophthalmologica Scandinavica*, *80*(4), 368–371.

Snodderly, D. M. (1995). Evidence for protection against age-related macular degeneration by carotenoids and antioxidant vitamins. *The American Journal of Clinical Nutrition*, *62*(6 Suppl), 1448S–1461S. https://doi.org/10.1093/ajcn/62.6.1448S

Snodderly, D. M., Auran, J. D., & Delori, F. C. (1984a). The macular pigment. II. Spatial distribution in primate retinasin primate retinas. *Investigative Ophthalmology & Visual Science*, *25*, 674–685.

Snodderly, D. M., Brown, P. K., Delori, F. C., & Auran, J. D. (1984b). The macular pigment. I. Absorbance spectra, localization, and discrimination from other yellow pigments in primate retinas. *Investigative Ophthalmology and Visual Science*, *25*(6), 660–673.

Sperduto, R. D., Hu, T.-S., Milton, R. C., Zhao, J.-L., Everett, D. F., Cheng, Q.-F., & Jun-Yao, L. (1993). The Linxian cataract studies: Two nutrition intervention trials. *Archives of Ophthalmology*, *111*(9), 1246–1253.

Stahl, W., & Sies, H. (1992). Uptake of lycopene and its geometrical isomers is greater from heat-processed than from unprocessed tomato juice in humans. *The Journal of Nutrition*, *122*(11), 2161–2166. https://doi.org/10.1093/jn/122.11.2161

Stahl, W., & Sies, H. (1996). Lycopene: A biologically important carotenoid for humans? *Archives of Biochemistry and Biophysics*, *336*(1), 1–9. https://doi.org/10.1006/abbi.1996.0525

Steinberg, W., Grashorn, M. A., Klünter, A. M., & Schierle, J. (2000). Comparative pigmentation efficiency of two products containing either apo-ester or tagetes extracts in egg yolks and liquid eggs. *Archiv für Geflügelkunde*, *64*(4), 180–187.

Steinmetz, J. D., Bourne, R. R., Briant, P. S., Flaxman, S. R., Taylor, H. R., Jonas, J. B., & Abu-Gharbieh, E. G. (2021). Causes of blindness and vision impairment in 2020 and trends over 30 years, and prevalence of avoidable blindness in relation to VISION 2020: The right to sight: An analysis for the global burden of disease study. *The Lancet Global Health*, *9*(2), e144–e160.

Storey, P., Hark, L., & Haller, J. A. (2014). Age-related macular degeneration: An overview. In V. R. Preedy (Ed.) *Handbook of nutrition, diet and the eye* (pp.11–20). Elsevier. https://doi.org/10.1016/B978-0-12-401717-7.00002-2

Straub, B. K., Stoeffel, P., Heid, H., Zimbelmann, R., & Schirmacher, P. (2008). Differential pattern of lipid droplet-associated proteins and de novo perilipin expression in hepatocyte steatogenesis. *Hepatology*, *47*(6), 1936–1946.

Stringham, J. M., & Stringham, N. T. (2016). Serum and retinal responses to three different doses of macular carotenoids over 12 weeks of supplementation. *Experimental Eye Research*, *151*, 1–8.

Suarez, D. (1969). Incorporation of lycopene in egg yolk. *Poultry Science*, *48*(2), 733–735. https://doi.org/10.3382/PS.0480733

Sujak, A. (2009). Interactions between canthaxanthin and lipid membranes—Possible mechanisms of canthaxanthin toxicity. *Cellular & Molecular Biology Letters*, *14*(3), 395–410. https://doi.org/10.2478/s11658-009-0010-8

Tan, J. S., Wang, J. J., Flood, V., Rochtchina, E., Smith, W., & Mitchell, P. (2008). Dietary antioxidants and the long-term incidence of age-related macular degeneration: The blue mountains eye study. *Ophthalmology*, *115*(2), 334–341.

Tavani, A., Negri, E., & La Vecchia, C. (1996). Food and nutrient intake and risk of cataract. *Annals of Epidemiology*, *6*(1), 41–46.

Taylor, A., Jacques, P. F., Chylack, L. T. Jr, Hankinson, S. E., Khu, P. M., Rogers, G., & Padhye, N. (2002). Long-term intake of vitamins and carotenoids and odds of early age-related cortical and posterior subcapsular lens opacities. *The American Journal of Clinical Nutrition*, *75*(3), 540–549.

Thürmann, P. A., Schalch, W., Aebischer, J.-C., Tenter, U., & Cohn, W. (2005). Plasma kinetics of lutein, zeaxanthin, and 3-dehydro-lutein after multiple oral doses of a lutein supplement. *The American Journal of Clinical Nutrition*, *82*(1), 88–97.

Thürmann, P. A., Steffen, J., Zwernemann, C., Aebischer, C.-P., Cohn, W., Wendt, G., & Schalch, W. (2002). Plasma concentration response to drinks containing β-carotene as carrot juice or formulated as a water dispersible powder. *European Journal of Nutrition*, *41*(5), 228–235.

Tikellis, G., Robman, L. D., Dimitrov, P., Nicolas, C., McCarty, C. A., & Guymer, R. H. (2007). Characteristics of progression of early age-related macular degeneration: The cardiovascular health and age-related maculopathy study. *Eye*, *21*(2), 169–176.

Titcomb, T. J., Kaeppler, M. S., Cook, M. E., Simon, P. W., & Tanumihardjo, S. A. (2019). Carrot leaves improve color and xanthophyll content of egg yolk in laying hens but are not as effective as commercially available marigold fortification. *Poultry Science*, *98*(10), 5208–5213. https://doi.org/10.3382/PS/PEZ257

Truscott, R. J. (2005). Age-related nuclear cataract—Oxidation is the key. *Experimental Eye Research*, *80*(5), 709–725. https://doi.org/10.1016/j.exer.2004.12.007

Tso, M. O., & Lam, T.-T. (1996). Method of retarding and ameliorating central nervous system and eye damage. Unites States Patent 5,527,533.

Tudor, C., & Pintea, A. (2020). A brief overview of dietary zeaxanthin occurrence and bioaccessibility. *Molecules*, *25*(18), 4067.

Vachali, P. P., Besch, B. M., Gonzalez-Fernandez, F., & Bernstein, P. S. (2013). Carotenoids as possible interphotoreceptor retinoid-binding protein (IRBP) ligands: A surface plasmon resonance (SPR) based study. *Archives of Biochemistry and Biophysics*, *539*(2), 181–186.

VandenLangenberg, G. M., Mares-Perlman, J. A., Klein, R., Klein, B. E., Brady, W. E., & Palta, M. (1998). Associations between antioxidant and zinc intake and the 5-year incidence of early age-related maculopathy in the beaver dam eye study. *American Journal of Epidemiology*, *148*(2), 204–214.

Vishwanathan, R., Goodrow-Kotyla, E. F., Wooten, B. R., Wilson, T. A., & Nicolosi, R. J. (2009). Consumption of 2 and 4 egg yolks/d for 5 wk increases macular pigment concentrations in older adults with low macular pigment taking cholesterol-lowering statins. *The American Journal of Clinical Nutrition*, *90*(5), 1272–1279. https://doi.org/10.3945/AJCN.2009.28013

Vu, H. T., Robman, L., Hodge, A., McCarty, C. A., & Taylor, H. R. (2006). Lutein and zeaxanthin and the risk of cataract:

The Melbourne visual impairment project. *Investigative Ophthalmology & Visual Science, 47*(9), 3783–3786.

Walker, L. A., Wang, T., Xin, H., & Dolde, D. (2012). Supplementation of laying-hen feed with palm tocos and algae astaxanthin for egg yolk nutrient enrichment. *Journal of Agricultural and Food Chemistry, 60*(8), 1989–1999.

Wang, J. J., Rochtchina, E., Smith, W., Klein, R., Klein, B. E., Joshi, T., & Mitchell, P. (2009). Combined effects of complement factor H genotypes, fish consumption, and inflammatory markers on long-term risk for age-related macular degeneration in a cohort. *American Journal of Epidemiology, 169*(5), 633–641.

Weigert, G., Kaya, S., Pemp, B., Sacu, S., Lasta, M., Werkmeister, R. M., & Schmetterer, L. (2011). Effects of lutein supplementation on macular pigment optical density and visual acuity in patients with age-related macular degeneration. *Investigative Ophthalmology & Visual Science, 52*(11), 8174–8178. https://doi.org/10.1167/iovs.11-7522

Wenzel, M., Seuss-Baum, I., & Schlich, E. (2011). Influences of storage time and temperature on the xanthophyll content of freeze-dried egg yolk. *Food Chemistry, 124*(4), 1343–1348. https://doi.org/10.1016/j.foodchem.2010.07.085

Widomska, J., & Subczynski, W. K. (2014). Why has nature chosen lutein and zeaxanthin to protect the retina? *Journal of Clinical & Experimental Ophthalmology, 5*(1), 326. https://doi.org/10.4172/2155-9570.1000326

Widomska, J., & Subczynski, W. K. (2019). Mechanisms enhancing the protective functions of macular xanthophylls in the retina during oxidative stress. *Experimental Eye Research, 178*, 238–246.

Wolf-Schnurrbusch, U. E. K., Zinkernagel, M. S., Munk, M. R., Ebneter, A., & Wolf, S. (2015). Oral lutein supplementation enhances macular pigment density and contrast sensitivity but not in combination with polyunsaturated fatty acids. *Investigative Ophthalmology & Visual Science, 56*(13), 8069–8074. https://doi.org/10.1167/iovs.15-17586

Wong, W. L., Su, X., Li, X., Cheung, C. M. G., Klein, R., Cheng, C.-Y., & Wong, T. Y. (2014). Global prevalence of age-related macular degeneration and disease burden projection for 2020 and 2040: A systematic review and meta-analysis. *The Lancet Global Health, 2*(2), e106–e116.

Yang, M., Chen, Y., Zhao, T., & Wang, Z. (2020). Effect of astaxanthin on metabolic cataract in rats with type 1 diabetes mellitus. *Experimental and Molecular Pathology, 113*, 104372.

Yu, C.-C., Nandrot, E. F., Dun, Y., & Finnemann, S. C. (2012). Dietary antioxidants prevent age-related retinal pigment epithelium actin damage and blindness in mice lacking αvβ5 integrin. *Free Radical Biology and Medicine, 52*(3), 660–670.

Zaheer, K. (2017). Hen egg carotenoids (lutein and zeaxanthin) and nutritional impacts on human health: A review. *CyTA - Journal of Food, 15*(3), 474–487. https://doi.org/10.1080/19476337.2016.1266033

Zeimer, M. B., Krömer, I., Spital, G., Lommatzsch, A., & Pauleikhoff, D. (2010). Macular telangiectasia: Patterns of distribution of macular pigment and response to supplementation. *Retina, 30*(8), 1282–1293. https://doi.org/10.1097/IAE.0b013e3181e096dd

Zhou, H., Zhao, X., Johnson, E. J., Lim, A., Sun, E., Yu, J., & Liu, N. (2011). Serum carotenoids and risk of age-related macular degeneration in a Chinese population sample. *Investigative Ophthalmology & Visual Science, 52*(7), 4338–4344. https://doi.org/10.1167/iovs.10-6519

38

Evidence-Based Nutrition-Fortified Egg—Omega-3 Egg, Folate Egg, Lutein Egg, VD Egg, and More

Hajime Hatta

CONTENTS

38.1 Introduction

Chicken eggs, like milk, have been called a complete nutritional food, but the biological roles of the two are decidedly different. Milk spoils when it is warmed for 21 days, but a chick is born if it is a fertilized egg. In other words, the egg is a capsule of life. Inside the egg, all ingredients necessary and sufficient for the development and growth of life are stored (Hatta et al., 2008). Chicken eggs are a typical food high in cholesterol, but even that cholesterol is stored without excess or deficiency as an essential constituent for the birth of life. Furthermore, consumers have become more healthy, food safety, and security conscious. Egg producers have developed their brand eggs that are not influenced by market egg prices and have increased sales by designing feed composition and rearing environments to enhance further the nutritional components of eggs, cage-free eggs, free-range eggs, organic-fed eggs, and fertilized eggs, among others.

Branded eggs, also known as designer eggs (Surai & Sparks, 2001), health-function eggs, or nutritionally enhanced eggs, have escalated to more than 1,500 varieties in Japan. The variety of labels and prices has led consumers to be confused in their egg selection, wondering if the health benefits of brand eggs have been scientifically proven or if price and function are correlated. Under these circumstances, the Consumer Affairs Agency was established in Japan in 2009, and scientific evidence is now required for the health function of branded eggs.

Currently, evidence-based information is attracting attention in pharmaceuticals and food selection, as new and reliable information is needed to make the right choices and decisions (Mah et al., 2019). In other words, selection and judgment are based on scientific data. Human clinical trial results are also required as evidence for specified health foods. In recent years, evidence-based medicine has gained attention in health and beauty foods, nutritional guidance, sports guidance, and nursing care and nursing. Nutritionally fortified eggs must also be scientifically proven to have health functions when consumed by humans. In other words, they are evidence-based, nutritionally fortified eggs.

38.2 Basic Nutritional Value of Chicken Eggs

We can obtain all the nutrients we need to maintain and improve our daily health from eggs. For this reason, eggs are also known as complete nutritional foods. Egg protein contains all the essential amino acids (leucine, valine, tryptophan, lysine, threonine, phenylalanine, methionine, isoleucine, and

DOI: 10.1201/9781003254430-46

histidine) our bodies cannot produce in sufficient amounts, and its amino acid score is the highest at 100. In growth tests with laboratory animals (rats), it is comparable to the nutritional value of breast milk protein, and whole egg protein is also used in animal growth tests as a replacement protein for breast milk (Horimoto & Hatta, 2016).

On the other hand, all egg lipids are contained in the egg yolk and exist as protein-bound forms (lipoproteins). About 65% of the lipids are triglycerides, about 30% are phospholipids, and about 5% are cholesterol. Phospholipids and cholesterol are essential components of the membranes of body cells and neurons. Egg yolk lipids are rich in essential fatty acids, which cannot be made in our bodies, and also contain lecithin (egg yolk phospholipids), which is expected to improve dementia. In addition, all fat-soluble vitamins (A, D, K, and E) and most water-soluble vitamins are present in egg yolks. Egg yolks are also rich in calcium and phosphorus, the building blocks of bone and red blood cells, and iron (Horimoto & Hatta, 2016).

The nutritional composition of a hen's egg and the nutrient intake ratios obtained from a single egg are shown in Table 38.1. The only nutrients missing from an egg are vitamin C and dietary fiber, so we can obtain many nutrients from the egg in a balanced and concentrated form. The average egg contains about 85 Kcal of energy, equivalent to 3.4% of Japanese estimated daily energy needs for men and 4.5% for women. The intake ratio of each nutrient is calculated from the same perspective, and the intake ratio of cholesterol is exceptionally high, at 34% for men and 42% for women.

Other nutrients that exceeded 20% of the total intake were vitamins D, B2, and B12, and nutrients that exceeded 10% of the total intake were protein and minerals such as phosphorus, iron, zinc, and zinc. The vitamins are A, K, folic acid, and pantothenic acid (Alagawany et al., 2018). Thus, a hen's egg is a food that contains an efficient concentration of the critical nutrients that make up our bodies, both quantitatively and qualitatively. We should recognize again that more than the energy obtained from a single egg (3.4% for men and 4.5% for women), its consumption provides us with an efficient intake of nutrients such as protein, vitamins, and minerals.

As described above, chicken eggs are capsules with various nutrients necessary for the maintenance and promotion of health, especially vitamins and minerals whose physiological functions have already been proven, are concentrated, and accumulated, and can be regarded as "natural supplements" in our diet.

38.3 Japan's Nutrition-Fortified Egg Certification System

Eggs are a nutritionally superior food that balances essential nutrients except for no dietary fiber and vitamin C. Furthermore, it is known that some nutrients are efficiently transferred to eggs when laying hens are raised on feed supplemented with various nutrients. Table 38.2 shows the variation in the chicken eggs' respective content when the feed's vitamin and mineral content changes (Okumura, 2002; Leeson & Gaston, 2003). It is generally said that nutrients with high variability are more likely to be transferred from feed to eggs.

Therefore, it is possible to produce high-value-fortified eggs with additional nutrients such as vitamins, minerals, and essential fatty acids in addition to regular eggs. Some nutrients are easily transferred to chicken eggs, while others are not. Minerals include iodine, fluorine, manganese, water-soluble and fat-soluble vitamins, and unsaturated fatty acids such as linoleic acid, α-linolenic acid, eicosapentaenoic acid (EPA), docosahexaenoic acid (DHA) are easily transferred (Sparks, 2006). In contrast, calcium, magnesium, iron, vitamin C, amino acids, etc. are not easily transferred. Fat-soluble nutrients added to feed are transferred to the egg yolk, while water-soluble nutrients are transferred to the yolk and egg white. Highly unsaturated fatty acids are incorporated into egg yolks as constituent fatty acids of phospholipids.

Nutritionally enriched eggs in Japan are defined in the "Fair Competition Code Concerning Labeling of Poultry and Eggs," which came into effect on March 27, 2009 (Label bank Co., Ltd., 2020), as eggs that meet the separately specified nutrient increase (Table 38.3) for the edible portion (egg yolk and egg white) by adding nutrients to chicken feed to increase the amount of nutrients in the eggs. The amount of nutrients is verified by periodic analysis of the ingredients. When labeling eggs as fortified eggs, the name of the nutrients increased or added, and the amount of nutrients per 100 g of the edible portion must be indicated so that the nutrients that meet the criteria for fortified eggs can be identified.

The labeling must also compare the amount of nutrients in the product with the amount in regular eggs so that consumers can easily compare them. In the case of a nutritional ingredient not contained in ordinary chicken eggs, the name of the nutritional ingredient shall be clearly stated with the unit of content per 100 g of edible portion, and the fact that it is a nutritional ingredient not contained in ordinary chicken eggs shall also be indicated. Currently, on a commercial basis, eggs fortified with iodine, folic acid, vitamin D, vitamin E, alpha-linolenic acid, docosahexaenoic acid (DHA), lutein, and other nutrients are commercially available.

Although the price of fortified eggs is higher than that of regular eggs, the advantage of consuming nutrients with various physiological functions from a familiar egg is significant from the viewpoint of food and health.

38.4 Omega-3 Fatty Acid (α-Linolenic Acid, EPA, DHA) Enriched Eggs

Mammals do not have enzymes that introduce double bonds at the ω-3 and ω-6 carbons (the third and sixth carbons from the methyl group end) of fatty acids. Therefore, the ω-3 fatty acid α-linolenic acid (C18:3) and the ω-6 fatty acid linoleic acid (C18:2) are called essential fatty acids because they cannot be synthesized in the human body and must be obtained from foods. Each of these essential fatty acids is metabolized in the body into ω-3 fatty acids (EPA, DHA, etc.) and ω-6 fatty acids (γ-linolenic acid, arachidonic acid, etc.), which are fatty acids in phospholipids that make up the lipid bilayer of cell membranes (Figure 38.1) (Juneja, 1997).

Linoleic acid is contained in large amounts as a constituent fatty acid of vegetable oils and fats, and its physiological

TABLE 38.1

Nutritional Composition of a Hen's Eggs and Nutrient Intake Ratios from an Egg

	Per 100 g			Per Egg			Intake Ratio Obtained from 1 Egg per Day (%)	Dietary Reference Intakes for Japanese Age 18 Years Old and Over with Normal Intensity of Life Activity
	of Edible Portion			Whole	EY	EW		
	Whole	EY	EW	(60 g)	(18 g)	(42 g)		
Enegry (Kcal)	142	336	44	85	60	18	3.4% (M) 4.5% (F)	2,490 (M)-1,900 (F)Kcal/D
Moisture (g)	75.0	49.6	88.3	45.0	8.9	37.1	—	—
Protein (g)	12.2	16.5	10.1	7.3	3.0	4.2	12% (M) 15% (F)	63 (M)-50 (F)g/D RDA
Fat (g)	10.2	34.3	Tr	6.1	6.2	-	8,9% (M) 11.7% (F)	20-30 (% Enagy) DG
Carbohydrates (g)	0.4	0.2	0.5	0.2	0.0	0.2	0.07% (M) 0.09% (F)	50-65 (% Enagy) DG
Ash (g)	1.0	1.7	0.7	0.6	0.3	0.3	—	
Na (mg)	140	53	180	84.0	9.5	75.6	14% (MF)	600 mg/D (MF) EAR
K (mg)	130	100	140	78.0	18.0	58.8	3.1% (M) 3.9% (F)	2,500 (M)2,000 (F)mg/D AI
Ca (mg)	46	140	5	28	25	2	3.7% (M) 4.3% (F)	750 (M)640 (F)mg/D RDA
Mg (mg)	10	11	10	6	2	4	1.7% (M) 2.1% (F)	350 (M)280 (F)mg/D RDA
P (mg)	170	540	11	102	97	5	9.7% (M) 12.7% (F)	1,050 (M)800 (F)mg/D AI
Fe (mg)	1.5	4.8	Tr	0.9	0.9	-	12% (M) 8.6% (F)	7.5 (M)10.5 (F)mg/D RDA
Zn (mg)	1.1	3.6	0.0	0.7	0.6	-	6.0% (M) 8.3% (F)	11.0 (M)8.0 (F)mg/D RDA
Cu (mg)	0.1	0.1	0.0	0.0	0.0	0.0	3.3% (M) 4.3% (F)	0.9 (M)0.7 (F)mg/D RDA
Vitamin A (μg)	210	690	0	126	124	0	16.5% (M)18.5% (F)	760 (M)680 (F)μg/D RDA
Vitamin D (μg)	3.8	12.0	0.0	2.3	2.2	0.0	26.8% (MF)	8.5 μg/D (MF) AI
Vitamin E (mg)	1.3	4.5	0.0	0.8	0.8	0.0	12.0% (M) 8.3% (F)	6.5 (M)6.0 (F)mg/D AI
Vitamin K (μg)	12.0	39.0	1.0	7.2	7.0	0.4	10% (M) 12% (F)	150 μg/D (MF) AI
Vitamin B1 (mg)	0.1	0.2	0.0	0.0	0.0	0.0	2.8% (M)3.6% (F)	1.3 (M)1.1 (F)mg/D RDA
Vitamin B2 (mg)	0.4	0.5	0.4	0.2	0.1	0.1	20% (M) 25% (F)	1.5 (M)1.2 (F)mg/D RDA
Niacin (mg)	0.1	0.0	0.1	0.1	0.0	0.0	0.7% (M) 0.8% (F)	14 (M)11 (F)mg当量/D RDA
Vitamin B6 (mg)	0.1	0.3	0.0	0.1	0.1	0.0	3.6% (M) 4.2% (F)	1.4 (M)1.1 (F)mg/D RDA
Vitamin B12 (μg)	1.1	3.5	Tr	0.7	0.6	-	21% (MF)	2.4 μg/D (MF) RDA
Folic Acid (μg)	49	150	0	29	27	0	11% (MF)	240 μg/D (MF) RDA
Pantothenic Acid (mg)	1.2	3.6	0.1	0.7	0.6	0.1	15% (M) 18% (F)	6 mg (M), 5 mg (F)/D AI
Vitamin C (mg)	0.0	0.0	0.0	0.0	0.0	0.0	0%	100 mg/D (MF) RDA
Saturated FAs (g)	3.1	9.4	Tr	1.9	1.7	-	—	7% Energy below DG
Monounsaturated FAs(g)	4.3	13.0	Tr	2.6	2.3	-	—	—
Polyunsaturated FAs (g)	1.4	4.5	Tr	0.9	0.8	-	—	—
Cholesterol (mg)	370	1200	1	222	216	0	34% (M) 42% (F)	750 (M), 600 (F)mg/D or less DG
Dietary fiber (g)	0.0	0.0	0.0	0.0	0.0	0.0	0%	20.5 (M), 17.6 (F)g/D AI
Salt equivalent (g)	0.4	0.1	0.5	0.2	0.0	0.2	2.7% (M)1.9% (F)	7.5 (M), 6.5 (F)mg/D DG

Source: Food Composition in Japan 2022 and Dietary Reference Intakes for Japanese (2020).

Note: Estimated Energy Requirement (Kcal/day): The amount of energy intake that minimizes the risk of energy deficiency and excess.

Estimated Average Requirement (EAR) The EAR is the amount that would meet the nutrient requirements of 50% of the population.

Recommended Daily Allowance (RDA): The daily intake that is estimated to meet the daily requirement for most (97–98%) of the human population of a given sex and age group.

Adequate Intake (AI): AI was developed for case in which the RDA cannot be set due to the insufficient scientific evidence. The AI indicates the amount that is adequate to maintain a certain nutritional status. Dietary intake of no less than the AI minimizes the risk of inadequacy.

Dietary Goal (DA): For the prevention of lifestyle-related diseases (LRDs), a tentative dietary goal for preventing LRDs (DG) was developed.

TABLE 38.2

The Range of Variation for Vitamins and Minerals in a 60 g Egg Due to Laying Hen Feed

Vitamins	Content	Minerals	Content
Vitamin A (I.U.)	0.5 – 416	Ca (mg)	32.0
Vitamin D (I.U.)	5.0 – 880	Phosphorus (mg)	118.0
Vitamin E (µg)	297 – 1,400	Mg (mg)	30.0
Vitamin K (µg)	44 – 300	Na (mg)	70.0
Vitamin B1 (µg)	38 – 340	K (mg)	76.0
Vitamin B2 (µg)	76 – 300	Chlorine (mg)	70.0
Niacin (µg)	32 – 38	Sulfate (mg)	98.0
Pantothenic Acid (µg)	464 – 810	Fe (mg)	1.0 – 5.0
Biotin (µg)	8.6 – 14.0	Cu (mg)	0.05 – 0.33
Vitamin B12 (µg)	130 – 730	I (mg)	0.004 – 0.01
Folic Acid (µg)	4.2 – 4.4	Mn (mg)	0.005 – 0.025
Vitamin B6 (µg)	130 – 150	Zn (mg)	0.7 – 1.0
Choline	238 – 300		

Source: Modified from Okumura (2002) and Leeson and Gaston (2003).

TABLE 38.3

Nutrients Fortified Eggs and Its Criterial Content for Labeling Nutritional Function

Nutrients	General Egg per 100 g	Nut. Fortified Egg (+ more)
Protein (g)	12.2	+8.1 g
Dietary fiber (g)	0.0	+3.0 g
Zn (mg)	1.1	+0.88 mg
K (mg)	130.0	+280 mg
Ca (mg)	46.0	+68 mg
Fe (mg)	1.5	+0.68 mg
Cu (mg)	0.1	+0.09 mg
Mg (mg)	10.0	+32 mg
Niacin (mg)	0.1	+1.3 mg
Pantothenic Acid (mg)	1.2	+0.48 mg
Biotin (µg)	24.0	+5 µg
Vitamin A (µg)	210.0	+77 µg
Vitamin B1 (mg)	0.1	+0.12 mg
Vitamin B2 (mg)	0.4	+0.14 mg
Vitamin B6 (mg)	0.1	+0.13 mg
Vitamin B12 (µg)	1.1	+0.24 µg
Vitamin C (mg)	0.0	+10 mg
Vitamin D (µg)	3.8	+0.55 µg
Vitamin E (mg)	1.3	+0.63 mg
Vitamin K (µg)	12.0	+15 µg
Folic Acid (µg)	49.0	+24 µg
Iodine (µg)	33.0	+240 µg
DHA (mg)	72.0	+60 mg
α-Linolenic acid (mg)	29.0	+22 mg

Source: Fair Competition Code and Enforcement Regulations Concerning Labeling of Chicken Eggs (2016) issued by the Poultry and Egg Fair Trade Council in Japan.

function of lowering cholesterol has attracted attention, but its effect is transient and does not last long. On the contrary, excessive intake of linoleic acid increases arachidonic acid (C20:4), a polyunsaturated fatty acid (PUFA) of the ω-6 family, in the body, which produces strong inflammatory mediators (such as prostaglandin E2) and is a cause of cancer, allergies, heart disease, stroke, and other This is a problematic factor in triggering diseases such as cancer, allergies, heart diseases, and strokes.

On the other hand, α-linolenic acid is an ω-3 polyunsaturated fatty acid (PUFA), which is converted to EPA and DHA of the same ω-3 family in vivo (Irawan et al., 2022). Moreover, its intake has a preventive or ameliorative effect on cancer, hypercholesterolemia, thrombosis, and immune hypersensitivity by producing weak inflammatory mediators that work on cells in antagonism to inflammatory mediators from arachidonic acid, resulting in the suppression of inflammation (Stupin et al., 2020).

The Dietary Guidelines for Americans 2015–2020 recommend 450–500 mg of omega-3 fatty acids per day (Vannice & Rasmussen, 2014). In Japan, the recommended amounts are 2.0–2.2 g for adult men and 1.6–1.9 g for adult women (18–64 years old). According to Commission Regulation (EU) No 116/2010, a product is an omega-3 fatty acid source if it contains at least 0.3 g ALA and at least 40 mg EPA + DHA per 100 g of product. It is also a high source of omega-3 fatty acids if it contains at least 0.6 g of ALA and at least 80 mg of EPA + DHA.

Currently, Japanese people tend to consume excessive amounts of linoleic acid due to the westernization of their diets, and there is concern about the resulting increase in lifestyle-related diseases. Developed countries have set lipid intake guidelines related to the maintenance and promotion of health. They instruct people to reduce the ω-6/ω-3 ratio of polyunsaturated fatty acid intake to 4 or less (Simopoulos, 2004). Eggs fortified with ω-3 fatty acids (α-linolenic acid, EPA, and DHA) have been developed from this perspective, and α-linolenic acid- and DHA-fortified eggs are spreading in

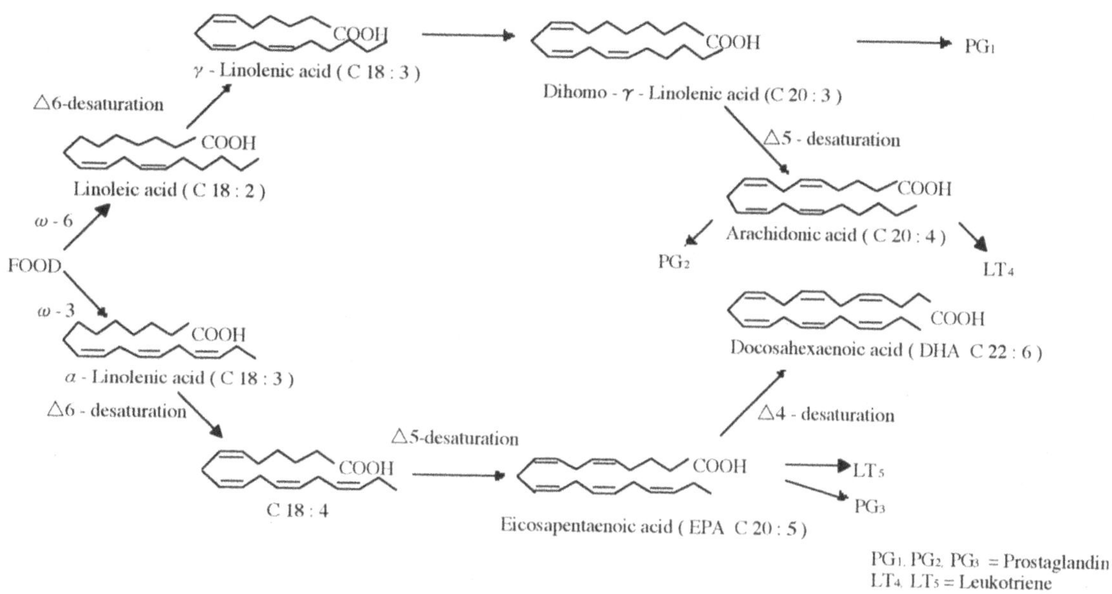

FIGURE 38.1 Metabolic pathway of ω-3 and ω-6 fatty acids. (Adapted from Juneja, 1997.)

the world egg market and attracting attention as high-value-added eggs.

38.4.1 α-Linolenic Acid-Fortified Eggs

Eggs fortified with α-linolenic acid are fed with plant seeds with high α-linolenic acid content or their extracted oils and fats (perilla oil, sesame oil, and linseed oil). As an example of the preparation of α-linolenic acid-fortified eggs, it was reported that adding 3% linseed oil to the feed resulted in fortified eggs containing about 7 times more α-linolenic acid than control eggs with fed 3% soybean oil (Hudečková et al., 2012). In this feeding trial, the α-linolenic acid concentration in the control and experimental group was 0.82 g/100 g of fat and 5.63 g/100 g of fat, respectively. The possibility of influencing the fatty acid profile in eggs is significant for developing value-added nutrition-fortified eggs.

In another study (Ferrier et al., 1995), eggs from Leghorn hens reared with 0%, 10%, and 20% ground flax seed in their feed contained 28 mg, 261 mg, and 527 mg/egg of α-linolenic acid and 51 mg, 81 mg, and 87 mg/egg of DHA, respectively. As a result, the ω-6/ω-3 ratio decreased to 9.3 for regular eggs, 2.2 for eggs containing 10% flax seed, and 1.4 for eggs containing 20% flax seed. On the other hand, the cholesterol content of each egg remained unchanged. In addition, 28 male volunteers were divided into 3 groups and fed 4 of each egg daily for 2 weeks. The results showed that the total cholesterol, HDL cholesterol, and triglycerides in the plasma of each group remained unchanged. On the other hand, total ω-3 fatty acids and DHA in platelet phospholipids increased significantly, and the ω-6/ω-3 ratio decreased considerably from 10.6 in the regular egg group to 8.3 and 8.1 in the egg groups containing 10% and 20% flax seed, respectively.

Furthermore, the fatty acid balance (ω-6/ω-3 ratio) of an average hen's egg is about 10. However, an α-linolenic acid-fortified egg with this ω-6/ω-3 ratio lowered to nearly 1 has been developed by Dr. Sim of the University of Alberta, Canada, under the name of Dr. Sim's Canadian Designer Eggs called Columbus Egg were marketed in about 30 countries around the world under the name of Dr. Sim of the University of Alberta, Canada, who developed them (Douny et al., 2014; Rogister & Scippo., 2015).

38.4.2 Canadian Designer Egg (CD Egg)

The first evidence-based fortified egg developed worldwide is probably the Canadian Designer Egg. It was designed by Dr. Sim of the University of Alberta, Canada, fortified with alpha-linolenic acid (Sim & Sunwoo, 2002). As a result of rearing laying hens by adding linseed oil (56% α-linolenic acid content) and tocopherol (antioxidant) to their feed and examining the fatty acid composition in their egg yolks in detail, eggs with a high content of ω-3 fatty acids in the yolk lipid in the quantitative order of α-linolenic acid > DHA > EPA were obtained.

The ω-3 fatty acid content is about 600 mg, and the tocopherol content is about 6 mg, equivalent to eating about 100 g of fish, and in countries where seafood consumption is low, α-linolenic acid-fortified eggs are attracting attention as an alternative to fish. Furthermore, human volunteers were recruited, and 24 healthy male students were divided into 2 groups (12 students in each group), one eating regular eggs and the other eating eggs fortified with α-linolenic acid and fed 2 eggs at breakfast every morning for 18 days. The total cholesterol level of the regular egg group increased by 6% before and after the test, while that of the α-linolenic acid-fortified egg group remained unchanged. On the other hand, HDL cholesterol levels did not change on the regular egg diet but increased by 8% on the fortified egg diet. In addition, neutral lipids were reduced by about 40% in the fortified egg group (Jiang & Sim, 1993).

In another clinical trial, 8 lactating mothers were fed 2 α-linolenic acid-fortified eggs (CD Eggs) daily for 6 weeks to determine changes in the fatty acid composition of breast milk and plasma. The ω-6/ω-3 ratio of CD Eggs is about 1, whereas the ω-6/ω-3 ratio of regular eggs is about 10. After eating 2 CD eggs daily for 6 weeks in addition to their diet, n-3 fatty acids in breast milk lipids increased from 1.9–3.6%, and the ω-6/ω-3 ratio decreased from 6.7 to 3.0 (Cherian & Sim, 1996).

As described above, CD eggs are the world's first evidence-based nutritionally fortified eggs whose physiological activities in humans, such as the effect of lowering triglycerides in the blood and modifying the fatty acid composition of lipids in breast milk when they are eaten, have been investigated and their effects published in academic journals (Surai et al., 2000).

38.4.3 EPA- and DHA-Fortified Eggs

Since 1979, when it was reported that the Inuit of Greenland had less thrombotic disease because they consumed much EPA and DHA from fish and other sources than Westerners, there have been many reports that EPA and DHA help prevent certain types of cancer, myocardial infarction, inflammatory diseases, and the like (Dyerberg & Bang, 1979; Anderson & Connor, 1989). EPA and DHA are highly unsaturated fatty acids, so these are easily oxidized and unstable. In chicken eggs, however, EPA and DHA are present as constituent fatty acids in egg yolk phospholipids, which are stable and not easily oxidized.

Eggs fortified with ω-3 fatty acids (mainly DHA) were prepared by adding 5% tuna oil to the feed of laying hens. The fatty acid composition of regular eggs vs. DHA eggs was 45% total saturated fatty acids vs. 35.60%, 14.5% total ω-6 fatty acids vs. 14.1%, 0.9% total ω-3 fatty acids vs. 7.8%, and the ω-6/ω-3 ratio was 16.0 vs. 1.80, respectively. In particular, DHA content was about 14 times higher in the DHA-fortified eggs (5.41%) than regular eggs (0.39%). Twenty-five healthy volunteers were recruited for a double-blind, crossover clinical trial (Bovet et al., 2007).

The volunteers were divided into Group A (11 volunteers) and Group B (14 volunteers). For the next 3 weeks, the groups were crossed over, with Group A receiving DHA eggs and Group B receiving regular eggs. Blood samples were taken from each group 3 times before, 3 weeks, and 6 weeks after the test. The blood test results for the first and second 3 weeks of the study showed a significant 16–18% decrease in blood triglyceride levels in the fish oil egg group but no significant change in the regular egg group. In both egg groups, serum LDL cholesterol levels increased slightly in the first 3 weeks but returned to the initial levels in the next 3 weeks.

The deficiency of DHA (C 22: 6) also causes loss of visual activity in monkeys and human infants, and DHA is physically essential for the normal function of neural tissue and brain membranes. Hoffman et al. (2004) administered baby food containing DHA-rich egg yolk (115 mg DHA/100 g) during the 6-month study period. As a result, DHA levels in red blood cells increased from 4.1–5.5% at 12 months in the supplemented infants. Formula feeding is a logical choice to provide DHA to younger infants who are not breastfed.

38.5 Vitamin-Fortified Eggs

Vitamins are essential in maintaining and improving health by assisting in the body's metabolism (chemical reactions) which is necessary for survival. Although the daily requirement for vitamins are small, we need to consume them daily because they are organic compounds that our bodies cannot synthesize. Currently, 13 vitamins are known, which are classified into water-soluble vitamins (B1, B2, B6, B12, niacin, pantothenic acid, folic acid, biotin, and C) and fat-soluble vitamins (A, E, D, and K). The deficiency of each vitamin causes specific deficiency symptoms. On the other hand, if taken in excess over a long period, water-soluble vitamins are quickly dissolved in urine and eliminated from the body. Fat-soluble vitamins accumulate in the body and cause excessive symptoms such as vomiting, headache, and nausea.

Both fat-soluble and water-soluble vitamins are easily transferred into chicken eggs, making it relatively easy to develop fortified eggs. Currently, eggs fortified with vitamins A, D, and E are commercially available on a commercial basis. Eggs containing vitamin D, pantothenic acid, folic acid, biotin, and vitamin B12 respond significantly to increases in the diet level of these vitamins. When fortifying 1–2 times vitamins to the National Research Council USA feed requirement levels, vitamin transfer efficiency to the egg is very high for vitamin A; it is high for riboflavin, pantothenic acid, biotin, and vitamin B12; it is medium for vitamin D3 and vitamin E; and low for vitamin K1, thiamine, and folic acid (Naber, 1993).

In addition, eggs that accumulate high concentrations of folic acid have recently been developed. Folic acid is a water-soluble B vitamin that is essential for cell differentiation. Therefore, if a pregnant woman suffers from a folic acid deficiency in the early stages of pregnancy, the fetus is more likely to develop neurological disorders, and supplementation with folic acid is actively recommended in the United States and Canada (Czeizel & Dudas, 1992; Stevenson et al., 2000).

38.5.1 Vitamin E-Fortified Eggs

Vitamin E is the generic name for the 4 tocopherols α, β, γ, and δ, as well as α, β, γ, and δ tocotrienols. They are ubiquitous and synthesized by photosynthetic organisms such as plants, algae, and cyanobacteria. D-alpha-tocopherol shows the most potent physiological activity and its content (mg) indicates vitamin E content. Vitamin E has strong antioxidant power, is widely distributed in cell membranes, prevents oxidation of polyunsaturated fatty acids and other fat-soluble components of phospholipids, which are components of cell membranes, keeps blood vessels healthy, inhibits oxidation of LDL cholesterol in the blood, and prevents cell aging.

In chicken eggs, vitamin E is mainly contained in the egg yolk. Hens efficiently transfer vitamin E in the feed to the egg yolk so that vitamin E-fortified eggs can be prepared. Furthermore, the functionality of vitamin E can be labeled as "vitamin E-fortified eggs" if the amount of vitamin E naturally contained per 100 g of table eggs is fortified from approximately 1.0 mg to an additional 1.2 mg or more, and vitamin E-fortified eggs containing 2.2 mg or more are prepared in Japan (refer to Table 38.3).

Vitamin E is an antioxidant for highly unsaturated fatty acids (PUFA) in preparing ω-3-fortified eggs. Elevated PUFA concentrations in eggs accelerate lipid peroxidation (Grashorn, 2005, Mohiti-Asli et al., 2008) and degrade the eggs' sensory characteristics and overall quality (Hayat et al., 2010). Many experiments have been conducted to prevent this, and lipid oxidation inhibitory effects have been obtained at vitamin E concentrations of 100–200 mg/kg or more (Hayat et al., 2010).

The antioxidant effect of vitamin E addition has also been evaluated in diets for laying hens with higher dietary PUFA levels; according to Barroeta (2007), at low dietary PUFA levels (15 g/kg PUFA), 60 mg of α-tocopherol per kg of feed is required to ensure lipid stability. Conversely, when dietary PUFA levels are high (30 g/kg of PUFA), 200 mg of α-tocopherol per kg of feed is required.

The amount of vitamin E in egg yolk is strictly related to the amount of α-tocopherol in the diet and increases linearly with increasing dietary dl-α-tocopheryl acetate. The groups with 100 ppm and 200 ppm (mg/Kg) of vitamin E in the diet had the highest vitamin E content of 3.98 mg and 4.70 mg per egg, respectively.

38.5.2 Vitamin D-Fortified Eggs

Vitamin D is obtained from the diet in two forms: vitamin D2 (Ergocalciferol) and vitamin D3 (Cholecalciferol). Vitamin D2 is mainly found in mushrooms, while vitamin D3 is in fish and meat. In addition, 7-dehydrocholesterol in the skin produces vitamin D3 in the blood due to the action of ultraviolet light. Both vitamins D2 and D3 are metabolized in the body by the liver and kidneys and undergo hydroxylation to activate vitamin D (1,25-dihydroxycholecalciferol (OH)2D), which has physiological activity. One of its well-known physiological activities is to promote the absorption of calcium (Ca) absorption from the intestinal tract and bone formation and strengthening. In addition, vitamin D has a wide range of critical biological effects, such as hormone-like effects on the immune system and reproductive systems (Jones et al., 1998).

Currently, many people are at risk for vitamin D deficiency worldwide due to inadequate sunlight exposure and dietary vitamin D intake. Data from the United States and Europe show that 8% and 13% of the population had serum 25-hydroxyvitamin D levels of 30 nmol/L, an indicator of vitamin D deficiency (Looker et al., 2008, Cashman et al., 2016). In particular, most people in the United Kingdom do not meet the standard nutrient intake of 10 μg/day, making vitamin D deficiency a significant public health issue. Eggs naturally contain vitamin D and are thought to contribute up to 10% of dietary vitamin D intake (Lanham-New et al., 2011). Each 100 g raw whole egg contains 3.15 μg of vitamin D. However, vitamin D in egg yolks can be enhanced by manipulating the vitamin D content of hen diets.

The Sunshine Eggs project, launched in 2015 in The United Kingdom is an initiative to prepare and offer vitamin D-fortified eggs to consumers (Clark et al., 2021). It was demonstrated that feeding egg-laying hens up to 75 μg of 25-hydroxyvitaminD3/kg of feed for 6 weeks significantly improved eggs' total vitamin D content. The vitamin D levels of regular eggs (3.15 μg/100 g eggs) were approximately 30% higher than those of standard eggs (4.2 μg/100 g) in the vitamin D3-enriched eggs. These British vitamin D-fortified eggs are currently marketed under the "Happy Eggs http://thehappyegg.co.uk/egg-nutrition/."

38.5.3 Folic Acid-Fortified Eggs

Folic acid, a water-soluble B vitamin discovered in spinach, is a pteridine compound bound to p-aminobenzoic acid and one or more glutamic acids. In natural foods, it is primarily found in the form of pteroyl polyglutamic acid with 3–7 glutamic acid linkages. Folic acid is widely distributed in various foods and is especially abundant in green and yellow vegetables and egg yolks (Ravanel & Rébeillé, 2012).

The physiological functions of folic acid include its involvement in the biosynthesis of nucleotides, methylation, and metabolism of amino acids in cells, and it is an essential component for cell differentiation. Therefore, if a pregnant woman suffers from a folic acid deficiency in early pregnancy, the risk of neural tube defects (anencephaly, spina bifida, and other fetal congenital disabilities) in the fetus is increased. For this reason, adding folic acid to cereal foods has been mandated in the United States and Canada, and the risk of developing neural tube closure disorder has been reported to be reduced (Honein et al., 2001; De Wals et al., 2007,). In addition, folic acid is involved in methyl group metabolism, transferring homocysteine to methionine. Therefore, plasma homocysteine increases when folic acid intake decreases, causing vascular damage and leading to vascular diseases such as atherosclerosis (Huo et al., 2012).

According to the Dietary Reference Intakes for Japanese 2020, women planning pregnancy and pregnant women in the early stages of pregnancy should consume 400 μg/day of folic acid in foods other than regular foods to reduce the risk of neural tube closure defects in the fetus. The recommended folic acid intake from regular foods is 240 μg/day for women aged 18–49 years, with an additional 240 μg/day for pregnant women and an additional 100 μg/day for nursing mothers. However, according to the Japanese National Health and Nutrition Survey in 2019, the daily intake of folic acid is 226–247 μg/day for women aged 20–49, 243 μg/day for pregnant women, and 220 μg/day for nursing mothers which does not meet the recommended intake.

Folic acid in food is unstable against light, heat, and oxidation, and nearly 50% of folic acid in food is decomposed during cooking or leached into boiled water due to its water solubility, quickly lost during cooking (Mckillop et al., 2002). Therefore, it is challenging to meet the intake of folic acid in the diet alone, and folic acid supplements are recommended, but it is essential to try to obtain folic acid from the diet daily if possible. Folic acid enrichment of eggs is possible when the folic acid concentration in the diet is high. Recently, eggs with a high concentration of folic acid have been developed. Adding 16 mg FA/kg to the feed results in a folic acid content of 75 mg per egg (32 mg for regular eggs) (Hoey et al., 2009).

38.6 Mineral-Fortified Eggs

Depending on the type of mineral, some are more easily transferred to the egg and some less easily. Iodine, fluorine, manganese, and selenium are highly transferred, while calcium, phosphorus, and iron are not altered in their transfer to the egg. By nature, eggs are rich in phosphorus, calcium, sodium, and potassium and contain all the essential trace elements, including copper, iron, magnesium, manganese, selenium, and zinc (Table 38.1). Iron and zinc are found only in egg yolks and are a supply source for the next generation. In recent years, these are reports have shown that deficiencies of some of these minerals (Iodine, Zn, Mg, and Se) are associated with anxiety, depression, and fatigue (Wang et al., 2018; Turan & Karaasian, 2020). Therefore, those micronutrients can be enriched in eggs by chicken feed, making use of mineral-fortified eggs available.

38.6.1 Iodine-Fortified Eggs

Iodine-fortified eggs were developed in Japan in 1976, a world pioneer in high-value-added functional eggs. These eggs are laid by egg-laying hens fed with feed fortified with iodine components such as seaweed powder and are still sold in Japan today.

Iodine (iodine) is an element with the element symbol I, atomic number 53, and atomic weight 127, and it is widely distributed in nature, mainly in seawater. About 90% of the iodine in the human body is localized in the thyroid gland, where it is an essential trace element as a component of the thyroid hormones thyroxine (T4) and triiodothyronine (T3). T3 and T4 are produced from the amino acid tyrosine and iodine in the thyroid gland's follicular cells and secreted into the blood by stimulating thyroid-stimulating hormone (TSH), which is secreted from the anterior pituitary gland (Chung, 2014).

The physiological function of iodine is the action of the thyroid hormones, which promote energy metabolism and enable daily activities in adults. Therefore, iodine deficiency or thyroid hormone deficiency causes hypothyroidism, decreasing motivation, thinking ability, and labor productivity. In addition, thyroid hormones are essential for developing the fetus's central nervous system and skeletal organs and are involved in the promotion of growth and the development of intelligence in infants and children. Thus, thyroid hormones, like growth hormones, are physiologically active substances essential for human development, growth, and development (Chung, 2014).

Iodine is found mainly in seaweeds. Most of the iodine ingested by humans is excreted in the urine, but that which is absorbed becomes a component of the thyroid hormone (thyroxine). The relationship between the amount of iodine excreted in urine and the frequency of goiter appearance has been studied, and it is known that goiter occurs when the daily excretion falls below 50 μg. Therefore, the recommended iodine intake is set at 0.13 mg/day to avoid deficiency. Japanese people are frequent seaweed eaters, and their average iodine intake is about 1.5 mg/day, so iodine deficiency is extremely rare. Globally, however, there are regions in Africa, South America, Asia, and Europe where iodine intake is 50 μg/day, and the World Health Organization (WHO) reports that 2.2 billion people, or 38% of the world's population, are at risk of iodine 2.2 billion people, or 38% of the world's population, are at risk of iodine deficiency (Vitti et al., 2001).

Iodine is also essential from a radiation protection perspective: volatile radioactive iodine (iodine-131) was dispersed into the surrounding environment in large amounts during the Chornobyl nuclear power plant accident in 1986 and the Fukushima nuclear power plant accident caused by 2011 Tohoku earthquake. When radioactive iodine enters the body, it collects in the thyroid gland, posing a risk of thyroid cancer due to internal exposure. Ordinary iodine (iodine-127) is non-radioactive, and its prophylactic intake can antagonistically inhibit the accumulation of radioactive iodine in the thyroid gland (Eng et al., 1999).

Adding iodine to the feed of laying hens with sodium iodide or seaweed increases the iodine content in the eggs. Suzuki and Tamura (1985) conducted a clinical trial in which 4 male university students in Japan consumed 2 iodine-fortified eggs daily. The average iodine concentration of 7.1 ± 0.6 μg in 100 ml serum and 218 ± 45 μg in daily urine increased to 12.0 ± 0.9 μg in 100 ml serum and 1593 ± 70 μ in daily urine, and the serum and urine iodine concentrations increased significantly ($p < 0.001$) by 1.7 and 7 times, respectively. The physiological functions of iodine are as follows. Iodine is taken up as organic iodine, and improved high cholesterol, dermatitis, adult diseases, allergic diseases, and the like have been reported.

38.6.2 Selenium-Fortified Eggs

A notable biological function of selenium is its abundance in the thyroid gland, where, along with iodine, it plays an essential role in thyroid hormone metabolism. Selenium is bound to the amino acid cysteine as selenoprotein, a component amino acid of enzymes such as glutathione peroxidase, thioreductase, and deiodinase (Bourre & Galea, 2006). These enzymes produce triiodothyronine (T3) from thyroxine (T4) in the thyroid gland. Glutathione peroxidase is also a potent antioxidant and protects cells in the body, including those in the brain, from free radical damage (Chen & Berry, 2003). In particular, sodium selenite directly activates natural killer cells and has anti-leukemic and anti-cancer effects. In other words, selenium deficiency affects the thymus gland, which is responsible for the production of macrophages and lymphocytes, and aggravates the immune system (Avery & Hoffmann, 2018).

Recently, eggs fortified with this selenium have been developed with the hope of improving human health. It has been shown that selenium content in eggs can be easily increased by adding selenium yeast to chicken feed at concentrations of 0.4–0.8 ppm. Each egg accumulates 30–35 μg of selenium, corresponding to about 50% of the recommended daily selenium intake. On the other hand, excess selenium is also toxic, with reports of alopecia, skin and nail lesions, bone fragility, nausea, and diarrhea at intakes of 400 μg/day or more (Kieliszek & Błażejak, 2016), leading to the recommendation that selenium-fortified eggs be used instead of taking supplements. Selenium-fortified eggs are available worldwide, including in Europe, Southeast Asia, and North America.

38.7 Carotenoid-Fortified Eggs

Carotenoids are fat-soluble pigments widely present in various plants, microalgae, bacteria, and fungi and are classified according to their structure and function into carotenes, such as lycopene, alpha-carotene, beta-carotene, and xanthophylls, such as lutein and zeaxanthin (Saini et al., 2015). Animals and birds cannot synthesize carotenoids in vivo, so these must be supplemented from the diet (O'Byrne & Blaner, 2013). Commonly known as precursors to vitamin A, carotenoids have been used in the poultry industry as pigments for egg yellows and chicken meat. Carotenoids also have intense free radical scavenging activity and reduce oxidative stress in vivo through mechanisms such as promoting the production of antioxidant enzymes (Lauretani et al., 2008). Therefore, their addition to poultry feed has been demonstrated to improve the oxidative stability of their eggs and meat and their productivity and fertility (Akdemir et al., 2012).

38.7.1 Lutein and Zeaxanthin-Fortified Eggs

The yellow pigments in egg yolks are lutein and zeaxanthin, derived from corn xanthophylls and transferred to eggs. Furthermore, when green and yellow plants such as kale, spinach, paprika, and marigold, rich in these pigments, are added to chicken feed, they are easily transferred into the egg yolk, making the egg yellow even brighter. Furthermore, this lutein in egg yolks is about 3 times more bioavailable than lutein in cooked spinach or crystal lutein in supplements (Chung et al., 2004).

In an experiment examining the transfer efficiency of lutein from feed to eggs, corn-soybean-based foundation diets supplemented with 0, 125, 250, 375, 500, 625, 750, or 1,000 ppm lutein were fed to laying hens for 30 days, and the lutein content of the eggs was determined (Leeson & Caston, 2003). Results showed that lutein migrated into the egg yolk, increasing from 0.3 mg/egg in the foundation feed to 1.5 mg/egg. However, there was no significant increase ($P > 0.05$) in egg yolk lutein when feeding diets above 375 ppm.

Lutein and zeaxanthin are also called macular pigments because they are very abundant in the macula, the darkest pigment in the most central part of the retina. These pigments accumulate in the macula and retina of the eye and have been shown to play an essential role in protecting the eye against oxidative stress and age-related retinal degeneration and cataracts (Walchuk et al., 2022).

Furthermore, it has been demonstrated that lutein- and DHA-fortified egg consumption can improve visual function in older people at risk for age-related eye diseases. Thirty elderly Caucasian subjects (64.0 ± 3.4 years old) consumed enriched eggs (0.87 mg lutein/day and 220 mg DHA/day) daily for 6 weeks and found that found to improve retinal function without adversely affecting plasma lipids (Walchuk et al., 2022).

Recently, lutein and zeaxanthin, as carotenoids that cross the cerebral blood barrier, have been studied for their ability to inhibit oxidative stress in the brain and maintain cognitive function, and expectations for lutein- and zeaxanthin-enriched eggs are increasing (García-Romera et al., 2022).

38.8 Conclusion

A chicken egg is a capsule of life containing all the nutrients necessary for the next generation's life. On the one hand, if a fertilized egg is incubated, chicks will be born. On the other hand, chicken eggs as food, while no vitamin C and dietary fiber, contain a good balance of nutrients such as protein, fat, vitamins, and minerals that are essential for maintaining and improving our health and are used in our daily diet as a complete nutritional food. Furthermore, utilizing the egg-laying mechanism of hens, nutrients to be fortified in eggs can be added to the feed, and these nutrients can be transferred to the eggs to obtain nutritionally fortified eggs. In this way, eggs fortified with ω-3 fatty acids, carotenoids, and nutrient-fortified eggs with even higher vitamin and mineral content have been developed, helping to supply nutrients often lacking in the everyday diet. In addition, evidence-based nutrient-enriched eggs attract attention because of their intended physiological effects have been demonstrated in human clinical trials.

Usually, the direct and common way to supplement nutrients is to take soft capsules or tablets of nutrients that tend to be deficient. Still, there are three advantages to taking these nutrients from eggs:

1. It prevents excessive intake of certain nutrients compared to the convenience and easy use of capsules and tablets.
2. A food chicken eggs are suitable for cooking and processing and are delicious.
3. Fat-soluble nutrients are stabilized in the egg-yolk emulsion and are more efficiently absorbed and utilized in the body.

This chapter summarizes the development and use of eggs fortified with nutrients from feed. However, it is also possible to develop eggs that can be immunized with antigens by injecting them into laying hens to provide eggs with IgY antibodies to prevent infectious diseases such as rotavirus, tooth decay, and *H. pylori* (see Chapter 27). In addition, it has recently become possible to use genome editing technology to create chicken eggs in which genes for peptides and proteins with pharmacological effects are inserted into genes encoding egg white and egg yolk proteins and expressed (Chapter 39). In this way, chicken eggs are expected to become not only food for our diet but also to contribute to medical care and welfare as eggs can be used as complete nutritional foods, eggs fortified with specific nutrients, eggs with specific IgY antibodies to prevent infectious diseases, and eggs are used as raw materials for pharmaceuticals.

REFERENCES

Akdemir, F., Orhan, C., Sahin, N., Sahin, K., & Hayirli, A. (2012). Tomato powder in laying hen diets: Effects on concentrations of yolk carotenoids and lipid peroxidation. *British Poultry Science*, *53*(5), 675–680. https://doi.org/10.1080/00071668.2012.729142

Alagawany, M., Farag, M.-N., Diiaala, K., & Patra, A. (2018). Nutritional significance and health benefits of designer eggs. *World's Poultry Science Journal*, *74*, 317–329. https://doi.org/10.1017/S0043933918000041

Anderson, G. J., & Connor, W. E. (1989). On the demonstration of ω-3 essential fatty acid deficiency in humans. *The American Journal of Clinical Nutrition*, *49*, 585–587. https://doi.org/10.1093/ajcn/49.4.585

Avery, J. C., & Hoffmann, P. R. (2018). Selenium, selenoproteins, and immunity. *Nutrients, 10*, 1203. http://doi.org/10.3390/nu10091203

Barroeta, A. C. (2007). Nutritive value of poultry meat: Relationship between vitamin E and PUFA. *World's Poultry Science Journal, 63*, 277–284. https://doi.org/10.1017/S0043933907001468

Bourre, J. M., & Galea, F. (2006). An important source of omega-3 fatty acids, vitamins D and E, carotenoids, iodine and selenium: A new natural multi-enriched egg. *Journal of Nutrition Health and Aging, 10*, 371–376.

Bovet, P., Faeh, D., Madeleine, G., Viswanathan, B., & Paccaud, F. (2007). Decrease in blood triglycerides associated with the consumption of eggs of hens fed with food supplemented with fish oil. *Nutrition, Metabolism and Cardiovascular Diseases, 17*, 280–287. https://doi.org/10.1016/j.numecd.2005.12.010

Cashman, K. D., Dowling, K. G., Skrabáková, Z., Gonzalez-Gross, M., Valtueña, J., De Henauw, S., Moreno, L., Damsgaard, C. T., Michaelsen, K. M., & Mølgaard, C., et al. (2016) Vitamin D deficiency in Europe: Pandemic? *The American Journal of Clinical Nutrition, 103*, 1033–1044. https://doi.org/10.3945/ajcn.115.120873

Chen, J., & Berry, M. J. (2003). Selenium and selenoproteins in the brain and brain diseases. *Journal of Neurochemistry, 86*, 1–12. https://doi.org/10.1046/j.1471-4159.2003.01854.x

Cherian, G., & Sim, J. S. (1996). Changes in the breast milk fatty acids and plasma lipids of nursing mothers following consumption of n-3 polyunsaturated fatty acid enriched eggs. *Nutrition, 12*, 8–12. https://doi.org/10.1016/0899-9007(95)00013-5

Chung, H. R. (2014). Iodine and thyroid function. *Annals of Pediatric Endocrinology & Metabolism, 19*, 8–12. http://dx.doi.org/10.6065/apem.2014.19.1.8

Chung, H.-Y., Rasmussen, H. M., & Johnson, E. J. (2004). Lutein bioavailability is higher from lutein-enriched eggs than from supplements and spinach in men. *Journal of Nutrition, 134*, 1887–1893. https://doi.org/10.1093/jn/134.8.1887

Clark, A., SKuznesof, H., Davies, S., Waller, A., Ritchie, A., Wilson, S., Harbord, L., & Hill, T. (2021). Egg enrichment with vitamin D: The sunshine eggs projects. *Nutrition Bulletin, 46*, 332–338. https://doi.org/10.1111/nbu.12509

Czeizel, A. E., & Dudas, I. (1992). Prevention of the first occurrence of neural-tube defect by periconceptional vitamin supplementation. *The New England Journal of Medicine, 327*, 1832–1835. https://doi.org/10.1056/NEJM199212243272602

De Wals, P., Tairou, F., & Van Allen, M. I., et al. (2007) Reduction in neural-tube defects after folic acid fortification in Canada. *The New England Journal of Medicine, 357*, 135–142. http://doi.org/10.1056/NEJMoa067103

Douny, C., Khoury, R., El Delmelle, J., Brose, F., Degand, G., Moula, N., Farnir, F., Clinquart, A., Rogister, G. M., & Scippo, M. L. (2014). Effect of storage and cooking on the fatty acid profile of omega-3 enriched eggs and pork meat marketed in Belgium. *Food Science & Nutrition, 3*, 140–152. https://doi.org/10.1002/fsn3.197

Dyerberg, J., & Bang, H. O. (1979). Haemostatic function and platelet polyunsaturated fatty acids in eskimos. *Lancet*, 2(8140), 433–435. https://doi.org/10.1016/s0140-6736(79)91490-9

Eng, P. H. K., Cardona, G. R., Fang, S., Previti, M., Alex, S., Carrasco, N., Chen, W. W., & Braverman, L. E. (1999). Escape from the acute wolff-chaikoff effect is associated with a decrease in thyroid Sodium/Iodide symporter messenger ribonucleic acid and protein. *Endocrinology, 140*, 3404–3410. https://doi.org/10.1210/endo.140.8.6893

Ferrier, L. K., Caston, L. J., Leeson, S., Squires, J., Weaver, B. J., & Holub, B. J. (1995). Alpha-linolenic acid- and docosahexaenoic acid-enriched eggs from hens fed flaxseed: Influence on blood lipids and platelet phospholipid fatty acids in humans. *The American Journal of Clinical Nutrition, 62*, 81–86. https://doi.org/10.1093/ajcn/62.1.81

García-Romera, M.-C., Silva-Viguera, M.-C., L'opez-Izquierdo, I., L'opez-Mu~noz, A., Capote-Puente, R., & Gargallo-Martínez, B. (2022). Effect of macular pigment carotenoids on cognitive functions: A systematic review. *Physiology & Behavior, 254*, 113–891. https://doi.org/10.1016/j.physbeh.2022.113891

Grashorn, M. A. (2005). Enrichment of eggs and poultry meat with biologically active substances by feed modifications and effects on the final quality of the product. *Polish Journal of Food and Nutrition Sciences, 14/55*(Special Issue), 15–20.

Hatta, H., Kapoor, M. P., & Juneja., L. R. (2008). Bioactive components in egg yolk. In Y. Mine (Ed.), *Egg bioscience and biotechnology* (pp. 185–237). Wiley & Sons, Inc.

Hayat, Z., Cherian, G., Pasha, T. N., Khattak, F. M., & Jabbar, M. A. (2010). Sensory evaluation and consumer acceptance of eggs from hens fed flax seed and 2 different antioxidants. *Poultry Science, 89*, 2293–2298. https://doi.org/10.3382/ps.2009-00575

Hoey, L., McNulty, H., McCann, E. M. E., McCracken, K. J., Scott, J. M., Marc, B. B., Molloy, A. M., Graham, C., & Pentieva, K. (2009). Laying hens can convert high doses of folic acid added to the feed into natural folates in eggs providing a novel source of food folate. *British Journal of Nutrition, 101*, 206–212. https://doi.org/10.1017/S0007114508995647

Hoffman, D. R., Theuer, R. C., Castañeda, Y. S., Wheaton, D. H., Bosworth, R. G., O'Connor, A. R., Morale, S. E., Wiedemann, L. E., & Birch, E. E. (2004). Maturation of visual acuity is accelerated in breast-fed term infants fed baby food containing DHA-enriched egg yolk. *Journal of Nutrition, 134*, 2307–2313. https://doi.org/10.1093/jn/134.9.2307

Honein, M. A., Paulozzi, L. J., Mathews, T. J., Erickson, J. D., & Wong, L.-Y. C. (2001). Impact of folic acid fortification of the US food supply on the occurrence of neural tube defects. *JAMA, 285*, 2981–2986. http://doi.org/10.1001/jama.285.23.2981

Horimoto, Y., & Hatta, H. (2016). The nutritional and physiological functions of egg yolk components. In J. Roberts (Ed.), *Achieving sustainable production of eggs volume 1 safety and quality* (pp. 49–96). Burleigh Dodds Science Publishing. https://doi.org/10.19103/AS.2016.0012.03

Hudečková, P., Rusníková, L., Straková, E., Suchý, P., Marada, P., & Macháček, M. (2012). The effect of linseed oil supplementation of the diet on the content of fatty acids in the egg yolk. *Acta Veterinaria Brno, 81*, 159–162. https://doi.org/10.2754/avb201281020159

Huo, Y., Qin, X., Wang, J., Sun, N., Zeng, Q., Xu, X., Liu, L., Xu, X., & Wang, X. (2012). Efficacy of folic acid supplementation in stroke prevention: New insight from a meta-analysis. *Journal of Clinical Practice*, *66*, 544–551. https://doi.org/10.1111/j.1742-1241.2012.02929.x

Irawan, A., Ningsih, N., Rusli, H. R. K., Suprayogi, W. P. S., Akhirini, N., Hadi, R. F., Setyono, W., & Jayanegara, A. (2022). Supplementary n-3 fatty acids sources on performance and formation of omega-3 in egg of laying hens: A meta-analysis. *Poultry Science*, *101*, 101566. https://doi.org/10.1016/j.psj.2021.101566

Jiang, Z., & Sim, J. S. (1993). Consumption of n-3 polyunsaturated fatty acid-enriched eggs and changes in plasma lipids of human subjects. *Nutrition*, *9*, 513–518.

Jones, G., Strugnell, S. A., & DeLuca, H. F. (1998). Current understanding of the molecular actions of vitamin D. *Physiological Reviews*, *78*, 1193–231. https://doi.org/10.1152/physrev.1998.78.4.1193

Juneja, L. R. (1997). Egg yolk lipids. In T. Yamamoto, L. R. Juneja, H. Hatta, & M. Kim (Eds.), *Hen eggs; Their basic and applied science* (pp. 179–191). CRC Press. https://doi.org/10.1201/9780203752081

Kieliszek, M., & Błażejak, S. (2016). Current knowledge on the importance of selenium in food for living organisms: A review. *Molecules*, *21*, 609. https://doi.org/10.3390/molecules21050609

Label bank Co., Ltd. (2020). About the importance and role of Japan's "Fair Competition Code" in food labeling, And the compliance to its rules. https://labelbank.com/newsletter/issues/202009_2.html

Lanham-New, S. A., Buttriss, J., l., Miles, L. M., Ashwell, M., Berry, J. l., & Boucher, B. J., et al. (2011). Proceedings of the rank forum on vitamin D. *The British Journal of Nutrition*, *105*, 144–156.

Lauretani, F., Semba, R. D., Dayhoff-Brannigan, M., Corsi, A. M., Di Iorio, A., Buiatti, E., Bandinelli, S., Guralink, J. M., & Ferrucci, L. (2008). Low total plasma carotenoids are independent predictors of mortality among older persons. *European Journal of Nutrition*, *47*(6), 335–340. https://doi.org/10.1007/s00394-008-0732-9

Leeson, S., & Caston, L. J. (2003). Vitamin enrichment of eggs. *Journal of Applied Poultry Research*, *12*, 24–26. https://doi.org/10.1093/japr/12.1.24

Looker, A. C., Pfeiffer, C. M., Lacher, D. A., Schleicher, R. L., Picciano, M. F., & Yetley, E. A. (2008). Serum 25-hydroxyvitamin D status of the US population: 1988–1994 compared with 2000–2004. *The American Journal of Clinical Nutrition*, *88*, 1519–1527. https://doi.org/10.3945/ajcn.2008.26182

Mah, E., Chen, C. Y. O., & Liska, D. J. (2019), The effect of egg consumption on cardiometabolic health outcomes: an umbrella review, *Public Health Nutrition*, *23*, 935–955 https://doi.org/10.1017/S1368980019002441

Mckillop, D. J., Pentieva, K., Daly, D., Mcpartlin, J. M., Hughes, J., Strain, J. J., Scott, J. M., & Mcnulty, H. (2002). The effect of different cooking methods on folate retention in various foods that are amongst the major contributors to folate intake in the UK diet. *British Journal of Nutrition*, *88*, 681–688. https://doi.org/10.1079/BJN2002733

Mohiti-Asli, M., Shariatmadari, F., Lotfollahian, H., & Mazuji, M. T. (2008). Effects of supplementing layer hen diets with selenium and vitamin E on egg quality, lipid oxidation and fatty acid composition during storage. *Canadian Journal of Animal Science*, *88*, 475–483. https://doi.org/10.4141/CJAS07102

Naber, E. C. (1993). Modifying vitamin composition of eggs: A review. *Journal of Applied Poultry Research*, *2*, 385–393. https://doi.org/10.1093/japr/2.4.385

O'Byrne, S. M., & Blaner, W. S. (2013). Retinol and retinyl esters: Biochemistry and physiology thematic review series: Fat-soluble vitamins: Vitamin A. *Journal of Lipid Research*, *54*(7), 1731–1743. https://doi.org/10.1194/jlr.R037648

Okumura, J. (2002). Current status and problems in developing special eggs (in Japanese). *Japan Poultry Science Association*, *39*, 63–66.

Outline of the Dietary Reference Intakes for Japanese. (2020). English version. https://www.mhlw.go.jp/content/10900000/000862500.pdf

Ravanel, S., & Rébeillé, F. (2012). Chapter6 folate. In A. Salter, H. Wiseman, & T. Gregoy (Ed.), *Phytonutrients* (pp. 173–174). Willy-Blackwell. https://doi.org/10.1002/9781118253649.ch6

Rogister, G. M., & Scippo, M. L. (2015). Effect of storage and cooking on the fatty acid profile of omega-3 enriched eggs and pork meat marketed in Belgium. *Food Science & Nutrition*, *3*, 140–152. https://doi.org/10.1002/fsn3.197

Saini, R. K., Nile, S. H., & Park, S. W. (2015). Carotenoids from fruits and vegetables: Chemistry, analysis, occurrence, bioavailability and biological activities. *Food Research International*, *76*(Pt 3), 735–750. https://doi.org/10.1016/j.foodres.2015.07.047

Sim, J. S., & Sunwoo, H. H. (2002). Designer eggs: Nutritional and functional significance. In R. Ross Watson ((Ed.), *Eggs and health promotion* (pp. 19–36). Iowa State Press.

Simopoulos, A. P. (2004). Omega-6/Omega-3 essential fatty acid ratio & chronic diseases. *Food Reviews International*, *20*, 77–90. https://doi.org/10.1081/FRI-120028831

Sparks, N. H. C. (2006). The hen's egg - is its role in human nutrition changing? *World's Poultry Science Journal*, *62*, 308–315. https://doi.org/10.1079/WPS200599

Stevenson, R. E., Allen, W. P., Pai, G. S., Best, R., Seaver, L. H., Dean, J., & Thompson, S. (2000) Decline in prevalence of neural tube defects in a high - risk region of the United States. *Pediatrics*, *106*, 677–683. https://doi.org/10.1542/peds.106.4.677

Stupin, A., Mihalj, M., Kolobari, N., Šušnjara, P., Kolar, L., Mihaljevi, Z., Mati, A., Stupin, M., Juki, I., Kralik, Z., Grˇcevi´c, M., Kralik, G., Šeri´c, V., & Drenjanˇcevi, I. (2020). Anti-inflammatory potential of n-3 polyunsaturated fatty acids enriched hen eggs consumption in improving microvascular endothelial function of healthy individuals-clinical trial. *International Journal of Molecular Sciences*, *21*, 4149. https://doi.org/10.3390/ijms21114149

Surai, P. G., MacPherson, A., Speake, B. K., & Sparks, N. H. (2000) Designer egg evaluation in a controlled trial. *European Journal of Clinical Nutrition*, *54*, 298–305. http://dx.doi.org/10.1038/sj.ejcn.1600939

Surai, P. F., & Sparks, N. H. C. (2001). Designer eggs: From improvement of egg composition to functional food. *Trends in Food Science & Technology*, *12*, 7–16. https://doi.org/10.1016/s0924-2244(01)00048-6

Suzuki, M., & Tamura, T. (1985). Iodine intake of Japanese male university students: Urinary iodine excretion of sedentary and physically active students and sweat iodine excretion during exercise. *Journal of Nutritional Science and*

Vitaminology (Tokyo), 31, 409–415. https://doi.org/10.3177/jnsv.31.409

Turan, E., & Karaasian, O. (2020). The relationship between iodine and selenium levels with anxiety and depression in patients with euthyroid nodular goiter. *Oman Medical Journal, 35*, e161. https://doi.org/10.5001/omj.2020.84

Vannice, G., & Rasmussen, H. (2014). Position of the academy of nutrition and dietetics. Dietary fatty acids for healthy adults. *Journal of the Academy of Nutrition and Dietetics, 114*, 136–153. https://doi.org/10.1016/j.jand.2013.11.001

Vitti, P., Rago, T., Aghini-Lombardi, F., & Pinchera, A. (2001). Iodine deficiency disorders in Europe. *Public Health Nutrition, 4*(2B), 529–535. http://doi.org/10.1079/PHN2001138

Walchuk, C., House, J. D., Aliani, M., & Suh, M. (2022). Lutein and docosahexaenoic acid enriched egg consumption improves retina function in healthy Caucasian older adults. *Journal of Functional Foods, 89*, 104913. https://doi.org/10.1016/j.jff.2021.104913

Wang, J., Um, P., Dickerman, B. A., & Liu, J. (2018). Zinc, magnesium, selenium and depression: A review of the evidence. *Potential Mechanisms and Implications. Nutrients, 10*, 584. https://doi.org/10.3390/nu10050584

Part IX

Egg Biotechnology

39

Genome Editing and Its Application to Egg Biotechnology

Jae Yong Han* **and Jin Se Park**

CONTENTS

39.1 Introduction

Poultry, especially chickens, are a major source of animal protein for humans. Importantly, chicken eggs are a good-quality protein source and an essential raw material in the food industry. Due to the high protein content and simple protein composition of egg white, chicken eggs are considered a cost-effective production platform for recombinant proteins. Furthermore, chickens do not produce non-human glycans, which can be immunogenic to humans, and chicken egg white proteins have a beneficial N-glycosylation pattern that can improve the efficacy of therapeutic proteins (Lillico et al., 2005; Zhu et al., 2005; Kim et al., 2018). Based on the characteristic N-glycosylation pattern of egg white proteins, recombinant enzymes for the treatment of lysosomal storage diseases (LSDs) have been produced from egg white of transgenic chickens and approved by the Food and Drug Administration (Sheridan, 2016). More recently, using genome editing technologies such as TALENs and CRISPR/Cas9, it has become possible to exchange endogenous egg white proteins with recombinant human therapeutic proteins or to remove major egg white allergens (Park et al., 2014; Oishi et al., 2018). Accordingly, these technologies are currently being used to produce various high-value poultry eggs. Additionally, genome editing in poultry can improve economic traits and disease resistance, which can accelerate poultry breeding programs to develop novel poultry lines with improved productivity (Koslova et al., 2020; Lee et al., 2020)

Genome engineering in poultry can produce valuable poultry species with improved economic traits, modulated egg components, and specific traits for basic biological research. Therefore, methods to introduce exogenous genes or remove endogenous genes have been studied during the past several decades. Although pronuclear injection and somatic cell nuclear transfer (SCNT) have been widely used for genome engineering in mammals, they are not feasible methods in poultry because of their unique physiological characteristics. Chicken oocytes ovulate with a large and opaque yolk; therefore, it is difficult to identify the oocyte nucleus, which makes it impractical to perform SCNT. Furthermore, multiple sperm penetrate poultry eggs during fertilization after ovulation. Therefore, it is difficult to identify the pronucleus engaged in fertilization. Established chicken embryonic stem cells (ESCs) exhibit very inefficient germline transmission, which is a major bottleneck for use of these cells in transgenic research (Pain et al., 1996; Petitte et al., 2004).

Therefore, distinct methods for genome engineering in poultry have been developed. Two main methods are widely used for genome engineering in poultry, namely, direct injection of viral vectors into blastoderm and cultured primordial germ cell (PGC)-mediated genome engineering. We will describe these two main strategies for poultry genome engineering and

* Corresponding author.

DOI: 10.1201/9781003254430-48

introduce several recently developed methods. Based on these established genome engineering technologies in poultry, we will introduce modulation of egg components by genome engineering and future use of egg bioreactors for the production of human recombinant proteins.

39.2 Genome Engineering Technologies in Poultry

39.2.1 Genome Engineering in Poultry by Injection of Viral Vectors into Embryos

To produce genome-engineered individuals, it is important to deliver exogenous genes into endogenous germ cells in embryos. These modulated germ cells will produce gametes and eventually genome-engineered progenies. PGCs, the early progenitor of the germ cell lineage that eventually differentiate into sperm or oocytes, were initially identified as having a scattered distribution in Eyal-Giladi and Kochav (EGK) stage X blastoderm in poultry (Ginsburg & Eyal-Giladi, 1987). During the formation of the primitive streak, PGCs passively move to the anterior region of the embryo and then actively migrate into the germinal crescent (Swift, 1914; Eyal-Giladi et al., 1981). After the formation of blood vessels, PGCs enter the bloodstream and migrate into embryonic gonads (Meyer, 1964). Therefore, if PGCs at the blastodermal stage are genetically engineered by introducing exogenous genes, the recipient embryo will be a germline chimera and eventually produce genome-engineered progenies after sexual maturation. Infection of PGCs in blastoderm with viral vectors is a traditional method to introduce foreign DNA into the poultry genome and has been widely used for transgenesis and genome editing in poultry. At the blastodermal stage, there is a hollow space called the subgerminal cavity between the blastodermal layer and yolk. When viral vectors are injected into the subgerminal cavity, they infect the upper blastodermal layer. During this process, some PGCs in the blastodermal layer are infected and transformed. The recipient embryo exhibits somatic and germline chimerism and produces transgenic progenies after sexual maturation and subsequent mating.

Researchers first used avian leucosis virus (ALV) vectors to infect PGCs in blastoderm and demonstrated that infection of blastoderm with ALV vectors induced viral genome insertion into blastodermal germ cells and transmission of the inserted viral genome to the next generations (Salter et al., 1987). Subsequently, replication-deficient ALV vectors encoding β-lactamase under the control of the cytomegalovirus (CMV) promoter were used to produce transgenic chickens that secrete foreign proteins in egg white (Harvey et al., 2002b). However, the germline transmission efficiency was very low (0.12–0.35%) when replication-deficient ALV vectors were used (Harvey et al., 2002a). Following the development of lentiviral vectors that can infect non-dividing cells and integrate their genome into that of target cells with high efficiency, lentiviral vectors were injected into blastoderm and showed a significantly enhanced germline transmission efficiency (4–45%) (McGrew et al., 2004). Thereafter, lentiviral vector systems were widely applied and various transgenic poultry models were produced (Kwon et al., 2010; Lyall et al., 2011). Furthermore, this system has been applied to produce transgenic models in other avian species such as zebra finch (*Taeniopygia guttata*) as animal models of vocal learning (Agate et al., 2009). However, lentiviral vector-mediated genome editing is hindered because the integration of the lentiviral genome induces constitutive expression of Cas9 and guides RNAs, which potentially increases off-target effects (Ortinski et al., 2017).

Recently, blastodermal injection of adenoviral vectors and genome editing have been performed in quails (*Coturnix japonica*). Adenoviral vectors efficiently infect both dividing and non-dividing cells and importantly are non-integrating viral vectors that enable episomal expression of Cas9 and guide RNA cassettes. Furthermore, they can deliver large DNA cargos, which can efficiently deliver Cas9 and guide RNA constructs. Based on these benefits of adenoviral vectors, adenoviral vectors encoding Cas9 and guide RNA targeting the *melanophilin* (*MLPH*) gene were constructed and microinjected into quail blastoderm. After mating of G0 chimeras with wild-type quails, it was reported that *MLPH* gene-edited progenies were produced and the germline transmission efficiency ranged from 2.4% to 10.0% (Lee et al., 2019).

These studies demonstrated that microinjection of viral vectors into blastoderm is a feasible strategy to modulate the poultry genome and that precise genome editing can be efficiently performed in poultry without PGC culture. However, the targeted insertion of long DNA constructs, by injecting viral vectors into blastoderm, has not been performed in poultry. If efficient targeted insertion can be performed via blastodermal injection, this will be a powerful tool to produce genome-edited poultry with valuable traits.

39.2.2 PGCs-Mediated Genome Engineering

PGCs are early progenitor cells that migrate to gonads and differentiate into sperm and oocytes. In poultry, especially chickens, PGCs are considered the most prominent germline-competent stem cells for transgenic research. Chicken PGCs can be isolated from the germinal crescent, embryonic blood vessels, and embryonic gonads. Most importantly, chicken PGCs maintain their germline competency during long-term *in vitro* culture, which significantly facilitates precise genome editing studies in chickens. Using cultured PGCs, numerous transgenic and genome-edited chickens with valuable purposes have been produced. In this section, we introduce the isolation and culture of chicken PGCs and the production of genome-engineered birds using cultured PGCs.

39.2.2.1 Isolation, Culture of PGCs and Production of Germline Chimera

As mentioned above, poultry PGCs migrate to the embryonic gonads by circulating in blood vessels, which is a unique characteristic compared to mammals. This unique characteristic facilitates research on the isolation of PGCs from embryos and analysis of the germline transmission efficiency following the injection of donor PGCs into the recipient's bloodstream. It was reported that chicken PGCs can be isolated from the blood of Hamburger and Hamilton (HH) stage 13–15 embryos by Ficoll density gradient centrifugation and that the germline transmission efficiency of isolated PGCs is very high up to 96% (Chang et al., 1992; Naito et al., 1994). Also, using monoclonal antibodies which specifically bind to PGC-specific markers, researchers performed experiments to isolate PGCs by magnetic-activated cell sorting and fluorescence-activated cell sorting in quail and chicken and efficiently isolated PGCs from embryonic blood and gonads (Ono & Machida, 1999; Kim et al., 2004; Mozdziak et al., 2005). More recently, simple methods to isolate avian PGCs from embryonic blood based on the size difference between PGCs and other blood cells have been developed (Jung et al., 2017). These studies suggest that chicken PGCs can be efficiently isolated from embryonic blood and gonads and that germline chimeras can be efficiently produced by transplanting PGCs. Furthermore, if chicken PGCs can be cultured *in vitro*, it will be possible to efficiently perform genome engineering including precise gene targeting in poultry species. With efforts to culture of chicken PGCs *in vitro*, it was reported that PGCs isolated from embryonic gonads have proliferative activity and that exogenous DNA can be introduced into cultured PGCs (Allioli et al., 1994). Furthermore, gonadal chicken PGCs can be cultured *in vitro* for 5 days, and germline chimera can be produced by transferring short-term cultured chicken PGCs (Chang et al., 1997). Also, mitomycin C-treated gonadal cells were used as feeder cells to culture blood PGCs and blood PGCs can be maintained *in vitro* for 4 days (Naito et al., 2001). Subsequently, it was demonstrated that chicken gonadal PGCs can be maintained *in vitro* for 2 months and showed that PGCs cultured for an extended duration have a germline transmission capacity (Han et al., 2002). These results demonstrate that chicken PGCs can be cultured *in vitro* and can be an optimal tool for genome engineering in poultry. In 2006, long-term *in vitro* culture methods for chicken blood PGCs were established and these long-term cultured PGCs showed efficient germline transmission efficiency (van de Lavoir et al., 2006). Furthermore, exogenous DNA can be introduced into cultured PGCs and these modified PGCs were able to produce transgenic progenies. Subsequently, several studies confirmed the feasibility of long term culture of chicken PGCs and demonstrated that the basic fibroblast growth factor (bFGF)-mediated MEK/ERK signaling pathway is essential for the proliferation of chicken PGCs *in vitro* (Choi et al., 2010; Macdonald et al., 2010). Recently, growth factors sufficient for self-renewal of germline-competent PGCs have been identified and defined culture condition for optimal chicken PGCs culture has been established. It was showed that bFGF, insulin, and Activin A signaling is sufficient to maintain self-renewal of germline-competent PGCs. Furthermore, the addition of ovotransferrin (TF) to cultures can remove all requirements for serum during PGC culture (Whyte et al., 2015). These studies of chicken PGC culture and germline chimera production have facilitated the production of genome-engineered poultry with valuable traits.

39.2.2.2 Genome Engineering in Chicken Using Cultured PGCs

After the establishment of long-term *in vitro* culture methods and validation of the high germline transmission efficiency of chicken PGCs, genome engineering studies, including those performing precise genome editing, have been mainly conducted using cultured PGCs in chicken. In 2012, two research groups introduced transgenes into cultured chicken PGCs using transposons and reported a high germline transmission efficiency of up to 95.2% and efficient production of transgenic progenies that ubiquitously expressed green fluorescent protein (GFP) (Macdonald et al., 2012; Park & Han, 2012). Furthermore, by delivering transposon vectors containing a synthetic ovalbumin (OVAL) promoter-controlled recombinant protein expression cassette into cultured PGCs, chicken bioreactors that accumulate recombinant proteins in egg white were produced (Park et al., 2015; Kim et al., 2018).

Precise genome editing and long-term selection can be performed with cultured PGCs. Consequently, the production of genome-edited chickens for various purposes was accelerated after the establishment of *in vitro* culture methods for PGCs. In 2013, by inducing homologous recombination in cultured PGCs, genome-edited chickens were produced in which the immunoglobulin heavy chain gene was knocked out, providing an avian B-cell depletion model for immunological research (Schusser et al., 2013). However, the targeting frequency was very low, with one targeted cell per 10^7 transfected cells. This problem can be solved by inducing double-strand breaks at the target site. After the development of precise genome editing tools such as TALENs and CRISPR/Cas9, these tools were rapidly applied to cultured chicken PGCs. In 2014, it was showed that TALENs can efficiently induce targeted knockout of the *OVAL* gene in cultured PGCs and produce genome-edited chickens (Park et al., 2014). Subsequently, targeted knockout of the *OVAL* and ovomucoid (*OVM*) genes in chicken using CRISPR/Cas9 was reported (Oishi et al., 2016). Furthermore, CRISPR/Cas9-mediated targeted insertion of exogenous DNA using homology-directed repair and non-homologous end joining was achieved in cultured chicken PGCs and genome-edited chickens were produced (Dimitrov et al., 2016; Lee et al., 2019).

This successful precise genome editing in chicken promoted research on the modulation of egg components and improvement of chicken productivity. The modulation of egg components by genome editing will be discussed in the following sections.

Furthermore, research about conferring viral resistance in chicken has been performed by applying precise genome editing technology to genes encoding certain host factors in order to inhibit viral infection and replication because viral diseases are a major threat to the poultry industry. For this purpose, precise deletion of tryptophan residue number 38 (W38) of chicken Na+/H+ exchanger type 1 (chNHE1), which are a critical host factor and amino acid residue for avian leucosis virus subgroup J (ALV-J) entry, was performed in cultured PGCs and then produced genome-edited chickens without visible abnormalities (Koslova et al., 2020). These chickens were resistant to the highly pathogenic ALV-J subtype, demonstrating that CRISPR/Cas9 and cultured PGCs can be used for precise breeding to induce disease resistance in poultry (Kheimar et al., 2021). Similarly, disruption of protein expression of Tva, which is a host factor required for ALV-A and E, was performed by inducing a frame-shift mutation in the *TVA* gene in cultured PGCs and produced genome-edited chickens. These chickens were resistant to ALV-A and E, but showed abnormalities in cobalamin/vitamin B_{12} metabolism (Koslova et al., 2021).

In conclusion, the culture of chicken PGCs significantly promotes and improves avian genome engineering technologies and contributes to production of novel chicken lines with traits valuable to the improvement of human welfare.

39.2.2.3 *Improvement of Production Efficiency of Germline Chimera Using Endogenous Germ Cell Sterilization System*

Although PGCs are an efficient tool for poultry genome editing, high variation in the germline transmission efficiency of PGCs cultured *in vitro* is a bottleneck for PGC-mediated genome editing (Song et al., 2014). To enhance the germline transmission efficiency of cultured PGCs, it is important to reduce or eliminate competitive endogenous germ cell populations in recipient embryos. Several studies have sought to reduce endogenous germ cell populations. Busulfan (1,4-butanediol-dimethanesulfonate), which is an alkylating agent, induced the depletion of PGCs when administered into mammalian embryos (Hemsworth & Jackson, 1962). Furthermore, busulfan induced depletion of endogenous PGCs in poultry embryos when administered into egg white or yolk (Aige-Gil & Simkiss, 1991). Due to its germ cell-selective activity, researchers have used busulfan to increase the efficiency of germline chimera production in poultry. In 2005, it was shown that chicken germline chimeras can be efficiently produced by transferring donor PGCs into busulfan-treated embryos; the efficiency was 8-fold higher in the busulfan-treated group than in the untreated group (Song et al., 2005). Also, the occupation efficiency of donor PGCs in gonads was about 28-fold higher in busulfan-treated embryos than in untreated embryos (Nakamura et al., 2008). Recently, the establishment of busulfan-resistant PGCs by introducing microsomal glutathione-S-transferase II (*MGSTII*) and the production of germline chimeras have been reported. In this model, when busulfan was injected into germline chimeras, the production efficiency of donor PGC-derived progenies and transgenic progenies was significantly improved (Kim et al., 2021). In addition to injection of chemical reagents, gamma irradiation of embryos also reduced the population of endogenous germ cells and enhanced the efficiency of germline chimera production (Carsience et al., 1993; Park et al., 2010). However, these methods cannot eliminate all endogenous germ cells and have detrimental effects on developing embryos, resulting in a low hatching rate.

In 2017, *DDX4* gene-targeted chickens mediated by TALENs were produced. Strikingly, *DDX4* null hens exhibited depletion of follicles in ovaries and did not start laying (Taylor et al., 2017). When introducing donor PGCs into *DDX4* null embryos, the recipient *DDX4* null hens laid eggs normally and progenies were mainly derived from donor PGCs, demonstrating that germ cell sterile models can be a powerful tool to produce chicken germline chimeras and genome-engineered chickens (Woodcock et al., 2019). Recently, the same research group developed novel germ-cell sterile chicken models in which inducible caspase-9 (iCaspase9) protein is specifically expressed in the germ cell lineage. iCaspase9 is a chimeric protein consisting of Caspase9 and the drug-dependent dimerization domain of FK506-binding protein. In the presence of B/B compound, iCaspase9 dimerizes and induces cellular apoptosis (Gargett & Brown, 2014). In this model, the iCaspase9 construct was specifically targeted to the 3'-end of the *DAZL* gene, and endogenous germ cells were eliminated when B/B compound was injected into embryos. Therefore, when donor PGCs and B/B compounds were co-injected into embryos, the endogenous germ cell population was specifically eliminated and injected donor PGCs occupied gonads, eventually leading to the production of gametes derived only from donor PGCs (Ballantyne et al., 2021). This conditional endogenous germ cell depletion system has been successfully used in studies about the maintenance of poultry genome resources (Hu et al., 2022). Also, it is anticipated that this germ cell sterile model will significantly accelerate genome editing studies in poultry, especially chickens.

39.2.3 Injection of Vectors into the Embryonic Bloodstream and Sperm Transfection-Assisted Gene Editing (STAGE)

Although cultured PGCs are a powerful tool for poultry genome editing, only chicken PGCs have been cultured for a long duration *in vitro*. Therefore, cultured PGC-mediated genome editing cannot be widely performed in poultry except chicken. Furthermore, the culture of PGCs is technically difficult and the production of genome-edited progenies using cultured PGCs is a time-consuming process that requires selection of genome-edited PGCs, transplantation of PGCs into recipient embryos, rearing of recipients to sexual maturation, and finally test crosses to produce genome-edited poultry. Therefore, to develop universal and convenient methods for genome editing in poultry, direct injection of vectors encoding genome editor into recipient embryos has been recently studied.

Avian PGCs migrate to gonads by circulating in the embryonic bloodstream. Therefore, it is expected that circulating PGCs can be modulated by directly injecting foreign DNA into embryonic blood vessels. In 2012, the production of transgenic quail by injecting lentiviral vectors encoding GFP into blood vessels of HH 13–15 stage embryos was reported (Zhang et al., 2012). Furthermore, in chicken, direct injection of Tol2 transposon and transposase plasmids into liposomes transformed circulating PGCs and produced transgenic progenies (Tyack et al., 2013). Furthermore, by introducing Cas9 and guide RNA constructs into Tol2 transposon vectors and injecting these Tol2 transposon vectors into embryonic bloodstream, genome-edited chickens were recently produced (Challagulla et al., 2020). Tol2 transposon-mediated transformation of circulating PGCs has also been performed in quails, and transgenic quail lines for use as developmental models have been established (Serralbo et al., 2020). Although the germline transmission efficiency is very low, these results demonstrate that direct injection of foreign DNA into embryonic blood vessels can transform circulating PGCs and efficiently produce transgenic and genome-edited offspring in various poultry species for which culture of PGCs is unfeasible.

In 2017, direct transfection of spermatozoa with Cas9 mRNA and guide RNA was reported. By artificially inseminating these transfected spermatozoa, genome-edited chickens were produced and the production efficiency of genome-edited progenies was 0–26% (Cooper et al., 2017). This STAGE method is a powerful method because it can be applied to various avian species and genome-edited progenies can be directly produced after insemination of transfected spermatozoa. However, this method requires further refinement to improve the editing efficiency in spermatozoa. Figure 39.1 shows the outline of genome engineering methods in poultry.

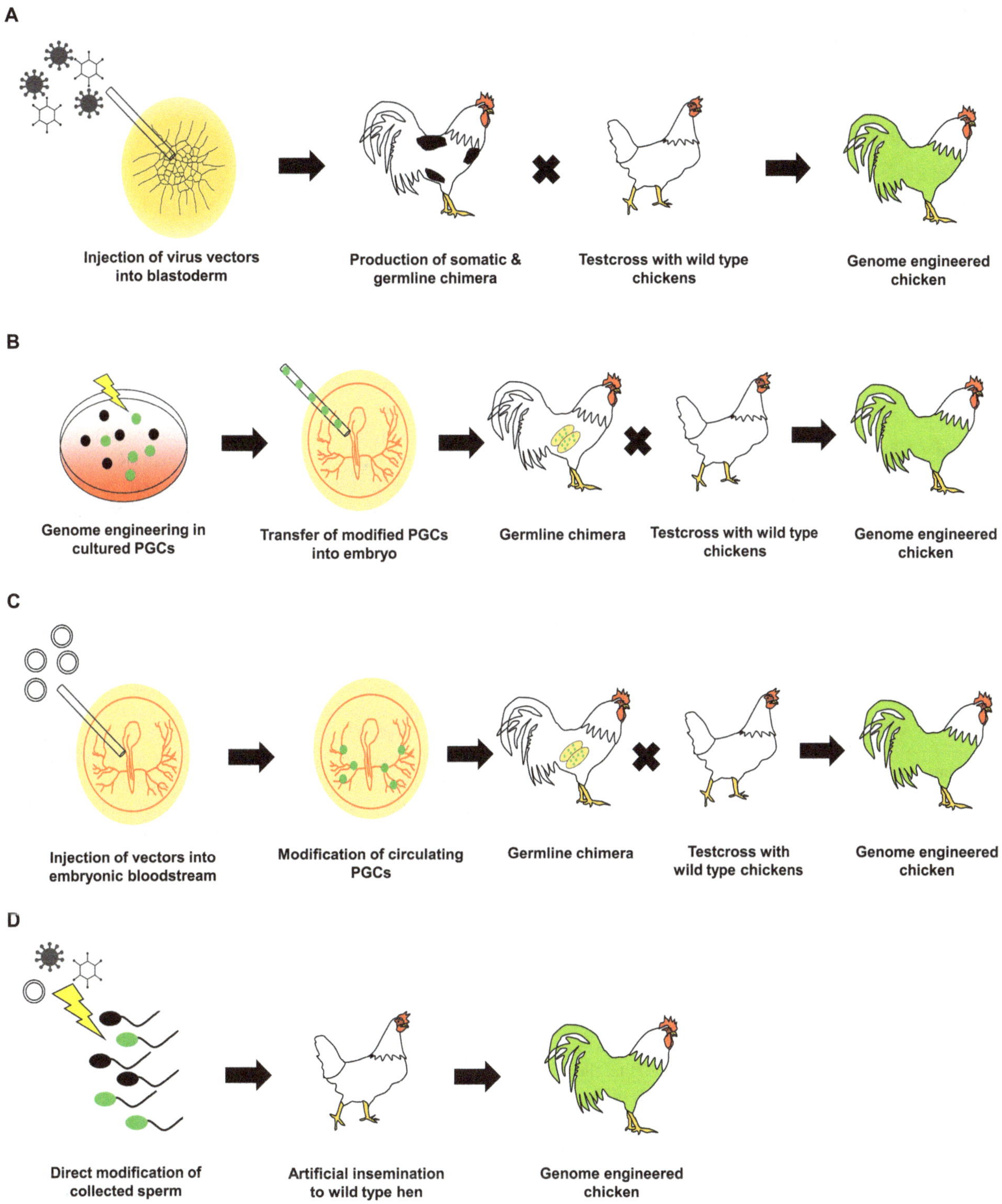

FIGURE 39.1 Schematic illustration of genome engineering methods in poultry. (A) Genome engineering in poultry by direct injection of viral vectors into blastoderm. PGCs in blastoderm can be modified by viral vector infection. Resultant germline chimera produce genome-engineered progeny by testcross with wild-type individuals. (B) Genome engineering in poultry by using cultured PGCs. PGCs can be modified in vitro and transferred into recipient embryos. Resultant germline chimera produce genome-engineered progeny by testcross with wild-type individuals. (C) *In vivo* PGCs modification by direct injection of vectors into the bloodstream of recipient embryos. When vectors are injected into the bloodstream, circulating PGCs are transfected and modified. Resultant germline chimera produce genome-engineered progeny by testcross with wild-type individuals. (D) STAGE methods. Direct modification of collected sperm can be performed *in vitro*. Modified sperm will be used in artificial insemination and producing genome-engineered progeny.

39.3 Egg Component Modifications by Genome Engineering

39.3.1 Modification of Egg White Components for Bioreactors

Chicken eggs are considered an efficient bioreactor for the production of recombinant human proteins because hens lay more than 300 eggs per year and chicken eggs contain abundant proteins. Furthermore, recombinant proteins with a humanized N-glycan pattern can be produced in eggs because egg proteins have a similar N-glycan pattern as human proteins and chickens do not produce non-human glycans such as α-galactose and Neu5Gc (Raju et al., 2000; Lillico et al., 2005). Therefore, various studies have sought to accumulate therapeutic human proteins in eggs, especially egg white.

Egg white is specifically synthesized and secreted by oviduct tubular gland cells of hens. It contains abundant proteins (around 3–4 g per egg), including mainly OVAL, TF, OVM, and lysozyme (LYZ). OVAL comprises almost 50% of egg white proteins and, consequently, the endogenous OVAL promoter is highly active in oviduct tubular gland cells. Therefore, researchers have constructed a recombinant protein expression cassette in which the expression of recombinant proteins is under the control of a synthetic OVAL promoter. In 2002, the accumulation of foreign proteins in egg white using the ubiquitous CMV promoter was reported (Harvey et al., 2002b). In 2005, the production of chimeric chickens in which a human monoclonal antibody accumulates in egg white was reported. In that report, expression vectors containing 7.5 or 15 kb of the OVAL 5' regulatory region and an antibody-encoding construct were transfected into cultured chicken ESCs and chimeric chickens were produced by transferring transformed ESCs into embryos, although germline transmission did not occur. The monoclonal antibody derived from egg white was glycosylated, with abundant terminal mannose and N-acetylglucosamine (GlcNAc) residues. Additionally, this antibody lacked core fucosylation. Based on this N-glycosylation, the antibody derived from egg white showed enhanced antibody-dependent cellular cytotoxicity (ADCC), although its serum half-life *in vivo* was compromised due to abundant terminal mannose residues. Importantly, the antibody efficiently accumulated in egg white (up to 3 mg per egg) (Zhu et al., 2005). This study demonstrated that the OVAL regulatory region can drive the specific synthesis of therapeutic proteins in oviducts that accumulate in egg white. Furthermore, eggs can be used as an efficient bioreactor for the production of therapeutic antibodies with enhanced efficacy. After that report was published, several studies sought to accumulate therapeutic proteins in egg white using the OVAL promoter. In 2007, transgenic chickens were produced by injection of lentiviral vectors containing 2.8 kb of the OVAL 5' regulatory region with an estrogen response element and a human therapeutic protein-encoding construct. The produced transgenic chickens showed oviduct-specific expression of human therapeutic proteins that accumulated in egg white. They also demonstrated that therapeutic proteins consistently accumulated in egg white after one generation without silencing effects, demonstrating that chicken eggs are a consistent production platform for human recombinant proteins (Lillico et al., 2007). Furthermore, transgenic quails that expressed human interleukin 1 receptor antagonist (rhIL1RN) in an oviduct-specific manner were produced using the chicken OVAL promoter and showed that rhIL1RN consistently accumulated in quail egg white after several generations (Kwon et al., 2010).

In 2015, using *piggyBac* transposon containing an OVAL mini-synthetic promoter and a human epidermal growth factor (hEGF) construct, transgenic chickens that accumulate hEGF in egg white were produced and hEGF derived from egg white showed tissue regeneration activity, demonstrating that egg white-derived recombinant proteins can be used as cosmetic ingredients or wound healing reagents (Park et al., 2015). Furthermore, By performing *piggyBac*-mediated transposition, transgenic chickens that accumulated an anti-CD20 monoclonal antibody in egg white were produced and showed that the Fc effector functions, such as ADCC and complement-dependent cytotoxicity, of egg white-derived anti-CD20 monoclonal antibody were better than those of the CHO cell-derived anti-CD20 monoclonal antibody Rituximab (Kim et al., 2018). This study demonstrated that chicken bioreactors are an efficient production platform for therapeutic monoclonal antibodies with improved Fc effector functions.

Following the development of precise genome editing technologies such as CRISPR/Cas9, it became possible to target a therapeutic protein construct to the endogenous OVAL locus, thereby allowing therapeutic proteins to be expressed under the control of the strong endogenous OVAL promoter. In 2018, OVAL-targeted hens accumulating a human interferon beta (hIFN-β) in egg white were reported. Eggs laid by these hens contained 1.86–4.42 mg/ml hIFN-β in egg white, which was significantly better than the accumulation efficiency in the previous bioreactor system in which expression of the protein construct was controlled by a synthetic OVAL promoter (Oishi et al., 2018). Subsequently, OVAL-targeted hens which accumulate anti-HER2 monoclonal antibody in egg white were also reported and this antibody accumulated in egg white at a concentration of 1.4–1.9 mg/ml (Mukae et al., 2020). These studies demonstrate that targeting the OVAL locus using genome editing technology can be a powerful strategy to develop chicken bioreactors for the production of therapeutic proteins in egg white.

39.3.2 Modification of Egg Yolk Components for Bioreactors

Proteins in egg yolk are transferred from the bloodstream when the yolk starts to mature in the ovary. In particular, chicken IgG is transported into egg yolk via receptor-mediated transcytosis. Furthermore, it was reported that human IgG can be transported into egg yolk and that the HEAL sequence in the human IgG Fc region is important for transportation (Morrison et al., 2002). Therefore, when human IgG Fc-containing proteins circulate in the chicken bloodstream, they accumulate in egg yolk (Kawabe et al., 2006). In 2005, the production of transgenic chickens by injecting retroviral vectors encoding anti-prion single-chain Fv fused to human IgG1 Fc (scFv-Fc) was reported. The ubiquitous chicken β-actin promoter regulated the expression of scFv-Fc, transgenic chickens ubiquitously expressed scFv-Fc in all organs, and this protein was secreted into serum, egg yolk, and egg white (Kamihira et al., 2005). Subsequently, transgenic chickens ubiquitously expressing a monoclonal antibody and human erythropoietin-Fc (hEPO-Fc) fusion protein were produced, and these target proteins circulated in serum and accumulated in egg yolk and white (Kamihira et al., 2009; Penno et al., 2010). However, a silencing effect of the ubiquitous promoter was observed in transgenic chickens after several generations, meaning that consistent expression and accumulation of target proteins in egg yolk are not guaranteed.

Recently, genome edited chickens that synthesize recombinant protein in liver specific manner has been reported (Park et al., 2023). In this report, recombinant human IgG1 Fc coding sequence was targeted in serum albumin (*ALB*) gene using CRISPR/Cas9 and human IgG1 Fc was secreted in bloodstream and eventually accumulated in egg yolk. The N-glycosylation pattern of human IgG1 Fc derived from chicken liver showed abundant terminal sialylation and low level of core fucosylation. Major serum proteins are dominantly synthesized and secreted by the liver; therefore, their N-glycosylation patterns and resultant biological activities will differ from those of recombinant proteins derived from oviducts. Further research is required to establish a consistent egg yolk bioreactor system using precise genome editing technologies.

39.3.3 Modification of Egg Components to Reduce Allergens

Egg allergy is a prevalent food allergy in children and affects approximately 1.6% of children (Eggesbo et al., 2001). Chicken eggs are a major food source and widely used in the food industry; therefore, it is desirable to develop hypoallergenic eggs. The major egg allergens are four egg white proteins, namely, OVAL, OVM, TF, and lysozyme (Mine & Yang, 2008). Among these, it was reported that OVM is the dominant allergen in egg white (Urisu et al., 1997). Following the development of precise genome editing technologies, it became possible to precisely and efficiently knockout the genes encoding these major egg allergens. In 2014, targeted disruption of *OVAL* gene using TALENs and the production of genome-edited chickens were reported (Park et al., 2014). Subsequently, using CRISPR/Cas9, knockout the *OVAL* and *OVM* genes and production of *OVAL*- and *OVM*-knockout chickens were reported (Oishi et al., 2016). A further study showed that eggs laid by these *OVM*-knockout hens lacked OVM protein, demonstrating that hypoallergenic eggs were produced using genome editing (Mukae et al., 2021). Recently, by applying base editing technology, a premature stop codon was introduced into the *TF* gene and showed that the quantity of TF was reduced in eggs laid by heterozygous *TF*-knockout hens (Lee et al., 2020). These studies show that genome editing can be efficiently performed in chicken to develop hypoallergenic egg products, which will contribute to related food industries (Figure 39.2).

39.4 Eggs as Potential Bioreactors for Production of Therapeutic Proteins

Eggs have beneficial characteristics for use as bioreactors to produce therapeutic proteins. As mentioned earlier, the high egg laying efficiency of hens and the high protein production capacity of hen oviducts are attractive characteristics for economical mass production of expensive biopharmaceuticals. In addition, the N-glycosylation pattern of egg white proteins is beneficial to improve the efficacy of certain biopharmaceuticals. Egg white proteins have an N-glycosylation pattern dominantly terminated with mannose and GlcNAc residues (Harvey et al., 2000). In general, terminal mannose and GlcNAc residues are not favorable because they can reduce the half-lives of therapeutic proteins via mannose receptor (MR)-mediated endocytosis (Lee et al., 2002). However, in the case of LSDs, in which certain lysosomal enzymes are depleted, an N-glycan terminated with mannose and GlcNAc has a strong benefit. To treat LSDs, the lacking enzyme should be supplied by enzyme replacement therapy and delivered to lysosomes. One effective way to deliver enzymes to lysosomes is to target MR and mannose-6-phosphate (M6P) receptor, which recognize terminal mannose, GlcNAc, and M6P (Kornfeld, 1987; Sato & Beutler, 1993). In mammalian cell culture systems, the N-glycosylation pattern is mainly terminated with galactose or sialic acids; therefore, an additional process is required to expose mannose or GlcNAc, which increases production costs (Friedman et al., 1999). When enzymes are produced in egg white, these additional processes are not required. Based on these characteristics, enzymes to treat LSDs have been produced and purified from chicken egg white and commercialized (Ratner, 2015). The use of genome-edited chickens in which enzymes to treat LSDs massively accumulate in egg white can significantly reduce production costs. It will be possible to produce various drugs for treatment of LSDs using genome-edited chickens as bioreactors in the near future.

Another beneficial characteristic of the N-glycosylation pattern of egg white proteins is the absence of core fucosylation (afucosylation). In the case of monoclonal antibodies, glycosylation of 297 asparagine residue is critical to maintain their binding affinity to Fc gamma receptors (FcγRs) and depletion of core fucosylation of Fc N-glycan significantly improves their binding affinity to FcγRIIIA, which is the main receptor to induce ADCC. Therefore, afucosylated monoclonal antibodies showed significantly improved ADCC (Shields et al., 2002). A major mode-of-action of anti-cancer monoclonal antibodies, such as anti-CD20, anti-HER2, and anti-EGFR antibodies, is to induce ADCC, and egg white-derived anti-cancer monoclonal antibodies show significantly improved anti-cancer activity (Kim et al., 2018; Pereira et al., 2018). Therefore, anti-cancer antibodies with increased efficacy can be economically produced using genome-edited chickens by targeting to a major egg white protein locus.

Also, it is possible to produce a chicken bioreactor system that accumulates recombinant proteins in egg yolk by targeting genes encoding major serum proteins. Serum proteins are mainly produced in the liver; therefore, their N-glycosylation pattern is expected to differ from that of egg white-derived recombinant proteins (Park et al., 2023). It was reported that α-2,6 sialyltransferase (ST6GAL1) activity is high and fucosyltransferase (FUT8) activity is low in the chicken liver (Kojima et al., 2015). Reflecting these characteristics, the N-glycosylation pattern of egg yolk proteins is characterized by efficient terminal sialylation and the absence of core fucosylation (Sumiyoshi et al., 2009). Therefore, it is

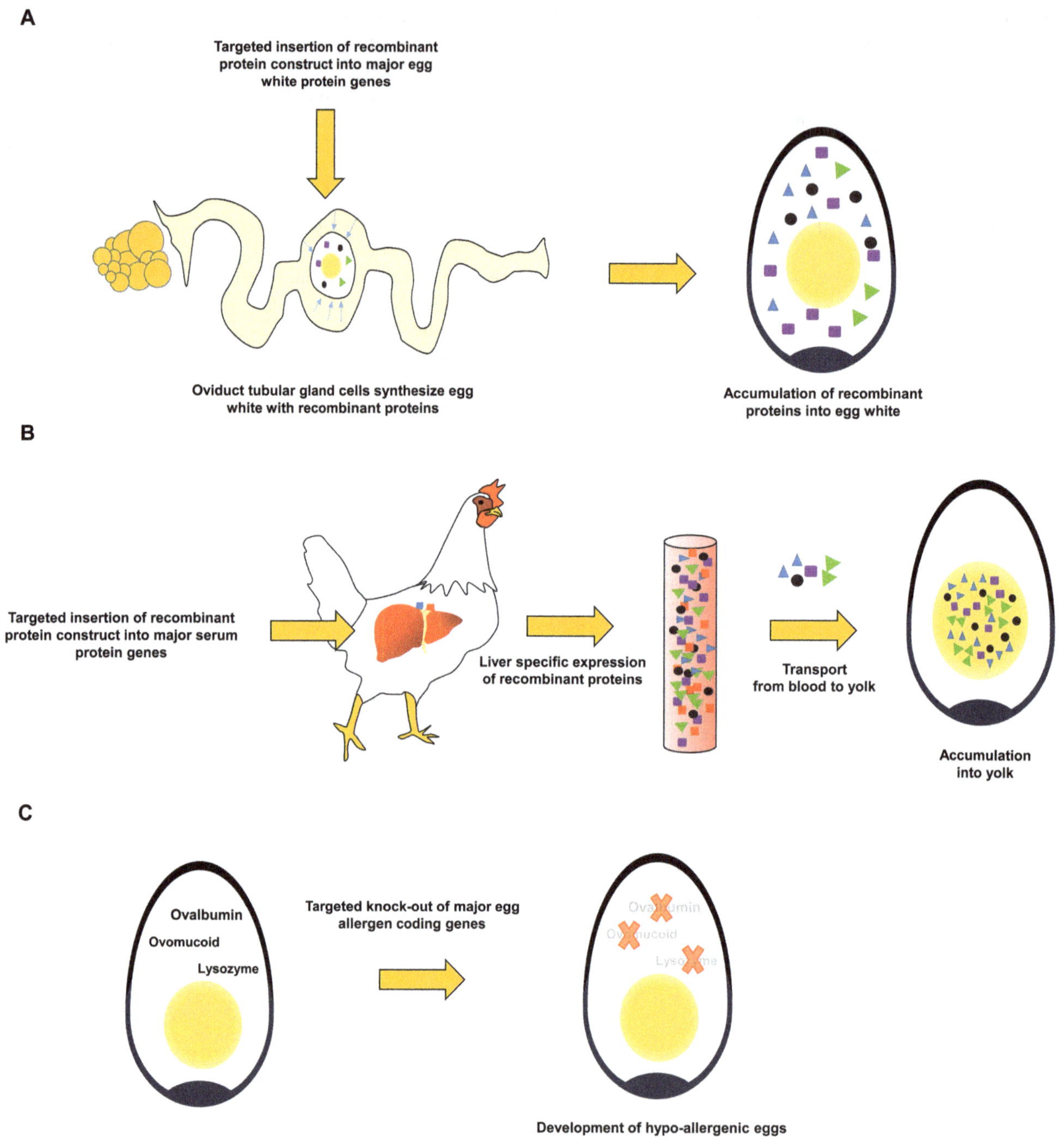

FIGURE 39.2 Modulation of egg components by genome engineering. (A) Accumulation of recombinant proteins in egg white. Targeted insertion of recombinant protein construct into egg white protein genes induces oviduct tubular gland cell-specific expression of recombinant proteins and accumulation of recombinant proteins in egg white. (B) Accumulation of recombinant proteins in egg yolk. Targeted insertion of recombinant protein construct into major serum protein genes induce liver-specific expression of recombinant proteins and accumulation of recombinant proteins in egg yolk. (C) Targeted disruption of major egg allergen coding genes can remove egg allergens in egg white.

expected that recombinant proteins derived from the liver will have an N-glycosylation pattern with abundant terminal α-2,6 sialylation and afucosylation. Terminal sialylation significantly increases the stability of proteins and their serum half-lives and will therefore be beneficial for most biopharmaceuticals, including hormones and cytokines (Ngantung et al., 2006). Furthermore, human serum proteins have an N-glycosylation pattern with abundant terminal α-2,6 sialylation and afucosylation (Clerc et al., 2016). Therefore, it will be possible to produce recombinant blood products with a humanized N-glycosylation pattern if human blood products are specifically synthesized in chicken liver. Thereafter, it will be possible to replace human blood with egg yolk as a source of human blood products, thereby solving problems related to human blood supply shortages. Furthermore, afucosylation of N-glycans is beneficial to produce monoclonal antibodies with enhanced efficacy, as mentioned above. Based on the developed genome editing technologies in chickens, a chicken bioreactor system that specifically synthesizes target proteins in a liver-specific manner and accumulates these proteins in egg yolk is expected to be developed in the near future.

39.5 Conclusion

Genome editing technologies in poultry have significantly developed in the last several decades and production of genome-edited poultry with valuable traits has become feasible in the last several years. These genome editing technologies can facilitate poultry breeding by selectively editing genes related to disease resistance and muscle growth. When used in combination with recently developed molecular breeding technologies, genome editing technologies will significantly accelerate poultry breeding programs and help to establish valuable poultry species with improved economic traits. More importantly, we can modulate egg protein components using genome editing technologies and develop high-valued eggs by removing major egg allergens or massively accumulating high-valued and expensive therapeutic proteins. In particular, the production of certain protein therapeutics in eggs can improve their *in vivo* efficacy due to the beneficial N-glycosylation pattern. Recently, mass accumulation of recombinant human proteins by genome editing was demonstrated in chickens, and various types of bioreactors will be produced. Therefore, genome-edited chickens are anticipated to be an optimal bioreactor system in pharmaceutical industries in the near future. Collectively, genome editing in poultry will contribute to the development of various industries related to poultry and improve human welfare in various respects.

REFERENCES

Agate, R. J., Scott, B. B., Haripal, B., Lois, C., & Nottebohm, F. (2009). Transgenic songbirds offer an opportunity to develop a genetic model for vocal learning. *Proceedings of the National Academy of Sciences of the United States of America*, *106*(42), 17963–17967. https://doi.org/10.1073/pnas.0909139106

Aige-Gil, V., & Simkiss, K. (1991). Sterilisation of avian embryos with busulphan. *Research in Veterinary Science*, *50*(2), 139–144. https://doi.org/10.1016/0034-5288(91)90096-7

Allioli, N., Thomas, J. L., Chebloune, Y., Nigon, V. M., Verdier, G., & Legras, C. (1994). Use of retroviral vectors to introduce and express the beta-galactosidase marker gene in cultured chicken primordial germ cells. *Developmental Biology*, *165*(1), 30–37. https://doi.org/10.1006/dbio.1994.1231

Ballantyne, M., Woodcock, M., Doddamani, D., Hu, T., Taylor, L., Hawken, R. J., & McGrew, M. J. (2021). Direct allele introgression into pure chicken breeds using sire dam surrogate (SDS) mating. *Nature Communications*, *12*(1), 659. https://doi.org/10.1038/s41467-020-20812-x

Carsience, R. S., Clark, M. E., Verrinder Gibbins, A. M., & Etches, R. J. (1993). Germline chimeric chickens from dispersed donor blastodermal cells and compromised recipient embryos. *Development*, *117*(2), 669–675. https://doi.org/10.1242/dev.117.2.669

Challagulla, A., Jenkins, K. A., O'Neil, T. E., Morris, K. R., Wise, T. G., Tizard, M. L., & Doran, T. J. (2020). Germline engineering of the chicken genome using CRISPR/Cas9 by in vivo transfection of PGCs. *Animal Biotechnology*, 1–10. https://doi.org/10.1080/10495398.2020.1789869

Chang, I. K., Jeong, D. K., Hong, Y. H., Park, T. S., Moon, Y. K., Ohno, T., & Han, J. Y. (1997). Production of germline chimeric chickens by transfer of cultured primordial germ cells. *Cell Biology International*, *21*(8), 495–499. https://doi.org/10.1006/cbir.1997.0173

Chang, I. K., Tajima, A., Yasuda, Y., Chikamune, T., & Ohno, T. (1992). Simple method for isolation of primordial germ cells from chick embryos. *Cell Biology International Reports*, *16*(9), 853–857. https://doi.org/10.1016/s0309-1651(06)80165-7

Choi, J. W., Kim, S., Kim, T. M., Kim, Y. M., Seo, H. W., Park, T. S., & Han, J. Y. (2010). Basic fibroblast growth factor activates MEK/ERK cell signaling pathway and stimulates the proliferation of chicken primordial germ cells. *PLoS One*, *5*(9), e12968. https://doi.org/10.1371/journal.pone.0012968

Clerc, F., Reiding, K. R., Jansen, B. C., Kammeijer, G. S., Bondt, A., & Wuhrer, M. (2016). Human plasma protein N-glycosylation. *Glycoconjugate Journal*, *33*(3), 309–343. https://doi.org/10.1007/s10719-015-9626-2

Cooper, C. A., Challagulla, A., Jenkins, K. A., Wise, T. G., O'Neil, T. E., Morris, K. R., & Doran, T. J. (2017). Generation of gene edited birds in one generation using sperm transfection assisted gene editing (STAGE). *Transgenic Research*, *26*(3), 331–347. https://doi.org/10.1007/s11248-016-0003-0

Dimitrov, L., Pedersen, D., Ching, K. H., Yi, H., Collarini, E. J., Izquierdo, S., & Leighton, P. A. (2016). Germline Gene editing in chickens by efficient CRISPR-mediated homologous recombination in primordial germ cells. *PLoS One*, *11*(4), e0154303. https://doi.org/10.1371/journal.pone.0154303

Eggesbo, M., Botten, G., Halvorsen, R., & Magnus, P. (2001). The prevalence of allergy to egg: A population-based study in young children. *Allergy*, *56*(5), 403–411. https://doi.org/10.1034/j.1398-9995.2001.056005403.x

Eyal-Giladi, H., Ginsburg, M., & Farbarov, A. (1981). Avian primordial germ cells are of epiblastic origin. *Journal of Embryology and Experimental Morphology*, *65*, 139–147. http://www.ncbi.nlm.nih.gov/pubmed/7334296

Friedman, B., Vaddi, K., Preston, C., Mahon, E., Cataldo, J. R., & McPherson, J. M. (1999). A comparison of the pharmacological properties of carbohydrate remodeled recombinant and placental-derived beta-glucocerebrosidase: Implications for clinical efficacy in treatment of Gaucher disease. *Blood*, *93*(9), 2807–2816. http://www.ncbi.nlm.nih.gov/pubmed/10216074

Gargett, T., & Brown, M. P. (2014). The inducible caspase-9 suicide gene system as a "safety switch" to limit on-target, off-tumor toxicities of chimeric antigen receptor T cells. *Front Pharmacol*, *5*, 235. https://doi.org/10.3389/fphar.2014.00235

Ginsburg, M., & Eyal-Giladi, H. (1987). Primordial germ cells of the young chick blastoderm originate from the central zone of the area pellucida irrespective of the embryo-forming process. *Development*, *101*(2), 209–219. https://doi.org/10.1242/dev.101.2.209

Han, J. Y., Park, T. S., Hong, Y. H., Jeong, D. K., Kim, J. N., Kim, K. D., & Lim, J. M. (2002). Production of germline chimeras by transfer of chicken gonadal primordial germ cells maintained in vitro for an extended period. *Theriogenology*, *58*(8), 1531–1539. https://doi.org/10.1016/s0093-691x(02)01061-0

Harvey, A. J., Speksnijder, G., Baugh, L. R., Morris, J. A., & Ivarie, R. (2002a). Consistent production of transgenic chickens using replication-deficient retroviral vectors and high-throughput screening procedures. *Poultry Science*, *81*(2), 202–212. https://doi.org/10.1093/ps/81.2.202

Harvey, A. J., Speksnijder, G., Baugh, L. R., Morris, J. A., & Ivarie, R. (2002b). Expression of exogenous protein in the egg white of transgenic chickens. *Nature Biotechnology*, *20*(4), 396–399. https://doi.org/10.1038/nbt0402-396

Harvey, D. J., Wing, D. R., Kuster, B., & Wilson, I. B. (2000). Composition of N-linked carbohydrates from ovalbumin and co-purified glycoproteins. *Journal of the American Society for Mass Spectrometry*, *11*(6), 564–571. https://doi.org/10.1016/S1044-0305(00)00122-7

Hemsworth, B. N., & Jackson, H. (1962). Effect of 'busulphan' on the foetal gonad. *Nature*, *195*, 816–817. https://doi.org/10.1038/195816a0

Hu, T., Taylor, L., Sherman, A., Keambou Tiambo, C., Kemp, S. J., Whitelaw, B., & McGrew, M. J. (2022). A low-tech, cost-effective and efficient method for safeguarding genetic diversity by direct cryopreservation of poultry embryonic reproductive cells. *Elife*, *11*. https://doi.org/10.7554/eLife.74036

Jung, K. M., Kim, Y. M., Ono, T., & Han, J. Y. (2017). Size-dependent isolation of primordial germ cells from avian species. *Molecular Reproduction and Development*, *84*(6), 508–516. https://doi.org/10.1002/mrd.22802

Kamihira, M., Kawabe, Y., Shindo, T., Ono, K., Esaka, K., Yamashita, T., & Iijima, S. (2009). Production of chimeric monoclonal antibodies by genetically manipulated chickens. *Journal of Biotechnology*, *141*(1–2), 18–25. https://doi.org/10.1016/j.jbiotec.2009.02.022

Kamihira, M., Ono, K., Esaka, K., Nishijima, K., Kigaku, R., Komatsu, H., & Iijima, S. (2005). High-level expression of single-chain Fv-Fc fusion protein in serum and egg white of genetically manipulated chickens by using a retroviral vector. *Journal of Virology*, *79*(17), 10864–10874. https://doi.org/10.1128/JVI.79.17.10864-10874.2005

Kawabe, Y., Naka, T., Ando-Noumi, N., Matsumoto, H., Ono, K., Nishijima, K., & Iijima, S. (2006). Transport of human immunoglobulin G and Fc-fusion proteins to chicken egg yolk. *Journal of Bioscience and Bioengineering*, *102*(6), 518–523. https://doi.org/10.1263/jbb.102.518

Kheimar, A., Klinger, R., Bertzbach, L. D., Sid, H., Yu, Y., Conradie, A. M., & Schusser, B. (2021). A genetically engineered commercial chicken line is resistant to highly pathogenic avian leukosis virus subgroup J. *Microorganisms*, *9*(5). https://doi.org/10.3390/microorganisms9051066

Kim, J. N., Kim, M. A., Park, T. S., Kim, D. K., Park, H. J., Ono, T., & Han, J. Y. (2004). Enriched gonadal migration of donor-derived gonadal primordial germ cells by immunomagnetic cell sorting in birds. *Molecular Reproduction and Development*, *68*(1), 81–87. https://doi.org/10.1002/mrd.20051

Kim, Y. M., Park, J. S., Kim, S. K., Jung, K. M., Hwang, Y. S., Han, M., & Han, J. Y. (2018). The transgenic chicken derived anti-CD20 monoclonal antibodies exhibits greater anti-cancer therapeutic potential with enhanced Fc effector functions. *Biomaterials*, *167*, 58–68. https://doi.org/10.1016/j.biomaterials.2018.03.021

Kim, Y. M., Park, K. J., Park, J. S., Jung, K. M., & Han, J. Y. (2021). In vivo enrichment of busulfan-resistant germ cells for efficient production of transgenic avian models. *Scientific Reports*, *11*(1), 9127. https://doi.org/10.1038/s41598-021-88706-6

Kojima, Y., Mizutani, A., Okuzaki, Y., Nishijima, K., Kaneoka, H., Sasamoto, T., & Iijima, S. (2015). Analyses of chicken sialyltransferases related to N-glycosylation. *Journal of Bioscience and Bioengineering*, *119*(6), 623–628. https://doi.org/10.1016/j.jbiosc.2014.11.009

Kornfeld, S. (1987). Trafficking of lysosomal enzymes. *The FASEB Journal*, *1*(6), 462–468. https://doi.org/10.1096/fasebj.1.6.3315809

Koslova, A., Trefil, P., Mucksova, J., Krchlikova, V., Plachy, J., Krijt, J., & Hejnar, J. (2021). Knock-out of retrovirus receptor Gene Tva in the chicken confers resistance to avian leukosis virus subgroups A and K and affects cobalamin (Vitamin B12)-dependent level of methylmalonic acid. *Viruses*, *13*(12). https://doi.org/10.3390/v13122504

Koslova, A., Trefil, P., Mucksova, J., Reinisova, M., Plachy, J., Kalina, J., & Hejnar, J. (2020). Precise CRISPR/Cas9 editing of the NHE1 gene renders chickens resistant to the J subgroup of avian leukosis virus. *Proceedings of the National Academy of Sciences of the United States of America*, *117*(4), 2108–2112. https://doi.org/10.1073/pnas.1913827117

Kwon, S. C., Choi, J. W., Jang, H. J., Shin, S. S., Lee, S. K., Park, T. S., & Han, J. Y. (2010). Production of bifunctional recombinant human interleukin 1 receptor antagonist (rhIL1RN) from transgenic quail egg white. *Biology of Reproduction*, *82*(6), 1057–1064. https://doi.org/10.1095/biolreprod.109.081687

Lee, H. J., Yoon, J. W., Jung, K. M., Kim, Y. M., Park, J. S., Lee, K. Y., & Han, J. Y. (2019). Targeted gene insertion into Z chromosome of chicken primordial germ cells for avian sexing model development. *The FASEB Journal*, *33*(7), 8519–8529. https://doi.org/10.1096/fj.201802671R

Lee, J., Ma, J., & Lee, K. (2019). Direct delivery of adenoviral CRISPR/Cas9 vector into the blastoderm for generation of targeted gene knockout in quail. *Proceedings of the National Academy of Sciences of the United States of America*, *116*(27), 13288–13292. https://doi.org/10.1073/pnas.1903230116

Lee, K. Y., Lee, H. J., Choi, H. J., Han, S. T., Lee, K. H., Park, K. J., & Han, J. Y. (2020). Highly elevated base excision repair pathway in primordial germ cells causes low base editing activity in chickens. *The FASEB Journal*, *34*(12), 15907–15921. https://doi.org/10.1096/fj.202001065RRR

Lee, J., Kim, D. H., & Lee, K. (2020). Muscle hyperplasia in Japanese quail by single amino acid deletion in MSTN propeptide. *International Journal of Molecular Sciences*, *21*(4). https://doi.org/10.3390/ijms21041504

Lee, S. J., Evers, S., Roeder, D., Parlow, A. F., Risteli, J., Risteli, L., & Nussenzweig, M. C. (2002). Mannose receptor-mediated regulation of serum glycoprotein homeostasis. *Science*, *295*(5561), 1898–1901. https://doi.org/10.1126/science.1069540

Lillico, S. G., McGrew, M. J., Sherman, A., & Sang, H. M. (2005). Transgenic chickens as bioreactors for protein-based drugs. *Drug Discovery Today*, *10*(3), 191–196. https://doi.org/10.1016/S1359-6446(04)03317-3

Lillico, S. G., Sherman, A., McGrew, M. J., Robertson, C. D., Smith, J., Haslam, C., & Sang, H. M. (2007). Oviduct-specific expression of two therapeutic proteins in transgenic hens. *Proceedings of the National Academy of Sciences of the United States of America*, *104*(6), 1771–1776. https://doi.org/10.1073/pnas.0610401104

Lyall, J., Irvine, R. M., Sherman, A., McKinley, T. J., Nunez, A., Purdie, A., & Tiley, L. (2011). Suppression of avian influenza transmission in genetically modified chickens. *Science*, *331*(6014), 223–226. https://doi.org/10.1126/science.1198020

Macdonald, J., Glover, J. D., Taylor, L., Sang, H. M., & McGrew, M. J. (2010). Characterisation and germline transmission of cultured avian primordial germ cells. *PLoS One*, *5*(11), e15518. https://doi.org/10.1371/journal.pone.0015518

Macdonald, J., Taylor, L., Sherman, A., Kawakami, K., Takahashi, Y., Sang, H. M., & McGrew, M. J. (2012). Efficient genetic modification and germ-line transmission of primordial germ cells using piggyBac and Tol2 transposons. *Proceedings of the National Academy of Sciences of the United States of America* *109*(23), E1466–1472. https://doi.org/10.1073/pnas.1118715109

McGrew, M. J., Sherman, A., Ellard, F. M., Lillico, S. G., Gilhooley, H. J., Kingsman, A. J., & Sang, H. (2004). Efficient production of germline transgenic chickens using lentiviral vectors. *EMBO Reports*, *5*(7), 728–733. https://doi.org/10.1038/sj.embor.7400171

Meyer, D. B. (1964). The migration of primordial germ cells in the chick embryo. *Developmental Biology*, *10*, 154–190. https://doi.org/10.1016/0012-1606(64)90009-0

Mine, Y., & Yang, M. (2008). Recent advances in the understanding of egg allergens: Basic, industrial, and clinical perspectives. *Journal of Agricultural and Food Chemistry*, *56*(13), 4874–4900. https://doi.org/10.1021/jf8001153

Morrison, S. L., Mohammed, M. S., Wims, L. A., Trinh, R., & Etches, R. (2002). Sequences in antibody molecules important for receptor-mediated transport into the chicken egg yolk. *Molecular Immunology*, *38*(8), 619–625. https://doi.org/10.1016/s0161-5890(01)00095-5

Mozdziak, P. E., Angerman-Stewart, J., Rushton, B., Pardue, S. L., & Petitte, J. N. (2005). Isolation of chicken primordial germ cells using fluorescence-activated cell sorting. *Poultry Science*, *84*(4), 594–600. https://doi.org/10.1093/ps/84.4.594

Mukae, T., Okumura, S., Watanobe, T., Yoshii, K., Tagami, T., & Oishi, I. (2020). Production of recombinant monoclonal antibodies in the egg white of gene-targeted transgenic chickens. *Genes (Basel)*, *12*(1). https://doi.org/10.3390/genes12010038

Mukae, T., Yoshii, K., Watanobe, T., Tagami, T., & Oishi, I. (2021). Production and characterization of eggs from hens with ovomucoid gene mutation. *Poultry Science*, *100*(2), 452–460. https://doi.org/10.1016/j.psj.2020.10.026

Naito, M., Sano, A., Tagami, T., Harumi, T., Matsubara, Y., & Kuwana, T. (2001). Effect of mytomycin c treatment on feeder cells and the proliferation of cultured chicken primordial germ cells isolated from embryonic blood. *Journal of Poultry Science*, *38*, 343–347.

Naito, M., Sano, A., Tajima, A., Yasuda, Y., & Kuwana, T. (1994). Production of germline chimeric chickens, with high transmission rate of donor-derived gametes, produced by transfer of primordial germ cells. *Molecular Reproduction and Development*, *39*(2), 153–161. https://doi.org/10.1002/mrd.1080390206

Nakamura, Y., Yamamoto, Y., Usui, F., Atsumi, Y., Ito, Y., Ono, T., & Tagami, T. (2008). Increased proportion of donor primordial germ cells in chimeric gonads by sterilisation of recipient embryos using busulfan sustained-release emulsion in chickens. *Reproduction, Fertility and Development*, *20*(8), 900–907. https://doi.org/10.1071/rd08138

Ngantung, F. A., Miller, P. G., Brushett, F. R., Tang, G. L., & Wang, D. I. (2006). RNA interference of sialidase improves glycoprotein sialic acid content consistency. *Biotechnology and Bioengineering*, *95*(1), 106–119. https://doi.org/10.1002/bit.20997

Oishi, I., Yoshii, K., Miyahara, D., & Tagami, T. (2018). Efficient production of human interferon beta in the white of eggs from ovalbumin gene-targeted hens. *Scientific Reports*, *8*(1), 10203. https://doi.org/10.1038/s41598-018-28438-2

Oishi, I., Yoshii, K., Miyahara, D., Kagami, H., & Tagami, T. (2016). Targeted mutagenesis in chicken using CRISPR/Cas9 system. *Scientific Reports*, *6*, 23980. https://doi.org/10.1038/srep23980

Ono, T., & Machida, Y. (1999). Immunomagnetic purification of viable primordial germ cells of Japanese quail (coturnix japonica). *Comparative Biochemistry and Physiology Part A: Molecular & Integrative Physiology*, *122*(2), 255–259. https://doi.org/10.1016/s1095-6433(99)00008-2

Ortinski, P. I., O'Donovan, B., Dong, X., & Kantor, B. (2017). Integrase-deficient lentiviral vector as an all-in-one platform for highly efficient CRISPR/Cas9-mediated Gene editing. *Molecular Therapy - Methods & Clinical Development*, *5*, 153–164. https://doi.org/10.1016/j.omtm.2017.04.002

Pain, B., Clark, M. E., Shen, M., Nakazawa, H., Sakurai, M., Samarut, J., & Etches, R. J. (1996). Long-term in vitro culture and characterisation of avian embryonic stem cells with multiple morphogenetic potentialities. *Development*, *122*(8), 2339–2348. https://doi.org/10.1242/dev.122.8.2339

Park, J. S., Choi, H. J., Jung, K. M., Lee, K. Y., Shim, J. H., Park, K. J., Kim, Y. M., & Han, J. Y. (2023). Production of recombinant human IgG1 Fc with beneficial N-glycosylation pattern for anti-inflammatory activity using genome-edited chickens. *Communications Biology*, *6*(1), 589. https://doi.org/10.1038/s42003-023-04937-5

Park, K. J., Kang, S. J., Kim, T. M., Lee, Y. M., Lee, H. C., Song, G., & Han, J. Y. (2010). Gamma-irradiation depletes endogenous germ cells and increases donor cell distribution in chimeric chickens. *In Vitro Cellular & Developmental Biology - Animal*, *46*(10), 828–833. https://doi.org/10.1007/s11626-010-9361-8

Park, T. S., & Han, J. Y. (2012). piggyBac transposition into primordial germ cells is an efficient tool for transgenesis in chickens. *Proceedings of the National Academy of Sciences of the United States of America*, *109*(24), 9337–9341. https://doi.org/10.1073/pnas.1203823109

Park, T. S., Lee, H. G., Moon, J. K., Lee, H. J., Yoon, J. W., Yun, B. N., & Han, B. K. (2015). Deposition of bioactive human epidermal growth factor in the egg white of transgenic hens using an oviduct-specific minisynthetic promoter. *The FASEB Journal*, *29*(6), 2386–2396. https://doi.org/10.1096/fj.14-264739

Park, T. S., Lee, H. J., Kim, K. H., Kim, J. S., & Han, J. Y. (2014). Targeted gene knockout in chickens mediated by TALENs. *Proceedings of the National Academy of Sciences of the United States of America*, *111*(35), 12716–12721. https://doi.org/10.1073/pnas.1410555111

Penno, C. A., Kawabe, Y., Ito, A., & Kamihira, M. (2010). Production of recombinant human erythropoietin/Fc fusion protein by genetically manipulated chickens. *Transgenic Research*, *19*(2), 187–195. https://doi.org/10.1007/s11248-009-9310-z

Pereira, N. A., Chan, K. F., Lin, P. C., & Song, Z. (2018). The "less-is-more" in therapeutic antibodies: Afucosylated anti-cancer antibodies with enhanced antibody-dependent cellular cytotoxicity. *MAbs*, *10*(5), 693–711. https://doi.org/10.1080/19420862.2018.1466767

Petitte, J. N., Liu, G., & Yang, Z. (2004). Avian pluripotent stem cells. *Mechanisms of Development*, *121*(9), 1159–1168. https://doi.org/10.1016/j.mod.2004.05.003

Raju, T. S., Briggs, J. B., Borge, S. M., & Jones, A. J. (2000). Species-specific variation in glycosylation of IgG: Evidence for the species-specific sialylation and branch-specific galactosylation and importance for engineering recombinant glycoprotein therapeutics. *Glycobiology*, *10*(5), 477–486. https://doi.org/10.1093/glycob/10.5.477

Ratner, M. (2015). Alexion pays big for Synageva's rare disease drug candidate. *Nature Biotechnology*, *33*(7), 679. https://doi.org/10.1038/nbt0715-679

Salter, D. W., Smith, E. J., Hughes, S. H., Wright, S. E., & Crittenden, L. B. (1987). Transgenic chickens: Insertion of retroviral genes into the chicken germ line. *Virology*, *157*(1), 236–240. https://doi.org/10.1016/0042-6822(87)90334-5

Sato, Y., & Beutler, E. (1993). Binding, internalization, and degradation of mannose-terminated glucocerebrosidase by macrophages. *Journal of Clinical Investigation*, *91*(5), 1909–1917. https://doi.org/10.1172/JCI116409

Schusser, B., Collarini, E. J., Yi, H., Izquierdo, S. M., Fesler, J., Pedersen, D., & Leighton, P. A. (2013). Immunoglobulin knockout chickens via efficient homologous recombination in primordial germ cells. *Proceedings of the National Academy of Sciences of the United States of America*, *110*(50), 20170–20175. https://doi.org/10.1073/pnas.1317106110

Serralbo, O., Salgado, D., Veron, N., Cooper, C., Dejardin, M. J., Doran, T., & Marcelle, C. (2020). Transgenesis and web resources in quail. *Elife*, *9*. https://doi.org/10.7554/eLife.56312

Sheridan, C. (2016). FDA approves 'farmaceutical' drug from transgenic chickens. *Nature Biotechnology*, *34*(2), 117–119. https://doi.org/10.1038/nbt0216-117

Shields, R. L., Lai, J., Keck, R., O'Connell, L. Y., Hong, K., Meng, Y. G., & Presta, L. G. (2002). Lack of fucose on human IgG1 N-linked oligosaccharide improves binding to human Fcgamma RIII and antibody-dependent cellular toxicity. *Journal of Biological Chemistry*, *277*(30), 26733–26740. https://doi.org/10.1074/jbc.M202069200

Song, Y., D'Costa, S., Pardue, S. L., & Petitte, J. N. (2005). Production of germline chimeric chickens following the administration of a busulfan emulsion. *Molecular Reproduction and Development*, *70*(4), 438–444. https://doi.org/10.1002/mrd.20218

Song, Y., Duraisamy, S., Ali, J., Kizhakkayil, J., Jacob, V. D., Mohammed, M. A., & de Lavoir, M. C. (2014). Characteristics of long-term cultures of avian primordial germ cells and gonocytes. *Biology of Reproduction*, *90*(1), 15. https://doi.org/10.1095/biolreprod.113.113381

Sumiyoshi, W., Nakakita, S., Miyanishi, N., & Hirabayashi, J. (2009). Strategic glycan elution map for the production of human-type n-linked oligosaccharides: The case of hen egg yolk and white. *Bioscience, Biotechnology, and Biochemistry*, *73*(3), 543–551. https://doi.org/10.1271/bbb.80598

Swift, C. H. (1914). Origin and early history of the primordial germ cells in the chick. *American Journal of Anatomy*, *18*, 483–516.

Taylor, L., Carlson, D. F., Nandi, S., Sherman, A., Fahrenkrug, S. C., & McGrew, M. J. (2017). Efficient TALEN-mediated gene targeting of chicken primordial germ cells. *Development*, *144*(5), 928–934. https://doi.org/10.1242/dev.145367

Tyack, S. G., Jenkins, K. A., O'Neil, T. E., Wise, T. G., Morris, K. R., Bruce, M. P., & Doran, T. J. (2013). A new method for producing transgenic birds via direct in vivo transfection of primordial germ cells. *Transgenic Research*, *22*(6), 1257–1264. https://doi.org/10.1007/s11248-013-9727-2

Urisu, A., Ando, H., Morita, Y., Wada, E., Yasaki, T., Yamada, K., & Wakamatsu, T. (1997). Allergenic activity of heated and ovomucoid-depleted egg white. *The Journal of Allergy and Clinical Immunology*, *100*(2), 171–176. https://doi.org/10.1016/s0091-6749(97)70220-3

van de Lavoir, M. C., Diamond, J. H., Leighton, P. A., Mather-Love, C., Heyer, B. S., Bradshaw, R., & Etches, R. J. (2006). Germline transmission of genetically modified primordial germ cells. *Nature*, *441*(7094), 766–769. https://doi.org/10.1038/nature04831

Whyte, J., Glover, J. D., Woodcock, M., Brzeszczynska, J., Taylor, L., Sherman, A., & McGrew, M. J. (2015). FGF, insulin, and SMAD signaling cooperate for avian primordial germ cell self-renewal. *Stem Cell Reports*, *5*(6), 1171–1182. https://doi.org/10.1016/j.stemcr.2015.10.008

Woodcock, M. E., Gheyas, A. A., Mason, A. S., Nandi, S., Taylor, L., Sherman, A., & McGrew, M. J. (2019). Reviving rare chicken breeds using genetically engineered sterility in surrogate host birds. *Proceedings of the National Academy of Sciences of the United States of America*, *116*(42), 20930–20937. https://doi.org/10.1073/pnas.1906316116

Zhang, Z., Sun, P., Yu, F., Yan, L., Yuan, F., Zhang, W., & Li, Z. (2012). Transgenic quail production by microinjection of lentiviral vector into the early embryo blood vessels. *PLoS One*, *7*(12), e50817. https://doi.org/10.1371/journal.pone.0050817

Zhu, L., van de Lavoir, M. C., Albanese, J., Beenhouwer, D. O., Cardarelli, P. M., Cuison, S., & Etches, R. J. (2005). Production of human monoclonal antibody in eggs of chimeric chickens. *Nature Biotechnology*, *23*(9), 1159–1169. https://doi.org/10.1038/nbt1132

40

Ex-Ovo Culture System of Egg

Yutaka Tahara* **and Yasumi Horimoto**

CONTENTS

40.1 Introduction

An avian embryo culture system enables the observation of embryonic development, which is usually enclosed in and obscured by an opaque eggshell. This avian embryo culture method can be applied to various fields of research, such as embryology, embryo manipulation, tumorigenesis, and basic research in regenerative medicine.

Widely used methods with potential hatchability in avian embryo culture include the eggshell windowing method, which facilitates observations by opening a hole in one part of the eggshell, and the surrogate eggshell method, which uses a surrogate eggshell as the culture vessel. On the other hand, a shell-less avian embryo culture system using a complete artificial culture vessel has also been developed. Kamihira et al. (1998) reported the first successful hatchings of quail and chick embryos in a complete artificial vessel using a gas-permeable porous polytetrafluoroethylene membrane. In addition, Tahara and Obara (2014) reported a novel shell-less culture system for chicken embryos using a transparent plastic artificial culture vessel with a gas-permeable polymethylpentene food-wrap film.

In the method developed by Tahara *et al.* (Tahara & Obara, 2014; Tahara, 2016; Tahara & Obara, 2021; Tahara et al., 2021)., embryonic development processes can be observed from various angles throughout the entire culture period; this method would be easily applied to various fields of studies, such as embryology, embryo manipulation, tumorigenesis, and basic research in regenerative medicine. The present chapter describes in detail Tahara *et al.*'s method of shell-less chick embryo culture.

40.2 Eggshell Windowing Method and Surrogate Eggshell Method

In the eggshell windowing method, a hole is made in an eggshell and then overlaid with a transparent film. This method facilitates the internal observation of the egg during the culturing period. Only a part of the eggshell is removed, and the rest of the eggshell is intact; therefore, various manipulations to the embryos can be performed relatively easily (Fisher & Schoenwolf, 1983; Andacht et al., 2004).

Regarding the *ex ovo* culture system, Rowlett and Simkiss (1987) succeeded in hatchings chicken embryos using a surrogate eggshell. Fertilized chicken eggs were cracked on the third day of incubation. Then the egg contents were transferred into surrogate eggshells and cultured continuously. Turkey and large chicken eggshells were used as surrogate eggshells.

Perry (1988) reported the successful hatching of chicken embryos using the *ex ovo* culture system. This culture system is comprised of three steps according to embryonic development from the single-cell fertilized egg stage to hatching. In Step 1, a fertilized egg taken out from the hen's oviduct is transferred to a glass vessel together with a culture solution and cultured for 24 hours. In Step 2, the embryo and water-soluble egg white are transferred to a surrogate eggshell and cultured without contact with air. Finally, in Step 3, on the third day of incubation, the embryo is transferred to a larger surrogate eggshell and cultured until hatching. The hatching rate by this method was less than or equal to 10%. With further improvement, the hatching rate reached 30–50% (Naito & Perry, 1989; Naito et al., 1990).

* Corresponding author.

DOI: 10.1201/9781003254430-49

With the development of methods using the eggshell windowing and the surrogate eggshells, the avian embryo culture system has been applied not only in basic research in developmental biology, but also in various fields of research, such as embryology, embryo manipulation, toxicology, tumorigenesis, and basic research in regenerative medicine (Boulland et al., 2010; Chiba et al., 2010; Kulesa et al., 2010; Funahashi & Nakamura, 2014; Haraguchi et al., 2016; Ribatti, 2016; Vu et al., 2018; Matheus et al., 2019).

40.3 Shell-Less Avian Embryo Culture

In avian embryo culture using windowed eggshells and surrogate eggshells, chick embryos can continue to develop after various treatments; however, it is still difficult to constantly observe the development of embryos from all angles because of the use of an opaque eggshell.

Shell-less artificial culture systems have been studied. Chicken embryos could not be hatched by the culture system using a polyethylene bag or food wrap film; however, the avian embryos transferred to the films could continue to develop for a certain period of time. Various problems such as limited oxygen and calcium availability to embryos were identified as the cause of the inability to hatch. (Elliott & Bennett, 1971; Rowlett & Simkiss, 1987).

Kamihira et al. (1998) reported the first successful hatching of quail embryos using an artificial culture vessel with a gas-permeable porous polytetrafluoroethylene (PTFE) membrane. In their method, fertilized quail eggs were broken on day 2 of incubation and the contents were transferred to an artificial vessel. As a calcium source, eggshell powder and calcium lactate were added to the culture vessel, and embryos were cultured while pure oxygen was supplied to the incubator. As a result, 42.8% of the quail embryos hatched. Moreover, the hatching rate in *ex ovo* culture using surrogate eggshells was improved by adding calcium lactate. These results indicate that the gas permeability of an artificial vessel and the addition of calcium supplements are important in *ex ovo* culture.

A chicken embryo culture system using a porous PTFE membrane is a novel technique that does not use an eggshell at all. However, the PTFE membrane is as opaque as the eggshells; therefore, an observation from any view other than from above is as difficult as the method using a surrogate eggshell.

Tahara and Obara (2014) developed a simpler plastic artificial culture vessel for the observation of the development of chicken embryos to allow omnidirectional visualization. A gas-permeable and transparent polymethylpentene food wrap film was used as a culture vessel. Fertilized chicken eggs on day 3 of incubation were broken, and the contents were transferred to an artificial vessel. At this time, calcium lactate was added to the culture vessel as calcium source. After day 17, humidified pure oxygen was directly supplied into the culture vessel for hatching. Using the same artificial culture system, Tahara et al. (2021) found that the addition of calcium was effective for the hatching of cultured chicken embryos. Calcium carbonate is the main component of eggshell. On day 10 of incubation, calcium carbonate was added. As a result, 40% of chicken embryos succeeded in hatching with the addition of only calcium carbonate.

Overall, in the method developed by Tahara *et al.* (Tahara & Obara, 2014; Tahara, 2016; Tahara & Obara, 2021; Tahara et al., 2021), a transparent polymethylpentene food wrap film was used as a culture vessel; therefore, embryos can be continuously observed from various angles throughout the entire culture period. Moreover, the easy access to the embryos in the artificial vessel will make it even easier to apply this method to various fields of research. In addition, no special techniques are required to create a plastic artificial culture vessel, and the production time and cost would be reduced (Tahara & Obara, 2014; Tahara, 2016; Tahara & Obara, 2021; Tahara et al., 2021).

40.4 Shell-Less Chick Embryo Culture Method Developed by Tahara et al.

40.4.1 Culture Vessel

A transparent polystyrene plastic cup (430 mL, ø86 mm × 118 mm) was used as a pod for the culture vessel. A hole with a diameter of 1 cm was made in the side of the cup approximately about 3 cm from the bottom, using a flame (e.g., a turbo lighter). The hole was plugged with a cotton pledget as a filter (Figure 40.1). For oxygen supply, a plastic tube with a diameter of approximately 2 mm was inserted through the space between the pledget and the hole. Then, 40 mL of 0.01% benzalkonium chloride solution diluted with distilled water was added to the cup.

Next, a polymethylpentene food wrap film (FOR-WRAP, Riken Technos, Tokyo, Japan) was stretched to avoid wrinkles. The film was then formed into a concave shape and installed along the edge of the plastic cup, which was used as a frame (Figure 40.2).

A porous polyethylene terephthalate plastic lid (ø86 mm × 7 mm) was placed on top of the culture vessel; the lid should fit tightly to the cup (Figure 40.3).

40.4.2 Embryo Culture

Fertilized chicken eggs were preincubated in an incubator at 38°C and 60% relative humidity for 55–56 hours with occasional rotation of eggs (stage 16 in Hamburger and Hamilton, 1951). Then, each eggshell was wiped using 70% ethanol and broken without using any tools. The egg contents were then transferred into the culture vessel (Figure 40.4).

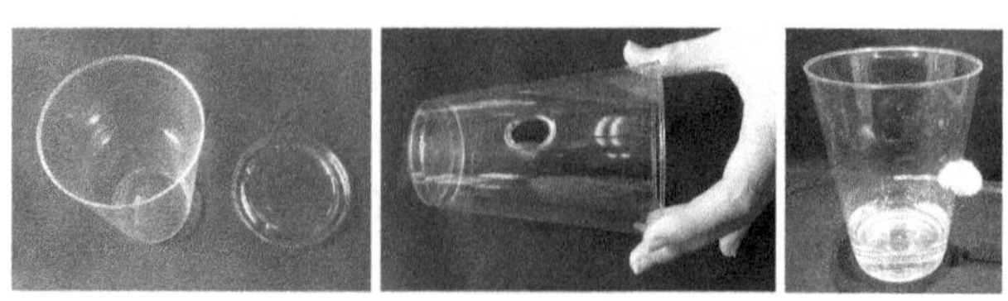

FIGURE 40.1 Photograph showing the preparation of the culture vessel using a plastic cup.

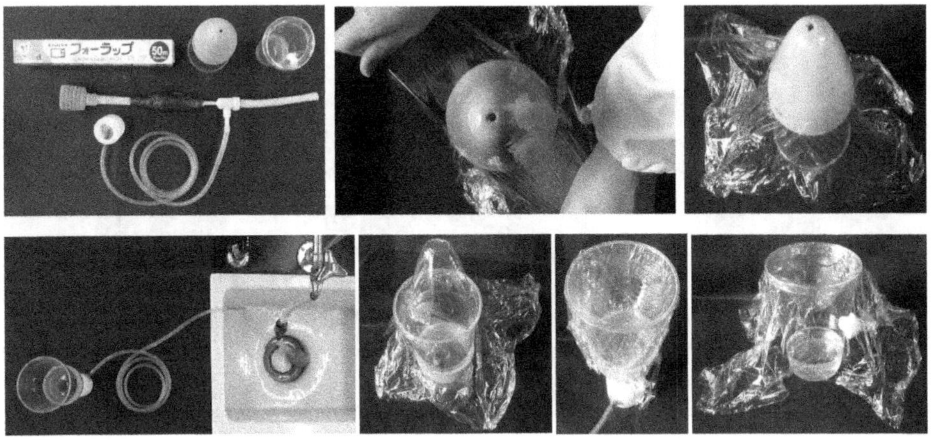

FIGURE 40.2 Photograph of the processing of the polymethylpentene film to make a concave shape and how to attach it to a plastic cup.

Next, the culture vessel was covered with a plastic lid and incubated at 38°C and a relative humidity of 80%. During the incubation, the culture vessel was placed at an angle of approximately 8° vertically. The shelf for the culture vessels was tilted by 8° in advance. Once a day, culture vessels were slowly rotated with 120° clockwise (Figure 40.5).

On day 14 of incubation, 10–14 ventilation holes (approximately 5–8 mm in diameter) were made in the upper surface of the polymethylpentene membrane using a 30W soldering iron, without contacting the egg contents (Figure 40.6).

From day 17 of incubation, humidified oxygen was supplied to the culture vessel at a flow rate of 500 mL/h through the plastic tube previously installed. The rotation was stopped after day 17 of incubation (Figure 40.7).

FIGURE 40.3 Photograph of the fully assembled plastic transparent culture vessel covered with a lid.

On day 21 of incubation, chicks were removed from the artificial vessel (Figure 40.8).

40.4.3 Calcium Addition Method

In order to hatch chicken embryos using an artificial vessel, it is necessary to add calcium supplements. As a calcium source, either calcium lactate or calcium carbonate was added to the embryos. For calcium lactate supplementation, eggs were precultured for 55–56 hours (stage 16 in Hamburger and Hamilton, 1951). Just before the contents of the eggs were transferred, 250 mg of calcium lactate powder and 2.5 mL of sterilized water were added to the culture vessel (Figure 40.9).

For calcium carbonate supplementation, on day 10 of incubation, 250 mg of calcium carbonate powder was directly added to the chorioallantois membrane. To avoid covering the top of the embryo, the calcium carbonate was added in a circle around the embryo (Figure 40.10).

The selection of either calcium lactate or calcium carbonate as a calcium source should be based on the purpose of an experiment. If you do not want to add any substances (except the egg contents) to the artificial vessel until day 10 of incubation, calcium carbonate is suitable. If it is important to observe the developmental process after day 10 of incubation, calcium lactate is a more suitable choice because the calcium carbonate powder interferes with the observation.

40.5 Conclusion

In the method developed by Tahara et al., embryonic development processes can be observed from various angles throughout the entire culture period; this method would be easily applied to various fields of studies. In the future, it will be possible to fully utilize this technology for birds other than chickens.

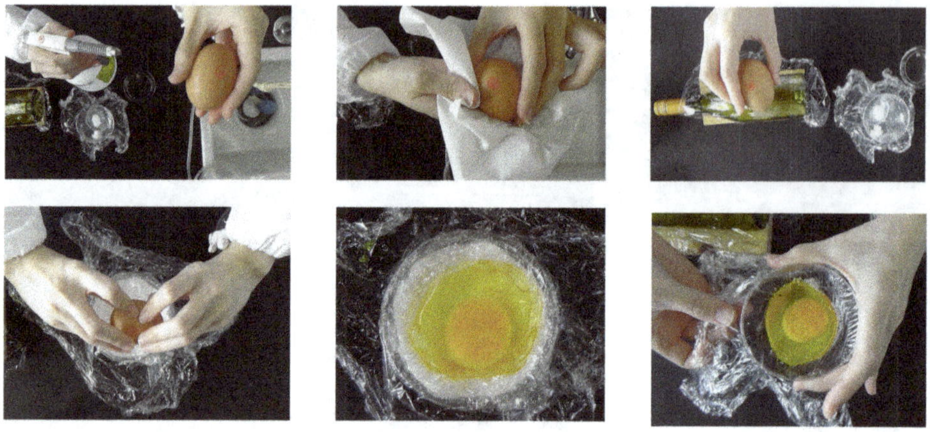

FIGURE 40.4 Photograph showing the procedure of breaking an egg and transferring the contents into an artificial vessel.

FIGURE 40.5 Photograph of an artificial culture vessel in an incubator.

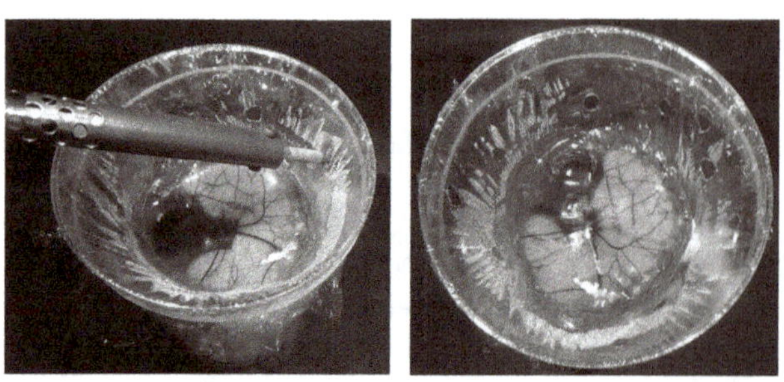

FIGURE 40.6 Photograph demonstrating how to make a hole in the upper part of the polymethylpentene membrane.

FIGURE 40.7 Photograph of an artificial culture vessel in an incubator with humidified pure oxygen supply.

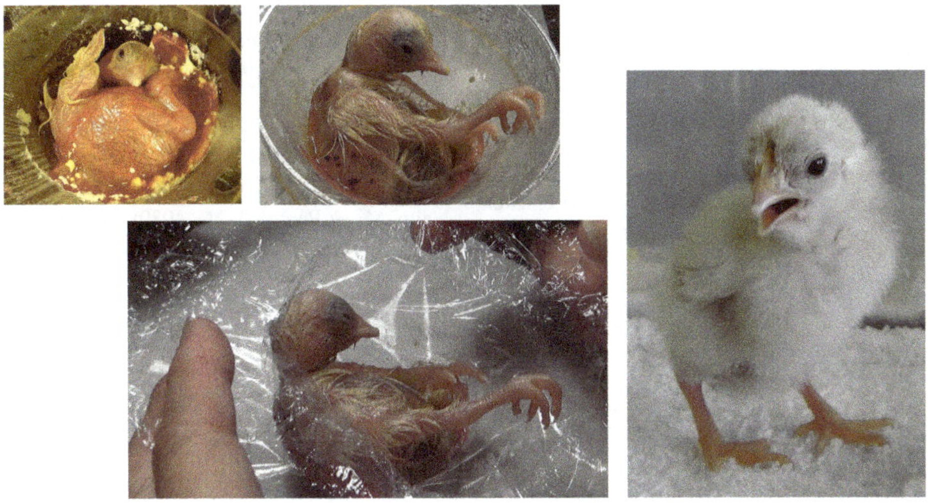

FIGURE 40.8 Photograph of the removal of the embryo after the blood pressure in the chorioallantois membrane has dropped and photograph of a successfully hatched chick standing.

FIGURE 40.9 Photograph of the addition of calcium lactate.

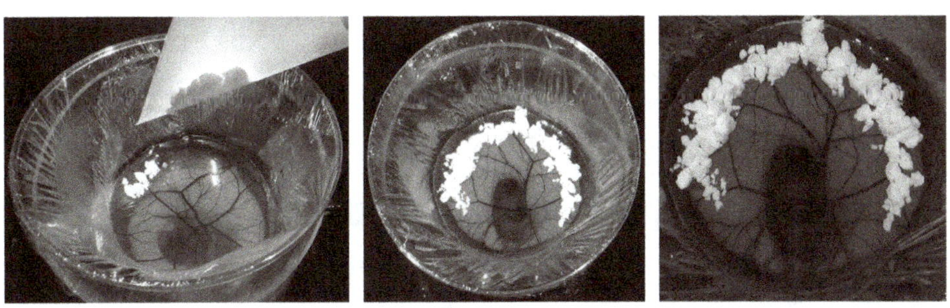

FIGURE 40.10 Photograph of the addition of calcium carbonate.

REFERENCES

Andacht, T., Hu, W., & Ivarie, R. (2004). Rapid and improved method for windowing eggs accessing the stage X chicken embryo. *Molecular Reproduction and Development*, *69*(1), 31–34. https://doi.org/10.1002/mrd.20155

Boulland, J., Halasi, G., Kasumacic, N., & Glover, J. (2010). Xenotransplantation of human stem cells into the chicken embryo. *Journal of Visualized Experiments*, *41*, e2071. 1–16. https://dx.doi.org/10.3791/2071

Chiba, A., Yui, C., & Hirano, S. (2010). Liver reconstruction on the chorioallantoic membrane of the chick embryo. *Archives of Histology and Cytology*, *73*(1), 45–53. https://doi.org/10.1679/aohc.73.45

Elliott, J., & Bennett, J. (1971). Growth of chick embryos in polyethylene bags. *Poultry Science*, *50*(3), 974–975. https://doi.org/10.3382/ps.0500974

Fisher, M., & Schoenwolf, G. C. (1983). The use of check embryos in experimental embryology and teratology: Improvements in standard procedures. *Teratology*, *27*(1), 65–72. https://doi.org/10.1002/tera.1420270110

Funahashi, J., & Nakamura, H. (2014). Time-lapse imaging system with shell-less culture chamber. *Development, Growth & Differentiation*, *56*, 305–309. https://doi.org/10.1111/dgd.12132

Hamburger, V., & Hamilton, H. L. (1951). A series of normal stages in the development of the chick embryo. *Journal of Morphology*, *88*(1), 49–92. https://doi.org/10.1002/jmor.1050880104

Haraguchi, S., Matsubara, Y., & Hosoe, M. (2016). Chick embryos can form teratomas from microinjected mouse embryonic stem cells. *Development, Growth & Differentiation*, *58*(2), 194–204. https://doi.org/10.1111/dgd.12260

Kamihira, M., Oguchi, S., Tachibana, A., Kitagawa, Y., & Iijima, S. (1998). Improved hatching for in vitro quail embryo culture using surrogate eggshell and artificial vessel. *Development, Growth & Differentiation*, *40*(4), 449–455. https://doi.org/10.1046/j.1440-169X.1998.t01-2-00010.x

Kulesa, P. M., Bailey, C. M., Coopere, C., & Fraser, S. E. (2010). *Ovo live imaging of avian embryos*. Cold Spring Harbor Protocols. https://doi.org/10.1101/pdb.prot5446

Matheus, F., Rusha, E., Rehimi, R., Molitor, L., Pertek, A., Modic, M., Feederle, R., Flatley, A., Kremmer, E., Geerlof, A., Rishko, V., Rada-Iglesias, A., & Drukker, M. (2019). Pathological ASXL1 mutations and protein variants impair neural crest development. *Stem Cell Reports*, *12*(5), 861–868. https://doi.org/10.1016/j.stemcr.2019.03.006

Naito, M., Nirasawa, K., & Oishi, T. (1990). Development in culture of the chick embryo from fertilized ovum to hatching. *Journal of Experimental Zoology*, *254*(3), 322–326. https://doi.org/10.1002/jez.1402540311

Naito, M., & Perry, M. M. (1989). Development in culture of the chick embryo from cleavage to hatch. *British Poultry Science*, *30*(2), 251–256. http://doi.org/10.1080/00071668908417145

Perry, M. M. (1988). A complete culture system for the chick embryo. *Nature*, *331*(6151), 70–72. https://doi.org/10.1038/331070a0

Ribatti, D. (2016). The chick embryo chorioallantoic membrane (CAM). A Multifaceted Experimental Model. *Mechanisms of Development*, 141, 70–77. https://doi.org/10.1016/j.mod.2016.05.003

Rowlett, K., & Simkiss, K. (1987). Explanted embryo culture: In vitro and in ovo techniques for domestic fowl. *British Poultry Science*, *28*, 91–101. https://doi.org/10.1080%2F00071668708416940

Tahara, Y. (2016) Method to hatch chicks without using egg-shell. Japan Patent JP5978523B2 (in Japanese).

Tahara, Y., & Obara, K. (2014). A novel shell-less culture system for chick embryos using a plastic film as culture vessels. *Journal of Poultry Science*, *51*, 307–312. https://doi.org/10.2141/jpsa.0130043

Tahara, Y., & Obara, K. (2021). Ex ovo culture system for avian embryos and its application. *Journal of Poultry Science*, *58*(1) 1–4. https://doi.org/10.2141/jpsa.0200016

Tahara, Y., Obara, K., & Kamihira, M. (2021) Calcium carbonate supplementation to chorioallantoic membranes improves hatchability in shell-less chick embryo culture. *Journal of Bioscience and Bioengineering*, *131*(3), 314–319. https://doi.org/10.1016/j.jbiosc.2020.11.001

Vu, B. T., Shahin, S. A., Croissant, J., Fatieiev, Y., Matsumoto, K., Doan, T. L., Yik, T., Simargi, S., Conteras, A., Ratliff, L., Jimenez, C. M., Raehm, L., Khashab, N., Durand, J., Glackin, C., & Tamanoi, F. (2018). Chick chorioallantoic membrane assay as an in vivo model to study the effect of nanoparticle-based anticancer drugs in ovarian cancer. *Scientific Reports*, *8*(1), 5254. https://doi.org/10.1038/s41598-018-25573-8

41

Egg and Embryonic Development

Yaqi Meng and Ning Qiu*

CONTENTS

41.1 Overview of the Dynamic Process during Chick Embryo Development

By definition, the biologically aseptic egg produced by laying hens is cleidoic, implying that the egg is isolated from the outside environment and must contain all the components required for its embryonic development from the stage of fertilized egg to a viable chick in 21 days. The chicken egg contains proteins, lipids, calcium salts, vitamins, and minerals, all necessary during the embryogenesis process, the most essential phase in the life of birds. The chicken embryo development can be divided into 3 main phases (Figure 41.1): the early embryogenesis (first week of the embryonic process), corresponding to the establishment of germ with the formation of some egg compartments (including allantois, chorion, amnion, and yolk sac (YS) to support the survival of the embryo. In the embryo, functional organs such as the reproductive organs, heart, limb buds for the legs, and wings develop between the third and seventh embryonic days (E3 and E7). The completion of the embryo during the second phase (from E8 to E14) relies on the recovery of nutrients from the egg white and yolk and entails the development of the highly vascularized chorio-allantoic membrane (CAM). The third phase, corresponding to the final preparation for hatching, includes intense metabolism as well as the oral absorption by the embryo of the mixture of nutrients remaining (Moran, 2007).

The chick embryo development is a dynamic process with various egg components comprehensively utilized as biological components during the embryogenic process. The chicken embryo is separated from the outside environment and therefore must derive during the developmental period all its nutrients only from within, at the risk of potential negative effects on hatchability, or long-term effects on the health and viability of chickens. For this reason, this closed environment represents an excellent model to study the mechanisms by which early embryonic nutrients are utilized. Moreover, with the substantial progress in the appreciation of similarities between tumorigenesis and embryogenesis, as well as the interaction between embryos and tumors (Ma et al., 2010), the fertilized chicken egg is also considered a proper model providing a theoretical basis for the developmental origin of certain diseases. Furthermore, by studying the changes in molecular composition between fertile and unfertile eggs, some insights may lead to the development of egg-based products optimized for specific applications (Qiu et al., 2013). The multi-omics data generated from various studies provide an integrative and holistic review of the dynamic molecular processes of the chicken egg components involved in embryo development.

41.2 Functions of the Different Egg Compartments during Embryo Development

Chicken eggs are basically composed of 3 compartments, the eggshell, egg white, and egg yolk, containing all the essential nutrients and required biological components to support the embryo development outside the hen's body. During the embryonic process, the egg white is mainly utilized to address the nutritional needs of the embryo. Many changes in the egg white will take place during the embryonic process, e.g., the pH of egg white will increase during the first 2 days of incubation and subsequently decrease with incubation time (Qiu et al., 2012). In addition, the conformational structure (e.g., ovalbumin) and enzymatic activity (e.g., lysozyme) of certain proteins in the egg white will change significantly over the course of the embryonic development. As for the egg yolk,

* Corresponding author.

DOI: 10.1201/9781003254430-50

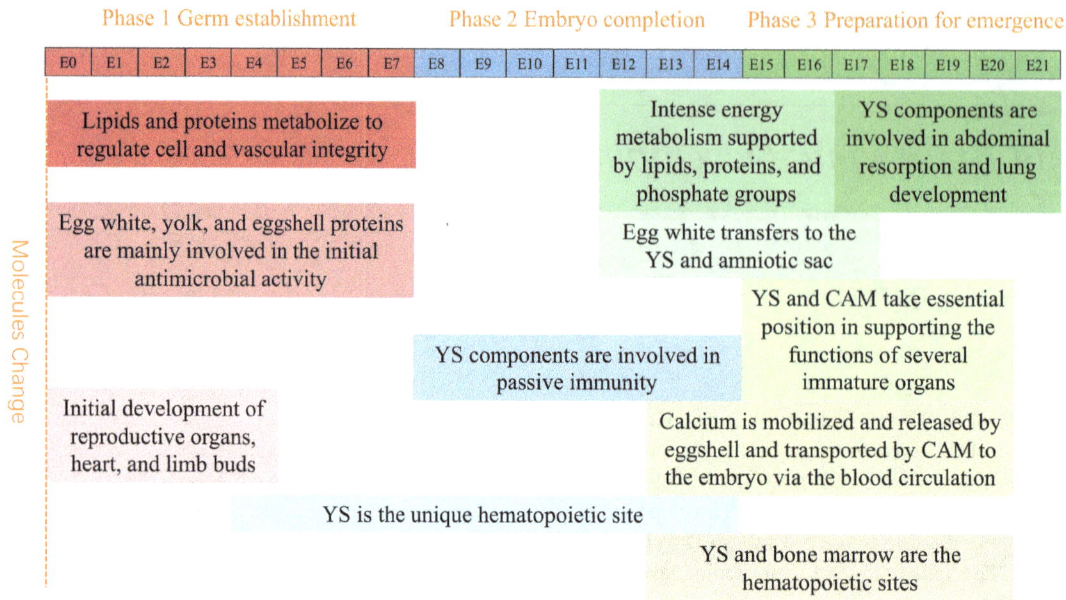

FIGURE 41.1 Egg and embryonic development.

its biofunction in avian reproduction is to satisfy energy needs and to provide numerous bioactive molecules to the embryo. Most egg yolk components are initially synthesized in the liver of the sexually mature hens, and then secreted into the bloodstream and captured by receptors or mediated endocytosis into the developing ovarian follicles. Once the embryogenesis process is activated, the egg yolk material is absorbed by cells lined inside the YS and digested products of the egg yolk are released into the blood circulation and delivered to the chick embryo to fuel its development (Starck, 2020). The chicken eggshell, representing about 10% of the weight of the egg, is made up of the membranes and shell. The shell is a calcareous structure mainly constituted of calcium carbonate (95%) with an organic matrix (3.5%), while the eggshell membranes are found between the calcified shell and the egg white, 2 smooth layers formed with homogeneous dense material, playing a role in the origin and function of the calcified shell (Shi et al., 2021). The shell membranes (later formed into CAM) also play a role in waste storage, respiration, and nutrient transport (especially calcium transport) during the late embryonic development.

Besides the above-mentioned common structural components of the egg, the avian embryo relies on additional extra-embryonic tissues to perform certain vital functions during its development, including digestion, breathing, and nutrient transport, since the embryo organs are not fully functional. These compartments, including the allantoic, amniotic, and chorioallantoic sacs, as well as the YS, develop at the early stage of embryogenesis and originate from embryonic tissues, but are not considered to be part of the embryo itself (Bellairs & Osmond, 2014). After E12, egg white proteins will massively transfer to the amniotic sac, where they will be orally absorbed by the embryo to support its rapid growth until hatching. In fact, before E11, the chicken amniotic fluid (AmF) is mainly composed of mineral elements, including sodium, chloride, phosphorus, potassium, calcium, magnesium, sulfur, and iron, and water similar to the human AmF. In amniotic vertebrates, the YS is the first organ mediating the nutrient transportation from the mother to the growing fetus (Ferner & Mess, 2011). Although developing embryos of oviparous species depend exclusively on the YS, representing the sole organ for transferring maternally-derived nutrients to the embryo, the structural and functional features of YS are highly conserved among species, including human, murine, and avian (Cindrova-Davies et al., 2017). Generally, the YS is formed during the first week of development stage and gradually covers the entire yolk by E11, which is regulated through the whole embryonic process. Its functions resemble that of the bone marrow for erythropoiesis, of the liver for the synthesis of plasma proteins, and of the intestine for the digestion and transport of nutrients to the embryo (Yadgary et al., 2014). Another chicken extra-embryo tissue, the allantoic sac, is the reservoir where the embryo discharged its waste products, and where chloride and sodium ions are stored (Mueller et al., 2015). More specifically, the allantois and chorion will fuse to generate the CAM around E6, undergoing rapid growth, and increasing its size by 20-fold. The allantois connects the chorion to the embryo, making the bloodstream run through the allantois, with the allantoic artery and veins connected to the heart of the embryo. At E14, the allantois completely surrounds the chick embryo and the vascular system reaches its final configuration on E18. CAM, as a highly vascularized structure, ensures many functions during the embryonic development, including respiratory exchanges, nitrogenous and excretory products collection, calcium transport, and acid–base homeostasis by reabsorbing ions and water from the allantoic fluid (AlF) (Bellairs & Osmond, 2014).

41.3 The Omics' View for Egg Embryonic Development

Since many structures in the fertilized egg play unique roles in support of the embryonic development, comprehensive studies are required to define their molecular profiles. Indeed, these molecules do not change independently in each structural component of the chicken egg, as there are some dynamic migration, transport, absorption, etc. throughout the egg. For example, the current hypothesis suggests the sequential transfer of egg white, first from the amniotic sac to the chicken embryo, then to the egg yolk and allantoic sacs. After being transported into the YS through the intestine and yolk stalk, egg white proteins as well as egg yolk proteins are digested by yolk proteases, and the digested products are redirected to the embryo via the chorioallantoic capillary plexus (Silva et al., 2017). In this section, we will introduce the applications of various omics platforms to define the molecular profiles of various molecules in a fertilized chicken egg during the different stages of the embryonic development (Table 41.1).

41.3.1 The First Stage of Embryo Development (E0-E7)

Nutrients initially supporting the germ are recovered from the adjacent yolk and albumen by an expanding vascular system. Low molecular weight (LMW) proteins are thought to be functionally important during embryo development, with many studies demonstrating that some specific domains within LMW proteins retain or even acquire some significant bioactivities. It was suggested that the digestion of egg white components facilitates their uptake by the cells during early embryogenesis and that these LMW proteins may play a crucial role in early embryo nutrition and immune defense. Since 2012, we have conducted a series of systematic proteomic analyses on the changes of egg white proteins during embryonic development. By using two-dimensional polyacrylamide gel electrophoresis-based proteomics, we firstly identified 8 proteins showing significant changes in abundance during embryo development (Qiu et al., 2012). The accelerating degradation of ovalbumin was noted with a wide range of molecular weights observed. Initially, the changed albumin proteins may mainly be functionalized for antibacterial activities and absorption preparation. This perspective was confirmed in a subsequent study characterizing the antibacterial potential of egg white proteins from fertilized eggs when compared to unfertilized eggs. The results indicated that the antibacterial potential of egg white proteins is effective at an early stage but progressively decreased during the first half of the embryonic period, possibly as a result of the alteration of specific antimicrobial proteins, while at the same time, the extra-embryonic structures progressively ensured some protective functions to the chick embryo (Guyot et al., 2016). As reported by Qiu et al., the apparent degradation of ovalbumin was observed through the identification of 33 protein spots, of which 22 showed a lower apparent molecular weight than the theoretical one. The degradation of ovalbumin may be caused mostly by the lack of specificity of glutamyl aminopeptidase and aminopeptidase under decreasing pH conditions during the embryonic process, or by its transformation into a less ordered structure facilitating its transport (Qiu et al., 2012). All these results and hypotheses suggest that the bioactivity of ovalbumin may be functionalized in the developing organic cells (Wang & Wu, 2014).

A vital function of the vascular system is the delivery of nutrients, mediated in adult animals by the mesoderm-derived vascularization together with the endoderm-derived hepatocytes and enterocytes. During the avian embryonic process, the YS endoderm is believed to play a similar role. A gene expression analysis of area vasculosa on YS endoderm tissues during the E2 to E4 period, covering the first 2 days of post-circulation development, was carried out to understand the roles of YS endoderm in the early development stage. The genes involved in lipid metabolism were highly enriched, indicating an active modification of lipid molecules during their transfer from the egg yolk to the circulatory system. The genes encoding key regulators of vascular integrity and major serum proteins were also uncovered, suggesting that the YS endoderm is an important source of protein and lipid constituents in the early embryonic serum and a vital site for regulating the vascular integrity of the developing embryo (Nakazawa et al., 2011). In a lipidomic study, some specific lipid species of acyl-carnitine (AcCa) present in the egg yolk, especially AcCa(18:0), AcCa(16:2), and AcCa(22:1), exhibited some significant decrease at E7 when compared to E0 ($P < 0.001$). AcCa is active in circulatory transport, delivering multiple acyl groups with a wide range of functions. On E7, the accumulation of cardiolipin in the egg yolk, a precursor step to its transport to the embryo and its integration into mitochondrial membranes, will enhance the speed of energy metabolism in the rapidly developing embryo at later stages (Meng et al., 2021a). In addition to egg white and egg yolk omics, a quantitative proteomic analysis was conducted for the eggshell membranes (ESM) during the embryonic development (Cordeiro & Hincke, 2015). During this first stage, the proteins are mainly modified to acquire protease inhibitory activities, inactivating any protease, a known factor of virulence, secreted by microorganisms. These modifications are important since the immune system has not fully developed at this initial stage of the embryonic development (Rose-Martel et al., 2012).

41.3.2 The Second Stage of Embryo Development (E8-E14)

The completion of the chick embryo structure continues through the second stage of embryogenesis process. The quantitative changes in yolk plasma proteins occurring at E12 were analyzed by comparing with the day of lay (E0) (Réhault-Godbert et al., 2014). Five up-regulated proteins (α-fetoprotein, vitronectin, one protein similar to thrombin, apovitellenin-1, and apolipoprotein B) and 15 downregulated proteins provided some clues to our understanding of the absorption kinetics of certain proteins by the embryo. Specifically, α-Fetoprotein, with an increased relative abundance after 12 days of development, may play an important role in the development of the avian immune system. The alterations of egg yolk proteome at different times (E0, E7, E14, and E18) were analyzed to further study the dynamics of egg yolk proteins during the entire incubation period. E14 is considered a key point in the embryonic process. Based on the alteration of egg yolk proteins shown on polyacrylamide gels, the accelerating transfer of ovalbumin from the albumen into the yolk was observed in fertilized eggs at E14, indicating the role of an unidentified absorption pathway for proteins/peptides in the developing embryo. After E14, the constantly decreasing abundance of yolk riboflavin-binding

TABLE 41.1

Applications of Omics in Egg Incubation Research

Omics Approach	Egg Compartment	Embryonic Days (E)	Main Findings
Transcriptome	Chorioallantoic membrane	E9, E13, E17, E20	• Genes of TRPV6, S100A10, RANKL, and OPG are involved in regulating calcium release and transport in chorioallantoic membrane • Chorioallantoic membrane calcium transport starts on E13
	Yolk sac	E13, E15, E17, E19, E21	• Several highest expressed lysosomal genes (CTSA, PSAP, NPC2) and apolipoproteins genes (apoA1, A2, B, C3) reflect the intensive digestion and re-synthesis of lipoproteins in yolk sac epithelial cells • Yolk sac is important to replace or support several immature organs that have not reached full functional maturity until the final week of incubation
Proteome	Egg white	E0, E2, E7	• An accelerating degradation of Ovalbumin was observed as well as its RNA-binding proteins
		E0, E1, E2, E3, E4, E5	• Low molecular weight proteins may be involved in early embryo nutrition and immunity
		E0, E1, E3, E5, E7, E9	• Ovomucoid is significantly increased ($P < 0.05$) while serum albumin precursor, hemopexin, Galline Ex-FABP, and Ex-FABP precursor are significantly decreased. • Provided a greater spectrum for the metabolic fate of egg white proteins during incubation
	Egg yolk	E0, E12	• Alpha-fetoprotein and 2 uncharacterized proteins (F1NMM2 and F1NHB8) are identified for the first time • Suggested a possible accumulation in the yolk during the incubation of proteins originating from other egg compartments
		E0, E7, E14, E18	• Accelerated transfer of ovalbumin from the egg white to the yolk is observed at E14, while riboflavin-binding protein is decreased at E14
		E0, E10, E18	• Differential proteins are involved in lipid transport activity, antigen binding, and nutrient reservoir activity at E10, and involved in substrate-specific transport activity and nutrient reservoir activity at E18
		24 hours post-lay	• The proteins increasing in abundance may contribute to angiogenesis (pleiotrophin, histidine-rich glycoprotein) and defense mechanisms against pathogens (mannose-binding lectin, β-defensin 11, serum amyloid P-component, ovostatin)
	Eggshell membrane	E0, E3, E7, E11, E15, E19	• Some proteins involved in oxygen transport and calcium binding (EGF-like repeats and discoidin I-like domains 3, Annexin 2) are only present in the second stage
	Egg white and yolk mixture	E0, E12	• Proteins are mainly implicated in ionic binding, antibacterial, embryonic development, and cell proliferation activities
		E0, E12	• Differentially expressed phosphorylated proteins are mainly associated with transport, molecular function regulation, and binding activity
	Amniotic fluid	E11, E16	• Proteins were mainly associated with the immune response mechanisms (e.g., lysozyme, avian betadefensin 11, vitelline membrane outer layer protein 1, and beta-microseminoprotein-like), developmental processes, and nutrients
Metabolome	Egg yolk	E7, E9, E11, E15, E17, E19	• Most metabolites including arginine, lysine, cysteine, and histidine are annotated in amino acid metabolism pathways from E11 to E15 and E17 to E19, probably due to their effects on the growth promotion and oxidative stress amelioration of the embryo • Ferroptosis was reported as the major process occurring after E15 owing to the upregulated expression of acyl-CoA synthetase long-chain family member 4
	Yolk sac and yolk content	E0, E7, E13, E18	• Lipids may be closely correlated with the development of organs in the embryo (e.g., phosphocholine and phosphatidylglycerol in supporting lungs, cardiolipin in supporting mitochondria membranes, ceramides on haematopoiesis)
	Chorioallantoic Membrane	Every day from E6 to E19	• The most active morphogenic process in the chorioallantoic membrane is the blood vessel development

Source: Meng et al. (2021b).

protein was also observed, likely contributing to meet the need for riboflavin, essential to support the embryo lipid metabolism (Meng et al., 2019). These basic proteomic studies have provided more knowledge about the modifications of egg proteins and gave some insights, from a molecular perspective, into the chick embryo development during incubation. With the benefit of this type of data and the development of omics technologies, more attention has been paid to protein post-translational modifications (PTMs) during embryogenesis. In honeybees, the phosphorylated and glycosylated modifications of the royal jelly proteins are essential for nutrient transport, cell growth and differentiation (Feng et al., 2015). Similar to the function of royal jelly, the nutrients required by the chick embryo are all derived from the egg. Therefore, our research group recently investigated the egg protein PTMs to explore their potential biological functions during the embryo development (Sun et al., 2019). The comparative and quantitative phosphoproteomic and N-glycoproteomic analysis of chicken egg yolk and egg white mixture (CEM) after 12 days of incubation was performed through tandem mass tag-labeling technology together with reversed-phase high-performance liquid chromatography (Zhu et al., 2019). Through gene ontology analysis, the differentially phosphorylated proteins were mainly associated with molecular functions in regulation, binding, and transport activities. For example, the increased phosphorylation degree of ovalbumin might enhance its ability to bind with minerals and facilitate a better absorption by the embryo. The increased phosphorylation of riboflavin binding protein may also enhance the transport efficiency of riboflavin to the embryo tissues. Moreover, the dephosphorylation process of vitellogenin may be triggered by temperature-regulated phosphatases, allowing to supply a lot of phosphate groups for the synthesis of nucleotides and nucleic acids as well as to support energy metabolism in the developing embryo (Sun et al., 2019). Phosphoproteins of Tibetan chicken eggs were shown to possess antibacterial, emulsifying, and antioxidative properties. In a N-glycoproteomic study, a total of 334 unique N-glycopeptides from 153 N-glycoproteins were identified and 82 N-glycopeptides showed some significant changes after 12 days of embryo development. For instance, we noted a significant decrease in dickkopf-related protein 3, suggesting a role in the embryo limb development and eye formation (Liu et al., 2020). Interestingly, some results suggested that a prior need for the chick embryos is for immune-defense-related proteins such as the complement proteins, produced from the CEM by the different combined and shifting effects of proteases and protease inhibitors. The reserved proteins supplied the exclusive nutrition and biological functions for the embryo may be degraded and reabsorbed at various priority levels for the subsequent embryonic development (Zhu et al., 2019).

From E4 to E12, the YS is the unique hematopoietic site while, hematopoiesis occurs in both the YS and bone marrow between E13 and E20. The developing chick embryo is mainly dependent for protection against pathogens on the maternally transmitted antibodies deposited into the egg yolk until the embryo can synthesize its own antibodies. The YS plays an essential role in passive immunity by allowing the uptake of the maternal antibodies and their transport into the embryo blood circulation (Meng et al., 2021b). At this stage, the accelerating transfer of egg white proteins to AmF and egg yolk around E13, driven by the needs of the developing embryo, is considered as the key point for the embryogenic process. The utilization of nutrients through YS to support the rapid growth of the embryo is considered as the transition point from the early embryogenesis stage to the second stage of development. Until the bone marrow can play its hematopoietic function, YS is the unique hematopoietic site during the second stage of embryonic development. It takes over the immune function performed by the egg white during the early embryogenesis period and plays a vital role in passive immunity by mobilizing egg yolk components, such as complement proteins, α-Fetoprotein, avian β-defensins. In addition to the prior need of immune-defense-related components for the chick embryo, the rest of the components are expected to supply the exclusive nutrition and facilitate provide biological functions, such as energy supplementation, organ formation, and lipid metabolism, at different priority levels for the subsequent embryonic development. Moreover, the CAM is active at this stage for calcium transport from the eggshell to the blood circulation, initiated around E13 and reaching its maximum by E17. In addition to proteins and lipids, carbohydrates and minerals are also utilized, but to a lesser extent.

41.3.3 The Third Stage of Embryonic Development (E15-E21)

The last period of development is mainly geared toward preparing the embryo for hatching. The YS, as an extra-embryonic tissue surrounding the yolk, possesses the ability to absorb, digest, and transport nutrients to the chick embryo during development, similar to the role of the YS during the early term mammalian embryonic development. The transcriptome analysis of YS tissues, as well as biochemical and histological analyses, on E13, E15, E17, E19, and E21 were performed to generate and analyze temporal large-scale patterns of gene expression data and gain insights into the processes and mechanisms of YS. The results demonstrated that in the final week of incubation, the YS supports the functioning of several immature organs, including the digestion of nutrients and transport in the intestine, the production of plasma carrier proteins in the liver, and the synthesis of blood cells in the bone marrow (Yadgary et al., 2014). Another proteomic study analyzed, through 2-dimensional gel electrophoresis, the egg yolk on E0, E10, and E18 (Zhu et al., 2020). At E10, differentially expressed proteins were mainly involved in lipid transport, nutrient reservoir, and antigen-binding ∡activities, while on E18, they were participating in substrate-specific transports and nutrient reservoir activities. The decreasing abundance of ovotransferrin on E10, for instance, suggested its contribution to the early cartilage formation, with its transfer from the egg yolk to the chick embryo. The significant decrease of vitellogenin on E18 is supposed to satisfy the embryo increasing metabolic needs and to provide the nutrients required for hatching. After E19, YS will resorb into the embryo abdomen, and the residual YS components will provide valuable nutrients to the newly hatched chick during its first few days of life. For this reason, the proteome of chick egg yolk within 24 hours post-lay was studied with a tandem mass tag labeling quantitative technology and compared with unfertilized chicken egg yolk. This study suggested that the yolk, recovered within 24 hours post-lay, may be an outstanding starting material to prepare dietary supplements or to isolate proteins with antimicrobial and pro-angiogenic properties (Padliya et al., 2015). Besides the YS, another active extra-embryo tissue is the CAM. The profile of metabolites present in the CAM and the change during maturation was investigated by the metabonomic 1H NMR spectroscopic analysis, which allows simultaneous measurements of various metabolites in an untargeted fashion. The study indicated the specific stages of blood vessel growth and identified decreasing energy-related substrates and waste product metabolites stored in the CAM during embryonic growth (Liu et al., 2021). As the energy substrate involved in fatty acyl decomposition, β-hydroxybutyrate increased during the development period, reflecting the intense lipid metabolism in the egg toward the end of development. The related CAM proteins involved in the binding and transport of lipids were also identified during the second and third phases of development. The lipid profile of the YS (including the yolk content) and the related effects in the embryonic process were investigated on E0, E7, E13, and E18 through UHPLC-MS/MS-based lipidomics to evaluate the changes in lipid metabolism during late development. With the evidence offered by the 2 non-operating de novo phosphocholine synthesis pathways in the embryo, phosphatidylglycerol and phosphocholine in YS potentially assist, during the late developmental stage, with the pre-hatching gas exchange and infection resistance once the embryo lungs start the respiration process. In addition, the decreasing cardiolipin may be transported to the embryo for its integration into mitochondrial membranes to accelerate the energy metabolism to meet the rapidly expanding needs of the developing embryo after E13 (Meng et al., 2021a). On E18, the accumulation of lysophosphatidylcholine is supposed to induce apoptosis, with the help of the expression of ACSL4 mRNA, and disturb the membrane structure to enable the best absorption conditions in the embryo abdomen (Liu et al., 2021). During this last stage of incubation, YS and CAM take essential roles in supporting the functions of several immature organs, releasing reserved nutrients, and favoring the intense lipid metabolism, to provide the required nutrients and satisfy the embryo increasing metabolic needs for hatching.

41.4 Conclusion

Chick embryos have a long and distinguished history as a major model system in developmental biology and have also made a major contribution to scientific fields such as immunology, genetics, virology, cancer, and cell biology. Omics approaches have revolutionized the egg biological sciences by providing the possibility to observe nutritional dynamics during incubation in a living system. They provide information related to DNA, RNAs, proteins, and metabolites to assemble the pieces required for our understanding of the dynamics of the incubation process. With the help of multi-omics data generated from the vast number of studies reported, this review may empower researchers to attempt to answer critical biological and nutritional questions from a broad perspective. In the early embryogenesis process, proteins in egg white are mainly mobilized for cell absorption and passive immunity. With the formation of blood vessels, the YS will gradually take a central role in the immune function with the intervention of some specific molecules (e.g., complement proteins, α-Fetoprotein, avian β-defensins). During the middle stage of incubation, proteins, lipids, and minerals experience significant transport, absorption, and relocation through the intervention of extra-embryo tissues, helping the gradually maturing organs and satisfying the rapid metabolism required in the last incubation stage. From the perspective of omics, we were able to gain a panoramic view of the dynamic utilization of the various egg components by the embryo, and several areas have been identified for further research, potentially leading to the identification of native or recombinant bio-molecules in the future.

REFERENCES

Bellairs, R., & Osmond, M. (2014). *The atlas of chick development* (3rd ed.). Academic Press. https://doi.org/10.1016/C2010-0-65149-2

Cindrova-Davies, T., Jauniaux, E., Elliot, M. G., Gong, S., Burton, G. J., & Charnock-Jones, D. S. (2017). RNA-seq reveals conservation of function among the yolk sacs of human, mouse, and chicken. *Proceedings of the National Academy of Sciences*, *114*(24), 4753–4761. https://doi.org/10.1073/pnas.1702560114

Cordeiro, C., & Hincke, M. T. (2015). Quantitative proteomics analysis of eggshell membrane proteins during chick embryonic development. *Journal of Proteomics*, *130*, 11–25. https://doi.org/10.1016/j.jprot.2015.08.014

Feng, M., Yu, F., Han, B., Xu, X., & Li, J. (2015). In-depth N-glycosylation reveals species-specific modifications and functions of the royal jelly protein from Western (*Apis mellifera*) and Eastern honeybees (*Apis cerana*). *Journal of Proteome Research*, *14*(12), 5327–5340. https://doi.org/10.1021/acs.jproteome.5b00829

Ferner, K., & Mess, A. (2011). Evolution and development of fetal membranes and placentation in amniote vertebrates. *Respiratory Physiology & Neurobiology*, *178*(1), 39–50. https://doi.org/10.1016/j.resp.2011.03.029

Guyot, N., Rehault-Godbert, S., Slugocki, C., Harichaux, G., Labas, V., Helloin, E., & Nys, Y. (2016). Characterization of egg white antibacterial properties during the first half of incubation: A comparative study between embryonated and unfertilized eggs. *Poultry Science*, *95*(12), 2956–2970. https://doi.org/10.3382/ps/pew271

Liu, H., Ding, P., Tong, Y., He, X., Yin, Y., Zhang, H., & Song, Z. (2021). Metabolomic analysis of the egg yolk during the embryonic development of broilers. *Poultry Science*, *100*(4), 101014. https://doi.org/10.1016/j.psj.2021.01.036

Liu, Y., Qiu, N., Geng, F., Sun, H., Wang, H., & Meng, Y. (2020). Quantitative phosphoproteomic analysis of fertilized egg derived from Tibetan and lowland chickens. *International Journal of Biological Macromolecules*, *149*, 522–531. https://doi.org/10.1016/j.ijbiomac.2020.01.211

Ma, Y., Zhang, P., Wang, F., Yang, J., Yang, Z., & Qin, H. (2010). The relationship between early embryo development and tumourigenesis. *Journal of Cellular and Molecular Medicine*, *14*(12), 2697–2701. https://doi.org/10.1111/j.1582-4934.2010.01191.x

Meng, Y., Qiu, N., Guyonnet, V., & Mine, Y. (2021b). Omics as a window to unravel the dynamic changes of egg components during chicken embryonic development. *Journal of Agricultural and Food Chemistry*, *69*(44), 12947–12955. https://doi.org/10.1021/acs.jafc.1c05883

Meng, Y., Qiu, N., Mine, Y., & Keast, R. (2021a). Comparative lipidomics of chick yolk sac during the embryogenesis provides insight into understanding the development-related lipid supply. *Journal of Agricultural and Food Chemistry*, *69*(26), 7467–7477. https://doi.org/10.1021/acs.jafc.1c01728

Meng, Y., Sun, H., Qiu, N., Geng, F., & Huo, Y. (2019). Comparative proteomic analysis of hen egg yolk plasma proteins during embryonic development. *Journal of Food Biochemistry*, *43*(12), e13045. https://doi.org/10.1111/jfbc.13045

Moran, E. T. Jr. (2007). Nutrition of the developing embryo and hatchling. *Poultry Science*, *86*(5), 1043–1049. https://doi.org/10.1093/ps/86.5.1043

Mueller, C. A., Burggren, W. W., & Tazawa, H. (2015). The physiology of the avian embryo. In *Sturkie's avian physiology* (6th ed., pp. 739–766). Elsevier. https://doi.org/10.1016/B978-0-12-407160-5.00032-4

Nakazawa, F., Alev, C., Jakt, L. M., & Sheng, G. (2011). Yolk sac endoderm is the major source of serum proteins and lipids and is involved in the regulation of vascular integrity in early chick development. *Developmental Dynamics*, *240*(8), 2002–2010. https://doi.org/10.1002/dvdy.22690

Padliya, N. D., Qian, M., Roy, S. M., Chu, P., Zheng, H., Tess, A., Dariani, M., & Hariri, R. J. (2015). The impact of fertilization on the chicken egg yolk plasma and granule proteome 24 hours post-lay at room temperature: Capitalizing on high-pH/low-pH reverse phase chromatography in conjunction with tandem mass tag (TMT) technology. *Food & Function*, *6*(7), 2303–2014. https://doi.org/10.1039/C5FO00304K

Qiu, N., Liu, W., Ma, M., Zhao, L., & Li, Y. (2013). Differences between fertilized and unfertilized chicken egg white proteins revealed by 2-dimensional gel electrophoresis-based proteomic analysis. *Poultry Science*, *92*(3), 782–786. https://doi.org/10.3382/ps.2012-02565

Qiu, N., Ma, M., Cai, Z., Jin, Y., Huang, X., Huang, Q., & Sun, S. (2012). Proteomic analysis of egg white proteins during the early phase of embryonic development. *Journal of Proteomics*, *75*(6), 1895–1905. https://doi.org/10.1016/j.jprot.2011.12.037

Réhault-Godbert, S., Mann, K., Bourin, M., Brionne, A., & Nys, Y. (2014). Effect of embryonic development on the chicken egg yolk plasma proteome after 12 days of incubation. *Journal of Agricultural and Food Chemistry*, *62*(12), 2531–2540. https://doi.org/10.1021/jf404512x

Rose-Martel, M., Du, J., & Hincke, M. T. (2012). Proteomic analysis provides new insight into the chicken eggshell cuticle. *Journal of Proteomics*, *75*(9), 2697–2706. https://doi.org/10.1016/j.jprot.2012.03.019

Shi, Y., Zhou, K., Li, D., Guyonnet, V., Hincke, M. T., & Mine, Y. (2021). Avian eggshell membrane as a novel biomaterial: A review. *Foods*, *10*(9), 2178. https://doi.org/10.3390/foods10092178

Silva, M., Labas, V., Nys, Y., & Réhault-Godbert, S. (2017). Investigating proteins and proteases composing amniotic and allantoic fluids during chicken embryonic development. *Poultry Science*, *96*(8), 2931–2941. https://doi.org/10.3382/ps/pex058

Starck, J. M. (2020). Morphology of the avian yolk sac. *Journal of Morphology*, *282*(7), 959–972. https://doi.org/10.1002/jmor.21262

Sun, H., Qiu, N., Keast, R., Wang, H., Li, B., Huang, Q., & Li, S. (2019). Comparative quantitative phosphoproteomic analysis of the chicken egg during incubation based on tandem mass tag labeling. *Journal of Agricultural and Food Chemistry*, *67*(48), 13353–13361. https://doi.org/10.1021/acs.jafc.9b04638

Wang, J., & Wu, J. (2014). Proteomic analysis of fertilized egg white during early incubation. *Eupa Open Proteomics*, *2*(C), 38–59. https://doi.org/10.1016/j.euprot.2013.11.001

Wu, J. (2019). *Eggs as functional foods and nutraceuticals for human health*. The Royal Society of Chemistry, pp. 22–39. https://doi.org/10.1039/9781788013833

Yadgary, L., Wong, E. A., & Uni, Z. (2014). Temporal transcriptome analysis of the chicken embryo yolk sac. *BMC Genomics*, *15*(1), 1–15. https://doi.org/10.1186/1471-2164-15-690

Zhu, F., Qiu, N., Sun, H., Meng, Y., & Zhou, Y. (2019). Integrated proteomic and N-glycoproteomic analyses of chicken egg during embryonic development. *Journal of Agricultural and Food Chemistry*, *67*(42), 11675–11683. https://doi.org/10.1021/acs.jafc.9b05133

Zhu, W., Zhang, J., He, K., Geng, Z., & Chen, X. (2020). Proteomic analysis of fertilized egg yolk proteins during embryonic development. *Poultry Science*, *99*(5), 2775–2784. https://doi.org/10.1016/j.psj.2019.12.056

Part X

Production and Sustainability

42

The Sustainability of the Egg Sector—Opportunities and Mitigation Strategies

Vincent Guyonnet* **and Audrey E. M. Guyonnet**

CONTENTS

42.1 Introduction

Sustainability has gained momentum since the report entitled "Our common future" (World Commission on Environment and Development, 1987) was issued by the United Nations in 1987. The term sustainability was used to describe a situation where the needs of current generations is met without compromising the ability of future generations to meet their own needs. The sustainability of egg production means that future generations will be able to enjoy the nutritional value of eggs in the same way we do now. With younger generations now acutely aware of the impact of human interventions and food choices on the planet and its ecosystem, this chapter will review the economic, environmental, and social sustainability of the egg sector, suggesting some avenues to further improve the overall sustainability of the egg sector.

42.2 Sustainability and Consumers

The definition of sustainability proposed by the United Nations is widely accepted but is often expressed in different manners. When applied to a business or economic sector, sustainability

* Corresponding author.

DOI: 10.1201/9781003254430-52

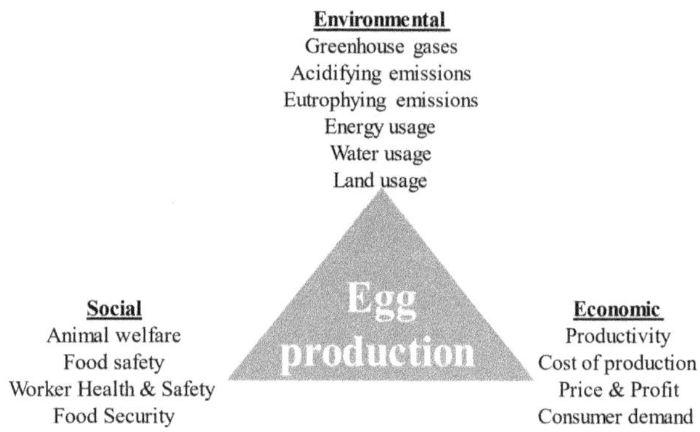

FIGURE 42.1 The 3 dimensions of sustainability.

is often divided into the following 3 components: economic, environmental, and social. While the intersection of these 3 components is viewed as the area in which a business or a sector will be the most sustainable, some compromises will have to be made (Guyonnet, 2018c; Pelletier et al., 2018). Another way to describe the sustainability for an egg company or the egg value chain is through its commitment to *People*, our *Planet*, and the *Prosperity* of the company or the value chain.

By 2027, the Gen Z, people born between 1998 and 2016 will be the largest demographic segment, representing 28% of the global population. Combined with Millennials born between 1981 and 1997, these generations will represent 52% of the global population and the largest consumer segment for food purchases. As expected, there are some differences between generations on several issues. Both Millennials and Gen Z reacted more positively to ethically-sourced brands than Baby boomers or Gen X (Donnan, 2018).

A survey of consumers was conducted in 2020 with 18,980 consumers from 28 different countries to identify the latest consumer trends and needs. The highest percentage of respondents in this survey (53%) believed that it was very important that products and services were designed to make their lives easier and simpler. But the availability of clean products (50% of respondents), the available of products that are sourced sustainably and/or environmentally responsible (45% of respondents) and the possibility to support recycling (44% of respondents) were also among the top demands of consumers. It is clear that consumers are more and more interested in the impact that their purchasing behaviors will have on the planet.

The survey probed consumer behaviors for 4 broad product categories: Foods & beverages, Apparel & Footwear, Personal care & beauty products, and Furniture and all house-related purchases. Purpose-driven consumers, or consumers willing to change their purchasing habits to reduce their impact on the environment, were observed with the highest percentage (44% of the respondents) for the purchase of foods and beverages. Likewise, the lowest percentage of consumers (35%) considered as value-driven consumers, or consumers primarily concerned with getting the best deal for their money, was related to the purchasing of foods and beverages. Within purpose-driven consumers, 57% were willing to change their purchasing habits to help reduce their environmental impact and 70% were willing to pay a premium, up to 35%, for the purchase of brands both sustainable and environmentally conscious (IBM, 2020).

A recent study evaluated the willingness of consumers to pay for a more sustainable egg supply chain, with interviews conducted with consumers from Brazil, China, India, Mexico, Spain and the United States of America (USA) (Sánchez-Bravo et al., 2021). Overall results showed that 47.8% of respondents were willing to pay at least a 5% premium for a more sustainable egg supply and 34.8% were willing to pay a premium of at least 20%. Only 17.4% of the respondents declared no interest in paying more for improvements in the sustainability of egg supply, although some differences were observed between countries, with at least 25% of the consumers in the USA and Spain unwilling to pay any premium.

Overall, consumers are now more concerned with the food chain sustainability and the consequences that their choices have on the planet.

42.3 Economic Sustainability

There are many ways to increase the economic sustainability of egg production and that of the overall egg sector. Some activities should focus on boosting the productivity and efficiency of egg production at the farm level, whereas other mechanisms should deal with consumers and the different marketing approaches that increase the overall value of the egg sector. The development of new products, the creation of new product categories and the growth of the egg processing sector are other mechanisms available to increase the economic sustainability of the egg sector (Guyonnet, 2019d; Guyonnet, 2020d).

42.3.1 Enhanced Productivity

Globally, egg producers have several options and tools available to enhance the productivity of egg farms. First, the design and implementation of strict biosecurity measures combined with efficient health programs ensure the health and well-being of laying hens, translating into longer production cycles. These programs also reduce any medication cost and extra labor required when birds are sick. The proper management of pullets also ensures that birds are prepared for a long and productive egg production cycle. Laying hen management such as feed formulation, temperature control, and ventilation must provide the best conditions for optimal egg production, which meets or exceeds the standards established by the genetics companies (Pelletier et al., 2018; Guyonnet, 2020e). The nutritional needs of laying hens vary throughout the laying cycle as older birds tend to produce larger eggs and will be less efficient at converting feeds into eggs (Guyonnet, 2021d). Since feed accounts for about 70% of egg production costs, it is essential that birds receive the right amount and quality feeds throughout their production cycle to lay eggs with a strong shell (see Chapter 15). A wider use of automated sensors and connected devices, or Internet-of-Things, on farms will also enhance productivity. Although egg farms are typically data-rich, most are not easy to access or compare, existing only as hard data (printed documents). While the use of Artificial Intelligence will offer many opportunities to assist farmers in optimizing many aspects of egg production, the sector must initially convert to digitalized data (Guyonnet, 2019e; Guyonnet, 2020c). With the assistance of Machine Learning, the digital data collected on farms will allow to build algorithms to predict pullet weight, egg weight, or a potential drop in egg production (Ramírez-Morales et al., 2017; Zimazile Sibanda et al., 2020; Guyonnet, 2021b). These predictions, sometimes 3–5 days in advance, will allow farm managers to make the appropriate changes to reach a desired outcome or prevent adverse impacts on productivity.

42.3.2 Growth in Egg Consumption

The growth in egg consumption was reviewed in Chapter 2. Between 2000 and 2020, in spite of the global population growing by 26.9% and reaching 7.795 billion in 2020, the global egg sector was able to make available an additional 46 eggs per person (+29.9%). National organizations of egg producers play a critical role in increasing egg consumption by conducting marketing and promotional campaigns that attract more consumers to the egg category and increase the frequency of consumption (Guyonnet, 2018a; Guyonnet, 2019f). Large-scale marketing research studies provide extremely valuable consumer insights: who they are, where they buy eggs, when and where they consume eggs, why they do eat eggs and, conversely, why do they not eat more eggs (Guyonnet, 2012). The responses to these many questions allow these organizations to segment the market into different categories of consumers, ranging from egg lovers to non-consumers, and to create the best messages for each consumer category. By integrating print media, social media, radio and TV advertising, these organizations can design very effective campaigns to address each consumer category (Guyonnet, 2020a). For instance, between 2010 and 2020, this holistic approach to the marketing and promoting of eggs has resulted in an increase in consumption of 119 eggs in Brazil, 111 eggs in Colombia and 59 eggs in Canada (IEC, 2021).

42.3.3 Branding of Eggs

At the global level, less than 50% of the eggs are sold as nicely packaged branded products (see Chapter 2). In major egg-producing countries like China, India, and Indonesia, the vast majority of shell eggs are sold in bulk, either by weight or by unit (Guyonnet, 2019a; Guyonnet, 2020b). One of the key roles of marketing will be to convert consumers from these "generic" eggs to branded eggs since the sale of branded products will typically generate more profits for egg producers and packers. As younger generations of consumers do no longer consider price as their main buying factor, their changing attitude will offer many opportunities to expand the global market share of nicely packaged, branded shell eggs. By focusing on marketing campaigns addressing the value and benefits of branded shell eggs, conducted either by national organizations of producers or by egg companies, the global sector will further enhance its economic sustainability while providing safe and nutritious foods to consumers (Guyonnet, 2018a; Guyonnet, 2019b).

42.3.4 Build New Product Categories

Another way to boosting the economic sustainability of egg production is to develop new categories of shell eggs and egg products that can be sold at a premium over the price of regular eggs (see Chapter 2).

One of the easiest ways to produce new categories of shell eggs is by modifying the diets given to laying hens and produce functional eggs. After adding a specific ingredient to the feed formulation, it usually takes between 10 and 21 days to reach a stable and predictable level for this ingredient in the egg, a characteristic that is essential to produce and sale functional eggs. Since more consumers will prefer natural foods, this natural enrichment of the egg composition is an important point of differentiation from other foods that cannot be enhanced without further processing (Guyonnet, 2011; Guyonnet, 2015). For this reason, functional eggs should also have a broader appeal with younger generations of consumers who tend to be adverse to highly processed foods (Guyonnet, 2019d).

The growing interest of younger generations of consumers for a more sustainable egg supply has convinced a growing number of egg producers to adopt a range of production systems claiming better welfare and quality of life for the laying hens (Rondoni et al., 2020; Sánchez-Bravo et al., 2021). While their cost of production is much higher than that of regular eggs (higher feed consumption, higher feed cost for organic production, etc.), the sale of "welfare" eggs will also increase profits across the egg value chain.

Globally, new categories of shell eggs have been developed based on criteria such as the date of lay and freshness, the age of hens, the guaranteed presence of double-yolk eggs, hens fed a vegetarian diet, the origin of production, the color of the shell or the production of eggs by rustic indigenous breeds. Recently, the support for a fair economic return to egg farmers has created a new product category in some countries (Guyonnet, 2019b).

42.3.5 Development of the Egg Processing Sector

New product categories can also be developed for egg products. Globally, the processing of shell eggs accounts for less than 10% of the total egg production (see Chapter 2). While the consumption of egg products account for about 50% of the total number of eggs consumed in Japan and about 25–30% in North America and Western European countries, the egg processing sector remains underdeveloped in many countries.

In countries with an established retail egg product segment, the development of more products addressing some of the key buying factors of consumers (convenience, authenticity, sustainability and taste) will provide more meal opportunities with eggs (see Chapter 14). The rapid expansion of Quick Service Restaurants (QSR) and food service companies from western countries, with specific requirements in terms of types of product and quality standards, will further expand globally the need for egg products. China represents a great example of this rapid development of the egg processing sector, with the production of liquid eggs increasing from 50,000 metric tonnes in 2010 to 230,000 metric tonnes in 2020 (IEC, 2021).

Overall, egg processing represents an excellent opportunity to further expand the consumption of eggs and boost the economic sustainability of the egg sector by developing more value-added egg products (Guyonnet, 2021a).

42.3.6 Research & Development

Although creating new shell egg categories is relatively simple, the development of new egg products requires a commitment to research and development (see Chapter 14). The focus of R&D should be on addressing the 5 main buying decisions (price, taste, convenience, authenticity and sustainability) and developing products and processes that increase sales and bring more consumers to the egg product category. An estimated 60% of the cost of production of egg products is related to the eggs, other ingredients used in the recipe and packaging, while 30% is related to labor cost and the remaining 10% accounts for energy such as electricity and water. Any R&D designed to identify ingredients and packaging materials of similar quality at a reduced price or to increase the yield of liquid egg during the breaking of shell eggs will have a significant impact on the cost of production of egg products. Manufacturing processes efficiency is extremely important in the calculation of the cost of production and egg processors are investing more into automated manufacturing processes, reducing both labor cost and human errors, a source of losses in time, money, product quality and company image. Further profit increases can also be obtained by implementing programs wherein engineers and food scientists will identify methods to reduce production losses, improve product quality and optimize the processing time of products. For instance, the use of a pasteurizer designed specifically for liquid egg products will allow for longer production runs that deliver optimum quality and safe products while requiring less energy for pasteurization and less usage of chemicals for the cleaning and sanitation of the processing equipment (see Chapter 10).

While the R&D for new egg products has initially concentrated on offering new breakfast solutions to accommodate for consumer lifestyles, there are numerous opportunities to expand the use of egg products to other meals. The success of hard-boiled eggs as a snack food item is an example of the many opportunities for a greater economic sustainability of the egg sector (Guyonnet, 2019c).

42.4 Environmental Sustainability

The environmental impact of products and services is often the only element considered by consumers when making sustainable purchasing decisions. A wide range of human-generated activities have an impact on the environmental sustainability of egg production. The increased emissions of various gases have an impact on air quality, the composition and quality of soils and water quality. In addition, some of the gases produced by human interventions have the potential to contribute to global warming. The challenge of feeding our growing global population also contributes to increased land use change and deforestation to produce more crops and foods. Therefore, the adoption of sustainable practices is essential to ensure the environmental sustainability of egg production.

42.4.1 Life Cycle Assessment (LCA)

The Environmental Footprint (EF) of products and services is defined using the Life Cycle Assessment (LCA) methodology. LCA, also known as the "From cradle to grave" analysis, investigates and assesses the EF of a product or service during all stages of its existence. It evaluates the potential impact of a product on the environment by quantifying its use of resources or "inputs" (e.g. energy, raw materials, water) and its environmental emissions or "outputs" in the air, water and soil, at each production stage. A procedure for implementing LCA was defined by the International Organization for Standardization, outlining 4 key phases: 1) definition of the goal and scope, 2) inventory analysis, 3) impact assessment, and 4) interpretation (ISO, 2006). Since the scope, including the system boundary and the depth and the breadth of the LCA could differ considerably between studies, most of the debate regarding LCA results within the poultry and livestock sectors did initially focus on the methodology used for these LCA studies and not on possible mitigation strategies. In 2012, the International Egg Commission (IEC) and the International Poultry Council (IPC), respectively representing the global egg and poultry meat sectors, joined the Food and Agriculture Organization of the United Nations (FAO) in the Alliance on Environmental Assessment and Ecological Performance of Livestock (LEAP), a multi-stakeholder initiative dedicated to improving the economic, social and environmental sustainability of livestock supply chains (LEAP, 2022). The LEAP alliance is composed of members from the livestock private sector, representatives of governments, the civil society, and various UN agencies. Since 2012, LEAP has published a series of guidelines offering some standardization to evaluate the EF of egg production (FAO, 2016a; FAO, 2016b; FAO, 2020). With producers, consumers and other stakeholders within the egg value chain increasingly in need of information on the ecological performance and sustainability of the supply chain, this standardized methodology is essential to identify the most promising mitigation strategies to reduce the EF of egg production.

42.4.2 Impact Categories

The EF of a product is evaluated for different impact categories, usually related to the use of resource (e.g. water, land) and the emissions of environmentally damaging substances (e.g. greenhouse gases and toxic chemicals) which may have an effect on the quality of water and soil or affect human health (Table 42.1). For each impact category, an indicator must be selected along with a method or model used for quantification. The results for each indicator provide some information on the environmental issues associated with the inputs and outputs of the product system (ISO, 2006; European Union, 2013).

42.4.3 Environmental Footprint of Egg Production

Although 14 impact categories (sometimes up to 19 by subdividing acidification and eutrophication into separate categories

TABLE 42.1

Environmental Footprint Impact Categories (with Relevant EF Impact Category Indicator)

EF Impact Category	Definition	EF Impact Category Indicator
Climate change	Relative effects of Greenhouse Gas emissions in terms of climate change over a fixed period compared to the same mass of carbon dioxide	kg CO_2 equivalent
Ozone depletion	Emissions causing the destruction of the stratospheric ozone layer	kg CFC-11 equivalent
Ecotoxicity for aquatic freshwater	Variety of toxic mechanisms caused by the release of substances with a direct effect on the health of the ecosystem	CTUe (Comparative Toxic Unit for ecosystems)
Human toxicity – Cancer effects	Adverse health effects caused by the intake of toxic substances insofar as they are related to cancer.	CTUh (Comparative Toxic Unit for humans)
Human toxicity – Non-cancer effects	Adverse health effects caused by the intake of toxic substances insofar as they are related to non-cancer effects	CTUh (Comparative Toxic Unit for humans)
Particulate Matter / Respiratory Inorganics	Adverse health effects on human health caused by emissions of Particulate Matter (PM) and its precursors (NO x, SO x, NH 3)	kg $PM_{2.5}$ equivalent
Ionizing radiation	Adverse effects on human health	kg U^{235} equivalent
Photochemical Ozone Formation	Formation of ozone at the ground level, damaging vegetation, human respiratory tracts, and man-made materials.	kg Non-Methane Volatile Organic Compounds
Acidifcation	Emissions of NO x, NH 3 and SO x lead to the release of H+ contributing to the acidification of soils and water	kg mol H+
Eutrophication – Terrestrial	Enrichment of the terrestrial ecosystem with nutritional elements, due to the emission of nitrogen-containing compounds.	mol N-equivalent
Eutrophication – Aquatic	Enrichment of aquatic ecosystems with nutritional elements, due to the emission of nitrogen or phosphor-containing compounds	kg PO_4-eq (fresh water) kg N-eq (marine water)
Resource depletion – Water	Relative use of water based on regionalized water scarcity factors	m^3 water use
Resource depletion – Mineral, fossil	Use of natural resources, either renewable or non-renewable, biotic or abiotic	kg Sb (antimony) equivalent
Land Use - Transformation	Use (Occupation) and conversion (Transformation) of land area by activities such as agriculture, roads, housing, mining, etc.	kg C (deficit) or kg CO_2 equivalent

Source: Adapted from European Union (2013).

for soil, fresh water and marine water) may be monitored to demonstrate the effects of human actions on the planet, most of the published LCA studies of egg production focused only on 3 to 5 of these impact categories (McClellan et al., 2018).

42.4.3.1 Climate Change

Climate change is the best-recognized impact category by consumers, although mostly known for its relation to the emissions of greenhouse gases (GHG). Solar energy is both absorbed at the surface of the earth and reflected into the atmosphere as infrared radiation (heat). While some of this energy travels back out into space, some of it is trapped in the atmosphere by GHG, contributing to the warming of the earth (greenhouse effect). GHG are naturally occurring and necessary to maintain a habitable atmosphere on Earth, but human interventions have caused an upregulation in GHG emissions responsible for the greenhouse effect. Primary greenhouse gases in the atmosphere include carbon dioxide (CO_2), methane (CH_4), nitrous oxide (N_2O), and ozone (O_3).

The Global Warming Potential (GWP) is defined by the Intergovernmental Panel on Climate Change (or IPCC) as an indicator reflecting the relative effect of the different greenhouse gases in terms of climate change considered over a fixed time period, usually 100 years, when compared to the effect of the same mass of carbon dioxide (IPCC, 2006). Therefore, carbon dioxide equivalent is used as the standard unit to compare the effects of emissions of different GHG. The total amount of GHG emissions associated with a product along its life is generally expressed in kilograms or tons of CO_2 equivalent (eq.) per unit of product, therefore the use of the term carbon footprint.

Since the mid-1800s, the global anthropogenic GHG emissions related to the use of forests as a source of wood for building and energy and to the conversion of land for farming or urbanization have remained at a similar level, less than 5 Giga tonnes of CO_2 equivalent. While remaining relatively low until the 1950s, the global anthropogenic emissions of GHG from fossil fuels have since increased almost 6-fold, reaching about 35 Giga tonnes of CO_2 eq. in 2010. In 2019, the concentration of GHG in the atmosphere was estimated at annual averages of 410 parts per million (ppm) for CO_2, 1866 ppb for CH_4, and 332 ppb for N_2O (IPCC, 2021). The global livestock sector contributes about 14.5% of the total anthropogenic GHG emissions, at par with the emissions of the transportation sector (14%) and less than the emissions from the production of electricity and heat (25%) and the global industrial sector (21%).

In 2011, the FAO estimated the GHG emissions from the livestock sector at 7.1 Giga tonnes of CO_2 eq. per year. The production and processing of animal feeds and the management

TABLE 42.2

Global Greenhouse Gas Emissions (Expressed in CO_2 Equivalent/Year) from Different Animal Productions

Animal Production	Global Greenhouse Gas Emissions (Million Tonnes of CO_2 Equivalent)
Beef	2,900
Cattle milk	1,400
Pork	700
Buffalo (milk & meat)	600
Small ruminants (milk & meat)	400
Chicken meat	400
Chicken eggs	200

Source: From Gerber et al. (2013).

of livestock were responsible for 47% and 50% of these emissions, respectively, while post-farm transportation and processing activities accounted for only 3% of emissions from the livestock sector (Gerber et al., 2013).

A breakdown of GHG emissions by animal production type is presented in Table 42.2. The production of beef (41%) and milk (29%) was responsible for most of the emissions from the global livestock sector whereas the production of eggs was responsible for only 2.8% of the total GHG emissions of the global livestock sector (Gerber et al., 2013).

The GHG emissions from the global egg value chain were estimated at 200 million CO_2 eq. per year. The production and processing of feeds is the main source of emissions (69.4%), due to the use of more land for crops, causing the oxidation of carbon in the soil and vegetation, the use of fossil fuels for the manufacture of fertilizers, and from the production and transport of poultry feeds. Manure management is responsible for 20% of the GHG emissions. Finally, Post-farm activities, including the packing of shell eggs and the processing of egg products are responsible for 10.6% of total emissions (Gerber et al., 2013).

A LCA study of egg production conducted in the United Kingdom (UK) showed 80.8% of the GHG emissions were related to the phase of egg production, while the pullet stage and the maintenance of breeding stocks accounted for 17.5% and 1.7%, respectively (Leinonen et al., 2012). Any strategy designed to reduce the emissions of GHG from the egg sector would therefore yield maximum impact at the egg production stage.

Numerous LCA studies have been conducted in different countries using national production statistics (Dekker et al., 2011; Wiedemann & McGahan, 2011; Leinonen et al., 2012; Pelletier, 2017; Pelletier et al., 2018) or data from individual farms (Mollenhorst et al., 2006; Ghasempour & Ahmadi, 2016; Abín et al., 2018; Estrada-González et al., 2020; González-Velandia et al., 2020) to calculate GHG emissions associated with egg production (Table 42.3). When laying hens were housed in conventional cage systems, the production of 1 kg of eggs was responsible for GHG emissions ranging from 1.3 kg CO_2 eq. to 5.58 kg CO_2 eq., depending on the studies. In enriched cage systems, the GHG emissions ranged between 2.31 and 3.40 kg CO_2 eq. per kilo of eggs produced. The differences in results between studies are due to variations in system boundaries used for the LCA study, the origin of feed ingredients (potentially contributing to a greater effect of land use change in the calculations) and in the breath of each dataset. Based on a limited number of studies, it appears that GHG emission from the production of eggs in conventional and enriched cage systems are comparable (Leinonen et al., 2014; Pelletier, 2017). In a cage-free production system, laying hens consume between 8% and 20% more feeds, translating into higher GHG emissions per kg of eggs produced. A study in the UK showed that the GHG emissions in an organic production system were comparable to those in a free-range system (3.42 vs. 3.38 kg CO_2 eq./kg of eggs) (Leinonen et al., 2012) whereas the LCA study of egg production in Canada showed that the organic production of eggs had the lowest level of GHG emissions (1.37 kg CO_2 eq./kg eggs) (Pelletier, 2017).

While GHG emissions from egg production are lower than those of most other animal-sourced foods, they are typically higher than those of plant-based foods, on the basis of weight (Gerber et al., 2013). However, it is important to consider the nutritional value delivered by the consumption of foods,

TABLE 42.3

Greenhouse Gas Emissions (Expressed in Kg CO_2 Equivalent/kg of Eggs) Depending on the Housing System Used for the Production of Eggs

Country	Conventional Cages	Enriched Cages	Barn/Aviary	Free-range	Reference
Australia	1.3	-	-	1.6	Wiedemann and McGahan (2011)
Canada	2.44	2.31	2.40	2.40	Pelletier (2017)
Colombia	-	-	-	2.77	González-Velandia et al. (2020)
Iran	4.07	-	-	-	Ghassempour and Ahmadi (2016)
Mexico	5.58	-	-	-	Estrada-González et al. (2020)
Netherlands	3.9	-	4.3	4.6	Mollenhorst et al. (2006)
Netherlands	2.23	-	2.68	2.75	Dekker et al. (2011)
UK	2.92	-	3.45	3.38	Leinonen et al. (2012)
UK	2.92	2.83			Leinonen et al. (2014)
USA	2.08	-	-	-	Pelletier et al. (2018)
Spain	-	3.40	-	-	Abín et al. (2018)

TABLE 42.4

Land Use (Occupation) for the Production of Eggs (Expressed in m^2/kg of Eggs)

Country	Conventional Cages	Barn/Aviary	Free-range	Organic	Reference
Canada	7.54	7.89	7.97	4.57	Pelletier (2017)
Netherlands	3.26	3.75	4.08	6.75	Dekker et al. (2011)
UK	4.0	4.2	5.1	16.9	Leinonen et al. (2012)

namely proteins and calories. While the GHG emissions of 1 kg of eggs is much higher than those for the same amount of rice or fresh produce, the GHG emissions expressed on the basis of protein content for eggs (24.4 kg of CO_2 eq./kg of proteins) are comparable to rice (21.2 kg CO_2 eq./kg protein) and much lower than for fresh produce (37.2 kg CO_2 eq./kg protein) (Clark & Tilman, 2017).

42.4.3.2 Land Use Change

Land use change (LUC) is a complex issue involving numerous factors, such as the conversion of forest areas for cultivation of crops or livestock ranching, logging activities or the urban encroachment on rural land. Altogether, these activities affect the ability of soils and vegetation to store carbon. This EF impact category is related to the way land areas are used (occupation) and potentially converted (transformation) by different activities. Land occupation considers the area involved and the duration of its occupation (changes in quality multiplied by area and duration), while Land transformation evaluates the extent of changes (changes in quality multiplied by area). The GHG emissions occurring as a result of land use change are allocated to the crops grown for the next 20 years following the change (European Union, 2013).

Depending on the LCA studies reported, feeds contribute between 60 and 75% of GHG emissions of the egg value chain. In 2013, the FAO estimated that feeds contributed to 70% of GHG emissions of the global egg sector. Direct emissions from feeds, accounting for about 57%, occur mostly during the production of crops, via the application of fertilizers, the use of fuel for farm machinery and for the transportation of all raw ingredients and finished feeds. The remaining 13% are related to LUC (Gerber et al., 2013). The most affected regions in terms of crop expansion are Latin America (soybean), Asia, and Africa (maize/corn). Many feed crops originating from these regions have higher LUC emissions than crops originating from Europe or North America (Guyonnet, 2021d). Although the expansion of soybean production in Brazil through deforestation is likely one of the most recognized drivers for LUC, a recent study in Brazil showed that the emission intensity of soybeans varies by more than 200-fold depending on their geographic origin within the country (Escobara et al., 2020).

LCA studies often report Land Use (Land Occupation) data or the amount of land occupied for the production of eggs. The data is related to pre-farm activities such as the production of crops and the area required to house birds (Table 42.4). Land use was lower for the production of eggs in cage systems when compared to those produced in barn or free-range systems. In both the Netherlands and the UK, organic egg production required significantly more land (up to 4 times) while in Canada, it was the lowest of all systems (Dekker et al., 2011; Leinonen et al., 2012; Pelletier, 2017).

42.4.3.3 Acidifying and Eutrophying Emissions

Acidifying emissions are a natural phenomenon that has been accelerated by recent human activities. Sulfur oxides (SO x) like sulfur dioxide, nitrogen oxides (NO x) like nitrogen dioxide and ammonia are gases with acidifying potential, meaning the ability to donate hydrogen ions (H+ ions). The release of H+ ions contributes to the acidification of soils and water, especially when their buffer capacity is low. Acidification also depends on the processes involved between emission and deposit and on the various processes in the soil and water. Although the impact indicator is the kg mol H+, acidifying emissions are combined and reported as acidification potential, expressed in g SO_2- eq./kg eggs produced. Depending on the LCA studies, the acidifying emissions related to the production of eggs in a cage system ranged from 32 to 78.4 g SO_2 eq. per kg of eggs (Table 42.5). Studies in the Netherlands and the UK showed lower acidifying emissions in cage systems compared to all other systems, including organic production while a Canadian study demonstrated the organic production of eggs had the lowest level of acidifying emissions (Dekker et al., 2011; Leinonen et al., 2012; Pelletier, 2017). Most acidifying emissions are related to the egg production stage (over

TABLE 42.5

Acidifying and Eutrophying Emissions from Egg Production in a Conventional Cage System

Country	Acidifying Emissions (g SO_2- eq. / kg eggs)	Eutrophying Emissions (g PO_4 eq. / kg eggs)	Reference
Canada	78.4	24.4	Pelletier (2017)
Iran	43.9	5.4	Ghasempour and Ahmadi (2016)
Netherlands	32	25	Mollenhorst et al. (2006)
UK	53.1	18.5	Leinonen et al. (2012)
USA	70	20	Pelletier et al. (2018)

80%), broken down into 3 main sources: manure (47%), housing and land (35%) and feeds (15%). There are little differences among housing systems used, except for the organic production system where manure contributed to 59% of these acidifying emissions (Leinonen et al., 2012).

In addition to their role as potentially acidifying substances, ammonia, nitrous oxides and sulfur dioxide also play a role in the formation of secondary fine particles that influence air quality.

Eutrophying emissions, evaluated for both fresh and marine water ecosystems as well for the terrestrial ecosystem, are linked to excessive amount of minerals and nutrients often rich in nitrate and phosphate in the environment and waterways. These minerals and nutrients often occur as runoff from fields after the use of fertilizers or the spreading of manure from poultry and livestock. Present in the water, these nutrients greatly facilitate the growth of algae and lead to the depletion of dissolved oxygen. With less oxygen available, water ecosystems change, often becoming inhabitable to plant and fish species. Although the Impact category indicators for these eutrophying emissions are reported in mol N-equivalent (terrestrial), kg N eq. (marine water) and kg PO_4 eq. (fresh water), most LCA studies of egg production reported these emissions in g PO_4 eq. per kilo of eggs produced, with values ranging from 5.4 to 24.4 g PO_4 eq. for production in a cage system (Table 42.5). Most of the eutrophying emissions are linked to the egg production stage (over 80%). Feeds (40–42%) and manure (39–41%) have an equal contribution to these emissions among housing systems used (cage, barn and free-range). The only exception was the organic production system wherein feeds and manure contributed 58.2% and 28.8%, respectively, to the acidifying emissions (Leinonen et al., 2012).

42.4.3.4 Water Footprint

Animal production accounts for nearly a third of water use in the agricultural sector related to animal production, meaning it has a significant impact on the fresh water resources of the planet. The Water Footprint (WF) is linked to the use of 3 water components: Blue water (surface and ground water), Green water (water from rainfall), and Grey water (fresh water required to dilute pollutants below an acceptable level). The WF for the production of eggs is calculated by adding water used for crops providing poultry feed ingredients and miscellaneous inputs (indirect needs) to the water used for production, such as drinking water for the birds and water required to clean the houses and equipment (direct needs) (Mekonnen & Hoekstra, 2012). The WF of egg production (2,872 liters of water/kg or about 170 liters/egg) is comparable to that of chicken meat, higher than that of milk production and much lower than that of beef and pork meat productions (Table 42.6). Depending on the type of plant-based foods, the WF of egg production also compares favorably. For instance, while egg production requires almost 9 times more water than the production of vegetables, it requires about 30% less water than pulse production. The WF of egg production can be broken down into 80% Green water, 7% Blue water, and 3% Grey (Mekonnen & Hoekstra, 2012). As previously mentioned for the Climate Change impact category, it is also important to compare WF based on nutritional values,

TABLE 42.6

Water Footprint of Different Animal and Plant Productions

Type of Production	Water Footprint (liters / kg product)	Water Footprint (liters / g proteins)
Beef meat	10,244	74.2
Pig meat	5,225	49.8
Chicken meat	2,873	22.6
Milk	1,207	36.6
Eggs	2,872	25.9
Vegetables	322	26.0
Pulses	4,055	19.0

Source: Adapted from Mekonnen and Hoekstra (2012)

i.e. calories or grams of proteins. When expressed in liters of water per gram of protein, eggs have a WF (25.9 liters/gram of protein) slightly higher than chicken meat and much lower than all other animal production types (Table 42.6). Although the WF of vegetables expressed on a kilogram of product basis was extremely low, it is fairly similar to that of eggs when compared on the basis of proteins provided for human nutrition. Conversely, pulses, extremely rich in proteins, have a lower WF based on protein content.

Production system types and geography will also affect the water footprint of egg production. In the Netherlands and the USA, the production of 1 kilogram of eggs in a cage system required 1,355 liters and 1,582 liters of water, respectively (Mekonnen & Hoekstra, 2012). In Australia, it was estimated that 2,187 liters of water were needed to produce 1 kilogram of eggs in a cage system whereas about 16% more water (2,543 liters) was required in a free-range production system (Wiedemann & McGahan, 2011). In China and India, with smaller farms and less automatization, the production of 1 kilogram of eggs required 2,920 liters and 4,483 liters of water, respectively (Mekonnen & Hoekstra, 2012). Due to the complexity of WF calculation, LCA studies on egg production have seldom reported WF as an impact category. Using only direct water use in their calculations (water as an inventory item), some LCA studies have reported the water use for 1 kg of eggs produced in a cage system: 5.11 liters in the UK and 16.5 liters in Canada (Leinonen et al., 2012; Pelletier, 2017). In Canada, these calculations for water use were comparable between conventional and enriched cage systems and about 12% higher in a cage-free indoor system (Pelletier, 2017). In each country, housing system types, management practices, and the origin of feed ingredients have an impact on the WF of egg production.

42.4.4 Mitigation Strategies

Although the impact of egg production on the environment is relatively low in comparison with other animal-sourced foods, consumers, large customers (food service and industrial customers), and ESG (Environmental, Social, and Governance) investors demand further reduction of its EF (Mench et al., 2011; Poore & Nemecek, 2018). LCA studies conducted in the USA and Canada have demonstrated the tremendous improvements made by the egg sector over a 50-year period

TABLE 42.7

Reduction (in %) over a 50-Year Period (USA; 1960–2010, Canada: 1962–2012) of the Level of Emissions and Water Use to Produce 1 Kg of Eggs

Country	GHG Emissions	Acidifying Emissions	Eutrophying Emissions	Water Use
Canada	72%	61%	68%	35%
USA	71%	65%	71%	32%

Source: From Pelletier et al. (2014) and Pelletier (2018a).

(Pelletier et al., 2014; Pelletier, 2018a). When considering the 3 main types of emissions (GHG, acidifying, and eutrophying emissions), the US and Canadian egg sectors were able to reduce emissions by at least 60% while increasing production between 30 and 50%. These improvements were attributed to several factors. Better nutrition with a more precise diet formulation, improved quality of feed ingredients and healthier hens mean that laying hens now requires half the amount of feed to produce a dozen eggs in comparison to 50 years ago. These improvements have also translated in a reduction in water use by more than 30% (Table 42.7). Better flock management with the use of effective vaccines and improved biosecurity measures has also contributed to a reduction in flock mortality, ranging from 57% (USA) to 75% (Canada). Overall, these improvements have allowed the egg sector to produce more eggs with less birds and lower land use.

Regardless of these great past achievements, the global egg sector must continue to identify and adopt new mitigation strategies. Under pressure mostly from NGOs, large customers (food service and food industry), and ESG investors, the egg sector is altering its practices by switching to cage-free production systems or by abandoning management methods that improved productivity (e.g. beak treatment). These changes may move the egg sector backward with regard to its EF and will make its sustainability journey much harder (Guyonnet, 2020e).

42.4.4.1 Crop Production, Feed Formulation and Feed Conversion

The intensity of GHG emissions of feedstuffs vary by crop, with wheat contributing less GHG emissions than corn, and corn itself less than corn Distiller's dried grains with solubles (DDGS) or soybean. Innovative formulation methodologies have suggested the incorporation of the environmental impacts of feed ingredients into the least-cost formulation of poultry feeds although there are still many issues to resolve. In many cases, the conversion of forest areas for the cultivation of crops will impact LUC and the overall environmental impact of feedstuffs. For instance, it was shown that the emission intensity of soybeans may vary by more than 200-fold depending on their geographic origin within Brazil (Escobara et al., 2020). Therefore, there will be need for data regionalization to accurately assess the environmental impact of crop production on the overall EF of egg production. In addition, the weighing method to reconcile the economic and environmental goals of these new approaches to feed formulation will have to be defined and agreed upon by all stakeholders (Pelletier et al., 2018).

In addition to the choice and quality of feed ingredients used, any decision made to improve the conversion of feeds into eggs will contribute to a reduction of emissions per kilogram of eggs produced. Laying hens typically digest and utilize about 80–90% of their grain-based diet. Therefore, feed enzymes, such as proteases, phytase, and non-starch polysaccharides degrading enzymes, are important tools to increase the nutritional value of feed ingredients, reduce feed costs, and improve egg production. The use of proteases will not only improve the conversion of feeds into eggs but will also reduce the protein level in layer rations, directly reducing the amount of nitrogen excreted by the birds (Leinonen & Kyriazakis, 2016). Beyond this traditional use of enzymes, it is now recognized that enzymes also have a functional role, reducing the presence of antinutritional and toxic compounds in the digestive system and promoting a more favorable gut microbiota. For instance, proteases act on allergenic proteins in the rations, reducing oxidative stress and gut inflammation, and promote a healthy gut for optimum nutrient absorption and conversion. Enzymes, like many other supplements in layer feeds, can reduce the GHG emissions per kilogram of eggs produced (Guyonnet, 2020e). Given the importance of feed to the overall profitability of the egg sector and influence on the egg production EF, future research will continue to focus on maximizing feed conversion with the addition of additives (ideally natural compounds) and on convincing more producers to adopt them.

42.4.4.2 Bird Management and Health

Feed efficiency will also be indirectly improved by better health and welfare practices, lower mortality rates, better bone strength, and improved genetics, all contributing to the reduction in the GHG emissions per kilograms of eggs produced. A meta-analysis, combining data from 10 studies with 2,823 free-range flocks of hens, evaluated the effects of cumulative mortality at 72 weeks of age on the EF of egg production. While cumulative mortality was as high as 69.3%, the average cumulative mortality was 10% across all flocks. The 4 impact categories (GHG, acidifying and eutrophying emissions, and land use) were assessed for cumulative mortalities ranging from 0% to 25% (Table 42.8). Reducing mortality from 20% to 5% translated into a significant reduction across all impact categories: GHG emissions (24.4%), acidifying emissions (17.9%), eutrophying emissions (17.9%), and land use (22.3%). A more modest reduction of the cumulative mortality from 10% to 5% also translated into significant reductions: GHG emissions (6.7%), acidifying emissions (6.1%), eutrophying emissions (6.0%), and land use (6.4%). For each percent reduction in mortality, these impact categories were reduced by 1–2 % (Weeks et al., 2016). Reducing mortality in flocks will not only improve the profitability of egg production (economic

TABLE 42.8

Effect of Flock Mortality in a Free-Range Production System on the EF of Egg Production (Cumulative Mortality at 72 Weeks of Age), Expressed per Kilogram of Eggs

Cumulative Mortality	GHG Emissions (kg CO_2 eq.)	Acidifying Emissions (g SO_2- eq.)	Eutrophying Emissions (g PO_4 eq.)	Land Use m^2 / year
25%	4.46	28.1	76.1	6.22
20%	4.13	26.2	70.9	5.76
15%	3.83	24.4	66.2	5.39
10%	3.56	22.9	61.9	5.03
5%	3.32	21.5	58.2	4.71
0%	3.11	20.3	54.9	4.44

Source: From Weeks et al. (2016).

sustainability) but also reduced the egg production EF, a true win-win situation for egg producers and for the planet.

The adoption of best practices, often referred to by the FAO as "bridging the gap", is a very important means of significantly reducing the EF of egg production. A survey conducted with 38 IEC member countries revealed the variability in the number of eggs produced between countries, despite similar production cycle length. For example, among 4 countries with a 60-week production cycle, there was a 28% difference between the lowest and highest production result. Ideally, the variability in productivity should be as low as possible and for similar length of production cycles, egg producers in a given country and across countries should aim to obtain as many eggs per hen housed as the best-performing country or producer. A recent study in Canada showed that the adoption of current best practices in terms of nutrition and management of laying hens could significantly affect the egg sector EF, reducing GHG emissions by 43%, acidifying emissions by 40%, eutrophying emissions by 44%, land use by 42% and water use by 32% (Pelletier, 2018a).

42.4.4.3 Manure Management

While egg farmers may have a limited influence on the emissions generated during the growing, harvesting, and processing of crops used for layer rations, the emissions produced by manure, the second largest source of emissions from the egg value chain (20% of total emissions), are under their direct control. Optimum manure management aims to reduce the fermentation processes by anaerobic bacteria before the application of manure onto fields. Poultry rations have a direct impact on manure chemistry and thereby the GHG emissions during manure storage and after application to soils. A decrease in the digestibility of nutrients in the feed will increase their availability as fermentable organic matters in manure, increasing in turn GHG emissions. Mitigation strategies will typically focus in 3 areas: manure composition, storage conditions, and methods used to spread manure onto fields (Hristov et al., 2013).

Depending on the feed composition, the manure excreted by laying hens will vary in terms of pH, humidity level, the presence of undigested organic substances, and its nitrogen content.

The composition of manure is rich in information, providing a better understanding on the true digestibility of feeds by the birds and allowing to appreciate its potential fertilizing value for soils and crops. Manure is not only a source of nitrogen and phosphorus but can also be rich in other minerals, including zinc, boron, and copper. By using ingredients or supplements that enhance gut absorption or promote a healthy digestive system, farmers can improve feed conversion and profitability on their farm as well as reduce the amount of organic matters favored by microorganisms in the manure.

The second mitigation strategy should focus on manure storage (form and length of storage). Manure will be typically stored in 1 of the 2 forms, dry or wet manure. The most common form, dry manure storage, involves gathering manure in piles left to dry for several months. Ideally, the manure is dried in a shed to protect it from the rain. In new housing systems, manure is partially dried when removed from the house, usually 2 or 3 times per week. Older housing systems may still use a deep pit under the cages where manure accumulates over the entire production cycle and is removed only once at the end of the egg production cycle (Meda et al., 2011). Wet manure involves the storage of waste material in outdoor tanks or lagoons where it is mixed with Grey water. In uncovered lagoons, the anaerobic fermentation of the manure leads to GHG production. The storage of manure in liquid form is more convenient for farmers to apply as a fertilizer onto fields. However, this type of storage produces more GHG emissions, regardless of the housing system used for egg production (Table 42.9). When using a wet manure storage system,

TABLE 42.9

Effects of Manure Storage Conditions on the Level of GHG Emissions (Kg CO_2 Eq. per Kilo of Eggs Produced)

Housing System	Dry Manure Storage	Wet Manure Storage	Increase
Conventional cages	1.66	3.35	101.8%
Aviary	1.7	3.49	105.3%

manure management may account for more than 50% of the overall GHG emissions linked to egg production.

While dry manure storage creates less GHG emission than the lagoon storage, some techniques can minimize the emissions from liquid manure storage. It is recommended that storage tanks be emptied completely, as "aged" manure left inside the tank accelerates the spread of bacteria and the fermentation process and emissions in the next batch of stored liquid manure. Studies have shown a reduction up to 55% in GHG emissions when slurry tanks were completely emptied. The acid treatment of manure also reduces methane and nitrous oxide emissions (Wood et al., 2014).

Finally, timing of manure application on fields also impacts the overall emissions from manure. Optimal manure application relies on the 4 "Rs" – the right rate of application, the right time, the right method of application and the right product. Manure should be analyzed and certain compounds such as nitrification inhibitors may be added. Spring application results in lower nitrogen losses and emissions than Fall spreading and should be the preferred period for spreading manure, coinciding also with the growing season and the increasing nutrient demand of plants. Applying manure on cloudy days should also be preferred to sunny days in order to limit ammonia volatilization (Chadwick et al., 2011).

Overall, reducing the nitrogen concentration in manure, preventing the formation of anaerobic conditions and reducing the presence of degradable carbon in manure are effective strategies to reduce GHG emissions during manure application (Loyon et al., 2016).

42.4.4.4 Housing Systems

In the previous section addressing Climate change, the influence of different housing systems on GHG emissions per kilograms of eggs produced was discussed and showed that the higher feed consumption and lower feed conversion in cage-free systems typically translated into higher GHG per kilo of eggs produced. A study in the USA, examining the potential impacts of changes in egg production from conventional cages to enriched cages or a cage-free aviary system, showed that while the feed conversion was comparable between conventional and enriched cages, laying hens in an aviary system consumed about 4% more feeds to produce 1 kg of eggs (Lin et al., 2016). Commercial data from the United Kingdom showed similar results, where the feed consumed by laying hens on a daily basis was 8% higher in the barn system and 13% higher in the free-range system when compared to the cage system. In addition, egg production was 4.8% lower in the barn system and 7% lower in the free-range system (Leinonen et al., 2012). These results not only impacted egg production cost in cage-free systems but also increased GHG emissions per kilogram of eggs produced (Table 42.3). However, as discussed in Chapter 2, the growth of cage-free production systems in Europe and North America is related to animal welfare and ethical considerations and does not consider the environmental impacts of this decision. When large retailers, food service and industrial users of eggs committed to a supply of eggs from cage-free systems, it is evident that they did not consider the results of the holistic approach designed by the Coalition for Sustainable Egg Supply Project to evaluate the impact of switching to a cage-free supply of eggs (CSES, 2022).

Unable to provide convincing data to select the housing system with the lowest GHG emissions, recent research has focused on the construction materials and equipment used in layer barns. Net zero energy (NZE) buildings are energy-efficient structures with net zero energy consumption, meaning that the total amount of energy used by the building is equal to the amount of renewable energy it can create on an annual basis. By design, these buildings rely on 3 approaches: structural considerations to reduce energy use, adoption of energy-efficient systems, and on-site generation of renewable energy. While government incentives, tax breaks, and savings on energy costs have increased the application of the NZE concept to industrial sites and private homes, it may be time to apply it to Poultry barns (Guyonnet, 2021).

Barns are typically built with timber, plywood, tin, and stone wool insulation material, all of which impact the environment and embodied energy, representing the sum of all the energy required to produce them. The addition of extra insulation, renewable energy systems, and more energy-efficient equipments in NZE barns will have a small incremental effect on the overall impact of the building itself but would easily be offset by the reduction in direct energy used for heating and ventilation during egg production. An LCA approach comparing a conventional and NZE barn reported the direct cumulative energy input was reduced by 36% per tonne of eggs in the NZE layer house. Depending on the origin of the electricity used (fossil fuel, gas, coal, or hydroelectricity), investments in NZE barns also provide significant benefits by reducing CO_2 emissions (Climate change) and other impact categories (Li et al., 2021). It is likely that NZE technology will be used more in the future when building new layer barns as an additional mitigation strategy to assist the egg sector to further reduce its EF (Guyonnet, 2021c).

42.4.4.5 Measuring to Mitigate

While consumers and producers both have roles in reducing the environmental impact of food choices, producers have a much harder task with the monitoring and management of multiple impact categories (Poore & Nemecek, 2018). As outlined previously, the EF of egg production is typically calculated at a national level using the LCA approach. However, this approach has its limitations and does not represent the situation on every farm. For instance, based on the LCA study conducted in Canada, the average mortality was 3.18% but ranged from 0.9% to 7.3%, representing a 10% difference in GHG emissions between the best and worst performing farms. Similar variations were also seen for feed conversion with 2.0 as the national average but values ranging from 1.64 to 2.82 (Pelletier, 2017). Such high variability implies egg producers cannot rely on average or national values to track and report the environmental impacts of their farms (Guyonnet, 2020e).

Some key differences between LCA studies and the latest tools available to assess the environmental impact of egg

TABLE 42.10

Comparison of LCA Studies versus Environmental Impact Assessment Tools

LCA Studies	Environmental Impact Assessment Tools
• Mostly conducted by academia • Time consuming to collect data • Concerns about data privacy • Mostly reporting average values at a macro level (national, state, province) • Little flexibility to test different scenarios	• Models developed by private companies • Built-in, user-friendly data collection • Controlled access to own data • Reporting at the level of company, farm site or specific layer house • Option to instantly evaluate the potential impact of various decisions made on farm

production are presented in Table 42.10. While LCA studies are complex and time-consuming studies typically conducted by academia, new tools have been developed by private enterprises to offer egg producers means of measuring the EF of their own production facilities. Through built-in simulation options, egg producers can also test and identify the best mitigation strategies for their farms.

The selection of the correct commercial tools is essential to accurately evaluate the EF of egg farms and several key characteristics must be considered. Models used by each tool must follow the set of international rules and guidelines (IPCC and LEAP) discussed previously. In addition, each tool has a number of built-in or default values used by its model and it is important to confirm that these default values are relevant to the local conditions. Since the contribution of the pullet stage to the total GHG emissions differed by 41% between data from Europe and North America in a model tested, it is clear that regional differences must be accounted for to formulate accurate mitigation strategies for each farm. Finally, these tools must be connected to relevant international databases to ensure consistent, relevant, and accurate calculations (Guyonnet, 2020e).

Overall, these new tools will allow egg producers to report accurately the EF of their own production and to take full credit for their investments to reduce the EF of eggs produced on their farms.

42.5 Social Sustainability

Social sustainability is often linked to the term of Social license, or the trust that consumers and the society at large confer to a sector to make ethical decisions when operating freely (Thompson et al., 2011). Similar to the evaluation of the environmental impact of a product through the LCA approach, formal guidelines also exist for the Social Life Cycle Assessment of products (UNEP, 2009) and ISO 26000:2010 standard provides further guidance to organizations on social responsibility (ISO, 2010). Various categories of stakeholders should be considered during this assessment: workers, consumers, local communities, society, and value chain actors. For each category, a range of items must be evaluated. For instance, the following elements must be considered with regard to workers: freedom of association and collective bargaining, child labor, fair salary, working hours, forced labor, equal opportunities, health & safety, and social benefits (UNEP, 2009). While the UNEP guidelines were used to evaluate the social sustainability of the Canadian egg production (Pelletier, 2018b), many evaluations at a national (Mench et al., 2011; de Olde et al., 2020; Australia Eggs, 2022) or company level (Cal-Maine Foods, 2021; Burnbrae Farms, 2022) have used their own terms of reference or incorporated certain elements of social sustainability into their sustainability reports. For the purpose of this section, the following components of social sustainability are presented: animal health and welfare, food safety, worker health & safety and food security, including the overall contribution of the eggs sector to the UN Sustainable Development Goals 2030.

42.5.1 Animal Health and Welfare

The general modern concept of animal welfare goes back to 1965 when the UK government appointed a technical committee to enquire on the welfare of farm animals raised in intensive production systems. This committee, led by Professor Brambell, issued a report outlining the Five Freedoms as the base to good welfare. In order to safeguard the welfare of farm animals, animals must be free from hunger and thirst, free from discomfort, free from pain, injury or disease, free to express natural behavior, and finally free from fear and distress (Brambell et al., 1965).

Any definition of animal welfare is complex and multifactorial, varying frequently depending on the interested party. For some, the welfare of laying hens is only about the ability of birds to express their natural behavior such as nesting and perching. For others, the welfare of hens includes not only the ability to express some natural behaviors but also functional criteria like nutrition, health and the conditions of the environment in which live the birds. The scientific evaluation of poultry welfare has made tremendous progress in recent years. Today, animal welfare specialists recognize the multifactorial aspect of animal welfare and the importance of positive experiences in these 4 fields: nutrition, environment, health, and behavior, all contributing to the welfare and well-being of laying hens.

Evaluating the quality of life of laying hens is likely the most challenging element to measure since it involves some ethical decisions made by people on behalf of the hens. It is obvious that there are many cultural and value differences around the world and that not everyone will necessarily share the same values (Widowski et al., 2017). The evaluation of the quality of life will certainly remain the most challenging issue for the egg sector over the next decade.

From the point of view of the birds, the enriched cages and cage-free systems offer several positive elements. A series of behavioral tests have clearly demonstrated that laying hens have the necessity to perform perching, nesting, and dustbathing behaviors (Weeks & Nicol, 2006; Hartcher & Jones, 2017; Widowski et al., 2017). Birds have a strong motivation to access perches and almost all hens will use perches at night if sufficient space is available. In aviary systems, the use of perches may contribute to reduce fearfulness, and a lower density of birds on the ground decreases the risk of piling and smothering. However, in cage-free systems, some birds with aggressive behavior will prevent other hens from accessing nests and perches. The use of perches along with exercise does improve bone strength and musculoskeletal health, but a higher incidence of keel bone deformities and fractures has been reported in cage-free systems when compared to cage systems (Petrik et al., 2015; Hartcher & Jones, 2017; Widowski et al., 2017).

Cage-free systems present more risks of diseases and parasitic infestations, feather pecking, and cannibalism than cage systems. A meta-analysis reviewed data from 10 studies in the United Kingdom, the Netherlands, and Sweden, assessing 3,851 commercial flocks and about 45 million birds housed in various housing systems. By analyzing the cumulated mortality rate between 60 and 80 weeks of age, it was shown that mortality was not only higher in cage-free systems compared to cage systems but also highly variable, especially for cage-free systems with access to the outdoors (Weeks et al., 2016). Cage-free systems are more complex to manage, with increased risks of diseases and parasitic infestations. The control of diseases, parasites, and overall mortality was identified as 1 of the 10 general principles for the welfare of animals in livestock production systems, guiding the World Organisation for Animal Health (WOAH, founded as OIE) in the development of its global animal welfare standards (Fraser et al., 2013).

Managing flocks in a cage-free environment, especially with access to the outdoors, requires more knowledge and higher skills to guarantee the welfare of laying hens. The review of data generated between 2000 and 2020, with more than 6,000 flocks and 176 million birds housed in the different housing systems, showed that with the exception of conventional cage systems, the mortality appeared to gradually drop over time, likely as egg producers acquired more experience with each new housing system (Schuck-Paim et al., 2021). Since 2000, each year of experience with cage-free systems (mostly aviary systems), was associated with a reduction in total mortality ranging between 0.35% and 0.65%. Mortality in alternative housing systems is likely to decline with more field experience. Therefore, it will be important for the egg sector to negotiate sufficient time for any effective production system transition, to test and adopt proper management practices.

It is clear that none of the current housing systems are perfect and some trade-offs are required. When US consumers were given a proper understanding of the different attributes of cage and cage-free aviary systems, they expressed a preference for the enriched cage system (Ochs et al., 2019).

Equating cage-free production to the ideal system is currently misleading consumers. The egg sector must continue to explain this scientific evidence for the benefit of consumers but also and foremost for the benefit of the laying hens (Guyonnet, 2021).

Other practices commonly used by the global egg sector, such as beak treatment and the culling of male chicks, are now contested by many stakeholders (Thompson et al., 2011; de Olde et al., 2020). Through innovations, the egg sector has recently developed new techniques to either minimize the pain of beak treatment or cull eggs at an early stage of the embryonic development (Dennis & Cheng, 2012; Gautron et al., 2021). However, the social sustainability of these mitigation strategies practices remains in question.

42.5.2 Food Safety

Consumers expect the provision of safe eggs. Over the years, the egg sector has developed and implemented a series of protocols to prevent the contamination of eggs with Salmonella and other pathogens, to prevent dirty eggs from being sold at retail, and to ensure that all eggs are free of chemical residues (see Chapters 18 and Chapter 20). In numerous countries, the mandatory candling or inspection of eggs prior to packing prevents the sale of cracked and/or leaking eggs as a measure to protect consumers from potential foodborne diseases (see Chapter 18).

With recent shifts to various alternative housing systems, a lot of research has been conducted to evaluate the effects of housing systems on the safety of eggs, comparing conventional to enriched cage systems (Mallet et al., 2006; Wall et al., 2008; Gast et al., 2015), cage systems to aviary or barn system (De Reu et al., 2005; De Reu et al., 2008; Hannah et al., 2011; Holt et al., 2011; Jones et al., 2015; Jones et al., 2016), and cage systems to free-range production systems (Huneau-Salaün et al., 2010; Jones et al., 2011; Jones et al., 2012). The results of these evaluations were extremely variable, likely due to additional factors besides the housing system. There seems however to be a correlation between the quality of the air in the barns and the level of contamination of the eggshell. Therefore, management practices on farms are likely more important than the housing system used to guarantee the safety of eggs. The age of the laying hens, the high incidence of eggs on the floor, and the manual handling of the eggs during collection and packaging are additional factors contributing to increased eggshell contamination. The use of reusable plastic containers to gather eggs can also increase the risk of contamination when improperly cleaned and disinfected.

In cage-free systems, the level of contamination of eggshells was also reduced when more laying hens were using the nest boxes. Without a doubt, the quality of eggs produced in cage-free systems is enhanced by reducing the percentage of floor eggs.

Programs at feed mills and on farms are also designed to prevent the cross-contamination of feeds provided to laying hens. Complying with local regulations will ensure the absence of drug and chemical residues in eggs. Since laying hens may also be able to pick up different pollutants from the soil, the risk of chemical contamination of eggs may be higher in a production system where laying hens are given access to the outdoors (Piskorska-Pliszczynska et al., 2019; Holt, 2021).

42.5.3 Worker Health and Safety

The health and safety of workers is an important part of the social commitment of the egg sector. As seen previously, the newer cage-free systems present potentially more risk to the health and safety of laying hens. The same may be true for farm workers, due to both the design of cage-free systems and the air quality inside the barn. Cage-free systems present many additional structures such as access ramps, perches, and other metal pieces on which workers can hurt themselves while walking through the houses, especially in a dimly lit environment. Fortunately, the manufacturers of housing systems are now including some elements of worker safety in their design, for example, specific structures on which farm workers can easily walk inside the houses will make the monitoring of birds more convenient and much safer.

The level of dust particles is typically higher in cage-free systems (barn and aviary systems) than in cage systems (Le Bouquin et al., 2013; Shepherd et al., 2015; Zhao et al., 2015). In a study comparing the air quality of conventional cages, enriched cages, and the aviary system in a commercial environment, the level of ammonia and the concentration of particulate matter (PM) were higher in the aviary system, especially when the outside temperature was below 10°C (Zhao et al., 2015). In the aviary system, the levels of $PM_{2.5}$ and PM_{10} were much higher (6.7 times and 6.4 times, respectively) than those in either cage system. Higher levels of PM_{10} and $PM_{2.5}$ in the aviary system result from an increased activity of laying hens combined with the presence of bedding and foraging materials (Shepherd et al., 2015). While the levels of airborne total bacteria concentration were above the recommended level for humans in all housing systems, a much higher concentration was observed in the cage-free system. The concentration and emission rate of airborne total bacteria were positively correlated with PM_{10}, and found to be higher in the afternoon when increased flock activity caused disruption to floor bedding (Zhao et al., 2016). With higher aerial dust concentrations in cage-free systems, it is necessary for workers to continuously wear protective masks to mitigate health risks (Le Bouquin et al., 2013; Pelletier, 2018b). Working in a safe environment is essential not only for current workers but also to attract future generations of farm workers.

42.5.4 Contribution to the UN Sustainable Development Goals

In 2015, members of the United Nations adopted the 2030 Agenda for Sustainable Development Goals. They committed to achieving sustainable development over the next 15 years while ending poverty and hunger (UN SDG, 2022). Among the 17 Sustainable Development Goals (SDGs) adopted, the global egg sector can contribute to the achievement of several, either directly or indirectly.

SDG1 – No Poverty: Globally, 1.3 billion people depend on livestock for their livelihood and Poultry represents the species mostly owned by smallholders. Activities to support and build capacity of poultry smallholders will further improve the livelihoods of about 600 million farmers living below the poverty line. National egg producer organizations like the South African Poultry Association provide dedicated services targeting the smallholder egg sector.

SDG2 – Zero Hunger: This SDG focuses on ensuring the access to all people, particularly infants, for safe, nutritious foods, and end all forms of malnutrition. It was clearly demonstrated that the daily consumption of 1 egg can contribute significantly to the reduction of stunting in young children (Iannotti et al., 2017). According to the FAO, in 2020, people in 55 countries (41 in Africa, 12 in Asia and 2 in the Americas) consumed less than 1 egg per week. The global egg sector must find ways to make eggs, a great source of high-quality proteins, more globally accessible and affordable. Currently, smallholder farmers, keeping only a few animals, produce up to 80% of the food consumed in Asia and Africa. By encouraging them to use inputs like vaccines and compounded feeds, share the knowledge available in other countries, and adopt more intensive production systems to improve production efficiencies, it is possible to boost the availability and reduce the price of eggs in developing countries (Guyonnet, 2022).

SDG3 – Good Health and Well-being: By producing an affordable food rich in high-quality proteins, an important component of a good functioning immune system, the egg sector provides for good nutrition, promoting good health and well-being (see Chapter 29 through Chapter 38).

SDG5 – Gender Equality: Agriculture employs 1.3 billion people globally, with 97% living in developing countries. Although women play a major role in the management of livestock species like Poultry, they do not have the same access as men to inputs such as feeds and vaccines. This lack of access leads to an estimated productivity reduction of 30%. The egg sector should ensure that women are given similar access as men to information and technology, knowledge and inputs, and promote their empowerment.

SDG12 – Responsible Consumption and Production: The egg sector contributes to this SDG by producing more eggs and reducing negative environmental impacts such as GHG emissions and degradation of ecosystems. The welfare and well-being of birds, a strong focus of egg producers, is part of the desire for a sustainable development in harmony with nature.

SDG13 – Climate Action: Egg production has the lowest total GHG emissions of all animal productions and the lowest emissions per kilogram of proteins produced. In addition, through the sharing and implementation of best practices, the egg sector can further reduce the emissions along its value chain and contribute significantly to this SDG.

SDG15 – Life on Land: The production of animal feeds uses 35% of all cultivable lands and 20% of the available water. Any improvement in feed conversion, meaning more eggs produced with the same amount of feeds, will have a direct benefit on the environment and will promote the responsible and sustainable use of natural resources by the egg sector (Guyonnet, 2018b).

Overall, the UN SDGs offer the global egg sector the opportunity to further demonstrate its commitment to sustainable development and to the welfare of the planet and its people.

42.6 Conclusions

The growth of global egg production over the past 20 years is a testimony of the work done by various stakeholders to improve the sustainability of the egg sector. Many innovations, delivering new products and creating new product categories, have consolidated the economic sustainability of the egg sector, providing more opportunities to consume eggs at any time of the day. While egg production has the lowest impact on the environment of all animal productions and has achieved significant gains over the past 50 years, the sector must further reduce its impact on the planet. Poultry feeds have the largest impact on the environmental footprint of eggs and the adoption of new feed ingredients, improved feeding practices and the overall effort to bridge the efficient gap will comfort the environmental sustainability of the egg sector. By providing a safe and affordable source of quality proteins, the egg sector will contribute to the challenge of feeding 9.7 billion by 2050. However, ensuring the social sustainability of some egg production systems and practices will likely be the biggest challenge for the egg sector over the next few decades. Future growth of the global egg sector will depend on its ability to identify new opportunities and mitigation strategies to further improve its economic, environmental, and social sustainability.

REFERENCES

Abín, R., Laca, A., Laca, A., & Díaz, M. (2018). Environmental assessment of intensive egg production: A Spanish case study. *Journal of Cleaner Production, 179*, 160–168. https://doi.org/10.1016/j.jclepro.2018.01.067

Australia Eggs (2022). Sustainability framework report 2022. Retrieved on April 29, 2022, from https://www.australianeggs.org.au/what-we-do/sustainable-production/sustainability-report

Brambell, F. W. R., Barbour, D. S., Barnett, L., Ewer, T. K., Hobson, A., Pitchforth, H., & Winship, F. J. W. (1965). Report of the technical committee to enquire into the welfare of animals kept under intensive livestock husbandry systems, HM stationary office, London, UK.

Burnbrae Farms (2022). Eggs for life. 2020/21Sustainability Report. Retrieved on April 29, 2021, from https://www.burnbraefarms.com/en/sustainability2021/

Cal-Maine Foods (2021). Scaling our sustainability in FY2020. Retrieved on April 29, 2022, from https://www.calmainefoods.com/media/1188/calm-fy2020-sustainability-report.pdf

Chadwick, D., Sommer, S., Thorman, R., Fangueiro, D., Cardenas, L., Amon, B., & Misselbrook, T. (2011). Manure management: Implications for greenhouse gas emissions. *Animal Feed Science and Technology, 66–167*, 514–531. https://doi.org/10.1016/j.anifeedsci.2011.04.036

Clark, M., & Tilman, D. (2017). Comparative analysis of environmental impacts of agricultural production systems, agricultural input efficiency, and food choice. *Environmental Research Letters, 12*, 064016. https://doi.org/10.1088/1748-9326/aa6cd5

CSES (2022, May 21). Coalition for sustainable egg supply - final results. https://www2.sustainableeggcoalition.org/research

de Olde, E. M., van der Linden, A., Bolhaar, L. D., & de Boer, I. J. M. (2020). Sustainability challenges and innovations in the Dutch egg sector. *Journal of Cleaner Production, 258*, 120974. https://doi.org/10.1016/j.jclepro.2020.120974

De Reu, K., Grijspeerdt, K., Heyndrickx, M., Zoons, J., De Baere, K., Uyttendaele, M., & Herman, L. (2005). Bacterial eggshell contamination in conventional cages, furnished cages and aviary housing systems for laying hens. *British Poultry Science, 46*(2), 149–155. https://doi.org/10.1080/00071660500065359

De Reu, K., Messens, W., Heyndrickx, M., Rodenburg, T. B., Uyttendaele, M., & Herman, L. (2008). Bacterial contamination of table eggs and the influence of housing systems. *World's Poultry Science Journal, 64*(1), 5–19. https://doi.org/10.1017/S0043933907001687

Dekker, S. E. M., de Boer, I. J. M., Vermeij, I., Aarnink, A. A., & Groot, P. W. G. (2011). Ecological and economic evaluation of Dutch egg production systems. *Livestock Science, 139*(1–2), 109–121. https://doi.org/10.1016/j.livsci.2011.03.011

Dennis, R. L., & Cheng, H. W. (2012). Effects of different infrared beak treatment protocols on chicken welfare and physiology. *Poultry Science, 91*(7), 1499–1505. https://doi.org/10.3382/ps.2011-01651

Donnan, D. (2018). Influence vs. affluence: The changing menu of food choices. Retrieved April 28, 2022, from https://www.kearney.com/consumer-retail/article/-/insights/influence-vs-affluence-the-changing-menu-of-food-choices-article

Escobara, N., Tizadoc, E. J., zu Ermgassend, E. K. H. J., Löfgrenf, P., Börnera, J., & Godarf, J. (2020). Spatially-explicit footprints of agricultural commodities: Mapping carbon emissions embodied in Brazil's soy exports. *Global Environmental Change, 62*(May 2020), 102067.https://doi.org/10.1016/j.gloenvcha.2020.102067

Estrada-González, I., Taboala- González, P. A., Guerrero-García-Rojas, H., & Márquez-Benavides, L. (2020). Decreasing the environmental impact in an egg-producing farm through the application of LCA and lean tools. *Applied Science, 10*, 1352–1366. https://doi.org/10.3390/app10041352

European Union (2013). 2013/179/EU: Commission Recommendation of 9 April 2013 on the use of common methods to measure and communicate the life cycle environmental performance of products and organisations. Retrieved May 12, 2022, from https://eur-lex.europa.eu/legal-content/EN/TXT/?uri=CELEX%3A32013H0179

FAO (2016a). Greenhouse gas emissions and fossil energy use from poultry supply chains: Guidelines for assessment. Livestock environmental assessment and performance partnership. FAO, Rome, Italy. Retrieved on May 12, 2022 from https://www.fao.org/3/i6421e/i6421e.pdf

FAO (2016b). Environmental performance of animal feeds supply chains: Guidelines for assessment. Livestock environmental assessment and performance partnership. FAO, Rome, Italy. Retrieved on May 12, 2022 from https://www.fao.org/3/i6433e/i6433e.pdf

FAO (2020). Environmental performance of feed additives in livestock supply chains – Guidelines for assessment – Version 1. Livestock environmental assessment and performance partnership (FAO LEAP). Rome. https://doi.org/10.4060/ca9744en

Fraser, D., Duncan, I. J. H., Edwards, S. A., Grandin, T., Gregory, N. G., Guyonnet, V., & Whay, H. R. (2013). General principles for the welfare of animals in production systems: The underlying science and its application. *The Veterinary Journal, 198*(1), 19–27. https://doi.org/10.1016/j.tvjl.2013.06.028

Gast, R. K., Guraya, R., Jones, D. R., & Anderson, K. E. (2015). Persistence of fecal shedding of salmonella enteritidis by experimentally infected laying hens housed in conventional or enriched cages. *Poultry Science, 94*(7), 1650–1656. https://doi.org/10.3382/ps/pev113

Gautron, J., Réhault-Godbert, S., Van de Braak, T. G. H., & Dunn, I. C. (2021). Review: What are the challenges facing the table egg industry in the next decades and what can be done to address them? *Animal*, *15*(Supplement 1), 100282. https://doi.org/10.1016/j.animal.2021.100282

Gerber, P. J., Steinfeld, H., Henderson, B., Mottet, A., Opio, C., Dijkman, J., & …Tempio, G. (2013). *Tackling climate change through livestock – A global assessment of emissions and mitigation opportunities*. FAO. https://www.fao.org/3/i3437e/i3437e.pdf

Ghasempour, A., & Ahmadi, E. (2016). Assessment of environment impacts of egg production chain using life cycle assessment. *Journal of Environmental Management*, *183*, 980–987. http://dx.doi.org/10.1016/j.jenvman.2016.09.054

González-Velandia, K. D., Landázury-Correa, A., & Chaparro, A. M. (2020). Evaluación de impactos ambientales en la cadena de producción de huevos agroecológicos con un enfoque de ciclo de vida. *Revista de Ciencias Ambientales*, *54*(2), 165–179. https://doi.org/10.15359/rca.54-2.9

Guyonnet, V. (2011). Strategic planning for the development of the egg nutraceutical industry. In Y. Nys, M. Bain, & F. Van Immerseel (Eds.), *Improving the safety and quality of eggs and egg products. Vol. 2: Egg safety and nutritional quality chemistry* (pp. 374–399) Woodhead Publishing. https://doi.org/10.1533/9780857093929.3.374

Guyonnet, V. (2012, August 5–9). *Eggs and egg products: Consumers' attitudes, perceptions and behaviours [conference invited lecture]*. World's Poultry Science Congress. http://www.facta.org.br/wpc2012-cd/pdfs/plenary/Vincent_Guyonnet.pdf

Guyonnet, V. (2015). Chapter 9. Strengths and weaknesses of functional eggs for human health. In R. R. Watson & F. De Meester (Eds.), *Handbook of eggs in human function* (pp. 153–165). Wageningen Academic Publishers https://doi.org/10.3920/978-90-8686-804-9

Guyonnet, V. (2018a). How egg marketing campaigns increase consumption. *Poultry International*, *57*(7), 10–11. https://www.wattagnet.com/articles/34529-how-egg-marketing-campaigns-increase-consumption

Guyonnet, V. (2018b). How the egg sector supports sustainable development goals. *Poultry International*, *57*(9), 22–23. https://www.wattagnet.com/articles/35004-how-the-egg-sector-supports-sustainable-development-goals

Guyonnet, V. (2018c). Is it time to reconsider caging laying hens? *Poultry International*, *57*(12), 14–20. https://www.wattagnet.com/articles/35828-is-it-time-to-reconsider-caging-laying-hens

Guyonnet, V. (2019a). Might China's egg industry be leading by example? *Poultry International*, *58*(1), 18–19. https://www.wattagnet.com/articles/36240-is-chinas-egg-industry-leading-by-example

Guyonnet, V. (2019b). Modern egg marketing geared toward younger generations. *Poultry International*, *58*(3), 14–15. https://www.wattagnet.com/articles/36582-modern-egg-marketing-geared-toward-younger-generations

Guyonnet, V. (2019c). Positioning hard-boiled eggs in the snacking market. *Poultry International*, *58*(6), 30–31. https://www.wattagnet.com/articles/37290-positioning-hard-boiled-eggs-in-the-snacking-market

Guyonnet, V. (2019d). Marketing poultry products to younger generations. *Poultry International*, *58*(7), 8–11. https://www.wattagnet.com/articles/38374-marketing-poultry-products-to-younger-generations

Guyonnet, V. (2019e). Big data: Connecting the chicken and the egg. *Poultry International*, *58*(8), 12–13. https://www.wattagnet.com/articles/38036-big-data-connecting-the-chicken-and-the-egg

Guyonnet, V.(2019f). Egg sector must play greater role in consumer education. *Poultry International*, *58*(12), 20–21. https://www.wattagnet.com/articles/38964-egg-sector-must-play-greater-role-in-consumer-education

Guyonnet, V. (2020a). Egg associations key for sector's sustainable growth. *Poultry International*, *59*(1), 30–31. https://www.wattagnet.com/articles/39096-egg-associations-key-for-sectors-sustainable-growth

Guyonnet, V. (2020b). Indian egg production, consumption to keep rising. *Poultry International*, *59*(3), 14–15. https://www.wattagnet.com/articles/39257-india—the-rise-of-a-new-egg-giant

Guyonnet, V. (2020c). Where artificial intelligence may aid the poultry sector. *Poultry International*, *59*(4), 27–29. https://www.wattagnet.com/articles/39650-how-artificial-intelligence-may-aid-the-poultry-industry

Guyonnet, V. (2020d). How the egg industry can grow its appeal to gen Z. *Poultry International*, *59*(7), 30–33. https://www.wattagnet.com/articles/40211-how-the-egg-industry-can-grow-its-appeal-to-gen-z

Guyonnet, V. (2020e). Egg produces need more sustainability guidance. *Poultry International*, *59*(12), 14–15. https://www.wattagnet.com/articles/41316-egg-producers-need-more-sustainability-guidance

Guyonnet, V. (2021, August 17). Time to rethink cage-free commitments for layer welfare? https://www.wattagnet.com/articles/43412-time-to-rethink-cage-free-commitments-for-layer-welfare

Guyonnet, V. (2021, August 2). Time to think about a net zero energy layer barn. https://www.wattagnet.com/articles/43316-time-to-think-about-a-net-zero-energy-layer-barn

Guyonnet, V. (2021a). Value through processing – As simple as boiling an egg! *Poultry International*, *60*(1), 18–19. https://www.wattagnet.com/articles/41640-value-through-processing—as-simple-as-boiling-an-egg

Guyonnet, V. (2021b). New technology redefining understanding of layer behavior. *Poultry International*, *60*(8), 32–33.

Guyonnet, V. (2021d). How nutrition can cut egg farms' environmental impact. *Poultry International*, *60*(12), 38–39. https://www.wattagnet.com/articles/43718-how-nutrition-can-cut-egg-farms-environmental-impact

Guyonnet, V. (2022, June 8–10). Development of egg production and consumption in low income countries in Africa. Poultry Economics session, Abstract 130226. pp. 4. 7th Mediterranean Poultry Summit of the Mediterranean Poultry Network of the WPSA, Cordoba, Spain

Hannah, J. F., Wilson, J. L., Cox, N. A., Cason, J. A., Bourassa, D. V., Musgrove, M. T., & Buhr, R. J. (2011). Comparison of shell bacteria from unwashed and washed table eggs harvested from cage laying hens and cage-free floor-housed laying hens. *Poult Science*, *90*(7), 1586–1593. https://doi.org/10.3382/ps.2010-01115

Hartcher, K. M., & Jones, B. (2017). The welfare of layer hens in cages and cage-free housing systems. *World's Poultry Science Journal*, *73*(4), 767–782. https://doi.org/10.1017/S0043933917000812

Holt, P. S. (2021). Centennial review: A revisiting of hen welfare and egg safety consequences of mandatory outdoor access for organic egg production. *Poultry Science*, *100*(12), 101436. https://doi.org/10.1016/j.psj.2021.101436

Holt, P. S., Davies, R. H., Dewulf, J., Gast, R. K., Huwe, J. K., Jones, D. R., & William, K. R. (2011). The impact of different housing systems on egg safety and quality. *Poultry Science*, *90*(1), 251–262. https://doi.org/10.3382/ps.2010-00794

Hristov, A. N., Oh, J., Lee, C., Meinen, R., Montes, F., Ott, T., & Oosting, S. (2013). Mitigation of greenhouse gas emissions in livestock production – A review of technical options for non-CO_2 emissions. In P. J. Gerber, B. Henderson, & H. P.S. Makkar (Eds.), *FAO animal production and health paper no. 177.* FAO. https://www.fao.org/3/i3288e/i3288e00.htm

Huneau-Salaün, A., Michel, V., Huonnic, D., Balaine, L., & Le Bouquin, S. (2010). Factors influencing bacterial eggshell contamination in conventional cages, furnished cages and free-range systems for laying hens under commercial conditions. *British Poultry Science*, 51(2), 163–169. https://doi.org/10.1080/00071668.2010.482462

Iannotti, L. L., Chessa, K., Lutter, C. K., Stewart, C. P., Gallegos Riofrío, C. A., Malo, C., & Waters, W. F. (2017). Eggs in early complementary feeding and child growth: A randomized controlled trial. *Pediatrics*, 140 (1), July 2017:e20163459. https://doi.org/10.1542/peds.2016-3459

IBM (2020). Research insights - Meet the 2020 consumers driving changes. Retrieved on March 3, 2022 from https://www.ibm.com/downloads/cas/EXK4XKX8

IEC (2021, November 9). International egg commission interactive statistics, Country data. https://marketdata.internationalegg.com/graphs/

IPCC (2006). 2006 IPCC guidelines for national greenhouse gas inventories. In H. S. Eggleston, L. Buendi, K. Miwa, T. Ngara, & K. Tanabe (Eds.), *Institute for global environmental strategies (IGES).* https://www.ipcc-nggip.iges.or.jp/public/2006gl/index.html

IPCC (2021). Climate change 2021 – The physical science basis – Summary for policymakers. Working group I, contribution to the 6th assessment report of the IPCC. IPCC, Switzerland. https://www.ipcc.ch/report/ar6/wg1/downloads/report/IPCC_AR6_WGI_SPM_final.pdf

ISO (2006). ISO 14040:2006 Environmental management - Life cycle assessment - Principles and framework. Retrieved on May 12, 2022 from https://www.iso.org/obp/ui/#iso:std:iso:14040:ed-2:v1:en

ISO (2010). ISO 26000:2010 Guidance on social responsibility. Retrieved on April 29, 2022 from https://www.iso.org/standard/42546.html

Jones, D. R., Anderson, K. E., & Guard, J. Y. (2012). Prevalence of coliforms, *Salmonella, Listeria* and *Campylobacter* associated with eggs and the environment of conventional and free-range egg production. *Poultry Science*, *91*(5), 1195–1202. https://doi.org/10.3382/ps.2011-01795

Jones, D. R., Anderson, K. E., & Musgrove, M. T. (2011). Comparison of environmental and egg microbiology associated with conventional and free-range laying hen management. *Poultry Science*, *90*(9), 2063–2068. https://doi.org/10.3382/ps.2010-01139

Jones, D. R., Cox, N. A., Guard, J., Fedorka-Cray, P. J., Buhr, R. J., Gast, R. K., & Makagon, M. M. (2015). Microbiological impact of three commercial laying hen housing systems. *Poultry Science*, *94*(3), 544–551. https://doi.org/10.3382/ps/peu010

Jones, D. R., Guard, J., Gast, R. K., Buhr, R. J., Fedorka-Cray, P. J., Abdo, Z., & …Karcher, D. M. (2016). Influence of commercial laying hen housing systems on the incidence and identification of *Salmonella* and *Campylobacter. Poultry Science*, *95*(5), 1116–1124. https://doi.org/10.3382/ps/pew036

Le Bouquin, S., Huneau-Salaün, A., Huonnic, D., Balaine, L., Martin, S., & Michel, V. (2013). Aerial dust concentration in cage-housed, floor-housed, and aviary facilities for laying hens, *Poultry Science*, 92(11), 2827–2833. https://doi.org/10.3382/ps.2013-03032

LEAP (2022, May 12). Livestock environmental assessment and performance partnership. https://www.fao.org/partnerships/leap/en/

Leinonen, I., & Kyriazakis, I. (2016). How can we improve the environmental sustainability of poultry production? *Proceedings of the Nutrition Society*, *75*(3), 265–273 https://doi.org/10.1017/S0029665116000094

Leinonen, I., Williams, A. G., & Kyriazakis, I. (2014). The effects of welfare-enhancing system changes on the environmental impacts of broiler and egg production. *Poultry Science*, *93*(2), 256–266. https://doi.org/10.3382/ps.2013-03252

Leinonen, I., Williams, A. G., Wiseman, J., Guy, J., & Kyriazakis, I. (2012). Predicting the environmental impacts of chicken systems in the United Kingdom through a life cycle assessment: Egg production systems. *Poultry Science*, *91*(1), 26–40. https://doi.org/10.3382/ps.2011-01635

Li, Y., Allacker, K., Feng, H., Davoud Heidari, M., & Pelletier, N. (2021). Net zero energy barns for industrial egg production: An effective sustainable intensification strategy? *Journal of Cleaner Production*, *316*, 128014. https://doi.org/10.1016/j.jclepro.2021.128014

Lin, X. J., Zhang, R., Jiang, S., Elmashad, H. M., & Mitloehner, F. (2016). Nutrient flow and distribution in conventional cage, enriched colony, and aviary layer houses. *Poultry Science*, *95*(1), 213–224. https://doi.org/10.3382/ps/pev307

Loyon, L., Burton, C. H., Misselbrook, T., Webb, J., Philippe, F. X., Aguilar, M., & Sommer, S. G. (2016). Best available technology for European livestock farms: Availability, effectiveness and uptake. *Journal of Environmental Management*, *166*, 1–11. https://doi.org/10.1016/j.jenvman.2015.09.046

Mallet, S., Guesdon, V., Ahmed, A. M. H., & Nys, Y. (2006). Comparison of eggshell hygiene in two housing systems : Standard and furnished cages. *British Poultry Science*, *47*(1), 30–35. https://doi.org/10.1080/00071660500468132

Meda, B., Hassouna, M., Aubert, C., Robin, P., & Dourmad, J.Y. (2011). Influence of rearing conditions and manure management practices on ammonia and greenhouse gas emissions from poultry houses. *World's Poultry Science Journal*, *67*(3), 441–455. https://doi.org/10.1017/S0043933911000493

Mekonnen, M. M., & Hoekstra, A. Y. (2012). A global assessment of the water footprint of farm animal products. *Ecosystems*, *15*, 401–415. https://doi.org/10.1007/s10021-011-9517-8

Mench, J. A., Summer, D. A., & Rosen-Molina, J. T. (2011). Sustainability of egg production in the United States – The policy and market context. *Poultry Science*, *90*(1), 229–240 https://doi.org/10.3382/ps.2010-00844

Mollenhorst, H., Berentsen, P. B. M., & De Boer, I. J. M. (2006). On-farm quantification of sustainability indicators: An application to egg production systems. *British Poultry Science Journal*, *47*(4), 405–417 https://doi.org/10.1080/00071660600829282

Ochs, D., Wolf, C. A., Widmar, N. J., & Bir, C. (2019). Is there a "Cage-free" lunch in U.S. Egg production? Public views of laying-hen housing attributes. *Journal of Agricultural and Resource Economics*, *44*(2), 345–361

Pelletier, N. (2017). Life cycle assessment of Canadian egg products, with differentiation by hen housing system type.

Journal of Cleaner Production, *152*, 167–180 https://doi.org/10.1016/j.jclepro.2017.03.050

Pelletier, N. (2018a). Changes in the life cycle environmental footprint of egg production in Canada from 1962 to 2012. *Journal of Cleaner Production*, *176*, 1144–1153. https://doi.org/10.1016/j.jclepro.2017.11.212

Pelletier, N. (2018b). Social sustainability assessment of Canadian egg production facilities: Methods, analysis and recommendations. *Sustainability*, *10*, 1601–1618. https://doi,org/10.3090/su10051601

Pelletier, N., Doyon, M., Muirhead, B., Widowski, T., Nurse-Gupta, J., & Hunniford, M. (2018). Sustainability in the Canadian industry – Learning from the past, navigating the present, planning for the future. *Sustainability*, *10*(10), 3524–3548. https://doi.org/10.3390/su10103524

Pelletier, N., Ibarburu, M., & Xin, H. (2014). Comparison of the environmental footprint of the egg industry in the United States in 1960 and 2010. *Poultry Science*, *93*(2), 241–255. https://doi.org/10.3382/ps.2013-03390

Petrik, M. T., Guerin, M. T., & Widowski, T. M. (2015). On-farm comparison of keel fracture prevalence and other welfare indicators in conventional cage and floor-housed laying hens in Ontario, Canada. *Poultry Science*, *94*(4), 579–585. https://doi.org/10.3382/ps/pev039

Piskorska-Pliszczynska, J., Struciński, P., & Kan, C. A. (2019). Contaminants in eggs: Dioxins/PCBs and other toxic substances and their possible health implications. In F. J. M. Smulders, I. M. C. M. Rietjens, & M. Rose (Eds.), *ECVPH food safety assurance*, *volume 7* (pp. 415–441). https://doi.org/10.3920/978-90-8686-877-3_17

Poore, J., & Nemecek, T. (2018). Reducing food's environmental impacts through produces and consumers. *Science*, *360*(6393), 987–992. https://doi.org/10.1126/science.aaq0216

Ramírez-Morales, I., Fernández-Blanco, E., Rivero, D., & Pazos, A. (2017). Automated early detection of drops in commercial egg production using neural networks. *British Poultry Science*, *58*(6), 739–747. https://doi.org/10.1080/00071668.2017.1379051

Rondoni, A., Asiolia, D., & Millana, E. (2020). Consumer behaviour, perceptions, and preferences towards eggs: A review of the literature and discussion of industry implications. *Trends in Food Science & Technology*, *106*, 391–401. https://doi.org/10.1016/j.tifs.2020.10.038

Sánchez-Bravo, P., Chambers V, E., Noguera-Artiaga, L., Sendra, E., Chambers, E., & Carbonell-Barrachina, A. (2021). Consumer understanding of sustainability concept in agricultural products. *Food Quality and Preference*, *89*, 104136. https://doi.org/10.1016/j.foodqual.2020.104136

Schuck-Paim, C., Negro-Calduch, E., & Alonso, W. J. (2021). Laying hen mortality in different indoor housing systems: A meta-analysis of data from commercial farms in 16 countries. *Scientific Reports*, *11*, 3052. https://doi.org/10.1038/s41598-021-81868-3

Shepherd, T. A., Zhao, Y., Li, H., Stinn, J. P., Hayes, M. D., & Xin, H. (2015). Environmental assessment of three egg production systems – Part II. Ammonia, greenhouse gas, and particulate matter emissions. *Poultry Science*, *94*(3), 534–543. https://doi.org/10.3382/ps/peu075

Thompson, P. B., Appleby, M., Busch, L., Kalof, L., Miele, M., Norwood, F., & Pajor, E. (2011). Values and public acceptability dimensions of sustainable egg production. *Poultry Science*, *90*(9), 2097–2109. https://doi.org/10.3382/ps.2010-0138

UN SDG (2022, May 24). 17 Goals to transform our world. Retrieved May 24, 2022, from https://www.un.org/sustainabledevelopment/sustainable-development-goals/

UNEP (2009). Guidelines for socila life cycle assessment of products. In The UNEP/ESTAC life cycle initiative Paris. https://www.unep.org/resources/report/guidelines-social-life-cycle-assessment-products

Wall, H., Tauson, R., & Sørgjerd, S. (2008). Bacterial contamination of eggshells in furnished and conventional cages. *Journal of Applied Poultry Research*, *17*(1), 11–16. https://doi.org/10.3382/japr.2006-00058

Weeks, C. A., Lambton, S. L., & Williams, A. G. (2016). Implications for welfare, productivity and sustainability of the variation in reported levels of mortality for laying hen flocks kept in different housing systems: A meta-analysis of ten studies. *PLoS One*, *11*(1), e0146394. https://doi.org/10.1371/journal.pone.0146394

Weeks, C. A., & Nicol, C. J. (2006). Behaviour needs, priorities and preferences of laying hens. *World's Poultry Science Journal*, *62*(2), 296–307. https://doi.org/10.1079/WPS200598

Widowski, T., Casey-Trott, T., Hunniford, M., & Morrissey, K. (2017). Chapter 3 - welfare of laying hens: An overview. In J. R. Roberts (Ed.), *Achieving sustainable production of eggs Vol. 2 animal welfare and sustainability* (pp. 1–28). Burleigh Dodds Science Publishing Limited http://dx.doi.org/10.19103/AS.2016.0012.22

Wiedemann, S. G., & McGahan, E. J. (2011). *Environmental assessment of an egg production supply chain using life cycle assessment*. Australian Egg Corporation Limited. https://www.freeranger.com.au/uploads/7/4/2/0/7420102/aecl_carbon_footprint.pdf

Wood, J. D., VanderZaag, A. C., Wagner-Riddle, C., Smith, E. L., & Gordon, R. J. (2014). Gas emissions from liquid dairy manure: Complete versus partial storage emptying. *Nutrient Cycling in Agroecosystems*, *99*, 95–105. https://doi.org/10.1007/s10705-014-9620-2

World Commission on Environment and Development. (1987). Our common future. Retrieved on November 15, 2021. https://sustainabledevelopment.un.org/content/documents/5987our-common-future.pdf

Zhao, Y., Shepherd, T. A., Li, H., & Xin, H. (2015). Environmental assessment of three egg production systems – Part I: Monitoring system and indoor air quality. *Poultry Science*, *94*(3), 518–533 https://doi.org/10.3382/ps/peu076

Zhao, Y., Zhao, D., Ma, H., Liu, K., Atilgan, A., & Xin, H. (2016). Environmental assessment of three egg production systems – Part III: Airborne bacteria concentrations and emissions. *Poultry Science*, 95(7), 1473–1481. https://doi.org/10.3382/ps/pew053

Zimazile Sibanda, T., Welch, M., Schneider, D., Kolakshyapati, D., & Ruhnke, I. (2020). Characterising free-range layer flocks using unsupervised cluster analysis. *Animals*, *10*, 855. https://doi.org/10.3390/ani10050855

43

List of Key Resources within the Egg Sector

Vincent Guyonnet and Yoshinori Mine*

CONTENTS

43.1 Introduction

While some information contained in this textbook may become obsolete over time, it is expected that this last chapter will provide readers with some valuable links to the most updated information on a wide range of topics, organized in four sections: Egg Production, Egg Science, Egg Nutrition and Egg Grading and Processing. The information provided in the following sections is far from exhaustive, focusing mostly on internet websites available in English. In most cases, the order of citation is based on the alphabetic order within each category. The listing of commercial companies does not mean any endorsement of the companies or their products.

43.2 Egg Production

43.2.1 Organizations of Egg Producers

<u>*International*</u>

Internation Egg Commission IEC): http://www.internationalegg.com

Asociación Latinoamericana de Avicultura (ALA): https://www.avicolatina.com/

<u>*North America*</u>

Canada

Egg Farmers of Canada (EFC): https://www.eggfarmers.ca/

Canadian Poultry and Egg Processors Council (CPEPC): https://www.cpepc.ca

* Corresponding author.

DOI: 10.1201/9781003254430-53

USA
American Egg Board (AEB): https://www.incredibleegg.org/
United Egg Producers (UEP): https://unitedegg.com/

Latin America

Argentina
Cámara Argentina de Productores Avícolas (CAPIA): https://www.capia.com.ar/

Brazil
Instituto Ovos Brasil: https://www.ovosbrasil.com.br/

Chile
Chile Huevos: https://www.chilehuevos.cl/

Colombia
Federación nacional de avicultores de Colombia (FENAVI): https://fenavi.org/

Guatemala
Asociación nacional de avicultores (ANAVI): http://www.anaviguatemala.org/

Honduras
Asociación Nacional de Avicultores de Honduras (ANAVIH): https://anavih.com/

Mexico
Unión nacional de avicultores (UNA): https://una.org.mx/

Uruguay
Asociación de Productores Avícolas Sur (APAS): http://www.apas.com.uy/

Europe

European Egg Processing Association (EEPA): https://www.eepa.info

France
Comité National pour la Promotion de l'Œuf (CNPO): https://oeuf-info.fr/

Germany
Zentralverband der Deutschen Geflügelwirtschaft e.V. (ZDG): https://zdg-online.de/

Spain
Organización Interprofesional del Huevo y Sus Productos (INPROVO): https://www.inprovo.com/que-es-inprovo/

United Kingdom
British Egg Industry Council (BEIC): https://britisheggindustrycouncil.com/

Africa/Middle-East

Morocco
Fédération Interprofessionnelle du Secteur Avicole du Maroc (FISA Maroc): https://www.fisamaroc.org.ma/index.php

Nigeria
Poultry Association of Nigeria (PAN): http://poultryassociationng.com/

South Africa
South Africa Poultry Association (SAPA): http://www.sapoultry.co.za/

Tunisia
Groupement Interprofessionnel des Produits Avicoles et Cunicoles (GIPAC): http://www.gipac.tn/index.php/la-filere-oeufs

Zambia
Poultry Association of Zambia (PAZ): http://www.paoz.org/

Asia/Oceania

Australia
Australian Eggs: https://www.australianeggs.org.au/

China
Wonderful Egg Alliance: http://www.wonderfulegg.com/

India
National Egg Coordination Committee (NECC): https://www.e2necc.com/

Japan
Japanese Poultry Association: https://www.jpa.or.jp/
Japan Poultry Science Association: https://jpn-psa.jp/en/home-en/

Korea
Korean Poultry Association: https://www.poultry.or.kr/

Malaysia
Federation of livestock farmers associations of Malaysia (FLFAM): http://www.flfam.org.my/

43.2.2 Genetics and Breeder Companies

Hendrix Genetics BV

Babcock: https://www.babcock-poultry.com/en/
Bovans: https://www.bovans.com/en/
Dekalb: https://www.dekalb-poultry.com/en/
Hisex: https://www.hisex.com/en/
ISA: https://www.isa-poultry.com/en/
Shaver: https://www.shaver-poultry.com/en/

PHW Group

Hy-Line: https://www.hyline.com/
Lohmann: https://lohmann-breeders.com/
Novogen: https://novogen-layers.com/en/

Yukou Poultry

https://m.ppxmw.com/

43.2.3 Housing Systems Manufacturers

AGCO/Tecno: https://www.poultryequipment.com/
Alaso: https://www.alaso.com/
Big Dutchman: https://www.bigdutchman.com/en/company/about-us/about-us/
Chore-Time: https://www.choretime.com/
FACCO Poultry Equipment: https://www.facco.net/
Hellmann Poultry Equipment: https://hellmannpoultry.com/en-us/
Henan Jinfeng Poultry Equipment Co.: https://www.jinfengpoultry.com/
Huabo (breeders only): http://www.hb-poultrysystem.com/
Huali: http://hualigf.com/en/
Kilbra: https://kilbra.com.br/
Little Giant Animal Husbandry Equipment Co. Ltd.: http://www.cdxjr.com/
SALMET GmbH & Co KG: https://www.salmet.de/
VALLI: https://www.valli-italy.com/
Vencomatic Group: https://www.vencomaticgroup.com/
Xian Qingan Zhihang: http://www.xianqzh.com/?mod=index
Zhengzhou Livi Machinery Manufacturing Co. Ltd.: https://poultryequipmentpro.com/about-us/
Zucami Poultry Equipment: https://zucami.com/en/

43.2.4 Feed Ingredients and Feed Suppliers

Adisseo: https://www.adisseo.com/en/
Alltech: https://www.alltech.com/

Cargill: https://www.cargill.com/animal-nutrition/species/poultry
DSM: https://www.dsm.com/anh/home.html
Evonik: https://corporate.evonik.com/en
Kemin: https://www.kemin.com/na/en-us/home
Novozymes: https://www.novozymes.com/en
Nutreco: https://www.nutreco.com/

43.2.5 Animal Health Companies

Boehringer-Ingelheim: https://www.boehringer-ingelheim.com/animal-health/products
CEVA: http://www.ceva.com/
Elanco: https://www.elanco.com/en-us
Globion: https://globionindia.com/wp/
Hester Biosciences: https://www.hester.in/
Joinsha Animal Health Products (Xiamen) Co. Ltd.: http://www.joinsha.com/en/
MSD Animal Health: https://www.msd-animal-health.com/
Phibro Animal Health: https://www.pahc.com/
UniFAHS: http://unifahs.com/
Zoetis: https://www.zoetis.com/

43.3 Egg Science

43.3.1 Key Universities/Research Institutes

International

World Poultry Science Association (WPSA): http://www.wpsa.com/
World Veterinary Education in Production Animal Health (WVEPAH): https://www.wvepah.net/

North America

Canada

University of Guelph, Department of Food Science, Guelph, ON: https://www.uoguelph.ca/foodscience/
University of Ottawa, Department of Cellular and Molecular Medicine, Faculty of Medicine, Ottawa, ON: https://www2.uottawa.ca/faculty-medicine/cellular-molecular
University of Alberta, Agricultural Food and Nutritional Science Department, Edmonton, AB: https://www.ualberta.ca/agricultural-food-nutritional-science/index.html

USA

Louisiana State University School of Nutrition and Food Sciences, Baton Rouge, LA: https://nfs.lsu.edu
Iowa State University Animal Science Department, Ames, IA; https://www.ans.iastate.edu
University of Illinois Urbana-Champlain Food Science & Human Nutrition, Urbana-Champlain, IL: https://www.fshn.illinois.edu

Europe

Belgium

KU Leuven Leuven Food science and Nutrition research centre, Leuven: https://www.biw.kuleuven.be/lforce/

Croatia

University of Zagreb Faculty of Agriculture, Division of Animal Sciences, Zagreb: https://www.agr.unizg.hr/en/group/345/Division+of+Animal+Science

Denmark

Aarhus University, Department of Animal Science & Department of Food Science, Aarhus: https://anivet.au.dk/en/

France

INRAE, Université de Tours, UMR BOA, Nouzilly: https://www6.val-de-loire.inrae.fr/boa/
INRAE, UR BIA, Nantes: https://www6.angers-nantes.inrae.fr/bia/
INRAE, Institut Agro Rennes-Angers, UMR STLO, Rennes: https://www6.rennes.inrae.fr/stlo
INRAE, Institut Agro Rennes-Angers, UMR PEGASE, Rennes: https://www6.rennes.inrae.fr/pegase/

Italy

University of Bologna, Department of Agricultural and Food Sciences, Bologna: https://distal.unibo.it/en/research/research-areas/animal-sciences

Netherlands

University of Wageningen, Food safety research group, Wageningen: https://www.wur.nl/en/research-results/research-institutes/food-safety-research.htm
Utrecht University, Faculty of Veterinary Medicine, Utrecht: https://www.uu.nl/en/organisation/faculty-of-veterinary-medicine

Spain

Institute of Food Science, Technology and Nutrition (ICTAN), CSIC, Madrid: https://www.csic.es/en/investigation/institutes-centres-units/institute-food-science-technology-and-nutrition
Universitat Autonoma de Barcelona, Department of Animal and Food Science, Barcelona: https://www.uab.cat/web/department-of-animal-and-food-science-1345715709728.html

United Kingdom

University of Edinburgh, The Roslin Institute, Midlothian https://www.ed.ac.uk/roslin/national-avian-research-facility
University of Reading, School of Biological Sciences, Reading: https://www.reading.ac.uk/biological-sciences/

Asia/Oceania:

China

China Agricultural University, College of Food Science & Nutritional Engineering, Beijing: http://spxy.cau.edu.cn/
Huazhong Agricultural University, College of Food Science and Technology, Wuhan, Hubei: http://shipin.hzau.edu.cn/index.htm
Jiangnan University, School of Food Science and Technology, Wuxi, Jiangsu: http://foodsci.jiangnan.edu.cn/
Jilin University, College of Food Science and Engineering, Changchun, Jilin: https://spxy.jlu.edu.cn/
Northeast Agricultural University, Food Science College, Harbin, Heilongjiang: http://english.neau.edu.cn/Schools___Departments/School_of_Food_Science.htm
South China Agriculture University, College of food science, Guangzhou, Guangdong: https://spxy.scau.edu.cn/spxy_en/

Japan

Hiroshima University, Graduate School of Integrated Sciences for Life, Hiroshima: https://www.hiroshima-u.ac.jp/en/ilife

Kyoto Women's University: https://www.kyoto-wu.ac.jp/english/
Nagoya University, Avian Bioscience Research Center, Nagoya: http://www.agr.nagoya-u.ac.jp/~abrc/abrc-E/index.html

Korea

Konkuk University, Department of Food Science and Biotechnology of Animal Resources, Seoul: http://foodbio.konkuk.ac.kr/main.do
National Institute of Animal Science, Wanju-gun: https://www.nias.go.kr/english/indexList.do
Seoul National University, Department of Food and Animal Biotechnology, Seoul: https://fabio.snu.ac.kr/

43.3.2 Main Scientific Journals

Advances in Nutrition: https://academic.oup.com/advances
Animal Production Journal: https://www.publish.csiro.au/an
Animal Science Journal: https://onlinelibrary.wiley.com/journal/17400929
Avian Diseases: https://meridian.allenpress.com/avian-diseases
Avian Pathology: https://www.tandfonline.com/toc/cavp20/current
Brazilian Journal of Poultry Science: https://www.scielo.br/j/rbca/
British Poultry Science: https://www.tandfonline.com/journals/cbps20
China Poultry (Zhongguo Jiaqin): http://www.zgjqzz.net/CN/1004-6364/home.shtml
Filières avicoles: https://www.filieres-avicoles.com/
Food Chemistry: https://www.journals.elsevier.com/food-chemistry
Food Science: https://www.spkx.net.cn/CN/1002-6630/home.shtml
Food Research International: https://www.sciencedirect.com/journal/food-research-international
Foods: https://www.mdpi.com/journal/foods
Japan Egg Science Society: http://japaneggscience.com/english.html
Journal of Agricultural and Food Chemistry: https://pubs.acs.org/journal/jafcau
Journal of Applied Poultry Research: https://www.journals.elsevier.com/journal-of-applied-poultry-research
Journal of Nutrition: https://academic.oup.com/jn
Journal of the Science of Food and Agriculture: https://onlinelibrary.wiley.com/journal/10970010
Korean Journal of Poultry Science: https://www.ekjps.org/
Poultry Science: https://www.journals.elsevier.com/poultry-science
Productions animales: https://productions-animales.org/
Science and Technology of Food Industry: http://www.spgykj.com/indexen.htm
The American Journal of Clinical Nutrition: https://academic.oup.com/ajcn
The Journal of Poultry Science: https://www.jstage.jst.go.jp/browse/jpsa
The Ornithological Society of Japan: https://ornithology.jp/en/publications.html
World Poultry Science Journal; https://www.tandfonline.com/journals/twps20

43.3.3 Egg and Poultry Related Conferences

International Poultry Forum China, Annual Conference: http://forum.guojixumu.com/eng/all.aspx
Japan Egg Science Society, Annual Egg Symposium: https://japaneggscience.com/english.html
Japan Egg Research Society, Annual Egg Science Forum, https://www.facebook.com/nihontamaken/
Japan Poultry Science Association, Conferences: https://jpn-psa.jp/en/home-en/
Middle-East North-Africa (MENA) Poultry Conferences: https://www.mepc-wpsa.com/
Poultry Association Association, Annual Meeting: https://poultryscience.org/PSA-Annual-Meeting
World's Poultry Science Association, World Poultry Congress,: http://www.wpsa.com/
World's Poultry Science Association, European Symposium on the Quality of Eggs and Egg Products.: http://www.wpsa.com/

43.3.4 Egg Genome Databases

BLAT server: https://birdbase.net/cgi-bin/webBlat
Chicken SNP data base: https://ngdc.cncb.ac.cn/chickensd/
Gallus GBrowse: https://birdbase.net/cgi-bin/gbrowse/gallus/
GenomeNet: https://www.genome.jp/
NIH: BLAST: https://blast.ncbi.nlm.nih.gov/Blast.cgi

43.4 Egg Nutrition

43.4.1 List of International and National Databases

International
FAO: https://www.fao.org/infoods/infoods/tables-and-databases/en/

North America
Canada: https://www.canada.ca/en/health-canada/services/food-nutrition/healthy-eating/nutrient-data.html
USA: https://fdc.nal.usda.gov/fdc-app.html#/food-details/748967/nutrients

Latin America
Argentina: http://www.unlu.edu.ar/~argenfoods/Tablas/Tabla.htm
Brazil: http://www.tbca.net.br/
Mexico: https://www.incmnsz.mx/2019/TABLAS_ALIMENTOS.pdf

Europe
France: https://ciqual.anses.fr/#/aliments/22000/egg-raw
Spain: https://www.bedca.net/bdpub/index.php
United Kingdom: https://www.gov.uk/government/publications/composition-of-foods-integrated-dataset-cofid

Africa
Africa (West): https://www.fao.org/3/ca7779b/CA7779B.PDF
South Africa: https://safoods.mrc.ac.za/database.html

Asia/Oceania
ASEAN: http://www.inmu.mahidol.ac.th/aseanfoods/composition_data.html

Australia: https://www.foodstandards.gov.au/science/monitoringnutrients/ausnut/foodnutrient/Pages/default.aspx

China: https://www.chinanutri.cn/yyzyxxpt/sjzx/swcfsjk/

Japan: https://www.mext.go.jp/en/policy/science_technology/policy/title01/detail01/sdetail01/sdetail01/1385122.htm

Korea: http://koreanfood.rda.go.kr/eng/fctFoodSrchEng/engMain

43.4.2 Key Organizations Focusing on Egg Nutrition

Egg Nutrition Center (ENC): https://www.incredibleegg.org/?site=b

Global Alliance for Improved Nutrition (GAIN): https://www.gainhealth.org/

International Egg Nutrition Centre (IENC): https://www.internationalegg.com/our-work/nutrition/

Instituto Latinoamericano del huevo (ILH): http://www.ilhala.org/

Korea Food Research Institute: https://www.kfri.re.kr/

National Institute of Health and Nutrition: https://www.nibiohn.go.jp/eiken/english/index.html

Sight and Life: https://sightandlife.org/

The Egg Safety Center: https://eggsafety.org

43.4.3 Intergovernmental and International Organizations

Codex Alimentarius: https://www.fao.org/fao-who-codexalimentarius/en/

Food and Agriculture Organization of the United Nations (FAO): https://www.fao.org/nutrition/en/

FAO STAT: https://www.fao.org/faostat/en/#data

International Food Policy Research Institute (IFPRI): https://www.ifpri.org/

Organisation for Economic Co-operation and Development (OECD) Agriculture Statistics: https://www.oecd-ilibrary.org/agriculture-and-food/data/oecd-agriculture-statistics_agr-data-en

World Health Organization (WHO): https://www.who.int/health-topics/nutrition#tab=tab_1

World Organisation for Animal Health (WOAH, founded as OIE): https://www.woah.org/en/home/

43.5 Egg Grading and Processing

43.5.1 Egg Grading Equipment Manufacturers

Eggtec Co. Ltd.: http://www.eggtec.com/eng/

Fuzhou Mintai Machinery Co., Ltd.: http://www.mtmt.cn/

MOBA: https://www.moba.net/page/en/Products/Grading

Nabel: https://nabel.com/

SANOVO Technology Group: https://www.sanovoegg.com/products/grading-solutions/?product-family=All+products

Seyang Co. Ltd.: http://www.seyang-kor.com/eng/main/

SIME-TEK: https://www.sime-tek.com/en/

Zenyer Egg Machinery Co., Ltd.: http://en.zenyer.cn/About-Zenyer/

43.5.2 Egg Processing Equipment Manufacturers

Actini: http://www.actini.com/en/food-en/expertise/

Fuzhou Mintai Machinery Co., Ltd.: http://www.mtmt.cn/

M-Mds Company: http://www.m-mds.com/html/main.html

MOBA: https://www.moba.net/page/en/Products/Processing

Ovobel: http://www.ovobel.com/

Ovoconcept: https://www.ovoconcept.eu/

SANOVO Technology Group: https://www.sanovoegg.com/products/processing-solutions/?product-family=All+products

43.5.3 National Regulations for Shell Eggs

International

Code of hygienic practice for eggs and egg products, CAC/RCP 15-1976: https://www.fao.org/fao-who-codexalimentarius/codex-texts/codes-of-practice/en/

North America

Canada

Safe Food for Canadians Regulations (SOR/2018-108): https://laws-lois.justice.gc.ca/eng/regulations/SOR-2018-108/index.html

USA

Agriculture Marketing Act. 7CFR Part 56, Voluntary Grading of shell eggs. *Code of Federal Regulations: https://www.ecfr.gov/current/title-7/subtitle-B/chapter-I/subchapter-C/part-56*

Egg Products Inspection Act. 7 CFR Part 57, Inspection of eggs. Code of Federal Regulations. https://www.ecfr.gov/current/title-7/subtitle-B/chapter-I/subchapter-C/part-57

USDA Egg grading manual https://www.ams.usda.gov/sites/default/files/media/Egg%20Grading%20Manual.pdf:

Latin America

Argentina

National AgriFood Safety and Quality, Resolución 1699/2019 (In Spanish): https://www.boletinoficial.gob.ar/detalleAviso/primera/223674/20191211

Brazil

New specifications for egg classification and inspection, Decree - 56.585, de 20/07/1965 (In Portuguese): https://www.defesa.agricultura.sp.gov.br/legislacoes/decreto-56585-de-20-07-1965,359.html

Mexico

Egg and its products. Sanitary measures and specifications. Testing procedures, Official Mexican Norm, NOM-159-SSA1- 2016 (In Spanish): https://www.snice.gob.mx/~oracle/SNICE_DOCS/NOM-159-SSA1-2016-NOM-159-DOF_20180719-20180116.pdf

Poultry products – Fresh egg from laying hen – Specifications and testing procedures, Official Mexican Norm, NMX-FF-127-SCFI-2016 (In Spanish): http://sitios1.dif.gob.mx/alimentacion/docs/NMX-FF-127-SCFI-2016_Huevo_fresco.pdf

Europe

European Union

Commission Regulation (EC) No 589/2008 of 23 June 2008 laying down detailed rules for implementing Council

Regulation (EC) No 1234/2007 as regards marketing standards for eggs: https://eur-lex.europa.eu/legal-content/EN/TXT/?uri=celex%3A32008R0589

Part VI – Eggs of hens of the Gallus gallus species. In Regulation (EU) No 1308/2013 of the European Parliament and of the Council of 17 December 2013 establishing a common organisation of the markets in agricultural products and repealing Council Regulations (EEC) No 922/72, (EEC) No 234/79, (EC) No 1037/2001 and (EC) No 1234/2007 (pp.818-819). https://eur-lex.europa.eu/legal-content/EN/TXT/?uri=CELEX:32013R1308

United Kingdom

Eggs marketing: guidance on the legislation: https://www.gov.uk/government/publications/eggs-marketing-guidance-on-the-legislation

Africa/Middle East

South Africa

Regulations regarding the grading, packing and marking of eggs intended for sales in the Republic of South Africa, No.R. 345, Government Gazette: https://www.gov.za/sites/default/files/gcis_document/202003/43108rg11055gon345.pdf

Asia/Oceania

China

Agriculture sector standard of the People's Republic of China, Technical specification for sorting, cleaning and grading chicken eggs, NY/T 1551-2007 (In Chinese): http://down.foodmate.net/wap/index.php?itemid=48415&moduleid=23

National standard of the People's Republic of China. Food safety national standard for Egg and egg products, GB 2749-2015 (In Chinese): http://down.foodmate.net/wap/index.php?itemid=48415&moduleid=23

Quality and safety control – Technical production of non-pollution agricultural products: Part 11: Fresh eggs, NY/T 2798.11-2016 (In Chinese): http://down.foodmate.net/wap/index.php?itemid=48415&moduleid=23

National standard of the People's Republic of China. Pack for shell eggs, GB/T 39438-2020 (In Chinese): http://down.foodmate.net/wap/index.php?itemid=48415&moduleid=23

India

Food safety and standards (Food products standards and Food additives) Regulations, version XII (08.09.2020): https://www.fssai.gov.in/upload/uploadfiles/files/Food_Additives_Regulations.pdf

Food Safety and Standards (Packaging and Labelling) Regulation, version III (27.08.2020): https://www.fssai.gov.in/upload/uploadfiles/files/Packaging_Labelling_Regulations.pdf

Japan

Hygiene management guidelines for egg sorting and packaging facilities (Ministry of Health and Welfare Notice No. 1674 (In Japanese): https://www.jz-tamago.co.jp/wp/wp-content/uploads/2020/03/E05_3_3.pdf

The illustrated egg handbook. Appendix (pp. 141-149). Packington, Leicestershire, United Kington: Context Publications: https://www.contextbookshop.com/user_uploads/files/booksamples/The-Illustrated-Egg-Handbook.pdf

Korea

Livestock Products Sanitation Control Act (Enforcement 2020. 10. 8.] [Act No. 17249, 2020. 4. 7., partially amended): https://www.law.go.kr/LSW/lsInfoP.do?efYd=20201008&lsiSeq=217419#0000

Enforcement Rule of Livestock Products Processing Act, Ordinance No.1478, August 4, 2004: https://www.qia.go.kr/getZipqiaEngNoticeWebAction.do?id=378

43.5.4 List of Major Egg Graders/Packers (Alphabetical Order, by Continent)

North America

Burnbrae Farms Ltd. (Canada): https://www.burnbraefarms.com/en/

Cal-Maine Foods (USA): https://www.calmainefoods.com/

Daybreak Foods, Inc. (USA): https://daybreakfoods.com/

Hillandale Farms (USA): https://www.hillandalefarms.com/

Rose Acres Farms (USA): https://www.goodegg.com/

Versova Holdings L.L.P. (USA): https://www.versova.com/

Latin America

Bachoco (Mexico): https://bachoco.com.mx/

Granja Mantiqueira (Brazil): https://www.ovosmantiqueira.com.br/

Huevo Guadalupe (Mexico): https://www.elhuevodehoy.com.mx/empresa/

Huevos Kikes (Colombia): https://www.huevoskikes.com/

La Calera (Peru): https://www.lacalera.pe/

PROAN (Mexico): https://proan.com/

Europe

Avangardco (Ukraine): https://avangardco.ua/

Dagu (Spain): https://www.dagu.es/

DANÆG A/S (Denmark): https://www.danaeg.com/

Eureden (France): https://www.eureden.com/fr

PEP (France): https://www.cocotine.com/en/

Prioskolye (Russian Federation): https://prioskol.ru/

Africa/Middle East

Arab Company for Livestock Development (ACOLID) (Saudi Arabia): http://www.acolid.com/

Irvine's (Zimbabwe): https://www.irvineschicken.co.zw/

Obasanjo Farms Nigeria Ltd. (Nigeria): https://obasanjo-farmsng.com/

Quantum Foods/Nulaid (South Africa): https://quantum-foods.co.za/

Sedima (Senegal): http://sedima.com/

Wadi Group (Egypt): https://www.wadigroup.com/

Asia/Oceania

CP Foods (Thailand): https://www.cpfworldwide.com/en/home

Deqingyuan (DQY) Agricultural Technology Co. Ltd. (China): http://www.dqy.com.cn/

Ise Inc. (Japan): https://www.ise-egg.co.jp/

Kazi Farms Group (Bangladesh): http://www.kazifarms.com/

Pace Farms (Australia): https://www.pacefarm.com/

Sundaily Farm Ecological Food Co: https://www.sundaily.cn/

43.5.5 National Regulations for Egg Products

Canada

Safe Food for Canadians Regulations (SOR/2018-108): https://laws-lois.justice.gc.ca/eng/regulations/SOR-2018-108/index.html

China

National standards for food safety (Egg and egg products) http://down.foodmate.net/wap/index.php?itemid=48415&moduleid=23

European Union

Regulation (EC) No 853/2004 of the European Parliament and of the Council of 29 April 2004 laying down specific hygiene rules for food of animal origin: https://eur-lex.europa.eu/eli/reg/2004/853/oj?locale=en

Japan

Food Sanitation Act (Act No. 233 of February 24, 1947): https://www.cas.go.jp/jp/seisaku/hourei/data/fsa.pdf

Mexico

Egg and its products. Sanitary measures and specifications. Testing procedures, Official Mexican Norm, NOM-159-SSA1- 2016 (In Spanish): https://www.snice.gob.mx/~oracle/SNICE_DOCS/NOM-159-SSA1-2016-NOM-159-DOF_20180719-20180116.pdf

USA

Egg Products Inspection Regulations: https://www.federalregister.gov/documents/2020/10/29/2020-20151/egg-products-inspection-regulations

43.5.6 List of Major Egg Processing Companies (Alphabetical Order, by Continent)

North America

Burnbrae Farms Ltd. (Canada): https://www.burnbraefarms.com/en/
EchoLake Foods (USA): http://www.echoforeggs.com/
Hillandale Farms (USA): https://www.hillandalefarms.com/
Michael Foods (USA): https://michaelfoods.com/
Rembrandt Foods (USA): https://rembrandtfoods.com/
Rose Acres Farms (USA): https://www.goodegg.com/

Latin America

Granja Mantiqueira (Brazil): https://www.ovosmantiqueira.com.br/
Granjazul (Guatemala): https://granjazul.com/
Huevos Kikes (Colombia): https://www.huevoskikes.com/
Ovobrand (Argentina): http://www.ovobrand.com/
Ovosur (Peru): https://www.ovosur.com/
PROAN (Mexico): https://proan.com/

Europe

AIA - Agricola Italiana Alimentare SpA (Italy): www.aiafood.com
Avangardco (Ukraine): https://avangardco.ua/
Cascina Italia (Italy): www.gruppomoretti.com
Copain (Spain): https://www.copain.co/
Derovo Group (Portugal): https://www.derovo.com/
Eurovo (Italy): https://www.eurovo.com/
Ferme du Pré (France): www.fermedupre.fr
Igreca S.A. (France): http://www.igreca.fr/
Lowrie Foods (UK): www.lowriefoods.co.uk
Monaldi Italia (Italy): http://monalditalia.it
Noble Egg Innovations (UK): https://www.noblefoods.co.uk/
Ovobest Eiprodukte (Germany): www.ovobest.de
Ovodan Foods A/S - former Sanovo Foods (Denmark): https://ovodan.com
Ovopol (Poland): www.ovopol.pl
Parmovo (Italy): http://parmovo.com
PEP/Groupe d'Aucy (France): www.eureden.com
Valoroeuf (France): www.valoroeuf.com
Suzhou Ovodan Foods (China): http://cn.ovodan.com/en/
Kewpie Co. Ltd (Japan): https://www.kewpie.com/en/

Asia/Oceania

Dalian Hanovo Foods Co., Ltd. (China): http://mobile.hanovofoods.com/3g/
DQY Ecological (China): https://en.dqy.com.cn/html/list-1415-1.html
Ganongbio, (Korea): http://ganongbio.co.kr
Hubei Shendan Health Food Co., Ltd. (China): http://en.shendan.com.cn/intro/1.html
Ise Food Inc. (Japan): https://www.iseegg.co.jp/eng/
Join (Korea): www.joinbio.co.kr
Kangde Egg Industry (China): http://www.jskangde.com/
KC Feed (Korea): http://www.kcfeed.co.kr/
Kenko Mayonnaise Co. Ltd. (Japan): https://www.kenkomayo.co.jp/en
Kewpie Egg Corp. (Japan): https://www.kewpie-egg.co.jp
Ovodan Foods (China): https://cn.ovodan.com/en/
Pace Farms (Australia): https://www.pacefarm.com/
Sanshu Foods (Japan): http://www.sansyu-egg.co.jp/index.html
Shendi Agriculture Branch Trade (China): http://www.hbsd.cn
Taiyo Kagaku Co. Ltd. (Japan): https://www.taiyokagaku.com/en/

43.6 Conclusion

The rapid pace of changes in many areas of the egg sector (e.g., production system, egg science, and research) means that continuing education is more than ever essential. New technologies are being developed to produce many new shell eggs and egg products, allowing consumers more opportunities to enjoy this very nutritious food. Egg science is identifying continuously newer molecules with some promising applications in many fields, ranging from the production of new biomaterials to that of semiconductors. The provision of these internet websites will give the reader the opportunity to keep abreast of the latest developments in Egg Science.

Index

Note: Locators in *italics* represent figures and **bold** indicate tables in the text.

D

J

K

L

M

N

O

P

T

For Product Safety Concerns and Information please contact our EU
representative GPSR@taylorandfrancis.com
Taylor & Francis Verlag GmbH, Kaufingerstraße 24, 80331 München, Germany

www.ingramcontent.com/pod-product-compliance
Lightning Source LLC
LaVergne TN
LVHW080104011026
849160LV00013B/1027

* 9 7 8 1 0 3 2 1 8 4 2 0 3 *